KT-521-304

THE ART OF ELECTRONICS

Second Edition

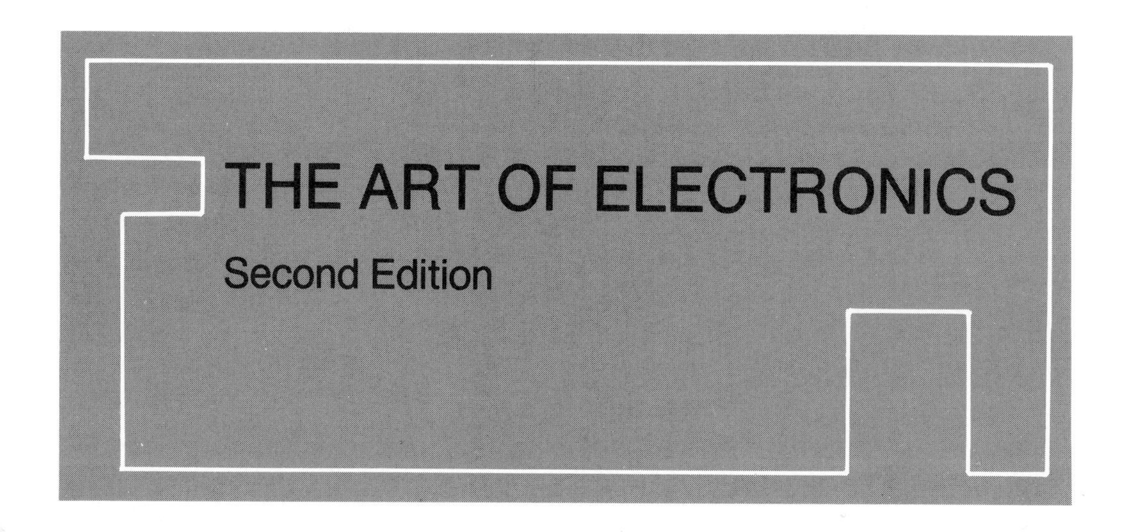
THE ART OF ELECTRONICS
Second Edition

Paul Horowitz HARVARD UNIVERSITY

Winfield Hill ROWLAND INSTITUTE FOR SCIENCE, CAMBRIDGE, MASSACHUSETTS

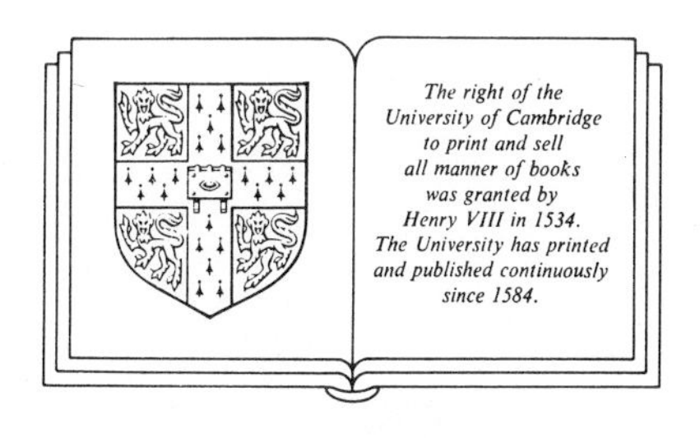
The right of the
University of Cambridge
to print and sell
all manner of books
was granted by
Henry VIII in 1534.
The University has printed
and published continuously
since 1584.

CAMBRIDGE UNIVERSITY PRESS

Cambridge

New York Port Chester Melbourne Sydney

Published by the Press Syndicate of the University of Cambridge
The Pitt Building, Trumpington Street, Cambridge CB2 1RP
40 West 20th Street, New York, NY 10011-4211, USA
10 Stamford Road, Oakleigh, Victoria 3166, Australia

First published 1980
Second edition 1989
Reprinted 1990 (twice), 1991

Printed in the United States of America

Library of Congress Cataloging-in-Publication Data

Horowitz, Paul, 1942–
The art of electronics / Paul Horowitz, Winfield Hill. – 2nd ed.
p. cm.
Bibliography: p.
Includes index.
ISBN 0-521-37095-7
1. Electronics. 2. Electronic circuit design. I. Hill,
Winfield. II. Title.
TK78155.H67 1989
621.381 – dc19 89-468
CIP

British Library Cataloguing in Publication Data

Horowitz, Paul, 1942–
The art of electronics. – 2nd ed.
1. Electronic equipment
I. Title. II. Hill, Winfield
621.381

ISBN 0-521-37095-7 hardback

TO CAROL, JACOB, MISHA, AND GINGER

CONTENTS

TABLES

PREFACE

Electronics, perhaps more than any other field of technology, has enjoyed an explosive development in the last four decades. Thus it was with some trepidation that we attempted, in 1980, to bring out a definitive volume teaching the art of the subject. By "art" we meant the kind of mastery that comes from an intimate familiarity with real circuits, actual devices, and the like, rather than the more abstract approach often favored in textbooks on electronics. Of course, in a rapidly evolving field, such a nuts-and-bolts approach has its hazards – most notably a frighteningly quick obsolescence.

The pace of electronics technology did not disappoint us! Hardly was the ink dry on the first edition before we felt foolish reading our words about "the classic [2Kbyte] 2716 EPROM ... with a price tag of about $25." They're so classic you can't even get them anymore, having been replaced by EPROMs 64 times as large, and costing less than half the price! Thus a major element of this revision responds to improved devices and methods – completely rewritten chapters on microcomputers and microprocessors (using the IBM PC and the 68008) and substantially revised chapters on digital electronics (including PLDs, and the new HC and AC logic families), on op-amps and precision design (reflecting the availability of excellent FET-input op-amps), and on construction techniques (including CAD/CAM). Every table has been revised, some substantially; for example, in Table 4.1 (operational amplifiers) only 65% of the original 120 entries survived, with 135 new op-amps added.

We have used this opportunity to respond to readers' suggestions and to our own experiences using and teaching from the first edition. Thus we have rewritten the chapter on FETs (it was too complicated) and repositioned it before the chapter on op-amps (which are increasingly of FET construction). We have added a new chapter on low-power and micropower design (both analog and digital), a field both important and neglected. Most of the remaining chapters have been extensively revised. We have added many new tables, including A/D and D/A converters, digital logic components, and low-power devices, and throughout the book we have expanded the number of figures. The book now contains 78 tables (available separately as *The Horowitz and Hill Component Selection Tables*) and over 1000 figures.

Throughout the revision we have strived to retain the feeling of informality and easy access that made the first edition so successful and popular, both as reference and text. We are aware of the difficulty students often experience when approaching electronics for the first time: The field is densely interwoven, and there is no path of learning that takes you, by logical steps, from neophyte to broadly competent designer. Thus we have added extensive cross-referencing throughout the text; in addition, we have expanded the separate *Laboratory Manual* into a *Student Manual* (*Student Manual for The Art of Electronics*, by Thomas C. Hayes and Paul Horowitz), complete with additional worked examples of circuit designs, explanatory material, reading assignments, laboratory exercises, and solutions to selected problems. By offering a student supplement, we have been able to keep this volume concise and rich with detail, as requested by our many readers who use the volume primarily as a reference work.

We hope this new edition responds to all our readers' needs – both students and practicing engineers. We welcome suggestions and corrections, which should be addressed directly to Paul Horowitz, Physics Department, Harvard University, Cambridge, MA 02138.

In preparing this new edition, we are appreciative of the help we received from Mike Aronson and Brian Matthews (AOX, Inc.), John Greene (University of Cape Town), Jeremy Avigad and Tom Hayes (Harvard University), Peter Horowitz (EVI, Inc.), Don Stern, and Owen Walker. We thank Jim Mobley for his excellent copyediting, Sophia Prybylski and David Tranah of Cambridge University Press for their encouragement and professional dedication, and the never-sleeping typesetters at Rosenlaui Publishing Services, Inc. for their masterful composition in TEX.

Finally, in the spirit of modern jurisprudence, we remind you to read the legal notice here appended.

Paul Horowitz
Winfield Hill

March 1989

LEGAL NOTICE

PREFACE TO FIRST EDITION

This volume is intended as an electronic circuit design textbook and reference book; it begins at a level suitable for those with no previous exposure to electronics and carries the reader through to a reasonable degree of proficiency in electronic circuit design. We have used a straightforward approach to the essential ideas of circuit design, coupled with an in-depth selection of topics. We have attempted to combine the pragmatic approach of the practicing physicist with the quantitative approach of the engineer, who wants a thoroughly evaluated circuit design.

This book evolved from a set of notes written to accompany a one-semester course in laboratory electronics at Harvard. That course has a varied enrollment – undergraduates picking up skills for their eventual work in science or industry, graduate students with a field of research clearly in mind, and advanced graduate students and postdoctoral researchers who suddenly find themselves hampered by their inability to "do electronics."

It soon became clear that existing textbooks were inadequate for such a course. Although there are excellent treatments of each electronics specialty, written for the planned sequence of a four-year engineering curriculum or for the practicing engineer, those books that attempt to address the whole field of electronics seem to suffer from excessive detail (the handbook syndrome), from oversimplification (the cookbook syndrome), or from poor balance of material. Much of the favorite pedagogy of beginning textbooks is quite unnecessary and, in fact, is not used by practicing engineers, while useful circuitry and methods of analysis in daily use by circuit designers lie hidden in application notes, engineering journals, and hard-to-get data books. In other words, there is a tendency among textbook writers to represent the theory, rather that the art, of electronics.

We collaborated in writing this book with the specific intention of combining the discipline of a circuit design engineer with the perspective of a practicing experimental physicist and teacher of electronics. Thus, the treatment in this book reflects our philosophy that electronics, as currently practiced, is basically a simple art, a combination of some basic laws, rules of thumb, and a large bag of tricks. For these reasons we have omitted entirely the usual discussions of solid-state physics, the h-parameter model of transistors, and complicated network theory, and reduced to a bare minimum the mention of load lines and the s-plane. The treatment is largely nonmathematical, with strong encouragement of circuit brainstorming and mental (or, at most, back-of-the-envelope) calculation of circuit values and performance.

In addition to the subjects usually treated in electronics books, we have included the following:

- an easy-to-use transistor model
- extensive discussion of useful subcircuits, such as current sources and current mirrors
- single-supply op-amp design
- easy-to-understand discussions of topics on which practical design information is often difficult to find: op-amp frequency

compensation, low-noise circuits, phase-locked loops, and precision linear design
- simplified design of active filters, with tables and graphs
- a section on noise, shielding, and grounding
- a unique graphical method for streamlined low-noise amplifier analysis
- a chapter on voltage references and regulators, including constant current supplies
- a discussion of monostable multivibrators and their idiosyncrasies
- a collection of digital logic pathology, and what to do about it
- an extensive discussion of interfacing to logic, with emphasis on the new NMOS and PMOS LSI
- a detailed discussion of A/D and D/A conversion techniques
- a section on digital noise generation
- a discussion of minicomputers and interfacing to data buses, with an introduction to assembly language
- a chapter on microprocessors, with actual design examples and discussion – how to design them into instruments, and how to make them do what you want
- a chapter on construction techniques: prototyping, printed circuit boards, instrument design
- a simplified way to evaluate high-speed switching circuits
- a chapter on scientific measurement and data processing: what you can measure and how accurately, and what to do with the data
- bandwidth narrowing methods made clear: signal averaging, multichannel scaling, lock-in amplifiers, and pulse-height analysis
- amusing collections of "bad circuits," and collections of "circuit ideas"
- useful appendixes on how to draw schematic diagrams, IC generic types, *LC* filter design, resistor values, oscilloscopes, mathematics review, and others
- tables of diodes, transistors, FETs, op-amps, comparators, regulators, voltage references, microprocessors, and other devices, generally listing the characteristics of both the most popular and the best types

Throughout we have adopted a philosophy of naming names, often comparing the characteristics of competing devices for use in any circuit, and the advantages of alternative circuit configurations. Example circuits are drawn with real device types, not black boxes. The overall intent is to bring the reader to the point of understanding clearly the choices one makes in designing a circuit – how to choose circuit configurations, device types, and parts values. The use of largely nonmathematical circuit design techniques does not result in circuits that cut corners or compromise performance or reliability. On the contrary, such techniques enhance one's understanding of the real choices and compromises faced in engineering a circuit and represent the best approach to good circuit design.

This book can be used for a full-year electronic circuit design course at the college level, with only a minimum mathematical prerequisite, namely, some acquaintance with trigonometric and exponential functions, and preferably a bit of differential calculus. (A short review of complex numbers and derivatives is included as an appendix.) If the less essential sections are omitted, it can serve as the text for a one-semester course (as it does at Harvard).

A separately available laboratory manual, *Laboratory Manual for the Art of Electronics* (Horowitz and Robinson, 1981), contains twenty-three lab exercises, together with reading and problem assignments keyed to the text.

To assist the reader in navigation, we have designated with open boxes in the margin those sections within each chapter that we feel can be safely passed over in an abbreviated reading. For a one-semester course it would probably be wise to omit, in addition, the materials of Chapter 5 (first half), 7, 12, 13, 14, and possibly 15, as explained in the introductory paragraphs of those chapters.

We would like to thank our colleagues for their thoughtful comments and assistance in the preparation of the manuscript, particularly Mike Aronson, Howard Berg, Dennis Crouse, Carol Davis, David Griesinger, John Hagen, Tom Hayes, Peter Horowitz, Bob Kline, Costas Papaliolios, Jay Sage, and Bill Vetterling. We are indebted to Eric Hieber and Jim Mobley, and to Rhona Johnson and Ken Werner of Cambridge University Press, for their imaginative and highly professional work.

Paul Horowitz
Winfield Hill

April 1980

INTRODUCTION

Developments in the field of electronics have constituted one of the great success stories of this century. Beginning with crude spark-gap transmitters and "cat's-whisker" detectors at the turn of the century, we have passed through a vacuum-tube era of considerable sophistication to a solid-state era in which the flood of stunning advances shows no signs of abating. Calculators, computers, and even talking machines with vocabularies of several hundred words are routinely manufactured on single chips of silicon as part of the technology of large-scale integration (LSI), and current developments in very large scale integration (VLSI) promise even more remarkable devices.

Perhaps as noteworthy is the pleasant trend toward increased performance per dollar. The cost of an electronic microcircuit routinely decreases to a fraction of its initial cost as the manufacturing process is perfected (see Fig. 8.87 for an example). In fact, it is often the case that the panel controls and cabinet hardware of an instrument cost more than the electronics inside.

On reading of these exciting new developments in electronics, you may get the impression that you should be able to construct powerful, elegant, yet inexpensive, little gadgets to do almost any conceivable task – all you need to know is how all these miracle devices work. If you've had that feeling, this book is for you. In it we have attempted to convey the excitement and know-how of the subject of electronics.

In this chapter we begin the study of the laws, rules of thumb, and tricks that constitute the art of electronics as we see it. It is necessary to begin at the beginning – with talk of voltage, current, power, and the components that make up electronic circuits. Because you can't touch, see, smell, or hear electricity, there will be a certain amount of abstraction (particularly in the first chapter), as well as some dependence on such visualizing instruments as oscilloscopes and voltmeters. In many ways the first chapter is also the most mathematical, in spite of our efforts to keep mathematics to a minimum in order to foster a good intuitive understanding of circuit design and behavior.

Once we have considered the foundations of electronics, we will quickly get into the "active" circuits (amplifiers, oscillators, logic circuits, etc.) that make electronics the exciting field it is. The reader with some background in electronics may wish to skip over this chapter, since it assumes no prior knowledge of electronics. Further generalizations at this time would be pointless, so let's just dive right in.

VOLTAGE, CURRENT, AND RESISTANCE

1.01 Voltage and current

There are two quantities that we like to keep track of in electronic circuits: voltage and current. These are usually changing with time; otherwise nothing interesting is happening.

Voltage (symbol: V, or sometimes E). The voltage between two points is the cost in energy (work done) required to move a unit of positive charge from the more negative point (lower potential) to the more positive point (higher potential). Equivalently, it is the energy released when a unit charge moves "downhill" from the higher potential to the lower. Voltage is also called *potential difference* or *electromotive force* (EMF). The unit of measure is the *volt*, with voltages usually expressed in volts (V), kilovolts ($1\text{kV} = 10^3\text{V}$), millivolts ($1\text{mV} = 10^{-3}\text{V}$), or microvolts ($1\mu\text{V} = 10^{-6}\text{V}$) (see the box on prefixes). A joule of work is needed to move a coulomb of charge through a potential difference of one volt. (The coulomb is the unit of electric charge, and it equals the charge of 6×10^{18} electrons, approximately.) For reasons that will become clear later, the opportunities to talk about nanovolts ($1\text{nV} = 10^{-9}\text{V}$) and megavolts ($1\text{MV} = 10^6\text{V}$) are rare.

Current (symbol: I). Current is the rate of flow of electric charge past a point. The unit of measure is the ampere, or amp, with currents usually expressed in amperes (A), milliamperes ($1\text{mA} = 10^{-3}\text{A}$), microamperes ($1\mu\text{A} = 10^{-6}\text{A}$), nanoamperes ($1\text{nA} = 10^{-9}\text{A}$), or occasionally picoamperes ($1\text{pA} = 10^{-12}\text{A}$). A current of one ampere equals a flow of one coulomb of charge per second. By convention, current in a circuit is considered to flow from a more positive point to a more negative point, even though the actual electron flow is in the opposite direction.

Important: Always refer to voltage *between* two points or *across* two points in a circuit. Always refer to current *through* a device or connection in a circuit.

To say something like "the voltage through a resistor . . . " is nonsense, or worse. However, we do frequently speak of the voltage *at a point* in a circuit. This is always understood to mean voltage between that point and "ground," a common point in the circuit that everyone seems to know about. Soon you will, too.

We *generate* voltages by doing work on charges in devices such as batteries (electrochemical), generators (magnetic forces), solar cells (photovoltaic conversion of the energy of photons), etc. We *get* currents by placing voltages across things.

At this point you may well wonder how to "see" voltages and currents. The single most useful electronic instrument is the oscilloscope, which allows you to look at voltages (or occasionally currents) in a circuit as a function of time. We will deal with oscilloscopes, and also voltmeters, when we discuss signals shortly; for a preview, see the oscilloscope appendix (Appendix A) and the multimeter box later in this chapter.

In real circuits we connect things together with wires, metallic conductors, each of which has the same voltage on it everywhere (with respect to ground, say). (In the domain of high frequencies or low impedances, that isn't strictly true, and we will have more to say about this later. For now, it's a good approximation.) We mention this now so that you will realize that an actual circuit doesn't have to look like its schematic diagram, because wires can be rearranged.

Here are some simple rules about voltage and current:

1. The sum of the currents into a point in a circuit equals the sum of the currents out (conservation of charge). This is sometimes called Kirchhoff's current law. Engineers like to refer to such a point as a *node*. From this, we get the following: For a series circuit (a bunch of two-terminal things all connected end-to-end) the current is the same everywhere.

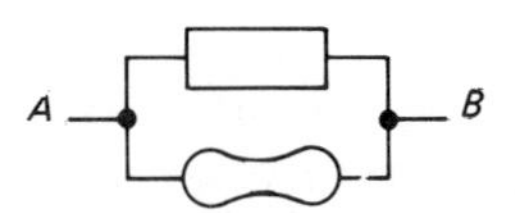

Figure 1.1

2. Things hooked in parallel (Fig. 1.1) have the same voltage across them. Restated, the sum of the "voltage drops" from A to B via one path through a circuit equals the sum by any other route equals the voltage between A and B. Sometimes this is stated as follows: The sum of the voltage drops around any closed circuit is zero. This is Kirchhoff's voltage law.

3. The power (work per unit time) consumed by a circuit device is

$$P = VI$$

This is simply (work/charge) × (charge/time). For V in volts and I in amps, P comes out in watts. Watts are joules per second (1W = 1J/s).

Power goes into heat (usually), or sometimes mechanical work (motors), radiated energy (lamps, transmitters), or stored energy (batteries, capacitors). Managing the heat load in a complicated system (e.g., a computer, in which many kilowatts of electrical energy are converted to heat, with the energetically insignificant by-product of a few pages of computational results) can be a crucial part of the system design.

PREFIXES

These prefixes are universally used to scale units in science and engineering.

Multiple	*Prefix*	*Symbol*
10^{12}	tera	T
10^{9}	giga	G
10^{6}	mega	M
10^{3}	kilo	k
10^{-3}	milli	m
10^{-6}	micro	μ
10^{-9}	nano	n
10^{-12}	pico	p
10^{-15}	femto	f

When abbreviating a unit with a prefix, the symbol for the unit follows the prefix without space. Be careful about upper-case and lower-case letters (especially m and M) in both prefix and unit: 1mW is a milliwatt, or one-thousandth of a watt; 1MHz is 1 million hertz. In general, units are spelled with lower-case letters, even when they are derived from proper names. The unit name is not capitalized when it is spelled out and used with a prefix, only when abbreviated. Thus: hertz and kilohertz, but Hz and kHz; watt, milliwatt, and megawatt, but W, mW, and MW.

Soon, when we deal with periodically varying voltages and currents, we will have to generalize the simple equation $P = VI$ to deal with *average* power, but it's correct as a statement of *instantaneous* power just as it stands.

Incidentally, don't call current "amperage"; that's strictly bush-league. The same caution will apply to the term "ohmage" when we get to resistance in the next section.

1.02 Relationship between voltage and current: resistors

This is a long and interesting story. It is the heart of electronics. Crudely speaking, the name of the game is to make and use gadgets that have interesting and useful I-versus-V characteristics. Resistors (I simply proportional to V), capacitors (I proportional to rate of change of V), diodes (I flows in only one direction), thermistors (temperature-dependent resistor), photoresistors (light-dependent resistor), strain gauges (strain-dependent resistor), etc., are examples. We will gradually get into some of these exotic devices; for now, we will start with the most mundane (and most widely used) circuit element, the resistor (Fig. 1.2).

Figure 1.2

Resistance and resistors

It is an interesting fact that the current through a metallic conductor (or other partially conducting material) is proportional to the voltage across it. (In the case of wire conductors used in circuits, we usually choose a thick enough gauge of wire so that these "voltage drops" will be negligible.) This is by no means a universal law for all objects. For instance, the current through a neon bulb is a highly nonlinear function of the applied voltage (it is zero up to a critical voltage, at which point it rises dramatically). The same goes for a variety of interesting special devices – diodes, transistors, light bulbs, etc. (If you are interested in understanding why metallic conductors behave this way, read sections 4.4–4.5 in the *Berkeley Physics Course*, Vol. II, see Bibliography). A resistor is made out of some conducting stuff (carbon, or a thin metal or carbon film, or wire of poor conductivity), with a wire coming out each end. It is characterized by its resistance:

$$R = V/I$$

R is in ohms for V in volts and I in amps. This is known as Ohm's law. Typical resistors of the most frequently used type (carbon composition) come in values from 1 ohm (1Ω) to about 22 megohms ($22\text{M}\Omega$). Resistors are also characterized by how

RESISTORS

Resistors are truly ubiquitous. There are almost as many types as there are applications. Resistors are used in amplifiers as loads for active devices, in bias networks, and as feedback elements. In combination with capacitors they establish time constants and act as filters. They are used to set operating currents and signal levels. Resistors are used in power circuits to reduce voltages by dissipating power, to measure currents, and to discharge capacitors after power is removed. They are used in precision circuits to establish currents, to provide accurate voltage ratios, and to set precise gain values. In logic circuits they act as bus and line terminators and as "pull-up" and "pull-down" resistors. In high-voltage circuits they are used to measure voltages and to equalize leakage currents among diodes or capacitors connected in series. In radiofrequency circuits they are even used as coil forms for inductors.

Resistors are available with resistances from 0.01 ohm through 10^{12} ohms, standard power ratings from 1/8 watt through 250 watts, and accuracies from 0.005% through 20%. Resistors can be made from carbon-composition moldings, from metal films, from wire wound on a form, or from semiconductor elements similar to field-effect transistors (FETs). But by far the most familiar resistor is the 1/4 or 1/2 watt carbon-composition resistor. These are available in a standard set of values ranging from 1 ohm to 100 megohms with twice as many values available for the 5% tolerance as for the 10% types (see Appendix C). We prefer the Allen-Bradley type AB (1/4 watt, 5%) resistor for general use because of its clear marking, secure lead seating, and stable properties.

Resistors are so easy to use that they're often taken for granted. They're not perfect, though, and it is worthwhile to look at some of their defects. The popular 5% composition type, in particular, although fine for nearly all noncritical circuit applications, is not stable enough for precision applications. You should know about its limitations so that you won't be surprised someday. Its principal defects are variations in resistance with temperature, voltage, time, and humidity. Other defects may relate to inductance (which may be serious at high frequencies), the development of thermal hot spots in power applications, or electrical noise generation in low-noise amplifiers. The following specifications are worst-case values; typically you'll do better, but don't count on it!

SPECIFICATIONS FOR ALLEN-BRADLEY AB SERIES TYPE CB

Standard tolerance is ±5% under nominal conditions. Maximum power for 70°C ambient temperature is 0.25 watt, which will raise the internal temperature to 150°C. The maximum applied voltage specification is $(0.25R)^{1/2}$ or 250 volts, whichever is less. They mean it! (See Fig. 6.53.) A single 5 second overvoltage to 400 volts can cause a permanent change in resistance by 2%.

	Resistance change		*Permanent?*
	(R = 1k)	(R = 10M)	
Soldering (350°C at 1/8 inch)	±2%	±2%	yes
Load cycling (500 ON/OFF cycles in 1000 hours)	+4%–6%	+4%–6%	yes
Vibration (20*g*) and shock (100*g*)	±2%	±2%	yes
Humidity (95% relative humidity at 40°C)	+6%	+10%	no
Voltage coefficient (10V change)	−0.15%	−0.3%	no
Temperature (25°C to −15°C)	+2.5%	+4.5%	no
Temperature (25°C to 85°C)	+3.3%	+5.9%	no

For applications that require any real accuracy or stability a 1% metal-film resistor (see Appendix D) should be used. They can be expected to have stability of better than 0.1% under normal conditions and better than 1% under worst-case treatment. Precision wire-wound resistors are available for the most demanding applications. For power dissipation above about 0.1 watt, a resistor of higher power rating should be used. Carbon-composition resistors are available with ratings up to 2 watts, and wire-wound power resistors are available for higher power. For demanding power applications, the conduction-cooled type of power resistor delivers better performance. These carefully designed resistors are available at 1% tolerance and can be operated at core temperatures up to 250°C with dependable long life. Allowable resistor power dissipation depends on air flow, thermal conduction via the resistor leads, and circuit density; thus, a resistor's power rating should be considered a rough guideline. Note also that resistor power ratings refer to *average* power dissipation and may be substantially exceeded for short periods of time (a few seconds or more, depending on the resistor's "thermal mass").

much power they can safely dissipate (the most commonly used ones are rated at 1/4 watt) and by other parameters such as tolerance (accuracy), temperature coefficient, noise, voltage coefficient (the extent to which R depends on applied V), stability with time, inductance, etc. See the box on resistors and Appendixes C and D for further details.

Roughly speaking, resistors are used to convert a voltage to a current, and vice versa. This may sound awfully trite, but you will soon see what we mean.

Resistors in series and parallel

From the definition of R, some simple results follow:

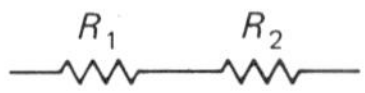

Figure 1.3

1. The resistance of two resistors in series (Fig. 1.3) is

$$R = R_1 + R_2$$

By putting resistors in series, you always get a *larger* resistor.

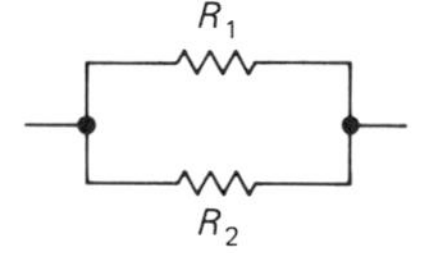

Figure 1.4

2. The resistance of two resistors in parallel (Fig. 1.4) is

$$R = \frac{R_1 R_2}{R_1 + R_2} \quad \text{or} \quad R = \frac{1}{\frac{1}{R_1} + \frac{1}{R_2}}$$

By putting resistors in parallel, you always get a *smaller* resistor. Resistance is measured in ohms (Ω), but in practice we frequently omit the Ω symbol when referring to resistors that are more than 1000Ω ($1k\Omega$). Thus, a $10k\Omega$ resistor is often referred to as a 10k resistor, and a $1M\Omega$ resistor as a 1M resistor (or 1 meg). On schematic diagrams the symbol Ω is often omitted altogether. If this bores you, please have patience – we'll soon get to numerous amusing applications.

EXERCISE 1.1

You have a 5k resistor and a 10k resistor. What is their combined resistance (a) in series and (b) in parallel?

EXERCISE 1.2

If you place a 1 ohm resistor across a 12 volt car battery, how much power will it dissipate?

EXERCISE 1.3

Prove the formulas for series and parallel resistors.

EXERCISE 1.4

Show that several resistors in parallel have resistance

$$R = \frac{1}{\frac{1}{R_1} + \frac{1}{R_2} + \frac{1}{R_3} + \cdots}$$

A trick for parallel resistors: Beginners tend to get carried away with complicated algebra in designing or trying to understand electronics. Now is the time to begin learning intuition and shortcuts.

Shortcut no. 1 A large resistor in series (parallel) with a small resistor has the resistance of the larger (smaller) one, roughly.

Shortcut no. 2 Suppose you want the resistance of 5k in parallel with 10k. If you think of the 5k as two 10k's in parallel, then the whole circuit is like three 10k's in parallel. Because the resistance of n equal resistors in parallel is $1/n$th the resistance of the individual resistors, the answer in this case is 10k/3, or 3.33k. This trick is handy because it allows you to analyze circuits quickly in your head, without distractions. We want to encourage mental designing, or at least "back of the envelope" designing, for idea brainstorming.

Some more home-grown philosophy: There is a tendency among beginners to want to compute resistor values and other circuit component values to many significant places, and the availability of inexpensive calculators has only made matters worse. There are two reasons you should try to avoid falling into this habit: (a) the components themselves are of finite precision (typical resistors are ±5%; the parameters that characterize transistors, say, frequently are known only to a factor of two); (b) one mark of a good circuit design is insensitivity of the finished circuit to precise values of the components (there are exceptions, of course). You'll also learn circuit intuition more quickly if you get into the habit of doing approximate calculations in your head, rather than watching meaningless numbers pop up on a calculator display.

In trying to develop intuition about resistance, some people find it helpful to think about *conductance*, $G = 1/R$. The current through a device of conductance G bridging a voltage V is then given by $I = GV$ (Ohm's law). A small resistance is a large conductance, with correspondingly large current under the influence of an applied voltage.

Viewed in this light, the formula for parallel resistors is obvious: When several resistors or conducting paths are connected across the same voltage, the total current is the sum of the individual currents. Therefore the net conductance is simply the sum of the individual conductances, $G = G_1 + G_2 + G_3 + \cdots$, which is the same as the formula for parallel resistors derived earlier.

Engineers are fond of defining reciprocal units, and they have designated the unit of conductance the siemens ($S = 1/\Omega$), also known as the mho (that's ohm spelled backward, given the symbol ℧). Although the concept of conductance is helpful in developing intuition, it is not used widely; most people prefer to talk about resistance instead.

Power in resistors

The power dissipated by a resistor (or any other device) is $P = IV$. Using Ohm's law, you can get the equivalent forms $P = I^2R$ and $P = V^2/R$.

EXERCISE 1.5

Show that it is not possible to exceed the power rating of a 1/4 watt resistor of resistance greater than 1k, no matter how you connect it, in a circuit operating from a 15 volt battery.

EXERCISE 1.6

Optional exercise: New York City requires about 10^{10} watts of electrical power, at 110 volts (this is plausible: 10 million people averaging 1 kilowatt each). A heavy power cable might be an inch in diameter. Let's calculate what will happen if we try to supply the power through a cable 1 foot in diameter made of pure copper. Its resistance is 0.05$\mu\Omega$ (5×10^{-8} ohms) per foot. Calculate (a) the power lost per foot from "I^2R losses," (b) the length of cable over which you will lose all 10^{10} watts, and (c) how hot the cable will get, if you know the physics involved ($\sigma = 6 \times 10^{-12}$W/°K^{4}cm^2).

If you have done your computations correctly, the result should seem preposterous. What is the solution to this puzzle?

Input and output

Nearly all electronic circuits accept some sort of applied *input* (usually a voltage) and produce some sort of corresponding *output* (which again is often a voltage). For example, an audio amplifier might produce a (varying) output voltage that is 100 times as large as a (similarly varying) input voltage. When describing such an amplifier, we imagine measuring the output voltage for a given applied input voltage. Engineers speak of the *transfer function* **H**, the ratio of (measured) output divided by (applied) input; for the audio amplifier above, **H** is simply a constant (**H** = 100). We'll get to amplifiers soon enough, in the next chapter. However, with just resistors we can already look at a very important circuit fragment, the *voltage divider* (which you might call a "de-amplifier").

1.03 Voltage dividers

We now come to the subject of the voltage divider, one of the most widespread electronic circuit fragments. Show us any real-life circuit and we'll show you half a dozen voltage dividers. To put it very simply, a voltage divider is a circuit that, given a certain voltage input, produces a predictable fraction of the input voltage as the output voltage. The simplest voltage divider is shown in Figure 1.5.

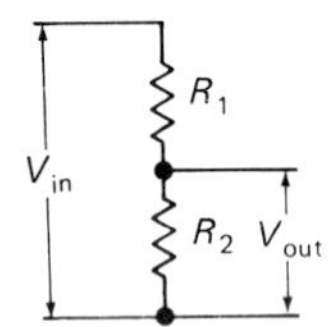

Figure 1.5. Voltage divider. An applied voltage V_{in} results in a (smaller) output voltage V_{out}.

What is V_{out}? Well, the current (same everywhere, assuming no "load" on the output) is

$$I = \frac{V_{\text{in}}}{R_1 + R_2}$$

(We've used the definition of resistance and the series law.) Then, for R_2,

$$V_{\text{out}} = IR_2 = \frac{R_2}{R_1 + R_2} V_{\text{in}}$$

Note that the output voltage is always less than (or equal to) the input voltage; that's why it's called a divider. You could get amplification (more output than input) if one of the resistances were negative. This isn't as crazy as it sounds; it is possible to make devices with negative "incremental" resistances (e.g., the tunnel diode) or even true negative resistances (e.g., the negative-impedance converter that we will talk about later in the book). However, these applications are rather specialized and need not concern you now.

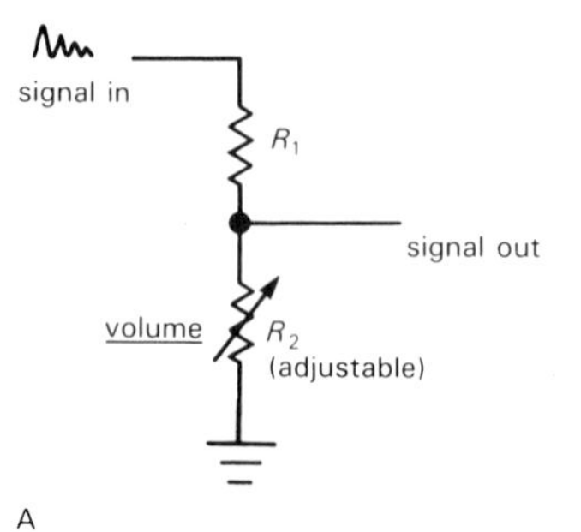

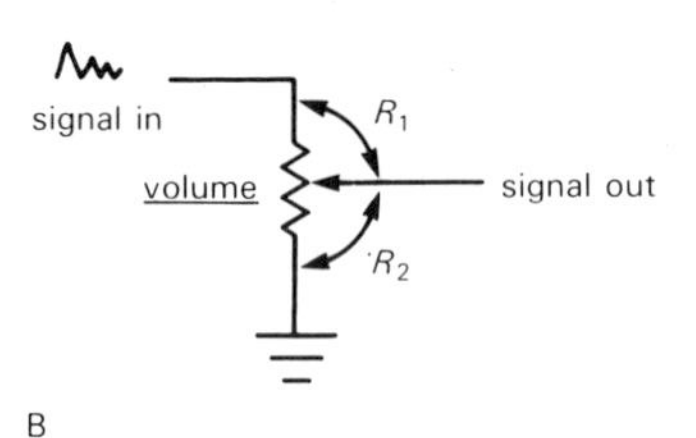

Figure 1.6. An adjustable voltage divider can be made from a fixed and variable resistor, or from a potentiometer.

Voltage dividers are often used in circuits to generate a particular voltage from a larger fixed (or varying) voltage. For instance, if V_{in} is a varying voltage and R_2 is an adjustable resistor (Fig. 1.6A), you have a "volume control"; more simply, the combination R_1R_2 can be made from a single variable resistor, or *potentiometer* (Fig. 1.6B). The humble voltage divider is even more useful, though, as a way of *thinking* about a circuit: the input voltage and upper resistance might represent the output of an amplifier, say, and the lower resistance might represent the input of the following stage. In this case the voltage-divider equation tells you how much signal gets to the input of that last stage. This will all become clearer after you know about a remarkable fact (Thévenin's theorem) that will be discussed later. First, though, a short aside on voltage sources and current sources.

1.04 Voltage and current sources

A perfect voltage source is a two-terminal *black box* that maintains a fixed voltage drop across its terminals, regardless of load resistance. For instance, this means that it must supply a current $I = V/R$ when a resistance R is attached to its terminals. A real voltage source can supply only a finite maximum current, and in addition it generally behaves like a perfect voltage source with a small resistance in series. Obviously, the smaller this series resistance, the better. For example, a standard 9 volt alkaline battery behaves like a perfect 9 volt voltage source in series with a 3 ohm resistor and can provide a maximum current (when shorted) of 3 amps (which, however, will kill the battery in a few minutes). A voltage source "likes" an open-circuit load and "hates" a short-circuit load, for obvious reasons. (The terms "open circuit" and "short circuit" mean the obvious: An open circuit has nothing connected to it, whereas a short circuit is a piece of wire bridging the output.) The symbols used to indicate a voltage source are shown in Figure 1.7.

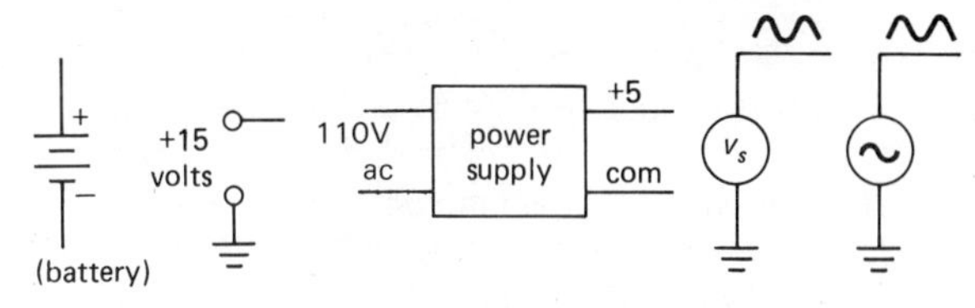

Figure 1.7. Voltage sources can be either steady (dc) or varying (ac).

A perfect current source is a two-terminal black box that maintains a constant current through the external circuit, regardless of load resistance or applied voltage. In order to do this it must be capable of supplying any necessary voltage across its terminals. Real current sources (a much-neglected subject in most textbooks) have a limit to the voltage they can provide (called the *output voltage compliance,* or just *compliance*), and in addition they do not provide absolutely constant output current. A current source "likes" a short-circuit load and "hates" an open-circuit load. The symbols used to indicate a current source are shown in Figure 1.8.

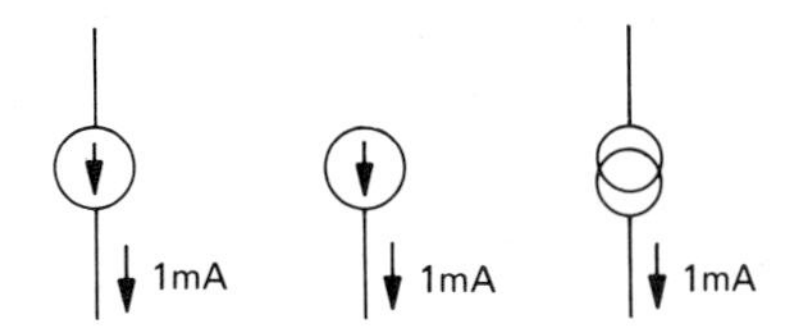

Figure 1.8. Current-source symbols.

A battery is a real-life approximation of a voltage source (there is no analog for a current source). A standard D-size flashlight cell, for instance, has a terminal voltage of 1.5 volts, an equivalent series resistance of about 1/4 ohm, and total energy capacity of about 10,000 watt-seconds (its characteristics gradually deteriorate with use; at the end of its life, the voltage may be about 1.0 volt, with an internal series resistance of several ohms). It is easy to construct voltage sources with far better characteristics, as you will learn when we come to the subject of feedback. Except in devices intended for portability, the use of batteries in electronic devices is rare. We will treat the interesting subject of low-power (battery-operated) design in Chapter 14.

MULTIMETERS

There are numerous instruments that let you measure voltages and currents in a circuit. The oscilloscope (see Appendix A) is the most versatile; it lets you "see" voltages versus time at one or more points in a circuit. Logic probes and logic analyzers are special-purpose instruments for troubleshooting digital circuits. The simple multimeter provides a good way to measure voltage, current, and resistance, often with good precision; however, it responds slowly, and thus it cannot replace the oscilloscope where changing voltages are of interest. Multimeters are of two varieties: those that indicate measurements on a conventional scale with a moving pointer, and those that use a digital display.

The standard VOM (volt-ohm-milliammeter) multimeter uses a meter movement that measures current (typically 50μA full scale). (See a less-design-oriented electronics book for pretty pictures of the innards of meter movements; for our purposes, it suffices to say that it uses coils and magnets.) To measure voltage, the VOM puts a resistor in series with the basic movement. For instance, one kind of VOM will generate a 1 volt (full-scale) range by putting a 20k resistor in series with the standard 50μA movement; higher voltage ranges use correspondingly larger resistors. Such a VOM is specified as 20,000 ohms/volt, meaning that it looks like a resistor whose value is 20k multiplied by the full-scale voltage of the particular range selected. Full scale on any voltage range is 1/20,000, or 50μA. It should be clear that one of these voltmeters disturbs a circuit less on a higher range, since it looks like a higher resistance (think of the voltmeter as the lower leg of a voltage divider, with the Thévenin resistance of the circuit you are measuring as the upper resistor). Ideally, a voltmeter should have infinite input resistance.

Nowadays there are various meters with some electronic amplification whose input resistance may be as large as 10^9 ohms. Most digital meters, and even a number of analog-reading meters that use FETs (field-effect transistors, see Chapter 3), are of this type. Warning: Sometimes the input resistance of FET-input meters is very high on the most sensitive ranges, dropping to a lower resistance for the higher ranges. For instance, an input resistance of 10^9 ohms on the 0.2 volt and 2 volt ranges, and 10^7 ohms on all higher ranges, is typical. Read the specifications carefully! For measurements on most transistor circuits, 20,000 ohms/volt is fine, and there will be little loading effect on the circuit by the meter. In any case, it is easy to calculate how serious the effect is by using the voltage-divider equation. Typically, multimeters provide voltage ranges from a volt (or less) to a kilovolt (or more), full scale.

A VOM can be used to measure current by simply using the bare meter movement (for our preceding example, this would give a range of 50μA full scale) or by shunting (paralleling) the movement with a small resistor. Because the meter movement itself requires a small voltage drop, typically 0.25 volt, to produce a full-scale deflection, the shunt is chosen by the meter manufacturer (all you do is set the range switch to the range you want) so that the full-scale current will produce that voltage drop through the parallel combination of the meter resistance and the shunt resistance. Ideally, a current-measuring meter should have zero resistance in order not to disturb the circuit under test, since it must be put in series with the circuit. In practice, you tolerate a few tenths of a volt drop (sometimes called "voltage burden") with both VOMs and digital multimeters. Typically, multimeters provide current ranges from 50μA (or less) to an amp (or more), full scale.

Multimeters also have one or more batteries in them to power the resistance measurement. By supplying a small current and measuring the voltage drop, they measure resistance, with several ranges to cover values from an ohm (or less) to 10 megohms (or more).

Important: Don't try to measure "the current of a voltage source," for instance by sticking the meter across the wall plug; the same applies for ohms. This is the leading cause of blown-out meters.

EXERCISE 1.7

What will a 20,000 ohms/volt meter read, on its 1 volt scale, when attached to a 1 volt source with an internal resistance of 10k? What will it read when attached to a 10k–10k voltage divider driven by a "stiff" (zero source resistance) 1 volt source?

EXERCISE 1.8

A 50μA meter movement has an internal resistance of 5k. What shunt resistance is needed to convert it to a 0–1 amp meter? What series resistance will convert it to a 0–10 volt meter?

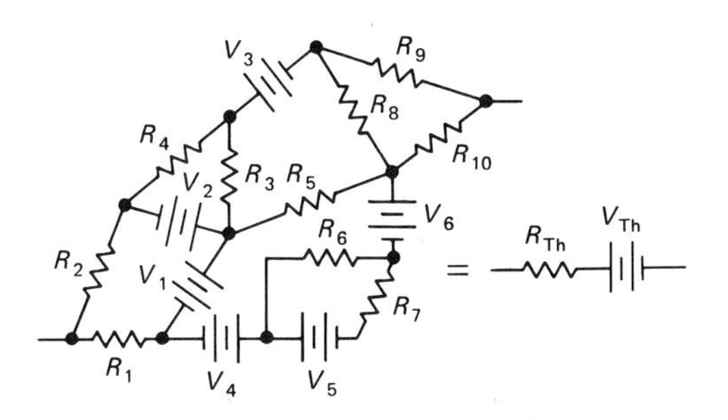

Figure 1.9

1.05 Thévenin's equivalent circuit

Thévenin's theorem states that any two-terminal network of resistors and voltage sources is equivalent to a single resistor R in series with a single voltage source V. This is remarkable. Any mess of batteries and resistors can be mimicked with one battery and one resistor (Fig. 1.9). (Incidentally, there's another theorem, Norton's theorem, that says you can do the same thing with a current source in parallel with a resistor.)

How do you figure out the Thévenin equivalent R_{Th} and V_{Th} for a given circuit? Easy! V_{Th} is the open-circuit voltage of the Thévenin equivalent circuit; so if the two circuits behave identically, it must also be the open-circuit voltage of the given circuit (which you get by calculation, if you know what the circuit is, or by measurement, if you don't). Then you find R_{Th} by noting that the short-circuit current of the equivalent circuit is $V_{\mathrm{Th}}/R_{\mathrm{Th}}$. In other words,

$$V_{\mathrm{Th}} = V \text{ (open circuit)}$$

$$R_{\mathrm{Th}} = \frac{V \text{ (open circuit)}}{I \text{ (short circuit)}}$$

Let's apply this method to the voltage divider, which must have a Thévenin equivalent:

1. The open-circuit voltage is

$$V = V_{\mathrm{in}} \frac{R_2}{R_1 + R_2}$$

2. The short-circuit current is

$$V_{\mathrm{in}}/R_1$$

So the Thévenin equivalent circuit is a voltage source

$$V_{\mathrm{Th}} = V_{\mathrm{in}} \frac{R_2}{R_1 + R_2}$$

in series with a resistor

$$R_{\mathrm{Th}} = \frac{R_1 R_2}{R_1 + R_2}$$

(It is not a coincidence that this happens to be the parallel resistance of R_1 and R_2. The reason will become clear later.)

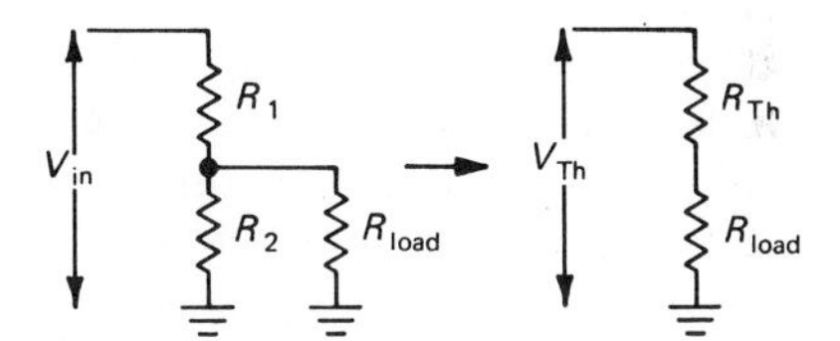

Figure 1.10

From this example it is easy to see that a voltage divider is not a very good battery, in the sense that its output voltage drops severely when a load is attached. As an example, consider Exercise 1.9. You now know everything you need to know to calculate exactly how much the output will drop for a given load resistance: Use the Thévenin equivalent circuit, attach a load, and calculate the new output, noting that the new circuit is nothing but a voltage divider (Fig. 1.10).

EXERCISE 1.9

For the circuit shown in Figure 1.10, with $V_{\mathrm{in}} = 30\mathrm{V}$ and $R_1 = R_2 = 10\mathrm{k}$, find (a) the output voltage with no load attached (the open-circuit voltage); (b) the output voltage with a 10k load (treat as voltage divider, with R_2 and R_{load} combined into a single resistor); (c) the Thévenin equivalent circuit; (d) the same as in part b, but using the Thévenin equivalent circuit (again, you wind up with a voltage divider; the answer should agree with the result in part b); (e) the power dissipated in each of the resistors.

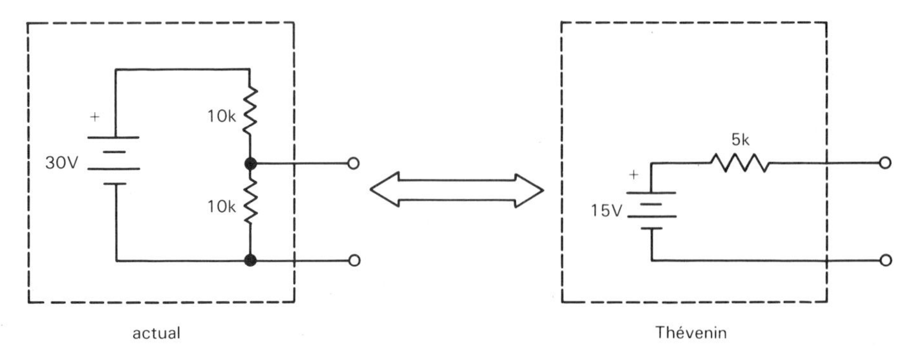

Figure 1.11

Equivalent source resistance and circuit loading

As you have just seen, a voltage divider powered from some fixed voltage is equivalent to some smaller voltage source in series with a resistor; for example, the output terminals of a 10k–10k voltage divider driven by a perfect 30 volt battery are precisely equivalent to a perfect 15 volt battery in series with a 5k resistor (Fig. 1.11). Attaching a load resistor causes the voltage divider's output to drop, owing to the finite *source resistance* (Thévenin equivalent resistance of the voltage divider output, viewed as a source of voltage). This is often undesirable. One solution to the problem of making a stiff voltage source ("stiff" is used in this context to describe something that doesn't bend under load) might be to use much smaller resistors in a voltage divider. Occasionally this brute-force approach is useful. However, it is usually best to construct a voltage source, or power supply, as it's commonly called, using active components like transistors or operational amplifiers, which we will treat in Chapters 2–4. In this way you can easily make a voltage source with internal (Thévenin equivalent) resistance measured in milliohms (thousandths of an ohm), without the large currents and dissipation of power characteristic of a low-resistance voltage divider delivering the same performance. In addition, with an active power supply it is easy to make the output voltage adjustable.

The concept of equivalent internal resistance applies to all sorts of sources, not just batteries and voltage dividers. Signal sources (e.g., oscillators, amplifiers, and sensing devices) all have an equivalent internal resistance. Attaching a load whose resistance is less than or even comparable to the internal resistance will reduce the output considerably. This undesirable reduction of the open-circuit voltage (or signal) by the load is called "circuit loading." Therefore, you should strive to make $R_{\text{load}} \gg R_{\text{internal}}$, because a high-resistance load has little attenuating effect on the source (Fig. 1.12). You will see numerous circuit examples in the chapters ahead. This high-resistance condition ideally

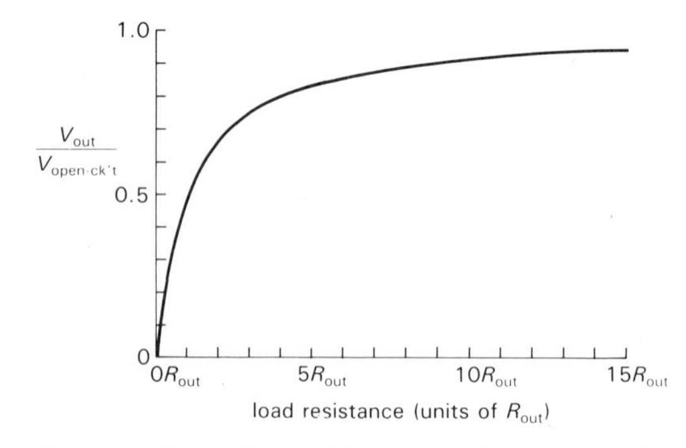

Figure 1.12. To avoid attenuating a signal source below its open-circuit voltage, keep the load resistance large compared with the output resistance.

characterizes measuring instruments such as voltmeters and oscilloscopes. (There are exceptions to this general principle; for example, we will talk about transmission lines and radiofrequency techniques, where you must "match impedances" in order to prevent the reflection and loss of power.)

A word on language: You frequently hear things like "the resistance looking into the voltage divider," or "the output sees a load of so-and-so many ohms," as if circuits had eyes. It's OK (in fact, it's a rather good way to keep straight which resistance you're talking about) to say what part of the circuit is doing the "looking."

Power transfer

Here is an interesting problem: What load resistance will result in maximum power being transferred to the load for a given source resistance? (The terms *source resistance, internal resistance*, and *Thévenin equivalent resistance* all mean the same thing.) It is easy to see that both $R_{\text{load}} = 0$ and $R_{\text{load}} = \infty$ result in zero power transferred, because $R_{\text{load}} = 0$ means that $V_{\text{load}} = 0$ and $I_{\text{load}} = V_{\text{source}}/R_{\text{source}}$, so that $P_{\text{load}} = V_{\text{load}}I_{\text{load}} = 0$. But $R_{\text{load}} = \infty$ means that $V_{\text{load}} = V_{\text{source}}$ and $I_{\text{load}} = 0$, so that $P_{\text{load}} = 0$. There has to be a maximum in between.

EXERCISE 1.10

Show that $R_{\text{load}} = R_{\text{source}}$ maximizes the power in the load for a given source resistance. Note: Skip this exercise if you don't know calculus, and take it on faith that the answer is true.

Lest this example leave the wrong impression, we would like to emphasize again that circuits are ordinarily designed so that the load resistance is much greater than the source resistance of the signal that drives the load.

1.06 Small-signal resistance

We often deal with electronic devices for which I is not proportional to V; in such cases there's not much point in talking about resistance, since the ratio V/I will depend on V, rather than being a nice constant, independent of V. For these devices it is useful to know the slope of the V–I curve, in other words, the ratio of a small change in applied voltage to the resulting change in current through the device, $\Delta V/\Delta I$ (or dV/dI). This quantity has the units of resistance (ohms) and substitutes for resistance in many calculations. It is called the small-signal resistance, incremental resistance, or dynamic resistance.

Zener diodes

As an example, consider the *zener diode*, which has the V–I curve shown in Figure 1.13. Zeners are used to create a constant voltage inside a circuit somewhere, simply by providing them with a (roughly constant) current derived from a higher voltage within the circuit. For example, the zener diode in Figure 1.13 will convert an applied current in the range shown to a corresponding (but narrower) range of voltages. It is important to know how the resulting zener voltage will change with applied current; this is a measure of its "regulation" against changes in the driving current provided to it. Included in the specifications of a zener will be its dynamic resistance, given at a certain current. (Useful fact: the dynamic resistance of a zener diode varies roughly in inverse proportion to current.) For example, a zener might have a dynamic resistance of 10 ohms at 10mA, at its zener voltage of 5 volts. Using the definition of dynamic resistance, we find that a 10% change in applied current will therefore result in a change in voltage of

$$\Delta V = R_{\text{dyn}}\Delta I = 10 \times 0.1 \times 0.01 = 10\text{mV}$$

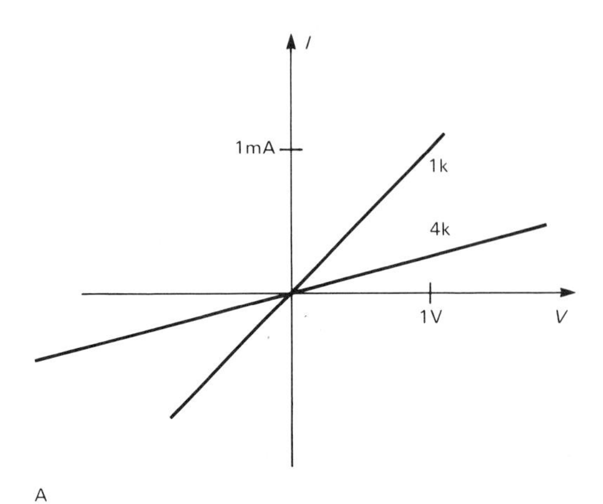

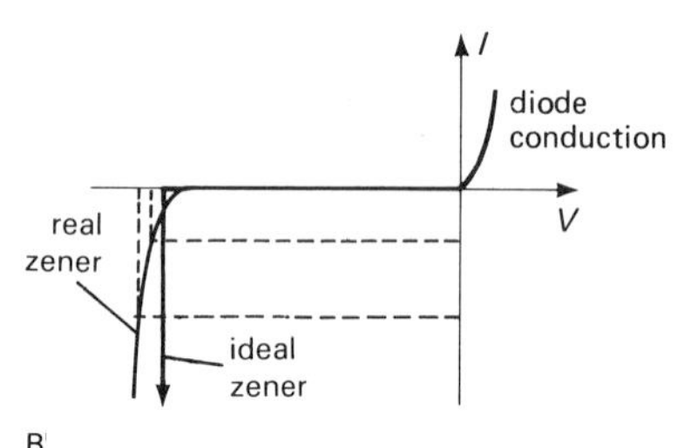

Figure 1.13. *V–I* curves.
A. Resistor (linear).
B. Zener diode (nonlinear).

or

$$\Delta V/V = 0.002 = 0.2\%$$

thus demonstrating good voltage-regulating ability. In this sort of application you frequently get the zener current through a resistor from a higher voltage available somewhere in the circuit, as in Figure 1.14.

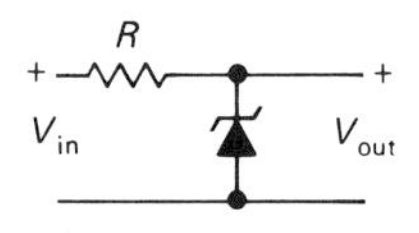

Figure 1.14. Zener regulator.

Then,

$$I = \frac{V_{\text{in}} - V_{\text{out}}}{R}$$

and

$$\Delta I = \frac{\Delta V_{\text{in}} - \Delta V_{\text{out}}}{R}$$

so

$$\Delta V_{\text{out}} = R_{\text{dyn}} \Delta I = \frac{R_{\text{dyn}}}{R}(\Delta V_{\text{in}} - \Delta V_{\text{out}})$$

and finally

$$\Delta V_{\text{out}} = \frac{R_{\text{dyn}}}{R + R_{\text{dyn}}} \Delta V_{\text{in}}$$

Thus, for *changes* in voltage, the circuit behaves like a voltage divider, with the zener replaced by a resistor equal to its dynamic resistance at the operating current. This is the utility of incremental resistance. For instance, suppose in the preceding circuit we have an input voltage ranging between 15 and 20 volts and use a 1N4733 (5.1V 1W zener diode) in order to generate a stable 5.1 volt power supply. We choose $R = 300$ ohms, for a maximum zener current of 50mA: (20 − 5.1)/300. We can now estimate the output voltage regulation (variation in output voltage), knowing that this particular zener has a specified maximum dynamic impedance of 7.0 ohms at 50mA. The zener current varies from 50mA to 33mA over the input voltage range; this 17mA change in current then produces a voltage change at the output of $\Delta V = R_{\text{dyn}} \Delta I$, or 0.12 volt. You will see more of zeners in Sections 2.04 and 6.14.

In real life, a zener will provide better regulation if driven by a current source, which has, by definition, $R_{\text{incr}} = \infty$ (same current regardless of voltage). But current sources are more complex, and therefore in practice we often resort to the humble resistor.

Tunnel diodes

Another interesting application of incremental resistance is the *tunnel diode*, sometimes called the Esaki diode. Its *V–I* curve is shown in Figure 1.15. In the region from *A* to *B* it has *negative* incremental resistance. This has a remarkable consequence: A voltage *divider* made with a resistor and

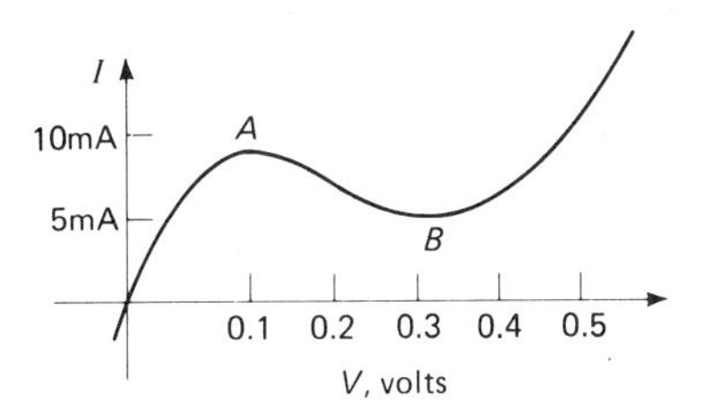

Figure 1.15

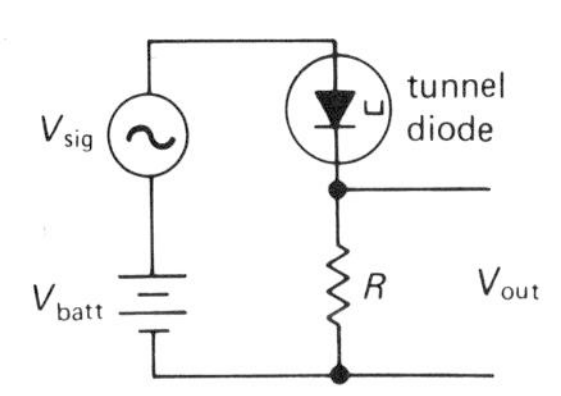

Figure 1.16

a tunnel diode can actually be an *amplifier* (Fig. 1.16). For a wiggly voltage v_{sig}, the voltage divider equation gives us

$$v_{out} = \frac{R}{R + r_t} v_{sig}$$

where r_t is the incremental resistance of the tunnel diode at the operating current, and the lower-case symbol v_{sig} stands for a small-signal variation, which we have been calling ΔV_{sig} up to now (we will adopt this widely used convention from now on). The tunnel diode has $r_{t(incr)} < 0$. That is,

$$\Delta V/\Delta I \quad (\text{or } v/i) \quad < \quad 0$$

from A to B on the characteristic curve. If $r_{t(incr)} \approx R$, the denominator is nearly zero, and the circuit amplifies. V_{batt} provides the steady current, or *bias*, to bring the operating point into the region of negative resistance. (Of course, it is always necessary to have a source of power in any device that amplifies.)

A postmortem on these fascinating devices: When tunnel diodes first appeared, late in the 1950s, they were hailed as the solution to a great variety of circuit problems. Because they were fast, they were supposed to revolutionize computers, for instance. Unfortunately, they are difficult devices to use; this fact, combined with stunning improvements in transistors, has made tunnel diodes almost obsolete.

The subject of negative resistance will come up again later, in connection with active filters. There you will see a circuit called a negative-impedance converter that can produce (among other things) a pure negative resistance (not just incremental). It is made with an operational amplifier and has very useful properties.

SIGNALS

A later section in this chapter will deal with capacitors, devices whose properties depend on the way the voltages and currents in a circuit are *changing*. Our analysis of dc circuits so far (Ohm's law, Thévenin equivalent circuits, etc.) still holds, even if the voltages and currents are changing in time. But for a proper understanding of alternating-current (ac) circuits, it is useful to have in mind certain common types of *signals*, voltages that change in time in a particular way.

1.07 Sinusoidal signals

Sinusoidal signals are the most popular signals around; they're what you get out of the wall plug. If someone says something like "take a 10 microvolt signal at 1 megahertz," he means a sine wave. Mathematically, what you have is a voltage described by

$$V = A \sin 2\pi f t$$

where A is called the amplitude, and f is the frequency in cycles per second, or hertz. A sine wave looks like the wave shown in Figure 1.17. Sometimes it is important to know the value of the signal at some arbitrary time $t = 0$, in which case you may see a *phase* ϕ in the expression:

$$V = A \sin(2\pi f t + \phi)$$

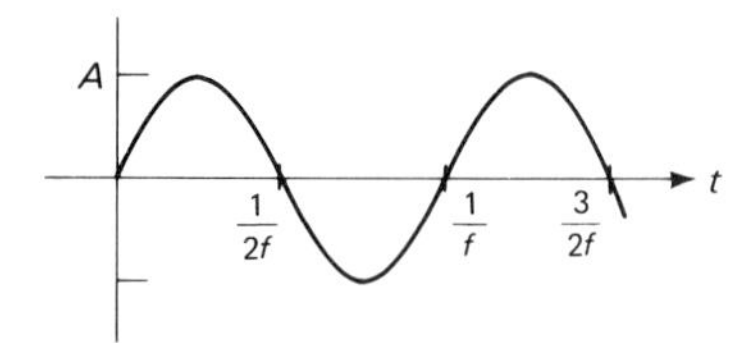

Figure 1.17. Sine wave of amplitude A and frequency f.

The other variation on this simple theme is the use of *angular frequency*, which looks like this:

$$V = A \sin \omega t$$

Here, ω is the angular frequency in radians per second. Just remember the important relation $\omega = 2\pi f$ and you won't go wrong.

The great merit of sine waves (and the cause of their perennial popularity) is the fact that they are the solutions to certain linear differential equations that happen to describe many phenomena in nature as well as the properties of linear circuits. A linear circuit has the property that its output, when driven by the sum of two input signals, equals the sum of its individual outputs when driven by each input signal in turn; i.e., if $O(A)$ represents the output when driven by signal A, then a circuit is linear if $O(A + B) = O(A) + O(B)$. A linear circuit driven by a sine wave always responds with a sine wave, although in general the phase and amplitude are changed. No other signal can make this statement. It is standard practice, in fact, to describe the behavior of a circuit by its *frequency response*, the way it alters the amplitude of an applied sine wave as a function of frequency. A high-fidelity amplifier, for instance, should be characterized by a "flat" frequency response over the range 20Hz to 20kHz, at least.

The sine-wave frequencies you will usually deal with range from a few hertz to a few megahertz. Lower frequencies, down to 0.0001Hz or lower, can be generated with carefully built circuits, if needed. Higher frequencies, e.g., up to 2000MHz, can be generated, but they require special transmission-line techniques. Above that, you're dealing with microwaves, where conventional wired circuits with lumped circuit elements become impractical, and exotic waveguides or "striplines" are used instead.

1.08 Signal amplitudes and decibels

In addition to its amplitude, there are several other ways to characterize the magnitude of a sine wave or any other signal. You sometimes see it specified by *peak-to-peak amplitude* (pp amplitude), which is just what you would guess, namely, twice the amplitude. The other method is to give the *root-mean-square amplitude* (rms amplitude), which is $V_{rms} = (1/\sqrt{2})A = 0.707A$ (this is for sine waves only; the ratio of pp to rms will be different for other waveforms). Odd as it may seem, this is the usual method, because rms voltage is what's used to compute power. The voltage across the terminals of a wall socket (in the United States) is 117 volts rms, 60Hz. The *amplitude* is 165 volts (330 volts pp).

Decibels

How do you compare the relative amplitudes of two signals? You could say, for instance, that signal X is twice as large as signal Y. That's fine, and useful for many purposes. But because we often deal with ratios as large as a million, it is easier to use a logarithmic measure, and for this we present the decibel (it's one-tenth as large as something called a bel, which no one ever uses). By definition, the ratio of two signals, in decibels, is

$$\text{dB} = 20 \log_{10} \frac{A_2}{A_1}$$

where A_1 and A_2 are the two signal amplitudes. So, for instance, one signal of twice the amplitude of another is +6dB relative to it, since $\log_{10}2 = 0.3010$. A signal 10 times as large is +20dB; a signal one-tenth as large is −20dB. It is also useful to express the ratio of two signals in terms of power levels:

$$dB = 10\log_{10}\frac{P_2}{P_1}$$

where P_1 and P_2 represent the power in the two signals. As long as the two signals have the same kind of waveform, e.g., sine waves, the two definitions give the same result. When comparing unlike waveforms, e.g., a sine wave versus "noise," the definition in terms of power (or the amplitude definition, with rms amplitudes substituted) must be used.

Although decibels are ordinarily used to specify the ratio of two signals, they are sometimes used as an absolute measure of amplitude. What is happening is that you are assuming some reference signal amplitude and expressing any other amplitude in decibels relative to it. There are several standard amplitudes (which are unstated, but understood) that are used in this way; the most common references are (a) dBV; 1 volt rms; (b) dBm; the voltage corresponding to 1mW into some assumed load impedance, which for radiofrequencies is usually 50 ohms, but for audio is often 600 ohms (the corresponding 0dBm amplitudes, when loaded by those impedances, are then 0.22V rms and 0.78V rms); and (c) the small noise voltage generated by a resistor at room temperature (this surprising fact is discussed in Section 7.11). In addition to these, there are reference amplitudes used for measurements in other fields. For instance, in acoustics, 0dB SPL is a wave whose rms pressure is 0.0002μbar (a bar is 10^6 dynes per square centimeter, approximately 1 atmosphere); in communications, levels can be stated in dBrnC (relative noise reference weighted in frequency by "curve C"). When stating amplitudes this way, it is best to be specific about the 0dB reference amplitude; say something like "an amplitude of 27 decibels relative to 1 volt rms," or abbreviate "27 dB re 1V rms," or define a term like "dBV."

EXERCISE 1.11

Determine the voltage and power ratios for a pair of signals with the following decibel ratios: (a) 3dB, (b) 6dB, (c) 10dB, (d) 20dB.

1.09 Other signals

The ramp is a signal that looks like the signal shown in Figure 1.18. It is simply a voltage rising (or falling) at a constant rate. That can't go on forever, of course, even in science fiction movies. It is sometimes approximated by a finite ramp (Fig. 1.19) or by a periodic ramp, or sawtooth (Fig. 1.20).

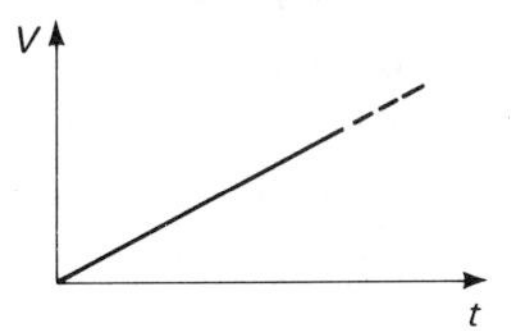

Figure 1.18. Voltage ramp waveform.

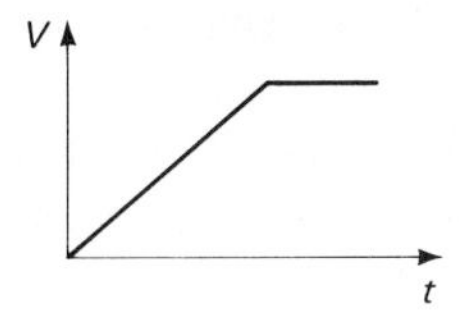

Figure 1.19. Ramp with limit.

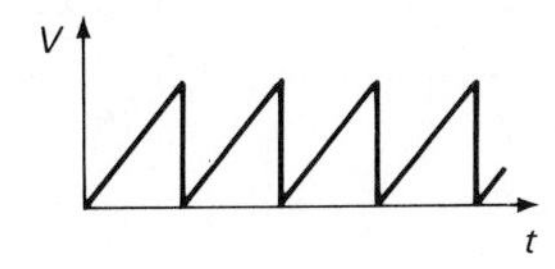

Figure 1.20. Sawtooth wave.

Triangle

The triangle wave is a close cousin of the ramp; it is simply a symmetrical ramp (Fig. 1.21).

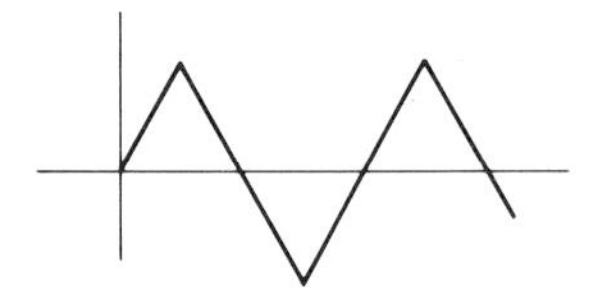

Figure 1.21. Triangle wave.

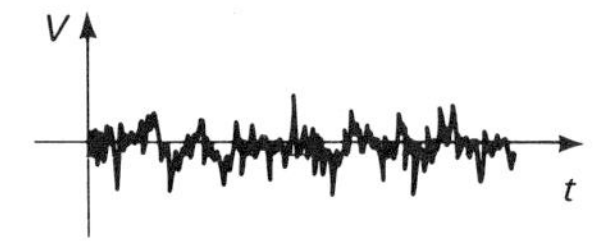

Figure 1.22. Noise.

Noise

Signals of interest are often mixed with *noise*; this is a catchall phrase that usually applies to random noise of thermal origin. Noise voltages can be specified by their frequency spectrum (power per hertz) or by their amplitude distribution. One of the most common kinds of noise is *band-limited white Gaussian noise*, which means a signal with equal power per hertz in some band of frequencies and a Gaussian (bell-shaped) distribution of amplitudes if large numbers of instantaneous measurements of its amplitude are made. This kind of noise is generated by a resistor (Johnson noise), and it plagues sensitive measurements of all kinds. On an oscilloscope it appears as shown in Figure 1.22. We will study noise and low-noise techniques in some detail in Chapter 7. Sections 9.32–9.37 deal with noise-generation techniques.

Square waves

A square wave is a signal that varies in time as shown in Figure 1.23. Like the sine wave, it is characterized by amplitude and frequency. A linear circuit driven by a square wave rarely responds with a square wave. For a square wave, the rms amplitude equals the amplitude.

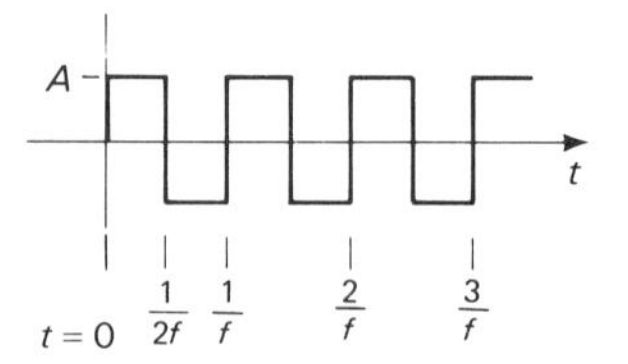

Figure 1.23. Square wave.

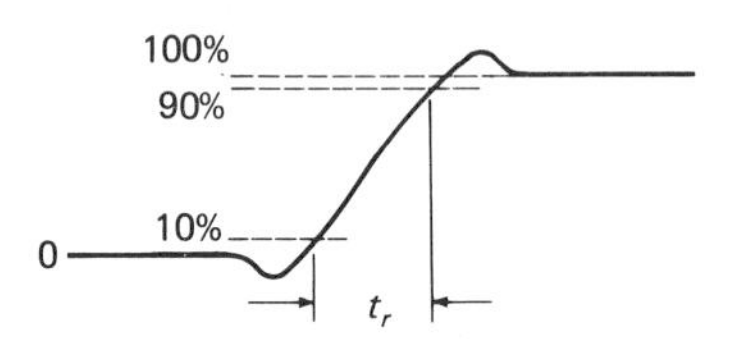

Figure 1.24. Rise time of a step waveform.

The edges of a square wave are not perfectly square; in typical electronic circuits the rise time t_r ranges from a few nanoseconds to a few microseconds. Figure 1.24 shows the sort of thing usually seen. The rise time is defined as the time required for the signal to go from 10% to 90% of its total transition.

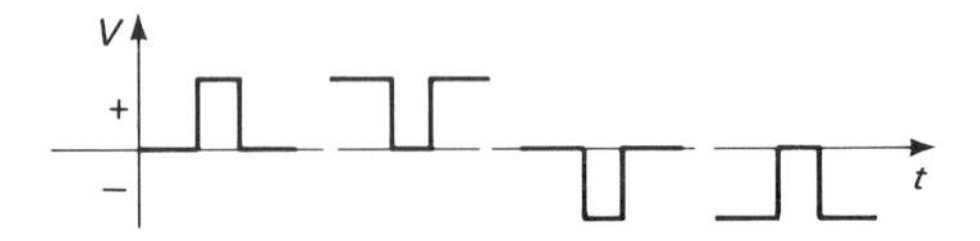

Figure 1.25. Positive- and negative-going pulses of both polarities.

Pulses

A pulse is a signal that looks as shown in Figure 1.25. It is defined by amplitude and pulse width. You can generate a train of periodic (equally spaced) pulses, in which case you can talk about the frequency, or pulse repetition rate, and the "duty cycle," the ratio of pulse width to repetition period (duty cycle ranges from zero to 100%). Pulses can have positive or negative polarity; in addition, they can be "positive-going" or "negative-going." For instance, the second pulse in Figure 1.25

is a negative-going pulse of positive polarity.

Figure 1.26

Steps and spikes

Steps and spikes are signals that are talked about a lot but are not often used. They provide a nice way of describing what happens in a circuit. If you could draw them, they would look something like the example in Figure 1.26. The step function is part of a square wave; the spike is simply a jump of vanishingly short duration.

1.10 Logic levels

Pulses and square waves are used extensively in digital electronics, where predefined voltage levels represent one of two possible states present at any point in the circuit. These states are called simply HIGH and LOW and correspond to the 0 (false) and 1 (true) states of Boolean logic (the algebra that describes such two-state systems).

Precise voltages are not necessary in digital electronics. You need only to distinguish which of the two possible states is present. Each digital logic family therefore specifies legal HIGH and LOW states. For example, the "74HC" digital logic family runs from a single +5 volt supply, with output levels that are typically 0 volts (LOW) and 5 volts (HIGH), and an input decision threshold of 2.5 volts. Actual outputs can be as much as a volt from ground or +5 volts without malfunction, however. We'll have much more to say about logic levels in Chapters 8 and 9.

1.11 Signal sources

Often the source of a signal is some part of the circuit you are working on. But for test purposes a flexible signal source is invaluable. They come in three flavors: signal generators, pulse generators, and function generators.

Signal generators

Signal generators are sine-wave oscillators, usually equipped to give a wide range of frequency coverage (50kHz to 50MHz is typical), with provision for precise control of amplitude (using a resistive divider network called an *attenuator*). Some units let you *modulate* the output (see Chapter 13). A variation on this theme is the *sweep generator*, a signal generator that can sweep its output frequency repeatedly over some range. These are handy for testing circuits whose properties vary with frequency in a particular way, e.g., "tuned circuits" or filters. Nowadays these devices, as well as many test instruments, are available in configurations that allow you to program the frequency, amplitude, etc., from a computer or other digital instrument.

A variation on the signal generator is the *frequency synthesizer*, a device that generates sine waves whose frequencies can be set precisely. The frequency is set digitally, often to eight significant figures or more, and is internally synthesized from a precise standard (a quartz-crystal oscillator) by digital methods we will discuss later (Sections 9.27–9.31). If your requirement is for no-nonsense accurate frequency generation, you can't beat a synthesizer.

Pulse generators

Pulse generators only make pulses, but what pulses! Pulse width, repetition rate, amplitude, polarity, rise time, etc., may all be adjustable. In addition, many units allow you to generate pulse pairs, with settable spacing and repetition rate, or even coded pulse trains. Most modern pulse generators are provided with logic-level outputs for easy connection to digital circuitry. Like signal generators, these come in the programmable variety.

Function generators

In many ways function generators are the most flexible signal sources of all. You can make sine, triangle, and square waves over an enormous frequency range (0.01 Hz to 10 MHz is typical), with control of amplitude and dc offset (a constant dc voltage added to the signal). Many of them have provision for frequency sweeping, often in several modes (linear or logarithmic frequency variation versus time). They are available with pulse outputs (although not with the flexibility you get with a pulse generator), and some of them have provision for modulation.

Like the other signal sources, function generators come in programmable versions and versions with digital readout of frequency (and sometimes amplitude). The most recent addition to the function-generator family is the synthesized function generator, a device that combines all the flexibility of a function generator with the stability and accuracy of a frequency synthesizer. An example is the HP 8116A, with sine, square, and triangle waves (as well as pulses, ramps, haversines, etc.) from 0.001Hz to 50MHz. Frequency and amplitude (10mV to 16V pp) are programmable, as are linear and logarithmic frequency sweeps. This unit also provides trigger, gate, burst, FM, AM, pulse-width modulation, voltage-controlled frequency, and single cycles. For general use, if you can have only one signal source, the function generator is for you.

CAPACITORS AND AC CIRCUITS

Once we enter the world of changing voltages and currents, or signals, we encounter two very interesting circuit elements that are useless in dc circuits: capacitors and inductors. As you will see, these humble devices, combined with resistors, complete the triad of passive linear circuit elements that form the basis of nearly all circuitry. Capacitors, in particular, are essential in nearly every circuit application. They are used for waveform generation, filtering, and blocking and bypass applications. They are used in integrators and differentiators. In combination with inductors, they make possible sharp filters for separating desired signals from background. You will see some of these applications as we continue with this chapter, and there will be numerous interesting examples in later chapters.

Let's proceed, then, to look at capacitors in detail. Portions of the treatment that follows are necessarily mathematical in nature; the reader with little mathematical preparation may find Appendix B helpful. In any case, an understanding of the details is less important in the long run than an understanding of the results.

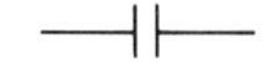

Figure 1.27. Capacitor.

1.12 Capacitors

A capacitor (Fig. 1.27) (the old-fashioned name was *condenser*) is a device that has two wires sticking out of it and has the property

$$Q = CV$$

A capacitor of C farads with V volts across its terminals has Q coulombs of stored charge on one plate, and $-Q$ on the other.

To a first approximation, capacitors are devices that might be considered simply frequency-dependent resistors. They allow you to make frequency-dependent voltage dividers, for instance. For some applications (bypass, coupling) this is almost all you need to know, but for other applications (filtering, energy storage, resonant circuits) a deeper understanding is needed. For example, capacitors cannot dissipate power, even though current can flow through them, because the voltage and current are 90° out of phase.

Taking the derivative of the defining equation above (see Appendix B), you get

$$I = C\frac{dV}{dt}$$

So a capacitor is more complicated than a resistor; the current is not simply proportional to the voltage, but rather to the rate of change of voltage. If you change the voltage across a farad by 1 volt per second, you are supplying an amp. Conversely, if you supply an amp, its voltage changes by 1 volt per second. A farad is very large, and you usually deal in microfarads (μF) or picofarads (pF). (To make matters confusing to the uninitiated, the units are often omitted on capacitor values specified in schematic diagrams. You have to figure it out from the context.) For instance, if you supply a current of 1mA to 1μF, the voltage will rise at 1000 volts per second. A 10ms pulse of this current will increase the voltage across the capacitor by 10 volts (Fig. 1.28).

Capacitors come in an amazing variety of shapes and sizes; with time, you will come to recognize their more common incarnations. The basic construction is simply two conductors near each other (but not touching); in fact, the simplest capacitors are just that. For greater capacitance, you need more area and closer spacing; the usual approach is to plate some conductor onto a thin insulating material (called a dielectric), for instance, aluminized Mylar film rolled up into a small cylindrical configuration. Other popular types are thin ceramic wafers (disc ceramics), metal foils with oxide insulators (electrolytics), and metallized mica. Each of these types has unique properties; for a brief rundown, see the box on capacitors. In general, ceramic and Mylar types are used for most noncritical circuit applications; tantalum capacitors are used where greater capacitance is needed, and electrolytics are used for power-supply filtering.

Capacitors in parallel and series

The capacitance of several capacitors in parallel is the sum of their individual capacitances. This is easy to see: Put voltage V across the parallel combination; then

$$\begin{aligned} C_{\text{total}}V &= Q_{\text{total}} = Q_1 + Q_2 + Q_3 + \ldots \\ &= C_1V + C_2V + C_3V + \ldots \\ &= (C_1 + C_2 + C_3 + \ldots)V \end{aligned}$$

or

$$C_{\text{total}} = C_1 + C_2 + C_3 + \ldots$$

For capacitors in series, the formula is like that for resistors in parallel:

$$C_{\text{total}} = \frac{1}{\frac{1}{C_1} + \frac{1}{C_2} + \frac{1}{C_3} + \ldots}$$

or (two capacitors only)

$$C_{\text{total}} = \frac{C_1C_2}{C_1 + C_2}$$

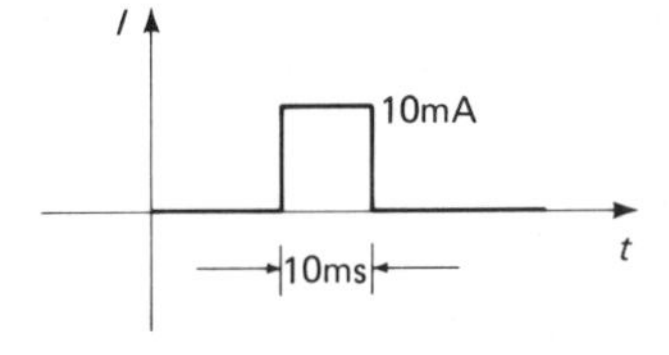

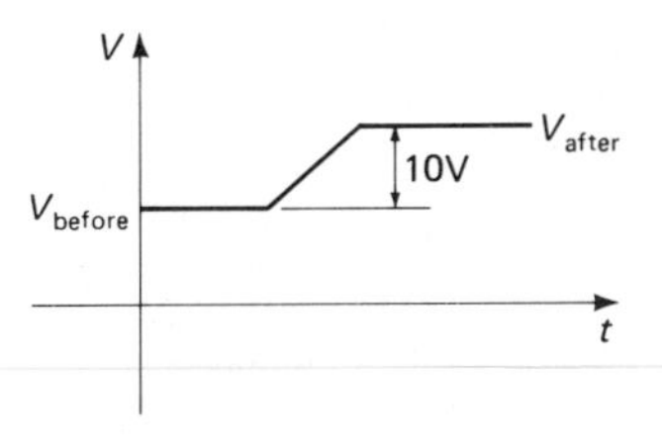

Figure 1.28. The voltage across a capacitor changes when a current flows through it.

CAPACITORS

There is wide variety among the capacitor types available. This is a quickie guide to point out their major advantages and disadvantages. Our judgments should be considered somewhat subjective:

Type	Capacitance range	Maximum voltage	Accuracy	Temperature stability	Leakage	Comments
Mica	1pF–0.01μF	100–600	Good		Good	Excellent; good at RF
Tubular ceramic	0.5pF–100pF	100–600		Selectable		Several tempcos (including zero)
Ceramic	10pF–1μF	50–30,000	Poor	Poor	Moderate	Small, inexpensive, very popular
Polyester (Mylar)	0.001μF–50μF	50–600	Good	Poor	Good	Inexpensive, good, popular
Polystyrene	10pF–2.7μF	100–600	Excellent	Good	Excellent	High quality, large; signal filters
Polycarbonate	100pF–30μF	50–800	Excellent	Excellent	Good	High quality, small
Polypropylene	100pF–50μF	100–800	Excellent	Good	Excellent	High quality, low dielectric absorption
Teflon	1000pF–2μF	50–200	Excellent	Best	Best	High quality, lowest dielectric absorption
Glass	10pF–1000pF	100–600	Good		Excellent	Long-term stability
Porcelain	100pF–0.1μF	50–400	Good	Good	Good	Good long-term stability
Tantalum	0.1μF–500μF	6–100	Poor	Poor		High capacitance; polarized, small; low inductance
Electrolytic	0.1μF–1.6F	3–600	Terrible	Ghastly	Awful	Power-supply filters; polarized; short life
Double layer	0.1F–10F	1.5–6	Poor	Poor	Good	Memory backup; high series resistance
Oil	0.1μF–20μF	200–10,000			Good	High-voltage filters; large, long life
Vacuum	1pF–5000pF	2000–36,000			Excellent	Transmitters

EXERCISE 1.12

Derive the formula for the capacitance of two capacitors in series. Hint: Because there is no external connection to the point where the two capacitors are connected together, they must have equal stored charges.

The current that flows in a capacitor during charging ($I = CdV/dt$) has some unusual features. Unlike resistive current, it's not proportional to voltage, but rather to the rate of change (the "time derivative") of voltage. Furthermore, unlike the situation in a resistor, the power (V times I) associated with capacitive current is not turned into heat, but is stored as energy in the capacitor's internal electric field. You get all that energy back when you discharge the capacitor. We'll see another way to look at these curious properties when we talk about *reactance*, beginning in Section 1.18.

1.13 *RC* circuits: *V* and *I* versus time

When dealing with ac circuits (or, in general, any circuits that have changing voltages and currents), there are two possible approaches. You can talk about V and I versus time, or you can talk about amplitude versus signal frequency. Both approaches have their merits, and you find yourself switching back and forth according to which description is most convenient in each situation. We will begin our study of ac circuits in the time domain. Beginning with Section 1.18, we will tackle the frequency domain.

What are some of the features of circuits with capacitors? To answer this question, let's begin with the simple RC circuit (Fig. 1.29). Application of the capacitor rules gives

$$C\frac{dV}{dt} = I = -\frac{V}{R}$$

This is a differential equation, and its solution is

$$V = Ae^{-t/RC}$$

So a charged capacitor placed across a resistor will discharge as in Figure 1.30.

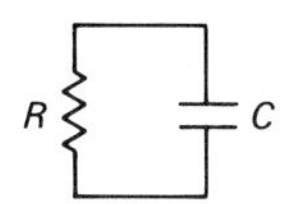

Figure 1.29

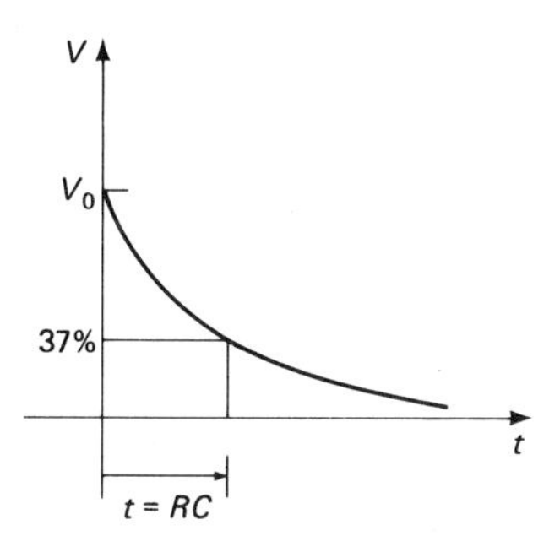

Figure 1.30. *RC* discharge waveform.

Time constant

The product RC is called the *time constant* of the circuit. For R in ohms and C in farads, the product RC is in seconds. A microfarad across 1.0k has a time constant of 1ms; if the capacitor is initially charged to 1.0 volt, the initial current is 1.0mA.

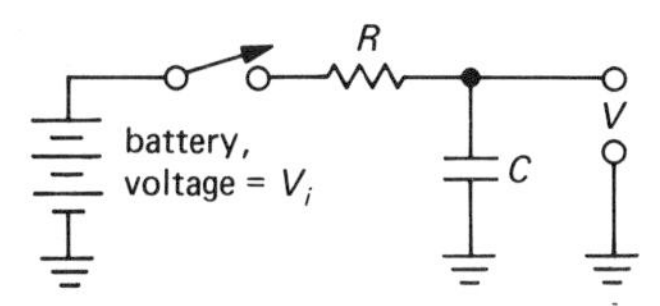

Figure 1.31

Figure 1.31 shows a slightly different circuit. At time $t = 0$, someone connects the battery. The equation for the circuit is then

$$I = C\frac{dV}{dt} = \frac{V_i - V}{R}$$

with the solution

$$V = V_i + Ae^{-t/RC}$$

(Please don't worry if you can't follow the mathematics. What we are doing is getting some important results, which you should remember. Later we will use the results often, with no further need for the mathematics used to derive them.) The constant A is determined by initial conditions (Fig. 1.32): $V = 0$ at $t = 0$; therefore, $A = -V_i$, and

$$V = V_i(1 - e^{-t/RC})$$

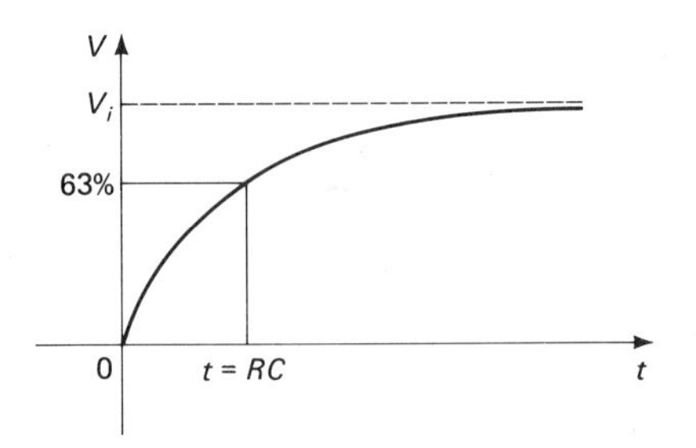

Figure 1.32

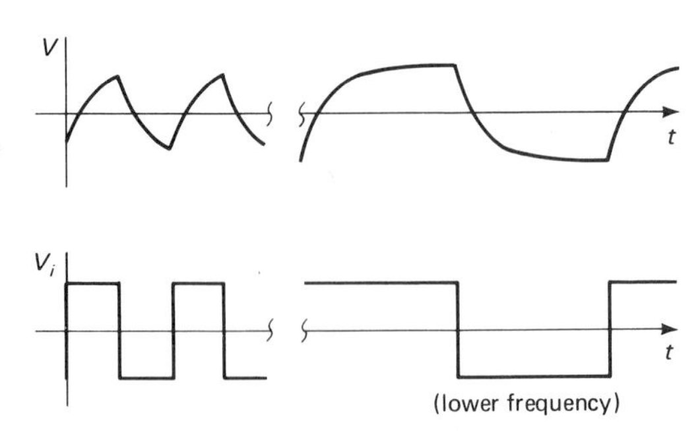

Figure 1.33. Output (top waveform) across a capacitor, when driven by square waves through a resistor.

Decay to equilibrium

Eventually (when $t \gg RC$), V reaches V_i. (Presenting the "$5RC$ rule of thumb": a capacitor charges or decays to within 1% of its final value in 5 time constants.) If we then change V_0 to some other value (say, 0), V will decay toward that new value with an exponential $e^{-t/RC}$. For example, a square-wave input for V_0 will produce the output shown in Figure 1.33.

EXERCISE 1.13

Show that the rise time (the time required to go from 10% to 90% of its final value) of this signal is $2.2RC$.

You might ask the obvious next question: What about $V(t)$ for arbitrary $V_i(t)$? The solution involves an inhomogeneous differential equation and can be solved by standard methods (which are, however, beyond the scope of this book). You would find

$$V(t) = \frac{1}{RC}\int_{-\infty}^{t} V_i(\tau)e^{-(t-\tau)/RC}d\tau$$

That is, the RC circuit averages past history at the input with a weighting factor

$$e^{-\Delta t/RC}$$

In practice, you seldom ask this question. Instead, you deal in the *frequency domain* and ask how much of each frequency component present in the input gets through. We will get to this important topic soon (Section 1.18). Before we do, though, there are a few other interesting circuits we can analyze simply with this time-domain approach.

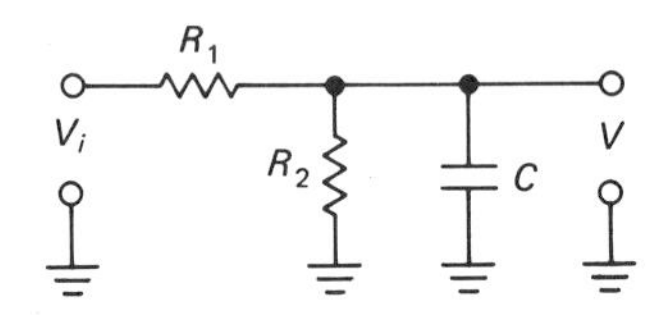

Figure 1.34

Simplification by Thévenin equivalents

We could go ahead and analyze more complicated circuits by similar methods, writing down the differential equations and trying to find solutions. For most purposes it simply isn't worth it. This is as complicated an RC circuit as we will need. Many other circuits can be reduced to it (e.g., Fig. 1.34). By just using the Thévenin equivalent of the voltage divider formed by R_1 and R_2, you can find the output

$V(t)$ produced by a step input for V_0.

EXERCISE 1.14

$R_1 = R_2 =$ 10k, and $C =$ 0.1μF in the circuit shown in Figure 1.34. Find $V(t)$ and sketch it.

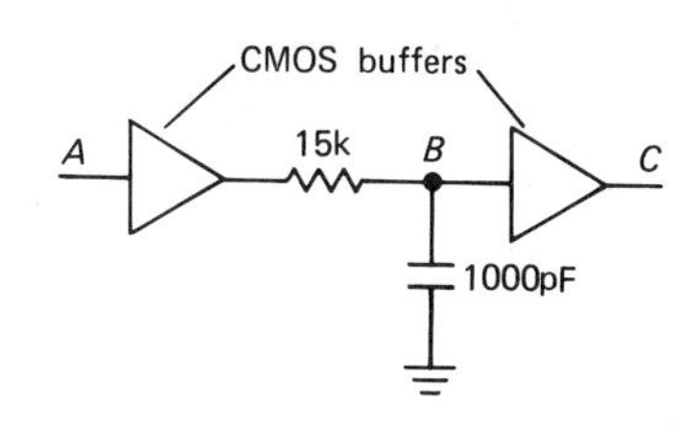

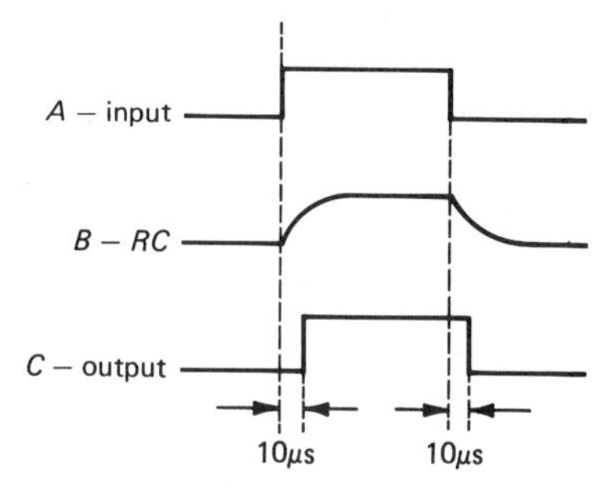

Figure 1.35. Producing a delayed digital waveform with the help of an RC.

Example: time-delay circuit

We have already mentioned logic levels, the voltages that digital circuits live on. Figure 1.35 shows an application of capacitors to produce a delayed pulse. The triangular symbols are "CMOS buffers." They give a HIGH output if the input is HIGH (more than one-half the dc power-supply voltage used to power them), and vice versa. The first buffer provides a replica of the input signal, but with low source resistance, and prevents input loading by the RC (recall our earlier discussion of circuit loading in Section 1.05). The RC output has the characteristic decays and causes the output buffer to switch 10μs after the input transitions (an RC reaches 50% output in $0.7RC$). In an actual application you would have to consider the effect of the buffer input threshold deviating from one-half the supply voltage, which would alter the delay and change the output pulse width. Such a circuit is sometimes used to delay a pulse so that something else can happen first. In designing circuits you try not to rely on tricks like this, but they're occasionally handy.

1.14 Differentiators

Look at the circuit in Figure 1.36. The voltage across C is $V_{\text{in}} - V$, so

$$I = C\frac{d}{dt}(V_{\text{in}} - V) = \frac{V}{R}$$

If we choose R and C small enough so that $dV/dt \ll dV_{\text{in}}/dt$, then

$$C\frac{dV_{\text{in}}}{dt} \approx \frac{V}{R}$$

or

$$V(t) = RC\frac{d}{dt}V_{\text{in}}(t)$$

That is, we get an output proportional to the rate of change of the input waveform.

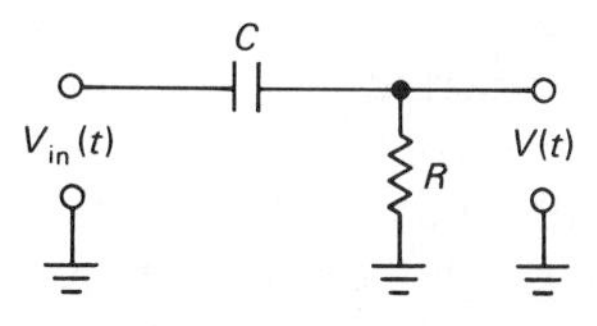

Figure 1.36

To keep $dV/dt \ll dV_{\text{in}}/dt$, we make the product RC small, taking care not to "load" the input by making R too small (at the transition the change in voltage across the capacitor is zero, so R is the load seen by the input). We will have a better criterion for this when we look at things in the frequency domain. If you drive this circuit with a square wave, the output will be as shown in Figure 1.37.

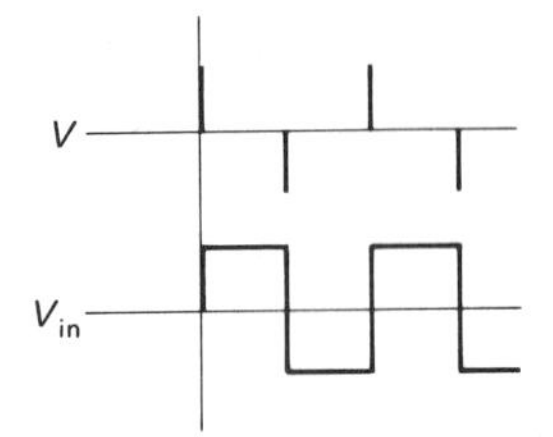

Figure 1.37. Output waveform (top) from differentiator driven by a square wave.

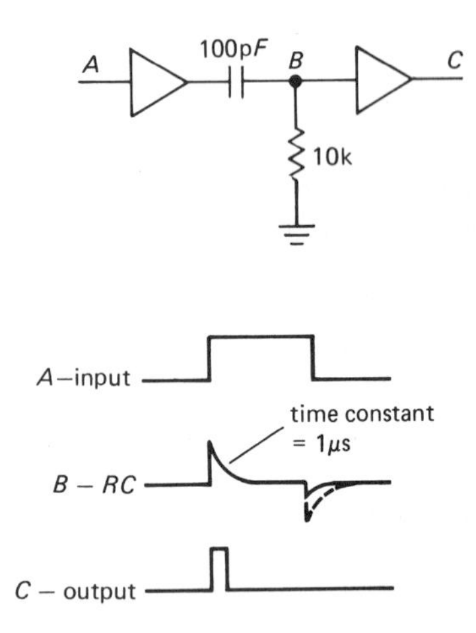

Figure 1.38. Leading-edge detector.

Differentiators are handy for detecting *leading edges* and *trailing edges* in pulse signals, and in digital circuitry you sometimes see things like those depicted in Figure 1.38. The RC differentiator generates spikes at the transitions of the input signal, and the output buffer converts the spikes to short square-topped pulses. In practice, the negative spike will be small because of a diode (a handy device discussed in Section 1.25) built into the buffer.

Unintentional capacitive coupling

Differentiators sometimes crop up unexpectedly, in situations where they're not welcome. You may see signals like those shown in Figure 1.39. The first case is caused by a square wave somewhere in the circuit coupling capacitively to the signal line you're looking at; that might indicate a missing resistor termination on your signal line. If not, you must either reduce the source resistance of the signal line or find a way to reduce capacitive coupling from the offending square wave. The second case is typical of what you might see when you look at a square wave, but have a broken connection somewhere, usually at the scope probe. The very small capacitance of the broken connection combines with the scope input resistance to form a differentiator. *Knowing that you've got a differentiated "something" can help you find the trouble and eliminate it.*

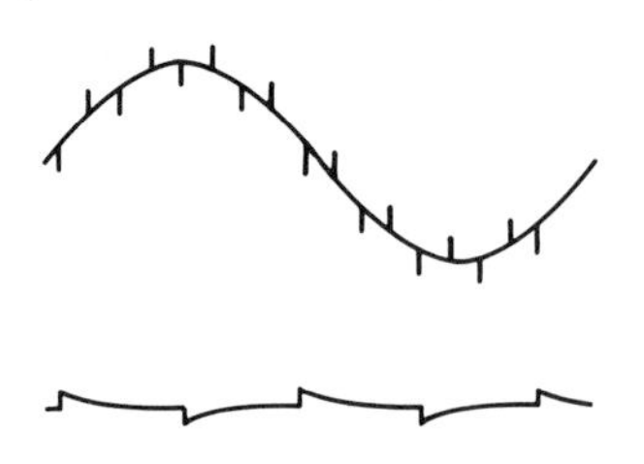

Figure 1.39

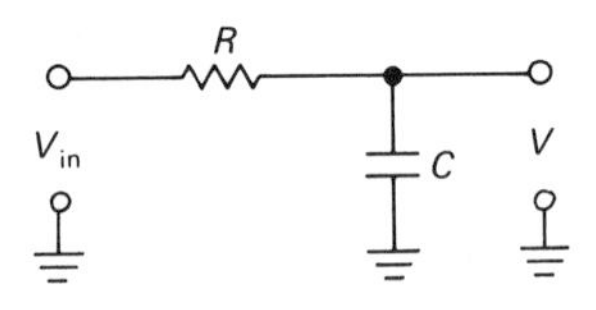

Figure 1.40

1.15 Integrators

Take a look at the circuit in Figure 1.40. The voltage across R is $V_{in} - V$, so

$$I = C\frac{dV}{dt} = \frac{V_{in} - V}{R}$$

If we manage to keep $V \ll V_{in}$, by keeping the product RC large, then

$$C\frac{dV}{dt} \approx \frac{V_{in}}{R}$$

or

$$V(t) = \frac{1}{RC}\int^{t} V_{in}(t)\,dt + \text{constant}$$

We have a circuit that performs the integral over time of an input signal! You can see how the approximation works for a square-wave input: $V(t)$ is then the exponential charging curve we saw earlier (Fig. 1.41). The first part of the exponential is a ramp, the integral of a constant; as we increase the time constant RC, we pick off a smaller part of the exponential, i.e., a better approximation to a perfect ramp.

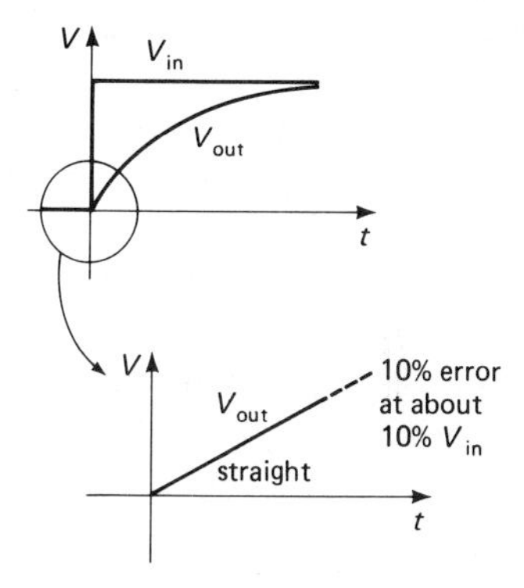

Figure 1.41

Note that the condition $V \ll V_{in}$ is just the same as saying that I is proportional to V_{in}. If we had as input a *current* $I(t)$, rather than a voltage, we would have an exact integrator. A large voltage across a large resistance approximates a current source and, in fact, is frequently used as one.

Later, when we get to operational amplifiers and feedback, we will be able to build integrators without the restriction $V_{out} \ll V_{in}$. They will work over large frequency and voltage ranges with negligible error.

The integrator is used extensively in analog computation. It is a useful subcircuit that finds application in control systems, feedback, analog/digital conversion, and waveform generation.

Ramp generators

At this point it is easy to understand how a ramp generator works. This nice circuit is extremely useful, for example in timing circuits, waveform and function generators, oscilloscope sweep circuits, and analog/digital conversion circuitry. The circuit uses a constant current to charge a capacitor (Fig. 1.42). From the capacitor equation $I = C(dV/dt)$, you get $V(t) = (I/C)t$. The output waveform is as shown in Figure 1.43. The ramp stops when the current source "runs out of voltage," i.e., reaches the limit of its compliance. The curve for a simple RC, with the resistor tied to a voltage source equal to the compliance of the current source, and with R chosen so that the current at zero output voltage is the same as that of the current source, is also drawn for comparison. (Real current sources generally have output compliances limited by the power-supply voltages used in making them, so the comparison is realistic.) In the next chapter, which deals with transistors, we will design some current sources, with some refinements to follow in the chapters on operational amplifiers (op-amps) and field-effect transistors (FETs). Exciting things to look forward to!

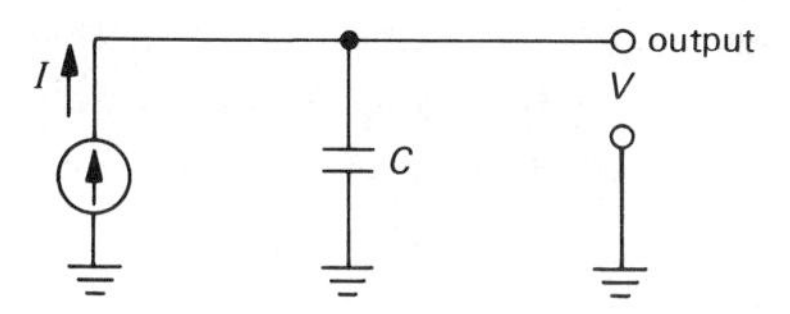

Figure 1.42. A constant current source charging a capacitor generates a ramp voltage waveform.

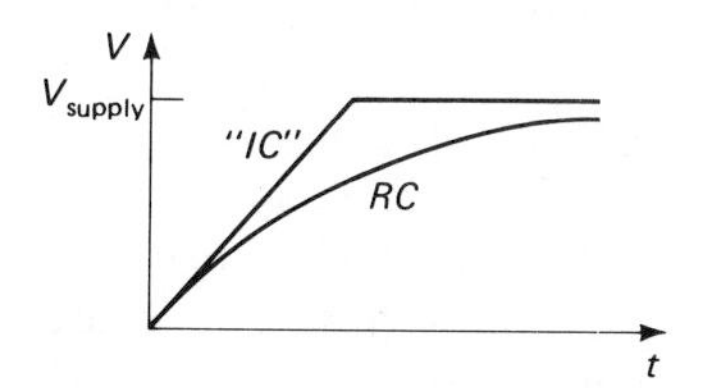

Figure 1.43

EXERCISE 1.15

A current of 1 mA charges a 1 μF capacitor. How long does it take the ramp to reach 10 volts?

INDUCTORS AND TRANSFORMERS

1.16 Inductors

If you understand capacitors, you won't have any trouble with inductors (Fig. 1.44). They're closely related to capacitors; the rate of current change in an inductor depends on the voltage applied across it, whereas the rate of voltage change in a capacitor depends on the current through it. The defining equation for an inductor is

$$V = L\frac{dI}{dt}$$

where L is called the *inductance* and is measured in henrys (or mH, μH, etc.). Putting a voltage across an inductor causes the current to rise as a ramp (for a capacitor, supplying a constant current causes the voltage to rise as a ramp); 1 volt across 1 henry produces a current that increases at 1 amp per second.

Figure 1.44. Inductor.

As with capacitive current, inductive current is not simply proportional to voltage. Furthermore, unlike the situation in a resistor, the power associated with inductive current (V times I) is not turned into heat, but is stored as energy in the inductor's magnetic field. You get all that energy back when you interrupt the inductor's current.

The symbol for an inductor looks like a coil of wire; that's because, in its simplest form, that's all it is. Variations include coils wound on various core materials, the most popular being iron (or iron alloys, laminations, or powder) and ferrite, a black, nonconductive, brittle magnetic material. These are all ploys to multiply the inductance of a given coil by the "permeability" of the core material. The core may be in the shape of a rod, a toroid (doughnut), or even more bizarre shapes, such as a "pot core" (which has to be seen to be understood; the best description we can think of is a doughnut mold split horizontally in half, if doughnuts were made in molds).

Inductors find heavy use in radio-frequency (RF) circuits, serving as RF "chokes" and as parts of tuned circuits (see Chapter 13). A pair of closely coupled inductors forms the interesting object known as a transformer. We will talk briefly about them in the next section.

An inductor is, in a real sense, the opposite of a capacitor. You will see how that works out in the next few sections of this chapter, which deal with the important subject of *impedance*.

1.17 Transformers

A transformer is a device consisting of two closely coupled coils (called primary and secondary). An ac voltage applied to the primary appears across the secondary, with a voltage multiplication proportional to the turns ratio of the transformer and a current multiplication inversely proportional to the turns ratio. Power is conserved. Figure 1.45 shows the circuit symbol for a laminated-core transformer (the kind used for 60Hz ac power conversion).

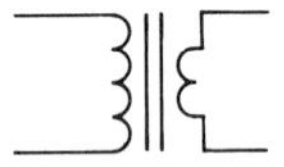

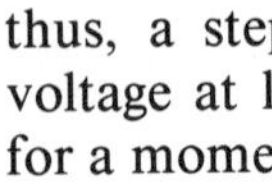

Figure 1.45. Transformer.

Transformers are quite efficient (output power is very nearly equal to input power); thus, a step-up transformer gives higher voltage at lower current. Jumping ahead for a moment, a transformer of turns ratio n increases the impedance by n^2. There is very little primary current if the secondary is unloaded.

Transformers serve two important functions in electronic instruments: They change the ac line voltage to a useful (usually lower) value that can be used by the circuit, and they "isolate" the electronic device from actual connection to the power line, because the windings of a transformer are electrically insulated from each other. *Power transformers* (meant for use from the 110V power line) come in an enormous variety of secondary voltages and currents: outputs as low as 1 volt or so up to several thousand volts, current ratings from a few milliamps to hundreds of amps. Typical transformers for use in electronic instruments might have secondary voltages from 10 to 50 volts, with current ratings of 0.1 to 5 amps or so.

Transformers for use at audiofrequencies and radiofrequencies are also available. At radiofrequencies you sometimes use tuned transformers, if only a narrow range of frequencies is present. There is also an interesting class of transmission-line transformer that we will discuss briefly in Section 13.10. In general, transformers for use at high frequencies must use special core materials or construction to minimize core losses, whereas low-frequency transformers (e.g., power transformers) are burdened instead by large and heavy cores. The two kinds of transformers are in general not interchangeable.

IMPEDANCE AND REACTANCE

Warning: This section is somewhat mathematical; you may wish to skip over the mathematics, but be sure to pay attention to the results and graphs.

Circuits with capacitors and inductors are more complicated than the resistive circuits we talked about earlier, in that their behavior depends on frequency: A "voltage divider" containing a capacitor or inductor will have a frequency-dependent division ratio. In addition, circuits containing these components (known collectively as *reactive* components) "corrupt" input waveforms such as square waves, as we just saw.

However, both capacitors and inductors are *linear* devices, meaning that the amplitude of the output waveform, whatever its shape, increases exactly in proportion to the input waveform's amplitude. This linearity has many consequences, the most important of which is probably the following: *The output of a linear circuit, driven with a sine wave at some frequency* f, *is itself a sine wave at the same frequency (with, at most, changed amplitude and phase).*

Because of this remarkable property of circuits containing resistors, capacitors, and inductors (and, later, linear amplifiers), it is particularly convenient to analyze any such circuit by asking how the output voltage (amplitude and phase) depends on the input voltage, *for sine-wave input at a single frequency*, even though this may not be the intended use. A graph of the resulting *frequency response*, in which the ratio of output to input is plotted for each sine-wave frequency, is useful for thinking about many kinds of waveforms. As an example, a certain "boom-box" loudspeaker might have the frequency response shown in Figure 1.46, where the "output" in this case is of course sound pressure, not voltage. It is desirable for a speaker to have a "flat" response, meaning that the graph of sound pressure versus frequency is constant over the band of audible frequencies. In this case the speaker's deficiencies can be corrected by introducing a passive filter with the inverse response (as shown) into the amplifiers of the radio.

As we will see, it is possible to generalize Ohm's law, replacing the word "resistance" with "impedance," in order to describe any circuit containing these linear passive devices (resistors, capacitors, and inductors). You could think of the subject of impedance and reactance as Ohm's law for circuits that include capacitors and inductors. Some important terminology: Impedance is the "generalized resistance"; inductors

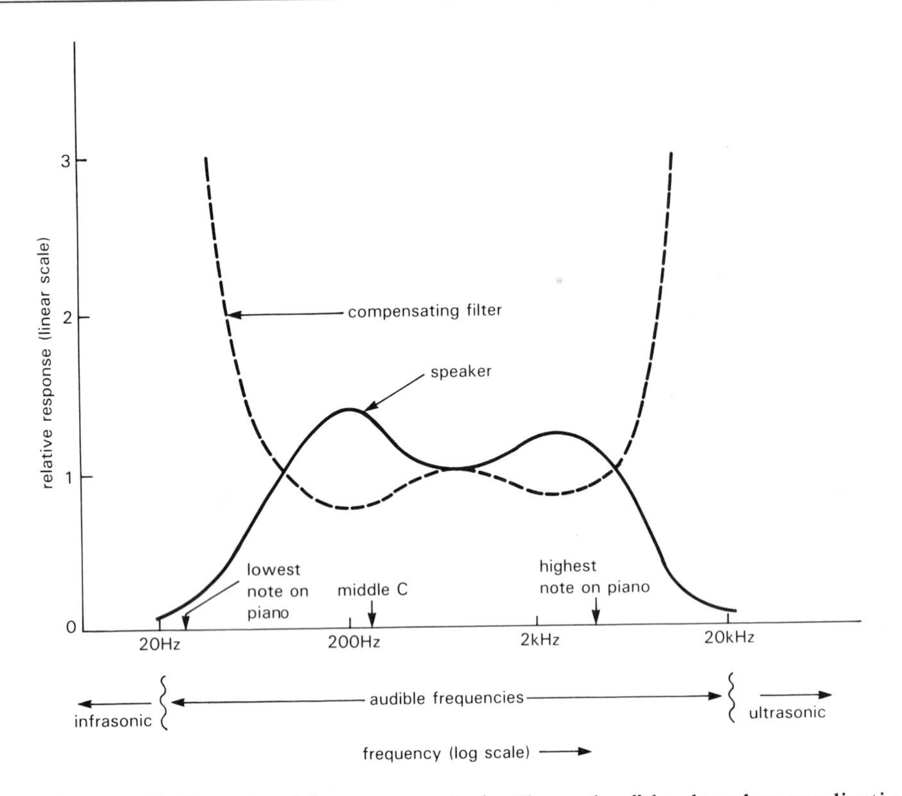

Figure 1.46. Example of frequency analysis: "boom box" loudspeaker equalization.

and capacitors have *reactance* (they are "reactive"); resistors have *resistance* (they are "resistive"). In other words, impedance = resistance + reactance (more about this later). However, you'll see statements like "the impedance of the capacitor at this frequency is ... " The reason you don't have to use the word "reactance" in such a case is that impedance covers everything. In fact, you frequently use the word "impedance" even when you know it's a resistance you're talking about; you say "the source impedance" or "the output impedance" when you mean the Thévenin equivalent resistance of some source. The same holds for "input impedance."

In all that follows, we will be talking about circuits driven by sine waves at a single frequency. Analysis of circuits driven by complicated waveforms is more elaborate, involving the methods we used earlier (differential equations) or decomposition of the waveform into sine waves (Fourier analysis). Fortunately, these methods are seldom necessary.

1.18 Frequency analysis of reactive circuits

Let's start by looking at a capacitor driven by a sine-wave voltage source (Fig. 1.47). The current is

$$I(t) = C\frac{dV}{dt} = C\omega V_0 \cos \omega t$$

i.e., a current of amplitude I, with the phase leading the input voltage by 90°. If we consider amplitudes only, and disregard phases, the current is

$$I = \frac{V}{1/\omega C}$$

(Recall that $\omega = 2\pi f$.) It behaves like a frequency-dependent resistance $R = 1/\omega C$, but in addition the current is 90° out of phase with the voltage (Fig. 1.48).

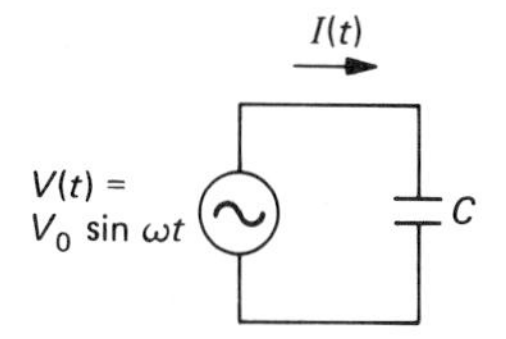

Figure 1.47

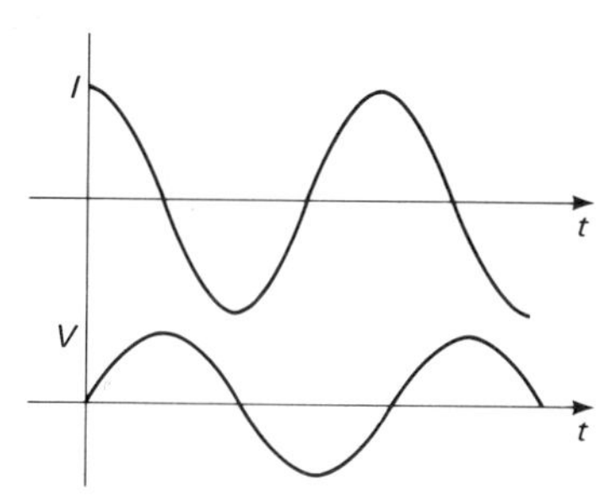

Figure 1.48

For example, a 1μF capacitor put across the 110 volt (rms) 60Hz power line draws a current of rms amplitude

$$I = \frac{110}{1/(2\pi \times 60 \times 10^{-6})} = 41.5\text{mA (rms)}$$

Note: At this point it is necessary to get into some complex algebra; you may wish to skip over the math in some of the following sections, taking note of the results as we derive them. A knowledge of the detailed mathematics is not necessary in order to understand the remainder of the book. Very little mathematics will be used in later chapters. The section ahead is easily the most difficult for the reader with little mathematical preparation. Don't be discouraged!

Voltages and currents as complex numbers

As you have just seen, there can be phase shifts between the voltage and current in an ac circuit being driven by a sine wave at some frequency. Nevertheless, as long as the circuit contains only *linear* elements (resistors, capacitors, inductors), the magnitudes of the currents everywhere in the circuit are still proportional to the magnitude of the driving voltage, so we might hope to find some generalization of voltage, current, and resistance in order to rescue Ohm's law. Obviously a single number won't suffice to specify the current, say, at some point in the circuit, because we must somehow have information about both the magnitude and phase shift.

Although we can imagine specifying the magnitudes and phase shifts of voltages and currents at any point in the circuit by writing them out explicitly, e.g., $V(t) = 23.7\sin(377t + 0.38)$, it turns out that our requirements can be met more simply by using the algebra of complex numbers to *represent* voltages and currents. Then we can simply add or subtract the complex number representations, rather than laboriously having to add or subtract the actual sinusoidal functions of time themselves. Because the actual voltages and currents are real quantities that vary with time, we must develop a rule for converting from actual quantities to their representations, and vice versa. Recalling once again that we are talking about a single sine-wave frequency, ω, we agree to use the following rules:

1. Voltages and currents are *represented* by the complex quantities $\mathbf{V}$ and $\mathbf{I}$. The voltage $V_0 \cos(\omega t + \phi)$ is to be represented by the complex number $V_0 e^{j\phi}$. Recall that $e^{j\theta} = \cos\theta + j\sin\theta$, where $j = \sqrt{-1}$.

2. *Actual* voltages and currents are obtained by multiplying their complex number representations by $e^{j\omega t}$ and then taking the real part: $V(t) = \mathcal{R}e(\mathbf{V}e^{j\omega t})$, $I(t) = \mathcal{R}e(\mathbf{I}e^{j\omega t})$

In other words,

circuit voltage versus time		complex number representation
$V_0 \cos(\omega t + \phi)$	⟷ multiply by $e^{j\omega t}$ and take real part	$V_0 e^{j\phi} = a + jb$

(In electronics, the symbol j is used instead of i in the exponential in order to avoid confusion with the symbol i meaning current.) Thus, in the general case the actual voltages and currents are given by

$$V(t) = \mathcal{R}e(\mathbf{V}e^{j\omega t})$$
$$= \mathcal{R}e(\mathbf{V})\cos\omega t - \mathcal{I}m(\mathbf{V})\sin\omega t$$
$$I(t) = \mathcal{R}e(\mathbf{I}e^{j\omega t})$$
$$= \mathcal{R}e(\mathbf{I})\cos\omega t - \mathcal{I}m(\mathbf{I})\sin\omega t$$

For example, a voltage whose complex representation is

$$\mathbf{V} = 5j$$

corresponds to a (real) voltage versus time of

$$V(t) = \mathcal{R}e[5j\cos\omega t + 5j(j)\sin\omega t]$$
$$= -5\sin\omega t \text{ volts}$$

Reactance of capacitors and inductors

With this convention we can apply complex Ohm's law to circuits containing capacitors and inductors, just as for resistors, once we know the reactance of a capacitor or inductor. Let's find out what these are. We have

$$V(t) = \mathcal{R}e(V_0 e^{j\omega t})$$

For a capacitor, using $I = C(dV/dt)$, we obtain

$$I(t) = -V_0 C\omega \sin\omega t = \mathcal{R}e\left(\frac{V_0 e^{j\omega t}}{-j/\omega C}\right)$$
$$= \mathcal{R}e\left(\frac{V_0 e^{j\omega t}}{X_C}\right)$$

i.e., for a capacitor

$$X_C = -j/\omega C$$

X_C is the reactance of a capacitor at frequency ω. As an example a 1μF capacitor has a reactance of $-2653j$ ohms at 60Hz and a reactance of $-0.16j$ ohms at 1MHz. Its reactance at dc is infinite.

If we did a similar analysis for an inductor, we would find

$$X_L = j\omega L$$

A circuit containing only capacitors and inductors always has a purely imaginary impedance, meaning that the voltage and current are always 90° out of phase – it is purely reactive. When the circuit contains resistors, there is also a real part to the impedance. The term "reactance" in that case means the imaginary part only.

Ohm's law generalized

With these conventions for representing voltages and currents, Ohm's law takes a simple form. It reads simply

$$\mathbf{I} = \mathbf{V}/\mathbf{Z}$$
$$\mathbf{V} = \mathbf{IZ}$$

where the voltage represented by $\mathbf{V}$ is applied across a circuit of impedance $\mathbf{Z}$, giving a current represented by $\mathbf{I}$. The complex impedance of devices in series or parallel obeys the same rules as resistance:

$$\mathbf{Z} = \mathbf{Z}_1 + \mathbf{Z}_2 + \mathbf{Z}_3 + \cdots \quad \text{(series)}$$
$$\mathbf{Z} = \frac{1}{\frac{1}{\mathbf{Z}_1} + \frac{1}{\mathbf{Z}_2} + \frac{1}{\mathbf{Z}_3} + \cdots} \quad \text{(parallel)}$$

Finally, for completeness we summarize here the formulas for the impedance of resistors, capacitors, and inductors:

$$\mathbf{Z}_R = R \quad \text{(resistor)}$$
$$\mathbf{Z}_C = -j/\omega C = 1/j\omega C \quad \text{(capacitor)}$$
$$\mathbf{Z}_L = j\omega L \quad \text{(inductor)}$$

With these rules we can analyze many ac circuits by the same general methods we used in handling dc circuits, i.e., application of the series and parallel formulas and Ohm's law. Our results for circuits such as voltage dividers will look nearly the same as before. For multiply connected networks we may have to use Kirchhoff's laws, just as with dc circuits, in this case using the complex representations for V and I: The sum of the (complex) voltage drops around a closed loop is zero, and the sum of the (complex) currents into a point is zero. The latter rule implies, as with dc circuits, that the (complex) current in a series circuit is the same everywhere.

EXERCISE 1.16

Use the preceding rules for the impedance of devices in parallel and in series to derive the formulas (Section 1.12) for the capacitance of two capacitors (a) in parallel and (b) in series. Hint: In each case, let the individual capacitors have capacitances C_1 and C_2. Write down the impedance of the parallel or series combination; then equate it to the impedance of a capacitor with capacitance C. Find C.

Let's try out these techniques on the simplest circuit imaginable, an ac voltage applied across a capacitor, which we considered just previously. Then, after a brief look at power in reactive circuits (to finish laying the groundwork), we'll analyze some simple but extremely important and useful RC filter circuits.

Imagine putting a 1μF capacitor across a 110 volt (rms) 60Hz power line. What current flows? Using complex Ohm's law, we have

$$\mathbf{Z} = -j/\omega C$$

Therefore, the current is given by

$$\mathbf{I} = \mathbf{V}/\mathbf{Z}$$

The phase of the voltage is arbitrary, so let us choose $\mathbf{V} = A$, i.e. $V(t) = A\cos\omega t$, where the amplitude $A = 110\sqrt{2} \approx 156$ volts. Then

$$\mathbf{I} = j\omega CA \approx 0.059\sin\omega t$$

The resulting current has an amplitude of 59mA (41.5mA rms) and leads the voltage by 90°. This agrees with our previous calculation. Note that if we just wanted to know the magnitude of the current, and didn't care what the relative phase was, we could have avoided doing any complex algebra: If

$$\mathbf{A} = \mathbf{B}/\mathbf{C}$$

then

$$A = B/C$$

where A, B, and C are the magnitudes of the respective complex numbers; this holds for multiplication, also (see Exercise 1.17). Thus, in this case,

$$I = V/Z = \omega CV$$

This trick is often useful.

Surprisingly, there is no power dissipated by the capacitor in this example. Such activity won't increase your electric bill; you'll see why in the next section. Then we will go on to look at circuits containing resistors and capacitors with our complex Ohm's law.

EXERCISE 1.17

Show that if $\mathbf{A}$=$\mathbf{BC}$, then A=BC, where A, B, and C are magnitudes. Hint: Represent each complex number in polar form, i.e., $\mathbf{A} = Ae^{i\theta}$.

Power in reactive circuits

The instantaneous power delivered to any circuit element is always given by the product $P = VI$. However, in reactive circuits where V and I are not simply proportional, you can't just multiply them together. Funny things can happen; for instance, the sign of the product can reverse over one cycle of the ac signal. Figure 1.49 shows an example. During time intervals A and C, power is being delivered to the capacitor (albeit at a variable rate), causing it to charge up; its stored energy is increasing (power is the rate of change of energy). During intervals B and D, the power delivered to the capacitor is negative; it is discharging. The average power over a whole cycle for this example is in fact exactly zero, a statement that is always true for any purely reactive circuit element (inductors, capacitors, or any combination thereof). If you know your trigonometric integrals, the next exercise will show you how to prove this.

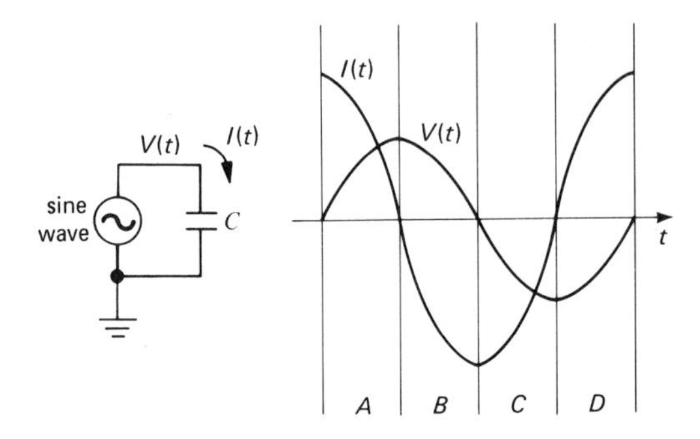

Figure 1.49. When driven by a sine wave, the current through a capacitor leads the voltage by 90°.

EXERCISE 1.18

Optional exercise: Prove that a circuit whose current is 90° out of phase with the driving voltage consumes no power, averaged over an entire cycle.

How do we find the average power consumed by an arbitrary circuit? In general, we can imagine adding up little pieces of VI product, then dividing by the elapsed time. In other words,

$$P = \frac{1}{T}\int_0^T V(t)I(t)\,dt$$

where T is the time for one complete cycle. Luckily, that's almost never necessary. Instead, it is easy to show that the average power is given by

$$P = \mathcal{R}e(\mathbf{VI}^*) = \mathcal{R}e(\mathbf{V}^*\mathbf{I})$$

where $\mathbf{V}$ and $\mathbf{I}$ are complex rms amplitudes.

Let's take an example. Consider the preceding circuit, with a 1 volt (rms) sine wave driving a capacitor. We'll do everything with rms amplitudes, for simplicity. We have

$$\mathbf{V} = 1$$

$$\mathbf{I} = \frac{\mathbf{V}}{-j/\omega C} = j\omega C$$

$$P = \mathcal{R}e(\mathbf{VI}^*) = \mathcal{R}e(-j\omega C) = 0$$

That is, the average power is zero, as stated earlier.

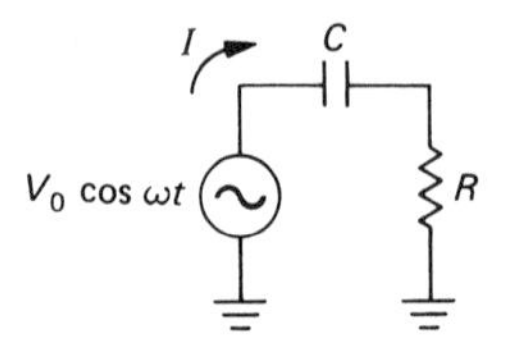

Figure 1.50

As another example, consider the circuit shown in Figure 1.50. Our calculations go like this:

$$\mathbf{Z} = R - \frac{j}{\omega C}$$

$$\mathbf{V} = V_0$$

$$\mathbf{I} = \frac{\mathbf{V}}{\mathbf{Z}} = \frac{V_0}{R - (j/\omega C)} = \frac{V_0[R + (j/\omega C)]}{R^2 + (1/\omega^2 C^2)}$$

$$P = \mathcal{R}e(\mathbf{VI}^*) = \frac{V_0^2 R}{R^2 + (1/\omega^2 C^2)}$$

(In the third line we multiplied numerator and denominator by the complex conjugate of the denominator, in order to make the denominator real.) This is less than the product of the magnitudes of $\mathbf{V}$ and $\mathbf{I}$. In fact, the ratio is called the *power factor*:

$$|\mathbf{V}|\,|\mathbf{I}| = \frac{V_0^2}{[R^2 + (1/\omega^2 C^2)]^{1/2}}$$

$$\text{power factor} = \frac{\text{power}}{|\mathbf{V}|\,|\mathbf{I}|} = \frac{R}{[R^2 + (1/\omega^2 C^2)]^{1/2}}$$

in this case. The power factor is the cosine of the phase angle between the voltage and the current, and it ranges from 0 (purely reactive circuit) to 1 (purely resistive). A power factor less than 1 indicates some component of reactive current.

EXERCISE 1.19
Show that all the average power delivered to the preceding circuit winds up in the resistor. Do this by computing the value of V_R^2/R. What is that power, in watts, for a series circuit of a 1μF capacitor and a 1.0k resistor placed across the 110 volt (rms), 60Hz power line?

Power factor is a serious matter in large-scale electrical power distribution, because reactive currents don't result in useful power being delivered to the load, but cost the power company plenty in terms of I^2R heating in the resistance of generators, transformers, and wiring. Although residential users are only billed for "real" power $[\mathcal{R}e(\mathbf{VI}^*)]$, the power company charges industrial users according to the power factor. This explains the capacitor yards that you see behind large factories, built to cancel the inductive reactance of industrial machinery (i.e., motors).

EXERCISE 1.20
Show that adding a series capacitor of value $C = 1/\omega^2 L$ makes the power factor equal 1.0 in a series RL circuit. Now do the same thing, but with the word "series" changed to "parallel."

Voltage dividers generalized

Our original voltage divider (Fig. 1.5) consisted of a pair of resistors in series to ground, input at the top and output at the junction. The generalization of that simple resistive divider is a similar circuit in which either or both resistors are replaced by a capacitor or inductor (or a more complicated network made from R, L, and C), as in Figure 1.51. In general, the division ratio $V_{\text{out}}/V_{\text{in}}$ of such a divider is not constant, but depends on frequency. The analysis is straightforward:

$$\mathbf{I} = \frac{\mathbf{V}_{\text{in}}}{\mathbf{Z}_{\text{total}}}$$

$$\mathbf{Z}_{\text{total}} = \mathbf{Z}_1 + \mathbf{Z}_2$$

$$\mathbf{V}_{\text{out}} = \mathbf{I}\,\mathbf{Z}_2 = \mathbf{V}_{\text{in}}\frac{\mathbf{Z}_2}{\mathbf{Z}_1 + \mathbf{Z}_2}$$

Rather than worrying about this result in general, let's look at some simple, but very important, examples.

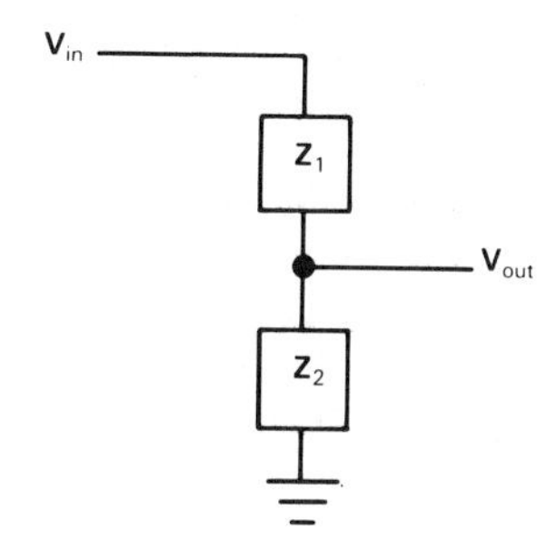

Figure 1.51. Generalized voltage divider: a pair of arbitrary impedances.

1.19 *RC* filters

By combining resistors with capacitors it is possible to make frequency-dependent voltage dividers, owing to the frequency dependence of a capacitor's impedance $\mathbf{Z}_C = -j/\omega C$. Such circuits can have the desirable property of passing signal frequencies of interest while rejecting undesired signal frequencies. In this section you will see examples of the simplest such RC filters, which we will be using frequently throughout the book. Chapter 5 and Appendix H describe filters of greater sophistication.

High-pass filters

Figure 1.52 shows a voltage divider made from a capacitor and a resistor. Complex Ohm's law gives

$$\mathbf{I} = \frac{\mathbf{V}_{\text{in}}}{\mathbf{Z}_{\text{total}}} = \frac{\mathbf{V}_{\text{in}}}{R - (j/\omega C)} = \frac{\mathbf{V}_{\text{in}}[R + (j/\omega C)]}{R^2 + 1/\omega^2 C^2}$$

(For the last step, multiply top and bottom by the complex conjugate of the denominator.) So the voltage across R is just

$$\mathbf{V}_{\text{out}} = \mathbf{I}\,\mathbf{Z}_R = \mathbf{I}\,R = \frac{\mathbf{V}_{\text{in}}[R + (j/\omega C)]R}{R^2 + (1/\omega^2 C^2)}$$

Most often we don't care about the phase of V_{out}, just its amplitude:

$$\mathbf{V}_{\text{out}} = (\mathbf{V}_{\text{out}}\mathbf{V}^*_{\text{out}})^{1/2} = \frac{R}{[R^2 + (1/\omega^2C^2)]^{1/2}}V_{\text{in}}$$

Note the analogy to a resistive divider, where

$$V_{\text{out}} = \frac{R_1}{R_1 + R_2}V_{\text{in}}$$

Here the impedance of the series RC combination (Fig. 1.53) is as shown in Figure 1.54. So the "response" of this circuit, ignoring phase shifts by taking magnitudes of the complex amplitudes, is given by

$$V_{\text{out}} = \frac{R}{[R^2 + (1/\omega^2C^2)]^{1/2}}V_{\text{in}} = \frac{2\pi fRC}{[1 + (2\pi fRC)^2]^{1/2}}V_{\text{in}}$$

and looks as shown in Figure 1.55. We could have gotten this result immediately by taking the ratio of the *magnitudes* of impedances, as in Exercise 1.17 and the example immediately preceding it; the numerator is the magnitude of the impedance of the lower leg of the divider (R), and the denominator is the magnitude of the impedance of the series combination of R and C.

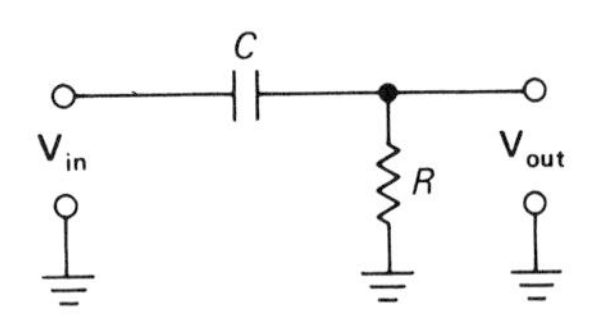

Figure 1.52. High-pass filter.

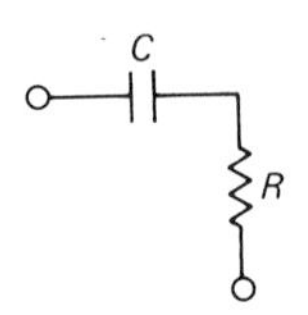

Figure 1.53

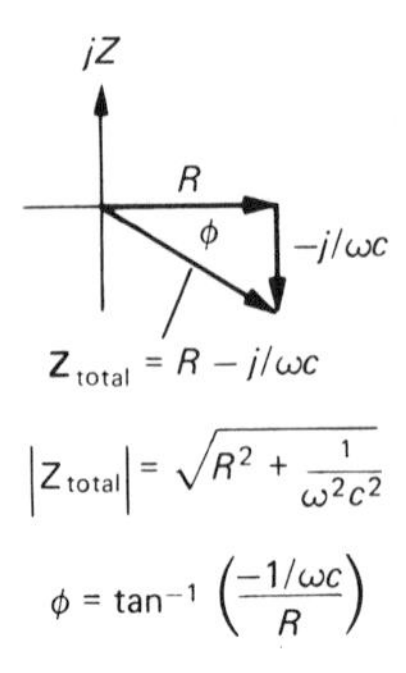

Figure 1.54

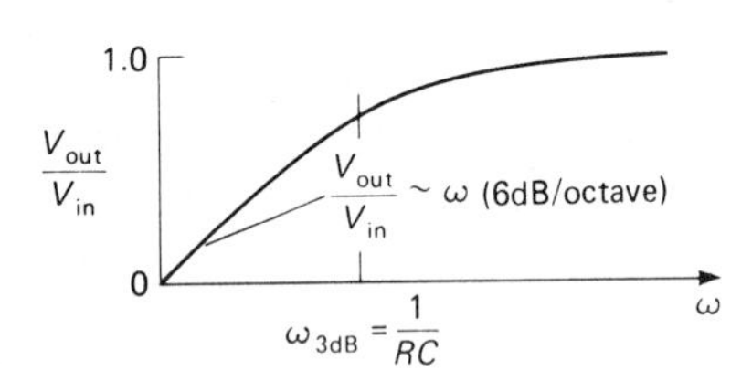

Figure 1.55. Frequency response of high-pass filter.

You can see that the output is approximately equal to the input at high frequencies (how high? $\gtrsim 1/RC$) and goes to zero at low frequencies. This is a very important result. Such a circuit is called a high-pass filter, for obvious reasons. It is very common. For instance, the input to the oscilloscope (Appendix A) can be switched to ac coupling. That's just an RC high-pass filter with the bend at about 10Hz (you would use ac coupling if you wanted to look at a small signal riding on a large dc voltage). Engineers like to refer to the −3dB "breakpoint" of a filter (or of any circuit that behaves like a filter). In the case of the simple RC high-pass filter, the −3dB breakpoint is given by

$$f_{3\text{dB}} = 1/2\pi RC$$

Note that the capacitor lets no steady current through ($f = 0$). This use as a dc *blocking capacitor* is one of its most frequent applications. Whenever you need to couple a signal from one amplifier to another, you almost invariably use a capacitor. For instance, every hi-fi audio amplifier has all its inputs capacitively coupled, because it doesn't know what dc level its input signals might be riding on. In such a coupling application you always pick R and C so that all frequencies of interest (in this case, 20Hz–20kHz) are passed without loss (attenuation).

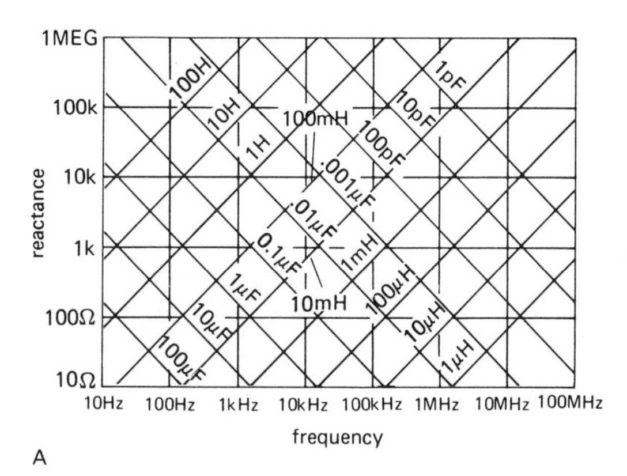

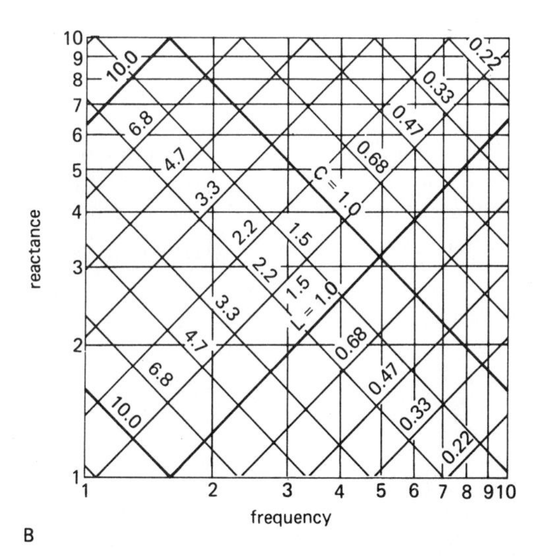

Figure 1.56. A. Reactance of inductors and capacitors versus frequency; all decades are identical, except for scale.
B. A single decade from part A expanded, with standard 20% component values shown.

You often need to know the impedance of a capacitor at a given frequency (e.g., for design of filters). Figure 1.56 provides a very useful graph covering large ranges of capacitance and frequency, giving the value of $|\mathbf{Z}| = 1/2\pi fC$.

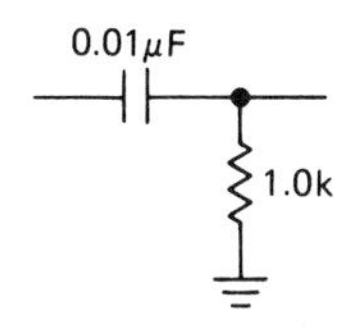

Figure 1.57

As an example, consider the filter shown in Figure 1.57. It is a high-pass filter with the 3dB point at 15.9kHz. The impedance of a load driven by it should be much larger than 1.0k in order to prevent circuit loading effects on the filter's output, and the driving source should be able to drive a 1.0k load without significant attenuation (loss of signal amplitude) in order to prevent circuit loading effects by the filter on the signal source.

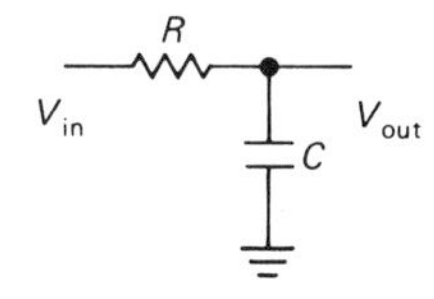

Figure 1.58. Low-pass filter.

Low-pass filters

You can get the opposite frequency behavior in a filter by interchanging R and C (Fig. 1.58). You will find

$$V_{\text{out}} = \frac{1}{(1 + \omega^2 R^2 C^2)^{1/2}} V_{\text{in}}$$

as seen in Figure 1.59. This is called a low-pass filter. The 3dB point is again at a frequency

$$f = 1/2\pi RC$$

Low-pass filters are quite handy in real life. For instance, a low-pass filter can be used to eliminate interference from nearby radio and television stations (550kHz–800MHz), a problem that plagues audio amplifiers and other sensitive electronic equipment.

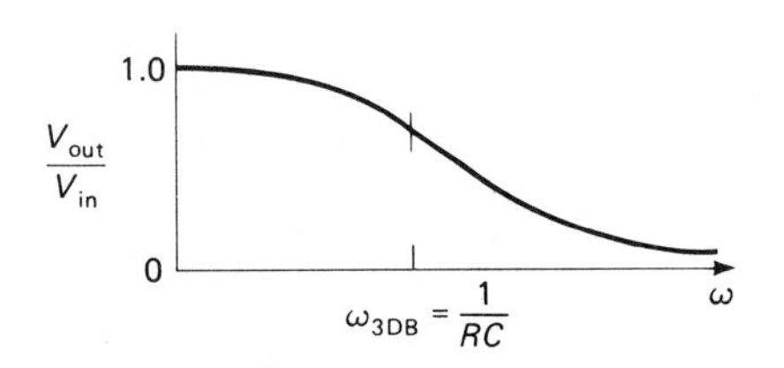

Figure 1.59. Frequency response of low-pass filter.

EXERCISE 1.21

Show that the preceding expression for the response of an RC low-pass filter is correct.

The low-pass filter's output can be viewed as a signal source in its own right. When driven by a perfect ac voltage (zero source impedance), the filter's output looks like R at low frequencies (the perfect signal source can be replaced by a short, i.e., by its small-signal source impedance, for the purpose of impedance calculations). It drops to zero impedance at high frequencies, where the capacitor dominates the output impedance. The signal driving the filter sees a load of R plus the load resistance at low frequencies, dropping to R at high frequencies.

In Figure 1.60, we've plotted the same low-pass filter response with *logarithmic* axes, which is a more usual way of doing it. You can think of the vertical axis as decibels, and the horizontal axis as octaves (or decades). On such a plot, equal distances correspond to equal ratios. We've also plotted the phase shift, using a linear

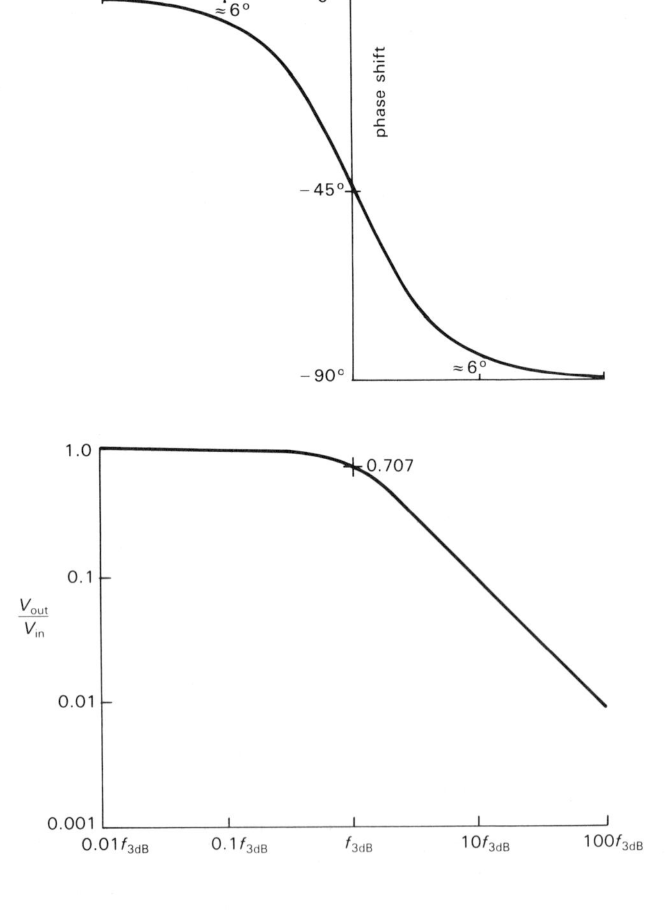

Figure 1.60. Frequency response (phase and amplitude) of low-pass filter, plotted on logarithmic axes. Note that the phase shift is 45° at the 3dB point and is within 6° of its asymptotic value for a decade of frequency change.

vertical axis (degrees) and the same logarithmic frequency axis. This sort of plot is good for seeing the detailed response even when it is greatly attenuated (as at right); we'll see a number of such plots in Chapter 5, when we treat active filters. Note that the filter curve plotted here becomes a straight line at large attenuations, with a slope of -20dB/decade (engineers prefer to say "-6dB/octave"). Note also that the phase shift goes smoothly from 0° (at frequencies well below the breakpoint) to 90° (well above it), with a value of 45° at the -3dB point. A rule of thumb for single-section RC filters is that the phase shift is $\approx 6°$ from its asymptotic value at $0.1f_{3\text{dB}}$ and $10f_{3\text{dB}}$.

EXERCISE 1.22
Prove the last assertion.

An interesting question is the following: Is it possible to make a filter with some arbitrary specified amplitude response and some other specified phase response? Surprisingly, the answer is no: The demands of causality (i.e., that response must follow cause, not precede it) force a relationship between phase and amplitude response of realizable analog filters (known officially as the Kramers-Kronig relation).

RC differentiators and integrators in the frequency domain

The RC differentiator that we saw in Section 1.14 is exactly the same circuit as the high-pass filter in this section. In fact, it can be considered as either, depending on whether you're thinking of waveforms in the time domain or response in the frequency domain. We can restate the earlier time-domain condition for its proper operation ($V_{\text{out}} \ll V_{\text{in}}$) in terms of the frequency response: For the output to be small compared with the input, the signal frequency (or frequencies) must be well below the 3dB point. This is easy to check. Suppose we have the input signal

$$V_{\text{in}} = \sin \omega t$$

Then, using the equation we obtained earlier for the differentiator output,

$$V_{\text{out}} = RC\frac{d}{dt}\sin \omega t = \omega RC \cos \omega t$$

and so $V_{\text{out}} \ll V_{\text{in}}$ if $\omega RC \ll 1$, i.e., $RC \ll 1/\omega$. If the input signal contains a range of frequencies, this must hold for the highest frequencies present in the input.

The RC integrator (Section 1.15) is the same circuit as the low-pass filter; by similar reasoning, the criterion for a good integrator is that the lowest signal frequencies must be well above the 3dB point.

Inductors versus capacitors

Inductors could be used, instead of capacitors, in combination with resistors to make low-pass (or high-pass) filters. In practice, however, you rarely see RL low- or high-pass filters. The reason is that inductors tend to be more bulky and expensive and perform less well (i.e., they depart further from the ideal) than capacitors. If you have a choice, use a capacitor. One exception to this general statement is the use of ferrite beads and chokes in high-frequency circuits. You just string a few beads here and there in the circuit; they make the wire interconnections slightly inductive, raising the impedance at very high frequencies and preventing "oscillations," without the added resistance you would get with an RC filter. An RF "choke" is an inductor, usually a few turns of wire wound on a ferrite core, used for the same purpose in RF circuits.

□ 1.20 Phasor diagrams

There's a nice graphic method that can be very helpful when trying to understand reactive circuits. Let's take an example, namely the fact that an RC filter attenuates 3dB at a frequency $f = 1/2\pi RC$, which we derived in Section 1.19. This is true for both high-pass and low-pass filters. It is easy to get a bit confused here, because at that frequency the reactance of the capacitor equals the resistance of the resistor; so you might at first expect 6dB attenuation. That is what you would get, for example, if you were to replace the capacitor by a resistor of the same impedance (recall that 6dB means half voltage). The confusion arises because the capacitor is reactive, but the matter is clarified by a phasor diagram (Fig. 1.61). The axes are the real (resistive) and imaginary (reactive) components of the impedance. In a series circuit like this, the axes also represent the (complex) voltage, because the current is the same everywhere. So for this circuit (think of it as an R–C voltage divider) the input voltage (applied across the series R–C pair) is proportional to the length of the hypotenuse, and the output voltage (across R only) is proportional to the length of the R leg of the triangle. The diagram represents the situation at the frequency where the magnitude of the capacitor's reactance equals R, i.e., $f = 1/2\pi RC$, and shows that the ratio of output voltage to input voltage is $1/\sqrt{2}$, i.e., -3dB.

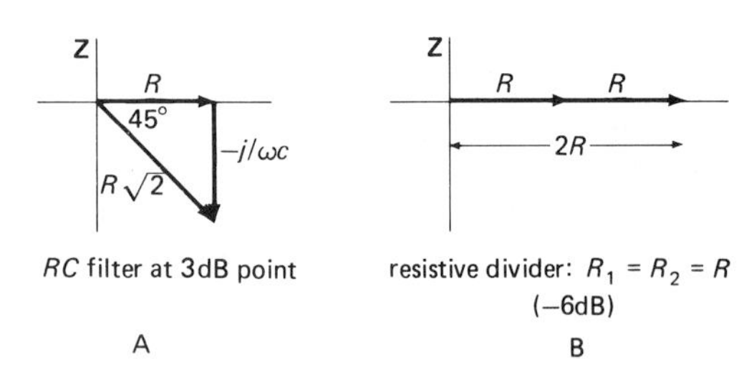

Figure 1.61

The angle between the vectors gives the phase shift from input to output. At the 3dB point, for instance, the output amplitude equals the input amplitude divided by the square root of 2, and it leads by 45° in phase. This graphic method makes it easy to read off amplitude and phase relationships in RLC circuits. For example, you can use it to get the response of the high-pass filter that we previously derived algebraically.

EXERCISE 1.23

Use a phasor diagram to derive the response of an RC high-pass filter:

$$V_{\text{out}} = \frac{R}{[R^2 + (1/\omega^2 C^2)]^{1/2}} V_{\text{in}}$$

EXERCISE 1.24

At what frequency does an RC low-pass filter attenuate by 6dB (output voltage equal to half the input voltage)? What is the phase shift at that frequency?

EXERCISE 1.25

Use a phasor diagram to obtain the low-pass filter response previously derived algebraically.

In the next chapter (Section 2.08) you will see a nice example of phasor diagrams in connection with a constant-amplitude phase-shifting circuit.

1.21 "Poles" and decibels per octave

Look again at the response of the RC low-pass filter (Fig. 1.59). Far to the right of the "knee" the output amplitude is dropping proportional to $1/f$. In one octave (as in music, one octave is twice the frequency) the output amplitude will drop to half, or -6dB; so a simple RC filter has a 6dB/octave falloff. You can make filters with several RC sections; then you get 12dB/octave (two RC sections), 18dB/octave (three sections), etc. This is the usual way of describing how a filter behaves beyond the cutoff. Another popular way is to say a "3-pole filter," for instance, meaning a filter with three RC sections (or one that behaves like one). (The word "pole" derives from a method of analysis that is beyond the scope of this book and that involves complex transfer functions in the complex frequency plane, known by engineers as the "s-plane.")

A caution on multistage filters: You can't simply cascade several identical filter sections in order to get a frequency response that is the concatenation of the individual responses. The reason is that each stage will load the previous one significantly (since they're identical), changing the overall response. Remember that the response function we derived for the simple RC filters was based on a zero-impedance driving source and an infinite-impedance load. One solution is to make each successive filter section have much higher impedance than the preceding one. A better solution involves active circuits like transistor or operational amplifier (op-amp) interstage "buffers," or active filters. These subjects will be treated in Chapters 2 through 5.

1.22 Resonant circuits and active filters

When capacitors are combined with inductors or are used in special circuits called active filters, it is possible to make circuits that have very sharp frequency characteristics (e.g., a large peak in the response at a particular frequency), as compared with the gradual characteristics of the RC filters we've seen so far. These circuits find applications in various audiofrequency and radiofrequency devices. Let's now take a quick look at LC circuits (there will be more on them, and active filters, in Chapter 5 and Appendix H).

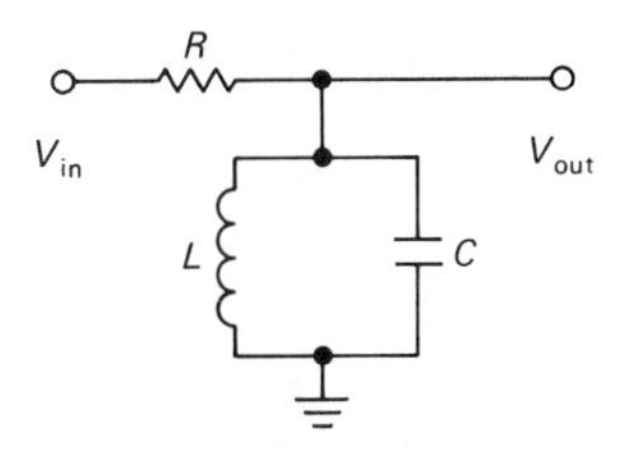

Figure 1.62. LC resonant circuit: bandpass filter.

First, consider the circuit shown in Figure 1.62. The reactance of the LC combination at frequency f is just

$$\frac{1}{\mathbf{Z}_{LC}} = \frac{1}{\mathbf{Z}_L} + \frac{1}{\mathbf{Z}_C} = \frac{1}{j\omega L} - \frac{\omega C}{j}$$
$$= j\left(\omega C - \frac{1}{\omega L}\right)$$

i.e.,

$$\mathbf{Z}_{LC} = \frac{j}{(1/\omega L) - \omega C}$$

In combination with R it forms a voltage divider; because of the opposite behaviors of inductors and capacitors, the impedance of the parallel LC goes to infinity at the *resonant frequency* $f_0 = 1/2\pi\sqrt{LC}$ (i.e., $\omega_0 = 1/\sqrt{LC}$), giving a peak in the response there. The overall response is as shown in Figure 1.63.

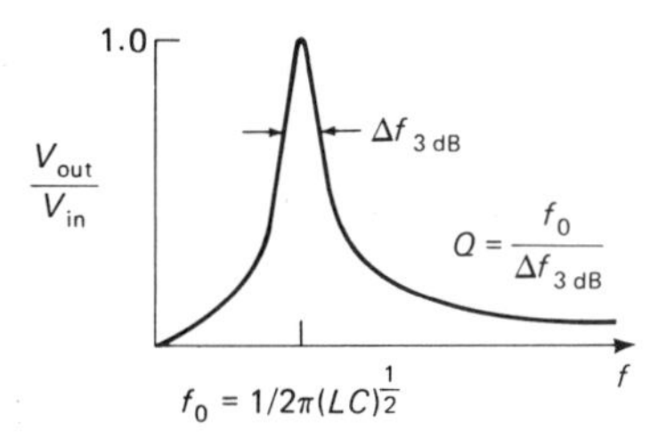

Figure 1.63

In practice, losses in the inductor and capacitor limit the sharpness of the peak, but with good design these losses can be made very small. Conversely, a Q-spoiling resistor is sometimes added intentionally to reduce the sharpness of the resonant peak. This circuit is known simply as a parallel LC resonant circuit or a tuned circuit and is used extensively in radiofrequency circuits to select a particular frequency for amplification (the L or C can be variable, so you can tune the resonant frequency). The higher the driving impedance, the sharper the peak; it is not uncommon to drive them with something approaching a current source, as you will see later. The *quality factor* Q is a measure of the sharpness of the peak. It equals the resonant frequency divided by the width at the −3dB points. For a parallel RLC circuit, $Q = \omega_0 RC$.

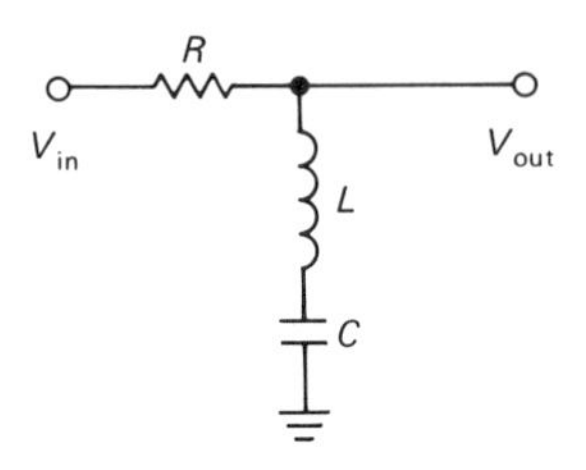

Figure 1.64. LC notch filter ("trap").

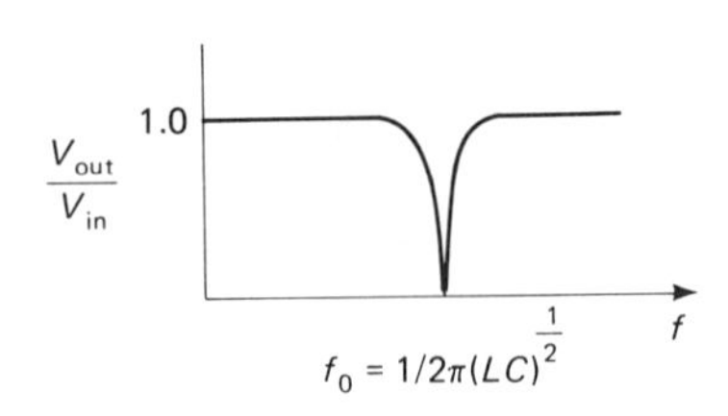

Figure 1.65

Another variety of LC circuit is the series LC (Fig. 1.64). By writing down the impedance formulas involved, you can convince yourself that the impedance of the LC goes to zero at resonance [$f_0 = 1/2\pi(LC)^{1/2}$]; such a circuit is a "trap" for signals at or near the resonant frequency, shorting them to ground. Again, this circuit finds application mainly in radiofrequency circuits. Figure 1.65 shows what the response looks like. The Q of a series RLC circuit is $Q = \omega_0 L/R$.

EXERCISE 1.26

Find the response (V_{out}/V_{in} versus frequency) for the series LC trap circuit in Figure 1.64.

1.23 Other capacitor applications

In addition to their uses in filters, resonant circuits, differentiators, and integrators, capacitors are needed for several other important applications. We will treat these in detail later in the book, mentioning them here only as a preview.

Bypassing

The impedance of a capacitor goes down with increasing frequency. This is the basis of another important application: bypassing. There are places in circuits where you want to allow a dc (or slowly varying) voltage, but don't want signals present. Placing a capacitor across that circuit element (usually a resistor) will help to kill any signals there. You choose the capacitor value so that its impedance at signal frequencies is small compared with what it is bypassing. You will see much more of this in later chapters.

Power-supply filtering

Power-supply filtering is really a form of bypassing, although we usually think of it as energy storage. The dc voltages used in electronics are usually generated from the ac line voltage by a process called *rectification* (which will be treated later in this chapter); some residue of the 60Hz input remains, and this can be reduced as much as desired by means of bypassing with suitably large capacitors. These capacitors really are large – they're the big shiny round things you see inside most electronic instruments. You will see how to design power supplies and filters later in this chapter and again in Chapter 6.

Timing and waveform generation

A capacitor supplied with a constant current charges up with a ramp waveform. This is the basis of ramp and sawtooth generators, used in function generators, oscilloscope sweep circuits, analog/digital converters, and timing circuits. RC circuits are also used for timing, and they form the basis of digital delay circuits (monostable multivibrators). These timing and waveform applications are important in many areas of electronics and will be covered in Chapters 3, 5, 8, and 9.

TABLE 1.1. DIODES

Type	$V_{R(max)}$[a] (V)	$I_{R(max)}$[b] (μA)	Continuous V_F (V)	@ I_F (mA)	Peak V_F (V)	@ I_F (A)	Reverse recovery (ns)	Capacitance (10V) (pF)	Class	Comments
PAD-1	45	1pA@20V	0.8	5	–	–	–	0.8	lowest I_R	Siliconix
FJT1100	30	0.001	–	–	1.1	0.05	–	1.2	very low I_R	1pA@5V, 10pA@15V
ID101	30	10pA@10V	0.8	1	1.1	0.03	–	0.8	very low I_R	Intersil; dual
1N3595	150	3	0.7	10	<1.0	0.2	3000	8.0	low I_R	1nA@125V
1N914	75	5	0.75	10	1.1	0.1	4	1.3	gen purp sig diode	indus std; same as 1N4148
1N6263	60	10	0.4	1	0.7	0.01	0	1.0	Schottky: low V_F	
1N3062	75	50	<1.0	20[b]	–	–	2	0.6	low cap, sig diode	1pF at 0 volts
1N4305	75	50	0.6	1	–	–	4	1.5	controlled V_F	
1N4002 }	100	50	0.9	1000	2.3	25	3500	15	1-amp rect	indus std; 7-member fam
1N4007 }	1000	50	0.9	1000	2.3	25	5000	10		
1N5819	40	10000	0.4	1000	1.1	20	–	50	pwr Schottky	lead mounted
1N5822	40	20000	0.45	3000	1.3	50	–	180	pwr Schottky	lead mounted
1N5625	400	50	1.1	5000	2.0	50	2500	45	5-amp rect	lead mounted
1N1183A	50	1000	1.1	40000	1.3	100	–	–	high curr rect	1N1183RA reverse

(a) $V_{R(max)}$ is repetitive peak reverse voltage, 25°C, 10μA leakage. (b) $I_{R(max)}$ is reverse leakage current at V_R and 100°C ambient temperature.

1.24 Thévenin's theorem generalized

When capacitors and inductors are included, Thévenin's theorem must be restated: Any two-terminal network of resistors, capacitors, inductors, and signal sources is equivalent to a single complex impedance in series with a single signal source. As before, you find the impedance and the signal source from the open-circuit output voltage and the short-circuit current.

DIODES AND DIODE CIRCUITS

1.25 Diodes

The circuit elements we've discussed so far (resistors, capacitors, and inductors) are all *linear*, meaning that a doubling of the applied signal (a voltage, say) produces a doubling of the response (a current, say). This is true even for the reactive devices (capacitors and inductors). These devices are also *passive*, meaning that they don't have a built-in source of power. And they are all two-terminal devices, which is self-explanatory.

Figure 1.66. Diode.

The diode (Fig. 1.66) is a very important and useful two-terminal passive *nonlinear* device. It has the *V–I* curve shown in Figure 1.67. (In keeping with the general philosophy of this book, we will not attempt to describe the solid-state physics that makes such devices possible.)

The diode's arrow (the anode terminal) points in the direction of forward current flow. For example, if the diode is in a circuit in which a current of 10mA is flowing from anode to cathode, then (from the graph) the anode is approximately 0.5 volt more positive than the cathode; this is called the "forward voltage drop." The reverse current, which is measured in the nanoamp range for a general-purpose diode (note the different scales in the graph for forward and reverse current), is almost never of any consequence until you reach the reverse breakdown voltage (also called the peak inverse voltage, PIV), typically 75 volts for a general-purpose diode like the 1N914. (Normally you never subject a diode to voltages large enough to cause reverse breakdown; the exception is the zener diode we mentioned earlier.) Frequently, also, the forward voltage drop of about 0.5 and 0.8 volt is of little concern, and the diode can be treated as a good approximation to an ideal one-way conductor. There are other important characteristics that distinguish the thousands of diode types available, e.g., maximum forward current, capacitance, leakage current, and reverse recovery time (see Table 1.1 for characteristics of some typical diodes).

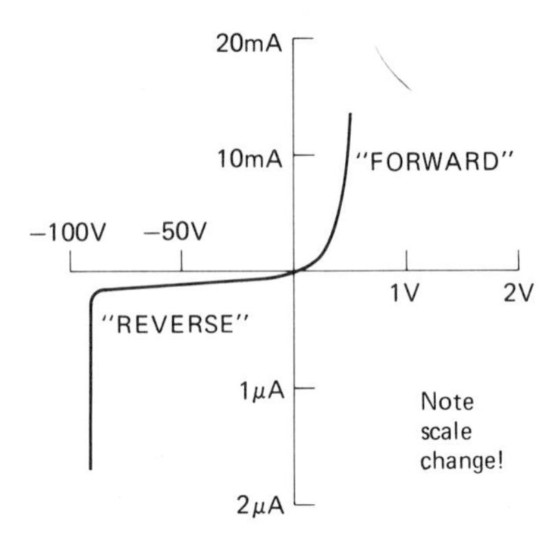

Figure 1.67. Diode *V–I* curve.

Before jumping into some circuits with diodes, we should point out two things: (a) A diode doesn't actually have a resistance (it doesn't obey Ohm's law). (b) If you put some diodes in a circuit, it won't have a Thévenin equivalent.

1.26 Rectification

A rectifier changes ac to dc; this is one of the simplest and most important applications of diodes (diodes are sometimes called rectifiers). The simplest circuit is shown in Figure 1.68. The "ac" symbol represents a source of ac voltage; in electronic circuits it is usually provided by a transformer, powered from the ac power line. For a sine-wave input that is much larger than the forward drop (about 0.6V for silicon diodes, the usual type), the output will look like that in Figure 1.69. If you think of the diode as a one-way conductor, you won't have any trouble understanding how the circuit works. This circuit is called a *half-wave rectifier*, because only half of the input waveform is used.

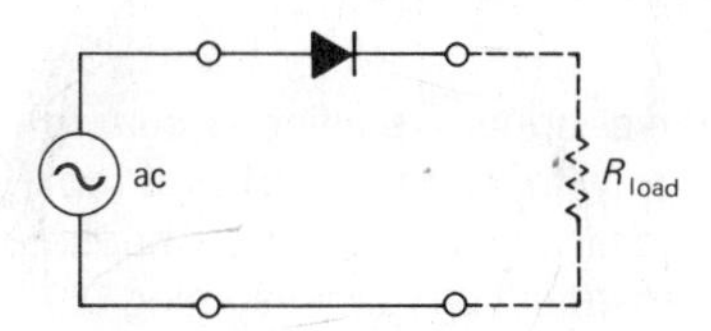

Figure 1.68. Half-wave rectifier.

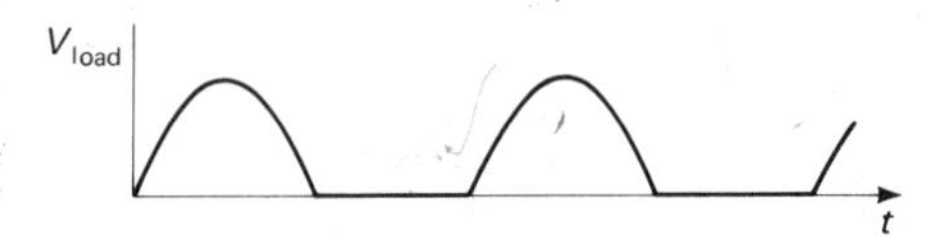

Figure 1.69

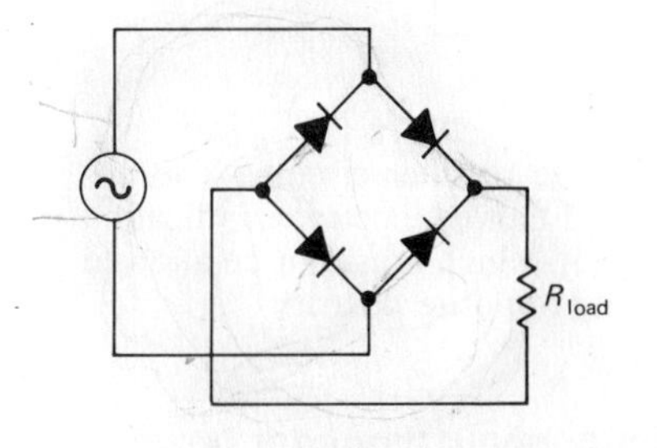

Figure 1.70. Full-wave bridge rectifier.

Figure 1.70 shows another rectifier circuit, a full-wave bridge. Figure 1.71 shows the voltage across the load for which the whole input waveform is used. The gaps at zero voltage occur because of the diodes' forward voltage drop. In this circuit, two diodes are always in series with the input; when you design low-voltage power supplies, you have to remember that.

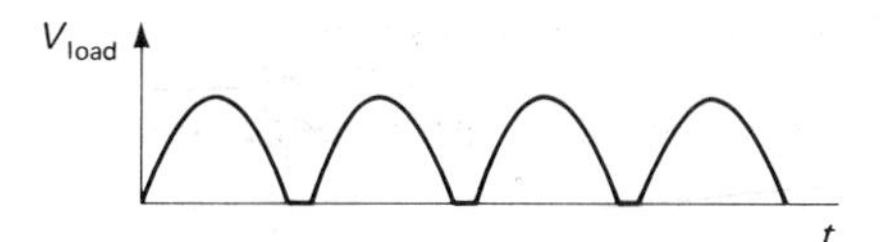

Figure 1.71

1.27 Power-supply filtering

The preceding rectified waveforms aren't good for much as they stand. They're dc only in the sense that they don't change polarity. But they still have a lot of "ripple" (periodic variations in voltage about the steady value) that has to be smoothed out in order to generate genuine dc. This we do by tacking on a low-pass filter (Fig. 1.72). Actually, the series resistor is unnecessary and is always omitted (although you sometimes see a very small resistor used to limit the peak rectifier current). The reason is that the diodes prevent flow of current back out of the capacitors, which are really serving more as energy-storage devices than as part of a classic low-pass filter. The energy stored in a capacitor is $U = \frac{1}{2}CV^2$. For C in farads and V in volts, U comes out in joules (watt-seconds).

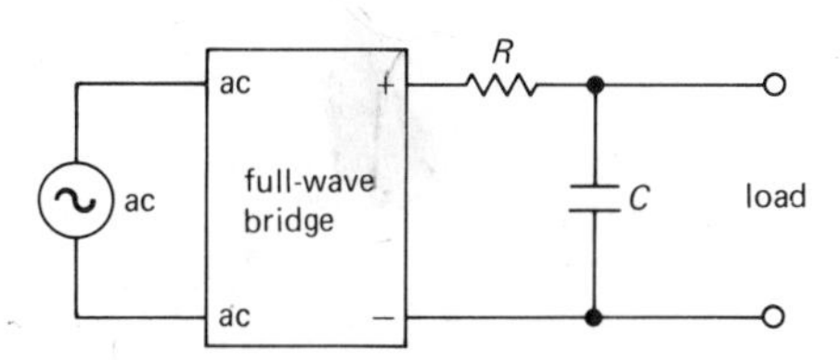

Figure 1.72

The capacitor value is chosen so that

$$R_{\text{load}}C \gg 1/f$$

(where f is the ripple frequency, here 120Hz) in order to ensure small ripple, by making the time constant for discharge much longer than the time between recharging. We will make this vague statement clearer in the next section.

Calculation of ripple voltage

It is easy to calculate the approximate ripple voltage, particularly if it is small compared with the dc (see Fig. 1.73). The load causes the capacitor to discharge somewhat between cycles (or half cycles, for full-wave rectification). If you assume that the load current stays constant (it will, for small ripple), you have

$$\Delta V = \frac{I}{C}\Delta t \qquad \left(\text{from } I = C\frac{dV}{dt}\right)$$

Just use $1/f$ (or $1/2f$ for full-wave rectification) for Δt (this estimate is a bit on the safe side, since the capacitor begins charging again in less than a half cycle). You get

$$\Delta V = \frac{I_{\text{load}}}{fC} \qquad \text{(half wave)}$$

$$\Delta V = \frac{I_{\text{load}}}{2fC} \qquad \text{(full wave)}$$

(While teaching electronics we've noticed that students love to memorize these equations! An informal poll of the authors showed that two out of two engineers don't memorize them. Please don't waste brain cells that way – instead, learn how to derive them.)

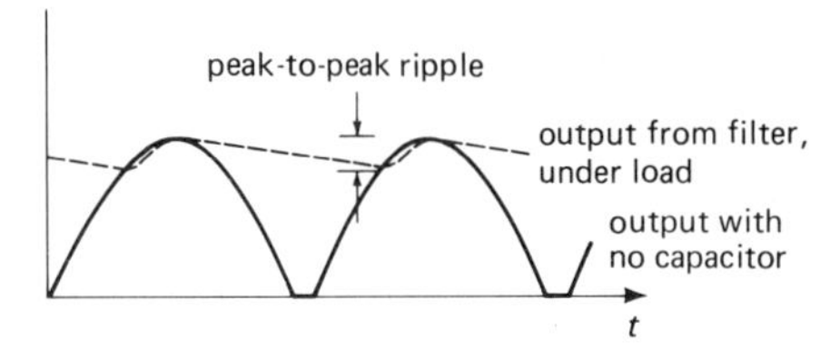

Figure 1.73. Power-supply ripple calculation.

If you wanted to do the calculation without any approximation, you would use the exact exponential discharge formula. You would be misguided in insisting on that kind of accuracy, though, for two reasons:

1. The discharge is an exponential only if the load is a resistance; many loads are not. In fact, the most common load, a *voltage regulator*, looks like a constant-current load.

2. Power supplies are built with capacitors with typical tolerances of 20% or more. Realizing the manufacturing spread, you design conservatively, allowing for the worst-case combination of component values.

In this case, viewing the initial part of the discharge as a ramp is in fact quite accurate, especially if the ripple is small, and in any case it errs in the direction of conservative design – it overestimates the ripple.

EXERCISE 1.27

Design a full-wave bridge rectifier circuit to deliver 10 volts dc with less than 0.1 volt (pp) ripple into a load drawing up to 10mA. Choose the appropriate ac input voltage, assuming 0.6 volt diode drops. Be sure to use the correct ripple frequency in your calculation.

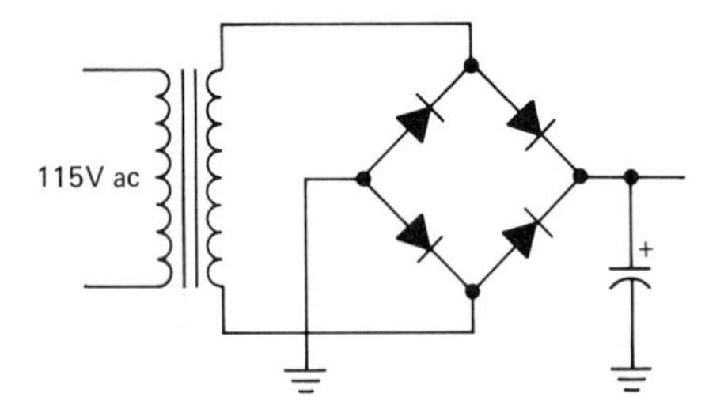

Figure 1.74. Bridge rectifier circuit. The polarity marking and curved electrode indicate a polarized capacitor, which must not be allowed to charge with the opposite polarity.

1.28 Rectifier configurations for power supplies

Full-wave bridge

A dc power supply using the bridge circuit we just discussed looks as shown in Figure 1.74. In practice, you generally buy the bridge as a prepackaged module. The smallest ones come with maximum current ratings of 1 amp average, with breakdown voltages going from 100 volts to 600 volts, or even 1000 volts. Giant bridge rectifiers are available with current ratings of 25 amps or more. Take a look at Table 6.4 for a few types.

Center-tapped full-wave rectifier

The circuit in Figure 1.75 is called a center-tapped full-wave rectifier. The output voltage is half what you get if you use a bridge rectifier. It is not the most efficient circuit in terms of transformer design, because each half of the secondary is used only half the time. Thus the current through the winding during that time is twice what it would be for a true full-wave circuit. Heating in the windings, calculated from Ohm's law, is I^2R, so you have four times the heating half the time, or twice the average heating of an equivalent full-wave bridge circuit. You would have to choose a transformer with a current rating 1.4 (square root of 2) times as large, as compared with the (better) bridge circuit; besides costing more, the resulting supply would be bulkier and heavier.

EXERCISE 1.28

This illustration of I^2R heating may help you understand the disadvantage of the center-tapped rectifier circuit. What fuse rating (minimum) is required to pass the current waveform shown in Figure 1.76, which has 1 amp average current? Hint: A fuse "blows out" by melting (I^2R heating) a metallic link, for steady currents larger than its rating. Assume for this problem that the thermal time constant of the fusible link is much longer than the time scale of the square wave, i.e., that the fuse responds to the value of I^2 averaged over many cycles.

Split supply

A popular variation of the center-tapped full-wave circuit is shown in Figure 1.77. It gives you split supplies (equal plus and minus voltages), which many circuits need. It is an efficient circuit, because both halves of the input waveform are used in each winding section.

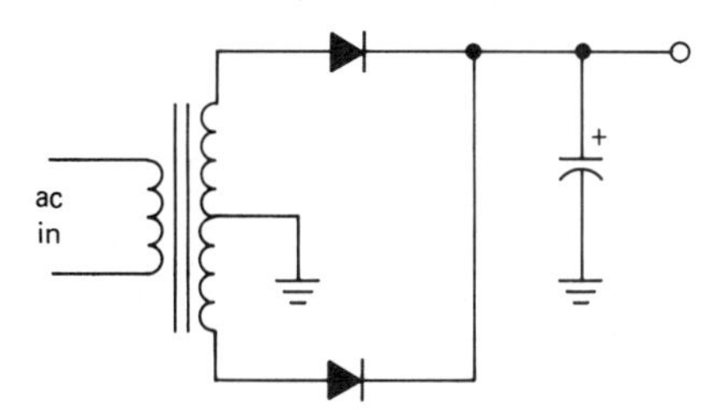

Figure 1.75. Full-wave rectifier using center-tapped transformer.

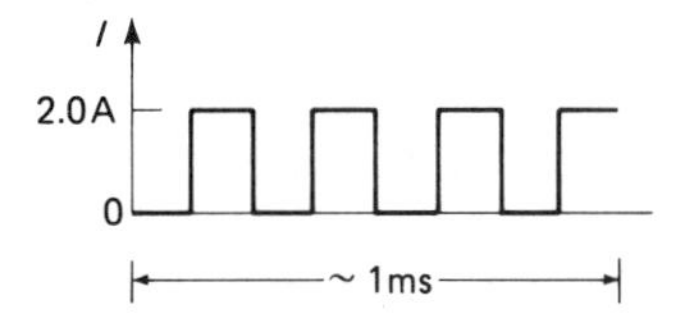

Figure 1.76

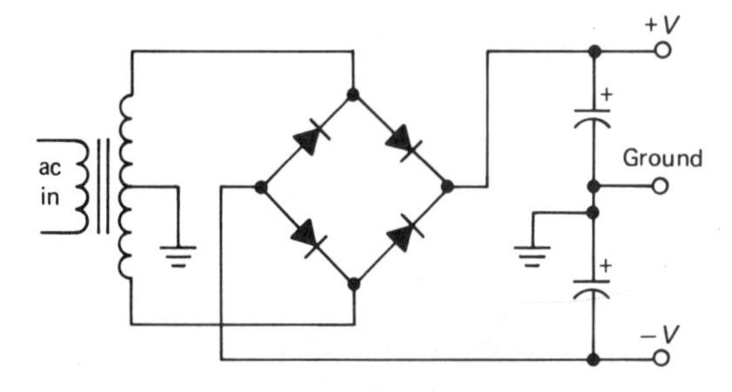

Figure 1.77. Dual-polarity (split) supply.

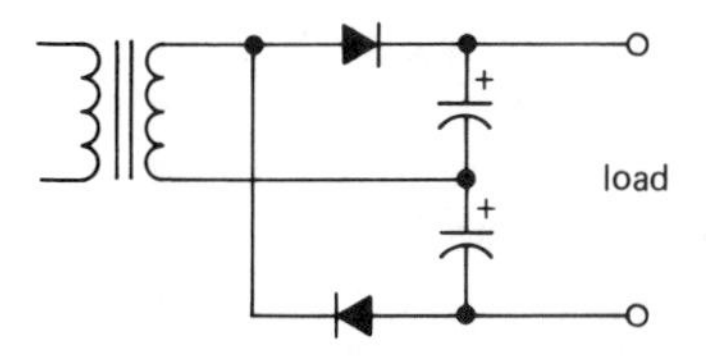

Figure 1.78. Voltage doubler.

□ *Voltage multipliers*

The circuit shown in Figure 1.78 is called a voltage doubler. Think of it as two half-wave rectifier circuits in series. It is officially a full-wave rectifier circuit, since

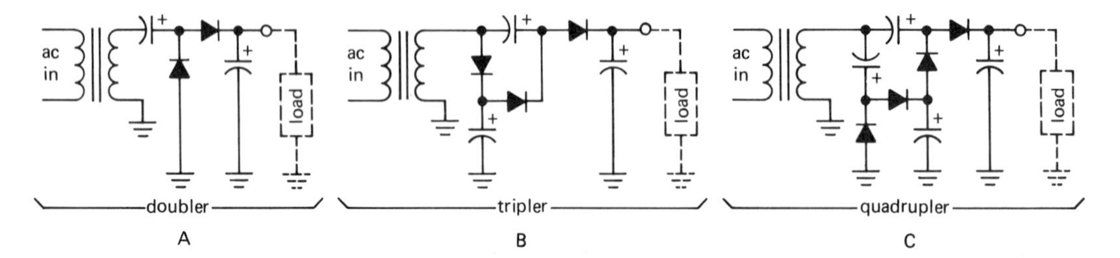

Figure 1.79. Voltage multipliers; these configurations don't require a floating voltage source.

both halves of the input waveform are used – the ripple frequency is twice the ac frequency (120Hz for the 60Hz line voltage in the United States).

Variations of this circuit exist for voltage triplers, quadruplers, etc. Figure 1.79 shows doubler, tripler, and quadrupler circuits that let you ground one side of the transformer.

1.29 Regulators

By choosing capacitors that are sufficiently large, you can reduce the ripple voltage to any desired level. This brute-force approach has two disadvantages:

1. The required capacitors may be prohibitively bulky and expensive.

2. Even with the ripple reduced to negligible levels, you still have variations of output voltage due to other causes, e.g., the dc output voltage will be roughly proportional to the ac input voltage, giving rise to fluctuations caused by input line voltage variations. In addition, changes in load current will cause the output voltage to change because of the finite internal resistances of the transformer, diode, etc. In other words, the Thévenin equivalent circuit of the dc power supply has $R > 0$.

A better approach to power-supply design is to use enough capacitance to reduce ripple to low levels (perhaps 10% of the dc voltage), then use an active *feedback circuit* to eliminate the remaining ripple. Such a feedback circuit "looks at" the output, making changes in a controllable series resistor (a transistor) as necessary to keep the output constant (Fig. 1.80).

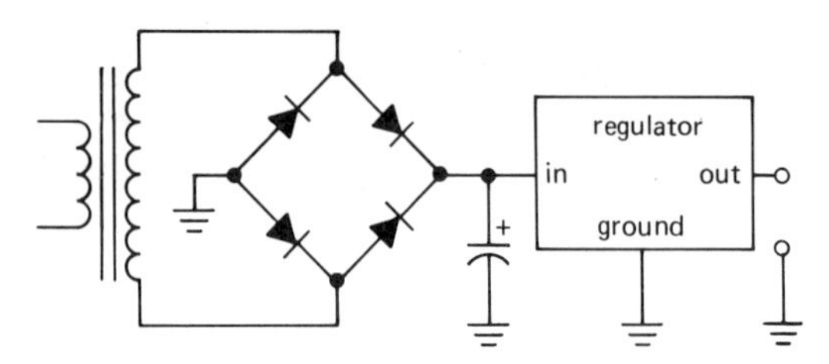

Figure 1.80. Regulated dc power supply.

These voltage regulators are used almost universally as power supplies for electronic circuits. Nowadays complete voltage regulators are available as inexpensive integrated circuits (priced under one dollar). A power supply built with a voltage regulator can be made easily adjustable and self-protecting (against short circuits, overheating, etc.), with excellent properties as a voltage source (e.g., internal resistance measured in milliohms). We will deal with regulated dc power supplies in Chapter 6.

1.30 Circuit applications of diodes

Signal rectifier

There are other occasions when you use a diode to make a waveform of one polarity only. If the input waveform isn't a sine wave, you usually don't think of it as a rectification in the sense of a power supply. For instance, you might want a train of pulses corresponding to the rising edge of a square wave. The easiest way is to rectify the differentiated wave (Fig. 1.81). Always keep in mind the 0.6 volt (approximately) forward drop of the diode. This circuit, for instance, gives no output for square waves smaller than 0.6 volt pp. If this is a problem, there are various tricks to circumvent this limitation. One possibility is to use *hot carrier diodes* (Schottky diodes), with a forward drop of about 0.25 volt (another device called a *back diode* has nearly zero forward drop, but its usefulness is limited by very low reverse breakdown voltage).

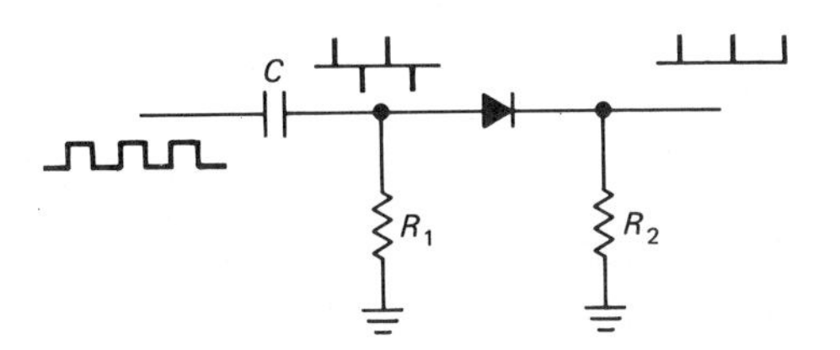

Figure 1.81

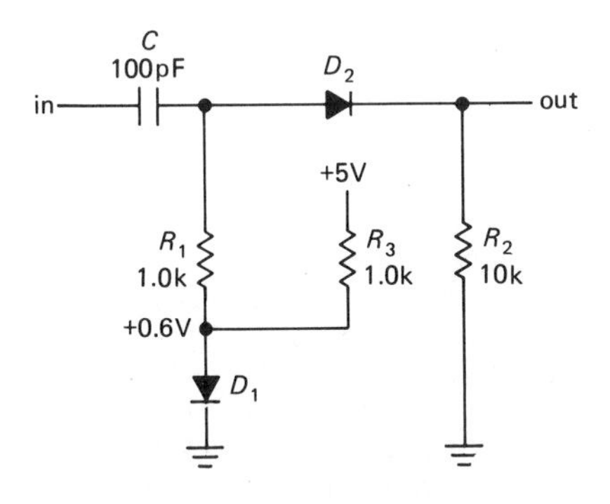

Figure 1.82. Compensating the forward voltage drop of a diode signal rectifier.

A possible circuit solution to this problem of finite diode drop is shown in Figure 1.82. Here D_1 compensates D_2's forward drop by providing 0.6 volt of *bias* to hold D_2 at the threshold of conduction. Using a diode (D_1) to provide the bias (rather than, say, a voltage divider) has several advantages: There is nothing to adjust, the compensation will be nearly perfect, and changes of the forward drop (e.g., with changing temperature) will be compensated properly. Later we will see other instances of matched-pair compensation of forward drops in diodes, transistors, and FETs: it is a simple and powerful trick.

Diode gates

Another application of diodes, which we will recognize later under the general heading of *logic*, is to pass the higher of two voltages without affecting the lower. A good example is *battery backup*, a method of keeping something running (e.g, a precision electronic clock) that must not stop when there is a power failure. Figure 1.83 shows a circuit that does the job. The battery does nothing until the power fails; then it takes over without interruption.

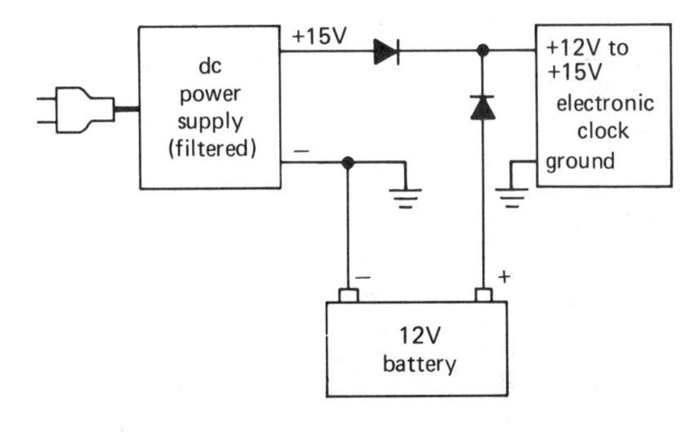

Figure 1.83. Diode OR gate: battery backup.

EXERCISE 1.29

Make a simple modification to the circuit so that the battery is charged by the dc supply (when power is on, of course) at a current of 10mA (such a circuit is necessary to maintain the battery's charge).

Diode clamps

Sometimes it is desirable to limit the range of a signal (i.e., prevent it from exceeding certain voltage limits) somewhere in a circuit. The circuit shown in Figure 1.84 will accomplish this. The diode prevents the output from exceeding about +5.6 volts, with no effect on voltages less than that (including negative voltages); the only limitation is that the input must not go so negative that the reverse breakdown voltage of the diode is exceeded (e.g., −70V for a 1N914). Diode clamps are standard equipment on all inputs in the CMOS family of digital logic. Without them, the delicate input circuits are easily destroyed by static electricity discharges during handling.

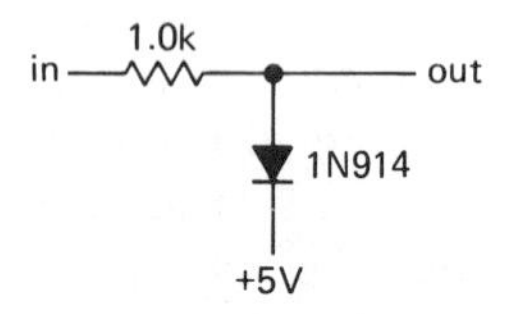

Figure 1.84. Diode voltage clamp.

EXERCISE 1.30

Design a symmetrical clamp, i.e., one that confines a signal to the range −5.6 volts to +5.6 volts.

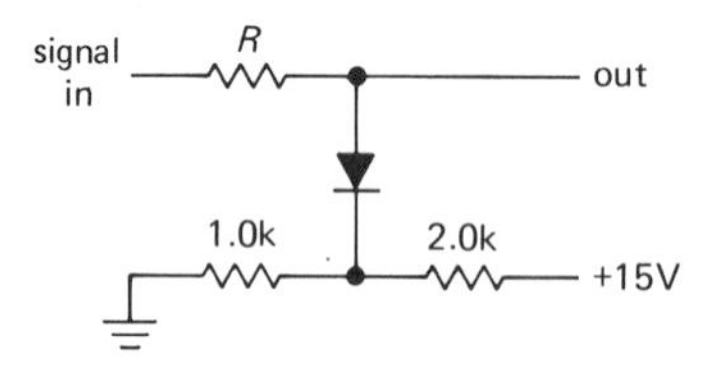

Figure 1.85

A voltage divider can provide the reference voltage for a clamp (Fig. 1.85). In this case you must ensure that the impedance looking into the voltage divider (R_{vd}) is small compared with R, because what you have looks as shown in Figure 1.86 when the voltage divider is replaced by its Thévenin equivalent circuit. When the diode conducts (input voltage exceeds clamp voltage), the output is really just the output of a voltage divider, with the Thévenin equivalent resistance of the voltage reference as the lower resistor (Fig. 1.87). So, for the values shown, the output of the clamp for a triangle-wave input would look as shown in Figure 1.88. The problem is that the voltage divider doesn't provide a stiff reference, in the language of electronics. A stiff voltage source is one that doesn't bend easily, i.e., it has low internal (Thévenin) impedance.

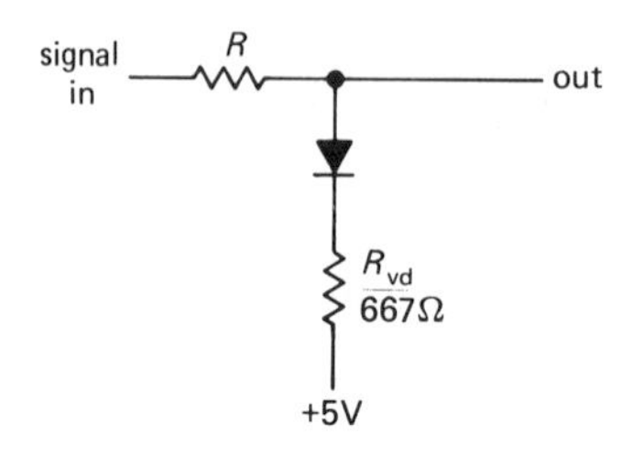

Figure 1.86

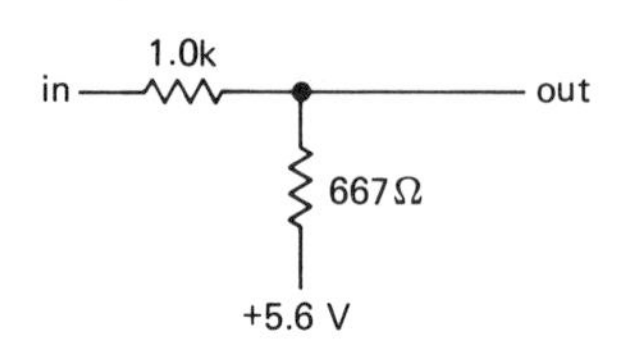

Figure 1.87

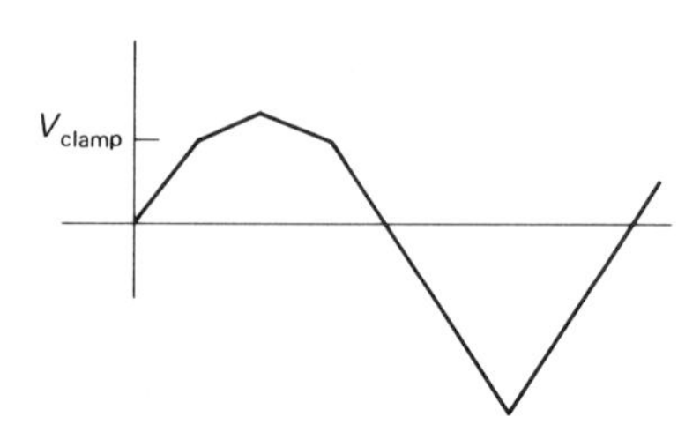

Figure 1.88

A simple way to stiffen the clamp circuit of Figure 1.85, at least for *high-frequency* signals, is to add a bypass capacitor across the 1k resistor. For example, a 15μF capacitor to ground reduces the impedance seen looking into the divider below 10 ohms for frequencies above 1kHz. (You could similarly add a bypass capacitor across D_1 in Fig. 1.82.) Of course, the effectiveness of this trick drops at low frequencies, and it does nothing at dc.

In practice, the problem of finite impedance of the voltage-divider reference can be easily solved using a transistor or operational amplifier (op-amp). This is usually a better solution than using very small resistor values, because it doesn't consume large currents, yet it provides impedances of a few ohms or less. Furthermore, there are other ways to construct a clamp, using an op-amp as part of the clamp circuit. You will see these methods in Chapter 4.

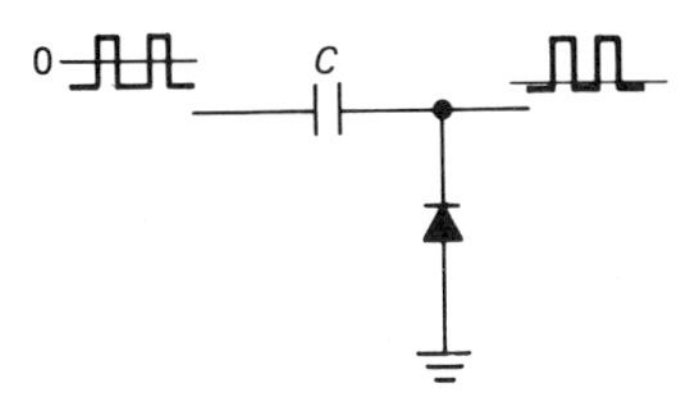

Figure 1.89. dc restoration.

One interesting clamp application is "dc restoration" of a signal that has been ac-coupled (capacitively coupled). Figure 1.89 shows the idea. This is particularly important for circuits whose inputs look like diodes (e.g., a transistor with grounded emitter); otherwise an ac-coupled signal will just fade away.

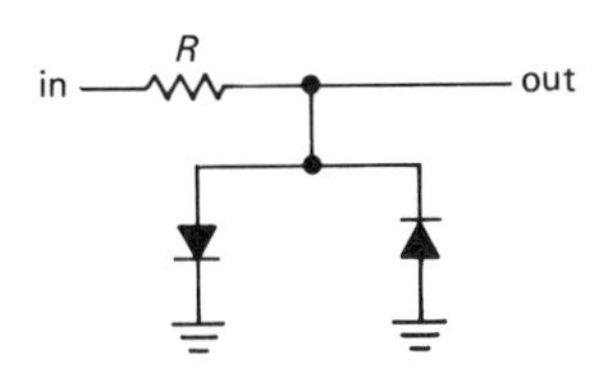

Figure 1.90. Diode limiter.

Limiter

One last clamp circuit is shown in Figure 1.90. This circuit limits the output "swing" (again, a common electronics term) to one diode drop, roughly 0.6 volt. That might seem awfully small, but if the next stage is an amplifier with large voltage amplification, its input will always be near zero volts; otherwise the output is in "saturation" (e.g., if the next stage has a gain of 1000 and operates from ±15V supplies, its input must stay in the range ±15mV in order for its output not to saturate). This clamp circuit is often used as input protection for a high-gain amplifier.

Diodes as nonlinear elements

To a good approximation the forward current through a diode is proportional to an exponential function of the voltage across it at a given temperature (for a discussion of the exact law, see Section 2.10). So you can use a diode to generate an output voltage proportional to the logarithm of a current (Fig. 1.91). Because V hovers in the region of 0.6 volt, with only small voltage changes that reflect input current variations, you can generate the input current with a resistor if the input voltage is much larger than a diode drop (Fig. 1.92).

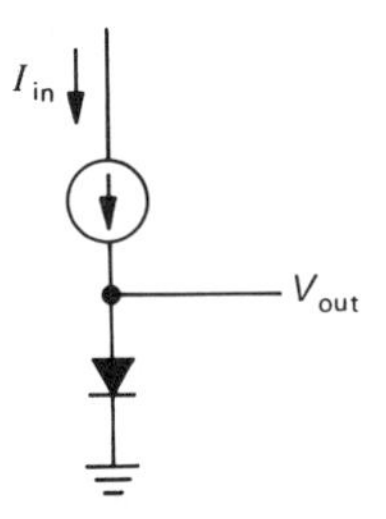

Figure 1.91. Exploiting the diode's nonlinear V–I curve: logarithmic converter.

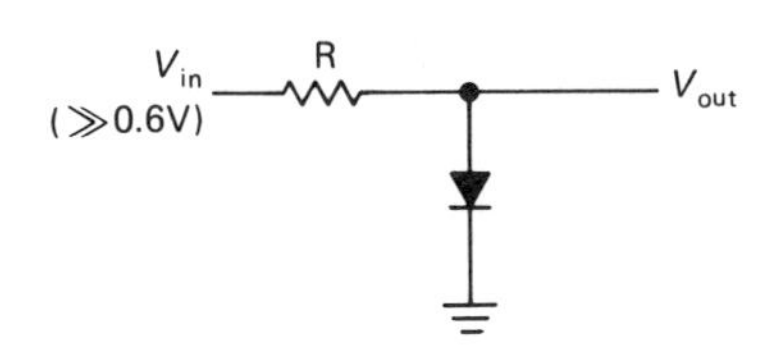

Figure 1.92

In practice, you may want an output voltage that isn't offset by the 0.6 volt diode drop. In addition, it would be nice to have a circuit that is insensitive to changes in temperature. The method of diode drop compensation is helpful here (Fig. 1.93). R_1 makes D_2 conduct, holding point A at about -0.6 volt. Point B is then near ground (making I_{in} accurately proportional to V_{in}, incidentally). As long as the two (identical) diodes are at the same temperature, there is good cancellation of the forward drops, except, of course, for the difference owing to input current through D_1, which produces the desired output. In this circuit, R_1 should be chosen so that the current through D_2 is much larger than the maximum input current, in order to keep D_2 in conduction.

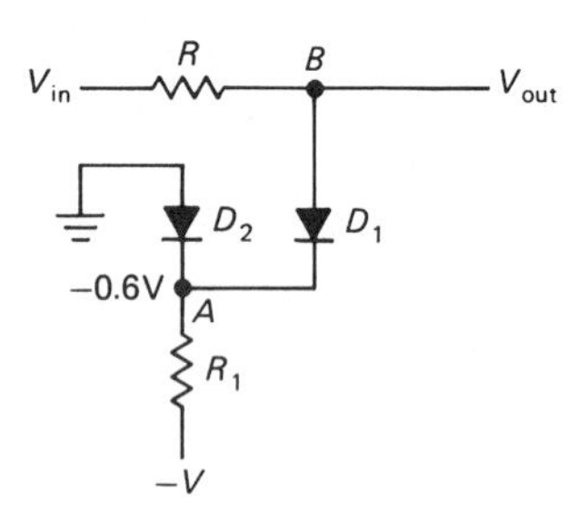

Figure 1.93. Diode drop compensation in the logarithmic converter.

In the chapter on op-amps we will examine better ways of constructing logarithmic converter circuits, along with careful methods of temperature compensation. With such methods it is possible to construct logarithmic converters accurate to a few percent over six decades or more of input current. A better understanding of diode and transistor characteristics, along with an understanding of op-amps, is necessary first. This section is meant to serve only as an introduction for things to come.

1.31 Inductive loads and diode protection

What happens if you open a switch that is providing current to an inductor? Because inductors have the property

$$V = L\frac{dI}{dt}$$

it is not possible to turn off the current suddenly, since that would imply an infinite voltage across the inductor's terminals. What happens instead is that the voltage across the inductor suddenly rises and keeps rising until it forces current to flow. Electronic devices controlling inductive loads can be easily damaged, especially the component that "breaks down" in order to satisfy the inductor's craving for continuity of current. Consider the circuit in Figure 1.94. The switch is initially closed, and current is flowing through the inductor (which might be a relay, as will be described later). When the switch is opened, the inductor "tries" to keep current flowing from A to B, as it had been. That means that terminal B goes positive relative to terminal A. In a case like this it may go 1000 volts positive before the switch contact "blows over." This shortens the life of the switch and also generates impulsive interference that may affect other circuits nearby. If the switch happens to be a transistor, it would be an understatement to say that its life is shortened; its life is ended!

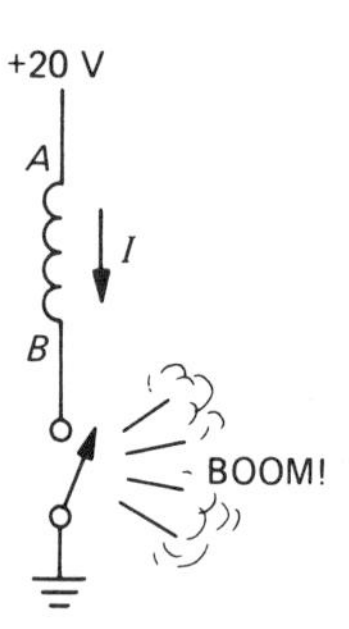

Figure 1.94. Inductive "kick."

The best solution is to put a diode across the inductor, as in Figure 1.95. When the switch is on, the diode is back-biased (from the dc drop across the inductor's winding resistance). At turn-off the diode goes into conduction, putting the switch terminal a diode drop above the positive supply voltage. The diode must be able to handle the initial diode current, which equals the steady current that had been flowing through the inductor; something like a 1N4004 is fine for nearly all cases.

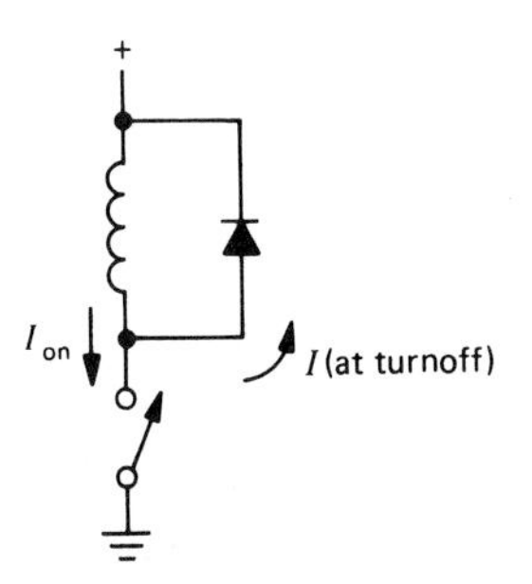

Figure 1.95. Blocking inductive kick.

The only disadvantage of this protection circuit is that it lengthens the decay of current through the inductor, since the rate of change of inductor current is proportional to the voltage across it. For applications where the current must decay quickly (high-speed impact printers, high-speed relays, etc.), it may be better to put a resistor across the inductor, choosing its value so that $V_{supply} + IR$ is less than the maximum allowed voltage across the switch. (For fastest decay with a given maximum voltage, a zener could be used instead, giving a ramp-down of current rather than an exponential decay.)

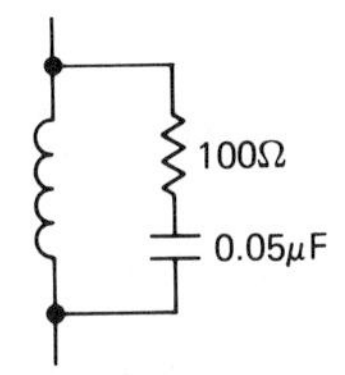

Figure 1.96. *RC* "snubber" for suppressing inductive kick.

For inductors driven from ac (transformers, ac relays), the diode protection just described will not work, since the diode will conduct on alternate half cycles when the switch is closed. In that case a good solution is an *RC* "snubber" network (Fig. 1.96). The values shown are typical for small inductive loads driven from the ac power line. Such a snubber should be included in all instruments that run from the ac power line, since a transformer is inductive. An alternative protection device is a metal-oxide varistor, or transient suppressor, an inexpensive device that looks something like a disc ceramic capacitor and behaves electrically like a bi-directional zener diode. They are available at voltage ratings from 10 to 1000 volts and can handle transient currents up to thousands of amperes (see Section 6.11 and Table 6.2). Putting a transient suppressor across the ac power-line terminals makes good sense in a piece of electronic equipment, not only to prevent inductive spike interference to other nearby instruments but also to prevent occasional large power-line spikes from damaging the instrument itself.

OTHER PASSIVE COMPONENTS

In the following sections we would like to introduce briefly an assortment of miscellaneous but essential components. If you are experienced in electronic construction, you may wish to proceed to the next chapter.

1.32 Electromechanical devices

Switches

These mundane but important devices seem to wind up in most electronic equipment. It is worth spending a few paragraphs on the subject. Figure 1.97 shows some common switch types.

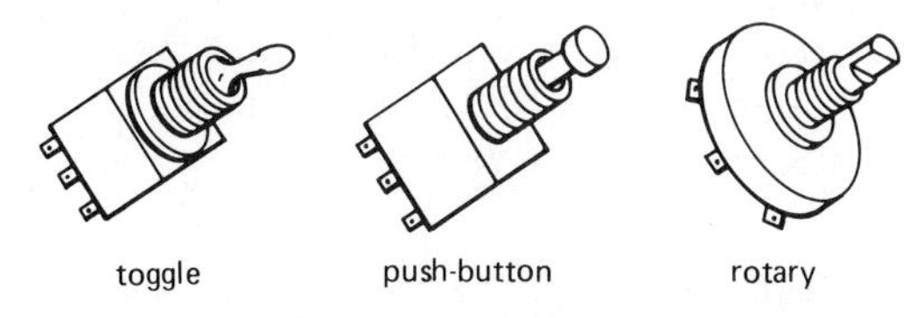

Figure 1.97. Panel switches.

Toggle switches. The simple toggle switch is available in various configurations, depending on the number of poles; Figure 1.98 shows the usual ones (SPDT indicates a single-pole double-throw switch, etc.). Toggle switches are also available with "center OFF" positions and with up to 4 poles switched simultaneously. Toggle switches are always "break before make," e.g., the moving contact never connects to both terminals in an SPDT switch.

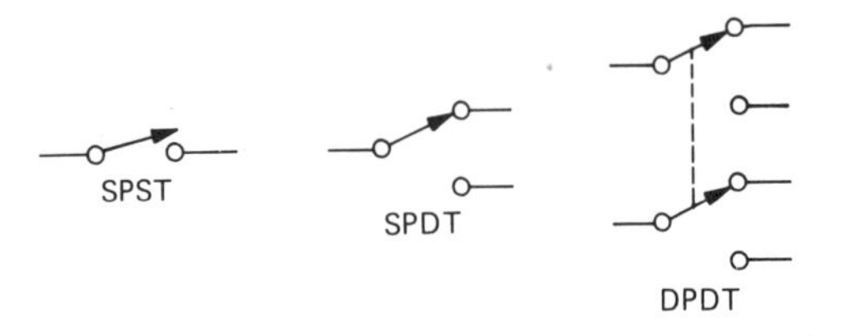

Figure 1.98. Fundamental switch types.

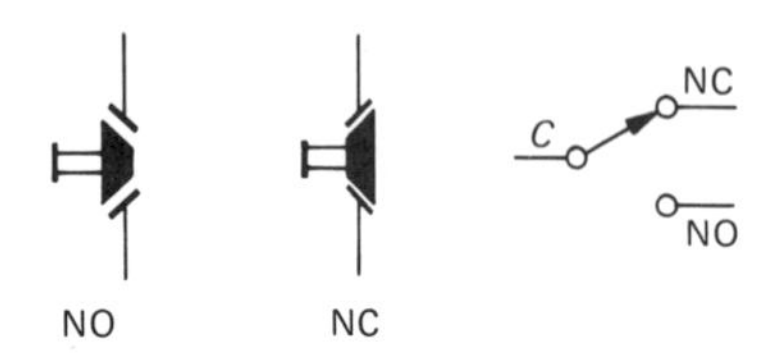

Figure 1.99. Momentary-contact (push-button) switches.

Push-button switches. Push-button switches are useful for momentary-contact applications; they are drawn schematically as shown in Figure 1.99 (NO and NC mean normally open and normally closed). For SPDT momentary-contact switches, the terminals must be labeled NO and NC, whereas for SPST types the symbol is self-explanatory. Momentary-contact switches are always "break before make." In the electrical (as opposed to electronic) industry, the terms form A, form B, and form C are used to mean SPST (NO), SPST (NC), and SPDT, respectively.

Rotary switches. Rotary switches are available with many poles and many positions, often as kits with individual wafers and shaft hardware. Both shorting (make before break) and nonshorting (break before make) types are available, and they can be mixed on the same switch. In many applications the shorting type is useful to prevent an open circuit between switch positions, because circuits can go amok with unconnected inputs. Nonshorting types are necessary if the separate lines being switched to one common line must not ever be connected to each other.

Other switch types. In addition to these basic switch types, there are available various exotic switches such as Hall-effect switches, reed switches, proximity switches, etc. All switches carry maximum current and voltage ratings; a small toggle switch might be rated at 150 volts and 5 amps. Operation with inductive loads drastically reduces switch life because of arcing during turn-off.

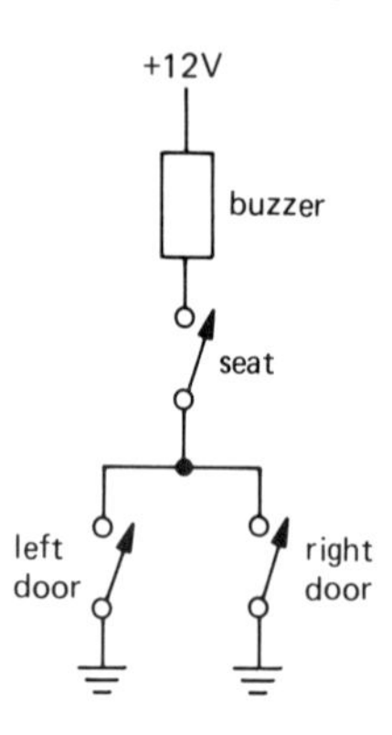

Figure 1.100

Switch examples. As an example of what can be done with simple switches, let's consider the following problem: Suppose you want to sound a warning buzzer if the driver of a car is seated and one of the car doors is open. Both doors and the driver's seat have switches, all normally open. Figure 1.100 shows a circuit that does what you want. If one OR the other door is open (switch closed) AND the seat switch is closed, the buzzer sounds. The words OR and AND are used in a logic sense here, and we will see this example again in Chapters 2 and 8 when we talk about transistors and digital logic.

Figure 1.101 shows a classic switch circuit used to turn a ceiling lamp on or off from a switch at either of two entrances to a room.

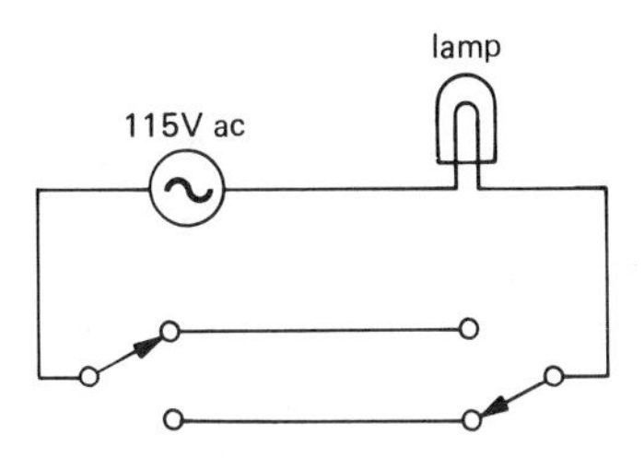

Figure 1.101. Electrician's "three-way" switch wiring.

EXERCISE 1.31

Although few electronic circuit designers know how, every *electrician* can wire up a light fixture so that any of N switches can turn it on or off. See if you can figure out this generalization of Figure 1.101. It requires two SPDT switches and $N-2$ DPDT switches. (Hint: First figure out how to use a DPDT switch to crisscross a pair of wires.)

Relays

Relays are electrically controlled switches. In the usual type, a coil pulls in an armature when sufficient coil current flows. Many varieties are available, including "latching" and "stepping" relays; the latter provided the cornerstone for telephone switching stations, and they're still popular in pinball machines. Relays are available for dc or ac excitation, and coil voltages from 5 volts up to 110 volts are common. "Mercury-wetted" and "reed" relays are intended for high-speed (~1ms) applications, and giant relays intended to switch thousands of amps are used by power companies. Many previous relay applications are now handled with transistor or FET switches, and devices known as solid-state relays are now available to handle ac switching applications. The primary uses of relays are in remote switching and high-voltage (or high-current) switching. Because it is important to keep electronic circuits electrically isolated from the ac power line, relays are useful to switch ac power while keeping the control signals electrically isolated.

Connectors

Bringing signals in and out of an instrument, routing signal and dc power around between the various parts of an instrument, providing flexibility by permitting circuit boards and larger modules of the instrument to be unplugged (and replaced) – these are the functions of the connector, an essential ingredient (and usually the most unreliable part) of any piece of electronic equipment. Connectors come in a bewildering variety of sizes and shapes.

Single-wire connectors. The simplest kind of connector is the simple pin jack or banana jack used on multimeters, power supplies, etc. It is handy and inexpensive, but not as useful as the shielded-cable or multiwire connectors you often need. The humble binding post is another form of single-wire connector, notable for the clumsiness it inspires in those who try to use it.

Shielded-cable connectors. In order to prevent capacitive pickup, and for other reasons we'll go into in Chapter 13, it is usually desirable to pipe signals around from one instrument to another in shielded coaxial cable. The most popular connector is the BNC ("baby N" connector) type that adorns most instrument front panels. It connects with a quarter-turn twist and completes both the shield (ground) circuit and inner conductor (signal) circuit simultaneously. Like all connectors used to mate a cable to an instrument, it comes in both panel-mounting and cable-terminating varieties (Fig. 1.102).

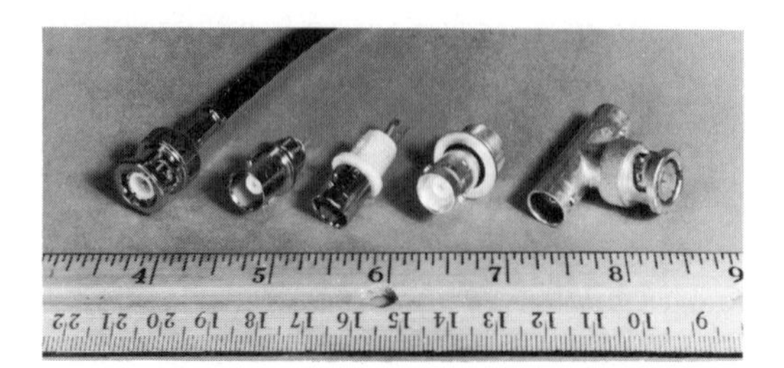

Figure 1.102. BNC connectors are the most popular type for use with shielded (coaxial) cable. From left to right: A male connector on a length of cable, a standard panel-mounted female connector, two varieties of insulated panel-mounted female connectors, and a BNC "T," a handy device to have in the laboratory.

Figure 1.103. A selection of popular multipin connectors. From left to right: *D* subminiature type, available in panel- and cable-mounting versions, with 9, 15, 25, 37, or 50 pins; the venerable MS-type connector, available in many (too many!) pin and mounting configurations, including types suitable for shielded cables; a miniature rectangular connector (Winchester MRA type) with integral securing jackscrews, available in several sizes; a circuit-board-mounting mass-termination connector with its mating female ribbon connector.

Among the other connectors for use with coaxial cable are the TNC (a close cousin of the BNC, but with threaded outer shell), the high-performance but bulky type N, the miniature SMA, the subminiature LEMO and SMC, and the MHV, a high-voltage version of the standard BNC connector. The so-called phono jack used in audio equipment is a nice lesson in bad design, because the inner conductor mates before the shield when you plug it in; furthermore, the design of the connector is such that both shield and center conductor tend to make poor contact. You've undoubtedly *heard* the results! Not to be outdone, the television industry has responded with its own bad standard, the type F coax "connector," which uses the unsupported inner wire of the coax as the pin of the male plug, and a shoddy arrangement to mate the shield.

Multipin connectors. Very frequently electronic instruments demand multiwire cables and connectors. There are literally dozens of different kinds. The simplest example is a 3-wire line cord connector. Among the more popular are the excellent type D subminiature, the Winchester MRA series, the venerable MS type, and the flat ribbon-cable mass-termination connectors (Fig. 1.103). Beware of connectors that can't tolerate being dropped on the floor (the miniature hexagon connectors are classic) or that don't provide a secure locking mechanism (e.g., the Jones 300 series).

Card-edge connectors. The most common method used to make connection to printed-circuit cards is the card-edge connector, which mates to a row of gold-plated contacts at the edge of the card. Card-edge connectors may have from 15 to 100 pins, and they come with different lug styles according to the method of connection. You can solder them to a "motherboard" or "backplane," which is itself just another printed-circuit board containing the interconnecting wiring between the individual circuit cards. Alternatively, you may want to use edge connectors with standard solder-lug terminations, particularly in a system with only a few cards (see Chapter 12 for some photographs).

1.33 Indicators

Meters

To read out the value of some voltage or current, you have a choice between the time-honored moving-pointer type of meter and digital-readout meters. The latter are more expensive and more accurate. Both types are available in a variety of voltage and current ranges. There are, in addition, exotic panel meters that read out such things as VU (volume units, an audio dB scale), expanded-scale ac volts (e.g., 105 to 130 volts), temperature (from a thermocouple), percentage motor load, frequency, etc. Digital panel meters often provide the option of logic-level outputs, in addition to the visible display, for internal use by the instrument.

Lamps and LEDs

Flashing lights, screens full of numbers and letters, eerie sounds - these are the stuff of science fiction movies, and except for the latter, they form the subject of lamps and displays (see Section 9.10). Small incandescent lamps used to be standard for front-panel indicators, but they have been replaced by light-emitting diodes (LEDs). The latter behave electrically like ordinary diodes, but with a forward voltage drop in the range of 1.5 to 2.5 volts. When current flows in the forward direction, they light up. Typically, 5mA to 20mA produces adequate brightness. LEDs are cheaper than incandescent lamps, they last forever, and they are even available in three colors (red, yellow, and green). They come in convenient panel-mounting packages; some even provide built-in current limiting.

LEDs are also used for digital displays, most often the familiar 7-segment numeric display you see in calculators. For displaying letters as well as numbers (alphanumeric display), you can get 16-segment displays or dot-matrix displays. For low power or outdoor use, liquid-crystal displays are superior.

1.34 Variable components

Resistors

Variable resistors (also called volume controls, potentiometers, pots, or trimmers) are useful as panel controls or internal adjustments in circuits. The most common panel type is known as a 2 watt type AB potentiometer; it uses the same basic material as the fixed carbon-composition resistor, with a rotatable "wiper" contact. Other panel types are available with ceramic or plastic resistance elements, with improved characteristics. Multiturn types (3, 5, or 10 turns) are available, with counting dials, for improved resolution and linearity. "Ganged" pots (several independent sections on one shaft) are also manufactured, although in limited variety, for applications that demand them.

For use inside an instrument, rather than on the front panel, *trimmer pots* come in single-turn and multiturn styles, most intended for printed-circuit mounting. These are handy for calibration adjustments of the "set-and-forget" type. Good advice: Resist the temptation to use lots of trimmers in your circuits. Use good design instead.

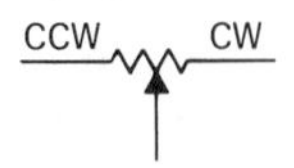

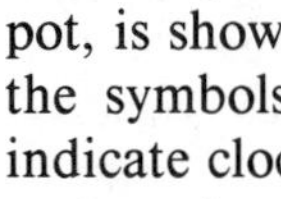

Figure 1.104. Potentiometer (3-terminal variable resistor).

The symbol for a variable resistor, or pot, is shown in Figure 1.104. Sometimes the symbols CW and CCW are used to indicate clockwise and counterclockwise.

One important point about variable resistors: Don't attempt to use a potentiometer as a substitute for a precise resistor value somewhere within a circuit. This is tempting, because you can trim the resistance to the value you want. The trouble is that potentiometers are not as stable as good (1%) resistors, and in addition they may not have good resolution (i.e., they can't be set to a precise value). If you must have a precise and settable resistor value somewhere, use a combination of a 1% (or better) precision resistor and a potentiometer, with the fixed resistor contributing most of the resistance. For example, if you need a 23.4k resistor, use a 22.6k (a 1% value) 1% fixed resistor in series with a 2k trimmer pot. Another possibility is to use a series combination of several precision resistors, choosing the last (and smallest) resistor to give the desired series resistance.

As you will see later, it is possible to use FETs as voltage-controlled variable resistors in some applications. Transistors can be used as variable-gain amplifiers, again controlled by a voltage. Keep an open mind when design brainstorming.

Figure 1.105. Variable capacitor.

Capacitors

Variable capacitors are primarily confined to the smaller capacitance values (up to about 1000pF) and are commonly used in radiofrequency (RF) circuits. Trimmers are available for in-circuit adjustments, in addition to the panel type for user tuning. Figure 1.105 shows the symbol for a variable capacitor.

Diodes operated with applied reverse voltage can be used as voltage-variable capacitors; in this application they're called varactors, or sometimes varicaps or epicaps. They're very important in RF applications, especially automatic frequency control (AFC), modulators, and parametric amplifiers.

Inductors

Variable inductors are usually made by arranging to move a piece of core material in a fixed coil. In this form they're available with inductances ranging from microhenrys to henrys, typically with a 2:1 tuning range for any given inductor. Also available are rotary inductors (coreless coils with a rolling contact).

Transformers

Variable transformers are very handy devices, especially the ones operated from the 115 volt ac line. They're usually "autoformers," which means that they have only one winding, with a sliding contact. They're also commonly called Variacs, and they are made by Technipower, Superior Electric, and others. Typically they provide 0 to 135 volts ac output when operated from 115 volts, and they come in current ratings from 1 amp to 20 amps or more. They're good for testing instruments that seem to be affected by powerline variations, and in any case to verify worst-case performance. Warning: Don't forget that the output is not electrically isolated from the power line, as it would be with a transformer!

ADDITIONAL EXERCISES

(1) Find the Norton equivalent circuit (a current source in parallel with a resistor) for the voltage divider in Figure 1.106. Show that the Norton equivalent gives the same output voltage as the actual circuit when loaded by a 5k resistor.

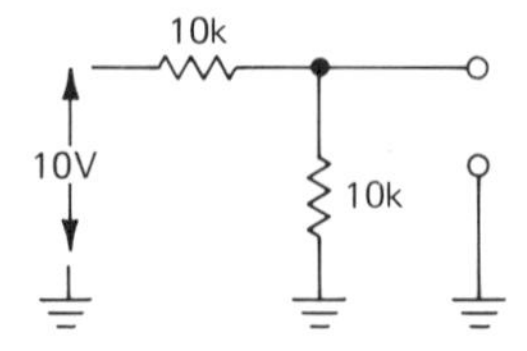

Figure 1.106

(2) Find the Thévenin equivalent for the circuit shown in Figure 1.107. Is it the

same as the Thévenin equivalent for exercise 1?

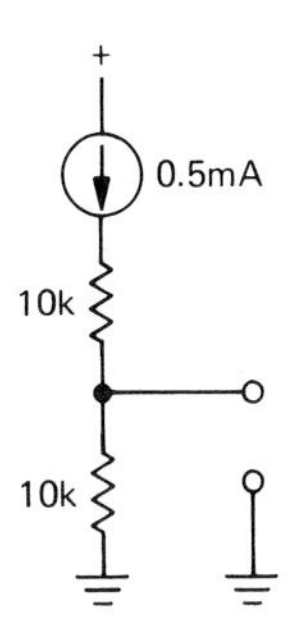

Figure 1.107

(3) Design a "rumble filter" for audio. It should pass frequencies greater than 20Hz (set the −3dB point at 10Hz). Assume zero source impedance (perfect voltage source) and 10k (minimum) load impedance (that's important so that you can choose R and C such that the load doesn't affect the filter operation significantly).

(4) Design a "scratch filter" for audio signals (3dB down at 10kHz). Use the same source and load impedances as in exercise 3.

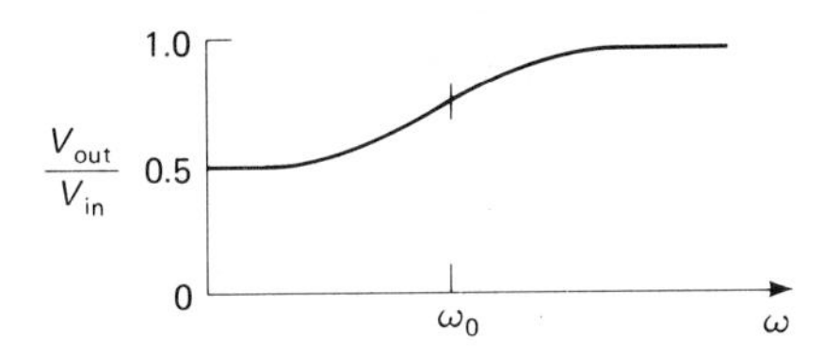

Figure 1.108

(5) How would you make a filter with Rs and Cs to give the response shown in Figure 1.108?

(6) Design a bandpass RC filter (as in Fig. 1.109); f_1 and f_2 are the 3dB points. Choose impedances so that the first stage isn't much affected by the loading of the second stage.

(7) Sketch the output for the circuit shown in Figure 1.110.

(8) Design an oscilloscope "×10 probe"

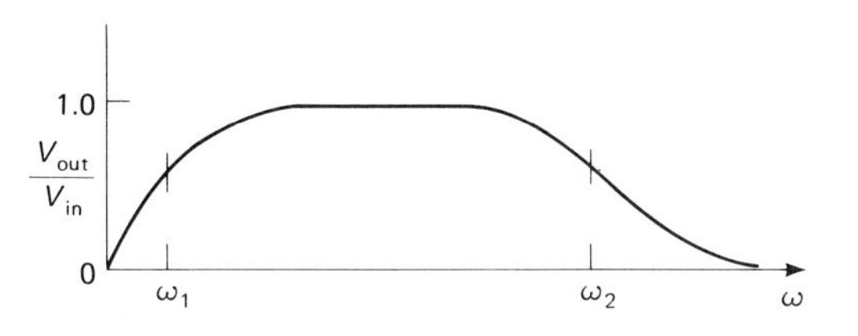

Figure 1.109

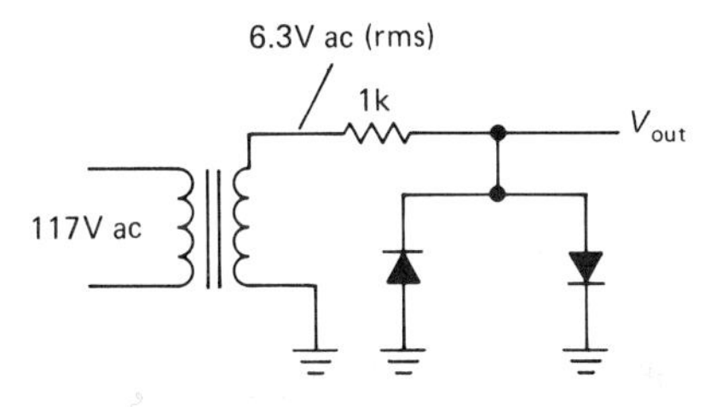

Figure 1.110

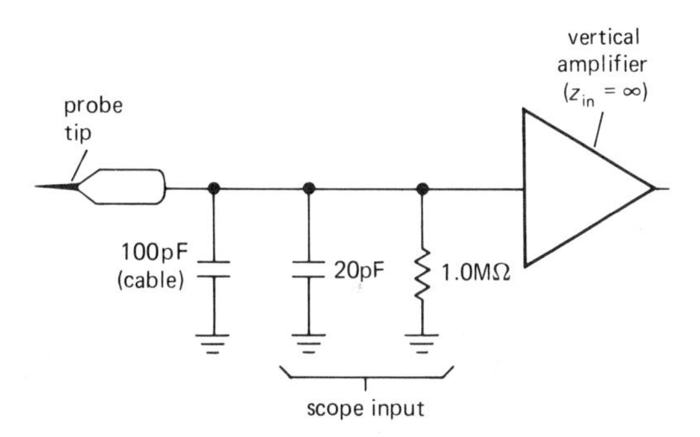

Figure 1.111

(see Appendix A) to use with a scope whose input impedance is 1MΩ in parallel with 20pF. Assume that the probe cable adds an additional 100pF and that the probe components are placed at the tip end (rather than at the scope end) of the cable (Fig. 1.111). The resultant network should have 20dB (×10) attenuation at all frequencies, including dc. The reason for using a ×10 probe is to increase the load impedance seen by the circuit under test, which reduces loading effects. What input impedance (R in parallel with C) does your ×10 probe present to the circuit under test, when used with the scope?

INTRODUCTION

The transistor is our most important example of an "active" component, a device that can amplify, producing an output signal with more power in it than the input signal. The additional power comes from an external source of power (the power supply, to be exact). Note that voltage amplification isn't what matters, since, for example, a step-up transformer, a "passive" component just like a resistor or capacitor, has voltage gain but no power gain. Devices with power gain are distinguishable by their ability to make oscillators, by feeding some output signal back into the input.

It is interesting to note that the property of power amplification seemed very important to the inventors of the transistor. Almost the first thing they did to convince themselves that they had really invented something was to power a loudspeaker from a transistor, observing that the output signal sounded louder than the input signal.

The transistor is the essential ingredient of every electronic circuit, from the simplest amplifier or oscillator to the most elaborate digital computer. Integrated circuits (ICs), which have largely replaced circuits constructed from discrete transistors, are themselves merely arrays of transistors and other components built from a single chip of semiconductor material.

A good understanding of transistors is very important, even if most of your circuits are made from ICs, because you need to understand the input and output properties of the IC in order to connect it to the rest of your circuit and to the outside world. In addition, the transistor is the single most powerful resource for interfacing, whether between ICs and other circuitry or between one subcircuit and another. Finally, there are frequent (some might say too frequent) situations where the right IC just doesn't exist, and you have to rely on discrete transistor circuitry to do the job. As you will see, transistors have an excitement all their own. Learning how they work can be great fun.

Our treatment of transistors is going to be quite different from that of many other books. It is common practice to use the h-parameter model and equivalent circuit. In our opinion that is unnecessarily complicated and unintuitive. Not only does circuit behavior tend to be revealed to you as something that drops out of elaborate equations, rather than deriving from a clear understanding in your own mind as to how the circuit functions; you also have the tendency to lose sight of which parameters of transistor behavior you can count on and, more important, which ones can vary over large ranges.

In this chapter we will build up instead a very simple introductory transistor model and immediately work out some circuits with it. Soon its limitations will become apparent; then we will expand the model to include the respected Ebers-Moll conventions. With the Ebers-Moll equations and a simple 3-terminal model, you will have a good understanding of transistors; you won't need to do a lot of calculations, and your designs will be first-rate. In particular, they will be largely independent of the poorly controlled transistor parameters such as current gain.

Some important engineering notation should be mentioned. Voltage at a transistor terminal (relative to ground) is indicated by a single subscript (C, B, or E): V_C is the collector voltage, for instance. Voltage between two terminals is indicated by a double subscript: V_{BE} is the base-to-emitter voltage drop, for instance. If the same letter is repeated, that means a power-supply voltage: V_{CC} is the (positive) power-supply voltage associated with the collector, and V_{EE} is the (negative) supply voltage associated with the emitter.

2.01 First transistor model: current amplifier

Let's begin. A transistor is a 3-terminal device (Fig. 2.1) available in 2 flavors (*npn* and *pnp*), with properties that meet the following rules for *npn* transistors (for *pnp* simply reverse all polarities):

1. The collector must be more positive than the emitter.

2. The base-emitter and base-collector circuits behave like diodes (Fig. 2.2). Normally the base-emitter diode is conducting and the base-collector diode is reverse-biased, i.e., the applied voltage is in the opposite direction to easy current flow.

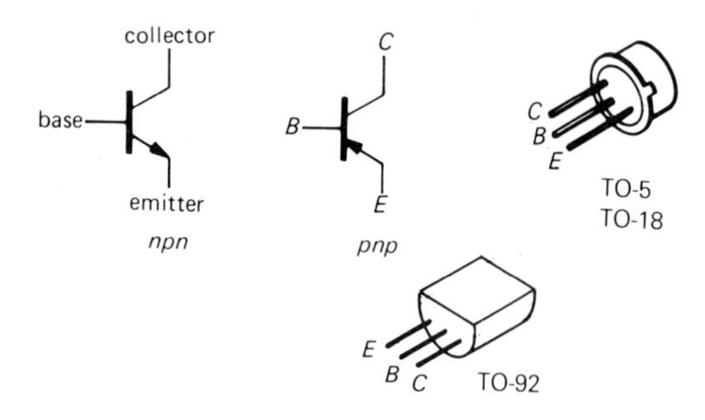

Figure 2.1. Transistor symbols, and small transistor packages.

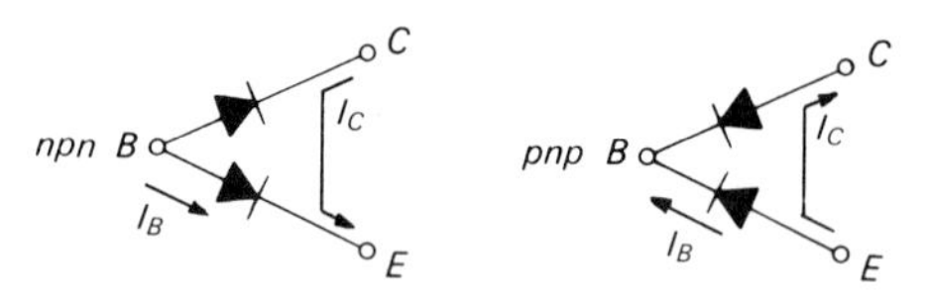

Figure 2.2. An ohmmeter's view of a transistor's terminals.

3. Any given transistor has maximum values of I_C, I_B, and V_{CE} that cannot be exceeded without costing the exceeder the price of a new transistor (for typical values, see Table 2.1). There are also other limits, such as power dissipation ($I_C V_{CE}$), temperature, V_{BE}, etc., that you must keep in mind.

4. When rules 1–3 are obeyed, I_C is roughly proportional to I_B and can be written as

$$I_C = h_{FE} I_B = \beta I_B$$

where h_{FE}, the current gain (also called beta), is typically about 100. Both I_C and I_E flow to the emitter. Note: The collector current is not due to forward conduction of the base-collector diode;

that diode is reverse-biased. Just think of it as "transistor action."

Property 4 gives the transistor its usefulness: A small current flowing into the base controls a much larger current flowing into the collector.

Warning: h_{FE} is not a "good" transistor parameter; for instance, its value can vary from 50 to 250 for different specimens of a given transistor type. It also depends upon the collector current, collector-to-emitter voltage, and temperature. *A circuit that depends on a particular value for h_{FE} is a bad circuit.*

Note particularly the effect of property 2. This means you can't go sticking a voltage across the base-emitter terminals, because an enormous current will flow if the base is more positive than the emitter by more than about 0.6 to 0.8 volt (forward diode drop). This rule also implies that an operating transistor has $V_B \approx V_E + 0.6$ volt ($V_B = V_E + V_{BE}$). Again, polarities are normally given for *npn* transistors; reverse them for *pnp*.

Let us emphasize again that you should not try to think of the collector current as diode conduction. It isn't, because the collector-base diode normally has voltages applied across it in the reverse direction. Furthermore, collector current varies very little with collector voltage (it behaves like a not-too-great current source), unlike forward diode conduction, where the current rises very rapidly with applied voltage.

SOME BASIC TRANSISTOR CIRCUITS

2.02 Transistor switch

Look at the circuit in Figure 2.3. This application, in which a small control current enables a much larger current to flow in another circuit, is called a transistor switch. From the preceding rules it is easy to understand. When the mechanical switch is open, there is no base current. So, from

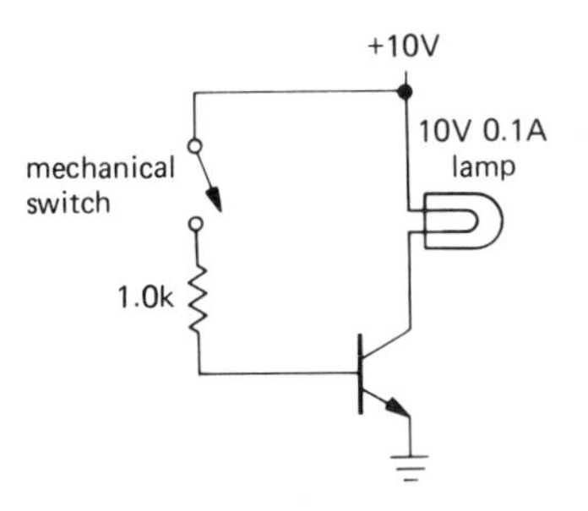

Figure 2.3. Transistor switch example.

rule 4, there is no collector current. The lamp is off.

When the switch is closed, the base rises to 0.6 volt (base-emitter diode is in forward conduction). The drop across the base resistor is 9.4 volts, so the base current is 9.4mA. Blind application of rule 4 gives $I_C = 940\text{mA}$ (for a typical beta of 100). That is wrong. Why? Because rule 4 holds only if rule 1 is obeyed; at a collector current of 100mA the lamp has 10 volts across it. To get a higher current you would have to pull the collector below ground. A transistor can't do this, and the result is what's called saturation – the collector goes as close to ground as it can (typical saturation voltages are about 0.05–0.2V, see Appendix G) and stays there. In this case, the lamp goes on, with its rated 10 volts across it.

Overdriving the base (we used 9.4mA when 1.0mA would have barely sufficed) makes the circuit conservative; in this particular case it is a good idea, since a lamp draws more current when cold (the resistance of a lamp when cold is 5 to 10 times lower than its resistance at operating current). Also transistor beta drops at low collector-to-base voltages, so some extra base current is necessary to bring a transistor into full saturation (see Appendix G). Incidentally, in a real circuit you would probably put a resistor from base to ground (perhaps 10k in this case) to make sure the base is at ground with the switch open. It wouldn't affect the "on" operation, because it would sink only 0.06mA from the base circuit.

There are certain cautions to be observed when designing transistor switches:
1. Choose the base resistor conservatively to get plenty of excess base current, especially when driving lamps, because of the reduced beta at low V_{CE}. This is also a good idea for high-speed switching, because of capacitive effects and reduced beta at very high frequencies (many megahertz). A small "speedup" capacitor is often connected across the base resistor to improve high-speed performance.
2. If the load swings below ground for some reason (e.g., it is driven from ac, or it is inductive), use a diode in series with the collector (or a diode in the reverse direction to ground) to prevent collector-base conduction on negative swings.
3. For inductive loads, protect the transistor with a diode across the load, as shown in Figure 2.4. Without the diode the inductor will swing the collector to a large positive voltage when the switch is opened, most likely exceeding the collector-emitter breakdown voltage, as the inductor tries to maintain its "on" current from V_{CC} to the collector (see the discussion of inductors in Section 1.31).

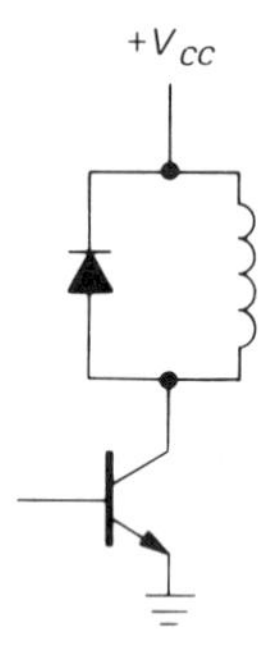

Figure 2.4. Always use a suppression diode when switching an inductive load.

Transistor switches enable you to switch very rapidly, typically in a small fraction of a microsecond. Also, you can switch many different circuits with a single control signal. One further advantage is the possibility of remote *cold switching*, in which only dc control voltages snake around through cables to reach front-panel switches, rather than the electronically inferior approach of having the signals themselves traveling through cables and switches (if you run lots of signals through cables, you're likely to get capacitive pickup as well as some signal degradation).

"Transistor man"

Figure 2.5 presents a cartoon that will help you understand some limits of transistor

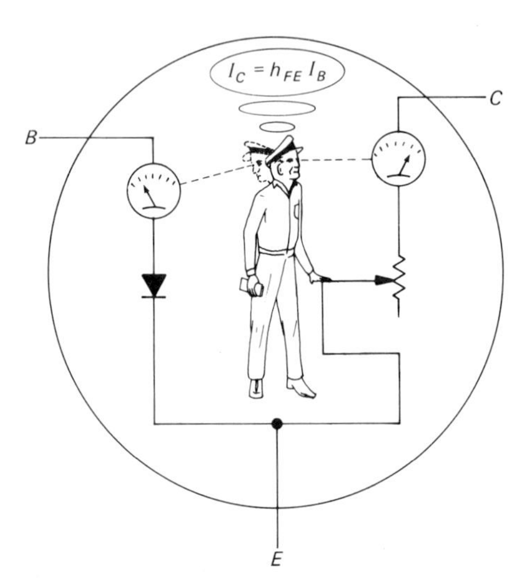

Figure 2.5. "Transistor man" observes the base current, and adjusts the output rheostat in an attempt to maintain the output current h_{FE} times larger.

behavior. The little man's perpetual task in life is to try to keep $I_C = h_{FE}I_B$; however, he is only allowed to turn the knob on the variable resistor. Thus he can go from a short circuit (saturation) to an open circuit (transistor in the "off" state), or anything in between, but he isn't allowed to use batteries, current sources, etc. One warning is in order here: Don't think that the collector of a transistor looks like a resistor. It doesn't. Rather, it looks approximately like a poor-quality constant-current sink (the value of current depending on the signal applied to the base), primarily because of this little man's efforts.

Another thing to keep in mind is that, at any given time, a transistor may be (a) cut off (no collector current), (b) in the active region (some collector current, and collector voltage more than a few tenths of a volt above the emitter), or (c) in saturation (collector within a few tenths of a volt of the emitter). See Appendix G on transistor saturation for more details.

2.03 Emitter follower

Figure 2.6 shows an example of an *emitter follower*. It is called that because the output terminal is the emitter, which follows the input (the base), less one diode drop:

$V_E \approx V_B - 0.6$ volt

The output is a replica of the input, but 0.6 to 0.7 volt less positive. For this circuit, V_{in} must stay at +0.6 volt or more, or else the output will sit at ground. By returning the emitter resistor to a negative supply voltage, you can permit negative voltage swings as well. Note that there is no collector resistor in an emitter follower.

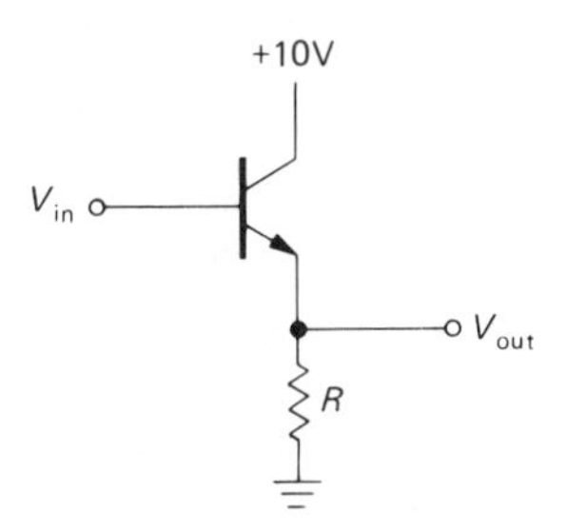

Figure 2.6. Emitter follower.

At first glance this circuit may appear useless, until you realize that the input impedance is much larger than the output impedance, as will be demonstrated shortly. This means that the circuit requires less power from the signal source to drive a given load than would be the case if the signal source were to drive the load directly. Or a signal of some internal impedance (in the Thévenin sense) can now drive a load of comparable or even lower impedance without loss of amplitude (from the usual voltage-divider effect). In other words, an emitter follower has current gain, even though it has no voltage gain. It has power gain. Voltage gain isn't everything!

Impedances of sources and loads

This last point is very important and is worth some more discussion before we calculate in detail the beneficial effects of emitter followers. In electronic circuits, you're always hooking the output of something to the input of something else, as suggested in Figure 2.7. The signal source might be the output of an amplifier stage (with Thévenin equivalent series impedance $\mathbf{Z}_{out}$), driving the next stage or perhaps a load (of some input impedance $\mathbf{Z}_{in}$). In general, the loading effect of the following stage causes a reduction of signal, as we discussed earlier in Section 1.05. For this reason it is usually best to keep $Z_{out} \ll Z_{in}$ (a factor of 10 is a comfortable rule of thumb).

In some situations it is OK to forgo this general goal of making the source stiff compared with the load. In particular, if the load is always connected (e.g., within a circuit) and if it presents a known and constant $\mathbf{Z}_{in}$, it is not too serious if it "loads" the source. However, it is always nicer if signal levels don't change when a load is connected. Also, if $\mathbf{Z}_{in}$ varies with signal level, then having a stiff source ($Z_{out} \ll Z_{in}$) assures linearity, where otherwise the level-dependent voltage divider would cause distortion.

Finally, there are two situations where $Z_{out} \ll Z_{in}$ is actually the wrong thing to

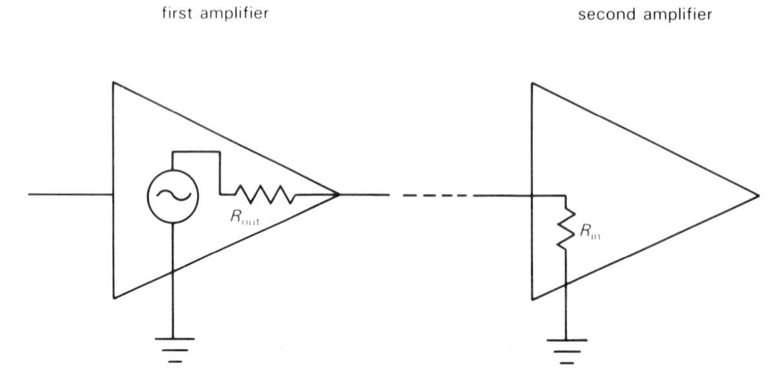

Figure 2.7. Illustrating circuit "loading" as a voltage divider.

do: In radiofrequency circuits we usually *match* impedances ($Z_{out} = Z_{in}$), for reasons we'll describe in Chapter 14. A second exception applies if the signal being coupled is a *current* rather than a voltage. In that case the situation is reversed, and one strives to make $Z_{in} \ll Z_{out}$ ($Z_{out} = \infty$, for a current source).

Input and output impedances of emitter followers

As you have just seen, the emitter follower is useful for changing impedances of signals or loads. To put it bluntly, that's the whole point of an emitter follower.

Let's calculate the input and output impedances of the emitter follower. In the preceding circuit we will consider R to be the load (in practice it sometimes is the load; otherwise the load is in parallel with R, but with R dominating the parallel resistance anyway). Make a voltage change ΔV_B at the base; the corresponding change at the emitter is $\Delta V_E = \Delta V_B$. Then the change in emitter current is

$$\Delta I_E = \Delta V_B / R$$

so

$$\Delta I_B = \frac{1}{h_{fe}+1}\Delta I_E = \frac{\Delta V_B}{R(h_{fe}+1)}$$

(using $I_E = I_C + I_B$). The input resistance is $\Delta V_B/\Delta I_B$. Therefore

$$r_{in} = (h_{fe}+1)R$$

The transistor beta (h_{fe}) is typically about 100, so a low-impedance load looks like a much higher impedance at the base; it is easier to drive.

In the preceding calculation, as in Chapter 1, we have used lower-case symbols such as h_{fe} to signify small-signal (incremental) quantities. Frequently one concentrates on the *changes* in voltages (or currents) in a circuit, rather than the steady (dc) values of those voltages (or currents). This is most common when these "small-signal" variations represent a possible signal, as in an audio amplifier, riding on a steady dc "bias" (see Section 2.05). The distinction between dc current gain (h_{FE}) and small-signal current gain (h_{fe}) isn't always made clear, and the term beta is used for both. That's alright, since $h_{fe} \approx h_{FE}$ (except at very high frequencies), and you never assume you know them accurately, anyway.

Although we used resistances in the preceding derivation, we could generalize to complex impedances by allowing ΔV_B, ΔI_B, etc., to become complex numbers. We would find that the same transformation rule applies for impedances: $\mathbf{Z}_{\text{in}} = (h_{fe} + 1)\mathbf{Z}_{\text{load}}$.

We could do a similar calculation to find that the output impedance $\mathbf{Z}_{\text{out}}$ of an emitter follower (the impedance looking into the emitter) driven from a source of internal impedance $\mathbf{Z}_{\text{source}}$ is given by

$$\mathbf{Z}_{\text{out}} = \frac{\mathbf{Z}_{\text{source}}}{h_{fe} + 1}$$

Strictly speaking, the output impedance of the circuit should also include the parallel resistance of R, but in practice $\mathbf{Z}_{\text{out}}$ (the impedance looking into the emitter) dominates.

EXERCISE 2.1

Show that the preceding relationship is correct. Hint: Hold the source voltage fixed, and find the change in output current for a given change in output voltage. Remember that the source voltage is connected to the base through a series resistor.

Because of these nice properties, emitter followers find application in many situations, e.g., making low-impedance signal sources within a circuit (or at outputs), making stiff voltage references from higher-impedance references (formed from voltage dividers, say), and generally isolating signal sources from the loading effects of subsequent stages.

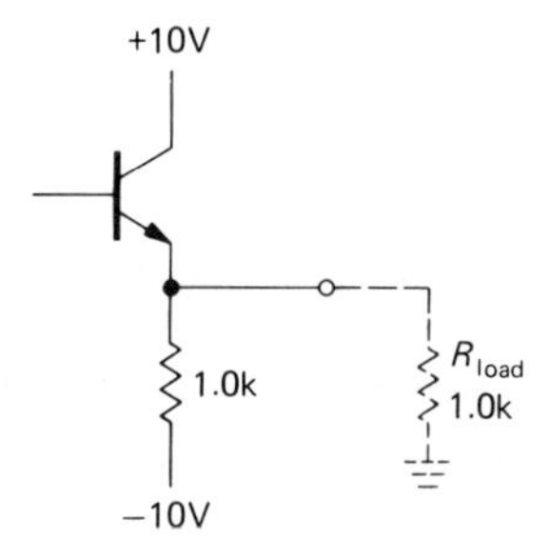

Figure 2.8. An *npn* emitter follower can source plenty of current through the transistor, but can sink limited current only through its emitter resistor.

EXERCISE 2.2

Use a follower with base driven from a voltage divider to provide a stiff source of +5 volts from an available regulated +15 volt supply. Load current (max) = 25mA. Choose your resistor values so that the output voltage doesn't drop more than 5% under full load.

Important points about followers

1. Notice (Section 2.01, rule 4) that in an emitter follower the *npn* transistor can only "source" current. For instance, in the loaded circuit shown in Figure 2.8 the output can swing to within a transistor saturation voltage drop of V_{CC} (about +9.9V), but it cannot go more negative than −5 volts. That is because on the extreme negative swing, the transistor can do no more than turn off, which it does at −4.4 volts input (−5V output). Further negative swing at the input results in backbiasing of the base-emitter junction, but no further change in output. The output, for a 10 volt amplitude sine-wave input, looks as shown in Figure 2.9.

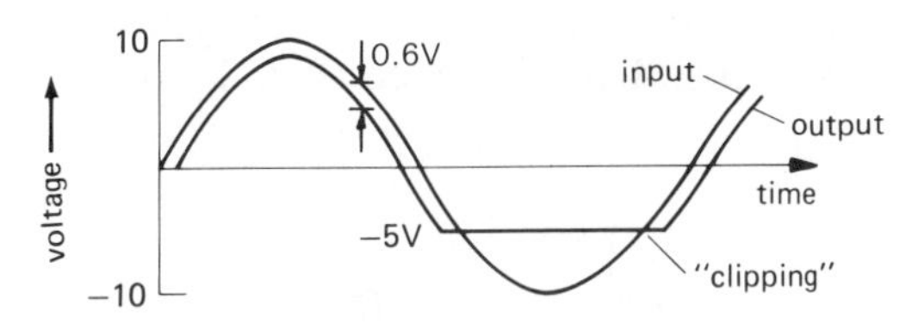

Figure 2.9. Illustrating the asymmetrical current drive capability of the *npn* emitter follower.

Another way to view the problem is to say that the emitter follower has low small-signal output impedance. Its large-signal output impedance is much larger (as large as R_E). The output impedance changes over from its small-signal value to its large-signal value at the point where the transistor goes out of the active region (in this case at an output voltage of -5V). To put this point another way, a low value of small-signal output impedance doesn't necessarily mean that the circuit can generate large signal swings into a low-resistance load. Low small-signal output impedance doesn't imply large output current capability.

Possible solutions to this problem involve either decreasing the value of the emitter resistor (with greater power dissipation in resistor and transistor), using a *pnp* transistor (if all signals are negative only), or using a "push-pull" configuration, in which two complementary transistors (one *npn*, one *pnp*), are used (Section 2.15). This sort of problem can also come up when the load of an emitter follower contains voltage or current sources of its own. This happens most often with regulated power supplies (the output is usually an emitter follower) driving a circuit that has other power supplies.

2. Always remember that the base-emitter reverse breakdown voltage for silicon transistors is small, quite often as little as 6 volts. Input swings large enough to take the transistor out of conduction can easily result in breakdown (with consequent degradation of h_{FE}) unless a protective diode is added (Fig. 2.10).

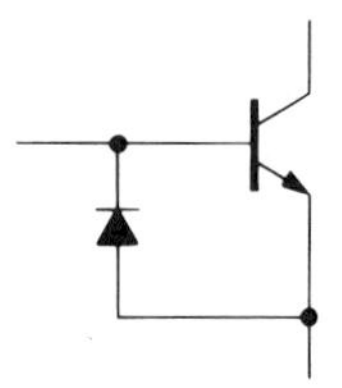

Figure 2.10. A diode prevents base-emitter reverse voltage breakdown.

3. The voltage gain of an emitter follower is actually slightly less than 1.0, because the base-emitter voltage drop is not really constant, but depends slightly on collector current. You will see how to handle that later in the chapter, when we have the Ebers-Moll equation.

2.04 Emitter followers as voltage regulators

The simplest regulated supply of voltage is simply a zener (Fig. 2.11). Some current must flow through the zener, so you choose

$$\frac{V_{in} - V_{out}}{R} > I_{out}\ (\max)$$

Because V_{in} isn't regulated, you use the lowest value of V_{in} that might occur for this formula. This is called worst-case design. In practice, you would also worry about component tolerances, line-voltage limits, etc., designing to accommodate the worst possible combination that would ever occur.

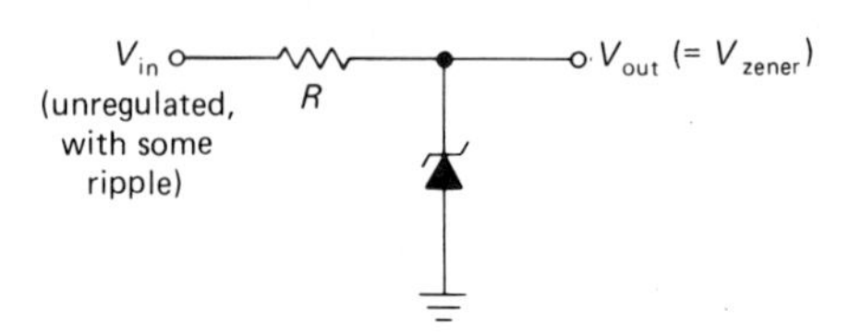

Figure 2.11. Simple zener voltage regulator.

The zener must be able to dissipate

$$P_{zener} = \left(\frac{V_{in} - V_{out}}{R} - I_{out}\right) V_{zener}$$

Again, for worst-case design, you would use V_{in} (max), R_{min}, and I_{out} (min).

EXERCISE 2.3

Design a +10 volt regulated supply for load currents from 0 to 100mA; the input voltage is +20 to +25 volts. Allow at least 10mA zener current under all (worst-case) conditions. What power rating must the zener have?

This simple zener-regulated supply is sometimes used for noncritical circuits, or circuits using little supply current. However, it has limited usefulness, for several reasons:

1. V_{out} isn't adjustable, or settable to a precise value.

2. Zener diodes give only moderate ripple rejection and regulation against changes of input or load, owing to their finite dynamic impedance.

3. For widely varying load currents a high-power zener is often necessary to handle the dissipation at low load current.

By using an emitter follower to isolate the zener, you get the improved circuit shown in Figure 2.12. Now the situation is much better. Zener current can be made relatively independent of load current, since the transistor base current is small, and far lower zener power dissipation is possible (reduced by as much as $1/h_{FE}$). The collector resistor R_C can be added to protect the transistor from momentary output short circuits by limiting the current, even though it is not essential to the emitter follower function. Choose R_C so that the voltage drop across it is less than the drop across R for the highest normal load current.

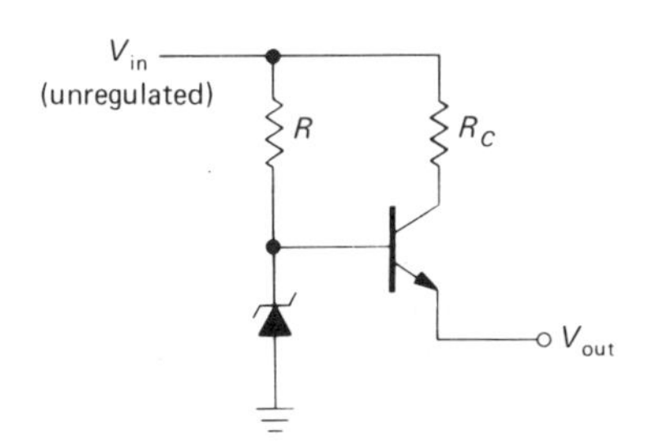

Figure 2.12. Zener regulator with follower, for increased output current. R_C protects the transistor by limiting maximum output current.

EXERCISE 2.4

Design a +10 volt supply with the same specifications as in Exercise 2.3. Use a zener and emitter follower. Calculate worst-case dissipation in transistor and zener. What is the percentage change in zener current from the no-load condition to full load? Compare with your previous circuit.

A nice variation of this circuit aims to eliminate the effect of ripple current (through R) on the zener voltage by supplying the zener current from a current source, which is the subject of Section 2.06. An alternative method uses a low-pass filter in the zener bias circuit (Fig. 2.13). R is chosen to provide sufficient zener current. Then C is chosen large enough so that $RC \gg 1/f_{\text{ripple}}$. (In a variation of this circuit, the upper resistor is replaced by a diode.)

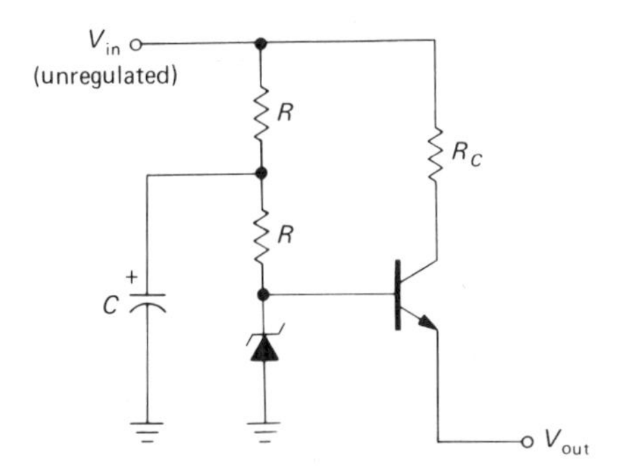

Figure 2.13. Reducing ripple in the zener regulator.

Later you will see better voltage regulators, ones in which you can vary the output easily and continuously, using feedback. They are also better voltage sources, with output impedances measured in milliohms, temperature coefficients of a few parts per million per degree centigrade, etc.

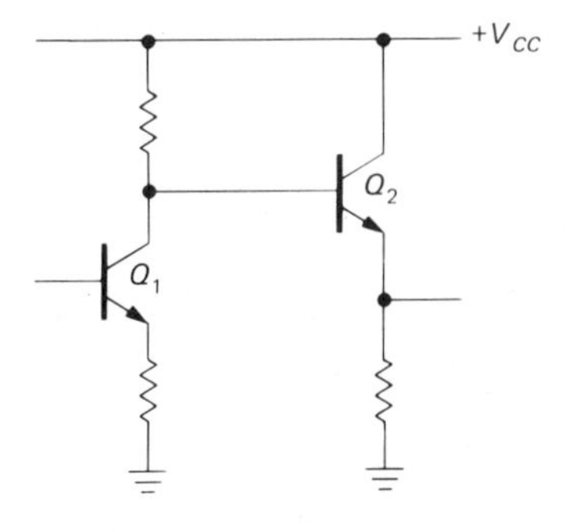

Figure 2.14

2.05 Emitter follower biasing

When an emitter follower is driven from a preceding stage in a circuit, it is usually OK to connect its base directly to the previous stage's output, as shown in Figure 2.14.

Because the signal on Q_1's collector is always within the range of the power supplies, Q_2's base will be between V_{CC} and ground, and therefore Q_2 is in the active region (neither cut off nor saturated), with its base-emitter diode in conduction and its collector at least a few tenths of a volt more positive than its emitter. Sometimes, though, the input to a follower may not be so conveniently situated with respect to the supply voltages. A typical example is a capacitively coupled (or ac-coupled) signal from some external source (e.g., an audio signal input to a high-fidelity amplifier). In that case the signal's average voltage is zero, and direct coupling to an emitter follower will give an output like that in Figure 2.15.

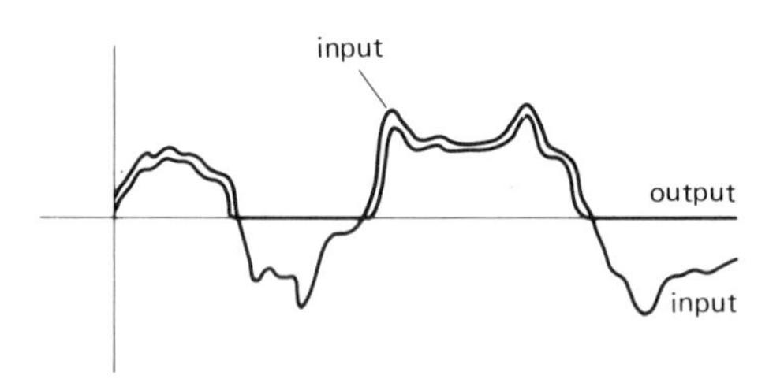

Figure 2.15. A transistor amplifier powered from a single positive supply cannot generate negative voltage swings at the transistor output terminal.

It is necessary to *bias* the follower (in fact, any transistor amplifier) so that collector current flows during the entire signal swing. In this case a voltage divider is the simplest way (Fig. 2.16). R_1 and R_2 are chosen to put the base halfway between ground and V_{CC} with no input signal, i.e., R_1 and R_2 are approximately equal. The process of selecting the operating voltages in a circuit, in the absence of applied signals, is known as setting the *quiescent point*. In this case, as in most cases, the quiescent point is chosen to allow maximum symmetrical signal swing of the output waveform without *clipping* (flattening of the top or bottom of the waveform). What values should R_1 and R_2 have? Applying our general principle (Section 1.05), we make the impedance of the dc bias source (the impedance looking into the voltage divider) small compared with the load it drives (the dc impedance looking into the base of the follower). In this case,

$$R_1 \| R_2 \ll h_{FE} R_E$$

This is approximately equivalent to saying that the current flowing in the voltage divider should be large compared with the current drawn by the base.

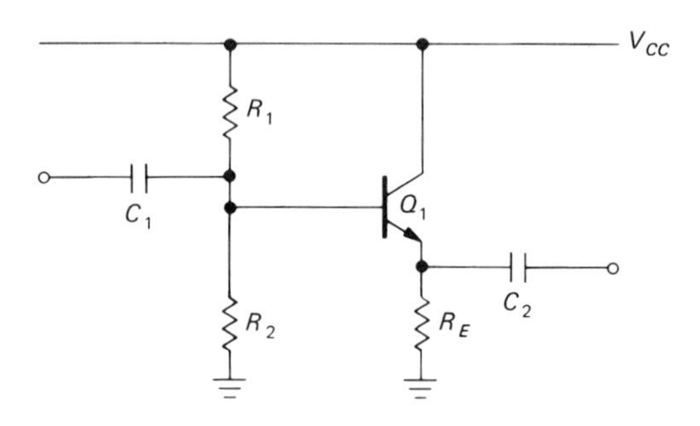

Figure 2.16. An ac-coupled emitter follower. Note base bias voltage divider.

Emitter follower design example

As an actual design example, let's make an emitter follower for audio signals (20Hz to 20kHz). V_{CC} is +15 volts, and quiescent current is to be 1mA.

Step 1. Choose V_E. For the largest possible symmetrical swing without clipping, $V_E = 0.5V_{CC}$, or +7.5 volts.

Step 2. Choose R_E. For a quiescent current of 1mA, $R_E = 7.5\text{k}$.

Step 3. Choose R_1 and R_2. V_B is V_E+ 0.6, or 8.1 volts. This determines the ratio of R_1 to R_2 as 1:1.17. The preceding loading criterion requires that the parallel resistance of R_1 and R_2 be about 75k or less (one-tenth of 7.5k times h_{FE}).

Suitable standard values are $R_1 = 130\text{k}$, $R_2 = 150\text{k}$.

Step 4. Choose C_1. C_1 forms a high-pass filter with the impedance it sees as a load, namely the impedance looking into the base in parallel with the impedance looking into the base voltage divider. If we assume that the load this circuit will drive is large compared with the emitter resistor, then the impedance looking into the base is $h_{FE}R_E$, about 750k. The divider looks like 70k. So the capacitor sees a load of about 63k, and it should have a value of at least $0.15\mu\text{F}$ so that the 3dB point will be below the lowest frequency of interest, 20Hz.

Step 5. Choose C_2. C_2 forms a high-pass filter in combination with the load impedance, which is unknown. However, it is safe to assume that the load impedance won't be smaller than R_E, which gives a value for C_2 of at least $1.0\mu\text{F}$ to put the 3dB point below 20Hz. Because there are now two cascaded high-pass filter sections, the capacitor values should be increased somewhat to prevent large attenuation (reduction of signal amplitude, in this case 6dB) at the lowest frequency of interest. $C_1 = 0.5\mu\text{F}$ and $C_2 = 3.3\mu\text{F}$ might be good choices.

Followers with split supplies

Because signals often are "near ground," it is convenient to use symmetrical positive and negative supplies. This simplifies biasing and eliminates coupling capacitors (Fig. 2.17).

Warning: You must always provide a dc path for base bias current, even if it goes only to ground. In the preceding circuit it is assumed that the signal source has a dc path to ground. If not (e.g., if the signal is capacitively coupled), you must provide a resistor to ground (Fig. 2.18). R_B could be about one-tenth of $h_{FE}R_E$, as before.

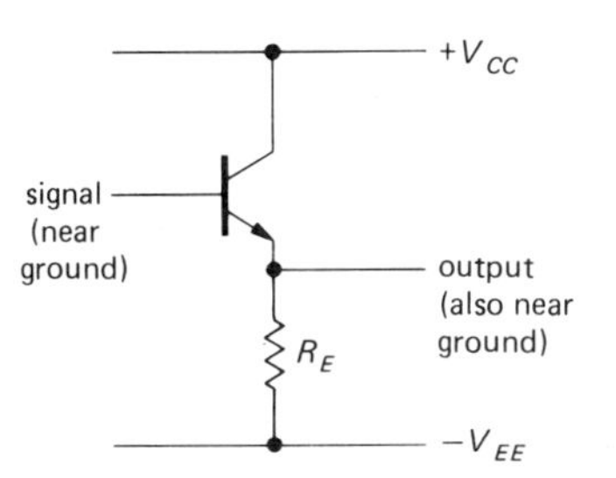

Figure 2.17. A dc-coupled emitter follower with split supply.

EXERCISE 2.5

Design an emitter follower with ±15 volt supplies to operate over the audio range (20Hz–20kHz). Use 5mA quiescent current and capacitive input coupling.

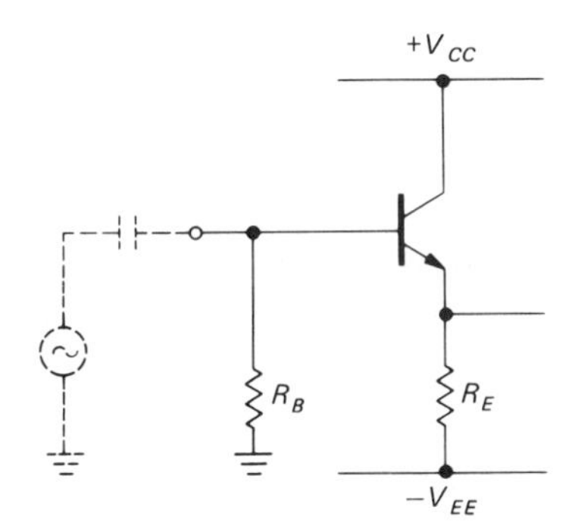

Figure 2.18

Bad biasing

Unfortunately, you sometimes see circuits like the disaster shown in Figure 2.19. R_B was chosen by assuming a particular value for h_{FE} (100), estimating the base current, and then hoping for a 7 volt drop across R_B. This is a bad design; h_{FE} is not a good parameter and will vary considerably. By using voltage biasing with a stiff voltage divider, as in the detailed example presented earlier, the quiescent point is insensitive to variations in transistor beta. For instance, in the previous design example the emitter voltage will increase by only 0.35 volt (5%) for a transistor with $h_{FE} = 200$ instead of the nominal

$h_{FE} = 100$. As with this emitter follower example, it is just as easy to fall into this trap and design bad transistor circuits in the other transistor configurations (e.g., the common-emitter amplifier, which we will treat later in this chapter).

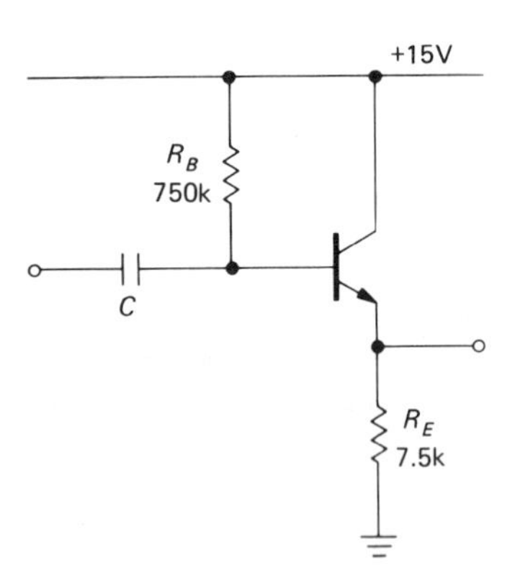

Figure 2.19. Don't do this!

2.06 Transistor current source

Current sources, although often neglected, are as important and as useful as voltage sources. They often provide an excellent way to bias transistors, and they are unequaled as "active loads" for super-gain amplifier stages and as emitter sources for differential amplifiers. Integrators, sawtooth generators, and ramp generators need current sources. They provide wide-voltage-range pull-ups within amplifier and regulator circuits. And, finally, there are applications in the outside world that require constant current sources, e.g., electrophoresis or electrochemistry.

Resistor plus voltage source

The simplest approximation to a current source is shown in Figure 2.20. As long as $R_{\text{load}} \ll R$ (in other words, $V_{\text{load}} \ll V$), the current is nearly constant and is approximately

$$I = V/R$$

The load doesn't have to be resistive. A capacitor will charge at a constant rate, as long as $V_{\text{capacitor}} \ll V$; this is just the first part of the exponential charging curve of an RC.

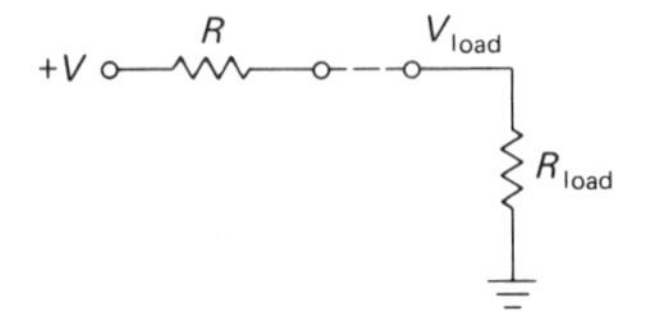

Figure 2.20

There are several drawbacks to a simple resistor current source. In order to make a good approximation to a current source, you must use large voltages, with lots of power dissipation in the resistor. In addition, the current isn't easily *programmable*, i.e., controllable over a large range via a voltage somewhere else in the circuit.

EXERCISE 2.6

If you want a current source constant to 1% over a load voltage range of 0 to +10 volts, how large a voltage source must you use in series with a single resistor?

EXERCISE 2.7

Suppose you want a 10mA current in the preceding problem. How much power is dissipated in the series resistor? How much gets to the load?

Transistor current source

Fortunately, it is possible to make a very good current source with a transistor (Fig. 2.21). It works like this : Applying V_B to the base, with $V_B > 0.6$ volt, ensures that the emitter is always conducting:

$$V_E = V_B - 0.6 \text{ volt}$$

So

$$I_E = V_E/R_E = (V_B - 0.6 \text{ volt})/R_E$$

But, since $I_E \approx I_C$ for large h_{FE},

$$I_C \approx (V_B - 0.6 \text{ volt})/R_E$$

independent of V_C, as long as the transistor is not saturated ($V_C > V_E + 0.2$ volt).

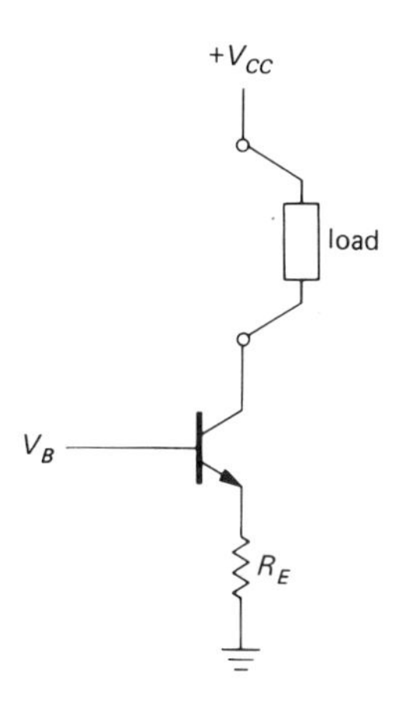

Figure 2.21. Transistor current source: basic concept.

Current-source biasing

The base voltage can be provided in a number of ways. A voltage divider is OK, as long as it is stiff enough. As before, the criterion is that its impedance should be much less than the dc impedance looking into the base ($h_{FE}R_E$). Or you can use a zener diode, biased from V_{CC}, or even a few forward-biased diodes in series from base to the corresponding emitter supply. Figure 2.22 shows some examples. In the last example (Fig. 2.22C), a *pnp* transistor *sources* current to a load returned to ground. The other examples (using *npn* transistors) should properly be called current *sinks*, but the usual practice is to call all of them current sources. ["Sink" and "source" simply refer to the direction of current flow: If a circuit *supplies* (positive) current to a point, it is a *source*, and vice versa.] In the first circuit, the voltage-divider impedance of ~1.3k is very stiff compared with the impedance looking into the base of about 100k (for $h_{FE} = 100$), so any changes in beta with collector voltage will not much affect the output current by causing the base voltage to change. In the other two circuits the biasing resistors are chosen to provide several milliamps to bring the diodes into conduction.

Compliance

A current source can provide constant current to the load only over some finite range of load voltage. To do otherwise would be equivalent to providing infinite power. The output voltage range over which a current source behaves well is called its output *compliance*. For the preceding transistor current sources, the compliance is set by the requirement that

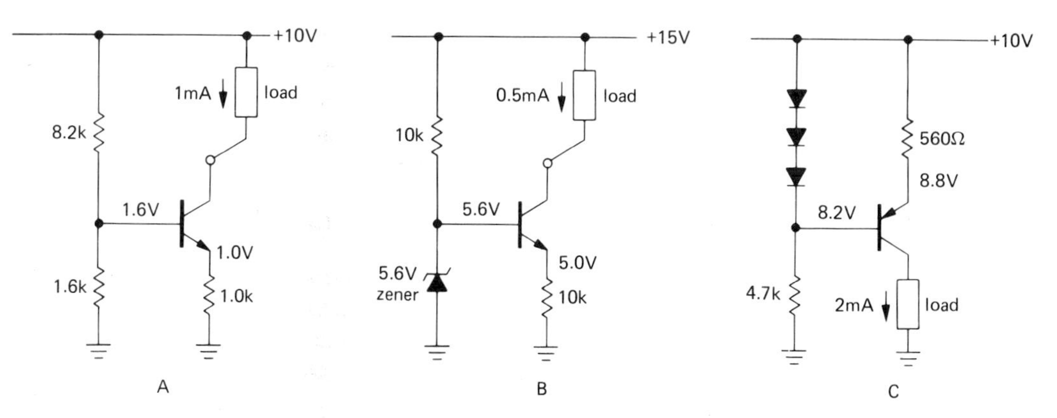

Figure 2.22. Transistor–current-source circuits, illustrating three methods of base biasing; *npn* transistors *sink* current, whereas *pnp* transistors *source* current. The circuit in C illustrates a load returned to ground.

the transistors stay in the active region. Thus in the first circuit the voltage at the collector can go down until the transistor is almost in saturation, perhaps +1.2 volts at the collector. The second circuit, with its higher emitter voltage, can sink current down to a collector voltage of about +5.2 volts.

In all cases the collector voltage can range from a value near saturation all the way up to the supply voltage. For example, the last circuit can source current to the load for any voltage between zero and about +8.6 volts across the load. In fact, the load might even contain batteries or power supplies of its own, carrying the collector beyond the supply voltage. That's OK, but you must watch out for transistor breakdown (V_{CE} must not exceed BV_{CEO}, the specified collector-emitter breakdown voltage) and also for excessive power dissipation (set by $I_C V_{CE}$). As you will see in Section 6.07, there is an additional safe-operating-area constraint on power transistors.

EXERCISE 2.8

You have +5 and +15 volt regulated supplies available in a circuit. Design a 5mA *npn* current source (sink) using the +5 volts on the base. What is the output compliance?

A current source doesn't have to have a fixed voltage at the base. By varying V_B you get a voltage-programmable current source. The input signal swing v_{in} (remember, lower-case symbols mean *variations*) must stay small enough so that the emitter voltage never drops to zero, if the output current is to reflect input voltage variations smoothly. The result will be a current source with variations in output current proportional to the variations in input voltage, $i_{\text{out}} = v_{\text{in}}/R_E$.

□ *Deficiencies of current sources*

To what extent does this kind of current source depart from the ideal? In other words, does the load current vary with voltage, i.e., have a finite ($R_{Th} < \infty$) Thévenin equivalent resistance, and if so why? There are two kinds of effects:

1. Both V_{BE} (Early effect) and h_{FE} vary slightly with collector-to-emitter voltage at a given collector current. The changes in V_{BE} produced by voltage swings across the load cause the output current to change, because the emitter voltage (and therefore the emitter current) changes, even with a fixed applied base voltage. Changes in h_{FE} produce small changes in output (collector) current for fixed emitter current, since $I_C = I_E - I_B$; in addition, there are small changes in applied base voltage produced by the variable loading of the nonzero bias source impedance as h_{FE} (and therefore the base current) changes. These effects are small. For instance, the current from the circuit in Figure 2.22A varied about 0.5% in actual measurements with a 2N3565 transistor. In particular, for load voltages varying from zero to 8 volts, the Early effect contributed 0.5%, and transistor heating effects contributed 0.2%. In addition, variations in h_{FE} contributed 0.05% (note the stiff divider). Thus these variations result in a less-than-perfect current source: The output current depends slightly on voltage and therefore has less than infinite impedance. Later you will see methods that get around this difficulty.

2. V_{BE} and also h_{FE} depend on temperature. This causes drifts in output current with changes in ambient temperature; in addition, the transistor junction temperature varies as the load voltage is changed (because of variation in transistor dissipation), resulting in departure from ideal current source behavior. The change of V_{BE} with ambient temperature can be compensated with a circuit like that shown in Figure 2.23, in which Q_2's base-emitter drop is compensated by the drop in emitter follower Q_1, with similar temperature dependence. R_3, incidentally, is a

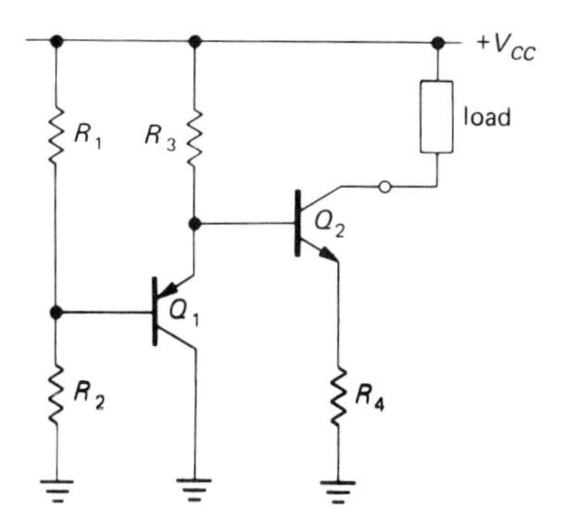

Figure 2.23. One method of temperature-compensating a current source.

pull-up resistor for Q_1, since Q_2's base sinks current, which Q_1 cannot source.

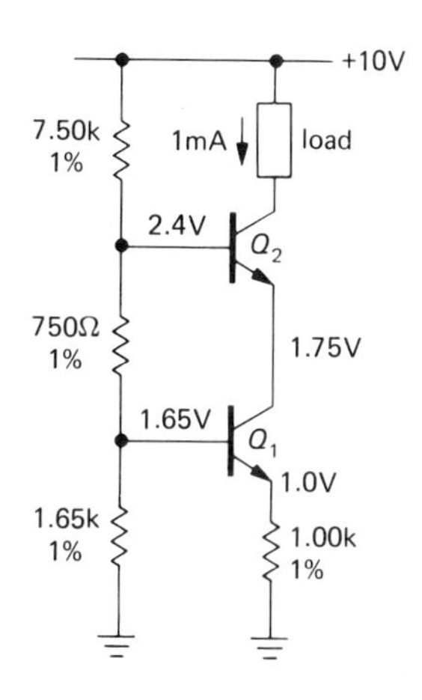

Figure 2.24. Cascode current source for improved current stability with load voltage variations.

□ *Improving current-source performance*

In general, the effects of variability in V_{BE}, whether caused by temperature dependence (approximately $-2\text{mV}/^\circ\text{C}$) or by dependence on V_{CE} (the Early effect, given roughly by $\Delta V_{BE} \approx -0.0001\,\Delta V_{CE}$), can be minimized by choosing the emitter voltage to be large enough (at least 1V, say) so that changes in V_{BE} of tens of millivolts will not result in large fractional changes in the voltage across the emitter resistor (remember that the *base* voltage is what is held constant by your circuit). For instance, choosing $V_E = 0.1$ volt (i.e., applying about 0.7V to the base) would cause 10% variations in output current for 10mV changes in V_{BE}, whereas the choice $V_E = 1.0$ volt would result in 1% current variations for the same V_{BE} changes. Don't get carried away, though. Remember that the lower limit of output compliance is set by the emitter voltage. Using a 5 volt emitter voltage for a current source running from a +10 volt supply limits the output compliance to slightly less than 5 volts (the collector can go from about V_E+ 0.2V to V_{CC}, i.e., from 5.2V to 10V).

Figure 2.24 shows a circuit modification that improves current-source performance significantly. Current source Q_1 functions as before, but with collector voltage held fixed by Q_2's emitter. The load sees the same current as before, since Q_2's collector and emitter currents are nearly equal (large h_{FE}). But with this circuit the V_{CE} of Q_1 doesn't change with load voltage, thus eliminating the small changes in V_{BE} from Early effect and dissipation-induced temperature changes. Measurements with 2N3565s gave 0.1% current variation for load voltages from 0 to 8 volts; to obtain performance of this accuracy it is important to use stable 1% resistors, as shown. (Incidentally, this circuit connection also finds use in high-frequency amplifiers, where it is known as the "cascode.") Later you will see current source techniques using op-amps and feedback that circumvent the problem of V_{BE} variation altogether.

The effects of variability of h_{FE} can be minimized by choosing transistors with large h_{FE}, so that the base current contribution to the emitter current is relatively small.

Figure 2.25 shows one last current source, whose output current doesn't depend on supply voltage. In this circuit, Q_1's V_{BE} across R_2 sets the output current, independent of V_{CC}:

$$I_{\text{out}} = V_{BE}/R_2$$

R_1 biases Q_2 and holds Q_1's collector at two diode drops below V_{CC}, eliminating Early effect as in the previous circuit. This circuit is not temperature-compensated; the voltage across R_2 decreases approximately 2.1mV/°C, causing the output current to decrease approximately 0.3%/°C.

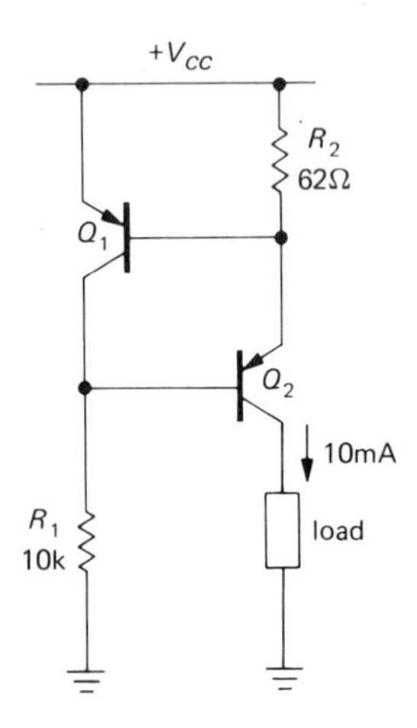

Figure 2.25. Transistor V_{BE}-referenced current source.

2.07 Common-emitter amplifier

Consider a current source with a resistor as load (Fig. 2.26). The collector voltage is

$$V_C = V_{CC} - I_C R_C$$

We could capacitively couple a signal to the base to cause the collector voltage to vary. Consider the example in Figure 2.27. C is chosen so that all frequencies of interest are passed by the high-pass filter it forms in combination with the parallel resistance of the base biasing resistors (the

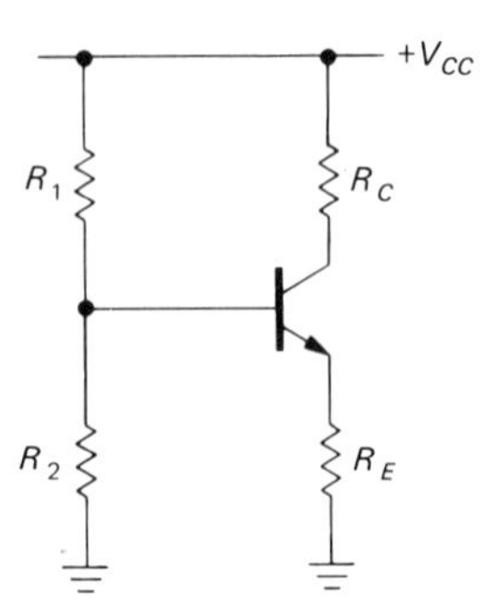

Figure 2.26

impedance looking into the base itself will usually be much larger because of the way the base resistors are chosen, and it can be ignored); that is,

$$C \geq \frac{1}{2\pi f(R_1 \| R_2)}$$

The quiescent collector current is 1.0mA because of the applied base bias and the 1.0k emitter resistor. That current puts the collector at +10 volts (+20V, minus 1.0mA through 10k). Now imagine an applied wiggle in base voltage v_B. The emitter follows with $v_E = v_B$, which causes a wiggle in emitter current

$$i_E = v_E/R_E = v_B/R_E$$

and nearly the same change in collector current (h_{fe} is large). So the initial wiggle in base voltage finally causes a collector voltage wiggle

$$v_C = -i_C R_C = -v_B(R_C/R_E)$$

Aha! It's a *voltage amplifier*, with a voltage amplification (or "gain") given by

$$\text{gain} = v_{\text{out}}/v_{\text{in}} = -R_C/R_E$$

In this case the gain is −10,000/1000, or −10. The minus sign means that a positive wiggle at the input gets turned into a negative wiggle (10 times as large) at the output. This is called a *common-emitter amplifier* with emitter degeneration.

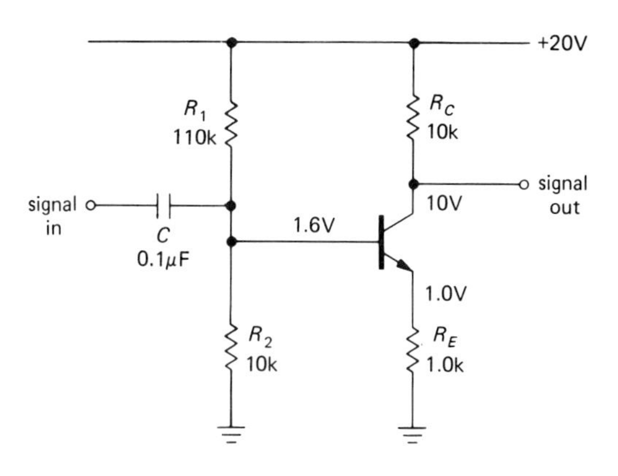

Figure 2.27. An ac common-emitter amplifier with emitter degeneration. Note that the output terminal is the collector, rather than the emitter.

Input and output impedance of the common-emitter amplifier

We can easily determine the input and output impedances of the amplifier. The input signal sees, in parallel, 110k, 10k, and the impedance looking into the base. The latter is about 100k (h_{fe} times R_E), so the input impedance (dominated by the 10k) is about 8k. The input coupling capacitor thus forms a high-pass filter, with the 3dB point at 200Hz. The signal driving the amplifier sees 0.1μF in series with 8k, which to signals of normal frequencies (well above the 3dB point) just looks like 8k.

The output impedance is 10k in parallel with the impedance looking into the collector. What is that? Well, remember that if you snip off the collector resistor, you're simply looking into a current source. The collector impedance is very large (measured in megohms), and so the output impedance is just the value of the collector resistor, 10k. It is worth remembering that the impedance looking into a transistor's collector is high, whereas the impedance looking into the emitter is low (as in the emitter follower). Although the output impedance of a common-emitter amplifier will be dominated by the collector load resistor, the output impedance of an emitter follower will not be dominated by the emitter load resistor, but rather by the impedance looking into the emitter.

2.08 Unity-gain phase splitter

Sometimes it is useful to generate a signal and its inverse, i.e., two signals 180° out of phase. That's easy to do – just use an emitter-degenerated amplifier with a gain of -1 (Fig. 2.28). The quiescent collector voltage is set to $0.75V_{CC}$, rather than the usual $0.5V_{CC}$, in order to achieve the same result – maximum symmetrical output swing without clipping at either output. The collector can swing from $0.5V_{CC}$ to V_{CC}, whereas the emitter can swing from ground to $0.5V_{CC}$.

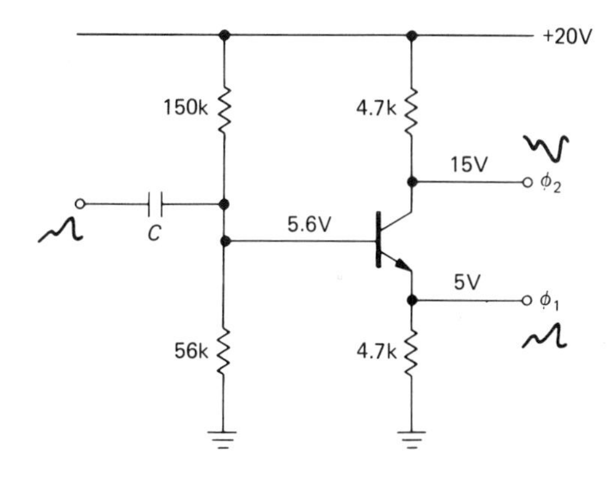

Figure 2.28. Unity-gain phase splitter.

Note that the phase-splitter outputs must be loaded with equal (or very high) impedances at the two outputs in order to maintain gain symmetry.

Phase shifter

A nice use of the phase splitter is shown in Figure 2.29. This circuit gives (for a sine wave input) an output sine wave of adjustable phase (from zero to 180°), but with constant amplitude. It can be best understood with a phasor diagram of voltages (see Chapter 1); representing the input signal by a unit vector along the real axis, the signals look as shown in Figure 2.30.

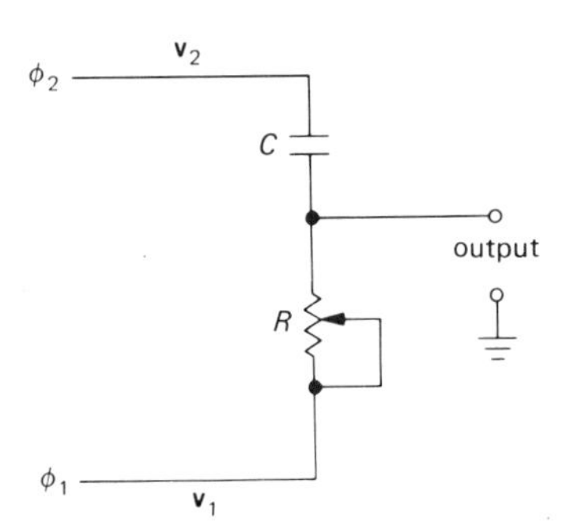

Figure 2.29. Constant-amplitude phase shifter.

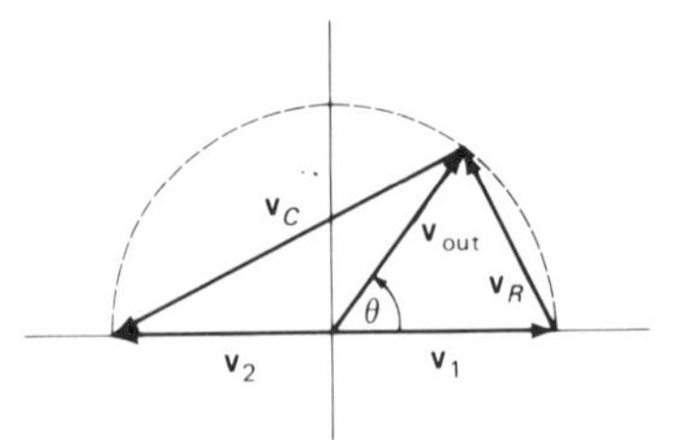

Figure 2.30. Phasor diagram for phase shifter.

Signal vectors $\mathbf{v}_R$ and $\mathbf{v}_C$ must be at right angles, and they must add to form a vector of constant length along the real axis. There is a theorem from geometry that says that the locus of such points is a circle. So the resultant vector (the output voltage) always has unit length, i.e., the same amplitude as the input, and its phase can vary from nearly zero to nearly 180° relative to the input wave as R is varied from nearly zero to a value much larger than Z_C at the operating frequency. However, note that the phase shift also depends on the frequency of the input signal for a given setting of the potentiometer R. It is worth noting that a simple RC high-pass (or low-pass) network could also be used as an adjustable phase shifter. However, its output amplitude would vary over an enormous range as the phase shift was adjusted.

An additional concern here is the ability of the phase-splitter circuit to drive the RC phase shifter as a load. Ideally, the load should present an impedance that is large compared with the collector and emitter resistors. As a result, this circuit is of limited utility where a wide range of phase shifts is required. You will see improved phase-splitter techniques in Chapter 4.

2.09 Transconductance

In the preceding section we figured out the operation of the emitter-degenerated amplifier by (a) imagining an applied base voltage swing and seeing that the emitter voltage had the same swing, then (b) calculating the emitter current swing; then, ignoring the small base current contribution, we got the collector current swing and thus (c) the collector voltage swing. The voltage gain was then simply the ratio of collector (output) voltage swing to base (input) voltage swing.

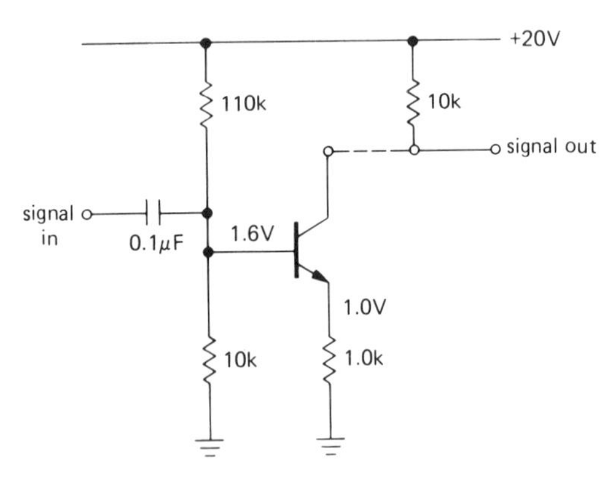

Figure 2.31. The common-emitter amplifier is a transconductance stage driving a (resistive) load.

There's another way to think about this kind of amplifier. Imagine breaking it apart, as in Figure 2.31. The first part is a voltage-controlled current source, with quiescent current of 1.0mA and gain of −1mA/V. Gain means the ratio output/input; in this case the gain has units of current/voltage, or 1/resistance. The inverse of resistance is called *conductance* (the inverse of reactance is *susceptance*, and the inverse of impedance is *admittance*) and has a special unit, the *siemens*, which used to be called the *mho* (ohm spelled backward). An amplifier whose gain has units of conductance is called a *transconductance* amplifier; the ratio I_{out}/V_{in} is called the transconductance, g_m.

Think of the first part of the circuit as a transconductance amplifier, i.e., a voltage-to-current amplifier with transconductance g_m (gain) of 1mA/V (1000μS, or 1mS, which is just $1/R_E$). The second part of the circuit is the load resistor, an "amplifier" that converts current to voltage. This resistor could be called a *transresistance* amplifier, and its gain (r_m) has units of voltage/current, or resistance. In this case its quiescent voltage is V_{CC}, and its gain (transresistance) is 10kV/A (10kΩ), which is just R_C. Connecting the two parts together gives you a voltage amplifier. You get the overall gain by multiplying the two gains. In this case $G = g_m R_C = R_C/R_E$, or −10, a unitless number equal to the ratio (output voltage)/(input voltage).

This is a useful way to think about an amplifier, because you can analyze performance of the sections independently. For example, you can analyze the transconductance part of the amplifier by evaluating g_m for different circuit configurations or even different devices, such as field-effect transistors (FETs). Then you can analyze the transresistance (or load) part by considering gain versus voltage swing trade-offs. If you are interested in the overall voltage gain, it is given by $G_V = g_m r_m$, where r_m is the transresistance of the load. Ultimately the substitution of an active load (current source), with its extremely high transresistance, can yield one-stage voltage gains of 10,000 or more. The *cascode* configuration, which we will discuss later, is another example easily understood with this approach.

In Chapter 4, which deals with operational amplifiers, you will see further examples of amplifiers with voltages or currents as inputs or outputs; voltage amplifiers (voltage to voltage), current amplifiers (current to current), and transresistance amplifiers (current to voltage).

Turning up the gain: limitations of the simple model

The voltage gain of the emitter-degenerated amplifier is $-R_C/R_E$, according to our model. What happens as R_E is reduced toward zero? The equation predicts that the gain will rise without limit. But if we made actual measurements of the preceding circuit, keeping the quiescent current constant at 1mA, we would find that the gain would level off at about 400 when R_E is zero, i.e., with the emitter grounded. We would also find that the amplifier would become significantly nonlinear (the output would not be a faithful replica of the input), the input impedance would become small and nonlinear, and the biasing would become critical and unstable with temperature. Clearly our transistor model is incomplete and needs to be modified in order to handle this circuit situation, as well as others we will talk about shortly. Our fixed-up model, which we will call the transconductance model, will be accurate enough for the remainder of the book.

EBERS-MOLL MODEL APPLIED TO BASIC TRANSISTOR CIRCUITS

2.10 Improved transistor model: transconductance amplifier

The important change is in property 4 (Section 2.01), where we said earlier that $I_C = h_{FE} I_B$. We thought of the transistor as a current amplifier whose input circuit behaved like a diode. That's roughly correct, and for some applications it's good enough. But to understand differential amplifiers, logarithmic converters, temperature compensation, and other important applications, you must think of the transistor as a *transconductance* device – collector current is determined by base-to-emitter *voltage*.

Here's the modified property 4:

4. When rules 1–3 (Section 2.01) are obeyed, I_C is related to V_{BE} by

$$I_C = I_S\left[\exp\left(\frac{V_{BE}}{V_T}\right) - 1\right]$$

where $V_T = kT/q = 25.3\text{mV}$ at room temperature (68°F, 20°C), q is the electron charge (1.60×10^{-19} coulombs), k is Boltzmann's constant (1.38×10^{-23} joules/°K), T is the absolute temperature in degrees Kelvin (°K =°C + 273.16), and I_S is the saturation current of the particular transistor (depends on T). Then the base current, which also depends on V_{BE}, can be approximated by

$$I_B = I_C/h_{FE}$$

where the "constant" h_{FE} is typically in the range 20 to 1000, but depends on transistor type, I_C, V_{CE}, and temperature. I_S represents the reverse leakage current. In the active region $I_C \gg I_S$, and therefore the -1 term can be neglected in comparison with the exponential.

The equation for I_C is known as the Ebers-Moll equation. It also approximately describes the current versus voltage for a diode, if V_T is multiplied by a correction factor m between 1 and 2. For transistors it is important to realize that the collector current is accurately determined by the base-emitter voltage, rather than by the base current (the base current is then roughly determined by h_{FE}), and that this exponential law is accurate over an enormous range of currents, typically from nanoamps to milliamps. Figure 2.32 makes the point graphically. If you measure the base current at various collector currents, you will get a graph of h_{FE} versus I_C like that in Figure 2.33.

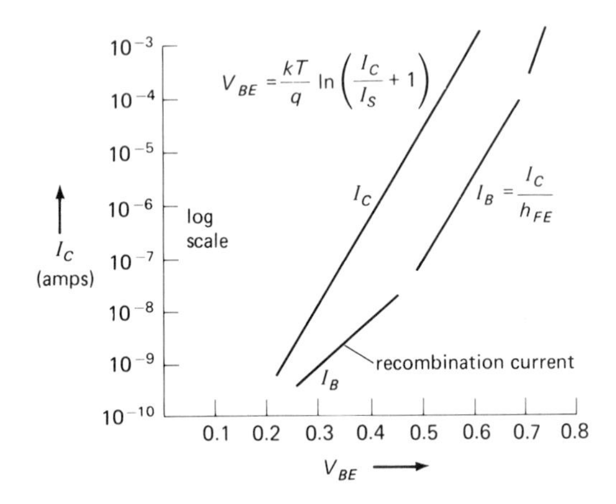

Figure 2.32. Transistor base and collector currents as functions of base-to-emitter voltage V_{BE}.

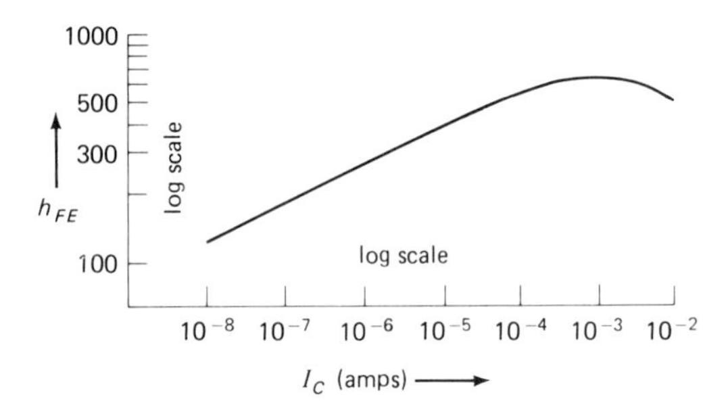

Figure 2.33. Typical transistor current gain (h_{FE}) versus collector current.

Although the Ebers-Moll equation tells us that the base-emitter voltage "programs" the collector current, this property may not be directly usable in practice (biasing a transistor by applying a base voltage) because of the large temperature coefficient of base-emitter voltage. You will see later how the Ebers-Moll equation provides insight and solutions to this problem.

Rules of thumb for transistor design

From the Ebers-Moll equation we can get

several important quantities we will be using often in circuit design:

1. The steepness of the diode curve. How much do we need to increase V_{BE} to increase I_C by a factor of 10? From the Ebers-Moll equation, that's just $V_T \log_e 10$, or 60mV at room temperature. *Base voltage increases 60mV per decade of collector current.* Equivalently, $I_C = I_{C0}e^{\Delta V/25}$, where ΔV is in millivolts.

2. The small-signal impedance looking into the emitter, for the base held at a fixed voltage. Taking the derivative of V_{BE} with respect to I_C, you get

$$r_e = V_T/I_C = 25/I_C \text{ ohms}$$

where I_C is in milliamps. The numerical value $25/I_C$ is for room temperature. This *intrinsic* emitter resistance, r_e, acts as if it is in series with the emitter in all transistor circuits. It limits the gain of a grounded emitter amplifier, causes an emitter follower to have a voltage gain of slightly less than unity, and prevents the output impedance of an emitter follower from reaching zero. Note that the transconductance of a grounded emitter amplifier is $g_m = 1/r_e$.

3. The temperature dependence of V_{BE}. A glance at the Ebers-Moll equation suggests that V_{BE} has a positive temperature coefficient. However, because of the temperature dependence of I_S, V_{BE} *decreases* about 2.1mV/°C. It is roughly proportional to $1/T_{abs}$, where T_{abs} is the absolute temperature.

There is one additional quantity we will need on occasion, although it is not derivable from the Ebers-Moll equation. It is the Early effect we described in Section 2.06, and it sets important limits on current-source and amplifier performance, for example:

4. Early effect. V_{BE} varies slightly with changing V_{CE} at constant I_C. This effect is caused by changing effective base width, and it is given, approximately, by

$$\Delta V_{BE} = -\alpha \Delta V_{CE}$$

where $\alpha \approx 0.0001$.

These are the essential quantities we need. With them we will be able to handle most problems of transistor circuit design, and we will have little need to refer to the Ebers-Moll equation itself.

2.11 The emitter follower revisited

Before looking again at the common-emitter amplifier with the benefit of our new transistor model, let's take a quick look at the humble emitter follower. The Ebers-Moll model predicts that an emitter follower should have nonzero output impedance, even when driven by a voltage source, because of finite r_e (item 2, above). The same effect also produces a voltage gain slightly less than unity, because r_e forms a voltage divider with the load resistor.

These effects are easy to calculate. With fixed base voltage, the impedance looking back into the emitter is just $R_{out} = dV_{BE}/dI_E$; but $I_E \approx I_C$, so $R_{out} \approx r_e$, the intrinsic emitter resistance $[r_e = 25/I_C(\text{mA})]$. For example, in Figure 2.34A, the load sees a driving impedance of $r_e = 25$ ohms, since $I_C = 1$mA. (This is paralleled by the emitter resistor R_E, if used; but in practice R_E will always be much larger than r_e.) Figure 2.34B shows a more typical situation, with finite source resistance R_S (for simplicity we've omitted the obligatory biasing components – base divider and blocking capacitor – which are shown in Fig. 2.34C). In this case the emitter follower's output impedance is just r_e in series with $R_s/(h_{fe}+1)$ (again paralleled by an unimportant R_E, if present). For example, if $R_s = 1$k and $I_C = 1$mA, $R_{out} = 35$ ohms (assuming $h_{fe} = 100$). It is easy to show that the intrinsic emitter r_e also figures into an emitter follower's *input* impedance, just as if it were in series with the load (actually, parallel combination of load resistor and

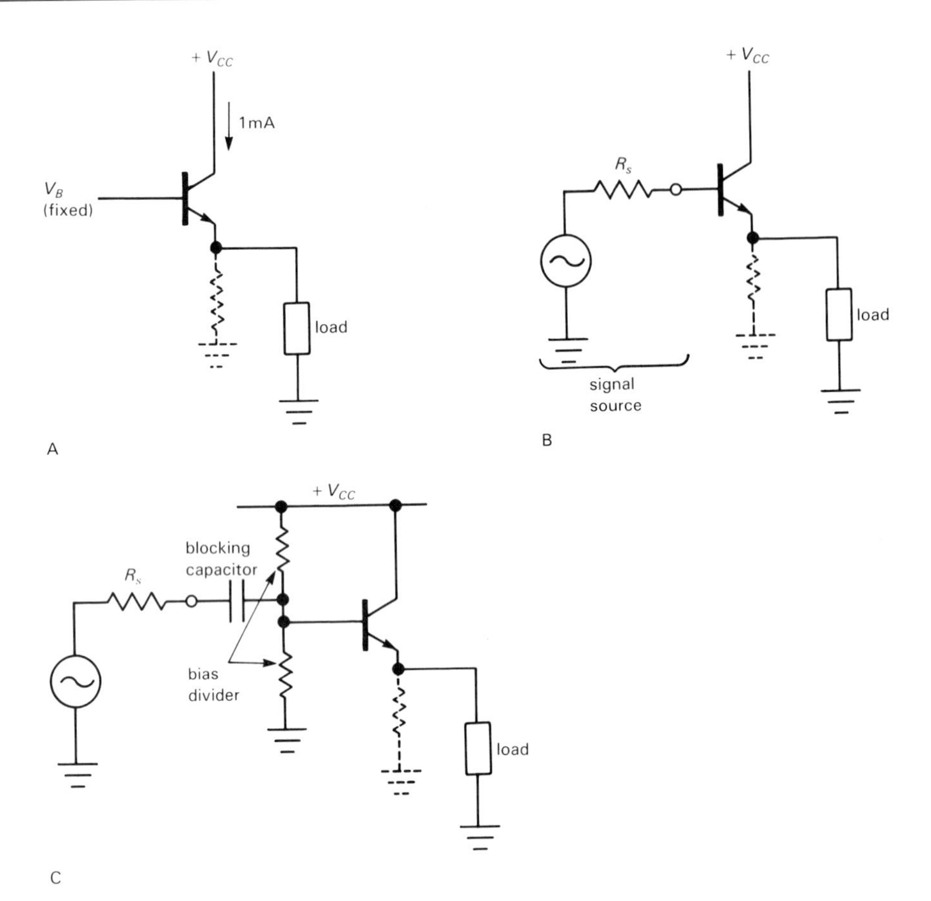

Figure 2.34

emitter resistor). In other words, for the emitter follower circuit the effect of the Ebers–Moll model is simply to add a series emitter resistance r_e to our earlier results.

The voltage gain of an emitter follower is slightly less than unity, owing to the voltage divider produced by r_e and the load. It is simple to calculate, because the output is at the junction of r_e and R_{load}: $G_V = v_{\text{out}}/v_{\text{in}} = R_L/(r_e + R_L)$. Thus, for example, a follower running at 1mA quiescent current, with 1k load, has a voltage gain of 0.976. Engineers sometimes like to write the gain in terms of the transconductance, to put it in a form that holds for FETs also (see Section 3.07); in that case (using $g_m = 1/r_e$) you get $G_V = R_L g_m/(1 + R_L g_m)$.

2.12 The common-emitter amplifier revisited

Previously we got wrong answers for the voltage gain of the common-emitter amplifier with emitter resistor (sometimes called emitter degeneration) when we set the emitter resistor equal to zero. The problem is that the transistor has $25/I_C$(mA) ohms of built-in (intrinsic) emitter resistance r_e that must be added to the actual external emitter resistor. This resistance is significant only when small emitter resistors (or none at all) are used. So, for instance, the amplifier we considered previously will have a voltage gain of $-10\text{k}/r_e$, or -400, when the external emitter resistor is zero. The input impedance is not zero, as we would have predicted earlier ($h_{fe}R_E$); it is approximately $h_{fe}r_e$, or in this case (1mA quiescent current) about 2.5k.

The terms "grounded emitter" and "common emitter" are sometimes used interchangeably, and they can be confusing. We will use the phrase "grounded emitter amplifier" to mean a common-emitter amplifier with $R_E = 0$. A common-emitter amplifier stage may have an emitter resistor; what matters is that the emitter circuit is common to the input circuit and the output circuit.

Shortcomings of the single-stage grounded emitter amplifier

The extra voltage gain you get by using $R_E = 0$ comes at the expense of other properties of the amplifier. In fact, the grounded emitter amplifier, in spite of its popularity in textbooks, should be avoided except in circuits with overall negative feedback. In order to see why, consider Figure 2.35.

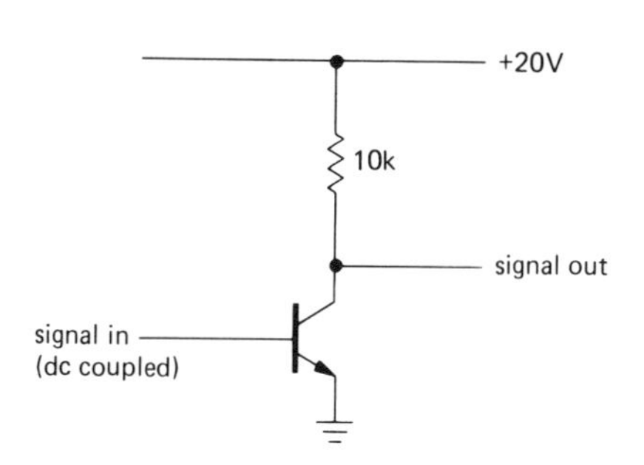

Figure 2.35. Common-emitter amplifier without emitter degeneration.

1. ***Nonlinearity.*** The gain is $G = -g_m R_C = -R_C/r_e = -R_C I_C(\text{mA})/25$, so for a quiescent current of 1mA, the gain is -400. But I_C varies as the output signal varies. For this example, the gain will vary from -800 ($V_{out} = 0$, $I_C = 2\text{mA}$) down to zero ($V_{out} = V_{CC}$, $I_C = 0$). For a triangle-wave input, the output will look like that in Figure 2.36. The amplifier has high distortion, or poor linearity. The grounded emitter amplifier without feedback is useful only for small signal swings about the quiescent point. By contrast, the emitter-degenerated amplifier has gain almost entirely independent of collector current, as long as $R_E \gg r_e$, and can be used for undistorted amplification even with large signal swings.

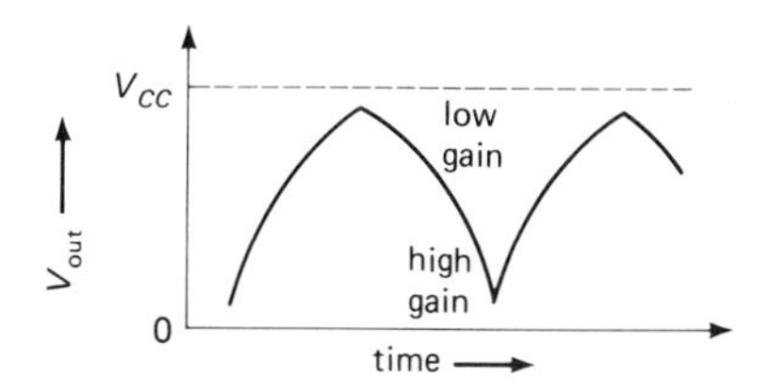

Figure 2.36. Nonlinear output waveform from grounded emitter amplifier.

2. ***Input impedance.*** The input impedance is roughly $Z_{in} = h_{fe}r_e = 25\ h_{fe}/I_C(\text{mA})$ ohms. Once again, I_C varies over the signal swing, giving a varying input impedance. Unless the signal source driving the base has low impedance, you will wind up with nonlinearity due to the nonlinear variable voltage divider formed from the signal source and the amplifier's input impedance. By contrast, the input impedance of an emitter-degenerated amplifier is constant and high.

3. ***Biasing.*** The grounded emitter amplifier is difficult to bias. It might be tempting just to apply a voltage (from a voltage divider) that gives the right quiescent current according to the Ebers-Moll equation. That won't work, because of the temperature dependence of V_{BE} (at fixed I_C), which varies about 2.1mV/°C (it actually decreases with increasing T because of the variation of I_S with T; as a result, V_{BE} is roughly proportional to $1/T$, the absolute temperature). This means that the collector current (for fixed V_{BE}) will increase by a factor of 10 for a 30°C rise in temperature. Such unstable biasing is useless, because even rather small changes in temperature will cause the amplifier to saturate. For example, a grounded emitter stage biased with the collector at half the supply voltage will go into saturation if the temperature rises by 8°C.

EXERCISE 2.9

Verify that an 8°C rise in ambient temperature will cause a base-voltage-biased grounded emitter stage to saturate, assuming that it was initially biased for $V_C = 0.5V_{CC}$.

Some solutions to the biasing problem will be discussed in the following sections. By contrast, the emitter-degenerated amplifier achieves stable biasing by applying a voltage to the base, most of which appears across the emitter resistor, thus determining the quiescent current.

Emitter resistor as feedback

Adding an external series resistor to the intrinsic emitter resistance r_e (emitter degeneration) improves many properties of the common-emitter amplifier, at the expense of gain. You will see the same thing happening in Chapters 4 and 5 , when we discuss *negative feedback*, an important technique for improving amplifier characteristics by feeding back some of the output signal to reduce the effective input signal. The similarity here is no coincidence; the emitter-degenerated amplifier itself uses a form of negative feedback. Think of the transistor as a transconductance device, determining collector current (and therefore output voltage) according to the voltage applied between the base and emitter; but the input to the amplifier is the voltage from base to ground. So the voltage from base to emitter is the input voltage, *minus a sample of the output* ($I_E R_E$). That's negative feedback, and that's why emitter degeneration improves most properties of the amplifier (improved linearity and stability and increased input impedance; also the output impedance would be reduced if the feedback were taken directly from the collector). Great things to look forward to in Chapters 4 and 5!

2.13 Biasing the common-emitter amplifier

If you must have the highest possible gain (or if the amplifier stage is inside a feedback loop), it is possible to arrange successful biasing of a common-emitter amplifier. There are three solutions that can be applied, separately or in combination: bypassed emitter resistor, matched biasing transistor, and dc feedback.

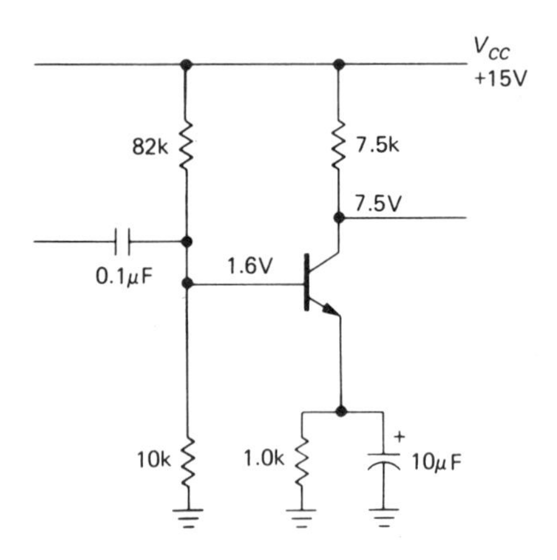

Figure 2.37. A bypassed emitter resistor can be used to improve the bias stability of a grounded emitter amplifier.

Bypassed emitter resistor

Use a bypassed emitter resistor, biasing as for the degenerated amplifier, as shown in Figure 2.37. In this case R_E has been chosen about $0.1R_C$, for ease of biasing; if R_E is too small, the emitter voltage will be much smaller than the base-emitter drop, leading to temperature instability of the quiescent point as V_{BE} varies with temperature. The emitter bypass capacitor is chosen by making its impedance small compared with r_e (not R_E) at the lowest frequency of interest. In this case its impedance is 25 ohms at 650Hz. At signal frequencies the input coupling capacitor sees an impedance of 10k in parallel with the base impedance, in this case h_{fe} times 25 ohms, or roughly 2.5k. At dc, the impedance looking into the base is much larger (h_{fe} times the emitter resistor, or about 100k), which is why stable biasing is possible.

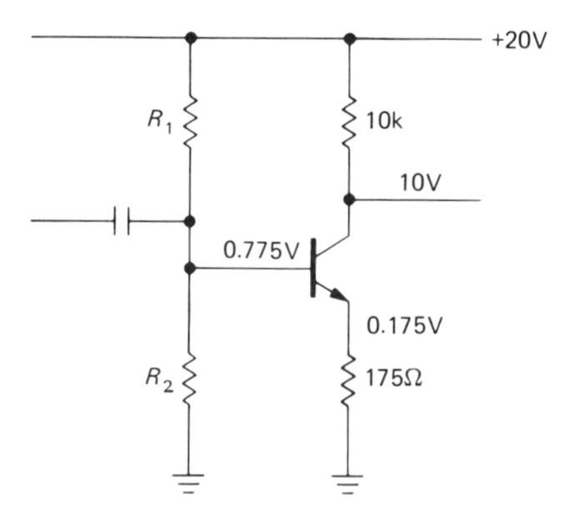

Figure 2.38

A variation on this circuit consists of using two emitter resistors in series, one of them bypassed. For instance, suppose you want an amplifier with a voltage gain of 50, quiescent current of 1mA, and V_{CC} of +20 volts, for signals from 20Hz to 20kHz. If you try to use the emitter-degenerated circuit, you will have the circuit shown in Figure 2.38. The collector resistor is chosen to put the quiescent collector voltage at $0.5V_{CC}$. Then the emitter resistor is chosen for the required gain, including the effects of the r_e of $25/I_C$(mA). The problem is that the emitter voltage of only 0.175 volt will vary significantly as the ~0.6 volt of base-emitter drop varies with temperature (−2.1mV/°C, approximately), since the base is held at constant voltage by R_1 and R_2; for instance, you can verify that an increase of 20°C will cause the collector current to increase by nearly 25%.

The solution here is to add some bypassed emitter resistance for stable biasing, with no change in gain at signal frequencies (Fig. 2.39). As before, the collector resistor is chosen to put the collector at 10 volts ($0.5V_{CC}$). Then the unbypassed emitter resistor is chosen to give a gain of 50, including the intrinsic emitter resistance $r_e = 25/I_C$(mA). Enough bypassed emitter resistance is added to make stable biasing possible (one-tenth of the collector resistance is a good rule). The base voltage is chosen to give 1mA of emitter current, with impedance about one-tenth the dc impedance looking into the base (in this case about 100k). The emitter bypass capacitor is chosen to have low impedance compared with 180+25 ohms at the lowest signal frequencies. Finally, the input coupling capacitor is chosen to have low impedance compared with the *signal-frequency* input impedance of the amplifier, which is equal to the voltage divider impedance in parallel with $(180 + 25)h_{fe}$ ohms (the 820Ω is bypassed, and looks like a short at signal frequencies).

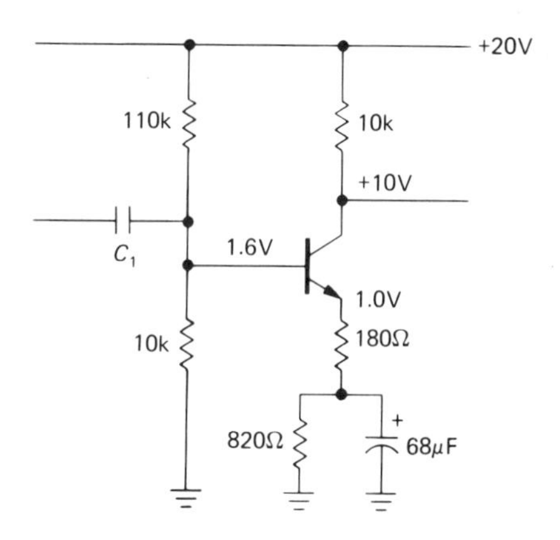

Figure 2.39. A common-emitter amplifier combining bias stability, linearity, and large voltage gain.

An alternative circuit splits the signal and dc paths (Fig. 2.40). This lets you vary the gain (by changing the 180Ω resistor) without bias change.

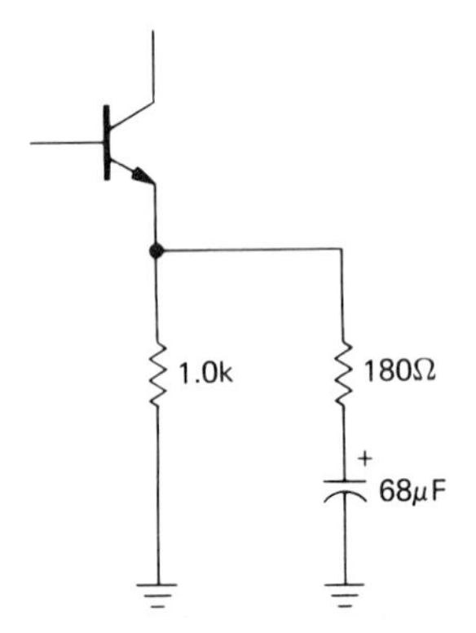

Figure 2.40. Equivalent emitter circuit for Figure 2.39.

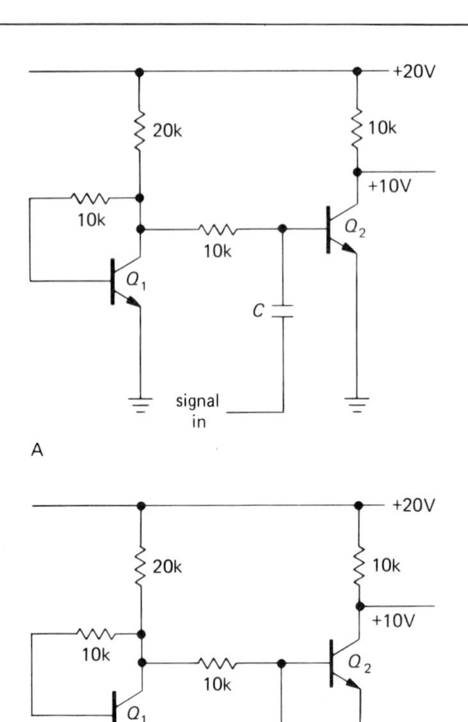

Figure 2.41. Biasing scheme with compensated V_{BE} drop.

□ *Matched biasing transistor*

Use a matched transistor to generate the correct base voltage for the required collector current; this ensures automatic temperature compensation (Fig. 2.41). Q_1's collector is drawing 1mA, since it is guaranteed to be near ground (about one V_{BE} drop above ground, to be exact); if Q_1 and Q_2 are a matched pair (available as a single device, with the two transistors on one piece of silicon), then Q_2 will also be biased to draw 1mA, putting its collector at +10 volts and allowing a full ±10 volt symmetrical swing on its collector. Changes in temperature are of no importance, as long as both transistors are at the same temperature. This is a good reason for using a "monolithic" dual transistor.

Feedback at dc

Use dc feedback to stabilize the quiescent point. Figure 2.42 shows one method. By taking the bias voltage from the collector, rather than from V_{CC}, you get some measure of bias stability. The base sits one diode drop above ground; since its bias comes from a 10:1 divider, the collector is at 11 diode drops above ground, or about 7 volts. Any tendency for the transistor

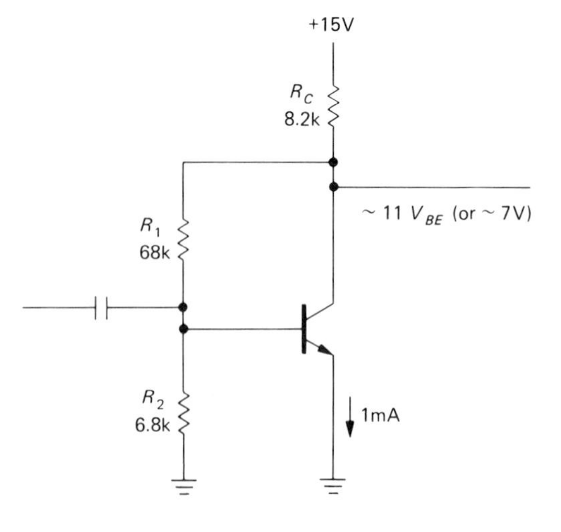

Figure 2.42. Bias stability is improved by feedback.

to saturate (e.g., if it happens to have unusually high beta) is stabilized, since the dropping collector voltage will reduce the base bias. This scheme is acceptable if great stability is not required. The quiescent point is liable to drift a volt or so as the ambient (surrounding) temperature changes, since the base-emitter voltage has a significant temperature coefficient. Better stability is possible if several stages of amplification are included within the feedback loop. You will see examples later in connection with feedback.

A better understanding of feedback is really necessary to understand this circuit. For instance, feedback acts to reduce the input and output impedances. The input signal sees R_1's resistance effectively reduced by the voltage gain of the stage. In this case it is equivalent to a resistor of about 300 ohms to ground. In Chapter 4 we will treat feedback in enough detail so that you will be able to figure the voltage gain and terminal impedance of this circuit.

Note that the base bias resistor values could be increased in order to raise the input impedance, but you should then take into account the non-negligible base current. Suitable values might be $R_1 = 220\text{k}$ and $R_2 = 33\text{k}$. An alternative approach might be to bypass the feedback resistance in order to eliminate feedback (and therefore lowered input impedance) at signal frequencies (Fig. 2.43).

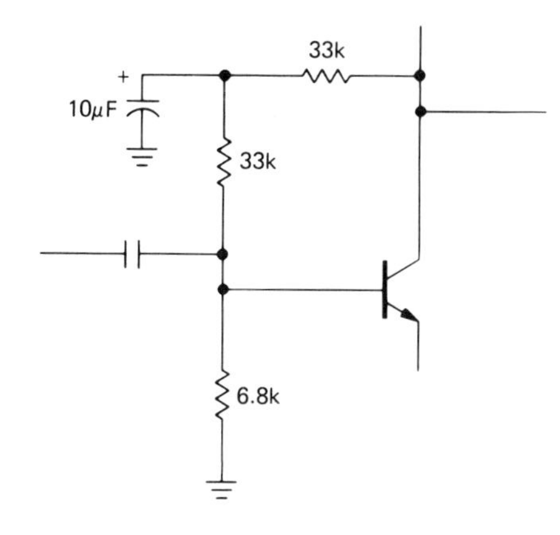

Figure 2.43. Eliminating feedback at signal frequencies.

Comments on biasing and gain

One important point about grounded emitter amplifier stages: You might think that the voltage gain can be raised by increasing the quiescent current, since the intrinsic emitter resistance r_e drops with rising current. Although r_e does go down with increasing collector current, the smaller collector resistor you need to obtain the same quiescent collector voltage just cancels the advantage. In fact, you can show that the small-signal voltage gain of a grounded emitter amplifier biased to $0.5V_{CC}$ is given by $G = 20V_{CC}$, independent of quiescent current.

EXERCISE 2.10
Show that the preceding statement is true.

If you need more voltage gain in one stage, one approach is to use a current source as an *active load*. Since its impedance is very high, single-stage voltage gains of 1000 or more are possible. Such an arrangement cannot be used with the biasing schemes we have discussed, but must be part of an overall dc feedback loop, a subject we will discuss in the next chapter. You should be sure such an amplifier looks into a high-impedance load; otherwise the gain obtained by high collector load impedance will be lost. Something like an emitter follower, a field-effect transistor (FET), or an op-amp presents a good load.

In radiofrequency amplifiers intended for use only over a narrow frequency range, it is common to use a parallel *LC* circuit as a collector load; in that case very high voltage gain is possible, since the *LC* circuit has high impedance (like a current source) at the signal frequency, with low impedance at dc. Since the *LC* is "tuned," out-of-band interfering signals (and distortion) are effectively rejected. Additional bonuses are the possibility of peak-to-peak output swings of $2V_{CC}$ and the use of transformer coupling from the inductor.

EXERCISE 2.11

Design a tuned common-emitter amplifier stage to operate at 100kHz. Use a bypassed emitter resistor, and set the quiescent current at 1.0mA. Assume $V_{CC} = +15$ volts and $L = 1.0$mH, and put a 6.2k resistor across the *LC* to set $Q = 10$ (to get a 10% bandpass; see Section 1.22). Use capacitive input coupling.

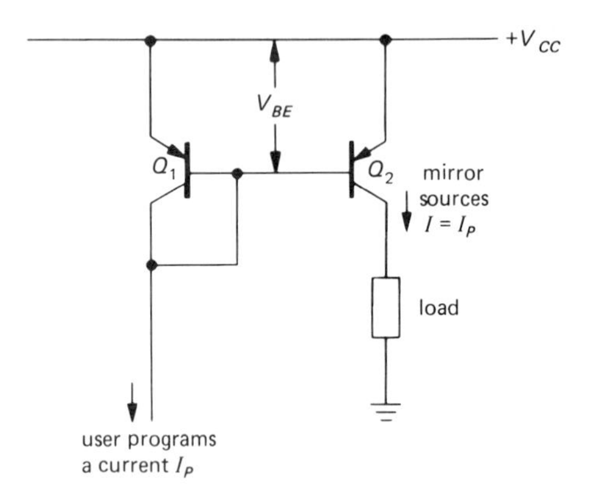

Figure 2.44. Classic bipolar-transistor matched-pair current mirror. Note the common convention of referring to the positive supply as V_{CC}, even when *pnp* transistors are used.

2.14 Current mirrors

The technique of matched base-emitter biasing can be used to make what is called a *current mirror* (Fig. 2.44). You "program" the mirror by sinking a current from Q_1's collector. That causes a V_{BE} for Q_1 appropriate to that current at the circuit temperature and for that transistor type. Q_2, matched to Q_1 (a monolithic dual transistor is ideal), is thereby programmed to source the same current to the load. The small base currents are unimportant.

One nice feature of this circuit is voltage compliance of the output transistor current source to within a few tenths of a volt of V_{CC}, since there is no emitter resistor drop to contend with. Also, in many applications it is handy to be able to program a current with a current. An easy way to generate the control current I_P is with a resistor (Fig. 2.45). Since the bases are a diode drop below V_{CC}, the 14.4k resistor produces a control current, and therefore an output current, of 1mA. Current mirrors can be used in transistor circuits whenever a current source is needed. They're very popular in integrated circuits, where (a) matched transistors abound and (b) the designer tries to make circuits that will work over a large range of supply voltages. There are even resistorless integrated circuit op-amps in which the operating current of the whole amplifier is set by one external resistor, with all the quiescent currents of the individual amplifier stages inside being determined by current mirrors.

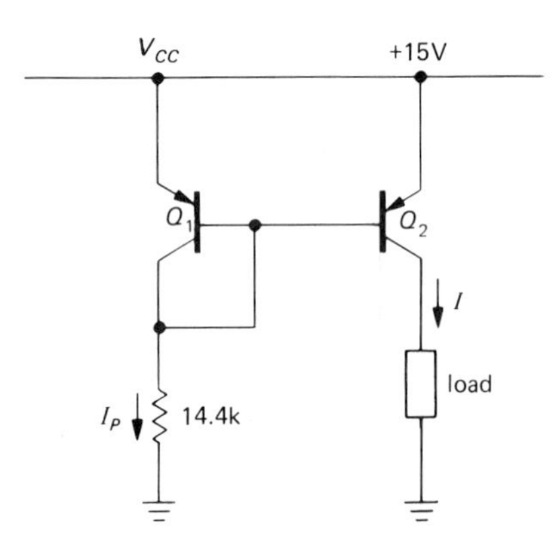

Figure 2.45

Current mirror limitations due to Early effect

One problem with the simple current mirror is that the output current varies a bit with changes in output voltage, i.e., the output impedance is not infinite. This is because of the slight variation of V_{BE} with collector voltage at a given current in Q_2 (due to Early effect); in other words, the curve of collector current versus collector-emitter voltage at a fixed base-emitter voltage is not flat (Fig. 2.46). In practice, the

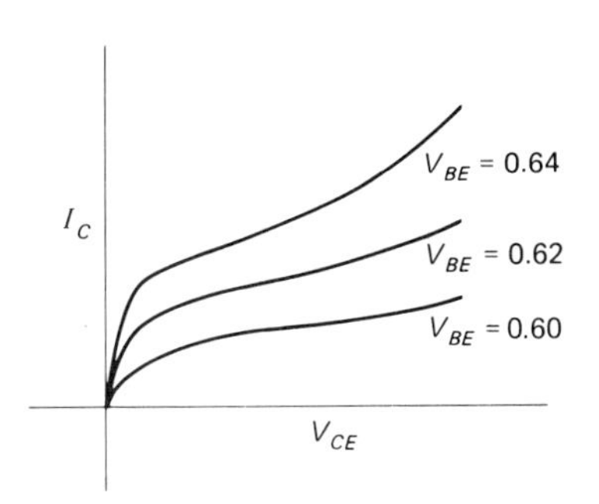

Figure 2.46

current might vary 25% or so over the output compliance range – much poorer performance than the current source with emitter resistor discussed earlier.

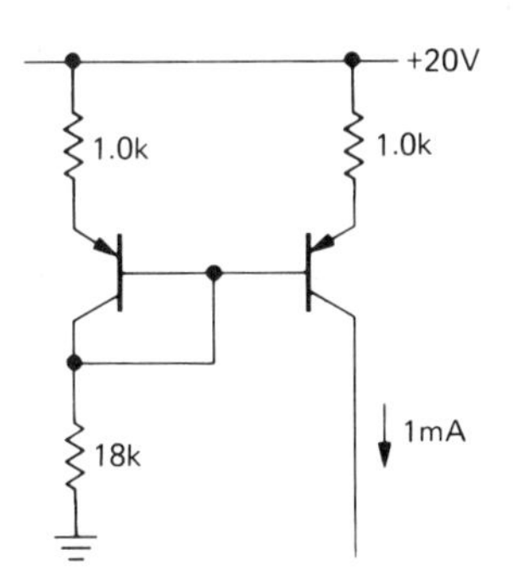

Figure 2.47. Improved current mirror.

One solution, if a better current source is needed (it often isn't), is the circuit shown in Figure 2.47. The emitter resistors are chosen to have at least a few tenths of a volt drop; this makes the circuit a far better current source, since the small variations of V_{BE} with V_{CE} are now negligible in determining the output current. Again, matched transistors should be used.

Wilson mirror

Another current mirror with very constant current is shown in the clever circuit of

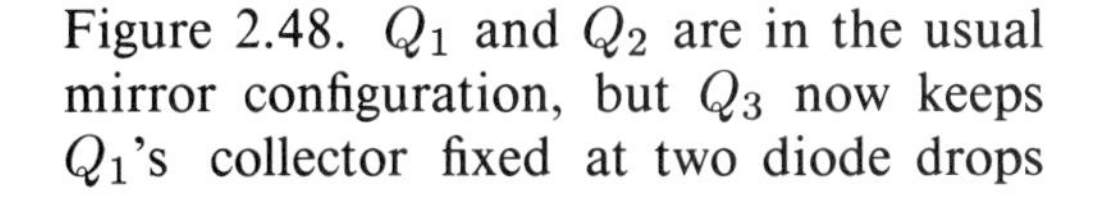

Figure 2.48. Q_1 and Q_2 are in the usual mirror configuration, but Q_3 now keeps Q_1's collector fixed at two diode drops

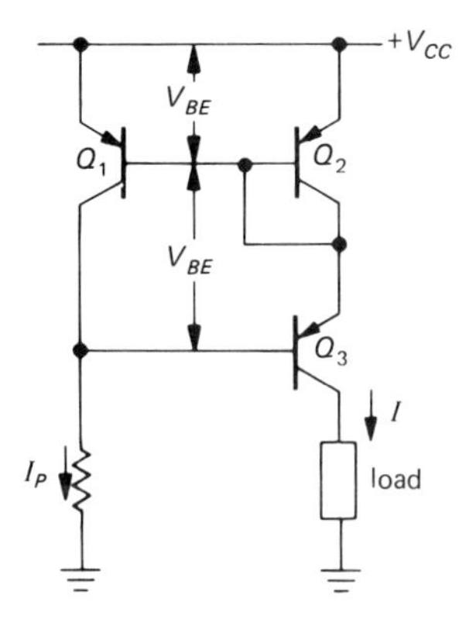

Figure 2.48. Wilson current mirror. Good stability with load variations is achieved through cascode transistor Q_3, which reduces voltage variations across Q_1.

below V_{CC}. That circumvents the Early effect in Q_1, whose collector is now the programming terminal, with Q_2 now sourcing the output current. Q_3 does not affect the balance of currents, since its base current is negligible; its only function is to pin Q_1's collector. The result is that both current-determining transistors (Q_1 and Q_2) have fixed collector-emitter drops; you can think of Q_3 as simply passing the output current through to a variable-voltage load (a similar trick is used in the cascode connection, which you will see later in the chapter). Q_3, by the way, does not have to be matched to Q_1 and Q_2.

Multiple outputs and current ratios

Current mirrors can be expanded to source (or sink, with *npn* transistors) current to several loads. Figure 2.49 shows the idea. Note that if one of the current source transistors saturates (e.g., if its load is disconnected), its base robs current from the shared base reference line, reducing the other output currents. The situation is

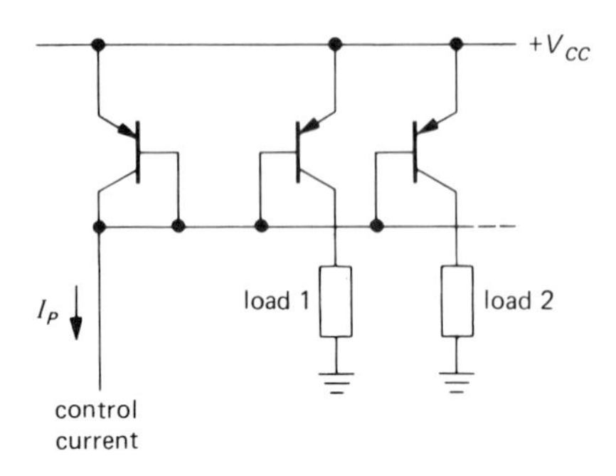

Figure 2.49. Current mirror with multiple outputs. This circuit is commonly used to obtain multiple programmable current sources.

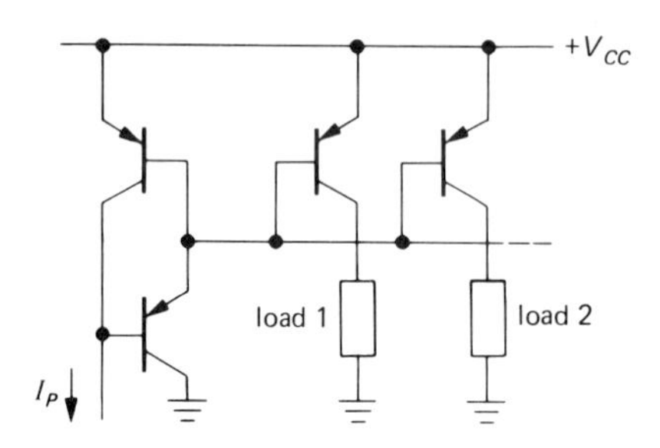

Figure 2.50

rescued by adding another transistor (Fig. 2.50).

Figure 2.51 shows two variations on the multiple-mirror idea. These circuits mirror twice (or half) the control current. In the design of integrated circuits, current mirrors with any desired current ratio can be made by adjusting the size of the emitter junctions appropriately.

Texas Instruments offers complete monolithic Wilson current mirrors in convenient TO-92 transistor packages. Their TL011 series includes 1:1, 1:2, 1:4, and 2:1 ratios, with output compliance from 1.2 to 40 volts. The Wilson configuration gives good current source performance – at constant programming current the output current increases by only 0.05% per volt – and they are very inexpensive (50 cents or less). Unfortunately, these useful devices are available in *npn* polarity only.

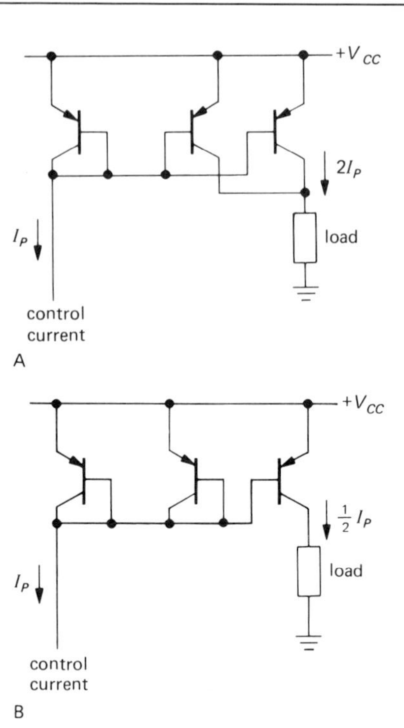

Figure 2.51. Current mirrors with current ratios other than 1:1.

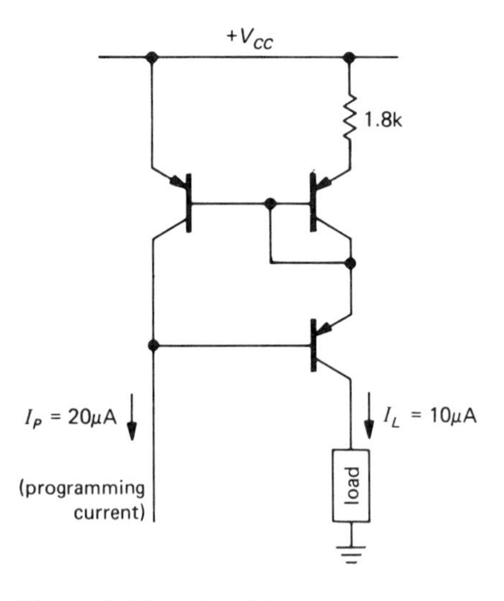

Figure 2.52. Modifying current-source output with an emitter resistor. Note that the output current is no longer a simple multiple of the progamming current.

Another way to generate an output current that is a fraction of the programming current is to add a resistor in the emitter circuit of the output transistor (Fig. 2.52). In any circuit where the transistors are operating at different current densities, the Ebers-Moll equation predicts that the difference in V_{BE} depends only on the ratio of the current densities. For matched transistors, the ratio of collector currents equals the ratio of current densities. The graph in Figure 2.53 is handy for determining the difference in base-emitter drops in such a situation. This makes it easy to design a "ratio mirror."

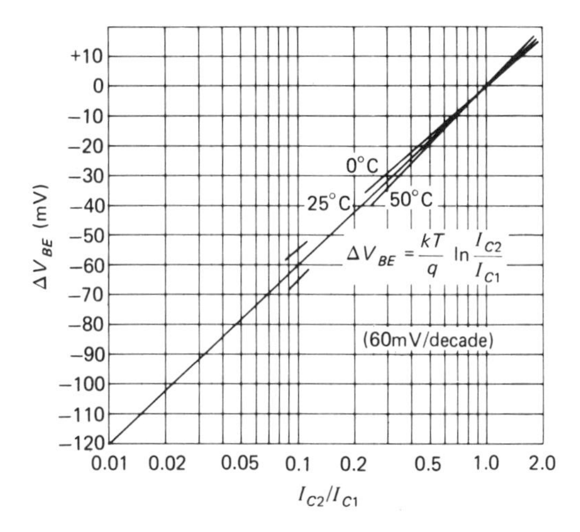

Figure 2.53. Collector current ratios for matched transistors as determined by the difference in applied base-emitter voltages.

EXERCISE 2.12

Show that the ratio mirror in Figure 2.52 works as advertised.

SOME AMPLIFIER BUILDING BLOCKS

□ 2.15 Push-pull output stages

As we mentioned earlier in the chapter, an *npn* emitter follower cannot sink current, and a *pnp* follower cannot source current. The result is that a single-ended follower operating between split supplies can drive a ground-returned load only if a high quiescent current is used (this is sometimes called a class A amplifier). The quiescent current must be at least as large as the maximum output current during peaks of the waveform, resulting in high quiescent power dissipation. For instance, Figure 2.54 shows a follower circuit to drive an 8 ohm load with up to 10 watts of audio. The *pnp* follower Q_1 is included to reduce drive requirements and to cancel Q_2's V_{BE} offset (zero volts input gives zero volts output). Q_1 could, of course, be omitted for simplicity. The hefty current source in Q_1's emitter load is used to ensure that there is sufficient base drive to Q_2 at the top of the signal swing. A resistor as emitter load would be inferior because it would have to be a rather low value (50Ω or less) in order to guarantee at least 50mA of base drive to Q_2 at the peak of the swing, when load current would be maximum and the drop across the resistor would be minimum; the resultant quiescent current in Q_1 would be excessive.

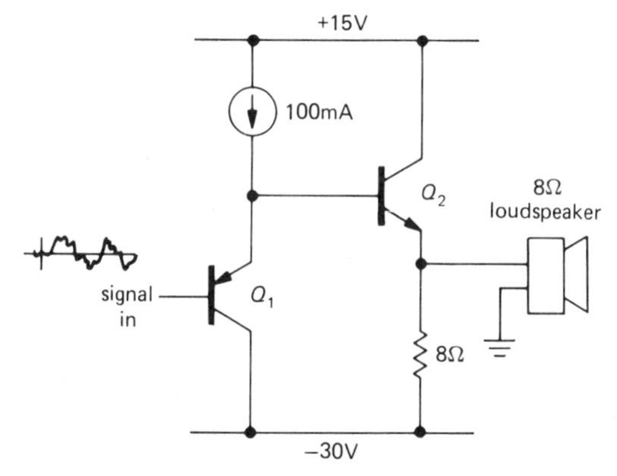

Figure 2.54. A 10 watt loudspeaker amplifier, built with a single-ended emitter follower, dissipates 165 watts of quiescent power!

The output of this example circuit can swing to nearly ±15 volts (peak) in both directions, giving the desired output power (9V rms across 8Ω). However, the output transistor dissipates 55 watts with no signal, and the emitter resistor dissipates another 110 watts. Quiescent power dissipation many times greater than the maximum output power is characteristic of this kind of class A circuit (transistor always in conduction); this obviously leaves a lot to be desired in applications where any significant amount of power is involved.

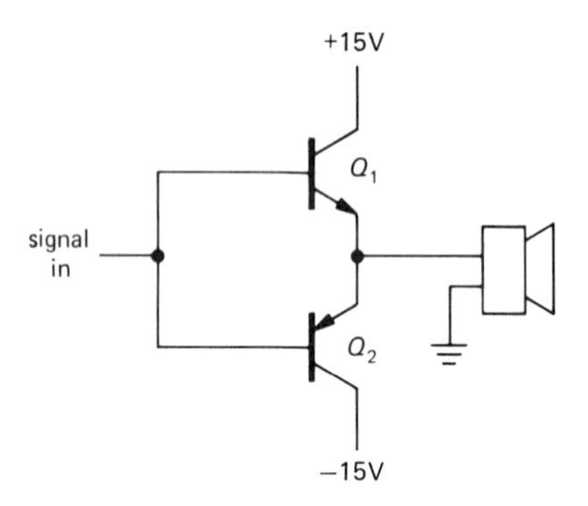

Figure 2.55. Push-pull emitter follower.

Figure 2.55 shows a push-pull follower to do the same job. Q_1 conducts on positive swings, Q_2 on negative swings. With zero input voltage, there is no collector current and no power dissipation. At 10 watts output power there is less than 10 watts dissipation in each transistor.

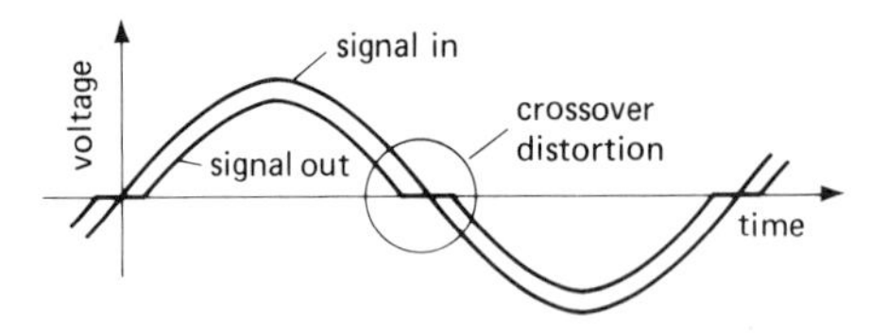

Figure 2.56. Crossover distortion in the push-pull follower.

□ *Crossover distortion in push-pull stages*

There is a problem with the preceding circuit as drawn. The output trails the

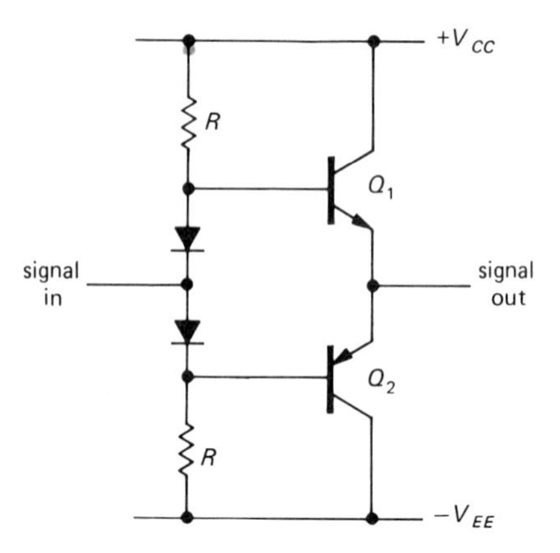

Figure 2.57. Biasing the push-pull follower to eliminate crossover distortion.

input by a V_{BE} drop; on positive swings the output is about 0.6 volt less positive than the input, and the reverse for negative swings. For an input sine wave, the output would look as shown in Figure 2.56. In the language of the audio business, this is called crossover distortion. The best cure (feedback offers another method, although it is not entirely satisfactory) is to bias the push-pull stage into slight conduction, as in Figure 2.57.

The bias resistors R bring the diodes into forward conduction, holding Q_1's base a diode drop above the input signal and Q_2's base a diode drop below the input signal. Now, as the input signal crosses through zero, conduction passes from Q_2 to Q_1; one of the output transistors is always on. R is chosen to provide enough base current for the output transistors at the peak output swing. For instance, with ± 20 volt supplies and an 8 ohm load running up to 10 watts sine-wave power, the peak base voltage is about 13.5 volts, and the peak load current is about 1.6 amps. Assuming a transistor beta of 50 (power transistors generally have lower current gain than small-signal transistors), the 32mA of necessary base current will require base resistors of about 220 ohms (6.5V from V_{CC} to base at peak swing).

Thermal stability in class B push-pull amplifiers

The preceding amplifier (sometimes called a class B amplifier, meaning that each transistor conducts over half the cycle) has one bad feature: It is not thermally stable. As the output transistors warm up (and they will get hot, because they are dissipating power when signal is applied), their V_{BE} drops, and quiescent collector current begins to flow. The added heat this produces causes the situation to get worse, with the strong possibility of what is called *thermal runaway* (whether it runs away or not depends on a number of factors, including how large a "heat sink" is used, how well the diode temperature tracks the transistor, etc.). Even without runaway, better control over the circuit is needed, usually with the sort of arrangement shown in Figure 2.58.

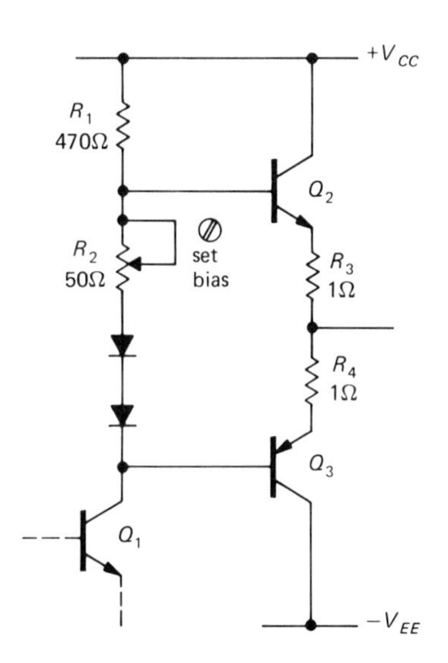

Figure 2.58. Small emitter resistors improve thermal stability in the push-pull follower.

For variety, the input is shown coming from the collector of the previous stage; R_1 now serves the dual purpose of being Q_1's collector resistor and providing current to bias the diodes and bias-setting resistor in the push-pull base circuit. Here R_3 and R_4, typically a few ohms or less, provide a "cushion" for the critical quiescent current biasing: The voltage between the bases of the output transistors must now be a bit greater than two diode drops, and you provide the extra with adjustable biasing resistor R_2 (often replaced by a third series diode). With a few tenths of a volt across R_3 and R_4, the temperature variation of V_{BE} doesn't cause the current to rise very rapidly (the larger the drop across R_3 and R_4, the less sensitive it is), and the circuit will be stable. Stability is improved by mounting the diodes in physical contact with the output transistors (or their heat sinks).

You can estimate the thermal stability of such a circuit by remembering that the base-emitter drop decreases by about 2.1mV for each 1°C rise and that the collector current increases by a factor of 10 for every 60mV increase in base-emitter voltage. For example, if R_2 were replaced by a diode, you would have three diode drops between the bases of Q_2 and Q_3, leaving about one diode drop across the series combination of R_3 and R_4. (The latter would then be chosen to give an appropriate quiescent current, perhaps 50mA for an audio power amplifier.) The worst case for thermal stability occurs if the biasing diodes are not thermally coupled to the output transistors.

Let us assume the worst and calculate the increase in output-stage quiescent current corresponding to a 30°C temperature rise in output transistor temperature. That's not a lot for a power amplifier, by the way. For that temperature rise, the V_{BE} of the output transistors will decrease by about 63mV at constant current, raising the voltage across R_3 and R_4 by about 20% (i.e., the quiescent current will rise by about 20%). The corresponding figure for the preceding amplifier circuit without emitter resistors (Fig. 2.57) will be a factor of 10 rise in quiescent current (recall that

I_C increases a decade per 60mV increase in V_{BE}), i.e., 1000%. The improved thermal stability of this biasing arrangement is evident.

This circuit has the additional advantage that by adjusting the quiescent current, you have some control over the amount of residual crossover distortion. A push-pull amplifier biased in this way to obtain substantial quiescent current at the crossover point is sometimes referred to as a class AB amplifier, meaning that both transistors conduct simultaneously during a portion of the cycle. In practice, you choose a quiescent current that is a good compromise between low distortion and excessive quiescent dissipation. Feedback, the subject of the next chapter, is almost always used to reduce distortion still further.

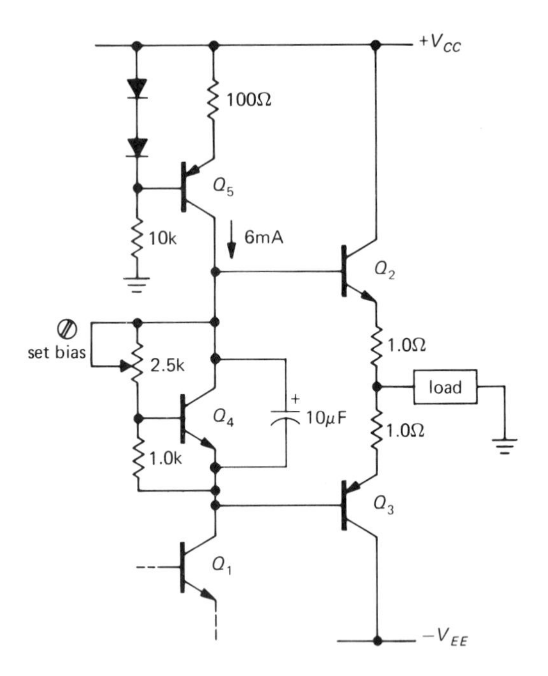

Figure 2.59. Biasing a push-pull output stage for low crossover distorion and good thermal stability.

An alternative method for biasing a push-pull follower is shown in Figure 2.59. Q_4 acts as an adjustable diode: The base resistors are a divider, and therefore Q_4's collector-emitter voltage will stabilize at a value that puts 1 diode drop from base to emitter, since any greater V_{CE} will bring it into heavy conduction. For instance, if both resistors were 1k, the transistor would turn on at 2 diode drops, collector to emitter. In this case, the bias adjustment lets you set the push-pull interbase voltage anywhere from 1 to 3.5 diode drops. The 10μF capacitor ensures that both output transistor bases see the same signal; such a bypass capacitor is a good idea for any biasing scheme you use. In this circuit, Q_1's collector resistor has been replaced by current source Q_5. That's a useful circuit variation, because with a resistor it is sometimes difficult to get enough base current to drive Q_2 near the top of the swing. A resistor small enough to drive Q_2 sufficiently results in high quiescent collector current in Q_1 (with high dissipation), and also reduced voltage gain (remember that $G = -R_{\text{collector}}/R_{\text{emitter}}$). Another solution to the problem of Q_2's base drive is the use of *bootstrapping*, a technique that will be discussed shortly.

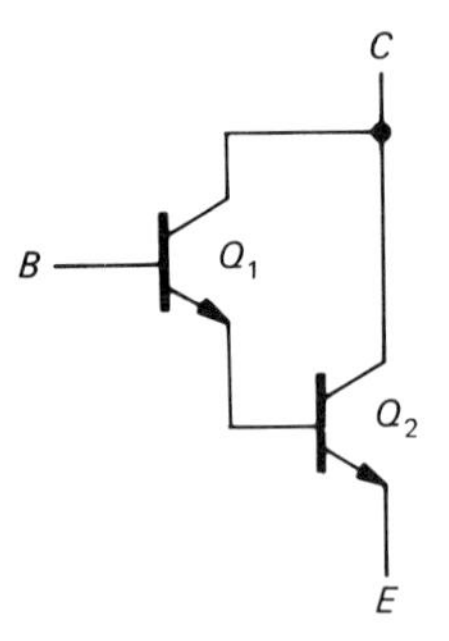

Figure 2.60. Darlington transistor configuration.

2.16 Darlington connection

If you hook two transistors together as in Figure 2.60, the result behaves like a single transistor with beta equal to the product of the two transistor betas. This can be very handy where high currents are involved (e.g., voltage regulators or power amplifier output stages), or for input stages of amplifiers where very high input impedance is necessary.

For a Darlington transistor the base-emitter drop is twice normal, and the saturation voltage is at least one diode drop (since Q_1's emitter must be a diode drop above Q_2's emitter). Also, the combination tends to act like a rather slow transistor because Q_1 cannot turn off Q_2 quickly. This problem is usually taken care of by including a resistor from base to emitter of Q_2 (Fig. 2.61). R also prevents leakage current through Q_1 from biasing

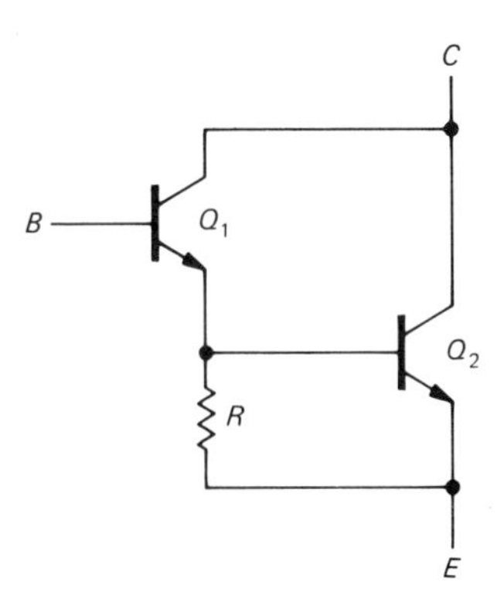

Figure 2.61. Improving turn-off speed in a Darlington pair.

Q_2 into conduction; its value is chosen so that Q_1's leakage current (nanoamps for small-signal transistors, as much as hundreds of microamps for power transistors) produces less than a diode drop across R and so that R doesn't sink a large proportion of Q_2's base current when it has a diode drop across it. Typically R might be a few hundred ohms in a power transistor Darlington, or a few thousand ohms for a small-signal Darlington.

Darlington transistors are available as single packages, usually with the base-emitter resistor included. A typical example is the *npn* power Darlington 2N6282, with current gain of 2400 (typically) at a collector current of 10 amps.

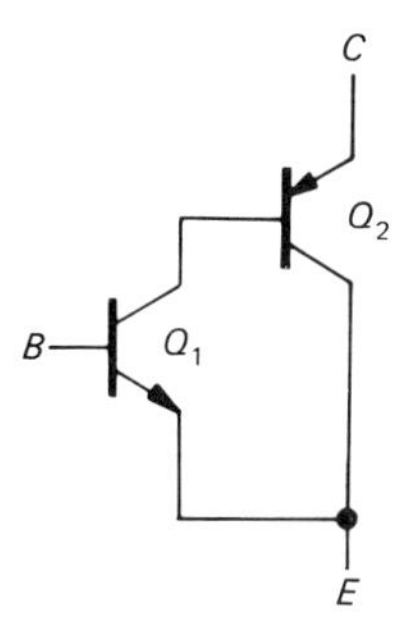

Figure 2.62. Sziklai connection ("complementary Darlington").

Sziklai connection

A similar beta-boosting configuration is the Sziklai connection, sometimes referred to as a complementary Darlington (Fig. 2.62). This combination behaves like an *npn* transistor, again with large beta. It has only a single base-emitter drop, but it also cannot saturate to less than a diode drop. A small resistor from base to emitter of Q_2 is advisable. This connection is common in push-pull power output stages where the designer wishes to use one polarity of output transistor only. Such a circuit is shown in Figure 2.63. As before, R_1 is Q_1's collector resistor. Darlington Q_2Q_3 behaves like a single *npn* transistor with high current gain. The Sziklai connected pair Q_4Q_5 behaves like a single high-gain *pnp* power transistor. As before, R_3 and R_4 are small. This circuit is sometimes called a pseudocomplementary push-pull follower. A true complementary stage would use a Darlington-connected *pnp* pair for Q_4Q_5.

Superbeta transistor

The Darlington connection and its near relatives should not be confused with the so-called superbeta transistor, a device with very high h_{FE} achieved through the manufacturing process. A typical superbeta transistor is the 2N5962, with a guaranteed minimum current gain of 450 at collector currents from 10μA to 10mA; it belongs to the 2N5961–2N5963 series; with a range of maximum V_{CE}s of 30 to 60 volts (if you need higher collector voltage, you have to settle for lower beta). Superbeta matched pairs are available for use in low-level amplifiers that require matched characteristics, a topic we will discuss in Section 2.18. Examples are the LM394 and MAT-01 series; these provide high-gain *npn* transistor pairs whose V_{BE}s are matched to a fraction of a millivolt (as little as 50μV in the best versions) and whose h_{FE}s are matched to about 1%. The MAT-03 is a *pnp* matched pair.

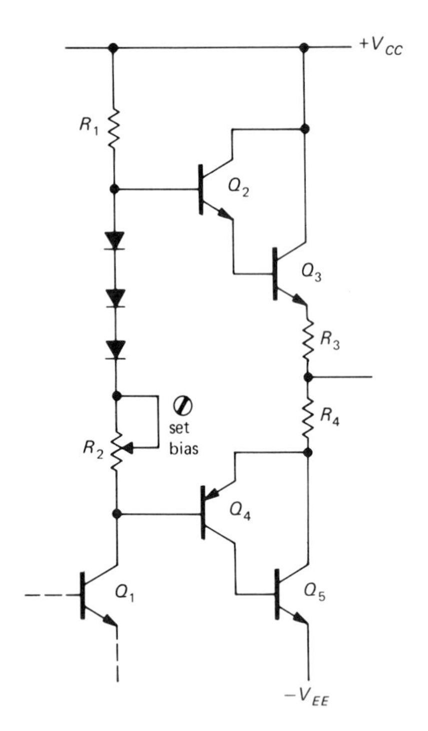

Figure 2.63. Push-pull power stage using only *npn* output transistors.

It is possible to combine superbeta transistors in a Darlington connection. Some commercial devices (e.g., the LM11 and LM316 op-amps) achieve base bias currents as low as 50 *pico*amps this way.

□ 2.17 Bootstrapping

When biasing an emitter follower, for instance, you choose the base voltage divider resistors so that the divider presents a stiff voltage source to the base, i.e., their parallel impedance is much less than the impedance looking into the base. For this reason the resulting circuit has an input impedance dominated by the voltage divider – the driving signal sees a much lower impedance than would otherwise be necessary. Figure 2.64 shows an example. The input

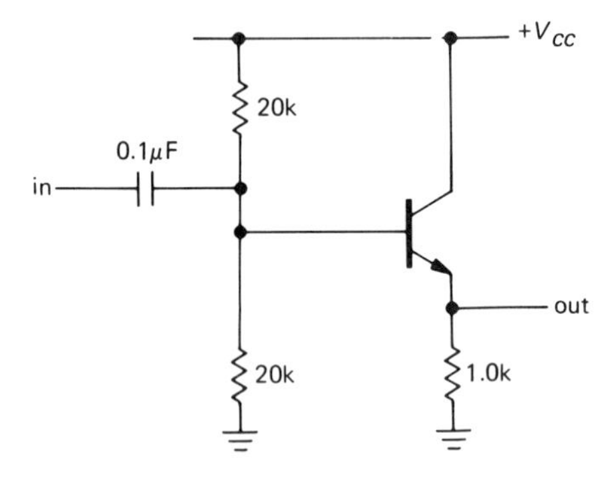

Figure 2.64

resistance of about 9k is mostly due to the voltage-divider impedance of 10k. It is always desirable to keep input impedances high, and anyway it's a shame to load the input with the divider, which, after all, is only there to bias the transistor. Bootstrapping is the colorful name given to a technique that circumvents this problem (Fig. 2.65). The transistor is biased by the divider R_1R_2 through series resistor R_3. C_2 is chosen to have low impedance at signal frequencies compared with the bias resistors. As always, bias is stable if the dc impedance seen from the base (in this case 9.7k) is much less than the dc impedance looking into the base (in this case approximately 100k). But now the signal-frequency input impedance is no longer the same as the dc impedance. Look at it this way: An input wiggle v_{in} results in an emitter wiggle $v_E \approx v_{\text{in}}$. So the change in current through bias resistor R_3 is $i = (v_{\text{in}} - v_E)/R_3 \approx 0$, i.e., Z_{in} (due to bias string) $= v_{\text{in}}/i_{\text{in}} \approx$ infinity. We've made the loading (shunt) impedance of the bias network very large *at signal frequencies.*

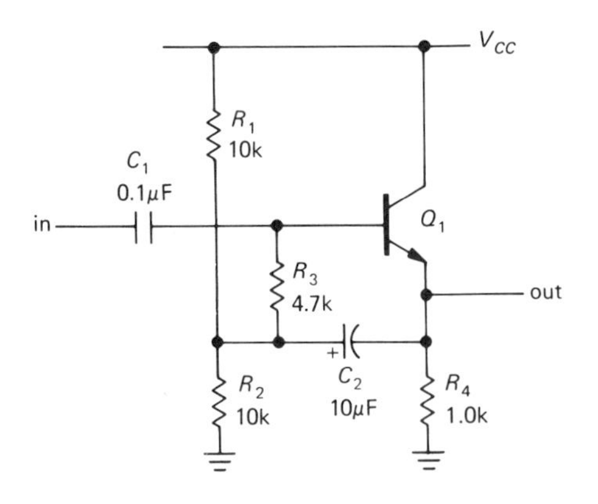

Figure 2.65. Raising the input impedance of an emitter follower at signal frequencies by bootstrapping the base bias divider.

Another way of seeing this is to notice that R_3 always has the same voltage across it at signal frequencies (since both ends of the resistor have the same voltage changes), i.e., it's a current source. But a current source has infinite impedance. Actually, the effective impedance is less than infinity because the gain of a follower is slightly less than 1. That is so because the base-emitter drop depends on collector current, which changes with the signal level. You could have predicted the same result from the voltage-dividing effect of the impedance looking into the emitter [$r_e = 25/I_C$(mA) ohms] combined with the emitter resistor. If the follower has voltage gain A ($A \approx 1$), the effective value of R_3 at signal frequencies is

$$R_3/(1 - A)$$

In practice the value of R_3 is effectively increased by a hundred or so, and the input impedance is then dominated by the transistor's base impedance. The emitter-degenerated amplifier can be bootstrapped in the same way, since the signal on the emitter follows the base. Note that the bias divider circuit is driven by the low-impedance emitter output at signal frequencies, thus isolating the input signal from this usual task.

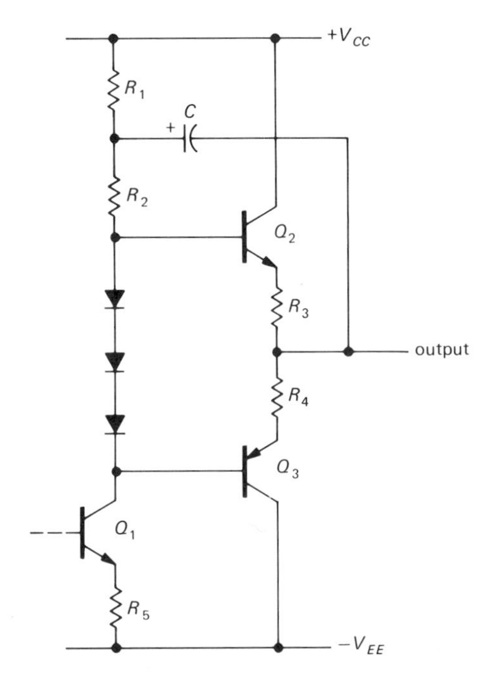

Figure 2.66. Bootstrapping driver-stage collector load resistor in a power amplifier.

□ *Bootstrapping collector load resistors*

The bootstrap principle can be used to increase the effective value of a transistor's collector load resistor, if that stage drives a follower. That can increase the voltage gain of the stage substantially [recall that $G_V = -g_m R_C$, with $g_m = 1/(R_E + r_e)$]. Figure 2.66 shows an example of a bootstrapped push-pull output stage similar to the push-pull follower circuit we saw earlier. Because the output follows Q_2's base signal, C bootstraps Q_1's collector load, keeping a constant voltage across R_2 as the signal varies (C must be chosen to have low impedance compared with R_1 and R_2 at all signal frequencies). That makes R_2 look like a current source, raising Q_1's voltage gain and maintaining good base drive to Q_2, even at the peaks of the signal swing. When the signal gets near V_{CC}, the junction of R_1 and R_2 actually rises above V_{CC} because of the stored charge in C. In this case, if $R_1 = R_2$ (not a bad choice) the junction between them rises to 1.5 times V_{CC} when the output reaches V_{CC}. This circuit has enjoyed considerable popularity in commercial audio amplifier design, although a simple current source in place of the bootstrap is superior, since it maintains the improvement at low frequencies and eliminates the undesirable electrolytic capacitor.

2.18 Differential amplifiers

The differential amplifier is a very common configuration used to amplify the difference voltage between two input signals. In the ideal case the output is entirely independent of the individual signal levels – only the difference matters. When both inputs change levels together, that's a *common-mode* input change. A differential change is called *normal mode*. A good differential amplifier has a high *common-mode rejection ratio* (CMRR), the ratio of response for a normal-mode signal to the response for a common-mode signal of the same amplitude. CMRR is usually specified in decibels. The common-mode input range is the voltage level over which the inputs may vary.

Differential amplifiers are important in applications where weak signals are contaminated by "pickup" and other miscellaneous noise. Examples include digital signals transferred over long cables (usually twisted pairs of wires), audio signals (the term "balanced" means differential, usually 600Ω impedance, in the audio business), radiofrequency signals (twin-lead cable is differential), electrocardiogram voltages, magnetic-core memory readout signals, and numerous other applications. A differential amplifier at the receiving end restores the original signal if the common-mode signals are not too large. Differential amplifiers are universally used in operational amplifiers, which we will come to soon. They're very important in dc amplifier design (amplifiers that amplify clear down to dc, i.e., have no coupling capacitors) because their symmetrical design is inherently compensated against thermal drifts.

Figure 2.67 shows the basic circuit. The output is taken off one collector with respect to ground; that is called a *single-ended output* and is the most common configuration. You can think of this amplifier as a device that amplifies a difference signal and converts it to a single-ended signal so that ordinary subcircuits (followers, current sources, etc.) can make use of the output. (If, instead, a differential output is desired, it is taken between the collectors.)

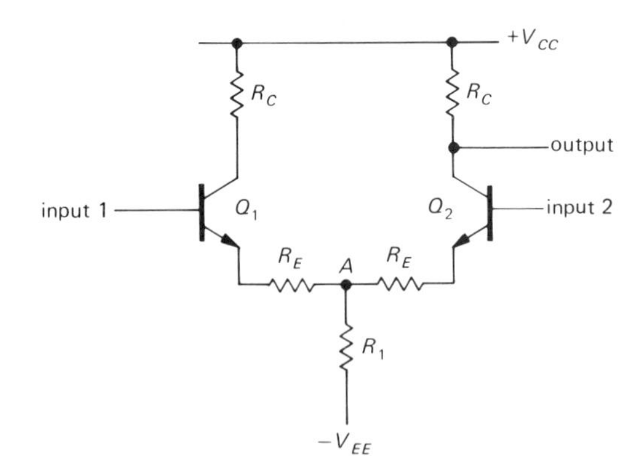

Figure 2.67. Classic transistor differential amplifier.

What is the gain? That's easy enough to calculate: Imagine a symmetrical input signal wiggle, in which input 1 rises by

v_{in} (a small-signal variation) and input 2 drops by the same amount. As long as both transistors stay in the active region, point A remains fixed. The gain is then determined as with the single transistor amplifier, remembering that the input change is actually twice the wiggle on either base: $G_{\text{diff}} = R_C/2(r_e + R_E)$. Typically R_E is small, 100 ohms or less, or it may be omitted entirely. Differential voltage gains of a few hundred are typical.

The common-mode gain can be determined by putting identical signals v_{in} on both inputs. If you think about it correctly (remembering that R_1 carries both emitter currents), you'll find $G_{CM} = -R_C/(2R_1 + R_E)$. Here we've ignored the small r_e, because R_1 is typically large, at least a few thousand ohms. We really could have ignored R_E as well. The CMRR is roughly $R_1/(r_e + R_E)$. Let's look at a typical example (Fig. 2.68) to get some familiarity with differential amplifiers.

R_C is chosen for a quiescent current of 100μA. As usual, we put the collector at $0.5V_{CC}$ for large dynamic range. Q_1's collector resistor can be omitted, since no output is taken there. R_1 is chosen to give total emitter current of 200μA, split equally between the two sides when the (differential) input is zero. From the formulas just derived, this amplifier has a differential gain of 30 and a common-mode gain of 0.5. Omitting the 1.0k resistors raises the differential gain to 150, but drops the (differential) input impedance from about 250k to about 50k (you can substitute Darlington transistors in the input stage to raise the impedance into the megohm range, if necessary).

Remember that the maximum gain of a single-ended grounded emitter amplifier biased to $0.5V_{CC}$ is $20V_{CC}$. In the case of a differential amplifier the maximum differential gain ($R_E = 0$) is half that figure, or (for arbitrary quiescent point) 20 times the voltage across the collector resistor. The corresponding maximum CMRR (again with $R_E = 0$) is equal to 20 times the voltage across R_1.

EXERCISE 2.13

Verify that these expressions are correct. Then design a differential amplifier to your own specifications.

The differential amplifier is sometimes called a "long-tailed pair," because if the length of a resistor symbol indicated its magnitude, the circuit would look like Figure 2.69. The long tail determines the

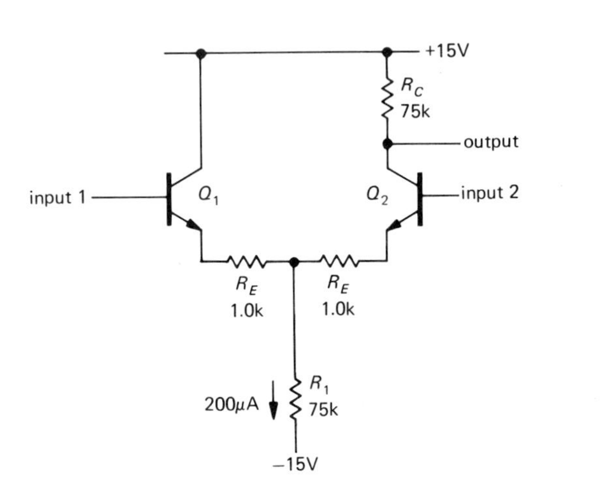

$$G_{\text{diff}} = \frac{v_{\text{out}}}{v_1 - v_2} = \frac{R_C}{2(R_E + r_e)}$$

$$G_{CM} = -\frac{R_C}{2R_1 + R_E + r_e}$$

$$\text{CMRR} \approx \frac{R_1}{R_E + r_e}$$

Figure 2.68. Calculating differential amplifier performance.

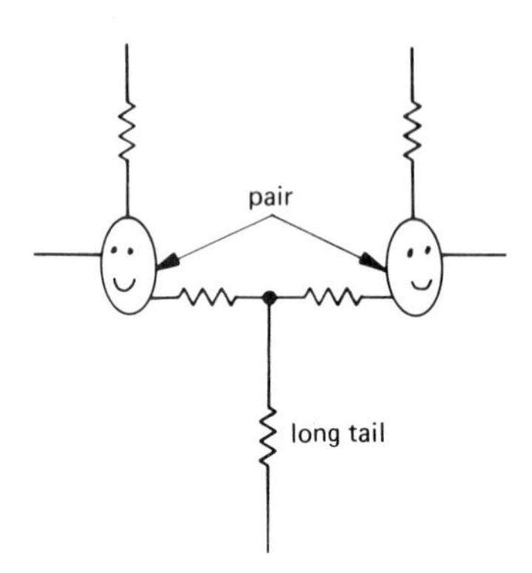

Figure 2.69

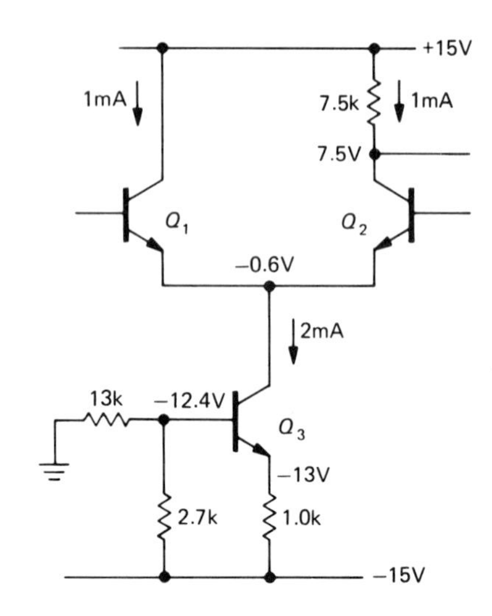

Figure 2.70. Improving CMRR of the differential amplifier with a current source.

common-mode gain, and the small inter-emitter resistance (including intrinsic emitter resistance r_e) determines the differential gain.

Current-source biasing

The common-mode gain of the differential amplifier can be reduced enormously by substituting a current source for R_1. Then R_1 effectively becomes very large, and the common-mode gain is nearly zero. If you prefer, just imagine a common-mode input swing; the emitter current source maintains a constant total emitter current, shared equally by the two collector circuits, by symmetry. The output is therefore unchanged. Figure 2.70 shows an example. The CMRR of this circuit, using an LM394 monolithic transistor pair for Q_1 and Q_2 and a 2N5963 current source is 100,000:1 (100dB). The common-mode input range for this circuit goes from −12 volts to +7 volts; it is limited at the low end by the compliance of the emitter current source and at the high end by the collector's quiescent voltage.

Be sure to remember that this amplifier, like all transistor amplifiers, must have a dc bias path to the bases. If the input is capacitively coupled, for instance, you must have base resistors to ground. An additional caution for differential amplifiers, particularly those without inter-emitter resistors: Bipolar transistors can tolerate only 6 volts of base-emitter reverse bias before breakdown; thus, applying a differential input voltage larger than this will destroy the input stage (if there is no inter-emitter resistor). An inter-emitter resistor limits the breakdown current and prevents destruction, but the transistors may be degraded (in h_{fe}, noise, etc.). In either case the input impedance drops drastically during reverse conduction.

Use in single-ended dc amplifiers

A differential amplifier makes an excellent dc amplifier, even for single-ended inputs. You just ground one of the inputs and connect the signal to the other (Fig. 2.71). You might think that the "unused" transistor could be eliminated. Not so! The differential configuration is inherently compensated for temperature drifts, and even when one input is at ground that transistor is still doing something: A temperature change causes both V_{BE}s to change the same amount, with no change in balance or output. That is, changes in V_{BE} are not amplified by G_{diff} (only by G_{CM}, which can be made essentially zero). Furthermore, the cancellation of V_{BE}s means that there are no 0.6 volt drops at the input to worry about. The quality of a dc amplifier constructed this way is limited only by mismatching of input V_{BE}s or their temperature coefficients. Commercial monolithic transistor pairs and commercial differential amplifier ICs are available with extremely good matching (e.g., the MAT-01 *npn* monolithic matched pair has a typical drift of V_{BE} between the two transistors of $0.15\mu V/°C$ and $0.2\mu V$ per month).

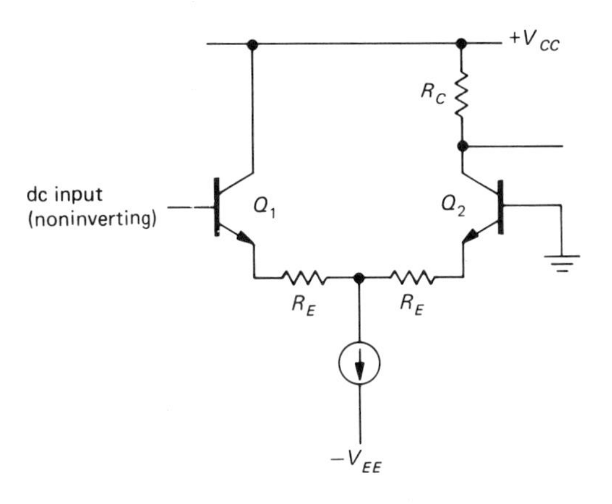

Figure 2.71. A differential amplifier can be used as a precision single-ended dc amplifier.

Either input could have been grounded in the preceding circuit example. The choice depends on whether or not the amplifier is supposed to invert the signal. (The configuration shown is preferable at high frequencies, however, because of *Miller effect*; see Section 2.19.) The connection shown is noninverting, and so the inverting input has been grounded. This terminology carries over to op-amps, which are simply high-gain differential amplifiers.

Current mirror active load

As with the simple grounded emitter amplifier, it is sometimes desirable to have a single-stage differential amplifier with very high gain. An elegant solution is a current mirror active load (Fig. 2.72). Q_1Q_2 is the differential pair with emitter current source. Q_3 and Q_4, a current mirror, form the collector load. The high effective collector load impedance provided by the mirror yields voltage gains of 5000 or more, assuming no load at the amplifier's output. Such an amplifier is usually used only within a feedback loop, or as a comparator (discussed in the next section). Be sure to load such an amplifier with a high impedance, or the gain will drop enormously.

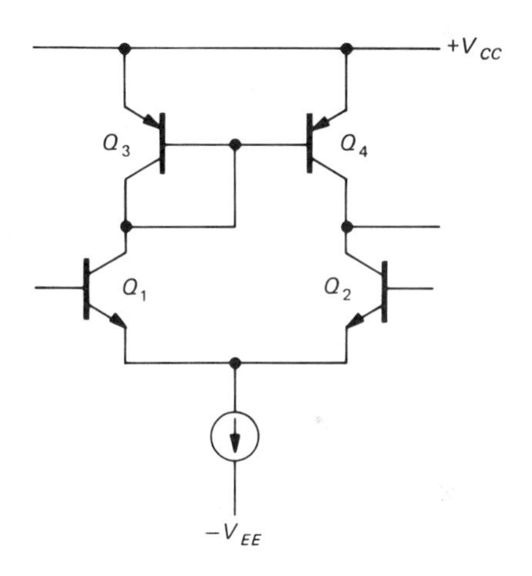

Figure 2.72. Differential amplifier with active current mirror load.

Differential amplifiers as phase splitters

The collectors of a symmetrical differential amplifier generate equal signal swings of opposite phase. By taking outputs from both collectors, you've got a phase splitter. Of course, you could also use a differential amplifier with both differential inputs and differential outputs. This differential output signal could then be used to drive an additional differential amplifier stage, with greatly improved overall common-mode rejection.

Differential amplifiers as comparators

Because of its high gain and stable characteristics, the differential amplifier is the main building block of the *comparator*, a circuit that tells which of two inputs is larger. They are used for all sorts of applications: switching on lights and heaters, generating square waves from triangles, detecting when a level in a circuit exceeds some particular threshold, class D amplifiers and pulse-code modulation, switching power supplies, etc. The basic idea is to connect a differential amplifier so that it turns a transistor switch on or off, depending on the relative levels of the input signals. The linear region of amplification is ignored, with one or the other of the two input transistors cut off at any time. A typical hookup is illustrated in the next section by a temperature-controlling circuit that uses a resistive temperature sensor (thermistor).

2.19 Capacitance and Miller effect

In our discussion so far we have used what amounts to a dc, or low-frequency, model of the transistor. Our simple current amplifier model and the more sophisticated Ebers-Moll transconductance model both deal with voltages, currents, and resistances seen at the various terminals. With these models alone we have managed to go quite far, and in fact these simple models contain nearly everything you will ever need to know to design transistor circuits. However, one important aspect that has serious impact on high-speed and high-frequency circuits has been neglected: the existence of capacitance in the external circuit and in the transistor junctions themselves. Indeed, at high frequencies the effects of capacitance often dominate circuit behavior; at 100 MHz a typical junction capacitance of 5pF has an impedance of 320 ohms!

We will deal with this important subject in detail in Chapter 13. At this point we would merely like to state the problem, illustrate some of its circuit incarnations, and suggest some methods of circumventing the problem. It would be a mistake to leave this chapter without realizing the nature of this problem. In the course of this brief discussion we will encounter the famous *Miller effect* and the use of configurations such as the cascode to overcome it.

Junction and circuit capacitance

Capacitance limits the speed at which the voltages within a circuit can swing ("slew rate"), owing to finite driving impedance or current. When a capacitance is driven by a finite source resistance, you see *RC* exponential charging behavior, whereas a capacitance driven by a current source leads to slew-rate-limited waveforms (ramps). As general guidance, reducing the source impedances and load capacitances and increasing the drive currents within a circuit will speed things up. However, there are some subtleties connected with feedback capacitance and input capacitance. Let's take a brief look.

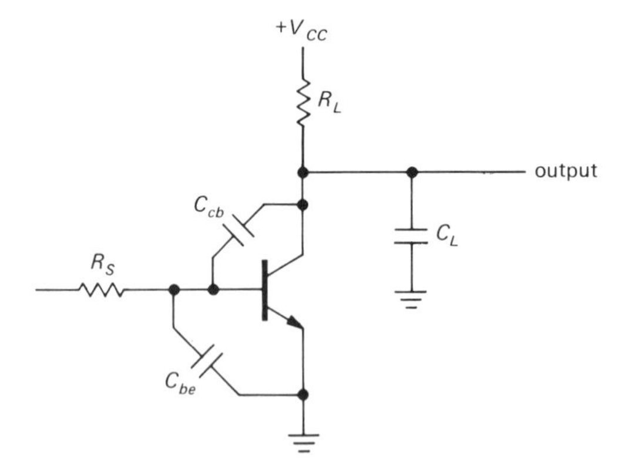

Figure 2.73. Junction and load capacitances in a transistor amplifier.

The circuit in Figure 2.73 illustrates most of the problems of junction capacitance. The output capacitance forms a time constant with the output resistance R_L (R_L includes both the collector and load resistances, and C_L includes both junction and load capacitances), giving a rolloff starting at some frequency $f = 1/2\pi R_L C_L$. The same is true for the input capacitance in combination with the source impedance R_S.

Miller effect

C_{cb} is another matter. The amplifier has some overall voltage gain G_V, so a small voltage wiggle at the input results in a wiggle G_V times larger (and inverted) at the collector. This means that the signal source sees a current through C_{cb} that is $G_V + 1$ times as large as if C_{cb} were connected from base to ground; i.e., for the purpose of input rolloff frequency calculations, the feedback capacitance behaves like a capacitor of value $C_{cb}(G_V + 1)$ from input to ground. This effective increase of C_{cb} is known as the Miller effect. It often dominates the rolloff characteristics of amplifiers, since a typical feedback capacitance of 4pF can look like several hundred picofarads to ground.

There are several methods available to beat the Miller effect. It is absent altogether in a grounded base stage. You can decrease the source impedance driving a grounded emitter stage by using an emitter follower. Figure 2.74 shows two other possibilities. The differential amplifier circuit (with no collector resistor in Q_1) has no Miller effect; you can think of it as an emitter follower driving a grounded base amplifier. The second circuit is the famous cascode configuration. Q_1 is a grounded emitter amplifier with R_L as its collector resistor. Q_2 is interposed in the collector path to prevent Q_1's collector from swinging (thereby eliminating the Miller effect) while passing the collector current through to the load resistor unchanged. V_+ is a fixed bias voltage, usually set a few volts above Q_1's emitter voltage to pin Q_1's collector and keep it in the active region. This fragment is incomplete as shown; you could either include a bypassed emitter resistor and base divider for biasing (as we did earlier in the chapter) or include it within an overall loop with feedback at dc. V_+ might be provided from a divider or zener, with bypassing to keep it stiff at signal frequencies.

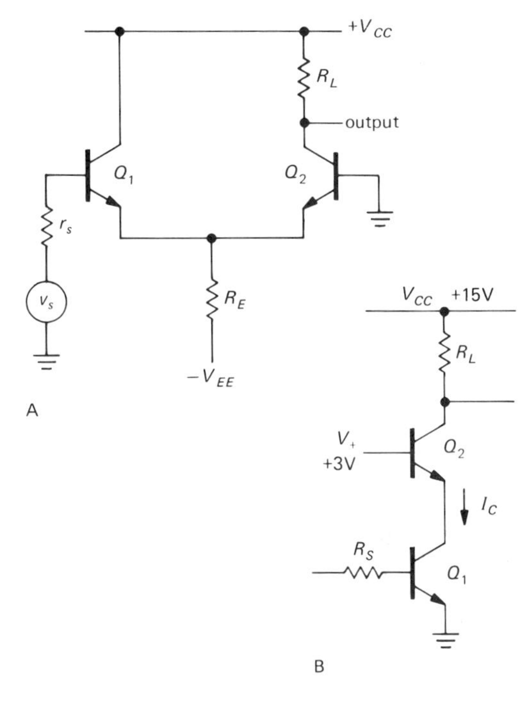

Figure 2.74. Two circuit configurations that avoid Miller effect. Circuit B is the cascode.

EXERCISE 2.14

Explain in detail why there is no Miller effect in either transistor in the preceding differential amplifier and cascode circuits.

Capacitive effects can be somewhat more complicated than this brief introduction might indicate. In particular: (a) The rolloffs due to feedback and output capacitances are not entirely independent; in the terminology of the trade there is pole splitting, an effect we will explain in the next chapter. (b) The input capacitance still has an effect, even with a stiff input signal source. In particular, current that flows through C_{be} is not amplified by the transistor. This base current "robbing" by the input capacitance causes the transistor's small-signal current gain h_{fe} to drop at high frequencies, eventually reaching unity at a frequency known as f_T. (c) To complicate matters, the junction capacitances depend on voltage. C_{be} changes so rapidly with base current that it is not even specified on transistor data sheets; f_T is given instead. (d) When a transistor is operated as a switch, effects associated with charge stored in the base region of a saturated transistor cause an additional loss of speed. We will take up these and other topics having to do with high-speed circuits in Chapter 13.

2.20 Field-effect transistors

In this chapter we have dealt exclusively with *bipolar junction transistors* (BJTs), characterized by the Ebers-Moll equation. BJTs were the original transistors, and they still dominate analog circuit design. However, it would be a mistake to continue without a few words of explanation about the other kind of transistor, the field-effect transistor (FET), which we will take up in detail in the next chapter.

The FET behaves in many ways like an ordinary bipolar transistor. It is a 3-terminal amplifying device, available in both polarities, with a terminal (the *gate*) that controls the current flow between the other two terminals (*source* and *drain*). It has a unique property, though: The gate draws no current, except for leakage. This means that extremely high input impedances are possible, limited only by capacitance and leakage effects. With FETs you don't have to worry about providing substantial base current, as was necessary with the BJT circuit design of this chapter. Input currents measured in *pico*amperes are commonplace. Yet the FET is a rugged and capable device, with voltage and current ratings comparable to those of bipolar transistors.

Most of the available devices fabricated with transistors (matched pairs, differential and operational amplifiers, comparators, high-current switches and amplifiers, radiofrequency amplifiers, and digital logic) are also available with FET construction, often with superior performance. Furthermore, microprocessors and memory (and other large-scale digital electronics) are built almost exclusively with FETs. Finally, the area of micropower design is dominated by FET circuits.

FETs are so important in electronic design that we will dèvote the next chapter to them, before treating operational amplifiers and feedback in Chapter 4. We urge the reader to be patient with us as we lay the groundwork in these first three difficult chapters; that patience will be rewarded many times over in the succeeding chapters, as we explore the enjoyable topics of circuit design with operational amplifiers and digital integrated circuits.

SOME TYPICAL TRANSISTOR CIRCUITS

To illustrate some of the ideas of this chapter, let's look at a few examples of circuits with transistors. The range of circuits we can cover is necessarily limited, since real-world circuits often use negative feedback , a subject we will cover in Chapter 4.

2.21 Regulated power supply

Figure 2.75 shows a very common configuration. R_1 normally holds Q_1 on; when the output reaches 10 volts, Q_2 goes into conduction (base at 5V), preventing further rise of output voltage by shunting base current from Q_1's base. The supply can be made adjustable by replacing R_2 and R_3

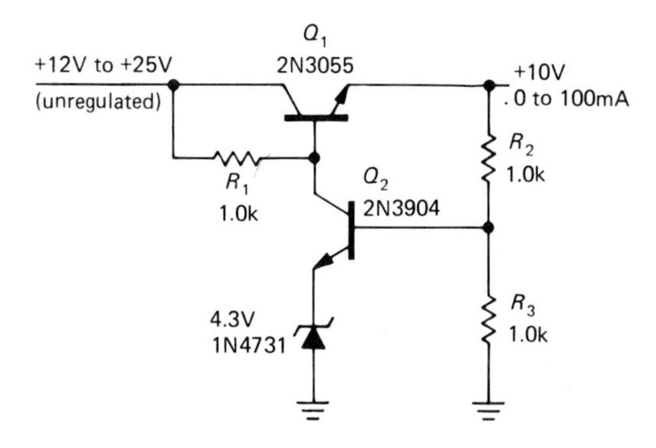

Figure 2.75. Feedback voltage regulator.

by a potentiometer. This is actually an example of negative feedback: Q_2 "looks at" the output and does something about it if the output isn't at the right voltage.

2.22 Temperature controller

The schematic diagram in Figure 2.76 shows a temperature controller based on a *thermistor* sensing element, a device that changes resistance with temperature. Differential Darlington $Q_1 - Q_4$ compares the voltage of the adjustable reference divider R_4–R_6 with the divider formed from the thermistor and R_2. (By comparing *ratios* from the same supply, the comparison becomes insensitive to supply variations; this particular configuration is called a Wheatstone bridge.) Current mirror Q_5Q_6 provides an active load to raise the gain, and mirror Q_7Q_8 provides emitter current. Q_9 compares the differential amplifier output with a fixed voltage, saturating Darlington $Q_{10}Q_{11}$, which supplies power to the heater, if the thermistor is too cold. R_9 is a current-sensing resistor that turns on protection transistor Q_{12} if the output current exceeds about 6 amps; that removes base drive from $Q_{10}Q_{11}$, preventing damage.

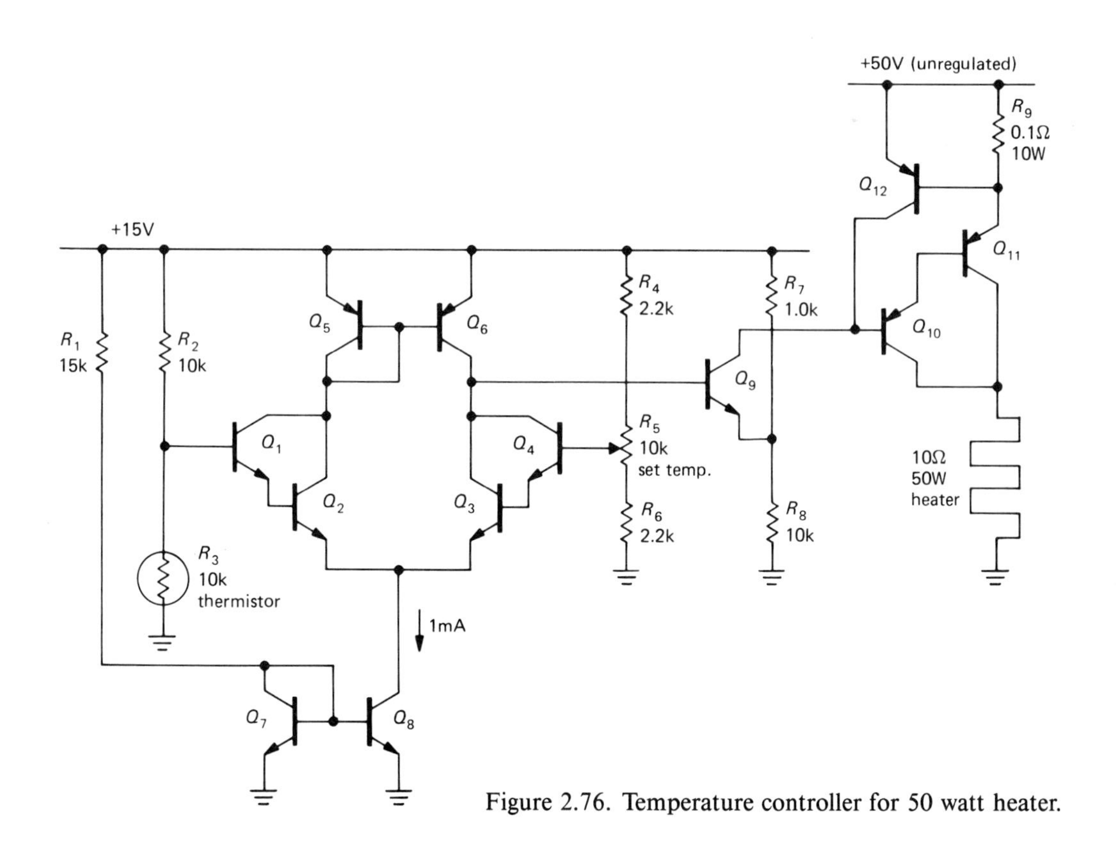

Figure 2.76. Temperature controller for 50 watt heater.

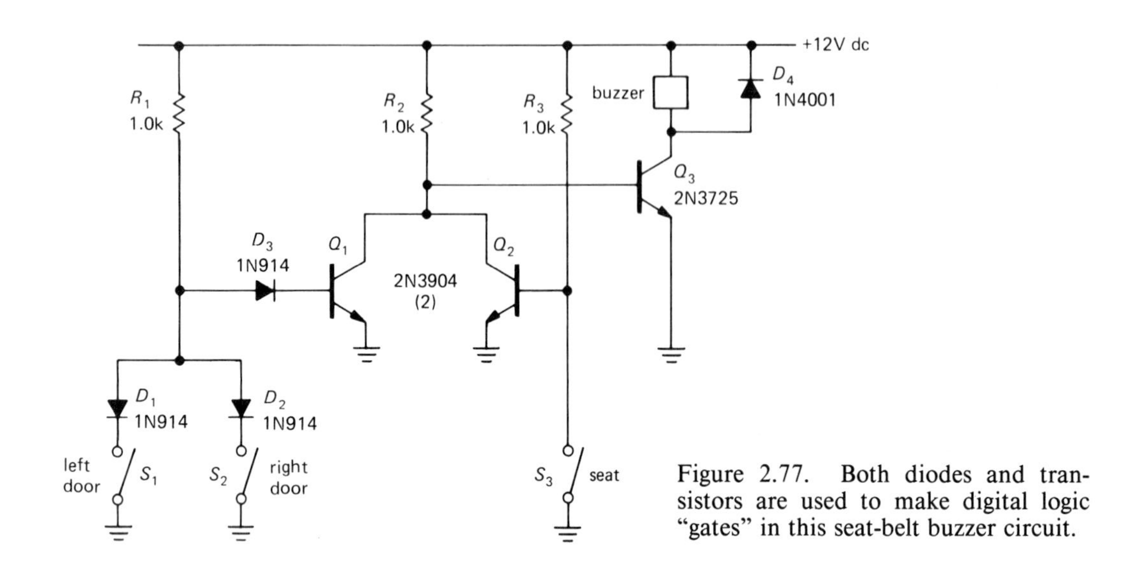

Figure 2.77. Both diodes and transistors are used to make digital logic "gates" in this seat-belt buzzer circuit.

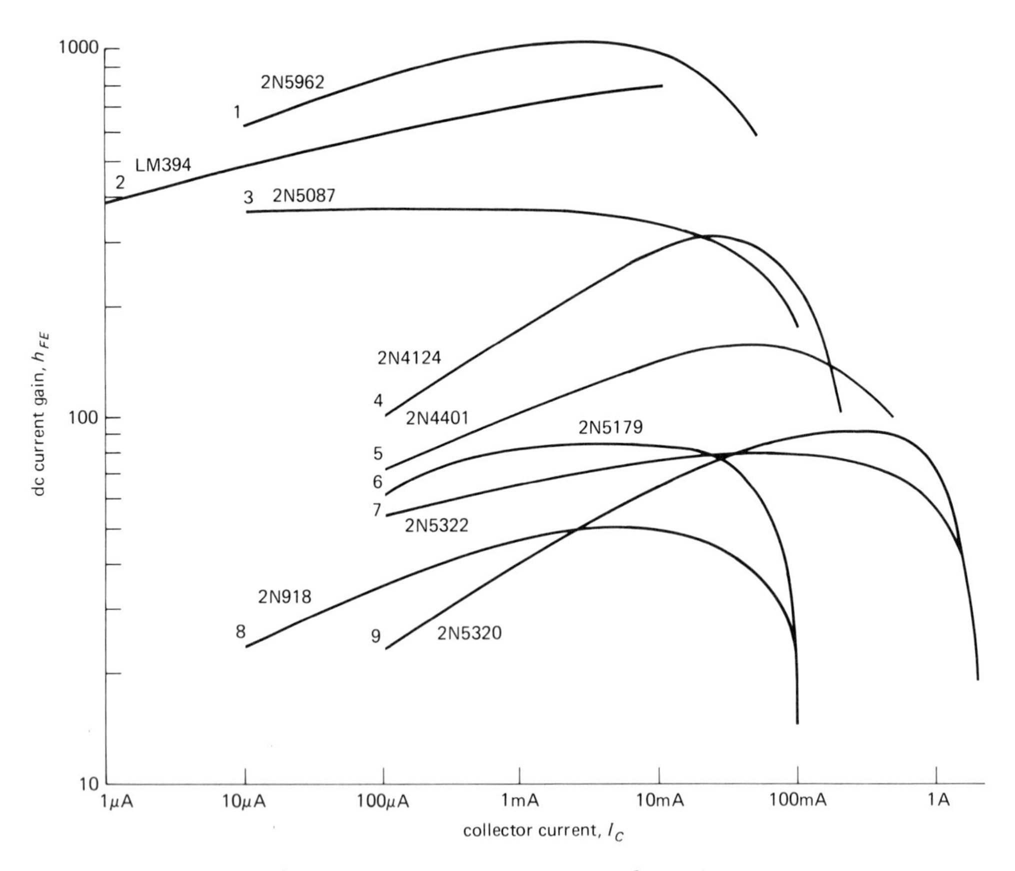

Figure 2.78. Curves of typical transistor current gain, h_{FE}, for a selection of transistors from Table 2.1. These curves are taken from manufacturers' literature. You can expect production spreads of +100%, −50% from the "typical" values graphed.

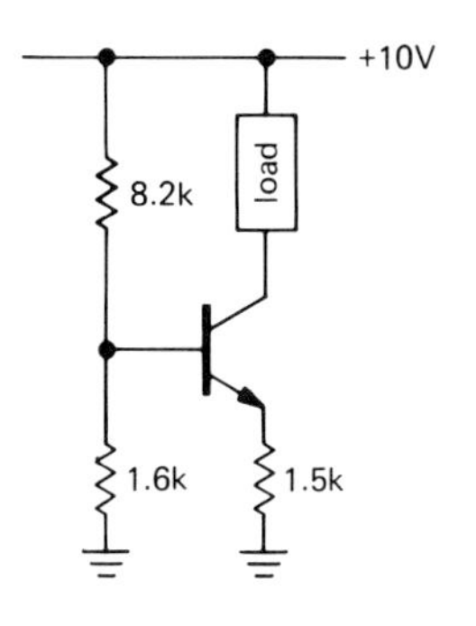

Figure 2.79

2.23 Simple logic with transistors and diodes

Figure 2.77 shows a circuit that performs a task we illustrated in Section 1.32: sounding a buzzer if either car door is open and the driver is seated. In this circuit the transistors all operate as switches (either off or saturated). Diodes D_1 and D_2 form what is called an OR gate, turning off Q_1 if either door is open (switch closed). However, the collector of Q_1 stays near ground, preventing the buzzer from sounding unless switch S_3 is also closed (driver seated); in that case R_2 turns Q_3 on, putting 12 volts across the buzzer. D_3 provides a diode drop so that Q_1 is off with S_1 or S_2 closed, and D_4 protects Q_3 from the buzzer's inductive turn-off transient. In Chapter 8 we will discuss logic circuitry in detail.

Table 2.1 presents a selection of useful and popular small-signal transistors; Figure 2.78 shows corresponding curves of current gain. See also Appendix K.

SELF-EXPLANATORY CIRCUITS

2.24 Good circuits

Figure 2.80 shows a couple of circuit ideas that use transistors.

2.25 Bad circuits

A lot can be learned from your own mistakes or someone else's mistakes. In this section we present a gallery of blunders (Fig. 2.81). You can amuse yourself by thinking of variations on these bad circuits, and then avoiding them!

ADDITIONAL EXERCISES

(1) Design a transistor switch circuit that allows you to switch two loads to ground via saturated *npn* transistors. Closing switch *A* should cause both loads to be powered, whereas closing switch *B* should power only one load. Hint: Use diodes.

Circuit ideas

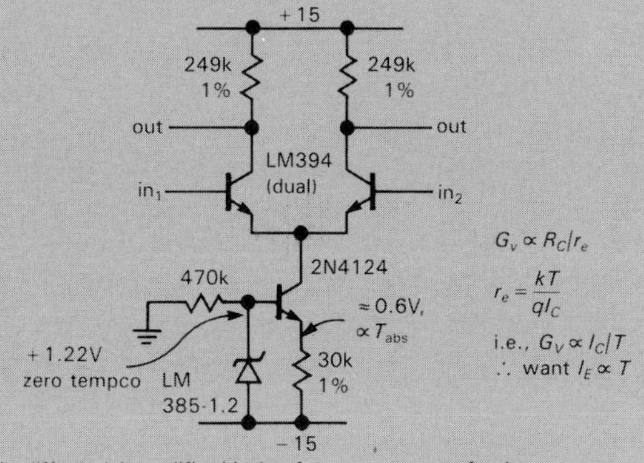

A. differential amplifier biasing for zero tempco of gain

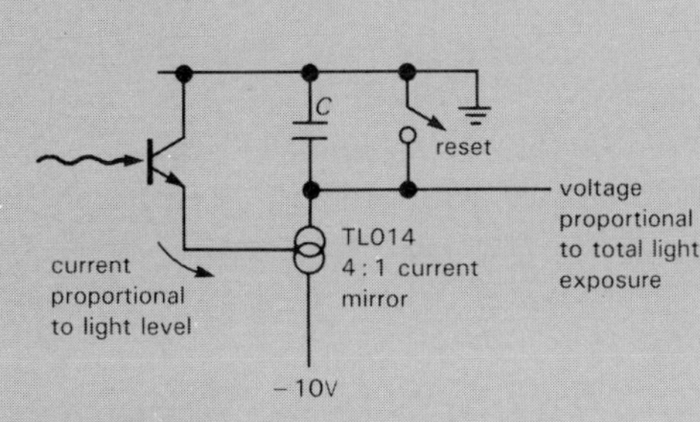

B. light integrator

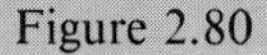

Figure 2.80

Bad circuits

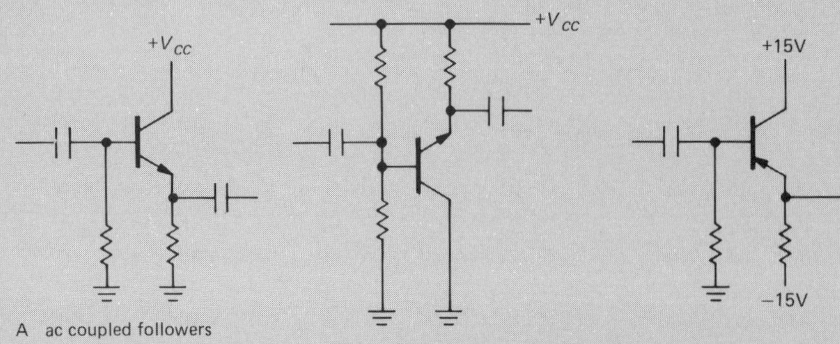

A ac coupled followers

unregulated
dc in (> 10V)
+5V out
5.6V zener

B +5V regulator

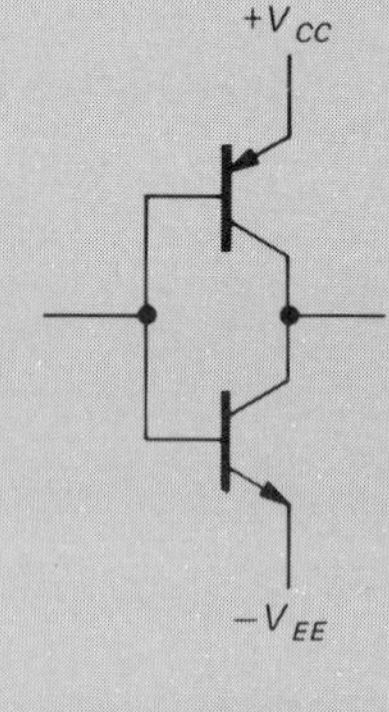

C push-pull follower

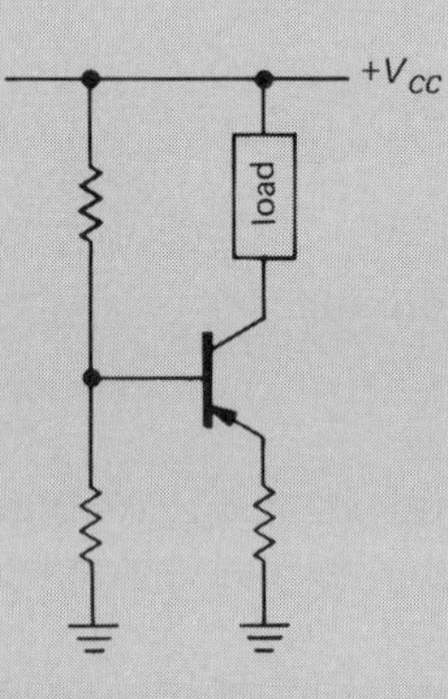

D current source

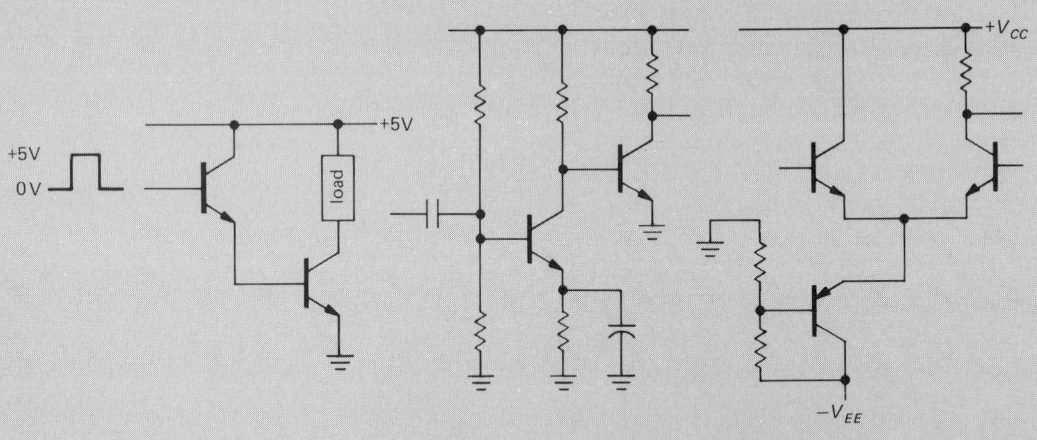

E high-current switch

F two-stage amplifier

G differential amplifier

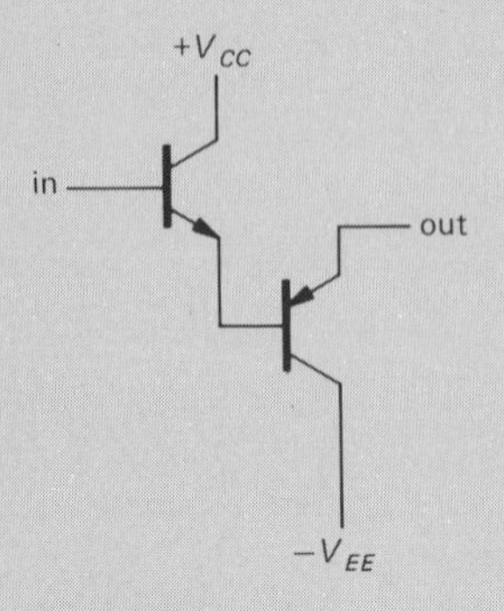

H zero-offset follower

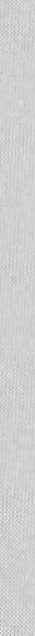

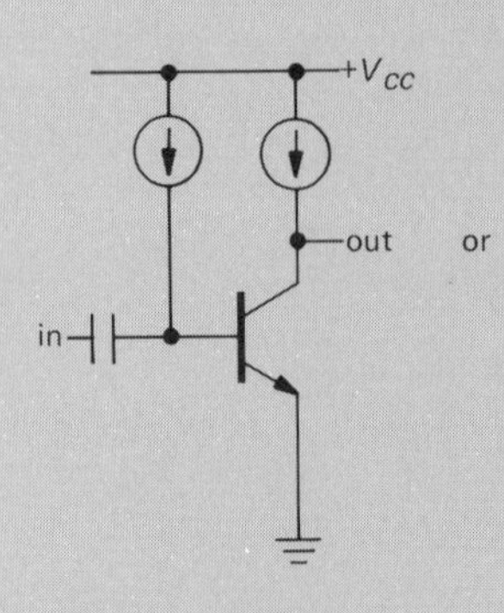

or

$+V_{CC}$
out
in

I high-gain ac amplifier

Figure 2.81

TABLE 2.1. SELECTED SMALL-SIGNAL TRANSISTORS[a]

	V_{CEO} (V)	I_C max (mA)	h_{FE} typ[b]	I_C (mA)	C_{cb} typ[c] (pF)	f_T typ[d] (MHz)	Gain curve	Metal				Plastic	
								TO-5[e]		TO-18[f]		TO-92[h]	
								npn	pnp	npn	pnp	npn	pnp
General purpose	20	500	100	150	16	200		–	–	–	–	–	–
	25	200	200	2	1.8-2.8	300	4	–	–	–	–	4124	4126
	40	200	200	10	1.8-2.8	300		–	–	3947	3251	3904	3906
High gain, low noise	25	50	300	10	2-7	150		–	–	–	–	3391A[h],3707[h]	4058[h]
	25	300	250	50	4	300		–	–	–	–	6008[h]	6009[h]
	25	50	500	5	1.5-4	500		–	–	–	–	5089	–
	40	20	700	1	14	200	2	LM394	–	–	–	–	–
	45	50	1000	10	1.5	300	1	–	–	–	–	5962	–
	50	50	350	5	1.8	400	3	–	–	2848	3965	4967,5210	4965,5087
High current	30-60	600	150	150	5	300	5	2219	2905	2222	2907,3251	4401	4403
	50	1000	100	200	7	450		3725	5022	4014	–	–	–
	60	1000	70	80	15	100		2102,3107	4036	–	–	–	–
	75	2000	70	500	20	60	7,9	5320	5322	–	–	–	–
High voltage	150	600	100	10	3-6	250		–	4929	–	–	5550	5401
	300	1000	50	50	10	50		3439	5416	–	–	–	–
High speed	12	50	80	3	0.7	1500	6	–	–	5179	–	3662[h]	–
	12	100	50	8	1.5	900	8	–	–	918	4208	5770	–
	12	200	75	25	3	500		–	–	2369	2894	5769	5771

(a) all transistors are 2Nxxx numbers, except for the LM394 dual transistor. Devices listed on a single row are similar in characteristics and in some cases are electrically identical. (b) see figure 2.76. (c) at V_{CB}=10V. (d) see figure 13.4. (e) or TO-39. (f) or TO-72, TO-46. (h) TO-92 and its variants have two basic pinouts: EBC and ECB. Transistors with superscript *h* are ECB; all others are EBC.

(2) Consider the current source in Figure 2.79. ***(a)*** What is I_{load}? What is the output compliance? Assume V_{BE} is 0.6 volt. ***(b)*** If h_{FE} varies from 50 to 100 for collector voltages within the output compliance range, how much will the output current vary? (There are two effects here.) ***(c)*** If V_{BE} varies according to $\Delta V_{BE} = -0.0001\,\Delta V_{CE}$ (Early effect), how much will the load current vary over the compliance range? ***(d)*** What is the temperature coefficient of output current assuming that h_{FE} does not vary with temperature? What is the temperature coefficient of output current assuming that h_{FE} increases from its nominal value of 100 by 0.4%/°C?

(3) Design a common-emitter *npn* amplifier with voltage gain of 15, V_{CC} of +15 volts, and I_C of 0.5mA. Bias the collector at $0.5V_{CC}$, and put the low-frequency 3dB point at 100Hz.

(4) Bootstrap the circuit in the preceding problem in order to raise the input impedance. Choose the rolloff of the bootstrap appropriately.

(5) Design a dc-coupled differential amplifier with voltage gain of 50 (to a single-ended output) for input signals near ground, supply voltages of ±15 volts, and quiescent currents of 0.1mA in each transistor. Use a current source in the emitter and an emitter follower output stage.

(6) In this problem you will ultimately design an amplifier whose gain is controlled by an externally applied voltage (in Chapter 3 you will see how to do the same thing with FETs). ***(a)*** Begin by designing a long-tailed pair differential amplifier with emitter current source and no emitter resistors (undegenerated). Use ±15 volt supplies. Set I_C (each transistor) at 1mA, and use R_C = 1.0k. Calculate the voltage gain from a single-ended input (other input grounded) to a single-ended output. ***(b)*** Now modify the circuit so that an externally applied voltage controls the emitter current source. Give an approximate formula for the gain as a function of controlling voltage. (In a real circuit you might arrange a second set of voltage-controlled current sources to cancel the quiescent-point shift that gain changes produce in this circuit, or a differential-input second stage could be added to your circuit.)

(7) Disregarding the lessons of this chapter, a disgruntled student builds the amplifier shown in Figure 2.82. He adjusts R until the quiescent point is $0.5V_{CC}$. ***(a)*** What is Z_{in} (at high frequencies where $Z_C \approx 0$)? ***(b)*** What is the small-signal voltage gain? ***(c)*** What rise in ambient temperature (roughly) will cause the transistor to saturate?

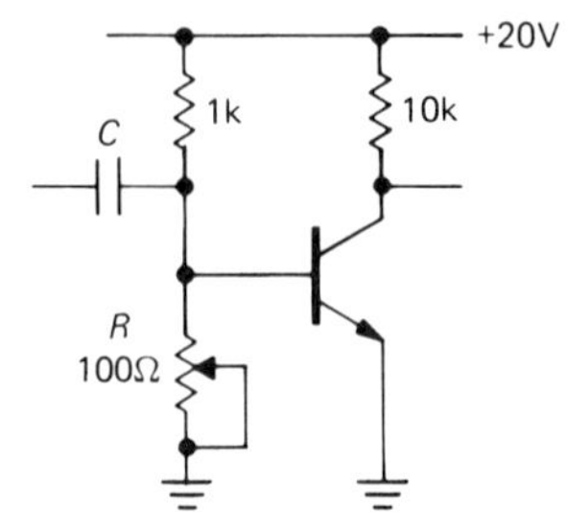

Figure 2.82

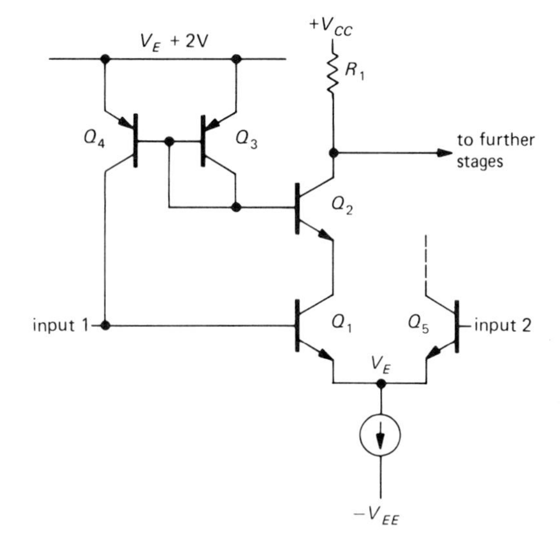

Figure 2.83. Base-current cancellation scheme, commonly used in high-quality operational amplifiers.

(8) Several commercially available precision op-amps (e.g., the venerable OP-07 and the recent LT1012) use the circuit in Figure 2.83 to cancel input bias current (only half of the symmetrical-input differential amplifier is shown in detail; the other half works the same way). Explain how the circuit works. Note: Q_1 and Q_2 are a beta-matched pair. Hint: It's all done with mirrors.

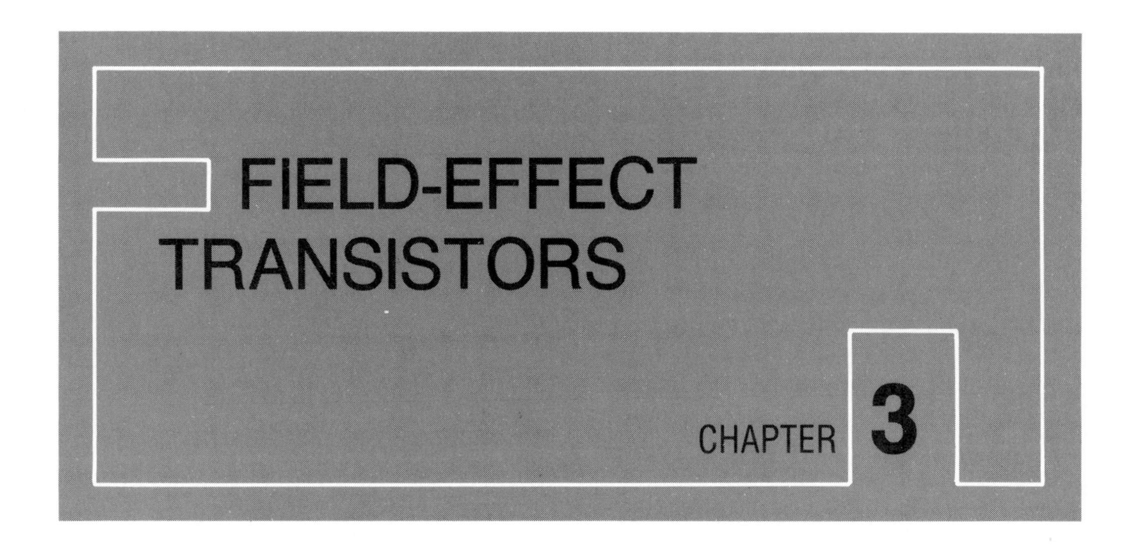

FIELD-EFFECT TRANSISTORS

CHAPTER 3

INTRODUCTION

Field-effect transistors (FETs) are different from the ordinary transistors (sometimes called "bipolar transistors," "bipolar junction transistors," or BJTs, to distinguish them from FETs) that we talked about in the last chapter. Broadly speaking, however, they are similar devices, which we might call *charge-control devices*: In both cases we have a 3-terminal device in which the conduction between two electrodes depends on the availability of charge carriers, which is controlled by a voltage applied to a third *control electrode*.

Here's how they differ: In an *npn* BJT the collector-base junction is back-biased, so no current normally flows. Forward-biasing the base-emitter junction by ≈0.6 volts overcomes its diode "contact potential barrier," causing electrons to enter the base region, where they are strongly attracted to the collector; although some base current results, most of these "minority carriers" are captured by the collector. This results in a collector current, controlled by a (smaller) base current. The collector current is proportional to the rate of injection of minority carriers into the base region, which is an exponential function of the *BE* potential difference (the Ebers–Moll equation). You can think of a bipolar transistor as a current amplifier (with roughly constant current gain, h_{FE}) or as a transconductance device (Ebers–Moll).

In a FET, as the name suggests, conduction in a *channel* is controlled by an *electric field*, produced by a voltage applied to the *gate* electrode. There are no forward-biased junctions, so the gate draws no current; this is perhaps the most important advantage of the FET. As with BJTs, there are two polarities, *n-channel* FETs (conduction by electrons) and *p-channel* FETs (conduction by holes). These two polarities are analogous to the familiar *npn* and *pnp* bipolar transistors, respectively. In addition, however, FETs tend to be confusing at first because they can be made with two different kinds of gates (thus JFETs and MOSFETs), and with two different kinds of channel doping (leading to *enhancement* and *depletion* modes). We'll sort out these possibilities shortly.

First, though, some motivation and perspective: The FET's nonexistent gate current is its most important characteristic. The resulting high input impedance (which can be greater than $10^{14}\Omega$) is essential in many applications, and in any case it makes circuit design simple and fun. For applications like analog switches and amplifiers of ultrahigh input impedance, FETs have no equal. They can be easily used by themselves or combined with bipolar transistors to make integrated circuits: In the next chapter we'll see how successful that process has been in making nearly perfect (and wonderfully easy to use) *operational amplifiers*, and in Chapters 8–11 we'll see how digital electronics has been revolutionized by MOSFET integrated circuits. Because many FETs using very low current can be constructed in a small area, they are especially useful for large-scale integration (LSI) digital circuits such as calculator chips, microprocessors, and memories. In addition, high-current MOSFETs (30A or more) of recent design have been replacing bipolar transistors in many applications, often providing simpler circuits with improved performance.

3.01 FET characteristics

Beginners sometimes become catatonic when directly confronted with the confusing variety of FET types (see, for example, the first edition of this book!), a variety that arises from the combined choices of polarity (*n-channel* or *p-channel*), form of gate insulation (semiconductor *junction* [JFET] or oxide *insulator* [MOSFET]), and channel doping (*enhancement* or *depletion* mode). Of the eight resulting possibilities, six *could* be made, and five actually are. Four of those five are of major importance.

It will aid understanding (and sanity), however, if we begin with one type only, just as we did with the *npn* bipolar transistor. Once comfortable with FETs, we'll have little trouble with their family tree.

FET V–I curves

Let's look first at the n-channel enhancement-mode MOSFET, which is analogous to the *npn* bipolar transistor (Fig. 3.1). In normal operation the drain (~collector) is more positive than the source (~emitter). No current flows from drain to source unless the gate (~base) is brought positive with respect to the source. Once the gate is thus "forward-biased" there will be drain current, all of which flows to the source. Figure 3.2 shows how the drain current I_D varies with drain-source voltage V_{DS}, for a few values of controlling gate-source voltage V_{GS}. For comparison, the corresponding "family" of curves of I_C versus V_{BE} for an ordinary *npn* bipolar transistor is shown. Obviously threre are a lot of similarities between n-channel MOSFETs and *npn* bipolar transistors.

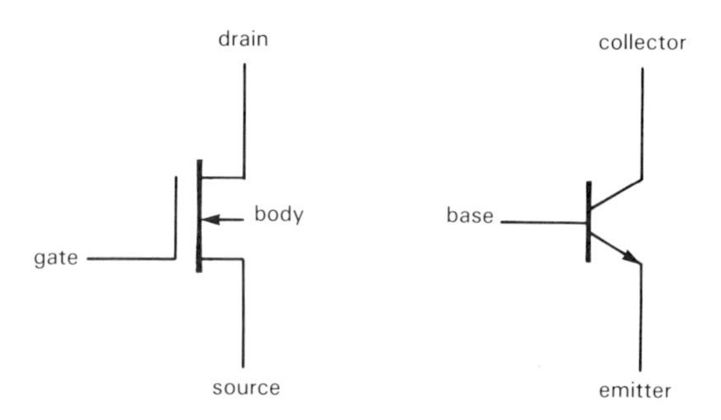

Figure 3.1

Like the *npn* transistor, the FET has a high incremental drain impedance, giving roughly constant current for V_{DS} greater than a volt or two. By an unfortunate choice of language, this is called the "saturation" region of the FET and corresponds to the "active" region of the bipolar transistor. Analogous to the bipolar transistor, larger gate-to-source bias produces larger drain current. If anything, FETs behave more nearly like ideal transconductance devices (constant drain current for constant gate-source voltage) than do bipolar

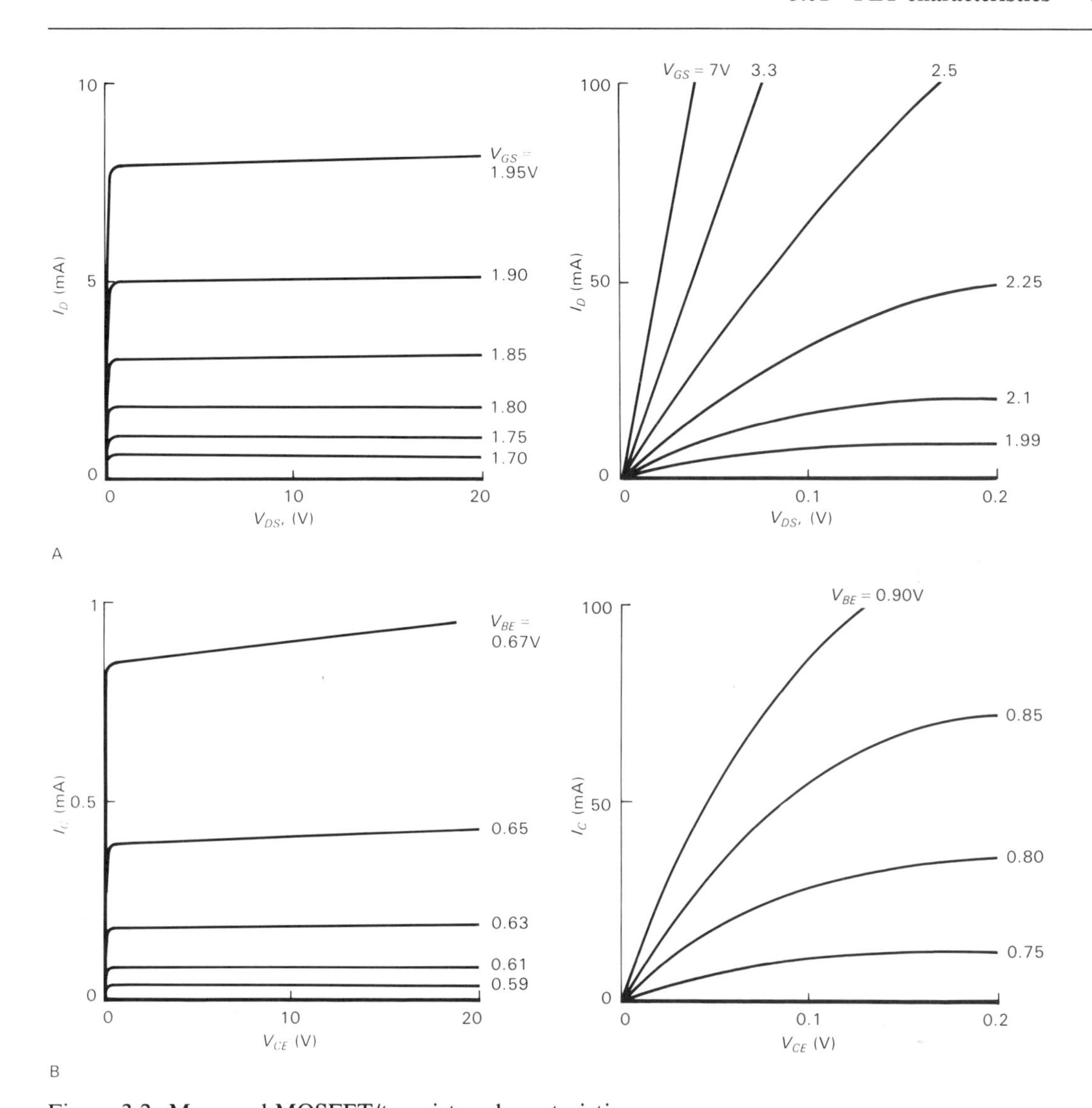

Figure 3.2. Measured MOSFET/transistor characteristic curves.
A. VN0106 n-channel MOSFET: I_D versus V_{DS} for various values of V_{GS}.
B. 2N3904 *npn* bipolar transistor: I_C versus V_{CE} for various values of V_{BE}.

transistors; the Ebers-Moll equation predicts perfect transconductance characteristics for bipolar transistors, but that ideal behavior is degraded by the Early effect (Section 2.10).

So far, the FET looks just like the *npn* transistor. Let's look closer, though. For one thing, over the normal range of currents the saturation drain current increases rather modestly with increasing gate voltage (V_{GS}). In fact, it is proportional to $(V_{GS} - V_T)^2$, where V_T is the "gate threshold voltage" at which drain current begins ($V_T \approx 1.63$V for the FET in Fig. 3.2); compare this mild quadratic law with the steep exponential transistor law, as given to us by Ebers and Moll. Second, there is *zero* dc gate current, so you mustn't think of the FET as a device with current gain (which would be infinite). Instead, think of the FET as a transconductance device, with gate-source voltage programming the drain current, as we did with the bipolar transistor in the Ebers-Moll treatment; recall that the transconductance g_m is simply the ratio i_d/v_{gs} (recall the convention of using lower-case letters to indicate "small-signal" changes in a parameter; e.g., $i_d/v_{gs} = \delta I_d/\delta V_{gs}$). Third, the gate of a MOSFET is truly insulated from the drain-source channel; thus, unlike the situation for bipolar transistors (or JFETs, as we'll see), you can bring it positive (or negative) at least 10 volts or more without worrying about diode conduction. Finally, the FET differs from the bipolar transistor in the so-called linear region of the graph, where it behaves rather accurately like a resistor, *even for negative* V_{DS}; this turns out to be quite useful because the equivalent drain-source resistance is, as you might guess, programmed by the gate-source voltage.

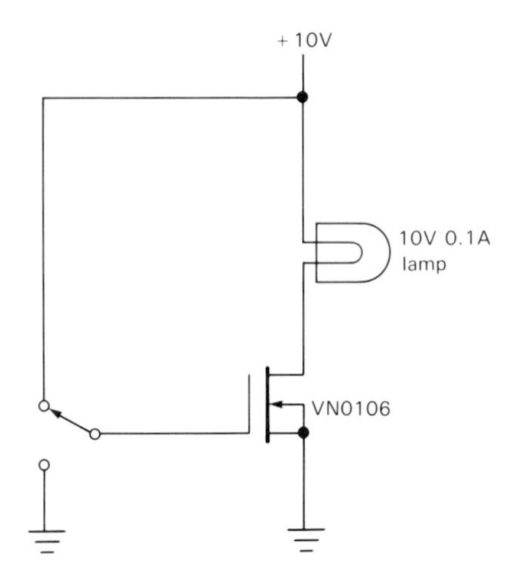

Figure 3.3. MOSFET switch.

Two examples

FETs have more surprises in store for us. Before getting into more details, though, let's look at two simple switching applications. Figure 3.3 shows the MOSFET equivalent of Figure 2.3, our first saturated transistor switch. The FET circuit is even simpler, because we don't have to concern ourselves with the inevitable compromise of providing adequate base drive current (considering worst-case minimum h_{FE} combined with the lamp's cold resistance) without squandering excessive power. Instead, we just apply a full-swing dc voltage drive to the cooperative high-impedance gate. As long as the switched-on FET behaves like a resistance small compared with the load, it will bring its drain close to ground; typical power MOSFETs have $R_{ON} < 0.2$ ohm, which is fine for this job.

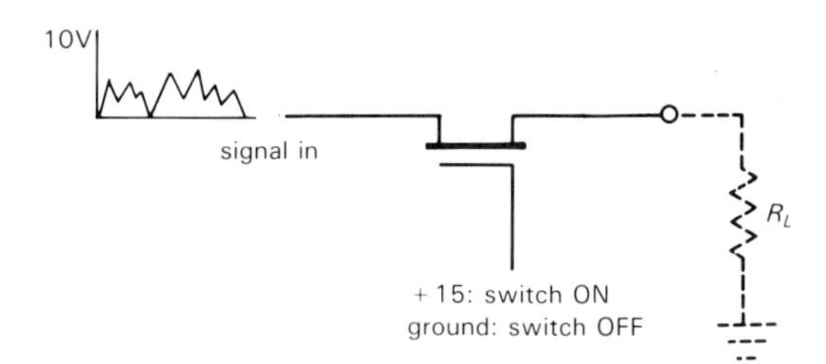

Figure 3.4

Figure 3.4 shows an "analog switch" application, which cannot be done at all with bipolar transistors. The idea here is to switch the conduction of a FET from open-circuit (gate reverse-biased) to short-circuit (gate forward-biased), thus blocking or passing the analog signal (we'll see plenty of reasons to do this sort of thing later). In this case we just arrange for the gate to be driven more negative than any input signal swing (switch *open*), or a few volts more positive than any input signal swing (switch *closed*). Bipolar transistors aren't suited to this application, because the base draws current and forms diodes with the emitter and collector, producing awkward clamping action. The MOSFET is delightfully simple by comparison, needing only a voltage swing into the (essentially open-circuit) gate. Warning:

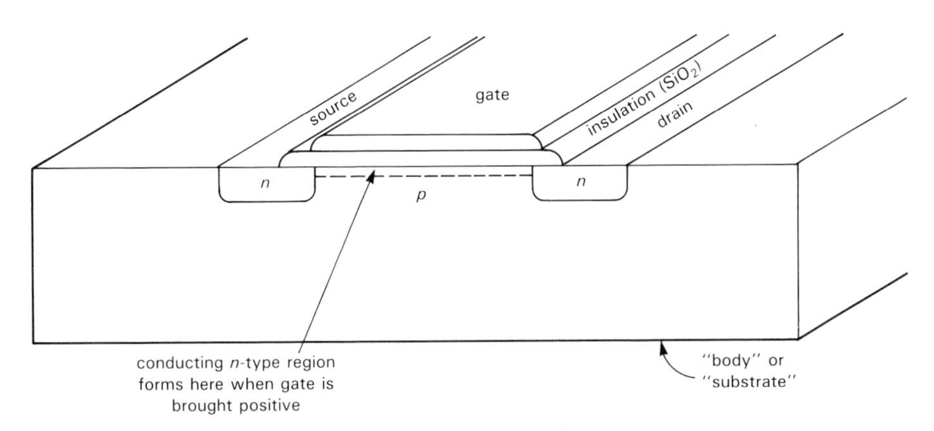

Figure 3.5. An n-channel MOSFET.

It's only fair to mention that our treatment of this circuit has been somewhat simplistic, for instance ignoring the effects of gate-channel capacitance and the variation of R_{ON} with signal swing. We'll have more to say about analog switches later.

3.02 FET types

n-channel, p-channel

Now for the family tree. First of all, FETs (like BJTs) can be fabricated in both polarities. Thus, the mirror twin of our n-channel MOSFET is a *p-channel* MOSFET. Its behavior is symmetrical, mimicking *pnp* transistors: The drain is normally negative with respect to the source, and drain current flows if the gate is brought at least a volt or two negative with respect to the source. The symmetry isn't perfect because the carriers are holes, rather than electrons, with lower "mobility" and "minority carrier lifetime." These are semiconductor parameters of importance in transistor performance. The consequence is worth remembering – p-channel FETs usually have poorer performance, manifested as higher gate threshold voltage, higher R_{ON}, and lower saturation current.

MOSFET, JFET

In a MOSFET ("Metal-Oxide-Semiconductor Field-Effect Transistor") the gate region is separated from the conducting channel by a thin layer of SiO_2 (glass) grown onto the channel (Fig. 3.5). The gate, which may be either metal or doped silicon, is truly insulated from the source-drain circuit, with characteristic input resistance $>10^{14}$ ohms. It affects channel conduction purely by its electric field. MOSFETs are sometimes called *insulated-gate* FETs, or IGFETs. The gate insulating layer is quite thin, typically less than a wavelength of light, and can withstand gate voltages up to ± 20 volts or more. MOSFETs are easy to use because the gate can swing either polarity relative to the source without any gate current flowing. They are, however, quite susceptible to damage from static electricity; you can destroy a MOSFET device literally by touching it.

The symbols for MOSFETs are shown in Figure 3.6. The extra terminal is the "body," or "substrate," the piece of silicon in which the FET is fabricated (see Fig. 3.5). Because the body forms a diode junction with the channel, it must be held at a nonconducting voltage. It can be tied to the source, or to a point in the

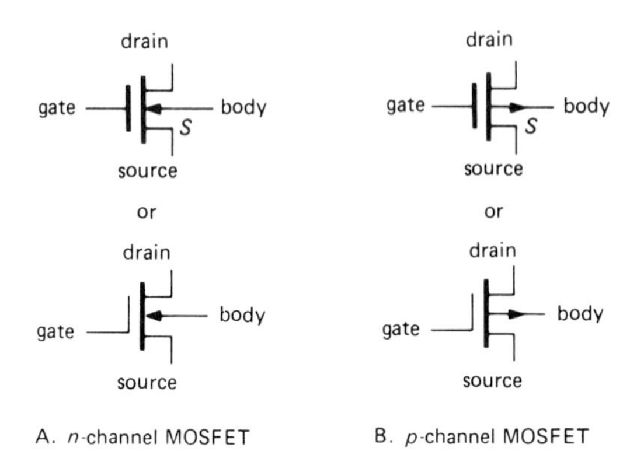

Figure 3.6

circuit more negative (positive) than the source for n-channel (p-channel) MOSFETs. It is common to see the body terminal omitted; furthermore, engineers often use the symbol with the symmetrical gate. Unfortunately, with what's left you can't tell source from drain; worse still, you can't tell n-channel from p-channel! We will use the lower set of schematic symbols exclusively in this book to avoid confusion, although we will often leave the body pin unconnected.

In a JFET ("Junction Field-Effect Transistor") the gate forms a semiconductor junction with the underlying channel. This has the important consequence that *a JFET gate should not be forward biased with respect to the channel, to prevent gate current.* For example, diode conduction will occur as the gate of an n-channel JFET approaches +0.6 volt with respect to the more negative end of the channel (which is usually the source). The gate is therefore operated reverse-biased with respect to the channel, and no current (except diode leakage) flows in the gate circuit. The circuit symbols for JFETs are shown in Figure 3.7. Once again, we favor the symbol with offset gate, to identify the source. As we'll see later, FETs (both JFET and MOSFET) are nearly symmetrical, but the gate-drain capacitance is usually designed to be less than the gate-source capacitance, making the drain the preferred output terminal.

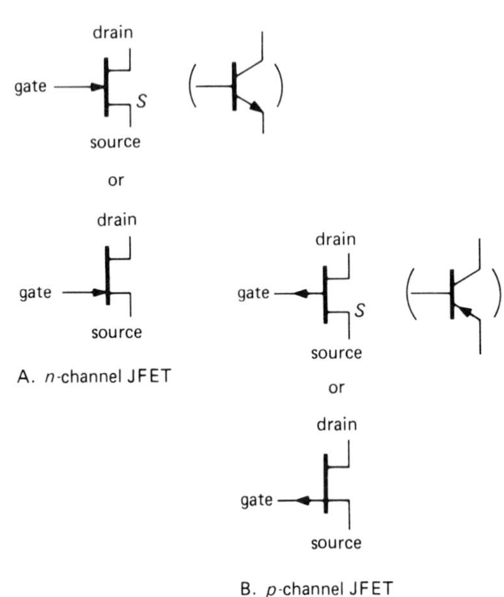

Figure 3.7

Enhancement, depletion

The n-channel MOSFETs with which we began the chapter were nonconducting, with zero (or negative) gate bias, and were driven into conduction by bringing the gate positive with respect to the source. This kind of FET is known as *enhancement mode.* The other possibility is to manufacture the n-channel FET with the channel semiconductor "doped" so that there is plenty of channel conduction even with zero gate bias, and the gate must be reverse-biased a few volts to cut off the drain current. Such a FET is known as *depletion mode.* MOSFETs can be made in either variety, since there is no restriction on gate polarity. But JFETs permit only reverse gate bias and therefore can be made only in depletion mode.

A graph of drain current versus gate-source voltage, at a fixed value of drain voltage, may help clarify this distinction (Fig. 3.8). The enhancement-mode device draws no drain current until the gate is brought positive (these are n-channel FETs) with respect to the source, whereas the depletion-mode device is operating at nearly its maximum value of drain current when the gate is at the same voltage as the source. In some sense the two categories are artificial, because the two curves are identical except for a shift along the V_{GS} axis. In fact, it is possible to manufacture "in-between" MOSFETs. Nevertheless, the distinction is an important one when it comes to circuit design.

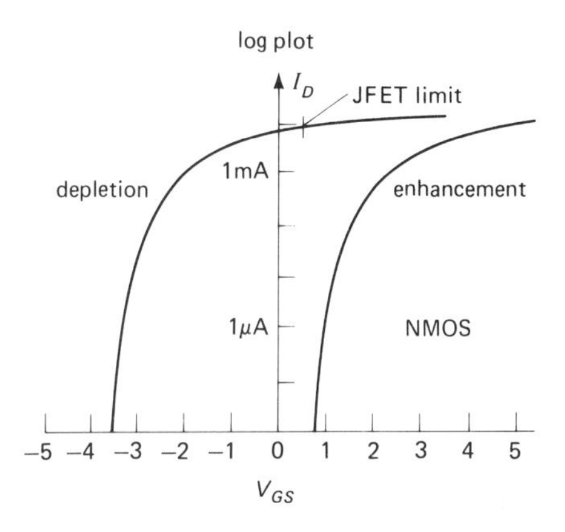

Figure 3.8

Note that JFETs are always depletion-mode devices and that the gate cannot be brought more than about 0.5 volt more positive (for n-channel) than the source, since the gate-channel diode will conduct. MOSFETs *could* be either enhancement or depletion, but in practice you rarely see depletion-mode MOSFETs (the exceptions being n-channel GaAs FETs and "dual-gate" cascodes for radiofrequency applications). For all practical purposes, then, you have to worry only about (a) depletion-mode JFETs and (b) enhancement-mode MOSFETs; they both come in the two polarities, n-channel and p-channel.

3.03 Universal FET characteristics

A family tree (Fig. 3.9) and a map (Fig. 3.10) of input/output voltage (source grounded) may help simplify things. The different devices (including garden-variety *npn* and *pnp* bipolar transistors) are drawn in the quadrant that characterizes their input and output voltages when they are in the active region with source (or emitter) grounded. You don't have to remember the properties of the five kinds of FETs, though, because they're all basically the same.

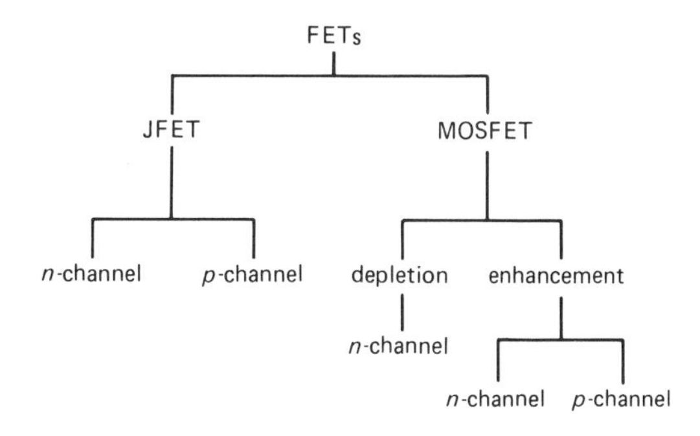

Figure 3.9

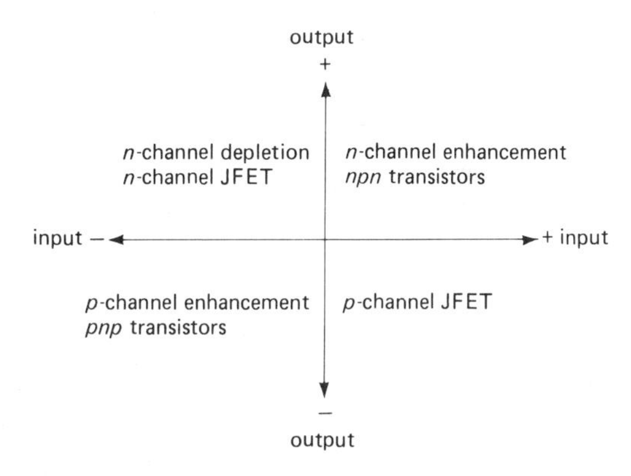

Figure 3.10

First, with the source grounded, a FET is turned on (brought into conduction) by bringing the gate voltage "toward" the active drain supply voltage. This is true for all five types of FETs, as well as the bipolar transistors. For example, an n-channel JFET (which is automatically depletion-mode) uses a positive drain supply, as do all n-type devices. Thus a positive-going gate voltage tends to turn on the JFET.

The subtlety for depletion-mode devices is that the gate must be (negatively) back-biased for zero drain current, whereas for enhancement-mode devices zero gate voltage is sufficient to give zero drain current.

Second, because of the near symmetry of source and drain, either terminal can act as the effective source (exception: not true for power MOSFETs, where the body is internally connected to the source). When thinking of FET action, and for purposes of calculation, the effective source terminal is always the one most "away" from the active drain supply. For example, suppose a FET is used to switch a line to ground, and both positive and negative signals are present on the switched line, which is usually selected to be the FET drain. If the switch is an n-channel MOSFET (therefore enhancement), and a negative voltage happens to be present on the (turned-off) drain terminal, then that terminal is actually the "source" for purposes of gate turn-on voltage calculation. Thus a negative gate voltage larger than the most negative signal, rather than ground, is needed to ensure turn-off.

The graph in Figure 3.11 may help you sort out all these confusing ideas. Again, the difference between enhancement and depletion is merely a question of displacement along the V_{GS} axis – i.e., whether there is a lot of drain current or no drain current at all when the gate is at the same potential as the source. The n-channel and p-channel FETs are complementary in the same way as *npn* and *pnp* bipolar transistors.

In Figure 3.11 we have used standard symbols for the important FET parameters of saturation current and cutoff voltage. For JFETs the value of drain current with gate shorted to source is specified on the data sheets as I_{DSS} and is nearly the maximum drain current possible. (I_{DSS} means current from *d*rain to *s*ource with the gate *s*horted to the source. Throughout the chapter you will see this notation, in which the first two subscripted letters designate the pair of terminals, and the third specifies the condition.) For enhancement-mode MOSFETs the analogous specification is $I_{D(ON)}$, given at some forward gate voltage ("I_{DSS}" would be zero for any enhancement-mode device).

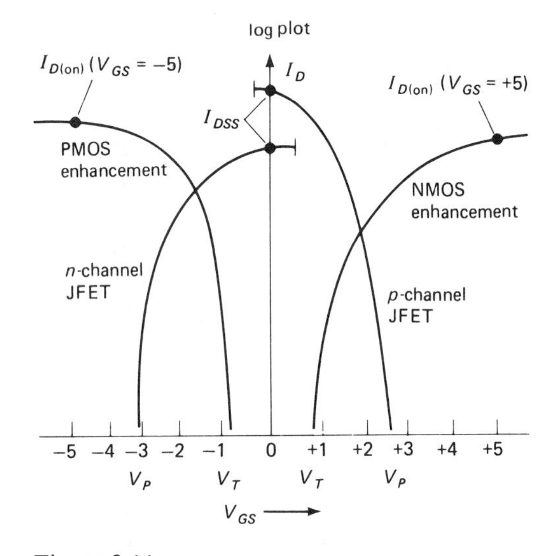

Figure 3.11

For JFETs the gate-source voltage at which drain current approaches zero is called the "gate-source cutoff voltage," $V_{GS(OFF)}$, or the "pinch-off voltage," V_P, and is typically in the range of -3 to -10 volts (positive for p-channel, of course). For enhancement-mode MOSFETs the analogous quantity is the "threshold voltage," V_T (or $V_{GS(th)}$), the gate-source voltage at which drain current begins to flow. V_T is typically in the range of 0.5 to 5 volts, in the "forward" direction, of course. Incidentally, don't confuse the MOSFET V_T with the V_T in the Ebers-Moll equation that describes bipolar transistor collector current; they have nothing to do with each other.

With FETs it is easy to get confused about polarities. For example, n-channel devices, which usually have the drain positive with respect to the source, can have positive or negative gate voltage, and positive (enhancement) or negative (depletion) threshold voltages. To make matters worse, the drain can be (and often is) operated negative with respect to the source. Of course, all these statements go in reverse for p-channel devices. In order to minimize confusion, we will always assume n-channel devices unless explicitly stated otherwise. Likewise, because MOSFETs are nearly always enhancement-mode, and JFETs are always depletion-mode, we'll omit those designations from now on.

3.04 FET drain characteristics

In Figure 3.2 we showed a family of curves of I_D versus V_{DS} that we measured for a VN0106, an n-channel enhancement-mode MOSFET. (The VN01 comes in various voltage ratings, indicated by the last two digits of the part number. For example, a VN0106 is rated at 60V.) We remarked that FETs behave like pretty good transconductance devices over most of the graph (i.e., I_D nearly constant for a given V_{GS}), except at small V_{DS}, where they approximate a resistance (i.e., I_D proportional to V_{DS}). In both cases the applied gate-source voltage controls the behavior, which can be well described by the FET analog of the Ebers-Moll equation. Let's look at these two regions a bit more closely.

Figure 3.12 shows the situation schematically. In both regions the drain current depends on $V_{GS} - V_T$, the amount by which the applied gate-source voltage exceeds the threshold (or pinch-off) voltage. The linear region, in which drain current is approximately proportional to V_{DS}, extends up to a voltage $V_{DS(\text{sat})}$, after which the drain current is approximately constant. The slope in the linear region, I_D/V_{DS}, is proportional to the gate bias, $V_{GS} - V_T$. Furthermore, the drain voltage at which the curves enter the "saturation region," $V_{DS(\text{sat})}$, equals $V_{GS} - V_T$, making the saturation drain current, $I_{D(\text{sat})}$, proportional to $(V_{GS} - V_T)^2$, the quadratic law we mentioned earlier. For reference, here are the universal FET drain-current formulas:

$$I_D = 2k[(V_{GS} - V_T)V_{DS} - V_{DS}^2/2] \qquad \text{(linear region)}$$

$$I_D = k(V_{GS} - V_T)^2 \qquad \text{(saturation region)}$$

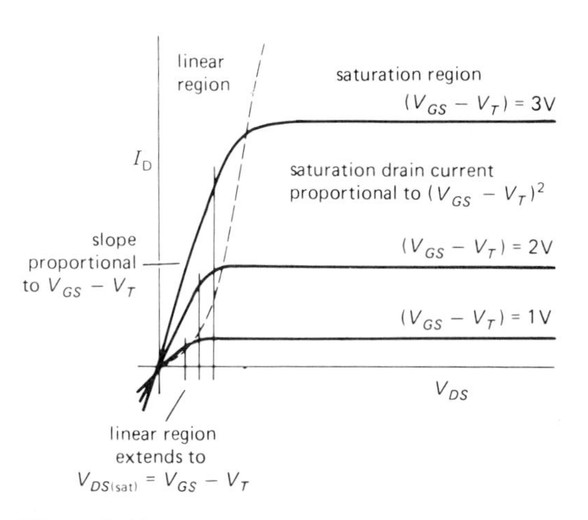

Figure 3.12

If we call $V_{GS} - V_T$ (the amount by which the gate-source voltage exceeds the threshold) the "gate drive," the important results are that (a) the resistance in the linear region is inversely proportional to gate drive, (b) the linear region extends to a voltage equal to the gate drive, and (c) saturation drain current is proportional to the square of the gate drive. These equations assume that the body is connected to the source. Note that the "linear region" is not really linear, because of the V_{DS}^2 term; we'll show a clever circuit fix later.

The scale factor k depends on particulars such as the geometry of the FET, oxide capacitance, and carrier mobility. It has a temperature dependence $k \propto T^{-3/2}$, which alone would cause I_D to decrease with increasing temperature. However, V_T also depends slightly on temperature (2–5mV/°C); the combined effect produces the curve of drain current versus temperature shown in Figure 3.13.

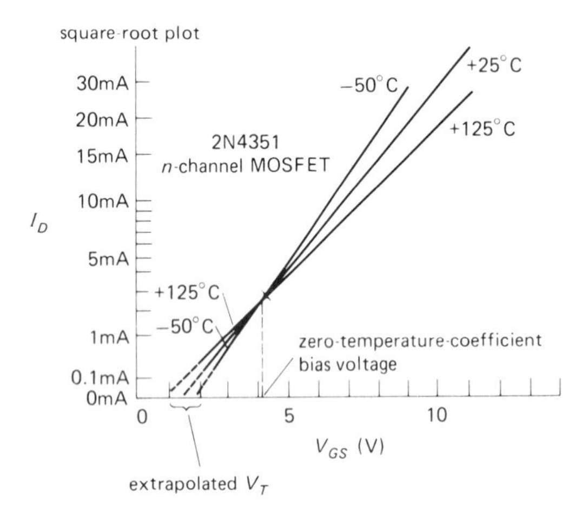

Figure 3.13

At large drain currents the negative temperature coefficient of k causes the drain current to decrease with increasing temperature – goodbye thermal runaway! As a consequence, FETs of a given type can be paralleled without the external current-equalizing ("emitter-ballasting") resistors that you must use with bipolar transistors (see Section 6.07). This same negative coefficient also prevents thermal runaway in local regions of the junction (an effect known as "current hogging"), which severely limits the power capability of large bipolar transistors, as we'll see when we discuss "second breakdown" and "safe operating area" in Chapter 6.

At small drain currents (where the temperature coefficient of V_T dominates), I_D has a positive tempco, with a point of zero temperature coefficient at some drain current in between. This effect is exploited in FET op-amps to minimize temperature drift, as we'll see in the next chapter.

Subthreshold region

Our expression given earlier for saturation drain current does not apply for very small drain currents. This is known as the "subthreshold" region, where the channel is below the threshold for conduction, but some current flows anyway because of a small population of thermally energetic electrons. If you've studied physics or chemistry, you probably know in your bones that the resulting current is exponential:

$$I_D = k\exp(V_{GS} - V_T)$$

We measured some MOSFETs over 9 decades of drain current (1nA to 1A) and plotted the result as a graph of I_D versus V_{GS} (Fig. 3.14). The region from 1nA to 1mA is quite precisely exponential; above this subthreshold region the curves enter the normal saturation region. For the n-channel MOSFET (type VN01) we checked out a sample of 20 transistors (from four different manufacturing runs spread over 2 years), plotting the extreme range to give you an idea of the variability (see next section). Note the somewhat poorer characteristics (V_T, $I_{D(\mathrm{ON})}$) of the "complementary" VP01.

3.05 Manufacturing spread of FET characteristics

Before we look at some circuits, let's take a look at the range of FET parameters (such as I_{DSS} and V_T), as well as their manufacturing "spread" among devices of the same nominal type, in order to get a better idea of the FET. Unfortunately, many of the characteristics of FETs show much greater process spread than the corresponding characteristics of bipolar transistors, a fact that the designer must keep in mind. For example, the VN01 (a typical n-channel MOSFET) has a specified V_T of 0.8 to 2.4 volts ($I_D = 1$mA), compared with the analogous V_{BE} spread of

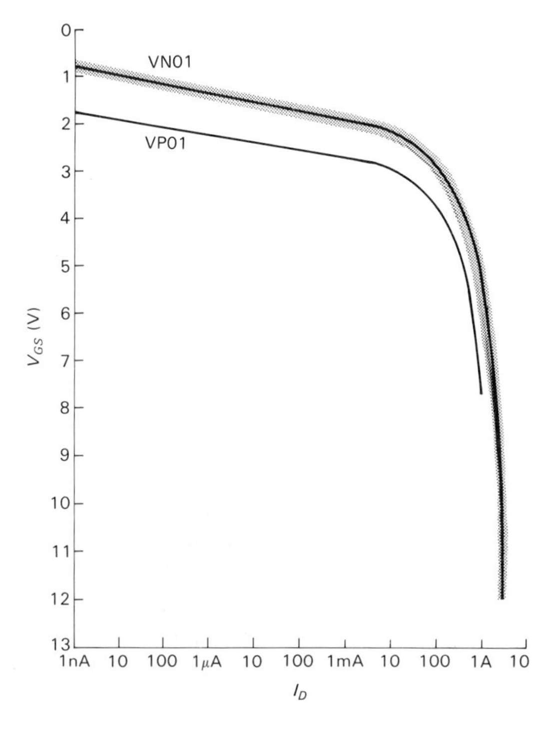

Figure 3.14. Measured MOSFET drain current versus gate-source voltage.

0.63 to 0.83 volt (also at $I_C = 1\text{mA}$) for an *npn* bipolar transistor. Here's what you can expect:

Characteristic	*Available range*	*Spread*
I_{DSS}, $I_{D(\text{ON})}$	1mA to 100A	×5
$R_{DS(\text{ON})}$	0.05Ω to 10k	×5
g_m @ 1mA	500–3000μS	×5
V_P (JFETs)	0.5–10V	5V
V_T (MOSFETs)	0.5–5V	2V
$BV_{DS(\text{OFF})}$	6–1000V	
$BV_{GS(\text{OFF})}$	6–125V	

$R_{D(\text{ON})}$ is the drain-source resistance (linear region, i.e., small V_{DS}) when the FET is conducting fully, e.g., with the gate grounded in the case of JFETs or with a large applied gate-source voltage (usually specified as 10V) for MOSFETs. I_{DSS} and $I_{D(\text{ON})}$ are the saturation-region (large V_{DS}) drain currents under the same turned-on gate drive conditions. V_P is the pinch-off voltage (JFETs), V_T is the turn-on gate threshold voltage (MOSFETs), and the BVs are breakdown voltages. As you can see, a JFET with grounded source may be a good current source, but you can't predict very well what the current will be. Likewise, the V_{GS} needed to produce some value of drain current can vary considerably, in contrast to the predictable (≈0.6V) V_{BE} of bipolar transistors.

Matching of characteristics

As you can see, FETs are inferior to bipolar transistors in V_{GS} predictability, i.e., they have a large spread in the V_{GS} required to produce a given I_D. Devices with a large process spread will, in general, have larger offset (voltage unbalance) when used as differential pairs. For instance, typical run-of-the-mill bipolar transistors might show a spread in V_{BE} of 50mV or so, at some collector current, for a selection of off-the-shelf transistors. The comparable figure for MOSFETs is more like 1 volt! Because FETs have some very desirable characteristics otherwise, it is worthwhile putting in some extra effort to reduce these offsets in specially manufactured matched pairs. IC designers use techniques like interdigitation (two devices sharing the same general piece of IC real estate) and thermal-gradient cancellation schemes to improve performance (Fig. 3.15).

The results are impressive. Although FET devices still cannot equal bipolar transistors in V_{GS} matching, their performance is adequate for most applications. For example, the best available matched FET has a voltage offset of 0.5mV and tempco of 5μV/°C (max), whereas the best bipolar pair has values of 25μV and 0.6μV/°C (max), roughly 10 times better. Operational amplifiers (the universal high-gain differential amplifiers we'll see in the next chapter) are available in both flavors; you would generally choose one with bipolar innards for high precision (because of its

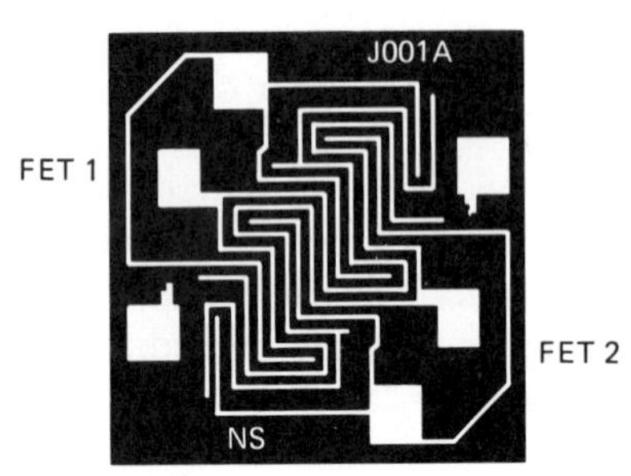

A. interdigitation

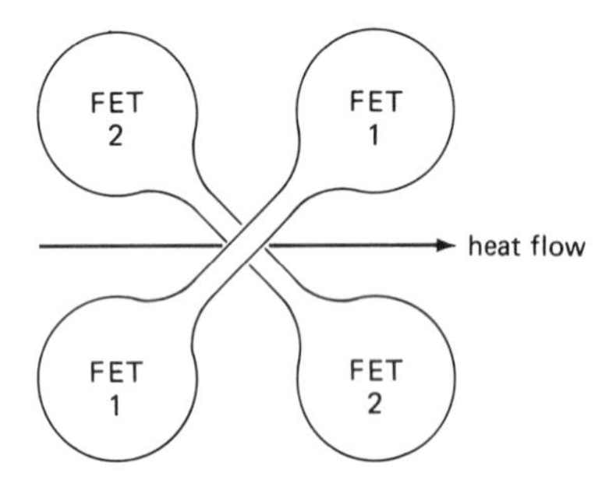

B. temperature-gradient cancellation

Figure 3.15

close input-transistor V_{BE} matching), whereas a FET-input op-amp is the obvious choice for high-impedance applications (because its inputs – FET gates – draw no current). For example, the inexpensive JFET-input LF411 that we will use as our all-around op-amp in the next chapter has a typical input current of 50pA and costs $0.60; the popular MOSFET-input TLC272 costs about the same and has a typical input current of only 1pA! Compare this with a common bipolar op-amp, the μA741, with typical input current of 80,000pA (80nA).

Tables 3.1–3.3 list a selection of typical JFETs (both single and dual) and small-signal MOSFETs. Power MOSFETs, which we will discuss in Section 3.14, are listed in Table 3.5.

BASIC FET CIRCUITS

Now we're ready to look at FET circuits. You can usually find a way to convert a circuit that uses BJTs into one using FETs. However, the new circuit may not be an improvement! For the remainder of the chapter we'd like to illustrate circuit situations that take advantage of the unique properties of FETs, i.e., circuits that work better with FETs, or that you can't build at all with bipolar transistors. For this purpose it may be helpful to group FET applications into categories; here are the most important, as we see it:

High-impedance/low-current. Buffers or amplifiers for applications where the base current and finite input impedance of BJTs limit performance. Although you *can* build such circuits with discrete FETs, current practice favors using integrated circuits built with FETs. Some of these use FETs as a high-impedance front-end for an otherwise bipolar design, whereas others use FETs throughout.

Analog switches. MOSFETs are excellent voltage-controlled analog switches, as we hinted in Section 3.01. We'll look briefly at this subject. Once again, you should generally use dedicated "analog switch" ICs, rather than building discrete circuits.

Digital logic. MOSFETs dominate microprocessors, memory, and most high-performance digital logic. They are used exclusively in micropower logic. Here, too, MOSFETs make their appearance in integrated circuits. We'll see why FETs are preferable to BJTs.

Power switching. Power MOSFETs are often preferable to ordinary bipolar power transistors for switching loads, as we suggested in our first circuit of the chapter. For this application you use *discrete* power FETs.

Variable resistors; current sources. In the "linear" region of the drain curves, FETs behave like voltage-controlled resistors; in the "saturation" region they are voltage-controlled current sources. You can exploit this intrinsic behavior of FETs in your circuits.

TABLE 3.1. JFETs

Type	BV_{GSS} (V)	I_{DSS} min (mA)	I_{DSS} max (mA)	$V_{GS(OFF)}$, V_P min (V)	$V_{GS(OFF)}$, V_P max (V)	C_{iss} max (pF)	C_{rss} max (pF)	Comments
n-channel								
2N4117A-	40	0.03	0.09	0.6	1.8	3	1.5	low leakage: 1pA (max)
2N4119A	40	0.24	0.6	2	6	4	1.5	
2N4338	50	0.2	0.6	0.3	1	6	2	0.5fA/√Hz @ 100Hz
2N4416	30	5	15	2.5	6	4	0.8	VHF low noise: <2dB@100MHz
2N4867A-	40	0.4	1.2	0.7	2	25	5	low freq, low noise:
2N4869A	40	2.5	7.5	1.8	5	25	5	10nV/√Hz(max)@10Hz
2N5265-	60	0.5	1	–	3	7	2	series of 6, tight I_{DSS} spec;
2N5270	60	7	14	–	8	7	2	2N5358-64 p-chan complement
2N5432	25	150	–	4	10	30	15	switch: R_{ON}=5Ω(max)
2N5457-	25	1	5	0.5	6	7	3	general purpose;
2N5459	25	4	16	2	8	7	3	2N5460-2 p-chan complement
2N5484-	25	1	5	0.3	3	5	1	low noise RF; inexpensive
2N5486	25	8	20	2	6	5	1	
2SK117	50	0.6	14	0.2	1.5	13^t	3^t	ultra low noise: 1nV/√Hz
2SK147	40	5	30	0.3	1.2	75^t	15^t	ultra low noise: 0.7nV/√Hz
p-channel								
2N5114	30	30	90	5	10	25	7	switch: R_{ON}=75Ω(max)
2N5358-	40	0.5	1	0.5	3	6	2	series of 7, tight I_{DSS} spec;
2N5364	40	9	18	2.5	8	6	2	2N5265-70 n-chan complement
2N5460-	40	1	5	0.75	6	7	2	general purpose;
2N5462	40	4	16	1.8	9	7	2	2N5457-9 n-chan complement
2SJ72	25	5	30	0.3	2	185^t	55^t	ultra low noise: 0.7nV/√Hz

$^{(t)}$ typical.

Generalized replacement for bipolar transistors. You can use FETs in oscillators, amplifiers, voltage regulators, and radio-frequency circuits (to name a few), where bipolar transistors are also normally used. FETs aren't *guaranteed* to make a better circuit – sometimes they will, sometimes they won't. You should keep them in mind as an alternative.

Now let's look at these subjects. We'll adopt a slightly different order, for clarity.

3.06 JFET current sources

JFETs are used as current sources within integrated circuits (particularly op-amps), and also sometimes in discrete designs. The simplest JFET current source is shown in Figure 3.16; we chose a JFET, rather than a MOSFET, because it needs no gate bias (it's depletion mode). From a graph of FET drain characteristics (Fig. 3.17) you can see that the current will be reasonably

TABLE 3.2. SELECTED MOSFETs

Type	Mfg[a]	Gate protec	$R_{DS(on)}$ max (Ω)	@V_{GS} (V)	$V_{GS(th)}$ min (V)	$V_{GS(th)}$ max (V)	$I_{D(on)}$ (V_{DS}=10V) min (mA)	C_{rss} max (pF)	BV_{DS} (V)	BV_{GS} (V)	I_{GSS} (nA)	Comments
n-channel												
3SK38A	TO	•	500	3	–	–	10	2.5	20	12	25	
3N170	IL	–	200	10	1.0	2	10	1.3	25	35	0.01	
SD210	SI	–	45	10	0.5	2	–	0.5	30	40	0.1	low R_{ON}
SD211	SI	•	45	10	0.5	2	–	0.5	30	15	10	low R_{ON}
VN1310	ST	–	8	10	0.8	2.4	500	5	100	20	0.1	small VMOS; D-S diode
IT1750	IL	–	50	20	0.5	3	10	1.6	25	25	0.01	
VN2222L	SI	–	8	5	0.6	2.5	750	5	60	40	0.1	small VMOS; D-S diode
CD3600	RC	•	500	10	1.5[t]	–	1.3	0.4	15	15	0.01	equiv to 4007 array
2N3796	MO	–	–	–	-4	–	14	0.8	25	10	0.001	depletion; I_{DSS}=1.5mA
2N4351	MO+	–	300	10	1.5	5	3	2.5	25	35	0.01	popular
p-channel												
3N163	IL	–	250	20	2	5	5	0.7	40	40	0.01	
VP1310	ST	–	25	10	1.5	3.5	250	5	100	20	0.1	small VMOS; D-S diode
IT1700	IL	–	400	10	2	5	2	1.2	40	40	0.01	
CD3600	RC	•	500	10	1.8[t]	–	1.3	0.8	15	15	0.02	equiv to 4007 array
2N4352	MO+	–	600	10	1.5	6	2	2.5	25	35	0.01	popular
3N172	IL	•	250	20	2	5	5	1	40	40	0.2	popular

(a) see footnotes to Table 4.1. (t) typical.

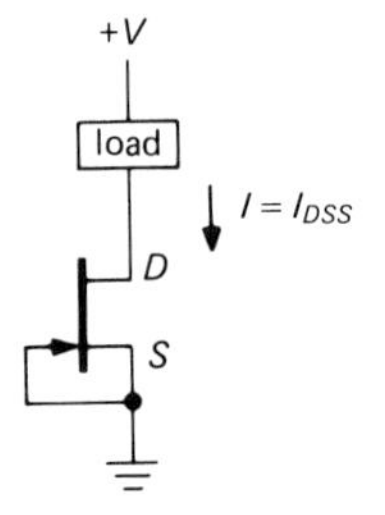

Figure 3.16

constant for V_{DS} larger than a couple of volts. However, because of I_{DSS} spread, the current is unpredictable. For example, the 2N5484 (a typical n-channel JFET) has a specified I_{DSS} of 1mA to 5mA. Still, the circuit is attractive because of the simplicity of a two-terminal constant-current device. If that appeals to you, you're in luck. You can buy "current-regulator diodes" that are nothing more than JFETs with gate tied to source, sorted according to current. They're the current analog of a zener (voltage regulator) diode. Here are the characteristics of the 1N5283–1N5314 series:

Currents available	0.22mA to 4.7mA
Tolerance	10%
Temperature coefficient	±0.4%/°C
Voltage range	1V–2.5V min, 100V max
Current regulation	5% typical
Impedance	1M typical (for 1mA device)

We plotted I versus V for a 1N5294 (rated at 0.75mA); Figure 3.18A shows

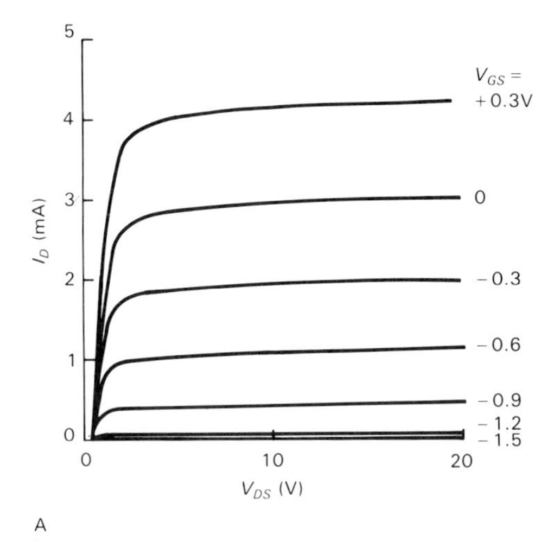

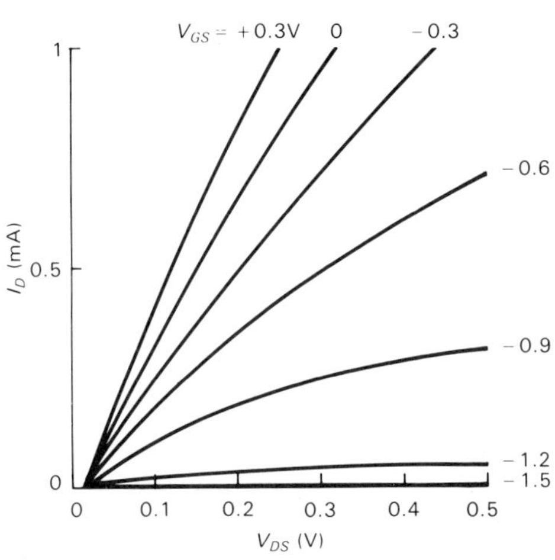

Figure 3.17. Measured JFET characteristic curves. 2N5484 n-channel JFET: I_D versus V_{DS} for various values of V_{GS}.

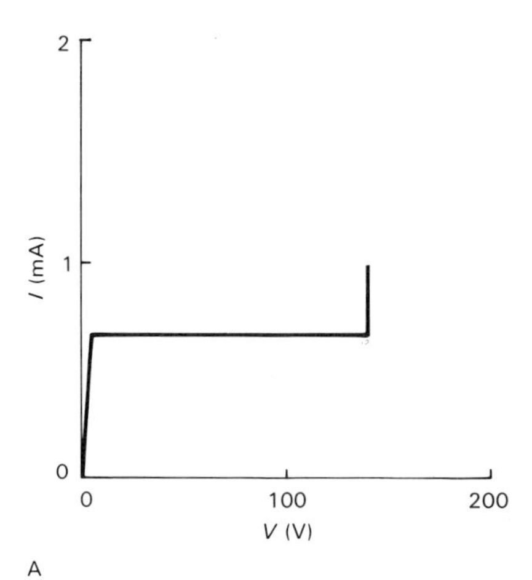

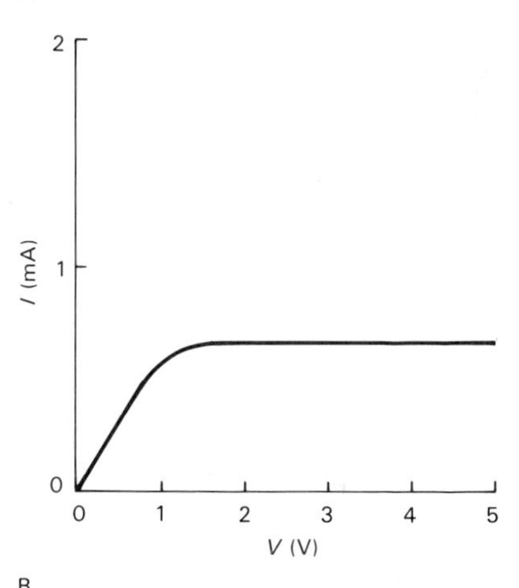

Figure 3.18. 1N5294 "current regulator diode."

good constancy of current up to the breakdown voltage (140V for this particular specimen), whereas Figure 3.18B shows that the device reaches full current with somewhat less than 1.5 volts across it. We'll show how to use these devices to make a cute triangle-wave generator in Section 5.13. Table 3.4 is a partial listing of the 1N5283 series.

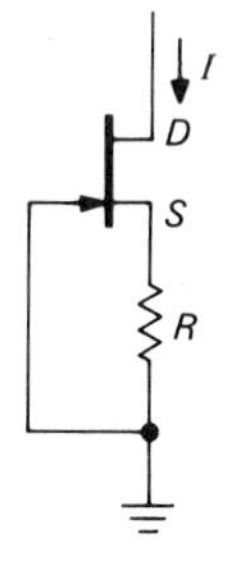

Figure 3.19

Source self-biasing

A variation of the previous circuit (Fig. 3.19) gives you an adjustable current source. The self-biasing resistor R back-biases the gate by I_DR, reducing I_D and bringing the JFET closer to pinch-off. You can calculate R from the drain curves for the particular JFET. This circuit allows you

TABLE 3.3. DUAL MATCHED n-CHANNEL JFETs

Type	V_{os} max (mV)	Drift max (μV/°C)	I_{GSS} (V_{DG}=20V) max (pA)	CMRR min (dB)	$V_{GS(OFF)}$, V_P min (V)	$V_{GS(OFF)}$, V_P max (V)	e_n (10Hz) max (nV/√Hz)	C_{rss} (V_{DG}=10V) max (pF)	Comments
U421	10	10	0.2	90	0.4	2	50	1.5	Siliconix
2N3954A	5	5	100	–	1	3	150[a]	1.2	gen purp, low drift
2N3955	5	25	100	–	1	4.5	150[a]	1.2	popular
2N3958	25	–	100	–	1	4.5	150[a]	1.2	
2N5196	5	5	15	–	0.7	4	20[b]	2	
2N5520	5	5	100	100	0.7	4	15	5	
2N5906	5	5	2	90[t]	0.6	4.5	70[t]	1.5	low gate leakage
2N5911	10	20	100	–	1	5	20[c]	1.2	low noise at high freq
2N6483	5	5	100	100	0.7	4	10	3.5	low noise at low freq
NDF9406	5	5	5	120	0.5	4	30	0.1	cascode: low C_{rss}
2N5452	5	5	100[d]	–	1	4.5	20[b]	1.2[e]	
2SK146	20	–	1000[d]	–	0.3	1.2	1.3	15[t]	ultra low noise

(a) at 100Hz. (b) at 1kHz. (c) at 10kHz. (d) at 30V. (e) at 20V. (t) typical.

to set the current (which must be less than I_{DSS}), as well as to make it more predictable. Furthermore, the circuit is a better current source (higher impedance) because the source resistor provides "current-sensing feedback" (which we'll learn about in Section 4.07), and also because FETs tend to be better current sources anyway when the gate is back-biased (as can perhaps be seen from the flatness of the lower drain-current curves in Figs. 3.2 and 3.17). Remember, though, that actual curves of I_D for some value of V_{GS} obtained with a real FET may differ markedly from the values read from a set of published curves, owing to manufacturing spread. You may therefore want to use an adjustable source resistor, if it is important to have a specific current.

EXERCISE 3.1

Use the 2N5484 measured curves in Figure 3.17 to design a JFET current source to deliver 1mA. Now ponder the fact that the specified I_{DSS} of a 2N5484 is 1mA (min), 5mA (max).

A JFET current source, even if built with source resistor, shows some variation of output current with output voltage; i.e., it has finite output impedance, rather than the desirable infinite Z_{out}. The measured curves of Figure 3.17, for example, suggest that over a drain voltage range of 5 to 20 volts, a 2N5484 shows a drain current variation of 5% when operated with gate tied to source (i.e., I_{DSS}). This might drop to 2% or so if you use a source resistor. The same trick used in Figure 2.24 can be used with JFET current sources and is shown in Figure 3.20. The idea (as with BJTs) is to use a second JFET to hold constant the drain-source voltage of the current source. Q_1 is an ordinary JFET current source, shown in this case with a source resistor. Q_2 is a JFET of larger I_{DSS}, connected "in series" with the current source. It passes Q_1's (constant) drain current through to the load, while holding Q_1's drain at a fixed voltage – namely the gate-source voltage that makes Q_2 operate at the same current as Q_1. Thus Q_2 shields Q_1 from voltage swings

TABLE 3.4. CURRENT-REGULATOR DIODES[a]

Type	I_P (mA)	Impedance (25V) min (MΩ)	V_{min} ($I > 0.8\ I_P$) (V)
1N5283	0.22	25	1.0
1N5285	0.27	14	1.0
1N5287	0.33	6.6	1.0
1N5288	0.39	4.1	1.1
1N5290	0.47	2.7	1.1
1N5291	0.56	1.9	1.1
1N5293	0.68	1.4	1.2
1N5294	0.75	1.2	1.2
1N5295	0.82	1.0	1.3
1N5296	0.91	0.9	1.3
1N5297	1.0	0.8	1.4
1N5299	1.2	0.6	1.5
1N5302	1.5	0.5	1.6
1N5304	1.8	0.4	1.8
1N5305	2.0	0.4	1.9
1N5306	2.2	0.4	2.0
1N5308	2.7	0.3	2.2
1N5309	3.0	0.3	2.3
1N5310	3.3	0.3	2.4
1N5312	3.9	0.3	2.6
1N5314	4.7	0.2	2.9

(a) all operate to 100V and 600mW, and look like diodes in the reverse direction

at its output; since Q_1 doesn't see drain voltage variations, it just sits there and provides constant current. If you look back at the Wilson mirror (Fig. 2.48), you'll see that it uses this same voltage clamping idea.

You may recognize this JFET circuit as the "cascode," which is normally used to circumvent Miller effect (Section 2.19). A JFET cascode is simpler than a BJT cascode, however, because you don't need a bias voltage for the gate of the upper FET: Because it's depletion-mode, you can simply ground the upper gate (compare with Fig. 2.74).

EXERCISE 3.2

Explain why the upper JFET in a cascode must have higher I_{DSS} than the lower JFET. It may help to consider a JFET cascode with no source resistor.

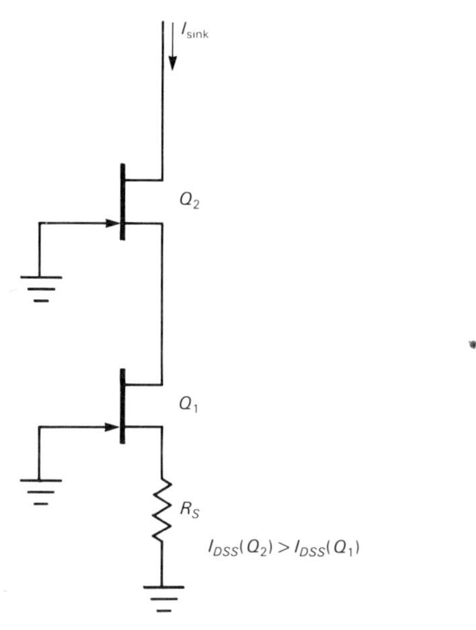

Figure 3.20. Cascode JFET current sink.

It is important to realize that a good bipolar transistor current source will give far better predictability and stability than a JFET current source. Furthermore, the op-amp-assisted current sources we'll see in the next chapter are better still. For example, a FET current source might vary 5% over a typical temperature range and load voltage variation, even after being set to the desired current by trimming the source resistor, whereas an op-amp/transistor (or op-amp/FET) current source is predictable and stable to better than 0.5% without great effort.

3.07 FET amplifiers

Source followers and common-source FET amplifiers are analogous to the emitter followers and common-emitter amplifiers made with bipolar transistors that we talked about in the last chapter. However, the absence of dc gate current makes it possible to realize very high input impedances. Such amplifiers are essential when dealing with the high-impedance signal sources encountered in measurement and instrumentation. For some specialized applications you may want to build followers or amplifiers with discrete FETs; most of the time, however, you can take advantage of FET-input op-amps. In either case it's worth knowing how they work.

With JFETs it is convenient to use the same self-biasing scheme as with JFET current sources (Section 3.06), with a single gate-biasing resistor to ground (Fig. 3.21); MOSFETs require a divider from the drain supply, or split supplies, just as we used with BJTs. The gate-biasing resistors can be quite large (a megohm or more), because the gate leakage current is measured in nanoamps.

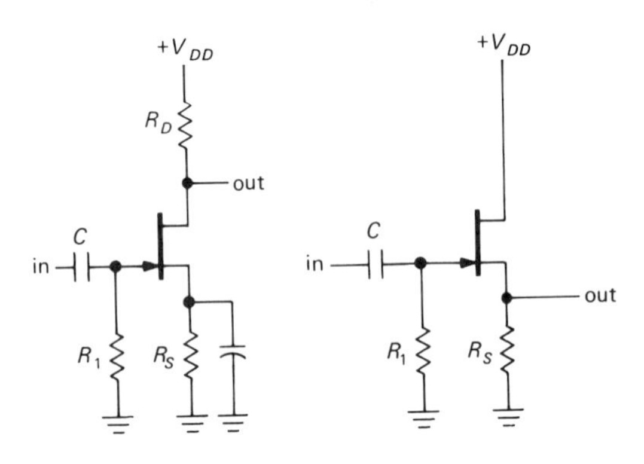

Figure 3.21

Transconductance

The absence of gate current makes *transconductance* (the ratio of output current to input voltage: $g_m = i_{out}/v_{in}$) the natural gain parameter for FETs. This is in contrast to bipolar transistors in the last chapter, where we at first flirted with the idea of current gain (i_{out}/i_{in}), then introduced the transconductance-oriented Ebers-Moll model: It's useful to think of BJTs either way, depending on the application.

FET transconductance can be estimated from the characteristic curves, either by looking at the increase in I_D from one gate-voltage curve to the next on the family of curves (Fig. 3.2 or 3.17), or, more simply, from the slope of the I_D–V_{GS} "transfer characteristics" curve (Fig. 3.14). The transconductance depends on drain current (we'll see how, shortly) and is, of course,

$$g_m(I_D) = i_d/v_{gs}$$

(Remember that lower-case letters indicate quantities that are small-signal variations.) From this we get the voltage gain

$$G_{\text{voltage}} = v_d/v_{gs} = -R_D i_d/v_{gs}$$
$$= -g_m R_D$$

just the same as the bipolar transistor result in Section 2.09, with load resistor R_C replaced by R_D. Typically, FETs have transconductances of a few thousand microsiemens (micromhos) at a few milliamps. Because g_m depends on drain current, there will be some variation of gain (nonlinearity) over the waveform as the drain current varies, just as we have with grounded emitter amplifiers (where $g_m = 1/r_e$, proportional to I_C). Furthermore, FETs in general have considerably lower transconductance than bipolar transistors, which makes them less suitable as amplifiers and followers. Let's look at this a little further.

Transconductance of FETs versus BJTs

To make our last remark quantitative, consider a JFET and a BJT, each operating at 1mA. Imagine they are connected as common source (emitter) amplifiers, with a drain (collector) resistor of 5k to a +10 volt supply (Fig. 3.22). Let's ignore details of biasing and concentrate on the gain. The BJT has an r_e of 25 ohms, hence a g_m of 40 mS, for a voltage gain of -200 (which you could have calculated directly as $-R_C/r_e$). A typical JFET (e.g., a

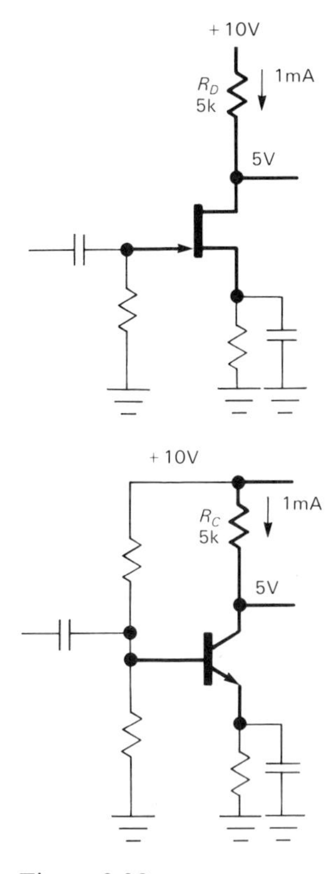

Figure 3.22

2N4220) has a g_m of 2mS at a drain current of 1mA, giving a voltage gain of −10. This seems discouraging by comparison. The low g_m also produces a relatively large Z_{out} in a follower configuration (Fig. 3.23): The JFET has

$$Z_{out} = 1/g_m$$

which in this case equals 500 ohms (independent of signal source impedance), to be compared with the BJT, which has

$$Z_{out} = R_s/h_{fe} + r_e = R_s/h_{fe} + 1/g_m$$

equal to R_s/h_{fe}+ 25 ohms (at 1mA). For typical transistor betas, say $h_{fe} = 100$, and reasonable signal sources, say with $R_s < 5k$, the BJT follower is an order of magnitude stiffer ($Z_{out} = 25\Omega$ to 75Ω). Note, however, that for $R_s > 50k$ the JFET follower will be better.

To see what is happening, let's look back at the expressions for FET drain current versus gate-source voltage and compare with the equivalent expression (Ebers-Moll) for BJT collector current versus base-emitter voltage.

BJT: The Ebers-Moll equation,

$$I_C = I_S\{\exp(V_{BE}/V_T) - 1\},$$
$$\text{with } V_T = kT/q = 25\,\text{mV}$$

predicts $g_m = dI_C/dV_{BE} = I_C/V_T$ for collector currents large compared with "leakage" current I_S. This is our familiar result $r_e(\text{ohms}) = 25/I_C(\text{mA})$, since $g_m = 1/r_e$.

FET: In the "subthreshold" region of very low drain current,

$$I_D \propto \exp(V_{GS})$$

which, being exponential like Ebers-Moll, also gives a transconductance proportional

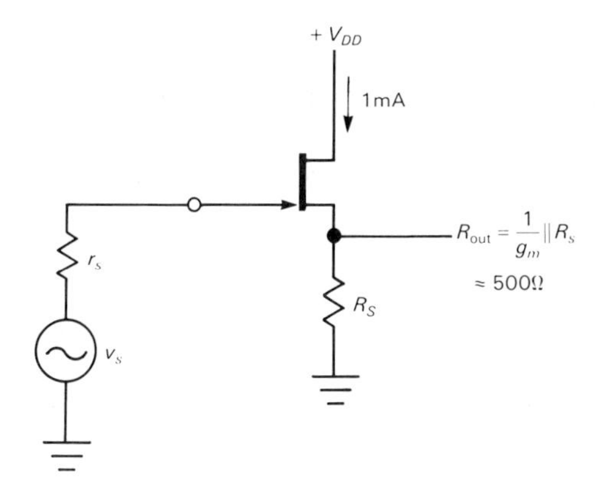

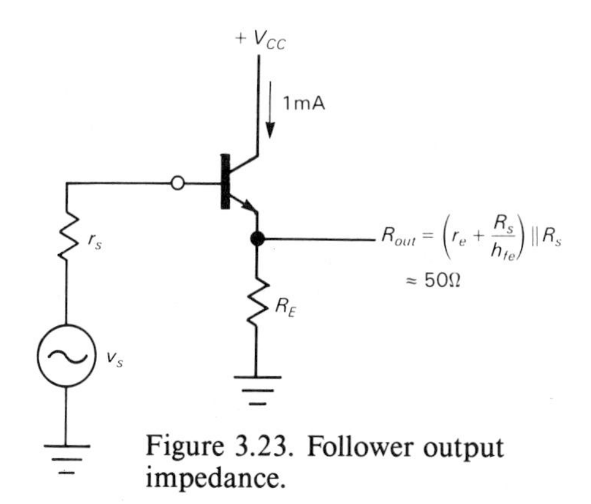

Figure 3.23. Follower output impedance.

to current. However, for real-world values of k (which is determined by FET geometry, carrier mobility, etc.) the FET's transconductance is somewhat lower than the BJT's, about $I/40\text{mV}$ for p-channel MOSFETs and $I/60\text{mV}$ for n-channel MOSFETs, as compared with $I/25\text{mV}$ for BJTs. As the current is increased, the FET enters the normal "saturation" region, where

$$I_D = k(V_{GS} - V_T)^2$$

which gives $g_m = 2(kI_D)^{1/2}$. That is, the transconductance increases only as the square root of I_D and is well below the transconductance of a bipolar transistor at the same operating current; see Figure 3.24. Increasing the constant k in our preceding equations (by raising the width/length ratio of the channel) increases the transconductance (and the drain current, for a given V_{GS}) in the region above threshold, but the transconductance still remains less than that of a bipolar transistor at the same current.

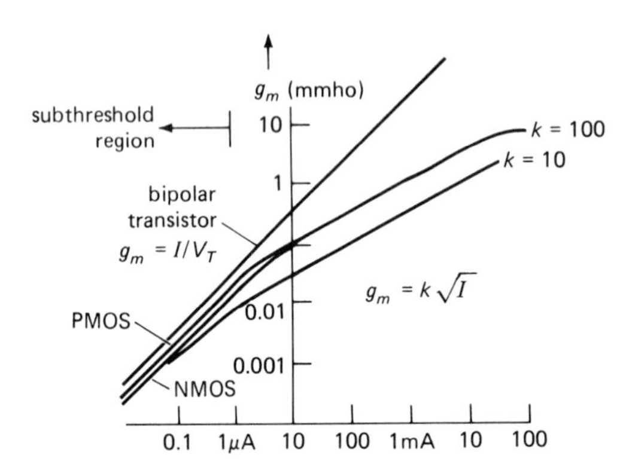

Figure 3.24. Comparison of g_m for bipolar transistors and FETs.

EXERCISE 3.3
Derive the foregoing expressions for g_m by differentiating I_{out} with respect to V_{in}.

The problem of low voltage gain in FET amplifiers can be circumvented by resorting to a current-source (active) load, but once again the bipolar transistor will be better in the same circuit. For this reason you seldom see FETs used as simple amplifiers, unless it's important to take advantage of their unique input properties (extremely high input resistance and low input current).

Note that FET transconductance in the saturation region is proportional to $V_{GS} - V_T$; thus, for example, a JFET with gate operated halfway to pinch-off has a transconductance approximately half that shown on the data sheet (where it is always given for $I_D = I_{DSS}$, i.e., $V_{GS} = 0$).

Differential amplifiers

Matched FETs can be used to construct high-input-impedance front-end stages for bipolar differential amplifiers, as well as the important op-amps and comparators we'll meet in the next chapter. As we mentioned earlier, the substantial V_{GS} offsets of FETs will generally result in larger input voltage offsets and offset drifts than with a comparable amplifier constructed entirely with bipolar transistors, but of course the input impedance will be raised enormously.

Oscillators

In general, FETs have characteristics that make them useful substitutes for bipolar transistors in almost any circuit that can benefit from their uniquely high input impedance and low bias current. A particular instance is their use in high-stability LC and crystal oscillators; we'll show examples in Sections 5.18, 5.19, and 13.11.

Active load

Just as with BJT amplifiers, it is possible to replace the drain-load resistor in a FET amplifier with an active load, i.e., a current source. The voltage gain you get that way can be very large:

$$G_V = -g_m R_D$$

(with a drain resistor as load)

$$G_V = -g_m R_0$$

(with a current source as load)

where R_0 is the impedance looking into the drain (called "g_{oss}"), typically in the range of 100k to 1M.

One possibility for an active load is a current mirror as the drain load for a differential FET pair (see Section 2.18); the circuit is not bias-stable, however, without overall feedback. The current mirror can be constructed with either FETs or BJTs. This configuration is often used in FET op-amps, as we'll see in the next chapter. You will see another nice example of the active load technique in Section 3.14 when we discuss the CMOS linear amplifier.

3.08 Source followers

Because of the relatively low transconductance of FETs, it's often better to use a FET "source follower" (analogous to an emitter follower) as an input buffer to a conventional BJT amplifier, rather than trying to make a common-source FET amplifier directly. You still get the high input impedance and zero dc input current of the FET, and the BJT's large transconductance lets you achieve high single-stage gain. Furthermore, discrete FETs (i.e., those that are not part of an integrated circuit) tend to have higher interelectrode capacitance than BJTs, leading to greater Miller effect (Section 2.19) in common-source amplifiers; the source follower configuration, like the emitter follower, has no Miller effect.

FET followers, with their high input impedance, are commonly used as input stages in oscilloscopes as well as other measuring instruments. There are many applications in which the signal source impedance is intrinsically high, e.g., capacitor microphones, pH probes, charged-particle detectors, or microelectrode signals in biology and medicine. In these cases a FET input stage (whether discrete or part of an integrated circuit) is a good solution. Within circuits there are situations where the following stage must draw little or no current. Common examples are analog "sample-and-hold" and "peak detector" circuits, in which the level is stored on a capacitor and will "droop" if the next amplifier draws significant input current. In all these applications the negligible input current of a FET is more important than its low transconductance, making source followers (or even common-source amplifiers) attractive alternatives to the bipolar emitter follower.

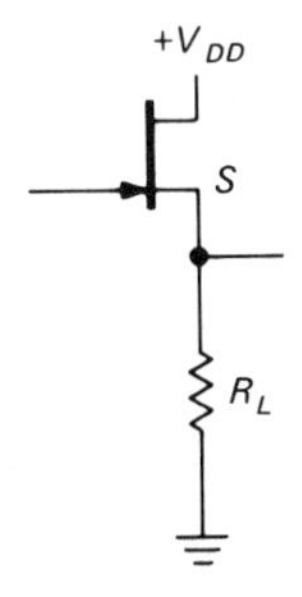

Figure 3.25

Figure 3.25 shows the simplest source follower. We can figure out the output amplitude, as we did for the emitter follower in Section 2.11, using the transconductance. We have

$$v_s = R_L i_d$$

since i_g is negligible; but

$$i_d = g_m v_{gs} = g_m(v_g - v_s)$$

so

$$v_s = \left[\frac{R_L g_m}{(1 + R_L g_m)}\right] v_g$$

For $R_L \gg 1/g_m$ it is a good follower ($v_s \approx v_g$), with gain approaching, but always less than, unity.

Output impedance

The preceding equation for v_s is precisely what you would predict if the source follower's output impedance were equal to $1/g_m$ (try the calculation, assuming a source voltage of v_g in series with $1/g_m$ driving a load of R_L). This is exactly analogous to the emitter follower situation, where the output impedance was $r_e = 25/I_C$, or $1/g_m$. It can be easily shown explicitly that a source follower has output impedance $1/g_m$ by figuring the source current for a signal applied to the output with grounded gate (Fig. 3.26). The drain current is

$$i_d = g_m v_{gs} = g_m v$$

so

$$r_{\text{out}} = v/i_d = 1/g_m$$

typically a few hundred ohms at currents of a few milliamps. As you can see, FET source followers aren't nearly as stiff as emitter followers.

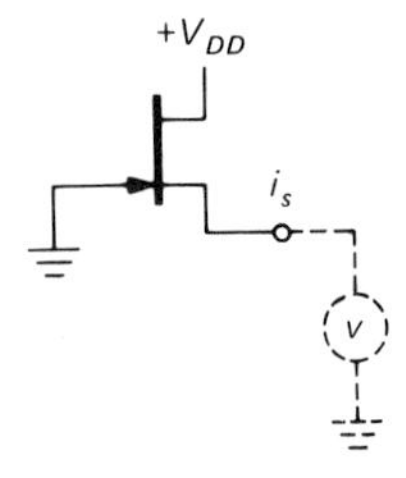

Figure 3.26

There are two drawbacks to this circuit:
1. The relatively high output impedance means that the output swing may be significantly less than the input swing, even with high load impedance, because R_L alone forms a divider with the source's output impedance. Furthermore, because the drain current is changing over the signal waveform, g_m and therefore the output impedance will vary, producing some nonlinearity (distortion) at the output. The situation is improved if FETs of high transconductance are used, of course, but a combination FET–bipolar follower is often a better solution.
2. Because the V_{GS} needed to produce a certain operating current is a poorly controlled parameter in FET manufacture, a source follower has an unpredictable dc offset, a serious drawback for dc-coupled circuits.

Active load

The addition of a few components improves the source follower enormously. Let's take it in stages:

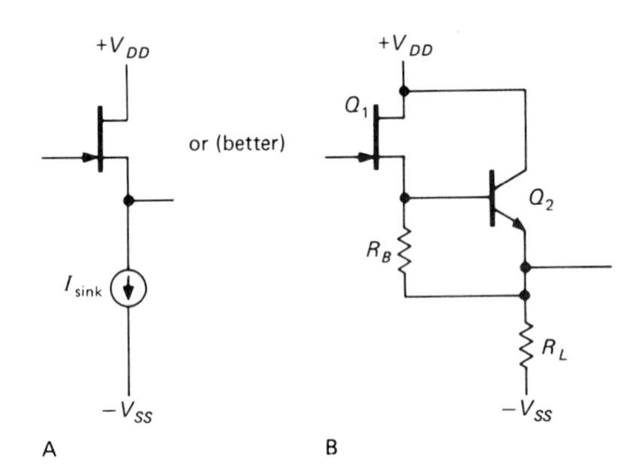

Figure 3.27

First, replace R_L with a (pull-down) current source (Fig. 3.27). The constant source current makes V_{GS} approximately constant, thus reducing nonlinearities. You can think of this as the previous case with infinite R_L, which is what a current source is. The circuit on the right has the advantage of providing low output impedance, while still providing a (roughly) constant source current of V_{BE}/R_B. We still have the problem of unpredictable (and therefore nonzero) offset voltage (from input to output) of V_{GS} ($V_{GS} + V_{BE}$ for the circuit on the right). Of course, we could simply adjust I_{sink} to the particular value of I_{DSS} for the given FET (in the first circuit) or adjust R_B (in the second). This is a poor solution, for two reasons: (a) It requires individual adjustment for each FET. (b) Even so, I_D may vary by a factor of two over the normal operating temperature range for a given V_{GS}.

A better circuit uses a matched FET pair to achieve zero offset (Fig. 3.28). Q_1 and Q_2 are a matched pair, on a single chip of silicon. Q_2 sinks a current exactly appropriate to the condition $V_{GS} = 0$. So, for both FETs, $V_{GS} = 0$, and Q_1 is therefore a follower with zero offset. Because Q_2 tracks Q_1 in temperature, the offset remains near zero independent of temperature.

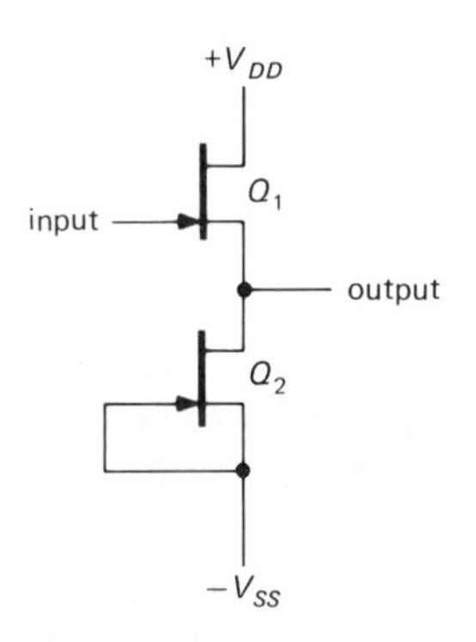

Figure 3.28

You usually see the preceding circuit with source resistors added (Fig. 3.29). A little thought should convince you that R_1 is necessary and that $R_1 = R_2$ guarantees that $V_{out} = V_{in}$ if Q_1 and Q_2 are matched. This circuit modification gives better I_D predictability, allows you to set the drain current to some value less than I_{DSS}, and gives improved linearity, since FETs are better current sources when operated below I_{DSS}. This follower circuit is popular as the input stage for oscilloscope vertical amplifiers.

For the utmost in performance you can add circuitry to bootstrap the drain (eliminating input capacitance) and use a bipolar output stage for low output impedance. That same output signal can then be used to drive an inner "guard" shield in order to effectively eliminate the effects of shielded-cable capacitance, which would otherwise be devastating for the high source impedances that you might see with this sort of high-impedance input buffer amplifier.

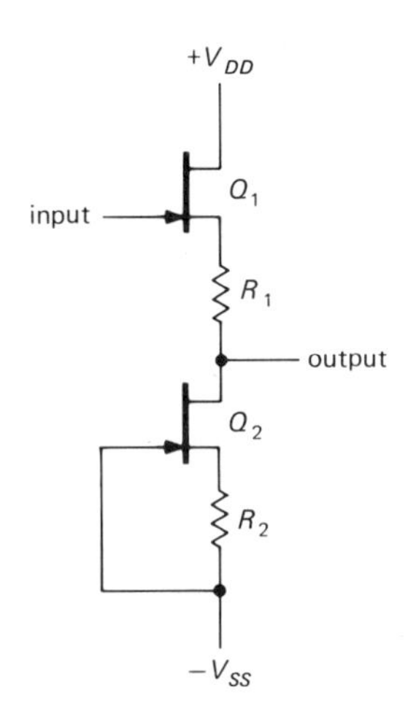

Figure 3.29

3.09 FET gate current

We said at the outset that FETs in general, and MOSFETs in particular, have essentially zero gate current. This is perhaps the most important property of FETs, and it was exploited in the high-impedance amplifiers and followers in the previous sections. It will prove essential, too, in applications to follow – most notably analog switches and digital logic.

Of course, at *some* level of scrutiny we might expect to see some gate current. It's important to know about gate current, because a naive zero-current model is guaranteed to get you in trouble sooner or later. In fact, finite gate current arises from

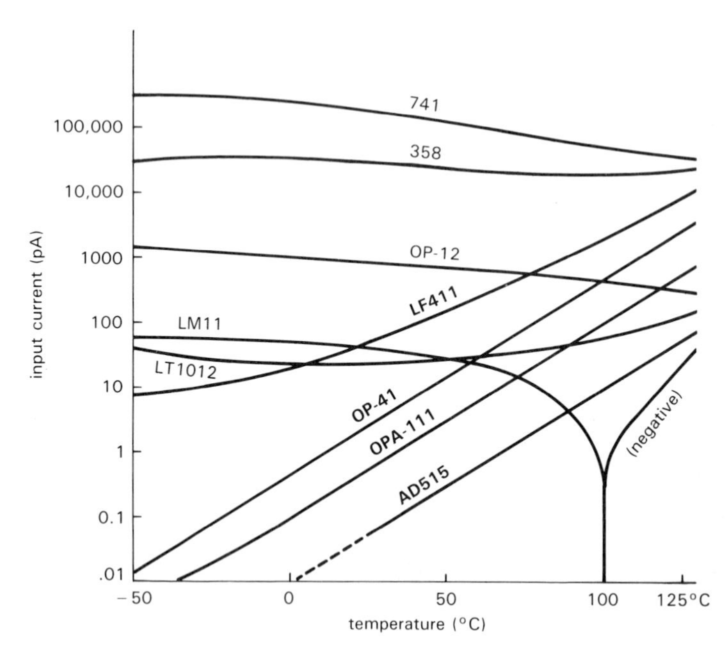

Figure 3.30. The input current of a FET amplifier is gate leakage, which doubles every 10°C.

several mechanisms: Even in MOSFETs the silicon dioxide gate insulation is not perfect, leading to leakage currents in the picoampere range. In JFETs the gate "insulation" is really a back-biased diode junction, with the same impurity and junction leakage current mechanisms as ordinary diodes. Furthermore, JFETs (n-channel in particular) suffer from an additional effect known as "impact-ionization" gate current, which can reach astounding levels. Finally, both JFETs and MOSFETs have *dynamic* gate current, caused by ac signals driving the gate capacitance; this can cause Miller effect, just as with bipolar transistors.

In most cases gate input currents are negligible in comparison with BJT base currents. However, there are situations in which a FET may actually have *higher* input current! Let's look at the numbers.

Gate leakage

The low-frequency input impedance of a FET amplifier (or follower) is limited by gate leakage. JFET data sheets usually specify a breakdown voltage, BV_{GSS}, defined as the voltage from gate to channel (source and drain connected together) at which the gate current reaches 1μA. For smaller applied gate-channel voltages, the gate leakage current, I_{GSS}, again measured with the source and drain connected together, is considerably smaller, dropping quickly to the picoampere range for gate-drain voltages well below breakdown. With MOSFETs you must never allow the gate insulation to break down; instead, gate leakage is specified as some maximum leakage current at a specified gate-channel voltage. Integrated circuit amplifiers with FETs (e.g., FET op-amps) use the misleading term "input bias current," I_B, to specify input leakage current; it's usually in the picoampere range.

The good news is that these leakage currents are in the picoampere range at room temperature. The bad news is that they increase rapidly (in fact, exponentially) with temperature, roughly doubling every 10°C. By contrast, BJT base currents aren't leakage, and in fact tend to *decrease* slightly with increasing temperature. The comparison is shown graphically in Figure 3.30, a plot of input current versus temperature for several IC amplifiers (op-amps).

The FET-input op-amps have the lowest input currents at room temperature (and below), but their input current rises rapidly with temperature, crossing over the curves for amplifiers with carefully designed BJT input stages like the LM11 and LT1012. These BJT op-amps, along with "premium" low-input-current JFET op-amps like the OPA111 and AD549, are fairly expensive. However, we also included everyday "jellybean" op-amps like the bipolar 358 and JFET LF411 in the figure to give an idea of input currents you can expect with inexpensive (less than a dollar) op-amps.

□ *JFET impact-ionization current*

In addition to conventional gate leakage effects, n-channel JFETs suffer from rather large gate leakage currents when operated with substantial V_{DS} and I_D (the gate leakage specified on data sheets is measured under the unrealistic conditions $V_{DS} = I_D = 0$!). Figure 3.31 shows what happens. The gate leakage current remains near the I_{GSS} value until you reach a critical drain-gate voltage, at which point it rises precipitously. This extra "impact-ionization" current is proportional to drain current, and it rises exponentially with voltage and temperature. The onset of this current occurs at drain-gate voltages of about 25% of BV_{GSS}, and it can reach gate currents of a microamp or more. Obviously a "high-impedance buffer" with a microamp of input current is worthless. That's what you would get if you used a 2N4868A as a follower, running 1mA of drain current from a 40 volt supply.

This extra gate leakage current afflicts primarily n-channel JFETs, and it occurs at higher values of drain-gate voltage. Some cures are to (a) operate at low drain-gate voltage, either with a low-voltage drain supply or with a cascode, (b) use a p-channel JFET, where the effect is much smaller, or (c) use a MOSFET. The most important thing is to be aware of the effect so that it doesn't catch you by surprise.

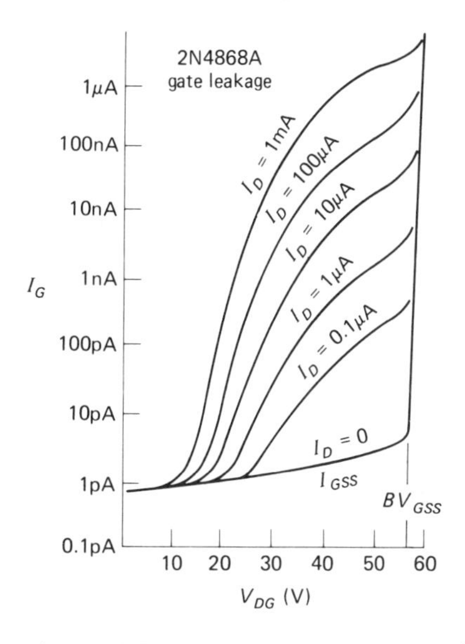

Figure 3.31. JFET gate leakage increases disastrously at higher drain-gate voltages and is proportional to drain current.

□ *Dynamic gate current*

Gate leakage is a dc effect. Whatever is driving the gate must also supply an *ac* current, because of gate capacitance. Consider a common-source amplifier. Just as with bipolar transistors, you can have the simple effect of input capacitance to ground (called C_{iss}), and you can have the capacitance-multiplying Miller effect (which acts on the feedback capacitance C_{rss}). There are two reasons why capacitive effects are more serious in FETs than in bipolar transistors: First, you use FETs (rather than BJTs) because you want very low input current; thus the capacitive currents loom relatively larger for the same capacitance. Second, FETs often have considerably larger capacitance than equivalent bipolar transistors.

To appreciate the effect of capacitance, consider a FET amplifier intended for a signal source of 100k source impedance.

At dc there's no problem, because the picoampere currents produce only microvolt drops across the signal source's internal impedance. But at 1MHz, say, an input capacitance of 5pF presents a shunt impedance of about 30k, seriously attenuating the signal. In fact, *any* amplifier is in trouble with a high-impedance signal at high frequencies, and the usual solution is to operate at low impedance (50Ω is typical) or use tuned *LC* circuits to resonate away the parasitic capacitance. The point to understand is that the FET amplifier doesn't look like a 10^{12} ohm load at signal frequencies.

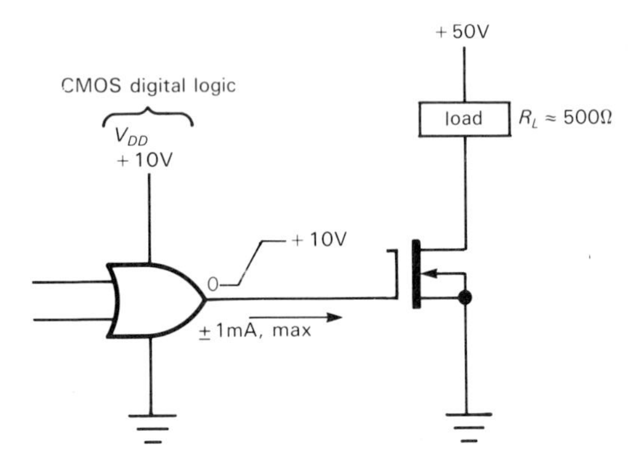

Figure 3.32

As another example, imagine switching a 10 amp load with a power MOSFET (there aren't any power JFETs), in the style of Figure 3.32. One might naively assume that the gate could be driven from a digital logic output with low current-sourcing capability, for example the so-called CMOS logic, which can supply output current on the order of 1mA with a swing from ground to +10 volts. In fact, such a circuit would be a disaster, since with 1mA of gate drive the 350pF feedback capacitance of the 2N6763 would stretch the output switching speed to a leisurely 20μs. Even worse, the dynamic gate currents ($I_{gate} = CdV_D/dt$) would force currents back into the logic device's output, possibly destroying it via a perverse effect known as "SCR latchup" (more of which in Chapters 8 and 9). Bipolar power transistors turn out to have comparable capacitances, and therefore comparable dynamic input currents; but when you design a circuit to drive a 10-amp power BJT, you're *expecting* to provide 500mA or so of base drive (via a Darlington or whatever), whereas with a FET you tend to take low input current for granted. In this example, once again, the ultra-high-impedance FET has lost some of its luster.

EXERCISE 3.4

Show that the circuit of Figure 3.32 switches in about 20μs, assuming 1mA of available gate drive.

3.10 FETs as variable resistors

Figure 3.17 showed the region of JFET characteristic curves (drain current versus V_{DS} for a small family of V_{GS} voltages), both in the normal ("saturated") regime and in the "linear" region of small V_{DS}. We showed the equivalent pair of graphs for a MOSFET at the beginning of the chapter (Fig. 3.2). The I_D-versus-V_{DS} curves are approximately straight lines for V_{DS} smaller than $V_{GS} - V_T$, and they extend in both directions through zero, i.e., the device can be used as a voltage-controlled resistor for small signals of either polarity. From our equation for I_D versus V_{GS} in the linear region (Section 3.04) we easily find the ratio (I_D/V_{DS}) to be

$$\frac{1}{R_{DS}} = 2k\left[(V_{GS} - V_T) - \frac{V_{DS}}{2}\right]$$

The last term represents a nonlinearity, i.e., a departure from resistive behavior (resistance shouldn't depend on voltage). However, for drain voltages substantially less than the amount by which the gate is above threshold ($V_{DS} \rightarrow 0$), the last term becomes unimportant, and the FET behaves approximately like a resistance:

$$R_{DS} \approx 1/[2k(V_{GS} - V_T)]$$

Because the device-dependent parameter k isn't a quantity you are likely to know, it's more useful to write R_{DS} as

$$R_{DS} \approx R_0(V_{G0} - V_T)/(V_G - V_T)$$

where the resistance R_{DS} at any gate voltage V_G is written in terms of the (known) resistance R_0 at some gate voltage V_{G0}.

EXERCISE 3.5
Derive the preceding "scaling" law.

From either formula you can see that the conductance ($= 1/R_{DS}$) is proportional to the amount by which the gate voltage exceeds threshold. Another useful fact is that $R_{DS} = 1/g_m$, i.e., the channel resistance in the *linear* region is the inverse of the transconductance in the *saturated* region. This is a handy thing to know, because g_m is a parameter nearly always specified on FET data sheets.

EXERCISE 3.6
Show that $R_{DS} = 1/g_m$ by finding the transconductance from the saturation drain-current formula in Section 3.04.

Typically, the values of resistance you can produce with FETs vary from a few tens of ohms (as low as 0.1Ω for power MOSFETs) all the way up to an open circuit. A typical application might be an automatic-gain-control (AGC) circuit in which the gain of an amplifier is adjusted (via feedback) to keep the output within the linear range. In such an AGC circuit you must be careful to put the variable-resistance FET at a place in the circuit where the signal swing is small, preferably less than 200mV or so.

The range of V_{DS} over which the FET behaves like a good resistor depends on the particular FET and is roughly proportional to the amount by which the gate voltage exceeds V_P (or V_T). Typically, you might have nonlinearities of about 2% for $V_{DS} < 0.1(V_{GS} - V_P)$, and perhaps 10% nonlinearity for $V_{DS} \approx 0.25(V_{GS} - V_P)$. Matched FETs make it easy to design a ganged variable resistor to control several signals at once. JFETs intended for use as variable resistors are available (Siliconix VCR series) with resistance tolerances of 30%, specified at some V_{GS}.

It is possible to improve the linearity, and simultaneously the range of V_{DS} over which a FET behaves like a resistor, by a simple compensation scheme. We'll illustrate with an application.

□ *Linearizing trick: electronic gain control*

By looking at the preceding equation for $1/R_{DS}$, you can see that the linearity will be nearly perfect if you can add to the gate voltage a voltage equal to one-half the drain-source voltage. Figure 3.33 shows two circuits that do exactly that. In the first, the JFET forms the lower half of a resistive voltage divider, thus forming a voltage-controlled attenuator (or "volume control"). R_1 and R_2 improve the linearity by adding a voltage of $0.5V_{DS}$ to V_{GS}, as just discussed. The JFET shown has an ON resistance (gate grounded) of 60 ohms (max), giving the circuit an attenuation range of 0 to 40dB.

The second circuit uses a MOSFET as a variable emitter resistance in an emitter-degenerated ac amplifier. Note the use of a constant-dc-current emitter pulldown (Wilson mirror or FET current-regulator diode); this (a) looks like a very high impedance at signal frequencies, thus letting the variable-resistance FET set the gain over a wide range (including $G_V \ll 1$), and (b) provides simple biasing. By using a blocking capacitor, we've arranged the circuit so that the FET affects only the

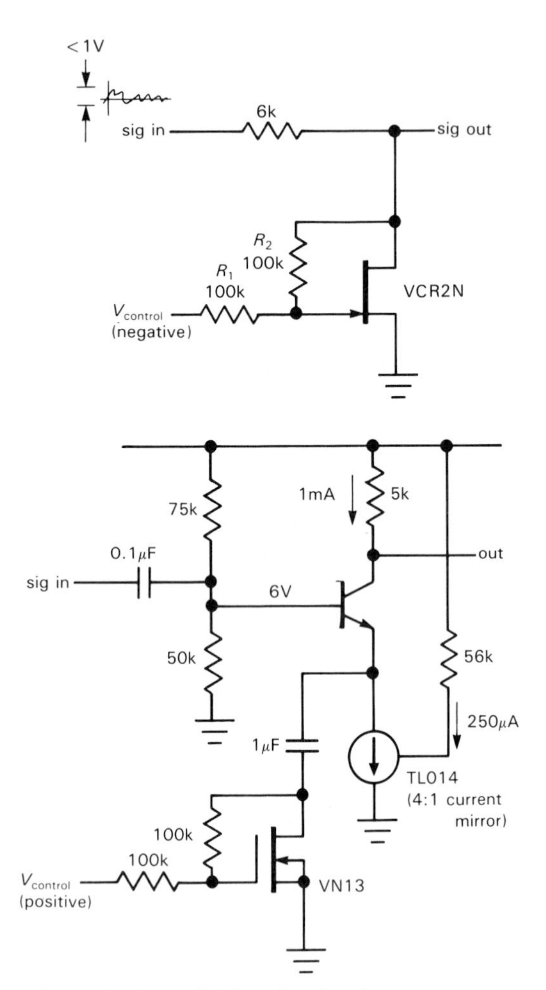

Figure 3.33. Variable-gain circuits.

ac (signal) gain. Without the capacitor, the transistor biasing would vary with FET resistance.

EXERCISE 3.7

The VN13 has an ON resistance ($V_{GS} = +5V$) of 15 ohms (max). What is the range of amplifier gain in the second circuit (assume that the current sink looks like 1MΩ)? What is the low-frequency 3dB point when the FET is biased so that the amplifier gain is (a) 40dB or (b) 20dB?

The linearization of R_{DS} with a resistive gate divider circuit, as above, is remarkably effective. In Figure 3.34 we've compared actual measured curves of I_D versus V_{DS} in the linear (low-V_{DS}) region for FETs with and without the linearizing circuit. The linearizing circuit is essential for low-distortion applications with signal swings of more than a few millivolts.

When considering FETs for an application requiring a gain control, e.g., an AGC or "modulator" (in which the amplitude of a high-frequency signal is varied at an audio rate, say), it is worthwhile to look also at "analog-multiplier" ICs. These are high-accuracy devices with good dynamic range that are normally used to form the product of two voltages. One of the voltages can be a dc control signal, setting the multiplication factor of the device for the other input signal, i.e., the gain. Analog multipliers exploit the g_m-versus-I_C characteristic of bipolar transistors [$g_m = I_C$(mA)/25 siemens], using matched arrays to circumvent problems of offsets and bias shifts. At very high frequencies (100MHz and above), passive "balanced mixers" (Section 13.12) are often the best devices to accomplish the same task.

It is important to remember that a FET in conduction at low V_{DS} behaves like a good resistance all the way down to zero volts from drain to source (there are no diode drops or the like to worry about). We will see op-amps and digital logic families (CMOS) that take advantage of this nice property, giving outputs that saturate cleanly to the power supplies.

FET SWITCHES

The two examples of FET circuits that we gave at the beginning of the chapter were both *switches*: a logic-switching application and a linear signal-switching circuit. These are among the most important FET applications and take advantage of the FET's unique characteristics: high gate impedance and bipolarity resistive conduction clear down to zero volts. In practice

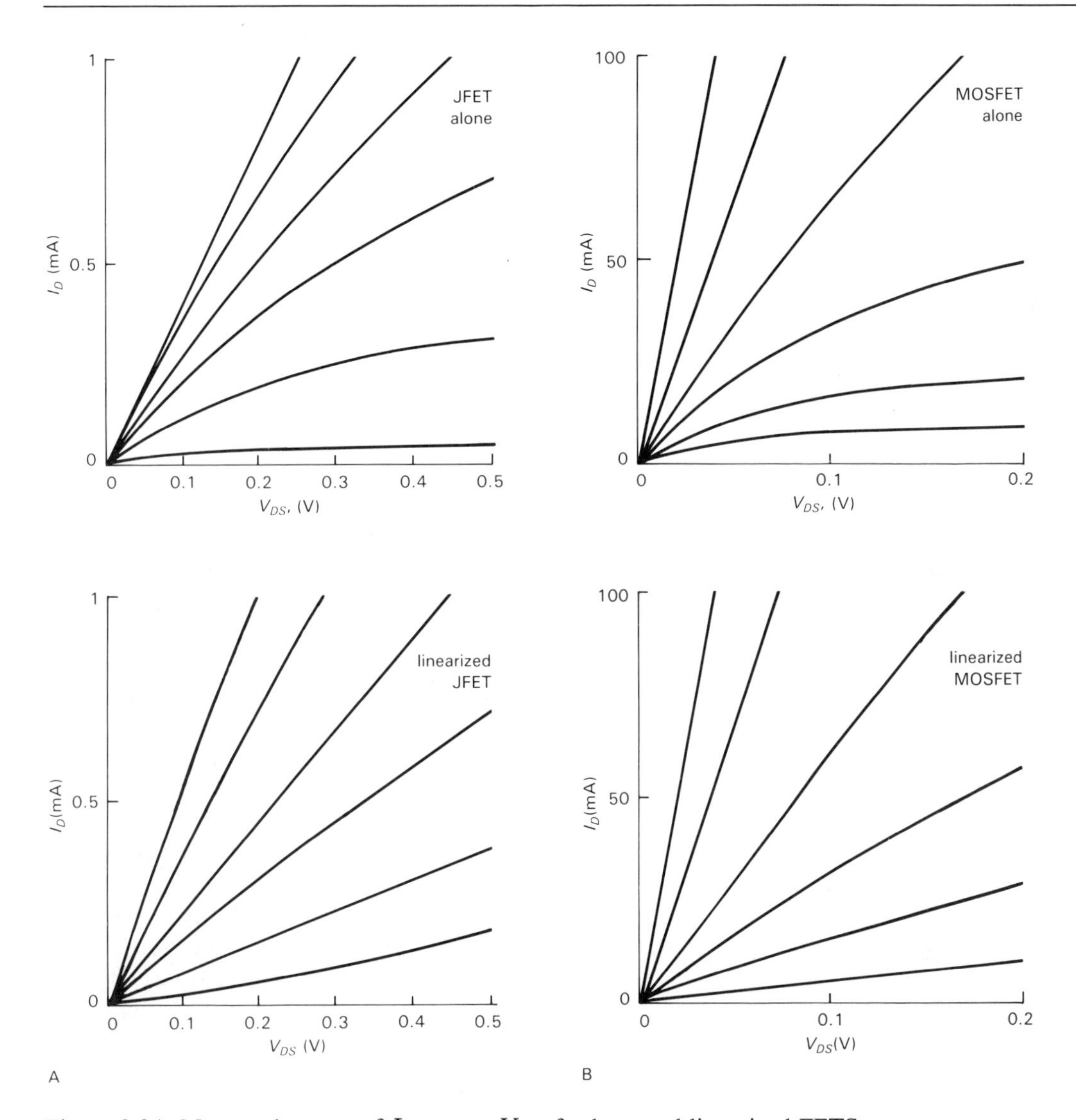

Figure 3.34. Measured curves of I_D versus V_{DS} for bare and linearized FETS.
A. 2N5484 JFET
B. VN0106 MOSFET

you usually use MOSFET integrated circuits (rather than discrete transistors) in all digital logic and linear switch applications, and it is only in power switching applications that you resort to discrete FETs. Even so, it is essential (and fun!) to understand the workings of these chips; otherwise you're almost guaranteed to fall prey to some mysterious circuit pathology.

3.11 FET analog switches

A common use of FETs, particularly MOSFETs, is as analog switches. Their combination of low ON resistance (all the way to zero volts), extremely high OFF resistance, low leakage currents, and low capacitance makes them ideal as voltage-controlled switch elements for analog signals. An ideal analog, or linear, switch behaves like a perfect mechanical switch: In the ON state it passes a signal through to a load without attenuation or nonlinearity; in the OFF state it is an open circuit. It should have negligible capacitance to ground and negligible coupling to the signal of the switching level applied to the control input.

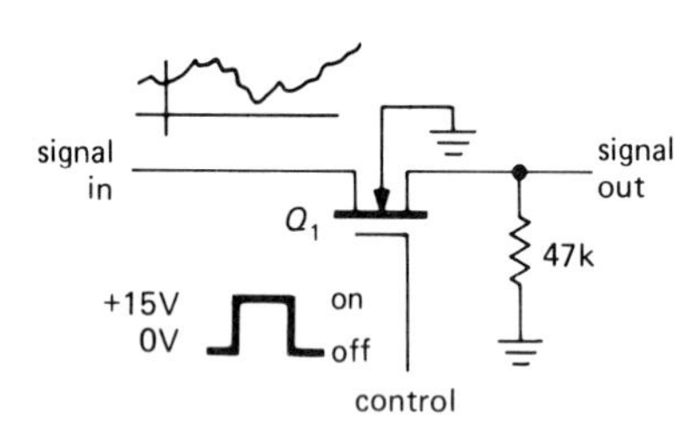

Figure 3.35

Let's look at an example (Fig. 3.35). Q_1 is an n-channel enhancement-mode MOSFET, and it is nonconducting when the gate is grounded or negative. In that state the drain-source resistance (R_{OFF}) is typically more than 10,000M, and no signal gets through (though at high frequencies there will be some coupling via drain-source *capacitance*; more on this later). Bringing the gate to +15 volts puts the drain-source channel into conduction, typically 25 to 100 ohms (R_{ON}) in FETs intended for use as analog switches. The gate signal level is not at all critical, as long as it is sufficiently more positive than the largest signal (to maintain R_{ON} low), and it could be provided from digital logic circuitry (perhaps using a FET or BJT to generate a full-supply swing) or even from an op-amp (whose ±13V output swing would do nicely, since gate breakdown voltages in MOSFETs are typically 20V or more). Swinging the gate negative (as from an op-amp output) doesn't hurt, and in fact has the added advantage of allowing the switching of analog signals of either polarity, as will be described later. Note that the FET switch is a bidirectional device; signals can go either way through it. Ordinary mechanical switches work that way, too, so it should be easy to understand.

The circuit as shown will work for positive signals up to about 10 volts; for larger signals the gate drive is insufficient to hold the FET in conduction (R_{ON} begins to rise), and negative signals would cause the FET to turn on with the gate grounded (it would also forward bias the channel-body junction; see Section 3.02). If you want to switch signals that are of both polarities (e.g., signals in the range −10V to +10V), you can use the same circuit, but with the gate driven from −15 volts (OFF) to +15 volts (ON); the body should then be tied to −15 volts.

With any FET switch it is important to provide a load resistance in the range of 1k to 100k in order to reduce capacitive feedthrough of the input signal that would otherwise occur during the OFF state. The value of the load resistance is a compromise: Low values reduce feedthrough, but they begin to attenuate the input signal because of the voltage divider formed by R_{ON} and the load. Because R_{ON} varies over the input signal swing (from changing V_{GS}), this attenuation also produces some undesirable nonlinearity. Excessively low load resistance appears at the switch input, of course, loading the signal source as well. Several possible solutions to this problem (multiple-stage switches, R_{ON} cancellation) are shown in Sections 3.12 and 4.30. An attractive alternative is to use a second FET switch section to connect the output to ground when the series FET is off, thus effectively forming an SPDT switch (more on this in the next section).

CMOS linear switches

Frequently it is necessary to switch signals that may go nearly to the supply voltages. In that case the simple n-channel switch circuit just described won't work, since the gate is not forward-biased at the peak of the signal swing. The solution is to use paralleled complementary MOSFET ("CMOS") switches (Fig. 3.36). The triangular symbol is a digital inverter, which we'll discuss shortly; it inverts a HIGH input to a LOW output, and vice versa. When the control input is high, Q_1 is held ON for signals from ground to within a few volts of V_{DD} (where R_{ON} starts increasing dramatically). Q_2 is likewise held ON (by its grounded gate) for signals from V_{DD} to within a few volts of ground (where its R_{ON} increases dramatically). Thus, signals anywhere between V_{DD} and ground are passed through with low series resistance (Fig. 3.37). Bringing the control signal to ground turns off both FETs, providing an open circuit. The result is an analog switch for signals between ground and V_{DD}. This is the basic construction of the 4066 CMOS "transmission gate." It is bidirectional, like the switches described earlier; either terminal can be the input.

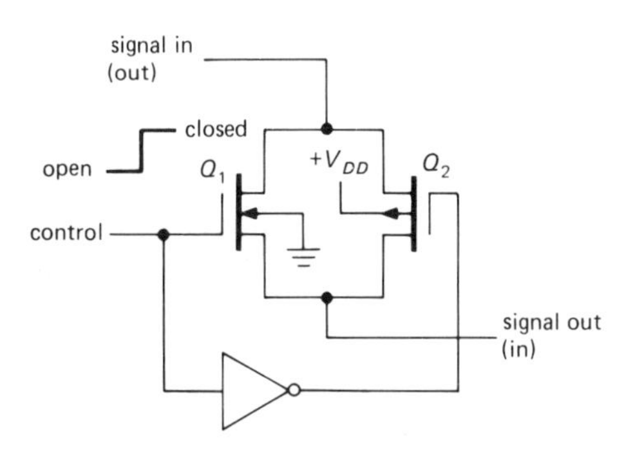

Figure 3.36. CMOS analog switch.

There is a variety of integrated circuit CMOS analog switches available, with various switch configurations (e.g., several independent sections with several poles each). The 4066 is the classic 4000-series CMOS "analog transmission gate," just another name for an analog switch for signals between ground and a single positive supply. The IH5040 and IH5140 series from Intersil and Harris and the DG305 and DG400 series from Siliconix are very convenient to use; they accept logic-level (0V = LOW, > 2.4V = HIGH) control signals, they will handle analog signals to ±15 volts (compared with only ±7.5V for the 4000 series), they come in a variety of configurations, and they have relatively low ON resistance (25Ω for some members of these families). Analog Devices, Maxim, and PMI also manufacture nice analog switches.

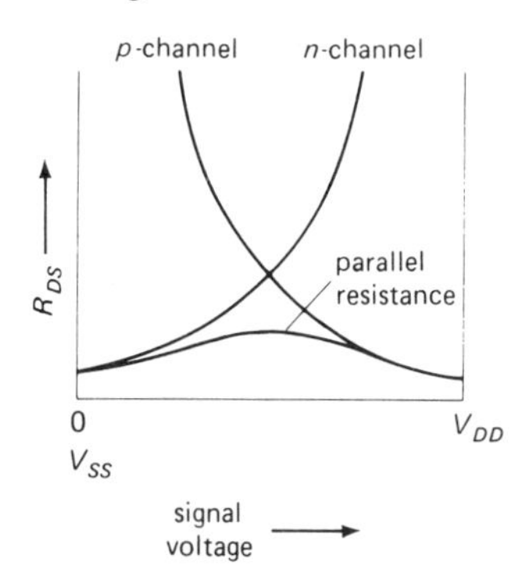

Figure 3.37

Multiplexers

A nice application of FET analog switches is the "multiplexer" (or MUX), a circuit that allows you to select any of several inputs, as specified by a digital control signal. The analog signal present on the selected input will be passed through to the (single) output. Figure 3.38 shows the basic scheme. Each of the switches SW0 through SW3 is a CMOS analog switch. The "select logic" decodes the address and *enables* (jargon for "turns on") the addressed switch only, disabling the remaining switches. Such a multiplexer is usually used in conjunction with digital circuitry that generates the appropriate addresses. A typical situation might involve a data-acquisition instrument in which a number of analog input voltages must be sampled in turn, converted to digital quantities, and used as input to some computation.

Because analog switches are bidirectional, an analog multiplexer such as this is

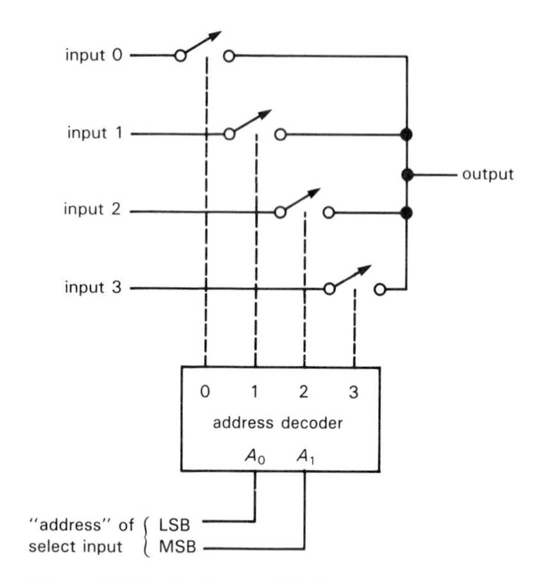

Figure 3.38. Analog multiplexer.

also a "demultiplexer": A signal can be fed into the "output" and will appear on the selected "input." When we discuss digital circuitry in Chapters 8 and 9, you will see that an analog multiplexer such as this can also be used as a "digital multiplexer/demultiplexer," because logic levels are, after all, nothing but voltages that happen to be interpreted as binary 1's and 0's.

Typical of analog multiplexers are the DG506–509 series and the IH6108 and 6116 types, 8- or 16-input MUX circuits that accept logic-level address inputs and operate with analog voltages up to ±15 volts. The 4051-4053 devices in the CMOS digital family are analog multiplexers/demultiplexers with up to 8 inputs, but with 15 volt pp maximum signal levels; they have a V_{EE} pin (and internal level shifting) so that you can use them with bipolarity analog signals and unipolarity (logic-level) control signals.

Other analog switch applications

Voltage-controlled analog switches form essential building blocks for op-amp circuits we'll see in the next chapter – integrators, sample-and-hold circuits, and peak detectors. For example, with op-amps we will be able to build a "true" integrator (unlike the approximation to an integrator we saw in Section 1.15): A constant input generates a linear ramp output (not an exponential), etc. With such an integrator you must have a method to reset the output; a FET switch across the integrating capacitor does the trick. We won't try to describe these applications here; because op-amps form essential parts of the circuits, they fit naturally into the next chapter. Great things to look forward to!

3.12 Limitations of FET switches

Speed

FET switches have ON resistances R_{ON} of 25 to 200 ohms. In combination with substrate and stray capacitances, this resistance forms a low-pass filter that limits operating speeds to frequencies of the order of 10MHz or less (Fig. 3.39). FETs with lower R_{ON} tend to have larger capacitance (up to 50pF with some MUX switches), so no gain in speed results. Much of the rolloff is due to protection components – current-limiting series resistance, and capacitance of shunt diodes. There are a few "RF/video" analog switches that obtain higher speeds, probably by eliminating some protection. For example, the

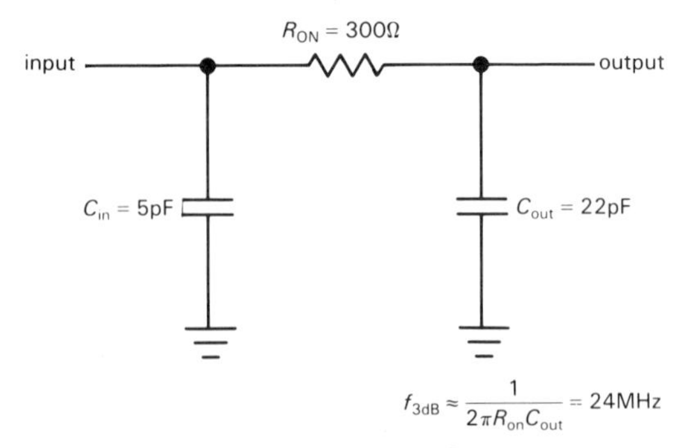

Figure 3.39

IH5341 and IH5352 switches handle analog signals over the usual ±15 volt range and have a bandwidth of 100MHz; the 74HC4051-53 series of "high-speed" CMOS multiplexers also provide a 3dB analog bandwidth of 100MHz, but handle signals only to ±5 volts. The MAX453–5 from Maxim combine a video multiplexer with an output video amplifier, so you can drive low-impedance cables or loads (usually 75Ω) directly; they have 50MHz typical bandwidth and are intended for ±1 volt low-impedance video signals.

ON resistance

CMOS switches operated from a relatively high supply voltage (15V, say) will have low R_{ON} over the entire signal swing, because one or the other of the transmission FETs will have a forward gate bias at least half the supply voltage. However, when operated with lower supply voltages, the switch's R_{ON} value will rise, the maximum occurring when the signal is about halfway between the supply and ground (or halfway between the supplies, for dual-supply voltages). Figure 3.40 shows why. As V_{DD} is reduced, the FETs begin to have significantly higher ON resistance (especially near $V_{GS} = V_{DD}/2$), since for enhancement-mode FETs V_T is at least a few volts, and a gate-source voltage of as much as 5 to 10 volts is required to achieve low R_{ON}. Not only will the parallel resistances of the two FETs rise for signal voltages between the supply voltage and ground, but also the peak resistance (at half V_{DD}) will rise as V_{DD} is reduced, and for sufficiently low V_{DD} the switch will become an open circuit for signals near $V_{DD}/2$.

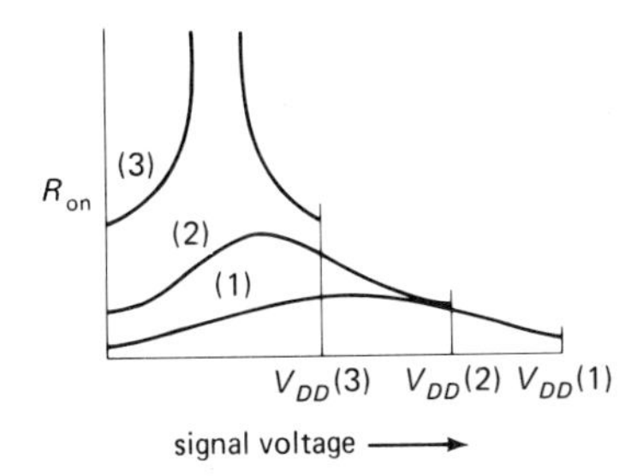

Figure 3.40

There are various tricks used by the designers of analog switch ICs to keep R_{ON} low, and approximately constant (for low distortion), over the signal swing. For example, the original 4016 analog switch used the simple circuit of Figure 3.36, producing R_{ON} curves that look like those in Figure 3.41. In the improved 4066 switch the designers added a few extra FETs so that the n-channel body voltage follows the signal voltage, producing the R_{ON} curves of Figure 3.42. The "volcano" shape, with

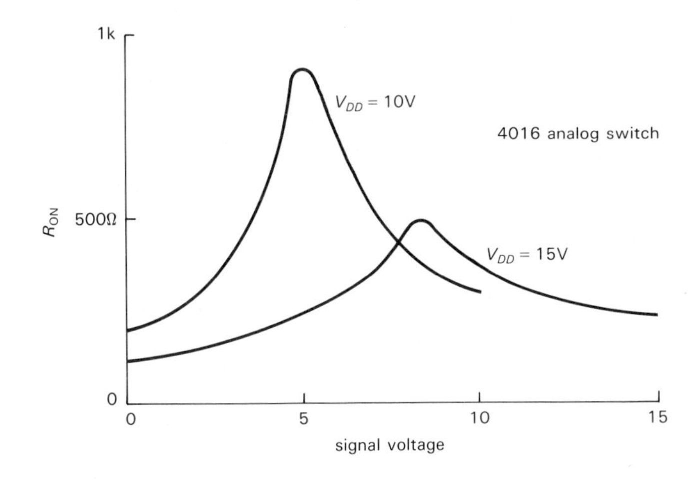

Figure 3.41. ON resistance for 4016 CMOS switch.

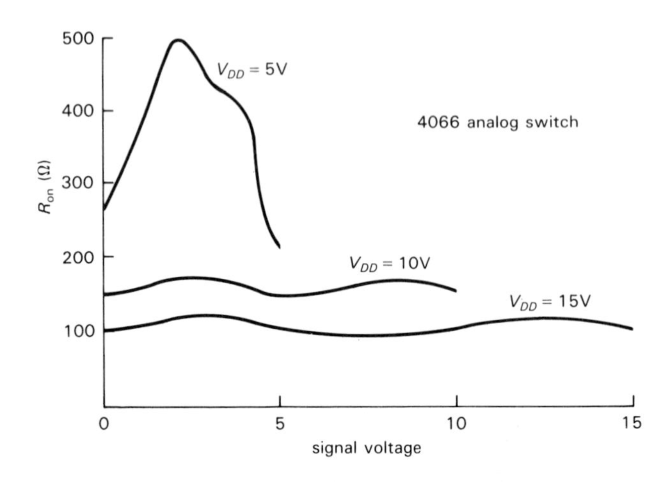

Figure 3.42. ON resistance for the improved 4066 CMOS switch; note change of scale from previous figure.

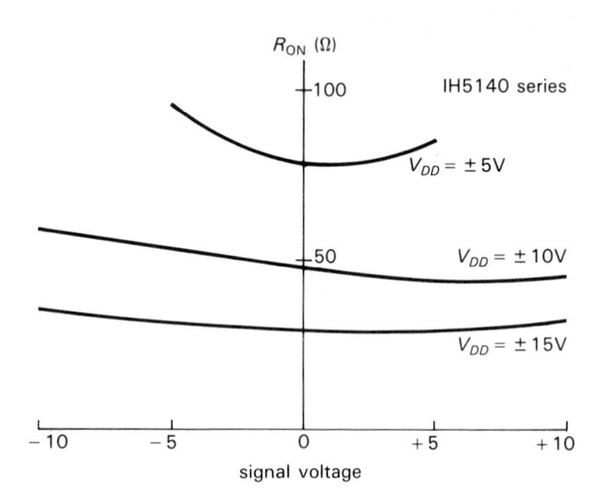

Figure 3.43. ON resistance for the IH5140-series bipolarity analog switches; note vertical scale.

its depressed central R_{ON}, replaces the "Everest" shape of the 4016. Sophisticated switches like the IH5140 series (or AD7510 series), intended for serious analog applications, succeed even better, with gentle R_{ON} curves like those shown in Figure 3.43. The recent DG400 series from Siliconix achieves an excellent R_{ON} of 20 ohms, at the expense of increased "charge transfer" (see the later section on *glitches*); this switch family (like the IH5140 series) has the additional advantage of zero quiescent current.

Capacitance

FET switches exhibit capacitance from input to output (C_{DS}), from channel to ground (C_D, C_S), from gate to channel, and from one FET to another within one IC package (C_{DD}, C_{SS}); see Figure 3.44. Let's look at the effects:

C_{DS}: Capacitance from input to output. Capacitance from input to output causes signal coupling in an OFF switch, rising at high frequencies. Figure 3.45 shows the effect for the IH5140 series. Note the use

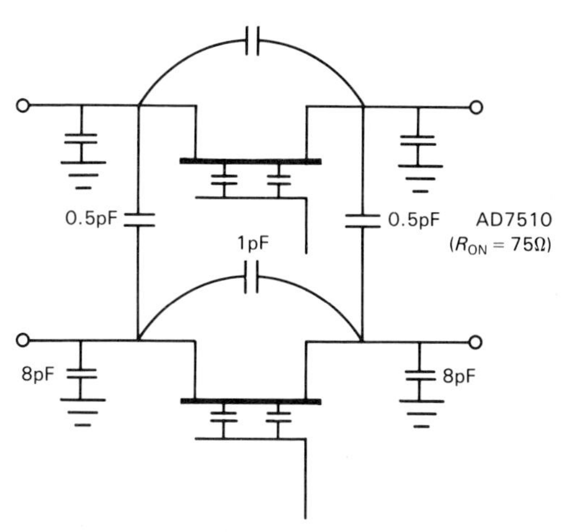

Figure 3.44. Analog switch capacitances – AD7510 4-channel switch.

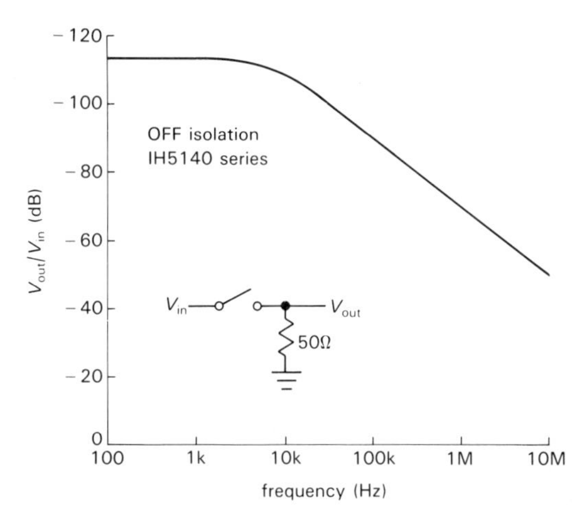

Figure 3.45

of a stiff 50 ohm load, common in radiofrequency circuits, but much lower than normal for low-frequency signals, where a typical load impedance is 10k or more. Even with a 50 ohm load, the feedthrough becomes significant at high frequencies (at 30MHz 1pF has a reactance of 5k, giving −40dB of feedthrough). And, of course, there is significant attenuation (and nonlinearity) driving a 50 ohm load, since R_{ON} is typically 30 ohms (75Ω worst-case). With a 10k load the feedthrough situation is much worse, of course.

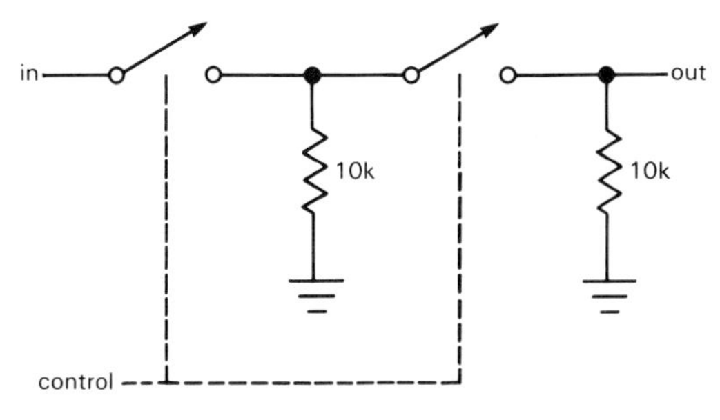

Figure 3.46

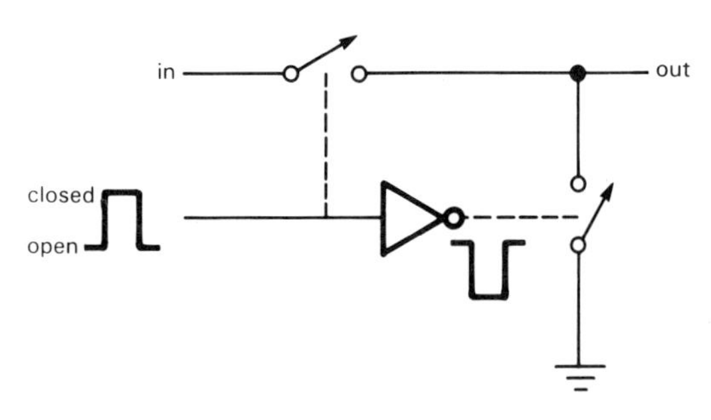

Figure 3.47

EXERCISE 3.8

Calculate the feedthrough into 10k at 1MHz, assuming $C_{DS} = 1$pF.

In most low-frequency applications capacitive feedthrough is not a problem. If it is, the best solution is to use a pair of cascaded switches (Fig. 3.46) or, better still, a combination of series and shunt switches, enabled alternately (Fig. 3.47). The series cascade doubles the attenuation (in decibels), at the expense of additional R_{ON}, whereas the series-shunt circuit (effectively an SPDT configuration) reduces feedthrough by dropping the effective load resistance to R_{ON} when the series switch is off.

EXERCISE 3.9

Recalculate switch feedthrough into 10k at 1MHz, assuming $C_{DS} = 1$pF and $R_{ON} = 50$ ohms, for the configuration of Fig. 3.47.

CMOS SPDT switches with controlled break-before-make are available commercially in single packages; in fact, you can get a pair of SPDT switches in a single package. Examples are the DG188 and IH5142, as well as the DG191, IH5143, and AD7512 (dual SPDT units). Because of the availability of such convenient CMOS switches, it is easy to use this SPDT configuration to achieve excellent performance. The RF/video switches mentioned earlier use a series-shunt circuit internally.

C_D, C_S: Capacitance to ground. Shunt capacitance to ground leads to the high frequency rolloff mentioned earlier. The situation is worst with a high-impedance signal source, but even with a stiff source the switch's R_{ON} combines with the shunt capacitance at the output to make a low-pass filter. The following problem shows how it goes.

EXERCISE 3.10

An AD7510 (here chosen for its complete capacitance specifications, shown in Fig. 3.44) is driven by a signal source of 10k, with a load impedance of 100k at the switch's output. Where is the high-frequency −3dB point? Now repeat the calculation, assuming a perfectly stiff signal source, and a switch R_{ON} of 75 ohms.

Capacitance from gate to channel. Capacitance from the controlling gate to the channel causes a different effect, namely the coupling of nasty little transients into your signal when the switch is turned on or off. This subject is worth some serious discussion, so we'll defer it to the next section on glitches.

C_{DD}, C_{SS}: Capacitance between switches. If you package several switches on a single piece of silicon the size of a kernel of corn, it shouldn't surprise you if there is some coupling between channels ("cross-talk"). The culprit, of course, is cross-channel capacitance. The effect increases with frequency and with signal impedance in the channel to which the signal is coupled. Here's a chance to work it out for yourself:

EXERCISE 3.11

Calculate the coupling, in decibels, between a pair of channels with $C_{DD} = C_{SS} = 0.5$pF (Fig. 3.44) for the source and load impedances of the last exercise. Assume that the interfering signal is 1MHz. In each case calculate the coupling for (a) OFF switch to OFF switch, (b) OFF switch to ON switch, (c) ON switch to OFF switch, and (d) ON switch to ON switch.

It should be obvious from this example why most broadband radiofrequency circuits use low signal impedances, usually 50 ohms. If cross-talk is a serious problem, don't put more than one signal on one chip.

Glitches

During turn-on and turn-off transients, FET analog switches can do nasty things. The control signal being applied to the gate(s) can couple capacitively to the channel(s), putting ugly transients on your signal. The situation is most serious if the signal is at high impedance levels. Multiplexers can show similar behavior during transitions of the input address, as well as momentary connection between inputs if turn-off delay exceeds turn-on delay. A related bad habit is the propensity of some switches (e.g., the 4066) to short the input to ground momentarily during changes of state.

Let's look at this problem in a bit more detail. Figure 3.48 shows a typical waveform you might see at the output of an n-channel MOSFET analog switch circuit similar to Figure 3.35, with an input

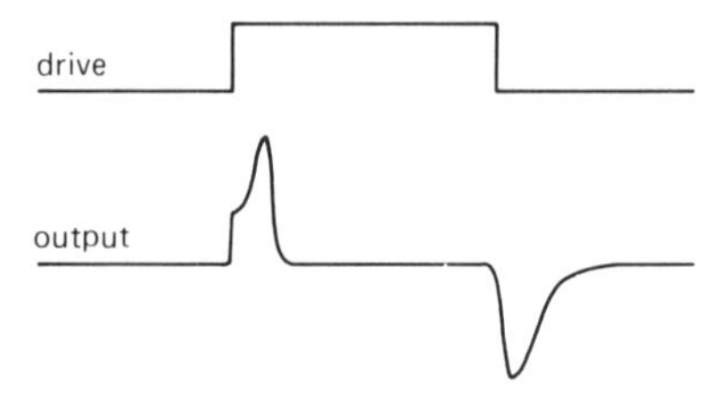

Figure 3.48

signal level of zero volts and an output load consisting of 10k in parallel with 20pF, realistic values for an analog switch circuit. The handsome transients are caused by charge transferred to the channel, through the gate-channel capacitance, at the transitions of the gate. The gate makes a sudden step from one supply voltage to the other, in this case between ±15 volt supplies, transferring a slug of charge

$$Q = C_{GC}[V_G(\text{finish}) - V_G(\text{start})]$$

C_{GC} is the gate-channel capacitance, typically around 5pF. Note that the amount of charge transferred to the channel depends only on the total voltage change at the gate, not on its rise time. Slowing down the gate signal gives rise to a smaller-amplitude glitch of longer duration, with the same total area under its graph. Lowpass filtering of the switch's output signal has the same effect. Such measures may help if the peak amplitude of the glitch must be kept small, but in general they are ineffective in eliminating gate feedthrough. In some cases the gate-channel capacitance may be predictable enough for you to cancel the spikes by coupling an inverted version of the gate signal through a small adjustable capacitor.

The gate-channel capacitance is distributed over the length of the channel, which means that some of the charge is coupled back to the switch's input. As a result, the size of the output glitch depends on the signal source impedance and is smallest when the switch is driven by a voltage source. Of course, reducing the size of the load impedance will reduce the size of the glitch, but this also loads the source and introduces error and nonlinearity due to finite R_{ON}. Finally, all other things being equal, a switch with smaller gate-channel capacitance will introduce smaller switching transients, although you pay a price in the form of increased R_{ON}.

Figure 3.49 shows an interesting comparison of gate-induced charge transfers for three kinds of analog switches, including JFETs. In all cases the gate signal is making a full swing, i.e., either 30 volts or the indicated supply voltage for MOSFETs, and a swing from −15 volts to the signal level for the n-channel JFET switch. The JFET switch shows a strong

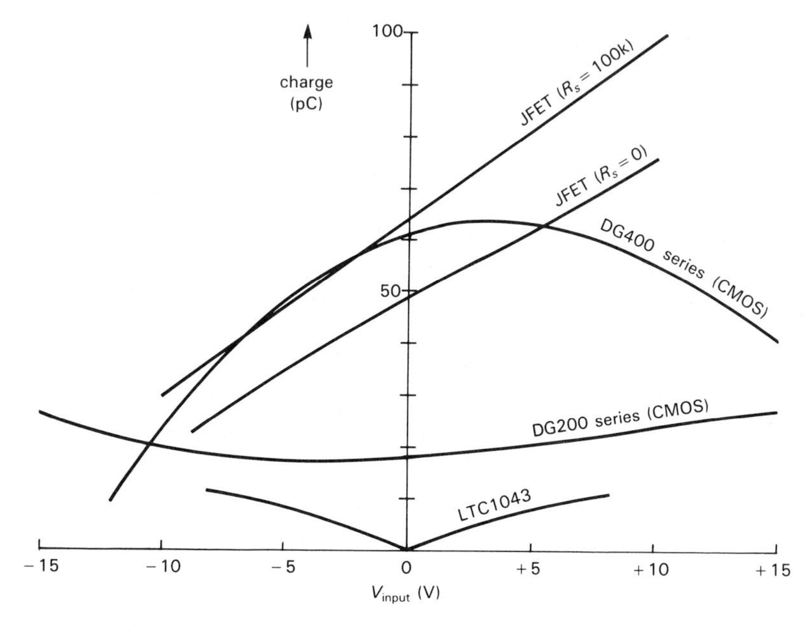

Figure 3.49. Charge transfer for various FET linear switches as a function of signal voltage.

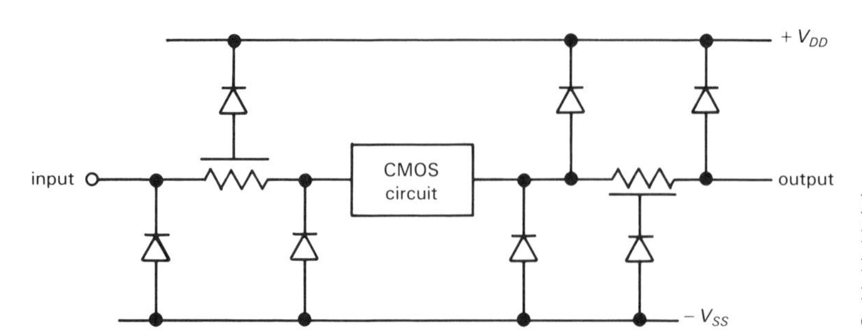

Figure 3.50. CMOS input/output protection networks. The series resistor at the output is often omitted.

dependence of glitch size on signal, because the gate swing is proportional to the level of the signal above −15 volts. Well-balanced CMOS switches have relatively low feedthrough because the charge contributions of the complementary MOSFETs tend to cancel out (one gate is rising while the other is falling). Just to give scale to these figures, it should be pointed out that 30pC corresponds to a 3mV step across a 0.01μF capacitor. That's a rather large filter capacitor, and you can see that this is a real problem, since a 3mV glitch is pretty large when dealing with low-level analog signals.

Latchup and input current

All CMOS integrated circuits have some form of input protection circuit, because otherwise the gate insulation is easily destroyed (see the later section on handling precautions). The usual protection network is shown in Figure 3.50: Although it may use distributed diodes, the network is equivalent to clamping diodes to V_{SS} and to V_{DD}, combined with resistive current limiting. If you drive the inputs (or outputs) more than a diode drop beyond the supply voltages, the diode clamps go into conduction, making the inputs (or outputs) look like a low impedance to the respective supplies. Worse still, the chip can be driven into "SCR latchup," a terrifying (and destructive) condition we'll describe in more detail in Section 14.16. For now, all you need to know about it is that you don't want it! SCR latchup is triggered by input currents (through the protection network) of roughly 20mA or more. Thus, you must be careful not to drive the analog inputs beyond the rails. This means, for instance, that you must be sure the power supply voltages are applied before any signals that have significant drive current capability. Incidentally, this prohibition goes for *digital* CMOS ICs as well as the analog switches we have been discussing.

The trouble with diode-resistor protection networks is that they compromise switch performance, by increasing $R_{\rm ON}$, shunt capacitance, and leakage. With clever chip design (making use of "dielectric isolation") it is possible to eliminate SCR latchup without the serious performance compromises inherent in traditional protection networks. Many of the newer analog switch designs are "fault protected"; for example, Intersil's IH5108 and IH5116 analog multiplexers claim you can drive the analog inputs to ±25 volts, even with the supply at zero (you pay for this robustness with an $R_{\rm ON}$ that is four times higher than that of the conventional IH6108/16). Watch out, though, because there are plenty of analog switch ICs around that are not forgiving!

You can get analog switches and multiplexers built with n-channel JFETs rather than complementary MOSFETs. They perform quite well, improving on CMOS switches in several characteristics. In particular, the series of JFET switches from PMI has superior constancy of $R_{\rm ON}$ versus analog voltage, complete absence of latchup, and low susceptibility to electrostatic damage.

Other switch limitations

Some additional characteristics of analog switches that may or may not be important in any given application are switching time, settling time, break-before-make delay, channel leakage current (both ON and OFF; see Section 4.15), R_{ON} matching, temperature coefficient of R_{ON}, and signal and power supply ranges. We'll show unusual restraint by ending the discussion at this point, leaving the reader to look into these details if the circuit application demands it.

3.13 Some FET analog switch examples

As we indicated earlier, many of the natural applications of FET analog switches are in op-amp circuits, which we will treat in the next chapter. In this section we will show a few switch applications that do not require op-amps, to give a feeling for the sorts of circuits you can use them in.

Switchable RC low-pass filter

Figure 3.51 shows how you could make a simple *RC* low-pass filter with selectable 3dB points. We've used a multiplexer to select one of four preset resistors, via a 2-bit (digital) address. We chose to put the switch at the input, rather than after the resistors, because there is less charge injection at a point of lower signal impedance. Another possibility, of course, is to use FET switches to select the *capacitor*. To generate a very wide range of time constants you might have to do that, but the switch's finite R_{ON} would limit attenuation at high frequencies, to a maximum of $R_{\mathrm{ON}}/R_{\mathrm{series}}$. We've also indicated a unity-gain buffer, following the filter, since the output impedance is high. You'll see how to make "perfect" followers (precise gain, high Z_{in}, low Z_{out}, and no V_{BE} offsets, etc.) in the next chapter. Of course, if the amplifier that follows the filter has high input impedance, you don't need the buffer.

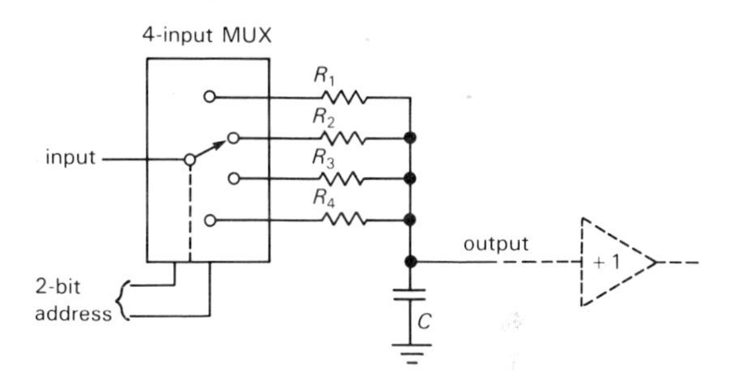

Figure 3.51

Figure 3.52 shows a simple variation in which we've used four independent switches, rather than a 4-input multiplexer. With the resistors scaled as shown, you can generate 16 equally spaced 3dB frequencies by turning on binary combinations of the switches.

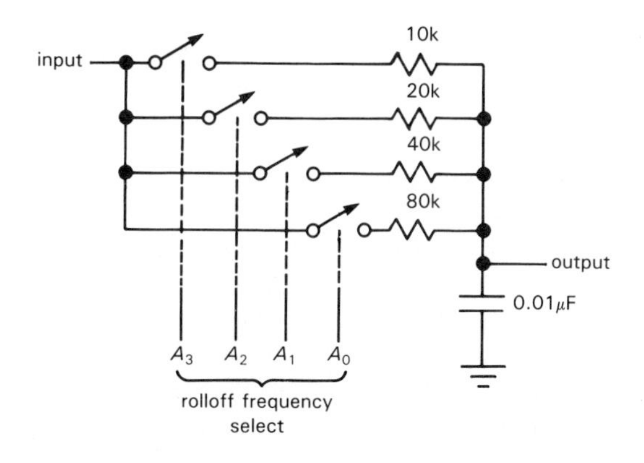

Figure 3.52. *RC* low-pass filter with choice of 15 equally spaced time constants.

EXERCISE 3.12
What are the 3dB points for this circuit?

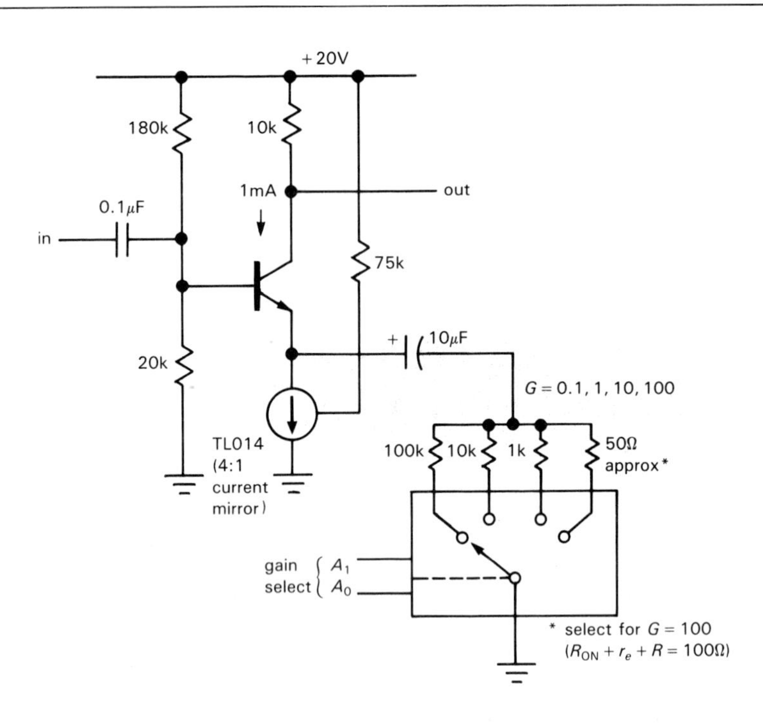

Figure 3.53. An analog multiplexer selects appropriate emitter degeneration resistors to achieve decade-switchable gain.

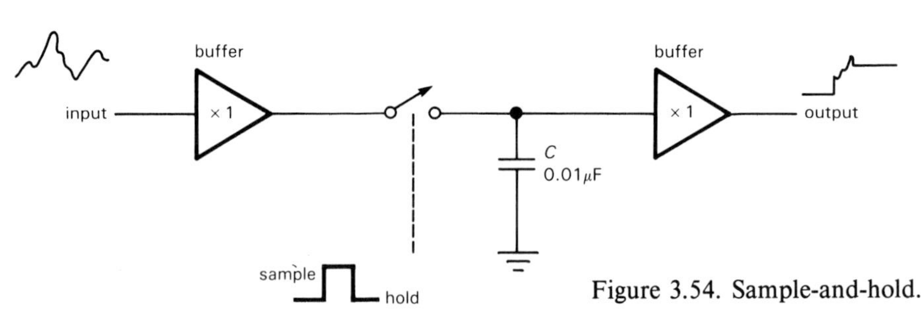

Figure 3.54. Sample-and-hold.

Switchable gain amplifier

Figure 3.53 shows how you can apply the same idea of switching resistors to produce an amplifier of selectable gain. Although this idea is a natural for op-amps, we can use it with the emitter-degenerated amplifier. We used a constant-current sink as emitter load, as in an earlier example, to permit gains much less than unity. We then used the multiplexer to select one of four emitter resistors. Note the blocking capacitor, needed to keep the quiescent current independent of gain.

Sample-and-hold

Figure 3.54 shows how to make a "sample-and-hold" circuit, which comes in handy when you want to convert an analog signal to a stream of digital quantities ("analog-to-digital conversion") – you've got to hold each analog level steady while you figure out how big it is. The circuit is simple: A unity-gain input buffer generates a low-impedance copy of the input signal, forcing it across a small capacitor. To hold the analog level at any moment, you simply open the switch. The high input impedance of the second buffer (which should have FET input transistors, to keep input current near zero) prevents loading of the capacitor, so it holds its voltage until the FET switch is again closed.

EXERCISE 3.13

The input buffer must supply current to keep the capacitor following a varying signal. Calculate the buffer's peak output current when the circuit is driven by an input sine wave of 1 volt amplitude at 10kHz.

Flying-capacitor voltage converter

Here's a nice way (Fig. 3.55) to generate a needed negative power-supply voltage in a circuit that is powered by a single positive supply. The pair of FET switches on the left connects C_1 across the positive supply, charging it to V_{in}, while the switches on the right are kept open. Then the input switches are opened, and the switches on the right are closed, connecting charged C_1 across the output, transferring some of its charge onto C_2. The switches are diabolically arranged so that C_1 gets turned upside down, generating a *negative* output! This particular circuit is available as the 7662 voltage converter chip, which we'll talk about in Sections 6.22 and 14.07. The device labeled "inverter" turns a HIGH voltage into a LOW voltage, and vice versa. We'll show you how to make one in the next section (and we'll really get you up to speed on them in Chapters 8–11!).

3.14 MOSFET logic and power switches

The *other* kinds of FET switch applications are *logic* and *power switching* circuits. The distinction is simple: In analog signal switching you use a FET as a series switch, passing or blocking a signal that has some range of analog voltage. The analog signal is usually a low-level signal, at insignificant power levels. In logic switching, on the other hand, MOSFET switches open and close to generate full swings between the power supply voltages. The "signals" here are really digital, rather than analog—they swing between the power supply voltages, representing the two states HIGH and LOW. In-between voltages are not useful or desirable; in fact, they're not even legal! Finally, "power switching" refers to turning on or off the power to a load such as a lamp, relay coil, or motor winding; in these applications, both voltages and currents tend to be large. We'll take logic switching first.

Logic switching

Figure 3.56 shows the simplest kind of logic switching with MOSFETs: Both circuits use a resistor as load and perform the logical function of *inversion* – a HIGH

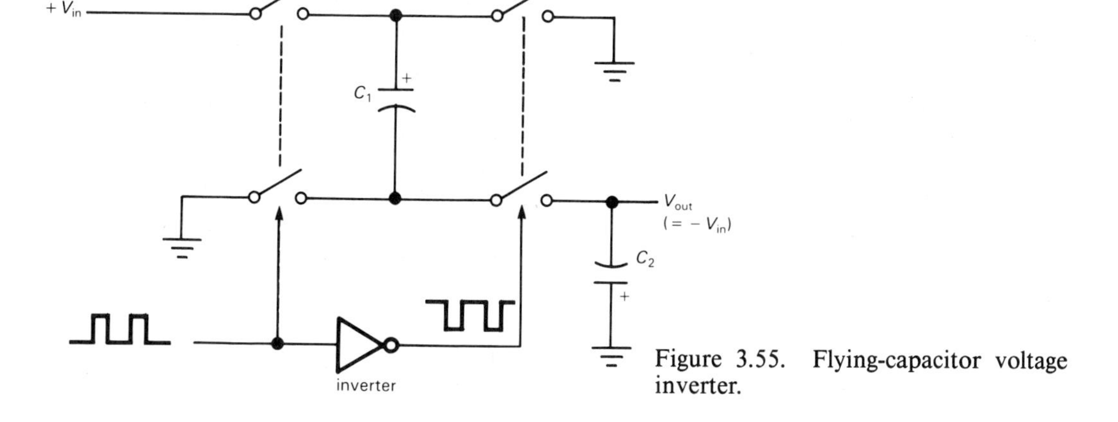

Figure 3.55. Flying-capacitor voltage inverter.

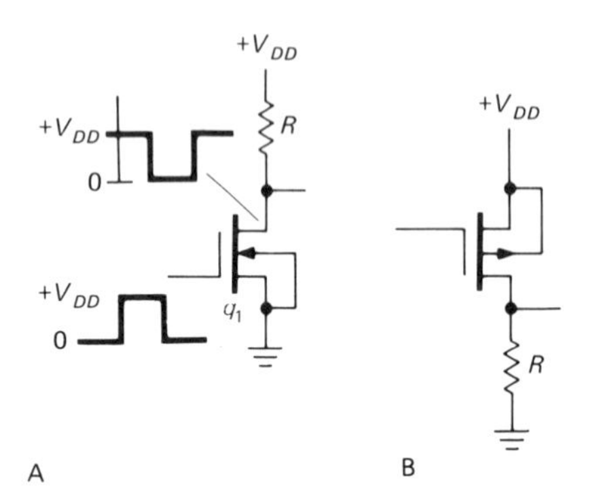

Figure 3.56. NMOS and PMOS logic inverters.

input generates a LOW output, and vice versa. The n-channel version pulls the output to ground when the gate goes HIGH, whereas the p-channel version pulls the resistor HIGH for grounded (LOW) input. Note that the MOSFETs in these circuits are used as common-source inverters, rather than as source followers. In digital logic circuits like these we are usually interested in the output voltage ("logic level") produced by a certain input voltage; the resistor serves merely as a passive drain load, to make the output swing to the drain supply when the FET is off. If, on the other hand, we replace the resistor by a light bulb, relay, printhead hammer, or some other hefty load, we've got a power-switching application (Fig. 3.3). Although we're using the same "inverter" circuit, in the power switching application we're interested instead in turning the load on and off.

CMOS inverter

The NMOS and PMOS inverters of the preceding circuits have the disadvantage of drawing current in the ON state and having relatively high output impedance in the OFF state. You can reduce the output impedance (by reducing R), but only at the expense of increased dissipation, and vice versa. Except for current sources, of course, it's never a good idea to have high output impedance. Even if the intended load is high impedance (another MOSFET gate, for example), you are inviting capacitive noise pickup problems, and you will suffer reduced switching speeds for the ON-to-OFF ("trailing") edge (because of stray loading capacitance). In this case, for example, the NMOS inverter with a compromise value of drain resistor, say 10k, would produce the waveform shown in Figure 3.57.

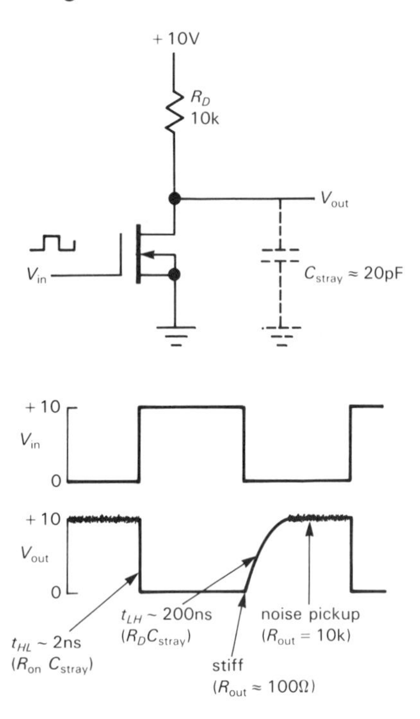

Figure 3.57

The situation is reminiscent of the single-ended emitter follower in Section 2.15, in which quiescent power dissipation and power delivered to the load were involved in a similar compromise. The solution there – the push-pull configuration – is particularly well suited to MOSFET switching. Look at Figure 3.58, which you might think of as a push-pull switch: Input grounded cuts off the bottom transistor and turns on the top transistor, pulling the output HIGH. A HIGH input ($+V_{DD}$) does the reverse, pulling the output to ground. It's an inverter with low output

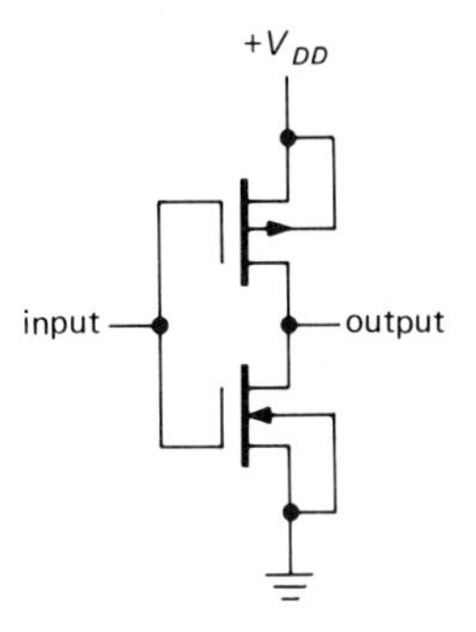

Figure 3.58. CMOS logic inverter.

impedance in *both* states, and no quiescent current whatsoever. It's called a CMOS (complementary MOS) inverter, and it is the basic structure of all digital CMOS logic, the logic family that has become dominant in large-scale integrated circuits (LSI), and seems destined to replace earlier logic families (with names like "TTL") based on bipolar transistors. Note that the CMOS inverter is two complementary MOSFET switches *in series*, alternately enabled, while the CMOS analog switch (treated earlier in the chapter) is two complementary MOSFET switches *in parallel*, enabled simultaneously.

EXERCISE 3.14

The complementary MOS transistors in the CMOS inverter are both operating as common-source inverters, whereas the complementary bipolar transistors in the push-pull circuits of Section 2.15 are (non-inverting) emitter followers. Try drawing a "complementary BJT inverter," analogous to the CMOS inverter. Why won't it work?

We'll be seeing much more of digital CMOS in the chapters on digital logic and microprocessors (Chapters 8-11). For now, it should be evident that CMOS is a low power logic family (with *zero* quiescent power) with high-impedance inputs, and with stiff outputs that swing the full supply range. Before leaving the subject, however, we can't resist the temptation to show you one additional CMOS circuit (Fig. 3.59). This is a logic *NAND gate*, whose output goes LOW only if input A AND input B are both HIGH. The operation is surprisingly easy to understand: If A and B are both HIGH, series NMOS switches Q_1 and Q_2 are both ON, pulling the output stiffly to ground; PMOS switches Q_3 and Q_4 cooperate by being OFF; thus, no current flows. However, if either A or B (or both) is LOW, the corresponding PMOS transistor is ON, pulling the output HIGH; since one (or both) of the series chain Q_1Q_2 is OFF, no current flows.

This is called a "NAND" gate because it performs the logical AND function, but

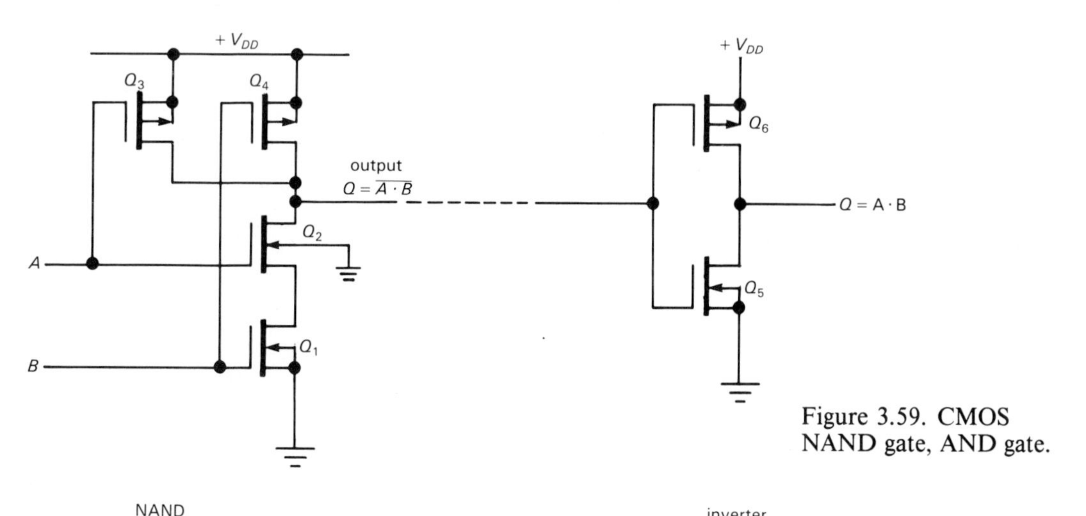

Figure 3.59. CMOS NAND gate, AND gate.

with inverted ("NOT") output – it's a NOT-AND, abbreviated NAND. Although gates and their variants are properly a subject for Chapter 8, you will enjoy trying your hand at the following problems.

EXERCISE 3.15
Draw a CMOS AND gate. Hint: AND = NOT-NAND.

EXERCISE 3.16
Now draw a NOR gate: The output is LOW if either A OR B (or both) is HIGH.

EXERCISE 3.17
You guessed it – draw a CMOS OR gate.

EXERCISE 3.18
Draw a 3-input CMOS NAND gate.

The CMOS digital logic we'll be seeing later is constructed from combinations of these basic gates. The combination of very low power dissipation and stiff rail-to-rail output swing makes CMOS logic the family of choice for most digital circuits, accounting for its popularity. Furthermore, for micropower circuits (such as wristwatches and small battery-powered instruments) it's the only game in town.

Lest we leave the wrong impression, however, it's worth noting that CMOS logic is not *zero*-power. There are two mechanisms of current drain: During transitions, a CMOS output must supply a transient current $I = CdV/dt$ to charge any capacitance it sees (Fig. 3.60). You get load capacitance both from wiring ("stray" capacitance) and from the input capacitance of additional logic that you are driving. In fact, because a complicated CMOS chip contains many internal gates, each driving some on-chip internal capacitance, there is some current drain in any CMOS circuit that is making transitions, even if the chip is not driving any external load. Not surprisingly, this "dynamic" current drain is proportional to the rate at which transitions take place. The second mechanism of CMOS current drain is shown in Figure 3.61: As the input jumps between the supply voltage and ground, there is a region where both MOSFETs are conducting, resulting in large current spikes from V_{DD} to ground. This is sometimes called "class-A current" or "power supply crowbarring." You will see some consequences of this in Chapters 8, 9, and 14. As long as we're dumping on CMOS, we should mention that an additional disadvantage of CMOS (and, in fact, of all MOSFETs) is its vulnerability to damage from static electricity. We'll have more to say about this in Section 3.15.

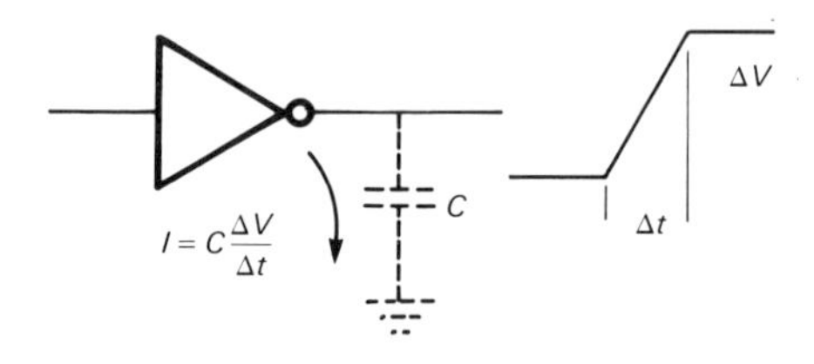

Figure 3.60. Capacitive charging current.

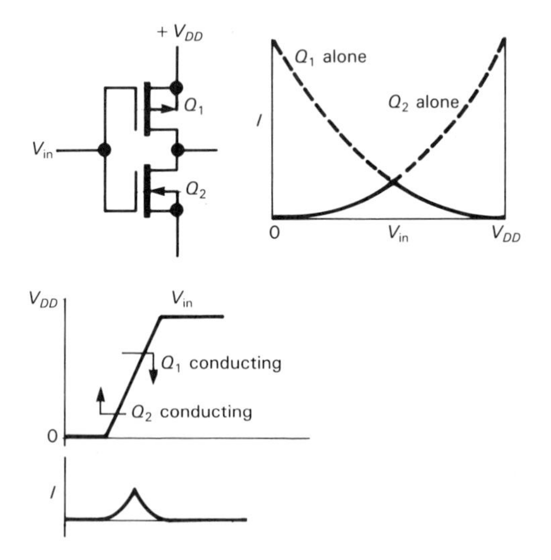

Figure 3.61. Class-A CMOS conduction.

□ CMOS linear amplifier

CMOS inverters – and indeed all CMOS digital logic circuits – are intended to be used with digital signal levels. Except during transitions between states, therefore, the inputs and outputs are close to ground or V_{DD} (usually +5V). And except during those transitions (with typical durations of a few nanoseconds), there is no quiescent current drain.

The CMOS inverter turns out to have some interesting properties when used with *analog* signals. Look again at Figure 3.61. You can think of Q_1 as an active (current-source) load for inverting amplifier Q_2, and vice versa. When the input is near V_{DD} or ground, the currents are grossly mismatched, and the amplifier is in saturation (or "clipping") at ground or V_{DD}, respectively. This is, of course, the normal situation with digital signals. However, when the input is near half the supply voltage, there is a small region where the drain currents of Q_1 and Q_2 are nearly equal; in this region the circuit is an inverting linear amplifier with high gain. Its transfer characteristic is shown in Figure 3.62. The variation of R_{load} and g_m with drain current is such that the highest voltage gain occurs for relatively low drain currents, i.e., at low supply voltages (say 5V).

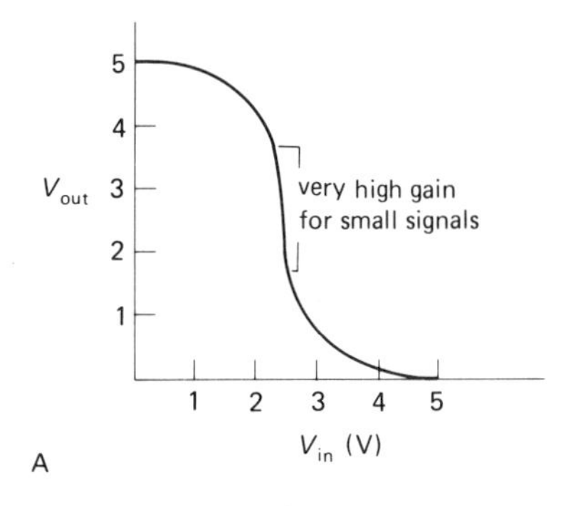

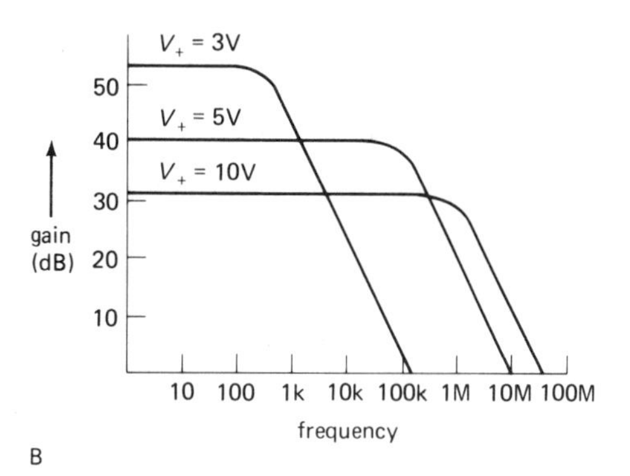

Figure 3.62

This circuit is not a good amplifier; it has the disadvantage of very high output impedance (particularly when operated at low voltage), poor linearity, and unpredictable gain. However, it is simple and inexpensive (CMOS inverters are available 6 to a package for under half a dollar), and it is sometimes used to amplify small input signals whose waveforms aren't important. Some examples are proximity switches (which amplify 60Hz capacitive pickup), crystal oscillators, and frequency-sensing input devices whose output is a frequency that goes to a frequency counter (see Chapter 15).

To use a CMOS inverter as a linear amplifier, it's necessary to bias the input so that the amplifier is in its active region. The usual method is with a large-value resistor from output to input (which we'll recognize as "dc feedback" in the next chapter), as shown in Figure 3.63. That puts us at the point $V_{\text{out}} = V_{\text{in}}$ in Figure 3.62. As we'll learn later, such a connection (circuit A) also acts to lower the input impedance, through "shunt feedback," making circuit B desirable if a high input impedance at signal frequencies is important. The third circuit is the classic CMOS crystal oscillator, discussed in Section 5.13. Figure 3.64 shows a variant of circuit A, used to generate a clean 10MHz full-swing square wave (to drive digital logic) from an input sine wave. The circuit works well for input amplitudes from 50mV rms to 5 volts rms. This is a good example of an "I don't know the gain, and I don't care" application. Note the input-protection network, consisting of a current-limiting series resistor and clamping diodes.

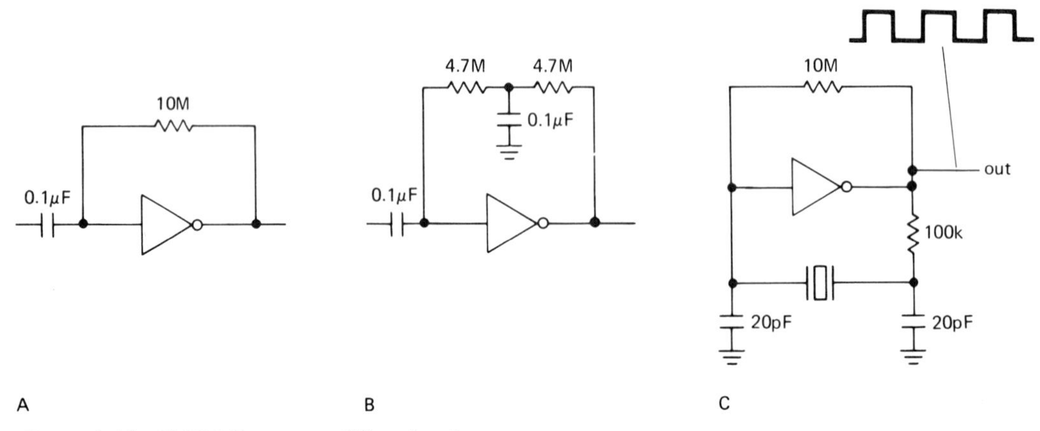

Figure 3.63. CMOS linear amplifier circuits.

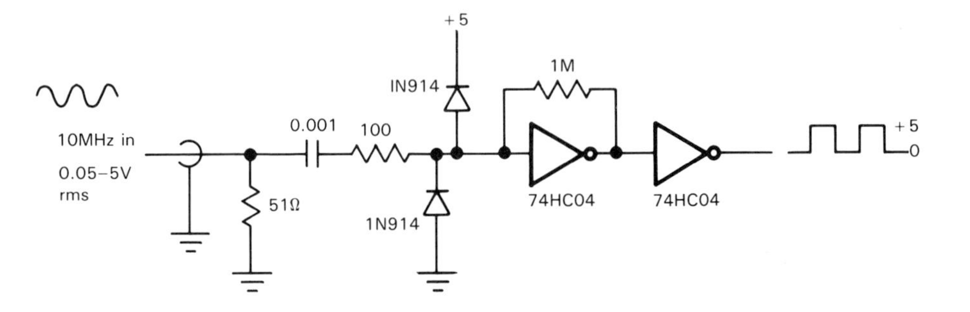

Figure 3.64

Power switching

MOSFETs work well as saturated switches, as we suggested with our simple circuit in Section 3.01. Power MOSFETs are now available from many manufacturers, making the advantages of MOSFETs (high input impedance, easy paralleling, absence of "second breakdown") applicable to power circuits. Generally speaking, power MOSFETs are easier to use than conventional bipolar power transistors. However, there are some subtle effects to consider, and cavalier substitution of MOSFETs in switching applications can lead to prompt disaster. We've visited the scenes of such disasters and hope to avert their repetition. Read on for our handy guided tour.

Power MOSFETs. FETs were feeble low-current devices, barely able to run more than a few tens of milliamps, until the late 1970s, when the Japanese introduced "vertical-groove" MOS transistors. Power MOSFETs are now manufactured by all the manufacturers of discrete semiconductors (e.g, GE, IR, Motorola, RCA, Siliconix, Supertex, TI, along with European companies like Amperex, Ferranti, Siemens, and SGS, and many of the Japanese companies), with names like VMOS, TMOS, vertical DMOS, and HEXFET. They can handle surprisingly high voltages (up to 1000V), and peak currents to 280 amps (continuous currents to 70A), with R_{ON} as low as 0.02 ohm. Small power MOSFETs sell for much less than a dollar, and they're available in all the usual transistor packages, as well as multiple transistors packaged in the convenient DIP (dual in-line package) that most integrated

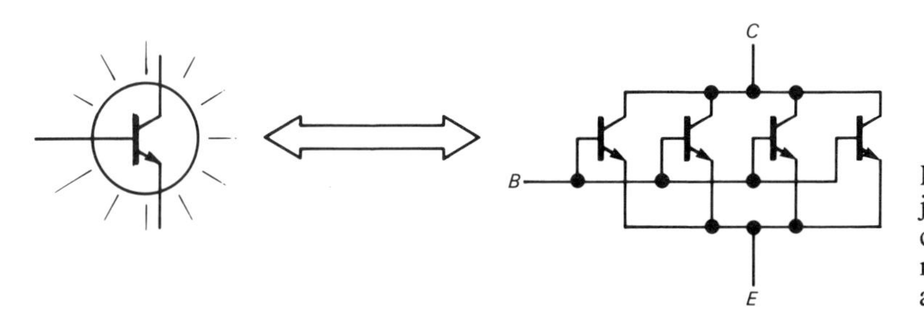

Figure 3.65. A large-junction-area transistor can be thought of as many paralleled small-area transistors.

circuits come in. Ironically, it is now discrete *low-level* MOSFETs that are hard to find, there being no shortage of power MOSFETs. See Table 3.5 for a listing of representative power MOSFETs.

High impedance, thermal stability. The two important advantages of the power MOSFET, compared with the bipolar power transistor, are its high input impedance (but watch out for high input capacitance, particularly with high-current devices; see below) and its complete absence of thermal runaway and second breakdown. This latter effect is very important in power circuits and is worth understanding: The large junction area of a power transistor (whether BJT or FET) can be thought of as a large number of small junctions in parallel (Fig. 3.65), all with the same applied voltages. In the case of a bipolar power transistor, the positive temperature coefficient of collector current at fixed V_{BE} (approximately +9%/°C, see Section 2.10) means that a local hot spot in the junction will have a higher current density, thus producing additional heating. At sufficiently high V_{CE} and I_C, this "current hogging" can cause local thermal runaway, known as *second breakdown*. As a result, bipolar power transistors are limited to a "safe operating area" (on a plot of collector current versus collector voltage) smaller than that allowed by transistor power dissipation alone (we'll see more of this in Chapter 6). The important point here is that the *negative* temperature coefficient of MOS drain current (Fig. 3.13) prevents these junction hot spots entirely. MOSFETs have no second breakdown, and their safe operating area (SOA) is limited only by power dissipation (see Fig. 3.66, where we've compared the SOAs of an *npn* and an NMOS power transistor of the same I_{max}, V_{max}, and P_{diss}). For the same reason, MOSFET power amplifiers don't have the nasty runaway tendencies that we've all grown to love in bipolar transistors (see Section 2.15), and as an added bonus, power MOSFETs can be paralleled without the current-equalizing "emitter-ballasting" resistors that are necessary with bipolar transistors (see Section 6.07).

Power switching examples and cautions. You often want to control a power MOSFET from the output of digital logic. Although there are logic families that generate swings of 10 volts or more ("4000-series CMOS"), the most common logic families use levels of +5 volts ("high-speed CMOS") or +2.4 volts ("TTL"). Figure 3.67 shows how to switch loads from these three logic families. In the first circuit, the +10 volt gate drive will fully turn on any MOSFET, so we chose the VN0106, an inexpensive transistor that specifies $R_{\text{ON}} < 5$ ohms at $V_{GS} = 5$ volts. The diode protects against inductive spike (Section 1.31); the series gate resistor, though not essential, is a good idea, because MOSFET drain-gate capacitance can couple the load's inductive transients back to the delicate CMOS logic (more on this soon). In the second circuit we have 5 volts of gate drive, still fine for the VN01/VP01 series; for variety we've

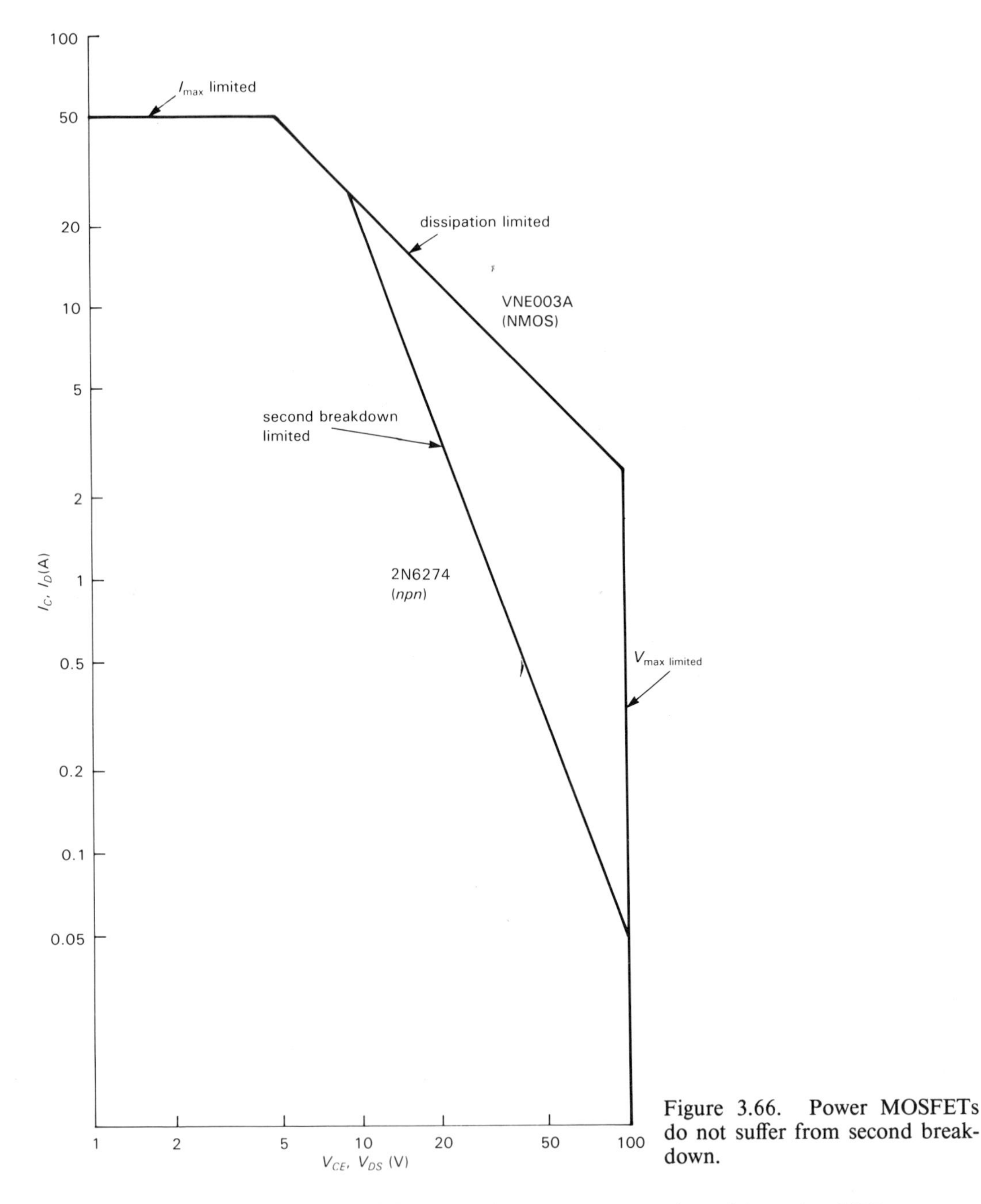

Figure 3.66. Power MOSFETs do not suffer from second breakdown.

used a p-channel MOSFET, driving a load returned to ground.

The last two circuits show two ways to handle the +2.4 volt (worst-case; it's usually around +3.5V) HIGH output from TTL digital logic: We can use a pullup resistor to +5 volts to generate a full +5 volt swing from the TTL output, which then drives a normal MOSFET; alternatively, we can use something like the TN0106, a "low-threshold" MOSFET designed for logic-level drive. Watch out, though, for misleading specifications. For example, the TN01 specifies "$V_{GS(\text{th})}$ =

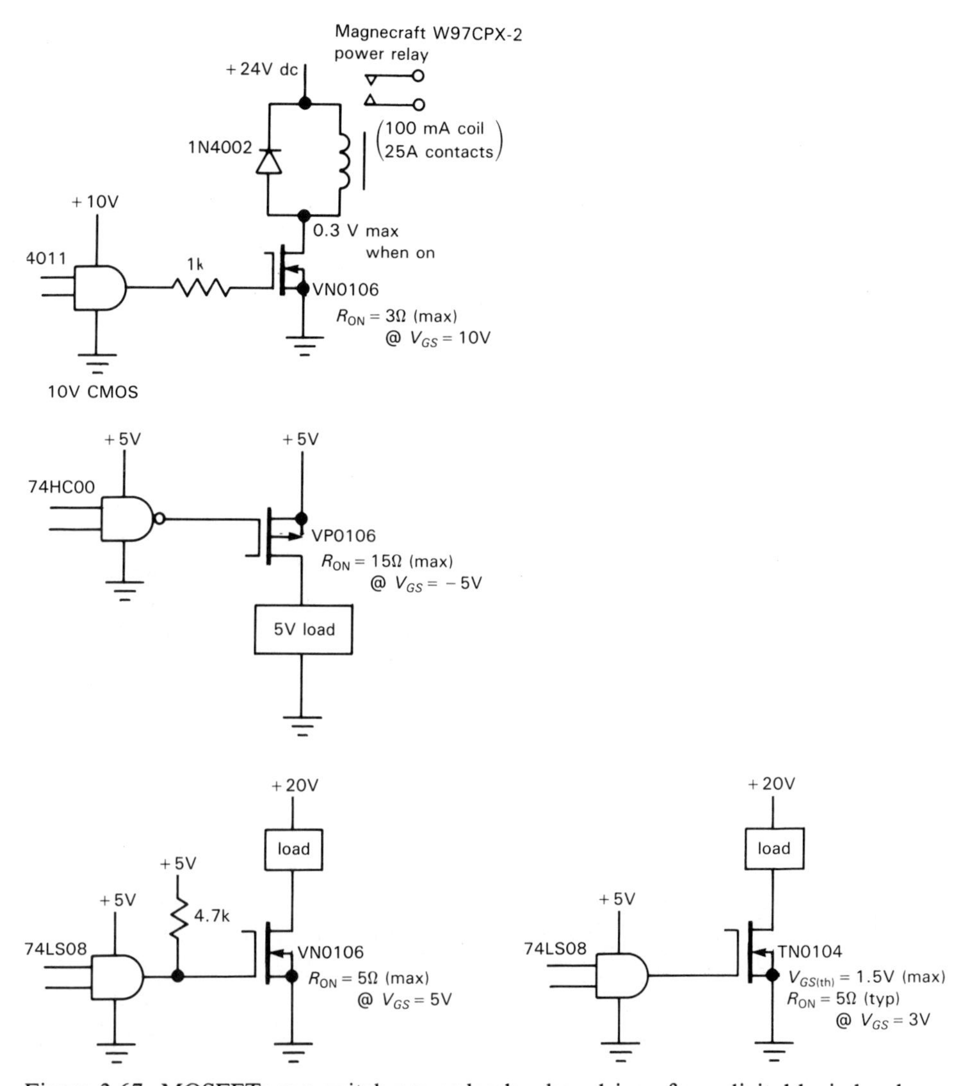

Figure 3.67. MOSFETs can switch power loads when driven from digital logic levels.

1.5 volts (max)," which sounds fine until you read the fine print ("at $I_D = 1\text{mA}$"). It takes considerably more gate voltage than $V_{GS(\text{th})}$ to turn the MOSFET on fully (Fig. 3.68). However, the circuit will probably work OK, because (a) a HIGH TTL output is rarely less than +3 volts, and typically more like +3.5 volts, and (b) the TN01 further specifies "$R_{ON}(\text{typ}) = 5\Omega$ at $V_{GS} = 3\text{V}$."

This example illustrates a frequent designer's quandary, namely a choice between a complicated circuit that meets the strict worst-case design criterion, and is therefore *guaranteed* to work, and a simple circuit that doesn't meet worst-case specifications, but is overwhelmingly likely to function without problems. There are times when you will find yourself choosing the latter, ignoring the little voice whispering into your ear.

□ *Capacitance.* In the preceding examples we put a resistor in series with the gate when there was an inductive load. As we mentioned earlier in the chapter (Section

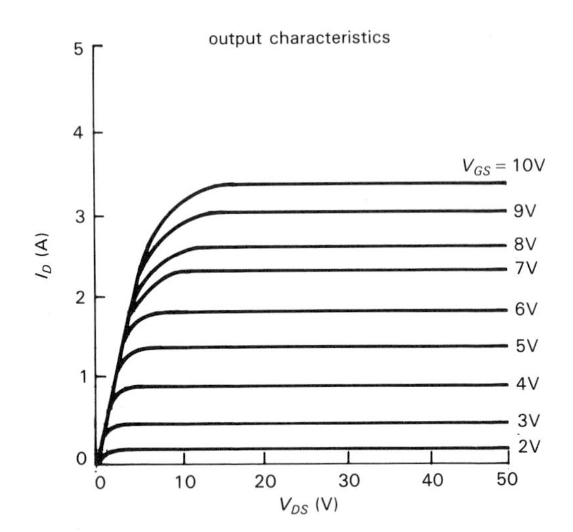

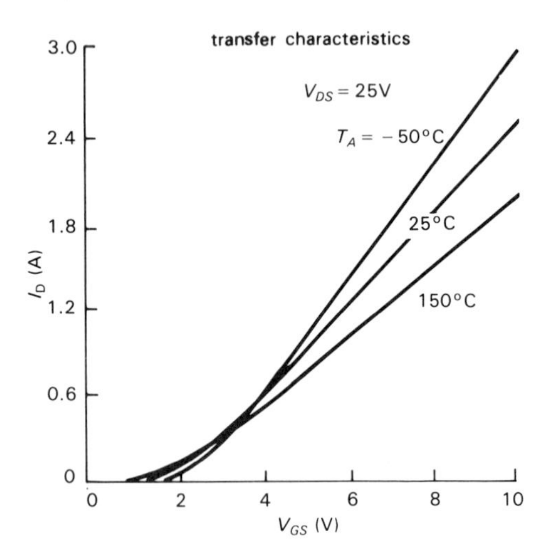

Figure 3.68. Drain characteristics of an n-channel low-threshold MOSFET (type TN0104).

3.09), MOSFETs have essentially infinite gate *resistance*, but finite *impedance* owing to gate-channel capacitance. With high-current MOSFETs the capacitance can be staggering: Compared with 45pF of input capacitance for the 1 amp VN01, the 10 amp IRF520 has $C_{\text{in}} = 450\text{pF}$, and the macho 70 amp SMM70N05 from Siliconix has $C_{\text{in}} = 4300\text{pF}$! A rapidly-changing drain voltage can produce milliamps of transient gate current, enough to overdrive (and even damage, via "SCR crowbarring") delicate CMOS driver chips.

The series resistance is a compromise between speed and protection, with values of 100 ohms to 10k being typical. Even without inductive loads there is dynamic gate current, of course: The capacitance to ground, C_{iss}, gives rise to $I = C_{iss}dV_{GS}/dt$, while the (smaller) feedback capacitance, C_{rss}, produces an input current $I = C_{rss}dV_{DG}/dt$. The latter may dominate in a common-source switch, because ΔV_{DG} is usually much larger than the ΔV_{GS} gate drive (Miller effect).

EXERCISE 3.19

An IRF520 MOSFET controlling a 2 amp load is switched off in 100ns (by bringing the gate from +10V to ground), during which the drain goes from 0 to 50 volts. What is the average gate current during the 100ns, assuming C_{GS} (also called C_{iss}) is 450pF, and C_{DG} (also called C_{rss}) is 50pF?

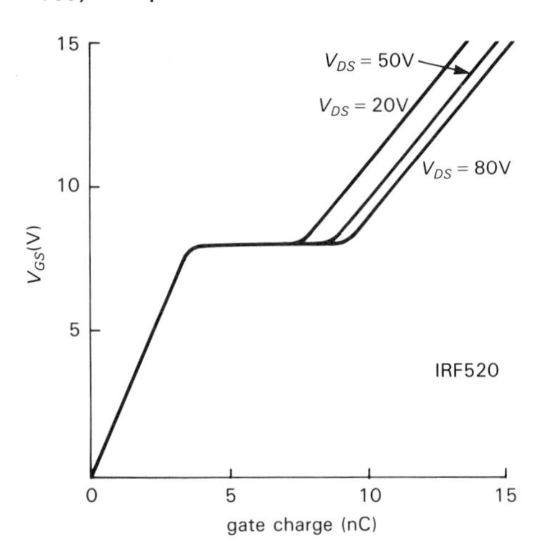

Figure 3.69. Gate charge versus V_{GS}.

In a common-source switch, the Miller-effect contribution to gate current occurs entirely during the drain transitions, whereas the gate-source capacitance causes current whenever the gate voltage is changing. These effects are often plotted as a graph of "gate charge versus gate-source voltage," as in Figure 3.69. The horizontal portion occurs at the turn-on voltage, where the rapidly falling drain forces the gate driver to supply additional charge to C_{rss} (Miller effect). If the feedback capacitance were independent of voltage, the horizontal portion would be proportional to drain voltage, after which the curve would continue at the previous slope. In fact, feedback capacitance C_{rss} rises rapidly at low voltage (Fig. 3.70), which means that most of the Miller effect occurs during the low-voltage portion of the drain waveform. This explains the change in slope of the gate charge curve, as well as the fact that the horizontal portion is almost independent of drain voltage.

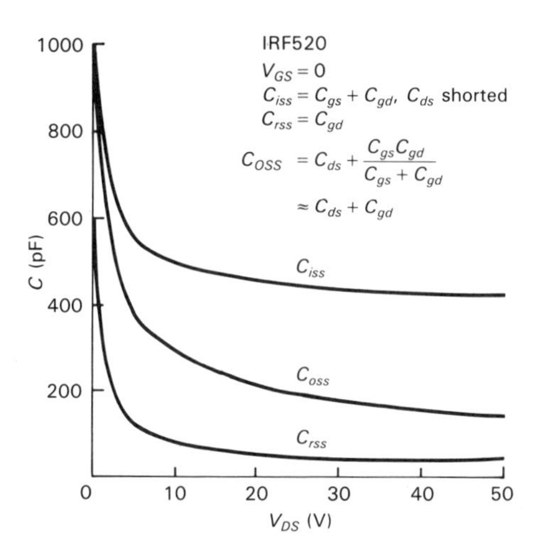

Figure 3.70. Power MOSFET capacitances.

EXERCISE 3.20

How does the voltage dependence of C_{rss} explain the change in slope of the gate charge curves?

Other cautions. Power MOSFETs have some additional idiosyncrasies you should know about. All manufacturers of power MOSFETs seem to connect the body internally to the source. Because the body forms a diode with the channel, this means that there is an effective diode from drain to source (Fig. 3.71); some manufacturers even draw the diode explicitly in their MOSFET symbol so that you won't forget. This means that you cannot use power MOSFETs bidirectionally, or at least not with more than a diode drop of reverse drain-source voltage. For example, you couldn't use a power MOSFET to zero an integrator driven with a bipolarity signal, and you couldn't use a power MOSFET as an analog switch for bipolarity signals. This problem does not occur with *integrated circuit* MOSFETs (analog switches, for example), where the body is connected to the most negative power-supply terminal.

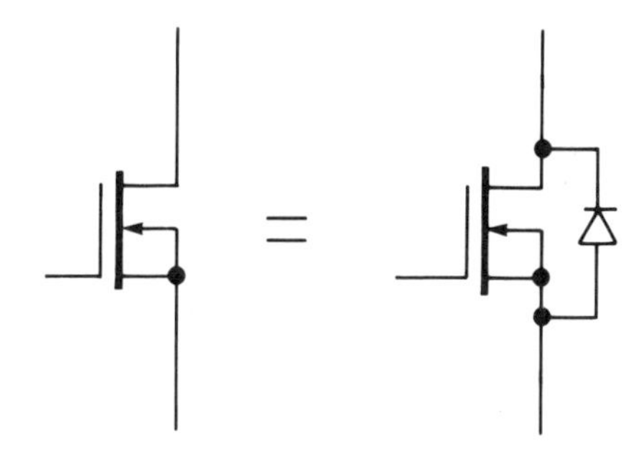

Figure 3.71. Power MOSFETs connect body to source, forming a drain-source diode.

Another trap for the unwary is the fact that gate-source breakdown voltages (±20V is a common figure) are lower than drain-source breakdown voltages (which range from 20V to 1000V). This doesn't matter if you're driving the gate from the small swings of digital logic, but you get into trouble immediately if you think you can use the drain swings of one MOSFET to drive the gate of another.

Finally, the issue of *gate protection*: As we discuss in the final section of this chapter, all MOSFET devices are extremely susceptible to gate oxide breakdown, caused by electrostatic discharge. Unlike JFETs or other junction devices, in which junction avalanche current can safely discharge the overvoltage, MOSFETs are damaged irreversibly by a single instance of gate breakdown. For this reason it is a very good idea to use gate series resistors of 1k–10k, particularly when the gate signal comes from another circuit board. This greatly reduces the chances of damage; it also prevents circuit loading if the gate is damaged, because the most common symptom of a damaged MOSFET is substantial dc gate current. Another precaution is to make sure you don't leave MOSFET gates unconnected, because they are much more susceptible to damage when floating (there is then no *circuit* path for

TABLE 3.5. POWER MOSFETs

BV_{DS}[a] (V)	Cont drain curr max (A)	$R_{DS(on)}$ max (Ω)	@ V_{GS} (V)	$V_{GS(th)}$ max (V)	C_{iss} typ (pF)	C_{rss} typ (pF)	Turn-on charge typ (nC)	Case[b]	Type/Comments[c]
n-channel									
30	0.8	1.8	5	2.5	110	35	-	DIP-14	VQ3001J[1]; 2N, 2P in DIP
40	4	2.5	5	1.5	60	5	0.8	TO-92	TN0104N3; low threshold
60	0.2	6	5	2.5	60	5	-	TO-92	VN0610L[3]; gate protec; sim to VN2222
60	0.4	5	5	2.5	60	10	-	DIP-14	VQ1004J[1]; quad in DIP
60	15	0.14	5	2	900	180	-	TO-220	RFP15N06L[2]; low threshold
100	0.25	15	5	2.4	27	3	0.6	TO-92	VN1310N3, BSS100
100	0.8	2.5	5	2.4	70	12	2.6	TO-92	VN0210N3
100	1.3	0.3	10	4	450	50	11	DIP-4	IRFD120
100	2	1	5	2	200	20	-	TO-220	RFP2N10L[2]; low threshold
100	4	0.6	10	4	180	15	5	TO-220	IRF510, MTP4N10, VN1110N5, 2SK295
100	8	0.25	10	4	350	24	10	TO-220	IRF520, BUZ72A, 2SK383, VN1210N5
100	25	0.08	10	4	1500	90	39	TO-220	IRF540, MTP25N10
100	40	0.06	10	4	2000	350	63	TO-3	IRF150, 2N6764
100	65	0.04	10	5	5200	640	-	TO-3	VNE003A[1]
120	0.2	10	2.5	2	125	20	-	TO-92	VN1206L[1]; low threshold
200	0.1	40	5	3.5	25	3	0.5	TO-92	VN1320N3
200	0.1	24	10	2	40	5	-	TO-92	VN2020L[1], BS107
200	0.25	15	5	3	40	5	1.0	TO-92	VN0120N3, BSS101
200	0.4	8	5	3	75	7	2.5	TO-92	VN0220N3, BSS89
200	3	1.5	10	4	140	9	6	TO-220	IRF610, VN1220N5
200	5	0.8	10	4	450	40	11	TO-220	IRF620, MTP5N20, BUZ30, 2SK440
200	9	0.4	10	4	600	80	19	TO-220	IRF630, MTP8N20, BUZ32
200	18	0.18	10	4	1300	93	43	TO-220	IRF640
200	30	0.09	10	4	2600	150	80	TO-3	IRF250, 2N6766, MTM40N20
500	0.05	85	5	4	45	2	-	TO-92	VN0550N3
500	0.2	20	5	4	75	10	-	TO-92	VN0650N3
500	2.5	3	10	4	350	10	13	TO-220	IRF820, BUZ74, MTP3N50
500	4	1.5	10	4	610	18	21	TO-220	IRF830, BUZ41A, VN5001D[1], MTP4N50
500	8	0.85	10	4	1300	45	42	TO-220	IRF840, MTP8N50, 2SK555[4]
500	12	0.4	10	4	2700	75	86	TO-3	IRF450, 2N6770, 2SK560[4]
500	20	0.3	10	5	4500	100	-	TO-3	VNP006A[1]
1000	1	10	10	4.5	1200[m]	80[m]	33	TO-220	MTP1N100, BUZ50B
1000	5	3	10	4.5	2600[m]	200[m]	110	TO-3	MTM5N100, BUZ54, IRFAG50

BV_{DS}[a] (V)	Cont drain curr max (A)	$R_{DS(on)}$ max (Ω)	@ V_{GS} (V)	$V_{GS(th)}$ max (V)	C_{iss} typ (pF)	C_{rss} typ (pF)	Turn-on charge typ (nC)	Case[b]	Type/Comments[c]
p-channel									
30	0.6	2	12	4.5	150	60	-	DIP-14	VQ3001J[1]; 2N, 2P in DIP
60	0.4	5	10	4.5	150	20	-	DIP-14	VQ2004J[1]; quad in DIP
100	0.15	40	5	3.5	20	3	0.4	TO-92	VP1310N3
100	0.4	8	5	3.5	90	15	3	TO-92	VP0210N3, VP1008L[1]
100	1	0.6	10	4	300	50	16	DIP-4	IRFD9120
100	6	0.6	10	4	300	50	16	TO-220	IRF9520, VP1210N5, MTP8P10
100	19	0.2	10	4	1100	250	70	TO-220	IRF9540, MTP12P10
200	0.06	100	5	3.5	35	2	0.5	TO-92	VP1320N3
200	0.1	40	5	3.5	50	5	1	TO-92	VP0120N3, BSS92
200	3.5	4	5	3.5	600	20	10	TO-220	VP1220N5, IRF9622
200	11	0.5	10	4	1100	150	70	TO-220	IRF9640
500	0.07	150	5	5	35	3	-	TO-92	VP0550N3
500	0.1	25	5	4	75	10	-	TO-92	VP0650N3
500	1	9	5	4.5	550	20	-	TO-220	VP0350N5
500	2	6	10	4.5	1000[m]	80[m]	20	TO-220	MTP2P50

(a) BV_{GS} is ±20V, except (1) ±40V, (2) ±10V, (3) +15, -0.3V, and (4) ±15V
(b) Θ_{JA}: DIP-4=120°C/W; DIP-14=100°C/W; TO-92=200°C/W; Θ_{JC}: TO-220=2.5°C/W; TO-3=0.8°C/W.
P_{diss} @ T_{amb}=75°C: DIP-4=0.6W; DIP-14=0.8W; TO-92=0.3W; P_{diss} @ T_{case}=75°C: TO-220=30W; TO-3=90W.
(c) expect variations in characteristics between manufacturers; those shown are typical. (m) maximum.

static discharge, which otherwise provides a measure of safety). This can happen unexpectedly if the gate is driven from another circuit board. The best practice is to connect a pulldown resistor (say 100k to 1M) from gate to source of any MOSFETs whose gates are driven from an off-card signal source.

☐ *MOSFETs versus BJTs as high-current switches.* Power MOSFETs are attractive alternatives to conventional power BJTs most of the time. They currently cost somewhat more, for the same capability; but they're simpler to drive, and they don't suffer from second breakdown and consequently reduced safe-operating-area (SOA) constraints (Fig. 3.66).

Keep in mind that an ON MOSFET behaves like a small resistance, rather than a saturated bipolar transistor, for small values of drain voltage. This can be an advantage, because the "saturation voltage" goes clear to zero for small drain currents. There is a general perception that MOSFETs don't saturate as well at high currents, but our research shows this to be largely false. In Table 3.6 we've chosen comparable pairs (*npn* versus n-channel MOSFET), for which we've looked up the specified $V_{CE(\text{sat})}$ or $R_{DS(\text{on})}$. The *low-current* MOSFET makes a poor showing when compared with its "small-signal" *npn* cousin, but in the range of 10–50 amps, 0–100 volts, the MOSFET does better. Note particularly the enormous base currents

TABLE 3.6. BJT-MOSFET COMPARISON

Class	Type	I_C, I_D	V_{sat}(max) (25°C) (V)	V_{sat}(max) (125°C) (V)	I_B, V_{GS}	C_{out} (10V) max	Price (100 pc)
60V, 0.5A	NPN – 2N4400	0.5A	0.75	0.8	50mA	8pF	$0.09
	NMOS – VN0610	0.5A	2.5	4.5	10V	25pF	$0.43
60V, 10A	NPN – 2N3055	10A	3	–	3.3A	600pF	$0.65
	NMOS – MTP3055A	10A	1.5	2.3	10V	300pF	$0.57
100V, 50A	NPN – 2N6274	20A	1	1.4	2A	600pF	$11.00
	NMOS – VNE003A	20A	0.7	1.1	10V	3000pF	$12.50
400V, 15A	NPN – 2N6547	15A	1.5	2.5	**2A**	500pF	$4.00
	NMOS – IRF350	15A	3	6	**10V**	900pF	$12.60

needed to bring the bipolar power transistor into good saturation – 10% or more of the collector current (!) – compared with the (zero-current) 10 volt bias at which MOSFETs are usually specified. Note also that high-voltage MOSFETs (say, $BV_{DS} >$ 200V) tend to have larger $R_{DS(on)}$, with larger temperature coefficients, than the lower-voltage units. Along with saturation data, we've listed capacitances in the table, because power MOSFETs often have more capacitance than BJTs of the same rated current; in some applications (particularly if switching speed is important) you might want to consider the product of capacitance and saturation voltage as a figure of merit.

Remember that power MOSFETs can be used as BJT substitutes for linear power circuits, for example audio amplifiers and voltage regulators (we'll treat the latter in Chapter 6). Power MOSFETs are also available as p-channel devices, although there tends to be a greater variety available among the (better performing) n-channel devices.

Some MOSFET power switching examples. Figure 3.72 shows three ways to use a MOSFET to control the dc power to some sub-circuit that you want to turn on and off. If you have a battery-operated instrument that needs to make some measurements occasionally, you might use circuit A to switch the power-hungry microprocessor off except during those intermittent measurements. Here we've used a PMOS switch, turned on by a 5 volt logic swing to ground. The "5V logic" is micropower CMOS digital circuitry, kept running even when the microprocessor is shut off (remember, CMOS logic has zero static dissipation). We'll have much more to say about this sort of "power-switching" scheme in Chapter 14.

In the second circuit (B), we're switching dc power to a load that needs +12 volts, at considerable current; maybe it's a radio transmitter, or whatever. Because we have only a 5 volt logic swing available, we've used a small n-channel switch to generate a full 12 volt swing, which then drives the PMOS gate. Note the high-value NMOS drain resistor, perfectly adequate here because the PMOS gate draws no dc current (even a beefy 10A brute), and we don't need high switching speed in an application like this.

The third circuit (C) is an elaboration of circuit B, with short-circuit current limiting via the *pnp* transistor. That's always a good idea in power supply design, because it's easy to slip with the oscilloscope probe. In this case, the current limiting

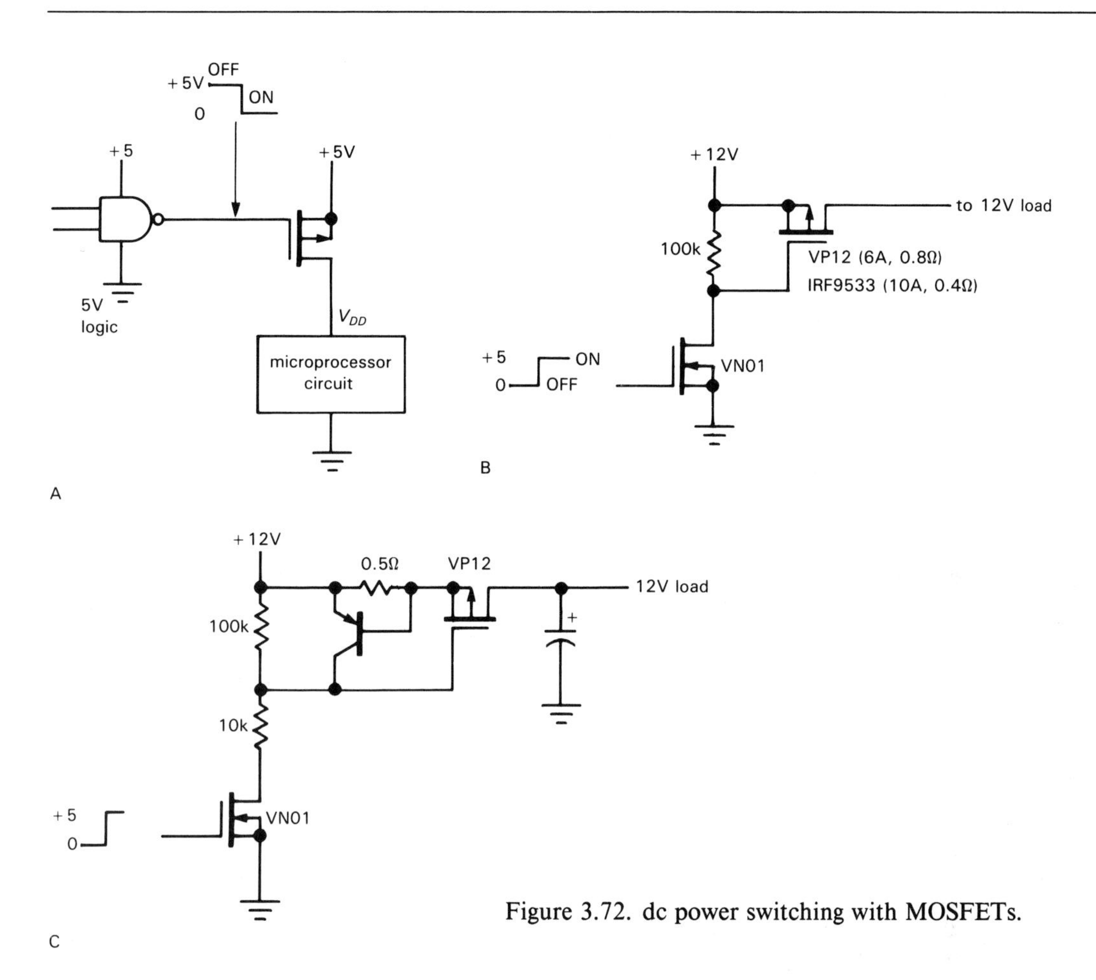

Figure 3.72. dc power switching with MOSFETs.

also prevents momentary short-circuiting of the +12 volt supply by the initially uncharged bypass capacitor. See if you can figure out how the current limiting circuit works.

EXERCISE 3.21

How does the current limiting circuit work? How much load current does it allow? Why is the NMOS drain resistor split in two?

The limited gate breakdown voltages of MOSFETs (usually ±20V) would create a real problem here if you attempted to operate the circuit from higher supply voltage. In that case you could replace the 100k resistor with 10k (allowing operation to 40V), or other appropriate ratio, always keeping the VP12 gate drive less than 20 volts.

Figure 3.73A shows a simple MOSFET switching example, one that takes advantage of the high gate impedance. You might want to turn on exterior lighting automatically at sunset. The photoresistor has low resistance in sunlight, high resistance in darkness. You make it part of a resistive divider, driving the gate directly (no dc loading!). The light goes on when the gate voltage reaches the value that produces enough drain current to close the relay. Sharp-eyed readers may have noticed that this circuit is not particularly precise or stable; that's OK, because the photoresistor undergoes an enormous change in resistance (from 10k to 10M, say) when it gets dark. The circuit's lack of a precise and stable threshold just means that the light may turn on a few minutes early or

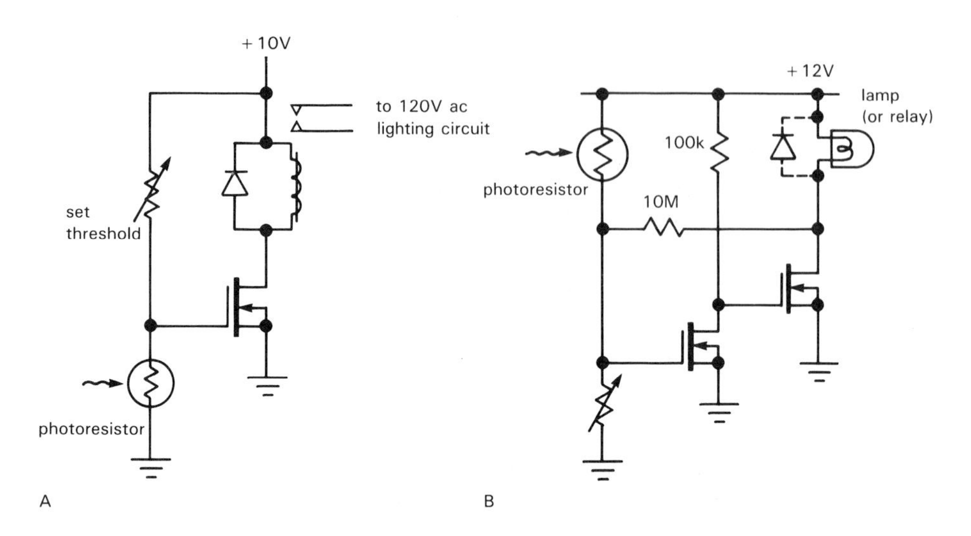

Figure 3.73. Ambient-light-controlled power switch.

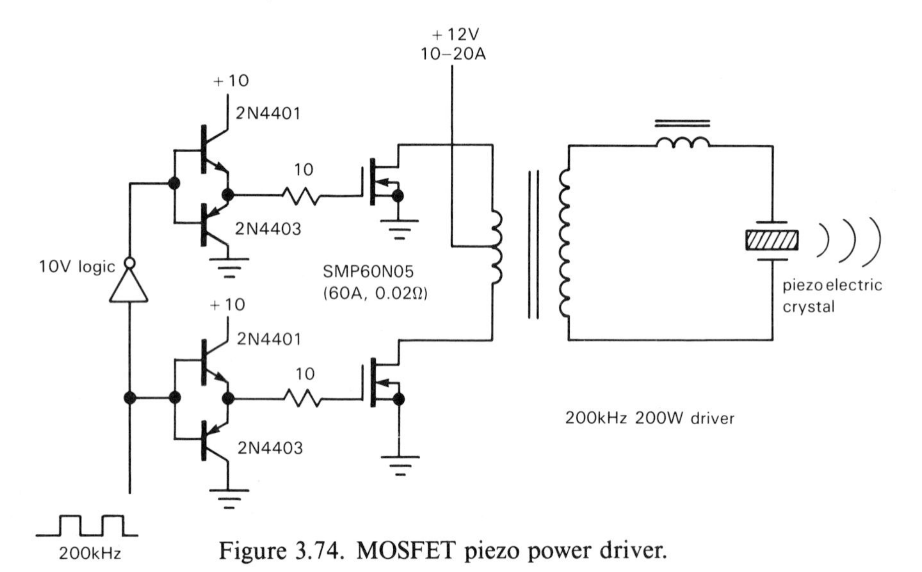

Figure 3.74. MOSFET piezo power driver.

late. Note that the MOSFET may have to dissipate some power during the time the gate bias is inching up, since we're operating in the linear region. That problem is remedied in Figure 3.73B, where a pair of cascaded MOSFETs delivers much higher gain, augmented by some positive feedback via the 10M resistor; the latter causes the circuit to snap on regeneratively as it reaches threshold.

Figure 3.74 shows a real power MOSFET job: A 200 watt amplifier to drive a piezoelectric underwater transducer at 200kHz. We've used a pair of hefty NMOS transistors, driven alternately to create ac drive in the (high-frequency) transformer primary. The bipolar push-pull gate drivers, with small gate resistors, are needed to overcome capacitive loading, since the FETs must be turned on fully in something less than a microsecond.

Finally, in Figure 3.75 we show a *linear* circuit example with power MOSFETs. Ceramic *piezoelectric* transducers are often used in optical systems to produce controlled small motions; for example,

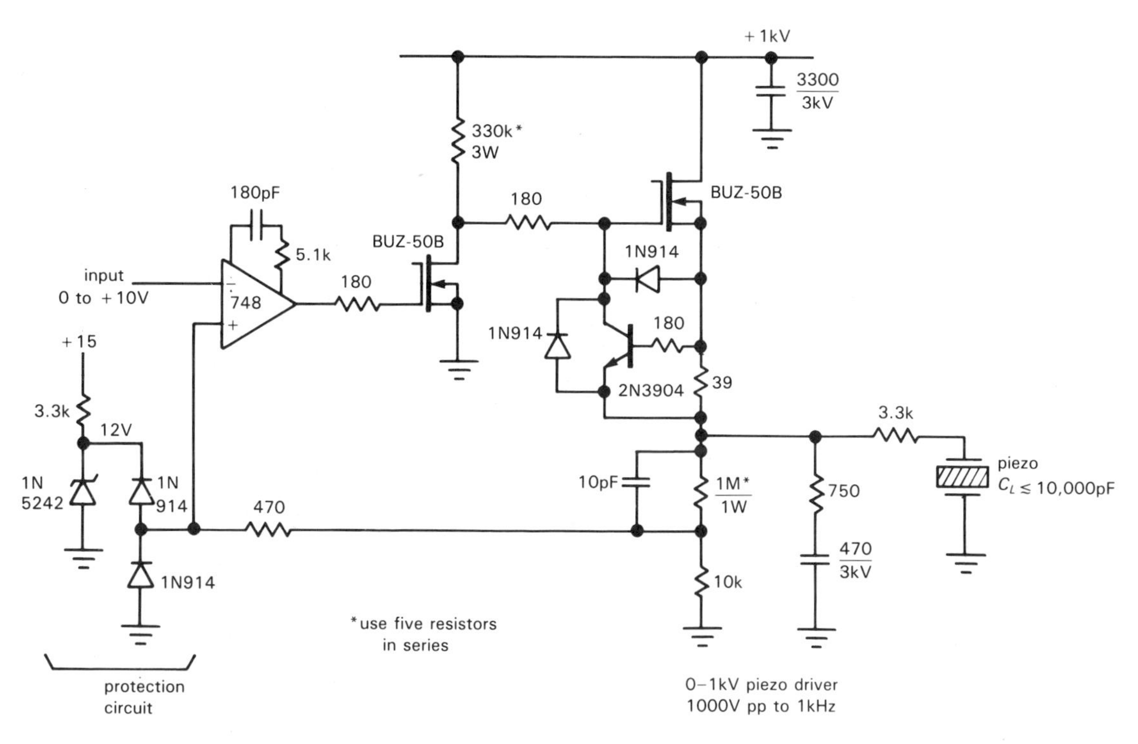

Figure 3.75. 1kV low-power piezo driver.

in *adaptive optics* you might use a piezoelectrically controlled "rubber mirror" to compensate for local variations in the index of refraction of the atmosphere. Piezo transducers are nice to use, because they're very stiff. Unfortunately, they require a kilovolt or more of voltage to produce significant motions. Furthermore, they're highly capacitive - typically 0.01μF or more - and have mechanical resonances in the kilohertz range, thus presenting a nasty load. We needed dozens of such driver amplifiers, which for some reason cost a few thousand dollars apiece if you buy them commercially. We solved our problem with the circuit shown. The BUZ-50B is an inexpensive ($4) MOSFET, good for 1kV and 2 amps. The first transistor is a common-source inverting amplifier, driving a source follower. The *npn* transistor is a current-limiter and can be a low-voltage unit, since it floats on the output. One subtle feature of the circuit is the fact that it's actually push-pull, even though it looks single-ended: You need plenty of current to push 10,000pF around at 2 volts per microsecond (how much?); the output transistor can *source* current, but the pulldown resistor can't sink enough (look back to Section 2.15, where we motivated push-pull with the same problem). In this circuit the driver transistor is the pulldown, via the gate-source diode! The rest of the circuit involves feedback (with an op-amp), a forbidden subject until the next chapter; in this case the magic of feedback makes the overall circuit linear (100V of output per volt of input), whereas without it the output voltage would depend on the (nonlinear) I_D-versus-V_{GS} characteristic of the input transistor.

3.15 MOSFET handling precautions

The MOSFET gate is insulated by a layer of glass (SiO_2) a few thousand angstroms (1Å = 0.1nm) thick. As a result it has very high resistance, and no resistive or junction-like path that can discharge static electricity as it is building up. In a classic situation you have a MOSFET (or MOSFET integrated circuit) in your hand. You walk over to your circuit, stick the device into its socket, and turn on the power, only to discover that the FET is dead. You killed it! You should have grabbed onto the circuit board with your other hand before inserting the device. This would have discharged your static voltage, which in winter can reach thousands of volts. MOS devices don't take kindly to "carpet shock," which is officially called *electrostatic discharge* (ESD). For purposes of static electricity, you are equivalent to 100pF in series with 1.5k; in winter your capacitor may charge to 10kV or more with a bit of shuffling about on a fluffy rug, and even a simple arm motion with shirt or sweater can generate a few kilovolts (see Table 3.7).

TABLE 3.7.
TYPICAL ELECTROSTATIC VOLTAGES[a]

Action	Electrostatic voltage 10%-20% humidity (V)	Electrostatic voltage 65%-90% humidity (V)
walk on carpet	35,000	1,500
walk on vinyl floor	12,000	250
work at bench	6,000	100
handle vinyl envelope	7,000	600
pick up poly bag	20,000	1,200
shift position on foam chair	18,000	1,500

(a) adapted from Motorola Power MOSFET Data Book.

Although any semiconductor device can be clobbered by a healthy spark, MOS devices are particularly susceptible because the energy stored in the gate-channel capacitance, when it has been brought up to breakdown voltage, is sufficient to blow a hole through the delicate gate oxide insulation. (If the spark comes from your finger, your additional 100pF only adds to the injury.) Figure 3.76 (from a series of ESD tests on a power MOSFET) shows the sort of mess this can make. Calling this "gate breakdown" gives the wrong idea; the colorful term "gate *rupture*" is closer to the mark!

high power (X1200)

Figure 3.76. Scanning electron micrograph of a 6 amp MOSFET destroyed by 1kV charge on "human body equivalent" (1.5k in series with 100pF) applied to its gate. (Courtesy of Motorola, Inc.)

The electronics industry takes ESD very seriously. It is probably the leading cause of nonfunctional semiconductors in instruments fresh off the assembly line. Books are published on the subject, and you can takes courses on it. MOS devices, as well as other susceptible semiconductors (which includes just about everything; e.g., it takes about 10 times as much voltage to zap a BJT), should be shipped in conductive foam or bags, and you have to be careful about voltages on soldering irons, etc., during fabrication. It is best to ground soldering irons, table tops, etc., and use conductive wrist straps. In addition, you can get "antistatic" carpets, upholstery, and even clothing (e.g., antistatic smocks containing 2% stainless steel fiber). A good antistatic workstation includes humidity control, air ionizers (to make the air slightly conductive, which keeps things from charging up), and educated workers. In spite of all this, failure rates increase dramatically in winter.

Once a semiconductor device is safely soldered into its circuit, the chances for damage are greatly reduced. In addition, most small-geometry MOS devices (e.g., CMOS logic devices, but not power MOSFETs) have protection diodes in the input gate circuits. Although the internal protection networks of resistors and clamping diodes (or sometimes zeners) compromise performance somewhat, it is often worthwhile to choose those devices because of the greatly reduced risk of damage by static electricity. In the case of unprotected devices, for example power MOSFETs, small-geometry (low current) devices tend to be the most troublesome, because their low input capacitance is easily brought to high voltage when it comes in contact with a charged 100pF human. Our personal experience with the small-geometry VN13 MOSFET has been so dismal in this regard that we no longer use it in production instruments.

It is hard to overstate the problem of gate damage due to breakdown in MOSFETs. Luckily, MOSFET designers realize the seriousness of the problem and are responding with new designs with higher BV_{GS}; for example, Motorola's new "TMOS IV" series features ±50 volt gate-source breakdown.

SELF-EXPLANATORY CIRCUITS

3.16 Circuit ideas

Figure 3.77 presents a sampling of FET circuit ideas.

3.17 Bad circuits

Figure 3.78 presents a collection of bad ideas, some of which involve a bit of subtlety. You'll learn a lot by figuring out why these circuits won't work.

Circuit ideas

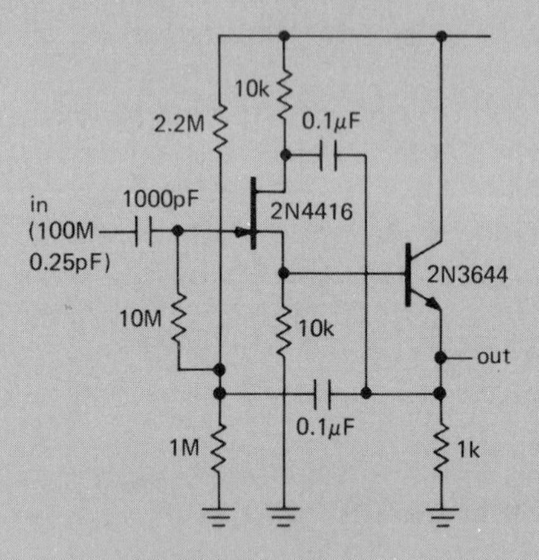

A. high-input-impedance follower

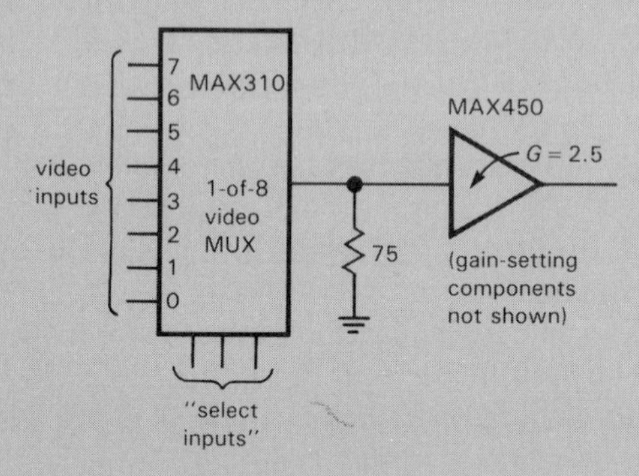

B. MOSFET video MUX with amplifier to compensate R_{on} losses.

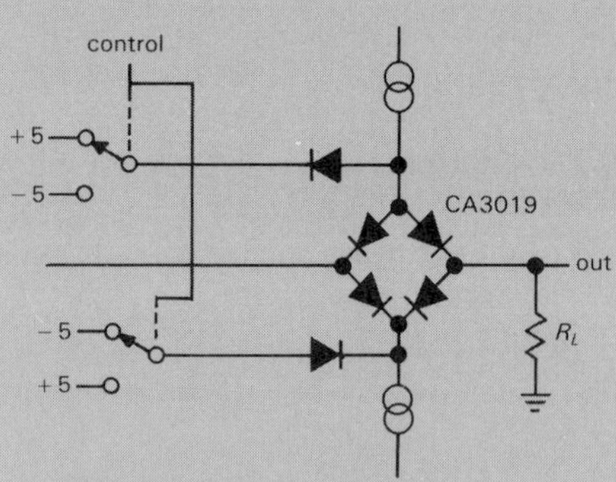

C. signal switching using a diode bridge; alternative to FETs.

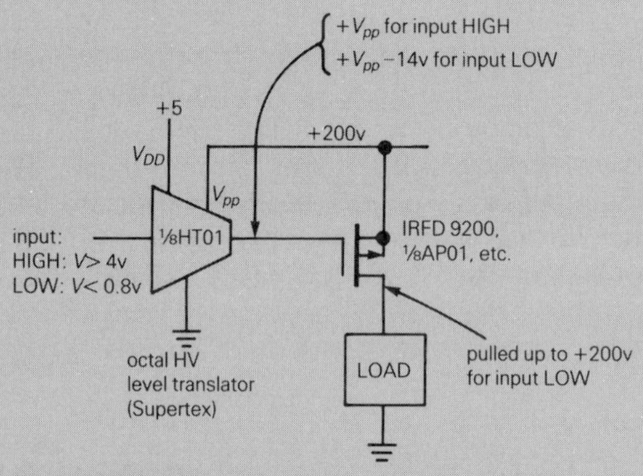

D. logic-level to *p*-channel HV switch

Figure 3.77

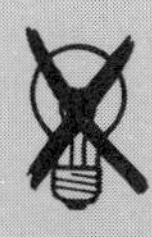

Bad circuits

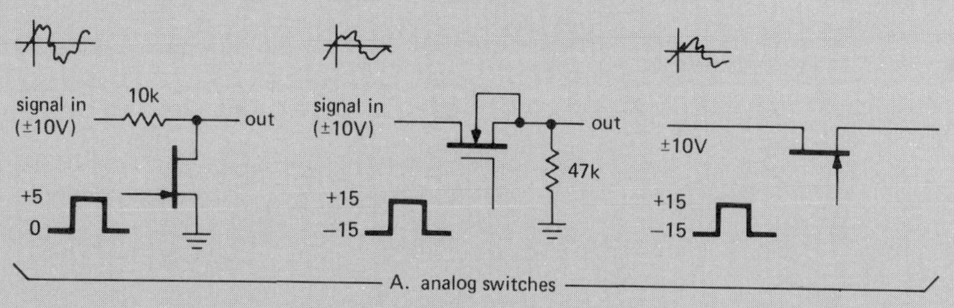

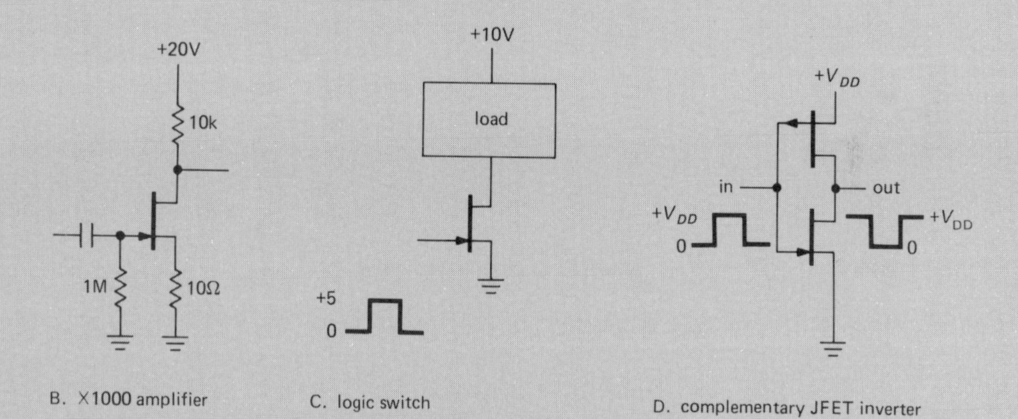

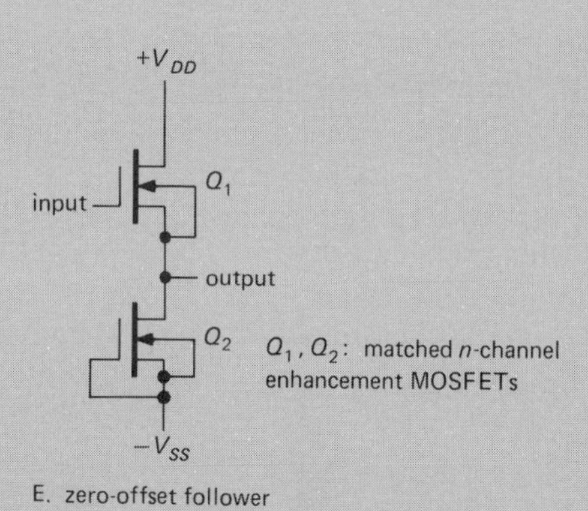

Figure 3.78

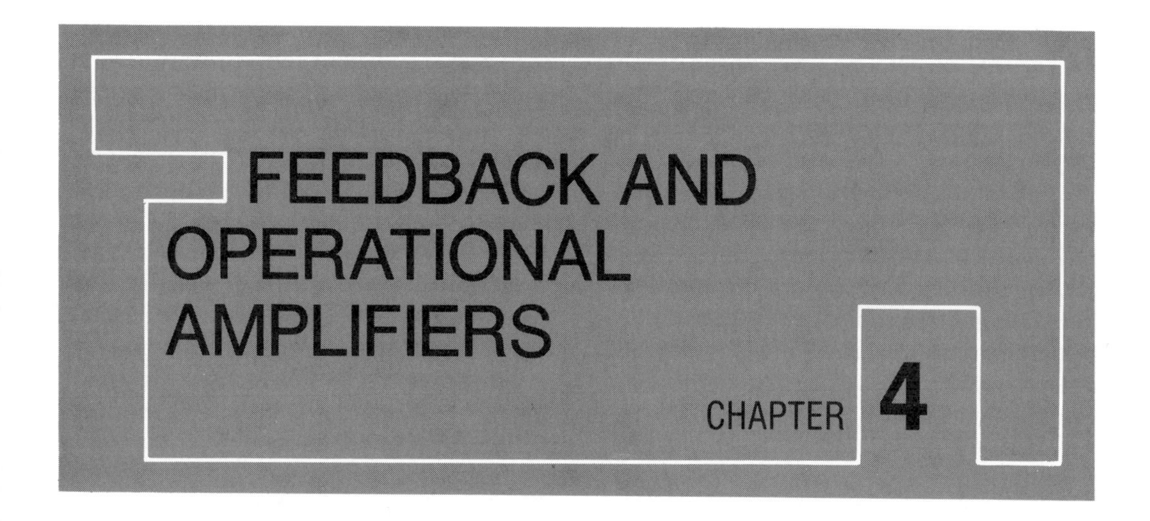

INTRODUCTION

Feedback has become such a well-known concept that the word has entered the general vocabulary. In control systems, feedback consists in comparing the output of the system with the desired output and making a correction accordingly. The "system" can be almost anything: for instance, the process of driving a car down the road, in which the output (the position and velocity of the car) is sensed by the driver, who compares it with expectations and makes corrections to the input (steering wheel, throttle, brake). In amplifier circuits the output should be a multiple of the input, so in a feedback amplifier the input is compared with an attenuated version of the output.

4.01 Introduction to feedback

Negative feedback is the process of coupling the output back in such a way as to cancel some of the input. You might think that this would only have the effect of reducing the amplifier's gain and would be a pretty stupid thing to do. Harold S. Black, who attempted to patent negative feedback in 1928, was greeted with the same response. In his words, "Our patent application was treated in the same manner as one for a perpetual-motion machine." (See the fascinating article in *IEEE Spectrum*, December 1977.) True, it does lower the gain, but in exchange it also improves other characteristics, most notably freedom from distortion and nonlinearity, flatness of response (or conformity to some desired frequency response), and predictability. In fact, as more negative feedback is used, the resultant amplifier characteristics become less dependent on the characteristics of the open-loop (no-feedback) amplifier and finally depend only on the properties of the feedback network itself. Operational amplifiers are typically used in this *high-loop-gain* limit, with *open-loop* voltage gain (no feedback) of a million or so.

A feedback network can be frequency-dependent, to produce an equalization amplifier (with specific gain-versus-frequency characteristics, an example being the famous RIAA phono amplifier characteristic), or it can be amplitude-dependent, producing a nonlinear amplifier (a popular example is a logarithmic amplifier, built with feedback that exploits the logarithmic V_{BE} versus I_C of a diode or transistor). It can be arranged to produce a current source (near-infinite output impedance) or a voltage source (near-zero output impedance), and it can be connected to generate very high or very low input impedance. Speaking in general terms, the property that is sampled to produce feedback is the property that is improved. Thus, if you feed back a signal proportional to the output current, you will generate a good current source.

Feedback can also be *positive*; that's how you make an oscillator, for instance. As much fun as that may sound, it simply isn't as important as negative feedback. More often it's a nuisance, since a negative-feedback circuit may have large enough phase shifts at some high frequency to produce positive feedback and oscillations. It is surprisingly easy to have this happen, and the prevention of unwanted oscillations is the object of what is called *compensation*, a subject we will treat briefly at the end of the chapter.

Having made these general comments, we will now look at a few feedback examples with operational amplifiers.

4.02 Operational amplifiers

Most of our work with feedback will involve operational amplifiers, very high gain dc-coupled differential amplifiers with single-ended outputs. You can think of the classic long-tailed pair (Section 2.18) with its two inputs and single output as a prototype, although real op-amps have much higher gain (typically 10^5 to 10^6) and lower output impedance and allow the output to swing through most of the supply range (you usually use a split supply, most often ±15V). Operational amplifiers are now available in literally hundreds of types, with the universal symbol shown in Figure 4.1, where the (+) and (−) inputs do as expected: The output goes positive when the noninverting input (+) goes more positive than the inverting input (−), and vice versa. The (+) and (−) symbols don't mean that you have to keep one positive with respect to the other, or anything like that; they just tell you the relative phase of the output (which is important to keep negative feedback negative). Using the words "noninverting" and "inverting," rather than "plus" and "minus," will help avoid confusion. Power-supply connections are frequently not displayed, and there is no ground terminal. Operational amplifiers have enormous voltage gain, and they are *never* (well, hardly ever) used without feedback. Think of an op-amp as fodder for feedback. The open-loop gain is so high that for any reasonable closed-loop gain, the characteristics depend only on the feedback network. Of course, at some level of scrutiny this generalization must fail. We will start with a naive view of op-amp behavior and fill in some of the finer points later, when we need to.

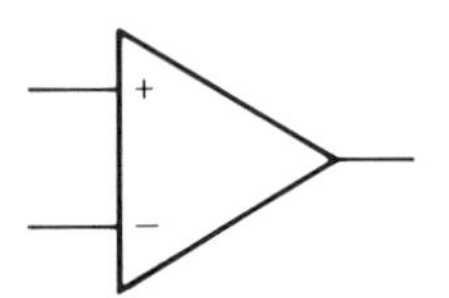

Figure 4.1

There are literally hundreds of different op-amps available, offering various performance trade-offs that we will explain later (look ahead to Table 4.1 if you want to be overwhelmed by what's available). A very good all-around performer is the popular LF411 ("411" for short), originally introduced by National Semiconductor. Like all op-amps, it is a wee beastie packaged in the so-called mini-DIP (dual in-line package), and it looks

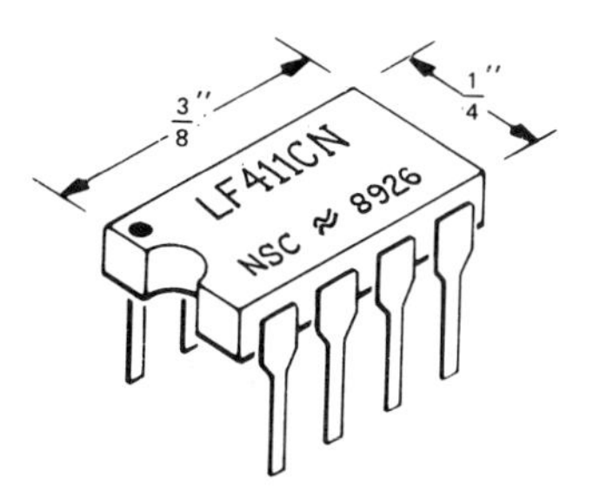

Figure 4.2. Mini-DIP integrated circuit.

as shown in Figure 4.2. It is inexpensive (about 60 cents) and easy to use; it comes in an improved grade (LF411A) and also in a mini-DIP containing two independent op-amps (LF412, called a "dual" op-amp). We will adopt the LF411 throughout this chapter as our "standard" op-amp, and we recommend it as a good starting point for your circuit designs.

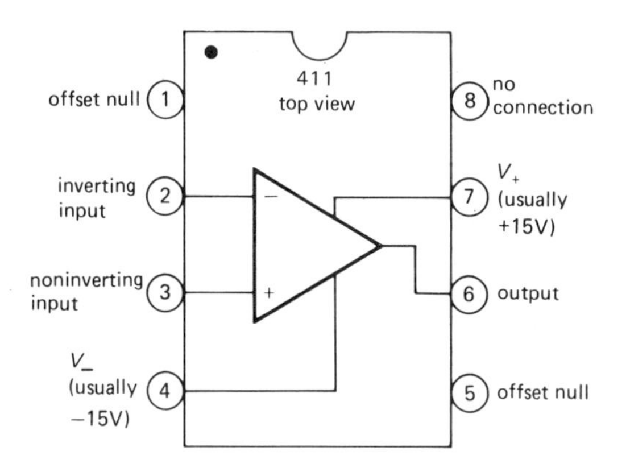

Figure 4.3

Inside the 411 is a piece of silicon containing 24 transistors (21 BJTs, 3 FETs), 11 resistors, and 1 capacitor. The pin connections are shown in Figure 4.3. The dot in the corner, or notch at the end of the package, identifies the end from which to begin counting the pin numbers. As with most electronic packages, you count pins counterclockwise, viewing from the top. The "offset null" terminals (also known as "balance" or "trim") have to do with correcting (externally) the small asymmetries that are unavoidable when making the op-amp. You will learn about this later in the chapter.

4.03 The golden rules

Here are the simple rules for working out op-amp behavior with external feedback. They're good enough for almost everything you'll ever do.

First, the op-amp voltage gain is so high that a fraction of a millivolt between the input terminals will swing the output over its full range, so we ignore that small voltage and state golden rule I:

I. The output attempts to do whatever is necessary to make the voltage difference between the inputs zero.

Second, op-amps draw very little input current (0.2nA for the LF411; picoamps for low-input-current types); we round this off, stating golden rule II:

II. The inputs draw no current.

One important note of explanation: Golden rule I doesn't mean that the op-amp actually changes the voltage at its *inputs*. It can't do that. (How could it, and be consistent with golden rule II?) What it does is "look" at its input terminals and swing its output terminal around so that the external feedback network brings the input differential to zero (if possible).

These two rules get you quite far. We will illustrate with some basic and important op-amp circuits, and these will prompt a few cautions listed in Section 4.08.

BASIC OP-AMP CIRCUITS

4.04 Inverting amplifier

Let's begin with the circuit shown in Figure 4.4. The analysis is simple, if you remember your golden rules:

1. Point B is at ground, so rule I implies that point A is also.

2. This means that (a) the voltage across R_2 is V_{out} and (b) the voltage across R_1 is V_{in}.

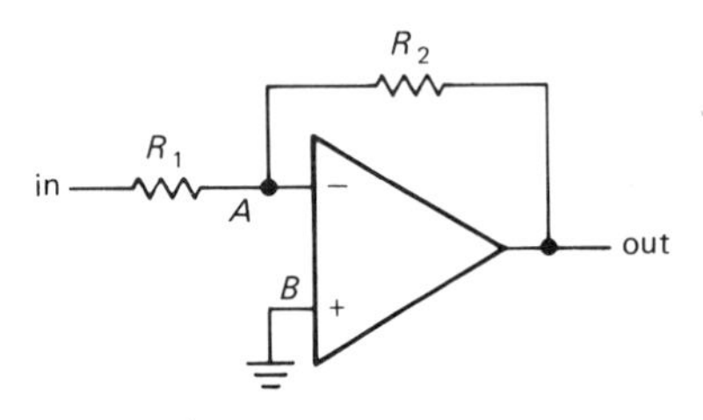

Figure 4.4. Inverting amplifier.

3. So, using rule II, we have

$$V_{out}/R_2 = -V_{in}/R_1$$

In other words,

$$\text{voltage gain} = V_{out}/V_{in} = -R_2/R_1$$

Later you will see that it's often better not to ground B directly, but through a resistor. However, don't worry about that now.

Our analysis seems almost too easy! In some ways it obscures what is actually happening. To understand how feedback works, just imagine some input level, say +1 volt. For concreteness, imagine that R_1 is 10k and R_2 is 100k. Now, suppose the output decides to be uncooperative, and sits at zero volts. What happens? R_1 and R_2 form a voltage divider, holding the inverting input at +0.91 volt. The op-amp sees an enormous input unbalance, forcing the output to go negative. This action continues until the output is at the required −10.0 volts, at which point both op-amp inputs are at the same voltage, namely ground. Similarly, any tendency for the output to go more negative than −10.0 volts will pull the inverting input below ground, forcing the output voltage to rise.

What is the input impedance? Simple. Point A is always at zero volts (it's called a *virtual ground*). So $Z_{in} = R_1$. At this point you don't yet know how to figure the output impedance; for this circuit, it's a fraction of an ohm.

Note that this analysis is true even for dc – it's a dc amplifier. So if you have a signal source offset from ground (collector of a previous stage, for instance), you may want to use a coupling capacitor (sometimes called a blocking capacitor, since it blocks dc but couples the signal). For reasons you will see later (having to do with departures of op-amp behavior from the ideal), it is usually a good idea to use a blocking capacitor if you're only interested in ac signals anyway.

This circuit is known as an *inverting amplifier*. Its one undesirable feature is the low input impedance, particularly for amplifiers with large (closed-loop) voltage gain, where R_1 tends to be rather small. That is remedied in the next circuit (Fig. 4.5).

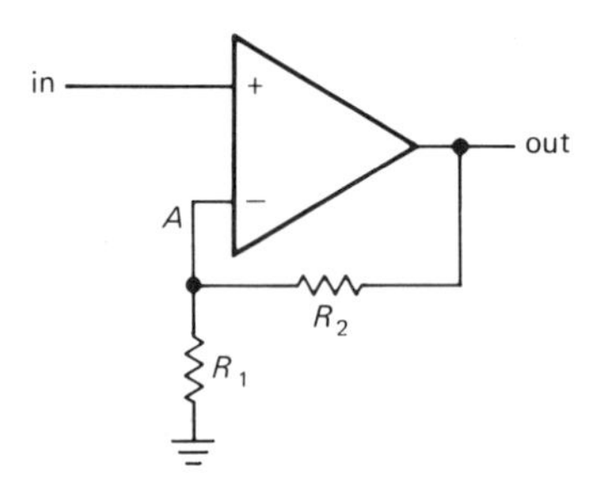

Figure 4.5. Noninverting amplifier.

4.05 Noninverting amplifier

Consider Figure 4.5. Again, the analysis is simplicity itself:

$$V_A = V_{in}$$

But V_A comes from a voltage divider:

$$V_A = V_{out}R_1/(R_1 + R_2)$$

Set $V_A = V_{in}$, and you get

$$\text{gain} = V_{out}/V_{in} = 1 + R_2/R_1$$

This is a *noninverting amplifier*. In the approximation we are using, the input impedance is infinite (with the 411 it would be $10^{12}\Omega$ or more; a bipolar op-amp will typically exceed $10^8\Omega$). The output impedance is still a fraction of an ohm. As with the inverting amplifier, a detailed look at the voltages at the inputs will persuade you that it works as advertised.

Once again we have a dc amplifier. If the signal source is ac-coupled, you must provide a return to ground for the (very small) input current, as in Figure 4.6. The component values shown give a voltage gain of 10 and a low-frequency 3dB point of 16Hz.

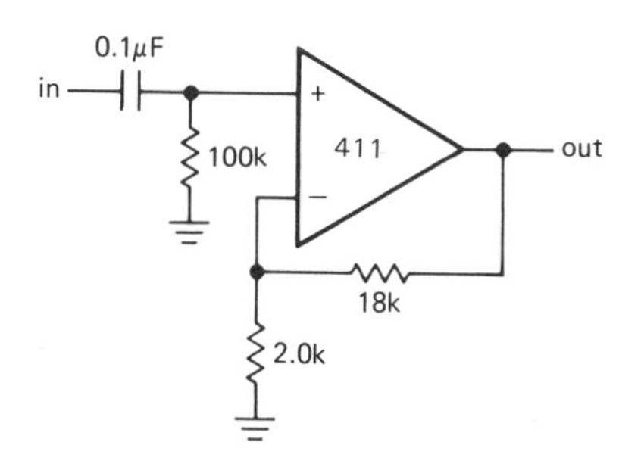

Figure 4.6

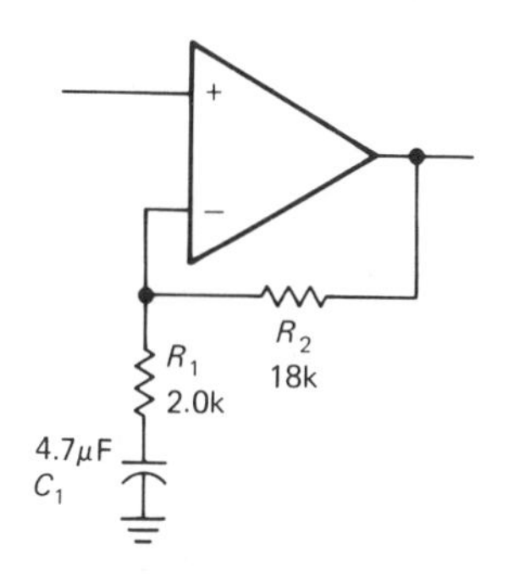

Figure 4.7

An ac amplifier

Again, if only ac signals are being amplified, it is often a good idea to "roll off" the gain to unity at dc, especially if the amplifier has large voltage gain, in order to reduce the effects of finite "input offset voltage." The circuit in Figure 4.7 has a low-frequency 3dB point of 17Hz, the frequency at which the impedance of the capacitor equals 2.0k. Note the large capacitor value required. For noninverting amplifiers with high gain, the capacitor in this ac amplifier configuration may be undesirably large. In that case it may be preferable to omit the capacitor and trim the offset voltage to zero, as we will discuss later (Section 4.12). An alternative is to raise R_1 and R_2, perhaps using a T network for the latter (Section 4.18).

In spite of its desirable high input impedance, the noninverting amplifier configuration is not necessarily to be preferred over the inverting amplifier configuration in all circumstances. As we will see later, the inverting amplifier puts less demand on the op-amp and therefore gives somewhat better performance. In addition, its virtual ground provides a handy way to combine several signals without interaction. Finally, if the circuit in question is driven from the (stiff) output of another op-amp, it makes no difference whether the input impedance is 10k (say) or infinity, because the previous stage has no trouble driving it in either case.

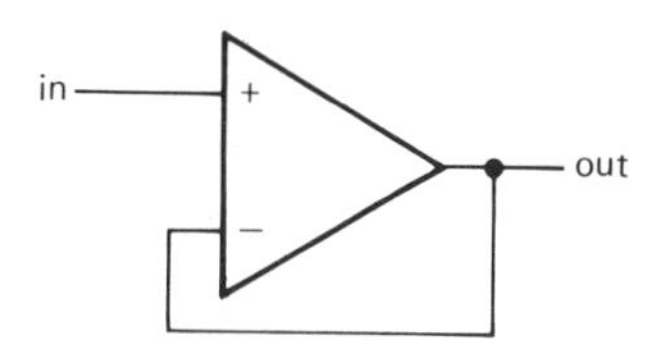

Figure 4.8. Follower.

4.06 Follower

Figure 4.8 shows the op-amp version of an emitter follower. It is simply a noninverting amplifier with R_1 infinite and R_2 zero (gain = 1). There are special op-amps, usable only as followers, with improved characteristics (mainly higher speed), e.g., the LM310 and the OPA633, or with simplified connections, e.g., the TL068 (which comes in a 3-pin transistor package).

An amplifier of unity gain is sometimes called a *buffer* because of its isolating properties (high input impedance, low output impedance).

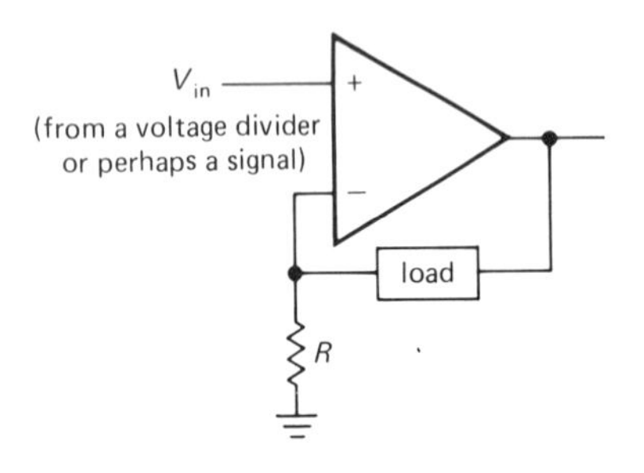

Figure 4.9

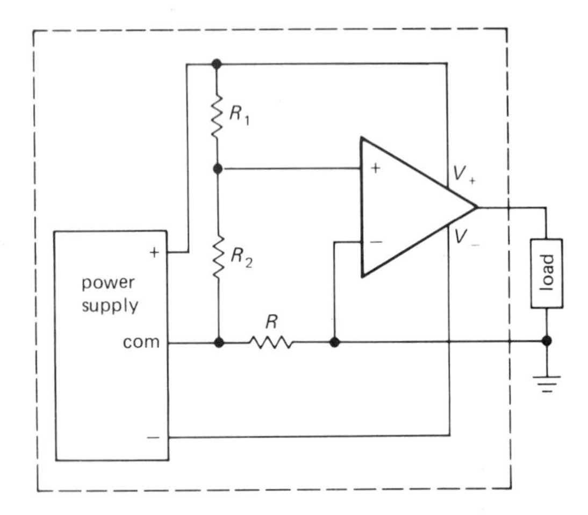

Figure 4.10. Current source with grounded load and floating power supply.

4.07 Current sources

The circuit in Figure 4.9 approximates an ideal current source, without the V_{BE} offset of a transistor current source. Negative feedback results in V_{in} at the inverting input, producing a current $I = V_{\text{in}}/R$ through the load. The major disadvantage of this circuit is the "floating" load (neither side grounded). You couldn't generate a usable sawtooth wave with respect to ground with this current source, for instance. One solution is to float the whole circuit (power supplies and all) so that you can ground one side of the load (Fig. 4.10). The circuit in the box is the previous current source, with its power supplies shown explicitly. R_1 and R_2 form a voltage divider to set the current. If this circuit seems confusing, it may help to remind yourself that "ground" is a relative concept. Any one point in a circuit could be called ground. This circuit is useful for generating currents into a load that is returned to ground, but it has the disadvantage that the control input is now floating, so you cannot program the output current with an input voltage referenced to ground. Some solutions to this problem are presented in Chapter 6 in the discussion of constant-current power supplies.

Current sources for loads returned to ground

With an op-amp and external transistor it is possible to make a simple high-quality current source for a load returned to ground; a little additional circuitry makes it possible to use a programming input referenced to ground (Fig. 4.11). In the first circuit, feedback forces a voltage $V_{CC} - V_{\text{in}}$ across R, giving an emitter current (and therefore an output current) $I_E = (V_{CC} - V_{\text{in}})/R$. There are no V_{BE} offsets, or their variations with temperature, I_C, V_{CE}, etc., to worry about. The current source is imperfect (ignoring op-amp errors: I_b, V_{os}) only insofar as the small base current may vary somewhat with V_{CE} (assuming the op-amp draws no input current), not too high a price to pay for the convenience of a grounded load; a Darlington for Q_1 would reduce this error considerably. This error comes about, of course, because the op-amp stabilizes the *emitter* current, whereas the load sees the *collector* current. A variation of this circuit, using a FET instead of a bipolar transistor, avoids this problem altogether, since FETs draw no gate current.

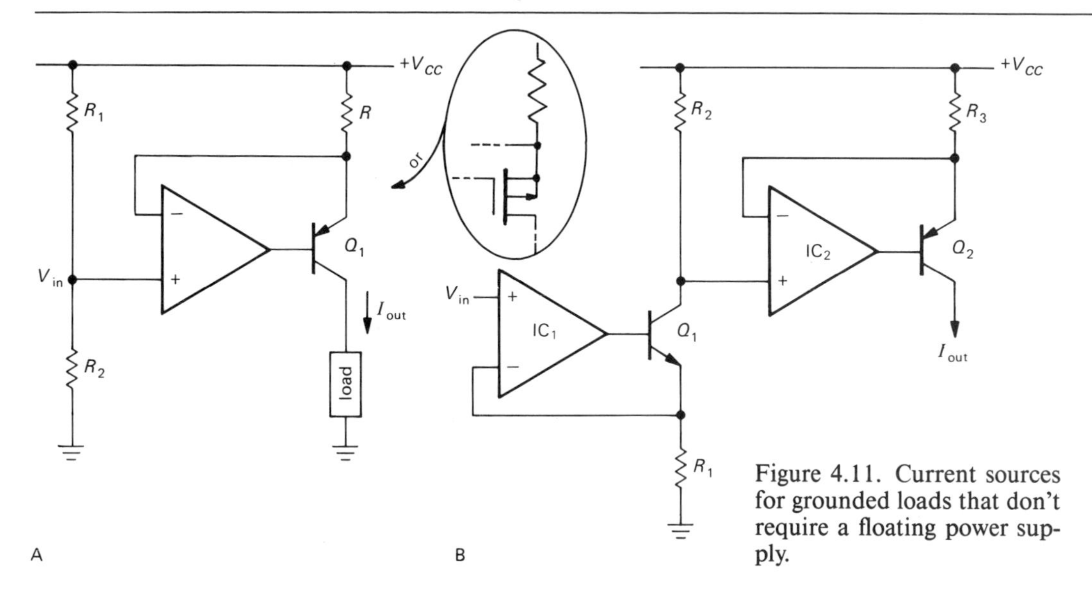

Figure 4.11. Current sources for grounded loads that don't require a floating power supply.

With this circuit the output current is proportional to the voltage drop below V_{CC} applied to the op-amp's noninverting input; in other words, the programming voltage is referenced to V_{CC}, which is fine if V_{in} is a fixed voltage generated by a voltage divider, but an awkward situation if an external input is to be used. This is remedied in the second circuit, in which a similar current source with *npn* transistor is used to convert an input voltage (referenced to ground) to a V_{CC}-referenced input to the final current source. Op-amps and transistors are inexpensive. Don't hesitate to use a few extra components to improve performance or convenience in circuit design.

One important note about the last circuit: The op-amp must be able to operate with its inputs near or at the positive supply voltage. An op-amp like the 307, 355, or OP-41 is good here. Alternatively, the op-amp could be powered from a separate V_+ voltage higher than V_{CC}.

EXERCISE 4.1

What is the output current in the last circuit for a given input voltage V_{in}?

Figure 4.12 shows an interesting variation on the op-amp/transistor current source. It has the advantage of zero base current error, which you get with FETs, without being restricted to output currents less than $I_{DS(ON)}$. In this circuit (actually a current *sink*), Q_2 begins to

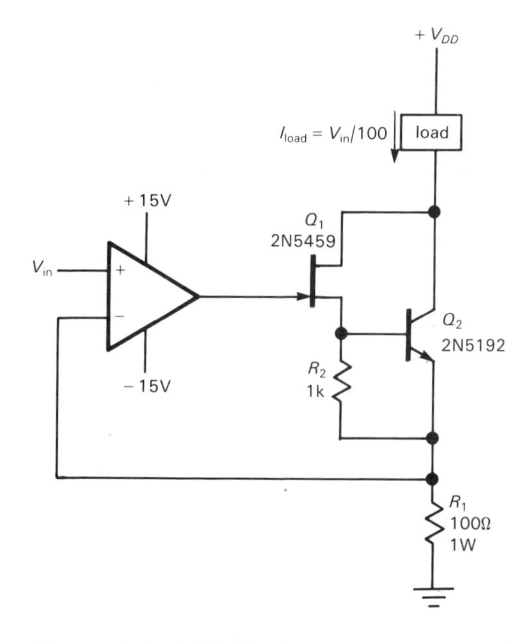

Figure 4.12. FET/bipolar current source suitable for high currents.

conduct when Q_1 is drawing about 0.6mA drain current. With Q_1's minimum I_{DSS}

of 4mA and a reasonable value for Q_2's beta, load currents of 100mA or more can be generated (Q_2 can be replaced by a Darlington for much higher currents, and in that case R_1 should be reduced accordingly). We've used a JFET in this particular circuit, although a MOSFET would be fine; in fact, it would be better, since with a JFET (which is a depletion-mode device) the op-amp must be run from split supplies to ensure a gate voltage range sufficient for pinch-off. It's worth noting that you can get plenty of current with a simple power MOSFET ("VMOS"); but the high interelectrode capacitances of power FETs may cause problems that you avoid with the hybrid circuit here.

Howland current source

Figure 4.13 shows a nice "textbook" current source. If the resistors are chosen so that $R_3/R_2 = R_4/R_1$, then it can be shown that $I_{\text{load}} = -V_{\text{in}}/R_2$.

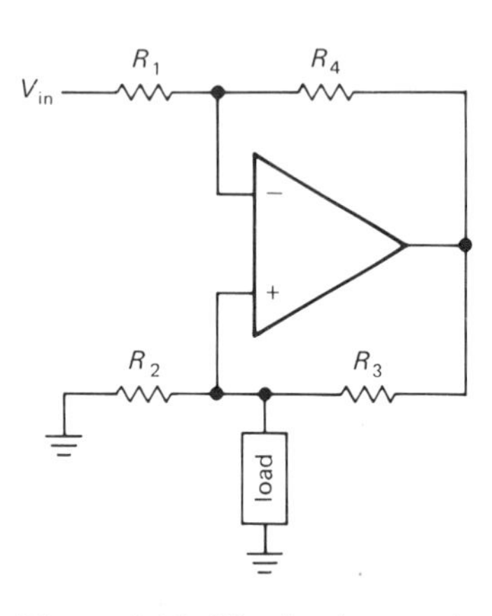

Figure 4.13. Howland current source.

EXERCISE 4.2
Show that the preceding result is correct.

This sounds great, but there's a hitch: The resistors must be matched exactly; otherwise it isn't a perfect current source. Even so, its performance is limited by the CMRR of the op-amp. For large output currents, the resistors must be small, and the compliance is limited. Also, at high frequencies (where the loop gain is low, as we'll learn shortly) the output impedance can drop from the desired value of infinity to as little as a few hundred ohms (the op-amp's open-loop output impedance). As clever as it looks, the Howland current source is not widely used.

4.08 Basic cautions for op-amp circuits

1. In all op-amp circuits, golden rules I and II (Section 4.03) will be obeyed only if the op-amp is in the active region, i.e., inputs and outputs not saturated at one of the supply voltages.

For instance, overdriving one of the amplifier configurations will cause output clipping at output swings near V_{CC} or V_{EE}. During clipping, the inputs will no longer be maintained at the same voltage. The op-amp output cannot swing beyond the supply voltages (typically it can swing only to within 2V of the supplies, though certain op-amps are designed to swing all the way to one supply or the other). Likewise, the output compliance of an op-amp current source is set by the same limitation. The current source with floating load, for instance, can put a maximum of $V_{CC} - V_{\text{in}}$ across the load in the "normal" direction (current in the same direction as applied voltage) and $V_{\text{in}} - V_{EE}$ in the reverse direction (the load could be rather strange, e.g., it might contain batteries, requiring the reverse sense of voltage to get a forward current; the same thing might happen with an inductive load driven by changing currents).

2. The feedback must be arranged so that it is negative. This means (among other things) that you must not mix up the inverting and noninverting inputs.

3. There must always be feedback at dc in an op-amp circuit. Otherwise the op-amp is guaranteed to go into saturation.

For instance, we were able to put a capacitor from the feedback network to ground in the noninverting amplifier (to reduce gain at dc to 1, Fig. 4.7), but we could not similarly put a capacitor in series between the output and the inverting input.

4. Many op-amps have a relatively small maximum differential input voltage limit. The maximum voltage difference between the inverting and noninverting inputs might be limited to as little as 5 volts in either polarity. Breaking this rule will cause large input currents to flow, with degradation or destruction of the op-amp.

We will take up some more issues of this type in Section 4.11 and again in Section 7.06 in connection with precision circuit design.

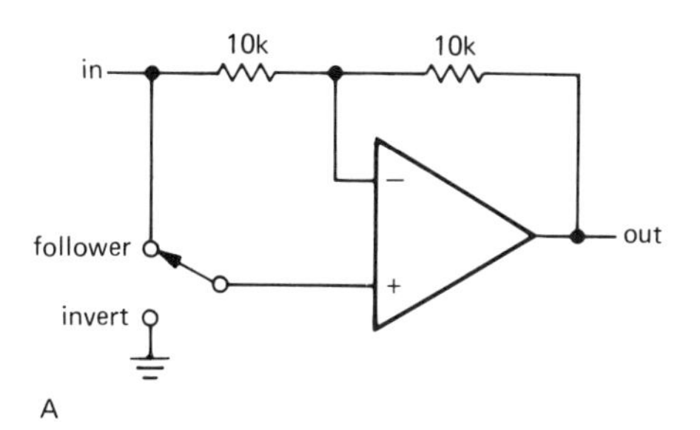

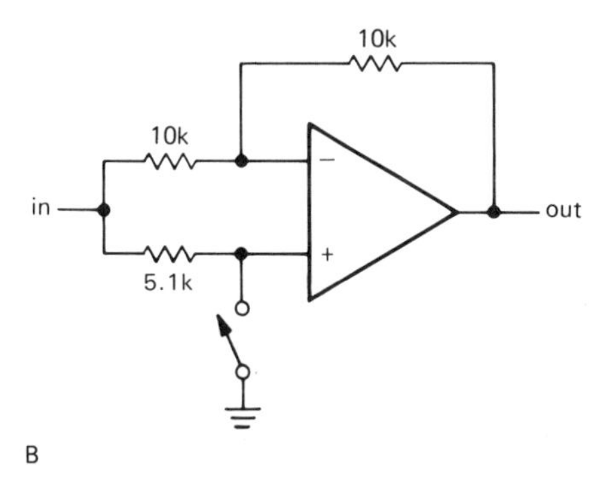

Figure 4.14

AN OP-AMP SMORGASBORD

In the following examples we will skip the detailed analysis, leaving that fun for you, the reader.

4.09 Linear circuits

Optional inverter

The circuits in Figure 4.14 let you invert, or amplify without inversion, by flipping a switch. The voltage gain is either +1 or −1, depending on the switch position.

EXERCISE 4.3

Show that the circuits in Figure 4.14 work as advertised.

Follower with bootstrap

As with transistor amplifiers, the bias path can compromise the high input impedance you would otherwise get with an op-amp,

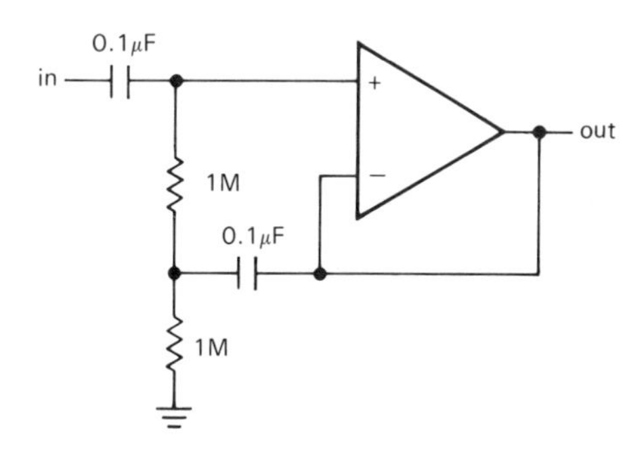

Figure 4.15

particularly with ac-coupled inputs, where a resistor to ground is mandatory. If that is a problem, the bootstrap circuit shown in Figure 4.15 is a possible solution. As in the transistor bootstrap circuit (Section 2.17), the 0.1μF capacitor makes the upper 1M resistor look like a high-impedance current source to input signals. The low-frequency rolloff for this circuit will begin at about 10Hz, dropping at 12dB per octave for frequencies somewhat below this. Note: You might be tempted to reduce the input coupling capacitor, since its load has been bootstrapped to high impedance. However, this can generate a peak in the frequency response, in the manner of an active filter (see Section 5.06).

Ideal current-to-voltage converter

Remember that the humble resistor is the simplest I-to-V converter. However, it has the disadvantage of presenting a nonzero impedance to the source of input current; this can be fatal if the device providing the input current has very little compliance or does not produce a constant current as the output voltage changes. A good example is a *photovoltaic cell,* a fancy name for a sun battery. Even the garden-variety signal diodes you use in circuits have a small photovoltaic effect (there are amusing stories of bizarre circuit behavior finally traced to this effect). Figure 4.16 shows the good way to convert current

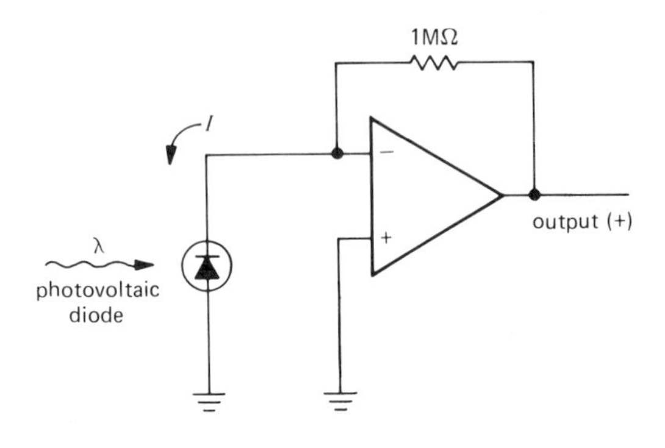

Figure 4.16

to voltage while holding the input strictly at ground. The inverting input is a virtual ground; this is fortunate, since a photovoltaic diode can generate only a few tenths of a volt. This particular circuit has an output of 1 volt per microamp of input current. (With BJT-input op-amps you sometimes see a resistor connected between the noninverting input and ground; its function will be explained shortly in connection with op-amp shortcomings.)

Of course, this *transresistance* configuration can be used equally well for devices that source their current via some positive excitation voltage, such as V_{CC}. Photomultiplier tubes and phototransistors (both devices that source current from a positive supply when exposed to light) are often used this way (Fig. 4.17).

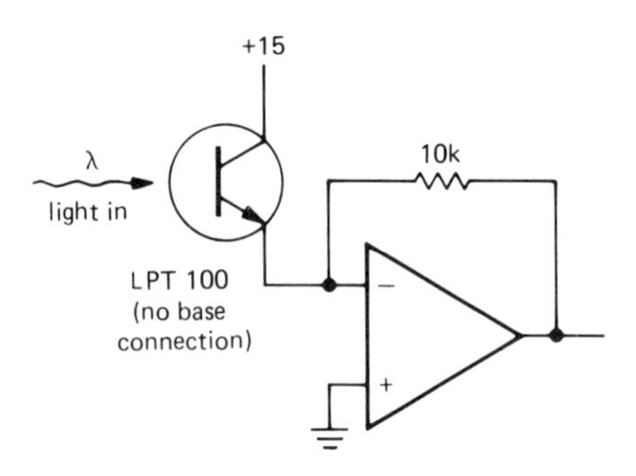

Figure 4.17

EXERCISE 4.4

Use a 411 and a 1mA (full scale) meter to construct a "perfect" current meter (i.e., one with zero input impedance) with 5mA full scale. Design the circuit so that the meter will never be driven more than ±150% full scale. Assume that the 411 output can swing to ±13 volts (±15V supplies) and that the meter has 500 ohms internal resistance.

Differential amplifier

The circuit in Figure 4.18 is a differential amplifier with gain R_2/R_1. As with the current source that used matched resistor ratios, this circuit requires precise resistor matching to achieve high common-mode rejection ratios. The best procedure is to stock up on a bunch of 100k 0.01% resistors next time you have a chance. All your differential amplifiers will have unity gain, but that's easily remedied with further (single-ended) stages of gain. We will treat differential amplifiers in more detail in Chapter 7.

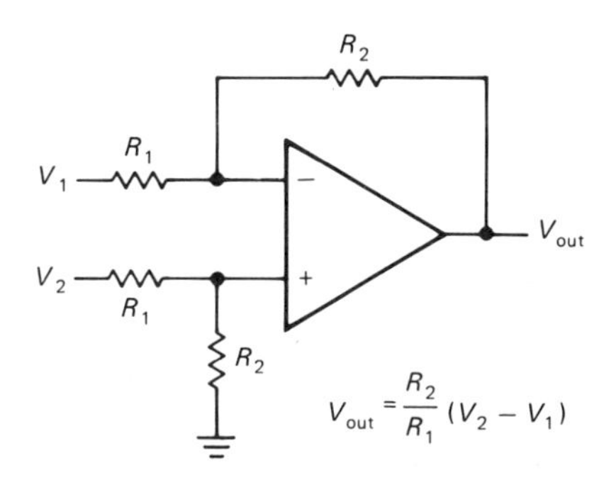

Figure 4.18. Classic differential amplifier.

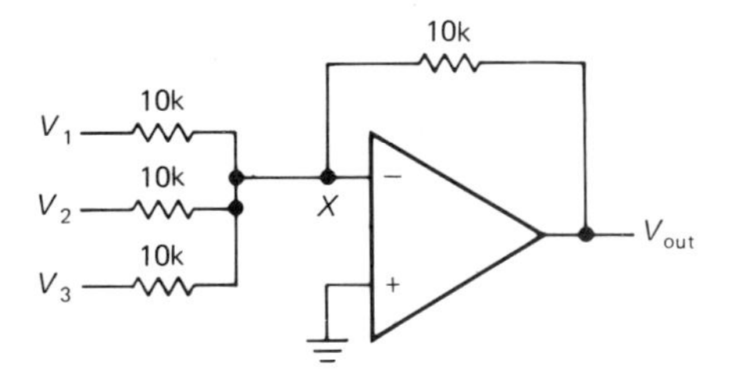

Figure 4.19

Summing amplifier

The circuit shown in Figure 4.19 is just a variation of the inverting amplifier. Point X is a virtual ground, so the input current is $V_1/R + V_2/R + V_3/R$. That gives $V_{\text{out}} = -(V_1 + V_2 + V_3)$. Note that the inputs can be positive or negative. Also, the input resistors need not be equal; if they're unequal, you get a weighted sum. For instance, you could have four inputs, each of which is +1 volt or zero, representing binary values 1, 2, 4, and 8. By using input resistors of 10k, 5k, 2.5k, and 1.25k you will get an output in volts equal to the binary count input. This scheme can be easily expanded to several digits. It is the basis of digital-to-analog conversion, although a different input circuit (an $R - 2R$ ladder) is usually used.

EXERCISE 4.5

Show how to make a two-digit digital-to-analog converter by appropriately scaling the input resistors in a summing amplifier. The digital input represents two digits, each consisting of four lines that represent the values 1, 2, 4, and 8 for the respective digits. An input line is either at +1 volt or at ground, i.e., the eight input lines represent 1, 2, 4, 8, 10, 20, 40, and 80. Because op-amp outputs generally cannot swing beyond ±13 volts, you will have to settle for an output in volts equal to one-tenth the value of the input number.

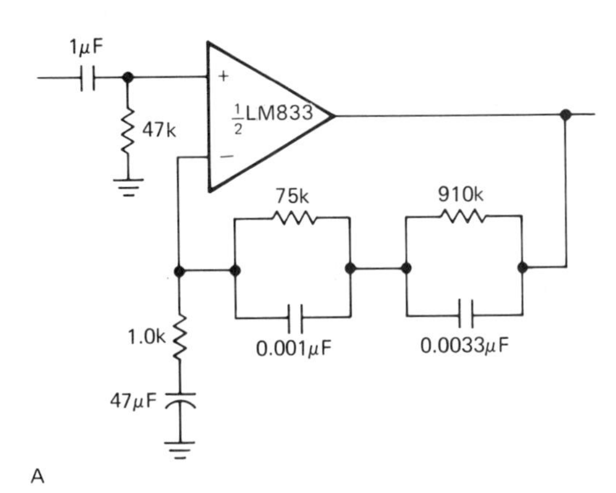

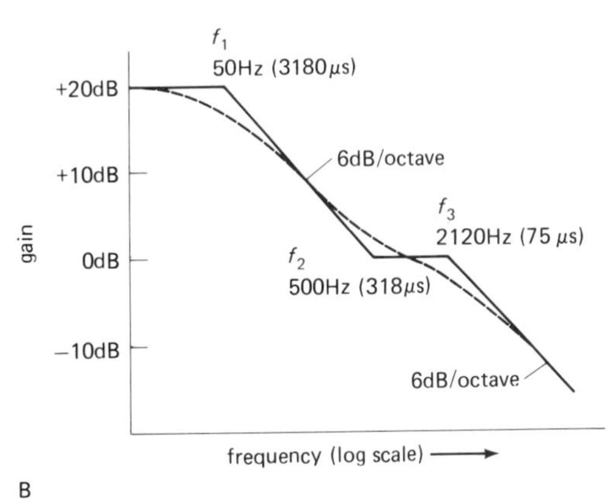

Figure 4.20. Op-amp RIAA phono playback amplifier.

RIAA preamp

The RIAA preamp is an example of an amplifier with a specifically tailored frequency response. Phonograph records are cut with approximately flat amplitude characteristics; magnetic pickups, on the other hand, respond to velocity, so a playback amplifier with rising bass response is required.

The circuit shown in Figure 4.20 produces the required response. The RIAA playback amplifier frequency response (relative to 0dB at 1kHz) is shown in the graph, with the breakpoints given in terms of time constants. The 47μF capacitor to ground rolls off the gain to unity at dc, where it would otherwise be about 1000; as we have hinted earlier, the reason is to avoid amplification of dc input "offsets." The LM833 is a low-noise dual op-amp intended for audio applications (a "gold-plated" op-amp for this application is the ultra-low-noise LT1028, which is 13dB quieter, and 10dB more expensive, than the 833!).

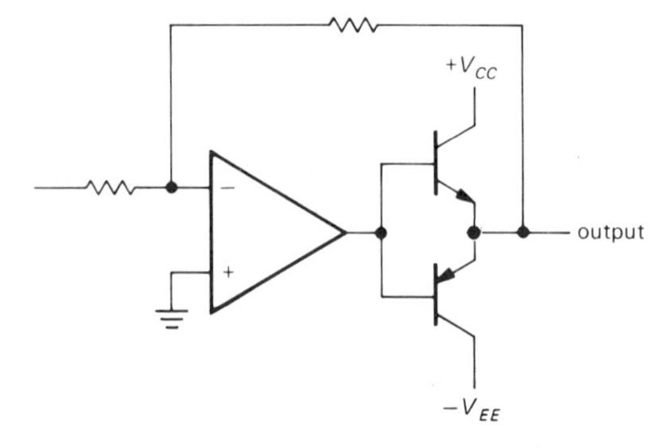

Figure 4.22

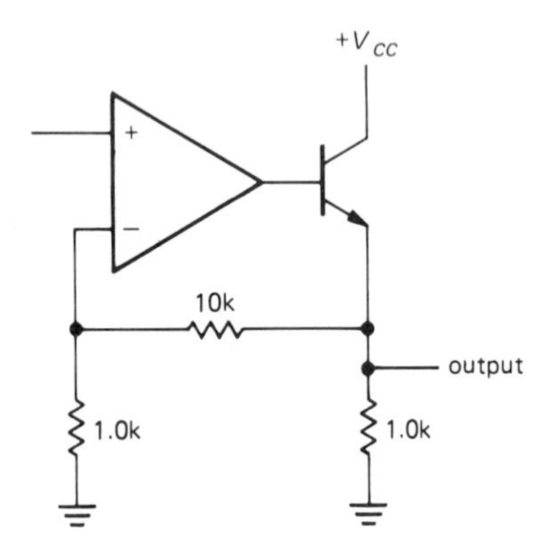

Figure 4.21

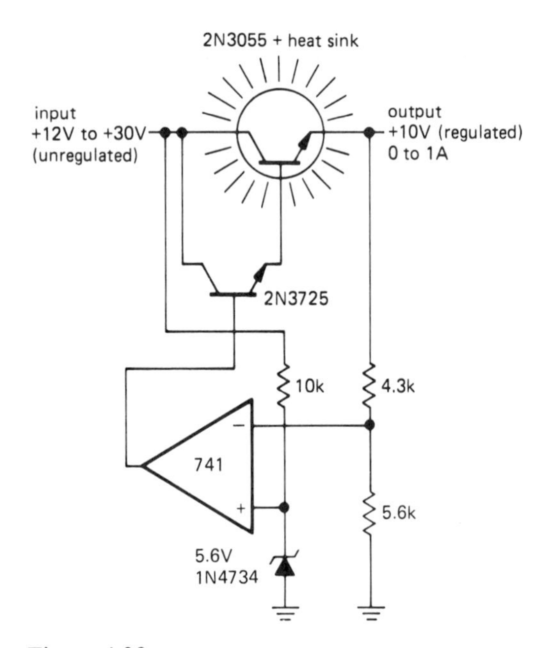

Figure 4.23

Power booster

For high output current, a power transistor follower can be hung on an op-amp output (Fig. 4.21). In this case a noninverting amplifier has been drawn; the follower can be added to any op-amp configuration. Notice that feedback is taken from the emitter; thus, feedback enforces the desired output voltage in spite of the V_{BE} drop. This circuit has the usual problem that the follower output can only *source* current. As with transistor circuits, the remedy is a push-pull booster (Fig. 4.22). You will see later that the limited speed with which the op-amp can move its output (slew rate) seriously limits the speed of this booster in the crossover region, creating distortion. For slow-speed applications you don't need to bias the push-pull pair into quiescent conduction, because feedback will take care of most of the crossover distortion. Commercial op-amp power boosters are available, e.g., the LT1010, OPA633, and 3553. These are unity-gain push-pull amplifiers capable of 200mA of output current and operation to 100MHz and above. You can include them inside the feedback loop without any worries (See Table 7.4).

Power supply

An op-amp can provide the gain for a feedback voltage regulator (Fig. 4.23). The op-amp compares a sample of the output with the zener reference, changing the drive to the Darlington "pass transistor" as needed. This circuit supplies 10 volts regulated, at up to 1 amp load current. Some notes about this circuit:

1. The voltage divider that samples the output could be a potentiometer, for adjustable output voltage.

2. For reduced ripple at the zener, the 10k resistor should be replaced by a current source. Another approach is to bias the zener from the output; that way you take advantage of the regulator you have built. Caution: When using this trick, you must analyze the circuit carefully to be sure it will start up when power is first applied.

3. The circuit as drawn could be damaged by a temporary short circuit across the output, because the op-amp would attempt to drive the Darlington pair into heavy conduction. Regulated power supplies should always have circuitry to limit "fault" current (see Section 6.05 for more details).

4. Integrated circuit voltage regulators are available in tremendous variety, from the time-honored 723 to the convenient 3-terminal adjustable regulators with internal current limit and thermal shutdown (see Tables 6.8–6.10). These devices, complete with temperature-compensated internal zener reference and pass transistor, are so easy to use that you will almost never use a general-purpose op-amp as a regulator. The exception might be to generate a stable voltage within a circuit that already has a stable power-supply voltage available.

In Chapter 6 we will discuss voltage regulators and power supplies in detail, including special ICs intended for use as voltage regulators.

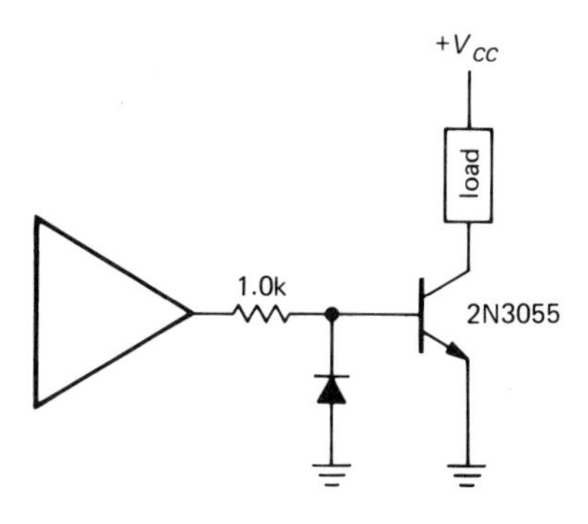

Figure 4.24

4.10 Nonlinear circuits

Power-switching driver

For loads that are either on or off, a switching transistor can be driven from an op-amp. Figure 4.24 shows how. Note the diode to prevent reverse base-emitter breakdown (op-amps easily swing more than −5V). The 2N3055 is everyone's power transistor for noncritical high-current applications. A Darlington (or power MOSFET) can be used if currents greater than about 1 amp need to be driven.

Active rectifier

Rectification of signals smaller than a diode drop cannot be done with a simple diode-resistor combination. As usual, op-amps come to the rescue, in this case by putting a diode in the feedback loop (Fig. 4.25). For V_{in} positive, the diode provides negative feedback; the output follows the input, coupled by the diode, but without a V_{BE} drop. For V_{in} negative, the op-amp goes into negative saturation and V_{out} is at ground. R could be chosen smaller for lower output impedance, with the tradeoff of higher op-amp output current. A better solution is to use an op-amp follower at the output, as shown, to produce very low output impedance regardless of the resistor value.

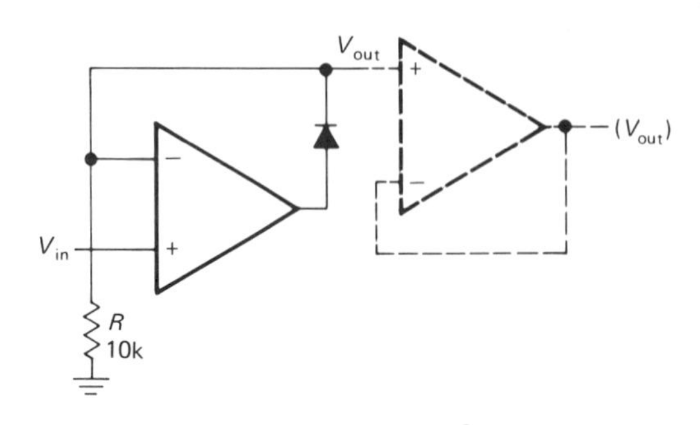

Figure 4.25. Simple active rectifier.

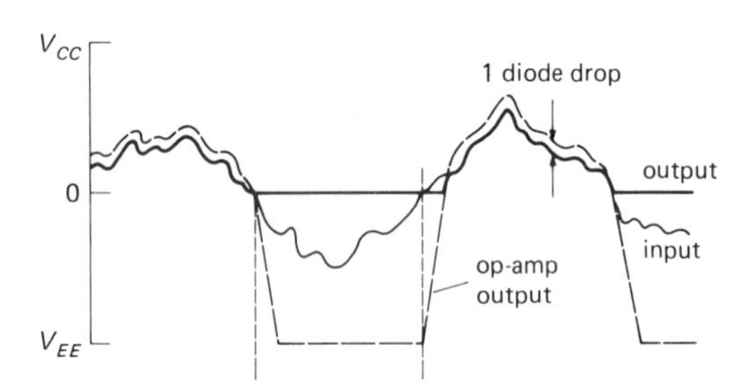

Figure 4.26. Effect of finite slew rate on the simple active rectifier.

There is a problem with this circuit that becomes serious with high-speed signals. Because an op-amp cannot swing its output infinitely fast, the recovery from negative saturation (as the input waveform passes through zero from below) takes some time, during which the output is incorrect. It looks something like the curve shown in Figure 4.26. The output (heavy line) is an accurate rectified version of the input (light line), except for a short time interval after the input rises through zero volts. During that interval the op-amp output is racing up from saturation near $-V_{EE}$, so the circuit's output is still at ground. A general-purpose op-amp like the 411 has a *slew rate* (maximum rate at which the output can change) of 15 volts per microsecond; recovery from negative saturation therefore takes about 1μs, which may introduce significant output error for fast signals. A circuit modification improves the situation considerably (Fig. 4.27).

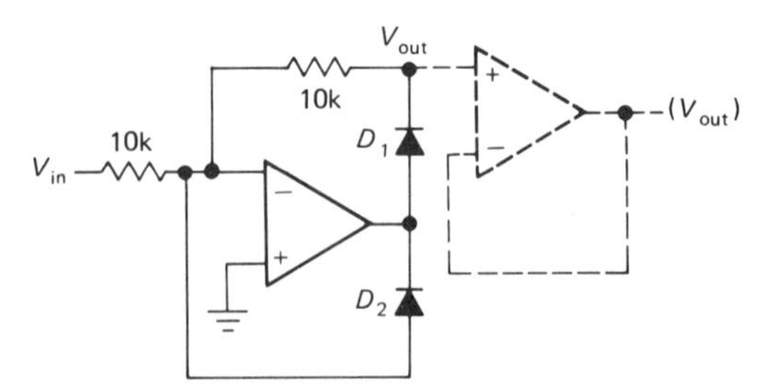

Figure 4.27. Improved active rectifier.

D_1 makes the circuit a unity-gain inverter for negative input signals. D_2 clamps the op-amp's output at one diode drop below ground for positive inputs, and since D_1 is then back-biased, V_{out} sits at ground. The improvement comes because the op-amp's output swings only two diode drops as the input signal passes through zero. Since the op-amp output has to slew only about 1.2 volts instead of V_{EE} volts, the "glitch" at zero crossings is reduced more than tenfold. This rectifier is inverting, incidentally. If you require a noninverted output, attach a unity-gain inverter to the output.

The performance of these circuits is improved if you choose an op-amp with a high slew rate. Slew rate also influences the performance of the other op-amp applications we've discussed, for instance the simple voltage amplifier circuits. At this point it is worth pausing for a while to see in what ways real op-amps depart from the ideal, since that influences circuit design, as we have hinted on several occasions. A good understanding of op-amp limitations and their influence on circuit design and performance will help you choose your op-amps wisely and design with them effectively.

A DETAILED LOOK AT OP-AMP BEHAVIOR

Figure 4.28 shows the schematic of the 741, a very popular op-amp. Its circuit is relatively straightforward, in terms of the kinds of transistor circuits we discussed in the last chapter. It has a differential input stage with current mirror load, followed by a common-emitter *npn* stage (again with active load) that provides most of the voltage gain. A *pnp* emitter follower drives the push-pull emitter follower output stage, which includes current-limiting circuitry. This circuit is typical of many op-amps now available. For many applications the properties of these amplifiers approach ideal op-amp performance characteristics. We will now take a look at the extent to which real op-amps depart from the ideal, what the consequences are for circuit design, and what to do about it.

4.11 Departure from ideal op-amp performance

The ideal op-amp has these characteristics:

1. Input impedance (differential or common mode) = infinity

2. Output impedance (open loop) = 0

3. Voltage gain = infinity

4. Common-mode voltage gain = 0

5. V_{out} = 0 when both inputs are at the same voltage (zero "offset voltage")

6. Output can change instantaneously (infinite slew rate)

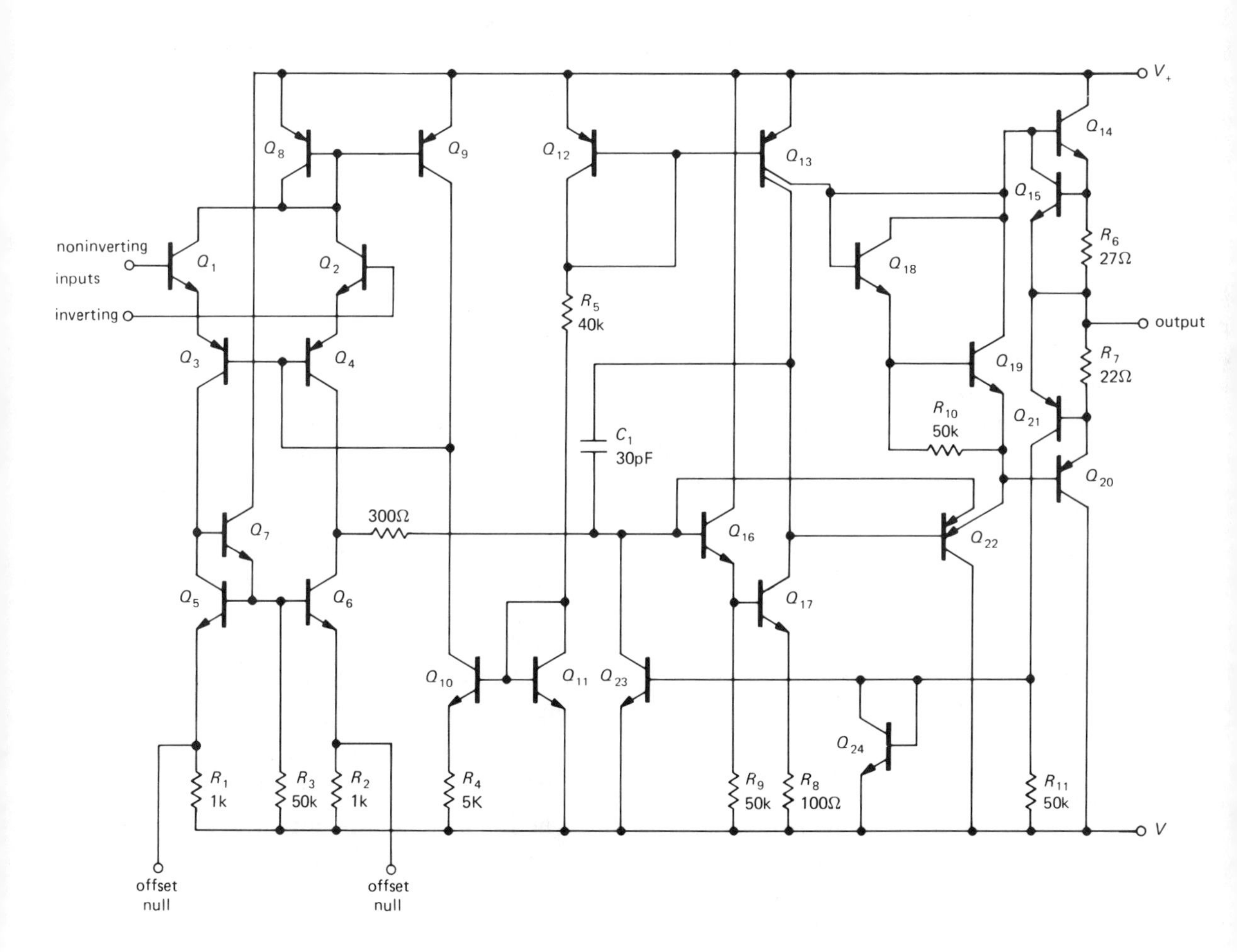

Figure 4.28. Schematic of the 741 op-amp. (Courtesy of Fairchild Camera and Instrument Corp.)

All of these characteristics are independent of temperature and supply voltage changes.

Real op-amps depart from these characteristics in the following ways (see Table 4.1 for some typical values).

Input current

The input terminals sink (or source, depending on the op-amp type) a small current called the input bias current, I_B, which is defined as half the sum of the input currents with the inputs tied together (the two input currents are approximately equal and are simply the base or gate currents of the input transistors). For the JFET-input 411 the bias current is typically 50pA at room temperature (but as much as 2nA at 70°C), while a typical BJT-input op-amp like the OP-27 has a typical bias current of 15nA, varying little with temperature. As a rough guide, BJT-input op-amps have bias currents in the tens of nanoamps, while FET-input op-amps have input currents in the tens of picoamps (i.e., 1000 times lower). Generally speaking, you can ignore input current with FET op-amps, but not with bipolar-input op-amps.

The significance of input bias current is that it causes a voltage drop across the resistors of the feedback network, bias network, or source impedance. How small a resistor this restricts you to depends on the dc gain of your circuit and how much output variation you can tolerate. You will see how this works later.

Op-amps are available with input bias currents down to a nanoamp or less for (bipolar) transistor-input circuit types or down to a few picoamps ($10^{-6}\mu$A) for FET-input circuit types. The very lowest bias currents are typified by the superbeta Darlington LM11, with a maximum input current of 50pA, the AD549, with an input current of 0.06pA, and the MOSFET ICH8500, with an input current of 0.01pA. In general, transistor op-amps intended for high-speed operation have higher bias currents.

Input offset current

Input offset current is a fancy name for the difference in input currents between the two inputs. Unlike input bias current, the offset current, I_{os}, is a result of manufacturing variations, since an op-amp's symmetrical input circuit would otherwise result in identical bias currents at the two inputs. The significance is that even when it is driven by identical source impedances, the op-amp will see unequal voltage drops and hence a difference voltage between its inputs. You will see shortly how this influences design.

Typically, the offset current is one-half to one-tenth the bias current. For the 411, $I_{\text{offset}} = 25\text{pA}$, typical.

□ *Input impedance*

Input impedance refers to the differential input resistance (impedance looking into one input, with the other input grounded), which is usually much less than the common-mode resistance (a typical input stage looks like a long-tailed pair with current source). For the FET-input 411 it is about 10^{12} ohms, while for BJT-input op-amps like the 741 it is about 2MΩ. Because of the input bootstrapping effect of negative feedback (it attempts to keep both inputs at the same voltage, thus eliminating most of the differential input signal), Z_{in} in practice is raised to very high values and usually is not as important a parameter as input bias current.

□ *Common-mode input range*

The inputs to an op-amp must stay within a certain voltage range, typically less than the full supply range, for proper operation. If the inputs go beyond this range, the gain of the op-amp may change drastically, even reversing sign! For a 411 operating from ±15 volt supplies, the guaranteed common-mode input range is ±11 volts minimum. However, the manufacturer claims that the 411 will operate with common-mode inputs all the way to the positive supply, though performance may be degraded. Bringing either input down to the negative supply voltage causes the amplifier to go berserk, with symptoms like phase reversal and output saturation to the positive supply.

There are op-amps available with common-mode input ranges down to the negative supply, e.g., the LM358 (a good dual op-amp) or the LM10, CA3440, or OP-22, and up to the positive supply, e.g., the 301, OP-41, or the 355 series. In addition to the *operating* common-mode range, there are maximum allowable input voltages beyond which damage will result. For the 411 they are ±15 volts (but not to exceed the negative supply voltage, if it is less).

Differential input range

Some bipolar op-amps allow only a limited voltage between the inputs, sometimes as small as ±0.5 volt, although most are more forgiving, permitting differential inputs nearly as large as the supply voltages. Exceeding the specified maximum can degrade or destroy the op-amp.

□ *Output impedance; output swing versus load resistance*

Output impedance R_0 means the op-amp's intrinsic output impedance *without feedback.* For the 411 it is about 40 ohms, but with some low-power op-amps it can be as high as several thousand ohms (see Fig. 7.16). Feedback lowers the output impedance into insignificance (or raises it, for a current source); so what usually matters more is the maximum output current, with typical values of 20mA or so. This is frequently given as a graph of output voltage swing V_{om} as a function of load resistance, or sometimes just a few values for typical load resistances. Many op-amps have asymmetrical output drive capability, with the ability to sink more current than they can source (or vice versa). For the 411, output swings to within about 2 volts of V_{CC} and V_{EE} are possible into load resistances greater than about 1k. Load resistances significantly less than that will permit only a small swing. Some op-amps can produce output swings all the way down to the negative supply (e.g., the LM358), a particularly useful feature for circuits operated from a single positive supply, since output swings all the way to ground are then possible. Finally, op-amps with MOS transistor outputs (e.g., the CA3130, 3160, ALD1701, and ICL761*x*) can swing all the way to both rails. The remarkable bipolar LM10 shares this property, without the limited supply voltage range of the MOS op-amps (usually ±8V max).

□ *Voltage gain and phase shift*

Typically the voltage gain A_{vo} at dc is 100,000 to 1,000,000 (often specified in decibels), dropping to unity gain at a frequency (called f_T) of 1MHz to 10MHz. This is usually given as a graph of open-loop voltage gain as a function of frequency. For *internally compensated* op-amps this graph is simply a 6dB/octave rolloff beginning at some fairly low frequency (for the 411 it begins at about 10Hz), an intentional characteristic necessary for stability, as you will see in Section 4.32. This rolloff (the same as a simple RC low-pass filter) results in a constant 90° lagging phase shift from input to output (open-loop) at all frequencies above the beginning of the rolloff, increasing to 120° to 160° as the open-loop gain approaches unity. Since a 180° phase shift at a frequency where the voltage gain equals 1 will result in positive feedback (oscillations), the term "phase margin" is used to specify the difference between the phase shift at f_T and 180°.

Input offset voltage

Op-amps don't have perfectly balanced input stages, owing to manufacturing variations. If you connect the two inputs together for zero input signal, the output will usually saturate at either V_{CC} or V_{EE} (you can't predict which). The difference in input voltages necessary to bring the output to zero is called the input offset voltage V_{os} (it's as if there were a battery of that voltage in series with one of the inputs). Usually op-amps make provision for trimming the input offset voltage to zero. For a 411 you use a 10k pot between pins 1 and 5, with the wiper connected to V_{EE}.

Of greater importance for precision applications is the drift of the input offset voltage with temperature and time, since any initial offset can be trimmed to zero. A 411 has a typical offset voltage of 0.8mV (2mV maximum), with temperature coefficient ("tempco") of 7μV/°C and unspecified coefficient of offset drift with time. The OP-77, a precision op-amp, is laser-trimmed for a typical offset of 10 *micro*volts, with temperature coefficient TCV_{os} of 0.2μV/°C and long-term drift of 0.2μV/month.

Slew rate

The op-amp "compensation" capacitance (discussed further in Section 4.32) and small internal drive currents act together to limit the rate at which the output can change, even when a large input unbalance occurs. This limiting speed is usually specified as *slew rate* or slewing rate (SR). For the 411 it is 15V/μs; low-power op-amps typically have slew rates less than 1V/μs, while a high-speed op-amp might slew at 100V/μs, and the LH0063C "damn fast buffer" slews at 6000V/μs. The slew rate limits the amplitude of an undistorted sine-wave output swing above some critical frequency (the frequency at which the full supply swing requires the maximum slew rate of the op-amp, Fig. 4.29), thus explaining the "output voltage swing as a function of frequency" graph. A sine wave of frequency f hertz and amplitude A volts requires a minimum slew rate of $2\pi Af$ volts per second.

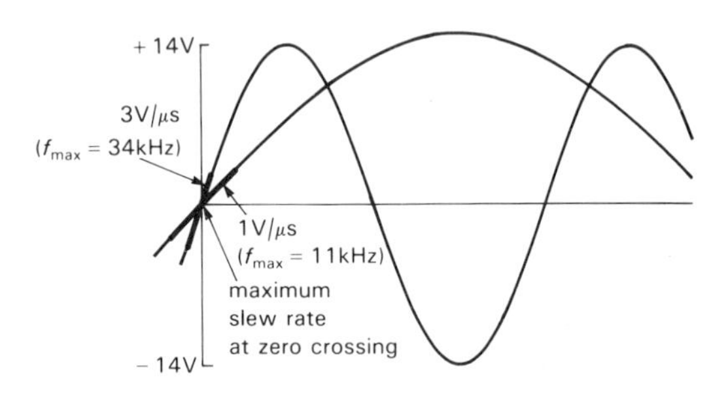

Figure 4.29. Slew-rate-induced distortion.

For externally compensated op-amps the slew rate depends on the compensation network used. In general, it will be lowest for "unity gain compensation," increasing to perhaps 30 times faster for ×100 gain compensation. This is discussed further in Section 4.32.

Temperature dependence

All these parameters have some temperature dependence. However, this usually doesn't make any difference, since small variations in gain, for example, are almost entirely compensated by feedback. Furthermore, the variations of these parameters with temperature are typically small compared with the variations from unit to unit.

The exceptions are input offset voltage and input offset current. This will matter, particularly if you've trimmed the offsets approximately to zero, and will appear as drifts in the output. When high precision is important, a low-drift "instrumentation" op-amp should be used, with external loads kept above 10k to minimize the horrendous effects on input-stage performance caused by temperature gradients. We will have much more to say about this subject in Chapter 7.

For completeness, we should mention here that op-amps are also limited in common-mode rejection ratio (CMRR), power-supply rejection ratio (PSRR), input noise voltage and current (e_n, i_n), and output crossover distortion. These become significant limitations only in connection with precision circuits and low-noise amplifiers, and they will be treated in Chapter 7.

4.12 Effects of op-amp limitations on circuit behavior

Let's go back and look at the inverting amplifier with these limitations in mind. You will see how they affect performance, and you will learn how to design effectively in spite of them. With the understanding you will get from this example, you should be able to handle other op-amp circuits. Figure 4.30 shows the circuit again.

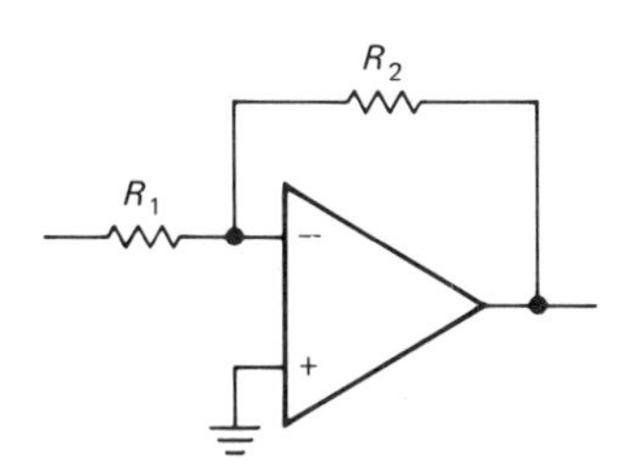

Figure 4.30

Open-loop gain

Because of finite open-loop gain, the voltage gain of the amplifier with feedback (closed-loop gain) will begin dropping at a frequency where the open-loop gain approaches R_2/R_1 (Fig. 4.31). For garden-variety op-amps like the 411, this means that you're dealing with a relatively low frequency amplifier; the open-loop gain is down to 100 at 50kHz, and f_T is 4MHz. Note that the closed-loop gain is always less than the open-loop gain; this means, for instance, that a ×100 amplifier built with a 411 will show a noticeable falloff of gain for frequencies approaching 50kHz. Later in the chapter (Section 4.25), when we deal with transistor feedback circuits with finite open-loop gains, we will have a more accurate statement of this behavior.

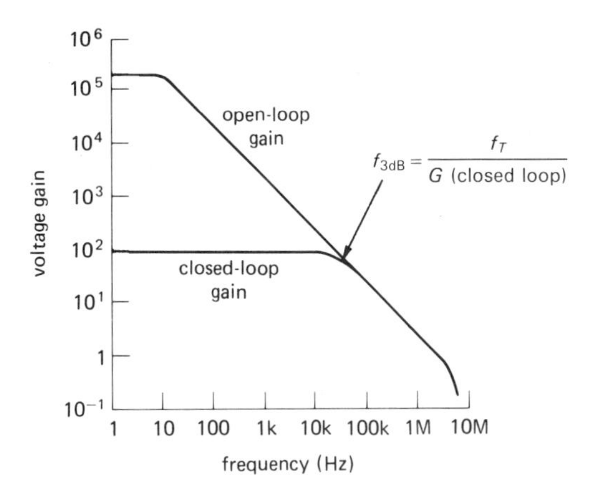

Figure 4.31. LF411 gain versus frequency ("Bode plot").

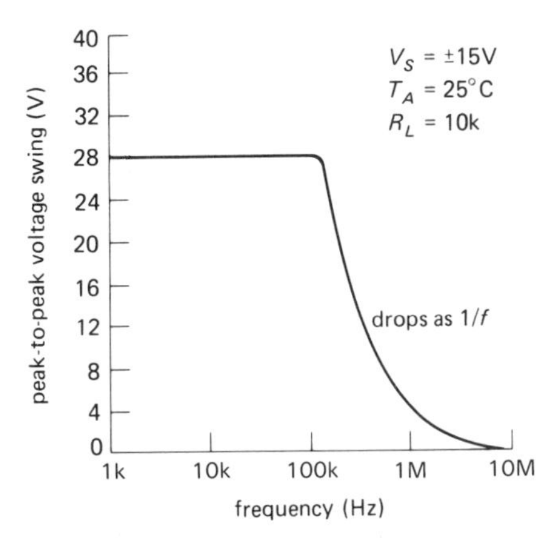

Figure 4.32. Output swing versus frequency (LF411).

Slew rate

Because of limited slew rate, the maximum undistorted sine-wave output swing drops above a certain frequency. Figure 4.32 shows the curve for a 411, with its 15V/μs slew rate. For slew rate S, the output amplitude is limited to $A(\text{pp}) \leq S/\pi f$ for a sine wave of frequency f, thus explaining the $1/f$ dropoff of the curve. The flat portion of the curve reflects the power-supply limits of output voltage swing.

As an aside, the slew-rate limitation of op-amps can be usefully exploited to filter sharp noise spikes from a desired signal, with a technique known as *nonlinear low-pass filtering*: By deliberately limiting the slew rate, the fast spikes can be dramatically reduced without any distortion of the underlying signal.

Output current

Because of limited output current capability, an op-amp's output swing is reduced for small load resistances. Figure 4.33 shows the graph for a 411. For precision applications it is a good idea to avoid large output currents in order to prevent on-chip thermal gradients produced by excessive power dissipation in the output stage.

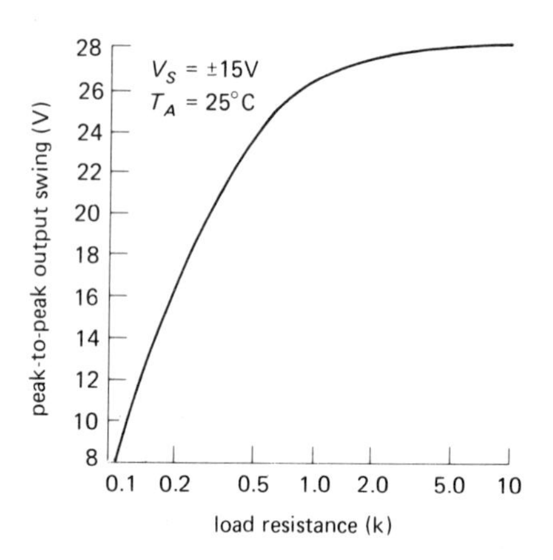

Figure 4.33. Output swing versus load (LF411).

Offset voltage

Because of input offset voltage, a zero input produces an output of $V_{\text{out}} = G_{dc}V_{os}$. For an inverting amplifier with voltage gain of 100 built with a 411, the output could be as large as ±0.2 volt when the input is grounded (V_{os} = 2mV max). Solutions: (a) If you don't need gain at dc, use a capacitor to drop the gain to unity at dc, as in Figure 4.7, as well as the RIAA amplifier circuit (Fig. 4.20). In this case you could do that by capacitively coupling the input signal. (b) Trim the voltage offset to zero using the manufacturer's recommended trimming network. (c) Use an op-amp with smaller V_{os}. (d) Trim the voltage offset to zero using an external trimming network as described in Section 7.06 (Fig. 7.5).

Input bias current

Even with a perfectly trimmed op-amp (i.e., $V_{os} = 0$), our inverting amplifier circuit will produce a non-zero output voltage when its input terminal is connected to ground. That is because the finite input bias current, I_B, produces a voltage drop across the resistors, which is then amplified by the circuit's voltage gain. In this circuit the inverting input sees a driving impedance of $R_1 \parallel R_2$, so the bias current produces a voltage $V_{\text{in}} = I_B(R_1 \parallel R_2)$, which is then amplified by the gain at dc, $-R_2/R_1$.

With FET-input op-amps the effect is usually negligible, but the substantial input current of bipolar op-amps can cause real problems. For example, consider an inverting amplifier with R_1 = 10k and R_2 = 1M; these are reasonable values for an inverting stage, where we might like to keep Z_{in} at least 10k. If we chose the low-noise bipolar LM833, the output (for grounded input) could be as large as 100×1000nA×9.9k, or 0.99 volt ($G_{dc}I_BR_{\text{unbalance}}$), which is unacceptable. By comparison, for our jellybean LF411 (JFET-input) op-amp the corresponding worst-case output (for grounded input) is 0.2mV; for most applications this is negligible, and in any case is dwarfed by the

V_{os}-produced output error (200mV, worst-case untrimmed, for the LF411).

There are several solutions to the problem of bias-current errors. If you must use an op-amp with large bias current, it is a good idea to ensure that both inputs see the same dc driving resistance, as in Figure 4.34. In this case, 9.1k is chosen as the parallel resistance of 10k and 100k. In addition, it is best to keep the resistance of the feedback network small enough so that bias current doesn't produce large offsets; typical values for the resistance seen from the op-amp inputs are 1k to 100k or so. A third cure involves reducing the gain to unity at dc, as in the RIAA amplifier earlier.

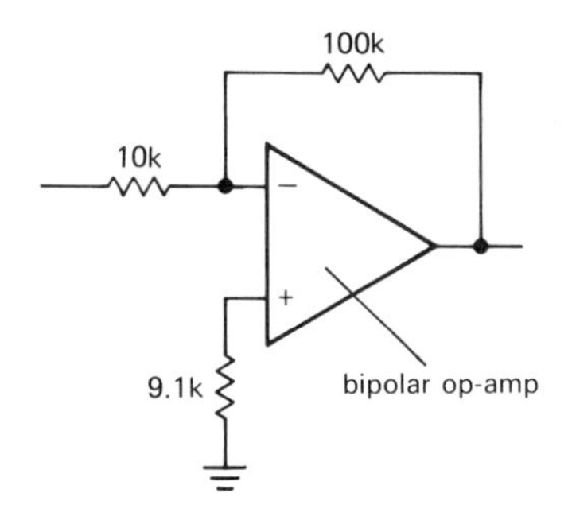

Figure 4.34. With bipolar op-amps, use a compensation resistor to reduce errors caused by input bias current.

In most cases, though, the simplest solution is to use op-amps with negligible input current. Op-amps with JFET or MOSFET input stages generally have input currents in the picoamp range (watch out for its rapid rise versus temperature, though, roughly doubling every 10°C), and many modern bipolar designs use superbeta transistors or bias-cancellation schemes to achieve bias currents nearly as low, *decreasing* slightly with temperature. With these op-amps, you can have the advantages of bipolar op-amps (precision, low noise) without the annoying problems caused by input current. For example, the precision low-noise bipolar OP-27 has I_B =10nA (typ), the inexpensive bipolar LM312 has I_B =1.5nA (typ), and its improved bipolar cousins (the LT1012 and LM11) have I_B = 30pA (typ). Among inexpensive FET op-amps, the JFET LF411 has I_B = 50pA (typ), and the MOSFET TLC270 series, priced under a dollar, have I_B = 1pA (typ).

Input offset current

As we just described, it is usually best to design circuits so that circuit impedances, combined with op-amp bias current, produce negligible errors. However, occasionally it may be necessary to use an op-amp with high bias current, or to deal with signals of extraordinarily high Thévenin impedances. In that case the best you can do is to balance the dc driving resistances seen by the op-amp at its input terminals. There will still be some error at the output ($G_{dc}I_{\text{offset}}R_{\text{source}}$), due to unavoidable asymmetry in the op-amp input currents. In general, I_{offset} is smaller than I_{bias} by a factor of 2 to 20 (with bipolar op-amps generally showing better matching than FET op-amps).

In the preceding paragraphs we have discussed the effects of op-amp limitations, taking the example of the simple inverting voltage amplifier circuit. Thus, for example, op-amp input current caused a *voltage* error at the output. In a different op-amp application you may get a different effect; for example, in an op-amp integrator circuit, finite input current produces an output *ramp* (rather than a constant) with zero applied input. As you become familiar with op-amp circuits you will be able to predict the effects of op-amp limitations in a given circuit and therefore choose which op-amp to use in a given application. In general, there is no "best" op-amp (even when price is no object): For example,

TABLE 4.1. OPERATIONAL AMPLIFIERS

Type	Mfg[a]	# per pkg[b] 1	2	4	Trim	Ext comp[c]	Min gain[d]	Total supply voltage min (V)	Total supply voltage max (V)	Supp curr max (mA)	Voltage Offset typ (mV)	Voltage Offset max (mV)	Voltage Drift typ (μV/°C)	Voltage Drift max (μV/°C)	Current Bias max (nA)	Current Offset max (nA)	e_n @1kHz typ nV/√Hz
BIPOLAR, PRECISION																	
OP-07A	PM+	•	A	–	•	–	1	6	44	4	0.01	0.025	0.2	0.6	2	2	9.6
OP-07E	PM+	•	A	–	•	–	1	6	44	4	0.03	0.08	0.3	1.3	4	3.8	9.6
OP-21A	PM	•	A	A	•	–	1	5	36	0.3	0.04	0.1	0.5	1	100	4	21
OP-27E	PM+	•	A	A	•	–	1	8	44	5	0.01	0.025	0.2	0.6	40	35	3.0
OP-27G	PM+	•	A	A	•	–	1	8	44	6	0.03	0.1	0.4	1.8	80	75	3.2
OP-37E	PM+	•	A	–	•	–	5	8	44	5	0.01	0.025	0.2	0.6	40	35	3.0
OP-50E	PM	•	–	–	•	•	5	10	36	4	0.01	0.025	0.15	0.3	5	1	4.5
OP-77E	PM	•	A	A	•	–	1	6	44	2	0.01	0.025	0.1	0.3	2	1.5	9.6
OP-90E	PM	•	A	A	•	–	1	1.6	36	0.02	0.05	0.15	0.3	2	15	3	60
OP-97E	PM	•	–	–	•	–	1	4.5	40	0.6	0.01	0.025	0.2	0.6	0.1	0.1	14
MAX400M	MA	•	–	–	•	–	1	6	44	4	0.004	0.01	0.2	0.3	2	2	9.6
LM607A	NS	•	–	–	•	•	1	6	44	1.5	0.015	0.025	0.2	0.3	2	2	6.5
AD707C	AD	•	A	–	•	–	1	6	36	3	0.005	0.015	0.03	0.1	1	1	9.6
AD846B	AD	•	–	–	–	•	2	10	36	6.5	0.025	0.075	0.8	3.5	250	(k)	2
LT1001A	LT	•	A	–	•	–	1	6	44	3.3	0.01	0.025	0.2	0.6	4	4	9.6
LT1007A	LT	•	–	–	•	–	1	5	44	4	0.01	0.025	0.2	0.6	35	30	2.5
LT1012C	LT+	•	A	–	•	•	1	4	40	0.6	0.01	0.05	0.2	1.5	0.15	0.15	14
LT1028A	LT	•	–	–	•	•	1	8	44	9.5	0.01	0.04	0.2	0.8	90	50	0.9
LT1037A	LT	•	–	–	•	–	5	5	44	4.5	0.01	0.025	0.2	0.6	35	30	2.5
RC4077A	RA	•	–	–	•	–	1	6	44	1.7	0.004	0.01	0.1	0.3	2	1.5	9.6
HA5134A	HA	–	–	•	–	–	1	10	40	8	0.05	0.1	0.3	1.2	25	25	7
HA5135	HA	•	–	–	•	–	1	8	40	1.7	0.01	0.08	0.4	1.3	4	4	9
HA5147A	HA	•	–	–	•	–	10	8	44	4	0.01	0.025	0.2	0.6	40	35	3.0
BIPOLAR, LOW-BIAS (see also "bipolar, precision")																	
OP-08E	PM	•	–	–	–	•	U	10	40	0.5	0.07	0.15	0.5	2.5	2	0.2	20
LM10	NS+	•	–	–	•	–	1	1	45	0.4	0.3	2	2	-	20	0.7	47
LM11	NS+	•	–	–	•	•	1	5	40	0.6	0.1	0.3	1	3	50pA	10pA	150
OP-12E	PM+	•	–	–	–	–	1	10	40	0.5	0.07	0.15	0.5	2.5	2	0.2	20
LM308	NS+	•	A	–	–	•	U	10	36	0.8	2	7.5	6	30	7	1	35
LM312	NS+	•	–	–	•	•	1	10	40	0.8	2	7.5	6	30	7	1	35
LP324	NS	–	–	•	–	–	1	4	32	0.25	2	4	10	-	10	2	-
BIPOLAR, SINGLE-SUPPLY																	
324A	NS+	A	A	•	–	–	1	3	32	3	2	3	7	30	100	30	-
LP324	NS	–	–	•	–	–	1	4	32	0.25	2	4	10	-	10	2	-
LT1013C	LT	–	•	A	–	–	1	4	44	1	0.06	0.3	0.4	2.5	50	2	22
HA5141A	HA	•	A	A	–	–	1	2	40	0.07	0.5	2	3	-	75	10	20
BIPOLAR, SINGLE-SUPPLY PRECISION																	
LT1006A	LT	•	–	–	•	–	1	2.7	44	0.5	0.02	0.05	0.2	1.3	15	0.5	22
LT1013A	LT	–	•	A	–	–	1	4	44	1	0.04	0.15	0.4	2	35	1.3	22

Type	Slew rate[e] typ (V/µs)	f_T typ (MHz)	CMRR min (dB)	PSRR min (dB)	Gain min (dB)	Max output curr (mA)	Max diff'l input[f] (V)	Swing to supplies?[g] In Out + − + −	Comments
OP-07A	0.17	0.6	110	100	110	10	30[h]	– – – –	
OP-07E	0.17	0.6	106	94	106	10	30[h]	– – – –	
OP-21A	0.25	0.6	100	104	120	-	30	– – – –	low power
OP-27E	2.8	8	114	100	120	20	0.5	– – – –	low noise
OP-27G	2.8	8	100	94	117	20	0.5	– – – –	cheap grade
OP-37E	17	63	114	100	120	20	0.5	– – – –	low noise, decomp OP-27
OP-50E	3	25	126	126	140	70	10[h]	– – – –	high current, low noise
OP-77E	0.3	0.6	120	110	134	12	30[h]	– – – –	improved OP-07
OP-90E	0.01	0.02	100	104	117	6	36	– • – •	micropower
OP-97E	0.2	0.9	114	114	110	10	0.5	– – – –	low power OP-77
MAX400M	0.3	0.6	114	100	114	12	30	– – – –	lowest non-chopper V_{os}
LM607A	0.7	1.8	124	100	134	10	0.5	– – – –	
AD707C	0.3	0.9	130	120	138	12	44	– – – –	improved OP-07; dual = 708
AD846B	450	310	110	110	-	50	18	– – – –	current feedback; fast
LT1001A	0.25	0.8	114	110	113	30	30	– – – –	
LT1007A	2.5	8	117	110	137	20	0.5	– – – –	low noise, ~OP-27
LT1012C	0.2	0.8	110	110	106	12	1	– – – –	improved 312; dual = 1024
LT1028A	15	75	114	117	137	20	1	– – – –	ultra low noise
LT1037A	15	60	117	110	137	20	0.5	– – – –	decomp 1007, ~OP-37
RC4077A	0.25	0.8	120	110	128	15	30	– – – –	lowest non-chopper V_{os}
HA5134A	1	4	94	100	108	20	40	– – – –	quad, low noise
HA5135	0.8	2.5	106	94	120	20	15[h]	– – – –	
HA5147A	35	140	114	80	120	15	0.5	– – – –	low noise, high speed, uncomp
OP-08E	0.12	0.8	104	104	98	5	0.5	– – – –	precision 308
LM10	0.12	0.1	93	90	102	20	40	– • • •	1V op-amp; precision; volt. ref.
LM11	0.3	0.5	110	100	100	2	0.5	– – – –	precision; lowest bias bipolar
OP-12E	0.12	0.8	104	104	98	5	0.5	– – – –	precision 312
LM308	0.15	0.3	80	80	88	5	0.5	– – – –	original low-bias (superbeta)
LM312	0.15	0.3	80	96	88	5	0.5	– – – –	compensated 308
LP324	0.05	0.1	80	90[t]	94	5	32	– • – •	low power, single supply
324A	0.5	1	65	65	88	20	30	– • – •	a classic; dual=358A
LP324	0.05	0.1	80	90[t]	94	5	32	– • – •	low power, low bias
LT1013C	0.4	0.8	97	100	122	25	30	– • – •	improved 358/324; quad = 1014
HA5141A	1.5	0.4	80	94	94	1	7	– • – •	micropower
LT1006A	0.4	1	100	106	120	20	30	– • – •	optional I_s = 90µA
LT1013A	0.4	0.8	100	103	124	25	30	– • – •	improved 358/324; quad = 1014

TABLE 4.1 (*cont'd*)

Type	Mfg[a]	# per pkg[b] 1 2 4	Trim	Ext comp[c]	Min gain[d]	Total supply voltage min (V)	Total supply voltage max (V)	Supp curr max (mA)	Voltage Offset typ (mV)	Voltage Offset max (mV)	Voltage Drift typ (μV/°C)	Voltage Drift max (μV/°C)	Current Bias max (nA)	Current Offset max (nA)	e_n @1kHz typ nV/√Hz
BIPOLAR, HIGH-SPEED															
OP-62E	PM	• – –	•	•	1	16	36	7	-	0.2	-	-	300	100	2.5
OP-63E	PM	• – –	•	•	1	16	36	7	-	0.75	-	-	300	100	7
OP-64E	PM	• – –	•	•	5	16	36	7	-	0.75	-	-	300	100	7
OP-65E	PM	• – –	•	–	1	9	14	25	-	2	-	-	3μA	1μA	-
CLC400	CL	• – –	•	–	1	-	7	15	2	5.5	20	40	25μA	(k)	12
AD509K	AD	• – –	•	•	3	10	40	6	4	8	-	40	200	25	19
SL541B	PL	• – –	•	–	10	-	24	21	-	5	15	-	25μA	10	-
VA705L	VT	• A A	•	–	1	8	12	10	1	2	20	-	900	25	-
VA706K	VT	• A A	•	–	1	8	12	10	4	10	20	-	1μA	120	-
VA707K	VT	• A A	•	–	12	8	12	10	3	6	20	-	1μA	120	-
LM837	NS	– – •	–	–	1	8	36	15	0.3	5	2	-	1μA	200	4.5
AD840K	AD	• – –	•	–	10	10	36	12	0.1	0.3	3	-	5μA	200	4
AD841K	AD	• – –	•	–	1	10	36	12	0.5	1	35	20	5μA	200	13
AD847J	AD	• – –	•	–	1	9	36	5.6	0.5	1	15	-	7μA	50[t]	15[i]
AD848J	AD	• – –	•	–	5	9	36	5.6	0.5	1	2	10	5μA	15[t]	4[i]
AD849J	AD	• – –	•	–	25	9	36	5.6	0.5	1	1	10	5μA	15[t]	4[i]
HA2539	HA	• – –	–	–	10	10	35	25	8	15	20	-	20μA	6μA	6
SL2541B	PL	• – –	•	–	1	14	30	25[t]	10	-	20	-	20μA	-	-
HA2541	HA	• – –	•	–	1	10	35	45	-	2	20	-	35μA	7μA	10
HA2542	HA	• – –	•	•	2	10	35	40	-	10	20	-	35μA	7μA	10
HA2544	HA	• – –	•	•	1	10	33	10	6	15	10	-	15μA	2μA	-
CA3450	RC	• – –	•	•	1	10	14	35	8	15	-	-	350	150	-
HA5101	HA	• A A	•	–	1	4	40	6	0.5	3	3	-	200	75	3.3
HA5111	HA	• A A	•	•	10	4	40	6	0.5	3	3	-	200	75	3.3
HA5147A	HA	• – –	•	–	10	8	44	4	0.01	0.025	0.2	0.6	40	35	3.0
HA5195	HA	• – –	–	–	5	20	35	25	3	6	20	-	15μA	4μA	6
LM6361	NS	• – –	•	–	1	5	36	6.5	5	20	10	-	5μA	2μA	15
LM6364	NS	• – –	•	–	5	5	36	6.5	2	9	6	-	5μA	2μA	8
LM6365	NS	• – –	•	–	25	5	36	6.8	1	6	3	-	5μA	2μA	5
BIPOLAR, OTHER															
OP-20B	PM	• A A	•	–	1	4	36	0.08	0.06	0.25	0.75	1.5	25	1.5	58
LM833	NS	– • –	–	–	1	10	36	8	0.3	5	2	-	1μA	200	4.5
CA3193A	RC	• – –	•	–	1	7	36	3.5	0.14	0.2	1	3	20	5	24
XR4560	XR	– • –	–	–	1	8	36	2	0.5	6	-	-	500	200	8
HA5151	HA	• A A	–	–	1	2	40	0.25	2	3	3	-	150	30	15
NE5534	SN+	• A –	•	•	3	6	44	8	0.5	4	-	-	2μA	300	4
MC33078	MO	– • A	–	–	1	10	36	5	0.15	2	2	-	750	150	4.5
MC33171	MO	• A A	•	–	1	3	44	0.25	2	4.5	10	-	100	20	32
MC34071A	MO	• A A	•	–	1	3	44	2.5	0.5	1.5	10	-	500	50	32

Type	Slew rate[e] typ (V/μs)	f_T typ (MHz)	CMRR min (dB)	PSRR min (dB)	Gain min (dB)	Max output curr (mA)	Max diff'l input[f] (V)	Swing to supplies?[g] In + − Out + −	Comments
OP-62E	15	50	110	105	111	20	5	– – – –	precision
OP-63E	50	50	110	105	100	20	5	– – – –	
OP-64E	200	200	110	105	100	20	5	– – – –	
OP-65E	200	150	85	90	100	50	5	– – – –	
CLC400	700	280	40	40	-	50	-	– – – –	transimpedance; decomp=401
AD509K	120	20	80	80	80	-	15	– – – –	fast
SL541B	175	100	60	46	46	6.5	9	– – – –	fast, video
VA705L	35	25	60	60	80	50	9	– – – –	video, drives 50Ω; fast settle
VA706K	42	25	60	60	66	50	9	– – – –	video, drives 50Ω; fast settle
VA707K	105	300	60	60	74	50	9	– – – –	decomp, fast, 50Ω
LM837	10	25	80	120	90	40	30	– – – –	low noise, low distortion
AD840K	400	400	100	94	104	50	6	– – – –	decomp 841; 842 has G>2
AD841K	300	40	90	90	88	50	6	– – – –	fast settle; decomp versions
AD847J	300	50	78	75	70	20	6	– – – –	fast settle; decomp versions
AD848J	300	250	104[t]	104[t]	82	25	6	– – – –	decomp 847
AD849J	300	725	110[t]	100[t]	90	25	6	– – – –	uncomp 847
HA2539	600	600	60	85	80	10	6	– – – –	low noise, sim to 2540
SL2541B	900	800	47	40[t]	45	10	10	– – – –	has uncommitted unity gain buf
HA2541	280	40	70	60	80	10	6	– – – –	fast settle, low distortion
HA2542	375	120	70	70	80	100	6	– – – –	fast settle, decomp
HA2544	150	33	75	70	70	35	6	– – – –	video
CA3450	420	190	50	60	96	75	5	– – – –	video amp/line driver
HA5101	10	10	100[t]	80	136[t]	30	7	– – – –	low noise
HA5111	50	100	100[t]	100[t]	136[t]	30	7	– – – –	low noise, uncomp
HA5147A	35	140	114	80	120	15	0.7	– – – –	low noise, precision, uncomp
HA5195	200	150	74	70	80	25	6	– – – –	Elantec EL2195 = improved
LM6361	300	50	70	72	52	30	8	– – – –	vertical PNP
LM6364	300	160	102[t]	70	66[t]	30	8	– – – –	vertical PNP
LM6365	300	725	80	104[t]	75	30	8	– – – –	vertical PNP
OP-20B	0.05	0.1	96	100	114	0.5	30	– • – –	accurate low power
LM833	7	15	80	80	90	10	30	– – – –	low noise, low distortion
CA3193A	0.25	1.2	110	100	110	7	5	– – – –	
XR4560	4	10	70	76	86	100	30	– – – –	intended for audio
HA5151	4.5	1.3	80	80	94	3	7	– – – •	low power
NE5534	6	10	70	80	88	20	0.5	– – – –	low noise, intended for audio
MC33078	7	16	80	80	90	20	36	– – – –	low noise, low distortion
MC33171	2.1	1.8	80	80	94	4	44	– • – –	
MC34071A	10	4.5	80	80	94	25	44	– • – –	drives 0.01μF

TABLE 4.1 (*cont'd*)

Type	Mfg[a]	# per pkg[b] 1	2	4	Trim	Ext comp[c]	Min gain[d]	Total supply voltage min (V)	Total supply voltage max (V)	Supp curr max (mA)	Voltage Offset typ (mV)	Voltage Offset max (mV)	Voltage Drift typ (μV/°C)	Voltage Drift max (μV/°C)	Current Bias max (nA)	Current Offset max (nA)	e_n @1kHz typ nV/√Hz
BIPOLAR, OBSOLESCENT																	
OP-01E	PM	•	–	–	•	–	1	10	44	3	1	2	3	10	50	5	-
OP-02E	PM	•	A	–	•	–	1	10	44	2	0.3	0.5	2	8	30	2	21
OP-05E	PM+	•	A	–	•	–	1	6	44	4	0.2	0.5	0.7	2	4	3.8	9.6
OP-11E	PM	–	–	•	–	–	1	10	44	6	0.3	0.5	2	10	300	20	12
307	NS+	•	–	–	–	–	1	10	44	2.5	2	7.5	6	30	250	50	16
LM318	NS+	•	–	–	•	•	1	10	40	10	4	10	-	-	500	200	14
349	NS	–	–	•	–	–	5	10	36	4.5	1	6	-	-	200	50	60
AD517L	AD	•	–	–	•	–	1	10	36	3	-	0.025	-	0.5	1	0.25	20
AD518J	AD	•	–	–	•	•	1	10	40	10	4	10	10	-	500	200	-
NE530	SN	•	A	–	•	–	1	10	36	3	2	5	6	-	150	40	30
NE531	SN	•	–	–	•	•	U	12	44	10	2	6	-	-	2μA	200	-
NE538	SN	•	A	–	•	–	5	10	36	2.8	2	5	6	-	150	40	18
μA725	FA+	•	–	–	•	•	U	6	44	3	0.5	1	2	5	100	20	-
μA739	FA	–	•	–	–	•	U	8	36	14	1	6	-	-	2μA	10μA	-
741C	FA+	•	A	A	•	–	1	10	36	2.8	2	6	-	-	500	200	-
748C	FA+	•	–	–	•	•	U	10	36	3.3	2	6	-	-	500	200	-
μA749	FA	–	•	–	–	•	U	8	36	10	1	3	3	-	750	400	-
1435	TP	•	–	–	•	•	10	24	32	30	2	5	5	25	20μA	-	-
1456	MO	•	–	–	•	–	1	10	36	3	5	10	-	-	30	10	45
HA2505	HA	•	–	–	•	•	1	20	40	6	4	8	20	-	250	50	-
HA2515	HA	•	–	–	•	•	1	20	40	6	5	10	30	-	250	50	-
HA2525	HA	•	–	–	•	•	3	20	40	6	5	10	30	-	250	50	-
HA2605	HA	•	–	–	•	•	1	10	45	4	3	5	10	-	25	25	-
HA2625	HA	•	–	–	•	•	5	10	45	4	3	5	10	-	25	25	-
CA3100	RC	•	–	–	•	•	10	13	36	11	1	5	-	-	2μA	400	-
4558	RA+	–	•	–	–	–	1	8	36	5.6	2	6	-	-	500	200	43
NE5535	SN	A	•	–	•	–	1	10	36	2.8	2	5	6	-	150	40	17
5539	SI+	•	–	–	–	•	7	6	24	15	2.5	5	5	10	20μA	-	4
JFET, PRECISION																	
OP-41E	PM	•	–	–	•	–	1	10	36	1	0.2	0.25	2.5	5	0.005	0.001	32
OP-43E	PM	•	–	–	•	–	1	10	36	1	0.2	0.25	2.5	5	0.005	0.001	32
OPA101B	BB	•	–	–	•	–	1	10	40	8	0.05	0.25	3	5	0.01	4pA	8
OPA111B	BB	•	A	–	•	–	1	10	36	3.5	0.05	0.25	0.5	1	1pA	0.7pA	7
AD547L	AD	•	A	–	•	–	1	5	36	1.5	-	0.25	-	1	0.025	2pA[t]	30
AD548C	AD	•	A	–	•	–	1	9	36	0.2	0.1	0.25	-	2	0.01	0.005	30
OPA627B	BB	•	–	–	•	–	1	9	36	8	0.04	0.1	0.5	0.8	0.02	0.02	5.2
AD711C	AD	•	A	A	•	–	1	9	36	2.8	0.1	0.25	2	3	0.025	0.01	18
AD845K	AD	•	–	–	•	–	1	9.5	36	12	0.1	0.25	1.5	5	1	0.1	25
LT1055A	LT	•	–	–	•	–	1	10	40	4	0.05	0.15	1.2	4	0.05	0.01	14
HA5170	HA	•	–	–	•	–	1	9	44	2.5	0.1	0.3	2	5	0.1	0.06	10

Type	Slew rate[e] typ (V/μs)	f_T typ (MHz)	CMRR min (dB)	PSRR min (dB)	Gain min (dB)	Max output curr (mA)	Max diff'l input[f] (V)	Swing to supplies?[g] In + − Out + −	Comments
OP-01E	18	2.5	80	80	94	6	30	– – – –	fast, precision
OP-02E	0.5	1.3	90	90	100	6	30	– – – –	precision, low current
OP-05E	0.17	0.6	110	94	106	10	30[h]	– – – –	
OP-11E	1	2	110	90	100	6	30	– – – –	precision quad
307	0.5	1	70	70	84	10	30	• – – –	a classic; uncomp=301
LM318	70	15	70	65	86	10	0.5	– – – –	was popular
349	2	4	70	77	88	15	36	– – – –	decomp 348 (quad 741)
AD517L	0.1	0.25	110	96	120	10	30	– – – –	
AD518J	70	12	70	65	88	15	-	– – – –	
NE530	35	3	70	76	94	10	30	• – – –	fast; dual=5530
NE531	35	1	70	76	86	-	15	• – – –	
NE538	60	5	70	76	94	10	30	• – – –	fast; dual=5538
μA725	0.005	0.08	110	100	108	15	5	– – – –	original precision op-amp
μA739	1	6	70	85[t]	76	1.5	5	– – – •	low noise, intended for audio
741C	0.5	1.2	70	76	86	20	30	– – – –	old classic; dual=1458, quad=348
748C	0.5	1.2	70	76	94	15	30	– – – –	uncomp 741
μA749	2	6	70	74	86	1.5	5	– – – •	sim to 739
1435	300	1GHz	80	75[t]	80	10	2	– – – –	fast settle
1456	2.5	1	70	74	97	5	40	– – – –	
HA2505	30	12	74	74	84	10	15	• – – –	
HA2515	60	12	74	74	78	10	15	• – – –	
HA2525	120	20	74	74	78	10	15	• – – –	
HA2605	7	12	74	74	98	10	12	– – – –	
HA2625	35	100	74	74	98	10	12	– – – –	
CA3100	25	30	76	60	58	15	12	– – – –	
4558	1	2.5	70	74	86	15	30	– – – –	fast 1458
NE5535	15	1	70	76	94	10	30	• – – –	fast
5539	600	1200	70	66	46	40	10	– – – •	small output swing
OP-41E	1.3	0.5	100	92	120	15	20	• – – –	low bias, low dist; OP-43 faster
OP-43E	6	2.4	100	92	120	15	20	• – – –	low bias, low dist; OP-41 stabler
OPA101B	7	20	80	86	96	45	20	– – – –	low noise; decomp = OPA102
OPA111B	2	2	100	100	120	10	36	– – – –	low noise, low bias
AD547L	3	1	80	80	108	20	20	– – – –	dual = AD642, 647
AD548C	1.8	1	86	86	110	20	20	– – – –	improved LF441; dual = AD648
OPA627B	55	16	106	106	110	30	-	– – – –	fast
AD711C	20	4	86	86	106	20	20	– – – –	improved LF411/2
AD845K	100	16	94	95	108	30	36	– – – –	fast
LT1055A	13	5	86	90	104	30	40	– – – –	LT1056 is 20% faster
HA5170	8	8	90	74	110	10	30	• – – –	low noise

TABLE 4.1 (*cont'd*)

Type	Mfg[a]	# per pkg[b] 1 2 4	Trim	Ext comp[c]	Min gain[d]	Total supply voltage min (V)	Total supply voltage max (V)	Supp curr max (mA)	Voltage Offset typ (mV)	Voltage Offset max (mV)	Voltage Drift typ (μV/°C)	Voltage Drift max (μV/°C)	Current Bias max (nA)	Current Offset max (nA)	e_n @1kHz typ nV/√Hz
JFET, HIGH-SPEED															
OP-42E	PM	• – –	•	–	1	15	40	6.5	0.3	0.75	4	10	0.2	0.04	13
OP-44E	PM	• – –	•	–	3	16	40	6	0.03	0.75	4	10	0.2	40pA	13
357B	NS+	• – –	•	–	5	10	36	7	3	5	5	-	100pA	0.02	12
AD380K	AD	• – –	•	•	U	12	40	15	-	1	-	10	0.1	5pA[t]	15
LF401A	NS	• – –	•	•	1	15	36	12	-	0.2	-	-	0.2	0.1	23
OPA404B	BB	– – •	–	–	1	10	36	10	0.26	0.75	3	-	0.004	4pA	15
LF457B	NS	• – –	•	–	5	10	36	10	0.18	0.4	3	4	50pA	20pA	10
OPA602C	BB	• – –	•	–	1	10	36	4	0.1	0.25	1	2	1pA	1pA	13
OPA605K	BB	• – –	•	•	50	10	40	9	0.25	0.5	-	5	0.035	2pA[t]	20
OPA606L	BB	• – –	•	–	1	10	36	9.5	0.1	0.5	3	5	0.01	5pA	13
AD744C	AD	• A –	•	•	2	9	36	4	0.1	0.25	2	3	0.05	0.02	18
AD843B	AD	• – –	•	–	1	9	36	12	0.5	1	15	-	1	0.1	13
AD845K	AD	• – –	•	–	1	9.5	36	10.2	0.1	0.25	1.5	3	0.4	0.05	25
LT1022A	LT	• – –	•	–	1	20	40	7	0.08	0.25	1.3	5	0.05	0.01	14
HA5160	HA	• – –	–	•	U	14	40	10	1	3	20	-	0.05	0.01	35
MC34080A	MO	• A A	•	–	2	6	44	3.4	0.3	0.5	10	-	0.2	0.1	30
MC34081A	MO	• A A	•	–	1	6	44	3.4	0.3	0.5	10	-	0.2	0.1	30
JFET, OTHER															
TL031C	TI	• A A	•	–	1	10	36	0.28	0.5	1.5	6	-	0.2	0.1	41
TL051C	TI	• A A	•	–	1	10	36	3.2	0.6	1.5	8	-	0.2	0.1	18
TL061C	TI+	• A A	•	–	1	4	36	0.25	3	15	10	-	0.4	0.2	42
TL071C	TI+	• A A	•	–	1	7	36	2.5	3	10	10	-	0.2	0.05	18
TL081B	TI+	• A A	•	–	1	7	36	2.8	2	3	10	-	0.2	0.01	18
OPA121	BB	• – –	•	–	1	10	36	4	0.5	2	3	10	0.005	4pA	8
OPA128L	BB	• – –	•	–	1	10	36	1.5	0.14	0.5	-	5	75fA	30fA[t]	27
LF351	NS+	• A A	•	–	1	10	36	3.4	5	10	10	-	0.2	0.1	25
355B	NS+	• – –	•	–	1	10	36	4	3	5	5	-	100pA	0.02	20
356B	NS+	• – –	•	–	1	10	36	7	3	5	5	-	100pA	0.02	12
LF411	NS+	• A –	•	–	1	10	36	3.4	0.8	2	7	20	0.2	0.1	25
LF*nnn*	NS	– • –	–	–	1	6	36	25	1	-	-	-	100pA	50pA	3.5
LF441	NS	• A A	•	–	1	10	36	0.25	1	5	10	20	0.1	0.05	35
LF455B	NS	• – –	•	–	1	10	36	4	0.18	0.4	3	4	50pA	20pA	12
LF456B	NS	• – –	•	–	1	10	36	8	0.18	0.4	3	4	50pA	20pA	10
AD549L	AD	• – –	•	–	1	10	36	0.7	0.3	0.5	5	10	60fA	20fA[t]	35
AD611K	AD	• – –	•	–	1	10	36	2.5	0.25	0.5	5	10	0.05	0.025	18
LT1057A	LT	– • A	–	–	1	20	40	3.8	0.15	0.45	1.8	7	0.05	0.04	13
HA5180	HA	• – –	•	–	1	10	40	1	0.1	0.5	5	-	0.001	200fA	70
MC34001A	MO	• A A	•	–	1	8	36	2.5	1	2	10	-	0.1	0.05	25
MC34181	MO	• A A	•	–	1	3	36	0.2	0.5	2	10	-	0.1	0.05	38

Type	Slew rate[e] typ (V/µs)	f_T typ (MHz)	CMRR min (dB)	PSRR min (dB)	Gain min (dB)	Max output curr (mA)	Max diff'l input[f] (V)	Swing to supplies?[g] In Out + − + −	Comments
OP-42E	58	10	88	86	114	25	40	– – – –	low Z_{out}
OP-44E	120	16	88	90	114	15	40	– – – –	
357B	50	20	85	85	94	20	30	• – – –	decomp 356
AD380K	330	300	60	60	92	60	20	– – – –	hybrid, fast, 50Ω
LF401A	30	16	90	80	100	50	32	– – – –	accurate
OPA404B	35	6.4	92	86	92	10	36	– – – –	accurate quad
LF457B	50	20	86	86	106	100[l]	40	• – – –	low noise; drives 0.01µF
OPA602C	35	6.5	92	86	92	20	36	– – – –	low bias, fast settle
OPA605K	94	20	80	74	104[t]	30	20	– – – –	uncomp
OPA606L	35	13	85	90	100	10	36	– – – –	improved LF356
AD744C	75	13	86	92	108	20	36	– – – –	very low dist (3ppm); fast settle
AD843B	250	35	100	95	88	50	-	– – – –	fast settle
AD845K	100	16	94	98	106	25	20	– – – –	fast settle
LT1022A	26	8.5	86	88	104	10	40	– – – –	
HA5160	120	100	74	108	98	22	40	– – – –	low bias
MC34080A	50	16	75	75	94	20	44	– – – –	$V_{in} > V_{-}+4V$; decomp 34081
MC34081A	25	8	75	75	94	20	44	– – – –	$V_{in} > V_{-}+4V$
TL031C	3	1	75	75	74	8	30	• – – –	low power; improved TL061
TL051C	24	3	75	75	94	30	30	• – – –	low dist; improved TL071/081
TL061C	3.5	1	70	70	70	5	30	– – – –	low power
TL071C	13	3	70	70	88	10	30	– – – –	lower noise
TL081B	13	3	80	80	94	10	30	– – – –	
OPA121	2	2	86	86	110	10	36	– – – –	low noise
OPA128L	3	1	90	90	110	10	36	– – – –	very low bias
LF351	13	4	70	70	88	10	30	• – – –	353=dual, 347=quad
355B	5	2.5	85	85	94	20	30	• – – –	popular
356B	12	5	85	85	94	20	30	• – – –	faster 355
LF411	15	4	70	70	88	20	30	• – – –	jellybean
LF*nnn*	20	10	80	80	100	15	2	• – – –	lowest noise JFET
LF441	1	1	70	70	88	4	30	• – – –	low current jellybean
LF455B	5	3	86	86	106	100[l]	40	• – – –	low noise; drives 0.01µF
LF456B	12.5	5	86	86	106	100[l]	40	• – – –	low noise; drives 0.01µF
AD549L	3	1	90	90	110	10	36	– – – –	electrometer; guard pin
AD611K	13	2	80	80	94	20	20	– – – –	low dist, gen purp JFET
LT1057A	13	5	86	88	104	10	40	– – – –	accurate dual/quad JFET
HA5180	7	2	90	90	106	15	40	– – – –	very low bias over temp; noisy
MC34001A	13	4	80	80	94	20	30	• – – –	
MC34181	10	4	70	70	88	8	36	– – – –	low power, fast, low dist.

TABLE 4.1 (*cont'd*)

Type	Mfg[a]	# per pkg[b] 1	2	4	Trim	Ext comp[c]	Min gain[d]	Total supply voltage min (V)	max (V)	Supp curr max (mA)	Voltage Offset typ (mV)	Offset max (mV)	Drift typ (μV/°C)	Drift max (μV/°C)	Current Bias max (nA)	Offset max (nA)	e_n @1kHz typ nV/√Hz
JFET, OBSOLESCENT																	
OP-15E	PM+	•	A	–	•	–	1	10	44	4	0.2	0.5	2	5	0.05	0.01	15
OP-16E	PM+	•	–	–	•	–	1	10	44	7	0.2	0.5	2	5	0.05	0.01	15
AD515L	AD	•	–	–	•	–	1	10	36	1.5	0.4	1	-	25	80fA	80fA	50
AD542L	AD	•	–	–	•	–	1	10	36	1.5	-	0.5	-	5	0.025	2pA[t]	30
AD544L	AD	•	–	–	•	–	1	10	36	2.5	-	0.5	-	5	0.05	0.5pA[t]	18
AD545L	AD	•	–	–	•	–	1	10	36	1.5	-	0.5	-	5	0.001	-	35
ICH8500A	IL	•	–	–	•	–	1	16	36	2.5	-	50	-	-	10fA	10fA	40
MOSFET																	
OP-80E	PM	•	–	–	•	–	1	4.5	16	0.2	0.4	1	-	-	60fA	10fA[t]	70
TLC27L2A	TI	A	•	A	–	–	1	3	18	0.04	-	5	0.7	-	1pA[t]	1pA[t]	70
TLC27M2A	TI	A	•	A	–	–	1	3	18	0.6	-	5	2	-	1pA[t]	1pA[t]	38
TLC272A	TI	A	•	A	–	–	1	3	18	4	-	5	5	-	1pA[t]	1pA[t]	25
TLC279C	TI	–	–	•	–	–	1	3	18	8	0.4	1.2	2	-	0.7pA[t]	0.1pA[t]	25
LMC660A	NS	–	–	•	–	–	1	5	16	2.2	1	2	1.3	5	20pA	20pA	22
TLC1078C	TI	–	•	A	–	–	1	1.4	16	0.05	0.18	0.6	1	-	0.7pA[t]	0.1pA[t]	68
ALD1701	AL	•	–	–	–	–	1	2	12	0.25	-	4.5	7	-	0.03	0.025	-
ALD1702	AL	•	–	–	–	–	1	2	12	2	-	4.5	7	-	0.03	0.025	100
CA3140A	RC	•	A	–	•	–	1	4	44	6	2	5	6	-	0.04	0.02	40
CA3160A	RC	•	A	–	•	•	1	5	16	15	2	5	10	-	0.03	0.02	72
CA3410A	RC	–	–	•	–	–	1	4	36	10	3	8	10	-	0.03	0.01	40
CA3420A	RC	•	–	–	•	•	1	2	22	1	2	5	4	-	0.005	0.004	62
CA5160A	RC	•	A	–	•	•	1	5	15	0.4	1.5	4	-	-	0.01	0.005	-
CA5420A	RC	•	–	–	•	•	1	2	20	0.5	1	5	-	-	0.001	0.5pA	-
CA5422	RC	•	–	–	•	–	1	2	20	0.7	1.8	10	20	-	0.005	0.004	-
ICL7612B	IL+	•	–	–	–	–	1	3	18	2.5	-	5	5	-	0.05	0.03	100
ICL7641B	IL+	A	A	•	–	–	1	1	18	2.5	-	5	5	-	0.05	0.03	100
CHOPPER STABILIZED																	
MAX420E	MA	•	–	–	–	–	1	6	33	2	0.001	0.005	0.02	0.05	0.03	0.06	1.1[j]
MAX422E	MA	•	–	–	–	–	1	6	33	0.5	0.001	0.005	0.02	0.05	0.03	0.06	1.1[j]
LMC668A	NS	•	–	–	–	–	1	5	16	3.5	0.001	0.005	0.05	-	0.06	-	2[j]
TSC900A	TS	•	–	–	–	–	1	4.5	16	0.2	-	0.005	0.02	0.05	0.05	0.5pA[t]	4[j]
TSC901	TS	•	A	A	–	–	1	5	32	0.6	0.007	0.015	0.05	0.15	0.05	0.1	5[j]
TSC911A	TS	•	A	A	–	–	1	4	16	0.6	0.005	0.015	0.05	0.15	0.07	0.02	11[j]
TSC915	TS	•	–	–	–	–	1	7	32	1.5		0.01	0.01	0.1	0.1	0.1	0.8[j]
TSC918	TS	•	–	–	–	–	1	4.5	16	0.8	-	0.05	0.4	0.8	0.1	0.5pA[t]	4[j]
LTC1050	LT	•	–	–	–	–	1	4.8	16	1.5	0.5μV	0.005	0.01	0.05	0.03	0.06	1.6[j]
LTC1052	LT	•	–	–	–	–	1	4.8	16	2	0.5μV	0.005	0.01	0.05	0.03	0.03	1.5[j]
ICL7650	IL+	•	–	–	–	–	1	4.5	16	3.5	0.002	0.005	0.1	-	0.01	5pA[t]	2[j]
ICL7650S	IL	•	–	–	–	–	1	4.5	16	3	0.7μV	0.005	0.02	0.1	0.01	0.02	2[j]
ICL7652	IL+	•	–	–	–	–	1	5	16	3.5	0.002	0.005	0.1	-	0.03	25pA[t]	0.7[j]
ICL7652S	IL	•	–	–	–	–	1	5	16	2.5	0.7μV	0.005	0.01	0.06	0.03	0.04	0.7[j]
TSC76HV52	TS	•	–	–	–	–	1	7	32	1.5	-	0.01	-	0.3	0.1	0.1	0.8[j]

Type	Slew rate[e] typ (V/μs)	f_T typ (MHz)	CMRR min (dB)	PSRR min (dB)	Gain min (dB)	Max output curr (mA)	Max diff'l input[f] (V)	Swing to supplies?[g] In +	In −	Out +	Out −	Comments
OP-15E	17	6	86	86	100	15	40	–	–	–	–	precision fast 355
OP-16E	25	8	86	86	100	20	40	–	–	–	–	precision fast 356 (OP-17=decomp)
AD515L	1	0.4	70	74	94	10	20	–	–	–	–	very low bias, precision
AD542L	3	1	80	80	110	10	20	–	–	–	–	precision
AD544L	13	2	80	80	94	15	20	–	–	–	–	precision, low noise
AD545L	1	0.7	76	74[t]	92	10	20	–	–	–	–	precision
ICH8500A	0.5	0.5	60	80[t]	100[t]	10	0.5	–	–	–	–	ultra low bias
OP-80E	0.4	0.3	60	60	100	10	16	–	•	–	•	electrometer; I_b<20pA @ 125°C
TLC27L2A	0.04	0.1	70	70	90	10	18	–	•	–	•	CMOS jellybeans
TLC27M2A	0.6	0.7	70	70	86	10	18	–	•	–	•	CMOS jellybeans
TLC272A	4.5	2.3	70	65	80	10	18	–	•	–	•	CMOS jellybeans
TLC279C	4.5	2.3	65	65	80	10	18	–	•	–	•	best V_{os} of 272-series
LMC660A	1.7	1.5	72	80	112	15	16	–	•	•	•	quad CMOS jellybean
TLC1078C	0.05	0.11	75	75	114	15	16	–	•	–	•	low offset
ALD1701	0.7	0.7	65	65	90	0.5	12	•	•	•	•	rail-to-rail; specs @ +5V supply
ALD1702	2.1	1.5	65	65	94	2	12	•	•	•	•	rail-to-rail; specs @ +5V supply
CA3140A	7	3.7	70	76	86	+10,-1	8	–	•	–	•	
CA3160A	10	4	80	76	94	12	8	–	•	•	•	MOS in/out (3130=uncomp)
CA3410A	10	5.4	80	80	86	6	16	–	•	–	–	high speed 324-type replacement
CA3420A	0.5	0.5	60	70	86	2	15	–	•	•	•	low I_b, good input protec.
CA5160A	10	4	-	-	90	1	?	–	•	•	•	CMOS output
CA5420A	0.5	0.5	-	-	85	0.5	15	–	•	•	•	similar to 3420
CA5422	1	1	60	60	80	2	15	–	•	•	•	unusual 2-section design
ICL7612B	1.6	1.4	60	70	80	5[m]	18	•	•	•	•	programmable; in/out to both rails
ICL7641B	1.6	1.4	60	70	80	5[m]	18	•	–	•	•	gen purp, low voltage
MAX420E	0.5	0.5	120	120	120	+2,-15	33	–	•	•	•	±15V V_s; 0.1μV/mo; 430 has C_{int}
MAX422E	0.13	0.13	120	120	120	+0.2,-8	33	–	•	•	•	±15V V_s; 0.1μV/mo; 432 has C_{int}
LMC668A	2.5	1	120	120	120	+5,-15	16	–	•	•	•	
TSC900A	0.2	0.7	110	120	120	2.5	16	–	•	–	•	low power
TSC901	2	0.8	120	120	120	-	36	–	•	–	•	±15V supply; int caps
TSC911A	2.5	1.5	110	112	116	3.5	16	–	•	–	•	int caps, noisy
TSC915	0.5	0.5	120	120	120	10	36	–	•	–	•	±15V supply
TSC918	0.2	0.7	98	105	100	-	16	–	•	–	•	inexpensive
LTC1050	4	2.5	120	125	130	+3,-20	16	–	•	•	•	int caps; 50nV/√month
LTC1052	4	1.2	120	120	120	+5,-15	16	–	•	•	•	improved 7652; 0.1μV/month
ICL7650	2.5	2	110	120	120	+5,-20	16	–	•	•	•	0.1μV/month
ICL7650S	2.5	2	120	120	136	+4,-20	16	–	•	•	•	improved 7650; 0.1μV/month
ICL7652	0.5	0.4	110	110	120	+5,-20	16	–	–	•	•	0.15μV/month
ICL7652S	1	0.5	120	120	136	+4,-20	16	–	–	•	•	improved 7652; 0.15μV/month
TSC76HV52	0.5	0.5	120	120	120	10	32	–	–	•	•	±15V supply

TABLE 4.1 (*cont'd*)

Type	Mfg[a]	# per pkg[b] 1	2	4	Trim	Ext comp[c]	Min gain[d]	Total supply voltage min (V)	max (V)	Supp curr max (mA)	Voltage Offset typ (mV)	Offset max (mV)	Drift typ (μV/°C)	Drift max (μV/°C)	Current Bias max (nA)	Current Offset max (nA)	e_n @1kHz typ nV/√Hz
HIGH VOLTAGE																	
LM343	NS	•	–	–	•	–	1	10	68	5	2	8	-	-	40	10	35
LM344	NS	•	–	–	•	•	U	10	68	5	2	8	-	-	40	10	35
OPA445B	BB	•	–	–	•	–	1	20	100	4.5	1	3	10	-	0.05	0.01	16
1436	MO+	•	–	–	•	–	1	10	80	5	5	10	-	-	40	10	50
HA2645	HA	•	–	–	•	•	1	20	80	4.5	2	6	15	-	30	30	30
3580	BB	•	–	–	•	–	1	30	70	10	-	10	-	30	0.05	-	15
3581	BB	•	–	–	•	–	1	64	150	8	-	3	-	25	0.02	0.02	25
3582	BB	•	–	–	•	–	1	140	300	6.5	-	3	-	25	0.02	-	25
3583	BB	•	–	–	•	–	1	100	300	8.5	-	3	-	25	0.1	0.1	50
3584	BB	•	–	–	•	•	U	140	300	6.5	-	3	-	25	0.1	0.1	50
MONOLITHIC POWER																	
LM12	NS	•	–	–	–	–	1	20	80	80	2	7	-	50	300	100	90
OPA541B	BB	•	–	–	–	–	1	20	80	25	0.1	1	15	30	0.05	0.03	50
LM675	NS	•	–	–	–	–	10	16	60	50	1	10	25	-	2μA	500	-
SG1173	SG	•	–	–	–	–	1	10	50	20	2	4	-	30	500	150	-

(a) manufacturers are as follows (a "+" suffix designates multiple sources):

AD - Analog Devices
AL - Advanced Linear Devices
AM - Advanced Micro Devices
AN - Analogic
AP - Apex
BB - Burr-Brown
BT - Brooktree
CL - Comlinear
CR - Crystal Semiconductor
CY - Cypress
DA - Datel
EL - Elantec
FA - Fairchild (National)
FE - Ferranti
GE - General Electric
GI - General Instrument
HA - Harris
HI - Hitachi
HO - Honeywell
HS - Hybrid Systems
ID - Integrated Device Technology
IL - GE/Intersil
IN - Intel
IR - International Rectifier
KE - M.S.Kennedy Corp
LT - Linear Technology Corp
MA - Maxim
MN - Micro Networks
MO - Motorola
MP - Micro Power Systems
NE - NEC
NS - National Semiconductor
OE - Optical Electronics Inc
PL - Plessey
PM - Precision Monolithics
RA - Raytheon
RC - GE/RCA
RO - Rockwell
SG - Silicon General
SI - Siliconix
SN - Signetics
SO - Sony
ST - Supertex
TI - Texas Instruments
TM - Telmos
TO - Toshiba
TP - Teledyne Philbrick
TQ - TriQuint
TR - TRW
TS - Teledyne Semiconductor
VT - VTC
XI - Xicor
XR - Exar
ZI - Zilog

Type	Slew rate[e] typ (V/μs)	f_T typ (MHz)	CMRR min (dB)	PSRR min (dB)	Gain min (dB)	Max output curr (mA)	Max diff'l input[f] (V)	Swing to supplies?[g] In Out + − + −	Comments
LM343	2.5	1	70	74	97	10	68	- - - -	monolithic
LM344	30	10	70	74	97	10	68	- - - -	uncomp 343
OPA445B	10	2	80	80	100	15	80	- - - -	low-bias, monolithic
1436	2	1	70	80	97	10	80	- - - -	monolithic
HA2645	5	4	74	74	100	10	37	- - - -	monolithic
3580	15	5	86[t]	87[t]	106[t]	60	70	- - - -	hybrid
3581	20	5	110[t]	105[t]	112[t]	30	150	- - - -	hybrid
3582	20	5	110[t]	105[t]	118[t]	15	300	- - - -	hybrid
3583	30	5	110[t]	84[t]	94	75	300	- - - -	fast JFET, hybrid
3584	150	20	110[t]	84[t]	100	15	300	- - - -	uncomp JFET, hybrid
LM12	9	0.7	75	80	94	10A	80	- - - -	full output protection
OPA541B	10	1.6	95	100	90	10A	80	- - - -	isolated case; no int. protec.
LM675	8	5.5	70	70	70	3000	60	- - - -	full output protection
SG1173	0.8	1	76	80	92	3500	50	- - - -	thermal shutdown

(b) the symbol • indicates the number of op-amps per package for the part number shown; an "A" indicates the availability of other quantities of op-amps per package from the same manufacturer; some electrical characteristics (particularly offset voltage) may be degraded somewhat in multiple packages.
(c) pins are provided for external compensation.
(d) a number gives the minimum closed-loop gain without instability. Op-amps with pins for external compensation can generally be operated at lower gain, if an appropriate ext comp network is used. The letter U means that the op-amp is uncompensated – external capacitance is necessary for any small value of closed-loop gain.
(e) at minimum stable closed-loop gain (usually unity gain), unless otherwise noted.
(f) the maximum value without damage to the chip; not to exceed the total supply voltage used, if that is less.
(g) a dot in an IN column means that the input operating common-mode range includes that supply rail; a dot in an OUT column means that the op-amp can swing its output all the way to the corresponding supply rail.
(h) resistor-diode network draws input current for input differential greater than ±1V.
(j) μV pp, 0.1-10Hz.
(k) current-sensing inverting input ("current feedback" configuration); the bias currents at the two inputs may differ widely. The listed bias current is for the non-inverting input.
(l) "raw" output (no current limit) available at pin 8, in addition to the conventional (protected) output at pin 6; the latter is limited to ±15mA.
(m) min/max (worst case).
(t) typical.

TABLE 4.2. RECOMMENDED OP-AMPS

Type	Mfg[a]	Amps per package[b] 1	2	4	Offset voltage max (mV)	Offset drift max (μV/°C)	Input curr max (nA)	Total supply voltage min (V)	max (V)	Supply curr max (mA)	e_n, typ 10Hz (nV/√Hz)	1kHz (nV/√Hz)	Slew rate typ (V/μs)	f_T typ (MHz)	Comments
LF411	NS	•	A	–	2	20	0.2	10	36	3.4	50	25	15	4	general purpose jellybean
AD711K	AD	•	A	–	0.5	10	0.05	9	36	3	45	18	20	4	improved LF411
LM358A	NS+	–	•	A	3	20	100	3	32	1.2	–	–	0.5	1	single supply jellybean
TLC27M2A	TI	A	•	A	5	2[t]	0.001[t]	3	18	0.6	–	–	0.6	0.7	CMOS jellybean
OP-27E	PM+	•	A	A	0.025	0.6	40	8	44	5	3.5	3.0	2.8	8	precision, low-noise
OP-37E	PM+	•	A	–	0.025	0.6	40	8	44	5	3.5	3.0	17[h]	63[h]	ditto, faster (decomp, min. gain = 5)
HA5147A	HA	•	–	–	0.025	0.6	40	8	44	4	3.5	3.0	35[c]	140[c]	ditto, still faster (min. gain = 10)
OP-77E	PM	•	A	A	0.025	0.3	2	6	44	2	10.3	9.6	0.3	0.6	precision
LT1028A	LT	•	–	–	0.04	0.8	90	8	44	9.5	1.0	0.85	15	75	precision ultra-low-noise
LT1013A	LT	–	•	A	0.15	2	35	4	44	1	24	22	0.4	0.8	precision single-supply
LT1055A	LT	•	–	–	0.15	4	0.05	10	40	4	28	14	13	5	precision JFET
LT1012C	LT+	•	A	–	0.05	1.5	0.15	4	40	0.6	17	14	0.2	0.8	precision low-bias
OPA111B	BB	•	A	–	0.25	1	0.001	10	36	3.5	30	7	2	2	precision low-bias JFET
AD744K	AD	•	–	–	0.5	10	0.1	9	36	4	45	18	75[f]	13[f]	ultra low dist, stable, fast settle
LTC1052	IL+	•	–	–	0.005	0.05	0.03	4.8	16	2	–	–	4	1.2	chopper
OP-90E	PM	•	A	A	0.15	2	15	1.6	36	0.02	60	60	0.012	0.02	precision micropower
CA3440A	RC	•	–	–	5	4[t]	0.04	4	15	(d)	250	110	0.003[e]	0.005[e]	nanopower (programmable)
AD549L	AD	•	–	–	0.5	10	60fA	10	36	0.7	90	35	3	1	ultra low input current JFET
LM10	NS+	•	–	–	2	2[t]	20	1.1	40	0.4	50	46	0.1	0.4	low supply voltage, rail-to-rail output

(a) see footnotes to Table 4.1. (b) • = this part number; A = available. (c) G>10. (d) programmable 0.02μA–10μA. (e) at I_s=1μA. (f) G>2. (h) G>5. (m) min/max. (t) typical.

op-amps with the very lowest input currents (MOSFET types) generally have poor voltage offsets, and vice versa. Good circuit designers choose their components with the right trade-offs to optimize performance, without going overboard on unnecessary "gold-plated" parts.

Limitations imply trade-offs

The limitations of op-amp performance we have talked about will have an influence on component values in nearly all circuits. For instance, the feedback resistors must be large enough so that they don't load the

---

"Here Yesterday, Gone Today"

In its untiring quest for better and fancier chips, the semiconductor industry can sometimes cause you great pain. It might go something like this: You've designed and prototyped a wonderful new gadget; debugging is complete, and you're ready to go into production. When you try to order the parts, you discover that a crucial IC has been discontinued by the manufacturer! An even worse nightmare goes like this: Customers have been complaining about late delivery on some instrument that you've been manufacturing for many years. When you go to the assembly area to find out what's wrong, you discover that a whole production run of boards is built, except for one IC that "hasn't come in yet." You then ask purchasing why they haven't expedited the order; turns out they have, just haven't received it. Then you learn from the distributor that the part was discontinued six months ago, and that none is available!

Why does this happen, and what do you do about it? We've generally found four reasons that ICs are discontinued:

1. Obsolescence: Much better parts come along, and it doesn't make much sense to keep making the old ones. This has been particularly true with digital memory chips (e.g., small static RAMs and EPROMs, which are superseded by denser and faster versions each year), though linear ICs have not entirely escaped the purge. In these cases there is often a pin-compatible improved version that you can plug into the old socket.

2. Not selling enough: Perfectly good ICs sometimes disappear. If you are persistent enough, you may get an explanation from the manufacturer – "there wasn't enough demand," or some such story. You might characterize this as a case of "discontinued for the convenience of the manufacturer." We've been particularly inconvenienced by Harris's discontinuation of their splendid HA4925 – a fine chip, the fastest quad comparator, now gone, with no replacement anything like it. Harris also discontinued the HA2705 – another great chip, the fastest low-power op-amp, now gone without a trace! Sometimes a good chip is discontinued when the wafer fabrication line changes over to a larger wafer size (e.g., from the original 3″ diameter wafer to a 5″ or 6″ wafer). We've noticed that Harris has a particular fondness for discontinuing excellent and unique chips; Intersil and GE have done the same thing.

3. Lost schematics: You might not believe it, but sometimes the semiconductor house loses track of the schematic diagram of some chip and can't make any more! This apparently happened with the Solid State Systems SSS-4404 CMOS 8-stage divider chip.

4. Manufacturer out of business: This also happened to the SSS-4404!

If you're stuck with a board and no available IC, you've got several choices. You can redesign the board (and perhaps the circuit) to use something that is available. This is probably best if you're going into production with a new design or if you are running a large production of an existing board. A cheap and dirty solution is to make a little "daughterboard" that plugs into the empty IC socket and includes whatever it takes to emulate the nonexistent chip. Although this latter solution isn't terribly elegant, it gets the job done.

---

output significantly, but they must not be so large that input bias current produces sizable offsets. High impedances in the feedback network also increase susceptibility to capacitive pickup of interfering signals and increase the loading effects of stray capacitance. These trade-offs typically dictate resistor values of 2k to 100k with general-purpose op-amps.

Similar sorts of trade-offs are involved in almost all electronic design, including the simplest circuits constructed with transistors. For instance, the choice of quiescent current in a transistor amplifier is limited at the high end by device dissipation, increased input current, excessive supply current, and reduced current gain, whereas the lower limit of operating current is limited by leakage current, reduced current gain, and reduced speed (from stray capacitance in combination with the high resistance values). For these reasons you typically wind up with collector currents in the range of a few tens of microamps to a few tens of milliamps (higher for power circuits, sometimes a bit lower in "micropower" applications), as mentioned in Chapter 2.

In the next three chapters we will look more carefully at some of these problems in order to give you a good understanding of the trade-offs involved.

EXERCISE 4.6

Draw a dc-coupled inverting amplifier with gain of 100 and $Z_{in} = 10$k. Include compensation for input bias current, and show offset voltage trimming network (10k pot between pins 1 and 5, wiper tied to V_-). Now add circuitry so that $Z_{in} \geq 10^8$ ohms.

4.13 Low-power and programmable op-amps

For battery-powered applications there is a popular group of op-amps known as "programmable op-amps," because all of the internal operating currents are set by an externally applied current at a bias programming pin. The internal quiescent currents are all related to this bias current by current mirrors, rather than by internal resistor-programmed current sources. As a consequence, such amplifiers can be programmed to operate over a wide range of supply currents, typically from a few

POPULAR OP-AMPS

Sometimes a new op-amp comes along at just the right time, filling a vacuum with its combination of performance, convenience, and price. Several companies begin to manufacture it (it becomes "second-sourced"), designers become familiar with it, and you have a hit. Here is a list of some popular favorites of recent times:

301 First easy-to-use op-amp; first use of "lateral *pnp*." External compensation. National.

741 The industry standard for many years. Internal compensation. Fairchild.

1458 Motorola's answer to the 741; two 741s in a mini-DIP, with no offset pins.

308 National's precision op-amp. Low power, superbeta, guaranteed drift specifications.

324 Popular quad op-amp (358=dual, mini-DIP). Single-supply operation. National.

355 All-purpose bi-FET op-amp (356, 357 faster). Practically as precise as bipolar, but faster and lower input current. National. (Fairchild tried to get the FET ball rolling with their 740, which flopped because of poor performance. Would you believe 0.1V input offset?)

TL081 Texas Instruments' answer to the 355 series. Low-cost comprehensive series of singles, duals, quads; low power, low noise, many package styles.

LF411 National's improved bi-FET series. Low offset, low bias, fast, low distortion, high output current, low cost. Dual (LF412) and low-power variants (LF441/2/4).

microamps to a few milliamps. The slew rate, gain-bandwidth product f_T, and input bias current are all roughly proportional to the programmed operating current. When programmed to operate at a few microamps, programmable op-amps are extremely useful in battery-powered circuits. We will treat micropower design in detail in Chapter 14.

The 4250 was the original programmable op-amp, and it is still a good unit for many applications. Developed by Union Carbide, this classic is now "second-sourced" by many manufacturers, and it even comes in duals and triples (the 8022 and 8023, respectively). As an example of the sort of performance you can expect for operation at low supply currents, let's look at the 4250 running at 10μA. To get that operating current, we have to supply a bias current of 1.5μA with an external resistor. When it is operated at that current, f_T is 75kHz, the slew rate is 0.05V/μs, and the input bias current I_B is 3nA. At low operating currents the output drive capability is reduced considerably, and the open-loop output impedance rises to astounding levels, in this case about 3.5k. At low

THE 741 AND ITS FRIENDS

Bob Widlar designed the first really successful monolithic op-amp back in 1965, the Fairchild μA709. It achieved great popularity, but it had some problems, in particular the tendency to go into a latch-up mode when the input was overdriven and its lack of output short-circuit protection. It also required external frequency compensation (two capacitors and one resistor) and had a clumsy offset trimming circuit (again requiring three external components). Finally, its differential input voltage was limited to 5 volts.

Widlar moved from Fairchild to National, where he went on to design the LM301, an improved op-amp with short-circuit protection, freedom from latch-up, and a 30-volt differential input range. Widlar didn't provide internal frequency compensation, however, because he liked the flexibility of user compensation. The 301 could be compensated with a single capacitor, but because there was only one unused pin remaining, it still required three external components for offset trimming.

Meanwhile, over at Fairchild the answer to the 301 (the now-famous 741) was taking shape. It had the advantages of the 301, but Fairchild engineers opted for internal frequency compensation, freeing two pins to allow simplified offset trimming with a single external trimmer. Since most circuit applications don't require offset trimming (Widlar was right), the 741 in normal use requires no components other than the feedback network itself. The rest is history – the 741 caught on like wildfire and became firmly entrenched as the industry standard.

There are now numerous 741-type amps, essentially similar in design and performance, but with various features such as FET inputs, dual or quad units, versions with improved specifications, decompensated and uncompensated versions, etc. We list some of them here for reference and as a demonstration of man's instinct to clutch onto the coattails of the famous (see Table 4.1 for a more complete listing).

Single units		*Dual units*		*Quad units*	
741S	fast (10V/μs)	747	dual 741	MC4741	quad 741(alias 348)
MC741N	low noise	OP-04	precision	OP-11	precision
OP-02	precision	1458	mini-DIP package	4136	fast (3MHz)
4132	low power (35μA)	4558	fast (15V/μs)	HA4605	fast (4V/μs)
LF13741	FET low input current	TL082	FET, fast (similar to LF353)	TL084	FET, fast (similar to LF347)
748	uncompensated				
NE530	fast (25V/μs)	LF412	FET, fast		
TL081	FET, fast (similar to LF351)				
LF411	FET, fast				

operating currents the input noise voltage rises, while the input noise current drops (see Chapter 7). The 4250 specifications claim that it can run from as little as 1 volt total supply voltage, but the claimed minimum supply voltages of op-amps may not be terribly relevant in an actual circuit, particularly where any significant output swing or drive capability is needed.

The 776 (or 3476) is an upgraded 4250, with better output-stage performance at lower currents. The 346 is a nice quad programmable op-amp, with three sections programmed by one of the programming inputs, and the fourth programmed by the other. Some other programmable op-amps constructed with ordinary bipolar transistors are the OP-22, OP-32, HA2725, and CA3078. Programmable CMOS op-amps include the ICL7612, TLC251, MC14573, and CA3440. These feature operation at very low supply voltage (down to 1V for the TLC251) and, for the astounding 3440, operation at quiescent currents down to 20 *nano*amps. The 7612 and 251 use a variation of the usual programming scheme; their quiescent current is pin-selectable (10μA, 100μA, or 1mA), according to whether the programming pin is connected to V_+ or V_- or is left open.

In addition to these op-amps, there are several nonprogrammable op-amps that have been designed for low supply currents and low-voltage operation and should be considered for low-power applications. Notable among these is the outstanding bipolar LM10, an op-amp that is fully specified at 1 volt total supply voltage ($\pm$0.5V, for example). This is extraordinary, considering that V_{BE} increases with decreasing temperature and is close to 1 volt at $-55°$C, the lower limit of the LM10's operating range. Some other excellent "micropower" op-amps (and their operating currents) are the precision OP-20 (40μA), OP-90 (12μA), and LT1006 (90μA), the inexpensive quad LP324 (20μA per amplifier), the JFET LF441/2/4 (150μA per amplifier), and the MOSFET TLC27L4 (10μA per amplifier).

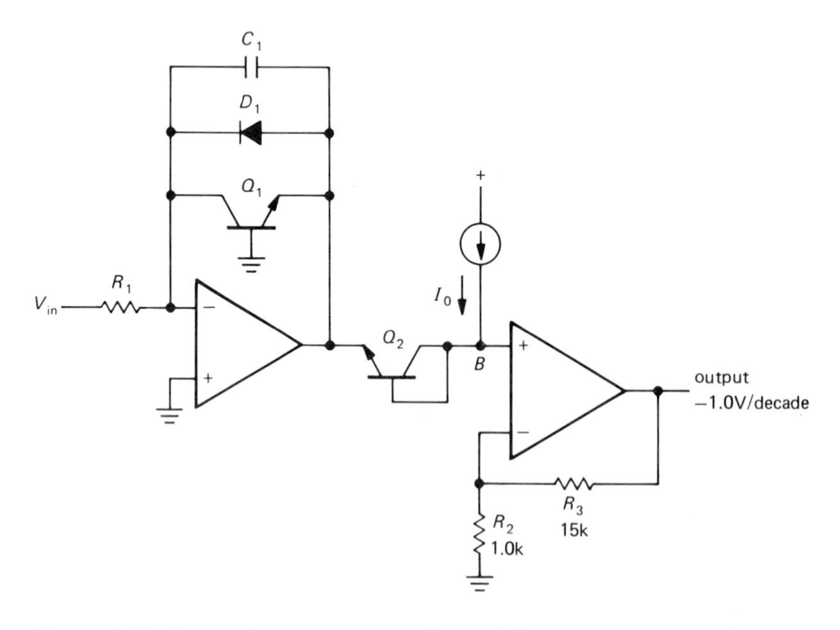

Figure 4.35. Logarithmic converter. Q_1 and Q_2 compose a monolithic matched pair.

TABLE 4.3. HIGH-VOLTAGE OP-AMPS

Type	Mfg[a]	Total supply min (V)	Total supply max (V)	Diff'l input[b] max (V)	FET	Ext comp	Trim	f_T typ (MHz)	Slew rate typ (V/μs)	Output current max (mA)	P_{diss} (50°C) max (W)	Therm lim	Case[c]	Comments
LM675	NS	20	60	60	–	–	–	5.5	8	3000	40	•	TO-220	monolithic pwr op-amp
LM343	NS	10	68	68	–	–	•	1	2	20	0.6	–	TO-99	superbeta
LM344	NS	10	68	68	–	•	–	1	30[d]	20	0.6	–	TO-99	superbeta
3580	BB	30	70	70	•	–	•	5	15	60	4.5	•	TO-3I	
LM12	NS	20	80	80	–	–	–	0.7	9	10000	90	•	TO-3	monolithic high-power
PA19	AP	30	80	40	•	•	–	100[e]	650[d]	5000	70	•	TO-3I	VMOS output
OPA541	BB	20	80	80	•	–	–	2	10	10000	90	–	TO-3I	monolithic high-pwr
MC1436	MO	10	80	80	–	–	•	1	2	10	0.6	–	TO-99	original, still good
1460	TP	30	80	6	–	•	•	1000[e]	300[e]	150	2.5	–	TO-3	VMOS output
1461	TP	30	80	25	•	•	•	1000[e]	1200[e]	750		–	P-DIP	VMOS output
1463	TP	30	80	25	•	–	•	17	165	1000	40	–	TO-3	fast unity-gain comp
HA2645	HA	20	80	74	–	•	•	4	5	10	0.6	•	TO-99	same as Philbrick 1332
OPA445	BB	20	100	80	•	–	•	2	10	15	0.6	–	TO-99	monolithic; miniDIP also
1481	TP	30	150	150	•	–	•	4.5	25	80	15	–	TO-3	current limit
3581	BB	65	150	150	•	–	•	5	20	30	4.5	•	TO-3I	
PA04	AP	30	200	20	•	•	–	2	50	20000	160	–	P-DIP	VMOS output; curr lim
1480	TP	30	300	450	•	–	•	20	100	80		•	TO-3	
3582	BB	140	300	300	•	–	•	5	20	15	4.5	•	TO-3I	
3583	BB	80	300	300	•	–	•	5	30	75	10	•	TO-3	
3584	BB	140	300	300	•	•	•	20[e]	150[e]	15	4.5	•	TO-3	
PA08V	AP	30	340	50	•	–	•	5	30	150	18	•	TO-3I	low V_{os}, low e_n
PA88	AP	30	450	25	•	•	–	1[d]	30[e]	100	12	•	TO-3I	low I_Q, V_{os}, e_n, VMOS
PA85	AP	30	450	25	•	•	–	20[d]	1000[e]	200	28	•	TO-3I	low V_{os}, low e_n, VMOS

(a) see notes to Table 4.1. (b) not to exceed total supply voltage. (c) "I" = isolated. (d) when comp for G>10. (e) when comp for G>100.

A DETAILED LOOK AT SELECTED OP-AMP CIRCUITS

The performance of the next few circuits is affected significantly by the limitations of op-amps; we will go into a bit more detail in their description.

4.14 Logarithmic amplifier

The circuit shown in Figure 4.35 exploits the logarithmic dependence of V_{BE} on I_C to produce an output proportional to the logarithm of a positive input voltage. R_1 converts V_{in} to a current, owing to the virtual ground at the inverting input. That current flows through Q_1, putting its emitter one V_{BE} drop below ground, according to the Ebers-Moll equation. Q_2, which operates at a fixed current, provides a diode drop of correction voltage, which is essential for temperature compensation. The current source (which can be a resistor, since point B is always within a few tenths of a volt of ground) sets the input current at which the output voltage is zero. The second op-amp is a noninverting amplifier with a voltage gain of 16, in order to give an output voltage of -1.0 volt per decade of input current (recall that V_{BE} increases 60mV per decade of collector current).

Some further details: Q_1's base could have been connected to its collector, but the base current would then have caused an error (remember that I_C is an accurate exponential function of V_{BE}). In this

TABLE 4.4. POWER OP-AMPS

Type	Mfg[a]	monolithic	FET	Trim?	Pkg[b]	I_{out} (A)	$\pm V_{supply}$ min (V)	$\pm V_{supply}$ max (V)	P_{diss} (W)	SR typ (V/μs)	f_T typ (MHz)	pwr BW (kHz)	V_{os}(max) (mV)	V_{os}(max) (μV/°C)	V_{os}(max) (μV/W)
PA03	AP	–	•	•	PD	30	15	75	500	10	5	70	3	30	20^t
PA04A	AP	–	•	–	PD	20	15	100	200	50	2	90	5	30	10^t
OPA512	BB	–	–	–	3I	15	10	50	125	4	4	20	3	40	20^t
LM12	NS	•	–	–	3	10	10	40	90	9	0.7	60	7	50	50
OPA501	BB	–	–	–	3I	10	10	40	80	1.4^m	1	16	5	40	35^t
OPA512B	BB	–	–	–	3I	10	10	50	125	4	4	20	6	65	20^t
OPA541B	BB	•	•	–	3I	10	10	40	90	10	2	55	2	30	60
1468	TP	–	–	–	3	10	10	50	125	4	4	20	6	65	20^t
PA19A	AP	–	•	–	3I	5	15	40	70	900	100	3500	0.5	10	20^t
OPA511	BB	–	–	–	3I	5	10	30	67	1.8	1	23	10	65	20^t
PA09A	AP	–	•	•	3I	4	10	40	78	400	75	2500	0.5	10	
SG1173	SG	•	–	–	220	3.5	5	25	20	0.8	1		4	30	-
LM675	NS	•	–	–	220	3	8	30	40	8	5.5		10	25^t	25^t
LH0101	NS	–	•	–	3	2	5	20	62	10	5	300	3	10^t	150^t
3572	BB	–	•	–	3I	2	15	40	60	3	0.5	16	2	40	20^t
3573	BB	–	–	–	3I	2	10	34	45	1.5	1	23	10	65	
LH0021	NS	–	–	–	3	1	5	15	23	3	1	20	3	25	15
MSK792	KE	–	–	•	3	1	5	22	5	2	1	11	0.1	2	-
1463	TP	–	•	•	3	1	15	40	40	165	17		5	20^t	
1461	TP	–	•	•	PD	0.75	15	40		1200^u	1000^u		5	50	-
LH0061	NS	–	–	•	3	0.5	-	15	20	70	-	1000	4	5^t	5^t
WA01A	AP	–	–	•	3I	0.4	12	16	10	4000	1000	150000	5	25	10^t
CLC203	CL	–	–	–	PD	0.2	9	20		6000	5000	60000	1.5	15	
1460	TP	–	–	•	3	0.15	15	40	2.5	300^u	1000^u	1500	5	50	-
3554B	BB	–	•	•	3I	0.15	5	18	5	1200	100	19000	1	15	-
HA2542	HA	•	–	•	D	0.1	5	15	1.6	375	120	4700^m	10^m	20	-
LH4101	NS	–	•	•	D	0.1	-	15	4	250	28	-	15	25^t	-
LH4104	NS	–	•	•	C	0.1	-	15	2.5	40	18	-	5	20^t	-
1480	TP	–	•	•	3	0.08	15	150		100	20	120	3	100	-
1481	TP	–	•	•	3	0.08	15	75	15	25	4.5	50	3	25	-
CA3450	RC	•	–	•	D	0.08	-	7	1.5	420	190	10000	15	-	-
3583	BB	–	•	•	3I	0.08	40	140	10	30	5	60	3	23	-
OP-50E	PM	•	–	•	D	0.07	5	18	0.5	3	25	20	0.03	0.3	-
3580	BB	–	•	•	3I	0.06	15	35	4.5	15	5	100	10	30	-
AMP-01E	PM	•	–	•	D	0.05	5	15	0.5	4.5	1	20	0.05	0.3	-
3581	BB	–	•	•	3I	0.03	32	75	4.5	20	5	60	3	25	-
3582/4	BB	–	•	•	3I	0.02	70	150	4.5	20/150	7	30/135	3	25	-

(a) see Table 4.1 notes. (b) 3 – TO-3; 220 – TO-220; PD – power DIP; D – DIP; I – isolated; C – metal can. (c) current limit: T – thermal limit; E – external adjust. (m) min or max. (t) typical. (u) uncompensated.

Type	I_b (max) @ 25°C (nA)	I_b (max) @ T_{max} (nA)	V_{sat} (V)	@ (A)	t_s (typ) (μs)	to (%)	I_{lim}[c] (A)	Thermal limit	Comments
PA03	0.05	50	7	30	2	0.1	T	•	a mighty brute
PA04A	0.02	-	7.5	15	2.5	0.1	E	–	high voltage brute
OPA512	20	15	7	15	2	0.1	E	–	PA-12 similar
LM12	300	150	8	10			13	•	
OPA501	20	15	7	10			E	–	PA-51 similar
OPA512B	30	-	6	10	2	0.1	E	–	
OPA541B	0.05	40	4.5	5	2	0.1	E	–	monolithic JFET
1468	30	-	6	10	2	0.1	E	–	
PA19A	0.05	50	5	4	1.2	0.01	E	•	VMOS output, wideband, prec
OPA511	40	30	8	5	2	0.1	E	–	PA-01 similar
PA09A	0.02	20	8	2	0.3	0.1	4.5	•	fast
SG1173	500	300	6	2			3.5	•	
LM675	2μA		10	3.5			4	•	
LH0101	0.3	300	5	2	2	0.01	E	–	PA-02 similar
3572	0.1	100	5	2			E	•	PA-07 similar; 3571 to 1A
3573	40	30	5	2			E	–	PA-73 similar
LH0021	100	35	4	1	4	0.1	E	–	ext comp
MSK792	100	100	3.5	1			E	–	
1463	0.2	200	8	1	0.25	0.1	E	–	VMOS output
1461	0.1	100	9	0.5	0.4	0.1	E	–	VMOS output; ext comp
LH0061	100	35	5	0.5	0.8	0.1	E	–	ext comp
WA01A	10μA		5	0.4	0.02	0.1			
CLC203	20μA	20μA	4	0.2	15ns	0.2	E	–	fast settle, wideband, prec
1460	10μA	-	6	0.15	1	0.1	0.25	–	VMOS output; ext comp
3554B	0.05	50	5	0.1	0.2	0.01	0.15	–	fast
HA2542	35μA	-	-	-	0.1	0.1	-	–	decomp (G>2)
LH4101	0.5	500	-	-	0.3	0.1	-	–	
LH4104	0.6	25	5	0.1	0.5	0.01	-	–	LH4105 has V_{os}<0.5mV
1480	0.2	200	10	0.08	1.5	0.01	0.13	–	high voltage
1481	0.1	100	5	0.08	7.5	0.1	0.13	–	
CA3450	350	-	2	0.08	35ns	0.1	-	–	video amp
3583	0.02	20	10	0.08	12	0.1	0.1	•	high voltage
OP-50E	5	7[f]	2	0.03	30	0.01	0.06	•	low noise, precision
3580	0.05	50	5	0.06	12	0.1	0.1	•	
AMP-01E	3	10[f]	2	0.03	15	0.01	0.06	•	low noise, prec inst amp
3581	0.02	20	5	0.03	12	0.1	0.05	•	
3582/4	0.02	20	5	0.02	12	0.1	0.03	•	high voltage

circuit the base is at the same voltage as the collector because of the virtual ground, but there is no base-current error. Q_1 and Q_2 should be a matched pair, thermally coupled (a matched monolithic pair like the LM394 or MAT-01 is ideal). This circuit will give accurate logarithmic output over seven decades of current or more (1nA to 10mA, approximately), providing that low-leakage transistors and a low-bias-current input op-amp are used. An op-amp like the 741 with 80nA of bias current is unsuitable, and a FET-input op-amp like the 411 is usually required to achieve the full seven decades of linearity. Furthermore, in order to give good performance at low input currents, the input op-amp must be accurately trimmed for zero offset voltage, since V_{in} may be as small as a few tens of microvolts at the lower limit of current. If possible, it is better to use a current input to this circuit, omitting R_1 altogether.

The capacitor C_1 is necessary to stabilize the feedback loop, since Q_1 contributes voltage gain inside the loop. Diode D_1 is necessary to prevent base-emitter breakdown (and destruction) of Q_1 in the event the input voltage goes negative, since Q_1 provides no feedback path for positive op-amp output voltage. Both these minor problems are avoided if Q_1 is wired as a diode, i.e., with its base tied to its collector.

Temperature compensation of gain

Q_2 compensates changes in Q_1's V_{BE} drop as the ambient temperature changes, but the changes in the slope of the curve of V_{BE} versus I_C are not compensated. In Section 2.10 we saw that the "60mV/decade" is proportional to absolute temperature. The output voltage of this circuit will look as shown in Figure 4.36. Compensation is perfect at an input current equal to I_0, Q_2's collector current. A change in temperature of 30°C causes a

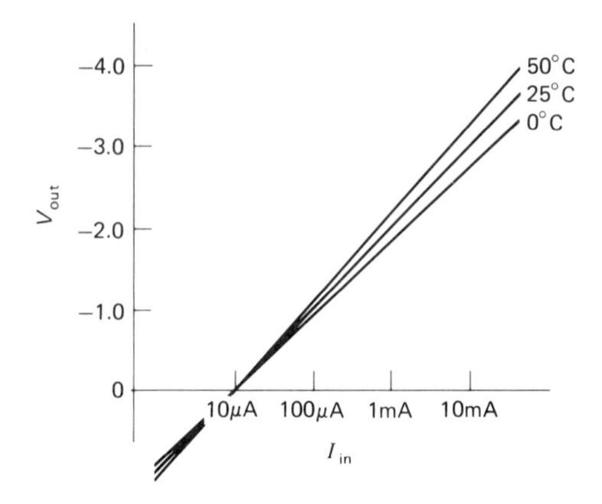

Figure 4.36

10% change in slope, with corresponding error in output voltage. The usual solution to this problem is to replace R_2 with a series combination of an ordinary resistor and a resistor of positive temperature coefficient. Knowing the temperature coefficient of the resistor (e.g., the TG 1/8 type manufactured by Texas Instruments has a coefficient of +0.67%/°C) allows you to calculate the value of the ordinary resistor to put in series in order to effect perfect compensation. For instance, with the 2.7k TG 1/8 type "sensistor" just mentioned, a 2.4k series resistor should be used.

There are several logarithmic converter modules available as complete integrated circuits. These offer very good performance, including internal temperature compensation. Some manufacturers are Analog Devices, Burr-Brown, Philbrick, Intersil, and National Semiconductor.

EXERCISE 4.7

Finish up the log converter circuit by (a) drawing the current source explicitly and (b) using a TG 1/8 resistor (+0.67%/°C tempco) for thermal slope compensation. Choose values so that V_{out} = +1 volt per decade, and provide an output offset control so that V_{out} can be set to zero for any desired input current (do this with an inverting amplifier offset circuit, not by adjusting I_0).

4.15 Active peak detector

There are numerous applications in which it is necessary to determine the peak value of some input waveform. The simplest method is a diode and capacitor (Fig. 4.37). The highest point of the input waveform charges up C, which holds that value while the diode is back-biased.

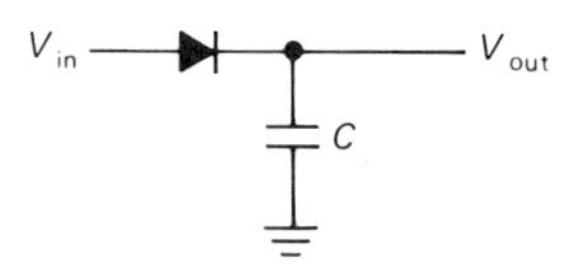

Figure 4.37

This method has some serious problems. The input impedance is variable and is very low during peaks of the input waveform. Also, the diode drop makes the circuit insensitive to peaks less than about 0.6 volt and inaccurate (by one diode drop) for larger peak voltages. Furthermore, since the diode drop depends on temperature and current, the circuit's inaccuracies depend on the ambient temperature and on the rate of change of output; recall that $I = C(dV/dt)$. An input emitter follower would improve the first problem only.

Figure 4.38 shows a better circuit, using feedback. By taking feedback from the voltage at the capacitor, the diode drop doesn't cause any problems. The sort of output waveform you might get is shown in Figure 4.39.

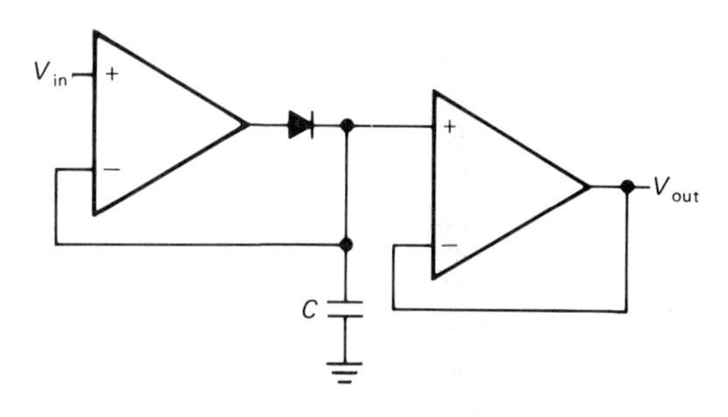

Figure 4.38. Op-amp peak detector.

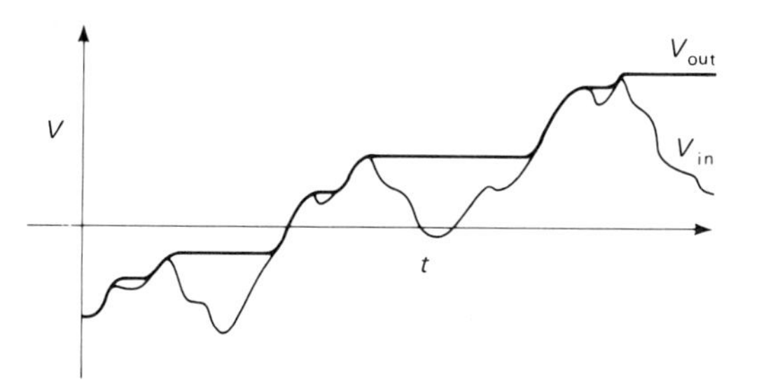

Figure 4.39

Op-amp limitations affect this circuit in three ways: (a) Finite op-amp slew rate causes a problem, even with relatively slow input waveforms. To understand this, note that the op-amp's output goes into negative saturation when the input is less positive than the output (try sketching the op-amp voltage on the graph; don't forget about diode forward drop). So the op-amp's output has to race back up to the output voltage (plus a diode drop) when the input waveform next exceeds the output. At slew rate S, this takes roughly $(V_0 - V_-)/S$, where V_- is the negative supply voltage and V_0 is the output voltage. (b) Input bias current causes a slow discharge (or charge, depending on the sign of the bias current) of the capacitor. This is sometimes called "droop," and it is best avoided by using op-amps with very low bias current. For the same reason, the diode must be a low-leakage type (e.g., the FJT1100, with less than 1pA reverse current at 20V, or a "FET diode" such as the PAD-1 from Siliconix or the ID101 from Intersil), and the following stage must also present high impedance (ideally it should also be a FET or FET-input op-amp). (c) The maximum op-amp output current limits the rate of change of voltage across the capacitor, i.e., the rate at which the output can follow a rising input. Thus, the choice of capacitor value is a compromise between low droop and high output slew rate.

For instance, a 1μF capacitor used in this circuit with the common 741 (which would be a poor choice because of its high bias current) would droop at $dV/dt = I_B/C = 0.08$V/s and would follow input changes only up to $dV/dt = I_{\text{output}}/C = 0.02$V/$\mu s$. This maximum follow rate is much less than the op-amp's slew rate of 0.5V/μs, being limited by the maximum output current of 20mA driving 1μF. By decreasing C you could achieve greater output slewing rate at the expense of greater droop. A more realistic choice of components would be the popular LF355 FET-input op-amp as driver and output follower (30pA typical bias current, 20mA output current) and a value of $C = 0.01\mu$F. With this combination you would get a droop of only 0.006V/s and an overall circuit slew rate of 2V/μs. For better performance, use a FET op-amp like the OPA111 or AD549, with input currents of 1pA or less. Capacitor leakage may then limit performance even if unusually good capacitors are used, e.g., polystyrene or polycarbonate (see Section 7.05).

□ *A circuit cure for diode leakage*

Quite often a clever circuit configuration can provide a solution to problems caused by nonideal behavior of circuit components. Such solutions are aesthetically pleasing as well as economical. At this point we yield to the temptation to take a closer look at such a high-performance design, rather than delaying until Chapter 7, where we treat such subjects under the heading of precision design.

Suppose we want the best possible performance in a peak detector, i.e., highest ratio of output slew rate to droop. If the lowest-input-current op-amps are used in a peak-detector circuit (some are available with bias currents as low as 0.01pA), the droop will be dominated by diode leakage; i.e., the best available diodes have higher leakage currents (see Table 1.1) than the op-amps' bias currents. Figure 4.40 shows a clever circuit solution. As before, the voltage on the capacitor follows a rising input waveform: IC_1 charges the capacitor through both diodes and is unaffected by IC_2's output. When the input drops below the peak value, IC_1 goes into negative saturation, but IC_2 holds point X at the capacitor voltage, eliminating leakage altogether in D_2. D_1's small leakage current flows through R_1, with negligible drop across the resistor. Of course, both op-amps must have low bias current. The OPA111B is a good choice here, with its combination of precision ($V_{os} = 250\mu$V, max) and low input current (1pA, max). This circuit is analogous to the so-called guard circuits used for high-impedance or small-signal measurements.

Note that the input op-amps in both peak-detector circuits spend most of their time in negative saturation, only popping up when the input level exceeds the peak voltage previously stored on the capacitor. However, as we saw in the active rectifier circuit (Section 4.10), the journey from negative saturation can take a while (e.g., 1μs–2μs for the LF411). This may restrict your choice to high-slew-rate op-amps.

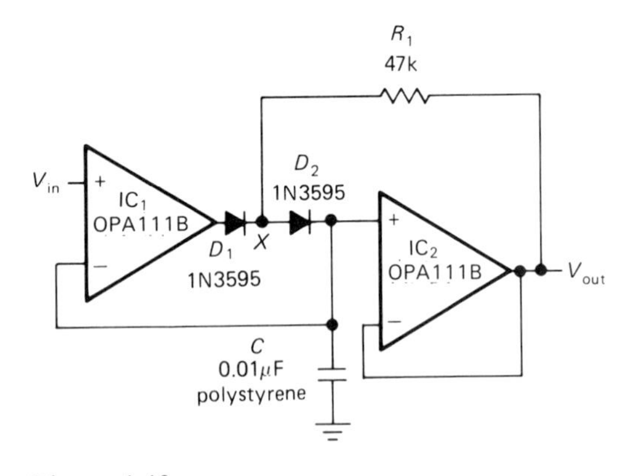

Figure 4.40

□ *Resetting a peak detector*

In practice it is usually desirable to reset the output of a peak detector in some way.

One possibility is to put a resistor across the output so that the circuit's output decays with a time constant RC. In this way it holds only the most recent peak values. A better method is to put a transistor switch across C; a short pulse to the base then zeros the output. A FET switch is often used instead. For example, in Figure 4.38 you could connect an n-channel MOSFET across C; bringing the gate momentarily positive then zeros the capacitor voltage.

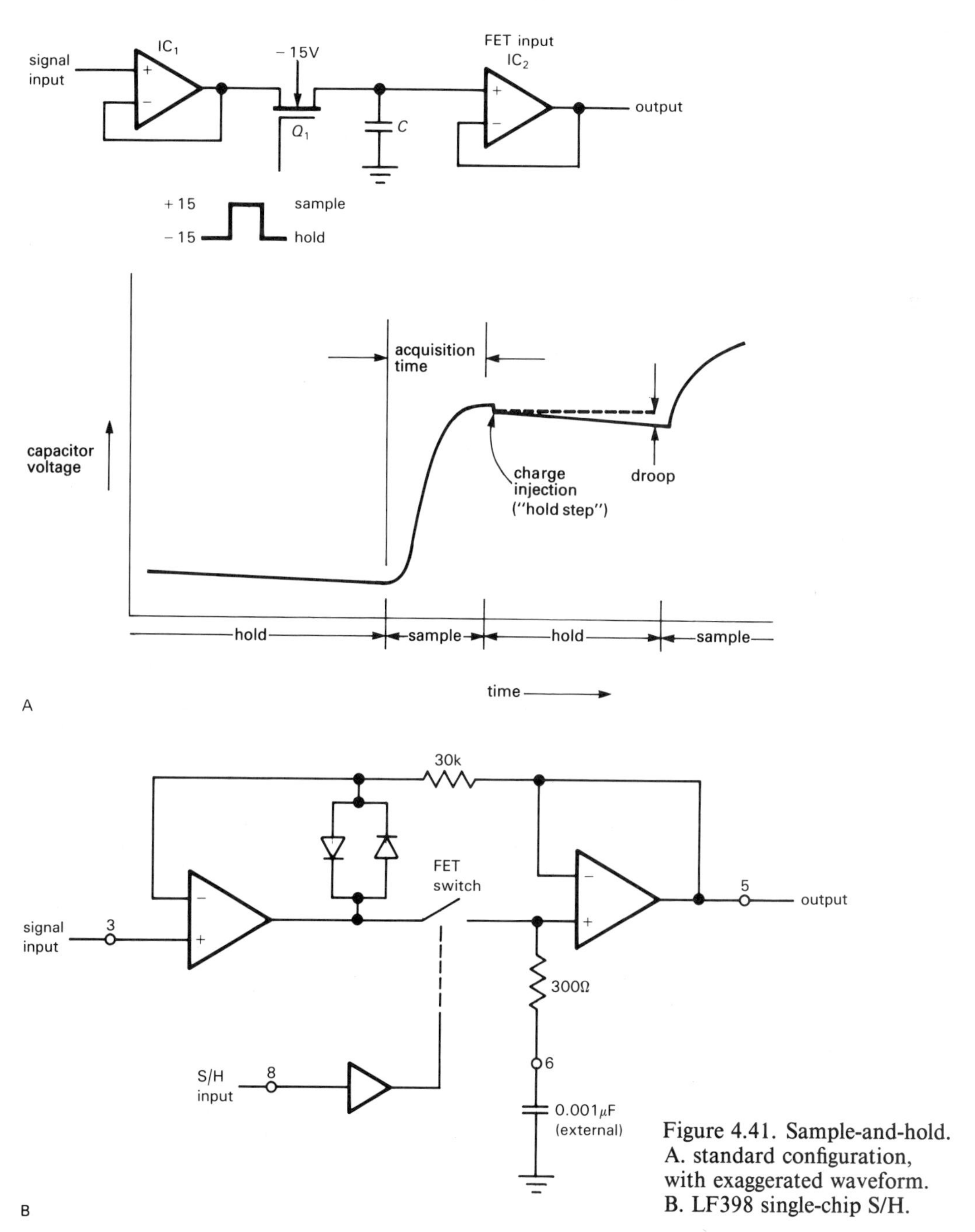

Figure 4.41. Sample-and-hold. A. standard configuration, with exaggerated waveform. B. LF398 single-chip S/H.

4.16 Sample-and-hold

Closely related to the peak detector is the "sample-and-hold" (S/H) circuit (sometimes called "follow-and-hold"). These are especially popular in digital systems, where you want to convert one or more analog voltages to numbers so that a computer can digest them: The favorite method is to grab and hold the voltage(s), then do the digital conversion at your leisure. The basic ingredients of a S/H circuit are an op-amp and a FET switch; Figure 4.41A shows the idea. IC_1 is a follower to provide a low-impedance replica of the input. Q_1 passes the signal through during "sample" and disconnects it during "hold." Whatever signal was present when Q_1 was turned OFF is held on capacitor C. IC_2 is a high-input-impedance follower (FET inputs), so that capacitor current during "hold" is minimized. The value of C is a compromise: Leakage currents in Q_1 and the follower cause C's voltage to "droop" during the hold interval, according to $dV/dt = I_{leakage}/C$. Thus C should be large to minimize droop. But Q_1's ON resistance forms a low-pass filter in combination with C, so C should be small if high-speed signals are to be followed accurately. IC_1 must be able to supply C's charging current $I = CdV/dt$ and must have sufficient slew rate to follow the input signal. In practice, the slew rate of the whole circuit will usually be limited by IC_1's output current and Q_1's ON resistance.

EXERCISE 4.8

Suppose IC_1 can supply 10mA of output current, and $C = 0.01\mu F$. What is the maximum input slew rate the circuit can accurately follow? If Q_1 has 50 ohms ON resistance, what will be the output error for an input signal slewing at 0.1V/μs? If the combined leakage of Q_1 and

☐ DIELECTRIC ABSORPTION

Capacitors are not perfect. The most commonly appreciated shortcomings are leakage (parallel resistance), series resistance and inductance, and nonzero temperature coefficient of capacitance. A more subtle problem is *dielectric absorption*, an effect that manifests itself clearly as follows: Take a large-value tantalum capacitor that is charged up to 10 volts or so, and rapidly discharge it by momentarily putting a 100 ohm resistor across it. Remove the resistor, and watch the capacitor's voltage on a high-impedance voltmeter. You will be amazed to see the capacitor *charge back up*, reaching perhaps a volt or so after a few seconds!

The origins of dielectric absorption (or dielectric *soakage*, dielectric *memory*) are not entirely understood, but the phenomenon is believed to be related to remnant polarization trapped on dielectric interfaces; mica, for example, with its layered structure, is particularly poor in this regard. From a circuit point of view, this extra polarization behaves like a set of additional series RCs across the capacitor (Fig. 4.42A), with time constants generally in the range of $\approx$100μs to several seconds. Dielectrics vary widely in their susceptibility to dielectric absorption; Figure 4.42B shows data for several high-quality dielectrics, plotted as voltage memory versus time after a 10 volt step of 100μs duration.

Dielectric absorption can cause significant errors in integrators and other analog circuits that rely on the ideal characteristics of capacitors. In the case of a sample/hold followed by precision analog-to-digital conversion, the effect can be devastating. In such situations the best approach is to choose your capacitors carefully (Teflon dielectric seems to be best), retaining a healthy skepticism until proven wrong. In extreme cases you may have to resort to tricks such as compensation networks that use carefully trimmed RCs to electrically cancel the capacitor's internal dielectric absorption. This approach is used in some high-quality sample/hold modules made by Hybrid Systems.

IC_2 is 1nA, what is the droop rate during the "hold" state?

For both the sample/hold circuit and the peak detector, an op-amp drives a capacitive load. When designing such circuits, make sure you choose an op-amp that is stable at unity gain when loaded by the capacitor C. Some op-amps, (e.g., the LF355/6) are specifically designed to drive large (0.01μF) capacitive loads directly. Some other tricks you can use are discussed in Section 7.07 (see Fig. 7.17).

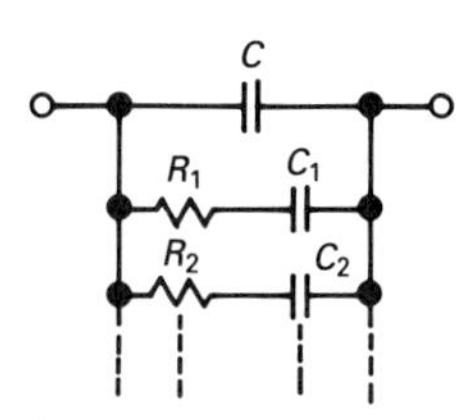

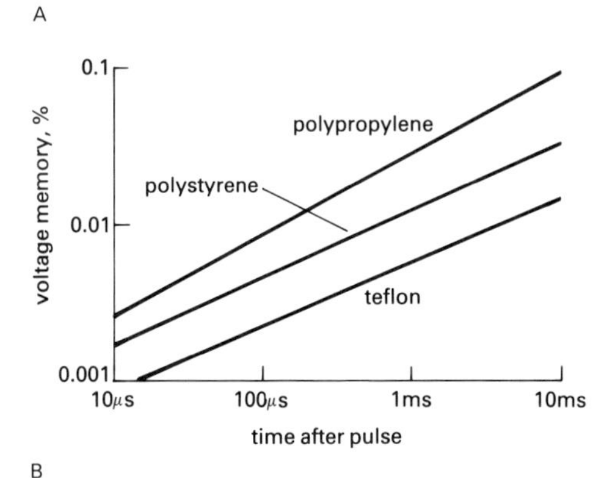

Figure 4.42. Dielectric absorption in capacitors. A. model B. measured properties for several dielectrics. (After Hybrid Systems HS9716 data sheet.)

You don't have to design S/H circuits from scratch, because there are nice monolithic ICs that contain all the parts you need except for the capacitor. National's LF398 is a popular part, containing the FET switch and two op-amps in an inexpensive ($2) 8-pin package. Figure 4.41B shows how to use it. Note how feedback closes the feedback loop around *both* op-amps. There are plenty of fancy S/H chips

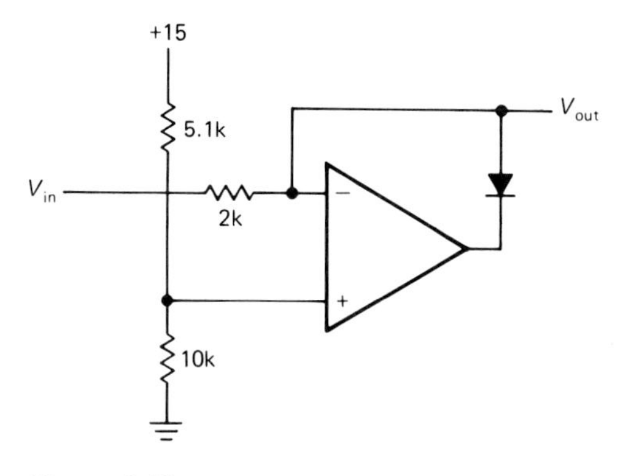

Figure 4.43

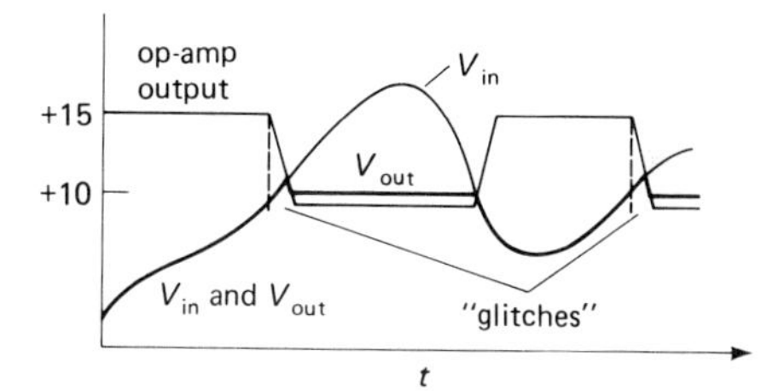

Figure 4.44

available, if you need better performance than the LF398 offers; for example, the AD585 from Analog Devices includes an internal capacitor and guarantees a maximum acquisition time of 3μs for 0.01% accuracy following a 10 volt step.

□ 4.17 Active clamp

Figure 4.43 shows a circuit that is an active version of the clamp function we discussed in Chapter 1. For the values shown, $V_{in} < +10$ volts puts the op-amp output at positive saturation, and $V_{out} = V_{in}$. When V_{in} exceeds +10 volts the diode closes the feedback loop, clamping the output at 10 volts. In this circuit, op-amp slew-rate limitations allow small glitches as the input reaches the clamp voltage from below (Fig. 4.44).

□ 4.18 Absolute-value circuit

The circuit shown in Figure 4.45 gives a positive output equal to the magnitude of the input signal; it is a full-wave rectifier. As usual, the use of op-amps and feedback eliminates the diode drops of a passive full-wave rectifier.

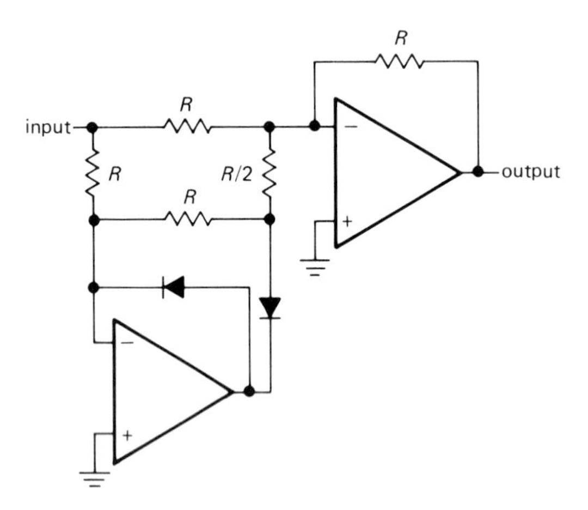

Figure 4.45. Active full-wave rectifier.

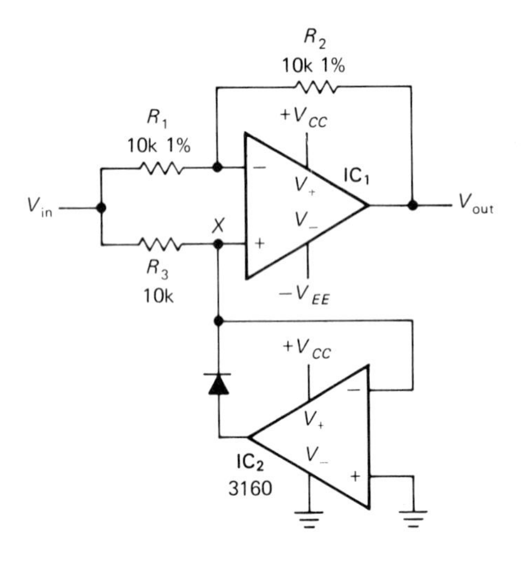

Figure 4.46

EXERCISE 4.9

Figure out how the circuit in Figure 4.45 works. Hint: Apply first a positive input voltage, and see what happens; then do negative.

Figure 4.46 shows another absolute-value circuit. It is readily understandable as a simple combination of an optional inverter (IC_1) and an active clamp (IC_2). For positive input levels the clamp is out of the circuit, with its output at negative saturation, making IC_1 a unity-gain inverter. Thus the output is equal to the absolute value of the input voltage. By running IC_2 from a single positive supply, you avoid problems of slew-rate limitations in the clamp, since its output moves over only one diode drop. Note that no great accuracy is required of R_3.

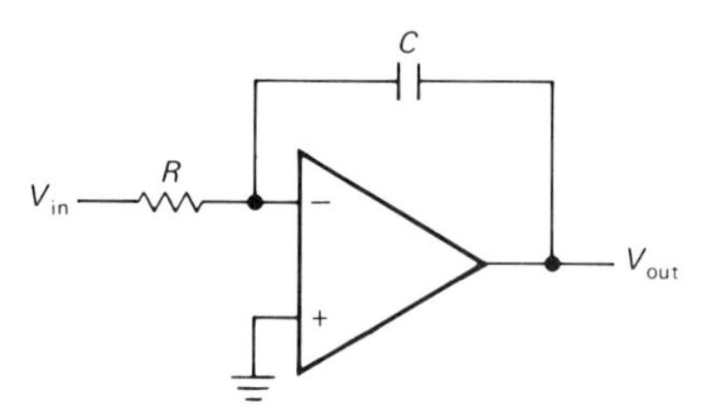

Figure 4.47. Integrator.

4.19 Integrators

Op-amps allow you to make nearly perfect integrators, without the restriction that $V_{out} \ll V_{in}$. Figure 4.47 shows how it's done. Input current V_{in}/R flows through C. Because the inverting input is a virtual ground, the output voltage is given by

$$V_{in}/R = -C(dV_{out}/dt)$$

or

$$V_{out} = \frac{1}{RC}\int V_{in}\,dt + \text{constant}$$

The input can, of course, be a current, in which case R will be omitted. One problem with this circuit as drawn is that the output tends to wander off, even with the input grounded, due to op-amp offsets and bias current (there's no feedback at dc, which violates rule 3 in Section 4.08). This problem can be minimized by using a FET op-amp for low input current and offset, trimming the op-amp input offset voltage, and using large R and C values. In addition, in many applications the integrator is zeroed periodically by closing a

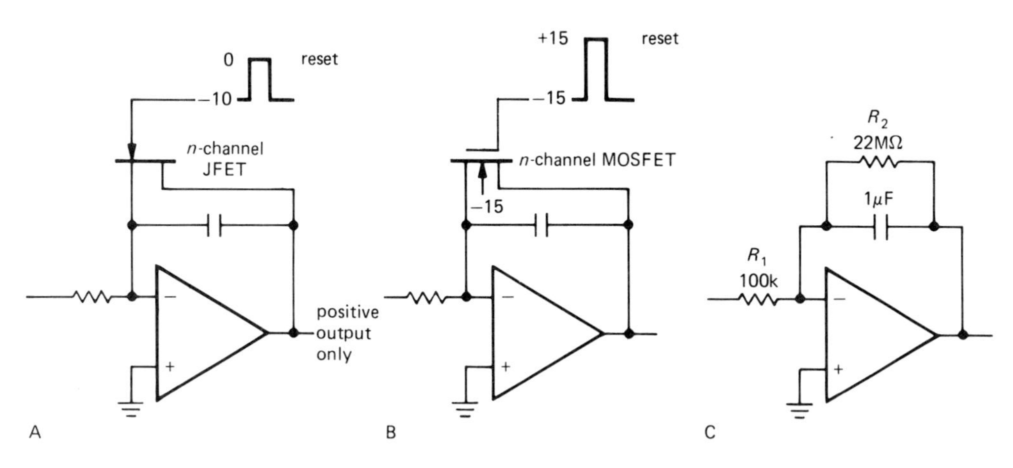

4.48. Op-amp integrators with reset switches.

switch placed across the capacitor (usually a FET), so only the drift over short time scales matters. As an example, an inexpensive FET op-amp like the LF411 (25pA typical bias current) trimmed to a voltage offset of 0.2mV and used in an integrator with R =10MΩ and C = 10μF will produce an output drift of less than 0.003 volt in 1000 seconds.

If the residual drift of the integrator is still too large for a given application, it may be necessary to put a large resistor R_2 across C to provide dc feedback for stable biasing. The effect is to roll off the integrator action at very low frequencies, $f < 1/R_2C$. Figure 4.48 shows integrators with FET zeroing switch and with resistor bias stabilization. The feedback resistor may become rather large in this sort of application. Figure 4.49 shows a trick for producing the effect of a large feedback resistor using smaller values. In this case the feedback network behaves like a single 10MΩ resistor in the standard inverting amplifier circuit giving a voltage gain of −100. This technique has the advantage of using resistors of convenient values without the problems of stray capacitance, etc., that occur with very large resistor values. Note that this "T-network" trick may increase the effective input offset voltage, if used in a transresistance configuration (Section 4.09). For example, the circuit of Figure 4.49, driven from a high-impedance source (e.g., the current from a photodiode, with the input resistor omitted), has an output offset of 100 times V_{os}, whereas the same circuit with a 10MΩ feedback resistor has an output equal to V_{os} (assuming the offset due to input current is negligible).

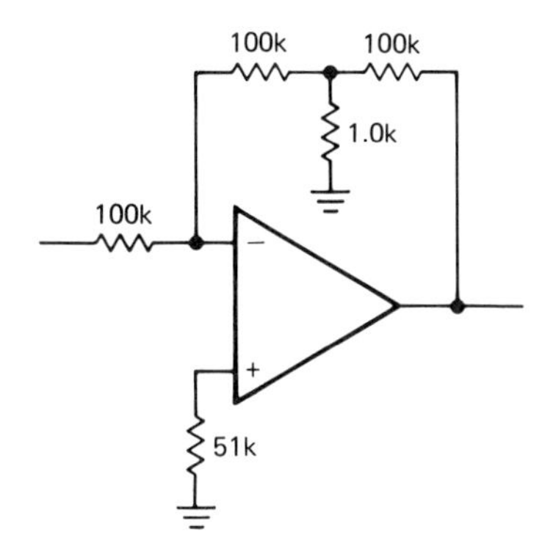

Figure 4.49

□ *A circuit cure for FET leakage*

In the integrator with a FET reset switch (Fig. 4.48), drain-source leakage sources a small current into the summing junction even when the FET is OFF. With an ultra-low-input-current op-amp and low-leakage capacitor, this can be the dominant error in the integrator. For example, the excellent AD549 JFET-input "electrometer" op-amp has a maximum input current of 0.06pA, and a high-quality 0.01μF metallized Teflon or polystyrene capacitor specifies leakage resistance as 10^7 *meg*ohms, minimum. Thus the integrator, exclusive of reset circuit, keeps stray currents at the summing junction below 1pA (for a worst-case 10V full-scale output), corresponding to an output dV/dt of less than 0.01mV/s. Compare this with the leakage contribution of a MOSFET such as the popular 2N4351 (enhancement mode), which specifies a maximum leakage current of 10nA at $V_{DS} = 10$V and $V_{GS} = 0$V! In other words, the FET contributes 10,000 times as much leakage as everything else combined.

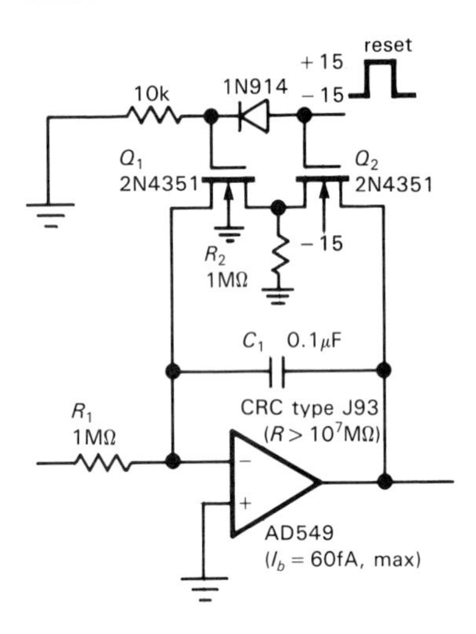

Figure 4.50

Figure 4.50 shows a clever circuit solution. Although both n-channel MOSFETs are switched together, Q_1 is switched with gate voltages of zero and +15 volts so that gate leakage (as well as drain-source leakage) is entirely eliminated during the OFF state (zero gate voltage). In the ON state the capacitor is discharged as before, but with twice R_{ON}. In the OFF state, Q_2's small leakage passes to ground through R_2 with negligible drop. There is no leakage current at the summing junction because Q_1's source, drain, and substrate are all at the same voltage. Compare this circuit with the zero-leakage peak-detector circuit of Figure 4.40.

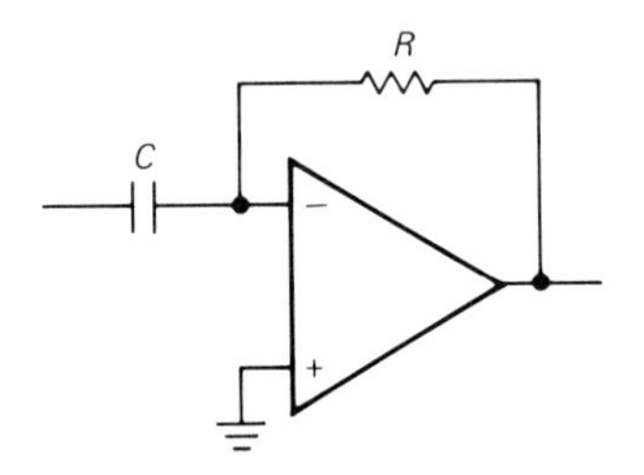

Figure 4.51

4.20 Differentiators

Differentiators are similar to integrators, but with R and C reversed (Fig. 4.51). Since the inverting input is at ground, the rate of change of input voltage produces a current $I = C(dV_{\text{in}}/dt)$ and hence an output voltage

$$V_{\text{out}} = -RC\frac{dV_{\text{in}}}{dt}$$

Differentiators are bias-stable, but they generally have problems with noise and instabilities at high frequencies because of the op-amp's high gain and internal phase shifts. For this reason it is necessary to roll off the differentiator action at some maximum frequency. The usual method is shown in Figure 4.52. The choice of the rolloff components R_1 and C_2 depends on the noise level of the signal and the bandwidth of the op-amp. At high frequencies this circuit becomes an integrator, due to R_1 and C_2.

OP-AMP OPERATION WITH A SINGLE POWER SUPPLY

Op-amps don't *require* ±15 volt regulated supplies. They can be operated from split

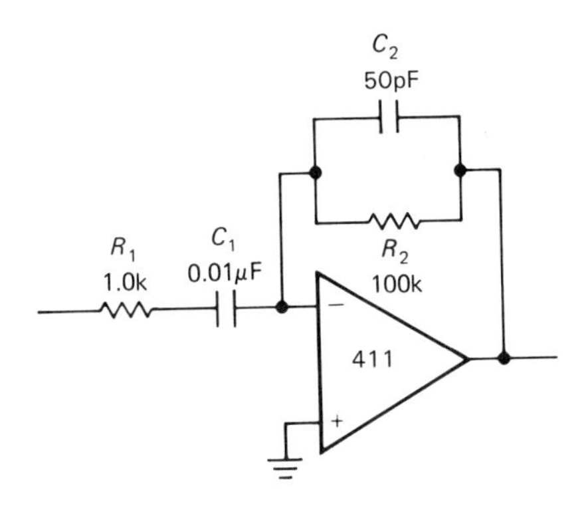

Figure 4.52

supplies of lower voltages, or from unsymmetrical supply voltages (e.g, +12 and −3), as long as the total supply voltage ($V_+ - V_-$) is within specifications (see Table 4.1). Unregulated supply voltages are often adequate because of the high "power-supply rejection ratio" you get from negative feedback (for the 411 it's 90dB typ). But there are many occasions when it would be nice to operate an op-amp from a single supply, say +12 volts. This can be done with ordinary op-amps by generating a "reference" voltage above ground, if you are careful about minimum supply voltages, output swing limitations, and maximum common-mode input range. With some of the more recent op-amps whose input and output ranges include the negative supply (i.e., ground, when run from a single positive supply), single-supply operation is attractive because of its simplicity. Keep in mind, though, that operation with symmetrical split supplies remains the usual technique for nearly all applications.

□ 4.21 Biasing single-supply ac amplifiers

For a general-purpose op-amp like the 411, the inputs and output can typically swing to within about 1.5 volts of either supply. With V_- connected to ground, you can't have either of the inputs or the output at ground. Instead, by generating a reference voltage (e.g., $0.5V_+$) you can bias the op-amp for successful operation (Fig. 4.53). This circuit is an audio amplifier with 40dB gain. $V_{ref} = 0.5V_+$ gives an output swing of about 17 volts pp before onset of clipping. Capacitive coupling is used at the input and output to block the dc level, which equals V_{ref}.

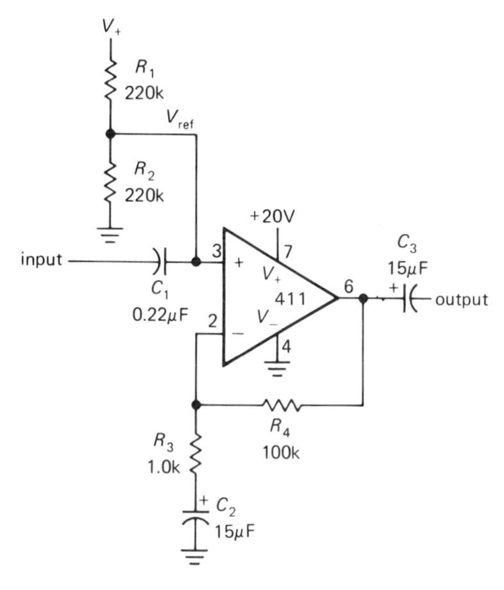

Figure 4.53

□ 4.22 Single-supply op-amps

There is a class of op-amps that permit simplified operation with a single positive supply, because they permit input voltages all the way down to the negative rail (normally tied to ground). They can be further divided into two types, according to the capability of the output stage: One type can swing all the way down to V_-, and the other type can swing all the way to *both* rails:

1. The LM324(quad)/LM358 (dual), LT1013, and TLC270 types. These have input common-mode ranges all the way down to 0.3 volt *below* V_-, and the output can swing down to V_-. Both inputs and output can go to within 1.5 volts of V_+. If instead you need an input range up to V_+, use something like an LM301/307, OP-41, or a 355; an example is illustrated in Section 6.24 in the discussion of constant-current supplies. In order to understand some of the subtleties of this sort of op-amp, it is helpful to look at the schematic (Fig. 4.54). It is a reasonably straightforward differential amplifier, with current mirror active load on the input stage and push-pull complementary output stage with current limiting. The special things to remember are these (calling V_- ground):

Inputs: The *pnp* input structure allows swings of 0.3 volt below ground; if that is exceeded by either input, weird things happen at the output (it may go negative, for instance).

Output: Q_{13} pulls the output down and can sink plenty of current, but it goes only to within a diode drop of ground. Outputs below that are provided by the 50μA current sink, which means you can't drive a load that sources more than 50μA and get closer than a diode drop above ground. Even for "nice" loads (an open circuit, say), the current source won't bring the output lower than a saturation voltage (0.1V) above ground. If you want the output to go clear down to ground, the load should sink a small current to ground; it could be a resistor to ground, for instance. Recent additions to the family of *pnp*-input single-supply op-amps include the precision LT1006 and LT1014 (single and quad, respectively) and the micropower OP-20 and OP-90 (both single), and LP324 (quad).

We will illustrate the use of these op-amps with some circuits, after mentioning the other kind of op-amp that lends itself well to single-supply operation.

2. The LM10 (bipolar) or CA5130/5160 (MOSFET) complementary-output-stage op-amps. When saturated, they look like a small resistance from the output to the supply (V_+ or V_-). Thus the output can swing all the way to either supply. In addition, the inputs can go 0.5 volt below V_-. Unlike the LM10, the CA5130

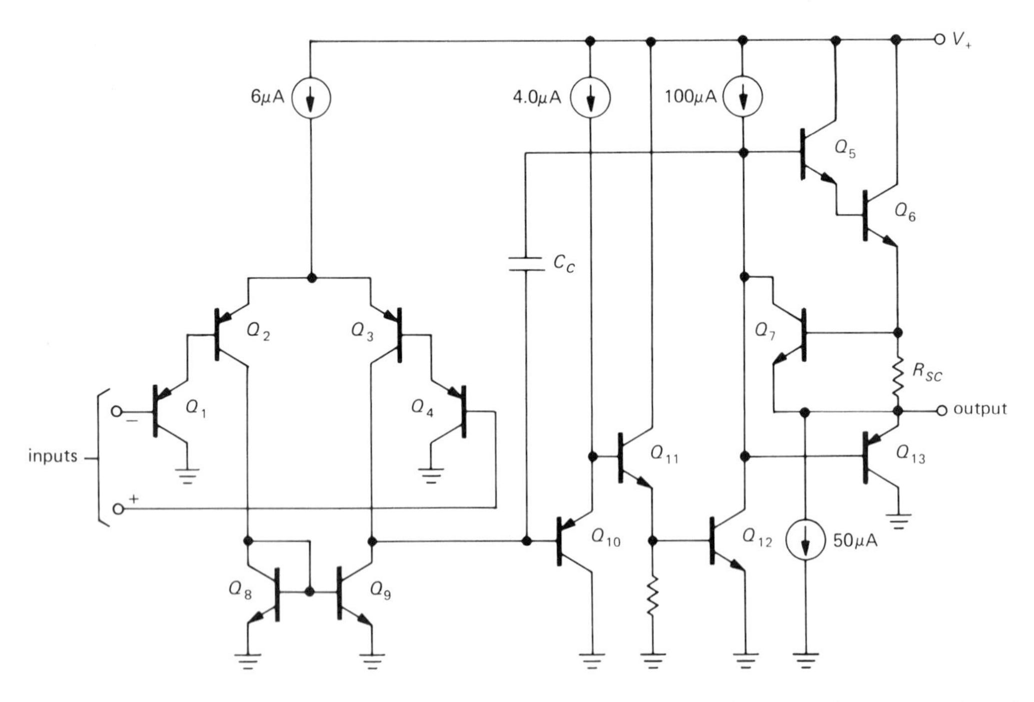

Figure 4.54. Schematic of the popular 324 and 358 op-amps. (National Semiconductor Corp.)

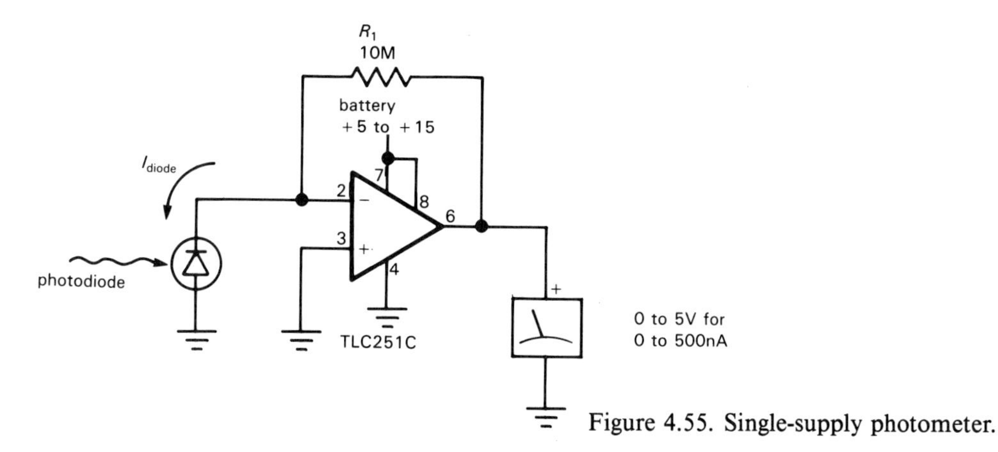

Figure 4.55. Single-supply photometer.

and 5160 are limited to 16 volts (max) total supply voltage and ±8 volts differential input voltage. Although most CMOS op-amps permit rail-to-rail output swings, watch out for some varieties that can only swing all the way to one rail; also note that the input common-mode range of most CMOS op-amps, like ordinary bipolar op-amps, includes at most one power-supply rail. For example, the popular TLC27*xx* series from TI has input and output capability to the negative rail only, whereas the LMC660 from National, along with the Intersil ICL76*xx* series and RCA's CMOS op-amps, has output swing to both rails (but input common-mode range only to the negative rail). Unique among op-amps are the CMOS ICL7612 and ALD1701/2, which claim both input and output operation to *both* rails.

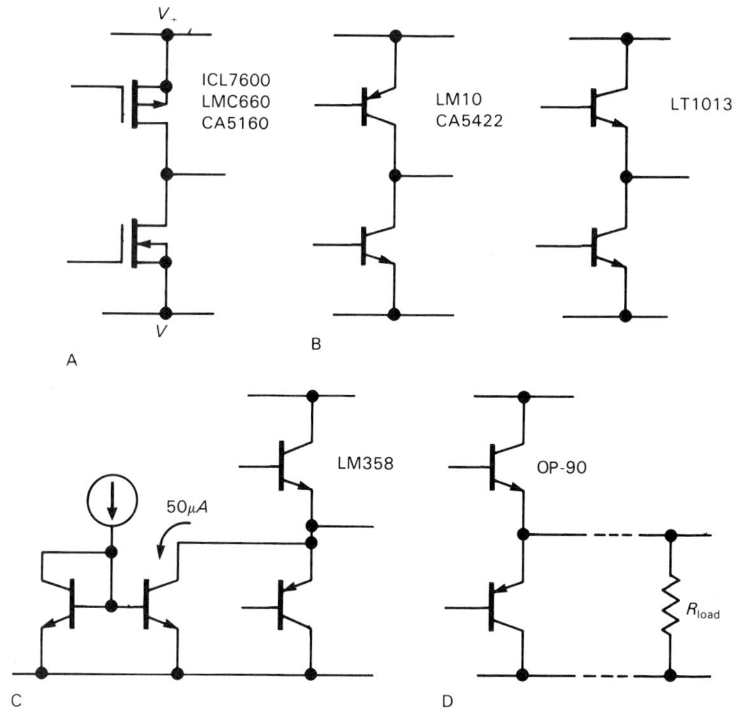

Figure 4.56. Output stages used in single-supply op-amps.

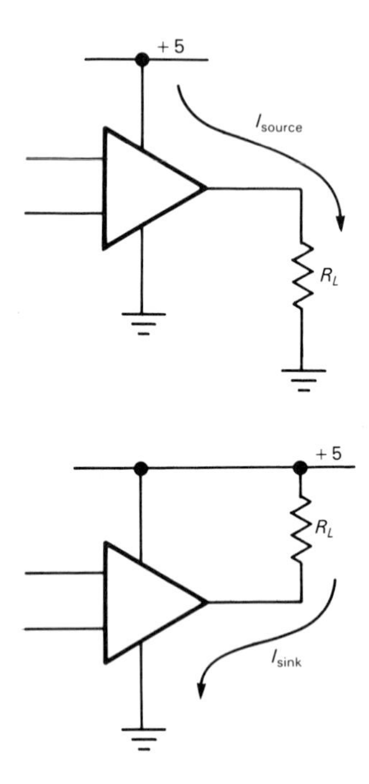

Figure 4.57. Connecting a load to a single-supply op-amp. All single-supply types (A–D) can swing all the way to ground while sourcing current. Some types (A and B) can swing nearly to ground while sinking moderate or substantial currents; type C can sink up to 50μA, and type D requires a load resistor returned to ground to operate near ground.

□ *Example: single-supply photometer*

Figure 4.55 shows a typical example of a circuit for which single-supply operation is convenient. We discussed a similar circuit earlier under the heading of current-to-voltage converters. Since a photocell circuit might well be used in a portable light-measuring instrument, and since the output is known to be positive only, this is a good candidate for a battery-operated single-supply circuit. R_1 sets the full- scale output at 5 volts for an input photocurrent of 0.5μA. No offset voltage trim is needed in this circuit, since the worst-case untrimmed offset of 10mV corresponds to a negligible 0.2% of full-scale meter indication. The TLC251 is an inexpensive micropower (10μA supply current) CMOS op-amp with input and output swings to the negative rail. Its low input current (1pA, typ, at room temperature) makes it good for low-current applications like this. Note that if you choose a bipolar op-amp for an application like this, better performance at low light levels results if the photodiode is connected as in the circuit shown in Figure 4.94J.

When using "single-supply" op-amps, watch out for misleading statements about output swing to the negative rail (ground). There are really four different kinds of output stages, all of which "swing down to ground," but they have very different properties (Fig. 4.56): (a) Op-amps with complementary MOS output transistors give true rail-to-rail swing; such a stage is

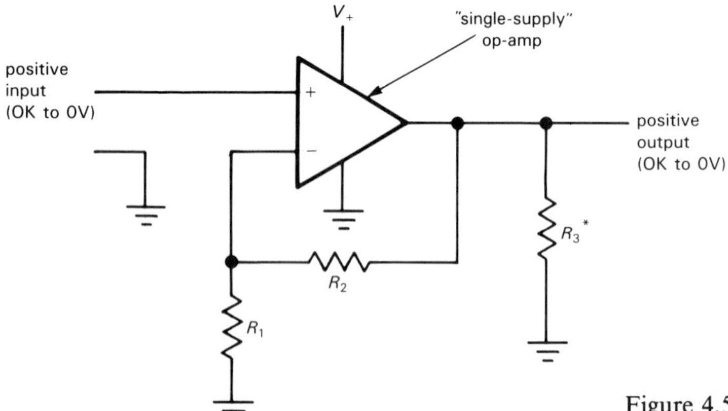

Figure 4.58. Single-supply dc amplifier.

capable of pulling its output to ground, *even when sinking moderate current.* Some examples are the ICL76*xx*, the LMC660, and CA5160. (b) Op-amps with an *npn* common-emitter transistor to ground behave similarly, i.e., they can pull their output to ground even while sinking current. Examples are the LM10, CA5422, and LT1013/14. Both kinds of output stages can, of course, handle an open circuit or a load that *sinks* current to ground. (c) Some op-amps, notably the 358 and 324, use a *pnp* follower to ground (which can only pull down to within a diode drop of ground), in parallel with an *npn* current sink (with compliance clear to ground). In the 358, the internal current sink is set at 50μA. Such a circuit can swing clear down to ground as long as it doesn't have to sink more than 50μA from the load. If the load sources more current, the output only works to within a diode drop of ground. As before, this kind of output circuit is happy *sourcing* current to a load that is returned to ground (as in the photometer example earlier). (d) Finally, some single-supply op-amps (e.g., the OP-90) use a *pnp* follower to ground, without the parallel current sink. Such an output stage can swing to ground only if the load helps out by sinking current, i.e., by being returned to ground. If you want to use such an op-amp with a load that sources current, you have to add an external resistor to ground (Fig. 4.57).

A note of caution: Don't make the mistake of assuming that you can make any op-amp's output work down to the negative rail simply by providing an external current sink. In most cases the circuitry driving the output stage does not permit that. Look for explicit permission in the data sheet!

Example: single-supply dc amplifier

Figure 4.58 shows a typical single-supply noninverting amplifier to amplify an input signal of known positive polarity. The input, output, and positive supply are all referenced to ground, which is the negative supply voltage for the op-amp. The output "pulldown" resistor may be needed with what we called type-1 amplifiers to ensure output swing all the way to ground; the feedback network or the load itself could perform this function. An important point: Remember that the output cannot go negative; thus you cannot use this amplifier with, say, ac-coupled audio signals.

Single-supply op-amps are indispensable in battery-operated equipment. We'll have more to say about this in Chapter 14.

COMPARATORS AND SCHMITT TRIGGER

It is quite common to want to know which of two signals is larger, or to know when a given signal exceeds a predetermined value. For instance, the usual method of generating triangle waves is to supply positive or negative currents into a capacitor, reversing the polarity of the current when the amplitude reaches a preset peak value. Another example is a digital voltmeter. In order to convert a voltage to a number, the unknown voltage is applied to one input of a comparator, with a linear ramp (capacitor + current source) applied to the other. A digital counter counts cycles of an oscillator while the ramp is less than the unknown voltage and displays the result when equality of amplitudes is reached. The resultant count is proportional to the input voltage. This is called single-slope integration; in most sophisticated instruments a dual-slope integration is used (see Section 9.21).

4.23 Comparators

The simplest form of comparator is a high-gain differential amplifier, made either with transistors or with an op-amp (Fig. 4.59). The op-amp goes into positive or

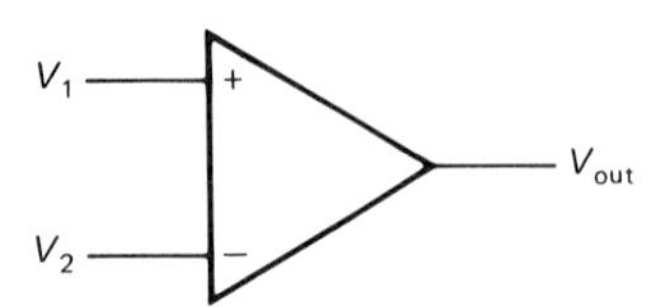

Figure 4.59

negative saturation according to the difference of the input voltages. Because the voltage gain typically exceeds 100,000, the inputs will have to be equal to within a fraction of a millivolt in order for the output not to be saturated. Although an ordinary op-amp can be used as a comparator (and frequently is), there are special integrated circuits intended for use as comparators. Some examples are the LM306, LM311, LM393, NE527, and TLC372. These chips are designed for very fast response and aren't even in the same league as op-amps. For example, the high-speed NE521 slews at several thousand volts per microsecond. With comparators, the term "slew rate" isn't usually used; you talk instead about "propagation delay versus input overdrive."

Comparators generally have more flexible output circuits than op-amps. Whereas an ordinary op-amp uses a push-pull output stage to swing between the supply voltages (±13V, say, for a 411 running from ±15V supplies), a comparator chip usually has an "open-collector" output with grounded emitter. By supplying an external "pullup" resistor (that's accepted terminology, believe it or not) connected to a voltage of your choice, you can have an output swing from +5 volts to ground, say. You will see later that logic circuits have well-defined voltages they like to operate between; the preceding example would be ideal for driving a TTL circuit, a popular type of digital logic. Figure 4.60 shows the circuit. The output switches from +5 volts to ground when the input signal goes negative. This use of a comparator is really an example of analog-to-digital conversion.

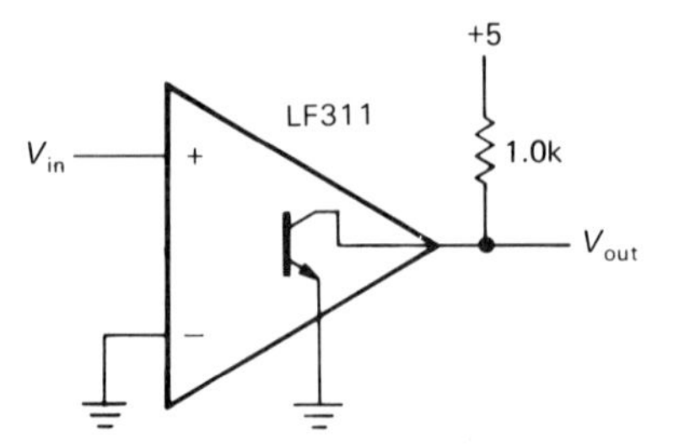

Figure 4.60

This is the first example we have presented of an *open-collector* output; this is a common configuration in logic circuits, as you will see throughout Chapters 8–11. If you like, you can think of the external pullup resistor as completing the comparator's internal circuit by providing a collector load resistor for an *npn* output transistor. Since the output transistor operates as a saturated switch, the resistor value is not at all critical, with values typically between a few hundred ohms and a few thousand ohms; small values yield improved switching speed and noise immunity at the expense of increased power dissipation. Incidentally, in spite of their superficial resemblance to op-amps, comparators are never used with negative feedback because they would not be stable (see Sections 4.32–4.34). However, some *positive* feedback is often used, as you will see in the next section.

Comments on comparators

Some points to remember: (a) Because there is no negative feedback, golden rule I is not obeyed. The inputs are not at the same voltage. (b) The absence of negative feedback means that the (differential) input impedance isn't bootstrapped to the high values characteristic of op-amp circuits. As a result, the input signal sees a changing load and changing (small) input current as the comparator switches; if the driving impedance is too high, strange things may happen. (c) Some comparators permit only limited differential input swings, as little as ±5 volts in some cases. Check the specs! See Table 9.3 and the discussion in Section 9.07 for the properties of some popular comparators.

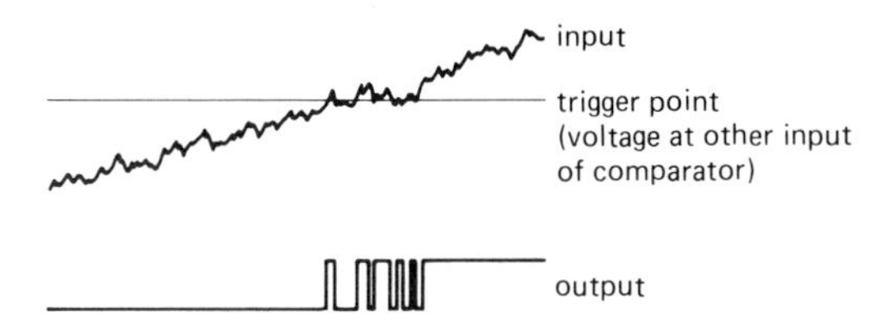

Figure 4.61

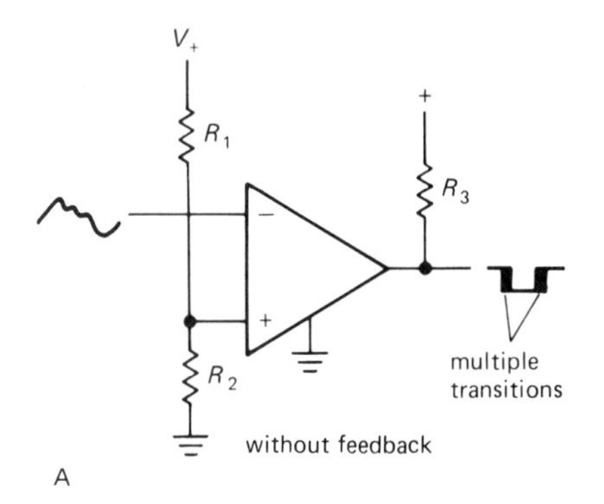

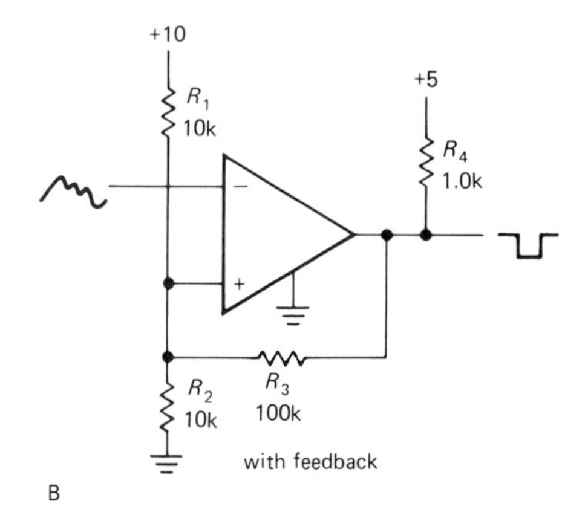

Figure 4.62

4.24 Schmitt trigger

The simple comparator circuit in Figure 4.60 has two disadvantages. For a very slowly varying input, the output swing can be rather slow. Worse still, if the input is noisy, the output may make several transitions as the input passes through the trigger point (Fig. 4.61). Both these problems can be remedied by the use of *positive* feedback (Fig. 4.62). The effect of R_3 is to make the circuit have two thresholds, depending on the output state. In the example shown, the threshold when the output is at ground (input high) is 4.76 volts, whereas the threshold with the output at +5 volts is 5.0 volts. A noisy input is less likely to produce multiple triggering (Fig. 4.63). Furthermore, the positive feedback ensures a rapid output transition, regardless of the speed of the input waveform. (A small "speedup" capacitor of 10–100pF is often connected across R_3 to enhance switching speed still further.) This configuration is known as a Schmitt trigger. (If an op-amp were used, the pullup would be omitted.)

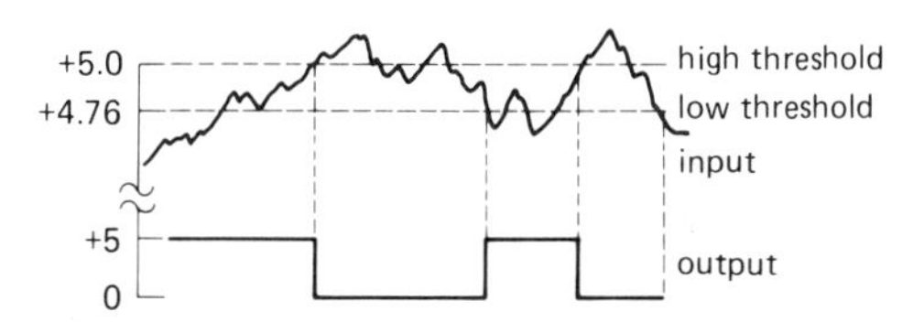

Figure 4.63

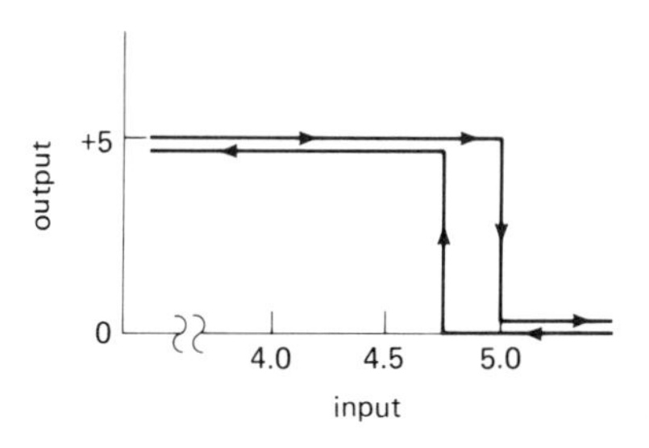

Figure 4.64

The output depends both on the input voltage and on its recent history, an effect called *hysteresis*. This can be illustrated with a diagram of output versus input, as in Figure 4.64. The design procedure is easy for Schmitt triggers that have a small amount of hysteresis. Use the circuit of Figure 4.62B. First choose a resistive divider (R_1, R_2) to put the threshold at approximately the right voltage; if you want the threshold near ground, just use a single resistor from noninverting input to ground. Next, choose the (positive) feedback resistor R_3 to produce the required hysteresis, noting that the hysteresis equals the output swing, attenuated by a resistive divider formed by R_3 and $R_1 \| R_2$. Finally, choose an output pullup resistor R_4 small enough to ensure nearly full supply swing, taking account of the loading by R_3. For the case where you want thresholds symmetrical about ground, connect an offsetting resistor of appropriate value from the noninverting input to the negative supply. You may wish to scale all resistor values in order to keep the output current and impedance levels within a reasonable range.

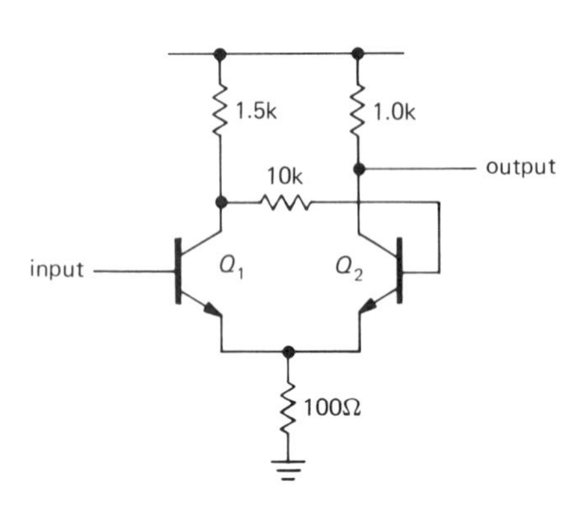

Figure 4.65

Discrete-transistor Schmitt trigger

A Schmitt trigger can also be made simply with transistors (Fig. 4.65). Q_1 and Q_2 share an emitter resistor. It is essential that Q_1's collector resistor be larger than Q_2's. In that way the threshold to turn on Q_1, which is one diode drop above the emitter voltage, rises when Q_1 is turned off, since the emitter current is higher with Q_2 conducting. This produces hysteresis in the trigger threshold, just as in the preceding integrated circuit Schmitt trigger.

EXERCISE 4.10

Design a Schmitt trigger using a 311 comparator (open-collector output) with thresholds at +1.0 volt and +1.5 volts. Use a 1.0k pullup resistor to +5 volts, and assume that the 311 is powered from ±15 volt supplies.

FEEDBACK WITH FINITE-GAIN AMPLIFIERS

We mentioned in Section 4.12 that the finite open-loop gain of an op-amp limits its performance in a feedback circuit. Specifically, the closed-loop gain can never exceed the open-loop gain, and as the open-loop gain approaches the closed-loop gain, the amplifier begins to depart from the ideal behavior we have come to expect. In this section we will quantify these statements so that you will be able to predict the performance of a feedback amplifier constructed with real (less than ideal) components. This is important also for feedback amplifiers constructed entirely with discrete components (transistors), where the open-loop gain is usually much less than with op-amps. In these cases the output impedance, for instance, will not be zero. Nonetheless, with a good understanding of feedback principles you will be able to achieve the performance required in any given circuit.

4.25 Gain equation

Let's begin by considering an amplifier of finite voltage gain, connected with feedback to form a noninverting amplifier (Fig. 4.66). The amplifier has open-loop voltage gain A, and the feedback network subtracts a fraction B of the output voltage from the input. (Later we will generalize things so that inputs and outputs can be currents or voltages.) The input to the gain block is then $V_{in} - BV_{out}$. But the output is just the input times A:

$$A(V_{in} - BV_{out}) = V_{out}$$

In other words,

$$V_{out} = \frac{A}{1 + AB} V_{in}$$

and the closed-loop voltage gain, V_{out}/V_{in}, is just

$$G = \frac{A}{1 + AB}$$

Some terminology: The standard designations for these quantities are as follows: G = closed-loop gain, A = open-loop gain, AB = loop gain, $1 + AB$ = return difference, or desensitivity. The feedback network is sometimes called the beta network (no relation to transistor beta, h_{fe}).

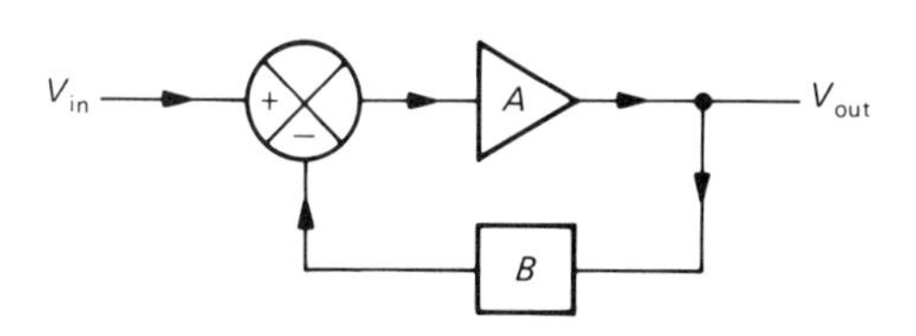

Figure 4.66

4.26 Effects of feedback on amplifier circuits

Let's look at the important effects of feedback. The most significant are predictability of gain (and reduction of distortion), changed input impedance, and changed output impedance.

Predictability of gain

The voltage gain is $A/(1 + AB)$. In the limit of infinite open-loop gain A, $G = 1/B$. We saw this result in the noninverting amplifier configuration, where a voltage divider on the output provided the signal to the inverting input (Fig. 4.69). The closed-loop voltage gain was just the inverse of the division ratio of the voltage divider. For finite gain A, feedback still acts to reduce the effects of variations of A (with frequency, temperature, amplitude, etc.). For instance, suppose A depends on frequency as in Figure 4.67. This

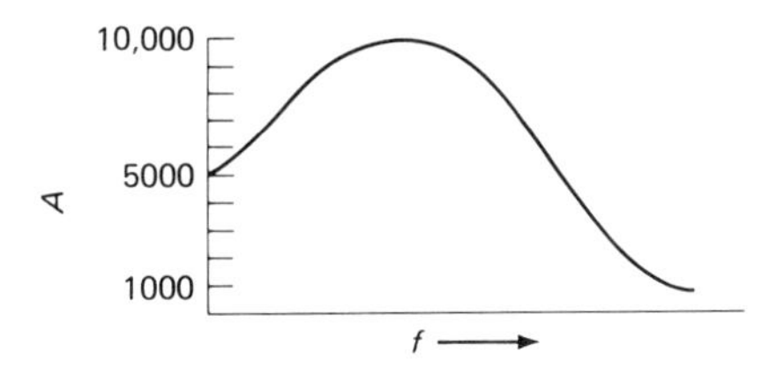

Figure 4.67

will surely satisfy anyone's definition of a poor amplifier (the gain varies over a factor of 10 with frequency). Now imagine we introduce feedback, with $B = 0.1$ (a simple voltage divider will do). The closed-loop voltage gain now varies from 1000/[1 +(1000×0.1)], or 9.90, to 10,000/[1 + (10,000×0.1)], or 9.99, a variation of just 1% over the same range of frequency! To put it in audio terms, the original amplifier is flat to ±10dB, whereas the feedback amplifier is flat to ±0.04dB. We can now recover the original gain of 1000 with nearly this linearity by just cascading three such stages. It was for just this reason (namely, the need for extremely flat telephone repeater amplifiers) that negative feedback was invented. As the inventor, Harold Black, described it in his first open publication on the invention (*Electrical Engineering*, 53:114, 1934), "by building an amplifier whose gain is made deliberately, say 40 decibels higher than necessary (10,000-fold excess on energy basis) and then feeding the output back to the input in such a way as to throw away the excess gain, it has been found possible to effect extraordinary improvement in constancy of amplification and freedom from nonlinearity."

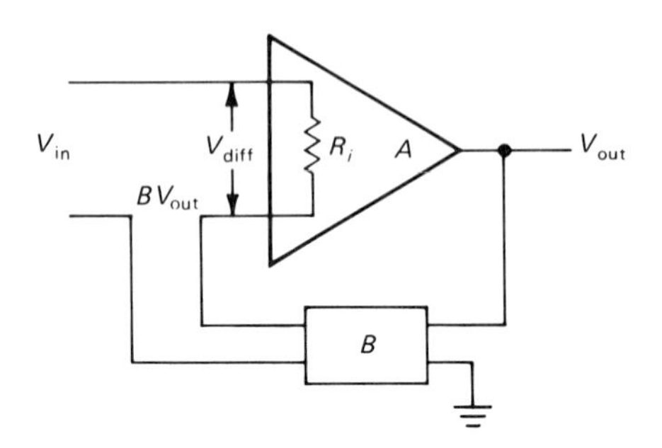

Figure 4.68

It is easy to show, by taking the partial derivative of G with respect to A ($\partial G/\partial A$), that relative variations in the open-loop gain are reduced by the desensitivity:

$$\frac{\Delta G}{G} = \frac{1}{1+AB}\frac{\Delta A}{A}$$

Thus, for good performance the loop gain AB should be much larger than 1. That's equivalent to saying that the open-loop gain should be much larger than the closed-loop gain.

A very important consequence of this is that nonlinearities, which are simply gain variations that depend on signal level, are reduced in exactly the same way.

Input impedance

Feedback can be arranged to subtract a voltage or a current from the input (these are sometimes called *series feedback* and *shunt feedback*, respectively). The noninverting op-amp configuration, for instance, subtracts a sample of the output voltage from the differential voltage appearing at the input, whereas in the inverting configuration a current is subtracted from the input. The effects on input impedance are opposite in the two cases: Voltage feedback multiplies the open-loop input impedance by $1 + AB$, whereas current feedback reduces it by the same factor. In the limit of infinite loop gain the input impedance (at the amplifier's input terminal) goes to infinity or zero, respectively. This is easy to understand, since voltage feedback tends to subtract signal from the input, resulting in a smaller change (by the factor AB) across the amplifier's input resistance; it's a form of bootstrapping. Current feedback reduces the input signal by bucking it with an equal current.

Let's see explicitly how the effective input impedance is changed by feedback. We will illustrate the case of voltage feedback only, since the derivations are similar for the two cases. We begin with an op-amp model with (finite) input resistance as shown in Figure 4.68. An input V_{in} is reduced by BV_{out}, putting a voltage $V_{diff} = V_{in} - BV_{out}$ across the inputs of the amplifier. The input current is therefore

$$I_{in} = \frac{V_{in} - BV_{out}}{R_i} = \frac{V_{in}\left(1 - B\frac{A}{1+AB}\right)}{R_i} = \frac{V_{in}}{(1+AB)R_i}$$

giving an effective input resistance

$$R_i' = V_{in}/I_{in} = (1+AB)R_i$$

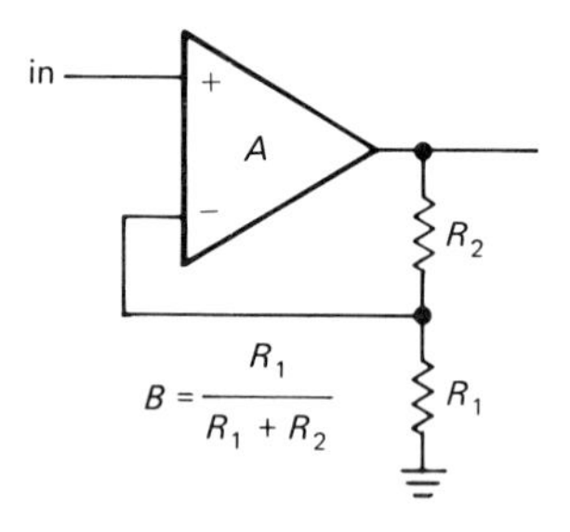

Figure 4.69

The classic op-amp noninverting amplifier is exactly this feedback configuration, as shown in Figure 4.69. In this circuit, $B = R_1/(R_1 + R_2)$, giving the usual voltage-gain expression $G_v = 1 + R_2/R_1$ and an infinite input impedance for the ideal case of infinite open-loop voltage gain

A. For finite loop gain, the equations as previously derived apply.

The op-amp *inverting* amplifier circuit is different from the noninverting circuit and has to be analyzed separately. It's best to think of it as a combination of an input resistor driving a shunt feedback stage (Fig. 4.70). The shunt stage alone has its input at the "summing junction" (the inverting input of the amplifier), where the currents from feedback and input signals are combined (this amplifier connection is really a "transresistance" configuration; it converts a current input to a voltage output). Feedback reduces the impedance looking into the summing junction, R_2, by a factor of $1 + A$ (see if you can prove this). In cases of very high loop gain (e.g, an op-amp) the input impedance is reduced to a fraction of an ohm, a good characteristic for a current-input amplifier. Some good examples are the photometer amplifier in Section 4.22 and the logarithmic converter in Section 4.14.

The classic op-amp inverting amplifier connection is a combination of a shunt feedback transresistance amplifier and a series input resistor, as in the figure. As a result, the input impedance equals the sum of R_1 and the impedance looking into the summing junction. For high loop gain, R_{in} approximately equals R_1.

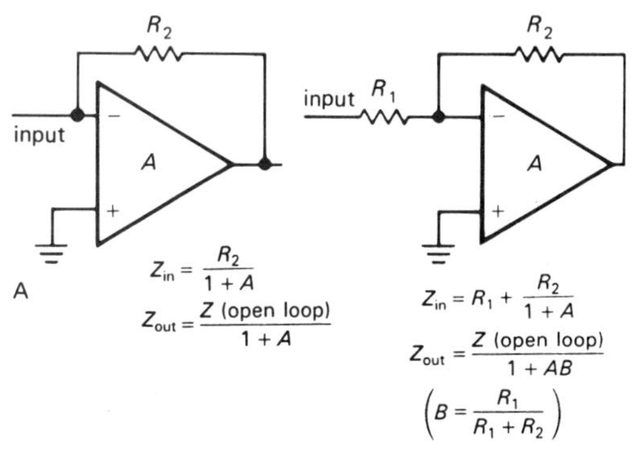

Figure 4.70. Input and output impedances for (A) transresistance amplifier and (B) inverting amplifier.

It is a straightforward exercise to derive an expression for the closed-loop voltage gain of the inverting amplifier with finite loop gain. The answer is

$$G = -A(1 - B)/(1 + AB)$$

where B is defined as before, $B = R_1/(R_1 + R_2)$. In the limit of large open-loop gain A, $G = -1/B + 1$ (i.e., $G = -R_2/R_1$).

EXERCISE 4.11

Derive the foregoing expressions for input impedance and gain of the inverting amplifier.

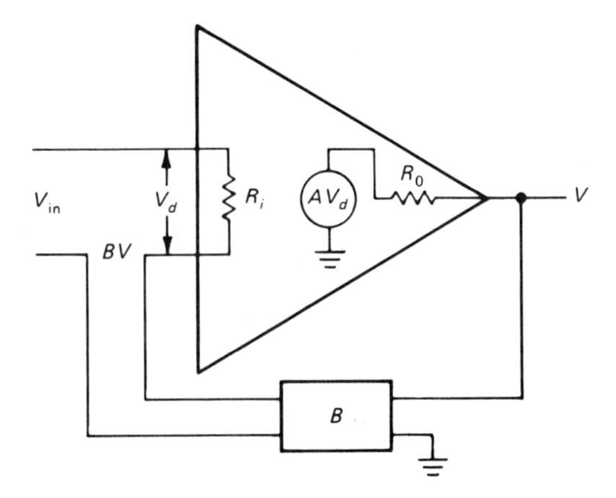

Figure 4.71

Output impedance

Again, feedback can extract a sample of the output voltage or the output current. In the first case the open-loop output impedance will be reduced by the factor $1 + AB$, whereas in the second case it will be increased by the same factor. We will illustrate this effect for the case of voltage sampling. We begin with the model shown in Figure 4.71. This time we have shown the output impedance explicitly. The calculation is simplified by a trick: Short the input, and apply a voltage V to the output; by calculating the output current I, we get the output impedance $R_0' = V/I$. Voltage V at the output puts a voltage $-BV$ across the amplifier's input, producing a voltage $-ABV$ in the amplifier's internal generator. The output current is therefore

$$I = \frac{V - (-ABV)}{R_O} = \frac{V(1 + AB)}{R_O}$$

giving an effective output impedance

$$R'_O = V/I = R_O/(1 + AB)$$

If feedback is connected instead to sample the output current, the expression becomes

$$R'_O = R_O(1 + AB)$$

It is possible to have multiple feedback paths, sampling both voltage and current. In the general case the output impedance is given by Blackman's impedance relation

$$R'_O = R_O \frac{1 + (AB)_{SC}}{1 + (AB)_{OC}}$$

where $(AB)_{SC}$ is the loop gain with the output shorted to ground and $(AB)_{OC}$ is the loop gain with no load attached. Thus, feedback can be used to generate a desired output impedance. This equation reduces to the previous results for the usual situation in which feedback is derived from either the output voltage or the output current.

□ *Loading by the feedback network*

In feedback computations, you usually assume that the beta network doesn't load the amplifier's output. If it does, that must be taken into account in computing the open-loop gain. Likewise, if the connection of the beta network at the amplifier's input affects the open-loop gain (feedback removed, but network still connected), you must use the modified open-loop gain. Finally, the preceding expressions assume that the beta network is unidirectional, i.e., it does not couple signal from the input to the output.

□ 4.27 Two examples of transistor amplifiers with feedback

Figure 4.72 shows a transistor amplifier with negative feedback.

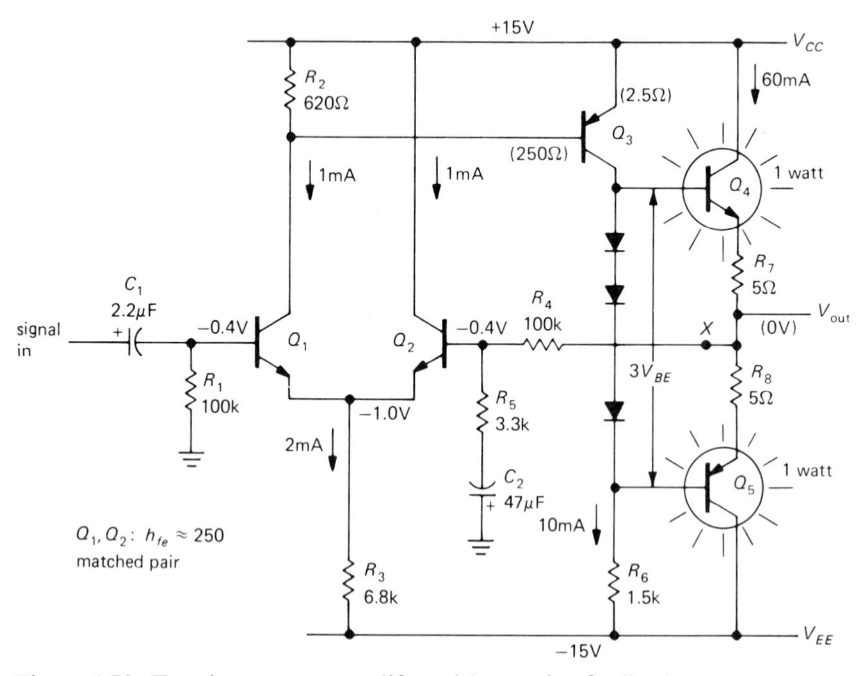

Figure 4.72. Transistor power amplifier with negative feedback.

□ ***Circuit description***

It may look complicated, but it is extremely straightforward in design and is relatively easy to analyze. Q_1 and Q_2 form a differential pair, with common-emitter amplifier Q_3 amplifying its output. R_6 is Q_3's collector load resistor, and push-pull pair Q_4 and Q_5 form the output emitter follower. The output voltage is sampled by the feedback network consisting of voltage divider R_4 and R_5, with C_2 included to reduce the gain to unity at dc for stable biasing. R_3 sets the quiescent current in the differential pair, and since overall feedback guarantees that the quiescent output voltage is at ground, Q_3's quiescent current is easily seen to be 10mA (V_{EE} across R_6, approximately). As we have discussed earlier (Section 2.15), the diodes bias the push-pull pair into conduction, leaving one diode drop across the series pair R_7 and R_8, i.e., 60mA quiescent current. That's class AB operation, good for minimizing crossover distortion, at the cost of 1 watt standby dissipation in each output transistor.

From the point of view of our earlier circuits, the only unusual feature is Q_1's quiescent collector voltage, one diode drop below V_{CC}. That is where it must sit in order to hold Q_3 in conduction, and the feedback path ensures that it will. (For instance, if Q_1 were to pull its collector closer to ground, Q_3 would conduct heavily, raising the output voltage, which in turn would force Q_2 to conduct more heavily, reducing Q_1's collector current and hence restoring the status quo.) R_2 was chosen to give a diode drop at Q_1's quiescent current in order to keep the collector currents in the differential pair approximately equal at the quiescent point. In this transistor circuit the input bias current is not negligible (4μA), resulting in a 0.4 volt drop across the 100k input resistors. In transistor amplifier circuits like this, in which the input currents are considerably larger than in op-amps, it is particularly important to make sure that the dc resistances seen from the inputs are equal, as shown (a Darlington input stage would probably be better here).

□ ***Analysis***

Let's analyze this circuit in detail, determining the gain, input and output impedances, and distortion. To illustrate the utility of feedback, we will find these parameters for both the open-loop and closed-loop situations (recognizing that biasing would be hopeless in the open-loop case). To get a feeling for the linearizing effect of the feedback, the gain will be calculated at +10 volts and −10 volts output, as well as the quiescent point (zero volts).

□ *Open loop.* Input impedance: We cut the feedback at point X and ground the right side of R_4. The input signal sees 100k in parallel with the impedance looking into the base. The latter is h_{fe} times twice the intrinsic emitter resistance plus the impedance seen at Q_2's emitter due to the feedback network at Q_2's base. For $h_{fe} \approx 250$, $Z_{in} \approx 250\times[(2\times25)+(3.3\text{k}/250)]$; i.e., $Z_{in} \approx 16\text{k}$.

Output impedance: Since the impedance looking back into Q_3's collector is high, the output transistors are driven by a 1.5k source (R_6). The output impedance is about 15 ohms ($h_{fe} \approx 100$) plus the 5 ohm emitter resistance, or 20 ohms. The intrinsic emitter resistance of 0.4 ohm is negligible.

Gain: The differential input stage sees a load of R_2 paralleled by Q_3's base resistance. Since Q_3 is running 10mA quiescent current, its intrinsic emitter resistance is 2.5 ohms, giving a base impedance of about 250 ohms (again, $h_{fe} \approx 100$). The differential pair thus has a gain of

$$\frac{250\|620}{2\times 25\Omega} \quad \text{or} \quad 3.5$$

The second stage, Q_3, has a voltage gain of 1.5k/2.5ohms, or 600. The overall voltage gain at the quiescent point is 3.5 × 600, or 2100. Since Q_3's gain depends on its collector current, there is substantial change of gain with signal swing, i.e., nonlinearity. The gain is tabulated in the following section for three values of output voltage.

☐ *Closed loop.* Input impedance: This circuit uses series feedback, so the input impedance is raised by (1 + loop gain). The feedback network is a voltage divider with $B = 1/30$ at signal frequencies, so the loop gain AB is 70. The input impedance is therefore 70 × 16k, still paralleled by the 100k bias resistor, i.e., about 92k. The bias resistor now dominates the input impedance.

Output impedance: Since the output *voltage* is sampled, the output impedance is reduced by (1 + loop gain). The output impedance is therefore 0.3 ohm. Note that this is a small-signal impedance and does not mean that a 1 ohm load could be driven to nearly full swing, for instance. The 5 ohm emitter resistors in the output stage limit the large signal swing. For instance, a 4 ohm load could be driven only to 10 volts pp, approximately.

Gain: The gain is $A/(1+AB)$. At the quiescent point, that equals 30.84, using the exact value for B. In order to illustrate the gain stability achieved with negative feedback, the overall voltage gain of the circuit with and without feedback is tabulated at three values of output level at the end of this paragraph. It should be obvious that negative feedback has brought about considerable improvement in the amplifier's characteristics, although in fairness it should be pointed out that the amplifier could have been designed for better open-loop performance, e.g., by using a current source for Q_3's collector load and degenerating its emitter, by using a current source for the differential-pair emitter circuit, etc. Even so, feedback would still make a large improvement.

	Open loop			*Closed loop*		
V_{out}	−10	0	+10	−10	0	+10
Z_{in}	16k	16k	16k	92k	92k	92k
Z_{out}	20Ω	20Ω	20Ω	0.3Ω	0.3Ω	0.3Ω
Gain	1360	2100	2400	30.60	30.84	30.90

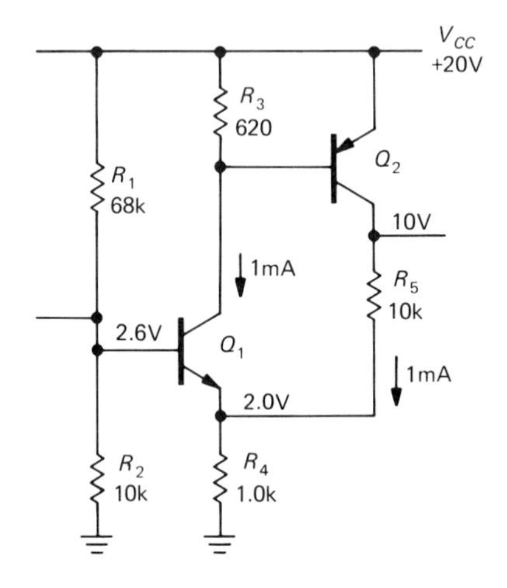

Figure 4.73

☐ **Series feedback pair**

Figure 4.73 shows another transistor amplifier with feedback. Thinking of Q_1 as an amplifier of its base-emitter voltage drop (thinking in the Ebers-Moll sense), the feedback samples the output voltage and subtracts a fraction of it from the input signal. This circuit is a bit tricky because Q_2's collector resistor doubles as the feedback network. Applying the techniques we used earlier, you should be able to show that G(open loop) ≈ 200, loop gain ≈ 20, Z_{out}(open loop) ≈10k, Z_{out}(closed loop) ≈ 500 ohms, and G(closed loop) ≈ 9.5.

SOME TYPICAL OP-AMP CIRCUITS

4.28 General-purpose lab amplifier

Figure 4.74 shows a dc-coupled "decade amplifier" with settable gain, bandwidth, and wide-range dc output offset. IC_1 is a FET-input op-amp with noninverting gain from unity (0dB) to ×100 (40dB) in accurately calibrated 10dB steps; a vernier is provided for variable gain. IC_2 is an inverting amplifier; it allows offsetting the output over a range of ±10 volts, accurately calibrated via R_{14}, by injecting current into the summing junction. $C_2 - C_4$ set the high-frequency rolloff, since it is often a nuisance to have excessive bandwidth (and noise). IC_5 is a power booster for driving low-impedance loads or cables; it can provide ±150mA output current.

Some interesting details: A 10MΩ input resistor is small enough, since the bias current of the 411 is 25pA (0.3mV error with open input). R_2, in combination with D_1 and D_2, limits the input voltage at the op-amp to the range V_- to $V_+ + 0.7$. D_3 is used to generate a clamp voltage at $V_- + 0.7$, since the input common mode range extends only to V_- (exceeding V_- causes the output to reverse phase). With the protection components shown, the input can go to ±150 volts without damage.

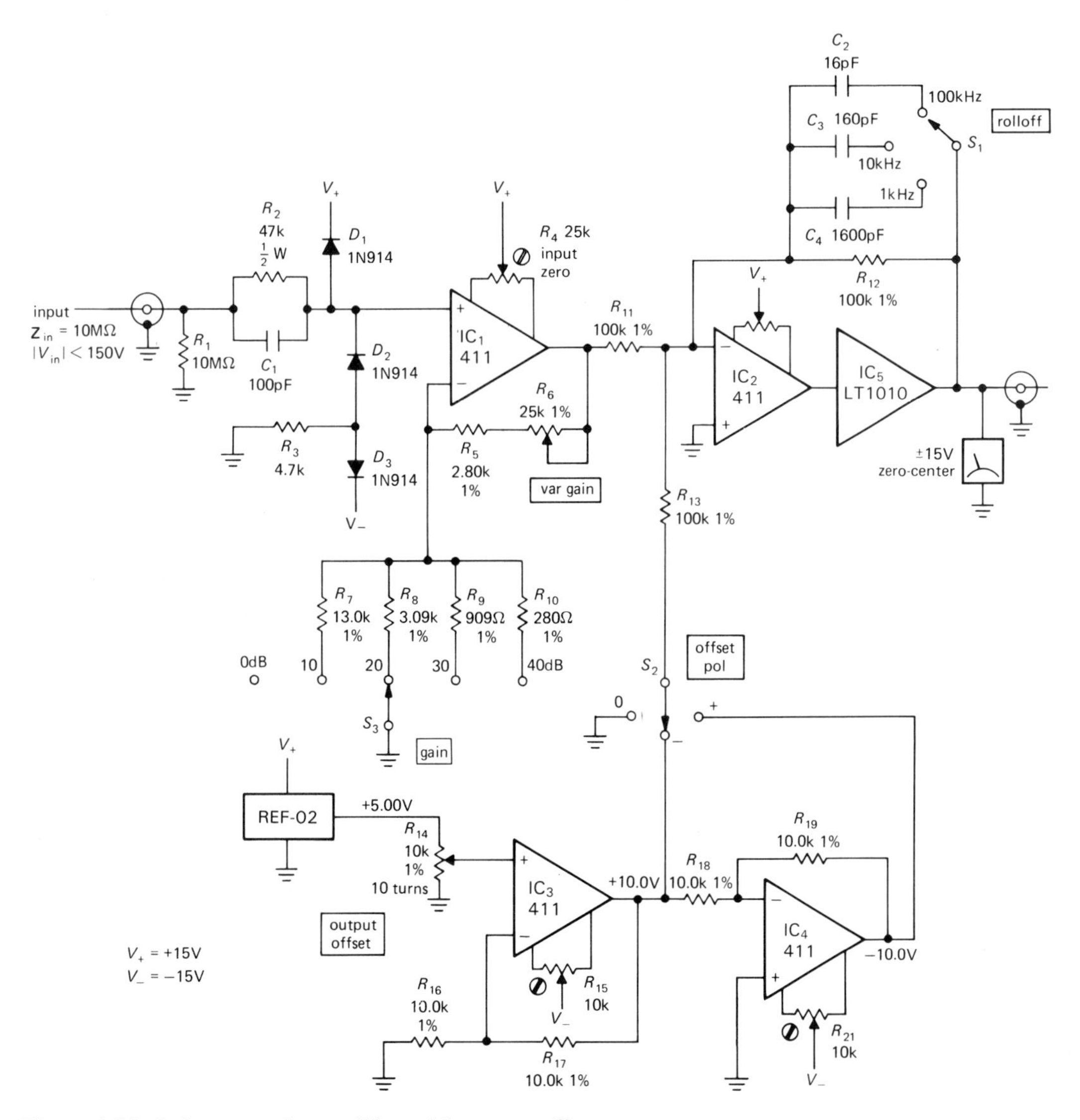

Figure 4.74. Laboratory dc amplifier with output offset.

EXERCISE 4.12

Check that the gain is as advertised. How does the variable offset circuitry work?

4.29 Voltage-controlled oscillator

Figure 4.75 shows a clever circuit, borrowed from the application notes of several manufacturers. IC_1 is an integrator, rigged up so that the capacitor current (V_{in}/200k) changes sign, but not magnitude, when Q_1 conducts. IC_2 is connected as a Schmitt trigger, with thresholds at one-third and two-thirds of V_+. Q_1 is an n-channel MOSFET, used here as a switch; it is simpler to use than bipolar transistors in this sort of application, but an alternative circuit using npn transistors is shown in addition. In either case, the bottom side of R_4 is pulled to ground when the output is HIGH and open-circuited when the output is LOW.

An unusual feature of this circuit is its operation from a single positive supply. The 3160 (internally compensated version of the 3130) has FETs as output transistors, guaranteeing a full swing between V_+ and ground at the output; this ensures that the thresholds of the Schmitt don't drift, as they would with an op-amp of conventional output-stage design, with its ill-defined limits of output swing. In this case this means that the frequency and amplitude of the triangle wave will be stable. Note that the frequency depends on the ratio V_{in}/V_+; this means that if V_{in} is generated from V_+ by a resistive divider (made from some sort of resistive transducer, say), the output frequency won't vary with V_+, only with changes in resistance.

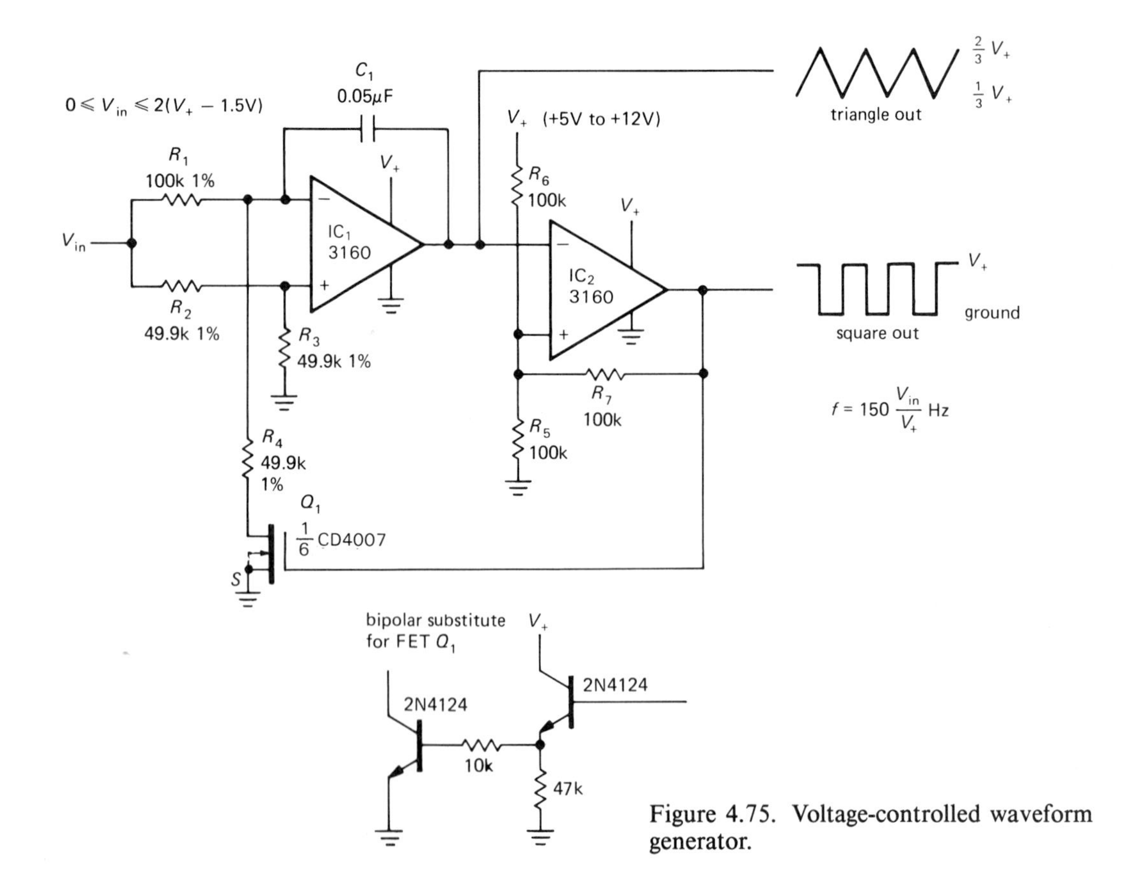

Figure 4.75. Voltage-controlled waveform generator.

EXERCISE 4.13

Show that the output frequency is given by $f(\text{Hz}) = 150V_{in}/V_{+}$. Along the way, verify that the Schmitt thresholds and integrator currents are as advertised.

□ 4.30 JFET linear switch with R_{ON} compensation

In Chapter 3 we considered MOSFET linear switches in some detail. It is also possible to use JFETs as linear switches. However, you have to be more careful about gate signals so that gate conduction doesn't occur. Figure 4.76 shows a typical arrangement. The gate is held well below ground to keep the JFET pinched off. This means that if the input signals go negative, the gate must be held at least V_P below the most negative input swing. To bring the FET into conduction, the control input is brought more positive than the most positive input excursion. The diode is then reverse-biased, and the gate rides at source voltage via the 1M resistor.

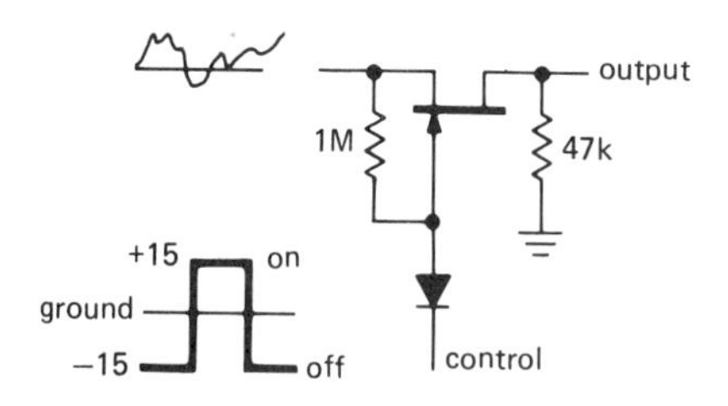

Figure 4.76

The awkwardness of this circuit probably accounts for much of the popularity of MOSFETs in linear switch applications. However, it is possible to devise an elegant JFET linear switch circuit if you use an op-amp, since you can tie the JFET source to the virtual ground at the summing junction of an inverting amplifier. Then you simply bring the gate to ground potential to turn the JFET on. This arrangement has the added advantage of providing a method of canceling precisely the errors caused by finite R_{ON} and its nonlinearity. Figure 4.77 shows the circuit.

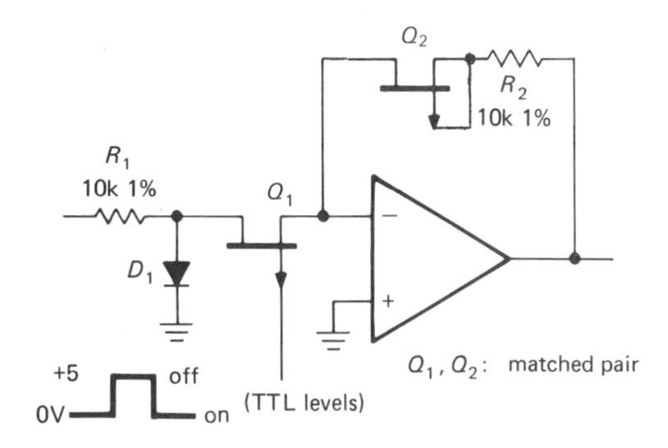

Figure 4.77. JFET-switched amplifier with R_{ON} cancellation.

There are two noteworthy features of this circuit: (a) When Q_1 is ON (gate grounded), the overall circuit is an inverter with identical impedances in the input and feedback circuits. That results in the cancellation of any effects of finite or nonlinear ON resistance, assuming the FETs are matched in R_{ON}. (b) Because of the low pinch-off voltage of JFETs, the circuit will work well with a control signal of zero to +5 volts, which is what you get with standard digital logic circuits (see Chapters 8 and 9). The inverting configuration, with Q_1's source connected to a virtual ground (the summing junction), simplifies circuit operation, since there are no signal swings on Q_1's source in the ON state; D_1 prevents FET turn-on for positive input swings when Q_1 is OFF, and it has no effect when the switch is closed.

There are *p*-channel JFETs with low pinch-off voltages available in useful configurations at low prices. For example, the IH5009–IH5024 family includes devices with four input FETs and one cancellation FET in a single DIP package, with R_{ON} of 100 ohms and a price less than two dollars. Add an op-amp and a few resistors and you've got a 4-input multiplexer. Note that the same R_{ON} cancellation trick can be used with MOSFET switches.

□ 4.31 TTL zero-crossing detector

The circuit shown in Figure 4.78 generates an output square wave for use with TTL logic (zero to +5V range) from an input wave of any amplitude up to 100 volts. R_1, combined with D_1 and D_2, limits the input swing to −0.6 volt to +5.6 volts, approximately. Resistive divider R_2R_3 is necessary to limit negative swing to less than 0.3 volt, the limit for a 393 comparator. R_5 and R_6 provide hysteresis, with R_4 setting the trigger points symmetrically about ground. The input impedance is nearly constant, because of the large R_1 value relative to the other resistors in the input attenuator. A 393 is used because its inputs can go all the way to ground, making single-supply operation simple.

EXERCISE 4.14

Verify that the trigger points are at ±25mV at the input signal.

□ 4.32 Load-current-sensing circuit

The circuit shown in Figure 4.79 provides a voltage output proportional to load current, for use with a current regulator, metering circuit, or whatever. The voltage across the 4-terminal resistor R_S goes from zero to 0.1 volt, with probable common-mode offset due to the effects of resistance in the ground lead (note that the power supply is grounded at the output). For that reason the op-amp is wired as a differential amplifier, with gain of 100. Voltage offset is trimmed externally with R_8, since the LT1013 doesn't have internal trimming circuitry (the single LT1006 does, however). A zener reference with a few percent stability is adequate for trimming, since the trimming is itself a small correction (you hope!). The venerable 358 could have been chosen because both inputs and output also go all the way to ground. V_+ could be unregulated, since the power-supply rejection of the op-amp is more than adequate, 100dB (typ) in this case.

FEEDBACK AMPLIFIER FREQUENCY COMPENSATION

If you look at a graph of open-loop voltage gain versus frequency for several op-amps,

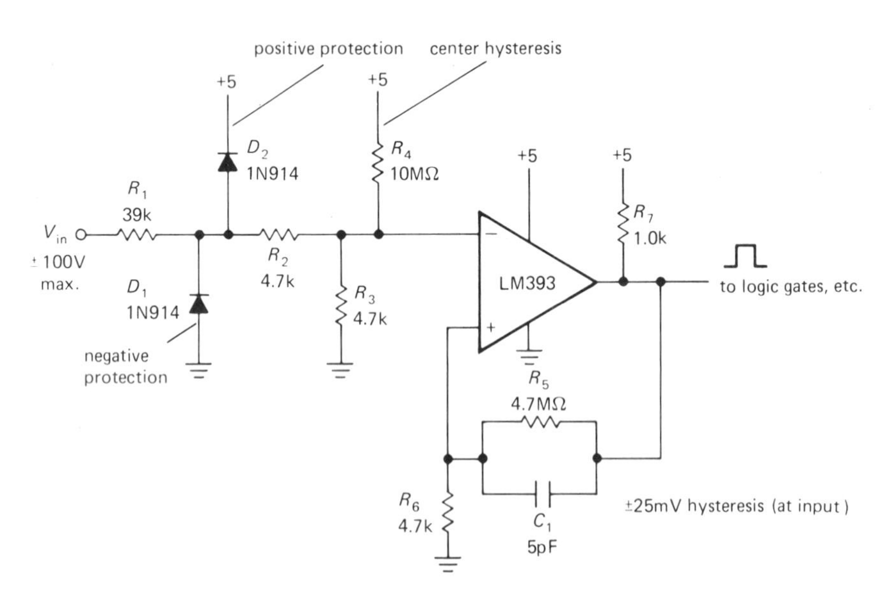

Figure 4.78. Zero-crossing level detector with input protection.

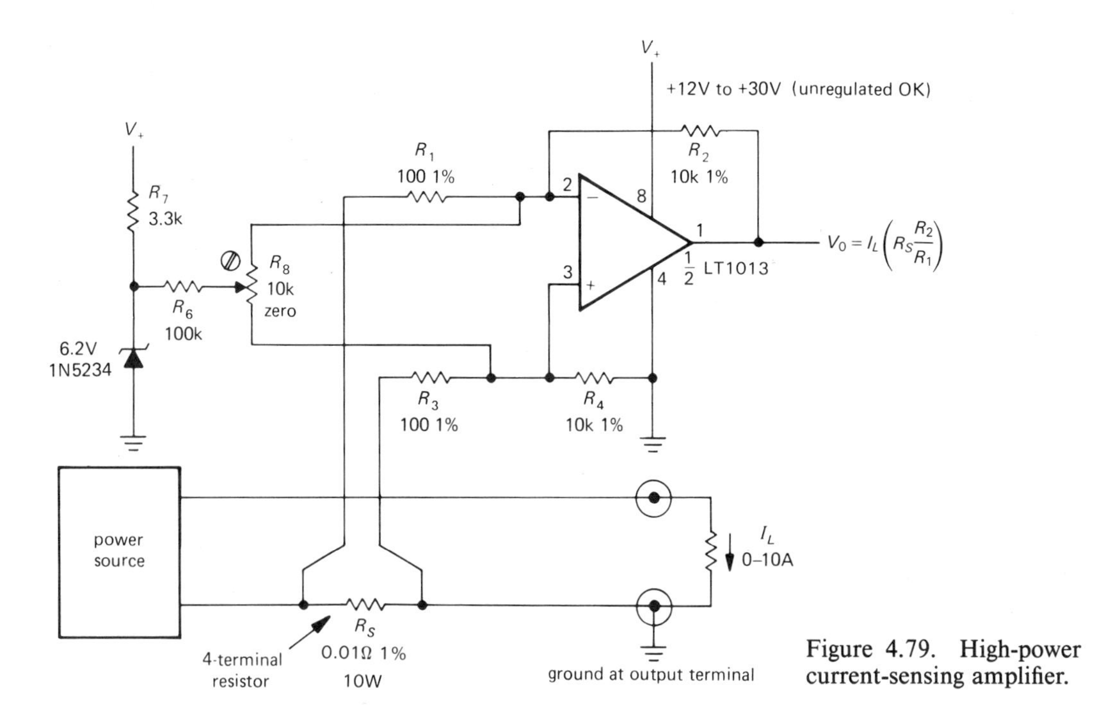

Figure 4.79. High-power current-sensing amplifier.

you'll see something like the curves in Figure 4.80. From a superficial look at such a *Bode plot* (a log-log plot of gain

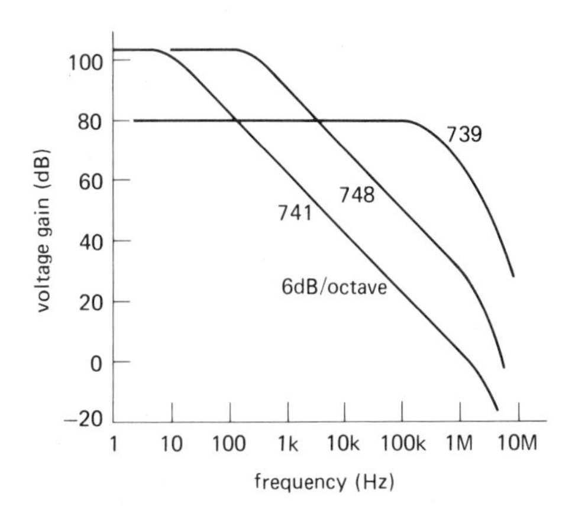

Figure 4.80

and phase versus frequency) you might conclude that the 741 is an inferior op-amp, since its open-loop gain drops off so rapidly with increasing frequency. In fact, that rolloff is built into the op-amp intentionally and is recognizable as the same −6db/octave curve characteristic of an *RC* low-pass filter. The 748, by comparison, is identical with the 741 except that it is *uncompensated* (as is the 739). Op-amps are generally available in internally compensated varieties and uncompensated varieties; let's take a look at this business of frequency compensation.

4.33 Gain and phase shift versus frequency

An op-amp (or, in general, any multistage amplifier) will begin to roll off at some frequency because of the low-pass filters formed by signals of finite source impedance driving capacitive loads within the amplifier stages. For instance, it is common to have an input stage consisting of a differential amplifier, perhaps with current mirror load (see the LM358 schematic in Fig. 4.54), driving a common-emitter second stage. For now, imagine that the capacitor labeled C_C in that circuit is removed. The high output impedance of the input stage, in combination with junction capacitance C_{ie} and feedback capacitance C_{cb} (Miller effect, see Sections 2.19 and 13.04) of the following stage, forms a low-pass filter whose 3dB point might fall somewhere in the range of 100Hz to 10kHz.

The decreasing reactance of the capacitor with increasing frequency gives rise to the characteristic 6db/octave rolloff: At sufficiently high frequencies (which may be below 1kHz), the capacitive loading dominates the collector load impedance, resulting in a voltage gain $G_V = g_m X_C$, i.e., the gain drops off as $1/f$. It also produces a 90° lagging phase shift at the output relative to the input signal. (You can think of this as the tail of an RC low-pass filter characteristic, where R represents the equivalent source impedance driving the capacitive load. However, it is not necessary to have any actual resistors in the circuit.)

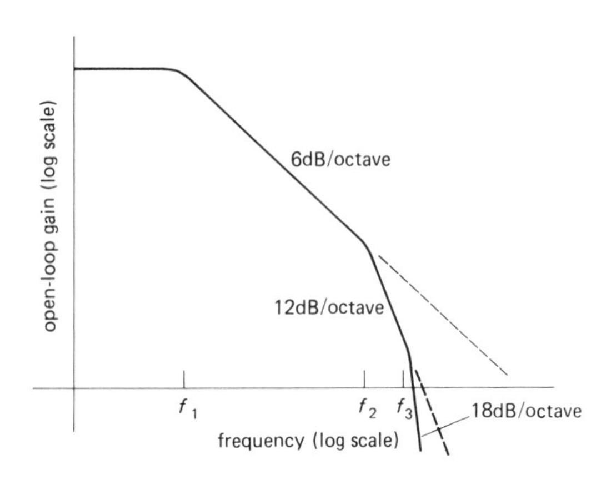

Figure 4.81

In a multistage amplifier there will be additional rolloffs at higher frequencies, caused by low-pass filter characteristics in the other amplifier stages, and the overall open-loop gain will look something like that shown in Figure 4.81. The open-loop gain begins dropping at 6dB/octave at some low frequency f_1, due to capacitive loading of the first-stage output. It continues dropping off with that slope until an internal RC of another stage rears its ugly head at frequency f_2, beyond which the rolloff goes at 12dB/octave, and so on.

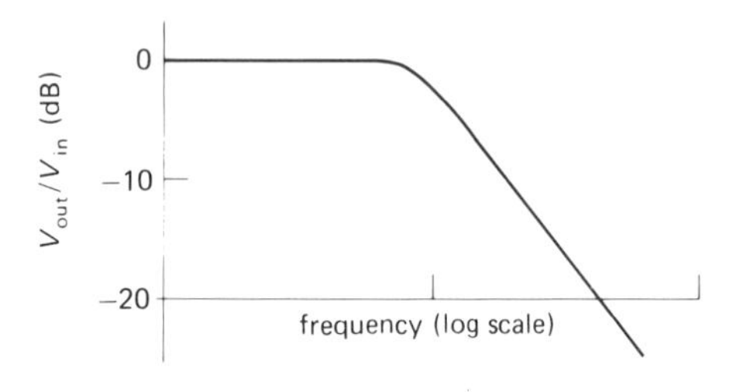

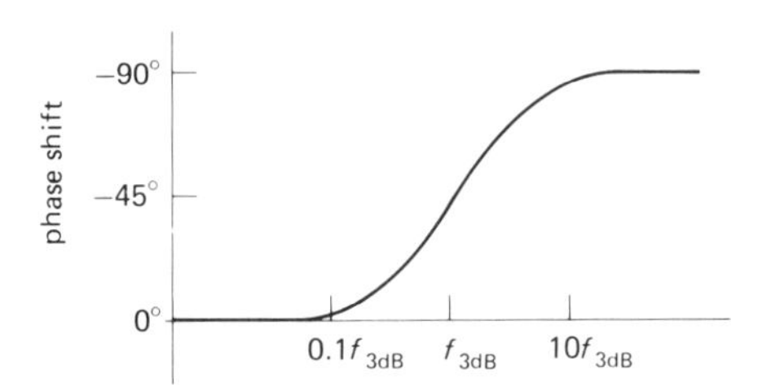

Figure 4.82. Bode plot: gain and phase versus frequency.

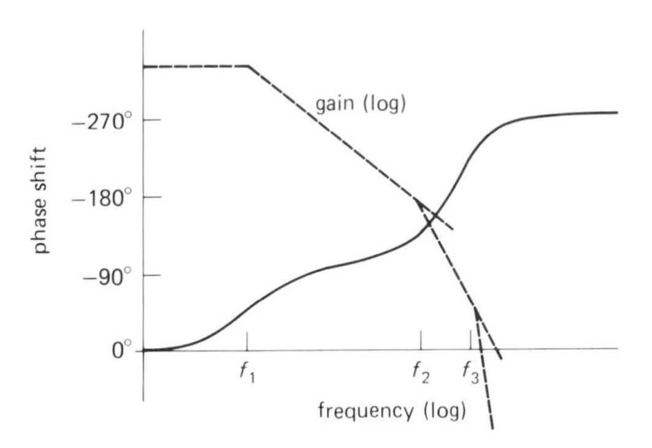

Figure 4.83

What is the significance of all this? Remember that an RC low-pass filter has a phase shift that looks as shown in Figure 4.82. Each low-pass filter within the amplifier has a similar phase-shift characteristic, so the overall phase shift of the hypothetical amplifier will be as shown in Figure 4.83.

Now here's the problem: If you were to connect this amplifier as an op-amp follower, for instance, it would oscillate. That's because the open-loop phase shift reaches 180° at some frequency at which the gain is still greater than 1 (negative feedback becomes positive feedback at that frequency). That's all you need to generate an oscillation, since any signal whatsoever at that frequency builds up each time around the feedback loop, just like a public address system with the gain turned up too far.

Stability criterion

The criterion for stability against oscillation for a feedback amplifier is that its open-loop phase shift must be less than 180° at the frequency at which the loop gain is unity. This criterion is hardest to satisfy when the amplifier is connected as a follower, since the loop gain then equals the open-loop gain, the highest it can be. Internally compensated op-amps are designed to satisfy the stability criterion even when connected as followers; thus they are stable when connected for any closed-loop gain with a simple resistive feedback network. As we hinted earlier, this is accomplished by deliberately modifying an existing internal rolloff in order to put the 3dB point at some low frequency, typically 1Hz to 20Hz. Let's see how that works.

4.34 Amplifier compensation methods

Dominant-pole compensation

The goal is to keep the open-loop phase shift much less than 180° at all frequencies for which the loop gain is greater than 1. Assuming that the op-amp may be used as a follower, the words "loop gain" in the last sentence can be replaced by "open-loop gain." The easiest way to do this is to add enough capacitance at the point in the circuit that produces the initial 6dB/octave rolloff, so that the open-loop gain drops to unity at about the 3dB frequency of the next "natural" RC filter. In this way the open-loop phase shift is held at a constant 90° over most of the passband, increasing toward 180° only as the gain approaches unity. Figure 4.84

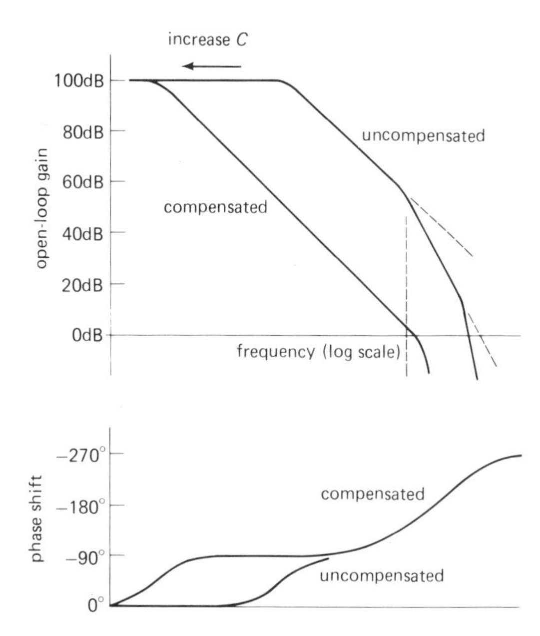

Figure 4.84

shows the idea. Without compensation the open-loop gain drops toward 1, first at 6dB/octave, then at 12dB/octave, etc., resulting in phase shifts of 180° or more before the gain has reached 1. By moving the first rolloff down in frequency (forming a "dominant pole"), the rolloff is controlled so that the phase shift begins to rise above 90° only as the open-loop gain approaches unity. Thus, by sacrificing open-loop gain, you buy stability. Since the natural rolloff of lowest frequency is usually caused by Miller effect in the stage driven by the input differential amplifier, the usual method of dominant-pole compensation consists simply of adding additional feedback capacitance around the second-stage transistor, so that the combined voltage gain of the two stages is $g_m X_C$ or $g_m/2\pi f C_{\text{comp}}$ over the compensated region of the amplifier's frequency response (Fig. 4.85). In practice, Darlington-connected transistors would probably be used for both stages.

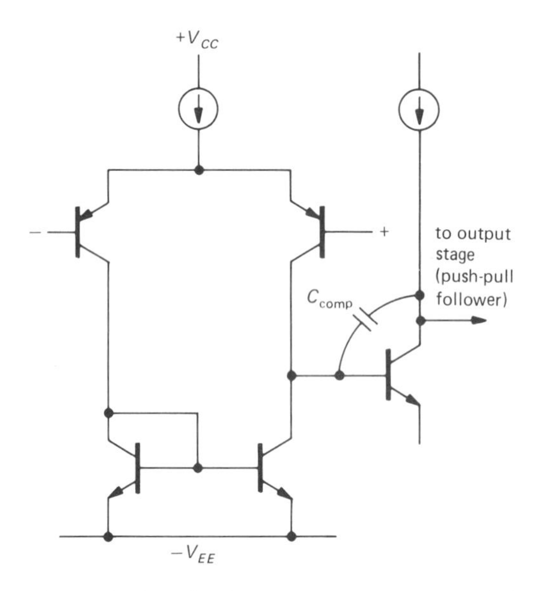

Figure 4.85. Classic op-amp input stage with compensation.

By putting the dominant-pole unity-gain crossing at the 3dB point of the next rolloff, you get a phase margin of about 45° in the worst case (follower), since a single RC filter has a 45° lagging phase shift at its 3dB frequency, i.e., the phase margin equals 180°− (90° + 45°), with the 90° coming from the dominant pole.

An additional advantage of using a Miller-effect pole for compensation is that the compensation is inherently insensitive to changes in voltage gain with temperature, or manufacturing spread of gain: Higher gain causes the feedback capacitance to look larger, moving the pole downward in frequency in exactly the right way to keep the unity-gain crossing frequency unchanged. In fact, the actual 3dB frequency of the compensation pole is quite irrelevant; what matters is the point at which it intersects the unity-gain axis (Fig. 4.86).

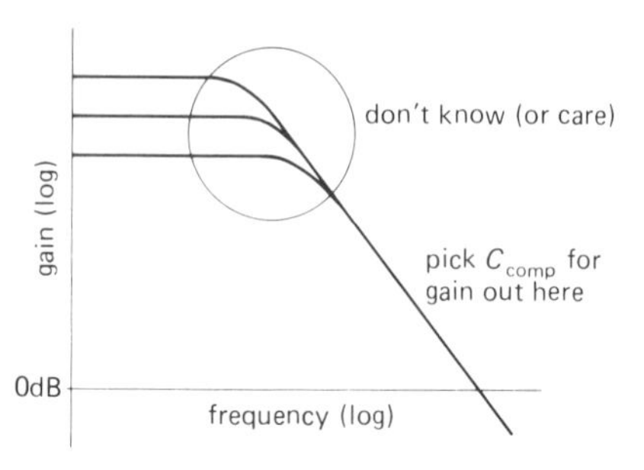

Figure 4.86

Uncompensated op-amps

If an op-amp is used in a circuit with closed-loop gain greater than 1 (i.e., not a follower), it is not necessary to put the pole (the term for the "corner frequency" of a low-pass filter) at such a low frequency, since the stability criterion is relaxed because of the lower loop gain. Figure 4.87 shows the situation graphically.

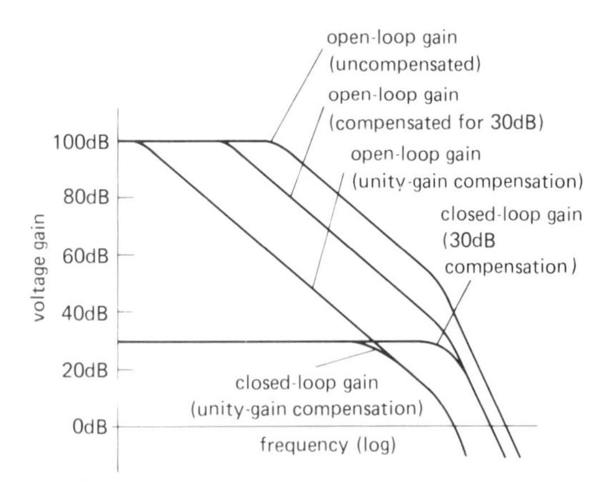

Figure 4.87

For a closed-loop gain of 30dB, the loop gain (which is the ratio of the open-loop gain to the closed-loop gain) is less than for a follower, so the dominant pole can be placed at a higher frequency. It is chosen so that the open-loop gain reaches 30dB (rather than 0dB) at the frequency of the next natural pole of the op-amp. As the graph shows, this means that the open-loop gain is higher over most of the frequency range, and the resultant amplifier will work at higher frequencies. Some op-amps are available in uncompensated versions [e.g., the 748 is an uncompensated 741; the same is true for the 308 (312), 3130 (3160), HA5102 (HA5112), etc.], with recommended external capacitance values for a selection of minimum closed-loop gains. They are worth using if you need the added bandwidth and your circuit operates at high gain. An alternative is to use "decompensated" (a better word might be "undercompensated") op-amps, such as the 357, which are internally compensated for closed-loop gains greater than some minimum ($A_V > 5$ in the case of the 357).

□ *Pole-zero compensation*

It is possible to do a bit better than with dominant-pole compensation by using a compensation network that begins dropping (6dB/octave, a "pole") at some low frequency, then flattens out again (it has a "zero") at the frequency of the second natural pole of the op-amp. In this way the amplifier's second pole is "canceled," giving a smooth 6dB/octave rolloff up to the amplifier's third pole. Figure 4.88 shows a frequency response plot. In practice, the zero is chosen to cancel the amplifier's second pole; then the position of the first pole is adjusted so that the overall response reaches unity gain at the frequency of the amplifier's third pole. A good set of data sheets will often give suggested component values (an R and a C) for pole-zero compensation, as well as the usual capacitor values for dominant-pole compensation.

As you will see in Section 13.06, moving the dominant pole downward in frequency actually causes the second pole of the amplifier to move upward somewhat in frequency, an effect known as "pole splitting." The frequency of the canceling zero will be chosen accordingly.

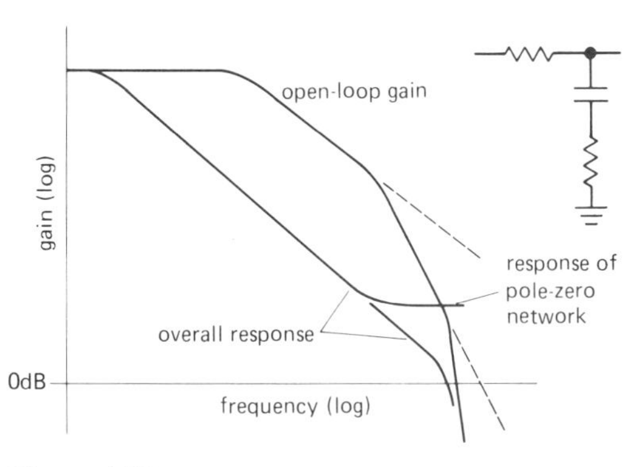

Figure 4.88

□ 4.35 Frequency response of the feedback network

In all of the discussion thus far we have assumed that the feedback network has a flat frequency response; this is usually the case, with the standard resistive voltage divider as a feedback network. However, there are occasions when some sort of equalization amplifier is desired (integrators and differentiators are in this category) or when the frequency response of the feedback network is modified to improve amplifier stability. In such cases it is important to remember that the Bode plot of loop gain versus frequency is what matters, rather than the curve of open-loop gain. To make a long story short, the curve of ideal closed-loop gain versus frequency should intersect the curve of open-loop gain, with a difference in slopes of 6dB/octave. As an example, it is common practice to put a small capacitor (a few picofarads) across the feedback resistor in the usual inverting or noninverting amplifier. Figure 4.89 shows the circuit and Bode plot.

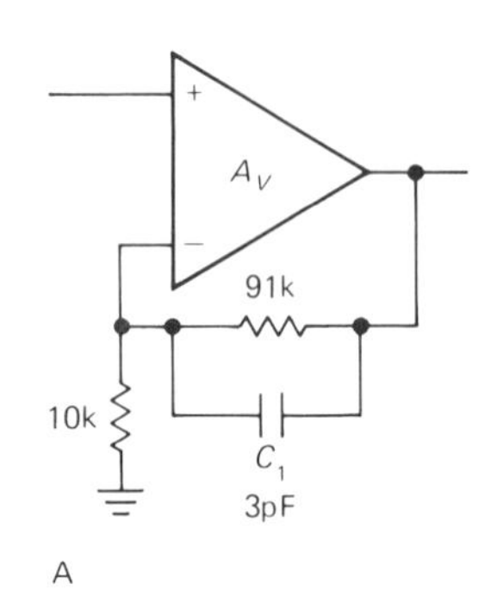

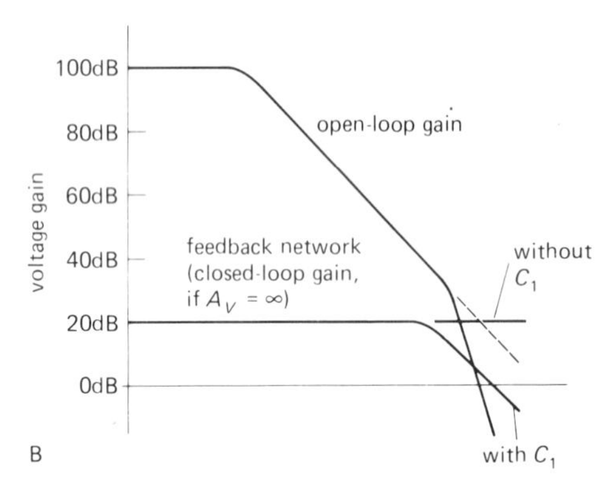

Figure 4.89

The amplifier would have been close to instability with a flat feedback network, since the loop gain would have been dropping at nearly 12dB/octave where the curves meet. The capacitor causes the loop gain to drop at 6dB/octave near the crossing, guaranteeing stability. This sort of consideration is very important when designing differentiators, since an ideal differentiator has a closed-loop gain that rises at 6dB/octave; it is necessary to roll off the differentiator action at some moderate frequency, preferably going over to a 6dB/octave rolloff at high frequencies. Integrators, by comparison, are very friendly in this respect, owing to their 6dB/octave closed-loop rolloff. It takes real talent to make a low-frequency integrator oscillate!

What to do

In summary, you are generally faced with the choice of internally compensated or uncompensated op-amps. It is simplest to use the compensated variety, and that's the usual choice. You might consider the internally compensated LF411 first. If you need greater bandwidth or slew rate, look for a faster compensated op-amp (see Table 4.1 or 7.3 for many choices). If it turns out that nothing is suitable, and the closed-loop gain is greater than unity (as it usually is), you can use an uncompensated op-amp, with an external capacitor as specified by the manufacturer for the gain you are using.

A number of op-amps offer another choice: a "decompensated" version, requiring no external compensation components, but only usable at some minimum gain greater than unity. For example, the popular OP-27 low-noise precision op-amp (unity-gain-compensated) is available as the decompensated OP-37 (minimum gain of 5), offering roughly seven times the speed, and also as the decompensated HA-5147 (minimum gain of 10), with 15 times the speed.

□ *Example: 60Hz power source*

Uncompensated op-amps also give you the flexibility of overcompensating, a simple solution to the problem of additional phase shifts introduced by other stuff in the feedback loop. Figure 4.90 shows a nice example. This is a low-frequency amplifier designed to generate a 115 volt ac power output from a variable 60Hz low-level sine-wave input (it goes with the 60Hz synthesizer circuit described in Section 8.31). The op-amp, together with R_2 and R_3, forms a $\times 100$ gain block; this is then used as the relatively low "open-loop gain" for overall feedback. The op-amp output drives the push-pull output stage, which in turn drives the transformer primary. Low-frequency feedback is taken from the transformer output via R_{10}, in order to generate low distortion and a stable output voltage under load variations. Because of

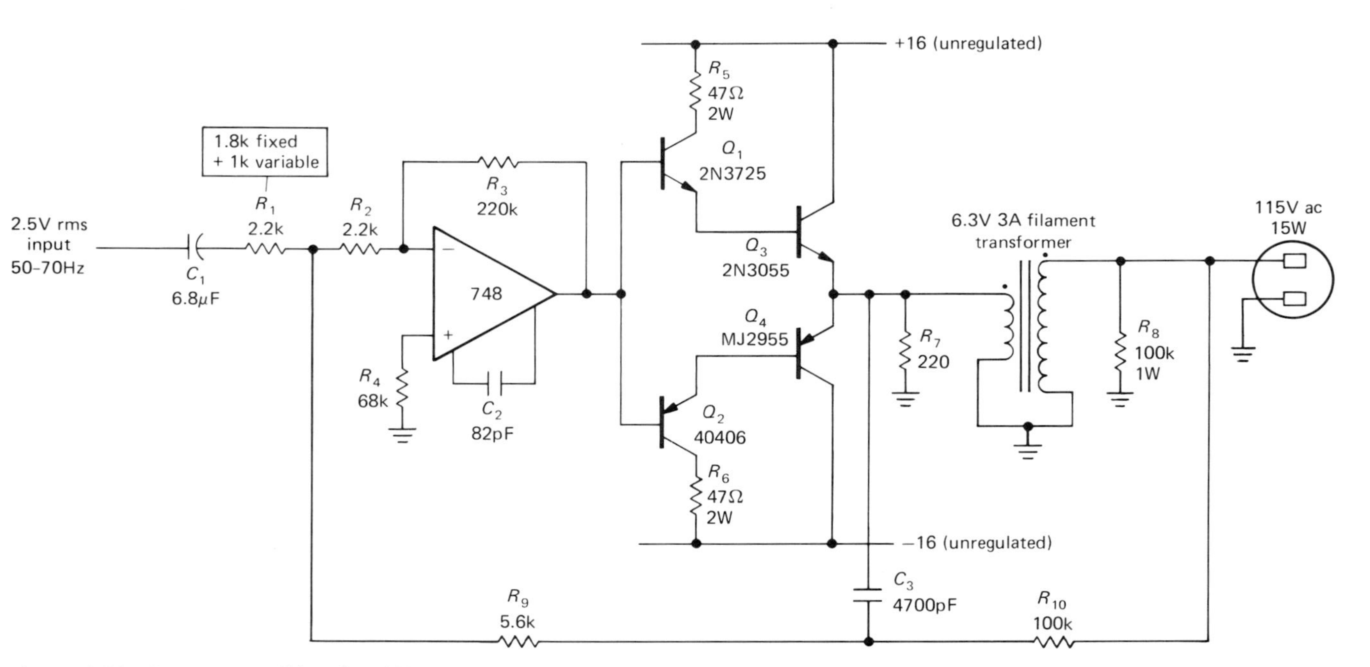

Figure 4.90. Output amplifier for 60Hz power source.

the unacceptably large phase shifts of such a transformer at high frequencies, the circuit is rigged up so that at higher frequencies the feedback comes from the low-voltage input to the transformer, via C_3. The relative sizes of R_9 and R_{10} are chosen to keep the amount of feedback constant at all frequencies. Even though high-frequency feedback is taken directly from the push-pull output, there are still phase shifts associated with the reactive load (the transformer primary) seen by the transistors. In order to ensure good stability, even with reactive loads at the 115 volt output, the op-amp has been overcompensated with an 82pF capacitor (30pF is the normal value for unity gain compensation). The loss of bandwidth that results is unimportant in a low-frequency application like this.

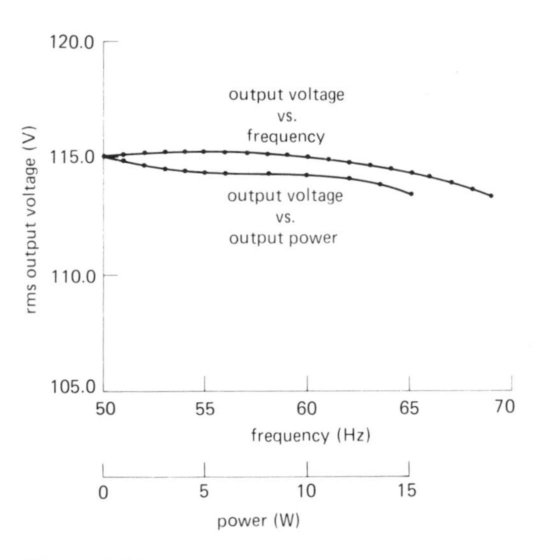

Figure 4.91

An application such as this represents a compromise, since ideally you would like to have plenty of loop gain to stabilize the output voltage against variations in load current. But a large loop gain increases the tendency of the amplifier to oscillate, especially if a reactive load is attached. This is because the reactive load, in combination with the transformer's finite output impedance, causes additional phase shifts within the low-frequency feedback loop. Since this circuit was built to derive a telescope's synchronous driving motors (highly inductive loads), the loop gain was intentionally kept low. Figure 4.91 shows a graph of the ac output voltage versus load, which illustrates good (but not great) regulation.

Motorboating

In ac-coupled feedback amplifiers, stability problems can also crop up at very low frequencies, due to the accumulated *leading* phase shifts caused by several capacitively coupled stages. Each blocking capacitor, in combination with the input resistance due to bias strings and the like, causes a leading phase shift that equals 45° at the low-frequency 3dB point and approaches 90° at lower frequencies. If there is enough loop gain, the system can go into a low-frequency oscillation picturesquely known as "motorboating." With the widespread use of dc-coupled amplifiers, motorboating is almost extinct. However, old-timers can tell you some good stories about it.

SELF-EXPLANATORY CIRCUITS

4.36 Circuit ideas

Some interesting circuit ideas, mostly lifted from manufacturers' data sheets, are shown in Figure 4.94.

4.37 Bad circuits

Figure 4.95 presents a zoo of intentional (mostly) blunders to amuse, amaze, and educate you. There are a few real howlers here this time. These circuits are guaranteed not to work. Figure out why. All op-amps run from ±15 volts unless shown otherwise.

ADDITIONAL EXERCISES

(1) Design a "sensitive voltmeter" to have $Z_{in} = 1M\Omega$ and full-scale sensitivities of 10mV to 10V in four ranges. Use a 1mA meter movement and an op-amp. Trim voltage offsets if necessary, and calculate what the meter will read with input open, assuming **(a)** $I_B = 25pA$ (typical for a 411) and **(b)** $I_B = 80nA$ (typical for a 741). Use some form of meter protection (e.g., keep its current less than 200% of full scale), and protect the amplifier inputs from voltages outside the supply voltages. What do you conclude about the suitability of the 741 for low-level high-impedance measurements?

(2) Design an audio amplifier, using an OP-27 op-amp (low noise, good for audio), with the following characteristics: gain = 20dB, Z_{in} = 10k, −3dB point = 20Hz. Use the noninverting configuration, and roll off the gain at low frequencies in such a way as to reduce the effects of input offset voltage. Use proper design to minimize the effects of input bias current on output offset. Assume that the signal source is capacitively coupled.

(3) Design a unity-gain phase splitter (see Chapter 2) using 411s. Strive for high input impedance and low output impedances. The circuit should be dc-coupled. At roughly what maximum frequency can you obtain full swing (27V pp, with ±15V supplies), owing to slew rate limitations?

(4) El Cheapo brand loudspeakers are found to have a treble boost, beginning at 2kHz (+3dB point) and rising 6dB/octave. Design a simple *RC* filter, buffered with AD611 op-amps (another good audio chip) as necessary, to be placed between preamp and amplifier to compensate this rise. Assume that the preamp has Z_{out} = 50k and that the amplifier has Z_{in} = 10k, approximately.

(5) A 741 is used as a simple comparator, with one input grounded; i.e., it is a zero-crossing detector. A 1 volt amplitude sine wave is fed into the other input (frequency 1kHz). What voltage(s) will the input be when the output passes through zero volts? Assume that the slew rate is 0.5V/μs and that the op-amp's saturated output is ±13 volts.

(6) The circuit in Figure 4.92 is an example of a "negative-impedance converter." **(a)** What is its input impedance? **(b)** If the op-amp's output range goes from V_+ to V_-, what range of input voltages will this circuit accommodate without saturation?

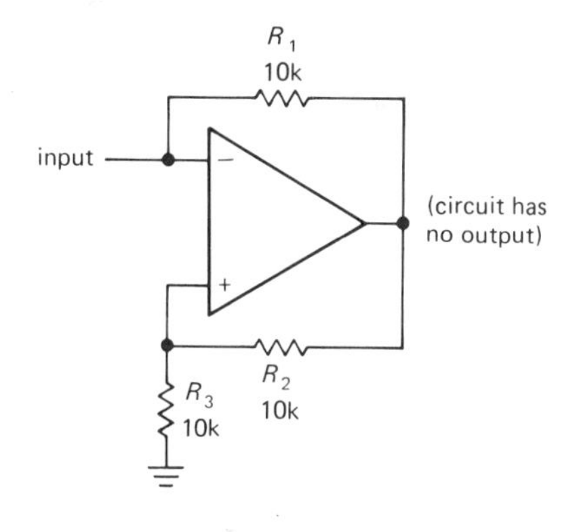

Figure 4.92

(7) Consider the circuit in the preceding problem as the 2-terminal black box (Fig. 4.93). Show how to make a dc amplifier with a gain of −10. Why can't you make a dc amplifier with a gain of +10? (Hint: The circuit is susceptible to a latchup condition for a certain range of source resistances. What is that range? Can you think of a remedy?)

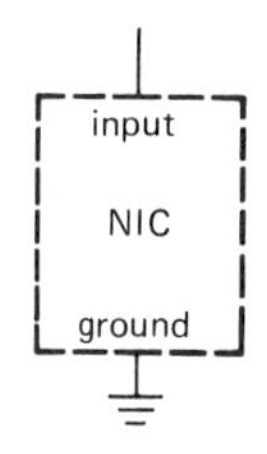

Figure 4.93

Circuit ideas

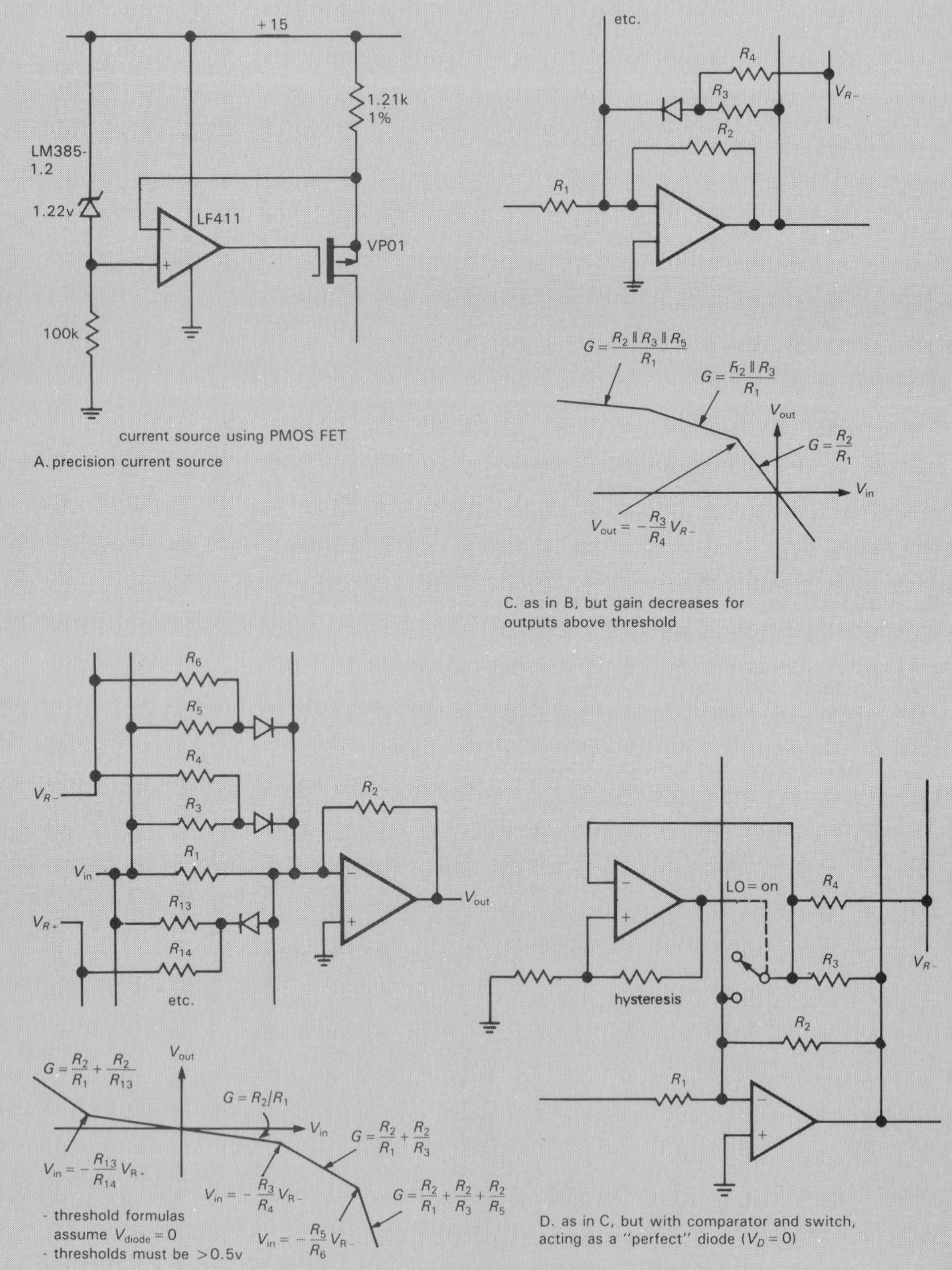

Figure 4.94

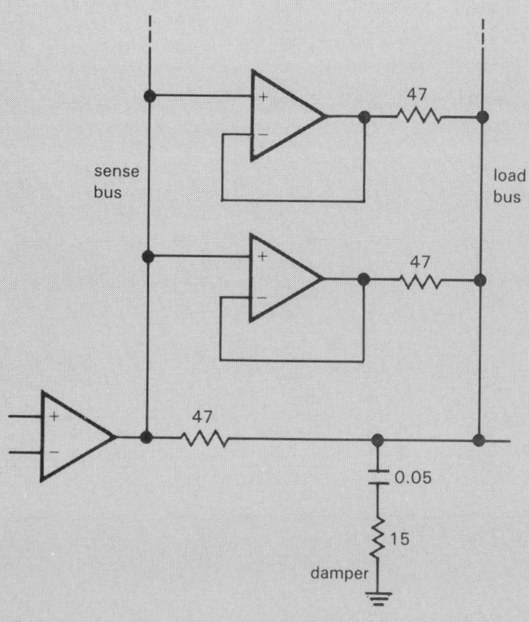

E. higher output current from additional op-amp sections; watch for excess heating

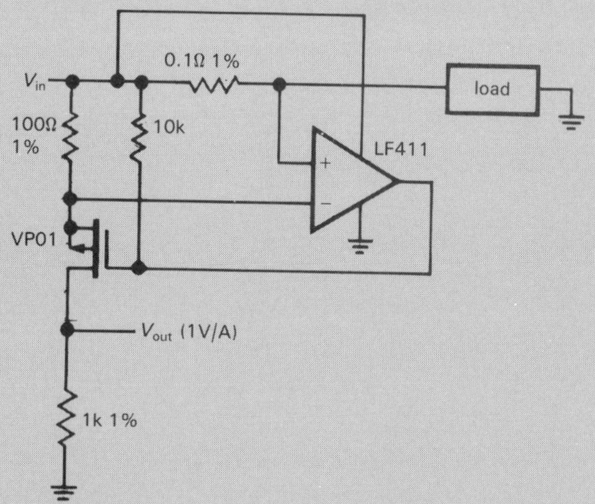

F. current monitor

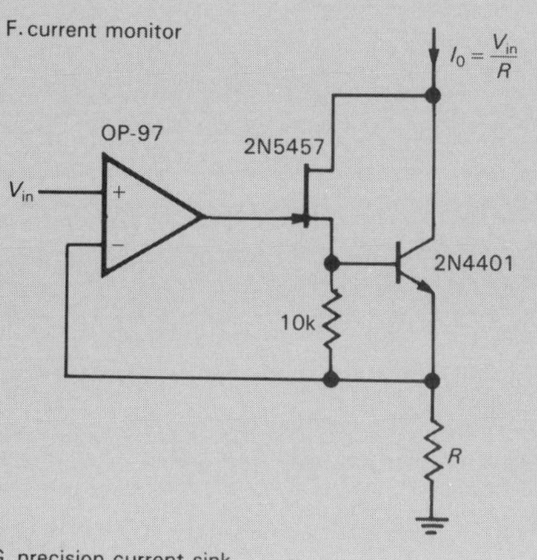

G. precision current sink

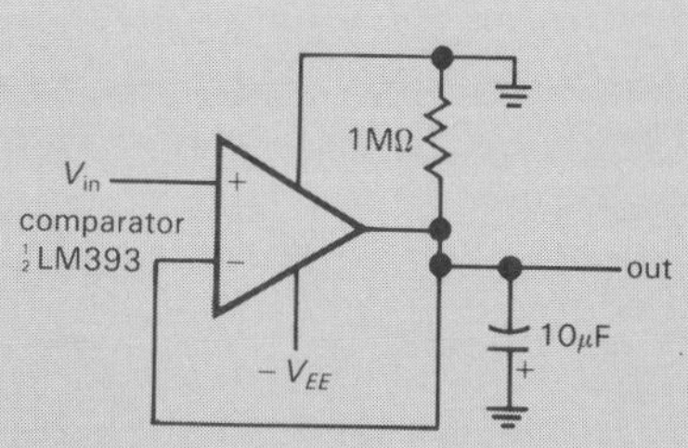

H. negative peak detector

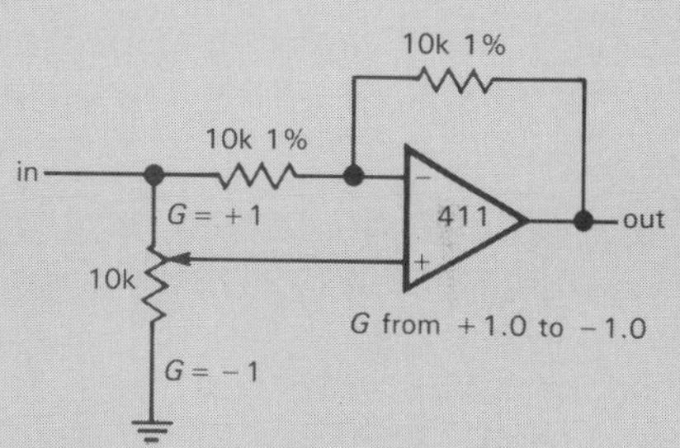

I. continuously variable gain

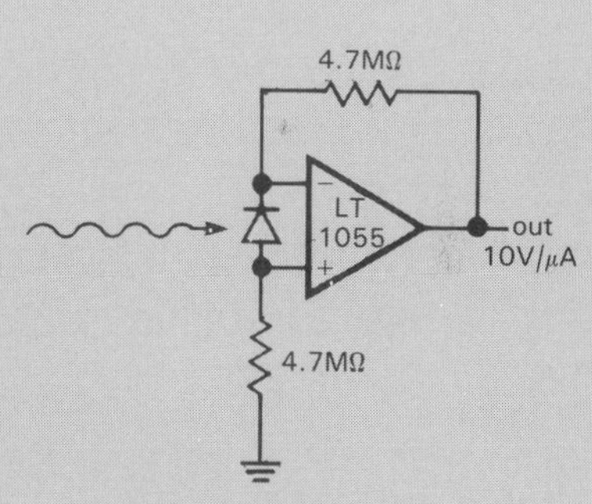

J. photodiode amplifier

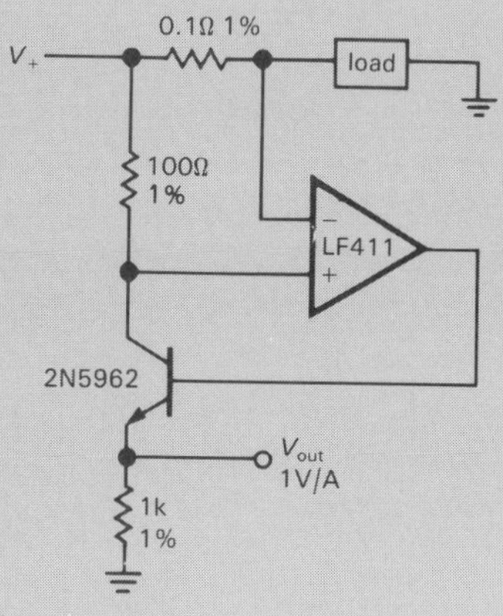

K. current monitor

Circuit ideas (*cont.*)

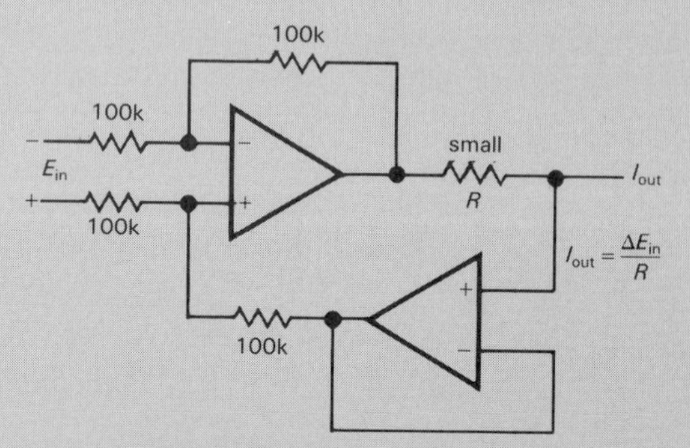

L. precision bipolarity current source

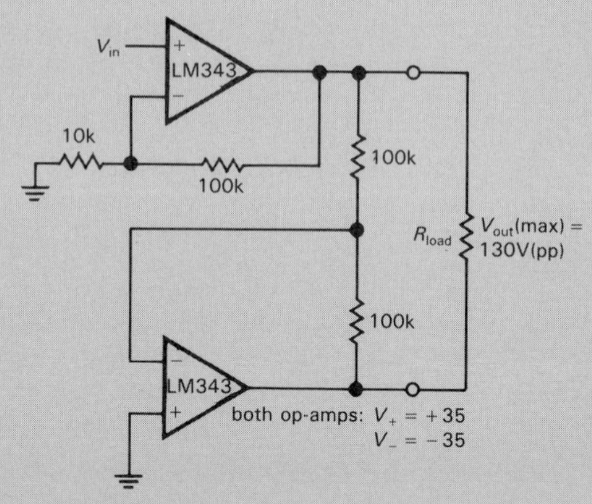

P. high-voltage (bridge) drive to floating load (gain = 22)

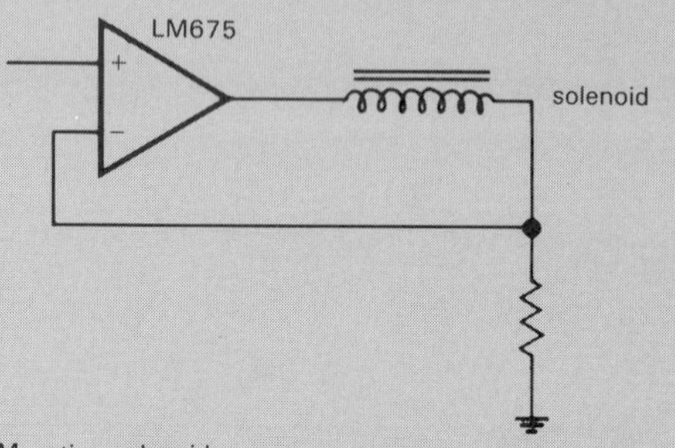

M. active solenoid

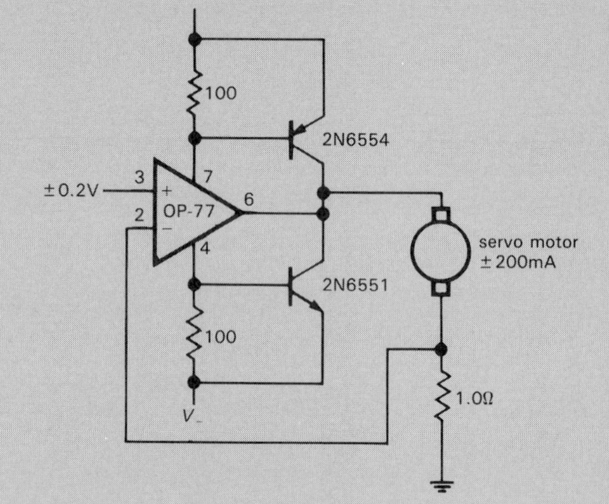

Q. 0.2A servo-motor amplifier

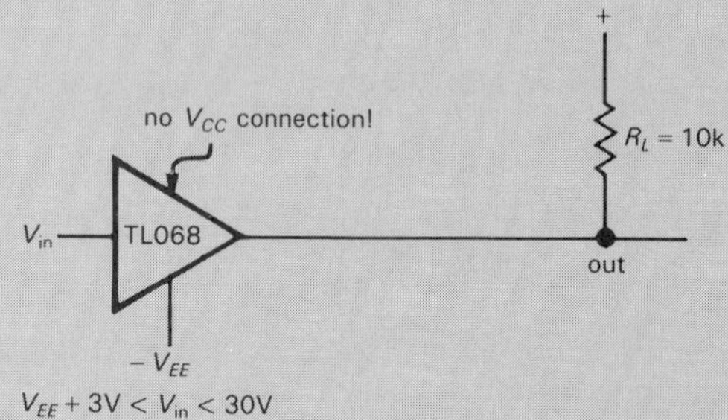

$V_{EE} + 3V < V_{in} < 30V$

N. unusual 3-pin JFET follower

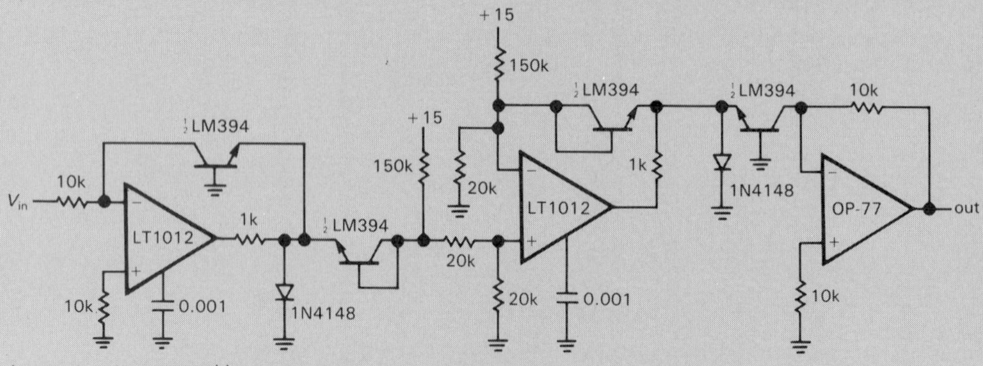

O. temperature-compensated log converter

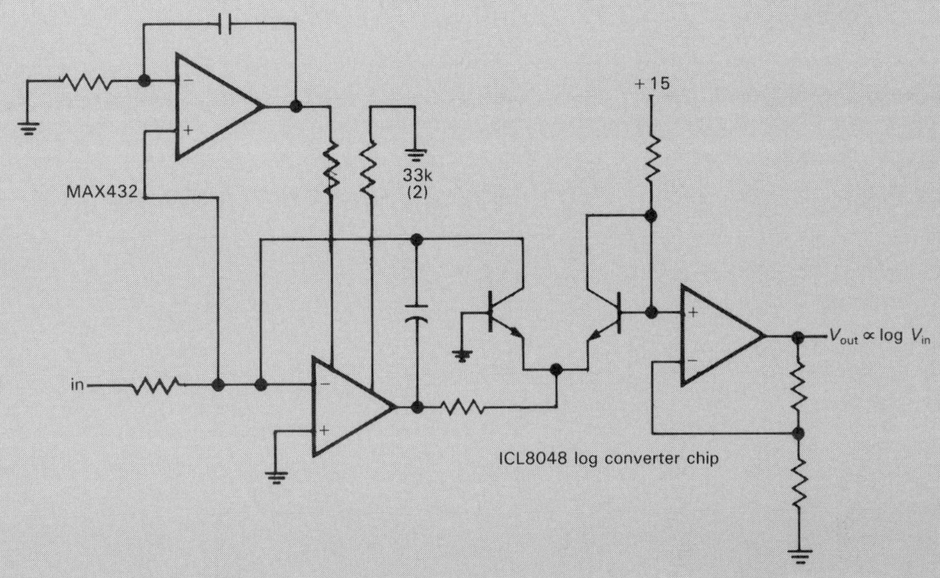

R. log converter with wide input range due to automatic nulling with a chopper amplifier

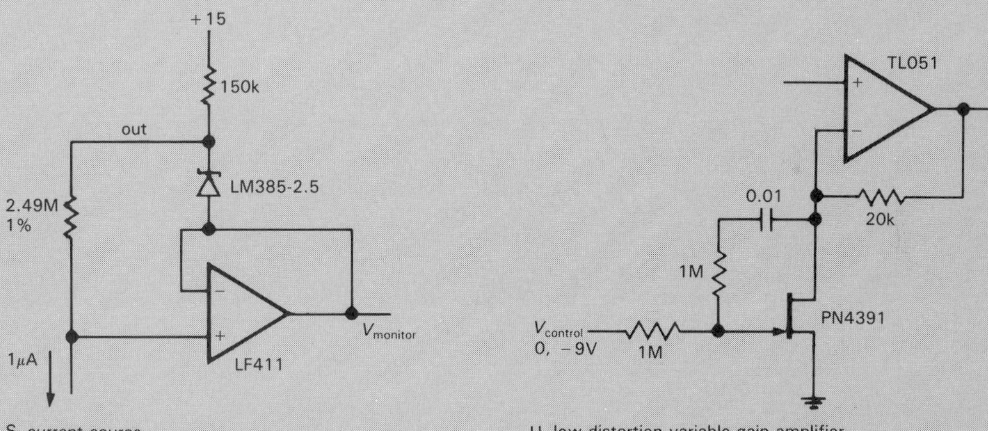

S. current source

U. low-distortion variable-gain amplifier

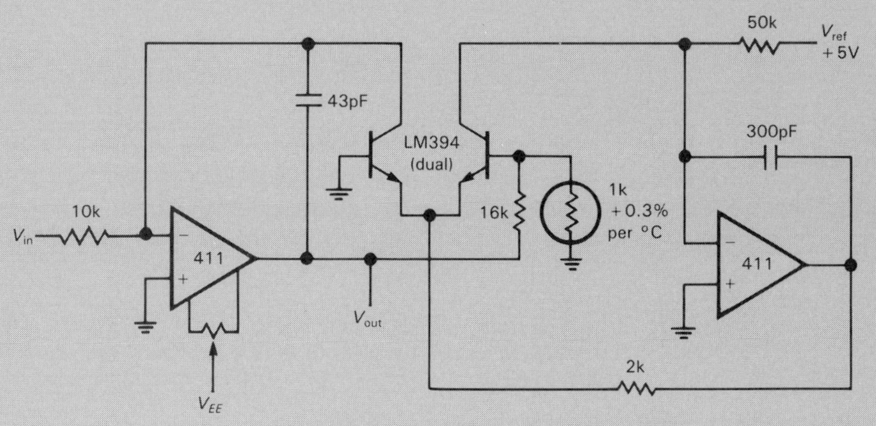

T. fast logarithmic converter

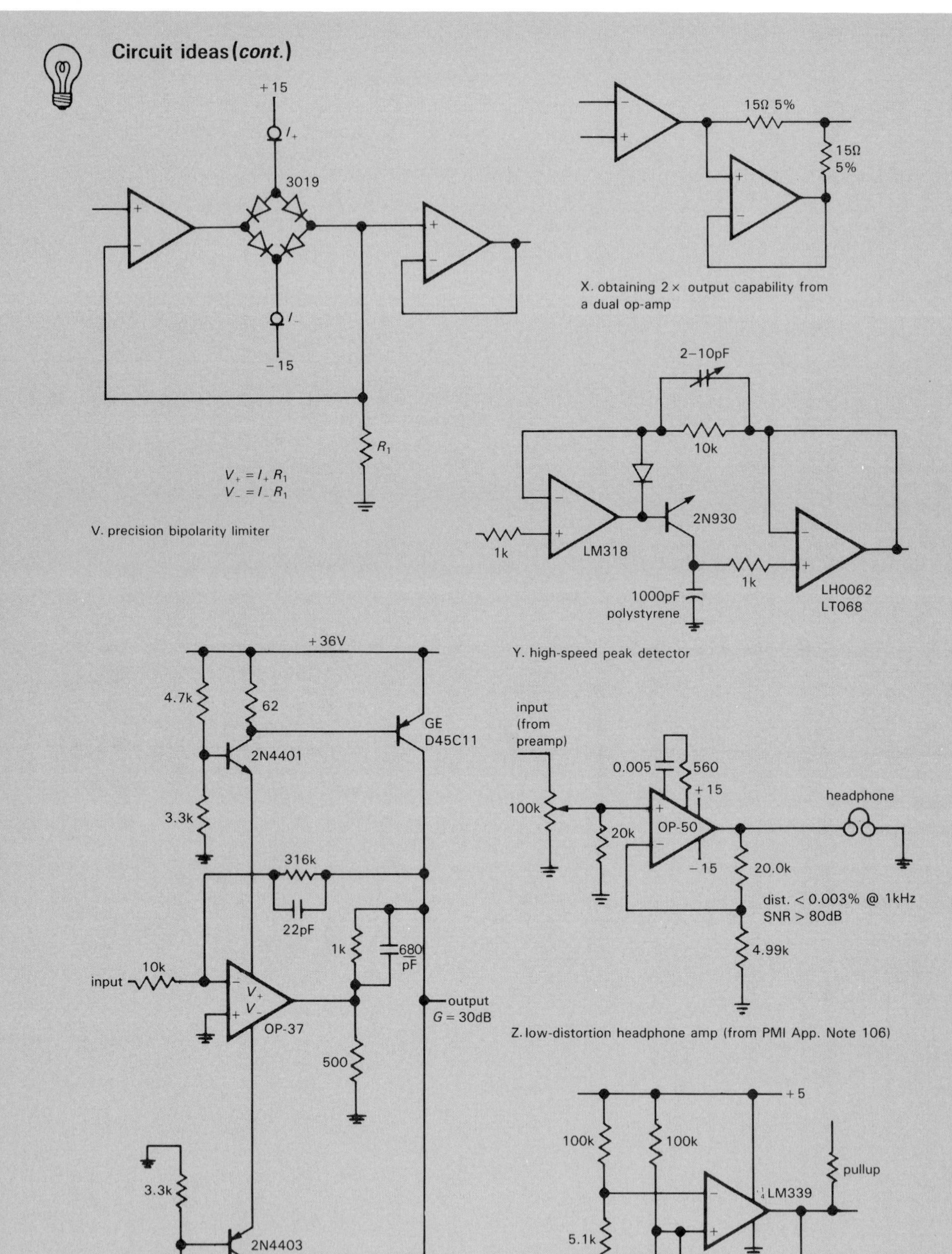

V. precision bipolarity limiter

X. obtaining 2× output capability from a dual op-amp

Y. high-speed peak detector

W. low-noise high-voltage op-amp (from PMI App. Note 106)

Z. low-distortion headphone amp (from PMI App. Note 106)

AA. zero-crossing detector, single supply

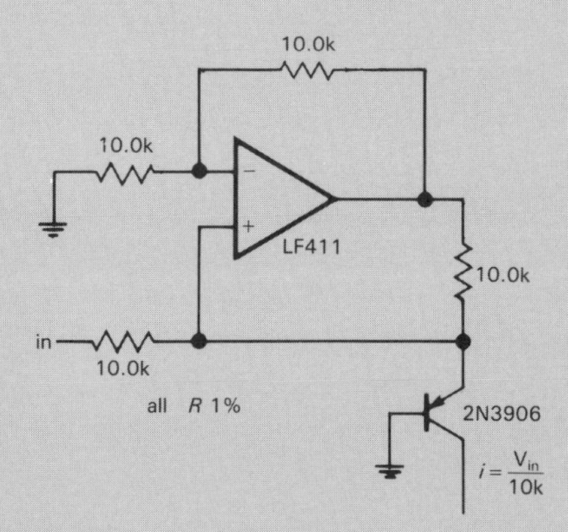

BB. Howland-style current source for transconductance voltage-current control circuits (1μA to 1mA)

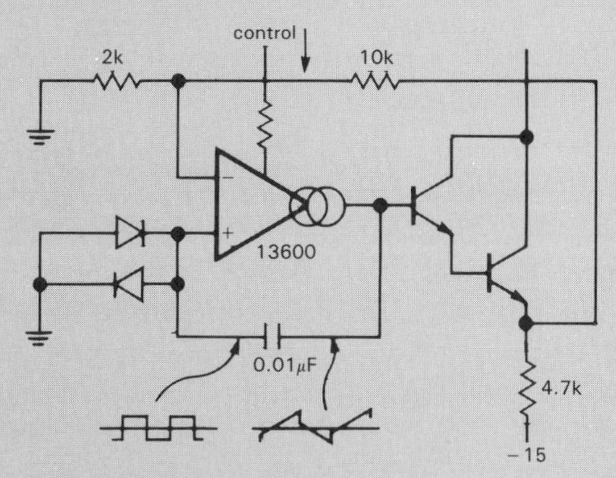

EE. voltage-contolled oscillator with transconductance amplifier

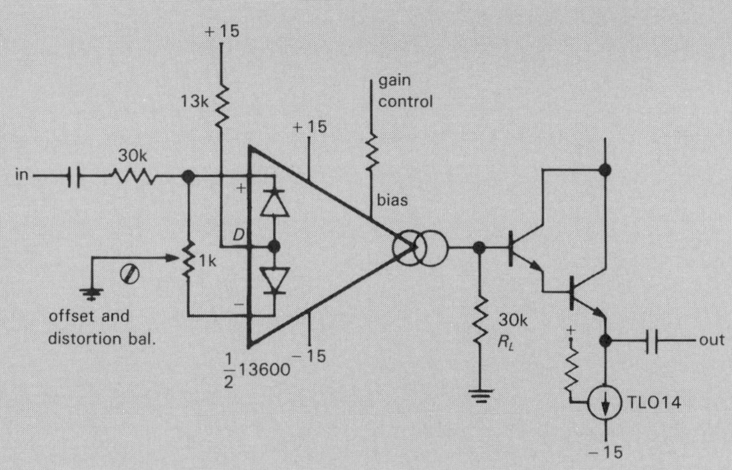

CC. transconductance linearized voltage-controlled amplifier (VCA)

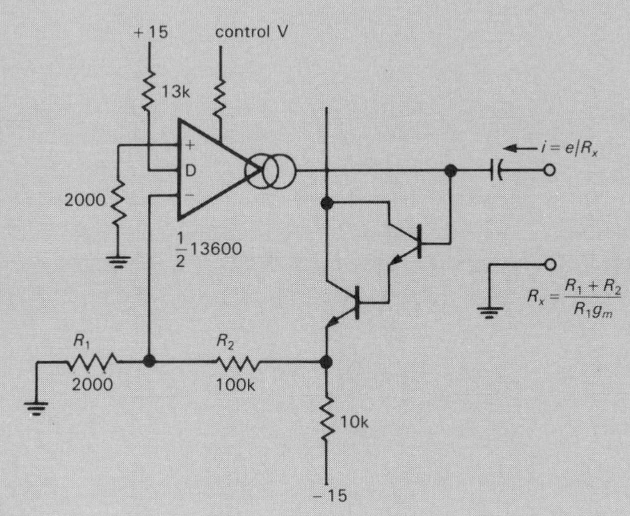

DD. voltage-controlled AC load resistor

Bad circuits

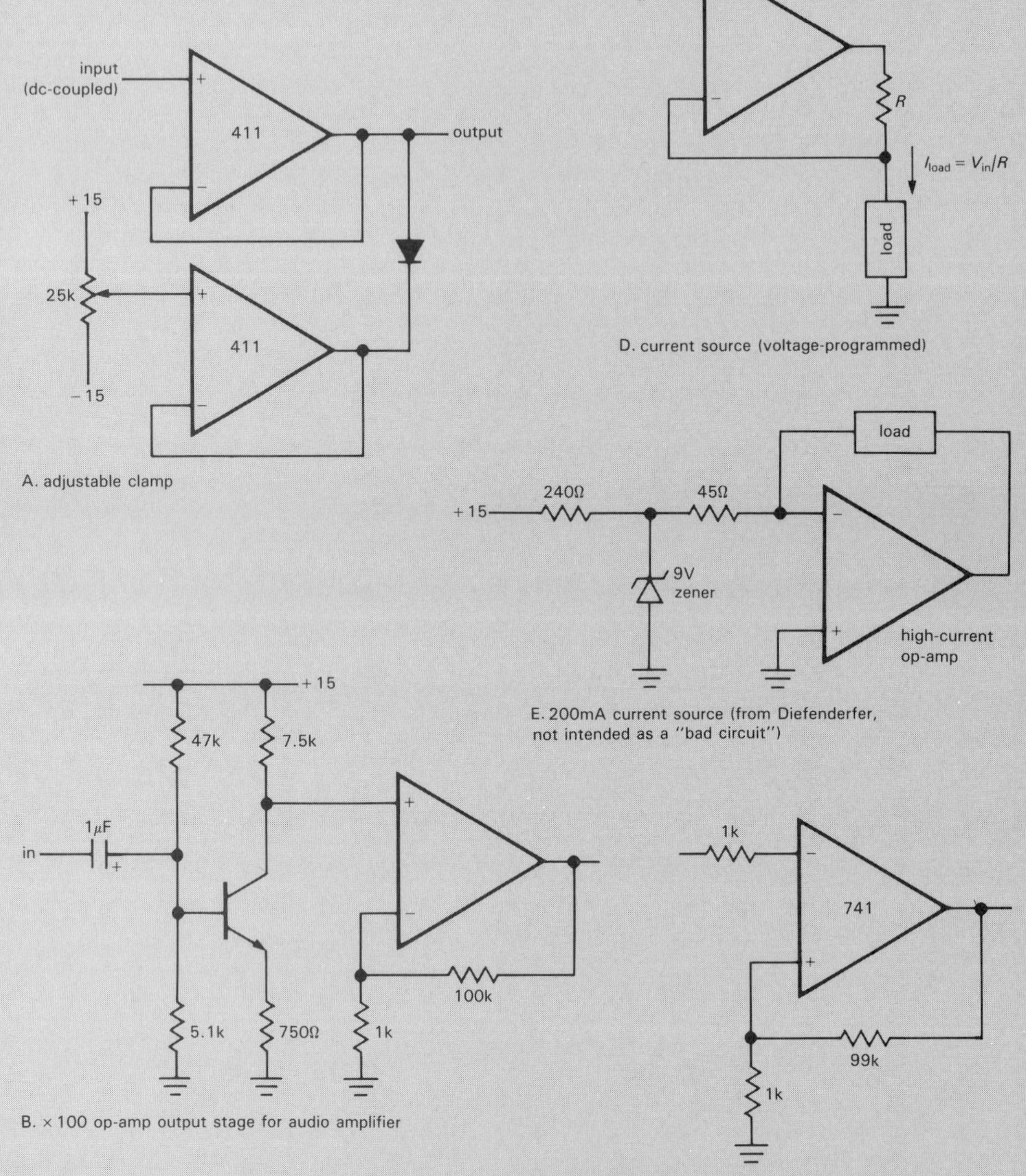

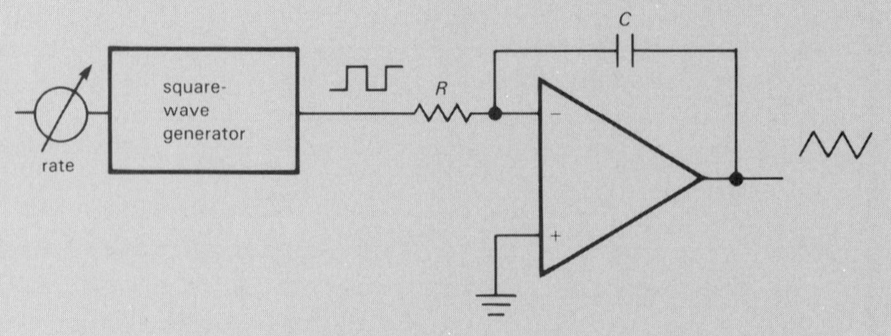

Figure 4.95

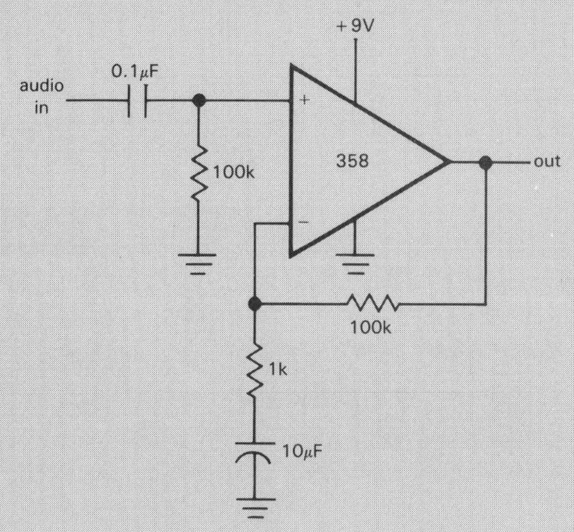

G. × 100 audio amplifier (single supply)

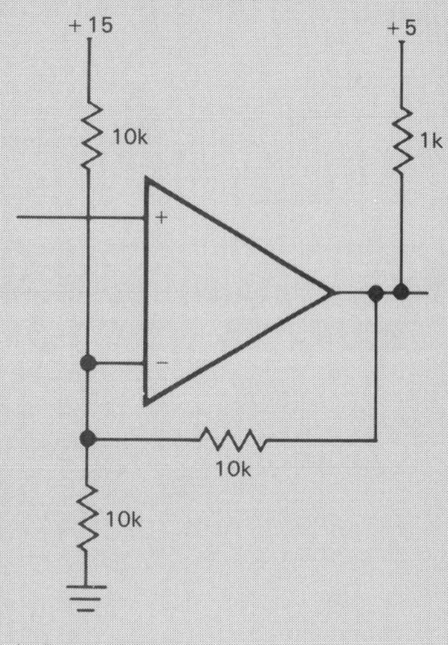

J. Schmitt trigger

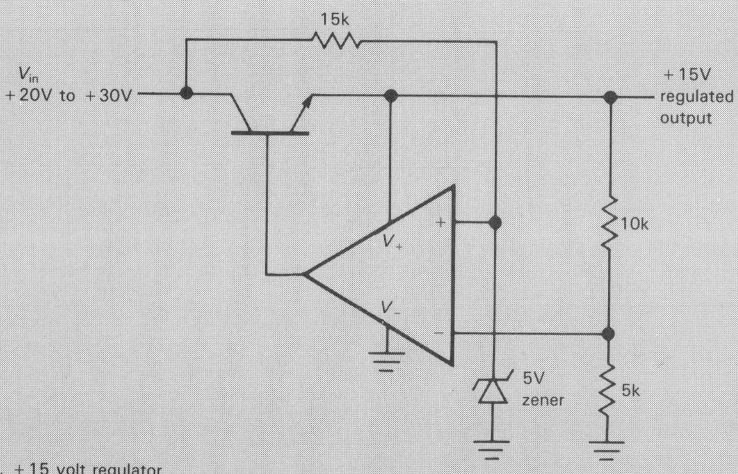

H. +15 volt regulator

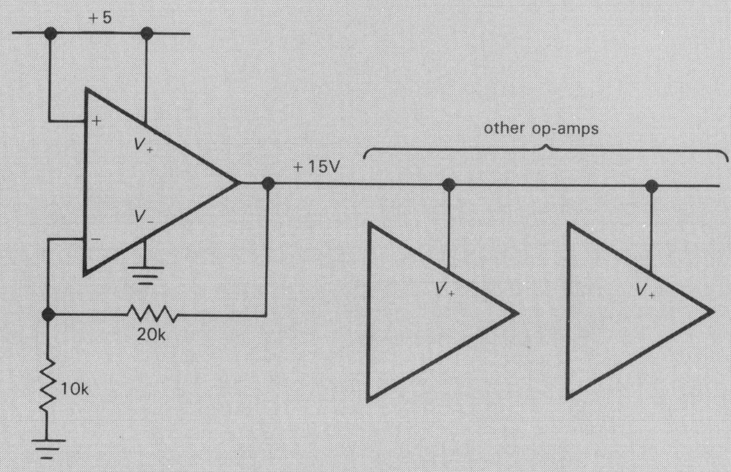

I. op-amp as +15V regulator

Bad circuits (*cont.*)

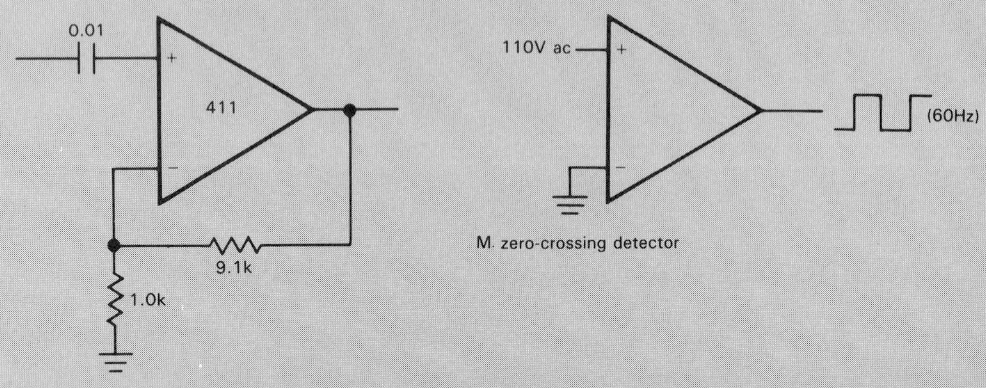

K. ac-coupled ×10 amplifier

M. zero-crossing detector

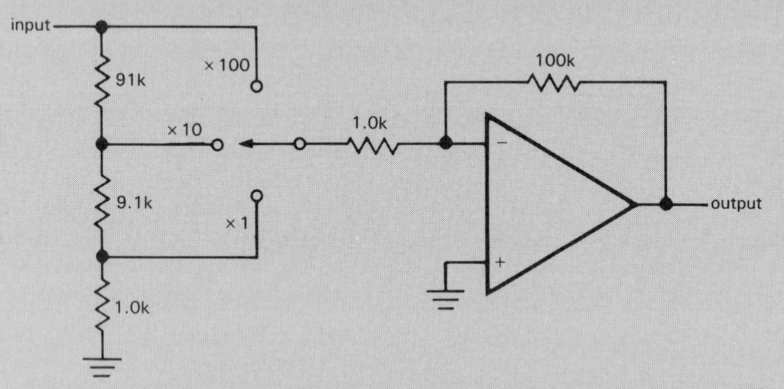

L. dc amplifier, selectable gains (×1 to ×100)

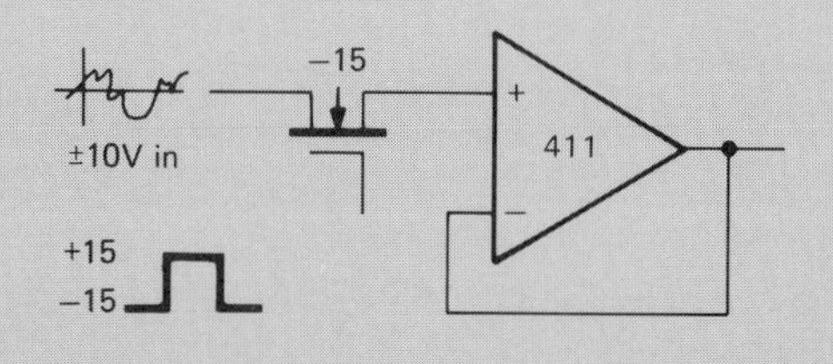

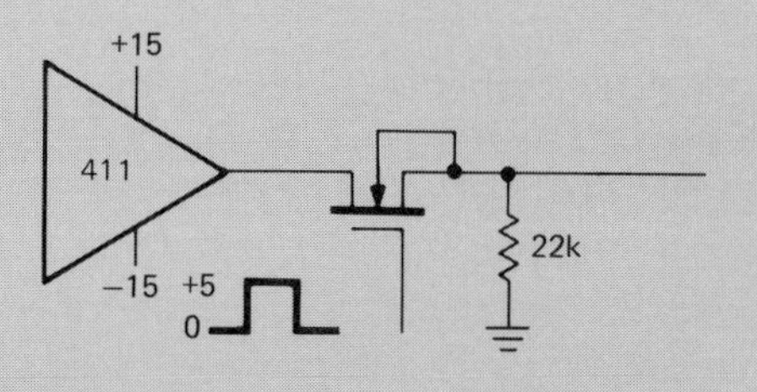

N. op-amps with switches

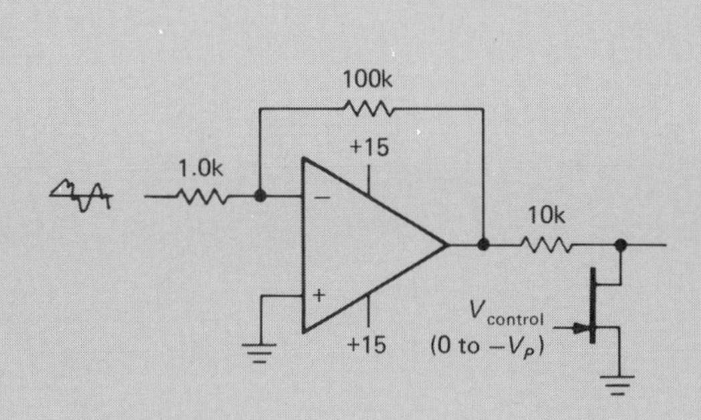

O. FET gain control

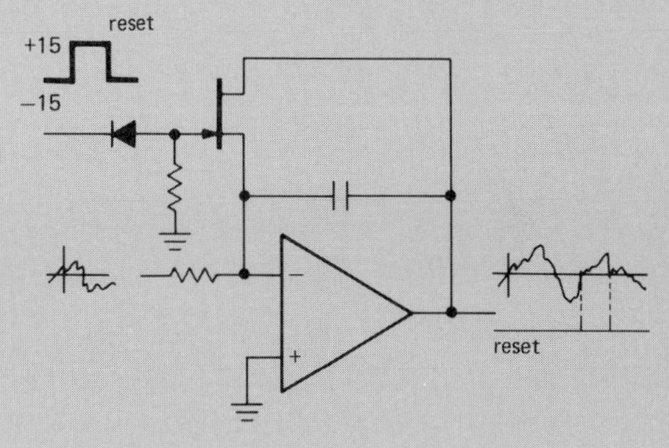

Q. integrator with reset

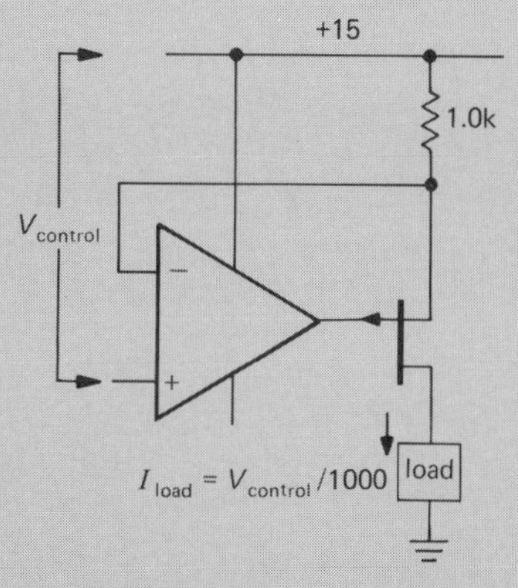

P. current source

ACTIVE FILTERS AND OSCILLATORS

CHAPTER 5

With only the techniques of transistors and op-amps it is possible to delve into a number of interesting areas of linear (as contrasted with digital) circuitry. We believe that it is important to spend some time doing this now, in order to strengthen your understanding of some of these difficult concepts (transistor behavior, feedback, op-amp limitations, etc.) before introducing more new devices and techniques and getting into the large area of digital electronics. In this chapter, therefore, we will treat briefly the areas of active filters and oscillators. Additional analog techniques are treated in Chapter 6 (voltage regulators and high-current design), Chapter 7 (precision circuits and low noise), Chapter 13 (radiofrequency techniques), Chapter 14 (low-power design), and Chapter 15 (measurements and signal processing). The first part of this chapter (active filters, Sections 5.01–5.11) describes techniques of a somewhat specialized nature, and it can be passed over in a first reading. However, the latter part of this chapter (oscillators, Sections 5.12–5.19) describes techniques of broad utility and should not be omitted.

ACTIVE FILTERS

In Chapter 1 we began a discussion of filters made from resistors and capacitors. Those simple *RC* filters produced gentle high-pass or low-pass gain characteristics, with a 6dB/octave falloff well beyond the −3dB point. By cascading high-pass and low-pass filters, we showed how to obtain bandpass filters, again with gentle 6dB/octave "skirts." Such filters are sufficient for many purposes, especially if the signal being rejected by the filter is far removed in frequency from the desired signal passband. Some examples are bypassing of radiofrequency signals in audio circuits, "blocking" capacitors for elimination of dc levels, and separation of modulation from a communications "carrier" (see Chapter 13).

5.01 Frequency response with *RC* filters

Often, however, filters with flatter passbands and steeper skirts are needed. This happens whenever signals must be filtered from other interfering signals nearby in frequency. The obvious next question is whether or not (by cascading a number of identical low-pass filters, say) we can generate an approximation to the ideal "brick-wall" low-pass frequency response, as in Figure 5.1.

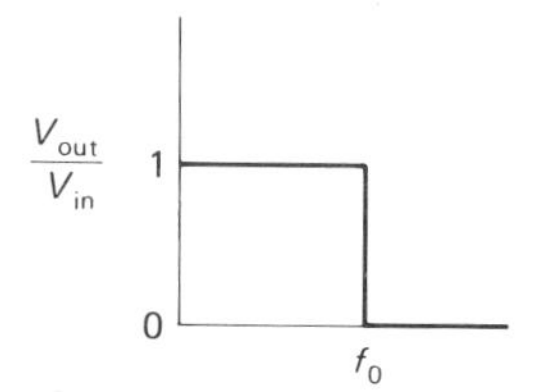

Figure 5.1

We know already that simple cascading won't work, since each section's input impedance will load the previous section seriously, degrading the response. But with buffers between each section (or by arranging to have each section of much higher impedance than the one preceding it), it would seem possible. Nonetheless, the answer is no. Cascaded *RC* filters do produce a steep *ultimate* falloff, but the "knee" of the curve of response versus frequency is not sharpened. We might restate this as "many soft knees do not a hard knee make." To make the point graphically, we have plotted some graphs of gain response (i.e., V_{out}/V_{in}) versus frequency for low-pass filters constructed from 1, 2, 4, 8, 16, and 32 identical *RC* sections, perfectly buffered (Fig. 5.2).

The first graph shows the effect of cascading several *RC* sections, each with its 3dB point at unit frequency. As more sections are added, the overall 3dB point is pushed downward in frequency, as you could easily have predicted. To compare filter characteristics fairly, the rolloff frequencies of the individual sections should be adjusted so that the overall 3dB point is always at the same frequency. The other graphs in Figure 5.2, as well as the next few graphs in this chapter, are all "normalized" in frequency, meaning that the −3dB point

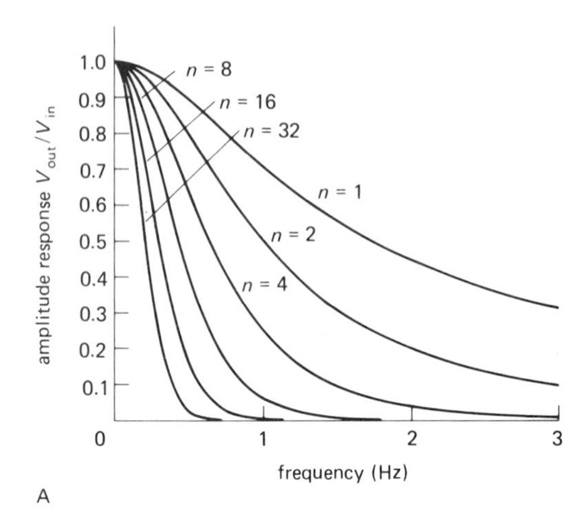

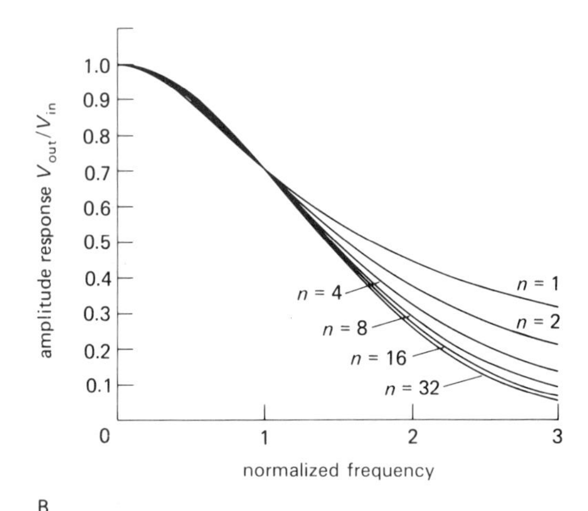

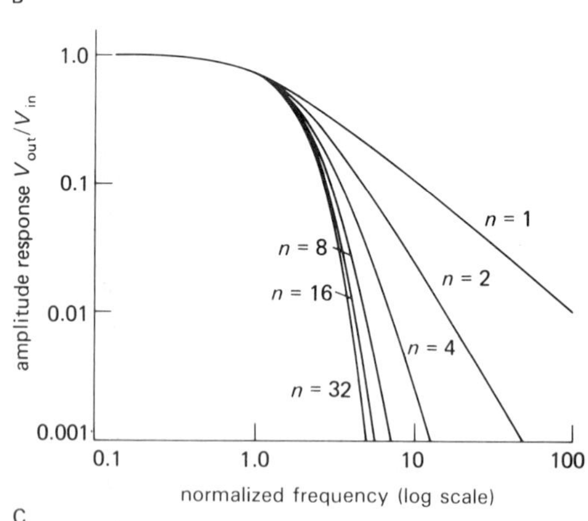

Figure 5.2. Frequency responses of multisection *RC* filters. Graphs A and B are linear plots, whereas C is logarithmic. The filter responses in B and C have been normalized (or scaled) for 3dB attenuation at unit frequency.

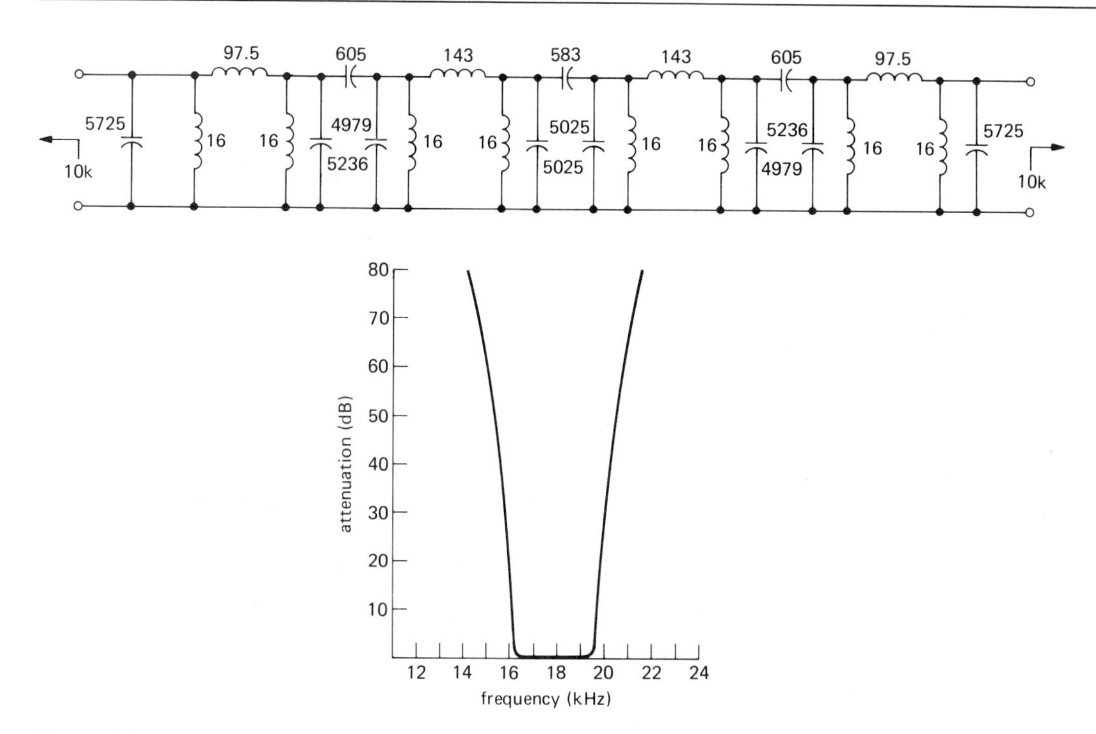

Figure 5.3. An unusually good passive bandpass filter implemented from inductors and capacitors (inductances in mH, capacitances in pF). Bottom: Measured response of the filter circuit. [Based on Figs. 11 and 12 from Orchard, H. J., and Sheahan, D. F., *IEEE Journal of Solid-State Circuits*, Vol. SC-5, No. 3 (1970).]

(or breakpoint, however defined) is at a frequency of 1 radian per second (or at 1 Hz). To determine the response of a filter whose breakpoint is set at some other frequency, simply multiply the values on the frequency axis by the actual breakpoint frequency f_c. In general, we will also stick to the log-log graph of frequency response when talking about filters, because it tells the most about the frequency response. It lets you see the approach to the ultimate rolloff slope, and it permits you to read off accurate values of attenuation. In this case (cascaded *RC* sections) the normalized graphs in Figures 5.2B and 5.2C demonstrate the soft knee characteristic of passive *RC* filters.

5.02 Ideal performance with *LC* filters

As we pointed out in Chapter 1, filters made with inductors and capacitors can have very sharp responses. The parallel *LC* resonant circuit is an example. By including inductors in the design, it is possible to create filters with any desired flatness of passband combined with sharpness of transition and steepness of falloff outside the band. Figure 5.3 shows an example of a telephone filter and its characteristics.

Obviously the inclusion of inductors into the design brings about some magic that cannot be performed without them. In the terminology of network analysis, that magic consists in the use of "off-axis poles." Even so, the complexity of the filter increases according to the required flatness of passband and steepness of falloff outside the band, accounting for the large number of components used in the preceding filter. The transient response and phase-shift characteristics are also generally degraded as the amplitude response is improved to approach the ideal brick-wall characteristic.

The synthesis of filters from passive components *(R, L, C)* is a highly developed subject, as typified by the authoritative handbook by Zverev (see chapter references at end of book). The only problem is that inductors as circuit elements frequently leave much to be desired. They are often bulky and expensive, and they depart from the ideal by being "lossy," i.e., by having significant series resistance, as well as other "pathologies" such as nonlinearity, distributed winding capacitance, and susceptibility to magnetic pickup of interference.

What is needed is a way to make inductorless filters with the characteristics of ideal *RLC* filters.

5.03 Enter active filters: an overview

By using op-amps as part of the filter design, it is possible to synthesize any *RLC* filter characteristic without using inductors. Such inductorless filters are known as active filters because of the inclusion of an active element (the amplifier).

Active filters can be used to make low-pass, high-pass, bandpass, and band-reject filters, with a choice of filter types according to the important features of the response, e.g., maximal flatness of passband, steepness of skirts, or uniformity of time delay versus frequency (more on this shortly). In addition, "all-pass filters" with flat amplitude response but tailored phase versus frequency can be made (they're also known as "delay equalizers"), as well as the opposite – a filter with constant phase shift but tailored amplitude response.

□ *Negative-impedance converters and gyrators*

Two interesting circuit elements that should be mentioned in any overview are the negative-impedance converter (NIC) and the gyrator. These devices can mimic the properties of inductors, while using only resistors and capacitors in addition to op-amps.

Once you can do that, you can build inductorless filters with the ideal properties of any *RLC* filter, thus providing at least one way to make active filters.

The NIC converts an impedance to its *negative*, whereas the gyrator converts an impedance to its *inverse*. The following exercises will help you discover for yourself how that works out.

EXERCISE 5.1
Show that the circuit in Figure 5.4 is a negative-impedance converter, in particular that $Z_{\text{in}} = -Z$. Hint: Apply some input voltage V, and compute the input current I. Then take the ratio to find $Z_{\text{in}} = V/I$.

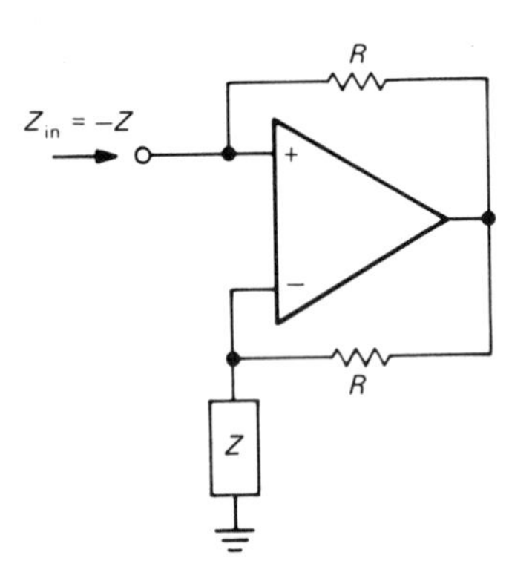

Figure 5.4. Negative-impedance converter.

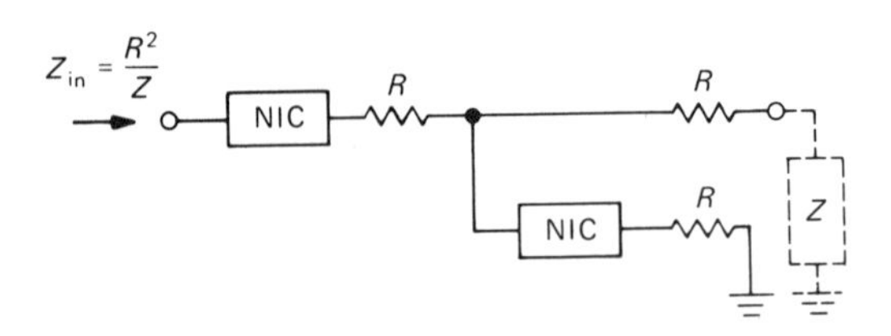

Figure 5.5

EXERCISE 5.2
Show that the circuit in Figure 5.5 is a gyrator, in particular that $Z_{\text{in}} = R^2/Z$. Hint: You can analyze it as a set of voltage dividers, beginning at the right.

The NIC therefore converts a capacitor to a "backward" inductor:

$$Z_C = 1/j\omega C \rightarrow Z_{\text{in}} = j/\omega C$$

i.e., it is inductive in the sense of generating a current that lags the applied voltage, but its impedance has the wrong frequency dependence (it goes down, instead of up, with increasing frequency). The gyrator, on the other hand, converts a capacitor to a true inductor:

$$Z_C = 1/j\omega C \rightarrow Z_{\text{in}} = j\omega C R^2$$

i.e., an inductor with inductance $L = CR^2$.

The existence of the gyrator makes it intuitively reasonable that inductorless filters can be built to mimic any filter using inductors: Simply replace each inductor by a gyrated capacitor. The use of gyrators in just that manner is perfectly OK, and in fact the telephone filter illustrated previously was built that way. In addition to simple gyrator substitution into preexisting *RLC* designs, it is possible to synthesize many other filter configurations. The field of inductorless filter design is extremely active, with new designs appearing in the journals every month.

Sallen-and-Key filter

Figure 5.6 shows an example of a simple and even partly intuitive filter. It is known as a Sallen-and-Key filter, after its inventors. The unity-gain amplifier can be an op-amp connected as a follower, or just an emitter follower. This particular filter is a 2-pole high-pass filter. Note that it would be simply two cascaded *RC* high-pass filters except for the fact that the bottom of the first resistor is bootstrapped by the output. It is easy to see that at very low frequencies it falls off just like a cascaded *RC*, since the output is essentially zero. As the output rises at increasing frequency, however, the bootstrap action tends to reduce the attenuation, giving a sharper knee. Of course, such hand-waving cannot substitute for honest analysis, which luckily has already been done for a prodigious variety of nice filters. We will come back to active filter circuits in Section 5.06.

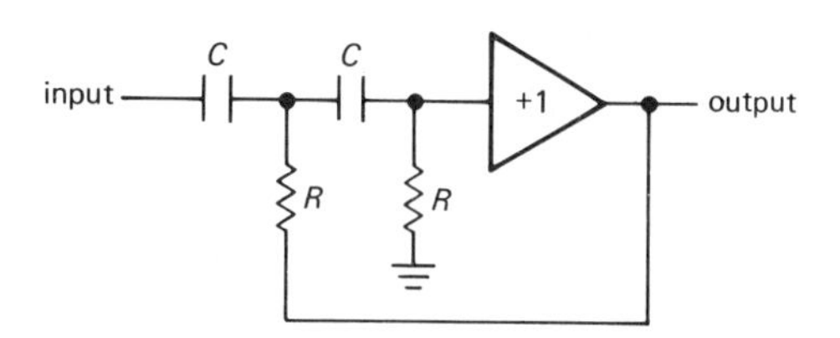

Figure 5.6

5.04 Key filter performance criteria

There are some standard terms that keep appearing when we talk about filters and try to specify their performance. It is worth getting it all straight at the beginning.

Frequency domain

The most obvious characteristic of a filter is its gain versus frequency, typified by the sort of low-pass characteristic shown in Figure 5.7.

The *passband* is the region of frequencies that are relatively unattenuated by the filter. Most often the passband is considered to extend to the −3dB point, but with certain filters (most notably the "equiripple" types) the end of the passband may be defined somewhat differently. Within the passband the response may show variations or *ripples*, defining a *ripple band*, as shown. The *cutoff frequency*, f_c, is the end of the passband. The response of the filter then drops off through a *transition region* (also colorfully known as the *skirt* of the filter's response) to a *stopband*, the region of significant attenuation. The stopband may be defined by some minimum attenuation, e.g., 40dB.

Along with the gain response, the other parameter of importance in the frequency

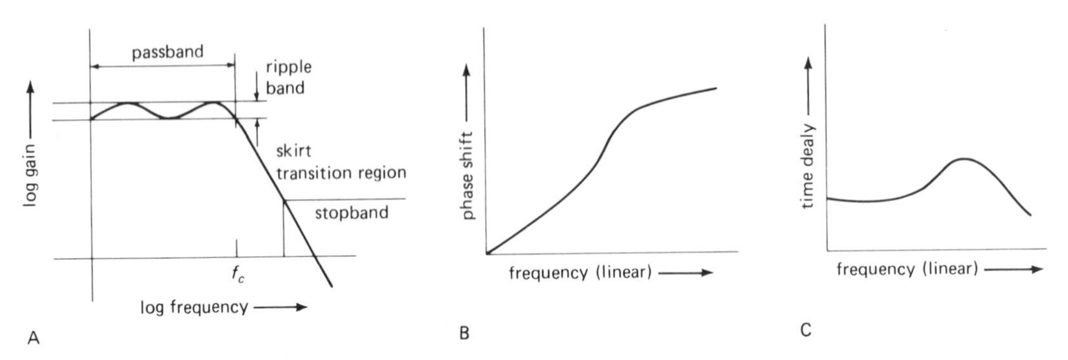

Figure 5.7. Filter characteristics versus frequency.

domain is the *phase shift* of the output signal relative to the input signal. In other words, we are interested in the *complex* response of the filter, which usually goes by the name of **H**(s), where s = $j\omega$, where **H**, s, and ω all are complex. Phase is important because a signal entirely within the passband of a filter will emerge with its waveform distorted if the time delay of different frequencies in going through the filter is not constant. Constant time delay corresponds to a phase shift increasing linearly with frequency; hence the term *linear-phase filter* applied to a filter ideal in this respect. Figure 5.8 shows a typical graph of phase shift and amplitude for a low-pass filter that is definitely not a linear-phase filter. Graphs of phase shift versus frequency are best plotted on a linear-frequency axis.

Time domain

As with any ac circuit, filters can be described in terms of their *time-domain* properties: rise time, overshoot, ringing, and settling time. This is of particular importance where steps or pulses may be used. Figure 5.9 shows a typical low-pass-filter step response. Here, *rise time* is the time required to reach 90% of the final value, whereas *settling time* is the time required to get within some specified amount of the final value and stay there. *Overshoot* and *ringing* are self-explanatory terms for some undesirable properties of filters.

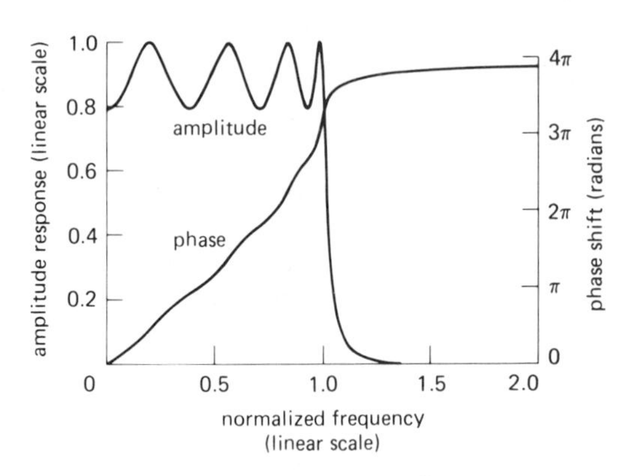

Figure 5.8. Phase and amplitude response for an 8-pole Chebyshev low-pass filter (2dB passband ripple).

5.05 Filter types

Suppose you want a low-pass filter with flat passband and sharp transition to the stopband. The ultimate rate of falloff, well into the stopband, will always be 6ndB/octave, where n is the number of "poles." You need one capacitor (or inductor) for each pole, so the required ultimate rate of falloff of filter response determines, roughly, the complexity of the filter.

Now, assume that you have decided to use a 6-pole low-pass filter. You are guaranteed an ultimate rolloff of 36dB/ octave at high frequencies. It turns out

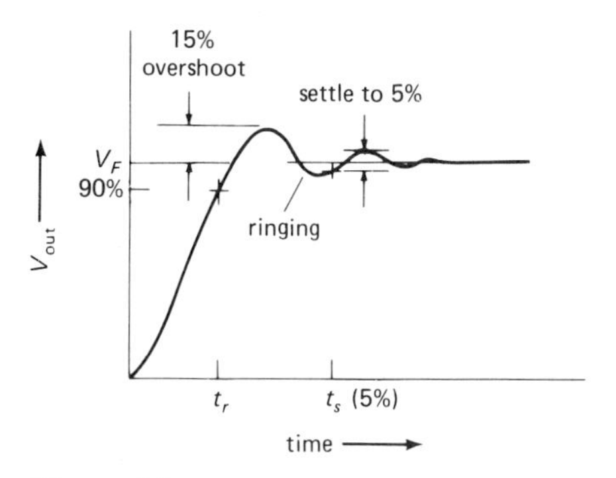

Figure 5.9

that the filter design can now be optimized for maximum flatness of passband response, at the expense of a slow transition from passband to stopband. Alternatively, by allowing some ripple in the passband characteristic, the transition from passband to stopband can be steepened considerably. A third criterion that may be important is the ability of the filter to pass signals within the passband without distortion of their waveforms caused by phase shifts. You may also care about rise time, overshoot, and settling time.

There are filter designs available to optimize each of these characteristics, or combinations of them. In fact, rational filter selection will not be carried out as just described; rather, it normally begins with a set of requirements on passband flatness, attenuation at some frequency outside the passband, and whatever else matters. You will then choose the best design for the job, using the number of poles necessary to meet the requirements. In the next few sections we will introduce the three popular favorites, the Butterworth filter (maximally flat passband), the Chebyshev filter (steepest transition from passband to stopband), and the Bessel filter (maximally flat time delay). Each of these filter responses can be produced with a variety of different filter circuits, some of which we will discuss later. They are all available in low-pass, high-pass, and bandpass versions.

Butterworth and Chebyshev filters

The Butterworth filter produces the flattest passband response, at the expense of steepness in the transition region from passband to stopband. As you will see later, it also has poor phase characteristics. The amplitude response is given by

$$\frac{V_{\text{out}}}{V_{\text{in}}} = \frac{1}{[1 + (f/f_c)^{2n}]^{\frac{1}{2}}}$$

where n is the order of the filter (number of poles). Increasing the number of poles flattens the passband response and steepens the stopband falloff, as shown in Figure 5.10.

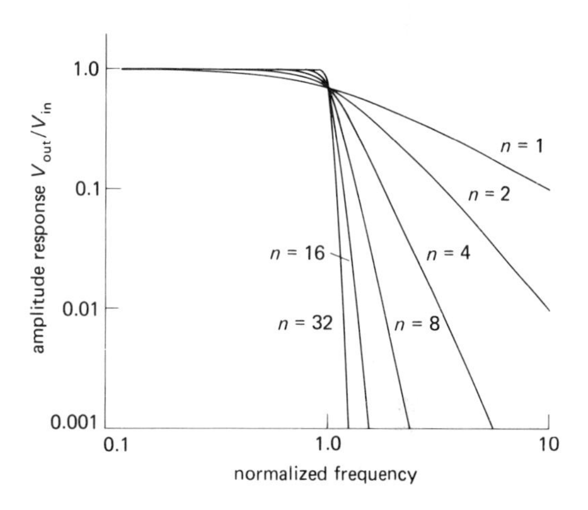

Figure 5.10. Normalized low-pass Butterworth-filter response curves. Note the improved attenuation characteristics for the higher-order filters.

The Butterworth filter trades off everything else for maximum flatness of response. It starts out extremely flat at zero frequency and bends over near the cutoff frequency f_c (f_c is usually the -3dB point).

In most applications, all that really matters is that the wiggles in the passband response be kept less than some amount, say 1dB. The Chebyshev filter responds to this reality by allowing some ripples throughout the passband, with greatly improved sharpness of the knee. A Chebyshev filter is specified in terms of its number of poles and passband ripple. By allowing greater passband ripple, you get a sharper knee. The amplitude is given by

$$\frac{V_{\text{out}}}{V_{\text{in}}} = \frac{1}{[1 + \epsilon^2 C_n^2(f/f_c)]^{\frac{1}{2}}}$$

where C_n is the Chebyshev polynomial of the first kind of degree n, and ϵ is a constant that sets the passband ripple. Like the Butterworth, the Chebyshev has phase characteristics that are less than ideal.

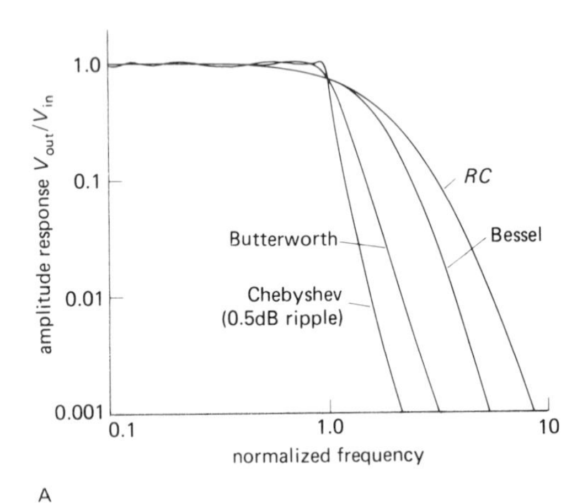

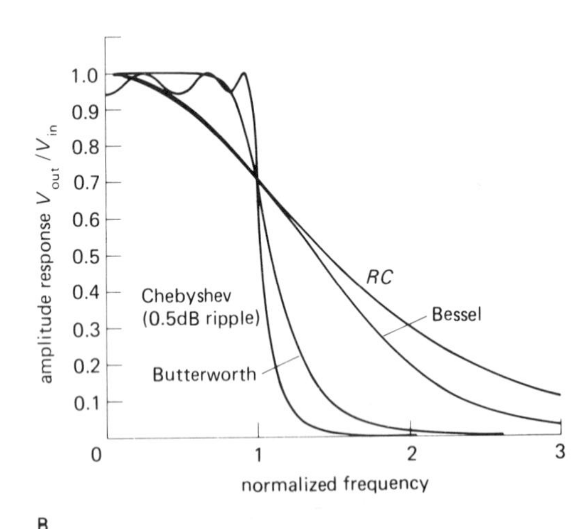

Figure 5.11. Comparison of some common 6-pole low-pass filters. The same filters are plotted on both linear and logarithmic scales.

Figure 5.11 presents graphs comparing the responses of Chebyshev and Butterworth 6-pole low-pass filters. As you can see, they're both tremendous improvements over a 6-pole RC filter.

Actually, the Butterworth, with its maximally flat passband, is not as attractive as it might appear, since you are always accepting some variation in passband response anyway (with the Butterworth it is a gradual rolloff near f_c, whereas with the Chebyshev it is a set of ripples spread throughout the passband). Furthermore, active filters constructed with components of finite tolerance will deviate from the predicted response, which means that a real Butterworth filter will exhibit some passband ripple anyway. The graph in Figure 5.12 illustrates the effects of worst-case variations in resistor and capacitor values on filter response.

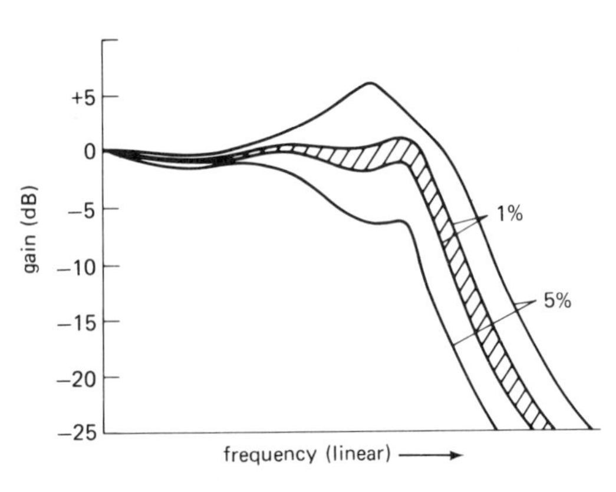

Figure 5.12. The effect of component tolerance on active filter performance.

Viewed in this light, the Chebyshev is a very rational filter design. It is sometimes called an equiripple filter: It manages to improve the situation in the transition region by spreading equal-size ripples throughout the passband, the number of ripples increasing with the order of the filter. Even with rather small ripples (as little as 0.1dB) the Chebyshev filter offers considerably improved sharpness of the knee

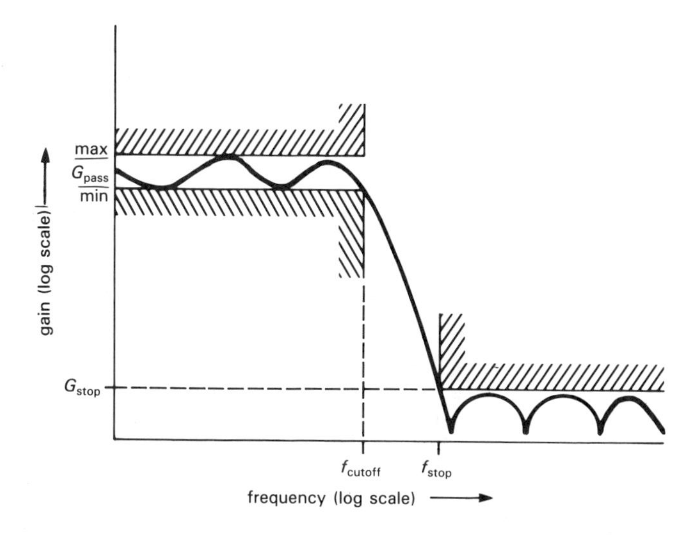

Figure 5.13. Specifying filter frequency response parameters.

as compared with the Butterworth. To make the improvement quantitative, suppose that you need a filter with flatness to 0.1dB within the passband and 20dB attenuation at a frequency 25% beyond the top of the passband. By actual calculation, that will require a 19-pole Butterworth, but only an 8-pole Chebyshev.

The idea of accepting some passband ripple in exchange for improved steepness in the transition region, as in the equiripple Chebyshev filter, is carried to its logical limit in the so-called elliptic (or Cauer) filter by trading ripple in both passband and stopband for an even steeper transition region than that of the Chebyshev filter. With computer-aided design techniques, the design of elliptic filters is as straightforward as for the classic Butterworth and Chebyshev filters.

Figure 5.13 shows how you specify filter frequency response graphically. In this case (a low-pass filter) you indicate the allowable range of filter gain (i.e., the ripple) in the passband, the minimum frequency at which the response leaves the passband, the maximum frequency at which the response enters the stopband, and the minimum attenuation in the stopband.

Bessel filter

As we hinted earlier, the amplitude response of a filter does not tell the whole story. A filter characterized by a flat amplitude response may have large phase shifts. The result is that a signal in the passband will suffer distortion of its waveform. In situations where the shape of the waveform is paramount, a linear-phase filter (or constant-time-delay filter) is desirable. A filter whose phase shift varies linearly with frequency is equivalent to a constant time delay for signals within the passband, i.e., the waveform is not distorted. The Bessel filter (also called the Thomson filter) had maximally flat time delay within its passband, in analogy with the Butterworth, which has maximally flat amplitude response. To see the kind of improvement in time-domain performance you get with the Bessel filter, look at Figure 5.14 for a comparison of time delay versus normalized frequency for 6-pole Bessel and Butterworth low-pass filters. The poor time-delay performance of the Butterworth gives rise to effects such as overshoot when driven with pulse signals. On the other hand, the price you pay for the Bessel's constancy of time delay is an amplitude response with even less steepness than that of the Butterworth in the transition region between passband and stopband.

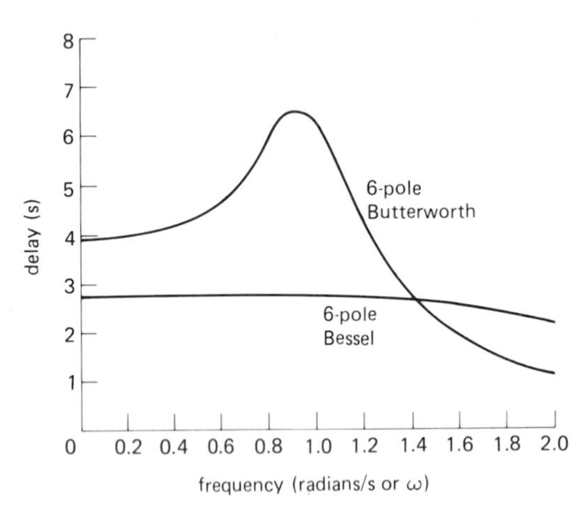

Figure 5.14. Comparison of time delays for 6-pole Bessel and Butterworth low-pass filters. The excellent time-domain performance of the Bessel filter minimizes waveform distortion.

There are numerous filter designs that attempt to improve on the Bessel's good time-domain performance by compromising some of the constancy of time delay for improved rise time and amplitude-versus-frequency characteristics. The Gaussian filter has phase characteristics nearly as good as those of the Bessel, with improved step response. In another class there are interesting filters that allow uniform ripples in the passband time delay (in analogy with the Chebyshev's ripples in its amplitude response) and yield approximately constant time delays even for signals well into the stopband. Another approach to the problem of getting filters with uniform time delays is to use all-pass filters, also known as delay equalizers. These have constant amplitude response with frequency, with a phase shift that can be tailored to individual requirements. Thus, they can be used to improve the time-delay constancy of any filter, including Butterworth and Chebyshev types.

Filter comparison

In spite of the preceding comments about the Bessel filter's transient response, it still has vastly superior properties in the time domain, as compared with the Butterworth and Chebyshev. The Chebyshev, with its highly desirable amplitude-versus-frequency characteristics, actually has the poorest time-domain performance of the three. The Butterworth is in between in both frequency and time-domain properties. Table 5.1 and Figure 5.15 give more information about time-domain performance for these three kinds of filters to complement the frequency-domain graphs presented earlier. They make it clear that the Bessel is a very desirable filter where performance in the time domain is important.

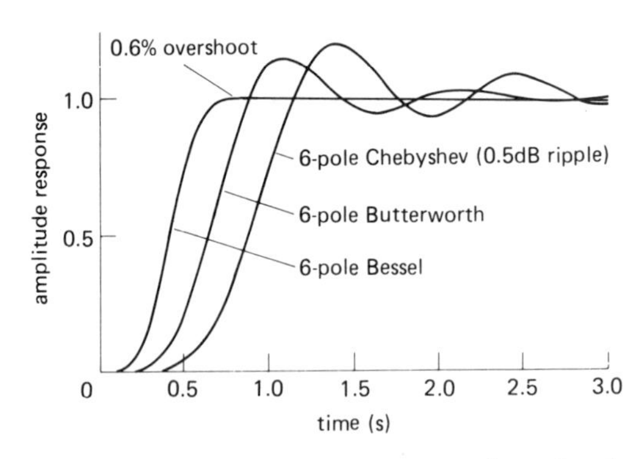

Figure 5.15. Step-response comparison for 6-pole low-pass filters normalized for 3dB attenuation at 1Hz.

ACTIVE FILTER CIRCUITS

A lot of ingenuity has been used in inventing clever active circuits, each of which can be used to generate response functions such as the Butterworth, Chebyshev, etc. You might wonder why the world needs more than one active filter circuit. The reason is that various circuit realizations excel in one or another desirable property, so there is no all-around best circuit.

Some of the features to look for in active filters are (a) small numbers of parts, both

TABLE 5.1. TIME-DOMAIN PERFORMANCE COMPARISON FOR LOW-PASS FILTERS[a]

Type	f_{3dB} (Hz)	Poles	Step rise time (0 to 90%) (s)	Over-shoot (%)	Settling time to 1% (s)	Settling time to 0.1% (s)	Stopband attenuation f = 2f_c (dB)	Stopband attenuation f = 10f_c (dB)
Bessel	1.0	2	0.4	0.4	0.6	1.1	10	36
(-3.0dB at	1.0	4	0.5	0.8	0.7	1.2	13	66
f_c = 1.0Hz)	1.0	6	0.6	0.6	0.7	1.2	14	92
	1.0	8	0.7	0.3	0.8	1.2	14	114
Butterworth	1.0	2	0.4	4	0.8	1.7	12	40
(-3.0dB at	1.0	4	0.6	11	1.0	2.8	24	80
f_c = 1.0Hz)	1.0	6	0.9	14	1.3	3.9	36	120
	1.0	8	1.1	16	1.6	5.1	48	160
Chebyshev	1.39	2	0.4	11	1.1	1.6	8	37
0.5dB ripple	1.09	4	0.7	18	3.0	5.4	31	89
(-0.5dB at	1.04	6	1.1	21	5.9	10.4	54	141
f_c = 1.0Hz)	1.02	8	1.4	23	8.4	16.4	76	193
Chebyshev	1.07	2	0.4	21	1.6	2.7	15	44
2.0dB ripple	1.02	4	0.7	28	4.8	8.4	37	96
(-2.0dB at	1.01	6	1.1	32	8.2	16.3	60	148
f_c = 1.0Hz)	1.01	8	1.4	34	11.6	24.8	83	200

(a) a design procedure for these filters is presented in Section 5.07.

active and passive, (b) ease of adjustability, (c) small spread of parts values, especially the capacitor values, (d) undemanding use of the op-amp, especially requirements on slew rate, bandwidth, and output impedance, (e) the ability to make high-Q filters, and (f) sensitivity of filter characteristics to component values and op-amp gain (in particular, the gain-bandwidth product, f_T). In many ways the last feature is one of the most important. A filter that requires parts of high precision is difficult to adjust, and it will drift as the components age; in addition, there is the nuisance that it requires components of good initial accuracy. The VCVS circuit probably owes most of its popularity to its simplicity and its low parts count, but it suffers from high sensitivity to component variations. By comparison, recent interest in more complicated filter realizations is motivated by the benefits of insensitivity of filter properties to small component variability.

In this section we will present several circuits for low-pass, high-pass, and bandpass active filters. We will begin with the popular VCVS, or controlled-source type, then show the state-variable designs available as integrated circuits from several manufacturers, and finally mention the twin-T sharp rejection filter and some interesting new directions in switched-capacitor realizations.

5.06 VCVS circuits

The voltage-controlled voltage-source (VCVS) filter, also known simply as a controlled-source filter, is a variation of the Sallen-and-Key circuit shown earlier. It replaces the unity-gain follower with a noninverting amplifier of gain greater than 1. Figure 5.16 shows the circuits for low-pass, high-pass, and bandpass realizations. The resistors at the outputs of the op-amps create a noninverting voltage amplifier

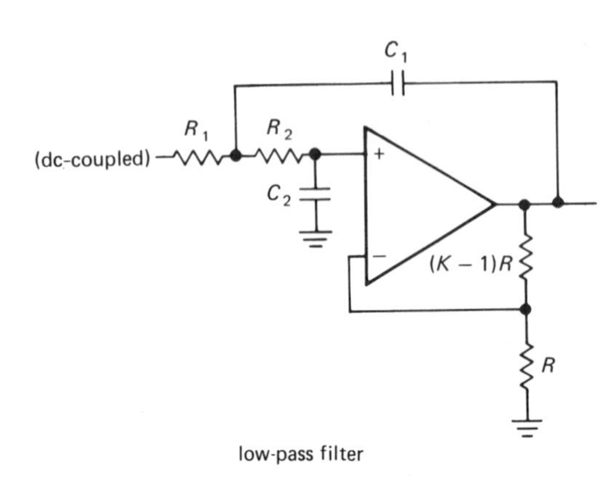

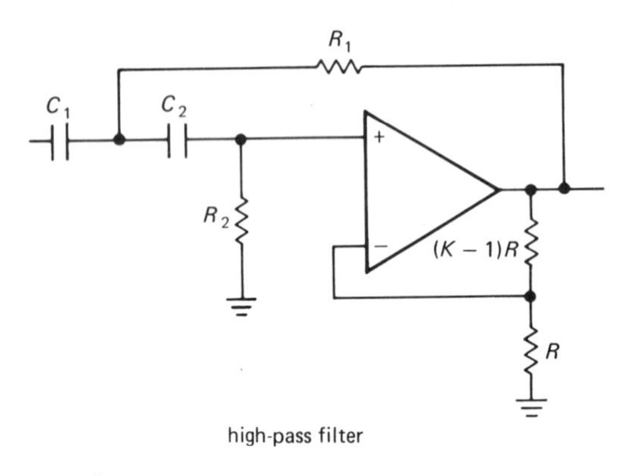

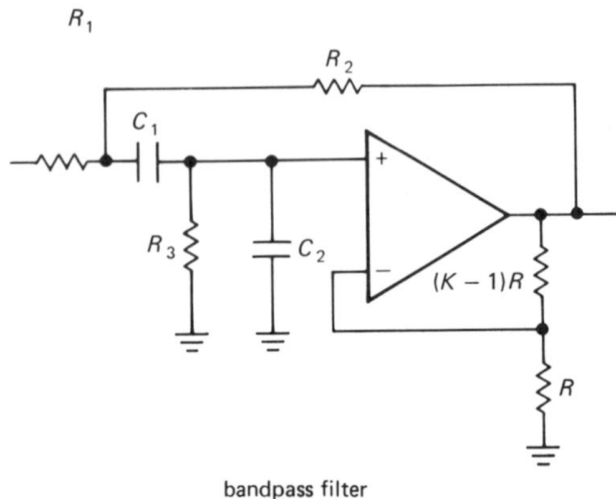

Figure 5.16. VCVS active filter circuits.

of voltage gain K, with the remaining Rs and Cs contributing the frequency response properties for the filter. These are 2-pole filters, and they can be Butterworth, Bessel, etc., by suitable choice of component values, as we will show later. Any number of VCVS 2-pole sections may be cascaded to generate higher-order filters. When that is done, the individual filter sections are, in general, not identical. In fact, each section represents a quadratic polynomial factor of the nth-order polynomial describing the overall filter.

There are design equations and tables in most standard filter handbooks for all the standard filter responses, usually including separate tables for each of a number of ripple amplitudes for Chebyshev filters. In the next section we will present an easy-to-use design table for VCVS filters of Butterworth, Bessel, and Chebyshev responses (0.5dB and 2dB passband ripple for Chebyshev filters) for use as low-pass or high-pass filters. Bandpass and band-reject filters can be easily made from combinations of these.

5.07 VCVS filter design using our simplified table

To use Table 5.2, begin by deciding which filter response you need. As we mentioned earlier, the Butterworth may be attractive if maximum flatness of passband is desired, the Chebyshev gives the fastest rolloff from passband to stopband (at the

TABLE 5.2. VCVS LOW-PASS FILTERS

Poles	Butter-worth K	Bessel f_n	Bessel K	Chebyshev (0.5dB) f_n	Chebyshev (0.5dB) K	Chebyshev (2.0dB) f_n	Chebyshev (2.0dB) K
2	1.586	1.272	1.268	1.231	1.842	0.907	2.114
4	1.152	1.432	1.084	0.597	1.582	0.471	1.924
	2.235	1.606	1.759	1.031	2.660	0.964	2.782
6	1.068	1.607	1.040	0.396	1.537	0.316	1.891
	1.586	1.692	1.364	0.768	2.448	0.730	2.648
	2.483	1.908	2.023	1.011	2.846	0.983	2.904
8	1.038	1.781	1.024	0.297	1.522	0.238	1.879
	1.337	1.835	1.213	0.599	2.379	0.572	2.605
	1.889	1.956	1.593	0.861	2.711	0.842	2.821
	2.610	2.192	2.184	1.006	2.913	0.990	2.946

expense of some ripple in the passband), and the Bessel provides the best phase characteristics, i.e., constant signal delay in the passband, with correspondingly good step response. The frequency responses for all types are shown in the accompanying graphs (Fig. 5.17).

To construct an n-pole filter (n is an even number), you will need to cascade $n/2$ VCVS sections. Only even-order filters are shown, since an odd-order filter requires as many op-amps as the next higher-order filter. Within each section, $R_1 = R_2 = R$, and $C_1 = C_2 = C$. As is usual in op-amp circuits, R will typically be chosen in the range 10k to 100k. (It is best to avoid small resistor values, because the rising open-loop output impedance of the op-amp at high frequencies adds to the resistor values and upsets calculations.) Then all you need to do is set the gain, K, of each stage according to the table entries. For an n-pole filter there are $n/2$ entries, one for each section.

Butterworth low-pass filters

If the filter is a Butterworth, all sections have the same values of R and C, given simply by $RC = 1/2\pi f_c$, where f_c is the desired -3dB frequency of the entire filter. To make a 6-pole low-pass Butterworth filter, for example, you cascade three of the low-pass sections shown previously, with gains of 1.07, 1.59, and 2.48 (preferably in that order, to avoid dynamic range problems), and with identical Rs and Cs to set the 3dB point. The telescope drive circuit in Section 8.31 shows such an example, with $f_c = 88.4$Hz ($R = 180$k, $C = 0.01\mu$F).

Bessel and Chebyshev low-pass filters

To make a Bessel or Chebyshev filter with the VCVS, the situation is only slightly more complicated. Again we cascade several 2-pole VCVS filters, with prescribed gains for each section. Within each section we again use $R_1 = R_2 = R$, and $C_1 = C_2 = C$. However, unlike the situation with the Butterworth, the RC products for the different sections are different and must be scaled by the normalizing factor f_n (given for each section in Table 5.2) according to $RC = 1/2\pi f_n f_c$. Here f_c is again the -3dB point for the Bessel filter, whereas for the Chebyshev filter it defines the end of the passband, i.e., it is the frequency at which the amplitude response falls out of the ripple band on its way into the stopband. For example, the response of a Chebyshev low-pass filter with 0.5dB ripple and $f_c = 100$Hz will be flat within $+0$dB to -0.5dB from dc to 100Hz, with 0.5dB attenuation at 100Hz and a rapid falloff for frequencies greater than 100Hz. Values are given for Chebyshev filters with 0.5dB and 2.0dB passband ripple; the latter have a somewhat steeper transition into the stopband (Fig. 5.17).

High-pass filters

To make a high-pass filter, use the high-pass configuration shown previously, i.e., with the Rs and Cs interchanged. For Butterworth filters, everything else remains unchanged (use the same values for R, C, and K). For the Bessel and Chebyshev filters, the K values remain the same, but the normalizing factors f_n must be inverted, i.e., for each section the new f_n equals $1/(f_n$ listed in Table 5.2).

A bandpass filter can be made by cascading overlapping low-pass and high-pass filters. A band-reject filter can be made by summing the outputs of nonoverlapping low-pass and high-pass filters. However, such cascaded filters won't work well for high-Q filters (extremely sharp bandpass filters) because there is great sensitivity to the component values in the individual (uncoupled) filter sections. In such cases a high-Q single-stage bandpass circuit (e.g., the VCVS bandpass circuit

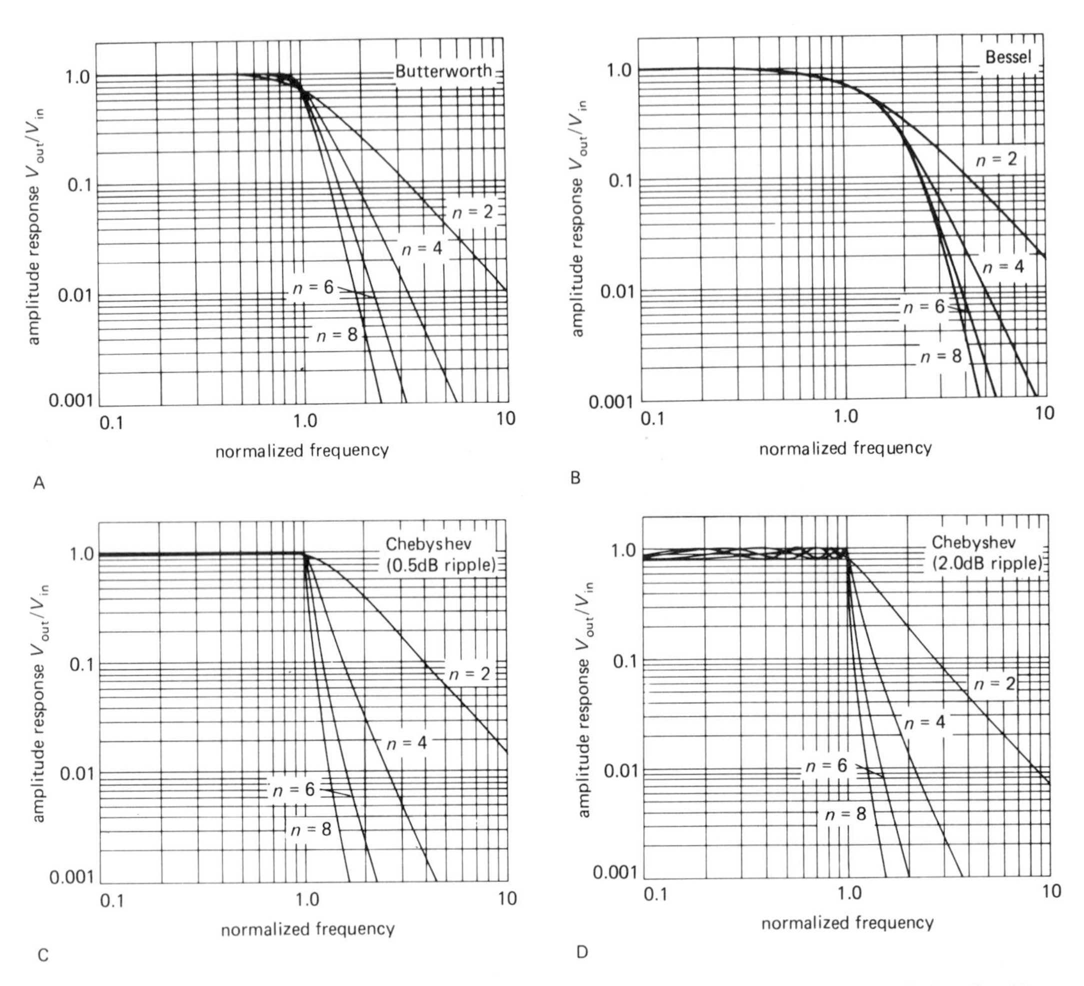

Figure 5.17. Normalized frequency response graphs for the 2-, 4-, 6-, and 8-pole filters in Table 5.2. The Butterworth and Bessel filters are normalized to 3dB attenuation at unit frequency, whereas the Chebyshev filters are normalized to 0.5dB and 2dB attenuations.

illustrated previously, or the state-variable and biquad filters in the next section) should be used instead. Even a single-stage 2-pole filter can produce a response with an extremely sharp peak. Information on such filter design is available in the standard references.

VCVS filters minimize the number of components needed (2 poles/op-amp) and offer the additional advantages of noninverting gain, low output impedance, small spread of component values, easy adjustability of gain, and the ability to operate at high gain or high Q. They suffer from high sensitivity to component values and amplifier gain, and they don't lend themselves well to applications where a tunable filter of stable characteristics is needed.

EXERCISE 5.3

Design a 6-pole Chebyshev low-pass VCVS filter with a 0.5dB passband ripple and 100Hz cutoff frequency f_C. What is the attenuation at $1.5f_C$?

5.08 State-variable filters

The 2-pole filter shown in Figure 5.18 is far more complex than the VCVS circuits,

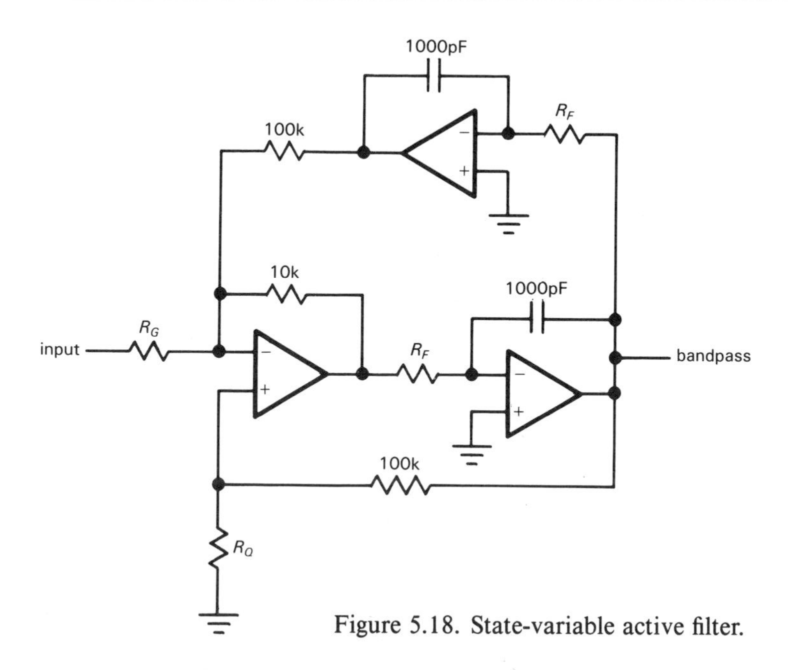

Figure 5.18. State-variable active filter.

but it is popular because of its improved stability and ease of adjustment. It is called a state-variable filter and is available as an IC from National (the AF100 and AF150), Burr-Brown (the UAF series), and others. Because it is a manufactured module, all components except R_G, R_Q, and the two R_Fs are built in. Among its nice properties is the availability of high-pass, low-pass, and bandpass outputs from the same circuit; also, its frequency can be tuned while maintaining constant Q (or, alternatively, constant bandwidth) in the bandpass characteristic. As with the VCVS realizations, multiple stages can be cascaded to generate higher-order filters.

Extensive design formulas and tables are provided by the manufacturers for the use of these convenient ICs. They show how to choose the external resistor values to make Butterworth, Bessel, and Chebyshev filters for a wide range of filter orders, for low-pass, high-pass, bandpass, and band-reject responses. Among the nice features of these hybrid ICs is integration of the capacitors into the module, so that only external resistors need be added.

Bandpass filters

The state-variable circuit, in spite of its large number of components, is a good choice for sharp (high-Q) bandpass filters. It has low component sensitivities, does not make great demands on op-amp bandwidth, and is easy to tune. For example, in the circuit of Figure 5.18, used as a bandpass filter, the two resistors R_F set the center frequency, while R_Q and R_G together determine the Q and band-center gain:

$$R_F = 5.03 \times 10^7/f_0 \text{ ohms}$$

$$R_Q = 10^5/(3.48Q + G - 1) \text{ ohms}$$

$$R_G = 3.16 \times 10^4 Q/G \text{ ohms}$$

So you could make a tunable-frequency, constant-Q filter by using a 2-section variable resistor (pot) for R_F. Alternatively, you could make R_Q adjustable, producing a fixed-frequency, variable-Q (and, unfortunately, variable-gain) filter.

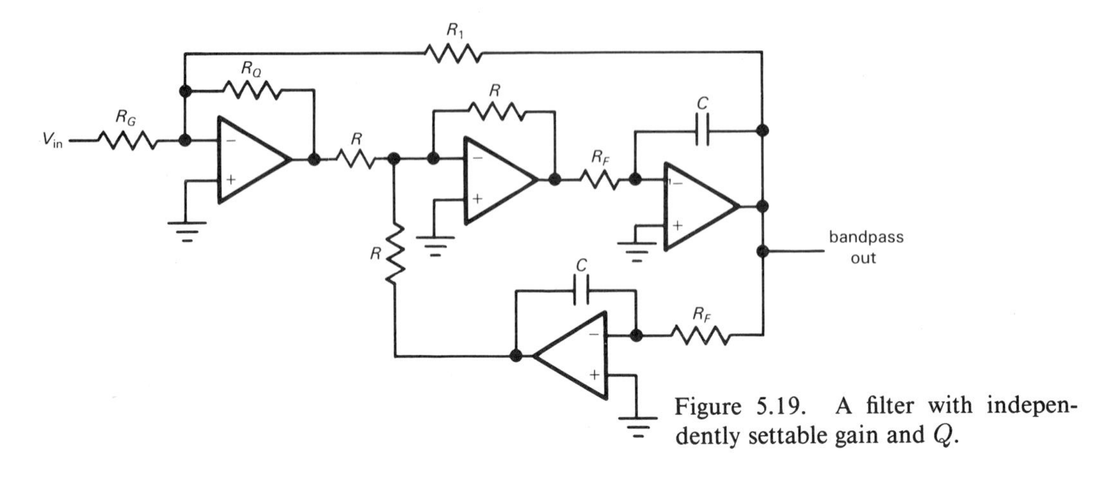

Figure 5.19. A filter with independently settable gain and Q.

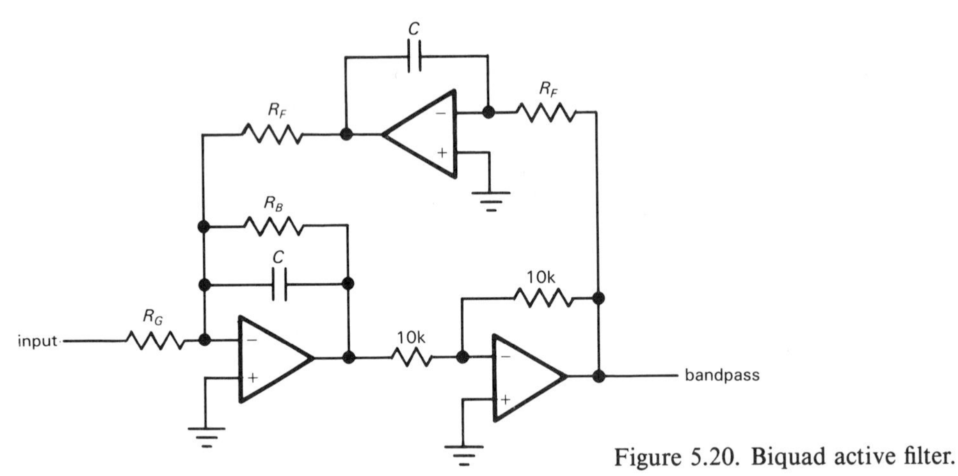

Figure 5.20. Biquad active filter.

EXERCISE 5.4

Calculate resistor values in Figure 5.18 to make a bandpass filter with $f_0 = 1\text{kHz}$, $Q = 50$, and $G = 10$.

Figure 5.19 shows a useful variant of the state-variable bandpass filter. The bad news is that it uses four op-amps; the good news is that you can adjust the bandwidth (i.e., Q) without affecting the midband gain. In fact, both Q and gain are set with a single resistor each. Q, gain, and center frequency are completely independent and are given by these simple equations:

$f_0 = 1/2\pi R_F C$
$Q = R_1/R_Q$
$G = R_1/R_G$
$R \approx 10\text{k}$ (noncritical, matched)

Biquad filter. A close relative of the state variable filter is the so-called biquad filter, shown in Figure 5.20. This circuit also uses three op-amps and can be constructed from the state-variable ICs mentioned earlier. It has the interesting property that you can tune its frequency (via R_F) while maintaining constant *bandwidth* (rather than constant Q). Here are the design equations:

$f_0 = 1/2\pi R_F C$
$\text{BW} = 1/2\pi R_B C$
$G = R_B/R_G$

The Q is given by f_0/BW and equals R_B/R_F. As the center frequency is varied (via R_F), the Q varies proportionately, keeping the bandwidth Qf_0 constant.

When you design a biquad filter from scratch (rather than with an active filter IC that already contains most of the parts), the general procedure goes something like this:

1. Choose an op-amp whose bandwidth f_T is at least 10 to 20 times Gf_0.

2. Pick a round-number capacitor value in the vicinity of

$$C = 10/f_0 \ \mu F$$

3. Use the desired center frequency to calculate the corresponding R_F from the first equation given earlier.

4. Use the desired bandwidth to calculate R_B from the second equation given earlier.

5. Use the desired band-center gain to calculate R_G from the third equation given earlier.

You may have to adjust the capacitor value if the resistor values become awkwardly large or small. For instance, in a high-Q filter you may need to increase C somewhat to keep R_B from becoming too large (or you can use the T-network trick described in Section 4.19). Note that R_F, R_B, and R_G each act as op-amp loads, and should not become less than, say, 5k. When juggling component values, you may find it easier to satisfy requirement 1 by decreasing integrator gain (increase R_F) and simultaneously increasing the inverter-stage gain (increase the 10k feedback resistor).

As an example, suppose we want to make a filter with the same characteristics as in the last exercise. We would begin by provisionally choosing $C = 0.01\mu$F. Then we find $R_F = 15.9$k ($f_0 = 1$kHz) and $R_B = 796$k ($Q = 50$; BW=20Hz). Finally, $R_G = 79.6$k ($G = 10$).

EXERCISE 5.5

Design a biquad bandpass filter with $f_0 = 60$Hz, BW=1Hz, and $G = 100$.

Higher order bandpass filters

As with our earlier low-pass and high-pass filters, it is possible to build higher order bandpass filters with approximately flat bandpass and steep transition to the stopband.

You do this by cascading several lower-order bandpass filters, the combination tailored to realize the desired filter type (Butterworth, Chebyshev, or whatever). As before, the Butterworth is "maximally flat," whereas the Chebyshev sacrifices passband flatness for steepness of skirts. Both the VCVS and state-variable/biquad bandpass filters just considered are second order (two pole). As you increase the filter sharpness by adding sections, you generally degrade the transient response and phase characteristics. The "bandwidth" of a bandpass filter is defined as the width between -3dB points, except for equiripple filters, for which it is the width between frequencies at which the response falls out of the passband ripple channel.

You can find tables and design procedures for constructing complex filters in standard books on active filters, or in the data sheets for active filter ICs. There are also some very nice filter design programs that run on inexpensive workstations (IBM PC, Macintosh).

□ 5.09 Twin-T notch filters

The passive RC network shown in Figure 5.21 has infinite attenuation at a frequency $f_c = 1/2\pi RC$. Infinite attenuation is

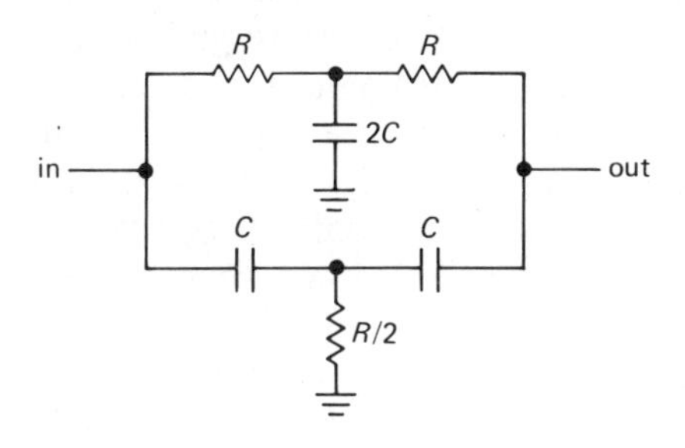

Figure 5.21. Passive twin-T notch filter.

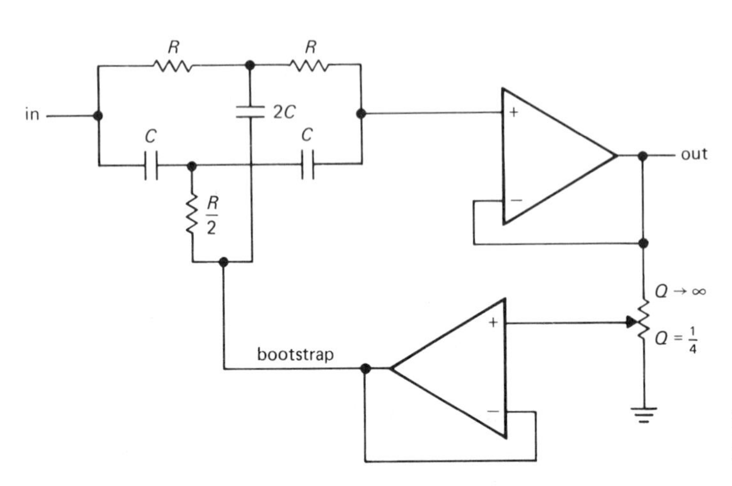

Figure 5.22. Bootstrapped twin-T.

uncharacteristic of *RC* filters in general; this one works by effectively adding two signals that have been shifted 180° out of phase at the cutoff frequency. It requires good matching of components in order to obtain a good null at f_c. It is called a twin-T, and it can be used to remove an interfering signal, such as 60Hz powerline pickup. The problem is that it has the same "soft" cutoff characteristics as all passive *RC* networks, except, of course, near f_c, where its response drops like a rock. For example, a twin-T driven by a perfect voltage source is down 10dB at twice (or half) the notch frequency and 3dB at four times (or one-fourth) the notch frequency. One trick to improve its notch characteristic is to "activate" it in the manner of a Sallen-and-Key filter (Fig. 5.22). This technique looks good in principle, but it is generally disappointing in practice, owing to the impossibility of maintaining a good filter null. As the filter notch becomes sharper (more gain in the bootstrap), its null becomes less deep.

Twin-T filters are available as prefab modules, going from 1Hz to 50kHz, with notch depths of about 60dB (with some deterioration at high and low temperatures). They are easy to make from components, but resistors and capacitors of good stability and low temperature coefficient should be used to get a deep and stable notch. One of the components should be made trimmable.

The twin-T filter works fine as a fixed-frequency notch, but it is a horror to make tunable, since three resistors must be simultaneously adjusted while maintaining constant ratio. However, the remarkably simple *RC* circuit of Figure 5.23A, which behaves just like the twin-T, can be adjusted over a significant range of frequency (at least two octaves) with a single potentiometer. Like the twin-T (and most active filters) it requires some matching of components; in this case the three capacitors must be identical, and the fixed resistor must be exactly six times the bottom (adjustable) resistor. The notch frequency is then given by

$$f_{\text{notch}} = 1/2\pi C\sqrt{3R_1R_2}$$

Figure 5.23B shows an implementation that is tunable from 25Hz to 100Hz. The 50k trimmer is adjusted (once) for maximum depth of notch.

As with the passive twin-T, this filter (known as a *bridged differentiator*) has a gently sloping attenuation away from the notch and infinite attenuation (assuming perfect matching of component values) at the notch frequency. It, too, can be "activated," by bootstrapping the wiper of the pot with a voltage gain somewhat less than unity (as in Fig. 5.22). Increasing

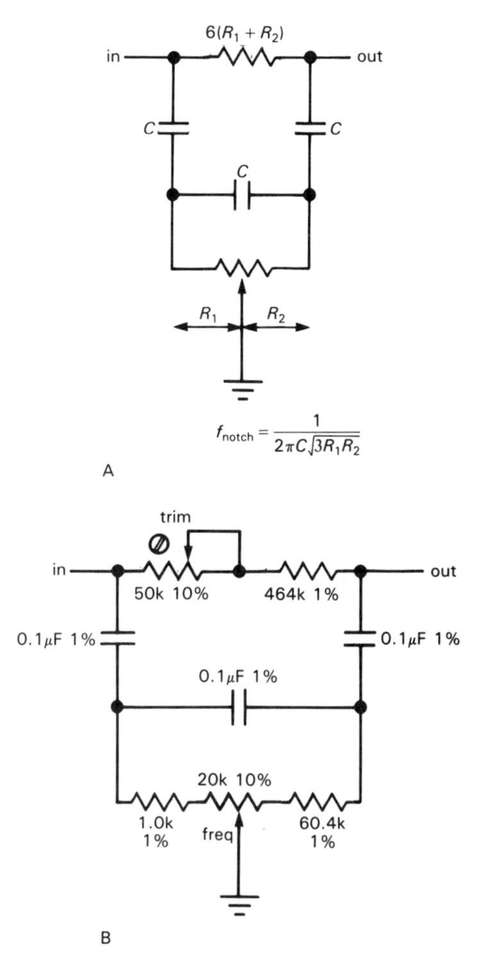

Figure 5.23. Bridged differentiator tunable-notch filter. The implementation in B tunes from 25Hz to 100Hz.

the bootstrap gain toward unity narrows the notch, but also leads to an undesirable response peak on the high frequency side of the notch, along with a reduction in ultimate attenuation.

□ 5.10 Gyrator filter realizations

An interesting type of active filter is made with gyrators; basically they are used to substitute for inductors in traditional filter designs. The gyrator circuit shown in Figure 5.24 is popular. Z_4 will ordinarily be a capacitor, with the other impedances

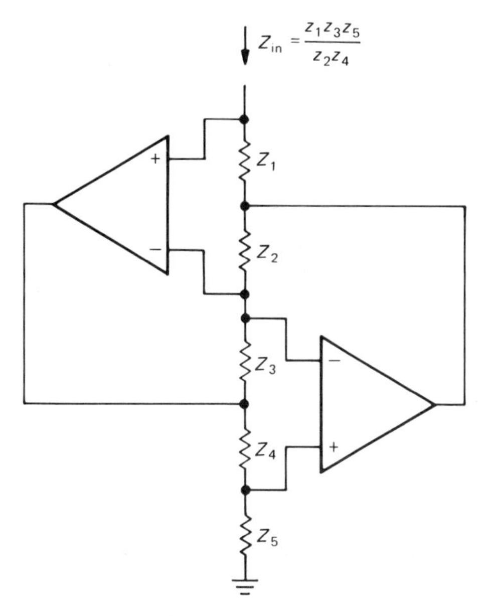

Figure 5.24. Gyrator.

being replaced by resistors, creating an inductor $L = kC$, where $k = R_1R_3R_5/R_2$. It is claimed that these gyrator-substituted filters have the lowest sensitivity to component variations, exactly analogous to their passive *RLC* prototypes.

5.11 Switched capacitor filters

One drawback to these state-variable or biquad filters is the need for accurately matched capacitors. If you build the circuit from op-amps, you've got to get pairs of stable capacitors (not ceramic or electrolytic), perhaps matched as closely as 2% for optimum performance. You also have to make a lot of connections, since the circuits use at least three op-amps and six resistors for each 2-pole section. Alternatively, you can buy a filter IC, letting the manufacturer figure out how to integrate matched 1000pF capacitors into his IC. IC manufacturers have solved those problems, but at a price: The AF100 "Universal Active Filter" IC from National is a hybrid IC and costs about $10 apiece.

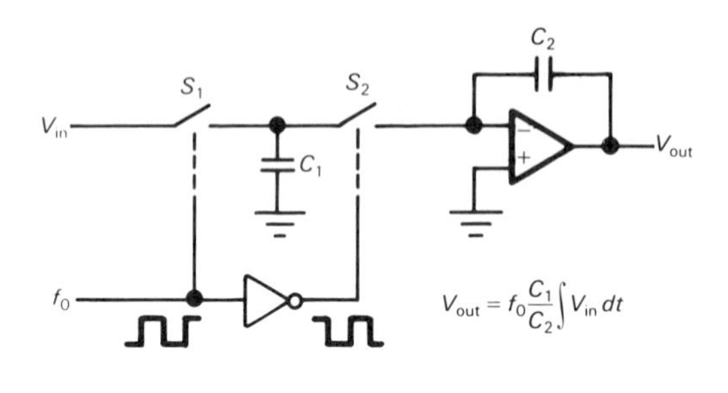

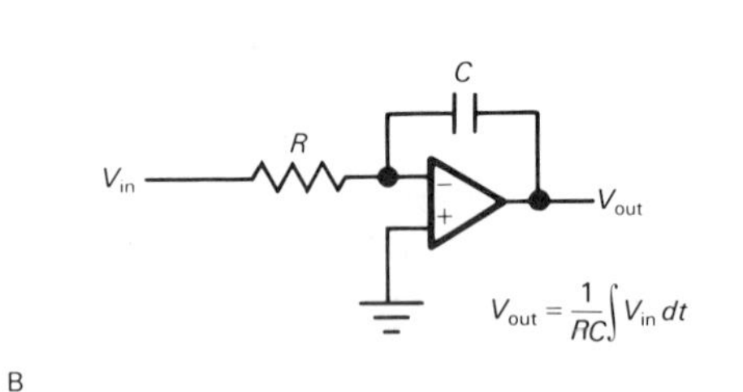

Figure 5.25. A. Switched-capacitor integrator B. conventional integrator.

There's another way to implement the integrators that are needed in the state-variable or biquad filter. The basic idea is to use MOS analog switches, clocked from an externally applied square wave at some high frequency (typically 100 times faster than the analog signals of interest), as shown in Figure 5.25. In the figure, the funny triangular object is a digital *inverter*, which turns the square wave upside down so that the two MOS switches are closed on opposite halves of the square wave. The circuit is easy to analyze: When S_1 is closed, C_1 charges to V_{in}, i.e., holding charge C_1V_{in}; on the alternate half of the cycle, C_1 discharges into the virtual ground, transferring its charge to C_2. The voltage across C_2 therefore changes by an amount $\Delta V = \Delta Q/C_2 = V_{in}C_1/C_2$. Note that the output voltage *change* during each cycle of the fast square wave is proportional to V_{in} (which we assume changes only a small amount during one cycle of square wave), i.e., the circuit is an integrator! It is easy to show that the integrators obey the equations in the figure.

EXERCISE 5.6
Derive the equations in Figure 5.25

There are two important advantages to using switched capacitors instead of conventional integrators. First, as hinted earlier, it can be less expensive to implement on silicon: The integrator gain depends only on the *ratio* of two capacitors, not on their individual values. In general it is easy to make a matched pair of anything on silicon, but very hard to make a similar component (resistor or capacitor) of precise value and high stability. As a result, monolithic switched-capacitor filter ICs are very inexpensive – National's universal switched-capacitor filter (the MF10) costs \$2 (compared with \$10 for the conventional AF100) and furthermore gives you *two* filters in one package!

The second advantage of switched-capacitor filters is the ability to tune the filter's frequency (e.g., the center frequency of a bandpass filter, or the −3dB point of a low-pass filter) by merely changing the frequency of the square wave ("clock") input. This is because the characteristic frequency of a state-variable or biquad filter is proportional to (and depends only on) the integrator gain.

Switched-capacitor filters are available in both dedicated and "universal" configurations. The former are prewired with on-chip components to form bandpass or low-pass filters, while the latter have various intermediate inputs and outputs brought out so you can connect external components to make anything you want. The price you pay for universality is a larger IC package and the need for external resistors. For example, National's self-contained MF4 Butterworth low-pass filter comes in an 8-pin DIP (\$1.30), while their MF5 universal filter comes in a 14-pin DIP (\$1.45), requiring 2 or 3 external resistors (depending on which filter configuration you choose). Figure 5.26 shows just how easy it is to use the dedicated type.

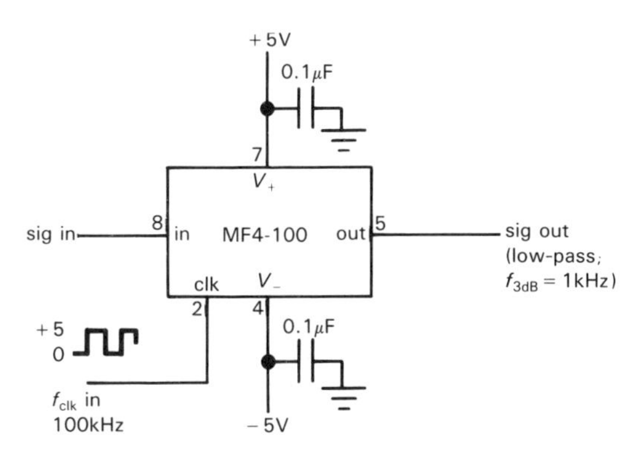

Figure 5.26

Now for the bombshell: Switched-capacitor filters have three annoying characteristics, all related and caused by the presence of the periodic clocking signal. First, there is *clock feedthrough*, the presence of some output signal (typically about 10mV to 25mV) at the clock frequency, independent of the input signal. Usually this doesn't matter, because it is far removed from the signal band of interest. If clock feedthrough is a problem, a simple *RC* filter usually gets rid of it. The second problem is more subtle: If the input signal has any frequency components near the clock frequency, they will be "aliased" down into the passband. To state it precisely, any input signal energy at a frequency that differs from the clock frequency by an amount corresponding to a frequency in the passband will appear (unattenuated!) in the passband. For example, if you use an MF4 as a 1kHz low-pass filter (i.e., set f_{clock} = 100kHz), any input signal energy in the range of 99kHz–101kHz will appear in the output band of dc–1kHz. No filter at the output can remove it! You must make sure the input signal doesn't have energy near the clock frequency. If this isn't naturally the case, you can usually use a simple *RC* filter, since the clock frequency is typically quite far removed from the passband. The third undesirable effect in switched-capacitor filters is a general reduction in signal dynamic range (an increase in the "noise floor") due to incomplete cancellation of MOS switch charge injection (see Section 3.12). Typical filter ICs have dynamic ranges of 80dB–90dB.

Like any linear circuit, switched-capacitor filters (and their op-amp analogs) suffer from amplifier errors such as input offset voltage and $1/f$ low-frequency noise. These can be a problem if, for example, you wish to low-pass filter some low-level signal without introducing errors or fluctuations in its average dc value. A nice solution is provided by the clever folks at Linear Technology, who dreamed up the LTC1062 "DC Accurate Low-Pass Filter" (or the MAX280, with improved offset voltage). Figure 5.27 shows how you use it. The basic idea is to put the filter outside the dc path, letting the low-frequency signal components couple passively to the output; the filter grabs onto the signal line only at higher frequencies, where it rolls off the response by shunting the signal to

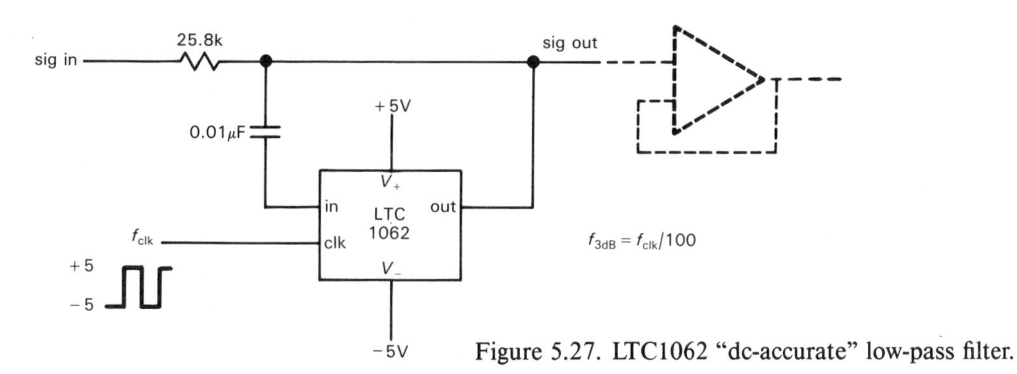

Figure 5.27. LTC1062 "dc-accurate" low-pass filter.

ground. The result is zero dc error, and switched-capacitor-type noise only in the vicinity of the rolloff (Fig. 5.28).

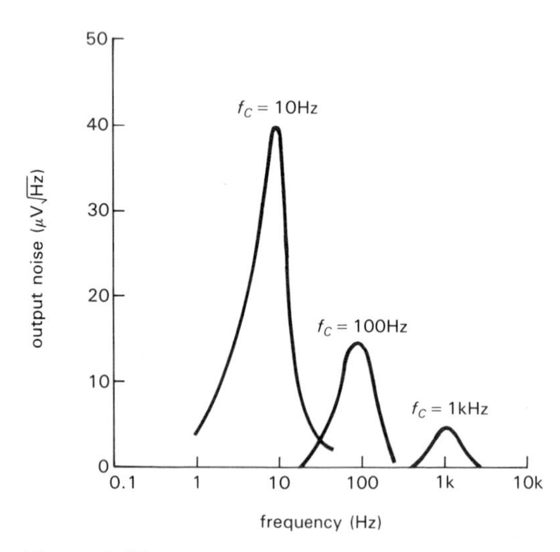

Figure 5.28

Switched-capacitor filter ICs are widely available, from manufacturers such as AMI-Gould, Exar, LTC, National, and EGG-Reticon. Typically you can put the cutoff (or band center) anywhere in the range of dc to a few tens of kilohertz, as set by the clock frequency. The characteristic frequency is a fixed multiple of the clock, usually $50f_{clk}$ or $100f_{clk}$. Most switched-capacitor filter ICs are intended for lowpass, bandpass, or notch (band-stop) use, though a few (e.g., the AMI 3529) are designed as high-pass filters. Note that clock feedthrough and discrete (clock frequency) output waveform quantization effects are particularly bothersome in the latter case, since they're both in-band.

OSCILLATORS

5.12 Introduction to oscillators

Within nearly every electronic instrument it is essential to have an oscillator or waveform generator of some sort. Apart from the obvious case of signal generators, function generators, and pulse generators themselves, a source of regular oscillations is necessary in any cyclical measuring instrument, in any instrument that initiates measurements or processes, and in any instrument whose function involves periodic states or periodic waveforms. That includes just about everything. For example, oscillators or waveform generators are used in digital multimeters, oscilloscopes, radiofrequency receivers, computers, every computer peripheral (tape, disk, printer, alphanumeric terminal), nearly every digital instrument (counters, timers, calculators, and anything with a "multiplexed display"), and a host of other devices too numerous to mention. A device without an oscillator either doesn't do anything or expects to be driven by something else (which probably contains an oscillator). It is not an exaggeration to say that an oscillator of some sort is as essential an ingredient in electronics as a regulated supply of dc power.

Depending on the application, an oscillator may be used simply as a source of regularly spaced pulses (e.g., a "clock" for a digital system), or demands may be made on its stability and accuracy (e.g., the time base for a frequency counter), its adjustability (e.g., the local oscillator in a transmitter or receiver), or its ability to produce accurate waveforms (e.g., the horizontal-sweep ramp generator in an oscilloscope).

In the following sections we will treat briefly the most popular oscillators, from the simple *RC* relaxation oscillators to the stable quartz-crystal oscillators. Our aim is not to survey everything in exhaustive detail, but simply to make you acquainted with what is available and what sorts of oscillators are suitable in various situations.

5.13 Relaxation oscillators

A very simple kind of oscillator can be made by charging a capacitor through a

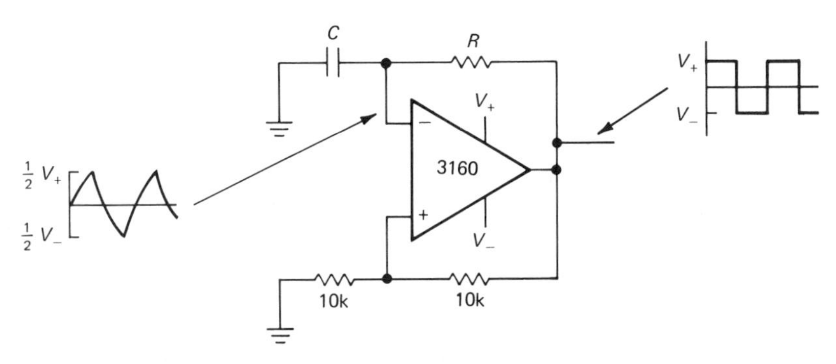

Figure 5.29. Op-amp relaxation oscillator.

resistor (or a current source), then discharging it rapidly when the voltage reaches some threshold, beginning the cycle anew. Alternatively, the external circuit may be arranged to reverse the polarity of the charging current when the threshold is reached, thus generating a triangle wave rather than a sawtooth. Oscillators based on this principle are known as relaxation oscillators. They are inexpensive and simple, and with careful design they can be made quite stable in frequency.

In the past, negative-resistance devices such as unijunction transistors and neon bulbs were used to make relaxation oscillators, but current practice favors op-amps or special timer ICs. Figure 5.29 shows a classic *RC* relaxation oscillator. The operation is simple: Assume that when power is first applied, the op-amp output goes to positive saturation (it's actually a toss-up which way it will go, but it doesn't matter). The capacitor begins charging up toward V_+, with time constant RC. When it reaches one-half the supply voltage, the op-amp switches into negative saturation (it's a Schmitt trigger), and the capacitor begins discharging toward V_- with the same time constant. The cycle repeats indefinitely, with period $2.2RC$, independent of supply voltage. A CMOS output-stage op-amp (see Sections 4.11 and 4.22) was chosen because its outputs saturate cleanly at the supply voltages. The bipolar LM10 also swings rail-to-rail and, unlike CMOS op-amps, allows operation at a full ±15 volts; however, it has a much lower f_T (0.1MHz).

EXERCISE 5.7

Show that the period is as stated.

By using current sources to charge the capacitor, a good triangle wave can be generated. A clever circuit using that principle was shown in Section 4.29.

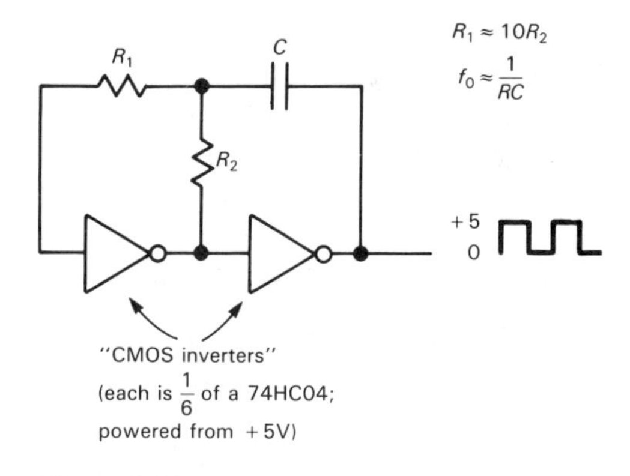

Figure 5.30

Sometimes you need an oscillator with very low noise content (also called "low sideband noise"). The simple circuit of Figure 5.30 is good in this respect. It uses a pair of CMOS inverters (a form of digital logic we'll use extensively in Chapters 8–11) connected together to form an *RC* relaxation oscillator with square wave output. Actual measurements

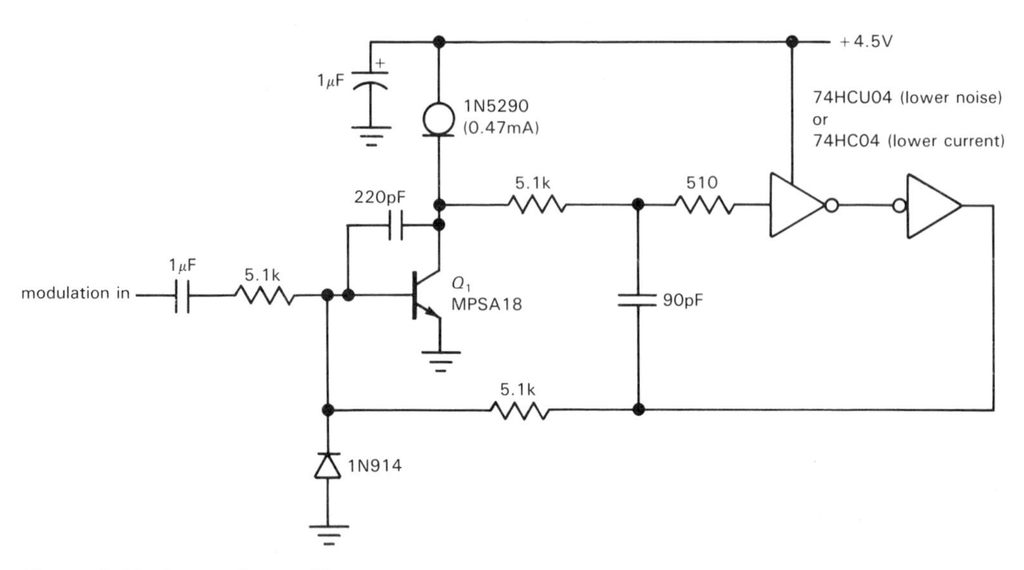

Figure 5.31. Low-noise oscillator.

on this circuit running at 100kHz show close-in sideband noise power density (power per square root hertz, measured 100Hz from the oscillator frequency), down at least 85dB relative to the carrier. You sometimes see a similar circuit, but with R_2 and C interchanged. Although it still oscillates fine, it is extremely noisy by comparison.

The circuit of Figure 5.31 has even lower noise and furthermore lets you modulate the output frequency via an external current applied to the base of Q_1. In this circuit Q_1 operates as an integrator, generating an asymmetrical triangle waveform at its collector. The inverters operate as a noninverting comparator, alternating the polarity of the base drive each half cycle. This circuit has close-in noise density of $-90\text{dBc}/\sqrt{\text{Hz}}$ measured 100Hz from the 150kHz carrier, and $-100\text{dBc}/\sqrt{\text{Hz}}$ measured at an offset of 300Hz. Although these circuits excel in low sideband noise, the oscillation frequency has more supply-voltage sensitivity than other oscillators discussed in this chapter.

5.14 The classic timer chip: the 555

The next level of sophistication involves the use of timer or waveform-generator ICs as relaxation oscillators. The most popular chip around is the 555 (and its successors). It is also a misunderstood chip, and we intend to set the record straight with the equivalent circuit shown in Figure 5.32. Some of the symbols belong to the digital world (Chapter 8 and following), so you won't become a 555 expert for a while yet. But the operation is simple enough: The output goes HIGH (near V_{CC}) when the 555 receives a TRIGGER' input, and it stays there until the THRESHOLD input is driven, at which time the output goes LOW (near ground) and the DISCHARGE transistor is turned on. The TRIGGER' input is activated by an input level below $\frac{1}{3}V_{CC}$, and the THRESHOLD is activated by an input level above $\frac{2}{3}V_{CC}$.

The easiest way to understand the workings of the 555 is to look at an example (Fig. 5.33). When power is applied, the capacitor is discharged; so the 555 is triggered, causing the output to go HIGH, the discharge transistor Q_1 to turn

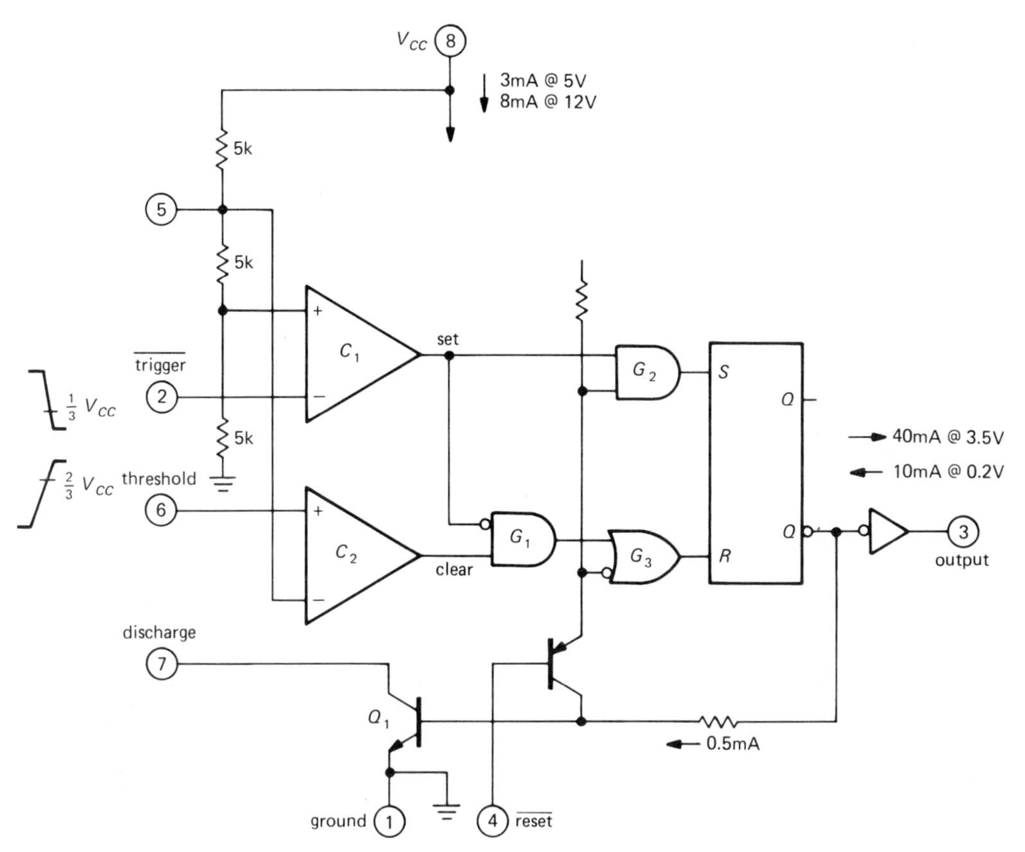

Figure 5.32. Simplified 555 schematic.

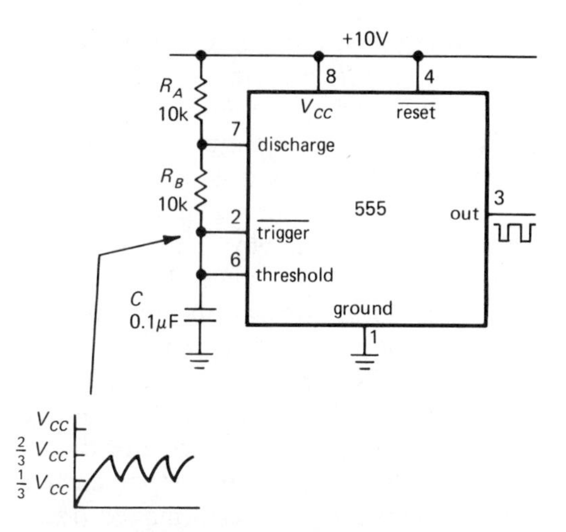

Figure 5.33. The 555 connected as an oscillator.

off, and the capacitor to begin charging toward 10 volts through $R_A + R_B$. When it has reached $\frac{2}{3}V_{CC}$, the THRESHOLD input is triggered, causing the output to go LOW and Q_1 to turn on, discharging C toward ground through R_B. Operation is now cyclic, with C's voltage going between $\frac{1}{3}V_{CC}$ and $\frac{2}{3}V_{CC}$, with period $T = 0.693(R_A + 2R_B)C$. The output you generally use is the square wave at the output.

EXERCISE 5.8

Show that the period is as advertised, independent of supply voltage.

The 555 makes a respectable oscillator, with stability approaching 1%. It can run from a single positive supply of 4.5 to 16 volts, maintaining good frequency stability with supply voltage variations because the thresholds track the supply fluctuations. The 555 can also be used to generate single pulses of arbitrary width, as well as a bunch of other things. It is really a small kit, containing comparators, gates, and flip-flops. It has become a game in the electronics industry to try to think of new uses for the 555. Suffice it to say that many succeed at this new form of entertainment.

A caution about the 555: The 555, along with some other timer chips, generates a big (≈150mA) supply-current glitch during each output transition. Be sure to use a hefty bypass capacitor near the chip. Even so, the 555 may have a tendency to generate double output transitions.

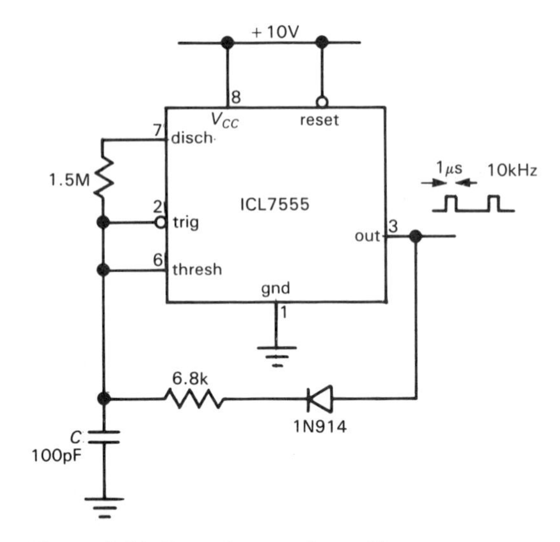

Figure 5.34. Low-duty-cycle oscillator.

CMOS 555s

Some of the less desirable properties of the 555 (high supply current, high trigger current, double output transitions, and inability to run with very low supply voltage) have been remedied in a collection of CMOS successors. You can recognize these by the telltale "555" somewhere in the part number. Table 5.3 lists most of these that we could find, along with their important specifications. Note particularly the ability to operate at very low supply voltage (down to 1V!) and the generally low supply current. These chips also can run at higher frequency than the original 555. The CMOS output stages give rail-to-rail swing, at least at low load currents (but note that these chips don't have the output-current muscle of the standard 555). All chips listed are CMOS except for the original 555 and the XR-L555. The latter is intended as a bipolar low-power 555 and reveals its pedigree by the hefty output sourcing capability and good tempco.

The 555 oscillator of Figure 5.33 generates a rectangular-wave output whose duty cycle (fraction of time the output is HIGH) is always greater than 50%. That is because the timing capacitor is charged through the series pair $R_A + R_B$, but discharged (more rapidly) through R_B alone. Figure 5.34 shows how to trick the 555 into giving you low duty-cycle positive pulses. The diode/resistor combination *charges* timing capacitor rapidly via the output, with slower discharge via the internal discharge transistor. You can only play this trick with a CMOS 555, because you need the full positive output swing.

By using a current source to charge the timing capacitor, you can make a ramp (or "sawtooth-wave") generator. Figure 5.35 shows how, using a simple *pnp* current source. The ramp charges to $\frac{2}{3}V_{CC}$, then discharges rapidly (through the 555's *npn* discharge transistor, pin 7) to $\frac{1}{3}V_{CC}$, beginning the ramp cycle anew. Note that the ramp waveform appears on the capacitor terminal and must be buffered with an op-amp since it is at high impedance. In this circuit you could simplify things somewhat by using a JFET "current-regulator diode" (Section 3.06) in place of the *pnp* current source; however, the performance (ramp linearity) would be slightly degraded, because a JFET operating at I_{DSS} is not as good a current source as the bipolar transistor circuit.

Figure 5.36 shows a simple way to

TABLE 5.3. 555-TYPE OSCILLATORS

Type	Mfg[a]	Qty per package	Supply voltage		Supply curr per osc (V_S = 5V)		Trig, thresh current		Max freq (V_S = 5V)		Tempco	V_{sat}, typ				Rail to rail?[b]	I_{out}, max (V_S=5V, V_O=2.5V)	
		1 2 4	min (V)	max (V)	typ (μA)	max (μA)	typ (nA)	max (nA)	min (MHz)	typ (MHz)	typ (ppm/°C)	V_{OH} (V)	@ I_{src} (mA)	V_{OL} (V)	@ I_{snk} (mA)		source (mA)	sink (mA)
555	SN+	• • –	4.5	18	3000	5000	100	500	–	0.5	30	1.4	2	0.1	10	–	200	200
ICL7555	IL	• • –	2	18	60	300	–	10	–	1	150	1	2	0.5	10	•	4	25
TLC551	TI	• • –	1	18	170	–	0.01	–	–	2.1[c]	–	1	2	0.2	10	•	–	–
TLC555	TI	• • –	2	18	170	–	0.01	–	–	2.1	–	1	2	0.2	10	•	–	–
LMC555	NS	• – –	1.5	15	100	250	0.01	–	–	3	75	0.3	2	0.3	10	•	–	–
ALD555-1	AL	• – –	1	12	100	180	0.001	0.2	1.4	2	300	0.4	2	0.2	10	•	3	100
ALD1504	AL	• • –	1	12	50	90	0.01	0.4	1.5	2.5	300	0.4	2	0.2	10	•	10	100
ALD4503	AL	– – •	1	12	35	70	0.01	0.4	–	2	300	0.4	2	0.2	10	•	3	100
XR-L555M	XR	• • –	2.7	15	150	300	500	–	–	–	30	1.7	10	0.3	2	–	100	–

(a) see footnotes to Table 4.1. (b) signifies that the output stage can swing to both rails. (c) at V_S=1.2V.

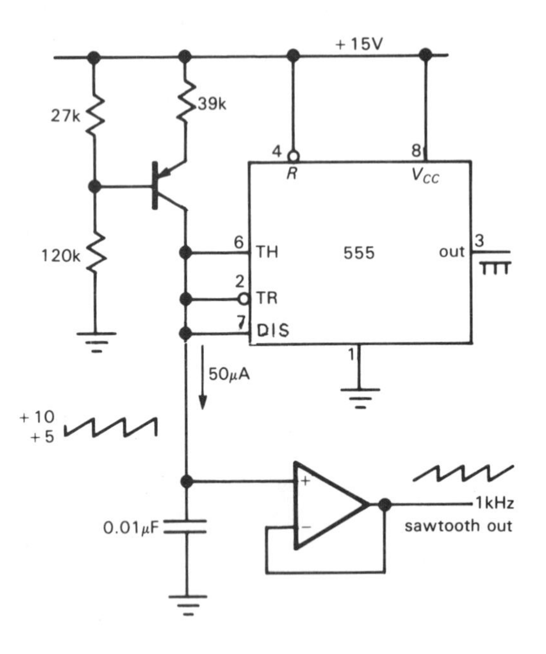

Figure 5.35. Sawtooth oscillator.

generate a *triangle* wave with a CMOS 555. Here we wired a pair of JFET current regulators in series to generate a bidirectional current regulator (each current regulator behaves like a normal diode in the reverse direction, owing to gate-drain conduction). The rail-to-rail output swing thus generates a constant current, of alternating polarity, producing a triangle waveform (going between the usual $\frac{1}{3}V_{CC}$ and $\frac{2}{3}V_{CC}$) at the capacitor. As before, you have to buffer the high-impedance waveform with an op-amp. Note that you must use a *CMOS* 555, particularly when operating the circuit from +5 volts, since the circuit depends on a full rail-to-rail output swing. For example, the HIGH output of a bipolar 555 is typically 2 diode drops below the positive rail (*npn* Darlington follower), or +3.8 volts with a 5 volt supply; this leaves only 0.5 volt across the series pair of current regulators at the top of the waveform, obviously insufficient to turn on the current regulator (approximately 1V) and the series JFET diode (0.6V).

EXERCISE 5.9

Demonstrate that you understand the circuits of Figures 5.35 and 5.36 by calculating the frequency of oscillation in each case.

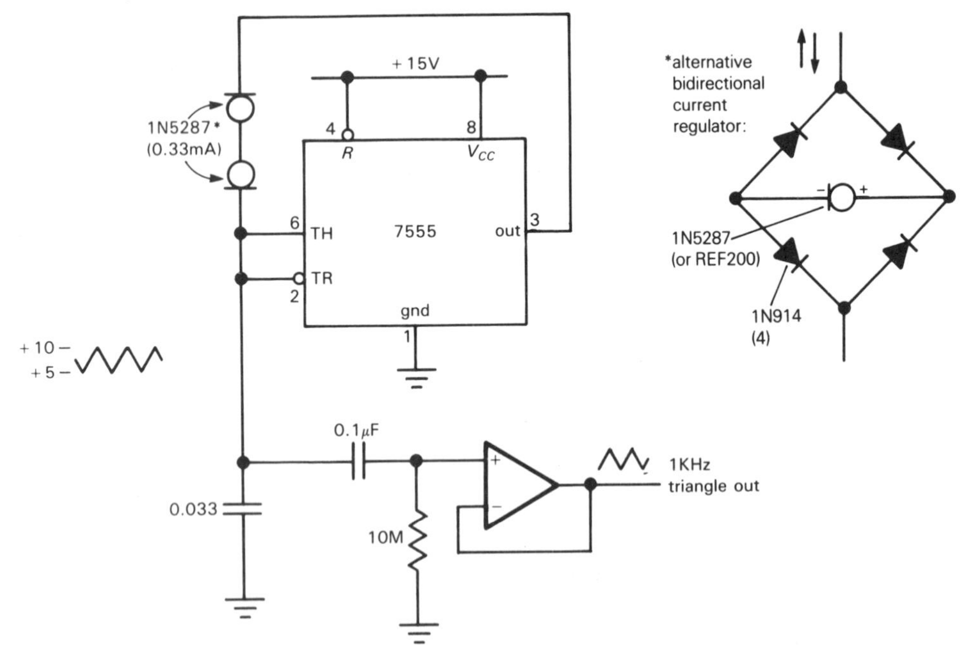

Figure 5.36. Triangle generator.

There are some other interesting timer chips available. The 322 timer from National includes its own internal precision voltage reference for determining the threshold. That makes it an excellent choice for generating a frequency proportional to an externally supplied current, as, for example, from a photodiode. Another class of timers uses a relaxation oscillator followed by a digital counter, in order to generate long delay times without resorting to large resistor and capacitor values. Examples of this are the 74HC4060, the Exar 2243, and the Intersil ICM7242 (also made by Maxim). The latter is CMOS, runs on a fraction of a milliamp, and generates an output pulse every 128 oscillator cycles. These timers (and their near relatives) are great for generating delays from a few seconds to a few minutes.

5.15 Voltage-controlled oscillators

Other IC oscillators are available as voltage-controlled oscillators (VCO's), with the output rate variable over some range according to an input control voltage. Some of these have frequency ranges exceeding 1000:1. Examples are the original NE566 and later designs like the LM331, 8038, 2206, and 74LS624–9 series.

The 74LS624 series, for example, generates digital-logic-level outputs up to 20 MHz and uses external *RC*s to set the nominal frequency. Faster VCOs like the 1648 can produce outputs to 200MHz, and in Chapter 13 we'll see how to make VCOs that operate in the *giga*hertz range. The LM331 is actually an example of a voltage-to-frequency (V/F) converter, designed for good linearity (see Sections 9.20 and 9.27). Where linearity is important, recent V/F converters like the AD650 really do the job, with linearity of 0.005%. Most VCOs use internal current sources to generate triangle-wave outputs, and the 8038 and 2206 even include a set of "soft" clamps to convert the triangle wave to a not-too-great sine wave. VCO chips sometimes have an awkward reference for the control voltage (e.g., the positive supply) and complicated symmetrizing schemes for sine-wave output. It is our opinion that the ideal VCO has yet to be developed. Many of these chips can be used with an external quartz crystal, as we will discuss shortly, for much higher accuracy and stability; in such cases the crystal simply replaces the capacitor. Figure 5.37 shows a VCO circuit with an output frequency range of 10Hz to 10kHz built with the LM331.

When shopping for VCO chips, don't overlook the ICs known as *phase-locked loops* (PLL), which contain both a VCO and a phase detector. An example is the popular CMOS 4046 (and its faster cousin, the 74HC4046). We will discuss PLLs in Sections 9.27–9.31. Table 5.4 lists most of the available VCOs.

5.16 Quadrature oscillators

There are times when you need an oscillator that generates a simultaneous *pair* of equal-amplitude sine waves, *90° out of phase*. You can think of the pair as sine and cosine. This is referred to as a *quadrature pair* (the signals are "in quadrature"). One important application is in radio communications circuits (quadrature mixers, single-sideband generation). Furthermore, as we'll explain below, a quadrature pair is all you need to generate any arbitrary phase.

The first idea you might invent is to apply a sine-wave signal to an integrator (or differentiator), thus generating a 90°-shifted cosine wave. The phase shift is right, but the amplitude is wrong (figure out why). Here are some methods that do work:

□ *Switched-capacitor resonator*

Figure 5.38 shows how to use an MF5 switched-capacitor filter IC as a self-excited

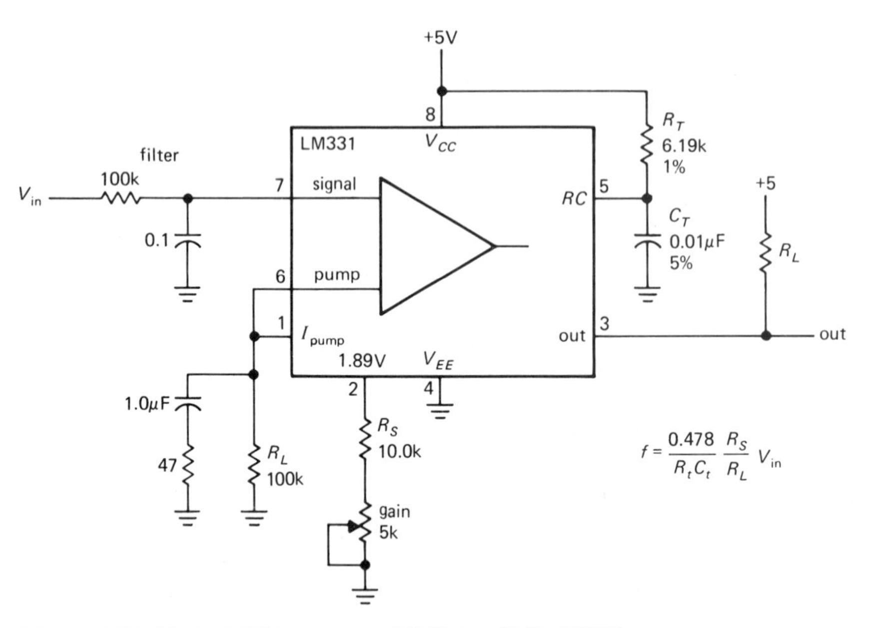

Figure 5.37. Typical V/F converter IC (0 to 10kHz VCO).

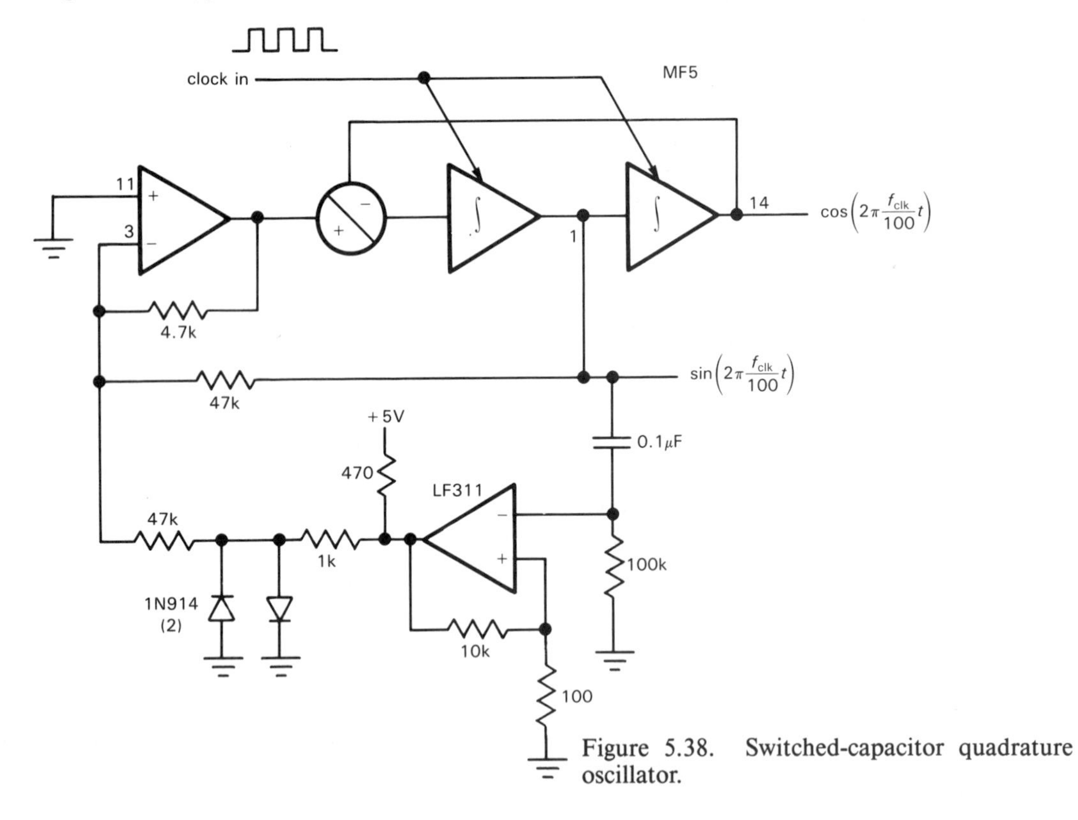

Figure 5.38. Switched-capacitor quadrature oscillator.

bandpass filter to generate a quadrature sine-wave pair. The easiest way to understand it is to assume there is already a sine-wave output present; the comparator

TABLE 5.4. SELECTED VCOs

Type	Mfg[a]	Family[b]	Max freq (MHz)	Outputs[c]	Supply voltage min (V)	Supply voltage max (V)	Linearity (at 10kHz)	Comments
VFC32	BB+	L	0.5	OC	±9	±18	0.01%	indus. st'd; good linearity
VFC62C	BB	L	1	OC	±13	±20	0.002%	excellent linearity
VFC110B	BB	L	4	OC	±8	±18	0.005%	fast, exc lin, int V_{ref}
74S124	TI	T	60	SQ	4.75	5.25		
74LS624-9	TI	T	20	SQ	4.75	5.25		
74LS724	TI	T	16	SQ	4.75	5.25		mini-DIP
215	XR	L	35	SQ	5	26		PLL
LM331	NS	L	0.1	OC	4	40		inexpensive, good linearity
AD537	AD	L	0.1	OC	4.5	36	0.07%	
566	SN	L	1	SQ,T	10	24		
AD650	AD	L	1	OC	±9	±18	0.005%	excellent linearity
AD654	AD	L	0.5	OC	4.5	36	0.1%[d]	inexpensive
1648	MO	E	200	P	-5.2			
1658	MO	E	130	P	-5.2			
XR2206	XR	L	0.5	SQ,T,SW	10	26	2%	0.5% sine dist (trimmed)
XR2207	XR	L	0.5	SQ,T	8	26	1%	
XR2209	XR	L	1	SQ,T	±4	±13	1%	
XR2212	XR	L	0.3	SQ	4.5	20		PLL
XR2213	XR	L	0.3	SQ	4.5	15		PLL
4024	MO	T	25	SQ	4.75	5.25		
4046	RC+	C	1	SQ	3	15		CMOS PLL
HC4046	RC+	C	15[t]	SQ	3	6		fast 4046
4151	RA	L	0.1	OC	8	22	0.013%	
4152	RA	L	0.1	OC	7	18	0.007%	
4153A	RA	L	0.5	OC	±12	±18	0.002%	excellent lin, easy to use
8038	IL	L	0.1	SQ,T,SW	10	30	0.2%	Exar 8038 to 1MHz
TSC9401	TP	L	0.1	OC	±4	±7.5	0.01%	V/F, linear, stable

(a) see footnotes to Table 4.1. (b) families: C – CMOS; E – ECL; L – linear; T – TTL. (c) outputs: OC – open collector, pulses; P – pulses; SQ – square waves; SW – sine waves; T – triangle. (d) at 250kHz.

converts this to a small-amplitude (1 diode drop) square wave, which is fed back as the filter's input. The filter has a narrow bandpass ($Q = 10$), so it converts the input square wave to a sine-wave output, sustaining the oscillation. A square-wave clock input (CLK) determines the bandpass center frequency, hence the frequency of oscillation, in this case $f_{\text{clk}}/100$. The circuit is usable over a frequency range of a few hertz to about 10kHz and generates a quadrature pair of sine waves of equal amplitude. Note that this circuit will actually have a "staircase" approximation to the desired sine-wave output, owing to the quantized output steps of the switched filter.

□ *Analog trigonometric-function generator*

Analog Devices makes an interesting nonlinear "function IC" that converts an input voltage to an output voltage proportional to $\sin(AV_{\text{in}})$, where the gain A is fixed at 50° per volt. In fact, this chip, the AD639, can actually do a lot more: It has four inputs, called X_1, X_2, Y_1, and Y_2, and generates as output the voltage $V_{\text{out}} = \sin(X_1 - X_2)/\sin(Y_1 - Y_2)$. Thus, for example, by setting $X_1 = Y_1 = 90°$

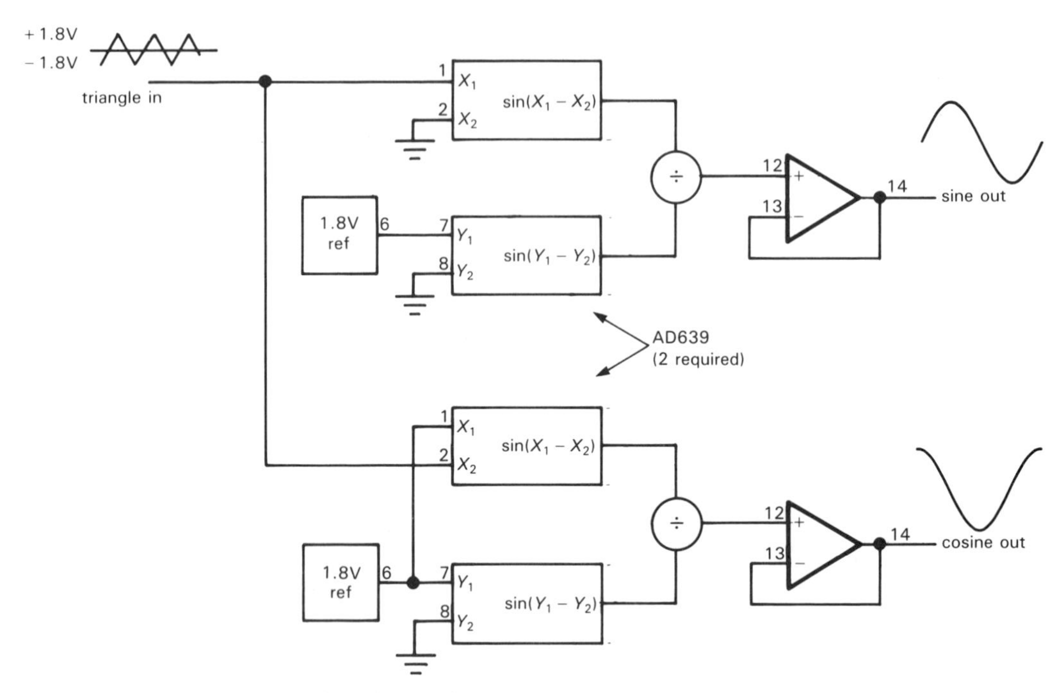

Figure 5.39. Trigonometric-function oscillator.

(i.e., +1.8 volt), $Y_2 = 0$ (ground), and applying an input voltage to X_2, we generate $\cos(X_2)$.

EXERCISE 5.10

Prove the last statement.

The AD639 even gives you a precise +1.8 volt output, to make life easy. Thus, a pair of AD639s, driven by a 1.8 volt amplitude triangle wave, generates a quadrature sine-wave pair, as shown in Figure 5.39. The AD639 operates from dc to about 1MHz.

□ *Lookup table*

This is a digital technique, which you will fully understand only after you've read Chapter 9. The idea is to program a digital memory with the numerical values of sine and cosine for a large set of equally spaced angle arguments (say for every 1°). You then make sine waves by rapidly generating the sequential addresses, reading the memory values for each address (i.e., each sequential angle), and applying the digital values to a pair of digital-to-analog (D/A) converters.

This method has some drawbacks. As with the switched-capacitor resonator, the output is actually a staircase wave, since it is constructed from a set of discrete voltages, one for each table entry. You can, of course, use a low-pass filter to smooth the output; but having done so, you cannot span a wide range of frequencies, since the low-pass filter must be chosen to pass the sine wave itself while blocking the (higher) angle step frequency (the same problem applies to the switched-capacitor resonator). Decreasing the angular step size helps, but reduces the maximum output frequency. With typical D/A converter speeds of something less than a microsecond, you can make sine waves up to a few tens of kilohertz or so, assuming you use angle steps of a degree or so. D/A converters also tend to generate large output spikes ("glitches") while jumping between diode limiter at $V_O = \frac{R_1}{R_2}(V_S + V_{diode})$

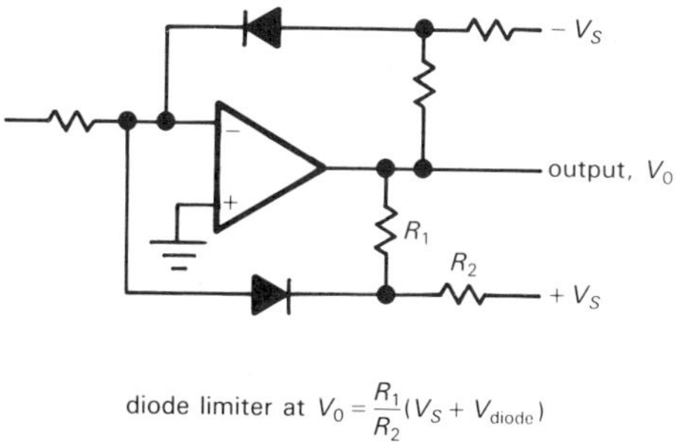

Figure 5.40

output voltages. You can get full-scale glitches even when jumping between adjacent (closest) output voltage levels! In Chapter 9 we'll see deglitching techniques to eliminate this problem. D/A converters are available with resolutions up to 16 bits (1 part in 65,536).

□ *State-variable oscillator*

The preceding methods all require some hard work. Luckily, the friendly folks at Burr-Brown have done their homework and have come up with the model 4423 "precision quadrature oscillator." It uses the standard 3-op-amp state-variable bandpass filter circuit (Figure 5.18), with the output diode-limited and fed back as input (see Fig. 5.40). It claims to operate from 0.002Hz to 20kHz, with good control of phase shift, amplitude, and frequency stability (100ppm/°C, max). The 4423 is a module (not a monolithic IC) in a 14-pin molded DIP; it costs $24 in small quantities.

□ *Phase sequence filters*

There are tricky *RC* filter circuits that have the property of accepting an input sine wave and producing as output a pair of sine-wave outputs whose phase *difference* is approximately 90°. The radio hams know this as the "phasing" method of single-sideband generation (due to Weaver), in which the input signal consists of the speech waveform that you want to transmit.

Unfortunately, this method works satisfactorily only over a rather limited range of frequencies and requires precision resistors and capacitors.

A better method for wideband quadrature generation uses "phase sequence networks," consisting of a cyclic repetitive structure of equal resistors and geometrically decreasing capacitors, as in Figure 5.41. You drive the network with a signal and its 180°-shifted cousin (that's easy, since all you need is a unity-gain inverter). The output is a fourfold set of quadrature signals, with a 6-section network giving ±0.5° error over a 100:1 frequency range.

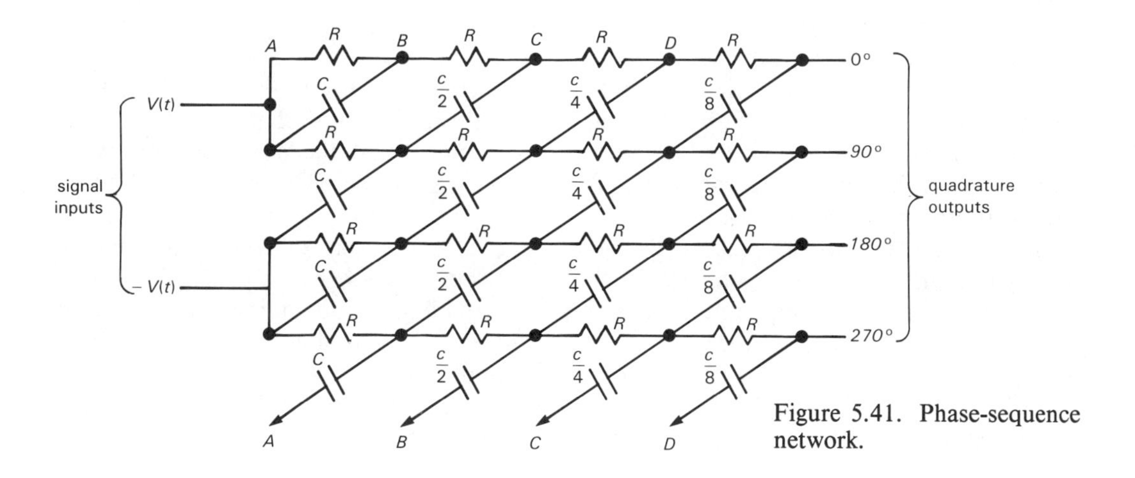

Figure 5.41. Phase-sequence network.

□ *Quadrature square waves*

For the special case of square waves, generating quadrature signals is a lead-pipe cinch. The basic idea is to generate twice the frequency you need, then divide by 2 with digital *flip-flops* (Chapter 8) and *decode* with *gates* (Chapter 8 again). This technique is essentially perfect from dc to at least 100MHz.

□ *Radiofrequency quadrature*

At radiofrequencies (upward of a few megahertz) the generation of quadrature sine-wave pairs again becomes easy, using devices known as *quadrature hybrids* (or quadrature *splitter/combiners*). At the low-frequency end of the radio spectrum (from a few megahertz to perhaps 1GHz) these take the form of small core-wound transformers, while at higher frequencies you find incarnations in the form of stripline (strips of foil insulated from an underlying ground plane) or waveguide (hollow rectangular tubing). We'll see these again in Chapter 13. These techniques tend to be fairly narrow-band, with typical operating bandwidths of an octave (i.e., ratio of 2:1).

□ *Generating a sine wave of arbitrary phase*

Once you have a quadrature pair, it's easy to make a sine wave of *arbitrary* phase. You simply combine the in-phase (I) and quadrature (Q) signals in a resistive combiner, made most easily with a potentiometer going between the I and Q signals. As you rotate the pot, you combine the I and Q in different proportions, taking you smoothly from 0° to 90° phase. If you think in terms of phasors, you'll see that the resulting phase is completely independent of frequency; however, the amplitude varies somewhat as you adjust the phase, dropping 3dB at 45°. You can extend this simple method to the full 360° by simply generating the inverted (180°-shifted) signals, I' and Q', with an inverting amplifier of gain $G_V = -1$.

□ 5.17 Wien bridge and *LC* oscillators

When a low-distortion sine wave is required, none of the preceding methods is generally adequate. Although wide-range function generators do use the technique of "corrupting" a triangle wave with diode clamps, the resulting distortion can rarely be reduced below 1%. By comparison, most hi-fi audiophiles insist on distortion levels below 0.1% for their amplifiers. To test such low-distortion audio components, pure sine-wave signal sources with residual distortion less than 0.05% or so are required.

At low to moderate frequencies the Wien bridge oscillator (Fig. 5.42) is a good source of low-distortion sinusoidal signals. The idea is to make a feedback amplifier with 180° phase shift at the desired output frequency, then adjust the loop gain so that a self-sustaining oscillation just barely takes place. For equal-value Rs and Cs as shown, the voltage gain from the noninverting input to op-amp output should be exactly +3.00. With less gain the oscillation will cease, and with more gain the output will saturate. The distortion is low if the amplitude of oscillation remains within the linear region of the amplifier, i.e., it must not be allowed to go into a full-swing oscillation. Without some trick to control the gain, that is exactly what will happen, with the amplifier's output increasing until the effective gain is reduced to 3.0 because of saturation. The tricks involve some sort of long-time-constant gain-setting feedback, as you will see.

In the first circuit, an incandescent lamp is used as a variable-resistance feedback element. As the output level rises, the lamp heats slightly, reducing the noninverting gain. The circuit shown has less

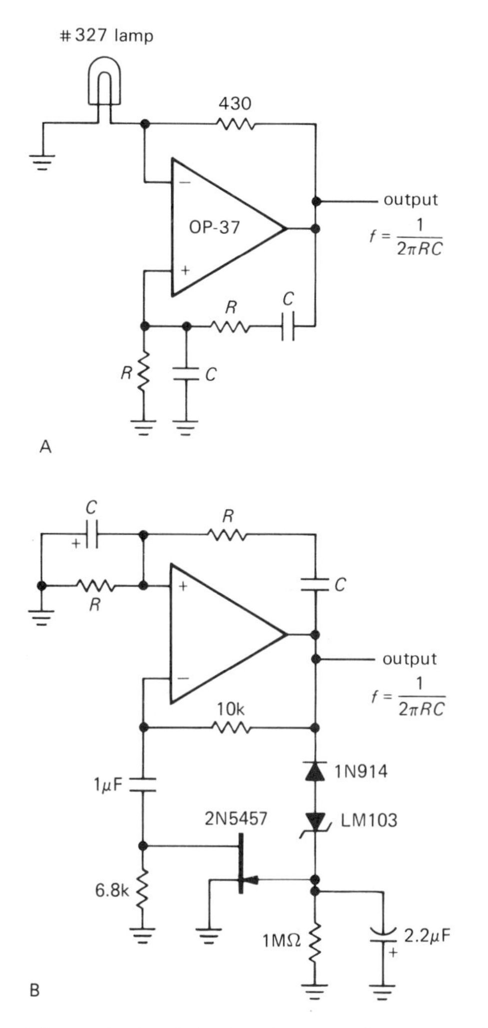

Figure 5.42. Wien-bridge low-distortion oscillators.

than 0.003% harmonic distortion for audiofrequencies above 1kHz; see LTC App. Note 5(12/84) for more details. In the second circuit, an amplitude discriminator consisting of the diodes and *RC* adjusts the ac gain by varying the resistance of the FET, which behaves like a voltage-variable resistance for small applied voltages (see Section 3.10). Note the long time constant used (2s); this is essential to avoid distortion, since fast feedback will distort the wave by attempting to control the amplitude within the time of one cycle.

□ 5.18 *LC* oscillators

At high frequencies the favorite method of sine-wave generation is an *LC*-controlled oscillator, in which a tuned *LC* is connected in an amplifier-like circuit to provide gain at its resonant frequency. Overall positive feedback is then used to cause a sustained oscillation to build up at the *LC*'s resonant frequency; such circuits are self-starting.

Figure 5.43 shows two popular configurations. The first circuit is the trusty Colpitts oscillator, a parallel tuned *LC* at the input, with positive feedback from the output. For this circuit it is claimed that its distortion is less than −60dB. The second circuit is a Hartley oscillator, built with an *npn* transistor. The variable capacitor is for frequency adjustment. Both circuits use *link coupling*, just a few turns of wire acting as a step-down transformer.

LC oscillators can be made *electrically* tunable over a modest range of frequency. The trick is to use a voltage-variable capacitor ("varactor") in the frequency-determining *LC* circuit. The physics of diode junctions provides the solution, in the form of a simple reverse-biased diode: The capacitance of a *pn* junction decreases with increasing reverse voltage (see Fig. 13.3). Although any diode acts as a varactor, you can get special varactor diodes designed for the purpose; Figure 5.44 shows some representative types. Figure 5.45 shows a simple JFET Colpitts oscillator (with feedback from the source) with ±1% tunability. In this circuit the tuning range has been made deliberately small, in order to achieve good stability, by using a relatively large fixed capacitor (100pF) shunted by a small tunable capacitor (maximum value of 15pF). Note the large biasing resistor (so the diode bias circuit doesn't load the oscillation) and the dc blocking capacitor. See also Section 13.11.

Varactors typically provide a maximum capacitance of a few picofarads to a few

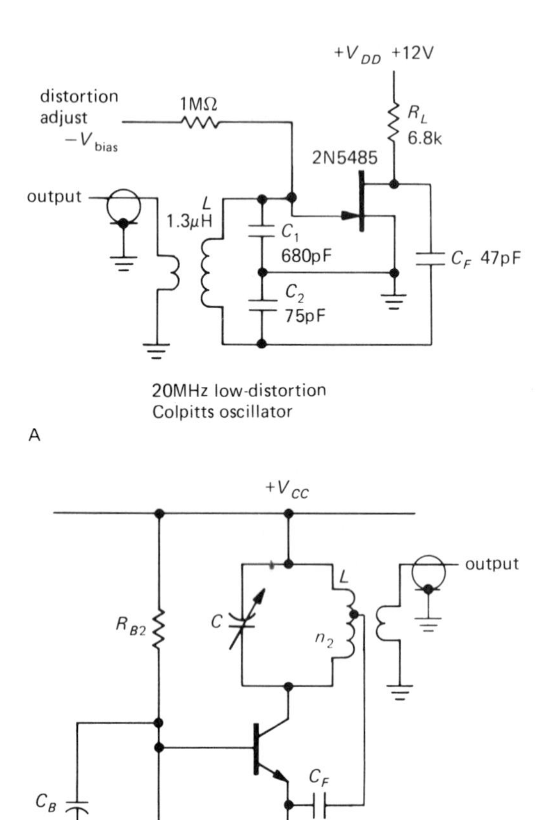

Figure 5.43

hundred picofarads, with a tuning range of about 3:1 (although there are wide range varactors with ratios as high as 15:1). Since the resonant frequency of an *LC* circuit is inversely proportional to the square root of capacitance, it is possible to achieve tuning ranges of up to 4:1 in frequency, though more typically you're talking about a tuning range of ±25% or so.

In varactor-tuned circuits the oscillation itself (as well as the externally applied dc tuning bias) appears across the varactor, causing its capacitance to vary at the signal frequency. This produces oscillator waveform distortion, and, more important, it causes the oscillator frequency to depend somewhat on the amplitude of oscillation. In order to minimize these effects, you should limit the amplitude of the oscillation (amplify in following stages, if you need more output); also, it's best to keep the dc varactor bias voltage above a volt or so, in order to make the oscillating voltage small by comparison.

Electrically tunable oscillators are used extensively to generate frequency modulation, as well as in radiofrequency phase-locked loops. We will treat these subjects in Chapters 9 and 13.

For historical reasons we should mention a close cousin of the *LC* oscillator, namely the tuning-fork oscillator. It used the high-*Q* oscillations of a tuning fork as the frequency-determining element of an oscillator, and it found use in low-frequency standards (stability of a few parts per million, if run in a constant-temperature oven) as well as wristwatches. These objects have been superseded by quartz oscillators, which are discussed in the next section.

□ *Parasitic oscillations*

Suppose you have just made a nice amplifier and are testing it out with a sine-wave input. You switch the input function generator to a square wave, but the output remains a sine wave! You don't have an amplifier; you've got trouble.

Parasitic oscillations aren't normally as blatant as this. They are normally observed as fuzziness on part of a waveform, erratic current-source operation, unexplained op-amp offsets, or circuits that behave normally with the oscilloscope probe applied, but go wild when the scope isn't looking. These are bizarre manifestations of untamed high-frequency parasitic oscillations caused by unintended Hartley or Colpitts oscillators employing lead inductance and interelectrode capacitances.

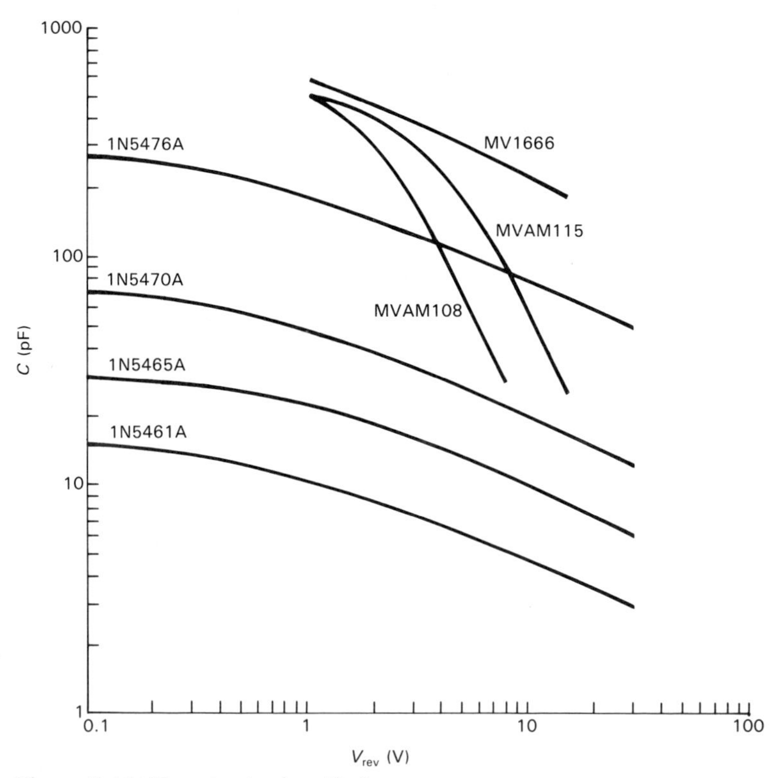

Figure 5.44. Varactor tuning diodes.

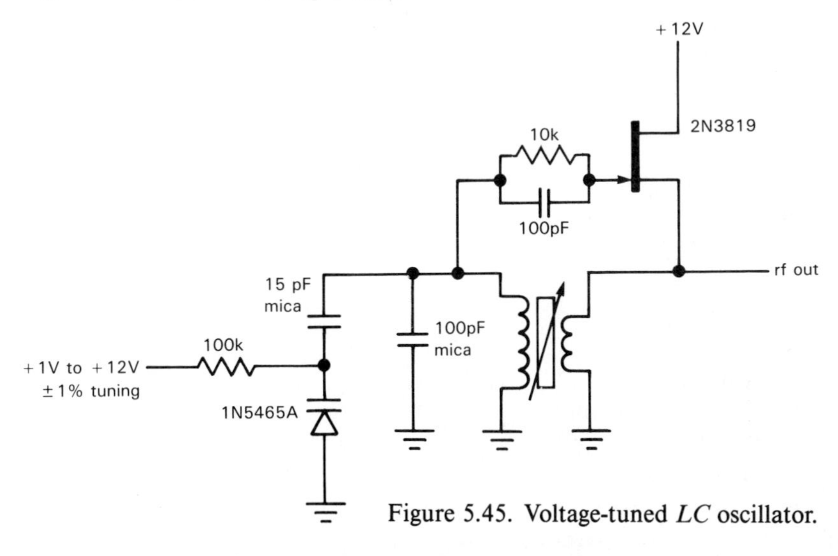

Figure 5.45. Voltage-tuned *LC* oscillator.

The circuit in Figure 5.46 shows an oscillating current source born in an electronics lab course where a VOM was used to measure the output compliance of a standard transistor current source. The current seemed to vary excessively (5% to 10%) with load voltage variations within its expected compliance range, a symptom that could be "cured" by sticking a finger on the collector lead! The collector-base

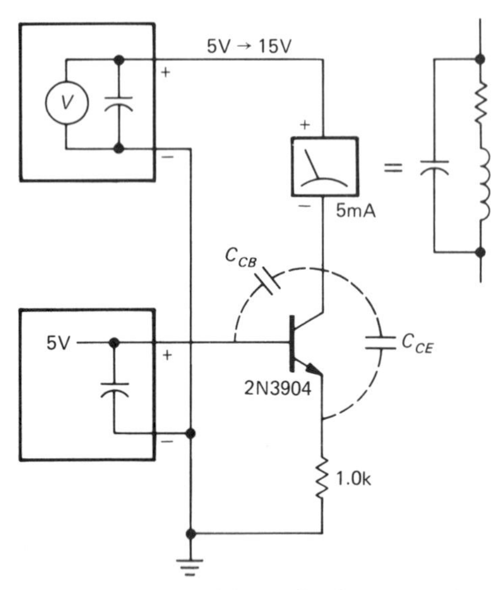

Figure 5.46. Parasitic oscillation example.

capacitance of the transistor and the meter capacitance resonated with the meter inductance in a classic Hartley oscillator circuit, with feedback provided by collector-emitter capacitance. Adding a small base resistor suppressed the oscillation by reducing the high-frequency common-base gain. This is one trick that often helps.

5.19 Quartz-crystal oscillators

RC oscillators can easily attain stabilities approaching 0.1%, with initial predictability of 5% to 10%. That's good enough for many applications, such as the *multiplexed display* in a pocket calculator, in which a multidigit numerical display is driven by lighting one digit after another in rapid succession (a 1kHz rate is typical). Only one digit is lit at any time, but your eye sees the whole display. In such an application the precise rate is quite irrelevant – you just want something in the ballpark. As stable sources of frequency, *LC* oscillators can do a bit better, with stabilities of 0.01% over reasonable periods of time. That's good enough for oscillators in radio-frequency receivers and television sets.

For real stability there's no substitute for a crystal oscillator. This uses a piece of quartz (same chemical as glass, silicon dioxide) that is cut and polished to vibrate at a certain frequency. Quartz is *piezoelectric* (a strain generates a voltage, and vice versa), so acoustic waves in the crystal can be driven by an applied electric field and in turn can generate a voltage at the surface of the crystal. By plating some contacts on the surface, you wind up with an honest circuit element that can be modeled by an *RLC* circuit, pretuned to some frequency. In fact, its equivalent circuit contains two capacitors, giving a pair of closely spaced (within 1%) series and parallel resonant frequencies (Fig. 5.47). The effect is to produce a rapidly changing reactance with frequency (Fig. 5.48). The quartz crystal's high Q (typically around 10,000) and good stability make it a natural for oscillator control, as well as for high-performance filters (see Section 13.12). As with *LC* oscillators, the crystal's equivalent circuit provides positive feedback and gain at the resonant frequency, leading to sustained oscillations.

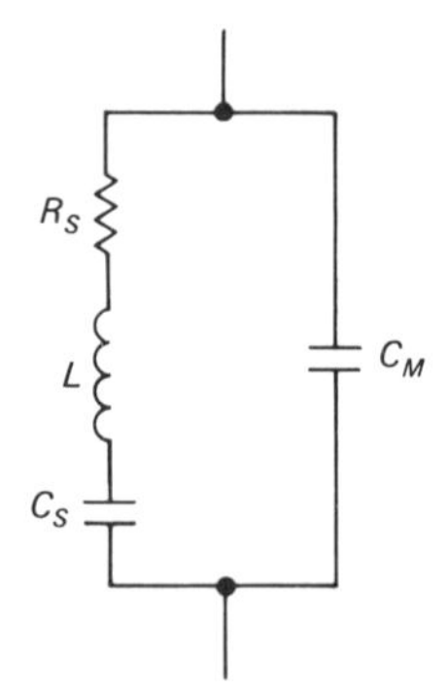

Figure 5.47

Figure 5.49 shows some crystal oscillator circuits. In A the classic Pierce oscillator is shown, using the versatile FET (see Chapter 3). The Colpitts oscillator, with a crystal instead of an *LC*, is shown in B. An *npn* bipolar transistor with the crystal as feedback element is used in C. The remaining circuits generate logic-level outputs using digital logic functions (D and E).

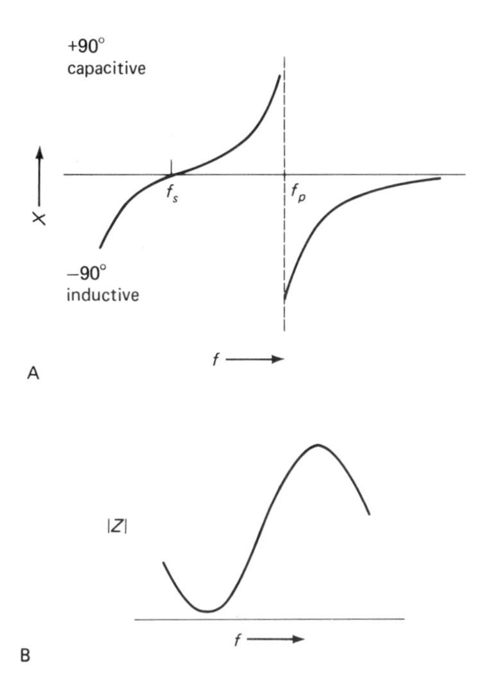

Figure 5.48

The last diagram uses the convenient MC12060/12061 series of crystal oscillator circuits from Motorola. These chips are intended for crystals in the range 100kHz to 20MHz and are designed to give excellent frequency stability by carefully limiting the amplitude of oscillation via internal amplitude discrimination and limiting circuitry. They provide sine-wave and square-wave outputs (both "TTL" and "ECL" logic levels).

An even more convenient alternative, if you're willing to accept a square wave output only, and if utmost stability isn't needed, is the use of complete crystal oscillator modules, usually provided as DIP IC-sized metal packages. They come in lots of standard frequencies (e.g., 1, 2, 4, 5, 6, 8, 10, 16, and 20MHz), as well as weird frequencies commonly used in microprocessor systems (e.g., 14.31818MHz, used for video boards). These "crystal clock modules" typically provide accuracies (over temperature, power supply voltage, and time) of only 0.01% (100ppm), but you get it cheap ($2 to $5), and you don't have to wire up any circuitry. Furthermore, they are guaranteed to oscillate, which isn't by any means assured when you wire your own oscillator: Crystal oscillator circuits depend on electrical properties of the crystal (such as series versus parallel mode, effective series resistance, and mount capacitance) that aren't always well specified. All too often you may find that your home-built crystal oscillator oscillates, but at a frequency unrelated to that stamped on the crystal! Our own experience with discrete crystal oscillator circuits has been, well, checkered.

Quartz crystals are available from about 10kHz to about 10MHz, with overtone-mode crystals going to about 250MHz. Although crystals have to be ordered for a given frequency, most of the commonly used frequencies are available off the shelf. Frequencies such as 100kHz, 1.0MHz, 2.0MHz, 4.0MHz, 5.0MHz, and 10.0MHz are always easy to get. A 3.579545MHz crystal (available for less than a dollar) is used in TV color-burst oscillators. Digital wristwatches use 32.768kHz (divide by 2^{15} to get 1Hz), and other powers of 2 are also common. A crystal oscillator can be adjusted slightly by varying a series or parallel capacitor, as shown in Figure 5.49D. Given the low cost of crystals (typically about 2 to 5 dollars), it is worth considering a crystal oscillator in any application where you would have to strain the capabilities of *RC* relaxation oscillators.

If you need a stable frequency with a very small amount of electrical tunability, you can use a varactor to "pull" the frequency of a quartz-crystal oscillator. The resulting circuit is called a "VCXO"

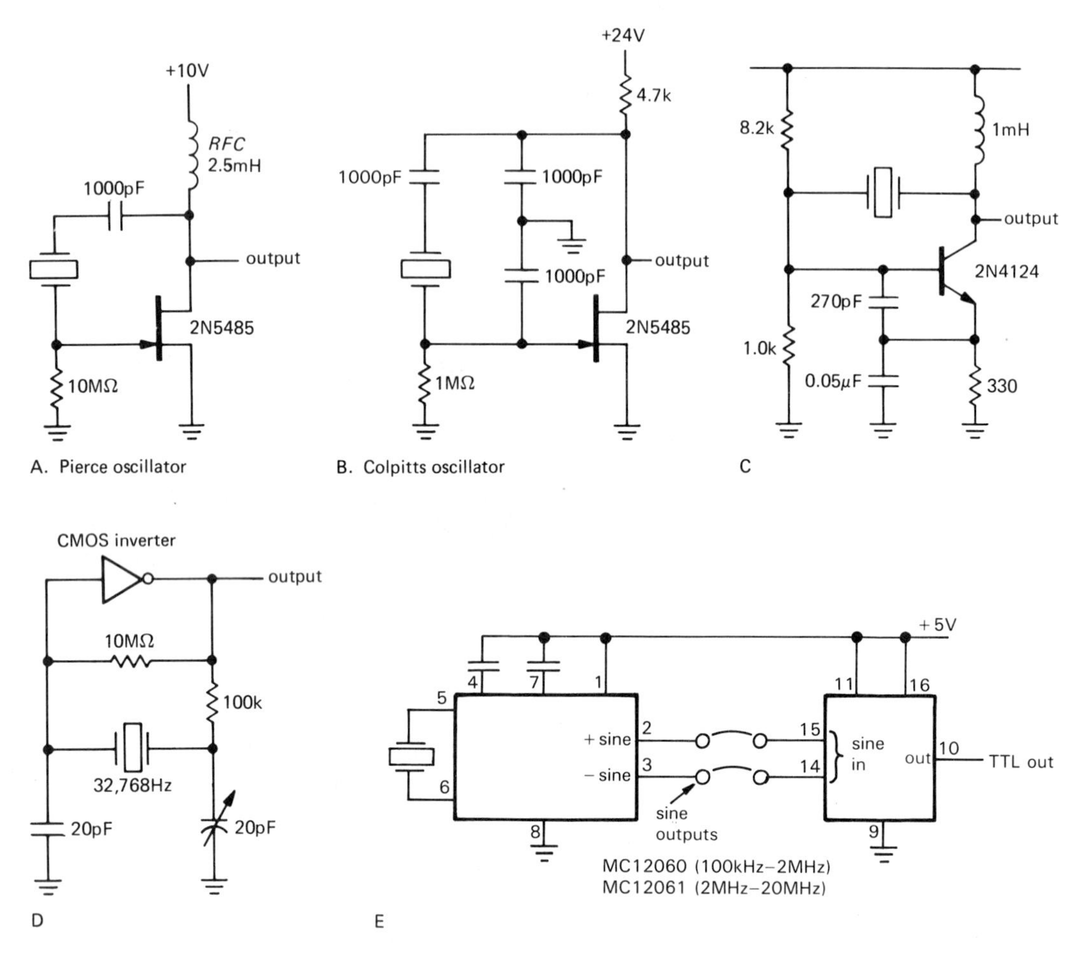

Figure 5.49. Various crystal oscillators.

(voltage-controlled crystal oscillator), and combines the good-to-excellent stability of crystal oscillators with the tunability of *LC* oscillators. The best approach is probably to buy a commercial VCXO, rather than attempt to design your own. Typically they produce maximum deviations of ±10ppm to ±100ppm from center frequency, though wide-deviation units (up to ±1000ppm) are also available.

Without great care you can obtain frequency stabilities of a few parts per million over normal temperature ranges with crystal oscillators. By using temperature-compensation schemes you can make a TCXO (temperature-compensated crystal oscillator) with somewhat better performance. Both TCXOs and uncompensated oscillators are available as complete modules from many manufacturers, e.g., Bliley, CTS Knights, Motorola, Reeves Hoffman, Statek, and Vectron. They come in various sizes, ranging down to DIP packages and TO-5 standard transistor cans. TCXOs deliver stabilities of 1ppm over the range 0°C to 50°C (inexpensive) down to 0.1ppm over the same range (expensive).

Temperature-stabilized oscillators

For the utmost in stability, you may need a crystal oscillator in a constant-temperature oven. A crystal with a zero temperature coefficient at some elevated temperature (80°C to 90°C) is used, with the thermostat set to maintain that temperature. Such oscillators are available as small modules for inclusion into an instrument or as complete frequency standards ready for rack mounting. The 10811 from Hewlett-Packard is typical of high-performance modular oscillators, delivering 10MHz with stabilities of a few parts in 10^{11} over periods of seconds to hours.

When thermal instabilities have been reduced to this level, the dominant effects become crystal "aging" (the frequency tends to decrease continuously with time), power-supply variations, and environmental influences such as shock and vibration (the latter are the most serious problems in quartz wristwatch design). To give an idea of the aging problem, the oscillator mentioned previously has a specified aging rate at delivery of 5 parts in 10^{10} per day, maximum. Aging effects are due in part to the gradual relief of strains, and they tend to settle down after a few months, particularly in a well-manufactured crystal. Our specimen of the 10811 oscillator ages about 1 part in 10^{11} per day.

Atomic frequency standards are used where the stability of ovenized-crystal standards is insufficient. These use a microwave absorption line in a rubidium gas cell, or atomic transitions in an atomic cesium beam, as the reference to which a quartz crystal is stabilized. Accuracy and stability of a few parts in 10^{12} can be obtained. Cesium-beam standards are the official timekeepers in this country, with timing transmissions from the National Bureau of Standards and the Naval Observatory. Atomic hydrogen masers have been suggested as the ultimate in stable clocks, with claimed stabilities approaching a few parts in 10^{14}. Recent research in stable clocks has centered on techniques using "cooled ions" to achieve even better stability. Many physicists believe that ultimate stabilities of parts in 10^{18} may be possible.

SELF-EXPLANATORY CIRCUITS

5.20 Circuit ideas

Figure 5.51 presents a variety of circuit ideas, mostly taken from manufacturers' data sheets and applications literature.

ADDITIONAL EXERCISES

1. Design a 6-pole high-pass Bessel filter with cutoff frequency 1kHz.
2. Design a 60Hz twin-T notch filter with op-amp input and output buffers.
3. Design a sawtooth-wave oscillator, to deliver 1kHz, by replacing the charging resistor in the 555 oscillator circuit with a transistor current source. Be sure to provide enough current-source compliance. What value should R_B (Fig. 5.33) have?
4. Make a triangle-wave oscillator with a 555. Use a pair of current sources I_0 (sourcing) and $2I_0$ (sinking). Use the 555's output to switch the $2I_0$ current sink on and off appropriately. The following figure shows one possibility.

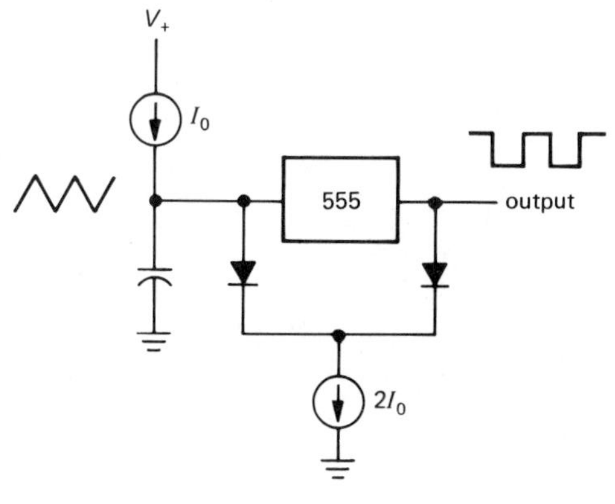

Figure 5.50

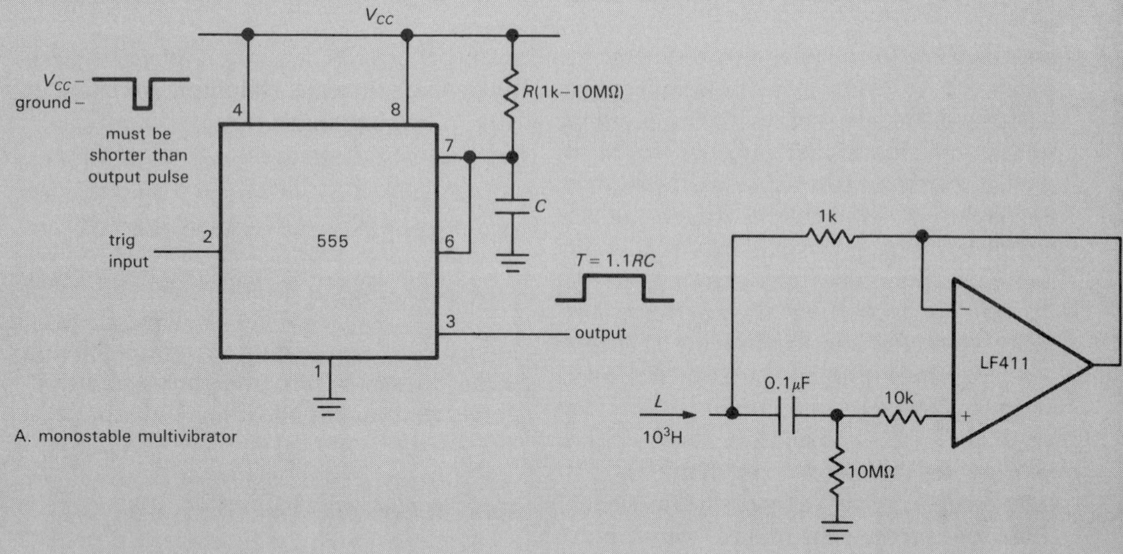

A. monostable multivibrator

B. active inductor

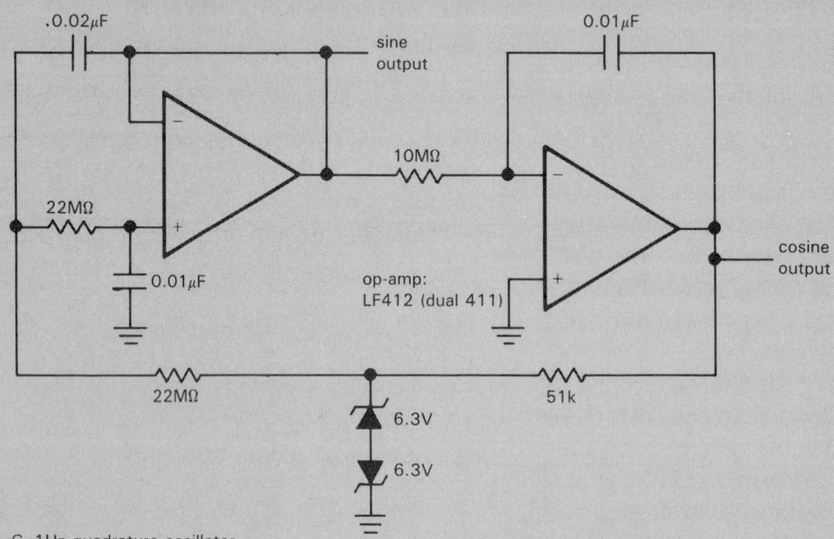

C. 1Hz quadrature oscillator

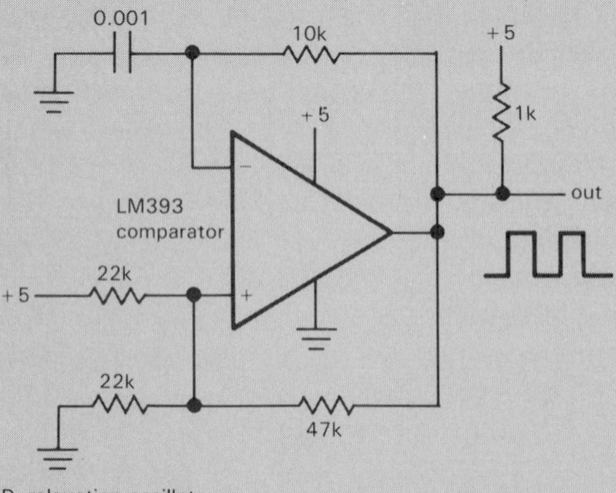

D. relaxation oscillator

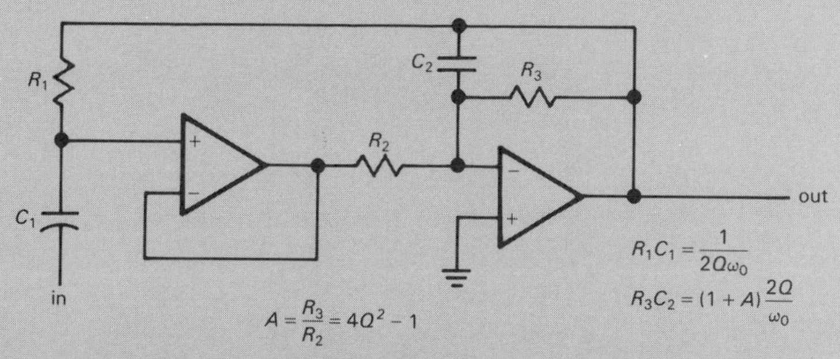

E. tuned amplifier; works for f_0 up to $f_T/2Q$. *Proc. IEEE 60*, 908 (1972)

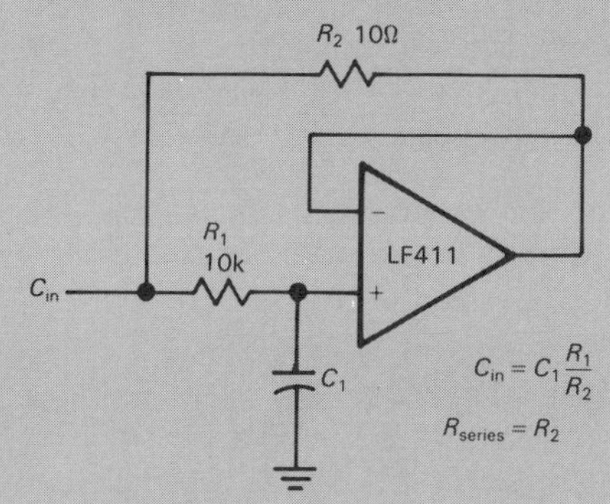

F. capacitance multiplier

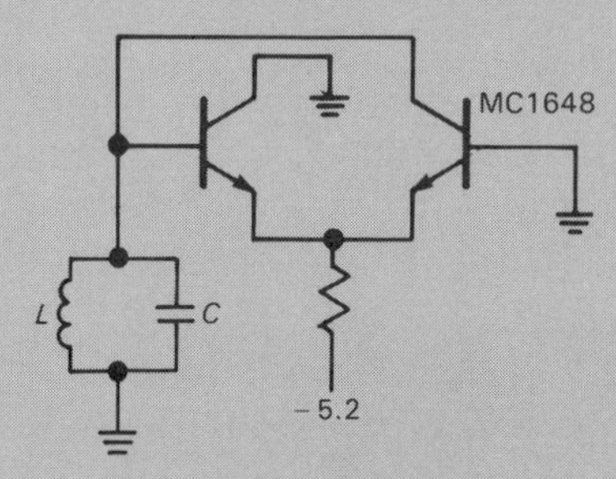

G. emitter-coupled *LC* oscillator

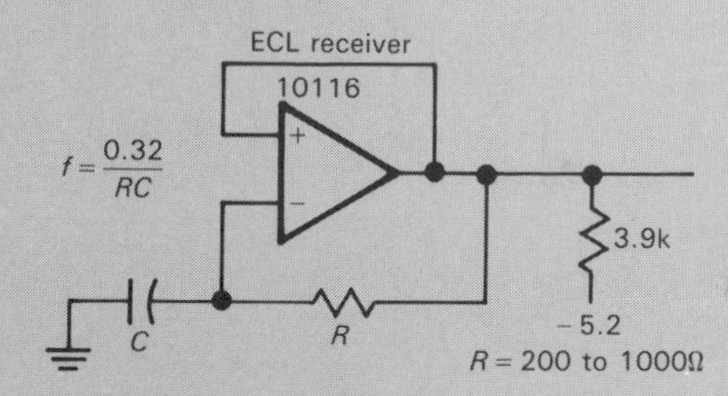

H. high-frequency ECL multivibrator

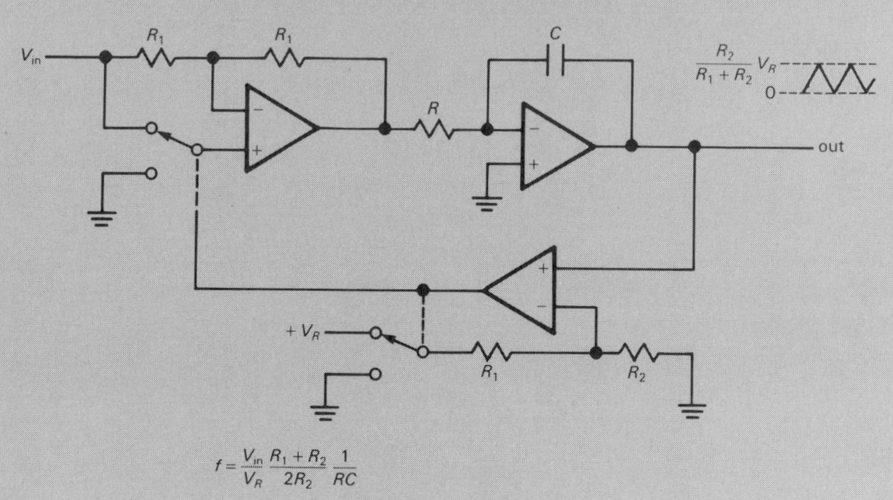

I. voltage-to-frequency converter

Circuit ideas (*cont.*)

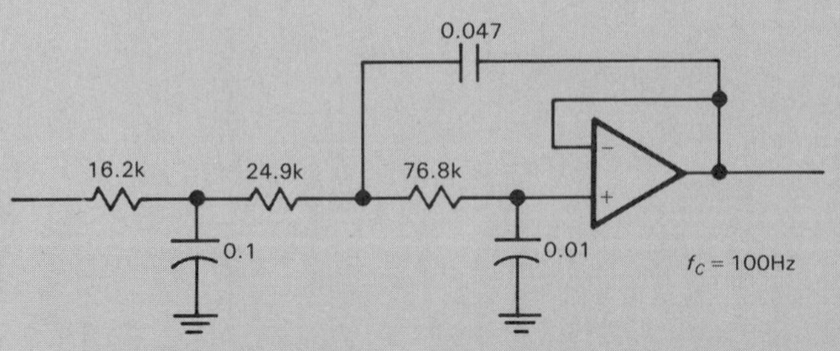

J. 3rd-order Bessel low-pass filter
scale values for other frequencies

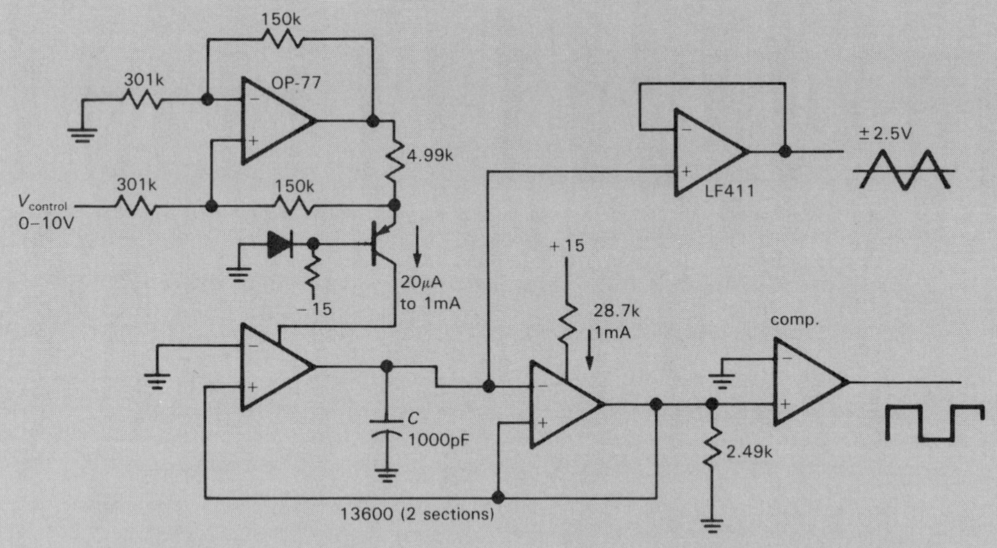

K. wide-range VCO (2Hz–100kHz) with
transconductance amplifiers

Nearly all electronic circuits, from simple transistor and op-amp circuits up to elaborate digital and microprocessor systems, require one or more sources of stable dc voltage. The simple transformer-bridge-capacitor unregulated power supplies we discussed in Chapter 1 are not generally adequate because their output voltages change with load current and line voltage and because they have significant amounts of 120Hz ripple. Fortunately, it is easy to construct stable power supplies using negative feedback to compare the dc output voltage with a stable voltage reference. Such regulated supplies are in universal use and can be simply constructed with integrated circuit voltage regulator chips, requiring only a source of unregulated dc input (from a transformer-rectifier-capacitor combination, a battery, or some other source of dc input) and a few other components.

In this chapter you will see how to construct voltage regulators using special-purpose integrated circuits. The same circuit techniques can be used to make regulators with discrete components (transistors, resistors, etc.), but because of the availability of inexpensive high-performance regulator chips, there is no advantage to using discrete components in new designs. Voltage regulators get us into the domain of high power dissipation, so we will be talking about heat sinking and techniques like "foldback limiting" to limit transistor operating temperatures and prevent circuit damage. These techniques can be used for all sorts of power circuits, including power amplifiers. With the knowledge of regulators you will have at that point, we will be able to go back and discuss the design of the unregulated supply in some detail. In this chapter we will also look at voltage references and voltage-reference ICs, devices with uses outside of power-supply design.

BASIC REGULATOR CIRCUITS WITH THE CLASSIC 723

6.01 The 723 regulator

The μA723 voltage regulator is a classic. Designed by Bob Widlar and first introduced in 1967, it is a flexible, easy-to-use regulator with excellent performance.

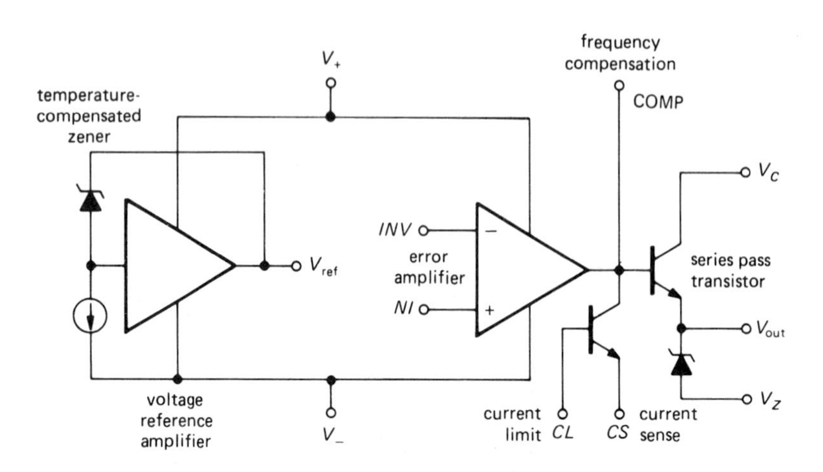

Figure 6.1. Simplified circuit of the 723 regulator. (Courtesy of Fairchild Camera and Instrument Corp.)

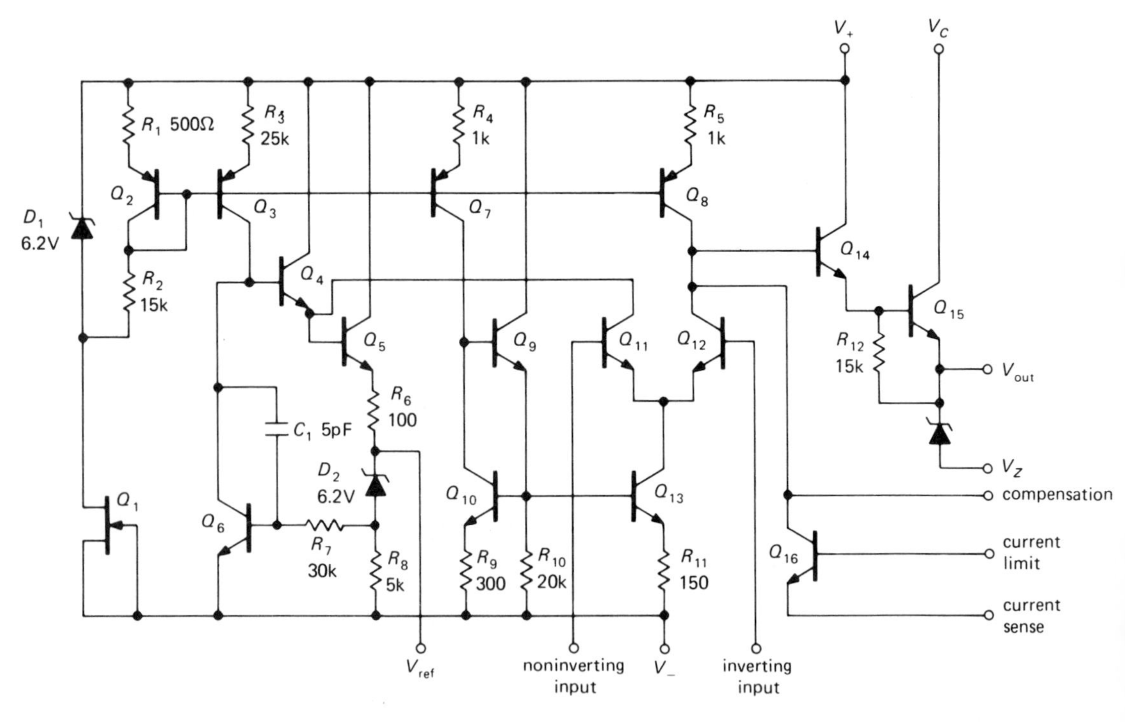

Figure 6.2. Schematic of the 723 regulator. (Courtesy of Fairchild Camera and Instrument Corp.)

Although you would not choose it for a new design nowadays, it is worth looking at in some detail, since more recent regulators work on the same principles. Its circuit is shown in Figures 6.1 and 6.2. As you can see, it is really a power-supply kit, containing a temperature-compensated voltage reference, differential amplifier, series pass transistor, and current-limiting protective circuit. As it comes, the 723 doesn't regulate anything. You have to hook up an external circuit to make it do what you want. Before going on to design regulators with it, let's look briefly at its internal circuit. It is straightforward and easy to understand (the innards of many ICs aren't).

The heart of the regulator is the temperature-compensated zener reference. Zener D_2 has a positive temperature coefficient, so its voltage is added to Q_6's base-emitter drop (remember, V_{BE} has a negative temperature coefficient of roughly -2mV/°C) to form a voltage reference (nominally 7.15V) of nearly zero temperature coefficient (typically 0.003%/°C). Q_4 through Q_6 are arranged to bias D_2 at $I = V_{BE}/R_8$ via negative feedback at dc, as indicated on the block diagram. Q_2 and Q_3 form an unsymmetrical current mirror to bias the reference; current to the mirror is set by D_1 and R_2 (their junction is fixed at 6.2V below V_+), which in turn is biased by Q_1 (the FET behaves roughly like a current source).

Q_{11} and Q_{12} form the differential amplifier (sometimes called the "error amplifier," thinking of the whole thing as an exercise in negative feedback), a classic long-tailed pair with emitter current source Q_{13}. The latter is half of a current mirror (Q_9, Q_{10}, and Q_{13}), driven in turn from current mirror Q_7 (Q_3, Q_7, and Q_8 all mirror the current generated by the D_1 reference, as we mentioned in Section 2.14). Q_{11}'s collector is tied to the fixed positive voltage at Q_4's emitter, and the error amplifier's output is taken from Q_{12}'s collector. Current mirror Q_8 supplies the latter's collector load. Q_{14} drives the pass transistor Q_{15}, in a not-quite-Darlington connection. Note that Q_{15}'s collector is brought out separately, to allow for separate positive supplies. By turning on Q_{16} you cut off drive to the pass transistors; this is used to limit output currents to nondestructive levels. Unlike many of the newer regulators, the 723 does not incorporate internal shut-down circuitry to protect against excessive load current or chip dissipation. The SG3532 and LAS1000 are improved 723-type regulators, with low-voltage bandgap reference (Section 6.15), internal current limiting, and thermal-overload shutdown circuitry.

6.02 Positive regulator

Figure 6.3 shows how to make a positive voltage regulator with the 723. All the components except the four resistors and the two capacitors are contained on the 723. Voltage divider R_1R_2 compares a fraction of the output with the voltage reference, and the 723 components do the rest; this circuit is identical with the op-amp noninverting amplifier with emitter follower, with V_{ref} as the "input." R_4 is chosen for about 0.5 volt drop at maximum desired output current, since a V_{BE} drop applied across the CL-CS inputs will turn on the current-limiting transistor (Q_{16} in Fig. 6.2), shutting off base drive to the output pass transistor. The 100pF capacitor stabilizes the loop. R_3 (sometimes omitted) is chosen so that the differential amplifier sees equal impedances at its inputs. This makes the output insensitive to changes in bias current (with changes in temperature, say), in the same way as we saw with op-amps (Section 4.12).

With this circuit, a regulated supply with output voltage ranging from V_{ref} to the maximum allowable output voltage (37V) can be made. Of course, the input voltage must stay a few volts more positive than the output at all times, including the effects of ripple on the unregulated supply. The "dropout voltage" (the amount by which the input voltage must exceed the regulated output voltage) is specified as 3 volts (minimum) for the 723, a value typical of most regulators. R_1 or R_2 is usually made adjustable, or trimmable, so the output voltage can be set precisely. The production spread in V_{ref} is 6.8 to 7.5 volts.

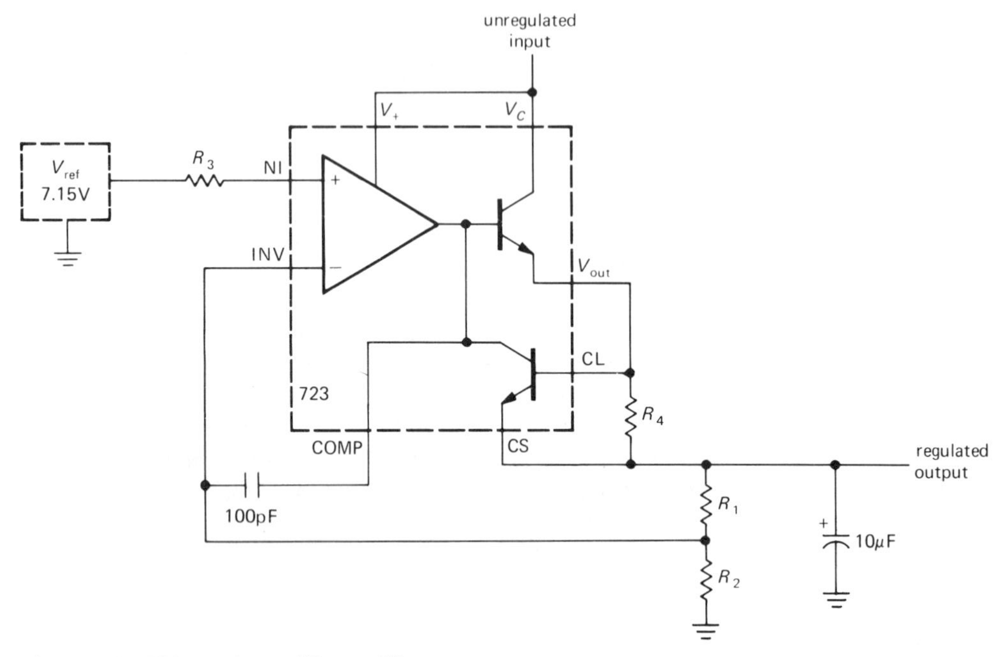

Figure 6.3. 723 regulator: $V_{out} > V_{ref}$.

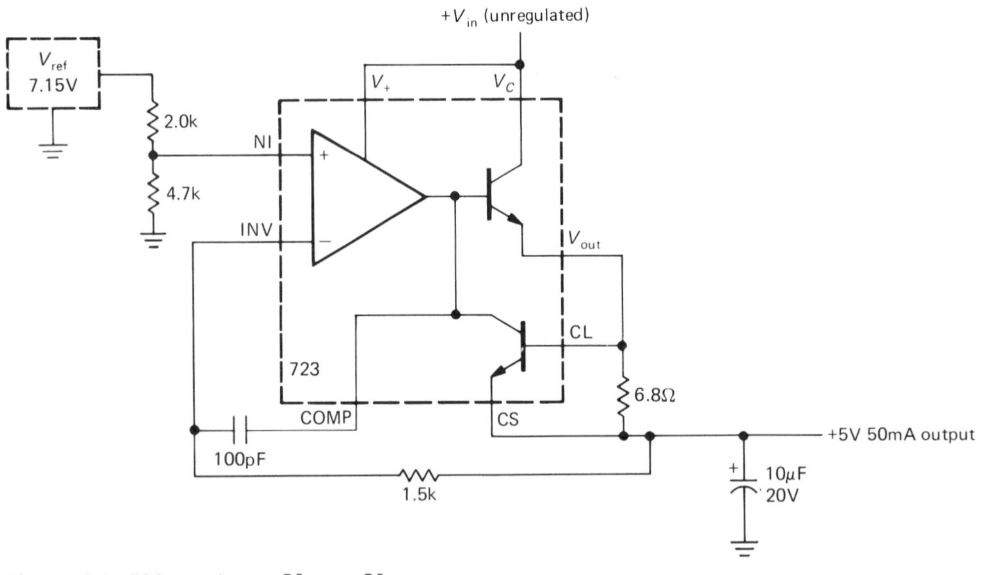

Figure 6.4. 723 regulator: $V_{out} < V_{ref}$.

It is usually a good idea to put a capacitor of a few microfarads across the output, as shown. This keeps the output impedance low even at high frequencies, where the feedback becomes less effective. It is best to use the output capacitor value recommended on the specification sheet, since oscillations can occur otherwise. In general, it is a good idea to bypass power-supply leads to ground liberally throughout a circuit, using a combination of ceramic types (0.01–0.1μF) and electrolytic or tantalum types (1–10μF).

For output voltages less than V_{ref}, you just put the voltage divider on the reference (Fig. 6.4). Now the full output voltage is compared with a fraction of the reference. The values shown are for +5 volts 50mA max. With this circuit configuration, output voltages from +2 volts to V_{ref} can be produced. The output cannot be adjusted down to zero volts because the differential amplifier will not operate below 2 volts input. This is given as a manufacturer's specification (see Table 6.9). With this circuit the unregulated input voltage must never drop below +9.5 volts, the voltage necessary to power the reference.

A third variation of this circuit is necessary if you want a regulator that is continuously adjustable through a range of output voltages around V_{ref}. In such cases, just compare a divided fraction of the output with a fraction of V_{ref} chosen to be less than the minimum output voltage desired.

EXERCISE 6.1

Design a regulator to deliver up to 50mA load current over an output voltage range of +5 to +10 volts, using a 723. Hint: Compare a fraction of the output voltage with $0.5V_{ref}$.

6.03 High-current regulator

The internal pass transistor in the 723 is rated at 150mA maximum; in addition, the power dissipation must not exceed 1 watt at 25°C (less at higher ambient temperatures; the 723 must be "derated" at 8.3mW/°C above 25°C in order to keep the junction temperature within safe limits). Thus, for instance, a 5 volt regulator with +15 volts input cannot deliver more than about 80mA to the load. To construct a higher-current supply, an external pass transistor must be used. It is easy to add one as a Darlington pair with the internal transistor (Fig. 6.5). Q_1 is the external pass transistor; it must be mounted on a *heat sink*, most often a finned metal plate designed to carry off heat (alternatively, the transistor can be mounted to one wall of the metal chassis housing the power supply). We will deal with thermal problems like these in the next section. A trimmer potentiometer has been used so that the output can be set accurately to +5 volts; its range of adjustment should be sufficient to allow for resistor tolerances as well as the maximum specified spread in V_{ref} (this is an example of worst-case design), and in this case it allows about ±1 volt adjustment from the nominal output voltage. Note the low-resistance high-power current-limiting resistor necessary for a 2 amp supply.

Pass transistor dropout voltage

One problem with this circuit is the high power dissipation in the pass transistor (at least 10W at full load current). This is unavoidable if the regulator chip is powered by the unregulated input, since it needs a few volts of "headroom" to operate (specified by the dropout voltage). With the use of a separate low-current supply for the 723 (e.g., +12V), the minimum unregulated input to the external pass transistor can be only a volt or so above the regulated output voltage (although you will always have to allow at least a few volts, since worst-case design dictates proper operation even at 105V ac line input).

Overvoltage protection

Also shown in this circuit is an *overvoltage crowbar* protection circuit consisting of

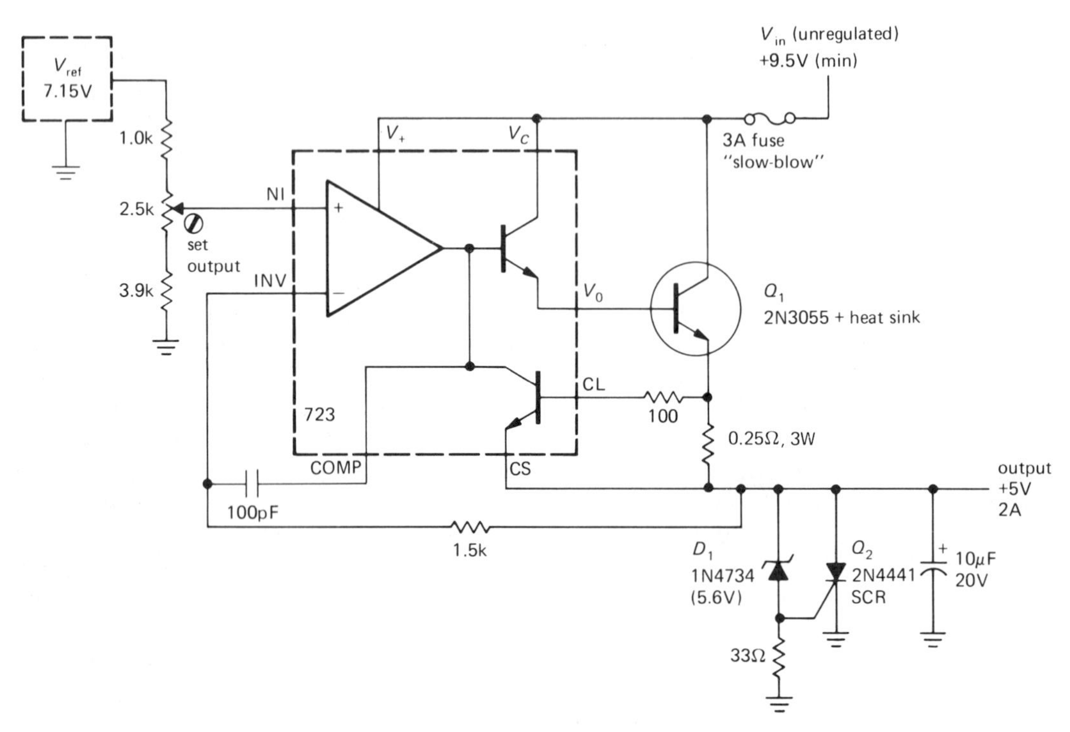

Figure 6.5. Five volt regulator with outboard pass transistor and crowbar.

D_1, Q_2, and the 33 ohm resistor. Its function is to short the output if some circuit fault causes the output voltage to exceed about 6.2 volts (this could happen if one of the resistors in the divider were to open up, for instance, or if some component in the 723 were to fail). Q_2 is an SCR (silicon-controlled rectifier), a device that is normally nonconducting but that goes into saturation when the gate-cathode junction is forward-biased. Once turned on, it will not turn off again until anode current is removed externally. In this case, gate current flows when the output exceeds D_1's zener voltage plus a diode drop. When that happens, the regulator will go into a current-limiting condition, with the output held near ground by the SCR. If the failure that produces the abnormally high output also disables the current-limiting circuit (e.g., a collector-to-emitter short in Q_1), then the crowbar will sink a very large current. For this reason it is a good idea to include a fuse somewhere in the power supply, as shown. We will treat overvoltage crowbar circuits in more detail in Section 6.06.

HEAT AND POWER DESIGN

6.04 Power transistors and heat sinking

As in the preceding circuit, it is often necessary to use power transistors or other high-current devices like SCRs or power rectifiers that can dissipate many watts. The 2N3055, an inexpensive power transistor of great popularity, can dissipate as much as 115 watts if properly mounted. All power devices are packaged in cases that permit contact between a metal surface and an external heat sink. In most cases the metal surface of the device is electrically connected to one terminal (e.g., for power transistors the case is always connected to the collector).

The whole point of heat sinking is to keep the transistor junction (or the junction of some other device) below some maximum specified operating temperature. For silicon transistors in metal packages the maximum junction temperature is usually 200°C, whereas for transistors in plastic packages it is usually 150°C. Table 6.1 lists some useful power transistors, along with their thermal properties. Heat sink design is then simple: Knowing the maximum power the device will dissipate in a given circuit, you calculate the junction temperature, allowing for the effects of heat conductivity in the transistor, heat sink, etc., and the maximum ambient temperature in which the circuit is expected to operate. You then choose a heat sink large enough to keep the junction temperature well below the maximum specified by the manufacturer. It is wise to be conservative in heat sink design, since transistor life drops rapidly at operating temperatures near or above maximum.

Thermal resistance

To carry out heat sink calculations, you use *thermal resistance*, θ, defined as heat rise (in degrees) divided by power transferred. For heat transferred entirely by conduction, the thermal resistance is a constant, independent of temperature, that depends only on the mechanical properties of the joint. For a succession of thermal joints in "series," the total thermal resistance is the sum of the thermal resistances of the individual joints. Thus, for a transistor mounted on a heat sink, the total thermal resistance from transistor junction to the outside (ambient) world is the sum of the thermal resistance from junction to case θ_{JC}, the thermal resistance from case to heat sink, θ_{CS}, and the thermal resistance from heat sink to ambient θ_{SA}. The temperature of the junction is therefore

$$T_J = T_A + (\theta_{JC} + \theta_{CS} + \theta_{SA})P$$

where P is the power being dissipated.

Let's take an example. The preceding power-supply circuit, with external pass transistor, has a maximum transistor dissipation of 20 watts for an unregulated input of +15 volts (10V drop, 2A). Let's assume that the power supply is to operate at ambient temperatures up to 50°C, not unreasonable for electronic equipment packaged together in close quarters. And let's try to keep the junction temperature below 150°C, well below its specified maximum of 200°C. The thermal resistance from junction to case is 1.5°C per watt. A TO-3 power transistor package mounted with an insulating washer and heat-conducting compound has a thermal resistance from case to heat sink of about 0.3°C per watt. Finally, a Wakefield model 641 heat sink (Fig. 6.6) has a thermal resistance from sink to ambient of about 2.3°C per watt. So the total thermal resistance from junction to ambient is about 4.1°C per watt. At 20 watts dissipation the junction will be 84°C above ambient, or 134°C (at maximum ambient temperature) in this example. The chosen heat sink will be adequate; in fact, a smaller one could be used if necessary to save space.

Comments on heat sinks

1. Where very high power dissipation (several hundred watts, say) is involved, forced air cooling may be necessary. Large heat sinks designed to be used with a blower are available with thermal resistances (sink to ambient) as small as 0.05°C to 0.2°C per watt.

2. When the transistor must be insulated from the heat sink, as is usually necessary (especially if several transistors are mounted on the same sink), a thin insulating washer is used between the transistor and sink, and insulating bushings are used around the mounting screws. Washers are available in standard

TABLE 6.1. SELECTED BIPOLAR POWER TRANSISTORS

npn	pnp	Pkg[a]	V_{CEO} max (V)	I_C max (A)	h_{FE} typ	@ I_C (A)	f_T min (MHz)	C_{cb}[b] typ (pF)	P_{diss} (T_C=25°C) (W)	Θ_{JC} (°C/W)	T_J max (°C)	Comments
Regular power: V_{CE}(sat) = 0.4V (typ); V_{BE}(on) = 0.8V (typ)												
2N5191	**2N5194**	A	60	4	100	0.2	2	80	40	3.1	150	low cost, gen purp
2N5979	**2N5976**	B	80	5	50	0.5	2	60	70	1.8	150	
2N3055	**MJ2955**	TO-3	60	15	50	2	2.5	125	115	1.5	200	metal, indus std
MJE3055	**MJE2955**	B	60	10	50	2	2.5	125	90	1.4	150	plastic, indus std
2N5886	**2N5884**	TO-3	80	25	50	10	4	400	200	0.9	200	
2N5686	**2N5684**	TO-3	80	50	30	25	2	700	300	0.6	200	for real power jobs
2N6338	**2N6437**	TO-3	100	25	50	8	40	200	200	0.9	200	premium audio
2N6275	**2N6379**	TO-3	120	50	50	20	30	400	250	0.7	200	premium audio
Darlington power: V_{CE}(sat) = 0.8V (typ); V_{BE}(on) = 1.4V (typ)												
2N6038	**2N6035**	A	60	4	2000	2	–	30	40	3.1	150	low cost
2N6044	**2N6041**	B	80	8	2500	4	4	80	75	1.7	150	
2N6059	**2N6052**	TO-3	100	12	3500	5	4	100	150	1.2	200	
2N6284	**2N6287**	TO-3	100	20	3000	10	4	150	160	1.1	200	high current

(a) A: small plastic pwr pkg (TO-126). B: large plastic pwr pkg (TO-127). (b) C_{cb} (npn) at V_{CB}=10V; C_{cb} (pnp) ≈ $2C_{cb}$ (npn).

transistor-shape cutouts made from mica, insulated aluminum, or beryllia (BeO). Used with heat-conducting grease, these add from 0.14°C per watt (beryllia) to about 0.5°C per watt.

An attractive alternative to the classic mica-washer-plus-grease is provided by greaseless silicone-based insulators that are loaded with a dispersion of thermally conductive compound, usually boron nitride or aluminum oxide. They're clean and dry, and easy to use; you don't get white slimy stuff all over your hands, your electronic device, and your clothes. You save lots of time. They have thermal resistances of about 0.2–0.4°C per watt, comparable to values with the messy method. Bergquist calls its product "Sil-Pad," Chomerics calls its "Cho-Therm," SPC calls it "Koolex," and Thermalloy calls its "Thermasil." We've been using these insulators, and we like them.

3. Small heat sinks are available that simply clip over the small transistor packages (like the standard TO-5). In situations of relatively low power dissipation (a watt or two) this often suffices, avoiding the nuisance of mounting the transistor remotely on a heat sink with its leads brought back to the circuit. An example is shown in Figure 6.6. In addition, there are various small heat sinks intended for use with the plastic power packages (many regulators, as well as power transistors, come in this package) that mount right on a printed-circuit board underneath the package. These are very handy in situations of a few watts dissipation; a typical unit is illustrated in Figure 6.6.

4. Sometimes it may be convenient to mount power transistors directly to the chassis or case of the instrument. In such cases it is wise to use conservative design (keep it cool), especially since a hot case will subject the other circuit components to high temperatures and shorten component life.

5. If a transistor is mounted to a heat sink without insulating hardware, the heat sink must be insulated from the chassis.

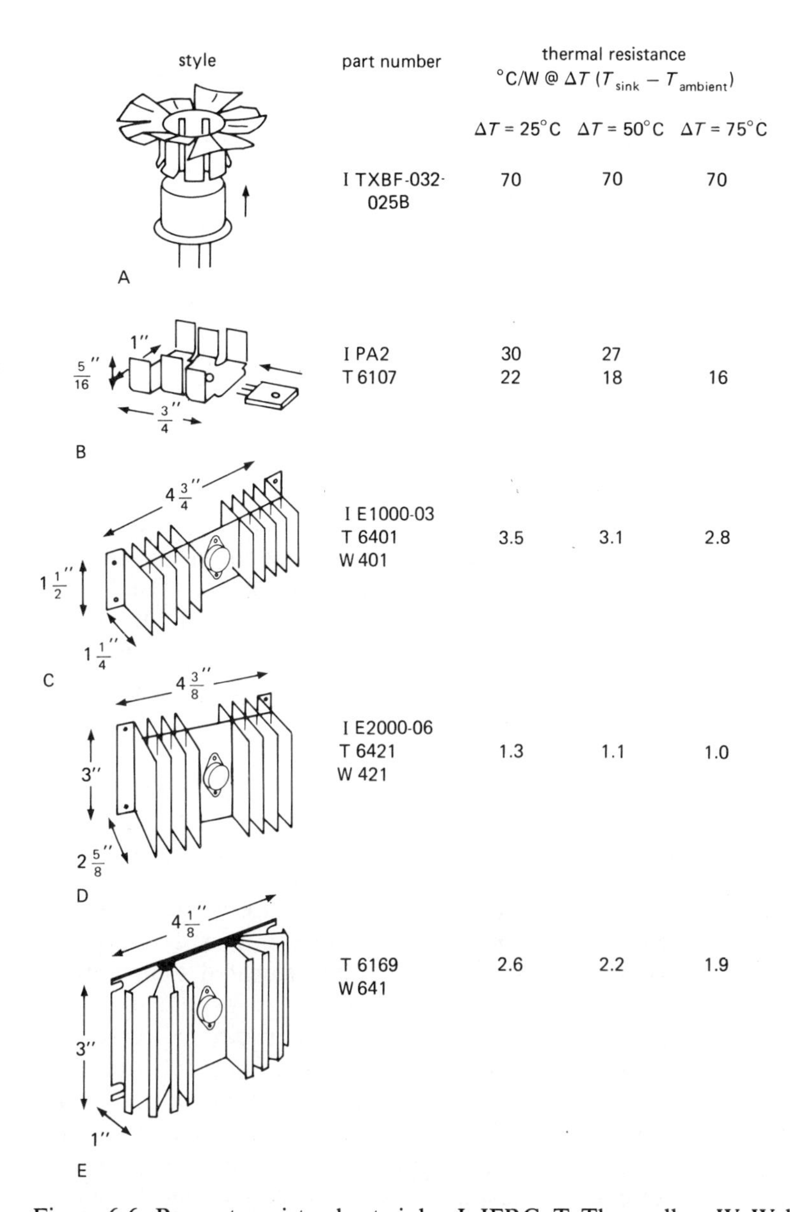

style	part number	thermal resistance °C/W @ ΔT ($T_{sink} - T_{ambient}$) ΔT = 25°C	ΔT = 50°C	ΔT = 75°C
A	I TXBF-032-025B	70	70	70
B	I PA2	30	27	
	T 6107	22	18	16
C	I E1000-03 T 6401 W 401	3.5	3.1	2.8
D	I E2000-06 T 6421 W 421	1.3	1.1	1.0
E	T 6169 W 641	2.6	2.2	1.9

Figure 6.6. Power transistor heat sinks. I, IERC; T, Thermalloy; W, Wakefield.

The use of insulating washers (e.g., Wakefield model 103) is recommended (unless, of course, the transistor case happens to be at ground). When the transistor is insulated from the sink, the heat sink may be attached directly to the chassis. But if the transistor is accessible from outside the instrument (e.g., if the heat sink is mounted externally on the rear wall of the box), it is a good idea to use an insulating cover over the transistor (e.g., Thermalloy 8903N) to prevent someone from accidentally coming in contact with it, or shorting it to ground.

6. The thermal resistance from heat sink to ambient is usually specified for the sink mounted with the fins vertical and with unobstructed flow of air. If the sink is mounted differently, or if the air flow is obstructed, the efficiency will be reduced (higher thermal resistance); usually it is best to mount it on the rear of the instrument with fins vertical.

EXERCISE 6.2

A 2N5320, with a thermal resistance from junction to case of 17.5°C per watt, is fitted with an IERC TXBF slip-on heat sink of the type shown in Figure 6.6. The maximum permissible junction temperature is 200°C. How much power can you dissipate with this combination at 25°C ambient temperature? How much must the dissipation be decreased per degree rise in ambient temperature?

□ 6.05 Foldback current limiting

For a regulator with simple current limiting, transistor dissipation is maximum when the output is shorted to ground (either accidentally or through some circuit malfunction), and it usually exceeds the maximum value of dissipation that would otherwise occur under normal load conditions. For instance, the pass transistor in the preceding +5 volt 2 amp regulator circuit will dissipate 30 watts with the output shorted (+15V input, current limit at 2A), whereas the worst-case dissipation under normal load conditions is 20 watts (10V drop at 2A). The situation is even worse in circuits in which the voltage normally dropped by the pass transistor is a smaller fraction of the output voltage. For instance, in a +15 volt 2 amp regulated supply with +25 volt unregulated input, the transistor dissipation rises from 20 watts (full load) to 50 watts (short circuit).

You get into a similar problem with push-pull power amplifiers. Under normal conditions you have maximum load current when the voltage across the transistors is minimum (near the extremes of output swing), and you have maximum voltage across the transistors when the current is nearly zero (zero output voltage). With a short-circuit load, on the other hand, you have maximum load current at the worst possible time, namely, with full supply voltage across the transistor. This results in much higher transistor dissipation than normal.

The brute-force solution to this problem is to use massive heat sinks and transistors of higher power rating (and safe operating area, see Section 6.07) than necessary. Even so, it isn't a good idea to have large currents flowing into the powered circuit under fault conditions, since other components in the circuit may then be damaged. The best solution is to use *foldback* current limiting, a circuit technique that reduces the output current under short-circuit or overload conditions. Figure 6.7 shows the basic configuration, again illustrated with a 723 with external pass transistor.

The divider at the base of the current-limiting transistor Q_L provides the foldback. At +15 volts output (the normal value) the circuit will limit at about 2 amps, since Q_L's base is then at +15.5 volts while its emitter is at +15 (V_{BE} is about 0.5V at the elevated temperatures at which regulator chips are normally run). But the short-circuit current is less; with the output shorted to ground, the output current is about 0.5 amp, holding Q_1's dissipation down to less than in the full-load case. This is highly desirable, since excessive heat sinking is not now required, and the thermal design need only satisfy the full-load requirements. The choice of the three resistors in the current-limiting circuit sets the short-circuit current, for a given full-load current limit. Warning: Use care in choosing the short-circuit current, since it is possible to be overzealous and design a supply that will not "start up" into a normal load. The short-circuit current should not be too small; as a

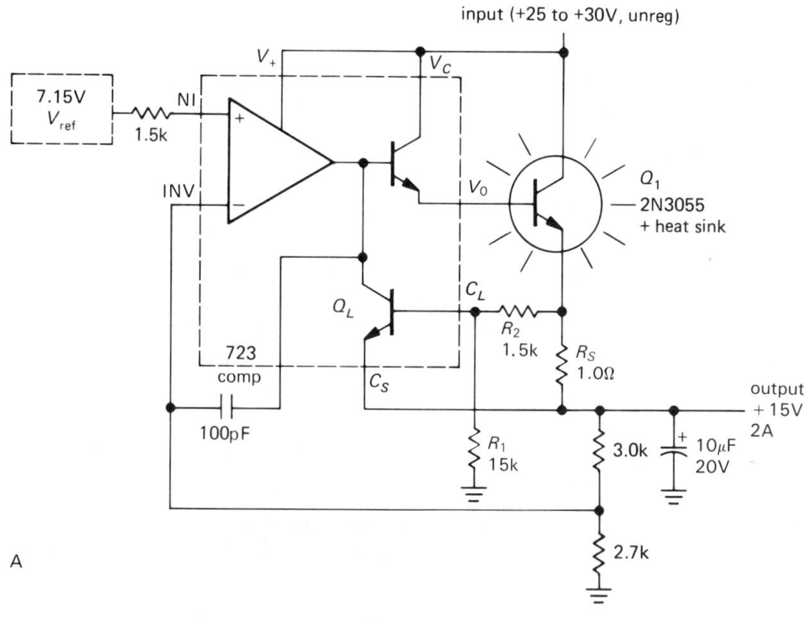

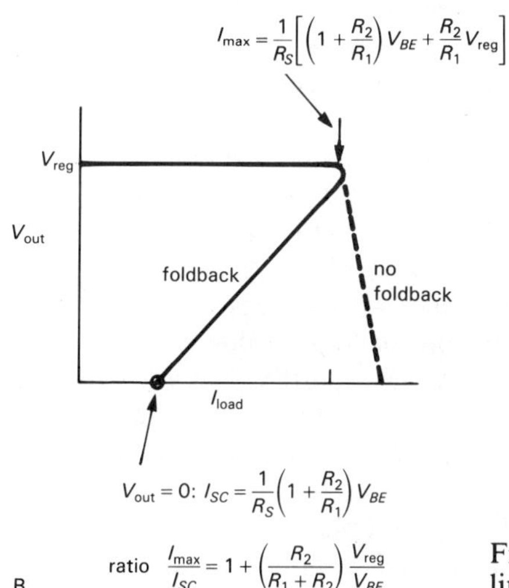

Figure 6.7. A. Power regulator with foldback current limiting. B. Output voltage versus load current.

rough guide, the short-circuit current limit should be set at about one-third the maximum load current at full output voltage.

EXERCISE 6.3

Design a 723 regulator with outboard pass transistor and foldback current limiting to provide up to 1.0 amp when the output is at its regulated value of +5.0 volts, but only 0.4 amp into a short-circuit load.

6.06 Overvoltage crowbars

As we remarked in Section 6.03, it is often a good idea to include some sort of overvoltage protection at the output of a regulated supply. Take, for instance, a +5 volt supply used to power a large digital system (you'll see lots of examples beginning in Chapter 8). The input to the regulator is probably in the range of +10 to +15 volts. If the series pass transistor fails by shorting its collector to emitter (a common failure mode), the full unregulated voltage will be applied to the circuit, with devastating results. Although a fuse probably will blow, what's involved is a race between the fuse and the "silicon fuse" that is constituted by the rest of the circuit; the rest of the circuit will probably respond first! This problem is most serious with TTL logic, which operates from a +5 volt supply, but cannot tolerate more than +7 volts without damage. Another situation with considerable disaster potential arises when you operate something from a wide-range "bench" supply, where the unregulated input may be 40 volts or more, regardless of the output voltage.

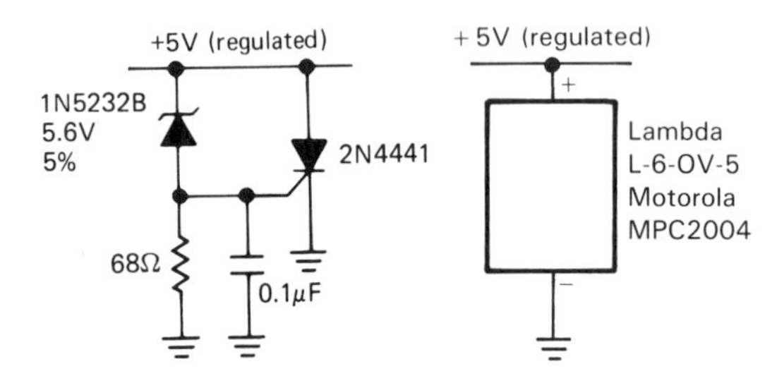

Figure 6.8. Overvoltage crowbars.

☐ *Zener sensing*

Figure 6.8 shows a popular crowbar circuit and a crowbar module. You hook the circuit between the regulated output terminal and ground. If the voltage exceeds the zener voltage plus a diode drop (about 6.2V for the zener shown), the SCR is turned on, and it remains in a conducting state until its anode current drops below a few milliamps. An inexpensive SCR like the 2N4441 can sink 5 amps continuously and withstand 80 amp surge currents; its voltage drop in the conducting state is typically 1.0 volt at 5 amps. The 68 ohm resistor is provided to generate a reasonable zener current (10mA) at SCR turn-on, and the capacitor is added to prevent crowbar triggering on harmless short spikes.

The preceding circuit, like all crowbars, puts an unrelenting 1 volt "short circuit" across the supply when triggered by an overvoltage condition, and it can be reset only by turning off the supply. Since the SCR maintains a low voltage while conducting, there isn't much problem with the crowbar itself failing from overheating. As a result, it is a reliable crowbar circuit. It is essential that the regulated supply have some sort of current limiting, or at least fusing, to handle the short. There may be overheating problems with the supply after the crowbar fires. In particular, if the supply includes internal current limiting, the fuse won't blow, and the supply will sit in the "crowbarred" state, with the output at low voltage, until someone notices. Foldback current limiting of the regulated supply would be a good solution here.

There are several problems with this simple crowbar circuit, mostly involving the choice of zener voltage. Zeners are available in discrete values only, with generally poor tolerances and (often) soft knees in the VI characteristic. The desired crowbar trigger voltage may involve rather tight tolerances. Consider a 5 volt supply used to power digital logic. There is typically a 5% or 10% tolerance on the supply voltage, meaning that the crowbar cannot be set less than 5.5 volts. The minimum permissible crowbar voltage is raised by the problem of transient response of a regulated supply: When the load current is changed quickly, the voltage can jump, creating a spike followed by some "ringing." This problem is exacerbated by remote sensing via long (inductive) sense leads. The resultant ringing puts glitches on the supply that we don't want to trigger the crowbar. The result is that the crowbar voltage should not be set less than about 6.0 volts, but it cannot exceed 7.0 volts without risk of damage to the logic circuits. When you fold in zener tolerance, the discrete voltages actually available, and SCR trigger voltage tolerances, you've got a tricky problem. In the example shown earlier, the crowbar threshold could lie between 5.9 volts and 6.6 volts, even using the relatively precise 5% zener indicated.

□ *IC sensing*

A nice solution to the problems of predictability and lack of adjustability in the simple zener/SCR crowbar circuit is to use a special crowbar trigger IC such as the MC3423–5, the TL431, or the MC34061–2. These inexpensive chips come in convenient packages (8-pin mini-DIP or 3-pin TO-92), they drive the SCR directly, and they're very easy to use. For example, the MC3425 has adjustable threshold and response time for its crowbar output, and in addition an *under*voltage sensor to signal your circuit that the supply voltage is low (very handy for circuits with microprocessors). It includes an internal reference and several comparators and drivers, and it requires only two external resistors, an optional capacitor, and an SCR to form a complete crowbar. These crowbar chips belong to a class of "power-supply supervisory circuits," which includes complex chips like the MAX691 that not only sense undervoltage but even switch over to battery backup when ac power fails, generate a power-on reset signal on return of normal power, and continually check for lockup conditions in microprocessor circuitry.

Modular crowbars

Why build it when you can buy it! From the designer's point of view the simplest crowbar of all is a 2-terminal gadget that says "crowbar" on top. You can buy just such a device from Lambda or Motorola, who offer a series of overvoltage protection modules in several current ranges. You just pick the voltage and current rating you need, and connect the crowbar across the regulated dc output. For example, the smallest units from Lambda are rated at 2 amps maximum, with the following set of fixed voltages (5V, 6V, 12V, 15V, 18V, 20V, and 24V). They're monolithic, come in a TO-66 package (small metal power transistor case), and cost $2.50 in small quantities. The Lambda monolithic 6 amp series comes in TO-3 packages (large metal power transistor case) and costs $5. They also make hybrid 12, 20, and 35 amp crowbars. Motorola's MPC2000 series are all monolithic (5V, 12V, and 15V only, rated at 7.5A, 15A, or 35A). The first two come in TO-220 (plastic power) packages, the last (available in 5V only) in TO-3 (metal power). The good news from Motorola is the incredibly low price: $1.96, $2.36, and $6.08 in small quantities for the three current ratings. One nice feature of these crowbars is the good accuracy; for example, the 5 volt units from Lambda have a specified trip point of 6.6 ± 0.2 volts.

□ *Clamps*

Another possible solution to overvoltage protection is to put a power zener, or its equivalent, across the supply terminals. This avoids the problems of false triggering on spikes, since the zener will stop drawing current when the overvoltage condition disappears (unlike an SCR, which has the memory of an elephant). Figure 6.9 shows

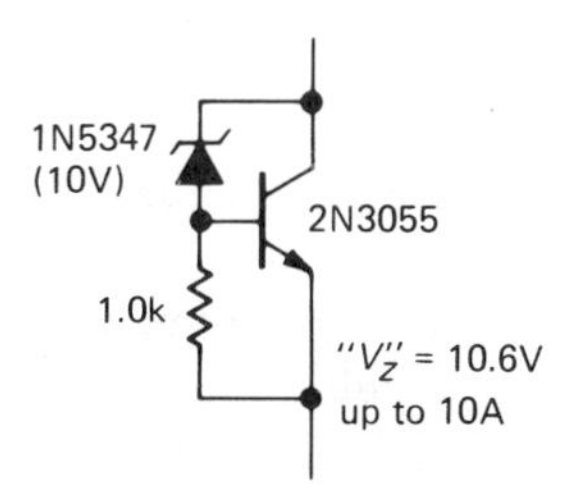

Figure 6.9. Active power zener.

the circuit of an "active zener." Unfortunately, a crowbar constructed from a power zener clamp has its own problems. If the regulator fails, the crowbar has to contend with high power dissipation ($V_{\text{zener}}I_{\text{limit}}$)

and may itself fail. We witnessed just such a failure in a commercial 15 volt 4 amp magnetic disc supply. When the pass transistor failed, the 16 volt 50 watt zener found itself dissipating more than rated power, and it proceeded to fail too.

□ 6.07 Further considerations in high-current power-supply design

□ *Separate high-current unregulated supply*

As we mentioned in Section 6.03, it is usually a good idea to use a separate supply to power the regulator in very high current supplies. In that way the dissipation in the pass transistors can be minimized, since the unregulated input to the pass transistor can then be chosen just high enough to allow sufficient "headroom" (regulators like the 723 have separate V_+ terminals for this purpose). For instance, a +5 volt 10 amp regulator might use a 10 volt unregulated input with a volt or two of ripple, with a separate low-current +15 volt supply for the regulator components (reference, error amplifier, etc.). As mentioned earlier, the unregulated input voltages must be chosen large enough to allow for worst-case ac power-line voltage (105V) as well as transformer and capacitor tolerances.

□ *Connection paths*

With high-current supplies, or supplies of highly precise output voltage, careful thought must be given to the connection paths, both within the regulator and between the regulator and its load. If several loads are run from the same supply, they should connect to the supply at the place where the output voltage is sensed; otherwise, fluctuations in the current of one load will affect the voltage seen by the other loads (Fig. 6.10).

In fact, it is a good idea to have one common ground point (a "mecca"), as shown, to which the unregulated supply, reference, etc., are all returned. The problem of unregulated voltage drops in the connecting leads from power supply to high-current load is sometimes solved by remote sensing: The connections back to the error amplifier and reference are brought out to the rear of the supply separately and may either be connected to the output terminals right there (the normal method) or brought out and connected to the load at a remote location along with the output voltage leads (this requires four wires, two of which must be able to handle

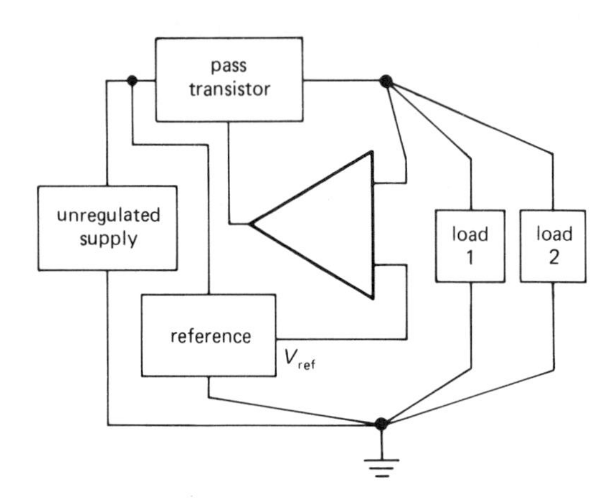

Figure 6.10. A power-supply ground "mecca."

the high load currents). Most commercially available power supplies come with jumpers at the rear that connect the sensing circuitry to the output and that may be removed for remote sensing. Four-wire resistors are used in an analogous manner to sense load currents accurately when constructing precision constant-current supplies. This will be discussed in greater detail in Section 6.24.

□ *Parallel pass transistors*

When very high output currents are needed, it may be necessary to use several pass transistors in parallel. Since there will be a spread of V_{BE}s, it is necessary to add a small resistance in series with each emitter, as in Figure 6.11. The *R*s ensure that the current is shared approximately equally among the pass transistors. *R* should be chosen for about 0.2 volt drop at maximum output current. Power FETs can be connected in parallel without any external components, owing to their negative temperature coefficient of drain current (Fig. 3.13).

Safe operating area

One last point about bipolar power transistors: A phenomenon known as "second breakdown" restricts the simultaneous

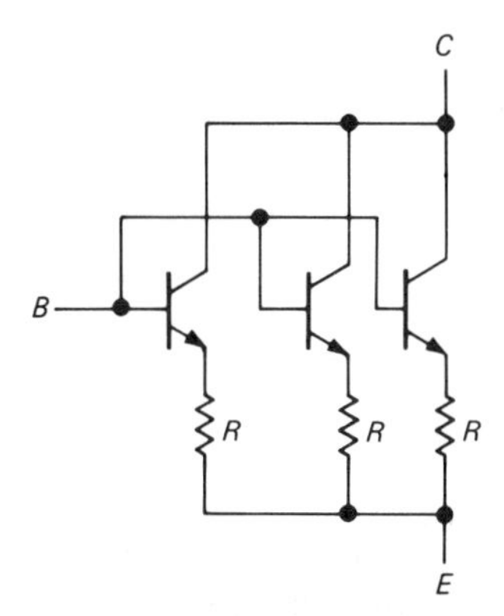

Figure 6.11. Use emitter ballasting resistors when paralleling bipolar power transistors.

voltage and current that may be applied for any given transistor, and it is specified on the data sheet as the safe operating area (SOA) (it's a family of safe voltage-versus-current regions, as a function of time duration). Second breakdown involves the formation of "hot spots" in the transistor junctions, with consequent uneven sharing of the total load. Except at low collector-to-emitter voltages, it sets a limit that is more restrictive than the maximum power dissipation specification. As an example, Figure 6.12 shows the SOA for the ever-popular 2N3055. For $V_{CE} > 40$ volts, second breakdown limits the dc collector current to values corresponding to less

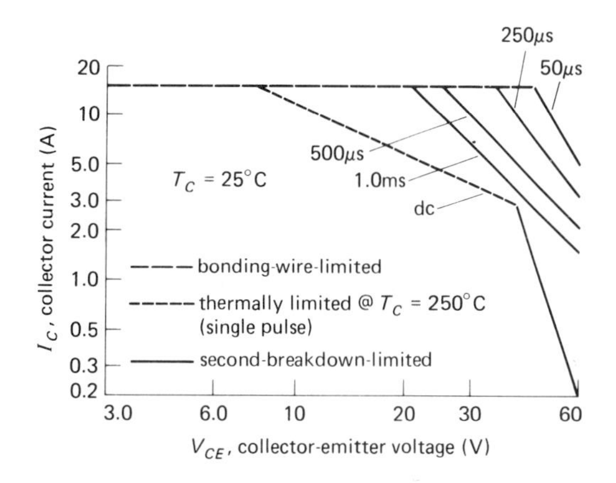

Figure 6.12. Safe operating area for 2N3055 bipolar power transistor. (Courtesy of Motorola, Inc.)

than the maximum allowable dissipation of 115 watts. Figure 6.13 shows the SOA for two similar high-performance power transistors: the 2N6274 *npn* bipolar transistor and the comparable Siliconix VNE003A *n*-channel MOSFET. For $V_{CE} > 10$ volts, second breakdown limits the *npn* transistor dc collector current to values corresponding to less than the maximum allowable dissipation of 250 watts. The problem is less severe for short pulses, and it effectively disappears for pulses of 1ms duration or less. Note that the MOSFET has no second breakdown; its SOA is bounded by maximum current (bonding-wire limited, therefore higher for short pulses), maximum dissipation, and maximum allowable drain-source voltage. See Chapter 3 for more details on power MOSFETs.

□ 6.08 Programmable supplies

There is frequently the need for power supplies that can be adjusted right down to zero volts, especially in bench applications where a flexible source of power is essential. In addition, it is often desirable to be able to "program" the output voltage with another voltage or with a digital input (via digital thumbwheel switches, for instance).

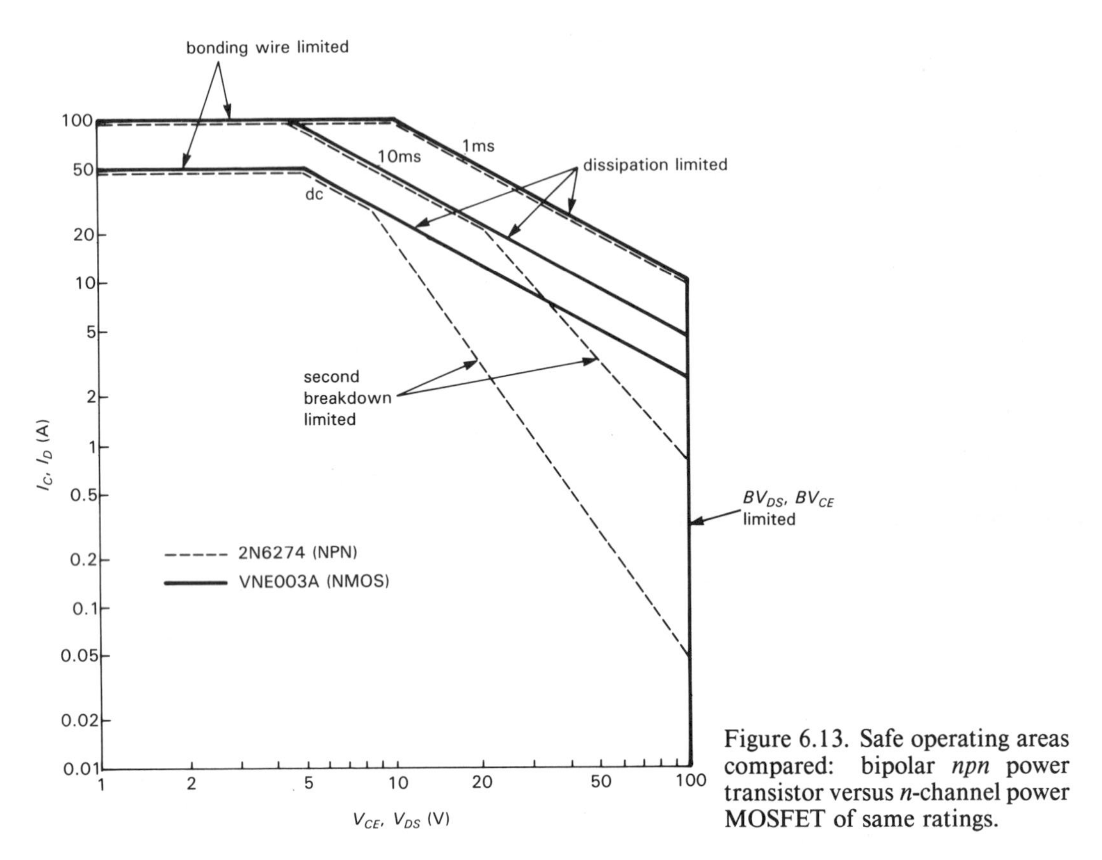

Figure 6.13. Safe operating areas compared: bipolar *npn* power transistor versus *n*-channel power MOSFET of same ratings.

Figure 6.14 shows the classic scheme for a supply that is adjustable down to zero output voltage (as our 723 circuits so far are not). A separate split supply provides power for the regulator and also generates an accurate negative reference voltage (more on references in Sections 6.14 and 6.15). R_1 sets the output voltage (since the inverting input will be at ground), which can be adjusted all the way down to zero (at zero resistance). When the regulator circuitry (which can be an integrated circuit or discrete components) is run from a split supply, no problems are encountered at low output voltages.

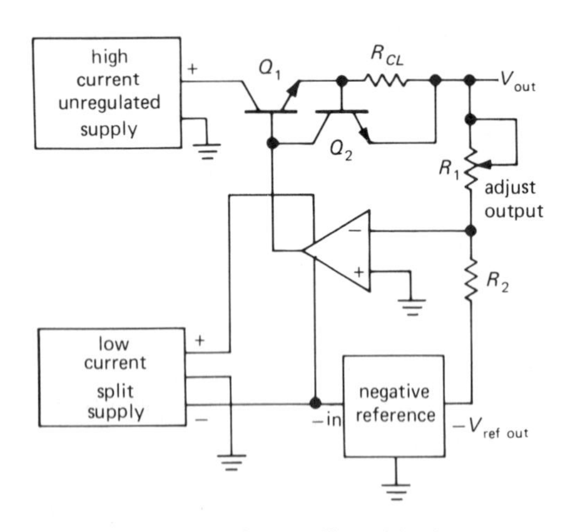

Figure 6.14. Regulator adjustable down to zero volts.

To make the supply programmable with an external voltage, just replace V_{ref} with an externally controlled voltage (Fig. 6.15). The rest of the circuit is unchanged. R_1 now sets the scale of $V_{control}$.

Digital programmability can be added by replacing V_{ref} with a device called a DAC (digital-to-analog converter) with current-sinking output. These devices, which we will discuss later, convert a binary input code to a proportional current (or voltage) output. A good choice here is the AD7548, a monolithic 12-bit DAC with current-sinking output and a price tag of about $9. By replacing R_2 with the DAC, you get a digitally programmed supply, with step size of 1/4096 (2^{-12}) of the full-scale output voltage. Since the inverting input is a virtual ground, the DAC doesn't even have to have any output compliance. In practice, R_1 would be adjusted to set a convenient scale for the output, say 1mV per input digit.

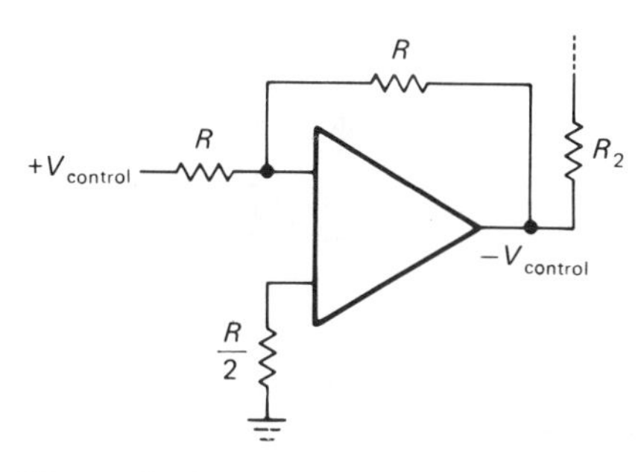

Figure 6.15

□ 6.09 Power-supply circuit example

The "laboratory" bench supply shown in Figure 6.16 should help pull all these design ideas together. It is important to be able to adjust the regulated output voltage right down to zero volts in a general-purpose bench supply, so an additional split supply is used to power the regulator. IC_1 is a high-voltage op-amp, which can operate with 80 volts total supply voltage. We used paralleled power MOSFETs as the output pass transistor, both because of its easy gate drive requirements and its excellent safe operating area (characteristic of all power MOSFETs). The combination can dissipate plenty of power (60W per transistor at 100°C case temperature), which is necessary even for moderate output current when such a wide range of output voltage is provided. This is because the unregulated input voltage has to be high enough for the maximum regulated output voltage, resulting in a large voltage drop across the pass transistors when the regulated output voltage is low. Some supplies solve this problem by having several ranges of output voltage, switching the unregulated input voltage accordingly. There are even supplies with the unregulated supply driven from a variable-voltage transformer ganged to the same control as the output voltage. In both cases you lose the capability of remote programmability.

EXERCISE 6.4

What is the maximum power dissipation in the pass transistors for this circuit?

R_1 is a precision multidecade potentiometer for precise and linear adjustment of the output voltage. The output voltage is referenced to the 1N829 precision zener (5ppm/°C tempco at 7.5mA zener current). The current-limiting circuitry is considerably better than the simple protective current limiters we have been discussing, since it is sometimes desirable to be able to set a precise and stable current limit when using a bench supply. Note the unusual (but convenient) method of current limiting by sinking current from the compensation pin of IC_1, which has unity gain to the output while operating at low current. By providing both precision-regulated voltage (all the way down to 0V) and current, the device becomes a flexible laboratory power supply. With this current-limit method, the supply becomes a flexible constant-current source. Q_4 provides a constant 100mA load, maintaining good performance near zero output voltage (or current) by keeping the pass transistors well into the active region. This current sink also allows the load to source some current into the supply without its output voltage rising. This is useful with the bizarre loads you sometimes encounter, e.g., an instrument that contains some additional supplies of its own capable of sourcing some current into the power-supply output terminal.

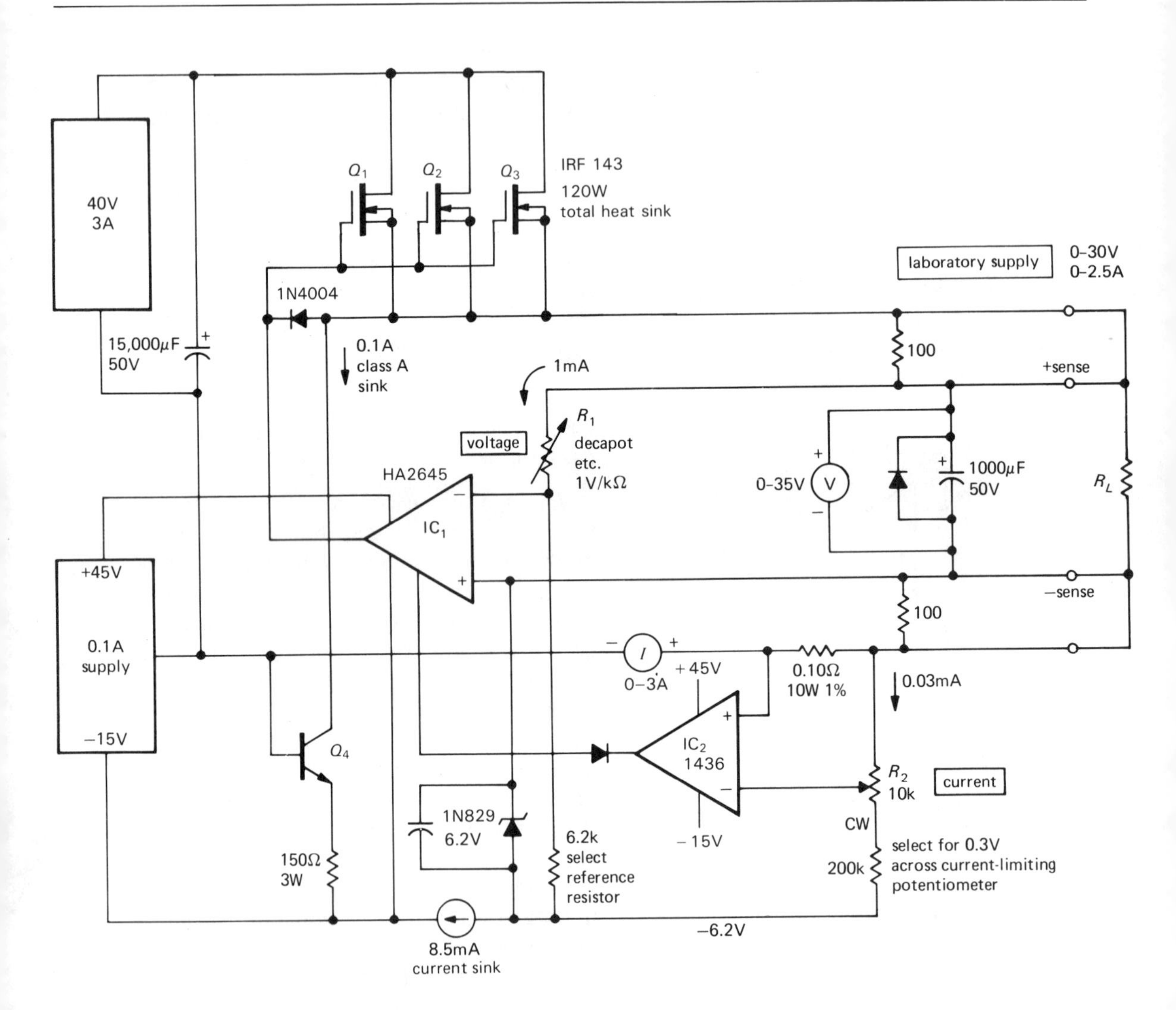

Figure 6.16. Laboratory bench supply.

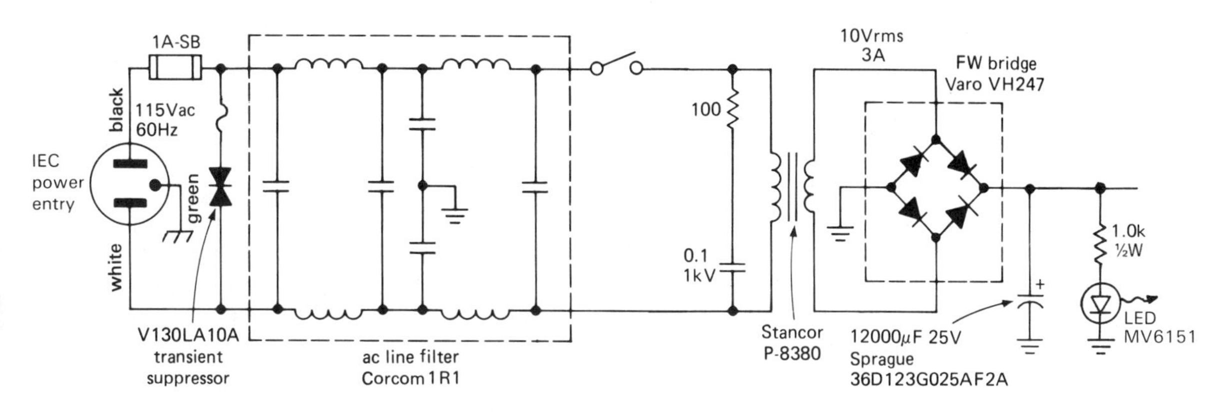

Figure 6.17. Unregulated supply with ac line connections. Note color convention of ac line cord.

Note the external sense leads, with default connection to the power-supply output terminals. For precise regulation of output voltage at the load, you would bring external sense leads to the load itself, eliminating (through feedback) voltage drops in the connecting leads.

6.10 Other regulator ICs

The 723 was the original voltage regulator IC, and it is still a useful chip. There are a few improved versions that work much the same way, however, and you should consider them when you design a regulated power supply. The LAS1000 and LAS1100 from Lambda and the SG3532 from Silicon General can operate down to 4.5 volts input voltage, because they use an internal 2.5 volt "bandgap reference" (see Section 6.15) rather than the 7.15 volt zener of the 723. They also have internal circuitry that shuts off the chip if it overheats; compare the 723's solution (burnout!). Although these regulators have the same pin names, you can't just plug these regulators into a socket intended for a 723, because (among other things) they assume a lower reference voltage. Another 723-like regulator is the MC1469 (and its negative twin, the MC1463) from Motorola.

If you look at modern power-supply circuits, you won't see many 723s, or even the improved versions we just mentioned. Instead, you'll see mostly ICs like the 7805 or 317, with a remarkable absence of external components (the 7805 requires none!). Most of the time you can get all the performance you need from these highly integrated and easy-to-use "three-terminal" regulators, including high output current (up to 10A) without external pass transistors, adjustable output voltage, excellent regulation, and internal current limiting and thermal shut-down. We'll talk about these shortly, but first an interlude on (a) the design of the unregulated supply and (b) voltage references.

THE UNREGULATED SUPPLY

All regulated supplies require a source of "unregulated" dc, a subject we introduced in Section 1.27 in connection with rectifiers and ripple calculations. Let's look at this subject in more detail, beginning with the circuit shown in Figure 6.17. This is an unregulated +13 volt (nominal) supply for use with a +5 volt 2 amp regulator. Let's go through it from left to right, pointing out some of the things to keep in mind when you do this sort of design.

6.11 ac line components

Three-wire connection

Always use a 3-wire line cord with neutral (green) connected to the instrument case. Instruments with ungrounded cases can become lethal devices in the event of transformer insulation failure or accidental connection of one side of the power line to the case. With a grounded case, such a failure simply blows a fuse. You often see instruments with the line cord attached to the chassis (permanently) using a plastic "strain relief," made by Heyco or Richco. A better way is to use an IEC three-prong male chassis-mounted connector, to mate with those popular line cords that have the three-prong IEC female molded onto the end. That way the line cord is conveniently removable. Better yet, you can get a combined "power entry module," containing IEC connector, fuse holder, line filter, and switch (as described later). Note that ac wiring uses a nonintuitive color convention: black = "hot," white = neutral, and green = ground.

Line filter and transient suppressor

In this supply we have used a simple *LC* line filter. Although they are often omitted, such filters are a good idea, since they serve the purpose of preventing possible radiation of radiofrequency interference (RFI) from the instrument via the power line, as well as filtering out incoming interference that may be present on the power line. Power-line filters with excellent performance characteristics are available from several manufacturers, e.g., Corcom, Cornell-Dubilier, and Sprague. Studies have shown that spikes as large as 1kV to 5kV are occasionally present on the power lines at most locations, and smaller spikes occur quite frequently. Line filters are reasonably effective in reducing such interference.

In many situations it is desirable to use a "transient suppressor," as shown, a device that conducts when its terminal voltage exceeds certain limits (it's like a bidirectional high-power zener). These are inexpensive and small and can short out hundreds of amperes of potentially harmful current in the form of spikes. Transient suppressors are made by a number of companies, e.g., GE and Siemens. Tables 6.2 and 6.3 list some useful RFI filters and transient suppressors.

Fuse

A fuse is essential in every piece of electronic equipment. The large wall fuses or circuit breakers (typically 15–20A) in house or lab won't protect electronic equipment, since they are chosen to blow only when the current rating of the wiring in the wall is exceeded. For instance, a

TABLE 6.2. 130 VOLT AC TRANSIENT SUPPRESSORS

Type	Manuf.	Diameter (in)	Energy (W-s)	Peak curr (A)	Capacitance (pF)
V130LA1	GE	0.34	4	500	180
S07K130	Siemens	0.35	6	500	130
V130LA10A	GE	0.65	30	4000	1000
S14K130	Siemens	0.67	22	2000	1000
V130LA20B	GE	0.89	50	6000	1900
S20K130	Siemens	0.91	44	4000	2300

TABLE 6.3. 115 VOLT AC POWER FILTERS (IEC CONNECTOR[a])

Manuf.	Part No.	Circuit	Current (A)	Attenuation[b] (line-to-gnd, 50Ω/50Ω) 150kHz (dB)	500kHz (dB)	1MHz (dB)	Comments
Corcom	**3EF1**	π	3	15	25	30	general purpose
	3EC1	π	3	20	30	37	higher attenuation
	3EDSC2-2	π	3	32	37	44	with fuse
	2EDL1S	π	2	14	–	24	with fuse and switch
Curtis	**F2100CA03**	π	3	15	25	30	general purpose
	F2400CA03	π	3	22	35	40	higher attenuation
	F2600FA03	π	3	21	35	41	with fuse
	PE810103	π	3	18	24	30	with fuse and switch
Delta	**03GEEG3H**	π	3	24	30	38	general purpose
	03SEEG3H	dual-π	3	42	65	70	higher attenuation
	04BEEG3H	π	4	26	35	40	with fuse
	03CK2	π	3	35	40	40	with fuse and switch
	03CR2	dual-π	3	50	60	55	same, higher attenuation
Schaffner	**FN323-3**	π	3	22	32	36	general purpose
	FN321-3	π	3	35	43	46	higher attenuation
	FN361-2	π	2	25	40	46	with fuse
	FN291-2.5	π	2.5	25	40	46	with fuse and switch
	FN1393-2.5	π	2.5	40	45	42	same, higher attenuation
Sprague	**3JX5421A**	π	3	15	25	30	general purpose
	3JX5425C	π	3	20	30	37	higher attenuation
	200JM6-2	π	6	12	25	–	with fuse

(a) these units are representative of a large selection, many of which do not include an IEC input connector. (b) rf attenuation figures are measured in a 50Ω system, and should not be relied upon to predict performance in an ac line circuit.

house wired with 14 gauge wire will have 15 amp breakers. Now, if the filter capacitor in the preceding supply becomes short-circuited someday (a typical failure mode), the transformer might then draw 5 amps primary current (instead of its usual 0.25A). The house breaker won't open, but your instrument becomes an incendiary device, with its transformer dissipating over 500 watts!

Some notes on fuses: (a) It is best to use a "slow-blow" type in the power-line circuit, because there is invariably a large current transient at turn-on (caused mostly by rapid charging of the power-supply filter capacitors). (b) You may think you know how to calculate the fuse current rating, but you're probably wrong. A dc power supply has a high ratio of rms current to average current, because of the small conduction angle (fraction of the cycle over which the diodes are conducting). The problem is worse if overly large filter capacitors are used. The result is an rms current considerably higher than you would estimate. The best procedure is to use a "true rms" ac current meter to measure the actual rms line current, then choose a fuse of at least 50% higher current rating (to allow for high line voltage, the effects of fuse "fatigue," etc.). (c) When wiring cartridge-type fuse holders (used with the popular 3AG fuse, which is almost universal in electronic equipment), be sure to connect the leads so that anyone changing the fuse cannot come in contact with the power line. This means connecting the "hot" lead to the rear terminal of the fuse holder (the authors learned this the hard way!). Commercial power-entry modules with integral fuse holders are usually arranged so that the fuse cannot be reached without removing the line cord.

Shock hazard

Incidentally, it is a good idea to insulate all exposed 110 volt power connections inside any instrument, using Teflon heat-shrink tubing, for instance (the use of "friction tape" or electrical tape inside electronic instruments is strictly bush-league). Since most transistorized circuits operate on relatively low dc voltages (±15V to ±30V or so), from which it is not possible to receive a shock, the power line wiring is the only place where any shock hazard exists in most electronic devices (there are exceptions, of course). The front-panel ON/OFF switch is particularly insidious in this respect, since it is close to other low-voltage wiring. Your test instruments (or, worse, your fingers) can easily come in contact with it when you go to pick up the instrument while testing it.

Miscellany

We favor "power-entry modules," combining a 3-prong IEC connector (use a removable line cord) and some combination of line filter, fuse holder, and power switch. For example, the Schaffner FN380 series (or Corcom L series) has all these features, and they are available with maximum currents from 2 to 6 amps. They give you options for fusing or switching either one or both sides of the line, and they offer several filter configurations. Some other manufacturers offering similar products are Curtis, Delta, and Power Dynamics (Table 6.3).

Our circuit shows an LED pilot light (with current-limiting resistor) running from the unregulated dc voltage. It is generally better practice to power the LED from the regulated dc, so that it doesn't flicker with load or power-line variations.

The series combination of 100 ohms and $0.1\mu F$ capacitor across the transformer primary prevents the large inductive transient that would otherwise occur at turn-off. This is often omitted, but it is highly desirable, particularly in equipment intended for use near computers or other digital devices. Sometimes this RC "snubber" network is wired across the switch, which is equivalent.

6.12 Transformer

Now for the transformer. Never build an instrument to run off the power line without a transformer! To do so is to flirt with disaster. Transformerless power supplies, which are popular in some consumer electronics (radios and televisions, particularly) because they're cheap, put the circuit at high voltage with respect to external ground (water pipes, etc.). This has no place in instruments intended to interconnect with any other equipment and should always be avoided. And use extreme caution when servicing any such equipment; just connecting your oscilloscope probe to the chassis can be a shocking experience.

The choice of transformer is more involved than you might at first expect. One problem is that manufacturers have been slow to introduce transformers with voltages and currents appropriate for transistorized circuitry (the catalogs are still cluttered with transformers designed for vacuum tubes), and you wind up making compromises you'd rather avoid. We have found the Signal Transformer Company unusual, with their nice selection of transformers and quick delivery. Don't overlook the possibility of having transformers custom-made if your application requires more than a few.

Even assuming that you can get the transformer you want, you still have to decide what voltage and current are best. The lower the input voltage to the regulator, the lower the dissipation in the pass transistors. But you must be absolutely certain the input to the regulator will never drop below the minimum necessary for regulation, typically 2 to 3 volts above the regulated output voltage, or you may encounter 120Hz dips in the regulated output. The amount of ripple in the unregulated output is involved here, since it is the minimum input to the regulator that must stay above some critical voltage, but it is the average input to the regulator that determines the transistor dissipation.

As an example, for a +5 volt regulator you might use an unregulated input of +10 volts at the minimum of the ripple, which itself might be a volt or two. From the secondary voltage rating you can make a pretty good guess of the dc output from the bridge, since the peak voltage (at the top of the ripple) is approximately 1.4 times the rms secondary voltage, less two diode drops. But it is essential to make actual measurements if you are designing a power supply with near-minimum drop across the regulator, because the actual output voltage of the unregulated supply depends on poorly specified parameters of the transformer, such as winding resistance and magnetic coupling, both of which contribute to voltage drop under load. Be sure to make measurements under worst-case conditions: full load and low power-line voltage (105V). Remember that large filter capacitors typically have loose tolerances: −30% to +100% about the nominal value is not unusual. It is a good idea to use transformers with multiple taps on the primary, when available, for final adjustment of output voltage. The Triad F-90X series and the Stancor TP series are very flexible this way.

One further note on transformers: Current ratings are sometimes given as rms secondary current, particularly for transformers intended for use into a resistive load (filament transformers, for instance). Since a rectifier circuit draws current only over a small part of the cycle (during the time the capacitor is actually charging), the rms current, and therefore the I^2R heating, is likely to exceed specifications for a load current approaching the rated rms current of the transformer. The situation gets worse as you increase capacitor size to reduce preregulator ripple; this simply requires a transformer of larger rating. Full-wave rectification is better in this respect, since a greater portion of the transformer waveform is used.

6.13 dc components

Filter capacitor

The filter capacitor is chosen large enough to provide acceptably low ripple voltage, with voltage rating sufficient to handle the worst-case combination of no load and high line voltage (125–130V rms). For the circuit shown in Figure 6.17, the ripple is about 1.5 volts pp at full load. Good design practice calls for the use of computer-type electrolytics (they come in a cylindrical package with screw terminals at one end), e.g., the Sprague 36D type. In smaller capacitance values most manufacturers provide capacitors of equivalent quality in an axial-lead package (one wire sticking out each end), e.g., the Sprague 39D type. Watch out for the loose capacitance tolerance!

At this point it may be helpful to look back at Section 1.27, where we first discussed the subject of ripple. With the exception of switching regulators (see Section 6.19 and following), you can always calculate ripple voltage by assuming a constant-current load equal to the maximum output load current. In fact, the input to a series regulator looks just like a constant-current sink. This simplifies your arithmetic, since the capacitor discharges with a ramp, and you don't have to worry about time constants or exponentials (Fig. 6.18).

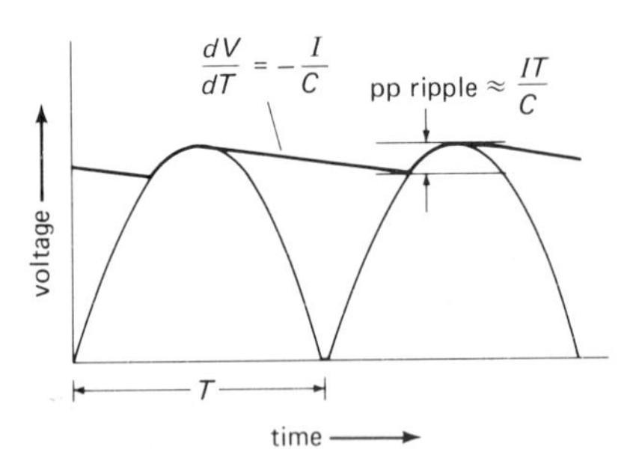

Figure 6.18

For example, suppose you want to choose a filter capacitor for the unregulated portion of a +5 volt 1 amp regulated supply, and suppose you have already chosen a transformer with a 10 volt rms secondary, to give an unregulated dc output of 12 volts (at the peak of the ripple) at full load current. With a typical regulator dropout voltage of 2 volts, the input to the regulator should never dip below +7 volts (the 723 will require +9.5V, but the convenient 3-terminal regulators discussed in Section 6.16 are more friendly). Since you have to contend with a ±10% worst-case line-voltage variation, you should keep ripple to less than 2 volts pp. Therefore,

$$2 = T(dV/dT) = TI/C = 0.008 \times 1.0/C$$

from which C=4000μF. A 5000μF 25 volt electrolytic would be a minimum choice, with allowance for a 20% tolerance in capacitor value. When choosing filter capacitors, don't get carried away: An oversize capacitor not only wastes space but also increases transformer heating (by reducing the conduction angle, hence increasing the ratio $I_{\rm rms}/I_{\rm avg}$). It also increases stress on the rectifiers.

The LED shown across the output in Figure 6.17 acts as a "bleeder" to discharge the capacitor in a few seconds under no-load conditions. This is a good feature, because power supplies that stay charged after things have been shut off can easily lead you to damage some circuit components if you mistakenly think that no voltage is present.

Rectifier

The first point to be made is that the diodes used in power supplies are quite different from the small 1N914-type signal diodes used in circuitry. Signal diodes are generally designed for high speed (a few nanoseconds), low leakage (a few nanoamps), and low capacitance (a few picofarads), and they can generally handle currents up to about 100mA, with breakdown voltages rarely exceeding 100 volts. By contrast, rectifier diodes and bridges for use in power supplies are hefty objects with current ratings going from 1 amp to 25 amps or more and breakdown voltages going from 100 volts to 1000 volts. They have relatively high leakage currents (in the range of microamps to milliamps) and plenty of junction capacitance. They are not intended for high speed. Table 6.4 lists a selection of popular types.

Typical of rectifiers is the popular 1N4001–1N4007 series, rated at 1 amp, with reverse-breakdown voltages ranging from 50 to 1000 volts. The 1N5625 series is rated at 3 amps, which is about the highest current available in a lead-mounted (cooled by conduction through the leads) package. The popular 1N1183A series typifies high-current stud-mounted rectifiers, with a current rating of 40 amps and breakdown voltages to 600 volts. Plastic-encapsulated bridge rectifiers are quite popular also, with lead-mounted 1 and 2 amp types and chassis-mounted packages in ratings up to 25 amps or more. For rectifier applications where high speed is important (e.g., dc-to-dc converters, see Section 6.19), fast-recovery diodes are

TABLE 6.4. RECTIFIERS

Type	Breakdown voltage V_{BR} (V)	Forward drop V_F typ (V)	@	Average current I_o (A)	Package	Comments
General purpose						
1N4001-07	50-1000	0.9		1	lead-mounted	popular
1N5059-62	200-800	1.0		2	lead-mounted	
1N5624-27	200-800	1.0		5	lead-mounted	
1N1183A-90A	50-600	0.9		40	stud-mounted	popular; -R for rev. pol.
Fast recovery (t_{rr} = 0.1μs typ)						
1N4933-37	50-600	1.0		1	lead-mounted	
1N5415-19	50-500	1.0		3	lead-mounted	
1N3879-83	50-400	1.2		6	stud-mounted	-R for reverse polarity
1N5832-34	50-400	1.0		20	stud-mounted	-R for reverse polarity
Schottky (low V_F, very fast)						
1N5817-19	20-40	0.6^m		1	lead-mounted	
1N5820-22	20-40	0.5^m		3	lead-mounted	
1N5826-28	20-40	0.5^m		15	stud-mounted	
1N5832-34	20-40	0.6^m		40	stud-mounted	
Full-wave bridge						
3N246-52	50-1000	0.9		1	plastic SIP	MDA100A
3N253-59	50-1000			2	plastic SIP	MDA200
MDA970A1-A5	50-400	0.85		8	chassis mtd	
MDA3500-10	50-1000			35	chassis mtd	
Exotic						
GE A570A-A640L	100-2000	1.0^m		1500	giant button	high current!
Semtech SCH5000-25000	5kV-25kV	$7\text{-}33^m$		0.5	lead-mounted	HV, curr; fast (0.2μs)
Varo VF25-5 to -40	5kV-40kV	$12\text{-}50^m$		0.025	lead-mounted	high voltage
Semtech SCKV100K3-200K3	100kV-200kV	150-300		0.1	plastic rod	very high voltage

(m) maximum.

available, e.g., the 1N4933 series of 1 amp diodes. For low-voltage applications it may be desirable to use Schottky barrier rectifiers, e.g., the 1N5823 series, with forward drops of less than 0.4 volt at 5 amps.

VOLTAGE REFERENCES

There is frequently the need for good voltage references within a circuit. For instance, you might wish to construct a precision regulated supply with characteristics better than those you can obtain using complete regulators like the 723 (since integrated voltage regulator chips usually dissipate considerable power because of the built-in pass transistor, they tend to heat up, with consequent drift). Or you might want to construct a precision constant-current supply. Another application that requires a precision reference, but not a precision power supply, is design of an accurate voltmeter, ohmmeter, or ammeter.

There are two kinds of voltage references – *zener diodes* and *bandgap references*; each can be used alone or as an internal part of an integrated circuit voltage reference.

□ 6.14 Zener diodes

The simplest form of voltage reference is the zener diode, a device we discussed in Section 1.06. Basically, it is a diode operated in the reverse-bias region, where current begins to flow at some voltage and increases dramatically with further increases in voltage. To use it as a reference, you simply provide a roughly constant current; this is often done with a resistor from a higher supply voltage, forming the most primitive kind of regulated supply.

Zeners are available in selected voltages from 2 to 200 volts (they come in the same series of values as standard 5% resistors), with power ratings from a fraction of a watt to 50 watts and tolerances of 1% to 20%. As attractive as they might seem for use as general-purpose voltage references, zeners are actually somewhat difficult to use, for a variety of reasons. It is necessary to stock a selection of values, the voltage tolerance is poor except in high-priced precision zeners, they are noisy, and the zener voltage depends on current and temperature. As an example of the last two effects, a 27 volt zener in the popular 1N5221 series of 500mW zeners has a temperature coefficient of +0.1%/°C, and it will change voltage by 1% when its current varies from 10% to 50% of maximum.

There is an exception to this generally poor performance of zeners. It turns out that in the neighborhood of 6 volts, zener diodes become very stiff against changes in current and simultaneously achieve a nearly zero temperature coefficient. The graphs in Figure 6.19, plotted from measurements on zeners with different voltages, illustrate the effects. This peculiar behavior comes about because "zener" diodes actually employ two different mechanisms: zener breakdown (low voltage) and avalanche breakdown (high voltage). If you need a zener for use as a stable voltage reference only, and you don't care what voltage it is, the best thing to use is one of the compensated zener references constructed from a 5.6 volts zener (approximately) in series with a forward-biased diode. The zener voltage is chosen to give a positive coefficient to cancel the diode's temperature coefficient of −2.1mV/°C.

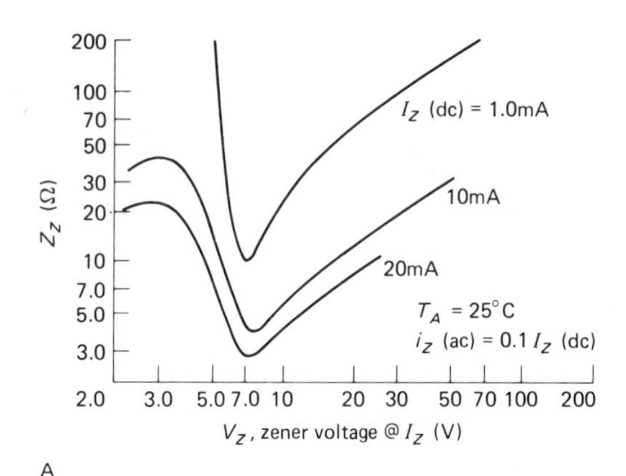

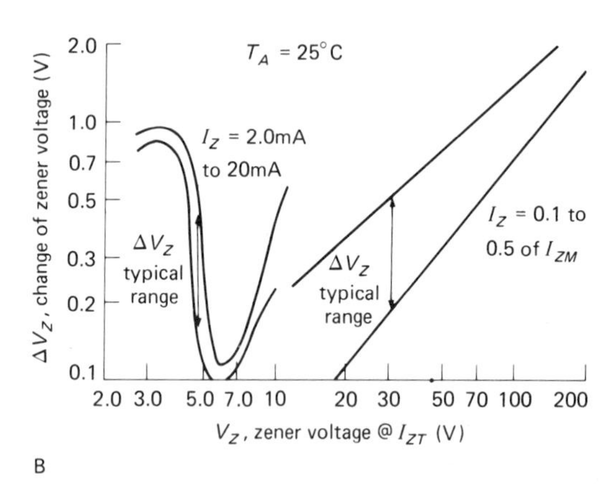

Figure 6.19. Zener diode impedance and regulation for zener diodes of various voltages. (Courtesy of Motorola, Inc.)

As you can see from the graph in Figure 6.20, the temperature coefficient depends on operating current and also on the zener voltage. Therefore, by choosing the zener current properly, you can "tune" the temperature coefficient somewhat. Such zeners with built-in series diodes make particularly good references. As an example, the 1N821 series of inexpensive 6.2 volt references offers temperature coefficients going from 100ppm/°C (1N821) down to 5ppm/°C (1N829); the 1N940 and 1N946 are 9 volt and 11.7 volt references with tempcos of 2ppm/°C.

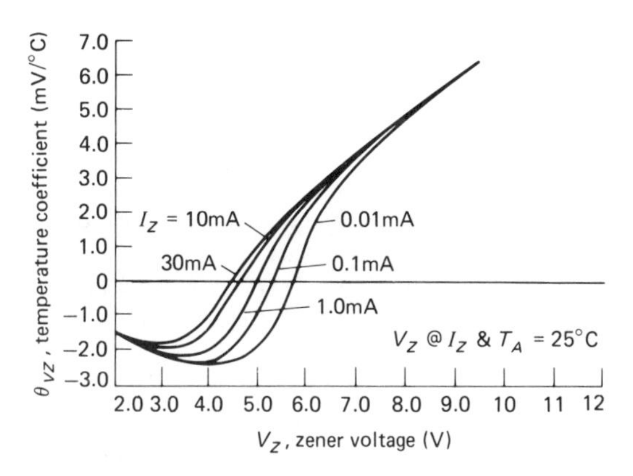

Figure 6.20. Temperature coefficient of zener diode breakdown voltage versus the voltage of the zener diode. (Courtesy of Motorola, Inc.)

□ *Providing operating current*

These compensated zeners can be used as stable voltage references within a circuit, but they must be provided with constant current. The 1N821 series is specified as 6.2 volts ±5% at 7.5mA, with an incremental resistance of about 15 ohms; thus, a change in current of 1mA changes the reference voltage three times as much as a change in temperature from −55°C to +100°C for the 1N829. Figure 6.21 shows a simple way to provide constant bias current for a precision zener. The op-amp is wired as a noninverting amplifier in order to generate an output of exactly +10.0 volts. That stable output is itself used to provide a precision 7.5mA bias current. This circuit is self-starting, but it can turn on with either polarity of output! For the "wrong" polarity, the zener operates as an ordinary forward-biased diode. Running the op-amp from a single supply, as shown, overcomes this bizarre problem. Be sure to use an op-amp that has common-mode input range to the negative rail ("single-supply" op-amps).

There are special compensated zeners available with guaranteed stability of zener voltage with *time*, a specification that normally tends to get left out. Examples are the 1N3501 and 1N4890 series. Zeners of this type are available with guaranteed stability of better than 5ppm/1000h. They're not cheap. Table 6.5 lists the characteristics of some useful zeners and reference diodes, and Table 6.6 shows part numbers for two popular 500mW general-purpose zener families.

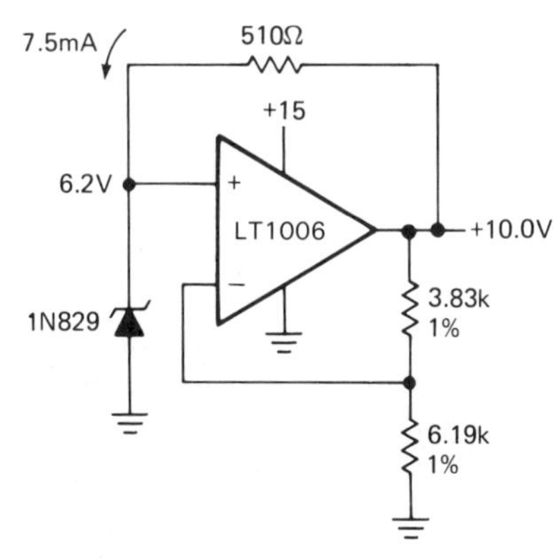

Figure 6.21

IC zeners

The 723 regulator uses a compensated zener reference to achieve its excellent performance (30ppm/°C stability of V_{ref}). The 723, in fact, is quite respectable as a voltage reference all by itself, and you can use the other components of the IC to generate a stable reference output at any desired voltage.

The 723 used as a voltage reference is an example of a *3-terminal reference*, meaning that it requires a power supply to

TABLE 6.5. ZENER AND REFERENCE DIODES[a]

Type	Zener voltage V_Z (V)	@	Test current I_{ZT} (mA)	Tolerance (±%)	Tempco max (ppm/°C)	Regulation ΔV for ±10% I_{ZT} max (mV)	P_{diss} max (W)	Comments
Reference zeners								
1N821A-	6.2		7.5	5	±100	7.5	0.4	5 member family, graded by
1N829A	6.2		7.5	5	±5	7.5	0.4	tempco; best and worst shown
1N4890-	6.35		7.5	5	±20		0.4	long-term stab < 100ppm/1000h
1N4895	6.35		7.5	5	±5		0.4	long-term stab < 10ppm/1000h
Regulator zeners								
1N5221A	2.4		20	10	-850	60	0.5	60 member family, 2.4V to 200V,
1N5231A	5.1		20	10	±300	34	0.5	in "5% resistor values," plus
1N5281A	200		0.65	10	+1100	160	0.5	some extras. -B = ±5%; popular[b]
1N4728A	3.3		76	10	-750	76	1.0	37 member family, 3.3V to 100V,
1N4735A	6.2		41	10	+500	8	1.0	in "5% resistor values."
1N4764A	100		2.5	10	+1100	88	1.0	-B = ±5%; popular

(a) see also Table 6.7 (IC Voltage References). (b) see Table 6.6 (500mW Zeners).

operate, and includes internal circuitry to bias the zener and buffer the output voltage. Improved 3-terminal IC zeners include the excellent LM369 from National (1.5ppm/°C typ), and the REF10KM from Burr-Brown (1ppm/°C max tempco); we've often used the inexpensive Motorola MC1404 (which is actually a bandgap reference, see below) in our circuits. We'll treat 3-terminal precision references in more detail shortly, after discussing the simpler 2-terminal types.

Precision temperature-compensated zener ICs are available as *2-terminal references* also; electrically they look just like zeners, although they actually include a number of active devices to give improved performance (most notably, constancy of "zener" voltage with applied current). An example is the inexpensive LM329, with a zener voltage of 6.9 volts. Its best version has a temperature coefficient of 6ppm/°C (typ), 10ppm/°C (max), when provided with a constant current of 1mA.

Some unusual IC zeners include the temperature-stabilized LM399 (0.3ppm/°C

TABLE 6.6. 500mW ZENER DIODES

1N5221 series	1N746 series	V_Z (V)	@	I_{ZT} (mA)
1N5230	1N750	4.7		20
1N5231	1N751	5.1		20
1N5232	1N752	5.6		20
1N5233	–	6.0		20
1N5235	1N754	6.8		20
1N5236	1N755	7.5		20
1N5237	1N756	8.2		20
1N5240	1N758	10		20
1N5242	1N759	12		20
1N5245	1N965	15		8.5
1N5248	1N967	18		7.0
1N5250	1N968	20		6.2
1N5253	–	25		5.0
1N5256	1N972	30		4.2
1N5259	1N975	39		3.2
1N5261	1N977	47		2.7
1N5267	1N982	75		1.7
1N5271	1N985	100		1.3
1N5276	1N989	150		0.85
1N5281	1N992	200		0.65

typ), the micropower LM385 (which operates down to 10μA), and the astounding LTZ1000 from Linear Technology, with its 0.05ppm/°C typical tempco, 0.3ppm per square-root-month drift, and 1.2μV low-frequency noise.

Zener diodes can be very noisy, and some IC zeners suffer from the same disease. The noise is related to surface effects, however, and *buried* (or *subsurface*) zener diodes are considerably quieter. In fact, the LTZ1000 buried zener just mentioned is the quietest reference of any kind. The LM369 and REF10KM also have very low noise.

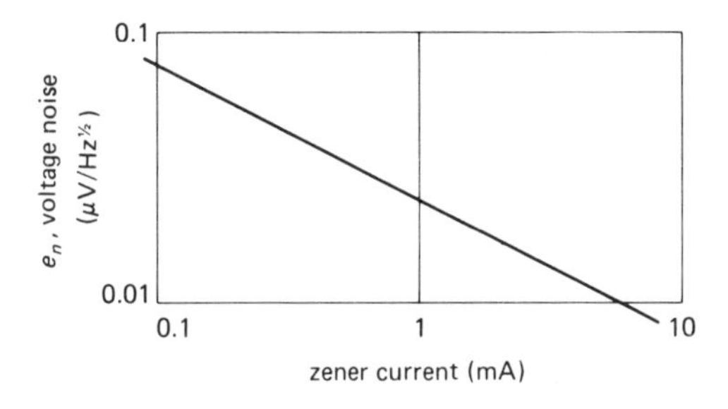

Figure 6.22. Voltage noise for a low-noise zener reference diode similar to the type used in the 723 regulator.

Table 6.7 lists the characteristics of nearly all available IC references, both zener and bandgap.

□ 6.15 Bandgap (V_{BE}) reference

More recently, a circuit known as a "bandgap" reference has become popular. It should properly be called a V_{BE} reference, and it is easily understandable using the Ebers-Moll diode equation. Basically, it involves the generation of a voltage with a positive temperature coefficient the same as V_{BE}'s negative coefficient; when added to a V_{BE}, the resultant voltage has zero tempco.

We start with a current mirror with two transistors operating at different emitter

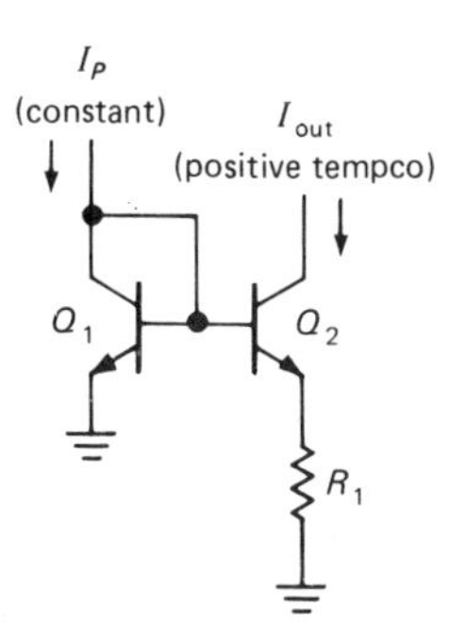

Figure 6.23

current densities (typically a ratio of 10:1) (see Fig. 6.23). Using the Ebers-Moll equation, it is easy to show that I_{out} has a positive temperature coefficient, since the difference in V_{BE}s is just $(kT/q)\log_e r$, where r is the ratio of current densities (see the graph in Fig. 2.53). You may wonder where we get the constant programming current I_P. Don't worry; you'll see the clever method at the end. Now all you do is convert that current to a voltage with a resistor and add a normal V_{BE}.

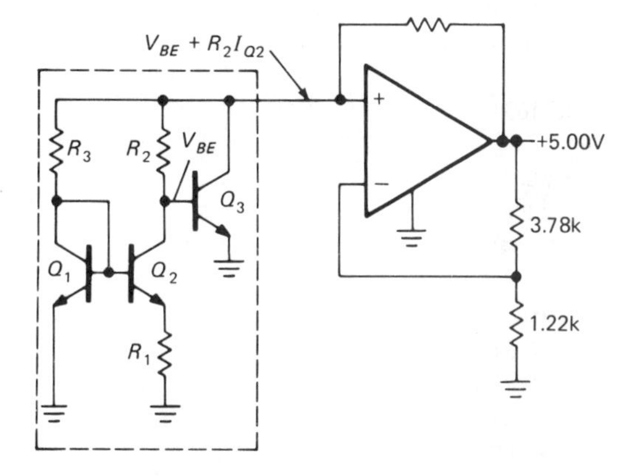

Figure 6.24. Classic V_{BE} bandgap voltage reference.

Figure 6.24 shows the circuit. R_2 sets the amount of positive-coefficient voltage you have added to V_{BE}, and by choosing it appropriately, you get zero overall temperature coefficient. It turns out that zero temperature coefficient occurs when the

TABLE 6.7. IC VOLTAGE REFERENCES

Type	Mfg[a]	B'gap/zener	Terminals	Trim	Voltage (V)	Acc'y (%)	Tempco typ (ppm/°C)	Min supply voltage (V)	Supply curr (mA)	Output curr max (mA)	Noise voltage 0.1-10Hz typ (μV pp)	Long-term stability typ (ppm/1000h)	Regulation: Line typ (%/V)	Regulation: Load 0-10mA typ (%)
Regulator type														
LM10C	NS+	B	8	•	0.20	5	30	1.1	0.3	20	–	–	0.001	0.01[a]
μA723C	FS+	Z	14	•	7.15	3	20	9.5	2.3	65	–	1000	0.003	0.03
SG3532J	SG+	B	10	•	2.50	4	50	4.5	1.6	150	–	300	0.005	0.02
Two-terminal (zener) type														
LM129A	NS	Z	2	–	6.9	5	6	–	1	15[b]	–	20	–	0.1
VR182C	DA	B	2	–	2.455	1.4	23	–	2	120[b]	10[e]	10	–	0.1
LM313	NS	B	2	–	1.22	5	100	–	1[l]	20[b]	5[f]	–	–	0.5[l]
LM329C	NS	Z	2	–	6.9	5	30	–	1	15[b]	–	20	–	0.1
LM336-2.5	NS	B	3	•	2.50	4	10	–	1	10[b]	–	20	–	0.1
LM336B-5	NS	B	3	•	5.0	1	15	–	1[o]	10[b]	–	20	–	0.1[o]
LM385B	NS	B	2	–	1.23	1	20	–	0.1[d]	20[b]	25	–	–	0.02[a]
LM385BX-1.2	NS	B	2	–	1.235	1	30[m]	–	0.1[d]	20[b]	60[f]	20	–	0.8[n]
LM385BX-2.5	NS	B	2	–	2.50	1.5	30[m]	–	0.1[p]	20[b]	120[f]	20	–	0.4[n]
LM299A	NS	Z	4	–	6.95	2	0.2	9	17	10[b]	–	20	–	0.1
LM399	NS	Z	4	–	6.95	5	0.3	9	17	10[b]	–	20	–	0.1
LM3999	NS	Z	3	–	6.95	5	2.0	9	17	10[b]	–	20	–	0.1
TL430	TI	B	3	•	2.75	5	120	–	10	100[b]	50	–	–	0.5
TL431	TI	B	3	•	2.75	2	10	–	10	100[b]	50	–	–	0.5
AD589M	AD	B	2	–	1.235	2	10[m]	–	0.1[h]	5[b]	5[f]	–	–	0.05[a]
LTZ1000	LT	Z	2	–	7.2	4	0.05	–	5		1.2	0.3[s]	–	1[r]
LT1004C-1.2	LT	B	2	–	1.235	0.3	20	–	0.1[d]	20[b]	60[f]	20	–	0.8[mr]
LT1009C	LT	B	3	•	2.50	0.2	15	–	1[o]	10[b]	–	20	–	0.1[o]
LT1029A	LT	B	3	•	5.0	0.2	8	–	1[o]	10[b]	–	20	–	0.04[o]
LT1034B	LT	B	3	–	1.225	1	10	–	0.1[p]	20[b]	4	–	–	0.3[p]
"	LT	Z			7.0	4	40	–	0.1[q]	20[b]	–	–	–	4[q]
HS5010N	HS	B	2	–	1.22	2	3	–	0.1[h]	5[b]	5[f]	–	–	0.05[a]
ICL8069A	IL	B	2	–	1.23	2	10	–	0.5	10[b]	–	–	–	0.2[a]
TSC9491	TS	B	2	–	1.22	2	30	–	0.1[k]	0.5[b]	–	–	–	1.2[k]
Three-terminal type														
REF-01A	PM	B	8	•	10.0	0.3	3	12	1	10	20	–	0.006	0.005
REF-02A	PM	B	8	•	5.0	0.3	3	7	1	10	10	–	0.006	0.005
REF-03E	PM	B	8	•	2.5	0.3	3	4.5	1	10	5	–	0.006	0.05
REF-05	PM	B	8	•	5.0	0.3	3	7	1	10	10	100[m]	0.006	0.05
REF-08G	PM	Z	8	•	-10.0	0.2	10[m]	-11.4	2[m]	10	10	–	0.02[m]	0.2[m]
REF-10	PM	B	8	•	10.0	0.3	3	12	1	10	20	50[m]	0.006	0.05
REF10KM	BB	Z	8	•	10.0	0.05	1[m]	13.5	4.5	10	6	10	0.001	0.01
REF-43E	PM	B	8	•	2.5	0.05	3[m]	4.5	0.2[m]	10	8[gm]	–	0.0002[m]	0.03[n]
LH0070-1	NS	Z	3	–	10.0	0.1	4	12.5	3	10	20	–	0.001	0.01
REF101KM	BB	Z	8	•	10.0	0.05	1[m]	13.5	4.5	10	6	25	0.0003	0.003
LM368Y-2.5	NS	B	8	•	2.5	0.2	11	4.9	0.35	10	12	–	0.0001	0.003
LM368-5	NS	B	4	•	5.0	0.1	15	7.5	0.25	10	16	–	0.0001	0.003
LM368-10	NS	B	4	•	10.0	0.1	15	12.5	0.25	10	30	–	0.0001	0.003
LM369B	NS	Z	3,8	•	10.0	0.05	1.5	13	1.4	10	4	6	0.0002	0.003
AD580M	AD	B	3	–	2.5	1	10	4.5	1	10	60	25	0.04	0.4

Type	Mfg[a]	B'gap/zener	Terminals	Trim	Voltage (V)	Acc'y (%)	Tempco typ (ppm/°C)	Min supply voltage (V)	Supply curr (mA)	Output curr max (mA)	Noise voltage 0.1-10Hz typ (µV pp)	Long-term stability typ (ppm/1000h)	Regulation: Line typ (%/V)	Regulation: Load 0-10mA typ (%)
Three-terminal type (cont'd)														
AD581L	AD+	B	3	–	10.0	0.05	5	12	0.75	10	50	25	0.005	0.002
AD584L	AD	B	8	•	2.5	0.05	10	5	0.75	18	50	25	0.005	0.002
"	AD				5.0	0.06	5	7.5	0.75	15	50	25	0.005	0.002
"	AD				7.5	0.06	5	10	0.75	13	50	25	0.005	0.002
"	AD				10.0	0.1	5	12.5	0.75	10	50	25	0.005	0.002
AD586L	AD	Z	8	•	5.0	0.05	5[m]	–	5[m]	10	–	15	–	–
AD587L	AD	Z	8	•	10.0	0.05	5[m]	–	5[m]	10	–	15	–	–
AD588B	AD	Z	14	•	±10.0	0.01	1.5[m]	±14	±10	±10	10	25[m]	0.002[m]	0.01[m]
MAX671C	MA	Z	14	•	10.0	0.01	1[m]	13.5	9	10	12	50	0.005[m]	0.01[m]
AD689L	AD	Z	8	•	8.192	0.05	5[m]	10.8	2	±10	2	15	0.002[m]	0.01[m]
R675C-3	HS	Z	14	•	±10.0	0.05	5	±13	+15,-3[m]	10	–	–	0.003[m]	0.02[m]
LT1019A-2.5	LT	B	8	•	2.5	0.002	3	4	0.7	10	6	–	0.00005	0.008
LT1021B-5	LT	Z	8	•	5.0	1	2	7	0.8	10	3	15	0.0004	0.01
LT1031B	LT	Z	3	–	10.0	0.05	3	11	1.2	10	6	15	0.00005	0.01
MC1403A	MO	B	8	–	2.5	1	10	4.5	1.2	10	–	–	0.002	0.06
MC1404AU5	MO	B	8	•	5.0	1	10	7.5	1.2	10	12	25	0.001	0.06
MC1404AU10	MO	B	8	•	10.0	1	10	12.5	1.2	10	12	25	0.0006	0.06
AD2702L[i]	AD+	Z	14	•	±10.0	0.05	5[m]	±13	+12,-2	±10	50	100	0.03[m]	0.05[m]
AD2712L[i]	AD+	Z	14	•	±10.0	0.01	1[m]	±13	+12,-2	±5	30	25	0.013	0.003[j]
LP2950ACZ	NS	B	3	–	5.0	0.5	20	5.4	0.08	100	–	–	0.002	0.004
ICL8212	IL	B	8	•	1.15	3	200	1.8	0.035	20	–	–	0.2	
TSC9495	TS	B	8	•	5.0	1	20	7	1	8	12	–	0.01	0.06
TSC9496	TS	B	8	•	10.0	1	20	12	1	8	25	–	0.01	0.06

(a) 0 to 1mA. (b) max zener curr. (c) on-chip heater/thermostat. (d) specified for 10µA to 20mA operating curr. (e) 1Hz to 10Hz. (f) 10Hz to 10kHz, rms. (g) 10Hz to 1kHz, rms. (h) spec'd for 50µA to 5mA. (i) 2700,2710: +10V; 2701: -10V; 2702,2712: ±10V. (j) 0 to 5mA. (k) spec'd for 50µA to 500µA. (l) spec'd for 0.5 to 20mA. (m) min or max. (n) 1 to 20mA, max. (o) specified for 0.5 to 10mA. (p) specified for 20µA to 20mA. (q) specified for 100µA to 20mA. (r) specified for 1mA to 5mA.

total voltage equals the silicon bandgap voltage (extrapolated to absolute zero), about 1.22 volts. The circuit in the box is the reference. Its own output is used (via R_3) to create the constant current we initially assumed.

Figure 6.25 shows another very popular bandgap reference circuit (it replaces the components in the box in Figure 6.24). Q_1 and Q_2 are a matched pair, forced to operate at a ratio of emitter currents of 10:1 by feedback from the collector voltages. The difference in V_{BE}s is $(kT/q)\log_e 10$, making Q_2's emitter current proportional to T (the preceding voltage applied across R_1). But since Q_1's collector current is larger by a factor of 10, it also is proportional to T. Thus, the total emitter current is proportional to T, and therefore it generates a positive-tempco voltage across R_2. That voltage can be used as a thermometer output, by the way, as will be discussed shortly. R_2's voltage is added to Q_1's V_{BE} to generate a stable reference of zero tempco at the base. Bandgap references appear in many variations, but they all feature the summation of V_{BE} with a voltage generated from a pair of transistors operated with some ratio of current densities.

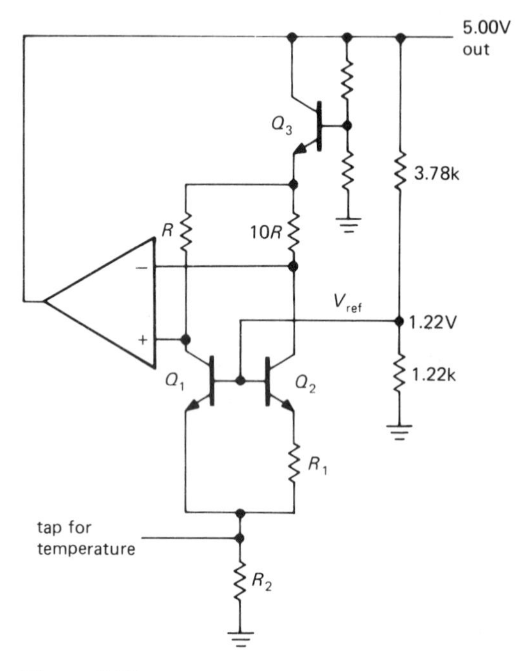

Figure 6.25

☐ *IC bandgap references*

An example of an IC bandgap reference is the inexpensive 2-terminal LM385-1.2, with a nominal operating voltage of 1.235 volts, ±1% (the companion LM385-2.5 uses internal circuitry to generate 2.50V), usable down to 10μA. That's much less than you can run any zener at, making these references excellent for micropower equipment (see Chapter 14). The low reference voltage (1.235V) is often much more convenient than the approximately 5 volt minimum usable voltage for zeners (you can get zeners rated at voltages as low as 3.3V, but they are pretty awful, with very soft knees). The best grade of LM385 guarantees 30ppm/°C maximum tempco and has a typical dynamic impedance of 1 ohm at 100μA. Compare this with the equivalent figures for a 1N4370 2.4 volt zener diode: tempco 800ppm/°C (typ), dynamic impedance ≈ 3000 ohms at 100μA, at which the "zener voltage" (specified as 2.4V at 20mA) is about 1.1 volt! When you need a precision stable voltage reference, these excellent bandgap ICs put conventional zener diodes to shame.

If you're willing to spend a bit more money, you can find bandgap references of excellent stability, for example, the 2-terminal LT1029, or the 3-terminal REF-43 (2.50V, 3ppm/°C max). The latter type, like the 3-terminal references based on zener technology, requires a dc supply. Table 6.7 lists most available bandgap (and zener) references, both 2-terminal and 3-terminal.

One other interesting voltage reference is the TL431C. It is an inexpensive "programmable zener" reference, and it is used as shown in Figure 6.26. The "zener" (made from a V_{BE} circuit) turns on when the control voltage reaches 2.75 volts; the device draws only a few microamps from the control terminal and gives a typical tempco of output voltage of 10ppm/°C. The circuit values shown give a zener voltage of 10.0 volts, for example. This device comes in a mini-DIP package and can handle currents to 100mA.

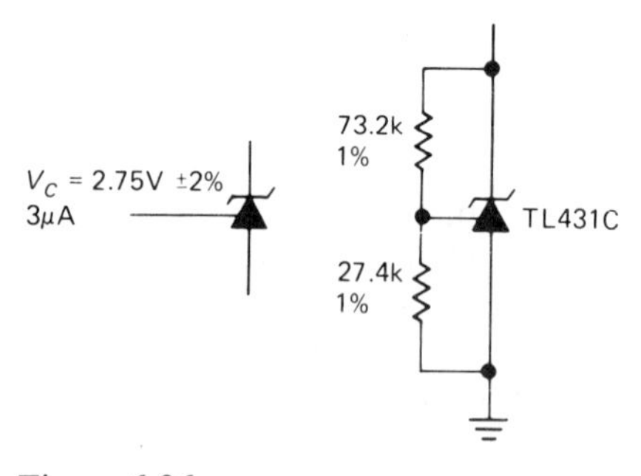

Figure 6.26

☐ *Bandgap temperature sensors*

The predictable V_{BE} variation with temperature can be exploited to make a temperature-measuring IC. The REF-02, for instance, generates an additional output voltage that varies linearly with temperature (see preceding discussion). With some simple external circuitry you can generate an output voltage that tells you the chip temperature, accurate to 1% over the full "military" temperature range (−55°C to +125°C). The AD590, intended for temperature measurement only, generates an accurate current of 1μA/°K. It's a 2-terminal device; you just put a voltage across it (4–30V) and measure the current. The LM334 can also be used in this manner. Other sensors, such as the LM35 and LM335, generate accurate voltage outputs with a slope of +10mV/°C. Section 15.01 has a detailed discussion on all these temperature "transducers."

Three-terminal precision references

As we remarked earlier, it is possible to make voltage references of remarkable temperature stability (down to 1ppm/°C or less). This is particularly impressive when you consider that the venerable Weston cell, the traditional voltage reference through the ages, has a temperature coefficient of 40ppm/°C (see Section 15.11). There are two techniques used to make such references.

□ *1. Temperature-stabilized references.* A good approach to achieving excellent temperature stability in a voltage reference circuit (or any other circuit, for that matter) is to hold the reference, and perhaps its associated electronics, at a constant elevated temperature. You will see simple techniques for doing this in Chapter 15 (one obvious method is to use a bandgap temperature sensor to control a heater). In this way the circuit can deliver equivalent performance with a greatly relaxed temperature coefficient, since the actual circuit components are isolated from external temperature fluctuations. Of greater interest for precision circuitry is the ability to deliver significantly improved performance by putting an already well-compensated reference circuit into a constant-temperature environment.

This technique of temperature-stabilized or "ovenized" circuits has been used for many years, particularly for ultrastable oscillator circuits. There are commercially available power supplies and precision voltage references that use ovenized reference circuits. This method works well, but it has the drawbacks of bulkiness, relatively large heater power consumption, and sluggish warm-up (typically 10min or more). These problems are effectively eliminated if the thermal stabilization is done at the chip level by integrating a heater circuit (with sensor) onto the integrated circuit itself. This approach was pioneered in the 1960s by Fairchild with the μA726 and μA727 temperature-stabilized differential pair and preamp, respectively.

More recently, temperature-stabilized voltage references such as the National LM199 series have appeared. It offers a temperature coefficient of 0.00002%/°C (typ), which is a mere 0.2ppm/°C. These references are packaged in standard metal transistor cans (TO-46); they consume about 0.25 watt of heater power and come up to temperature in 3 seconds. Users should be aware that the subsequent op-amp circuitry, and even precision wire-wound resistors with their ±2.5ppm/°C tempco, may degrade performance considerably, unless extreme care is used in design. In particular, low-drift precision op-amps such as the OP-07, with 0.2μV/°C (typ) input-stage drift, are essential. These aspects of precision circuit design are discussed in Sections 7.01 to 7.06.

One caution when using the LM399: The chip can be damaged if the heater supply hovers below 7.5 volts for any length of time.

The LT1019 bandgap reference, though normally operated unheated, has an on-chip heater and temperature sensor. So

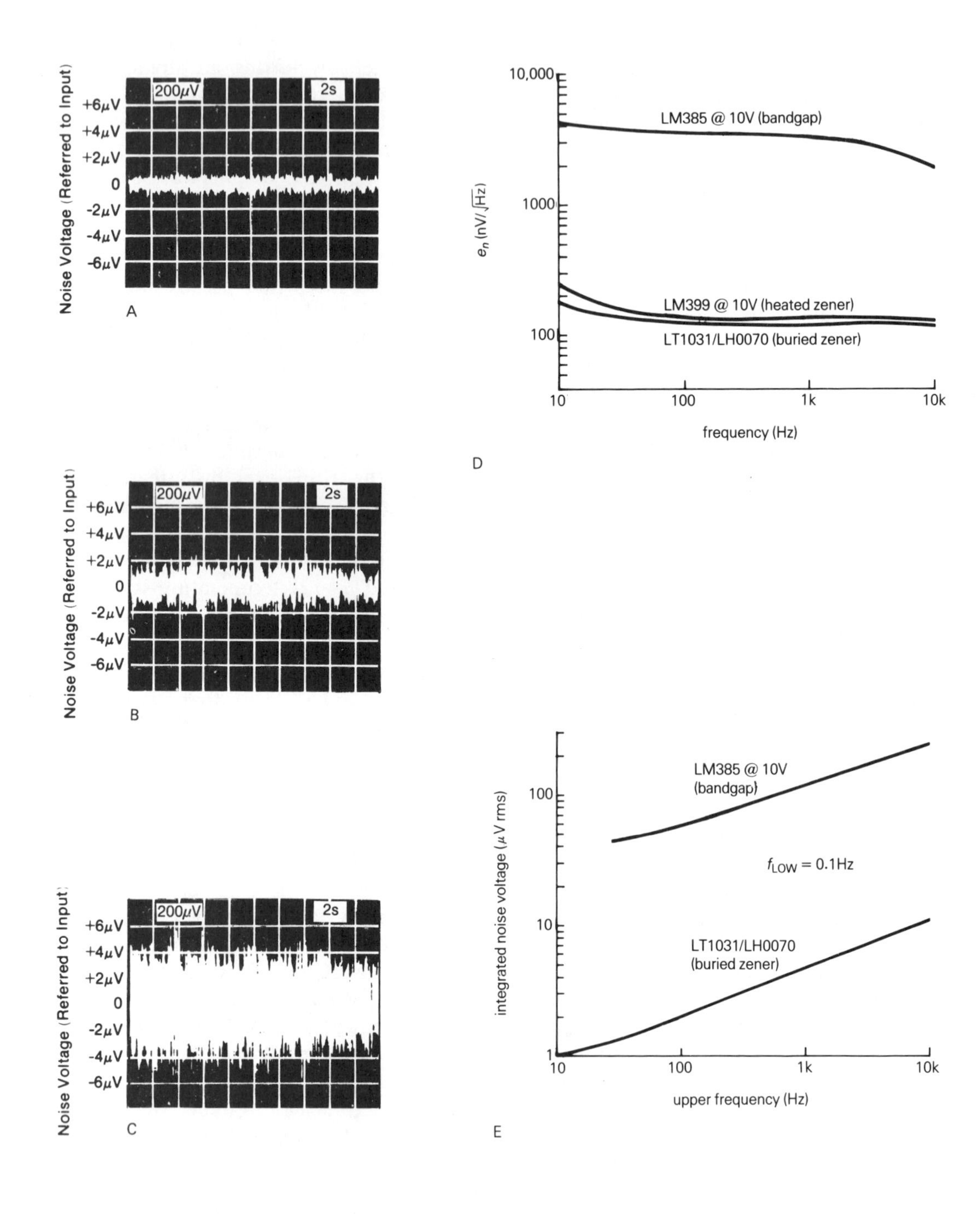

Figure 6.27. Buried zener references (A) have lower noise than either heated zeners (B) or bandgap references (C). (Courtesy of Burr-Brown Corporation.) (D) Noise density (e_n) comparison; (E) integrated noise voltage comparison.

you can use it like the LM399, to get tempcos less than 2ppm/°C. However, unlike the LM399, the LT1019 requires some external circuitry to implement the thermostat (an op-amp and a half dozen components).

□ *2. Precision unheated references.* The thermostated LM399 has excellent tempco, but it does not exhibit extraordinary noise or long-term drift specs (see Table 6.7). The chip also takes a few seconds to heat up, and it uses plenty of power (4W at start-up, 250mW stabilized).

Clever chip design has made possible unheated references of equivalent stability. The REF10KM and REF101KM from Burr-Brown have tempcos of 1ppm/°C (max), with no heater power or warm-up delays. Furthermore, they exhibit lower long-term drift and noise than the LM399-style references. Other 3-terminal references with 1ppm/°C maximum tempco are the MAX671 from Maxim and the AD2710/2712 references from Analog Devices. In 2-terminal configurations the only contender is the magnificent LTZ1000 from Linear Technology, with its claimed 0.05ppm/°C tempco. It also claims long-term drift and noise specs that are a factor of 10 better than any other reference of any kind. The LTZ1000 does require a good external biasing circuit, which you can make with an op-amp and a few parts. All of these high stability references (including the heated LM399) use buried zeners, which additionally provide much lower noise than ordinary zener or bandgap references (Figure 6.27).

THREE-TERMINAL AND FOUR-TERMINAL REGULATORS

6.16 Three-terminal regulators

For most noncritical applications the best choice for a voltage regulator is the simple 3-terminal type. It has only three connections (input, output, and ground) and is factory-trimmed to provide a fixed output. Typical of this type is the 78*xx*. The voltage is specified by the last two digits of the part number and can be any of the following: 05, 06, 08, 10, 12, 15, 18, or 24. Figure 6.28 shows how easy it is to make a +5 volt regulator, for instance, with one of these regulators. The capacitor across the output improves transient response and keeps the impedance low at high frequencies (an input capacitor of at least 0.33μF should be used in addition if the regulator is located a considerable distance from the filter capacitors). The 7800 series is available in plastic or metal power packages (same as power transistors). A low-power version, the 78L*xx*, comes in the same plastic and metal packages as small-signal transistors (see Table 6.8). The 7900 series of negative regulators works the same way (with negative input voltage, of course). The 7800 series can provide up to 1 amp load current and has on-chip circuitry to prevent damage in the event of overheating or excessive load current; the chip simply shuts down, rather than blowing out. In addition, on-chip circuitry prevents operation outside the transistor safe operating area (see Section 6.07) by reducing available output current for large input–output voltage differential. These regulators are inexpensive and easy to use, and they make it practical to design a system with many printed-circuit boards in which the unregulated dc is brought to each board and regulation is done locally on each circuit card.

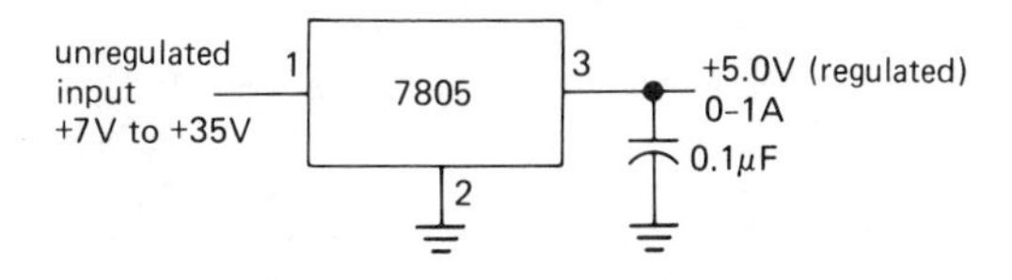

Figure 6.28

Three-terminal fixed regulators come in some highly useful variants. The LP2950 works just like a 7805, but draws only

TABLE 6.8. FIXED VOLTAGE REGULATORS

Type	Pkg	V_{out} (V)	Accuracy (%)	Output current (max)[a] @75°C case I_{out} (A)	No heatsink[b] I_{out} (A)	No heatsink[b] P_{diss} (W)	Regulation (typ) Load[c] (mV)	Regulation (typ) Line[d] (mV)	Θ_{JC} (°C/W)	Input voltage min[i] (V)	Input voltage max (V)
Positive											
LM2950CZ-5.0	TO-92	5	1	0.08	0.1	0.5	2	1.5	160	5.4	30
LM2931Z-5.0	TO-92	5	5	0.1	0.1	0.5	14	3	160	5.3	26
LM78L05ACZ	TO-92	5	4	0.1	0.1	0.6	5	50	160	7	35
LM330T-5.0[g]	TO-220	5	4	0.15	0.15	1.5	14	20	4	5.3	26
TL750L05	TO-92	5	4	0.15	0.15	0.6	20	6	160	5.6	26
LM2984CT	TO-220[h]	5	3	0.5	0.5	2	12	4	3	5.5	26
LM2925T	TO-220	5	5	0.75	0.5	2	10	8	3	5.6	26
LM2935T	TO-220	5	5	0.75	0.5	2	10	8	3	5.5	26
LM309K	TO-3	5	4	1	0.6	2.2	20	4	3	7	35
LT1005CT	TO-220	5	2	1	0.5	2	5	5	3	7	20
LM2940T-5.0	TO-220	5	3	1	0.5	2	35	20	3	5.5	26
LM7805CK	TO-3	5	4	1	0.6	2.2	10	3	3.5	7	35
LM7805CT	TO-220	5	4	1	0.45	1.7	10	3	3	7	35
LM7815CT	TO-220	15	4	1	0.15	1.7	12	4	3	17	35
LT1086-5CT	TO-220	5	1	1.5	0.5	2	5	0.5	3	6.3	30
LAS16A05	TO-3	5	2	2	0.75	2.8	30[m]	100[m]	2.5	7.6	30
LM323K	TO-3	5	4	3	0.6	2	25	5	2	7	20
LT1035CK	TO-3	5	2	3	0.8	3	10	5	1.5	7.3	20
LT1085-5CT	TO-220	5	1	3	0.5	2	5	0.5	3	6.3	30
LAS14A05	TO-3	5	2	3	0.8	3	30[m]	50[m]	2.3	7.5	35
LT1003CK	TO-3	5	2	5	0.8	3	25	5	1	7.3	20
LT1084-5CK	TO-3	5	1	5	0.8	3	5	0.5	1.6	6.3	30
LAS19A05	TO-3	5	2	5	0.8	3	30[m]	50[m]	0.9	7.6	30
LT1083-5CK	TO-3	5	1	7.5	0.8	3	5	0.5	1.6	6.3	30
LAS3905	TO-3	5	5	8	0.8	3	20[m]	100[m]	0.7	7.6	25
Negative											
LM79L15ACZ	TO-92	-15	4	0.1	0.05	0.6	75[m]	45[m]	160	-17	-35
LM7915CK	TO-3	-15	4	1	0.2	2.2	4	3	3.5	-16.5	-35
LM7915CT	TO-220	-15	4	1	0.15	1.7	4	3	3	-16.5	-35
LM345K-5.0	TO-3	-5	4	3	0.2	2.1	10	5	2	-7.5	-20

(a) with V_{in}=1.75V_{out}. (b) 50°C ambient. (c) 0 to I_{max}. (d) ΔV_{in}=15V. (e) ΔV_{out} for 0°C to 100°C junc temp. (f) 1000 hours. (g) similar to LM2930T-5.0, LM2931T-5.0. (h) wide TO-220. (i) at I_{max}. (m) min or max. (t) typical. All include internal thermal shutdown and current-limiting circuitry. Most are available in ±5, 6, 8, 10, 12, 15, 18, and 24V units; a few are available in -2, -3, -4, -5.2, -9, +2.6, +9, and +17V units.

Type	120Hz ripple reject typ (dB)	Temp stab[e] typ (mV)	Long-term stab[f] max (%)	Output impedance 10Hz (Ω)	Output impedance 10kHz (Ω)	Comments
LM2950CZ-5.0	70	10	–	0.01	0.5	micropower, 1%
LM2931Z-5.0	80	–	0.4[t]	0.1	0.2	low dropout, low power
LM78L05ACZ	50	–	0.25	0.2	0.2	small; LM240LAZ-5.0
LM330T-5.0[g]	56	25	0.4[t]	0.1	0.2	low dropout; 2930
TL750L05	65	50	–	–	–	TL751 has enable
LM2984CT	70	3	0.4[t]	0.01	0.02	dual outputs (μP); reset, on/off
LM2925T	66	–	0.4[t]	0.2	0.2	microprocessor; reset
LM2935T	66	–	0.4[t]	0.02	0.02	dual outputs (μP); reset, on/off
LM309K	80	50	0.4	0.04	0.05	original +5V regulator
LT1005CT	70	25	–	0.003	0.01	dual outputs (μP)
LM2940T-5.0	72	20	0.4[t]	0.03	0.03	
LM7805CK	80	30	0.4	0.01	0.03	LM340K-5
LM7805CT	80	30	0.4	0.01	0.03	popular; LM340T-5
LM7815CT	70	100	0.4	0.02	0.05	LM340T-15
LT1086-5CT	63	25	1	–	–	low dropout
LAS16A05	75	–	–	0.002	0.02	Lambda, monolithic
LM323K	70	30	0.7	0.01	0.02	
LT1035CK	70	25	–	0.003	0.01	dual +5; 1036 is +12/+5
LT1085-5CT	63	25	1	–	–	low dropout
LAS14A05	70	100[m]	–	0.001	0.003	Lambda, monolithic
LT1003CK	66	25	0.7	0.003	0.02	
LT1084-5CK	63	25	1	–	–	low dropout
LAS19A05	70	150[m]	–	0.01	0.2	lambda, monolithic
LT1083-5CK	63	25	1	–	–	low dropout
LAS3905	60[m]	100	–	0.004	0.01	Lambda, monolithic
LM79L15ACZ	40	–	0.4[t]	0.05	0.05	small; LM320LZ-15
LM7915CK	60	60	0.4	0.06	0.07	LM320KC-15
LM7915CT	60	60	0.4	0.06	0.07	LM320T-15
LM345K-5.0	65	25	1.0	0.02	0.04	

75μA of quiescent current (compared with the 7805's 5mA, or the 78L05's 3mA); it also regulates with as little as a 0.4 volt drop from unregulated input to regulated output (called the "dropout voltage"), compared with 2 volts dropout for the classic 7805. The LM2931 is also low-dropout, but you might call it *milli*power (0.4mA quiescent current), compared with the "micropower" LP2950. Low-dropout regulators also come in high-current versions – for example, the LT1085/4/3 series from

LTC (3A, 5A, and 7.5A, respectively, with both +5V and +12V available in each type). Regulators like the LM2984 are basically 3-terminal fixed regulators, but with extra outputs to signal a microprocessor that power has failed, or resumed. Finally, regulators like the 4195 contain a pair of 3-terminal 15 volt regulators, one positive and one negative. We'll say a bit more about these special regulators shortly.

6.17 Three-terminal adjustable regulators

Sometimes you want a nonstandard regulated voltage (say +9V, to emulate a battery) and can't use a 78*xx*-type fixed regulator. Or perhaps you want a standard voltage, but set more accurately than the ±3% accuracy typical of fixed regulators. By now you're spoiled by the simplicity of 3-terminal fixed regulators, and therefore you can't imagine using a 723-type regulator circuit, with all its required external components. What to do? Get an "adjustable 3-terminal regulator"! Table 6.8 lists the characteristics of a representative selection of 3-terminal fixed regulators.

These wonderful ICs are typified by the classic LM317 from National. This regulator has no ground terminal; instead, it adjusts V_{out} to maintain a constant 1.25 volts (bandgap) from the output terminal to the "adjustment" terminal. Figure 6.29 shows the easiest way to use it. The regulator puts 1.25 volts across R_1, so 5mA flows through it. The adjustment terminal draws very little current (50–100μA), so the output voltage is just

$$V_{out} = 1.25(1 + R_2/R_1) \text{ volts}$$

In this case the output voltage is adjustable from 1.25 volts to 25 volts. For a fixed-output-voltage application, R_2 will normally be adjustable only over a narrow range, to improve settability (use a fixed resistor in series with a trimmer). Choose your resistive divider values low enough to allow for a 50μA change in adjustment current with temperature. Because the loop compensation for the regulator is the output capacitor, larger values must be used compared with other designs. At least a 1μF tantalum is required, but we recommend something more like 6.8μF.

The 317 is available in several packages, including the plastic power package (TO-220), the metal power package (TO-3), and the small transistor packages (metal, TO-5; plastic, TO-92). In the power packages it can deliver up to 1.5 amps, with proper heat sinking. Because it doesn't "see" ground, it can be used for high-voltage regulators, as long as the input–output voltage differential doesn't exceed the rated maximum of 40 volts (60V for the LM317HV high-voltage variant).

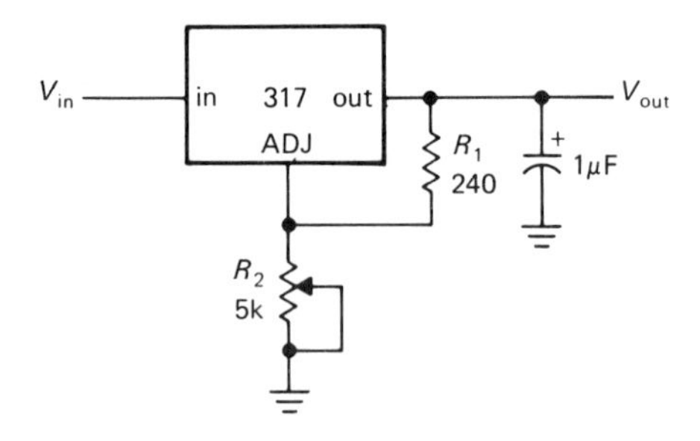

Figure 6.29. Three-terminal adjustable regulator.

EXERCISE 6.5

Design a +5 volt regulator with the 317. Provide ±20% voltage adjustment range with a trimmer pot.

Three-terminal adjustable regulators are available with higher current ratings, e.g., the LM350 (3A), the LM338 (5A), and the LM396 (10A), and also with higher voltage ratings, e.g., the LM317H (60V) and the TL783 (125V). Read the data sheets carefully before using these parts, noting bypass capacitor requirements and safety diode suggestions. As with the fixed

3-terminal regulators, you can get low-dropout versions (e.g., the LT1085, with 1.3V dropout at 3.5A), and you can get micropower versions (e.g., the LP2951, the adjustable variant of the fixed 5V LP2950; both have $I_Q = 75\mu A$). You can also get *negative* versions, although there's less variety: The LM337 is the negative cousin of the LM317 (1.5A), and the LM333 is a negative LM350 (3A).

□ *Four-terminal regulators*

Three-terminal adjustable regulators are the favorite for noncritical requirements. Historically they were preceded by four-terminal adjustable regulators, which you connect as shown in Figure 6.30. You drive the "control" terminal with a sample of the output; the regulator adjusts the output to keep the control terminal at a fixed voltage (+3.8V for the Lambda regulators in Table 6.9, +5V for the μA79G, and −2.2V for the negative regulators). Four-terminal regulators aren't any better than the simpler 3-terminal variety (but they aren't any worse, either), and we mention them here for completeness.

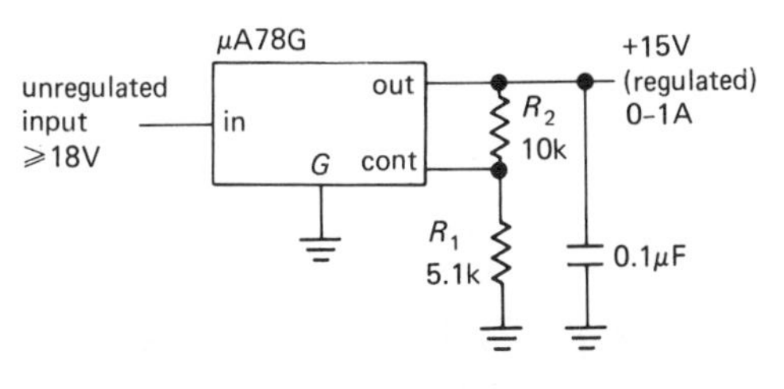

Figure 6.30

6.18 Additional comments about 3-terminal regulators

General characteristics of 3- and 4-terminal regulators

The following specifications are typical for most 3- and 4-terminal regulators, both fixed and adjustable, and they may be useful as a rough guide to the performance you can expect:

Output voltage tolerance:	1–2%
Dropout voltage:	0.5–2 volts
Maximum input voltage:	35 volts (except TL783 to +125V)
Ripple rejection:	0.01–0.1%
Spike rejection:	0.1–0.3%
Load regulation:	0.1–0.5%, full load change
dc input rejection:	0.2%
Temperature stability:	0.5%, over full temp range

Improving ripple rejection

The circuit of Figure 6.29 is the standard 3-terminal regulator, and it works fine. However, the addition of a 10μF bypass capacitor from the adjust (ADJ) terminal to ground (Fig. 6.31) improves the ripple (and spike) rejection by about 15dB (factor of 5 in voltage). For example, the LM317 ripple rejection factor goes from 65dB to 80dB (the latter is 0.1mV output ripple when supplied with 1V input ripple, a typical value). Be sure to include the safety discharge diode; look at the specification sheet of the particular regulator for more details.

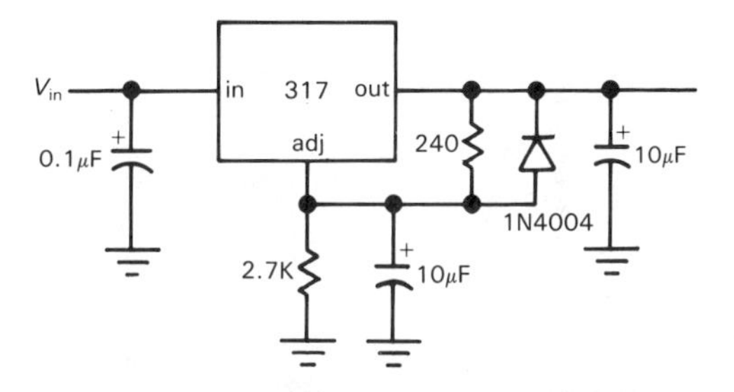

Figure 6.31. The ADJ pin may be bypassed for lower noise and ripple, but a safety discharge diode must be included.

Low-dropout regulators

As we mentioned earlier, most series regulators need at least 2 volts of "headroom" to function; that's because the base of the *npn* pass transistor is a V_{BE} drop above

TABLE 6.9. ADJUSTABLE VOLTAGE REGULATORS

Type	Polarity	Pkg	Output voltage min (V)	Output voltage max (V)	I_{max} (A)	Regulation (typ) Load[a] (%)	Regulation (typ) Line[b] (%)	Θ_{JC} (°C/W)	Input voltage min (V)	Input voltage max (V)	Dropout voltage @I_{max} max (V)	120Hz ripple reject typ (dB)	Temp stab[c] typ (%)	Long-term stab[d] max (%)	Output impedance 10Hz (Ω)	Output impedance 10kHz (Ω)	Therm lim	Curr lim[i]	Comments
Three-terminal																			
LM317L	+	TO-92	1.2	37	0.1	0.1	0.15	160[h]	–	40[e]	2.5[t]	65	0.5	1	0.07	4	•	I	miniature
LM337L	–	TO-92	1.2	37	0.1	0.1	0.15	160[h]	–	-40[e]	2.5[t]	65	0.5	–	–	–	•	I	miniature (neg 317L)
LM317H	+	TO-39	1.2	37	0.5	0.1	0.2	12	–	40[e]	2[t]	80	0.6	0.3	0.01	0.03	•	I	317 in TO-39
LM337H	–	TO-39	-1.2	-37	0.5	0.3	0.2	12	–	-40[e]	2[t]	75	0.5	0.3	0.02	0.02	•	I	negative 317H
TL783C	+	TO-220	1.3	125	0.7	0.2[f]	0.02	4	–	125[e]	10	50	0.3	0.2	0.05	0.3	•	I	MOSFET, high voltage
LM317T	+	TO-220	1.2	37	1.5	0.1	0.2	4	–	40[e]	2.5[t]	80	0.6	0.3	0.01	0.03	•	I	popular
LM317HVK	+	TO-3	1.2	57	1.5	0.1	0.2	2.3	–	60[e]	2.5[t]	80	0.6	0.3	0.01	0.03	•	I	high voltage 317
LM337T	–	TO-220	-1.2	-37	1.5	0.3	0.2	4	–	-40[e]	2.5[t]	75	0.5	0.3	0.02	0.02	•	I	negative 317
LM337HVK	–	TO-3	-1.2	-47	1.5	0.3	0.2	2.3	–	-50[e]	2.5[t]	75	0.5	0.3	0.02	0.02	•	I	high voltage 337
LT1086CP	+	TO-220	1.3	30	1.5	0.1	0.02	–	–	30[e]	1.5	75	0.5	1	–	–	•	I	low dropout
LM350K	+	TO-3	1.2	32	3	0.1	0.1	2	–	35[e]	2.5[t]	80	0.6	0.3	0.005	0.02	•	I	3A monolithic
IP3R07T	+	TO-220	1.2	37	3	0.1	0.08	2.3	–	15[e]	0.8[t]	65	–	–	–	–	•	I	two unreg inputs
LM333T	–	TO-220	-1.2	-32	3	0.2	0.02	50	–	-35[e]	2.5[t]	60	0.5	0.2	–	–	•	I	neg 350; LT1033 is imprvd
LT1085CT	+	TO-220	1.3	30	3	0.1	0.02	3	–	30[e]	1.5	75	0.5	1	–	–	•	I	low dropout
LM338K	+	TO-3	1.2	32	5	0.1	0.1	2	–	35[e]	2.5[t]	80	0.6	0.3	–	–	•	I	5A monolithic
LT1084CP	+	TO-247	1.3	30	5	0.1	0.02	2.3	–	30[e]	1.5	75	0.5	1	–	–	•	I	low dropout
LT1083CP	+	TO-247	1.3	30	7.5	0.1	0.02	1.6	–	30[e]	1.5	75	0.5	1	–	–	•	I	low dropout
LM396K	+	TO-3	1.2	15	10	0.4[m]	0.08	1	–	20[e]	2.1[t]	74	0.3	1	0.01	0.02	•	I	10A monolithic
LT1038CK	+	TO-3	1.2	32	10	0.1	0.08	1	–	35[e]	2.5[t]	60	1	1	0.005	0.1	•	I	10A monolithic, 1% acc'y

Four-terminal																			
μA78GU1C	+	TO-220	5	30	1	1^{m}	1^{m}	7.5	7.5	40	2.5	80	3^{m}	–	–	–	•	I	TO-39 pkg avail
μA79GU1C	–	TO-220	-2.5	-30	1	1^{m}	1^{m}	7.5	-7	-40	2^{t}	60	3^{m}	–	–	–	•	I	TO-39 pkg avail
LAS15U	+	TO-3	4	30	1.5	0.6^{m}	2^{m}	3	6.5	40	2.4	70	3^{m}	–	0.003	0.02	•	I	Lambda
LAS18U	–	TO-3	-2.6	-30	1.5	0.6^{m}	2^{m}	3	-5	-40	2.1	60	3^{m}	–	0.02	0.04	•	I	Lambda
LAS16U	+	TO-3	4	30	2	0.6^{m}	2^{m}	2.5	6.5	35	2.6	70	2^{m}	–	0.002	0.02	•	I	Lambda
LAS14AU	+	TO-3	4	35	3	0.6^{m}	1^{m}	1.5	6.5	40	2.3	70	2^{m}	–	0.001	0.01	•	I	Lambda
LAS19U	+	TO-3	4	30	5	0.6^{m}	1^{m}	0.9	6.5	35	2.6	65	2^{m}	–	0.0005	0.004	•	I	Lambda
LAS39U	+	TO-3	4	16	8	0.6^{m}	2^{m}	0.7	6.6	25	2.6	60^{m}	3^{m}	–	0.002	0.01	•	I	Lambda
Multiterminal																			
LM376N	+	DIP-8	5	37	0.03	0.2^{m}	0.6^{m}	190^{h}	9	40	3	60^{m}	1^{m}	–	–	–	–	E	
LM304H	–	TO-5	0	-40	0.03	1mV	0.2	45	-8	-40	2	65	0.3	0.01	–	–	–	E	orig neg reg
ICL7663S	+	DIP-8	1.3	16	0.04	0.4^{f}	0.5	200^{h}	1.5	16	1^{t}	20	1	–	–	–	–	E	micropower; also MAX663
MAX664	–	DIP-8	-1.3	-16	0.04	0.8^{f}	0.5	120^{h}	-2	-16	0.3^{t}	15	1	–	–	–	–	E	μpwr, impr 7664; low dropout
LM305AH	+	TO-5	4.5	40	0.05	0.03	0.3	45	8.5	50	3	80	0.3	0.1	–	–	–	E	
LM2931CT	+	TO-220	3	24	0.1	0.3	0.06^{f}	3	3.6	26	0.3^{t}	60^{f}	–	0.4	0.1^{f}	0.2^{f}	•	I	low dropout, low power
LP2951CN	+	DIP-8	1.3	29	0.1	0.1^{f}	0.03^{f}	105	1.7	30	0.4^{t}	70^{f}	0.5	–	0.01	0.5	•	I	low dropout, micropower
LT1020CN	+	DIP-14	2.5	35	0.13	0.2	0.15	60?	4.5	36	0.4^{t}	60	1	–	–	–	•	I	micropower
NE550N	+	DIP-14	2	40	0.15	0.03	0.08	150^{h}	8.5	40	3	90	0.2	0.1^{t}	0.1	0.1	–	E	
μA723PC	+	DIP-14	2	37	0.15	0.03	0.1	150^{h}	9.5	40	3	75	0.3	0.1	0.05	0.1	–	E	classic
LAS1000	+	TO-5	3	38	0.15	0.1^{m}	0.2	150^{h}	5	40	2	60^{m}	1.5^{m}	–	0.004	0.05	•	E	Lambda, improved 723
LAS1100	+	TO-5	3	48	0.15	0.1^{m}	0.2	150^{h}	5	50	2	60^{m}	1.5^{m}	–	0.004	0.05	•	E	high voltage LAS1000
SG3532J	+	DIP-14	2	38	0.17	0.1	0.1	125^{h}	4.7	40	2	66	0.5	0.3	–	–	•	E	improved 723
MC1469R	+	TO-66	2.5	32	0.6	0.005	0.05	7	9	35	3	100	0.2	–	0.05	0.1	–	E	precision, may oscillate
MC1463R	–	TO-66	-3.8	-32	0.6	0.005	0.05	17	-9	-35	3	90	0.2	–	0.02	0.03	–	E	neg MC1469
LM2941CT	+	TO-220	1.3	25	1			3		26	1	74	0.4	0.4^{t}	0.04	0.1	–	I	low dropout
LAS2200	+	module	2.5	28	5	0.2^{m}	0.15^{m}	2	9.6	40	2.5	60^{m}	0.7^{m}	–	–	–	•	I	
LAS3000	+	module	2.7	29	10	0.2^{m}	0.15^{m}	1.3	7.9	40	2.5	60^{m}	1.5^{m}	–	–	–	•	I	Lambda hybrid; 2 unreg inputs
LAS5000	+	module	4.8	29	20	0.2^{m}	0.2^{m}	0.7	11.9	40	2.5	60^{m}	1.5^{m}	–	–	–	•	I	Lambda hybrid; 2 unreg inputs
LAS7000	+	module	4.8	29	30	0.2^{m}	0.2^{m}	0.4	12.3	40	2.5	60^{m}	1.5^{m}	–	–	–	•	I	Lambda hybrid; 2 unreg inputs
MC1466L	+	DIP-14	0	1000	–	0.02	0.05	170^{h}	–	–	2^{t}	70	0.4	–	–	–	–	E	lab supply; good curr lim
LAS3700	+	TO-5	0	1000	–	0.003	0.15	220	–	–	–	65	0.5^{m}	–	–	–	–	E	floating reg with on-chip heater

(a) 10% to 50% I_{max}. (b) for ΔV_{in}=15V. (c) ΔV_{out} for 0°C to 100°C T_J. (d) 1000 hrs. (e) max V_{in}-V_{out}. (f) at 5V. (h) Θ_{JA}. (i) E – external; I – internal. (m) min/max. (t) typ.

the output, and it has to be driven by a driver transistor, usually another *npn* whose base is pulled up with a current mirror. That's already two V_{BE} drops. Furthermore, you need to allow another V_{BE} drop across the current-sensing resistor for short-circuit protection; see Figure 6.32A, a simplified schematic of the 78L*xx*. The three V_{BE}s add up to about 2 volts, below which the regulator drops out of regulation at full current.

By using a *pnp* (or *p*-channel MOSFET) pass transistor, the dropout voltage can be reduced from the three V_{BE} drop of the conventional *npn* scheme, down nearly to the transistor saturation voltage. Figure 6.32B shows a simplified schematic of the LM330 low-dropout fixed +5 volt

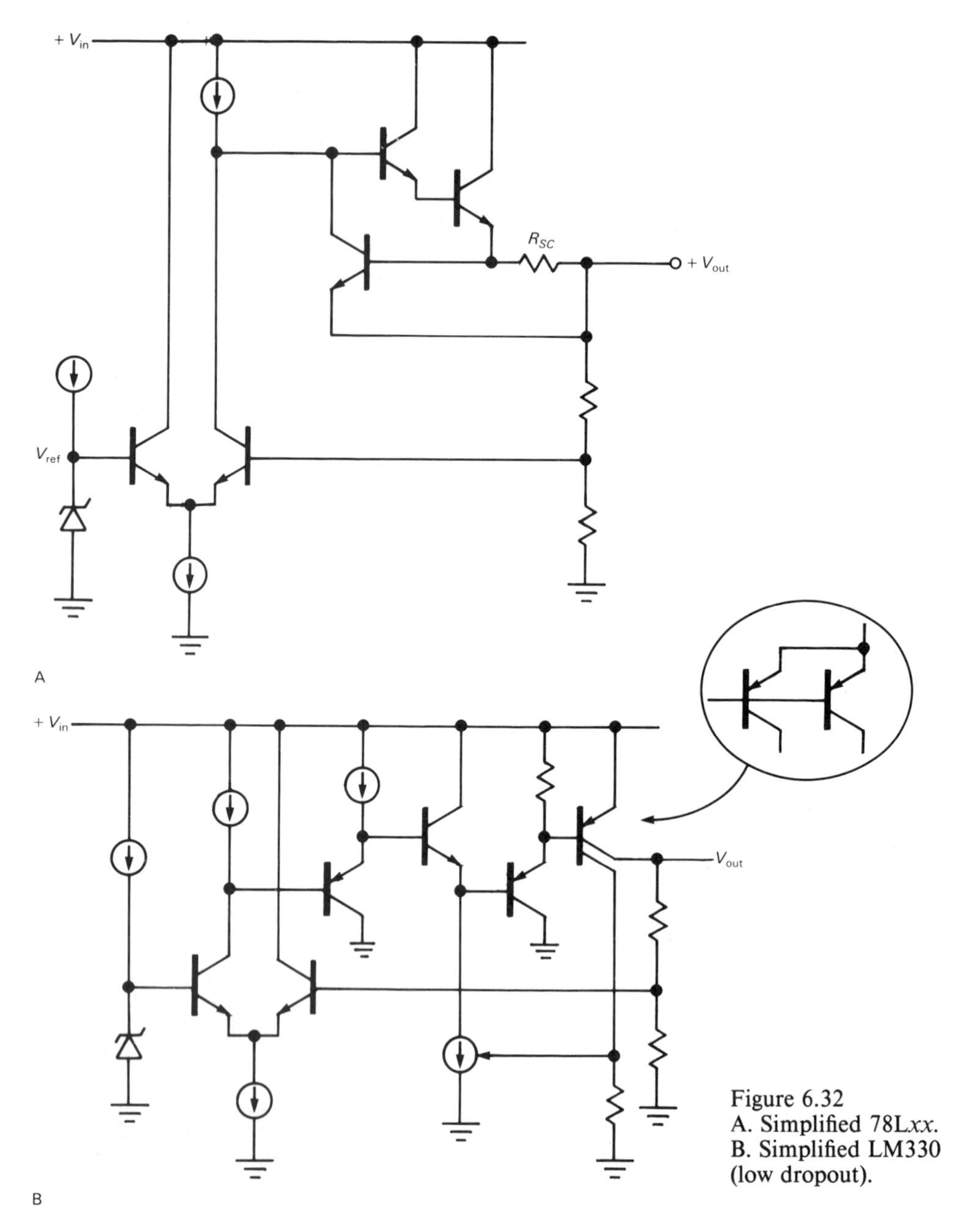

Figure 6.32
A. Simplified 78L*xx*.
B. Simplified LM330 (low dropout).

(150mA) regulator. The output can be brought within a saturation voltage of the unregulated input voltage by the *pnp* pass transistor. Having thus eliminated the Darlington V_{BE} drops of the *npn* regulator circuit, the designers weren't about to waste a diode drop with the usual (series resistor) short-circuit protection scheme. So they used a clever trick, deriving a sample of the output current via a second collector: That current is a fixed fraction of the output current and is used to shut off base drive as shown. This current-limit scheme is not particularly precise (I_L is specified as 150mA min, 700mA max), but it's good enough to protect the regulator, which also has internal thermal limiting.

Low-dropout regulators are available in most of the popular types, for example 3-terminal fixed voltage [LM2931, LM330, LT1083/4/5 (5V and 12V), TL750], 3-terminal adjustable (LT1083/4/5, LM2931), and micropower (LP2950/1, MAX664, LT1020). Tables 6.8 and 6.9 include all low-dropout regulators available at the time of writing.

Processor-oriented regulators

Electronic devices that include microprocessors (the subject of Chapters 10 and 11) require more than a simple regulated voltage. In order to retain the contents of volatile memory (and in order to keep track of elapsed time) they need a separate source of low-current dc when the regular power source is off; this may happen because the device is shut off, or because of a power failure. They also need to know when ordinary power has resumed, so they can "wake up" in a known state. Furthermore, a microprocessor-based device may need to have a few milliseconds warning that normal power is about to fail, so it can put data into safe memory.

Until recently you had to design extra circuitry to do these things. Now life is easy – you can get "(micro)processor-oriented" regulator ICs, with various combinations of these functions built in. These ICs sometimes go by the name of "power supply supervisory chips" or "watchdog" chips. An example is the LM2984, which has two high-current +5 volt outputs (one for the microprocessor, one for other circuitry) and a low-current +5 volt output (for memory), a delayed RESET flag output to initialize your microprocessor when power resumes, and an ON/OFF controlling input for the high-current outputs. It also has an input that monitors microprocessor activity, resetting the processor if it grinds to a halt. An example of a watchdog chip *without* regulator is the MAX691 from Maxim, which monitors the regulated supply voltage and microprocessor activity, and provides reset (and "interrupt") signals to the microprocessor, just like the LM2984. However, it adds both power-fail *warning* and battery switchover circuitry to the other capabilities of the LM2984. Used with an ordinary +5 volt regulator, the MAX691 does everything you need to keep a microprocessor happy. We'll learn much more about the care and feeding of microprocessors in Chapters 10 and 11.

□ Micropower regulators

As we suggested earlier, most regulator chips draw a few milliamps of quiescent current to run the internal voltage reference and error amplifiers. That's no problem for an instrument powered from the ac mains, but it's undesirable in a battery-operated instrument, powered by a 400mA-hour 9 volt alkaline battery, and it's intolerable in a micropower instrument that must run a thousand hours, say, on a battery.

The solution is a micropower regulator. The most miserly of these are the ICL7663/4, positive and negative adjustable regulators with quiescent currents of 4μA. At that current a 9 volt battery would last 100,000 hours (more than 10 years), which exceeds the "shelf life" (self-discharge time) of all batteries except some lithium-based types. Micropower design is challenging and fun, and we'll tell you all about it in Chapter 14.

Dual-polarity regulated supplies

Most of our op-amp circuits in Chapter 4 ran from symmetrical bipolarity supplies, typically ±15 volts. That's a common requirement in analog circuit design, where you often deal with signals near ground, and the simplest way to generate symmetrical split supplies is to use a pair of 3-terminal regulators. For example, to generate regulated ±15 volts, you could use a 7815 and a 7915, as in Figure 6.33A. We tend to favor the use of adjustable 3-terminal regulators, because (a) you only need to stock one type for each polarity and current range, and (b) you can trim the voltage exactly, if needed; Figure 6.33B shows how the circuit looks with a 317 and 337.

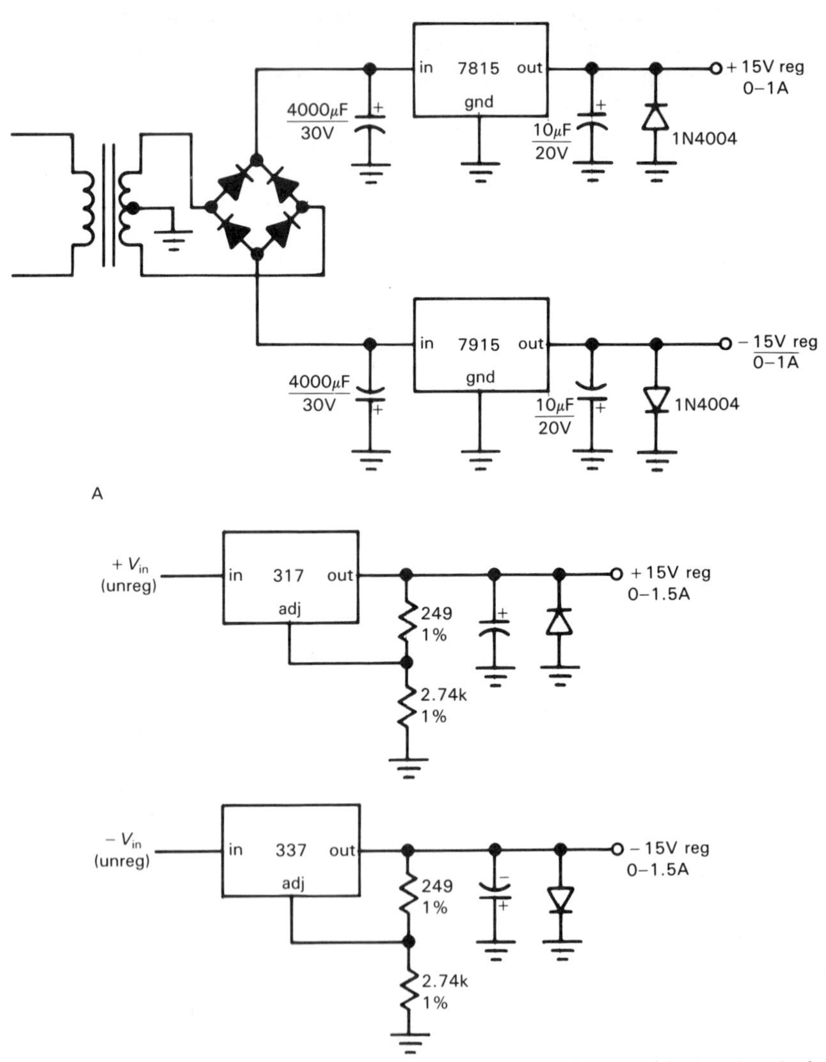

Figure 6.33. Dual-polarity regulated supplies.

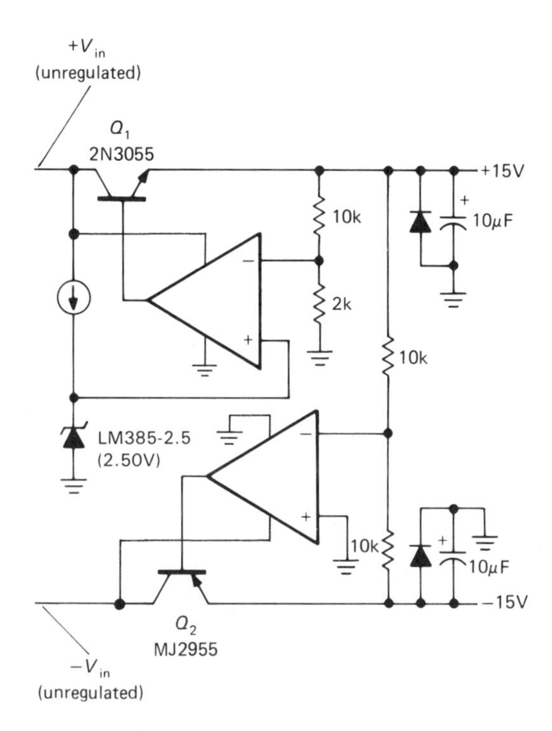

Figure 6.34. Dual-tracking regulator.

□ *Dual tracking regulators.* Given the need for regulated split supplies, you might wonder why there aren't "dual 3-terminal regulators." Wonder no more – they exist and are known as "dual tracking regulators." To understand why they have such a complicated name, take a look at Figure 6.34, which shows the classic dual tracking regulator circuit. Q_1 is the pass transistor for a conventional positive regulated supply. The positive regulated output is then simply used as the reference for a negative supply. The lower error amplifier controls the negative output by comparing the average of the two output voltages with ground, thus giving equal 15 volt positive and negative regulated outputs. The positive supply can be any of the configurations we have already talked about; if it is an adjustable regulator, the negative output follows any changes in the positive regulated output. In practice, it is wise to include current-limiting circuitry, not shown in Figure 6.34 for simplicity.

As with single-polarity regulators, dual-tracking regulators are available as complete integrated circuits in both fixed and adjustable versions, though in considerably less variety. Table 6.10 lists the characteristics of most types now available. Typical are the 4194 and 4195 regulators from Raytheon, which are used as shown in Figure 6.35. The 4195 is factory-trimmed for ±15 volt outputs, whereas the 4194's symmetrical outputs are adjustable via the single resistor R_1. Both regulators are available in power packages as well as the small DIP packages, and both have internal thermal shutdown and current limiting. For higher output currents you can add outboard pass transistors (see below).

Many of the preceding regulators (e.g., the 4-terminal adjustable regulators) can be connected as dual-tracking regulators. The manufacturers' data sheets often give suggested circuit configurations. It is worth keeping in mind that the idea of referencing one supply's output to another supply can be used even if the two supplies are not of equal and opposite voltages. For instance, once you have a stable +15 volt supply, you can use it to generate a regulated +5 volt output, or even a regulated −12 volt output.

EXERCISE 6.6

Design a ±12 volt regulator using the 4194.

Reverse-polarity protection. An additional caution with dual supplies: Almost any electronic circuit will be damaged extensively if the supply voltages are reversed. The only way that can happen with a single supply is if you connect the wires backward; sometimes you see a high-current rectifier connected across the circuit in the reverse direction to protect against this error. With circuits that use several supply voltages (a split supply, for instance), extensive damage can result if there is a component failure that shorts the two supplies together; a common

TABLE 6.10. DUAL-TRACKING REGULATORS

Type	Pkg	V_{out} (V)	Adj. output	Balance trim	Adj. cur. limit	Thermal limit	Max V_+-V_- Input (V)	Maximum output current[a] (each supply) @75°C case I_{out} (mA)	No sink[b] I_{out} (mA)	No sink[b] P_{diss} (W)	Regulation typ Load[c] (mV)	Regulation typ Line[d] (mV)	Θ_{JC} (°C/W)	120Hz ripple reject typ (dB)	Temp stab[e] typ (mV)	Noise[f] (μV rms)
Motorola																
MC1468L	DIP	±15	•	•	•	–	60	55	30	0.5	10[m]	10[m]	50	75	45	100
MC1468R	TO-66	±15	•	•	•	–	60	100	65	1.2	10[m]	10[m]	17	75	45	100
National																
LM325H[g]	TO-5	±15	–	–	•	•	60	100	50	0.5	6	2	12	75	45	150
LM325N[g]	DIP	±15	–	–	•	•	60	–	50	0.5	6	2	90[h]	75	45	150
LM326H[g]	TO-5	adj	–	–	•	•	60	100	70	0.5	6	2	12	75	35	100
LM326N[g]	DIP	adj	–	–	•	•	60	–	70	0.5	6	2	90[h]	75	35	100
Raytheon																
RC4194DB	DIP	adj	•	•	–	•	70	30[i]	25[i]	0.5	0.1%	0.2%	160[h]	70	0.2%	250[j]
RC4194TK	TO-66	adj	•	•	–	•	70	250[i]	90[i]	1.8	0.2%	0.2%	7	70	0.2%	250[j]
RC4195NB	miniDIP	±15	–	•	–	•	60	–	20	0.35	2	2	210[h]	75	75	60
RC4195TK	TO-66	±15	–	•	–	•	60	150	70	1.2	3	2	11	75	75	60
Silicon General																
SG3501AN	DIP	±15	–	•	–	•	60	60	30	0.6	30	20	125[h]	75	150	50
SG3502N	DIP	adj	•	•	•	•	50	50[i]	30	0.6	0.3%	0.2%	125[h]	75	1%	50

(a) V_{in}=1.6V_{out} (each supply). (b) for 50°C ambient. (c) 10% to 50% I_{max}. (d) for ΔV_{in}=15V. (e) ΔV_{out} for 0°C to 100°C T_J. (f) 100Hz to 10kHz. (g) intended for use with a pair of external pass transistors. (h) Θ_{JA}. (i) 10V drop (each supply). (j) 10Hz to 100kHz. (m) max.

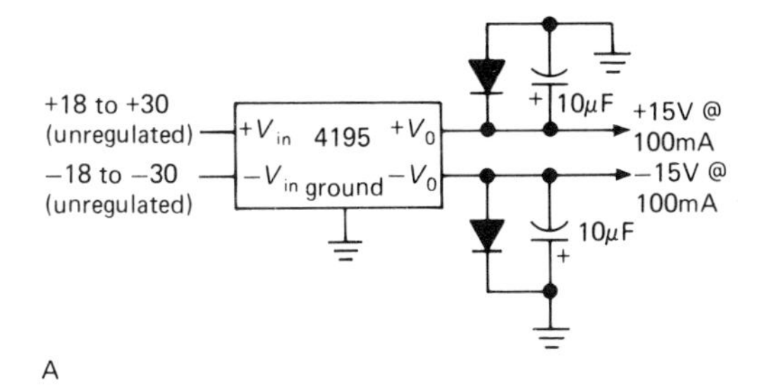

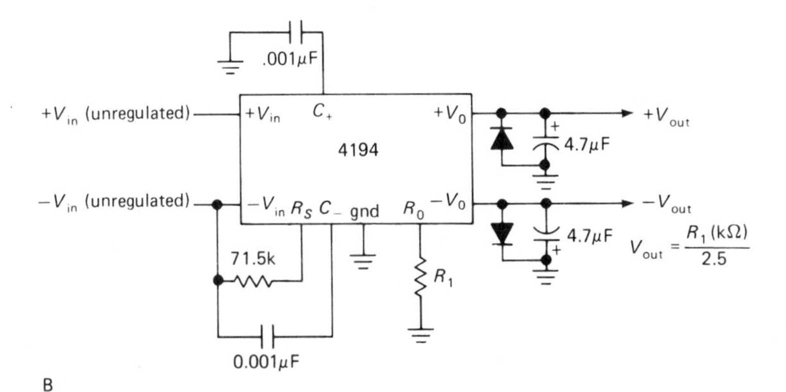

Figure 6.35

situation is a collector-to-emitter short in one transistor of a push-pull pair operating between the supplies. In that case the two supplies find themselves tied together, and one of the regulators will win out. The opposite supply voltage is then reversed in polarity, and the circuit starts to smoke. For this reason it is wise to connect a power rectifier (e.g., a 1N4004) in the reverse direction from each regulated output to ground, as we drew in Figure 6.33.

Outboard pass transistors

Three-terminal fixed regulators are available with 5 amps or more of output current, for example the adjustable 10 amp LM396. However, such high current operation may be undesirable, since the maximum chip operating temperature for these regulators is lower than for power transistors, mandating oversize heat sinks. Also, they are expensive. An alternative solution is the use of external pass transistors, which can be added to the 3- and 4-terminal regulators (and dual-tracking regulators) just as with the classic 723. Figure 6.36 shows the basic circuit.

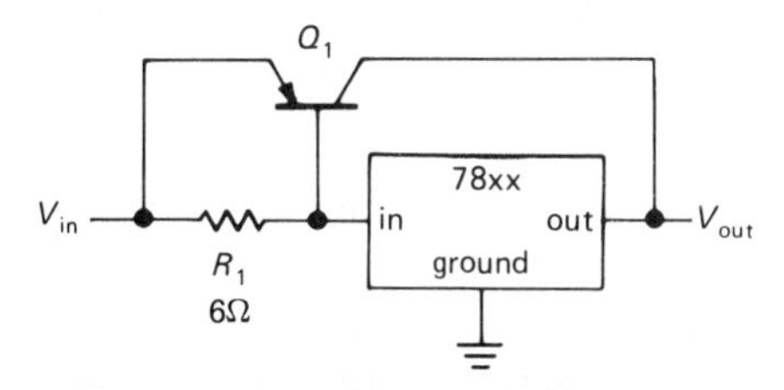

Figure 6.36. Three-terminal regulator with current-boosting outboard transistor.

The circuit works normally for load currents less than 100mA. For greater load currents, the drop across R_1 turns on Q_1, limiting the actual current through the 3-terminal regulator to about 100mA. The 3-terminal regulator maintains the output at the correct voltage, as usual, by reducing input current and hence drive to Q_1 if the output voltage rises, and vice versa. It never even realizes the load is drawing more than 100mA! With this circuit the input voltage must exceed the output voltage by the dropout voltage of the 78xx (2V) plus a V_{BE} drop.

In practice, the circuit must be modified to provide current limiting for Q_1, which could otherwise supply an output current equal to h_{FE} times the regulator's internal current limit, i.e., 20 amps or more! That's enough to destroy Q_1, as well as the unfortunate load that happens to be connected at the time. Figure 6.37 shows two methods of current limiting.

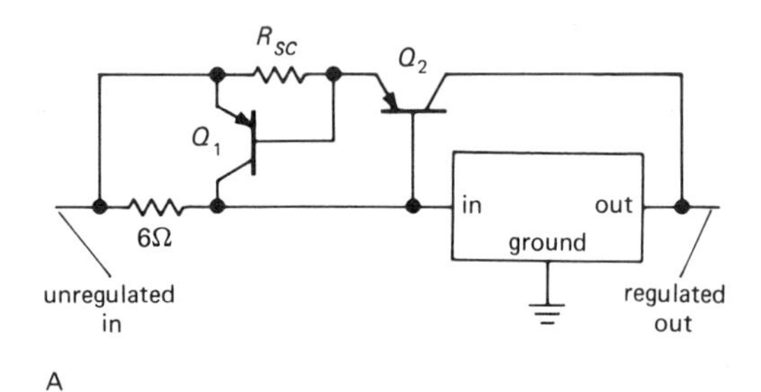

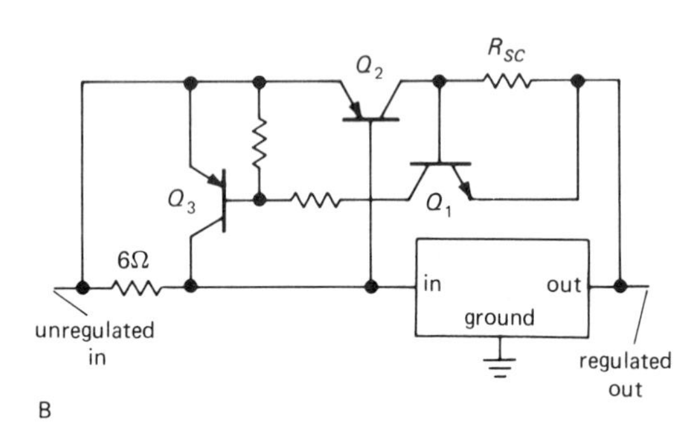

Figure 6.37. Current-limit circuits for outboard transistor booster.

In both circuits, Q_2 is the high-current pass transistor, and its emitter-to-base resistor has been chosen to turn it on at 100mA load current. In the first circuit, Q_1 senses the load current via the drop across R_{SC}, cutting off Q_2's drive when the drop exceeds a diode drop. There are a couple of drawbacks to this circuit: The input voltage must now exceed the regulated output voltage by the dropout voltage of the 3-terminal regulator plus two diode drops, for load currents near the current limit. Also, Q_1 must be capable of handling high currents (equal to the current limit of the regulator), and it is difficult to add foldback limiting because of the small resistor values required in Q_1's base.

The second circuit helps solve these problems, at the expense of some additional complexity. With high-current regulators, a low dropout voltage is often important to reduce power dissipation to acceptable levels. To add foldback limiting to the latter circuit, just tie Q_1's base to a divider from Q_2's collector to ground, rather than directly to Q_2's collector.

External pass transistors can be added to the adjustable 3- and 4-terminal regulators in exactly the same way. See the manufacturers' data sheets for further details.

Current source

A 3-terminal adjustable regulator makes a handy high-power constant-current source. Figure 6.38 shows one to source 1 amp. The addition of an op-amp follower, as in the second circuit, is necessary if the circuit is used to source small currents, since the "ADJ" (adjust) input contributes a current error of about 50μA. As with the previous regulators, there is on-chip current limit, thermal-overload protection, and safe operating area protection.

EXERCISE 6.7

Design an adjustable current source for output currents from 10μA to 1mA using a 317. If $V_{\text{in}} = +15\text{V}$, what is the output compliance? Assume a dropout voltage of 2 volts.

Note that the current source in Figure 6.38A is a 2-terminal device. Thus, the load can be connected on either side. The figure shows how you might connect things

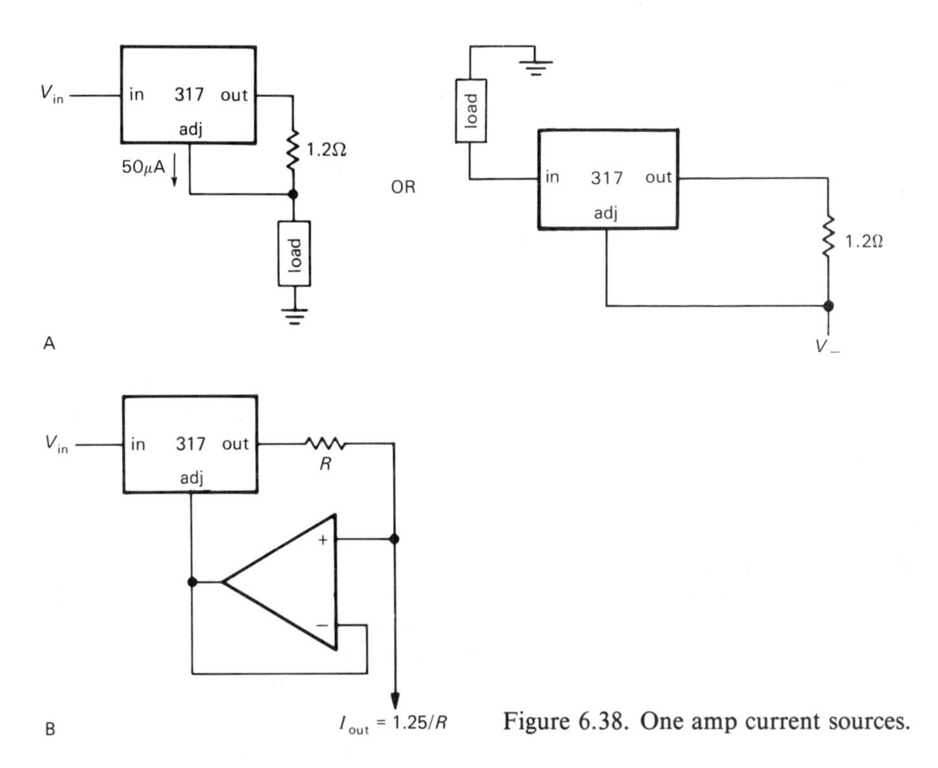

Figure 6.38. One amp current sources.

to *sink* current from a load returned to ground (of course, you could always use the negative-polarity 337, in the configuration analogous to Fig. 6.38A).

National makes a special 3-terminal device, the LM334, optimized for use as a low-power current source. It comes in the small plastic transistor package (TO-92), as well as the standard DIP IC package. You can use it all the way down to 1 μA, because the "adj" current is a small fraction of the total current. It has one peculiarity, however: The output current is temperature-dependent, in fact, precisely proportional to absolute temperature. So although it is not the world's most stable current source, you can use it (Section 15.01) as a temperature sensor!

6.19 Switching regulators and dc-dc converters

All the voltage regulator circuits we have discussed so far work the same way: A linear control element (the "pass transistor") in series with the unregulated dc is used, with feedback, to maintain constant output voltage (or perhaps constant current). The output voltage is always lower in voltage than the unregulated input voltage, and some power is dissipated in the control element [the average value of $I_{out}(V_{in} - V_{out})$, to be precise]. A minor variation on this theme is the *shunt regulator*, in which the control element is tied from the output to ground, rather than in series with the load; the simple resistor-plus-zener is an example.

There is another way to generate a regulated dc voltage, fundamentally different from what we've seen so far; look at Figure 6.39. In such a *switching regulator* a transistor operated as a saturated switch periodically applies the full unregulated voltage across an inductor for short intervals. The inductor's current builds up during each pulse, storing $\frac{1}{2}LI^2$ of energy in its magnetic field; the stored energy is transferred to a filter capacitor at the output,

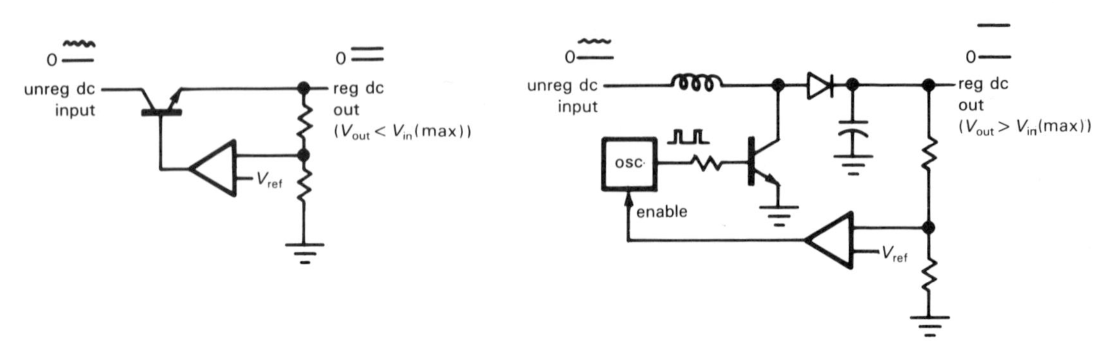

Figure 6.39. Two kinds of regulators.
A. Linear (series).
B. Step-up switcher.

which also smooths the output (to carry the output load between charging pulses). As with a linear regulator, feedback compares the output with a voltage reference – but in a switching regulator it controls the output by changing the oscillator's pulse width or switching frequency, rather than by linearly controlling the base or gate drive.

Switching regulators have unusual properties that have made them very popular: Since the control element is either off or saturated, there is very little power dissipation; switching supplies are thus very efficient, even if there is a large drop from input to output. Switchers (slang for "switching power supplies") can generate output voltages *higher* than the unregulated input, as in Figure 6.39B; they can just as easily generate outputs *opposite in polarity* to the input! Finally, switchers can be designed with no dc path from input to output; that means they can run directly from the rectified power line, with no ac power transformer! The result is a very small, lightweight, and efficient dc supply. For these reasons, switching supplies are used almost universally in computers.

Switching supplies have their problems, too. The dc output has some switching "noise," and they can put hash back onto the power line. They used to have a bad reputation for reliability, with occasional spectacular pyrotechnic displays during episodes of catastrophic failure. Most of these problems have been solved, however, and the switching supply is now firmly entrenched in electronic instruments and computers.

In this section we'll tell you all about switching supplies, in two steps: First, we'll describe the basic switching regulator, operating from a conventional unregulated dc supply. There are three circuits, used for (a) step-down (output voltage less than input), (b) step-up (output voltage greater than input), and (c) inverting (output polarity opposite to input). Then we'll take a radical step, describing the heretical (and most widely used!) designs that run straight from the rectified ac power line, without an isolation transformer. Both kinds of power supplies are in wide use, so our treatment is practical (not just pedagogically pleasing). Finally, we'll give you plenty of advice on the subject: When to use switchers, when to avoid them; when to design your own, when to buy them. With characteristic overstatement, we won't leave you in any doubt!

□ *Step-down regulator*

Figure 6.40 shows the basic step-down (or "bucking") switching circuit, with feedback omitted for simplicity. When the MOS

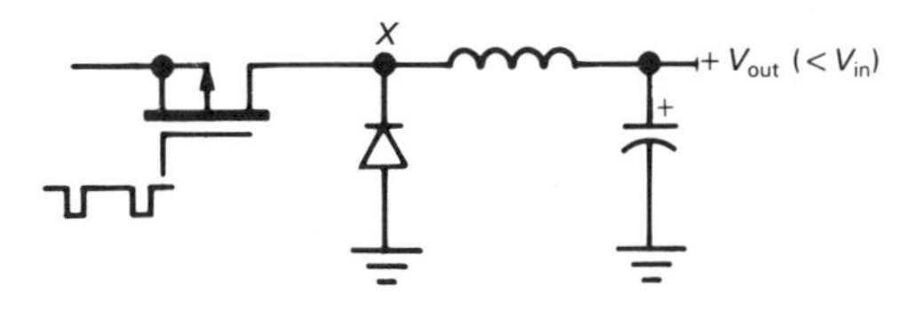

Figure 6.40. Step-down switcher.

switch is closed, $V_{out} - V_{in}$ is applied across the inductor, causing a linearly increasing current (recall $dI/dt = V/L$) to flow through the inductor. (This current flows to the load and capacitor, of course.) When the switch opens, inductor current continues to flow in the same direction (remember that inductors don't like to change their current suddenly, according to the last equation), with the "catch diode" now

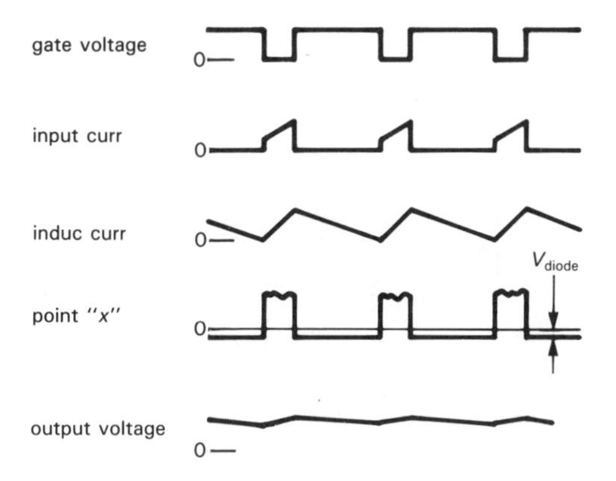

Figure 6.41

conducting to complete the circuit. The output capacitor acts as an energy "flywheel," smoothing the inevitable sawtooth ripple (the larger the capacitor, the less the ripple). The inductor current now finds fixed voltage $V_{out} - 0.6V$ across it, causing its current to decrease linearly. Figure 6.41 shows the corresponding voltage and current waveforms. To complete the circuit as a *regulator*, you would of course add feedback, controlling either the pulse width (at constant pulse repetition rate) or the repetition rate (with constant pulse width) from an error amplifier that compares the output voltage with a reference.

Figure 6.42 shows a low-current +5 volt regulator using the MAX638 from Maxim. This nice chip gives you a choice of fixed +5 volt output (no external divider needed) or adjustable positive output, with external resistive divider. It includes nearly all components in a convenient mini-DIP package. In the MAX638 the oscillator runs at a constant 65kHz, with the error amplifier either permitting or cutting off gate drive pulses, according to the output voltage. The circuit shown gives about 85% efficiency, pretty much independent of the input voltage. Compare that with a linear regulator by doing the next problem:

EXERCISE 6.8

What is the maximum theoretical efficiency of a linear (series pass) regulator, when used to generate regulated +5 volt from a +12 volt unregulated input?

EXERCISE 6.9

What does a step-down regulator's high efficiency imply about the ratio of output current to input current? What is the corresponding ratio of currents, for a linear regulator?

□ *Step-up regulator; inverting regulator*

Apart from its high efficiency, the step-down switching regulator of the previous paragraph has no significant advantage (and some significant disadvantages – component count, switching noise) over a linear regulator. However, when there is a need for an output voltage greater than the unregulated input, or for an output voltage of opposite polarity to the unregulated input, switching supplies become very attractive indeed. Figure 6.43 shows the basic step-up (or "boosting") and inverting (sometimes called "flyback") circuits.

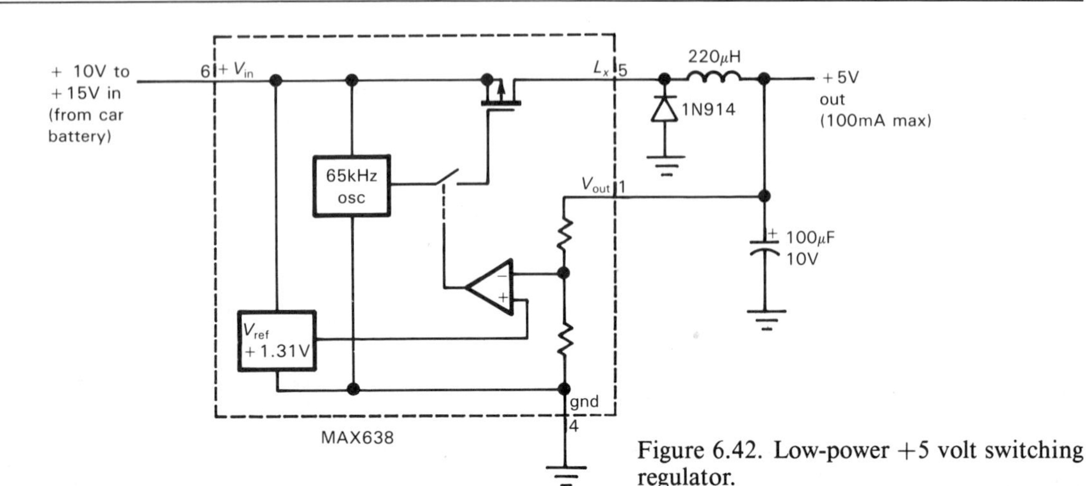

Figure 6.42. Low-power +5 volt switching regulator.

We showed the step-up circuit of Figure 6.39A earlier, in comparison with the linear regulator. The inductor current ramps up during switch conduction (point X near ground); when the switch is turned off, the voltage at point X rises rapidly as the inductor attempts to maintain constant current. The diode turns on, and the inductor dumps current into the capacitor. The output voltage can be much larger than the input voltage.

EXERCISE 6.10

Draw waveforms for the step-up switcher, showing voltage at point X, inductor current, and output voltage.

EXERCISE 6.11

Why can't the step-up circuit be used as a step-down regulator?

The inverting circuit is shown in Figure 6.43B. During switch conduction, a linearly increasing current flows from point X to ground. To maintain the current when the switch is open, the inductor pulls point X negative, as much as needed to maintain current flow. Now, however, the current is flowing *into* the inductor from the filter capacitor. The output is thus negative, and its average value can be larger or smaller in magnitude than the input (as determined by feedback); in other words, the inverting regulator can be either step-up or step-down.

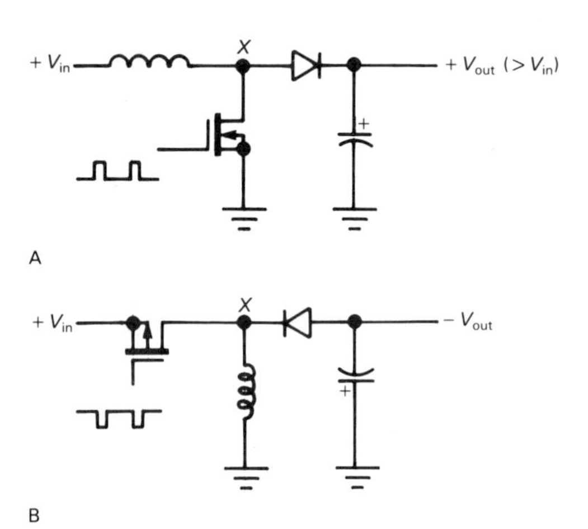

Figure 6.43. Two switching-element configurations.
A. Step-up ("boost").
B. Inverting.

EXERCISE 6.12

Draw waveforms for the inverting switcher, showing voltage at point X, inductor current, and output voltage.

Figure 6.44 shows how you might use low-power switching regulators to generate ±15 volt op-amp supply voltages from a single +12 volt automotive battery, a trick that is impossible with linear regulators. Here we've again used low-power

This leads to a radical idea: We could eliminate the heavy 60Hz step-down transformer if we ran the regulator directly from rectified (unregulated) and filtered ac power. Two immediate comments: (a) The dc input voltage will be approximately 160 volts (for 115V ac power) – this is a dangerous circuit to tinker with! (b) The absence of a transformer means that the dc input is not isolated from the power line. Thus, the switching circuit itself must be modified to provide isolation.

The usual way to isolate the switching circuit is to wind a secondary onto the energy-storage inductor and use an isolation device (either transformer or opto-isolator) to couple the feedback to the switching oscillator; see the simplified block diagram in Figure 6.45. Note that the oscillator circuitry is powered from the high-voltage unregulated dc, whereas the feedback control circuitry (error amplifier, reference) is powered from the regulated dc output. Sometimes an auxiliary low-current unregulated supply (with its own 60Hz low-voltage transformer) is used to power the control elements. The box labeled "isolation" is often a small pulse transformer, although optical isolation can also be used (more on this later).

Figure 6.45. Direct ac-line-powered switching supply.

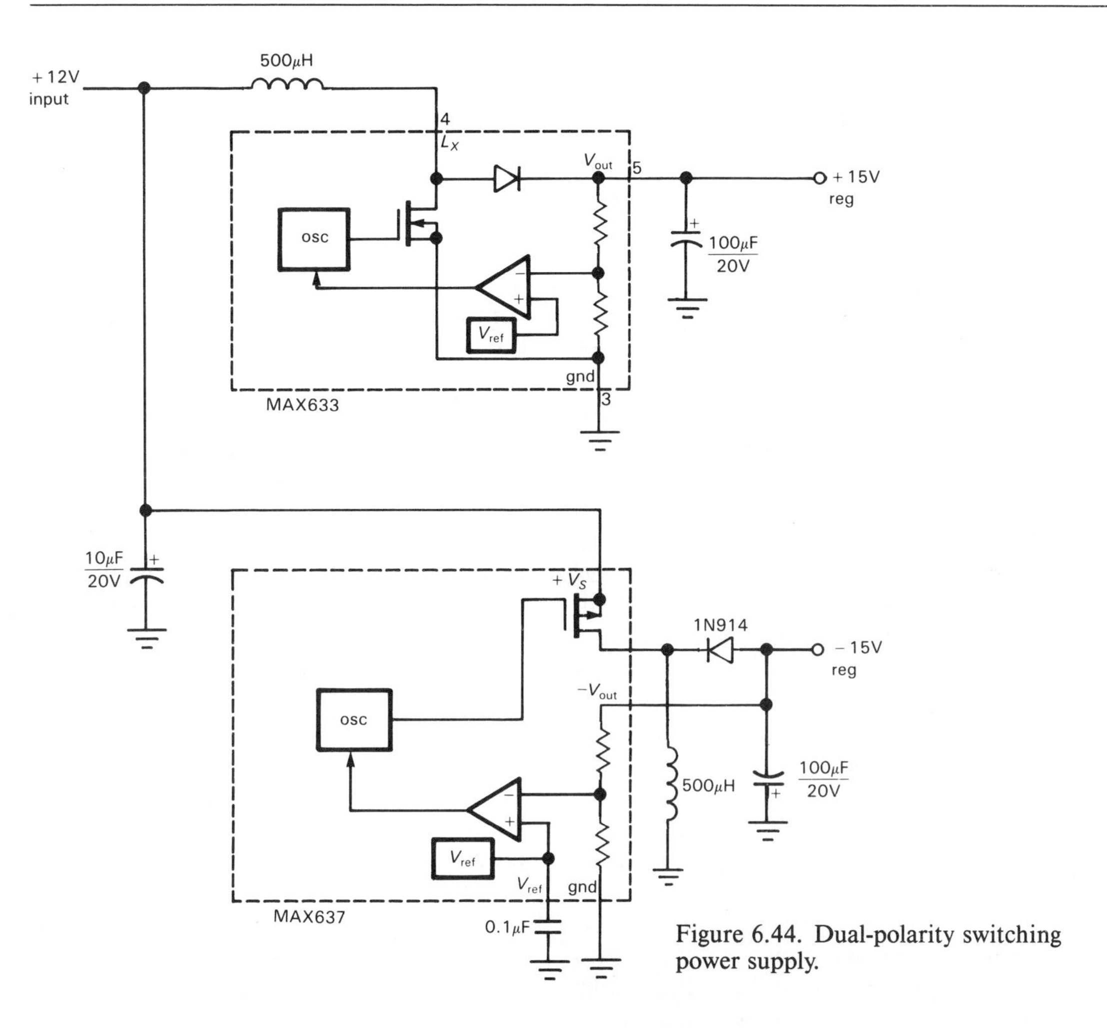

Figure 6.44. Dual-polarity switching power supply.

fixed-output ICs from Maxim, in this case the step-up MAX633 and the inverting MAX637. The external components shown were chosen according to the manufacturer's data sheets. They're not critical, but, as always in electronic design, there are trade-offs. For example, a larger value of inductor lowers the peak currents and increases the efficiency, at the expense of maximum available output current. This circuit is rather insensitive to input voltage, as long as it doesn't exceed the output voltage, and will work all the way down to +2 volts input, with greatly reduced maximum output current.

Before leaving the subject of inverting and step-up regulators, we should mention that there is one other way to accomplish the same goal, namely with "flying capacitors." The basic idea is to use MOS switches to (a) charge a capacitor from the dc input, and then (b) change the switches to connect the now-charged capacitor in series with another (step-up), or with reversed polarity across the output (inverting). Flying-capacitor voltage converters (e.g., the popular 7662) have some advantages (no inductors) and some disadvantages (low power, poor regulation, limited voltage). We'll discuss them later in the chapter.

General comments on switching regulators

As we've seen, the ability of switchers to generate stepped-up or inverted outputs makes them quite handy for making, say, low-current ± 12 volt supplies on an otherwise all-digital $+5$ volt circuit board. You'll often need such bipolarity supplies to power "serial ports" (more in Chapters 10 and 11) or linear circuitry using op-amps or A/D (analog-to-digital) and D/A (digital-to-analog) converters. Another good use for step-up switchers is to power displays that require relatively high voltage, for example using fluorescent or plasma technology. In these applications, where the dc input (typically $+5$V) is already regulated, you often use the phrase "dc-to-dc converter," rather than "switching regulator," although it's really the same circuit. Finally, in battery-operated equipment you often want high efficiency over a wide range of battery voltage; for example, a 9 volt alkaline "transistor" battery begins life at about 9.5 volts, dropping steadily to about 6 volts at the end of its useful life. A $+5$ volt low-power step-down regulator maintains high efficiency, with current step-up over most of the battery's life.

Note that the inductor and capacitor in a switching regulator are not functioning as an *LC* filter. In the simple step-down regulator, that might seem to be true, but obviously a circuit that *inverts* a dc level is hardly a filter! The inductor is a lossless energy-storage device (stored energy $= \frac{1}{2}LI^2$), able to transform impedance in order to conserve energy. This is an accurate statement from a physicist's point of view, in which the magnetic field contains stored energy. We're more used to thinking of capacitors as energy storage devices (stored energy $= \frac{1}{2}CV^2$), which is their role in switching supplies, as in conventional series regulators.

A bit of nomenclature: You sometimes see the phrases "PWM switch-mode

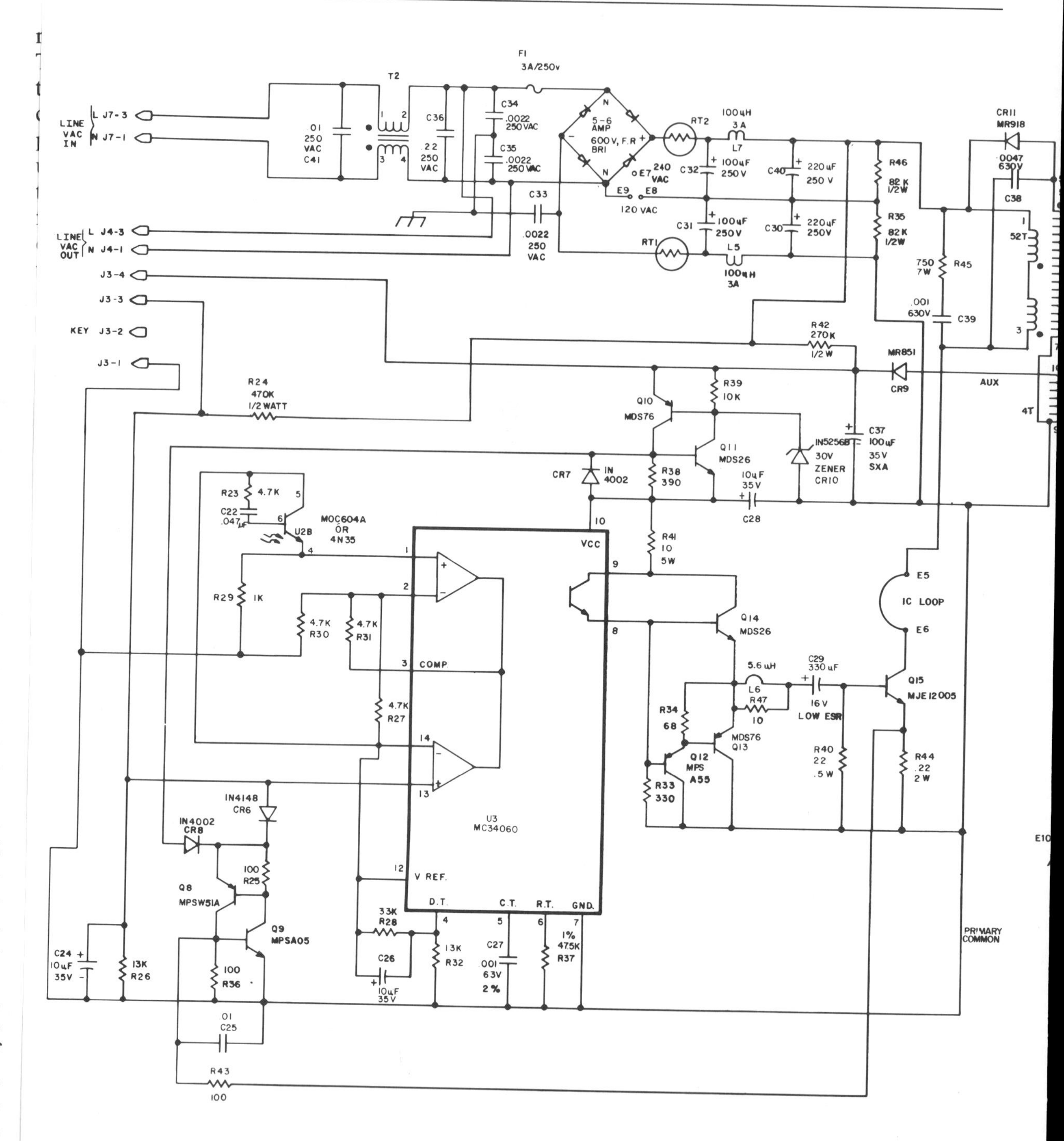

Figure 6.46. Switching power supply used in the Tandy model 2000 personal computer. Feedback from the $+5$ volt output is provided via opto-isolator U_{2A}-U_{2B}. (Courtesy of Tandy Corporation. Copyright 1984.)

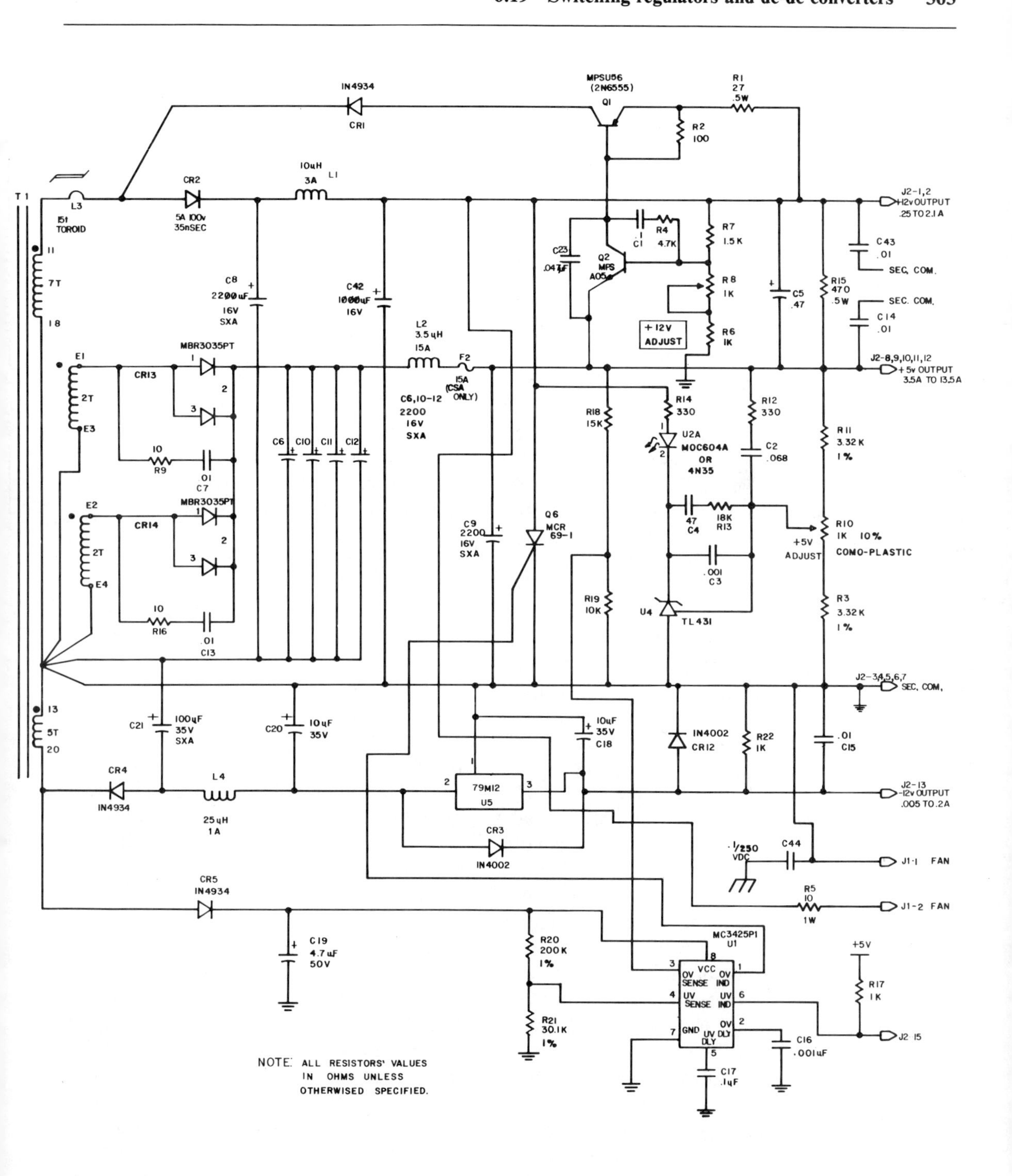

J2-1,2 +12v OUTPUT .25 TO 2.1 A
J2-8,9,10,11,12 +5v OUTPUT 3.5A TO 13.5A
J2-3,4,5,6,7 SEC. COM.
J2-13 -12v OUTPUT .005 TO .2A
J1-1 FAN
J1-2 FAN
J2 15
MPSU56 (2N6555) Q1
79M12 U5
MC3425P1 U1
U2A MOC604A OR 4N35
U4 TL431
Q6 MCR 69-1
+12V ADJUST
+5V ADJUST
COMO-PLASTIC
NOTE: ALL RESISTORS' VALUES IN OHMS UNLESS OTHERWISED SPECIFIED.

(Radio Shack) to power their model 2000 personal computer. (We tried to get power-supply schematics from both IBM and Apple, but were ignored or haughtily rebuffed. Tandy, by comparison, publishes excellent documentation, with complete schematics and extensive circuit description.) It provides regulated outputs of +5 volts at 13 amps, +12 volts at 2.5 amps, and −12 volts at 0.2 amp (95W total), which are used to power the logic circuits and floppy-disk drives in the computer.

Let's take a walk through Figure 6.46 to see how a line-powered switcher copes with real-world problems. The circuit topology chosen by Tandy's designers is precisely that shown in Figure 6.45, though there are a few more components! Begin by comparing the figures: The line-powered bridge rectifier (BR_1) charges filter capacitors C_{30}, C_{31}, C_{32}, and C_{40} (T_2 is not a transformer – note the connections – but rather an interference filter). The charged capacitors are switched across the transformer primary (pins 1 and 3) by power transistor Q_{15}, whose switching waveform (a fixed-frequency square wave of variable pulse width) is provided by IC U_3 (a "PWM switch-mode regulator"). The secondary winding (there are actually three windings, one for each output voltage) is half-wave-rectified to generate the dc output: The +12 volts are produced by CR_2 from the 7-turn winding of pins 11 and 18, the −12 volts by CR_4 from the 5-turn winding of pins 13 and 20, and the +5 volts by the paralleled combination of CR_{13} and CR_{14}, each powered from its own (2-turn!) winding.

With multioutput switchers, only one output can be used for voltage-regulating feedback. It is conventional to use the +5 volts logic supply for this purpose, as Tandy has done here: R_{10} selects a fraction (nominally 50%) of the +5 volt output to compare with U_4's internal +2.5 volt reference, turning on photodiode U_{2a} if the output is too high. This photodiode couples optically to phototransistor U_{2b}, which varies the pulse width of U_3 to maintain +5 volts output. Thus the block labeled "isolation" in Figure 6.45 is an opto-coupler (see Section 9.10).

At this point we have accounted for perhaps 25% of the components in Figure 6.46. The rest are needed to cope with problems such as (a) short-circuit protection, (b) overvoltage and undervoltage shut-down, (c) auxiliary power for the regulating circuitry, (d) ac power filtering, and (e) linear post-regulation of the (tracking) ±12 volt supplies. Let's explore the circuit in some more detail.

Beginning at the ac line input, we find four capacitors and a series inductor pair, together forming an RFI filter. It's always a good idea, of course, to clean up the ac power entering an instrument (see Section 6.11); here, however, the careful filtering is additionally needed to keep radiofrequency hash generated *inside* the machine (mostly from the switching action in the power supply) from radiating *out* through the power line. Note next the optional jumper at E_8E_9, which converts the input from full-wave bridge (jumper open) to full-wave doubler (jumper shorted); manufacturers who wish to export their electronic wares must provide 110/220 volt compatibility, which is remarkably simple in the case of switching supplies.

Thermistors RT_1 and RT_2 are used to limit the high inrush current when the supply is first switched on, at which point the power line sees a few hundred microfarads of uncharged capacitor. Without the thermistors (or some other trick) the inrush current can easily exceed 100 amperes! The thermistors provide an ohm or two of series resistance, dropping to near zero when they warm up. Even with thermistors, the inrush current is impressive: The power supply has a specified "Input Surge Current" of 70 amps, maximum.

The 100μH series inductors L_5 and L_7 in the unregulated supply further clean

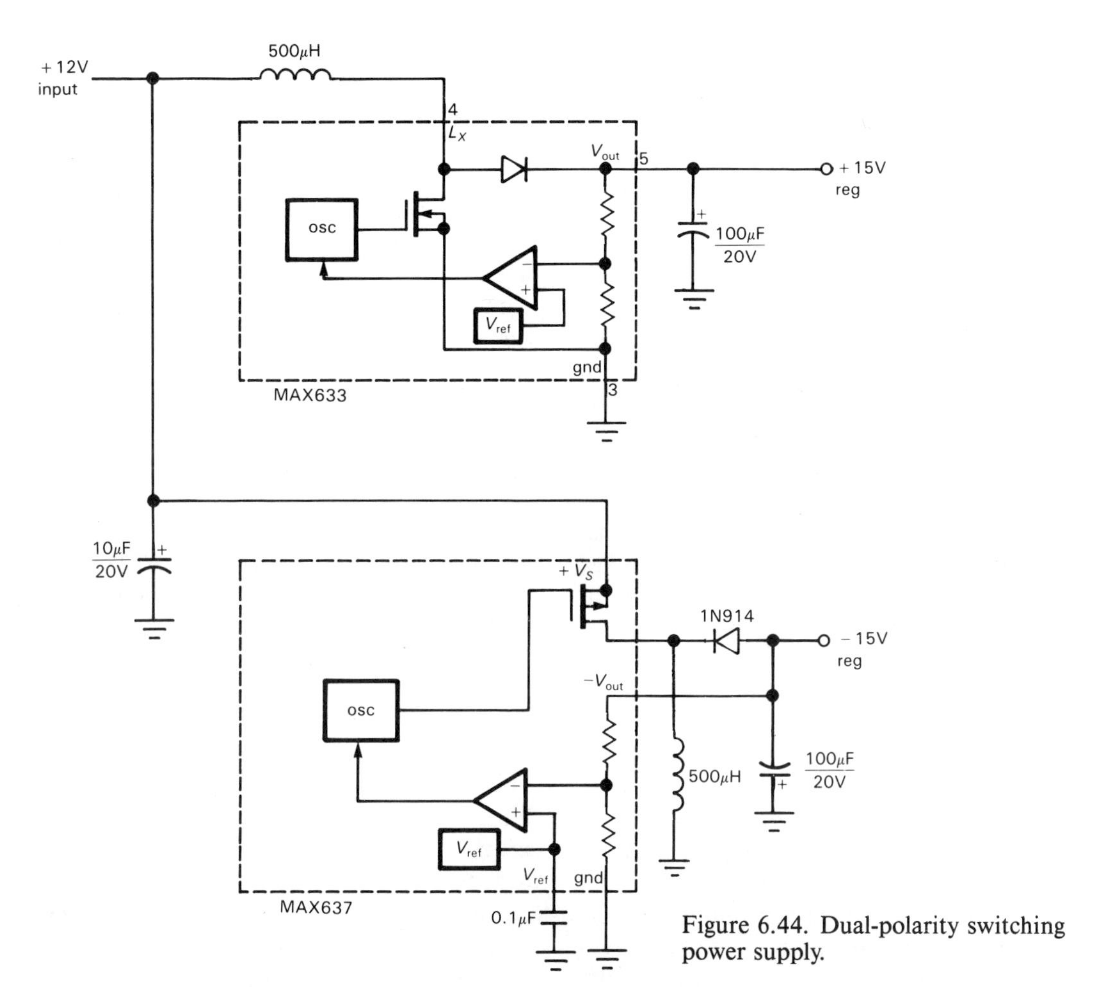

Figure 6.44. Dual-polarity switching power supply.

fixed-output ICs from Maxim, in this case the step-up MAX633 and the inverting MAX637. The external components shown were chosen according to the manufacturer's data sheets. They're not critical, but, as always in electronic design, there are trade-offs. For example, a larger value of inductor lowers the peak currents and increases the efficiency, at the expense of maximum available output current. This circuit is rather insensitive to input voltage, as long as it doesn't exceed the output voltage, and will work all the way down to +2 volts input, with greatly reduced maximum output current.

Before leaving the subject of inverting and step-up regulators, we should mention that there is one other way to accomplish the same goal, namely with "flying capacitors." The basic idea is to use MOS switches to (a) charge a capacitor from the dc input, and then (b) change the switches to connect the now-charged capacitor in series with another (step-up), or with reversed polarity across the output (inverting). Flying-capacitor voltage converters (e.g., the popular 7662) have some advantages (no inductors) and some disadvantages (low power, poor regulation, limited voltage). We'll discuss them later in the chapter.

General comments on switching regulators

As we've seen, the ability of switchers to generate stepped-up or inverted outputs makes them quite handy for making, say, low-current ±12 volt supplies on an otherwise all-digital +5 volt circuit board. You'll often need such bipolarity supplies to power "serial ports" (more in Chapters 10 and 11) or linear circuitry using op-amps or A/D (analog-to-digital) and D/A (digital-to-analog) converters. Another good use for step-up switchers is to power displays that require relatively high voltage, for example using fluorescent or plasma technology. In these applications, where the dc input (typically +5V) is already regulated, you often use the phrase "dc-to-dc converter," rather than "switching regulator," although it's really the same circuit. Finally, in battery-operated equipment you often want high efficiency over a wide range of battery voltage; for example, a 9 volt alkaline "transistor" battery begins life at about 9.5 volts, dropping steadily to about 6 volts at the end of its useful life. A +5 volt low-power step-down regulator maintains high efficiency, with current step-up over most of the battery's life.

Note that the inductor and capacitor in a switching regulator are not functioning as an LC filter. In the simple step-down regulator, that might seem to be true, but obviously a circuit that *inverts* a dc level is hardly a filter! The inductor is a lossless energy-storage device (stored energy = $\frac{1}{2}LI^2$), able to transform impedance in order to conserve energy. This is an accurate statement from a physicist's point of view, in which the magnetic field contains stored energy. We're more used to thinking of capacitors as energy storage devices (stored energy = $\frac{1}{2}CV^2$), which is their role in switching supplies, as in conventional series regulators.

A bit of nomenclature: You sometimes see the phrases "PWM switch-mode regulator" and "current-mode regulator." They refer to the particular way in which the switching waveform is modified according to the feedback (error) signal. In particular, PWM means pulse-width modulation, in which the error signal is used to control the conduction pulse width (at fixed frequency), whereas in current-mode control the error voltage is used to control the peak inductor current (as sensed by a resistor) via width on a pulse-by-pulse basis. Current-mode regulators have some significant advantages and are becoming more popular now that good current-mode controller ICs have become available.

Keep in mind, when considering any switching supply, the noise generated by the switching process. This takes three forms, namely (a) output ripple, at the switching frequency, typically of order 10mV–100mV peak-to-peak, (b) ripple, again at the switching frequency, impressed onto the input supply, and (c) radiated noise, at the switching frequency and its harmonics, from switching currents in the inductor and leads. You can get into plenty of trouble with switching supplies in a circuit that has low-level signals (say 100μV or less). Although an aggressive job of shielding and filtering may solve such problems, you're probably better off with linear regulators from the outset.

Line-powered switching supplies

As we have seen, switching supplies have high efficiency even when the output voltage is nowhere near the input voltage. It may help our understanding to think of the inductor as an "impedance converter," since the average dc output current can be larger (step-down) or smaller (step-up) than the average dc input current. This is in stark contrast to linear series regulators, where the average values of input and output currents are always equal (ignoring the quiescent current of the regulator circuitry, of course).

This leads to a radical idea: We could eliminate the heavy 60Hz step-down transformer if we ran the regulator directly from rectified (unregulated) and filtered ac power. Two immediate comments: (a) The dc input voltage will be approximately 160 volts (for 115V ac power) – this is a dangerous circuit to tinker with! (b) The absence of a transformer means that the dc input is not isolated from the power line. Thus, the switching circuit itself must be modified to provide isolation.

The usual way to isolate the switching circuit is to wind a secondary onto the energy-storage inductor and use an isolation device (either transformer or opto-isolator) to couple the feedback to the switching oscillator; see the simplified block diagram in Figure 6.45. Note that the oscillator circuitry is powered from the high-voltage unregulated dc, whereas the feedback control circuitry (error amplifier, reference) is powered from the regulated dc output. Sometimes an auxiliary low-current unregulated supply (with its own 60Hz low-voltage transformer) is used to power the control elements. The box labeled "isolation" is often a small pulse transformer, although optical isolation can also be used (more on this later).

It might seem as if the advantage of a transformerless unregulated supply is more than overcome by the need for at least *two* other transformers! Not so. The size of a transformer is determined by the core size, which decreases dramatically at high frequencies. As a result, line-powered switching supplies are much smaller and lighter than the equivalent linear supply; they also run cooler, owing to higher efficiency. For example, Power-One manufactures both kinds of supplies. Comparing their model F5-25 (5V, 25A) linear supply with their comparably priced SPL130-1005 (5V, 26A) switcher, we find that the switcher weighs 2.5 pounds, compared with 19 pounds for the linear, and occupies just one-fourth the volume. Furthermore, the switcher will run cool, while the 19-pound linear will run hot, dissipating up to 75 watts at full load.

□ *Real-world switcher example*

In order to give you a feel for the real complexity of line-powered switching supplies, we've reproduced in Figure 6.46 the complete schematic of a commercial switcher, in fact the power supply used by Tandy

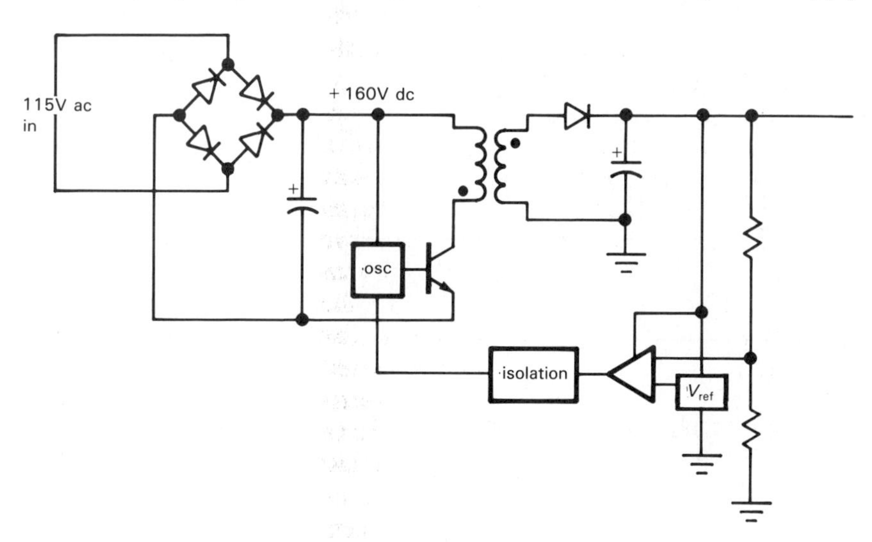

Figure 6.45. Direct ac-line-powered switching supply.

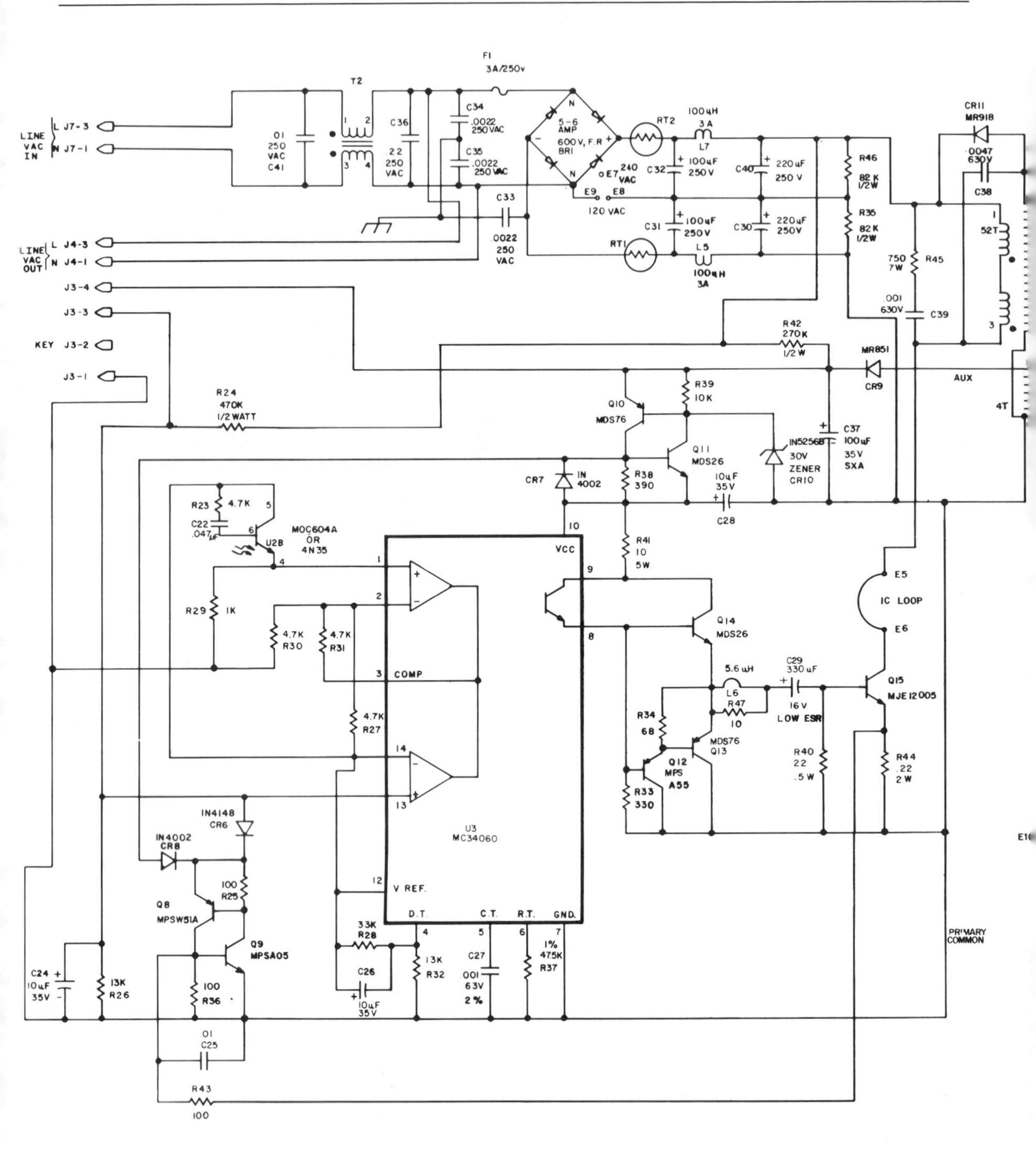

Figure 6.46. Switching power supply used in the Tandy model 2000 personal computer. Feedback from the +5 volt output is provided via opto-isolator U_{2A}-U_{2B}. (Courtesy of Tandy Corporation. Copyright 1984.)

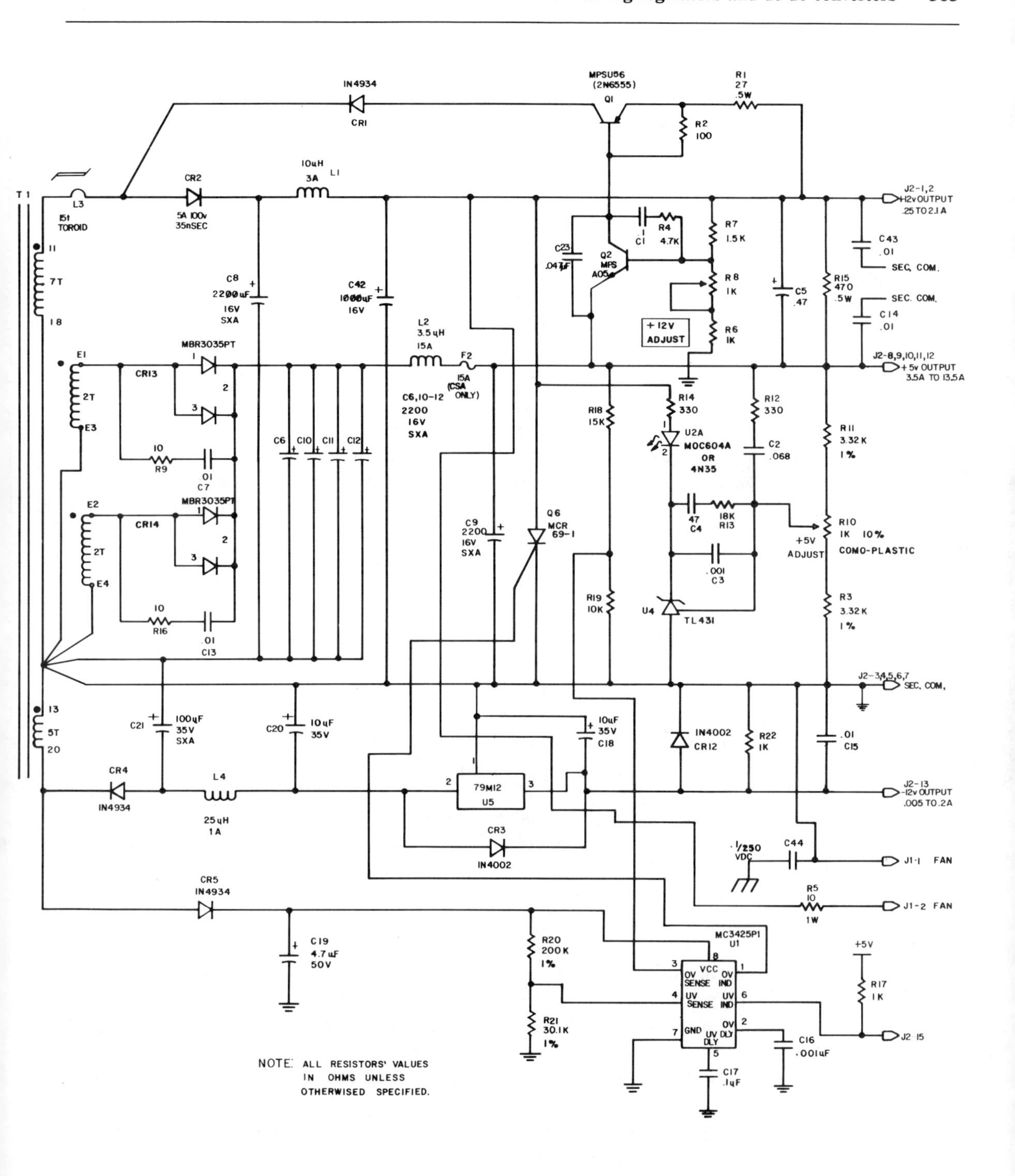
IN4934
CR1
MPSU56
(2N6555)
Q1
R1
27
.5W
R2
100
T1
L3
15t
TOROID
CR2
5A 100v
35nSEC
10uH
3A
L1
J2-1,2
+12v OUTPUT
.25 TO 2.1 A
C1
.1
R4
4.7K
R7
1.5K
C43
.01
SEC. COM.
C23
.047uF
Q2
MPS
A05
R8
1K
R15
470
.5W
C5
.47
C14
.01
C8
2200uF
16V
SXA
C42
1000uF
16V
L2
3.5uH
15A
+12V
ADJUST
R6
1K
J2-8,9,10,11,12
+5v OUTPUT
3.5A TO 13.5A
MBR3035PT
E1
CR13
2T
E3
F2
15A
(CSA
ONLY)
C6,10-12
2200
16V
SXA
R14
330
R12
330
R18
15K
U2A
MOC604A
OR
4N35
C2
.068
R11
3.32 K
1%
R9
10
C7
.01
C6
C10
C11
C12
E2
CR14
E4
C9
2200
16V
SXA
Q6
MCR
69-1
47
C4
18K
R13
R10
1K 10%
COMO-PLASTIC
+5V
ADJUST
.001
C3
R19
10K
U4
TL431
R3
3.32K
1%
R16
C13
J2-3,4,5,6,7
SEC. COM.
13
5T
20
C21
100uF
35V
SXA
C20
10uF
35V
C18
IN4002
CR12
R22
1K
C15
CR4
IN4934
L4
25uH
1A
79M12
U5
J2-13
-12v OUTPUT
.005 TO .2A
CR3
IN4002
1/250
VDC
C44
J1-1 FAN
CR5
IN4934
R5
10
1W
J1-2 FAN
C19
4.7uF
50V
R20
200K
1%
MC3425P1
U1
+5V
OV SENSE
VCC
OV IND
UV SENSE
UV IND
GND
UV DLY
OV DLY
R17
1K
J2 15
C16
.001uF
R21
30.1K
1%
C17
.1uF
NOTE: ALL RESISTORS' VALUES
IN OHMS UNLESS
OTHERWISED SPECIFIED.

(Radio Shack) to power their model 2000 personal computer. (We tried to get power-supply schematics from both IBM and Apple, but were ignored or haughtily rebuffed. Tandy, by comparison, publishes excellent documentation, with complete schematics and extensive circuit description.) It provides regulated outputs of +5 volts at 13 amps, +12 volts at 2.5 amps, and −12 volts at 0.2 amp (95W total), which are used to power the logic circuits and floppy-disk drives in the computer.

Let's take a walk through Figure 6.46 to see how a line-powered switcher copes with real-world problems. The circuit topology chosen by Tandy's designers is precisely that shown in Figure 6.45, though there are a few more components! Begin by comparing the figures: The line-powered bridge rectifier (BR_1) charges filter capacitors C_{30}, C_{31}, C_{32}, and C_{40} (T_2 is not a transformer - note the connections - but rather an interference filter). The charged capacitors are switched across the transformer primary (pins 1 and 3) by power transistor Q_{15}, whose switching waveform (a fixed-frequency square wave of variable pulse width) is provided by IC U_3 (a "PWM switch-mode regulator"). The secondary winding (there are actually three windings, one for each output voltage) is half-wave-rectified to generate the dc output: The +12 volts are produced by CR_2 from the 7-turn winding of pins 11 and 18, the −12 volts by CR_4 from the 5-turn winding of pins 13 and 20, and the +5 volts by the paralleled combination of CR_{13} and CR_{14}, each powered from its own (2-turn!) winding.

With multioutput switchers, only one output can be used for voltage-regulating feedback. It is conventional to use the +5 volts logic supply for this purpose, as Tandy has done here: R_{10} selects a fraction (nominally 50%) of the +5 volt output to compare with U_4's internal +2.5 volt reference, turning on photodiode U_{2a} if the output is too high. This photodiode couples optically to phototransistor U_{2b}, which varies the pulse width of U_3 to maintain +5 volts output. Thus the block labeled "isolation" in Figure 6.45 is an opto-coupler (see Section 9.10).

At this point we have accounted for perhaps 25% of the components in Figure 6.46. The rest are needed to cope with problems such as (a) short-circuit protection, (b) overvoltage and undervoltage shut-down, (c) auxiliary power for the regulating circuitry, (d) ac power filtering, and (e) linear post-regulation of the (tracking) ±12 volt supplies. Let's explore the circuit in some more detail.

Beginning at the ac line input, we find four capacitors and a series inductor pair, together forming an RFI filter. It's always a good idea, of course, to clean up the ac power entering an instrument (see Section 6.11); here, however, the careful filtering is additionally needed to keep radiofrequency hash generated *inside* the machine (mostly from the switching action in the power supply) from radiating *out* through the power line. Note next the optional jumper at E_8E_9, which converts the input from full-wave bridge (jumper open) to full-wave doubler (jumper shorted); manufacturers who wish to export their electronic wares must provide 110/220 volt compatibility, which is remarkably simple in the case of switching supplies.

Thermistors RT_1 and RT_2 are used to limit the high inrush current when the supply is first switched on, at which point the power line sees a few hundred microfarads of uncharged capacitor. Without the thermistors (or some other trick) the inrush current can easily exceed 100 amperes! The thermistors provide an ohm or two of series resistance, dropping to near zero when they warm up. Even with thermistors, the inrush current is impressive: The power supply has a specified "Input Surge Current" of 70 amps, maximum.

The 100μH series inductors L_5 and L_7 in the unregulated supply further clean up transmitted switching hash, and the 82k shunt resistors (R_{35} and R_{46}) are "bleeders," to make sure the power-supply filter capacitors discharge fully after power is turned off. Some additional passive "snubber" components (C_{38}, C_{39}, and R_{45}) are used to damp the large voltage spikes that otherwise might destroy the switching transistor Q_{15}. CR_{11}'s function is more subtle – it cleverly returns unused transformer energy to filter capacitors C_{30} and C_{40}.

Moving down the page, we encounter some real trickery, namely the "auxiliary supply." The circuits need some low-voltage, low-current dc to run the PWM controller chip and associated circuits. One possibility is to use a separate little linear supply, with its own line-powered transformer, etc. However, the temptation is overwhelming to hang another small winding (with half-wave rectifier) on T_1, thus saving a separate transformer. That's what the designer has done here, with a 4-turn winding (pins 9 and 10), rectified and filtered by CR_9 and C_{37}. This simple supply generates a nominal 15 volt output.

Sharp-eyed readers will have noticed a flaw in this scheme: The circuit cannot start itself, since the auxiliary power is only present if the supply is already running! This turns out to be an old problem: Designers of television sets love to play the same trick, deriving all their low-voltage supplies from auxiliary windings on the high-frequency horizontal drive transformer. The solution is the so-called kick-start circuit, in which some of the unregulated dc is brought over to start the circuit; once going, the supply keeps itself going from its transformer-derived dc power. In this circuit the kick-start comes via R_{42}, which begins charging up C_{37} at power-on. Nothing happens until the capacitor reaches a diode drop above CR_{10}'s zener voltage, at which point the SCR-like combination of Q_{10} and Q_{11} is switched into conduction (figure out how that works), dumping C_{37}'s charge across C_{28}, thus momentarily powering the control circuitry (U_3 and all components to its left). Once the oscillation starts, CR_9 provides 15 volts with enough current to power the control circuitry continuously (which R_{42} cannot do).

Most of the components surrounding U_3 pander to its needs (C_{27} and R_{37}, for example, set the pulse repetition rate at 25kHz). At the input side, U_{2b} provides overall feedback to maintain the output at +5 volts, as described earlier. Q_8 and Q_9 are another SCR-like latch, this time triggered to shut down the oscillator (and the series latching switch $Q_{10}Q_{11}$) if driver Q_{15}'s emitter current (sensed by R_{44}) is excessive, for example if the power supply sees a short-circuit load. The series combination $R_{43}C_{25}$ provides a 1μs time constant so that the circuit is not triggered by switching spikes. The shutdown circuitry also derives an input from divider $R_{26}R_{24}$, quenching oscillation if the ac input drops below 90 volts ac. At the output side of the controller U_3, Q_{12} – Q_{14} provide high-current push-pull drive to Q_{15}'s base from the single-ended on-chip *npn* driver transistor (figure out how). Note the "I_C loop," an accessible length of wire in Q_{15}'s collector, which lets you observe the current waveform on a 'scope by using a clip-on current probe (see, for example, the Tektronix catalog).

Things are considerably simpler on the output side of T_1. The +5 volt supply uses paralleled power Schottky diodes (CR_{13} and CR_{14}) for fast recovery and low forward drop (the MBR3035PT is rated at 30A average current with 20kHz drive, 35V reverse breakdown, and 0.5V typical forward drop at 10A), with "snubber networks" (10Ω/0.01μF) to protect the diodes from high-voltage spikes. The "π-section" filter consists of 8800μF of input capacitance, a 3.5μH series inductor, and a 2200μF output capacitor. (The lower-current ±12V outputs also use half-wave

Schottky rectifiers and π-section filters, with smaller-valued components.) This degree of filtering might seem extreme by linear regulator standards, but remember that there is no post-regulation – what comes out of the filter is the "regulated dc" – therefore lots of filtering is needed to reduce ripple, predominantly at the switching frequency, to the requisite $\approx$50mV or so at the output.

The +5 volt output is sensed via divider $R_3R_{10}R_{11}$, driving TI's TL431 "3-terminal zener" (U_4), which, in combination with a few resistors and capacitors for feedback compensation, provides isolated feedback via opto-coupler U_{2ab}. The +5 volt output is also sensed, via $R_{18}R_{19}$, to trigger the overvoltage-sensor IC (U_1: $V_{\text{thresh}} = +2.5\text{V}$); the latter drives the gate of SCR Q_6, which crowbars the +12 volt supply, shutting things down via current limiting in the primary side, as described earlier. U_1 is also wired to sense an *under*voltage condition, via its dedicated auxiliary power from CR_5 and C_{19}; the undervoltage signal (a saturated *npn* transistor to ground) is sent to the microprocessor, alerting the system to imminent power failure so that the program can be brought to an orderly shut-down during the few remaining milliseconds without loss of data.

The power-supply designers used a bit of trickery to improve regulation in the ±12 volt supplies, which otherwise ride virtually open-loop on what is basically a +5 volt supply. For the +12 volt supply they used the +5 volt output as a reference for error amplifier Q_2, which controls a "magnetic amplifier." The latter consists of series saturable reactor L_3, provided with an opposing "reset current" via Q_1. The reset current determines how many volt-seconds the inductor will block before reaching the state of magnetic saturation, in which it acts as a good conductor. A magnetic amplifier deserves its name, because a small control current modifies a large output current. Mag-amp controllers are available as complete integrated circuits, for example the UC3838 from Unitrode.

For the lower-current −12 volt supply the designers opted for the simpler solution of a linear 7912-type post-regulator, complete with diodes for protection against reverse polarity. Throughout, the designers have used bypass capacitors and bleeders on the dc outputs.

This power-supply circuit illustrates most of the details that seldom get mentioned in textbooks, but are essential in the real world. The extra component count in this circuit pays handsomely in ensuring a power supply that is robust under field conditions. Although this extra care in design might appear to be a display of unnecessary compulsiveness, in fact it is hard-nosed cost-effectiveness – each field failure under warranty costs the manufacturer at least a hundred dollars in real shipping and repair costs, not to mention the tarnished reputation produced by persistent failure.

General comments on line-powered switching power supplies

1. Line-powered switchers (also called "off-line" switchers, though we don't like the term) make excellent *high-power* supplies. Their high efficiency keeps them cool, and the absence of a low-frequency transformer makes them considerably lighter and smaller than the equivalent linear supply. As a result, they are used almost exclusively to power computers, even desktop personal computers. They are finding their way into other portable instruments, too, even such noise-sensitive applications as oscilloscopes.

2. Switchers are noisy! Their outputs have tens of millivolts of switching ripple at their outputs, they put garbage onto the power line, and they can even scream audibly! One cure for output ripple, if that's a problem, is to add an external high-current LC low-pass filter; alternatively, you can add a low-dropout linear post-regulator. Some dc-dc converters include this feature, as well as complete shielding and extensive input filtering.

3. Switchers with multiple outputs are available and are popular in computer systems. However, the separate outputs are generated from additional windings on a common transformer. Typically, feedback is taken from the highest current output (usually the +5V output), which means that the other outputs are not particularly well regulated. There is usually a "cross-regulation" specification, which tells, for example, how much the +12 volt output, say, changes when you vary the load on the +5 volt output from 75% of full load to either 50% or 100% of full load; a typical cross-regulation specification is 5%. Some multiple-output switchers achieve excellent regulation by using linear post-regulators on the auxiliary outputs, but this is the exception. Check the specs!

4. Line-powered switchers may have a minimum load current requirement. If your load-current may drop below the minimum, you'll have to add some resistive loading; otherwise the output may soar or oscillate. For example, the +5 volt, 26 amp switching supply above has a minimum load current of 1.3 amps.

5. When working on a line-powered switcher, *watch out!* Many components are at line potential and can be lethal. You can't clip the ground of your scope probe to the circuit without catastrophic consequences.

6. When you first turn on the power, the ac line sees a large discharged electrolytic filter capacitor across it (through a diode bridge, of course). The resulting "inrush" current can be enormous; for our Power One switcher it's specified as 17 amps, maximum (compared with a full-load input current of 1.6A). Commercial switchers use various "soft-start" tricks to keep the inrush current within civilized bounds. One method is to put a negative-tempco resistor (a low-resistance thermistor) in series with the input; another method is to actively switch out a small (10Ω) series resistor a fraction of a second after the supply is turned on.

7. Switchers usually include overvoltage "shut-down" circuitry, analogous to our SCR crowbar circuits, in case something goes wrong. However, this circuit often is simply a zener sensing circuit at the output that shuts off the oscillator if the dc output exceeds the trip point. There are imaginable failure modes in which such a "crowbar" wouldn't crowbar anything. For maximum safety you may want to add an autonomous outboard SCR-type crowbar.

8. Switchers used to have a bad reputation for reliability, but recent designs seem much better. However, when they decide to blow out, they sometimes do it with great panache! We had one blow its guts out in a "catastrophic deconstruction," spewing black crud all over its innards and innocent electronic bystanders as well.

9. Line-powered switchers are definitely complex and tricky to design reliably. You need special inductors and transformers (and lots of them; Fig. 6.46). Our advice is to avoid the design phase entirely, by *buying* what you need! After all, why build what you can buy?

10. A switching supply presents a peculiar load to the power line that drives it. In particular, an increase in line voltage results in a *decrease* in average current, because the switcher operates at roughly constant efficiency: That's a negative resistance load (averaged over the 60Hz wave), and it can cause some crazy effects. If there's a lot of inductance in the power line, the system may oscillate.

Advice

Luckily for you, we're not bashful about giving advice! Here it is:

1. For *digital* systems, you usually need +5 volts, often at high current (10A or more).

Advice: (a) Use a line-powered switcher. (b) Buy it (perhaps adding filtering, if needed).

2. For analog circuits with low-level signals (small-signal amplifiers, signals less than 100μV, etc.). *Advice:* Use a linear regulator; switchers are too noisy – they will ruin your life. *Exception:* For some battery-operated circuits it may be better to use a low-power dc-dc switching converter.

3. For high-power anything. *Advice:* Use a line-powered switcher. It's smaller, lighter, and cooler.

4. For high-voltage, low-power applications (photomultiplier tubes, flash tubes, image intensifiers, plasma displays). *Advice:* Use a low-power step-up converter.

In general, low-power dc-dc converters are easy to design and require few components, thanks to handy chips like the Maxim series we saw earlier. Don't hesitate to build your own. By contrast, high-power switchers (generally line-powered) are complex and tricky and extremely trouble-prone. If you must design your own, be careful, and test your design very thoroughly. Better yet, swallow your pride and buy the best switcher you can find.

SPECIAL-PURPOSE POWER-SUPPLY CIRCUITS

□ 6.20 High-voltage regulators

Some special problems arise when you design linear regulators to deliver high voltages. Since ordinary transistors typically have breakdown voltages of less than 100 volts, supplies to deliver voltages higher than that require some clever circuit trickery. This section will present a collection of such techniques.

□ *Brute force: high-voltage components*

Power transistors, both bipolar and MOSFET, are available with breakdown voltages to 1000 volts and higher, and they're not even very expensive. Motorola's MJ12005, for example, is an 8 amp *npn* power transistor with conventional (V_{CEO}) collector-to-emitter breakdown of 750 volts, and base back-biased breakdown (V_{CEX}) of 1500 volts; it costs less than 5 dollars in single quantities. Their MTP1N100 (similar to the European BUZ-50) is a 1 amp n-channel power MOSFET with 1000 volt breakdown and a price tag of a few dollars. Power MOSFETs in particular are often excellent choices for high-voltage regulators, owing to their excellent safe operating area (absence of thermally induced second breakdown).

By running the error amplifier near ground (the output-voltage-sensing divider gives a low-voltage sample of the output), you can build a high-voltage regulator with only the pass transistor and its driver seeing high voltage. Figure 6.47 shows the idea, in this case a +100 to +500 volt regulated supply using NMOS pass transistor and driver. Q_2 is the series pass transistor, driven by inverting amplifier Q_1. The op-amp serves as error amplifier, comparing an adjustable fraction of the output with a precision +5 volt reference. Q_3 provides current limiting by shutting off drive to Q_2 when the drop across the 33 ohm resistor equals a V_{BE} drop. The remaining components serve more subtle, but necessary, functions: The diode protects Q_2 from reverse gate breakdown if Q_1 decides to pull its drain down rapidly (while the output capacitor holds up Q_2's source). The various small capacitors in the circuit provide compensation, which is needed because Q_1 is operated as an inverting amplifier with voltage gain, thus making the op-amp loop unstable (especially considering the circuit's capacitive load). This circuit is an exception to the general rule that transistor circuits do not present a shock hazard!

We can't resist an aside here: In slightly modified form (reference replaced by

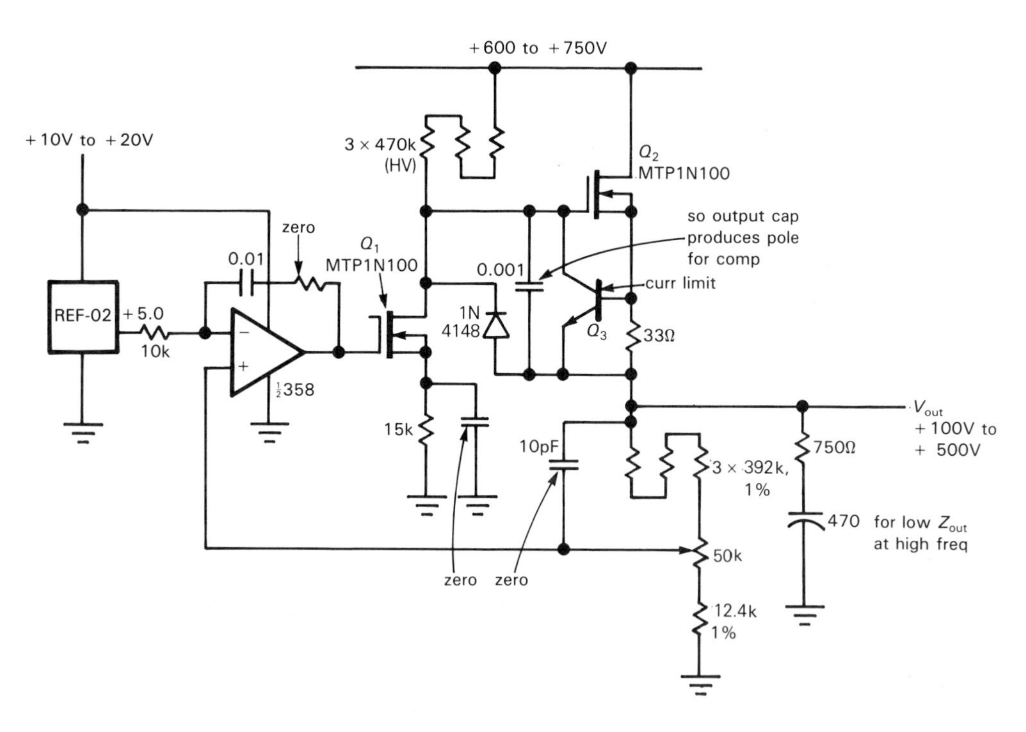

Figure 6.47. High-voltage regulated supply.

signal input) this circuit makes a very nice high-voltage amplifier, useful for driving crazy loads such as piezoelectric transducers. For that particular application the circuit must be able both to sink and to source current into the capacitive load. Oddly enough, the circuit acts like a "pseudo-push-pull" output, with Q_2 sourcing current and Q_1 sinking current (via the diode), as needed; see Section 3.14.

If a high-voltage regulator is designed to provide a fixed output only, the pass transistor may have a breakdown voltage less than the output voltage. In the preceding circuit, replacing the voltage-adjustment resistors with a fixed 12.4k resistor results in a fixed +500 volt regulator. A 300 volt pass transistor will then be fine, provided that the circuit ensures that the voltage across it never exceeds 300 volts, even during turn-on, turn-off, and output short-circuit conditions. The latter condition presents a challenge, but bridging Q_2 with a 300 volt zener may solve the problem. If the zener can handle high current, it can also protect the pass transistor against short-circuit loads, if suitable fusing is provided ahead of the regulator. The active zener circuit mentioned in Section 6.06 would be a good choice here.

□ *Regulating the ground return*

Figure 6.48 shows another way to regulate high voltages with low-voltage components. Q_1 is a series pass transistor, but it is connected in the low side of the supply; its "output" goes to ground. It has only a fraction of the output voltage across it, and it sits near ground, simplifying the driver circuitry. As before, protection must be provided during on/off transients and overloads. The simple zener protection shown is adequate, but remember that the zener must be able to handle the full short-circuit current.

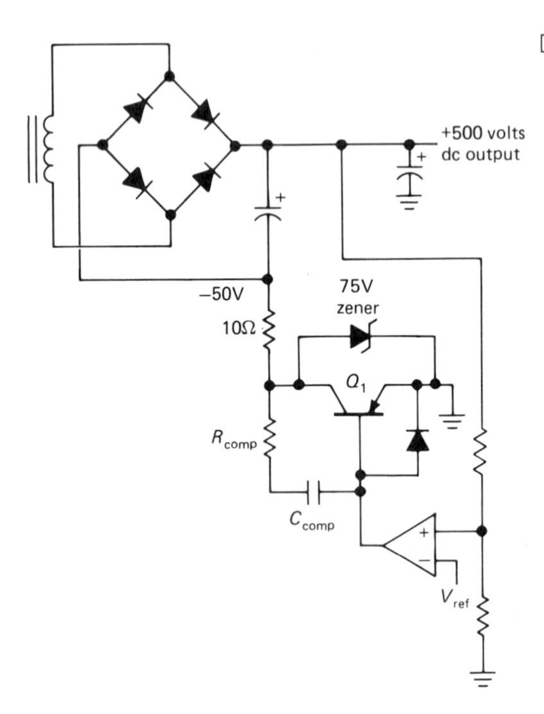

Figure 6.48. Regulating the ground return.

☐ *Lifting the regulator above ground*

Another method sometimes used to extend the voltage range of regulators, including the simple 3-terminal type, is to raise the common terminal off ground with a zener (Fig. 6.49). In this circuit D_1 adds

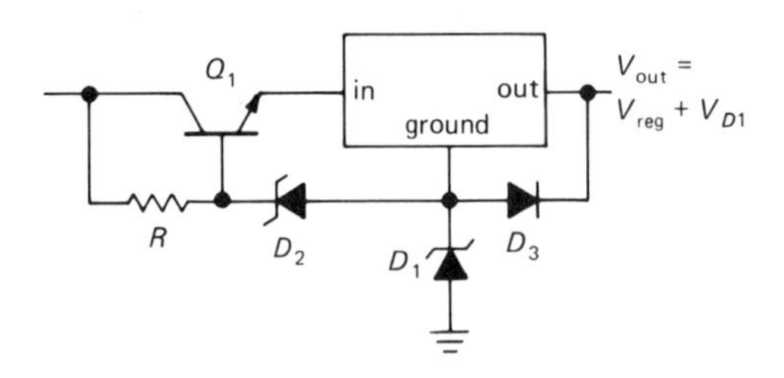

Figure 6.49

its voltage to the normal output of the regulator. D_2 sets the drop across the regulator via follower Q_1 and provides protection during short circuit because of D_3.

☐ *Optically coupled transistor*

There is another way to handle the problem of transistor breakdown ratings in high-voltage supplies, especially if the pass transistor can be a relatively low voltage unit because of fixed (known) output voltage. In such cases only the driver transistor has to withstand high voltage, and even that can be avoided by using optically coupled transistors. These devices, which we will talk about further in connection with digital interfacing in Chapter 9, actually consist of two units electrically isolated from each other: a light-emitting diode (LED), which lights up when current flows through it in the forward direction, and a phototransistor (or photo-Darlington) mounted in close proximity in an opaque package. Running current through the diode causes the transistor to conduct, just as if there were base current. As with an ordinary transistor, you apply collector voltage to put the phototransistor in the active region. In many cases no separate base lead is actually brought out. Optically coupled devices are typically insulated to withstand several thousand volts between input and output.

Figure 6.50 shows a couple of ways to use an optically coupled transistor in a high-voltage supply. In the first circuit, phototransistor Q_2 shuts off pass transistor Q_3 when the output rises too high. In the second version, for which only the pass-transistor circuitry is shown, phototransistor Q_2 increases the output voltage when driven, so the error-amplifier inputs should be reversed. Both circuits generate some output current through the pass-transistor biasing circuit, so some load from output to ground is needed to keep the output voltage from rising under no-load conditions. The output-sensing voltage divider can do the job, or a separate "bleeder" resistor can be connected across the output, which is always a good idea anyway in a high-voltage supply.

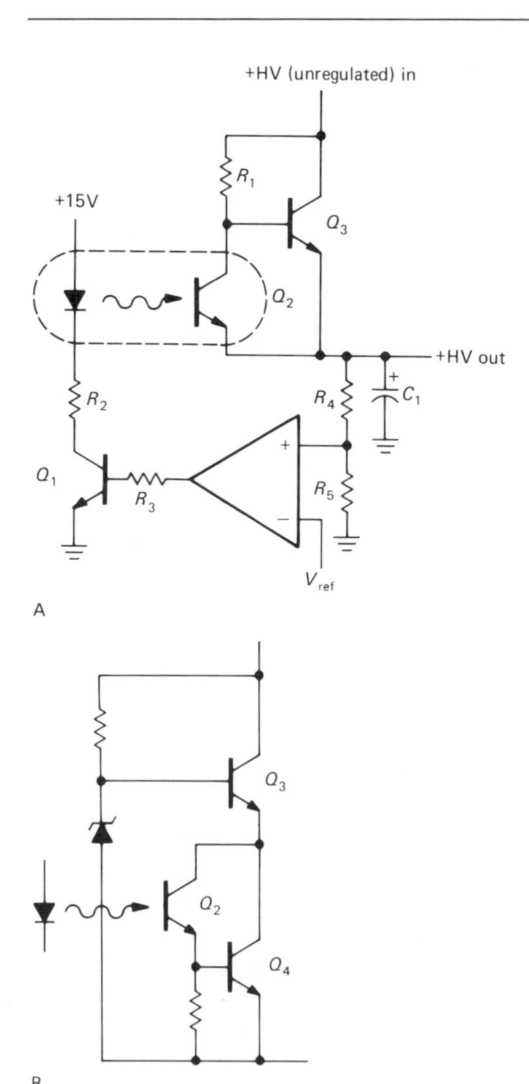

Figure 6.50. Opto-isolated high-voltage regulator.

□ *Floating regulator*

Another way to avoid applying large voltages to the control components of a high-voltage power supply is to "float" the control circuitry at the pass-transistor potential, comparing the drop across its own voltage reference with the drop down to ground. The excellent MC1466 regulator chip is intended for this kind of application, which normally requires an auxiliary low-current floating power supply to provide dc (20–30V) for the chip itself. The output voltage is limited only by the pass transistors and the isolation (transformer insulation breakdown voltage) of the auxiliary supply. The MC1466 features very good regulation and precise current-limiting circuitry and is well suited for accurate "laboratory" power supplies. A warning, however: The MC1466, unlike more recent regulator designs, does not include on-chip thermal protection.

An elegant way to rig up a floating regulator is provided by the LM10 op-amp plus voltage-reference combination, a remarkable breakthrough in chip technology from the legendary Widlar (see Section 4.13) that will operate from a single 1.2 volt supply. Such a chip can be powered from the base-emitter drop of a Darlington pass transistor! Figure 6.51 shows an example. If you like analogies, think of a giraffe who measures his height by looking at the distance to the ground, then stabilizes it by craning his neck accordingly. The TL783 from Texas Instruments is a 125 volt IC regulator that works this way; for lower-current applications it replaces the discrete circuitry of Figure 6.51.

□ *Transistors in series*

Figure 6.52 shows a trick for connecting transistors in series to increase the breakdown voltage. Driver Q_1 drives series-connected transistors Q_2 – Q_4, which share the large voltage from Q_2's collector to the output. The equal base resistors are chosen small enough to drive the transistors to full output current. The same circuit works with MOSFETs as well, but be sure to provide reverse-gate-protection diodes, as shown (you don't have to worry about *forward* gate breakdown, because the MOSFETs should turn on vigorously long before gate-channel breakdown). Note that the bias resistors produce some output current even when the transistors are cut off,

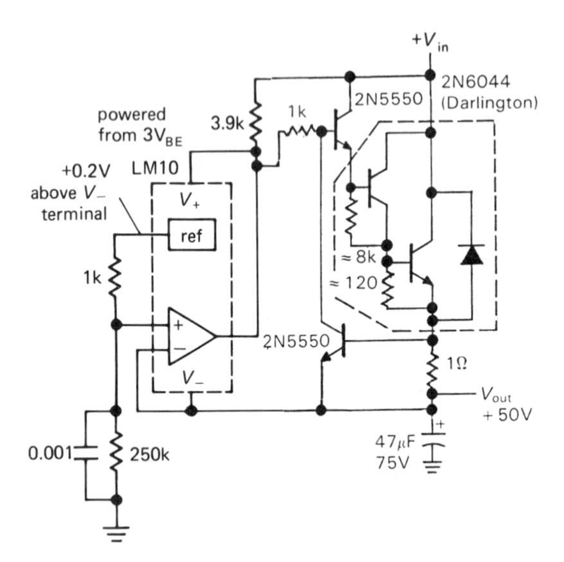

Figure 6.51. High-voltage floating regulator.

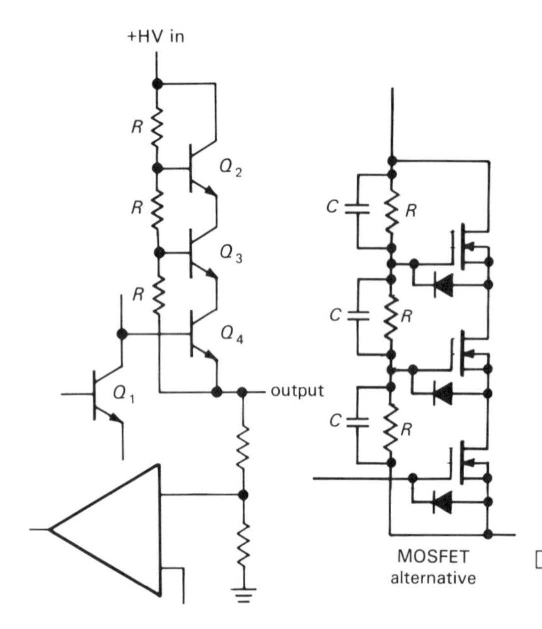

Figure 6.52. Connecting transistors in series to raise breakdown voltage.

so there must be a minimum load to ground to prevent the output from rising above its regulated voltage. It's often a good idea to parallel the divider resistors with small capacitors, as indicated, in order to maintain the divider action at high frequencies; choose a capacitor value large enough to swamp differences in transistor input capacitance, which otherwise cause unequal division, reducing overall breakdown voltage.

Series-connected transistors can, of course, be used in circuits other than power supplies. You'll sometimes see them in high-voltage amplifiers, although the availability of high-voltage MOSFETs often makes it unnecessary to resort to the series connection at all.

In high voltage circuits like this, it's easy to overlook the fact that you may need to use 1 watt (or larger) resistors, rather than the standard $\frac{1}{4}$ watt type. A more subtle trap awaits the unwary, namely the maximum *voltage* rating of 250 volts for standard $\frac{1}{4}$ watt composition ("carbon") resistors, regardless of power dissipation. Carbon resistors run at higher voltages show astounding voltage coefficients, not to mention permanent changes of resistance. For example, in an actual measurement (Fig. 6.53) a 1000:1 divider (10Meg, 10k) produced a division ratio of 775:1 (29% error!) when driven with 1kV; note that the *power* was well within ratings. This non-ohmic effect is particularly important in the output-voltage-sensing divider of high-voltage supplies and amplifiers – beware! Companies like Victoreen make resistors in many styles designed for high-voltage applications like this.

□ *Regulating the input*

Another technique sometimes used in high-voltage supplies, particularly those intended for low currents, is to regulate the input rather than the output. This is usually done with high-frequency dc-to-dc switching supplies, since attempting to regulate the 60Hz ac input will result in poor regulation and plenty of residual ripple. Figure 6.54 shows the general idea. T_1 and associated circuitry generate unregulated dc at some manageable voltage, say 24

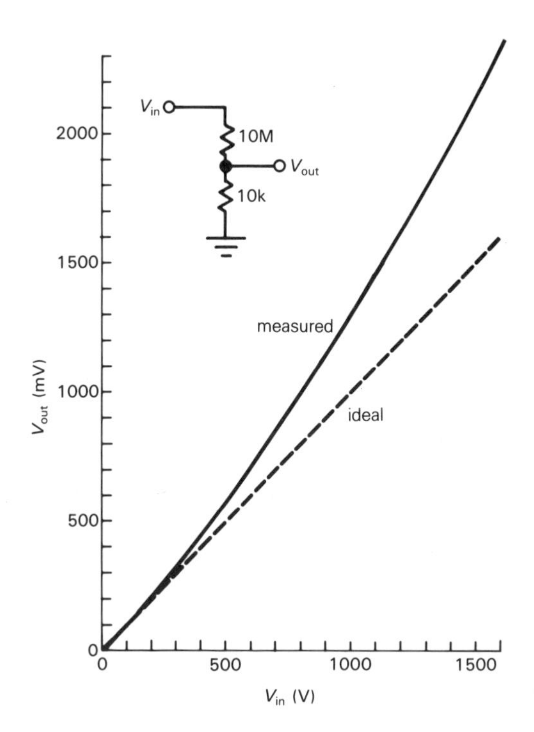

Figure 6.53. Carbon composition resistors exhibit a reduction in resistance above 250 volts.

volts; alternatively, batteries might provide the dc input. This powers a high-frequency square-wave power oscillator, with its output full-wave-rectified and filtered. This filtered dc is the output, a sample of which is fed back to control the oscillator's duty cycle or amplitude in response to the output voltage. Since the oscillator runs at high frequency, the response is rapid, and its rectified waveform is easy to filter, especially since it is a full-wave-rectified square wave. T_2 must be designed for high-frequency operation, since ordinary laminated-core power transformers will have excessive core losses. Suitable transformers are built with iron powder, ferrite, or "tape-wound" toroidal cores and are much smaller and lighter than conventional power transformers of the same power rating. No high-voltage components are used, except, of course, for the output bridge rectifier and capacitor.

The astute reader may experience a sense of *déjà vu* while reading the last paragraph. In fact, it describes switching regulators (Section 6.19) in nearly all respects. The one significant difference is that switching supplies usually use inductors as energy-storage devices, whereas the input-regulated high-voltage supply uses T_2 as a "normal" (albeit high-frequency) transformer. In common with switching supplies, these high-voltage supplies display high-frequency ripple and noise.

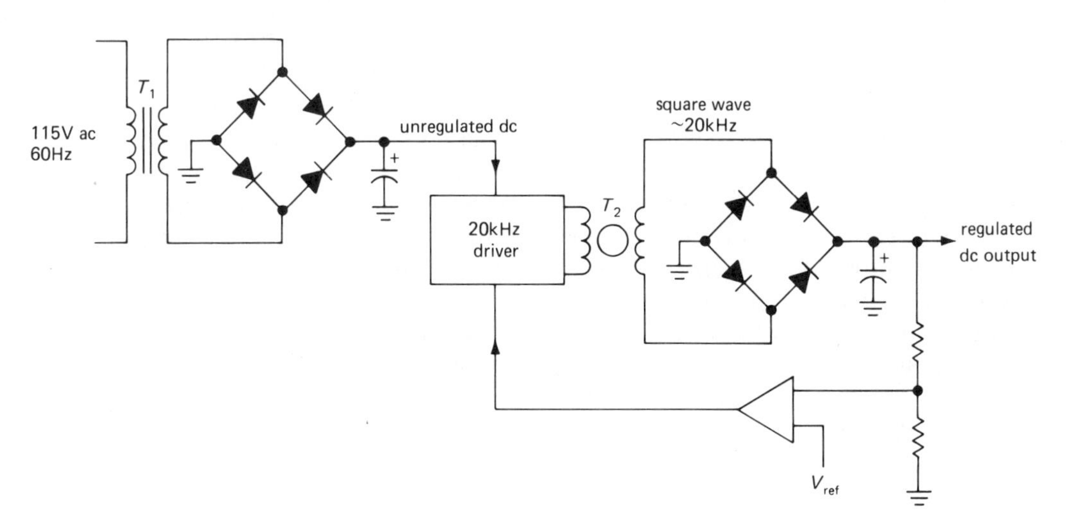

Figure 6.54. High-voltage switching supply.

Video flyback supplies

A variation on the conventional flyback switching regulator (Fig. 6.43A) is commonly used to generate the high dc voltages (10kV or more) needed in television and cathode-ray-tube (CRT) video displays. As we'll see, this circuit is especially clever, because it also generates the horizontal sweep signal used to drive the deflection coils.

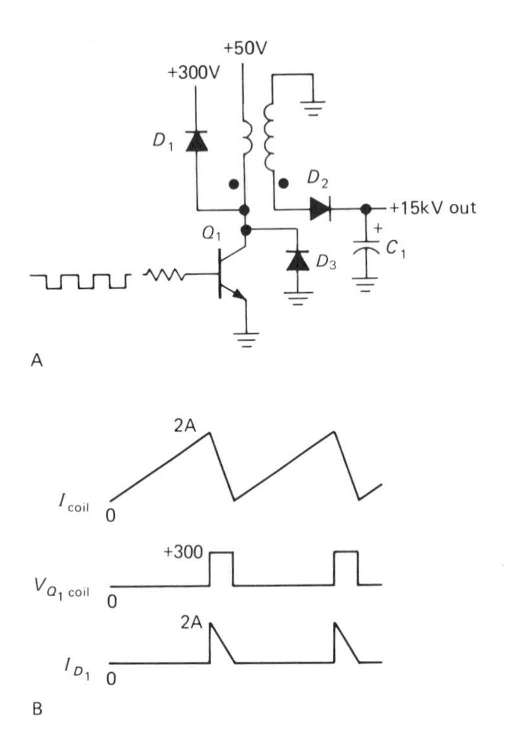

Figure 6.55. Video flyback high-voltage supply.

The basic idea is to use a transformer with a large turns ratio, driving the primary with a saturated transistor, just like a conventional flyback circuit. The output is taken from the secondary, rectified to generate high-voltage dc; see Figure 6.55. Q_1 is driven by wide pulses, pulling the primary to ground. It may be self-excited or driven by an oscillator. D_1 is a "damper" diode that prevents Q_1's collector from rising too high during the flyback. D_2, connected to the high-voltage secondary winding, rectifies the output, typically 10kV–20kV at a few microamps. The circuit is operated at frequencies of 15kHz or more, which means that filter capacitor C_1 can be as small as a few hundred picofarads (check this for yourself, by calculating the ripple).

Note that the collector-current waveform is a linearly rising ramp, which is often used to drive the magnetic deflection coils (called the "yoke") of the CRT, thus producing the linear horizontal raster scan. In such cases the oscillator frequency sets the horizontal scan rate. A related circuit is the so-called blocking oscillator, which generates its own excitation pulses.

□ 6.21 Low-noise, low-drift supplies

The regulated supplies we have described thus far are pretty good – they typically have ripple and noise below a millivolt, and drift with temperature of 100ppm/°C or so. This is more than adequate for just about everything you will ever need to power. However, there are times when you may need better performance, and you can't get it with any available regulator ICs. The solution is to design your own regulator circuit, using the best available IC references (in terms of stability and noise; see, e.g., the REF101KM in Table 6.7). This kind of stability (<1ppm/°C) is far better than the tempco of ordinary metal-film resistors (50ppm/°C), for example; so you must use great care to select op-amps and passive components whose errors and drifts do not degrade overall performance.

Figure 6.56 shows a complete design of an exceptional low-noise, low-drift dc regulated supply. It begins with the excellent REF10KM from Burr-Brown, which guarantees better than 1ppm/°C tempco, along with very low noise (6μV pp, 0.1–10Hz). Furthermore, it achieves this without thermostatic control, which helps keep the subsurface zener noise low. The reference

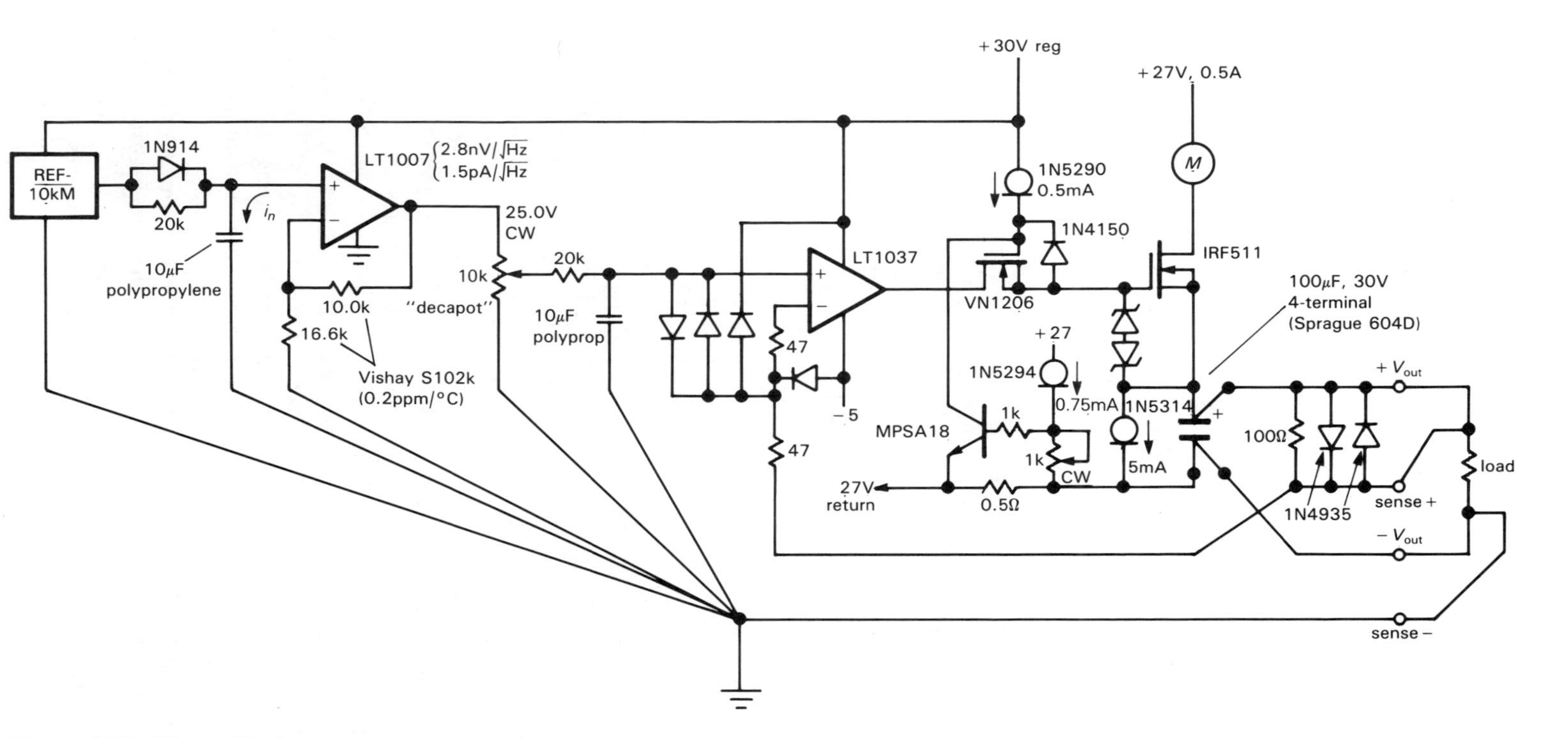

Figure 6.56. Ultrastable low-noise power supply.

is followed by a low-pass filter, to reduce the noise further. The large capacitor value is needed to suppress current noise from the op-amp: the value shown converts the current noise (1.5pA/$\sqrt{\text{Hz}}$ at 10Hz) to a voltage noise of 2.4nV$\sqrt{\text{Hz}}$, comparable with the op-amp's e_n. A polypropylene capacitor is used because the capacitor leakage (more precisely, *changes* in leakage over time and temperature) must be less than 0.1nA in order to avoid microvolt drifts in output voltage. The reference is boosted to +25 volts by the op-amp, whose feedback resistors have ultra-low tempco (0.2ppm/°C, max); note the +30 volt supply voltage. The resultant +25.0 volt reference drives a voltage divider to produce the desired output voltage, which is then low-pass-filtered a second time, again using a low-leakage capacitor. Because a potentiometer is used to divide the reference voltage, resistor tempco isn't as critical here – it's a *ratiometric* measurement.

The rest of the circuit is simply a follower, using a precision low-noise error amplifier to compare the output voltage from a power MOSFET series pass transistor. A decompensated op-amp has been used, since the large output capacitor provides the dominant pole for compensation. Note the unusual current-limit circuit and the liberal use of constant-current "diodes" (really JFETs) to provide operating bias. Note also the use of "sense" wires to sample the voltage across the load. In a precision circuit like this it is important to pay careful attention to ground paths, since, for example, a 100mA load current flowing through 1 inch of #20 wire produces a voltage drop of 100μV – which is a 100ppm error for a 1 volt output! The circuit shown has excellent performance and surpasses the typical noise and drift figures given earlier by at least a factor of 100. According to EVI, Inc. (Columbia, MD), which kindly provided the circuit, it produces noise and hum below 1μV, tempco below 1ppm/°C, output impedance below 1$\mu\Omega$, and drift below 1ppm/working day.

We will talk more about such precision and low-noise design in the next chapter.

□ 6.22 Micropower regulators

As we've hinted earlier, it's possible to design battery operated circuits that use very low quiescent current, often as little as tens of microamps. That's what's needed, of course, to make the circuit run for months or years on a small battery, as it must if it is a wristwatch or calculator. For example, an alkaline 9 volt transistor battery is exhausted after supplying about 400mA-hours; thus you can run a 50μA circuit with it for about a year (8800 hours). If such *micropower* circuits need regulated voltages, you clearly can't afford to squander the 3mA quiescent current of a 78L05, since that would degrade battery life to less than a week!

The solution is either to design a micropower regulator from discrete components or use one of the ICs intended for micropower applications. Luckily, some good ICs have come along in recent years. One of the best is the LP2950 series from National, available as a TO-92 (small transistor package) 3-terminal fixed 5 volt regulator (LP2950ACZ-5.0) or as a multiterminal adjustable 1.2–30 volt regulator (LP2951). Both versions have a quiescent current of 75μA. For even lower quiescent currents there are the ICL7663/4 (or MAX663/4), adjustable regulators of both polarities with 4μA quiescent current. We will discuss micropower regulators, along with all aspects of battery-powered circuit design, in Chapter 14.

As an example of what you can do with discrete design, we show in Figure 6.57 a micropower circuit, designed for possible use in a lithium-battery-powered heart pacemaker, that converts an input voltage in the range +5 volts down to +3 volts (as

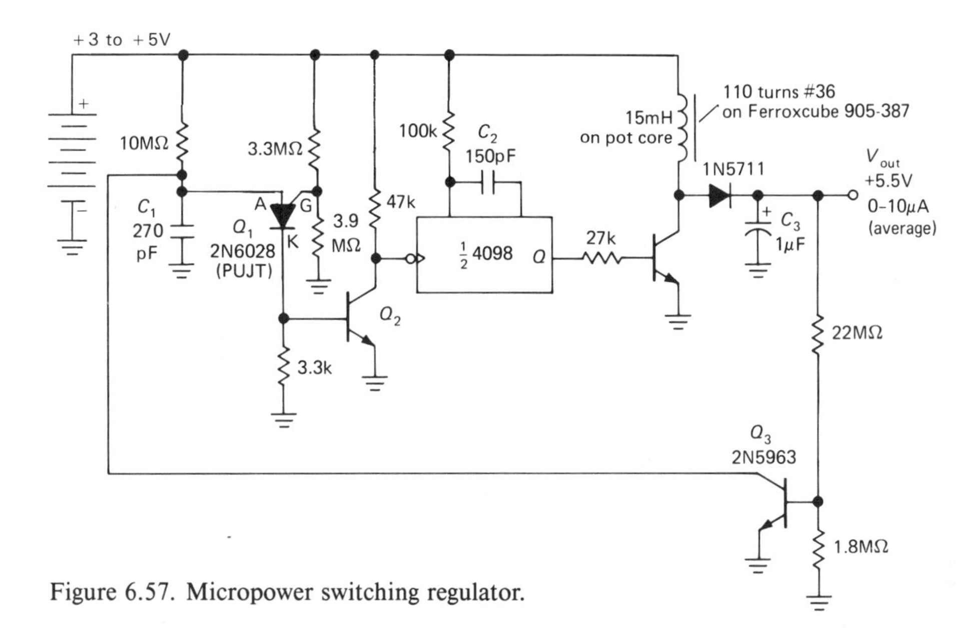

Figure 6.57. Micropower switching regulator.

the battery ages) to a regulated +5.5 volt supply. The power supply has a quiescent current of 1μA and provides line and load regulation of about 5%, with 85% conversion efficiency under full load for all battery voltages. As we remarked when discussing switching supplies, a conventional linear supply using an oscillator, doubler, and series pass regulator would be far less efficient because of regulator losses following the higher unregulated dc voltage. The flyback technique is effectively like a variable-ratio voltage multiplier, which yields extremely high efficiency, making it an attractive technique for micropower applications.

The 2N6028 programmable unijunction transistor (PUJT) is a versatile relaxation oscillator component. Its sense terminal (the anode) draws no current until its voltage exceeds the gate programming voltage by a diode drop, at which point it goes into heavy conduction from anode to cathode, discharging the capacitor. The resulting positive pulse at Q_2's base pulls Q_2's collector to ground, triggering the 4098, a device known as a "monostable multivibrator" (see Section 8.20), which generates positive pulses of constant width at its output terminal labeled Q.

In this circuit, Q_3 senses the output voltage and robs charging current from C_1, reducing the energy-transfer pulsing rate of the inductor as necessary to maintain the desired output voltage. Note the large resistor values throughout the circuit. Temperature compensation is not an issue here because the circuit operates in a stable 98.6°F mobile oven. (Warning: We remind the reader to look again at the "Legal notice" in the Preface.)

6.23 Flying-capacitor (charge pump) voltage converters

In Section 6.19 we discussed switching supplies, with their bizarre ability to produce a dc output voltage *larger* than the dc input, or even of opposite polarity. We mentioned there that flying-capacitor voltage converters let you do some of the same

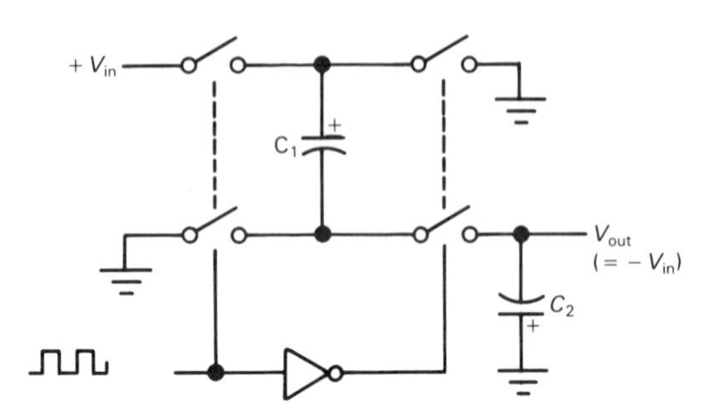

Figure 6.58. Flying-capacitor voltage inverter. C_1 and C_2 are external 10μF tantalum capacitors.

things. What is this strange "flying capacitor"?

Figure 6.58 shows a simplified circuit of the 7662 CMOS IC introduced by Intersil, and widely second-sourced. It has an internal oscillator and some CMOS switches, and it requires a pair of external capacitors to do its job. When the input pair of switches are closed (conducting), C_1 charges to V_{in}; then, during the second half cycle, C_1 is disconnected from the input and connected, upside-down, across the output. It thus transfers its charge to C_2 (and the load), producing an output of approximately $-V_{in}$. Alternatively, to use the 7662 to create an output of $2V_{in}$ you can arrange things so that C_1 charges as before, but then gets hooked in series with V_{in} during the second (transfer) half cycle.

This flying-capacitor technique is simple and efficient and requires few parts and no inductors. However, the output is not regulated, and it drops significantly under load currents greater than a few milliamps (Fig. 6.59). Also, like most CMOS devices, it has a limited supply voltage range; for the 7662, V_{in} can only range from 4.5 to 20 volts (1.5V to 10V for its predecessor, the 7660). Finally, unlike the inductive step-up or inverter circuits, which can generate any output voltage at all, the flying capacitor voltage converter can only generate discrete multiples of the input voltage. In spite of these drawbacks, flying-capacitor voltage converters can be very useful in some circumstances, for example to power a bipolarity op-amp or serial port (see Chapters 10 and 11) on a circuit board that has only +5 volts available.

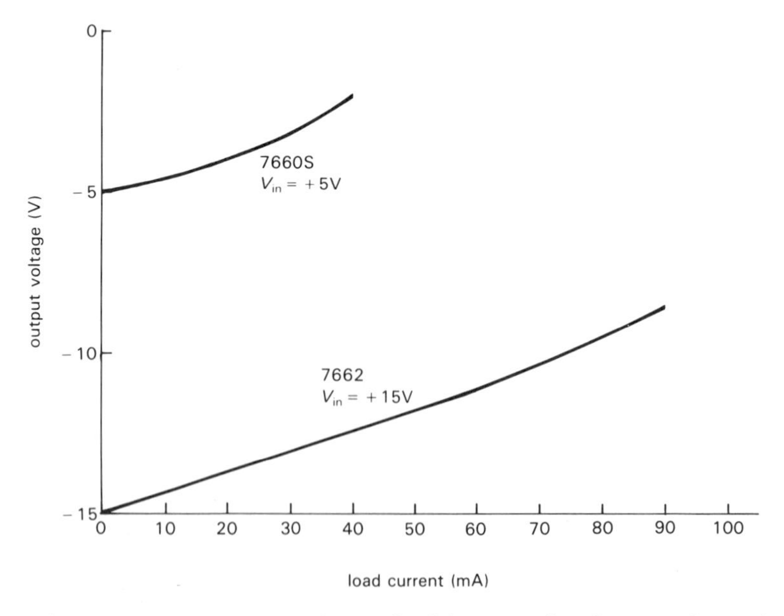

Figure 6.59. The output voltage of a flying-capacitor inverter drops significantly under load.

There are some other interesting flying-capacitor chips. The MAX680 from Maxim is a dual supply that generates ±10 volts (up to 10mA) from +5 volts (Fig. 6.60). The similar LT1026 from LTC operates to ±20 volts output (up to 20mA) and uses smaller capacitors (1μF instead of 20μF). The LT1054 from LTC combines a flying-capacitor converter with a linear regulator to provide a stiff regulated output up to 100mA (at lower efficiency, of course). The MAX232 series and the LT1080 combine a ±10 volt switched-capacitor supply with an RS-232C digital serial port (see Chapter 11), eliminating the need for bipolarity supplies in many computer boards; some chips in the MAX232 series even have built-in capacitors. And the LTC1043 is an uncommitted flying-capacitor building block, which you can use to do all kinds of magic. For example, you can use a flying capacitor to transfer a voltage drop measured at an inconvenient potential (e.g., a current-sensing resistor at the positive supply voltage) down to ground, where you can easily use it. The LTC1043 data sheet has 8 pages of similar clever applications.

6.24 Constant-current supplies

In Sections 2.06 and 2.14 we described some methods for generating constant currents within a circuit, including voltage-programmed currents with floating or grounded loads and various forms of current mirrors. In Section 3.06 we showed how to use FETs to construct some simple current-source circuits, including "current-regulator diodes" (a JFET with gate tied to source) such as the 1N5283 series. In Section 4.07 we showed how to get improved performance (at low frequencies, anyway) by using op-amps to construct current sources. And in Section 6.15 we mentioned the convenient LM334 3-terminal current source IC. There is often a need, however, for a flexible constant-current supply, which can supply substantial voltage and current, as a complete instrument. In this section we will look at some of the more successful circuit techniques.

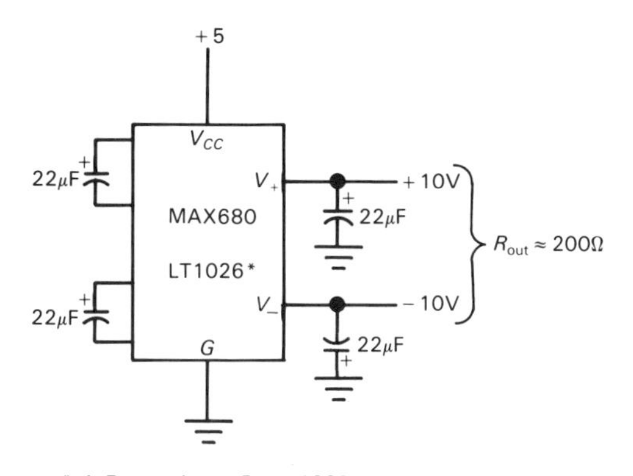

Figure 6.60. Flying-capacitor dual supply. The LT1026 is similar, but has $R_{out} \approx 100$ ohms and requires only 1μF capacitors.

□ *Three-terminal regulator*

In Section 6.18 we showed how you can use a 3-terminal adjustable regulator to make a delightfully simple current source. The 317-type regulator, for example, maintains a constant 1.25 volts (bandgap) between its output and its "ADJ" pin; by putting a resistor across these pins, you form a 2-terminal constant-current device (Fig. 6.38), which can be used as a sink or source. Performance degrades with less than about 3 volts across the circuit, since the regulator itself has a dropout voltage near 2 volts.

This type of current source is suitable for moderate to high currents: The LM317 has a maximum current of 1.5 amps and can operate with up to 37 volts drop. Its high-voltage cousin, the LM317HVK, can withstand 57 volts drop. Higher-current versions are available, e.g., the LM338 (5A) and LM396 (10A), although these have lower voltage ratings. Three-terminal regulators won't work as current sources below about 10mA, the worst-case quiescent current. However, note that the latter is not a source of current error, since it flows from input pin to output pin; the much smaller current that flows out of the ADJ pin (50μA, nominal) varies about 20% over the operating temperature range and is negligible by comparison.

In ancient times, before 3-terminal adjustable regulators were available, people sometimes used 5 volt *fixed* regulators (e.g., the 7805) as current sources in a similar arrangement (substituting "GND" for "ADJ"). This is an inferior circuit, because at low output currents the regulator's quiescent current (8mA max) contributes a large error, and at high currents the 5 volt drop across the current-setting resistor results in unnecessary power dissipation.

□ *Supply-line sensing*

A simple technique that yields good performance involves constructing a conventional series pass regulator, with current sensing at the input to the pass transistor (Fig. 6.61). R_2 is the current-sensing

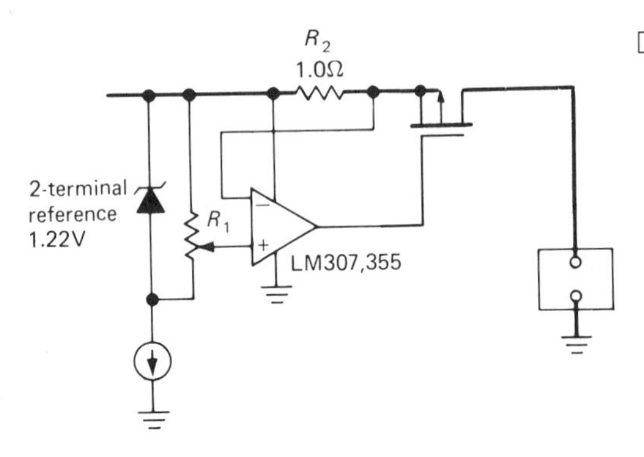

Figure 6.61. Input-rail current sensing.

resistor, preferably a low-temperature-coefficient type. For very high current or high-precision applications, you should use a 4-wire resistor, intended for current-sensing applications, in which the sensing leads are connected internally. The sensing voltage does not then depend on the connection resistance of the joints to the current-carrying leads, which for clarity are drawn with heavy lines in this schematic.

For this circuit you must use an op-amp that has an input common-mode range all the way to the positive supply (the 307, 355, and 441 have this virtue), unless, of course, you power the op-amp with a more positive auxiliary supply. The MOSFET in this circuit could be replaced by a *pnp* pass transistor; however, since the output current would then include the base current, you should use a Darlington connection to minimize that error. Note that an n-channel output transistor (connected as a follower) can be used instead of the p-channel shown, if the input connections to the op-amp are reversed. However, the current source will then have an undesirably low output impedance at frequencies approaching f_T of the op-amp loop, since the output is actually a source follower. This is a common error in current-source design, since the dc analysis shows correct performance.

□ *Return-line current sensing*

A good way to make a precise current source is to sense the voltage across a precision resistor directly in series with the load, since this makes it easier to meet the simple criterion for eliminating current-source errors due to base drive currents; the base drive current must either pass through both the load and sense amplifier, or pass through neither. However, to meet this criterion it is necessary to "float" either the load or the power supply by at least the voltage drop across the current-sensing resistor. Figure 6.62 shows a couple of circuits that use floating loads.

The first circuit is a conventional series pass circuit, with the error signal derived from the drop across the small resistor in the load's return path to ground. The high-current path is again drawn with bold lines. The Darlington connection is used here

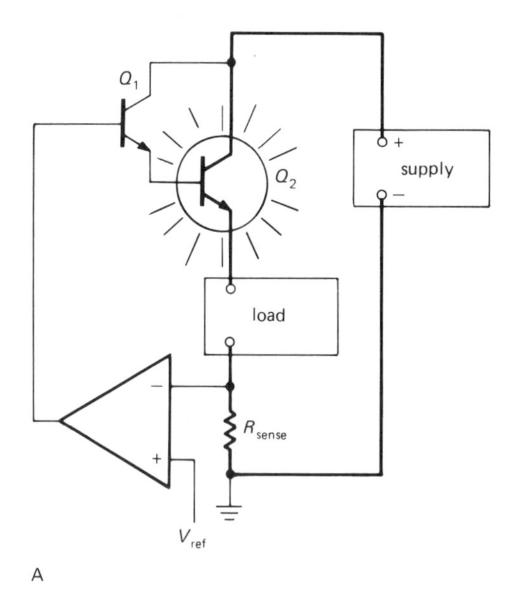

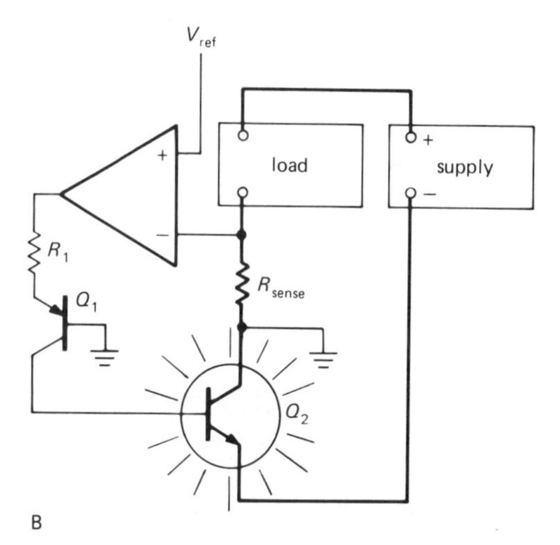

Figure 6.62. Return-line current sensing.

not to avoid base-current error, since the actual current through the load is sensed, but rather to keep the drive current down to a few milliamps so that ordinary op-amps can be used for the error amplifier. The sensing resistor should be a precision power resistor of low temperature coefficient, preferably a 4-wire resistor. In the second circuit the regulating transistor Q_2 is in the ground return of the high-current supply. The advantage here is that its collector is at ground, so you don't have to worry about insulating the transistor case from the heat sink.

In both circuits, R_{sense} will normally be chosen to drop a volt or so at typical operating currents; its value is a compromise between op-amp input offset errors, at one extreme, and a combination of reduced current-source compliance and increased dissipation at the other. If the circuit is meant to operate over large ranges of output current, R_{sense} should probably be a set of precision power resistors, with the appropriate resistor selected by a range switch.

□ *Grounded load*

If it is important for the load to be returned to circuit ground, a circuit with floating supply can be used. Figure 6.63 shows two examples. In the first circuit, the funny-looking op-amp represents an error amplifier with a high-current buffer output, run from a single split supply; it could be something as simple as a 723 (for currents up to 150mA) or one of the high-current op-amps listed in Table 4.4. The high-current supply has a common terminal that floats relative to circuit ground, and it is important that the error amplifier (or at least its buffer output) be powered from the floating supply so that base drive currents return through R_{sense}. An additional low-current supply with grounded common would be needed if other op-amps, etc., were in the same instrument. A negative reference (relative to circuit ground) programs the output current. Note the polarity at the error-amplifier inputs.

The second circuit illustrates the use of a second low-power supply when an ordinary low-current op-amp is used as error amplifier. Q_1 is the outboard pass transistor, which must be a Darlington (or MOSFET), since the base current returns through the load, but not through the sense resistor. The error amplifier is now

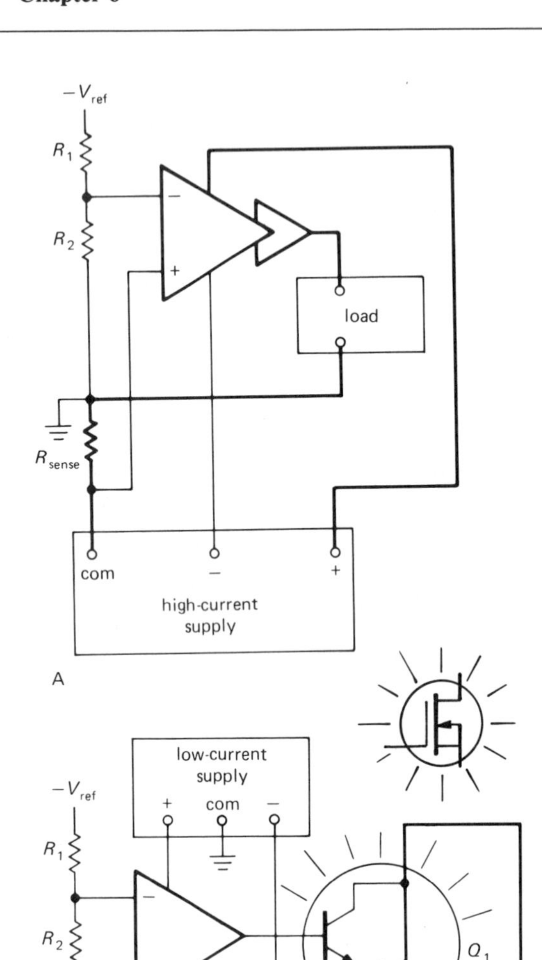

Figure 6.63. Current sources for grounded loads, employing floating high-current supplies.

powered from the same split supply with grounded common that powers the rest of the instrument. This circuit is well suited as a simple bench-instrument current source, with the low-current split supply built in and the high-current supply connected externally. You would choose the latter's voltage and current capability to fit each application.

6.25 Commercial power-supply modules

Throughout the chapter we have described how to design your own regulated power supply, implicitly assuming that is the best thing to do. Only in the discussion of line-operated switching power supplies did we suggest that the better part of valor is to swallow your pride and buy a commercial power supply.

As the economic realities of life would have it, however, the best approach is often to use one of the many commercial power supplies sold by companies such as ACDC, Acopian, Computer Products Inc., Lambda, Power-One, and literally hundreds more. They offer both switching and linear supplies, and they come in four basic packages (Figure 6.64):

1. Modular "potted" supplies: These are low-power supplies, often dual (±15) or triple (+5, ±15), packaged in "potted" modules that are usually 2.5"×3.5", and about 1" thick. The most common package has stiff wire leads on the bottom, so you can mount it directly on a circuit board; you can also screw it to a panel, or plug it into a socket. They are also available with terminal-strip screw connections along one side, for chassis mounting. A typical linear triple supply provides +5 volts at 0.5 amp and ±15 volts at 0.1 amp and costs about $100 in small quantity. Linear modular supplies fall in the 1–10 watt range, switchers in the 15–25 watt range.

2. "Open-frame" supplies: These consist of a sheet-metal chassis, with circuit board, transformer, and power transistors mounted in full view. They're meant to go

Figure 6.64. Commercial power supplies come in a variety of shapes and sizes, including potted modules, open-frame units, and fully enclosed boxes. (Courtesy of Computer Products, Inc.)

inside a larger instrument. They come in a wide range of voltages and currents and include dual and triple units as well as single-output supplies. For example, a popular triple open-frame linear provides +5 volts at 3 amps and ±15 volts at 0.8 amp and costs $75 in small quantity. Open-frame supplies are larger than potted modules, and you always screw them to the chassis. Open-frame linears fall in the 10–200 watt range, switchers in the 20–400 watt range. Open-frame supplies at the low end of the power range may have all components mounted directly on a circuit board, with no metal frame at all. As with the potted supplies, you are expected to provide switches, filters, and fuses for the ac line voltage circuits.

3. Fully enclosed supplies: These supplies have a full metal enclosure, usually perforated for cooling, and usually free of the protruding power transistors, etc., that you find on an open-frame supply. They can be mounted externally, because their full enclosure keeps fingers out; you can also mount them inside an instrument, if you want. They come with single and multiple outputs, in both linear and switchers. Fully enclosed linears fall in the 15–750 watt range, switchers in the 25–1500 watt range.

4. Wall plug-in power supplies: These are the familiar black plastic boxes that come with small consumer electronic gadgets and plug directly into the wall. They actually come in three varieties, namely (a) step-down ac transformer only, (b) unregulated dc supply, and (c) complete regulated supply; the latter can be either linear or switcher. For example, Ault has a nice series of dual (±12V or ±15V) and triple (+5V and ±12V or ±15V) linear regulated wall-plug-in supplies. These save you the trouble of bringing the ac line power into your instrument, and keep it light and small. Some of us think that these convenient supplies are getting a bit *too* popular, though, as measured by the cluster of wall plug-ins found feeding at the outlets in our house! Some "desktop" models have two cords, one each for the ac input and dc output. Some of the switching units allow a full 95 to 252 volts ac input range, useful for traveling instruments. We'll have more to say about wall plug-ins in Section 14.03, when we deal with low-power design.

SELF-EXPLANATORY CIRCUITS

6.26 Circuit ideas

Figure 6.65 presents a variety of current ideas, mostly taken from manufacturers' data sheets.

6.27 Bad circuits

Figure 6.66 presents some circuits that are guaranteed not to work. Figure them out, and you will avoid these pitfalls.

ADDITIONAL EXERCISES

(1) Design a regulated supply to deliver exactly +10.0 volts at currents up to 10mA using a 723. You have available a 15 volt (rms) 100mA transformer, diodes by the bucketful, various capacitors, a 723, resistors, and a 1k trimmer pot. Choose your resistors so that they are standard (5%) values and so that the range of adjustment of the trimmer will be sufficient to accommodate the production spread of internal reference voltages (6.80V to 7.50V).

(2) Design +5 volt 50mA voltage regulators, assuming +10 volt unregulated input, using the following: **(a)** zener diode plus emitter follower; **(b)** 7805 3-terminal regulator; **(c)** 723 regulator; **(d)** 723 plus outboard *npn* pass transistor; use foldback current limiting with 100mA onset (full-voltage current limit) and 25mA short-circuit current limit; **(e)** a 317 3-terminal adjustable positive regulator; **(f)** discrete components, with zener reference and feedback. Be sure to show component values; provide 100mA current limiting for (a), (c), and (f).

(3) Design a complete +5 volt 500mA power supply for use with digital logic. Begin at the beginning (the 115V ac wall socket), specifying such things as transformer voltage and current ratings, capacitor values, etc. To make your job easy, use a 7805 3-terminal regulator. Don't squander excess capacitance, but make your design conservative by allowing for ±10% variation in all parameters (power-line voltage, transformer and capacitor tolerances, etc.). When you're finished, calculate worst-case dissipation in the regulator.

Next, modify the circuit for 2 amp load capability by incorporating an outboard pass transistor. Include a 3 amp current limit.

Circuit ideas

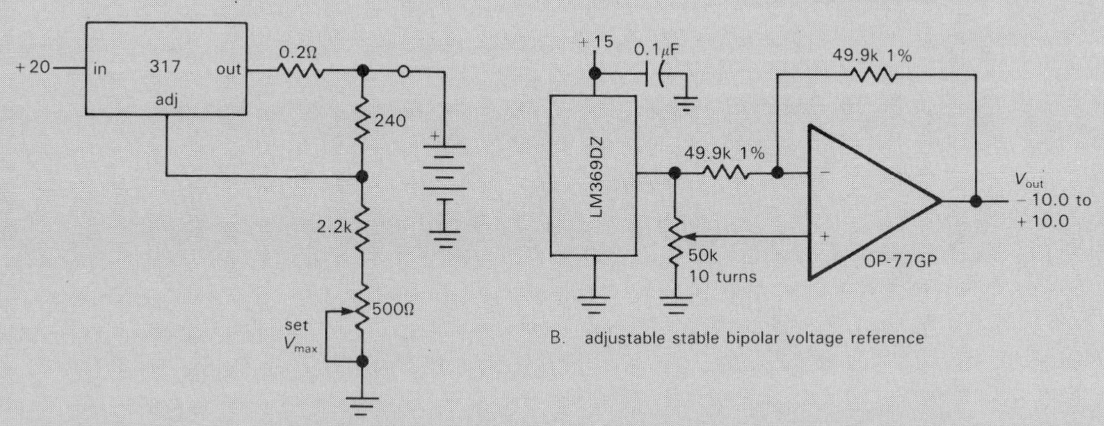

B. adjustable stable bipolar voltage reference

A. 12V battery charger

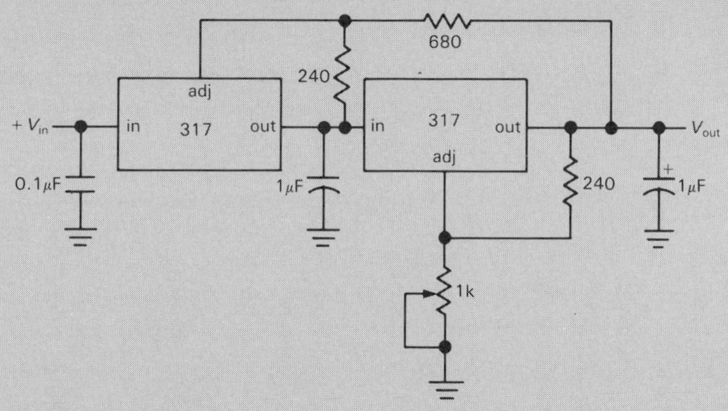

C. tracking preregulator

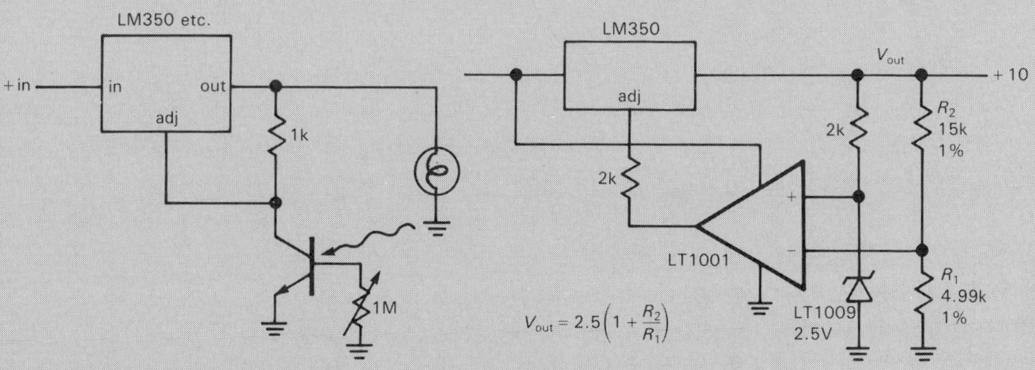

D. automatic incandescent bulb light regulator

E. precision power voltage source

Figure 6.65

Circuit ideas (*cont.*)

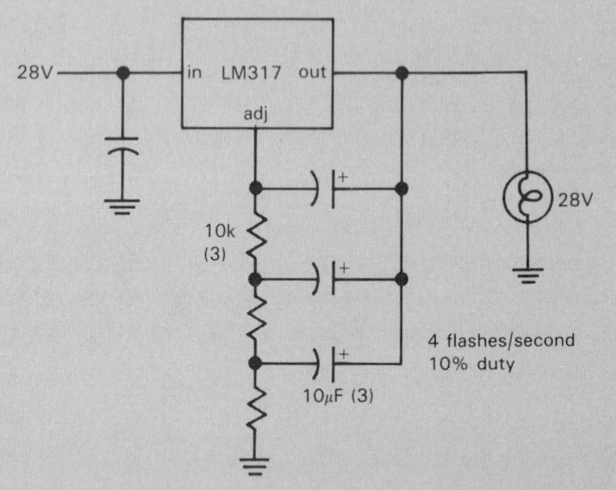

F. lamp flasher (from NSC317 data sheet)

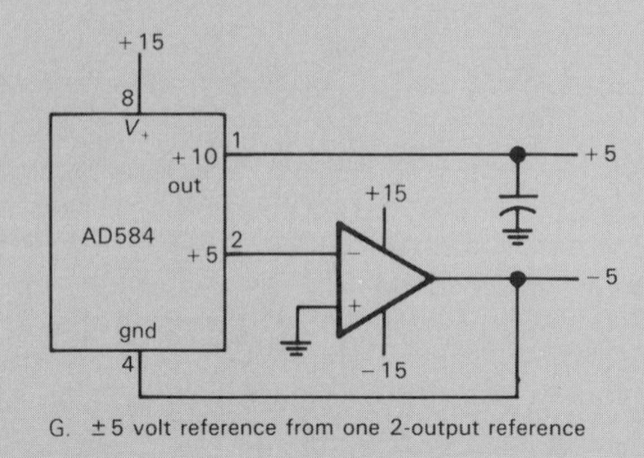

G. ±5 volt reference from one 2-output reference

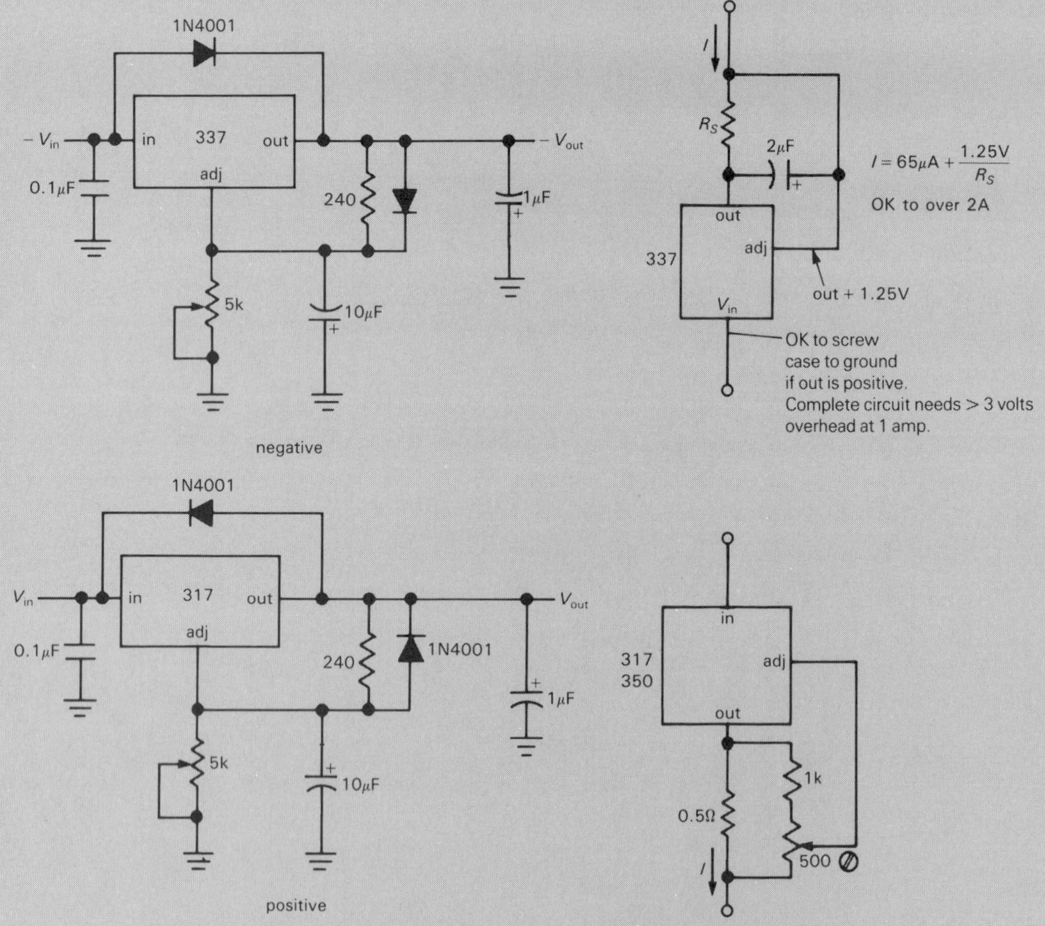

H. 3-terminal regulators with improved ripple rejection (diodes protect against input/output shorts)

I. power current sources

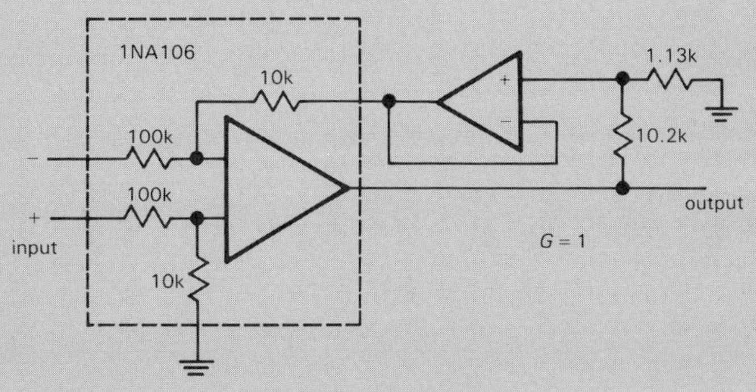

J. ±100V common-mode range differential follower

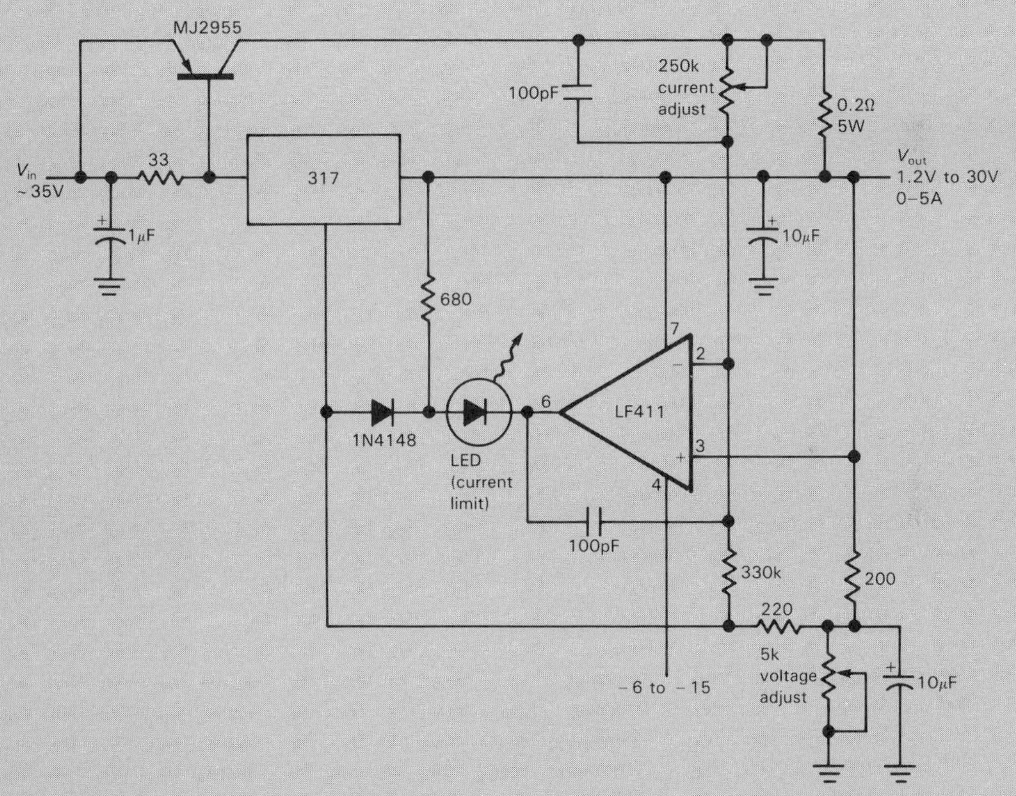

K. constant voltage/constant current supply

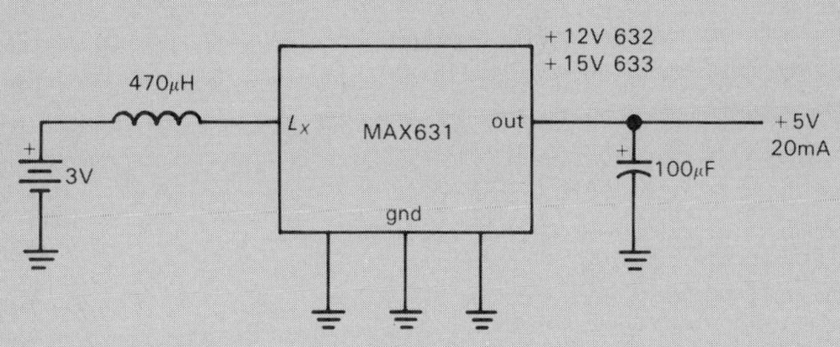

L. "world's simplest" dc–dc converter

Circuit ideas (*cont.*)

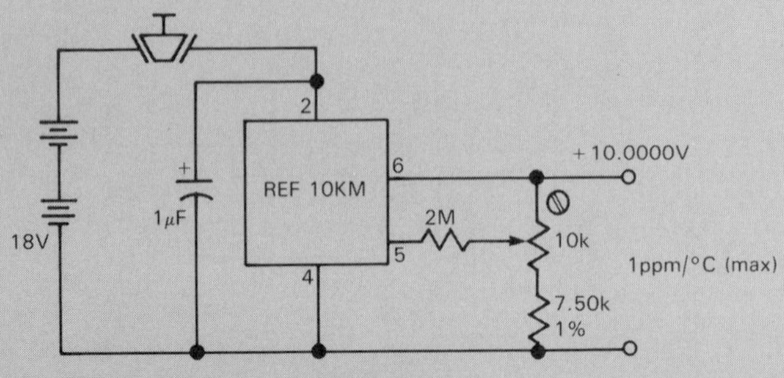

M. portable 10ppm voltage reference

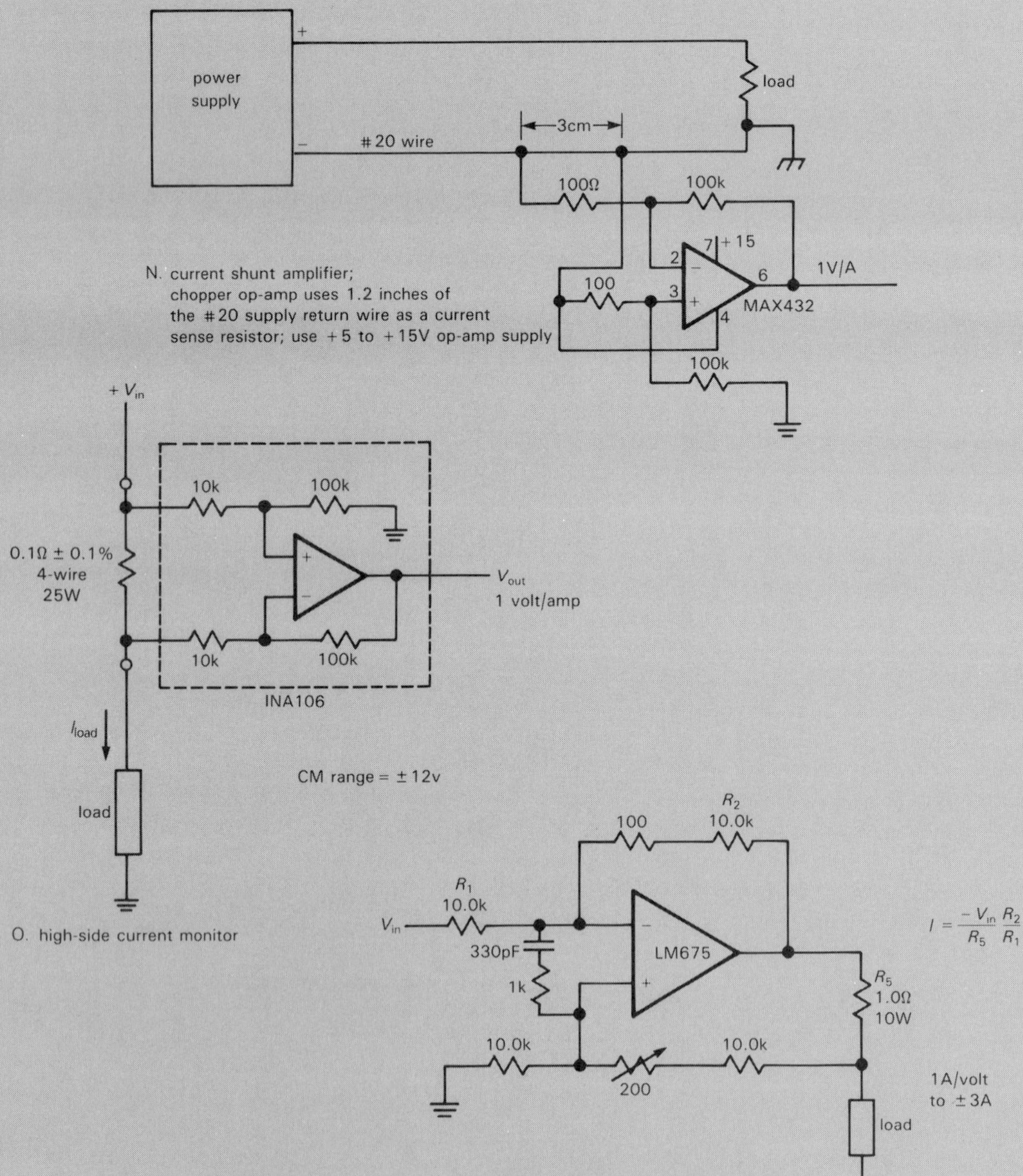

N. current shunt amplifier; chopper op-amp uses 1.2 inches of the #20 supply return wire as a current sense resistor; use +5 to +15V op-amp supply

O. high-side current monitor

P. high-current bipolarity current source

Bad circuits

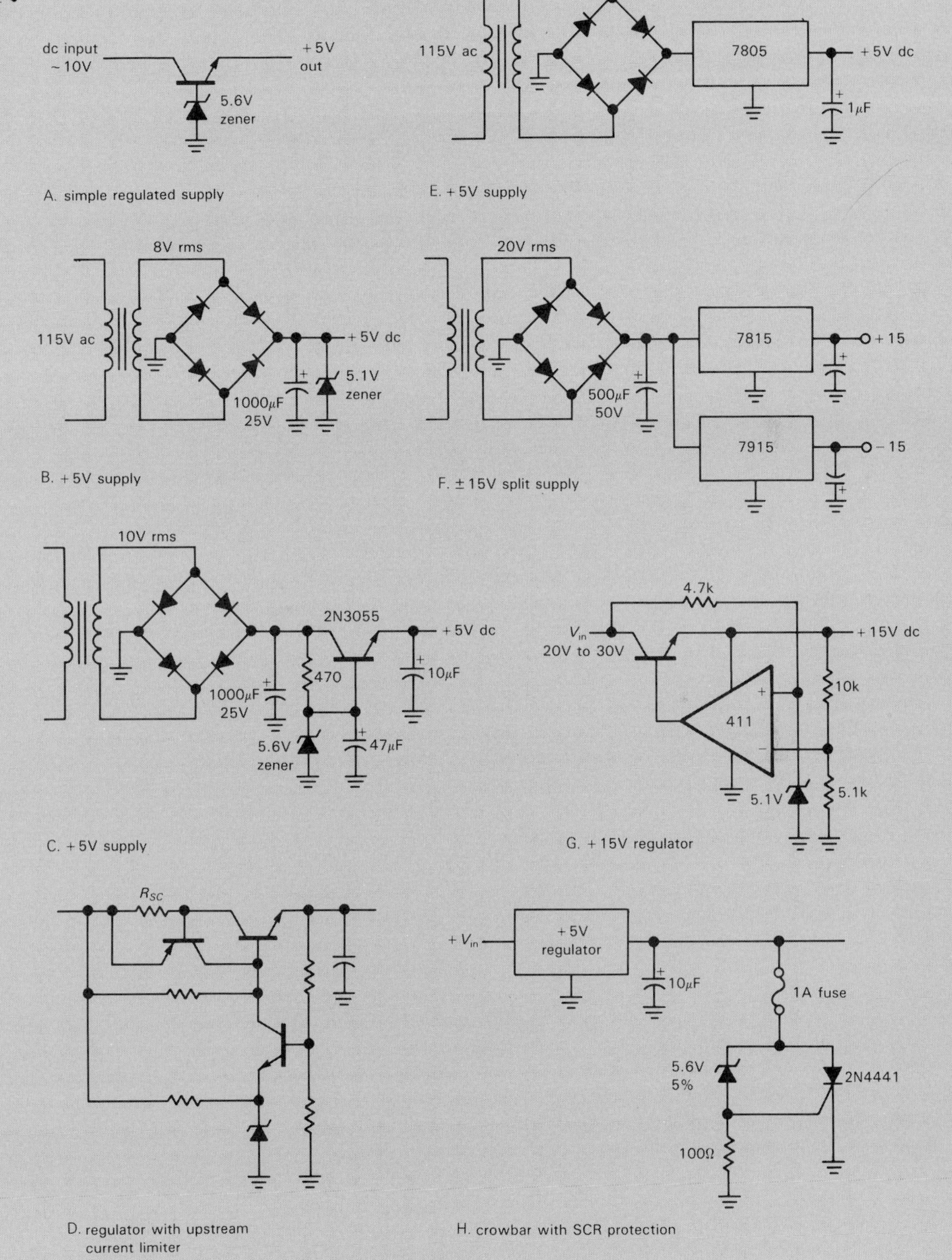

Figure 6.66

PRECISION CIRCUITS AND LOW-NOISE TECHNIQUES

CHAPTER 7

In the preceding chapters we have dealt with many aspects of analog circuit design, including the circuit properties of passive devices, transistors, FETs, and op-amps, the subject of feedback, and numerous applications of these devices and circuit methods. In all our discussions, however, we have not yet addressed the question of the best that can be done, for example, in minimizing amplifier errors (nonlinearities, drifts, etc.) and in amplifying weak signals with minimum degradation by amplifier "noise." In many applications these are the most important issues, and they form an important part of the art of electronics. In this chapter, therefore, we will look at methods of precision circuit design and the issue of noise in amplifiers. With the exception of the introduction to noise in Section 7.11, this chapter can be skipped over in a first reading. This material is not essential for an understanding of later chapters.

PRECISION OP-AMP DESIGN TECHNIQUES

In the field of measurement and control there is often a need for circuits of high precision. Control circuits should be accurate, stable with time and temperature, and predictable. The usefulness of measuring instruments likewise depends on their accuracy and stability. In almost all electronic subspecialties we always have the desire to do things more accurately – you might call it the joy of perfection. Even if you don't always actually *need* the highest precision, you can still delight in the joy of fully understanding what's going on.

7.01 Precision versus dynamic range

It is easy to get confused between the concepts of *precision* and *dynamic range*, especially since some of the same techniques are used to achieve both. Perhaps the difference can best be clarified by some examples: A 5-digit multimeter has high precision; voltage measurements are accurate to 0.01% or better. Such a device also has wide dynamic range; it can measure millivolts and volts on the same scale. A precision decade amplifier (one with selectable gains of 1, 10, and 100, say) and a precision voltage reference may have plenty of precision, but not necessarily much dynamic range. An example of a device with wide dynamic range but only moderate accuracy might be a 6-decade logarithmic amplifier (log amp) built with carefully trimmed op-amps but with components of only 5% accuracy; even with accurate components a log amp might have limited accuracy because of lack of log conformity (at the extremes of current) of the transistor junction used for the conversion. Another example of a wide-dynamic-range instrument (greater than 10,000:1 range of input currents) with only moderate accuracy (1%) is the coulomb meter described in Section 9.26. It was originally designed to keep track of the total charge put through an electrochemical cell, a quantity that needs to be known only to approximately 5% but that may be the cumulative result of a current that varies over a wide range. It is a general characteristic of wide-dynamic-range design that input offsets must be carefully trimmed in order to maintain good proportionality for signal levels near zero; this is also necessary in precision design, but, in addition, precise components, stable references, and careful attention to every possible source of error must be used to keep the sum total of all errors within the so-called error budget.

7.02 Error budget

A few words on *error budgets*. There is a tendency for the beginner to fall into the trap of thinking that a few strategically placed precision components will result in a device with precision performance. On rare occasions this will be true. But even a circuit peppered with 0.01% resistors and expensive op-amps won't perform to expectations if somewhere in the circuit there is an input offset current multiplied by a source resistance that gives a voltage error of 10mV, say. With almost any circuit there will be errors arising all over the place, and it is essential to tally them up, if for no other reason than to locate problem areas where better devices or a circuit change might be needed. Such an error budget results in rational design, in many cases revealing where an inexpensive component will suffice, and eventually permitting a careful estimate of performance.

7.03 Example circuit: precision amplifier with automatic null offset

In order to motivate the discussion of precision circuits, we have designed an extremely precise decade amplifier with automatic offset. This gadget lets you "freeze" the value of the input signal, amplifying any subsequent changes from that level by gains of exactly 10, 100, or 1000. This might come in particularly handy in an experiment in which you wish to measure a small change in some quantity (e.g., light transmission or radiofrequency absorption) as some condition of the experiment is varied. It is ordinarily difficult to get accurate measurements of small changes in a large dc signal, owing to drifts and instabilities in the amplifier. In such a situation a circuit of extreme precision and stability is required. We will describe the design choices and errors of this particular circuit in the framework of precision design in general, thus rendering painless what could otherwise become a tedious exercise. A note at the outset: Digital techniques offer an attractive alternative to the purely analog circuitry used here. Look forward to exciting revelations in chapters to come! Figure 7.1 shows the circuit.

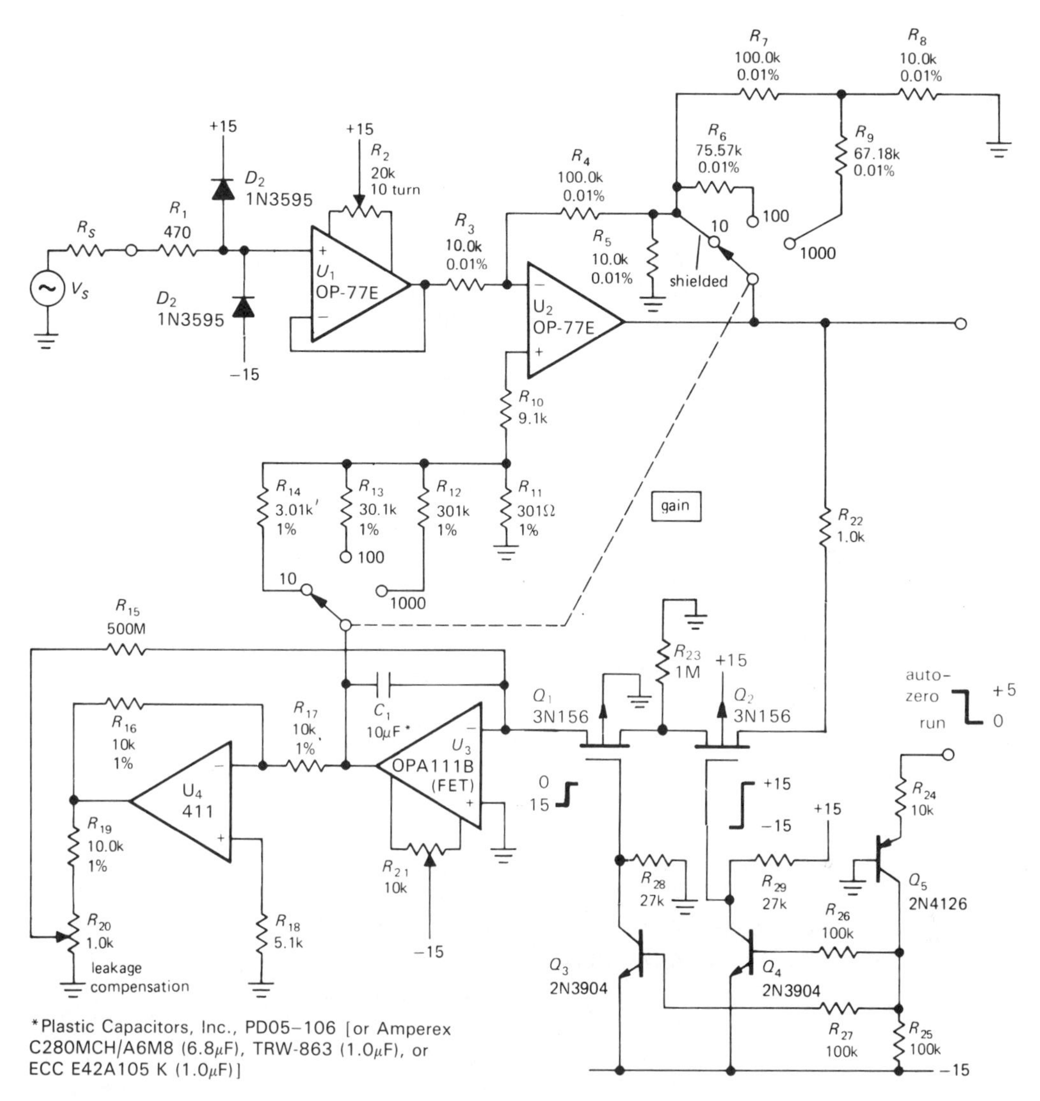

*Plastic Capacitors, Inc., PD05–106 [or Amperex C280MCH/A6M8 (6.8μF), TRW-863 (1.0μF), or ECC E42A105 K (1.0μF)]

Figure 7.1. Autonulling dc laboratory amplifier.

Circuit description

The basic circuit is a follower (U_1) driving an inverting amplifier of selectable gain (U_2), the latter offsettable by a signal applied to its noninverting input. Q_1 and Q_2 are FETs, used in this application as simple analog switches; Q_3–Q_5 generate suitable levels, from a logic-level input, to activate the switches. Q_1 through Q_5 and their associated circuitry could all be replaced by a relay, or even a switch, if desired. For now, just think of it as a simple SPST switch.

When the logic input is HIGH ("auto-zero"), the switch is closed, and U_3 charges the analog "memory" capacitor (C_1) as necessary to maintain zero output. No attempt is made to follow rapidly changing signals, since in the sort of application for which this was designed the signals are essentially dc, and some averaging is a desirable feature. When the switch is opened, the voltage on the capacitor remains stable, resulting in an output signal proportional to the wanderings of the input thereafter.

There are a few additional features that should be described before going on to explain in detail the principles of precision design as applied here: (a) U_4 participates in a first-order leakage-current compensation scheme, whereby the tendency of C_1 to discharge slowly through its own leakage (100,000M, minimum, corresponding to a time constant of 2 weeks!) is compensated by a small charging current through R_{15} proportional to the voltage across C_1. (b) Instead of a single FET switch, two are used in series in a "guarded leakage-cancellation" arrangement. The small leakage current through Q_2, when switched OFF, flows to ground through R_{23}, keeping all terminals of Q_1 within millivolts of ground. Without any appreciable voltage drops, Q_1 hasn't any appreciable leakage! (See Section 4.15 and Fig. 4.50 for similar circuit tricks.) (c) The offsetting voltage generated at the output of U_3 is attenuated by R_{11}–R_{14}, according to the gain setting. This is done to avoid problems with dynamic range and accuracy in U_3, since drifts or errors in the offset holding circuitry are not amplified by U_2 (more on this later).

7.04 A precision-design error budget

For each category of circuit error and design strategy we will devote a few paragraphs to a general discussion, followed by illustrations from the preceding circuit. Circuit errors can be divided into the categories of (a) errors in the external network components, (b) op-amp (or amplifier) errors associated with the input circuitry, and (c) op-amp errors associated with the output circuitry. Examples of the three are resistor tolerances, input offset voltage, and errors due to finite slew rate, respectively.

Let's start by setting out our error budget. It is based on a desire to keep input errors down to the 10μV level, output drift (from capacitor "droop") below 1mV in 10 minutes, and gain accuracy in the neighborhood of 0.01%. As with any budget, the individual items are arrived at by a process of trade-offs, based on what can be done with available technology. In a sense the budget represents the end result of the design, rather than the starting point. However, it will aid our discussion to have it now.

Error budget (worst-case values)

1. Buffer amplifier (U_1)
Voltage errors referred to input:

Temperature	1.2μV/4°C
Time	1.0μV/month
Power supply	0.3μV/100mV change
Bias current $\times R_S$	2.0μV/1k of R_S
Load-current heating	0.3μV @ full scale (10V)

2. Gain amplifier (U_2)
Voltage errors referred to input:

Temperature	1.2μV/4°C
Time	1.0μV/month
Power supply	0.3μV/100mV change
Bias offset current drift	1.6μV/4°C/1k
Load-current heating	0.3μV @ full scale ($R_L \geq$ 10k)

3. Hold amplifier (U_3)
Voltage errors referred to output:

U_3 offset tempco	60μV/4°C
Power supply	10μV/100mV change
Capacitor droop (see current error budget)	100μV/min
Charge transfer	10μV

Current errors applied to C_1 (needed for preceding voltage error budget):

Capacitor leakage	
Maximum (uncompensated)	(100pA)
Typical (compensated)	10pA
U_3 input current	0.2pA
U_3 & U_4 offset voltage across R_{15}	1.0pA
FET switch OFF leakage	0.5pA
Printed-circuit-board leakage	5.0pA

The various items in the budget will make sense as we discuss the choices faced in this particular design. We will organize by the categories of circuit errors listed earlier: network components, amplifier input errors, and amplifier output errors.

7.05 Component errors

The degrees of precision of reference voltages, current sources, amplifier gains, etc., all depend on the accuracy and stability of the resistors used in the external networks. Even where precision is not involved directly, component accuracy can have significant effects, e.g., in the common-mode rejection of a differential amplifier made from an op-amp (see Section 4.09), where the ratios of two pairs of resistors must be accurately matched. The accuracy and linearity of integrators and ramp generators depend on the properties of the capacitors used, as do the performances of filters, tuned circuits, etc. As you will see shortly, there are places where component accuracy is crucial,and there are other places where the particular component value hardly matters at all.

Components are generally specified with an initial accuracy, as well as the changes in value with time (stability) and temperature. In addition, there are specifications of voltage coefficient (nonlinearity) and bizarre effects such as "memory" and dielectric absorption (for capacitors). Complete specifications will also include the effects of temperature cycling and soldering, shock and vibration, short-term overloads, and moisture, with well-defined conditions of measurement. In general, components of greater initial accuracy will have their other specifications correspondingly better, in order to provide an overall stability comparable with the initial accuracy. However, the overall error due to all other effects combined can exceed the initial accuracy specification. Beware!

As an example, RN55C 1% tolerance metal-film resistors have the following specifications: temperature coefficient (tempco), 50ppm/°C over the range −55°C to +175°C; soldering, temperature, and load cycling, 0.25%; shock and vibration, 0.1%; moisture, 0.5%. By way of comparison, ordinary 5% carbon-composition resistors (Allen-Bradley type CB) have these specifications: tempco, 3.3% over the range 25°C to 85°C; soldering and load cycling, +4%, −6%; shock and vibration, ±2%; moisture, +6%. From these specs it should be obvious why you can't just select (using an accurate digital ohmmeter) carbon resistors that happen to be within 1% of their marked value for use in a precise circuit, but are obliged to use 1% resistors (or better) designed for long-term stability as well as initial accuracy. For the utmost in precision it is necessary to use an ultra-precise metal-film resistor, such as Mepco 5023Z (5ppm/°C and 0.025%), or wire-wound resistors, available with tolerances of 0.01%. See Appendix D for more information on precision resistors.

Nulling amplifier: component errors

In the preceding circuit (Fig 7.1), 0.01% resistors are used in the gain-setting network, R_3–R_9, giving highly predictable gain. As you will see shortly, the value of R_3 is a compromise, with small values reducing offset current error in U_2 but increasing heating and thermal offsets in U_1. Given the value of R_3, the feedback network is forced to take on its complicated form to keep the resistor values below 301k, the maximum value generally available in 1% precision resistors. This trick is discussed in Section 4.19. Note that 1% resistors are used in the offset attenuator network, R_{11}–R_{14}; here accuracy is irrelevant, and metal-film resistors are used only for their good stability.

The largest error term in this circuit, as the error budget shows, is capacitor leakage in the holding capacitor, C_1. Capacitors intended for low-leakage applications give a leakage specification, sometimes as a leakage resistance, sometimes as a time constant (megohm-microfarads). In this circuit C_1 must have a value of at least a few microfarads in order to keep the charging rate from other current error terms small (see budget). In that range of capacitance, polystyrene, polycarbonate, and polysulfone capacitors have the lowest leakage. The unit chosen has a leakage specified as 1,000,000 megohm-microfarads maximum, i.e., a parallel leakage resistance of at least 100,000M. Even so, that's equivalent to a leakage current of 100pA at full output (10V), corresponding to a droop rate of nearly 1mV/min at the output, the largest error term by far. For that reason we have added the leakage-cancellation scheme described earlier. It is fair to assume that the effective leakage can be reduced to 0.1 of the capacitor's worst-case leakage specification (in practice, we can probably do much better). No great stability is required in the cancellation circuit, given the modest demands made of it. As you will see later when we discuss voltage offsets, R_{15} is kept intentionally large so that input voltage offsets in U_3 aren't converted to a significant current error.

While on the subject of errors produced by components external to the amplifiers themselves, it should be pointed out that leakage in FET switches is normally in the range of 1nA, a value completely unacceptable in this circuit. The trick of using a pair of series-connected FETs, with Q_2's leakage resulting in only 1mV across Q_1 (with negligible leakage into U_3's summing junction), is elegant and powerful; it is sometimes used in integrator circuits, as discussed in Section 4.19. We have also used it in a novel peak-detector circuit in Section 4.15. As you will see shortly, U_3 is chosen carefully to keep currents through C_1 down in the picoampere range. The philosophy is the same everywhere: Choose circuit configurations and component types as necessary to meet the error budget. At times this involves hard work and circuit trickery, but at other times it falls easily within standard practice.

One subtle source of error in any circuit using FET switches is *charge transfer* from the controlling gate to the signal-carrying channel: The full-swing transitions at the gate couple capacitively to the drain and source. As we remarked in Chapter 3, the total charge transferred is independent of the transition time and depends only on the gate swing and gate-channel capacitance: $\Delta Q = C_{GC}\Delta V_G$. In this circuit, charge transfer results in a simple voltage error of the auto-zero, because the charge is converted to a voltage in the holding capacitor C_1. It's easy to estimate the error. The 3N156 specifies a C_{rss} (drain-gate capacitance) of 1.3pF maximum, and a C_{iss} (gate-channel capacitance, mostly to the source) of 5pF maximum. The 15 volt gate swing therefore produces a maximum charge transfer of 75pC, corresponding to a voltage step of $\Delta V_C = \Delta Q/C_1 = 7.5\mu$V across the 10$\mu$F capacitor C_1. This is within our error budget; in fact, we may have overestimated the effect, since we included the capacitance to source as well as drain, whereas during a portion of the gate step the channel is cut off, decoupling the source from the drain.

7.06 Amplifier input errors

The deviations of op-amp input characteristics from the ideal that we discussed in Chapter 4 (finite values of input impedance and input current, voltage offset, common-mode rejection ratio, and power-supply rejection ratio, and their drifts with time and temperature) generally constitute serious obstacles to precision circuit design and force trade-offs in circuit configuration, component selection, and the choice of a particular op-amp. The point is best made with examples, as we will do shortly. Note that these errors, or their analogs, exist for amplifiers of discrete design as well.

Input impedance

Let's discuss briefly the error terms just listed. The effect of finite input impedance is to form voltage dividers in combination with the source impedance driving the amplifier, reducing the gain from the calculated value. Most often this isn't a problem, because the input impedance is bootstrapped by feedback, raising its value enormously. As an example, the OP-77E precision op-amp (with transistor, not FET, input stage) has a typical "differential-mode input impedance," of 45M. In a circuit with plenty of loop gain, feedback raises the input impedance to the "common-mode input impedance" 200,000M. In any case, some FET-input op-amps have astronomical values of R_{in}, if there's still a problem.

Input bias current

More serious is the input bias current. Here we're talking about currents measured in nanoamps, and this already produces voltage errors of microvolts for source impedances as small as 1k. Again, FET op-amps come to the rescue, but with generally increased voltage offsets as part of the bargain. Bipolar superbeta op-amps such as the LT1012, 312, and LM11 can also have surprisingly low input currents. As an example, compare the OP-77 precision bipolar op-amp with the LT1012 (bipolar, optimized for low bias current), the OPA111 (JFET, precision and low bias), the AD549 (ultra-low-bias JFET), and the ICH8500 (MOSFET, lowest-bias op-amp); these are the best you can get at the time of writing, and we've chosen the best grade of each one:

	Bias current @ 25°C I_B max	Offset voltage @ 25°C V_{os} max	Tempco of V_{os} ΔV_{os} max
OP-77E (bipolar)	2000pA	25μV	0.3μV/°C
LT1012C (superbeta)	150pA	50μV	1.5μV/°C
OPA111B (JFET)	1pA	250μV	1μV/°C
AD549L (JFET)	0.06pA	500μV	10μV/°C
ICH8500A (MOSFET)	0.01pA	50,000μV	2000μV/°C

Well-designed FET amplifiers have extremely low bias current, but with much larger offset voltage, as compared with the precision OP-77. Since the offset voltage can always be trimmed, what matters more is the drift with temperature. In this case the FET amplifiers are 3 to 6000 times worse. The op-amp with the lowest input current uses MOSFETs for the input stage. MOSFET op-amps are becoming popular because of the availability of inexpensive units like the 3440, 3160, the TLC270 series, and the ICL7610 series, as well as the ultra-low-bias-current devices like the 8500A listed earlier. However, unlike JFETs or bipolar transistors, MOSFETs can have very large drifts of offset voltage with time, an effect that will be discussed shortly. So the improvement in current errors you buy with a FET op-amp can be wiped out by the larger voltage error terms. With any circuit in which bias current can contribute significant error, it is always wise to ensure that both op-amp input terminals see the same dc source resistance, as we discussed in Section 4.12; then the op-amp's *offset current* becomes the relevant specification. A note on bias-current compensation: A number of precision op-amps use a "bias compensation" scheme to cancel (approximately) the input current, in order to make that error

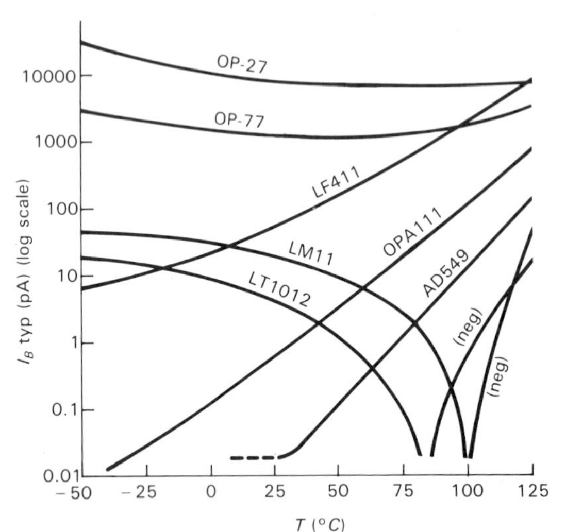

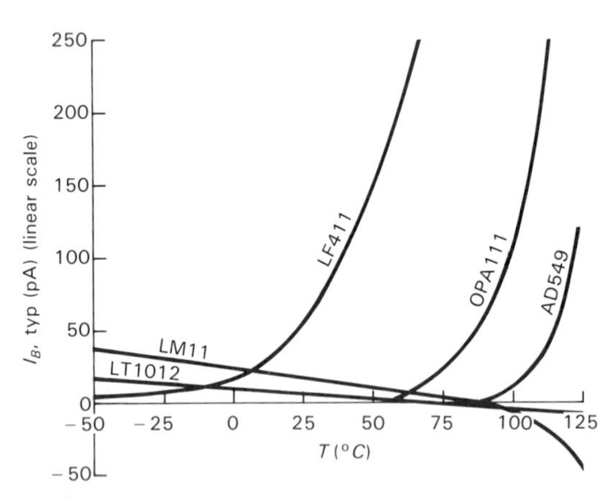

Figure 7.2. Op-amp input current versus temperature.
A. Logarithmic scale
B. Linear scale

term smaller; look back at Additional Exercise 8 at the end of Chapter 2 to see how it's done. With op-amps of this type you generally don't gain anything by matching the dc resistances seen by the two inputs, since for a bias-compensated op-amp the residual bias current and the offset current are comparable.

One additional point to keep in mind when using FET-input op-amps is that the input 'bias" current is actually gate leakage current, and it rises dramatically with increasing temperature (it roughly doubles for every 10°C increase in chip temperature; see Fig. 3.30). Since FET op-amps often run warm (the popular 356 dissipates 150mW quiescent power), the actual input current may be considerably higher than the 25°C figures you see on the data sheet. The input current of a bipolar-transistor-input op-amp, by comparison, is actual base current, and it *drops* with rising temperature (Fig. 7.2). So a FET-input op-amp with impressive input-current specs on paper may not give such an improvement over a good superbeta bipolar unit. As an example, the OPA111 with its 1pA input current (at 25°C) will have an input current of about 10pA at 65°C chip temperature, which is higher than the input current of the superbeta LT1012 at the same temperature. The popular 355 series of FET op-amps has an input current that is comparable to that of the LT1012 or LM11 at 25°C and is many times higher at elevated temperatures. Finally, when comparing op-amp input currents, watch out for some FET types whose I_B depends on the input voltage. The spec sheet usually lists I_B only at 0 volts (mid-supply), but a good data sheet will show curves as well. See Figure 7.3 for some typical I_B–V_{in} behavior. Note the excellent performance of the OPA111, due in part to its cascode input stage.

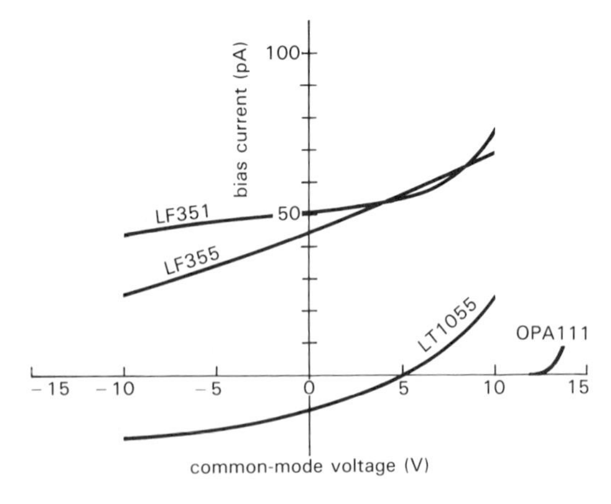

Figure 7.3. FET op-amp input current versus common-mode voltage.

Voltage offset

Voltage offsets at the amplifier input are obvious sources of error. Op-amps differ widely in this parameter, ranging from "precision" op-amps offering worst-case V_{os} values generally in the tens of microvolts to ordinary jellybean op-amps like the LF411 with V_{os} values of 2 to 5mV. At the time of writing, the champion in the (non-chopper, see below) world of low offsets is the MAX400M ($V_{os} = 10\mu$V, max). We expect to see further incremental improvements in this area.

Although most good single op-amps (but not duals or quads) have offset-adjustment terminals, it is still wise to choose an amplifier with inherently low initial offset V_{os} max, for several reasons. First, op-amps designed for low initial offset tend to have correspondingly low offset drift with temperature and time. Second, a sufficiently precise op-amp eliminates the need for external trimming components (a trimmer takes up space, needs to be adjusted initially, and may change with time). Third, offset voltage drift and common-mode rejection are degraded by the unbalance caused by an offset-adjustment trimmer.

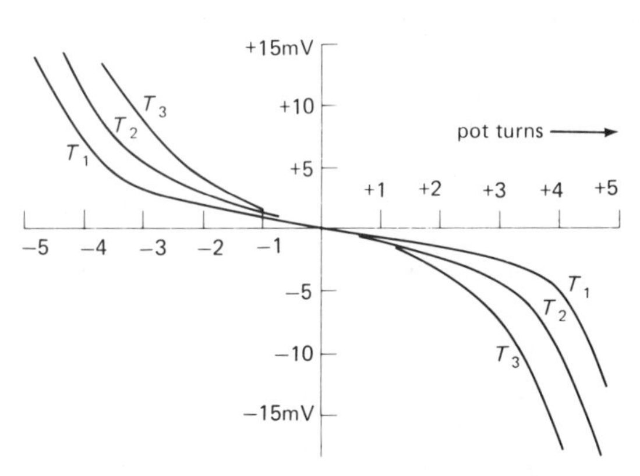

Figure 7.4. Typical op-amp offset versus offset-adjustment potentiometer rotation for several temperatures.

Figure 7.4 illustrates how a trimmed offset has larger drifts with temperature. We have also shown how the offset adjustment is spread over the trimmer pot rotation, with best resolution near the center, especially for large values of trimmer resistance. Finally, you'll generally find that the recommended external trimming network provides far too much range, making it nearly impossible to trim V_{os} down to a few microvolts; even if you succeed, the adjustment is so critical it won't stay trimmed for long. Another way to think about it is to realize that the manufacturer of a precision op-amp has *already* trimmed the offset voltage, in a custom test jig using "laser-zapping" techniques; you may be unable to do any better yourself. Our advice is (a) to use precision op-amps for precision circuits, and (b) if you must trim them further, arrange a narrow-range trim circuit like the ones shown in Figure 7.5, which have a full-scale range of $\pm 50\mu$V, linear in trimmer rotation.

Because voltage offsets can be trimmed to zero, what ultimately matters is the drift of offset voltage with time, temperature, and power-supply voltage. Designers of precision op-amps work hard to minimize these errors. You get the best performance from bipolar (as opposed to FET) op-amps in this regard, but input current effects may then dominate the error budget. The best op-amps keep drifts below 1μV/°C; at the time of writing, the AD707 claims the smallest drift (for a non-chopper op-amp) – $\Delta V_{os} = 0.1\mu$V/°C, max.

Another factor to keep in mind is the drift caused by self-heating of the op-amp when it drives a low-impedance load. It is often necessary to keep the load impedance above 10k to prevent large errors from this effect. As usual, that may compromise the next stage's error budget from the effects of bias current! You will see just such a problem in this design example. For applications where drifts of a few microvolts are important, the related effects of thermal gradients (from nearby heat-producing components) and thermal emf's

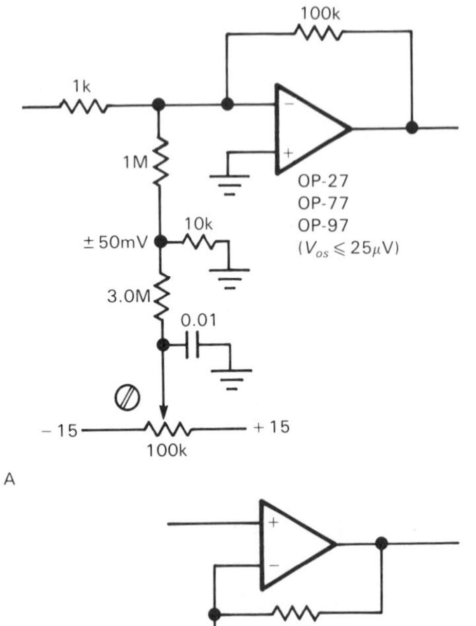

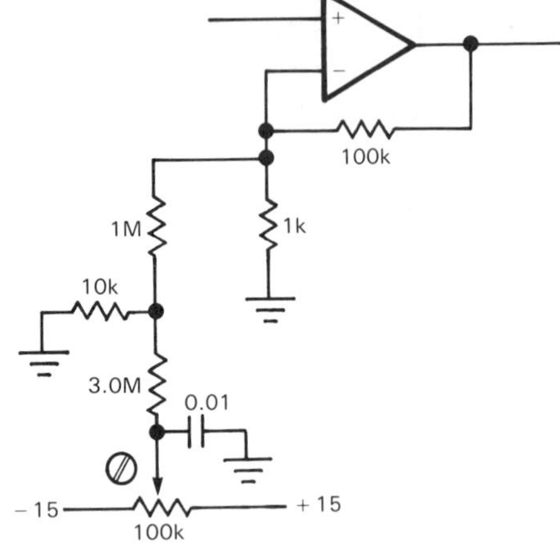

Figure 7.5. External trimming networks for precision op-amps.
A. Inverting.
B. Noninverting.

(from junctions of dissimilar metals) become important. This will come up again when we discuss the ultraprecise *chopper-stabilized* amplifier in Section 7.08.

Table 7.1 compares the important specifications for seven of our favorite precision op-amps. Spend some time with it – it will give you a good feeling for the trade-offs you face in high-performance design with op-amps. Note particularly the trade-offs of offset voltage (and drift) versus input current for the best bipolar and JFET op-amps. You also get the lowest noise voltage from bipolar op-amps, dropping with increasing bias current; we'll see why that happens later in the chapter, when we discuss noise. The awards for low-noise *current*, however, always go to the FET op-amps, again for reasons that will become clear later. In general, choose FET op-amps for low input current and current noise; choose bipolar op-amps for low input voltage offset, drift, and voltage noise.

Among FET-input op-amps, those using JFETs dominate the scene, particularly where precision is needed. MOSFETs, in particular, are subject to a unique debilitating effect that neither FETs nor bipolar transistors have. It turns out that sodium-ion impurities in the gate insulating layer migrate slowly under the influence of the gate's $V_{GS(\mathrm{ON})}$ electric field, resulting in a drift of the offset voltage under closed-loop conditions of as much as 0.5mV over a period of years. The effect is increased for elevated temperatures and for a large applied differential input signal. For example, the data sheet for the CA3420 MOSFET-input op-amp shows a typical 5mV change of V_{os} over 3000 hours of operation at 125°C with 2 volts across the input. This sodium-ion disease can be cured by introducing phosphorus into the gate region. Texas Instruments, for example, uses a phosphorus-doped polysilicon gate in its "LinCMOS" series of op-amps (TLC270 series) and comparators (TLC339 and TLC370 series). These popular inexpensive parts come in a variety of packages and speed/power selections and maintain respectable offset voltages with time (50μV eventual offset drift per volt of differential input).

There is an important exception to the generalization that FET op-amps, particularly MOSFET types, suffer from larger initial offsets and much larger drifts of V_{os} with temperature and time than do bipolar-transistor op-amps. That exception is the so-called auto-zero (or "chopper-stabilized") amplifier, which uses MOSFET analog switches and amplifiers

TABLE 7.1. SEVEN PRECISION OP-AMPS

Parameter	Symbol	bipolar OP-77E typ	min/max	low-noise OP-27E typ	min/max	low-bias LT1012C typ	min/max	micropower OP-90E typ	min/max	fast JFET LT1055A typ	min/max	low-bias JFET OPA111B typ	min/max	chopper MAX430C typ	min/max	units
($V_s = \pm15V$; $T_A = 25^\circ C$)																
Offset voltage	V_{os}	10	25	10	25	10	50	50	150	50	150	50	250	1	5	μV
Offset voltage drift	ΔV_{os}	0.2	-	0.2	1	0.3	-	0.3	2	-	-	-	-	0.1	-	μV/month
Bias current	I_b	1.2	2	10	40	0.03	0.15	4	15	0.01	0.05	0.0005	0.001	0.02	0.2	nA
Offset current	I_{os}	0.1	1.5	7	35	0.02	0.15	0.4	3	0.002	0.01	0.0003	0.0008	0.01	0.1	nA
Input resistance – diff'l	R_{in}	45	25	6	1.3	-	-	30	-	10^6	-	10^7	-	10^6	-	MΩ
Input resistance – common mode	R_{inCM}	200	-	3	-	-	-	20	-	10^3	-	10^5	-	10^3	-	GΩ
Input noise voltage (0.1-10Hz)	e_n pp	0.4	0.6	0.1	0.2	0.5	-	3	-	1.8	-	1.2	2.5	1.1	-	μV, pp
Input noise voltage density (10Hz)	e_n	10	18	3.5	5.5	17	30	60	-	28	50	30	60	-	-	nV/√Hz
Input noise voltage density (1kHz)	e_n	10	11	3	4	14	22	60	-	14	20	7	12	-	-	nV/√Hz
Input noise current density (10Hz)	i_n	0.3	0.8	1.7	4	0.02	-	1.5	-	0.002	0.004	0.0004	-	0.01	-	pA/√Hz
Input noise current density (1kHz)	i_n	0.1	0.2	0.4	0.6	0.006	-	0.7	-	0.002	0.004	0.0004	-	-	-	pA/√Hz
Large signal voltage gain	A_{vo}	12	5	1.8	1	0.2	2	1.2	0.7	0.4	0.015	2	1	30	1	V/μV
Common-mode rejection ratio	CMRR	140	120	126	114	132	110	130	100	100	86	110	100	140	120	dB
Power supply rejection ratio	PSRR	120	110	120	100	132	110	120	105	106	90	110	100	140	120	dB
Slew rate	SR	0.3	0.1	2.8	1.7	0.2	0.1	0.01	0.005	13	10	2	1	0.5	-	V/μs
Gain-bandwidth product	GBW	0.6	0.2	8	5	1	-	0.02	-	5	-	2	-	0.5	-	MHz
Supply current	I_s	1.7	2	3	4.7	0.4	0.6	0.014	0.02	2.8	4	2.5	3.5	1.3	2	mA
(Over temperature range)	T	-25 to +85		-25 to +85		0 to +70		-25 to +85		0 to +70		-25 to +85		0 to +70		°C
Offset voltage	V_{os}	10	45	20	50	20	120	70	270	100	330	100	500	-	-	μV
Offset voltage tempco	TCV_{os}	0.1	0.3	0.2	0.6	0.2	1.5	0.3	2	1.2	4	0.5	1	0.02	0.05	μV/°C
Bias current	I_b	2.4	4	14	60	0.035	0.23	4	15	0.03	0.15	0.03	0.13	0.05	-	nA
Offset current	I_{os}	0.1	2.2	10	50	0.02	0.23	0.8	3	0.01	0.05	0.02	0.1	0.04	-	nA

to sense, and correct, the residual offset error of an ordinary op-amp (which itself is often built with MOSFETs, on the same chip). Chopper-stabilized op-amps deliver lower voltage offsets and drifts than even the best precision bipolar op-amps – 5μV (max), 0.05μV/°C (max) – but not without cost. They have some unpleasant characteristics that make them unsuited for many applications. We will discuss them in detail in Section 7.08.

Common-mode rejection

Insufficient common-mode rejection ratio (CMRR) degrades circuit precision by effectively introducing a voltage offset as a function of dc level at the input. This effect is usually negligible, since it is equivalent to a small gain change, and in any case it can be overcome by choice of configuration: An inverting amplifier is insensitive to op-amp CMRR, in contrast with a noninverting amplifier. However, in "instrumentation amplifier" applications you are looking at a small differential signal riding on a large dc offset, and a high CMRR is essential. In such cases you have to be careful about circuit configurations and, in addition, choose an op-amp with a high CMRR specification. Once again, a superior op-amp like the OP-77 can solve your problems, with a CMRR (min) of 120dB, compared with the 411's meager specification of 70dB. We will discuss high-gain differential and instrumentation amplifiers shortly.

Power-supply rejection

Changes in power-supply voltage cause small op-amp errors. As with most op-amp specifications, the power-supply rejection ration (PSRR) is referred to a signal at the *input*. For example, the OP-77 has a specified PSRR of 110dB at dc, meaning that a 0.3 volt change in one of the power-supply voltages causes a change at the output equivalent to a change in differential input signal of 1μV.

The PSRR drops drastically with increasing frequency, and a graph documenting this scurrilous behavior is often given on the data sheet. For example, the PSRR of our favorite OP-77 begins dropping at 0.3Hz and is down to 83dB at 60Hz and 42dB at 10kHz. This actually doesn't present much of a problem, since power-supply noise is also decreasing at higher frequencies if you have used good bypassing. However, 120Hz ripple could present a problem if an unregulated supply is used.

It is worth noting that the PSRR will not, in general, be the same for the positive and negative supplies. Thus, the use of dual-tracking regulators (Section 6.19) doesn't necessarily bring any benefits.

Nulling amplifier: input errors

The amplifier circuit in Figure 7.1 begins with a follower, to keep a high input impedance. It is tempting to consider a FET type, but the poor V_{os} specification more than offsets the advantage of low input current, except with sources of very high impedance. The OP-77's 2nA bias current gives an error of 2μV/1k source impedance; a JFET LT1055A, although giving negligible current error, would give voltage offset drifts of 16μV/4°C (4°C is considered a typical laboratory ambient temperature range). The input follower is provided with offset trimming, since the initial 25μV spec is too large. As mentioned earlier, feedback bootstraps the input impedance to 200,000M and eliminates any errors from finite source impedance, up to 20M (for gain error less than 0.01%). D_1 and D_2 are included for input overvoltage protection and are low-leakage types (less than 1nA).

U_1 drives an inverting amplifier (U_2), with R_3 being a compromise between heat-produced thermal offsets in U_1 and bias-current offset errors in U_2. The value chosen keeps heating down to 5.6mW (at 7.5V output, the worst case), which works out to a temperature rise of 0.8°C (the op-amp has a thermal resistance of about 0.14°C/mW, see Section 6.04), with a consequent voltage offset of 0.3μV. The resultant 10k source impedance seen by U_2 results in an error due to bias-current offset, but since U_2 is inside a feedback loop with U_3 trimming the overall offset to zero, all that matters is the drift in the current error term. The OP-77 has a specification for bias offset drift with temperature (not often specified by manufacturers, incidentally), from which the error result of 1.6μV/4°C in the error budget is calculated. Reducing the value of R_3 would improve this term, at the expense of the heating term in U_1.

As explained in the overall circuit description earlier, the value of R_3 forces the bizarre feedback T network in order to keep the feedback resistor values in the range where precision wire-wound resistors can be manufactured. Using the ordinary inverting amplifier configuration, for example, you would need resistors of 100.0k, 1.0M, and 10.0M for gains of 10, 100, and 1000, respectively.

The dc input impedance of U_2 comes closer to presenting a problem. At a gain of 1000 its differential input impedance of 25M is bootstrapped by a factor of $A_{vol}/1000$ to 125,000M. Fortunately this exceeds the 9.4k impedance of the gain-setting network by a factor of more than a million, contributing much less than 0.01% error. This is one of the toughest examples we could think of, and even so the op-amp input impedance presents no problem, thus demonstrating that, in general, you can ignore the effects of op-amp input impedances.

Drifts in offset voltage in both U_1 and U_2 over time, temperature, and power-supply variations affect the final error equally and are tabulated in the budget. It is worth pointing out that they are all automatically cancelled at each "zeroing" cycle, and only short-term drifts matter anyway. These errors are all in the microvolt range, thanks to a good op-amp. U_3 has larger drifts, but it must be a FET type to keep capacitor current small, as already explained. Since U_3's output is attenuated according to the gain selected, its error, *referred to the input*, is reduced at high gain. This is an important point, since high gains are used with small input signal levels where high accuracy is needed. U_3's errors are always the same *at the output*, and they are therefore specified as output errors in the error budget.

Note the general philosophy of design that emerges from this example: You work at the problem areas, choosing configurations and components as necessary to reduce errors to acceptable values. Trade-offs and compromises are involved, with some choices depending on external factors (e.g., the use of a FET-input follower for U_1 would be preferable for source impedances greater than about 50k).

Table 7.2 compares the specifications of op-amps you might choose for precision circuit design.

7.07 Amplifier output errors

As we discussed in Chapter 4, op-amps have some serious limitations associated with the output stage. Limited slew rate, output crossover distortion (see Section 2.15), and finite open-loop output impedance can all cause trouble, and they can cause precision circuits to display astoundingly large errors if not taken into account.

Slew rate: general considerations

As we mentioned in Section 4.11, an op-amp can swing its output voltage only at some maximum rate. This effect originates in the frequency-compensation circuitry of the op-amp, as we will explain in a bit more detail shortly. One consequence

TABLE 7.2. PRECISION OP-AMPS

		Voltage					Current												Settling time, typ		
		Offset		Drift			Bias		Offset		e_n, typ		i_n, typ				Slew				
Type	Mfg[a]	typ (μV)	max (μV)	typ (μV/°C)	max (μV/°C)	typ (μV/mo)	typ (nA)	max (nA)	typ (nA)	max (nA)	@10Hz (nV/√Hz)	@1kHz (nV/√Hz)	@10Hz (fA/√Hz)	@1kHz (fA/√Hz)	PSRR min (dB)	Gain min (x1000)	rate typ (V/μs)	f_T typ (MHz)	0.1% (μs)	0.01% (μs)	Comments
BIPOLAR																					
OP-07A	PM+	10	25	0.2	0.6	0.2	0.7	2	0.3	2	10.3	9.6	320	120	100	300	0.17	0.6	-	-	classic prec op-amp
OP-08E	PM	70	150	0.5	2.5	-	0.8	2	0.05	0.2	22	20	150	130	104	80	0.12	0.8	-	-	improved 308
LM11	NS+	100	300	1	3	1	0.025	0.05	0.5pA	0.01	180	150	15	4	100	100	0.3	0.5	-	70	lowest bias bipolar
OP-12E	PM+	70	150	0.5	2.5	-	0.8	2	0.05	0.2	22	20	150	130	104	80	0.12	0.8	-	-	improved 312
OP-20B	PM	60	250	0.75	1.5	-	12	25	0.15	1.5	58	58	140	90	100	500	0.05	0.1	-	-	micropower
OP-21A	PM	40	100	0.5	1	-	50	100	0.6	4	21	21	380	210	104	1000	0.25	0.6	-	-	low power
OP-27E	PM+	10	25	0.2	0.6	0.5	10	40	7	35	3.5	3.0	1700	400	100	1000	2.8	8	-	-	low noise
OP-37E	PM+	10	25	0.2	0.6	0.5	10	40	7	35	3.5	3.0	1700	400	100	1000	17	63	-	-	low noise, decomp OP-27 (G>5)
OP-50E	PM	10	25	0.15	0.3	-	1	5	0.1	1	5.5	4.5	300	230	126	10M	3	25[b]	-	30	hi curr, low noise, decomp (G>5)
OP-62E	PM	-	200	-	-	-	-	300	-	100	-	2.5	-	-	105	350	15	50	-	-	
OP-77E	PM	10	25	0.1	0.3	0.3	1.2	2	0.1	1.5	10.3	9.6	320	120	110	5000	0.3	0.6	-	-	improved OP-07
OP-90E	PM	50	150	0.3	2	-	4	15	0.3	3	60	60	1600	700	104	700	0.01	0.02	-	-	micropower
OP-97E	PM	10	25	0.2	0.6	0.3	0.03	0.1	0.03	0.1	17	14	20	6	114	300	0.2	0.9	-	-	low power OP-77
MAX400M	MA	4	10	0.2	0.3	0.2	0.7	2	0.3	2	10.3	9.6	320	120	100	500	0.3	0.6	-	-	lowest non-chopper V_{os}
LM607A	NS	15	25	0.2	0.3	0.2	1	2	0.5	2	9	6.5	320	120	100	5000	0.7	1.8	-	-	
AD707C	AD	5	15	0.03	0.1	0.2	0.5	1	0.1	1	10.3	9.6	320	120	120	8000	0.3	0.9	-	-	improved OP-07; dual=708
LT1001A	LT	10	25	0.2	0.6	0.2	1	4	0.8	4	10.3	9.6	320	120	110	450	0.25	0.8	-	-	
LT1006A	LT	20	50	0.2	1.3	0.4	9	15	0.12	0.5	23	22	70	30	106	1000	0.4		-	-	single supply; optional $I_s = 90\mu A$
LT1007A	LT	10	25	0.2	0.6	0.2	10	35	7	30	2.8	2.5	1500	400	110	7000	2.5	8	-	-	low noise, ~OP-27
LT1012C	LT+	10	50	0.2	1.5	0.3	0.03	0.15	0.02	0.15	17	14	20	6	110	200	0.2	0.8	-	-	low bias, imprvd 312; PM1012[f]
LT1013A	LT	40	150	0.4	2	0.4	15	35	0.2	1.3	24	22	70	15	103	1500	0.4	0.8	-	-	improved 358/324; sing supply[g]
LT1028A	LT	10	40	0.2	0.8	0.3	25	90	12	50	1	0.9	4700	1000	117	7000	15	75	-	-	ultra low noise
LT1037A	LT	10	25	0.2	0.6	0.2	10	35	7	30	2.8	2.5	1500	400	110	7000	15	60	-	-	decomp 1007 (G>5), ~OP-37
RC4077A	RA	4	10	0.1	0.3	0.2	0.3	2	0.1	1.5	10.3	9.6	320	120	110	2500	0.25	0.8	-	-	lowest non-chopper V_{os}
HA5134	HA	25	250	-	5	-			-	25	-	7	-	2000	-	250		4	-	-	quad, low noise
HA5135	HA	10	80	0.4	1.3	-	1	4	-	4	13	9	400	140	94	1000	0.8	2.5	11	13	
HA5147A	HA	10	25	0.2	0.6	-	10	40	7	35	3.5	3.0	1700	400	80	1000	35	140	0.4	-	low noise, fast, uncomp (G>10)

JFET																					
OPA101B	BB	50	250	3	5	-	6pA	0.01	1.5pA	4pA	25	8	1.4	1.4	86	60	7	20	2.5	10	low noise; decomp = OPA102
OPA111B	BB	50	250	0.5	1	-	0.5pA	1pA	0.3pA	0.7pA	30	7	0.4	0.4	100	1000	2	2	6	10	low noise, low bias
LF*nnn*	NS	1000	–	–	–	-	0.05	0.01	0.01	0.05	14	3.5	10	10	80	100	20	10	-	-	lowest noise JFET, no popcorn
LF455A	NS	75	180	3	4	-	7pA	0.05	3pA	0.02	100	12	10	10	86	200	5	3	-	4	456 and 457 faster
AD547L	AD	-	250	-	1	-	0.01	0.025	2pA	-	70	30	-	-	80	250	3	1	3.5	4.5	dual = AD642, 647
AD548C	AD	100	250	-	2	15	3pA	0.01	2pA	5pA	80	30	-	2	86	300	1.8	1	6	7	improved LF441; dual = AD648
AD711C	AD	100	250	2	3	15	0.015	0.025	5pA	0.01	45	18	-	10	86	200	20	4	0.9	1	improved LF411/2
LT1055A	LT	50	150	1.2	4	5	0.01	0.05	2pA	0.01	28	14	2	2	90	150	13	5	1.2	1.8	LT1056 is 20% faster
HA5170	HA	100	300	2	5	-	0.02	0.1	3pA	0.06	20	10	50	10	74	300	8	8	1	1.1	low noise
JFET, HIGH-SPEED																					
OP-44E	PM	30	750	4	10	5	0.08	0.2	4pA	0.04	38	13	-	7	90	500	120[h]	16[h]	0.2	-	decomp (G>3)
LF401A	NS	-	200	-	-	-	-	0.2		0.1	60	23	-	10	80	100	30	16	0.2	0.3	fast settle
OPA404B	BB	260	750	3	-	-	1pA	4pA	0.5pA	4pA	32	15	0.6	0.6	86	40	35	6.4	0.6	1.5	quad
OPA602C	BB	100	250	1	2	-	0.5pA	1pA	0.5pA	1pA	23	13	0.6	0.6	86	40	35	6.5	0.7	1	low bias, fast settle
OPA605K	BB	250	500	-	5	-	0.01	0.035	2pA	-	80	20	-	-	74	-	94	20	0.3	0.4	uncomp (G>50)
OPA606L	BB	100	500	3	5	-	5pA	0.01	0.4pA	5pA	30	13	1.3	1.3	90	100	35	13	1	2.1	improved LF356
AD744C	AD	100	250	2	3	15	0.03	0.05	0.01	0.02	45	18	-	10	92	250	75	13	0.4	0.5	low dist (3ppm); decomp (G>2)
AD845K	AD	100	250	1.5	3	-	0.25	0.4	0.015	0.05	80	25	-	100	98	200	100	16	0.3	0.3	fast settle
LT1022A	LT	80	250	1.3	5	-	0.01	0.05	2pA	0.01	28	14	2	2	88	150	26	8.5	0.8	1.8	
CHOPPER STABILIZED[e]																					
MAX420E	MA	1	5	0.02	0.05	0.1	0.01	0.03	0.015	0.06	1.1[c]	-	10	-	120	1000	0.5	0.5	-	-	±15V V_s; 430 has C_{int}
MAX422E	MA	1	5	0.02	0.05	0.1	0.01	0.03	0.015	0.06	1.1[c]	-	10	-	120	1000	0.13	0.13	-	-	±15V V_s; 432 has C_{int}
LMC668A	NS	1	5	0.05	-	0.1	0.02	0.06	-	-	2[c]	-	10	-	120	1000	2.5	1	-	-	
TSC900A	TS	-	5	0.02	0.05	-	-	0.05	0.5pA	-	4[c]	-	-	-	120	1000	0.2	0.7	-	-	low power
TSC901	TS	7	15	0.05	0.15	-	0.03	0.05	0.05	0.1	5[c]	-	-	-	120	1000	2	0.8	-	-	±15V supply; int caps
TSC911A	TS	5	15	0.05	0.15	-	-	0.07	5pA	0.02	11[c]	-	-	-	112	600	2.5	1.5	-	-	int caps, noisy
TSC915	TS		10	0.01	0.1	-	0.03	0.1	0.05	0.1	0.8[c]	-	-	-	120	1000	0.5	0.5	-	-	±15V supply
TSC918	TS	-	50	0.4	0.8	-	-	0.1	0.5pA	-	4[c]	-	-	-	105	100	0.2	0.7	-	-	inexpensive
LTC1050	LT	0.5	5	0.01	0.05	0.05[d]	0.01	0.03	0.02	0.06	1.6[c]	-	2.2	-	125	300	4	2.5	-	-	int caps
LTC1052	LT	0.5	5	0.01	0.05	0.1[d]	1pA	0.03	5pA	0.03	1.5[c]	-	0.6	-	120	1000	4	1.2	-	-	improved 7652
ICL7650S	IL+	0.7	5	0.02	0.1	0.1	4pA	0.01	8pA	0.02	2[c]	-	10	-	120	6000	2.5	2	-	-	improved 7650
ICL7652S	IL+	0.7	5	0.01	0.06	0.2	3pA	0.03	0.015	0.04	0.7[c]	-	10	-	120	6000	1	0.5	-	-	improved 7652
TSC76HV52	TS	-	10	-	0.3	-	0.03	0.1	0.05	0.1	0.8[c]	-	-	-	120	1000	0.5	0.5	-	-	±15V 7652

(a) see footnotes to Table 4.1. (b) at G=50. (c) μV pp, 0.1-10Hz. (d) μV per square root month. (e) total supply=18V unless noted. (f) dual=1024. (g) quad=1014.

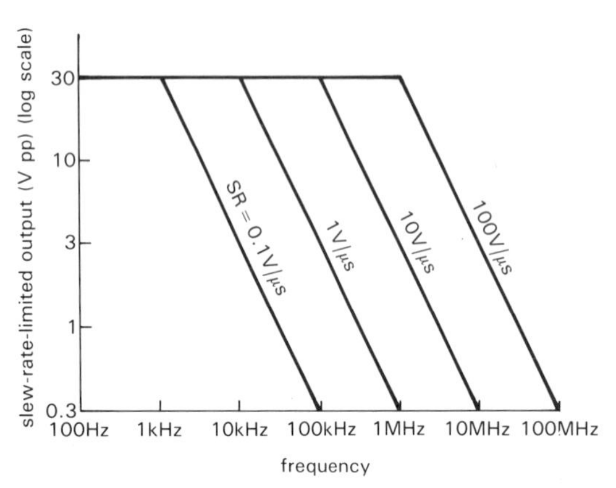

Figure 7.6. Maximum output swing versus frequency.

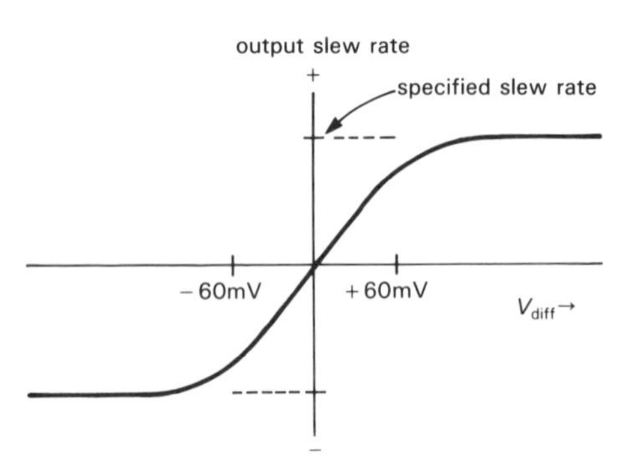

Figure 7.7. A substantial differential input voltage is required to produce full op-amp slew rate.

of finite slew rate is to limit the output swing at high frequencies, as we showed in Section 4.12 and illustrated in Figure 7.6:

$V_{\text{pp}} = S/\pi f$

A second consequence is best explained with the help of a graph of slew rate versus differential input signal (Fig. 7.7). The point to be made here is that a circuit that demands a substantial slew rate operates with a substantial voltage error across the op-amp's input terminals. This can be disastrous for a circuit that pretends to be highly precise.

Let's look at the innards of an op-amp in order to get some understanding of the origin of slew rate. The vast majority of op-amps can be summarized with the circuit

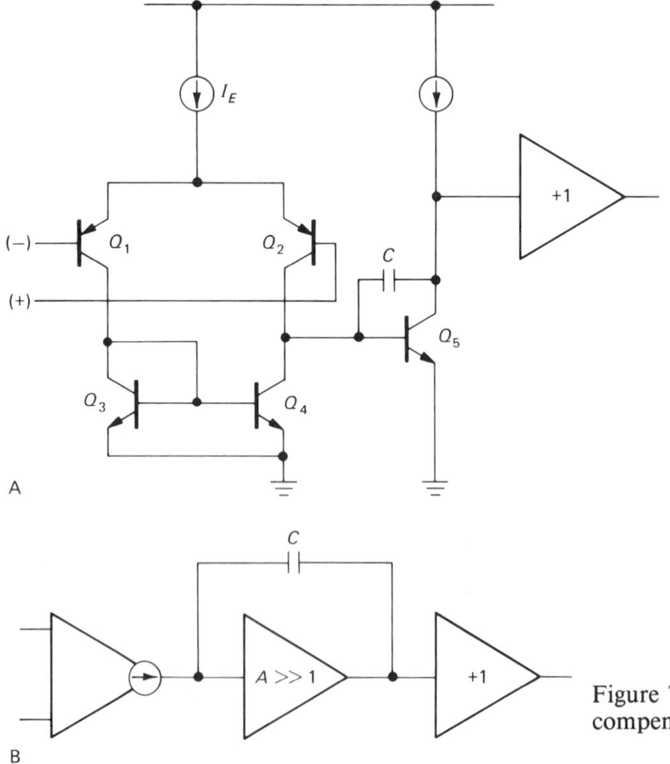

Figure 7.8. Typical op-amp internal compensation scheme.

shown in Figure 7.8. A differential input stage, loaded with a current mirror, drives a stage of large voltage gain with a compensation capacitor from output to input.The output stage is a unity-gain push-pull follower. The compensation capacitor C is chosen to bring the open-loop gain of the amplifier to unity before the phase shifts caused by the other amplifier stages have become significant. That is, C is chosen to put f_T, the unity-gain bandwidth, near the frequency of the next amplifier rolloff pole, as described in Section 4.34. The input stage has very high output impedance, and it looks like a current source to the next stage.

The op-amp is slew-rate-limited when the input signal drives one of the differential-stage transistors nearly to cutoff, driving the second stage with the total emitter current I_E of the differential pair. This occurs for a differential input voltage of about 60mV, at which point the ratio of currents in the differential stage is 10:1. At this point Q_5 is slewing its collector as rapidly as possible, with all of I_E going into charging C. Q_5 and C thus form an integrator, with a slew-rate-limited ramp as output. Read the accompanying section "Slew rate: a detailed look" to see how to derive an expression for the slew rate, knowing how bipolar transistors work.

☐ SLEW RATE: A DETAILED LOOK

First, let us write an expression for the open-loop small-signal ac voltage gain, ignoring phase shifts:

$$A_V = g_m X_C = g_m/2\pi f C$$

from which the unity-gain bandwidth product (the frequency at which $A_V = 1$) is

$$f_T = \frac{1}{2\pi}\frac{g_m}{C}$$

Now, the slew rate is determined by a current I_E charging a capacitance C:

$$S = \frac{dV}{dt} = \frac{I_E}{C}$$

For the usual case of a differential amplifier with no emitter resistors, g_m is related to I_E by

$$g_m = \frac{1}{r_e} = \frac{I_E}{2V_T} = \frac{I_E}{50\text{mV}}$$

By substituting this into the slew-rate formula, we find

$$S = 2V_T\frac{g_m}{C}$$

i.e., the slew rate is proportional to g_m/C, just the same as the unity-gain bandwidth! In fact,

$$S = 4\pi V_T f_T = 0.3 f_T$$

with f_T in MHz and S in V/μs. This is independent of the particular values of C, g_m, I_E, etc., and it gives a good estimate of slew rate (e.g., the classic 741, with $f_T \approx$ 1.5MHz, has a slew rate of 0.5V/μs). It shows that an op-amp with greater gain-bandwidth product f_T will have a higher slew rate. You can't improve matters in a slow op-amp by merely increasing input-stage current I_E, because the increased gain (from increased g_m) then requires a correspondingly increased value of C for compensation. Adding gain anywhere else in the op-amp doesn't help either.

The preceding result shows that increasing f_T (by raising collector currents, using faster transistors, etc.) will increase the slew rate. A high f_T is, of course, always desirable, a fact not lost on the IC designer, who has already done the best he can with what's on the chip. However, there is a way to get around the restriction that $S = 0.3f_T$. That result depended on the fact that the transconductance was determined by I_E (through $g_m = I_E/2V_T$). You can use simple tricks to raise I_E (and therefore the slew rate) while keeping f_T (and therefore compensation) fixed. The easiest is to add some emitter resistance to the input differential amplifier. Let's imagine we do something like that, causing I_E to increase by a factor m while holding g_m constant. Then, by going through the preceding derivation, you would find

$$S = 0.3mf_T$$

EXERCISE 7.1

Prove that such a trick does what we claim.

☐ *Increasing slew rate*

Here, then, are some ways to obtain a high slew rate: (a) Use an op-amp with high f_T. (b) Increase f_T by using a smaller compensation capacitor; of course, this is possible only in applications where the closed-loop gain is greater than unity. (c) Reduce the input-stage transconductance g_m by adding emitter resistors; then reduce C or raise I_E proportionately. (d) Use a different input-stage circuit.

The third technique (reduced g_m) is used in many op-amps. As an example, the HA2605 and HA2505 op-amps are nearly identical, except for the inclusion of emitter resistors in the input stage of the HA2505. The emitter resistors increase the slew rate, at the expense of open-loop gain. The following data demonstrate this trade-off. FET op-amps, with their lower input-stage g_m, tend to have higher slew rates for the same reason.

	HA2605	HA2505
f_T	12MHz	12MHz
Slew rate	7V/μs	30V/μs
Open-loop gain	150,000	25,000

The fourth technique generally uses the method of "cross-coupled transconductance reduction," which involves having a second set of transistors available at the input stage, biding their time during small signal swings, but ready to help out with some extra current when needed. This has the advantage of improved noise and offset performance, at the expense of some complexity, as compared with the simple emitter resistor scheme. This technique is used in the Harris HA5141 and HA5151, Raytheon 4531, and Signetics 535 and 538 to boost the slew rate for large differential input signals. The resultant graph of slew rate versus input error signal is shown in Figure 7.9.

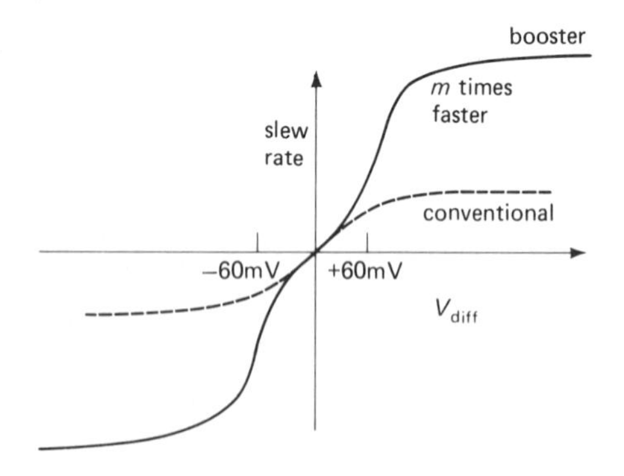

Figure 7.9

Bandwidth and settling time

Slew rate measures how rapidly the output voltage can change. The op-amp slew-rate specification usually assumes a large differential input voltage (60mV or more), which is realistic, since an op-amp whose output isn't where it's supposed to be will have its input driven hard by feedback, assuming a reasonable amount of loop gain. Of perhaps equal importance in high-speed precision applications is the time required for the output to get where it's going following an input change. This *settling-time* specification (the time required to get within the specified accuracy of the final value and stay there, Fig. 7.10) is always given for devices such as digital-to-analog converters, where precision is the name of the game, but it is not normally specified for op-amps.

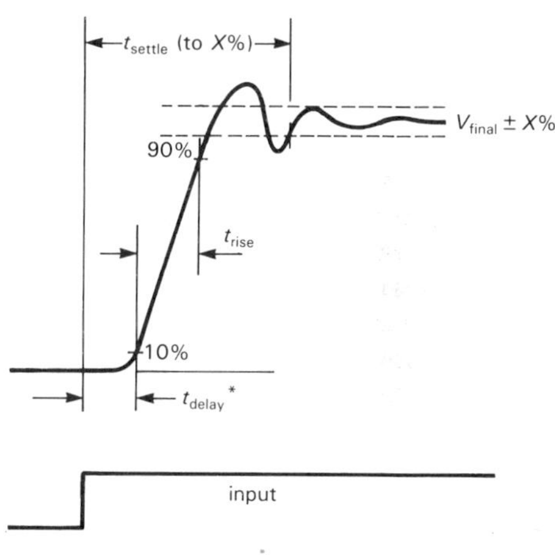

*Sometimes defined to V_{out} = logic threshold, or $V_{out} = 0.5V_{final}$

Figure 7.10. Settling time defined.

We can estimate op-amp settling time by considering first a different problem, namely what would happen to a perfect voltage step somewhere in a circuit if it were followed by a simple *RC* low-pass filter (Fig. 7.11). It is a simple exercise to show that the filtered waveform has the settling times shown. This is actually an important result, since you often limit bandwidth with a filter to reduce noise (more on that later in the chapter). To extend this simple result to an op-amp, just remember that a compensated op-amp has a 6dB/octave rolloff, just like a low-pass filter. When connected for closed-loop gain G_{CL}, its "bandwidth" (the frequency at which the loop gain drops to unity) is approximately given by

$$f_{3dB} = f_T/G_{CL}$$

As a general result, a system of bandwidth B has response time $\tau = 1/2\pi B$; thus, the equivalent "time constant" of the op-amp is

$$\tau \approx G_{CL}/2\pi f_T$$

The settling time is then roughly $5\tau - 10\tau$.

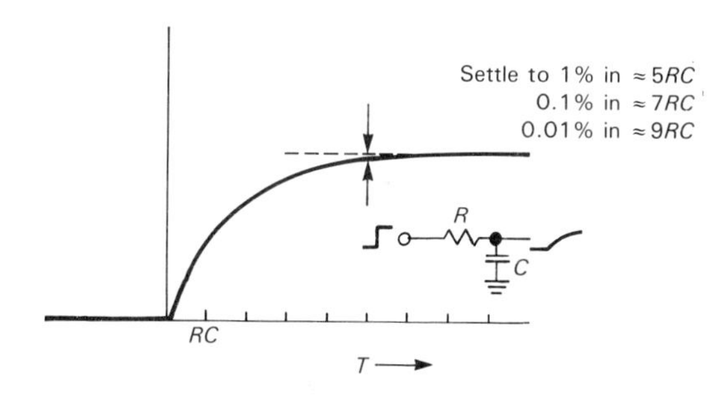

Figure 7.11. Settling time of an *RC* low-pass filter.

Let's try our prediction on a real case. The OP-44 from PMI is a precision fast-settling decompensated ($G_{CL} \geq 3$) op-amp, with an f_T of 23MHz (typ). Our simple formula then estimates the response time to be 21ns, which implies a settling time of 0.15μs (7τ) to 0.1%. This is in pretty good agreement with the actual value, which the data sheet gives as 0.2μs (typ) to 0.1%.

There are several points worth making: (a) Our simple model only gives us a lower bound for the actual settling time in a real circuit; you should always check the slew-rate-limited rise time, which may dominate. (b) Even if slew rate is not a problem, the settling time may be much longer than our idealized "single-pole" model, depending on the op-amp's compensation and phase margin. (c) The op-amp will settle more quickly if the frequency compensation scheme used gives a plot of open-loop phase shift versus frequency that is a nice straight line on a log-log graph (e.g., the OP-42, Fig. 7.12); op-amps with wiggles in the phase-shift graph are more likely to exhibit overshoot and ringing, as in the upper waveform shown in Figure 7.10. (d) A fast settling time to 1%, say, doesn't necessarily guarantee a fast settling time to 0.01%, since there may be a long tail (Fig. 7.13). (e) There's no substitute for an actual settling-time specification from the manufacturer.

Table 7.3 lists a selection of high-speed op-amps suitable for applications that demand high f_T, high slew rate, fast settling time, and reasonably low offset voltage.

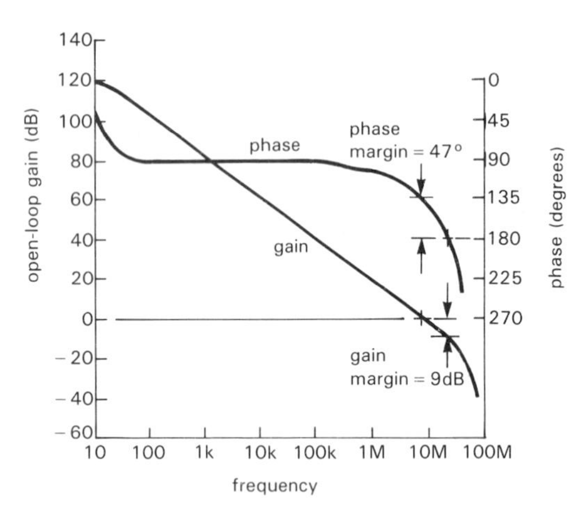

Figure 7.12. OP-42 gain and phase versus frequency.

Gain error

There's one more error that arises from finite open-loop gain, namely an error in closed-loop gain owing to finite loop gain.

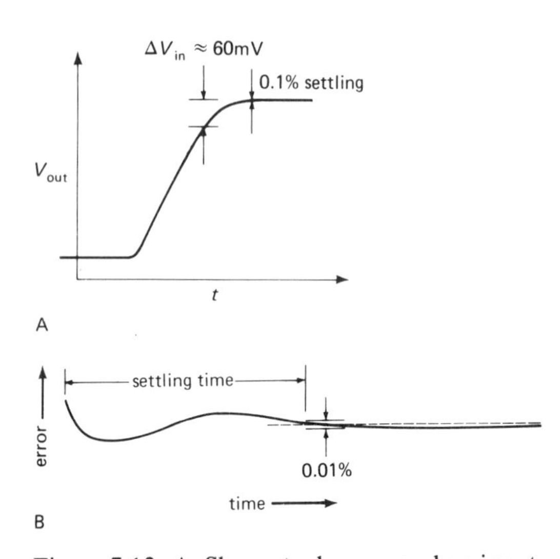

Figure 7.13. A. Slew rate decreases when input error approaches 60mV.
B. Settling to high precision can be surprisingly lengthy.

We calculated in Chapter 3 the expression for closed-loop gain in a feedback amplifier, $G = A/(1+AB)$, where A is the open-loop gain and B is the "gain" of the feedback network. You might think that the $A \geq 100$dB of op-amp open-loop gain is plenty, but when you try to construct extremely precise circuits you are in for a surprise. From the preceding gain equation it is easy to show that the "gain error," defined as

$$\delta_G = \text{gain error} \equiv \frac{G_{\text{ideal}} - G_{\text{actual}}}{G_{\text{ideal}}}$$

is just equal to $1/(1+AB)$ and ranges from 0 for $A = \infty$ to 1 (100%) for A = 0.

EXERCISE 7.2

Derive the foregoing expression for gain error.

The resulting frequency-dependent gain error is far from negligible. For instance, a 411, with 106dB of low-frequency open-loop gain will have a gain error of 0.5% when configured for a closed-loop gain of 1000. Worse yet, the open-loop gain drops 6dB/octave above 20Hz, so our amplifier would have a gain error of 10% at

500Hz! Figure 7.14 plots gain error versus frequency, for closed-loop gains of 100 and 1000, for the OP-77, with its extraordinary 140dB of low-frequency gain. It should be obvious that you need plenty of gain and a high f_T to maintain accuracy at even moderate frequencies.

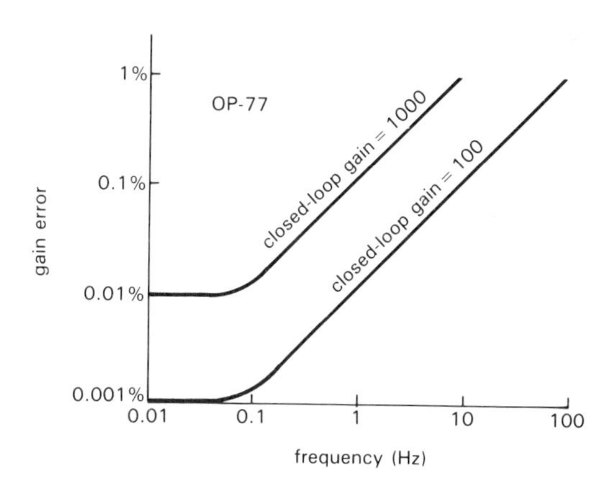

Figure 7.14. OP-77 gain error.

We plotted these curves using the graph of open-loop gain versus frequency given in the data sheet. Even if your op-amp data sheet provides a curve, it's best to work backward from the specified f_T and dc open-loop gain, figuring the open-loop gain at the frequency of interest, hence the gain error (as above) as a function of frequency. This procedure yields

$$\delta_G = \frac{1}{1 + Bf_T/f}$$

where B is, as usual, the gain of the feedback network. Of course, in some applications, such as filters, B may also depend on frequency.

EXERCISE 7.3

Derive the foregoing result for $\delta_G(f)$.

Crossover distortion and output impedance

Some op-amps use a simple push-pull output stage, without biasing the bases two

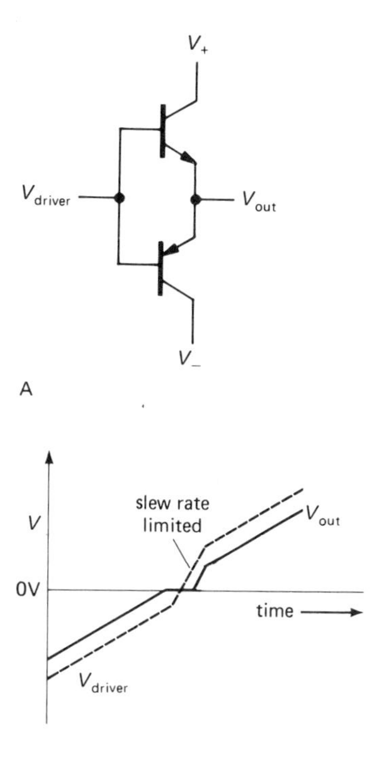

Figure 7.15. Crossover distortion in class B push-pull output stage.

diode drops apart, as we discussed in Section 2.15. This leads to class B distortion near zero output, since the driver stage has to slew the bases through $2V_{BE}$ as the output current passes through zero (Fig. 7.15). This crossover distortion can be substantial, particularly at higher frequencies where the loop gain is reduced. It is greatly reduced in op-amp designs that bias the output push-pull pair into slight conduction (class AB). The popular 741 is an example of the latter, whereas its predecessor, the 709, uses the simple class B output-stage biasing. The otherwise admirable 324 can exhibit large crossover distortion for this reason. The right choice of op-amp can have enormous impact on the performance of low-distortion audio amplifiers. Perhaps this problem has contributed to what the audiophiles refer to

TABLE 7.3. HIGH-SPEED PRECISION OP-AMPS

Type	Mfg[a]	FET	Trim	Ext comp	Min gain	V_{os} max (mV)	ΔV_{os} max (μV/°C)	I_{os} max (nA)	I_b max (nA)	e_n @1kHz typ (nV/√Hz)	Input cap (pF)	Slew rate typ (V/μs)	f_T typ (MHz)	Settle, typ 0.1% (ns)	Settle, typ 0.01% (ns)	R_{out} typ (Ω)	Over-shoot @G_{min} typ (%)	Phase margin @G_{min} (deg)	Diff'l input max (V)	Swing into load ±V_o (±V)	Swing into load R_L (Ω)	Max output curr (mA)	Comments
OP-37E	PM+	–	•	–	5	0.025	0.6	35	40	3.0	-	17	63	1000	-	70	10	71	0.7	11	600	5	low-noise (decomp OP-27)
OP-42E	PM	•	•	–	1	0.75	10	0.04	0.2	13	6	58	10	450	700	50	15	47	40	12	600	6	stable into 300pF
OP-44E	PM	•	•	–	3	0.75	10	0.04	0.2	12	-	122	20	200	300?	-	25	53	40	-	-	7	decomp OP-42
OP-62E	PM	–	•	•	1	0.2	-	100	300	2.5	-	15	50	-	-	-	-	-	5	12	600	-	
OP-63E	PM	–	•	•	1	0.75	-	100	300	7	-	50	50	-	-	-	-	-	5	12	600	-	
OP-64E	PM	–	•	•	5	0.75	-	100	300	7	-	200	200	-	-	-	-	-	5	12	600	-	
OP-65E	PM	–	•	–	1	2	-	2000	2.5μA	-	-	200	150	-	-	-	-	-	5	3.5	500	50	±5V supply
CLC221	CL	–	–	–	1	1	15	-	10μA[b]	4	2.4	6500	-	15	18[c]	8	12	-	-	-	-	30	hybrid
CLC400	CL	–	•	–	1	5.5	40	-	25μA[b]	12	0.5	700	280	10	15	-	0	-	-	3.5	100	70	monolithic; transresistance
CLC401	CL	–	–	–	7	6.5	50	-	35μA[b]	12	0.5	1200	2100	10	13	-	0	-	-	3.5	100	70	monolithic; transresistance
LF401A	NS	•	•	–	1	0.2	10	0.1	0.2	20	7	70	16	200	340	50	-	55	32	12	600	12	
OPA602C	BB	•	•	–	1	0.25	2	0.001	0.001	13	3	35	6.5	700	1000	80	20	40	36	10	600	4	low-bias
AD711C	AD	•	•	–	1	0.25	5	0.01	0.025	18	6	20	4		1000	-	10	-	20	10	600	2.8	inexpensive
MS738B	KE	–	•	–	2	0.075	1	20	40	3.8	3	3500	1500	30	200	2	5	-	12	12	100	120	
MS739	KE	•	–	–	1	0.025[t]	0.3[t]	0.03[t]	0.08[t]	-	-	5500	-	15	30[c]	-	-	-	-	12	100	120	
AD744C	AD	•	•	•	2	0.25	3	0.02	0.05	18	5.5	75	13	400	500	-	4	70	36	12	600	4	ultra low dist (3ppm)[f]
AD840K	AD	–	•	–	10	0.5	5	200	5μA	3	1	400	400	80	110	15	20	-	-	10	500	50	decomp 841
AD841K	AD	–	•	–	1	1	20	200	5μA	13	1	300	40	80	110	5	20	-	-	10	500	12	vert PNP, decomp avail
AD842K	AD	–	•	–	2	1	10	200	5μA	9	1	375	80	80	110	5	20	-	-	10	500	100	decomp 841
AD845K	AD	•	•	–	1	0.25	3	0.05	0.4	25	4	100	16	250	300	-	-	-	20	12.5	500	10.2	
AD846K	AD	–	–	•	2	0.2	2	-	150[b]	1.3	2	450	40	80	110	16	20	-	18	10	500	7	low-noise
AD847J	AD	–	•	–	1	1	30	15[t]	5μA	15[d]	1.5	300	50	80	-	-	-	50	6	10	500	5.6	vert PNP, decomp avail
AD848J	AD	–	•	–	1	1	10	15[t]	5μA	4[d]	1.5	300	250	80	-	-	-	50	6	2.5	150	20	vert PNP
LT1028A	LT	–	•	•	2	0.04	0.8	50	90	0.85	5	15	75	-	-	80	5	50	0.7	12	600	10.5	ultra low noise
LT1055A	LT	•	•	–	1	0.15	4	0.01	0.05	14	4	13	5	1200	1800	60	10	-	40	10	600	4	LT1056 is faster
1435	TP	–	•	•	2	5	25	300[t]	20μA	16	2	300	1000	40	70	-	1	25	4	7	500	30	hybrid
LH4105C	NS	•	–	–	1	0.5	20[t]	0.4	0.6	-		40	18	-	500	-	-	-	30	10	100	25	no current limit
HA5147A	HA	–	•	–	10	0.025	0.6	35	40	3.0	-	35	140	400	-	70	20	-	0.7	11	600	4	low-noise (decomp OP-27)
AD9611B	AD	–	–	–	1	3	20	-	5	1[e]	3	1900	280	13	-	0.03	4	-	-	3	100	50	cur fdbk; no protec; hybrid

(a) see notes to Table 4.1. (b) current-sensing inv input; bias current shown is for non-inv input only. (c) to 0.02%. (d) at 10kHz. (e) 5MHz to 280MHz. (f) stable into 1nF. (t) typical.

as "transistor sound." Some modern op-amps, particularly those intended for audio applications, are designed to produce extremely low crossover distortion. Examples are the LT1028, the LT1037, and the LM833. The LM833, for example, has less than 0.002% distortion over the full audio band of 20Hz–20kHz. (That's the claim, anyway; we may be overly gullible!) These amplifiers all have very low noise voltage, as well; in fact, the LT1028 is currently the world noise-voltage champion, with $e_n = 1.7\text{nV}/\sqrt{\text{Hz}}$ (max) at 10Hz.

The open-loop output impedance of an op-amp is highest near zero output voltage, because the output transistors are operating at their lowest current. The output impedance also rises at high frequency as the transistor gain drops off, and it may rise slightly at very low frequencies due to thermal feedback on the chip.

It is easy to neglect the effects of finite open-loop output impedance, thinking that feedback will cure everything. But when you consider that some op-amps have open-loop output impedances of a few hundred ohms, it becomes clear that the effects may not be negligible, especially at low to moderate loop gains. Figure 7.16 shows some typical graphs of op-amp output impedance, both with and without feedback.

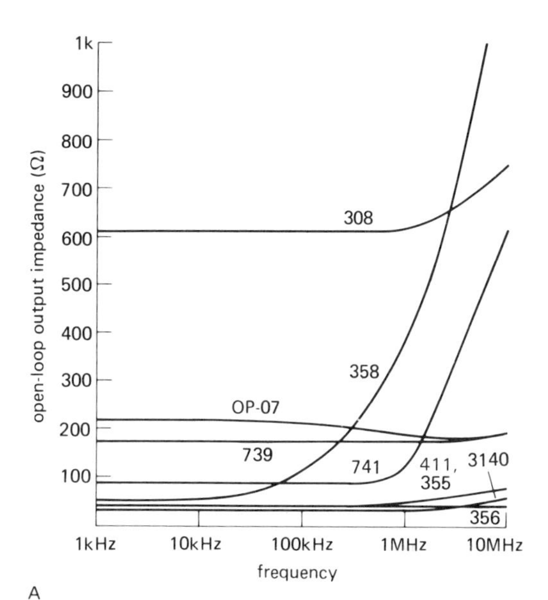

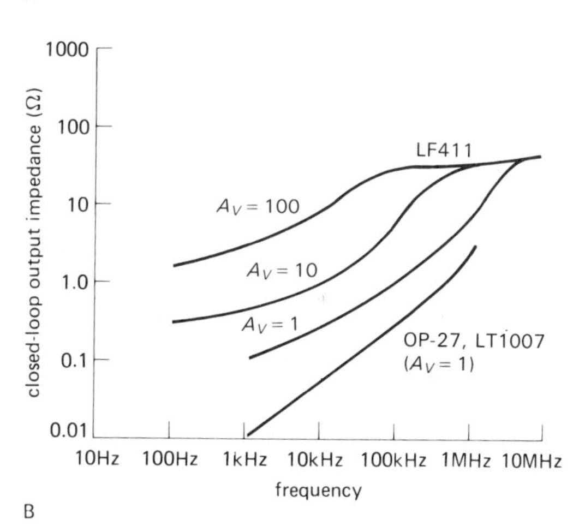

Figure 7.16. A. Measured open-loop output impedance for some popular op-amps.
B. Closed-loop output impedance for the 411 and OP-27 op-amps.

Driving capacitive loads

The finite open-loop output impedance of op-amps leads to serious difficulties when you attempt to drive a capacitive load, owing to lagging phase shifts produced by the output impedance in combination with the load capacitance to ground. These can lead to feedback instabilities if the 3dB frequency is low enough, since it adds to the 90° phase shift already present with frequency compensation. As an example, imagine driving a hundred feet of coaxial cable from an op-amp with 200 ohms output impedance. The unterminated coax line acts like a 3000pF capacitor, generating a low-pass *RC* with a 3dB point of 270kHz. This is well below the unity-gain frequency of a typical op-amp, so oscillations are likely at high loop gain (a follower, for example).

There are a couple of solutions to this problem. One is to add a series resistor, taking feedback at high frequencies from the op-amp output and feedback at low frequencies and dc from the cable (Fig. 7.17).

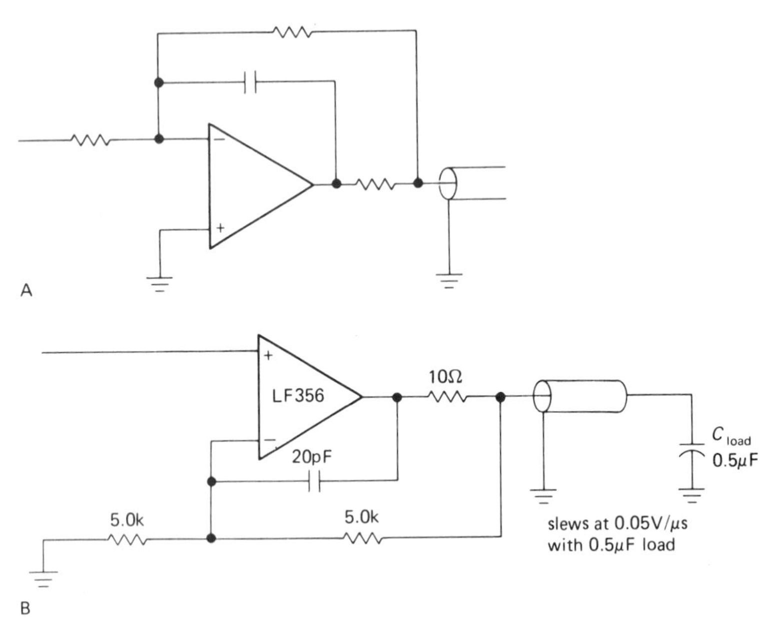

Figure 7.17

The parts values shown in the second circuit are specific for that op-amp and circuit configuration, and they give an idea of how large a capacitance can be driven. Of course, this technique degrades the high-frequency performance, since feedback isn't operative at high frequencies on the signal at the cable.

Unity-gain power buffers

If this technique of split feedback paths is unacceptable, the best thing to do is add a unity-gain high-current buffer inside the loop (Fig. 7.18). The devices listed have voltage gain near unity and low output impedance, and they can supply up to 250mA output current. They have no significant phase shifts up to the unity-gain frequency (f_T) of most op-amps, and they can be included in the feedback loop without any additional frequency compensation. Table 7.4 presents a brief listing of buffer amplifiers. These "power boosters" can, of course, be used for loads that require high current, regardless of whether or not there are problems with capacitance. Unfortunately, most buffer amplifiers do not contain either internal current limiting or thermal shutdown circuits and must therefore

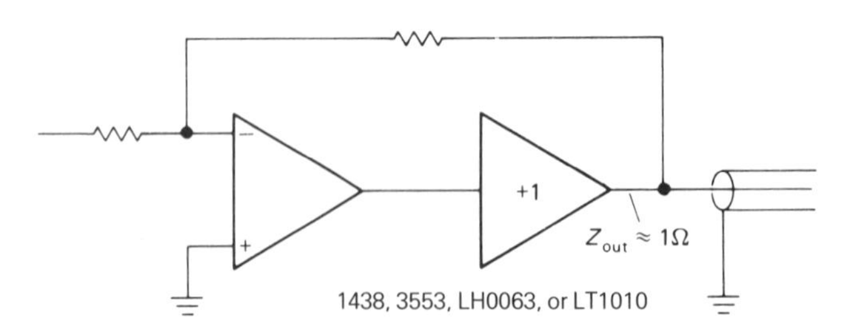

Figure 7.18

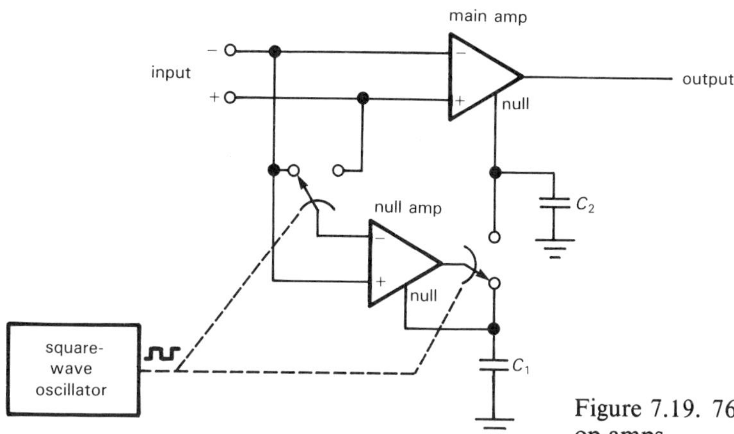

Figure 7.19. 7650-type chopper-stabilized op-amps.

be used carefully. The exceptional devices that include on-chip protection are noted in Table 7.4, for example the LT1010.

Note also that the preceding example would be changed if the cable were terminated in its characteristic impedance. In that case it would look like a pure resistance, somewhere in the range of 50 to 100 ohms, depending on the type of cable. In such a case a buffer would be mandatory, with ±200mA drive capability in order to drive ±10 volt signals into the 50 ohm load impedance. This subject is discussed in greater detail in Section 13.09.

The preceding circuit example does not suffer from any op-amp output-related errors, since it operates essentially at dc.

7.08 Auto-zeroing (chopper-stabilized) amplifiers

Even the best of precision low-offset op-amps cannot match the stunning V_{os} performance of the so-called "chopper-stabilized" or "auto-zero" op-amps. Ironically, these interesting amplifiers are built with CMOS, otherwise famous for its mediocrity when it comes to offset voltage or drift. The trick here is to put a second *nulling* op-amp on the chip, along with some MOS analog switches and offset-error storage capacitors (Fig. 7.19). The main op-amp functions as a conventional imperfect amplifier. The nulling op-amp's job is to monitor the input offset of the main amplifier, adjusting a slow correction signal as needed in an attempt to bring the input offset exactly to zero. Since the nulling amplifier has an offset error of its own, there is an alternating cycle of operation in which the nulling amplifier corrects its own offset voltage.

Thus, the auto-zeroing cycle goes like this: (a) Disconnect nulling amplifier from input, short its inputs together, and couple its output back to C_1, the holding capacitor for its correction signal; the nulling amplifier now has zero offset. (b) Now connect nulling amplifier across input, and couple its output to C_2, the holding capacitor for the main amplifier's correction signal; the main amplifier now has zero offset. The MOS analog switches are controlled by an on-board oscillator, typically at a rate of a few hundred hertz. The error-voltage holding capacitors are typically $0.1\mu F$ and in most cases must be supplied externally; LTC, Maxim, and Teledyne make some convenient auto-zero amplifiers with discrete capacitors encapsulated right into the IC package.

Auto-zero op-amps do best what they are optimized for, namely delivering V_{os} values (and tempcos) five times better than the best precision bipolar op-amp (see Table 7.2). What's more, they do this while

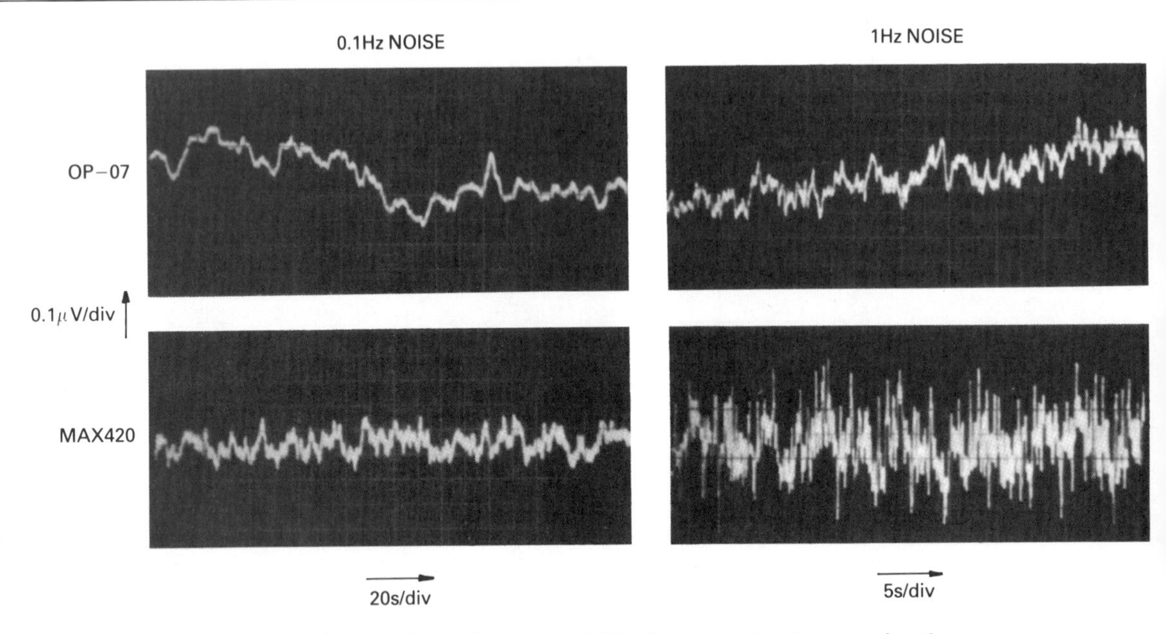

Figure 7.20. At very low frequencies a chopper-stabilized op-amp has lower noise than a conventional low-noise op-amp.
A. dc to 0.1Hz
B. dc to 1Hz (Courtesy of Maxim Integrated Products, Inc.)

delivering full op-amp speed and bandwidth, unlike earlier synchronous amplifiers that were also called "chopper amplifiers," but which had bandwidth limited to a fraction of the chopping clock frequency (see below).

That's the good news. The bad news is that auto-zero amplifiers have a number of diseases that you must watch out for. First of all, being CMOS devices, most of them have a severely limited supply voltage (typically 15V total supply) and thus cannot run from conventional ±15 volt supplies. The Maxim MAX430/2 and Teledyne TSC915 and TSC76HV52 "high-voltage" auto-zero op-amps are exceptions and will operate from ±15 volt supplies. Second, most auto-zero op-amps require external capacitors (exceptions: LTC1050, Maxim MAX430/2, Teledyne TSC911/13/14). A third problem with many auto-zero amplifiers (particularly those with limited supply voltage) is the rather restricted common-mode input range; for example, the popular ICL7650 has a guaranteed common-mode input range of −5 to +1.5 volts, when running from its usual ±5 volt supplies (for the improved ICL7652 the range is −4.3V to +3.5V; that's a wider range, but it doesn't include the negative rail, so you can't use it as a "single-supply" op-amp). The high-voltage amplifiers are much better – for example, the MAX432 has a guaranteed common-mode range of −15 to +12 volts, when running from ±15 volt supplies. The op-amp table (Table 4.1) shows which chopper amps have common-mode range to the negative rail; although the popular ICL7652 doesn't, improved versions from LTC (LTC1052) and Maxim (ICL7652B) do, permitting convenient single-supply operation.

A fourth drawback is the tendency of these CMOS op-amps to have poor output sourcing capability, sometimes as little as 1–2mA in the sourcing (positive-output) direction. The otherwise admirable MAX432 can only source 0.5mA! Fifth on the list of drawbacks, but often first in importance, is the problem of clock-induced noise. This is caused by charge coupling from the MOS switches (see Section 3.12) and can cause wicked spikes at the output. The specifications are often misleading here, because it is conventional to quote input-referred noise with R_S = 100 ohms and also to give the specification only for very low frequencies; for example, a typical input-referred noise voltage is 0.2μV (dc to 1Hz, with $R_S = 100\Omega$). However, with zero input signal the output waveform might consist of a train of 5μs-wide 15mV spikes of alternating polarity! In low-frequency applications you can (and should) *RC*-filter the output to a bandwidth of a few hundred hertz, which will make these spikes disappear. This spiky noise is also of no importance in integrating applications (e.g., integrating A/D converters, see Section 9.21) or in applications where the output is intrinsically slow (e.g., a thermocouple circuit with a meter at the output). In fact, if you only want very slow output response, and therefore low-pass-filter the output to extremely low frequencies (below 1Hz), a chopper amplifier will actually have *less* noise than a conventional low-noise op-amp; see Figure 7.20.

A final problem with auto-zero amplifiers is their disastrous saturation characteristic. What happens is this: The auto-zeroing circuit, in attempting to bring the input difference voltage to zero, implicitly assumes there is overall feedback operating. If the amplifier's output saturates (or if there is no external circuit to provide feedback), there will be a large differential input voltage, which the nulling amplifier sees as an input offset error; it therefore blindly generates a large correction voltage that charges up the correction capacitors to a large voltage before the nulling amplifier itself finally saturates. Recovery is incredibly slow – up to a *second*! The "cure" is to sense when the output is approaching saturation, and clamp the input to prevent it. Most auto-zeroing op-amps provide a "clamp" output for this purpose, which you tie back to the inverting input to prevent saturation. You can prevent saturation in chopper amplifiers without a "clamp" pin (and in ordinary op-amps, as well) by bridging the feedback network with a bidirectional zener (two zeners in series), which clamps the output at the zener voltage, rather than letting it limit at the supply rail; this works best in the inverting configuration.

□ Chopper miscellany

□ *ac-coupled "chopper amp."* When considering auto-zeroing chopper amplifiers, be sure you don't confuse this technique with another "chopper" technique, namely the traditional low-bandwidth chopper amplifier in which a small dc signal is converted to ac ("chopped") at a known frequency, amplified in ac-coupled amplifiers, then finally demodulated by multiplying with the same waveform used to chop the signal initially (Fig. 7.21). This scheme is quite different from the full-bandwidth auto-zeroing technique we've been considering, in that it rolls off at signal frequencies approaching the clock frequency, typically just a few hundred hertz. You sometimes see it used in chart recorders and other low-frequency instrumentation.

□ *Thermal offsets.* When you build dc amplifiers with submicrovolt offset voltages, you should be aware of *thermal offsets*, which are little thermally driven batteries produced by the junction of dissimilar metals (see Section 15.01). You get a Seebeck-effect "thermal emf" when you have a pair of such junctions at different temperatures. In practice you usually have joints between wires with different plating; a thermal gradient, or even a little draft, can easily produce thermal voltages of a few microvolts. Even similar wires from different manufacturers can produce thermal emf's of 0.2μV/°C, four times the drift

TABLE 7.4. FAST BUFFERS

		Small signal			Supply voltages		Large signal					
		Rolloff freq							Output swing			
Type	Mfg[a]	-3dB (MHz)	-40° (MHz)	Z_{out} (Ω)	min (±V)	max (±V)	Slew rate (V/μs)	Maximum output current (±mA)	V_{out} (±V)	R_{load} (Ω)	V_{os} max (mV)	Comments
LT1010	LT	40	15	7	2.5	20	200	150	12	80	150	thermal limit; monolithic
LH0002	NS	50	60	6	6	22	200	100	10	50	30	
LH4001	NS	50	-	6	5	22	125	200	10	50	50	10-pin DIP
LM6321	NS	50	40	5	5	16	800	300	10	50	50	mini-DIP; therm lim; monolithic
AH0010	OE	60	-	20	6	18	1500	100	10	100	20	mini-DIP; alias 9910
BUF03	PM	65	20	2	6	18	250	70	10	150	6	monolithic
EL2001	EL	70	-	-	5	15	500	100	-	-	-	mini-DIP; monolithic
LH0033	NS+	100	80	6	5	20	1400	100	10	50	10	also EL2033, and others
1490	TP	100	-	20	12	18	500	100	-	-	20	FET input
HA5002	HA	110	-	3	-	20	1300	200	11	100	20	monolithic
HOS100	AD	125	-	8	5	20	1500	100	10	100	10	
MAX460	MA	140	65	4	5	20	1500	100	10	100	5	monolithic
LH4004	NS	140	-	-	4	15	1500	-	10	50	15	FET input; ext feedback
EL2005	EL	140	60	4	5	15	1500	100	10	100	5	FET input, precision
EL2002	EL	180	-	-	5	15	1000	100	-	-	-	mini-DIP; monolithic
LH0063	NS+	200	30	1	5	20	4000	250	10	50	25	"damn fast" buffer
MSK330	KE	200	-	2	-	18	3000	200	13	100	25	
LH4002	NS	200	150	6	4	6	1250	60	2.2	50	50	video
9911	OE	200	-	6	11	18	1000	500	10	20	20	
9963	OE	200	-	3	6	18	3000	200	10	50	50	FET input
1359	TP	250	60	5	12[b]	18	1300	100	10	100	15	
LH4003	NS	250	-	-	5	8	1200	100	3	50	15	video; ext feedback
HA5033	HA	250	80	5	5	20	1300	100	10	100	15	mini-DIP; monolithic; also AH001
OPA633	BB	275	150	5	5	16	2500	100	11	50	15	monolithic
3553	BB	300	60	1	5	20	2000	200	10	50	50[t]	insulated metal case
MP2004	MP	350	280	4	5	20	2500	100	10	100	10	FET input; also EL2004
LH4006	NS	350	-	-	4	8	1200	100	3	50	15	video; ext feedback
EL2031	EL	500	-	-	-	-	5000	100	-	-	-	FET input
CLC110	CL	730	200	2	5?	7	800	70	4	100	8	monolithic

(a) see footnotes to Table 4.1. (b) nominal. (t) typical.

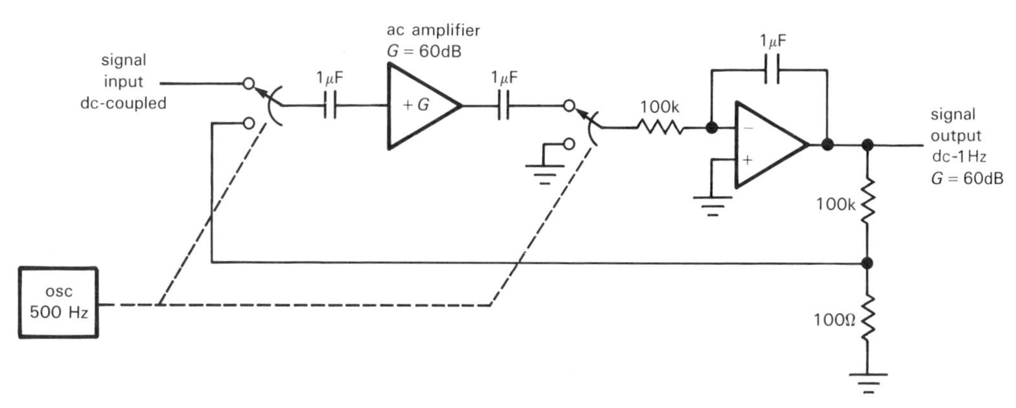

Figure 7.21. An ac-coupled chopper amplifier.

spec of a MAX432! The best approach is to strive for symmetrical wiring and component layouts, and then avoid drafts and gradients.

□ *External auto-zero.* National makes a nice "auto-zero" chip (the LMC669) that can be used as an outboard nulling amplifier to make any op-amp of your choosing into an auto-zeroing amplifier (Fig. 7.22). It works most naturally with the inverting configuration, as shown, generating an error voltage to the noninverting input to maintain zero input offset. It doesn't do as well as the dedicated auto-zero amplifiers we've been considering: The V_{os} specification is 5μV (typ), 25μV (max). However, it does let you use the auto-zero technique with any op-amp. You might, for instance, use it to zero an imprecise but high-power or high-speed op-amp. The circuits shown are good examples. The LM675 is a nice high-power op-amp (3A output current, with sophisticated on-chip safe-operating-area and thermal protection), but with a maximum offset voltage of 10mV. The auto-zero reduces that by nearly a factor of 1000. Likewise, the LM6364 is a fast op-amp (f_T = 175MHz, SR = 350V/μs) with maximum offset voltage of 9mV, here reduced by a factor of 400. Note the *RC* filter components at both input and output of the auto-zero: These are necessary to suppress chopper noise in the (slow) correction loop, when this technique is used with small signals and low-noise parts like the LM6364 (8nV/$\sqrt{\text{Hz}}$).

□ *Instrumentation amplifier.* Another "chopper" technique in use is the so-called "commutating auto-zeroing" (or CAZ) amplifier, originated by Intersil. In this technique, typified by the ICL7605 flying-capacitor instrumentation amplifier, MOSFET switches enable you to store the differential input signal across a capacitor, then amplify it with a single-ended chopper-stabilized amplifier (Fig. 7.23). You get charge-coupled spikes at the clock rate, just as with the standard auto-zeroing amplifier, which puts the same sort of limitations on the CAZ technique as we saw earlier. Although we raved about CAZ amplifiers in our first edition ("... stands a good chance of revolutionizing precision op-amp and instrumentation amplifier technology"), they've been finessed by the better auto-zeroing technique in which the signal always passes through a single amplifier.

However, in fairness to the CAZ-amp, we should point out that the flying-capacitor technique used in the 7605 has

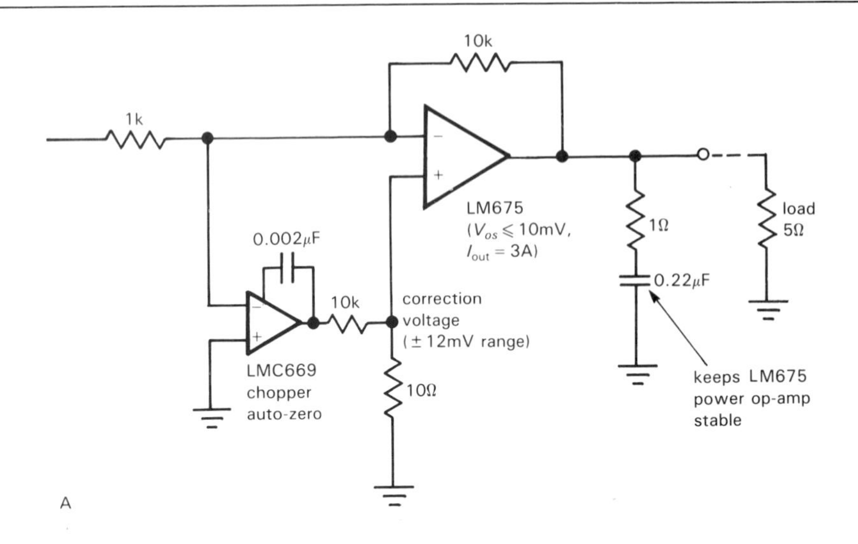

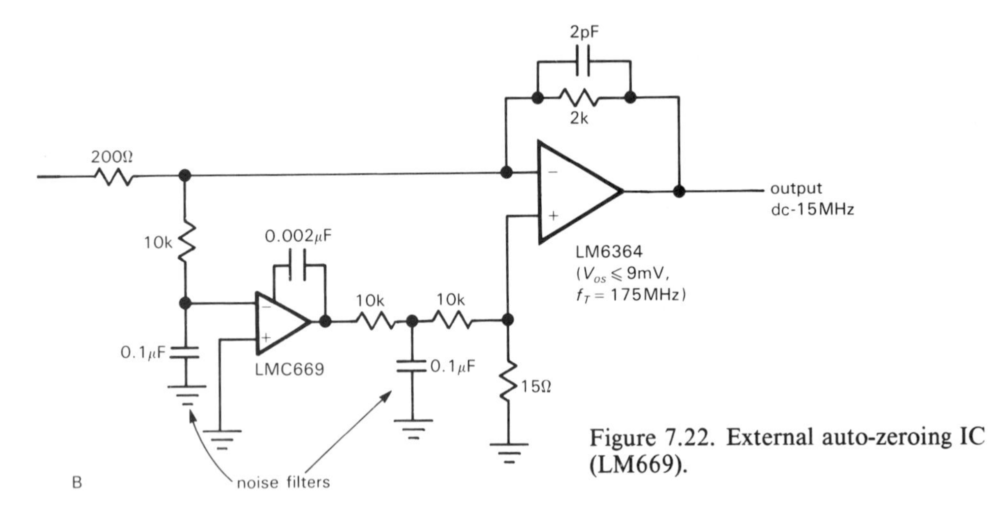

Figure 7.22. External auto-zeroing IC (LM669).

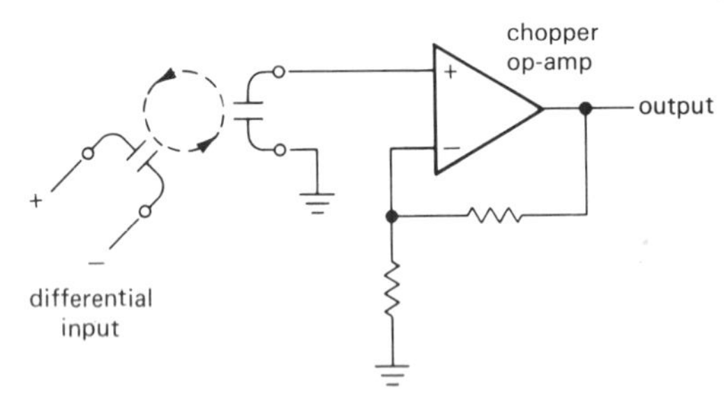

Figure 7.23. ICL7605 flying-capacitor differential amplifier with high CMRR.

some unique advantages, including input common-mode operation 0.3 volt beyond both supply rails, 100dB CMRR (min), even at unity gain, and the lowest offset voltage of any monolithic amplifier. If you use these amplifiers, however, don't forget the required output noise filter, the limited supply voltage (±8V max), and the requirement of a high-impedance load, since the output impedance rises periodically at the clock rate.

The LTC1043 flying-capacitor building block lets you make your own high-CMRR differential amplifier. Instrumentation amplifiers are discussed in detail in the next section. The precision op-amp table (Table 7.2) includes most of the currently available auto-zeroing op-amps.

DIFFERENTIAL AND INSTRUMENTATION AMPLIFIERS

The term *instrumentation amplifier* is used to denote a high-gain dc-coupled differential amplifier with single-ended output, high input impedance, and high CMRR. They are used to amplify small differential signals coming from transducers in which there may be a large common-mode signal or level.

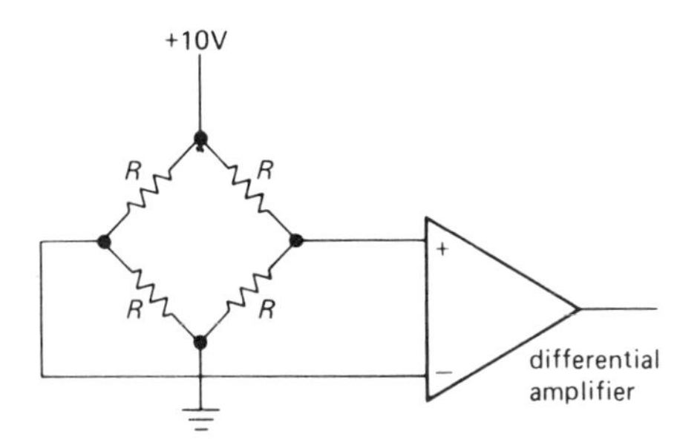

Figure 7.24. Strain gauge with amplifier.

An example of such a transducer is a strain gauge, a bridge arrangement of resistors that converts strain (elongation) of the material to which it is attached into resistance changes (see Section 15.03); the net result is a small change in differential output voltage when driven by a fixed dc bias voltage (Fig. 7.24). The resistors all have roughly the same resistance, typically 350 ohms, but they are subjected to differing strains. The full-scale sensitivity is typically 2mV per volt, so that the full-scale output is 20mV for 10 volts dc excitation. This small differential output voltage proportional to strain rides on a 5 volt dc level. The differential amplifier must have extremely good CMRR in order to amplify the millivolt differential signals while rejecting variations in the ~5 volt common-mode signal. For example, suppose that a maximum error of 0.1% is desired. Since 0.1% of full scale is 0.02mV, riding on 5000mV, the CMRR would have to exceed 250,000 to 1, i.e., about 108dB.

The tricks involved in making good instrumentation amplifiers and, more generally, high-gain differential amplifiers are similar to the techniques just discussed. Bias current, offsets, and CMRR errors are all important. Let's begin by discussing the design of differential amplifiers for noncritical applications first, working up to the most demanding instrumentation requirements and their circuit solutions.

7.09 Differencing amplifier

Figure 7.25 shows a typical circuit situation requiring only modest common-mode rejection. This is a current-sensing circuit used as part of a constant-current power supply to generate a constant current in the load. The drop across the precision 4-wire 0.01 ohm power resistor is proportional to load current. Even though one side of R_5 is connected to ground, it would be unwise to use a single-ended amplifier, since connection resistances of a milliohm would contribute 10% error! A differential amplifier is obviously required, but it need not have particularly good CMRR, since only very small common-mode signals are expected.

The op-amp is connected in the standard differencing amplifier configuration, as discussed in Section 4.09. R_1, R_2, and R_5 are precision wire-wound types for extreme stability of gain, whereas R_3 and R_4, which set CMRR, can be mere 1% metal-film types. The overall circuit thus has a gain accuracy approaching that of the current-sensing resistor and a CMRR of about 40dB.

Precision differential amplifier

For applications such as strain gauges, thermocouples, and the like, 40dB of common-mode rejection is totally inadequate, and figures more like 100dB to 120dB are often needed. In the preceding example of the strain gauge, for instance, you might have a full-scale differential (unbalance) signal

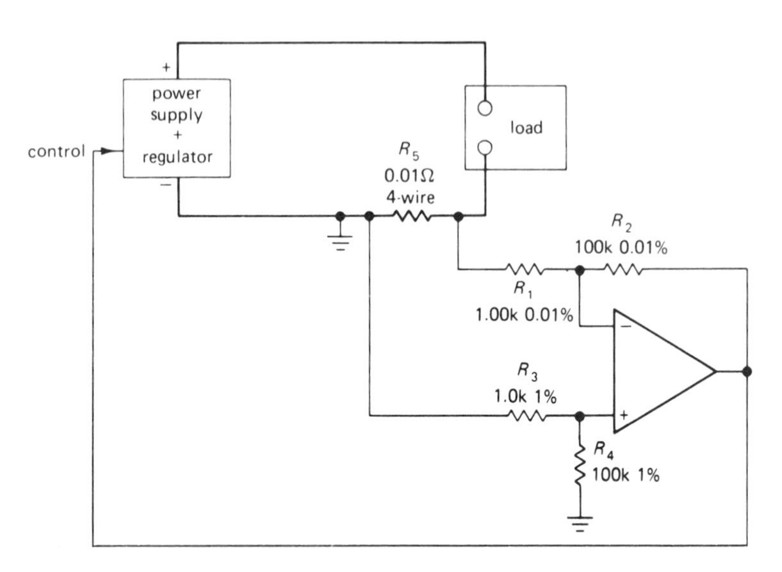

Figure 7.25. Current regulator.

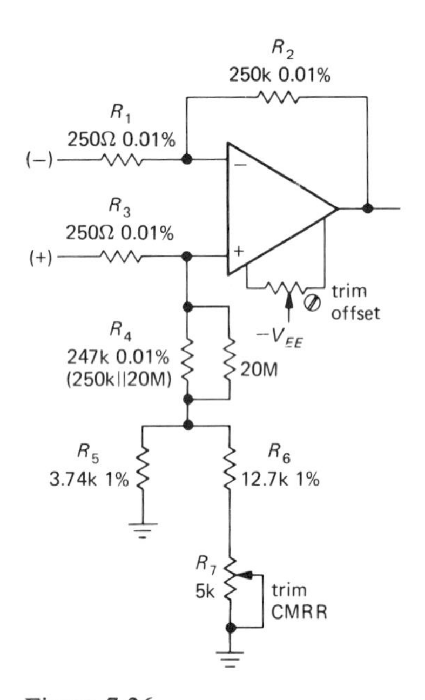

Figure 7.26

of 2mV per volt. If you want accuracy of 0.05%, you need a common-mode rejection of 114dB, minimum. (Note that this requirement can be relaxed considerably in the special case that the amplifier is zeroed with the common-mode voltage present, as might be done in a laboratory situation.)

The obvious first approach to improved CMRR is to beef up the resistor precision in the differencing circuit (Fig. 7.26). The resistor values are chosen to keep the large feedback resistors within the range of available precision wire-wound resistors. With 0.01% resistors, the common-mode rejection is in the range of 80dB (68dB worst case), assuming the op-amp has high CMRR. It takes only one trimmer to null the common-mode sensitivity, as shown. With the values shown, you can trim out an accumulated error up to 0.05%, i.e., a bit more than the worst-case resistor error. The fancy network shown is used because small-value trimmer resistors tend to be somewhat unstable with time and are best avoided.

A point about ac common-mode rejection: With good op-amps and careful trimming, you can achieve 100dB or better CMRR at dc. However, the wire-wound resistors you need for the best stability have some inductance, causing degradation of CMRR with frequency. Noninductive wire-wound resistors (Aryton-Perry type) are available to reduce this effect, which is common to all the circuits we

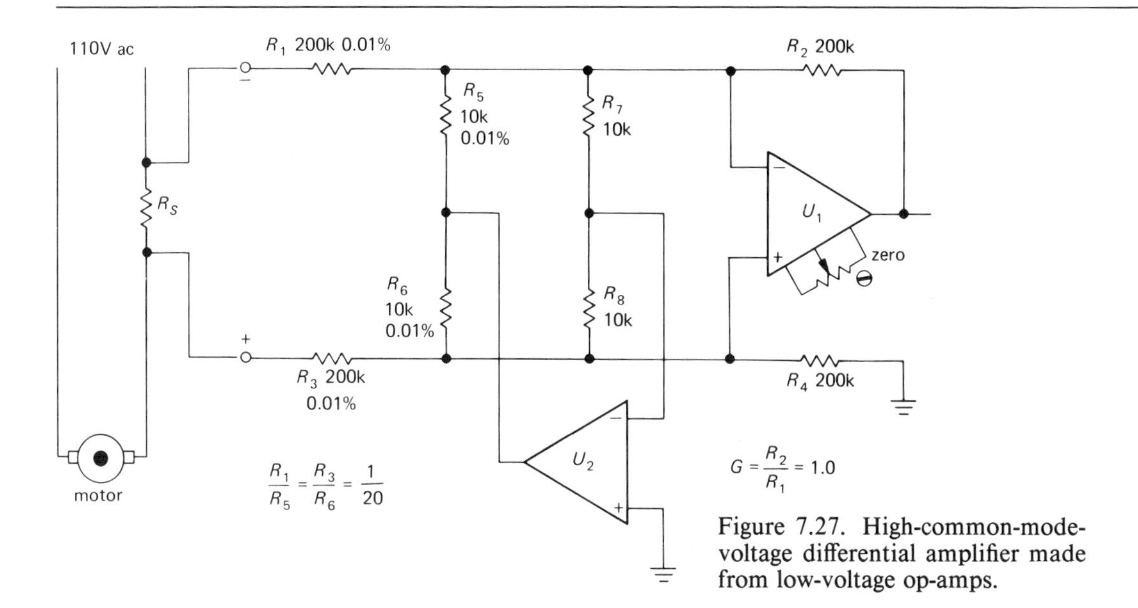

Figure 7.27. High-common-mode-voltage differential amplifier made from low-voltage op-amps.

will be talking about. Note also that it is necessary to balance the circuit capacitances to achieve good CMRR at high frequencies. This may require careful mirror-image placement of components.

Burr-Brown offers a series of precision differential amplifiers, complete with matched resistors, in a nice mini-DIP (8-pin) package. The INA105 is unity-gain (±0.01% maximum gain error), with input impedance of 25kΩ, and the INA106 has a gain of 10, with the same accuracy and an input impedance of 10kΩ. The latter has a minimum CMRR of 94dB and maximum V_{os} of 100μV and is stable into 1000pF. Burr-Brown also offers a version with high-input common-mode voltage range (±200V), as described later.

□ *High-voltage differential amplifier*

Figure 7.27 shows a clever method for increasing the common-mode input voltage range of the differencing amplifier circuit beyond the supply voltages without a corresponding reduction in differential gain. U_2 looks at the common-mode input signal at U_1's input and removes it via R_5 and R_6. Since there is no common-mode signal left at either U_1 or U_2, the CMRR of the op-amps is unimportant. The ultimate CMRR of this circuit is thus set by the matching of resistor ratios $R_1/R_5 = R_3/R_6$, with no great demands made on the accuracy of R_2 and R_4. The circuit shown has a common-mode input range of ±200 volts, a CMRR of 80dB, and a differential gain of 1.0.

Burr-Brown's unity-gain INA117 uses a different trick to achieve large common-mode voltage range, namely a 200:1 resistive voltage divider to bring the ±200 volt signal within the op-amp's common-mode range of ±10 volt (Fig. 7.28). This scheme is simpler than Figure 7.27, but suffers from degraded offset and noise specs: V_{os} is 1000μV (versus 250μV for the INA105), and output noise voltage is 25μV pp (0.01–10Hz) versus 2.4μV for the INA105.

□ *Raising input impedance*

The differencing circuit with carefully trimmed resistor values would seem to give the performance you want, until you look at the restrictions it puts on allowable source resistances. To get a gain accuracy of 0.1% with the circuit of Figure 7.26, you have to keep the source impedance below

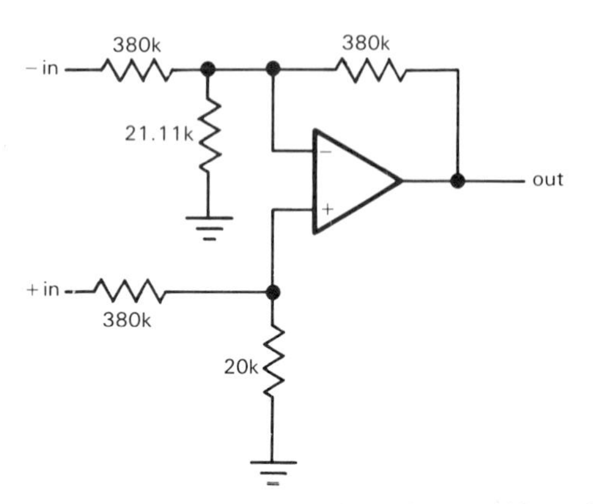

Figure 7.28. INA117 differential amplifier with ±200 volts common-mode input range.

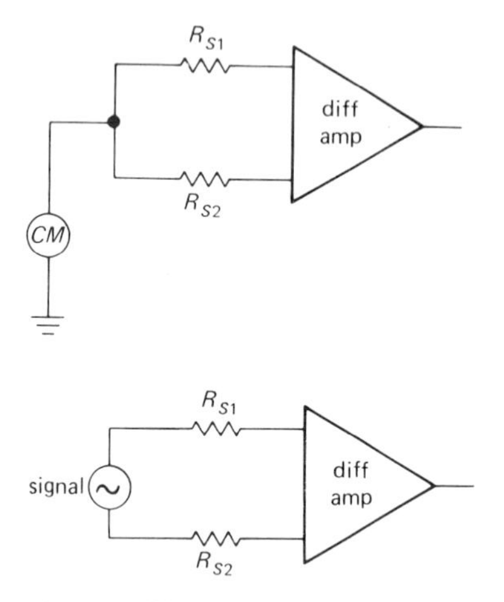

Figure 7.29

0.25 ohm! Furthermore, the source impedance seen at the two terminals has to be matched to 0.0025 ohm in order to attain a CMRR of 100dB. This last result follows from a look at the equivalent circuit (Fig. 7.29). The triangles represent the whole differential amplifier circuit or, in general, any differential or instrumentation amplifier, and R_{S1} and R_{S2} represent the Thévenin source resistances in each leg. For common-mode signals, the overall amplifier circuit includes the two source impedances in series with the input resistors R_1 and R_3, and so the CMRR now depends on the matching of $R_{S1}+R_1$ with $R_{S2}+R_3$. Obviously the demands this circuit makes on the source impedances as calculated earlier are unreasonable.

Some improvement can be had by increasing the resistor values, using the trick of a T network for the feedback resistors, as in Figure 7.30. This is the differential amplifier version of the T network discussed in Sections 7.06 and 4.19. With the values shown, you get a differential voltage gain of 1000 (60dB). For a gain accuracy of 0.1%, the source impedance must be less than 25 ohms and must be matched to 0.25 ohm for 100dB CMRR. This is still an unacceptable demand on the source

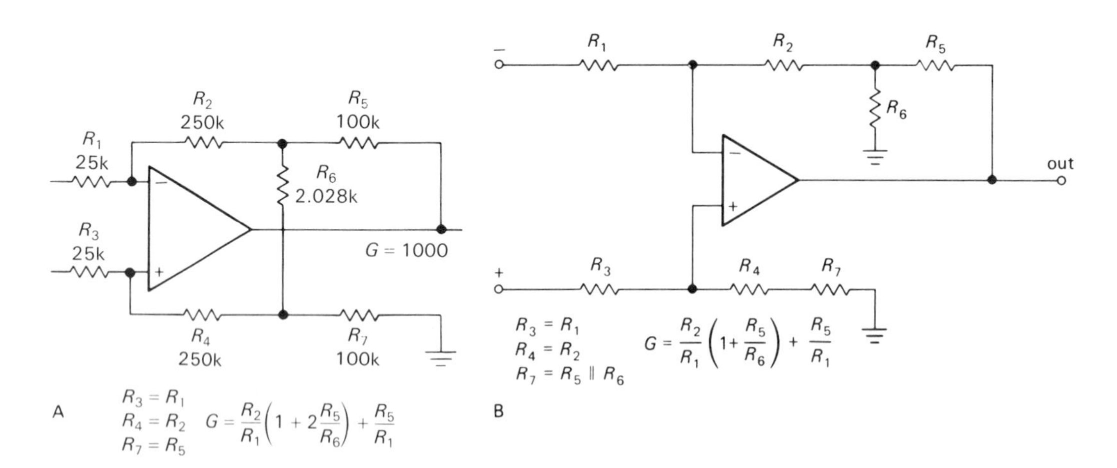

Figure 7.30. Differential amplifiers with T networks allow higher input impedances with smaller feedback resistors.

in most applications. A strain gauge, for instance, typically has a source impedance of about 350 ohms.

The general solution to this problem involves followers, or noninverting amplifiers, to attain high input impedance. The simplest method would be to add followers to the conventional differential amplifier (Fig. 7.31). With the enormous input impedances you get, there is no longer any problem with any reasonable source impedance, at least at dc. At higher frequencies it again becomes important to have matched source impedances relative to the common-mode signal, because the input capacitance of the circuit forms a voltage divider in combination with the source resistance. By "high frequencies" we often mean 60Hz, since common-mode ac power-line pickup is a common nuisance; at that frequency the effect of a few picofarads of input capacitance isn't serious.

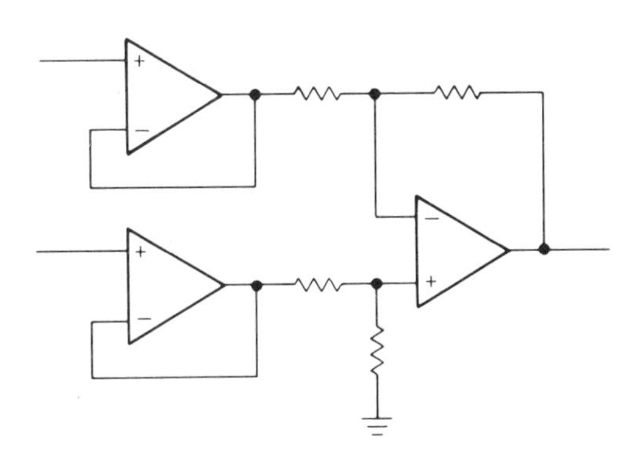
Figure 7.31. Differential amplifier with high Z_{in}.

7.10 Standard three-op-amp instrumentation amplifier

One disadvantage of the previous follower circuit (Fig. 7.31) is that it requires high CMRR both in the followers and in the final op-amp. Since the input buffers operate at unity gain, all the common-mode rejection must come in the output amplifier, requiring precise resistor matching, as we discussed. The circuit in Figure 7.32 is a significant improvement in this respect. It constitutes the standard instrumentation amplifier configuration. The input stage is a clever configuration of two op-amps that provides high differential gain and unity common-mode gain without any close resistor matching. Its differential output represents a signal with substantial reduction in the comparative common-mode signal, and it is used to drive a conventional differential amplifier circuit. The latter is often arranged for unity gain and is used to generate a single-ended output and polish off any remaining common-mode signal. As a result, the output op-amp, U_3, needn't have exceptional CMRR itself, and resistor matching in U_3's circuit is not terribly critical. Offset trimming for the whole circuit can be done at one of the input op-amps, as shown. The input op-amps must still have high CMRR, and they should be chosen carefully.

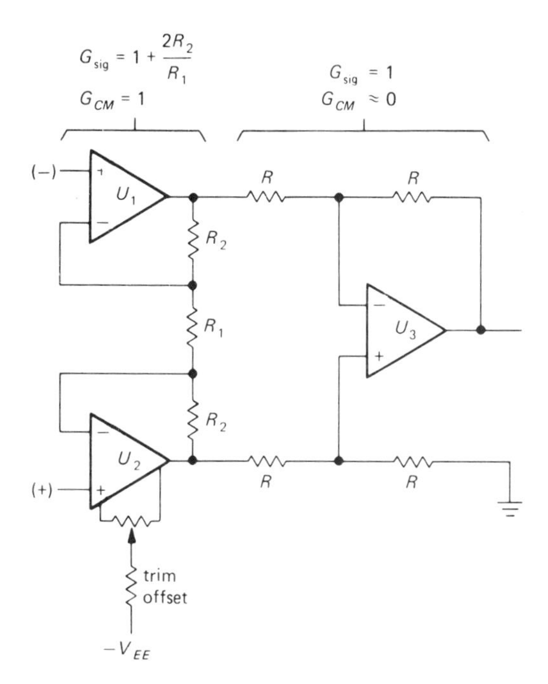

Figure 7.32. Classic instrumentation amplifier.

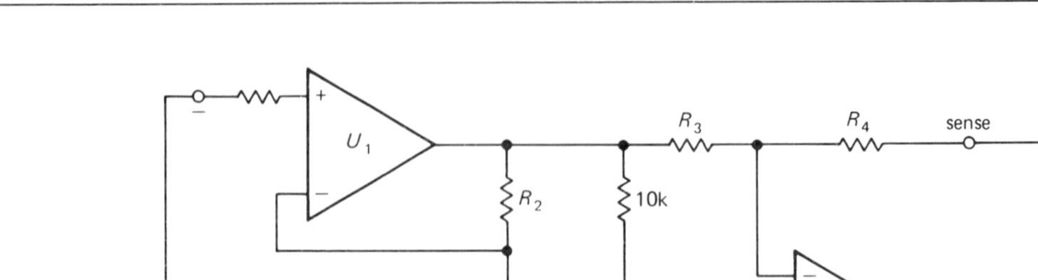

Figure 7.33. Instrumentation amplifier with guard, sense, and reference terminals.

Complete instrumentation amplifier ICs containing this standard configuration are available from several manufacturers. All components except R_1 are internal, with gain set by the single external resistor R_1. Typical examples are the micropower INA102, high-speed INA110, and the high-accuracy AD624. All of these amplifiers offer a gain range of 1 to 1000, CMRR in the neighborhood of 100dB, and input impedances greater than 100M. The micropower hybrid LH0036 can run from supply voltages as low as ±1 volt. The AD624 offers gain linearity of 0.001%, initial offset voltage of 25μV, and offset drift of 0.25μV/°C, with provision for external trimming of offset voltage. Some instrumentation amplifiers (e.g., the high-accuracy INA104) have provision for CMRR trimming. Don't confuse these with the 725 "instrumentation operational amplifier," which is nothing more than a good op-amp intended as a building block for instrumentation amplifiers. Figure 7.33 shows the complete instrumentation amplifier circuit that is usually used.

A few comments about these instrumentation amplifier circuits (Fig. 7.33): (a) The buffered common-mode signal at U_4's output can be used as a "guard" voltage to reduce the effects of cable capacitance and leakage. When used this way, the guard output will be tied to the shield of the input cables. If the gain-setting resistor (R_1) is not immediately adjacent to the amplifier (e.g., if it is a panel adjustment, a configuration that should usually be avoided), its connections should be shielded and guarded also. (b) The SENSE and REF terminals allow sensing of output voltage *at the load* so that feedback can operate to eliminate losses in the wiring or external circuit. In addition, the REF terminal also allows you to offset the output signal by a dc level (or by another signal); however, the impedance from the ref terminal to ground must be kept small, or the CMRR will be degraded. (c) With any of these instrumentation amplifiers there must be a bias path for input current; for example, you can't just connect a thermocouple across the input. Figure 7.34 shows the simple application of an IC instrumentation amplifier with guard, sense, and reference terminals.

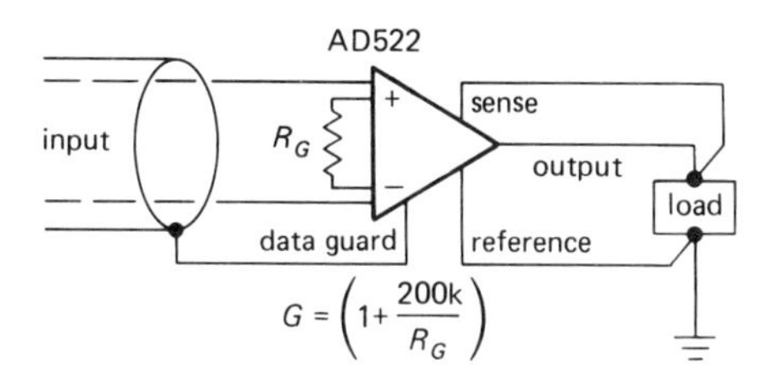

Figure 7.34. IC instrumentation amplifier.

□ *Bootstrapped power supply*

The CMRR of the input op-amps may be the limiting factor in the ultimate common-mode rejection of this circuit. If CMRRs greater than about 120dB are needed, the trick shown in Figure 7.35 can be used. U_4 buffers the common-mode signal level, driving the common terminal of a small floating split supply for U_1 and U_2. This bootstrapping scheme effectively eliminates the input common-mode signal from U_1 and U_2, because they see no swing (due to common-mode signals) at their inputs relative to their power supplies. U_3 and U_4 are powered by the system power supply, as usual. This scheme

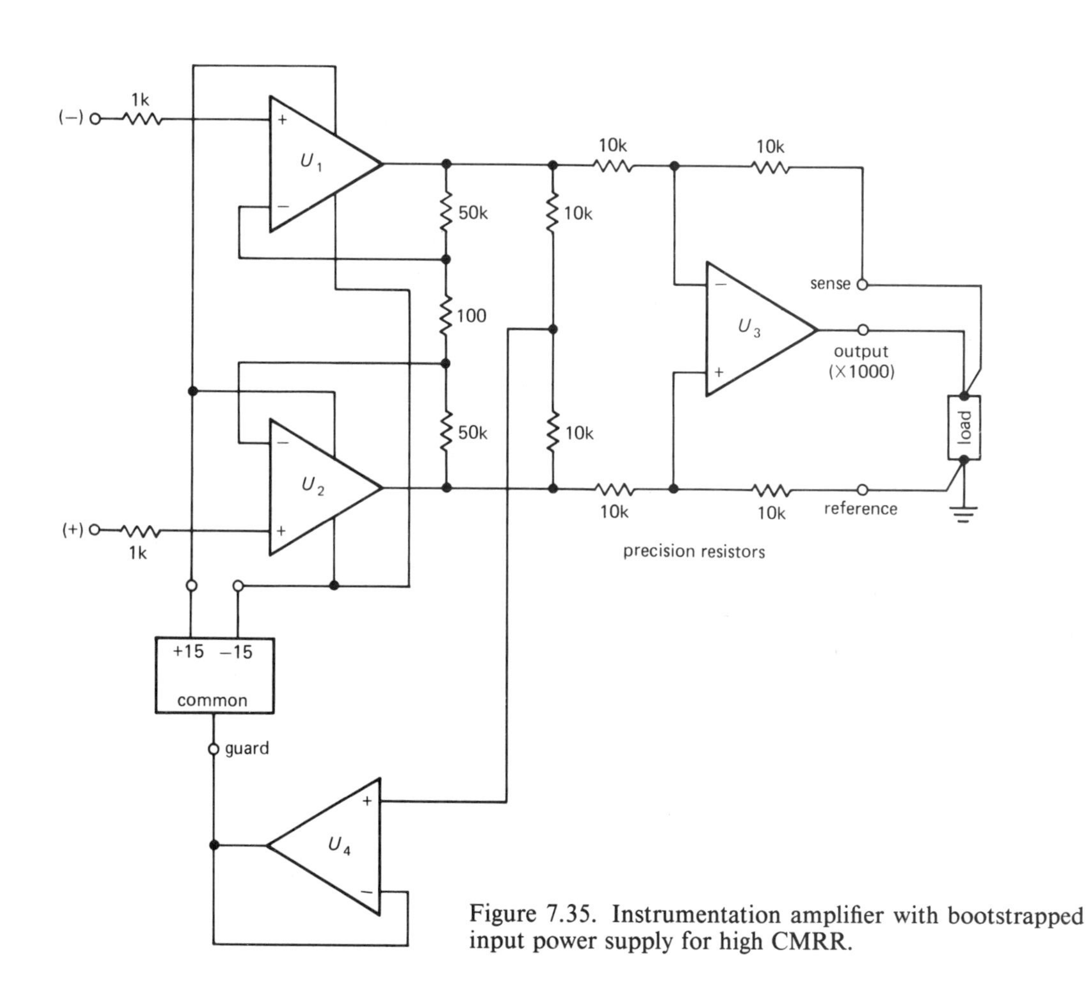

Figure 7.35. Instrumentation amplifier with bootstrapped input power supply for high CMRR.

can do wonders for the CMRR, at least at dc. At increasing frequencies you have the usual problems of presenting matched impedances to the input capacitances.

Two-op-amp configuration

Figure 7.36 shows another configuration that offers high input impedance with only two op-amps. Since it doesn't accomplish the common-mode rejection in two stages, as in the three-op-amp circuit, it requires precise resistor matching for good CMRR, in a manner similar to that of the standard differencing amplifier circuit.

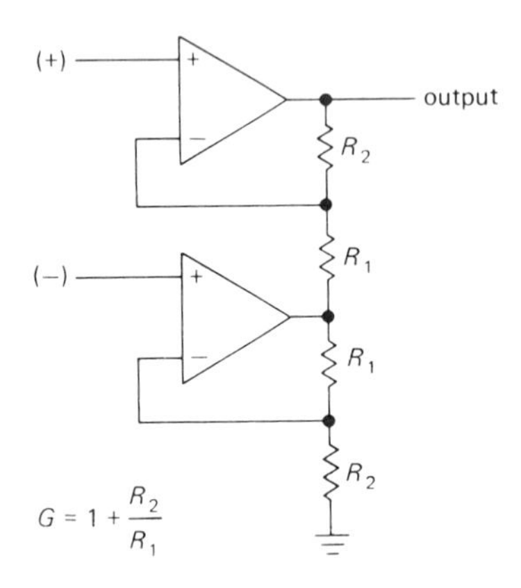

Figure 7.36. Instrumentation amplifier circuit with two op-amps.

Special IC instrumentation amplifiers

There are several interesting instrumentation amplifier configurations available as monolithic (and therefore inexpensive) ICs, some with extremely good performance. They use methods unrelated to the preceding circuits.

□ *Current-feedback amplifier technique.* This technique, typified by the LM363, AD521, and JFET AMP-05, achieves high CMRR without the need for matched external resistors. In fact, the gain is set by the ratio of a pair of external resistors. Figure 7.37 shows a block diagram of the AMP-01. The circuit employs two differential transconductance amplifier pairs, with a single external resistor setting the gain in each case. One pair is driven by the input signal, and the other is driven by the output signal, relative to the ref terminal. The AMP-05 uses FETs to keep input currents low, whereas the AMP-01 uses bipolar technology to achieve low offset voltage and drift (Table 7.5).

Computer-aided design methods can be extremely useful in precision circuit design; see Section 13.24.

AMPLIFIER NOISE

In almost every area of measurement the ultimate limit of detectability of weak signals is set by noise – unwanted signals that obscure the desired signal. Even if the quantity being measured is not weak, the presence of noise degrades the accuracy of the measurement. Some forms of noise are unavoidable (e.g., real fluctuations in the quantity being measured), and they can be overcome only with the techniques of *signal averaging* and *bandwidth narrowing*, which we will discuss in Chapter 15. Other forms of noise (e.g., radiofrequency interference and "ground loops") can be reduced or eliminated by a variety of tricks, including filtering and careful attention to wiring configuration and parts location. Finally, there is noise that arises in the amplification process itself, and it can be reduced through the techniques of low-noise amplifier design. Although the techniques of signal averaging can often be used to rescue a signal buried in noise, it always pays to begin with a system that is free of preventable interference and that possesses the lowest amplifier noise practicable.

We will begin by talking about the origins and characteristics of the different

TABLE 7.5. INSTRUMENTATION AMPLIFIERS

Type	FET	Curr fdbk	Total supply: Voltage min (V)	Total supply: Voltage max (V)	Total supply: Curr max (mA)	Max input errors[b]: Offset voltage RTI[a] (mV)	Offset voltage RTI[a] (μV/°C)	Offset voltage RTO[a] (mV)	Offset voltage RTO[a] (μV/°C)	Current: Bias (nA)	Current: Offset (nA)	Noise voltage 0.1-10Hz RTI[a] (μV, pp)	0.1-10Hz RTO[a] (μV, pp)	10Hz-10kHz RTI[a] (μV, rms)	10Hz-10kHz RTO[a] (μV, rms)	Noise Current 10Hz-10kHz (pA, rms)	CMRR (@dc, min) G=1 (dB)	CMRR G=1k (dB)	Slew rate (V/μs)	-3dB bandwidth G=1 (kHz)	-3dB bandwidth G=1k (kHz)	Bandwidth for 1% error G=1 (kHz)	Bandwidth for 1% error G=1k (kHz)	Settling time to 1% G=1 (μs)	Settling time to 1% G=1k (μs)
AMP-01A	–	•	9	36	5	0.05	0.3	3	50	3	1	0.1	13	0.5	-	-	85	125	4.5			570	26	12	50
AMP-05A	•	•	10	36	10	1	10	15	100	0.05	0.025	4	7	3	-	1	90	110	7.5	3000	120	-	-	5	5
LH0036	–	–	2	36	0.6	1	10[t]	5	15[t]	100	40	-	-	5	-	-	50	100	0.3	350	0.35	-	-	8	600
LH0038[c]	–	–	10	36	2	0.1	0.25	10	25[t]	100	5	0.2	-	0.6	-	10	-	114	0.3	-	1.6	-	-	-	80[d]
INA101C	–	–	10	40	8.5	0.025	0.25	0.2	10	20	20	0.8	-	1.5	-	50	80	106	0.4	300	2.5	20	0.2	30	500
INA102C	–	–	7	36	0.8	0.1	2	0.2	5	30	10	0.1	-	2.5	8	20	90	90	0.2	300	0.3	30	0.03	50	3300
INA104C	–	–	10	40	10	0.025	0.25	0.2	10	20	20	0.8	-	1.5	-	50	80	106	0.4	300	2.5	20	0.2	30	350
INA110B	•	–	12	36	4.5	0.25	2	3	50	0.05	0.025	1	8	5	10	-	80	106	17	2500	100[f]	-	-	4	11[f]
LM363A	–	•	10	36	2	0.05	0.5	10	250	5	2	0.4[h]	100[h]	1.2	100	15	100[g]	126	0.4	200[g]	30	30[g]	5	20[g]	70
AD521	–	•	10	36	5	3	15	400	400	80	20	0.5	150	1.2	30	-	70	100	10	2000	40	75	6	7	35
AD522	–	–	10	36	10	-	6	0.4	50	25	20	1.5	15	-	15	-	75	100	0.1	300	0.3	-	-	500[d]	20000[d]
AD524C	–	–	12	36	5	0.05	0.5	2	25	15	10	0.3	15	0.5	10	40	80	120	5	1000	25			10	50
AD624C	–	–	10	36	5	0.025	0.25	2	10	15	10	0.2	10	0.5	10	40	80	130	5	1000	25	-	-	10	50
AD625C	–	–	10	36	5	0.025	0.25	2	15	15	5	0.2	-	0.4	7	30	80	120	5	650	25	-	-	15[d]	75[d]
ICL7605[e]	•	–	4	18	5	0.005	0.2	-	-	1.5	-	1.7	-	-	-	-	100[t]	100[t]	0.5	0.01	0.01	slow		slow	

(a) RTI: referred to the input; RTO: referred to the output. Noise and errors can be separated into components generated at both the input and output. The total input-referred noise (or error) is thus given by *RTI+RTO/G*. (b) diff'l input impedance > 1GΩ except LH0038 (5MΩ), AMP-05 (1TΩ), and INA110B (5TΩ). (c) gain range 10-2000. (d) to 0.01%. (e) CAZ type (see section 7.10); 7606 is uncomp. (f) G = 500. (g) G = 10. (h) 0.01Hz to 10Hz. (t) typical.

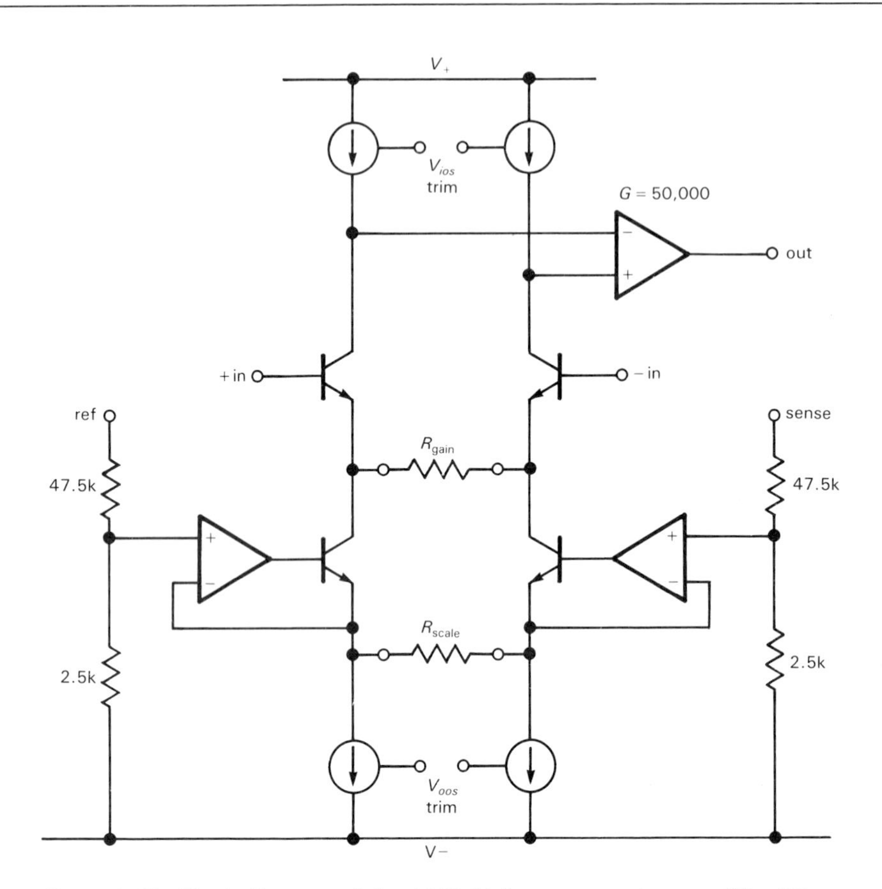

Figure 7.37. Block diagram of the AMP-01 instrumentation amplifier IC.

kinds of noise that afflict electronic circuits. Then we will launch into a discussion of transistor and FET noise, including methods for low-noise design with a given signal source, and will present some design examples. After a short discussion of noise in differential and feedback amplifiers, we will conclude with a section on proper grounding and shielding and the elimination of interference and pickup. See also Section 13.24 (Analog modeling tools).

7.11 Origins and kinds of noise

Since the term *noise* can be applied to anything that obscures a desired signal, noise can itself be another signal ("interference"); most often, however, we use the term to describe "random" noise of a physical (often thermal) origin. Noise can be characterized by its frequency spectrum, its amplitude distribution, and the physical mechanism responsible for its generation. Let's next look at the chief offenders.

Johnson noise

Any old resistor just sitting on the table generates a noise voltage across its terminals known as Johnson noise. It has a flat frequency spectrum, meaning that there is the same noise power in each hertz of frequency (up to some limit, of course). Noise with a flat spectrum is also called "white noise." The actual open-circuit noise voltage generated by a resistance R at temperature T is given by

$$V_{\text{noise}}(\text{rms}) = V_{nR} = (4kTRB)^{\frac{1}{2}}$$

where k is Boltzmann's constant, T is the absolute temperature in degrees Kelvin (°K = °C + 273.16), and B is the bandwidth in hertz. Thus, $V_{\text{noise}}(\text{rms})$ is what you would measure at the output if you drove a perfect noiseless bandpass filter (of bandwidth B) with the voltage generated by a resistor at temperature T. At room temperature (68°F = 20°C = 293°K),

$$4kT = 1.62 \times 10^{-20}\text{V}^2/\text{Hz} - \Omega$$

$$(4kTR)^{\frac{1}{2}} = 1.27 \times 10^{-10} R^{\frac{1}{2}} \quad \text{V/Hz}^{\frac{1}{2}}$$

$$= 1.27 \times 10^{-4} R^{\frac{1}{2}} \quad \mu\text{V/Hz}^{\frac{1}{2}}$$

For example, a 10k resistor at room temperature has an open-circuit rms voltage of 1.3μV, measured with a bandwidth of 10kHz (e.g., by placing it across the input of a high-fidelity amplifier and measuring the output with a voltmeter). The source resistance of this noise voltage is just R. Figure 7.38 plots the simple relationship between Johnson-noise voltage density (rms voltage per square root bandwidth) and source resistance.

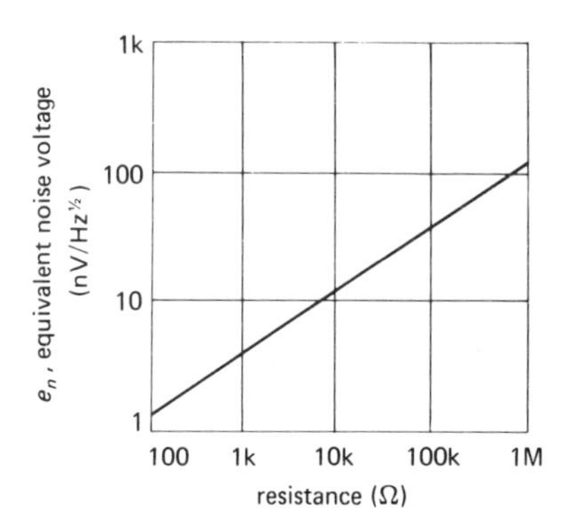

Figure 7.38. Thermal noise voltage versus resistance.

The amplitude of the Johnson-noise voltage at any instant is, in general, unpredictable, but it obeys a Gaussian amplitude distribution (Fig. 7.39),

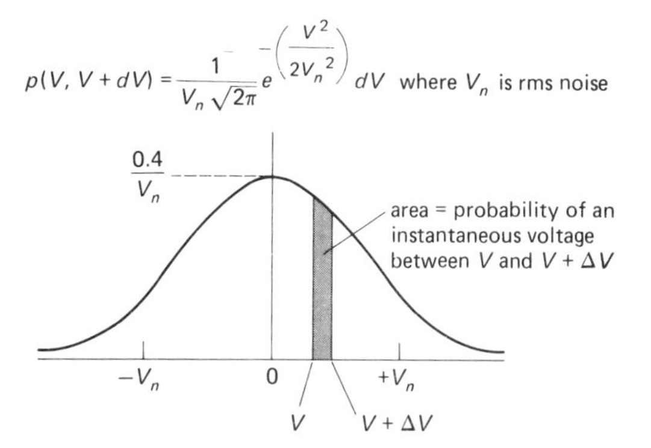

Figure 7.39

where $p(V)dV$ is the probability that the instantaneous voltage lies between V and $V + dV$, and V_n is the rms noise voltage, given earlier.

The significance of Johnson noise is that it sets a lower limit on the noise voltage in any detector, signal source, or amplifier having resistance. The resistive part of a source impedance generates Johnson noise, as do the bias and load resistors of an amplifier. You will see how it all works out shortly.

It is interesting to note that the physical analog of resistance (any mechanism of energy loss in a physical system, e.g., viscous friction acting on small particles in a liquid) has associated with it fluctuations in the associated physical quantity (in this case, the particles' velocity, manifest as the chaotic Brownian motion). Johnson noise is just a special case of this fluctuation-dissipation phenomenon.

Johnson noise should not be confused with the additional noise voltage created by the effect of resistance fluctuations when an externally applied current flows through a resistor. This "excess noise" has a $1/f$ spectrum (approximately) and is heavily dependent on the actual construction of the resistor. We will talk about it later.

Shot noise

An electric current is the flow of discrete electric charges, not a smooth fluidlike flow. The finiteness of the charge quantum results in statistical fluctuations of the current. If the charges act independent of each other, the fluctuating current is given by

$$I_{\text{noise}}(\text{rms}) = I_{nR} = (2qI_{dc}B)^{\frac{1}{2}}$$

where q is the electron charge (1.60×10^{-19} coulomb) and B is the measurement bandwidth. For example, a "steady" current of 1 amp actually has an rms fluctuation of 57nA, measured in a 10kHz bandwidth; i.e., it fluctuates by about 0.000006%. The relative fluctuations are larger for smaller currents: A "steady" current of 1μA actually has an rms current noise fluctuation, measured over a 10kHz bandwidth, of 0.006%, i.e., −85dB. At 1pA dc, the rms current fluctuation (same bandwidth) is 56fA, i.e., a 5.6% variation! Shot noise is "rain on a tin roof." This noise, like resistor Johnson noise, is Gaussian and white.

The shot-noise formula given earlier assumes that the charge carriers making up the current act independently. That is indeed the case for charges crossing a barrier, as for example the current in a junction diode, where the charges move by diffusion; but it is not true for the important case of metallic conductors, where there are long-range correlations between charge carriers. Thus, the current in a simple resistive circuit has far less noise than is predicted by the shot-noise formula. Another important exception to the shot-noise formula is provided by our standard transistor current-source circuit (Fig. 2.21), in which negative feedback acts to quiet the shot noise.

EXERCISE 7.4

A resistor is used as the collector load in a low-noise amplifier; the collector current I_C is accompanied by shot noise. Show that the output noise voltage is dominated by shot noise (rather than Johnson noise in the resistor) as long as the quiescent voltage drop across the load resistor is greater than $2kT/q$ (50mV, at room temperature).

1/f noise (flicker noise)

Shot noise and Johnson noise are irreducible forms of noise generated according to physical principles. The most expensive and most carefully made resistor has exactly the same Johnson noise as the cheapest carbon resistor (of the same resistance). Real devices have, in addition, various sources of "excess noise." Real resistors suffer from fluctuations in resistance, generating an additional noise voltage (which adds to the ever-present Johnson noise) proportional to the dc current flowing through them. This noise depends on many factors having to do with the construction of the particular resistor, including the resistive material and especially the end-cap connections. Here is a listing of typical excess noise for various resistor types, given as rms microvolts per volt applied across the resistor, measured over one decade of frequency:

Carbon-composition	0.10μV to 3.0μV
Carbon-film	0.05μV to 0.3μV
Metal-film	0.02μV to 0.2μV
Wire-wound	0.01μV to 0.2μV

This noise has approximately a $1/f$ spectrum (equal power per decade of frequency) and is sometimes called "pink noise." Other noise-generating mechanisms often produce $1/f$ noise, examples being base current noise in transistors and cathode current noise in vacuum tubes. Curiously enough, $1/f$ noise is present in nature in unexpected places, e.g., the speed of ocean currents, the flow of sand in an hourglass, the flow of traffic on Japanese expressways, and the yearly flow of the Nile measured over the last 2000 years. If you plot the loudness of a piece of classical music versus time, you get a $1/f$ spectrum! No unifying principle has been found for all the $1/f$ noise that seems to be swirling around us, although particular sources can often be identified in each instance.

Interference

As we mentioned earlier, an interfering signal or stray pickup constitutes a form of noise. Here the spectrum and amplitude characteristics depend on the interfering signal. For example, 60Hz pickup has a sharp spectrum and relatively constant amplitude, whereas car ignition noise, lightning, and other impulsive interferences are broad in spectrum and spiky in amplitude. Other sources of interference are radio and television stations (a particularly serious problem near large cities), nearby electrical equipment, motors and elevators, subways, switching regulators, and television sets. In a slightly different guise you have the same sort of problem generated by anything that puts a signal into the parameter you are measuring. For example, an optical interferometer is susceptible to vibration, and a sensitive radiofrequency measurement (e.g., NMR) can be affected by ambient radiofrequency signals. Many circuits, as well as detectors and even cables, are sensitive to vibration and sound; they are *microphonic*, in the terminology of the trade.

Many of these noise sources can be controlled by careful shielding and filtering, as we will discuss later in the chapter. At other times you are forced to take draconian measures, involving massive stone tables (for vibration isolation), constant-temperature rooms, anechoic chambers, and electrically shielded rooms.

7.12 Signal-to-noise ratio and noise figure

Before getting into the details of amplifier noise and low-noise design, we need to define a few terms that are often used to describe amplifier performance. These involve ratios of noise voltages, measured at the same place in the circuit. It is conventional to refer noise voltages to the input of an amplifier (although the measurements are usually made at the output), i.e., to describe source noise and amplifier noise in terms of microvolts *at the input* that would generate the observed output noise. This makes sense when you want to think of the relative noise added by the amplifier to a given signal, independent of amplifier gain; it's also realistic, because most of the amplifier noise is usually contributed by the input stage. Unless we state otherwise, noise voltages are referred to the input.

Noise power density and bandwidth

In the preceding examples of Johnson noise and shot noise, the noise voltage you measure depends both on the measurement bandwidth B (i.e., how much noise you see depends on how fast you look) and on the variables (R and I) of the noise source itself. So it's convenient to talk about an rms noise-voltage "density" v_n:

$$V_n(\text{rms}) = v_n B^{\frac{1}{2}} = (4kTR)^{\frac{1}{2}} B^{\frac{1}{2}}$$

where V_n is the rms noise voltage you would measure in a bandwidth B. White-noise sources have a v_n that doesn't depend on frequency, whereas pink noise, for instance, has a v_n that drops off at 3dB/octave. You'll often see v_n^2, too, the mean squared noise density. Since v_n always refers to rms, and v_n^2 always refers to mean square, you can just square v_n to get v_n^2! Sounds simple (and it is), but we want to make sure you don't get confused.

Note that B and the square root of B keep popping up. Thus, for example, for Johnson noise from a resistor R

$$v_{nR}(\text{rms}) = (4kTR)^{\frac{1}{2}} \qquad \text{V/Hz}^{\frac{1}{2}}$$

$$v_{nR}^2 = 4kTR \qquad \text{V}^2\text{/Hz}$$

$$V_n(\text{rms}) = v_{nR} B^{\frac{1}{2}} = (4kTRB)^{\frac{1}{2}} \qquad \text{V}$$

$$V_n^2 = v_{nR}^2 B = 4kTRB \qquad \text{V}^2$$

On data sheets you may see graphs of v_n or v_n^2, with units like "nanovolts per root Hz" or "volts squared per Hz." The quantities e_n and i_n that will soon appear work just the same way.

When you add two signals that are uncorrelated (two noise signals, or noise plus a real signal), the *squared* amplitudes add:

$$v = (v_s^2 + v_n^2)^{\frac{1}{2}}$$

where v is the rms signal obtained by adding together a signal of rms amplitude v_s and a noise signal of rms amplitude v_n. The rms amplitudes *don't* add.

Signal-to-noise ratio

Signal-to-noise ratio (SNR) is simply defined as

$$\text{SNR} = 10\log_{10}\left(\frac{V_s^2}{V_n^2}\right) \quad \text{dB}$$

where the voltages are rms values, and some bandwidth and center frequency are specified; i.e., it is the ratio, in decibels, of the rms voltage of the desired signal to the rms voltage of the noise that is also present. The "signal" itself may be sinusoidal or a modulated information-carrying waveform or even a noiselike signal itself. It is particularly important to specify the bandwidth if the signal has some sort of narrowband spectrum, since the SNR will drop as the bandwidth is increased beyond that of the signal: The amplifier keeps adding noise power, while the signal power remains constant.

Noise figure

Any real signal source or measuring device generates noise because of Johnson noise in its source resistance (the real part of its complex source impedance). There may be additional noise, of course, from other causes. The *noise figure* (NF) of an amplifier is simply the ratio, in decibels, of the output of the real amplifier to the output of a "perfect" (noiseless) amplifier of the same gain, with a resistor of value R_s connected across the amplifier's input terminals in each case. That is, the Johnson noise of R_s is the "input signal."

$$\text{NF} = 10\log_{10}\left(\frac{4kTR_s + v_n^2}{4kTR_s}\right)$$

$$= 10\log_{10}\left(1 + \frac{v_n^2}{4kTR_s}\right) \quad \text{dB}$$

where v_n^2 is the mean squared noise voltage per hertz contributed by the amplifier, with a noiseless (cold) resistor of value R_s connected across its input. This latter restriction is important, as you will see shortly, because the noise voltage contributed by an amplifier depends very much on the source impedance (Fig. 7.40).

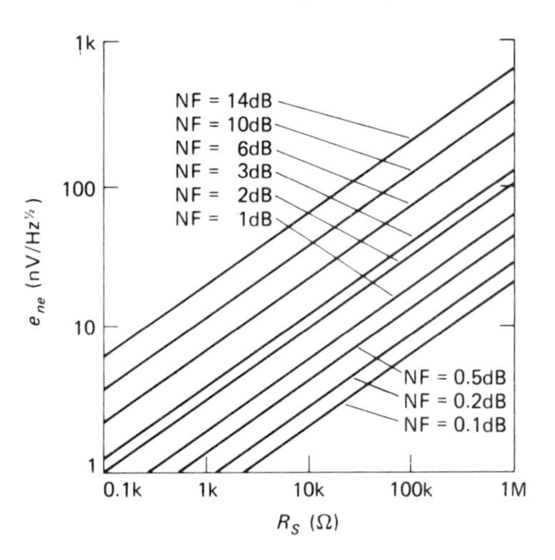

Figure 7.40. Effective noise voltage versus noise figure and source resistance. (National Semiconductor Corp.)

Noise figure is handy as a figure of merit for an amplifier when you have a signal source of a given source impedance and want to compare amplifiers (or transistors, for which NF is often specified). NF varies with frequency and source impedance, and it is often given as a set of contours of constant NF versus frequency and R_s. It may also be given as a set of graphs of NF versus frequency, one curve for each collector current, or a similar set of graphs of NF versus R_s, one for each collector current. Note: The foregoing expressions for NF assume that the amplifier's input impedance is much larger than the source impedance, i.e., $Z_{\text{in}} \gg R_s$. However, in the special case of radiofrequency amplifiers, you usually have $R_s = Z_{\text{in}} = 50$ ohms, with NF defined accordingly. For this special case of matched impedances, simply remove the factors "4" from the foregoing equations.

Big fallacy: Don't try to improve things by adding a resistor in series with a signal source to reach a region of minimum NF. All you're doing is making the source noisier to make the amplifier look better! Noise figure can be very deceptive for this reason. To add to the deception, the NF specification (e.g., NF = 2dB) for a transistor or FET will always be for the optimum combination of R_s and I_C. It doesn't tell you much about actual performance, except that the manufacturer thinks the noise figure is worth bragging about.

In general, when evaluating the performance of some amplifier, you're probably least likely to get confused if you stick with SNR calculated for that source voltage and impedance. Here's how to convert from NF to SNR:

$$\text{SNR} = 10 \log_{10}\left(\frac{v_s^2}{4kTR_s}\right) - \text{NF(dB) (at } R_s) \quad \text{dB}$$

where v_s is the rms signal amplitude, R_s is the source impedance, and NF is the noise figure of the amplifier for source impedance R_s.

□ *Noise temperature*

Rather than noise *figure*, you sometimes see noise *temperature* used to express the noise performance of an amplifier. Both methods give the same information, namely the excess noise contribution of the amplifier when driven by a signal source of impedance R_s; they are equivalent ways of expressing the same thing.

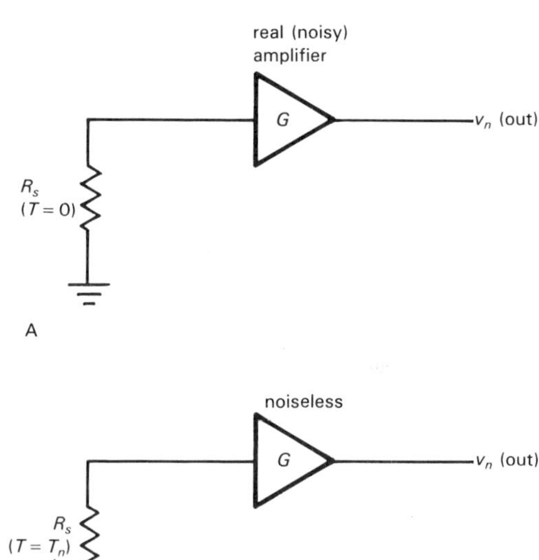

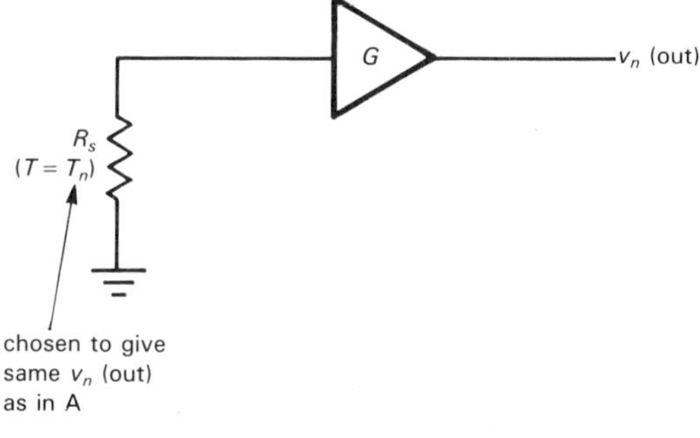

Figure 7.41

Look at Figure 7.41 to see how noise temperature works: We first imagine the actual (noisy) amplifier connected to a *noiseless* source of impedance R_s (Fig. 7.41A). If you have trouble imagining a noiseless source, think of a resistor of value R_s cooled to absolute zero. There will be some noise at the output, even though the source is noiseless, because the amplifier has noise. Now imagine constructing Figure 7.41B, where we magically make the amplifier noiseless, and bring the source R_s up to some temperature T_n such that *the output noise voltage is the same as in Figure 7.41A*. T_n is called the noise temperature of the amplifier, for source impedance R_s.

As we remarked earlier, noise figure and noise temperature are simply different ways of conveying the same information. In fact, you can show that they are related by the following expressions:

$$T_n = T\left(10^{\mathrm{NF(dB)}/10} - 1\right)$$

$$\mathrm{NF(dB)} = 10\log_{10}\left(\frac{T_n}{T} + 1\right)$$

where T is the ambient temperature, usually taken as 290°K.

Generally speaking, good low-noise amplifiers have noise temperatures far below room temperature (or, equivalently, they have noise figures far less than 3dB). Later in the chapter we will explain how you go about measuring the noise figure (or temperature) of an amplifier. First, however, we need to understand noise in transistors and the techniques of low-noise design. We hope the discussion that follows will clarify what is often a murky subject!

After reading the next two sections, we trust you won't ever be confused about noise figure again!

7.13 Transistor amplifier voltage and current noise

The noise generated by an amplifier is easily described by a simple noise model that is accurate enough for most purposes. In Figure 7.42, e_n represents a noise voltage source in series with the input, and i_n represents an input noise current. The transistor (or amplifier, in general) is assumed noiseless, and it simply amplifies the input noise voltage it sees. That is, the amplifier contributes a total noise voltage e_a, referred to the input, of

$$e_a(\mathrm{rms}) = [e_n^2 + (R_s i_n)^2]^{\frac{1}{2}} \quad \mathrm{V/Hz}^{\frac{1}{2}}$$

The two terms are simply the amplifier input noise voltage and the noise voltage generated by the amplifier's input noise current passing through the source resistance. Since the two noise terms are usually uncorrelated, their squared amplitudes add to produce the effective noise voltage seen by the amplifier. For low source resistances the noise voltage e_n dominates, whereas for high source impedances the noise current i_n generally dominates.

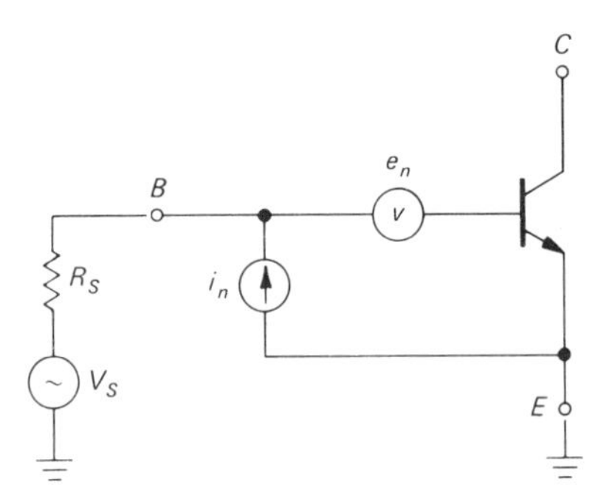

Figure 7.42. Noise model of a transistor.

Just to give an idea of what these look like, Figure 7.43 shows a graph of e_n and i_n versus I_C and f, for a 2N5087. We'll go into some detail now, describing these and showing how to design for minimum noise. It is worth noting that voltage noise and current noise for a transistor are in the range of nanovolts and picoamps per root hertz ($\mathrm{Hz}^{\frac{1}{2}}$).

Voltage noise, e_n

The equivalent voltage noise looking in series with the base of a transistor arises from Johnson noise in the base spreading resistance, r_{bb}, and collector current shot noise generating a noise voltage across the intrinsic emitter resistance r_e. These two terms look like this:

$$e_n^2 = 4kTr_{bb} + 2qI_C r_e^2$$

$$= 4kTr_{bb} + \frac{2(kT)^2}{qI_C} \quad \mathrm{V^2/Hz}$$

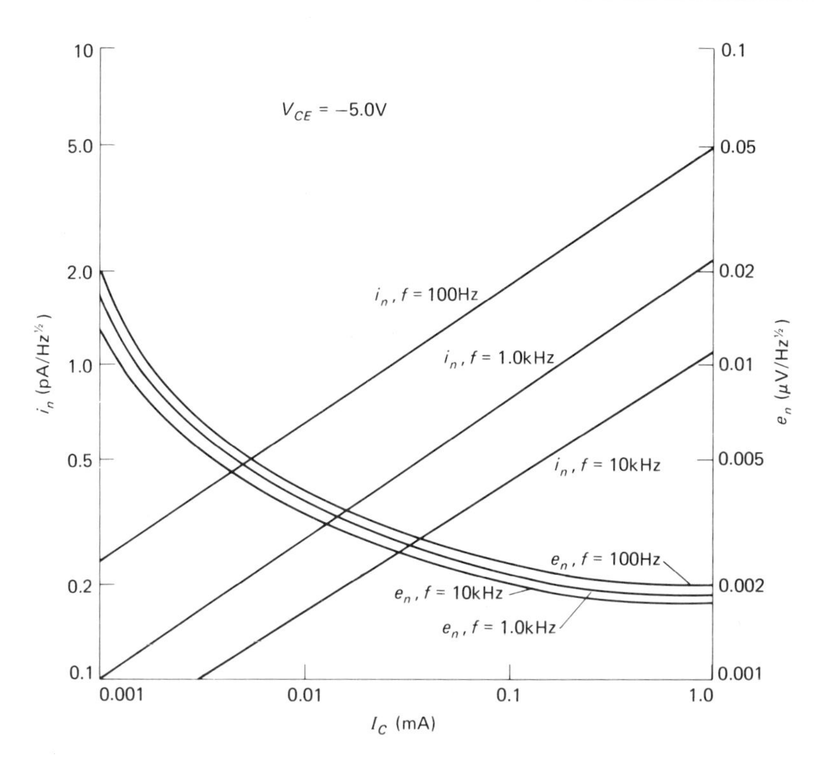

Figure 7.43. Equivalent rms input noise voltage (e_n) and noise current (i_n) versus collector current for a 2N5087 *npn* transistor. (Courtesy of Fairchild Camera and Instrument Corp.)

Both of these are Gaussian white noise. In addition, there is some flicker noise generated by base current flowing through r_{bb}. This last term is significant only at high base current, i.e., at high collector current. The result is that e_n is constant over a wide range of collector currents, rising at low currents (shot noise through an increasing r_e) and at sufficiently high currents (flicker noise from I_B through r_{bb}). This latter rise is present only at low frequencies, because of its $1/f$ character. As an example, at frequencies above 10kHz the 2N5087 has an e_n of $5\text{nV/Hz}^{\frac{1}{2}}$ at $I_C = 10\mu\text{A}$ and $2\text{nV/Hz}^{\frac{1}{2}}$ at $I_C = 100\mu\text{A}$. Figure 7.44 shows graphs of e_n versus frequency and current for the low-noise LM394 *npn* differential pair, and the low-noise 2SD786 from Toyo-Rohm. The latter uses special geometry to achieve an unusually low r_{bb} of 4 ohms, which is needed to realize the lowest values of e_n.

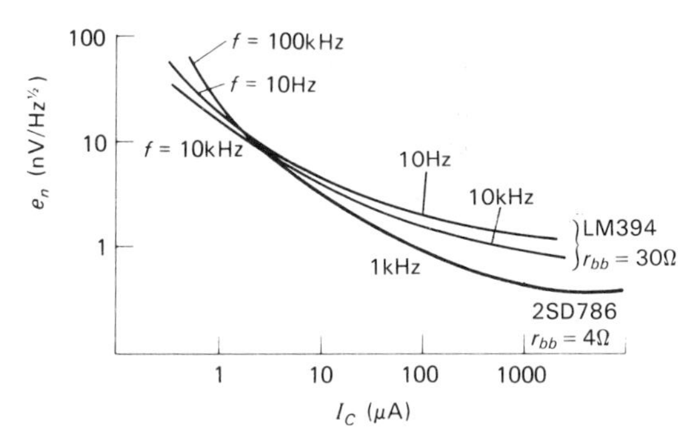

Figure 7.44. Input noise voltage (e_n) versus collector current for two low-noise bipolar transistors.

Current noise, i_n

Noise current is important, because it generates an additional noise voltage across the input signal source impedance. The main source of current noise is shot-noise fluctuation in the steady base current, added to the fluctuations caused by flicker

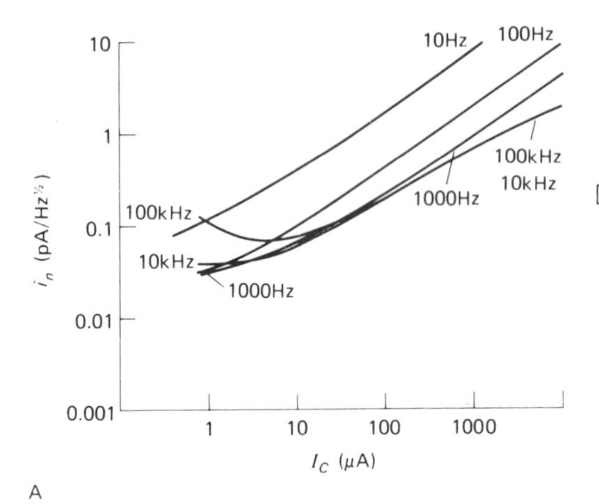

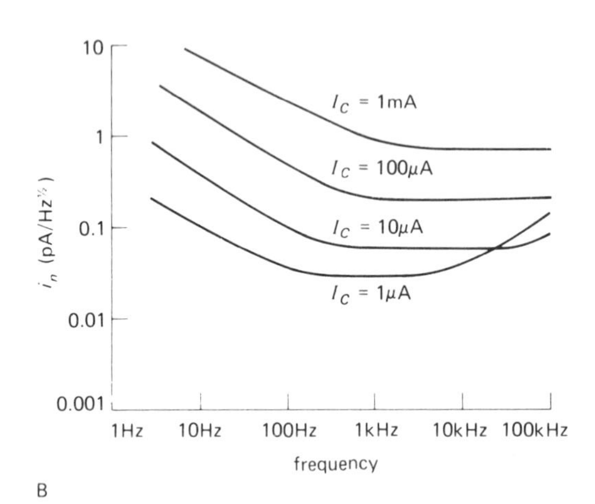

Figure 7.45. Input noise current for the LM394 bipolar transistor.
A. Noise current (i_n) versus collector current.
B. Noise current (i_n) versus frequency.

noise in r_{bb}. The shot-noise contribution is a noise current that increases proportional to the square root of I_B (or I_C) and is flat with frequency, whereas the flicker-noise component rises more rapidly with I_C and shows the usual $1/f$ frequency dependence. Taking the example of the 2N5087 again, above 10kHz i_n is about 0.1pA/Hz$^{1/2}$ at $I_C = 10\mu$A and 0.4pA/Hz$^{1/2}$ at $I_C = 100\mu$A. The noise current increases, and the noise voltage drops, as I_C is increased. In the next section you will see how this dictates operating current in low-noise design.

Figure 7.45 shows graphs of i_n versus frequency and current, again for the low-noise LM394.

□ 7.14 Low-noise design with transistors

The fact that e_n drops and i_n rises with increasing I_C provides a simple way to optimize transistor operating current to give lowest noise with a given source. Look at the model again (Fig. 7.46). The noiseless signal source v_s has added to it an irreducible noise voltage from the Johnson noise of its source resistance.

$$e_R^2(\text{source}) = 4kTR_s \qquad \text{V}^2/\text{Hz}$$

The amplifier adds noise of its own, namely,

$$e_a^2(\text{amplifier}) = e_n^2 + (i_nR_s)^2 \qquad \text{V}^2/\text{Hz}$$

Thus the amplifier's noise voltage is added to the input signal, and in addition, its noise current generates a noise voltage across the source impedance. These two are uncorrelated (except at very high frequencies), so you add their squares. The idea is to reduce the amplifier's total noise contribution as much as possible. That's easy, once you know R_s, because you just look at a graph of e_n and i_n versus I_C, in the region of the signal frequency, picking I_C to minimize $e_n^2 + (i_nR_s)^2$. Alternatively, if you are lucky and have a plot of noise-figure contours versus I_C and R_s, you can quickly locate the optimum value of I_C.

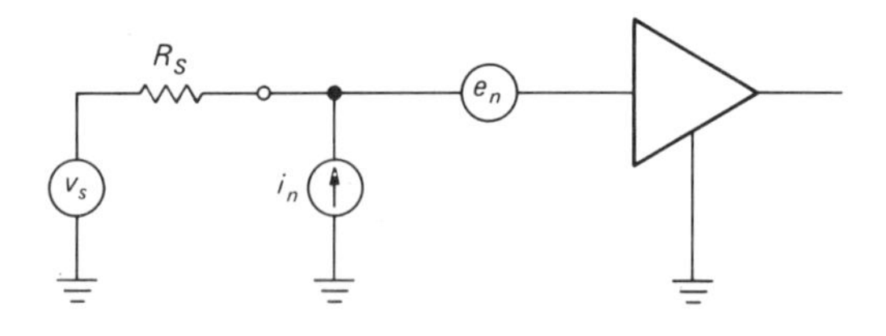

Figure 7.46. Amplifier noise model.

□ *Noise figure example*

As an example, suppose we have a small signal in the region of 1kHz with source

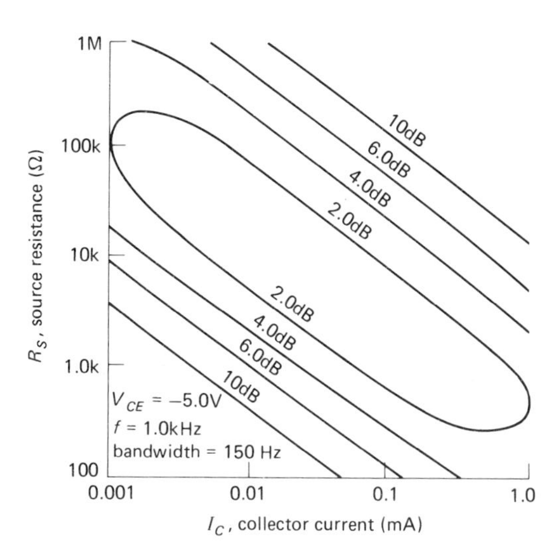

Figure 7.47. Contours of constant narrowband noise figure for the 2N5087 transistor. (Courtesy of Fairchild Camera and Instrument Corp.)

resistance of 10k, and we wish to make an amplifier with a 2N5087. From the e_n-i_n graph (Fig. 7.47) we see that the sum of voltage and current terms (with 10k source) is minimized for a collector current of about 10–20μA. Since the current noise is dropping faster than the voltage noise is rising as I_C is reduced, it might be a good idea to use slightly less collector current, especially if operation at a lower frequency is anticipated (i_n rises rapidly with decreasing frequency). We can estimate the noise figure using i_n and e_n at 1kHz:

$$\text{NF} = 10\log_{10}\left(1 + \frac{e_n^2 + (i_n R_s)^2}{4kTR_s}\right) \text{ dB}$$

For $I_C = 10\mu\text{A}$, $e_n = 3.8\text{nV/Hz}^{1/2}$, $i_n = 0.29\text{pA/Hz}^{1/2}$, and $4kTR_s = 1.65 \times 10^{-16}\text{V}^2/\text{Hz}$ for the 10k source resistance. The calculated noise figure is therefore 0.6dB. This is consistent with the graph (Fig. 7.48) showing NF versus frequency, in which they have chosen $I_C = 20\mu\text{A}$ for $R_s = 10\text{k}$. This choice of collector current is also roughly what you would get from the graph in Figure 7.47 of noise-figure contours at 1kHz, although the actual noise figure can be estimated only approximately from that plot as being less than 2dB.

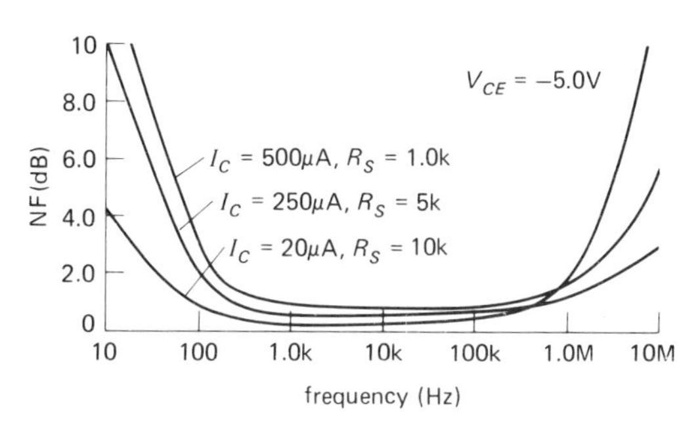

Figure 7.48. Noise figure (NF) versus frequency, for three choices of I_C and R_S, for the 2N5087. (Courtesy of Fairchild Camera and Instrument Corp.)

EXERCISE 7.5

Find the optimum I_C and corresponding noise figure for $R_s = 100\text{k}$ and $f = 1\text{kHz}$, using the graph in Figure 7.43 of e_n and i_n. Check your answer from the noise-figure contours (Fig. 7.47).

For the other amplifier configurations (follower, grounded base) the noise figure is essentially the same, for given R_s and I_C, since e_n and i_n are unchanged. Of course, a stage with unity voltage gain (a follower) may just pass the problem along to the next stage, since the signal level hasn't been increased to the point that low-noise design can be ignored in subsequent stages.

□ *Charting amplifier noise with e_n and i_n*

The noise calculations just presented, although straightforward, make the whole subject of amplifier design appear somewhat formidable. If you misplace a factor of Boltzmann's constant, you suddenly get an amplifier with 10,000dB noise figure! In this section we will present a simplified noise-estimation technique of great utility.

The method consists of first choosing some frequency of interest in order to get

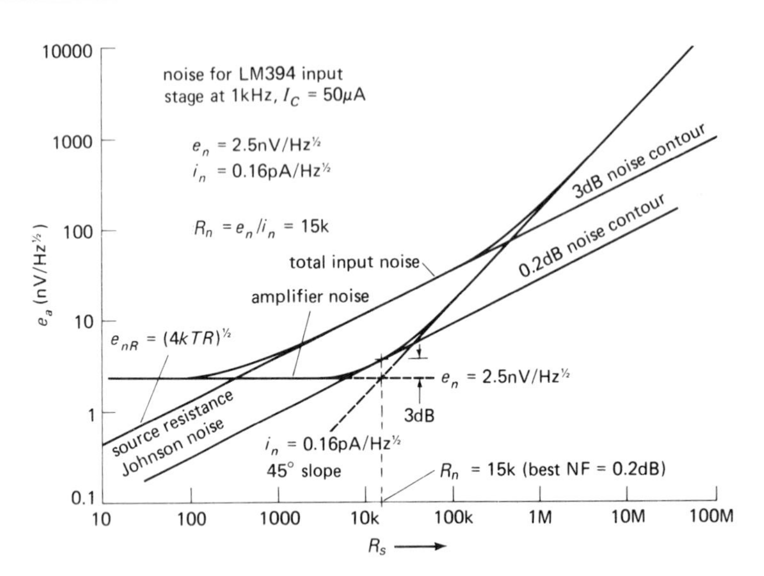

Figure 7.49. Total amplifier input voltage noise (e_a) plotted from the e_n and i_n parameters.

values for e_n and i_n versus I_C from the transistor data sheets. Then, for a given collector current, you can plot the total noise contributions from e_n and i_n as a graph of e_a versus source resistance R_s. Figure 7.49 shows what that looks like at 1kHz for a differential input stage using an LM394 matched superbeta transistor running 50μA of collector current. The e_n noise voltage is constant, and the $i_n R_s$ voltage increases proportional to R_s, i.e., with a 45° slope. The amplifier noise curve is drawn as shown, with care being taken to ensure that it passes through a point 3dB (voltage ratio of 1.4) above the crossing point of individual voltage and current noise contributions. Also plotted is the noise voltage of the source resistance, which also happens to be the 3dB NF contour. The other lines of constant noise figure are simply straight lines parallel to this line, as you will see in the examples that follow.

The best noise figure (0.2dB) at this collector current and frequency occurs for a source resistance of 15k, and the noise figure is easily seen to be less than 3dB for all source resistances between 300 ohms and 500k, the points at which the 3dB NF contour intersects the amplifier noise curve.

The next step is to draw a few of these noise curves on the same graph, using different collector currents or frequencies, or maybe a selection of transistor types, in order to evaluate amplifier performance. Before we go on to do that, let's show how we can talk about this same amplifier using a different pair of noise parameters, the noise resistance R_n and the noise figure NF(R_N), both of which pop right out of the graph.

□ *Noise resistance*

The lowest noise figure in this example occurs for a source resistance R_s = 15k, which equals the ratio of e_n to i_n. That defines the noise resistance

$$R_n = \frac{e_n}{i_n}$$

You can find the noise figure for a source of that resistance from our earlier expression for noise figure. It is

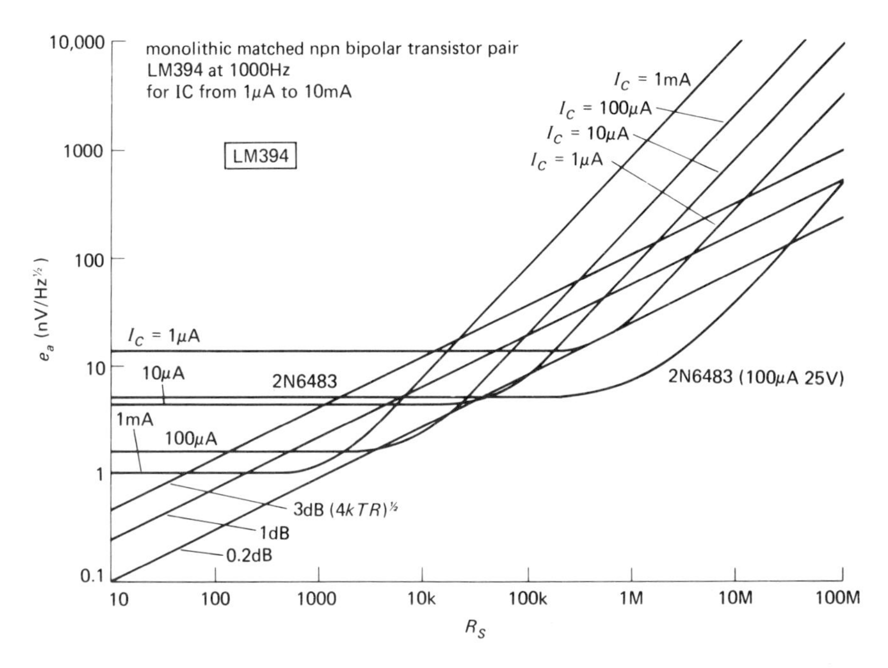

Figure 7.50. Total amplifier input voltage noise (e_a) for the LM394 bipolar transistor under various conditions, compared with the 2N6483 JFET.

$$\text{NF (at } R_n) = 10\log_{10}\left(1 + 1.23\times 10^{20}\frac{e_n^2}{R_n}\right) \quad \text{dB}$$
$$\approx 0.2 \quad \text{dB}$$

Noise resistance isn't actually a real resistance in the transistor, or anything like that. It is a tool to help you quickly find the value of source resistance for minimum noise figure, ideally so that you can vary the collector current to shift R_n close to the value of source resistance you're actually using. R_n corresponds to the point where the e_n and i_n lines cross.

The noise figure for a source resistance equal to R_n then follows simply from the preceding equation.

□ *Charting the bipolar/FET shootout*

Let's have some fun with this technique. A perennial bone of contention among engineers is whether FETs or bipolar transistors are "better." We will dispose of this issue with characteristic humility by matching two of the best contenders and letting them deliver their best punches. In the interest of fairness, we'll let National Semiconductor intramural teams compete, choosing two game fighters.

In the bipolar corner we have the magnificent LM394 superbeta monolithic matched pair, already warmed up, as described earlier. We'll run it at 1kHz, with collector currents from 1μA to 1mA (Fig. 7.50).

The FET entry is the 2N6483 monolithic n-channel JFET matched pair, known far and wide for its stunning low-noise performance, reputed to exceed that of bipolar transistors. According to its data sheet, it was trained only for 100μA and 400μA drain currents (Fig. 7.51).

And the winner? Well, it's a split decision. The FET won points on lowest minimum noise figure, NF(R_n), reaching a phenomenal 0.05dB noise figure, and dipping well below 0.2dB from 100k to 100M source impedance. For high source

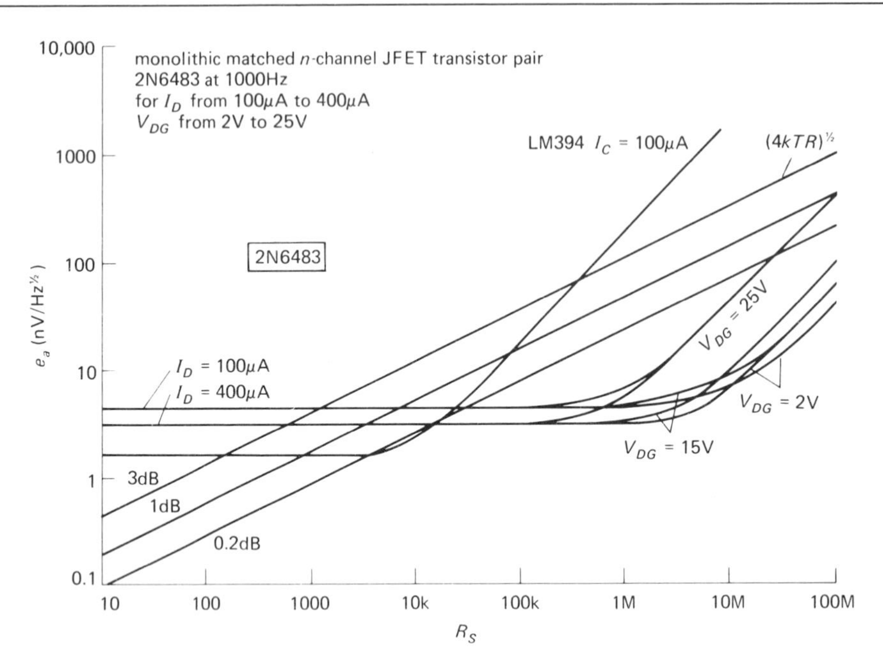

Figure 7.51. Total amplifier input voltage noise (e_a) for the 2N6483 JFET compared with the LM394 bipolar transistor.

impedances, FETs remain unbeaten. The bipolar transistor is best at low source impedances, particularly below 5k, and it can reach a 0.3dB noise figure at $R_s = 1$k, with suitable choice of collector current. By comparison, the FET cannot do better than 2dB with a 1k source resistance, owing to larger voltage noise e_n.

Just as in boxing, where the best fighters haven't yet had a chance to compete in a world championship, there are some younger contenders for the best low-noise transistor. For example, the 2SJ72 and 2SK147 complementary JFETs from Toshiba use a meshed-gate geometry to achieve a phenomenal e_n of $0.7\text{nV}/\sqrt{\text{Hz}}$ at $I_D = 10$mA (equivalent to Johnson noise from a 30Ω resistor!). But these are *JFETs*, with their low input current (hence low i_n), and thus the noise resistance is about 10k. When used as an amplifier with a source impedance equal to their noise resistance (i.e., $R_s = 10$k), their performance is unbeatable – the noise temperature is just 2°K!

Before you go out and buy a bushel of these remarkable JFETs, consider the remarks of the critics, who claim they are muscle-bound – they have high input and feedback capacitance (85pF and 15pF, respectively), which limits their usefulness at high frequencies. Their relative, the 2SK117, is better in this regard, at the expense of higher e_n. These same critics argue that the Toyo-Rohm bipolar complementary pair, the 2SD786 and 2SB737, with e_n as low as $0.55\text{nV}/\sqrt{\text{Hz}}$, can offer even better performance at moderate source impedances and frequencies.

□ *Low source impedances*

Bipolar transistor amplifiers can provide very good noise performance over the range of source impedances from about 200 ohms to 1M; corresponding optimum collector currents are generally in the range of several milliamps down to a microamp. That is, collector currents used for the input stage of low-noise amplifiers generally tend to be lower than in amplifier stages not optimized for low-noise performance.

For very low source impedances (say 50Ω), transistor voltage noise will always dominate, and noise figures will be poor. The best approach in such cases is to use a transformer to raise the signal level (and impedance), treating the signal on the secondary as before. High-quality signal transformers are available from companies such as James and Princeton Applied Research. As an example, the latter's model 116 FET preamp has voltage and current noise such that the lowest noise figure occurs for signals of source impedance around 1M. A signal around 1kHz with source impedance of 100 ohms would be a poor match for this amplifier, since the amplifier's voltage noisc is much larger than the signal source's Johnson noise; the resultant noise figure for that signal connected directly to the amplifier would be 11dB. By using the optional internal step-up transformer, the signal level is raised (along with its source impedance), thus overriding amplifier noise voltage and giving a noise figure of about 1.0dB.

At radiofrequencies (e.g., beginning around 100kHz) it is extremely easy to make good transformers, both for tuned (narrowband) and broadband signals. At these frequencies it is possible to make broadband "transmission-line transformers" of very good performance. We will treat some of these methods in Chapter 13. It is at the very low frequencies (audio and below) that transformers become problematic.

Three comments: (a) The voltage rises proportional to the turns ratio of the transformer, whereas the impedance rises proportional to the square of the ratio. Thus a 2:1 voltage step-up transformer has an output impedance four times the input impedance (this is mandated by conservation of energy). (b) Transformers aren't perfect. They have trouble at low frequencies (magnetic saturation) and at high frequencies (winding inductance and capacitance), as well as losses from the magnetic properties of the core and from winding resistance. The latter is a source of Johnson noise, as well. Nevertheless, when dealing with a signal of very low source impedance, you may have no choice, and transformer coupling can be very beneficial, as the preceding example demonstrates. Exotic techniques such as cooled transformers, superconducting transformers, and SQUIDs (superconducting quantum interference devices) can provide good noise performance at low impedance and voltage levels. With SQUIDs you can measure voltages of 10^{-15} volt! (c) Again, a warning: Don't attempt to improve performance by adding a resistor in series with a low source impedance. If you do that, you're just another victim of the noise-figure fallacy.

□ *High source impedances*

If the source impedance is high, say greater than 100k or so, transistor current noise dominates, and the best device for low-noise amplification is a FET. Although their voltage noise is usually greater than that of bipolar transistors, the gate current (and its noise) can be exceedingly small, making them ideally suited for low-noise high-impedance amplifiers. Incidentally, it is sometimes useful to think of Johnson noise as a current noise $i_n = v_n/R_s$. This lets you compare source noise contributions with amplifier current noise (Fig. 7.52).

7.15 FET noise

We can use the same amplifier noise model for FETs, namely a series noise voltage source and a parallel noise current source. You can analyze the noise performance with exactly the same methods used for bipolar transistors. For example, see the graphs in the section on bipolar/FET shootout.

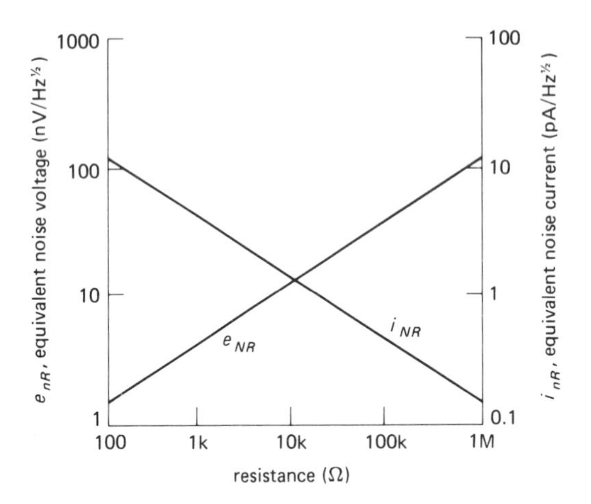

Figure 7.52. Thermal noise voltage density versus resistance at 25° C. The equivalent short-circuit current noise density is also shown.

Voltage noise of JFETs

For JFETs the voltage noise e_n is essentially the Johnson noise of the channel resistance, given approximately by

$$e_n^2 = 4kT\left(\frac{2}{3}\frac{1}{g_m}\right) \qquad \mathrm{V^2/Hz}$$

where the inverse transconductance takes the place of resistance in the Johnson-noise formula. Since the transconductance rises with increasing drain current (as $\sqrt{I_D}$), it is generally best to operate FETs at high drain current for lowest voltage noise. However, since the e_n is Johnson noise, which goes only as $1/\sqrt{g_m}$, and that in turn goes as $\sqrt{I_D}$, e_n is finally proportional to $I_D^{-1/4}$. With such a mild dependence of e_n on I_D it doesn't pay to run at a drain current so high that other properties of the amplifier are degraded. In particular, a FET running at high current gets hot, which (a) decreases g_m, (b) increases offset voltage drift and CMRR, and (c) raises gate leakage dramatically; the latter effect can actually *increase* voltage noise, since there is some contribution to e_n from flicker noise associated with the gate leakage current.

There is another way to increase g_m, and therefore decrease JFET voltage noise: By paralleling a pair of JFETs you get twice the g_m, but of course this is at twice the I_D. But now if you run the combination at the previous I_D, you still improve g_m by a factor of $\sqrt{2}$ over the single-JFET value, without increasing total drain current. In practice you can simply parallel a number of matched JFETs, or look for a large-geometry JFET like the 2SJ72 and 2SK147 mentioned earlier.

There is a price to pay, however. All the capacitances scale with the number of paralleled JFETs. As a result, high-frequency performance (including noise figure) is degraded. In practice you should stop paralleling additional transistors when the circuit's input capacitance roughly matches the source's capacitance. If you care about performance at high frequencies, choose JFETs with high g_m and low C_{rss}; you might consider the ratio g_m/C_{rss} a high-frequency figure of merit. Note that circuit configurations can also play an important role; e.g., the cascode circuit can be used to eliminate the Miller effect (gain multiplication) on C_{rss}.

MOSFETs tend to have much higher voltage noise than JFETs, with $1/f$ noise predominating, since the $1/f$ knee is as high as 10kHz to 100kHz. For this reason you wouldn't normally choose a MOSFET for low-noise amplifiers below 1MHz.

Current noise of JFETs

At low frequencies the current noise i_n is extremely small, arising from the shot noise in the gate leakage current (Fig. 7.53):

$$i_n = (3.2 \times 10^{-19} I_G B)^{\frac{1}{2}} \qquad \mathrm{A(rms)}$$

In addition, there is a flicker-noise component in some FETs. The noise current rises with increasing temperature, as the gate leakage current rises. Watch out for the rapidly increasing gate leakage in n-channel JFETs that occurs for operation at high V_{DG} (see Section 3.09).

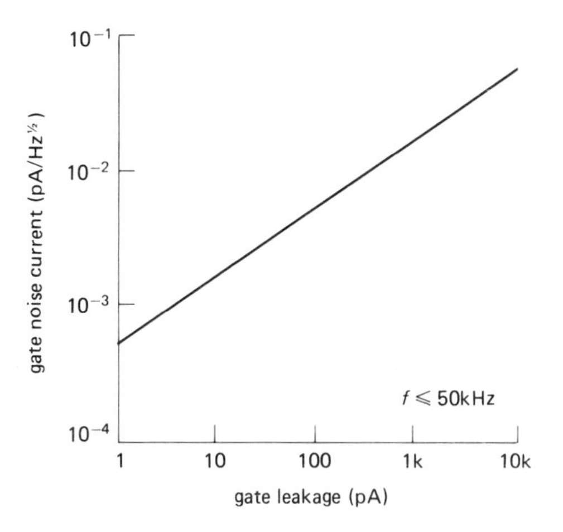

Figure 7.53. Input noise current versus gate leakage current for JFETs. (Courtesy of National Semiconductor Corp.)

At moderate to high frequencies there is an additional noise term, namely the real part of the input impedance seen looking into the gate. This comes from the effect of feedback capacitance (Miller effect) when there is a phase shift at the output due to load capacitance; i.e., the part of the output signal that is shifted 90° couples through the feedback capacitance C_{rss} to produce an effective resistance at the input, given by

$$R = \frac{1 + \omega C_L R_L}{\omega^2 g_m C_{rss} C_L R_L^2} \quad \text{ohms}$$

As an example, the 2N5266 p-channel JFET has a noise current of $0.005\text{pA/Hz}^{\frac{1}{2}}$ and a noise voltage e_n of $12\text{nV/Hz}^{\frac{1}{2}}$, both at I_{DSS} and 10kHz. The noise current begins climbing at about 50kHz. These figures are roughly 100 times better in i_n and 5 times worse in e_n than the corresponding figures for the 2N5087 used earlier.

With FETs you can achieve good noise performance for input impedances in the range of 10k to 100M. The PAR model 116 preamp has a noise figure of 1dB or better for source impedances from 5k to 10M in the frequency range from 1kHz to 10kHz. Its performance at moderate frequencies corresponds to a noise voltage of $4\text{nV/Hz}^{\frac{1}{2}}$ and a noise current of $0.013\text{pA/Hz}^{\frac{1}{2}}$.

7.16 Selecting low-noise transistors

As we mentioned earlier, bipolar transistors offer the best noise performance with low source impedances, owing to their lower input voltage noise. Voltage noise, e_n, is reduced by choosing a transistor with low base spreading resistance, r_{bb}, and operating at high collector current (as long as h_{FE} remains high). For higher source impedances the current noise can be minimized instead by operating at lower collector current.

At high values of source impedance, FETs are the best choices. Their voltage noise can be reduced by operating at higher drain currents, where the transconductance is highest. FETs intended for low-noise applications have high k values (see Section 3.04), which usually means high input capacitance. For example, the low-noise 2N6483 has C_{iss} = 20pF, whereas the 2N5902 low-current FET has C_{iss} = 2pF.

Figures 7.54 and 7.55 show comparisons of the noise characteristics of a number of popular and useful transistors.

□ 7.17 Noise in differential and feedback amplifiers

Low-noise amplifiers are often differential, to obtain the usual benefits of low drift and good common-mode rejection. When you calculate the noise performance of a differential amplifier, there are three points to keep in mind: (a) Be sure to use the individual collector currents, not the sum, to get e_n and i_n from data sheets. (b) The i_n seen at each input terminal is the same as for a single-ended amplifier configuration. (c) The e_n seen at one input, with the other input grounded, say, is 3dB

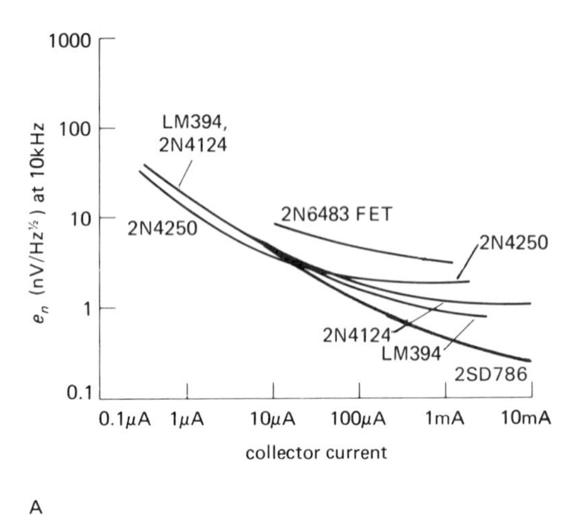

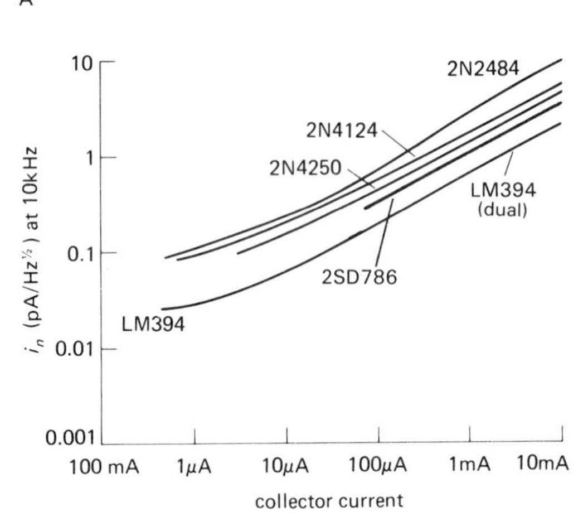

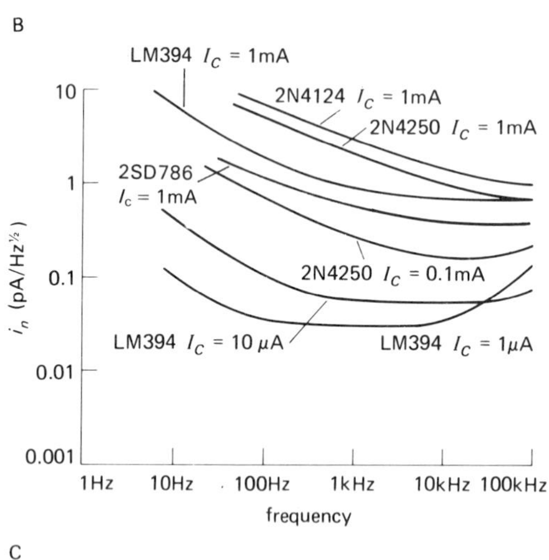

Figure 7.54. Input noise for some popular transistors.
A. Input noise voltage (e_n) versus collector current.
B. Input noise current (i_n) versus collector current.
C. Input noise current (i_n) versus frequency.

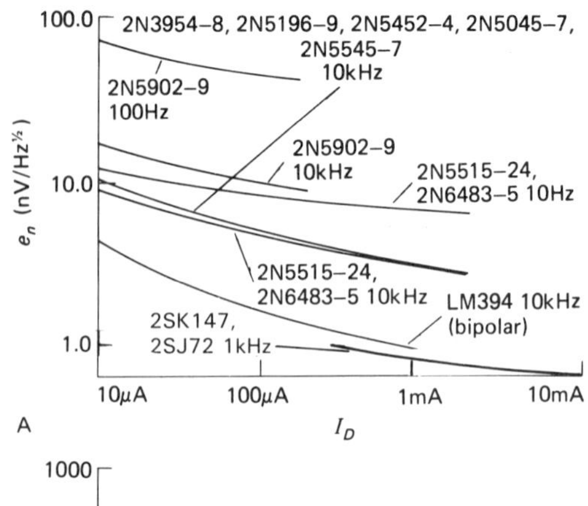

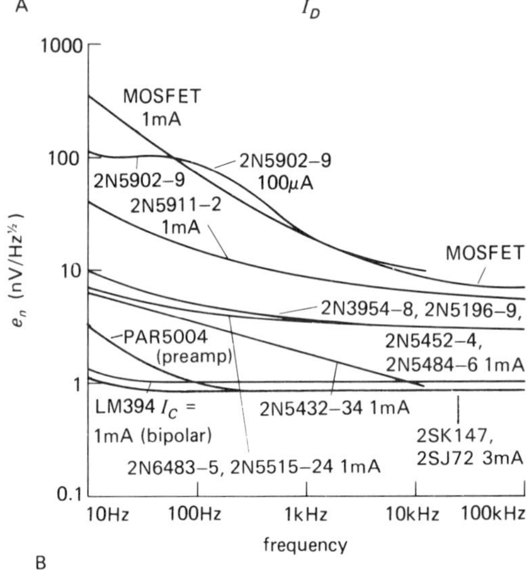

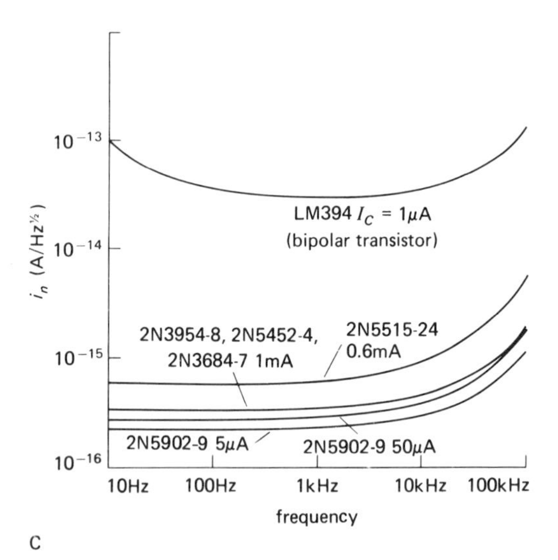

Figure 7.55. Input noise for some popular FETs.
A. Input noise voltage (e_n) versus drain current (I_D).
B. Input noise voltage (e_n) versus frequency.
C. Input noise current (i_n) versus frequency.

larger than the single-transistor case, i.e., it is multiplied by $\sqrt{2}$.

In amplifiers with feedback, you want to take the equivalent noise sources e_n and i_n out of the feedback loop, so you can use them as previously described when calculating noise performance with a given signal source. Let's call the noise terms brought out of the feedback loop e_A and i_A, for *amplifier* noise terms. Thus the amplifier's noise contribution to a signal with source resistance R_s is

$$e^2 = e_A^2 + (R_s I_A)^2 \quad \text{V}^2/\text{Hz}$$

Let's take the two feedback configurations separately.

□ *Noninverting*

For the noninverting amplifier (Fig. 7.56) the input noise sources become

$$i_A^2 = i_n^2$$
$$e_A^2 = e_n^2 + 4kTR_{\|} + (i_n R_{\|})^2$$

where e_n is the "adjusted" noise voltage for the differential configuration, i.e., 3dB larger than for a single-transistor stage. The additional noise voltage terms arise from Johnson noise and input-stage noise current in the feedback resistors. Note that the effective noise voltage and current are now not completely uncorrelated, so calculations in which their squares are added can be in error by a maximum factor of 1.4.

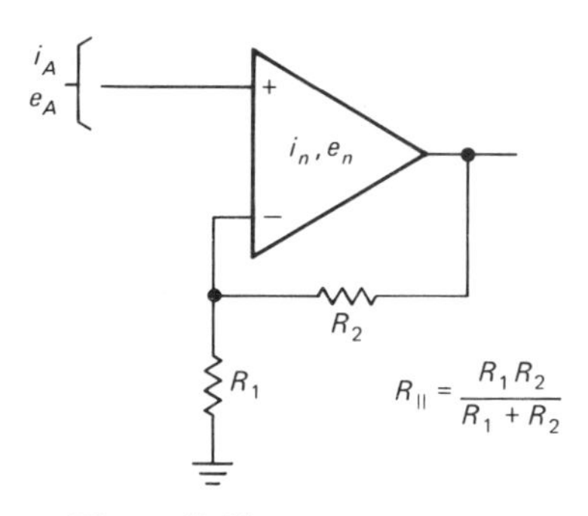

Figure 7.56

For a follower, R_2 is zero, and the effective noise sources are just those of the differential amplifier alone.

□ *Inverting*

For the inverting amplifier (Fig. 7.57) the input noise sources become

$$i_A^2 = i_n^2 + 4kT\frac{1}{R_2}$$
$$e_A^2 = e_n^2 + R_1^2\left(i_n^2 + 4kT\frac{1}{R_2}\right)$$
$$= e_n^2 + R_1^2 i_A^2$$

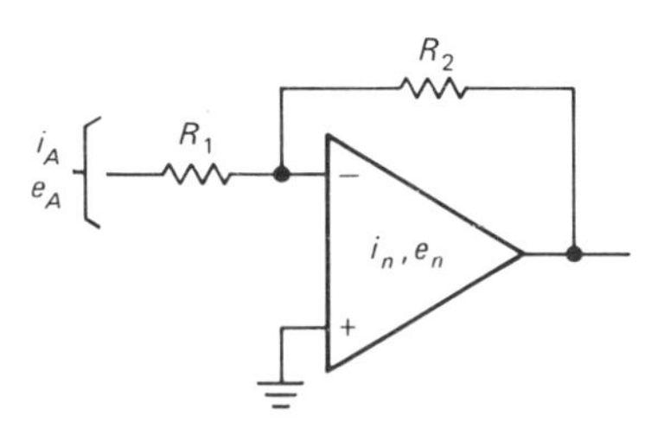

Figure 7.57

Op-amp selection curves

You now have all the tools necessary to analyze op-amp input circuits. Their noise is specified in terms of e_n and i_n, just as with transistors and FETs. You don't get to adjust anything, though; you only get to use them. The data sheets may need to be taken with a grain of salt. For example, "popcorn noise" is typified by jumps in offset at random times and duration. It is rarely mentioned in polite company. Figure 7.58 summarizes the noise performance of some popular op-amps.

Wideband noise

Op-amp circuits are generally dc-coupled and extend to some upper frequency limit f_{cutoff}. Therefore it is of interest to know the total noise voltage over this band, not merely the noise power density. Figure 7.59 presents some graphs showing the rms noise voltage in a band extending

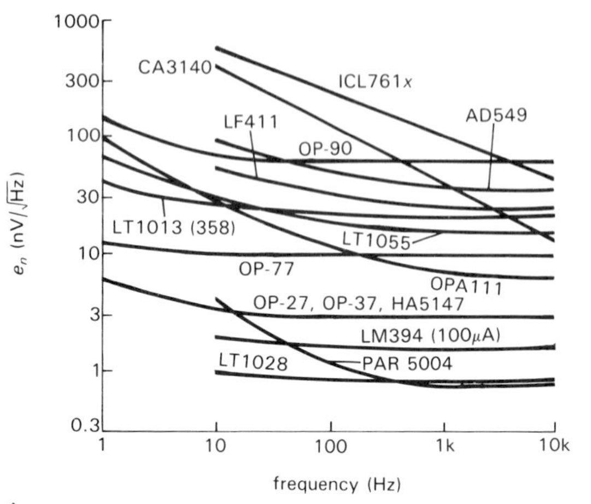

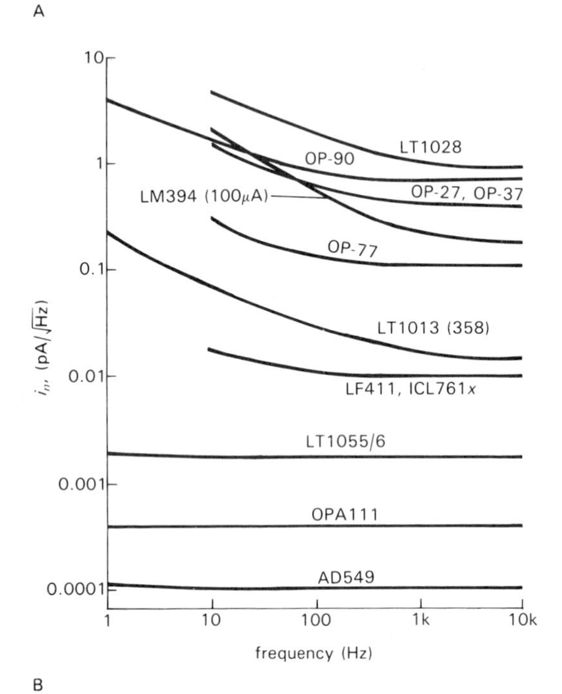

Figure 7.58. Input noise for some popular op-amps.
A. Input noise voltage (e_n) versus frequency.
B. Input noise current (i_n) versus frequency.

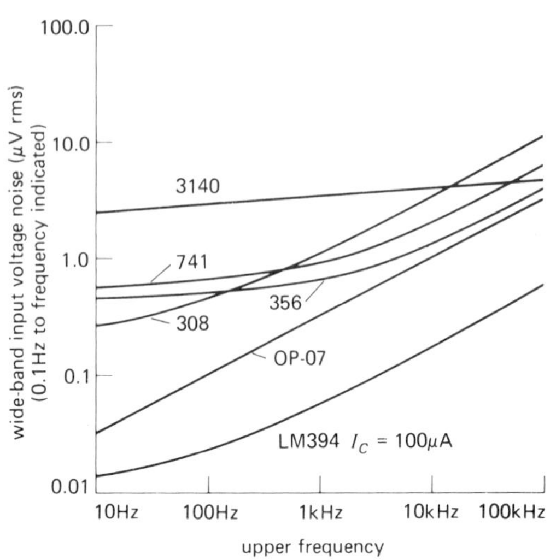

Figure 7.59. Wideband noise voltage for some popular op-amps.

from dc to the indicated frequency; they were calculated by integrating the noise power curves for the various op-amps.

Choosing a low-noise op-amp

It is simple to choose an op-amp to minimize noise in some frequency range, given the signal impedance seen from the op-amp, R_{sig} (which includes the effects of feedback components, as given in the foregoing expressions). Generally speaking, you want op-amps with low i_n for high signal impedances, and op-amps with low e_n for low signal impedances. Assuming the signal source is at room temperature, the total input-referred squared noise voltage density is just

$$e_A^2 = 4kTR_{sig} + {e_n}^2 + (i_n R_{sig})^2$$

where the first term is due to Johnson noise, and the last two terms are due to op-amp noise voltage and current. Obviously the Johnson noise sets a lower bound to the input-referred noise. In Figure 7.60 we've plotted the quantity e_A (at 10Hz) as a function of R_{sig} for the quietest op-amps we could find. For comparison we included our jellybean JFET LF411 and the micropower bipolar OP-90. The latter, although an excellent micropower op-amp, has high noise voltage (because the front end operates at low collector current, hence high r_e and therefore high Johnson noise) and also high noise current (because the bipolar input has substantial base current); it shows just how good

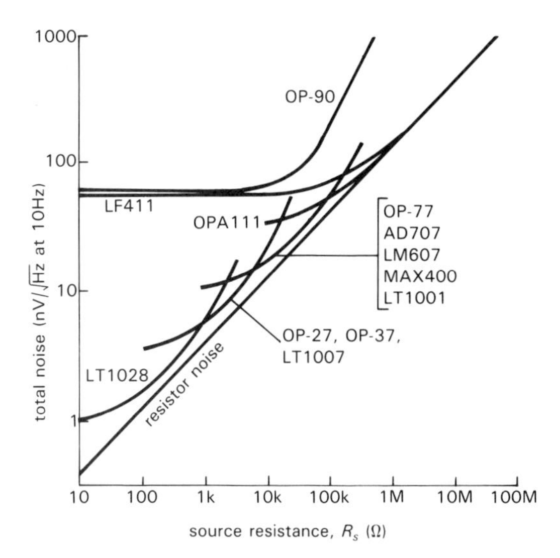

Figure 7.60. Total noise (source resistor plus amplifier, at 10Hz) for high-performance op-amps.

the premium low-noise op-amps really are.

□ *Low-noise preamps*

In addition to the low-noise op-amps, there are some nice low-noise IC *preamplifiers.* Unlike op-amps, these generally have fixed voltage gain, though in some models you can attach an external gain-setting resistor. People sometimes call these "video amplifiers" because they often have bandwidths into the tens of megahertz, though they can be used for low-frequency applications as well. Examples are the Plessey SL561B and several models from Analog Systems. These amplifiers typically have e_n less than $1\text{nV}/\sqrt{\text{Hz}}$, achieved (at the expense of high input noise *current,* i_n) by running the input transistor at relatively high collector current.

□ NOISE MEASUREMENTS AND NOISE SOURCES

It is a relatively straightforward process to determine the equivalent noise voltage and current of an amplifier, and from these the noise figure and signal-to-noise ratio for any given signal source. That's all you ever need to know about the noise performance of an amplifier. Basically the process consists of putting known noise signals across the input, then measuring the output noise signal amplitudes within a certain bandwidth. In some cases (e.g., a matched input impedance device such as a radio-frequency amplifier) an oscillator of accurately known and controllable amplitude is substituted as the input signal source.

Later we will discuss the techniques you need to do the output voltage measurement and bandwidth limiting. For now, let's assume you can make rms measurements of the output signal, with a measurement bandwidth of your choice.

□ 7.18 Measurement without a noise source

For an amplifier stage made from a FET or transistor and intended for use at low to moderate frequencies, the input impedance is likely to be very high. You want to know e_n and i_n so that you can predict the SNR with a signal source of arbitrary source impedance and signal level, as we discussed earlier. The procedure is simple:

First, determine the amplifier's voltage gain G_V by actual measurement with a signal in the frequency range of interest. The amplitude should be large enough to override amplifier noise, but not so large as to cause amplifier saturation.

Second, short the input and measure the rms noise output voltage, e_s. From this you get the input noise voltage per root hertz from

$$e_n = \frac{e_s}{G_V B^{\frac{1}{2}}} \qquad \text{V/Hz}^{\frac{1}{2}}$$

where B is the bandwidth of the measurement (see Section 7.21).

Third, put a resistor R across the input, and measure the new rms noise output voltage, e_r. The resistor value should be large enough to add significant amounts of current noise, but not so large that the input impedance of the amplifier begins to dominate. (If this is impractical, you can leave the input open and use the amplifier's input impedance as R.) The output you measure is just

$$e_r^2 = [e_n^2 + 4kTR + (i_nR)^2]BG_V^2$$

from which you can determine i_n to be

$$i_n = \frac{1}{R_s}\left[\frac{e_r^2}{BG_V^2} - (e_n^2 + 4kTR)\right]^{\frac{1}{2}}$$

With some luck, only the first term in the square root will matter (i.e., if current noise dominates both amplifier voltage noise and source resistor Johnson noise).

Now you can determine the SNR for a signal V_s of source impedance R_s, namely

$$\text{SNR} = 10\log_{10}\left(\frac{V_s^2}{V_n^2}\right)$$

$$= 10\log_{10}\left[\frac{V_s^2}{[e_n^2 + (i_nR_s)^2 + 4kTR_s]B}\right]$$

where the numerator is the signal voltage (presumed to lie within the bandwidth B) and the terms in the denominator are the amplifier noise voltage, amplifier noise current applied to R_s, and Johnson noise in R_s. Note that increasing the amplifier bandwidth beyond what is necessary to pass the signal V_s only decreases the final SNR. However, if V_s is broadband (e.g., a noise signal itself), the final SNR is independent of amplifier bandwidth. In many cases the noise will be dominated by one of the terms in the preceding equation.

□ 7.19 Measurement with noise source

The preceding technique of measuring the noise performance of an amplifier has the advantage that you don't need an accurate and adjustable noise source, but it requires an accurate voltmeter and filter, and it assumes that you know the gain versus frequency of the amplifier, with the actual source resistance applied. An alternative method of noise measurement involves applying broadband noise signals of known amplitude to the amplifier's input and observing the relative increase of output noise voltage. Although this technique requires an accurately calibrated noise source, it makes no assumptions about the properties of the amplifier, since it measures the noise properties right at the point of interest, at the input.

Again, it is relatively straightforward to make the requisite measurements. You connect the noise generator to the amplifier's input, making sure that its source impedance R_g equals the source impedance of the signal you ultimately plan to use with the amplifier. You first note the amplifier's output rms noise voltage, with the noise source attenuated to zero output signal. Then you increase the noise source rms amplitude V_g until the amplifier's output rises 3dB (a factor of 1.414 in rms output voltage). The amplifier's input noise voltage in the measurement bandwidth, for this source impedance, equals this value of added signal. The amplifier therefore has a noise figure

$$\text{NF} = 10\log_{10}\left(\frac{V_g^2}{4kTR_g}\right)$$

From this you can figure out the SNR for a signal of any amplitude with this same source impedance, using the formula from Section 7.12

$$\text{SNR} = 10\log_{10}\left(\frac{V_s^2}{4kTR_s}\right) - \text{NF}(R_s)\text{ dB}$$

There are nice calibrated noise sources available, most of which provide means for attenuation to precise levels in the microvolt range. Note: Once again, the preceding formulas assume $R_{\text{in}} \gg R_s$. If, on the other hand, the noise-figure measurement is made with a *matched* signal source, i.e., if $R_s = Z_{\text{in}}$, then omit the factors "4" in the preceding expressions.

Note that this technique does not tell you e_n and i_n directly, just the appropriate combination for a source of impedance equal to the driving impedance you used in the measurement. Of course, by making several such measurements with different noise source impedances, you could infer the values of e_n and i_n.

A nice variation on this technique is to use resistor Johnson noise as the "noise source." This is a favorite technique used by designers of very low noise radiofrequency amplifiers (in which, incidentally, the signal source impedance is usually 50Ω and matches the amplifier's input impedance). It is usually done the following way: A dewar of liquid nitrogen holds a 50 ohm "termination" (a fancy name for a well-designed resistor that has negligible inductance or capacitance) at the temperature of boiling nitrogen, 77°K; a second 50 ohm termination is kept at room temperature. The amplifier's input is connected alternately to the two resistors (usually with a high-quality coax relay), while the output noise power (at some center frequency, with some measurement bandwidth) is measured with an RF power meter. Call the results of the two measurements P_C and P_H, the output noise power corresponding to cold and hot source resistors, respectively. It is then easy to show that the amplifier's noise temperature, at the frequency of the measurement, is just

$$T_n = \frac{T_H - YT_C}{Y - 1}$$

where $Y = P_H/P_C$, the ratio of noise powers. Noise figure is then given by the formula of Section 7.12, namely

$$\text{NF(dB)} = 10\log_{10}\left(\frac{T_n}{290} + 1\right)$$

EXERCISE 7.6

Derive the foregoing expression for noise temperature. Hint: Begin by noting that $P_H = \alpha(T_n + T_H)$ and $P_C = \alpha(T_n + T_C)$, where α is a constant that will shortly disappear. Then note that the noise contribution of the amplifier, stated as a noise temperature, *adds* to the noise temperature of the source resistor. Take it from there.

EXERCISE 7.7

Amplifier noise temperature (or noise figure) depends on the value of signal source impedance, R_s. Show that an amplifier characterized by e_n and i_n (as in Fig. 7.46) has minimum noise temperature for a source impedance $R_s = e_n/i_n$. Then show that the noise temperature, for that value of R_s, is given by $T_n = e_n i_n/2\text{k}$.

□ *Amplifiers with matched input impedance*

This last technique is ideal for noise measurements of amplifiers designed for a matched signal source impedance. The most common examples are in radiofrequency amplifiers or receivers, usually meant to be driven with a signal source impedance of 50 ohms, and which themselves have an input impedance of 50 ohms. We will discuss in Chapter 13 the reasons for this departure from our usual criterion that a signal source should have a small source impedance compared with the load it drives. In this situation e_n and i_n are irrelevant as separate quantities; what matters is the overall noise figure (with matched source) or some specification of SNR with a matched signal source of specified amplitude.

Sometimes the noise performance is explicitly stated in terms of the *narrowband* input signal amplitude required to obtain a certain output SNR. A typical radiofrequency receiver might specify a 10dB SNR with a 0.25μV rms input signal and 2kHz receiver bandwidth. In this case the procedure consists of measuring the rms receiver output with the input driven by a

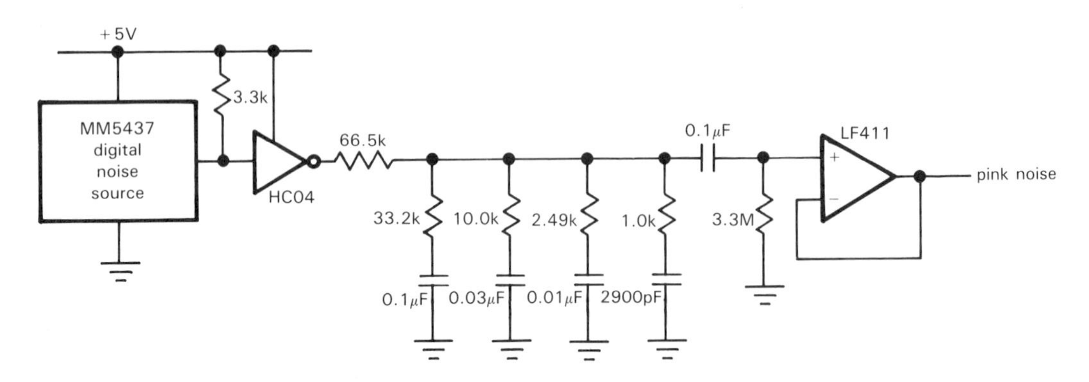

Figure 7.61. Pink noise source (−3dB/octave, ±0.25dB from 10Hz to 40kHz).

matched sine-wave source initially attenuated to zero, then increasing the (sine-wave) input signal until the rms output rises 10dB, in both cases with the receiver bandwidth set to 2kHz. It is important to use a meter that reads true rms voltages for a measurement where noise and signal are combined (more about this later). Note that radiofrequency noise measurements often involve output signals that are in the audiofrequency range.

□ 7.20 Noise and signal sources

Broadband noise can be generated from the effects we discussed earlier, namely Johnson noise and shot noise. The shot noise in a vacuum diode is a classic source of broadband noise that is especially useful because the noise voltage can be predicted exactly. More recently, zener diode noise has been used in noise sources. Both of these extend from dc to very high frequencies, making them useful in audiofrequency and radiofrequency measurements.

An interesting noise source can be made using digital techniques, in particular by connecting long shift registers with their input derived from a modulo-2 addition of several of the later bits (see Section 9.33). The resultant output is a pseudorandom sequence of 1's and 0's that after low-pass filtering generates an analog signal of white spectrum up to the low-pass filter's break-point, which must be well below the frequency at which the register is shifted. These things can be run at very high frequencies, generating noise up to 100kHz or more. The "noise" has the interesting property that it repeats itself exactly after a time interval that depends on the register length (an n-bit maximal-length register goes through $2^n - 1$ states before repeating). Without much difficulty that time can be made to be very long (months or years), although most often a period of a second is long enough. For example, a 50-bit register shifted at 10MHz will generate white noise up to 100kHz or so, with a repeat time of 3.6 years. A design for a pseudorandom noise source based on this technique is shown in Section 9.36.

Some noise sources can generate pink noise as well as white noise. Pink noise has equal noise power per *octave*, rather than equal power per hertz. Its power density (power per hertz) drops off at 3dB/octave. Since an *RC* filter drops off at 6dB/octave, a more complicated filter is necessary to generate a pink spectrum from a white noise input. The circuit shown in Figure 7.61 uses a 23-bit pseudorandom white noise generator chip to generate pink noise, accurate to ±0.25dB from 10Hz to 40kHz.

Versatile signal sources are available with precisely controlled output amplitude (down to the microvolt range and below) over frequencies from a fraction of a hertz to gigahertz. Some can even be programmed via a digital "bus." An example is the Hewlett-Packard model 8660 synthesized signal generator, with output frequencies from 0.01 to 110MHz, calibrated amplitudes from 10nV to 1 volt rms, handsome digital display and bus interface, and nifty accessories that extend the frequency range to 2.6GHz and provide modulation and frequency sweeping. This is a bit more than you usually need to do the job.

□ 7.21 Bandwidth limiting and rms voltage measurement

□ *Limiting the bandwidth*

All the measurements we have been talking about assume that you are looking at the noise output only in a limited frequency band. In a few cases the amplifier may have provision for this, making your job easier. If not, you have to hang some sort of filter on the amplifier output before measuring the output noise voltage.

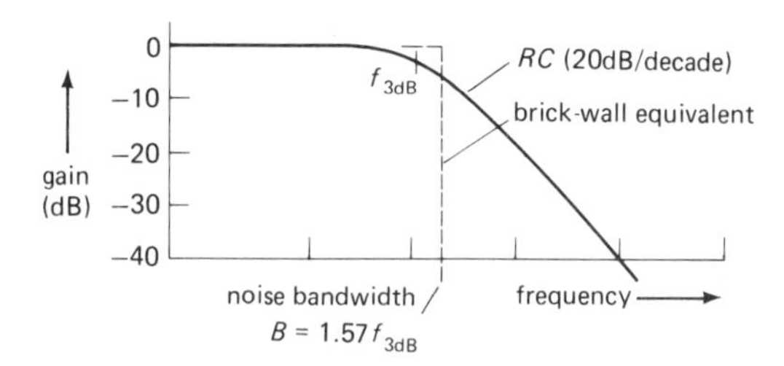

Figure 7.62. Equivalent "brick-wall" noise bandwidth for *RC* low-pass filter.

The easiest thing to use is a simple *RC* low-pass filter, with 3dB point set at roughly the bandwidth you want. For accurate noise measurements, you need to know the equivalent "noise bandwidth," i.e., the width of a perfect "brick-wall" low-pass filter that lets through the same noise voltage (Fig. 7.62). This noise bandwidth is what should be used for B in all the preceding formulas. It is not terribly difficult to do the mathematics, and you find

$$B = \frac{\pi}{2} f_{3dB} = 1.57 f_{3dB}$$

For a pair of cascaded *RC*s (buffered so they don't load each other), the magic formula becomes $B = 1.22 f_{3dB}$. For the Butterworth filters discussed in Section 5.05, the noise bandwidth is

$B = 1.57\ f_{3dB}$ (1 pole)
$B = 1.11\ f_{3dB}$ (2 poles)
$B = 1.05\ f_{3dB}$ (3 poles)
$B = 1.025\ f_{3dB}$ (4 poles)

If you want to make band-limited measurements up at some center frequency, you can just use a pair of *RC* filters (Fig. 7.63), in which case the noise bandwidth is as indicated. If you have had experience with contour integration, you may wish to try the following exercise.

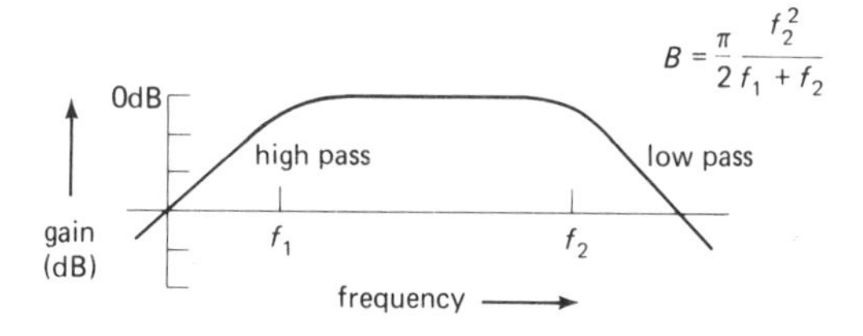

Figure 7.63. Equivalent "brick-wall" noise bandwidth for *RC* bandpass filter.

EXERCISE 7.8

Optional exercise: Derive the preceding result, beginning with the response functions of *RC* filters. Assume unit power per hertz input signal, and integrate the output power from zero to infinity. A contour integral then gets you the answer.

Another way to make a bandpass filter for noise measurements is to use an *RLC* circuit. This is better than a pair of cascaded high-pass and low-pass *RC* filters if you want your measurement over a bandpass that is narrow compared with the center frequency (i.e., high Q). Figure 7.64 shows both parallel and series *RLC* circuits and their exact noise bandwidths. In both cases the resonant frequency is given by $f_0 = 1/2\pi\sqrt{LC}$. You might arrange the bandpass filter circuit as a parallel *RLC* collector (or drain) load, in which case you use the expression as given. Alternatively, you might interpose the filter as shown in Figure 7.65; for noise bandwidth purposes the circuit is exactly equivalent to the parallel *RLC*, with $R = R_1 \| R_2$.

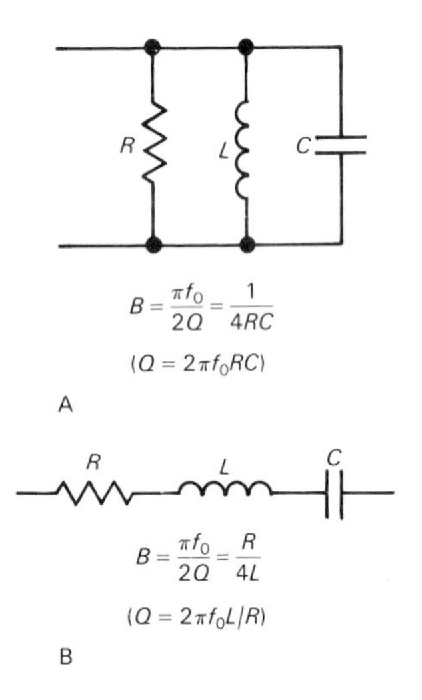

Figure 7.64. Eqivalent "brick-wall" noise bandwidth for *RLC* bandpass filter.

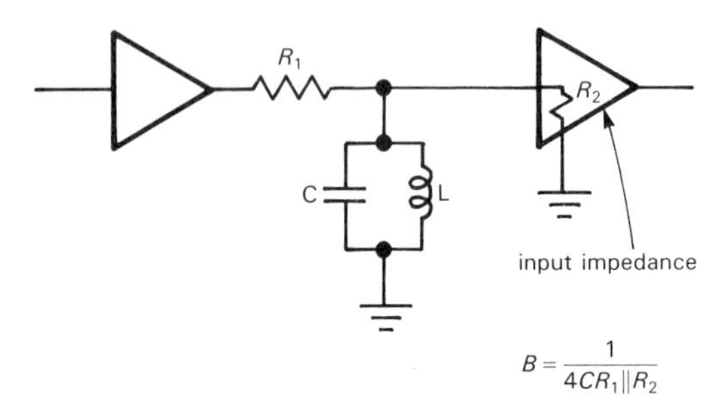

Figure 7.65

Measuring the noise voltage

The most accurate way to make output noise measurements is to use a true rms voltmeter. These operate either by measuring the heating produced by the signal waveform (suitably amplified) or by using an analog squaring circuit followed by averaging. If you use a true rms meter, make sure it has response at the frequencies you are measuring; some of them only go up to a few kilohertz. True rms meters also specify a "crest factor," the ratio of peak voltage to rms that they can handle without great loss of accuracy. For Gaussian noise, a crest factor of 3 to 5 is adequate.

You can use a simple averaging-type ac voltmeter instead, if a true rms meter is unavailable. In that case, the values read off the scale must be corrected. As it turns out, all averaging meters (VOMs, DMMs, etc.) already have their scales adjusted, so what you read isn't actually the *average*, but rather the rms voltage *assuming a sinewave signal*. For example, if you measure the power-line voltage in the United States, your meter will read something close to 117 volts. That's fine, but if the signal you're reading is Gaussian noise, you have to apply an additional correction. The rule is as follows: To get the rms voltage of Gaussian noise, multiply the "rms" value you read on an averaging ac voltmeter by 1.13 (or add 1dB). Warning: This works fine if the signal you are measuring is pure noise (e.g., the output of an amplifier with a resistor or noise source as input), but it won't give accurate results if the signal consists of a sine wave added to noise.

A third method, not exactly world-famous for its accuracy, consists of looking at the noise waveform on an oscilloscope: The rms voltage is 1/6 to 1/8 of the peak-to-peak value (depending on your subjective reading of the pp amplitude). It isn't very accurate, but at least there's no problem getting enough measurement bandwidth.

7.22 Noise potpourri

Herewith a collection of interesting, and possibly useful, facts.

1. The averaging time required in an indicating device to reduce the fluctuations of a rectified noise signal to a desired level for a given noise bandwidth is

$$\tau \approx \frac{1600}{B\sigma_2} \text{ seconds}$$

where τ is the required time constant of the indicating device to produce fluctuations of standard deviation σ percent at the output of a linear detector whose input is noise of bandwidth B.

2. For band-limited white noise the expected number of maxima per second is

$$N = \sqrt{\frac{3(f_2^5 - f_1^5)}{5(f_2^3 - f_1^3)}}$$

where f_1 and f_2 are the lower and upper band limits. For $f_1 = 0$, $N = 0.77f_2$; for narrowband noise ($f_1 \approx f_2$), $N \approx (f_1 + f_2)/2$.

3. rms-to-average (i.e., average magnitude) ratios:

Gaussian noise: $\text{rms/avg} = \sqrt{\pi/2} = 1.25 = 1.96\text{dB}$

Sine wave: $\text{rms/avg} = \pi/2^{\frac{3}{2}} = 1.11 = 0.91\text{dB}$

Square wave: $\text{rms/avg} = 1 = 0\text{dB}$

4. Relative occurrence of amplitudes in Gaussian noise. Figure 7.66 gives the fractional time that a given amplitude level is exceeded by a Gaussian noise waveform of amplitude 1 volt rms.

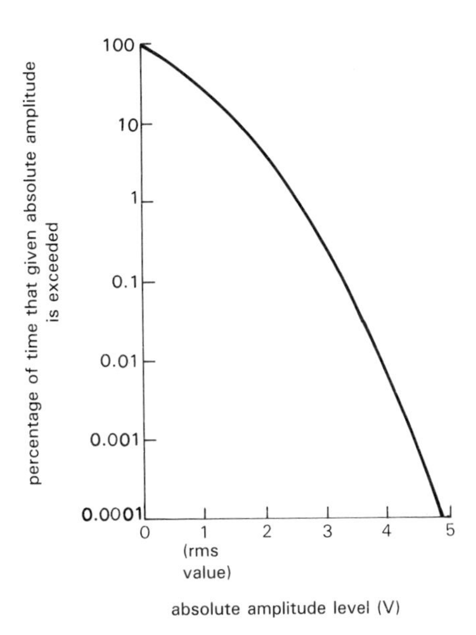

Figure 7.66. Relative occurrence of amplitudes in Gaussian noise.

INTERFERENCE: SHIELDING AND GROUNDING

"Noise" in the form of interfering signals, 60Hz pickup, and signal coupling via power supplies and ground paths can turn out to be of far greater practical importance than the intrinsic noise sources we've just discussed. These interfering signals can all be reduced to an insignificant level (unlike thermal noise) with proper layout and construction. In stubborn cases the cure may involve a combination of filtration of input and output lines, careful layout and grounding, and extensive electrostatic and magnetic shielding. In these sections we would like to offer some suggestions that may help illuminate this dark area of the electronic art.

7.23 Interference

Interfering signals can enter an electronic instrument through the power-line inputs or through signal input and output lines. In addition, signals can be capacitively coupled (electrostatic coupling) onto wires in the circuit (the effect is more serious for high-impedance points within the circuit), magnetically coupled to closed loops in the circuit (independent of impedance level), or electromagnetically coupled to wires acting as small antennas for electromagnetic radiation. Any of these can become a mechanism for coupling of signals from one part of a circuit to another. Finally, signal currents from one part of the circuit can couple to other parts through voltage drops on ground lines or power-supply lines.

Eliminating interference

Numerous effective tricks have been evolved to handle most of these commonly occurring interference problems. Keep in mind the fact that these techniques are all aimed at reducing the interfering signal or signals to an acceptable level; they rarely eliminate them altogether. Consequently, it often pays to raise signal levels, just to improve the signal-to-interference ratio. Also, it is important to realize that some environments are much worse than others; an instrument that works just dandy on the bench may perform miserably on location. Some environments worth avoiding are those (a) near a radio or television station (RF interference), (b) near a subway (impulsive interference and power-line garbage), (c) near high-voltage lines (radio interference, frying sounds), (d) near motors and elevators (power-line spikes), (e) in a building with triac lamp and heater controllers (power-line spikes), (f) near equipment with large transformers (magnetic pickup), and (g) near arc welders (unbelievable pickup of all sorts). Herewith a gathering of advice, techniques, and black magic:

Signals coupled through inputs, outputs, and power line

The best bet for power-line noise is to use a combination of RF line filters and transient suppressors on the ac power line. You can achieve 60dB or better attenuation of interference above a few hundred kilohertz this way, as well as effective elimination of damaging spikes.

Inputs and outputs are more difficult, because of impedance levels and the need to couple desired signals that may lie in the frequency range of interference. In devices like audio amplifiers you can use low-pass filters on inputs *and outputs* (much interference from nearby radio stations enters via the speaker wires, acting as antennas). In other situations shielded lines are often necessary. Low-level signals, particularly at high impedance levels, should always be shielded. So should the instrument cabinet.

Capacitive coupling

Signals within an instrument can get around handsomely via electrostatic coupling: Some point within the instrument has a 10 volt signal jumping around; a high-Z input nearby does some sympathetic jumping, too. The best things to do are to reduce the capacitance between the offending points (move them apart), add shielding (a complete metal enclosure, or even close-knit metal screening, eliminates this form of coupling altogether), move the wires close to a ground plane (which "swallows" the electrostatic fringing fields, reducing coupling enormously), and lower the impedance levels at susceptible points, if possible. Op-amp outputs don't pick up interference easily, whereas inputs do. More on this later.

Magnetic coupling

Unfortunately, low-frequency magnetic fields are not significantly reduced by metal enclosures. A turntable, microphone, tape recorder, or other sensitive circuit placed in close proximity to an instrument with a large power transformer will display astounding amounts of 60Hz pickup. The best therapy here is to avoid large enclosed areas within circuit paths and try to keep the circuit from closing around in a loop. Twisted pairs of wires are quite effective in reducing magnetic pickup, because the enclosed area is small, and the signals induced in successive twists cancel.

When dealing with very low level signals, or devices particularly susceptible to magnetic pickup (tape heads, inductors, wire-wound resistors), it may be desirable to use magnetic shielding. "Mu-metal shielding" is available in preformed pieces and flexible sheets. If the ambient magnetic field is large, it is best to use shielding of high permeability (high mu) on the inside, surrounded by an outer shield of lower permeability (which can be ordinary iron, or low-mu shielding material), to prevent magnetic saturation in the inner shield. Of course, moving the offending source of magnetic field is often a simpler solution. It may be necessary to exile large power transformers to the hinterlands, so to speak. Toroidal transformers have smaller fringing fields than the standard frame types.

Radiofrequency coupling

RF pickup can be particularly insidious, because innocent-looking parts of the circuit can act as resonant circuits, displaying enormous effective cross section for pickup. Aside from overall shielding, it is best to keep leads short and avoid loops that can resonate. Ferrite beads may help, if the problem involves very high frequencies. A classic situation is the use of a pair of bypass capacitors (one tantalum, one disc ceramic), often recommended to improve bypassing. The pair can form a lovely parasitic tuned circuit somewhere in the HF to VHF region (tens to hundreds of megahertz), with self-oscillations!

7.24 Signal grounds

Ground leads and shields can cause plenty of trouble, and there is a lot of misunderstanding on this subject. The problem, in a nutshell, is that currents you forgot about flowing through a ground line can generate a signal seen by another part of the circuit sharing the same ground. The technique of a ground "mecca" (a common point in the circuit to which all ground connections are tied) is often seen, but it's a crutch; with a little understanding of the problem you can handle most situations intelligently.

Common grounding blunders

Figure 7.67 shows a common situation. Here a low-level amplifier and a high-current driver are in the same instrument. The first circuit is done correctly: Both amplifiers tie to the supply voltages at the regulator (right at the sensing leads), so *IR* drops along the leads to the power stage don't appear on the low-level amplifier's supply voltages. In addition, the load current returning to ground does not appear at the low-level input; no current flows from the ground side of the low-level amplifier's input to the circuit mecca (which might be the connection to the case near the BNC input connector).

In the second circuit there are two blunders. Supply voltage fluctuations caused by load currents at the high-level stage are impressed on the low-level supply voltages. Unless the input stage has very good supply rejection, this can lead to oscillations. Even worse, the load current returning to the supply makes the case "ground" fluctuate with respect to power-supply ground. The input stage ties to this fluctuating ground, a very bad idea. The general idea is to look at where the large signal currents are flowing and make sure their *IR* drops don't wind up at the input. In some cases it may be a good idea to decouple the supply voltages to the low-level stages with a small *RC* network (Fig. 7.68). In stubborn cases of supply coupling it may pay to put a zener or 3-terminal regulator on the low-level-stage supply for additional decoupling.

□ 7.25 Grounding between instruments

The idea of a controlled ground point within one instrument is fine, but what do you do when a signal has to go from one instrument to another, each with its own idea of "ground"? Some suggestions follow.

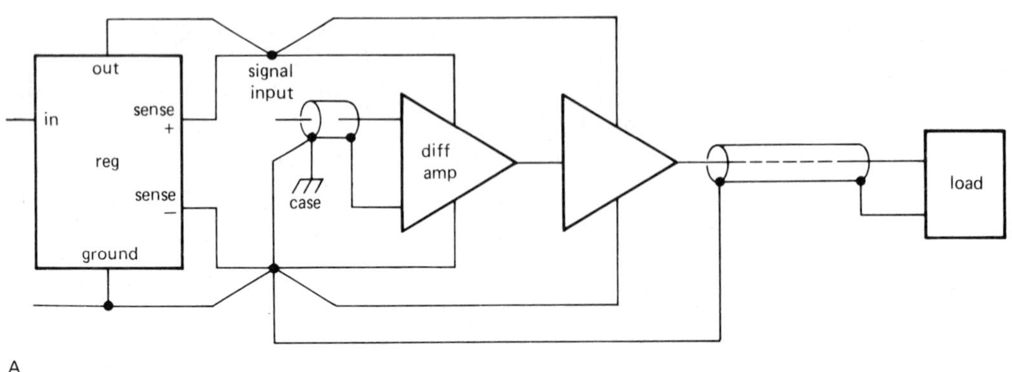

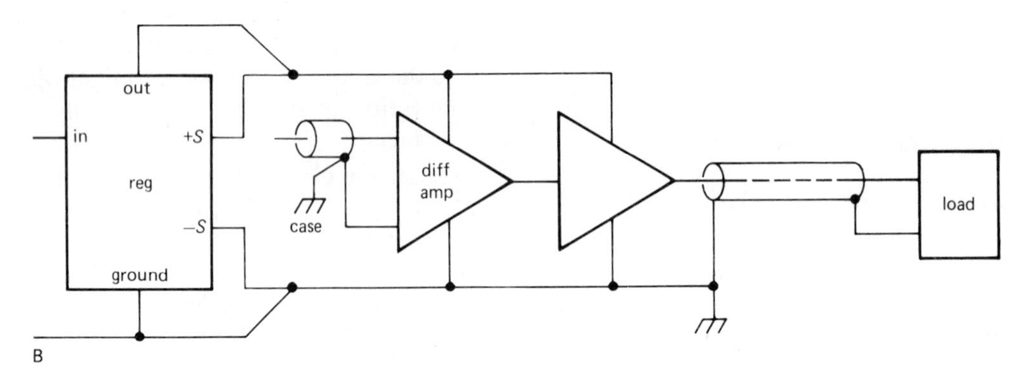

Figure 7.67. Ground paths for low-level signals.
A. Right
B. Wrong

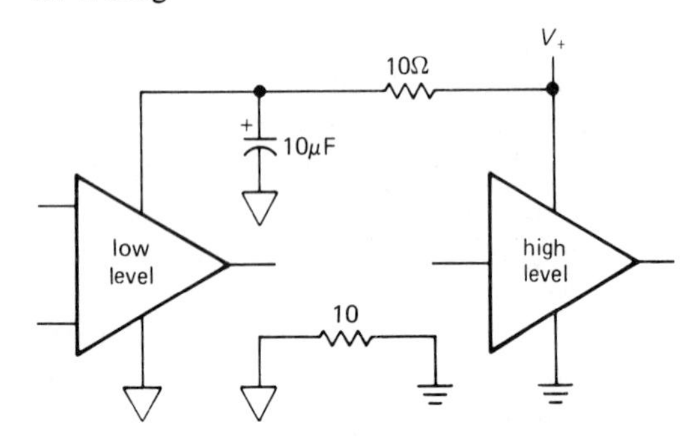

Figure 7.68

□ High-level signals

If the signals are several volts, or large logic swings, just tie things together and forget about it (Fig. 7.69). The voltage source shown between the two grounds represents the variations in local grounds you'll find on different power-line outlets in the same room or (worse) in different rooms or buildings. It consists of some 60Hz voltage, harmonics of the line frequency, some radiofrequency signals (the power line makes a good antenna), and assorted spikes and other garbage. If your signals are large enough, you can live with this.

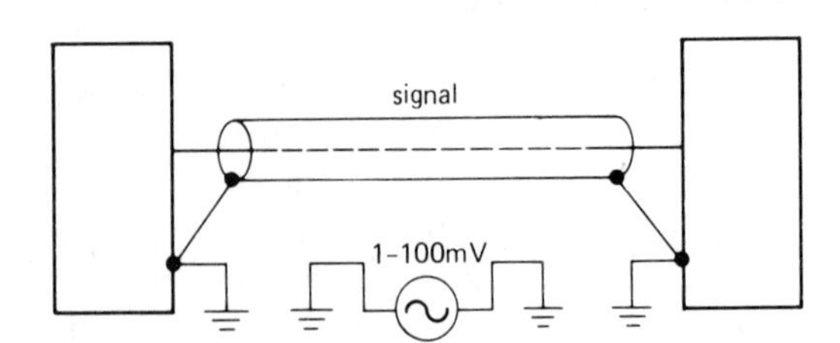

Figure 7.69

□ Small signals and long wires

For small signals this situation is intolerable, and you will have to go to some effort

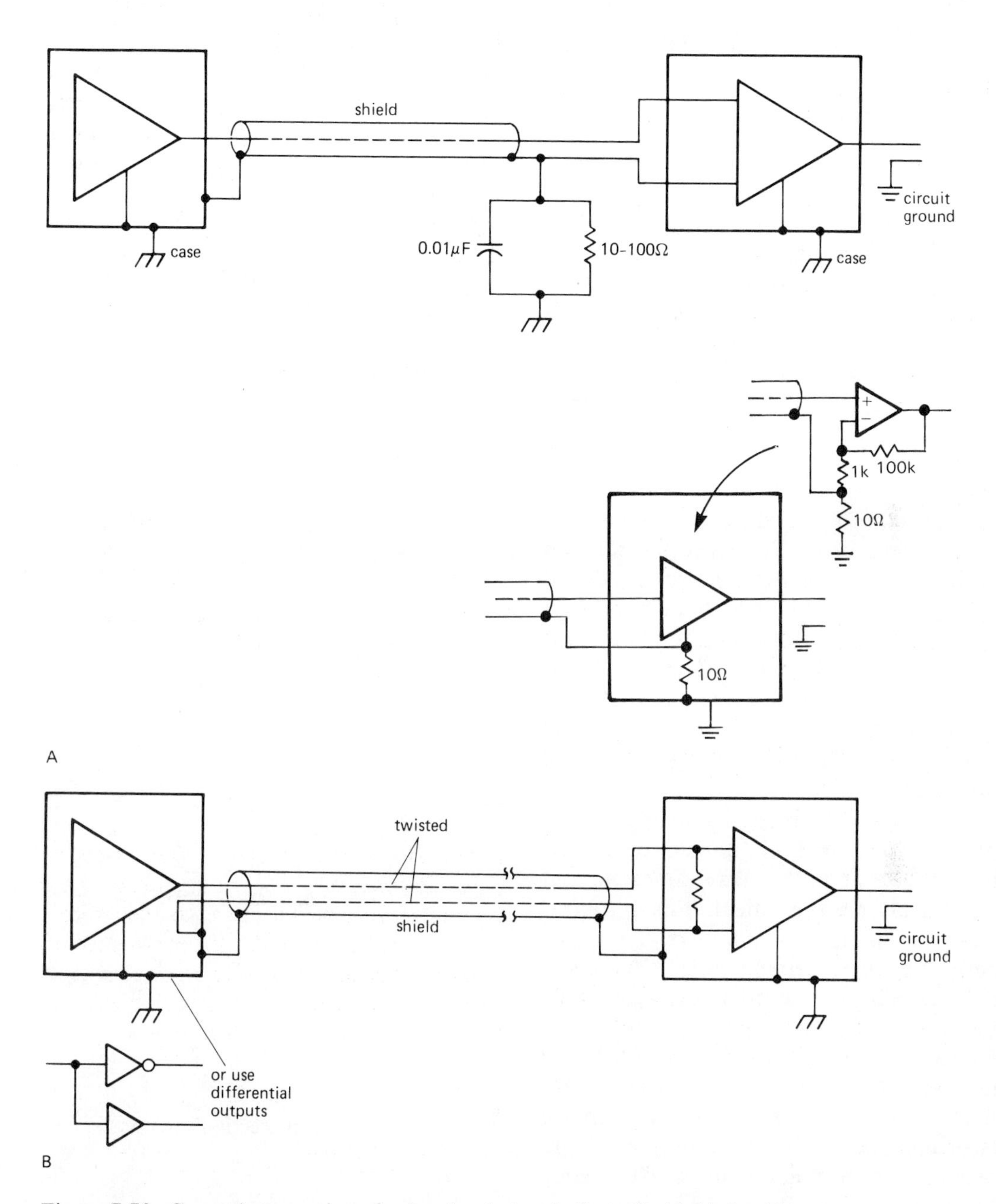

Figure 7.70. Ground connections for low-level signals through shielded cables.

to remedy the situation. Figure 7.70 shows some ideas. In the first circuit, a coaxial shielded cable is tied to the case and circuit ground at the driving end, but it is kept isolated from the case at the receiving end (use a Bendix 4890-1 or Amphenol 31-010 insulated BNC connector). A differential amplifier is used to buffer the input signal, thus ignoring the small amount of "ground signal" appearing on the shield. A small resistor and bypass capacitor to ground is a good idea to limit ground swing and prevent damage to the input stage. The alternate receiver circuit in Figure 7.70 shows the use of a "pseudodifferential" input connection for a single-ended amplifier stage (which might, for example, be a standard non-inverting op-amp connection, as indicated). The 10 ohm resistor between amplifier common and circuit ground is large enough to let the signal source's reference ground set the potential at that point, since it is much larger than the impedance of the source's ground. Any noise present at that node, of course, appears also at the output. However, this becomes unimportant if the stage has sufficiently high voltage gain, G_V, since the ratio of desired signal to ground noise is reduced by G_V. Thus, although this circuit isn't truly differential (with infinite CMRR), it works well enough (with effective CMRR $= G_V$). This pseudo-differential ground-sensing trick can be used also for low-level signals *within* an instrument, when ground noise is a problem.

In the second circuit, a shielded twisted pair is used, with the shield connected to the case at both ends. Since no signal travels on the shield, this is harmless. A differential amplifier is used as before on the receiving end. If logic signals are being transmitted, it is a good idea to send a differential signal (the signal and its inverted form), as indicated. Ordinary differential amplifiers can be used as input stages, or if the ground interference is severe, special "isolated amplifiers" are available from manufacturers like Analog Devices and Burr-Brown. The latter permit kilovolts of common-mode signals. So do opto-isolator modules, a handy solution for digital signals in some situations.

At radiofrequencies, transformer coupling offers a convenient way of removing common-mode signal at the receiving end; this also makes it easy to generate a differential bipolarity signal at the driving end. Transformers are popular in audio applications as well, although they tend to be bulky and lead to some signal degradation.

For very long cable runs (measured in miles) it is useful to prevent large ground currents flowing in the shield at radiofrequencies. Figure 7.71 suggests a method. As before, a differential amplifier looks at the twisted pair, ignoring the voltage on the shield. By tying the shield to the case through a small inductor, the dc voltage is kept small while preventing large radiofrequency currents. This circuit also shows protection circuitry to prevent common-mode excursions beyond ±10 volts.

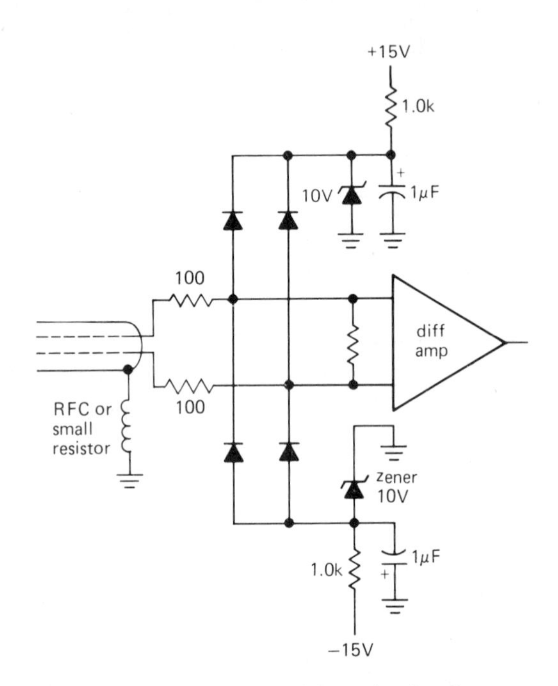

Figure 7.71. Input-protection circuits for use with very long lines.

Figure 7.72 shows a nice scheme to save wires in a multiwire cable in which the common-mode pickup has to be eliminated. Since all the signals suffer the same common-mode pickup, a single wire tied to ground at the sending end serves to cancel the common-mode signals on each of the n signal lines. Just buffer its signal

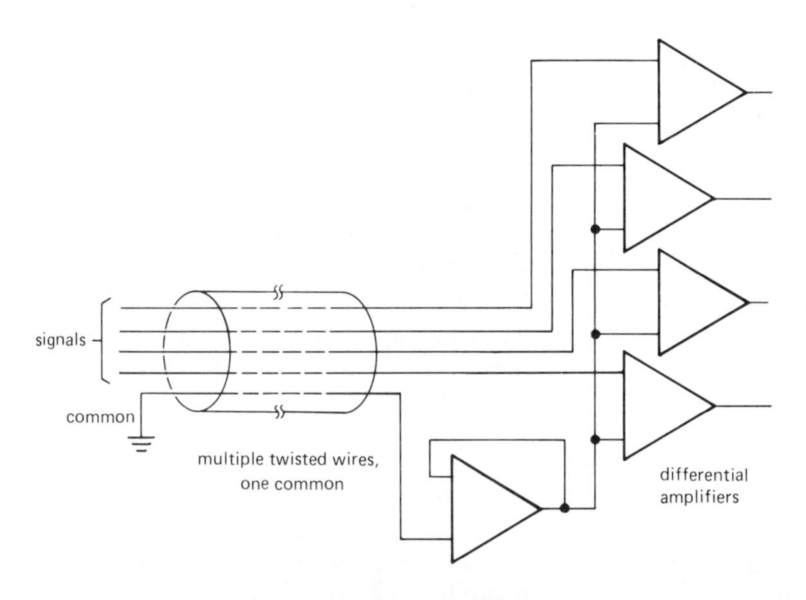

Figure 7.72. Common-mode interference rejection with long multiwire cables.

(with respect to ground at the receiving end), and use it as the comparison input for each of n differential amplifiers looking at the other signal lines.

The preceding schemes work well to eliminate common-mode interference at low to moderate frequencies, but they can be ineffective against radiofrequency interference, owing to poor common-mode rejection in the receiving differential amplifier.

One possibility here is to wrap the whole cable around a ferrite toroid (Fig. 7.73). That increases the series inductance of the whole cable, raising the impedance to common-mode signals of high frequency and making it easy to bypass them at the far end with a pair of small bypass capacitors to ground. The equivalent circuit shows why this works without attenuating the differential signal: You have a series inductance inserted into both signal lines and the shield, but since they form a tightly coupled transformer of unit turns ratio, the differential signal is unaffected. This is actually a "1:1 transmission-line transformer," as discussed in Section 13.10.

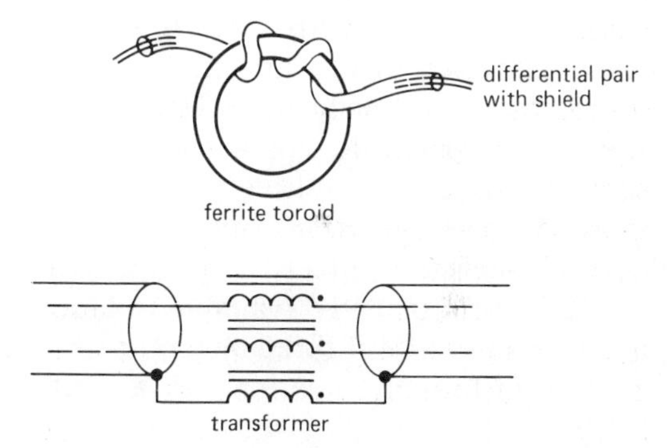

Figure 7.73

□ *Floating signal sources*

The same sort of disagreement about the voltage of "ground" at separated locations enters in an even more serious way at low-level inputs, just because the signals are so small. An example is a magnetic tape head or other signal transducer that requires a shielded signal line. If you ground the shield at both ends, differences in ground potential will appear as signal at the amplifier input. The best approach is to lift the shield off ground *at the transducer* (Fig. 7.74).

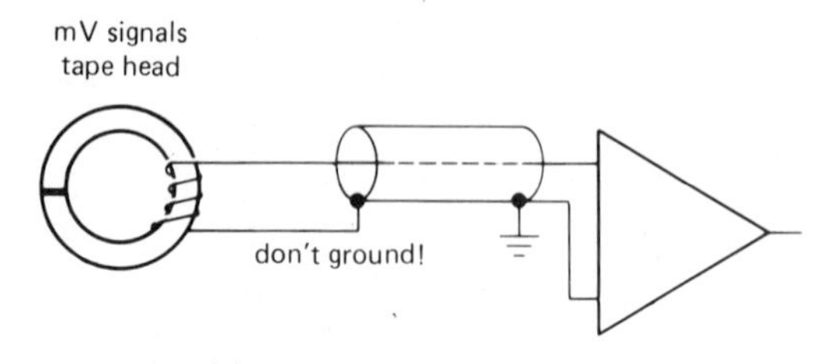

Figure 7.74

Isolation amplifiers

Another solution to serious ground-contention problems is the use of an "isolation amplifier." Isolation amplifiers (iso-amps) are commercial devices intended for coupling an analog signal (with bandwidth clear down to dc) from a circuit with one ground reference to another circuit with a completely different ground (Fig. 7.75). In fact, in some bizarre situations the "grounds" can differ by many kilovolts! Isolation amplifiers are mandatory for medical electronics in which electrodes are applied to human subjects, in order to isolate completely those connections from any instrument circuits powered directly from the ac power lines. Currently available isolation amplifiers use one of three methods:

1. Transformer isolation of a high-frequency carrier signal, which is either frequency-modulated or pulse-width-modulated with the relatively low bandwidth signal (dc to 10kHz or so) to be isolated (Fig. 7.76). This method is used in all of the isolation amplifiers from Analog Devices, as well as some units from Burr-Brown. Transformer-isolated iso-amps have the convenient feature of requiring dc power only on one side; they all include a transformer-coupled dc-to-dc converter in the package. Transformer-coupled iso-amps can isolate up to 3.5kV and have typical bandwidths of 2kHz, though some units go to 20kHz.

2. Optically coupled signal transmission via an LED at the sending end and photodiode at the receiving end. This technique is typified by the ISO100 from Burr-Brown. No high-frequency carrier is needed, since signals all the way to dc can be transmitted optically. However, to achieve good linearity, Burr-Brown uses a cute trick: A second matched photodiode at the transmitting side receives light from the LED, in a feedback arrangement that cancels nonlinearities in both LED and photodiode; see Figure 7.77. The ISO100 requires power supplies at both ends, isolates to 750 volts, and has 60kHz bandwidth.

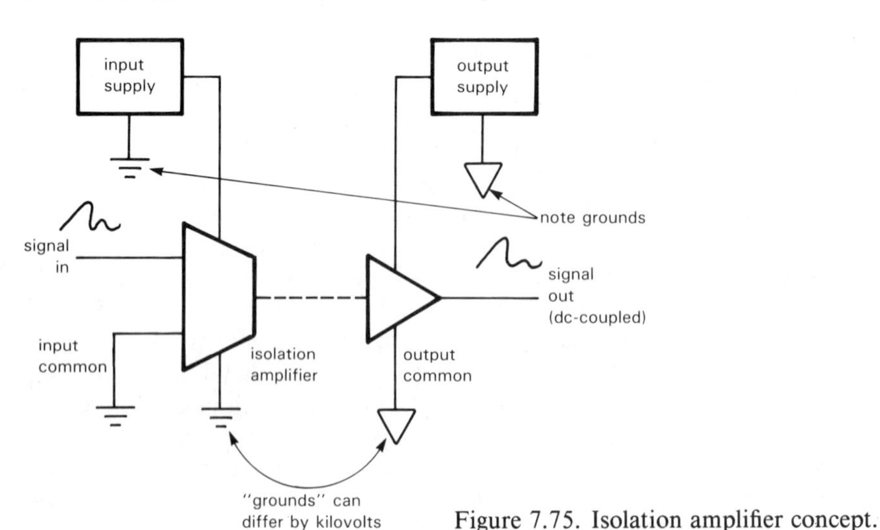

Figure 7.75. Isolation amplifier concept.

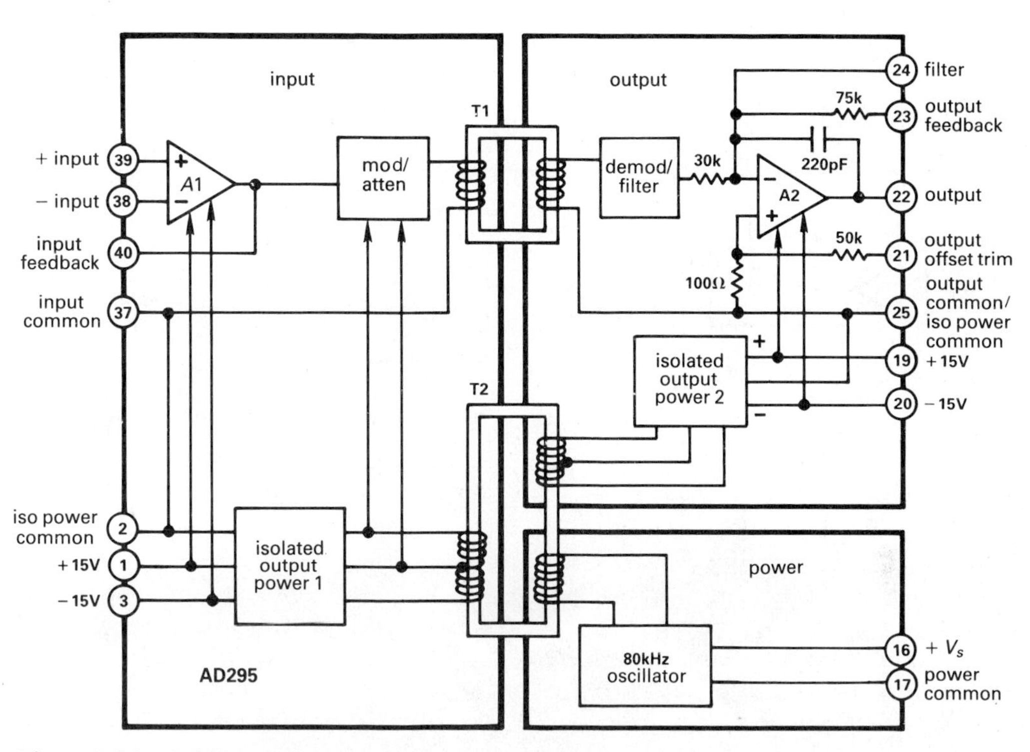

Figure 7.76. AD295 transformer-coupled isolation amplifier. (Courtesy of Analog Devices.)

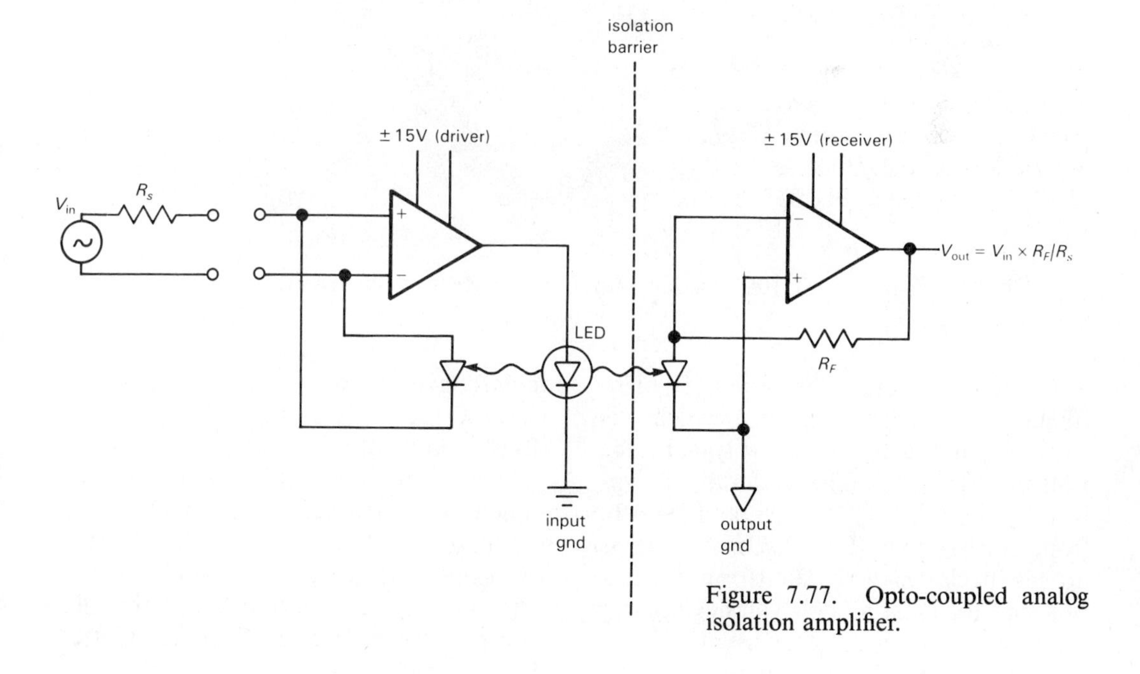

Figure 7.77. Opto-coupled analog isolation amplifier.

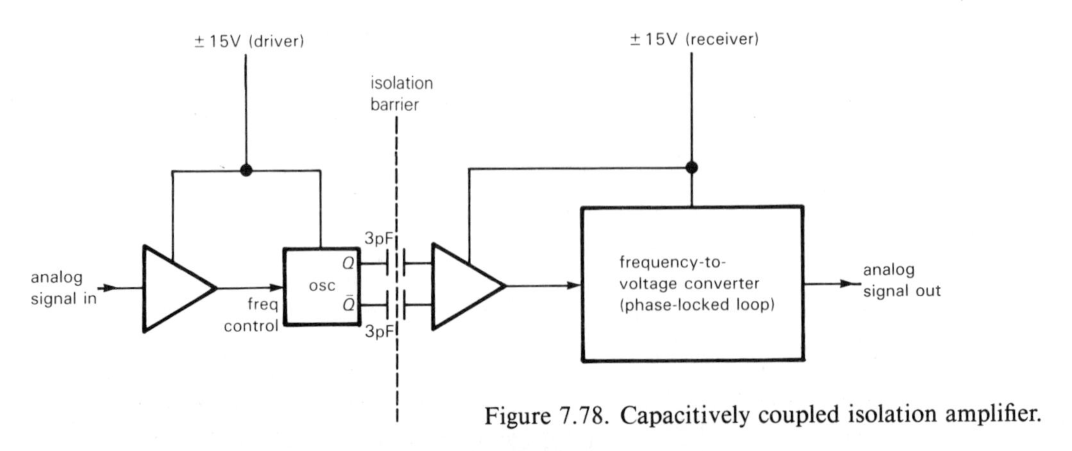

Figure 7.78. Capacitively coupled isolation amplifier.

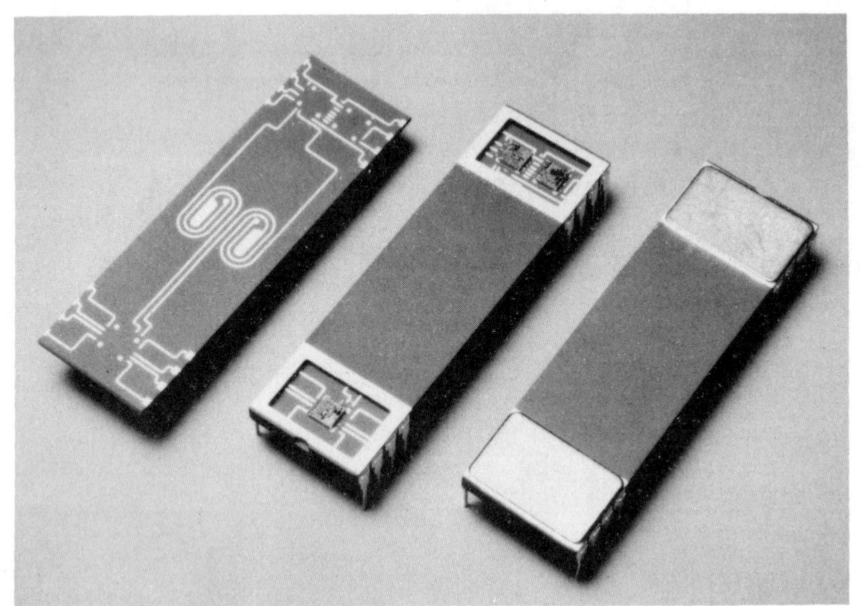

Figure 7.79. Burr-Brown ISO106 isolation amplifier. (Courtesy of Burr-Brown Corporation.)

3. Capacitively coupled isolation of a high-frequency carrier signal, which is frequency-modulated with the signal to be isolated (Fig. 7.78). This technique is typified by the ISO102, ISO106, and ISO122 from Burr-Brown (Fig. 7.79). There is no feedback, as with transformer isolation, but for most models you need power supplies at both ends. This usually isn't a problem, since you are likely to have electronics at both ends, generating and using the signal. If not, you can get an isolated dc-dc converter to use with the iso-amp. The ISO106 isolates to 3.5kV and has 70kHz bandwidth.

These isolation amplifiers are all intended for *analog* signals, of modest bandwidth; they cost from $25 to $100 each. The same sorts of ground problems can arise in digital electronics, where the solution is simple and effective: Optically coupled isolators ("opto-isolators") are available, with plenty of bandwidth (10MHz or more), isolation of several

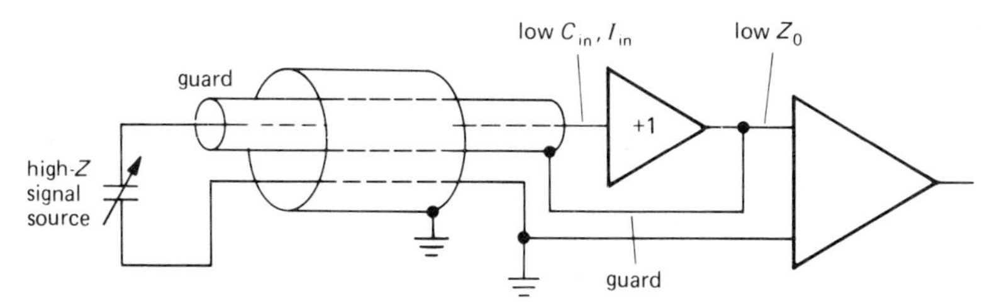

Figure 7.80. Using a guard to raise input impedance.

kilovolts, and low cost (a dollar or two). We'll see them in Chapter 9.

☐ *Signal guarding*

A closely related issue is signal *guarding*, an elegant technique to reduce the effects of input capacitance and leakage for small signals at high impedance levels. You may be dealing with signals from a microelectrode or a capacitive transducer, with source impedances of hundreds of megohms. Just a few picofarads input capacitance can form a low-pass filter, with rolloffs beginning at a few hertz! In addition, the effects of insulation resistance in the connecting cables can easily degrade the performance of an ultra-low input current amplifier (bias currents less than a picoamp) by orders of magnitude. The solution to both these problems is a *guard electrode* (Fig. 7.80).

A follower bootstraps the inner shield, effectively eliminating leakage current and capacitive attenuation by keeping zero voltage difference between the signal and its surrounding. An outer grounded shield is a good idea, to keep interference off the guard electrode; the follower has no trouble driving that capacitance and leakage, of course, since it has low output impedance.

You shouldn't use these tricks more than you need to; it would be a good idea to put the follower as close to the signal source as possible, guarding only the short section of cable connecting them. Ordinary shielded cable can then carry the low-impedance output signal out to the remote amplifier. We will discuss signal guards in Section 15.08 in connection with high-impedance microelectrodes.

☐ *Coupling to outputs*

Ordinarily the output impedance of an op-amp is low enough that you don't have to worry about capacitive signal coupling. In the case of high-frequency or fast-switching interference, however, you have just cause for alarm, particularly if the desired output signal involves some degree of precision. Consider the example in Figure 7.81. A precision signal is buffered by an op-amp and passes through a region containing digital logic signals jumping around with slew rates of 0.5V/ns. The op-amp's closed-loop output impedance rises with frequency, typically reaching values of 10 to 100 ohms or more at 1MHz (see Section 7.07). How large a coupling capacitance is permissible, keeping coupled interference less than the analog signal's resolution of 0.1mV? The surprising answer is 0.02pF.

There are some solutions. The best thing is to keep your small analog waveforms out of the reach of fast-switching signals. A moderate bypass capacitor across the op-amp's output (with perhaps a small series resistor, to maintain op-amp stability) will help, although it degrades the slew rate. You can think of the action of this capacitor as lowering the frequency of the coupled charge bundles to the point where the op-amp's feedback can swallow them. A

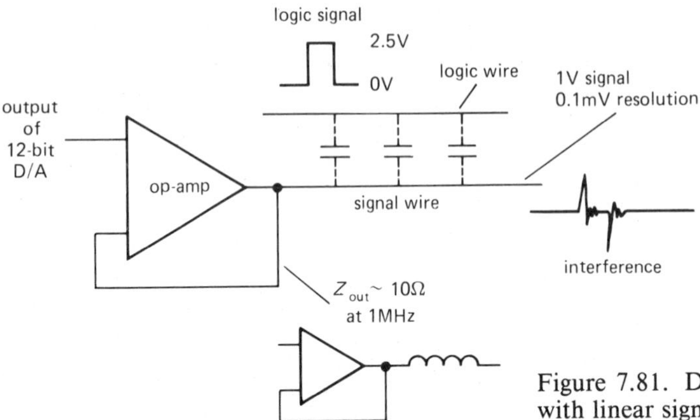

Figure 7.81. Digital cross-coupling interference with linear signals.

few hundred picofarads to ground will adequately stiffen the analog signal at high frequencies (think of it as a capacitive voltage divider). Another possibility is to use a low-impedance buffer, such as the LT1010, or a power op-amp such as the LM675. Don't neglect the opportunity to use shielding, twisted pairs, and proximity to ground planes to reduce coupling.

SELF-EXPLANATORY CIRCUITS

7.26 Circuit ideas

Figure 7.82 presents some circuit ideas relevant to the subjects of this chapter.

ADDITIONAL EXERCISES

(1) Prove that SNR $= 10\log_{10}(v_s^2/4kTR_s) -$ NF(dB) (at R_s).

(2) A 10μV (rms) sine wave at 100Hz is in series with a 1M resistor at room temperature. What is the SNR of the resultant signal **(a)** in a 10Hz band centered at 100Hz and **(b)** in a 1MHz band going from dc to 1MHz?

(3) A transistor amplifier using a 2N5087 is operated at 100μA collector current and is driven by a signal source of impedance 2000 ohms. **(a)** Find the noise figure at 100Hz, 1kHz, and 10kHz. **(b)** Find the SNR (at each of listed frequencies) for an input signal of 50nV (rms) and an amplifier bandwidth of 10Hz.

(4) Measurements are made on a commercial amplifier (with $Z_{\text{in}} = 1\text{M}$) in order to determine its equivalent input noise e_n and i_n at 1kHz. The amplifier's output is passed through a sharp-skirted filter of bandwidth 100Hz: A 10μV input signal results in a 0.1 volt output. At this level the amplifier's noise contribution is negligible. With the input shorted, the noise output is 0.4mV rms. With the input open, the noise output rises to 50mV rms. **(a)** Find e_n and i_n for this amplifier at 1kHz. **(b)** Find the noise figure of this amplifier at 1kHz for source resistances of 100 ohms, 10k, and 100k.

(5) Noise measurements are made on an amplifier using a calibrated noise source whose output impedance is 50 ohms. The generator output must be raised to 2nV/Hz$^{1/2}$ in order to double the output noise power of the amplifier. What is the amplifier's noise figure for a source impedance of 50 ohms?

(6) The output noise voltage of a white noise generator is measured with the circuit shown in Figure 7.83. At a particular setting of the noise generator output level, the ac voltmeter reads 1.5 volts "rms." What is the noise generator's output noise density (i.e., rms volts per root hertz)?

Circuit ideas

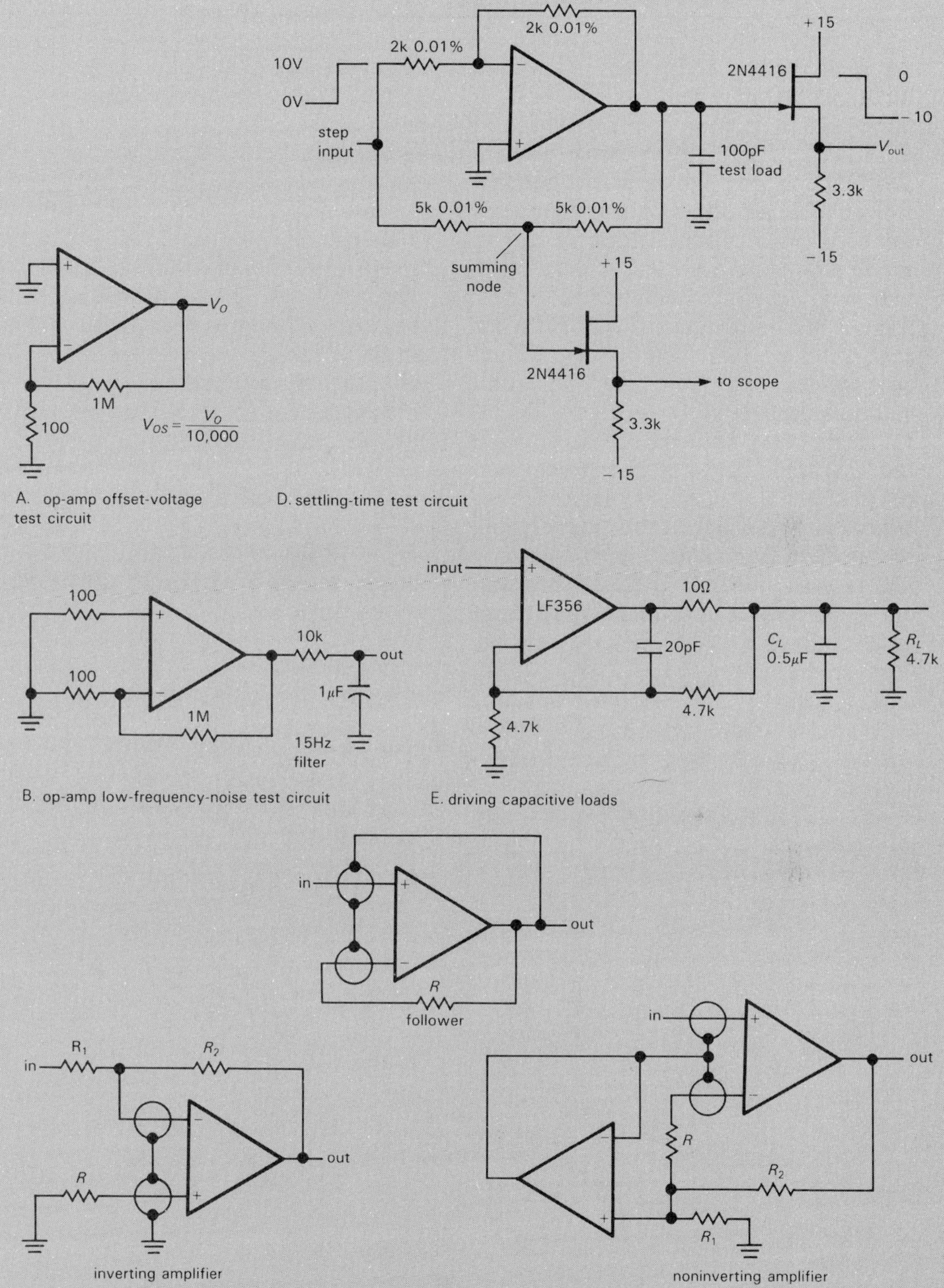

Figure 7.82

Circuit ideas (*cont.*)

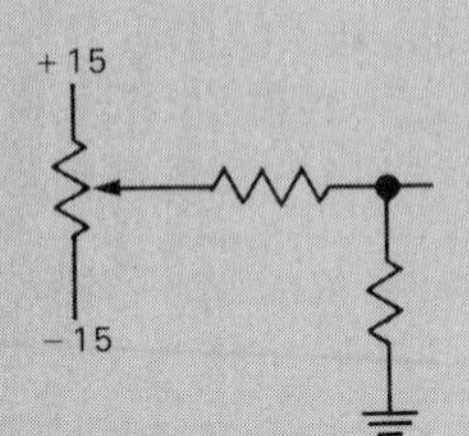

A

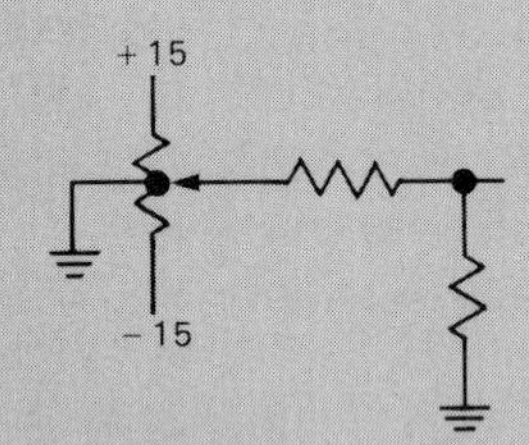

B

F. A. traditional offset trimming circuit
B. with improved trimpot (available from Bourns), circuit has less dependance on unbalanced supply voltage changes

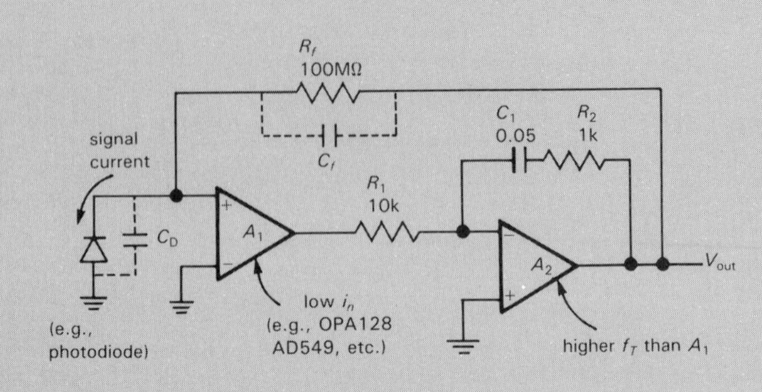

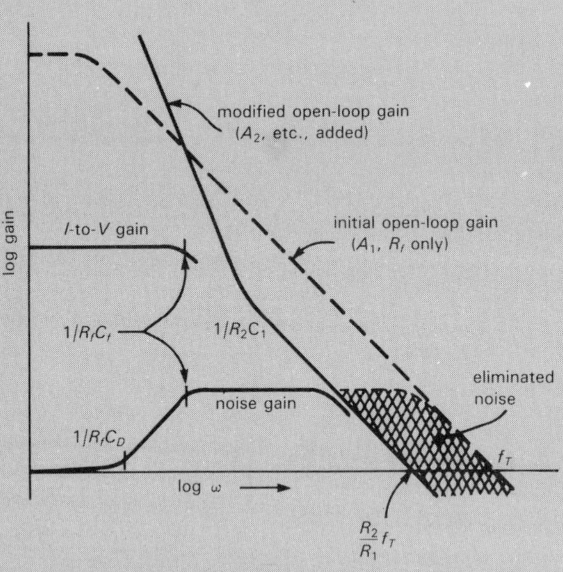

I. reducing "noise gain" in low-level transimpedance amplifier (from Burr-Brown App. Note)

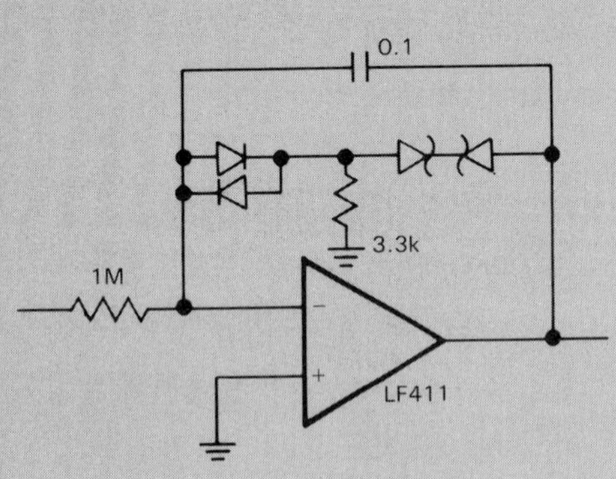

G. integrator with low-leakage zener clamp

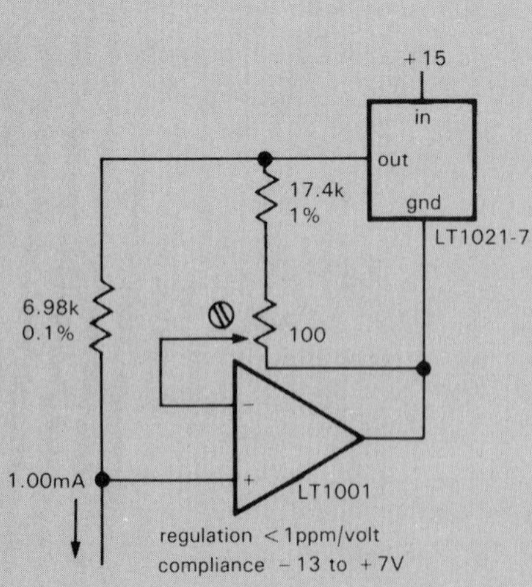

H. ultraprecise current source

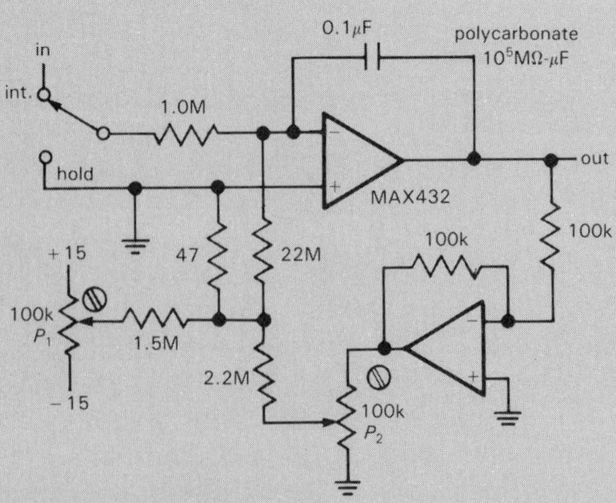

J. precision integrator with capacitor leakage compensation; use P_1 to null hold drift while output is near zero and P_2 while output is near +10V

$$\text{gain} = \left[\frac{R_2 + R_3}{R_5} + \frac{1}{2}\left(\frac{R_2}{R_1} + \frac{R_3}{R_4}\right) + 1\right]\frac{R_4}{R_3}$$

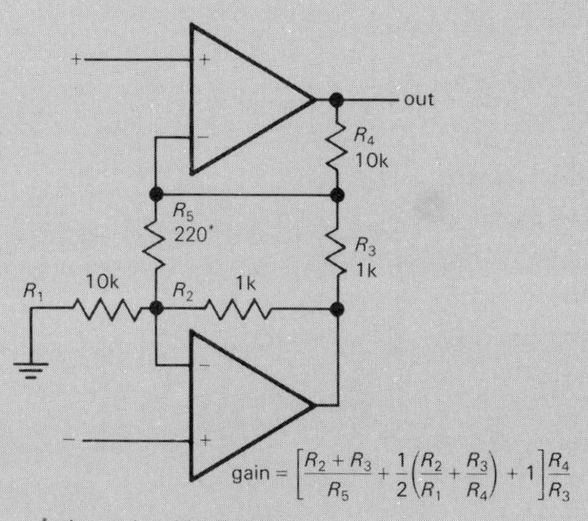

*select gain with changes in R_5

K. instrumentation amplifier

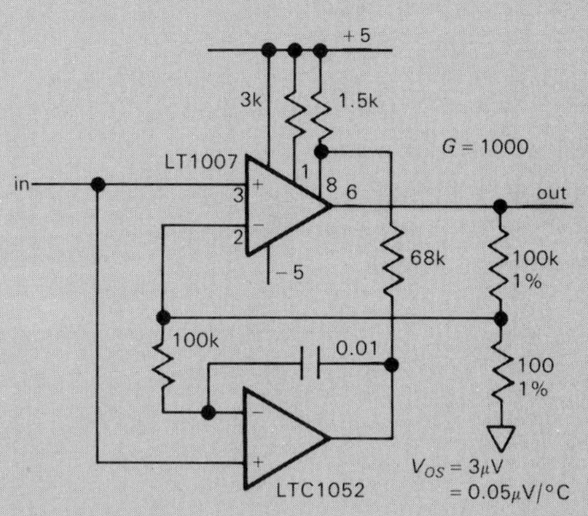

N. low-noise, low-drift amplifier; noise 60nV p-p, 0.1 to 10Hz, rather than the larger value of 1.5μ V p-p for the chopper amplifier

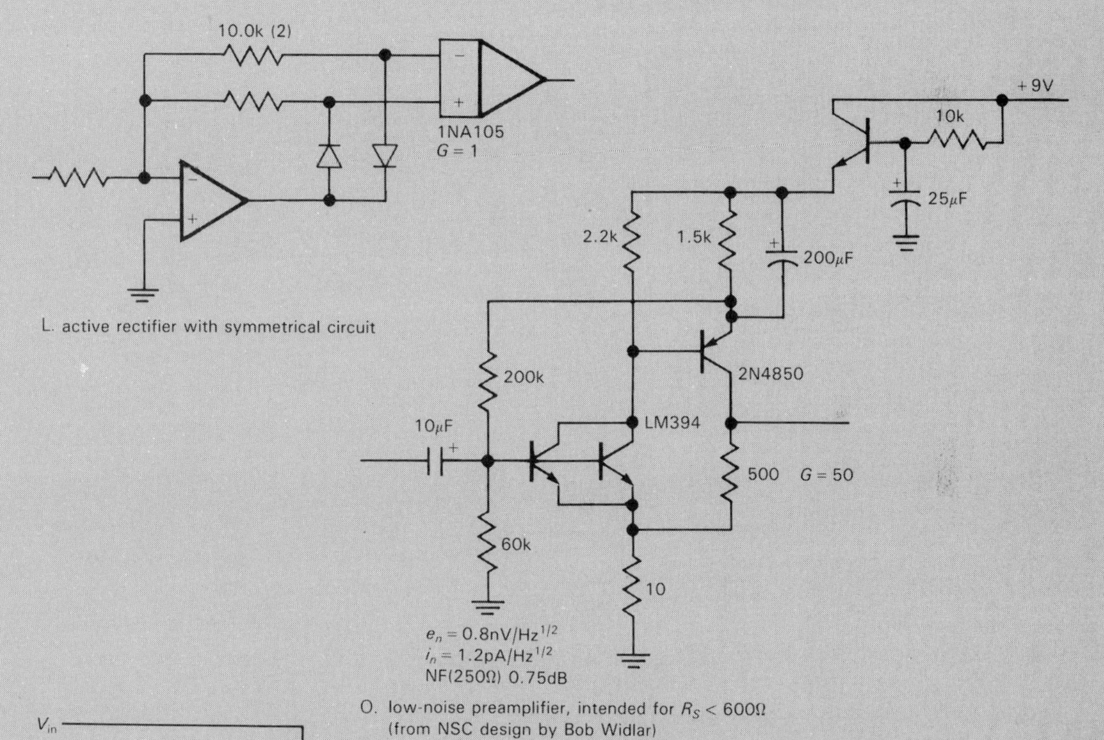

L. active rectifier with symmetrical circuit

$e_n = 0.8\text{nV/Hz}^{1/2}$
$i_n = 1.2\text{pA/Hz}^{1/2}$
NF(250Ω) 0.75dB

O. low-noise preamplifier, intended for $R_S < 600\Omega$ (from NSC design by Bob Widlar)

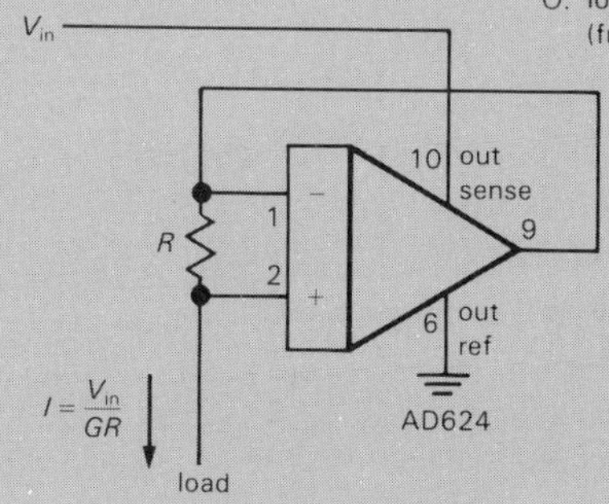

M. programmable current source using instrumentation amplifier

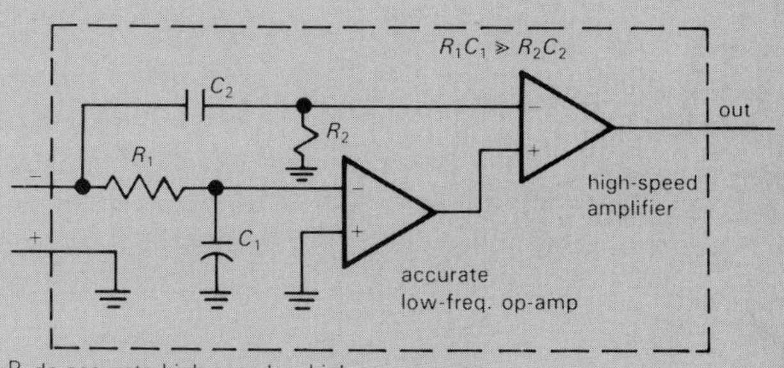

P. dc-accurate high-speed or high-power op-amp

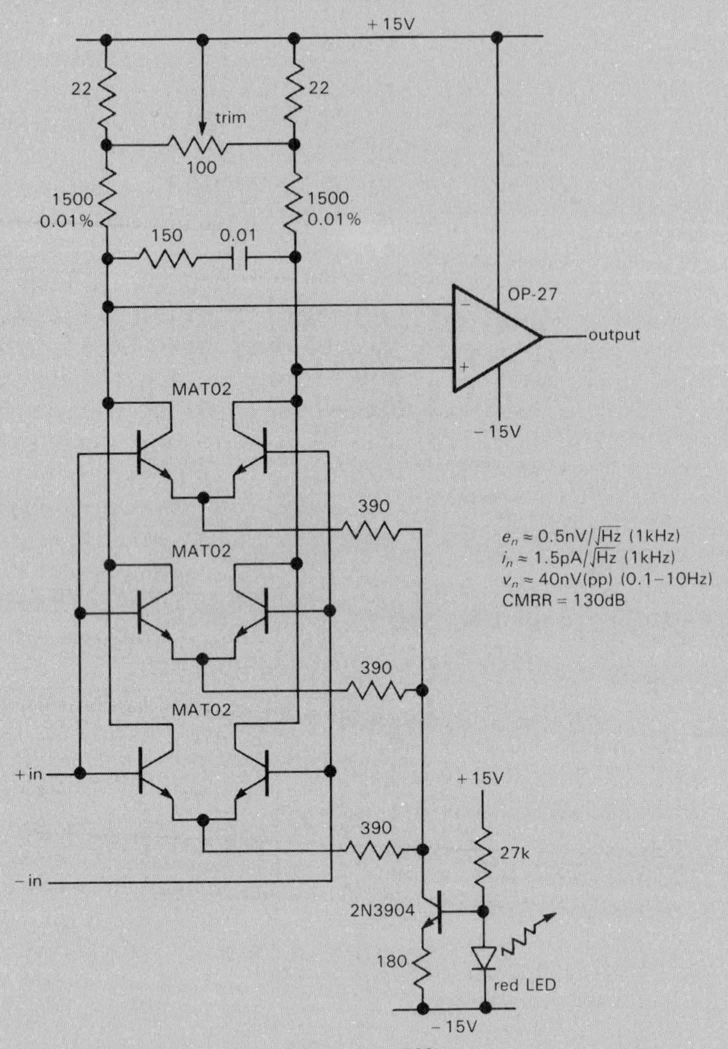

Q. ultra-low-noise op-amp (from PMI App. Note 102)

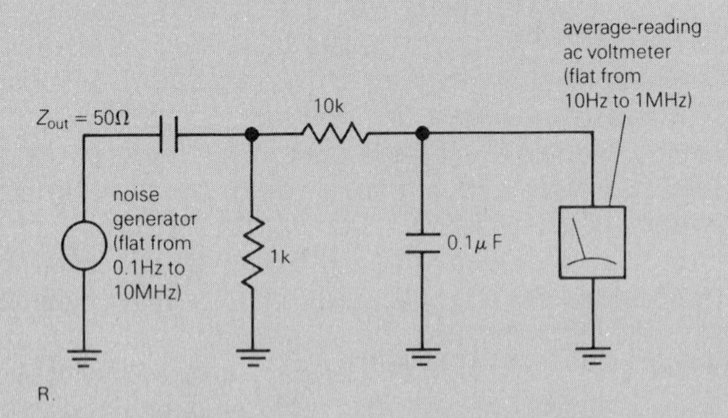

R.

Figure 7.83

BASIC LOGIC CONCEPTS

8.01 Digital versus analog

Thus far we have been dealing mainly with circuits in which the input and output voltages have varied over a range of values: *RC* circuits, amplifiers, integrators, rectifiers, op-amps, etc. This is natural when dealing with signals that are continuous (e.g., audio signals) or continuously varying voltages from measuring instruments (e.g., temperature-reading or light-detecting devices, or biological or chemical probes).

However, there are instances in which the input signal is naturally discrete in form, e.g., pulses from a particle detector, or "bits" of data from a switch, keyboard, or computer. In such cases the use of digital electronics (circuits that deal with data made of 1's and 0's) is natural and convenient. Furthermore, it is often desirable to convert continuous (analog) data to digital form, and vice versa (using D/A and A/D converters), in order to perform calculations on the data with a calculator or computer or to store large quantities of data as numbers. In a typical situation a microprocessor or computer might monitor signals from an experiment or industrial process, control the experimental parameters on the basis of the data obtained, and store for future use the results collected or computed while the experiment is running.

Another interesting example of the power of digital techniques is the transmission of analog signals without degradation by noise: An audio or video signal, for instance, picks up "noise" while being transmitted by cable or radio that cannot be removed. If, instead, the signal is converted to a series of numbers representing its amplitude at successive instants of time, and these numbers are transmitted as digital signals, the analog signal reconstruction at the receiving end (done with D/A converters) will be without error, providing the noise level on the transmission channel isn't high enough to prevent accurate recognition of 1's and 0's. This technique, known as PCM (pulse-code modulation), is particularly attractive where a signal must pass through a series of "repeaters," as in the case of a transcontinental telephone call, since digital regeneration at each stage guarantees noiseless transmission. The information and pictures sent back by recent deep space probes were done with PCM. Digital audio is now commonplace in the home, in the form of 12cm optical "compact discs" (CDs); these store a piece of music in the form of a stereo pair of 16-bit numbers every 23 microseconds, 6 billion bits of information in all.

In fact, digital hardware has become so powerful that tasks that seem well suited to analog techniques are often better solved with digital methods. As an example, an analog temperature meter might incorporate a microprocessor and memory in order to improve accuracy by compensating the instrument's departure from perfect linearity. Because of the wide availability of microprocessors, such applications are becoming commonplace. Rather than attempt to enumerate what can be done with digital electronics, let's just start learning about it. Applications will emerge naturally as we go along.

8.02 Logic states

By "digital electronics" we mean circuits in which there are only two (usually) states possible at any point, e.g., a transistor that can either be in saturation or be nonconducting. We usually choose to talk about voltages rather than currents, calling a level HIGH or LOW. The two states can represent any of a variety of "bits" (binary digits) of information, such as the following:

one bit of a number
whether a switch is opened or closed
whether a signal is present or absent
whether some analog level is above or below some preset limit
whether or not some event has happened yet
whether or not some action should be taken
etc.

HIGH and LOW

The HIGH and LOW states represent the TRUE and FALSE states of Boolean logic, in some predefined way. If at some point HIGH represents TRUE, that signal line is called "positive true," and vice versa. This can be confusing at first. Figure 8.1 shows an example. SWITCH CLOSED is true when the output is LOW; that's a negative-true signal ("LOW-true" might be a better label, since no negative voltages are involved), and you might label the lead as shown (a bar over a symbol means NOT; that line is HIGH when the switch is *not* closed). Just remember that the presence or absence of the negation bar over the label tells whether the wire is LOW or HIGH when the stated condition (SWITCH CLOSED) is true.

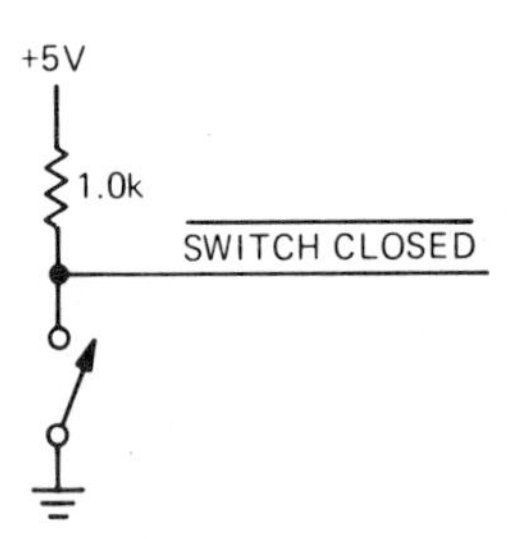

Figure 8.1

A digital circuit "knows" what a signal represents by where it comes from, just as an analog circuit might "know" what the output of some op-amp represents. However, added flexibility is possible in digital circuits; sometimes the same signal lines are used to carry different kinds of information, or even to send it in different directions, at different times. In order to do this "multiplexing," additional information must also be sent (address bits, or status bits). You will see many examples of this very useful ability later. For now, imagine that any given circuit is wired up to perform a predetermined function and that it knows what that function is, where its inputs are coming from, and where the outputs are going.

To lend a bit of confusion to a basically simple situation, we introduce 1 and 0. These symbols are used in Boolean logic to mean TRUE and FALSE, respectively, and are sometimes used in electronics in exactly that way. Unfortunately, they are also used in another way, in which 1 = HIGH and 0 = LOW! In this book we will try to avoid any ambiguity by using the word HIGH (or the symbol H) and the word LOW (or the symbol L) to represent logic states, a method that is in wide use in the electronics industry. We will use 1 and 0 only in situations where there can be no ambiguity.

Voltage range of HIGH and LOW

In digital circuitry, the voltage levels corresponding to HIGH and LOW are allowed to fall in some range. For example, with high-speed CMOS ("HC") logic, *input* voltages within about 1.5 volts of ground are interpreted as LOW, while voltages within 1.5 volts of the +5 volt supply are interpreted as HIGH. In fact, typical LOW- and HIGH-state *output* voltages are usually within a tenth of a volt of 0 and +5 volts, respectively (the output is a saturated MOS transistor to one of the rails; see Fig. 8.17). This allows for manufacturing spread, variations of the circuits with temperature, loading, supply voltage, etc., and the presence of "noise," the miscellaneous garbage that gets added to the signal in its journey through the circuit (from capacitive coupling, external interference, etc.). The circuit receiving the signal decides if it is HIGH or LOW and acts accordingly. As long as noise does not change 1's to 0's, or vice versa, all is well, and any noise is eliminated at each stage, since "clean" 0's and 1's are regenerated. In that sense digital electronics is noiseless and perfect.

The term *noise immunity* is used to describe the maximum noise level that can be added to logic levels (in the worst case) while still maintaining error-free operation. For instance, TTL has 0.4 volt noise immunity, since a TTL *input* is guaranteed to interpret anything less than +0.8 volt as LOW and anything greater than +2.0 volts as HIGH, whereas the worst-case *output* levels are +0.4 volt and +2.4 volts, respectively (see the accompanying box on logic levels). In practice, noise immunity is considerably better than this figure, with typical LOW and HIGH voltages of +0.2 and +3.4 volts and an input decision threshold near +1.3 volts. But always remember that if you are doing good circuit design, you use worst-case values. It is worth keeping in mind that different logic families have different amounts of noise immunity. CMOS has greater voltage noise immunity than TTL, whereas the speedy ECL family has less. Of course, susceptibility to noise in a digital system depends also on the amplitude of noise that is present, which in turn depends on factors such as output-stage stiffness, inductance in the ground leads, the existence of long "bus" lines, and output slew rates during logic transitions (which produces transient currents, and therefore voltage spikes on the ground line, due to capacitive loading). We will worry about some of these problems in Sections 9.11–9.13.

8.03 Number codes

Most of the conditions we listed earlier that can be represented by a digital level are self-explanatory. How a digital level can represent part of a number is a more involved, and very interesting, question.

A decimal (base-10) number is simply a string of integers that are understood to multiply successive powers of 10, the individual products then being added together.

For instance,

$$137.06 = 1 \times 10^2 + 3 \times 10^1 + 7 \times 10^0 + 0 \times 10^{-1} + 6 \times 10^{-2}$$

Ten symbols (0 through 9) are needed, and the power of 10 each multiplies is determined by its position relative to the decimal point. If we want to represent a number using two symbols only (0 and 1), we use the *binary*, or base-2, number system. Each 1 or 0 then multiplies a successive power of 2. For instance,

$$1101_2 = 1 \times 2^3 + 1 \times 2^2 + 0 \times 2^1 + 1 \times 2^0 = 13_{10}$$

The individual 1's and 0's are called "bits" (binary digits). The subscript (always given in base 10) tells what number system we are using, and often it is essential in order to avoid confusion, since the symbols all look the same.

We convert a number from binary to decimal by the method just described. To convert the other way, we keep dividing the number by 2, and write down the remainders. To convert 13_{10} to binary,

$$13/2 = 6 \quad \text{remainder } 1$$
$$6/2 = 3 \quad \text{remainder } 0$$
$$3/2 = 1 \quad \text{remainder } 1$$
$$1/2 = 0 \quad \text{remainder } 1$$

from which $13_{10} = 1101_2$. Note that the answer comes out in the order LSB (least significant bit) to MSB (most significant bit).

Hexadecimal ("hex") representation

The binary number representation is the natural choice for two-state systems (although it is not the only way; you'll see some others soon). Since the numbers tend to get rather long, it is common to write them in hexadecimal (base-16) representation: Each position represents successive powers of 16, with each hex symbol having a value from 0 to 15. Since you want a single symbol for each hex position, the symbols A–F are assigned to the values 10–15. To write a binary number in hexadecimal, just group it in 4-bit groups, beginning with the LSB, and write the hexadecimal equivalent of each group:

$$707_{10} = 1011000011_2 \ (= 10\,1100\,0011_2) = 2C3_{16}$$

Hexadecimal representation is well suited to the popular "byte" (8-bit) organization of computers, which are most often organized as 16-bit or 32-bit computer "words"; a word is then 2 or 4 bytes. An alphanumeric character (letter, number, or symbol) is one byte. So in hexadecimal, each byte is 2 hex digits, a 16-bit word is 4 hex digits, etc.

For example, in the widely used ASCII representation (more on that in Section 10.19), lower-case "a" is ASCII value 01100001 (61 hex, which we will write as 61_H), "b" is 62_H, etc. Thus the word "nerd" could be stored in a pair of 16-bit words, whose hex values are $6E65_H$ and 7264_H. As another example, the memory locations in a computer with 65,536 ("64K") bytes of memory can be identified by a 2-byte address, since $2^{16} = 65{,}536$; the lowest address is 0000_H, the highest address is $FFFF_H$, the second half of memory begins at 8000_H, and the fourth quarter of memory begins at $C000_H$.

You occasionally see "octal" (base-8) notation, a relic of an earlier era when computers used 12-bit and 36-bit words, with 6-bit alphanumeric representation. Although octal has the comfortable feature of using only familiar symbols (0–7), it is extremely awkward when applied to byte-organized words. Exercise 8.1 shows you why.

EXERCISE 8.1

Begin by writing down the octal representation for ASCII "a" and "b" using the hex values

LOGIC LEVELS

The diagram shows the ranges of voltages that correspond to the two logic states (HIGH and LOW) for the most popular families of digital logic. For each logic family it is necessary to specify legal values of both output and input voltages corresponding to the two states HIGH and LOW. The shaded areas above the line show the specified range of output voltages that a logic LOW or HIGH is guaranteed to fall within, with the pair of arrows indicating typical output values (LOW, HIGH) encountered in practice. The shaded areas below the line show the range of input voltages guaranteed to be interpreted as LOW or HIGH, with the arrow indicating the typical *logic threshold* voltage, i.e., the dividing line between LOW and HIGH. In all cases a logic HIGH is more positive than a logic LOW.

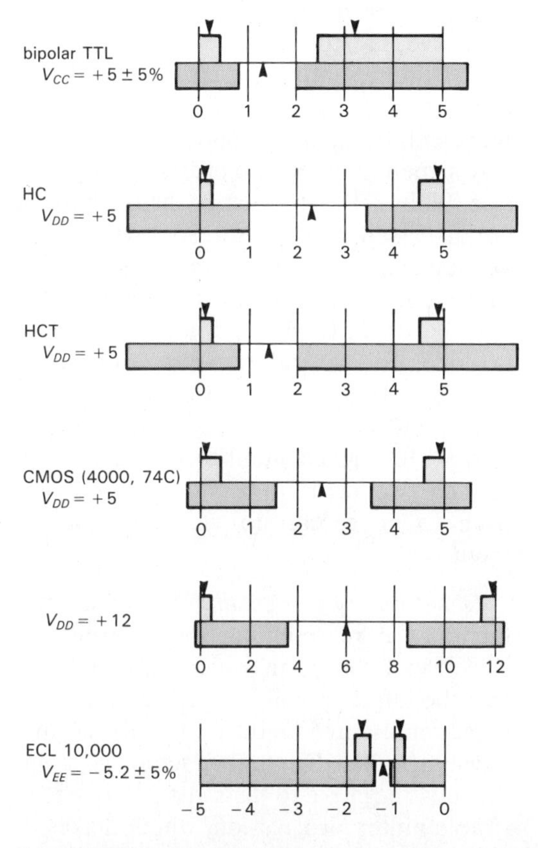

The meanings of "minimum," "typical,"and "maximum," in electronic specifications are worth a few words of explanation. Most simply, the manufacturer guarantees that the components will fall in the range minimum–maximum, with many close to "typical." What this means is that typical specifications are what you use when designing circuits; however, those circuits must work properly over the whole range of specifications from minimum to maximum (the extremes of manufacturing variability). In particular, a well-designed circuit must function under the worst possible combination of minimum and maximum values. This is known as *worst-case design*, and it is essential for any instrument produced from off-the-shelf (i.e., not specially selected) components.

given earlier. Then write down the octal representation of the 16-bit word formed by putting the two bytes for "ab" together. Why are the individual identities of the characters lost? What is "ba" in octal? Now do the same things, but using hexadecimal notation.

EXERCISE 8.2

Convert to decimal: (a) 1110101.0110_2, (b) $11.01010101\ldots_2$, (c) $2A_H$. Convert to binary: (a) 1023_{10}, (b) 1023_H. Convert to hexadecimal: (a) 1023_{10}, (b) 101110101101_2, (c) 61453_{10}.

BCD

Another way to represent a number is to encode each decimal digit into binary. This is called BCD (binary-coded decimal), and it requires a 4-bit group for each digit. For instance,

$$137_{10} = 0001\,0011\,0111 \qquad \text{(BCD)}$$

Note that BCD representation is *not* the same as binary representation, which in this case would be $137_{10} = 10001001_2$. You can think of the bit positions (starting from the right) as representing 1, 2, 4, 8, 10, 20, 40, 80, 100, 200, 400, 800, etc. It is clear that BCD is wasteful of bits, since each 4-bit group could represent numbers 0 through 15, but in BCD never represents numbers greater than 9. However, BCD is ideal if you want to display a number in decimal, since all you do is convert each BCD character to the appropriate decimal number and display it. (There are many devices that do exactly this, e.g., a "BCD decoder, driver, and display," which is a little IC with a transparent top. You apply logic levels for your BCD character, and it lights up with the digit.) For this reason, BCD is commonly used for input and output of numeric information. Unfortunately, the conversion between pure binary and BCD is complicated, since *each* decimal digit depends on the state of almost every binary bit, and vice versa. Nevertheless, binary arithmetic is so efficient that most computers convert all input data to binary, converting back only when data need to be output. Think how much effort and bother would have been saved if *Homo sapiens* had evolved with 8 (or 16) fingers!

Signed numbers

Sign magnitude representation. Sooner or later it becomes necessary to represent negative numbers in binary, particularly in devices where some computation is done. The simplest method is to devote one bit (the MSB, say) to the sign, with the remaining bits representing the magnitude of the number. This is called "sign magnitude representation," and it corresponds to the way signed numbers are ordinarily written (see Table 8.1). It is used when numbers are displayed, as well as in some A/D conversion schemes. In general, it is not the best method for representing signed numbers, particularly where some computation is done, for several reasons: Computation is awkward; subtraction is different from addition (i.e., addition doesn't "work" for signed numbers). Also, there can be two zeros (+0 and −0), so you have to be careful to use only one of them.

Offset binary representation. A second method for representing signed numbers is "offset binary," in which you subtract half the largest possible number to get the value represented (Table 8.1). This has the advantage that the number sequence from the most negative to the most positive is a simple binary progression, which makes it a natural for binary "counters." The MSB still carries the sign information, and zero appears only once. Offset binary is popular in A/D and D/A conversions, but it is still awkward for computation.

2's complement representation. The method most widely used for integer computation is called "2's complement."

TABLE 8.1. 4-BIT SIGNED INTEGERS IN THREE SYSTEMS OF REPRESENTATION

Integer	Sign-magnitude	Offset binary	2's comp
+7	0111	1111	0111
+6	0110	1110	0110
+5	0101	1101	0101
+4	0100	1100	0100
+3	0011	1011	0011
+2	0010	1010	0010
+1	0001	1001	0001
0	0000	1000	0000
–1	1001	0111	1111
–2	1010	0110	1110
–3	1011	0101	1101
–4	1100	0100	1100
–5	1101	0011	1011
–6	1110	0010	1010
–7	1111	0001	1001
–8	–	0000	1000
(–0)	1000	–	–

In this system, positive numbers are represented as simple unsigned binary. The system is rigged up so that a negative number is then simply represented as the binary number you add to a positive number of the same magnitude to get zero. To form a negative number, first complement each of the bits of the positive number (i.e., write 1 for 0, and vice versa; this is called the "1's complement"), then add 1 (that's the "2's complement"). As you can see from Table 8.1, 2's complement numbers are related to offset binary numbers by having the MSB complemented. As with the other signed number representations, the MSB carries the sign information. There's only one zero, conveniently represented by all bits 0 ("clearing" a counter or register sets its value to zero).

Arithmetic in 2's complement

Arithmetic is simple in 2's complement. To add two numbers, just add bitwise (with carry), like this:

$$\begin{array}{lll} 5+(-2): & 0101 & (+5) \\ & \underline{1110} & (-2) \\ & 0011 & (+3) \end{array}$$

To subtract B from A, take the 2's complement of B and add (i.e., add the negative):

$$\begin{array}{llll} 2-5: & 0010 & (+2) & \\ & \underline{1011} & (-5) & (+5 = 0101 : \text{1's comp} = 1010, \text{so 2's comp} = 1011) \\ & 1101 & (-3) & \end{array}$$

Multiplication also "works right" in 2's complement representation. Try the following exercise.

EXERCISE 8.3

Multiply $+2$ by -3 in 3-bit 2's complement binary arithmetic. Hint: The answer is -6.

EXERCISE 8.4

Show that the 2's complement of -5 is $+5$.

Because the 2's complement system is natural for computation, it is universally used for integer arithmetic in computers (note, however, that "floating point" numbers are usually represented in a form of "sign magnitude," namely sign-exponent-mantissa).

□ *Gray code*

The following code is used for mechanical shaft-angle encoders, among other things. It is called a Gray code, and it has the property that only one bit changes in going from one state to the next. This prevents errors, since there is no way of guaranteeing that all bits will change simultaneously at the boundary between two encoded values. If straight binary were used, it would be possible to generate an output of 15 in going from 7 to 8, for instance. Here is a simple rule for generating Gray-code states: Begin with a state of all zeros. To get to the next state, always change the single least significant bit that brings you to a new state.

0000
0001
0011
0010
0110
0111
0101
0100
1100
1101
1111
1110
1010
1011
1001
1000

Gray codes can be generated with any number of bits. They find use also in "parallel encoding," a technique of high-speed A/D conversion that you will see later. We will talk about translation between Gray-code and binary-code representations in the next section.

8.04 Gates and truth tables

Combinational versus sequential logic

In digital electronics the name of the game is generating digital outputs from digital inputs. For instance, an *adder* might take two 16-bit numbers as inputs and generate a 16-bit (plus carry) sum. Or you might build a circuit to multiply two numbers. These are the kinds of operations a computer's processing unit should be able to do. Another task might be to compare two numbers to see which is larger or to compare a set of inputs with the desired input to make sure that "all systems are go." Or you might want to attach a "parity bit" to a number to make the total number of 1's even, say, before transmission over a data link; then the parity could be checked on receipt as a simple check of correct transmission. Another typical task is to take some numbers expressed in binary and display, print, or punch them as decimal characters. All of these are tasks in which the output or outputs are predetermined functions of the input or inputs. As a class, they are known as "combinational" tasks. They can all be performed with devices called *gates*, which perform the operations of Boolean algebra applied to two-state (binary) systems.

There is a second class of problems that cannot be solved by forming a combinational function of the inputs alone, but require knowledge of past inputs as well. Their solution requires the use of "sequential" networks. Typical tasks of this type might be converting a string of bits in serial form (one after another) into a parallel set of bits, or keeping count of the number of 1's in a sequence, or recognizing a certain pattern in a sequence, or giving one output pulse for each four input pulses. All these tasks require digital memory of some sort. The basic device here is the "flip-flop" (the fancy name is "bistable multivibrator").

We will begin with gates and combinational logic, since they're basic to everything. Digital life will become more interesting when we get to sequential devices, but there will be no lack of fun and games with gates alone.

OR gate

The output of an OR gate is HIGH if either input (or both) is HIGH. This can be expressed in a "truth table," as shown in Figure 8.2. The gate illustrated is a 2-input OR gate. In general, gates can have any number of inputs, but the standard packages usually contain four 2-input gates, three 3-input gates, or two 4-input gates. For instance, a 4-input OR gate will have a HIGH output if any one input (or more) is HIGH.

The Boolean symbol for OR is $+$. "A OR B" is written $A+B$.

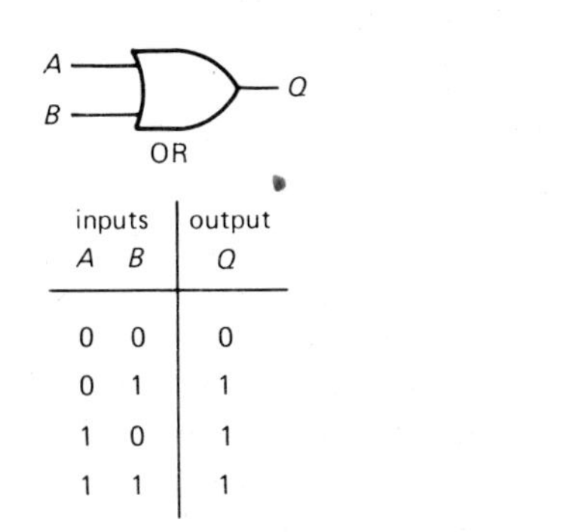

inputs		output
A	B	Q
0	0	0
0	1	1
1	0	1
1	1	1

Figure 8.2

AND gate

The output of an AND gate is HIGH only if both inputs are HIGH. The logic symbol and truth table are as shown in Figure 8.3. As with OR gates, AND gates are available with 3 or 4 (sometimes more) inputs. For instance, an 8-input AND gate will have a HIGH output only if *all* inputs are HIGH.

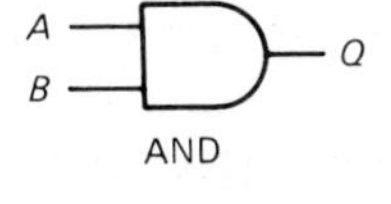

inputs		output
0	0	0
0	1	0
1	0	0
1	1	1

Figure 8.3

The Boolean symbol for AND is a dot ($\cdot$); this can be omitted, and usually is. "A AND B" is written $A \cdot B$, or simply AB.

Inverter (the NOT function)

Frequently we need the complement of a logic level. That is the function of an inverter, a "gate" with only one input (Fig. 8.4).

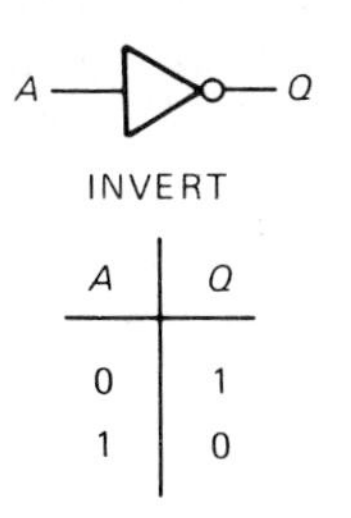

A	Q
0	1
1	0

Figure 8.4

The Boolean symbol for NOT is a bar over the symbol, or sometimes a prime symbol. "NOT A" is written $\overline{A}$, or A'. For the convenience of typesetters, the symbols /, *, –, and ′ are often used, in place of the overbar, to indicate NOT; thus, "NOT A" might be written as any of the following: A', $-A$, $^*A$, $/A$, A^*, $A/$. A given document will usually pick one of these alternatives and stick with it throughout. We have chosen the form A' for this book.

NAND and NOR

The INVERT function can be combined with gates, forming NAND and NOR (Fig. 8.5). These are actually more popular than AND and OR, as you will see shortly.

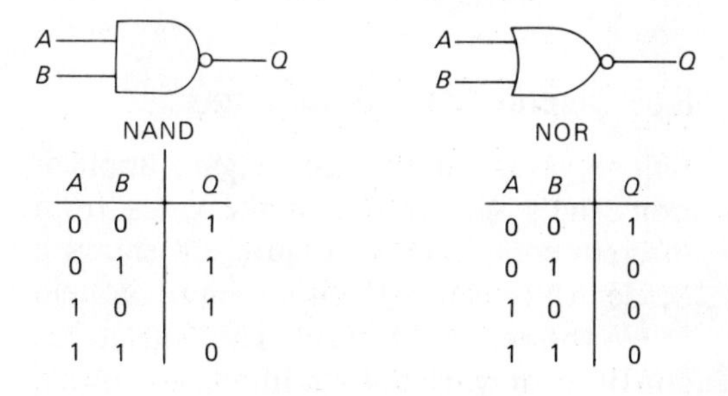

NAND:

A	B	Q
0	0	1
0	1	1
1	0	1
1	1	0

NOR:

A	B	Q
0	0	1
0	1	0
1	0	0
1	1	0

Figure 8.5

Exclusive-OR

Exclusive-OR is an interesting function, although less fundamental than AND and

OR (Fig. 8.6). The output of an exclusive-OR gate is HIGH if one or the other (but not both) input is HIGH (it never has more than two inputs). Another way to say it is that the output is HIGH if the inputs are different. The exclusive-OR gate is identical with modulo-2 addition of two bits.

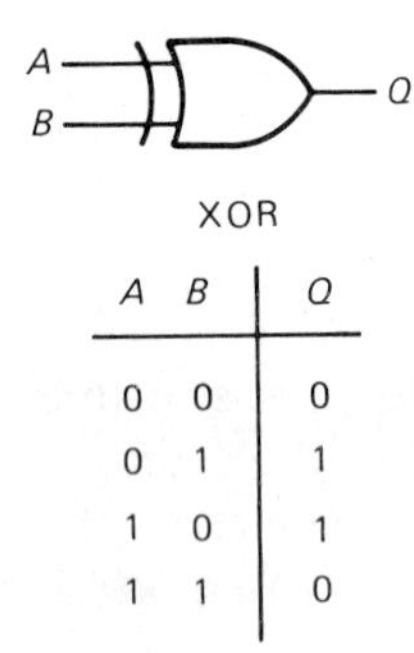

A	B	Q
0	0	0
0	1	1
1	0	1
1	1	0

Figure 8.6

EXERCISE 8.5

Show how to use the exclusive-OR gate as an "optional inverter," i.e., it inverts an input signal or buffers it without inversion, depending on the level at a control input.

EXERCISE 8.6

Verify that the circuits in Figure 8.7 convert binary code to Gray code, and vice versa.

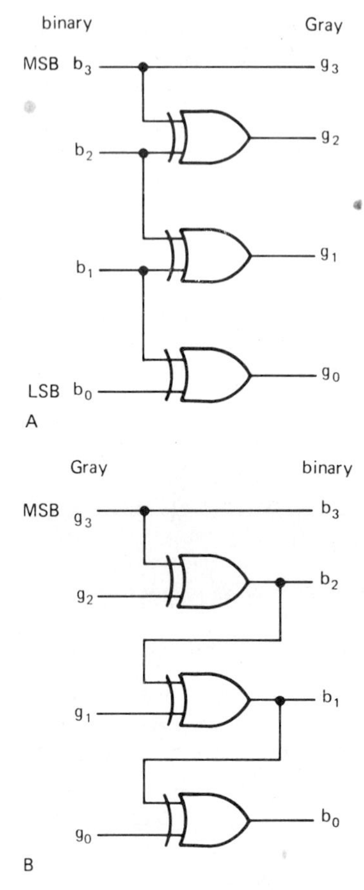

Figure 8.7. Parallel code converters: binary to Gray and Gray to binary.

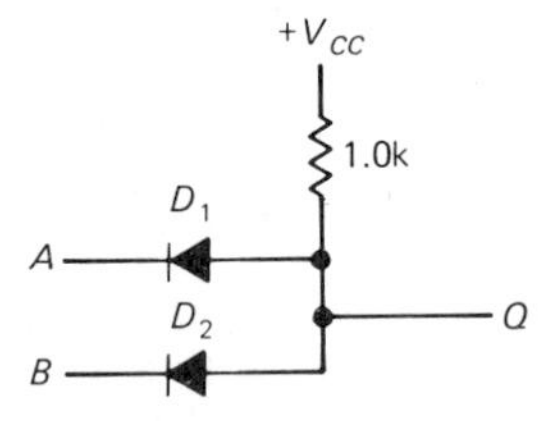

Figure 8.8

□ 8.05 Discrete circuits for gates

Before going on to discuss gate applications, let's see how to make gates from discrete components. Figure 8.8 shows a diode AND gate. If either input is held LOW, the output is LOW. The output can go HIGH only when both inputs go HIGH. This circuit has many disadvantages. In particular: (a) Its LOW output is a diode drop above the signal holding the input LOW. Obviously you couldn't use very many of these in a row! (b) There is no "fanout" (the ability of one output to drive several inputs), since any load at the output is seen by the signal at the input. (c) It is slow, because of resistive pull-up. As a general rule, you cannot do as well with logic constructed from discrete components as with IC logic. Part of the superiority of IC logic lies in the use of special techniques (e.g., ion implantation) to achieve excellent performance.

Figure 8.9 shows the simplest form of transistor NOR gate. This circuit was used in the family of logic known as RTL (resistor-transistor logic), which was popular in the 1960s because of its low price, but is now obsolete. A HIGH at either input (or both) turns on at least one transistor, pulling the output LOW. Since this gate is intrinsically inverting, you would have to add an inverter, as shown, to make an OR gate.

Although the discrete gate circuits just illustrated are simple to understand, you wouldn't use them in practice because of their disadvantages. In fact, except in rare circumstances you would never construct gates (or any other logic) from discrete components, since a full range of excellent logic is available as inexpensive and compact integrated circuits, as we will see shortly. Currently the most popular IC logic circuits are built with complementary MOSFETs ("CMOS"). Look back at Figure 3.59 to remind yourself how you would make a CMOS NAND gate.

8.06 Gate circuit example

Let's work out a circuit to perform the logic we gave as an example in Chapters 1 and 2: the task to sound a buzzer if either car door is open and the driver is seated. The answer is obvious if you restate the problem as "output HIGH if either the left door OR the right door is open, AND driver is seated," i.e., $Q = (L + R)S$. Figure 8.10 shows it with gates. The output of the OR gate is HIGH if one OR the other door (or both) is open. If that is so, AND the driver is seated, Q goes HIGH. With an additional transistor, this could be made to sound a buzzer or close a relay.

In practice, the switches generating the inputs will probably close a circuit to ground, to save extra wiring (there are additional reasons, particularly in the case of the popular TTL logic, and we will get to them shortly). This means that the inputs will go LOW when a door is opened, for example. In other words, we have "negative-true" inputs. Let's rework the example with this in mind, calling the inputs L', R', and S'.

First, we need to know if either door input (L', R') is LOW; i.e., we must distinguish the state "both inputs HIGH" from all others. That's an AND gate. So we make L' and R' the inputs to an AND gate. The output will be LOW if either input is LOW; call that EITHER$'$. Now we need to know when EITHER$'$ is LOW and S'

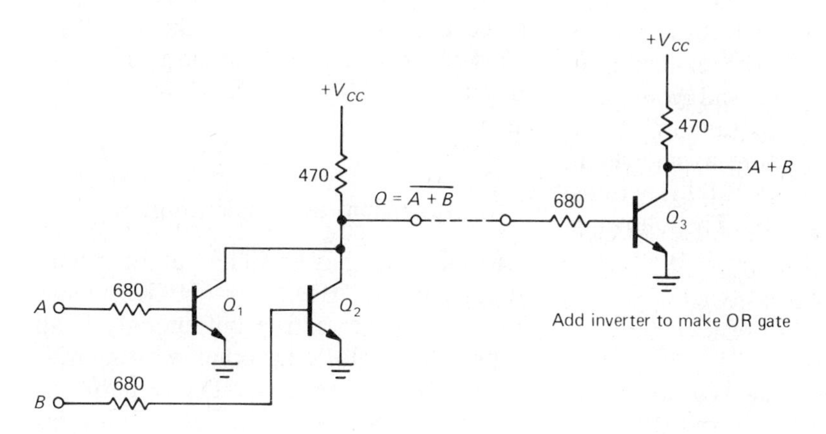

Figure 8.9

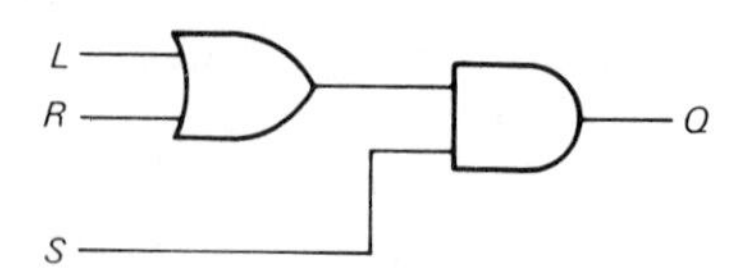

Figure 8.10

is LOW; i.e., we must distinguish the state "both inputs LOW" from all others. That's an OR gate. Figure 8.11 shows the circuit. We have used a NOR gate, instead of an OR gate, to get the same output as earlier: Q HIGH when the desired condition is present. Something strange seems to be going on here, though. We have used AND instead of OR (and vice versa), as compared with the earlier circuit. Section 8.07 should clarify the matter. First, consider the following exercise.

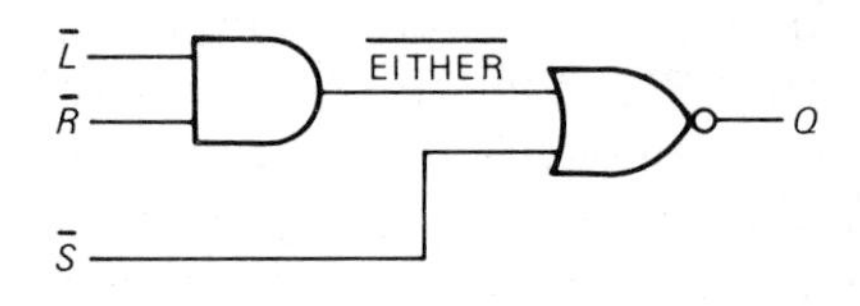

Figure 8.11

EXERCISE 8.7
What do the circuits shown in Figure 8.12 do?

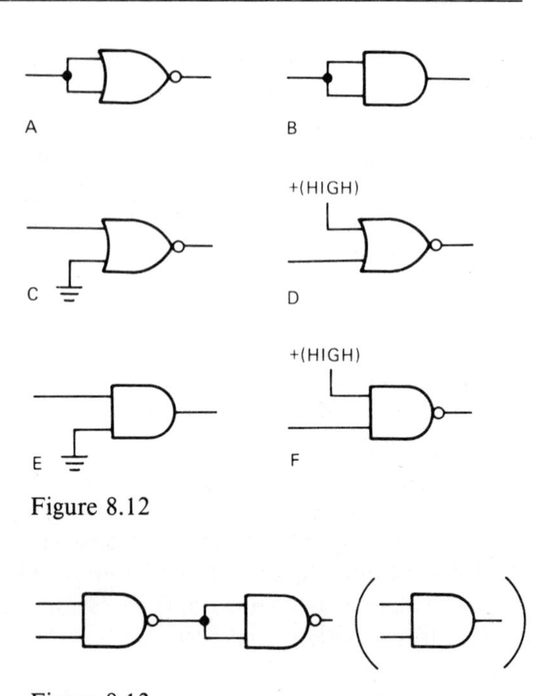

Figure 8.12

Figure 8.13

Gate interchangeability

When designing digital circuits, keep in mind that it is possible to form one kind of gate from another. For example, if you need an AND gate, and you have half of a 7400 available (quad 2-input NAND), you can substitute as shown in Figure 8.13. The second NAND functions as an inverter, making AND. The following exercises should help you explore this idea.

EXERCISE 8.8
Using 2-input gates, show how to make (a) INVERT from NOR, (b) OR from NORs, and (c) OR from NANDs.

EXERCISE 8.9
Show how to make (a) a 3-input AND from 2-input ANDs, (b) a 3-input OR from 2-input ORs, (c) 3-input NOR from 2-input NORs, and (d) a 3-input AND from 2-input NANDs.

In general, multiple use of one kind of inverting gate (e.g., NAND) is enough to make any combinational function. However, this isn't true for a noninverting gate, since there's no way to make INVERT. This probably accounts for the greater popularity of NAND and NOR in logic design.

8.07 Assertion-level logic notation

An AND gate has a HIGH output if both inputs are HIGH. So, if HIGH means "true," you get a true output only if all inputs are true. In other words, with positive-true logic, an AND gate performs the AND function. The same holds for OR.

What happens if LOW means "true," as in the last example? An AND gate gives a LOW if either input is true (LOW): It's an OR function! Similarly an OR gate gives a LOW only if both inputs are true (LOW). It's an AND function! Very confusing.

There are two ways to handle this problem. The first way is to think through any digital design problem as we did earlier, choosing the kind of gate that gives the needed output. For instance, if you need to know if any of three inputs is LOW, use a 3-input NAND gate. This method is still used by some misguided designers. When designing this way, you would draw a NAND gate, even though the gate is performing a NOR function on the (negative-true) inputs. You would probably label the inputs as in Figure 8.14. In this example, CLEAR′, MR′ (master reset), and RESET′ might be negative-true levels coming from various places in a circuit. The output, CLR, is positive-true and will go to the devices that are to be cleared if *any* of the reset signals goes LOW (true).

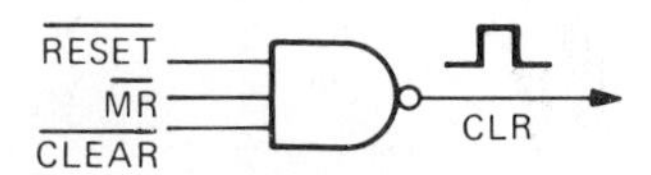

Figure 8.14

The second way to handle the problem of negative-true signals is to use "assertion-level logic." If a gate performs an OR function on negative-true inputs, draw it that way, as in Figure 8.15. The 3-input OR gate with negated inputs is functionally identical with the preceding 3-input NAND. That equivalence turns out to be an important logical identity, as stated in DeMorgan's theorem, and we will spell out a number of such useful identities shortly. For now, it is enough to know that you can change AND to OR (and

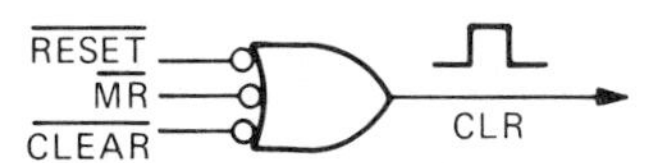

Figure 8.15

vice versa) if you negate the output and all inputs (see Table 8.2). Assertion-level logic looks forbidding at first, because of the proliferation of funny-looking gates. It is better, though, because the logical functions of the gates in the circuit stand out clearly. You'll find it "friendly" after you've used it for a while, and you won't want to use anything else. Let's rework the car door example again with assertion-level logic (Fig. 8.16). The gate on the left determines if L or R is true, i.e., LOW, giving a negative-true output. The second gate gives a HIGH output if both $(L + R)$ and S are true, i.e., LOW. From DeMorgan's theorem (after a while you won't even need that, you'll recognize these gates as equivalent) the first gate is AND, the second is NOR, just as in the circuit drawn earlier. Two important points:

1. Negative-true doesn't mean that the logic levels are negative *polarity*. It means that the lower of the two states (LOW) stands for TRUE.

2. The symbol used to draw the gate itself assumes positive-true logic. A NAND gate used as an OR for negative-true signals can be drawn as a NAND or, using assertion-level logic, as an OR with negation symbols (little circles) at the inputs. In the latter case you think of the circles as indicating inversion of the input signals, followed by an OR gate operating on positive-true logic as originally defined.

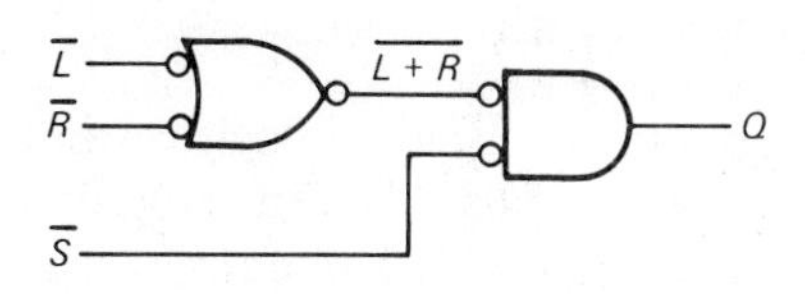

Figure 8.16

TABLE 8.2 COMMON GATES IN THE TTL AND CMOS FAMILIES

Name	Expression	Symbol	Negative true symbol	Type	No. per chip	Part number 4000B CMOS	74xx	→ ALS	AS	F	LS	C	AC(T)	HC(T)
AND	AB			2-input	4	4081	7408	✓	✓	✓	✓	✓	✓	✓
				3-input	3	4073	7411	✓	✓	✓	✓		✓	✓
				4-input	2	4082	7421	✓	✓	✓	✓			✓
NAND	$\overline{AB}$			2-input	4	4011	7400	✓	✓	✓	✓	✓	✓	✓
				3-input	3	4023	7410	✓	✓	✓	✓	✓	✓	✓
				4-input	2	4012	7420	✓	✓	✓	✓	✓	✓	✓
				8-input	1	4068	7430	✓	✓	✓	✓	✓		✓
				13-input	1	—	74133	✓		✓				✓
OR	$A + B$			2-input	4	4071	7432	✓	✓	✓	✓	✓	✓	✓
				3-input	3	4075	—							✓
				4-input	2	4072	74802		✓					
NOR	$\overline{A + B}$			2-input	4	4001	7402	✓	✓	✓	✓	✓	✓	✓
				3-input	3	4025	7427	✓	✓	✓	✓			✓
				4-input	2	4002	7425							✓
				5-input	2	—	74260			✓				✓
				8-input	1	4078	—							
INVERT	$\overline{A}$				6	4069/4049	7404	✓	✓	✓	✓	✓	✓	✓
					8		74240	✓		✓	✓	✓	✓	✓
BUFFER	A				6	4503/4050	74365	✓	✓	✓	✓			✓
					8		74241/244 (−541/−544)	✓		✓	✓	✓	✓	✓
XOR	$A \oplus B$			2-input	4	4070	7486/386 (−135)	✓		✓	✓	✓	✓	✓
XNOR	$\overline{A \oplus B}$			2-input	4	4077	74266 (−135)				✓			✓
AOI				2-2-input	2	4085	7450/51			✓	✓			✓
				2-2-2-2-input	1	4086	7453/54				✓			

Postscript: Logical AND and OR shouldn't be confused with the *legal* equivalents. The weighty legal tome known as *Words and Phrases* has over 40 pages of situations in which AND can be construed as OR. For example: "OR will be construed AND, and AND will be construed OR, as the necessities of the case may require. . . ." This isn't the same as DeMorgan's theorem!

TTL AND CMOS

TTL (transistor-transistor logic) and CMOS (complementary MOS) are the two most popular logic families in current use, with at least 10 manufacturers of integrated circuits offering an enormous variety of functions in both families. These families should satisfy your needs for all digital design, with the exceptions of some large-scale integration (LSI), which uses either CMOS or NMOS logic, and ultra-high-speed logic, where GaAs devices and emitter-coupled logic (ECL) reign supreme. Throughout the rest of the book we will rely heavily on these families.

8.08 Catalog of common gates

Table 8.2 shows the common gates you can get in the TTL and CMOS families of digital logic. Each gate is drawn in its normal (positive-true) incarnation, and also the way it looks for negative-true logic. The last entry in the table is an AND-OR-INVERT gate, sometimes abbreviated AOI.

A word of explanation: Digital logic is available in 10 popular subfamilies (*CMOS*: 4000B, 74C, 74HC, 74HCT, 74AC, 74ACT; and *TTL*: 74LS, 74ALS, 74AS, 74F), all offering the same functions and with a pretty good degree of compatibility between them. The differences have to do with speed, power dissipation, output drive capability, and logic levels (see Sections 8.09 and 9.02). The best type for most applications is currently "high-speed CMOS," specified by adding the letters HC after the digits 74, e.g., 74HC00. Where compatibility with existing bipolar TTL is required, however, you should use the HCT (or possibly LS) subfamily. For simplicity we will routinely omit such letters (and the 74- prefix) in this book, indicating digital IC types with an apostrophe, e.g., '00 for a 2-input NAND. Note that the original TTL ("7400 family" - no letters after the "74") is obsolete. We'll describe the interesting history of these families in Section 9.01.

8.09 IC gate circuits

Although a NAND gate, for instance, performs identical logic operations in the various TTL and CMOS versions, the logic levels and other characteristics (speed, power, input current, etc.) are quite different. In general, you have to be careful when mixing logic family types. To understand the differences, look at the schematics of a NAND gate in Figure 8.17.

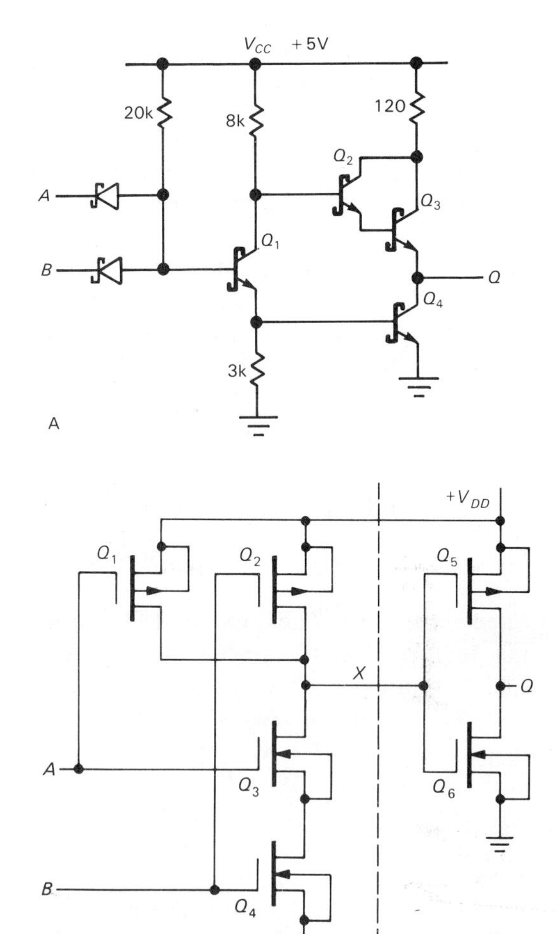

Figure 8.17. A. LS TTL NAND gate.
B. CMOS AND gate.

The CMOS gate is constructed from enhancement-mode MOSFETs of both polarities, connected as switches rather than followers. An ON FET looks like a low resistance to whichever supply rail it is connected. Both inputs must be HIGH to turn on the series pair Q_3Q_4 and to turn off both of the pull-up transistors Q_1Q_2. That produces a LOW at the output, i.e., it is a NAND gate. Q_5 and Q_6 constitute the standard CMOS inverter, to generate an AND gate. From this example it should be evident how to generalize to AND, NAND, OR, and NOR with any number of inputs.

EXERCISE 8.10

Draw the circuit of a 3-input CMOS OR gate.

The bipolar LS (low-power Schottky) TTL NAND basically consists of the diode-resistor logic of Figure 8.8 driving a transistor inverter followed by a push-pull output. If both inputs are HIGH, the 20k resistor holds Q_1 on, thus producing a LOW output by saturating Q_4 and shutting off Darlington Q_2Q_3. If at least one input is LOW, Q_1 is held off, thus producing a HIGH output by follower action of Q_2Q_3 combined with Q_4 being held off. Schottky diodes and Schottky-clamped transistors are used throughout for enhanced speed.

Note that both CMOS and bipolar TTL gates have an output circuit with "active pullup" to the positive supply rail, unlike our discrete gate examples.

8.10 TTL and CMOS characteristics

Let's compare family characteristics:

Supply voltage. The bipolar TTL families require +5 volts, ±5%, whereas the CMOS families have a wider range: +2 to +6 volts for HC and AC, +3 to +15 volts for 4000B and 74C. The HCT and ACT CMOS families, designed for compatibility with bipolar TTL (see below), require +5 volts.

Input. A TTL input held in the LOW state sources current into whatever drives it (0.25mA typ for LS), so to pull it LOW you must sink current. Since the TTL output circuit (a saturated *npn* transition) is good at sinking current, this presents no problem when TTL logic is wired together, but you must keep it in mind when driving TTL with other circuitry. By contrast, CMOS has no input current.

The TTL input logic threshold is about two diode drops above ground (about 1.3V), whereas most CMOS families have their threshold nominally at half the supply voltage (though with considerable spread, typically 1/3 to 2/3 the supply voltage). The HCT and ACT CMOS families are designed with a low threshold similar to bipolar TTL for compatibility, since a bipolar TTL output does not swing all the way to +5 volts (see below).

CMOS inputs are susceptible to damage from static electricity during handling. In both families, unused inputs should be tied HIGH or LOW, as necessary (more on this later).

Output. The TTL output stage is a saturated transistor to ground in the LOW state, and a (Darlington) follower in the HIGH state (two diode drops below V_+). For all CMOS families (including HCT and ACT) the output is a turned-on MOSFET, either to ground or to V_+; i.e., rail-to-rail output swings. In general, the faster families (F, AS; AC, ACT) have greater output drive capability than the slower families (LS; 4000B, 74C, HC, HCT).

Speed and power. The bipolar TTL families consume considerable quiescent current, more for the faster families (AS and F); the corresponding speeds go from about 25MHz (for LS) to about 100MHz (for AS and F). All CMOS families consume zero quiescent current. However, their power consumption rises linearly with increasing frequency (switching capacitive loads requires current), and CMOS operated near its upper frequency limit often dissipates as much power as the equivalent bipolar TTL family (Fig. 8.18). The speed range of CMOS goes from about 2MHz (for 4000B/74C at 5V) to about 100MHz (for AC/ACT).

In general, the nice characteristics of CMOS (zero quiescent current, rail-to-rail output swings, good noise immunity) make it the logic of choice, and we recommend the HC family for most new designs. However, for greater speed, use AC; for wide supply range where high speed is not needed, use 74C or 4000B; use HCT (or perhaps LS) for compatibility with bipolar TTL outputs, unless you need the speed of ACT (or AS or F). In some

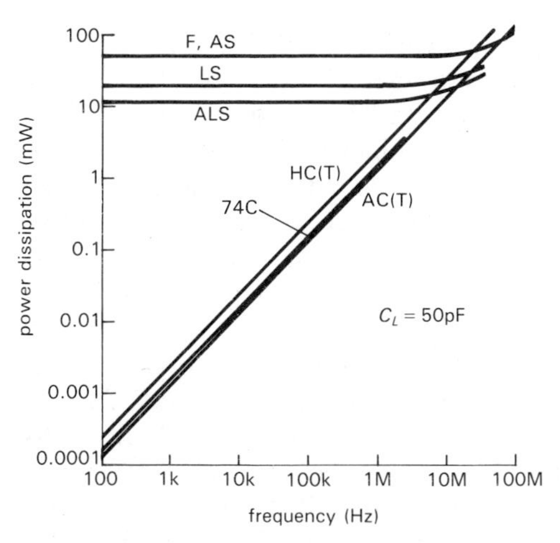

Figure 8.18. Gate power dissipation versus frequency.

high-density applications (memory, microprocessor), NMOS devices are preferred, in spite of their relatively high power dissipation. And for the highest-speed applications (above 100MHz) you are forced to use either ECL, which goes up to about 500MHz, or logic based on GaAs, which is usable to about 4GHz. See Section 14.15 and Table 9.1 for further discussion of CMOS logic families.

Within any one logic family, outputs are designed to drive other inputs easily, so you don't often have to worry about thresholds, input current, etc. For instance, with TTL or CMOS, any output can drive at least 10 other inputs (the official term for this is *fanout*: TTL has a fanout of 10), so you don't have to do anything special to ensure compatibility. In the next chapter we will go into the issue of interfacing between logic families and between logic circuits and the outside world.

8.11 Three-state and open-collector devices

The TTL and CMOS gates we have just discussed have push-pull output circuits: The output is held either HIGH or LOW by an ON transistor or MOSFET. Nearly all digital logic uses this sort of circuit (called active pullup; in TTL it's also called a totem-pole output) because it provides low output impedance in both states, giving faster switching time and better noise immunity, as compared with an alternative such as a single transistor with passive collector pullup resistor. In the case of CMOS, it also results in lower power dissipation.

However, there are a few situations for which active pullup output is unsuitable. As an example, imagine a computer system in which several functional units have to exchange data. The central processor (CPU), memory, and various peripherals all need to be able to send and receive 16-bit words. It would be awkward (to put it mildly) to have separate 16-wire cables connecting each device to all others. The solution is the so-called *data bus*, a single set of 16 wires accessible to all devices. It's like a telephone party line: Only one device at a time may "talk" (assert data), but all may "listen" (receive data). With a bus system there must be agreement as to who may talk, and words like "bus arbitrator," "bus master," and "control bus" pop up.

You can't use gates (or any other devices) with active pullup outputs to drive a bus, since you couldn't disconnect your output from the shared data lines (you're holding it either HIGH or LOW at all times). What's needed is a gate whose output can be "open." Such devices are available, and they come in two varieties, "three-state devices" and "open-collector devices."

Three-state logic

Three-state logic, also called TRI-STATE logic (a trademark of National Semiconductor Corporation, who invented it), provides an elegant solution. The name is misleading; it is not digital logic with three voltage levels. It's just ordinary logic, with a third output state: open circuit (Fig. 8.19). A separate *enable* input determines whether the output behaves like an ordinary active pullup output or goes into the "third" (open) state, regardless of the logic levels present at the other inputs. Three-state outputs are available on many digital chips, including counters, latches, registers, etc., as well as on gates and inverters. A device with three-state output behaves exactly like ordinary active pullup logic when enabled, always driving its output either HIGH or LOW; when disabled, it effectively disconnects its output, so another logic device can drive the same line. Let's look at an example.

A look ahead: data buses

Three-state drivers are widely used to drive computer data buses. Every device (memory, peripherals, etc.) that needs to put data on the (shared) bus ties onto it with three-state gates (or more complex functions such as "registers"). Things are cleverly arranged so that at most one device has its drivers enabled at any instant, all other devices being disabled into the open (third) state. In a typical situation the selected device "knows" to assert data onto the bus by recognizing its particular address on a set of address and control lines (Fig. 8.20). In this simplified case the device is wired as port 6: It looks at address lines A_0–A_2 and asserts data onto data bus D_0 through D_3 when it sees its particular address (i.e., 6) on the address lines *and* it sees a READ pulse. Such a bus protocol is adequate for many simple systems. Something like this is used in most microcomputers, as you will see in Chapters 10 and 11.

Note that there must be some external logic to make sure that three-state devices sharing the same output lines don't try to talk at the same time (that undesirable condition is officially called "bus contention"). In this case all is well as long as each device responds to a unique address.

Open-collector logic

The predecessor to three-state logic was "open-collector" logic, which allows you to share a single line among the outputs of several drivers. An open-collector (or open-drain) output simply omits the active pullup transistor of the output stage (Fig. 8.21). The name "open-collector" is a good one. When you use such gates, you must supply an external pullup resistor somewhere. Its value isn't critical; a small-value resistor gives increased speed and improved noise immunity, at the expense of increased power dissipation and loading

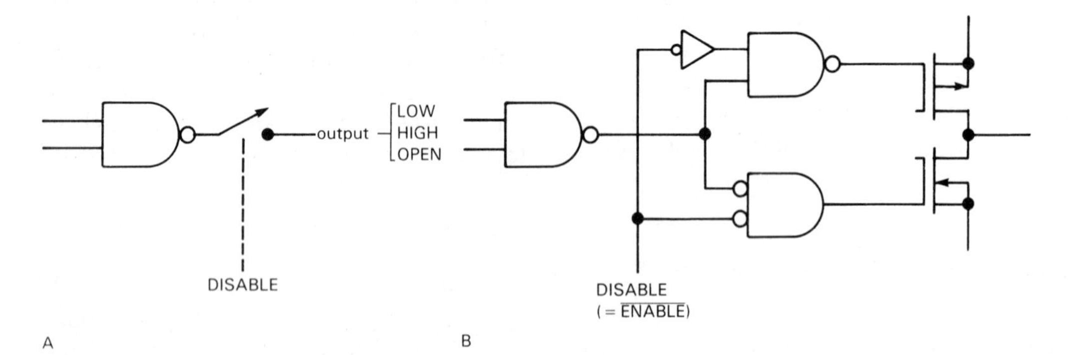

Figure 8.19. Three-state CMOS NAND gate.
A. Conceptual diagram.
B. Realization with internal CMOS gates.

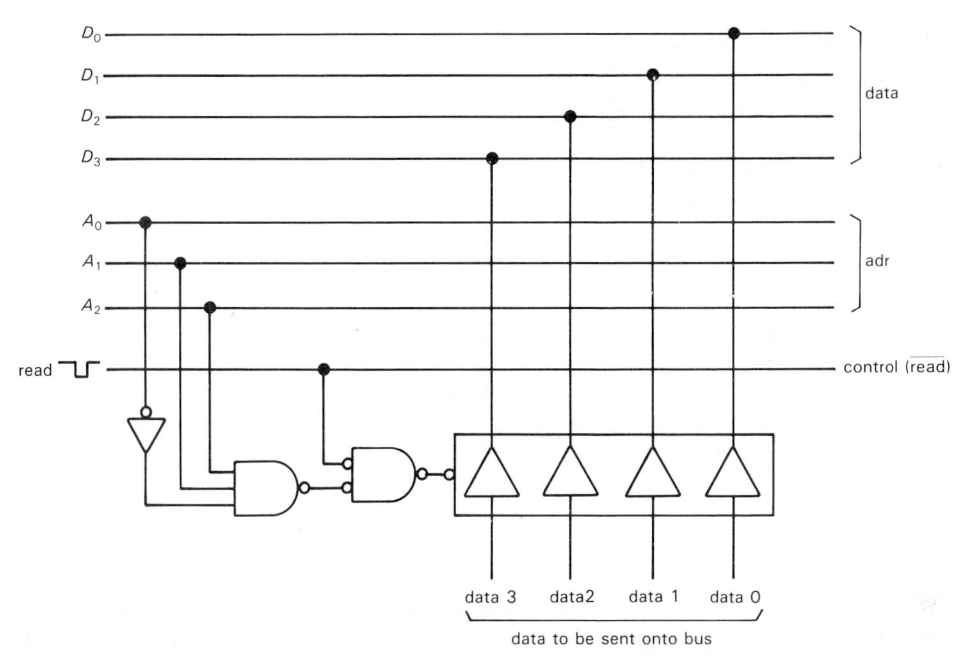

Figure 8.20. Data bus.

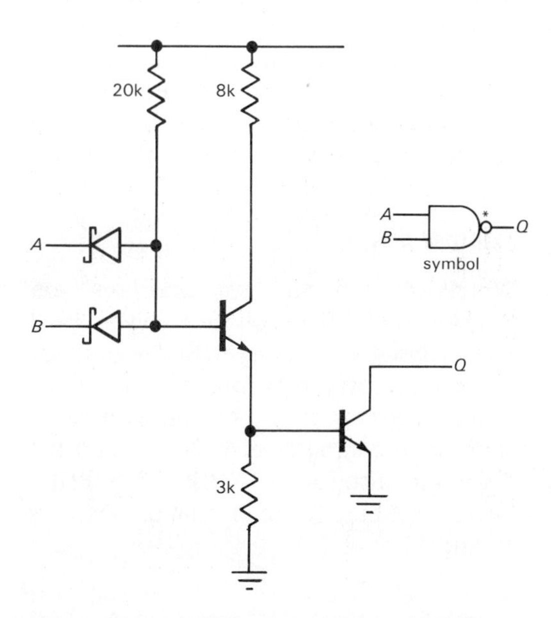

Figure 8.21. LS TTL open-collector NAND.

of the driver. Values of a few hundred to a few thousand ohms are typical. If you wanted to drive a bus with open-collector gates, you would substitute 2-input open-collector NANDs for the three-state drivers of Figure 8.20, bringing one input of each gate HIGH to enable the gates onto the bus; note that the data then asserted onto the bus are inverted. Each bus line would need a resistive pullup to +5 volts.

The disadvantage of open-collector logic is that speed and noise immunity are degraded, when compared with logic constructed with active pullup devices, because of the resistive pullup circuit. That's why three-state drivers are nearly universally favored for computer bus applications. However, there are three situations in which you would choose open-collector (or open-drain) devices: driving external loads, "wired-OR," and external buses. Let's look at them briefly.

Driving external loads

Open-collector logic is good for driving external loads that are returned to a higher-voltage positive supply. You might want to drive a low-current lamp that requires 12 volts, or perhaps just generate a 15 volt logic swing by running a resistor from a gate's output to +15 volts, as in Figure 8.22. For example, the '06 is an open-collector hex inverter with 30 volt breakdown rating, and the CMOS 40107 is a dual-NAND open-drain buffer with up to 120mA sink capability. The 75450 series of "dual peripheral drivers" can sink up to 300mA from loads returned to +30 volts, and the UHP/UDN series from Sprague extends this to more than 1 amp and 80 volts. More on these subjects in the next chapter.

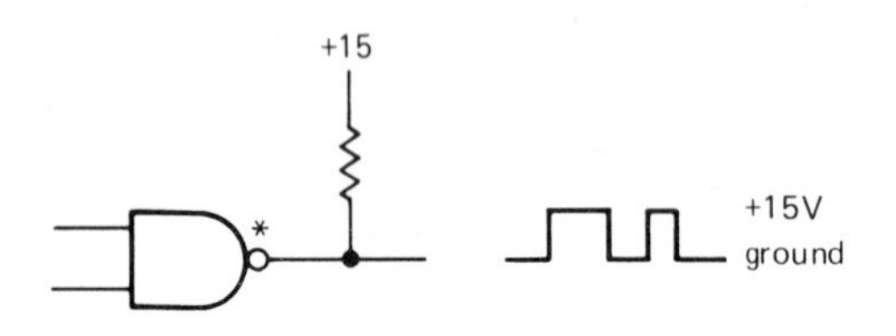

Figure 8.22

Wired-OR

If you wire together some open-collector gates as shown in Figure 8.23, you get what's called "wired-OR" – the combination behaves in this case like a larger NOR gate, with the output going LOW if any input is HIGH. You can't do this with active pullup outputs, because there would be a contest of wills, if all the gates didn't agree on what the output should be. You can combine NORs, NANDs, etc., with this kind of connection, and the output will be LOW if any gate asserts a LOW output. This connection is sometimes called "wired-AND," since the output is HIGH only if all gates have HIGH (open) outputs. Both names are describing the same thing: It's wired-AND for positive-true logic and wired-OR for negative-true logic. This will make more sense to you after you've seen DeMorgan's theorem, in the next section.

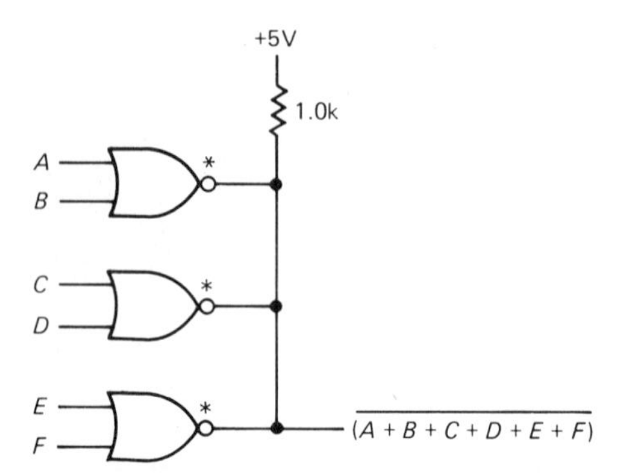

Figure 8.23. Wired-OR.

Wired-OR enjoyed some brief popularity in the early days of digital electronics, but it is not much used today, with two exceptions: (a) In the family of logic known as ECL (emitter-coupled logic) the outputs are what you might call "open-emitter," and can be wired-OR'ed painlessly, and (b) there are some shared lines in computer buses (most notably the line called *interrupt*) whose function is not to transfer data bits, but merely to indicate if *at least one* device is requesting attention; in that case you use wired-OR, since it does what you want and doesn't require external logic to prevent contention.

External buses

Where speed is not too important, you sometimes see open-collector drivers used to drive buses. This is often the case for buses that carry data out of computers; common examples are the buses used to connect to computer disk drives, and the IEEE-488 (also called "HPIB" or "GPIB") instrument bus. More on this in Chapters 10 and 11.

COMBINATIONAL LOGIC

As we discussed earlier in Section 8.04, digital logic can be divided into *combinational* and *sequential*. Combinational circuits are those in which the output state depends only on the present input states in some predetermined fashion, whereas in sequential circuits the output state depends both on the input states and on the previous history. Combinational circuits can be constructed with gates alone, whereas sequential circuits require some form of memory (flip-flops). In these sections we will explore the possibilities of combinational logic before entering the turbulent world of sequential circuits.

8.12 Logic identities

No discussion of combinational logic is complete without the identities shown in Table 8.3. Most of these are obvious. The last two compose DeMorgan's theorem, the most important for circuit design.

TABLE 8.3. LOGIC IDENTITIES

$ABC = (AB)C = A(BC)$
$AB = BA$
$AA = A$
$A1 = A$
$A0 = 0$
$A(B+C) = AB + AC$
$A + AB = A$
$A + BC = (A + B)(A + C)$
$A + B + C = (A + B) + C = A + (B + C)$
$A + B = B + A$
$A + A = A$
$A + 1 = 1$
$A + 0 = A$
$1' = 0$
$0' = 1$
$A + A' = 1$
$AA' = 0$
$(A')' = A$
$A + A'B = A + B$
$(A + B)' = A'B'$
$(AB)' = A' + B'$

Example: exclusive-OR gate

We will illustrate the use of the identities with an example: making the exclusive-OR function from ordinary gates. Figure 8.24 shows the XOR truth table. From studying this, and by realizing that the output is 1 only when $(A, B) = (0,1)$ or $(1,0)$, we can write

$$A \oplus B = \overline{A}B + A\overline{B}$$

from which we have the realization shown in Figure 8.25. However, this realization is not unique. Applying the identities, we find

$$\begin{aligned} A \oplus B &= A\overline{A} + A\overline{B} + B\overline{A} + B\overline{B} \\ &\qquad\qquad (A\overline{A} = B\overline{B} = 0) \\ &= A(\overline{A} + \overline{B}) + B(\overline{A} + \overline{B}) \\ &= A(\overline{A\,B}) + B(\overline{A\,B}) \\ &= (A + B)(\overline{A\,B}) \end{aligned}$$

A	B	$A \oplus B$
0	0	0
0	1	1
1	0	1
1	1	0

Figure 8.24. XOR.

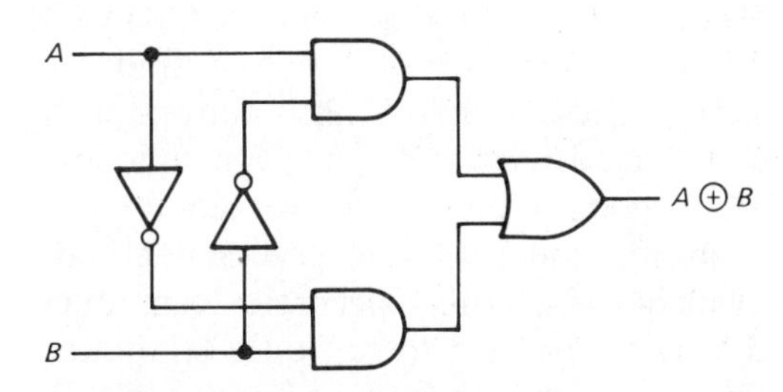

Figure 8.25. XOR realization.

(In the first step we used the trick of adding two quantities that equal zero; in the third step we used DeMorgan's theorem.) This has the realization shown in Figure 8.26. There are still other ways to construct XOR. Consider the following exercise.

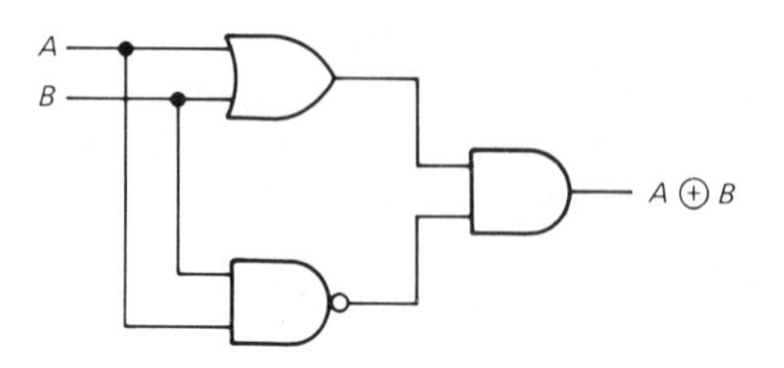

Figure 8.26. XOR realization.

EXERCISE 8.11

Show that

$$A \oplus B = \overline{AB + \overline{A}\,\overline{B}}$$

$$A \oplus B = (A + B)(\overline{A} + \overline{B})$$

by logic manipulation. You should be able to convince yourself that these are true by inspection of the truth table, combined with suitable hand-waving.

EXERCISE 8.12

What are the following: (a) 0 · 1, (b) 0 + 1, (c) 1 · 1, (d) 1 + 1, (e) $A(A + B)$, (f) $A(A' + B)$, (g) A XOR A, (h) A XOR A'?

8.13 Minimization and Karnaugh maps

Since a realization of a logic function (even one as simple as exclusive-OR) isn't unique, it is often desirable to find the simplest, or perhaps most conveniently constructed, circuit for a given function. Many good minds have worked on this problem, and there are several methods available, including algebraic techniques that can be coded to run on a computer. For problems with four or fewer inputs, a Karnaugh map provides one of the nicest methods; it also enables you to find a logic expression (if you don't know it) once you can write down the truth table.

We will illustrate the method with an example. Suppose you want to generate a logic circuit to count votes. Imagine that you have three positive-true inputs (each either 1 or 0) and an output (0 or 1). The output is to be 1 if at least two of the inputs are 1.

Step 1. Make a truth table:

A	B	C	Q
0	0	0	0
0	0	1	0
0	1	0	0
0	1	1	1
1	0	0	0
1	0	1	1
1	1	0	1
1	1	1	1

All possible permutations must be represented, with corresponding output(s). Write an X (= "don't care") if either output state is OK.

Step 2. Make a Karnaugh map. This is somewhat akin to a truth table, but the variables are represented along two axes. Furthermore, they are arranged in such a way that only one input bit changes in going from one square to an adjacent square (Fig. 8.27).

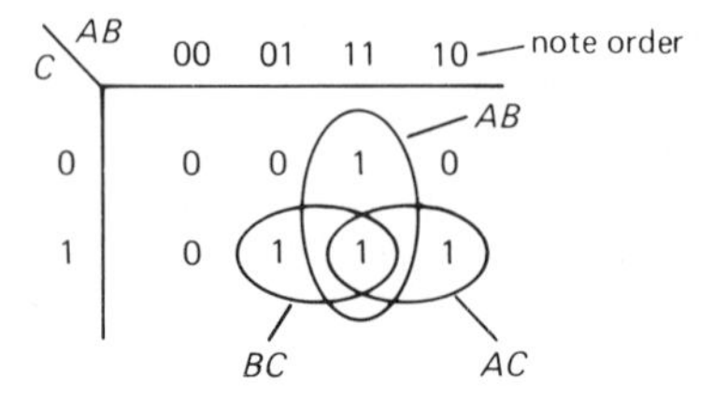

Figure 8.27. Karnaugh map.

Step 3. Identify on the map groups of 1's (alternatively, you could use groups of 0's): The three blobs enclose the logic expressions AB, AC, and BC. Finally, read off the required function, in this case

$$Q = AB + AC + BC$$

with the realization shown in Figure 8.28. The result seems obvious, in retrospect. We could have read off the pattern of 0's to get instead

$$Q' = A'B' + A'C' + B'C'$$

which might be useful if the complements A', B', and C' already exist somewhere in the circuit.

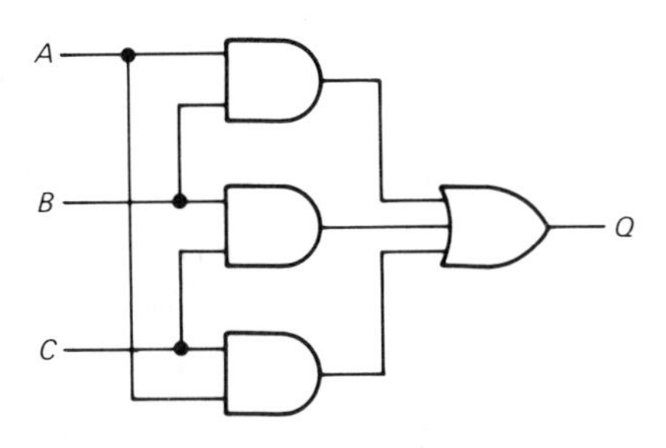

Figure 8.28

Some comments on Karnaugh maps

1. Look for groups of 2, 4, 8, etc., squares; they have the simple logic expressions.
2. The larger the block you describe, the simpler the logic.
3. The edges of the Karnaugh map connect up. For instance, the map in Figure 8.29 is described by $Q = B'C$.

CD \ AB	00	01	11	10
00	0	0	0	0
01	0	0	0	0
11	1	0	0	1
10	1	0	0	1

Figure 8.29

4. A block of 1's with only one or two 0's may be best described by the grouping illustrated in Figure 8.30, which corresponds to the logic expression

$$Q = A(BCD)'$$

5. Xs (don't care) are "wild cards." Use them as 1's or 0's to generate the simplest logic.

CD \ AB	00	01	11	10
00	0	0	1	1
01	0	0	1	1
11	0	0	(0)	1
10	0	0	1	1

Figure 8.30

6. A Karnaugh map may not lead directly to the best solution: A more complicated logic expression may sometimes have a simpler realization in gates, e.g., if some of its terms already exist as logic in your circuit, and you can exploit intermediate outputs (from other terms) as inputs. Furthermore, exclusive-OR realizations are not always obvious from Karnaugh maps. Finally, package constraints (e.g., the fact that four 2-input gates come in a single IC) also figure into the choice of logic used in the final circuit realization. When programmable logic devices such as PALs (Section 8.15) are used to construct logic functions, the internal structure (programmable AND; fixed OR) constrains the realization that can be used.

EXERCISE 8.13

Draw a Karnaugh map for logic to determine if a 3-bit integer (0 to 7) is prime (assume that 0,1, and 2 are not primes). Show a realization with 2-input gates.

EXERCISE 8.14

Find logic to perform multiplication of two 2-bit unsigned numbers (i.e., each 0 to 3), producing a 4-bit result. Hint: Use a separate Karnaugh map for each output bit.

8.14 Combinational functions available as ICs

Using Karnaugh maps, you can construct logic to perform rather complicated functions such as binary addition or magnitude comparison, parity checking, multiplexing (selecting one of several inputs, as determined by a binary address), etc. In the real world the most frequently used complex functions are available as single MSI functions (medium-scale integration, upward of 100 gates on one chip). Although many of the MSI functions involve flip-flops, which we will get to shortly, lots of them are combinational functions involving only gates. Let's see what animals live in the MSI combinational zoo.

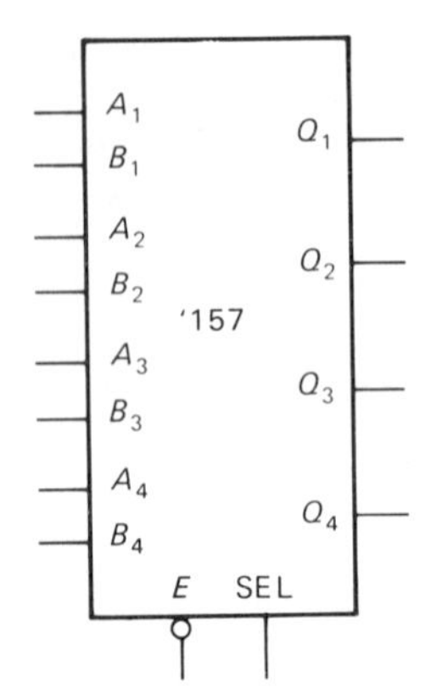

Figure 8.31. Quad 2-input select.

Quad 2-input select

The quad 2-input select is a very useful chip. It is basically a 4-pole 2-position switch for logic signals. Figure 8.31 shows the basic idea. When SELECT is LOW, the A inputs are passed through to their respective Q outputs. For SELECT HIGH, the B inputs appear at the output. ENABLE′ HIGH disables the device by forcing all outputs LOW. This is an important concept you will see more of later. Here's the truth table, which illustrates the X (don't care) entry:

Inputs				*Outputs*
E'	SEL	A_n	B_n	Q_n
H	*X*	*X*	*X*	*L*
L	*L*	*L*	*X*	*L*
L	*L*	*H*	*X*	*H*
L	*H*	*X*	*L*	*L*
L	*H*	*X*	*H*	*H*

Figure 8.31 and the preceding table correspond to the '157 quad 2-input select chip. The same function is also available with inverted output ('158) and with 3-state outputs (true: '257; inverted: '258).

EXERCISE 8.15

Show how to make a 2-input select from an AND-OR-INVERT gate.

Although the function of a select gate can be performed by a mechanical switch in some cases, the gate is a far better solution, for several reasons: (a) it is cheaper; (b) all channels are switched simultaneously and rapidly; (c) it can be switched, nearly instantaneously, by a logic level generated elsewhere in the circuit; (d) even if the select function is to be controlled by a front-panel switch, it is better not to run logic signals around through cables and switches, to avoid capacitive signal degradation and noise pickup. With a select gate actuated by a dc level, you keep logic signals on the circuit board and get the bonus of simpler off-board wiring (a single line with pullup switched to ground by an SPST switch). Controlling circuit functions with externally generated dc levels in this manner is known as "cold switching," and it is a much better approach than controlling the signals themselves with switches, potentiometers, etc. Besides its other advantages, cold switching lets you bypass control lines with capacitors to eliminate interference, whereas signal lines cannot generally be bypassed. You will see some examples of cold switching later.

Transmission gates

As we discussed in Sections 3.11 and 3.12, with CMOS it is possible to make "transmission gates," simply a pair of complementary MOSFET switches in parallel, so that an input (analog) signal between ground and V_{DD} is either connected through to the output through a low resistance (a few hundred ohms) or open-circuited (essentially infinite resistance). As you may remember, such a device is bidirectional and doesn't know (or care) which end is input and which end is output. Transmission gates work perfectly well with digital CMOS levels and are used extensively in CMOS design. Figure 8.32 shows the layout of the popular 4066 CMOS "quad bilateral switch." Each switch has a separate "control" input; input HIGH closes the switch, and LOW opens it. Note that transmission gates are merely switches, and therefore have no fanout; i.e., they simply pass input logic levels through to the output, without providing additional drive capability.

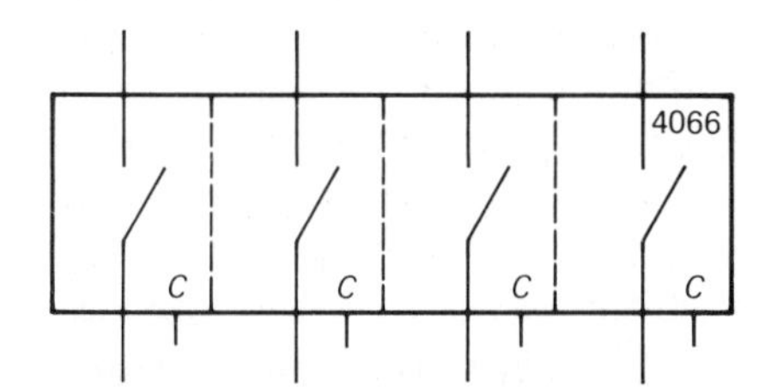

Figure 8.32. Quad transmission gate.

With transmission gates you can make 2-input (or more) select functions, usable with CMOS digital levels or analog signals. To select among a number of inputs, you can use a bunch of transmission gates (generating the control signals with a "decoder," as will be explained later). This is such a useful logic function that it has been institutionalized as the "multiplexer," which we will discuss next.

EXERCISE 8.16
Show how to make a 2-input select with transmission gates. You will need an inverter.

Multiplexers

The 2-input select gate is also known as a 2-input multiplexer. Multiplexers are also available with 4, 8, and 16 inputs (the 4-input variety comes as a dual unit, 2 in one package). A binary address is used to select which of the input signals appears at the output. For instance, an 8-input multiplexer has a 3-bit address input to address the selected data input (Fig. 8.33). The digital multiplexer illustrated is a '151. It has a STROBE (another name for ENABLE) input (negative-true), and it provides true and complemented outputs. When the chip is disabled (STROBE held HIGH), Q is LOW and Q' is HIGH, independent of the states of the address and data inputs.

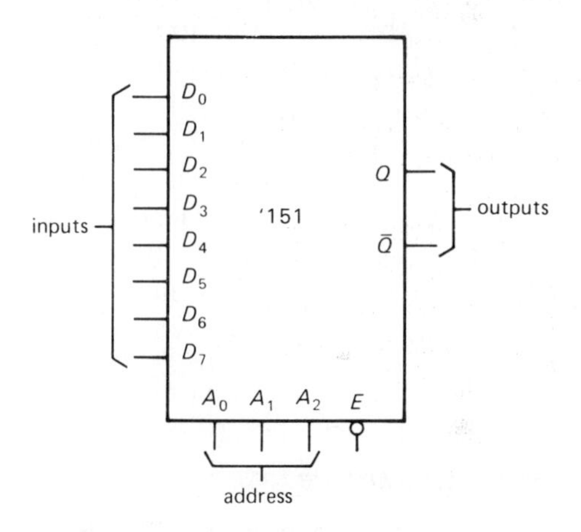

Figure 8.33. 8-input multiplexer.

In CMOS, two varieties of multiplexers are available. One type is for digital levels only, with an input threshold and "clean" regeneration of output levels according to the input state; that's also the way all TTL functions work. An example is the '153 TTL multiplexer. The other kind of CMOS multiplexer is analog and bidirectional; it's really just an array of transmission gates. The 4051–4053 CMOS multiplexers work this way (remember that logic made from transmission gates has no fanout). Since transmission gates are bidirectional, these multiplexers can be used as "demultiplexers," or decoders. We will discuss them next.

EXERCISE 8.17

Show how to make a 4-input muitipiexer using (a) ordinary gates, (b) gates with 3-state outputs, and (c) transmission gates. Under what circumstances would (c) be preferable?

You might wonder what to do if you want to select among more inputs than are provided in a multiplexer. This question comes under the general category of chip "expansion" (using several chips that have small individual capabilities to generate a larger capability), and it applies to decoders, memories, shift registers, arithmetic logic, and many other functions as well. In this case the job is easy (Fig. 8.34). Here we have expanded two 74LS151 8-input multiplexers into a 16-input multiplexer. There's an additional address bit, of course, and you use it to enable one chip or the other. The disabled chip holds its Q LOW, so an OR gate at the output completes the expansion. With three-state outputs the job is even simpler, since you can connect the outputs directly together.

Demultiplexers and decoders

A demultiplexer takes an input and routes it to one of several possible outputs, according to an input binary address. The other outputs are either held in the inactive state or open-circuited, depending on the type of demultiplexer.

A decoder is similar, except that the address is the only input, and it is "decoded" to assert one of n possible outputs. Figure 8.35 shows an example. This is the '138 "1-of-8 decoder." The output corresponding to (addressed by) the 3-bit input data is LOW; all others are HIGH. This particular decoder has three ENABLE inputs, all of which must be asserted (two LOW, one HIGH); otherwise all outputs are HIGH. A favorite use of the decoder is to cause different things to happen, depending on the state of a "counter" chip that drives it (more on this, soon). Decoders are commonly used when interfacing to microprocessor, to trigger different actions depending on the address; we'll treat this subject in detail in Chapter 10. Another common use of a decoder is to enable a sequence of actions in turn, according to an advancing address given by the output of a binary *counter* (Section 8.25). A close cousin of the '138 is the '139, a dual 1-of-4 decoder with a single LOW-true enable per section. Figure 8.36 shows how to use a pair of '138 1-of-8 decoders to generate a 1-of-16 decoder. No external gates are necessary,

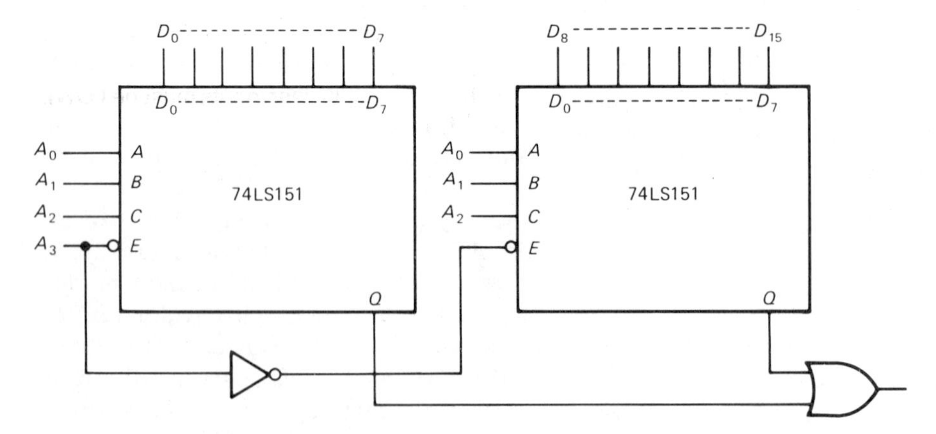

Figure 8.34. Multiplexer expansion.

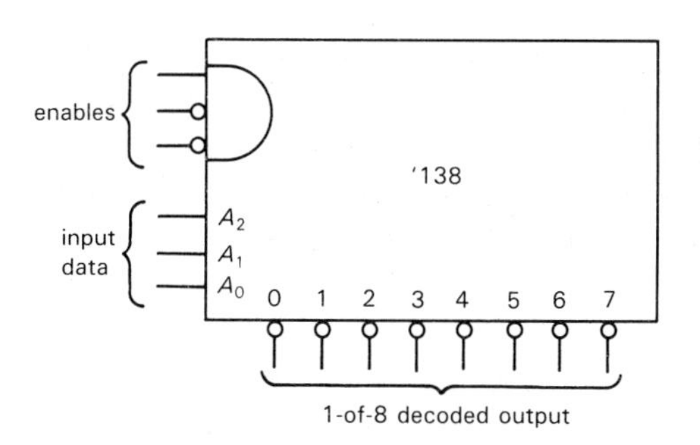

Figure 8.35. 1-of-8 decoder.

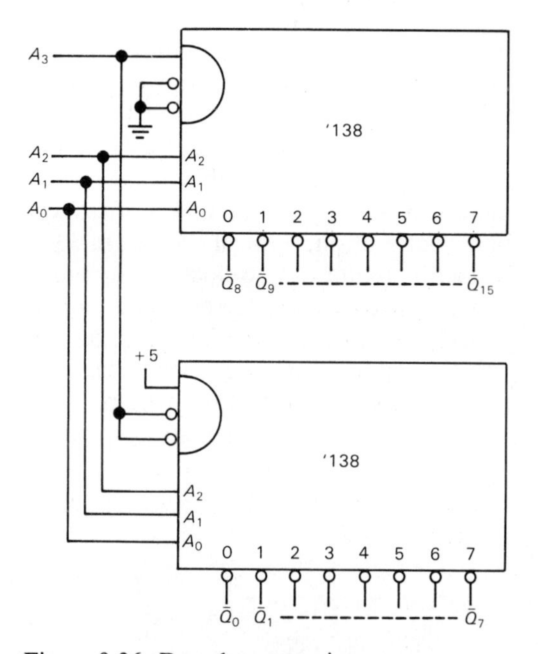

Figure 8.36. Decoder expansion.

since the '138 has enable inputs of both polarities.

EXERCISE 8.18
More expansion: Make a 1-of-64 decoder from nine '138s. Hint: Use one of them as an enabling switchyard for the others.

In CMOS logic, the multiplexers that use transmission gates are also demultiplexers, since transmission gates are bidirectional. When they are used that way, it is important to realize that the outputs that aren't selected are open-circuited. A pullup resistor, or equivalent, must be used to assert a well-defined logic level on those outputs (the same requirement as with TTL open-collector gates).

There is another kind of decoder generally available in all logic families. An example is the '47 "BCD-to-7-segment decoder/driver." It takes a BCD input and generates outputs on 7 lines corresponding to the segments of a "7-segment display" that have to be lit to display the decimal character. This type of decoder is really an example of a "code converter," but in common usage it is called a decoder. Table 8.6 at the end of the chapter lists most available decoders.

EXERCISE 8.19
Design a BCD-to-decimal (1-of-10) decoder using gates.

Priority encoder

The priority encoder generates a binary code giving the address of the highest-numbered input that is asserted. It is particularly useful in "parallel-conversion" A/D converters (see next chapter) and in microprocessor system design. An example is the '148 8-input (3 output bits) priority encoder. The '147 encodes 10 inputs.

EXERCISE 8.20
Design a "simple" encoder: a circuit that outputs the (2-bit) address telling which of 4 inputs is HIGH (all other inputs must be LOW).

Adders and other arithmetic chips

Figure 8.37 shows a "4-bit full adder." It adds the 4-bit number A_i to the 4-bit number B_i, generating a 4-bit sum S_i plus carry bit C_o. Adders can be "expanded" to add larger numbers: The "carry-in" input C_i is provided to accept the carry out of the next lower adder. Figure 8.38 shows how you would add two 8-bit numbers.

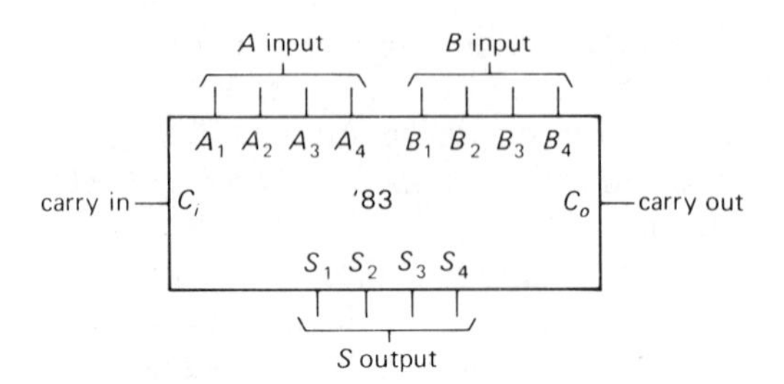

Figure 8.37. 4-bit full adder.

A device known as an arithmetic logic unit (ALU) is often used as an adder. It actually has the capability of performing a number of different functions. For instance, the '181 4-bit ALU (expandable to larger word lengths) can do addition, subtraction, bit shifts, magnitude comparison, and a few other functions. Adders and ALUs do their arithmetic in times measured in nanoseconds to tens of nanoseconds, depending on logic family.

Dedicated integer multiplier chips are available in configurations such as 8 bits times 8 bits, or 16 bits times 16 bits. A variation that is particularly suited to digital signal processing is the so-called MAC (multiplier-accumulator), which accumulates a sum of products; these, too, are available in sizes up to 32×32, with 64-bit product plus a few additional bits to keep the sum from overflowing. CMOS MACs and multipliers are available with typical speeds of 25–50ns; ECL multipliers are much faster – 5ns (typ) for 16×16 multiply.

Another arithmetic chip that's handy in digital signal processing is the *correlator*, which compares the corresponding bits of a pair of bit strings, calculating the number of bits that agree. A typical corrclator chip compares a pair of 64-bit words, which can be shifted in internal shift registers. Any pattern of bits can be ignored ("masked") in the correlation. Typical speeds are 30ns, i.e., a bit stream can be clocked through at 35MHz, with a 7-bit correlation available at each clock tick. A variation (known as an FIR digital filter) calculates instead the true sum (with carry) of the true pairwise product of a pair of integer strings; typical

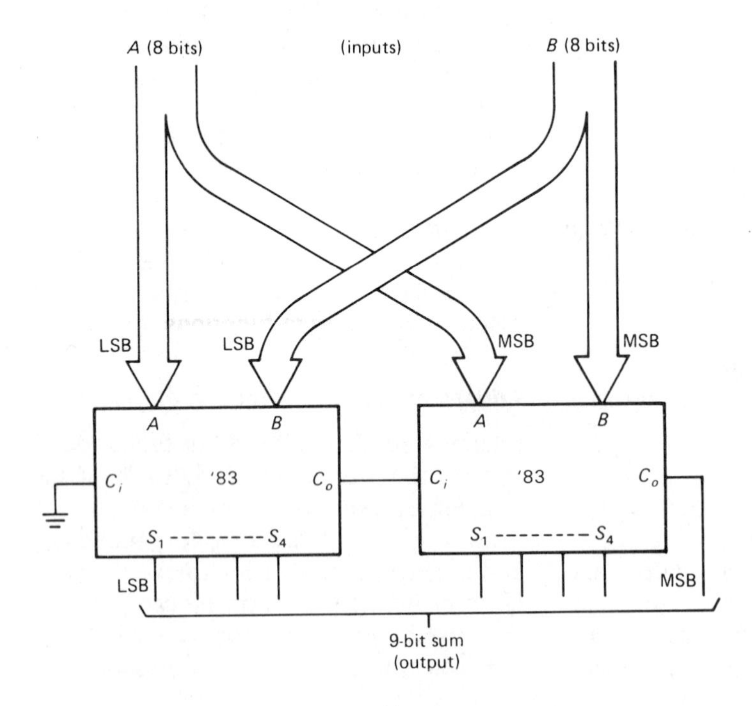

Figure 8.38. Adder expansion.

sizes are 4- to 10-bit integers, with string lengths of 3 to 8 words (expandable to greater length, of course).

The most complex arithmetic chips are the *floating-point processors*, which perform compares, sums, and products, as well as trigonometric functions, exponentials, and square roots. These are usually designed to work closely with particular microprocessors, and they generally conform to a nice standard known as IEEE P754, which specifies word size (up to 80 bits), format, etc. Examples (plus *matching microprocessors*) are the 8087 (*8086/8*), 80387 (*80386*), and 68881 (*68020*). These are truly stunning performers, with speeds of 10Mflops (million floating-point operations per second) or more.

Magnitude comparators

Figure 8.39 shows a 4-bit magnitude comparator. It determines the relative sizes of the 4-bit input numbers A and B and tells you via outputs whether $A < B$, $A = B$, or $A > B$. Inputs are provided for expansion to numbers larger than 4 bits.

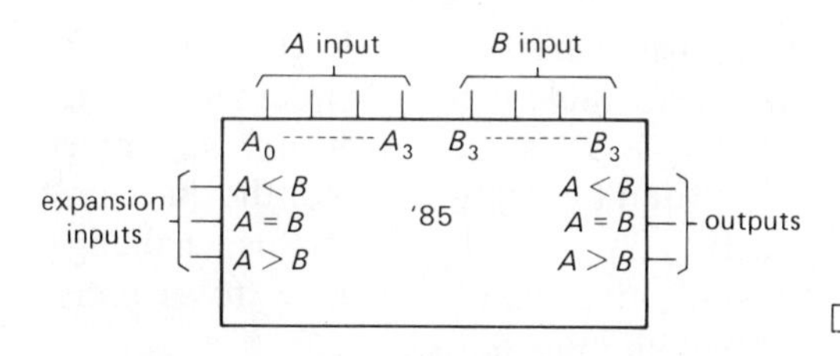

Figure 8.39. Magnitude comparator.

EXERCISE 8.21
Construct a magnitude comparator, using XOR gates, that tells whether or not $A = B$, where A and B are 4-bit numbers.

Table 8.7 at the end of the chapter lists most available magnitude comparators.

Parity generator/checker

This chip is used to generate a parity bit to be attached to a "word" when transmitting (or recording) data and to check the received parity when such data are recovered. Parity can be even or odd (e.g., with odd parity the number of 1 bits in each character is odd). The '280 parity generator, for instance, accepts a 9-bit input word, giving an even and an odd parity bit output. The basic construction is an array of exclusive-OR gates.

EXERCISE 8.22
Figure out how to make a parity generator using XOR gates.

Programmable logic devices

You can build your own custom combinational (and even sequential) logic on a chip, using ICs that contain an array of gates with programmable interconnections. There are several varieties, of which the most popular are PALs (programmable array logic) and PLAs (programmable logic arrays). PALs, in particular, have become extremely inexpensive and flexible and should form a part of every designer's toolbox of tricks. We will describe combinational PALs in the next section.

□ Some other strange functions

There are many other interesting MSI combinational chips worth knowing about. For example, in CMOS you can get a "majority logic" IC that tells you whether or not a majority of n inputs are asserted. Also available is a BCD "9's complementer," whose function is obvious. A "barrel-shifter" IC shifts an input word over by n (selectable) bits and can be expanded to any width.

8.15 Implementing arbitrary truth tables

Luckily, most of digital circuit design does not consist of cooking up crazy arrangements of gates to implement some complex logic function. However, there are times when you do need to wire up some complicated truth table, and the number of gates can become awfully large. You may begin to ask yourself if there isn't some other way. Fortunately, there are several. In this section we will look briefly at the use of multiplexers and demultiplexers to implement arbitrary truth tables. Then we will discuss the generally more powerful methods using programmable logic chips, particularly ROMs and PALs.

☐ *Multiplexers as generalized truth tables*

It should be obvious that an n-input multiplexer can be used to generate any n-entry truth table, without any external components, by simply connecting its inputs to HIGH or LOW as required. For example, Figure 8.40 shows a circuit that tells if a 3-bit binary input is prime.

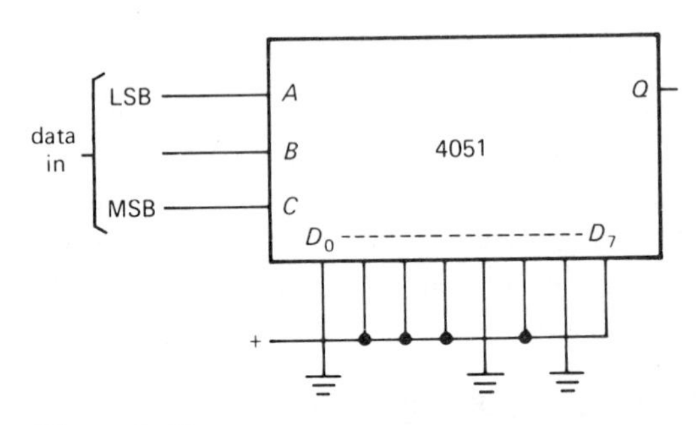

Figure 8.40

What is not so obvious is that an n-input multiplier can be used to generate any $2n$-entry truth table, with at most one external inverter. For example, Figure 8.41 shows a circuit that tells whether or not a given month of the year has 31 days, where the month (1 to 12) is specified by a 4-bit input. The trick is to notice that for a given state of address bits applied to the multiplexer, the output (as a function of the remaining input bit B) must be one of the four choices H, L, B, or B'; the corresponding multiplexer input is therefore tied to logic HIGH, logic LOW, B or inverted B.

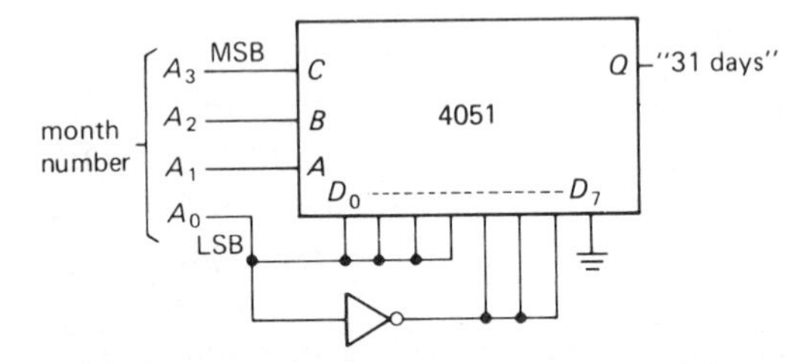

Figure 8.41

EXERCISE 8.23

Design of Figure 8.41. Make a table showing whether or not a given month has 31 days, with the month addressed in binary. Group the months in pairs, according to the most significant 3 bits of address. For each pair, figure out how Q ("31-ness") depends on the least significant address bit A_0. Compare with Figure 8.41. Finally, verify (using your knuckles, or by reciting poetry) that the circuit does indeed tell you if a given month has 31 days.

Amusing postscript: It turns out that this truth table can be implemented with a single XOR gate, if you take advantage of Xs (don't care) for the months that don't exist! Try your hand at this challenge. It will give you a chance to exercise Karnaugh map skills.

☐ *Decoders as generalized truth tables*

Decoders also provide a nice shortcut for combinational logic, particularly in situations where you need several simultaneous outputs. As an example, let's generate a circuit to convert BCD to excess-3. Here's the truth table:

Decimal	BCD	XS3	Decimal	BCD	XS3
0	0000	0011	5	0101	1000
1	0001	0100	6	0110	1001
2	0010	0101	7	0111	1010
3	0011	0110	8	1000	1011
4	0100	0111	9	1001	1100

We use the 4-bit (BCD) input as an address to the decoder, then use the (negative-true) decoded outputs as inputs to several OR gates, one for each output bit, as shown in Figure 8.42. Note that with this scheme the output bits don't have to be mutually exclusive. You might use something like this as a cycle controller for a washing machine, in which you turn on several functions (pump out water, fill, spin, etc.) at each input state. You will see shortly how to generate equally timed consecutive binary codes. The individual outputs from the decoder are known as *minterms*, and they correspond to positions on a Karnaugh map.

ROM and programmable logic

These are ICs that let you program their internal connections, roughly speaking. In that sense they are really devices with memory and should probably be discussed later, along with flip-flops, registers, etc. However, once programmed they are strictly combinational (although there are also sequential programmable logic devices;

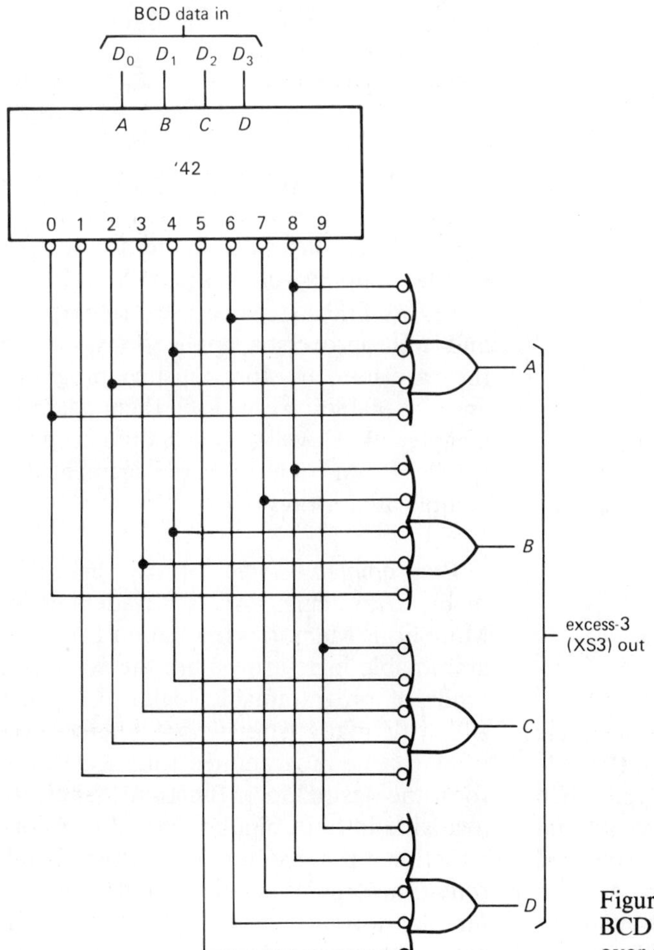

Figure 8.42. Minterm code conversion: BCD to "excess-3" (an obsolete code, left over from the first edition).

see Section 8.27), and they are so useful that it would be unforgivable not to discuss them now.

ROM. A ROM, or read-only memory, holds a bit pattern (typically 4 or 8 bits, parallel output) for each distinct address applied to its input. For example, a 1K×8 ROM gives eight output bits for each of 1024 input states, specified by a 10-bit input address (Fig. 8.43). Any combinational truth table can therefore be programmed into a ROM, provided there are enough input (address) lines. For example, the 1K×8 ROM above could be used to implement a 4-bit by 4-bit multiplier; in this case the limitation is the "width" (8 bits), not the "depth" (10 bits).

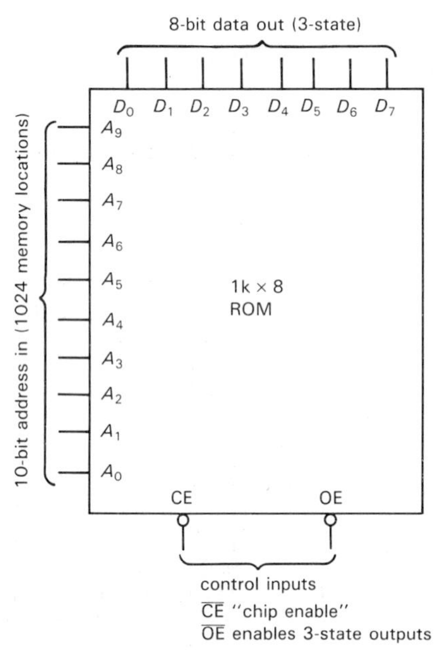

Figure 8.43

ROMs (and also programmable logic devices) are *nonvolatile*, meaning that the stored information is retained even when power is removed. There are several basic varieties, according to their method of programming: (a) "Mask-programmed ROMs" have their bit pattern built in at the time of manufacture. (b) "Programmable ROMs" (PROMs) are programmable by the user: PROMs use tiny interconnections that can be blown (like fuses) by applying appropriate address and control signals; they tend to be very fast (25–50ns), of relatively high power (bipolar: 0.5–1W), and small to medium in size (32×8 up to 8K×8). (c) "Erasable programmable ROMs" (EPROMs) store their bits as charge held on floating MOS gates, and hence can be erased by exposing them to intense ultraviolet (UV) light for some tens of minutes (they have a transparent quartz window); they are available in NMOS and CMOS varieties and are rather slow (200ns), of low power (particularly in standby mode), and large (8K×8 to 128K×8). Recent CMOS EPROMs are approaching bipolar speeds (35ns). A variant known as "one-time-programmable" (OTP) is an identical chip, but omits the quartz window for economy and ruggedness. (d) "Electrically erasable programmable ROMs" (EEPROMs) behave like EPROMs, but can be programmed and erased electrically, while in the circuit, with standard supply voltages (+5V).

ROMs find extensive use in computer and microprocessor applications, where they are used to store finished programs and data tables; we will see them again in Chapter 11. However, you should keep the smaller ROMs in mind as replacements for complicated arrays of gates.

Programmable logic. PALs (programmable array logic; PAL is a trademark of Monolithic Memories Inc.) and PLAs (programmable logic arrays) are the two basic kinds of programmable logic. They are ICs with many gates whose interconnections can be programmed (like ROMs) to form the desired logic functions. They are available in both bipolar and CMOS construction, the former using fusible-link (one-time-programmable), and the latter floating-gate MOS (UV or electrically erasable). You can't program *any*

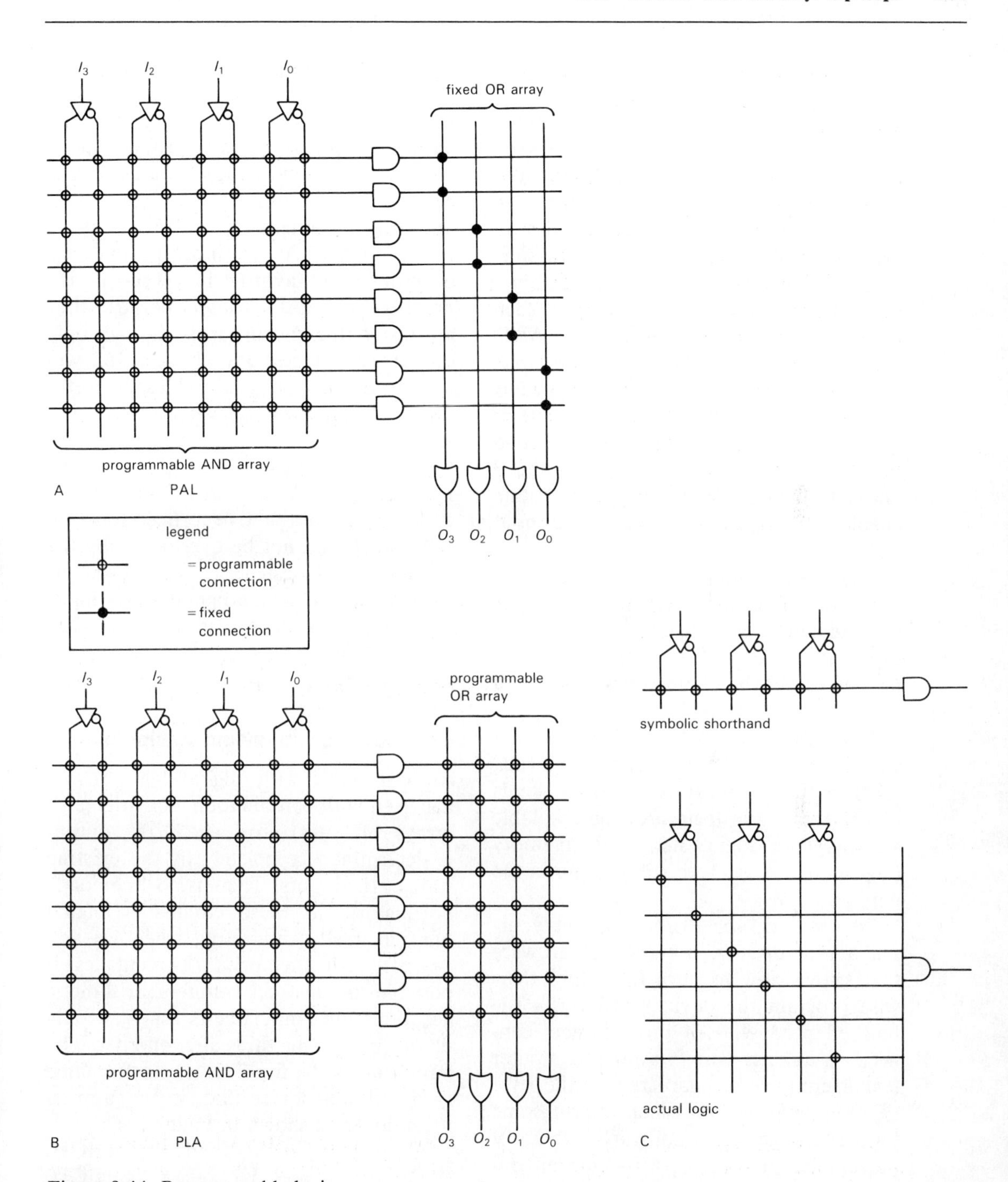

Figure 8.44. Programmable logic.
A. PAL
B. PLA
C. Detail of programmable connections to many-input AND gate; the circles are fusible links or other programmable connections.

interconnection you want – you're limited by the built-in structure. Figure 8.44 shows the basic design of combinational (no registers) PALs and PLAs. To keep this figure simple, the AND and OR gates, though drawn with a single input line, are actually multiple-input gates, with an input implied at every connected crossing.

Each (three-state) output of a combinational PAL comes from an OR gate, each of whose inputs is prewired to an AND gate with dozens of inputs. For example, the 16L8 (Fig. 8.45) has eight 7-input OR gates; every possible signal is available at each AND gate, including the 10 dedicated input pins (and their inverts) and the 8 outputs (and their inverts). Each tristate enable is also derived from a 32-input AND gate.

PLAs are similar to PALs, but with the added flexibility that the AND-gate outputs can be connected to the OR-gate inputs in any combination (i.e., programmable), rather than being prewired as in a PAL.

Note that the PALs and PLAs that we have described are *combinational* (i.e., gates only, no memory). Both kinds of programmable logic are also available as *sequential* logic, i.e., with memory (registers), a subject we will take up in the next section.

To use PALs or PLAs, you get yourself a *programmer*, a piece of hardware that knows how to burn fuses (or otherwise program the device) and verify the finished product. All programmers connect via a serial port to a microcomputer (engineers have standardized on the IBM PC or compatible), on which you run some form of programmer software. Some of the fancier programmers include an onboard computer that runs its own software. The simplest kind of software simply lets you select the fuses to burn; you figure that out by deciding what logic you want, at the gate level, then listing (or marking on a graphics display) those fuses. Figure 8.46 shows a trivial example, forming an exclusive-OR of two inputs as one of the outputs. Better programmers let you specify Boolean expressions (if you know them) or truth tables; the software does the rest, including minimization, simulation, and programming.

Although PLAs are more flexible, the overwhelming favorite in recent design has been the PAL. That is because they are faster (the signal passes through only one array of fuses) and cheaper and will usually do the job. As we'll see shortly, sophisticated new PALs using "macrocells" and "folded architecture" give you some additional flexibility within the fixed-OR PAL design. PALs provide a flexible and compact alternative to fixed-function ICs and should not be overlooked by the serious circuit designer. We'll show how (and when) to use programmable logic, along with useful tricks, in Section 8.27.

SEQUENTIAL LOGIC

8.16 Devices with memory: flip-flops

All our work with digital logic so far has been with combinational circuits (e.g., arrays of gates), for which the output is determined completely by the existing state of the inputs. There is no "memory," no history, in these circuits. Digital life gets really interesting when we add devices with memory. This makes it possible to construct counters, arithmetic accumulators, and circuits that generally do one interesting thing after another. The basic unit is the flip-flop, a colorful name to describe a device that, in its simplest form, looks as shown in Figure 8.47.

Assume that both A and B are HIGH. What are X and Y? If X is HIGH, then both inputs of G_2 are HIGH, making Y LOW. This is consistent with X being HIGH, so we're finished. Right?

$X = \text{HIGH}$
$Y = \text{LOW}$

16L8

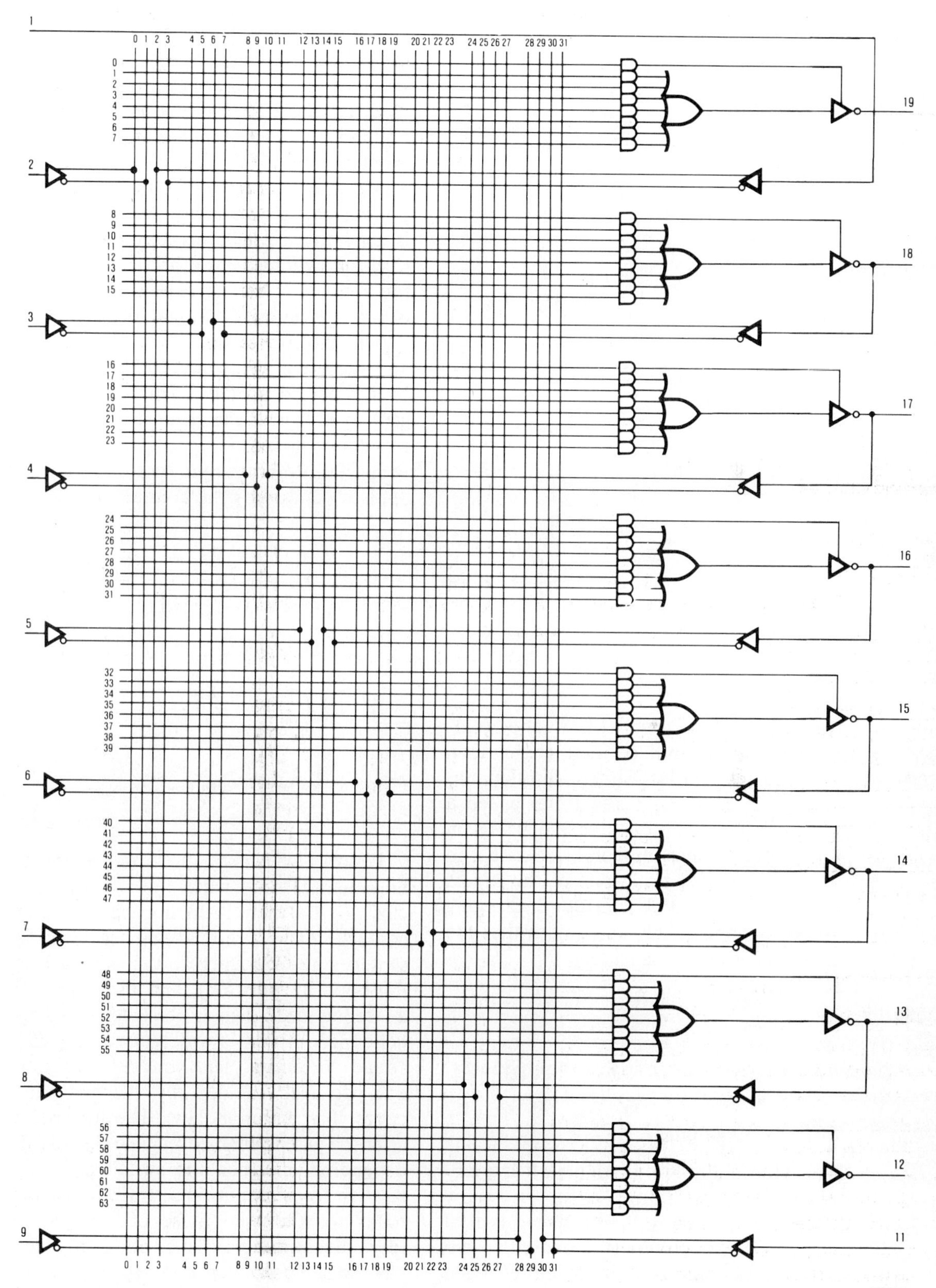

Figure 8.45. The 16L8 combinational PAL® has 10 dedicated inputs, 2 dedicated outputs, and 6 bidirectional (three-state) input/output lines; "16L8" means 16 (max) inputs, 8 (max) outputs (LOW-true). (Diagrams courtesy of Advanced Micro Devices of Sunnyvale, California.)

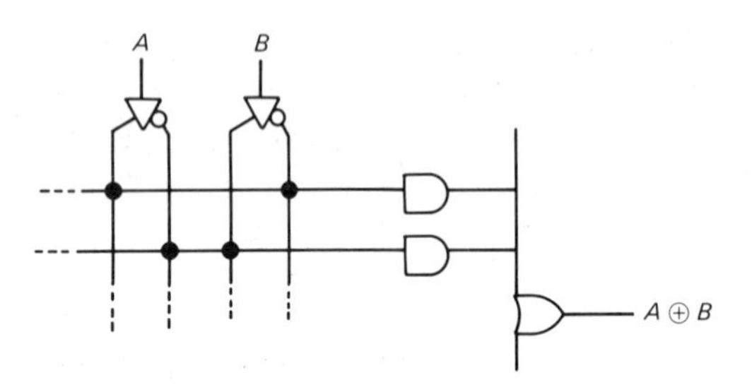

Figure 8.46. PAL exclusive OR.

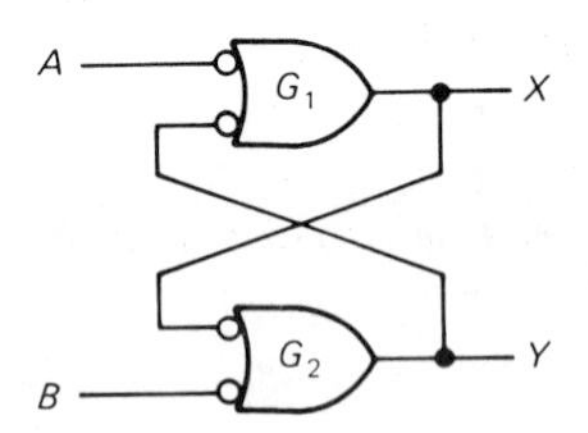

Figure 8.47. Flip-flop ("set-reset" type).

Wrong! The circuit is symmetrical, so an equally good state is

$X = \text{LOW}$

$Y = \text{HIGH}$

The states X, Y both LOW and X, Y both HIGH are not possible (remember, $A = B = \text{HIGH}$). So the flip-flop has two stable states (it's sometimes called a "bistable"). Which state it is in depends on past history. It has memory! To write into the memory, just bring one of the inputs momentarily LOW. For instance, bringing A LOW momentarily guarantees that the flip-flop goes into the state

$X = \text{HIGH}$

$Y = \text{LOW}$

no matter what state it was in previously.

Switch debouncing

This kind of flip-flop (with a SET and RESET input) is quite useful in many applications. Figure 8.48 shows a typical example. This circuit is supposed to enable the gate and pass input pulses when the switch is opened. The switch is tied to ground (not +5V), because of a peculiarity of bipolar TTL (as opposed to CMOS): You must *sink* substantial current from an input in the LOW state (0.25mA for LS TTL), whereas in the HIGH state the input current is near zero. Besides, ground is generally available as a convenient return for switches and other controls. The problem with this circuit is that switch contacts "bounce." When the switch is closed, the two contacts actually separate and reconnect, typically 10 to 100 times over a period of about 1ms. You would get waveforms as sketched; if there were a counter or shift register using the output, it would faithfully respond to all those extra "pulses" caused by the bounce.

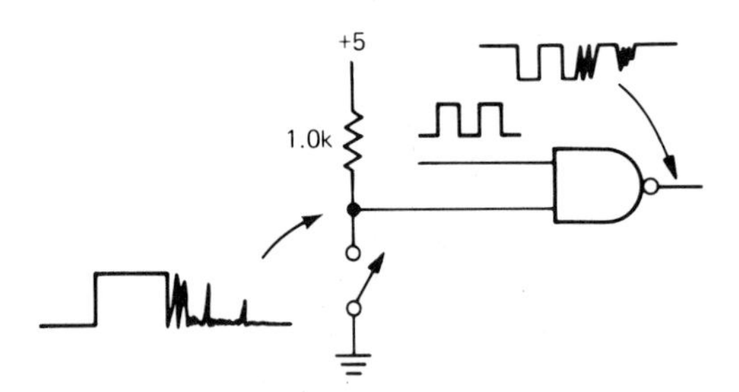

Figure 8.48. Switch "bounce."

Figure 8.49 shows the cure. The flip-flop changes state when the contacts first close. Further bouncing against that contact makes no difference (SPDT switches never bounce all the way back to the opposite position), and the output is a "debounced" signal, as sketched. This debouncer circuit is widely used; the '279 "quad *SR* latch" lets you get four into one package. Incidentally, the preceding circuit has a minor flaw: The first pulse after the gate is enabled may be shortened, depending on when the switch is closed relative to the input pulse train; the same holds for the final pulse of a sequence (of course, a switch that is not debounced has the same problem). A "synchronizer" circuit (see Section 8.19) can be used to prevent this from happening, for applications where it makes a difference.

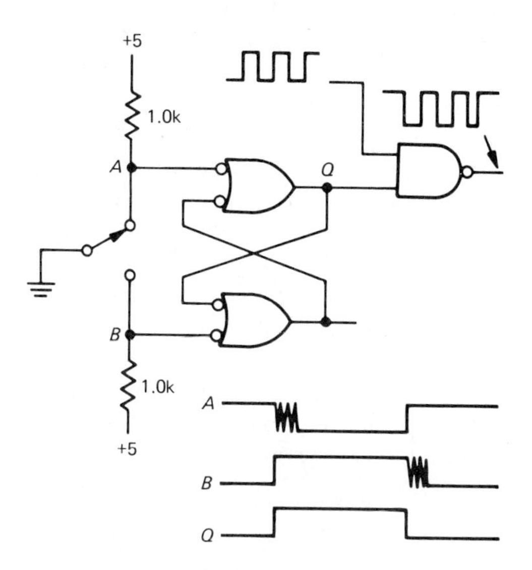

Figure 8.49. Switch debouncer.

Multiple-input flip-flop

Figure 8.50 shows another simple flip-flop. Here NOR gates have been used; a HIGH input forces the corresponding output LOW. Multiple inputs allow various signals to set or clear the flip-flop. In this circuit fragment, no pullups are used, since logic signals generated elsewhere (by standard active pullup outputs) are used as inputs.

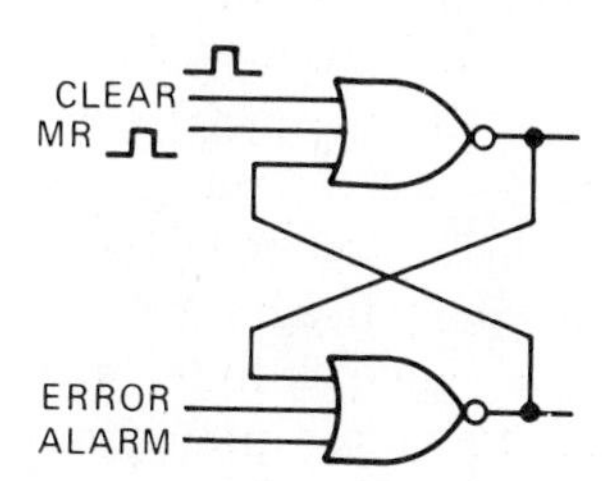

Figure 8.50

8.17 Clocked flip-flops

Flip-flops made with two gates, as in Figures 8.47 and 8.50, are known generically as *SR* (set-reset), or jam-loaded, flip-flops. You can force them into one state or the other whenever you want by just generating the right input signal. They're handy for switch debouncing and many other applications. But the most widely used form of flip-flop looks a little different. Instead of a pair of jam inputs, it has one or two "data" inputs and a single "clock" input. The outputs can change state or stay the same, depending on the levels at the data inputs when the clock pulse arrives.

The simplest clocked flip-flop looks as shown in Figure 8.51. It's just our original flip-flop, with a pair of gates (controlled by the clock) to enable the SET and RESET inputs. It is easy to verify that the truth table is

S	R	Q_{n+1}
0	0	Q_n
0	1	0
1	0	1
1	1	indeterminate

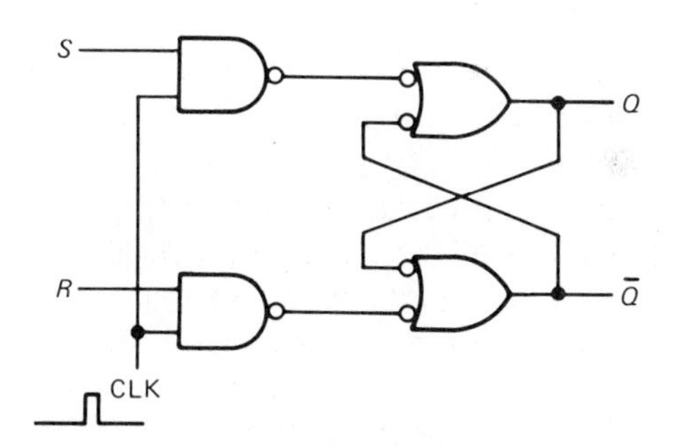

Figure 8.51. Clocked flip-flop.

where Q_{n+1} is the Q output after the clock pulse and Q_n is the output before the clock pulse. The basic difference between this and the previous flip-flops is that R and S should now be thought of as data inputs. What is present on R and S when a clock pulse comes along determines what happens to Q.

This flip-flop has one awkward property, however. The output can change in response to the inputs during the time the clock is HIGH. In that sense it is still like the jam-loaded SR flip-flop (it's also known as a "transparent latch," since the output "sees through" to the input when the clock is HIGH). The full utility of clocked flip-flops comes with the introduction of slightly different configurations, the master-slave flip-flop and the edge-triggered flip-flop.

Master-slave and edge-triggered flip-flops

These are by far the most popular flip-flops. The data present on the input lines just before a clock transition, or "edge," determines the output state after the clock has changed. These flip-flops are available as inexpensive packaged ICs and are always used in that form. But it is worth looking at their innards in order to understand what is going on. Figure 8.52 shows the schematics. Both are known as type D flip-flops. Data present at the D input will be transferred to the Q output after the clock pulse. The master-slave configuration is probably easier to understand. Here's how it works:

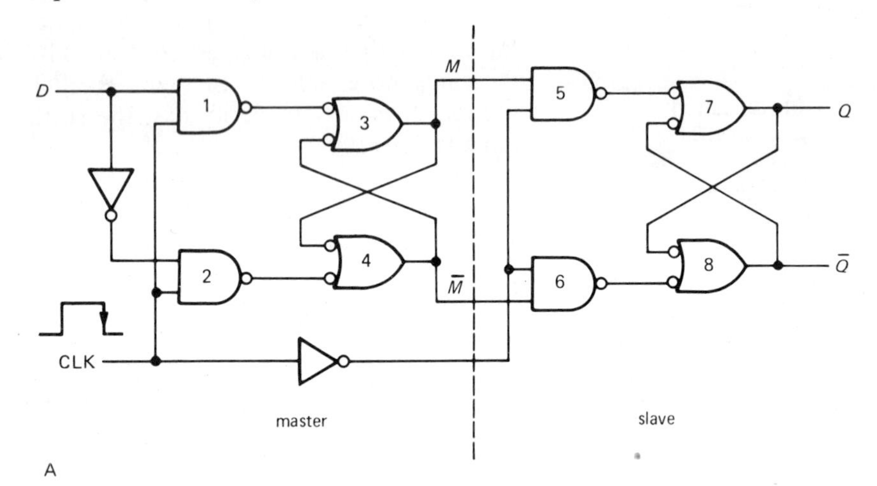

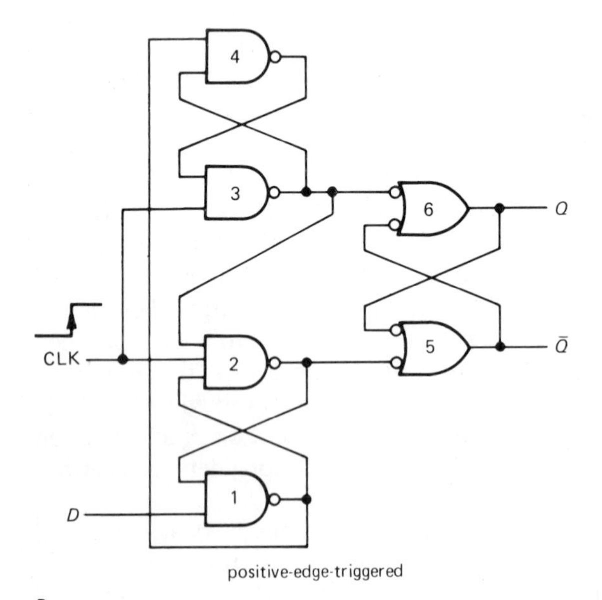

Figure 8.52. Edge-triggered type D flip-flops.

While the clock is HIGH, gates 1 and 2 are enabled, forcing the master flip-flop (gates 3 and 4) to the same state as the D input: $M = D, M' = D'$. Gates 5 and 6 are disabled, so the slave flip-flop (gates 7 and 8) retains its previous state. When the clock goes LOW, the inputs to the master are disconnected from the D input, while the inputs of the slave are simultaneously coupled to the outputs of the master. The master thus transfers its state to the slave. No further changes can occur at the output, because the master is now stuck. At the next rising edge of the clock, the slave will be decoupled from the master and will retain its state, while the master will once again follow the input.

The edge-triggered circuit behaves the same externally, but the inner workings are different. It is not difficult to figure it out. The particular circuit shown happens to be the popular '74 positive-edge-triggered type D flip-flop. The preceding master-slave circuit transfers data to the output on the negative edge. Flip-flops are available with either positive or negative edge triggering. In addition, most flip-flops also have SET and CLEAR jam-type inputs. They may be set and cleared on HIGH or on LOW, depending on the type of flip-flop. Figure 8.53 shows a few popular flip-flops. The wedge means "edge-triggered," and the little circles mean "negation," or complement. Thus, the '74 is a dual type D positive-edge-triggered flip-flop with active LOW jam-type SET and CLEAR inputs. The 4013 is a CMOS dual type D positive-edge-triggered flip-flop with active HIGH jam-type SET and CLEAR inputs. The '112 is a dual JK master-slave flip-flop with data transfer on the negative edge and with active LOW jam-type SET and CLEAR inputs.

The JK flip-flop. The JK flip-flop works on principles similar to those of the type D flip-flop, but it has two data inputs. Here's the truth table:

J	K	Q_{n+1}
0	0	Q_n
0	1	0
1	0	1
1	1	Q_n'

Thus, if J and K are complements, Q will go to the value of the J input at the next clock edge. If J and K are both LOW, the output won't change. If J and K are both HIGH, the output will "toggle" (reverse its state after each clock pulse).

Warning: Some older JK flip-flops are "ones-catching," a term you won't find in the data sheet, but an effect that can have dire consequences for the unsuspecting. This means that if either J or K (or both) changes state momentarily while the slave is enabled by the clock, then returns to its previous state before the clock makes its transition, the flip-flop will "remember" that momentary state and behave as if that state had persisted. Thus, the flip-flop may change state at the next clock transition even if the J and K inputs existing at that transition should cause the flip-flop to remain in its current state. This can

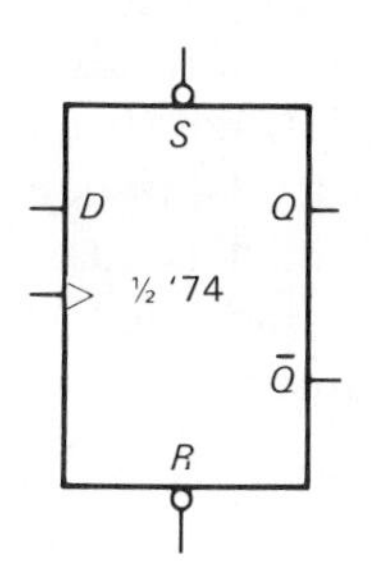

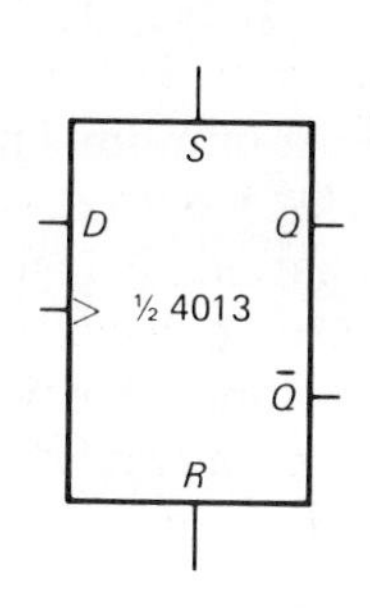

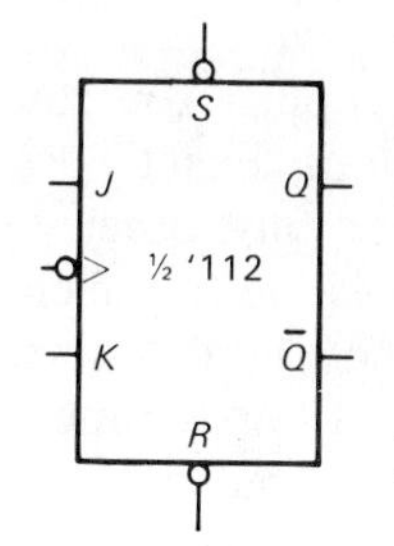

Figure 8.53. D-type and JK flip-flops.

lead to peculiar behavior, to put it mildly. The problem arises because such flip-flops were designed with short clock pulses in mind, whereas in common usage you clock flip-flops with just about anything. Be careful when using master-slave flip-flops, or avoid them altogether and use true edge-triggered flip-flops instead.

Two good choices that employ true edge triggering are the '112 and the '109. Both are dual (two per package) *JK* flip-flops with (negative-true) SET and CLEAR jam-type inputs; the '112 clocks on the negative edge, the '109 on the positive edge. The '109 has an interesting quirk, namely the K input is complemented (it's sometimes called a "*JK*-bar" flip-flop). Thus, if you tie the J and K inputs together, you've got a D flip-flop; to make it toggle, you ground K' and tie J HIGH.

Divide by 2

It is easy to make a divide-by-2 circuit by just exploiting the toggling capability of flip-flops. Figure 8.54 shows two ways. The *JK* flip-flop toggles when both inputs are HIGH, producing the output shown. The second circuit also toggles, since with the D input tied to its own Q' output, the D flip-flop always sees the complement of its existing output at its D input at the time of the clock pulse. The output signal in either case is at half the frequency of the input.

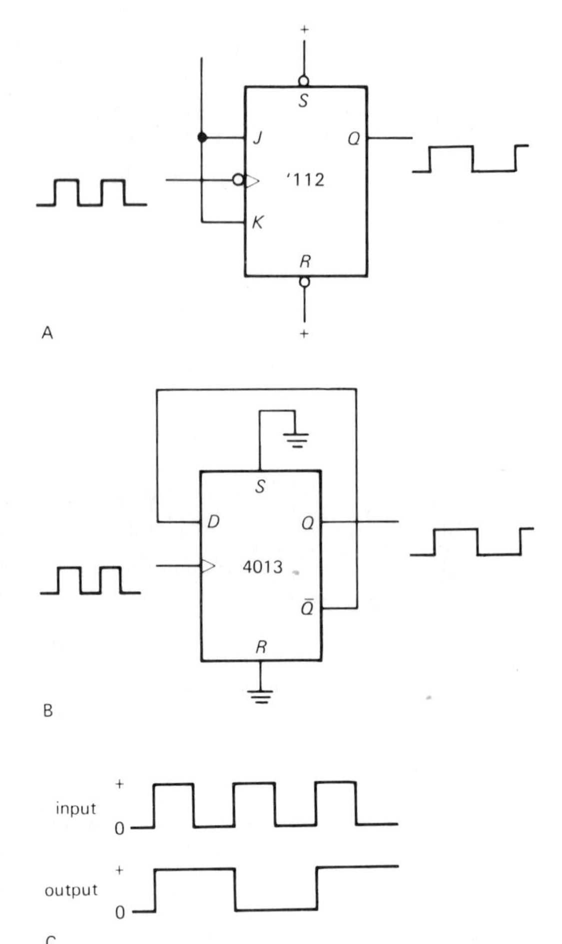

Figure 8.54. Toggling flip-flops.

Data and clock timing

This last circuit raises an interesting question: Will the circuit fail to toggle, since the D input changes almost immediately after the clock pulse? In other words, will the circuit get confused, with such crazy things happening at its input? You could, instead, ask this question: Exactly *when* does the D flip-flop (or any other flip-flop) look at its input, relative to the clock pulse? The answer is that there is a specified "setup time" t_s and "hold time" t_h for any clocked device. Input data must be present and stable from at least t_s before the clock transition until at least t_h after it, for proper operation. For the 74HC74, for instance, $t_s = 20\text{ns}$ and $t_h = 3\text{ns}$ (Fig. 8.55). So, for the preceding toggling connection, the setup-time requirement is met if the output has been stable for at least 20ns before the next clock rising edge. It may look as if the hold-time requirement is violated, but that's OK, also. The minimum "propagation time" from clock to output is 10ns, so a D flip-flop connected to toggle as described is guaranteed to have its D input stable for at least 10ns after the clock transition. Most devices nowadays have a zero hold-time requirement.

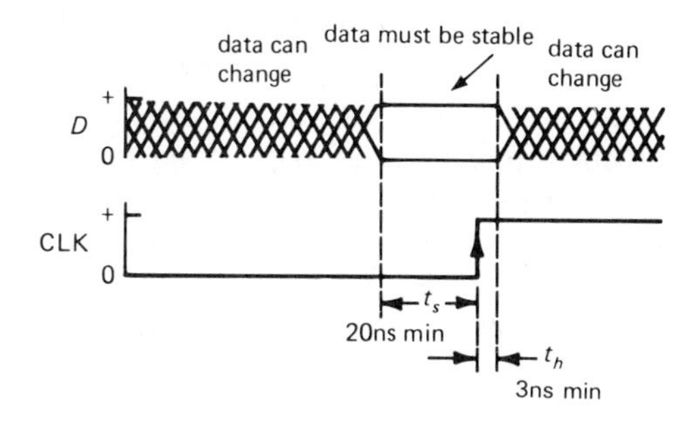

Figure 8.55. Data setup and hold times.

An interesting thing can happen if the level at the D input changes during the setup-time interval, namely a so-called *metastable* state in which the flip-flop can't make up its mind which state to go into. We will have more to say about this shortly.

Divide by more

By cascading several toggling flip-flops (connect each Q output to the next clock input), it is easy to make a divide-by-2^n, or binary, counter. Figure 8.56 shows a four-stage "ripple counter" and its waveforms. Note that flip-flops that clock on the falling edge (indicated by the negation circle) must be used if each Q output drives the next clock input. This circuit is a divide-by-16 counter: The output waveform from the last flip-flop is a square wave whose frequency is 1/16 of the circuit's input clock frequency. Such a circuit is called a *counter* because the data present at the four Q outputs, considered as a single 4-bit binary number, go through a binary sequence from 0 to 15, incrementing after each input pulse. The waveforms in Figure 8.56 demonstrate this fact. In the

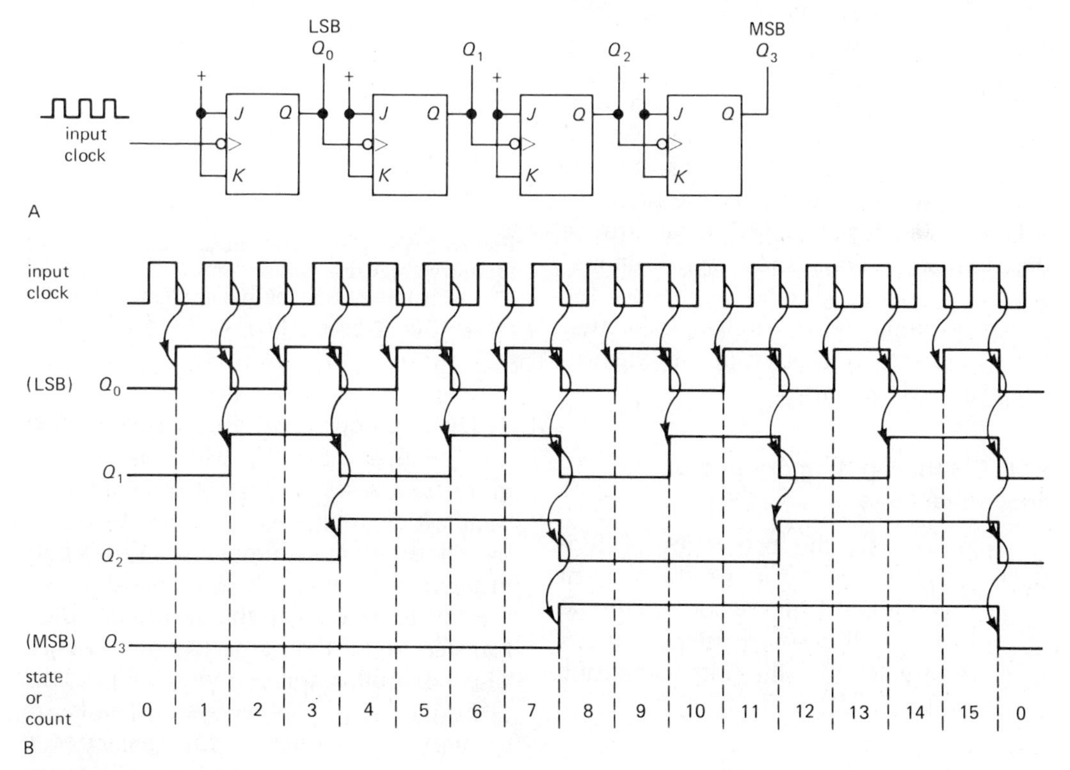

Figure 8.56. Four-bit counter.
A. Schematic.
B. Timing diagram.

figure the abbreviation MSB is used to mean "most significant bit," and LSB means "least significant bit"; the curved arrows are used to indicate what causes what, to aid in understanding.

As you will see in Section 8.25, the counter is such a useful function that many versions are available integrated onto single chips, including 4-bit, BCD, and multidigit counting formats. By cascading several such counters and displaying the count on a numeric display device (e.g., an LED digital display) you can easily construct an event counter. If the input pulse train to such a counter is gated for exactly 1 second, you've got a frequency counter, which displays frequency (cycles per second) by actually counting the number of cycles in a second. Section 15.10 shows diagrams of this simple and highly useful scheme. In fact, single-chip frequency counters are available, complete with oscillator, counter, control, and display circuitry; see Figure 8.71 for an example.

In practice, the simple scheme of cascading counters by connecting each Q output to the next clock input has some interesting problems related to the cascaded delays as the signal "ripples" down through the chain of flip-flops, and a "synchronous" scheme (in which all clock inputs see the same clocking signal) is usually better. Let's look into this question of synchronous clocked systems.

8.18 Combining memory and gates: sequential logic

Having explored the properties of flip-flops, let's see what can be done when they are combined with the combinational (gate) logic we discussed earlier. Circuits made with gates and flip-flops constitute the most general form of digital logic.

Synchronously clocked systems

As we hinted in the last section, sequential logic circuits in which there is a common source of clock pulses driving all flip-flops have some very desirable properties. In such a *synchronous system* all action takes place just after each clocking pulse, based on the levels present just before each clock pulse. Figure 8.57 shows the general scheme.

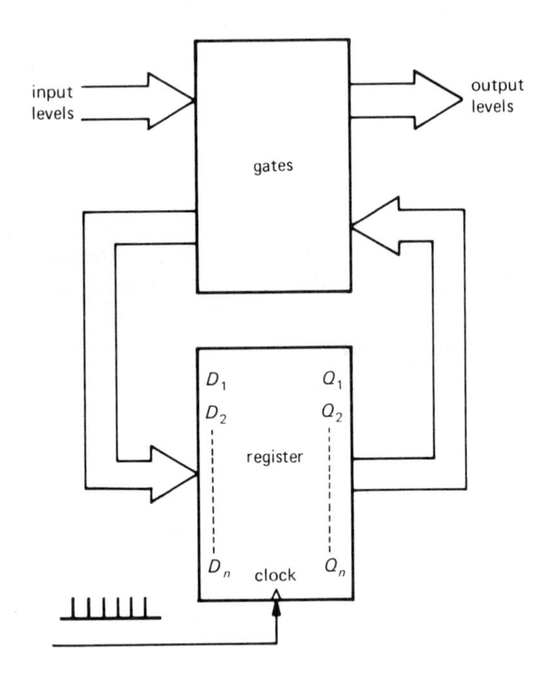

Figure 8.57. The classical sequential circuit: memory registers plus combinational logic. This scheme can be easily implemented with single-chip "registered PALs" (Section 8.27).

The flip-flops have all been combined into a single *register*, which is nothing more than a set of type D flip-flops with their clock inputs all tied together and their individual D inputs and Q outputs brought out; i.e., each clock pulse causes the levels present at the D inputs to be transferred to the respective Q outputs. The box full of gates looks at both the Q outputs and whatever input levels are applied to the circuit and generates a new set of D inputs and logic outputs. This simple-looking scheme is extremely powerful. Let's look at an example.

Example: divide-by-3

Let's design a synchronous divide-by-3 circuit with two type D flip-flops, both clocked from the input signal. In this case, D_1 and D_2 are the register inputs, Q_1 and Q_2 are the outputs, and the common clock line is the master clocking input (Fig. 8.58).

1. Choose the three states. Let's use

Q_1	Q_2	
0	0	
0	1	
1	0	
0	0	(i.e., first state)

2. Find the combinational logic network outputs necessary to generate this sequence of states, i.e., figure out what the D inputs have to be to get those outputs:

Q_1	Q_2	D_1	D_2
0	0	0	1
0	1	1	0
1	0	0	0

3. Concoct suitable gating (combinational logic), using available outputs, to produce those D inputs. In general, you can use a Karnaugh map. In this simple case you can see by inspection that

$$D_1 = Q_2$$
$$D_2 = (Q_1 + Q_2)'$$

from which the circuit of Figure 8.59 follows.

It is easy to verify that the circuit works as planned. Since it is a synchronous counter, all outputs change simultaneously (when you feed one output to the next clock, you've got a *ripple* counter instead). In general, synchronous (or "clocked") systems are desirable, since susceptibility to noise is improved: Things have settled down by the time of the clock pulse, so circuits that only look at their inputs at clock edges aren't troubled by capacitively coupled interference from other flip-flops, etc. A further advantage of clocked systems is that transient states (caused by

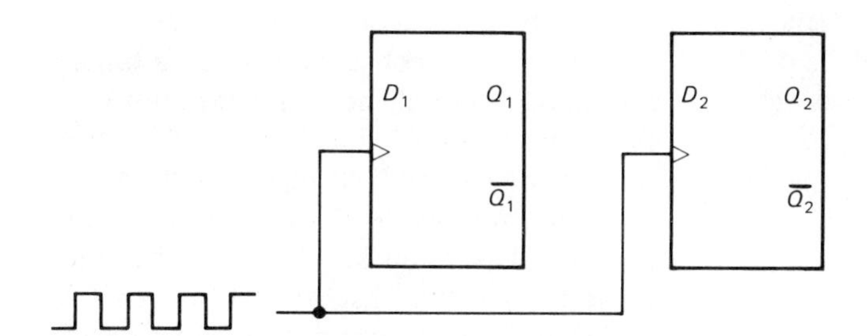

Figure 8.58

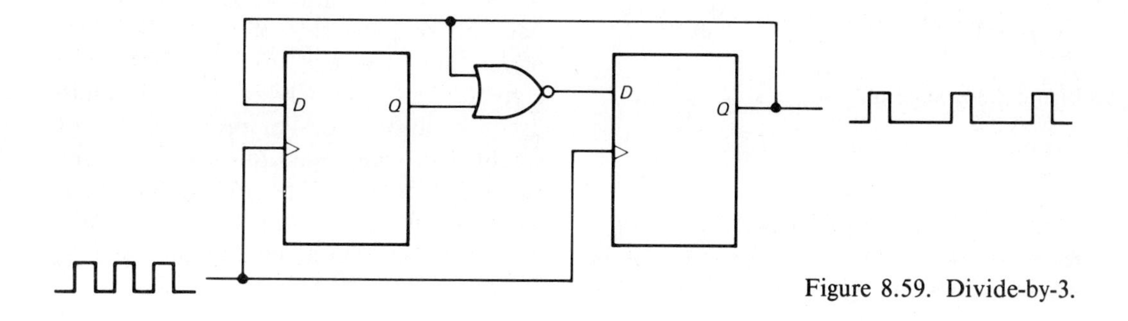

Figure 8.59. Divide-by-3.

delays, so that all outputs don't change simultaneously) don't produce false output, since the system is insensitive to what happens just *after* a clock pulse. You will see some examples later.

Excluded states

What happens to the divide-by-3 circuit if the flip-flops somehow get into the state $(Q_1, Q_2) = (1,1)$? This can easily happen when the circuit is first turned on, since the initial state of a flip-flop is anyone's guess. From the diagram, it is clear that the first clock pulse will cause it to go to the state (1,0), from which it will function as before. It is important to check the excluded states of a circuit like this, since it is possible to be unlucky and have it get stuck in one of those states. (Alternatively, the initial design procedure can include a specification of all possible states.) A useful diagnostic tool is the *state diagram*, which for this example looks like Figure 8.60. Usually you write the conditions for each transition next to the arrows, if other variables of the system are involved. Arrows may go in both directions between states, or from one state to several others.

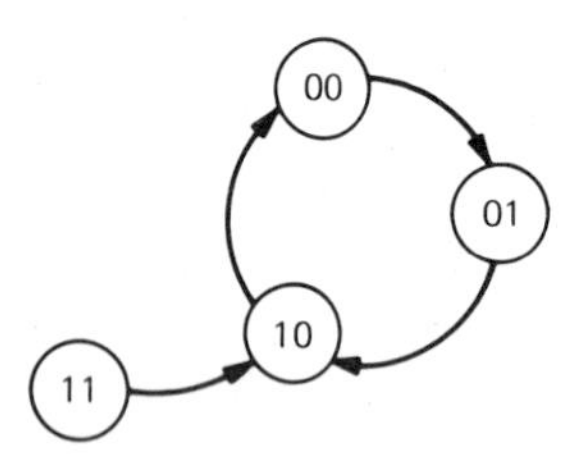

Figure 8.60. State diagram: divide-by-3.

EXERCISE 8.24

Design a synchronous divide-by-3 circuit using two JK flip-flops. It can be done (in 16 different ways!) without any gates or inverters. One hint: When you construct the table of required J_1, K_1 and J_2, K_2 inputs, keep in mind that there are two possibilities for J, K at each point. For instance, if a flip-flop output is to go from 0 to 1, J, K = 1,X (X = don't care). Finally, check to see if the circuit will get stuck in the excluded state (of the 16 distinct solutions to this problem, 4 will get stuck and 12 won't).

EXERCISE 8.25

Design a synchronous 2-bit UP/DOWN counter: It has a clock input, and a control input (U/D′); the outputs are the two flip-flop outputs Q_1 and Q_2. If U/D′ is HIGH, it goes through a normal binary counting sequence; if LOW, it counts backward – Q_2Q_1 = 00,11,10,01,00. . . .

State diagrams as design tools

The state diagram can be very useful when designing sequential logic, particularly if the states are connected together by several paths. In this design approach, you begin by selecting a set of unique states of the system, giving each a name (i.e., a binary address). You will need a minimum of n flip-flops, or bits, where n is the smallest integer for which 2^n is equal to or greater than the number of distinct states in the system. Then you set down all the rules for moving between states, i.e., all possible conditions for entering and leaving each state. From there it is a straightforward (but perhaps tedious) job to generate the necessary combinational logic, since you have all possible sets of Qs and the set of Ds that each leads to. Thus you have converted a sequential design problem into a combinational design problem, always soluble through techniques such as the Karnaugh map. Figure 8.61 shows a real-world example. Note that there may be states that don't lead to others, e.g., "receive diploma."

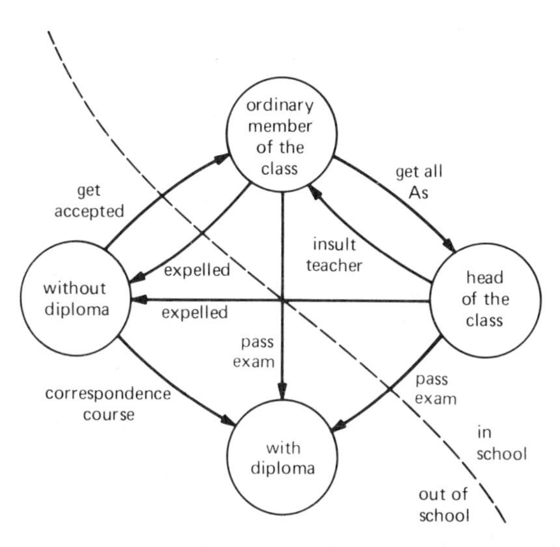

Figure 8.61. State diagram: going to school.

Registered PALs

Programmable logic (PALs and PLAs, see Section 8.15) is available with both gates and synchronously clocked *D* flip-flops on the same chip; these are known as *registered* PALs and PLAs, and they are ideal for implementing custom sequential circuits. We'll show how in Section 8.27.

8.19 Synchronizer

An interesting application of flip-flops in sequential circuits is their use in a *synchronizer*. Suppose you have some external control signal coming into a synchronous system that has clocks, flip-flops, etc., and you want to use the state of that input signal to control some action. For example, a signal from an instrument or experiment might signify that data are ready to be sent to a computer. Since the experiment and the computer march to the beats of different drummers, so to speak (in fancy language you would say they are *asynchronous* processes), you need a method to restore order between the two systems.

Example: pulse synchronizer

As an example, let's reconsider the circuit in which a debouncer flip-flop gated a pulse train (Section 8.16). That circuit enables the gate whenever the switch is closed, regardless of the phase of the pulse train being gated, so that the first or last pulse may be shortened. The problem is that the switch closure is asynchronous with the pulse train. In some applications it is important to have only *complete* clock cycles, and that requires a synchronizer circuit like that in Figure 8.62. Pushing START brings the output of gate 1 HIGH, but *Q* stays LOW until the next falling edge of the input pulse train. In that way, only complete pulses are passed by NAND gate 3. Figure 8.62 shows some waveforms. The curved arrows are drawn to show exactly what causes what. You can see, for instance, that the transitions of *Q* occur slightly *after* the falling edges of the input.

Logic races and glitches

This example brings up a subtle but extremely important point: What would happen if a positive-edge-triggered flip-flop were used instead? If you analyze it carefully, you'll find that START still works OK, but if STOP is pushed while the input is LOW, a bad thing happens (Fig. 8.63). A short spike, or "glitch," gets through because the final NAND gate isn't disabled until the flip-flop output has a chance to go LOW, a delay of about 20ns for HC or LS TTL. This is a classic example of a "logic race." With some care these situations can be avoided, as the example shows. Glitches are terrible things to have running through your circuits. Among other things, they're hard to see on an oscilloscope, and you may not know they are present. They can clock subsequent flip-flops erratically, and they may be widened – or narrowed to extinction – by passage through gates and inverters.

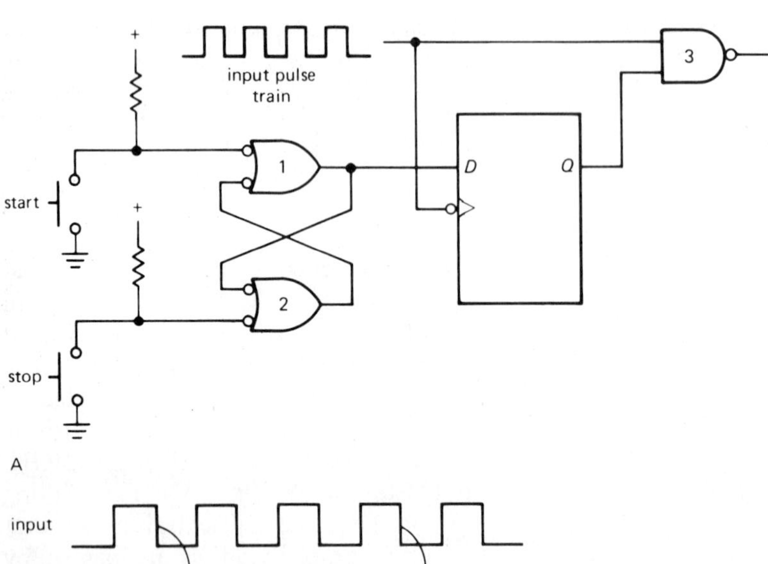

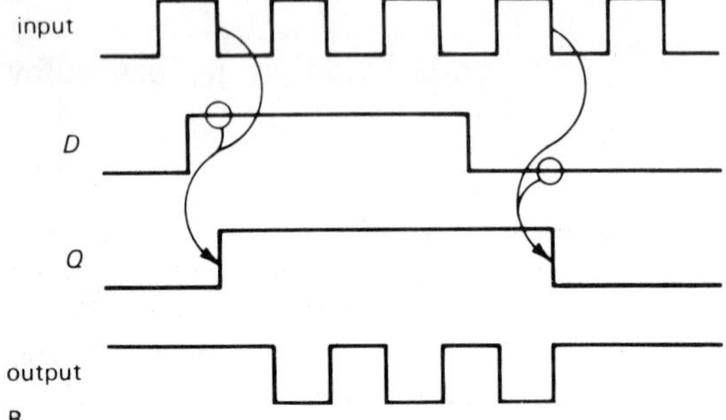

Figure 8.62. Pulse-train synchronizer.

EXERCISE 8.26

Demonstrate that the preceding pulse synchronizer circuit (Fig. 8.62) does not generate glitches.

EXERCISE 8.27

Design a circuit that lets exactly one full negative pulse (from an input train of pulses) pass through to the output, after a button is pushed.

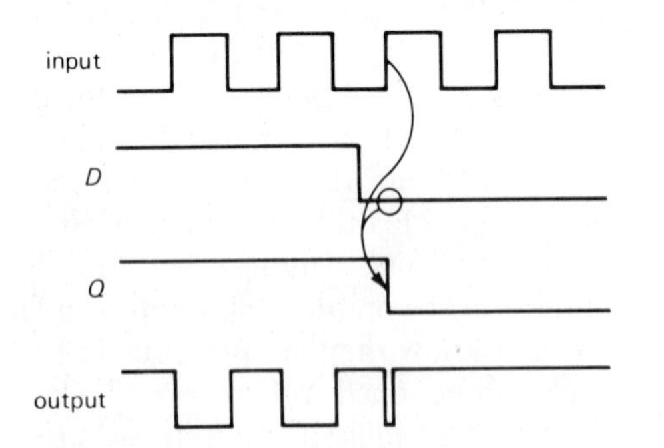

Figure 8.63. A logic race can generate a "runt pulse."

A few comments about synchronizers: The input to the *D* flip-flop can come from other logic circuitry, rather than from a debounced switch. There are applications in computer interfacing, etc., where an asynchronous signal must communicate with a clocked device; in such cases clocked flip-flops or synchronizers are ideal. In this circuit, as in all logic, unused inputs must be handled properly. For instance, SET and CLEAR must be connected so that they are not asserted (for a '74, tie them HIGH; for a 4013, they are grounded). Unused inputs that have no influence on the outputs can be left unconnected (e.g., inputs to unused gates), except in CMOS, where they should be grounded to prevent output-stage current (more on that in Chapter 9). A dual synchronizer is available as the 74120, although it has not been widely used.

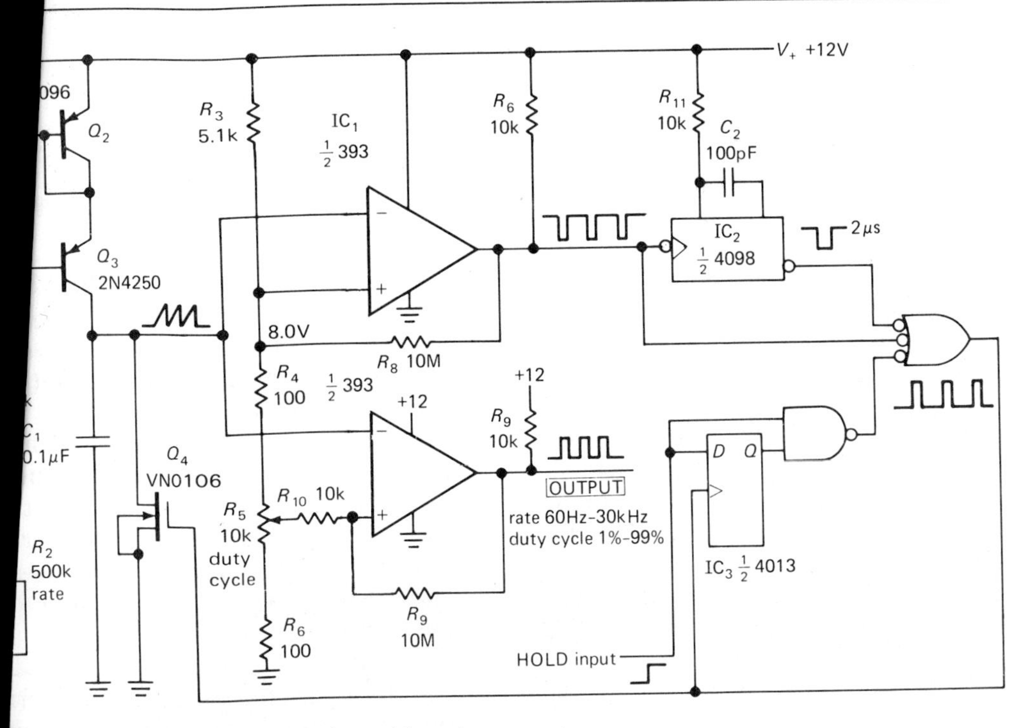

re 8.66. Autosynchronizing triggerable pulse generator.

linear circuits have the usual problems V_{BE} and h_{FE} variation with tempera-, etc., one-shots tend to exhibit tem-ature and supply voltage sensitivity of put pulse width. A typical unit like '4538 will show pulse-width variations a few percent over a 0–50°C tempera-e range and over a ±5% supply voltage nge. In addition, unit-to-unit variations e you a ±10% prediction accuracy for y given circuit. When looking at temper-ure and voltage sensitivity, it is impor-nt to remember that the chip may exhibit lf-heating effects and that supply volt-e variations *during the pulse* (e.g., small litches on the V_+ line) may affect the pulse idth seriously.

Long pulses. When generating long pulses, the capacitor value may be a few microfarads or more; in that case elec-trolytic capacitors are necessary. You have to worry about leakage current (which is insignificant with the smaller capacitor types), especially since most monostable types apply voltage of both polarities across the capacitor during the pulse. It may be necessary to add a diode or tran-sistor to prevent this problem, or to use a digital delay method instead (involving a clock and many cascaded flip-flop stages, as in Section 8.23). The use of an exter-nal diode or transistor will degrade tem-perature and voltage sensitivity and pulse-width predictability; it may also degrade retriggerable operation.

Duty cycle. With some one-shots the pulse width is shortened at high duty cycle. A typical example is the TTL 9600–9602 series, which has constant pulse width up to 60% duty cycle, decreasing about 5% at 100% duty cycle. The otherwise admirable '121 is considerably worse in this respect, with erratic behavior at high duty cycles.

Triggering. One-shots can produce sub-standard or jittery output pulses when trig-gered by too short an input pulse. There is

MONOSTABLE MULTIVIBRATORS

The monostable multivibrator, or "one-shot" (emphasis on the word "one"), is a variation of the flip-flop (which is sometimes called a bistable multivibrator) in which the output of one of the gates is capacitively coupled to the input of the other gate. The result is that the circuit sits in one state. If it is forced to the other state by a momentary input pulse, it will return to the original state after a delay time de-termined by the capacitor value and the circuit parameters (input current, etc.). It is very useful (some would say *too* useful!) for generating pulses of selectable width and polarity. Making one-shots with gates and *RC*s is tricky, and it depends on the details of the gate's input circuit, since, for instance, you wind up with voltage swings beyond the supply voltages. Rather than encourage bad habits by illustrating such circuits, we will just treat the one-shot as an available functional unit. In actual cir-cuits it is best to use a packaged one-shot; you construct your own only if absolutely necessary, e.g., if you have a gate available and no room for an additional IC package (even then, maybe you shouldn't).

8.20 One-shot characteristics

Inputs

One-shots are triggered by a rising or falling edge at the appropriate inputs. The only requirement on the triggering signal is that it have some minimum width, typically 25ns to 100ns. It can be shorter or longer than the output pulse. In general, several inputs are provided so that several signals can trigger the one-shot, some on positive edges and some on negative edges (remember, a negative edge means a HIGH-to-LOW transition, not a negative polarity). The extra inputs can also be used to inhibit triggering. Figure 8.64 shows four examples.

Each horizontal row of the table rep-resents a valid input triggering transition. For example, the '121 will trigger when one of the A inputs makes a HIGH-to-LOW transition, if the B input and the other A input are both HIGH. The '4538 is a dual CMOS monostable with OR gating at the input; if only one input is used, the other must be disabled, as shown. The '121 has three inputs, with a combination of OR and AND gating (and triggering), as shown. Its B input is a Schmitt trigger, more forgiving with slowly rising or noisy input signals. This monostable also in-cludes a not-too-good internal timing resistor you can use instead of R, if you're feeling lazy. The '221 is a dual '121; CMOS users can get only the dual version. The popular '123 is a dual monostable with AND input gating; unused inputs must be enabled. Note particularly that it triggers when RESET is disabled if both trigger inputs are already asserted. This is not a universal property of monostables, and it may or may not be desirable in a given application (it's usually not). The '423 is the same as the '123, but without this "feature."

When drawing monostables in a circuit diagram, the input gating is usually omitted, saving space and creating a bit of confusion.

Retriggerability

Most monostables, e.g., the 4538, '123, and '423 mentioned earlier, will begin a new timing cycle if the input triggers again during the duration of the output pulse. They are known as retriggerable mono-stables. The output pulse will be longer than usual if they are retriggered during the pulse, finally terminating one pulse width after the last trigger. The '121 and '221 are nonretriggerable; they ignore in-put transitions during the output pulse. Most retriggerable one-shots can be connected as nonretriggerable one-shots.

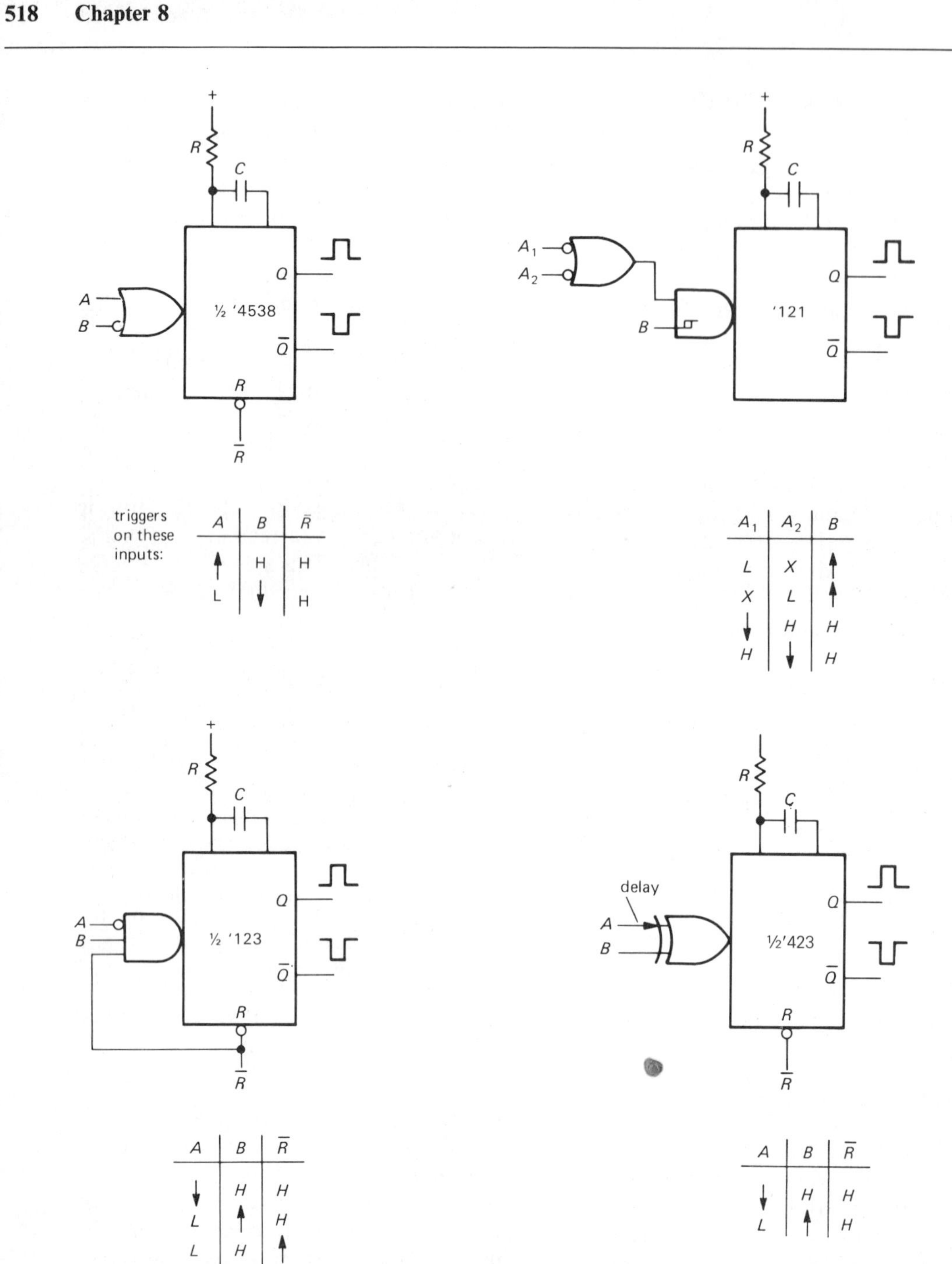

Figure 8.64. Four popular one-shots with their truth tables.

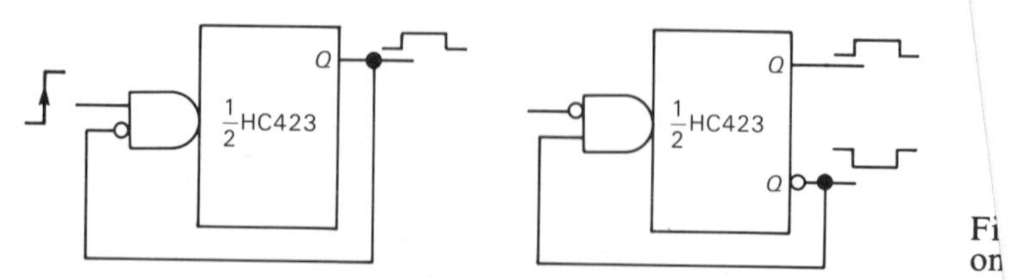

Figure 8.65 shows an example that's easy to understand.

Resettability

Most monostables have a jam RESET input that overrides all other functions. A momentary input to the RESET terminal terminates the output pulse. The RESET input can be used to prevent a pulse during power-up of the logic system; however, see the preceding comment about the '123.

Pulse width

Pulse widths from 40ns up to milliseconds (or even seconds) are attainable with standard monostables, set by an external capacitor and (usually) resistor combination. A device like the 555 (Section 5.14) can be used to generate longer pulses, but its input properties are sometimes inconvenient. Very long delays are best generated digitally (see Section 8.23).

Table 8.8 at the end of the chapter lists most available monostables.

8.21 Monostable circuit example

Figure 8.66 shows a square-wave generator with independently settable rate and duty cycle (ratio of HIGH to LOW) and an input that permits an external signal to "hold" the output following a negative edge. Current mirror Q_1–Q_3 generates a ramp at C_1. When it reaches the threshold of the upper comparator at two-thirds V_+, the one-shot is triggered and generates a 2μs positive pulse, putting n-channel VFET Q_4 into conduction and discharging the capacitor. C_1 therefore has a sawtooth

waveform going
with rate set by
lower comparato
square wave from
cycle adjustable li
99% via R_5. Bo
few millivolts of h
to prevent noise-in
tions. The LM39
comparator with
range right down
collector outputs.

A feature of this
to synchronize (start/
applied control level.
lets the driven circuit
the next negative tran
When HOLD is again
oscillator immediately
as if a falling edge had o
HOLD was released. Th
to the 3-input NAND fro
output ensures that the
stuck with C_1 charged u
the one-shot pulse width
long enough to ensure
discharged during the pul

8.22 Cautionary notes ab monostables

Monostables have some
don't see in other digital ci
dition, there are some gene
involved in their use. First, a
monostable pathology.

Some problems with monost

Timing. One-shots involve
tion of linear and digital techni

CA3
Q_1
R_1
1.0
Figu

the
of
tur
per
out
the
of
tu
ra
gi
ar
at
ta
s
a
g

a minimum trigger pulse width specified, e.g., 50ns for the 'LS121, 140ns for the 4098 with +5 volt supply, and 40ns for the 4098 with +15 volt supply (CMOS is faster and has more output drive capability when operated at higher supply voltages).

Noise immunity. Because of the linear circuits in a monostable, the noise immunity is generally poorer than in other digital circuits. One-shots are particularly susceptible to capacitive coupling near the external R and C used to set the pulse width. In addition, some one-shots are prone to false triggering from glitches on the V_+ line or ground.

Specsmanship. Be aware that monostable performance (predictability of pulse width, temperature and voltage coefficients, etc.) may degrade considerably at the extremes of its pulse-width range. Specifications are usually given in the range of pulse widths where performance is good, which can be misleading. In addition, there can be a lot of difference from manufacturer to manufacturer in the performance of monostables of the same part number. Read the data sheets carefully!

Output isolation. Finally, as with any digital device containing flip-flops, outputs should be buffered (by a gate, inverter, or perhaps an interface component like a line driver) before going through cables or to devices external to the instrument. If a device like a one-shot drives a cable directly, the load capacitance and cable reflections may cause erratic operation to occur.

General considerations for using monostables

Be careful, when using one-shots to generate a train of pulses, that an extra pulse doesn't get generated at the "ends." That is, make sure that the signals that enable the one-shot inputs don't themselves trigger a pulse. This is easy to do by looking carefully at the one-shot truth table, if you take the time.

Don't overuse one-shots. It is tempting to put them everywhere, with pulses running all over the place. Circuits with lots of one-shots are the mark of the neophyte designer. Besides the sort of problems just mentioned, you have the added complication that a circuit full of monostables doesn't allow much adjustment of the clock rate, since all the time delays are "tuned" to make things happen in the right order. In many cases there is a way to accomplish the same job without a one-shot, and that is to be preferred. Figure 8.67 shows an example.

The idea is to generate a pulse and then a second delayed pulse following the falling edge of an input signal. These might be used to set up and initiate operations that require that some previous operation be completed, as signaled by the input falling edge. Since the rest of the circuit is probably controlled by a "clock" square wave, let's assume that the signal at the D input falls synchronous with a clock rising edge. In the first circuit the input triggers the first one-shot, which then triggers the second one-shot at the end of its pulse.

The second circuit does the same thing with type D flip-flops, generating output pulses with width equal to one clock cycle. This is a synchronous circuit, as opposed to the asynchronous circuit using cascaded flip-flops. The use of synchronous methods is generally preferable from several standpoints, including noise immunity. If you wanted to generate shorter pulses, you could use the same kind of circuit, with the system clock divided down (via several toggling flip-flops) from a master clock of higher frequency. The master clock would then be used to clock the D flip-flops in this circuit. The use of several subdivided

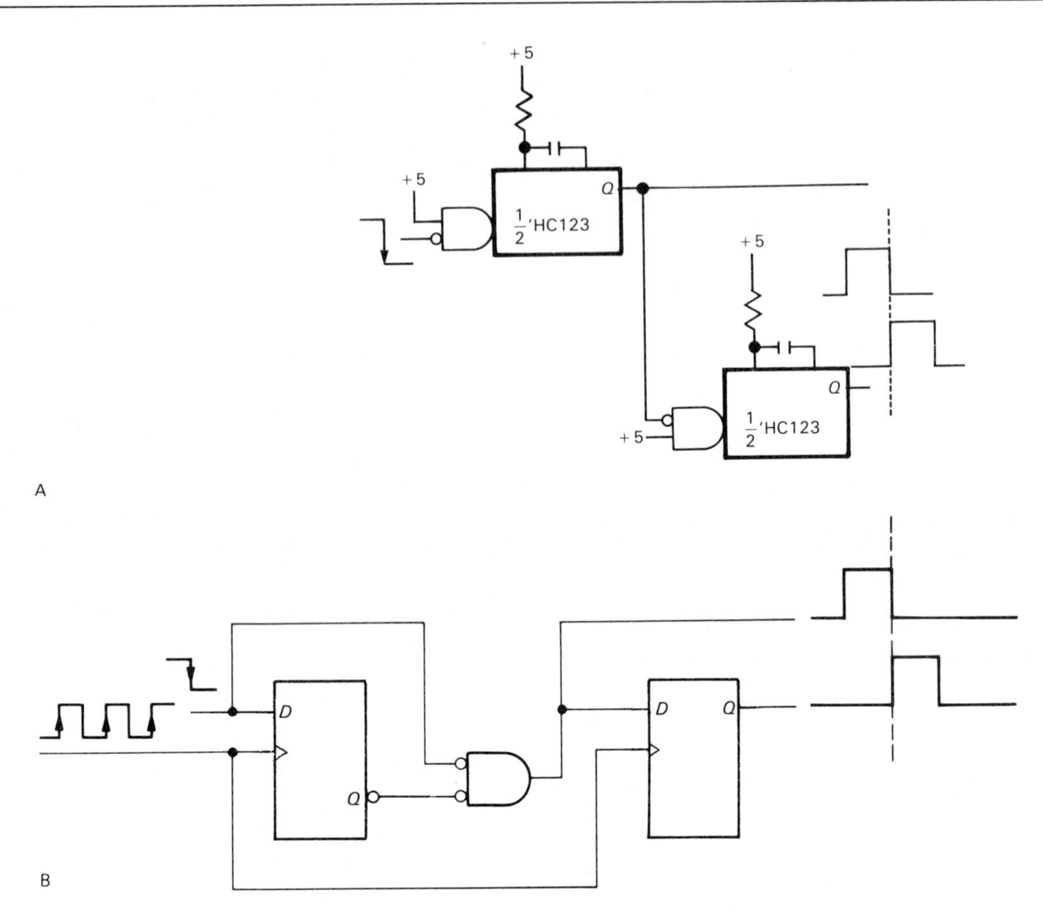

Figure 8.67. A digital delay can replace one-shot delays.

system clocks is common in synchronous circuits.

8.23 Timing with counters

As we have just emphasized, there are many good reasons for avoiding the use of monostables in logic design. Figure 8.68 shows another case where flip-flops and counters (cascaded toggling flip-flops) can be used in place of a monostable to generate a long output pulse. The '4060 is a 14-stage CMOS binary counter (14 cascaded flip-flops). A rising edge at the input brings Q HIGH, enabling the counter. After 2^{n-1} clock pulses, Q_n goes HIGH, clearing the flip-flop and the counter. This circuit generates an accurate long pulse whose length may be varied by factors of 2. The '4060 also includes internal oscillator circuitry that can substitute for the external clock reference. Our experience is that the internal oscillator has poor frequency tolerance and (in some HC versions) may malfunction.

You can get complete integrated circuits to implement timing with counters. The ICM7240/50/60 (Intersil, Maxim) have 8-bit or 2-digit internal counters and the necessary logic to make delays equal to an integral number of counts (1–255 or 1–99 counts); you can set the number either with "hardwired" connections or with external thumbwheel switches. The ICM7242 is similar, but with prewired divide-by-128 counter. Exar makes a close cousin, called the XR2243, which has a fixed divide-by-1024 counter.

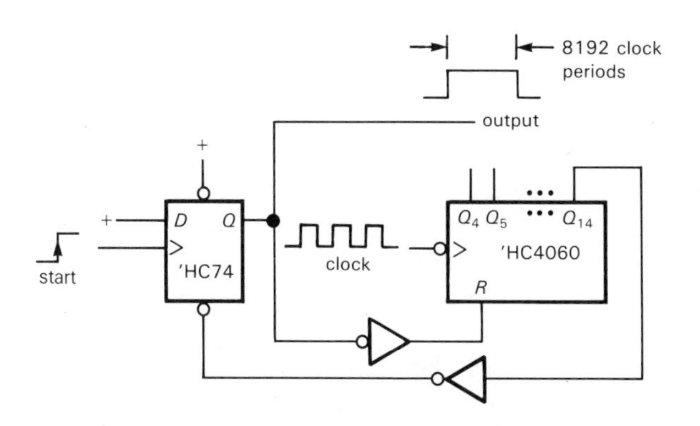

Figure 8.68. Digital generation of long pulses.

SEQUENTIAL FUNCTIONS AVAILABLE AS ICs

As with the combinational functions we described earlier, it is possible to integrate various combinations of flip-flops and gates onto a single chip. In the following sections we will present a survey of the most useful types, listed according to function.

As with pure combinational logic, *programmable* logic (PALs and GALs in particular) provides an attractive alternative to the use of prewired sequential functions. We'll talk about them, also, after looking at the standard functions.

8.24 Latches and registers

Latches and registers are used to "hold" a set of bits, even if the inputs change. A set of D flip-flops constitutes a register, but it has more inputs and outputs than necessary. Since you don't need separate clocks, or SET and CLEAR inputs, those lines can be tied together, requiring fewer pins and therefore allowing 8 flip-flops to fit in a 20-pin package. The popular '574 is an octal D register with positive clock edge and three-state outputs; the '273 is similar, but has a reset instead of three-state outputs. Figure 8.69 shows a quad D register with both true and complemented outputs.

The term "latch" is usually reserved for a special kind of register: one in which the outputs follow the inputs when enabled, and hold the last value when disabled. Since the term "latch" has become ambiguous with use, the terms "transparent latch" and "type D register" are often used to distinguish these closely related devices. As an example, the '573 is the octal transparent-latch equivalent of the '574 D register.

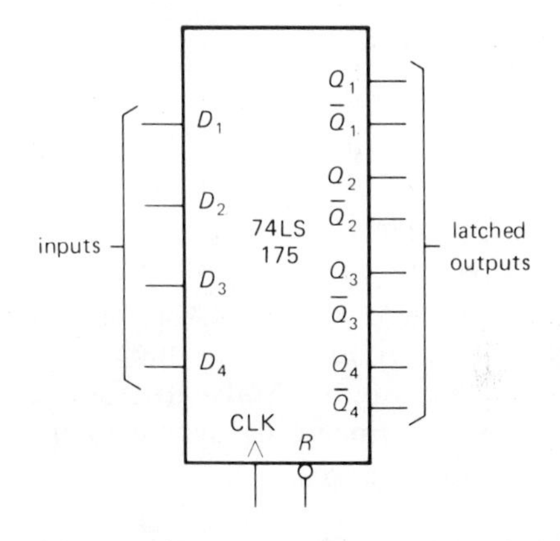

Figure 8.69. '175 4-bit D register.

Some variations on the latch/register are as follows: (a) random-access memories (RAMs), which let you write to, and read from, a (usually large) set of registers, but only one (or at most a few) at a time; RAMs come in sizes from a handful of bytes up to 1M bytes or more and are used primarily for memory in microprocessor systems (see Chapters 10 and 11); (b) addressable latches, a multibit latch that lets you update individual bits while keeping the others unchanged; (c) a latch or register built into a larger chip, for example a digital-to-analog converter; such a device only needs the input applied momentarily (with appropriate clocking edge), since an internal register can hold the data.

Table 8.9 at the end of the chapter lists most of the useful registers and latches. Note features such as input enable, reset, three-state outputs, and "broadside" pinout (inputs on one side of the chip, outputs on the other); the latter is very convenient when you are laying out a printed-circuit board.

8.25 Counters

As we mentioned earlier, it is possible to make a "counter" by connecting flip-flops together. There is available an amazing variety of such devices as single chips. Here are some of the features to look for:

Size

You can get BCD (divide-by-10) and binary (or *hexadecimal*, divide-by-16) counters in the popular 4-bit category. There are larger counters, up to 24 bits (not all available as outputs), and there are *modulo-n* counters that divide by an integer n, specified as an input. You can always cascade counters (including synchronous types) to get more stages.

Clocking

An important distinction is whether the counter is a "ripple" counter or a "synchronous" counter. The latter clocks all flip-flops simultaneously, whereas in a ripple counter each stage is clocked by the output of the previous stage. Ripple counters generate transient states, since the earlier stages toggle slightly before the later ones. For instance, a ripple counter going from a count of 7 (0111) to 8 (1000) goes through the states 6, 4, and 0 along the way. This doesn't cause trouble in well-designed circuits, but it would in a circuit that used gates to look for a particular state (this is a good place to use something like a D flip-flop, so that the state is examined only at the clock edge). Ripple counters are slower than synchronous counters, because of the accumulated propagation delays. Ripple counters clock on negative-going edges for easy expandability (by connecting the Q output of one counter directly to the clock input of the next); synchronous counters clock on the positive edge.

We favor the '160–'163 family of 4-bit synchronous counters for most applications that don't require some special feature. The '590 and '592 are good 8-bit synchronous counters. Figure 8.70 shows the '390 dual BCD ripple counter.

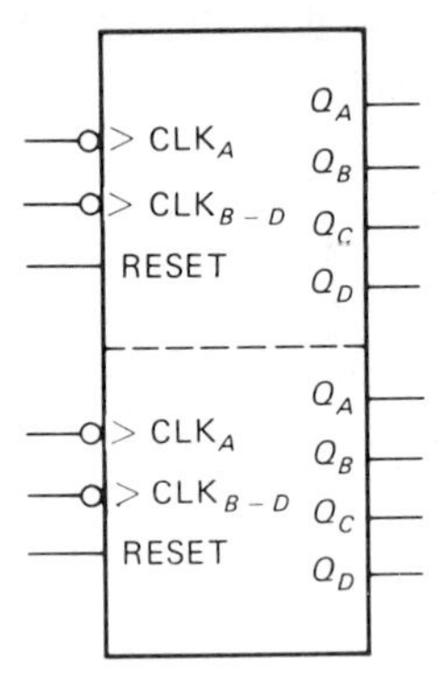

Figure 8.70. '390 dual BCD ripple counter.

Up/down

Some counters can count in either direction, under control of some inputs. The two possibilities are (a) an U/D′ input that sets the direction of count and (b) a pair of clocking inputs, one for UP, one for DOWN. Examples are the '191 and '193, respectively. The '569 and '579 are useful 8-bit up/down counters.

Load and clear

Most counters have data inputs so that they can be preset to a given count. This is handy if you want to make a modulo-n counter, for example. The load function can be either synchronous or asynchronous: the '160–'163 have synchronous load, which means that data on the input lines are transferred to the counter coincident with the next clock edge, if the LOAD′ line is also asserted LOW; the '190–'193 are asynchronous, or *jam*-load, which means that input data are transferred to the counter when LOAD′ is asserted, independent of the clock. The term "parallel load" is sometimes used, since all bits are loaded at the same time.

The CLEAR (or RESET) function is a form of presetting. The majority of counters have a jam-type CLEAR function, though some have synchronous CLEAR; for example, the '160/161 are jam CLEAR, while the '162/163 are synchronous CLEAR.

Other counter features

Some counters feature latches on the output lines; these are always of the transparent type, so the counter can be used as if no latch were present. (Keep in mind that any counter with parallel-load inputs can function as latch, but you can't count at the same time as data are held, as you can with a counter/latch chip.) The combination of counter plus latch is sometimes very convenient, e.g., if you want to display or output the previous count while beginning a new counting cycle. In a frequency counter this would allow a stable display, with updating after each counting cycle, rather than a display that repeatedly gets reset to zero and then counts up.

There are counters with three-state outputs. These are great for applications where the digits (or 4-bit groups) are multiplexed onto a bus for display or transfer to some other device. An example is the '779, an 8-bit synchronous binary counter whose three-state outputs also serve as parallel inputs; by sharing input/output lines, the counter fits in a 16-pin package. The '593 is similar, but in a 20-pin package.

If you want a counter to use with a display, there are several that combine counter, latch, 7-segment decoder, and driver on one chip. An example is the 74C925–74C928 series of 4-digit counters. Another amusing chip is the TIL306/7, a counter *with display* on one chip: You just look at the IC, which lights up with a digit telling the count! Figure 8.71 shows a nice LSI (large-scale-integration) counter circuit that doesn't require a lot of support circuits.

Table 8.10 at the end of the chapter lists most of the counter chips that you might want to use. Many of them are only available in one family (e.g., LS or F), so be sure to check the data books before you design with them.

8.26 Shift registers

If you connect a series of flip-flops so that each Q output drives the next D input, and all clock inputs are driven simultaneously, you get what's called a "shift register." At each clock pulse the pattern of 0's and 1's in the register shifts to the right, with the data at the first D input entering from the left. As with flip-flops, the data present at the serial input just prior to the clock pulse are entered, and there is the usual propagation delay to the outputs. Thus they may be cascaded without fear of a logic race. Shift registers are very useful for conversion of parallel data (n bits present simultaneously, on n separate lines) to serial data (one bit after another, on a single data line), and vice versa. They're also handy as memories, particularly if the data are always read and written in order. As with counters and latches, shift registers come in a pleasant variety of prefab styles. The important things to look for are the following:

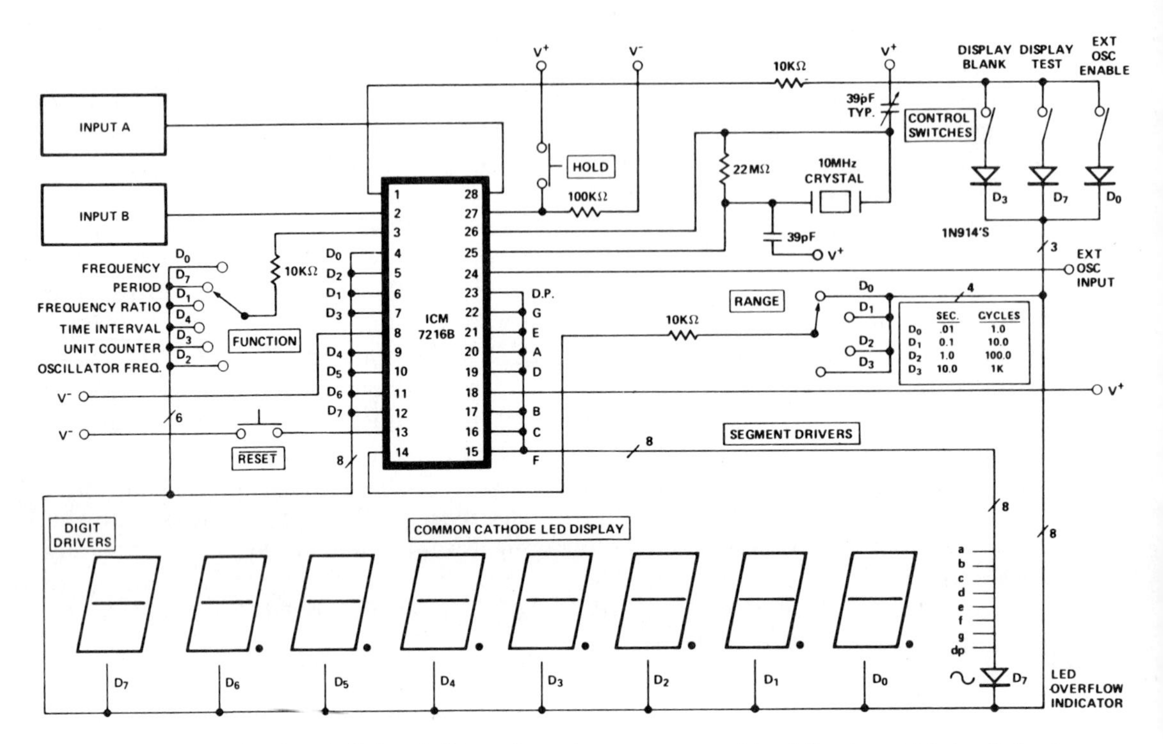

Figure 8.71. Intersil 7216 8-digit 10MHz universal counter on a chip. (Courtesy of Intersil, Inc.)

Size

The 4-bit and 8-bit registers are standard, with some larger sizes available (up to 64 bits or more). There are even variable-length registers (e.g., the 4557: 1 to 64 stages, set by a 6-bit input).

Organization

Shift registers are usually 1 bit wide, but there are also dual-, quad-, and hex-width registers. Most shift registers only shift right, but there are bidirectional registers like the '194 and '323 that have a "direction" input (Fig. 8.72). Watch out for trickery like the "bidirectional" '95, which can shift left only by tying each output bit to the previous input, then doing a parallel load.

Inputs and outputs

Small shift registers can provide parallel inputs or outputs, and usually do; an example is the '395, a 4-bit parallel-in, parallel-out (PI/PO) shift register with three-state outputs. Larger registers may only provide *serial* input or output, i.e., only the input to the first flip-flop or the output from the last is accessible. In some cases a few selected intermediate taps are provided. One way to provide both parallel input and output in a small package is to share input and output (three-state) on the same pins, e.g., the '299, an 8-bit bidirectional PI/PO register in a 20-pin package. Some shift registers include a latch at the input or output, so shifting can go on while data are being loaded or unloaded.

As with counters, parallel LOAD and

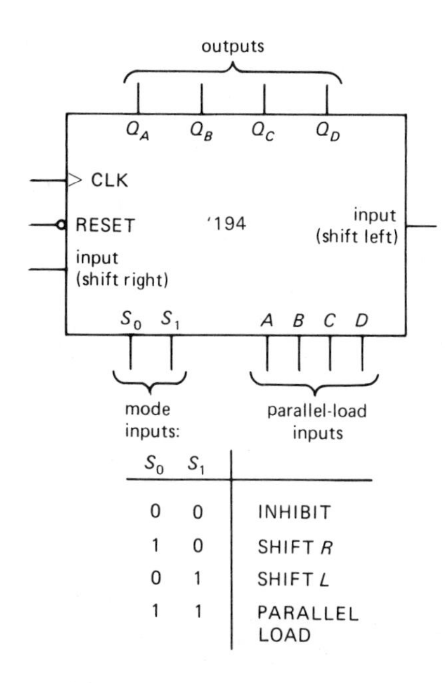

Figure 8.72. '194 4-bit bidirectional shift register.

CLEAR can be either synchronous or jam-load; for example, the '323 is the same as the '299, but with synchronous clear.

Table 8.11 at the end of the chapter lists the shift registers you're likely to use. As always, not all types are available in all logic families; be sure to check the data books.

RAMs as shift registers

A random-access memory can always be used as a shift register (but not vice versa) by using an external counter to generate successive addresses. Figure 8.73 shows the idea. An 8-bit synchronous up/down counter generates successive addresses for a 256-word×4-bit CMOS RAM. The combination behaves like a quad 256-bit shift register, with left/right direction of shift selected by the counter's UP/DOWN′ control line. The other inputs of the counter are shown enabled for counting. By choosing a fast counter and memory, we were able to achieve a maximum clocking rate of 30MHz (see timing diagram), which is the same as that of an integrated (but much smaller) HC-type shift register. This technique can be used to produce very large shift registers, if desired.

EXERCISE 8.28

In the circuit of Figure 8.73, input data seem to go into the same location that output data are read from. Nevertheless, the circuit behaves identically to a classic 256-word shift register. Explain why.

8.27 Sequential PALs

The combinational (gates-only) PALs we talked about in Section 8.15 belong to a larger family that includes devices with various numbers of on-chip D-type registers (called "registered PALs"). Typical of these PALs is the 16R8, shown in Figure 8.74. The programmable-AND/fixed-OR array typical of combinational PALs generates the input levels for 8 synchronously clocked D-type registers with three-state outputs; the register outputs (and their inverts) are available, along with the standard input pins, as inputs to the logic array. If you look back at Figure 8.57, you'll see that a registered PAL is a general-purpose sequential circuit element; within limits set by the number of registers and gates available, you can construct just about anything you want. For instance, you could make a shift register or counter, or some of both! In practice, you're more likely to make some custom piece of logic that is part of a larger circuit, for which the alternative is "discrete" logic built with gates and flip-flops. Let's look at some examples.

Hand-generated fuse maps

Simple designs can be implemented in PALs by figuring out the logic, then burning the appropriate pattern into the fuse array with a "PAL programmer." As an

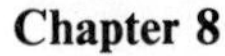

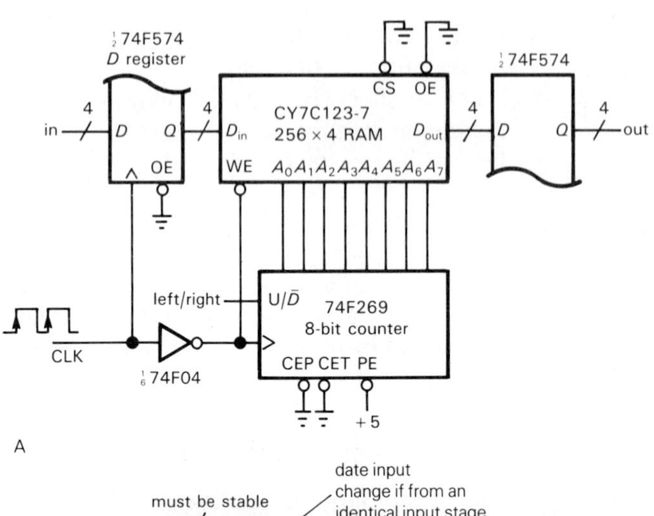

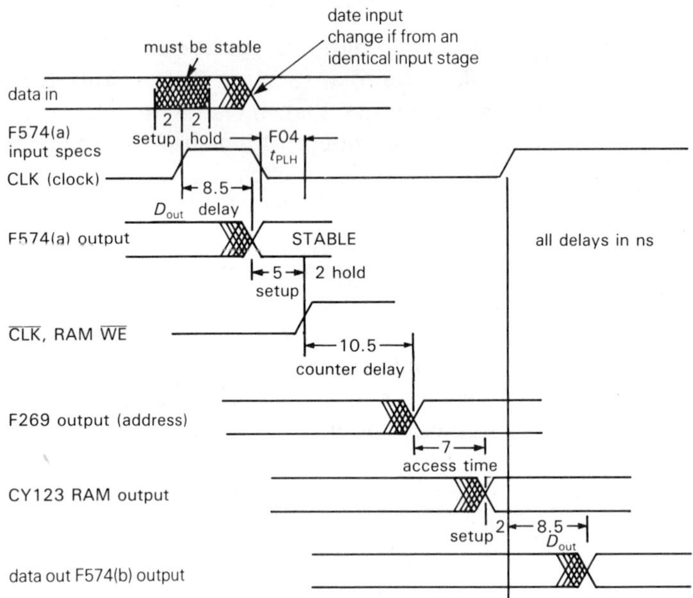

F04 t_{PLH} delay	3.7ns	
F374 D_{out} delay	8.5ns	} 13.5 − 3.7 = 9.8ns HI
CYC123 setup time	5	
F269 counter out delay	10.5	
CYC123 access time	7	} 19.5 + 3.7 = 23.2ns LO
F374 D_{IN} setup time	2	
shift clock period(min)	33 ns	

C

Figure 8.73. A. Large shift register made from RAM plus counter; the slash indicates multiple lines, in this case a 4-bit-wide data path.
B. Timing diagram to calculate maximum clocking rate, assuming worst-case timing specifications.
C. Calculation showing worst-case sequence of delays in a single clock cycle.

example, let's imagine we want a 4-input multiplexer with latched outputs. We can write the logic equation for the multiplexer portion (i.e., the D-input to the flip-flop) as

$$Q.d = I_0 * S_0' * S_1' + I_1 * S_0 * S_1' + I_2 * S_0' * S_1 + I_3 * S_0 * S_1$$

where the inputs S_0 and S_1 address the selected input I_0–I_3 and "*" and "+" represent AND and OR. A registered PAL makes it easy to latch the result. Note that we have used the OR of 3-input ANDs, rather than first decoding the select address in 2-input ANDs, because we

Logic Diagram 16R8

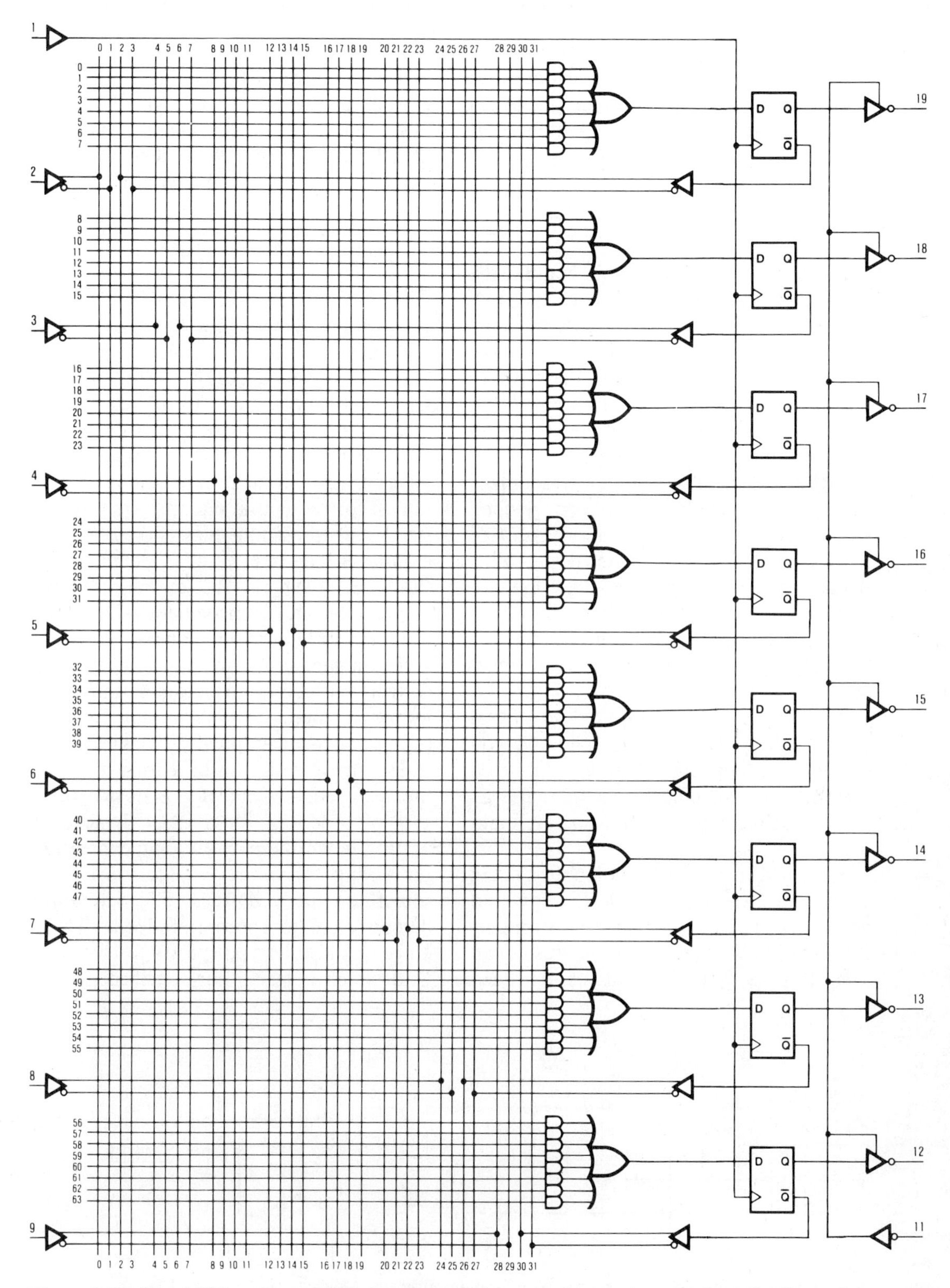

Figure 8.74. The 16R8 registered PAL has 8 external inputs, 8 outputs, a clock, and a three-state control line. The register outputs are also available as input to the AND array. (Diagram courtesy of Advanced Mircro Devices of Sunnyvale, California.)

are constrained to use a sum of products (it's also faster). The finished design is shown in Figure 8.75. (Note: There's a subtlety to this circuit; see "Glitches" at the end of this section.)

PALASM

For circuits of any reasonable complexity, some PAL logic design aids are needed. For example, a 16L8 PAL has 2048 fuses; a complex design might require you to blow a few hundred of them, and unless you're unusually compulsive you're unlikely to succeed in manually specifying them all correctly to a PAL programmer.

PALASM (PAL Assembler; trademark of Monolithic Memories Inc.), developed by Monolithic Memories (the inventor of the PAL), was one of the first design aids. It takes Boolean expressions, like the one we wrote above, and converts them to fuse maps. No logic minimization is performed by the program, so you have to do most of the hard work yourself. However, PALASM does let you input a set of test states (called "test vectors"), giving you back the outputs that would result from your Boolean specification. In that way you can debug your equations before making the PAL.

PALASM is widely available. There are FORTRAN source code listings, and versions to run on popular microcomputers, which can then send the fuse map to a PAL programmer (in standardized "JEDEC" format) via a serial port. Many PAL programmers with built-in microprocessors include a resident PALASM; some examples are Data I/O, Digilec, Stag, and Structured Design. With these you just hook up a terminal and you're in business.

ABEL and CUPL

PALASM is a help, but the serious PAL user needs more. High-level programmable-logic languages like ABEL (from Data I/O) and CUPL (from Logical Devices, Inc.) make programming PALs (and PLAs) easy work. They let you specify logic either by Boolean expressions or truth tables; for sequential circuits you give the states and transition rules. Like any good high-level language, you can define arrays (for a set of signals, e.g., an address bus), expressions, and intermediate values, then use them in later expressions.

These languages are smart enough to convert truth tables to logic equations, then minimize them (as well as any Boolean expressions you supplied) via logical identities, finally resulting in a form that fits the logic constraints of the device (e.g., sum of products, for a PAL). Instead of writing down the explicit logic expression for a range of values, you can just write something like ADDR:[10..FF], which will be converted to appropriate logic. These languages also let you specify test vectors, with which it tests your specified design;

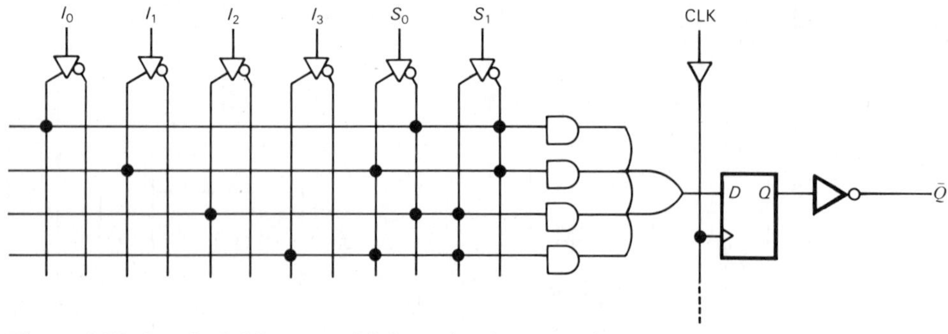

Figure 8.75. Latched 4-input multiplexer implemented in a PAL.

the test vectors can also be sent to the programmer to test the actual programmed chip. Finally, these languages provide standardized documentation of the finished chip, which is essential when you've got to debug a gadget with these mystery devices in it.

Just to make these ideas concrete, let's look at both combinational and sequential design examples using CUPL.

CUPL example: 7-segment-to-hex encoder (combinational). There are times when you might like to use an available LSI chip that performs a convenient function (e.g., a stopwatch or calculator) as part of an instrument you're building. The trouble is that these LSI chips usually provide outputs to drive 7-segment displays directly, rather than the hexadecimal (or BCD) outputs that you want. Therefore let's design an "encoder" chip that converts from 7-segment back to 4-bit binary, a function that is not available as a standard IC (although there is a 7-segment-to-BCD encoder, the 74C915).

The inputs are the individual segment signals, which are always labeled a–g; see Figure 8.76, which also shows how the digits A–F are represented with 7-segment displays. Note that it is possible to represent 9 and C in two ways, both of which should be correctly recognized by our logic. For the PAL, we choose a 16L8,

Figure 8.76. 7-segment display codes.

a 20-pin combinational part whose logic was shown in Figure 8.45.

Figure 8.77 is the input to CUPL. We assigned the (positive-true) segment drive signals a–g as inputs, and the (negative-true) hexadecimal bits D0–D3 as outputs. CUPL lets you define intermediate variables that can be used in later expressions; in this case it is convenient to define the obvious variables *zero* through *hexf*, the possible displayed digits in terms of the segment inputs. These are simply large product (AND) terms of the input segment variables, which you can read from the digit shapes in Figure 8.76. Finally, each binary output bit is written as the sum (OR) of the digit variables in which that bit is set. We've used negative-true levels because the 16L8 is an AND-NOR array, thus minimizing logic. This completes the logic specification to CUPL.

EXERCISE 8.29

Check for yourself that we did our work right by drawing some of the displayed characters as specified by our intermediate variables *zero* through *hexf*.

CUPL first uses the intermediate variable definitions to write the D0–D3 expressions directly in terms of the input variables a–g, a job that an assembler like PALASM would force us to do initially. At this point the logic equations are in the desirable AND-NOR form. However, we're not finished yet, because the 16L8 (and all other combinational PALs) permit at most 7 product terms in each sum, whereas we have 9, 8, 9, and 10, respectively, for the outputs D0–D3. One solution would be to string each output through a second OR gate, in order to get enough product terms in the sum; this is generally considered poor form because it doubles the propagation delay, though it wouldn't matter in a slow application like this. The better solution is to perform a logic minimization, using logic identities, DeMorgan's formula, etc.

```
/**  Inputs  **/

PIN 1   = a   ;     /*  segment a  */
PIN 2   = b   ;     /*  segment b  */
PIN 3   = c   ;     /*  segment c  */
PIN 4   = d   ;     /*  segment d  */
PIN 5   = e   ;     /*  segment e  */
PIN 6   = f   ;     /*  segment f  */
PIN 7   = g   ;     /*  segment g  */


/**  Outputs  **/

PIN 19  = !D3   ;  /* msb of hex encode     */
PIN 18  = !D2   ;  /*                       */
PIN 17  = !D1   ;  /*                       */
PIN 16  = !D0   ;  /* lsb                   */


/** Declarations and Intermediate Variable Definitions **/

zero = a & b & c & d & e & f & !g ;
one = !a & b & c & !d & !e & !f & !g ;
two = a & b & !c & d & e & !f & g ;
three = a & b & c & d & !e & !f & g ;
four = !a & b & c & !d & !e & f & g ;
five = a & !b & c & d & !e & f & g ;
six = a & !b & c & d & e & f & g ;
seven =  a & b & c & !d & !e & !f & !g ;
eight = a & b & c & d & e & f & g ;
nine = a & b & c & !d & !e & f & g
     # a & b & c & d & !e & f & g ;     /* two ways */
hexa = a & b & c & !d & e & f & g ;
hexb = !a & !b & c & d & e & f & g ;
hexc = !a & ! b & !c & d & e & !f & g
     # a & !b & !c & d & e & f & !g ;   /* two ways */
hexd = !a & b & c & d & e & !f & g ;
hexe = a & !b & !c & d & e & f & g ;
hexf = a & !b & !c & !d & e & f & g ;


/**  Logic Equations  **/

D3 = eight # nine # hexa # hexb # hexc # hexd # hexe # hexf ;
D2 = four # five # six # seven # hexc # hexd # hexe # hexf ;
D1 = two # three # six # seven # hexa # hexb # hexe # hexf ;
D0 = one # three # five # seven # nine # hexb # hexd # hexf ;
```

Figure 8.77. 7-segment-to-hex encoder specification, written in CUPL language.

We ran CUPL's minimizer, producing the product terms shown in Figure 8.78. Luckily, all fit within the 7-product constraint. CUPL also draws a fuse map for you (Fig. 8.79). Of course, you don't program the PAL from that, but use instead a direct download of the universal JEDEC programming format. In this example, CUPL has obviously made a laborious problem simple.

```
**  Expanded Product Terms **

D0 =>
     a & b & c & d & !e & g
   # a & b & c & !e & f & g
   # a & c & d & !e & f & g
   # b & c & !d & !e & !f & !g
   # !a & !b & c & d & e & f & g
   # a & !b & !c & !d & e & f & g
   # !a & b & c & d & e & !f & g

D1 =>
     a & !b & !c & e & f & g
   # !b & c & d & e & f & g
   # a & b & c & !d & e & f & g
   # a & b & !c & d & e & !f & g
   # a & b & c & d & !e & !f & g
   # a & b & c & !d & !e & !f & !g

D2 =>
     a & !b & !c & d & e & f
   # a & !b & c & d & f & g
   # a & !b & !c & e & f & g
   # !a & b & c & !d & !e & f & g
   # !a & b & c & d & e & !f & g
   # !a & !b & !c & d & e & !f & g
   # a & b & c & !d & !e & !f & !g

D3 =>
     a & b & c & f & g
   # a & !b & !c & d & e & f
   # a & !b & !c & e & f & g
   # !a & !b & c & d & e & f & g
   # !a & b & c & d & e & !f & g
   # !a & !b & !c & d & e & !f & g

D0.oe  => 1
D1.oe  => 1
D2.oe  => 1
D3.oe  => 1
```

Figure 8.78. 7-segment-to-hex encoder: minimized product terms.

CUPL example: vending machine (sequential). An arbitrary state machine (Section 8.18) has a set of *states* and a set of *transition rules* for moving between those states at each clock edge. In general, the transition rules depend both on the present state and on the particular combination of input levels present at the next clock edge. You can implement a state machine in programmable logic containing registers if (a) there are enough register bits to represent all possible states (e.g., with 4 registers you could have up to 16 states) and (b) there are enough inputs and logic gates to implement the transition rules.

As an example, let's design a registered PAL for the state diagram of Figure 8.80. It's a vending machine, and it is supposed to disgorge a bottle of fizzy sweet liquid when 25 cents (or more) has been deposited. There is some sort of coin interface that gobbles up and recognizes money and sends to our PAL a 2-bit input (C1, C0), valid for one clock edge, indicating the coin just deposited (01 = nickel, 10 = dime, 11 = quarter, 00 = slug or no coin).

```
** Fuse Plot **

Pin #19
 0000 --------------------------------
 0032 x-x-x-----------x---x-----------
 0064 -xx--x--x---x---x---------------
 0096 -xx--x------x---x---x-----------
 0128 -x-xx---x---x---x---x-----------
 0160 x--xx---x---x----x--x-----------
 0192 -x-x-x--x---x----x--x-----------
 0224 xxxxxxxxxxxxxxxxxxxxxxxxxxxxxxxx

Pin #18
 0256 --------------------------------
 0288 -xx--x--x---x---x---------------
 0320 -xx-x---x-------x---x-----------
 0352 -xx--x------x---x---x-----------
 0384 x--xx----x---x--x---x-----------
 0416 x--xx---x---x----x--x-----------
 0448 -x-x-x--x---x----x--x-----------
 0480 x-x-x----x---x---x---x----------

Pin #17
 0512 --------------------------------
 0544 -xx--x------x---x---x-----------
 0576 -x--x---x---x---x---x-----------
 0608 x-x-x----x--x---x---x-----------
 0640 x-x--x--x---x----x--x-----------
 0672 x-x-x---x----x---x--x-----------
 0704 x-x-x----x---x---x---x----------
 0736 xxxxxxxxxxxxxxxxxxxxxxxxxxxxxxxx

Pin #16
 0768 --------------------------------
 0800 x-x-x---x----x------x-----------
 0832 x-x-x--------x--x---x-----------
 0864 --x-x---x----x--x---x-----------
 0896 x---x----x---x---x---x----------
 0928 -x-xx---x---x---x---x-----------
 0960 -xx--x---x--x---x---x-----------
 0992 x--xx---x---x----x--x-----------

LEGEND     X : fuse not blown
           - : fuse blown
```

Figure 8.79. 7-segment-to-hex encoder: fuse map.

The state machine's job is to add up the total deposited and generate an output called *bottle* when there's enough money.

Figure 8.81 shows the specification, in CUPL's state-machine syntax. As before, we begin by defining input and output pins. Note that we've added a *reset* input so that you can initialize to the state S0 (no money). Next we define the states, then the rules for moving between them. If any outputs, either registered or combinational, need to be generated during states or transitions between states, they are specified at the same time. In this example, for instance, the output *bottle* has been specified as a separate output register, so that no output state decoding is needed. In fact, this is the only output needed, and the state-machine bits Q0–2 could by implemented in internal registers that don't generate outputs directly; some programmable logic devices have such

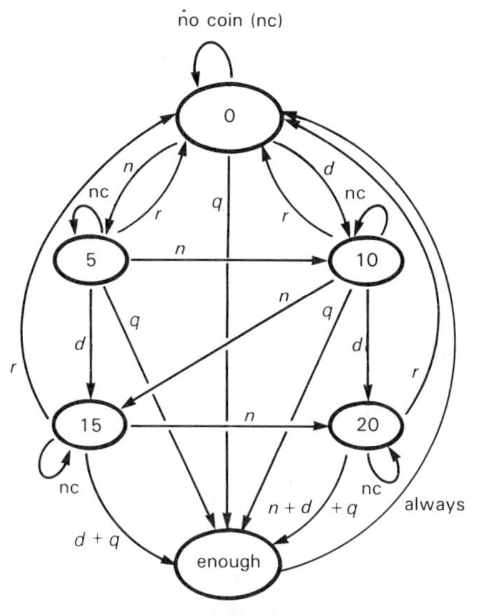

Figure 8.80. Vending machine state diagram.

"buried" registers, in addition to the usual output registers.

Note that you have to specify explicitly the transition from a state to itself, as we have done here for the input *nocoin.* An unspecified condition implicitly resets the state to all zeros. That is because these conditionals are compiled into combinational logic to assert the *D* inputs of the registers, and thus if the condition is not met, the corresponding *D* input is not asserted.

Figure 8.82 shows the output from CUPL. There is nothing obvious or simple about the logic, because both the machine state (S0–S5) and the input variable (C0–1) are specified as binary numbers, whereas the logic operates on individual

```
/**  Inputs  **/

Pin  1  = clk    ;  /* clock -- positive edge  */
Pin  3  = c0     ;  /* coin type -- low bit    */
Pin  4  = c1     ;  /* coin type -- high bit   */
Pin  6  = reset  ;  /* reset input             */


/**  Outputs  **/

Pin  18  = !Q0     ;  /* bit 0 of state variable    */
Pin  17  = !Q1     ;  /* bit 1                      */
Pin  16  = !Q2     ;  /* bit 2                      */
Pin  15  = !bottle ;  /* bottle disgorge command    */

/* Define machine states with symbolic names;
     "enough" = 25 cents or more */

$define S0      'b'000
$define S5      'b'001
$define S10     'b'010
$define S15     'b'011
$define S20     'b'100
$define ENOUGH 'b'101

/* define intermediate variables */

nocoin = !c0 & !c1 & !reset;
nickel =  c0 & !c1 & !reset;
dime =   !c0 &  c1 & !reset;
quarter = c0 &  c1 & !reset;

/* Define state bit variable field */

field statebit = [Q2..0] ;
```

Figure 8.81. Vending machine specification (CUPL). (*Continued on next page.*)

```
/* Transition rules for vending machine */

sequence statebit {
   present S0    if nocoin     next S0;
                 if nickel     next S5;
                 if dime       next S10;
                 if quarter    next ENOUGH    out bottle;

   present S5    if nocoin     next S5;
                 if nickel     next S10;
                 if dime       next S15;
                 if quarter    next ENOUGH    out bottle;

   present S10   if nocoin     next S10;
                 if nickel     next S15;
                 if dime       next S20;
                 if quarter    next ENOUGH    out bottle;

   present S15   if nocoin     next S15;
                 if nickel     next S20;
                 if dime       next ENOUGH    out bottle;
                 if quarter    next ENOUGH    out bottle;

   present S20   if nocoin     next S20;
                 if nickel     next ENOUGH    out bottle;
                 if dime       next ENOUGH    out bottlc;
                 if quarter    next ENOUGH    out bottle;

   present ENOUGH              next S0;    }
```

Figure 8.81. (*cont.*)

bits. Thus, the resulting logic does not bear much relation to the original state description (Fig. 8.81). In fact, the particular choice of states (ascending binary, 0–5) could have been chosen differently, completely changing the resulting logic. In this case, this example fits easily within the constraints of the 16R6 PAL (8 product terms per register); if it had not, we could have tried redefining the states, which often leads to a simpler logic realization. Note that the *reset* input acts by an overriding disassertion of all *D* inputs, which we forced by our definition of the intermediate variables *nocoin*, *nickel*, etc.

EXERCISE 8.30

Verify that the finished logic equations are correct, by checking a few transition rules. You might try all the transitions from 00, or a nickel or dime from some other state.

EXERCISE 8.31

Good vending machines give change. Redraw the state diagram (Fig. 8.80) so that there are states (how many?) for each possible amount of change. Modify the transition rules accordingly. Make sure your modified vending machine still does its primary job – dispensing bottles.

EXERCISE 8.32

Draw a state diagram and transition rules for an electronic combination lock: It should open only if four digits are entered in the correct order. Any error should reset it.

Hints for using programmable logic devices (PLDs)

For certain jobs, PLDs really hit the spot. Here are the most important uses and advantages of PLDs:

```
** Expanded Product Terms **

Q0.d  =>
    !Q0 & !Q1 & c0 & !reset
  # !Q0 & !Q2 & c0 & !reset
  # Q0 & !Q2 & !c0 & !reset
  # Q0 & !Q2 & c1 & !reset
  # !Q0 & !Q1 & Q2 & c1 & !reset

Q1.d  =>
    !Q1 & !Q2 & !c0 & c1 & !reset
  # !Q0 & Q1 & !Q2 & !c1 & !reset
  # Q1 & !Q2 & !c0 & !c1 & !reset
  # Q0 & !Q1 & !Q2 & c0 & !c1 & !reset

Q2.d  =>
    !Q0 & !Q1 & Q2 & !reset
  # Q1 & !Q2 & c1 & !reset
  # !Q2 & c0 & c1 & !reset
  # Q0 & Q1 & !Q2 & c0 & !reset

bottle.d  =>
    !Q2 & c0 & c1 & !reset
  # !Q0 & !Q1 & Q2 & c0 & !reset
  # !Q0 & !Q1 & Q2 & c1 & !reset
  # Q0 & Q1 & !Q2 & c1 & !reset

statebit =>
    Q2 , Q1 , Q0
```

Figure 8.82. Vending machine: CUPL output.

State machines. As in the foregoing example, the programmable logic device is a natural for an arbitrary synchronous state machine. You would be foolish to use an array of D flip-flops and discrete combinational logic when a PAL does the job in one inexpensive and powerful package.

Replacing "random" logic. Within many circuits you find little knots and tangles of gates, inverters, and flip-flops, scornfully referred to as *random logic*, or *glue*. A PLD will generally cut the package count by a factor of 4 or more. It also generally results in cleaner design, because the enormous number of gates available means that you can do all your gating at the *inputs* to the registers (resulting in strictly synchronous outputs), instead of the gate-conserving method of also combining register outputs via gates. With the latter the outputs are not strictly synchronous; Figure 8.83.

Flexibility. Sometimes you're not quite sure how you ultimately want some circuit to work, yet you must finish the design so that you can play around with it. PLDs are great here, because you can substitute one with different programming at some later stage, without the rewiring you'd have to do if you had used discrete logic. With PLDs, the *circuit* is a form of software!

Multiple versions. PLDs make it possible to design a single circuit, then produce several different versions of the instrument by populating the board with different PLDs. For example, you could have a computer that could accept either 256K or 1M memory chips, with just a change of PAL.

Speed and inventory. With PLDs you can generally get the design job done more quickly (once you've learned the ropes). Furthermore, you only need to stock a few PLD types, rather than dozens of

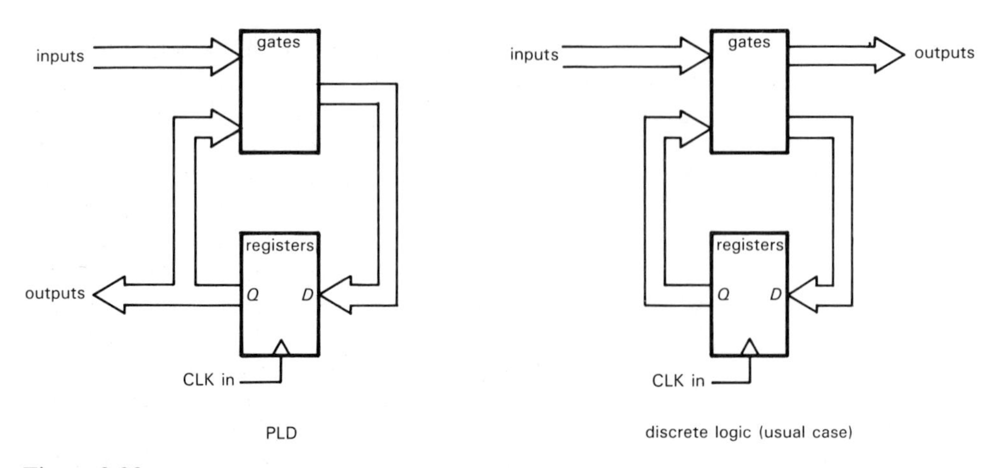

Figure 8.83

standard-function MSI logic types. In fact, just two of the newer GALs (generic array logic) mimic a whole set of PALs, by making their internal architecture (as well as the logic connections) programmable. In particular, the 20-pin GAL16V8 and 24-pin GAL20V8 can each mimic 21 standard PALs. What's more, they can be programmed as mongrel in-between PALs (e.g., an odd number of registers).

PAL loose ends

I/O pins. Three-state output pins that are internally wired as inputs to the AND array can be used as *inputs*. For example, the 16L8 shown in Figure 8.45 has 16 inputs (each true and complemented) to the AND array; 10 of these are dedicated as input pins, and 6 are fed back from three-state outputs. The latter can be converted to "permanent" inputs by disabling the corresponding output (connect a true/complement pair to its AND control); alternatively, those outputs can operate bidirectionally, by enabling the three-state driver according to some logic variables.

"Super-PALs." As we remarked earlier, more flexible programmable logic has followed the original PALs. Notable among these are erasable CMOS variants from Lattice, VTI, Altera, and others.

For example, the "generic array logic" (GAL) from Lattice uses electrically erasable programmable logic, so you can reprogram the chip. Even better, the output structure (called a "macrocell") is itself programmable – each output can be either registered or combinational, true or complemented; there's similar programmable flexibility for the three-state enable line and the feedback line (the latter can come before or after the three-state buffer, or from the adjacent output); see Figure 8.84. The result is that you can mimic any of the common 20-pin PALs with the single GAL16V8 (and any of the common 24-pin PALs with the GAL20V8). This kind of flexibility helps keep your inventory within manageable bounds.

Altera has a line of programmable CMOS logic that can be erased with UV light, just like EPROMs (the IC has a quartz window over the chip). They call their devices EPLDs, for "erasable programmable logic devices." Their smallest chip (the EP320) has output macrocells, and it mimics all 20-pin PALs, just like the GAL16V8. Furthermore, it runs at very low power, unlike the original power-hungry PALs (see below). Finally, Altera makes a number of larger EPLDs, as well as programmable microsequencers, etc. Cypress and VTI also make erasable CMOS programmable logic, complete with macrocells.

Another approach to programmable logic is typified by the Xilinx RAM-configurable gate arrays. These impressive chips contain huge blocks of configurable logic, holding the connection configuration in on-chip RAM (volatile memory). This memory gets loaded from external memory after power has been applied, either by downloading from a microprocessor or by loading itself from nonvolatile ROM-type memory.

Speed and power. The original bipolar transistor PALs introduced by Monolithic Memories (and quickly copied by National and AMD) consumed considerable current – about 200mA for the 16L8/16R8 – and had propagation delays of 40ns. The subsequent "half-power" bipolar PALs were more reasonable, running 90mA with speeds of 35ns. However, the fastest PALs still burn plenty of power; for example, AMD's 16R8D and 16R8-7 have propagation times of 10 and 7.5ns, respectively, but require 180mA (max). The CMOS devices are significantly better: Lattice's "quarter-power" GALs (GAL20V8-15Q) draw 45mA, with a delay time of 15ns, and Altera's EP320-1 delivers 25ns delays with 5mA of current. More important for low-power design, the Altera chips (and AMD's Z-series PALs) can be put into a "zero-power" (10μA typical) standby mode. The designers of future programmable logic will surely continue this healthy trend toward high speed and low power; the days of watt-guzzler programmable logic are behind us!

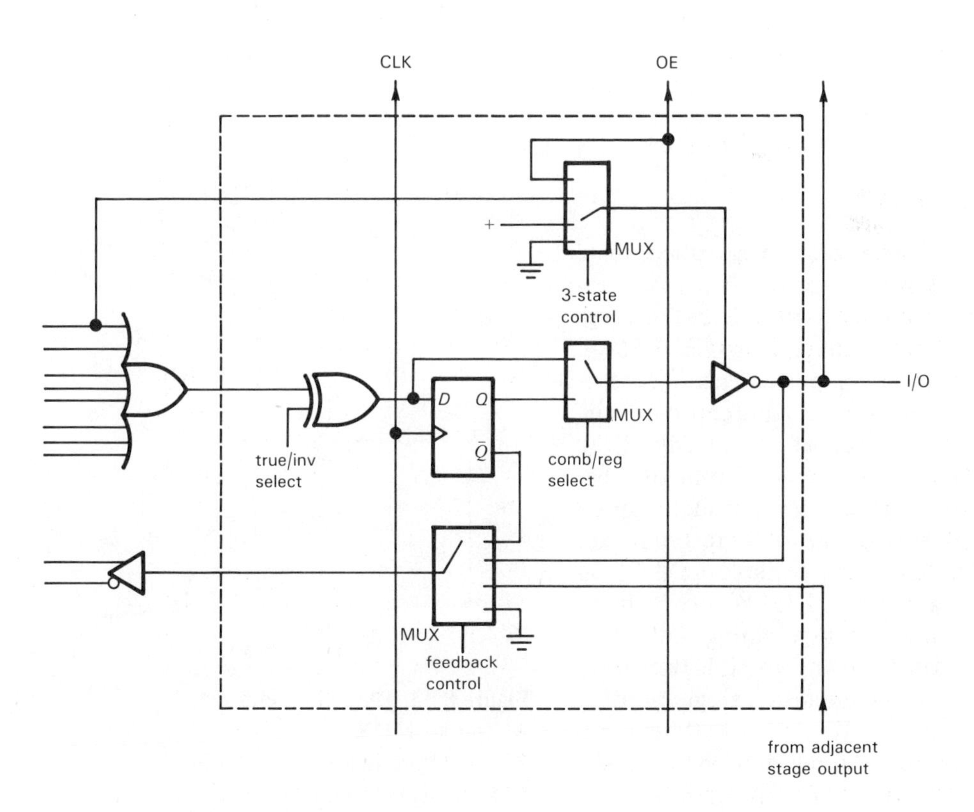

Figure 8.84. GAL programmable-output macrocell.

Glitches. Programmable logic devices are wonderful things. But you can occasionally get into trouble if you forget about the possibility of logic races. Figure 8.85 shows a 2-input multiplexer implemented in the obvious way; both the PAL diagram and the equivalent circuit are shown. It looks fine, the logic is fine, but the circuit has a flaw: If both data inputs (A and B) are HIGH, and the SELECT line changes state, the output may produce a glitch, as shown. That is because the internal gate delays in S and S' can be unequal, causing a transient state in which both AND gates have one LOW input. The solution in this case (Fig. 8.85B) is to add a redundant term, $A * B$, which you can easily prove to yourself will guarantee no output glitches.

EXERCISE 8.33

Prove that the additional term eliminates all possibility of glitches.

EXERCISE 8.34

What logic terms need to be added to the 4-input multiplexer example (Fig. 8.75) to eliminate glitches?

You can visualize this so-called logic hazard in terms of Karnaugh maps: Diagram the 2-input multiplexer of Figure 8.85A as a Karnaugh map (Fig. 8.85C). Each group in the map is one product term that forms an input to the common OR gate. The OR output is true if any of the product terms is true; but a transition between product groups can produce a glitch if the variables of the starting group are disasserted before the variables of the final group are asserted. The cure (which we used earlier) consists of adding redundant terms to ensure that any possible transition between 1's is included in a single product term; in other words, any 1's that lie in adjacent rows or columns must be enclosed by a product group. This prescription can be cast into a generalized form that applies

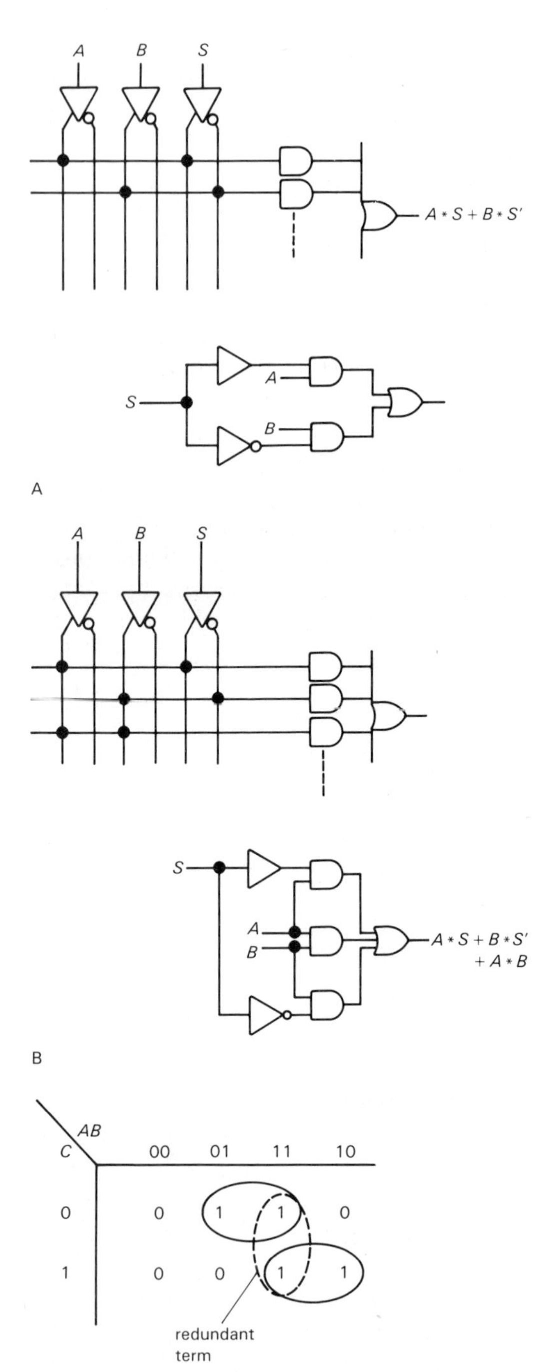

Figure 8.85. PAL glitch elimination.
A. 2-input MUX.
B. Added product term to eliminate glitch.
C. Karnaugh map showing necessary redundant term.

to the Boolean logic expression directly, rather than the Karnaugh map. That's useful for logic with more than four variables, for which Karnaugh maps become awkward.

The foregoing example is called a *static* glitch, because the output should have remained static. There are also *dynamic* glitches, in which an output that should make a single transition makes multiple transitions instead. It is important to be aware of these internal race conditions when you use programmable logic. In general, design aids like PALASM, ABEL, and CUPL do not attempt to identify such problems. If anything, they usually make things worse, because their logic optimizers work zealously to eliminate just such redundant terms needed to prevent glitches!

8.28 Miscellaneous sequential functions

With the widespread availability of large-scale integration (LSI, equivalent to 1000 gates or more on a chip) and very large scale integration (VLSI), you can get weird and wonderful gadgets all on one chip. This brief section will present just a sampling.

□ *First-in/first-out memory*

A first-in/first-out (FIFO) memory is somewhat akin to a shift register in that data entered at the input appear at the output in the same order. The important difference is that with a shift register the data get "pushed along" as additional data are entered and clocked, but with a FIFO the data "fall through" to the output queue with only a small delay. Input and output are controlled by separate clocks, and the FIFO keeps track of what data have been entered and what data have been removed. A helpful analogy might be a bowling alley, in which black and white bowling balls (bits) are returned to the bowling station: The bits are input by the pin-setting machine, and the time it takes for a ball to roll the length of the alley is the "ripple-through delay time" of the FIFO (typically $1\mu s$–$25\mu s$), whereupon the bits are available at the output to be removed at the whim (asynchronously) of the user.

FIFOs are useful for buffering asynchronous data. The classic application is buffering a keyboard (or other input device, such as magnetic tape) to a computer or sluggish instrument. By this method, no data are lost if the computer isn't ready for each word as it is generated, provided the FIFO isn't allowed to fill up completely. Some typical FIFOs are the 74F433 (TTL, 64 words of 4 bits each, 10MHz, $4\mu s$ fall-through) and the IDT7202 (CMOS, 4096×9, 15MHz, zero fall-through).

A FIFO is unnecessary if the device to which you are sending data can always get it before the next data arrive. In computer language, you must ensure that the maximum *latency* is less than the minimum time between data words. Note that a FIFO will not help if the data recipient is not able, *on the average*, to keep up with the incoming data.

□ *Rate multiplier*

Rate multipliers are used to generate output pulses at a frequency that is related to the clock frequency by a rational fraction. For instance, a 3-decade BCD rate multiplier allows you to generate output frequencies of $nnn/1000$ of the input frequency, where nnn is a 3-digit number specified as three BCD input characters. This isn't the same as a modulo-n counter, since, for instance, you cannot generate an output frequency of $3/10$ of the input frequency with a modulo-n divider. One important note: The output pulses generated by a rate multiplier are not, in general, equally spaced. They coincide with input pulses, and therefore they come in funny patterns whose *average* rate is as above. Examples are the '97 (6-bit binary) and the '167 (BCD) rate multipliers.

Frequency counters

Intersil has a good selection of integrated frequency counters. These include provision to gate the input signal for accurately known intervals, up to 8 digits of BCD counter, display drivers, auto-decimal-point and leading-zero blanking, etc. These chips generally require very little external circuitry.

Digital voltmeters

You can get complete digital voltmeters (DVMs) on a single chip. They include analog/digital conversion circuitry and the necessary timing, counting, and display circuitry. Examples are the low-power $3\frac{1}{2}$-digit ICL7136 and the $4\frac{1}{2}$-digit ICL7129; both use LCD 7-segment displays and run from a single 9-volt battery.

□ *Special-purpose circuits*

There are nice collections of LSI chips for arcane jobs like radio communications (e.g., frequency synthesizers), digital signal processing (multiplier/accumulators, digital filters, correlators, arithmetic units), data communications (UARTs, modems, network interfaces, data encryption/decryption ICs, serial format converters), and the like. Often these chips are used in conjunction with microprocessor-based devices, and many of them cannot stand alone.

□ *Consumer chips*

The semiconductor industry loves to develop ICs for use in large-market consumer products. You can get single chips to make digital (or "analog") watches, clocks, locks, calculators, smoke detectors, telephone dialers, music synthesizers, rhythm and accompaniment generators, etc. The guts of radios, TVs, and compact discs are nearly empty these days, thanks to large-scale integration. Speech synthesis (and ultimately speech recognition) has seen a lot of work recently; that's why elevators, Coke machines, automobiles, and even kitchen appliances now speak to us in those sci-fi voices we've all come to love. The development of effective automobile circuits (for engine functions, collision-avoidance systems, etc.) seems to be the next big frontier.

Microprocessors

The most stunning example of the wonders of LSI is the microprocessor, a computer on a chip. At one extreme there are powerful number crunchers like the 68020/30 and 80386/486 (32-bit fast processors with prefetch and cache, large address space, virtual memory, and powerful numeric coprocessors) and chips like the MicroVAX that emulate existing mainframe computers. At the other extreme are single-chip processors with various input, output, and memory functions included on the same chip, for stand-alone use. An example of the latter is the Toshiba TLCS-90 (Fig. 8.86), a CMOS low-power microcontroller with 6 channels of 8-bit A/D converter, internal timers, RAM and ROM, 20 bidirectional digital I/O lines, a serial port, and two stepping-motor ports. This latter type is intended as a dedicated controller in an instrument, rather than as a versatile computation device.

The microprocessor revolution hasn't begun to slow, and we have seen a doubling of computer power and memory size (now 1Mbit per chip, compared with 16kbit/chip at the time the previous edition of this book was written) each year; at the same time, prices have dropped dramatically (Fig. 8.87). Along with bigger and better

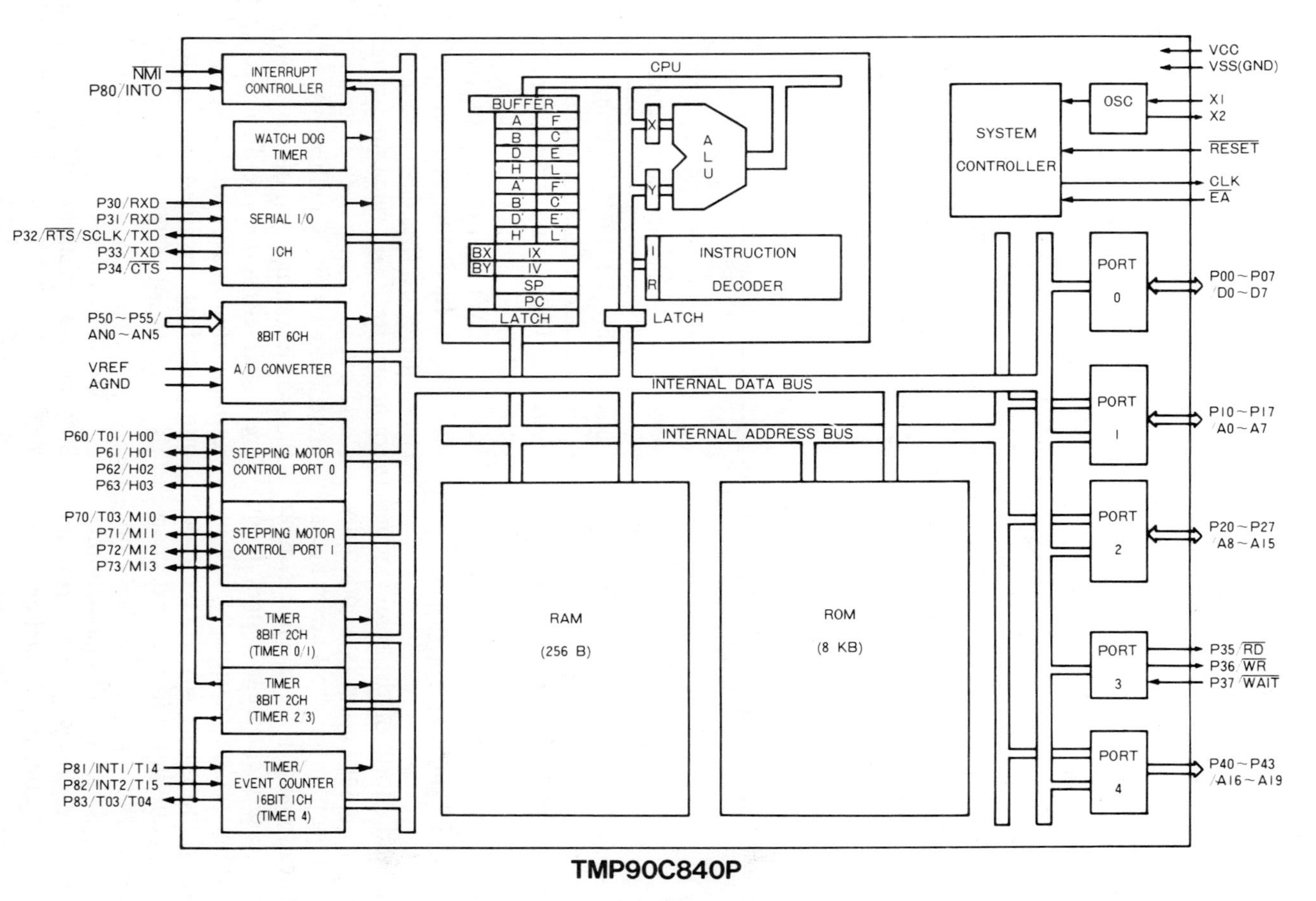

Figure 8.86. Single-chip microprocessor with lots of on-chip I/O. (Courtesy of Toshiba America, Inc.)

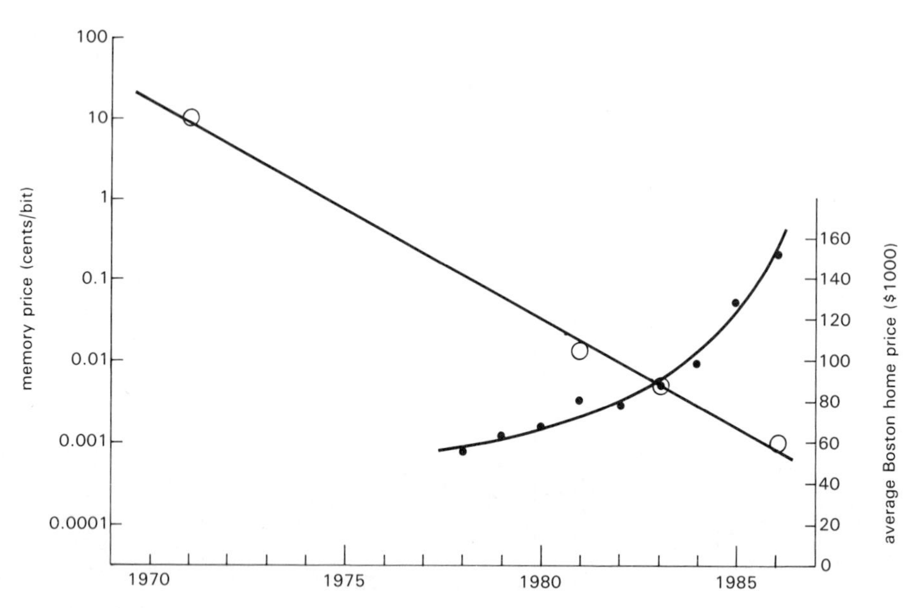

Figure 8.87. The law of Silicon Valley: learning curve.

processors and memory, recent activity in very high speed devices and large parallel architectures promises more excitement in coming years.

SOME TYPICAL DIGITAL CIRCUITS

Thanks to the efforts of the semiconductor industry, digital design is wonderfully easy and pleasant. It's almost never necessary to "breadboard" a digital circuit, as so often is the case with linear design. Generally speaking, the only serious pitfalls involve timing and noise. We'll have more the say about the latter in the next chapter. This is a good place to illustrate timing with some sequential design examples. Some of these functions can be performed with LSI circuits, but the implementations shown are reasonably efficient and illustrate the kind of circuit design being done with what's now available.

8.29 Modulo-n counter: a timing example

The circuit in Figure 8.88 produces one output pulse for every $n+1$ input clock pulses, where n is the 8-bit number you've set on the pair of hexadecimal thumbwheel switches. The '163s are 4-bit synchronous up-counters, with synchronous load (when LD$'$ is LOW) via the D inputs. The idea is to load the counters with the *complement* of the desired count, then count up to FF_H, reloading at the next clock pulse. Since we've generated the preload levels with pullups to +5 (with the switch common grounded), those levels are negative-true for the displayed switch settings; that makes the preload values, interpreted as positive-true, equal to the 1's complement of the switch settings.

EXERCISE 8.35

Show that the last statement is true, by figuring out the positive-true value that will be loaded for the switch settings in Figure 8.88.

Circuit operation is entirely straightforward. To cascade synchronous counter, you tie all clocks together, then tie a "maximum-count" output of each counter to an enable of the successive counter. For an enabled '163, the RCO (ripple-clock output) goes HIGH at maximum count,

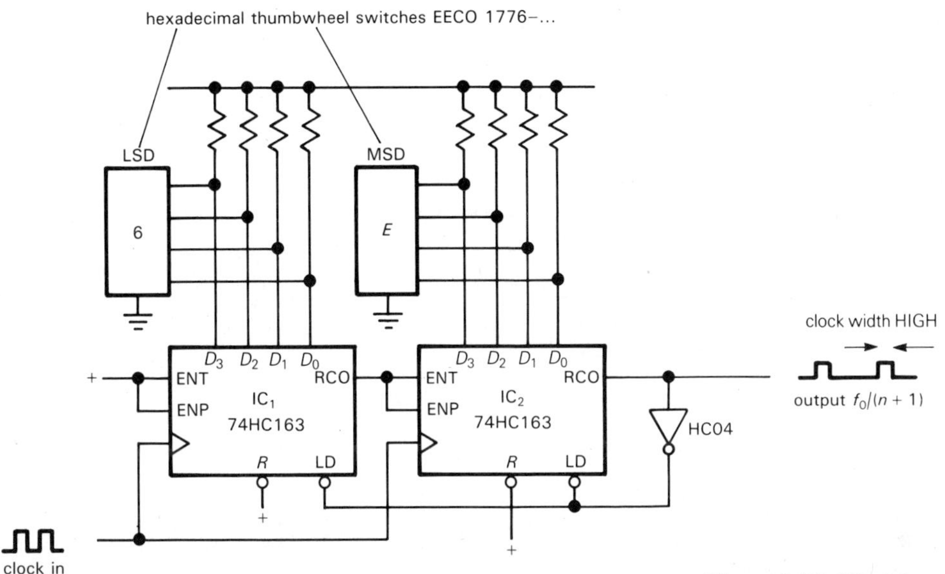

Figure 8.88. Modulo-n counter.

enabling the second counter via the enabling inputs ENT and ENP. Thus IC_1 advances at each clock, and IC_2 advances at the clock after IC_1 reaches F_H. The pair thus counts in binary until the state FF_H, at which point the LD′ input is asserted. This causes synchronous preload at the next clock. In this example we've chosen counters with *synchronous* load in order to avoid the logic race (and runt pulse RCO) that you would get with a *jam*-loaded counter. Unfortunately, this makes the counter divide by $n+1$, rather than n.

EXERCISE 8.36

Explain what would happen if jam-load counters (e.g. '191s) were substituted for the synchronous-load '163s. In particular, show how a runt pulse would be created. Demonstrate also that the foregoing circuit divides by $n+1$, whereas the asynchronous-load would divide by n (if it worked at all!).

Timing

How fast can our modulo-n counter count? The 74HC163 specifies a guaranteed f_{max} of 27MHz. However, in our circuit there are additional time delays associated with the cascading connection (IC_2 has to know that IC_1 has reached maximum count in time for the next clock pulse), and also the load-on-overflow connection. To figure the maximum frequency at which the circuit is guaranteed to work, we have to add up the worst-case delays and make sure there is enough setup time remaining. Look at Figure 8.89, where we've drawn a timing diagram showing the load sequence that occurs at maximum count.

A LOW-to-HIGH change on any Q output follows the positive edge of CLK by 34ns max. That's interesting, but not relevant, because the load sequence uses the RCO output; IC_1's RCO follows the rising edge of the CLK pulse that brings it to maximum count by 35ns max, and IC_2's RCO follows its input enable (assuming, of course, that it is at maximum count) by 32ns max. The 74HC04 adds a delay of 19ns max to generate LD′, which must precede CLK (t_{setup}) by 30ns min. That brings us to the next CLK; therefore, $1/f_{max} = (35 + 32 + 19 + 30)$ns, or $f_{max} = 8.6$MHz. This is considerably less than the maximum guaranteed counting frequency of a single 74HC163.

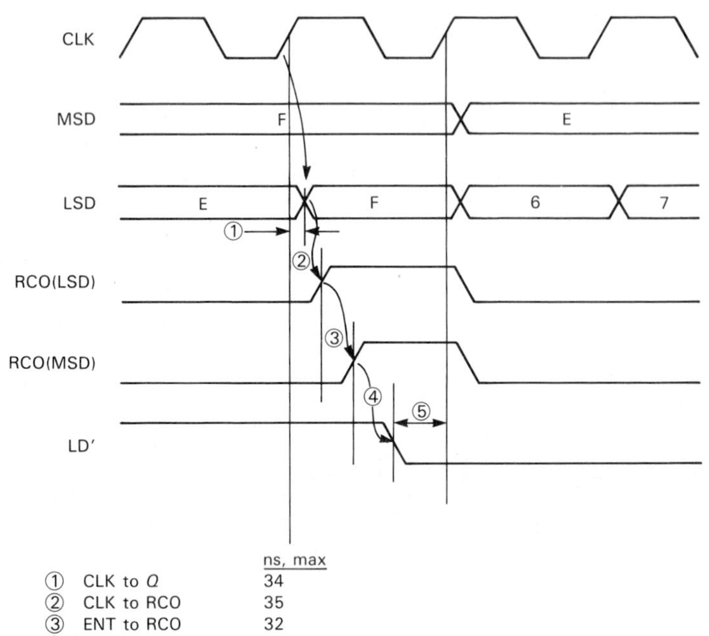

Figure 8.89. Timing diagram and calculation for modulo-n counter.

EXERCISE 8.37

Show by a similar calculation that a pair of synchronously cascaded 74HC163s (without load on overflow) have a maximum guaranteed counting rate of 15.4MHz.

Of course, if greater speed is needed, you can always use faster logic. Doing the same calculation for 74F logic (for which the maximum count rate of a single 74F163 is 100MHz), we find $f_{max} = 29$MHz.

Modulo-n counter devotees should take note of the 'HC40103, an 8-bit synchronous *down*-counter with parallel load (synchronous *or* jam load!), decoded zero-state output, and reset-to-maximum input. It has a close cousin, the 'HC40102, identical except organized as 2-digit BCD.

8.30 Multiplexed LED digital display

This example illustrates the technique of display multiplexing: displaying an n-digit number by displaying successive digits rapidly on successive 7-segment LED displays (of course, the characters need not be numbers, and the displays can have a different organization than the popular 7-segment arrangement). Display multiplexing is done for reasons of economy and simplicity: Displaying each digit continuously requires separate decoders, drivers, and current-limiting resistors for each digit, as well as separate connections from each register to its corresponding decoder (4 lines) and from each driver to its corresponding display (7 wires); it's a mess!

With multiplexing, there's only one decoder/driver and one set of current-limiting resistors. Furthermore, since LED displays come in n-character "sticks" with the corresponding segments of all characters tied together, the number of interconnections is enormously reduced. An 8-digit display requires 15 connections when multiplexed (7 segment inputs, common to all digits, plus one cathode or anode return for each digit), rather than the 57 required for continuous display. An interesting bonus of multiplexing is that the subjective brightness perceived by the eye is greater than if all digits were illuminated continuously with the same average brightness.

Figure 8.90 shows the schematic diagram. The digits to be displayed are resident in register IC_1–IC_4; they could be counters, if the device happened to be a frequency counter, or perhaps a set of latches receiving data from a computer, or possibly the output of an A/D converter, etc. In any case, the technique is to assert each digit successively onto an internal 4-bit "bus" (in this case with 4503 CMOS three-state buffers) and decode and display it while on the bus (4511 BCD-to-7-segment decoder/driver).

In this circuit a pair of inverters is used to form a classic CMOS oscillator operating at about 1kHz, driving a 4022 octal counter/decoder. As each successive output of the counter goes HIGH, it enables one digit onto the bus and simultaneously pulls the corresponding digit's cathode LOW via the high-current open-drain 40107 buffers. The 4022 is rigged up to

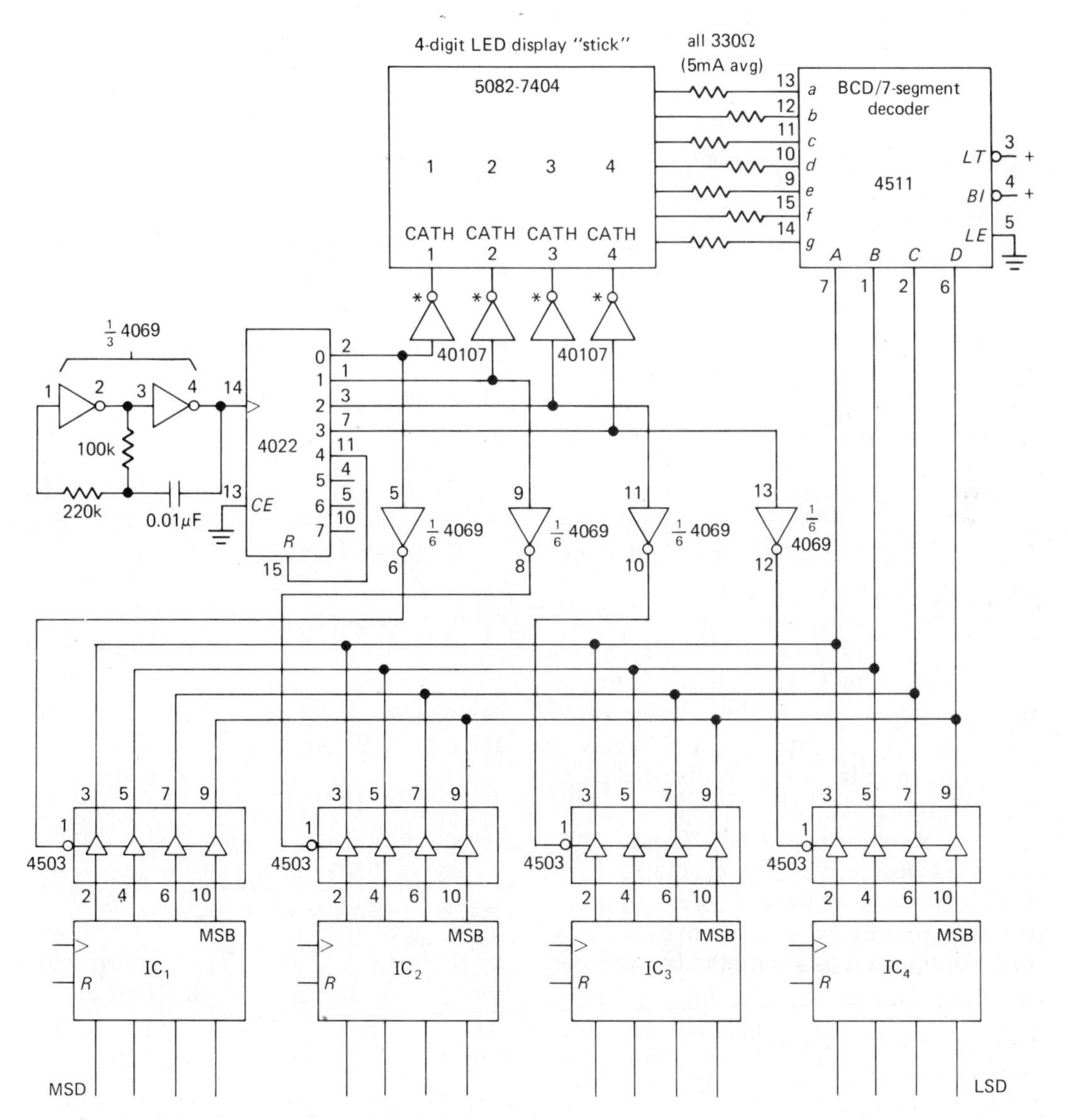

Figure 8.90. Four-digit multiplexed display. Numbers outside symbols are IC pin numbers.

cycle through the states 0–3 by resetting when the count reaches 4. Display multiplexing will work with greater numbers of digits, and it is universally used in instruments with multidigit LED displays. Try waving a calculator around in front of your eyes – you get numeric alphabet soup!

Many LSI display-oriented chips, such as counters, timers, and watches, include on-chip display multiplexing (and even driver) circuitry. In addition, you can get LSI display controller chips (e.g., the 74C911 and 74C912) for handling the kind of job done earlier with MSI circuits.

□ 8.31 Sidereal telescope drive

The circuit in Figure 8.91 was designed to drive Harvard's 61-inch optical telescope. We needed a 60Hz power source for the equatorial drive motor (1 revolution/day), accurately settable to any frequency near 60Hz (55Hz to 65Hz, say). You wouldn't want exactly 60Hz, for several reasons: (a) stars move at the sidereal rate, not the solar rate, so you would want 60.1643Hz, approximately; (b) starlight gets bent, traveling obliquely through the atmosphere; this "refraction" depends on zenith angle, so the apparent motion is at a slightly different rate; (c) sometimes you want to look at the moon, planets, or comets, which have different rates. The solution here was to use a 5-digit rate multiplier to generate output pulses at a rate $f_{\text{in}}n/10^5$, where n is a 5-digit number set by front-panel BCD thumbwheel switches.

The output is then near 600kHz, since f_{in} is an accurate 1MHz generated by a stable crystal oscillator. The output of the rate multiplier is divided by 10^4 by four decade counters, with the last counter arranged as a divide-by-5 followed by a divide-by-2 for symmetrical square waves at 60Hz. The output is clamped by a zener for stable square-wave amplitude and filtered by a 6-pole Butterworth low-pass filter (f_0 = 90Hz) to generate a good sine wave (you can think of the filter as stripping away the higher Fourier components, or "overtones," of the square wave). Then 115 volts ac is generated by the "overcompensated" amplifier illustrated in Section 4.35. The output of the Butterworth looks "perfect" on a scope, as it should, since in this case a 6-pole Butterworth reduces the largest overtone to 1.5% of its unfiltered amplitude; this means that the distortion is more than 35dB down. Note that this technique of sine-wave generation is convenient only if the input frequency is confined to a narrow range.

The ±10% guiding inputs alter the synthesized output frequency 10% by changing the third divider to divide-by-9 or divide-by-11. That stage is a modulo-n divider constructed along the lines of Figurc 8.88.

□ 8.32 An n-pulse generator

The n-pulse generator is a useful little test instrument. It generates a burst of n output pulses following an input trigger signal (or you can push a button), with a set of selectable pulse repetition rates. Figure 8.92 shows the circuit. The 'HC40102s are high-speed CMOS 2-decade downcounters, clocked continuously by a selected power-of-10 subdivision of the fixed 10MHz crystal oscillator, but disabled by having both APE (asynchronous preset enable) asserted and CI (carry in) disasserted. When a trigger pulse comes along (note the use of 'HCT logic at this input, for compatibility with bipolar TTL), flip-flop 1 enables the counter, and flip-flop 2 synchronizes counting following the next rising edge of the clock. Pulses are passed by NAND gate 3 until the counter reaches zero, at which time both flip-flops are reset; this parallel-loads the counter to n from the BCD switches, disables counting, and readies the circuit for another trigger. Note that the use of pulldown

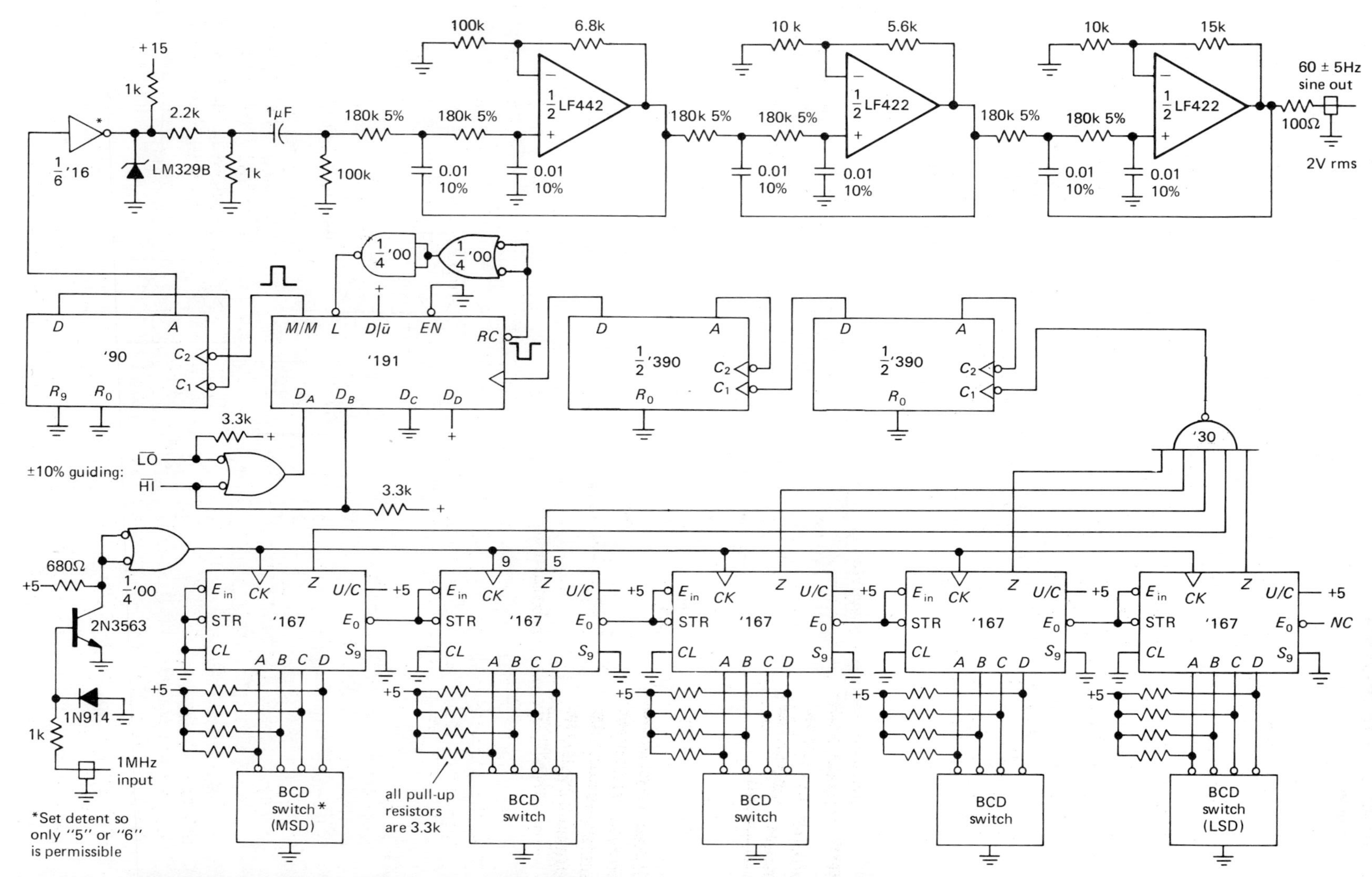

Figure 8.91. Precision 60Hz ac signal source. *Output* frequency = $xx.xxx$; e.g., to generate sidereal rate, set switches to 60165.

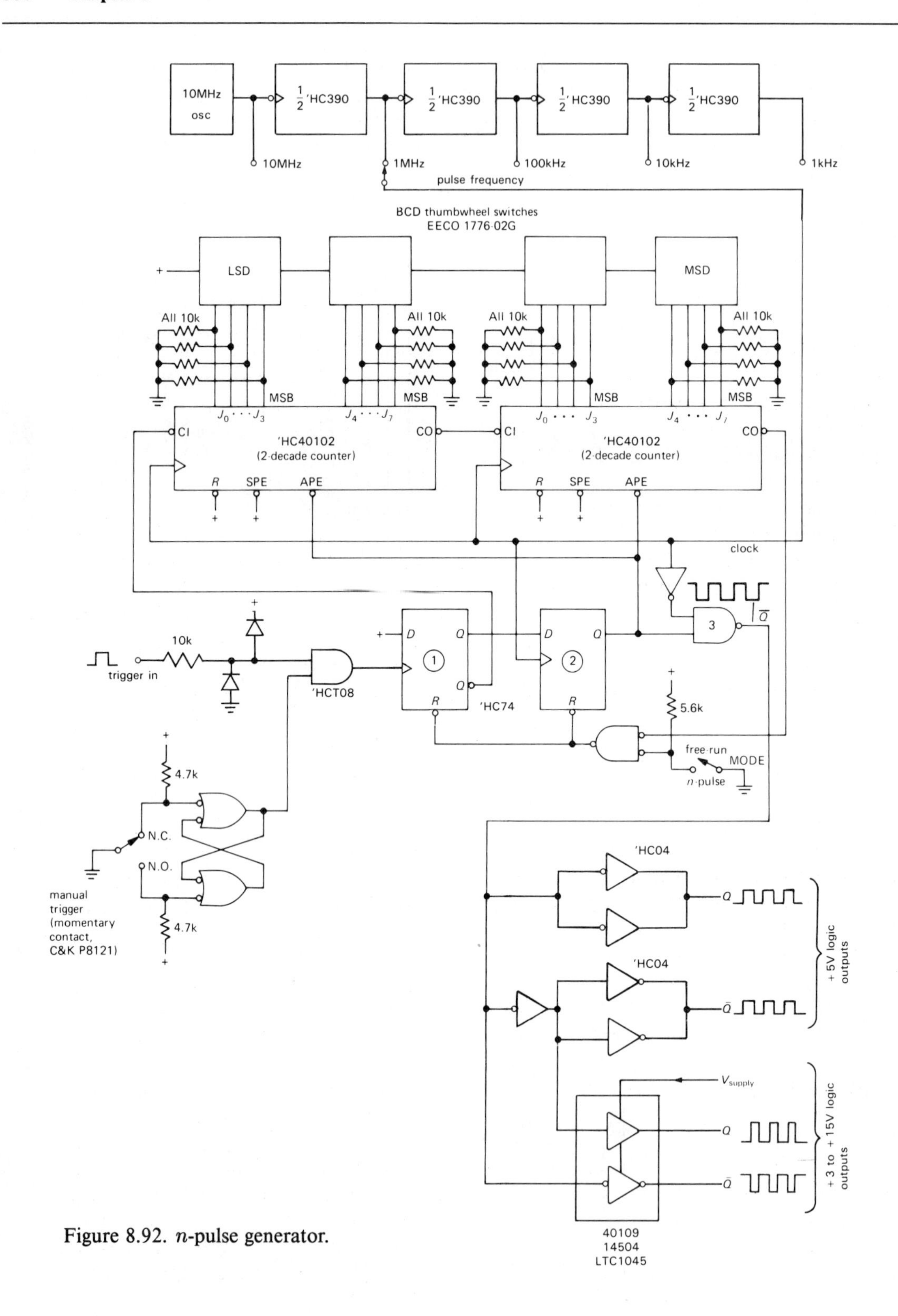

Figure 8.92. n-pulse generator.

resistors in this circuit means that true (rather than complemented) BCD switches must be used. Note also that the manual trigger input must be debounced, since it clocks a flip-flop. That is not necessary for the free-run/n-pulse switch, which simply enables a continuous stream of output pulses.

The output stage delivers two pairs of true/complement signals. The paralleled 'HC04 inverters give normal +5 volt logic swings, saturating cleanly at the rails because we're using CMOS. We paralleled them to increase drive capability; the circuit as drawn can drive at least ±10mA with logic levels within 0.3 volt of the rails. If you need more output current, you can replace them with 'AC04s, a paralleled pair of which will give you ±50mA with logic levels within 0.3 volt of the rails. We added the boxed driver pair so that you can drive logic that runs on voltages other than +5 volts. For example, low-power design often uses 4000B or 74C series CMOS running directly from 9 volt batteries (it's rated for 3V to 15V operation); 'HC logic accepts 2 to 6 volt supply voltage; and there have been suggestions recently that 'AC-style CMOS be operated at +3.3 volts (JEDEC Standard No. 8). The 40109, 14504, and LTC1045 are all level shifters, chips with active pullups to a second supply terminal connected (in this case) to the driven circuit's V_{DD} line, which can be higher or lower than the pulse generator's +5 volt supply. In that way you get clean CMOS output levels at the right voltage for the circuit under test.

LOGIC PATHOLOGY

There are interesting, and sometimes amusing, pitfalls awaiting the unsuspecting digital logician. Some of these, such as logic races and lockup conditions, can occur regardless of the logic family in use. Others (e.g., "SCR latchup" in CMOS chips) are "genetic abnormalities" of one logic family or another. In the following sections we have collected our bad experiences in the hope that such anecdotes can help others avoid such problems.

8.33 dc problems

Lockup

It is easy to fall into the trap of designing a circuit with a lockup state. Suppose you have some gadget with a number of flip-flops, all going through their proper states. Everything seems to work fine. Then one day it just stops dead. The only way you can get it to work is to turn the power off and back on again. The problem is that there is a lockup state (an excluded state of the system that you can't escape from), and you got into it because of some power-line transient that sent the system into the forbidden state. It is very important to look for such states when you design the circuit and rig up logic so that the circuit recovers automatically. At a minimum, things should be arranged so that a RESET signal (generated manually, at start-up, etc.) brings the system to a good state. This may not require any additional components (e.g., Exercise 8.24).

Start-up clear

A related issue is the state of the system at start-up. It is always a good idea to provide some sort of RESET signal at start-up. Otherwise the system may do weird things when first turned on. Figure 8.93 shows a suitable circuit. The series resistor is necessary with CMOS to prevent damage when power is removed from the circuit, since otherwise the electrolytic capacitor will try to power the system via the CMOS input-gate protection diode. A Schmitt trigger (4093, '14) may be a good idea, to make the RESET signal switch off cleanly. The hysteresis symbol shown in the figure indicates an inverter with Schmitt trigger input, e.g., the TTL 74LS14 (hex inverter) or CMOS 40106 or 74HC14.

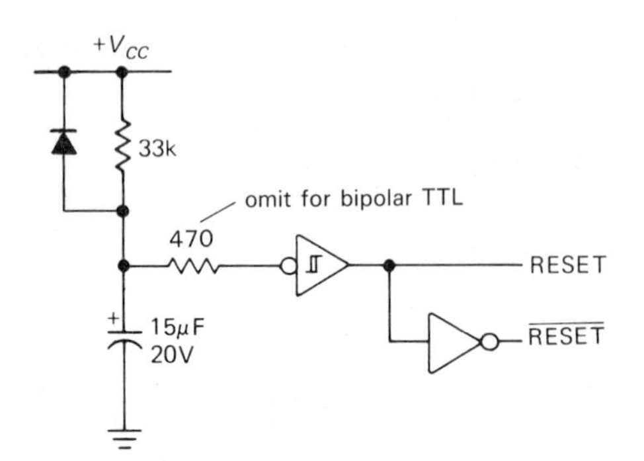

Figure 8.93. Power-on clear circuit.

8.34 Switching problems

Logic races

Lots of subtle traps lurk here. The classic race was illustrated with the pulse synchronizer in Section 8.19. Basically, in any situation where gates are enabled by signals coming from flip-flops (or any clocked device), you must be sure that a gate doesn't get enabled and then disabled a flip-flop delay time later. Likewise, make sure that signals appearing at flip-flop inputs aren't delayed with respect to the clock (another plus for synchronous systems!). In general, delay the clock rather than the data. It is surprisingly easy to overlook a race condition.

□ *Metastable states*

As we remarked earlier, a flip-flop (or any clocked device) can get confused if the data input changes during the setup-time interval preceding a clock pulse. As long as the flip-flop makes *some* decision in this ambiguous case, all is well. However, there is a chance that the input may have changed at just the wrong time, at exactly the "moment of truth," such that the flip-flop can't make up its mind; its output can hover at the logic threshold literally for microseconds (by comparison, normal propagation delays are around 20ns for HC or LS TTL), or (even worse) go into one state, then change its mind and switch back to the other state! (Carver Mead compares the metastable condition with Jean Buridan's paradox – a dog could starve if placed midway between two equal amounts of food.)

This problem does not arise in properly designed synchronous systems, where setup times are always satisfied (by using logic fast enough so that inputs to flip-flops are stable by t_{setup} before the next clocking pulse). However, it can create problems in situations where asynchronous signals (e.g., going from device A, with its own clock, to device B, with a separate clock) must be synchronized. In these cases you cannot guarantee that input transitions do not occur during the setup interval; in fact, you can calculate how often they do! The metastability problem has been blamed for mysterious computer crashes, although we are skeptical. The cure generally involves a set of concatenated synchronizers, or a "metastable state detector" that resets the flip-flop. Awareness of this problem is growing, and there are even "metastable-hardened" logic families, e.g., the AMD 29800 bus-interface series, which claims a maximum metastable delay of 6ns before finally making up its mind.

Clock skew

Clock skew primarily affects slow CMOS logic. The problem arises when you have a clocking signal of slow rise time driving several interconnected devices (Fig. 8.94). In this case two shift registers are clocked by a slowly rising edge, caused by capacitive loading of a relatively high impedance CMOS output (around 500Ω, when operating from +5V). The problem is that the first register may have its threshold at a

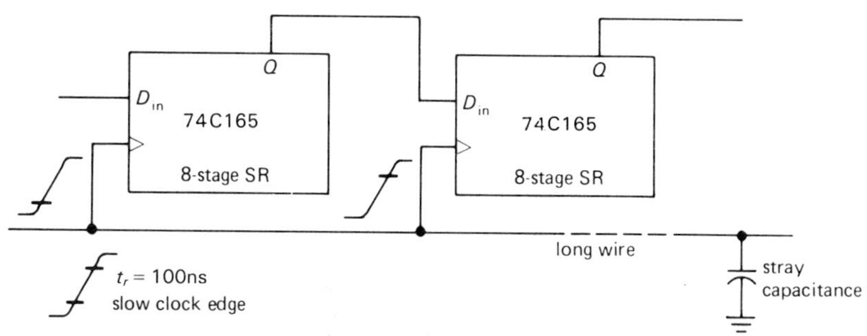

Figure 8.94. Slow rise times cause clocking skew.

lower voltage than that of the second register, and this causes it to shift earlier than the second register. The last bit of the first register is then lost. CMOS devices can display quite a spread of input threshold voltages, which compounds the problem (the threshold is specified only to be between one-third and two-thirds of V_{DD}, and they mean it!). The best cure is to use a nearby chip without much capacitive loading to drive clock inputs in this sort of situation.

Speaking generally, edge-triggered clock inputs on any digital IC should always be treated with respect. For example, clock lines with noise or ringing should always be cleaned up with a gate (perhaps one with input hysteresis) before driving the clocked chip. You're especially likely to have problems with clock lines that come from another board, or from a different family of logic. For example, slow 4000B or 74C logic driving the faster HC or AC families is likely to exhibit problems of clock skew or multiple transitions.

Runt pulses

In Section 8.29 (modulo-n counter) we remarked that some delay should be added if a counter's output clears itself, in order to prevent a pulse of substandard width. The same comment goes for LOAD pulses when using counters or shift registers. Runt pulses will make your life miserable, since you may have marginal operation or intermittent failures. Use the worst-case propagation delay specifications when designing.

Unspecified rules

As the semiconductor industry was finding its way, beginning with the simplest RTL integrated circuits of the 1960s (see Section 9.01 for a brief chronology), then the improved TTL and Schottky families, to the modern high-performance CMOS families, there was an understandable lack of standardization of pinouts, specifications, and functionality. As examples, the 7400 (NAND) had its gates pointing "down," but the 7401 (open-collector NOR) was built with the gates going the other way. This created so much confusion that it had to be mutated into the 7403, which is a 7401 with 7400-style pinouts; a similar disaster happened with the 7490 (BCD ripple counter), with power-supply pins in the middle instead of the corners. (Ironically, mid-chip power-supply pins are making a comeback in fast CMOS, because of reduced inductance and better isolation.)

An important legacy of this early anarchy is the hodge-podge of "unspecified rules" that we're stuck with. For example, the ever-popular '74 D flip-flop exists in every logic family; asserting both SET and CLEAR makes both outputs HIGH in every family except 74C, where it makes both outputs LOW! That's not exactly an unspecified rule, since if you look carefully in the fine print you'll find the inconsistency; the technical term for it is a *gotcha*. Another of our favorite gotchas is the '96, a 5-bit shift register with tricky jam-load inputs: They can SET, but not CLEAR!

A genuine unspecified rule, and in fact a very important one, is "removal time": That's the amount of time you must wait after disasserting a jam-type input before a clocked device is guaranteed to clock properly. Chip designers didn't bother specifying this (though circuit designers always wanted to know it) until the logic families of the early 1980s, specifically the advanced Schottky and fast CMOS families. If you're designing with earlier logic (e.g., 74C), our advice is to be conservative; for example, assume that the removal time is the same as the data setup time [it's usually less; for example, the 74HC74 *D*-type flip-flop specifies a minimum removal time (preset or clear to clock) of 5ns, while the minimum data setup time is 20ns].

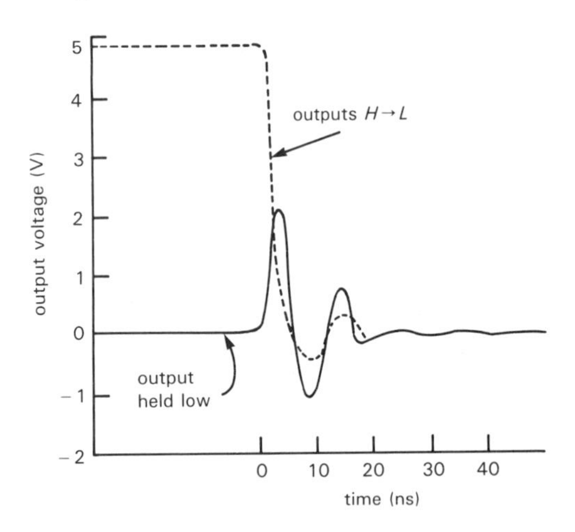

Figure 8.95. 74AC244 octal buffer, driving seven 50pF loads H→ L, holding eighth output LOW. "Ground" is a copper plate (1 oz/ft^2). (After Fig. 1.1-4, TI Advanced CMOS Logic Designer's Handbook.)

8.35 Congenital weaknesses of TTL and CMOS

We will divide this section into nuisance problems and really bizarre behavior.

Nuisance problems

Bipolar TTL. You have to remember that TTL inputs *source* current when held in the LOW state (e.g., 0.25mA for LS, 0.5mA for F). That makes it difficult to use *RC* delays, etc., because of the low impedances necessary, and in general you have to give some thought when interfacing linear levels to TTL inputs.

The TTL threshold (and that of its imitators, HCT and ACT) is too close to ground, making the whole logic family somewhat noise-prone (more on this in Chapter 9). The high speed of these logic families makes them recognize short spikes on the ground line; those spikes, in turn, are generated by the fast output transition speeds, making the problem worse.

Bipolar TTL makes demands on the power supply (+5V, ±5%, with relatively high quiescent power dissipation). Power-supply current spikes generated by the active pullup output circuitry generally require liberal use of power-supply bypassing, ideally one 0.1μF capacitor per chip (Fig. 8.96).

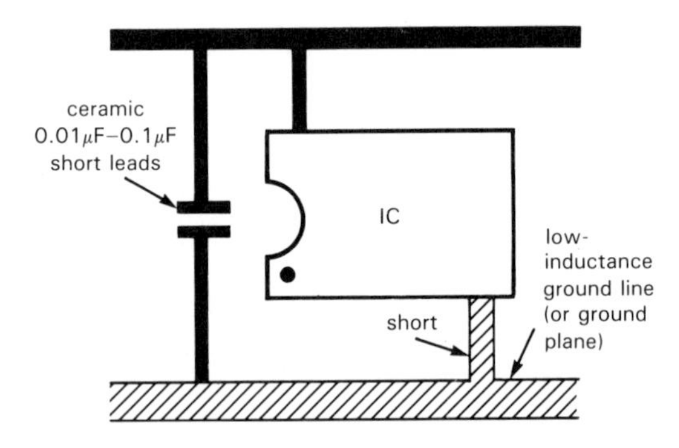

Figure 8.96. It's always a good idea to use robust low-inductance ground wiring, with liberal use of bypass capacitors.

CMOS. CMOS inputs are prone to damage from static electricity. The mortality rate really climbs in winter! Recent families with polysilicon gates [HC(T), AC(T)] are much more rugged than their metal-gate ancestors. CMOS inputs show a large spread in logic threshold, which can lead to problems of clock skew (Section 8.34), particularly with the slow CMOS families (4000B, 74C) owing to their high output impedance (200–500Ω). These slow families can even exhibit double output transitions when driven with slowly rising inputs. CMOS requires *all* unused inputs, even those of unused gate sections, to be connected to HIGH or LOW.

An interesting congenital problem with the newer fast CMOS families, particularly AC and ACT, is the presence of "ground bounce": A fast CMOS chip driving its capacitive load generates enormous transient ground currents, causing the chip's ground line to jump momentarily, and thereby carrying with it LOW outputs that happen to be innocent bystanders on the same chip. Figure 8.95 shows the sort of thing you see. Notice particularly the magnitude of the effect – 1 to 2 volts is not uncommon! When you consider that a 3ns, 5 volt transition into 50pF amounts to a transient current $I = C\,dV/dt = 83\text{mA}$, and that an octal buffer might drive eight such loads simultaneously (total current of $\frac{2}{3}$A!), this behavior isn't surprising. This problem is turning out to be harder to solve than anyone imagined, leading to a controversial new set of AC/ACT circuits with "center-pin" power and grounds (for lower inductance). At the time of writing, the major players have taken sides, with TI championing the new pinouts, and RCA and Fairchild defending the conventional corner-pin layout. We have received handsomely bound polemics from both sides, pointing out the inferiority of both the opposing arguments and the opposing chips. At the very least, users should be aware of this serious problem and take measures to keep ground inductance as low as possible when using AC/ACT. It's best to use circuit boards with a dedicated ground plane and plenty of low-inductance bypass capacitors. Better still, if you don't need the speed, stay away from AC/ACT entirely, and use HC/HCT instead.

Bizarre behavior

Bipolar TTL. TTL doesn't do many really weird things. However, some TTL monostables will trigger on a glitch on the supply (or ground) line, and they generally behave somewhat fidgety. A circuit that works well with LS TTL may malfunction when replaced with AS TTL, because of faster edge times and consequently larger ground-line currents and ringing (74F TTL seems better in this regard). Most weird TTL operation can be traced to noise problems.

CMOS. CMOS can drive you crazy! For example, the chip can go into "SCR latchup" if the input (or output) is driven beyond the supply momentarily. The resultant current (50mA or so) through the input-protection diodes turns on a pair of parasitic cross-connected transistors that are a side effect of the "junction-isolated" CMOS process (see Fig. 3.50 and Section 14.16). This effectively shorts V_{DD} to ground; the chip gets hot, and you have to turn off the power supply before it will behave itself again. If you let this happen for more than a few seconds, you'll have to replace the chip. Some of the newer CMOS designs (the polysilicon-gate HC processes from GE/RCA and National) claim to be immune to latchup even with input swings *5 volts* beyond the rails and to operate properly for input swings 1.5 volts beyond the rails.

CMOS has some strange and subtle failure modes. One of the output FETs can open up, giving pattern-sensitive failures that are difficult to detect. An input may begin to sink or source current. Or the whole chip may start drawing substantial supply current. Putting a 10 ohm resistor in series with each chip's V_{DD} lead makes it easy to locate faulty CMOS chips that are drawing quiescent supply current (for chips driving many outputs, or power drivers such as the AC series, use 1Ω sense resistors).

Besides the input threshold variation between chips, a single chip can exhibit different thresholds for several functions driven from a single input. For example, the RESET input of a 4013 can bring Q' HIGH before it brings Q LOW. This means that you should not terminate a reset pulse based on the output at Q', since the runt pulse that will be generated may actually fail to clear the flip-flop.

Open inputs on CMOS chips are bad news! You might have a circuit that intermittently misbehaves. You put you scope probe onto a point in the circuit, and it shows zero volts, as it should. Then the circuit works fine for a few minutes – before malfunctioning again! What happened was that the scope discharged the open input, and it took a long time to charge back up to the logic threshold.

Here's the craziest of them all: You forgot to wire up the V_{DD} pin on a CMOS chip, but the circuit works just fine! That's because it is being powered by one of its logic inputs (via the input-protection diodes from the input to the V_{DD} line in the chip). You might get away with this for a long time, but suddenly the circuit reaches a state where all the logic inputs to the chip are simultaneously LOW; the chip loses its power and forgets its state. Of course, this is a bad situation anyway, since the output stage isn't adequately powered and can't source much current. The trouble is that this situation may produce symptoms only occasionally, and it can have you running around in circles until you figure out what's going on.

SELF-EXPLANATORY CIRCUITS

8.36 Circuit ideas

Figure 8.97 shows some digital circuit ideas.

8.37 Bad circuits

Figure 8.98 shows some classic digital circuit blunders.

ADDITIONAL EXERCISES

(1) Show how to make a *JK* flip-flop using a type *D* flip-flop and a 4-input multiplexer. Hint: Use the address inputs for *J* and *K*.

(2) Design a circuit that reads out, on 7-segment digits, how many milliseconds you've held a button down. The device should be smart enough to reset itself each time. Use a 1.0MHz oscillator.

(3) Design a reaction timer. "A" pushes his button; an LED goes on, and a counter begins counting. When "B" pushes her button, the light goes out and an LED display reads the time, in milliseconds. Be sure to design the circuit so that it will function properly even if A's button is still held down when B's button is pushed.

(4) Design a period counter: a device to measure the number of microseconds in one period of an input waveform. Use a Schmitt comparator to generate TTL; use a 1.0MHz clock frequency. Make it work so that pushing a button initiates the next measurement.

(5) Add latches to the period counter, if you haven't already.

(6) Now make it measure the time for 10 periods. Also, have it light an LED while it's counting.

(7) Design a true electronic stopwatch. Button *A* starts and stops the count. Button *B* resets the count. The output should be of the form $xx.x$ (seconds and tenths); assume that you have a 1.0MHz square wave.

(8) Some stopwatches use a single button (start, stop, reset, start, etc., each time it is pushed). Design an electronic equivalent.

(9) Design a nice frequency counter to measure the number of cycles per second of an input waveform. Include lots of

Circuit ideas

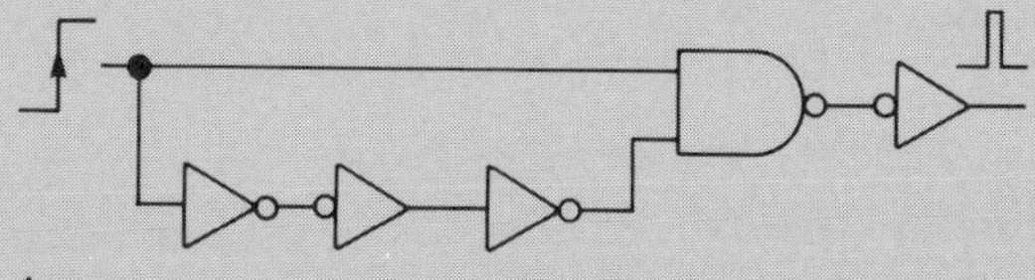

A. rising edge pulse generator

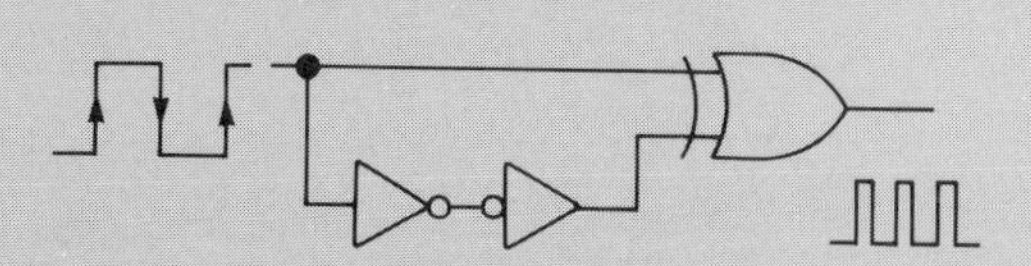

E. both-edge pulse generator

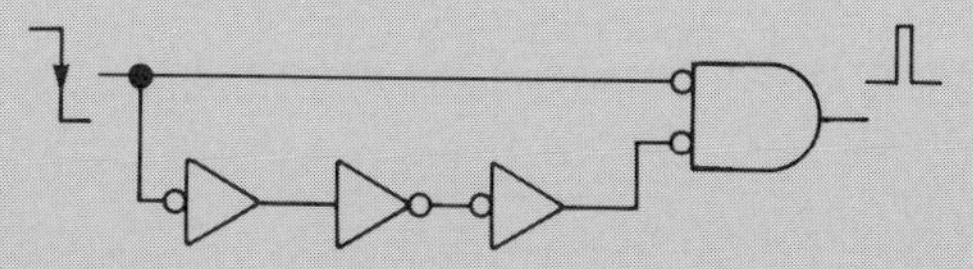

B. falling-edge pulse generator

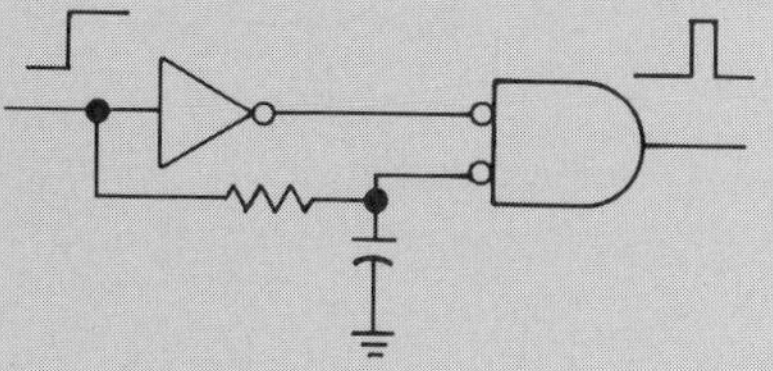

F. adjustable rising-edge pulse generator

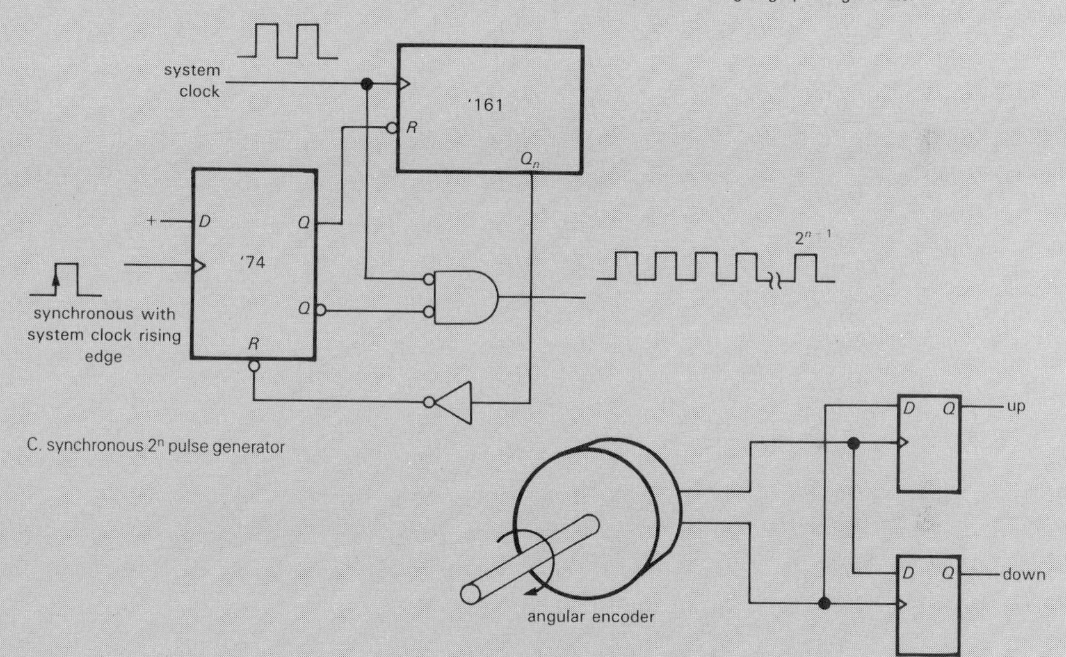

C. synchronous 2^n pulse generator

G. quadrature to up-down-counter conversion; use to obtain rotary position from angular encoder

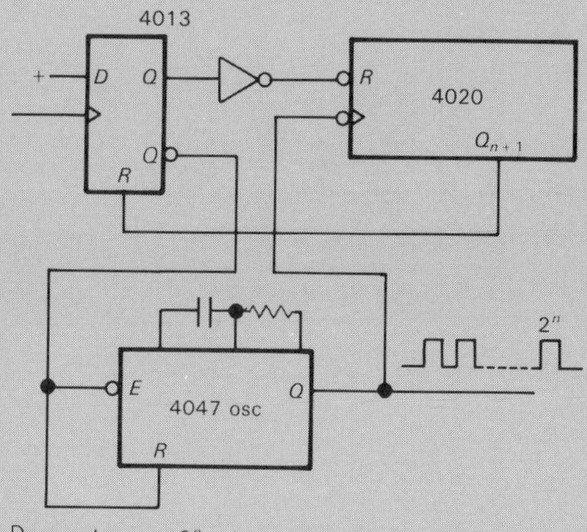

D. asynchronous 2^n pulse generator

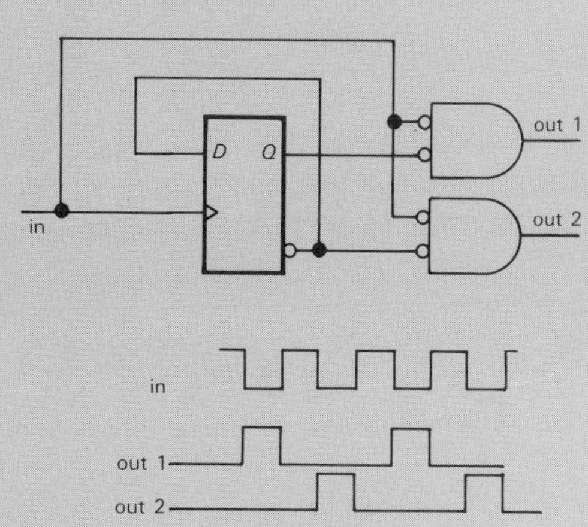

H. quadrature clock pulse generator

Figure 8.97

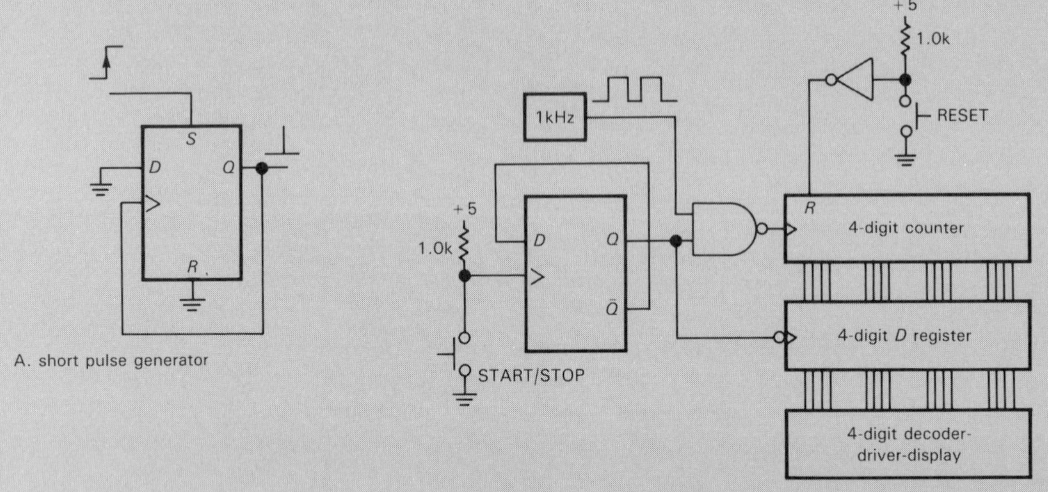

A. short pulse generator

B. stopwatch (with single START/STOP button)

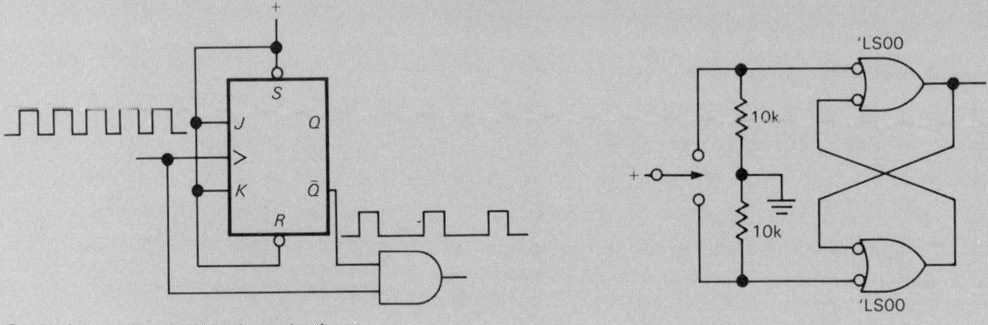

C. circuit to remove every other pulse from an input train of 1μs pulses (subtle error)

D. switch debouncer

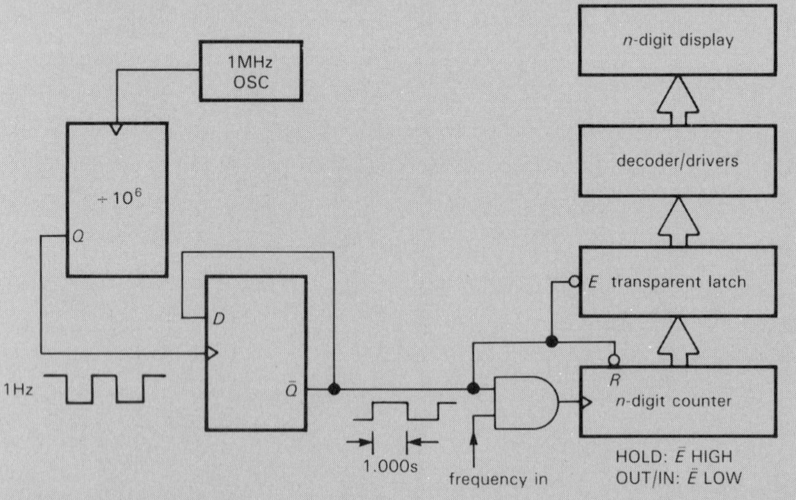

E. frequency counter with latched display

Figure 8.98

digits, latched count while counting the next interval, and choice of 1 second, 0.1 second, or 0.01 second counting interval. You might add a good input circuit with several sensitivities, a Schmitt trigger with adjustable hysteresis and trigger point (use a fast comparator), and a logic signal input for TTL signals. How about a BCD output? Multiplex the digits on output, as well as parallel output? Spend some time on this one.

(10) Design a circuit, using HC logic, to time a speeding bullet. The projectile breaks a thin wire stretched across its path; then, some measured distance farther along its path, it breaks a second wire. Beware of problems like "contact bounce." Assume that you have a 10MHz logic square wave, and design your circuit to read out, in microseconds (4 digits), the time interval between breaking the two wires. A push button should reset the circuit for the next shot.

(11) Make a 1-of-16 decoder from two 74HC42s (1-of-10). The input is a 4-bit binary number. Output will be negative-true (as with the 74HC42). Hint: MSB input to the 74HC42 can be used as an "ENABLE."

(12) Imagine that you have four 256-bit ROMs, TTL style, each of which has an 8-bit parallel address input, a three-state positive-true output, and an output enable (negative-true); i.e., the ROM asserts the selected data bit at its output if the enable is LOW. Show how to "expand" these into a 1024-bit ROM, using whatever else you need. (A 74HC138 might be handy, or you can do it with gates. Try both.)

(13) Invent a circuit to keep a running sum of successive 4-bit numbers that are input to it. Keep your result to only 4 bits (i.e., perform a sum modulo-16). (Such a sum is useful as a "check sum" to be written on inherently error-prone data-recording media, e.g., punched paper tape.) Assume that a positive 1μs TTL pulse occurs once during the time that each input number is valid. Provide a reset input. Thus, your circuit is as shown in Figure 8.99.

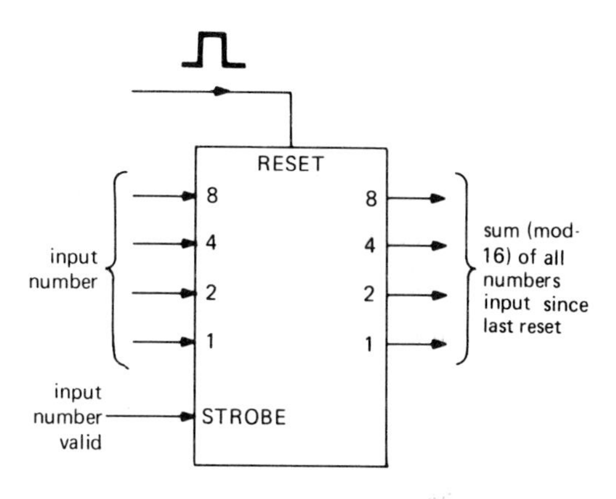

Figure 8.99

Now add another feature to the circuit, namely an output bit that is 1 if the *total number* of 1 bits in all the input numbers (since the last reset input) is odd, 0 if even. Hint: An XOR "parity tree" will tell you if the sum of 1's in each number is odd; figure it out from there.

(14) In Exercise 8.14 you designed a 2×2 multiplier by using a Karnaugh map for each output bit. In this problem you are to accomplish the same task by the process of "shift and add." Begin by writing out the product the way you would in elementary school:

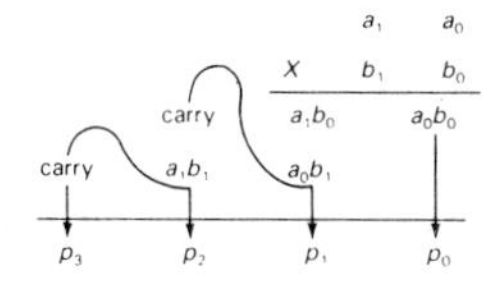

This process has a simple repeating pattern, requiring 2-input gates (what kind?) to generate the intermediate terms a_0b_0, etc., and 1-bit "half adders" (adders that have a carry out but no carry in) to sum the intermediate terms.

(15) Now design a 4×4 multiplier along the same lines, this time using three 4-bit full adders (74HC83) and sixteen 2-input gates.

TABLE 8.4. BUFFERS

Type: true	Type: inv	Bits	Pins	Output drive[a]: sink (mA)	Output drive[a]: source (mA)	Family	Enable inputs[b]	Output[c]	Broadside[d]	Comments
'125	–	4	14	24	2.6	LS	4 L	3S	–	each enable 1 bit
'126	–	4	14	24	2.6	LS	4 H	3S	–	each enable 1 bit
'365	'366	6	16	24	2.6	LS	2 L	3S	–	
'367	'368	6	16	24	2.6	LS	2 L	3S	–	enable 2,4 bits
'1034	'1004	6	14	24	15	LS	–	2S	–	74AS is 48/48 mA
'230	–	8	20	64	15	AS	2 L	3S	–	4 true, 4 inv; each enable 4 bits
–	'231	8	20	64	15	AS	L,H	3S	–	each enable 4 bits
'241	–	8	20	24	15	LS	L,H	3S	–	each enable 4 bits
'244	'240	8	20	24	15	LS	2 L	3S	–	each enable 4 bits
'465	'466	8	20	24	2.6	LS	2 L	3S	–	
'467	'468	8	20	24	2.6	LS	2 L	3S	–	each enable 4 bits
'541	'540	8	20	24	15	LS	2 L	3S	•	
'656	'655	8	24	64	15	F	3 L	3S	•	+ parity gen
'2966	'2965	8	20	12	1		2 L	3S	–	25Ω R_{out}; each enable 4 bits
'827	'828	10	24	64	15	F	2 L	3S	•	Am29827/8

(a) output drive capability for members of the indicated family. (b) ENABLE inputs, HIGH or LOW true.
(c) 2S = 2-state; 3S = 3-state. (d) broadside (flow-through) pinout.

TABLE 8.5. TRANSCEIVERS

Type: true	Type: inv	Bits	Pins	Registers	Output drive[a]: sink (mA)	Output drive[a]: source (mA)	Enables: pol	Enables: mode[b]	Broadside[c]	Comments[d]
'243	'242	4	14	–	24	15	L,H	LR	•	
'245	–	8	20	–	24	15	L	DE	•	
'543	'544	8	24	2			L	LR	•	
'545	–	8	20	–			L	DE	•	
'550	'551	8	28	2			L	LR	–	Am2950/1
'552	–	8	28	2			L	LR	–	+ parity gen
'588	–	8	20	–			L	DE	•	IEEE-488 term resistors
'623	'620	8	20	–	24	15	L,H	LR	•	'621/2 are OC
'639	'638	8	20	–	24	15	L	DE	•	3-state in one dir, OC in other
'643	–	8	20	–	24	15	L	DE	•	Q and Q' outputs; '644 is OC
'645	'640	8	20	–	24	15	L	DE	•	'641/2 are OC
'646	'648	8	24	2	24	15	L	DE	•	'647/9 are OC
'652	'651	8	24	2	24	15	L,H	LR	•	'654/3 are 3S one dir, OC other
'657	–	8	24	–			L,H	DE	•	+ parity gen
'2623	'2620	8	20	–	12	2	L,H	LR	•	25Ω output resistors
'2645	'2640	8	20	–	12	2	L	DE	•	25Ω output resistors
2952	'2953	8	24	2	24	6.5	L	LR	•	'2950/1 have handshake

(a) for LS-family devices. (b) DE - DIRection and ENable inputs; LR - separate Left and Right enables.
(c) broadside (flow-through) pinout. (d) all are 3-state, unless stated otherwise.

TABLE 8.6. DECODERS

Type	Bits[a]	Pins	E/E' [b]	Outputs qty[a]	Outputs pol[c]	Output enables[a]	Comments
'42	4	16	0/0	10	L	–	'156 is OC
'131	3	16	1/1	8	L	–	
'137	3	16	1/1	8	L	–	input D flip-flop
'138	3	16	1/2	8	L	–	
'139	2+2	16	0/1+1	4+4	L	–	
'155	2	16	0+1/2+1	4+4	L	–	input latch
'538	3	20	2/2	8	H	2	'537 is 1 of 10
'539	2+2	20	0/1+1	4+4	H	1+1	
'547	3	20	2/1	8	L	–	input latch; ACK output when enabled
'548	3	20	2/2	8	L	–	ACK output when enabled

(a) multiple sections are indicated by "*X+Y*," where *X* and *Y* are the number of bits in each section.
(b) "*H/L*" indicates the number of active-H/L enable inputs; for multiple sections see note (a).
(c) H = active HIGH; L = active LOW.

TABLE 8.7. MAGNITUDE COMPARATORS

Type	Bits	Pins	Pullup	Enable	Latch	Outputs Q	Outputs Q'	Outputs <	Outputs >	Comments
'85	4	16	–	•	–	•	–	•	•	
'518	8	20	Q	•	–	•	–	–	–	OC; '519 lacks pullups
'520	8	20	Q	•	–	–	•	–	–	'521 lacks pullups; '522 is OC
'524	8	20	–	•	•	•	–	•	•	latch is ser/par load SR with 3S outputs
'682	8	20	Q	–	–	–	•	–	•	'683 has OC
'684	8	20	–	–	–	–	•	–	•	'685 has OC
'686	8	24	–	•	–	–	•	–	•	'687 has OC
'688	8	20	–	•	–	–	•	–	–	'689 has OC
'866	8	28	–	•	•	•	–	•	•	P,Q output latches; Q reset; logical or arith comp
'885	8	24	–	–	P	–	–	•	•	input latch for P; logical or arith comp

TABLE 8.8. MONOSTABLE MULTIVIBRATORS

Type	# sect	Pins	Trig logic	Reset	Retrig	Int R[a]	PRD[b]	Comments
'121	1	14	(L+L)•H	–	–	•	–	precision
'221	2	16	L•H	L	–	–	Y	precision
'122	1	14	(L+L)•H•H	L	•	•	Y	
'422	1	14	(L+L)•H•H	L	•	•	N	'122 w/o trigger on clear
'123	2	16	L•H	L	•	–	Y	
'423	2	16	L•H	L	•	–	N	'123 w/o trigger on clear
'4098	2	16	L+H	L	•	–		
'4538	2	16	L+H	L	•	–	N	'4528
'9601	1	14	(L+L)•H•H	–	•	–	–	
'9602	2	16	L+H	L	•	–	N	

(a) internal timing resistor. (b) generates pulse on RESET disassert, if triggering logic is asserted.

TABLE 8.9. D-TYPE REGISTERS AND LATCHES[a]

Type											
true		inverted									
D-type	latch	D-type	latch	Bits	Pins	Enable	Reset[b]	Broadside[c]	Q/Q'[d]	Output[e]	Comments
'173	–	–	–	4	16	-	A	•	-	3S	
'175	**'375**	**'175**	**'375**	4	16	-	A	-	•	2S	
'298	–	–	–	4	16	-	-	•	-	2S	MUXed input
'379	–	**'379**	–	4	16	•	-	-	•	2S	sim to '175, but CLR→EN
'398	–	**'398**	–	4	20	-	-	-	•	2S	MUXed input
'399	–	–	–	4	16	-	-	-	-	2S	MUXed input
'174	–	–	–	6	16	-	A	-	-	2S	
'378	–	–	–	6	16	•	-	-	-	2S	sim to '174, but CLR→EN
'273	–	–	–	8	20	-	A	-	-	2S	
'374	**'373**	**'534**	**'533**	8	20	-	-	-	-	3S	
'377	–	–	–	8	20	•	-	-	-	2S	sim to '374, but 3S→EN
–	**'412**	–	**'432**	8	24	•	A	-	-	3S	Intel 8212; status bit
'574	**'573**	**'564**	**'563**	8	20	-	-	•	-	3S	broadside '374/3; also '576/'580
'575	–	**'577**	–	8	24	-	S	•	-	3S	'574 with synch CLR
'825	**'845**	**'826**	**'846**	8	24	•	A	•	-	3S	Am29825
'823	**'843**	**'824**	**'844**	9	24	•	A	•	-	3S	Am29823
'821	**'841**	**'822**	**'842**	10	24	-	-	•	-	3S	Am29821; 10-bit '374
'396	–	–	–	4+4	16	-	-	-	-	2S	cascaded 4-bit reg; 8-bit out
'874	**'873**	**'876**	**'880**	4+4	24	-	A	•	-	3S	
'878	–	**'879**	–	4+4	24	-	S	•	-	3S	'874 with synch CLR
–	**'604**	–	–	8+8	28	-	-	-	-	2S	MUXed input; '605 is OC
–	**'606**	–	–	8+8	28	-	-	-	-	2S	glitch-free '604; '607 is OC

(a) see also "TRANSCEIVERS," some of which have latches. (b) A – asynchronous; S – synchronous. (c) broadside (flow-through) pinout. (d) both true and complemented outputs. (e) 2S – push-pull (2-state) output; 3S – 3-state output.

TABLE 8.10. COUNTERS

Type binary	Type BCD	Bits	Pins	Clock[a]	Load[a]	Reset[a]	U/D[b]	Direct/ latch[c]	Output[d]	Comments
'93	'90	4	14	A	-	A	-	D	2S	non-st'd Vcc, gnd; '92 is modulo-12
'161	'160	4	16	S	S	A	-	D	2S	
'163	'162	4	16	S	S	S	-	D	2S	
'169	'168	4	16	S	S	-	•	D	2S	
'191	'190	4	16	S	A	-	•	D	2S	
'193	'192	4	16	S	A	-	•	D	2S	separate U/D clock inputs
'197	'196	4	14	A	A	A	-	D	2S	
'293	'290	4	14	A	-	A	-	D	2S	'93 with st'd Vcc, gnd
'561	'560	4	20	S	B	B	-	D	3S	
'569	'568	4	20	S	S	B	•	D	3S	25LS2569/8
'669	'668	4	16	S	S	-	•	D	2S	improved '169
'691	'690	4	20	S	S	A	-	B	3S	
'693	'692	4	20	S	S	S	-	B	3S	
'697	'696	4	20	S	S	A	•	B	3S	
'699	'698	4	20	S	S	S	•	B	3S	
'4516	'4510	4	16	S	A	A	•	D	2S	
–	'4017	5	16	S	-	A	-	D	2S	decoded 1-of-10 outputs
'4024	–	7	14	A	-	A	-	D	2S	
'69	'68	8	16	A	-	A	-	D	2S	
'269	–	8	24	S	S	-	•	D	2S	skinny-DIP
'393	'390	8	14/16	A	-	A	-	D	2S	dual '93/'90
'461	–	8	24	S	S	S	-	D	3S	PAL
'469	–	8	24	S	S	-	•	D	3S	PAL
'579	–	8	20	S	S	B	•	D	3S	8 bidirectional in/out lines
'590	–	8	16	S	-	A	-	L	3S	
'591	–	8	16	S	-	A	-	L	OC	
'592	–	8	16	S	A	A	-	L	2S	8 inputs, 1 output (MAX CNT)
'593	–	8	16	S	A	A	-	L	3S	8 bidirectional in/out lines
'779	–	8	16	S	S	-	•	D	3S	8 bidirectional in/out lines
'867	–	8	24	S	S	A	•	D	2S	skinny-DIP
'869	–	8	24	S	S	S	•	D	2S	skinny-DIP
'4520	'4518	8	16	S	-	A	-	D	2S	pos or neg edge clk
'40103	'40102	8	16	S	B	A	D	D	2S	
'4040	–	12	16	A	-	A	-	D	2S	
'4020	–	14	16	A	-	A	-	D	2S	
'4060	–	14	16	A	-	A	-	D	2S	

(a) A - asynchronous; all A clock inputs count on neg edge. S - synchronous; all S clock inputs count on positive edge. B - both. (b) D - count down only. (c) B - both. (d) 2S - 2-state (totem-pole); 3S - 3-state.

TABLE 8.11. SHIFT REGISTERS

Type	Bits	Pins	Serial/parallel input	Serial/parallel output	Direction	Latches[a]	Reset[b]	Output[c]	Comments[d]
'95	4	14	P/S	P	R	-	-	2S	
'194	4	16	P/S	P	R/L	-	A	2S	
'195	4	16	P/JK	P	R	-	A	2S	
'295	4	14	P/S	P	R	-	-	3S	
'395	4	16	P/S	P/S	R	-	A	3S	
'671	4	20	P/S	P	R/L	O	A	3S	output MUXed: SR or latch; reset SR only
'672	4	20	P/S	P	R/L	O	S	3S	'671 with synch CLR
'96	5	16	P/S	P	R	-	A	2S	high load only
'91	8	14	2S	2S	R	-	-	2S	
'164	8	14	2S	P	R	-	A	2S	
'165	8	16	P/S	2S	R	-	-	2S	
'198	8	24	P/S	P	R/L	-	A	2S	
'299	8	20	P/S	P/S	R/L	-	A	3S	common I/O pins
'322	8	20	P/S	P/S	R	-	A	3S	common parallel I/O
'323	8	20	P/S	P/S	R/L	-	S	3S	'299 with synch reset
'589	8	16	P/S	S	R	I	-	3S	power-up clear (SR only)
'594	8	16	S	P/S	R	O	2A	2S	'599 is O/C; separate resets
'595	8	16	S	P/S	R	O	A	3S	'596 is O/C; reset SR only
'597	8	16	P/S	S	R	I	A	2S	reset SR only
'598	8	20	P/2S	P/S	R	I	A	3S	common parallel I/O; reset SR only
'673	16	24	S	P/S	R	O	A	2S	common serial 3S I/O; reset latch only; CS, R/W
'674	16	24	P/S	S	R	-	-	3S	common serial 3S I/O; CS, R/W
'675	16	24	S	P/S	R	O	-	2S	CS, R/W
'676	16	24	P	S	R	-	-	2S	CS

(a) O - at output; I - at input. (b) A - asynchronous; S - synchronous. (c) 2S - 2-state (totem pole); 3S - 3-state. (d) CS - chip select input; R/W - read/write input.

DIGITAL MEETS ANALOG

CHAPTER 9

Although sheer "number crunching" is an important application of digital electronics, the real power of digital techniques is seen when digital methods are applied to analog (or "linear") signals and processes. In this chapter we will begin with a brief chronology of the rise and fall of digital logic families and a review of the input and output properties of the surviving TTL and CMOS families that you are likely to use in circuit design. This is essential to understand how to interface logic families to each other and to digital input devices (switches, keyboards, comparators, etc.) and output devices (indicator lamps, relays, etc.). We will also look at n-channel MOS logic, since it is widely used in LSI functions. We will continue with the important subject of bringing digital signals on and off circuit boards, in and out of instruments, and through cables. Then we will discuss the major subject of conversion between analog and digital signals. Finally, with an understanding of these techniques, we will look at a number of applications in which combined analog and digital techniques provide powerful solutions to interesting problems.

CMOS AND TTL LOGIC INTERFACING

□ 9.01 Logic family chronology

In the prehistoric early 1960s, adventurous people who didn't want to build their logic from discrete transistors struggled with RTL (resistor-transistor logic), a simple logic family introduced by Fairchild and characterized by poor fanout and poor noise immunity. Figure 9.1 shows the problem, namely a logic threshold at one V_{BE} above ground, and miserable fanout (in some cases one output could drive only one input!) caused by passive pullup and a low-impedance current-sinking load. Those were the days of small integration, and the most complicated function you could get was a dual flip-flop, which would operate to 4MHz. We bravely built circuits with RTL; they sometimes malfunctioned when you switched on a soldering iron in the same room.

The death knell for RTL came a few years later with the introduction by Signetics of DTL (diode-transistor logic) and, soon thereafter, Sylvania's SUHL – "Sylvania Universal High Speed Logic" –

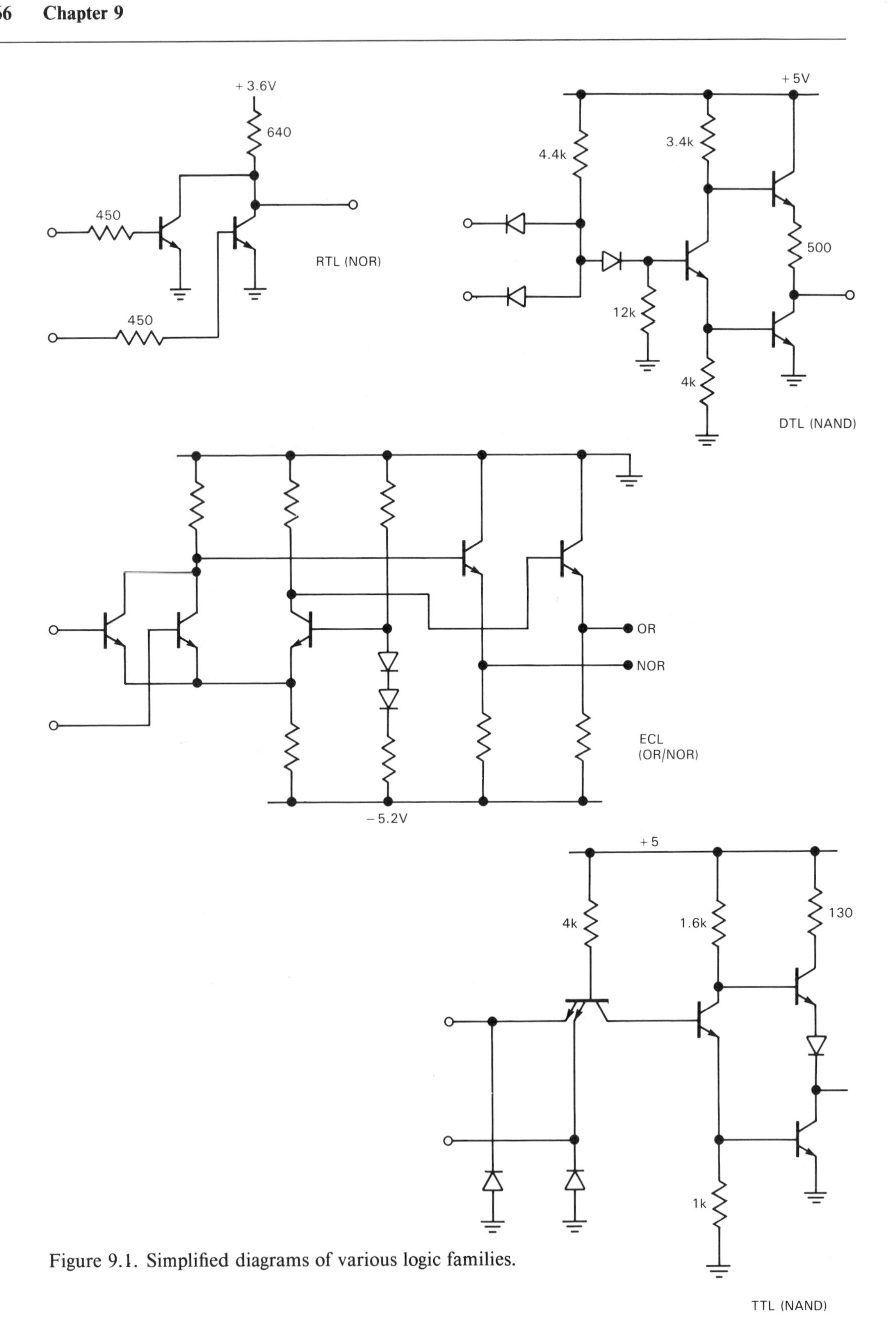

Figure 9.1. Simplified diagrams of various logic families.

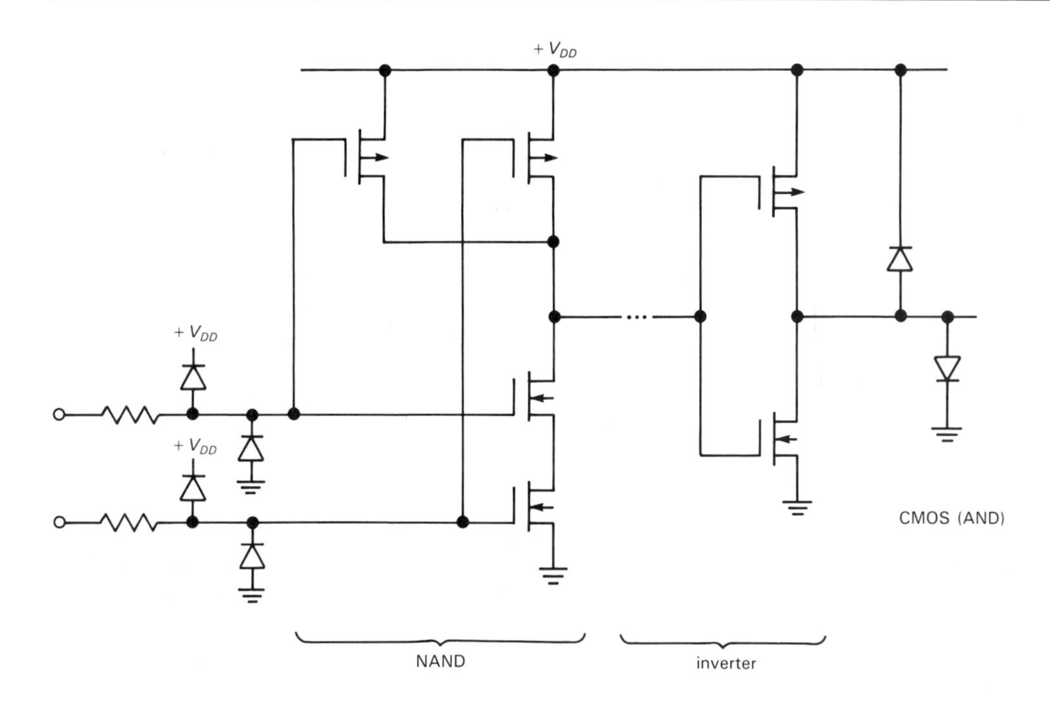

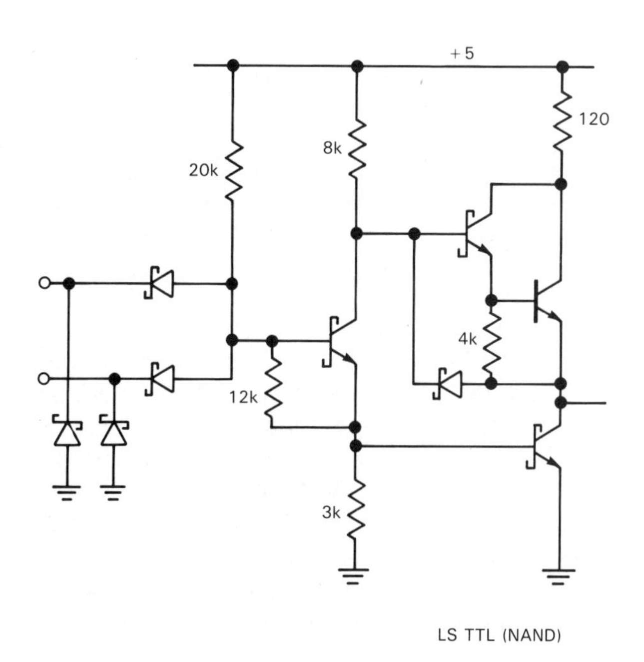

Figure 9.1 (*cont*).

which is now called TTL (transistor-transistor logic). Signetics had a popular mixture of the two, called 8000-series DCL Utilogic ("Designer's Choice Logic"). TTL caught on quickly, particularly in the "74xx" numbering system originated by Texas Instruments. These families used current-*sourcing* inputs with logic threshold at two V_{BE}s and (usually) push-pull "totem-pole" outputs (Fig. 9.1). DTL and TTL began the era of +5 volt logic (RTL used +3.6V), and offered speeds of 25MHz and fanouts of 10 (i.e., an output could drive 10 inputs). Designers rejoiced at the speed, reliability, and complex functions (divide-by-10 counters, for example) of these families. It seemed to us that you couldn't ask for more; TTL would live forever.

People are greedy, though. They wanted more speed. They wanted less power consumption. They soon got both, sort of. In the high-speed arena, a souped-up TTL (74H series; "High-speed" TTL) delivered roughly twice the speed, at twice the power! (It accomplished this underwhelming feat by cutting all resistor values in half.) Another family, ECL (emitter-coupled logic), delivered real speed (30MHz in its original version), using a negative power supply and rather closely spaced logic levels (−0.9V and −1.75V); it consumed lots of power (30mW/gate) and only came in small integration. For low power there was a souped-down TTL (74L series; "Low-power" TTL) with 1/4 the speed at 1/10 the power of the corresponding 7400 "standard" TTL.

Back at RCA the first of the MOSFET logic families was developed, 4000-series CMOS. It had zero quiescent power consumption and wide supply range (+3V to +12V). The outputs swung rail-to-rail, and the inputs drew no current. That was the good news. The bad news was the speed (1MHz at 10V supply) and the price (about $20 for a package of 4 gates). In spite of the price, a whole generation of battery-powered instrument designers grew up on the micropower CMOS, simply because there was no alternative. They learned the true meaning of static electricity as they worked with the easily damaged inputs.

This, then, was the situation at the beginning of the 1970s – two main lines of bipolar logic (TTL and ECL) and the extraordinary CMOS. The TTL variants were essentially compatible, except that 74L TTL had feeble output drive (3.6mA sink) and could drive only two standard (74-series) TTL loads (whose inputs sourced 1.6mA when held LOW). There was almost no compatibility between the major families (though a pulled-up TTL could drive CMOS, and 5V CMOS could just barely drive a single 74L TTL load).

During the 1970s there were steady improvements on all fronts. TTL sprouted the non-saturating "Schottky-clamped" (Section 13.23) families: first the 74S (Schottky) series, which made 74H obsolete by giving three times the speed at twice the power; then the 74LS ("Low-power Schottky"), which made the "standard" 74-series TTL obsolete by delivering slightly improved speed at 1/5 the power. Life with 74LS and 74S was nice, but then Fairchild came up with 74F (F for FAST: "Fairchild Advanced Schottky TTL"), which is up to 50% faster than 74S at 1/3 the power; it also has other improved properties that make it extremely nice to design with. Texas Instruments, the originators of many 74xx lines, came up with a pair of advanced Schottky families, 74AS ("Advanced Schottky") and 74ALS ("Advanced Low-power Schottky"); the former is intended to replace 74S, and the latter is intended to replace 74LS. All these TTL families have the same logic levels and plenty of output drive, and so they can be mixed within a circuit. Table 9.1 and Figure 9.2 compare the speed and power of these families.

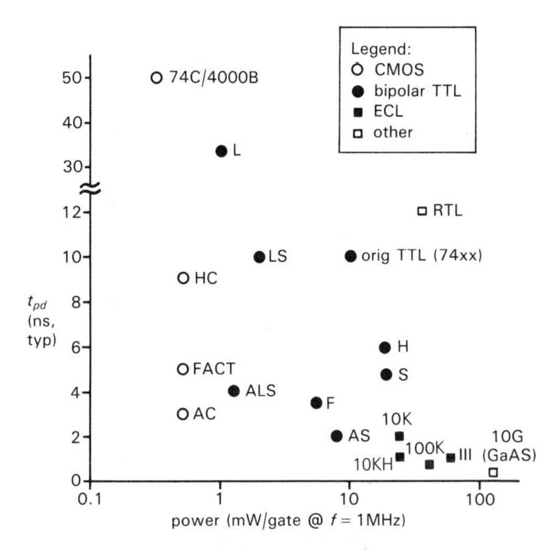

Figure 9.2. Speed versus power for various logic families.

Meanwhile, the 4000-series CMOS evolved into the improved 4000B series, with wider supply range (3V to 18V), better input protection, and higher speed (3.5MHz at 5V). The 74C series is essentially the same, with 74-family functions and pinouts to take advantage of the tremendous success of the 74-family bipolar logic. ECL sprouted the ECL II, ECL III, ECL 10,000, and ECL 100,000 series, with speeds to 500MHz.

The situation in 1980, then, was this: Most design was done with 74LS, with 74F (or 74AS) mixed in where higher speed was needed. This same TTL was used as "glue" to hold together NMOS microprocessor circuits, whose inputs and outputs are TTL-compatible. Micropower design was always done with 4000B or 74C CMOS, equivalent and compatible with each other. And for the highest speeds (100MHz–500MHz), ECL was used. There wasn't much mixing between families, except occasional combination of CMOS and TTL, or perhaps TTL interfaces to an ECL high-speed circuit.

During the 1980s came the remarkable development of CMOS logic with the speed and output drive of TTL: first 74HC ("High-speed CMOS"), with the same speed as 74LS, and, of course, zero quiescent current; then 74AC ("Advanced CMOS"), with the same speed as 74F or 74AS. With rail-to-rail output swings, and input threshold at half the supply voltage, this logic combines the best features of previous TTL and CMOS and should gradually replace bipolar TTL. There is an incompatibility, though, since the logic HIGH output of TTL or NMOS (2.4V min) is insufficient to drive the HC and AC input. Since there is likely to be a period in which you want to use some of the older bipolar TTL or NMOS, each CMOS family is offered in a variant with the lower input threshold. These are named 74HCT and 74ACT ("High-speed CMOS with TTL threshold"). Don't be tempted to use them universally – the CMOS-threshold devices (HC, AC) have better noise immunity and are the families of choice. During the 1980s, also, LSI and VLSI (microprocessors, memory, etc.) have been gradually switching over from NMOS to CMOS (with consequent low power and CMOS compatibility), at the same time increasing speed and complexity. And at the extreme high-speed end, GaAs (gallium arsenide) devices are delivering speeds of several gigahertz.

Note that all CMOS families (which includes 4000B, 74C, HC, HCT, AC, and ACT) have the pleasant characteristic of zero "static" (i.e., nothing happening) power dissipation, with typical quiescent currents less than a microamp. But CMOS does draw "dynamic" current when logic levels are switching, because of the combined effects of (a) transient rail-to-rail conduction of internal push-pull pairs during the middle of the logic swing and (b) dynamic current needed to charge and discharge internal and load capacitances. Dynamic supply currents are proportional to switching frequency and can rival bipolar logic as you reach maximum operating frequencies. See Section 8.10 (Fig. 8.18)

TABLE 9.1. LOGIC FAMILIES

Family	t_{pd}('00) (C_L=50pF) typ (ns)	t_{pd}('00) (C_L=50pF) max (ns)	f_{clk}('74) max (MHz)	P_{diss} (C_L=0) @1MHz (mW/gate)	I_{OL} @0.5V max (mA)	I_{IL} max (mA)	V_{th} typ (V)	V_{supply} min (V)	V_{supply} nom (V)	V_{supply} max (V)	Date of intro
CMOS											
AC	3	5.1	125	0.5	24	0	$V_+/2$	2	5 or 3.3	6	1985
ACT							1.4	4.5	5	5.5	
HC	9	18	30	0.5	8	0	$V_+/2$	2	5	6	1982
HCT							1.4	4.5	5	5.5	
4000B/74C @10V	30	60	5	1.2	1.3	0	$V_+/2$	3	5-15	18	1970
4000B/74C @5V	50	90	2	0.3	0.5	0					
TTL											
AS	2	4.5	105	8	20	0.5	1.5	4.5	5	5.5	1980
F	3.5	5	100	5.4	20	0.6	1.6	4.75	5	5.25	1979
ALS	4	11	34	1.3	8	0.1	1.4	4.5	5	5.5	1980
LS	10	15	25	2	8	0.4	1.1	4.75	5	5.25	1976
ECL											
ECL III	1.0	1.5	500	60	–	–	-1.3	-5.19	-5.2	-5.21	1968
ECL 100K	0.75	1.0	350	40	–	–	-1.32	-4.2	-4.5	-5.2	1981
ECL 100KH	1.0	1.5	250	25	–	–	-1.29	-4.9	-5.2	-5.5	1981
ECL 10K	2.0	2.9	125	25	–	–	-1.29	-5.19	-5.2	-5.21	1971
GaAs											
10G	0.3	0.32	2700	125	–	–	-1.3	-3.3 -5.1	-3.4 -5.2	-3.5 -5.5	1986

and Section 14.16 (Fig. 14.38) for more detail.

We wrap up our brief history with a recommendation: Use 74HC logic for nearly all your new designs, mixing in (a) 74HCT for compatibility with existing NMOS or TTL and (b) 74AC(T) for speed. Bipolar TTL (74LS/ALS and 74F/AS) are OK to use, but probably are outclassed by CMOS. Choose the older 4000B/74C where the extended supply voltage range is needed and speed is unimportant (e.g., a portable device powered by an unregulated 9V battery).

9.02 Input and output characteristics

Digital logic families are designed so that the output from a chip can properly drive many inputs within the same logic family. A typical fanout capability is 10 loads, meaning that an output from a gate or flip-flop, for example, can be connected to 10 inputs and still perform within specifications. In other words, in normal digital design practice you can get by without knowing anything about the electrical properties of the chips you're using, as long as your circuit consists only of digital logic driving more digital logic of the same type. In practice, this means that you don't often have to worry about what's actually going on at logic inputs and outputs.

However, as soon as you attempt to drive digital circuitry with externally generated signals, whether digital or analog, or whenever you use digital logic outputs to drive other devices, you must face the realities of what it takes to drive a logic input and what a logic output can drive. Furthermore, when mixing logic families it is essential to understand the circuit properties of logic inputs and outputs. Interfacing between logic families is not an academic question. In order to take advantage of advanced LSI chips, or special functions that are available in only one logic family, you must know how to mix logic types. In the next few sections we will consider the circuit properties of logic inputs and outputs in detail, with examples of interfacing between logic families and between logic devices and the outside world.

Input characteristics

The graphs in Figure 9.3 show the important properties of CMOS and TTL inputs: input current and output voltage (for an inverter) as a function of input voltage. We have extended the graphs to input voltages beyond the range normally encountered in digital circuits, since in interface situations the input signals might easily exceed the power-supply voltages. As the graphs imply, both CMOS and TTL are normally operated with the negative supply pin connected to ground.

A TTL input sources a sizable current when held LOW, and it draws only a small current when HIGH (typically a few microamperes, never more than 20μA). To drive a TTL input, you must be able to sink a milliamp or so (see Table 9.1 for exact values) while holding the input below 0.4 volt. Failure to understand this may lead to widespread circuit malfunction in interfacing situations! For input voltages below ground, a TTL input looks like a clamping diode to ground; for inputs above +5 volts, the current is set by breakdown of a diode (LS, F) or base-emitter junction (ALS, AS), with onset of current somewhere above 10 volts.

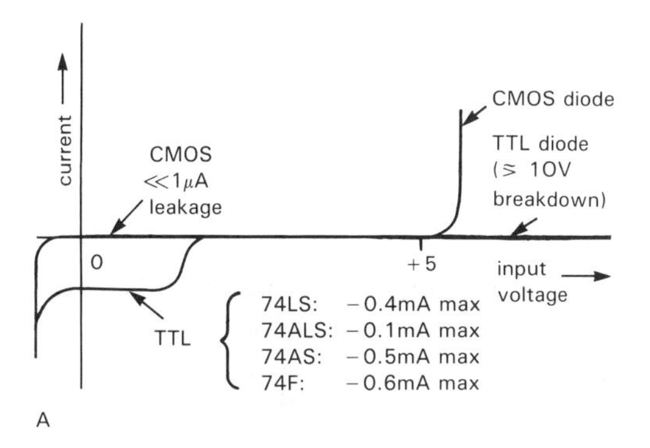

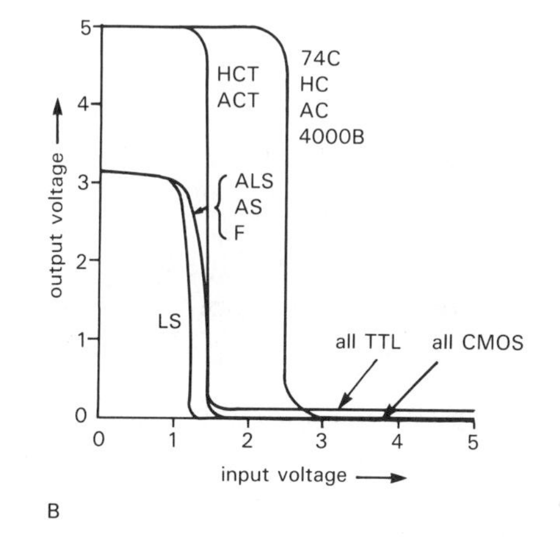

Figure 9.3. Logic gate characteristics.
A. Input current.
B. Transfer function.

The TTL input threshold is typically around +1.3 volts, although the specifications only guarantee that it is between +0.8 and +2.0 volts. TTL gates with Schmitt-trigger inputs are available ('13, '14, '132), with ±0.4 volt hysteresis; they are indicated by a hysteresis symbol within the gate outline (e.g., Fig. 9.9). V_{supply} (usually called V_{CC}) is +5.0 volts ±5%.

A CMOS input draws no current (except for leakage current, typically $10^{-5}\mu$A) for input voltages between ground and the supply voltage. For voltages beyond the supply range, the input looks like a pair of clamping diodes to the positive supply and ground (Fig. 9.1). Momentary currents greater than about 10mA through these diodes is all it takes to put many CMOS devices into SCR latchup (see Section 8.35; newer designs withstand higher currents and tend to be resistant, or immune, to this disease; for example, the inputs of HC and HCT families can be driven 1.5V beyond the supply rails without malfunction or damage). These are the famous input-protection diodes, without which CMOS would be extremely susceptible to damage from static electricity during handling (as it is, CMOS is still relatively fragile). For 4000B, 74C, 74HC, and 74AC, the input logic threshold is typically at half the supply voltage, but it can range from about one-third to two-thirds of V_+ (V_+ is usually called V_{DD}); for 74HCT and 74ACT types the input threshold is around +1.5 volts, for compatibility with TTL. As with TTL, CMOS gates with Schmitt-trigger inputs are available. CMOS supply voltages are +2 volts to +6 volts (HC, AC), +5 volts ±10% (HCT, ACT), and +3 volts to +18 volts (4000B, 74C).

Output characteristics

The TTL output circuit is an *npn* transistor to ground and an *npn* follower (or Darlington) to V_+ with a current-limiting resistor in its collector. One transistor is saturated and the other is off. As a result, a TTL device can sink a large current (8mA for 74LS, 24mA for 74F) to ground with a small (saturation) voltage drop and can source at least a few milliamps with its output HIGH (about +3.5V). The output circuit is designed to drive TTL inputs, with a fanout of 10.

The CMOS output circuit is a push-pull pair of complementary MOSFETs, one ON and the other OFF (Fig. 9.1). The output looks like a MOSFET r_{ON} to ground or to V_+ when it is within a volt or so of the respective rail, becoming something like a current source when you draw so much current that the output is forced more than a volt or two away from the supply rails. Typical values of r_{ON} are 200 ohms to 1k (4000B/74C), 50 ohms [74HC(T)], and 10 ohms [74AC(T)]. Figure 9.4 shows a summary of the output characteristics of CMOS and TTL.

We have plotted the typical output voltage, for both HIGH and LOW output states, against output current. To simplify the graph, output current is always drawn positive. Note that CMOS pulls its outputs all the way to V_+ or ground, generating a full swing unless heavily loaded; when driving only CMOS loads (zero dc current), the swing is fully rail-to-rail. TTL levels, by comparison, are typically 50 – 200mV (LOW) or +3.5 volts (HIGH) when driving other TTL devices as loads. With a pullup resistor (discussed later), HIGH TTL outputs go all the way to +5 volts.

9.03 Interfacing between logic families

It is important to know how to make different logic families talk to each other, because there are situations when you must mix logic types. For example, many desirable LSI chips are built with NMOS, which has TTL-like output levels (i.e., a HIGH output around +3V) that cannot directly drive 74HC. Another example is the nice 74C9*xx* series of counters, which you might want to use within an existing circuit built with 74LS. Or you might want to put some 5 volt logic around the edges of a 12 volt CMOS system, for easy connection to external TTL-compatible signals and for driving cables.

The three things that can keep you from connecting any pair of logic chips together

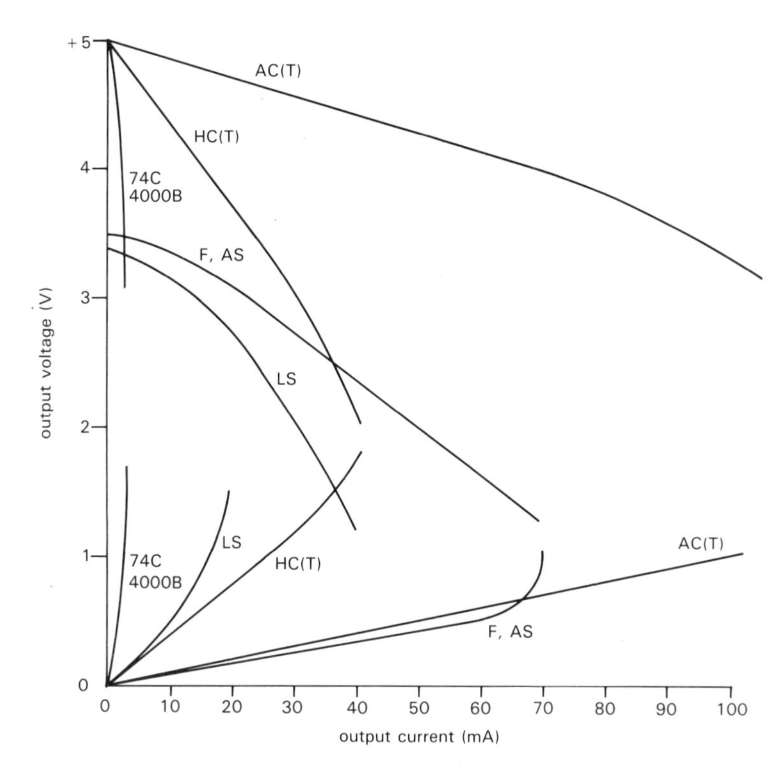

Figure 9.4. Logic gate output characteristics.

are (a) input logic-level incompatibility, (b) output drive capability, and (c) supply voltages. Rather than bore you with pages of explanations of what works and what doesn't, we've boiled down the interface problem to a simple table (Table 9.2). Let's take a quick tour.

TTL requires a +5 volt supply and typically pulls HIGH only to about +3.5 volts; it has good sinking, almost to ground. So it can drive logic with low threshold, namely TTL, HCT, ACT, and NMOS (which is routinely designed to be compatible). You need full swing to +5 volts to drive HC, AC, and 4000B/74C running at 5 volts, which you can do with a resistor pullup to +5 volts, or by interposing an HCT buffer (recall that HCT and ACT have full-swing outputs). If you use a pullup, note that the resistor value is a compromise – smaller is faster, but uses more power. A value of 4.7k is typical. A pullup resistor brings the TTL HIGH output all the way to +5 volts, though the last part of the rising waveform (during which the resistor is doing all the work) is slow. To drive high-voltage CMOS, use a level translator like the 40109, 14504, or LTC1045; they are very slow, but who cares, since you're driving slow CMOS anyway. NMOS outputs are similar to TTL, but generally with less drive capability. You therefore use the same techniques.

All CMOS families swing their outputs rail-to-rail, which means you can drive TTL, NMOS, and 5 volts CMOS directly from 5V-powered CMOS. Note, however, that old-fashioned CMOS (4000B/74C) has feeble drive when running from 5 volts (0.5mA sinking), crippling its ability to drive TTL. For all these families, use a level up-translator to drive high-voltage CMOS.

A nice solution to the CMOS-to-TTL/NMOS interface problem is to run the CMOS at reduced supply voltage; JEDEC Standard No. 8 specifies +3.3 volts, which puts the input threshold near the usual TTL value of 1.4 volts. Thus, TTL can drive 3.3V-powered HC/AC

TABLE 9.2. LOGIC FAMILY CONNECTIONS

TO → FROM ↓	TTL	HCT ACT	HC AC	HC, AC @3.3V	NMOS LSI	4000B, 74C @5V	4000B, 74C @10V
TTL	OK	OK	A	OK	OK	A	B
HCT ACT	OK	OK	OK	NO	OK	OK	B
HC AC	OK	OK	OK	NO	OK	OK	B
HC, AC @3.3V	OK	OK	NO	OK	OK	B	B
NMOS LSI	OK	OK	A	OK	OK	A	B
4000B, 74C @5V	OK[a]	OK	OK	NO	OK	OK	B
4000B, 74C @10V	C	C	C	C	C	C	OK

(a) with limited fanout. A – pullup to +5V, or use HCT as interface.
B – use *i)* OC pullup to +10V, or *ii)* 40109, 14504, or LTC1045 level translator.
C – use 74C901/2, 4049/50, 14504, or LTC1045 level translator.

directly, and vice versa. As an added bonus, 3.3 volt operation decreases dynamic power consumption (see Sections 8.10 and 14.16 and Figs. 8.18 and 14.38) by 55% from its 5 volt value, while increasing propagation delays by about 40%. Note, however, that you cannot interconnect (in either direction) 3.3 volt CMOS from other CMOS running at 5 volts.

EXERCISE 9.1.
Explain why the last statement is true.

Finally, high-voltage CMOS can drive 5 volt logic if you interpose a level translator (74C901/2, 14504, LTC1045, or 4049/4050) to generate 5 volt output swings. It is probably safe to drive LS TTL directly from high-voltage CMOS, since there are no input-protection diodes, and input breakdown usually exceeds 10 volts; however, a strict interpretation of LS specifications (7V absolute maximum input voltage) requires the translator.

Warning: Although the static logic levels are OK, there is an interesting *dynamic* incompatibility that sometimes crops up when you try to drive edge-sensitive inputs (e.g., clocking inputs of counters) of

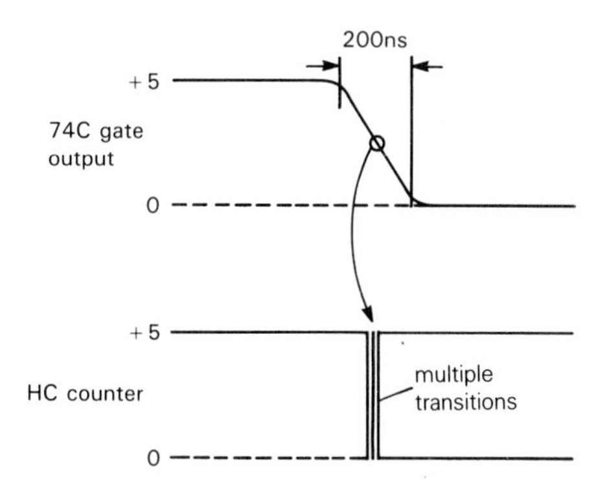

Figure 9.5. Edge-sensitive fast logic should not be driven with slow signals (e.g., slow logic parts).

HC or AC logic from the outputs of slower 4000B- or 74C-type logic. Figure 9.5 shows the multiple logic transitions you often see; sometimes the HC chip refuses to count at all, unless an oscilloscope probe (or small capacitor) is attached! Apparently the combination of slow transition time and relatively high output impedance from the sluggish CMOS is the culprit.

Figure 9.6 illustrates most of the cross-family hookups that you're likely to encounter.

9.04 Driving CMOS and TTL inputs

Switches as input devices

It is easy to drive digital inputs from switches, keyboards, comparators, etc., if you keep in mind the input characteristics of the logic you're driving. The simplest way is with a pullup resistor (Fig. 9.7). When driving TTL, a pullup resistor with switch closure to ground is by far the best method, because of the input properties

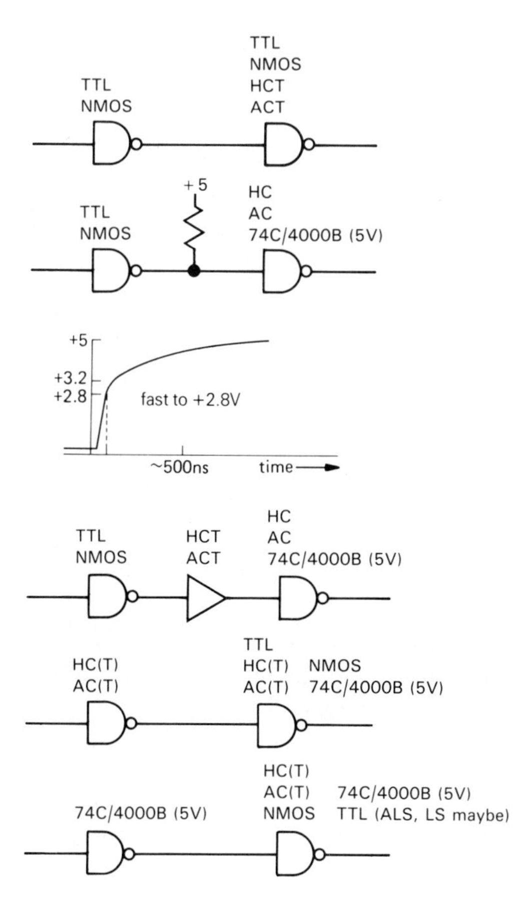

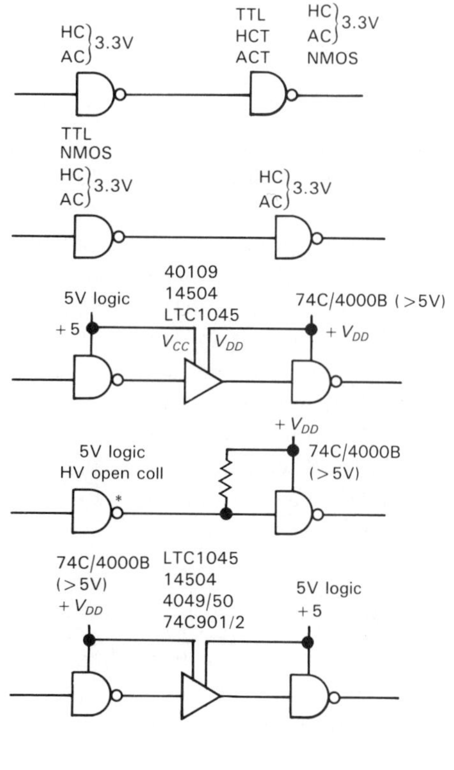

Figure 9.6. Logic family interconnections.

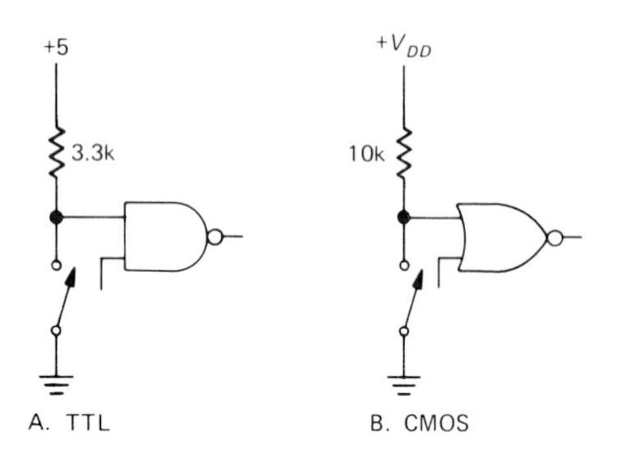

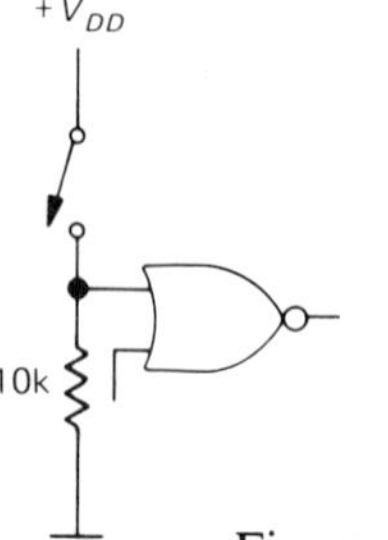

Figure 9.7. Mechanical switch to logic-level circuits (not debounced).

of TTL. The switch easily sinks the LOW-state input current, and the pullup brings the HIGH state to +5 volts, giving good noise immunity; in addition, it is convenient to have the switch return to ground.

The alternative possibility, a pulldown resistor to ground with switch closure to +5 volts, is undesirable because it requires a small pulldown resistor (such as 220 Ω) to guarantee a TTL LOW of a few tenths of a volt, which means you have rather large currents with the switch closed. In the pullup circuit, the noise immunity with the switch open (the worst case, from the standpoint of noise pickup) is at least 3 volts, whereas in the undesirable pulldown circuit it could be as little as 0.6 volt (FAST TTL, −0.6mA input current, LOW threshold at +0.8V).

Either pullup or pulldown with CMOS is fine, since the inputs draw no current and the threshold is typically halfway from V_{DD} to ground. It is usually more convenient to ground one side of the switch, but if the circuit is simplified by having a HIGH input when the switch is closed, the method with pulldown resistor will be perfectly OK. Figure 9.7 shows these three methods.

Switch bounce

As we remarked in Chapter 8, mechanical switch contacts usually bounce for about a millisecond after closure. With large switches the bounce can last for as much as 50ms. This can wreak havoc with circuits that are sensitive to changes of state, or "edges" (a flip-flop or counter would toggle many times if clocked directly from a switch input, for instance). In such cases it is essential to debounce the switch electronically. Here are a few methods:

1. Use a pair of gates to make a jam-type *SR* flip-flop. Use pullups at the inputs to the debouncer, of course (Fig. 9.8). You could use a flip-flop with SET and CLEAR inputs instead (e.g., a '74); in that case, ground the clocking input.

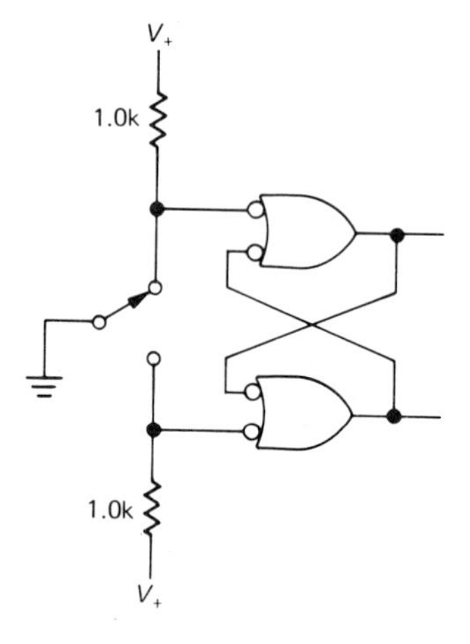

Figure 9.8. Switch debouncer: *SR* flip-flop.

2. Use an integrated version of the preceding arrangement. The '279, 4043, and 4044 are quad *SR* latches.

3. Use an *RC* slowdown network to drive a CMOS Schmitt trigger (Fig. 9.9). The low-pass filter R_2C_1 smooths the bouncy waveform so that the Schmitt-trigger gate

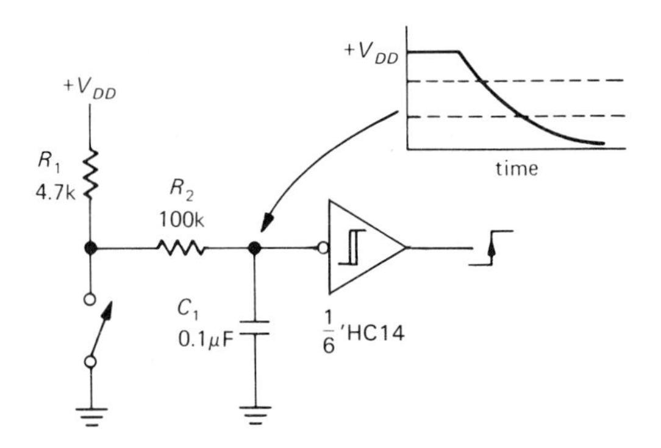

Figure 9.9. Switch debouncer: *RC* plus Schmitt trigger.

makes only one transition. A 10ms to 25ms *RC* time constant is generally long enough. This method isn't well suited to TTL because of the low driving impedance that TTL inputs require.

4. Use a chip like the 4490 "hex contact bounce eliminator," a nifty device that uses a digital delay (a 5-bit shift register for each switch) as a sort of digital low-pass filter. It includes internal pullup resistors and clock circuitry. The user supplies a timing capacitor that sets the oscillator frequency, thus determining the delay time.

5. Use the circuit shown in Figure 9.10, with either a noninverting gate or buffer.

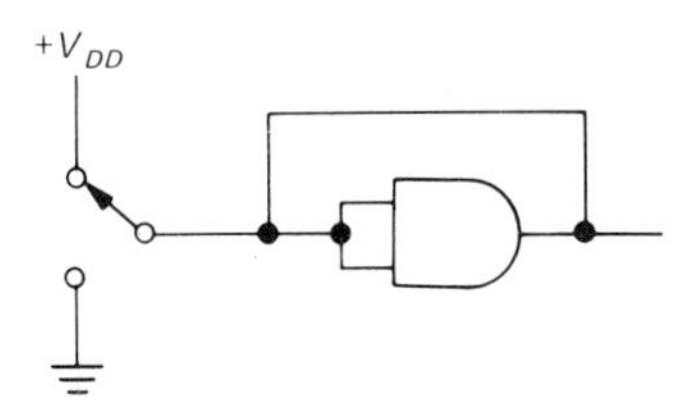

Figure 9.10. Switch debouncer: noninverting gate with feedback.

It's always OK to override a logic output by shorting it to V_+ or ground, providing the duration is kept short. In this circuit there is no problem, because the output is forced only for a gate propagation delay, after which it holds itself in the new state.

6. Use a device with built-in debouncer. Keyboard encoders, for instance, are designed with mechanical switches in mind as input devices, and they usually include debouncing circuitry.

7. Use a Hall-effect switch. These are magnetically operated solid-state switches, available as panel switches or keyboard switches. In either case the magnet and switch come as a complete assembly. They require +5 volts and produce a debounced logic output suitable for driving TTL or CMOS (operated from +5V). Since they have no mechanical contact assembly to wear out, Hall-effect switches last just about forever (although we once had an epidemic of progressive magnet anemia in a Hall-effect keyboard; we trust this disease has now been eradicated).

A few general comments on switches as input devices: Note that SPST switches (sometimes called "form A") can be used with methods 3 and 4 (and usually 6), whereas SPDT switches (form B) are necessary with the other methods. Keep in mind, also, that often it isn't necessary to debounce switch inputs, since they aren't always used to drive edge-sensitive circuitry. Another point: Well-designed switches are usually "self-wiping" to maintain a clean contact surface (take one apart to see what that means), but it's a good idea to choose circuit values so that a current of at least a few milliamps flows through the switch contacts to clean them. With suitable choice of contact material (e.g., gold) and mechanical design, switches can be designed to avoid this "dry-switching" problem and will work properly even when switching zero current.

9.05 Driving digital logic from comparators and op-amps

Comparators and op-amps, together with analog-to-digital converters, are the common input devices by which analog signals can drive digital circuits. Figure 9.11

shows some examples. In the first circuit, a comparator drives TTL directly. Since most comparators have an *npn* output transistor with open collector and grounded emitter, all that's needed is a pullup to +5 volts. The same scheme works for CMOS as well with pullup to $+V_{DD}$. The comparator may not have to run from a split supply, since many are designed for single-supply operation (V_- grounded), and some will even operate with a single 5 volt supply (e.g., the 311, 339, 393, or 372/4).

The second circuit shows an op-amp driving CMOS with only a series current-limiting resistor. The input-protection diodes of CMOS device form effective clamps to V_{DD} and ground, provided that input current is kept below 10mA. In the third circuit, an op-amp switches an *npn* transistor into saturation to drive a TTL load; the diode prevents base-emitter reverse breakdown (~6V). In this circuit you can eliminate R_1 and D_1 by replacing the *npn* transistor with an *n*-channel MOSFET. The final circuit is not highly recommended, but it does work. The input clamping diode of the TTL device limits its negative swing to a diode drop below ground, and the external diode limits positive swing. The series resistor prevents damage if base-emitter reverse breakdown occurs in the TTL input transistor. The series resistor is chosen small enough to sink the TTL LOW-state input current when the op-amp output is a few volts negative.

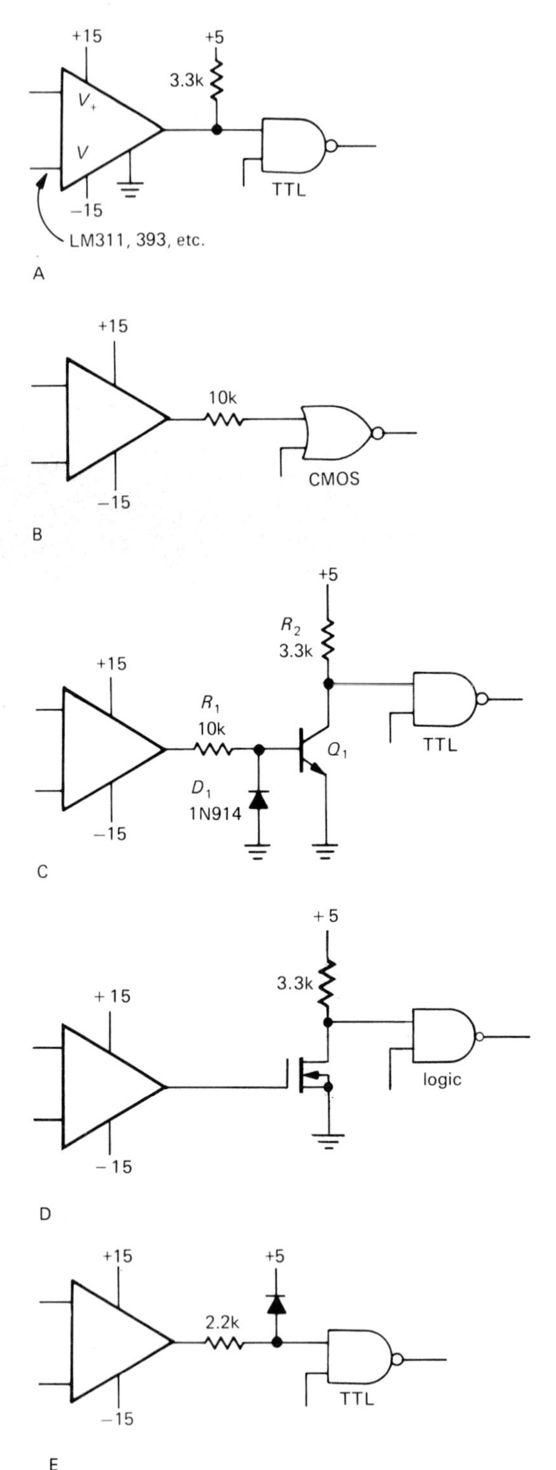

Figure 9.11. Driving logic from comparators and op-amps.

Clock inputs: hysteresis

A general comment about driving digital logic from op-amps: Don't try to drive clock inputs from these op-amp interfaces; the transition times are too long, and you may get glitches as the input signal passes through the logic threshold voltage. If you intend to drive clocking inputs (of flip-flops, shift registers, counters, monostables, etc.), it is best to use a comparator with hysteresis or to buffer the input with a gate (or other logic device) with Schmitt-trigger input. The same comment goes for signals derived from transistor analog circuitry. Figure 9.12 shows the idea. R_2 is chosen for 50mV hysteresis. A small capacitor (C_2) across the feedback resistor is a good idea to ensure fast transitions and to prevent multiple pulses as the input passes through the threshold (the 311, in particular, has a tendency to do this). Bypass capacitor C_1 is also important in preventing glitches on the reference supply during transitions. In many cases the reference voltage will be ground, in which case C_1 is omitted.

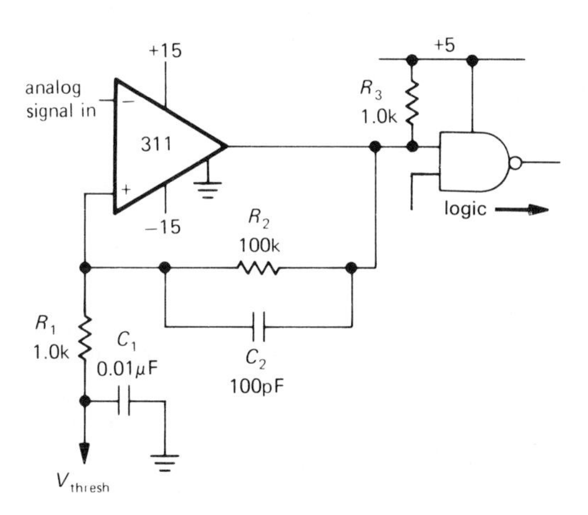

Figure 9.12. Threshold detector with hysteresis.

9.06 Some comments about logic inputs

TTL chips (and their HC and AC counterparts) tend to be designed with active LOW control inputs; for instance, flip-flops generally have SET and RESET inputs that are enabled when LOW. External digital signals used as inputs will therefore nearly always have a pullup resistor and will pull LOW (sinking current) when active, a convenient arrangement, because switches, etc., can use a common ground return. It also leads to greater noise immunity, since a line held near +5 volts has 3 volts of noise immunity, as compared with the 0.8 volt of noise immunity of a line held near ground. This intrinsic weakness of TTL logic (poor noise immunity in the LOW state) is particularly apparent when you realize that a negative 0.5 volt spike on the chip's ground line can be interpreted as a HIGH input by the chip. Such spikes are not uncommon, since they can be generated by short current spikes through the ground-line inductance. See Section 9.11 for further discussion of this troublesome problem.

With CMOS, the noise immunity is the same for either state, so you can use pullup or pulldown resistors as input terminations when driving from devices that have an open state. Pulldown resistors tend to be used more often, although you will see pullups when the driving device is something like a switch, with a ground return.

An open TTL input is "barely HIGH." It sits at the logic threshold (~1.3V), but since no current is being sunk, it does not turn on the input transistor. You may occasionally see "designs" in which inputs that should be tied HIGH on TTL devices are left open. Don't *ever* do this! It is foolish and dangerous: There is zero noise immunity on an open input, so capacitive coupling from nearby signals can generate short LOW spikes at the input. This generates glitches at the outputs of combinational devices (gates), which is bad enough, but in the case of a flip-flop or a register it can be devastating, since an open RESET′ input can clear the device at unpredictable times. The offending glitches may be impossible to see on a scope, since they may be "single-event" pulses of about 20ns duration. Although you may be able to get away with this unsavory practice most of the time, especially if there is low capacitance between the open pin and neighboring pins, it is still a very poor idea; if you try to troubleshoot the circuit by clipping a logic analyzer or test clip over the IC, you've got a new circuit, since the additional capacitance of the test device is almost certain to cause transient LOWs at the open pin(s). Besides, why build unreliable circuits when you know how to make them reliable with a few simple connections? (End of tirade.)

Unused inputs

Unused inputs that affect the logic state of a chip (e.g., a RESET input of a flip-flop) must be tied HIGH or LOW, as appropriate. Inputs that have no effect (e.g., inputs of unused gate sections in the same package) may be left unconnected in TTL, but not in CMOS. An open input of an unused CMOS gate, for instance, can float up to the logic threshold, causing the output to go to half the supply voltage, with both MOS output transistors conducting, thus drawing considerable class A current. This results in excessive supply current and can even lead to failure in devices with hefty output stages. It is best to ground all inputs of unused functions on every CMOS chip.

With TTL you can ignore unused sections of a chip, as well as irrelevant inputs of circuits you are using. For instance, you can leave the parallel-load data lines of a counter unconnected if you never enable the LOAD line.

9.07 Comparators

Comparators were introduced briefly in Section 4.23 to illustrate the use of positive feedback (Schmitt trigger) and to show that special-purpose comparator ICs deliver considerably better performance than general-purpose op-amps used as comparators. These improvements (short delay times, high output slew rate, and relative immunity to large overdrive) come at the expense of the properties that make op-amps useful (in particular, careful control of phase shift versus frequency). Comparators are not frequency-compensated (Section 4.33) and cannot be used as linear amplifiers.

Comparators provide an important interface between analog (linear) input signals and the digital world. In this section we would like to look at comparators in some detail, with emphasis on their output properties, their flexibility regarding power-supply voltages, and the care and feeding of input stages.

Supply voltages and outputs

Most comparators have open-collector outputs, suitable for driving logic inputs (with a pullup resistor, of course) or higher current/voltage loads. The 311, for instance, can drive loads connected to supplies up to 40 volts with currents up to 50mA, and the 306 can handle even more current. These comparators have a ground pin in addition to the negative and positive supply pins, so the load is pulled to ground regardless of supply voltages. Comparators intended for very high speed (521, 527, 529, 360, 361, Am686, CMP-05, LT1016, and VC7695/7) often come with active pullup output circuits. These are meant to drive 5 volt digital logic, and they usually have 4 power-supply pins: V_+, V_-, V_{CC} (+5), and ground.

One thing to watch for is the fact that many comparators require positive and negative supplies, even if the inputs never go negative. Examples are the 306, 710, and 711, in addition to the comparators with active pullup listed previously. It is inconvenient to have to generate a negative supply just to run a comparator in an instrument that otherwise uses only positive supplies, so it is important to know about comparators that will run on a single positive supply (e.g., the 311, 319, 339, 393, 365, CA3290, HA4905,

CMP-01, CMP-02, LT1016, AD790, and TLC372/4). In fact, these will all operate with a single 5 volt supply, a great advantage in a digital system. When operated from a single +5 volt supply, the 339, 393, 365, CA3290, HA4905, LT1017/18, AD790, and TLC372/4 have an input common-mode range all the way down to ground. These last comparators are intended specifically for single-supply operation and, with the exception of the 4905, 365, and 790, have only two supply pins (V_+ and ground); if operated from split supplies, the output will be pulled to V_-. Several of these also have the unusual property of being able to operate from a single supply of as little as +2 volts.

While on the subject of power supplies, it should be mentioned that some comparators are designed to run on low supply current, generally less than 0.5mA; examples are the LP311, LP339, TLC373/4, TLC339/393, TLC3702/4, CMP-04, LT1017/8, MC14574, and LP365. The last two are quad comparators with programmable operating current. The price you pay for low power is low speed, with response times measured in microseconds. See Chapter 14 for the full story on low-power electronics; Table 14.8 lists low-power comparators.

Inputs

There are some general cautions concerning input circuits of comparators. Hysteresis (Section 4.24) should be used whenever possible, because erratic switching is otherwise likely to result. To see why, imagine a comparator without hysteresis in which the differential input voltage has just passed through zero volts, slewing relatively slowly, since it is an analog waveform. A mere 2mV input differential causes the output to change state, with switching times of 50ns or less. Suddenly you have 3000mV fast digital logic transitions in your system, with current pulses impressed on the power supplies, etc. It would be a miracle if some of these fast waveforms didn't couple into the input signal, at least to the extent of a few millivolts, overcoming the 2mV input differential and thus causing multiple transitions and oscillations. This is why generous amounts of hysteresis (including a small capacitor across the feedback resistor), combined with careful layout and bypassing, are generally required to get sensitive comparator circuits to function well. It is generally a good idea to avoid driving comparator inputs directly from high-impedance signals; use an op-amp output instead. It is also a good idea to avoid high-speed comparators, which only aggravate these problems, if speed is not needed. Then, too, some comparators are more troublesome in this regard than others; we have had plenty of headaches using the otherwise admirable 311.

Another caution on inputs: Some comparators have a very limited differential input voltage range, as little as 5 volts for some types (e.g., the CMP-05, 685-7, and VT9695/7). In such cases it may be necessary to use diode clamps to protect the inputs, since excessive differential input voltage will degrade h_{FE} and cause permanent input offset errors and may even destroy the base-emitter junctions of the input stage. General-purpose comparators are generally better in this respect, with typical differential input voltage ranges of ±30 volts (e.g., the 311, 393, LT1011, etc.).

An important feature of comparator inputs is the bias current at the input terminals and the way it changes with differential input voltage. Most comparators use bipolar transistors for their input stages, with input bias currents ranging from tens of nanoamps to tens of microamps. Since the input stage is just a high-gain differential amplifier, the bias current changes as the input signal takes the comparator through its threshold. In addition, internal protection circuitry may cause a larger change in bias current a few volts away from threshold. The curve shown in Figure 9.13 (for the CMP-02) is typical. The small current "step" at zero volts (differential) is actually a smooth transition taking place over 100mV or so, and represents the voltage change necessary to switch the input differential amplifier stage fully from one state to the other.

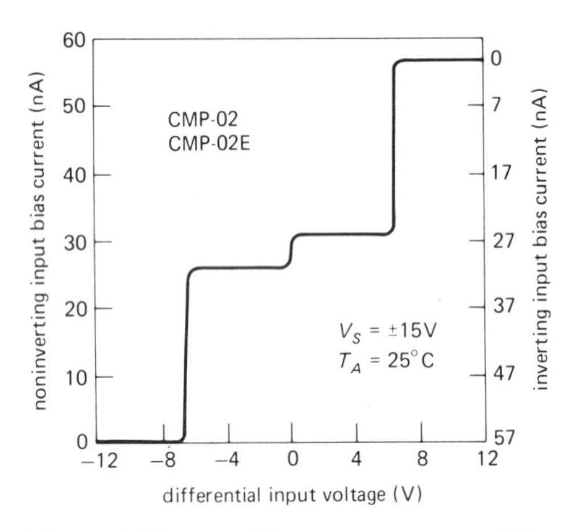

Figure 9.13. Input bias current versus differential input voltage for the CMP-02 comparator. (Courtesy of Precision Monolithics, Inc.)

For comparator applications where extremely low input current is necessary, there are FET-input comparators available. Examples are the CA3290 and TLC372, TLC3702, and TLC393 MOSFET-input dual comparators and the JFET LF311, a version of the popular 311. The latter offers maximum input currents of 50pA, compared with 250nA for the 311, with essentially no sacrifice in offset voltage or speed. In situations where the properties of a particular comparator are needed, but with lower input current, one solution is to add a matched-pair FET follower at the input.

One last comment on input properties: Thermal gradients set up on the chip from dissipation in the output stage can degrade input offset voltage specifications. In particular, it is possible to have "motorboating" (a slow oscillation of the output state) take place for input signals near zero volts (differential), since the state-dependent heat generated at the output can cause the input to switch.

Overall speed

It is convenient to think of a comparator as an ideal switching circuit for which any reversal in the differential input voltage, however small, results in a sudden change at the output. In reality, a comparator behaves like an amplifier, for small input signals, and the switching performance depends on the gain properties at high frequencies. As a result, a smaller input "overdrive" (i.e., more than enough signal to cause saturation at dc) causes a greater propagation delay and (often) a slower rise or fall time at the output. Comparator specifications usually include a graph of "response times for various input overdrives." Figure 9.14 shows some for the 311. Note particularly the reduced performance in the configuration where the output transistor is used as a follower, i.e., with less gain. Increased input drive speeds things up because the amplifier's reduced gain at high frequencies is overcome by a larger signal. In addition, larger internal amplifier currents cause internal capacitances to charge faster.

Table 9.3 compares the properties of most of the comparators currently available.

9.08 Driving external digital loads from CMOS and TTL

It is easy to drive on/off devices like lamps (LEDs), relays, displays, and even ac loads from the outputs of CMOS or TTL logic. Figure 9.15 shows some methods. Circuit A shows the standard method of driving LED indicator lamps from 5 volt logic. For TTL, which is better at sinking than sourcing, the LED is returned to +5 volts; for CMOS you can return the LED either

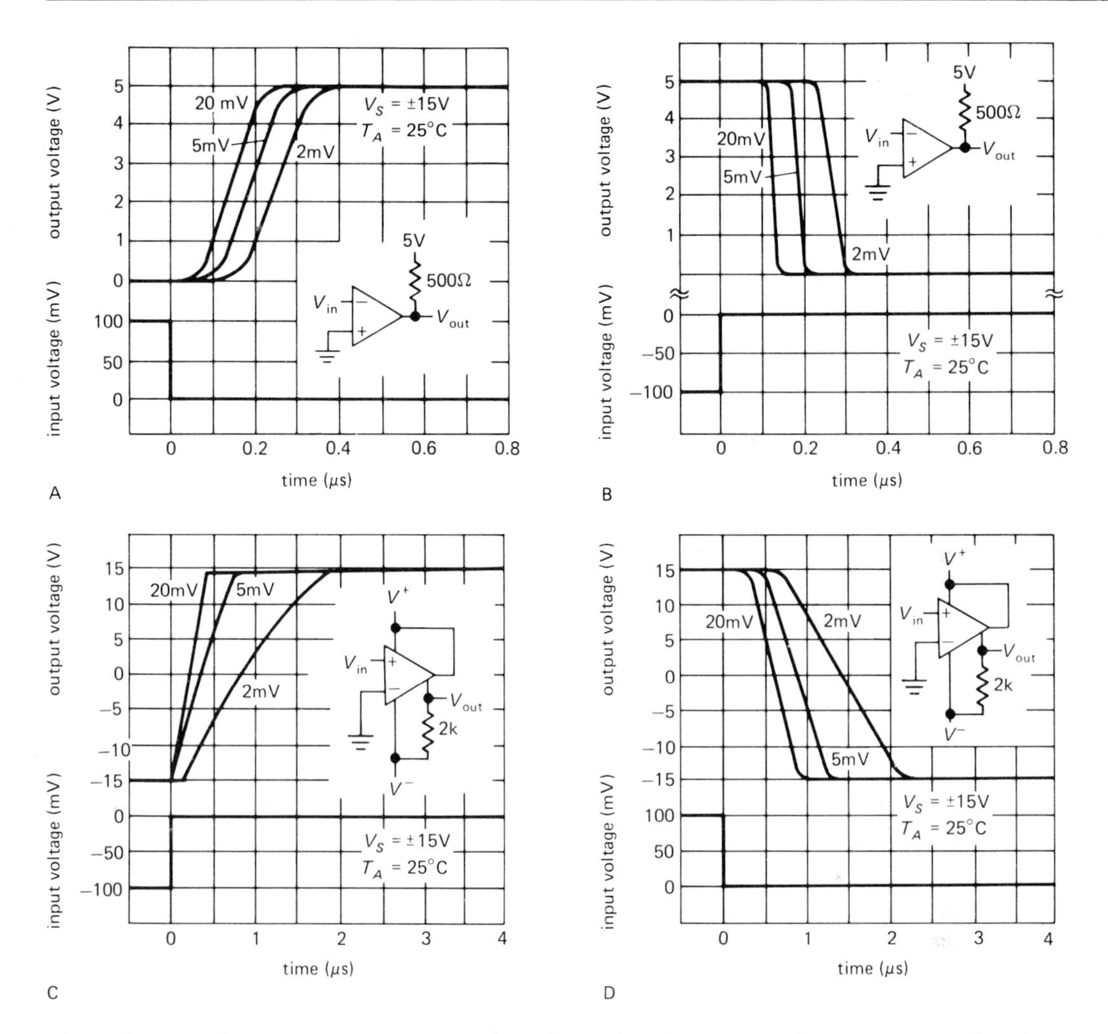

Figure 9.14. LM311 comparator response times for various input overdrives. (Courtesy of National Semiconductor Corp.)

to V_+ or to ground. The LED behaves like a diode with a forward drop of 1.5 to 2.5 volts at typical operating currents of 5mA to 20mA; with some of the newer high-efficiency LEDs (you can get some real dazzlers from Stanley) you get plenty of light with just a few milliamps. Instead of using a discrete LED plus resistor, you can get LEDs with integral current-limiting resistor (or current regulator) from most manufacturers; check the catalogs of Dialight, General Instrument, Siemens, and Hewlett-Packard.

Circuit B shows how to drive a 5 volt low-current relay directly from logic, sinking current as in circuit A. The diode is essential to clamp the inductive spike. The relay shown comes in a standard DIP IC package and has 500 ohms coil resistance (10mA, within the drive capability of most 5 volt logic). For driving higher-voltage loads, the schemes shown in circuits C through E are effective. In circuit C, a 74LS26 open-collector gate with 15 volt rating drives a 12 volt relay, and in circuit D a 75451 "dual peripheral driver" chip rated at up to 300mA and 30 volts drives an unspecified load. Open-collector

TABLE 9.3. COMPARATORS

							CM input range abs max[d]		Abs max diff'l input[e]	Power supplies			
										Positive		Negative	
Type	Mfg[a]	Qty per pkg	t_r typ[b] (ns)	V_{os} max (mV)	I_b max (μA)	CM to V_-[c]	min (V)	max (V)	(V)	min (V)	max (V)	min (V)	max (V)
CMP01C	PM	1	110	2.8	0.9	–	V_-	V_+	11	5	30	0	-30
CMP02	PM	1	190	0.8	0.003	–	V_-	V_+	11	5	30	0	-30
CMP04	PM	4	1300	1	0.1	•	-0.3	30	36	3	36	0	-30
CMP05	PM	1	40	0.6	1.2	–	V_-	V_+	5	5[i]	6	-5.2[i]	-18
LM306	NS	1	28	6.5	5	–	-7	7	5	12[i]	-	-3	-12
LM311	NS	1/2	200	3	0.1	–	V_+-30	V_-+30	30	5	30	0	-30
LF311	NS	1	200	4	0.00005	–	V_+-30	V_-+30	30	5	30	0	-30
LP311	NS	1	2000	7.5	0.1	–	V_-	V_-+30	30	3	30	0	-30
LM319	NS	2	80	4	0.5	–	V_-	V_+	5	5	30	0	-30
LM339	NS	4/2	1300	5	0.25	•	-0.3	36	36	2	36	-	-
LP339	NS	4	10000	5	0.025	•	-0.3	36	36	2	36	-	-
TLC339	TI	4/2	2500	5	0.005[t]	•	-0.3	V_+	18	3	16	-	-
TLC3702	TI	2/4	2500	5	0.005[t]	•	-0.3	V_+	18	3	16	-	-
LM360	NS	1	14	5	20	–	V_-	V_+	5	4.5	6.5	-4.5	-6.5
LM361	NS	1	14	5	30	–	-6	6	5	5	15[j]	-6	-15
LP365	NS	4	2000	6	0.075	•	-0.3	36	36	2	36	0	-36
TLC372	TI	2/4	650	10	1pA[t]	•	-0.3	18	18	2	18	-	-
CMP404	PM	4	3500	1	0.05	•	-0.3	V_+	V_+	5	30	-	-
TL510C	TI	1/2	30	3.5	20	–	-7	7	5	10	14	-5	-7
NE521	SN	2	11	7.5	20	–	-5	5	6	5[i]	-	-5[i]	-
NE522	SN	2	14	7.5	20	–	-5	5	6	5[i]	-	-5[i]	-
NE527	SN	1	33	6	2	–	-6	6	5	5	10	-6	-10
NE529	SN	1	20	6	20	–	-6	6	5	5	10	-6	-10
Am685	AM	1	6	2	10	–	-4	4	6	6[i]	-	-5.2[i]	-
Am686	AM	1	9	2	10	–	-4	4	6	5[i]	-	-6[i]	-
Am687	AM	2	7	2	10	–	-4	4	6	5[i]	-	-5.2[i]	-
Bt687	BT	2	1.8			–	-3.3	3.3		5[i]	-	-5.2[i]	-
AD790K	AD	1	35	0.3	3	•	-18	18	15	3.5	18	0	-18
TL810C	TI	1/2	30	3.5	20	–	-7	7	5	10	14	-5	-7
LT1011	LT	1	150	0.5	0.025	–	V_--0.6	V_++0.6	36	3	36	0	-36
LT1016	LT	1	10	3	10	–	V_-	V_+	5	4.5	7	0	-7
LT1017	LT	2	20000	1	0.015	•	-0.3	40	40	1.1	40	-	-
LT1018	LT	2	6000	1	0.075	•	-0.3	40	40	1.1	40	-	-
LT1040	LT	1	80000	0.5	0.0003[t]	•	V_-+0.3	V_+-0.3		2.8	16	-	-
SP1650B	PL	2	3.5[k]	20	10	–	-3	2.5	5	5[i]	-	-5.2[i]	-
EL2018C	EL	1	20	3	0.3	–	V_-	V_+	36	4.5	18	-4.5	-18
EL2019C	EL	1	6[l]	5	0.3	–	V_-	V_+	36	4.5	18	-4.5	-18
CA3290A	RC	2	1000	10	0.00004	•	V_--5	V_++5	36	4	36	-	-
RC4805A	RA	1	22	0.3	1.2	–	-4	4	3	4.5	5.5	-4.5	-16.5
HA4905	HA	4	150	7.5	0.15	•	V_-	V_+	15	5	30	0	-30
VC7695	VT	1	1.5	5	5[n]	–	-5	5	3.5	-	6	-	-6
VC7697	VT	2	2	5	5[n]	–	-5	5	3.5	-	6	-	-6
SP9685	PL	1	2.3	5	20	–	-5	3	5	5[i]	-	-5.2[i]	-
SP9687	PL	2	2.8	5	20	–	-5	5	5	5[i]	-	-5.2[i]	-
MC14574	MO	4	20000	30	0.00005	•	-0.5	V_++0.5	V_+	3	15	-	-
SP93808	PL	8	1.0	3.5	9	–	V_-	V_+	3.8	1.5	7.3	-4.9	-5.5

(a) see footnote to Table 4.1. (b) 100mV step with 5mV overdrive. (c) operating input common-mode range includes negative supply. (d) maximum range without input breakdown; will not operate properly over entire range. (e) maximum allowable voltage between input terminals. (f) ability to accept signals of both polarities and drive unipolar logic. (g) E - output (open NPN emitter) is designed to drive ECL logic; G - output pulled

Type	Total supply min (V)	Total supply max (V)	Gain typ	Output parameters: Gnd pin?	Single +5 OK?	Drive TTL?	±Sig to logic?[f]	OC output?	Active pullup?	Q and Q' out?	Strobe?	Latch?	V_{OL}[g]	Pullup V_{max}[h]	Comments
CMP01C	5	36	500k	–	•	•	•	–	•	–	–	–	R	32	
CMP02	5	36	500k	–	•	•	•	–	•	–	–	–	R	32	precision
CMP04	3	36	200k	•	•	•	–	•	–	–	–	–	G	30	precision 339
CMP05	9.5	24	16k	•	–	•	•	–	•	–	–	•	G	-	fast precision
LM306	-	30	40k	•	–	•	•	–	•	–	•	–	G	24	high output current
LM311	4.5	36	200k	•	•	•	•	•	–	–	–	–	R	40	may osc; popular; dual=2311
LF311	4.5	36	200k	•	•	•	•	•	–	–	–	–	R	40	JFET 311
LP311	3	36	200k	•	•	•	•	•	–	–	–	–	R	40	low power 311
LM319	4.5	36	40k	–	•	•	•	•	–	–	–	–	R	36	
LM339	2	36	200k	–	•	•	–	•	–	–	–	–	G	30	favorite; low pwr; dual=393
LP339	2	36	500k	–	•	•	–	•	–	–	–	–	G	30	low power 339
TLC339	3	16	-	–	•	•	–	•	–	–	–	–	G	18	MOSFET; dual=393
TLC3702	3	16	-	–	•	•	–	–	•	–	–	–	G	V_+	MOSFET; quad=3704
LM360	9	13	3k	•	–	•	•	–	•	•	–	–	G	-	similar to 760
LM361	11	30	3k	•	–	•	•	–	•	•	•	–	G	7	similar to 529
LP365	4	36	300k	•	•	•	•	•	–	–	–	–	R	36	prog; specs @ I_{set} = 10μA
TLC372	2	18	200k	•	•	•	–	•	–	–	–	–	G	18	MOSFET; quad=374
CMP404	5	30	400k	•	•	•	–	•	–	–	–	–	G	-	low power
TL510C	15	21	33k	•	–	•	•	–	•	–	•	–	G	. -	TL514C=dual
NE521	9.5	10.5	-	•	–	•	•	–	•	–	•	–	G	-	
NE522	9.5	10.5	-	•	–	•	•	–	•	–	•	–	G	-	
NE527	10	20	-	•	–	•	•	–	•	•	•	–	G	15	529 with Darlington
NE529	10	20	-	•	–	•	•	–	•	•	•	–	G	15	
Am685	9.7	14	1600	•	–	–	•	–	–	•	•	•	E	-	ECL; also CMP-07
Am686	9.7	14	-	•	–	•	•	–	•	–	•	–	G	-	fastest TTL comp
Am687	9.7	14	-	•	–	–	•	–	–	•	•	•	E	-	ECL
Bt687	-	12	100	•	–	–	•	–	–	•	•	•	E	-	ECL; fastest 687-type
AD790K	3.5	36	10k	•	•	•	•	–	•	–	–	•	G	-	fast single +5V
TL810C	15	21	33k	•	–	•	•	–	•	–	–	–	G	-	510 w/o STB; 820C=dual
LT1011	3	36	500k	–	•	•	•	•	–	–	•	–	R	50	improved 311
LT1016	5	14	3k	•	•	•	•	–	•	•	–	•	G	-	fastest single +5V
LT1017	1.1	40	500k	•	•	•	–	–	•	–	–	–	G	-	low power
LT1018	1.1	40	2M	•	•	•	–	–	•	–	–	–	G	-	low power
LT1040	2.8	16	-	•	•	•	•	–	•	•	•	–	G	-	micropower, sampling
SP1650B	-	-	-	•	–	–	•	–	–	–	•	•	E	-	ECL; 1651 is faster
EL2018C	9	36	40k	•	–	•	•	–	•	–	•	•	G	-	fast, accurate, HV
EL2019C	9	36	-	•	–	•	•	–	•	–	•	•	G	-	fast, HV, clocked
CA3290A	4	36	150k	•	•	•	–	•	–	–	–	–	G	36	MOSFET
RC4805A	-	22	20k	•	–	•	•	–	•	–	–	•	G	-	fastest precision
HA4905	5	33	400k	–	•	•	•	–	•	–	–	–	R	-	flexible output stage
VC7695	-	12	-	•	–	–	•	–	–	•	–	•	E	-	ultra-fast
VC7697	-	12	-	•	–	–	•	–	–	•	–	•	E	-	fastest dual
SP9685	-	12	300	•	–	–	•	–	–	•	•	•	E	-	ECL; fast Am685
SP9687	-	12	300	•	–	–	•	–	–	•	•	•	E	-	ECL; fast Am687
MC14574	3	18	100k	•	•	•	–	–	•	–	–	–	G	V_+	CMOS, prog; specs@100μA
SP93808	6.5	13	20	•	–	–	•	–	•	•	–	•	E[o]	-	ultra-fast octal

to GND; R - low output is saturated NPN transistor to terminal that may be operated at voltages other than GND. (h) maximum voltage to which output can be externally pulled up. (i) nominal. (j) and additional +5V logic supply. (k) 100mV overdrive. (l) t_{setup}. (n) offset current. (o) use -5.2 and -10V supplies for ECL output. (t) typical.

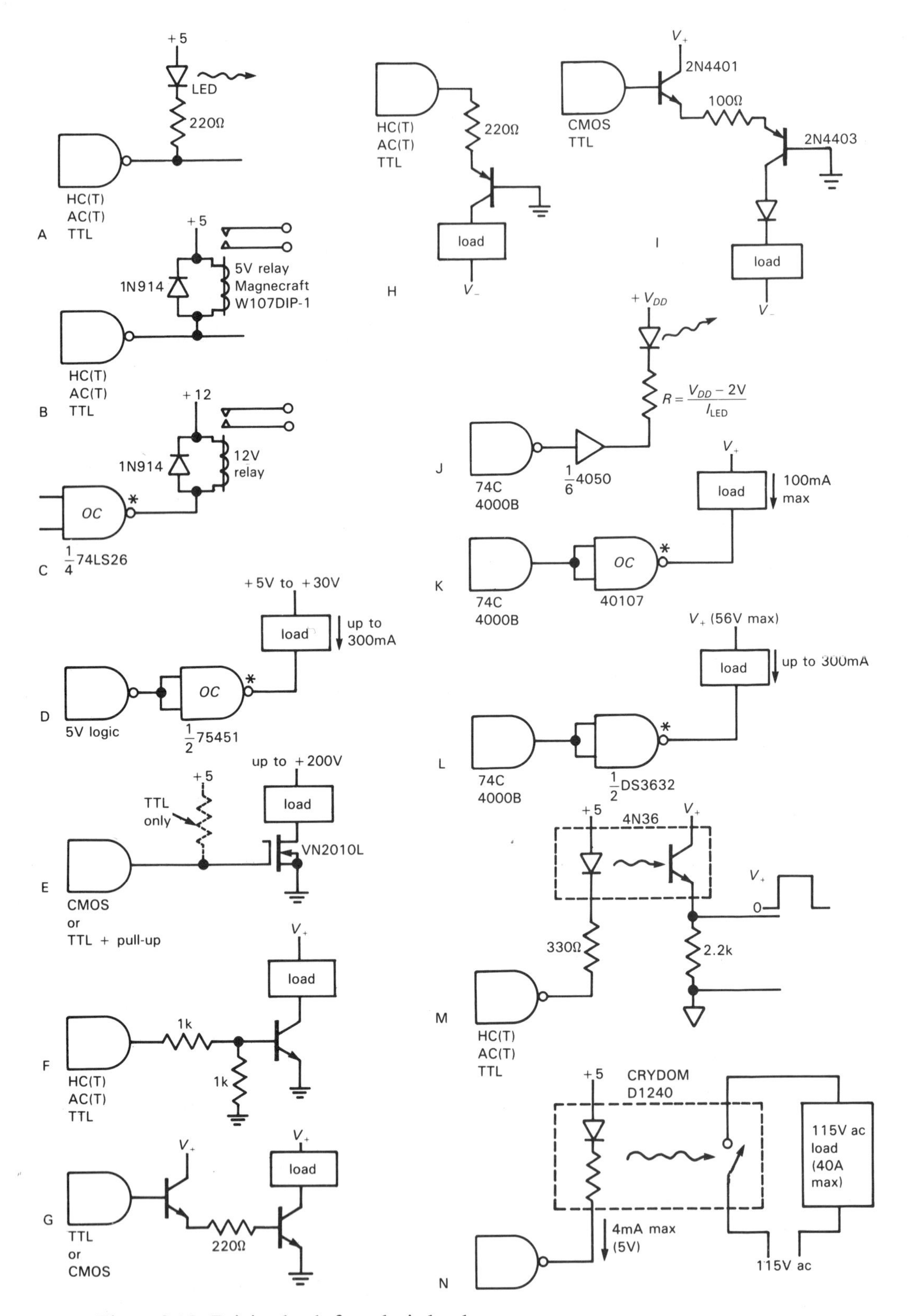

Figure 9.15. Driving loads from logic levels.

devices like these are available with 80 volt ratings and above, and with even higher current capability; look at the DS3600 series from National and the series of power drivers from Sprague (UCN/UDN/ULN), which include some very nice octal drivers in DIPs. In circuit E, we've used a low-threshold *n*-channel power MOSFET, especially convenient because of its high input impedance; if driven with TTL output levels, it's best to use a pullup to guarantee good conduction, since TTL's minimum guaranteed HIGH output (2.4V) is otherwise too low.

The schemes mentioned previously may not be practical with HC, LS, or 74C logic, because of their limited output drive (5, 8, and 3.5mA sinking, respectively). It may be a good idea to set aside a chip like the 74AS1004, a robust hex inverter (48mA sink or source), for driving large LEDs. When driving high-current loads directly from logic chips, take particular care to use a hefty ground wire to the chip, since the current from the load returns through the chip into the logic supply ground. In some cases a separate ground return path may be desirable.

Circuit F shows how to use an *npn* transistor to switch a high-current load from 5 volt logic. To switch higher currents, use a second transistor, as in G. The circuits H and I show one way to drive loads returned to a negative supply. A HIGH output state turns on the *pnp* transistor, pulling the collector into saturation at one diode drop above ground. In circuit H, the resistor (or positive current limit of the gate) sets the emitter current, and therefore the maximum collector (load) current, whereas in the improved circuit I, an *npn* follower is used as a buffer, while a diode in series with the output keeps the load from swinging above ground. In both cases the maximum load current is equal to the drive current to the *pnp* transistor's emitter. Similar IC circuits are available, with CMOS/TTL-compatible inputs and high-voltage outputs with up to a few hundred milliamps drive capability; check out National's DS3687 (300mA, −56V) and the ever-popular UDN series from Sprague.

If you're using the low-current 4000B/74C logic, with barely a milliamp of output drive, it is generally necessary to use some sort of power driver, even to light LEDs. Circuit J shows the trusty 4050 hex buffer driving an LED. It can typically sink 5mA to 50mA with 5 to 15 volt supplies, respectively (output drive capability increases with increasing supply voltage). The circuits K and L show the application of even heftier drivers: The 40107 has a large "open-drain" *n*-channel MOS output transistor good for sinking 16mA to 50mA (5-15V supply voltages, respectively), and the DS3632 uses an *npn* Darlington driver, good for 300mA. Of course, discrete external transistors can always be used, as in circuits G and I, but they will be limited to less than a milliamp of base drive current. The discrete *n*-channel MOSFET solution used in circuit E works particularly well with gutless CMOS.

For driving remote loads, or loads with an independent ground system, the best method is an opto-coupler. This is a device with an LED (on the driver side) illuminating a photodetector (on the load side). Opto-couplers are available in various speeds, and with different input/output configurations [logic-compatible input or bare LED; logic output, saturated transistor (or Darlington) output, MOSFET output, or SCR or triac output; see Fig. 9.26]. A typical example is the popular 4N36 shown in circuit M, with bare LED input, *npn* transistor output, isolation of 2500 volt, and 4μs switching speed. It has a minimum current transfer ratio of 1.0, so you only need to drive the LED with a current equal to the maximum output current, as shown. There are some opto-couplers that conveniently use logic levels at both input and output. An example is the 74OL6000 from General Instrument;

it has LS inputs and outputs, 60ns propagation times (15MHz), and 2500 volt isolation. It costs $3 in quantity.

For driving ac loads, the easiest method is to use a "solid-state relay," as in circuit N. It is an optically coupled triac with logic-compatible input and 1 to 40 amp load-current capability when switching a 115 volt ac load. The low-current variety are available in DIP packaging (e.g., the "chipswitch" series from International Rectifier), while the heftier ones come as a chassis-mounting rectangular block roughly 2 inches square. Alternatively, you can switch ac loads with an ordinary relay, energized from logic. However, be sure to check the specifications, because most small logic-driven relays cannot drive heavy ac loads, and you may have to use the logic relay to drive a second larger relay. Most solid-state relays use "zero-crossing" (or "zero-voltage") switching, which is actually a combination of zero-voltage turn-on and zero-current turn-off; it is a desirable feature that prevents spikes and noise from being impressed onto the power line. Much of the "garbage" on the ac power line comes from triac controllers that don't switch at zero crossings, e.g., phase-controlled dimmers used on lamps, thermostatic baths, motors, etc. As an alternative to the optical coupling used internally in circuit N, you sometimes see a pulse transformer used to couple triggering pulses to a triac or SCR.

When driving 7-segment numeric displays, it is usually simplest to take advantage of one of the decoder/driver combinations available. They come in a bewildering variety, including LED and LCD drivers, sinking or sourcing polarities, etc. Typical examples are the 74HC4511 (common cathode LED) and 74HC4543 (LCD) latch/decoder/drivers. More on this in the opto-electronics section (Section 9.10).

9.09 NMOS LSI interfacing

Most large scale and very large scale integrated circuits (LSI, VLSI) are now being designed as CMOS, with pretty much the same interfacing properties as the 5 volt CMOS logic gates and other MSI (medium-scale integration) functions we just discussed. However, for a long time, LSI and VLSI chips were made from *n*-channel enhancement-mode MOSFETs alone, because of simpler processing and higher density. Because of the widespread use of this NMOS logic, it is important to know how to interface between NMOS and CMOS/TTL and how to connect between NMOS inputs/outputs and external discrete circuitry. Most NMOS LSI chips are designed for TTL compatibility; however, there are some subtle points to be considered.

NMOS inputs

The input circuit of an *n*-channel MOS IC intended for use with TTL driving circuitry is shown in Figure 9.16. Q_1 is

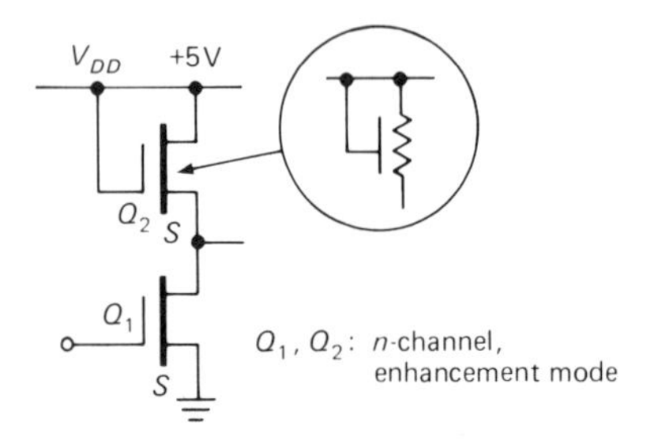

Figure 9.16. NMOS logic input circuit.

an inverter, and Q_2 is a small-geometry source follower providing pullup current (resistors take up too much space, so MOSFETs are universally used as drain loads); the alternative symbol shown for Q_2 is widely used. The threshold voltage of the input transistor is in the range of 1 to 1.5 volts in modern silicon-gate designs, so it can be driven directly from either

TTL or CMOS. In some older designs the threshold can be in the range of 2 to 3 volts, for which it is best to use a 1k–10k pullup when driving from TTL; as usual, no pullup is needed with a CMOS driver.

NMOS outputs

The output stage of 5 volt NMOS logic is shown in Figure 9.17. Q_1 is a switch, and Q_2 is a source follower. To put the output in the LOW state, Q_1's gate is brought to +5 volts, forcing the output below 0.5 volt even while sinking a few milliamps. The situation in the HIGH output state is dismal by comparison: At a minimum TTL HIGH output of +2.4 volts, Q_2 finds only 2.6 volts from gate to source, resulting in relatively high R_{ON}, with the situation rapidly deteriorating for higher output voltages.

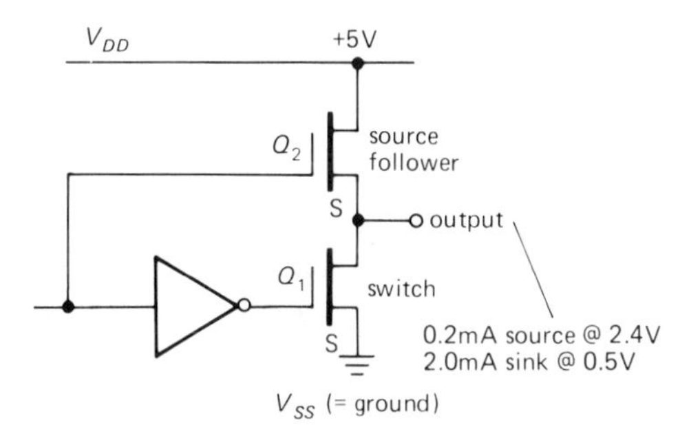

Figure 9.17. NMOS logic output circuit.

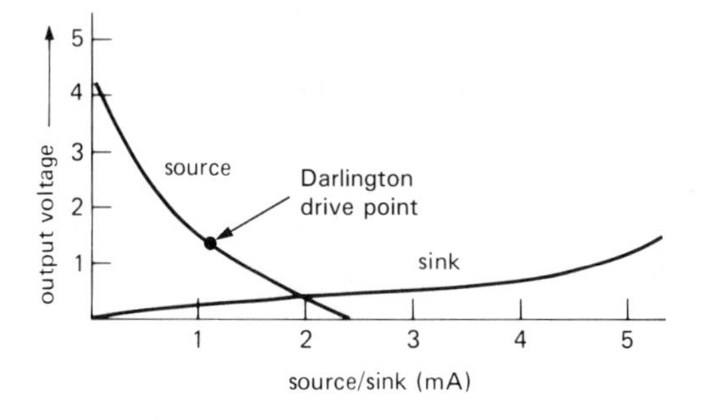

Figure 9.18. Typical NMOS output source and sink current capability.

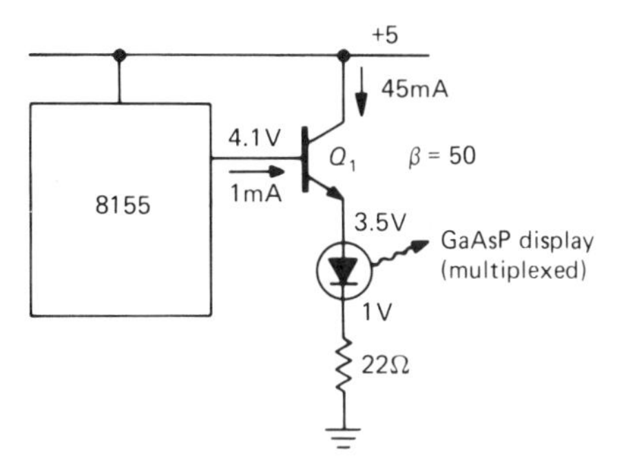

Figure 9.19

The graph in Figure 9.18 illustrates the situation. As a result, the NMOS output drive capability might be only 0.2mA (sourcing) at +2.4 volts output. This is fine for driving TTL inputs, but it is marginal for 5 volt CMOS (use a pullup, or interpose an HCT or ACT gate), and it is a disaster in a circuit like that shown in Figure 9.19.

To operate the LED with multiplexed-display current levels (25-50mA while ON), the NMOS output would have to put out about 1mA at +4.1 volts. That's impossible, since V_{GS} would then be only 0.9 volt, possibly even below the threshold voltage for the FET. Remember, too, that all 5 volt logic circuits should work with supply voltages as low as +4.5 volts to allow for ±10% tolerance of supply voltage. To drive LEDs (or other high-current devices) from NMOS, the circuits in Figure 9.20 are good.

In the first circuit, the NMOS output LOW sinks 2mA, driving the *pnp* transistor into hard conduction. In the second circuit, a Darlington *npn* transistor is switched ON by the small output current of the NMOS device while in the HIGH state. This circuit clamps the HIGH output two diode drops above ground, which might seem a bit unfriendly, but it turns out that NMOS outputs are designed so that they can be shorted to ground in this manner, with output currents controlled at small enough values so that they can drive

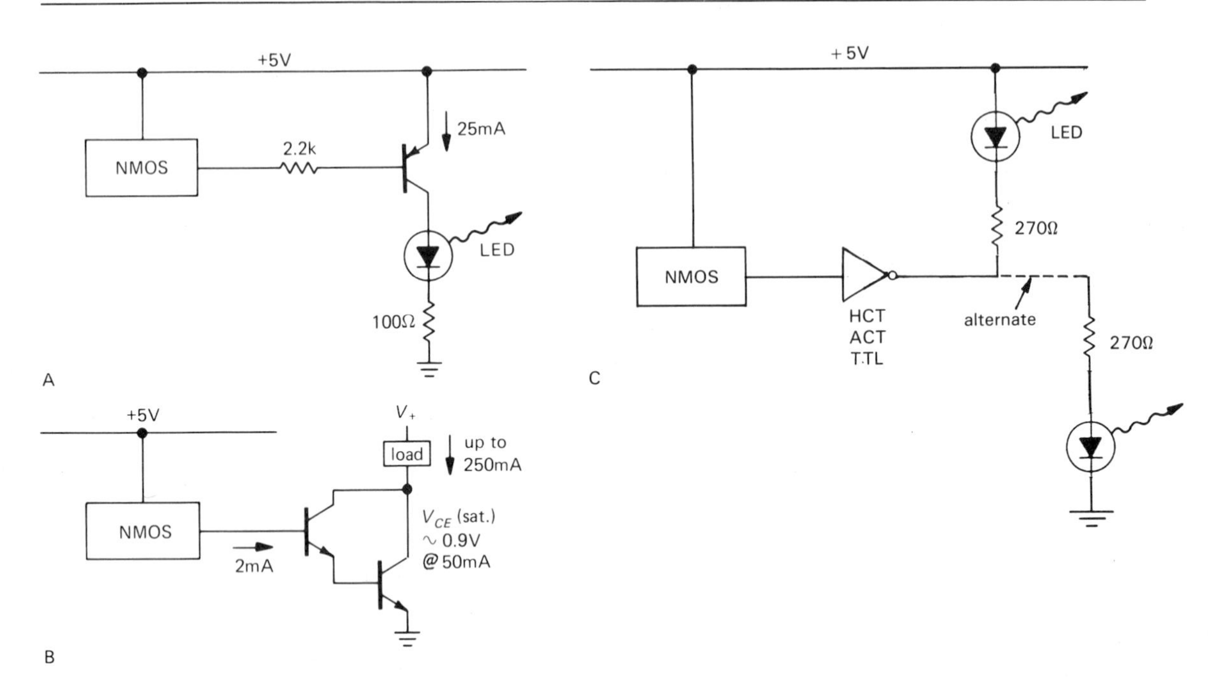

Figure 9.20. NMOS logic outputs driving loads.

the base of a grounded emitter Darlington without damage. A typical NMOS output would source 2mA into the Darlington's base at +1.5 volts, giving an output sink capability of 250mA at 1 volt with an IC like the 75492 hex Darlington array. The ULN series from Sprague has several hex and octal Darlington arrays in DIP packages.

Finally, you can always buffer NMOS outputs with an HCT or ACT (or even TTL) buffer, since NMOS outputs are fully compatible with those families' inputs. The buffered outputs can then sink current from a load; with HCT or ACT you can equally well *source* current, since the CMOS families have the same output drive capability whether sinking or sourcing.

9.10 Opto-electronics

In the last two chapters we have been using LED indicators and LED numeric display devices in various circuit contexts, as we needed them. LEDs belong to the general area of *opto-electronics*, which includes displays based on other technologies as well, notably liquid crystals, fluorescence, and gas discharge. It also includes optical electronics used for purposes other than indicators and displays: light-coupled isolators ("opto-isolators"), solid-state relays, position sensors ("interrupters" and "reflective sensors"), diode lasers, array detectors ("charge-coupled devices," CCDs), image intensifiers, and a variety of components used in fiber optics.

Although we will continue to conjure up miscellaneous magic devices as we need them, this seems like a good place to pull together the area of opto-electronics, since it is related to the logic interface problems we have just been discussing.

Indicators

Electronic instruments look nicer, and are more fun to use, if they have pretty little colored lights on them. LEDs have replaced all earlier technologies for this purpose. You can get red, yellow, and green indicators, and you can get them in many packages, the most useful of which are (a) panel-mounting lights, and (b) printed-circuit board (PCB) mounting types. The catalogs present you with a bewildering variety of them, differing mostly in size, color, efficiency, and illumination angle. The latter deserves some explanation: A "flooded" LED has some diffusing stuff mixed in, so the lamp looks uniformly bright over a range of viewing angles; that's usually best, but you pay a price in brightness.

An LED looks electrically like a diode, with a forward drop of about 2 volts (they're built with gallium arsenide phosphide, which has a larger band gap, and hence a larger forward drop, than silicon). Typical flooded panel-type LED indicators look good at 10mA forward current; on a board inside an instrument you can usually get away with 2–5mA, particularly if you use a narrow-angle LED.

Figure 9.21 shows how to drive LED indicators. Most of the circuits are obvious, but note that bipolar TTL is poor at *sourcing* current, so you always arrange the circuit so that logic LOW turns on the LED; by comparison, CMOS families have symmetrical drive capability. NMOS circuits not only share the feeble sourcing of bipolar TTL but also tend to have rather limited current-sinking capability; it's best to interpose a buffer (an HCT gate is good), or perhaps a discrete MOSFET. Note also that some LED indicators come with internal current-limiting resistors (or even internal constant-current circuits) – with these you omit the external resistor.

You can get little arrays of indicators – sticks of 2, 4, or 10 LEDs in a row – designed for PCB mounting. The latter are actually intended for linear "bar-graph" readouts. They come in upright or right-angle mounting. You can also get panel-mounting indicators that combine a red LED and green LED in one uncolored package. These make an impressive panel, with lights changing color to indicate good/bad conditions.

We've used LED indicators from manufacturers such as Dialight, General Instrument, HP, Panasonic, Siemens, and Stanley. The latter specializes in lamps of unusually high efficiency; you can usually locate their exhibit at electronics shows by the dazed look (and incipient sunburn) of recent visitors.

Displays

A *display* means an opto-electronic device that can show a number ("numeric" display); a hexadecimal digit, namely 0–9 and A–F ("hexadecimal display"), or any letter or number ("alphanumeric display"). The dominant display technologies today are LEDs and LCDs (liquid-crystal displays). LCDs are the newer technology, with significant advantages for (a) battery-operated equipment, owing to its very low power dissipation, (b) equipment for use outdoors or in high ambient light levels, (c) displays that require custom shapes and symbols, and (d) displays with many digits or characters. LEDs, by comparison, are somewhat simpler to use, particularly if you only need a few digits or characters. They also come in three colors, and they look good in subdued light, where their good contrast makes them easier to read than LCD displays.

For displays of many characters – say a line or two of text – gas discharge ("plasma") display panels compete with LCDs, particularly if you care about clarity and contrast. They do require significant power, however, so LCD displays are usually preferred for battery-powered applications.

LED displays. Figure 9.22 shows the choices you have in LED displays. The original 7-segment display is the simplest and can display the digits 0–9 and the hex extension (A–F), albeit somewhat crudely (the hex letters are displayed as

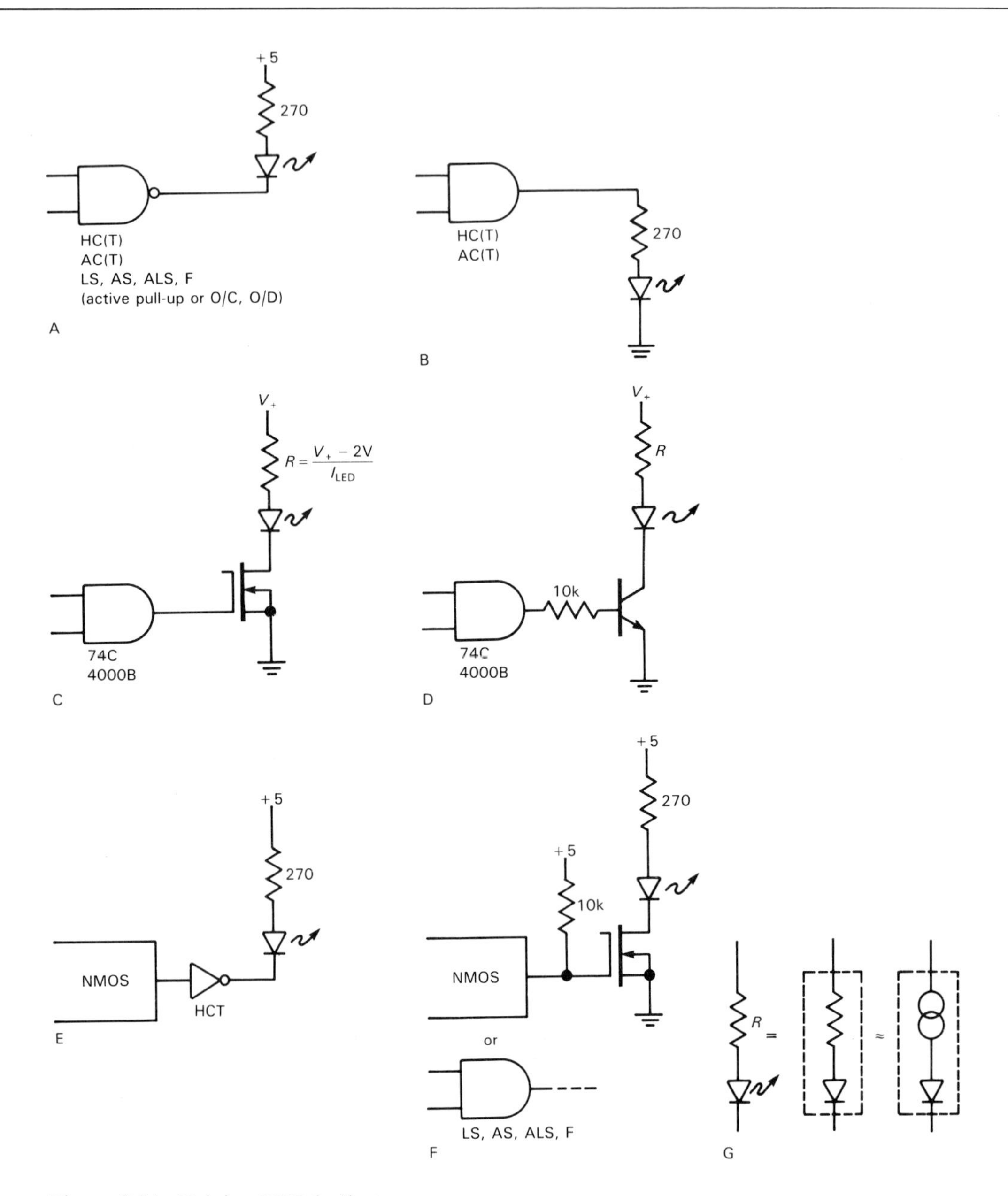

Figure 9.21. Driving LED indicators.

"AbcdEF"). You can get single-character 7-segment displays in many sizes, and in "sticks" containing 2, 3, 4, or 8 characters (generally intended to be "multiplexed" – the characters displayed one at a time in rapid sequence). Single-character displays bring out leads for the 7 segments and the common electrode; the two flavors are thus "common cathode" and "common anode." Multiple-character sticks bring each character's common electrode out, but tie the corresponding segments together, which is what you want for multiplexing.

Sixteen-segment and 5×7 dot-matrix displays are available in two varieties: "dumb" displays that bring out the

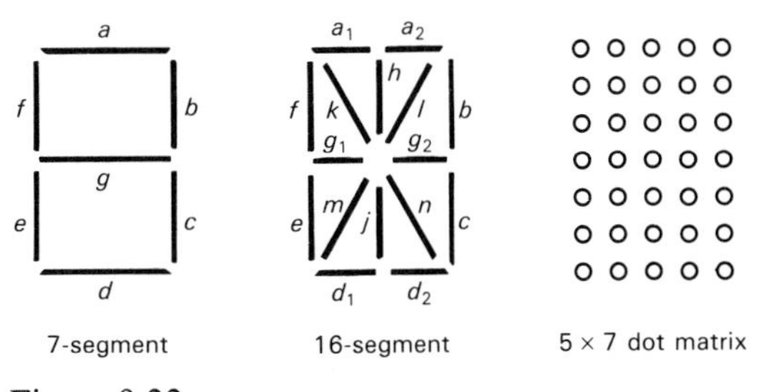

Figure 9.22

segments and common leads (the same as with 7-segment displays) and "smart" displays that do the hard work of decoding and driving for you.

Rather than speak further in generalities, let's look at some examples (Fig. 9.23). The first circuit shows how to drive a single-digit 7-segment common-cathode LED display. The 'HC4511 is a "BCD-to-7-segment latch-decoder-driver," able to source about 15mA while holding its active outputs at +4.5 volts. The series resistors make sure the segment current is limited to that value, with a forward diode drop of 2 volts. You can get arrays of equal-value resistors in convenient SIPs (single in-line package).

You only need a single decoder-driver chip, even if you are displaying multiple digits, as long as you multiplex the display, i.e., illuminate only one displayed digit at a time. Figure 9.23B shows the idea, in this case using an LSI 4-digit counter chip with built-in 7-segment multiplexed drivers. The 74C925 asserts its segment drivers (active HIGH, with plenty of drive capability) for each digit in turn, simultaneously asserting an active HIGH on the corresponding digit output A–D. The rest of the circuitry is self-evident, except perhaps for the unsavory manner in which the digit outputs are pinned a diode drop above ground; luckily, the 74C925 specifies proper operation with this circuit, since the digit outputs are buffered and current-limited.

Figure 9.24A shows how to drive a single hexadecimal display, implemented with a 5×7 dot matrix. The HP 5082-7340 is an example of a "smart" display, with built-in latch, decoder, and driver. All you have to do is assert the 4-bit data, wait at least 50ns, then bring the latch enable HIGH. If you don't want to use its latch, just keep the enable LOW. In Figure 9.24B we've shown one of Siemens' "intelligent" (smarter than "smart"?) displays, in this case a 4-character stick that uses 16-segment displays. This display is intended to look like memory to a microprocessor, something we'll learn about in the next two chapters. To make a long story short, you just assert any 7-bit character and its position ("address," 2 bits), then assert WR′ (write) while making sure the chip is enabled. The data then get stored

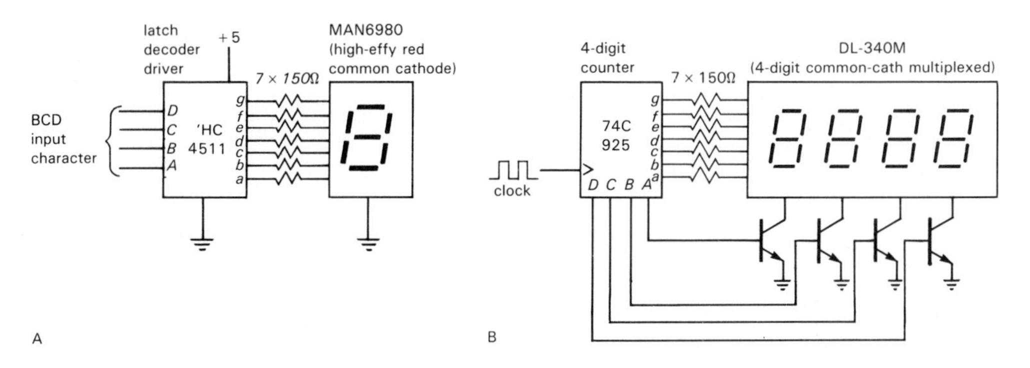

Figure 9.23. Driving 7-segment LED displays.
A. Single-digit.
B. Multiplexed.

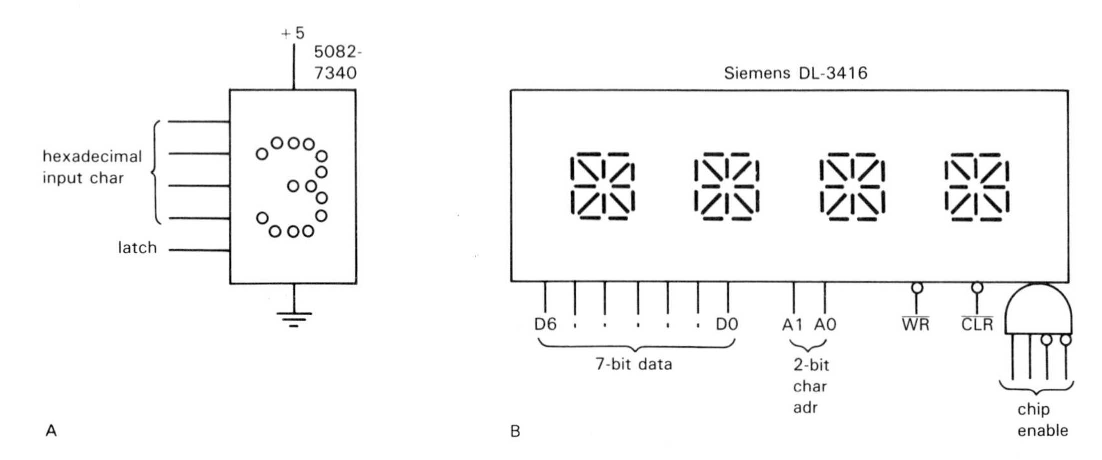

Figure 9.24. Integrated displays.
A. Single-character, dot matrix.
B. 4-character, 16-segment, addressable.

internally, and the corresponding display position changes to the new character. Figure 9.25 shows the character set that can be displayed.

If you want to use a dumb display (perhaps it isn't available with intelligence), but you're spoiled by the simplicity of these intelligent displays, you can simply interpose a chip like the 8-digit Intersil ICM7218/28, which looks like memory to the microprocessor, and which drives a dumb LED display stick with appropriate segment and digit drives. Another alternative is to let the microprocessor do all the work of figuring out which segments and digits to drive, using bits of its "parallel ports" to drive the appropriate lines. This will make more sense to you after you've digested the two microprocessor chapters (Chapters 10 and 11).

LCD and gas-discharge displays. Much of what we have said about LED displays applies also to LCD displays. However, there are some important differences. For one thing, LCDs must be driven by an *ac* waveform; otherwise their liquid guts are ruined. So LCD driver chips usually have

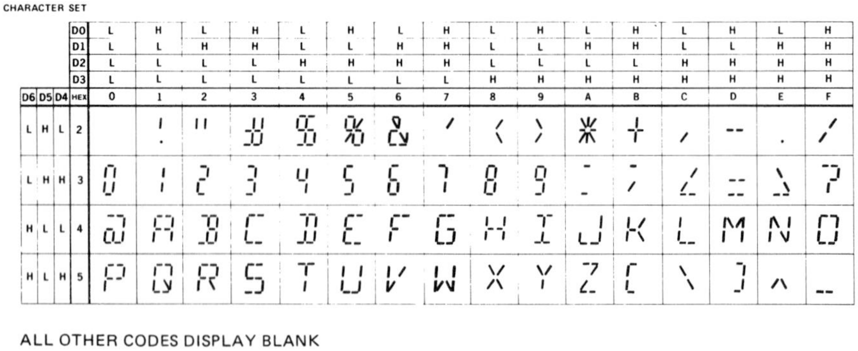

CHARACTER SET

			D0	L	H	L	H	L	H	L	H	L	H	L	H	L	H	L	H
			D1	L	L	H	H	L	L	H	H	L	L	H	H	L	L	H	H
			D2	L	L	L	L	H	H	H	H	L	L	L	L	H	H	H	H
			D3	L	L	L	L	L	L	L	L	H	H	H	H	H	H	H	H
D6	D5	D4	HEX	0	1	2	3	4	5	6	7	8	9	A	B	C	D	E	F
L	H	L	2		!	"	#	$	%	&	'	(	)	*	+	,	-	.	/
L	H	H	3	0	1	2	3	4	5	6	7	8	9	:	;	<	=	>	?
H	L	L	4	@	A	B	C	D	E	F	G	H	I	J	K	L	M	N	O
H	L	H	5	P	Q	R	S	T	U	V	W	X	Y	Z	[	\	]	^	_

ALL OTHER CODES DISPLAY BLANK

Figure 9.25. Display codes for Siemens DL-3416 16-segment display. (Courtesy of Siemens Components, Inc.)

some way to generate a square-wave segment drive, synchronized to the LCD backplane waveform. An example is the 'HC4543, the LCD cousin of the 'HC4511 LED latch/decoder/driver.

Another difference with LCD displays is that you don't often see single-character displays. Instead, they come in rather large panels that can display a line or two of text. Luckily, the manufacturers realize how complicated things can get, and therefore they provide complete displays that are more than intelligent, they're positively at the genius level. In general, you're talking to these things with a microprocessor, so (as with the display in Figure 9.24) the display is configured to look like a block of memory. Whatever you write gets displayed. Some of the fancier displays go even further, with the ability to store several messages and communicate via serial ports. Look in the EEM (see bibliography) for manufacturers.

Gas-discharge displays feature those handsome red-orange characters you see on some higher-priced portable computers. They require high-voltage drivers, and the manufacturers generally provide the driver circuitry. You can get single- and multiple-digit displays, and you can get large multi-character panels complete with memory and convenient interface. Examples of the latter are the multi-line displays by Cherry, with battery-backed memory that can hold 512 messages, interleave real-time data, and let you edit its memory. You probably shouldn't think of these as *displays*, but as computers that happen to have a display attached!

Opto-couplers and relays

An LED emitter, combined with a photodetector in close proximity, forms a very useful object known as an *opto-coupler* or *opto-isolator*. In a nutshell, opto-couplers let you send digital (and sometimes analog) signals between circuits with separate grounds. This "galvanic isolation" is a good way to prevent ground loops in equipment that drives a remote load. It is essential in circuits that interact with the ac power mains. For example, you might want to turn a heater on and off from a digital signal provided by a microprocessor; in this case you would probably use a "solid-state" relay, which consists of an LED coupled to a high-current triac. Some ac-operated switching supplies (e.g., the supply used in the IBM PC-AT) use opto-couplers for the isolated feedback path (see Section 6.19). Similarly, designers of high-voltage power supplies sometimes use opto-couplers to get a signal up to a circuit floating at high voltage.

Even in less exotic situations you can take advantage of opto-isolators. For example, an opto-FET lets you switch analog signals with no charge injection whatsoever; the same goes for sample-and-hold circuits and integrators. Opto-isolators can keep you out of trouble when driving industrial current loops, hammer drivers, etc. Finally, the galvanic isolation of opto-isolators comes in handy in high-precision or low-level circuits: For example, it is difficult to take full advantage of a 16-bit analog-to-digital converter, because the digital output signals (and noise on the digital ground to which you connect the converter's output) get back into the analog front end. You can extricate yourself from "noise city" with optical isolation of the digital half.

Opto-couplers typically provide 2500 volts (rms) isolation, 10^{12} ohms insulation resistance, and less than a picofarad coupling between input and output.

Before looking at actual opto-couplers, let's take a quick look at photodiodes and phototransistors. Visible light causes ionization in silicon, producing charge pairs in the exposed base region; this mimics exactly the effect of an externally applied base current. There are two ways to use a phototransistor: (a) You can use it as a *photodiode*, connecting only to the base and collector leads; in that case the detected photocurrent will typically be a few percent of the LED drive current. A photodiode generates its photocurrent whether or not you apply a bias voltage; thus you can hook it directly into an op-amp summing junction (a virtual short circuit), or you can back-bias it (Fig. 9.26A,B). (b) If you let the photodiode current act as a base current, you get normal transistor current multiplication, with a resulting I_{CE} that is typically 100 times larger; in this case you must bias the transistor, as in Figure 9.26C. You pay for the increased current with slower response, because of the open base circuit. You can add a resistor from base to emitter to improve the speed; however, this produces a threshold effect, since the phototransistor doesn't begin to conduct until the photodiode current is large enough to produce a V_{BE} across the external base resistor. In digital applications the threshold can be useful, but in analog applications it is an undesirable nonlinearity.

Figure 9.26D–S shows typical examples of nearly every kind of opto-coupler you are likely to encounter. The earliest (and simplest) is typified by the 4N35, an LED-phototransistor pair with 40% (min) current transfer ratio (CTR) as a phototransistor, and sluggish 5μs turn-off time (t_{OFF}) into a 100 ohm load. The figure shows how to use it: A gate and resistor generate current-limited 8mA drive, and a relatively large collector resistor guarantees saturated switching of the output between logic levels. Note the use of a Schmitt-trigger inverter, a good idea here because of the long switching times. You can get LED-phototransistor pairs with CTRs of 100% or more (e.g., the MCT2201, with CTR=100% min), and you can get LED-photo-Darlingtons, as shown; they're even slower than phototransistors! To get improved speed, the manufacturers sometimes use separate photodiode and transistor, as shown in the 6N136 and 6N139 opto-transistor and opto-Darlingtons.

These opto-couplers are nice, but somewhat annoying to use because you have to supply discrete components at both input and output. Furthermore, the input loads ordinary logic gates to their maximum capability, and the passive pullup output suffers from slow switching and mediocre noise immunity. To remedy these deficiencies the silicon wizards bring us "logic" opto-couplers. The 6N137 in Figure 9.26I goes halfway, with diode input and logic output; you still need plenty of input current (specified as 6.3mA, min, to guarantee output switching), but you get clean logic swings at the output (albeit open-collector), and speeds to 10Mbit/sec. Note that you must supply +5 volts to the internal output circuitry. The newer 74OL6000 series from General Instrument (Fig. 9.26J) does what you really want: It accepts logic-level inputs, and produces logic-level outputs, with both totem-pole and open-collector types available; these opto-couplers operate to 15Mbit/sec. Because of the internal logic circuitry at both input and output, both sides of the chip require logic supply voltages.

Figure 9.26 continues with some variations on the LED-phototransistor theme. The IL252 hooks a pair of LEDs back-to-back, so you can drive it with ac. The IL11 uses a long isolation gap (and package) to obtain 10kV rms voltage standoff, compared with the usual 2.5kV rms value for all the other couplers shown. The H11C4 is an opto-*SCR*, useful for switching high voltages and currents. The MCP3023 replaces the unidirectional SCR with a *triac*, which is a bidirectional SCR; it can drive an ac load directly, as in Figure 9.15N. When driving ac loads, it's best to switch on the load during a zero crossing of the ac waveform, in order to avoid putting spikes onto the power line. This is easily done, with an opto-triac containing

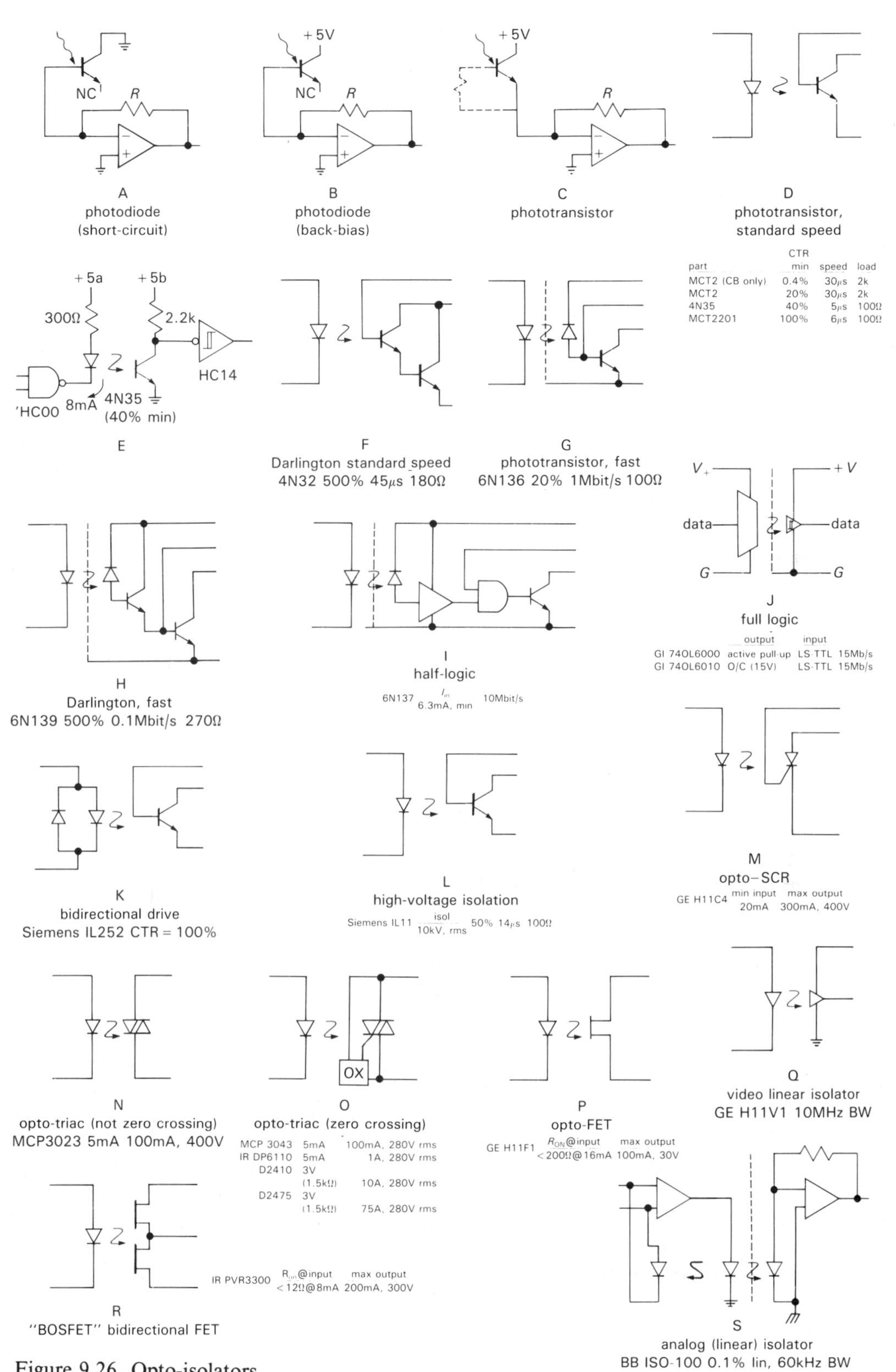

Figure 9.26. Opto-isolators.

internal "zero-voltage-switching" circuitry (which blocks triac drive until the next zero crossing); the small MCP3043 shown uses such circuitry, as do the higher-current "solid-state relays" shown. The DP6110 "ChipSwitch" from IR comes in a 16-pin DIP (with all but 4 pins missing), while the mighty D2410 and D2475 are packaged in 1.75″×2.25″×1" power modules intended for mounting to a heat sink.

The remaining opto-couplers in Figure 9.26 can be used for linear signals. The H11F series of opto-FETs can be used as an isolated variable resistor or isolated analog switch. There are no problems of compatible voltage levels, SCR latchup, or charge injection. You might use one of these in a sample/hold or integrator. The PVR series of "BOSFETs" are similar, but with a pair of power MOSFETs hooked in series as the output element. They are intended primarily to switch ac loads directly, in the manner of an opto-triac. The H11V1 is a linear video isolator, with 10MHz bandwidth. And the ISO-100 from Burr-Brown is a clever analog isolation element, in which the LED couples to *two* matched photodiodes; one is used in a feedback loop to linearize the coupled response to the second photodiode.

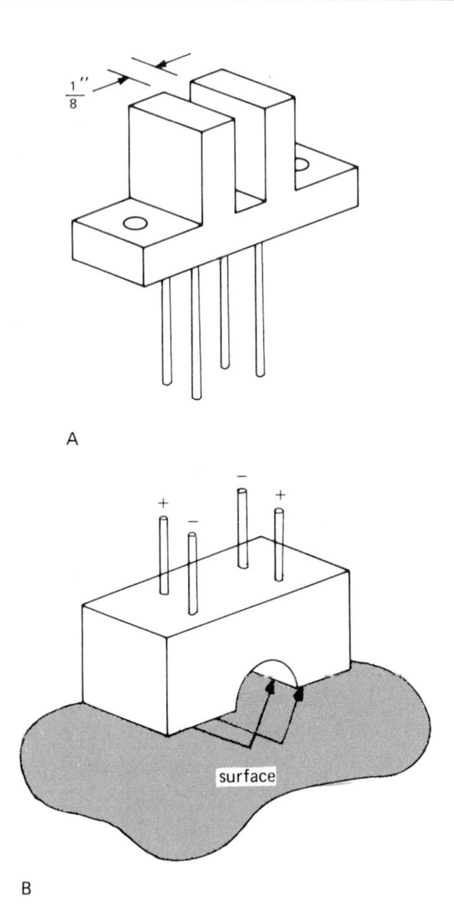

Figure 9.27. A. Optical interrupter.
B. Reflective object sensor.

□ *Interrupters*

You can use LED-phototransistor technology to sense proximity or motion. An "optical interrupter" consists of an LED coupled to a phototransistor across a 1/8 inch slot. It can sense the presence of an opaque strip, for example, or the rotation of a slotted disk. An alternative form has the LED and photodetector looking in the same direction, and it senses the presence of a reflective object nearby (most of the time, anyway!). Take a look at Figure 9.27. Optical interrupters are used in disk drives and printers, to sense the end of travel of the moving assembly. You can get optical "rotary encoders" that generate a quadrature pulse train (two outputs, 90° out of phase) as the shaft is rotated. These provide a nice alternative to resistive panel controls (potentiometers). See Section 11.09. In any application where you're considering an optical interrupter or reflective sensor, take a look at Hall-effect sensors as an alternative; they use solid-state magnetic-field sensors to indicate proximity. They're commonly used in automobile ignition systems as an alternative to mechanical breaker points.

□ *Emitters and detectors*

We've already mentioned LEDs, both for displays and opto-couplers. A recent development in opto-electronics is the availability of the inexpensive solid-state diode *laser*, a true coherent source of light, unlike the diffuse LED. You can see one of these when you open the top door of a portable compact-disk player. Diode lasers cost about $20, sold by the companies that manufacture consumer electronics (Matsushita, Mitsubishi, Sharp, and Sony). A typical diode laser generates 10mW of light output at 800nm (invisible, in the "near infrared") when powered by 80mA through its forward diode drop of 2 volts. The output beam emerges directly from a tiny spot on the GaAlAs chip, diverging with an angle of about 10°–20°; it can be collimated with a lens to form a parallel beam or a very small focal spot. LED lasers are used extensively in optical fiber transmission.

Another recent emitter technology is embodied in high-density linear LED arrays, 300 emitters/inch or more, intended for LED printers. If the semiconductor technology is successful, these are likely to replace laser printers, because they are simpler and more reliable and ultimately of higher resolution.

In the detector arena there are several alternatives to the simple photodiode and phototransistor that we discussed earlier, particularly when speed or sensitivity is important. We will discuss PIN diodes, CCDs, and intensifiers in Section 15.02.

DIGITAL SIGNALS AND LONG WIRES

Special problems arise when you try to send digital signals through cables or between instruments. Effects such as capacitive loading of the fast signals, common-mode interference pickup, and "transmission-line" effects (reflections from impedance mismatching, see Section 13.09) become important, and special techniques and interface ICs are often necessary to ensure reliable transmission of digital signals. Some of these problems arise even on a single circuit board, so a knowledge of digital transmission techniques is generally handy. We will begin by considering on-card problems. Then we will go on to consider the problems that arise when signals are sent between cards, on data buses, and finally between instruments via twisted-pair or coaxial cables.

9.11 On-board interconnections

Output-stage current transient

The push-pull output circuit for TTL and CMOS ICs consists of a pair of transistors going from V_+ to ground. When the output changes state, there is a brief interval during which both transistors are ON; during that time, a pulse of current flows from V_+ to ground, putting a short negative-going spike on the V_+ line and a short positive spike on the ground line. The situation is shown in Figure 9.28. Suppose that IC_1 makes a transition, with a momentary large current from +5 to ground along the paths as indicated [with 74Fxx or 74AC(T)xx circuits the current might reach 100mA]. This current, in combination with the inductance of the ground and V_+ leads, causes short voltage spikes relative to the reference point, as shown. These spikes may be only 5ns to 20ns long, but they can cause plenty of trouble: Suppose that IC_2, an innocent bystander located near the offending chip, has a steady LOW output that drives IC_3 located some distance away. The positive spike at IC_2's ground line appears also at its output, and if it is large enough, it gets interpreted by IC_3 as a short HIGH spike. Thus, at IC_3, some distance away from the troublemaker IC_1, a full-size bona fide logic output pulse appears, ready to mess up an otherwise well-behaved circuit. It doesn't take very much to toggle or reset a flip-flop, and this sort of ground-current spike can do the job nicely.

The best therapy for this situation consists of (a) using hefty ground lines throughout the circuit, even to the extent of

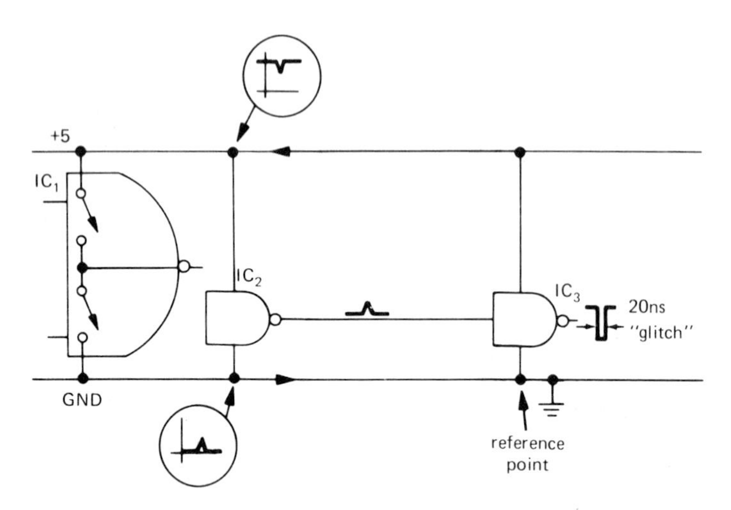

Figure 9.28. Ground current noise.

using a large "ground plane" (one whole side of a double-sided printed-circuit board devoted to ground), and (b) using bypass capacitors liberally throughout the circuit. Large ground lines mean smaller current-induced spikes (lower inductance and resistance), and bypass capacitors from V_+ to ground sprinkled throughout the circuit mean that current spikes travel only over short paths, with the reduced inductance resulting in much smaller spikes (the capacitor acts as a local voltage source, since its voltage does not change appreciably during the brief current spikes). It is best to use a 0.05μF to 0.1μF capacitor near each IC, although one capacitor for every two or three ICs may suffice. In addition, a few larger tantalum capacitors (10μF, 20V is a good value) scattered throughout the circuit for energy storage is a good idea. Incidentally, bypass capacitors from power-supply lines to ground are recommended in any circuit, digital or linear. They help make the supply lines low-impedance voltage sources at high frequencies, and they prevent signal coupling between circuits via the power supply. Unbypassed power-supply lines can cause peculiar circuit behavior, oscillations, and headaches.

Spikes caused by driving capacitive loads

Even with the supplies bypassed, your problems aren't over. Figure 9.29 shows why. A digital output sees the stray wiring capacitance and the input capacitance of the chip it drives (5–10pF, typically) as part of its overall load. To make a fast transition between states it must sink or source a large current into such a load, according to $I = C(dV/dt)$. For instance, consider a 74ACxx chip (5V output swing in 3ns) driving a total load capacitance of 25pF (equivalent to three or four logic loads connected with short leads). The current during the logic transition is 40mA, nearly the maximum output capability of the driving chip! This current returns through ground (HIGH-to-LOW transition) or the +5 volt line (LOW-to-HIGH transition), producing those cute little spikes at the receiving end, as before. [Just to get an idea of the magnitude of this effect, consider the fact that wiring inductance is roughly 5nH/cm. An inch of ground wire carrying this logic transition current would have a spike of $V = L(dI/dt) = 0.2$V.] If the chip happened to be an octal buffer, with simultaneous transitions on a half dozen outputs, the ground spike would be over a volt; see Figure 8.95. A similar (though generally smaller) ground spike is generated near the driven chip, where the drive-current spikes return to ground through the input capacitance of the driven device.

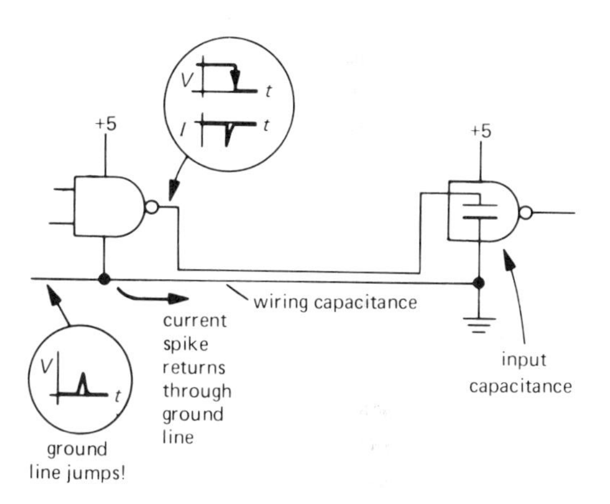

Figure 9.29. Capacitive-load ground current noise.

In a synchronous system, with a number of devices making output transitions simultaneously, the noise-spike situation can become so serious that the circuit will not function reliably. This is especially true in a large printed-circuit card, with long interconnections and long ground runs. The circuit may fail only occasionally, when a whole group of data lines unluckily happen to make a simultaneous HIGH-to-LOW transition, generating a momentary very large ground current. This kind of pattern sensitivity is characteristic of noise-induced error and is a good reason for running extensive memory tests on microprocessor systems (where you typically have 16 data lines and 24 address lines bouncing around in crazy patterns).

The best design approach is to use massive ground runs (for low inductance), preferably in the form of an internal ground-plane layer on a multilayer circuit board (see Chapter 12), or at least a perpendicular "gridded" arrangement of grounds on both sides of a simpler two-sided board. Copious use of bypass capacitors is mandatory. These problems are much less severe with high-voltage CMOS (because of slow edge times); conversely, they are most severe with F, AS, and AC(T) logic families. In fact, the AC(T) families are so prone to dynamic current spikes that some manufacturers (led by TI) have abandoned the popular "corner-pin" power/ground arrangement in favor of a "center-pin" arrangement with lower internal lead inductance; to improve things still further, they use up to four adjacent pins for a lower-inductance ground. Because of these noise problems, it is best not to use a faster logic family than you need; that is why we recommend HC logic, rather than AC, for general-purpose use.

9.12 Intercard connections

With logic signals going between circuit cards, the opportunities for trouble multiply rapidly. There is greater wiring capacitance, as well as longer ground paths through cables, connectors, card extenders, etc., so the ground spikes induced by drive currents during logic transitions are generally larger and more troublesome. It is best to avoid sending clocking signals with large fanout between boards, if possible, and the ground connections to the individual cards should be robust. If clocking signals are sent between boards, it is important to use a gate as input buffer on each board. In extreme cases it may be necessary to use line driver and receiver chips, as we will discuss later. In any case, it is best to try to keep critical circuits together on one card, where you can control the inductance of the ground paths and keep wiring capacitance at a minimum. The problems you will encounter in sending fast signals around through several cards should not be underestimated; they can turn out to be the major headache of an entire project!

□ 9.13 Data buses

Where many subcircuits are connected together by a data bus (more on this in Chapters 10 and 11), the sort of problems already mentioned become more severe. In addition, a new factor comes into the picture: transmission-line effects due to the length and inductance of the signal lines themselves. With the fastest ECL chips ("ECL III," "ECL100K," with rise times of less than 1ns), these effects are so severe that signal runs of more than 1 inch must be treated as transmission lines and properly terminated!

The best approach with data buses of any substantial length (a foot or more) is probably the use of a "motherboard" with ground plane. As we will explain in Chapter 12, a motherboard is simply a printed-circuit board containing a row of printed-circuit edge connectors to accept the individual circuit cards that make up the logic circuit. Motherboards are economical solutions to the problem of interconnecting cards, and if they are properly laid out, they are electrically superior as well. Wires that run close to a ground plane have lower inductance and less tendency to couple capacitively to adjacent signal lines, so a good way to arrange a simple motherboard is with all the signal lines on one side, with the other side being a solid ground plane (though double-sided printed-circuit boards are standard, multiple-layer boards are being increasingly used for complicated circuits).

One last note on this problem: In desperation you may be tempted to use the common trick of putting a capacitor directly across the input of a gate driven from a long line, when transmission-line effects like "ringing" or ground spikes are driving you up the wall. In spite of the fact that we have done it ourselves, we don't recommend this inelegant fix, since it only compounds the problem of large ground currents during logic transitions (Section 9.11).

□ *Bus termination*

With buses of substantial length, it is common to "terminate" the signal lines at the far end with a resistive pullup or pulldown. As we will discuss in Chapter 13, long pairs of wires or coaxial cables have a "characteristic impedance" Z_0. For a cable terminated with that impedance (which is always a resistance), a signal traveling down the cable will be entirely absorbed, with no reflection whatsoever. Any other value of termination, including an open circuit, will produce reflected waves, with amplitude and phase depending on the impedance mismatch. The sorts of conductor width and spacing used on printed-circuit boards give a characteristic impedance in the neighborhood of 100 ohms, which is also close to the characteristic impedance of twisted pairs made from ordinary insulated wire of 24 gauge or so.

A popular method of terminating a TTL bus is to use a voltage divider from +5 volts to ground. In that way the logic HIGH is kept around +3 volts, which means that less swing (and therefore less current into the load capacitance) is necessary during logic transitions. The combination of a 180 ohm resistor to +5 and a 390 ohm resistor to ground is typical (Fig. 9.32). Another method that works well with either TTL or CMOS is to use an ac termination, consisting of a series resistor/capacitor from the data line to ground (Fig. 9.30). The resistor value should be close to the characteristic impedance of the bus (typically 100Ω); the capacitor value should be chosen for a low capacitive reactance at a frequency equal to the inverse of the edge time (100pF generally works well).

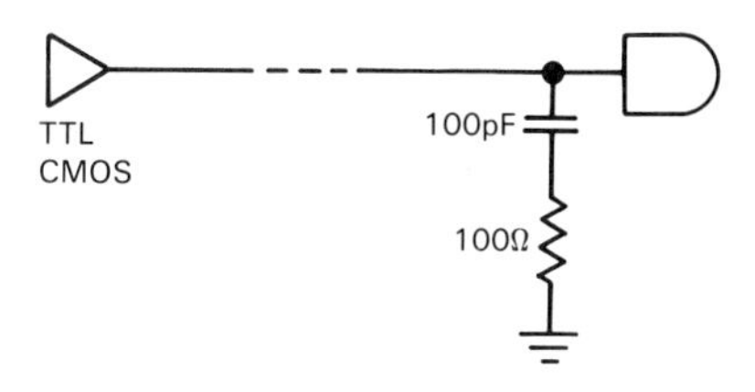

Figure 9.30. ac termination.

Bus drivers

When driving bus lines of substantial length or fanout, it is necessary to use special logic functions with high current capability. Here are the most popular types:

Type	Description
'365–8	hex 3S
'1004/34	hex 2S
'11004/34	hex 2S broadside
'240–4	octal 3S
'540–1	octal 3S broadside
'827–8	10-bit 3S broadside

Family	Sink/source (mA)
LS	24/15
AS, F	64/15
AS1xxx	48/48
HC(T)	6/6
AC(T)	24/24
Am298xx	48/24

"Broadside" means that the chip's pins are arranged with corresponding inputs and outputs on opposite sides. There are also *transceiver* chips of comparable drive capability; those can buffer data in either direction by using a parallel pair of three-state buffers back-to-back for each data line, with a "direction" input determining which way data are going. See Tables 8.4 and 8.5 for additional device types.

9.14 Driving cables

You can't run digital signals from one instrument to another by simply stringing a single conductor between them, because such an arrangement is prone to pickup of interference (as well as generating interference of its own). Digital signals are generally piped through coaxial cables, twisted pairs, flat ribbon cables (sometimes with ground plane or shield), multiwire bundled cables, and, increasingly, fiber-optic cables. We will see coaxial cables (affectionately called "coax") again in Chapter 13 in connection with radiofrequency techniques; here we would like to describe some of the methods used to send digital signals between boxes of electronics, since these methods constitute an important part of digital interfacing. In most cases there are special-purpose driver/receiver chips available to make your job easier.

RS-232

For relatively slow transmission (thousands of bits per second) through simple multiwire cables, the popular RS-232C (or newer RS-232D) signal standard is convenient. It specifies bipolarity levels of ±5 volts to ±15 volts (positive and negative supply voltages are needed for the driver, but not usually for the receiver), and the receivers usually allow control of hysteresis and response time for particularly noisy environments. With RS-232 you can use a multiwire cable without any special shielding, because the drivers are deliberately limited to a maximum slewing rate of 30V/μs to minimize cross-talk. In addition to the original TTL-compatible quad driver/receiver pair (the 1488/1489), there is now a selection of improved chips, including low-power versions (LT1032, 1039, and MC145406, see Section 14.17) and versions that operate from a single +5 volt supply (MAX-232 and LT1130 series, LT1080); the latter have on-board flying-capacitor voltage converters to generate the necessary negative voltage. Figure 9.31 shows a typical setup.

RS-232 is used extensively for communication between computers and terminals,

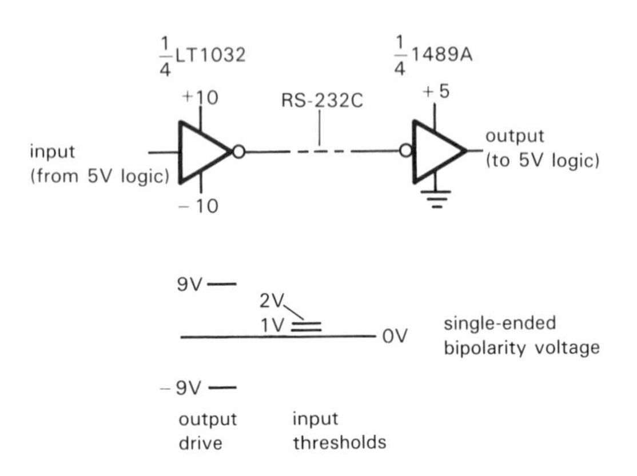

Figure 9.31. RS-232 high-noise-immunity cable driver and receiver.

at standardized data rates ranging from 110 to 38,400 bits per second. The full standard even specifies the pin connections to be used with a 25-pin type D subminiature connector and is usually used for transferring ASCII-encoded data, as we will discuss in Section 10.19.

Direct 5 volt logic drive

Just as with data buses, it is possible to drive lines of moderate length directly with logic levels; in general, gates of high current capability should be used (see list under "bus drivers," above). Figure 9.32 shows some methods. In the first circuit, a buffer (which could be open-collector) drives a terminated line, with a TTL Schmitt trigger as a receiver for improved noise immunity. If noise is a serious problem, the *RC* slowdown network shown in the second circuit can be used, with the *RC* time constant (and the transmitted bit rate!) adjusted to give good noise immunity; in this circuit the Schmitt trigger is essential. In the last circuit a robust CMOS buffer drives a line with ac termination and a CMOS Schmitt-trigger receiver.

Direct logic drive will work over modest lengths (say 10 feet) of twisted-pair, flat

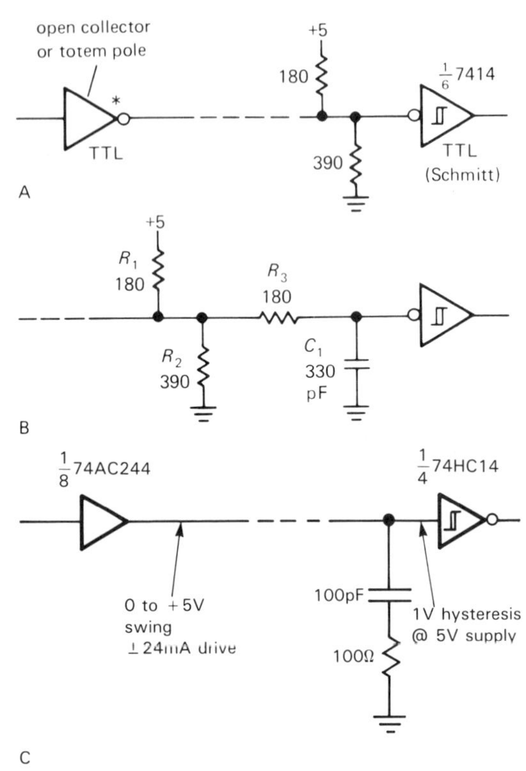

Figure 9.32. Use of termination with logic-level drive.

ribbon cable, or coax. Because of the fast edge times, capacitive pickup on adjacent lines is a problem; the usual cure is to alternate ground lines (ribbon cable), or to pair signal lines with ground (twisted-pair). This signal pickup problem makes direct logic drive with bundled multiwire cables almost hopeless. In the next section we've got some interesting scope photos that illustrate the problem, and another effective cure, namely *differential* logic drive.

Important note: *Never* attempt to drive long lines from the output of unbuffered clocked devices (flip-flops, monostables, counters, some shift registers); the capacitive loading and transmission-line effects can cause erratic or erroneous behavior. A "buffered" device interposes output drivers between the internal registers and

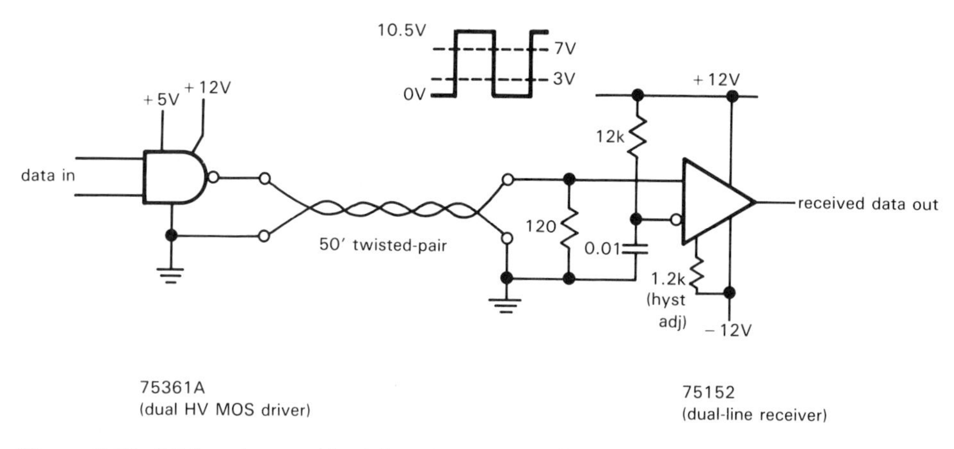

Figure 9.33. High-voltage cable driver improves noise immunity.

the output pins and thus, because the chip doesn't see the (degraded) signals actually present on the output lines, does not have this problem.

□ *High-voltage logic drive*

When you use direct logic drive to send signals through cables, you can improve the noise immunity by increasing the signal swing. Figure 9.33 shows an example, using the 75361A "TTL-to-MOS driver" as a generator of 12 volt logic swings for a twisted-pair line. The receiver is a 75152, which lets you set the input threshold (the input resistance is about 9k, hence the 12k biasing resistor to set the threshold at +5 volt) and the hysteresis (here set to ±2 volt). The 120 ohm line termination matches the characteristic impedance of the twisted-pair.

□ *Trapezoidal drive*

National makes a line driver/receiver (DS3662; DS3890 series) with controlled driver transition time combined with controlled receiver response time, to reduce the problem of capacitive coupling from adjacent lines. It is meant to drive lines terminated with the network of Figure 9.32.

Differential drive; RS-422

Much better noise immunity can be obtained by using differential signals, i.e., Q and Q' on a twisted-pair, with a differential receiver (Fig. 9.34). In this example, paired TTL inverters drive a terminated twisted-pair with true and complemented signals, and a 75115 differential line receiver regenerates clean TTL levels from the mess. We selected bipolar TTL drivers, rather than CMOS, because of their greater resistance to damage from static electricity, and to SCR latchup from transmission-line reflections. This scheme gives very good common-mode rejection and generates clean logic levels from line signals that can look just awful. The waveform shown gives an idea of what you might see on the individual signal lines in a relatively clean system; the individual signals tend to acquire bumps and wiggles, although they generally remain monotonic through the transition (they don't reverse direction).

The 75115 is an example of a line receiver with adjustable response time; the 75152, another differential receiver, allows control of hysteresis instead. For peace of mind, it is a nice idea to use a receiver with hysteresis (and adjustable time constant), given the bizarre waveforms these receivers are called on to recognize.

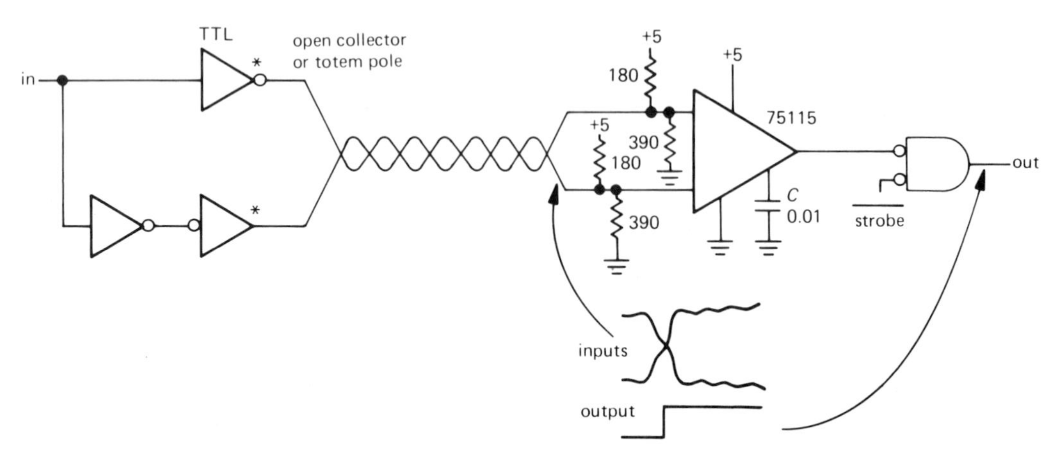

Figure 9.34. High-speed differential TTL cable driver and receiver.

□ *Current-sinking drivers.* Chips like the 75S110 and MC3453 have switched current-sinking outputs, which can be used as single-ended outputs or in a differential mode, as in Figure 9.35. The 75107 is a companion differential receiver chip, normally used with a termination as shown. Several drivers can share a common differential line in a "party line" configuration, since their outputs can be disabled via a "three-state" enable; in that case the terminations at each driver are omitted, and moved to the end of the line farthest from the receiver.

In our experience, differential current-sinking drivers allow really impressive data rates, probably because the high-impedance current-source drive ensures that the cable can be terminated in its characteristic impedance, for both states of the driver. The data sheets claim rates of more than 1 megabit per second (Mb/s) over line lengths of 2000 feet, and rates up to 10Mb/s over lengths of several hundred feet or less.

The actual scope photos in Figure 9.36 show how effective differential current-sinking drive can be in eliminating problems of common-mode noise. In the example shown, a 50mV pp signal is contaminated by 4 volts pp of common-mode noise.

RS-422/423. This is a data-transmission standard intended to supplant the popular RS-232 and is intended primarily for use with twisted-pair or flat cable. It provides for both unbalanced ("RS-423,"100kb/s max) and balanced

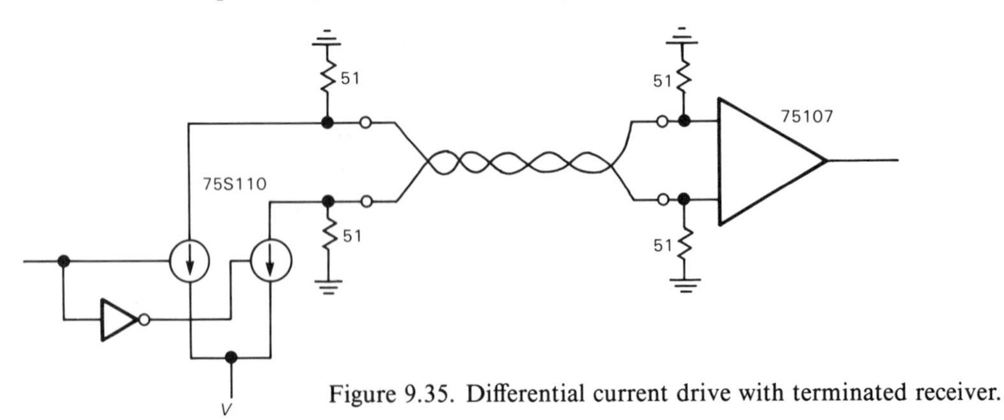

Figure 9.35. Differential current drive with terminated receiver.

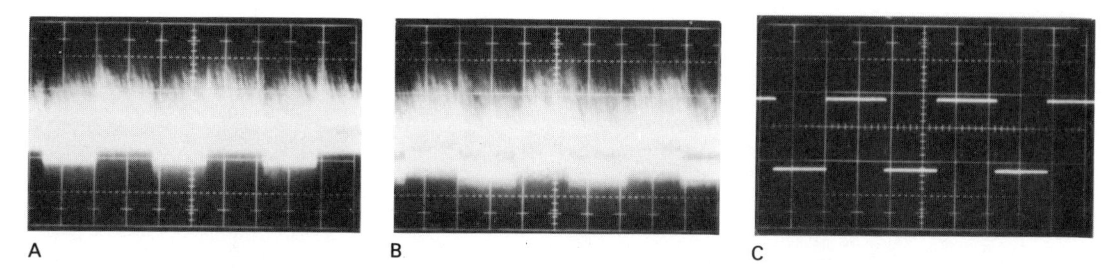

Figure 9.36. Scope photos showing excellent noise immunity of differential data transmission (75108 differential receiver). (Courtesy of Texas Instruments, Inc., Dallas, Texas.)
A. (+) receiver input.
B. (−) receiver input.
C. Receiver output.

("RS-422," 10Mb/s max) schemes. In the unbalanced mode you use bipolarity signal levels (±5V supplies) with controlled slew rates, analogous to RS-232. In the balanced mode you use unipolarity TTL-like levels (and thus single +5V supplies at both ends), with no slew-rate limiting. Figure 9.37 shows the trade-off of practical data rate versus line length.

A popular series of RS-422/3 driver/receivers is the 26LS30–34 from AMD, widely second-sourced by other manufacturers; the recent 75ALS192/4 and DS34F30/80 series claim improved performance and less power. We used RS-422 with twisted-pair flat cable in an application where we wanted to tie together the parallel ports and control signals of a set

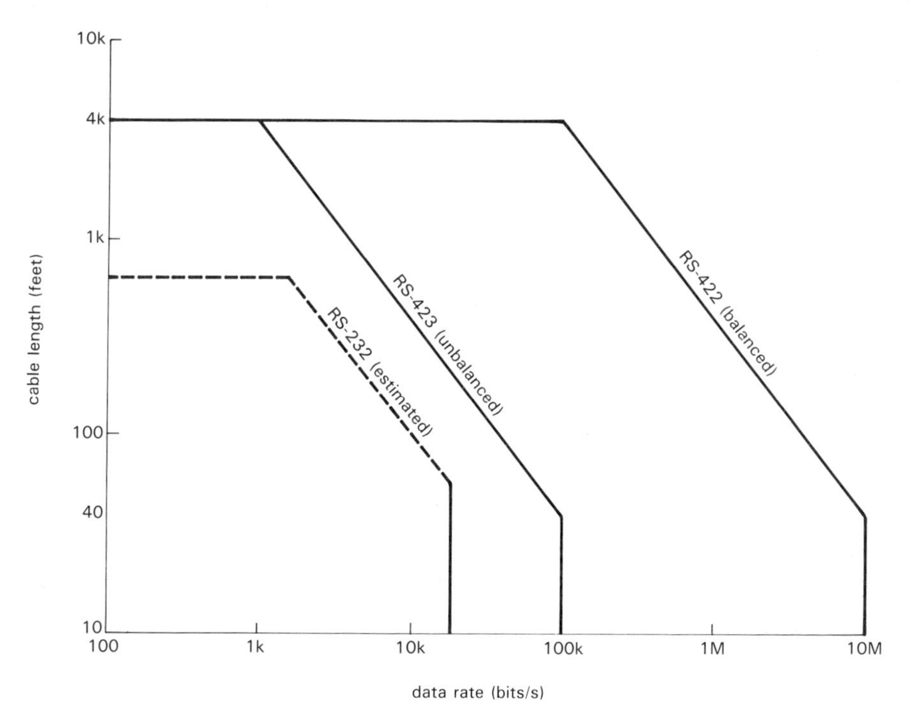

Figure 9.37. Serial communication data rate versus length.

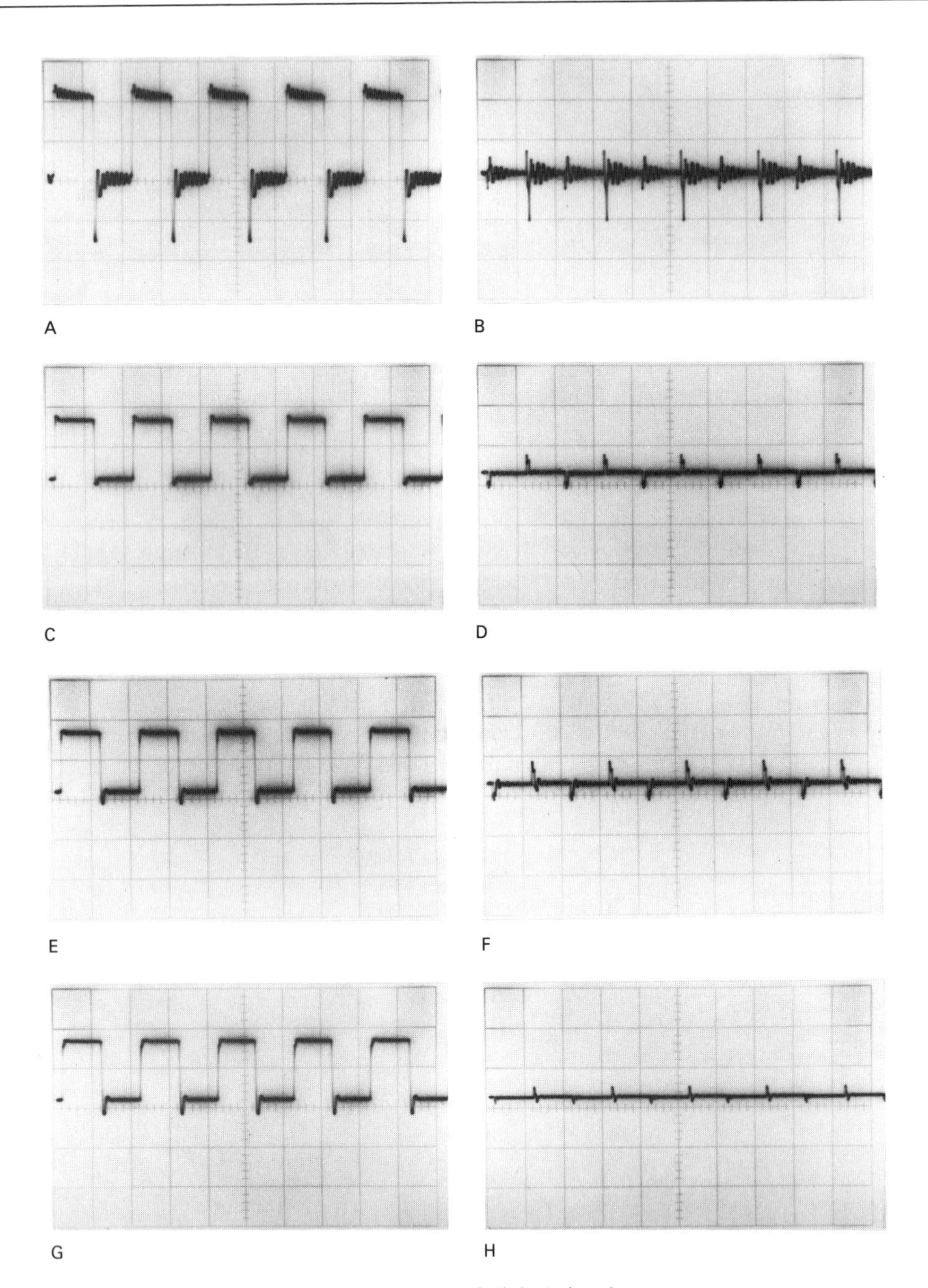

Figure 9.38. Degradation and cross-talk of digital signals.
A. 1MHz TTL square wave, through 10 feet of flat ribbon cable, alternating grounds, unterminated; 1V/div.
B. Pair next to A, but driven by TTL LOW.
C,D. Same as A,B, but terminated with 220Ω/330Ω network to +5 volts.
E,F. Same as C,D, but twisted-pair instead of flat ribbon cable.
G,H. Same as C,D, but flat ribbon with ground plane.

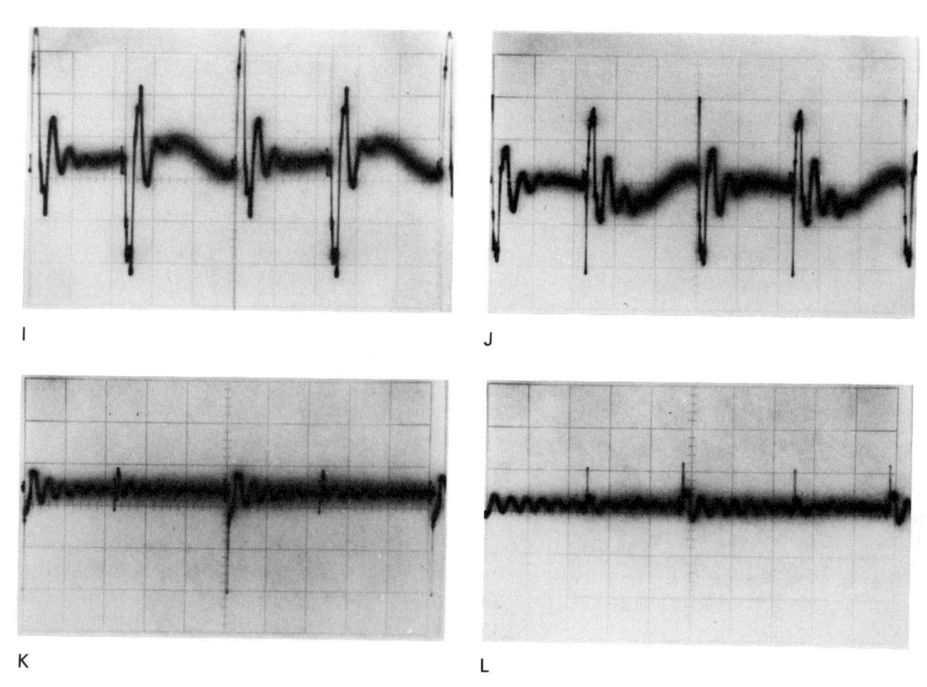

Figure 9.38. (*cont.*)
I. LOW RS-422 level pair, through 100 feet of flat ribbon cable adjacent to pair driven by 100kHz RS-422 differential square wave; 0.1V/div (note scale change).
J. Same as I, but signals separated by a grounded pair.
K,L. Same as I,J, but "twist-and-flat" pair instead of flat ribbon cable.

of 144 microprocessor boards in a "star" network. We made 9 groups of 16 processor boards, each group also including one interface board, using TTL signals between the processors and interface; the 9 interface boards were then strung together with an external computer using (differential) RS-422. The total cable length was about 25 feet, terminated by bridging 100 ohm resistors across each pair at both ends. The whole system is extremely clean and reliable at our data rates of about 1Mb/s.

We favor differential signal transmission where reliability and good noise immunity are important. The differential signal results in very small signal coupling ("crosstalk"), because of cancellation effects. The use of twisted-pair, rather than flat cable, makes things even better. Figure 9.38 shows some measured waveforms, for both RS-422 and direct logic drive, using both flat cable and flat twisted-pair (the latter was actually a variant known as "twist-and-flat," consisting of twisted-pair cable interrupted for 2 inches every 20 inches to make a flat termination area). For RS-422 we used 100 feet of cable, driving the 6 volt pp 100kHz signal on one pair and looking at cross-talk on the adjacent pair; both were terminated. For the direct logic drive we used 74LS244 drivers at 1MHz through 10 feet of cable, with or without termination as indicated. From the waveforms it should be clear that RS-422 is very safe, even with long cable runs, whereas direct logic drive is marginal even with modest cable runs, though improved by termination and the use of flat cable with integral ground plane. Unexpectedly, twisted-pair did not fare any better than flat cable for direct logic drive.

Differential line receivers work well, as long as the received signal stays within the allowable common-mode voltage range, usually a few volts (±3V for the 75108). With long lines, however, you can get into a situation where either (a) high-frequency common-mode noise or (b) low-frequency (power-line) ground-voltage differences exceed the receiver's common-mode range. If the problem isn't too severe, you can use a pair of resistive dividers at the receiver's input, or use a receiver with built-in attenuator, for example the 26LS33 RS-422 receiver, with ±15 volt common-mode range.

For really long cable runs, or very noisy environments, the usual solution is to use transformer coupling. With transformers, of course, you can't send dc logic levels: You are forced to encode the data in some fashion, for example with a "carrier" signal. Local-area networks (Section 10.21) usually use transformer coupling.

□ *AMD "TAXI chip®."* AMD has introduced an interesting differential transmitter/receiver pair, the Am7968/9, with all sorts of on-chip registers to make it easy to use (Fig. 9.39). For example, you can treat the transmitter chip like an 8-bit latch with strobe and handshake; in that mode the chip converts the bytes to serial data, appends appropriate synchronization bits, transmits the data over a serial link, and reconstructs the byte at the other end. To the user the link looks like a simple parallel register. These chips include the cable drivers and receivers for 50 ohm cable, they run on a single +5 volt supply, and they go *fast*: Data throughput rates are 32–100 megabits/s (4 to 12.5 megabytes/s). The TAXI chips are intended as a high-speed general-purpose data link, with dc or ac coupling. The actual transmission medium can be direct connection via wire, twisted-pair, coax, transformer-coupled cable, or even a fiber-optic link.

Coaxial-cable drivers

Coaxial cable gives very good protection against interference because of its completely shielded geometry. In addition, its uniformity of diameter and spacing (compared with a certain randomness in

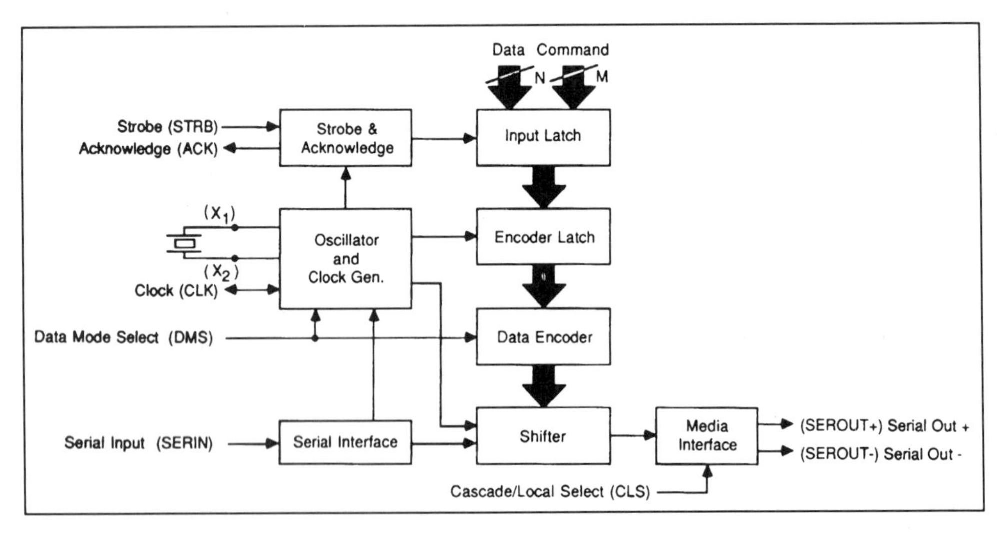

Figure 9.39. AMD "TAXI chip®" set: transparent high-speed serial link. (Diagram courtesy of Advanced Micro Devices of Sunnyvale, California.)
A. Am7968 transmitter.

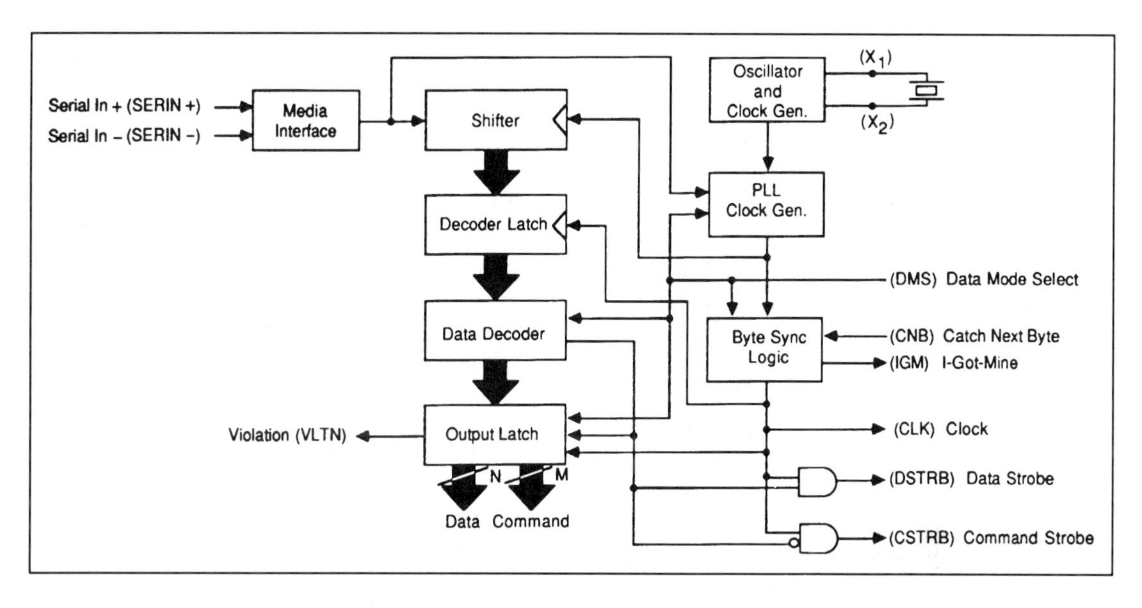

Figure 9.39. B. Am7969 receiver.

the case of bundled cables or twisted-pair) leads to predictable characteristic impedance and therefore excellent transmission properties; that is why it is used exclusively for analog radiofrequency signals.

There are several driver/receiver pairs that are suited for digital transmission with coax; Figure 9.40 shows an example. The cable is terminated in its characteristic impedance, in this case 51 ohms. The 8T23 can drive a 50 ohm load directly, and the 8T24 has a fixed amount of hysteresis for noise immunity and fast output transition time. This circuit allows bit rates of 100kb/s over a mile of cable and up to 20Mb/s over shorter lines. Other driver/receiver pairs are available in the 8T*xx* and 75*xxx* interface families. The 74F3037 (quad) and 74F30244 (octal) drivers are meant to drive cables with impedances down to 30 ohms (e.g., a cable terminated at both ends). Be sure to use the specified receivers when driving 50 ohm coaxial lines, since the voltage levels on a terminated cable may be less than ordinary logic levels.

The various ECL logic families provide a number of 50 ohm coax driver/receiver pairs, for example the 10128/10129 pair. A very nice coax driver chip is the 10194: It's actually a bus transceiver, designed to

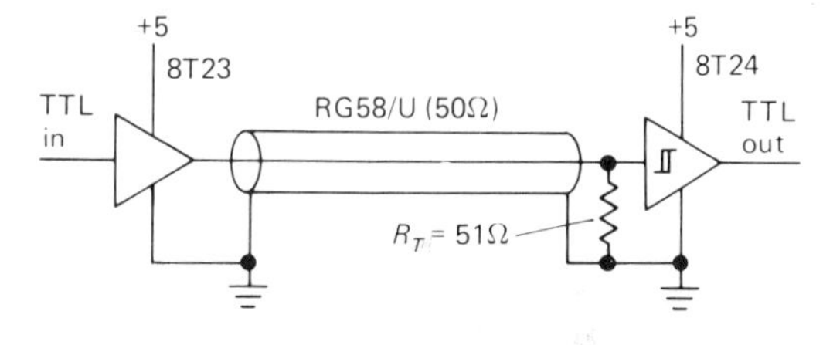

Figure 9.40. 50 ohm cable driver and receiver.

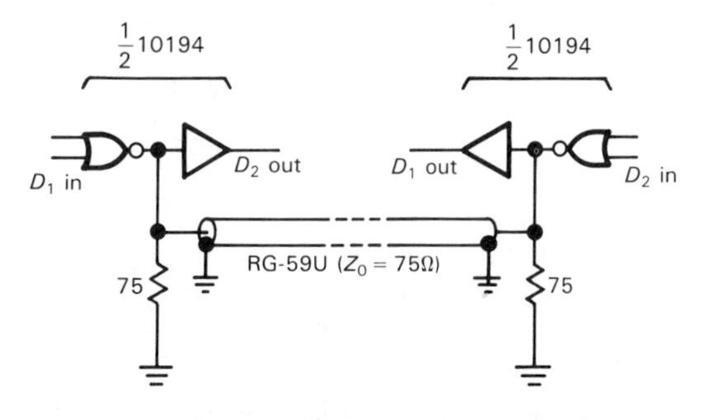

Figure 9.41. ECL current-sourcing transceiver (full duplex).

transmit and receive *simultaneously* ("full duplex") on the same line (Fig. 9.41). When used this way, each chip can be transmitting to the other transceiver while receiving asynchronously from it at the same time, without cross-talk, at speeds of 100MHz or more.

You can drive a length of coax directly from +5 volt logic if you use a single *npn* emitter follower (Fig. 9.42). The 2N4401 is a husky little transistor with plenty of beta at high currents ($h_{FE} > 100$ at $I_C = 150\text{mA}$). The 10 ohm resistor is for short-circuit protection. In comparison with the carefully designed and expensive 50 ohm cable driver chips that are available, this circuit is embarrassingly simple. Note that the "open-emitter" output must be loaded with a low-resistance return to ground in order to work, which is also true of some IC cable driver chips.

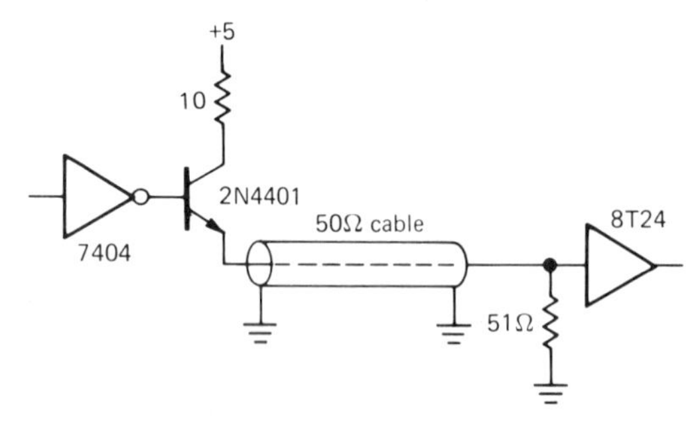

Figure 9.42. Simple 50 ohm cable driver.

Fiber-optics

An exciting new method of signal transmission is provided by fiber-optic cables. These come as nice plastic-coated cables, with matching connectors, emitters, and detectors. High-performance fiber-optic cables can transmit bandwidths of several gigahertz over distances of tens or hundreds of kilometers, with losses of fractional decibels per kilometer. Compared with coaxial cables, which are dispersive (different frequencies travel at different speeds, and with different amounts of loss, thereby corrupting waveforms), fiber-optic cables can be designed so that dispersion is negligible. And fiber-optic cables are insulators, so they can be used to send signals between devices with separate grounds, or at different voltages. Unlike ordinary cables, they do not act as antennas for radio interference or impulsive noise. They are lighter, safer, and more rugged than conventional cables, and they are potentially cheaper, too.

There are several types of fiber-optic cables, the choice involving a trade-off between cost and performance (length times bandwidth). The cheapest is the so-called step-index, multimode fiber, typically a plastic fiber 1mm in diameter; you would normally drive it with an infrared LED (as opposed to a laser diode) and use a phototransistor or pin-diode detector. Motorola makes an inexpensive set of driver/receivers (less than $1 each) that mate directly with jacketed fiber-optic cable (the MFOE71/MFOD71-73 series); with them you can send data at rates up to 1Mb/s over 30 feet of the plastic cable just described (see Fig. 9.43). Higher-performance cables use glass fibers – step-index multimode, graded-index multimode (better), or single-mode (best). With 200μm glass step-index fiber you can run 5Mb/s over a 1km path with standard off-the-shelf fiber-optic components, which include connectors, couplers, splitter/combiners, and detectors with built-in amplifiers. The current record for high-bandwidth long-haul fiber-optic communication is 4GHz over a 120km path, using no repeaters.

ANALOG/DIGITAL CONVERSION

9.15 Introduction to A/D conversion

In addition to applying the purely "digital" interfacing (switches, lights, etc.) discussed in the last few sections, it is

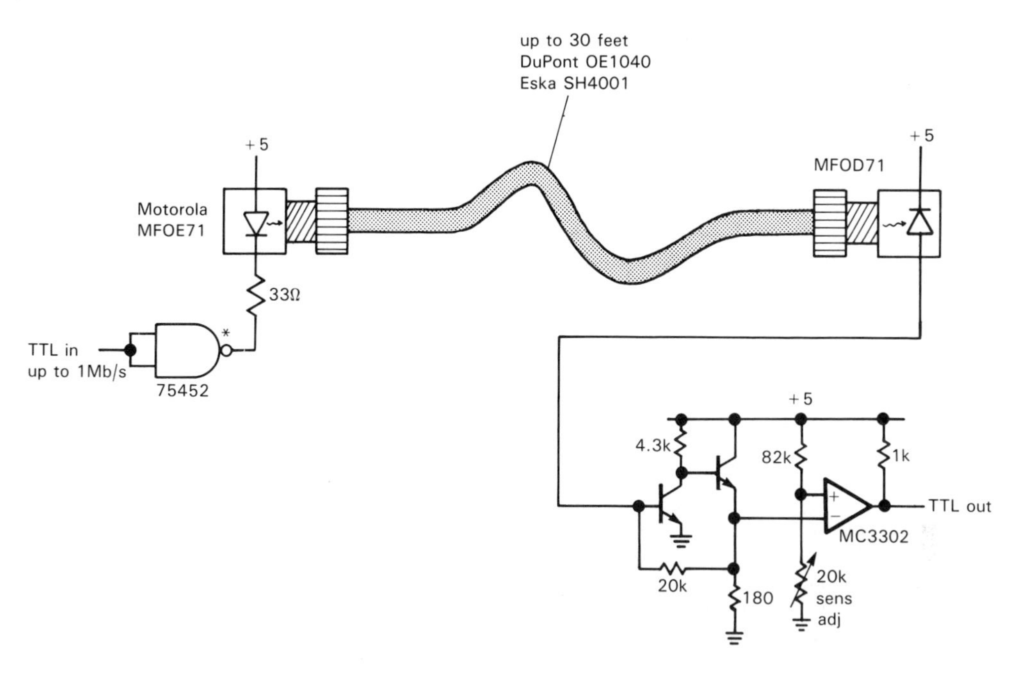

Figure 9.43. Inexpensive fiber-optic link. (Adapted from Fig. 7, MFOD71 Data Sheet, Motorola, Inc.)

often necessary to convert an analog signal to an accurate digital number proportional to its amplitude, and vice versa. This is essential in any application in which a computer or processor is logging or con trolling an experiment or process, or whenever digital techniques are used to do a "normally" analog job. Applications in which analog information is converted to an intermediate digital form for error- and noise-free transmission [e.g., "digital audio" or pulse-code modulation (PCM)] make heavy use of analog/digital conversion. It is necessary in a wide variety of measurement instruments (including ordinary bench instruments such as digital multimeters and more exotic instruments such as transient averagers, "glitch catchers," and digital memory oscilloscopes), as well as signal-generation and processing instruments such as digital waveform synthesizers and data encryption devices.

Finally, conversion techniques are essential ingredients for the generation of analog displays by a digital instrument, e.g., a meter indication or xy display (or plot) created by a computer. Even in a relatively unsophisticated electronic apparatus there are plenty of nice applications that call for A/D and D/A conversion, and it is worth developing familiarity with the various techniques and available modules used in analog/digital conversion, especially since the day of the $5 A/D and D/A converter is here.

Our treatment of the various conversion techniques will not be aimed at developing skill in converter design itself. Rather, we will try to point out the advantages and disadvantages of each method, since in most cases the sensible thing is to buy commercially available chips or modules, rather than build the converter from scratch. An understanding of conversion techniques and idiosyncrasies will guide you in choosing among the hundreds of available units.

Codes

At this point you should review Section 8.03 on the various number codes used to represent signed numbers. Offset binary and 2's complement are commonly used in A/D conversion schemes, with sign-magnitude and Gray codes also popping up from time to time. Here is a reminder:

	Offset binary	*2's Complement*
+Full scale	11111111	01111111
+Full scale−1	11111110	01111110
↓	↓	↓
0+1 LSB	10000001	00000001
0	10000000	00000000
0−1LSB	01111111	11111111
↓	↓	↓
−Full scale+1	00000001	10000001
−Full scale	00000000	10000000

Converter errors

The subject of A/D and D/A errors is a complicated one, about which whole volumes could be written. According to Bernie Gordon at Analogic, if you think a high-accuracy converter system lives up to its claimed specifications, you probably haven't looked closely enough. We won't go into the application scenarios necessary to support Bernie's claim, but we will show the four most common types of converter errors. Rather than bore you with a lot of complicated talk, we'll just present self-explanatory graphs of the four most common errors: offset error, scale error, nonlinearity, and nonmonotonicity (Fig. 9.44).

9.16 Digital-to-analog converters (DACs)

The goal is to convert a quantity specified as a binary (or multidigit BCD) number to a voltage or current proportional to the value of the digital input. There are several popular methods:

Scaled resistors into summing junction

As you saw in Section 4.09, by connecting a set of resistors to an op-amp summing junction you get an output proportional to a weighted sum of the input voltages. (Fig. 9.45). This circuit generates an output from zero to −10 volts, with full output corresponding to an input count of 64. Actually, the maximum input count is always 2^n-1, i.e., all bits set to 1. In this case the maximum input count is 63, with output voltage of $-10 \times 63/64$. By changing the feedback resistor, you can generate an output of zero to −6.3 volts (i.e., output in volts numerically equal to −1/10 of input count), or you can add an inverting amplifier, or dc offset at the summing junction, to get positive outputs. By changing the input resistor values you can properly convert a multidigit BCD input code, or any other weighted code. The input voltages must be clamped to an accurate reference, and the smaller input resistors must be of correspondingly higher precision. Of course, the switch resistance must be smaller than $1/2^n$ of the smallest resistor, an important consideration since in actual circuits the switching is done with transistor or FET switches. This conversion technique is used only in fast, low-precision converters.

EXERCISE 9.2

Design a 2-digit BCD D/A converter. Assume that the inputs are zero or +1 volt; the output should go from zero to 9.9 volts.

R-2R ladder

The scaled resistor technique becomes awkward when you have more than a few bits. For example, in a 12-bit converter you would need a range of resistor values of 2000:1, with corresponding precision in the small resistor values. An elegant solution is provided by the *R-2R* ladder,

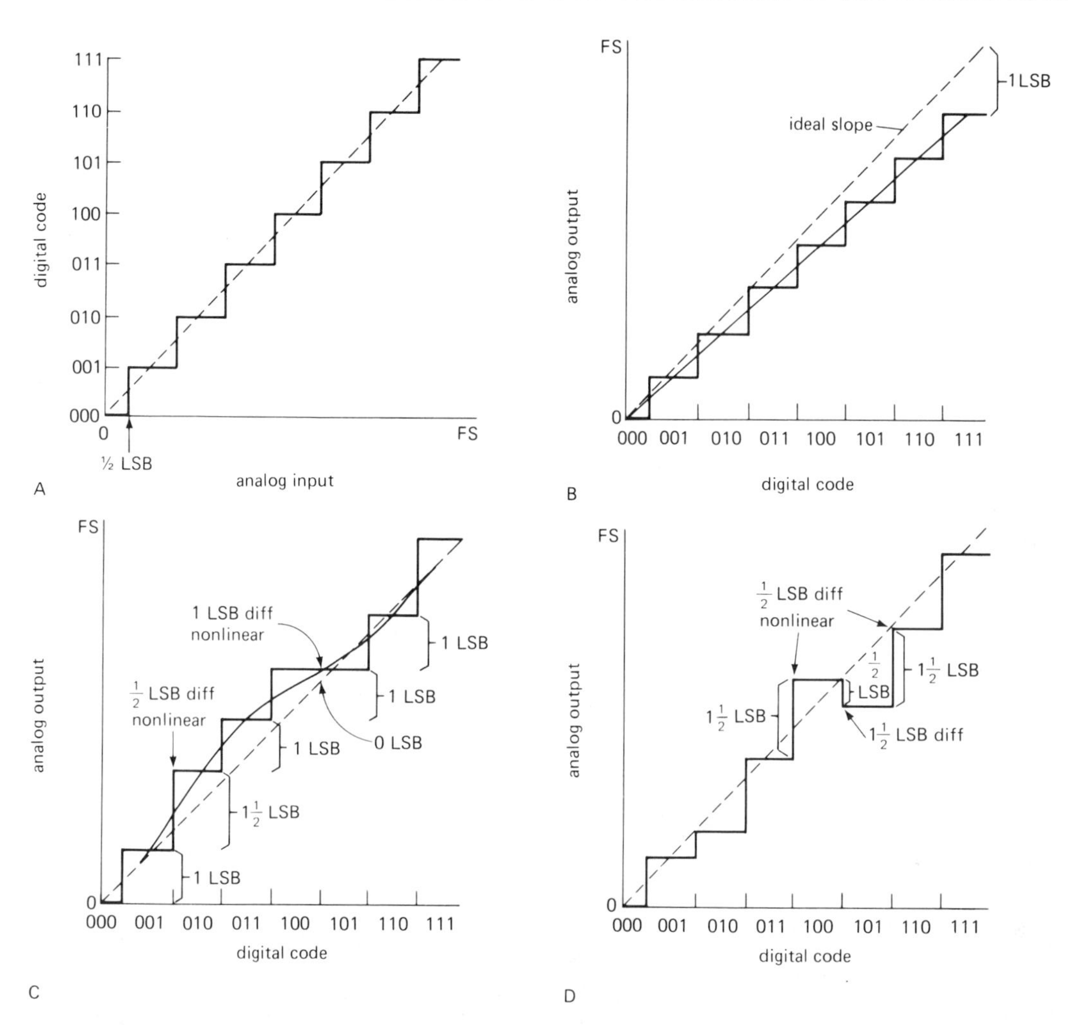

Figure 9.44. Graphs illustrating the definitions of four common digital conversion errors. (Courtesy of National Semiconductor Corp.).
A. ADC transfer curve, $\frac{1}{2}$ LBS offset at zero.
B. Linear, 1LSB scale error.
C. $\pm\frac{1}{2}$ LSB nonlinearity (implies 1 LSB possible error); 1 LSB differential nonlinearity (implies monotonicity).
D. Nonmonotonic (must be $> \pm\frac{1}{2}$ LSB nonlinear).

shown in Figure 9.46. Only two resistor values are needed, from which the *R-2R* network generates binary scaled currents. Of course, the resistors must be precisely matched, though the actual resistor values are not critical. The circuit shown generates an output of zero to -10 volts, with "full output" corresponding to an input count of 16 (again, the maximum input count is 15, with output voltage of $10 \times 15/16$). With a few modifications, an *R-2R* scheme can be used for BCD conversion.

EXERCISE 9.3

Prove that the foregoing *R-2R* ladder above works as advertised.

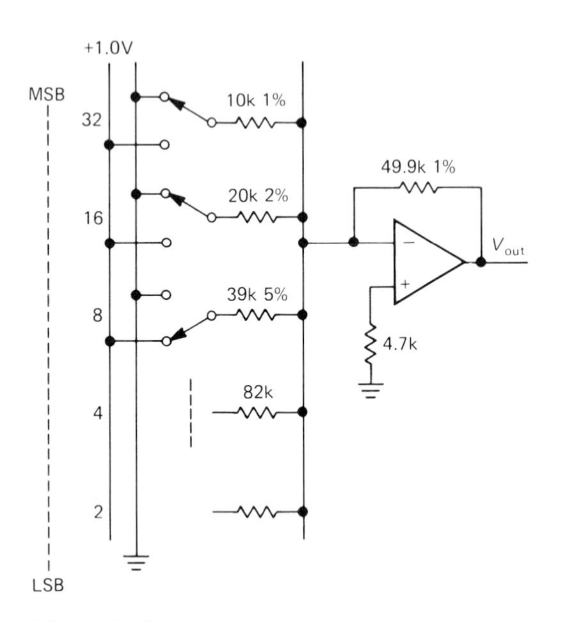

Figure 9.45

Scaled current sources

In the *R-2R* converter in the preceding paragraph, the op-amp converts binary scaled *currents* to an output *voltage*. Although an output voltage is often most convenient, the op-amp tends to be the slowest part of the converter circuit. In situations where you can use a converter with current output, you will get better performance, usually at lower price. Figure 9.47 shows the general idea. The currents can be generated by an array of transistor current sources with scaled emitter resistors, although IC designers usually use instead an *R-2R* ladder of emitter resistors. In most converters of this type, the current sources are ON all the time, and their output current is switched to the output terminal or to ground, under control of the digital input code. Watch out for limited output compliance in current-output DACs; it can be as little as 0.5 volt, though values of a few volts are typical.

Generating a voltage output

There are a few ways to generate an output voltage from a current DAC. Figure 9.48 shows some ideas. If the load capacitance is low, and large voltage swings aren't needed, a simple resistor to ground will do nicely. With the usual 1mA full-scale output current, a 100 ohm load resistor will give 100mV full-scale output with 100 ohms output impedance. If the capacitance of the DAC's output combined with the load capacitance doesn't exceed 100pF, you will get 100ns settling time in the preceding example, assuming the DAC is that fast. When worrying about the effect of *RC* time constants on DAC output response, don't forget that it takes quite a few *RC* time constants for the output to settle to within

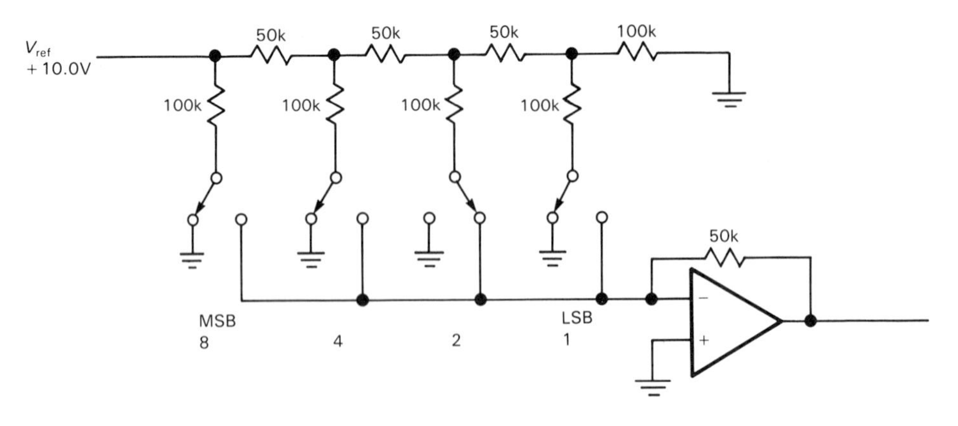

Figure 9.46. *R-2R* ladder.

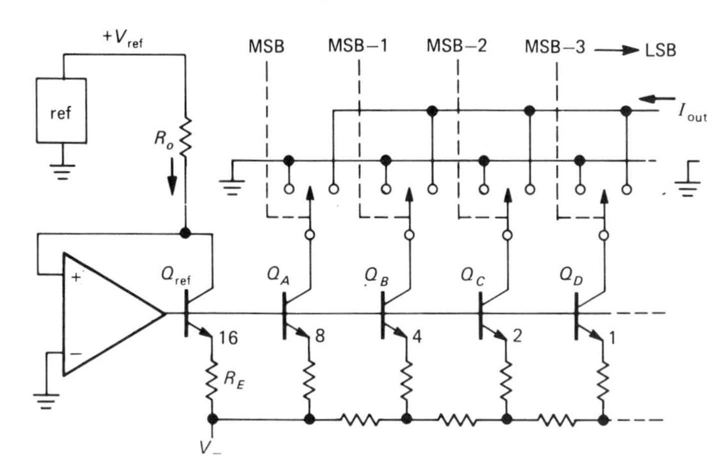

Figure 9.47. Classic current-switched DAC.

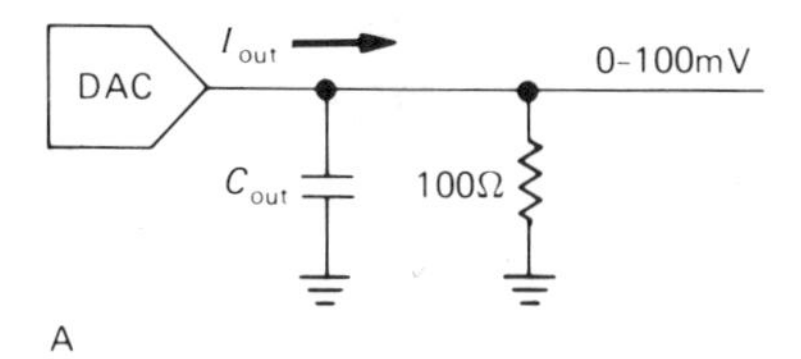

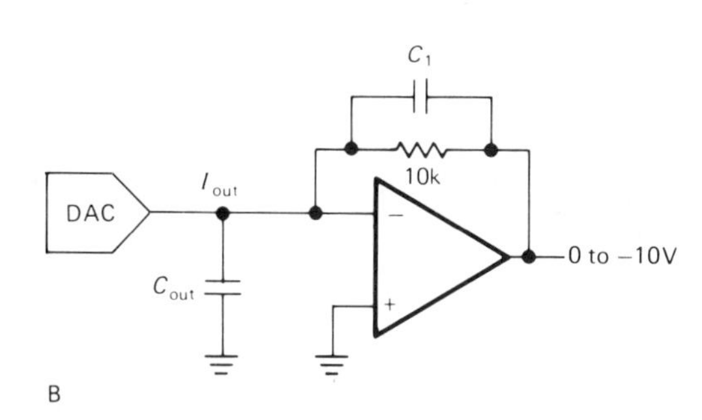

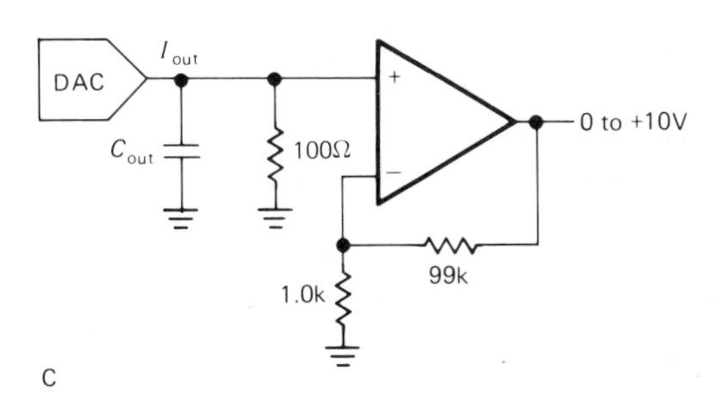

Figure 9.48. Generating voltages from current-output DACs.

$\frac{1}{2}$ LSB of the final voltage. It takes 7.6 RC time constants, for instance, to settle to within 1 part in 2048, which is what you would want for a 10-bit converter output.

To generate large swings, or to buffer into small load resistances or large load capacitances, an op-amp can be used in the transresistance configuration (current-to-voltage amplifier), as shown. The capacitor across the feedback resistor may be necessary for stability, because the DAC's output capacitance in combination with the feedback resistance introduces a lagging phase shift; unfortunately, that compromises the speed of the amplifier. It is an irony of this circuit that a relatively expensive high-speed (fast-settling) op-amp may be necessary to maintain the high speed of even an inexpensive DAC. In practice, the last circuit may give better performance, since no compensation capacitor is needed. Watch out for offset voltage error, since the op-amp's input offset voltage is amplified 100 times.

Commercial D/A converter modules are available with precision ranging from 6 bits up to 18 bits, with settling times of 25ns to 100μs (for converters of the highest precision). Prices range from a few dollars to several hundred dollars. A typical popular unit is the AD7248, a

12-bit converter with latch and internal reference, settling time of 5μs for voltage output, and a price of about \$10.

9.17 Time-domain (averaging) DACs

Frequency-to-voltage converters

In conversion applications the "digital" input may be a train of pulses or other waveform of some frequency; in that case, direct conversion to a voltage is sometimes more convenient than the alternative of counting for predetermined time, then converting the binary count via the preceding methods. In direct F/V conversion, a standard pulse is generated for each input cycle; it may be a voltage pulse or a pulse of current (i.e., a fixed amount of charge).

An RC low-pass filter or integrator then averages the pulse train, giving an output voltage proportional to the average input frequency. Of course, some output ripple results, and the low-pass filter necessary to keep this ripple less than the D/A precision (e.g., $\frac{1}{2}$ LSB) will, in general, cause a slow output response. To ensure less than $\frac{1}{2}$ LSB output ripple, the time constant T of a simple RC low-pass filter must be at least $T = 0.69(n+1)T_o$, where T_o is the output period of the n-bit F/V converter corresponding to maximum input frequency. The output of this RC network will settle to $\frac{1}{2}$ LSB, following a full scale change at the input, in $0.69(n+1)$ filter time constants. In other words, the output settling time to $\frac{1}{2}$ LSB will be approximately $t = 0.5(n+1)^2 T_o$. A 10-bit F/V converter with 100kHz maximum input frequency, smoothed with an RC filter, will have an output voltage settling time of 0.6ms. With more complicated low-pass filters (sharp cutoff) you can get improved performance. Before you get carried away with fancy filter design, however, you should remember that F/V techniques are most often used when a voltage output is not what's needed. For some perspective, we will next comment about intrinsically slow loads in connection with pulse-width modulation.

Pulse-width modulation

In this technique the digital input code is used to generate a train of pulses of fixed frequency, with width proportional to the input count; this can be easily done with a counter, magnitude comparator, and high-frequency clock (see Exercise 9.4). Once again, a simple low-pass filter can be used to generate an output voltage proportional to the average time spent in the HIGH state, i.e., proportional to the digital input code. More often, this sort of D/A conversion is used when the load is itself a slowly responding system; the pulse-width modulator then generates precise parcels of energy, averaged by the system connected as a load. For example, the load might be capacitive (as in a switching regulator, see Chapter 6), thermal (a thermostated bath with heater), mechanical (a tape speed servo), or electromagnetic (a large electromagnet controller).

EXERCISE 9.4

Design a circuit to generate a 10kHz train of pulses of width proportional to an 8-bit binary input code. Use counters and magnitude comparators (suitably expanded).

Averaged rate multiplier

The rate-multiplier circuit described in Section 8.28 can be used to make a simple D/A converter. A parallel binary or BCD input code is converted to a train of output pulses of average frequency proportional to the digital input; simple averaging, as in the preceding F/V converter, can be used to generate a dc output proportional to the digital input code, although in this case the resulting output time constant may be intolerably long, since the rate-multiplier output will have to be averaged for a time equal to the longest output period it can generate. Rate multipliers are most advantageously used as D/A converters when the output is intrinsically averaged by the slowly responding characteristics of the load itself, as described earlier.

Perhaps the nicest application is in digital temperature control, where complete cycles of ac power are switched across the heater for each rate-multiplier output pulse. In this application the rate multiplier is arranged so that its lowest output frequency is an integral submultiple of 120Hz, and a solid-state relay (or triac) is used to switch the ac power (at zero crossings of its waveform) from logic signals.

Note that the last three conversion techniques involve some time averaging, whereas the resistor-ladder and current-source methods are "instantaneous," a distinction that also exists in the various methods of analog-to-digital conversion. Whether a converter averages the input signal or converts an instantaneous sample of it can make an important difference, as you will see shortly in some examples.

9.18 Multiplying DACs

Most of the preceding techniques can be used to construct a "multiplying DAC," in which the output equals the product of an input voltage (or current) and an input digital code. For instance, in a scaled current-source DAC, you can scale all the internal current sources by an input programming current. Multiplying DACs can be made from DACs that have no internal reference by using the reference input for the analog input signal. However, not all DACs are optimized for use in this way, so it is best to check the data sheets of the converters you're considering for details. A DAC with good multiplying properties (wide analog input range, high speed, etc.) will usually be called a "multiplying DAC" right at the top of the data sheet. The AD7541, 7548, 7845, and DAC1230 are all examples of 12-bit multiplying DACs, priced around $10 to $20.

Multiplying DACs (and the A/D equivalent) open the possibility of *ratiometric* measurements and conversions. If a sensor of some sort (e.g., a variable-resistance transducer like a thermistor) is powered by a reference voltage that also supplies the reference for the A/D or D/A converter, then variations in the reference voltage will not affect the measurement. This concept is very powerful, since it permits measurement and control with accuracy greater than the stability of voltage references or power supplies; conversely, it relaxes the requirements on supply stability and accuracy. The ratiometric principle is used in its simplest form in the classic *bridge* circuit, in which two ratios are adjusted to equality by nulling the differential signal taken between the two voltage-divider outputs (see Section 15.02). Devices like the 555 (see Section 5.14) achieve good stability of output frequency with large variations of supply voltage by using essentially a ratiometric scheme: The capacitor voltage, generated by an RC network from the supply, is compared with a fixed fraction of the supply voltage ($\frac{1}{3}V_{CC}$ and $\frac{1}{3}V_{CC}$), giving an output frequency that depends only on the RC time constant. We will have more to say on this important subject in connection with A/D converters later in this chapter, as well as in Chapter 15, when we discuss scientific measurement techniques.

9.19 Choosing a DAC

To guide you in choosing a DAC for a particular application, we've assembled in Table 9.4 a representative selection of DACs of various precisions and speeds. This listing is by no means exhaustive, but it does include many of the more popular converters and some more recent

TABLE 9.4. DIGITAL-TO-ANALOG CONVERTERS

Type	Mfg[a]	# per pkg	Bits	Latch	V or I ?	Speed to MSB/2 (ns)	Pol[b]	Input org	Pkg[c]	V_{supply} (V)	I_{supply} (mA)	Reference	Trim[d]	Multiplying?	Ext'd compl?[e]	Low glitch?	Price	Comments
AD9702	AD	3	4	1	I	5	–	12	24	±6	1.8W	ext	–	–	4.2	•	$45	color video; ECL or TTL
AD7225	AD	4	8	2	V	5000	+	4x8	24S	+15	10	ext	NT	–	–	–	$18	double buffered; 7226 is single buff.
AD558	AD	1	8	1	V	1000	+	8	16	5 to 15	15	int	NT	–	–	–	$6	complete, easy to use
DAC0830	NS	1	8	2	I	1000	M	8	20	5 to 15	2	0 to ±25	NT	•	N	–		same pinout as 12-bit DAC1230
AD7528	AD	2	8	1	I	350	M	8	20	5 to 15	0.1	0 to ±25	NT	•	N	–	$6	dual, easy to use
DAC8408	PM	4	8	1	I	190	M	4x8	28	+5	0.05	ext	NT	•	N	–		can read back buffers
Bt-110	BT	8	8	1	I	100	+	8	40	+5	30	int	•	–	N	–		octal
AD7524	AD	1	8	1	I	100	M	8	16	5 to 15	0.1	0 to ±10	NT	•	N	–	$5	indus std multiplying DAC
DAC-08	AD	1	8	-	I	85	–	8	16	±15	+2,-6	ext	–	–	28	–	$1	obsolete; indus std
Bt-453	BT	3	8	1	I[f]	15	+	8	40	+5	160	ext	–	–	N	–	$58	256x24 palette, for 256 of 16M colors
HDG0807	AD	1	8	-	V[f]	14	+	8	24	+5	185	int	NT	–	–	–	$43	video, 75Ω output
TDC1018	TR	1	8	1	I	10	–	8	24	-5.2	100	ext	–	–	4	•		video
AD9768	AD	1	8	-	I	5	–	8	18?	±5	+15,-70	int	NT	–	N	–	$40	ECL inputs
TDC1318	TR	3	8	1	I	5	–	3x8	40	-5.2	200	int	NT	–	N	–		high res'n color video; many 2nd sources
TQ6112	TQ	1	8	1	V	1	–	8	44	-3.5,-9	3.5W	ext	–	–	N	•		GaAs: ultra-fast
IDT75C29	ID	1	9	1	I	8	–	9	24	+5	80	ext	–	–	N	–		
DAC1000	NS	1	10	2	I	500	M	8+2	20	5 to 12	0.5	0 to ±25	NT	•	N	–		double buffered; cheap
AD7248	AD	1	12	2	V	5000	±	8+4	20	±15	5	int	NT	–	–	–	$10	single supply possible; +5V ref out
AD7537	AD	2	12	2	I	1500	M	8+4	24S	+15	5	ext	NT	–	N	–	$15	
AD7548	AD	1	12	2	I	1000	M	8+4	20	5 to 15	1	0 to ±25	NT	•	N	–	$10	
DAC1230	NS	1	12	2	I	1000	M	8+4	20	+15	1.2	0 to ±25	NT	•	N	–		same pinout as DAC0830
AD568	AD	1	12	-	I	35	±	12	24S	±15	+30,-8	int	•	–	N	•	$42	
AD7534	AD	1	14	2	I	1500	M	8+4	20	+15,-0.3	0.5	0 to ±25	•	•	N	–	$17	7535,6,8 have 14-bit bus
AD569	AD	1	16	2	V	6000	M	8+8	28	±12	±6	±5	NT	•	–	–	$28	
DAC71/72	all	1	16	-	I	1000	±	16	24	+5,±15	10,+10,-30	int	•	–	11	–	$45	indus std; V_{out} also: 10μs
PCM54	BB	1	16	-	I	350	±	16	28	±5 to ±15	±13	int	NT	–	N	•	$11	digital audio, cheap; V_{out} also: 3μs
DAC729	BB	1	18	-	I	300	–	18	40	+5, +15	18,+30,-40	int	•	–	6	–	$141	V_{out} also: 4μs; 18 bits = 4ppm!

(a) see footnote to Table 4.1. (b) M - multiplying. (c) all are DIP; S - skinny-DIP. (d) NT - no trim required. (e) extended compliance, in volts. (f) to drive 75Ω.

entries that were intended as improved replacements.

When looking for a DAC for some application, some issues to keep in mind are (a) precision, (b) speed, (c) accuracy (external trimming required?), (d) input structure (latched? CMOS/TTL/ECL-compatible?), (e) reference (internal, or externally supplied?), (f) output structure (current output? compliance? voltage output? range?), (g) required supply voltages and power dissipation, (h) package (low pin count and 0.3″ wide "skinny-DIP" packaging are desirable), and (i) price.

9.20 Analog-to-digital converters

There are half a dozen basic techniques of A/D conversion, each with its peculiar advantages and limitations. Since you usually use a commercial A/D module or chip rather than build your own, we will describe the various conversion techniques somewhat briefly, mainly to serve as a guide for intelligent selection in a given application. The next section of the chapter will illustrate some typical A/D applications. In Chapter 11 we will discuss some A/D converters that use exactly the same conversion methods, but that have outputs designed for simplified interfacing to microprocessors.

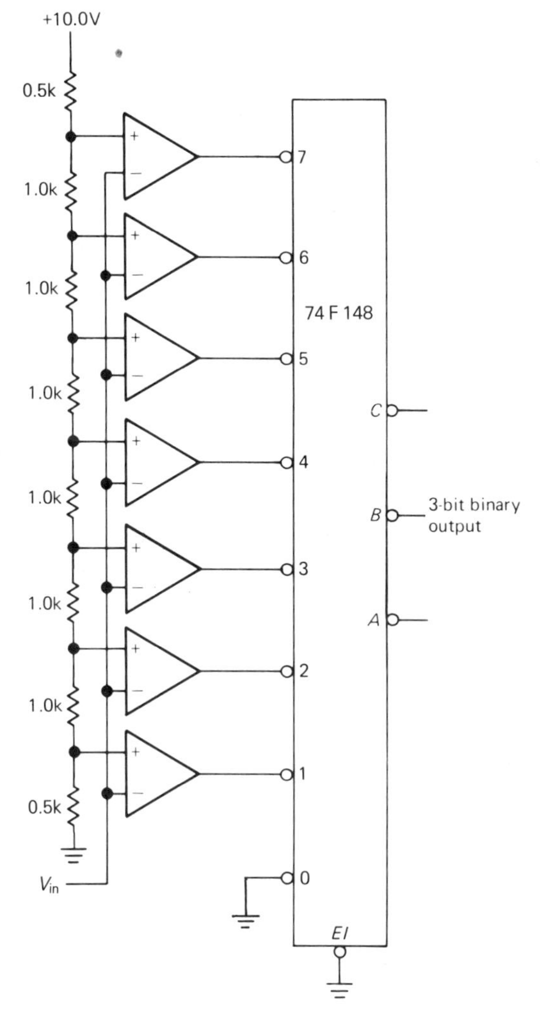

Figure 9.49. Parallel-encoded ("flash") A/D converter (ADC).

Parallel encoder

In this method the input signal voltage is fed simultaneously to one input of each of n comparators, the other inputs of which are connected to n equally spaced reference voltages. A priority encoder generates a digital output corresponding to the highest comparator activated by the input voltage (Fig. 9.49).

Parallel encoding (also called "flash" encoding) is the fastest method of A/D conversion. The delay time from input to output equals the sum of comparator plus encoder delays. Commercial parallel encoders are available with 16 to 1024 levels (4-bit to 10-bit outputs). Beyond that they become prohibitively expensive and bulky. Speeds range from 15 to 300 megasamples per second (MSPS). A typical flash ADC is the TRW TDC1048, a bipolar 8-bit 20MSPS converter in a 28-pin package, which sells for about $100; IDT makes the 75C48, a CMOS equivalent with improved specs.

A variant on the simple parallel encoder is the so-called half-flash technique, a two-step process in which the input is flash-converted to half the final precision; an internal DAC converts this approximation back to analog, where the difference "error" between it and the input is flash-converted to obtain the least significant bits (Fig. 9.50). This technique yields low-cost converters that are faster than anything else except full flash converters. It is used in inexpensive converters like the 8-bit ADC0820 (National) and AD7820/4/8 (Analog Devices).

Flash encoders are worth considering in waveform digitizing applications even when the conversion rate is relatively slow, because their high speed (or, more precisely, their short *aperture* interval, during which the comparator outputs are latched) ensures that the input signal is effectively not changing during the conversion. The alternative – the slower converters we'll describe next – usually requires an analog sample-and-hold circuit to freeze the input waveform while conversion is going on.

Successive approximation

In this popular technique you try various output codes by feeding them into a D/A converter and comparing the result with the analog input via a comparator (Fig. 9.51). The way it's usually done is to set all bits initially to 0. Then, beginning with the most significant bit, each bit in turn is set provisionally to 1. If the D/A output does not exceed the input signal voltage, the bit is left as a 1; otherwise it is set back to 0. For an n-bit A/D, n such steps are required. What you're doing could be described as a binary search, beginning at the middle. A successive-approximation A/D module has a BEGIN CONVERSION input and a CONVERSION DONE output. The digital output is always provided in parallel format (all bits at once, on n separate output lines) and usually in serial format as well (n successive output bits, MSB first, on a single output line).

In our electronics course the students construct a successive-approximation ADC, complete with DAC, comparator, and control logic. Figure 9.52A shows the

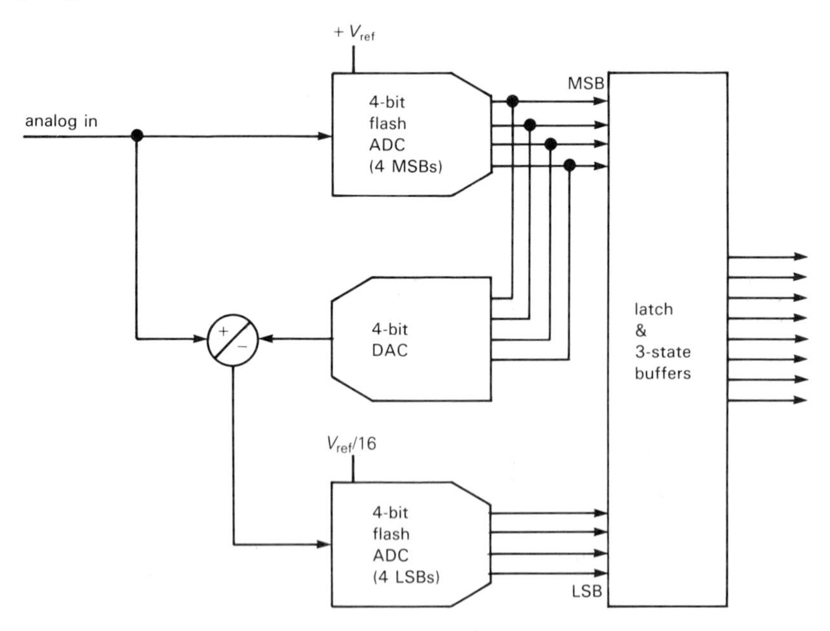

Figure 9.50. Half-flash ADC.

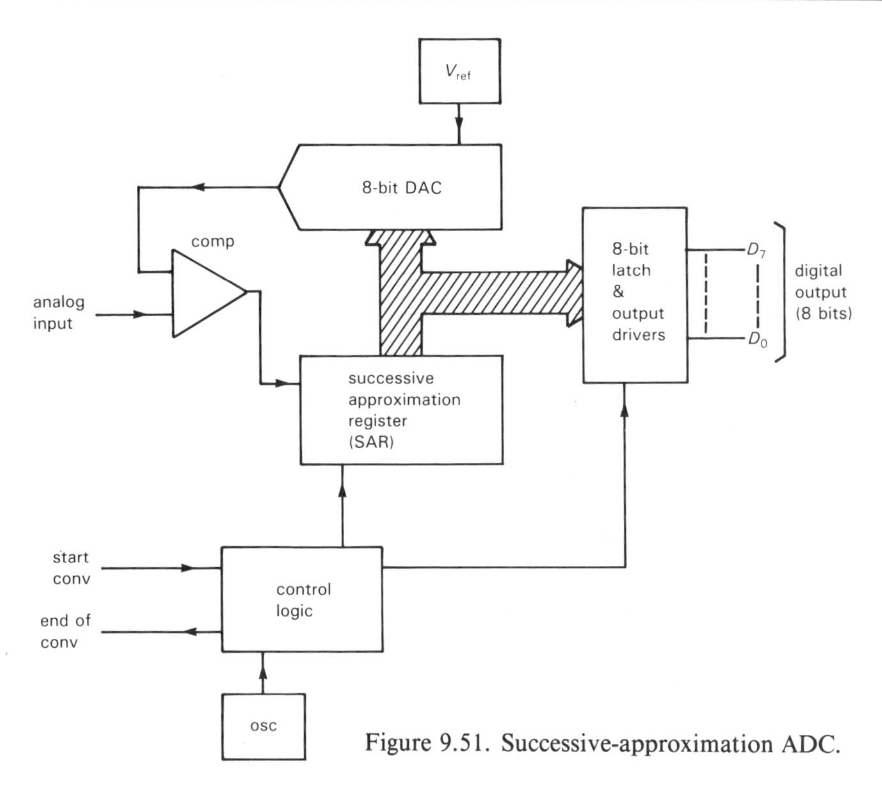

Figure 9.51. Successive-approximation ADC.

successive outputs from the DAC, along with the eight clock pulses, as the trial analog output converges to the input voltage. Figure 9.52B shows the full 8-bit "tree," a pretty picture you can generate by watching the DAC output while driving the input with a slow ramp that runs over the full analog input range.

Successive-approximation A/D converters are relatively accurate and fast,

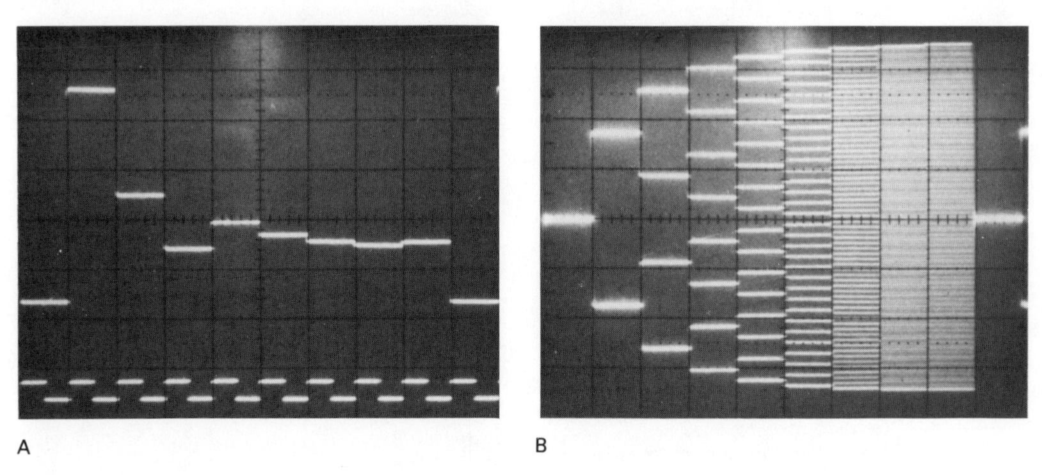

Figure 9.52. 8-bit successive-approximation waveforms. (Courtesy of P. Emery, R. Lovett, and K. Rudin.).
A. Analog output converging to final value; note clock waveform.
B. Full "tree".

requiring only n settling times of the DAC for n-bit precision. Typical conversion times range from $1\mu s$ to $50\mu s$, with accuracies of 8 to 12 bits commonly available; prices range from about \$10 to upwards of \$400. This type of converter operates on a brief sample of the input voltage, and if the input is changing during the conversion, the error is no greater than the change during that time; however, spikes on the input are disastrous. Although generally quite accurate, these converters can have strange nonlinearities and "missing codes."

National Semiconductor uses a nice trick in its ADC0800-series converters to eliminate missing codes: Instead of using a conventional *R-2R* ladder DAC, they use a chain of 2^n resistors and analog switches to generate the trial analog voltages (Fig. 9.53), in the manner of a flash encoder.

A variation known as a "tracking A/D converter" uses an up/down counter to generate successive trial codes; it is slow in responding to jumps in the input signal, but it follows smooth changes somewhat more rapidly than a successive-approximation converter. For large changes its slew rate is proportional to its internal clocking rate.

□ *Voltage-to-frequency conversion*

In this method an analog input voltage is converted to an output pulse train whose

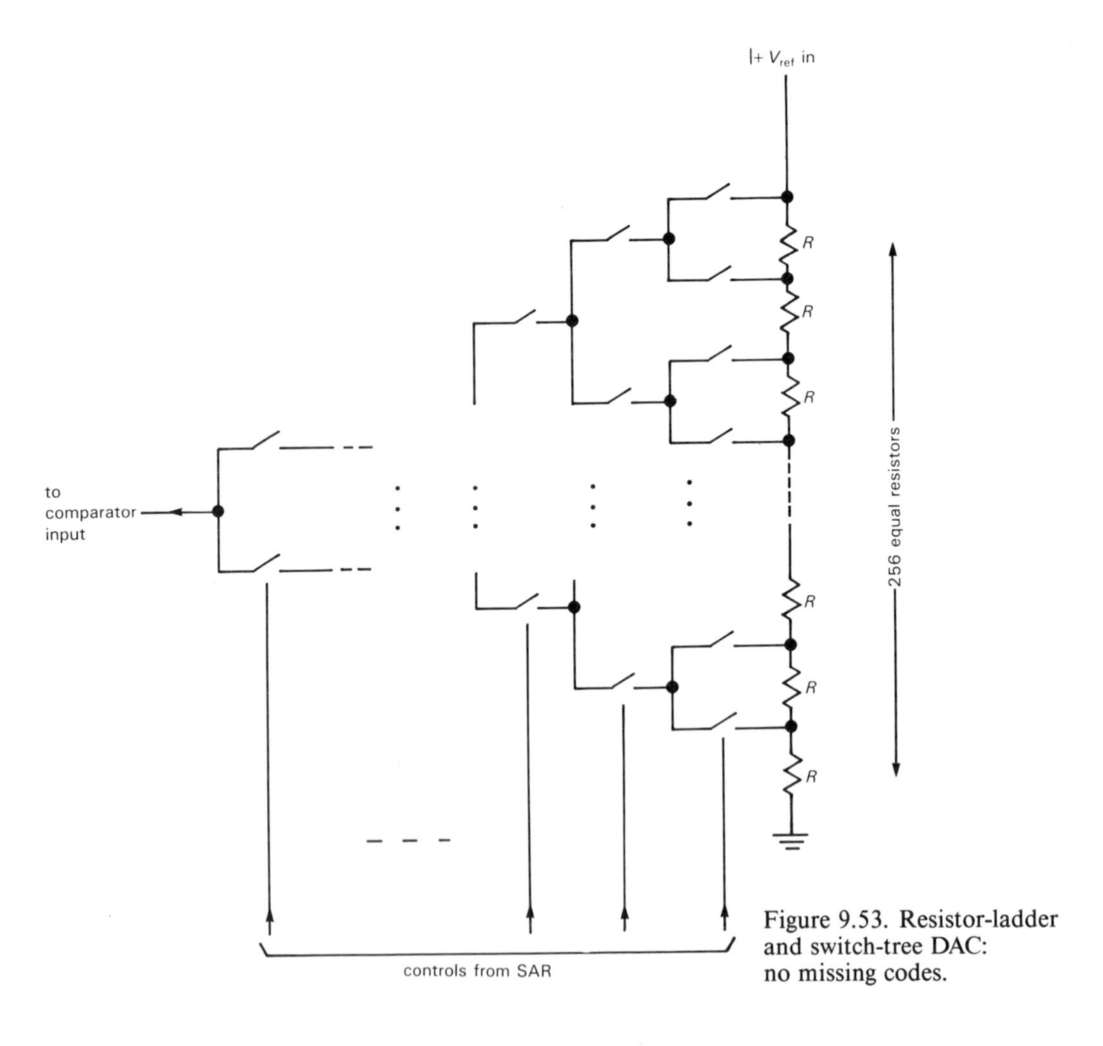

Figure 9.53. Resistor-ladder and switch-tree DAC: no missing codes.

frequency is proportional to the input level. This can be done simply by charging a capacitor with a current proportional to the input level and discharging it when the ramp reaches a preset threshold. For greater accuracy, a feedback method is generally used. In one technique you compare the output of an F/V circuit with the analog input level and generate pulses at a rate sufficient to bring the comparator inputs to the same level. In the more popular methods, a "charge-balancing" technique is used, as will be described in greater detail later (in particular, the "capacitor-stored charge-dispensing" method).

Typical V/F output frequencies are in the range 10kHz to 1MHz for maximum input voltage. Commercial V/F converters are available with the equivalent of 12-bit resolution (0.01% accuracy). For example, the excellent AD650 from Analog Devices (Section 5.15) has a typical nonlinearity of 0.002% when operating from 0 to 10kHz. They are inexpensive, and they are handy when the output is to be transmitted digitally over cables or when an output frequency (rather than digital code) is desired. If speed isn't important, you can get a digital count proportional to the average input level by counting the output frequency for a fixed time interval. This technique is popular in moderate-accuracy (3-digit) digital panel meters.

□ *Single-slope integration*

In this technique an internal ramp generator (current source + capacitor) is started to begin conversion, and at the same time a counter is enabled to count pulses from a stable clock. When the ramp voltage equals the input level, a comparator stops the counter; the count is proportional to the input level, i.e., it's the digital output. Figure 9.54 shows the idea.

At the end of the conversion the circuit discharges the capacitor and resets the counter, and the converter is ready for another cycle. Single-slope integration is simple, but it is not used where high accuracy is required because it puts severe requirements on the stability and accuracy of the capacitor and comparator. The method of "dual-slope integration" eliminates that problem (and several others as well) and is now generally used where precision is required.

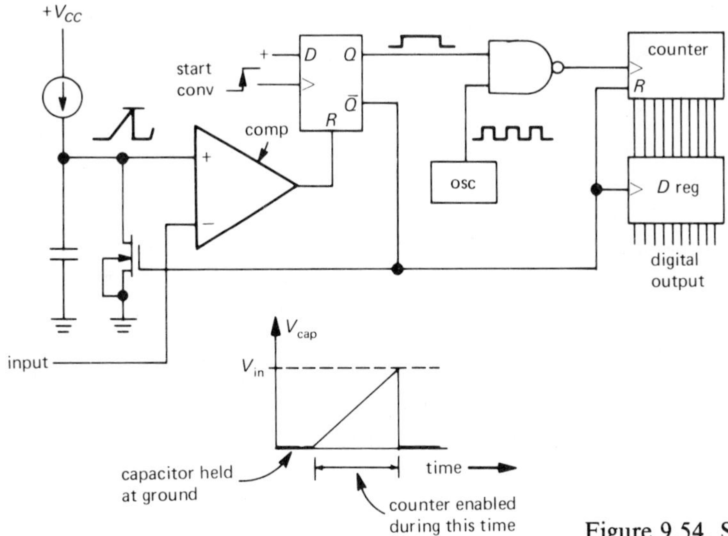

Figure 9.54. Single-slope ADC.

Single-slope integration is still alive and well, particularly in applications that don't require absolute accuracy, but rather need conversion with good resolution and uniform spacing of adjacent levels. A good example is pulse-height analysis (see Section 15.16), where the amplitude of a pulse is held (peak detector) and converted to an address. Channel width equality is essential for this application, for which a successive-approximation converter would be totally unsuitable. The technique of single-slope integration is also used in time-to-amplitude conversion (TAC).

9.21 Charge-balancing techniques

There are several techniques that have in common the use of a capacitor to keep track of the ratio of an input signal level to a reference. These methods all average (integrate) the input signal for a fixed time interval for a single measurement. There are two important advantages:

1. Because these methods use the same capacitor for the signal and reference, they are relatively forgiving of capacitor stability and accuracy. These methods also make fewer demands on the comparator. The result is better accuracy for equivalent-quality components, or equivalent accuracy at reduced cost.

2. The output is proportional to the *average* input voltage over the (fixed) integration time. By choosing that time interval to be a multiple of the power-line period, the converter becomes insensitive to 60Hz "hum" (and its harmonics) on the input signal. As a result, the sensitivity to interfering signals as a function of frequency is as shown in Figure 9.55 (0.1s integration).

This nulling of 60Hz interference requires accurate control of the integration time, since an error of a fraction of a percent in the clock timing will result in incomplete cancellation of hum. One possibility is to use a crystal oscillator. You will see in Section 9.29 an elegant method of synchronizing the workings of an integrating converter to a multiple of the line frequency in order to make this rejection perfect.

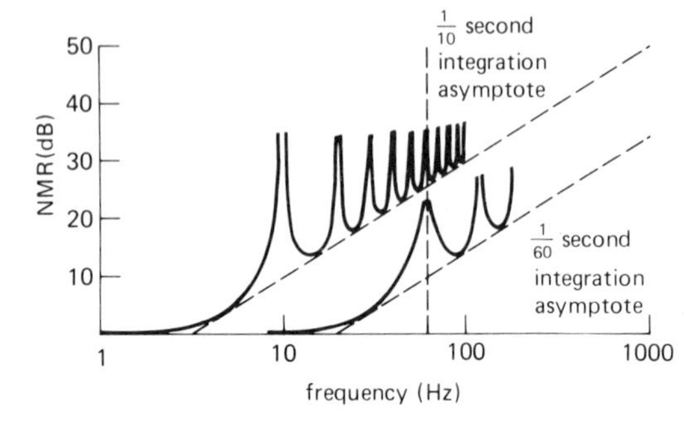

Figure 9.55. Normal-mode rejection with integrating A/D converters.

Integrating techniques have the disadvantage of slow speed, as compared with successive approximation.

Dual-slope integration

This elegant and very popular technique eliminates most of the capacitor and comparator problems inherent in single-slope integration. Figure 9.56 shows the idea. First, a current accurately proportional to the input level charges a capacitor for a fixed time interval; then the capacitor is discharged by a constant current until the voltage reaches zero again. The time to discharge the capacitor is proportional to the input level and is used to enable a counter driven from a clock running at a fixed frequency. The final count is proportional to the input level, i.e., it's the digital output.

Dual-slope integration achieves very good accuracy without putting extreme requirements on component stability. In particular, the capacitor value doesn't have to be particularly stable, since the charge cycle and the discharge cycle both go at a rate inversely proportional to C. Likewise, drifts or scale errors in the comparator are canceled out by beginning and ending each conversion cycle at the same voltage and, in some cases, at the same slope. In the most accurate converters, the conversion cycle is preceded by an "auto-zeroing" cycle in which the input is held at zero. Since the same integrator and comparator are used during this phase, subtracting the resulting "zero-error" output from the subsequent measurement results in effective cancellation of errors associated with measurements near zero; however, it does not correct for errors in overall scale.

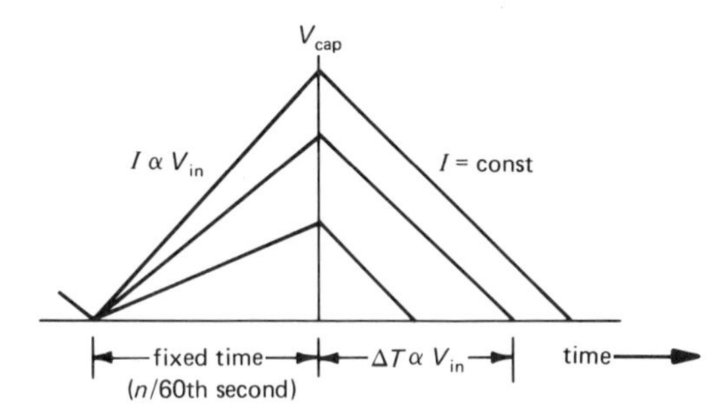

Figure 9.56. Dual-slope conversion cycle.

Note that even the clock frequency does not have to have high stability in dual-slope conversion, because the fixed integration time during the first phase of the measurement is generated by subdivision from the same clock used to increment the counter. If the clock slows down by 10%, the initial ramp will go 10% higher than normal, requiring 10% longer ramp-down time. Since that's measured in clock ticks that are 10% slower than normal, the final count will be the same! Only the discharge current has to be of high stability in a dual-slope converter with internal auto-zeroing. Precision voltage and current references are relatively easy to produce, and the (adjustable) reference current sets the scale factor in this type of converter.

When choosing components for dual-slope conversion, be sure to use a high-quality capacitor with minimum dielectric absorption ("memory" effect; see the model in Fig. 4.42) – polypropylene, polystyrene, or Teflon capacitors work best. Although these capacitors are not polarized, you should connect the outside foil (indicated with a band) to the low-impedance point (the output of the integrator op-amp). To minimize errors, choose integrator R and C values to use nearly the full analog range of the integrator. A high clock frequency improves resolution, although you gain little once the clock period becomes shorter than the comparator response time.

When using precision dual-slope converters (and, indeed, any kind of precision converter) it is essential to keep digital noise out of the analog signal path. Converters usually provide separate "analog ground" and "digital ground" pins for this purpose. It is often wise to buffer the digital outputs (say with a '244 three-state octal driver, asserted only when reading the output) in order to decouple the converter from the digital roar of a microprocessor bus (see next chapter). In extreme cases you might use opto-couplers to quarantine the noise of a particularly dirty bus. Be sure to use liberal power-supply bypassing right at the converter chip. And be careful not to introduce noise during the critical endpoint of the integration, as the ramp reaches the comparator trip point: For example, some converters let you check for end-of-conversion by reading the output word; don't do it! Instead, use the separate BUSY line, suitably isolated.

Dual-slope integration is used extensively in precision digital multimeters, as well as in conversion modules of 10-bit to 18-bit resolution. It offers good accuracy and high stability at lowest cost, combined with excellent rejection of power-line (and other) interference, for applications where speed is not important. For a fixed amount of money, you will get greatest precision with a module that uses this technique. The digital output codes are strictly monotonic with increasing input.

□ *Delta-sigma converters*

There are several methods of A/D conversion that involve cancellation of the (average) signal input current with a switched internal source of current or charge. Figure 9.57 shows a functional diagram of a "delta-sigma" converter.

The input voltage drives an integrator, whose output is compared with any fixed voltage, such as ground. Depending on the comparator output, pulses of current of fixed length (i.e., fixed increments of charge) are switched into the summing junction or to ground at each clock transition, with the effect of maintaining zero average current into the summing junction. This is the balancing concept. A counter keeps track of the number of charge pulses switched into the summing junction for a given number of clock pulses, say 4096. That count is proportional to the average input level during the 4096 clock pulses, i.e., it's the output.

Delta-sigma converters can also be constructed with the current pulses generated with a resistor from a stable reference voltage, since the summing junction is a virtual ground. In that case you have to make sure that the switch ON resistance is small compared with the series resistor, so that variations of R_{ON} don't cause drifts.

□ *Switched-capacitor A/D*

A closely related charge-balancing method uses the "capacitor-stored charge-dispensing A/D," or "switched-capacitor" A/D. In this technique, fixed amounts of charge are created by repeatedly charging a capacitor from a stable reference voltage, then discharging it into the summing junction. The comparator looks at the integrator output, as previously, and controls the rate at which the capacitor is switched. That rate is counted for a fixed time interval to generate the digital output. This method has advantages for circuits that are meant to operate from a single supply voltage, since the effective polarity of charge transferred from the capacitor to the summing junction can be reversed by suitably connected FET switches (i.e., by switching both sides of the capacitor).

An example of this technique is the LM331 voltage/frequency converter, which

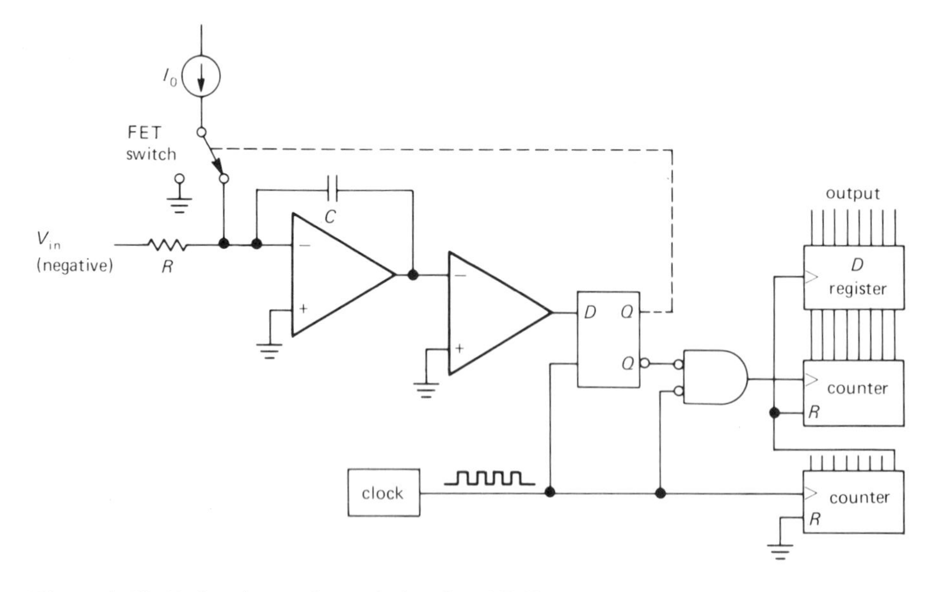

Figure 9.57. Delta-sigma charge-balancing ADC.

has the unique advantage of operation from a single +5 volt supply. We showed its application as a VCO in Section 5.14.

Comments on integrating A/D converters

As with the dual-slope integration A/D, these charge-balancing methods all average the input over fixed time intervals, so they can be made insensitive to 60Hz pickup and its harmonics. In general, charge-balancing methods are inexpensive (they don't require a particularly good comparator, for instance) and accurate, and they produce a strictly monotonic output. However, they are slow, as compared with successive approximation. The AD1170 gives 18-bit resolution with 66ms conversion time, with a price of about $100. By comparison, the AD76 is a 16-bit successive-approximation converter with 15μs conversion time and $120 price tag. Compared with dual-slope converters, delta-sigma and switched-capacitor methods can be characterized by low-accuracy comparators following the integrators, but with precise charge-switching circuits, whereas dual-slope methods use highly repeatable endpoint comparators, with somewhat simpler switch requirements, at least from the viewpoint of speed and charge injection. Continuing our comparison of actual devices, the AD1175K is a multiple-slope 22-bit converter with 50ms conversion time and $800 price (Section 9.22).

One interesting point to keep in mind with any of the integrating techniques (single- and dual-slope integration and charge balancing) is that the input to the integrator can be either a current, or a voltage in series with a resistor. In fact, some converters have two input terminals, one tied directly to the summing junction for use with devices that source a current directly. When used with a current input, the offset voltage of the integrator becomes unimportant, whereas with a voltage input (with internal series resistor) the integrator op-amp produces an error equal to its input offset voltage. Current input is useful, therefore, to get wide dynamic range, particularly if the A/D is used with a device that has a current output anyway; examples are photomultipliers and photodiodes. Beware of this specmanship "gotcha," though: A/D accuracy may be specified assuming input current, for converters that accept either current or voltage input; don't expect as good performance for small inputs when using such a converter with voltage input.

Note that these charge-balancing methods all include a highly accurate V/F converter and that they can be used as such if an output *frequency* is desired, as indicated in Figure 9.58.

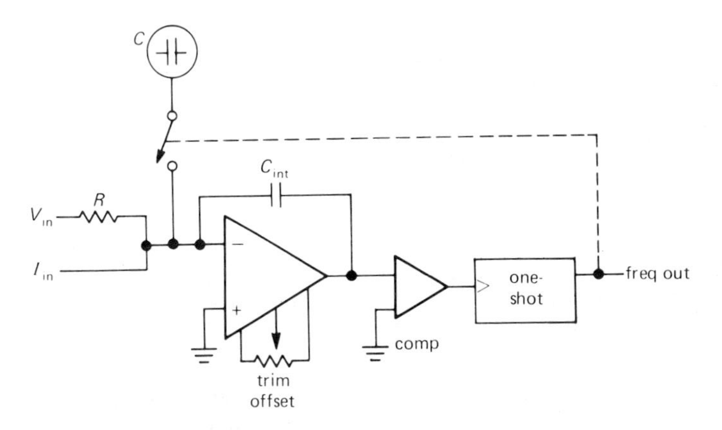

Figure 9.58. Charge-balancing V/F converter.

☐ 9.22 Some unusual A/D and D/A converters

Four recent introductions by Analog Devices, a traditional leader in conversion ICs and modules, are too interesting to pass by without comment.

☐ *AD7569 combined DAC/ADC*

The monolithic AD7569 combines 8-bit A/D and D/A on a single chip, complete with sample/hold, internal clock, and reference (Fig. 9.59). The successive-approximation ADC converts in 2μs, and the DAC provides a voltage output with 1μs typical settling time. This chip is great for microprocessor systems: The converters share a single 8-bit digital port, with convenient control signals and fast timing (unlike many sluggish converter chips that require extra "wait" states and awkward setup times; see Chapters 10 and 11), and the chip can run from a single +5 volt supply. Furthermore, it needs no external components or trimming, the chip is packaged in a convenient 24-pin skinny-DIP, it runs on low power (60mW), and the price is right ($6 in 100 piece quantities).

☐ *AD1175 22-bit integrating ADC*

This impressive module (Fig. 9.60) uses auto-zeroing multiple-slope integration to obtain 22-bit ($6\frac{1}{2}$-digit) accuracy, with unusual conversion speed (20 conversions per second). To get an idea what this means, consider the alternative – a benchtop (or rack-sized) meter, typically costing $4000, and delivering perhaps 2 conversions per second. The AD1175, by comparison, occupies 10 cubic inches, uses 3 watts, and costs $800 in single quantities. It contains an internal microprocessor and lets you set coarse gain and offset via its digital bus (which is used both for command inputs and converted data output).

☐ *HDG0807 and AD9502 video converters*

These converters (Figs. 9.61 and 9.62) are just what you need for digital video. The HDG0807 is an 8-bit DAC with standard video signal levels and corresponding 75 ohm output impedance. Furthermore, it even generates the proper "composite sync" pulses, overlaid with the analog video signal, to create a complete video output signal. This converter is ready to use, fast (to 50MHz), and affordable. The AD9700 is the monolithic core of this hybrid converter; it operates to 100MHz.

The AD9502 video ADC does the opposite job, namely digitizing a video input signal. It has internal circuitry that strips the sync pulses from the analog composite

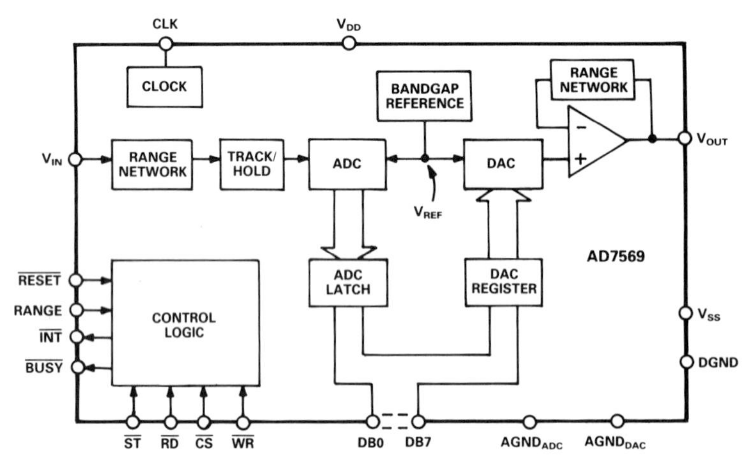

Figure 9.59. AD7569 combined 8-bit DAC/ADC. (Courtesy of Analog Devices.)

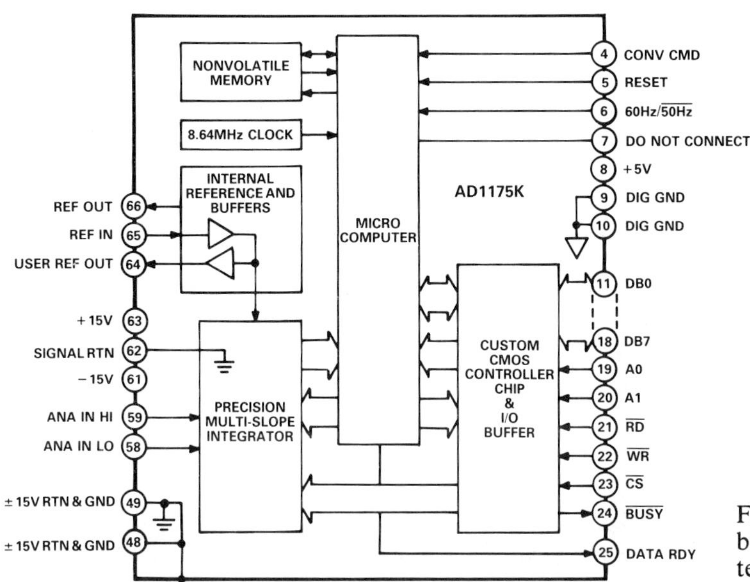

Figure 9.60. AD1175K 22-bit integrating ADC. (Courtesy of Analog Devices.)

video input, uses an internal phase-locked loop to generate a "pixel clock" synchronized to the raster sweep, then converts the analog voltage to 8-bit digital quantities. It can convert at a 13MHz rate, generating digitized outputs with 512×512 screen resolution.

9.23 Choosing an ADC

As with DACs, we've pulled together (Tables 9.5 and 9.6) a selection of ADCs that span a useful range of performance and price. We've tried to include both the most popular units and some of the recent

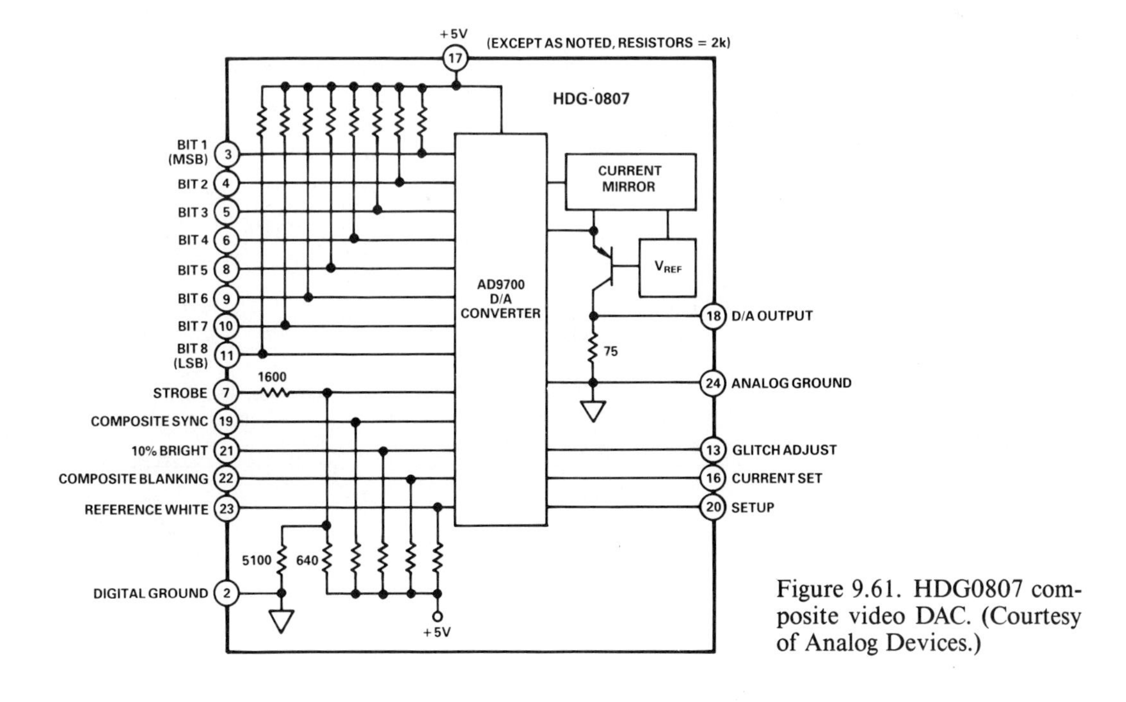

Figure 9.61. HDG0807 composite video DAC. (Courtesy of Analog Devices.)

TABLE 9.5. ANALOG-TO-DIGITAL CONVERTERS

Type	Mfg[a]	Bits	Method[b]	Speed max (µs)	Pkg[c]	Supply voltage (V)	Supply curr typ (mA)	V_{ref}	+5 ref OK?	Input range (V)
HS9582	HS	6	F	0.07	18	+5	30	ext 1-5	•	0-V_{ref}
TDC1047	TR	7	F	0.05	24	+5,-5.2	+20,-140	ext −1	–	0-V_{ref}
ADC0844	NS	8	SA	40	20	+5	1	ext 1-5	•	0-V_{ref}; 2V_{ref}
ADC0831	NS	8	SA	32	8	+5	1	ext 1-5	•	0-V_{ref}; 2V_{ref}
TLC548	TI	8	SA	22	8	+5	1.9	ext	•	0-5
AD670	AD	8	SA	10	20	+5	30	int	–	0.25; 2.5
AD7575	AD	8	SA	5	18	+5	3	ext 1.2	–	0-2V_{ref}
ADC0820	NS	8	HF	2.5	20	+5	8	ext 1-5	•	0-V_{ref}; 2V_{ref}
AD7820	AD	8	HF	1.6	20	+5	8	ext 1-5	•	0-V_{ref}
HS9583	HS	8	F	0.2	24	+5	20	ext +5V	•	0-5
AD9002	AD	8	F	0.007	28	-5.2	150	ext	–	-2 to 0
AD770	AD	8	F	0.005	40	+5,-5.2	270,125	ext	–	-1 to 1
CXA1176K	SO	8	F	0.003	68LCC	-5.2	300	ext	–	-2 to 0
TDC1049	TR	9	F	0.03	64	-5.2	950	ext −2	–	-2 to 0
HADC77600	HO	10	F	0.02	72PGA	+5,-5.2	+440,-380	ext ±2	–	±0.5; ±2
ADC1001	NS	10	SA	200	20	+5	1.5	ext 1-5	•	0-V_{ref}; 2V_{ref}
AD573	AD	10	SA	20	20	+5,-12	15,-9	int	–	0-10; -5 to 5
AD7578	AD	12	SA	100	24S	+5,±12	0.1,±3	ext +5V	•	0-5
AD574A	AD	12	SA	25	28	+5,±12	30,+2,-18	int 10.0	–	0-10; ±5; ±10
ADC80	BB	12	SA	25	32	+5,±12	11,+5,-21	int 6.3	–	0-5;0-10;±5;±10
AD7572	AD	12	SA	5	24S	+5,-15	5,-10	int −5.2	–	0-5
AD7672	AD	12	SA	3	24S	+5,-12	7,-12	ext −5	–	0-5; 0-10; ±5
AD578	AD+	12	SA	3	32	+5,±12	100,+3,-22	int 10.0	–	0-10;0-20;±5;±10
ADC511	DA	12	HF	1	24	+5,±15	65, ±25	int	–	0-10; ±5
AD9003	AD	12	HF	1	40	+5,±15	2.5W	int	–	0-5
THC1201	TR	12	HF?	0.1	46	±15	160	int	–	-1 to 1
CAV1220	AD	12	HF	0.05	PCB	±5,±15	20W	int	–	±1
TLC1205B	TI	13	SA	10	24	±5	3	ext	•	±5
ICL7115	IL	14	SA	40	40	±5	±2	ext +5V	•	0-5
ADC71	AD	16	SA	50	32HY	+5,±15	70,±20	int 6.3V	–	0-5;0-10;±5;±10
ADC76	BB+	16	SA	15	32	+5,±12	10,+14,-17	int	–	0-5;0-10;±5;±10
CX20018	SO	16	DS	9	28	±5	10,100	ext		±10
ADAM-826-3	AN	16	SA	1.5	PCB	+5,-6,±15	3W	int	–	0-10; ±10
MN5420	MN	20[h]	SA	3	40HY	+5,±15	+400,±150	int	–	±5

(a) see footnote to Table 4.1. (b) DS - dual slope; F - flash; HF - half-flash; SA - successive approx.
(c) all are DIP, unless otherwise noted; HY - hybrid; S - skinny-DIP. (d) AZ - auto-zero; NT - no trim req'd.
(e) 20k with internal amp. (f) 1000-pc qty. (g) needs external op-amps. (h) floating point.

Type	Input				Output			Clock	Price (plastic 100 pcs) ($)	Comments
	Z_{in}	Diff'l?	S/H?	Trim?[d]	Bits	3-state?	Serial?			
HS9582	∞	–	–	NT	6	•	–	ext	15	fast
TDC1047	100k	–	–	NT	7	•	–	–	40	latched output
ADC0844	∞	–	–	AZ	8	•	–	int	3.85	4-input MUX; easy to use; +5V
ADC0831	∞	•	–	AZ	–	–	•	ext	2.70	small, no glue
TLC548	1μA	–	•	AZ	–	–	•	int		easy-to-use miniDIP
AD670	∞	•	–	NT	–	•	–	int	6	instrumentation amp input
AD7575	∞	–	•	NT	–	•	–	extRC	5.50	fast
ADC0820	∞	–	•	NT	–	•	–	int		latched output; overflow
AD7820	∞	–	•	NT	–	•	–	int	10	fast; 4 & 8-ch versions; overflow
HS9583	100k	–	–	NT	–	•	–	ext	44	fast
AD9002	20k	–	–	NT	–	–	–	–	90	fast, lo-pwr flash, w latch & ovflw
AD770	3.3k	–	–	NT	8	–	–	–	175	fast, w latch & ovflw
CXA1176K		–	–	NT	–		–	–		world speed champion
TDC1049	16k	–	–	•	10	–	–	–		9-bit flash
HADC77600	1k[e]	–	–	•	10	–	–	–		highest resolution flash
ADC1001	∞	•	–	AZ	8+2	•	–	extRC		5V, minimum glue
AD573	5k	–	–	NT	8+2	•	–	int	14	fast, no glue
AD7578	∞	–	–	AZ	8	•	–	extRC	20	low power, cheap
AD574A	5k	–	–	•	8+4	•	–	int	28	V_{ref} out; classic
ADC80	5k	–	–	•	12	–	•	int	33	V_{ref} out, classic, clk out
AD7572	2.5k	–	–	NT	8+4	•	–	int	46	fast, no glue, clk out
AD7672	5k	–	–	–	8+4	•	–	extRC	75	fast
AD578	5k	–	–	NT	12	–	•	int	100	fast, no glue, V_{ref} out, clk out
ADC511	2.5k	–	–	NT	12	•	–	int	99[f]	no missing codes over temp
AD9003	1k	–	•	NT	12	–	–	int	250	
THC1201		–	•	NT	12	•	–	–		very fast, easy to use
CAV1220	1k	–	•	NT	12	–	–	int	2500	fastest
TLC1205B	1μA	•	•	–	8+5	•	–	ext	30	
ICL7115	5k	–	–	AZ	8+6	•	–	ext	50	radix 1.85 + internal ROM
ADC71	5k	–	–	•	16	–	•	int	63	indus std, clk out, ref out
ADC76	2.5k	–	–	•	16	–	•	int	100	indus std; AD376
CX20018	10k[g]		•	•	–	–	•	ext	18	digital audio, 2 ch
ADAM-826-3	1.4k	–	–	NT	8+8	•	–	int		module; versions w S/H, buffer
MN5420	5k	–	–	–	16	•	–	ext		float pt: 4-bit exp, 12-bit mantissa

TABLE 9.6. INTEGRATING A/D CONVERTERS

Type	Mfg[a]	Chips	Bits	Method[b]	Conv/sec	Pkg[c]	V_{supply} (V)	V_{ref} (V)	5V ref OK?	Input range (V)	Input			Output				
											$Z_{in}(\Omega)$	Ext trim?	Differential	Bits parallel	3-state?	UART compat?	Clock	Comments
AD7552	AD	1	12[d]	QS	6	40	+12,±5	1 to 5	•	0 to ±0.5V_{ref}	1M	–	–	8+6	•	–	ext	
TSC804	TS	1	12[d]	QS	30	60	±5	int	–	±4	∞	–	•	8+6	•	•	xtal	8-channel MUX
ICL7109	IL	1	12[d]	DS	30	40	±5	0.2 to 2	–	0 to ±2V_{ref}	∞	–	•	8+6	•	•	xtal	industry standard
AD7550	AD	1	13	QS	25	40	+12,±5	1 to 5	•	0 to 0.5V_{ref}	1M	–	–	8+6	•	–	ext	
TSC800	TS	1	15[d]	DS	2.5	40	±5	0.2 to 2	–	0 to ±2V_{ref}	∞	–	•	8+8	•	•	xtal	
TSC850	TS	1	15[d]	TS	40	40	±5	1.6,0.025	–	0 to 2V_{ref}	∞	V_{ref}	•	8+8	•	–	ext	fast, low pwr (2mA)
ICL7104-16[e]	IL	2	16[d]	DS	3	40	+5,±15	int	–	0 to ±4	∞	V_{ref}	–	8+8+2	•	–	xtal	use 8052 for low I_{in}
CSZ5316	CR	1	16	CB	20k	18	±5	int	–	±2.75	30k	–	–	serial	•	–	ext	int S/H
AD1170	AD	1	18	CB	1000[f]	40	+5,±15	int	•	±5	100M	–	–	8+8+6	•	–	xtal	software trims
AD1175K	AD	1	22	QS	20	66M	+5,±15	int	–	±5	1000M	–	•	8+8+6	•	–	int	accurate, stable; software trims

(a) see footnote to Table 4.1. (b) CB - charge balancing; DS - dual slope; TS - triple slope; QS - quad slope. (c) M - module. (d) plus additional sign bit. (e) plus 8068A companion chip. (f) prog conversion rate; value shown is max, at which resolution is degraded. Full resolution can be obtained at 50 conv/sec.

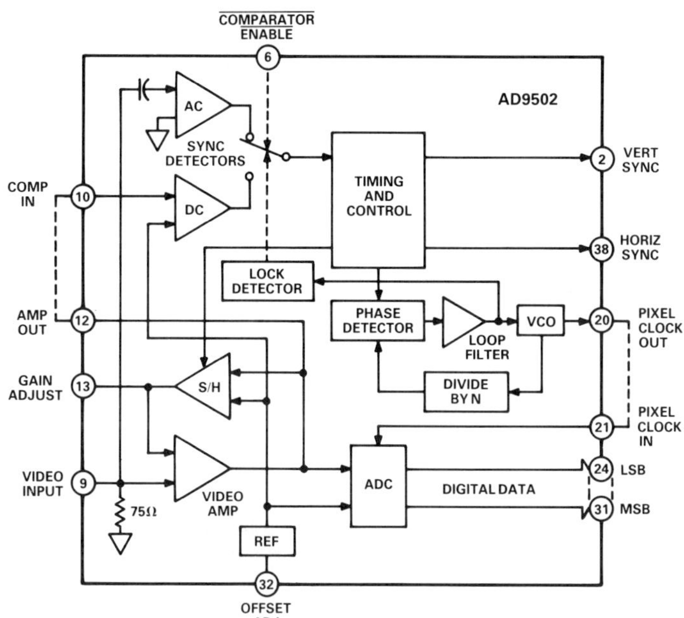

Figure 9.62. AD9502 composite video ADC. (Courtesy of Analog Devices.)

entries that should be next year's winners.

When looking for an ADC, some of the factors to consider are (a) precision, (b) speed, (c) accuracy (external trimming required? guaranteed monotonicity?), (d) required supply voltages (some work with +5V alone) and power dissipation, (e) small package, (f) reference and clock (internal or externally supplied? if external reference, is +5V OK? if internal, is it accessible externally, e.g. for ratiometric measurements? how good is it? can you override it?), (g) input impedance and analog voltage range (unipolar, bipolar, or both?), (h) input structure (differential? internal multiplexer or sample/hold? inverted polarity, i.e., more negative for larger output?), (i) output structure (parallel, serial, or both? is the parallel output "microprocessor-compatible," as a pair of individually enabled byte-wide groups?), and, of course, (j) price.

Complete A/D subsystems

If you need a high-accuracy A/D converter, particularly with an input multiplexer and sample/hold, you should look carefully at the "A/D subsystems" offered by several manufacturers. These are modules, rather than ICs, usually supplied as a metal box typically 0.4" thick and 2"×4" (or 3"×5") in size, with pin connections that mate with a special socket (or solder directly into a printed circuit board). Although these converters aren't cheap, they are delightfully easy to use. Better still, the manufacturers have solved the really thorny problems that plague high-resolution conversion, such as noise pickup, isolation of digital and analog sections, stable references and amplifier offsets, etc.

The DT-5716A from Data Translation (Fig. 9.63) is typical of these modules. It accepts 16 single-ended (or 8 differential) inputs, with an analog multiplexer at the input, followed by a sample/hold, programmable gain amplifier, and 16-bit A/D converter. It can convert at a 20kHz throughput rate, with the output organized as a pair of bytes for easy connection to a microprocessor bus (see Chapters 10 and 11).

A/D subsystem modules are available with resolution of 12 to 16 bits, both with

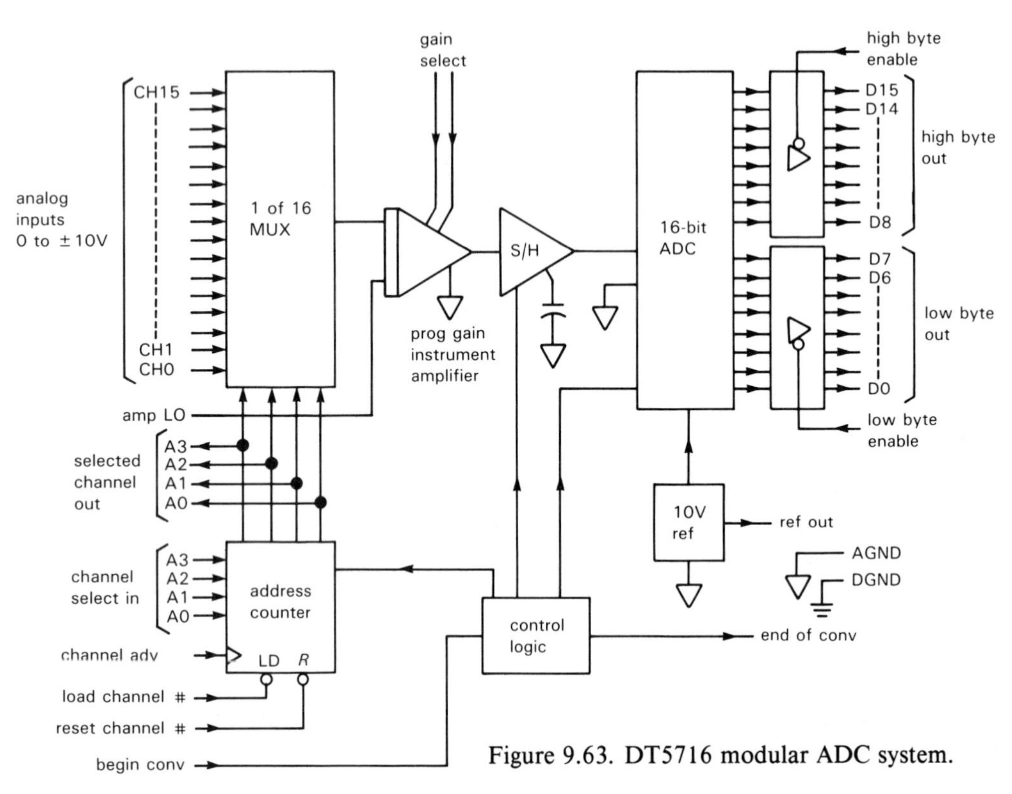

Figure 9.63. DT5716 modular ADC system.

and without input multiplexers. You pay a real premium for high accuracy and high speed, and most of the available modules are considerably cheaper than the units cited earlier. For example, the DAS1157-9 series from Analog Devices are single-channel converters with 14 to 16 bits of resolution and 18kHz throughput; they cost less than $300 in 100-pc quantities. You can get conversion modules from several companies, including Analog Devices, Analogic, Data Translation, and Intech.

SOME A/D CONVERSION EXAMPLES

9.24 16-Channel A/D data-acquisition system

Figure 9.64 shows a circuit to digitize any of 16 analog inputs on command, with a 12-bit digital output. This could form the "front end" of a microprocessor-controlled data-taking experiment.

The HI-506 is a 16-channel CMOS analog multiplexer with CMOS-compatible digital inputs. This particular multiplexer has some very nice properties. In particular, its switches are of the "break-before-make" variety, which means that the various input channels don't find themselves shorted together during address changes in the MUX. In addition, the inputs may swing beyond the supply rails, and when they do there is no "SCR latchup" or cross-talk between inputs. Watch out for considerations of this type when shopping for linear switches. They sometimes involve a compromise. For example, "break-before-make" results in a slower switching-time specification, since the "make" must be delayed to allow the switch to open.

The multiplexer's single analog output drives an LF398, a $2 monolithic sample/hold amplifier (Fig. 4.41) in convenient 8-pin DIP. The latter is used as a "track-and-hold," freezing the analog waveform only when conversion begins. With a

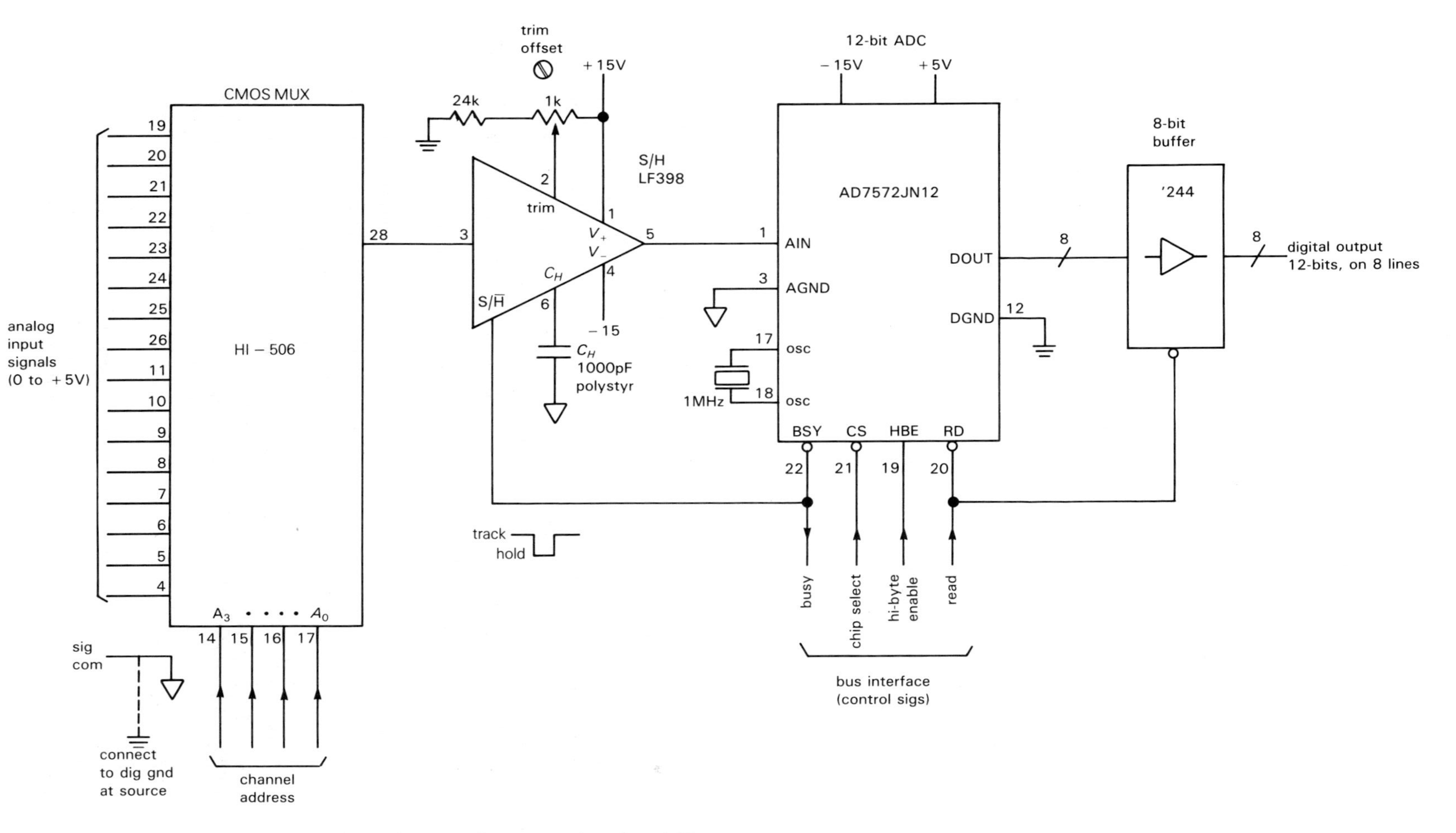

Figure 9.64. A 12-bit 16-channel successive-approximation A/D converter system (12μs/conversion).

1000pF capacitor the LF398 settles to 1 LSB in 0.5μs following "HOLD" and droops less than 3μV during the subsequent 12μs conversion. The AD7572 is a very nice low-power 12-bit ADC with on-chip voltage reference and clock; it has convenient controls for microprocessor interfacing, including the option of multiplexing the 12-bit result onto 8 lines (a "byte-wide" data bus) in two successive cycles.

The device controlling this circuit normally asserts an address to the multiplexer, then initiates a conversion by asserting CS′ and RD′. The ADC responds by asserting BUSY′, which freezes the analog input. The conversion is complete 12μs later, at which time BUSY′ goes HIGH. The full 12-bit result is then available, if you want to use all 12 D_{out} lines; however, if you have an 8-bit bus, you can read the least significant 8 bits first, then assert HBEN ("high-byte enable") to put the most significant 4 bits on $D_0 - D_3$.

After initiating conversion, the device controlling the converter can check BUSY′ to see when conversion is complete. A simpler alternative is just to wait 12μs (a software "timing loop" will waste the requisite time; see Chapter 11). The controlling device is obliged to wait 4μs after conversion is complete before initiating another conversion; this is the LF398's "acquisition time" – the time for the output to track the input again to within 0.1%. During that time the controller could be reading the digital output, of course. The overall conversion cycle is thus 16μs minimum, or 60,000 conversions per second.

Several points are worth making about this circuit: (a) For full 12-bit accuracy you have to provide an offset trim, to compensate for three errors: *(i)* the S/H has a V_{os} of 7mV (max), *(ii)* the S/H FET introduces a small voltage step at HOLD, owing to FET gate charge injection (Section 3.12), in this case a relatively stable 10mV negative step, and *(iii)* the ADC itself has a V_{os} specified as 4 LSB (equivalent to 5mV for a 0–5V input range). We put the trimming circuit on the LF398, using the manufacturer's recommended network. (b) The hold capacitor value is a compromise. Small values yield a short acquisition time, but introduce a larger droop and charge-injection step. The value we chose produces negligible droop and produces a HOLD step equivalent to 8 LSB; the step is relatively stable and can be canceled by proper adjustment of the offset trim control. (c) The circuit is wired for unipolar (0–5V) input; if you wanted bipolarity input signal range, you would have to add an op-amp offsetting circuit, taking care to keep errors less than 1 LSB (1 part in 4000). The AD7572 provides a precision reference voltage out, to make your job easier, but it still takes a bunch of components. (d) A nice addition to this sort of circuit is a programmable-gain amplifier, so the controlling microprocessor can command a change of gain to accommodate a range of input signal amplitudes. The AD526 from Analog Devices is a single-chip solution, providing programmable gains of 1, 2, 4, 8, and 16, with gain accuracy of 0.02% (12-bit accuracy); the LF13006/7 from National is an alternative, containing the resistors and FET switches (but not the amplifier itself) to set gains of 1 to 128 (by factors of 2) or 1 to 100 (1-2-5 sequence); these parts have a gain accuracy of 0.5% (8-bit accuracy).

A successive-approximation A/D is the natural choice here, since speed is important when jumping from one input to another. We picked components to minimize cost; the circuit shown will cost about $50 at today's prices, dominated by the $35 converter.

9.25 $3\frac{1}{2}$-Digit voltmeter

Figure 9.65 shows a circuit that exploits the advantages of dual-slope integration. Almost the entire circuit is included in the

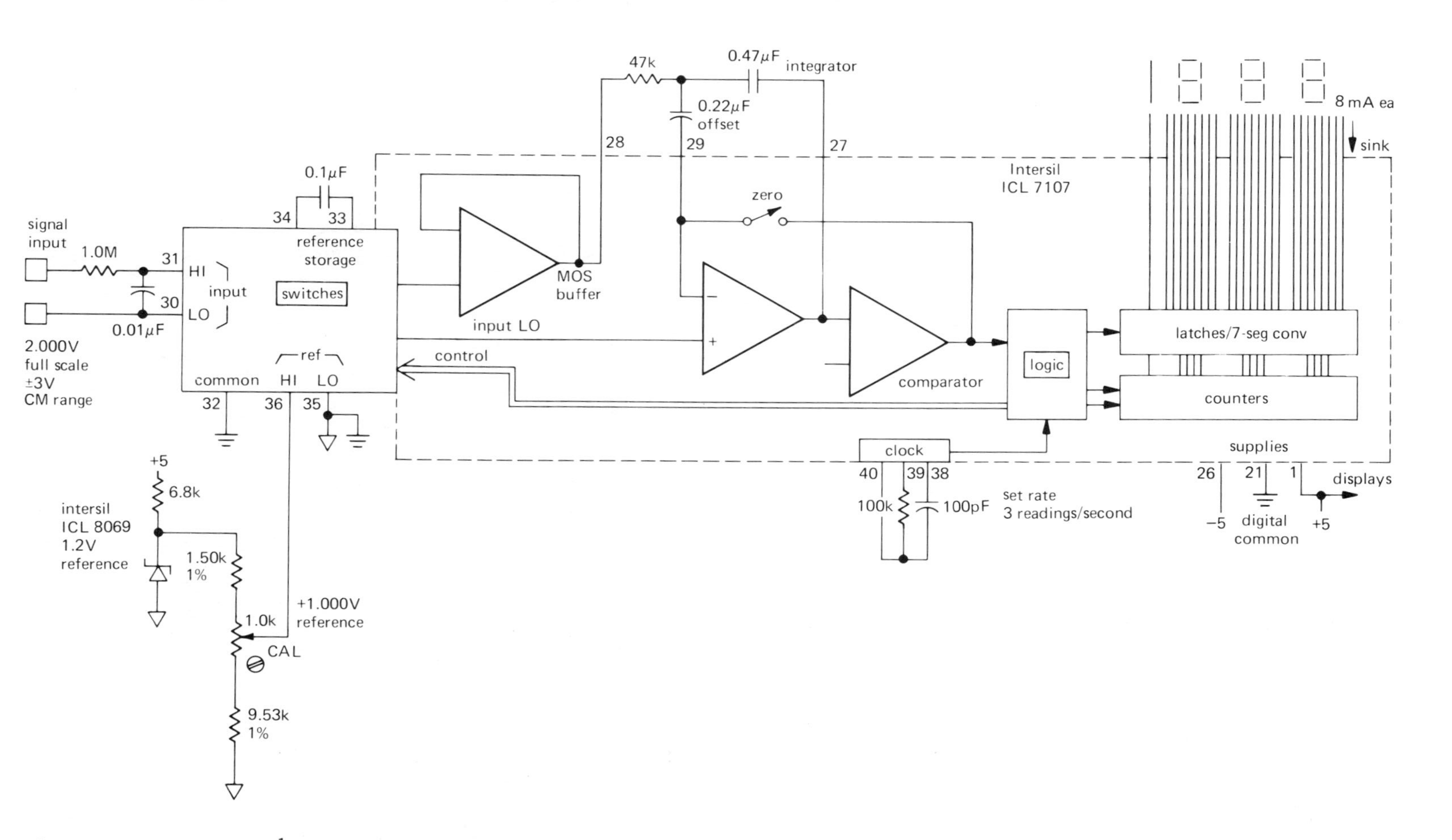

Figure 9.65. Single-chip $3\frac{1}{2}$-digit dual-slope DVM.

digital-voltmeter CMOS LSI chip, the only external components being the integrator and clock RCs, an accurate voltage reference, and the display itself. The ICL 7107 includes an automatic zeroing cycle in its operation, and it even generates all the 7-segment multiplexed outputs to drive a 4-digit LED display directly. By using an external attenuator at the input (or a reference of a different voltage), you can generate other full-scale voltage ranges. Dual-slope conversion is well suited to DVM operation because it provides good accuracy (including auto-zeroing) and 60Hz rejection in an averaging instrument at low cost; the converter chip used here costs less than $20.

□ 9.26 Coulomb meter

The circuit in Figure 9.66 is a charge-balancing current integrator, or "coulomb meter." This instrument can be used to measure the integrated current (total charge) over some time period; it might find application in electrophoresis or electrochemistry. The action begins in the lower left-hand corner, where the current to be integrated flows through a precision 4-wire power resistor, generating a proportional voltage. IC_2 is a relatively low cost (under $5) precision single-supply op-amp with low initial voltage offset (80μV max) and low drift of offset with time and temperature (less than 2μV per degree and 0.5μV per month). It generates an output current, programmed by the current being measured, to drive the charge-balancing integrator, IC_3. Five decades of input sensitivity are selectable via the rotary switch at the input, with 200μA collector current in Q_1 corresponding to full-scale input in any range. Q_1 is a MOSFET (rather than a BJT), to eliminate control current error.

The charge-balancing circuitry is a standard delta-sigma scheme, with p-channel enhancement-mode MOSFET Q_2 doling out parcels of charge as directed by the state of flip-flop IC_{5a} after each clock cycle. IC_{5b} acts as a monostable, incrementing binary scaler IC_7 for each clock cycle during which Q_2 conducts. This circuit doesn't count for a fixed number of clock cycles, but simply integrates until it is stopped. The 4-digit counters IC_9 and IC_{10} deep track of the total charge, driving an 8-digit LED display.

If the current being measured ever exceeds the full-scale current of the range selected, Q_2 will be unable to balance Q_1's current even if it is ON continuously, and the measured charge registered in the counters will be in error. IC_{4a} checks for this overrange condition, lighting the LED if the integrator output rises past a fixed reference voltage (chosen comfortably larger than the output excursions of the integrator under normal conditions).

□ *Design calculations*

There are several interesting decisions that have to be made in designing a circuit like this. For instance, most of the CMOS logic is operated from +15 volts in order to simplify switching of Q_2. Since the 4-digit counters require +5 volts, a 4049 is used to interface the high-level CMOS logic signals to the counters. IC_4 is operated with single supply so that its output goes between ground and +15, for simple connection to IC_{5a}. The reference voltage for the integrator and comparator is put at about +4.7 volts by zener D_2, in order to allow headroom for Q_1; a simple zener is fine, since no accuracy is needed here. Note that a precision reference rides on the +4.7 volt level used to scale the current switched into the integrator. The REF-02's operating current is conveniently used to bias the zener.

The choice of switch (Q_2) can critically affect the overall precision of the instrument. If it has too much capacitance, the additional charge residing on its drain will cause error. The scheme used in the previous circuit example (switching to ground during current-OFF cycles) is not used here because offset voltage errors in IC_3 will cause a fixed error at very low currents. You get increased dynamic range at the expense of some accuracy (owing to residual charge on Q_2's drain that gets integrated each cycle) by using an SPST switch configuration as shown. Note that the integrator op-amp chosen is a low-bias-current MOSFET type for negligible current error (10pA typ). Since FET op-amps tend to have larger voltage offsets than bipolar transistor types, this choice of op-amp will aggravate the dynamic-range problem just discussed if SPDT switching is used.

□ *Dynamic range*

It is important to understand that this instrument is designed to have a large dynamic range, accurately integrating a current that may vary over several orders of magnitude during the measurement. That is why great care has been exercised in the design of the "front end," using a precision op-amp with an offset trimming scheme capable of precise adjustment (the normal trimming scheme usually has a total range of a few millivolts, making it difficult to adjust the offset precisely to zero). With IC_2 trimmed to $10\mu V$ offset or better, the instrument's dynamic range exceeds 10,000:1.

PHASE-LOCKED LOOPS

9.27 Introduction to phase-locked loops

The phase-locked loop (PLL) is a very interesting and useful building block, available from several manufacturers as a single integrated circuit. A PLL contains a phase detector, amplifier, and voltage-controlled oscillator (VCO) and represents a blend of digital and analog techniques all in one package. A few of its applications, which we will discuss shortly, are tone decoding, demodulation of AM and FM signals, frequency multiplication, frequency synthesis, pulse synchronization of signals from noisy sources (e.g., magnetic tape), and regeneration of "clean" signals.

There has traditionally been some reluctance to use PLLs, partly because of the complexity of discrete PLL circuits and partly because of a feeling that they cannot be counted on to work reliably. With inexpensive and easy-to-use PLLs now widely available, the first barrier to their acceptance has vanished. And with proper design and conservative application, the PLL is as reliable a circuit element as an op-amp or flip-flop.

Figure 9.67 shows the classic PLL configuration. The phase detector is a device that compares two input frequencies, generating an output that is a measure of their phase difference (if, for example, they differ in frequency, it gives a periodic output at the difference frequency). If f_{IN} doesn't equal f_{VCO}, the phase-error signal, after being filtered and amplified, causes the VCO frequency to deviate in the direction of f_{IN}. If conditions are right (lots more on that soon), the VCO will quickly "lock" to f_{IN}, maintaining a fixed phase relationship with the input signal.

At that point the filtered output of the phase detector is a dc signal, and the control input to the VCO is a measure of the input frequency, with obvious applications to tone decoding (used in digital transmission over telephone lines) and FM detection. The VCO output is a locally generated frequency equal to f_{IN}, thus providing a clean replica of f_{IN}, which may itself be noisy. Since the VCO output can be a triangle wave, sine wave, or whatever, this provides a nice method of generating a sine wave, say, locked to a train of input pulses.

In one of the most common applications of PLLs, a modulo-n counter is hooked

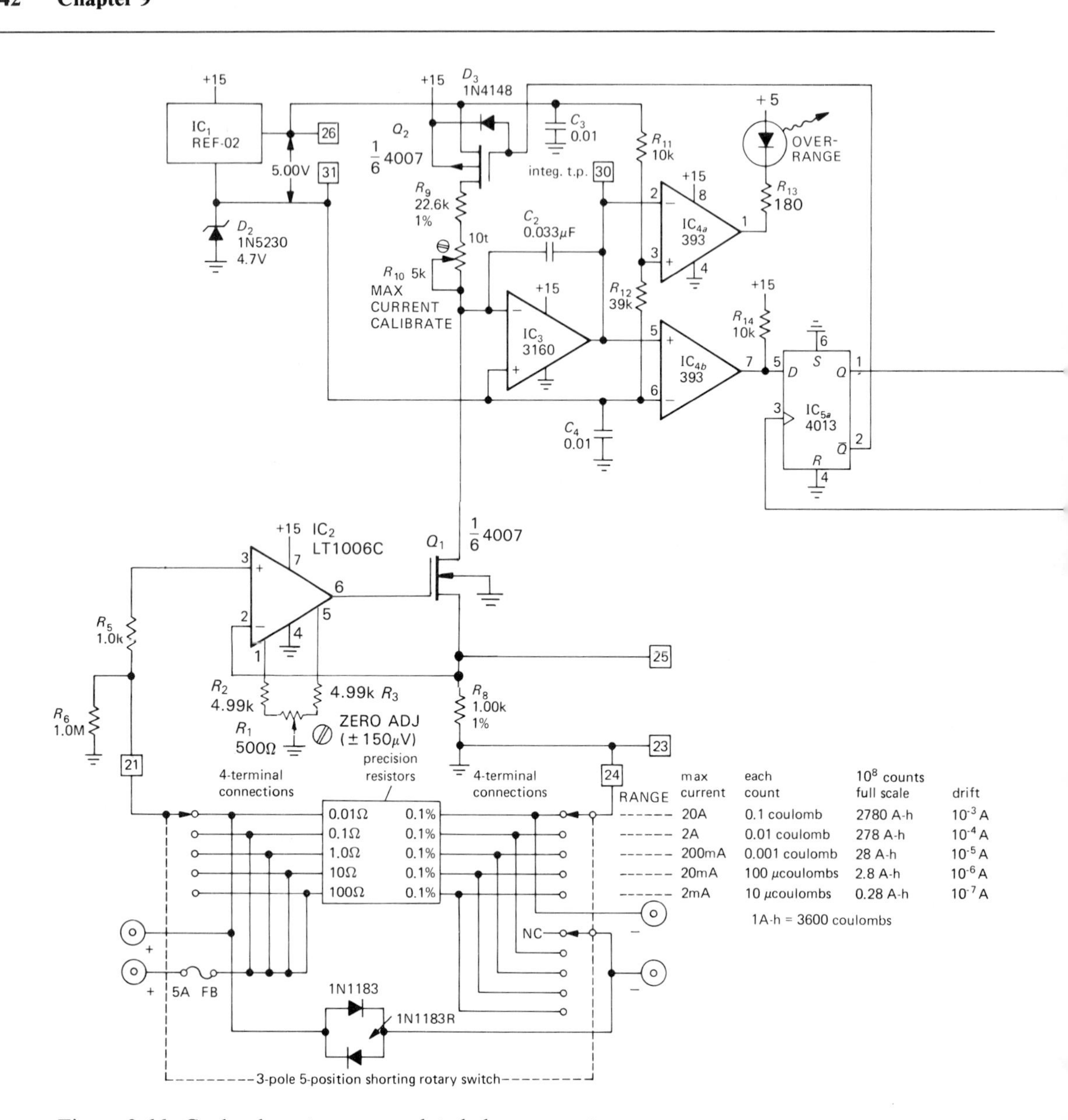

Figure 9.66. Coulomb meter: accumulated charge counter.

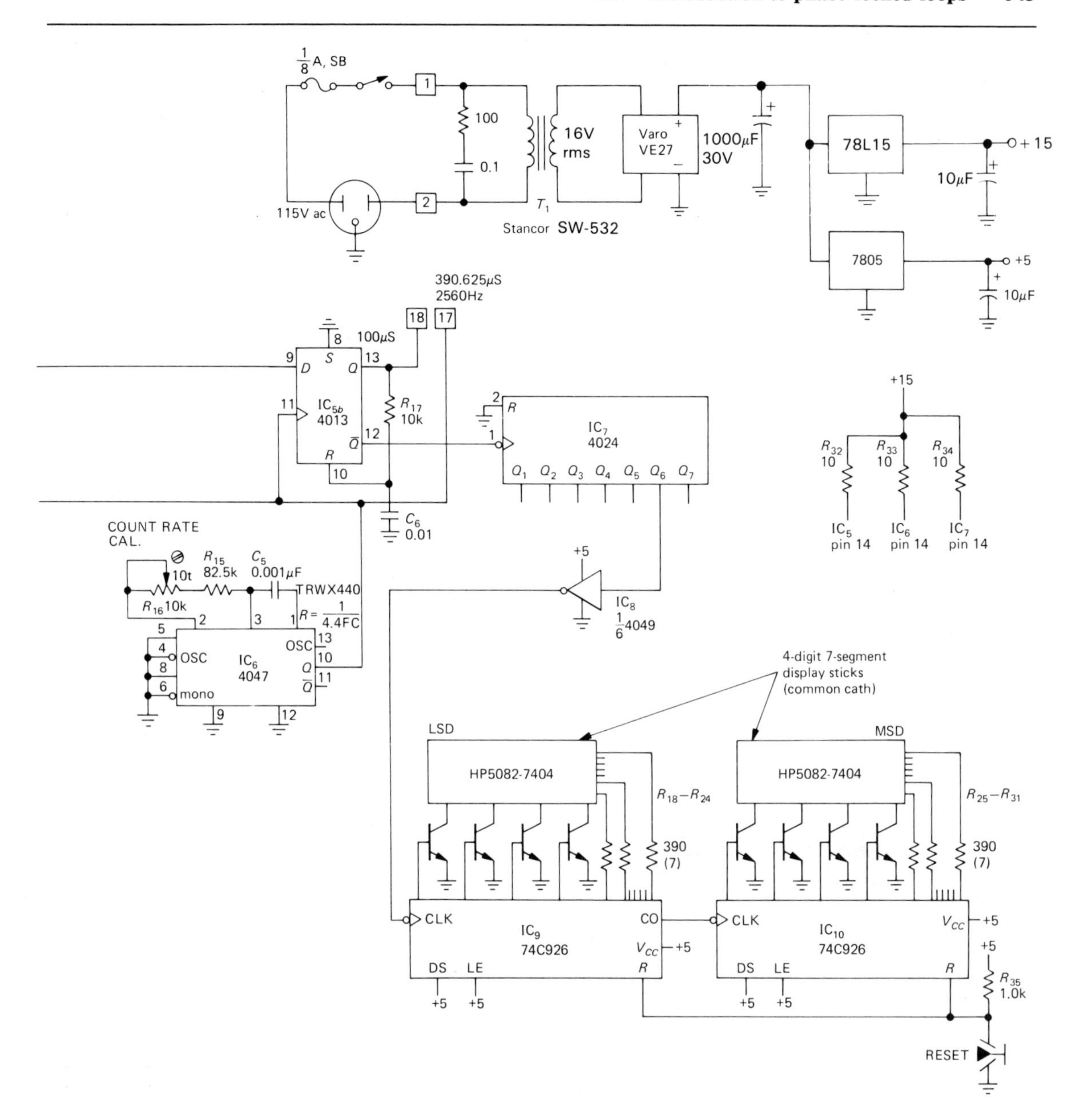

Figure 9.66. (*cont.*)

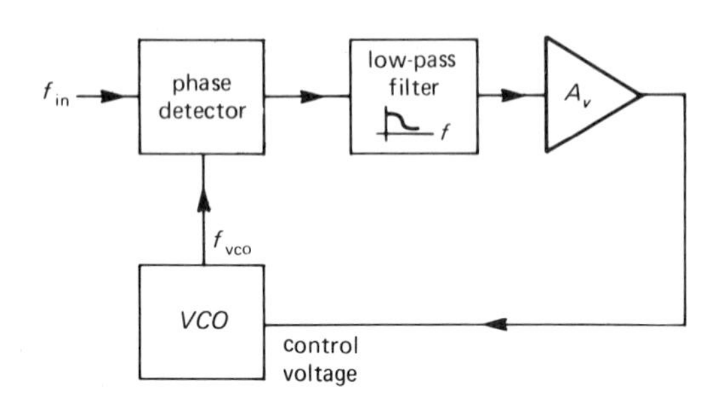

Figure 9.67. Phase-locked loop.

between the VCO output and the phase detector, thus generating a multiple of the input reference frequency f_{IN}. This is an ideal method for generating clocking pulses at a multiple of the power-line frequency for integrating A/D converters (dual-slope, charge-balancing), in order to have infinite rejection of interference at the power-line frequency and its harmonics. It also provides the basic technique of frequency synthesizers.

□ *PLL components*

□ *Phase detector.* Let's begin with a look at the phase detector. There are actually two basic types, sometimes referred to as type I and type II. The type I phase detector is designed to be driven by analog signals or digital square-wave signals, whereas the type II phase detector is driven by digital transitions (edges). They are typified by the 565 (linear, type I) and the CMOS 4046, which contains both.

The simplest phase detector is the type I (digital), which is simply an exclusive-OR gate (Fig. 9.68). With low-pass filtering, the graph of the output voltage versus phase difference is as shown, for input square waves of 50% duty cycle. The type I (linear) phase detector has similar output-voltage-versus-phase characteristics, although its internal circuitry is actually a "four-quadrant multiplier," also known as a "balanced mixer." Highly linear phase detectors of this type are essential for *lock-in detection*, a lovely technique we will discuss in Section 15.15.

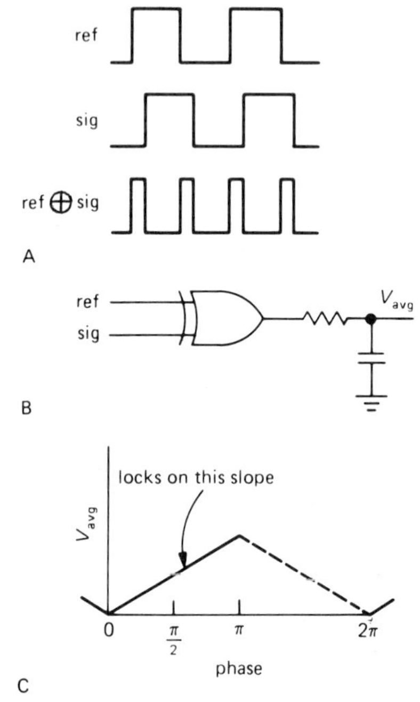

Figure 9.68. Exclusive-OR-gate phase detector (type I).

The type II phase detector is sensitive only to the relative timing of *edges* between the signal and VCO input, as shown in Figure 9.69. The phase comparator circuitry generates either *lead* or *lag* output pulses, depending on whether the transitions of the VCO output occur before or after the transitions of the reference signal, respectively. The width of these pulses is equal to the time between the respective edges, as shown. The output circuitry then either sinks or sources current (respectively) during those pulses and is otherwise open-circuited, generating an average output-voltage-versus-phase difference like that in Figure 9.70. This is completely independent of the duty cycle of the input signals, unlike the situation with the type I phase comparator discussed earlier. Another nice feature of this phase detector is the fact that the output pulses disappear entirely when the two signals are in lock. This means that there is no "ripple" present at the output to generate periodic phase modulation in the loop, as there is with the type I phase detector.

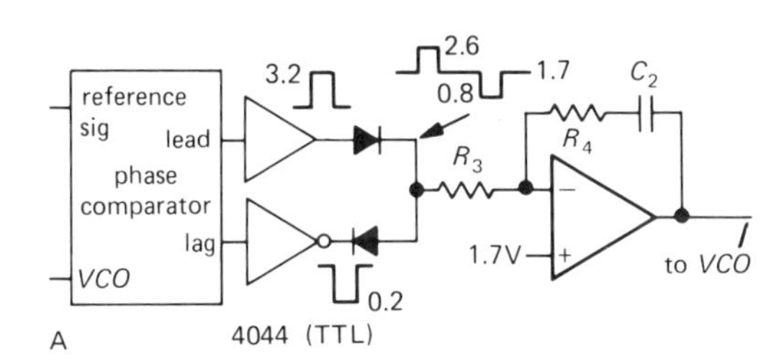

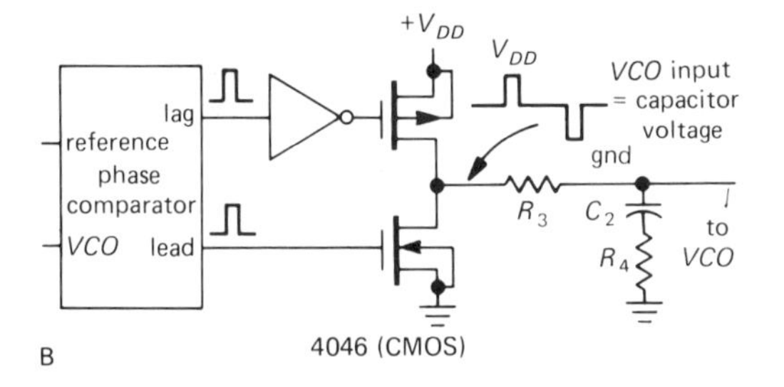

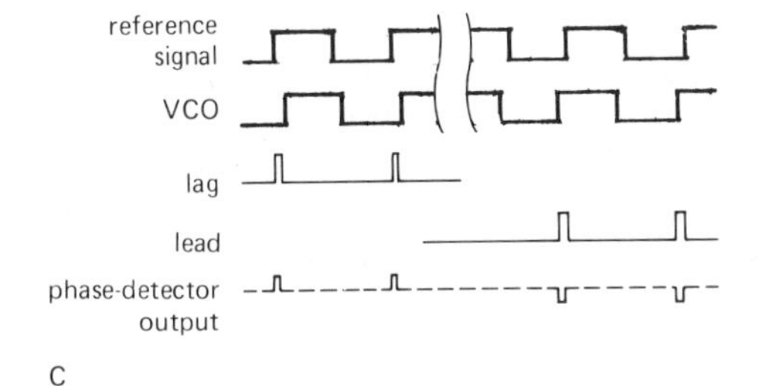

Figure 9.69. Edge-sensitive lead-lag phase detector (type II).

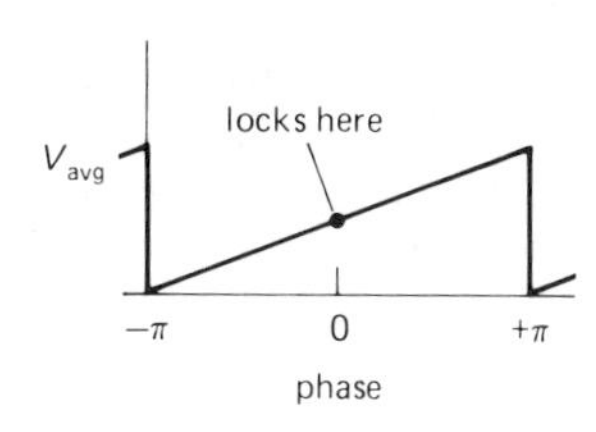

Figure 9.70

Here is a comparison of the properties of the two basic types of phase detector:

	Type I exclusive-OR	*Type II edge-triggered ("charge pump")*
Input duty cycle	50% optimum	irrelevant
Lock on harmonic?	yes	no
Rejection of noise	good	poor
Residual ripple at $2f_{IN}$	high	low
Lock range (L)	full VCO range	full VCO range
Capture range	$fL(f < 1)$	L
Output frequency when out of lock	f_{center}	f_{min}

There is one additional point of difference between the two kinds of phase detectors. The type I detector is always generating an output wave, which must then be filtered by the loop filter (much more on this later). Thus in a PLL with type I phase detector, the loop filter acts as a low-pass filter, smoothing this full-swing logic-output signal. There will always be residual ripple, and consequent periodic phase variations, in such a loop. In circuits where phase-locked loops are used for frequency multiplication or synthesis, this adds "phase-modulation sidebands" to the output signal (see Section 13.18).

By contrast, the type II phase detector generates output pulses only when there is a phase error between the reference and VCO signal. Since the phase detector output otherwise looks like an open circuit, the loop filter capacitor then acts as a voltage-storage device, holding the voltage that gives the right VCO frequency. If the reference signal moves away in frequency, the phase detector generates a train of short pulses, charging (or discharging) the capacitor to the new voltage needed to put the VCO back into lock.

□ *VCOs.* An essential component of a phase-locked loop is an oscillator whose frequency can be controlled by the phase detector output. Some PLL ICs include a VCO (e.g., the linear 565 and the CMOS 4046). Then there are separate VCO chips, listed in Table 5.4. An interesting class of VCOs is composed of the sine-wave-output types (8038, 2206, etc.), since they let you generate a clean sine wave locked to some horrendous input waveform. Don't overlook yet another class of VCOs, namely "voltage-to-frequency converters" (V/F), which are generally designed to optimize linearity; they tend to have a modest maximum frequency (1MHz or less), and logic-level pulses as output (see Section 5.15).

One thing to keep in mind is the fact that the VCO doesn't have to be restricted to logic speeds. You could, for instance, use a radiofrequency oscillator tuned with a varactor (variable capacitor) diode (Fig. 9.71).

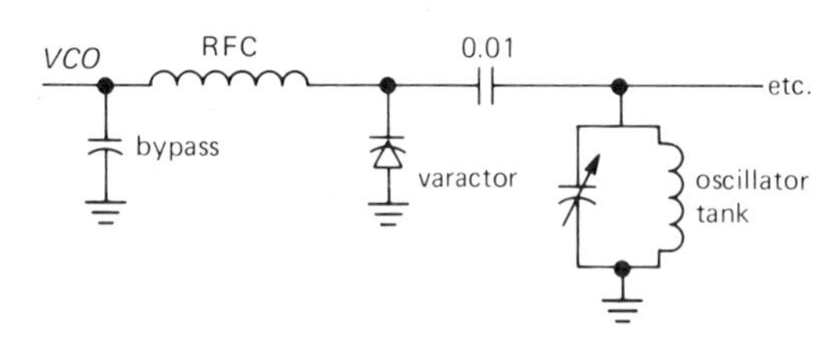

Figure 9.71

Carrying this idea one step further, you could even use something like a reflex klystron, a microwave (gigahertz) oscillator that is electrically tuned by varying the voltage on the *repeller*. Of course, a phase-locked loop built with such oscillators would require a radiofrequency phase detector.

A VCO for use in a phase-locked loop doesn't have to be particularly linear in its frequency-versus-control-voltage characteristic, but if it is highly nonlinear, the loop gain will vary according to the signal frequency, requiring better loop stability.

□ 9.28 PLL design

□ *Closing the loop*

The phase detector gives us an error signal related to the phase difference between the signal and reference inputs. The VCO allows us to control its frequency with a voltage input. It would seem straightforward to treat this like any other feedback amplifier, closing the loop with some gain, just as we did with op-amp circuits.

However, there is one essential difference. Previously, the quantity adjusted by feedback was the same quantity measured to generate the error signal, or at least a proportional quantity. For example, in a voltage amplifier we measured output voltage and adjusted input voltage accordingly. In a PLL there's an integration; we measure *phase*, but adjust *frequency*, and phase is the integral of frequency. This introduces a 90° phase shift in the loop.

This integrator included within the feedback loop has important consequences, since an additional 90° of lagging phase shift at a frequency where the loop gain is unity can produce oscillations. A simple solution is to avoid any further lagging components within the loop, at least at frequencies where the loop gain is close to unity. After all, op-amps have a 90° lagging phase shift over most of their frequency range, and they work quite nicely. This is one approach, and it produces what is known as a "first-order loop." It looks just like the PLL block diagram shown earlier, with the low-pass filter omitted.

Although they are useful in many circumstances, first-order loops don't have the desirable property of acting as a "flywheel," allowing the VCO to smooth out noise or fluctuations in the input signal. Furthermore, a first-order loop will not maintain a fixed phase relationship between the reference and VCO signals, since the phase detector output drives the VCO directly. A "second-order loop" has additional low-pass filtering within the feedback loop (as drawn earlier), carefully designed to prevent instabilities. This provides flywheel action and also reduces the "capture range" and increases the capture time. Furthermore, with type II phase detectors, a second-order loop guarantees phase lock with zero phase difference between reference and VCO, as will be explained later. Second-order loops are used almost universally, since the applications of phase-locked loops usually demand an output frequency with low phase noise and some "memory," or flywheel action. Second-order loops permit high loop gain at low frequencies, resulting in high stability (in analogy with the virtues of high loop gain in feedback amplifiers). Let's get right down to business, illustrating the use of phase-locked loops with a design example.

□ 9.29 Design example: frequency multiplier

Generating a fixed multiple of an input frequency is one of the most common applications of PLLs. This is done in frequency synthesizers, where an integer multiple n of a stable low-frequency reference signal (1Hz, say) is generated as an output; n is settable digitally, giving you a flexible signal source that can even be controlled by a computer. In more mundane applications, you might use a PLL to generate a clock frequency locked to some other reference frequency already available in the instrument. For example, suppose we want to generate a 61,440Hz clock signal for a dual-slope A/D converter. That particular choice of frequency permits 7.5 measurement cycles per second, allowing 4096 clock periods for the ramp-up (remember that dual-slope conversion uses a constant time interval) and 4096 counts full scale for the constant-current ramp-down. The unique virtue of a PLL scheme is that the 61.440kHz clock can be locked to the 60Hz power line ($61{,}440 = 60 \times 1024$), giving infinite rejection of 60Hz pickup present on any signal input to the converter, as we discussed in Section 9.21.

We begin with the standard PLL scheme, with a divide-by-n counter added between the VCO output and the phase detector (Fig. 9.72). In this diagram we have indicated the units of gain for each function in the loop. That will be important in our stability calculations. Note particularly that the phase detector converts phase to voltage and that the VCO converts voltage to the time derivative of phase (i.e., frequency). This has the important consequence that the VCO is actually an integrator, thinking of phase as the variable in the lower part of the diagram; a fixed input voltage error produces a linearly rising phase error at the VCO output. The low-pass filter and the divide-by-n counter both have unitless gain.

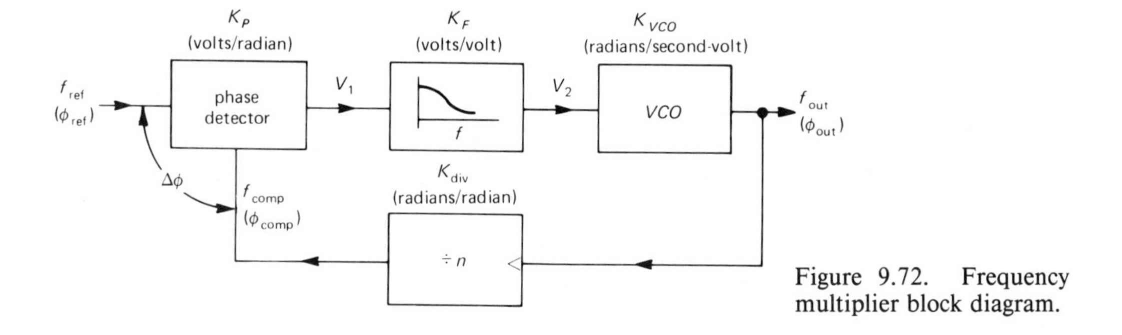

Figure 9.72. Frequency multiplier block diagram.

Stability and phase shifts

The trick to a stable second-order phase-locked loop is shown in the Bode plots of loop gain in Figure 9.73. The VCO acts as an integrator, with $1/f$ response and 90° lagging phase shift (i.e., its response is proportional to $1/j\omega$, a current source driving a capacitor). In order to have a respectable phase margin (the difference between 180° and the phase shift around the loop at the frequency of unity loop gain), the low-pass filter has an additional resistor in series with the capacitor to stop the rolloff at some frequency (fancy name: a "zero"). The combination of these two responses produces the loop gain shown. As long as the loop gain rolls off at 6dB/octave in the neighborhood of unity loop gain, the loop will be stable. The "lead-lag" low-pass filter does the trick, if you choose its properties correctly (this is exactly the same as lead-lag compensation in op-amps). Next you will see how it is done.

Loop gain calculations

Figure 9.74 shows the schematic of the 61.440Hz PLL synthesizer. Both the phase detector and VCO are parts of a 4046 CMOS PLL. We have used the edge-triggered type of phase detector in this circuit (the 4046 actually contains both kinds). Its output comes from a pair of CMOS transistors generating saturated pulses to V_{DD} or ground. It is really a three-state output as explained earlier, since it is in the high-impedance state except during actual phase-error pulses.

The VCO allows you to set the minimum and maximum frequencies corresponding to control voltages of zero and V_{DD}, respectively, by choosing R_1, R_2, and C_1 according to some design graphs. We have made the choices shown. Note: The 4046 has chronic severe supply sensitivity disease; check the graphs on the data sheet. The rest of the loop is standard PLL procedure.

Having rigged up the VCO range, the remaining task is the low-pass-filter design. This part is crucial. We begin by writing down the loop gain, as in the "PLL gain calculation" box, considering each component (refer to Fig. 9.72). Take special pains here to keep your units consistent; don't switch from f to ω or (worse) from hertz to kilohertz. The only gain term still to be decided is K_F. We do this by writing down the overall loop gain, remembering that the VCO is an integrator:

$$\phi_{\text{out}} = \int V_2 K_{\text{VCO}} dt$$

PLL GAIN CALCULATION

Component	*Function*	*Gain*	*Gain calculation* ($V_{DD} = +10$V)
Phase detector	$V_i = K_P \Delta\phi$	K_P	0 to $V_{DD} \leftrightarrow 0°$ to 360°
Low-pass filter	$V_2 = K_F V_1$	K_F	$K_F = \dfrac{1 + j\omega R_4 C_2}{1 + j\omega(R_3 C_2 + R_4 C_2)}$ volts/volt
VCO	$\dfrac{d\phi_{\text{out}}}{dt} = K_{\text{VCO}} V_2$	K_{VCO}	20kHz ($V_2 = 0$) to 200kHz ($V_2 = 10$V) $\rightarrow K_{\text{VCO}} =$ 18kHz/volt $= 1.13 \times 10^5$ radians/second-volt
Divide-by-n	$\phi_{\text{comp}} = \frac{1}{n}\phi_{\text{out}}$	K_{div}	$K_{\text{div}} = \frac{1}{n} = \frac{1}{1024}$

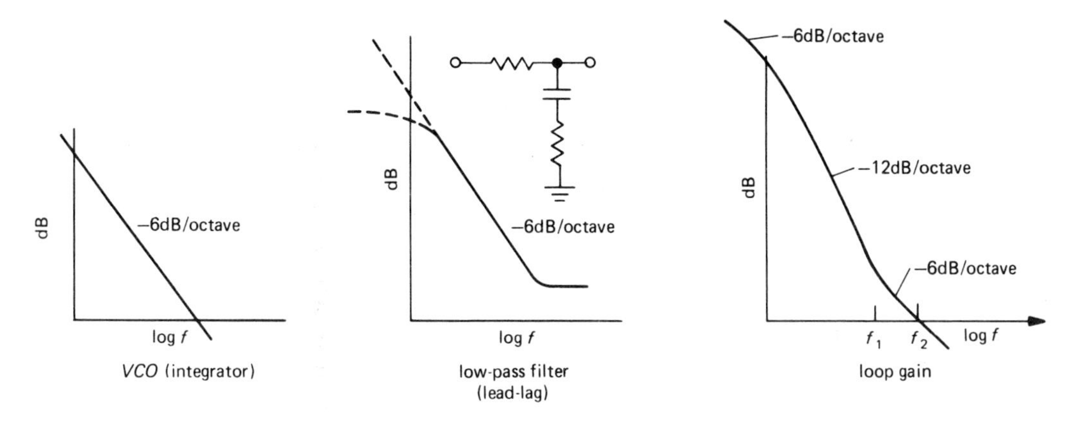

Figure 9.73. PLL Bode plots.

The loop gain is therefore given by

$$\text{Loop gain} = K_P K_F \frac{K_{\text{VCO}}}{j\omega} K_{\text{div}}$$

$$= 1.59 \times \frac{1 + j\omega R_4 C_2}{1 + j\omega(R_3 C_2 + R_4 C_2)} \times \frac{1.13 \times 10^5}{j\omega} \times \frac{1}{1024}$$

Now comes the choice of frequency at which the loop gain should pass through unity. The idea is to pick a unity-gain frequency high enough so that the loop can follow input frequency variations you want to follow, but low enough to provide flywheel action to smooth over noise and jumps in the input frequency. For example, a PLL designed to demodulate an FM input signal, or decode a rapid sequence of input tones, needs to have rapid response (for the FM input signal, the loop should have as much bandwidth as the input signal, i.e., response up to the maximum modulating frequency, while to decode input tones, its response time must be short compared with the time duration of the tones). On the other hand, a loop such as this one, designed to generate a fixed multiple of a stable and slowly varying input frequency, should have a low

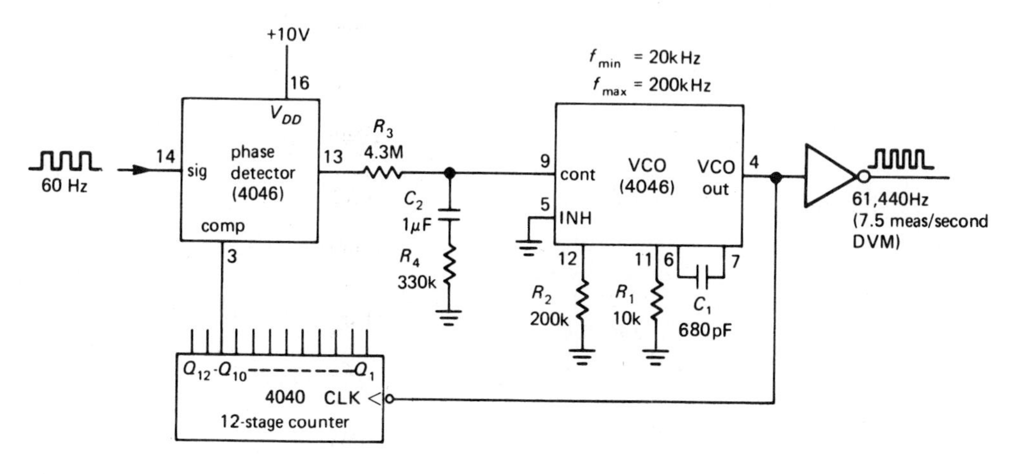

Figure 9.74. Using a PLL multiplier to generate a clock locked to the 60Hz ac line.

unity-gain frequency. That will reduce phase noise at the output and make the PLL insensitive to noise and glitches on the input. It will hardly even notice a short dropout of input signal, because the voltage held on the filter capacitor will instruct the VCO to continue producing the same output frequency.

In this case, we choose the unity-gain frequency f_2 to be 2Hz, or 12.6 radians per second. This is well below the reference frequency, and you wouldn't expect genuine power-line frequency variations on a scale shorter than this (remember that the 60Hz power is generated by enormous generators with lots of mechanical inertia). As a rule of thumb, the breakpoint of the low-pass filter (its "zero") should be lower by a factor of at least 3 to 5, for comfortable phase margin. Remember that the phase shift of a simple RC goes from 0° to 90° over a frequency range of roughly 0.1 to 10 times the −3dB frequency (its "pole"), with a 45° phase shift at the −3dB frequency. In this case we put the frequency of the zero, f_1, at 0.5Hz, or 3.1 radians per second (Fig. 9.75). The breakpoint f_1 determines the time constant R_4C_2 : $R_4C_2 = 1/2\pi f_1$. Tentatively, take $C_2 = 1\mu\text{F}$ and $R_4 = 330\text{k}$. Now all we do is choose R_3 so that the magnitude of the loop gain equals 1 at f_2. In this case, that works out to $R_3 = 4.3\text{M}\Omega$.

EXERCISE 9.5

Show that these choices of filter components actually give a loop gain of magnitude 1.0 at $f_2 = 2.0\text{Hz}$

Sometimes the filter values are inconvenient, so you have to readjust them, or move the unity-gain frequency somewhat. With a CMOS phase-locked loop these values are acceptable (the VCO input terminal has a typical input impedance of $10^{12}\Omega$). With bipolar transistor PLLs (the 4044, for example), you might want to use an external op-amp to buffer the impedances.

We used an edge-triggered (type II) phase detector in this circuit example because of its simplified loop filter; in practice that might not be the best choice for a PLL locked to the 60Hz power line because of the relatively high noise level present on the 60Hz signal. With careful design of the analog input circuit (e.g., a low-pass filter followed by a Schmitt trigger) it would probably perform well; otherwise an exclusive-OR (type I) phase detector should be used.

☐ *"Cut and try"*

For some people, the art of electronics consists in fiddling with filter component values until the loop "works." If you are one of those, we will oblige you by looking the other way. We have presented these loop calculations in detail because we suspect that much of the PLL's bad reputation is the result of too many people "looking the other way." Nevertheless, we can't resist supplying a hot tip for cut-and-try addicts: R_3C_2 sets the smoothing (response) time of the loop, and R_4/R_3 determines the damping, i.e., absence of overshoot for step changes in frequency. You might begin with $R_4 = 0.2R_3$.

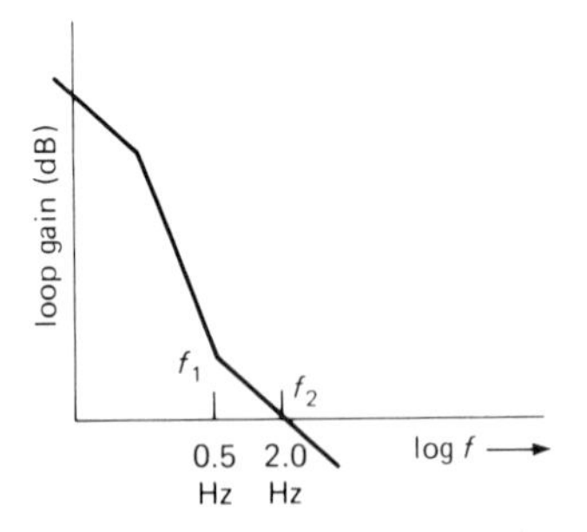

Figure 9.75

☐ *Video clock generation*

Another nice application of a high-frequency oscillator locked to the 60Hz power line is in video signal generation, as in alphanumeric computer terminals. The standard video display rate is 30 pictures per second. Since a small amount of 60Hz pickup is almost inevitable, the picture will "weave" slowly sideways unless the vertical video sync rate is locked exactly to the power-line frequency. PLLs provide a nice way. You just lock a high-frequency VCO (around 15MHz) to a predetermined multiple of 60Hz, so that subdivisions of that high-frequency clock generate (successively) the dots for each displayed character, the number of characters on each line, and the number of horizontal lines in each picture.

□ 9.30 PLL capture and lock

Once locked, it is clear that a PLL will stay locked as long as the input frequency doesn't wander outside the range of the feedback signal. An interesting question to ask is how PLLs get locked in the first place. After all, an initial frequency error results in a periodic output from the phase detector at the difference frequency. After filtering by the low-pass filter, it is reduced to small-amplitude wiggles, rather than a nice clean dc error signal.

□ *Capture transient*

The answer is a little complicated. First-order loops will always lock, because there is no low-pass attenuation of the error signal. Second-order loops may or may not lock, depending on the type of phase detector and the bandpass of the low-pass filter. In addition, the exclusive-OR (type I) phase detector has a limited *capture* range that depends on the filter time constant (this fact can be used to advantage, if you want a PLL that will lock to signals only within a certain frequency range).

The capture transient goes like this: As the (phase) error signal brings the VCO frequency closer to the reference frequency, the error-signal waveform varies more slowly, and vice versa. So the error signal is asymmetric, varying more slowly over that part of the cycle during which f_{VCO} is closer to f_{ref}. The net result is a nonzero average, i.e., a dc component that brings the PLL into lock. If you look carefully at the VCO control voltage during this *capture transient*, you'll see something like what is shown in Figure 9.76. That final

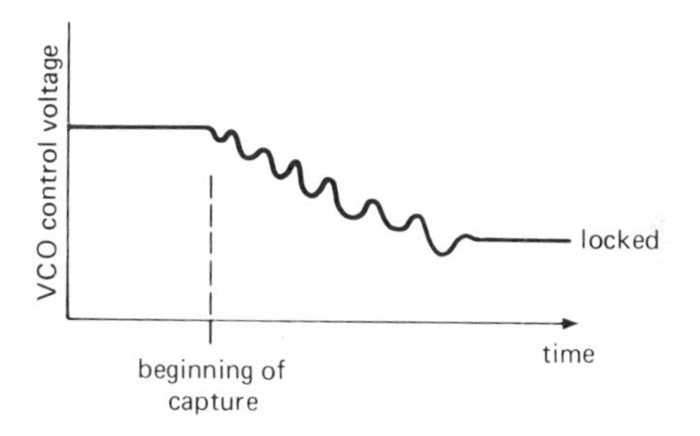

Figure 9.76

overshoot has an interesting cause. Even when the VCO *frequency* reaches its correct value (as indicated by correct VCO control voltage), the loop isn't necessarily in lock, since the *phase* may be wrong. So it may overshoot. Each capture transient is an individual – it looks a bit different each time!

□ *Capture and lock range*

For the exclusive-OR (type I) phase detector, the capture range is limited by the low-pass-filter time constant. This makes sense, because if you begin sufficiently far away in frequency, the error signal will be attenuated so much by the filter that the loop will never lock. It should be evident that a longer filter time constant results in narrower capture range, as does reduced loop gain. It turns out that the edge-triggered phase detector does not have this limitation. Both types have a lock range extending to the limits of the VCO, given the available control input voltage.

□ 9.31 Some PLL applications

We have spoken already of the common use of phase-locked loops in frequency multiplication. The latter application, as in the preceding example, is so straightforward that there should be no hesitation about using these mysterious PLLs. In simple frequency-multiplication applications (e.g., the generation of higher clock frequencies in a digital system) there isn't even any problem of noise on the reference signal, and a first-order loop may suffice.

We would like to point out some other interesting applications, just to give an idea of the diversity of PLL uses.

□ *FM detection*

In frequency modulation, information is encoded onto a "carrier" signal by varying its frequency proportional to the information waveform. We will talk about FM and other modulation techniques in some detail in Chapter 13. There are two methods of recovering the modulating information using phase detectors or PLLs. The word *detection* is used to mean a technique of demodulation.

In the simplest method, a PLL is locked to the incoming signal. The voltage setting the VCO frequency is proportional to the input frequency and is therefore the desired modulating signal (Fig. 9.77). In such a system you would choose the filter bandwidth wide enough to pass the modulating signal, i.e., the response time of the PLL must be short compared with the time scale of variations in the signal being recovered. As you will see in Chapter 13, the signal applied to the PLL does not have to be the actual transmitted waveform; it can be an "intermediate frequency" (IF) generated in the receiving system by the process of *mixing*. A high degree of linearity in the VCO is desirable in this method of FM detection, for low audio distortion.

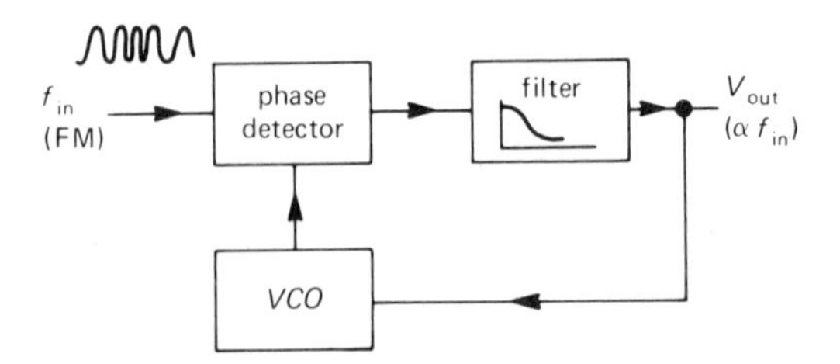

Figure 9.77. PLL FM discriminator.

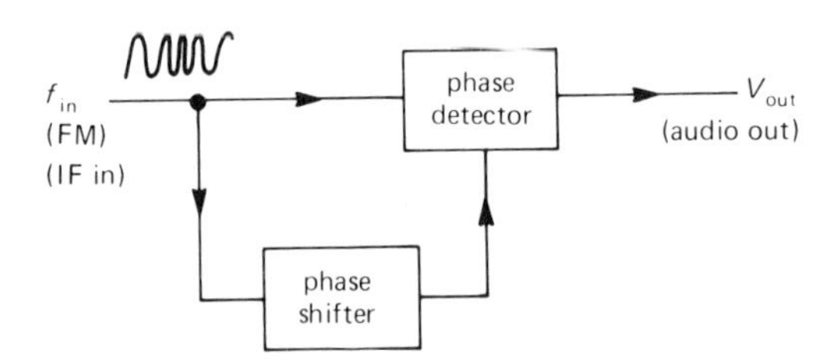

Figure 9.78. Quadrature FM detection.

The second method of FM detection involves a phase detector, although not in a phase-locked loop. Figure 9.78 shows the idea. Both the input signal and a phase-shifted version of the signal are applied to a phase detector, generating some output voltage. The phase-shifting network is diabolically arranged to have a phase shift varying linearly with frequency in the region of the input frequency (this is usually done with resonant *LC* networks), thus generating an output voltage with linear dependence on input frequency. That is the demodulated output. This method is called doubly balanced quadrature FM detection, and it is used in many IF amplifier/detector integrated circuits (e.g., the CA3189).

□ *AM detection*

Wanted: a technique to give an output signal proportional to the instantaneous *amplitude* of a high-frequency signal. The usual method involves rectification (Fig. 9.79). Figure 9.80 shows a fancy method ("homodyne detection") using PLLs. The PLL generates a square wave at the same frequency as the modulated carrier.

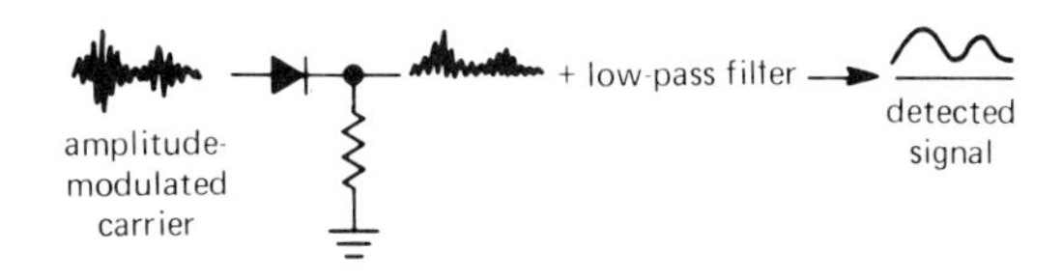

Figure 9.79. AM detection.

Multiplying the input signal by this square wave generates a full-wave-rectified signal that only needs some low-pass filtering to remove the remnants of the carrier frequency, leaving the modulation *envelope*. If you use the exclusive-OR type of phase detector in the PLL, the output is shifted 90° relative to the reference signal, so a 90° phase shift would have to be inserted in the signal path to the multiplier.

□ *Pulse synchronization and clean-signal regeneration*

In digital signal transmission, a string of bits containing the information is sent over a communications channel. The information may be intrinsically digital, or it may be digitized analog signals, as in "pulse-code modulation" (PCM, see Section 13.20). A closely related situation is the decoding of digital information from magnetic tape or disk. In both cases there may be noise or variations in pulse rate (e.g., from tape stretch), and it is desirable to have a clean clock signal at the same rate as the bits you are trying to read. PLLs work very nicely here. The low-pass filter would be chosen to eliminate the jitter and noise in the input synchronizing signal while following slow variations in tape speed, for example.

Another example of signal synchronization might be the circuit in Section 8.31, in which an accurate digitally generated "60Hz" signal (actually, anything from 50Hz to 70Hz) is used to generate a nice output sine wave. In that circuit we used a 6-pole Butterworth low-pass filter to convert the square wave to a sine. An attractive alternative would be to use a sine-wave VCO chip (e.g., the 8038) phase-locked to the precision 60Hz square wave. That would guarantee a constant sine-wave amplitude, permit a wide range of frequency variation, and allow you to eliminate jitter in the rate-multiplier output.

LC oscillator

Figure 9.81 shows an example of a PLL that uses an *LC* oscillator, with digital phase comparison at a lower frequency. We needed a stable and precise source of 14.4MHz, locked to the master observatory oscillator at 10MHz. A varactor (tuning diode, see Section 5.18) fine tunes the JFET *LC* oscillator, according to the type II phase detector output of the 'HC4046. Note that the varactor's tuning range of 18pF–30pF (5V to 1V, respectively) translates to a change in *LC* shunt capacitance of about 2pF (8.2pF to 10pF), for a tuning range of ±0.5% in oscillator frequency. We intentionally made the tuning range narrow to achieve good oscillator stability.

Both reference and output are digitally divided to 400kHz, where the phase detector works well. Note the use of an 'HC-style gate, biased at its logic threshold with a large feedback resistor, to convert the input sine wave to logic levels. Note also the simple emitter follower output stage (with

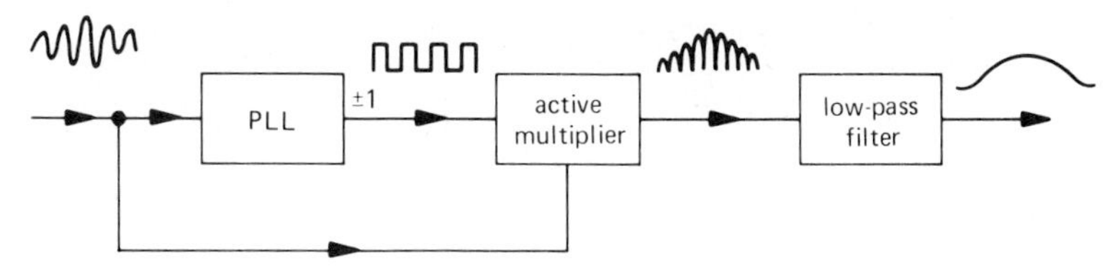

Figure 9.80. Homodyne detection.

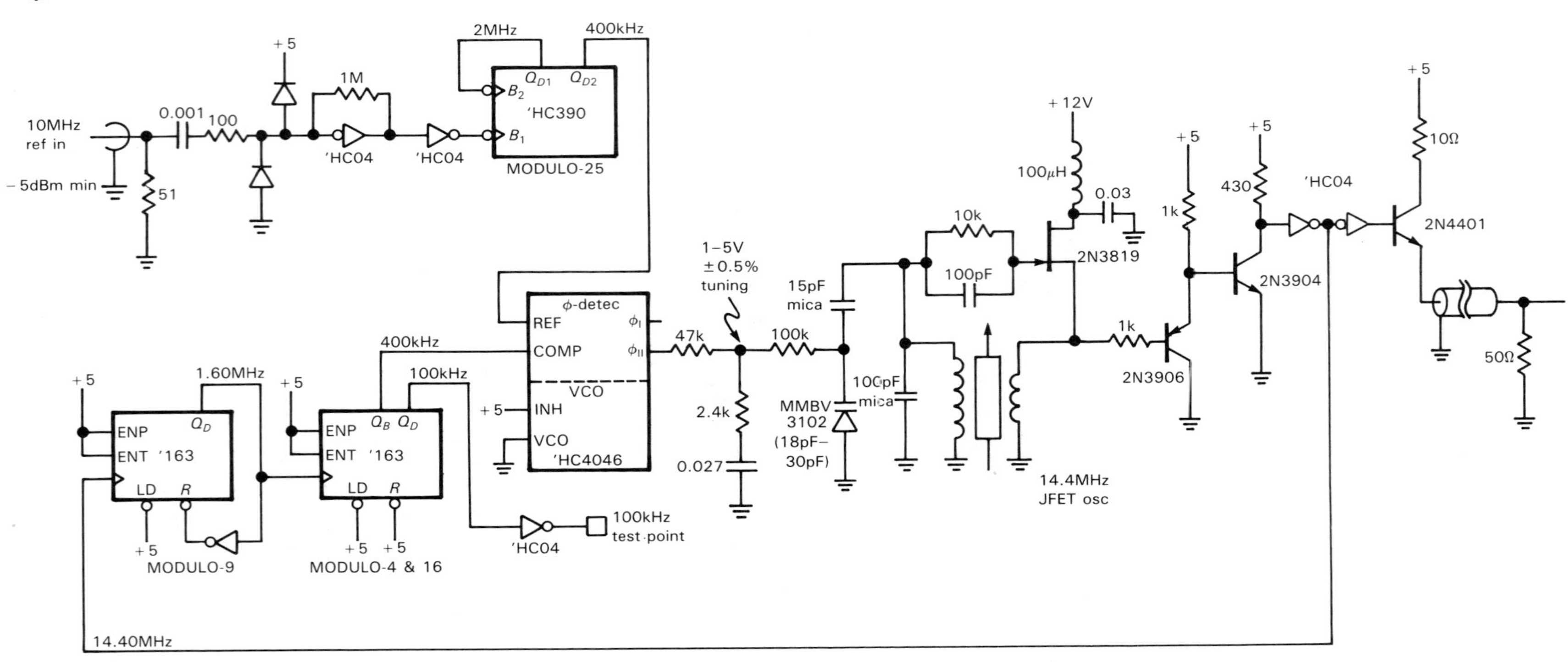

Figure 9.81. Varactor-tuned PLL.

current limiting), used to drive 50 ohm terminated cable, as in Figure 9.42. To align this circuit, you tune the ferrite core in the oscillator until the phase detector filter output swings through its range; adjust the core for mid-range output.

Motorola makes a nice series of inexpensive "PLL Frequency Synthesizer" chips, the MC145145–59, that contain type II phase detectors and on-chip modulo-n dividers for both the signal and reference signals; both dividers are programmable, with 14-bit (or more) precision. Keep these in mind next time you need to synthesize weird frequencies.

PSEUDO-RANDOM BIT SEQUENCES AND NOISE GENERATION

□ 9.32 Digital noise generation

An interesting blend of digital and analog techniques is embodied in the subject of pseudo-random bit sequences (PRBS). It turns out to be remarkably easy to generate sequences of bits (or words) that have good randomness properties, i.e., a sequence that has the same sort of probability and correlation properties as an ideal coin-flipping machine. Since these sequences are generated by standard deterministic logic elements (shift registers, to be exact), the bit sequences generated are in fact predictable and repeatable, although any portion of such a sequence looks for all the world just like a random string of 0's and 1's. With just a few chips you can generate sequences that literally go on for centuries without repeating, making this a very accessible and attractive technique for the generation of digital bit sequences or analog noise waveforms. In fact, there is even an inexpensive "digital noise source" chip available in a mini-DIP package (National MM5437), and shift register noise generators are included in the innards of many sound effects chips.

□ *Analog noise*

Simple low-pass filtering of the output bit pattern of a PRBS generates band-limited white Gaussian noise, i.e., a noise voltage with a flat power spectrum up to some cutoff frequency (see Chapter 7 for more on noise). Alternatively, a weighted sum of the shift register contents (via a set of resistors) performs *digital filtering*, with the same result. Flat noise spectra out to several megahertz can easily be made this way. As you will see later, such digitally synthesized analog noise sources have many advantages over purely analog techniques such as noise diodes or resistors.

□ *Other applications*

Besides their obvious applications as analog or digital noise sources, pseudo-random bit sequences are useful in a number of applications that have nothing to do with noise. They can be used for encipherment of messages or data, since an identical PRBS generator at the receiving end provides the key. They are used extensively in error-detecting and error-correcting codes, since they allow the transcription of blocks of data in such a way that valid messages are separated by the greatest "Hamming distance" (measured by the number of bit errors). Their good autocorrelation properties make them ideal for radar ranging codes, in which the returned echo is compared (cross-correlated, to be exact) with the transmitted bit string. They can even be used as compact modulo-n dividers.

□ 9.33 Feedback shift register sequences

The most popular (and the simplest) PRBS generator is the feedback shift register (Fig. 9.82). A shift register of length m bits is clocked at some fixed rate, f_0. An exclusive-OR gate generates the serial input signal from the exclusive-OR combination of the nth bit and the last (mth) bit of the shift register. Such a circuit goes through a set of states (defined by the set of bits in the register after each clock pulse), eventually repeating itself after K clock pulses; i.e., it is cyclic with period K.

The maximum number of conceivable states of an m-bit register is $K = 2^m$, i.e., the number of binary combinations of m bits. However, the state of all 0's would get "stuck" in this circuit, since the exclusive-OR would regenerate a 0 at the input. Thus, the maximum-length sequence you can possibly generate with this scheme is $2^m - 1$. It turns out that you can make such "maximal-length shift register sequences" if m and n are chosen correctly, and the resultant bit sequence is pseudo-random. (The criterion for maximal length is that the polynomial $1 + x^n + x^m$ be irreducible and prime over the Galois field.) As an example, consider the 4-bit feedback shift register in Figure 9.83. Beginning with the state 1111 (we could start anywhere except 0000), we can write down the states it goes through:

1111
0111
0011
0001
1000
0100
0010
1001
1100
0110
1011
0101
1010
1101
1110

We have written down the states as 4-bit numbers $Q_A Q_B Q_C Q_D$. There are 15 distinct states $(2^4 - 1)$, after which it begins again; therefore it is a maximal-length register.

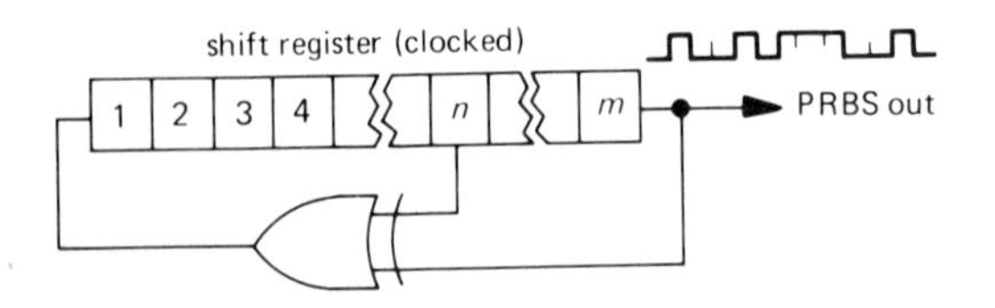

Figure 9.82. Pseudo-random bit sequence generator.

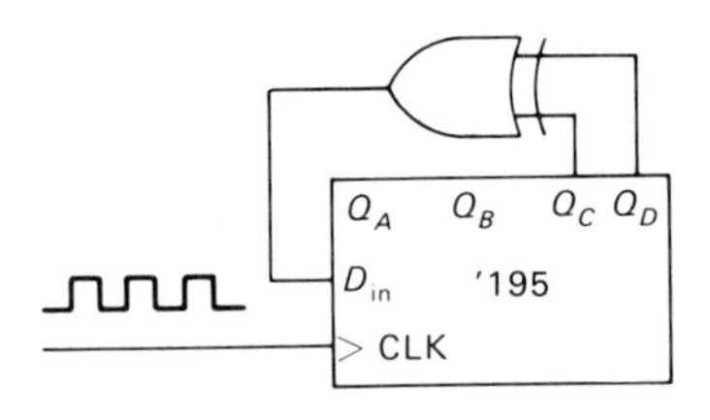

Figure 9.83

EXERCISE 9.6

Demonstrate that a 4-bit register with feedback taps at the second and fourth bits is not maximal length. How many distinct sequences are there? How many states within each sequence?

☐ *Feedback taps*

Maximal-length shift registers can be made with exclusive-OR feedback from more than two taps (in these cases you use several exclusive-OR gates in the standard parity-tree configuration, i.e., modulo-2 addition of several bits). In fact, for some values of m, a maximal-length register can only be made with more than two taps. Here is a listing of all values of m up to 40 for which maximal-length registers can be made with just two taps, i.e., feedback from the nth bit and the mth (last) bit, as previously. A value is given for n and for the cycle length K, in clock cycles. In some cases there is more than one possibility for n, and in every case the value $m - n$ can be used instead of n; thus, the earlier 4-bit example could have used taps at $n = 1$ and $m = 4$.

m	*n*	*Length*
3	2	7
4	3	15
5	3	31
6	5	63
7	6	127
9	5	511
10	7	1023
11	9	2047
15	14	32767
17	14	131071
18	11	262143
20	17	1048575
21	19	2097151
22	21	4194303
23	18	8388607
25	22	33554431
28	25	268435455
29	27	536870911
31	28	2147483647
33	20	8589934591
35	33	34359738367
36	25	68719476735
39	35	549755813887

Since shift register lengths of multiples of 8 are common, you may want to use one of those lengths. In that case, more than two taps are necessary. Here are the magic numbers:

m	*Taps*	*Length*
8	4, 5, 6	255
16	4, 13, 15	65535
24	17, 22, 23	16777215

The MM5437 IC noise-generator chip uses a 23-bit register with a tap at stage 18. Its internal clock runs at about 160kHz, generating white noise output up to about 70kHz (3dB down) with a cycle time of about 1 minute; Figure 7.61 showed it in a "pink noise" generator circuit. With a 33-bit register clocked at 1MHz, the cycle time would be over 2 hours. A 100-bit register clocked at 10MHz would have a cycle time a million times longer than the age of the universe!

□ *Properties of maximal-length shift register sequences*

We generate a string of pseudo-random bits from one of these registers by clocking it and looking at successive output bits. The output can be taken from any position of the register; it is conventional to use the last (mth) bit as the output. Maximal-length shift register sequences have the following properties:

1. In one full cycle (K clock cycles), the number of 1's is one greater than the number of 0's. The extra 1 comes about because of the excluded state of all 0's. This says that heads and tails are equally likely (the extra 1 is totally insignificant for any reasonable-length register; a 17-bit register will produce 65,536 1's and 65,535 0's in one of its cycles).

2. In one full cycle (K clock cycles), half the runs of consecutive 1's have length 1, one-fourth the runs have length 2, one-eighth have length 3, etc. There are the same numbers of runs of 0's as of 1's, again with the exception of a missing 0. This says that the probability of heads and tails does not depend on the outcome of past flips, and therefore the chance of terminating a run of successive 1's or 0's on the next flip is $\frac{1}{2}$ (contrary to the man-in-the-street's understanding of the "law of averages").

3. If one full cycle (K clock cycles) of 1's and 0's is compared with the same sequence shifted cyclically by any number of bits n (where n is not 0 or a multiple of K), the number of disagreements will be one greater than the number of agreements. In fancy language, the autocorrelation function is a Kronecker delta at zero delay, and $-1/K$ everywhere else. This absence of "side lobes" in the autocorrelation function is what makes PRBSs so useful for radar ranging.

EXERCISE 9.7

Show that the 4-bit shift register sequence listed earlier (taps at $n = 3$, $m = 4$) satisfies these properties, considering the Q_A bit as the "output": 100010011010111.

□ 9.34 Analog noise generation from maximal-length sequences

Advantages of digitally generated noise

As we remarked earlier, the digital output of a maximal-length feedback shift register can be converted to band-limited white noise with a low-pass filter whose cutoff frequency is well below the clock frequency of the register. Before getting into the details, we will point out some of the advantages of digitally generated analog noise. Among other things, it allows you to generate noise of known spectrum and amplitude, with adjustable bandwidth (via clock frequency adjustment), using reliable and easily maintained digital circuitry. There is none of the variability of diode noise generators, nor are there interference and pickup problems that plague the sensitive low-level analog circuitry used with diode or resistor noise generators. Finally, it generates repeatable "noise" and, when filtered with a weighted digital filter (more about this later), repeatable noise waveforms independent of clocking rate (output noise bandwidth).

□ 9.35 Power spectrum of shift register sequences

The output spectrum generated by maximal-length shift registers consists of noise extending from the repeat frequency of the entire sequence, f_{clock}/K, up to the clock frequency and beyond. It is flat within ± 0.1dB up to 12% of the clock frequency (f_{clock}), dropping rather rapidly beyond its -3dB point of 44% f_{clock}. Thus a low-pass filter with a high-frequency cutoff of 5%–10% of the clock frequency will convert the unfiltered shift register output to a band-limited analog noise voltage. Even a simple *RC* filter will suffice, although it may be desirable to use active filters with sharp cutoff characteristics (see Chapter 5) if a precise frequency band of noise is needed.

To make these statements more precise, let's look at the shift register output and its power spectrum. It is usually desirable to eliminate the dc offset characteristic of digital logic levels, generating a bipolarity output with 1 corresponding to $+a$ volts and 0 corresponding to $-a$ volts (Fig. 9.84). This can be easily done with the

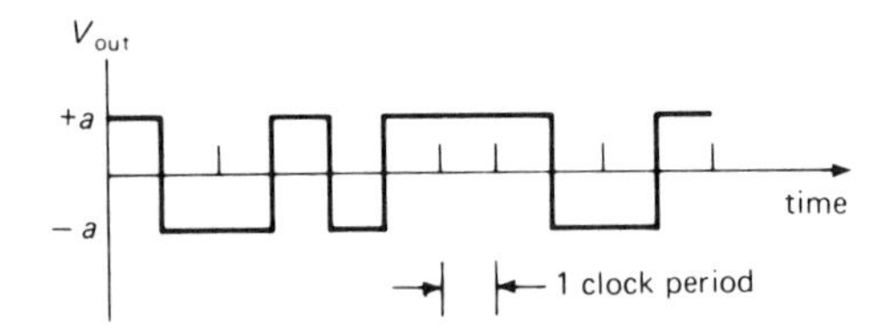

Figure 9.84

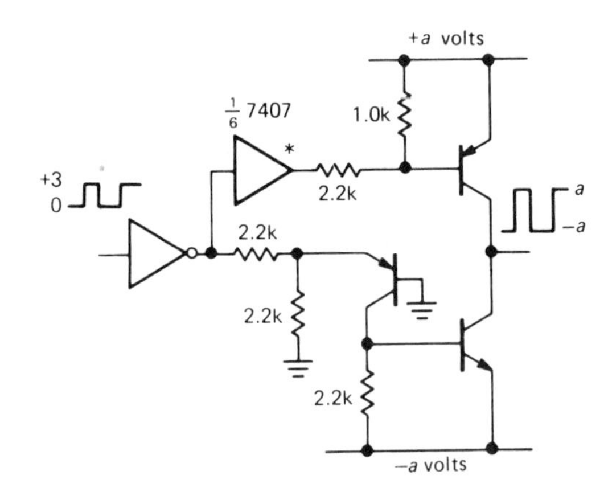

Figure 9.85. Precise bipolarity output stage with low Z_{out}.

sort of transistor push-pull stage shown in Figure 9.85. Alternatively, you can use MOS transistors, a circuit with clamping diodes to stable reference voltages, a fast op-amp with adjustable dc offset current into a summing junction, or a '4053 CMOS switch, running from $\pm a$ volts, with a pair of inputs tied to the supplies.

As we remarked earlier, the string of output bits has a single peak in its autocorrelation. If the output states represent $+1$ and -1, the digital autocorrelation (the sum of the product of corresponding bits, when the bit string is compared with a shifted version of itself) is as shown in Figure 9.86.

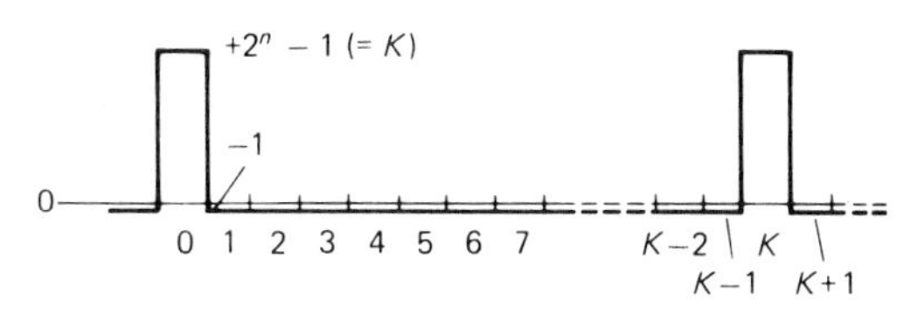

Figure 9.86. Full-cycle discrete autocorrelation for a maximal-length shift register sequence.

Don't confuse this with a *continuous* autocorrelation function, which we will consider later. This graph is defined only for shifts corresponding to a whole number of clock cycles. For all shifts that aren't zero or a multiple of the overall period K, the autocorrelation function has a constant -1 value (because there is an extra 1 in the sequence), negligible when compared with the zero-offset value of K. Likewise, if we consider the unfiltered shift register output as an *analog* signal (whose waveform happens to take on values of $+a$ and $-a$ volts only), the normalized autocorrelation becomes a continuous function, as shown in Figure 9.87. In other words, the waveform is totally uncorrelated with itself when shifted more than one clock period forward or backward.

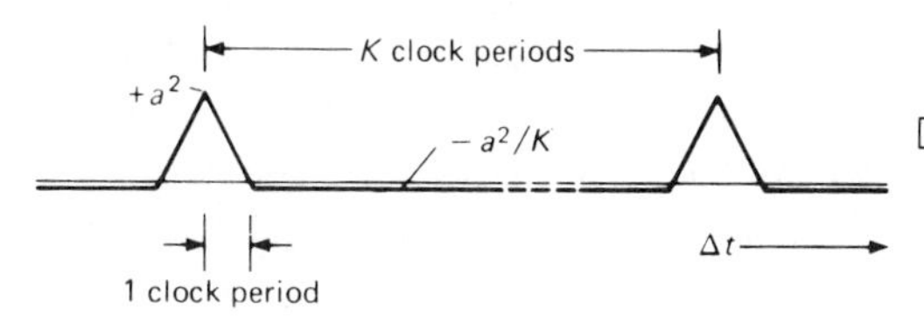

Figure 9.87. Full-cycle continuous autocorrelation for a maximal-length shift register sequence.

The power spectrum of the unfiltered digital output can be obtained form the autocorrelation by standard mathematical techniques. The result is a set of equally spaced series of spikes (delta functions), beginning at the frequency at which the whole sequence repeats, f_{clock}/K, and going up in frequency by equal intervals f_{clock}/K. The fact that the spectrum consists of a set of discrete spectral lines reflects the fact that the shift register sequence eventually (and periodically) repeats itself. Don't be alarmed by this funny spectrum; it will look continuous for any measurement or application that takes less time than the cycle time of the register. The envelope of the spectrum of the unfiltered output is shown in Figure 9.88. The envelope is proportional to the square of $(\sin x)/x$. Note the peculiar property that there is *no* noise power at the clock frequency or its harmonics.

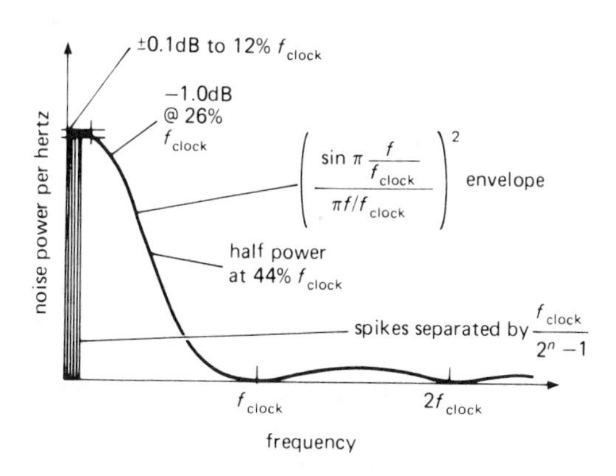

Figure 9.88. Power spectrum of unfiltered digital shift register output signal.

□ *Noise voltage*

Of course, for analog noise generation you use only a portion of the low-frequency end of the spectrum. It turns out to be easy to calculate the noise power per hertz in terms of the half-amplitude *(a)* and the clock frequency (f_{clock}). Expressed as an rms noise *voltage*, the answer is

$$V_{rms} = a\left(\frac{2}{f_{clock}}\right)^{1/2} \qquad \mathrm{V/Hz}^{1/2}$$
$$(f \leq 0.2 f_{clock})$$

This is for the bottom end of the spectrum, the part you usually use (you can use the envelope function to find the power density elsewhere).

For example, suppose we run a maximal-length shift register at 1.0MHz and arrange it so that the output voltage swings between +10.0 and −10.0 volts. The output is passed through a simple RC low-pass filter with 3dB point at 1kHz (Fig. 9.89). We can calculate the rms noise voltage at the output exactly. We know from the preceding equation that the output from the level shifter has an rms noise voltage of 14.14mV per root hertz. From Section 7.21 we know that the noise bandwidth of the low-pass filter is (π/2)(1.0kHz), or 1.57kHz. So the output noise voltage is

$$V_{\text{rms}} = 0.01414(1570)^{1/2} = 560\,\text{mV}$$

with the spectrum of a single-section RC low-pass filter.

9.36 Low-pass filtering

Analog filtering

The spectrum of useful noise from a pseudo-random-sequence generator extends from a low-frequency limit of the reciprocal repeat period (f_{clock}/K) up to a high-frequency limit of perhaps 20% of the clock frequency (at that frequency the noise power per hertz is down by 0.6dB). Simple low-pass filtering with RC sections, as illustrated in the earlier example, is adequate provided that its 3dB point is set far below the clock frequency (e.g., less than 1% of f_{clock}). In order to use the spectrum closer to the clock frequency, it is advisable to use a filter with sharper cutoff, e.g., a Butterworth or Chebyshev. In that case the flatness of the resultant spectrum depends on the filter characteristics, which should be measured, since component variations can produce ripples in the passband gain. Likewise, the filter's actual voltage gain should be measured if the precise value of noise voltage per root hertz is important.

Digital filtering

A disadvantage of analog filtering is the need to readjust the filter cutoff if the clock frequency is changed by large factors. In situations where that is desirable, an elegant solution is provided by digital filtering, in this case performed by taking an analog weighted sum of successive output bits (nonrecursive digital filtering). In this way the effective filter cutoff frequency changes to match changes in the clock frequency. In addition, digital filtering lets you go to extremely low cutoff frequencies (fractions of a hertz) where analog filtering becomes awkward.

In order to perform a weighted sum of successive output bits simultaneously, you can simply look at the various parallel outputs of successive shift register bits, using resistors of various values into an op-amp summing junction. For a low-pass filter the weights should be proportional to $(\sin x)/x$; note that some levels will have to be inverted, since the weights are of both signs. Since no capacitors are used in

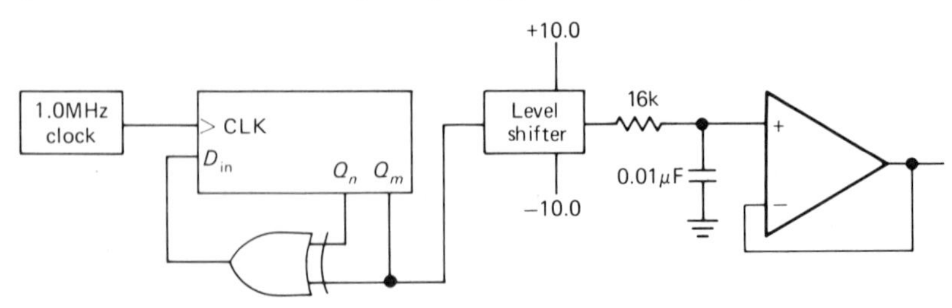

Figure 9.89. Simple pseudo-random noise source.

this scheme, the output waveform consists of a set of discrete output voltages.

The approximation to Gaussian noise is improved by using a weighting function over many bits of the sequence. In addition, the analog output then becomes essentially a continuous waveform. For this reason it is desirable to use as many shift register stages as possible, adding additional shift register stages outside the exclusive-OR feedback if necessary. As before, pull-ups or MOS switches should be used to set stable digital voltage levels (CMOS logic is ideal for this application, since the outputs saturate cleanly at V_{DD} and ground).

The circuit in Figure 9.90 generates pseudo-random analog noise, with bandwidth selectable over an enormous range, using this technique. A 2.0MHz crystal oscillator drives a 14536 24-stage programmable divider, generating clock frequencies going from 1.0MHz down to 0.12Hz by factors going from 1.0MHz down to 0.12Hz by factors of 2. A 32-bit shift register is connected with feedback from stages 31 and 18, generating a maximal-length sequence with 2 billion states (at the maximum clock frequency the register completes one cycle in a half hour). In this case we have used a $(\sin x)/x$ weighted sum over 32 successive values of the sequence. U_1 and U_2 amplify the inverted and noninverted terms, respectively, driving differential amplifier U_3. The gains are chosen to generate a 1.0 volt rms output with no dc offset into a 50 ohm load impedance (2.0V rms open circuit). Note that this noise amplitude is independent of the clock rate, i.e., the total bandwidth. This digital filter has a cutoff at about $0.05f_{\text{clock}}$, giving a white noise output spectrum extending from dc to 50kHz (maximum clock frequency) down to dc to 0.006Hz (minimum clock frequency), in 24 steps of bandwidth. The circuit also provides an unfiltered output waveform, going between +1.0 volt and −1.0 volt.

There are a few interesting points about this circuit. Note that an exclusive-NOR gate is used for feedback so that the register can be simply initialized by bringing it to the state of all zeros. This trick of inverting the serial input signal makes the excluded state the state of all 1's (rather than all 0's as with the usual exclusive-OR feedback), but it leaves all other properties unaffected.

A weighted sum of a finite number of bits cannot ever produce truly Gaussian noise, since the peak amplitude is limited. In this case it can be calculated that the peak output amplitude (into 50Ω) is ±4.34 volts, giving a "crest factor" of 4.34. That calculation is important, by the way, because you must keep the gain of U_1 through U_3 low enough to prevent clipping. Look carefully at the methods used to generate an output of zero dc offset from the CMOS levels of +6.0 volts average value (LOW=0V, HIGH=12.0V).

This method of digital low-pass filtering of maximal-length shift register sequences is used in many commercial noise generators.

□ 9.37 Wrap-up

A few comments about shift register sequences as analog noise sources: You might be tempted to conclude from the three properties of maximal-length shift registers listed earlier that the output is "too random," in the sense of having exactly the right number of runs of a given length, etc. A genuine random coin-flipping machine would not generate exactly one more head than tail, nor would the autocorrelation be absolutely flat for a finite sequence. To put it another way, if you used the 1's and 0's that emerge from the shift register to control a "random walk," moving forward one step for a 1 and back one step for a 0, you would wind up exactly one step away from your beginning point after the register had gone

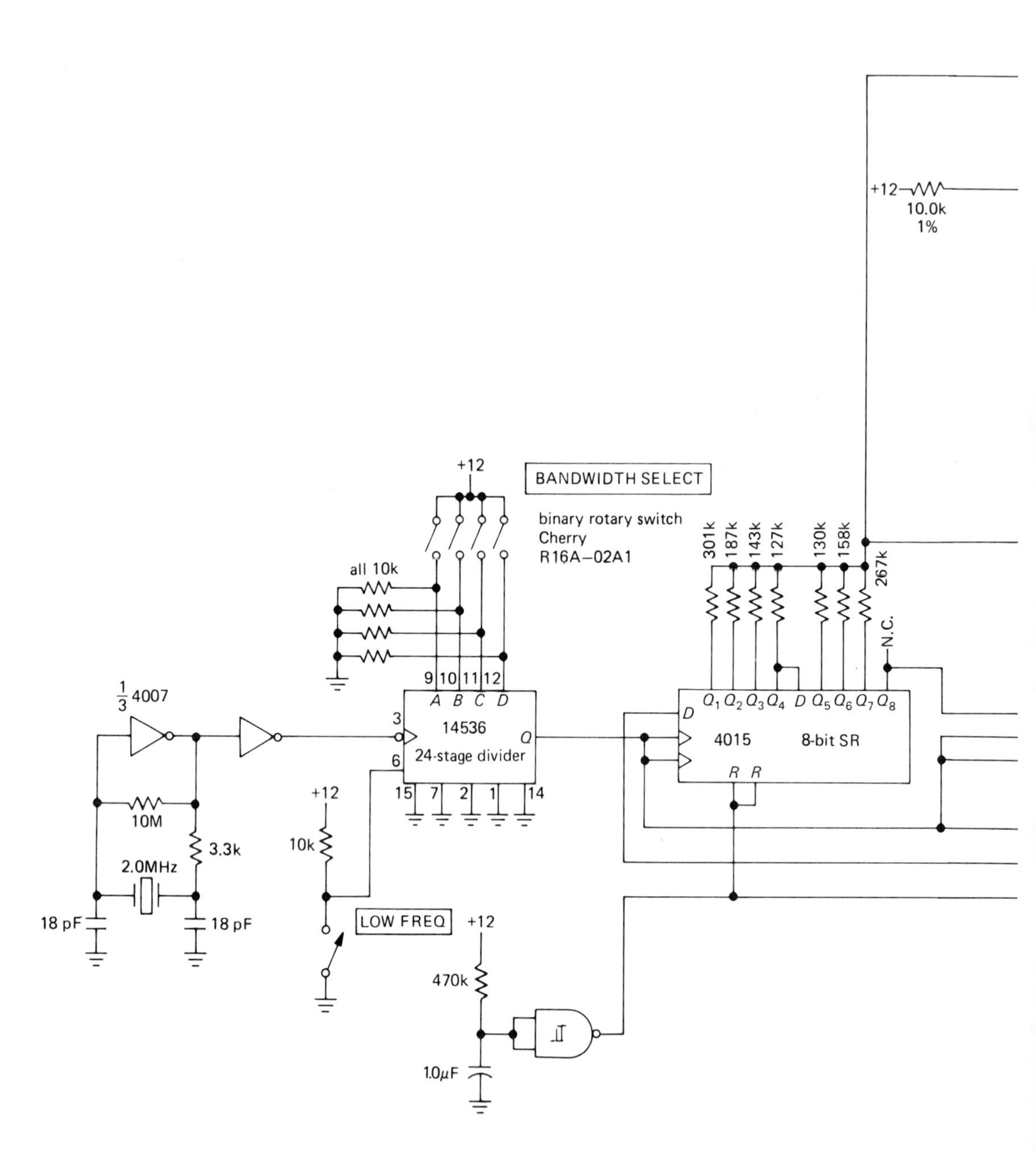

Figure 9.90. Wide-frequency-range laboratory noise source.

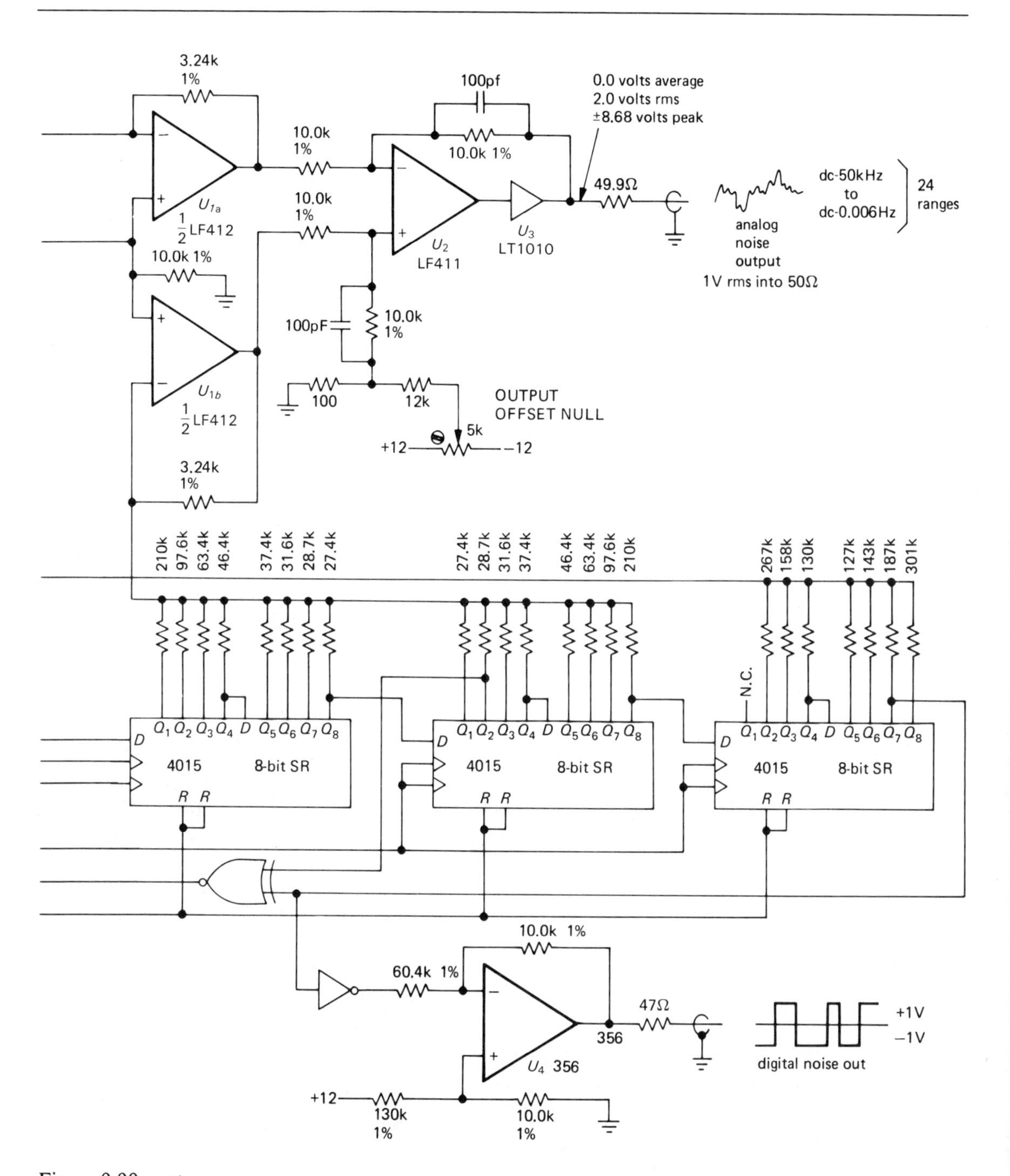

Figure 9.90 *cont.*

through its entire cycle, a result that is anything but "random"!

However, the shift register properties mentioned earlier are only true of the entire sequence of $2^n - 1$ bits, *taken as a whole*. If you use only a section of the entire bit sequence, the randomness properties closely approximate a random coin-flipper. To make an analogy, it is as if you were drawing red balls and blue balls at random from an urn initially containing K balls in all, half red and half blue. If you do this *without replacing them*, you would expect to find approximately random statistics at first. As the urn becomes depleted, the statistics are modified by the requirement that the total numbers of red and blue balls must come out the same.

You can get an idea how this goes by thinking again about the random walk. If we assume that the only nonrandom property of the shift sequence is the exact equality of 1's and 0's (ignoring the single excess 1), it can be shown that the random walk as described should reach an average distance from the starting point of

$$X = [r(K-r)/(K-1)]^{1/2}$$

after r draws from a total population of $K/2$ 1's and $K/2$ 0's. (We are indebted to E. M. Purcell for this calculation.) Since in a completely random walk X equals the square root of r, the factor $(K-r)/(K-1)$ expresses the effect of finite urn contents. As long as $r \ll K$, the randomness of the walk is only slightly reduced from the completely random case (infinite urn contents), and the pseudo-random-sequence generator is indistinguishable from the real thing. We tested this with a few thousand PRBS-mediated random walks, each a few thousand steps in length, and found that the randomness was essentially perfect, as measured by this simple criterion.

Of course, the fact that PRBS generators pass this simple test does not guarantee that they would satisfy some of the more sophisticated tests of randomness, e.g., as measured by higher-order correlations. Such correlations also affect the properties of analog noise generated from such a sequence by filtering. Although the noise amplitude distribution is Gaussian, there may be higher-order amplitude correlations uncharacteristic of true random noise. Current thinking on this subject is that the use of many (preferably about $m/2$) feedback taps (using an exclusive-OR parity-tree operation to generate the serial input) generates "better" noise in this respect.

Noise generator builders should be aware of the 4557 CMOS variable-length shift register (1 to 64 stages); of course, you have to use it in combination with a parallel-output register (such as the '4015 or '164) in order to get at the n tap.

In Section 7.20 there is a discussion of noise, with an example of a "pink noise" generator using the MM5437 maximal-length shift register IC.

□ 9.38 Digital filters

The last example brought up the interesting topic of digital filtering, in that case the generation of a low-pass-filtered analog output signal by taking a weighted sum of 32 samples of pseudo-random bits, each corresponding to a voltage level of 0 volts or +12 volts. The "filter" accepted as its input a waveform that happened to have only two voltage levels. In general, the same thing can be done with an analog waveform as input, forming a weighted sum over its values (x_i) at equally spaced times

$$Y_i \sum_{k=-\infty}^{\infty} h_k x_{i-k}$$

The x_is are the discrete input signal samples, the h_ks are the weights, and the y_is are the output of the filter. In real life a digital filter will sum only over a finite set of input values, as for example in the noise generator circuit where we used 32 terms.

Figure 9.91 shows schematically what is happening.

Note that such a filter can have the interesting property of being symmetrical in time, i.e., averaging past and future "history" to arrive at its present output. Of course, real analog filters can only look backward in time, corresponding to a digital filter with nonzero weighting factors h_k only for $k \geq 0$.

□ *Symmetrical filter frequency response*

For a symmetrical filter ($h_k = h_{-k}$) it can be shown that the frequency response is given by

$$H(f) = h_0 + 2\sum_{k=1}^{\infty} h_k \cos 2\pi k f t_s$$

where t_s is the time between samples. Thus the individual $h_k s$ are recognizable (to those who know about such things!) as the Fourier series components of the desired frequency response curve, which explains why the weights in the noise-generator circuit shown earlier were chosen proportional to $(\sin x)/x$, the Fourier components of a "brick-wall" low-pass filter. For such a symmetrical filter the phase shift at any frequency is either zero or 180°.

□ *Recursive filters*

An interesting class of digital filters can be made if you allow the filter to use as its inputs the value of its own output, in addition to the value of the signal being filtered. You can think of the filter as having "feedback." The fancy name for such a filter is a *recursive* (or *infinite impulse response*, IIR) digital filter, as opposed to the nonrecursive (or *finite impulse response*, FIR) filters just discussed. For example, you could form the outputs y_i according to

$$y_i = Ay_{i-1} + (1 - A)x_i$$

This happens to give you a low-pass response, equivalent to that of a simple RC low-pass filter, according to

$$A = e^{-t_s/RC}$$

where t_s is the time between successive samples x_i of the input waveform. Of course, the situation is not *identical* with an analog low-pass filter operating on an analog waveform because of the discrete nature of the sampled waveform.

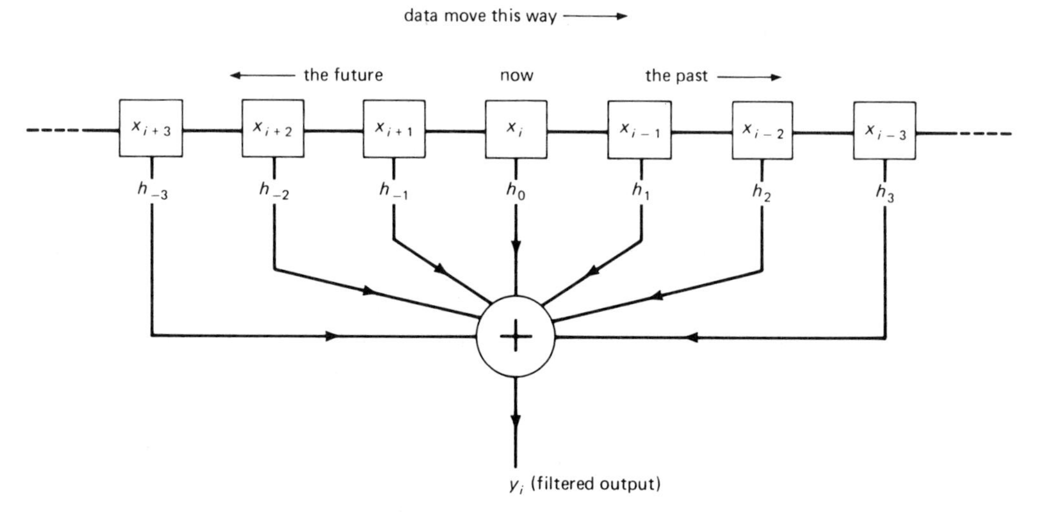

Figure 9.91. Nonrecursive digital filter.

□ *Low-pass filter example*

As a numerical example, suppose you want to filter a set of numbers representing a signal, with a low-pass 3dB point at $f_{3\text{dB}} = 1/20t_s$. Thus the time constant equals the time for 20 successive samples. Then $A = 0.95123$, and so the output is given by

$$y_i = 0.95123y_{i-1} + 0.04877x_i$$

The approximation to a real low-pass filter becomes better as the time constant becomes long compared with the time between samples, t_s.

You would probably use a filter like this to process data that are already in the form of discrete samples, e.g., an array of data in a computer. In that case the recursive filter becomes a trivial arithmetic pass once through the data. Here is what the low-pass filter would look like in a FORTRAN program:

```
      A = EXP(-TS/TC)
      B = 1. - A
      DO 10 I = 2,N
10    X(I) = A * X(I - 1) + B * X(I)
```

where X is the array of data, TS is the time between samples of the data (i.e., TS $= 1/f_s$), and TC is the desired filter time constant. Ideally, TC $\gg$ TS. This little program does the filtering *in place*, i.e., it replaces the original data by the filtered version. You could, of course, put the filtered data into a separate array.

□ *Low-pass commutating filter*

This same filter can be built with hardware, using the circuit shown in Figure 9.92. The FET switches S_1 and S_2 are toggled at some clock rate f_s, repeatedly charging C_1 to the input voltage, then transferring its charge to C_2. If C_2 has voltage V_2, and C_1 charges up to the input level V_1, then when C_1 is connected to C_2, the new voltage will be

$$V = \frac{C_1V_1 + C_2V_2}{C_1 + C_2}$$

i.e., it is identical with the preceding low-pass recursive filter, with

$$y_i = \frac{C_2}{C_1 + C_2}y_{i-1} + \frac{C_1}{C_1 + C_2}x_i$$

Equating these coefficients to the value of A, given earlier, gives us

$$f_{3\text{dB}} = \frac{1}{2\pi}f_s \log_e\left(\frac{C_1 + C_2}{C_2}\right)$$

EXERCISE 9.8

Prove that this result is correct.

This filter is perfectly practicable, and it offers the nice feature of electronic tuning via the clock rate f_s. In practice, you would use CMOS switches, and C_1 would probably be much larger than C_2. Consequently, the switch-driving waveform should be unsymmetrical, spending most of its time closing S_1.

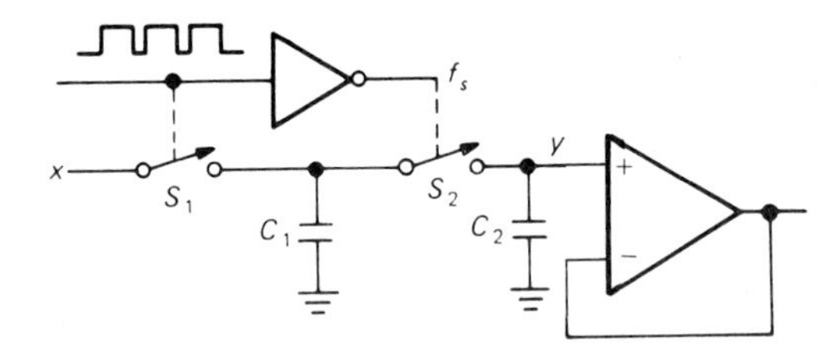

Figure 9.92. Recursive switched-capacitor filter.

The preceding circuit is a simple example of a *commutating filter*, which includes filters made from arrays of switched capacitors. They have periodic frequency response properties that make them particularly suitable for "comb" and notch filters.

It is possible to synthesize discrete approximations to all the classic filters (Butterworth, Chebyshev, etc.) in high-pass, low-pass, bandpass, and band-reject forms, either symmetrical in time or with the genuine "lagging" time response. Such filters are extremely useful when processing digitally quantized data, which is clearly the way of the future.

Inexpensive switched-capacitor filter ICs are now widely available. For example, the MF4 from National is a 4-pole Butterworth low-pass filter in a mini-DIP; it requires no external components and runs from a single +5 volt to +14 volt supply. The filter's cutoff frequency (0.1Hz min, 20kHz max) is set by the externally supplied clock, $f_{\rm clk} = 100 f_{\rm 3dB}$. The MF5 and MF10 "universal switched capacitor filters" work somewhat differently. A few external resistors set the type of filter (high-pass, low-pass, bandpass, notch) and filter characteristic (Butterworth, Chebyshev, etc.), with the clock rate setting the cutoff frequency as before. Some other manufacturers of switched-capacitor filters are American Microsystems (AMI), Linear Technology (LTC), and Reticon. LTC, as usual, has found some clever tricks to improve performance. Their LTC1062 (or MAX280) is similar to the MF4, but with 5 poles and with *zero* dc error! They accomplish the latter by putting the filter outside the dc path (Fig. 9.94). The flexible MAX260 series allows microprocessor control of important filter parameters.

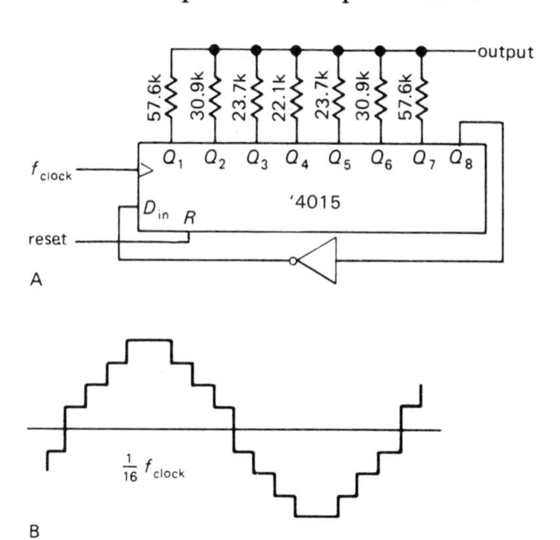

Figure 9.93. Digital sine-wave generation.

In general, these switched-capacitor filters operate only to the top of the audio range. They also tend to suffer from significant clock feedthrough to the output, typically in the range of 10–25mV. This limits their dynamic range in applications where the clock frequency is within the bandpass (e.g., high-pass filters). They can have significant amounts of noise, sometimes limiting their dynamic range to 80dB or less (compared with 140dB or more for a good op-amp). On the positive side, switched-capacitor filters are easy to use and let you tune your filter response effortlessly (via $f_{\rm clk}$). They are finding widespread use in modems (data communication via audio telephone circuits) and other communications applications. See Section 5.11.

□ *Digital sine-wave generation*

An interesting technique related to nonrecursive digital filtering is the synthesis of sine waves by taking weighted sums from the outputs of a Johnson counter (or "twisted-ring" or "walking-ring" counter). The circuit in Figure 9.93 shows the way. The 4015 is an 8-stage parallel-output shift register. By driving the input with the complement of the last stage, you get a Johnson counter, which goes through 16 states ($2n$ states, in general, for an n-stage shift register). Beginning with the state of all 0's, 1's begin marching in from the left until the register contains all 1's, at which point 0's march in again, and so on. The weighting shown generates an 8-level approximation to a sine wave, as shown, with a frequency 1/16th that of the clock, and with the first nonzero distortion term (assuming perfect resistor values) being the 15th harmonic, which is down by 24dB.

SELF-EXPLANATORY CIRCUITS

9.39 Circuit ideas

Figure 9.95 shows a few examples of interfacing between logic and linear signals.

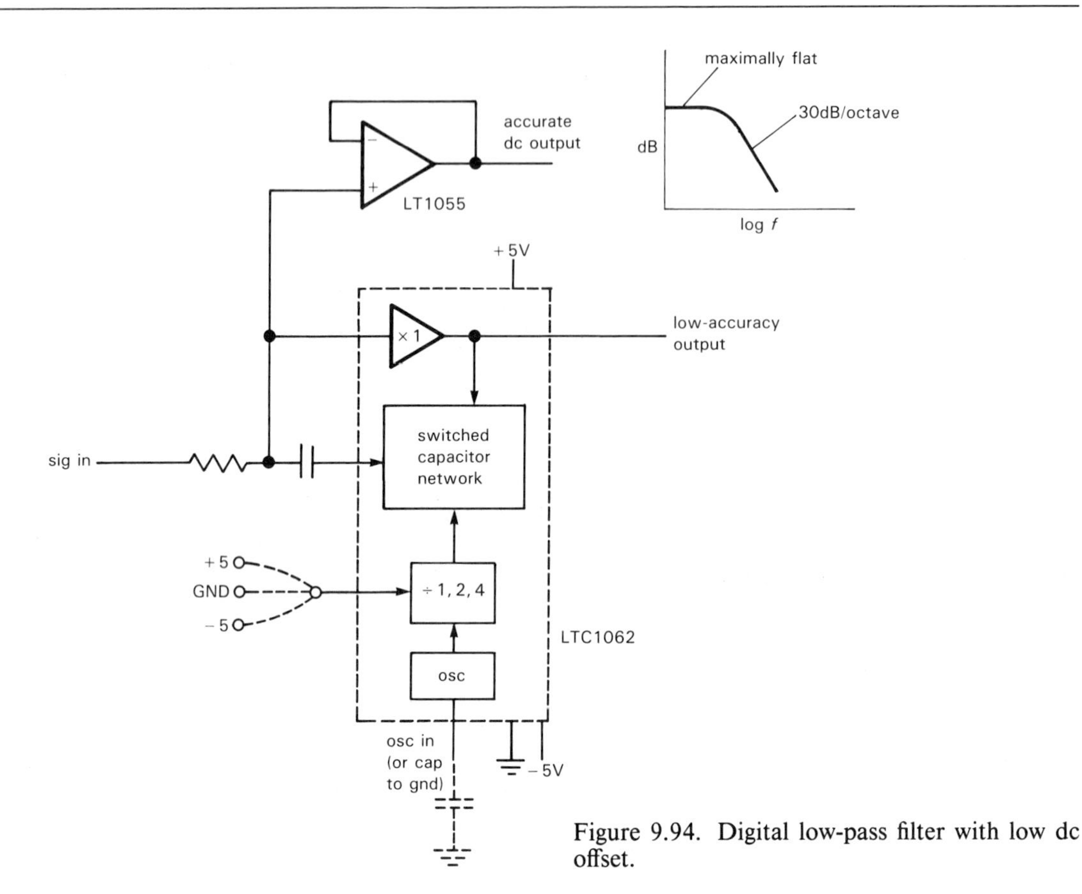

Figure 9.94. Digital low-pass filter with low dc offset.

9.40 Bad circuits

The circuits in Figure 9.96 illustrate some basic interfacing blunders; in each case try to figure out what is wrong and how to fix it.

ADDITIONAL EXERCISES

1. Design a circuit to indicate if the logic power (+5V) has failed momentarily. It should have a push button to RESET it and a LED lamp that indicates CONTINUOUS POWER; make it operate from the +5 volt logic supply.

2. Why can't two n-bit DACs be used to make a $2n$-bit DAC by just summing their outputs proportionally ($\mathrm{OUT}_1 + \mathrm{OUT}_2/2^n$)?

3. Verify that the peak output of the pseudo-random-noise generator (Fig. 9.90) is ±8.68 volts.

4. An experiment is being controlled by a programmable calculator interfaced to various stimulus and measurement devices. The calculator increments a variable under its control (e.g., the wavelength of light coming from a monochromator) and processes the corresponding measurement (e.g., the amount of transmitted light, corrected for the known sensitivity curve of the detector). The result is a set of x, y pairs. Your job is to design a circuit so that they can be plotted on analog x, y plotter.

The calculator outputs each x, y pair as two 3-digit BCD characters. To reduce the number of connections necessary, the numbers are presented one digit at a time ("bit parallel, character serial"),

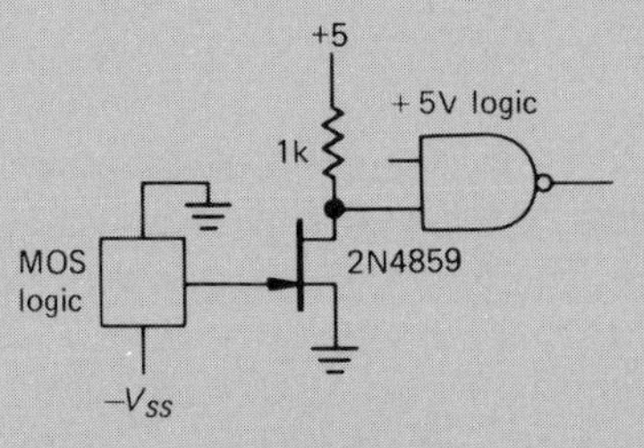

A. negative logic to TTL shifter

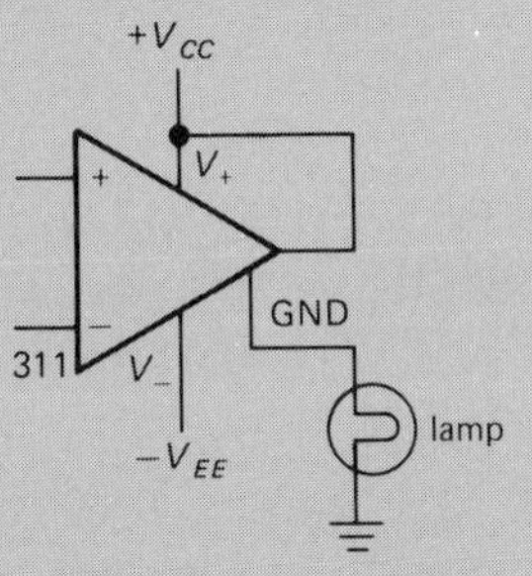

B. driving ground-returned load

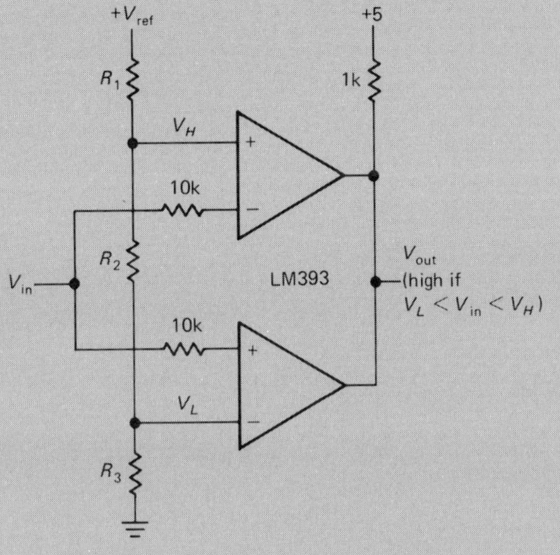

C. window discriminator

+15
1N4002
solenoid
15V 1A
8
1.5k
2N5194
1
311
4

D. solenoid driver

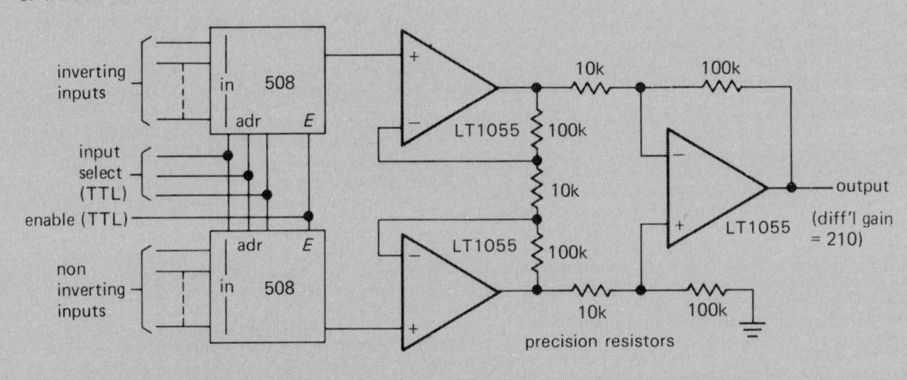

E. 8-channel multiplexer with differential input

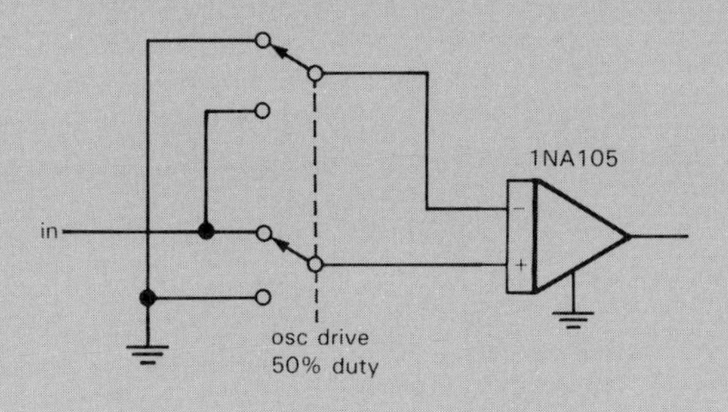

F. phase detector with 60dB osc rejection

+5
$\frac{1}{3}$4053
$+V_{ref}$
$-V_{ref}$

G. precision Schmitt trigger

Figure 9.95

Bad circuits

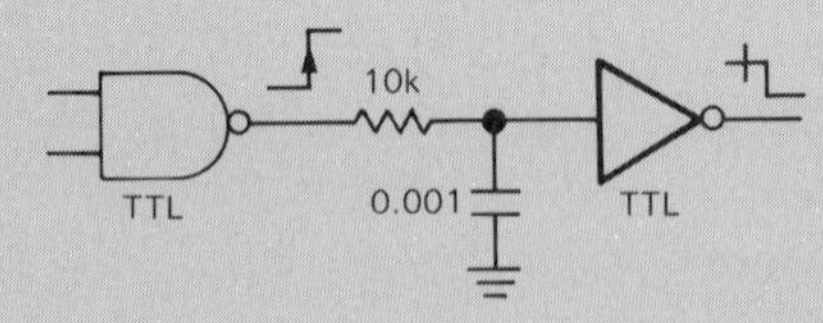

A. delayed edge generator

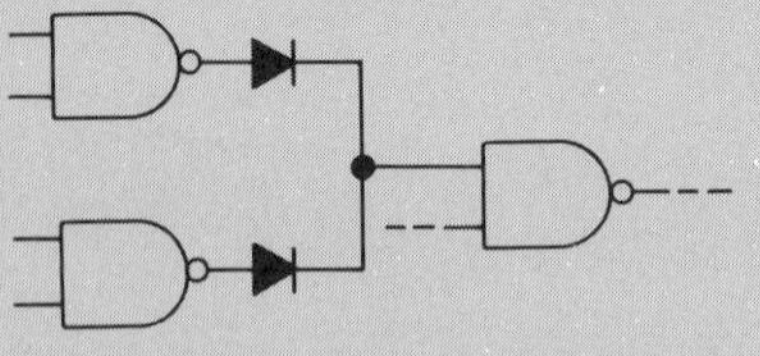

E. wired-OR from active pull-up gates

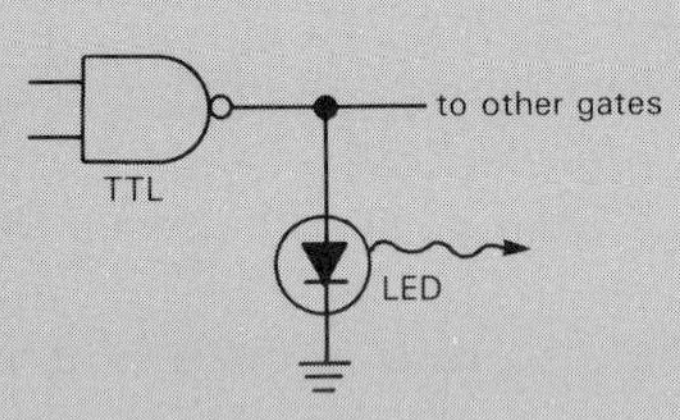

B. logic-state indicator

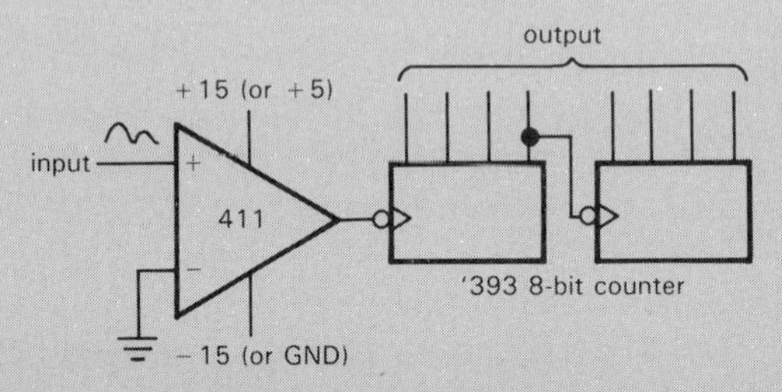

F. zero-crossing counter

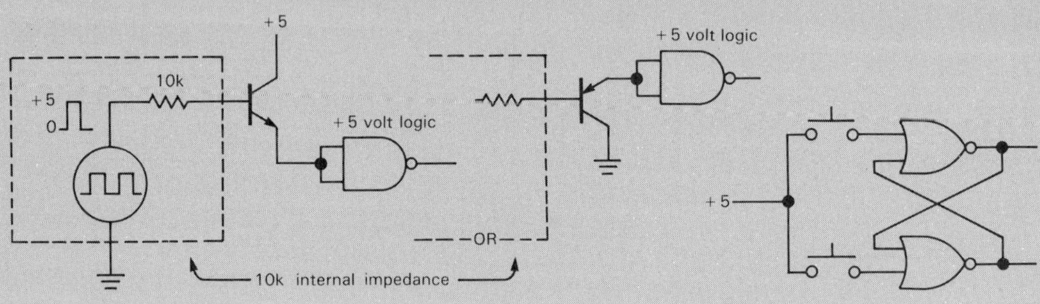

C. high-Z-to-logic interface (2 bad circuits)

G. SR flip-flop

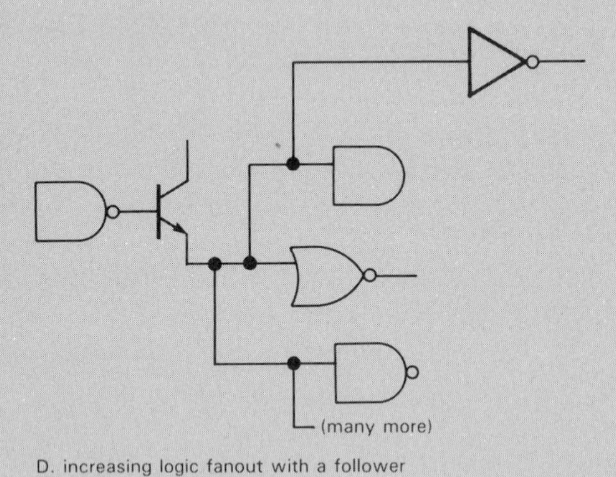

D. increasing logic fanout with a follower

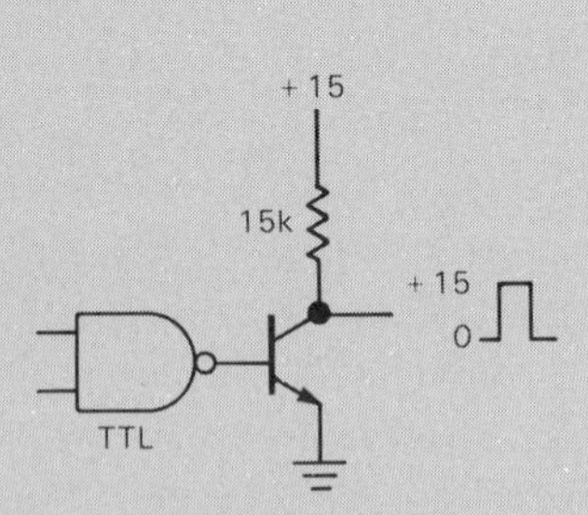

H. TTL-to-high-level interface

Figure 9.96

along with an address (2 bits). A CHARACTER VALID pulse signifies that the data and address are valid and can be latched, for example. An x'/y level tells whether the character being presented belongs to the x or y number. Figure 9.97 presents a summary.

The data are sent in the order x_n (LSD) ... x_n (MSD), y_n (LSD) ... y_n (MSD), so you know you've got one complete x, y pair after receiving the MSD of a y value ($A_1 = 0$, $A_2 = 1$, $x'/y = 1$). At *that* point you should update the digits seen by your D/A converters (don't update them one at a time).

In your circuit you needn't use particular device numbers; just label them generically, e.g., a type D flip-flop, or a 1-of-10 decoder. Be sure to indicate where inputs or outputs are inverted (by showing small circles). Assume that you have some D/A converters that accept 3-digit logic-level BCD inputs, and whose outputs are currents, zero to 1 mA corresponding to inputs of 000 to 999. Since the x, y plotter has 10 volt full-scale sensitivity, you'll have to convert the current to a voltage. As an additional obstacle, so that you can exercise your ingenuity, assume that the D/A converters have an output compliance of only 1 volt.

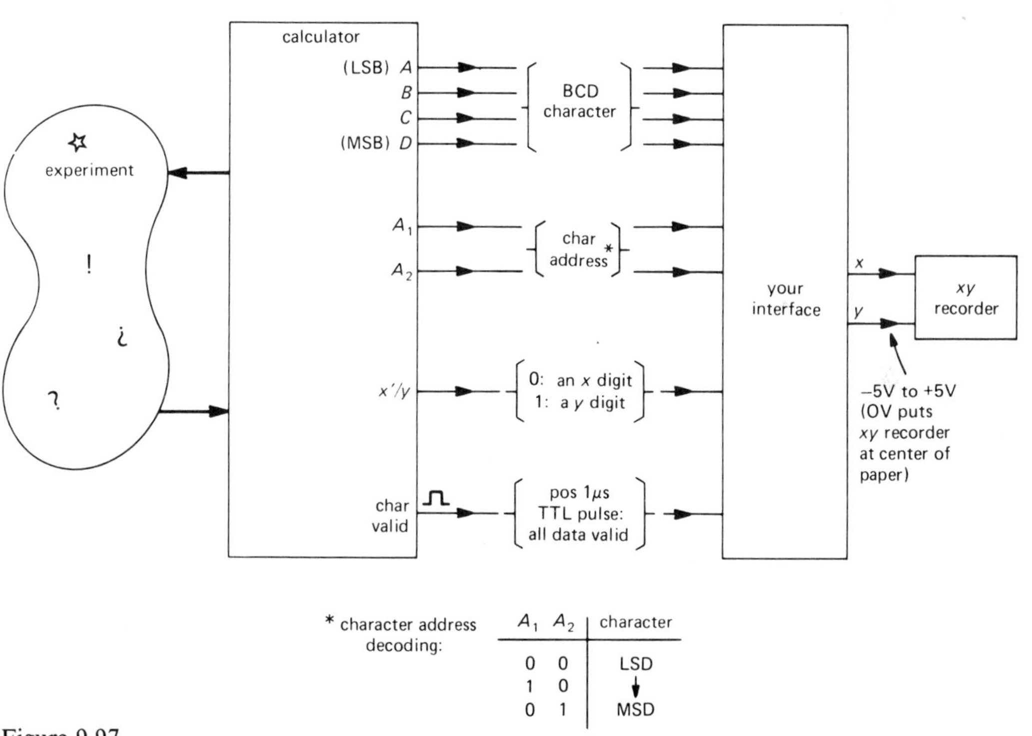

A_1	A_2	character
0	0	LSD
1	0	↓
0	1	MSD

Figure 9.97

MINICOMPUTERS, MICROCOMPUTERS, AND MICROPROCESSORS

The availability of inexpensive ($1k) small computers has made it attractive to control experiments and processes, collect data, and perform computation directly under the control of a computer. Small computers are commonly used in laboratory and industrial settings, and knowledge of their capabilities, program languages, and interfacing requirements is an essential part of electronics know-how.

The microcomputer evolved from the earlier *minicomputer*, a small machine whose central processing unit (CPU) was constructed from SSI and MSI ICs, usually occupying one or more large printed-circuit boards. As large-scale integration improved, it became possible to put minicomputer CPU performance into a single LSI chip; thus a *microcomputer* is a computer whose CPU is constructed from just a few (often only one) LSI microcircuits; the CPU chip (or chip set) constitutes a *microprocessor*. For example, DEC's popular PDP-11 minicomputers (CPU on several interconnected boards) were succeeded by a family of similarly named computers whose CPUs were built from a few LSI chips in place of many SSI/MSI chips; at about the same time, Motorola introduced a high-performance microprocessor (the 68000) that has many similarities to the PDP-11 and was obviously influenced by it.

Most modern small computers are in fact microcomputers, relying on the impressive performance of the present generation of microprocessors. The phrase "superminicomputer" has recently surfaced and seems to signify a class of machines that achieve higher performance, in some cases rivaling the large and expensive "mainframe" computers. In some cases the distinction refers more to physical size or number of peripherals than to the scale of integration used in the construction of the CPU.

A more important distinction separates microcomputers from *microcontrollers*, a term used to describe the use of a microprocessor, along with a small amount of memory and other support chips, for dedicated control of a process or instrument. In this role a microprocessor plus a few assorted chips and some ROM (read-only memory) can flexibly replace a complicated logic circuit of gates, flip-flops, and analog/digital conversion functions and should be considered whenever embarking on a large design project. There are microprocessors optimized for this kind of application, generally characterized by on-chip timers, ports, and other functions that usually require extra ICs, at the expense of the computational power and large address space that characterizes microprocessors intended for microcomputer-based computational tasks.

In this chapter we will describe microcomputer architecture, programming, and interfacing, with some examples of useful and simple interfacing of peripherals to the IBM PC/XT (here we refer to the original PC bus and its derivatives such as the PC/AT and compatibles, and the low end of the PS/2 line). Most of the ideas introduced in this chapter will carry over to the next chapter, where we will get into a detailed discussion of the selection and construction of microprocessor-based circuits and systems; for those examples we will use the 68008 microprocessor, a member of the Motorola 68000 family that, together with the Intel 8086 family, dominates small computers. Generally speaking, with microcomputers the design of the computer itself, including the integration of memory, disks, and I/O control, as well as system programming and utility program development, is taken care of by the manufacturer (and suppliers of compatible hardware and software). The user need only worry about special-purpose interfaces and the job of user programming. By contrast, in a dedicated microprocessor system, the choices of memory types, system interconnection, and programming generally have to be made by the designer. Microcomputer manufacturers are generally committed to providing system and utility software as part of a complete computing system (often including peripherals), whereas the microprocessor manufacturers (semiconductor companies) generally see the design and marketing of microprocessor and support chips as their central tasks. In this chapter, then, we will describe computer architecture and programming and will concentrate on the details of internal communication and interfacing.

10.01 Computer architecture

Figure 10.1 summarizes the organization typical of most computers. Let's take it from left to right:

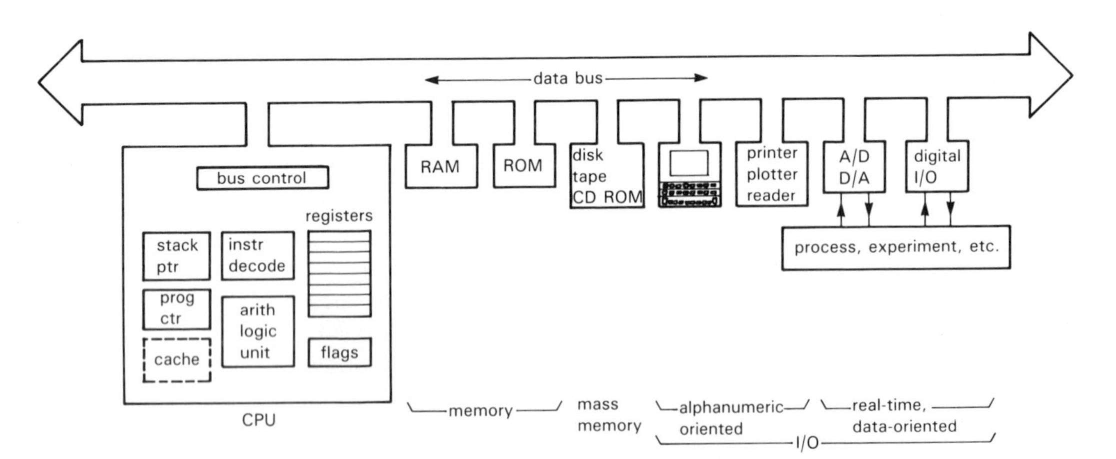

Figure 10.1. Block diagram of a computer.

CPU

The central processing unit, or CPU, is the heart of the machine. Computers do their computation in the CPU on chunks of data organized as computer *words*. Word size can range from 4 bits to 32 bits or more, with a 16-bit word size being the most popular in current microcomputers. A *byte* is 8 bits (half a byte, or 4 bits, is sometimes called a "nybble"). A portion of the CPU called the *instruction decoder* interprets the successive instructions (fetched from memory), figuring out what should be done in each case. The CPU has an *arithmetic unit*, which can perform the instructed operations, such as add, complement, compare, shift, move, etc., on quantities contained in *registers* (and sometimes in *memory*). The *program counter* keeps track of the current location in the executing program. It normally increments after each instruction, but it can take on a new value after a "jump" or "branch" instruction. The *bus control circuitry* handles communication with memory and I/O. Most computers also have a *stack pointer register* (more on that later) and a few *flags* (carry, zero, sign) that get tested for conditional branching. Many high-performance processors also include *cache* memory, which holds values recently fetched from memory for quicker access.

There has been a lot of activity in the experimental field of "parallel processing," in which you interconnect many CPUs to get tremendous computational power. With time this trend may become dominant in high-performance processing. For the time being, however, our single-CPU machine, executing instructions serially, represents the standard microcomputer architecture.

Memory

All computers have some fast random-access memory, called RAM (it used to be called "core," because tiny magnetic cores held the data, one bit per core). In a large microcomputer this may include 10 megabytes or more, although a megabyte is more typical, and as little as 16K may be used in a microcontroller. (When used to describe memory sizes, K doesn't mean 1000, but rather 1024, or 2^{10}; thus, 16K bytes is actually 16,384 bytes. We employ the lower-case symbol k to mean 1000.) This memory can typically be read and written in about 100ns. RAM is almost always *volatile*, which means that its information evaporates when power is removed (maybe it should be called "forgettory"!). All computers therefore include some nonvolatile memory, usually ROM (read-only memory), to "bootstrap" the computer, i.e., get it started from a state of total amnesia when power is first turned on. Additional ROM is often programmed with system routines, graphics routines, and other programs that you want to be there all the time.

To get or store information in memory, the CPU "addresses" the desired word. Most computers address memory by bytes, beginning at byte 0 and going sequentially through to the last byte in memory. Since most computer words are several bytes long, you are usually storing or fetching a group of bytes at a time; this is usually expedited by having a data bus that is several bytes wide. For example, microcomputers that use the 80386 or 68020 use a bus 32 bits (4 bytes) wide, so that a 32-bit word can be moved to/from memory in one memory fetch. (There are control signals to specify how many contiguous bytes are being moved, since even with a large bus you may want only 1 or 2 bytes.)

In a computer with lots of memory, it takes three or four bytes to specify an arbitrary memory address anywhere in the machine. Since most memory references in an actual program are usually "nearby," all computers provide for simplified addressing modes: "Relative" addressing specifies an address by its distance from the present instruction; "indirect" addressing uses the contents of a CPU register to point to a location in memory; "paged" addressing uses a shortened address to refer to a memory location within a small area (a page); "direct" or "absolute" addressing uses the next few bytes in memory to specify an address. A modern CPU embellishes this short list with additional "indexed," "autoincrementing," and other useful addressing modes, which we'll learn about in the next chapter.

Both programs and data are kept in memory during program execution. The CPU fetches instructions from memory, figures out what they mean, and does the appropriate things, often involving data stored somewhere else in memory. General-purpose computers usually store programs and data in the same memory, and in fact the computer doesn't even know one from the other. Amusing things start to happen if a program goes awry and you "execute" data!

Since computer programs spend most of their time looping through a relatively short sequence of instructions, you can enhance performance by providing a small, but fast, *cache* memory, in which you routinely store copies of the most recently used memory locations. A cached CPU checks its local cache first, before fetching from (slower) main memory; when looping through familiar territory, you often achieve a cache "hit" rate of 95% or better, dramatically improving execution speed.

Mass memory

Computers intended for program development or computation, as opposed to dedicated control processors, usually have one or more mass-storage devices. "Hard" disks (also called "Winchester") and "floppy" disks ("diskettes") are the usual ones, with storage capacities going from a few hundred kilobytes to a few megabytes (floppy disks), and from a few tens of megabytes to a few hundred megabytes (hard disks). Most well-endowed computers also have a tape drive or two, ranging from a simple cartridge-tape "streaming" drive to a full-fledged 9-track, half-inch, large-reel tape (the kind that are always spinning in the background in science fiction movies). A newer technology uses 8mm videotape (the kind that lives in those little hand-held video cameras) to store a gigabyte on a small tape cartridge. And the latest in mass storage is the "CD ROM," which uses the same optical disk technology as audio CDs (compact discs); they store 600 megabytes on one side of a 5 inch plastic disk, with much faster access than any tape medium. Unlike audio CDs, there are CD ROM drives that let you write as well as read, by laser-burning pits in a blank CD; they're called "WORM," for "write once, read many." Furthermore, fully erasable read/write magneto-optic disk memories are also available.

Compared with RAM, mass-storage media are generally slow, magnetic tape being the slowest, with access times of many seconds, and hard disks being the fastest (and most expensive), with average access times of tens of milliseconds. With all mass-storage devices, data transfer is rapid (10K to 100K bytes per second or more) once the data has been located. You generally keep programs, data files, plot files, etc., on some sort of mass-storage device and bring these into RAM only when doing computation. Many users can simultaneously fit their programs on one disk; a moderate-size optical disk can hold the contents of the *Encyclopaedia Britannica* several times over.

If your computer has lots of RAM, a nice way to speed up computer operations that make heavy use of disk is to form a "RAM disk" by loading all the relevant disk files into RAM when you start. Thus you might put a text editor, compiler, and linker/loader into RAM; then you can switch back and forth without waiting for the disk. Be careful, though; because none of your work is being saved on nonvolatile disk, you lose all your work if the computer crashes.

Alphanumeric and graphic I/O

It is nice to have a powerful computer, capable of millions of smart computations per second, but it doesn't do you any good if it keeps all its results to itself. Peripherals such as a keyboard and screen (the combination is a "terminal"), "mouse," printer, etc., let man and machine communicate, and these are essential in any "friendly" computer system. These peripherals are mostly oriented toward programming, word processing, spreadsheets, and graphics; you use them when writing programs, debugging, listing, writing and printing documents, manipulating quantities and objects, and playing flight simulator. These sorts of peripherals, together with suitable interfaces, are available from many sources, including the microcomputer manufacturer.

Real-time I/O

For experiment or process control and data logging, or for exotic applications such as speech or music synthesis, you need A/D and D/A devices that can communicate with the computer in "real time," i.e., while things are happening. The possibilities are almost endless here, although a general-purpose set of multiplexed A/D converters, a few fast D/As, and some digital "ports" (serial or parallel) for exchange of digital data will permit many interesting applications. Such general-purpose peripherals are commercially available for most popular computer buses. If you want something fancier, such as improved performance (higher speed, more channels) or special-purpose functions (tone generation, frequency synthesis, time-interval generation, etc.), you may have to build it yourself. This is where a knowledge of bus interfacing and programming techniques is essential, though it's helpful in any case.

Network interface

Powerful desktop computers become even more powerful when they can exchange files with other computers. One way to do this is to "log on" to a remote computer via telephone lines, then use the features of the remote computer that you need. That might include access to a large data base or special programs, a powerful supercomputer, computer "mail," or a colleague's text or data file. For these purposes you need a "modem" (modulator/demodulator), which either plugs directly into your computer's bus or hooks onto a serial data port. We'll have more to say about this later.

Another way to extend the scope of your machine is to use a local area network (LAN) to link a group of computers together. An example is Ethernet, which provides communication at rates up to 10Mb/s among linked machines, via a single coaxial cable. A LAN lets you access files on anyone's machine; in fact, with a good LAN you would probably pool your resources, sharing a fast large disk, high-priced plotters and printers, etc. Each "workstation" would then have only limited mass storage, but enough computational and display capability for the work you want to do with it. Such a setup is ideal for a publishing house or newspaper, for example, where different people work on manuscripts as they are readied for publication. You can get Ethernet (and other LAN) interfaces for most microcomputers.

Data bus

For communication within the computer between the CPU and memory or peripherals, all computers use a *bus*, a set of shared lines for exchange of digital words. (Many buses also allow communication *between* peripherals, though this capability is used less often.) The use of a shared bus vastly simplifies interconnections, since otherwise you would need multiwire cables connecting every pair of communicating devices. With a little care in bus design and implementation, everything works fine.

The bus contains a set of DATA lines (generally the same number as bits in a word – 8 for microcontrollers and low-performance PCs, 16 or 32 for more sophisticated microcomputers), some ADDRESS lines for determining who should "talk" or "listen" on the line, and a bunch of CONTROL lines that specify what action is going on [data going to or from the CPU, interrupt handling, DMA (direct memory access) transfers, etc.]. All the DATA lines, as well as a number of others, are *bidirectional* – they're driven by three-state devices, or in some cases by open-collector gates with resistor pullups somewhere (usually at the end of the bus, where they also serve as terminators to minimize reflections, see Section 13.09); pullups may be necessary with three-state drivers also, if the bus is physically long.

Three-state or open-collector devices are used so that devices connected to the bus can disable their bus drivers, since in normal operation only one device is asserting data onto the bus at any time. Each computer has a well-defined protocol for determining who asserts data, and when. If it didn't, total chaos would result, with everyone shouting at once (so to speak). (Computer people can't resist personalizing their machines, peripherals, etc. Engineers are even worse, with flip-flops and even gates coming to life. Naturally, we follow the trend.)

There is one interesting distinction in computer buses. They can be either *synchronous* or *asynchronous*, with examples of each in currently popular microcomputers. You will see what this means when we get into the details of communication via the bus.

We'll return to the bus in detail, with interface examples, using the example of the popular IBM PC/XT family. First, though, we need to look at the CPU's instruction set.

A COMPUTER INSTRUCTION SET

10.02 Assembly language and machine language

In order to understand bus signals and computer interfacing, you've got to understand what the CPU does when it executes various instructions. At this point, therefore, we would like to introduce the instruction set that goes with the IBM PC/XT family. Unfortunately, the instruction sets of most real-world microprocessors tend to be rich with complexities and extra features, and the Intel 8086 series is no exception. However, since our purpose is only to illustrate bus signals and interfacing (not fancy programming), we'll take a shortcut by laying out a subset of 8086 instructions. By leaving out the "extra" instructions we'll wind up with a compact set of instructions that is both understandable and complete enough to do any programming task. We'll then use it to show some examples of interfacing and programming. These examples will help convey the idea of programming at the "machine-language" level, something quite different from programming in a high-level language like FORTRAN or C.

First, a word on "machine language" and "assembly language." As we mentioned earlier, the computer's CPU is designed to interpret certain words as instructions and carry out the appointed tasks. This "machine language" consists of a set of binary instructions, each of which may occupy one or more bytes. Incrementing (increasing by one) the contents of a CPU register would be a single-byte instruction, for example, whereas loading a register with the contents of a memory location would usually require at least two bytes, perhaps as many as five (the first would specify the operation and register destination, and four more would be necessary to specify an arbitrary memory location in a large machine). It is a sad fact of life that different computers have different machine languages, and there is no standard whatsoever.

Programming directly in machine language is extremely tedious, since you wind up dealing with columns of binary numbers, each bit of which has to be bit-perfect, so to speak. For this reason you invariably use a program called an *assembler*; it allows you to write programs using easily remembered mnemonics for the instructions, and symbolic names of your own choosing for memory locations and variables. This *assembly-language* program, really nothing more than a number of cryptic-looking lines of letters and numbers, is massaged by a program called an *assembler* to produce as its output a finished program in machine-language *object code* that the computer can execute. Each line of assembly code gets turned into a few machine-language bytes (1 to 6 bytes, for the 8086). The computer cannot execute assembly-language instructions directly. To make these ideas concrete, let's look at our subset of the 8086/8 assembly language and do a few examples.

10.03 Simplified 8086/8 instruction set

The 8086 is a 16-bit processor with a rich, and somewhat idiosyncratic, instruction set; part of its complexity stems from the designers' objective to maintain compatibility with the earlier 8080 8-bit processor. Newer CPUs, such as the 80286 and 80386, can still execute the full 8086 instruction set. We've gone through the instructions with a machete, keeping 10 arithmetic operations and 11 others. Here they are:

Instruction	What you call it	What it does
arithmetic		
MOV *b,a*	move	*a*→*b*; *a* unchanged
ADD *b,a*	add	*a*+*b*→*b*; *a* unchanged
SUB *b,a*	subtract	*b*–*a*→*b*; *a* unchanged
AND *b,a*	and	*a* AND *b*→*b* bitwise; *a* unchanged
OR *b,a*	or	*a* OR *b*→*b* bitwise; *a* unchanged
CMP *b,a*	compare	set flags as if *b*–*a*; *a,b* unchanged
INC *rm*	increment	*rm*+1→*rm*
DEC *rm*	decrement	*rm*–1→*rm*
NOT *rm*	not	1's complement of *rm*→*rm*
NEG *rm*	negate	negative (2's comp) of *rm*→*rm*
stack		
PUSH *rm*	push	push *rm* onto stack (2 bytes)
POP *rm*	pop	pop 2 bytes from stack to *rm*
control		
JMP *label*	jump	jump to instr *label*
J*cc* *label*	jump conditional	jump to instr *label* if *cc* true
CALL *label*	call	push next adr, jump to instr *label*
RET	return	pop stack, jump to that adr
IRET	return from int	pop stack, restore flags, return
STI	set interrupt	enable interrupts
CLI	clear interrupt	disable interrupts
input/output		
IN AX(AL),*port*	input	*port*→AX (or AL)
OUT *port*,AX(AL)	output	AX (or AL)→*port*

notes

b,a: any of *m,r* *r,m* *r,r* *m,imm* *r,imm*
rm: *r* or *m*, via various addressing modes
cc: any of Z NZ G GE LE L C NC
label: via various addressing modes
port: byte (via *imm*) or word (via DX)

A quick tour

Some explanations: The first six arithmetic instructions operate on pairs of numbers ("2-operand" instructions), which we've abbreviated as *b,a*, and which can be any of the 5 pairs listed in the notes; *m* means the contents of a memory location, *r* means the contents of a CPU register (there are 8), and *imm* means an *immediate* argument, which is a number stored in the next 1 to 4 bytes of memory following the

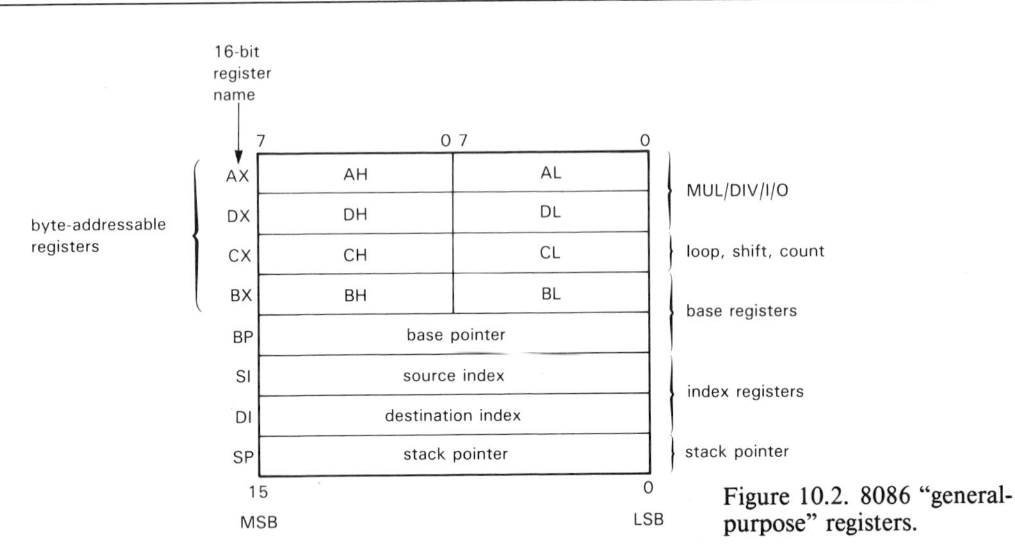

Figure 10.2. 8086 "general-purpose" registers.

instruction. Thus, for example, the instructions

```
MOV   count,CX
ADD   small,02H
AND   AX,007FH
```

have arguments of the form *m,r*, *m,imm*, and *r,imm*, respectively. The first copies the contents of register CX to a memory location that we've named "count"; the second adds 2 to the contents of another memory location called "small"; the third clears the top 9 bits of 16-bit register AX, while preserving the bottom 7 bits unchanged (a so-called masking operation). Note Intel's argument convention: The first argument is replaced or modified by the second argument. (In the next chapter we'll learn that Motorola decided to do it the other way around!)

The last four arithmetic operations take only a single operand, which can be either the contents of a register or memory. Here are two examples:

```
INC   count
NEG   AL
```

The first adds 1 to the contents of memory location "count," while the second changes the sign of register AL.

A detour: addressing

Before continuing, a word on registers and memory addressing. The 8086 claims to have 8 "general-purpose" registers, but after reading the fine print you'll realize that most of them have special uses (Fig. 10.2). Four of them (A–D) can be used either as single 16-bit registers (e.g., AX; think of "X" as "extended") or as a pair of byte registers (AH, AL; "high" and "low" halves). The BX and BP registers can hold addresses, as can the SI and DI registers, and tend to be used for addressing (see below). Special looping instructions (which we omitted from our short list) use register C, while multiply/divide and I/O instructions make analogous use of registers A and D.

Data used in instructions can be an immediate constant, a value held in a register, or a value in memory. You specify immediates by value, and registers by name, as in the examples above. To address memory, the 8086 provides six addressing modes, three of which are described by the diagrams in Figure 10.3. You can just name the variable *directly*, in which case its address gets assembled as a pair of bytes immediately following the instruction; you can put the variable's address in an addressing register (BX, BP, SI, or DI), then use an instruction that specifies addressing *indirectly* through the register; or you can combine the above, adding an immediate *displacement* to the value in a designated addressing register to get the variable's address. The indirect

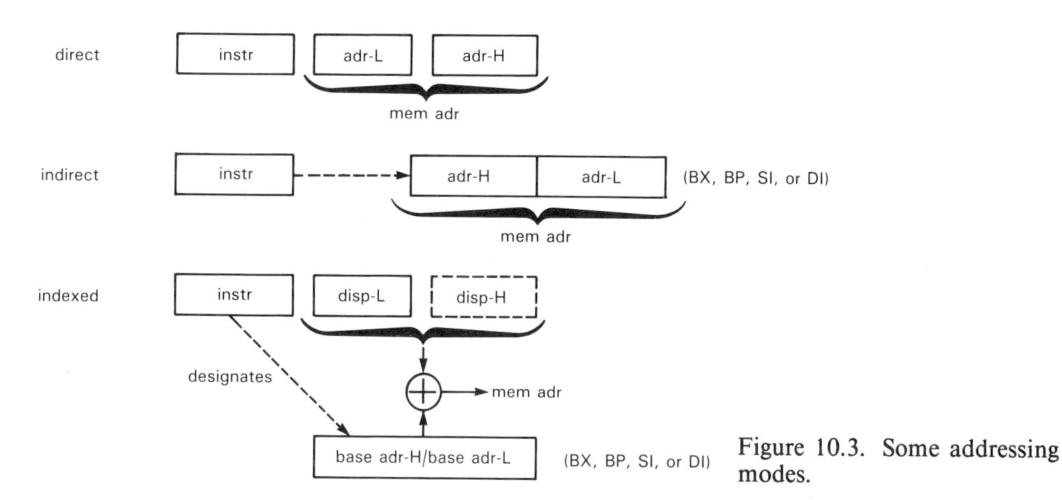

Figure 10.3. Some addressing modes.

mode is faster (assuming the address has already been loaded into an addressing register) and much better if you want to do something to a whole set of numbers (a *string* or *array*). Here are a few addressing examples:

```
MOV   count,100H        (direct,immediate)
MOV   [BX],100H         (indirect,immediate)
MOV   [BX+1000H],AX     (indexed,register)
```

The last two assume you've already put an address into BX. The last instruction copies the contents of AX to a memory location 4K (1000 hex) higher than BX points in memory; we'll give an example shortly showing how you could use this to copy an array.

There's one other complexity of 8086 memory addressing that we've swept under the rug: The "address" generated by any of the above addressing modes is not actually the final address, as should be obvious from the fact that it has only 16 bits (which can address only 64K bytes of memory). In fact, it's called an *offset*; to get an actual address, you add to the offset a 20-bit *base* formed by shifting left 4 bits the contents of a 16-bit *segment register* (there are four such registers). In other words, the 8086 lets you access groups of 64K bytes of memory at a time, with the location of those "segments" within a total memory size of 1Mbyte set by the contents of the segment registers. The use of 16-bit addressing in the 8086 was basically a big mistake, inherited from earlier generations of microprocessors. Newer processors (80386 onward, and the 68000 series) are done right, with 32-bit addressing throughout. Rather than complicate our examples, we'll simply ignore segments entirely; in real life you would, of course, have to worry about them.

Instruction set tour (continued)

The *stack* instructions PUSH and POP come next. A stack is a portion of memory, organized in a special way: When you put data onto the stack (a *push*), it goes into the next available spot ("top" of the stack); and when you retrieve data (a *pop*), it is taken from the top, i.e., it is the item last pushed onto the stack. Thus a stack is a consecutive list of data, stored last-in, first-out (LIFO). It may help to think of a bus driver's coin dispenser (or a lunchroom tray dispenser).

Figure 10.4 shows how it works. The stack lives in ordinary RAM, with the CPU's *stack pointer* (SP) keeping track of the location of the current "top" of the stack. The 8086 stack holds 16-bit words and grows *down* in memory as you push data onto it. The SP is automatically decremented by 2 before each PUSH, and incremented by 2 after each POP. Thus, in

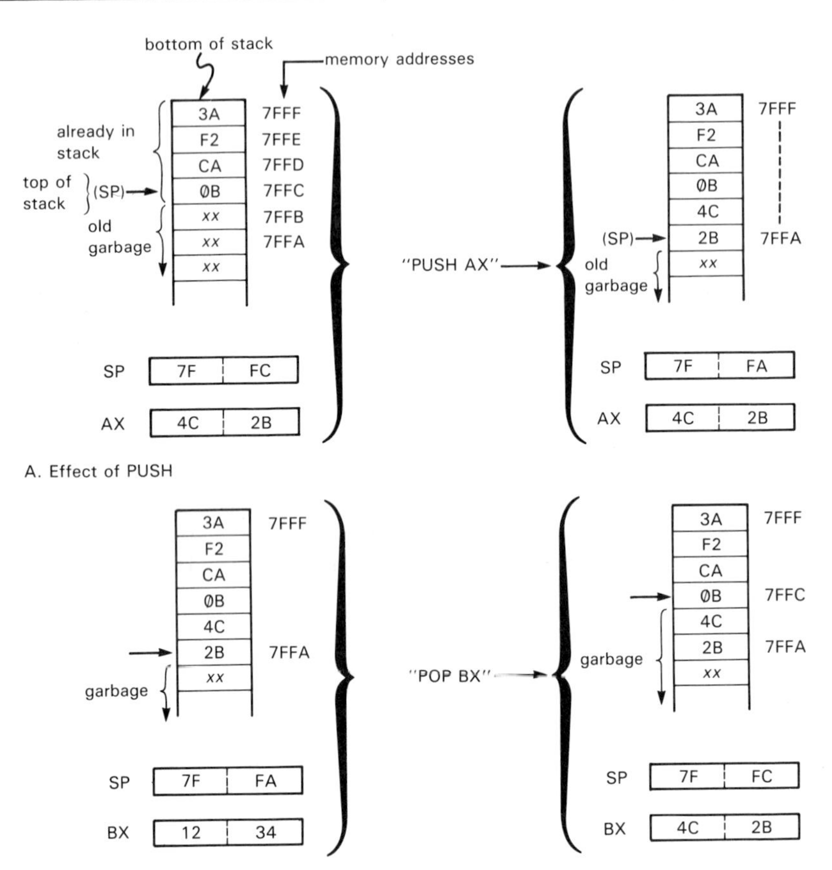

Figure 10.4. Stack operation.

the example, the 16-bit data in register AX is copied onto the top of the stack by the instruction PUSH AX; the SP is left pointing at the last byte pushed. POP reverses the process, as shown. As we will see, the stack plays a central role in subroutine calls and interrupts.

JMP causes the CPU to depart from its usual habit of executing instructions in sequential order, detouring instead to the instruction that you jump to. Conditional jumps (there are eight possibilities, indicated generically as J*cc*) test the flag register (which lives in the CPU, and whose bits are set according to the result of the most recent arithmetic operation), then either jump (if the condition is true) or execute the next instruction in sequence (if the condition is not true). Program 10.1 shows an example. It copies 100 words from the array beginning at 1000 hex to a new array beginning 1K bytes (400H) higher.

Program 10.1

```
        MOV  BX,1000H           ;put array address in BX
        MOV  CL,100             ;initialize loop counter
LOOP:   MOV  AX,[BX]            ;copy array element to AX
        MOV  [BX+400H],AX       ;then to new array
        ADD  BX,2               ;increment array pointer
        DEC  CL                 ;decrement counter
        JNZ  LOOP               ;loop if count not zero
NEXT:   (next statement)        ;exit here when done
```

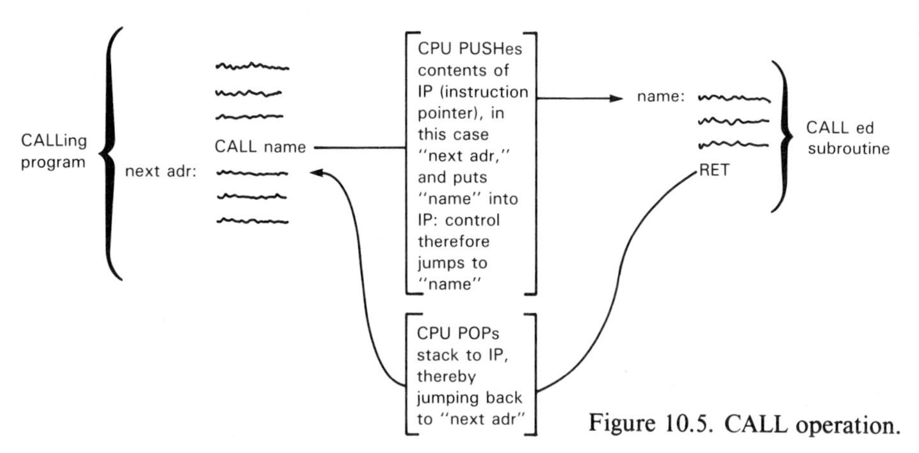

Figure 10.5. CALL operation.

Note the explicit loading of the pointer (to address register BX) and the loop count (to CL). The actual array of words had to move through a register (we chose AX) because the 8086 does not permit memory-to-memory operations (see the instruction set notes). At the end of the 100th pass through the loop, CZ = 0, and the jump nonzero (JNZ) instruction no longer jumps. This example will work, but in practice you would probably use one of the 8086's faster *string move* instructions. Also, it's good programming practice to use symbolic names for sizes and arrays, rather than constants like 400H and 1000H.

The CALL statement is a subroutine call; it's like a jump, except that the return address (the address of the instruction that would have come next, except for the intervening CALL) is pushed onto the stack. At the end of the subroutine you execute a RET statement, which pops the stack so the program can find its way home (Fig. 10.5). The three statements STI, CLI, and IRET have to do with interrupts, which we'll illustrate with a circuit example later in the chapter. Finally, the I/O instructions IN and OUT move a word or byte between the A register and the addressed port; more on this shortly.

10.04 A programming example

As the example above suggests, assembly language tends to verbosity, with a lot of little steps needed to do a basically simple thing. Here's another example: Suppose you want to increment a number, N, if it equals another number, M. This will typically be a tiny step in a larger program, and in higher level languages it will be a single instruction:

```
if (n==m) ++n;          (C)
IF (N.EQ.M) N=N+1       (Fortran)
if n=m then n:=n+1;     (Pascal), etc.
```

In 8086 assembler, it looks like Program 10.2. The assembler program will

Program 10.2

```
n   DW 0                        ;n (a "word") lives here, and
m   DW 0                        ;m lives here, both initialized to 0

        MOV  AX,n               ;get n
        CMP  AX,m               ;compare
        JNZ  NEXT               ;unequal, do nothing
        INC  n                  ;equal, increment n
NEXT:   (next statement)
          o
          o
          o
```

convert this set of mnemonics to machine language, generally translating each line of assembler *source code* to several machine-language bytes, and the resultant machine-language code will get loaded into successive locations in memory before being executed. Note that it is necessary to tell the assembler to assign some storage space for variables. This you do with the assembler *pseudo-op* "`DW`" (define word) (*pseudo*-op because it doesn't produce executable code). Unique symbolic labels (e.g., `NEXT`) can be used to tag instructions; this is usually done only if there is a jump to that location (`JNZ NEXT`). Giving some locations understandable (to you!) names and adding comments (separated by a semicolon) make the job of programming easier; it also means that you have a chance of understanding what you've written a few weeks later. Programming in assembly language can still be a nuisance, but it is often necessary to write short routines in it, callable from a higher-level language, to handle I/O. Assembly-language programs run faster than programs compiled from a higher language, so it is often used where speed is crucial (e.g., the innermost loop of a long numerical calculation). To some extent the development of the powerful C programming language has minimized the occasions when you must use assembly code. In any case, you can't really understand computer interfacing without understanding the nature of assembly-language I/O. The correspondence between mnemonic assembly language and executable machine language is explored further in Section 11.03, in that case illustrated by 68000 microprocessor programming.

BUS SIGNALS AND INTERFACING

A typical microcomputer data bus has about 50–100 signal lines, devoted to the transfer of data, addresses, and control signals. The IBM PC/XT is typical of a small machine, with 53 signal lines and 8 power/ground lines. Rather than throw them all at you at once, we will approach the subject by building up the bus, beginning with the signal lines necessary for the simplest kind of data interchange (programmed I/O) and adding additional signal lines as they become necessary. We will give some useful interface examples as we go along, to keep things comprehensible and interesting.

10.05 Fundamental bus signals: data, address, strobe

To move data on a shared (party-line) bus, you have to be able to specify the data, the recipient, and the moment when data is valid. Thus, a minimum bus must have DATA lines (for the data to be transferred), ADDRESS lines (to identify the I/O device or memory address), and some STROBE lines (which tell when data is being transferred). There are usually as many DATA lines as bits in the computer word, so a whole word can be transferred at once. In the PC, however, there are only 8 DATA lines (D0–D7); you can move a byte in one transfer, but to move a 16-bit word you have to do two transfers. The number of ADDRESS lines determines the number of addressable devices: If the bus is used for both I/O and memory (the usual situation) there will be 16 to 32 ADDRESS lines (corresponding to a 64Kbyte to 4Gbyte address space); a bus used for I/O only might have 8 to 16 ADDRESS bits (256 to 64K I/O devices). [The IBM PC talks to both memory and I/O on its bus, and has 20 ADDRESS lines (A0–A19), corresponding to a 1Mbyte address space.] Finally, data transfer itself is synchronized by pulses on additional "strobing" bus lines. There are two ways in which this can be done: by having separate READ and WRITE lines, with a pulse on one or the other synchronizing data transfer; or by having one STROBE line and one READ/WRITE′ line, with a pulse on STROBE synchronizing data transfer in a direction specified by the level on the READ/WRITE′ line. The IBM PC uses the first scheme, with (active-LOW) read/write lines called IOR′, IOW′, MEMR′, and MEMW′; there are four because the PC distinguishes between memory and I/O, with individual pairs of read/write strobes for each.

These bus signals – DATA, ADDRESS, and the four strobes – would normally be all you need to do the simplest kind of data transfers. However, on the PC bus you need one more, called ADDRESS ENABLE (AEN), to distinguish normal I/O transfers from what's called "direct memory access" (DMA). We'll get to DMA in Section 10.12; for now, all you need to know is that AEN is LOW for normal I/O, and HIGH for DMA. We now have 33 bus signals: D0–D7, A0–A19, IOR′, IOW′, MEMR′, MEMW′, and AEN. Let's see how they work.

10.06 Programmed I/O: data out

The simplest method of data exchange on a computer bus is known as "programmed I/O," meaning that data is transferred via an `IN` or `OUT` statement in the program (the directions for `IN` and `OUT` are among the few things on which all computer manufacturers agree: `IN` always means *toward* the CPU, and `OUT` always means *from* the CPU). The whole process of data `OUT` (and memory write) is extremely simple and logical (Fig. 10.6). The ADDRESS of the recipient and the DATA to be sent are put onto the respective bus lines by the CPU. A write strobe (IOW′ or MEMW′) is asserted (LOW) by the CPU to signal the recipient that data is good. On the PC's bus the address is guaranteed valid beginning about 100ns before IOW′, and the data are guaranteed valid at least 500ns before the end of IOW′ (and for another 185ns thereafter). To play the game, the peripheral (in this case, an XY "vector" scope display) looks at the ADDRESS and DATA lines. When it sees its own address, it latches the information on the DATA lines, using the trailing edge of the IOW′ pulse as a clocking signal. That's all there is to it.

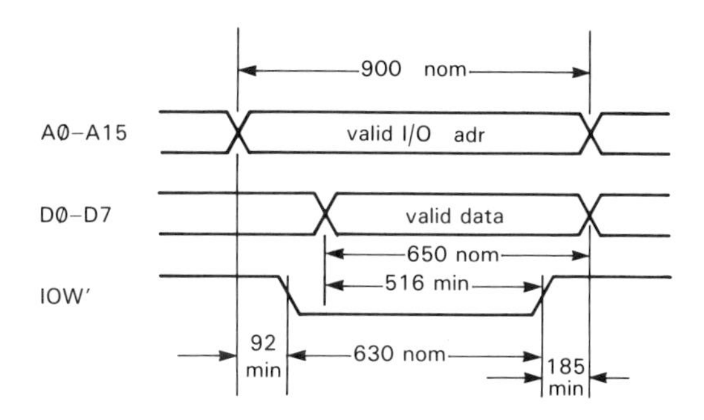

Figure 10.6. I/O WRITE cycle.

Let's look at the example shown in Figure 10.7. Here we have designed an XY scope display; you send it successive X, Y pairs of numbers, and it plots each point in turn on an XY display oscilloscope. First we have to pick an I/O address. Figure 10.8 shows the reserved and available I/O addresses on the IBM PC; we've chosen $3C0_H$ for the X register, and $3C1_H$ for the Y register. The '688 is an octal comparator with enable and LOW-true output on equality, giving a LOW output when the eight high-order bits A2–A9 match the fixed comparison bits, in this case when the address bus contains addresses 3C0–3C3 (you could use a bunch of gates, but an address comparator is more compact). We've also required AEN to be LOW, as explained earlier. The 3-input NANDs complete the address decoding, using A0 and A1, to give LOW outputs on individual addresses 3C0 and 3C1 (another method will be described shortly). Finally, these outputs are ANDed with IOW′ to get the clocks for the X and Y registers, which are '574 octal *D* flip-flops. These latch bytes from the data bus when (a) the correct address is present, (b) AEN is LOW, and (c) an IOW′ is sent. The 8-bit DACs convert the latched bytes to analog voltages, to drive the X and Y inputs

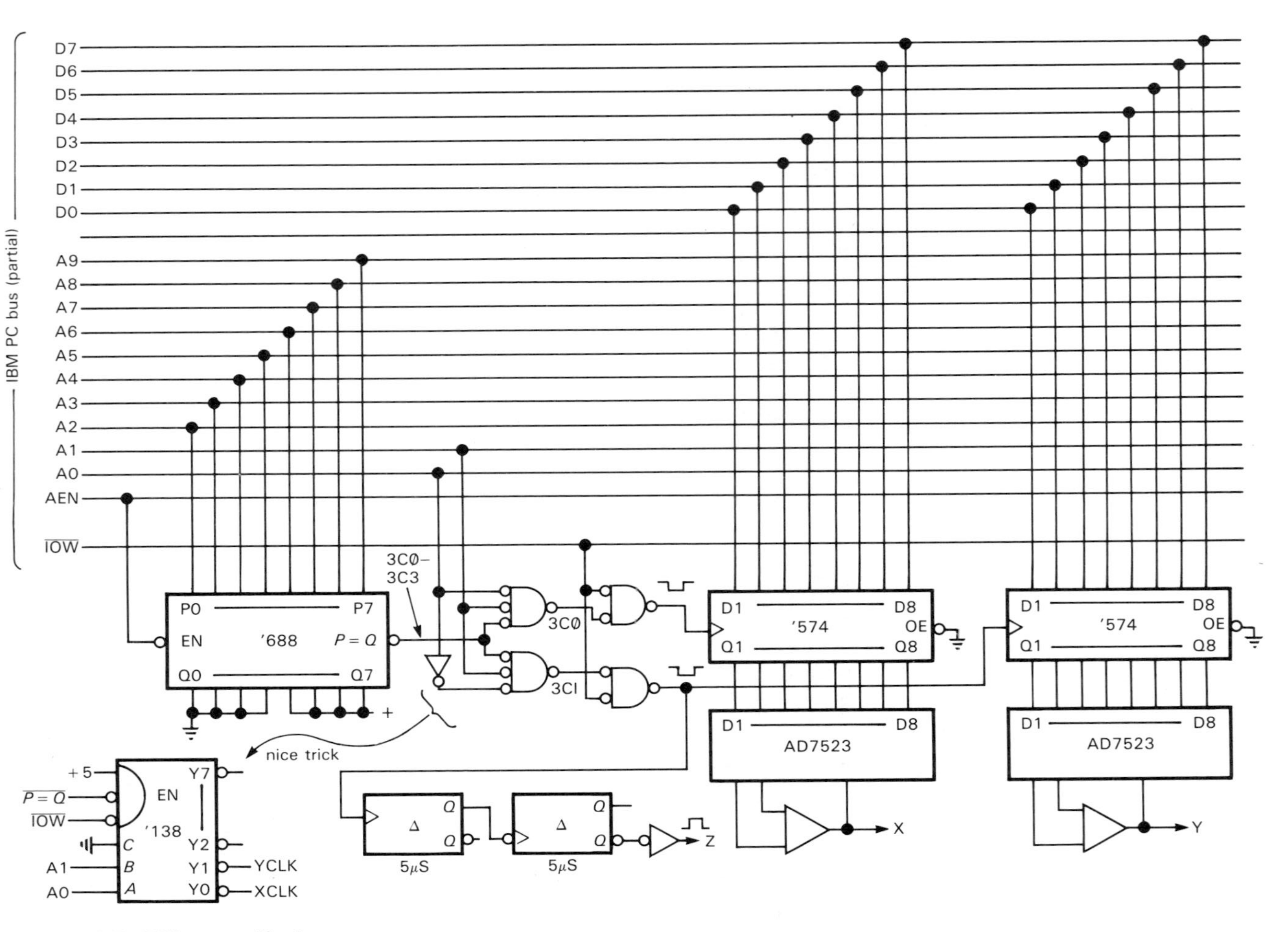

Figure 10.7. XY scope display.

of a display scope. A pair of monostables generates a 5μs "unblanking" pulse a few microseconds after the Y coordinate has been latched, to intensify the selected spot on the scope (all scopes have a "Z input" for that purpose). To draw a graph or set of characters on the screen, all you do is output successive XY coordinates repetitively (send X, then Y), fast enough so the eye doesn't see the flicker. Microcomputers are fast enough to display a few thousand XY pairs repetitively without annoying flicker. Given the fact that video (raster-scanned) displays for viewing are commonplace on microcomputers, this example might be more useful as an ultra-high-resolution plotter for photographic "hard copy," using 14-bit DACs and a micro-spot-size hard-copy scope display (see the next exercise).

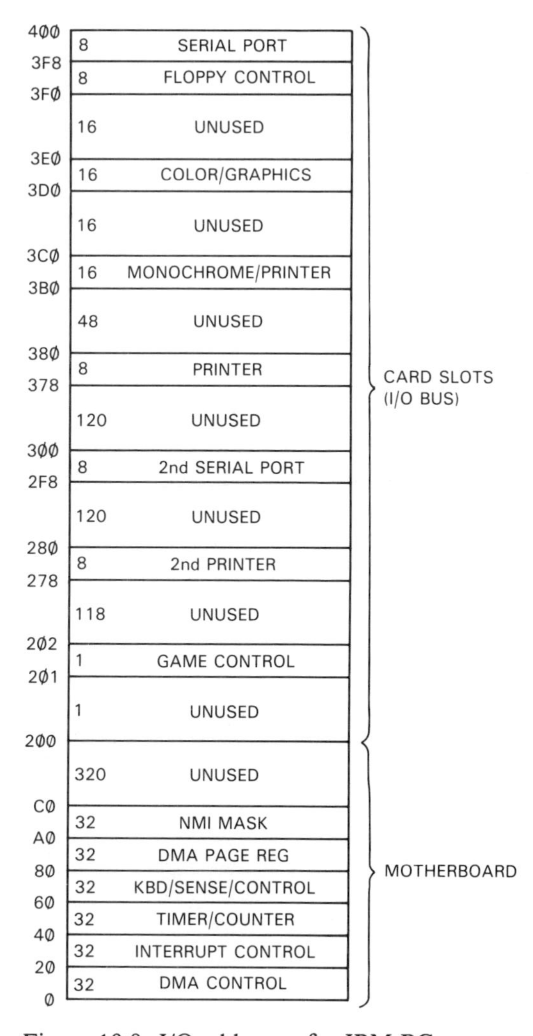

Figure 10.8. I/O addresses for IBM PC.

Some useful comments: (a) Note that we've arranged polarities so that the *trailing* edge of IOW′ clocks the *D* flip-flops; that is essential, since the data isn't yet valid on the leading edge of IOW′. If we were being very careful, we would check to see that required setup and hold times are satisfied for the '574s; in fact, for a slow bus like the PC's, you can't go wrong, since there are more than 500ns from valid data to trailing edge of IOW′. (b) You can save a few parts by using a strobed decoder in the address decoding circuitry, as indicated. Decoders like the '138 (3-line to 8-line) and '139 (dual 2-line to 4-line) include one or more enable inputs, and they are handy in this sort of application. (c) Note also that we could have combined the 3-input and 2-input NANDs into 4-input NANDs; we kept them separate only for clarity, decoding the addresses first, then AND-ing with the IOW′ strobe. (d) In fact, we could have ignored A1 entirely, and the circuit would work just the same! However, it would then respond also to addresses 3C2 and 3C3 (as X and Y, respectively), in effect "wasting" two I/O locations. In practice you often cheat in this way, incompletely decoding the address, because it saves parts (and there is plenty of I/O space, even if you waste some). In this example we could then have connected IOW′ where A1 is now connected, and omitted the 2-input NANDs entirely. (e) An interface like this is more flexible if its address can be set using a DIP switch (or DIP jumper block); then you can always make sure its address doesn't conflict with that of another interface you've got somewhere else. In this case the change is simple – replace the "hardwired" address lines to the comparator with eight lines that have switches to ground and pullups to +5 volts. (f) We used separate octal registers and DACs in this example for clarity. In real life you would probably choose a DAC with built-in latch (e.g., the "microprocessor-compatible" AD7528, a dual DAC with input latches); these even come in quad versions, (e.g., the AD7226) and in "double-buffered" versions with two cascaded latches for each DAC (e.g., the AD7225 quad).

EXERCISE 10.1

Redraw the address comparator logic with selectable I/O address.

EXERCISE 10.2

Redraw the XY display interface, using 16-bit DACs for both X and Y. You'll need four consecutive addresses: Assign the first two to the X register, and the last two to the Y register; use DIP-selectable I/O base address, of course. In each case the even address is the low-order byte, and the odd address is the high-order byte; that's the good choice, because that's how the 8086 stores 16-bit words, so you can use *word* I/O instructions to send data to your interface.

Programming the scope display

The programming to run this interface is straightforward. Program 10.3 shows what you do. The addresses of the first X and Y, and the number of points to be plotted, have to be available to the program. The display program will probably be a subroutine, with those parameters passed as arguments in the subroutine call. The program puts the addresses of the X and Y arrays (i.e., the address of the first X and Y) into address pointer registers SI and DI, and the byte count into CX. It then enters a loop in which successive XY pairs are sent to I/O ports 3C0 and 3C1. The X and Y pointers are advanced each time around, and the counter is decremented and tested for zero, which means the last point has been displayed; the pointers and counter are then reinitialized, and the process begins again.

A couple of important points: Once started, this program displays the XY array forever. In real life the program would probably check the keyboard to see if the operator wants the plot terminated. Alternatively, the display could be terminated after a specified time had elapsed, or by an "interrupt," which we will discuss shortly. With this sort of "refreshed" display, there usually isn't time to do much computing while displaying. A display device refreshed from its own memory takes that burden off the computer, and this is generally a better method. Nevertheless, if the objective is to make a precision plot for photographic hard copy, this program and interface (souped up as in Exercise 10.2) will do the job nicely.

Program 10.3

```
                                  ;routine to drive XY display
INIT:   MOV   SI,xpoint           ;initialize x pointer
        MOV   DI,ypoint           ;initialize y pointer
        MOV   CX,npoint           ;initialize counter

PLOT:   MOV   AL,[SI]             ;get x byte
        OUT   3C0H,AL             ;send it out
        MOV   AL,[DI]             ;get y byte
        OUT   3C1H,AL             ;send it out
        INC   SI                  ;advance x pointer
        INC   DI                  ;advance y pointer
        DEC   CX                  ;decrement counter
        JNZ   PLOT                ;not done, plot more stuff
        JMP   INIT                ;done, start over
```

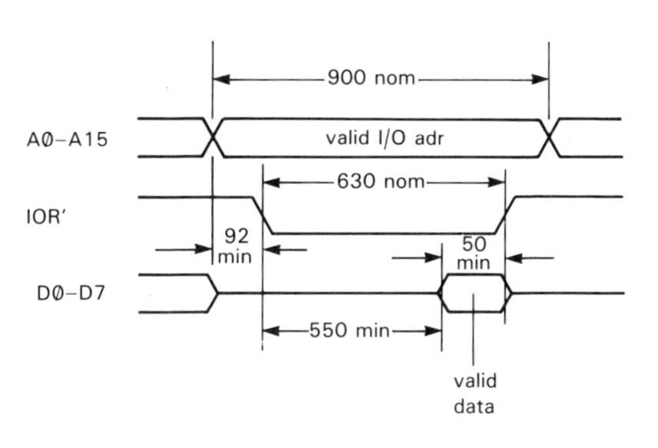

Figure 10.9. I/O READ cycle.

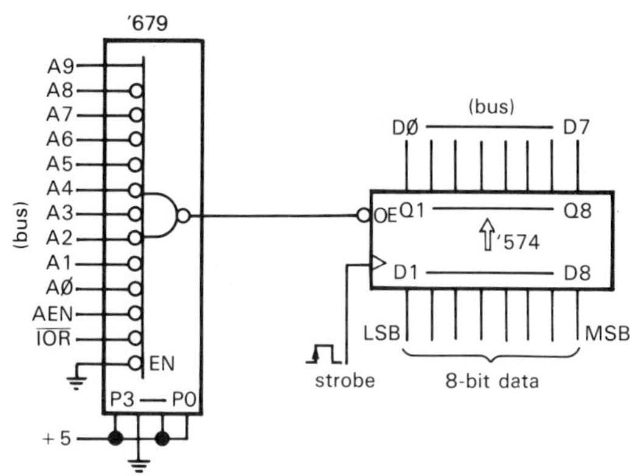

Figure 10.10. Parallel input port.

10.07 Programmed I/O: data in

The other direction of programmed I/O is equally simple. The interface looks at the ADDRESS lines as before. If it sees its own address (and AEN is LOW), it puts data onto the DATA lines coincident with the IOR′ pulse (Fig. 10.9). Figure 10.10 shows an example. This interface lets the PC read a byte latched in the '574 *D*-type register. Since the clock input and data inputs of the register are accessible to an external device, the register could hold just about any sort of digital information (the output of a digital instrument, A/D converter, etc.). For variety, we've eliminated all gates by using a '679 "12-bit address decoder" IC. It's a clever chip with 12 address inputs, an enable, and 4 "programming" inputs. If you want to decode a fixed address, it does the trick: It's functionally a 12-input NAND gate, for which a programmable number of the inputs can be inverted; the inverted inputs are always the lowest-numbered ones, and the number of them is the number you have asserted at the (4-bit) programming inputs.

In this case we've decided to plunder that lonely unused port at I/O address 200_H (Fig. 10.8). We need to recognize the state A9 = HIGH, A0–A8 = LOW. We might as well use the '679 to qualify the decoded address with AEN = LOW and IOR′ = LOW. So altogether we need a NAND with 11 inverted inputs and 1 noninverted input, which we get by hard-wiring the programming inputs with 11 in binary (1011). Then we connect the address lines and strobe as shown. When an

```
IN  AL,200H
```

instruction is executed, the CPU asserts 200_H on A0–A9, waits a while, then asserts IOR′ for 630ns. The CPU latches what it sees on the DATA bus (D0–D7) at the trailing edge of IOR′, then disasserts A0–A9. The peripheral's responsibility is to get the data onto D0–D7 at least 50ns before the end of IOR′; that's pretty relaxed timing, since it has known that data is being requested from it for at least 600ns. With typical HC or LS gate propagation times of 10ns, 600ns looks like forever.

Beginning with this example, we will omit the tangle of bus lines and simply call them out by name.

Bus signals: bidirectional versus one-way

From the two examples we've done so far you can see that some bus lines are *bidirectional*, for example the DATA lines: They are asserted by the CPU during write, but asserted by the peripheral during read. Both CPU and peripheral use three-state drivers for these lines. Others, like IOW′ and IOR′, are always driven by the CPU,

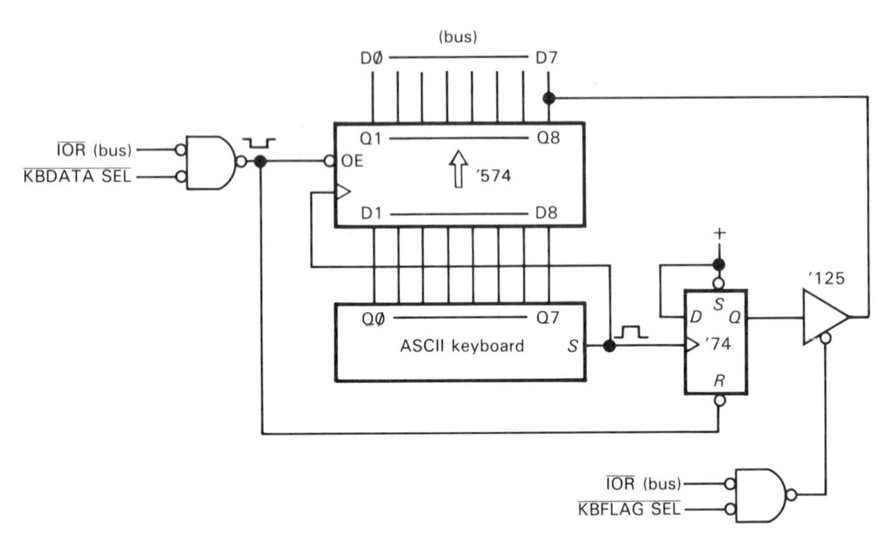

Figure 10.11. Keyboard interface with status bit.

with standard totem-pole driver chips. It is typical of computer buses to have both kinds of lines, using bidirectional lines for data that goes both ways, and one-way lines for signals that are always generated by the CPU (or, more accurately, generated by the associated bus control logic). There is always some clear protocol, like our rules for asserting/reading according to IOW′, IOR′, and ADDRESS, to prevent "bus contention" on these shared lines.

Of the signals so far, only the DATA lines are bidirectional; the ADDRESS lines, AEN, and strobes are one-way from the CPU. (Lest we give the wrong impression, we should point out that more complex computer systems permit other riders on the bus to become bus "masters"; obviously in such a system nearly all bus signals must be shared and bidirectional. The PC is unusually simple.)

10.08 Programmed I/O: status registers

In our last example, the computer can read a byte from the interface any time it wants to. That's nice, but how does it know when there's something worth reading? In some situations you may want the computer to read data at equally spaced intervals, as determined by its "real-time clock." Perhaps the computer instructs an A/D converter to begin conversions at regular intervals (via an `OUT` command), then reads the result a few microseconds later (via an `IN` command). That might suffice in a data-logging application. However, it is often the case that the external device has a mind of its own, and it would be nice if it could communicate what's happening to the computer without having to wait around.

A classic example is an alphanumeric input terminal, with someone banging away at a keyboard. You don't want characters to get lost; the computer has to get every character, and without much delay. With a fast storage device like disk or tape the situation is even more serious; data must be moved at rates up to 100,000 bytes per second without delay. There are actually three ways to handle this general problem: status registers, interrupts, and direct memory access. Let's begin with the simplest method – status registers – illustrated by the keyboard interface in Figure 10.11.

In this example, an ASCII keyboard drives a '574 octal *D*-type register, clocking in a character via the keyboard's STB (strobe) output pulse when a key is struck. We rig up the standard programmed data-incircuit, as shown, using the three-state outputs of the '574 to drive the DATA bus directly. The input labeled KBDATA SEL′ comes from an address decoding circuit of the sort shown explicitly in the previous examples, and it goes LOW when the particular address chosen for this interface appears on the ADDRESS lines of the bus (in combination with AEN asserted LOW).

What's new in this example is the flip-flop, which gets set when a character is struck, and cleared when a character is read by the computer. It's a 1-bit *status* register, HIGH if there's a new character available, LOW otherwise. The computer can query the status bit by doing a data IN from the other address of this device, decoded (with gates, decoders, or whatever) as KBFLAG SEL′. You need only one bit to convey the status information, so the interface drives only the most significant bit, in this case with a '125 three-state buffer. (*Never* drive a bidirectional line with a totem-pole output!) The line coming into the side of the buffer symbol is the three-state output enable, asserted when LOW, as indicated by the negation bubble.

Program example: keyboard terminal

The computer now has a way to find out when new data is ready. Program 10.4 shows how. This is a routine to get characters from keyboard terminal, whose data port address is `KBDATA` (it's good programming style to define the actual numerical port addresses - which correspond to what the hardware decodes as KBDATA SEL, etc. - in some statements near the beginning of the program, as shown); each character is "echoed" on the computer's display device (port address = `OUTBYTE`). When it has gotten a whole line, it transfers control to a line-handling routine, which might do just about anything, based on what the line says. When it's ready for another line, it types an asterisk. This sort of function should make sense to you if you've had some experience with computers.

The program begins by initializing the character buffer pointer, by moving the *address* of the buffer that we just allocated to the address register BP. Note we can't just say

```
MOV  BP,charbuf
```

because that would load the *contents* of charbuf, not its address; in 8086 assembly language you use the word "offset" in front of a memory label to signify its address. The program then reads the keyboard status bit via an `IN` instruction, `AND`s it with 80_H to keep only the status bit (this is called "masking"), and tests for zero. Zero means the bit isn't set, so the program loops. When a nonzero status bit is detected, it reads the keyboard data port (which clears the status flag flip-flop), stores it consecutively in the line buffer, increments the pointer (`BP`), and calls the routine that echoes the character to the screen. Finally, it checks to see if the line was terminated by a carriage return: If it wasn't, it goes back and loops on the keyboard status flag again; if it was a CR, it transfers control to the line handler, after which it types an asterisk and begins the entire process anew.

A subroutine has been used to display a character, since even that simple operation requires some flag checking and masking. The routine first saves the byte into `AH`, then reads and masks the screen's busy flag. Nonzero means the screen is busy, so it keeps checking; otherwise it restores the character to `AL`, sends it to the screen's data port, and returns.

Some notes on the program: (a) We could have omitted the keyboard flag masking step, since the MSB (where we put the flag bit in our hardware) is the sign bit; thus we could have used the instruction `JPL KFCHK`. However, this trick works only for testing the MSB and thus is

Program 10.4

```
                                    ;keyboard handler -- uses flags
KBDATA  equ ***H                    ;put kbd data port adr here
KBFLAG  equ ***H                    ;ditto for kbd flag
KBMASK  equ 80H                     ;kbd flag mask
OUTBYTE equ ***H                    ;put disp port adr here
OUTFLAG equ ***H                    ;ditto for disp port flag
OUTMASK equ ***H                    ;disp port busy mask

charbuf  DB 100 dup(0)              ;allocates buffer of 100 bytes

INIT:  MOV  BP,offset charbuf ;initialize char buffer pointer
KFCHK: IN   AL,KBFLAG               ;read kbd flag
       AND  AL,KBMASK               ;mask unused bits
       JZ   KFCHK                   ;flag not set -- no data
       IN   AL,KBDATA               ;get new kbd byte
       MOV [BP],AL                  ;store it in line buffer
       INC  BP                      ;advance pointer
       CALL TYPE                    ;echo last char to display
       CMP  AL,0DH                  ;was it carriage return?
       JNZ  KFCHK                   ;if not, get next char
LINE:    o                          ;if so, do something with line
         o                          ;keep at it
         o                          ;don't quit now
         o                          ;done at last!
       MOV  AL,'*'
       CALL TYPE                    ;type a "prompt" -- asterisk
       JMP  INIT                    ;get another line

                                    ;routine to type character
                                    ;types and preserves AL
TYPE:  MOV  AH,AL                   ;save the char in AH
PCHK:  IN   AL,OUTFLAG              ;check printer busy?
       AND  AL,OUTMASK              ;printer flag mask
       JNZ  PCHK                    ;if busy check again
       MOV  AL,AH                   ;restore char to AL
       OUT  OUTBYTE,AL              ;type it
       RET                          ;return
```

somewhat specialized. (b) In keeping with good programming practice, the carriage return symbol (0DH) and asterisk probably should be defined constants, similar to KBMASK. (c) The line handler probably should be a subroutine, also. (d) Characters will be lost if the line handler takes too long; this leads us to the more elegant approach of *interrupts*, which we'll take up shortly. (e) Keyboard and terminal handlers are used so often that the PC provides built-in handlers, accessed through "software interrupts" (we'll see them later); thus, our program isn't even needed!

Status bits generalized

This keyboard example illustrates status bit protocol; but it's so simple that you may come away with the wrong idea. In an actual peripheral interface of some complexity, there will usually be several flags to signal various conditions. For example, in a magnetic-tape interface you will normally have status bits for beginning of tape, end of reel, parity error, tape in motion, etc. The usual procedure is to put all the status bits into one byte or word, so that a data IN command from the status register gets all bits at once. Typically you would have a bit indicating any of a set of error conditions as the MSB of the status word, so a simple check of sign tells if there are *any* errors; if there are, you test specific bits of the word (by ANDing with masks) to find out what's wrong. Furthermore, in a complex interface you probably wouldn't have the status bits reset "automatically," as we did with our single bit; instead, a data OUT statement might be used, each bit of which clears a specific flag.

EXERCISE 10.3

With our keyboard interface there is no way for the computer to know if it missed a character. Modify the circuit so there are two status bits: CHAR READY (that's what we have already) and LOST DATA. The LOST DATA flag should be readable as D6 on the same status port as CHAR READY; it is 1 if a key was struck before the previous character was fetched by the computer, zero otherwise.

EXERCISE 10.4

Add a program segment to Program 10.4 that checks for lost data. It should call a subroutine called LOST if it detects lost data; otherwise it should work as before.

10.09 Interrupts

The use of status flags just illustrated is one of three ways for a peripheral device to "tell" the computer when some action needs to be taken. Although it will suffice in many simple situations, it has the serious drawback that the peripheral cannot "announce" that some action needs to be taken – it has to wait to be "asked" by the CPU, via a data IN command from its status register. Devices that need quick action (such as disks or latency-sensitive real-time I/O) would have to have their status flags queried often, and with a few such devices in a computer system the CPU would soon find itself spending most of its time checking status flags, as in the last example.

Furthermore, even with continual status flag checking you can still get in trouble: In the last example, for instance, the CPU will have no trouble keeping up with someone typing at the keyboard when it is in the main (flag checking) loop. But what if it spends 1/10 second in the line-handling portion? Or what if the display device is a slow one, making the program wait for its busy flag to clear?

What is needed is a mechanism for a peripheral to *interrupt* the normal action of the CPU when something needs to be done. The CPU can then check the status register to find out what the trouble is, take care of what needs to be done, and go back to its normal business.

To add interrupt capability to a computer, it is necessary to add a few new bus signals: At least one shared line for peripherals to signal an interrupt, and (usually) a pair of lines by which the CPU can determine who interrupted. As luck would have it, the IBM PC is not a very instructive example, because it does not implement a full interrupt capability. What it lacks in power, though, it more than makes up for in simplicity; implementing hardware interrupts in a PC peripheral interface is like falling off a log.

Here's how it works: The PC bus has a set of 6 *interrupt request* lines, called IRQ2–IRQ7. They are positive-true inputs to the CPU's support circuitry (specifically, to the 8259 interrupt controller). To make an interrupt, you simply bring one of the lines HIGH. If interrupts are enabled in general (along with the particular IRQ you assert), the CPU will break off after its next instruction, then (after saving its flags and current location onto the stack) jump to an "interrupt-handler" program somewhere in memory. You write the handler to do what you want (e.g., get keyboard data), and you can put the handler anywhere you wish, because the CPU figures out where to jump by

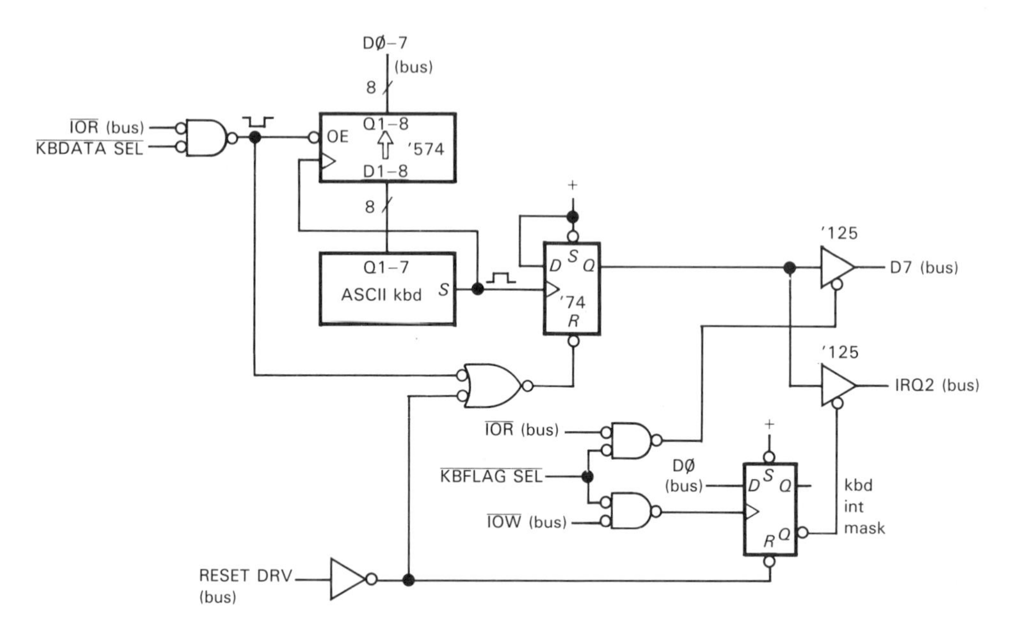

Figure 10.12. Keyboard interface with interrupts.

looking for the handler's 4-byte address in a special location in low memory. That location depends on which IRQ you've asserted; for the 8086 it is given in hex by $20 + 4n$, where n is the interrupt level. For example, the CPU would respond to an interrupt on IRQ2 by jumping to the (4-byte) address stored in locations 28_H through $2B_H$ (it's just like indirect addressing, except that the address is found in memory rather than in a register); of course, you would have cleverly arranged for the starting address of your handler to be there. At the end of your handler you execute an `IRET` instruction, which causes the CPU to restore the preexisting flag register and jump back to wherever it was when the interrupt happened.

Let's illustrate by adding interrupts to the keyboard interface (Fig. 10.12). We've left the flag bit ("character ready") and programmed I/O circuitry essentially as before, except that we've ORed the flag clear with a new bus line, RESET DRV, a CPU output that is momentarily asserted HIGH when the computer is turned on. This signal is generally used to force your flip-flops and other sequential logic into a known state at power-up. Obviously it should reset a flag that indicates a valid byte is ready to be claimed (and that, in our new interface, will even cause an interrupt). The only other change we've made is to use a compact notation for the byte-wide data paths, to make the diagram easy to read.

The new interrupt circuitry consists of a driver to assert IRQ2 when a character is ready. That's all the new hardware you need. Although not strictly necessary, we've added the capability to disable the interrupt driver (it's a three-state buffer) by sending a byte with D0 LOW to the `KBFLAG` port address. This would be used if you wanted to plug in another peripheral with interrupts at the same IRQ level, allowing only one peripheral to use its interrupts at any given time (later we'll have further explanation on this awkward point).

10.10 Interrupt handling

The IBM PC/XT family makes interrupt handling easy (though limited in flexibility) by using an 8259 interrupt-controller IC on the motherboard. This chip does most of the hard work, which consists of prioritizing, masking, and asserting vectors (we'll describe these after finishing the example). The CPU, for its part, recognizes the interrupt and responds by saving the instruction pointer and flag register, disabling further interrupts, then making a jump via the corresponding address stored in the low-memory vector area. Your handler program does the rest, namely: (a) save (push) any registers you'll be using (remember that the interrupted program can't prepare for the interrupt, since it can happen anywhere in the running program; it's a bolt out of the blue), (b) figure out what needs to be done, by reading status register(s) if necessary, (c) do it, (d) restore the saved registers from the stack, (e) tell the 8259 you're done (by sending an "end-of-interrupt" byte 20_H to its register at I/O address 20_H), and finally (f) execute a return from interrupt instruction `IRET`; this causes the CPU to restore the old flag register that it saved on the stack, and jump (via the old instruction pointer it saved on the stack) back to the program that was interrupted. Somewhere in the program, you must have (g) loaded the handler's address into the vector location corresponding to the IRQ level used by the hardware, and told the 8259 to enable interrupts at that level.

Program 10.5 shows the code for the keyboard with interrupt. Here's the overall scheme: The main program sets things up, then loops on a flag (in memory, not hardware) that the interrupt handler sets when it recognizes a carriage return; when the main program sees that flag set, it goes off and does something with the line, then returns to the flag-checking loop. The handler, entered at each interrupt, puts a character into the line buffer, sets the flag if it was a carriage return, then returns.

Let's look at the program in some detail. After defining port addresses and the all-critical vector location for IRQ2, it allocates 100 bytes (initially filled with zeros) for the character buffer. The actual program execution begins by putting the buffer address in address register SI, zero in the end-of-line flag, and the address of the handler (which begins with `KBINT`) in location 28_H. To enable level-2 interrupts in the 8259, we clear bit 2 of its existing mask (`IN`, `AND`, `OUT`); then we enable CPU interrupts and send a 1 to `KBFLAG`, which enables the three-state driver. Now we're running. The program then loops, with interrupts secretly happening right under the main program's nose, until it mysteriously finds "buflg" set. It resets the pointer and flag immediately (in case another interrupt occurs soon), then gobbles up the line. It would be well advised either to move quickly or to copy the line to another buffer, since another interrupt (with a new byte to go in the buffer) could come along in a few milliseconds; in that time you can execute a few thousand instructions, however, more than enough to copy the line.

The interrupt handler is a separate little piece of code, with no entry from the main program. It gets entered upon a level-2 interrupt, via its address that we initially loaded into 28_H. It knows exactly what it has to do, and it does it without complaining: It saves `AX` (since it plans to clobber it), reads the character from the keyboard data port, puts it in the buffer, increments the pointer, echoes the character to the screen, sets the flag if it was a carriage return, sends end-of-interrupt to the 8259, restores `AX`, and returns.

If you look back at our list of handler tasks above, you'll see that we omitted just

Program 10.5

```
                                  ;keyboard handler -- uses interrupts
KBVECT equ word pntr 0028H        ;INT2 vector
KBDATA equ ***H                   ;put kbd data port adr here
KBFLAG equ ***H                   ;put kbd flag port adr here

buflg    DB   0                   ;allocates "buffer-full" flag
charbuf  DB 100 dup(0)            ;allocates 100-byte character buffer

       CLI                        ;disable interrupts
       MOV   SI,offset charbuf    ;initialize buffer pointer
       MOV   buflg,0              ;and end-of-line flag
       MOV   KBVECT,offset KBINT  ;handler adr to vector area
       IN    AL,21H               ;existing 8259 int mask
       AND   AL,0FBH              ;clear bit 2 to enable INT2
       OUT   21H,AL               ;and send to 8259 OCW1
       STI                        ;enable interrupts
       MOV   AL,1
       OUT   KBFLAG,AL            ;enable hardware 3-state driver

LNCHK: MOV   AL,buflg
       JZ    LNCHK                ;loop until end-of-line flag set

LINE:  MOV   SI,offset charbuf    ;reset pointer
       MOV   buflg,0              ;and line flag
       MOV   AL,'*'
       CALL  TYPE                 ;type prompt "*"
         o                        ;do something with line
         o
         o
       JMP   LNCHK                ;and wait for another line

                                  ;keyboard interrupt handler
                                  ;an INT2 lands you here, via vector we loaded
KBINT: PUSH  AX                   ;save AX register, used here
       IN    AL,KBDATA            ;get data byte from keyboard
       MOV   [SI],AL              ;put it in line buffer
       INC   SI                   ;and advance pointer
       CALL  TYPE                 ;echo to screen
       CMP   AL,0DH               ;check for carriage return
       JNZ   HOME                 ;not a CR -- return
       MOV   buflg,0FFH           ;CR -- set end-of-line flag
HOME:  MOV   AL,20H
       OUT   20H,AL               ;end-of-interrupt signal to 8259
       POP   AX                   ;restore old AX
       IRET                       ;and return
```

one step, namely reading status flags to figure out which of several actions needed to be done. That's unnecessary here, though, because there's only one reason to interrupt, namely a new keyboard character needs to be read. (The programmer obviously has to understand under what conditions the hardware makes an interrupt, and what is required to service the interrupt.)

A few notes on this program: First, even though we're using interrupts, the program seems as dumb as before – it loops continually on the end-of-line flag.

However, it could be doing other things, if there were things to do. In fact, it does just that beginning at statement `LINE`, where it processes the finished line; during that time, interrupts make sure that new characters are put into the buffer, whereas they would have been lost in our previous example without interrupts.

This brings up a second point, namely, even with interrupts we're still in trouble if the program is doing things with the previous line when the next line has been completely entered. Of course, *on the average* the program simply has to keep up with keyboard entry; but you could have a situation in which the line user occasionally spends a lot of time, and you need to buffer more than one line temporarily. One solution to this is to make a copy to a second buffer, or to alternate between two buffers. An elegant alternative is to organize input as a queue, implemented as a "ring buffer" (or "circular buffer"), in which a pair of pointers keep track of where the next input character goes, and where the next character is removed. The interrupt handler advances the input pointer, and the line user advances the output pointer. Such a ring buffer might typically be 256 bytes long, permitting the line user to get behind by a few lines.

A third point concerns the interrupt handler itself. It's usually best to keep it short and simple, perhaps setting flags to signal the need for complicated operations in the main program. If the handler does become long-winded, you risk losing data from other interrupting devices, because interrupts are disabled when the CPU jumps into the handler. The solution in this case is to re-enable interrupts *within* your handler with an `STI` instruction, after doing the critical things that have to be done first. Then if an interrupt occurs, your interrupt handler will itself be interrupted. Since flags and return addresses are stored on the stack, the program will find its way back, first to your handler, finally to the main program.

10.11 Interrupts in general

Our keyboard example illustrates the essence of interrupts – a spontaneous hardware request for attention by a peripheral, producing a program jump to a dedicated handler routine (usually resulting in some programmed I/O), followed by a return to the code that was interrupted. Other examples of interrupting devices are real-time clocks, in which a periodic interrupt (often 10 per second, but 18.2 per second in the PC) signals a timekeeping routine to advance the current time; another example is a parallel printer port, which interrupts each time it is ready for a new character. By using interrupts, these peripherals let the computer interleave other tasks simultaneously; that's why you can be doing word processing while your PC is printing a file (and, of course, keeping proper time throughout).

The IBM PC does not, however, illustrate the full generality of interrupts. As we saw, it has a set of six IRQ lines on the bus, each one of which can be used only for a single interrupting device. The IRQ lines are numbered according to priority; in the event of multiple interrupts, the lowest-numbered interrupt is serviced first. Four of the IRQ lines are preassigned to essential peripherals, namely the serial port (IRQ4), hard disk (IRQ5), floppy disk (IRQ6), and printer port (IRQ7), leaving only IRQ2 and IRQ3 available [lines for two additional IRQ levels recognized in the IBM PC are not even brought out on the bus, being used on the motherboard for the 18.2Hz clock (IRQ0) and the keyboard (IRQ1)]. If you were to add a streaming tape backup or local area network, you would have to use IRQ2 and IRQ3. Furthermore, the interrupt is *edge-triggered*, which frustrates any reasonable possibility of using wired-OR to combine several peripherals on a single IRQ line.

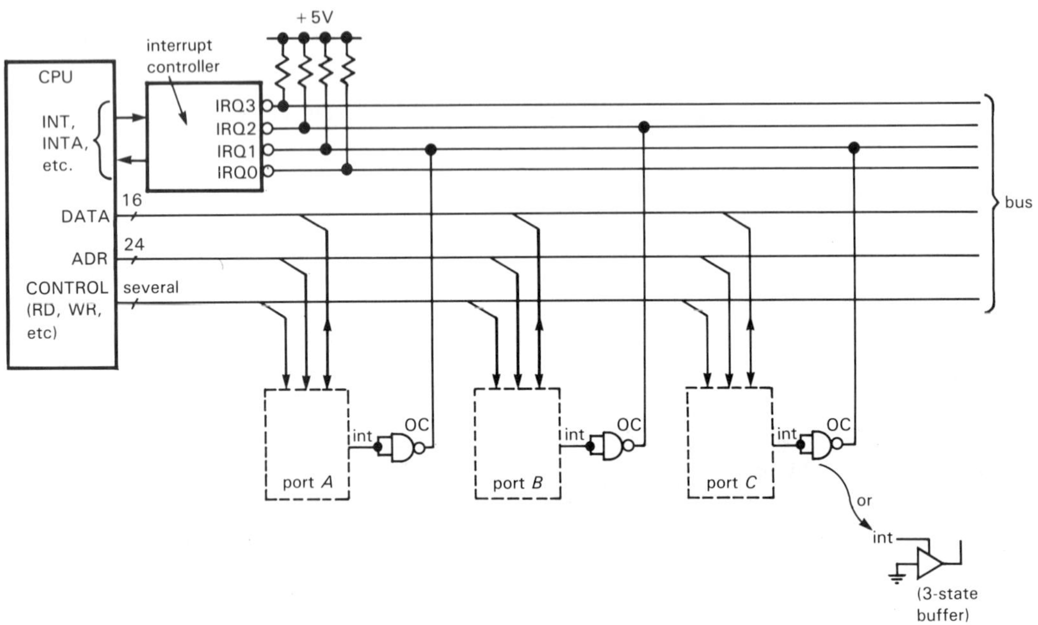

Figure 10.13. Shared interrupt lines.

□ *Shared interrupt lines*

The usual interrupt protocol, as implemented on many microcomputers, circumvents these limitations. Look at Figure 10.13. There are several (prioritized) IRQ-type lines; these are negative-true inputs to the CPU (or its immediate support circuitry). To request an interrupt, you pull one of the IRQ′ lines LOW, using an open-collector (or three-state) gate, as shown (note the trick for using a three-state gate to mimic an open-collector gate). The IRQ′ lines are shared, with a single resistive pullup, so you can put as many devices on each IRQ′ line as you want; in our example two ports share IRQ1. You would generally connect a latency-sensitive (impatient) device to a higher-priority IRQ′ line.

Since the IRQ′ lines are shared, there could always be another device interrupting on the same line at the same time. The CPU needs to know who interrupted, so it can jump to the appropriate handler. There is a simple way, and a complicated way, to do this. The simple way is called *autovectored polling* and is used nearly universally (though not on the IBM PC). Here's how it works.

□ *Autovectored polling.* Some circuitry on the CPU board (we'll have an example in Chapter 11) instructs the microprocessor that it is to use autovectoring, which works just like the IBM PC – each level of interrupt forces a jump through a corresponding vectoring location in low memory. For example, the 68000 microprocessor family we'll meet in Chapter 11 has seven levels of prioritized interrupt, which autovector through 4-byte pointers stored in the 28 (7×4) locations 64_H through $7F_H$. You put the addresses of the handlers in those locations, just as in our example above. For example, you would put the (4-byte) address of the handler for a level-3 interrupt in hex locations 6C through 6F.

Once in the handler, you know which level of interrupt you're servicing; you just don't know which particular device caused the interrupt. To find out, you simply check the status registers of each of the devices connected to that level of interrupt (a device *never* requests an interrupt without also indicating its need by setting one or more readable status bits). If a bit is set indicating that something needs to be done, you do it, including whatever it takes to cause the device to disassert its IRQ′: Some devices (like our keyboard) clear their interrupt when read, whereas others may need a particular byte sent to some I/O port address.

If the device you serviced was the only one interrupting at that level, that IRQ′ will now be HIGH upon returning to the interrupted program, and execution will continue. However, if there had been a second interrupting device at the same level, that IRQ′ line will still be held LOW (by the wired-OR action of the shared IRQ′ line) upon return from the service routine, so the CPU will immediately autovector back to the same handler. This time the polling will find the other interrupting device, do its thing, and return. Note that the order in which you poll status registers effectively sets up a "software priority," in addition to the hardware priority of the multiple IRQ′ levels.

☐ *Interrupt acknowledgment.* We shouldn't leave the subject of interrupts without mentioning a more sophisticated procedure for identifying who interrupted – *interrupt acknowledgment.* In this method the CPU doesn't need to poll the status registers of possible interrupters, because the interrupting device *tells* the CPU its name, when asked. The interrupter does this by putting an "interrupt vector" (usually a unique 8-bit quantity) onto the DATA lines in response to an "interrupt acknowledge" signal that the CPU generates during the interrupt processing.

Nearly every microprocessor generates the needed signals. The sequence of events goes like this: (a) The CPU notices a pending interrupt. (b) The CPU finishes the current instruction, then asserts (i) bus signals that announce an interrupt, (ii) the interrupt level being serviced (on the low-order ADDRESS lines), and (iii) READ-like strobes that invite the interrupting device to identify itself. (c) The interrupting device responds to this bus activity by asserting its identity (interrupt vector) onto the DATA lines. (d) The CPU reads the vector and jumps into the corresponding unique handler for the interrupting device. (e) The handler software, as in our last example, reads flags, gets and sends data, etc., as needed; among its other duties, it must make sure the interrupting device disasserts its interrupt. (f) Finally, the interrupt handler software returns control to the program that was interrupted.

Sharp-eyed readers may have noticed a flaw in the procedure just outlined. In particular, there has to be a protocol to ensure that only one device asserts its vector, since there may be several simultaneous interrupting devices at the same IRQ level. The usual way to handle this is to have a bus signal (call it INTP, "interrupt priority") that is unusual in not being shared by devices on the bus, but rather is passed along *through* each device's interface circuit, beginning as a HIGH level at the device closest to the CPU and threading along through each interface. That's called a "daisy chain" in the colorful language of electronics. The rule for INTP hardware logic is as follows: If you have not requested an interrupt at the level being acknowledged, pass INTP through to the next device unchanged; if you *have* interrupted at that level, hold your INTP output LOW. Now the rule for asserting your vector goes like this: Put your vector number onto the data bus when requested by the CPU only if (a) you have an interrupt pending at the level being acknowledged and (b) your input INTP is HIGH. This guarantees that only one device asserts its vector; it also establishes a "serial priority" chain within each IRQ level, with devices electrically closest to the CPU getting serviced first. Computers that implement this scheme have little jumper plugs to pass INTP over unused motherboard slots. Don't forget to remove these jumpers when you plug in a new interface card (and put them back when you take one out!).

There is a nice alternative to the serial daisy-chain method of interrupt acknowledgment: Instead of threading a line through each possible interrupter, you bring individual lines back from each one to a priority encoder (Section 8.14), which in turn acknowledges the interrupt by asserting the identity of the highest-priority interrupting device. This scheme avoids the nuisance of daisy-chain jumpers. We will describe it in detail in Section 11.4 (Fig. 11.8).

In most microcomputer systems it isn't worth implementing the full-blown interrupt acknowledgment just described. After all, with 8-level autovectoring you can handle up to 8 interrupting devices without polling, and several times that number with polling. Only in large computer systems, in which you demand fast response with dozens of interrupting devices present, might you succumb to the complexity of the interrupt acknowledgment protocol, whether with serial daisy-chained hardware priority or with parallel priority encoding.

However, it is important to realize that even simple computers may be using vectored interrupt acknowledgment *internally*. For example, the simple 6-level autovector interrupt scheme of the IBM PC seen by the bus user is actually generated by an 8259 "programmable interrupt controller" chip that lives close to the CPU and generates the proper interrupt acknowledgment sequence just described (see below). This is necessary because the 8086 (and successors) can't implement autovectoring by themselves. On the other hand, the popular 68000 series of CPU chips can implement autovectoring internally, with just a single external gate (see Chapter 11).

□ *Interrupt masks*

We put a flip-flop in our simple keyboard example so that its interrupts could be disabled, even though the 8259 controller lets you turn off ("mask") each level of interrupt individually. We did that so that some other device could then use IRQ2. For a bus with shared (*level*-sensitive) IRQ′ lines, it is especially important to make each interrupt source maskable, again with an I/O output port bit. For example, a printer port normally interrupts each time its output buffer is empty ("give me more data"); when you've finished printing, though, you don't care. The obvious solution is to turn off printer interrupts. Since there might be other devices hooked to the same interrupt level, you must not mask that whole level; instead, you just send a bit to the printer port to disable its interrupts.

□ *How the IBM PC got the way it is*

The 8086/8 microprocessor used in the IBM PC actually implements the full vectored interrupt acknowledgment protocol. To keep things simple, however, the PC designers used an 8259 interrupt controller IC on the motherboard. The way it is used in the PC, it has a set of IRQ inputs from the I/O bus card slots (that's where you make your interrupt requests), and it connects to the microprocessor's data bus and signal lines. When it gets a request on an IRQ line from a peripheral, it figures out priority and goes through the whole business of asserting the corresponding vector onto the data bus. It has a mask register (accessible as I/O port 21_H) so that you can disable any specified group of interrupts.

The 8259 lets you select (through software) either *level-* or *edge*-triggered interrupts on its IRQ input lines, according to a byte sent to a control register (I/O port 20_H). Unfortunately, the PC designers decided to use edge triggering, probably because that makes it a little easier to implement interrupts (for example, you can just connect the real-time clock's square-wave output directly to IRQ0). If they had selected level-sensitive interrupts instead, you would be able to hang multiple interrupting devices on each IRQ′ line, with software polling as illustrated above. Unfortunately, the PC's ROM BIOS (Basic Input/Output System) and operating system (not to mention the hardware) assume edge triggering, so the choice is irrevocable. (Nearly all other computers, including even the successors to the PC and PC/AT, use level-sensitive interrupts.)

There is a partial solution to this problem. As long as there is an IRQ line available, you *can* combine several interrupting devices on a single PC board, with logic to generate edge-triggered interrupts on that single IRQ line; in fact, you could use your own 8259 (with its I/O ports accessible to the CPU) to do the job. But, since the interrupting devices have to know about each other, you can't use this scheme for independent plug-in peripherals. Furthermore, you still use up an IRQ line per card, and in a complicated system, given that there are only two IRQ levels free in the IBM PC, there will not be enough to go around.

Software interrupts

The Intel 8086-series of CPUs have an instruction ("`INT n`," where n is 0–255) that allows you to produce the same kind of vectored jump as an actual hardware interrupt. In fact, among its 256 possible jump vectors are duplicates of the 8 levels of IRQ-requested hardware interrupts (`INT 8` through `INT 15`, to be exact). Thus, you can make a "software interrupt" from a program statement. The IBM PC uses these software interrupts to let you communicate with the operating system and its various ROM-based utilities. For example, `INT 5` sends a replica of the screen to the printer. `INT 21H` turns out to be particularly important, because it is a function call to the operating system; you tell the system which DOS function you want by putting the corresponding number in register `AH` before executing the `INT 21H`.

Don't confuse these software interrupts with the externally triggered hardware interrupts we've been talking about. Software interrupts turn out to be a handy way of implementing vectored jumps from user code into system software. But they are not real interrupts, in the sense of a hardware call for attention from an external autonomous device. On the contrary, you build these into your software, you know when they are coming (that's why you can pass arguments through registers), and they are merely the response (albeit identical with what follows a true interrupt) of the CPU to its own code. You might think of software interrupts as a clever way to extend the instruction set.

10.12 Direct memory access

There are situations in which data must be moved very rapidly to or from a device. The classic examples are fast mass-storage devices like disk or tape, and on-line data-acquisition applications such as multichannel pulse-height analysis. Interrupt-initiated programmed processing of each data transfer in these examples would be awkward, and probably too slow. For example, data come from a "high-density" floppy disk at about 500kbit/s, or one byte every 16μs. With all the bookkeeping involved in handling an interrupt, data would almost certainly be lost, even if the floppy were the only interrupting device in the system; with a few such devices the situation becomes hopeless. Even worse, a typical hard disk transfers a byte every $2\mu s$, completely beyond the capability of programmed I/O. Devices like disks and tapes (not to mention real-time signals and data) can't stop in mid-stream, so a method must be provided for reliably fast response and high overall byte transfer rates. Even with peripherals with low average data transfer rates, there are sometimes requirements for short *latency* time, the time from initial request to actual movement of data.

The solution to these problems is direct memory access, or DMA, a method for direct communication from peripheral to memory. In some microcomputers (e.g., the IBM PC) the communication is actually handled by the CPU hardware, but that doesn't really matter. The important point is that no programming is involved in the actual transfer of data; bytes are moved between memory and peripheral via the bus, without program intervention. The only effect on the executing program is some slowing down of execution time, because DMA activity "steals" bus cycles that would otherwise be used to access memory for program execution. DMA usually involves more hardware complexity in the interface itself, and it should not be used unless necessary. However, it is useful to know what can be done, so we will describe briefly what you need to make a DMA interface. As with interrupts, the IBM PC designers streamlined their DMA protocol; a "DMA controller" chip on the motherboard does the hard work for you, making a DMA interface relatively straightforward. In general, though, DMA interfaces tend to be machine-dependent and complicated. We'll first explain the more usual "bus mastership" method of DMA, then the PC's simplified DMA protocol.

A typical DMA protocol

In DMA transfers, the peripheral requests access to the bus via special "bus request" lines (prioritized like the IRQ lines) that are part of the bus. The CPU gives permission and releases control of address, data, and strobing lines. The peripheral then asserts memory addresses onto the bus and either sends or receives data, one byte at a time, according to the strobes it asserts; in other words, it takes over the bus (it becomes "bus master") and acts like a CPU, directing data transfers. The DMA bus master is responsible for generating addresses (usually a block of successive addresses, generated with a binary counter) and keeping track of the number of bytes moved. The usual way to do this is to have a byte counter and an address counter in the interface. These are initially loaded from the CPU, via programmed I/O, to set up the DMA transfer desired. On command from the CPU (via a command bit, written with programmed I/O), the interface makes its DMA request and begins to move its data. It may release the bus between each byte (allowing the CPU to sneak a few instructions in), or it may take the more antisocial approach of keeping the bus for a block of transfers. When all transfers are complete, it releases the bus for the last time and notifies the program that it is finished by setting a status bit and requesting an interrupt, whereupon the CPU can decide what to do next.

Getting data or programs from disk is a common example of DMA transfer: The executing program asks for some "file" by name; the "operating system" (more about this soon) translates this into a set of programmed data OUT commands to the disk interface's control (or "command") register, byte count register, and address register (specifying where to go on the disk, how many bytes to read, and where to put them in memory). Then the disk interface finds the right place on the disk, makes a

DMA request, and begins moving blocks of data to the specified place in memory. When it's done, it sets bits in its status register to signify completion and then makes an interrupt. The CPU, which has meanwhile been executing other instructions (or possibly just waiting for data from the disk), responds to the interrupt, finds out from the status register of the disk interface that the data are now in memory, and then goes on to the next task. Thus, programmed I/O to the interface (the simplest kind of I/O) was used to set up the DMA transfer, DMA itself (stealing bus cycles from the CPU) was used for rapid transfer of data, and an interrupt was used to let the computer know the task was done. This sort of I/O hierarchy is extremely common, especially with mass-storage devices; you can expect maximum DMA transfer rates of one to ten million words per second on a typical microcomputer bus.

□ *DMA on the IBM PC*

The IBM PC, which is basically a simple microcomputer, has a simpler DMA protocol. The motherboard has a DMA controller (the Intel 8237) with built-in address and byte counters, along with the logic to disable the CPU and take over the bus. A peripheral that wants to do DMA, therefore, doesn't have to generate addresses and drive the bus. Instead, it signals the controller (via one of the three DRQ1–DRQ3 "DMA request" lines), which in turn responds by returning the corresponding DACK0–3′ ("DMA acknowledge"). The controller then controls the transfers, asserting address and strobing lines, with the peripheral asserting (or receiving) data to (or from) memory. In this whole process the memory sees nothing unusual going on, since addresses and memory strobes (MEMW′ or MEMR′), normally supplied by the CPU, are supplied by the 8237 controller, and if it's DMA *to* memory, data are supplied by the peripheral. The peripheral, on the other hand, knows something special is happening since it requested DMA access (and received confirmation via DACK′); so when the DMA controller asserts IOR′ (or IOW′), the peripheral supplies (or accepts) successive bytes. You might wonder why some innocent bystander peripheral doesn't get hurt in the DMA process, since both I/O strobes and addresses are being asserted, whereas the addresses are in fact the *memory* addresses that go with the memory strobes MEMW′ or MEMR′ asserted by the controller; they have nothing to do with I/O port addresses. The secret is our old friend AEN, specifically added to the bus just to solve this problem. AEN is asserted HIGH during DMA transfers, and all I/O port addressing must be qualified by ANDing with AEN LOW to prevent spurious responses to DMA memory addresses.

Even with the use of a separate controller chip, you still have to set up the starting address, byte count, and direction for the impending DMA transfer. These data go to the 8237, which is obliging, having a set of registers that you write (via programmed I/O) from the CPU. It's pretty straightforward (see Eggebrecht's book for clear guidance), except that, as with most peripheral LSI chips, there is a confusing variety of choices of "modes" (single transfer, block transfer, etc.). Luckily, the PC is sufficiently primitive that you're allowed to use only "single transfer," which transfers only one byte per DRQ request. If you insist on transferring a whole block of data by holding DRQ high, the 8237 releases the bus for one CPU cycle between each DMA cycle; that keeps the computer alive, even if you have a greedy peripheral that tries to hog the bus. The standard PC has a rather modest DMA capacity, about 2μs per byte transferred. As with interrupts, the PC is sparse on DMA channels: Three

TABLE 10.1. IBM PC BUS SIGNALS

Signal name	Number	Active	Type[a]	Direction CPU↔I/O	Pin #	Function
A0–A19	20	H	2S	→	A31-A12	address (A0-A15 for I/O)
D0–D7	8	H	3S	↔	A9-A2	data
IOR'	1	L	2S	→	B14	I/O read strobe
IOW'	1	L	2S	→	B13	I/O write strobe
MEMR'	1	L	2S	→	B12	memory read strobe
MEMW'	1	L	2S	→	B11	memory write strobe
AEN	1	H	2S	→	A11	DMA address signal
IRQ2–IRQ7	6	↑	2S	←	B4,B25-B21	interrupt request
RESET DRV	1	H	2S	→	B2	power-on reset
DRQ1–DRQ3	3	H	2S	←	B18,B6,B16	DMA request
DACK0'–DACK3'	4	L	2S	→	B19,B17,B26,B15	DMA acknowledge
ALE	1	H	2S	→	B28	"address latch enable"
CLK	1	–	2S	→	B20	CPU clock (4.77MHz)
I/O CH CK'	1	L	OC	←	A1	I/O error - makes NMI
I/O CH RDY	1	H	OC	←	A10	pull LOW for wait states
OSC	1	–	2S	→	B30	14.31818MHz (3×CPU clk)
T/C	1	H	2S	→	B27	DMA terminal count
GND	3	–	PS	→	B1,B10,B31	signal & power gnd
+5V DC	2	–	PS	→	B3,B29	+5V supply
+12V DC	1	–	PS	→	B9	+12V supply
–5V DC	1	–	PS	→	B5	-5V supply
–12V DC	1	–	PS	→	B7	-12V supply

[a] OC - open-collector; PS - power supply; 2S - 2-state (totem-pole); 3S - 3-state.

channels (DRQ1–DRQ3) are accessible on the I/O bus (DRQ0 is already used internally to refresh dynamic memory): DRQ1 is used for hard disk, and DRQ2 is used for floppy disk. That leaves DRQ3 for everything else.

10.13 Summary of the IBM PC's bus signals

Through our examples – programmed I/O, interrupts, and DMA – we've seen most of the bus signals that go to the card slots in the IBM PC. Table 10.1 (and Fig. 10.14) lists the full bus, with pin connections. For completeness we'll summarize them all here, beginning with the ones we've already met.

A0–A19

Address bus. Two-state, output only, active-HIGH. All 20 lines are used to address memory (with MEMR′ and MEMW′ as strobes, analogous to IOR′ and IOW′), but only the 16 least significant lines are used during I/O access

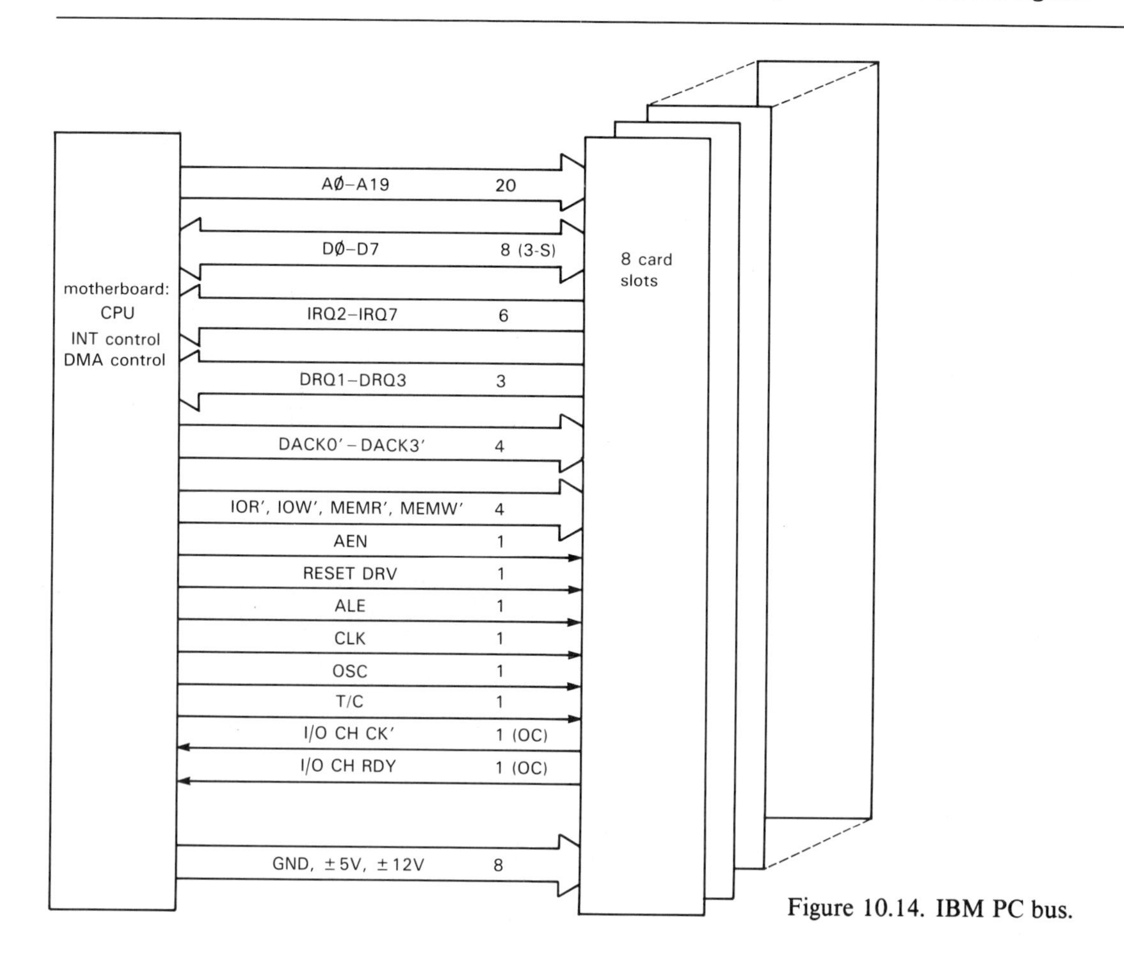

Figure 10.14. IBM PC bus.

(64K port addresses); I/O devices should qualify address with AEN LOW. Important note: I/O on the motherboard only looks at A0–A9, and uses 000_H–$1FF_H$; so external I/O must have its low ten bits in the range 200_H–$3FF_H$. You can be clever, though, by roosting in an unused 10-bit I/O address, then using the top 6 bits to create 64 I/O port addresses.

D0–D7

Data bus. Three-state, bidirectional, active-HIGH. Asserted by CPU during memory or I/O write; asserted by memory during memory read or DMA from memory; asserted by I/O port during I/O read or DMA to memory.

IOR′, IOW′, MEMR′, MEMW′

Data strobes. Two-state, output only, active-LOW. Asserted by CPU during read or write. On writes, data should be latched on trailing (rising) edge, qualified by address; on reads, data should be asserted coincident with strobe, qualified by address.

AEN

Address enable. Two-state, output only, active-HIGH. Asserted by CPU during DMA cycles. I/O ports must not respond with normal address decoding to IOR′ and IOW′; instead, I/O port that received DACK uses IOR′ or IOW′ to strobe DMA data bytes.

IRQ2–IRQ7

Interrupt request. Two-state, input only, rising-edge-triggered. Asserted by interrupting device. Prioritized, with IRQ2 highest, IRQ7 lowest. Maskable in 8259 interrupt controller, via CPU write to port 21_H. Each IRQ level can be used by only one device at a time.

RESET DRV

Reset driver. Two-state, output only, active-HIGH. Asserted by CPU during power-on. Used to initialize I/O devices to known start-up state.

DRQ1–DRQ3

DMA request. Two-state, input only, active-HIGH. Asserted by I/O device requesting DMA channel. Prioritized, with DRQ1 highest, DRQ3 lowest. Acknowledged by DACK1′-DACK3′.

DACK0′–DACK3′

DMA acknowledge. Two-state, output only, active-LOW. Asserted by CPU (DMA controller) to indicate grant of corresponding DMA request.

ALE

Address latch enable. Two-state, output only, active-HIGH. The 8088 used a multiplexed data/address bus, and this signal corresponds to the 8088's strobing signal, used by latches on the motherboard to latch the address. Can be used to signal beginning of a CPU cycle; usually ignored in I/O design.

CLK

Clock. Two-state, output only. This is the CPU's clocking signal; it's asymmetrical, 1/3 HIGH and 2/3 LOW. The original PC used a 4.77MHz clock, but higher speeds are common. CLK is used to synchronize wait-state requests (via I/O CH RDY), in order to stretch an I/O cycle for slow devices.

OSC

Oscillator. Two-state, output only. This is a 14.31818MHz square wave, which can be used (when divided by 4) as a color-burst oscillator for color display.

T/C

Terminal count. Two-state, output only, active-HIGH. Tells I/O port that a DMA block data transfer is complete. A DMA device must qualify it with DACK′ for the channel in use, since T/C is asserted when any of the DMA channels finishes a block transfer.

I/O CH CK′

I/O channel check. Open-collector, input only, active-LOW. Generates highest-priority interrupt (NMI, "nonmaskable interrupt"); used to signal error condition from some peripheral. CPU figures out who's in trouble by device polling (Section 10.11); each peripheral that can assert I/O CH CK′ must therefore have a status bit that can be read by the CPU.

I/O CH RDY

I/O channel ready. Open-collector, input only, active-HIGH. Generates "wait states" if disasserted (i.e., pulled LOW) before the second CLK rising edge of a processor cycle (normally 4 CLKs). Used to extend bus cycle for slow I/O or memory.

GND, +5VDC, −5VDC, +12VDC, −12VDC

Ground and dc supplies. Regulated dc voltages that are bused for use by peripheral interface cards. Check the specifications of your computer for power limitations, which are machine-dependent. Generally speaking, there should be enough power to run anything you can fit into the I/O slots.

□ 10.14 Synchronous versus asynchronous bus communication

The data IN/OUT protocol we described earlier constitutes a *synchronous* exchange of data; data are asserted onto or retrieved from the bus synchronously with strobing signals generated in the CPU (or DMA controller). Such a scheme has the virtue of simplicity, but it does open the possibility of problems with long buses, since the long propagation delays you get mean that data may not be asserted soon enough during a data `IN` operation for reliable transmission. In fact, with a synchronous bus the device sending the data never even knows if it was received! This sounds like a serious disadvantage, but in reality computer systems with synchronous buses seem to work just fine.

The alternative is an *asynchronous* bus, in which a data `IN` transfer, for example, goes something like this: The CPU asserts the port ADDRESS and a *level* (not a pulse) on a strobing line (call it IOR′, as before) that signifies data `IN` from the addressed device. The addressed device then asserts the DATA and a level signifying that the DATA is valid (call it DTACK′, "data transfer acknowledged"). When the CPU sees DTACK, it latches the DATA and then releases its IOR′ level. When the interface sees the IOR′ line go HIGH, it releases the DTACK′ and DATA lines. In other words, the CPU says "Give me data." The peripheral then says "OK, here it is." The CPU then says "OK, got it." And the peripheral finally says "Great! I'll go back to sleep again." This is sometimes referred to as "interlocked communication," or "handshaking."

Asynchronous bus protocol allows long buses and gives the communicating devices assurance that data are being moved. If a remote device is switched off, the CPU will know about it! Actually, that information is available via status registers with any kind of bus, and the chief advantage of asynchronous protocol is the flexibility of using any length of bus, bought at the expense of slightly greater hardware complexity.

There are situations where you want to attach relatively slow interface ICs to a bus; an example is slow-access ROM, or even RAM. All buses provide some way to stretch a bus cycle: With an asynchronous bus it's automatic, since the bus cycle goes on until the DTACK′ handshake is returned. With synchronous buses there is always some sort of HOLD′ line (it's called I/O CH RDY on the PC) to create wait states, effectively stretching the strobes and thereby delaying the end of the cycle. The overall bus cycle is always lengthened by an integral number of CPU clock cycles; that is the number of "wait states" inserted. For example, a standard IBM PC has a clock frequency of 4.77MHz (period of 210ns), and a standard memory access bus cycle is 4 clock periods (840ns). If I/O CH RDY is brought LOW before the second rising edge of CLK during a memory access, and brought HIGH again before the third, one wait state is generated, stretching the bus cycle (and MEMW′ or MEMR′) to 5 clocks (1050ns). If you hold I/O CH RDY LOW for additional clocks, you get additional wait states, up to a maximum of 10 clock periods.

Now we can reveal a well-kept secret about synchronous versus asynchronous buses: In reality, all single-processor (or, more precisely, single-bus-master) microcomputer buses are really synchronous, because all timing is slaved to a single CPU oscillator (like the 4.77MHz clock signal for the original IBM PC). Thus, if a peripheral device delays its handshake on an "asynchronous" bus, the cycle is always stretched by an integral number of CPU clocks. The distinction that is usually called synchronous versus asynchronous is really this: On an "asynchronous" bus, wait states are inserted by default unless a wired-OR line (DTACK′) is asserted LOW, whereas on a "synchronous" bus the default bus cycle has no wait states, which are generated only if a wired-OR line (HOLD′) is asserted LOW. The difference is more than semantic – you can't have a physically long bus with "synchronous" protocol, because the HOLD′ signal gets back too late to stretch the cycle, whereas with an "asynchronous" bus the CPU won't terminate the bus cycle without your permission (DTACK′). In our usually humble manner, we offer the following suggestive terminology to clear up this confusion: If the bus makes wait states by default ("asynchronous"), call it *default-wait*; if the bus makes wait states only when you ask ("synchronous"), call it *request-wait*. The IBM PC is request-wait, whereas the VME bus (see below) is default-wait.

This whole bus situation becomes more complicated with multiprocessor systems, in which bus mastership changes hands. A synchronous bus with multiple masters requires all masters to use a single clock, whereas an asynchronous bus permits different clock rates. Luckily for you, multiprocessor systems are beyond the scope of this book!

A possible point of confusion: You don't add wait states because you have a slow *peripheral* (a printer, for example); you do it only because you have a slow IC (ROM, say, with 250ns access time, or a slow LSI peripheral chip), which cannot latch (or produce) the data within the normal bus access time. A slow peripheral is usually hopelessly slow (*milli*seconds, not nanoseconds); the solution is to send (or receive) a byte at full bus speed, latching it in a byte-wide register chip, then wait for an interrupt (or possibly a status flag) before doing another full-speed transfer.

10.15 Other microcomputer buses

We chose the IBM PC to illustrate microcomputer bus architecture – bus signals, memory and programmed I/O, interrupts, and DMA. Since the PC is widely copied and widely used in engineering and data acquisition/control, this is a good illustrative choice for an electronics book. Furthermore, the PC bus is extraordinarily simple and easy to use.

However, simplicity has its costs. The original PC bus is seriously limited in important ways, some of which we have already mentioned (e.g., scarcity of interrupt and DMA channels). Even more seriously, by today's standards the PC bus has too little address space (20 bits, only 640K usable), too narrow a data path (8 bits), insufficient data transfer rate (1.2Mbyte/s, max), and no provision for multiple bus masters. IBM has evolved improved buses in subsequent PC generations, first the PC/AT (a compatible enhancement of the original PC), then the all-new (= incompatible!) "microchannel" bus of the PS/2 series. Outside the IBM world there are competing buses peculiar to a given manufacturer (e.g., DEC's Q-bus and VAXBI bus), and generic buses (Multibus, NuBus, VME bus). Let's take a quick tour of the computer buses listed in Table 10.2.

PC/AT and Micro Channel

IBM's PC/AT (for "Advanced Technology") was introduced in 1984, and discontinued in 1987, at the peak of its popularity, to make way for IBM's PS/2 series of "clone-killer" computers, which use the improved "Micro Channel" bus. [The PC/AT continued to thrive, however, since the clone manufacturers (and many buyers) initially ignored IBM's newer machines, whose advanced features required nonexistent software.] The PC/AT uses the 80286 CPU and a compatible

TABLE 10.2. COMPUTER BUSES

Bus	Raw bandwidth (Mbyte/s)	Data width	Address width	Block xfer?	MUXed data/adr?	Multimaster?	Sync/Async	IRQ lines[a]	Drivers	Connector[b]	Comments
STD bus		8	16	–	–	–	S	1	TTL	CE	controller-type applications
PC/XT	1.2	8	20	–	–	–	S	5E	TTL	CE	original IBM PC & compatibles
PC/AT	5.3	8,16	20,24	–	–	(c)	S	10E	TTL	CE	accepts PC/XT cards
EISA	33	8,16,32	20,24,32	•	–	•	S	11P	TTL	CE	enhanced PC/AT; auto-configure
MicroChannel	20	8,16,(32)	24,(32)	•	–	•	A	11	TTL	CE	IBM PS/2; auto-configure
Q-bus	2	16	22	•	•	•	A	4	(d)	CE	LSI-11, μVAX-I,II; daisy-chained IACK
Multibus I	10	8,16	20,24	–	–	•	A	8	TTL	CE	Intel; SUN-I and others
CAMAC	3	24	9	•	–	–	S	L	TTL/OC	CE	data acquisition & control bus
VAX BI	13.3	8,16,24,32	32	•	•	•	S	4	TTL	ZIF	VAX 780, 8600 series; parity
Multibus II	40	8,16,24,32	16,32	•	•	•	S	M	TTL	DIN	parity; 40MB/s for blk xfer, 20M otherwise
NuBus	40	32	32	•	•	•	S	M	TTL	DIN	Macintosh II adds 1 dedicated INT per slot; ""
VME	40	8,16,32	16,24,32	•	–	•	A	7	TTL	DIN	daisy-chained IACK; SUN-3
Futurebus	120					•	A	–	(d)		
Fastbus	160	32	32	•	•	•	A	M	ECL	H	communication across many crates

(a) E - edge-sensitive; L - LAM ("look at me"); M - "int" via bus mastership; P - programmable edge- or level-sensitive interrupts.
(b) CE - card-edge; DIN - 2-part "Eurocard" 96-pin connector; H - high density 2-part conn. (c) almost. (d) National Semi special.

enhancement of the original PC bus: An additional (and optional) connector carries an extra 8 bits of data, 4 bits of address, and 5 additional IRQ lines (edge-triggered, as before). The resulting 16-bit data path and higher CPU clock speed raise the maximum bandwidth to 5.3Mbyte/s, which, with the additional address space and interrupts, makes the PC/AT a serious microcomputer. The PC/AT bus (sometimes called Industry Standard Architecture, or ISA) even supports multiple bus masters, though its abilities here are limited. Cards that work on the original PC bus will work in the PC/AT (if they are fast enough), because you can ignore the bus enhancement carried on the extra connector; in that case, of course, you revert to an 8-bit data path and 20-bit address space. AT-compatible computers generally run their I/O bus at higher speeds, which can create additional timing problems with older plug-in boards.

The Micro Channel bus was first used in IBM's PS/2 series of second-generation personal computers, introduced in 1987. It allows for data and address paths up to 32 bits wide (in the high-end 80386-based machines), 11 levels of shared (level-sensitive) interrupts, multiple bus masters, and asynchronous protocol. Cards that plug into the Micro Channel don't have hardwired I/O port addresses; instead, the CPU assigns an address (and other configuration choices) at start-up, based on information it reads from ROM on the card. This pleasant feature means that you don't have to set little switches on each card, and you don't have to worry about cards using overlapping address space. Micro Channel cards have tight dimensional tolerances, owing to the daring use of 0.050 inch spacing between pads on the edge connectors.

EISA

The Extended Industry Standard Architecture (EISA) is the clone-makers' answer to the Micro Channel. It was introduced in 1988 by nine manufacturers of AT-compatible computers. By adding an extra connector to the AT bus, the EISA's designers implemented many of the desirable features of the Micro Channel, *while maintaining compatibility with existing AT plug-in cards*. Thus, you can plug standard AT boards into EISA and get normal AT functionality. Moreover, when used with boards designed specifically for it, the EISA supports 32-bit data transfers (with peak transfer rates of 33Mbyte/s), 32-bit memory addressing, multiple bus masters, programmable level or edge-triggered interrupts, and automatic board configuration.

Multibus I and II

Originally introduced by Intel, the Multibus formats have found their way into many computers. The original Multibus I is a capable bus with 16-bit data path and 24-bit address space, and it allows multiple bus masters. Multibus II is intended for high-performance multiprocessor systems, with 32-bit data and address paths, parity checking, distributed arbitration, and message-passing protocols. It uses a synchronous 10MHz clock and can transfer up to 40Mbyte/s to sequential addresses in "block transfer" mode. In common with some other large buses (NuBus, Fastbus), Multibus II saves pins by multiplexing data and address on a common set of 32 lines. It also uses a 96-pin card-mounted DIN connector, rather than the simple gold-plated "card-edge" connector: By using a well-designed card-mounted ("2-part") connector, you get better reliability and a connection system that is insensitive to card warp and rough handling.

Although Multibus II seems to have all the advantages, its flexibility can make your work hard. For example, it doesn't have conventional interrupts; instead, you "interrupt" by requesting bus mastership, then sending a message to the processor you want to interrupt! For simple systems, the simpler Multibus I (or some other simple bus) may be better.

NuBus

This is another high-performance synchronous multiprocessor bus with multiplexed 32-bit data and address paths, DIN connectors, and high data-transfer rates (to 40Mbyte/s in "block transfer" mode). In common with Multibus II, it forces you to go through a bus mastership protocol to interrupt. It is used in the high-end Macintosh computers where, thankfully, Apple added a dedicated interrupt line to each slot. Thus, each card slot has a unique vector assigned; the corresponding software handler knows which card interrupted without polling and has to poll only if that card has more than one possible interrupting device.

VME bus

The VME bus, like NuBus and Multibus II, is intended for multiprocessor 32-bit systems. Unlike those buses, however, it does not use multiplexed data/address lines. Nor does it use a synchronous master clock, preferring asynchronous protocol; this lets you mix processors of varying speeds without pain. The VME bus also implements conventional multilevel IRQ-type interrupts, with full interrupt acknowledgment (complete with daisy-chained INTP line). The VME bus is often viewed as an alternative to Multibus; for example, the original Sun computer from Sun Microsystems used Multibus, whereas their more recent Sun 2 and Sun 3 use VME. VME bus and Multibus II are currently slugging it out in the trade press, cheered on by Motorola and Intel, complete with diatribes and name-calling.

Fastbus and Futurebus

These are *very* high performance buses, with blazing speed. The Fastbus uses large cards (14×16 inches), ECL drivers, and arbitration protocols to support multiple bus masters. In fact, bus communication is one of its strong points, with capability for sophisticated "geographic" communication beyond the immediate crate of cards.

Q-bus and VAXBI

These are proprietary buses used in DEC computers. The Q-bus, used in the LSI-11 and early MicroVAX computers, evolved from DEC's original PDP-11 "Unibus." It supports 16-bit data and 22-bit addressing, asynchronous protocol with multiple masters, and multilevel IRQ-type interrupts. The VAXBI is a high-performance multiplexed 32-bit data/address bus used in the larger VAX 8600-series machines.

10.16 Connecting peripherals to the computer

Interfaces are usually built on printed-circuit cards or Wire-Wrap cards (see Chapter 12) designed to plug into the microcomputer's card slots. Microcomputers generally contain a number of unused slots for just this purpose (or they can be "expanded" to accommodate extra cards), with power-supply voltages and bus signals distributed to the card slots. Some machines use a "proprietary" bus (e.g., the IBM PC), others use a standardized microcomputer bus (e.g., the Sun 3 workstation, which rides on the VME bus), and some have no bus slots at all (e.g., the original Macintosh). Each bus has a standard card size (or sizes), ranging from the small 3.2×11.5 inch IBM PS/2 cards to the giant 14.4×15.9 inch Fastbus cards. Depending on the particular bus, each card has 50 to 300 connections along one edge, either in the form of a set of gold-plated printed-circuit edge connections or as a set of multipin connectors that are soldered to the board; the latter are known as "two-part" connectors and are generally more reliable than PCB edge connectors.

Commercially available interfaces for common tasks (disk, graphics, communications, analog I/O) are usually built on cards that plug into unused bus slots. Cables then go from connectors on the interface card to the peripheral (if any); if the interface involves many inputs or outputs (e.g., a digital logic analyzer), it may connect by cable to an external panel or box where there is more room for connectors (and additional circuitry). In either case it is common to use flat *ribbon cable*, with some care being taken to prevent cross-coupling of strobing signals with data. One method is to ground every other wire in the ribbon; another technique uses ribbon cable bonded to a flexible metal groundplane to reduce inductance and coupling, at the same time maintaining a nearly constant cable impedance. In both cases you can get nice multipin "mass-termination" connectors that attach to the cable with one simple crimping operation; check the catalogs of AMP, Berg, T& B Ansley, 3M, etc. An alternative to ribbon cable is a cable made of multiple twisted-pairs, each pair consisting of one signal line and one ground wire. Twisted-pair cable is available in many configurations, including a nifty ribbonlike flat cable (Allied/Spectra "Twist-'n-flat") in which there is a flat untwisted region every 20 inches for easy connection to crimp-on connectors of the type used for ordinary ribbon cable. Because of the strobed data-transfer protocol used between an interface card and the device it controls, it generally isn't necessary to use signal/ground pairs for *all* signal lines, just for the synchronizing pulses and other strobing or enabling lines. Suitable terminations and driver/receiver combinations should be used for long lines, as described in Section 9.14.

Custom interfaces are best handled in the same way, either laying out printed-circuit boards or using one of the general-purpose interfacing cards available commercially from companies such as Douglas, Electronic Solutions, and Vector. These blank cards have places for ICs and other components (including mass-termination connectors for external cables), and they come in solder and Wire-Wrap styles (more in Chapter 12). Some of them include built-in circuitry to handle bus communication, including interrupts and even DMA.

In some cases the best plan may be to build an interface that resides partly in the computer and partly outside, as suggested in Figure 10.15. In such cases the interface circuitry that goes in the computer will probably be a simple parallel input/output port, either a commercially available parallel port card or a custom card you design. The cable connecting the two parts of the interface is simple and could use one of the high-performance driver/receiver combinations we discussed in Section 9.14 if high-speed communication over long cable runs is needed (for example, RS-422, or the differential current-sinking 75S110 IC, or even fiber optics). This sort of scheme may be particularly useful for interfaces that handle low-level analog signals, since the noise-susceptible linear circuitry can be kept away from the general roar of digital interference present in the computer (and close to its analog signal source); this also allows you to pay careful attention to maintaining clean analog signal ground lines.

SCSI, IEEE-488, and other interfaces

There are literally hundreds of plug-in boards, performing an incredible variety of functions, available for common buses

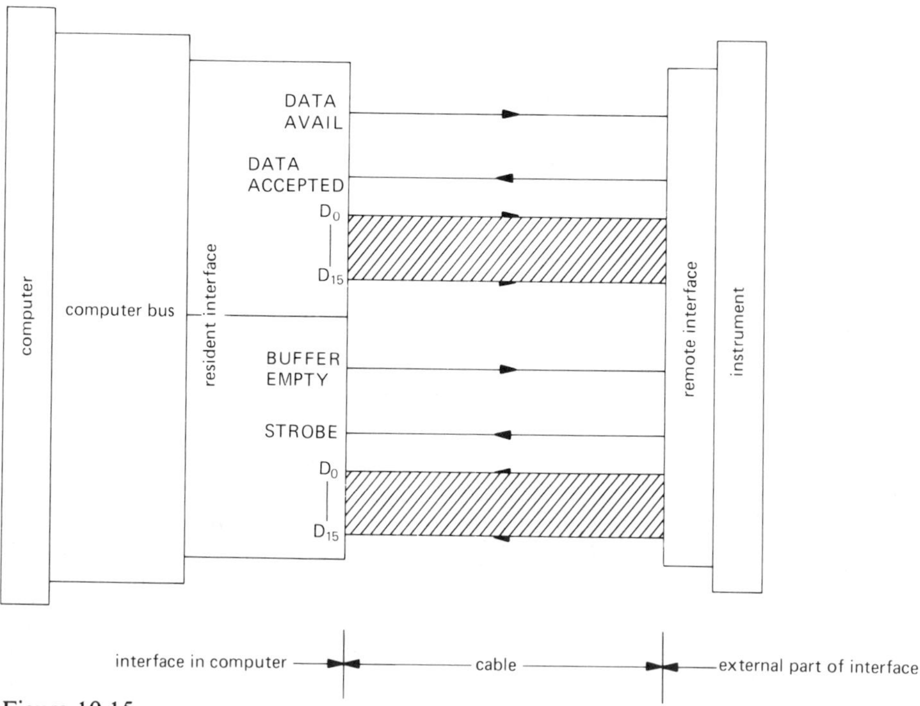

Figure 10.15

such as the IBM PC, Multibus, VME, and Q-bus. These are so inexpensive and easy to use that you should always check out the possibility that either (a) the board you are designing already exists or (b) you can use a simple parallel-port card as a computer-resident part of your interface, as described in the last section. Another possibility is to use a standard built-in "Centronics" parallel port, or an RS-232 serial port (see Sections 10.19 and 10.20), to couple a custom gadget to a microcomputer. This scheme has the virtue of making your gadget portable, even to a microcomputer with a different bus (or no bus at all!), since these ports look the same on all computers. Such a gadget for connection to a serial port will probably have its own microprocessor, so you might tend to think of it as a computer rather than a peripheral. But, as we'll explain in the next chapter, it's fun and easy (and cheap) to build little microprocessor-based instruments; there's really no good reason to treat a microprocessor differently from any other LSI chip, which you wouldn't hesitate to include in a custom circuit.

Following this idea a step further, there are a few "cable interface" standards that have become popular recently. They have names like SCSI ("small computer system interface"), IPI ("intelligent peripherals interface"), ESDI ("enhanced small-disk interface"), and IEEE-488 (also known as HPIB and GPIB, "general-purpose interface bus"). SCSI (pronounced "skuzzy") in particular is now standard equipment on many microcomputers, thanks to the availability of disks and other peripherals that connect directly to a SCSI port. You can get add-in SCSI interface cards for computers without built-in SCSI ports. SCSI is actually a descendant of SASI (Shugart Associates System Interface, a simple parallel bus that Shugart cooked up for their hard-disk drives) and in its simplest incarnation is a byte-wide bidirectional parallel protocol with handshaking. It allows several modes, including synchronous or asynchronous transfer, with single-ended or differential drivers; although it originally was used to connect a single CPU to a single disk, it can be used to couple multiple CPUs to multiple disks. Typical transfer rates are 1.5Mbyte/s (asynch) and 4Mbyte/s (synch); asynchronous protocol is slower because the handshakes are bouncing back and forth during each transfer. SCSI can go 20 feet with single-ended drivers, and 80 feet with differential drivers.

The IEEE-488 bus, originated by Hewlett-Packard as the HPIB, was designed for connecting laboratory instruments to computers. There is a full protocol for connecting multiple instruments on a bus, with phrases like "talkers" and "listeners." IEEE-488 is firmly entrenched in the instrumentation field, with manufacturers like Hewlett-Packard, Keithley, Philips/Fluke, Tektronix, and Wavetek offering it on most of their instrumentation. You can get 488 interfaces for nearly all microcomputers. We'll have more to say on SCSI and IEEE-488 in Section 10.20.

SOFTWARE SYSTEM CONCEPTS

In this section we will discuss some general aspects of small-computer programming, since a knowledge of computer interfacing is of limited value without an understanding of the hierarchy of programs that actually make the computer come to life. In particular, we would like to discuss the important areas of programming, operating systems, files, and use of memory. It is easy to get carried away admiring the beauty of computer hardware and underestimate the importance of good software. Software is what makes the computer fly, and a good operating system and package of "utilities" can make all the difference.

Following our discussion of software and systems, we will end the chapter with a section on communications concepts, in particular the standardized RS-232 serial ASCII protocol, the "Centronics" parallel port, other parallel communications schemes (SCSI, IPI, GPIB), and local area networks.

10.17 Programming

Assembly language

As we mentioned earlier in the chapter, the computer's CPU recognizes certain groupings of bits as valid instructions, which it then acts upon. It is extremely rare to program directly in this binary machine language. Instead, you write programs in a mnemonic assembly language (like our interfacing examples earlier), which a program called an *assembler* converts into relocatable machine code. Assembly language is very close to machine language; each instruction is converted directly into one line or a few lines of machine code (the first line is usually the operation code, with the extra lines generally completing the addressing of the variables, or furnishing constants). Assembly-language programming produces the most efficient code and allows you to get at flags and registers that are inaccessible from higher-level languages. But it is tedious programming, as the examples illustrate, and for most computing jobs (especially those involving plenty of numerical computation) it pays to use a compiled or interpreted high-level language, such as C or FORTRAN, with calls to assembly-language routines only where necessary.

Compilers and interpreters

C, FORTRAN, PASCAL, and BASIC are popular examples of high-level languages. You write a program with algebraic types of commands, for instance

$$x = (-b + sqrt(b * b - 4 * a * c))/(2 * a)$$

and with control structures like *if... elseif ... else*, *for ...* , *while ...* , and *do ...* . You don't have to shuttle your little bytes hither and thither, or worry about addressing, saving registers, etc.; you just declare variables and arrays by type and size and use them in arithmetic or logical expressions. Everything is chocolate-coated.

This is called *source* code, from which there are two routes to a running program. Languages like C and FORTRAN are *compiled*, a process in which a language compiler converts the source-code statements to assembly code; from there it's business as usual, with the assembler converting that intermediate assembly language into machine language. Languages like BASIC and APL have traditionally been *interpreted*; instead of compiling an assembly-language program from the source program, an *interpreter* program "looks at" the statements and executes appropriate computer instructions.

In general, interpreted languages run much more slowly than compiled languages. However, since there's no compilation, assembly, or linking (discussed next), there's no delay after entering a program before it can run. Interpreters often include a simple editor, convenient for quick modification and retrial of a program you're debugging. Interpreted BASIC gained popularity in the early days of microcomputers, when hard disks were a rarity, since it ran entirely in memory; this contrasted with the tedious multipass compilation process. With today's fast disks and efficient compilers, there's not much to complain about. In fact, recent compilers have followed the lead of Borland's interpreted "Turbo Pascal" by providing a "total environment" in which you can hop around effortlessly between editor and running program: If there's a bug, the system puts you back into the editor, pointing to the bad statement; these compilers include debuggers, provision for making "libraries," and other pleasant features.

The current all-around favorite among heavy-duty programmers seems to be C, which combines the power of high-level languages with the beauty of structured languages and the bit-pushing flexibility of assembly code. However, FORTRAN still claims the lion's share of scientific computing.

Linkers and libraries

The assembler produces machine code (well, almost; it's actually called "relocatable machine code") from the assembly code produced by the compiler and from separate subroutines written in (or compiled to) assembly code. In addition, there are usually routines needed by particular commands in the high-level program. For example, a C program might need a math function like *sqrt*, or a host of I/O functions like *printf* or *fopen*. A program called a "linker" handles the bureaucratic nightmare of getting the appropriate subroutines (in relocatable form) from a "library," then rigging up all the linking jumps and addressing so the whole mess fits together in memory. It is the linker's job to put final numerical values into the memory references and variable addresses of the assembled code, and it can do this only when it knows which program calls which, and how long each program is. That's why the code produced by the assembler must be in relocatable form, as must the assembled subroutines that sit in the various libraries [there are usually several – a library of compiler functions, an I/O library, a math library, a library of system calls, and perhaps a home-grown (or store-bought) library of useful subroutines].

Editors and formatters

In prehistoric days (before 1970) you could find card-carrying computer programmers, literally: You wrote your programs by hand on coding forms, then punched them (or paid someone else to punch them) onto those handsome "IBM cards" that had rows of numbers printed on pastel cardboard. Nowadays even toddlers know how to use computer editors, the universal program entry method. Old-timers (those over 30) can still remember the first awkward "interactive" computer editors, with which you could create and modify a text file that, for some reason, the editor never let you see much of. Don Lancaster teased us with his "TV Typewriter," a build-it-yourself project that let you display a line of text on a television. That's all it did. No editing, no storage, no nothing. Our joy was truly unbounded, therefore, when we first used "full-screen" editors.

A good editor (and they're all good, now) lets you type and correct as you go, search for words, change text, move blocks of text around, open multiple windows on multiple files, and write "macro" definitions that do complex manipulations. The screen should redraw quickly, even if you add text near the beginning of a large file. Very large files shouldn't slow things down.

A general-purpose editor doesn't know, or care, what you are writing; it could be a program, a sonnet, or a book. It just creates the text file according to your keyboard instructions. If the file consists of statements in a programming language, the compiler, interpreter, or assembler reads it directly. If, on the other hand, the file is text that you want to print, you have two choices: You can send it directly to a printer, or you can mark it up with formatting information and send it to a formatter program that tells the printer how it should be printed. A good text formatter takes care of margins and line justification, proportional spacing, changes of font, italics, boldface, underline, and so on. The editor and formatter are often combined, sometimes with a screen display showing what the printed page will look like (that's called WYSIWYG, pronounced "wizzywig": what you see is what you get), but more often with the screen display only partially faithful to the final page. The most advanced formatters are capable of typesetting mathematical and scientific formulas. For "camera-ready" quality, you do your printing on a typesetting machine, which exposes photographic paper or film directly; laser or LED printers offer quite good quality at moderate cost and high speed; "dot-matrix" impact printers are the cheapest, as the result shows.

Editor/formatters go by names like MacWrite, Manuscript, Microsoft Word, Sprint, and WordPerfect. Popular technical formatters (which do both text and equations) are TEX and Troff. One caution: When creating text (as opposed to programs), most editor/formatters insert unusual characters in the edited text stream, for example to indicate italics, or temporary end-of-line. These characters are unacceptable to compilers and assemblers. Thus, you've got to force the editor to run in a "vanilla" mode, in order to create unadorned source code that the compiler, etc., won't choke on.

Here's some free advice: (a) Find a good editor and stick with it, and (b) Don't try to persuade others that your editor is better than theirs.

10.18 Operating systems, files, and use of memory

Operating systems

As you might guess from the preceding discussion, you frequently want to run different programs at different times, trading data back and forth between them. For instance, in writing and running a program you begin by running the editor program, creating a text file from the keyboard (good programmers never set pencil to paper, as far as we can tell). After temporarily storing the text file, you bring in the compiler program and compile the stored text file to form an assembly-language file. You store that, bring in the assembler, and produce a relocatable machine-language file from the stored assembly-language file. Finally, the linker combines the relocatable machine code with other assembled subroutines and library routines to produce the executable machine-language program, which (at last!) you run. For all these operations you need some sort of super program to juggle things around, getting programs from disk, putting them into memory, and transferring control to the relevant programs. In addition, it would be nice if each program didn't have to contain all the commands necessary to do disk reads and writes (including interrupt handling, loading of status and command registers, etc.), or, indeed, any of the other detailed data communications tasks.

These are some of the tasks of the *operating system*, a vast program that oversees the loading and running of user programs (the ones you write) and utility programs (editor, compiler, assembler, linker, debugger, etc.), as well as the handling of I/O and interrupts, and file creation and manipulation. The operating system includes a *monitor* for user interface (you tell it to run the editor, compile a program, or run a program) and many "system calls" that permit a running program to read or write a line of text from some device, find out the time of day, swap control to another program, let several multitasking "processes" share CPU time and communicate among themselves, bring in a program "overlay," etc. Good operating systems handle all the busywork of I/O handling, including "spooling" (the buffering of input or output data so that the program can run at the same time that data are being read or written to some device). When running under an operating system, a user program doesn't have to worry about interrupts; an interrupt is taken care of by the system, and it affects the running program only if it wants to take part in the handling of a particular device's interrupts. The whole business of successful "time sharing" (using one computer to handle many users at once), with the disk providing "virtual memory" for unlimited program size, is system programming at its finest.

Some popular microcomputer operating systems are MS-DOS (used on the IBM PC and its imitators), OS/2 (used on the PS/2, IBM's successor to the PC), UNIX (created at Bell Labs, widely used on VAX and 68000-based machines), MacOS, and VMS (company-supplied VAX operating system).

Files

The mass-storage medium in widest use currently is magnetic disk, either flexible ("floppy"), with contacting read/write heads, or rigid ("hard" disk, or "Winchester"), with flying heads. Typical storage capacities are in the range of 1Mbyte for floppies and 20–500Mbyte for small Winchesters. The data are organized into *files*. Text, user programs, utility programs (e.g., editor, assembler, compiler), libraries, etc., are all stored in similar ways, and all constitute files. Although the mass-storage medium is divided into physical blocks or sectors of well-defined size (512 bytes/sector is common), the files themselves may have any length. The operating system mercifully takes care of track/sector addressing, etc.; it gets the data you want, if you know the file name. There are all sorts of interesting details having to do with file organization that we don't have space to describe here. What is important is to understand that all those programs (editor, compiler, etc., as well as user source text, compiled programs, and even data) reside on some mass-storage device as named files, and the system can get them for you (read the next subsection, however, on "ramdisk"). In the normal course of its duties, the system does enormous amounts of file handling.

Recent additions to the mass-storage stable are based on consumer electronics media and provide very high density storage in small packages: (a) Optical disks of the kind used in audio CD players store nearly a gigabyte, as prerecorded "read-only" memory, as *WORM* memory ("Write Once, Read Many"), or (as with magnetic media) as fully erasable read/write memory. (b) Videotape, in both VHS and 8mm formats, lets you store a gigabyte of read/write memory on inexpensive tape; the major drawback is the long access time. Both storage systems use sophisticated error-correction schemes to overcome errors due to media blemishes, etc., which are a minor nuisance in the original audio/video applications of these media, but would be devastating for data or program storage if uncorrected.

Use of memory

Files are stored in some mass-storage device, but a program must reside in memory while being executed. A simple stand-alone program of the sort we'll talk about in the next chapter can be loaded almost anywhere in memory. But in a microcomputer with an operating system there are special areas reserved for special functions. For example, the MS-DOS operating system itself, along with its command interpreter, disk buffers, stack, etc., is usually loaded at the bottom of memory, taking care to put its interrupt vectors in the specific locations in low memory that the CPU requires, while the portion of MS-DOS that is in ROM is located high in memory, above the portion of memory reserved for video display buffers. When operating under an operating system, the allocation of memory for user programs will be handled by the system. This is particularly important to understand if you intend to use DMA; in that case you have to let the system figure out where your data buffer wound up, and use that as the starting address for the DMA block transfer.

The situation is even more complicated if programs are being swapped in and out of memory, or moved around in memory. There may be several programs in memory simultaneously, sharing "time slices" of the CPU in a multitasking mode. To add to the complexity, most microcomputers use "memory mapping," in which *physical* memory addresses (what's actually on the bus lines) are mapped to different *logical* addresses (where your program thinks it is). If that isn't enough to confuse you, consider "virtual memory," a feature of advanced microcomputers in which your program is diced up into little "pages," any of which may or may not be in memory at any instant; the program "pages" them in and out in a crazy quilt of frenzied activity.

No discussion of memory use is complete without mentioning *ramdisk*, which can be invoked even on relatively simple machines, if they have enough memory. The basic idea is to make memory look like disk, from the operating-system point of view; you then load into this ramdisk memory the programs that you need frequently. This can be handy during program development, when you need to keep using the editor, compiler, assembler, and linker. With ramdisk, things move along quickly, since no actual disk access is required. It does have the hazard that you can lose all your work if the computer crashes, since files are not automatically saved on disk. A related concept is a disk *cache*, in which an area of RAM holds the results of recent disk accesses.

Drivers

The computer world is rich with diversity – each month we see products using novel technologies in data storage (magnetic, optical), printers (laser, LED), networks, etc. Different hardware requires different controlling signals, with different timing requirements, etc. This would appear to create real programming problems, since publishing software designed for a dot-matrix printer, for example, would appear to be totally inappropriate for a laser typesetter.

The solution consists of software *drivers*, which are special programs designed to create a uniform programming interface to each particular piece of hardware. Thus, for example, the typesetting language TEX creates output in the form of *dvi* (device-independent) files; a printer driver (specific to the particular printer you are using) eats the *dvi* file and spits out the corresponding idiosyncratic printer codes to instruct the printer. TEX works with any printer, once you have the *dvi*-translating driver. The same sort of device independence goes for mass-storage devices such as disk drives, so that you can attach any of a variety of disks to UNIX, PC-type, or Macintosh computers.

Drivers are really part of the overall system software, and the average computer user is unaware of their operation. If you are designing new computer hardware, however, you will probably find yourself quickly becoming an expert on these essential software modules, since you will have to write your own drivers to make your hardware play with the rest of the team.

DATA COMMUNICATIONS CONCEPTS

A small computer system will usually be configured with some mass-storage devices, such as disks and tape, and some "hardcopy" or interactive devices, such as alphanumeric terminals, printers, plotters, etc. In addition, it may have a modem (modulator-demodulator) so that it can dial up other computers through ordinary telephone lines. Finally, local area networks (LANs) are becoming increasingly popular. With a LAN you can have access to files stored in other computers on the network, as well as the ability to share expensive resources (for example, large disks, tape drives, printers, and typesetters). In each case your CPU has to communicate data. Let's see how it works.

Incompatibility

In the dark "middle ages" of computers (say, up to 1975) the situation was pretty bleak. Each brand of computer had its own bus structure and interfacing protocol (not to mention programming language). You bought (or sometimes built) interfacing cards to fit the particular computer, with custom cables going from the interface to the peripheral itself. This general lack of compatibility extended to the peripherals themselves: You couldn't hook a tape drive to a disk interface, or a terminal to a plotter interface, etc. To make matters worse, the peripherals offered by different manufacturers generally used different signals and data-transfer conventions and were not "plug-compatible."

Compatibility

Some of this incompatibility was unavoidable, since to maximize performance different peripherals transfer their data to and from the interface differently. For example, a magnetic disk moves words in parallel byte-wide format for high speed, and the corresponding interface must use DMA transfer, as we explained earlier; by contrast, a keyboard terminal uses a standardized alphanumeric bit-serial format, with the interface using simpler interrupt-driven programmed I/O. Although some of this incompatibility is still with us, the situation is vastly improved, with most of the industry standardizing on a few agreed-upon data communications standards. The introduction of the IBM PC defined a much-needed small-machine format and data bus, while nonproprietary high-performance buses like VME and Multibus became the backplane for a number of other computers. You can get interface cards for these buses (and others, like DEC's Q-bus) from many manufacturers, which simplifies things enormously. Even more important, the manufacturers of peripherals have agreed on a few standardized "cable interfaces." The most important of these are (a) RS-232 serial format, usually used with alphanumeric ASCII data, (b) Centronics' parallel printer format, (c) SCSI parallel bus, (d) IPI bus, and (e) IEEE-488 (GPIB) instrument bus. Let's take a look at these, and then finish the chapter with a brief description of two popular kinds of local area networks, Ethernet and token-ring networks.

10.19 Serial communication and ASCII

As mentioned earlier, alphanumeric communication between a computer and devices of moderate speed is most frequently done using the 7-bit ASCII code (American Standard Code for Information Interchange), with bit-serial transmission over a single line. Table 10.3 presents a listing of the 7-bit codes. Devices communicating via serial ASCII almost always send an 8th bit, but it is not part of the ASCII code; it is most often a hardware parity bit (sometimes odd parity, sometimes even, but most often set to 0 and ignored), but it is occasionally used as a "meta" shift key to generate an additional 128 characters, which may be Greek symbols, alternate fonts, etc. There are no standards for these extra symbols. (The 8th bit also gets used when you ship *binary* data via a serial connection; this doesn't always work, though, because serial data links are so used to getting rid of the 8th bit during ASCII transfer that they may not permit you to retain it as data.)

A few notes on the ASCII table. The upper-case alphabet begins at 41_H; setting bit 5 to a 1 generates the corresponding lower-case character. The ASCII value for a digit is just the digit plus 30_H. The first 32 ASCII characters are nonprinting, or "control" characters. Some of them are important enough to have earned their own keys on keyboards, for example CR (which may be labeled "return," since keyboards don't have carriages), BS ("backspace"), HT ("tab"), and ESC ("escape"). You can generate any control character (including the above) by holding down CTRL and typing the corresponding letter from the upper-case alphabet; for example, CR is CTRL-M (try it on your computer). The control characters are used to control printing or program execution, or they can be used by programs that otherwise expect to receive alphanumeric characters, e.g., text editors. Some other important control characters, besides the ones listed above, are NUL (null), a character of all zeros often used to delimit character strings; FF (form feed), used to begin a new page; ETX (end of text, affectionately called "control C"), which many operating systems interpret as a command to abort a running program; DC3 (control S), used as a "soft handshake" to stop serial transmission; and DC1 (control Q), the complementary character to resume transmission.

Unfortunately, ASCII doesn't provide for subscripts, exponents, or any Greek or scientific characters. As a minimum, it would be nice to have π, μ, Ω, and the degree symbol (°), which crop up frequently in technical writing. Of course, it is possible to use a control character (or sequence of characters) to indicate a change of font or alphabet. This is

TABLE 10.3. ASCII CODES

non-printing					printing			printing			printing		
Name	**Control char**	**Char**	**Hex**	**Dec**	**Char**	**Hex**	**Dec**	**Char**	**Hex**	**Dec**	**Char**	**Hex**	**Dec**
null	ctrl-@	**NUL**	00	00	**SP**	20	32	@	40	64	‘	60	96
start of heading	ctrl-A	**SOH**	01	01	**!**	21	33	**A**	41	65	**a**	61	97
start of text	ctrl-B	**STX**	02	02	"	22	34	**B**	42	66	**b**	62	98
end of text	ctrl-C	**ETX**	03	03	**#**	23	35	**C**	43	67	**c**	63	99
end of xmit	ctrl-D	**EOT**	04	04	**$**	24	36	**D**	44	68	**d**	64	100
enquiry	ctrl-E	**ENQ**	05	05	**%**	25	37	**E**	45	69	**e**	65	101
acknowledge	ctrl-F	**ACK**	06	06	**&**	26	38	**F**	46	70	**f**	66	102
bell	ctrl-G	**BEL**	07	07	’	27	39	**G**	47	71	**g**	67	103
backspace	ctrl-H	**BS**	08	08	(	28	40	**H**	48	72	**h**	68	104
horizontal tab	ctrl-I	**HT**	09	09	)	29	41	**I**	49	73	**i**	69	105
line feed	ctrl-J	**LF**	0A	10	*	2A	42	**J**	4A	74	**j**	6A	106
vertical tab	ctrl-K	**VT**	0B	11	**+**	2B	43	**K**	4B	75	**k**	6B	107
form feed	ctrl-L	**FF**	0C	12	,	2C	44	**L**	4C	76	**l**	6C	108
carriage return	ctrl-M	**CR**	0D	13	**-**	2D	45	**M**	4D	77	**m**	6D	109
shift out	ctrl-N	**SO**	0E	14	**.**	2E	46	**N**	4E	78	**n**	6E	110
shift in	ctrl-O	**SI**	0F	15	/	2F	47	**O**	4F	79	**o**	6F	111
data line escape	ctrl-P	**DLE**	10	16	**0**	30	48	**P**	50	80	**p**	70	112
device control 1	ctrl-Q	**DC1**	11	17	**1**	31	49	**Q**	51	81	**q**	71	113
device control 2	ctrl-R	**DC2**	12	18	**2**	32	50	**R**	52	82	**r**	72	114
device control 3	ctrl-S	**DC3**	13	19	**3**	33	51	**S**	53	83	**s**	73	115
device control 4	ctrl-T	**DC4**	14	20	**4**	34	52	**T**	54	84	**t**	74	116
neg acknowledge	ctrl-U	**NAK**	15	21	**5**	35	53	**U**	55	85	**u**	75	117
synchronous idle	ctrl-V	**SYN**	16	22	**6**	36	54	**V**	56	86	**v**	76	118
end of xmit block	ctrl-W	**ETB**	17	23	**7**	37	55	**W**	57	87	**w**	77	119
cancel	ctrl-X	**CAN**	18	24	**8**	38	56	**X**	58	88	**x**	78	120
end of medium	ctrl-Y	**EM**	19	25	**9**	39	57	**Y**	59	89	**y**	79	121
substitute	ctrl-Z	**SUB**	1A	26	:	3A	58	**Z**	5A	90	**z**	7A	122
escape	ctrl-[	**ESC**	1B	27	;	3B	59	[	5B	91	{	7B	123
file separator	ctrl-\	**FS**	1C	28	<	3C	60	\	5C	92	|	7C	124
group separator	ctrl-]	**GS**	1D	29	=	3D	61	]	5D	93	}	7D	125
record separator	ctrl-^	**RS**	1E	30	>	3E	62	^	5E	94	~	7E	126
unit separator	ctrl-_	**US**	1F	31	**?**	3F	63	_	5F	95	**DEL**	7F	127

the usual method used in technical word processing, where the formatter interprets subsequent ASCII characters differently. This is probably the best solution anyway, since, given the variety of symbols needed for any serious technical writing, you wouldn't be happy for long even with a very large fixed ASCII alphabet.

Note that computer keyboards are often implemented not simply as ASCII code generators, one code per keystroke; instead, recent practice is to generate unique "key down" and "key up" codes for each key. Special system software (a "keyboard driver," see Section 10.18) may then translate the keystrokes into vanilla

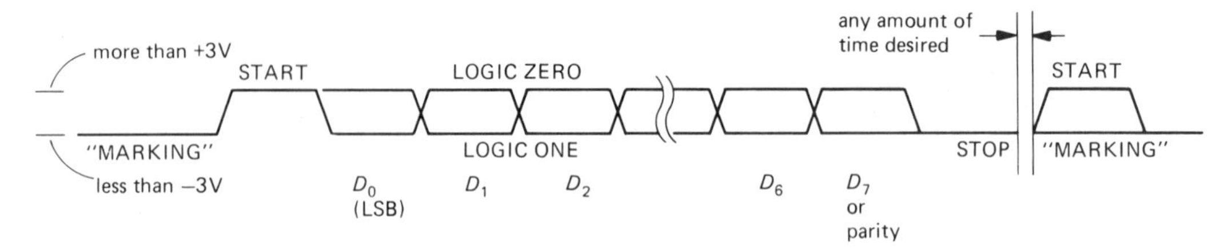

Figure 10.16. RS-232 serial data-byte timing waveform.

ASCII. However, this implementation allows much greater flexibility, since you can configure the keyboard driver to have auto-repeat keys, multiple shifts, keyboard remapping (e.g., a Dvorak keyboard), "hot keys," etc.

Bit-serial transmission

ASCII (or any other alphanumeric code) can be transmitted either as a parallel 8-bit group (8 separate wires) or as a serial string of 8 bits, one after the other. For transmission at low to moderate speeds it is most convenient to use serial transmission, to simplify wiring. A modem (discussed later in this section) converts a serial bit stream to an audio signal, and vice versa (e.g., by using one audio tone for "1," another for "0"), which can then be sent via telephone lines; serial transmission is a natural here, too. Serial transmission has a standard bit-transmission protocol and fixed bit rates: With *asynchronous* transmission, a *start* bit and a *stop* bit (sometimes two) are attached to the ends of each 8-bit character, forming a 10-bit group. The sender and receiver use a fixed bit rate, the most popular of which are 300, 1200, 2400, 4800, 9600, and 19,200 baud (=clock periods per second). Figure 10.16 shows the idea.

When no information is being sent, the transmitter sits in the "marking" state (the language comes from the teletypewriter days, with "mark" and "space"). Every character begins with a START bit, followed by the 8 ASCII bits, least significant bit first (usually organized as 7 data bits, plus 1 optional parity bit), and a final STOP bit; the latter must be at least one clock period, but may extend any amount longer. At the receiving end, a UART ("universal synchronous/asynchronous receiver/transmitter," see Section 11.11) operating at the same baud rate synchronizes to each 10-bit group, generating successive 8-bit parallel data groups from the input serial string. By resynchronizing on the START and STOP bits of each character, the receiver doesn't require a highly accurate clock; it only has to be accurate and stable enough for the transmitter and receiver to stay synchronized to a fraction of a bit period over the time of one character, i.e., an accuracy of a few percent. The receive UART is triggered by the transition at the beginning of the START bit, waits for half a bit cell to be sure the START bit is still present, and then examines the data value at the middle of each data cell. The STOP bit terminates the character and is the resting state if no new characters are sent immediately. The receive UART looks for the STOP bit level 10.5 bit-cell intervals after the START transition, to help verify a correctly sent character. "Break" is a continuous space, which cannot occur during normal character transmission. Programmable baud-rate generators (i.e., programmable dividers) are available that generate any of the standard baud rates from a single oscillator input frequency, with the output baud rate selected by a binary input code. Most modern UARTs (for example the dual-channel synchronous/asynchronous 8530 from Zilog) include internal software-programmable baud-rate generators.

RS-232

The actual serial ASCII signals can be sent in one of several ways. The original method, which dates back many decades, consists of switching a 20mA (or sometimes 60mA) current at the selected baud rate. This is known as "current-loop" signaling. It is sometimes available as an option, but has been superseded for moderate baud rates by the EIA RS-232C standard of 1969 (and subsequent RS-232D standard of 1986), which uses bipolarity *voltage* signaling. The RS-232 standard specifies the properties of both drivers and receivers: A driver must generate voltage levels of +5 to +15 volts (logic LOW input), and −5 to −15 volts (logic HIGH input), into a load of 3k to 7k, with a slew rate of less than 30V/μs, and the ability to withstand a short to any other output (which can be as inhospitable as ±5V@500mA); a receiver must present a 3k to 7k load resistance, converting an input of +3 to +25 volts to logic LOW, and an input of −3 to −25 volts to logic HIGH. Note that logic 1 gets inverted by the RS-232 driver to a negative level, called "mark"; logic 0 is a positive level ("space"). In current-loop transmission, current flows during logic 1 (mark), and ceases during logic 0 (space).

RS-232 receivers usually have voltage hysteresis at the input, and some types let you limit the response speed with a capacitor, to reduce susceptibility to noise pulses. Look at Sections 9.14 and 14.17 for a discussion of official RS-232 driver and receiver ICs. RS-232 works well up to 38,400 baud over distances of tens of feet, even with unshielded bundled multiwire cable; for short links it is sometimes used at 115,200 baud.

RS-232 also specifies the connector type and pin assignments. Unfortunately, it doesn't specify enough! This is an eternal source of confusion because, in general, two RS-232 devices, when connected together, won't work. The problem is so annoying that readers of the previous edition of this book have even complained to *us*, because we didn't tell them what to do about it. Luckily for you, you're reading the second edition. Here's the story:

There are two basic problems in this business: (a) There are two flavors of device defined, with input pins of one type corresponding to output pins of the other; you may want to connect two similar devices together, or you may want to connect two complementary types together. (b) There are five "handshaking" signals; some devices send them out, and expect to receive them back, while others ignore their inputs (and don't drive their outputs). To make things work, you've got to understand these in detail. Let's plunge in.

RS-232 was designed for connecting DTEs ("data terminal equipment") to DCEs ("data communication equipment"). A terminal always looks like a DTE, and a modem always looks like a DCE; but other devices, including microcomputers, can be either. The IBM PC looks like a DTE with a male connector, although most large computers are DCE-like. When you connect a DTE to a DCE, you just connect corresponding pins of their DB-25 connectors (which can be either male or female, at either end!), and, with some luck, it may work. We say *may*, because it still depends on which handshaking lines each device expects from the other, and bothers to drive itself. (Of course, even when the cable is right, you still have to agree on baud rate, parity, and a few other software parameters!) When you want to connect two *similar* devices, on the other hand, you can't connect corresponding pins, because that would connect the two outputs together: A DTE transmits on pin 2 and receives on pin 3, while a DCE does the reverse. So you have to connect them with a cable (called a "null modem") that criss-crosses

TABLE 10.4. RS-232 SIGNALS

Name	Pin number 25-pin	9-pin	Direction (DTE↔DCE)	Function (as seen by DTE)	
TD	2	3	→	transmitted data	data pair
RD	3	2	←	received data	
RTS	4	7	→	request to send (= DTE ready)	handshake pair
CTS	5	8	←	clear to send (= DCE ready)	
DTR	20	4	→	data terminal ready	handshake pair
DSR	6	6	←	data set ready	
DCD	8	1	←	data carrier detect	enable DTE input
RI	22	9	←	ring indicator	
FG	1	–		frame ground (= chassis)	
SG	7	5		signal ground	

pins 2 and 3. Unfortunately, that's not all there is to it.

Table 10.4 shows all the important lines. TD and RD are the data transmit and receive lines; RTS and CTS are "ready to send" and "clear to send"; DTR, DSR, and DCD are "data terminal ready," "data set ready," and "data carrier detect." There are, in addition, two grounds: a "frame ground" (or chassis, pin 1) and a "signal ground" (pin 7); most machines just tie them together. The five signals that aren't data are *handshaking*-type control signals: A DTE asserts RTS and DTR when it's ready to receive, and a DCE asserts CTS and DSR when it is ready to receive. Some DTEs also expect their DCD input to be asserted before they will do anything. All signal lines are RS-232 bipolarity levels, with data (TD, RD) asserted *negative*, but control lines (RTS, CTS, DSR, DTR, DCD) asserted *positive*.

Note that the signal names make sense only as viewed by the DTE: For instance, pin 2 is called TD ("transmitted data") by *both* sides, even though the DTE asserts it and the DCE receives it. Thus, the name of a pin isn't enough to tell you if it's an input or output – you also need to know whether the device thinks it's a DTE or a DCE (or you can cheat and use a voltmeter!).

If all RS-232 devices asserted everything they are supposed to and listened to everything they are supposed to, then you could just connect corresponding pins (for DTE ↔ DCE), or cross corresponding pairs (for DCE ↔ DCE, or DTE ↔ DTE). However, when you connect a device that ignores all handshaking lines to one that expects them, nothing happens. So you have to tailor your strategy to the reality; this sometimes involves trickery. Figure 10.17 shows how to make cables that actually work, for all (well, *nearly* all) situations. In part A we show the connection for DTE ↔ DCE when both devices use full handshaking. RTS/CTS is one pair of handshakes, and DTR/DSR is the other. In C we show the same thing, but with a "null modem" cable to cross inputs and outputs for a DTE ↔ DTE pair. The same cable works for a DCE ↔ DCE pair, but you should reverse the arrows in the picture, and omit the connections to pin 8. These cables won't work, though, if one device is looking for handshaking and the other isn't providing it. In that case the easiest thing

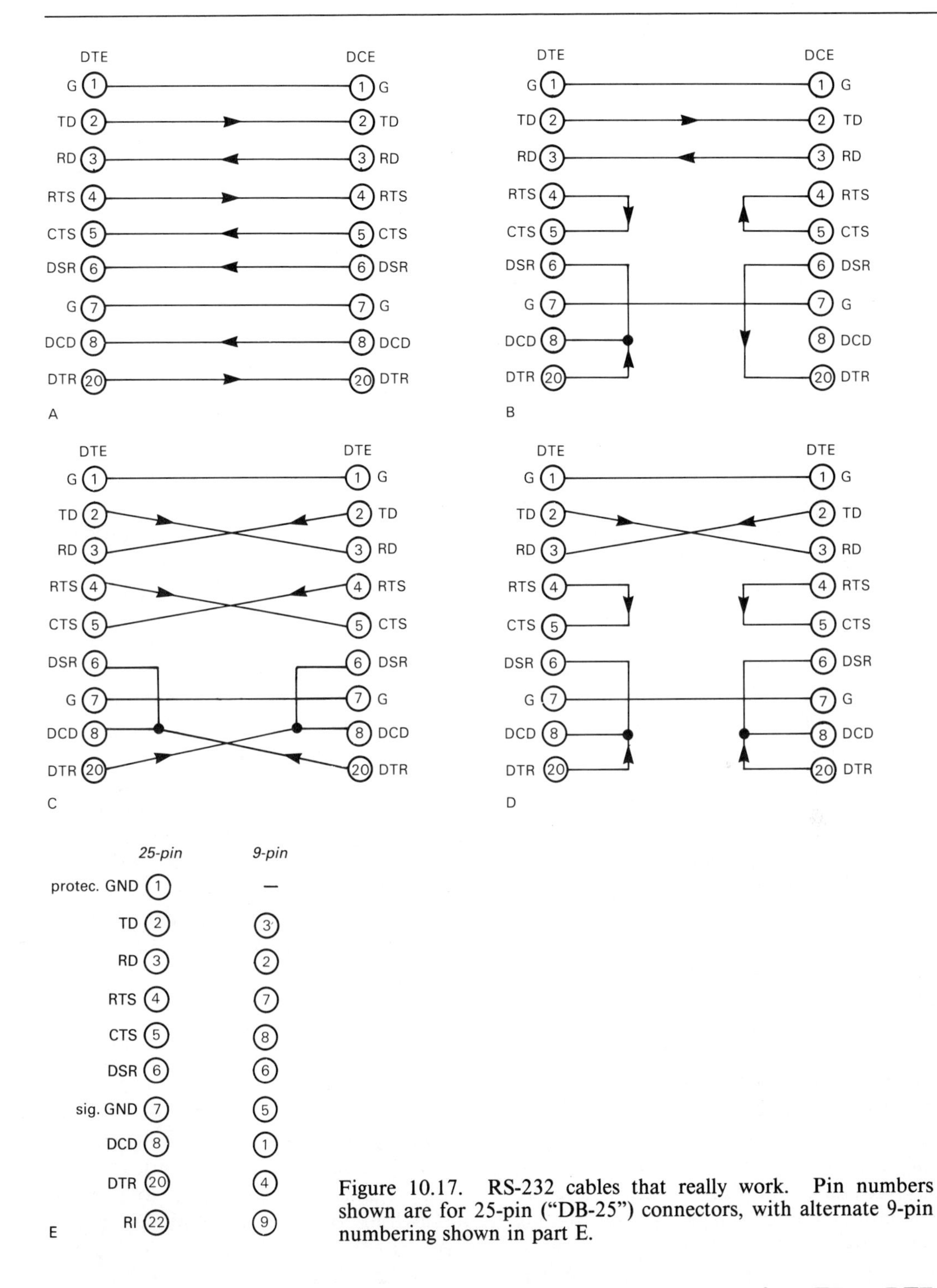

Figure 10.17. RS-232 cables that really work. Pin numbers shown are for 25-pin ("DB-25") connectors, with alternate 9-pin numbering shown in part E.

is to wire the cable so that each device provides its own handshakes, i.e., tells itself to go ahead. That's shown in B for DTE ↔ DCE and in D for DTE ↔ DTE (works also for DCE ↔ DCE, but you should omit the connections to pin 8).

How to become an RS-232 genius. If you make up these four cables, with a male *and* female connector at each end, you can make anything work with anything else (almost). Your colleagues will swear you're a genius. They will, that is, unless they've discovered the *real* professional's gimmick, an "RS-232 breakout box." It has LEDs for each line, so that you can see who is asserting what, and it has little jumpers so that you can connect any given pin to any other pin. Instructions: Look at the lights to get TD and RD connected right, then look again to see who asserts the handshakes. If a device asserts RTS, it probably looks at CTS. If both do, connect them together; otherwise, loop its RTS back to its CTS. Play the same game with DTR and DSR. If only one pair of handshakes is implemented, it is usually DTR/DSR. In general, the DTR/DSR pair is used to make sure the other side is connected and turned on, while the RTS/CTS pair is used to start and stop transmission as one side gets ahead of the other.

If you're too cheap to buy a breakout box, use a voltmeter to check for implemented signals: Any line with a large (>4V) negative or positive level is asserted; any line floating near ground is not.

Software handshaking. Some devices use the RTS/CTS hardware handshakes to start and stop data transmission while the slower device (e.g., a printer) catches up. Others transmit a "software handshake": CTRL-S (to stop) and CTRL-Q (to resume). If you're lucky, you'll have a choice. The software method means you can use a simpler cable, and if the devices ignore the the hardware lines altogether, your cable is extremely simple, with only pins 1, 2, 3, and 7 connected (all you have to figure out is whether or not to cross pins 2 and 3). The devices may still expect the hardware handshakes to be connected to enable the link, even if they use CTRL-S and CTRL-Q for detailed handshaking. In that case you can get away with the scheme of Figure 10.17B,D. Just make sure you remember to turn on the power at both ends, because neither side has any way to know that the other is alive, or even connected!

Other serial standards: RS-422, RS-423, and RS-485

The RS-232C standard was frozen in 1969, when serial data communication was a relatively leisurely occupation. It works well up to 50 feet, at speeds up to 19,200 baud. But computer and peripheral speeds have been doubling every year or two, and a better standard for serial communication was needed.

As we discussed in Section 9.14, RS-423 is an improved bipolarity single-ended protocol, good to 100kbaud and to 4000 feet (not at the same time); it is essentially compatible with RS-232. RS-422 is a unipolarity differential protocol good to 10Mbaud and to 4000 feet (see Fig. 9.37 for the speed/length tradeoff). RS-485 is similar to RS-422, but with additional specifications so that many drivers and receivers can share a single line. Table 10.5 summarizes the characteristics of these four standards.

Modems

As we remarked earlier, a *modem* ("*mo*dulator/*dem*odulator") is used to convert bit-serial digital quantities into analog signals that can be sent over telephone lines or other transmission paths (Fig. 10.18). An *internal* modem plugs into a slot in your computer (or comes built-in), whereas an *external* modem is a stand-alone box, powered from the ac power line, with RS-232 connection to

TABLE 10.5. SERIAL DATA STANDARDS

	RS-232C/D	RS-423A	RS-422A	RS-485
Mode	single-ended	single-ended	differential	differential
Maximum number				
drivers	1	1	1	32
receivers	1	10	10	32
Maximum cable length	15m	1200m	1200m	1200m
Maximum data rate (bits/s)	20k	100k	10M	10M
Transmit levels	±5V min ±15V max	±3.6V min ±6.0V max	±2V min (diff'l)	±1.5V min
Receive sensitivity	±3V	±0.2V	±0.2V	±0.2V
Load impedance	3k to 7k	450Ω min	100Ω min	60Ω min
Output current limit	500mA to V_{cc} or gnd	150mA to gnd	150mA to gnd	150mA to gnd 250mA to -8V or +12V
Driver Z_{out}, min (pwr off)	300Ω	60k	60k	120k

your computer's serial port. In either case the modem communicates with the telephone line, in one of two ways: (a) direct connection, via a telephone-type "modular jack," or (b) "acoustically coupled," by seating the telephone handset into a rubbery cradle containing microphone and speaker. Acoustically coupled modems are pretty much out of style these days, although they can be handy in hotel rooms where you may not want to crawl around under the beds looking for a modular jack (which may not even exist!).

In most situations you want to be able to send data on a single telephone channel in both directions simultaneously ("full duplex"), sharing the telephone audio bandwidth, which is roughly 300Hz–3kHz. There are three full-duplex formats in common use: 300 baud FSK (Bell 103), 1200 baud dibit PSK (Bell 212A), and 2400 baud dibit PSK (FSK stands for "frequency-shift keying," and PSK stands for "phase-shift keying"). A modem designed for 1200 baud, say, generally also supports 300 baud communication, etc. Although you don't need to understand how the modem encodes its data in order

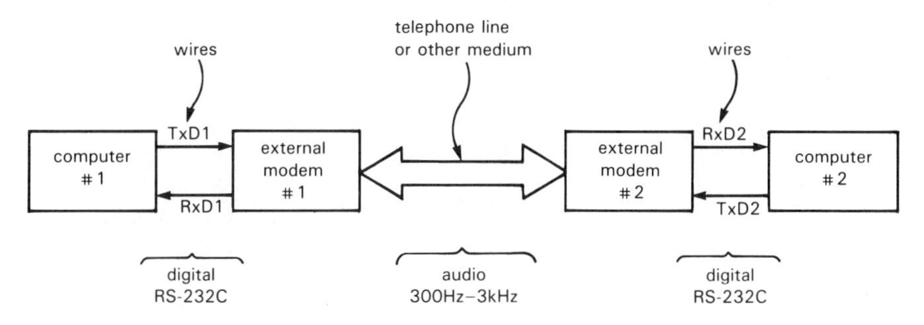

Figure 10.18. Modem communication.

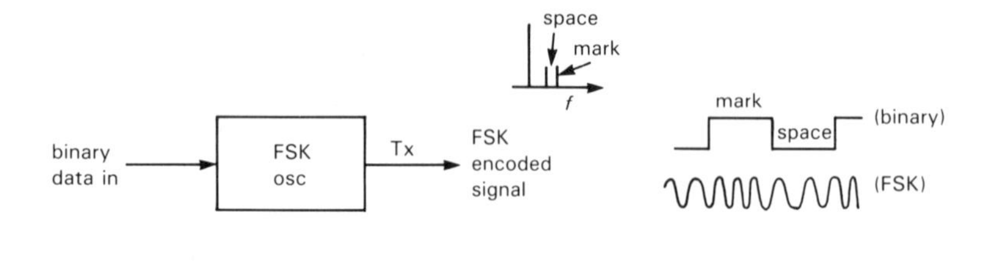

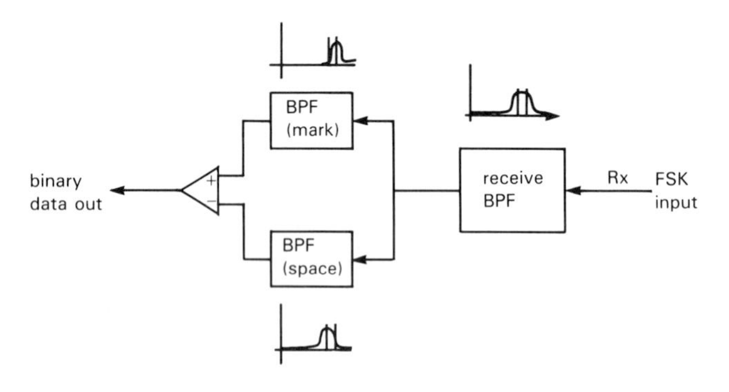

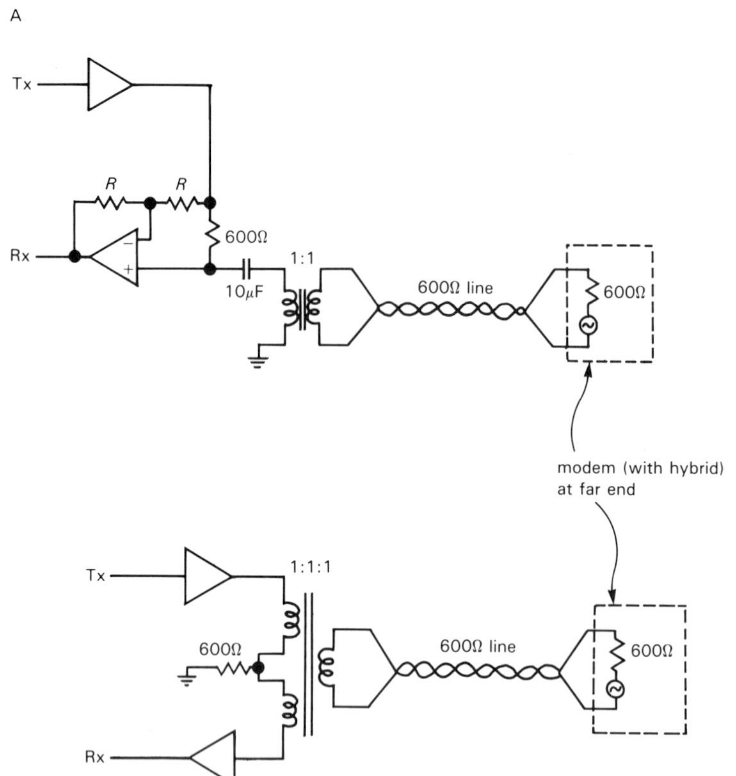

Figure 10.19. A. FSK modem.
B. Hybrid couplers

to use it, the methods are interesting in their own right, and we can't resist describing them briefly.

The 300 baud standard (Bell 103) uses frequency-shift keying (FSK), in which a designated pair of audio tones represents mark and space: 1270Hz (mark) and 1070Hz (space) in one direction, 2225Hz and 2025Hz in the other. A Bell 103 modem is very simple, with a switchable oscillator for transmitting, and a pair of audio filters for receiving (Fig. 10.19A). Note also the use of a *hybrid* circuit (Fig. 10.19B) to isolate the outgoing signal from the received signal: Assuming the telephone line is close to its nominal 600 ohms impedance, none of the modem's own transmitted signal (Tx) appears back at its received-signal (Rx) output. In practice, hybrids don't work that well, because the telephone line impedance can deviate substantially from the nominal 600 ohms (see Section 14.5). Thus, it is important to have a very sharp receive filter, which adds some complexity to the modem circuit.

EXERCISE 10.5

Figure out how the hybrid circuits in Figure 10.19 work. Then impress your friends with your new knowledge.

The 1200 baud standard (Bell 212A) works differently. The digital data stream is grouped into bit *pairs* ("dibits"); each of the four possible dibits is transmitted as a designated phase shift of a fixed-frequency carrier (00: +90°, 01: 0°, 10: 180°, and 11: −90°), with smooth transitions of phase from each transmitted dibit to the next. Thus, dibits are transmitted at a 600Hz rate. The (phase-modulated) carrier frequency is 1200Hz in one direction, 2400Hz in the other. The receiving modem decodes by looking at the *difference* in phase of adjacent dibits. This clever idea has one pitfall, namely that the receiver loses track of relative phase if there is a long run of similar dibits. Therefore, in order to prevent long runs of constant phase, the transmitted data stream is randomized by exclusive-ORing it with a pseudo-random sequence (generated by a 17-bit shift register with XOR feedback from the 14th bit, see Section 9.32), with an identical descrambling process at the receiving end.

The 2400 baud full-duplex modems also use phase-encoded dibits, though with a different set of phases. These sophisticated modems tend to use real-time adaptive equalizers to correct the frequency and time-delay errors of the telephone line, and highly optimized filters for both transmitted and received signals. The end result is that the error rate is not significantly degraded when compared with the earlier 300 baud FSK modems.

You don't have to construct a modem from scratch, because complete modem chips and modules are made by AMI/Gould, Exar, National, Rockwell, Silicon Systems, and TI. Your life is made even easier, however, if you buy a complete modem, whether in the form of an internal plug-in card or an external box with RS-232 connection to your computer. Modems cost \$100–\$300, depending on features. Look for "Hayes-compatible" modems, which accept standardized commands for dialing, etc., that are now the de facto standard used by all communications software.

Some good advice: When using a modem to transfer data files between computers, be sure to use a block-checking modem protocol such as Kermit or XMODEM. These send the data in fixed-length blocks, each with error-checking checksums. The receiving modem compares the checksums, automatically insisting on retransmission of bad blocks. Files received this way are guaranteed error-free; files sent with plain unformatted ASCII transmission, by contrast, can almost be guaranteed to have errors!

10.20 Parallel communication: Centronics, SCSI, IPI, GPIB (488)

For cable communications with high-speed peripherals, parallel transmission is generally better than serial. Here are the popular favorites.

Centronics

This is a simple byte-wide unidirectional parallel port with handshaking, originated by Centronics and now widely used for printers. Unlike RS-232, *it always works!* Table 10.6 lists the signals, which are supposed to be sent with twisted-pairs and terminated in a 36-pin connector. Figure 10.20 shows the corresponding timing.

The basic signals are listed in the first group: D0–D7, STROBE′, ACKNLG′, and BUSY. BUSY is a flag: When LOW, the printer is not "busy," i.e., it's ready to accept data; the data source (computer) therefore asserts DATA, then a STROBE′ (with data guaranteed valid on both sides).

TABLE 10.6. CENTRONICS (PRINTER) SIGNALS

	Pin number			
Name	sig	com	Direction	Description
STROBE'	1	19	OUT	data strobe
D0	2	20	OUT	data LSB
D1	3	21	OUT	•
D2	4	22	OUT	•
D3	5	23	OUT	•
D4	6	24	OUT	•
D5	7	25	OUT	•
D6	8	26	OUT	•
D7	9	27	OUT	data MSB
ACKNLG'	10	28	IN	finished with last char; pulse
BUSY	11	29	IN	not ready (note 1)
PE'	12	30	IN	HIGH = no paper
SLCT	13	–	IN	pulled HIGH
AUTO FEED XT'	14	–	OUT	auto LF
INIT'	31	16	OUT	initialize printer
ERROR'	32	–	IN	can't print (note 2)
SLCT IN'	36	–	OUT	deselect protocol (note 3)
GND	–	33	–	additional ground
CHASSIS GND	17	–	–	chassis ground

note 1: BUSY = HIGH
- *i)* during each char transfer
- *ii)* if buffer full
- *iii)* if off-line
- *iv)* if error state

note 2: ERROR' = LOW
- *i)* if out-of-paper
- *ii)* if off-line
- *iii)* if error state

note 3: normally LOW
- *i)* sending DC3 when SLCT IN' = HIGH deselects printer
- *ii)* can only re-select by sending DC1 when SLCT IN' = HIGH

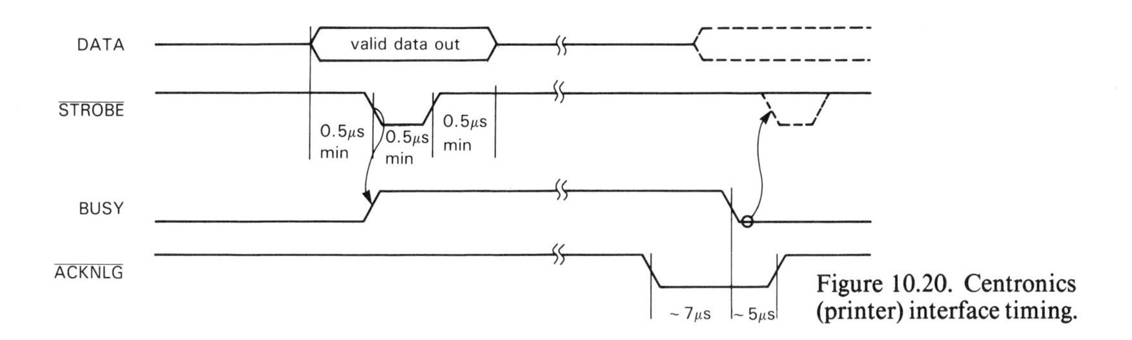

Figure 10.20. Centronics (printer) interface timing.

BUSY then goes HIGH, and it comes LOW again only when the printer is ready for another byte. The computer should look at BUSY, as shown, in order to know when it can send another byte. ACKNLG′ (which is a pulse, not a level) can be used to trigger an interrupt; don't try to use it instead of BUSY, though, because it may be gone by the time you look, and you'll wait forever.

There are several other signals, to indicate that the printer is out of paper (PE′), or off-line (ERROR′ or BUSY); the computer can initialize the printer (INIT′), ask for automatic line feed (AUTO FEED XT′), or send a byte to deselect the printer (set SLCT IN′ HIGH, then send an ASCII DC3). Note the relaxed timings, obviously intended for a slow (mechanical) device that can't accept data at a high rate. Most printers have some buffer memory, so they can accept data at a high rate initially; on the average, though, you can send bytes only at the printing rate. For a dot-matrix printer you're talking 100–300 bytes per second.

If you need to design a Centronics interface to go on some computer's bus, the easiest thing is to drive all the output lines from latched data via programmed I/O: Make D0–D7 one port, and the remaining lines (including STROBE′) a second port. For the input signals (BUSY, etc.), don't latch anything, just enable them onto the bus for programmed IN. A nice touch is to use ACKNLG′ to make an interrupt. Figure 10.21 shows the idea, for the IBM PC bus. Note that interrupts are easy here, because the PC uses edge triggering; just use the trailing edge of ACKNLG′, as shown. We've used one of the latched output bits to disable the interrupt line, as discussed in Sections 10.09 and 10.11. Note also the use of the bus signal RESET DRV to disassert all outputs (and also interrupts) at power-on; that's why we chose the '273 octal D register (which has a RESET′ input).

To use this interface, you assert and disassert output control lines selectively by sending OUT bytes to port B, with appropriate bits set to 1 or 0. With a latched output arrangement like this you can always safely change the state of one output bit without introducing glitches on the unchanged outputs. For this purpose, keep a copy in memory of the current byte latched in port B, so you can send out a new byte to port B with only one bit changed (by using AND and OR, see example below). To generate a STROBE′ pulse you must use software, since the interface has no ugly monostables. Program 10.6 shows how you make a "software pulse" on the STROBE′ line. Note the use of AND and OR, to clear and set a single bit, respectively. In this example we didn't bother updating the byte stored in "current," because at the end it was unchanged. If instead we had changed (and left changed) one of the other control bits, we would have saved the new byte with a "`MOV current,AL`" instruction at the end.

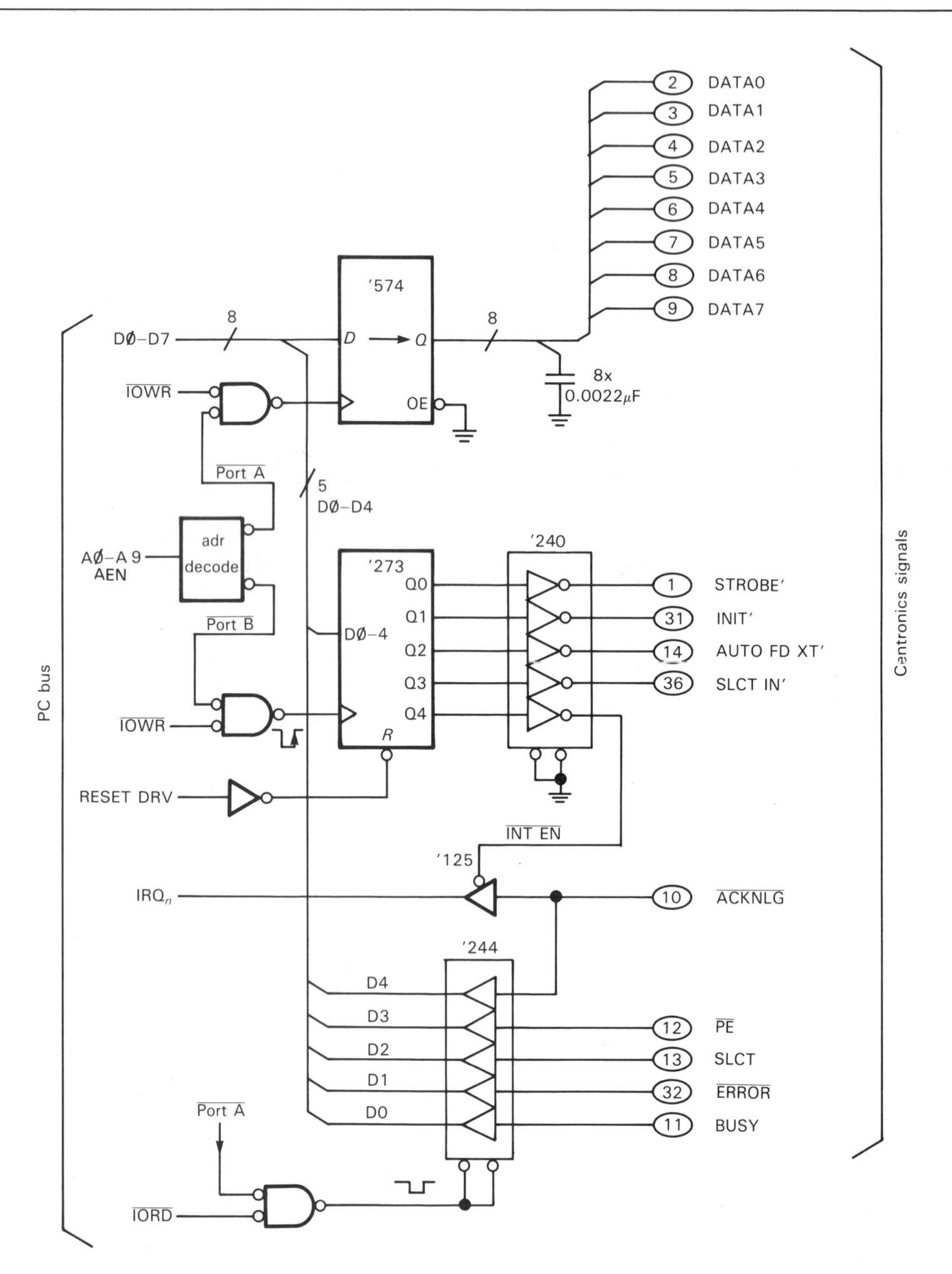

Figure 10.21. Centronics port for PC.

A hardware alternative to keeping a copy of the port byte in memory is to add a "readback" port to the interface, so a programmed IN lets you look at what's actually latched. The next example lets you discover how.

EXERCISE 10.6
Imagine that you are feeling energetic and want to add a readback port to the Centronics interface circuit. Make an IN from port B do the job. You should be pleasantly surprised at how little hardware is required.

Program 10.6

```
                                    ;make a software pulse
                                    ;assume Cenronics "port B" address is in DX
                                    ;assume strobe bit (bit D0) is initially "1"
current DB 0                        ;copy of port B kept here
        o
        o
      MOV  AL,current               ;current value of control byte
      AND  AL,0FEH                  ;clear D0
      OUT  DX,AL                    ;send to port B
      OR   AL,1                     ;set D0
      OUT  DX,AL                    ;and send it out again
        o
        o
```

EXERCISE 10.7

Now rewrite Program 10.6, using your new port and omitting the use of "current."

Centronics ports are standard on nearly all microcomputers; don't hesitate to take advantage of it, if you need a quick and simple parallel output port. In many cases (but not on the IBM PC) the microcomputer will even let you use the port bidirectionally; the usual way that's done is by sending a control bit to the port to reverse the direction of the single 8-bit data path.

SCSI and IPI

These are universal parallel interface standards for connecting disks and other high-performance peripherals to microcomputers, as mentioned briefly in Section 10.16. SCSI ("Small Computer System Interface") is an 8-bit parallel cable interface with handshakes and protocols for handling multiple hosts and multiple peripherals. It has both asynchronous and synchronous modes, and defined software protocols. You can get SCSI interface cards to plug into most popular microcomputer buses, including VME and Multibus I and II; you then connect this SCSI "host adapter" to the peripheral's controller card via a flat-cable SCSI bus (Fig. 10.22). The controller card is often part of the peripheral itself (e.g., it may be attached to a hard-disk drive) and communicates with the drive by a "device-level interface," which will have a name like "ST-506/412," ESDI, or SMD.

SCSI has the advantage of effectively making all microcomputers compatible with all peripherals. Everyone's rushing to adopt SCSI, and new microcomputer designs incorporate it right on the CPU motherboard. At the peripheral end, manufacturers are eliminating the controller by going to an "embedded-SCSI" architecture, in which the SCSI bus becomes also the device-level interface. In other

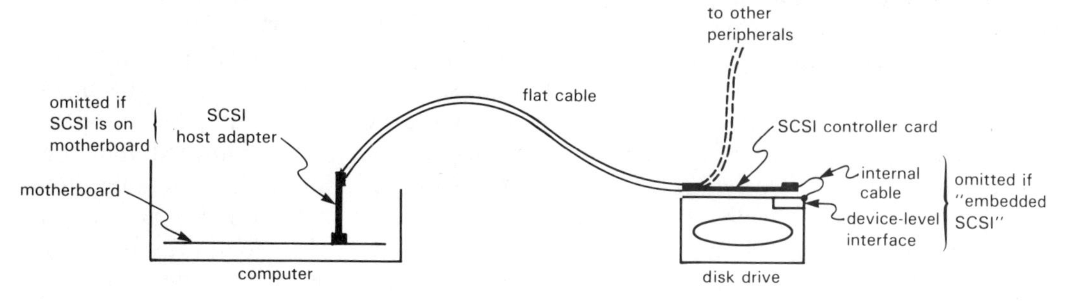

Figure 10.22. SCSI bus with single peripheral.

words, you just hook a cable from the microcomputer's motherboard to the disk drive. SCSI supports data rates to 1.5Mbyte/s (asynchronous) or 4Mbyte/s (synchronous), with cable lengths to 20 feet (single-ended) or 80 feet (differential).

SCSI is complicated enough that we don't have room here to define all its signals, modes, command protocols, and interfacing possibilities. However, because of its popularity, there are single-IC interface chips (e.g., the NCR 5380 series, Western Digital 33C90 series, and others from Fujitsu, Ferranti, etc.) to make your life easy.

SCSI works well with current-generation disks. However, in order to increase data-transfer rates, the industry is considering going to a 16-bit wide interface bus. For this the IPI ("Intelligent Peripheral Interface") may be the next interface bus of choice. IPI specifies a 16-bit parallel bus operating to 10Mbyte/s (5MHz transfer rate); like SCSI, it also works with multiple hosts and peripherals. Hard-disk drives have been getting denser and faster at an amazing pace lately; given the increasing transfer rates, the world is heading rapidly toward universal embedded-bus interfaces (SCSI or IPI). In a few years you probably won't see any other formats.

□ *IEEE-488 (GPIB, HPIB)*

When laboratory instruments first became available with actual data outputs on the back, it was a case of "each company for itself." There were nearly as many interface protocols as there were instruments, with parallel and serial modes, positive and negative polarities, and all sorts of crazy handshakes. It was total pandemonium. We remember vividly designing a huge-digit (6 inches high) display for use in Harvard's lecture halls: It had separate input circuitry for each instrument we owned!

Hewlett-Packard decided in the mid-1960s to end this craziness by defining a universal instrument interface. They modestly called it the Hewlett-Packard Interface Bus (HPIB) and implemented it as the only option on all new designs. It permits up to 15 instruments on a single bus cable up to 20 meters long, with a cleverly designed connector that you can stack at each node. The HPIB bus protocol is byte-wide with handshakes, and it allows data-transfer rates to 1Mbyte/s; it includes software commands to enable any connected device to become a "talker" (source of data), and any combination of the remaining devices to be "listeners" (recipients of data). A "controller" (dictator) tells everybody what to do.

HPIB worked so well that a standards committee was set up by the IEEE to make it official. The resulting standard is known as IEEE-488-1975/ANSI MC1.1, which everyone except HP refers to as "GPIB" ("general-purpose interface bus") or "488-bus." It has become the universal digital interface for laboratory instrumentation. The instruments of all companies can be strung together on the same GPIB, with a microcomputer (or fancy desk calculator) giving the orders. For example, you can set the waveform, frequency, and amplitude of a frequency synthesizer, then take voltage measurements from the same experiment or process.

10.21 Local area networks

In prehistoric times, computing was done in "batch mode" on large centralized computers. They were powerful (slower than the least powerful of today's personal computers, with a tiny fraction of the memory) and expensive (comparable to today's supercomputers). You punched your programs on decks of cards, then submitted the job. With luck, your aborted output was available by the end of the day, so you could resubmit the job the next morning to find the next bug.

Nowadays we're all spoiled by incredible desktop horsepower, fast disks, beautiful graphics. We want more. We want to be able to exchange files with the guy down the hall without getting out of our chairs. We want instant access to everyone's data bases, printers, and fancy peripherals. The way we get it is with networking – both the worldwide networks like BITNET and DECNET and "local area networks" (LANs) like Ethernet and LocalTalk.

The field of networking is still in its infancy, and we expect dramatic changes in the next decade. A few trends have emerged, however, and it's worth describing the kinds of LANs in use today.

CSMA/CD (Ethernet)

Ethernet typifies "carrier-sense multiple-access/collision-detection" (CSMA/CD) networks. It uses coaxial line to transmit 10Mbit/s signals to the addressed recipient. An Ethernet message is sent in "packets," with a preamble and error-checking. The sending protocol goes like this: (a) wait until you see no activity on the network; (b) begin sending your message packet (see below); (c) while sending, check simultaneously for interference (a "collision"); (d) (i) as long as all is clear, continue sending your message, but (ii) if you detect interference, jam the network intentionally (to ensure that everyone else sees the collision!), then abort your transmission, wait a random length of time, and try again; wait a longer "random" time after each successive failure.

Ethernet messages are organized into relatively short packets (≈1kbyte maximum), each of which includes a *header* (identifying recipient and sender), a few bytes telling the packet's length, type, and sequence number, the actual group of data bytes, and finally a "cyclical redundancy checksum" (CRC), from which the recipient can verify error-free transmission. Note that a collision can occur only during the beginning of transmission of a packet, since [by rule (a) above] a transmission in progress for twice the network travel time will not be interfered with.

Ethernet was invented by Xerox and is widely used. It has ample bandwidth for most local area networks, and its performance degrades somewhat gracefully under heavy use, owing to the random retry protocol. You can get Ethernet controllers for most serious microcomputers (VAX, IBM PC, etc.) and buses (Multibus, VME), and it's the official network for the popular Sun and NeXT workstations. An Ethernet network can go up to 1km per segment, with up to 2 repeaters; you can also have fiber-optic "bridges" of greater length. A number of desktop computers can share a multiple-port RS-232 "server," tied into one node on the Ethernet coax. Servers can also tie into shared resources such as printers and large disks.

Token-ring networks

A token-ring network visits a closed set of nodes, in a ring configuration. Collisions are not allowed here, and the rules of the game go like this: Imagine some token object; whoever has it is permitted to send messages, while all others can only listen. In a token ring, the token is a short message that can be passed around when the owner is finished. At any time, one node owns the token and is free to send messages. As with Ethernet (and any other sensible network), the messages are packetized, often using the SDLC format ("Synchronous Data Link Control": one packet = flag + address + header + message + checksum + flag). The message packets circulate around the ring until the addressed recipient receives them. When the sender is finished sending the full message (normally many packets), he sends the token. It circulates around until some other node in the ring, desiring to send a message, swallows it, becoming the new token owner.

LocalTalk

LocalTalk (formerly Appletalk) is a simplified collision network, designed by (guess who) Apple Computer. It is a linear network, not a ring. One node can transmit, while all listen. The cable is a single differential pair, with RS-422 signals transformer-coupled at each node. The packet format is SDLC. Maximum network length is 1000 feet, with up to 32 nodes attached. The network bandwidth is 230.4kbit/s. A compatible variant known as PhoneNET (Farallon Computing Inc.) uses standard telephone cable and connectors and claims to work up to 4000 feet.

The protocol is similar to Ethernet, but simpler: If you hear no activity, you may send a packet. The network hardware doesn't attempt to detect collisions; it just forwards received packets with valid checksums up to the next higher level of software. A collision generally clobbers the colliding packets, rendering both their checksums invalid; thus, the software never gets the message at all! It is the software's job to notice this: For example, the sender of a message expects a reply; if he doesn't get one after a while, he initiates an identical message and tries again. LocalTalk is a "CSMA/CA" network; the "CA" stands for collision *avoidance*, rather than Ethernet's collision *detection*.

LocalTalk has defined protocols for sharing of files and resources (printers, modems, etc.), and it has a method for naming devices connected to the network. You can even get LocalTalk interfaces for non-Apple computers, letting you ship files between Macintoshes, IBM-compatibles, and UNIX computers, and to shared resources such as laser printers.

□ 10.22 Interface example: hardware data packing

If all your instruments connect to a standardized interface bus (such as the GPIB), you're in great shape: Just buy the interface card for your computer, buy some cables, string things together, and hire a programmer. It doesn't take much talent, only money. However, this chapter is about bus interfacing, so we would like to conclude with a complete design example.

If you're like us, you probably don't throw out all your functioning instruments when something new comes along. Some extremely capable measurement instruments were made before the era of GPIB; you can bring new life to them by cooking up an interface to your lab computer. As an example, an 8-digit frequency counter with multiplexed display is likely to have a rear-panel output that gives you one digit after another ("digit-serial, bit-parallel"), encoded as 4-bit BCD, and probably presented at the display's internal refresh rate. You have no control over the timing; each valid digit, along with its 3-bit digit-position address, is signaled with a strobe. Such an instrument most likely uses TTL output levels.

Figure 10.23 shows how to interface such an instrument to an IBM PC. This is a complete interface, including a status flag, interrupt, and selectable I/O port address. The action begins at the lower left, where the counter is busy putting out successive digits, their addresses (0–7), and a STROBE′ pulse when the data is valid. The counter goes from the least significant digit (LSD) to the most significant digit (MSD), so a complete output cycle ends with the receipt of the MSD (digit 7). The eight '173 registers (4-bit D registers with three-state outputs) latch the successive digits, being driven in parallel and separately clocked via the decoded digit addresses. Note the use of a '138 strobed 1-of-8 decoder to generate

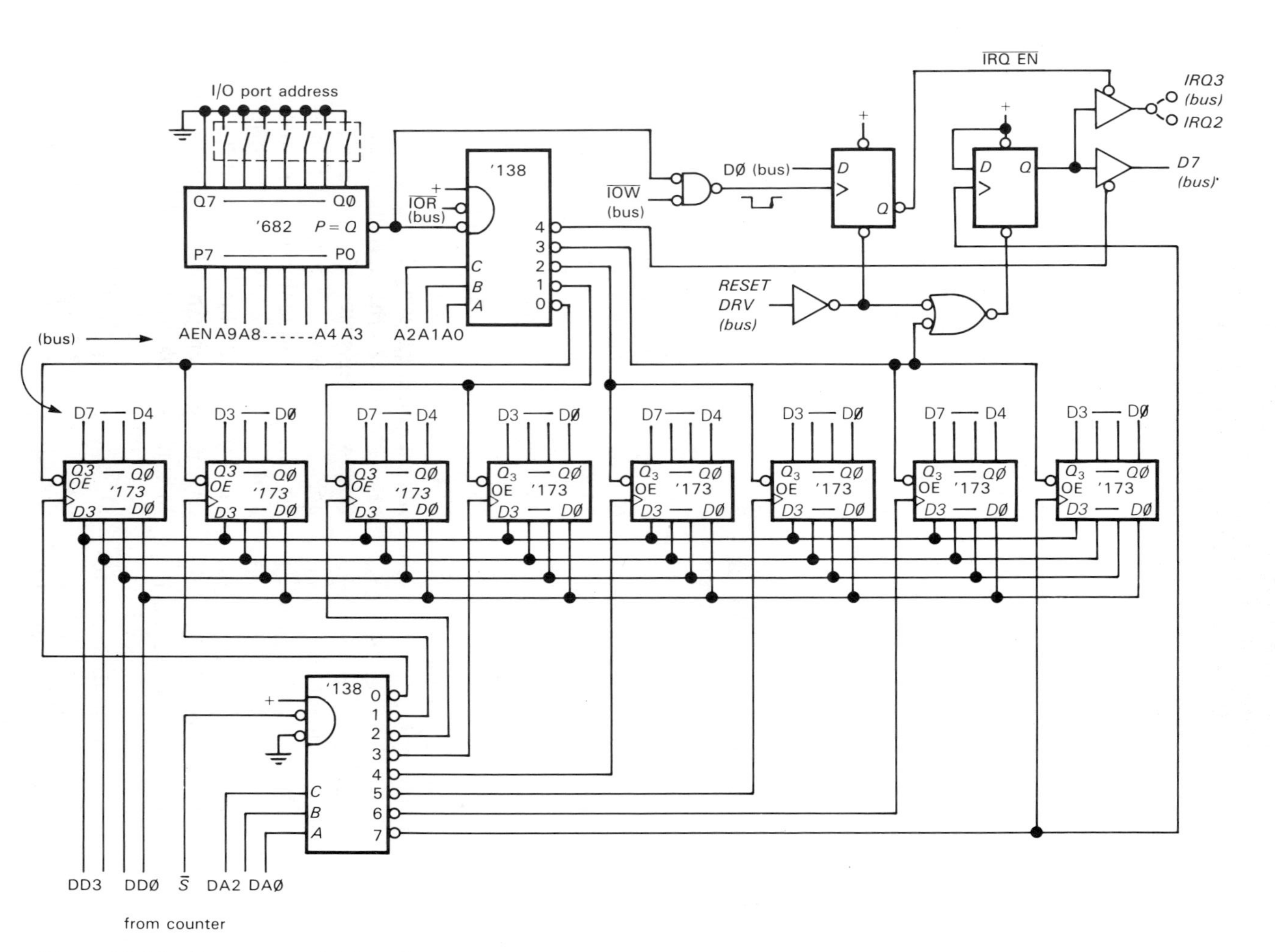

Figure 10.23. Character-serial interface.

the digit clocking signals from the address and strobe.

The counter output is thus latched in the eight 4-bit registers, with the outputs connected as four 2-digit groups (8 bits each). The PC can thus bring in all eight digits with four byte-wide data IN commands, from four successive I/O port addresses (beginning with the one set on the DIP switch). In fact, it can do even better by reading from a 16-bit register (i.e., doing an "IN AX,DX," rather than an "IN AL,DX"), which causes two successive byte reads from consecutive I/O port addresses.

Note the simple address decoding scheme: A '682 octal comparator generates a LOW output when the 7 high-order address bits match the switch settings (and also the nuisance AEN is LOW); this "base address" enables a '138 1-of-8 decoder, strobed by IOR′, which decodes the low-order three address bits to generate the separate data IN enabling pulses corresponding to successive port addresses. This is a common method of handling address decoding, since you usually assign a few contiguous port addresses to the various registers of a single interface.

The status flag is set when the last digit of each group is received from the counter; it can be read with a data IN from PORT+4, where PORT is the address set with the DIP switch. The flag is cleared when the CPU reads the last (most significant) data byte (from PORT+3). This interface also has provision to make interrupts, jumper-selectable on either IRQ2 or IRQ3, and enabled by sending a 1 to PORT (and disabled by sending a 0); note the lazy address decoding we've used for OUT, to save a gate. In a spirit of good citizenship, both the status flag and interrupt enable flip-flops are cleared at power-on.

This interface is an example of "packing" data, the process by which several numbers are stuffed into one computer word. If the "numbers" happen to consist of single bits, you can pack 16 of them into each 16-bit word. This isn't as crazy as it sounds: In digital signal processing you sometimes deal with periodically sampled "hard-clipped" waveforms (which you can think of as 1-bit A/D conversion); for highest I/O throughput rate you pack in hardware (as we did in this example) and read in bus-wide words. Of course, if speed is not important, the simplest thing is to bring in the data with the least hardware and then do the packing and conversion in software. In the preceding example, for instance, you might latch and transfer to the CPU one digit at a time if you can be sure that the latency time of the computer is short enough that no digits will be lost.

EXERCISE 10.8

Modify the interface circuit so that the IRQ line used by the interface is *programmable*: Sending 01_H to PORT enables interrupts on IRQ2, and sending 02_H to PORT enables interrupts on IRQ3; both are disabled by sending 0 to PORT, and also at power-on.

A practical note about this interface circuit: In general it is best to avoid loading bus lines excessively. Our circuit ties each Dn line to the outputs of four '173 three-state registers, which is an undesirably large capacitive load. Although our circuit would undoubtedly work properly, it might limit the number of additional cards you could plug into the bus (particularly if the others sinned in the same way!). In this example, a single '244 three-state octal buffer, interposed between the D0–D7 outputs and the PC data bus, would be a good solution. It should be enabled with the AND of the decoded port address and IOR.

10.23 Number formats

In the preceding example, the bytes (or words) brought in are not in the computer's internal binary-number format; they're really BCD, packed two digits per byte (or four per word). To do meaningful computation, it is best to convert them into an integer or a floating-point number (although there are "decimal-adjust" operations that let you do arithmetic directly on packed BCD numbers). Let's take a look at the usual number formats used in computers (Fig. 10.24), a subject we touched on briefly at the beginning of Chapter 8.

Integers

Signed integers are always represented in 2's complement, using either 1, 2, or 4 bytes, as shown. Thus, the most significant bit (MSB) tells the sign, even though 2's complement is not the same as sign/magnitude representation (e.g., -1 is 11111111, not 10000001; see Section 8.03). You can think of 2's complement as offset binary with inverted MSB; alternatively, you can think of it as an integer with the bit values as shown in the figure. Many computers let you declare variables as *unsigned* integers, in addition to 2's complement signed integers. A 2-byte unsigned integer can have values from 0 to 65535.

Floating-point numbers

Floating-point numbers, also called *real* numbers, are usually 32-bit ("single precision") or 64-bit ("double precision"), with an additional 80-bit format sometimes used for temporary values during calculations. Unfortunately there are several common representations in use. The most popular is the recently completed IEEE standard (officially known as ANSI/IEEE Std 754-1985), which has been implemented by nearly all floating-point chip sets (including Intel's 8087/287/387, Motorola's 68881, and chip sets from AMD, Weitek, et al.) and is therefore universal in microcomputers that accept those chips (this includes the IBM PC).

Figure 10.24 shows the IEEE 32-bit and 64-bit formats. The 32-bit single-precision format has 1 sign bit, 8 exponent bits, and 23 bits of fraction. The exponent tells the power of 2 that the fraction (see below) should be multiplied by. The exponent is "biased" by adding 127, so that the exponent field 01111111 corresponds to an exponent of 0; exponents thus go from -127 to $+128$. The fraction itself uses an interesting trick, originated by DEC in their floating-point format. A floating-point number in binary can always be written in the form $\text{f.fff}\times 2^e$, where f.fff is the (base-2) mantissa ("significand"), and e is the (power-of-2) exponent. In order to maximize the precision you get with a given number of mantissa bits, you "normalize" it by shifting the mantissa left (and decrementing the exponent) until the leading bit is non-zero, thus casting it in the form $1.\text{fff}\times 2^e$. Now, here's the "hidden-bit" trick: Since the resulting normalized significand always has a nonzero MSB, it would be redundant to display it; i.e., you don't put 1fff in the number, just the fff, with the leading 1 assumed. The resulting number gains one bit of precision, and has a range of $\pm 1.2\times 10^{-38}$ to $\pm 3.4\times 10^{38}$.

EXERCISE 10.9

Show that the range of normalized floating-point numbers is as claimed, by constructing the smallest and largest numbers.

The IEEE double-precision format is similar, but with the significand precision more than doubled (by attaching 29 more bits) and with the exponent fortified by an additional 3 bits. The range of numbers is as shown in the figure. There is also a whopping "extended-precision" (80-bit) format, as shown. The IEEE format allows non-normalized numbers also, to give some additional range at the small end (at the expense of precision); these "denormalized" numbers go down to $\pm 1.4\times 10^{-45}$. The standard also defines zero (e = fff = 0; thus there are two

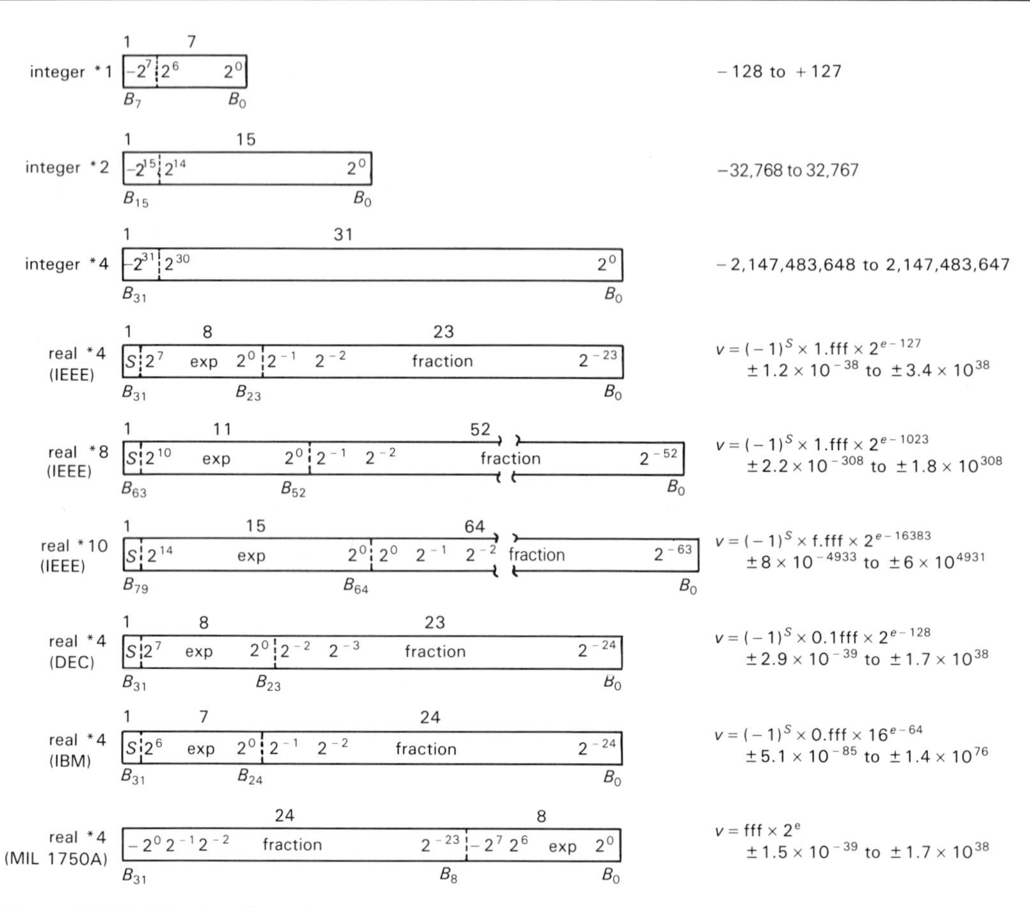

Figure 10.24. Number formats.

zeros, +0 and −0), infinity (e = all 1s, fff = 0; therefore both signs), and a curious class of reserved quantities known officially as NANs (NAN = "not a number")

The other important microcomputer floating-point format is DEC's, used in the MicroVAX and LSI-11 computers (and their ancestors, the VAX and PDP-11 minicomputers). It is very close to the IEEE standard, with the same number of exponent and mantissa bits (including the use of a hidden bit) used for single-precision numbers. In fact, the only differences are the exponent bias (128 instead of 127) and the fact that the mantissa has no leading bits, being instead of the form .1fff (with the "1" hidden). DEC defines only one zero (all bits zero), and does not permit non-normalized numbers or infinity; there are, however, analogs of the IEEE NANs. DEC also has a 64-bit double-precision format.

The last two formats in Figure 10.24 are used in large or special-purpose computers, but not in microcomputers. The "IBM" format has been used in mainframe IBM computers for some time and even in minicomputers like the Nova line from Data General. The 7-bit biased exponent tells the power of 16, rather than 2, giving greater exponent range. The mantissa therefore may have up to three leading zeros; i.e., a normalized fraction has a nonzero most significant hex digit.

EXERCISE 10.10

In order to understand the meaning of this last statement, write out the IBM representation of the number 1.0. Now write the next smaller number that can be represented in this format.

By its choice of exponent radix the IBM format sacrifices some precision for dynamic range. Furthermore, the precision varies somewhat from one number to another, owing to the variable number of leading binary zeros; this is known as "wobble." IBM format has no infinities or NANs, and only one zero (all bits zero); it does permit non-normalized numbers. IBM also has a 64-bit double-precision format.

The last format in the figure is MIL-STD-1750A, used in military systems. It is unusual in departing from the "sign/magnitude" convention of the previous formats, using instead a 2's complement mantissa with a 2's complement exponent. (Actually, the previous formats are more accurately described as sign/magnitude mantissa with offset-binary exponents.) It has no infinities, NANs, or non-normalized numbers; it, too, has a double-precision version.

Number storage in memory

Microprocessor designers like to express their individuality by storing numbers in memory in peculiar orders. The 8086/8 (therefore the IBM PC and compatibles) stores numbers beginning with the least significant byte in the lowest-numbered memory byte; the 68000 family does it the other way around. Lots of luck!

I/O data conversion

We detoured earlier to discuss number formats in the context of our hardware interface with its packed-BCD format. What is the best way to handle the kind of 8-digit data you would get from such an interface? Depending on the type of input data, the number of significant digits, its range of variation, etc., it may be best to convert the incoming data to floating-point (for greatest dynamic range) or to integers (for best resolution) or to do some other sort of numerical massaging (e.g., taking differences from the average value, or between successive data). This might be done in the particular device's software "driver," the section of program that handles the actual input of data. In this sense the software cannot be optimized without an understanding of the hardware and what its data means. Just another reason why it is important to know your way around the wonderful world of electronic hardware!

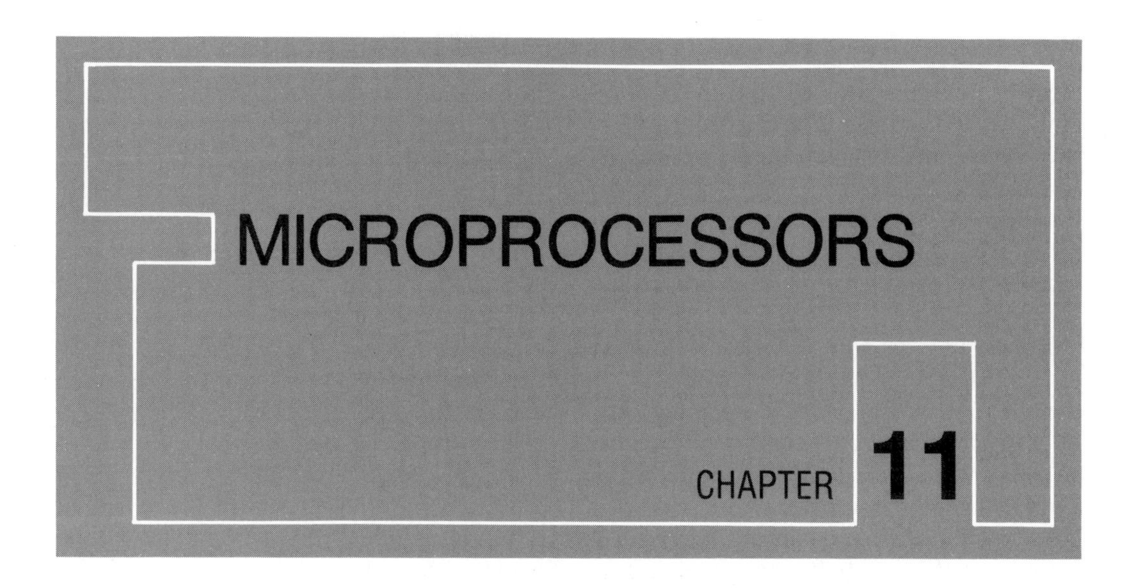

The microcomputers we discussed in the last chapter are stand-alone computational systems built around a microprocessor, usually including mass storage (disks), graphics, printers, and perhaps some networking. They come complete with terminal, memory, and I/O ports, and with plug-in boards you can augment their native capabilities. They're wonderful machines for computation, word processing, computer-aided design (CAD), and even computer-aided manufacture (CAM). With commercially available hardware enhancements you can use them to program logic devices, perform as logic analyzers, or in fact serve as the front end for a variety of engineering tools.

You can have the power of a microcomputer in the instruments and systems you design, by incorporating a microprocessor and some associated circuitry. In this sort of "dedicated" application the processor executes a fixed program you've burned into ROM ("firmware"), and there generally is no mass storage (disks, tape), terminals, etc. From the outside the instrument may look quite ordinary, although a keypad often reveals its superior IQ. Microprocessor-based instruments generally deliver better performance at lower cost and with simpler construction than do equivalent instruments implemented with discrete logic chips. Furthermore, making changes and improvements is often as simple as writing new firmware. As a consequence, no competent designer can afford to ignore these versatile devices. If any further incentive is needed, we might point out that microprocessors are fun; designing an instrument with them gives you an exhilarating sense of power.

When microprocessors are designed into an instrument as dedicated devices, the designer must play a greater role in the design and programming than he would with a microcomputer system. In particular, design with microprocessors includes tasks such as choosing the type of memory (static or dynamic RAM, EPROM, EEPROM) and deciding where in "memory space" to put it, deciding the form that input/output takes (including the choice of I/O hardware, whether constructed from the conventional MSI functions we dealt with in Chapters 8 and 9, or constructed from custom LSI "peripheral support" chips), and writing and debugging the dedicated software (firmware) in the context of the instrument it controls. In general, designers of microprocessor-based instruments must have a thorough knowledge of both hardware and assembly-language software techniques in order to be successful.

Most of the bus interfacing and programming concepts we introduced in the previous chapter in connection with microcomputers are directly applicable to circuits with dedicated microprocessors, and this chapter assumes a familiarity with the contents of Chapter 10. In this chapter we will begin by looking in detail at a small microprocessor with an elegant instruction set, the Motorola 68008, which is really a 32-bit processor (the 68000) with an 8-bit external data bus. After looking at its architecture and instruction set, we will show a complete design example – an analog "signal averager," complete with graphic XY display, digital serial/parallel ports, and some other niceties. No microprocessor system is complete without software, and we will show the programming necessary for this example. Along the way we will have used LSI peripheral and memory chips, and we will have some additional advice about them. The chapter will continue with a discussion of timing, data buses, and other popular processors, including the highly integrated "microcontroller" chips. Finally, we will step back and look at the overall process of electronic design with microprocessors – development systems, evaluation boards, and emulators.

A DETAILED LOOK AT THE 68008

The abundance of different microprocessor types can present a real problem to the circuit designer. Incompatibility between different microprocessor chips, both in their hardware implementation (signal lines, interfacing protocol, etc.) and in their instruction sets, is the rule in the microprocessor world. Rather than attempt to choose precisely the best microprocessor for each job, it is perhaps best to settle on a sufficiently good microprocessor, then build a good development system and gain expertise with it. This is especially true in view of the fact that software development costs and effort often exceed those of the hardware design in microprocessor-based designs.

In this chapter we will concentrate on the Motorola 68008, a junior member of the elegant and popular 68000 family of processors that are used in microcomputers like the Macintosh, NeXT, Sun, and Apollo. It is essentially identical with the 68000 (16-bit data bus, 24-bit address bus), but packaged in a 48-pin DIP with an 8-bit data bus and a 20-bit address bus. It executes exactly the same code as the 68000; the smaller data bus is entirely transparent to the user.

11.01 Registers, memory, and I/O

Registers

Figure 11.1 shows the internal registers of the 68000 (we will use "68000" to refer to features that are common to the 68000 and 68008). There are 8 *data registers* and 7 *address registers*, all of which are completely general-purpose; this contrasts with the 8086/8, in which `AX` (`AL`) must be used for I/O, `DX` for port addressing, etc. The *data registers* can hold bytes (8 bits), "words" (16 bits), or "longs" (32 bits), with the data type in any given operation specified in the assembly-language instruction itself (see below). You do your computation and byte-pushing in the data registers.

The *address registers* are used as pointers into memory or I/O in 5 of the 68000's 12 possible addressing modes; only a few arithmetic operations are allowed on address registers (add, subtract, compare, move). There are no segments or segment registers; with the 68000 family you always have access to the full address space (1 or 4Mbyte for the 68008 in DIP or quad package, 16Mbyte for the 68000, 4Gbyte for the 68020/30).

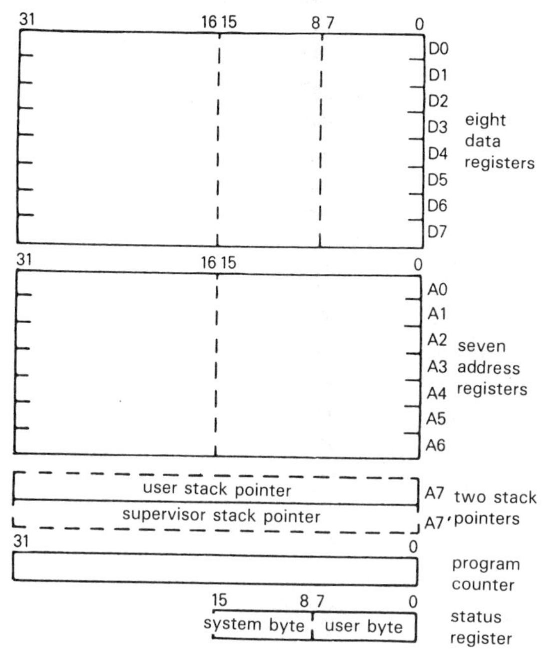

Figure 11.1. 68000/8 registers.

There is a dedicated stack pointer (actually there are two, with only one active at any given time), program counter (or "PC," called the "instruction pointer" in the 8086/8), and status register (SR). The latter holds the flags (zero, carry, overflow, etc.) and also the interrupt mask and mode bits.

Memory and I/O

Unlike the 8086/8, the 68000 processors don't have separate I/O bus signals, and they don't have I/O instructions like `IN` and `OUT`. Instead, they treat I/O just like memory, asserting a full-length address and a strobing signal. To attach an I/O port, you decode the address lines and make your port registers look like nonexistent memory. This is called "memory-mapped I/O" and is used in many microprocessors. (Of course, even with the 8086's separate I/O protocol, you could always have put I/O into memory space, using the MEMR′ and MEMW′ strobes.)

The advantage of memory-mapped I/O is that you can operate on port registers with all the instructions that normally operate on memory: `MOVE` replaces the solitary `IN/OUT` of the 8086/8; but you can also do arithmetic operations (add, subtract, rotate, compare, test), logical operations (and, or), and bit-manipulation operations (bit set, bit test) directly on ports. The only real disadvantage of memory-mapped I/O is that you have to decode a large number of address lines; in practice this is not a problem, because the large address space and small number of I/O ports mean that you can incompletely qualify the address decoding (examples later).

11.02 Instruction set and addressing

Table 11.1 lists the complete set of instruction op-codes for the 68000. To form a valid assembly-language instruction you must specify the operands (via one of 12 addressing modes) and the data type (byte, word, or long). In Motorola's assembly-language syntax, an instruction looks like

`OPCODE`.size source,destination

where OPCODE is from Table 11.1, *size* is either B, W, or L (byte, word, or long, respectively), and the *source* and *destination* can be registers, immediate constants, or memory. Here are some examples:

```
MOVE.W #$FFFF,D0   (immediate, register)
MOVE.B (A0),(A1)   (indirect, indirect)
ADD.L  D5,(A2)+    (direct, indirect with
                      postincrement)
BTST.B #2,$C0000   (immediate, absolute
                      long)
```

TABLE 11.1. 68000/8 INSTRUCTION SET

Op-code	Description
Arithmetic	
ABCD	add packed BCD
ADD	add
ASL	arith shift left
ASR	arith shift right
CLR	clear operand
DIVS	divide (signed)
DIVU	divide (unsigned)
EXT	sign extend
LSL	logical shift left
LSR	logical shift right
MOVE	move
MULS	multiply (signed)
MULU	multiply (unsigned)
NBCD	negate packed BCD
NEG	negate
SBCD	subtract packed BCD
SUB	subtract
Logical	
AND	logical AND
BCHG	bit test and change
BCLR	bit test and clear
BSET	bit test and set
BTST	bit test
CHK	check reg against bounds
CMP	compare
EOR	exclusive OR
EXG	exchange registers
NOT	bitwise complement
OR	logical OR
RESET	pulse RESET line
ROL	rotate left w/o extend
ROR	rotate right w/o extend
ROXL	rotate left with extend
ROXR	rotate right with extend
S*cc*[a]	set a byte conditional
SWAP	swap register halves
TAS	test and set operand
TST	test operand, set flags

Op-code	Description
Control	
B*cc*[a]	branch conditionally
BRA	branch always (relative)
BSR	branch to subr (relative)
DB*cc*[a]	test, decrement, and branch
JMP	branch always (7 modes)
JSR	branch to subr (7 modes)
LEA	load effective address
LINK	link stack
NOP	no operation
PEA	push effective address
RTE	return from exception
RTR	return, restore cc's
RTS	return from subroutine
STOP	stop
TRAP	trap (vectored exception)
TRAPV	trap on overflow
UNLK	unlink stack

Condition codes ("cc")

CC	carry clear
CS	carry set
EQ	equal to zero
F	never true[b]
GE	≥ zero
GT	greater than zero
HI	high
LE	≤ zero
LS	low or same
LT	less than zero
MI	minus
NE	not equal to zero
PL	plus
T	always true[b]
VC	overflow clear
VS	overflow set

(a) see "Condition codes." (b) not available for the B*cc* instruction; use BRA for "BT."

The first example sets 16 bits of register D0 to all 1's (the symbol "$" means hexadecimal, and the "#" signifies an "immediate" constant); the second instruction copies a byte from the memory location whose address is in A0 to the memory location whose address is in A1; the third adds the 32-bit signed integer in D5 to the 4-byte ("long") integer that begins in the memory byte addressed by A2, then increments A2 by 4; the last instruction tests bit #2 in memory location $C0000_H$, setting the Z (zero) flag (for a subsequent branch instruction) accordingly. Note that the operands are in the order *src,dest*, which is the reverse of the 8086.

In general, the 68000 permits nearly all addressing modes and operand sizes with any instruction (the 68000 Programmer's Reference Manual tells precisely what you can do; we've condensed the most useful information into Table 11.2). As a result, it is relatively painless to write good and efficient assembly code. For example, with the 8086 you have to clear out the AL register in order to test an I/O port flag, requiring 5 instructions (PUSH, IN, TEST, POP, Jcc). By contrast, with the 68000 the BTST instruction, above, followed by a Bcc test, does the same job; *no* registers are needed, because the 68000 lets you test a memory bit (and therefore a port register) directly. Furthermore, "auto-incrementing" addressing modes like "(A2)+" facilitate array operations. Even though we haven't explained all the addressing modes and instructions yet, you should be able to do this one yourself:

EXERCISE 11.1

Copy an array of $100 bytes from a table beginning at location $A0000 to a table beginning at location $A8000. You might find the instruction BGT label (branch if greater than zero) useful.

Addressing modes

In the examples above, instructions operate on *constants*, values held in *registers*, and values held in *memory* (or ports). The 68000 provides a nice selection of "addressing modes" by which you specify these operands. Table 11.3 lists the 12 addressing modes, which Motorola counts as 14. Here's what they mean:

Register direct

Syntax: Dn (or An)
Example: `MOVE.W D0,D1`
The operand is the contents of the named register

Immediate

Syntax: # xxxx
Example: `MOVE.B #$FF,D0`
The operand is the specified constant

Memory absolute

Syntax: xxx.W or xxx.L
Example: `ADD.W D0,$B000.W`
The operand's address is given as an immediate constant

Indirect

Syntax: (An)
Example: `SUB.W D0,(A0)`
The specified register contains the operand's address

Indirect postincrement

Syntax: (An)+
Example: `MOVE.B (A0)+,(A1)+`
Same as *indirect*, then An is incremented by *size*

Indirect predecrement

Syntax: −(An)
Example: `MOVE.W D0,-(A7)`
An is first decremented by *size*, then same as indirect

TABLE 11.2. ALLOWABLE ADDRESSING MODES[a]

Operation	Size			Source, count, or bit							Destination					
	B	W	L	Dn	An[b]	( )[c]	abs	PC rel	imm	SR	Dn	An[b]	( )	abs	PC rel	SR
ADD	•	•	•	•	-	-	-	-	•	-	•	•	•	•	-	-
"	•	•	•	•	•	•	•	•	•	-	•	•	-	-	-	-
ADDQ	•	•	•	-	-	-	-	-	3	-	•	•	•	•	-	-
AND	•	•	•	•	-	-	-	-	•	-	•	-	•	•	-	-
"	•	•	•	•	-	•	•	•	•	-	•	-	-	-	-	-
"	-	•	-	-	-	-	-	-	•	-	-	-	-	-	-	•
ASL, ASR	•	•	•	•	-	-	-	-	3	-	•	-	-	-	-	-
"	•	•	•	-	-	-	-	-	(d)	-	-	-	•	•	-	-
B*cc*, BSR	•	•	-	-	-	-	-	-	-	-	-	-	-	-	-	-
BCHG, BCLR, BSET	•	-	-	•	-	-	-	-	8	-	-	-	•	•	-	-
	-	-	•	•	-	-	-	-	8	-	•	-	-	-	-	-
BTST	•	-	-	•	-	-	-	-	8	-	-	-	•	•	•	-
"	-	-	•	•	-	-	-	-	8	-	•	-	-	-	-	-
CLR	•	•	•	-	-	-	-	-	-	-	•	-	•	•	-	-
CMP	•	•	•	•	•	•	•	•	•	-	•	•	-	-	-	-
"	•	•	•	-	-	-	-	-	•	-	•	-	•	•	-	-
DB*cc*	-	•	-	-	-	-	-	-	-	-	•	-	-	-	-	-
DIVS, DIVU	-	•	-	•	-	•	•	•	•	-	•	-	-	-	-	-
EOR	•	•	•	•	-	-	-	-	•	-	•	-	•	•	-	-
"	-	•	-	-	-	-	-	-	•	-	-	-	-	-	-	•
EXT	-	•	•	-	-	-	-	-	-	-	•	-	-	-	-	-
LEA	-	-	•	-	-	(e)	•	•	-	-	-	•	-	-	-	-
LSL, LSR	*(see ASL, ASR)*															
MOVE	•	•	•	•	•	•	•	•	•	-	•	•	•	•	-	-
"	-	•	-	•	-	•	•	•	•	-	-	-	-	-	-	•
"	-	•	-	-	-	-	-	-	-	•	•	-	•	•	-	-
MOVEM	-	•	•	•	•	-	-	-	-	-	-	-	(f)	•	-	-
"	-	•	•	-	-	(g)	•	•	-	-	•	•	-	-	-	-
MOVEQ	-	-	•	-	-	-	-	-	-	8	•	-	-	-	-	-
MULS, MULU	-	•	-	•	-	•	•	•	•	-	•	-	-	-	-	-
NEG, NOT	•	•	•	-	-	-	-	-	-	-	•	-	•	•	-	-
OR	*(see AND)*															
PEA	-	-	•	-	-	(e)	•	•	-	-	-	-	-	-	-	-
ROXL, ROXR	*(see ASL, ASR)*															
S*cc*	•	-	-	-	-	-	-	-	-	-	•	-	•	•	-	-
SUB, SUBQ	*(see ADD, ADDQ)*															
SWAP	-	•	-	-	-	-	-	-	-	-	•	-	-	-	-	-
TAS	•	-	-	-	-	-	-	-	-	-	•	-	•	•	-	-
TST	•	•	•	-	-	-	-	-	-	-	•	-	•	•	-	-

(a) of the most-used instructions. (b) W or L size only. (c) "( )" = all adr reg indirect modes: (A_n), $(A_n)+$, $-(A_n)$, $d_{16}(A_n)$, $d_8(A_n,X_n)$. (d) 1-bit shift. (e) except $-(A_n)$ and $(A_n)+$. (f) except $(A_n)+$. (g) except $-(A_n)$.

TABLE 11.3. 68000/8 ADDRESSING MODES

Mode	Syntax	Address generation
REGISTER DIRECT		
Data register direct	Dn	EA = Dn
Address register direct	An	EA = An
ABSOLUTE		
Absolute short	xxx.W	EA = (next word)
Absolute long	xxx.L	EA = (next two words)
PROGRAM COUNTER RELATIVE		
PC relative with offset	d.W(PC)	EA = (PC) + d_{16}
PC relative with index and offset	{ d.B(PC, Xn.W) / d.B(PC, Xn.L) }	EA = (PC) + (Xn) + d_8
REGISTER INDIRECT		
Register indirect	(An)	EA = (An)
Register indirect with postincrement	(An)+	EA = (An); An ← An + N
Register indirect with predecrement	–(An)	An ← An – N; EA = (An)
Register indirect with offset	d.W(An)	EA = (An) + d_{16}
Indexed register indirect with offset	{ d.B(An, Xn.W) / d.B(An, Xn.L) }	EA = (An) + (Xn) + d_8
IMMEDIATE		
Immediate	#xxxx	DATA = next word(s)
Immediate quick	#x	inherent data

Notes:
EA = effective address
An = address register (A0 – A6)
Dn = data register (D0 – D7)
Xn = address or data register used as index register
SR = status register
PC = program counter
( ) = "contents of"
d_8 = 8-bit offset ("displacement")
d_{16} = 16-bit offset ("displacement")
N = 1 for Byte, 2 for Word, and 4 for Long. If An is the stack pointer, and the operand size is Byte, N = 2 to keep the stack pointer on a word boundary.
← = "replaces"

Indirect with offset

Syntax: d_{16}(An)
Example: `MOVE.L (A0),100(A0)`
The operand's address is (An) plus the 16-bit signed displacement, d_{16}

Indexed indirect with offset

Syntax: d_8(An,Xn.W [or .L]) (Xn can be either Dn or An)
Example: `MOVE.L 100(A0),100(A0,D7)`
The operand's address is (An) plus (Xn) plus the 8-bit signed displacement, d_8

PC-relative with offset

Syntax: d_{16}(PC)
Example: `LEA 100(PC),A3`
The operand address differs from the address of this instruction by the 16-bit signed displacement

PC-relative with index and offset

Syntax: d_8(PC,Xn.W [or .L])
Example: `MOVE.W 100(PC,D0.W),D1`
The operand address differs from the address of this instruction by the sum of the 8-bit signed displacement and the contents of Xn

A few explanations: The first two modes don't address memory; they address *registers* or *immediate constants* (constants embedded in the instruction stream; they can never be destinations, only sources). All the rest are memory-addressing modes. Memory absolute is good for I/O ports or individual memory accesses. Indirect (particularly with postincrement/predecrement) is good for arrays or the stack; furthermore, if the address is already in an address register, it's faster than absolute addressing, because the (absolute) address doesn't have to be fetched during instruction execution. The PC-relative addressing modes are particularly useful if you want to make "position independent" code, since all addressing is relative to the code itself; note that the 8- or 16-bit displacements are 2's complement (signed) integers, allowing displacements of ±127 or ±32767, respectively. Note that you cannot modify immediate or PC-relative operands (they are "nonalterable").

11.03 Machine-language representation

As we mentioned earlier, the assembly language that we have been using is not the "object code" actually executed by the microprocessor, but rather a mnemonic representation convenient for writing programs. The set of assembly-language instructions that constitutes a program must be converted to a set of binary bytes that the processor actually executes. As with the 8086, each 68000 assembly-language instruction assembles into several bytes of machine code. The op-code is always 2 bytes long, with additional words (2 bytes) following only if needed to complete the addressing mode. Depending on the instruction and addressing modes, a single instruction may be from 2 bytes to 10 bytes in length. For example, the instruction

```
ADD.W (A1)+,D3
```

assembles to the minimum length of 2 bytes, namely $(D6\ 59)_H$, with the register numbers and addressing modes encoded (along with the operation) into the 2-byte instruction. However, the instruction

```
MOVE.W #$FFFF,$A0000
```

assembles to an 8-byte instruction, namely $(33\ FC\ FF\ FF\ 00\ 0A\ 00\ 00)_H$, with the operation and addressing modes specified in the first 2 bytes, the immediate constant in the next 2 bytes, and the absolute long destination address in the last 4 bytes.

The CPU, of course, is genetically programmed to know how to interpret this resulting machine code. Looking at the construction of a particular instruction op-code may help you, too, to get a glimmer of how a CPU thinks. Figure 11.2 shows the anatomy of the 68000's best-selling instruction, "`MOVE`." Let's walk through it. The two leading zeros identify the instruction as a `MOVE` operation (almost), with the next two bits defining the operand size (as listed in the figure). It's interesting to note that, since the bit pair 00 is not a legal size, 0000xxx..xx is not a `MOVE` (don't worry, it's not wasted – Motorola used this combination for some other instructions). The next 6 bits tell the addressing mode and register (if any) for the destination, and the last 6 do the same for the source; the figure shows you how to form those bit

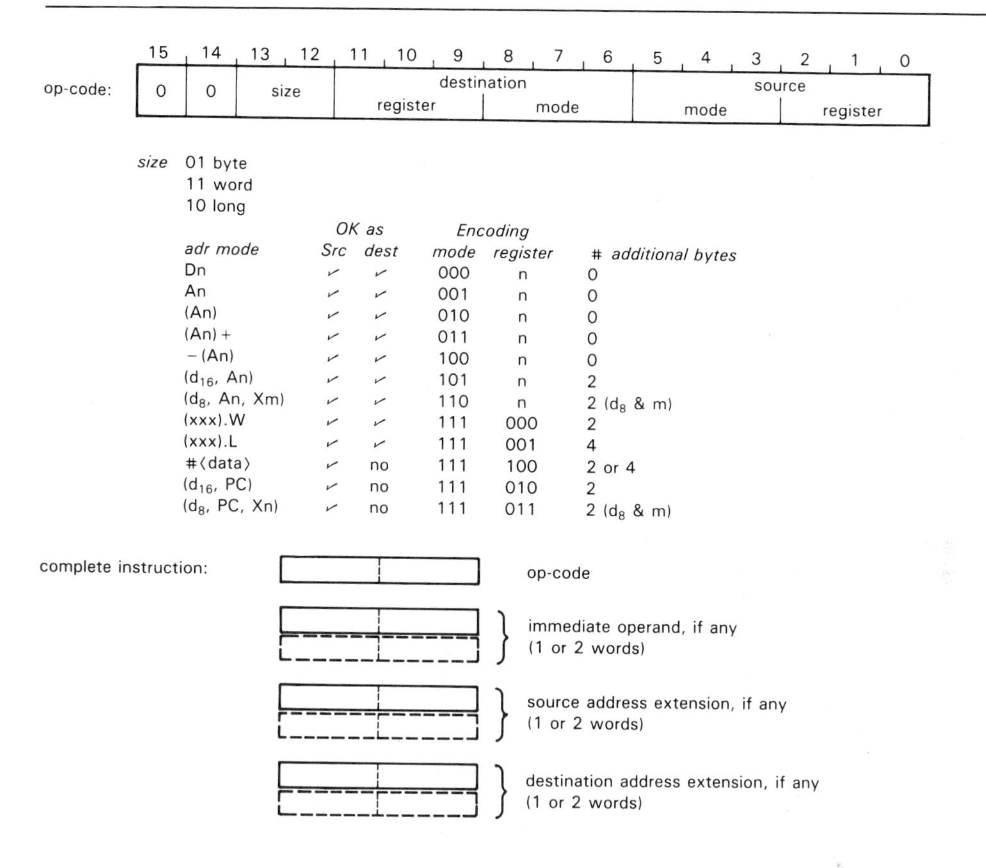

adr mode	OK as Src	OK as dest	Encoding mode	Encoding register	# additional bytes
Dn	✓	✓	000	n	0
An	✓	✓	001	n	0
(An)	✓	✓	010	n	0
(An)+	✓	✓	011	n	0
−(An)	✓	✓	100	n	0
(d_{16}, An)	✓	✓	101	n	2
(d_8, An, Xm)	✓	✓	110	n	2 (d_8 & m)
(xxx).W	✓	✓	111	000	2
(xxx).L	✓	✓	111	001	4
#⟨data⟩	✓	no	111	100	2 or 4
(d_{16}, PC)	✓	no	111	010	2
(d_8, PC, Xn)	✓	no	111	011	2 (d_8 & m)

Figure 11.2. The **MOVE** instruction, dissected.

encodings. Note that the last 5 addressing modes, which don't actually use a register, all share the only remaining mode number (111) and are distinguished by fake "register" numbers. If the addressing mode of either operand requires additional information (immediate data, absolute addresses, or displacements), extra bytes get attached to the instruction, as indicated in the table.

It's interesting to note that the 68000 squanders $1/4 \times 3/4 = 19\%$ of its instruction set on MOVE, in order to accommodate all addressing modes for both source and destination. Motorola couldn't afford to be that profligate with the remaining 50-odd instructions in Table 11.1, so they limited the addressing mode options. For example, using Motorola's term ⟨ea⟩ for the full set of addressing modes, you can form the instructions

```
ADD <ea>,Dn
```

or

```
ADD Dn,<ea>
```

but you can't do the fully general

```
ADD <ea>,<ea>
```

In real life you always use an assembler (running on a computer or a microprocessor "development system") to do the dirty work of constructing these instructions. In order to be able to say we really understand it, however, let's try our luck at "hand assembly": Let's do a

```
MOVE.W #$3FFF,(A1)+
```

It's easy – size is 11 (word); destination is mode 011, register 001; source is mode 111, "register" 100. So the op-code is

00 11 001 011 111 100, or $32FC_H$

and the complete instruction is therefore

32 FC 3F FF

It is our belief that if we persist in abstract discussions of the instruction set and addressing modes, you may well shut the book and never open it again! So let's show a simple programming example instead, then move on to the 68008 bus signals. At that point we'll be able to do a complete 68008 circuit design, with software.

As an illustration of 68000 instructions and addressing, Program 11.1 shows two ways to copy a table of 100_H bytes beginning at location $8000 to the memory space just above it (beginning at $8100). In the first program, we used indirect addressing with offset to do the memory-to-memory move (a nice feature of the 68000 not present in the 8086), followed by pointer increment and counter decrement, and finally a test. With a 10MHz clock, the loop takes 6.2μs, and the table is moved in 1.6ms. The second program uses a second address register to point to the destination, which executes faster and also lets us use postincrement, thus eliminating the `ADDQ` instruction. We also used the more efficient (but tricky) "decrement and branch" (`DBcc`) instruction. This loop executes almost twice as fast (3.4μs/loop, 0.87ms total).

EXERCISE 11.2

Write a program to calculate the sum of 16-bit words in a table that begins at $10000. Assume that the length of the table, in words, is given as the first table entry (which should not be part of the sum); assume also that the sum will not overflow.

EXERCISE 11.3

Write a program to reverse the order of bytes in a table of $100 bytes beginning at $1000. A straightforward (but slow) way is to reverse the order while copying to a temporary array, then copy the reversed version back. A faster method does the reversal "in place" (but be careful not to step on your own feet as you go). Program both methods.

Program 11.1

```
                                      ;move $100 byte table
                                      ;first method
       MOVE.L  #$8000, A0             ;address of table
       MOVE.W  #$100, D0              ;size of table
LOOP:  MOVE.B  (A0), $100(A0)         ;move byte
       ADDQ.L  #1, A0                 ;increment pointer
       SUBQ.W  #1, D0                 ;decrement counter
       BHI     LOOP                   ;loop until done
         o
         o
         o

                                      ;second method
       MOVE.L  #$8000, A0             ;source table
       MOVE.L  #$8100, A1             ;destination table
       MOVE.W  #$FF, D0               ;size-1
LOOP:  MOVE.B  (A0)+, (A1)+           ;move byte
       DBF     D0, LOOP               ;loop until done
         o
         o
         o
```

11.04 Bus signals

If you understood the IBM PC's bus signals, you won't have any trouble with the 68008's, which are similar. We've collected them in Table 11.4 (same format as Table 10.1) and Figure 11.3. The best way to learn about them is to look at the same progression as before: programmed data transfer ("programmed I/O"), interrupts, and DMA. It won't take long.

Programmed data transfer

Programmed data transfer is shown in Figure 11.4; we have also included some signals (CLK and AS′) that you can generally ignore in your circuit design. The 68008 uses a single data strobe (DS′) and a read/write direction line (R/W′), rather than the PC's pair of strobes (IOR′, IOW′).

For WRITE, the CPU asserts R/W′ LOW, and asserts ADDRESS and DATA, then (allowing some setup time) asserts DS′ LOW. The recipient (memory or I/O) latches the data, which (unlike the PC) is guaranteed valid before the leading edge of DS′, and acknowledges by asserting DTACK′. The CPU completes the cycle by disasserting DS′ and then (allowing some hold time, in case the recipient is using transparent latches) disasserting ADDRESS and DATA. Thus the data is guaranteed good during DS′, and for a short time on either side. For READ, the only differences are that the CPU holds the R/W′ line HIGH (to indicate a read cycle), and also asserts DS′ one CPU clock earlier, to give the data source more time to respond to the call for data. That data must be valid before the end of DS′, with actual timing as shown.

TABLE 11.4. 68008 BUS SIGNALS

Signal name	Number	Active	Type[a]	Direction CPU↔BUS	Function
A0–A19	20	H	2S[b]	→	address
D0–D7	8	H	3S	↔	data
AS'	1	L	2S[b]	→	address strobe
DS'	1	L	2S[b]	→	data strobe
R/W'	1	–	2S[b]	→	direction (read/write')
DTACK'	1	L	OC	←	data xfer acknowledge (handshake)
IPL0–IPL2'	2	L	in	←	interrupt request inputs
FC0–FC2	3	H	2S[b]	→	indicates type of cycle
VPA'	1	L	in	←	autovector (or 6800-type I/O)
BERR'	1	L	in	←	bus error signal to CPU
RESET'	1	L	2S[c]	↔	reset
HALT'	1	L	2S[c]	↔	stop
BR'	1	L	OC	←	bus master request
BG'	1	L	2S	→	bus master grant
E	1	H	2S	→	6800-type I/O enable
CLK	1	–	in	←	CPU clock (10MHz typ)

(a) 2S - 2-state (totem-pole); 3S - 3-state. (b) released if not bus master. (c) dual-function: asserted (2-state) by CPU (as output); can be overdriven externally (as input).

The business of DTACK′ deserves some explanation. The 68008 bus is what we called "default-wait" (asynchronous) in Section 10.14: Having asserted DS′, the CPU waits for the (wired-OR) acknowledge signal DTACK′ from the addressed device before completing the cycle. If DTACK′ comes back before the end of S4, no wait states are inserted, and the timing is as shown in Figure 11.4; but if DTACK′

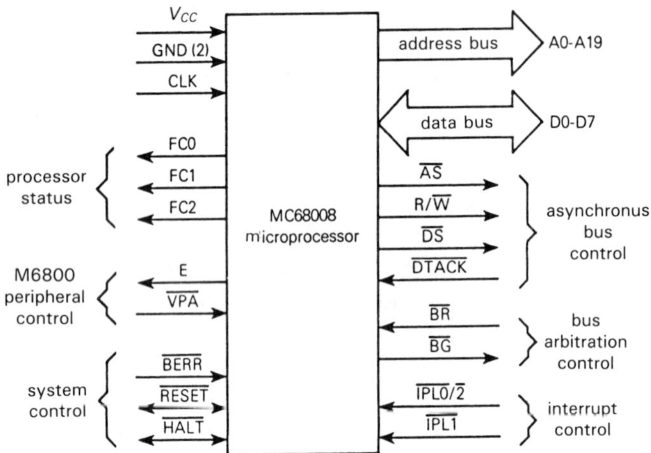

Figure 11.3. 68008 bus.

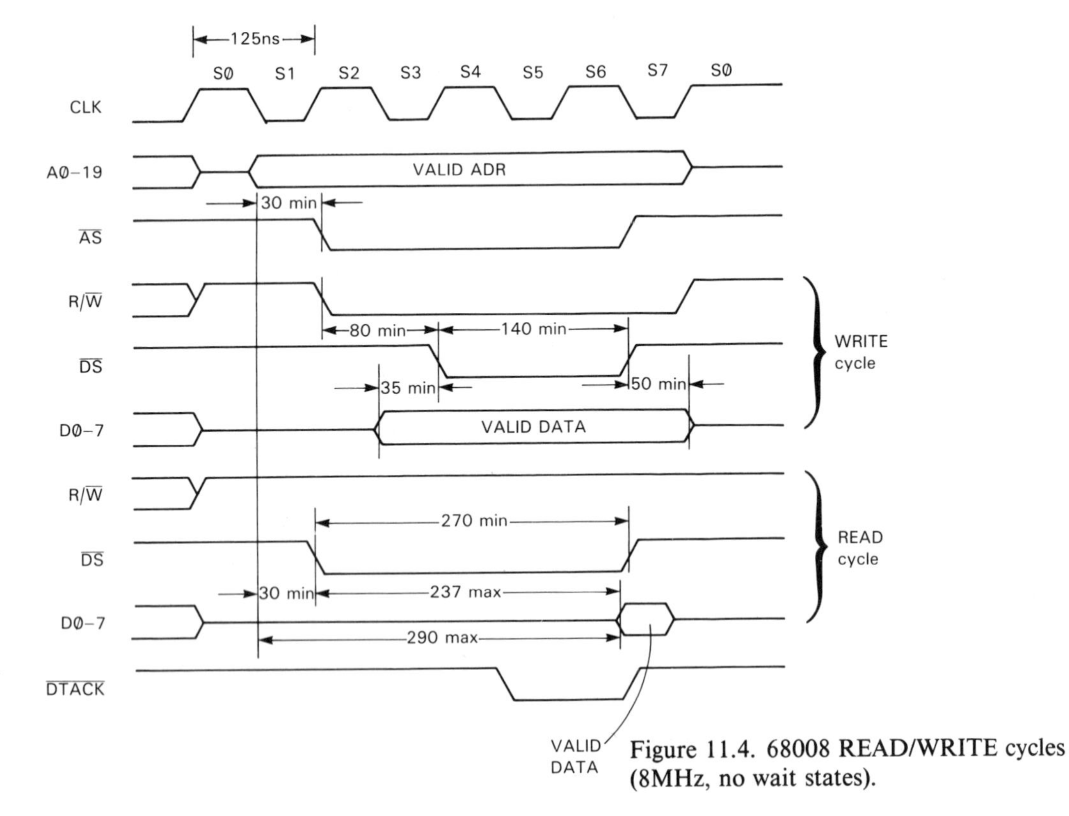

Figure 11.4. 68008 READ/WRITE cycles (8MHz, no wait states).

is delayed, the CPU just holds all its outputs stable (inserting "wait states" after S4) until it sees DTACK′, after which it finishes the cycle with S5–S7. Unless the addressed device is very slow, no wait states are needed, so the addressed device should assert DTACK′ as soon as it recognizes its address (it can just assert DTACK′ based on its address decoding, or, if it's compulsive, AND that with AS′, which signals a valid address). In fact, if all devices on the bus are fast, you can live dangerously and tie DTACK′ permanently LOW, thus disabling wait states entirely; this explains the title of a magazine devoted to high-performance 68000-family applications: *DTACK Grounded.*

All this sounds complicated, but in fact interfacing to the 68008 is simple. Figure 11.5 shows the simplest sort of read/write port. The address lines are decoded and qualified with DS′ and R/W′ to generate the *D*-register clock (write) and three-state enable (read). DTACK′ is asserted whenever this port is addressed, since you never need wait states for devices as fast as octal registers; we've used the usual trick of making a three-state driver into an open-collector driver. Note that we've used the trailing edge of DS′ to clock the *D* register; that is because the 35ns minimum "valid data" to the leading edge of DS′ (see Fig. 11.4) is uncomfortably close to the setup time requirement of many octal registers (e.g., both LS and HCT families have $t_{\text{setup}} = 20\text{ns}$, min). In fact, if the data bus is *buffered* (for example with '245 octal bidirectional buffers), the additional delay of DATA relative to DS′ might violate minimum setup time for the

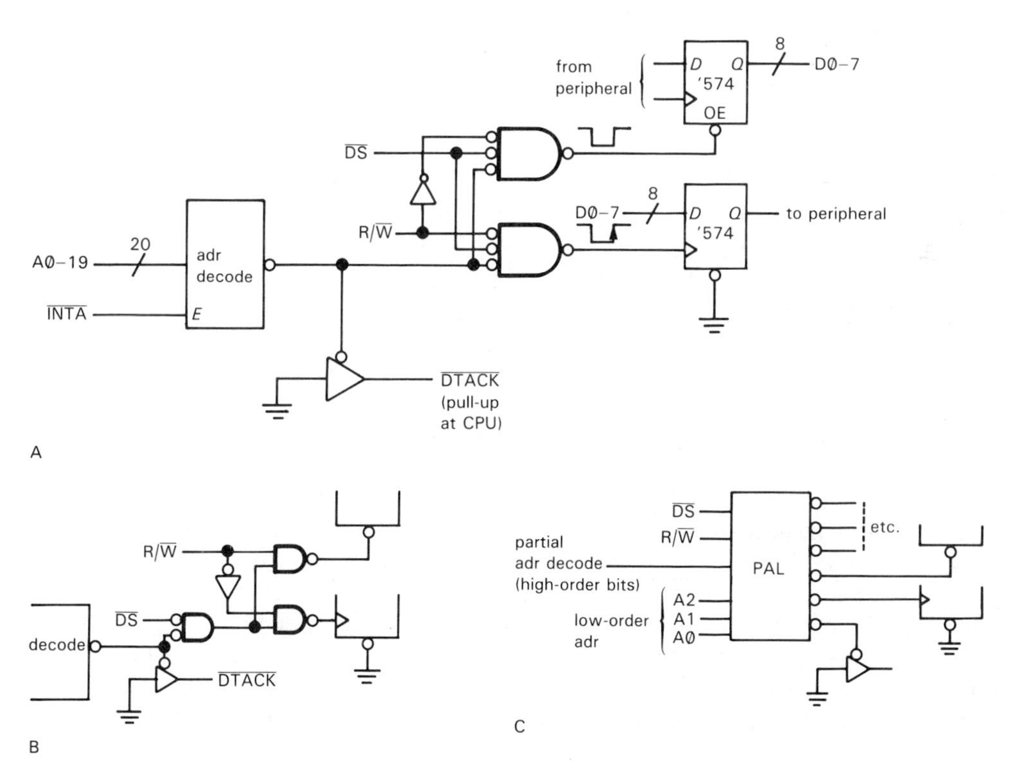

Figure 11.5. Parallel I/O port.
A. Basic implementation.
B. As constrained by available gate types.
C. Implementation with a PAL.

'574. By using the trailing edge, we gain an extravagant 140ns of additional setup time. (Alternatively, we could have used a transparent latch, e.g., a '573, which freezes the data present at the trailing edge of its latch enable; it requires setup and hold times of about 15ns, min.)

The alternative logic forms shown in Figure 11.5 deal with some practical realities. As it turns out (see Table 8.2), in most logic families you can't get 3-input OR gates! One solution is to use 2-input gates, as shown. A more modern approach is to swallow all the gate logic into a combinational PAL; in addition to reducing the chip count, the PAL's extravagant gate content lets you generate clocking and enabling signals for additional peripheral ports, all with one chip.

In this example we have anticipated the next section by showing one minor detail: During interrupts (discussed next), the 68008 executes an interrupt acknowledge cycle, which looks like a READ from the very top of memory (A4–A19 all 1's). If you actually place any memory or registers there, you must disable it during the interrupt cycle, signaled by the "function code" bits FC0–2. The following section explains all.

Interrupts

The 68008 allows both autovectored and fully vectored (acknowledged) interrupts (see Section 10.11 if you've forgotten how these work), using the bus signals in the second group in Table 11.4. In either case you request an interrupt by pulling some combination of the two prioritized request (IPL′) lines LOW. The two lines define three levels of interrupt (the fourth state – both lines HIGH – corresponds to no interrupt at all). These lines are similar to the IRQ lines of the PC bus, but since they are *level*-sensitive, you can put multiple interrupting devices on each level. (It's worth noting that the 68000, and some versions of the 68008, have 3 IPL′ lines, defining 7 levels of interrupt.)

When the CPU sees an interrupt request (at least one IPL′ line LOW), it executes an acknowledge cycle (Fig. 11.6) to identify the source of the interrupt: It asserts the interrupt level on A1–A3 and asserts A4–A19 HIGH; it also asserts the function code lines FC0–2 all HIGH. Then it does a read cycle (i.e., R/W′ HIGH). Your external circuitry now determines the kind of acknowledge cycle: either autovector (jump according to the IPL′ level) or acknowledge (jump according to a vector asserted onto D0–D7 by the interrupting device).

Autovectoring is the simplest (Fig. 11.7). External circuitry detects the acknowledge cycle by looking at FC0–2, and asserts the VPA′ input, coincident with AS′. The CPU then jumps to the service routine corresponding to the IPL level of the interrupt, using vectors (i.e., 32-bit addresses of the routines) stored in absolute locations $68, $74, or $7C. If you have no more than three devices that can interrupt, autovectoring is fine. In fact, even with more interrupters, all is well as long as you poll the status registers of all eligible devices (i.e., all known devices hooked to the interrupt level being serviced) to find the culprit. It's only when you have many possible interrupters (an unlikely situation in a small gadget with a dedicated 68008), and also insist on minimum latency, that you should resort to vector acknowledgment.

Full interrupt acknowledgment works like this: First, leave the processor's VPA′ permanently disasserted (HIGH). Then arrange circuitry so that each interrupt-capable device can assert a unique vector onto the data lines when the processor (a) does a READ with FC0–2 HIGH, (b) at the IPL level (as seen on A1–3) at which the device interrupted. Your circuitry must ensure that no more than one interrupting device asserts its vector, even if several devices interrupted

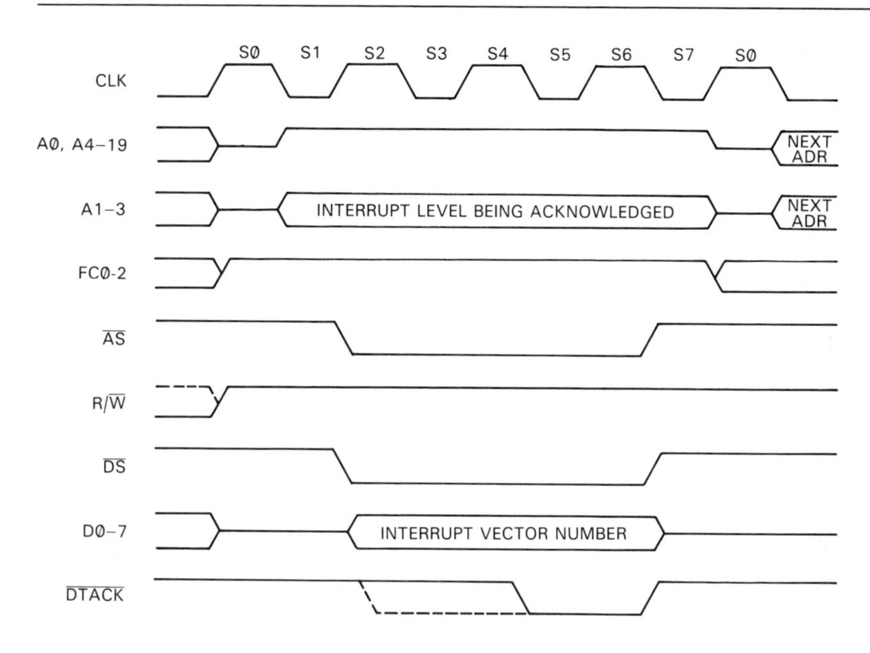

Figure 11.6. Interrupt acknowledge cycle.

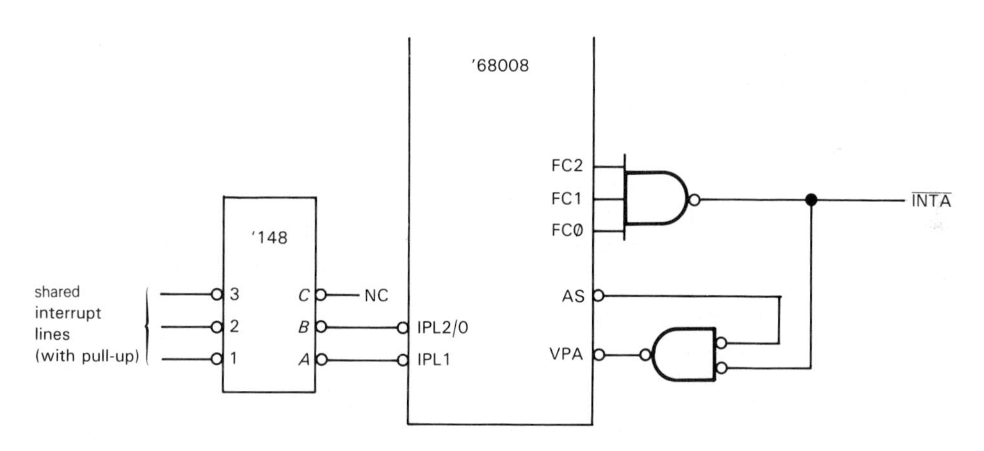

Figure 11.7. Autovectored interrupt.

simultaneously. One method for doing this involves a daisy-chained serial interrupt priority signal, INTP, as discussed in Section 10.11; it guarantees that only the device (of proper IPL level) electrically closest to the CPU acknowledges the interrupt, even if more than one device interrupted on the same IPL level.

An elegant alternative is shown in Figure 11.8. It avoids the need for awkward daisy-chaining, requiring instead a request line from each interrupt-capable device. The state of those request lines is latched at the start of each bus cycle (via the leading edge of AS) and fed to a priority encoder (which generates the binary address of the highest-numbered asserted input, see Section 8.14). The encoder also generates an output (GS′) if *any* inputs are asserted; we use this to initiate a CPU interrupt. For simplicity we put all device interrupts on a single IPL level. The CPU responds to the interrupt by stacking the return address, then initiating the acknowledge cycle of Figure 11.6, during which our circuit asserts both the priority-encoded vector and DTACK′. The CPU now does a vectored jump to the appropriate handler.

This scheme is simple to implement, and for the 68000 family it is actually faster than autovectoring. Furthermore, it is relatively painless to expand the number of interrupters in multiples of 8 with additional '574s and '148s. This scheme does require a dedicated (not bused) line from each peripheral device; although that spoils the symmetry of the data bus, it is probably preferable to the serial daisy-chained scheme, which malfunctions completely if you forget to install jumpers over unused slots! In fact, recent computer buses (e.g. the NuBus in the Macintosh II) seem to be moving toward slot-dedicated interrupt lines.

One interesting (and important) point about this circuit: You might wonder why we need to latch the individual request inputs, since each interrupting device generates its request as a latched bit already (see, e.g., Fig. 10.12). The reason is subtle. Interrupting devices are, in general, asynchronous with respect to the CPU clock and can generate interrupts at any time. If we omitted the latch, and a second peripheral device decided to request an interrupt at precisely the moment the CPU was reading the interrupt vector from the original interrupter, the asserted vector would change in midstream (during the CPU's interrupt acknowledge vector fetch), with unpredictable results. You might object that this scenario is unlikely, and indeed it is; but it *can* happen, and in fact you can estimate about how often. By forcing a decision time for everyone's "intention to interrupt" at the start of each bus cycle, we eliminate the problem (actually, there is still a tiny possibility for an error, due to the "metastability problem"; see Section 8.17 if you need something to worry about).

EXERCISE 11.4

Assume that we choose to live dangerously by omitting the '574 register and that we have two asynchronous devices interrupting at the rate of 1000 interrupts per second each. Assume that the vector fetch cycle has a critical time window of one ns during which a change of asserted vector will lead to an incorrect fetch (i.e., the CPU will read a vector number different from both asserted vectors). Estimate how often the CPU will vector into the wild blue yonder (i.e., crash).

One last point about our circuit. The 68000 series of CPUs has an instruction called `HALT`, which kills all bus activity but permits restarting via an interrupt (or, of course, a complete reboot). Unfortunately, our circuit won't let you restart via an interrupt (why not?). Thus, you must either abstain from `HALT`ing or use some other timing information (perhaps a variation on CLK) to latch the interrupt requests.

The 68000 allows devices to assert 192 distinct interrupt vectors, numbered 40_H–FF_H; the corresponding jump addresses (i.e., the addresses of the corresponding service routines) are stored in memory locations 100_H–$3FF_H$.

Direct memory access

With the 68000, DMA is not mediated by an on-board DMA controller, with address counters, etc., as it is on the PC bus. Instead, the 68000 is willing to relinquish the bus entirely, in an orderly transfer of bus mastership; the new master (which might be another 68000, or just a humble peripheral interface) can then do anything it wants, including (but not limited to) the classic DMA function of sending data to and from memory.

To become master, any device can make a "bus request" by pulling the wired-OR line BR′ LOW. The CPU really takes this

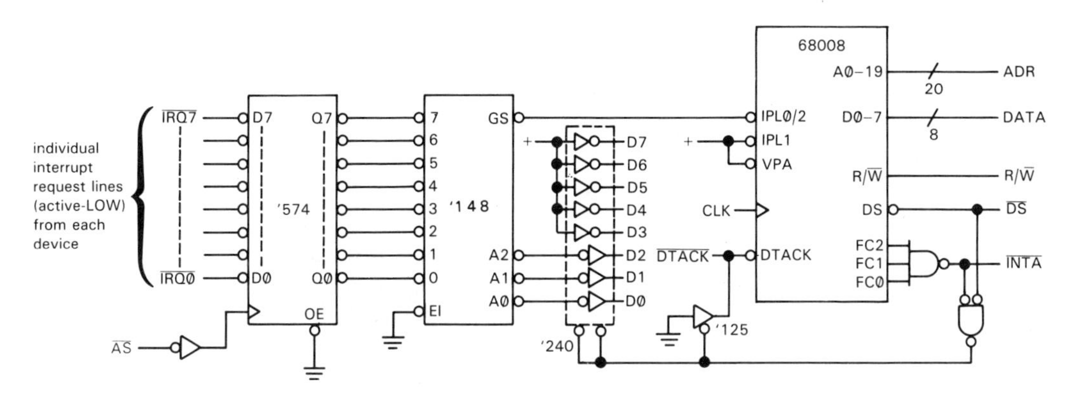

Figure 11.8. Fully vectored interrupt.

seriously, responding as soon as humanly(?) possible by asserting the "bus grant" line BG′ LOW. The CPU also releases control of all bus lines (except BG′), including address, strobes, and other control lines indicated by a superscript "b" in Table 11.4. The external device is now in charge, and it stays that way until it releases BR′, whereupon the CPU regains mastership. The external master is obliged to run the bus under the same rules the CPU normally does, so other bus members don't get confused. In fact, they won't even know anything unusual is happening, unless they happen to look at BR′/BG′.

If there is more than one external bus-master candidate, they have to sort things out ("arbitrate") among themselves. Note, too, that the CPU holds onto control of BG′, thereby retaining a measure of dominance.

Remaining bus signals

Here's what the remaining signals in Table 11.4 are for:

CLK. This is the CPU clock input, which we showed in Figures 11.3 and 11.4. We recommend using inexpensive commercial crystal oscillators in DIP packages for microprocessor clocks, available from companies like CTS, Dale, Motorola, Statek, and Vectron. The 68008 likes a symmetrical clock waveform, which is best generated by dividing the oscillator output with a toggling flip-flop. Maximum clock speeds are usually indicated in microprocessor part numbers (memory, too): Current versions of the 68008 go to 10MHz (MC68008P10). Two-byte instructions generally execute in four clock periods (as in Fig. 11.4), while instructions with the fancier addressing modes have to do more memory fetches and can take up to 70 clock periods or so.

BERR′. You assert this input to tell the CPU that something has gone wrong on the bus. For example, the CPU will be happy to wait forever if no one asserts DTACK′ in response to a data strobe; this could happen if a program tries to access nonexistent memory. Asserting BERR′ causes an interrupt-like jump (officially called an "exception") to a software handler. Look ahead to Figure 11.10 for a simple BERR′ circuit implementation.

RESET′ and HALT′. These signal lines are unusual in serving both as inputs (to reset or halt the processor) and as outputs (by which the CPU can initialize the system). Look again at Figure 11.10 for a straightforward way to handle these lines.

E (Enable). This output signal sounds important, but isn't. It (in combination with VPA′) makes it easy to hook older 6800-series peripheral chips (designed for the synchronous, and relatively slow, 6800 8-bit microprocessor) directly to the 68008. Otherwise you just ignore it.

A COMPLETE DESIGN EXAMPLE: ANALOG SIGNAL AVERAGER

In the following sections we will design a complete 68008-based instrument – an analog "signal averager" (a subject we'll discuss further in Section 15.13). Our example will include the CPU circuitry (with its DTACKs, BERRs, and so on), memory (both RAM and ROM), and plenty of interfacing: A DIP-switch and LED array, serial and parallel ports, calendar clock/timer, A/D and D/A converters, and a solid-state relay for switching ac loads. In fact, we decided to put a little bit of everything into our design. So it's really a general-purpose microprocessor module, and the software makes it into whatever instrument you want.

We'll take you through the hardware design, pointing out how we chose the parts and designed the circuit. You'll learn how to choose and connect memory and peripherals and put everything in memory space in a sensible way. Once done with the hardware, we'll talk about the overall programming and write some software "modules" (sections of code) to handle the interesting tasks. We won't bore you with every line of code, though, since a design like this includes tedious (but essential) routines, for example to get setup commands from a keyboard. Finally, we'll analyze the instrument's performance – the flexibility that we got by using a microprocessor in the design, and the speed limitations that it implies.

11.05 Circuit design

Block diagram

Figure 11.9 is a block diagram, and Figure 11.10 the schematic, of the microprocessor instrument. Look first at the block diagram, which shows the devices connected to the bus. Viewed in terms of micro*computers*, the memory at first seems lopsided, with four times as much ROM as RAM. But it makes sense when you realize that for a dedicated instrument all the programming and tables reside in ROM, not RAM, which is used only to buffer data and hold temporary results of computations. Besides, the manufacturers of UV-erasable programmable ROMs ("EPROMs") have been discontinuing the smaller ROMs as they get better at their trade; it's hard to find EPROMs smaller than 8K×8 nowadays. In any case, the amount of memory shown is the minimum you could use (one ROM chip and one RAM chip); it's easy to add more, as you'll find out by working the problems.

Next in line on the bus: A calendar clock. *Calendar clock??!!* Isn't that just a luxury, for people too lazy to look at their watch when logging on? No, it's essential for any instrument that might be called upon to make periodic measurements, or keep track of when data came in, or a host of other timekeeping jobs. You can program the calendar clock to make periodic interrupts, at rates from 100 per second to once per day; you can also use it as an alarm clock (no bell, of course; it interrupts instead!), settable to any time in the next hundred years. For our signal averager we'll use the timers in the 8536 parallel port, but it's nice to have the calendar handy, anyway.

The 8530 serial port chip is a high-performance 2-channel USART (universal synchronous/asynchronous receiver/transmitter; see Section 10.19), complete with a pair of on-chip baud-rate generators. It's a very nice chip that can do all the usual

RS-232 asynchronous tricks and also has full capabilities for "SDLC/HDLC" synchronous protocols, which include error checking, clock recovery, frame synchronization, etc.; probably overkill for the job, but what the heck. The 8536 is a companion timer and parallel port from the friendly folks at Zilog; it is a powerhouse, also, with a zillion modes of operation. For example, each of its 20 lines can be programmed as either input or output, normal or inverted; each output can be either open-drain or totem-pole, while each input can be normal or "ones-catching" (a momentary HIGH pulse sets an input register). The data sheet goes on and on (26 pages), astounding you with the miracles it can perform.

Moving to the top row in the block diagram, the LEDs are a simple array of 8 lights that are handy for indicating what's going on; they're also great during debugging, if all else fails. We've piggybacked a solid-state ac relay onto one of the latched LED outputs, so you can control some big machine. For example, you could make a temperature-stabilized bath by using an A/D input to sense temperature, and the ac relay to turn on a heater. We'll give you an opportunity to show your stuff with an exercise along these lines. The DIP switch is one of those little 8-station switches, useful for indicating configuration information, e.g., you can tell the processor which serial port (and what baud rate) to use at power-on. Finally, we've hooked up a pair each of A/D and D/A converters so that the instrument can live in an analog world.

Circuit details

Now for the fun part. Let's go through the circuit design (Fig. 11.10).

CPU

CLK. The 68008 needs a CPU clocking signal (CLK), a logic-level square wave in the range 2MHz to 10MHz. The upper limit is set by the intrinsic speed of internal gates and registers; currently you can get 68008s specified for 8, 10, or 12.5MHz maximum clock frequency. The lower limit is dictated by the fact that the CPU uses *dynamic* registers, which have to be periodically "refreshed" because the data is stored on charged capacitors, not flip-flops. The speed of computation is proportional to the clock frequency,

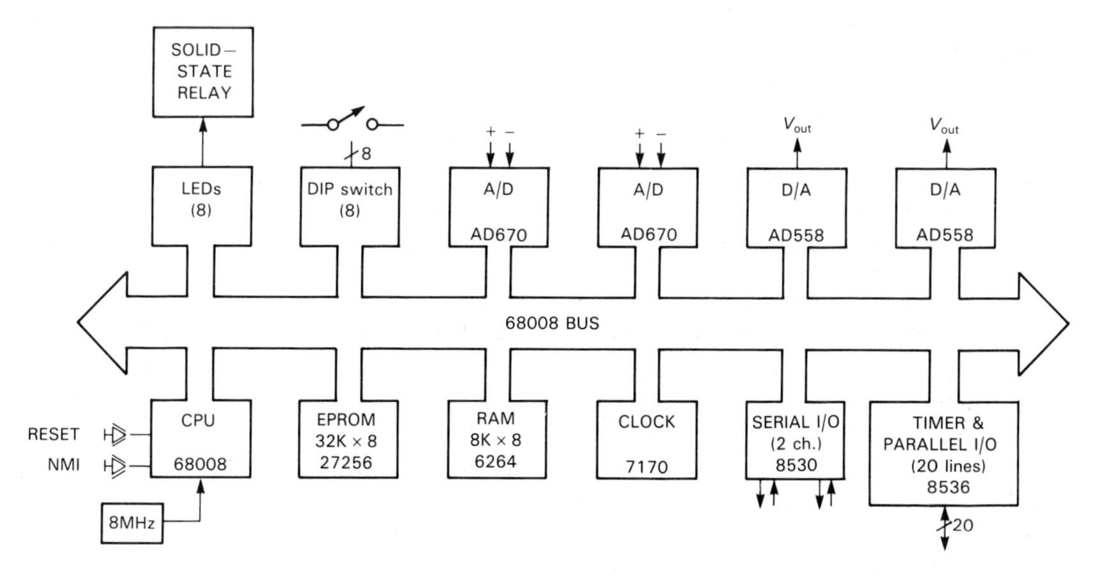

Figure 11.9. General-purpose microprocessor-based instrument – block diagram.

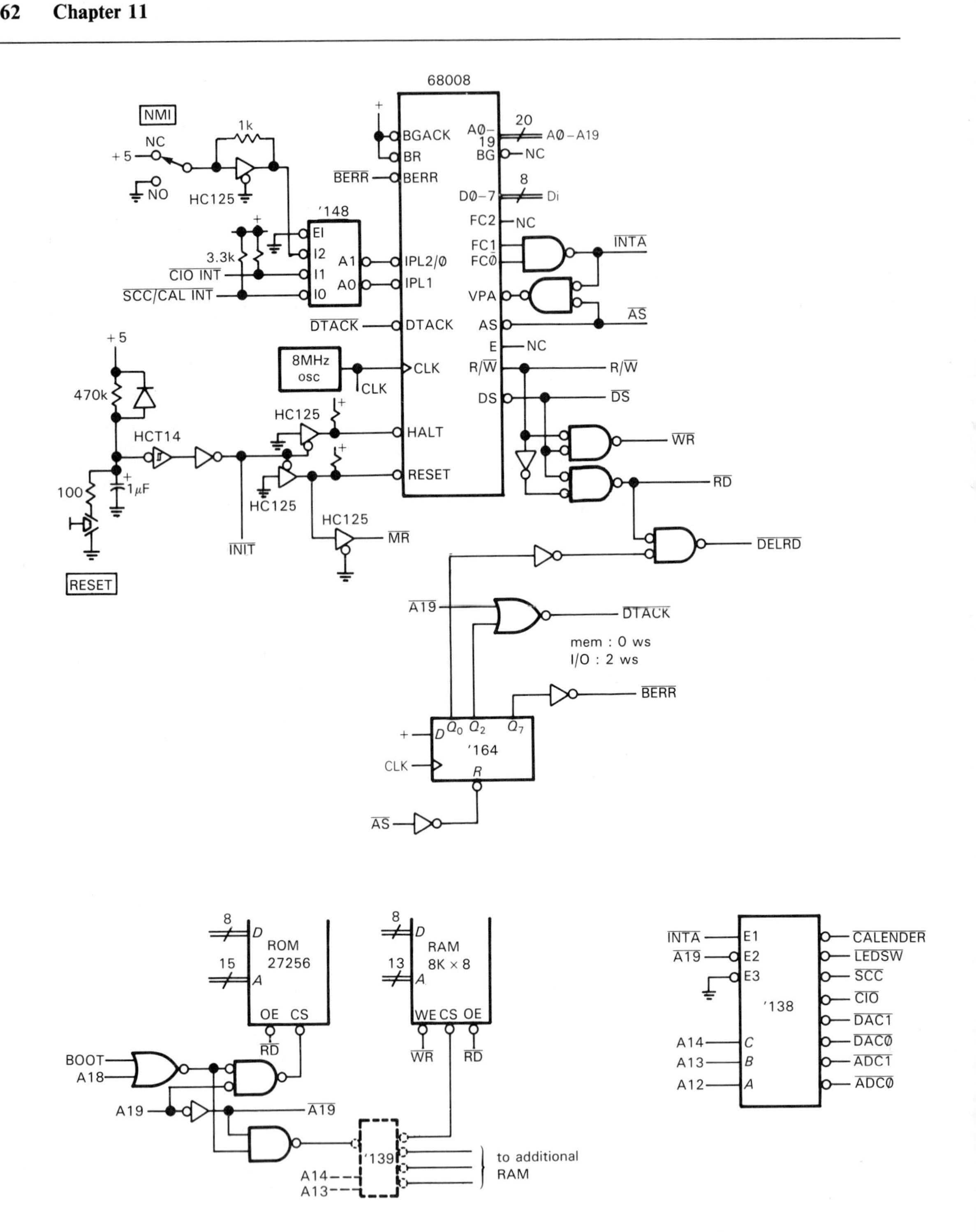

Figure 11.10. General-purpose microprocessor-based instrument – schematic.

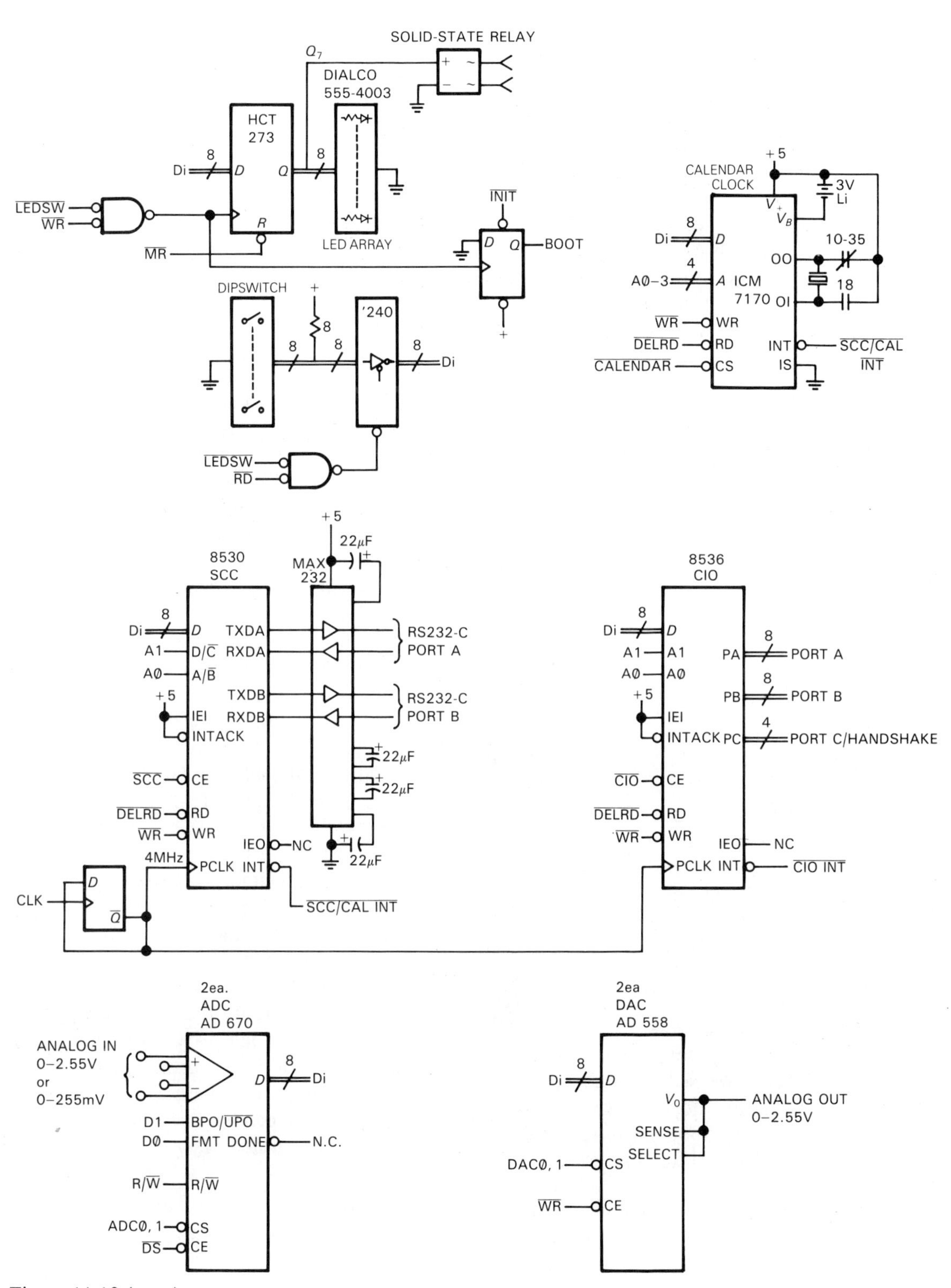

Figure 11.10 (*cont.*)

so you generally want to go as fast as you can; the disadvantages of high clock speeds are (a) greater timing demands on memory and peripherals, (b) higher price, and (c) greater power dissipation, particularly for low-power CMOS CPUs and peripherals. You usually don't care about power, except in battery-powered instruments; see Chapter 14. We used an 8MHz clock frequency because with that choice we can use the same clock (divided by two) for the serial port chip; otherwise we would have to use a separate oscillator for the USART, or run at reduced baud rates.

RESET, interrupts, strobes. To boot the 68008 ("boot" means "bootstrap," as in "pull yourself up by your bootstraps"; i.e., start it up from nothing) you assert both RESET′ and HALT′ (these are bidirectional; you must use open-collector with pull-up). We implemented a simple power-on boot with an RC and Schmitt trigger, paralleled with a push button for manual reboot. Note the diode for fast discharge during short supply glitches; a better power-on reset circuit would use something like the MAX692 "microprocessor supervisory" circuit, with well controlled reset output. The line we named MR′ is asserted both at boot and (for 128 clock cycles) when the CPU executes a RESET instruction, while our INIT′ line gets asserted only at boot.

We opted for autovectoring in this simple system; the AND of FC0 and FC1 indicates an interrupt acknowledgment cycle, during which we are therefore obliged to assert VPA′, coincident with address strobe AS′. We will also use our INTA′ signal to inhibit normal I/O decoding (see below). The 68008 permits three levels of autovectored interrupt: We therefore combined, via wired-OR, the slow serial port and calendar interrupts at the lowest level (IPL1) and put the latency-sensitive timer (called a "CIO" chip) interrupt at the next higher level (IPL0/2). We reserved the highest "nonmaskable" level (asserting both IPL lines) for a push-button ("NMI," nonmaskable interrupt), so you can always get the board out of a hung state during program debugging.

We used a few gates to generate a (RD′, WR′) pair of strobes from the (DS′,R/W′) strobe/direction pair that the 68008 provides. They will come in handy for several of the Intel-style peripherals that prefer separate strobes.

DTACK′, BERR′, and slow peripherals. Finally, we used an 8-bit parallel-out shift register ('164) as a state machine to generate several necessary signals in sequence. The shift register is held reset until the CPU asserts AS′, which signals the beginning of a bus cycle (Fig. 11.4), whereupon 1's begin propagating down the register, one stage per CPU-clock rising edge. The Q_0 output is used to generate a delayed RD′ signal for two fussy peripherals (SCC and CIO), as we'll explain shortly. Several I/O devices are slow and require wait states; we therefore used shift register output Q_2 to generate a delayed DTACK, resulting in two wait states for all I/O ports (in our scheme, all I/O is memory-mapped above $80000, i.e., A19 asserted), and no wait states for memory (A19 not asserted). We're in trouble if a 1 ever propagates as far as the last stage of the shift register, since all bus cycles should end (with disassertion of AS) long before that. Therefore we've used the last stage (Q_7) to assert BERR, which forces a vectored jump (through $08) rather than letting the CPU hang forever. Such a bus "time-out" signal is particularly important in a general-purpose computer, where a CPU access to a nonexistent peripheral would otherwise crash the machine.

Memory

When you boot a 68008 (by asserting RESET and HALT), it goes to the bottom of memory to find two crucial addresses:

the 32-bit *starting address*, stored in memory bytes \$04–\$07, and the initial value for the *stack pointer*, stored in bytes \$00–\$03. Having read these addresses, it initializes the stack pointer, then jumps to the starting address.

Because these low-memory locations are read by the CPU before any other program activity, you've got to have nonvolatile memory down there, i.e., memory that retains its information when power is shut off completely. The most obvious choice is EPROM ("erasable programmable read-only memory"; see Section 11.12), the inexpensive, UV-erasable, non-volatile byte-wide memory that you can recognize by the little glass (actually, quartz) window on the top of each package. EPROMs take a half hour to erase, and a minute or so to program. They come in sizes up to a megabit, and they retain their data longer than the lifetime of the instruments you put them in. The only snag with putting EPROM in low memory is that the various vectors (from interrupts, bus errors, and other "exceptions") also live there, and you like to be able to change them under program control.

You could do that by using a variant of EPROM called "EEPROM" ("electrically erasable programmable read-only memory"). You could also use a two-stage process in which the vectors reside permanently in EPROM, pointing to a set of jumps (a "jump table") in writable memory (RAM, see next). But there's a cleverer way: You arrange things so that at bootup there is ROM at the bottom of memory, but later it is replaced (under program control) by ordinary writable memory, called RAM ("random-access memory," which always means addressable read/write memory).

Look again at Figure 11.10. We've used a 27256 EPROM, which is moderate-sized by present standards, and organized as 32K×8. It has 15 address inputs, 8 three-state data outputs, a chip-select input (CS′), and an output-enable input (OE′). Each addressed byte (written onto the chip previously by an EPROM programmer, and now unchangeable) is asserted onto the data lines only when both enables are asserted. The usual scheme is to assert CS′ as soon as possible with address-decoding logic, then strobe OE′ with the subsequent read pulse. In our case, memory (ROM or RAM) is enabled only if A19 is disasserted; i.e., memory resides in the lower half of address space. In addition, ROM is enabled only when (a) A18 is asserted or (b) the BOOT bit (set at start-up, cleared under program control) is set. RAM also resides in the lower half of address space, but is enabled only when ROM is not. So when you boot the CPU, the BOOT flip-flop is set, and ROM temporarily lives in address space \$0000–\$7FFF; RAM is nonexistent. The ROM also has a dual existence up in its real home at \$40000–\$47FFF. The first 8 bytes of ROM are cleverly coded to make a jump into its continuing boot-up code in the higher address space, where (among other things) it clears the LED port (address \$86000). Writing to the LED port has the side effect of clearing the BOOT flip-flop, causing RAM to replace the temporary low-memory image of ROM. To make this precise, here are the first 16 bytes of ROM needed to work this trick:

```
0000: 00 00 20 00                ;initial stack ptr
0004: 00 04 00 08                ;starting address,
                                   in "real" ROM
0008: 13 FC 00 00 00 08 60 00    ;MOVE.B #0,$86000
                                   - clear boot
```

Note that the last instruction executes at address \$40008, due to the starting address fetched from \$0004.

The wiring of the 8K×8 RAM chip is simple: It receives the lower 13 bits of address (8K) and is enabled when A19 is disasserted and ROM is disasserted. The RD′ and WR′ strobes are tied to output enable (OE′) and write enable (WE′), respectively. For the time being, imagine that the extra decoding logic shown in dotted outline is omitted. Then RAM resides at the bottom of address space, except during the boot-up, when it is replaced by the temporary image of ROM.

Now there's a peculiar thing happening with our address decoding. Look at RAM. We've *ignored* address bits A13–A17! So, for instance, the memory byte addressed at $0000 has many aliases – you can find it at $2000, $4000, and in fact at any address that has zeros in bits A0–A12 and A18–A19. It makes multiple appearances in address space. To "fix" this we could qualify our RAM CS′ enable by insisting that all those bits also be zero; but it would serve no purpose. It may seem sloppy to have ghost images of memory all over address space, but it is harmless, and in fact saves gates. The same thing is happening with ROM (and with I/O). Figure 11.11 shows the circuit's *memory map*, with the repeating memory shown explicitly.

You do have to look at additional address lines if you want additional memory in the system, of course. The circuit shows the natural way to do that – simply add a 1-of-4 decoder ('139) on the next two address lines, enabled by our existing RAM enable, and, presto, you can add 3 more RAMs. The extension to even larger memory should be evident.

EXERCISE 11.5

Use a 1-of-8 decoder ('138) to expand our original circuit to accommodate eight 8K×8 RAMs.

EXERCISE 11.6

Modify the original circuit to accommodate four 32K×8 RAMs.

EXERCISE 11.7

Now change the circuit to accept a pair of 64K×8 ROMs (27512).

EXERCISE 11.8

Redraw the memory map for each of the previous exercises.

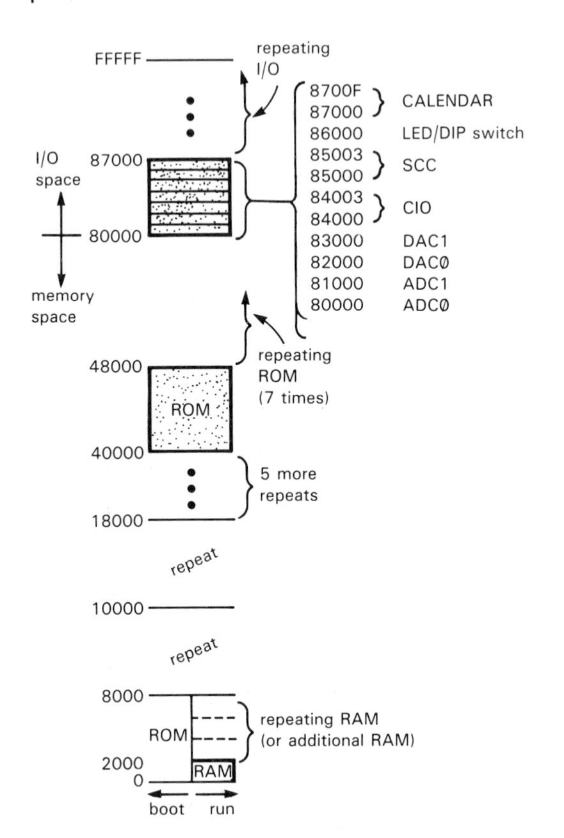

Figure 11.11. Memory map.

Memory timing. Before going on to I/O, it's worth looking at the memory timing situation. We said earlier that our DTACK circuit generated no wait states for memory accesses. That's OK as long as the memory is in fact fast enough to satisfy the timing constraints of the read and write cycles in Figure 11.4. But is it? The way to find out is to start with the 68008 no-wait-state timing diagram, then subtract worst-case delays of the supporting "glue" circuitry to see how much time is left for the memory to respond. Let's try it.

Figure 11.12 shows the situation for a read cycle, which usually presents the more difficult timing. We began with the CPU

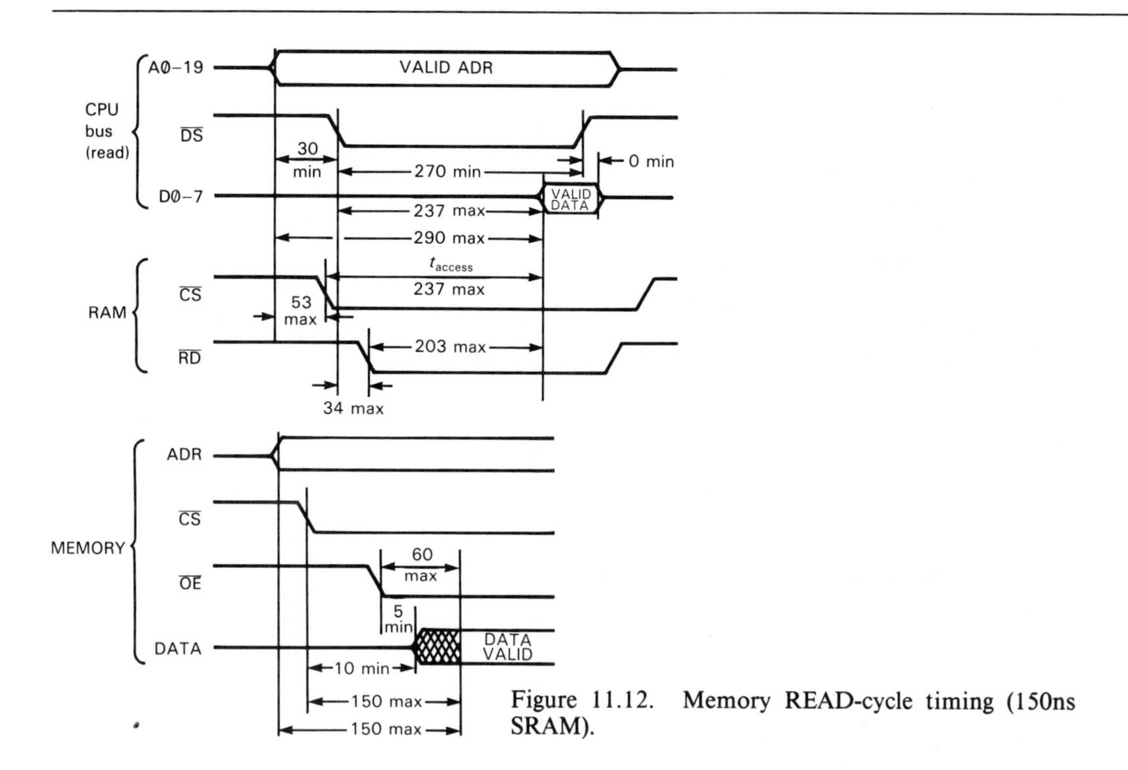

Figure 11.12. Memory READ-cycle timing (150ns SRAM).

timing specifications for the 8MHz chip, since we chose that clock frequency for our circuit. The most important timing is from CPU address valid to memory data valid, because that determines the maximum "address access time" of the memory that you can tolerate. In this case the CPU asserts a valid address at least 290ns before data must be valid; the corresponding figure for DS′ is 237ns. Our RAM CS′ circuit has two cascaded gates, which we'll assume are 74HCT02 and 74HCT00; the worst-case delays are 28ns and 25ns, for a total of 53ns maximum. That leaves 237ns (290−53) for the memory's access time from CS′. By a similar calculation (assuming a single 74HCT32 to generate RD′) the memory has to provide data no later than 203ns after OE′ is asserted. The figure also shows the worst-case timings for the slowest grade (150ns) of 8K×8 static RAM: 150ns access time from address, 150ns access time from CS′, and 60ns access time from OE′. Since our circuit allows 290ns, 237ns, and 203ns, respectively, we have nearly 100ns of margin in the closest case (access from CS′).

We won't go through the analogous calculation for a write cycle, which turns out to be even more relaxed. Evidently there is no RAM timing problem, even with the slowest grade of memory and no CPU wait states.

Unfortunately the same cannot be said of ROM, which tends to be slower than RAM. For example, 32K×8 EPROM comes in standard speeds (address-to-data or CS′-to-data) of 150ns, 200ns, and 250ns. The calculation goes as before, but with an additional 6ns of CS′ delay because of different gating to CS′. Thus, only the two fastest grades of EPROM meet the 231ns CS′-to-data requirement and thus are guaranteed to work without wait states. Instead of using the fast ROMs, we could use faster glue logic, either 74ACT or 74F, which would let us use the 250ns grade of

ROM. In reality, they would probably work fine in our circuit anyway, since worst-case calculations tend to be very conservative. They take into account the worst-case combinations of power-supply voltage, temperature, capacitive loading, and batch variation; for example, our worst-case conditions assumed a V_{CC} of 4.5 volts, a temperature range of −40°C to +85°C, and an unrealistically high load capacitance of 50pF. However, if you want to be *sure* your circuits work reliably, particularly if you are shipping production quantities of instruments, you should adhere to worst-case design.

Peripheral circuits

We have nine peripherals in this circuit, so we used a 1-of-8 decoder ('138) as an address decoding switchyard (with the LED indicator and DIP switch sharing one read/write port). The decoder is enabled on A19 asserted, which puts us in the I/O portion (the top half) of address space; it is also disabled during interrupt acknowledge, as we explained earlier. We used address lines A12–A14 into the decoder, which puts the successive peripherals at addresses $80000, $81000, $82000, etc.; we ignored the remaining high-order address lines, as we did with memory, thus producing a "lazy" address decoding, with each peripheral making multiple appearances in address space. In fact, *every* address above $80000, right up to the top of memory at $FFFFF (that's a half million addresses), has some peripheral living in it!

EXERCISE 11.9

Elaborate on this last statement by calculating exactly how many times one of these peripherals appears in memory. Then write down the general form for the LED indicator's address, using x's for "don't cares."

EXERCISE 11.10

The only real disadvantage of our incomplete address decoding is that it wastes a half megabyte of address space (most of which could hold memory) on a few paltry peripherals. Show how a designer who wants to use most of the 1 Mbyte address space for memory might decode I/O addresses so that the eight ports are memory-mapped at $FF000, $FF100, ..., $FF700 and do not respond to lower addresses. You could now install a megabyte of RAM, except for the fact that both memory and I/O would respond to those high I/O port addresses. Figure out a way to solve this problem.

Note that the address decoding assigns a whole set of contiguous addresses to each peripheral, since the low-order address lines are also ignored by the '138. Some peripherals have multiple internal registers, and we'll use a few low-order address lines to address them. Think of the decoder as responding to the peripheral's *base address*. Let's now look at the particular I/O devices in the circuit.

LED and DIP switch. These are the simplest of ports. For output, we drive a set of LEDs with an octal D register, clocked by qualifying the address-decoded LEDSW′ with WR′. Note the use of trailing-edge triggering for relaxed setup timing constraints. We used a '273 octal register with two-state outputs, rather than the usual three-state '574, because it has a RESET′ input, which we assert during boot or processor reset; that's useful here so that the LED register starts up with no lights lit. HCT logic has rail-to-rail saturation and good output sourcing capability (8mA at +4.5V out), so we can drive an LED array that returns to ground (with LS logic you have to return the LEDs to +5V); this is nice because the LEDs then indicate 1's, not 0's. The LED array shown has built-in resistors, limiting the current to 6mA. Note the use of one LED port bit to drive a solid-state ac relay.

These relays are easily driven by logic levels (3V guaranteed turn-on, 1.5k load resistance), and they use "zero-voltage" switching (Sections 9.08 and 9.10). Note also that the LED port's WRITE strobe does double duty, by clearing the BOOT flip-flop the first time it is asserted; once you've cleared the flip-flop, you can use the LED port as you wish.

The DIP-switch input port is also easy, with a '240 octal inverting three-state buffer driven by the pulled-up switch levels, and enabled by the same LEDSW′ port decoding, this time qualified by RD′. In other words, if you write to $86000 it shows on the LEDs; if you read, you get a byte indicating the DIP-switch settings. We used an inverting buffer so that a switch that is closed reads as a 1, not a 0.

ADCs and DACs. These are easy ports, also. Both converters are advertised as "complete" converters, with internal references and clocks. The AD670 ADC matches the (R/W′, DS′) strobing convention, since it has a direction input and a chip enable input. A write (chip enabled with R/W′ LOW) begins conversion, while a read gets the resulting byte. During the write, the ADC latches the two data bits: BPO/UPO′ controls the input range (HIGH = bipolar, LOW = unipolar), and FMT controls the digital output format (H = 2's complement, L = unsigned binary). The DONE output tells you the ADC has finished a conversion; we ignored it in our design because we think it's easier to execute a few no-ops during the $10\mu s$ (max) conversion time than it is to hang on a flag. The AD670, like many peripheral chips, has a sluggish processor interface. This one, in particular, requires a 300ns minimum CE′ strobe during write and has a 250ns access time from assertion of CE′ during read. If you look back at Figure 11.4, you'll see that this violates the fast timing requirements of the 68008 during a normal (0 wait state) bus cycle; but with two wait states (which our circuit generates for all addresses $80000 and up), all is well: the DS′ for a write becomes 390ns wide, and the DS′ setup on read becomes 487ns.

The AD558 DAC is also a complete converter, with convenient single +5 volt supply and *voltage* output. It is write-only, so we used the WR′ strobe for chip enable, with the decoded address used for chip select. Once again the timing is violated without wait states: The AD558 needs 200ns of data setup to the trailing edge of CE′, and 150ns minimum CE′ pulse width. With no wait states you have only 180ns and 140ns, respectively; two wait states increases that to a comfortable 430ns and 390ns.

Serial and parallel ports. The Zilog 8530 SCC (serial port) and 8536 CIO (parallel port and timer) typify large-scale-integration (LSI) peripheral support chips. These tend to be designed with lots of flexibility and a dizzying choice of operating modes, programmed by sending particular bytes to one or more internal registers. Many of these chips are as complicated as microprocessors (see Fig. 11.13), and you should figure on spending up to a half day learning how to program their operation.

Although LSI peripheral chips are usually designed for specific microprocessors, their generality allows you to use one manufacturer's support chips with another's CPU. The Zilog 85*xx* chips actually claim to be "universal" bus-independent peripherals, although in the case of the 68008 there is a minor bus incompatibility involving the RD′ strobe that we will fix by generating a delayed RD′ strobe.

Look first at the 8536 parallel port/timer. It uses the (RD′,WR′) strobing pair, along with a chip enable input, CE′ (which we assert, as usual, from the I/O address decoder). In addition, it requires a clocking input, both to clock its timers and

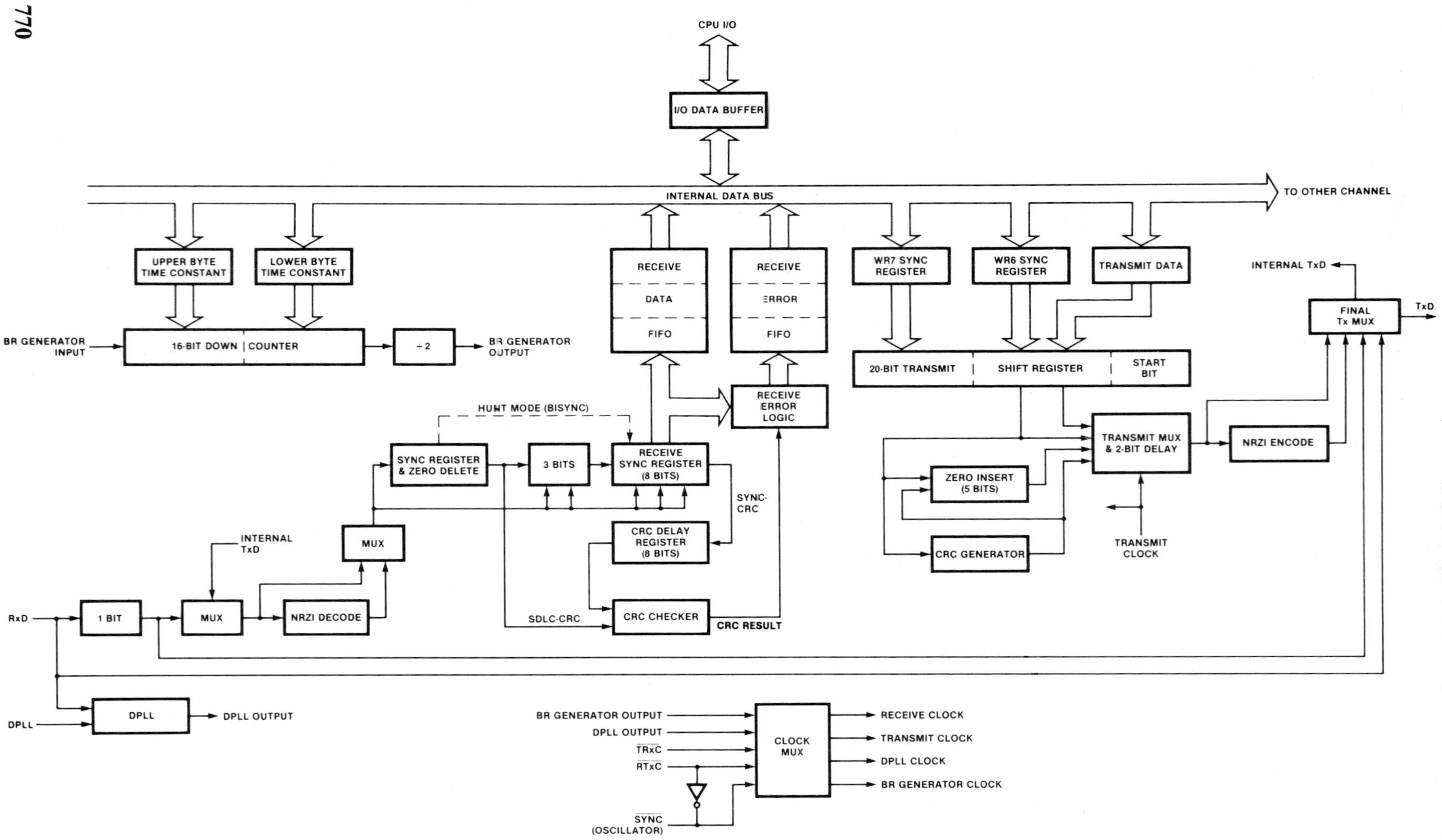

Figure 11.13. Zilog 8530 serial port block diagram. (Courtesy of Zilog, Inc.)

to control internal logic. The 8536 has full interrupt acknowledge circuitry, and it can assert a vector onto the data lines during an interrupt acknowledge cycle. We won't be using most of that capability, which includes a daisy-chained priority line (IEI input, IEO output) and the INTACK′ input that signals it to assert its (programmed) vector; we will use the INT′ output, however, which is used to request an interrupt. The only other bus interface lines, besides the data lines D0–D7, are the two inputs (A0, A1) used to address internal registers; you always connect them to the low-order bus address lines, which puts the internal registers into the CPU's address space beginning at the base address. In this case, for example, the internal registers are located in memory space at $84000–$84003. The addressing pins suggest that there are only 4 internal registers, which is quite incorrect: There are actually 41 writable registers and 48 readable registers! (We warned you these chips are hell to program!) You get to them in a two-step process by first writing to the "control" register at BASE+3 ($85003) a byte containing the target register address, then reading or writing that register. The parallel port data registers are a special case, with the addressing streamlined so that you can read/write them directly (at addresses BASE, BASE+1, and BASE+2).

Figure 11.14 shows the RD′ strobe timing problem: The 8536 specifies an 80ns minimum setup time of the A0–A1 address lines before the RD′ strobe leading edge; it also specifies the usual sluggish response – in this case it takes the poor 8536 255ns to produce the data, and it insists on a 390ns (min) width for the RD′ pulse. The slow *response* is by now a familiar problem, solved by wait states. But the address *setup* to RD′ isn't helped by wait states (see Fig. 11.4, which shows that DS′ can come as soon as 30ns after valid address). To solve this problem we have to delay RD′ by a CPU clock cycle; we can do this conveniently with the same shift register circuit we built to generate DTACK′. In this case we simply AND the "prompt" RD′ strobe with the (inverted) Q_0 shift register output, which isn't asserted until the S3–S4 CPU clock edge. This produces a delayed RD′ strobe, which we've called DELRD′, which begins one clock late (at the same time as a normal *write*-cycle DS′). The result is to give the 8536 an extra 125ns of address setup time (i.e., 155ns total). The wait-state generator still introduces 2 wait states, which makes the overall cycle long enough for the slow peripheral.

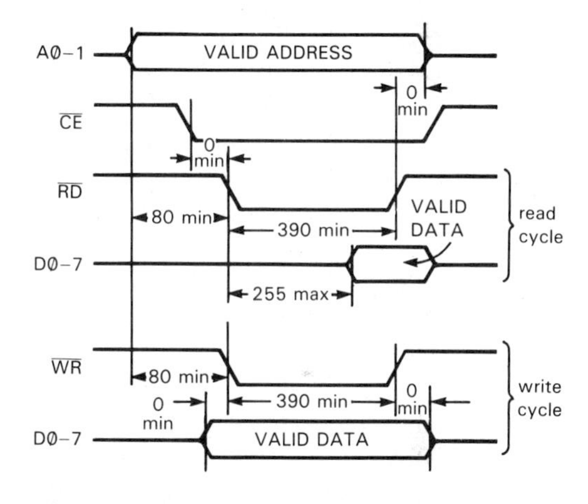

Figure 11.14. Zilog 8536 parallel port timing.

As luck would have it, a similar circuit isn't needed for WR′, because the 68008 considerately provides an extra clock cycle of setup during write cycles (note the delayed DS′ in Fig. 11.4 during write), while the 8536 still requires the same 80ns of setup (Fig. 11.14).

The 8530 serial port interfaces almost identically. The only difference is that the address lines to the internal registers are named differently, with A0 driving A/B′ (which selects channel *A* or *B* of the dual port), and A1 driving D/C′ (which selects data or control registers). This chip is also well endowed with registers: There are 16 writable registers and 9 readable registers for each of the two channels, with the same two-step access as the 8536.

The 8530 accepts clock inputs to 6MHz; we chose a 4MHz clock because that permits asynchronous baud rates up to 9600 baud. The asynchronous data lines TxD and RxD are TTL logic levels, not bipolarity RS-232 (see Sections 9.14 and 10.19). Most RS-232 driver chips (typified by the classic 1488) require dual supplies, which would be a real show-stopper in an otherwise single +5 volt system like this. Luckily you can now get clever chips with on-board flying-capacitor voltage converters, originated by Maxim with their MAX232 series of chips (or the LT1080 from LTC). Note that the open-drain interrupt request line from the 8530 is wired-ORed with the corresponding calendar clock line, so either makes an autovectored interrupt at level IPL1. The interrupt handler for this level has to find out which interrupted, by polling each device's status register via programmed read, as we will illustrate.

Calendar clock. The last peripheral is the calendar clock, another LSI chip with some tricky features. Its bus interface is very similar to the Zilog parts, with the Intel-style (RD′,WR′) strobe pair, and five bits of internal register addressing. It may even have the same RD′ timing problem. We say "may" because the data sheet is ambiguous: It specifies a "typical" setup time from address to RD′ (100ns), but doesn't give a *minimum*. We don't know what it is they're trying to say, but why take chances. We've already got the DELRD′ strobe, so let's use it. The ICM7170 is a modern calendar clock chip with internal battery switchover circuitry; you just hang a 3 volt (lithium) cell on, as shown. Earlier calendar chips forced you to worry about disasserting control lines in an orderly sequence at shut-down, but the 7170 also takes care of all that. Of course, you can't talk to the chip when the +5 volts is gone; the battery just keeps the clock ticking during hibernation, so it wakes up with a clear head and a keen sense of time.

Circuit power

Having finished the circuit design, it's tempting to ignore minor details such as power and ground. You shouldn't. Our circuit uses "5 volt logic," which often means 5 volts ±5% (in this example, the CPU and several peripherals require V_{CC} between +4.75V and +5.25V). Furthermore, the supply voltage should be free of large spikes, which is best managed by liberal use of 0.1μF ceramic capacitors, with a sprinkling of larger tantalum electrolytics. The extreme case of a "large spike" is rampant overvoltage caused by outright failure of the +5 volt series regulator in the power supply. For these occasions you should provide an overvoltage crowbar (Section 6.06), either on the motherboard or at the power supply. When thinking about power supplies, keep in mind that you may easily have a dc current drain of an amp or more in a microprocessor card, and perhaps many amps on the motherboard traces that power plug-in cards. Plan ahead, with hefty printed circuit traces and ample connector current capability.

For reasons explained in Section 9.11, ground wiring on and between PC boards is especially important and should be made low-inductance. The best way is to devote a layer of a multilayer PC board, though a "gridded" ground arrangement on a two-layer board often works well enough (more on this in the next chapter). Finally, a word of advice about power-on reset circuits: It's tempting to use a simple *RC* (+ diode) circuit of the kind we showed in Figure 11.10. The problem is that such a circuit may not respond to a short power dip that is sufficient to disrupt the running program. If the microprocessor circuit is part of an instrument, the effect is that the instrument starts malfunctioning, and the only way you may be able to make it regain sanity is by switching the ac power off and back on again! We've had this happen to commercial instruments, as well as to our own designs. The safest solution is to use a good supervisory circuit, for example the MAX690 series from Maxim.

Having disposed of the circuit design, which fell together pretty easily, we'll next proceed to the really hard part, programming.

Warm-up: boiled egg

Our signal-averager example will turn out to be a complex programming task, as in fact most real-world problems are. Close reading of the following sections will reward those readers intending to design their own microprocessor systems, since it will illustrate in detail most of the techniques needed to build serious microprocessor-based instruments.

Instead of jumping immediately into the shark-infested waters, however, let's dip a toe in gently with a simple (and frivolous) example of software for our general-purpose microprocessor circuit. Computers are supposed to spare us unnecessary drudgery in our lives. Let's make this one cook our 5 minute boiled egg each morning at 8AM!

Imagine that the solid-state relay in Figure 11.10 is hooked to our coffee-cup immersion heater, sitting in a cupful of water in which a raw egg is waiting. Look at the code in Program 11.2.

To keep this program simple, we assume the computer is running, and the calendar clock is set. (In the signal-averager example we'll see how to handle those uninteresting, but important, details!) The code begins by sending a zero byte to the LED port, to turn off the heater; it also sets the calendar clock to 24 hour mode and turns off interrupts. It then goes into a loop ("WAKE") that continually tests the hours digit from the calendar clock, until it reads an "8," whereupon it sends a byte of all 1's to the LED port; that turns on the heater and lights all the LEDs.

Now the program falls through to a second loop ("COOK"), which continually tests the *minutes* digit of the calendar clock, until it reads a "5," whereupon it sends a zero byte to shut off the heat (and LEDs). Finally, the program falls

Program 11.2

```
    ;cook egg 5 minutes at 8AM every day
    ;assume calendar clock set, computer running
        CLR.B    $86000              ;clear LED and ac relay
        MOVE.B   #$0C, $87011        ;24-hour calendar mode, no int

wake:   MOVE.B   $87001, D0          ;get time -- hours digit
        CMP.B    D0, #8              ;is it 8AM?
        BNE      wake
        MOVE.B   #$FF, $86000        ;yes. start egg, light all LEDs

cook:   MOVE.B   $87002, D0          ;get minutes digit
        CMP.B    D0, #5              ;cooked 5 minutes?
        BNE      cook
        CLR.B    $86000              ;yes. stop egg, turn off LEDs

wait:   MOVE.B   $87001, D0          ;get hours digit
        CMP.B    D0, #8              ;9AM yet?
        BEQ      wait
        BRA      wake                ;yes. look forward to tomorrow's egg
```

through to a third loop ("WAIT"), which continually tests the hours digit again, until it *stops* being "8." At that point it branches unconditionally back to the first loop, waiting for 8AM, tomorrow.

We wrote this rough-and-ready program to show how simple it can be. Don't copy our style, though – we did many things crudely, to save space and explanation. If we were willing to expand the code, we could make the timer "smart," for example by using one of the ADC ports to sense when the water reaches boiling; that could initiate the actual egg timing, and even thermostat the heater to save energy! We could allow you to set the wake-up time, the cooking interval, etc., via the NMI button. A DAC port could display the time, perhaps as "walking digits," while the other DAC port (attached to a pillow speaker) could speak to you in a quiet voice, gently waking you with soothing thoughts and memorized music ... but we digress!

OK, it's time to dive in.

11.06 Programming: defining the task

The surest way to waste a lot of time, and get yourself totally confused in the process, is to begin programming before you've figured out what you want to do. This is especially true when you are programming a dedicated controller in assembly language, because assembly code itself does not have the clean modularity and control flow of a structured high-level language; furthermore, to optimize real-time performance you often resort to tricks and less-than-transparent ways of doing things. The resultant code – full of branches, tasks divided between interrupt handlers and main code, software flags that get modified in unpredictable places, command bytes going to peripherals, etc. – can quickly become a patchwork horror if halfway through you are still discovering how you should have viewed the job at the outset. Just as in painting a house, the preparation may take longer than the actual work, but it's well worth the effort.

The signal averager we're designing is a good example. It is not a very sophisticated instrument, yet a glance ahead at the interrupt handler's flow chart (Fig. 11.21) should convince you that there is considerable complexity in the setting and reading of flags and signals, the real-time alteration of interrupt vectors, and the overall control flow. So it's worth spending the time now to understand how the instrument is intended to work.

What is a signal averager?

A signal averager, sometimes called a multichannel scaler, is an instrument used to improve the signal quality (signal/noise ratio) of a periodic analog input signal that is unavoidably combined with nonperiodic noise (or interfering signals). It does this by measuring the signal amplitude many times during each period, storing these amplitude samples into a set of sequential "bins," then adding to each bin the corresponding samples from many additional periods of the input waveform. In other words, the signal waveform is co-added *modulo its period*. As we explain in detail in Section 15.13, this improves the signal/noise ratio of the signal accumulating in the bins because the (periodic) signal grows linearly with time, while the (random) noise fluctuations grow only as the square root of time. We call each successive periodic accumulation through the bins a "sweep"; a typical data run might consist of a few thousand sweeps.

A good signal averager continually shows you the accumulated waveform (stored in its 1000 or so bins) on a CRT display and gives you a wide choice of dwell time per bin, trigger modes, display scales, etc. We'll put many of these features in ours, but we'll stop short of implementing everything that is possible, in order to fit the example into this chapter. In deciding what to include, we've picked a nice set of functions that lets us illustrate the full range of programming tricks and trade-offs, without getting carried away with gold-plated bells and whistles.

Features

For reasons we'll explain shortly, we chose to use a conventional control panel with labeled switches, rather than the contemporary fashionable keyboard with CRT-screen menus. Our signal averager therefore looks just like a conventional instrument, and the controls have to have predefined functions and ranges. When we planned the chapter, we actually began, as we're beginning here, by deciding what functions to implement, and over what ranges.

We decided to use a fixed number of bins (256), with a large set of selectable dwell times per bin. Since signal averagers get used for two kinds of periodic phenomena – those that have an intrinsic period of their own (e.g., ocean tides) and those that you can trigger or drive periodically (e.g., nerve impulses or resonance scans), we provided two sweeping modes: "triggered," in which the instrument waits for an external trigger before beginning each sweep, and "autolooping," in which it just cycles around periodically. We provided two ways to terminate the signal average: a set of preprogrammed total sweeps, and a "stop" switch that terminates with the next complete sweep. Our design includes analog X and Y outputs (with Z-unblanking pulses) for a continuously refreshed display of the averaged signal on an XY CRT: We provided a set of display scales (progressing by factors of 2) and a smart "autoscale" mode, in which the data are continually rescaled (normalized) according to the number of sweeps completed. Finally, there are LED status indicators (standby, sweeping) and logic-level outputs indicating sweep-in-progress and end-of-sweep. Here are the specifications of our signal averager:

Analog input range: ±5 volts

Number of bins: 256

Internal representation: 32-bit signed integer

Dwell time per bin: 100μs to 1 second, in a 1-2-5 sequence

Sampling: Finite integration (sum of 100μs samples)

Preprogrammed number of sweeps: 1 to 20,000, in a 1-2-5 sequence

Sweep modes: Wait-for-trigger; or autoloop

Display modes: Selectable scale, with wraparound; or autoscale

Display scales: 1 to 16K input range, by factors of 2

Inputs: Analog signal, ext trig

Outputs: X, Y, Z (to CRT), SWEEP, END

Additional controls: START, STOP, RESET (reboot)

Figure 11.15 shows the signals coming in and going out of the microprocessor board. We've used the 8536 parallel port for all the digital signals, with direction and polarity programmed appropriately; all digital inputs that come from switches should have resistor pullups, with switch closure to ground. No debouncing circuitry is needed, because we will do software debouncing.

The analog filter is important and deserves some explanation. If you make periodic brief samples of the amplitude of a continuous analog waveform of finite bandwidth (where the maximum frequency present is $f_{\max}$), you preserve the input information as long as you sample at $2f_{\max}$ or faster. If you don't meet this *Nyquist* criterion, funny things can happen; in particular, Figure 11.16 shows the phenomenon of *aliasing* due to undersampling, in this case with frequencies close to the sampling frequency appearing folded back down near zero frequency. In order to prevent aliasing, you must low-pass-filter the incoming waveform at $f_{\text{sample}}/2$ or less.

This is simple enough, but how do we handle the fact that the signal averager has adjustable dwell time per bin, i.e., adjustable sampling time? One possibility is to build an adjustable low-pass filter (perhaps a switched-capacitor filter with programmed clock rate) at the input to match the dwell time; this makes sense, because if you choose a long dwell time, you're not interested in high frequencies anyway. But a simpler solution is to notice that if you *integrate* (or average) the signal during the dwell time, you've got an automatically tracking low-pass filter. For this reason, signal averagers sometimes use V/F converters (which are intrinsically integrating) at the input. Our solution amounts to the same thing: We always sample the analog input at 10kHz (which is filtered with a matching 5kHz anti-aliasing filter); for longer dwell times we effectively integrate by adding the appropriate number of successive samples. Figure 11.17 shows what the front panel might look like.

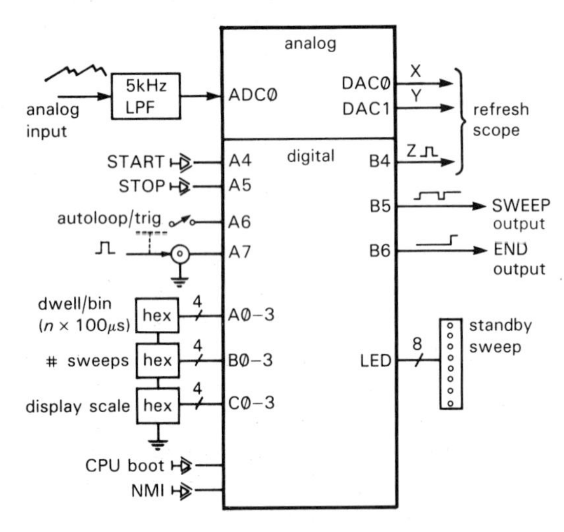

Figure 11.15. Signal-averager controls and inputs/outputs. "LED" is the indicator port in Fig. 11.10; "A," "B," and "C" refer to parallel port bits of the 8536.

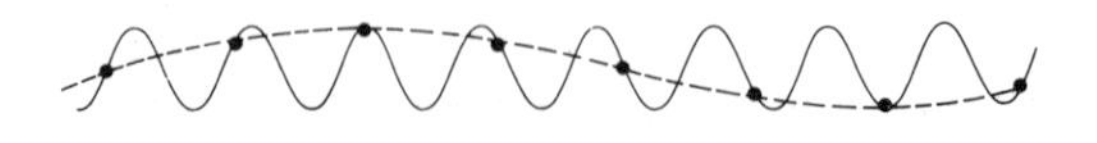

Figure 11.16. Aliasing due to undersampling.

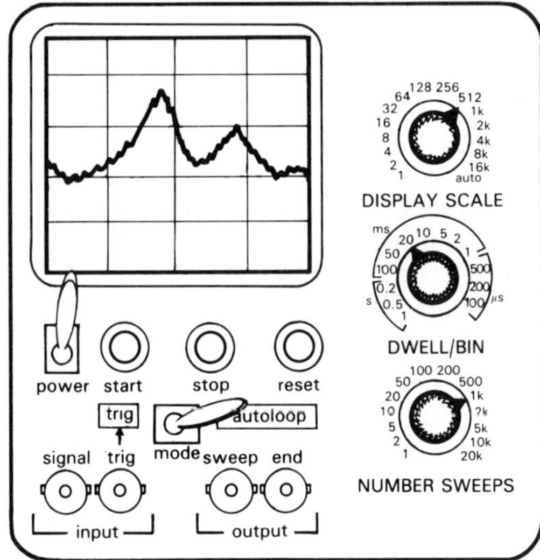

Figure 11.17. Signal-averager panel.

Non-features

It's worth explaining what we left out, and why. We decided not to succumb to the keypad/menu mania, for a few reasons. First, we often find keypad instruments annoying, because each one works differently, and you have to keep relearning how to use them. Second, they're often slower to use; you have to do some *typing* just to change scales! Last, a keypad/menu scheme requires tedious parsing and screen-driving software routines, with no redeeming pedagogical value. So we decided to use the parallel port to read some panel controls, which is quick and simple, even though it sacrifices flexibility by limiting your choices.

There are quite a few other features that would make a better instrument, but we left them out to keep the program from getting too large for this chapter. We could have had a second analog input channel (there are two ADCs), selectable numbers of bins per sweep, multiple "banks" of data memory, digital outputs telling current bin number, and analog outputs proportional to bin number (for controlling analog quantities). These are all in the nature of desirable enhancements; an *essential* feature that we omitted is the capability to send the final data to a microcomputer, best done through one of the serial ports.

Even the features we included in our design could have additional modes. In particular, microprocessor systems with both A/D and D/A converters can benefit from built-in analog multiplexers that let you loop the DAC outputs back into the ADCs. That way, you can test all the converters with software, as part of a power-on test sequence (that also includes memory and port tests, etc.). You can even check the power-supply voltage(s), using the ADCs.

EXERCISE 11.11

Show how to do this, by connecting analog multiplexers at the input of the ADCs. You'll need to arrange a port address to which the CPU can ship its MUX-select commands.

11.07 Programming: details

Overview

Programs are usually complicated. Time-critical assembly-language programs always are. Rather than show you some fake code that wouldn't do the job, we are going to show you finished code that would work in our hardware microprocessor circuit, in all its beauty (and ugliness). You'll never understand it if we don't ease you in, layer by layer. The best way to get started is in a "top-down" fashion, which means that we begin by identifying the major tasks of the job, the order in which they will be handled, and the protocol for interchange of information between major modules of the program. Having done this, we can proceed with flow charts of the individual program modules. Finally, we write the actual assembly-language code.

Note: The material that follows is necessarily detailed. Readers who simply want to get the "general idea" may wish to skim the discussion of programming, rejoining us in Section 11.08 or 11.09.

Figure 11.18 shows the big picture, considerably simplified. We have assigned three arrays in RAM: a block of 256 32-bit ("long") integers, to hold the current DATA for each bin; a block of 256 16-bit integers ("words"), to hold the number of sweeps completed for each bin, used to NORMalize the data in "autoscale" mode; and an array of 256 bytes, to hold the data that are continually sent to the DISPLAY. Our jobs are to add fresh data from the ADC into the DATA array, while updating the NORM array accordingly; scale those long integers to bytes in the DISPLAY array (using the NORM array, if autoscale mode; otherwise just by shifting); and display those bytes continually.

Here's how the program works, in broad brush strokes: At power-on, the CPU does a complete boot, and begins executing from ROM (whose ghost image at the bottom of memory cleverly provides start-up vectors to its real self, as we described in Section 11.05). The various peripheral chips must be initialized (by sending the right bytes to their command registers), and the program has to initialize arrays, pointers, starting values, etc. An instrument like this has several possible states (waiting for START, taking data, etc.), and we begin in a standby state, waiting for someone to push the START button. The machine isn't totally dead, though – it should also be displaying its data, which we have initialized to all zeros.

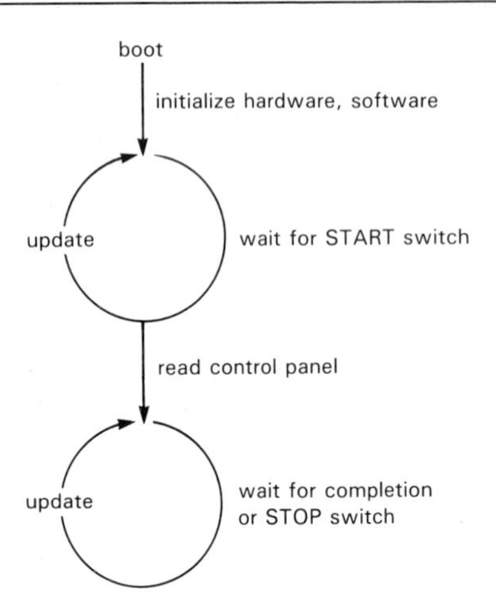

Figure 11.18. Programming: the big picture.

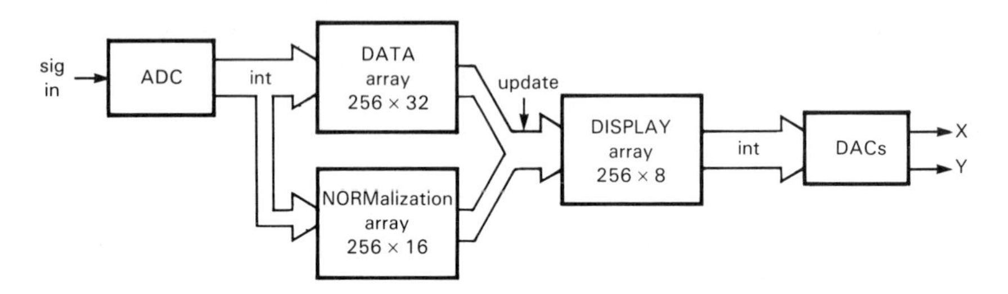

When the START button is pressed, the program proceeds to read the rest of the control panel, to get operating parameters. Then it goes into a data-taking mode (with parameters as specified by the control panel), while displaying the ongoing signal averaging. It jumps back into standby mode when it has completed the specified number of sweeps, or sooner if it senses the STOP button.

Timing; interrupt-driven versus programmed tasks

The most important task is to ensure that the ADC makes conversions every $100\mu s$ and that these data get integrated and added into the DATA array. The next most important task is to keep the display refreshed, at a rate of at least 40Hz or so to prevent flicker. Of lesser importance is the job of keeping the DISPLAY memory up to date with what is happening in the DATA array.

You might at first think that the most important task should be tightly coded in the main program loop, with nuisance interrupts occasionally stealing time for minor tasks. You would be wrong. Interrupts take priority over running code, so the time-critical tasks should be in the interrupt handler, and the less important tasks should be in the "main" code, which will get executed with whatever time is left over. (This assumes, of course, that the processor handles interrupts quickly. As we will see, the extravagant number of registers in the 68008 makes it an extremely efficient interrupt handler; the 68008 is well suited to this job.)

As we've indicated in Figure 11.18, ADC data go into the DATA array via interrupts, created at $100\mu s$ intervals by a programmable timer in the 8536. The main program keeps busy just continually updating the DISPLAY array, using a routine called "***update***" (since it is oblivious to all the interesting data collection going on under its nose, it is obliged to check a flag that the interrupt routine sets to know when the run is complete). We had originally planned to do the display refreshing in the main loop also. But there was an interesting snag: With each X, Y pair you send to the DACs for display, you have to send a "Z-axis" (trace intensifying) pulse. We showed in Section 10.20 how to make a "software pulse" by sending a 1 followed by a 0 to some port bit. The Z-axis pulses must all have the same duration; otherwise the display will have some points brighter than others. But you can't safely make constant-duration software pulses if interrupts are going on underneath your code.

EXERCISE 11.12
Why not?

One solution is to turn off interrupts, make the pulse, then turn interrupts back on. This is ugly, because it creates undesirable latency for our most important task, the regular collection of ADC samples. Then we hit on the better way: Display one point as an auxiliary task in the interrupt handler. It executes every 100μs, so the complete 256-point graph will be displayed 40 times per second. What's even better, the interrupts go on even when the "main" program is in its other state ("standby" – waiting for START), so the screen display doesn't go away. Finally, there's a wonderful serendipity here, since you have to wait 10μs after starting the ADC before you can read its conversion; that's just enough time to send an X, Y pair to the DACs. In other words, putting the display refresh in the interrupt handler doesn't cost any execution time at all!

☐ *Main program: setup*

We've beat around the bush long enough. Let's look at the detailed tasks done by the program. Look first at the main program, which is diagrammed in the unconventional flow chart of Figure 11.19. The chart corresponds very closely to the assembly code, which is shown in Program 11.3.

The program listing begins with RAM address definitions (including our interrupt vector, space for variables and arrays) and port address (and bit) definitions. These definitions will later be used as operands in memory- and port-addressing instructions, causing the assembler to substitute the actual addresses. Even though the result is equivalent, you should always use definitions (rather than the alternative of using the addresses directly in the code), because it not only makes the subsequent code more readable but also makes it easy to change port and bit assignments in later revisions. The port addresses correspond to our schematic diagram and include the peripherals' internal registers, addressable via the low-order address bits or by two-byte transfers.

The program listing also shows how we will use the 68008's registers. At each interrupt we'll be getting data from the ADC, adding it to the current bin count and checking to see if the bin or sweep is complete. We could store the pointers and counters in memory (as you would have to when using a less capable processor like the 8086), but by reserving enough registers for the interrupt handler's needs, we can make the interrupts very efficient. Therefore, we've assigned data registers for the current bin accumulation (`D7`), remaining dwell count within bin (`D6`) and bin count within sweep (`D5`), offset into the display array (`D4`), and a scratch register (`D3`). Likewise, we've reserved address registers for the three arrays (NORM, `A6`; DATA, `A5`; DISPLAY, `A4`), and for the most-used ports [ADC0, `A3`; CIO (parallel port), `A2`]. The main program promises not to use these whenever interrupts are enabled.

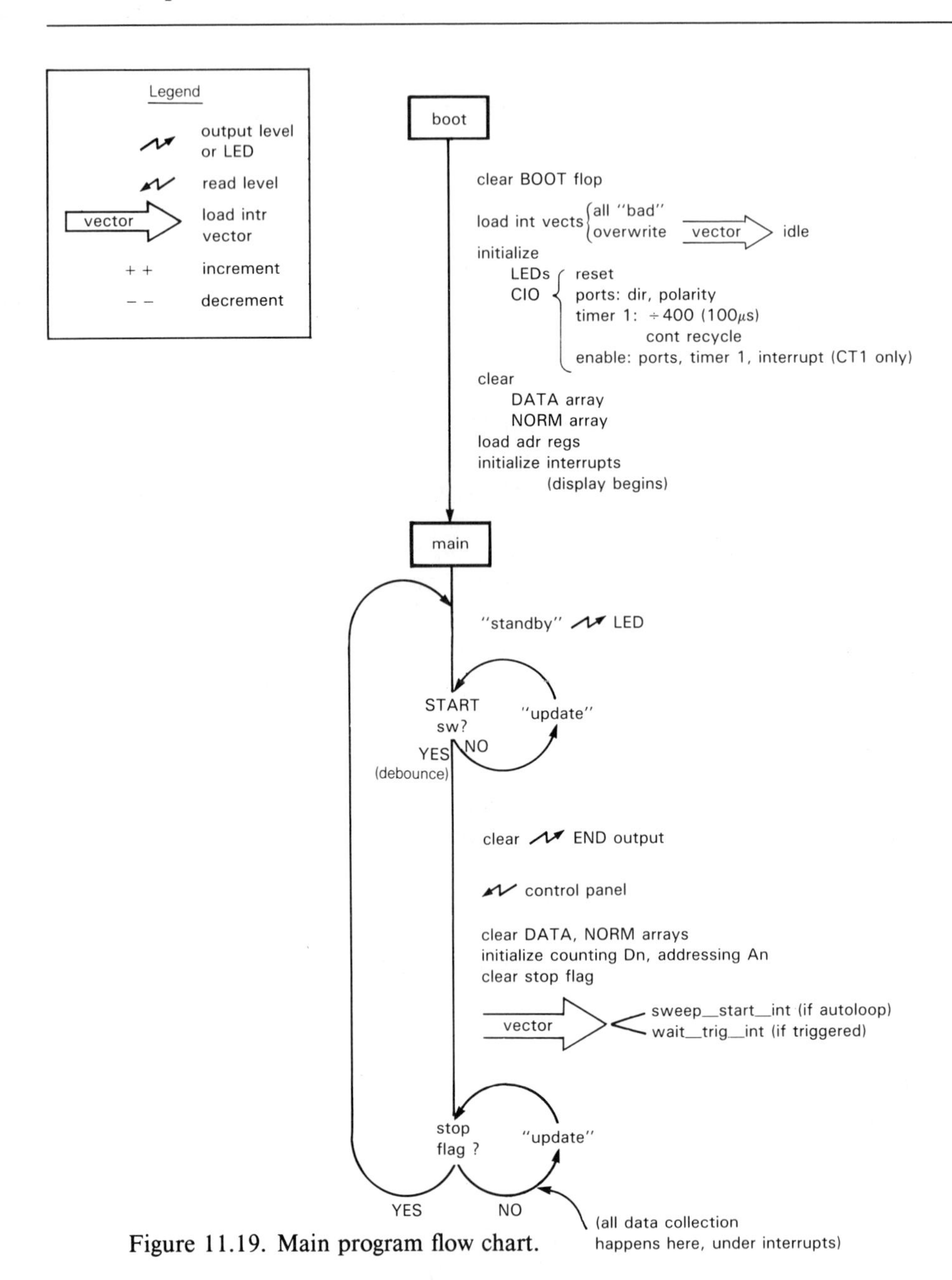

Figure 11.19. Main program flow chart.

You might wonder why we waste addressing registers (with all their fancy array-oriented auto-incrementing features) on fixed port addresses, when absolute addressing will do. The reason is speed. The absolute-addressing instruction

```
MOVE.B ADC0,D0
```

where ADC0 is a long absolute address ($80000 here), takes 28 clocks (3.5μs in our processor), whereas the instruction

```
MOVE.B (A3),D0
```

which uses indirect addressing through A3, takes only 12. The speed difference is caused entirely by the bus traffic, with the

Program 11.3

```
;RAM location definitions
   ;vectors
init_stack_top  EQU     $000     ;initial SSP
reset_vect      EQU     $004     ;startup vector
int5_vect       EQU     $074     ;level 5 int
   ;parameters
dwell_per_bin   EQU     $400     ;# ticks per bin
num_sweeps      EQU     $402     ;via decode_table
auto_loop_flag  EQU     $404     ;1=autoloop, 0=trigger
   ;internal flags
stop_flag       EQU     $405     ;set by handler when sweep done
   ;variables
led_store       EQU     $407     ;LED image in memory
update_offset   EQU     $408     ;indexes next point
   ;arrays
data_array      EQU     $1000    ;unscaled data (long)
norm_array      EQU     $1400    ;#sweeps/bin (word)
display_array   EQU     $1600    ;offset binary, scaled (byte)


;Port definitions
ADC0            EQU     $80000   ;kept in A3
DAC0_OFFSET     EQU     $2000    ;offset from ADC0 (index from A3)
DAC1_OFFSET     EQU     $3000
LED             EQU     $86000
   ;parallel port addresses
CIO_CNTRL       EQU     $84003   ;control reg
CIO_PA_DATA     EQU     $84002   ;port A data
CIO_PB_DATA     EQU     $84001   ;port B data -- kept in A2
CIO_PC_DATA     EQU     $84000   ;port C data
CIO_CNTRL_OFFSET EQU    2        ;index from A2
CIO_PA_OFFSET   EQU     1
CIO_PC_OFFSET   EQU     -1
   ;parallel port internal registers (access via CIO_CNTRL)
MAST_CNTRL      EQU     $00
MAST_CONFIG     EQU     $01
PA_CMDSTAT      EQU     $08
PA_MODE         EQU     $20
PA_POLARITY     EQU     $22
PA_DIRECTION    EQU     $23
PA_SPECIAL      EQU     $24
PB_CMDSTAT      EQU     $09
PB_MODE         EQU     $28
PB_POLARITY     EQU     $2A
PB_DIRECTION    EQU     $2B
PB_SPECIAL      EQU     $2C
PC_POLARITY     EQU     $05
PC_DIRECTION    EQU     $06
PC_SPECIAL      EQU     $07
   ;timer1 internal registers (access via CIO_CNTRL)
CT1_CMDSTAT     EQU     $0A
CT1_MODE        EQU     $1C
CT1_FROM_MSB    EQU     $26
CT1_FROM_LSB    EQU     $27
CT3_CMDSTAT     EQU     $0C
```

```
   ;parallel port bit assignments
   ;inputs -- port A
START_BIT       EQU     4         ;starts sweep
STOP_BIT        EQU     5         ;stops sweep at end
AUTO_LOOP       EQU     6         ;1=autoloop, 0=ext trig
EXT_TRIGGER     EQU     7         ;trig input if not autoloop
   ;outputs -- port B
Z_BLANK         EQU     4         ;display scope unblank
SWEEP_BIT       EQU     5         ;HIGH while sweeping
END             EQU     6         ;HIGH after last sweep
   ;LED bit assignments
LED_STAND_BY    EQU     7
LED_SWEEP       EQU     6
BOOT_BIT        EQU     0         ;rising edge kills ghost ROM
   ;cal clk and serial port
CAL_CNTRL       EQU     $87011    ;cal clk control reg
SCC             EQU     $85000    ;serial port base address

;Global register usage, for fast interrupt handling
   ;data registers
;D7 - currently accumulating value for one dwell time
;D6 - remaining dwell count (0 is terminal count)
;D5 - remaining bin count (0 is terminal count)
;D4 - index (offset) for X,Y display value each tick
;D3 - temp. reg. for integration; also START debounce
   ;address registers
;A6 - pointer into NORM array
;A5 - pointer into DATA array
;A4 - base pointer into DISPLAY array
;A3 - points at ADC; use offset for DACs
;A2 - points at CIO_PB_DATA port; use offset for other CIO ports

;ROM code begins here
        .ORG $40000               ;assembler directive, defines ROM location
        .long $2000               ;initial SSP -- top of RAM
        .long reset_entry         ;boot-up vector, should be $40008

           ;first executible statement next
reset_entry:
        MOVE.B  #0, LED           ;make sure LED reg is cleared, and turn off
                                  ;BOOT flip-flop, since we're in real ROM
           ;now initialize vector table
        MOVE.W  #255, D0                  ;table size minus 1
        MOVE.L  #bad_int, D1              ;the vector to load
        MOVE.L  #0, A0                    ;first vector location

vect_init_loop:
        MOVE.L  D1,(A0)+                  ;load the vector
        DBF     D0, vect_init_loop        ;and loop

           ;now load int5 with the initial handler entry
        MOVE.L  #idle_int, int5_vect

           ;now initialize ports
           ;LED port first
        MOVE.B  #0, led_store             ;clear LED memory image
        BSET    #LED_STAND_BY, led_store  ;and set standby bit
        MOVE.B  led_store, LED            ;ship to LED port
```

```
    ;kill calendar clk and serial port
CLR.B   CAL_CNTRL               ;shut off cal & interrupts
MOVE.B  #09, SCC
MOVE.B  #$C0, SCC               ; shut off SCC & interrupts
    ;parallel port (CIO)
MOVE.B  CIO_CNTRL, D0           ;read to force state 0
MOVE.B  #MAST_CNTRL, CIO_CNTRL  ;prepare to reset
MOVE.B  #$01, CIO_CNTRL         ;in reset state
MOVE.B  #$00, CIO_CNTRL         ;out of reset state
MOVE.B  #MAST_CNTRL, CIO_CNTRL  ;master int cntrl
MOVE.B  #$00, CIO_CNTRL         ;dont enable yet
    ;port A
MOVE.B  #PA_DIRECTION, CIO_CNTRL ;port A direction
MOVE.B  #$FF, CIO_CNTRL         ;all inputs
MOVE.B  #PA_POLARITY, CIO_CNTRL ;port A polarity
MOVE.B  #$7F, CIO_CNTRL         ;invert all switch inputs
MOVE.B  #PA_SPECIAL, CIO_CNTRL  ;port A mode
MOVE.B  #$20, CIO_CNTRL         ;STOP switch is "1's catching"
MOVE.B  #PA_CMDSTAT, CIO_CNTRL
MOVE.B  #$E0, CIO_CNTRL         ;disable port A interrupts
    ;port B
MOVE.B  #PB_DIRECTION, CIO_CNTRL ;port B direction
MOVE.B  #$0F, CIO_CNTRL         ;top 4 bits are output
MOVE.B  #PB_POLARITY, CIO_CNTRL ;port B polarity
MOVE.B  #$0F, CIO_CNTRL         ;invert inputs
MOVE.B  #PB_SPECIAL, CIO_CNTRL  ;port B mode
MOVE.B  #$00, CIO_CNTRL         ;all unlatched
MOVE.B  #PB_CMDSTAT, CIO_CNTRL
MOVE.B  #$E0, CIO_CNTRL         ;disable port B interrupts
MOVE.B  #$00, CIO_PB_DATA       ;clear all outputs
    ;port C
MOVE.B  #PC_DIRECTION, CIO_CNTRL ;port C direction
MOVE.B  #$0F, CIO_CNTRL         ;4 input bits only
MOVE.B  #PC_POLARITY, CIO_CNTRL ;port C polarity
MOVE.B  #$0F, CIO_CNTRL         ;inverted
MOVE.B  #PC_SPECIAL, CIO_CNTRL  ;port C mode
MOVE.B  #$00, CIO_CNTRL         ;all unlatched
    ;timer
MOVE.B  #CT1_FROM_MSB, CIO_CNTRL ;MSByte of counting modulus
MOVE.B  #1, CIO_CNTRL           ;MSByte of 400
MOVE.B  #CT1_FROM_LSB, CIO_CNTRL ;LSByte of modulus
MOVE.B  #144, CIO_CNTRL         ;LSByte of 400
MOVE.B  #CT1_MODE, CIO_CNTRL    ;timer mode
MOVE.B  #$83, CIO_CNTRL         ;continuous, no in/outputs
MOVE.B  #CT1_CMDSTAT, CIO_CNTRL ;interrupts
MOVE.B  #$20, CIO_CNTRL         ;clear interrupts
    ;final doings
MOVE.B  #MAST_CONFIG, CIO_CNTRL
MOVE.B  #$D4, CIO_CNTRL         ;port A,B,C, & timer1 enabled
MOVE.B  #MAST_CNTRL, CIO_CNTRL
MOVE.B  #$80, CIO_CNTRL         ;enable chip's interrupts
MOVE.B  #CT1_CMDSTAT, CIO_CNTRL
MOVE.B  #$23, CIO_CNTRL         ;timer 1 start + interrruptss

    ;port initialization done
    ;set up arrays, registers, pointers, etc.
BSR     clear_arrays            ;zero DATA and NORM arrays
MOVE.L  #display_array, A4      ;initialize screen data pointer,
```

```
        MOVE.L  #ADC0, A3               ;ADC pointer,
        MOVE.L  #CIO_PB_DATA, A2        ;port B pointer,
        CLR.B   D5                      ;remaining scan count,
        CLR.L   D4                      ;display offset,
        CLR.W   update_offset           ;and update index to DISPLAY array

        AND.W   #$F8FF, SR              ;enable interrupts
main_loop:                              ;and PLAY BALL!
        BCLR    #LED_SWEEP, led_store
        BSET    #LED_STAND_BY, led_store
        MOVE.B  led_store, LED          ;do good things to LEDs
           ;now wait for START switch
           ;must stay open a while, then close
wait_for_zero:
        BSR     update                  ;update screen, waste 40us
        BTST    #START_BIT, CIO_PA_DATA
        BNE     wait_for_zero           ;loop until START is open
        MOVE.W  #1024, D3               ;debounce delay (update uses D0-2)
check_debounce:
        BSR     update                  ;update screen (40us) while looping
        BTST    #START_BIT, CIO_PA_DATA
        BNE     wait_for_zero           ;if it bounces start over
        DBF     D3, check_debounce      ;must stay open 1K loops
           ;now wait for it to be pressed
wait_press:
        BSR     update                  ;update screen while waiting
        BTST    #START_BIT, CIO_PA_DATA
        BEQ     wait_press              ;loop until START is pressed

           ;START has been pressed. Now read control panel, etc.
        BCLR    #END, CIO_PB_DATA       ;clear END output signal
        MOVE.L  #decode_tbl, A0         ;table for dwell/bin and #sweeps
        MOVE.B  CIO_PA_DATA, D0         ;read autoloop and dwell/bin
        BTST    #AUTO_LOOP, D0          ;autoloop switch set?
        SNE     auto_loop_flag          ;if so set flag
        AND.B   #$0F, D0                ;mask hex switch
        ASL.W   #1, D0                  ;convert to a word offset
        MOVE.W  (A0,D0.W), dwell_per_bin ;get table value and save

        MOVE.B  CIO_PB_DATA, D0         ;read #sweeps hex switch
        AND.B   #$0F, D0                ;mask
        ASL.W   #1, D0
        MOVE.W  (A0, D0.W), num_sweeps  ;get table value and save

           ;final setup before data collection
        BSR     clear_arrays            ;clear DATA and NORM arrays
        MOVE.L  #norm_array, A6         ;load dedicated registers
        MOVE.L  #data_array, A5
        CLR.L   D7                      ;clear accumulating register
        MOVE.W  dwell_per_bin, D6
        CLR.B   D5                      ;clear (=256) bin count
        CLR.B   stop_flag               ;it would be premature
           ;setup which interrupt handler entry is being used
        TST.B   auto_loop_flag          ;check for autoloop
        BEQ     free_run_int
        MOVE.L  #wait_trig_int, int5_vect ;handler entry for triggered
        BRA     update_loop
```

```
free_run_int:
        MOVE.L  #sweep_start_int, int5_vect ;handler entry for autoloop
           ;enter "main" loop
update_loop:
        BSR     update                  ;update screen
        TST.B   stop_flag               ;see if handler says it's done
        BEQ     update_loop             ;still sweeping
        BRA     main_loop               ;sweep is done
           ;LOOKUP TABLE for decoding rotary hex switches
           ;used for #sweeps and dwell/bin
decode_tbl:
        .word   1, 2, 5, 10, 20, 50, 100, 200, 500, 1000
        .word   2000, 5000, 10000, 20000, 30000, 0

           ;SUBROUTINES
           ;"clear-arrays" -- clears DATA and NORM arrays
clear_arrays:
        CLR.L   D0                      ;zero lives here
        MOVE.L  #data_array, A0         ;pointers
        MOVE.L  #norm_array, A1
        MOVE.W  #$FF, D1                ;counter
clr_loop:
        MOVE.L  D0, (A0)+
        MOVE.W  D0, (A1)+
        DBF     D1, clr_loop            ;fastest looping primitive
        RTS

           ;"update" -- update one more value to DISPLAY array
           ;checks panel control for scale
           ;also updates LED port from memory image (led_store)
           ;registers -- D0 - update offset  D1 - current data value
           ; D2 - scale factor, storage for normalizers  A0 - array ptr
update:
        MOVE.B  led_store, LED          ;update LEDs
        MOVE.L  #data_array, A0         ;raw data base pointer
        MOVE.W  update_offset, D0       ;indexes next point to update
        ASL.W   #2, D0                  ;make into a long offset
        MOVE.L  (A0, D0.W), D1          ;get data
        ASR.L   #2, D0                  ;restore integer offset
           ;get scale factor
        MOVE.B  CIO_PC_DATA, D2         ;read switch
        AND.B   #$0F, D2                ;mask hex value
        CMP.B   #$0F, D2                ;check for autoscale (n=$F)
        BEQ     auto_scale
        ASR.W   D2, D1                  ;else shift right n bits
        BRA     comp_and_save           ;and convert to offset binary byte
auto_scale:
        MOVE.L  #norm_array, A0
        ASL.W   #1, D0                  ;make a word offset
        MOVE.W  (A0, D0.W), D2          ;get normalizer
        ASR.W   #1, D0                  ;restore integer offset
           ;divide by norm and by dwell/bin
        TST.W   D2                      ;test before dividing
        BEQ     comp_and_save           ;dont divide, data is zero anyway
        DIVS    D2, D1                  ;data/norm
        MOVE.W  dwell_per_bin, D2
        BEQ     comp_and_save           ;data is zero anyway
        DIVS    D2, D1
```

```
comp_and_save:
        BCHG    #7, D1                  ;complement MSB of display byte
        MOVE.L  #display_array, A0      ;screen array base pointer
        MOVE.B  D1, (A0, D0.W)          ;store finished value in array
        ADDQ.B  #1, D0                  ;increment index, modulo-256
        MOVE.W  D0, update_offset       ;and save
        RTS


            ;INTERRUPT HANDLER(s)
            ;entered every timer-1 tick (100us)
            ;always refreshes display
        ;five entry points --
            ;"bad" - display LED walking bit forever
            ;"idle" - display refresh only
            ;"get_data" - get ADC data, check for end of bin & sweep
            ;"sweep_start" - initialize, load get_data vector, get_data
            ;"wait_trig" - sweep_start if triggered, else idle

bad_int:
        BCLR    #Z_BLANK, CIO_PB_DATA   ;turn off z-axis
        BCLR    #SWEEP_BIT, CIO_PB_DATA ;and SWEEP output
        BSET    #END, CIO_PB_DATA       ;set END output
            ;now make "walking bit" pattern
            ;and loop forever
        MOVE.B  #$01, D0                ;initialize LED value
flsh_loop:
        MOVE.B  D0, LED                 ;send to LED
        ROL.B   #1, D0                  ;circular shift left
        MOVE.L  #$8000, D1              ;reset delay counter
flsh_delay:
        SUBQ.L  #1, D1
        BNE     flsh_delay              ;make a delay
        BRA     flsh_loop

idle_int:
        MOVE.B  D4, DAC0_OFFSET(A3)     ;send X position
        MOVE.B  (A4, D4.W), DAC1_OFFSET(A3)  ;send Y position
        ADDQ.B  #1, D4                  ;incr index
        BRA     z_pulse                 ;make unblank pulse

wait_trig_int:
        BTST    #STOP_BIT, (A2)         ;check STOP switch
        BNE     stop_sweep
        BTST    #EXT_TRIGGER, (A2)      ;check trigger signal
        BEQ     idle_int

sweep_start_int:
        BSET    #LED_SWEEP, led_store   ;gets displayed by "update"
        BSET    #SWEEP_BIT, (A2)
        BCLR    #STOP_BIT, (A2)         ;clear STOP 1's catcher
        MOVE.L  #get_data_int, int5_vect  ;load "getdata" vector

get_data_int:
        MOVE.B  #$03, (A3)              ;start ADC (bipolar, 2's comp)
            ;update display
        MOVE.B  D4, DAC0_OFFSET(A3)     ;send X
        MOVE.B  (A4, D4.W), DAC1_OFFSET(A3)  ;send Y
```

```
        ADDQ.B  #1, D4                      ;incr index
        NOP                                 ;waste time so ADC can finish
           ;done sending XY pair
        MOVE.B  (A3), D3                    ;read ADC, which must be done
        EXT.W   D3                          ;extend byte->word
        EXT.L   D3                          ;extend word->long
        ADD.L   D3, D7                      ;add data to accumulating register
        SUBQ.W  #1, D6                      ;decrement dwell counter
        BNE     z_pulse                     ;we're still in this bin
           ;next bin
        MOVE.W  dwell_per_bin, D6           ;reset dwell counter
        ADD.L   D7, (A5)+                   ;add new value to DATA bin
        ADDQ.W  #1, (A6)+                   ;incr normalizer
        CLR.L   D7                          ;clear accumulating register
        SUBQ.B  #1, D5                      ;decr remaining bin count
        BNE     z_pulse                     ;if still in sweep
           ;end of this sweep. bin count is zero already
        MOVE.L  #data_array, (A5)           ;reset array pointers
        MOVE.L  #norm_array, (A6)
        BCLR    #LED_SWEEP, led_store       ;turn off sweep LED
        BCLR    #SWEEP_BIT, (A2)            ;and sweep signal
        BTST    #STOP_BIT, (A2)             ;check for manual abort
        BNE     stop_sweep
           ;now check if at end of sweep count
        MOVE.W  num_sweeps, D3
        BEQ     re_trigger                  ;already zero -> go forever
        SUBQ.W  #1, D3                      ;otherwise decr and test
        MOVE.W  D3, num_sweeps
        BEQ     stop_sweep
           ;now check if autoloop
re_trigger:
        TST.B   auto_loop_flag
        BNE     re_trig_auto                ;autoloop, needs vector loaded
        MOVE.L  #wait_trig_int, int5_vect  ;trig -- load vector
        BRA     z_pulse
re_trig_auto:
        MOVE.L  #sweep_start_int, int5_vect  ;load autoloop vector
        BRA     z_pulse
           ;stop sweep, manual or end of sweeps
stop_sweep:
        BSET    #END, (A2)                  ;set END output signal
        MOVE.L  #idle_int, int5_vect        ;load idle vector
        ST      stop_flag                   ;tell main prog we're done
           ;fall through to Z-axis unblanking pulse
z_pulse:
        BSET    #Z_BLANK, (A2)              ;start unblank software-pulse
        MOVE.B  #CT1_CMDSTAT, CIO_CNTRL_OFFSET(A2)  ;need to clear int
        MOVE.B  #$23, CIO_CNTRL_OFFSET(A2)  ;this does it
        BCLR    #Z_BLANK, (A2)              ;end unblank pulse
        RTE                                 ;return from exception (int)
           ;end of program
```

68008 using 4 clocks for each byte moved: The first instruction requires the CPU to fetch a 2-byte opcode, 4 bytes of extension for the (long) address, and finally the data byte requested; that's 7 bytes, or 28 clocks. The second instruction is a 2-byte fetch, followed by the data byte requested, for 3 bytes total, or 12 clocks. In general, devices with narrow buses (like our 68008, whose internal 32-bit architecture must feel positively claustrophobic in its 8-bit bus outerwear) suffer the most from fetch-intensive catatonia.

Finally the program begins! The first 8 bytes of ROM are the all-important start-up vectors, for the stack pointer and program entry point, respectively. The entry point is up in "true" ROM (it should be at $40008), so we can immediately clear the BOOT bit, which makes RAM replace the phantom ROM image we needed in order to boot. Now we can load interrupt vectors into low RAM, in the specific locations the 68008 dictates (Table 11.5 shows the full vector area): $68 (INT2), $74 (INT5), and $7C (NMI = INT7). We've used only INT5 (from the 100μs timer in the parallel port chip), which we load with the address of our interrupt handler. Different things need to be done by the handler, depending on the overall state of the machine (standby, waiting for trigger, beginning new sweep, or in the middle of a sweep), so we've written one grand handler, with separate entry points according to what needs to be done. At this point we're not ready to take data yet, so we load the address of the ***idle_int*** entry point into the INT5 vector location. It's a good idea to load all the unused interrupts and other vectors with a ***bad_int*** vector(s), in case something goes wrong (divide-by-zero, spurious interrupt, etc.); we loaded them all with a pointer to a routine that will flash something distinctive on the LED display (you'll find out what later).

Now comes the tedious but essential task of initializing the ports. The price you pay for the wonderful flexibility of LSI peripherals like the 8536 is the need to figure out precisely what control bytes to send to which register, and in what order, in order to make it do what you want. This includes choices of direction, polarity, mode, and interrupts, in the case of simple parallel ports, and choices of modulus, cascading, triggering mode, interrupts, etc., for the timers. We've shown the full initializing code for the parallel port/timer in Program 11.3: Our initialization enables

TABLE 11.5. 68000/8 VECTORS

Vector number	Address[a] (hex)	Assignment
0	000	initial SSP } reset
	004	initial PC } reset
2	008	bus error
3	00C	address error
4	010	illegal instruction
5	014	divide-by-zero
6	018	CHK instruction
7	01C	TRAPV instruction
8	020	privilege violation
9	024	trace
10	028	1010 emulator
11	02C	1111 emulator (fl point)
12–14	030–038	reserved
15	03C	uninitialized intr vector
16–23	040–05C	reserved
24	060	spurious interrupt
25[b]	064	level-1 autovector
26	068	level-2 autovector
27[b]	06C	level-3 autovector
28[b]	070	level-4 autovector
29	074	level-5 autovector
30[b]	078	level-6 autovector
31	07C	level-7 autovector
32–47	080–0BC	TRAP vectors
48–63	0C0–0FC	reserved
64–255	100–3FC	user (acknowledged) intr vectors

[a] all are 4-byte ("long") addresses. [b] not available on the 48-pin 68008, in which IPL0' and IPL2' are combined onto one pin.

parallel ports A, B, and C, and it makes bits 4–6 of port B outputs, and all the rest inputs (see Fig. 11.15). It initializes timer-0 to divide its 4MHz clock by 400, then retrigger continuously, generating interrupts (on INT5) every 100μs cycle. Note that we've made all switch inputs *inverting*, so that a switch (pulled up to +5V, with common to ground) that is closed reads as a 1, not a 0. We took advantage of the "1's catching" input option for the STOP switch, so a momentary closure is latched until we check it at the end of a sweep.

Finally, we clear the arrays in RAM (note use of a subroutine), initialize registers, enable interrupts, and jump into the "main" loop.

□ *Main program: main loop*

Once things are initialized, we enter the main loop forever. It actually consists of two loops: a loop that waits for the START button to be pressed, and then a second loop that continually updates the display memory while the actual data collection goes on underneath via interrupts. When the interrupt routine has finished its last sweep, it sets a software "stop flag," which the second main loop is continually checking. This signals the main routine to loop back to its first loop, its resting state while waiting for START again. Let's look at the flow chart and code.

The main loop (Fig. 11.19) begins by setting the LEDs to standby. Then it looks for the START button to be pressed, i.e., to go from open to closed. This is trickier than it seems, because the switch is not debounced in hardware, and therefore you typically get a few dozen closely spaced closures, extending over perhaps 25ms. That might be enough time to finish the shortest possible sweep (e.g., if you had selected 1 sweep at 100μs dwell/bin) and then erroneously begin again because the switch is still making transitions from open to closed. So we wrote a simple debouncing routine, which waits until the switch has been open continuously for about 50ms (it keeps busy with the ***update*** routine meanwhile), and then changes to a closed state. Now we've got our marching orders! The routine clears the END output signal, and then reads the control panel and uses the values accordingly (setting software flags like "`auto_loop`," and parameters like "`dwell_per_bin`" and "`num_sweeps`"). Note the use of the *decode_tbl* (and indirect addressing with index) to assign values corresponding to the switch positions.

The program then clears the DATA and NORM arrays, initializes some address and data registers, and clears the stop flag. The last step is to change the INT5 vector location (which currently points to the ***idle_int*** handler entry) to either the "***wait_trigger***" entry or the "***sweep_start***" entry, depending on whether the panel controls specified triggered or autoloop mode.

The main program then enters a tight loop in which it alternately calls ***update*** (update the DISPLAY memory from the DATA memory) and checks the *stop_flag*. Of course, interrupts are secretly making everything interesting happen during this insipid loop.

□ *Main loop: subroutines*

As an interlude before attacking the interrupt handler, which is the most complex code of all, let's look at the two subroutines called by the main program (Fig. 11.20). The ***clear_arrays*** routine sets both DATA and NORM to all zeros; it doesn't even bother with DISPLAY, because DATA's zeros will be rapidly copied to DISPLAY by ***update***. This routine updates one DISPLAY value, using current display parameters from the control panel, and input values from DATA and NORM; it also updates the LED port by copying the memory byte `led_store`.

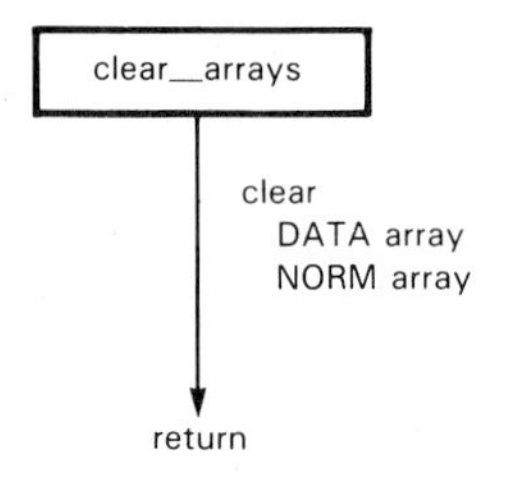

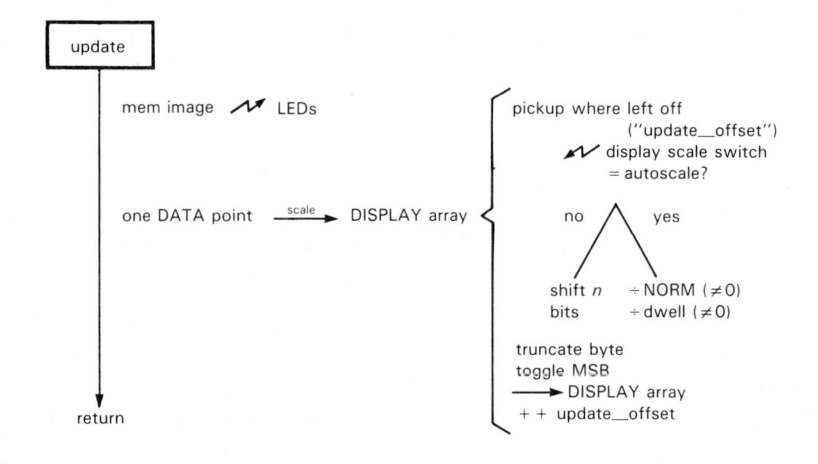

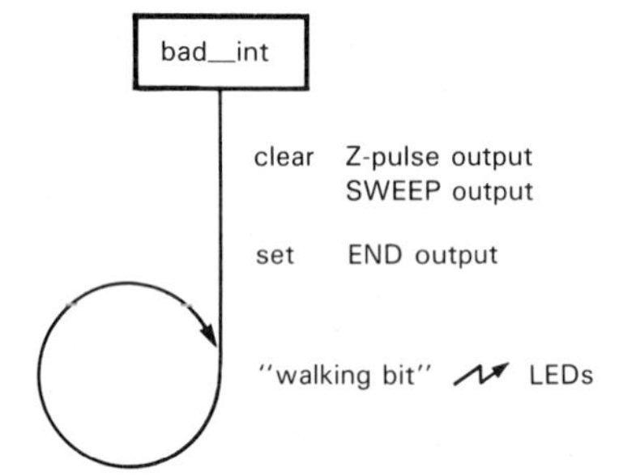

Figure 11.20. Subroutine flow charts.

Look first at the straightforward ***clear_arrays*** subroutine in Program 11.3. Pointers to the two arrays are initialized in A0 and A1, and all 32 bits of D0 are zeroed. D1 is a counter, initialized to the array size *minus one*; we'll see why in a moment. The loop moves either word- or long-sized zeros to the arrays, using indirect addressing (with postincrement); remember that postincrement is smart, and increments the address register by the right amount, in this case adding 2 for a word operation, 4 for a long operation. The `DBF` instruction is tricky and deserves explanation: It is really the `DBcc` instruction, with the condition code cc = False. The general `DBcc Dn,label` actually tests two conditions: First it checks the condition `cc` (i.e., flags set by the previous operation), doing nothing (i.e., falling through to the next instruction) if `cc` = True; however, if `cc` = False, it decrements the named register (as a word) and branches to `label` unless the register now equals −1, in which case it falls through. We forced `DBcc` to become a simple looping primitive by setting `cc` = F (always false, see Table 11.1), so it always decrements D1, looping until D1 = -1.

In spite of this trickery, and also the restriction that the counter must be word-length, DBcc is handy because it replaces two instructions (SUBQ, Bcc) and is very fast. Since it checks for -1, the counter register must be initialized to one less than the number of loops you want, which explains our initialization. The subroutine ends with the usual RTS (return from subroutine), which restores the old PC (program counter), thus returning to the calling program.

Note that we didn't need to save any registers at the beginning, because the calling program didn't leave anything valuable in D0–D1 or A0–A1. Note also the use of MOVE rather than CLR to zero the arrays; it turns out that MOVE is faster than CLR, because of an idiosyncrasy of the 68000 – to do a CLR, the 68000 does a read followed by a write. The designers did this strange thing in order to simplify the CPU logic.

EXERCISE 11.13

Rewrite *clear_ arrays* using SUBQ and Bcc in place of DBF. Rewrite it again using CLR in place of MOVE.

Update is a busier subroutine. Its job is to keep the DISPLAY memory (and also the LEDs) up to date, and it is repeatedly called in both loops of the main program. Since interrupts get priority, their work gets done on schedule (every 100μs), with ***update*** taking essentially all the remaining time. It begins by sending the memory image of the LEDs to the physical LED port. Even this simple operation involves some subtlety. The obvious question you should be asking yourself is why we don't just update the LEDs directly when we want to set or clear a bit (an LED). The answer is twofold: First, we can't just write a new byte to the LED port, because we'll clobber all the other bits; either we need an LED port that we can read as well as write or we need to keep a memory image of the LED. Since we didn't make the LED port readable, we use a memory location ("led_store") to keep a copy of what the LED port was last sent. Second, having been forced to use a memory location anyway, we can save time inside critical loops in the interrupt handler by updating only led_store there, letting the persistent ***update*** get the message out to the panel LEDs in its own good time. When we look at the interrupt handler this will make more sense.

EXERCISE 11.14

Add hardware (very little required) to make the LED port readable. Use cleverness to keep the additional address decoding simple.

The rest of ***update*** updates the DISPLAY memory. First it retrieves from memory the offset (number of points from the beginning of the array) for the next point to update. (It would be nice to use a dedicated address register here, but the speedy interrupt handler got first priority when registers were being given out.) The offset is multiplied by 4 (shift left 2 bits) so that it can be used for indexed addressing within the DATA array of longs. Having fetched the DATA value to D1, we read the panel for current display scale factor, and mask to obtain a number between 0 and 15. The value 15 ($0F) signifies autoscale, while smaller numbers signify fixed scaling by the specified power of 2: We either shift accordingly or branch to the autoscaling code.

For autoscaling we need to divide the current (indexed by update_offset) DATA value by the current NORM value (which tells how many scans are included in the DATA value), and then divide again by dwell-per-bin (which tells how many samples were included in each scan). Before you divide anything, *always test for zero!* Finally, whether we've autoscaled or shifted, we have to convert the resulting signed long to an offset-binary byte.

In the case of autoscaling, the final long is always in the range ±128; in the case of fixed scaling there will be overflow if you choose a scale smaller than the largest bin; in that case the best thing is to make the points at the top of the display roll over to the bottom on overflow, and vice versa. If you write some numbers down and play around a while, you can convince yourself that the right thing to do is always to truncate to 8 bits, then complement the MSB. We did this by using `BCHG` (bit change), then doing a byte-length `MOVE` to the DISPLAY array. Finally, we incremented and saved the `update_offset` index, and executed the essential RTS.

□ *Interrupt handler*

Now we come to the interrupt handler, which is really the centerpiece of this program. There are four entries to the timer-initiated interrupt; there is also the simple ***bad_int*** handler for spurious interrupts and all manner of other error vectors (Table 11.5). Let's warm up with ***bad_int***, then, with no more excuses left, we'll tackle the timer interrupt handler.

The 68008, as we described earlier, is genetically programmed to recognize both interrupts and the other various "exceptions" listed in the table, and after saving the current PC and status register on the stack, it jumps to the instruction whose address it finds in the vector location corresponding to the exception. Thus, if you try to divide by zero, the CPU will stack the PC and status register, then jump to the instruction whose (32-bit) address is stored in absolute locations \$014–\$017. Interrupts work the same, with the vectors for fully acknowledged interrupts in locations \$100–\$3FF, and the vectors for autovectored interrupts in locations \$064–\$07F. You can do whatever you want in the interrupt handler, and when you're done you execute an RTE instruction (return from exception). In order to prevent pandemonium, the CPU disables interrupts when it jumps to the handler, and reenables them when it executes RTE. If you have a long-winded handler, you may want to re-enable interrupts (at higher-priority levels only) within the handler, which you can do by sending the appropriate byte to the status register.

□ *Bad_int*

In our example, Figure 11.20 and Program 11.3 show the operation of ***bad_int***, whose job is to bring about an orderly shutdown of output signals, then make the LED display do something eye-catching. Its starting address, known to the linker after all the relocatable assembly code has been hooked together, is loaded (by our main program during the boot-up sequence) into all those reserved vector locations (in low RAM) shown in the table. An exception or spurious interrupt (i.e., any except level 5) causes the CPU to do the sequence described above, putting us into the code at ***bad_int***. We first shut off the Z-axis signal, because if we're unlucky enough the exception could have happened right in the middle of the Z-axis software pulse, causing the XY display to stay at full intensity (and in one spot) after the crash. While we're at it, we might as well terminate the SWEEP output and assert the END output, since whatever we were doing is now surely a hopeless cause.

Now for the fun. We send 01_H to the LED, then enter a loop that does a left circular shift, wastes a biologically significant amount of time, then sends the shifted byte out, etc. The net effect is a dazzling "walking bit" display, guaranteed to arouse even the most jaded button pusher. We never do an RTE, so the thing just goes on forever. The operator is obliged to do a total RESET to get things going again.

EXERCISE 11.15

Think of something smarter, so the operator can find out which exception caused the trouble. Hint: There are slightly fewer than 256 possible exceptions; there are also 8 LED bits. Can you write the code for your brainchild?

□ *Timer interrupts: four entries*

OK, we've got no more excuses. Let's plunge in. Figure 11.21 and Program 11.3 show the interrupt handler. It has four entry points, according to what state the machine is in. They are called ***idle***, ***wait_trig***, ***sweep_start***, and ***get_data***. The program cleverly causes interrupts to vector to the right entry by changing the interrupt vector (at $074) according to the overall machine state. You enter at ***idle*** if you don't want to collect data; it just displays one point on the screen, then returns. If you enter at ***get_data*** it reads the ADC, checks for end-of-bin or end-of-sweep (processing accordingly), and refreshes the display. ***Sweep_start*** sets appropriate LED and output signals, then falls through to ***get_data***. Finally, ***wait_trig*** checks for a trigger input, entering ***sweep_start*** or ***idle*** accordingly. There are some other labels in the interrupt handler (e.g., ***z_pulse***), but they are not entry points, just targets for loops or branches.

□ *Timer interrupts:* ***idle***. Let's go through the handler in detail, since it is important code. Early in the main program the vector is set to the ***idle*** entry, to create a screen display while waiting to start. Thus, execution commences at the ***idle_int*** label. With our reserved registers, everything is simple. `D4` holds the index of the next screen point to be refreshed, so we send it to the X-coordinate converter, DAC0 (addressed indirectly with offset, which is faster than absolute). Then we send the data value (using `D4` to index from `A4`, the DISPLAY base pointer) to the Y-coordinate converter, DAC1. `D4` is incremented (but not checked for end-of-array), and control jumps to the Z-axis software pulser.

EXERCISE 11.16

Why can we get away without checking after incrementing the DISPLAY index, `D4`?

By this time, the X and Y DACs have settled ($1\mu s$ settling time), so the Z-axis pulse routine uses `BSET` (bit set) to set the `Z_BLANK` bit (bit 4, see definitions) of parallel port B, an often-used port whose address we keep at the ready in `A2`. We could clear the bit next, but the rather short pulse that would result ($3\mu s$) would give you a dim display ($3\mu s$ unblanking every $100\mu s$). However, since all interrupts return via this code, we can use this opportunity to do a good deed while killing some time, namely telling the timer that it can now disassert its interrupt request. Writing to the timer-1 command/status register is a two-step process (as it was in the main program's initialization code), in which we first send the register's internal address ($0A) to the chip's control register ($83000), and then we send the actual controlling byte ($20), which the 8536 interprets as a command to disassert its timer-1 interrupt request. Now there's nothing left to do before returning from the interrupt, so we clear the Z-axis pulse (with a `BCLR`) and execute an `RTE` (return from exception). By putting the awkward interrupt acknowledge code within the Z-axis loop, we've managed to stretch the unblanking pulse out to $10\mu s$, repeated every $100\mu s$. Since we had to acknowledge the interrupt anyway, this is a great place to do it. We'll see a similar serendipity when we send the X,Y pairs out during the ADC conversion, coming next.

□ *Timer interrupts:* ***get_data***. This is the entry point used most of the time, namely when the signal averager is in the midst

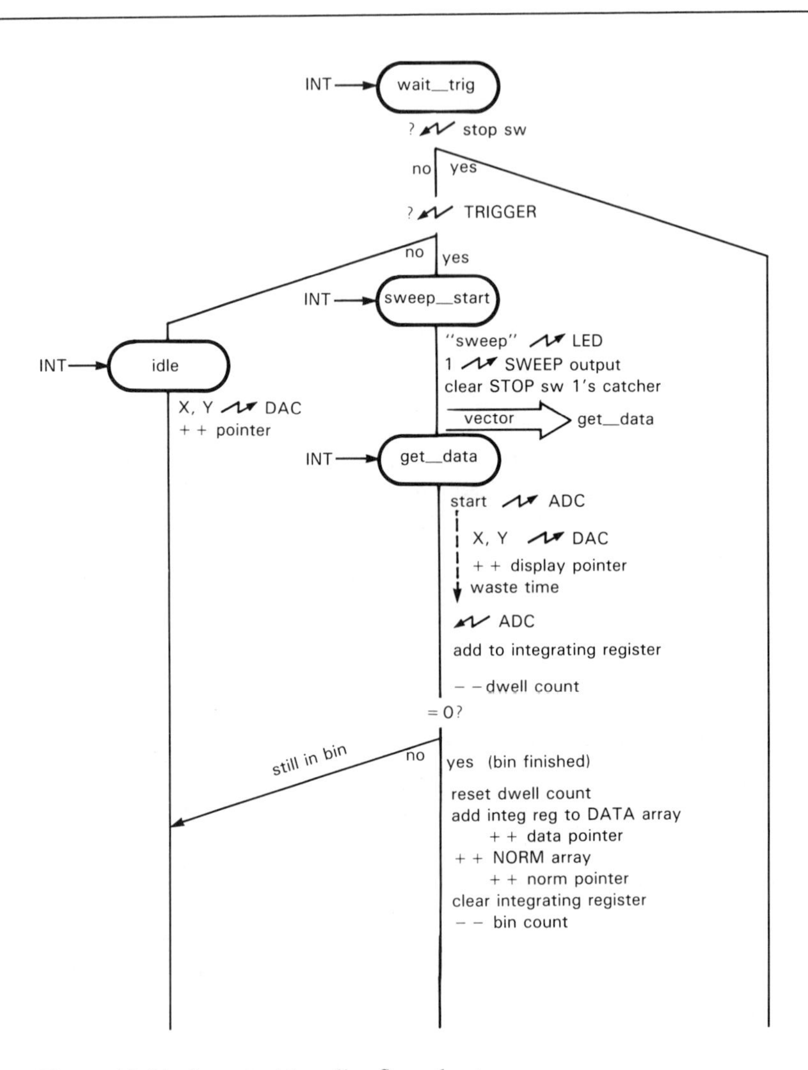

Figure 11.21. Interrupt handler flow chart.

of a sweep. We start the ADC by sending a mode byte ($03) to it; $03 specifies bipolarity 2's-complement conversion. As before, for maximum speed we use indirect addressing through A3 (which holds the ADC's address). Now we have to wait 10μs for the conversion, a perfect opportunity to send a new X,Y pair to the display DACs, with identical code as used in ***idle***. We're done a microsecond early, so we waste time with a `NOP` (no operation), then read the ADC. Notice how much better this is than having a readable hardware status bit signaling ADC conversion done, as we remarked in the circuit design discussion (Section 11.05); note, however, that we'd have to remember to add some more `NOP`s if we later decided to use a faster CPU clock.

We've read a 2's-complement byte from the ADC, but our DATA array and bin accumulator (`D7`) use 2's-complement *longs*. So we use the `EXT` (sign extend) instruction twice to make a long integer. Sign extension is simply the replication of the MSB leftward until the larger integer word is full; it preserves the value of a signed integer (simple zero-filling does not).

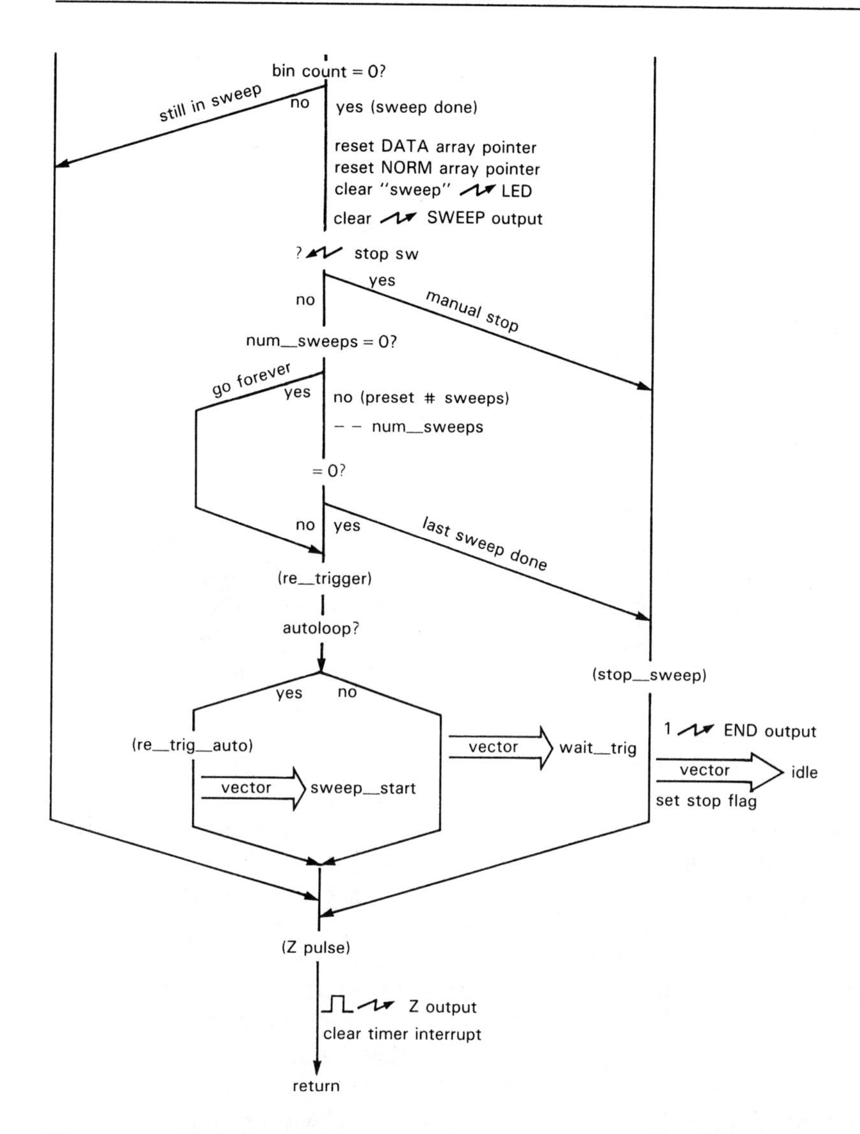

The extended integer is now added to the accumulating bin value in D7, and the dwell_per_bin counter (D5) is decremented. If it is still nonzero, we return via ***z_pulse***, described above. The total time in the handler in this case is 32.3μs, plus 9μs for the CPU interrupt processing and 5μs more for the RTE, for a grand total of 46.3μs. Thus, the main program gets more than half the CPU processing power for its simple DISPLAY update task.

If the bin accumulation is done, the handler resets the dwell count, adds the accumulated value in D7 to the DATA array (as pointed to by A5), increments the NORM value (through A6), clears the accumulating register (D7), decrements the bin counter, and (if nonzero, i.e., scan not done) jumps to ***z_pulse***. Note the use of auto-incrementing addressing modes. The additional time spent in the handler is 14.8μs.

If the *sweep* is also done, as indicated by zero in the remaining-bin-count register D5, the handler resets pointers, LEDs, and signal outputs. Then it checks to see if the STOP switch has been pushed; you always do the checking at the end (or beginning) of a sweep, so that all data are always averaged for the same number of sweeps. If STOP has been pushed, the routine jumps to ***stop_sweep***, which sets the END output and stop flag, and loads the ***idle*** entry point into the INT5 vector.

If the STOP button hasn't been pushed, the routine checks to see if it should quit anyway, since the panel-selected number of sweeps may have been completed (the number remaining is kept in the memory variable `num_sweeps`). Since the value 0 means "sweep forever," we first check for zero; if `num_sweeps` is zero, we leave it alone and jump to ***re_trigger***, otherwise we decrement and again check for zero. If it is now zero, we have done the last preprogrammed sweep, so we jump to ***stop_sweep***, otherwise we jump to ***re_trigger***.

Re_trigger's job is to worry about how to start the next sweep. If the variable `autoloop`, set by ***main*** after reading the control panel, is true, it loads the ***sweep_start*** entry into the INT5 vector, otherwise it loads the ***wait_trig*** entry. Note that there's no danger of an interrupt occurring while the vector is being changed, because interrupts are turned off by the CPU during interrupt processing; since we don't turn them on in the handler, they stay off.

□ *Timer interrupts:* ***sweep_start*** *and* ***wait_trig***. These entries are used if the next interrupt should begin a sweep, or if instead we're waiting for an external trigger pulse (at least 100μs duration!). The corresponding vector will have been loaded into the INT5 vector location, either by the main program when the START button is pushed or by the interrupt handler when it finishes a sweep that is not its last (at ***re_trigger***); look at the flow charts to remind yourself where this happens.

Sweep_start begins sweeping immediately, and is easy: It sets the SWEEPing LED and output signal, clears the latching (1's catching) STOP-switch bit, loads the ***get_data*** vector, and falls through to ***get_data***. Subsequent interrupts cause entry at ***get_data***.

Wait_trig is the entry point if the next sweep should not begin until a trigger signal is received (parallel port A, bit 7). Since STOPping should override starting, it looks first at the STOP switch (which causes a jump to ***stop_sweep***), then at the trigger input; if there is no trigger, it jumps to ***idle***, otherwise it falls through to ***sweep_start***.

□ 11.08 Performance

You can design a microprocessor-based instrument so that all the essential data-taking is handled entirely by fast hardware, with the microprocessor taking a role only in setup, display, and readout. In that case the instrument will run as fast as the hardware permits, with the microprocessor adding convenience and flexibility. You pay a price, of course, in hardware complexity and cost; you may also reduce the instrument's flexibility, owing to the fixed hardware architecture. If instead you simplify the hardware and use the processor to handle the data in real time, as we have done in our example, you can reduce the hardware complexity and maintain flexibility, at the probable expense of speed. In many cases speed is not important, though, so the choice is simple.

In our case the basic sampling rate, and therefore the maximum channel-advance (bin-stepping) rate, is limited by the processor speed. Each interrupt service has to be completed before the next one comes along. When we designed this example, we looked at what had to be done and estimated (based on a gut feeling) that 100μs was probably enough time. We didn't know for sure, though, but we decided that we could tolerate a slower sampling rate if that were necessary. Let's look at the numbers.

The 68008 data "sheet," which is really a 100-page book, includes tables of the instruction execution times (in clock cycles). We've used those tables to calculate the execution speeds we've quoted above. Here are the times (including interrupt vectoring and return) for the critical interrupt handler:

Entry	*Time (μs)*
idle	37
get_data	46.3 (within bin)
	61 (end of bin)
	92 (end of sweep, manual STOP)
	105 (end of sweep, programmed stop)
	113 (end of sweep, wait for trig)
	114 (end of sweep, autoloop)
sweep_start	61
wait_trig	46 (no trigger)
	69 (trigger)

Most of the times are less than the $100\mu s$ "heartbeat" of the signal averager, which is fine. There are three conditions that lead to a $>100\mu s$ interrupt service time, however. For the first one (end of sweep, programmed stop) it doesn't matter, since you don't care about an extra few microseconds once all the data have been collected. Likewise, you probably don't care about the second bad case (waiting for an external trigger), since such a trigger would always allow some resting time before beginning each sweep. However, the last bad case (end of sweep, autoloop) could be serious, since in autoloop mode you expect the overall period to be exactly $256 \times$ (dwell/bin). However, the situation is actually OK, for the following reason: When you use a signal averager in autoloop mode, you always trigger the external experiment from the averager (that's why we provided a SWEEP output signal), so you don't care if the period is a fraction of a percent different from what you expect. However, if you insist on getting done in $<100\mu s$, just substitute a 68008 running at 10MHz, which will multiply all the times by 0.8, making even the slowest time less than $100\mu s$ ($91\mu s$, in fact). If you do use a faster processor, be sure to fiddle with the interrupt handler code so that the ADC still has enough time to convert.

In summary, our guess that the 68008 would let us run at a 10kHz sampling rate was correct. The authors are particularly pleased, because we had written everything up to the last paragraph before we learned the good news! It's also obvious that you could not hope to speed up this instrument to a 20kHz rate without going to an all-hardware data collection scheme.

11.09 Some afterthoughts

In this design example we made a number of choices, both in the way we designed the hardware and in the way we implemented the software. In many places we could have done things differently. Most of the time the "best" choice is pretty obvious, but there were some cases where another method would have worked as well; in such situations we generally made our choice to maximize clarity, illustrate the most generally useful methods (avoiding fancy tricks and hardware idiosyncrasies), and minimize the length of assembly code. In real life (as opposed to book-writing) it's OK to use tricks, taking advantage of special features in the hardware, and it's also OK to write a lot of code. Here are some places where a different choice makes sense.

Switch-reading via table

Our program used a block of "in-line" code to read and test the various bits from the control panel, setting software parameters accordingly. That's a perfectly good way to do it. But a nice alternative, and one that is particularly easy to modify, uses instead a short *loop* to cycle through the panel's bits, with the port addresses, bit locations, and corresponding affected software variables specified by *tables*. Because such a scheme requires more explanation, and in our example would probably involve more lines of code, we opted for the simpler preprogrammed in-line switch-reading code. You should consider a table-driven scheme, though, in any application with a large number of parameters, particularly if you are likely to change the assignments or values of the input bits.

□ *One-shot Z-unblank*

We used a parallel port "software pulse" for the scope unblanking, because we wanted to illustrate this important technique. We particularly wanted to make the point that you cannot safely make software pulses when interrupts are turned on. An alternative is to use a hardware pulse generator, for example a monostable IC (rather than a parallel port bit). Monostables generally leave us with a bad feeling. However, in this case the powerful Zilog 8536 CIO comes to the rescue, with a built-in monostable that can be used to drive any output bit. Its "monostable" is actually one of the three on-board timers, which you program for pulse length (you can even cascade two timers to get longer pulse lengths). In our application there are timers to spare, so this method is ideally suited: By using the 8536 one-shot, you reduce the code inside the interrupt handler, and you can also choose the longest possible Z-unblanking pulse width.

□ *STOP-switch 1's catcher*

To read the STOP switch, we took advantage of a unique feature of the 8536, namely a built-in "1's catcher" flip-flop. When the 8536 is initialized, any input port bit can be made 1's catching; such a bit is then set by a momentary switch closure and holds that state until reset by a software WRITE to that port bit. This is perfect for our application because we only want to know *at the end of a sweep* if the STOP button has been pushed since the completion of the last sweep. Because a sweep can take many seconds, the memory feature saves us the trouble of repetitively reading the STOP switch; in our code, therefore, we look at the STOP bit only at the end of a sweep (see Fig. 11.21, the interrupt handler flow chart).

Since most parallel port chips don't have input memory, you may have to do the hard work we avoided. Here's what you would do: First, define an internal software flag, which you might call "`stop_at_end`"; in our code you would put the definition next to `stop_flag`. Be sure to clear this software flag before entering the data-taking loop; a good place would be after reading the control panel. Then add a few instructions in ***update_loop*** to check the STOP_BIT input repeatedly, setting `stop_at_end` if the STOP button is pressed. Finally, change the code in the interrupt handler to check this software flag, rather than the STOP switch, at the end of each sweep.

EXERCISE 11.17

Pencil in these changes to the assembly-code listing we've provided.

□ *Interrupt handler: multiple entries versus flags*

We used multiple entry points into our interrupt handler, one for each of four possible machine states (idle, waiting for trigger, beginning sweep, getting data). Since the handler is not a called subroutine, but rather a vectored entry at each interrupt, the program altered the entry point by loading the corresponding interrupt vector (in low RAM) each time it changed state. An obvious alternative is to have a single handler entry point, with flag-checking code in the handler to decide what to do. The program then tells the handler what to do by changing this software flag (rather than by fiddling with the handler entry vector). This scheme has the virtue of simplicity, but it takes longer to execute, since every entry involves tests and branches.

The difference isn't all that significant, though, so you should feel free to use flags to alter handler function if you like that method better.

□ *Serial port: data dump and slave control*

As we remarked in the "non-features" section (11.06), our signal averager lacks the essential capability of sending its averaged data out to an external computer! The code for this isn't terribly difficult, just tedious, involving both an initialization procedure analogous to that for the 8536 and a parsing and handshaking procedure so that the data recipient can trigger and acknowledge the transfer of data.

Assuming that the serial port has been set up for connection to a computer, it makes sense to use the same port as an alternative control panel, so that the external computer can set parameters and trigger data collection. To do this, the parsing routine would look for certain bytes that the computer sends to tell the averager that it wants control. Additional bytes would specify parameters (dwell per bin, number of sweeps, etc.), unconstrained by the limited number of switch positions in our panel-reading design. The software should be designed, of course, so that the panel switches take over if the computer does not ask for control. This gives us the best of both worlds, namely the simplicity of front-panel switches with the optional flexibility of computer control.

Reading panel knobs

In our microprocessor instrument we managed to avoid the complexity of panel *knobs*, sticking with the simpler option of panel *switches*, each of which drove a single parallel port bit. This kind of laziness on the part of instrument designers has led to an unfortunate tendency to omit knobs entirely, substituting instead (for example on a microprocessor-controlled oscillator) a pair of "up" and "down" buttons. If you're like us, you miss the nostalgic feel of a real rotary control. We could make our signal averager more civilized by having a knob that selects a bin on the screen, causing a display of the address and accumulated count.

The easiest way to regain control in a microprocessor instrument is to use an ADC input to convert the voltage from a panel potentiometer that is connected between +5 volts (or some better reference) and ground. There are inexpensive and small 8-bit ADCs with on-chip 8-input multiplexers and S/H; typically you have a few inputs left over, and you can use them to read several panel controls. In fact, you can even use an ADC input to read out the state of an n-position rotary switch – just string a resistive divider chain of n-1 equal resistors along the switch contacts, and use the ADC to read out the voltage!

If you need more analog readout resolution than a simple 8-bit ADC provides, consider instead a *rotary encoder*. This typically consists of a panel-mounting control no larger than an ordinary panel potentiometer, containing a pair of optical interrupters that provide quadrature pulses (90° out of phase) as the knob is turned. By providing pulses in quadrature, the encoder lets you determine which way the knob is being turned (see Fig. 8.97). Unlike an ordinary potentiometer, a rotary encoder has no limit stop, so you can turn the shaft through multiple rotations. A typical unit like the Bourns EN series produces 256 pulses per revolution.

MICROPROCESSOR SUPPORT CHIPS

In our microprocessor circuit there are 22 ICs, of which 10 are powerful LSI functions (CPU, memory, parallel and serial ports, calendar clock, and ADC/DAC converters), with the remaining 12 (a clear majority) consisting of lowly gates, flip-flops, buffers, and latches. The latter small- and medium-scale chips are needed to stick the big chips together and are sometimes called "glue logic." You can reduce the glue substantially with PALs, or, in large-production situations, with custom or semicustom chips designed for the task. Nevertheless, latches and three-state buffers, in particular, are used extensively in all microprocessor systems, and so it is worth taking a quick look at some of the choices you have. After that we'll deal briefly with some common LSI support chips (serial and parallel ports, and converters), then conclude the section with a discussion of memory.

11.10 Medium-scale integration

Latches and D registers

We mentioned latches and registers briefly in Section 8.24. The term "latch" strictly refers only to a *transparent latch*, whose outputs follow the respective inputs while enabled. A so-called *edge-triggered latch* is properly called a *D*-type register and consists of an array of *D* flip-flops with common clock. The difference has important consequences when latching data from a bus, because of the relative timing of DATA and its corresponding WRITE strobe. In particular, with some microprocessor buses (e.g., the IBM PC) the DATA is not necessarily valid at the leading edge of the write strobe, but it is guaranteed to be valid (and to have been valid for some minimum setup time) by the trailing edge of the strobe; see Figures 10.6 and 11.22. If you use a transparent latch, enabled during the entire strobe, you will most likely get transient states at the output, as shown. By comparison, the outputs of a *D*-type register (clocked on the trailing edge in this case) change state at the clock edge and are guaranteed not to have any glitches. It's an important fact that an output bit that has not changed state since the previous WRITE will have no momentary spikes or transients; this means you can safely use the various output lines of a latched byte to generate data and strobes for following circuitry.

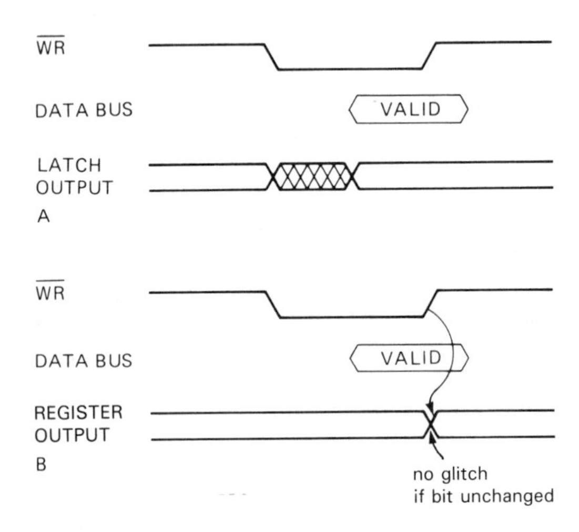

Figure 11.22. WRITE-cycle timing.
A. Transparent latch.
B. Edge-triggered *D* register.

When choosing between latches and registers, note that valid output data is available sooner with a transparent latch, which is sometimes important. Note also that many buses (e.g., the 68008) provide valid data throughout the strobe, including some setup and hold time, and with these you can clock a *D* register on the leading edge if there is enough setup time. On such a bus, of course, a transparent latch will not have transient output states.

There is a nice variety available in both D registers and transparent latches, with features such as RESET input, "broadside" pinouts (all inputs on one side, outputs on the other), inverted outputs, three-state outputs (useful for driving buses), and separate input enable. The latter lets you simplify external gating by driving the clock input with the write strobe and the input enable with the address decoding logic. Look back to Table 8.9 for a listing of features. Some perennial favorites are the octal three-state '373 (latch) and '374 (*D*-type), now modernized into the broadside '573 and '574. These are all packaged in 20-pin DIPs. In that same package the '273 is a '374 with RESET (but not three-state), and the '377 is a '374 with ENABLE (but not three-state). The newer 24-pin "skinny-DIP" packaging maintains the convenient 0.3 inch width while freeing up some extra pins. Thus, the modern '821 series includes 8- and 9-bit registers and latches with ENABLE, RESET, *and* three-state outputs, all in broadside pinout.

Note that a few of these 20-pin or 24-pin narrow chips may be a better choice than a fancy 40-pin (0.6 inch wide) LSI parallel port chip in many applications. For example, in our microprocessor design we used a '273 octal register for the LED port, and a '240 octal buffer for the DIP-switch port. The alternative would be a parallel port LSI chip like the Zilog 8536 (though we'd probably use the simpler Intel 8255 for such a trivial application), which costs more, uses up more space and power, and requires extra programming. In fact, the LSI chip also has poorer output drive (1.7mA sink and negligible source for the 8255, compared with 8mA source/sink for the 'HCT273). Some of the MSI latch/register chips are real brutes when it comes to output drive: The 'AC(T) family can sink/source 24mA (with rail-to-rail swings), and the 'AS821 series can source 24mA and sink 48mA. The LSI chips show their true colors, on the other hand, when you need fancy operating modes (interrupts, tricky input and output modes, etc.) or lots of programmable flexibility, rather than sheer muscle.

Buffers

Another chip used by the bucketful in microprocessor system design is the three-state buffer. You use it for asserting data and address information onto the bus. Most often you're simply sending data to the CPU, as with our simple DIP-switch input. As with latches, there are 20- and 24-pin versions with 8-bit (or more) width. Various features (see Table 8.4) include input hysteresis (to suppress noise), inverted output, broadside pinout, and separate enable inputs for bidirectional use. Special bidirectional buffers with a DIRECTION input and an ENABLE input (rather than a pair of enable inputs) are also available and are usually called *transceivers*; see Table 8.5 for a listing. Figure 11.23 shows a bidirectional buffer used to fortify the relatively feeble (~5mA) microprocessor data bus so it can drive the wiring and input capacitance of a board full of chips; such buffers are mandatory in a micro*computer*, in which a CPU board has to drive relatively high currents into a system bus (backplane) of high capacitance.

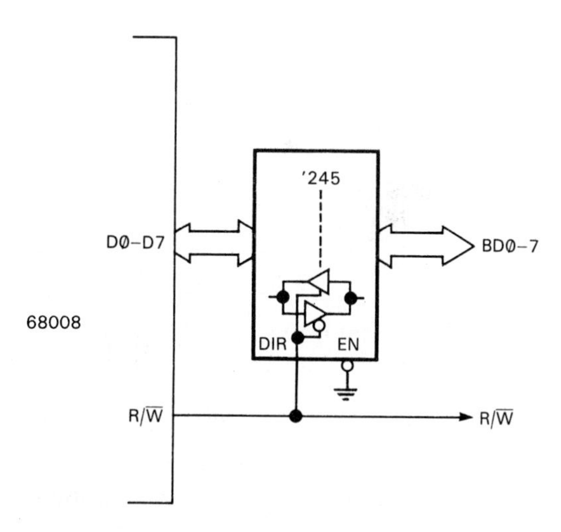

Figure 11.23. Data bus transceiver.

In many cases you can find MSI chips that combine a three-state output buffer with some other function you need; for example, you can connect a counter or latch, or even an A/D converter with three-state outputs, directly to a microprocessor bus. Our microprocessor example illustrated that method with its A/D converters.

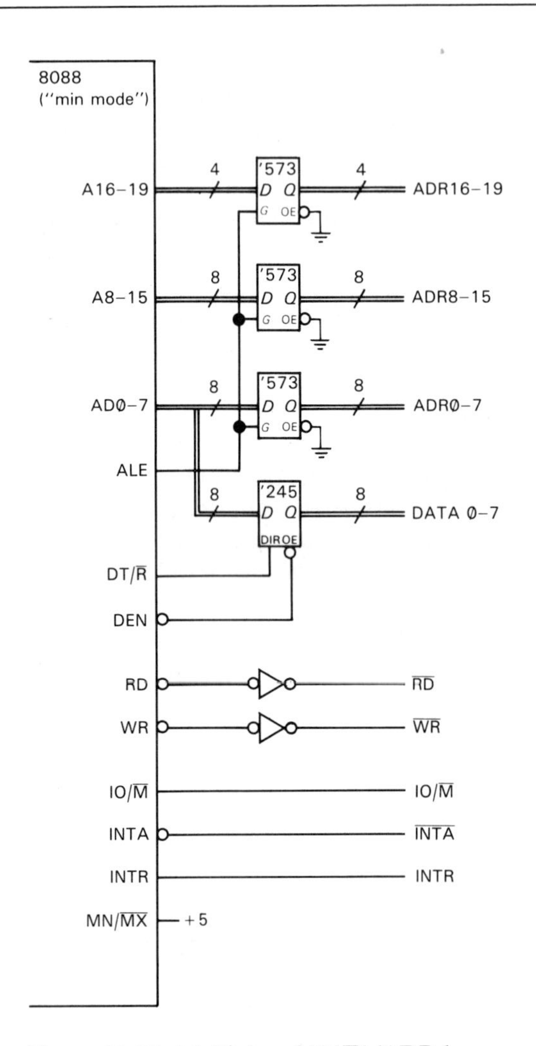

Figure 11.24. Multiplexed DATA/ADR bus.

In Figure 11.24 we show another example of bus buffering: Some microprocessors (e.g., the 8086 and 8088) combine pin functions to save space, multiplexing both DATA and low-order ADDRESS lines onto the same pins. An output called *address latch enable* (ALE) signals valid address and is used to enable a set of latches, as shown. You don't need to latch the DATA, however, because valid RD′ and WR′ strobes occur only when DATA is valid anyway. Note the use of a '245 transceiver to buffer the bidirectional DATA lines.

11.11 Peripheral LSI chips

General characteristics

As we mentioned earlier, LSI chips intended for microprocessor support are usually constructed with NMOS or CMOS technology, and they are usually supplied in large packages with 28 or 40 pins. They tend to be designed with lots of flexibility, often with programmable parameters of operation. Although they are usually designed for specific microprocessors, their generality allows you to use one manufacturer's support chips with another's CPU; thus we combined an Intersil calendar clock and two Zilog ports with a Motorola CPU in our microprocessor circuit. LSI peripherals tend to be expensive at the time of their introduction (the 8530 costs about $25, for example), often costing more than the CPU itself, but they show the usual exponential decay in price that is characteristic of IC technology (and precious little else in this world!). Figure 8.87 illustrated this apparently universal law of "Silicon Valley" (a section of the San Andreas Fault between San Francisco and San Jose).

Although we cast some disparaging remarks in their direction, many LSI support chips are virtually indispensable: Disk and video controllers are obvious examples. Another widely used type of support chip is the universal synchronous/asynchronous receiver/transmitter, USART for short.

How to use a USART

A USART is a microprocessor-controlled serial port chip, for example the Zilog 8530 in our design. Good USARTs include a programmable baud-rate generator, flexible control of bit format (number of bits, parity, etc.), and, in the most advanced USARTs, powerful synchronous modes of operation (with names like HDLC and SDLC), along with a choice of modulation methods (NRZ, FM, Manchester), clock recovery, error checking, etc. Most USARTs include interrupt hardware, and some even support block data transfer to the CPU via DMA. Most CPU families have their own USART, though only a small effort is involved in adapting a different USART. For example, IBM chose the National 8250 UART (asynchronous only), rather than Intel's 8251 USART, to go with the Intel 8088 in their PC. We chose Zilog's 8530 (also used in the Macintosh computer) because of flexibility, availability, and price, and we will use it to illustrate USART interfacing and programming.

USARTs are most often used to send data to and from terminals, modems, or hard-copy devices (printers, plotters), or directly between computers, where the major requirement is universal compatibility and simplicity of interconnection. The usual method is to use serial ASCII transmitted via bipolarity RS-232 levels, as described in Section 9.14 or 10.19. For this simplest sort of communication the USART is operated in the asynchronous mode, with each 8-bit character sandwiched between a START and STOP bit, and transmitted as a 10-bit serial string at a standard baud rate; for this job the 8530 is overqualified.

The 8530 comes in a 40-pin package (Fig. 11.25), and it communicates with the CPU via a set of *processor interface* lines, while independently communicating with the outside via a set of *communication interface* lines.

Processor interface. The 8530 uses 8 bidirectional DATA lines to connect to the CPU's bus, and the usual pair of strobes (RD′, WR′) and chip enable (CE′) for programmed I/O. The A/B′ input tells which of the two USART channels is being accessed, and the D/C′ input tells if the transfer is data (D/C′ HIGH) or control/status information (D/C′ LOW). As with the 8536, there are actually multiple control/status registers, accessed via a pair of successive transfers (look back at the 8536 initialization code). In the usual circuit implementation, the A/B′ and D/C′ lines are simply connected to the low-order CPU address lines, which map them into address space starting at the base address (as determined by the device address decode logic) of the USART. Finally, the processor interface includes four interrupt lines.

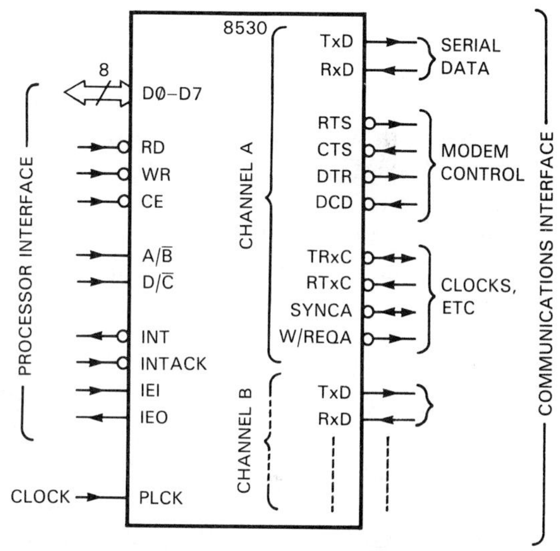

Figure 11.25. Zilog 8530 "serial communications controller" (UART) signals.

Communication interface. Each USART channel (labeled A and B) has transmitted and received serial data (TxD, RxD), and "modem control" handshaking lines (RTS, CTS, etc.), just like you find on the back of a computer. In addition, there are less familiar clock lines used for synchronous communication only (TRxC, RTxC). Finally, the USART requires an external oscillator signal at an integral multiple of 32 times the highest baud rate.

The USART doesn't know anything about bipolarity RS-232 levels, so you have to use RS-232C drivers and receivers on all these lines. For decades the classic RS-232 interface chips have been the bipolar 1488 (quad driver) and 1489 (quad receiver); however, in our microprocessor design we used the CMOS MAX233 (dual driver + receiver) because it conveniently has on-chip flying-capacitor voltage doublers and inverters, therefore running from a single +5 volt supply. Note that we didn't bother with the modem control lines (RTS, CTS, DSR, DTR); they're ignored by most of the world anyway, which uses "soft" handshaking (ctrl-S, ctrl-Q) embedded into the data stream instead.

Software. As we said at the outset, the operating modes of the USART are controlled by software commands. A byte sent to the USART in command mode (D/C′ LOW) is interpreted by the USART as a control command, setting the mode of operation. You can choose, for example, synchronous versus asynchronous operation, the number of STOP bits, type of parity, and so on. The simple USARTs of yesteryear were easy to program, with a single control register; fancy chips like the 8530 have literally dozens of registers and require the services of a PhD to program. Unfortunately this sort of complexity is the price you pay for the extreme flexibility of powerful LSI microprocessor support chips.

To illustrate, let's look at the command sequence to initialize an 8530 for asynchronous serial communication on channel A, at 1200 baud, 8 bits, no parity, and one stop bit; we'll also disable interrupts. The full initialization is somewhat long-winded; we'll therefore display the full sequence, but only show in gory detail how we figured out one or two of the critical command bytes. Table 11.6 lists the writable and readable registers in the 8530, which, as we've explained, are accessed by first writing the register *number* as a command (D/C′ LOW), then writing (or reading) from that register. This two-step process is waived for the transmit/receive buffers (WR8 and RR8), because they are needed for each transmitted byte; for these, a simple read or write with D/C′ HIGH does the trick. Likewise, the buffer status byte needs streamlined access, since you probably want to read its flag for each transmitted or received byte; the 8530 accommodates you by letting you read RR0 as a simple command/status read (D/C′ LOW). We'll see how this all works shortly, with some simple assembly language routines.

TABLE 11.6. ZILOG 8530 REGISTERS

Register	Function
Read registers	
RR0	xmit/recv buff status and extnl status
RR1	special receive condition status
RR2	unmodified intr vector (ch A); modified intr vector (ch B)
RR3	interrupt pending bits (ch A)
RR8	receive buffer
RR10	miscellaneous status
RR12	baud rate gen count (lower byte)
RR13	baud rate gen count (upper byte)
RR15	external/status interrupt info
Write registers	
WR0	initialization, pointers
WR1	intr and xfer mode definition
WR2	interrupt vector
WR3	receive parameters and control
WR4	misc parameters and modes
WR5	xmit parameters and control
WR6	sync chars or SDLC address field
WR7	sync char or SDLC flag
WR8	xmit buffer
WR9	master intr control and reset
WR10	misc xmit/recv control bits
WR11	clock mode control
WR12	baud rate gen count (lower byte)
WR13	baud rate gen count (upper byte)
WR14	misc control bits
WR15	external/status interrupt control

Each bit of each register signifies something. For example, Figure 11.26 shows WR3 and WR4, which are used to set up various communication options. The only bits of WR3 that affect asynchronous operation are D0, which enables the receiver, D5, which enables hardware

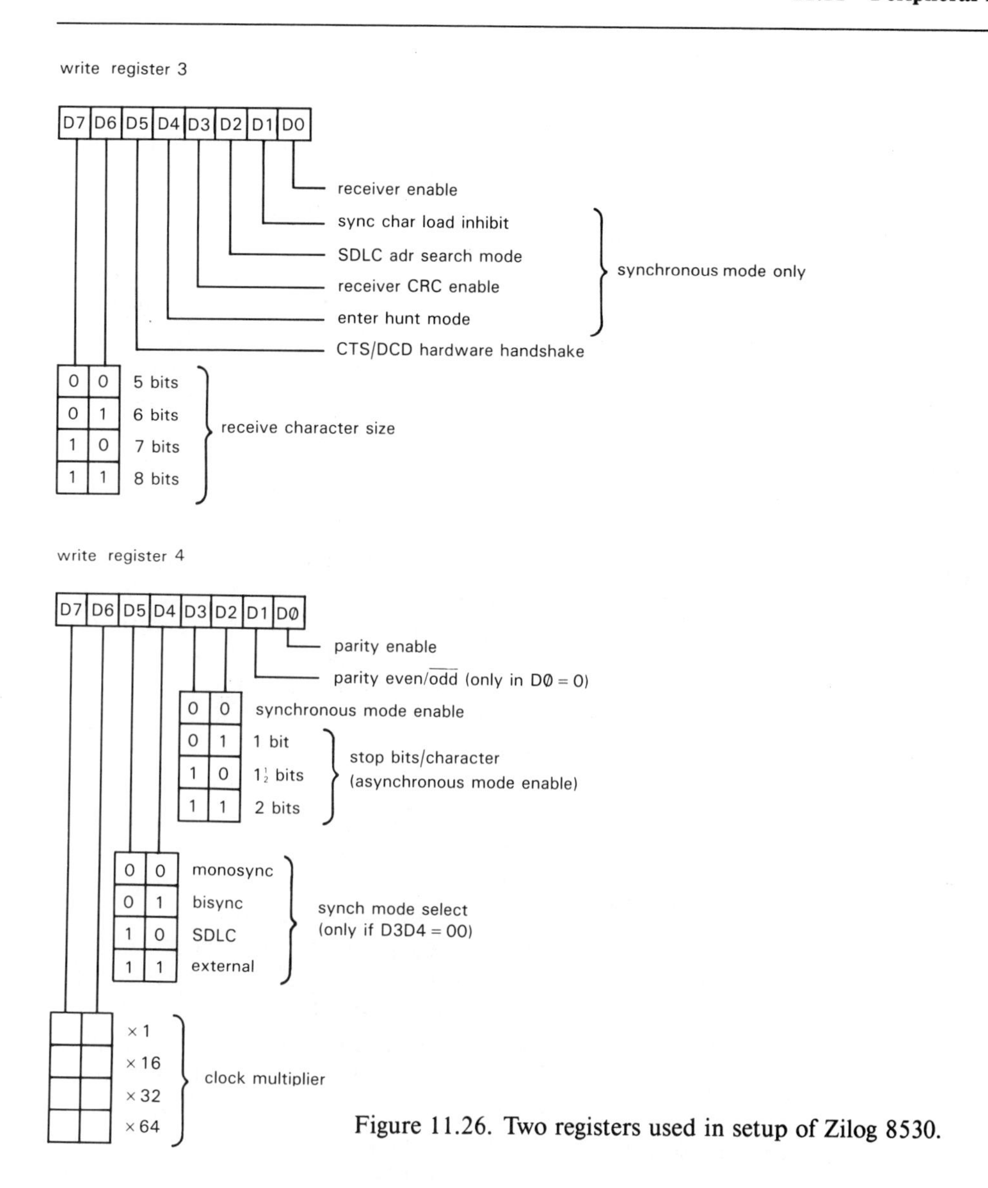

Figure 11.26. Two registers used in setup of Zilog 8530.

handshaking (see next section) via the "modem control" signals CTS and DCD, and the two high-order bits, which select the number of bits/character. The rest have to do with synchronous modes, which we're going to disable by appropriate choice of D2–D3 in WR4. So we make (D7,D6) = (1,1), D5 = 0, and D0 = 1, i.e., we send the hex byte C1 to WR3. For WR4 we select ×16 clock mode (the minimum divisor for asynchronous operation – the USART has to sample in the middle of each bit cell, so it needs the input clock to be a multiple of the baud rate), 1 stop bit/character (the standard choice, except for the obsolete 110 baud teletype standard), and no parity; thus we send the hex byte 44. Note that D5–D4 don't matter, having to do with synchronous options; also, D1 (parity choice) is irrelevant if D0 (parity enable) is 0. Note also that choosing 1 stop bit/character (which makes sense only in asynchronous mode) automatically disabled the synchronous mode, and with it all control register bits that specify synchronous options (e.g., D4–D1 of WR3).

In similar manner you navigate the remaining control registers. It's pretty boring work, and it's easy to make a mistake. Table 11.7 lists what we believe are the correct bytes (we tried it out to be sure!). Note the XXs ("don't care") for registers that are ignored because we aren't using interrupts or synchronous mode. WR9 does a total chip reset, which must precede all other commands. WR12 and WR13 set the 16-bit divisor for the baud rate generator, which for the 8530 is given by

$$\text{baud rate} = f_{\text{clock}} / [2(\text{clock mode})(\text{divisor} + 2)]$$

Thus with a 4MHz clock and ×16 clock mode, we need a divisor of 102_{10} for 1200 baud (this actually produces 1201.92 baud, which is close enough). Our choice of clock frequency allows any of the standard baud rates up to 9600 baud (for which the divisor should be set to 13).

TABLE 11.7. 8530 SERIAL PORT INITIALIZATION[a]

Register	Byte[b]	Result
WR0	(reg pntr)	use to access WR1-WR15
WR1	00	disable interrupts
WR2	XX	interrupt vector (if enabled)
WR3	C1	8 bits, Rx enable
WR4	44	x16 clk, 1 stop bit, no parity
WR5	68	8 bits, Tx enable
WR6	XX	synch char (sync mode only)
WR7	XX	synch char (sync mode only)
WR8	(xmit buf)	Tx data (direct access via D/C'=1)
WR9	C0	reset
WR10	XX	misc sync mode options
WR11	50	Tx, Rx clock from baud generator
WR12	102_{10}	baud rate divider – high byte
WR13	00	baud rate divider – low byte
WR14	03	enable baud rate generator
WR15	XX	interrupt options (if enabled)

(a) for 1200 baud asynchronous, 8 bits, no parity.
(b) XX = "don't care"; all are hex except WR12.

Note that there are numerous other possible operating modes under your control, set by corresponding sets of initialization control bytes. The 8530, like all USARTs, also permits synchronous communication in various modes at rates to 1Mbit/s; this is particularly useful for communication between a pair of processors. Such an application doesn't make much sense for our processor design, because of its small RAM, though it might come in handy for attaching a hard-disk drive.

Note that the correct initialization bytes must be sent to the USART by the CPU before any serial data are transmitted, just as with the 8536 parallel port chip; since we didn't use the serial port in our example earlier, we omitted the initialization. Program 11.4 shows how the initialization would go in this case. This sort of complexity (including deciphering the data sheet in order to figure out those crucial bytes) is the price you pay for the extreme flexibility of these microprocessor support chips.

After setting up the USART operating mode via the control registers, actual bytes of data are sent and received by CPU writes and reads with D/C′ HIGH. The status register must also be interrogated (D/C′ LOW) to determine when the USART has a new received data byte to be picked up by the CPU (D0 of RR0 set) or when it can accept a new byte to be transmitted (D2 of RR0 set). In addition, other bits of the status registers tell if a parity error was detected, if incoming data were lost, etc. You often ignore these latter dire indications of doom and plunge boldly ahead. Program 11.5 is an example, with the same register mapping into address space as above.

Note that these are the simplest forms of handlers, using programmed I/O to check for status information (see Sections 10.06–10.08). These hold up CPU operation by looping on the status flags. Input, in particular, would benefit from an

Program 11.4

```
    ;serial port initialization
    ;serial port addresses
CTRL_A  EQU     $85001  ;channel A control
CTRL_B  EQU     $85000  ;channel B control
DATA_A  EQU     $85003  ;channel A data
DATA_B  EQU     $85002  ;channel B data
    ;initialize port A (but see warning in text)
        MOVE.L  #CTRL_A, A0     ;port address, used often

        MOVE.B  #9, (A0)        ;first WR9, to reset chip
        MOVE.B  #$C0, (A0)      ;reset both channels

        MOVE.B  #4, (A0)
        MOVE.B  #$44, (A0)      ;x16 clk, 1 stop bit, no parity

        MOVE.B  #1, (A0)
        MOVE.B  #0, (A0)        ;disable interrupts

        MOVE.B  #3, (A0)
        MOVE.B  #$C1, (A0)      ;Rx 8 bits/char, enable Rx

        MOVE.B  #5, (A0)
        MOVE.B  #$68, (A0)      ;Tx 8 bits/char, enable Tx

        MOVE.B  #11, (A0)
        MOVE.B  #$50, (A0)      ;Tx and Rx clock from baud gen

        MOVE.B  #12, (A0)
        MOVE.B  #102, (A0)      ;baud rate divider, low byte

        MOVE.B  #13, (A0)
        MOVE.B  #0, (A0)        ;baud rate divider, high byte

        MOVE.B  #14, (A0)
        MOVE.B  #$03, (A0)      ;enable baud rate generator
    ;all remaining WR registers affect synchronous operation only
```

Program 11.5

```
    ;transmit routine
    ;enter here, with outgoing data byte in D0
trans:  BTST.B  #2, CTRL_A      ;Tx buffer empty?
        BEQ     trans           ;if not, keep checking
        MOVE.B  D0, DATA_A      ;if so, send byte
        RTS                     ;and return

    ;receive routine
    ;incoming data byte returned in D0
recv:   BTST.B  #0, CTRL_A      ;Rx character available?
        BEQ     recv            ;if not, keep checking
        MOVE.B  DATA_A, D0      ;if so, get byte
        RTS                     ;and return
```

interrupt-driven routine. The 8530 would be happy to oblige by making interrupts on any specified condition; it can even acknowledge with an 8-bit vector of your choosing, if you connect the interrupt acknowledgment lines (IEI, IEO, INTACK). These options are, of course, specified through initialization control bytes.

Parallel I/O chips

We showed an example of these versatile parallel port chips (often combined with one or more timers) in our microprocessor example. The 8536 we used there is a companion to the 8530 serial port chip, and it uses a similar processor interface and setup protocol. Good parallel port chips let you individually program the direction and mode (latching, open-drain, inverting, etc.) of each bit. The data-transfer protocol can also be programmed; for example, with the 8536 you can enable vectored interrupts on any *pattern* of input bits. You can also select one of four *handshake* modes, as we'll describe shortly.

PIO chips, like all LSI peripheral chips, are constructed with NMOS or CMOS technology, with the latter being favored in new designs. Outputs can generally sink a few milliamps, but NMOS outputs, unlike CMOS, can source only a fraction of a milliamp. As a result, they are usually used in conjunction with power driver chips to drive loads requiring significant amounts of current. Don't try to turn on a relay directly with a PIO output (but see our comments on MSI ports under "Latches and *D* registers," above).

Figure 11.27 suggests the sort of circuit you might use, in this case to refresh a 6-digit display. You would, of course, have to write software to repetitively output the successive digit values, along with a "walking bit" in the A port, taking care that interrupts are disabled during display to prevent flicker. An easier way to handle multidigit LED displays in microprocessor systems is to use something like the Siemens "intelligent display" series of memory-mapped display sticks, which conveniently look like memory to the

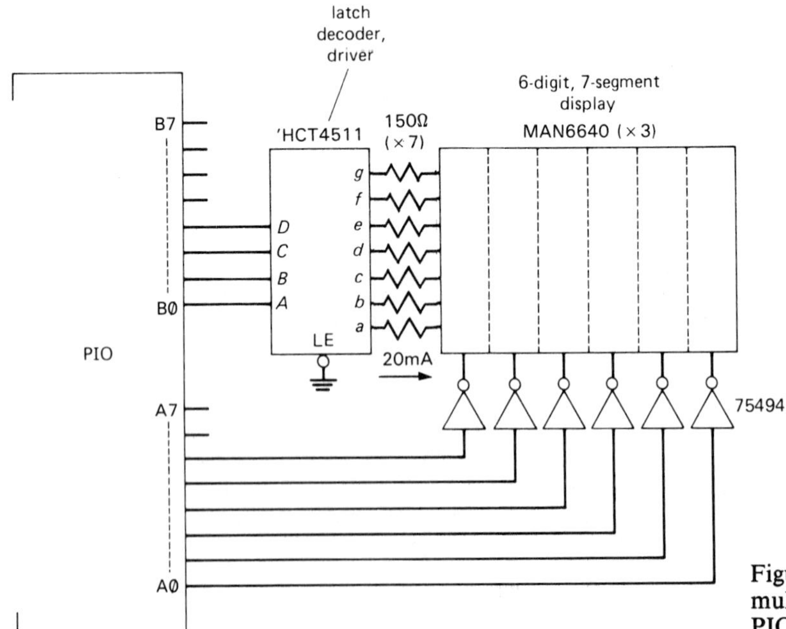

Figure 11.27. Driving a multiplexed display with a PIO.

CPU (Fig. 9.24); since they latch their displayed values, you need to write to them only when you want to change the display.

Handshaking

The business of handshaking deserves a bit more explanation. Suppose you are an external device sending bytes of data to a processor via its PIO port. You want to know when the PIO is ready to accept the next byte, i.e., when the previous byte has been picked up by the processor. The natural way to handle this is to have a "ready for data" (RFD) output from the PIO, which is disasserted by the PIO when you give it a byte, and reasserted after that byte has been picked up by the CPU. In other words, you can strobe your data at any time RFD is asserted.

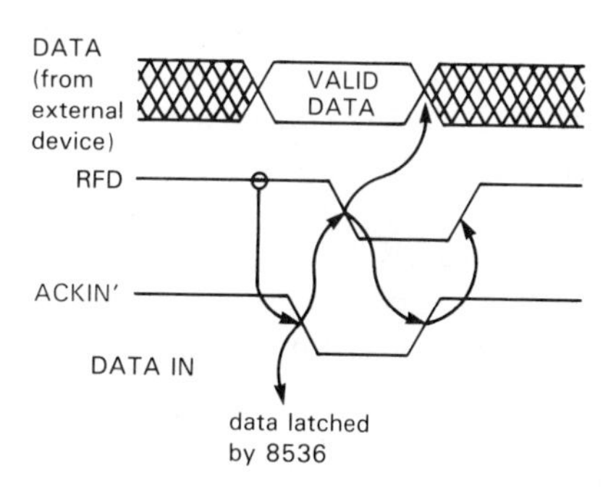

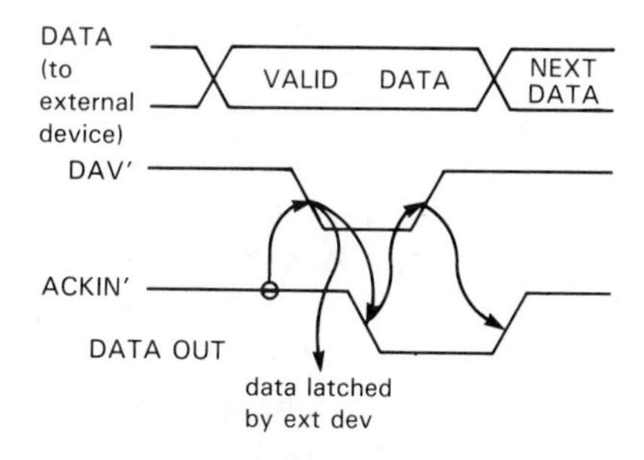

Figure 11.28. Handshaking.

Figure 11.28 shows how you use this "interlocked handshake," which is actually one of four possible handshaking modes provided by the 8536. For data IN, the external source may assert data, but waits for RFD true before asserting the strobe ACKIN′ (that's Zilog's name for this pin!). It releases the strobe (and may disassert data) when it sees RFD disasserted, then waits for RFD again. For data OUT the procedure is similar, with the PIO asserting valid data, then asserting DAV′ (data available). The external device latches the data, then asserts the acknowledgment ACKIN′. This causes the PIO to disassert both the data and DAV′. The latter is a signal for the external device to disassert ACKIN′, beginning a new cycle. Note that the handshake is fully interlocked, in that each side in the transaction waits for the other to complete each step. This fully interlocked data exchange guarantees that no data are lost. It's sometimes nice to simplify things, however. The 8536 therefore has a "pulsed handshake" mode in which the ACKIN′ signal doesn't have to wait for permission to disassert; instead, it is a pulse of at least 250ns duration, asserted when RFD or DAV′ is true, as above.

As you might expect, the handshaking modes are selected by sending those all-important control bytes during initialization. When handshaking is selected, some or all of the four port-C lines are used for ACKIN′, DAV′, and RFD. If you don't tell the chip to use handshaking, you can use the port-C lines instead for ordinary I/O bits, just like ports A and B.

Warning: A general disease of many LSI peripheral chips, particularly those with a clock input, is that they can have a surprisingly large latency between allowed accesses. Our experience with the 8530 is instructive: It was necessary to put some NOPs between successive WRITEs when using a 68000 running at 10MHz, because the speed with which the CPU could send successive bytes ($0.8\mu s$) was faster than the $1.7\mu s$ minimum "valid access recovery time" of the 8530. Watch out, also, for peculiar processor-interface timing requirements. Recall that in our microprocessor circuit, for example, the Zilog peripherals had an unusual requirement of 80ns setup time from address to leading edge of the *read* strobe, forcing us to add circuitry to generate a delayed RD′ pulse. These idiosyncrasies, which you never see in plain MSI digital logic, are the result of the chip's clocked state-machine logic, the relatively slow NMOS speeds, or both.

The only safe cure for these LSI pathologies is careful design, which implies studying the data sheets carefully and completely, talking to other users, and exhaustive testing. If you notice an anomaly while testing a prototype, don't assume that it will go away in the more carefully constructed final version. In fact, these are the very clues you are seeking. Often you can test a circuit more thoroughly by artificially varying the clock rate and supply voltage, and perhaps also the temperature (use a heat gun), in the prototype.

Microprocessor-bus A/D and D/A converters

With the increasing use of microprocessor-controlled (as opposed to stand-alone) data acquisition, newer A/D and D/A converters have been designed with data buses in mind. "Microprocessor-compatible" D/A converters, for example, have byte-wide input data paths and double buffering so that you can latch a 12-bit quantity from an 8-bit bus in two write cycles; the double buffering ensures that the entire 12-bit quantity is presented to the converter simultaneously, to minimize output glitches. Likewise, microprocessor-compatible A/D converters have three-state output drivers, organized in 8-bit-wide groups. Both A/D and D/A converters tend to have familiar bus-oriented signals such as RD′, WR′, and CS′. These converters can generally be connected directly to the microprocessor bus, with only a small amount of "glue" needed for address decoding. You don't need to fool around with the usual interface paraphernalia such as external latches and three-state drivers. Take a look back at our microprocessor circuit, which has both ADCs and DACs, to see how easy it is.

As examples, the AD7537 from Analog Devices is a dual 12-bit DAC with 8-bit loading (i.e., each 12-bit quantity is loaded as $8+4$, in two bus cycles), with provision for simultaneous updating of the two double-buffered 12-bit channels (Fig. 11.29); the companion AD7547 has a 12-bit-wide input, for 16-bit buses. The AD7572 from the same manufacturer is a no-glue 12-bit ADC with choice of byte-wide or 12-bit-wide parallel output via three-state drivers; it can thus be used efficiently with microprocessor buses of 8-bit or 16-bit width. Some bus-compatible converters even allow left- or right-justification when the word is smaller than the bus. When choosing a bus-compatible converter, watch out for devices with sluggish processor-interface timing (which has nothing to do with the *conversion* speed), which can force you to insert wait states, delayed strobes, etc. For example, the AD558, an otherwise admirable self-contained 8-bit ADC, has a 200ns minimum data setup to trailing edge of WR′, requiring a wait state with the 68008.

In any ADC microprocessor interface with resolution of 12 bits or more, consider isolating the ADC chip outputs from the system bus with buffers (or even opto-isolators); otherwise the digital current transients and microprocessor noise are likely to degrade the resolution. For the highest resolutions (16 bits or more), it may be best to put the converter outside the box containing the digital electronics. As an example of what can happen, we offer our experience with a commercial 16-bit ADC board designed for the IBM PC: The converter module is on the PC board, inside the computer. We were skeptical that it would attain a full 16 bits of resolution, so before buying one we asked what

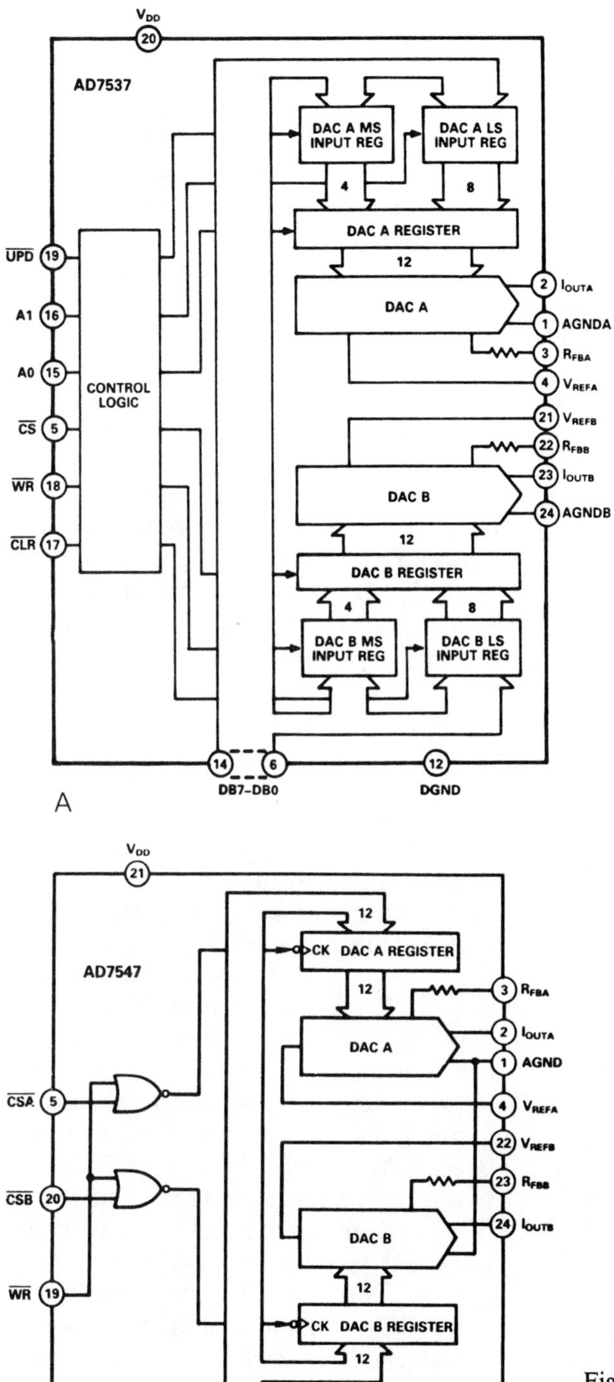

Figure 11.29. 12-bit dual DACs. (Courtesy of Analog Devices.)
A. 7537 byte-wide loading.
B. 7547 12-bit-wide loading.

would happen if a fixed voltage were applied to the board's analog inputs. The manufacturer's technical support department assured us that the result would be "at most two adjacent digital codes." In fact, the board actually bounces around among seven consecutive codes, which amounts to a 14-bit conversion. At their suggestion we brought the board back for tests, which confirmed the noise problem. When we asked about this, we were told that the particular employee no longer worked there. They told us all their boards work that way; and, to add insult to injury, they threatened to charge us for the "service" they had performed in testing the board.

11.12 Memory

In a commercial microcomputer it's easy to add memory – all you have to decide is how many megabytes you want to add, and from whom to buy it. You get to have more fun in a dedicated microprocessor application, where the assignment of memory is part of the design, with blocks of nonvolatile (permanent) ROM being used for program storage, and read/write volatile RAM being used for temporary storage of data, stacks, and program workspace. Nonvolatile program storage is universally used in dedicated instruments so that it isn't necessary to load a program each time the instrument is turned on.

In this section we will discuss the various kinds of memory: static RAM, dynamic RAM, EPROM, and EEPROM. Once you get the hang of it, the choices are pretty clear. You may want to refer ahead to Figure 11.35, where we summarize memory families.

Static and dynamic RAM

Static RAM, or SRAM, stores bits in an array of flip-flops, whereas dynamic RAM, or DRAM, stores bits as charged capacitors. A bit once written in a static RAM stays there until rewritten or until the power is turned off. In a dynamic RAM the data will disappear in less than a second, typically, unless "refreshed." In other words, a dynamic RAM is always busy forgetting data, and it is rescued only by periodic clocking through the "rows" of the two-dimensional pattern of bits in the chip. For example, you have to access each of 256 row addresses in a 256K-bit RAM every 4ms.

You might wonder what would possess anyone to choose a dynamic RAM. By not using flip-flops, the DRAM saves space, giving you more data on a chip, at lower cost. For example, the currently popular 32K×8 (256kbit) SRAMs cost about $10, twice the current price of 1Mbit DRAMs. Thus, you can get four times as much memory on a large memory board, at half the price, by using dynamic RAM.

Now you might wonder why anyone would choose *static* RAM (fickle, aren't you?). The major virtue of SRAM is its simplicity, with no refresh clocks or timing complexity to worry about (the refresh cycle competes with normal memory access cycles and must be properly synchronized). Thus, for a small system with only a few memory chips, the natural choice is SRAM. Furthermore, most SRAM these days is CMOS, essential for battery-powered devices. In fact, CMOS static RAM, backed up with a battery when main power is off (using a power control chip like the MAX690), forms a good alternative to ROM for nonvolatile memory. A further advantage of SRAM is its availability in high-speed versions, down to 25ns or less, and convenient byte-wide packaging. Let's take a closer look at both SRAM and DRAM.

Static RAM. We saw SRAM in our microprocessor design, where we used a single 32K×8 SRAM for data, stack, and workspace (the program was in EPROM).

Using SRAM is like falling off a log: You assert the address, chip select (CS′), and output enable (OE′) for READ; the requested data appears on the three-state data lines a maximum of t_{aa} (address access time) later. For WRITE, you assert address, data, and CS′, then follow (after an address setup time t_{as}) with a write enable (WE′) pulse; valid data are written at the end of the WE′ pulse. The actual timing constraints for 120ns SRAM are shown in Figure 11.30, from which you can see that the "speed" of the memory is the time from assertion of valid address to valid data (read) or to completion of the write cycle (write). For SRAM, the time from one memory access to the next (the "cycle time") equals the access time; this isn't true for DRAM, as we'll see.

Static RAM is available in sizes from 1Kbit or less up to 1Mbit, organized as 1 bit, 4 bits, or 8 bits in width. Speeds range from a leisurely 150ns access time down to 10ns or so. Currently popular types include the inexpensive 8K×8 and 32K×8 CMOS SRAM with 80ns access times, and smaller high-speed (<30ns) CMOS SRAM for cache memory. Variations include separate input/output pins, dual-port access, and SIP (single-row) packaging.

For whatever it's worth, note that you don't have to hook up the CPU's data lines to the corresponding memory data pins – you can scramble them any way you want, since they get unscrambled when you read back what you wrote! The same goes for addresses. Don't fool around with ROM in this way, though.

EXERCISE 11.18
Why not?

Dynamic RAM. By comparison with SRAM, DRAM is a royal headache. Figure 11.31 shows a normal cycle. The address (e.g., 20 bits for 1M DRAM) is split into two groups and multiplexed onto half the number of pins, first the "row address," signaled by a "row address strobe" (RAS′); then the "column address," signaled by a "column address strobe" (CAS′). The data are written (or read, as determined by the direction input R/W′) following assertion of CAS′. Some additional "RAS-precharge" time must elapse before the next memory cycle, so the cycle time is greater than the access time; for example, a particular DRAM with 100ns access time has 200ns cycle time. A refresh cycle looks similar, but without the assertion of CAS′. Actually, normal memory accesses would refresh perfectly well, if you could count on accessing all required combinations of row addresses!

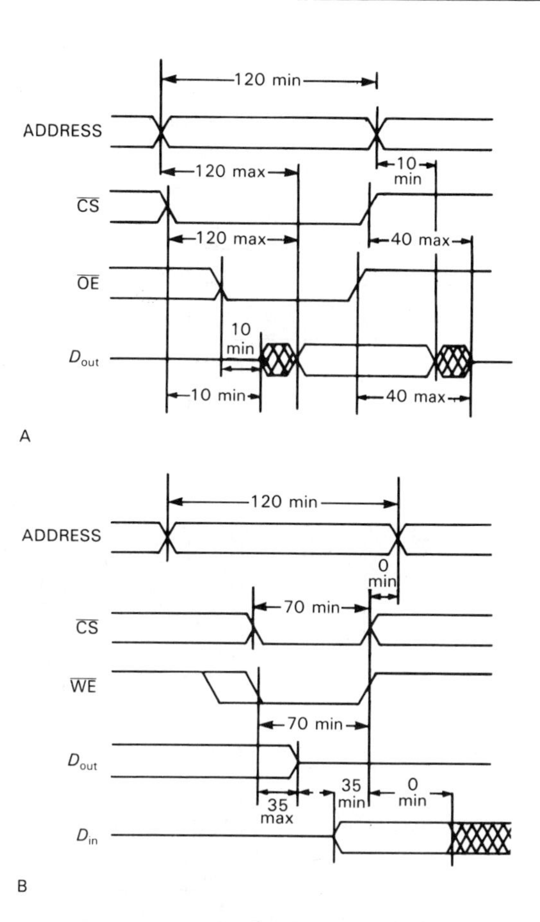

Figure 11.30. 120ns SRAM timing.
A. READ cycle.
B. WRITE cycle.

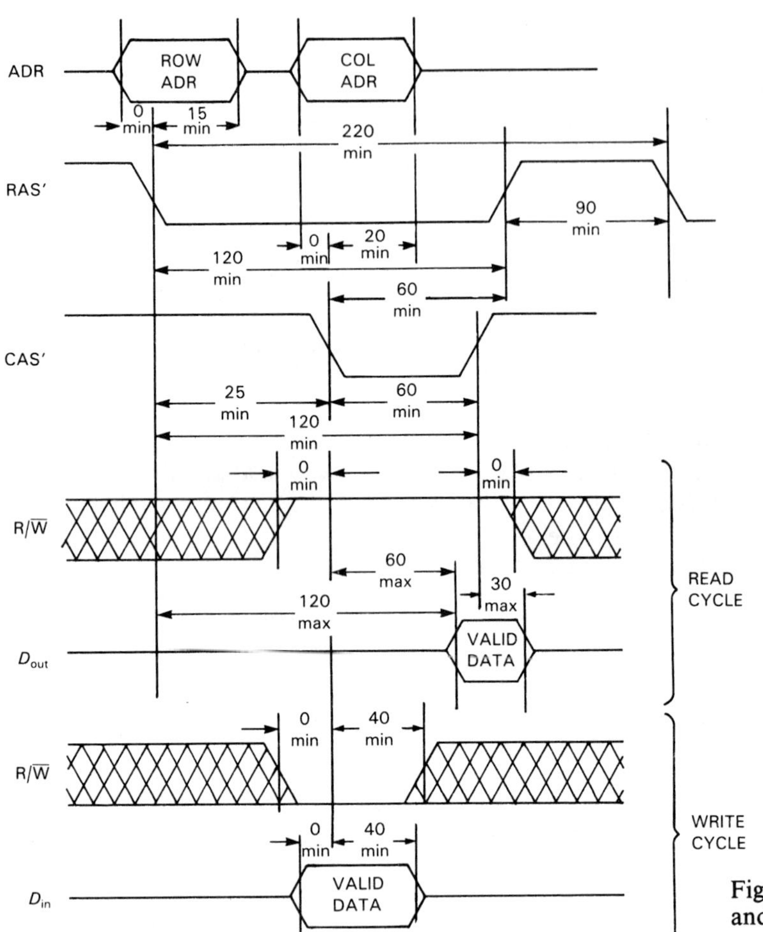

Figure 11.31. DRAM READ and WRITE cycles (Hitachi, 120ns).

DRAMs, like SRAMs, come in data widths of 1, 4, and 8 bits, in sizes from 64Kbit to 4Mbit, and in speeds from about 70ns to 150ns. The most popular are the large 1-bit types, which makes sense: If you are building a large memory array, say 16 bits wide and 4Mbyte total, and you have a choice of 1Mbit DRAMs, organized either as 1M×1 or 128K×8, it pays to use the 1-bit chips because (a) each data line will be tied to only two chips (rather than 16), which results in far less capacitive loading, and (b) the chips are smaller, since fewer data pins more than makes up for the extra address pins. Furthermore, the 1-bit chips will generally be cheaper. This argument assumes you're building a large memory, of course, and, for example, doesn't apply for our simple 32K×8 microprocessor. Note, however, that improved high-density packaging technologies ("ZIPs" and "SIMMs") are reducing the importance of minimizing pin count.

There are several ways to generate the proper sequence of multiplexed addresses, RAS, CAS, and R/W′ signals that you need to use dynamic RAM. Since the DRAM is invariably connected to a microprocessor bus, you start things going when you see an AS′ (or equivalent) signaling a valid address in DRAM memory space (as determined by the high-order address lines). The traditional method uses discrete MSI components to do the address multiplexing (a few '257 quad 2-input MUXs) and generates the RAS, CAS, and MUX control signals either with a shift register clocked from a multiple of the processor clock or, better, from a tapped delay line. You need some logic to interpose RAS-only refresh cycles periodically, with a counter to generate successive row addresses. Figure on at least 10 chips if you design your circuit this way.

PALs offer an attractive alternative to "discrete" DRAM control circuitry, and you can usually get most of the logic you need in one or two chips this way. An even easier way is to use one of the special-purpose "DRAM support" chips, for example the AM2968. These chips handle not only the address MUX and RAS/CAS generation but also the refresh arbitration and row address generation; furthermore, they even include the robust drivers and damping resistors you need to drive a large array of memory chips, as we'll explain in a moment. These DRAM controller chips usually include companion chips for timing and error detection/correction; thus, a small chip set forms a complete solution to your DRAM design.

Well, *almost* a complete solution! The real agony of DRAM memory comes when you try to keep all those bused strobes and address lines noise-free. The basic problem is that you typically have a few dozen MOS ICs spread over a substantial area of PC board, with the address lines and other control signals bused to all the chips. To drive a few dozen chips you need high-current Schottky output stages; but the long lines and distributed input capacitances, in combination with fast output-stage edge times, result in severe ringing. It's not unusual to see −2 volts undershoot at DRAM address pins! The usual solution (not always completely successful) is to put series damping resistors, typically about 33 ohms, at the output of each driver. A related problem is caused by the enormous transient currents, which can easily reach 100mA per line. Imagine an octal driver chip in which most of the outputs happen to make a HIGH-to-LOW transition. This causes a transient current of nearly an ampere, which makes the ground pin rise momentarily above ground, and with it any outputs that were supposed to stay LOW. This problem is far from academic – we once had memory errors because just such a ground transient, caused by CAS-driver current spikes, let the RAS drivers on the same chip spike HIGH enough to terminate the memory cycle!

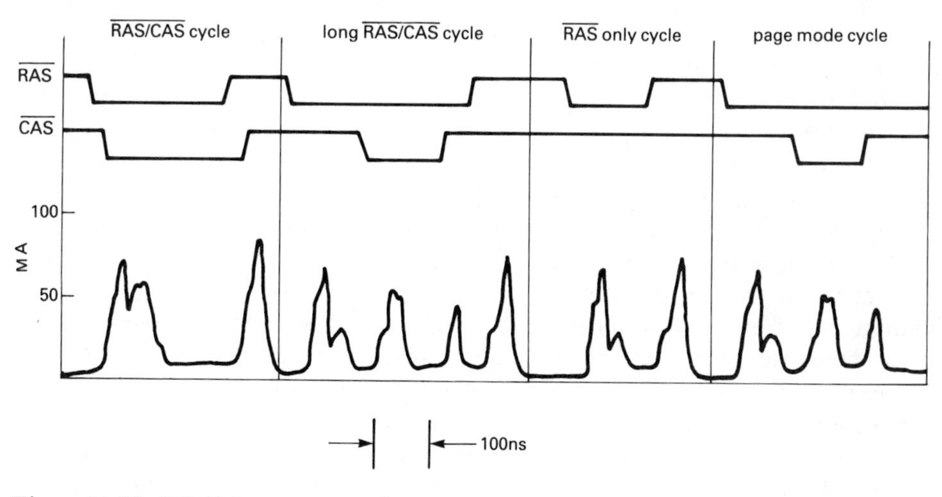

Figure 11.32. DRAM current transients.

An additional noise source in DRAM memories is the large rail-to-rail transient current produced by the memory chips themselves, which some manufacturers are even honest enough to show in their data sheets (Fig. 11.32). The usual solution is to use extensive bypassing to a low-inductance groundplane; current practice favors one $0.1\mu F$ ceramic capacitor per memory chip.

We have found that 74F logic drivers with external resistors work well with DRAM, as do special drivers, such as the Am2966, that include integral damping resistors. The Am2968 DRAM controller mentioned above claims to drive up to 88 memory chips without external components, with −0.5 volts maximum undershoot. Perhaps more important than the particular driver is the use of a low-inductance groundplane and liberal V_+ bypassing. The best PC-board memories we've seen are invariably 4-layer (or more), with layers devoted entirely to ground and V_+. Two-sided boards with skimpy grounds are certain invitations to disaster; Wire-Wrap panels often don't do much better.

It's important to realize that noise-induced memory errors tend to be highly pattern-sensitive and may not show up in a simple read/write memory test. Conservative design and exhaustive memory testing (with oscilloscope examination of waveforms) are the best ways to ensure reliable memory operation.

Read-only memory (ROM)

These are nonvolatile memories, which you need in nearly all computer systems. In microcomputers, for example, you need as a minimum some nonvolatile memory to hold the boot-up sequence of instructions, which includes not only the stack allocation and port and interrupt initializations but also enough code to get the operating system read from disk storage. When your PC does its memory test and then comes up in DOS, it's following orders from some ROM. In addition, microcomputers usually keep a portion of the operating system (usually the more hardware-specific portions) in ROM; this is called the "ROM-BIOS" (basic I/O system), and it makes the operating system feel at home by providing universal "hooks" to its particular ports. ROMs are widely used for lookup tables, for example for character generators for a display device. In the most extreme case, a microcomputer can put the entire operating system, and even compilers and graphics routines, in ROM. The Macintosh, for example, provides almost all system software in ROM, liberating nearly 256Kbyte of RAM for user programs. However, this ROM-based approach has enjoyed only limited popularity in microcomputers, because it limits flexibility; note, however, that software "fixes" and modest upgrades can usually be accommodated by placing patches in RAM.

In dedicated microcontroller applications, ROM is used more extensively. As in our signal-averaging example, the entire stand-alone program lives in ROM, and volatile RAM is used only for arrays and temporary storage. Finally, ROM is sometimes useful in discrete digital hardware, for example to construct arbitrary sequential state machines, or as lookup tables for linearizing the response of some measuring system. Let's take a look at nonvolatile memory choices: UV-erasable EPROMs, fusible-link ROMs, mask-programmed ROMs, and electrically erasable EEPROMs.

EPROM. Those large chips with the quartz windows are EPROMs, erasable programmable read-only memories. They are by far the most popular form of nonvolatile memory for computers. They are available in CMOS and NMOS and consist of a large array of MOSFETs with floating gates that can be charged by "avalanche injection," a fancy name for breakdown of the gate insulating layer by an applied pulse of more than 20 volts. These memories store data by retaining indefinitely a tiny charge (about 10^6 electrons) on these insulated "buried" gates, which you can think of as capacitors with time constants of centuries. You read out the state of an individual capacitor by allowing it to be the gate of an associated MOSFET channel. Since the gate is not electrically accessible, it can be erased only by exposing it to intense ultraviolet radiation for 10 to 30 minutes, which causes the stored charge to leak off by photoconduction. Thus, the individual stored bytes cannot be erased selectively.

In the first edition of this book we mentioned the "classic 2716" 2K×8 EPROM, with its price of $25. It's now so classic that you can't even get it! The current crop of EPROMs range from 8K×8 up to 128K×8, with prices of a few dollars. Typical access times are in the range of 150–300ns, though companies like Cypress offer parts as fast as 25ns in the small sizes. To program EPROMs, you have to apply elevated programming voltages (usually 12.5V or 21V) while asserting the desired bytes. The original algorithms specified 50ms programming time per byte (which is 100 seconds for the 2716, but would add up to a half hour for a modest-sized 32K×8 EPROM). Thus, the larger ROMs forced the designers to concoct newer "smart" algorithms, in which for each byte you try a succession of 1ms programming pulses, checking the result by reading after each write attempt; when the byte reads back correctly, you hit it with a final write pulse equal to three times the previous total. Typically most of the bytes program the first time, so you spend about 4ms per byte, or 2 minutes for a 32K×8 ROM.

EPROMs are great for prototyping, since you can re-use them after erasure. They're fine also for small production runs. You can get them in a cheaper no-window variant, sometimes called "one-time-programmable" (OTP) EPROM. Although the latter shouldn't be called *E*PROM, engineers persist in this misnomer. EPROM manufacturers are conservative, only guaranteeing a 10 year data retention. That figure assumes worst-case conditions (especially high temperature, which causes charge leakage); in reality they don't seem to lose data, unless you happen to get a defective part.

EPROMs have a limited *endurance*, the number of times the memory can be erased and reprogrammed. The manufacturers are notoriously shy about revealing these numbers, but you can generally expect to get 100 or so erase/program cycles before the chip degrades seriously.

□ *Mask ROM and fusible-link ROM.* Mask-programmed ROMs are essentially custom chips born with your bit pattern built in. The semiconductor house converts your bit specification into a custom metallization mask used to process the chip. These are for large production jobs, and you wouldn't dream of having a mask-programmed ROM designed for prototyping. Typical costs are $1k to $3k setup charge, with the manufacturer strongly discouraging you from buying fewer than a thousand ROMs at a time. In those quantities the chip might cost a few dollars.

Many single-chip microcontrollers include a few kilobytes of on-chip RAM and ROM, so that a finished instrument needn't have any external memory chips. In most cases the micro-controller family includes versions that take external ROM and, in some cases, versions with on-chip EPROM (Fig. 11.33). The idea is that you develop the instrument and write the code using the EPROM (or external ROM) version, then go into production with the cheaper mask-programmed controller.

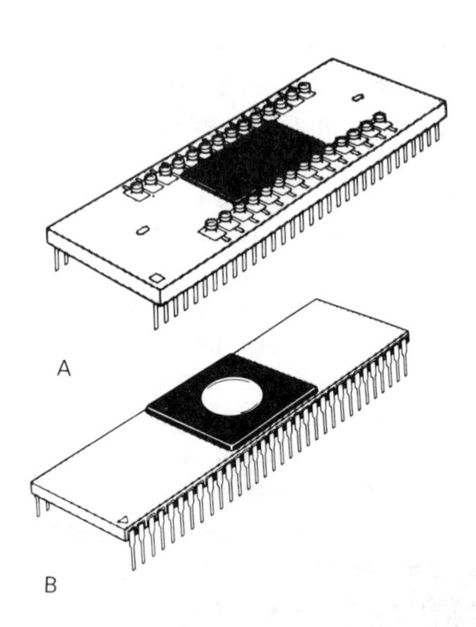

Figure 11.33. Microcontroller/EPROM combinations.
A. 8-bit microcontroller with "piggyback" EPROM.
B. 8-bit microcontroller with on-chip EPROM.

Another kind of one-time-programmable ROM is the *fusible-link* ROM. These begin life with all bits set, and you give them electrical shock treatments until you've blown out the offending bits. A typical part is the Harris HM6617, a 2K×8 CMOS PROM. Fusible-link PROMs are also available in bipolar (TTL) technology.

EEPROM. These ROMs can be selectively erased and reprogrammed electrically, while in-circuit. They are therefore ideal for holding configuration information, calibration parameters, etc., that cannot be frozen before the computer or instrument is used. They use the same MOS floating-gate technique as EPROMS.

First-generation EEPROMs were difficult to use and required elevated voltages and lengthy programming procedures, just like EPROMs. More recent EEPROMs run from a single +5 volt supply and work nearly the same as SRAMs – in other words, you can reprogram any byte with a single bus WRITE cycle. The EEPROM uses internal circuitry to generate the higher programming voltages, and internal logic latches the data and generates the several-millisecond programming sequence, either setting a BUSY′ flag during the process, or producing complemented DATA on a READ to indicate WRITE-in-progress. Some EEPROMs implement *both* protocols, which are commonly called "RDY/BUSY′" and "DATA′-polling." It's easy to interface these ROMs – just hook them up like SRAMs, and use the BUSY′ line to make an interrupt (or read it or the DATA as a status flag); see Figure 11.34. DATA′-polling has the advantage of working in a standard SRAM socket without any circuit changes (you must, of course, alter your programs to test the read-back data, waiting until it agrees with what you wrote). Since you don't write often to EEPROM, you don't really need the interrupt capability of RDY/BUSY′.

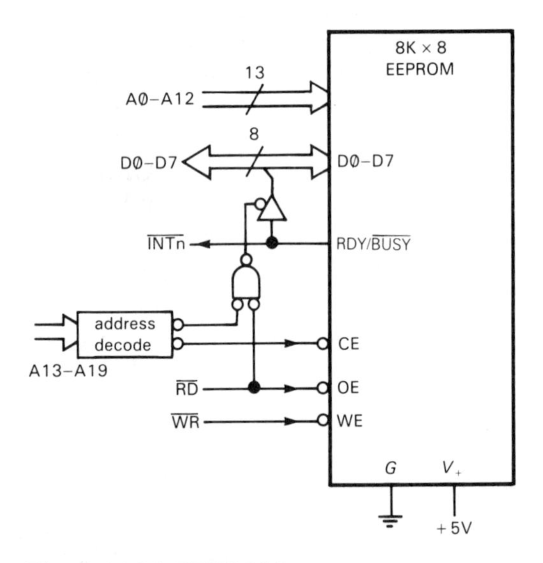

Figure 11.34. EEPROM.

CMOS EEPROMs are currently available in 2K×8, 8K×8, and 32K×8 sizes, with prices around $10–$50. Access times (200ns–300ns) and programming speeds (~2ms/byte, with internal "smart" algorithm) are comparable with standard

EPROMs. EEPROMs, like EPROMs, have limited read/write endurance. Although the manufacturers avoid specifying actual endurance figures, you see numbers like 100,000 read/write cycles at 25°C operating temperature.

Note: Although EEPROMs are unique in allowing *in-circuit* reprogramming, they can perfectly well be programmed externally, in an EPROM-type programmer. This makes them ideal during firmware development, because you don't have to wait around while your old EPROMs cook for a half hour in the UV hot seat.

There are two interesting variations on these EEPROMs. From National, Xicor, etc., you can get little 8-pin mini-DIP EEPROMs, with memory sizes from 16×16 to 2K×8 bits; they use serial access, with a clock input and single data line. You can't easily use these memories without a microprocessor around to access them; but in microprocessor instrumentation applications they're great for storing a few setup parameters, etc. And from Xicor comes the "EEPOT," a clever use of electrically erasable memory to hold the position of a "digital wiper": These chips have a string of 99 equal resistors, with a tap position you can set digitally and store in on-chip nonvolatile memory. You can imagine applications where it would be nice to have automatic or remote calibration of some instrument, without the need for adjusting a mechanical trimmer pot.

A recent variety of EEPROM known as *flash* combines the high density of EPROM with the in-circuit reprogrammability of EEPROM. However, with flash EEPROMs you generally are not able to erase single bytes, as you can with conventional EEPROM: Intel's current flash EEPROM permits only *complete* erasure (like EPROM), whereas Seeq's product permits erasure either by sector (512 bytes) or in bulk. Furthermore, most available flash EEPROMs require an additional switched +12 volt supply during erase/write, a significant price to pay when compared with the single +5 volt supply of conventional EEPROMs. Flash EEPROMs claim endurances of 100 to 10,000 programming cycles.

These PROM technologies are still evolving, and we're watching to see what happens; you should, too!

Nonvolatile RAM

EPROM is fine for nonvolatile read-only storage, but you often want nonvolatile read/write memory. EEPROM achieves that, but with sluggish (10ms) write cycle times (and limited total number of read/write cycles). There are two ways to get full RAM read/write speeds (~100ns) with unlimited read/write endurance: Use either battery backed-up CMOS static RAM or the unusual "NOVRAM" from Xicor, which combines SRAM and EEPROM on the same chip.

We talked about battery backed-up SRAM earlier. In many ways this gives you the best of both worlds, namely the low price and full read/write speed of RAM combined with the nonvolatility of ROM. You must, of course, use CMOS RAM with specified micropower shut-down current. Some manufacturers are producing "nonvolatile RAM" by packaging a lithium battery and shut-down logic together with the CMOS RAM chip in an ordinary DIP package. An example is Dallas Semiconductor's 8K×8 DS1225 and 32K×8 DS1230; they also make a line of "smart-sockets" containing battery and logic, which magically render your ordinary CMOS RAM nonvolatile. Note that nonvolatile RAM constructed this way is not exactly immortal; the battery, and therefore the data, has a life of about 10 years. As with ordinary SRAM, there is no limit on the number of read/write cycles the memory can sustain.

NOVRAM (nonvolatile RAM) from Xicor combines ordinary static RAM with some on-chip "shadow" EEPROM. A STORE′ input preserves the contents of SRAM on the EEPROM, with a total write cycle time of 10ms; the data are retrieved more quickly – about 1μs. With a power supervisory chip like the MAX690 series, you get enough warning to save the SRAM contents before the +5 volt power drops too far. The NOVRAM is specified to withstand 10,000 storage operations, and, like ordinary SRAM, it has unlimited RAM read/write endurance.

Among these two options in nonvolatile RAM, the battery-backed SRAM seems generally preferable, because it lets you use any SRAM you can find, as long as it has a zero current shut-down mode. That means you can use the latest large RAMs, and you can, for example, shop for memory with quick access time, if that is what you need. Although battery life is finite, it's long enough for most applications. For short retention times (a day or less), you can substitute a high-capacitance double-layer capacitor for the lithium battery; these are available from Panasonic, Sohio, and others, and they offer capacitances up to a farad or more in a very small package.

Memory wrap-up

Figure 11.35 summarizes the important attributes of the various memory types. Among the types shown, we would recommend 1-bit-wide DRAM for the largest read/write memory arrays, byte-wide SRAM for smaller microprocessor memory arrays, EPROM for read-only parameters and programs, and either EEPROM (if write speed is unimportant) or battery-backed SRAM (for full-speed read/write) for nonvolatile read/write storage.

11.13 Other microprocessors

Like any Darwinian process, the evolution of microprocessors has proceeded in several divergent directions. In the contest for survival, some of the less fit have become endangered species. As an example of different evolutionary paths, there is the division between processors with separate I/O instructions and those that require "memory-mapped I/O," in which the peripheral device registers simply look like individual locations in memory (examples are the 8086 and 68000 families, respectively). Then there is the division created by machines that use memory instead of registers for most arithmetic operations. Another choice involves the use of available pinouts: Some CPUs make multiple use of pins, to allow more flexibility within the package constraint. Then there is the question of word size (4, 8, 16, or 32 bits), stacks, and elegance (or

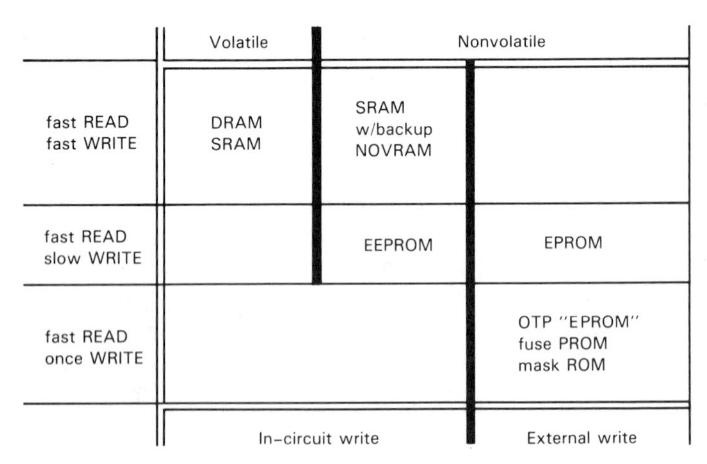

Figure 11.35. Memory types.

richness) of the instruction set. Each microprocessor family has its own assembly language, another stumbling block for the beginner. Microprocessors are usually fabricated with CMOS technology, but you'll see some built with NMOS or bipolar logic.

There are even greater differences among processors that have to do with their intended applications. There are the "single-chip" processors with on-chip RAM and ROM, parallel ports, UARTs, timers, and even analog/digital converters. At the other extreme, powerful 32-bit CPUs like the 80486, 68040, and AM29000 overlap the computational finesse of large computers, but they require extensive hardware and software support to take full advantage of their advanced features. "High-integration" chips try to strike a balance; for example, the H16 is a CMOS 68000 with 2 UARTs, timers, and DMA channels on-chip.

We have used the 68008 for all our examples in this chapter, but we don't want to leave the impression that other microprocessors are less useful. Table 11.8 lists a selection of the most popular microprocessors now available. This table is not intended to be a comprehensive listing of everything that is available.

11.14 Emulators, development systems, logic analyzers, and evaluation boards

How are you going to get a program written, assembled, debugged, and loaded into a ROM for use in some instrument you've designed? This is a major problem, especially for someone beginning to work with microprocessors. There are a few techniques available, involving methods that range from the simple stand-alone "burn and crash" up to elaborate development systems and high-level language emulators. In this section we will try to describe what is available and how it can be useful in designing instruments with microprocessors.

"Burn and crash"

This colorful term describes the simplest method of code development for microprocessor-based hardware. It goes as follows. You first use an assembler or compiler in some general-purpose computer (perhaps a desktop personal computer) to generate executable code that can be burned in EPROM. If the target processor – the one in your custom instrument – is different from the processor in the development computer (which we'll call a PC), you'll need a "cross-assembler"; otherwise just use the PC's "native assembler." Now you program an EPROM (that's the "burn") and try it in your instrument (that's the "crash"). Debugging now consists of looking at the symptoms created by the faulty code (or faulty hardware), making corrections or inserting diagnostic tests, and trying again. There's plenty of room for cleverness as you go about finding bugs. For example, if you put the right statements in your trial programs, you can make diagnostic use of LED indicators or other ports in the instrument to see what is happening in the program. Don't forget about the traditional tools of the electronic trade – logic probe, oscilloscope, and (if you're desperate) logic analyzer (see below).

ROM substitutes. Burn-and-crash, as described above, is a slow procedure. Although it's sometimes adequate to the task, you quickly become impatient with the need to cycle EPROMs through the ultraviolet eraser and ROM programmer. There are some shortcuts: (a) Use EEPROMs instead of EPROMs. They don't program any faster, but at least you don't have to fool around erasing them. (b) Use a battery-backed-up (nonvolatile) CMOS RAM, instead of EPROM, during the development cycle. These are as fast as conventional RAM (i.e., they "program" instantly) and have EPROM-compatible pinouts so that you can plug them into

TABLE 11.8. MICROPROCESSORS

Type	Reg size (bits)	Bus width (bits)	Adr space	Fastest instr (μs)	Clock @ freq (MHz)	CMOS?	Queue?	Segments?	MMU?	Virtual mem?	Cache?	FPU?	Pipelined?	Comments
Traditional 8-bit														
6502	8	8	64kB			A	–	–	–	–	–	–	–	Apple II; 65C802=upgrade
6800	8	8	64kB			–	–	–	–	–	–	–	–	
8085	8	8	64kB	0.8	10	A	–	–	–	–	–	–	–	replaced 8080
Z80	8/16	8	64kB	0.5	8	A	–	–	–	–	–	–	–	popular CP/M μP
6809	8/16	8	64kB			A	–	•	–	–	–	–	–	
High Integration 8-bit														
50740, 65C124			64kB			•	–	–	–	–	–	–	–	6502 inst set
6801, 6301, 68HC11			64kB			A	–	•	•	–	–	–	–	6800 inst set
64180, Z180	8	8	1MB		10	•	–	•	•	–	–	–	–	Z80 inst set, on-chip peripherals
Z280	8/16	16	16MB		25	•	–	•	•	–	•	E	•	Z80 inst set
Traditional 16-bit														
8088	16	8	1MB			A	•	•	–	–	–	E	–	
8086	16	16	1MB	0.3	10	A	•	•	–	–	–	E	–	
Z8000	16	16	8MB	0.3	10	–	•	•	E	–	–	E	–	fast interrupts
F9450	16	16	4MB	0.2	20	–	•	–	•	•		•	•	MIL std 1750A
Advanced 16-bit														
65C816	16	16	16MB	0.13	8	•	–	–	–	–	E	–	–	AppleIIGS, has 6502 subset
80286	16/32	16	16MB	0.16	12.5	–	•	•	•	•	–	E	•	Harris 80C286 to 20MHz
V20,V30	16	8/16	1MB		10	•	•	•	–	–	–	E	–	has 8080 subset; CMOS
V60	32	16	4GB		16	•	•	•	•	–	•	•	•	1.5 MIPS average; CMOS
RTX2000	16	16	1MB	0.1	10	•	–	•	–	–	–	–	–	Forth at 10 MIPS
High-integration 16-bit														
80188	16	8	1MB			A	•	•	–	–	–	E	–	
80186	16	16	1MB			A	•	•	–	–	–	E	–	popular as a controller
V40,V50	16	8/16	1MB		8	•	•	•	–	–	–	E	–	8086 inst set controller
Traditional 32-bit														
68008	32	8	4MB			–	•	–	–	–	–	–	–	
68000	32	16	8MB			–	•	–	–	–	–	–	–	
68010	32	16	16MB			–	•	–	•	•	–	–	•	
68012	32	16	4GB			–	•	–	•	•	–	–	•	
32008	32	8	16MB		10	•	•	–	•	•	–	•	•	
32C016	32	16	16MB		10	•	•	–	•	•	–	E	•	
32C032	32	32	16MB		10	•		–	E	•		E	•	
Advanced 32-bit														
V70	32	32	4GB		20	•		•	•	•	•	•	•	2.5 MIPS avg; CMOS
32332	32	32	4GB	0.2	15	•	•	–	E	•	•	E	•	1st 32-bit data+adr; NMOS
68020	32	32	4GB	0.08	25	•	–	–	E	•	•	E	•	5 MIPS sustained
80386	32	32	4GB	0.03	33	•	•	•	•	•	E	E	•	has 8086 subset (SX=16-bit bus)
WE32100	32	32	4GB	0.13	18	•	•	–	E	•	•	E	•	AT&T "3B" computer; 4 MIPS avg
Clipper	32	32	4GB	0.03	33	•	•	–	E	•	•	•	•	RISC (Harv arch), 5 MIPS avg (C100)
Z80000	32	32	4GB	0.08	25	–	–	•	•	•	•	E	•	5 MIPS avg (has Z8000 subset)
68030	32	32	4GB	0.06	30	•	–	–	•	•	•	E	•	has data cache, twice 68020 speed
32C532	32	32	4GB	0.06	30	•	•	–	•	•	•	E	•	UNIX top performer, 10 MIPS
T414	32	32	4GB	0.05	20		•	–	–	–	E	E	•	transputer; 4 multiprocessing links
86C010	32	32	64MB	0.08	12	•	•	–	E	•	E	E	–	Acorn RISC (44 instructions)
80960	32	32	4GB	0.06	16	•	–	–	•	•	•	•	•	RISC, multitasking, 8MIPS, 0.5K cache
68040	32	32	4GB	0.06	33	•	–	–	•	•	•	•	•	17 MIPS, 3Mflops, 8K cache
80486	32	32	4GB	0.03	33	•	•	•	•	•	•	•	•	20 MIPS, 8Mflops, 8K cache
29000	32	32	4GB	0.04	25	•	•	–	•	•	•	E	•	RISC, 17 MIPS sustained
88000	32	32	4GB				–	–	E	E	E	•	•	RISC
T800	32	32	4GB	0.05	20		•	–	–	–	E	•	•	Transputer
86900	32	32	4GB	0.07	17		–	–	–	–	E	E	•	Sun μP (107 instrtuctions) SPARC
WE32200[a]	32	32	4GB	0.13	30	•	•	–	E	•	•	E	•	8 MIPS avg (28MHz)
SPARC	32	32	4GB	0.05	33	•	–	–	•	•	–	•	•	Cypress, Fujitsu, etc, >20 MIPS

(a) AT&T may discourage OEM sales. A - CMOS version available. E - external chip can provide this function.

the target instrument carrying trial ROM code. You can get battery backed-up RAM from Dallas, Thomson-Mostek, and others, or you can make your own. (c) Use a "ROM-emulator": This is a clever little box you buy, with a cable and 28-pin DIP plug at the end. You plug the latter into the EPROM socket of your instrument, and it emulates a ROM. However, the little box actually contains a dual-ported RAM, which you download from your PC via a serial port. The ROM-emulator technique is the fastest of all, because you don't have to plug and unplug memory chips each time you change the software – you just download the new code to the emulator, which stays plugged into the target instrument. ROM emulators go by names like "Memulator" or "Romulator" (the latter from Onset Computer Corp., N. Falmouth, MA).

Monitor ROM. If your target instrument has a serial port in it, you can make the software development job easier by coding a simple "monitor" EPROM whose task is not to run the instrument but only to serve as a communication link between the port and the target instrument's memory. The simplest monitor lets you load code into RAM, then begin program execution. That lets you speed up code development, because you can load trial programs into the target instrument's RAM directly from your PC. With just a little more effort you can add additional features to the monitor, for example, the ability to look at the contents of specified locations in RAM. With that addition, your trial programs can tell you what's happening, for example, by depositing numbers in specified locations in RAM, then jumping back to the monitor (through which the PC can examine those RAM locations). You might exploit this capability by coding diagnostic "software breakpoints" into your trial code: At such a breakpoint the contents of designated registers or memory locations would be copied into an unused area of RAM, there to be subsequently read by your PC via the monitor. This sort of tool really shortens the development cycle, because it lets you quickly locate the source of trouble in the program.

In-circuit emulator

The burn-and-crash technique, and its elaborations above, can usually succeed in getting the job done. But it's not an ideal method. For one thing, it requires the use of resources on the target instrument, such as serial ports. And, more seriously, it doesn't provide a way to get at faulty operation at the *hardware* level. To understand this last statement, imagine you have an instrument that is crashing because it is improperly trying to write to EPROM. You've figured that out, somehow, but you're stuck because there's no easy way to code a *software* breakpoint to trap a fault known only from its hardware symptom. In this case, for example, an address register may have been overwritten earlier. It's a subtle error: The actual crash occurs long after the initiating event, and there's just no way to find the bug by looking at the software code. What you need is a way to set a "hardware breakpoint."

The solution is an *in-circuit emulator* (ICE): This is a box of hardware (or plug-in card) you buy, this time emulating the *microprocessor* in the target system via a cable with CPU-like plug on the end. The in-circuit emulator can execute code in the target system's memory (EPROM or RAM), or it can execute code you have downloaded to the emulator. Either way, it knows everything that is happening in the CPU – it can monitor the contents of registers, and it can set *hardware* breakpoints. For example, to solve our hypothetical problem we could ask it to check for a write cycle to the EPROM's area of memory space, providing us with a register dump and listing of the last 100 instructions executed before the doomed write-to-EPROM.

In-circuit emulation is the best way to develop code – it's the fastest, and it's the most powerful. The disadvantages are cost (several thousand dollars, sometimes much more) and the need to buy a new emulator for each type of microprocessor you use. The ROM-emulator, by comparison, is completely general, but not as powerful. It's the poor man's in-circuit emulator.

Development systems

"Development system" is the collective term for the paraphernalia of cross-assembler, EPROM burner, and hardware in-circuit emulator. Development systems traditionally were packaged as impressive stand-alone systems, but recent trends are favoring plug-in cards that use a PC as host, perhaps controlling an external box containing additional circuitry. Whatever the format, you should invest in a development system if you are in the business of developing instruments based on a particular microprocessor. Development systems are generally offered by the manufacturer of the microprocessor family you're using. There are, in addition, companies making "universal emulators" that accept plug-in boards for additional processor types. Hewlett-Packard, Tektronix, Microcosm, and Applied Microsystems are some companies offering development systems.

Logic analyzers

Logic analyzers are the "super-oscilloscopes" of digital hardware development. The logic analyzer was a major character in Tracy Kidder's book *Soul of a New Machine*. These marvellous devices look like fancy oscilloscopes, but they have dozens of channels, lots of memory, complex "word-recognizer" triggers, and the ability to disassemble executing instructions and display them on the screen. Logic analyzers work in two modes, *state analysis* and *timing analysis*. Here's how you use them.

State analysis. In state-analysis mode, you provide the logic analyzer with a synchronous clock signal from your circuit (usually the CPU clock), and you clip clusters of leads (which come from "pods") onto data and address bus lines, and any other signals you're interested in. Good logic analyzers can handle 60 or 80 channels, and clock rates to 25MHz or more. You then set the triggering to catch the software event you're interested in: There are usually several *word recognizers*, which you set up with the pattern of 0's, 1's, and X's (don't care) you want to trigger on; these are, of course, address and data bits. Good logic analyzers let you combine the outputs of the word recognizers in Boolean and state expressions; you could, for example, trigger on the 10th pass through a certain subroutine.

The logic analyzer waits for the trigger, then records successive states on all input lines. You can display them as digital waveforms, or as lists of 1's and 0's (or hex or octal), with disassembled code shown alongside. You can walk through the array of recorded states (typically 4K or more), and, most important, you can look "backward in time" before the triggering event. This lets you see the few thousand states that preceded the triggering event, which usually provides the cause of the trouble.

Timing analysis. In timing-analysis mode, the logic analyzer runs from a fast *asynchronous* clock, typically 100MHz, logging the logic states of a smaller number of input lines (typically 16). The triggering logic is necessarily simpler, usually a single word recognizer. The analyzer waits for the trigger condition, then fills up its memory with the fast samples. In timing mode you can see fast glitches and other waveform aberrations that might be missed in state-analysis mode. You can also turn on a "glitch-detection" mode, in which the analyzer looks for two edges occurring during a 10ns sample period.

Cross-triggering. You can combine state analysis and timing analysis with a powerful technique called *cross-triggering.* In this combined mode, state analysis can arm timing analysis, or vice versa. Thus, you might set the state triggering logic to catch a problem that occurs within some particular software loop, arming the timing logic to store a fast burst on the next occurrence of its trigger word. In that way you could discover a short logic glitch that happens only rarely.

Logic analyzers with cross-triggering capability provide split-screen display, so you can watch the fast timing waveform go by as you scroll through the state display. Some of the bigger names in logic analyzers are Gould, Hewlett-Packard, Philips, and Tektronix.

Evaluation boards

During the 1970s, when 8-bit microprocessors like the 6800 and Z80 were becoming popular, there was an *evaluation board* offered by the manufacturer for each new microprocessor. These boards had a small keypad, hex display, RAM, monitor EPROM, some parallel and serial ports, and a breadboard area for adding custom circuitry of your own choosing. You could hand-assemble little programs, enter them through the keypad, and delight in the results. For a world feeling its way into microprocessors, it was an easy way to learn the ropes.

It's a more sophisticated world now, and these breadboards have become almost extinct. However, you'll still see evaluation boards offered for specialized processors, for example, large signal-processing ICs, or complex video processors. These boards contain the processor itself, surrounded by "glue" logic, analog signal components, and often a conventional microprocessor used for control. These evaluation boards are frequently made in the form of plug-in cards for personal computers, complete with driver software. Given the complexity of many of these new special-purpose processors, the evaluation board saves lots of time and makes good sense.

ELECTRONIC CONSTRUCTION TECHNIQUES

CHAPTER 12

Between completing a circuit design and testing a finished product, you've got a number of decisions to make: Is the instrument going in a benchtop case, a "relay rack" enclosure, or perhaps some sort of modular "bin" chassis? Should the circuit itself be constructed with point-to-point wiring, on a "breadboard" card, with Wire-Wrap connections, or on a printed-circuit board? Are connections to the circuit board made with solder lugs, flat ribbon-cable connectors, or edge connectors? Should individual circuit cards be housed in a card "cage," plug into a motherboard, or what? Does it pay to design a printed-circuit motherboard or use hand-wired backplane connections? Which adjustments should be on the circuit board, and which on the front (or rear) panel? Decisions like these are important for the appearance, reliability, and serviceability of the finished product, not to mention cost and ease of construction and testing. In this chapter we will try to give some information and guidance on this important subject, one that tends to be overlooked in electronics courses where circuit work is usually done on plug-in breadboarding gadgets. We will begin with the circuit construction itself, treating interconnections, controls, and enclosures last.

Because this chapter does not deal with circuit design, it could be skipped in an abbreviated reading.

PROTOTYPING METHODS

12.01 Breadboards

The unusual name "breadboard" seems to have arisen from the early practice of building radios on handsome slabs of varnished wood, with tubes, coils, condensers, etc., and the interconnecting wires all fastened to the topside of the board. Later, radios of greater refinement and elegance (to be used by stiff-skirted ladies in the parlor) were built with holes near each component so that the wiring would be hidden from view underneath the board. The practice of testing circuits by constructing trial versions on some sort of board or jig is still called breadboarding.

Wooden breadboards are no longer used (except in the kitchen). Instead, you can get handy plastic blocks with rows of holes spaced to accommodate ICs or other components and (usually) some extra rows for distributing the power-supply voltages. These breadboards are typified by those manufactured by AP and Global Specialties, and more elaborate breadboarding boxes are made by these companies and by E&L Instruments, among others. These are intended for *testing* circuits, not for constructing permanent (even semipermanent) versions.

12.02 PC prototyping boards

To construct one-of-a-kind circuits of some permanence, probably the best approach is to use a different kind of breadboard, one of the many printed-circuit (PC) prototyping cards available with predrilled pads for ICs and other components, but with no interconnections laid out on the board itself. Instead, each component pad is connected to two or three uncommitted pads nearby, and you wire the circuit by soldering lengths of insulated wire from pad to pad. There are usually some additional lines running around the board for power-supply distribution and ground. These boards, made by companies such as Douglas Electronics, Artronics, Vector, Triad, Radio Shack, and many others, usually include a card-edge connector – gold plated "fingers" of copper board material aligned at one end of the card for easy plug-in connection to a mating PC card-edge socket.

There are several standard connector configurations, a very common one being 22 connections on each side of the card, with 0.156 inch spacing (spacings of 0.125 and 0.1 inch are also common). A connector to mate with such a card is called a 44-pin "dual-readout" PC card-edge connector. Prototyping cards are available in various sizes, accommodating from 12 to 36 or more IC packages, with some larger computer-compatible cards available that will accept 100 or more ICs and will plug directly into small-computer mainframes. Some of these boards are single-sided, and others are double-sided with plated-through holes, a subject we will discuss shortly in connection with custom PC design. Figure 12.1 shows a small PC prototyping card (Douglas Electronics 11-DE-3) plugged into a 44-pin dual-readout socket that includes card-supporting guides (Elco 6022).

Another form of breadboard that has enjoyed considerable popularity is the so-called perfboard, a thin sheet of laminated insulating material manufactured with regularly spaced holes (3/16 inch is a common spacing) designed to accept little metal pins. To wire up a circuit, you shove in dozens of the little pins wherever you want, then stick in the components. You then solder wires from pin to pin to complete the circuit. Perfboard works OK, but it is awkward to use with the tight spacings of IC packages (0.1 inch between pins). Figure 12.2 shows an example.

12.03 Wire-Wrap panels

A variation on the PC breadboard is the Wire-Wrap (a registered trademark of Gardner-Denver) panel, a circuit card festooned with IC sockets (or pads), with a pin 0.3 to 0.6 inch long protruding from each socket connection (Fig. 12.3). These pins are square in cross section, typically 0.025 inch on a side, and made of a hard metal with sharp corners, plated with gold or tin. Instead of soldering to the pin, you wrap an inch of bare wire tightly around it, using an electric wire-wrapping tool (there's an inexpensive variation known euphemistically as a "hand-operated Wire-Wrap tool"). Wire-wrapping is very fast. You just stick the stripped end of the wire into the Wire-Wrap tool, put the tool over the Wire-Wrap pin, and zip, you're done. The standard wire used for this purpose is

Figure 12.1. A "solder breadboard" prototyping card is useful for wiring up small circuits, especially those involving both discrete parts and ICs. This particular specimen will accommodate 12 dual in-line (DIP) packages, and it includes common lines for ground and power supply voltages. An "edge-connector" foil pattern is standard, so the card can be plugged into a card cage or supporting connector, as shown. This particular circuit uses a variety of components, including single-turn and multiturn trimmers, inductors, a crystal, a DIP switch, a miniature relay, and a logic-state indicator, in addition to both transistor and IC circuitry.

silver-plated copper wire of 26 or 30 gauge, with Kynar insulation. There are special tools available for stripping the insulation from the thin wire without nicking it. The wire is stretched tightly around the sharp corners during the wrapping process, forming a few dozen gas-tight cold welds. As a result, wire-wrapped connections are as reliable as well-soldered connections, and it is extremely easy to do them rapidly. For logic circuits, where you have few discrete components, wire-wrapping is probably the best technique for constructing one or two custom circuits of reasonable complexity. Because Wire-Wrap panels are laid out primarily for IC packages, the technique is less convenient for linear circuits with many resistors, capacitors, etc., and the soldered prototype breadboard technique described earlier will usually be better.

It is possible to use discrete components with a Wire-Wrap panel. You just mount them on little "headers" that plug into IC sockets, then do the wire-wrapping as usual from the socket pins. Some Wire-Wrap panels have extra solderable pads (rather than IC sockets) available for discrete components. A very nice kind

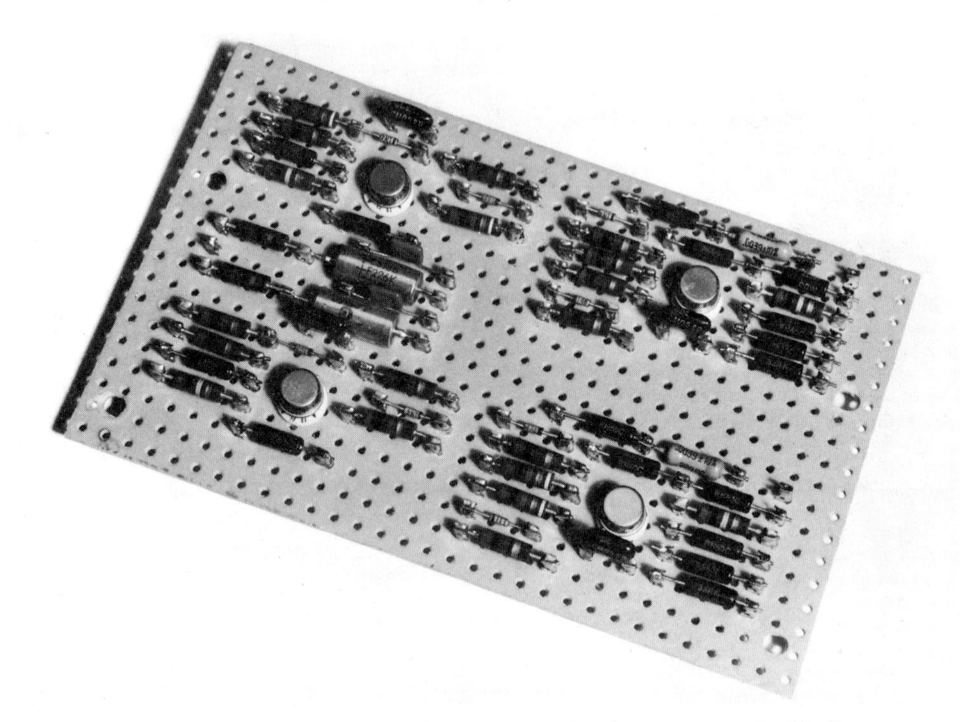

Figure 12.2. "Perfboard" can be handy for prototype circuits constructed with discrete components, although it is not particularly good for ICs. The terminals are press-fit into the holes (or flared with a special tool) and wired underneath.

of Wire-Wrap board is available with the pins on the *component* side of the board (the usual procedure is to have the pins stick out the other side). Although this type of board is less dense (fewer ICs per square inch), it is easier to use with discrete components, since you can see them while wrapping, and it allows closer spacing between adjacent circuit boards, since both components and Wire-Wrap pins take space on the same side. This kind of board without sockets is actually quite convenient for construction of linear or digital circuits. Figure 12.4 shows an example. In Figure 12.5 we have compared a prototype circuit built on a Wire-Wrap panel with the final printed-circuit version used in production. Printed circuits are much easier to produce in quantity; they are superior electrically and less cluttered than Wire-Wrap panels. We will talk about PC cards next.

PRINTED CIRCUITS

12.04 PC board fabrication

The best method of constructing any electronic circuit in quantity is to use a printed circuit, a stable insulating sheet of material with thin plated copper lines bonded to the sheet forming the circuit paths. Although early printed circuits were associated with poor reliability (Remember the advertisements stressing the superior quality that only handcrafted television sets without printed circuits could provide?), the process of manufacturing board material and producing finished boards has been perfected to the point that printed-circuit boards now have very few problems. In fact, PC boards offer the most reliable fabrication technique. They are routinely used in computers, spacecraft, and military electronics where high reliability is essential.

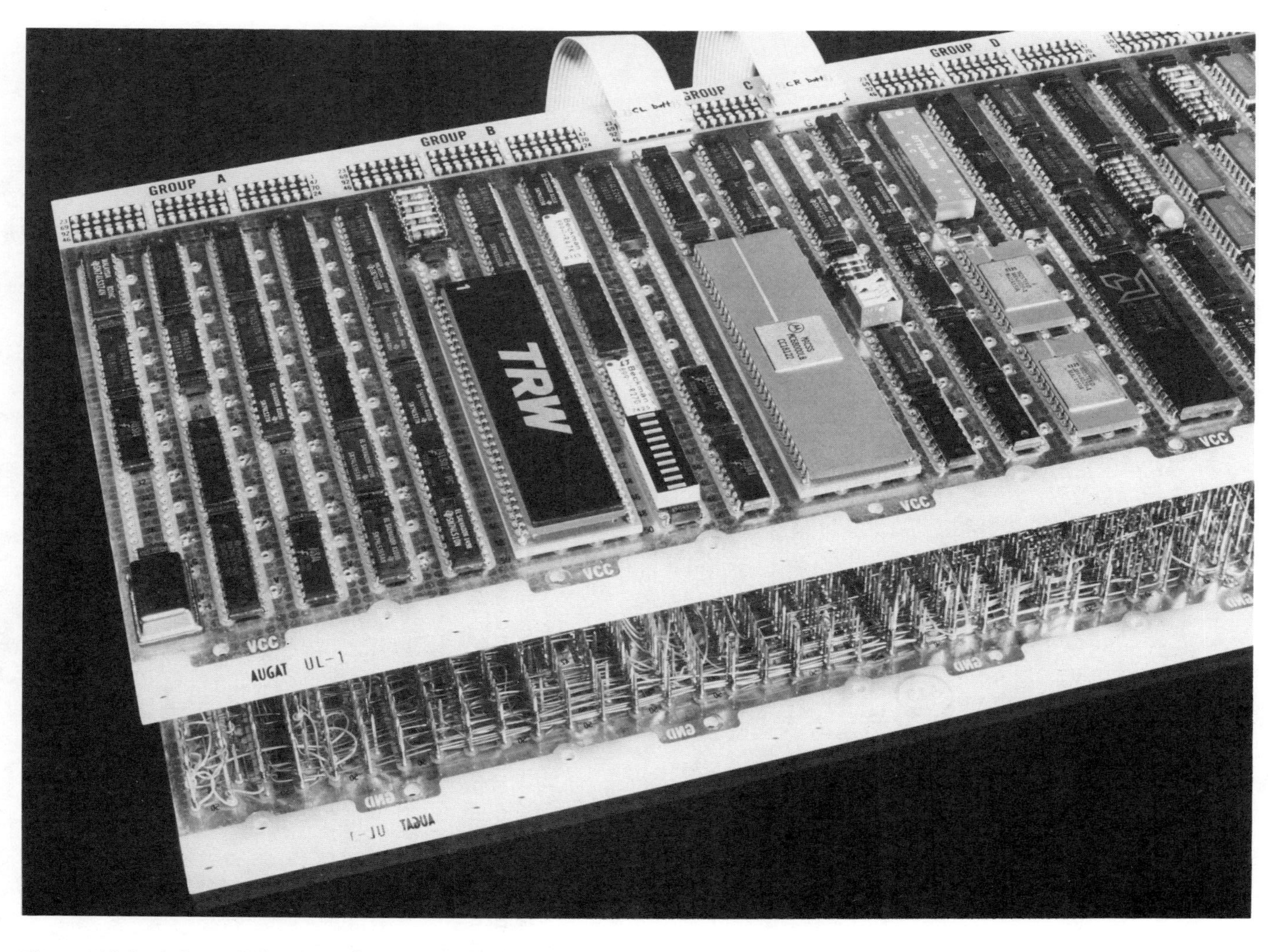

Figure 12.3. A large Wire-Wrap® panel and its underside (visible in a mirror). This microcomputer circuit was wired by machine ("CAD/CAM," see Section 12.08). Chips of different sizes, as well as component-carrying "headers," are accommodated by this general-purpose panel, which also includes an area for off-board connections via flat ribbon cable with 14-pin DIP plugs.

Figure 12.4. Wire-Wrap boards provide a neat and fast construction method particularly good for circuits made with digital ICs. This board uses a printed-circuit pattern to bring out the Wire-Wrap pins on the component side, an alternative to the usual underside pin configuration. Its peculiar shape is dictated by the interior of the oceanographic pressure cell into which it fits.

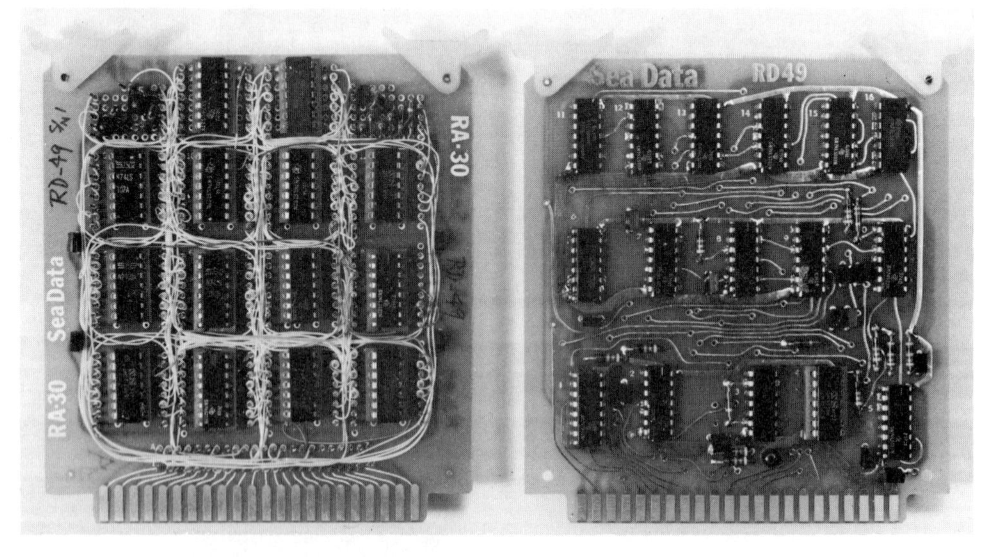

Figure 12.5. A Wire-Wrap prototype board and its printed-circuit successor. PC boards are less cluttered and far easier to fabricate in quantity. They eliminate wiring errors, too.

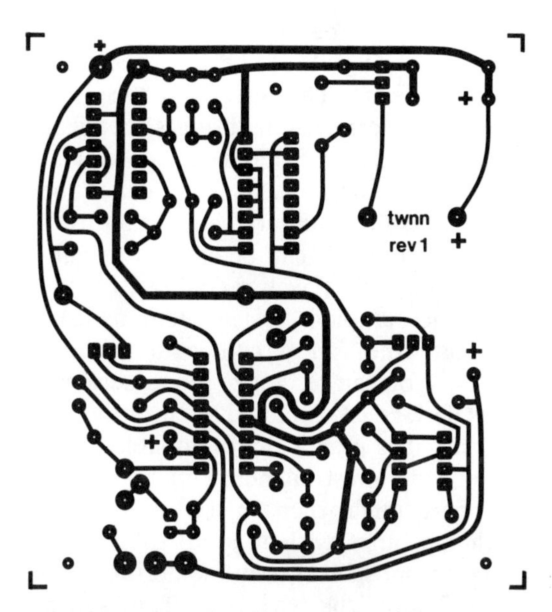

Figure 12.6. Foil pattern for a simple single-sided PC board. This "layer positive" was photographically reduced from a double-sized hand-taped Mylar pattern. The circuit board is 3.25×3.5 inches and has 4 ICs, 24 resistors, 11 capacitors, 5 diodes, 1 trimmer, and 1 piezo buzzer.

The "Mylar" or photoplot

Printed-circuit-board production begins with a set of actual-size transparencies on which an opaque pattern delineates the desired circuit traces and "pads" (Fig. 12.6). There are lots of rules and tricks in this business, but the basic idea is to figure out how to make all the interconnections the circuit demands by running lines around a board. As we will describe shortly (Section 12.08), these transparencies are usually produced directly on film by computer-driven photoplotters or laser plotters, working from a description of the circuit that you produce on a CAD (computer-aided design) system. However, for simple circuits, you may choose to do the layout manually, sticking opaque tape and patterns onto clear Mylar film. In the latter case you usually make the taped Mylar pattern twice actual size, from which you photographically produce an actual-size transparency.

No matter what the parentage of the final 1:1 transparency, the result is a set of conductor patterns on film. For the simplest boards you may be able to make all the needed connections (perhaps aided by a few wire "jumpers") with a *single-sided* board, which has all its traces on the underside (officially called the "solder" side; the top is called the "component" side). Most often, however, you need traces on both sides of the board. Such *double-sided* boards invariably use *plated-through* holes (the holes in the board are lined with foil, connecting corresponding pads on both sides of the board). This really makes a difference when you are attempting to route traces, because you can always switch sides (using a "via" hole) when a trace runs into a dead end, whereas with a single-sided board you often get hopelessly boxed in. As an important side benefit, the use of plated-through holes ensures a superior solder joint to the component leads, because the solder wicks up through the hole.

For complex digital circuits you often see laminated multilayer PC boards, in which interior layers (called the "core") are used to carry ground and power-supply voltages, and sometimes signal lines as well. Four-layer and six-layer boards are pretty routine these days, with occasional use of more layers (up to 40!) in desperate situations.

Manufacturing

The board material (usually 1/16 inch of so-called FR-4 board, a fire-resistant epoxy-bonded fiberglass) comes clad on both sides with copper ("2 ounce" thickness is standard; the copper is 0.0027 inch thick). The first step is to drill the holes, using a template or automated drilling machine keyed to the full-size photopositive from the photoplotter or the Mylar pattern. The holes are then "plated through" by a tricky multistep copper plating process, creating continuous conducting paths from one side of the board to the other.

The next step is to create a tough "resist" material, adhering to both sides of the board everywhere except where the foil for the circuit is to remain. This is done by (a) coating the board with a light-sensitive film (usually a thin adhesive "dry film"), then (b) exposing the board to light with the full-size photopositive accurately sandwiched on top, and (c) chemically "developing" the film (as in conventional photography) to make the exposed areas permanent. A step analogous to photographic "fixing" then removes the unexposed film, in precisely the pattern that will ultimately become the circuit traces. Then the board, with the pattern of resist now masking the areas where the copper foil is ultimately to be removed, is immersed into a solder-plating bath. The result is to plate solder (a tin/lead alloy) everywhere that the foil pattern is to remain, including the insides of the holes.

Next the resist is removed chemically, exposing the copper that is to be removed, and the board is treated with a copper-etching compound. That leaves the desired pattern of solder-plated copper, complete with plated-through holes. At this point it is important to carry out a step known as "reflow soldering," which consists of heating the board to make the thin solder plating flow. This prevents the formation of tiny slivers of metal (from the undercutting action of the etching bath) that could otherwise cause conductive bridges. Reflow soldering also improves the solderability of the finished board; a reflow-soldered board is a delight to "stuff" with components.

The next step in board manufacture is to electroplate the edge-connector fingers with gold. The final process in board manufacture consists of applying a tough "soldermask" coating over the entire board, covering all of the foil except for the pad areas. This greatly reduces the tendency to form "solder bridges" between closely spaced traces during subsequent soldering operations. It also makes the board moisture- and scratch-resistant. Soldermask materials can be applied by silk-screen methods ("wet mask") or by the same photographic "resist" methods used to create the foil circuit pattern ("dry mask"). You can recognize the soldermask by its dark green color and by the observed fact that it is practically impossible to remove. In industrial board manufacture, the board might then be stuffed by automatic machines, with all joints soldered in a few seconds in a "wave-soldering" machine. The alternative is to stuff and solder by hand.

There is a simpler process of board manufacture that is sometimes used, especially in small production situations or for single-sided boards, where plated-through holes aren't needed. In this method you begin by coating the board with photo-resist, then expose it through a full-size *negative* (rather than positive) of the desired pattern, i.e., a photographic film that is transparent wherever you want foil to *remain*. The resist is developed, and then the unexposed resist is dissolved away. This board then has a layer of tough resist covering the copper that is to remain, so you simply expose it directly to the etching compound (omitting the solder-plating step, above). After the superfluous copper has been etched away, the remaining resist is washed off with solvent, leaving the desired pattern in copper. At this point it is best to treat the board with an "electroless" tin-plating bath in order to cover the copper with a metal less susceptible to corrosion. As before, the edge-connector fingers will then be gold-plated. The final step in this process consists of drilling the holes by hand, using the actual conductive pattern as a guide (each "donut" pad has a small opening in the center to aid in drilling the finished board).

12.05 PC board design

There are several important decisions you have to make during PC board design, during component "stuffing," and finally when the board is used in an instrument. In this section we will try to touch on the most important of these.

PC board layout

To make a printed-circuit board, you must ultimately convert your schematic diagram into a corresponding pattern of desired copper-foil traces that will compose the finished board. There are basically two ways: (a) Working from the manually drawn schematics, you can use pencil and paper (and lots of erasers!) to figure out a set of interconnection paths ("routes") that does the job, then manually apply opaque tape and preformed connection patterns accurately onto clear polyester film to make the finished "Mylar" masters, or (b) you can convert your hand-drawn schematic to a connection "netlist," then use a CAD (computer-aided design) program to figure the trace routing, producing a set of precision machine-drawn "photoplots" directly; better still, you can replace the manual schematic drafting with CAD-based "schematic capture," in which you draw the diagrams directly on a graphics workstation, using a graphic tablet or mouse.

CAD-based circuit drawing and layout (option b, above) has many advantages, including automatic netlist extraction, painless documentation, the ability to check for design and layout rule errors, the ability to make changes relatively easily, and the ability to produce complex multilayer boards with precise alignment of pads and traces. It is the method of choice for nearly all complex and high-density PC board production. However, we would like to begin simply, by briefly describing manual methods of PC board layout (option a, above). Once you know how to do a PC board layout by hand, you will understand what you're trying to make with the complex computer-based CAD tools. Furthermore, for simple jobs the manual methods may be all you need, and they are cheaper (and sometimes quicker) than CAD-based methods. They are particularly appropriate for the uncrowded boards you might use in simple unsophisticated instruments, of the sort you might build in small quantities for research laboratory use. They are also well adapted to boards containing parts with unusual shapes and lead spacings. Later, in Section 12.08, we will describe the CAD methods that are mandatory for high-density digital multilayer boards intended for large production.

There are several stages along the way from a schematic diagram to a final printed circuit. Beginning with the diagram, you generally work out trial pencil sketches of component layouts and interconnections, eventually working these together into a final pencil layout drawing. From this you make the "Mylar," consisting of accurately aligned "pads" (terminal areas for component connections) and taped interconnections. Precut patterns are used for IC and transistor pads and for ribbon and edge connectors, since these have standard spacings and dimensions. The pencil sketch and Mylar are usually made double size to allow greater accuracy (and to keep your eyes from popping out!). When the Mylar (two Mylars for double-sided boards) is completed, it is photographically reduced to an actual-size negative, from which a trial board is made as described previously. You generally "stuff" the prototype board with components, turn on the power, and then hunt down the errors; this lets you correct the Mylar artwork to produce final boards. The following subsections provide some further details and hints.

□ *Initial sketch*

We recommend doing the initial layout with pencil on grid paper (5 lines/inch), with two colors to indicate foil patterns on the top and bottom (assuming it is a double-sided board). We usually use black pencil for runs on the bottom and green or red for the top (component) side. Since you're likely to do plenty of erasing, it is best to use vellum graph paper. The 0.2 inch gridding corresponds to 0.1 inch final size, the universal measure for IC pin spacings, transistor pinouts, edge connectors, etc. Your drawing should be the view from the component side; i.e., the sketch of the component-side (top-side) foil pattern looks like the final pattern, and the sketch of the bottom foil pattern is what you would see looking down through the finished board with x-ray eyes. While working on the layout, indicate component outlines with a pencil of a third color. All this work should be freehand. Don't waste time with an outline template; just use the grid lines as a guide to draw IC and component pinouts.

It is generally best to work up some trial layouts on a piece of scratch paper, particularly for sections of the circuit that may require special layout to minimize long lines or capacitive coupling. It may take some experimentation to arrive at good component arrangements. A trial layout might consist of a block of the circuit with two or three op-amps, or perhaps the input or output section of the circuit. These blocks should then be worked together onto the large gridded vellum, with adjustments being made as you go. Don't hesitate to do lots of erasing!

□ *Layout dimensions and hints*

Try to have all ICs pointing in the same direction, preferably in straight rows. Likewise, resistors should be in even rows, not staggered. We use 0.030 or 0.040 black tape for signal runs, with wider tape for power supplies (0.05 or 0.062 inch) and very wide ground runs (0.1 to 0.2 inch, or even wider; it's common to broaden the ground runs with lots of tape). Be sure to include plenty of bypass capacitors, one $0.1\mu F$ for every two to four ICs. As you scratch your head, trying to juggle the tangled maze of interconnections, don't forget that components act as "jumpers" – they can hop over runs on the board.

Dimensions and spacings: On the actual-size PC board, we recommend holes spaced 0.4 inch for resistors ($\frac{1}{4}$W size), with spacing of 0.1 or 0.15 inch between resistors (with 0.15 inch spacing you can get a tape run between adjacent pads). We favor CK05 and CK06 types of ceramic capacitors, with their controlled 0.2 inch lead spacing, or the "DIP" 0.3 inch types (AVX type MD01, or Kemet C630C104M5U); they can also be spaced 0.1 inch from other capacitors or resistors. Leave some room around ICs for logic clips: a minimum of 0.2 inch to the next IC pads and a minimum of 0.15 inch to the nearest resistor or capacitor pads. Leave 0.030 inch spacing between tape runs, and don't run anything closer than 0.25 inch from the edge of the board, to allow room for card lifters, guides, standoffs, etc. Avoid running lines between the 0.1 inch spaced pads of an IC, unless necessary. You can fit up to six tape runs lengthwise between the pads of a standard DIP IC pattern (they're spaced 0.3 inch).

Recent PC design practice generally favors much higher line densities (both thinner lines and closer spacing) than the values recommended above; the latter would be called "15-15" design rules, signifying 0.015 inch minimum line width with 0.015 inch minimum conductor clearance. In the PC board industry, 15-15 rules are considered quite relaxed, with 12-12 pretty standard; 12-12 rules let you run one trace between adjacent IC pads on 0.1 inch centers (check for yourself that this is permitted, providing the pads aren't larger than 0.064 inch in diameter). High-density boards often use 10-10 or 8-8 rules, with which you can squeeze *two* traces between adjacent IC pads (the maximum pad diameters are then 0.050 and 0.060 inch, respectively). Occasionally you see daring boards with 0.006 inch or thinner traces; the designers of such boards are trading the increased circuit density (*three* traces between adjacent pads, 20 down the center channel!) against smaller plated-through holes and other compromises that result in poorer production yield and board robustness.

□ *Connections to the board*

For the majority of boards it is probably best to bring out all connections through "edge-connector" contacts, which mate directly with sockets available in a variety of contact configurations. The most commonly used spacings are 0.156 inch, 0.125 inch, and 0.100 inch between fingers. Generally you'll put an edge-connector pattern at one end of the card, bringing power-supply voltages and signals through that connector. The card is mechanically supported, and it plugs in at that end (more on that shortly).

Often you see an edge-connector pattern at the other end of the card also, used instead of a flat ribbon connector to bring some other signals off the board or to other boards. Another method for bringing out signals is to use flat ribbon cable terminated in DIP plugs; such cables plug right into IC sockets on the board. You can buy these cables prefabricated in various lengths, or you can make them yourself with a kit consisting of flat cable, unassembled DIP plugs, and a crimping tool. Ribbon cables can also connect to the board via in-line or "mass-termination" connectors, which use one or two rows of pins on 0.1 inch centers.

For simple boards the best method of connection may be to use swage-solder terminals or PC-type barrier strips with screw terminals. Avoid the use of large pads alone for connection of external wires to PC boards.

Figure 12.7 illustrates a variety of PC board connection techniques.

□ *Odds and ends*

With plated-through boards, use several holes to join ground foils on opposite sides of the board. Try to avoid using multiple passes through the board to reach your destination, since plated-through connections where no component is mounted are more likely to give trouble. The layout of a double-sided board generally winds up with most tape runs going horizontally on one side, vertically on the other.

General philosophy: Use smooth curves or 45° turns, rather than right-angle turns, for hand-taped layouts. Bring lines into pads as if heading for the center of the pad, rather than coming in at an oblique angle. Don't mount heavy components on boards (a couple of ounces ought to be the limit); assume that the instrument will be dropped 6 feet onto a hard surface sometime during its life! Put polarity markings on the component side for diodes and electrolytic capacitors, and label IC numbers and pin 1 location (if there's room). It is always nice to label test points, trimmer functions (e.g., "ZERO ADJ"), inputs and outputs, indicator light functions, etc., if you have room.

□ *Taping the Mylar*

General advice: Use an illuminated "light table" with a piece of precision gridded Mylar taped to it. Don't confuse this with the inexpensive gridded plastic films that are neither accurate nor dimensionally stable; a piece of precision gridded film will set you back at least $20. Put your clear Mylar over, and stick down the IC pads accurately on it. Use the pencil sketch for guidance while taping. Wash

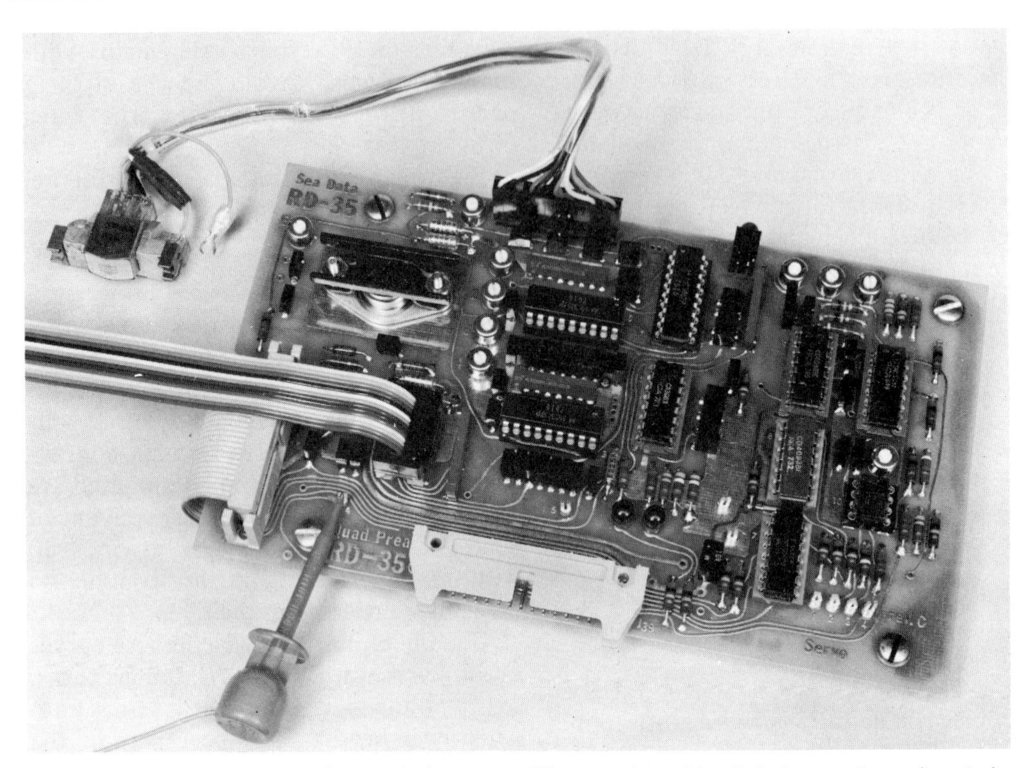

Figure 12.7. Several connection techniques are illustrated in this digital recorder printed-circuit card. The tape head connects via an in-line connector (which mates with a row of Wire-Wrap-type pins), and other signals are brought out with "mass-termination" ribbon connectors and a dual in-line ribbon connector. A test lead is shown clipped onto a "test point" terminal. This board also illustrates PC board heat sinking (upper left), a logic-state indicator (upper right), miniature single-turn trimmers, and single in-line (SIP) resistor networks.

your hands often to prevent deposition of oily film on the Mylar, and use alcohol to wipe any areas that might become oily. Use an Xacto knife with curved blade for tape and outline cutting, and learn not to cut through the Mylar. Press the tape down firmly after positioning; otherwise it will eventually curl up. Allow generous overlap where tape meets pad, etc. When laying out tape, don't hold it under tension; it will shrink and pull away from pads. Use precut bends and circles for the larger tape widths (0.062 inch or wider) when navigating tight turns. After the Mylar is completely taped, check it against the schematic by going over each connection on the diagram with a red pencil. When all seems OK, seal up flaws on the Mylar with an indelible black felt-tip pen.

Precut PC graphics patterns are available from several manufacturers. Table 12.1 shows some recommended types. The Bishop Graphics catalog (5388 Sterling Center Drive, Westlake Village, CA 91359) includes extensive information on PC board layout and execution.

12.06 Stuffing PC boards

Your worries aren't over when you've got a finished board. You've got some decisions to make (e.g., whether or not to use IC sockets) and some important things to do (e.g., defluxing and lead trimming). Herewith, some thoughts on these subjects.

TABLE 12.1. SELECTED PC GRAPHIC PATTERNS

Pattern[a]	Bishop	Datak
Small pads (0.150" OD)	D203	JD-145
Medium pads (0.187" OD)	D104	JD-146
Large pads (0.250" OD)	D108	JD-150
Giant pads (0.300" OD)	D293	JD-343
0.150" thermal relief – pos	5272	JDS-532
0.150" thermal relief – neg	5278	–
0.187" thermal relief – pos	5232	–
0.187" thermal relief – neg	5238	–
16-pin DIP	6109	JD-64
16-pin DIP with in-betweens	6946	JD-179
20-pin DIP	6999	JD-575
20-pin DIP with in-betweens	–	JD-585
28-pin DIP	6904	JDS-398
28-pin DIP with in-betweens	–	JDS-591
TO-5 transistor	6077	–
TO-18 transistor	6274	JD-88
TO-92 transistor	–	JD-91
0.100" connector pads	5004	JD-145
0.100" edge-connector strip	6714	JD-123
0.156" edge-connector strip	6722	JD-121
0.031" black tape	201-031-11	–
0.040" black tape	201-040-11	–
0.050" black tape	201-050-11	–
0.062" black tape	201-062-11	–
0.100" black tape	201-100-11	–
0.200" black tape	201-200-11	–
0.062" universal corners	CU601	–
0.100" universal corners	CU607	–
0.200" universal corners	CU609	–

(a) for 2:1 artwork originals.

Sockets

There is great temptation to use IC sockets everywhere, for ease of troubleshooting. However, if you're not careful, the sockets may well cause more trouble than they prevent. In general, sockets are a good idea at the prototyping phase, where IC substitution may be necessary to convince you that the trouble you're having is a *design* error, not a bad component. They should also be used for expensive ICs (e.g., a D/A converter, microprocessor, or the like), ICs that you're likely to want to change from time to time (e.g., a program ROM), and ICs that have a good chance of being damaged sooner or later (e.g., chips that buffer input or output signals from outside the instrument).

The problem is that a poorly designed socket may prove unreliable over extended time periods. A nonsoldered joint must have a gas-tight seal, such as that created by a mechanical metal-to-metal wiping action, with the seal then being left undisturbed. PC edge connectors, for example, used to be somewhat unreliable; with time, manufacturers learned some good tricks: bifurcated contacts (two independently sprung contacts for each finger), gold plating on the socket and on the edge fingers, and good mechanical design to ensure firm contact pressure during wiping and afterward. Joints that aren't gas-tight can be expected to fail after some time, perhaps a year or so. This sometimes happens inadvertently, e.g., by inserting a component in a PC board and then forgetting to solder it. Such connections have the maddening property of working fine at first, then becoming intermittent months or years later, owing to the formation of corrosion. A different problem can arise when heavy ICs (24 pins or more) are held in sockets. They can work their way out of the sockets after repeated vibration or shock.

We have found that the pin-and-jack type of IC socket (popularized by the Augat $5xx$-AG series), although expensive compared with many other socket types, gives good reliability.

Soldering and defluxing

The usual procedure is to insert some components, turn the board over and bend the leads aside to hold the components in place, then solder them using a thermostated soldering iron and fine solder. ICs can be inserted easily with an insertion tool (highly recommended), and it is best to use a lead bender on resistor leads, etc., in order to prevent slivers of solder being shaved off the leads during insertion. We recommend the adjustable "flip-over" board holders (marketed by OK Industries), which hold the components firmly in place with foam rubber during soldering. After soldering, the leads should be trimmed with a snipper.

Now comes a very important step: Solder flux should be removed from the board. If it isn't, the board will look just terrible in a few years, when you're not around to defend it! Rules for defluxing:

1. Do it.
2. Do it soon. The stuff gets much harder to remove with time.
3. Use a solvent such as Freon, alcohol, or some other organic solvent recommended for this purpose. Use a small brush to help dislodge stubborn globs of flux.

Commercial board manufacturers clean their boards with *vapor-phase degreasers*, in which hot vapor from a bath of boiling solvent condenses on the board (suspended above), dissolves the flux, and drips back into the bath. This method works well, especially because the distillation process continually bathes the board with clean (and hot) solvent. Because organic solvents aren't exactly good for your health, people have experimented with water-based cleaning processes. One method uses a water-based "saponifier" to emulsify the rosin so it will wash away; alternatively, you can use a water-soluble solder flux (rather than the usual flux, which is made from insoluble tree rosin). As nice as they sound, these water-based cleaning methods turn out to be tricky to do correctly; if you're not careful you can leave corrosive residue that ruins the board, in time. For the do-it-yourselfer we recommend organic solvent cleaning; take care not to breathe the stuff or get it on your hands.

When defluxing boards, keep in mind that mere removal of the rosin residues is not enough – you have to get rid of the ionic "activators" also. Incomplete defluxing may do more harm than good, since it can release the activators from the area of the solder joints and spread them all over the board. Note also that solder flux comes in several grades of aggressiveness. We routinely use "RA" (rosin, activated), which is the most active flux for electronic soldering, because it produces good solder joints even when surface oxidation is present; RA flux residues should be thoroughly removed after soldering. The alternative, "RMA" flux (rosin, mildly activated), is less active and is often specified for government contracts or for applications where defluxing is not possible.

12.07 Some further thoughts on PC boards

The solderability of PC boards tends to decrease with time, owing to oxide formation, so it is best to stuff the components soon after the board is made. For the same reason, you should keep unstuffed boards in plastic bags, away from corrosive fumes. Good circuit boards should be made from 1/16 inch FR-4-type board material (sometimes referred to as "epoxy fiberglass") clad with 2 ounce copper. Remember that a circuit on a PC board is basically sitting on a piece of glued-together stuff; the board can absorb moisture and develop electrical leakage. Another pathology of PC board material is "hook," the variation of dielectric constant with frequency; the consequent variation of stray capacitance can make it impossible to build an amplifier with flat frequency response, for instance. Oscilloscope manufacturers are very aware of this bizarre effect.

PC runs with large currents passing through them have to be widened to prevent excessive heating and voltage drops. As a rough guide, here is a table of approximate conductor widths that give temperature rises of 10°C or 30°C for the currents listed, for 2 ounce copper PC boards. For other foil thicknesses, just scale the widths accordingly.

	0.5A	1.0A	2.0A
10°C rise	0.004"	0.008"	0.020"
30°C rise	0.002"	0.004"	0.010"

	5.0A	10.0A	20.0A
10°C rise	0.070"	0.170"	0.425"
30°C rise	0.030"	0.080"	0.200"

PC runs with high voltage need correspondingly wider spacing – a good rule is 5 volts per mil (0.001"). It is also a good idea to avoid sharp corners and points, in favor of rounded contours.

Tools

As a starting point, we lined up the most heavily used tools on our bench and came up with the following part numbers:

Long-nose pliers	Erem 11d, Utica 321-4½, C.K 3772H, Xcelite 72CG
Snippers	Erem 90E, C.K 3786HF
Soldering iron	Weller WCTP-N, Ungar "Ungarmatic"
Solder	Ersin Multicore 22ga, Sn63 alloy, RA flux
IC inserter	Solder Removal 880
Lead bender	Production Devices PD801
Solvent dispenser	Menda 613
Solder sucker	Edsyn Soldapullt DS017

For the most effective desoldering of PC boards, it's worth investing in a controlled-vacuum desoldering station. These work well even on solder-filled plated-through holes, with less tendency to damage the delicate foil pads (when compared with the simple spring-powered plunger types). These desoldering stations do have a tendency to clog with solder, however. They are manufactured by several companies, including Edsyn, OK, Pace, Ungar, and Weller. Lots of useful gimmicks for PC assembly are listed in the Contact East catalog (335 Willow Street South, N. Andover, MA 01845) and the Marshall Claude Michael catalog (9674 Telstar Avenue, El Monte, CA 91731).

12.08 Advanced techniques

Hand-drawn schematics, manually converted first to PC board layouts and then to hand-taped Mylars, are the traditional steps in PC board design technique, and they were used nearly universally through the mid-1970s. It still makes sense to do simple boards this way, particularly if you aren't trying to break new records in board component density. With a little kit of double-sized patterns, some opaque tape, a gridded table, and some clear Mylar, you're ready to go. And you don't have to buy expensive CAD software (and learn how to use it!), pay photoplotting charges, etc.

However, as soon as you strive for high-density boards, paved wall to wall with fifty or a hundred IC's, often requiring four or six layers laid out to 10-10 design rules, the honeymoon is over. Even with extraordinary effort and quadruple-sized Mylar, you can hardly achieve the precision interlayer alignment required. Furthermore, you'll need a month's vacation to recover from the concentration required to work out the routing. The first board you get back from the PC house is unlikely to be error-free, and any serious changes to the taped Mylars are sheer hell, often requiring massive areas of wiring to be pulled up and repositioned (which often introduces new errors). We aren't kidding.

CAD/CAM

The solution is CAD/CAM (computer-aided design/computer-aided manufacture). These software packages are powerful, and with the advent of desktop workstations with multimegabyte memory, dazzling graphics, and processing speeds of tens of MIPS (million instructions per second), you don't need special hardware to run them. Some popular PC CAD systems come from Valid Logic, Mentor Graphics, and Daisy Systems. They aren't cheap, but at least they're not getting any more expensive. Here's a quick tour of board design using CAD/CAM.

☐ *Schematic capture.* You begin by "capturing" your schematic directly into the graphics workstation. You can draw and edit diagrams, using standard electronic symbols from a "library." Just as with word processing, you can pull up old work, extract sections you want to reuse (e.g., a DRAM memory control circuit, an active filter, etc.). With a mouse (or joystick, trackball, or graphic tablet) you can move things around, with the wiring following you around (not always totally successfully!). You name signals, give the ICs part numbers, and so on. Good CAD systems assist you by looking up the ICs and assigning pin numbers. And good CAD systems are *hierarchical*: You can, for example, do a top-level diagram in which the major subcircuits are indicated by big boxes; each box opens up to reveal its sub-boxes, eventually displaying the lowest (gate-level) description. At any stage of the design you can get handsome laser-printed schematics. See Appendix E for further thoughts on schematic diagrams.

☐ *Checking and simulation.* The output from the schematic capture consists of a set of drawings (Fig. 12.8) and a *netlist*, which is simply a list of every signal, telling every "node" (component pin) it is connected to.

At this stage in the design you should, of course, spend plenty of time with the schematics, making sure they're what you want. A good CAD system can help you out, by flagging obvious errors, for example an output tied to ground, or to another output. You can ask the system to simulate digital circuit operation, but you must provide it with test "vectors" that describe machine states, and you must describe the function of each chip in your circuit that is not already described in the library. Since a complex chip like a microprocessor requires lots of description, some CAD systems have a socket into which you can plug such a chip; it then uses the chip itself for hardware modeling.

At this point, too, you can ask the CAD system to disgorge voluminous documentation showing (a) for each named signal, all the pins ("nodes") it visits and (b) for each pin of each component, all other nodes to which it is connected. Good systems will even inform you if you have violated loading rules for digital outputs, etc.

☐ *Placement and routing.* The next step is component placement and routing. One of the great beauties of an integrated CAD/CAM system is that it automatically extracts the netlist from its representation of the schematic – if the schematic is correct, the final PC board will be too. Although some CAD systems claim to be able to juggle the component placements into a near-ideal configuration, it's usually best to do component placement manually. As with schematic capture, this is done with mouse, joystick, or graphic tablet. You begin by setting up the board outline (it's often the same as some previous board you have designed), within which you now position the parts. Since the component library has outline and pin information, you get to play with the characteristic IC and component shapes. Good CAD systems can flag design-rule errors as you make them, so you can see if the parts are too close together, etc.

Once the parts have been placed, the wiring must be routed. At this point it is conventional to display what's called the "rat's nest," a display of the board with all the connections shown as straight lines

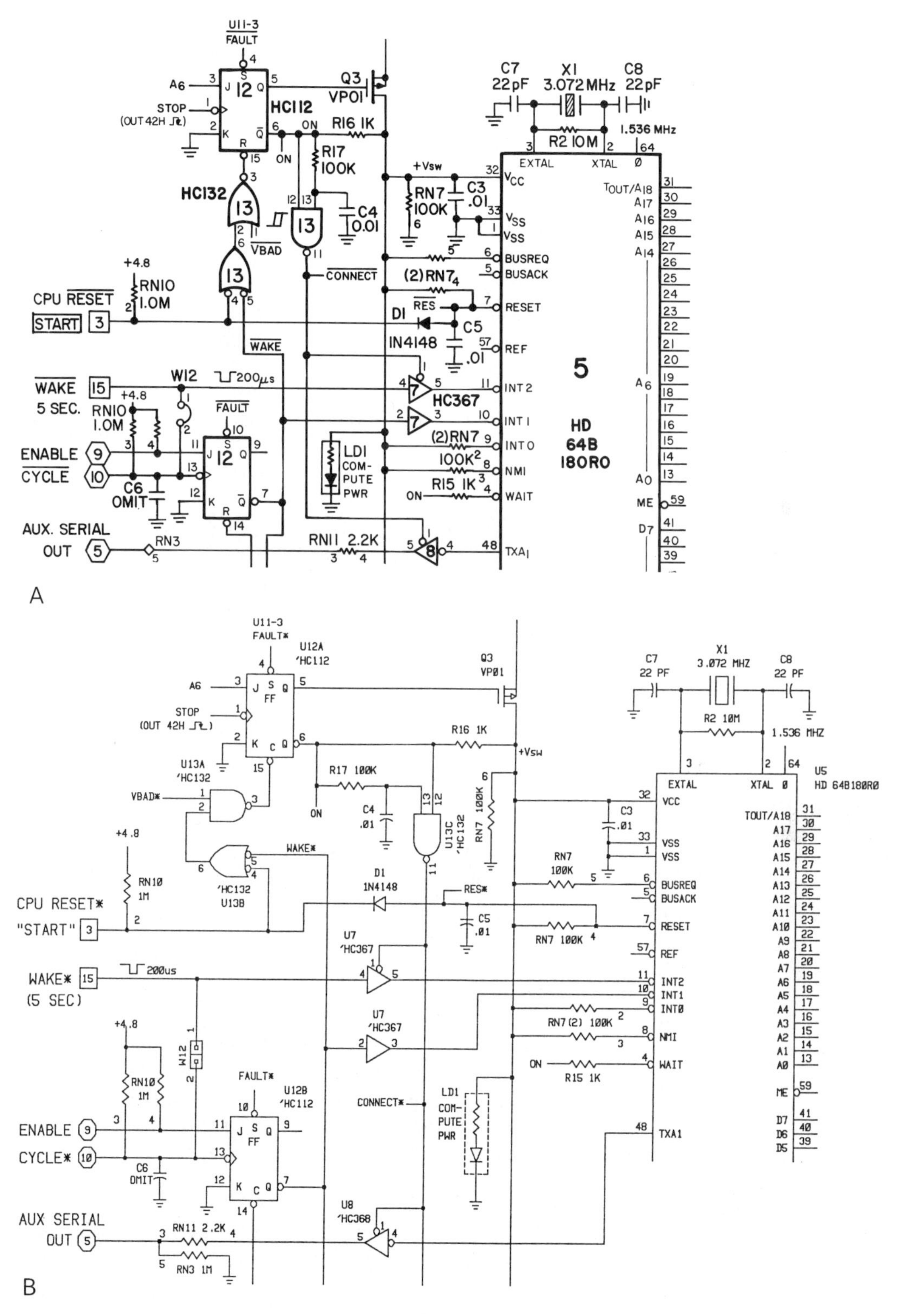

Figure 12.8. Portion of a complex schematic diagram from which the PC board in Figure 12.11 was produced.
A. Manually inked on a drafting table.
B. Machine-drawn, from Case Technologies schematic capture.

connecting their respective pins. It's a tangled mess. With creative use of color you can make some sense of it, for example by selectively turning on only power and ground connections.

Early CAD systems made you work hard at this point. You were forced to route every connection by hand, with just a little help from the computer. Current systems provide *autorouting*, which means what you hope it does: The software finds connection paths, taking care not to violate the design rules that specify not only line widths and clearances but also parameters such as the maximum number of "vias" (plated-through holes used only for a signal trace, not a component lead). The best CAD machines can do 100% autorouting, although the result often lacks the full elegance of hand-routed lines. For example, it may put a via daringly close to a component pad; although its placement meets all the rules, you can create a board that's more easily manufactured and soldered by shifting the via away from the pad. You usually spend a few hours cleaning up the machine's work in this way, pushing and pulling traces around a bit (Fig. 12.9).

In complex digital circuits you can often simplify the routing considerably by reassigning gates or sections within chips, or even swapping gates between IC packages. As an example, you might have a byte-wide data bus connected to an octal D register. Your initial assignment of bus bits to corresponding register bits might force each line to leapfrog the next, whereas by reversing the order of bits in the register the resulting routing requires no crossings. Good CAD systems can do this automatically (the library specification of each chip identifies duplicated functions), even updating ("back-annotating") the schematic via the modified netlist.

The final step is to check the proposed routing for any rule violations, and also for exact equivalence to the schematic. The latter step is important because you may have made manual changes to the routing netlist that altered the circuit. At this point all CAD systems will produce paper pen-plots of the routing, showing any designated combination of components, wiring, labels, etc., in a dizzying array of colors.

□ *Photoplot and drill tape.* If all has gone well, the last stage of the process consists of producing a machine-readable board description suitable for the PC house. For this you need two things: a tape that tells a photoplotter how to draw a precision plot of the "artwork" (the individual foil patterns for each layer, a pattern for the "solder mask," and a silkscreen pattern for printed legends) and a "drill tape," which tells each hole size and its precise position. The artwork tape is usually written in "Gerber format," named after a brand of photoplotter that exposes film by moving a large easel under a stationary projector in response to commands written on tape. (A newer form of photoplotter uses laser scanning, producing large plots in minutes rather than the hours required by the early Gerber machines.) Some PC board fabricators want you to supply the finished photoplots (Fig. 12.10), whereas others ask for the Gerber tape itself. A surprising number of board houses ignore the drill tape, preferring to figure out where the holes go by manually digitizing the photoplotted artwork (if you ask, they'll tell you that customer-supplied drill tapes sometimes have crazy commands that break their drill bits). Drill tapes, believe it or not, are not magnetic tapes; they are *paper* tapes!

Board production. In this world you don't manufacture your own PC boards; you go to a PC board house. They're all over the place, and they just love to make boards (Fig. 12.11), in exchange for money of course. Some specialize in

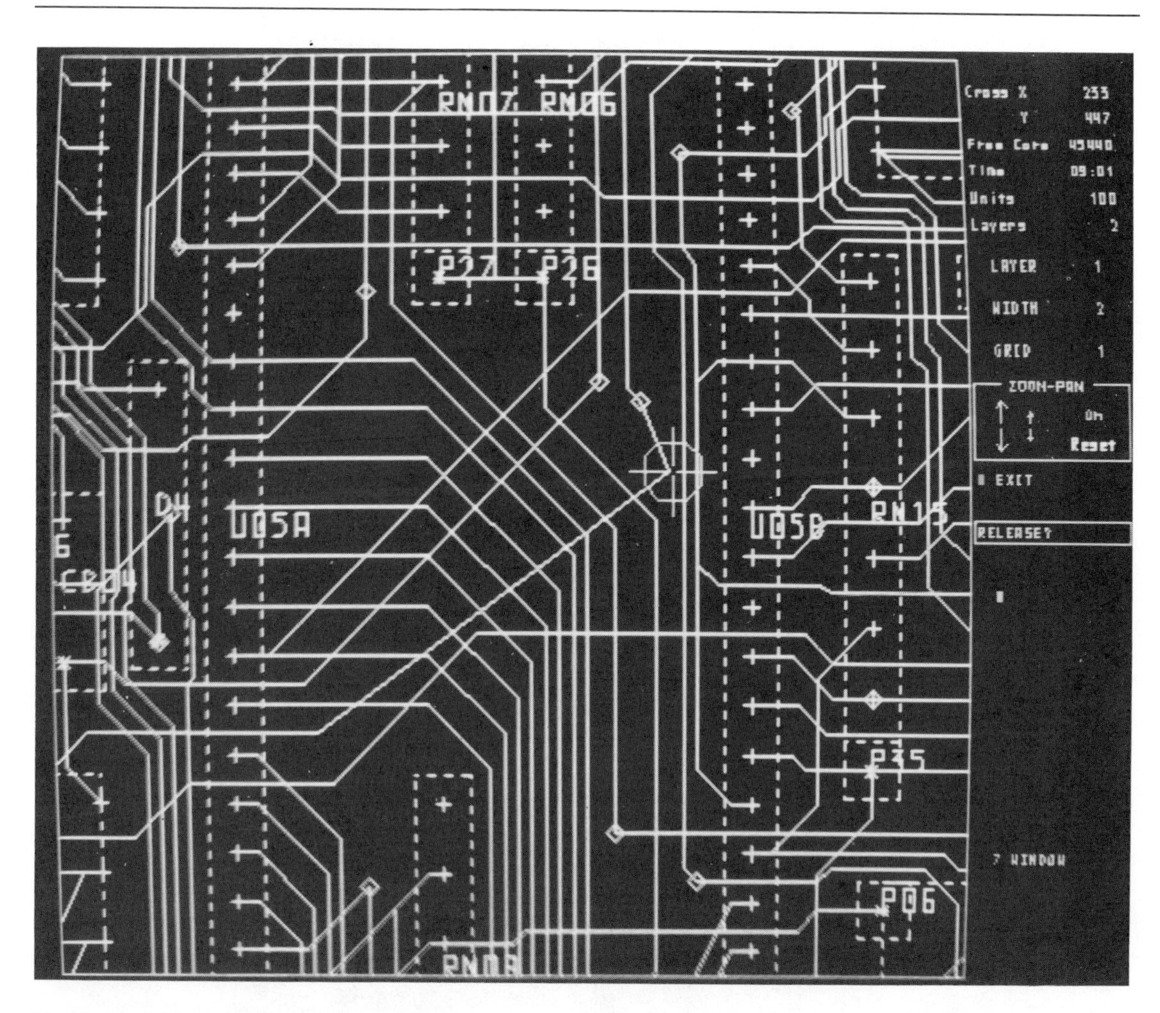

Figure 12.9. Photograph of PC board design system screen display (Racal-Redac), showing manual "cleanup" of trace routing. The cursor, steered via a "graphics tablet" stylus, shows the circuit trace that is being modified. This display is in color, with different colors and intensities designating the different layers, component outlines, legends, etc.

small-quantity, fast-turnaround prototyping, whereas others may do only production quantities. Your first task is to find out which companies do a good job. This is harder than it sounds. You can't tell much on the telephone, because it is possible to *know* how to make good boards, but still make bad ones. Here are some tricks we've tried: (a) Look at the little logos on good boards you've seen (e.g., boards you find in instruments or computers sold by major companies, who, like you, usually buy their boards from external board houses) – companies like IBM and Apple are extremely careful about their board suppliers, (b) Ask everyone you know whom they use to make their boards, (c) Go and visit the board house; while there, look for cleanliness, high morale, native intelligence, diligence, and pride in workmanship; you might also ask to see some boards, (d) Finally, if you know

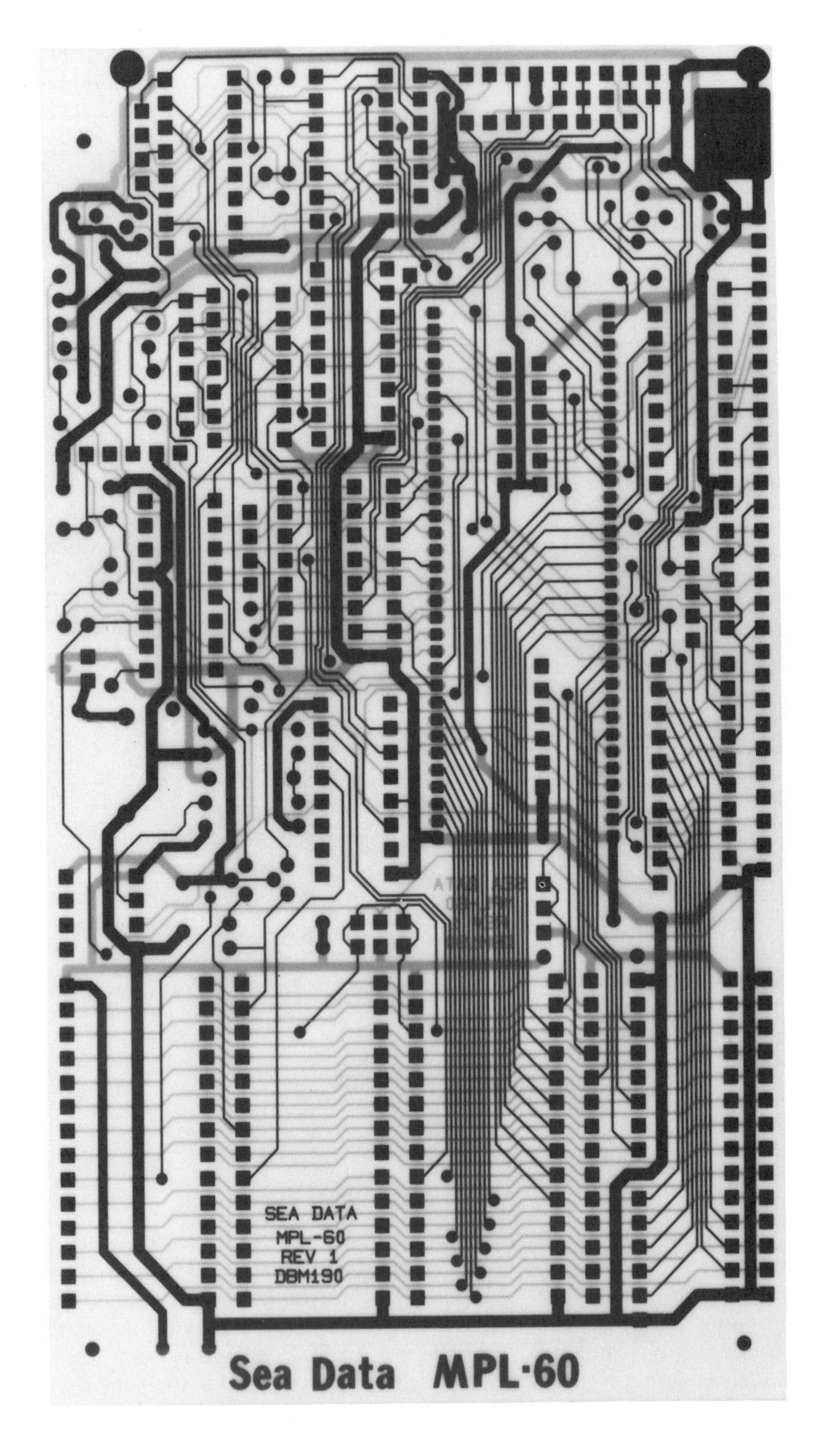

Figure 12.10. Photoplot for the double-sided board shown in Figure 12.11. Both layers are overlaid in this photograph, showing the view from the solder side.

Figure 12.11. Finished microprocessor board, designed to double-sided 12-12 rules. Higher component density can be achieved only by going to tighter design rules, or, preferably, a multilayer board.

someone who works at a bare-board testing house, cajole him into telling you who makes good boards: Those guys really know, but they're not supposed to tell!

Next you have to get a quote for price and delivery. Be prepared to answer the following set of questions, which every board house will ask you:

1. board dimensions?
2. number of layers (sides)?
3. design rules (minimum conductor width and spacing)?
4. any gold-plated edge-connectors?
5. any fancy board cutouts or shapes?
6. solder mask? wet or dry? to cover the vias ("tented") or not?
7. board and copper thickness?
8. number of holes?
9. number of hole sizes?
10. silkscreen 2 sides, 1 side, or none?
11. any unusual options, e.g. "dry mask over bare copper"?

Having agreed on terms, you must now supply the following things:

1. actual-size layer positive transparencies (i.e., opaque = copper)
2. soldermask patterns(s)
3. silkscreen pattern(s)
4. drill drawing (hole pattern, coded by sizes)
5. mechanical drawing (precise outline, aligned to specific holes)
6. drill tape (a paper tape; often optional)
7. additional specifications and notes

The last item is important. You should strictly specify the minimum trace width (an over-etched board has thin traces), the minimum conductor spacing (under-etched board), the minimum annular ring (caused by misaligned holes or misaligned layers), the tolerances for finished hole diameters, minimum thickness of copper (and solder) plating, the maximum number of repairs permitted per board, and so on. The industry generally adheres to some typical numbers (e.g., the IPC-600 standards), but it's best to have it clearly in writing, in case you get some boards you're unhappy with. Furthermore, your board may have unusual requirements; for example, if you've used unusually small pads for your vias, your minimum annular ring specification for vias will be less forgiving than usual.

Testing. For any new PC board design, *always get a few prototypes made first.* Stuff them with sockets, and test the dickens out of the circuit. You can expect several kinds of bugs, namely (a) your circuit had a design error, (b) the circuit was OK, but you made an error in schematic capture (which you should have caught earlier), (c) an error crept in during routing (which also should have been caught earlier), (d) the board has flaws, usually shorts or opens, due to blemishes, misalignments, or other problems during manufacture (see below), or, finally, (e) there are some bad components or solder joints.

Assuming circuit changes are needed during debugging, the usual procedure is to cut existing foil traces with a small knife, then solder in wires to make the new connections. This may be impossible in a multilayer board with buried signal layers, however. In that case you have to make your changes at the final destinations – the IC or component pins. A good way is to stack up two IC sockets, with a pin removed from the upper one.

During circuit testing, mark up your schematics with circuit fixes, for later entry into the schematic capture (or hand-drawn circuit). The second pass through the CAD system is usually easy and fast; but be *very* careful, because you probably will go directly to a production run, and any errors remaining in the production board will have to be fixed on each unit. If that becomes necessary, you use the same procedures of cutting and jumpering as above, taking care to secure the wiring (a hot-wax gun seems to work well). These extra wire patches are sometimes called "roach wires"; the official term is "ECO," for "engineering change order."

There are plenty of things that can go wrong in board production (see the fine handbook by Coombs for gory details). With high-density boards (Fig. 12.12) even a minor blemish can create a short or open circuit somewhere. And plated-through holes have all sorts of pathologies of their own that can lead to open circuits. It's an axiom in electronics that the cost to repair a flaw multiplies at each stage of production – a bad board connection is easily fixed at the board house, fixed with greater investment of time during circuit checkout, and repaired with considerable expense once shipped into the field. So it makes good sense to have the board house do a "bare-board" test on all your production boards. This they do on a curious little setup that jabs the board with a set of pins that align with the pads (they fabricate the "bed of nails" from your drill tape, of course). Oddly enough, none of the testing houses compares the bare board with your *netlist*; instead, they just test a bunch of boards, and assume that if a number of boards are identical, they are also correct. Although sharp-eyed readers will see a flaw in the logic, they can rest assured that the system usually works anyway.

□ *CAD-based Wire-Wrap.* The netlist output from the schematic-capture phase contains all that is necessary to construct the circuit. Many CAD systems offer output formats compatible with automated Wire-Wrap fabrication. You begin by choosing component placements on a standard Wire-Wrap panel (see the Augat catalog for numerous specimens), after which the CAD system constructs a special form of netlist called a "From-To" list. Each entry tells the coordinates of the two pins connected by a wire, and the wrap level (height above board) of each end's wrap. If you send this tape to a wire-wrapping house such as DataCon, they will produce a machine-wrapped board. It typically takes a few weeks and costs about 6 cents per wire plus the price of the Wire-Wrap panel (about $2 per IC position). As with PC boards, there can be flaws. Our experience is that you might find one or two problems per board, usually in the form of a broken wire that looks OK from the outside.

A more serious question about Wire-Wrap prototypes is "Why bother?" If your objective is to make PC boards, you're going to have to do a PC board routing eventually, so why not do the work for the prototype? Furthermore, a wire-wrapped board behaves differently than a PC board, owing to lead inductance and differences in ground-pin connection length and ground-plane inductance. The effects are most noticeable if you have fast logic such as 74F, 74AS, or 74AC(T), or memory arrays with widely bused signals. In general, a good multilayer PC board performs far better than a comparable wire-wrapped board, so you do yourself a favor by building the prototype with a PC board.

Of course, if you are building only a handful of units, you may want to avoid the one-time ("nonrecurring") expenses associated with PC board production and just build wrapped boards as the finished product. In that case the machine-wrapped boards are a rational choice. Don't overlook the alternative, however, which is the following.

□ *CAD-based Multiwire.* "Multiwire" is a trademark of the Kollmorgen Corporation, which manufactures a clever hybrid of custom-routed PC board. The basic board has pads in standard or custom groups to hold your ICs and to provide power and ground. The signal interconnections are made not with PC traces, however, but with fine insulated wire (#34 or #38) that

Figure 12.12. Surface-mounted devices (SMDs), with their 0.050 inch pad spacing, permit high component density, as shown in this photograph of a portion of a 4-layer board with 10-10 design rules. Both SMD and conventional through-board-mounting components are seen. The edge connector provides a scale, with its pads on 0.1 inch centers.

is machine-placed onto a sticky coating on the board. The wire ends are attached to the IC pads with a hole plating process.

Multiwire has good signal properties because of its complete groundplane, and it lets you make circuit changes in subsequent production runs relatively painlessly. Since the (insulated) wires can cross over each other, you can achieve very high component density, comparable with that of a 10-layer PC board. Kollmorgen claims that Multiwire is thus a cheaper solution than custom multilayer PC boards, particularly in small production runs and in situations where your circuit is evolving.

□ *CAD-base ASIC.* We finish our CAD narrative by pointing out that you would not dream of designing custom or semicustom (gate-array) ICs by hand (Fig. 12.13). These belong in the general class of "application-specific integrated circuits," or ASICs. Once again, CAD comes to the rescue, with capture and simulation software that gives you some assurance that your chip has a chance of working. ASICs are rational in the context of substantial production runs, and we confidently assume that the situation will only improve with time. By the time of the next revision of this book it may have become

Figure 12.13. At the *chip*-design level, CAD/CAM isn't a luxury – it's a necessity. (Courtesy of Performance Semiconductor Corporation, Sunnyvale, California.)

commonplace to design custom ASICs, promptly delivered by your local silicon foundry, for just the sort of circuits for which you now design a PC board filled with standard ICs. We see no fundamental reason to prevent the ASIC solution from becoming the less expensive (and better performing) alternative.

INSTRUMENT CONSTRUCTION

12.09 Housing circuit boards in an instrument

Circuit cards, whether printed circuits, Wire-Wrap panes, or breadboarding cards, have to be mounted in some sort of enclosure and connected to power supplies, panel controls and connectors, and other circuitry. In this section we will discuss some of the popular methods of putting instruments together so that circuits are neatly mounted and accessible for testing and repair. We will begin with methods for holding the circuit cards themselves, then discuss the business of cabinets, front and rear controls, power-supply mounting, etc.

Circuit card mounting

In simple instruments you may have only a single circuit card, whether printed circuit, Wire-Wrap, breadboard card, or whatever. In that case a simple solution is to drill holes near the corners and mount the card with screws (and standoff bushings) to a flat surface, component side up. Connections can then be made with a card-edge connector socket (if the card has plated fingers), with flat cable terminated with a connector to mate with a plug on the board, or with individual soldered connections to swaged terminals on the board. With edge or ribbon connectors, the card will support the connector adequately, so no extra connector supports have to be used. Whatever the method of connection, it is wise to arrange the wiring such that the board can be tipped upward for access to the underside so that you can make modifications or repairs.

In a system with several circuit boards, the best way to arrange things is with some sort of card "cage," a rigid assembly with guides for individual cards to slide into and aligned holes along the rear so that you can mount edge connectors to mate with the cards. There's lots of flexibility in card width, spacing, and number of cards that can fit into a card cage. A very common size accommodates cards 4.50 inches wide with 44-pin dual-readout (22 pins each side) edge connections on 0.156 inch centers. There are plenty of other formats, often with tighter pin spacing (0.1 inch is common), multiple groups of edge connectors on a single card, or the more reliable "2-part" connectors (in which one mating portion solders to the board as a component); the most popular example of the latter is the so-called VME connector, which has 64 or 96 pins. Cards can be spaced as little as 0.5 inch apart, if necessary, although 0.6 inch is a more comfortable spacing; if space is no problem, 0.75 inch spacing will allow plenty of room, even with Wire-Wrap pins and bulky components. It is best to look at some catalogs to see what's available. You can get variations with plastic card guides or just dimples in the metal sides to align the cards, and there are various types of card ejectors (attached to the PC cards) to help remove a card.

Card cages are available with simple flange mounting to a flat surface parallel to the cards, as well as in various configurations that fit nicely into rack enclosures, etc. You can even get modular enclosures that include an integral card cage, with some additional room for power supplies, panel controls, etc.

Warning: Instruments built with the circuit spread over a number of plug-in circuit boards achieve a nice modularity and ease of repair. But this method of construction can cause difficulty in circuits with low-level signals (less than a millivolt), and in circuits with high-frequency signals (above a few megahertz). The basic problem is that you often cannot provide a sufficiently stable and low-inductance grounding system to the set of boards by way of the connectors at the ends of the boards. Combining low-level analog circuitry with digital switching waveforms is particularly dangerous. This problem is worst with a hand-wired backplane, in which the ground distribution depends on a few wires running between the connectors. Typical symptoms are low-level 60Hz or 120Hz pickup, miscellaneous "fuzz" at the level of fractions of a millivolt, and coupling of radiofrequency signals between circuits that are supposed to be isolated. If the boards have a bare groundplane and are plugged into a metal card cage, the symptoms often change as you press on the cards, altering the uncontrolled ground contact.

We've suffered with such problems on more than one occasion, and we offer the following advice. First, it may be best to avoid interconnected small boards entirely, building all the critical circuitry on one larger board with integral groundplane. On that board you may need to use coaxial lines or twisted-pairs to make connections between separated portions of the circuit. Second, if you do use interconnected boards, you will achieve better ground distribution with a PC motherboard provided with wide ground traces, rather than a hand-wired backplane. In radiofrequency systems you sometimes see springy metallic fingers used along the card guides, to provide a continuous and stable ground connection. Third, the use of coaxial line or twisted-pairs, combined with differential inputs (or "pseudo-differential" ground-sensing inputs, see Fig. 7.70), is often the best way to handle microvolt signals, which are otherwise prone to ground-loop noise and pickup. Finally, except for the ground-sensing method just described, we suggest as many redundant ground connections as possible – multiple connections to the chassis, doubled connector pins and wires, etc. – to reduce the inductance through which ground currents must flow. Don't worry about "ground loops" in a digital or RF circuit; that's a microvolt audio circuit issue. Refer to Section 7.24 for additional ground philosophy.

Backplane connections

Card-edge connectors are available with lugs for solder connections, with Wire-Wrap pins, and with small pins for insertion into PC boards. In many cases it is best to wire up the intercard connections with point-to-point wiring between card-edge connector pins, using the edge connectors with solder lugs. A neat job requires some cabling of wire bundles, with the wires running in straight lines parallel to the card cage dimensions. In other cases it may be preferable to use Wire-Wrap connections on this backplane, especially if there are many connections between backplane pins with relatively few connections to other points in the instrument, and if there is no need for shielded-cable connections to the backplane.

The third possibility is to use a motherboard backplane, a PC board designed just to hold the card-edge sockets. Motherboards are popular in bused systems (they are nearly universal in computers) and should be considered in any case if the instrument is intended for production in significant quantities. With double-sided motherboards, you can have the advantage of a groundplane (lower inductance and coupling of signal lines), or you can use both sides for signals if the intercard wiring is complicated. With bused systems the backplane is usually simple, with lines connecting corresponding bus pins on all cards. In computer backplanes you sometimes see a motherboard used with Wire-Wrap pins sticking through. This is very handy if you want the motherboard to do all the bus and power-supply connections, leaving the unbused pins to be connected in a custom configuration by wire-wrapped connections. Figure 12.14 shows a simple PC motherboard.

12.10 Cabinets

Depending on the intended use, an electronic instrument might be housed in a benchtop cabinet (complete with rubber feet and hinged front "bail"), in a cabinet or panel designed for mounting in a standard "relay rack" 19 inches wide (either screwed directly to the rack flanges that run vertically up from the floor or mounted on ball-bearing rack slides for simplified access), in a modular instrument case designed to plug into slots in a larger rack-mounted "bin," "cage," or "crate" (the latter usually provide dc power connections through standardized connectors at the rear), or perhaps in some other format such as a free-standing pedestal-mounted case.

There are many cabinet configurations available in both benchtop and rack-mounting formats. Among the most popular are cabinets 17 inches wide, available in various heights (always multiples of 1.75 inch) and depths, that accept optional rack-mounting flanges or slides (a rack 19 inches wide has about 17 1/2 inches of clearance between the flanges). That way you can convert an instrument from rack-mounted to benchtop format, or vice versa, by just changing a bit of cabinet hardware.

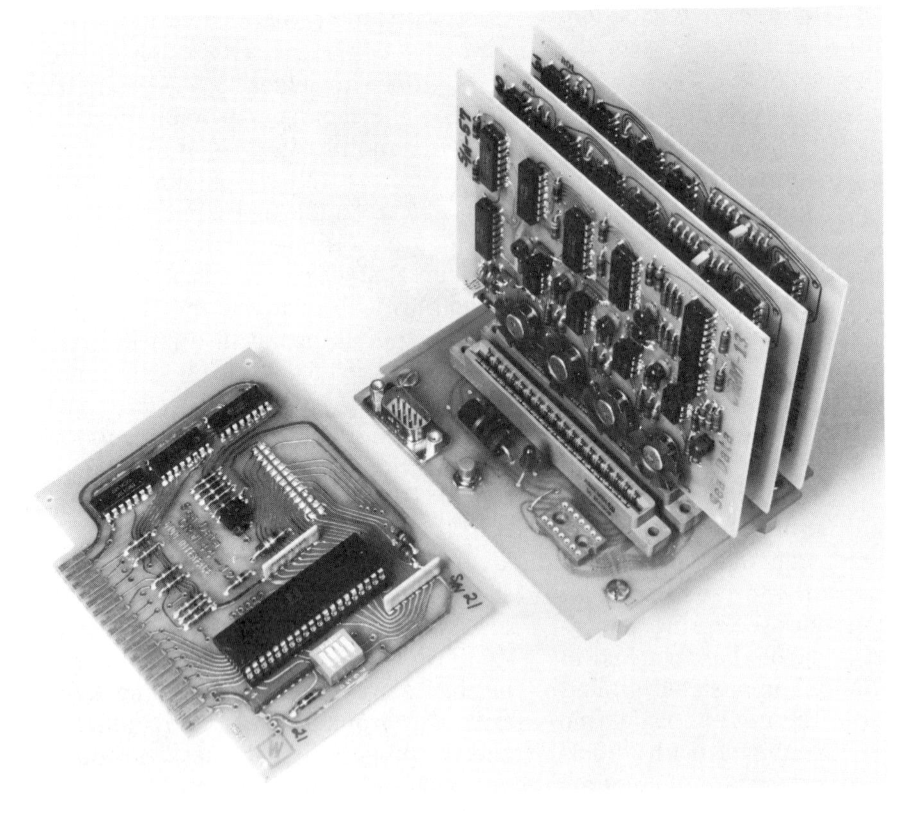

Figure 12.14. A "motherboard" provides a convenient method of interboard connection, reducing hand wiring and the possibilities for error considerably, while simultaneously providing superior electrical performance. In large systems, the motherboard and its connectors would probably be mounted rigidly at the rear of a card cage.

A point to check: Some of these convertible cabinets require removal of the outer case for rack mounting, whereas others let you keep the cabinet intact.

In the category of modular instrumentation, the NIM bin is popular in nuclear and atomic measurements, the CAMAC crate has established itself in some computer interfacing, and several manufacturers have defined modules and bins (e.g., the TM500 series from Tektronix and the EFP series of blank modules from Vector). Blank chassis are available in each of these formats, complete with rear connectors to mate with the dc power receptacle of the mother bin.

12.11 Construction hints

Rather than attempt to list the enormous variety of manufactured cabinets by name or style, we will simply offer some general comments on construction of instruments. These suggestions, together with the figures in this chapter, should help you choose electronics enclosures wisely and fill them up with circuitry in a sensible way.

In general, you use the front panel for indicators, meters, displays, keypads, etc., as well as controls and frequently used connectors. It is common to put seldom-used adjustments and connectors that don't require frequent access on the rear panel, along with large connectors, line cord, fuses, etc. (Fig. 12.15).

The traditional way to lay out a professional-looking front panel is by silkscreening the legends onto the painted or anodized aluminum surface. Although the result doesn't look bad, it tends to erode with frequent rough handling (although a clear overcoat can help considerably). If you look at a recent Fluke, Tektronix, or Hewlett-Packard instrument, you'll see the modern alternative – an adhesive Lexan panel of pleasing appearance and unbelievable toughness. These are made by silkscreening the legends on the *back* side of a matte-textured 0.010 inch Lexan film, to which a tough adhesive is then applied. You can apply lettering and patterns in several colors, and you can specify colored plastic windows or cutouts. You can get these custom laminated panel coverings in quantity from many labeling services; you just provide them with the "artwork" (usually an actual-size positive or negative). You might prepare the artwork with dry-transfer rub-on lettering or, better yet, use some desktop publishing software and a laser printer.

Perhaps the most important thing to remember when laying out an instrument is the need for good accessibility to circuit cards and controls. It should be possible to replace any component in the instrument without great pain. This means neat cabling of wiring, so that modular units can be raised up without having to unsolder anything, and careful planning, so that circuit cards can be tested while operating in the instrument. For instance, a card cage might be mounted with the cards vertical; to get at them, you remove the top panel from the cabinet, then plug in extender cards to make the circuit cards accessible. If the cards are mounted horizontally, you might make the front panel removable, or hinged, to provide access. At all costs, fight the temptation to lay down the circuit in "layers," with circuitry nicely covering other circuitry. Figure 12.16 shows an example of neat and accessible front-panel cabling built in a cabinet with a removable front panel.

12.12 Cooling

Instruments that consume more than a few watts will usually require some sort of forced air cooling. As a rule of thumb, a small instrument running more than 10 watts, or a larger (rack-width) instrument consuming more than about 25 watts, will probably benefit from a blower. It is important to keep in mind that a box full of electronics may run at

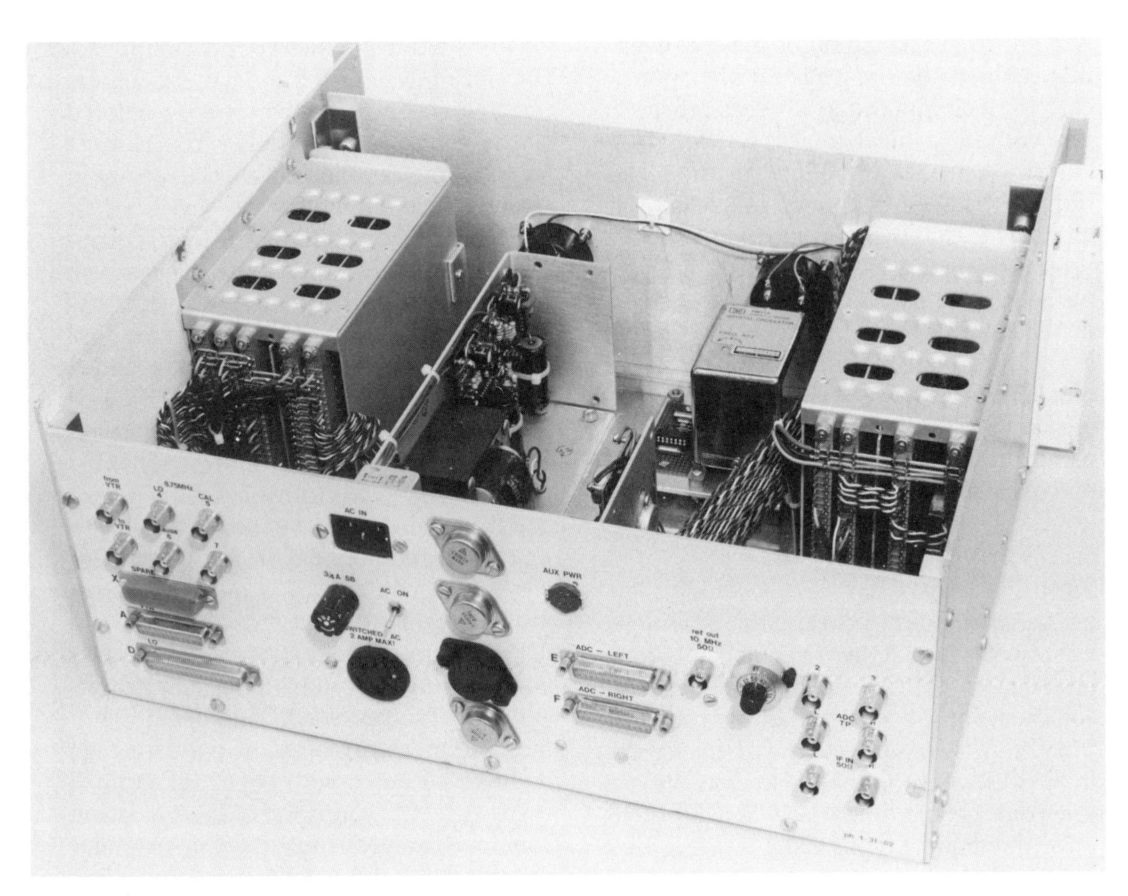

Figure 12.15. In this instrument we used the rear panel for heat-sinking power components, and for seldom-changed controls and connectors. The front panel detaches for access to the nine circuit boards housed in card cages.

a nice temperature when sitting on the bench with the top cover removed, but when installed in a rack with other heat-producing equipment (where the ambient temperature may reach 50°C), complete with its outer cover, it is likely to run very hot, leading to early failure of components and generally unsatisfactory operation.

Instruments running at moderately low power, say at the figures mentioned earlier, can often be cooled adequately with simple convective cooling. In such cases you might perforate the top and bottom covers, paying attention to the location of major heat-producing components (power resistors and transistors). It may be best to mount the high-power components on the rear panel, using heat sinks with their fins aligned vertically (see Section 6.04). Circuit boards will also be better ventilated if mounted vertically, although heat dissipation in circuit cards is often negligible. If simple convective cooling doesn't keep things cool enough, you have to resort to a blower.

The simple "Muffin-type" venturi instrument blower, with flow rate of about 100 cubic feet per minute (CFM) assuming relatively unimpeded airflow, will adequately cool instruments running 100 watts or more. Here's the relevant formula:

$$\text{air temperature rise}(^\circ\text{C}) = \frac{1.6 \times P(\text{watts})}{\text{airflow(CFM)}}$$

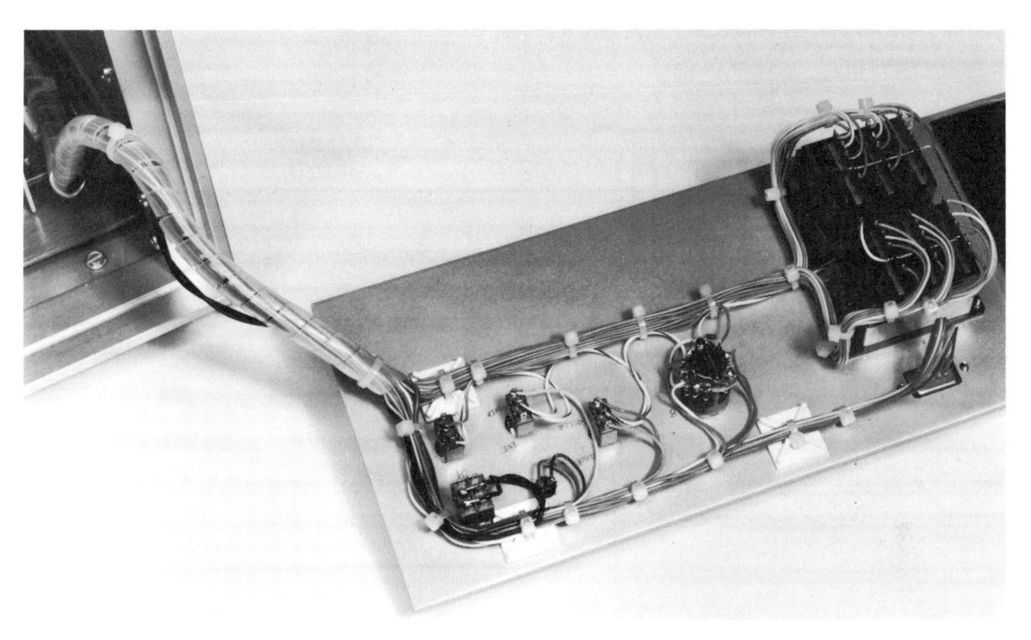

Figure 12.16. One way to ensure good accessibility to panel controls is to bring all wiring away at one end, so the panel can be hinged or otherwise detached from the instrument. In this example the panel slides into a slotted instrument case. Note the use of cable "ties" and self-adhesive supports to keep the wiring tidy.

If less airflow will suffice, a quieter version of the venturi fan is available from most manufacturers. Table 12.2 shows some part numbers. The airflow of these fans is greatly reduced when operating against high back-pressure. Figure 12.17 shows a graph.

All the blower manufacturers now sell brushless dc fans, in addition to the traditional line-voltage-operated ac fans. Running the fan from low-voltage dc (usually 12V or 24V) makes it easy to control the fan speed electronically, in response to the actual temperature inside the instrument. You can either do the temperature sensing and speed-control feedback yourself (perhaps using an outboard module like the "SmartFan" controllers made by Control Resources) or buy a fan with built-in "thermal speed control" (e.g., the Rotron "ThermaPro-V" series). Either way, using a variable-speed blower makes good sense, because under most conditions the fan noise will be far less than the worst-case (full-speed) condition, which is needed only when operating an instrument at its maximum ambient temperature.

When laying out an instrument designed for forced air cooling, try to arrange things so that the air enters the box at one end, flows around the components, and exits at the far end. In an instrument with an interior horizontal chassis partition, for example, you might punch some inlet perforations at the bottom rear, perforate the internal chassis near the front of the instrument, and mount the exhaust fan at the top rear, thus forcing the airflow to pass through all parts of the instrument. Keep in mind that a circuit board will block airflow, and plan accordingly. If there is significant impedance to the flow of air (high back-pressure), a centrifugal blower will work better than the propeller type. The blades of the latter go into "stall" when the back-pressure exceeds about 0.3 inch of water, rendering the fan totally ineffective. Finally, in any

TABLE 12.2. VENTURI FANS

Manuf	Standard 4.7" square 105–120 CFM	Quiet 4.7" square 70 CFM	Very quiet 4.7" square 50 CFM	Mini 3.1" square 35 CFM
Rotron	MU2A1	WR2H1	WR2A1	SU2A1
IMC	4715FS-12T-B50	4715FS-12T-B20	4715FS-12T-B00	3115FS-12T-B30
Pamotor	4600X	–	4800X	8500D
Torin	A30108	A30390	A30769	A30473

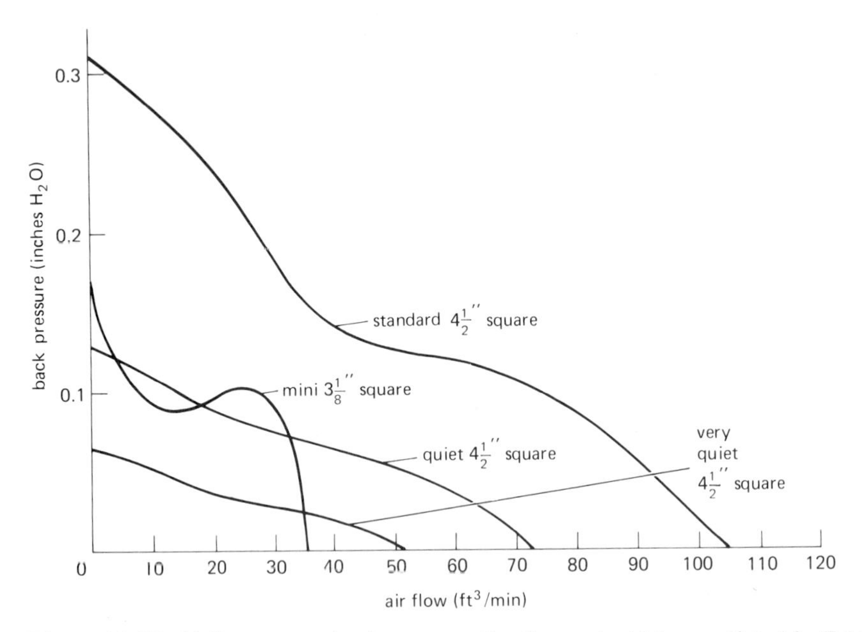

Figure 12.17. Airflow versus back-pressure for the venturi blowers listed in Table 12.2.

cooling situation it is a good idea to design conservatively; failure rates for electronic instruments rise dramatically when equipment is operated hot. Figure 12.18 illustrates good instrument design in regard to cooling and accessibility.

12.13 Some electrical hints

Unreliable components

The most unreliable components in any electronic system will be the following (worst first):

1. Connectors and cables
2. Switches
3. Potentiometers and trimmers

Keep them in mind as your brainchild proliferates in complexity.

RF line filters

As we mentioned earlier, it is a good idea to use RF filters on the ac power-line inputs. These are manufactured by a number of companies, including Corcom, Cornell-Dubilier, and Sprague. They are

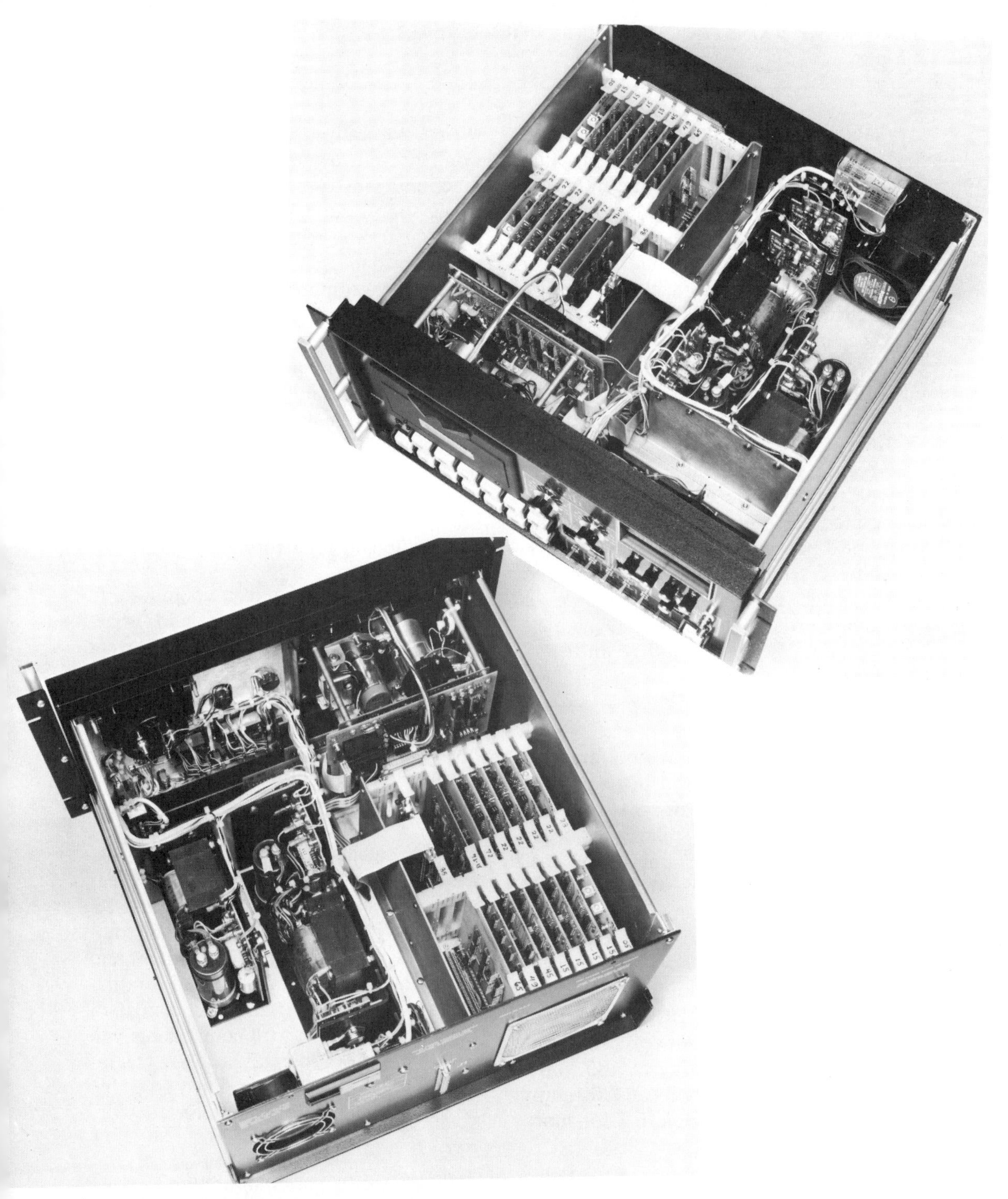

Figure 12.18. Views of a complete instrument (a digital cassette tape reader) illustrating several techniques of support and interconnection. Most of the electronics are housed in a card cage (with hand-wired backplane and mass-termination connections), and the electronics associated with the tape drive are on two boards near the drive motor (with mass-termination, in-line, and DIP plug connections). Adjustments and test points are accessible near the edges of each circuit card. Note the cooling path: Air is sucked in behind the card cage; it flows between the cards, then around the central partition and back over the power supplies before being blown out by the exhaust fan at the right rear.

available as simple modules with solder-lug terminals or in configurations that include an integral chassis-mounting ac line plug to mate with the standard IEC line cord. These filters provide excellent rejection of RF signals on the power line (as well as preventing their emission by the instrument itself), and they are also partially effective in reducing line transients. As an example, the Corcom 3R1 filter (rated at 3A, 115V) has 50dB rejection of RF at 200kHz and more than 70dB rejection for signals above 0.5MHz (see Section 6.11 and Table 6.3).

We favor the use of complete "power entry modules," rear-mounted units that combine 3-prong IEC power connector, RF line filter, fuse(s), line-voltage selector, and power switch. Look in the catalog pages of Corcom, Curtis, Delta, Power Dynamics, and Schaffner for representative units.

Transient suppressors

Power-line transient suppressors are also a good idea in any instrument to prevent malfunction (or even damage) from the occasional 1kV to 5kV spikes that occur on everyone's ac power line. You just put them across the power-line terminals downstream of the fuse; they act like bidirectional zeners with enormous peak current capability. They come in packages that look like disc ceramic capacitors or power diodes; the inexpensive and small GE V130LA10A, for instance, costs about a dollar, begins conducting at 185 volts, and can handle peak currents of 4000 amps (see Section 6.11 and Table 6.2 for more details).

Fusing

A line fuse is *mandatory* in every line-operated electronic instrument, without exception. As we indicated in Section 6.11, the wall socket is fused at a current designed to prevent fire hazard in the wall wiring, typically 15 or 20 amps. That won't prevent a major disaster in a malfunctioning instrument, such as one with a power-supply capacitor failure, where the instrument may begin to draw 10 amps or so (more than 1kW heating in the power transformer). Important fact (learned the hard way by the authors): The lead from the power line goes to the innermost terminal of the fuse holder, so when you insert a new fuse, you can't get your fingers onto the "hot" terminal. Use a slow-blow fuse with rating 50%–100% greater than worst-case current drain of the instrument.

Cold-switching philosophy

Whenever possible, it is a good idea to avoid running logic signals or analog waveforms to panel controls; this is to prevent cross-coupling and signal degradation that otherwise might occur. Instead, run dc control levels to the panel switches and pots, and use on-board circuitry to handle the actual signal switching, etc. This is especially important in noisy environments, or when dealing with high-speed or low-level signals, since the dc control signals can be thoroughly bypassed, whereas fast signals cannot. For example, use select gates (multiplexers) rather than routing the logic signals through a switch, and use a voltage-controlled oscillator rather than an *RC* oscillator for panel control of frequency. The few extra components you need to do things this way will buy you increased reliability and simplified assembly (no shielded cable, for example).

12.14 Where to get components

Getting the parts you need to build some piece of electronic equipment can present some real difficulties. Most of the large electronics distributors have abandoned over-the-counter sales, making it nearly impossible for the small purchaser to go down to the store and buy a few parts. Fortunately, the large (well-stocked) distributors will still take an order by telephone, with cash pickup at the "will call" counter. When playing this game, it is essential to know exactly what you want, by part number and manufacturer (for ICs, you may have to know the full part number, prefixes, suffixes, and all).

Many distributors are extremely hesitant to sell in small quantities, so you're often forced to buy at least five or ten of each item. Add to this the fact that a given distributor handles only a fraction of the brands you may need, and you're faced with a major chore. The consumer-oriented electronics stores (Radio Shack, etc.) will deal in small quantities and do have counter sales, but they tend to stock an extremely limited range of parts. The parts distribution system seems to be aimed at the industrial user, with his quantity ordering. Electronics manufacturers are treated well by the distributors, who pay them frequent visits, give them handsome catalogs and data books, and engage in competitive pricing.

Some special cautions are in order for buying ICs. Many kinds of ICs are manufactured without 100% testing; instead, a sample of each batch is tested, with the whole batch being rejected if the sample shows excessive failure rate. As a result, you can, on occasion, get a perfectly worthless chip straight from a reputable manufacturer's production line. As a rough guide, you can expect up to 0.1% of new chips to be defective. That's not too serious, and in any case you can have your chips tested if a lower reject rate is necessary. Furthermore, all manufacturers test their LSI chips, and some manufacturers (AMD, for example) perform 100% testing of all their ICs.

A more serious problem arises when those rejected batches find their way into the hands of a "relabeler," a kind term for a junk peddler. Labeling machines are inexpensive; consequently, "counterfeit ICs" are all too common. Our experience has been that the large distributors (Arrow, Hamilton/Avnet, Newark, Schweber, and Wyle, to name a few) are reliable, at least for the brands for which they are authorized distributors. Surprisingly, most mail-order houses seem to ship good merchandise, often at very good prices, but there is an element of risk involved. Be suspicious of any IC without a date code. Because of the extra time and annoyance involved in finding bad ICs in a circuit, we generally recommend that you play it safe by using regular distributors for all IC buying, in spite of the generally higher prices. However, two mail-order houses that we have used successfully are Digi-Key (full-range catalog; Thief River Falls, MN) and Microprocessors Unlimited (computer chips; Beggs, OK).

HIGH FREQUENCY AND HIGH-SPEED TECHNIQUES

CHAPTER 13

HIGH-FREQUENCY AMPLIFIERS

In this chapter we will discuss the important subject of high-frequency and radio-frequency techniques, as well as the digital equivalent, high-speed switching. High-frequency techniques find wide application in communications and broadcasting and in the domain of radiofrequency laboratory measurements (resonance, plasmas, particle accelerators, etc.), whereas high-speed switching techniques are essential for the fast digital instrumentation used in computers and other digital applications. High-frequency and high-speed techniques are extensions of our ordinary linear and digital techniques into the domain where the effects of interelectrode capacitance, wiring inductance, stored charge, and short wavelength begin to dominate circuit behavior. As a result, circuit techniques depart radically from those used at lower frequencies, with such bizarre incarnations as stripline and waveguide and devices like Gunn diodes, klystrons, and traveling-wave tubes. To give an idea of what is possible, there are now commercially available digital ICs (counters, etc.) that operate at pulse rates of 3GHz and higher and linear circuit elements (amplifiers, etc.) that operate at frequencies in excess of 100GHz.

We will begin with a discussion of high-frequency transistor amplifiers, complete with simple transistor and FET models. After a few examples, we will move to the important subject of radiofrequency techniques, followed by a discussion of communications concepts and methods, including modulation and detection. Finally, we will look at high-speed switching techniques in some detail. Because of the specialized nature of these subjects, this chapter could be passed over in a first reading.

13.01 Transistor amplifiers at high frequencies: first look

Amplifiers of the type we discussed earlier (e.g.,common-emitter amplifiers with resistive collector load) show a rolloff of gain with increasing signal frequency, mostly owing to the effects of load capacitance and junction capacitance. Figure 13.1 shows the situation in its

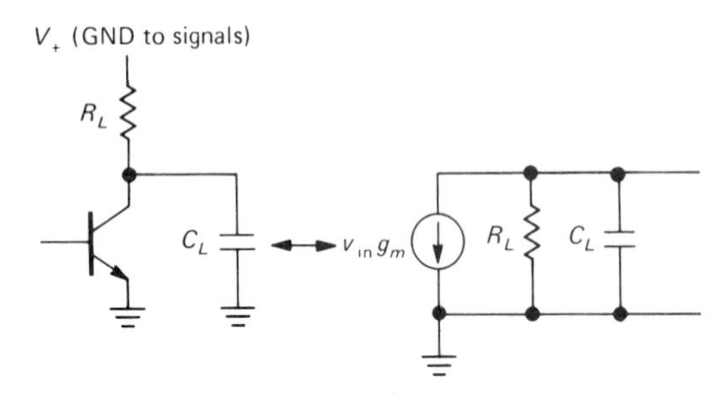

Figure 13.1

simplest form (we'll complicate things soon enough!). C_L represents the effective capacitance from collector to ground and forms a low-pass filter of time constant R_LC_L in combination with the amplifier's collector load resistance R_L. Remember that at signal frequencies V_+ is the same as ground; hence the equivalent circuit shown. C_L includes collector-to-emitter and collector-to-base capacitances, as well as load capacitance. At frequencies approaching $f \approx 1/R_LC_L$ the amplifier's gain begins dropping rapidly.

Reducing load capacitance effects

The simplest therapy consists of measures to reduce the product R_LC_L. For example:
1. Choose a transistor (or FET) with low interelectrode (junction and lead) capacitance; these are usually designated as RF or switching transistors.
.2 Isolate the load with an emitter follower, thus reducing the capacitive load seen at the collector.

3. Reduce R_L. If you keep I_C constant, the gain drops, owing to reduced g_mR_L. Remember that for a transistor, $g_m = 1/r_e$, or $I_C(mA)/25$ for an amplifier with bypassed emitter. To keep the gain constant with decreasing R_L, you have to raise the collector current by keeping V_+ constant. Thus,

$$f_{\max} \approx 1/R_LC_L \propto I_C/C_L$$

which accounts for the rather high currents often used in high-frequency circuits.

□ 13.02 High-frequency amplifiers: the ac model

Load capacitance is not the only effect reducing amplifier gain at high frequencies. As we mentioned earlier (see the discussion of Miller effect in Chapter 2), the feedback capacitance (C_{cb}) from output to input can dominate the high-frequency rolloff, especially if the input signal source impedance is not low. In order to determine where an amplifier will roll off, and what to do about it, it is necessary to introduce a relatively simple ac model of transistors and FETs. We will do that now, with a worked example of a high-frequency amplifier to illustrate how to use it.

□ ac model

The common-emitter (or source) models diagrammed in Figure 13.2 are just about

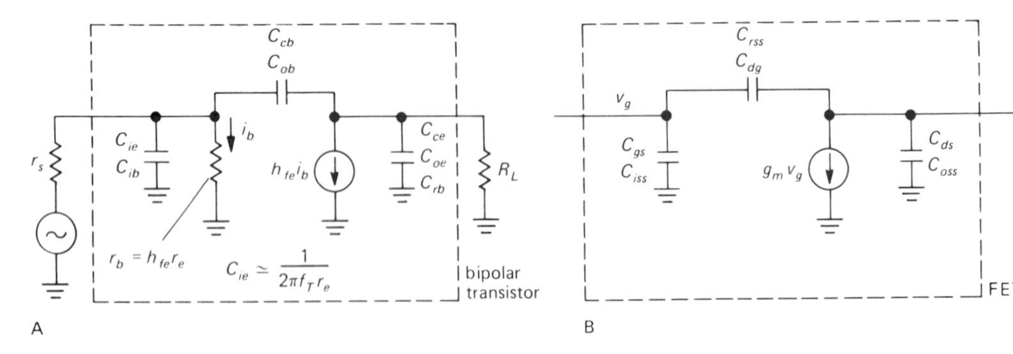

Figure 13.2. Bipolar transistor and FET high-frequency equivalent circuits.

the simplest possible; yet they are reasonably useful in estimating the performance of high-speed circuits. Both models are straightforward. In the bipolar transistor model, C_{ie} (also called C_{ib} or C_{be}; note the alternative naming of input and output capacitances) is the input junction capacitance, r_b is the impedance looking into the base, C_{cb} is the feedback (Miller) capacitance, and C_{ce} is the capacitance from collector to emitter. The current source models the transistor's gain at signal frequencies. The FET model is similar, but with different names for the capacitances and with the simplification of infinite input resistance.

☐ *Effects of collector voltage and current on transistor capacitances*

The feedback and output capacitances (C_{cb}, C_{rss}, C_{ce}) consist of a combination of the small capacitance of the transistor leads and the larger capacitances of the semiconductor junctions. The latter behave like reverse-biased diodes, with a capacitance that decreases gradually with increasing back-bias, as shown in Figure 13.3 (this effect is exploited in the voltage-variable capacitors known as "varactors"). The capacitance varies with voltage approximately as $C = k(V - V_d)^n$, where n is in the range of $-1/2$ to $-1/3$ for transistors and V_d is a "built-in" voltage of about 0.6 volt.

The input capacitance C_{ie} is different, since you're dealing with a forward-biased junction. In this case the effective capacitance rises dramatically with increasing base current, since V is near V_d, and it would make little sense to specify a value for C_{ie} on a transistor data sheet. However, it turns out that the effective C_{ie} increases with increasing I_E (and therefore decreasing r_e) in such a way that the RC product ($r_b C_{ie}$)

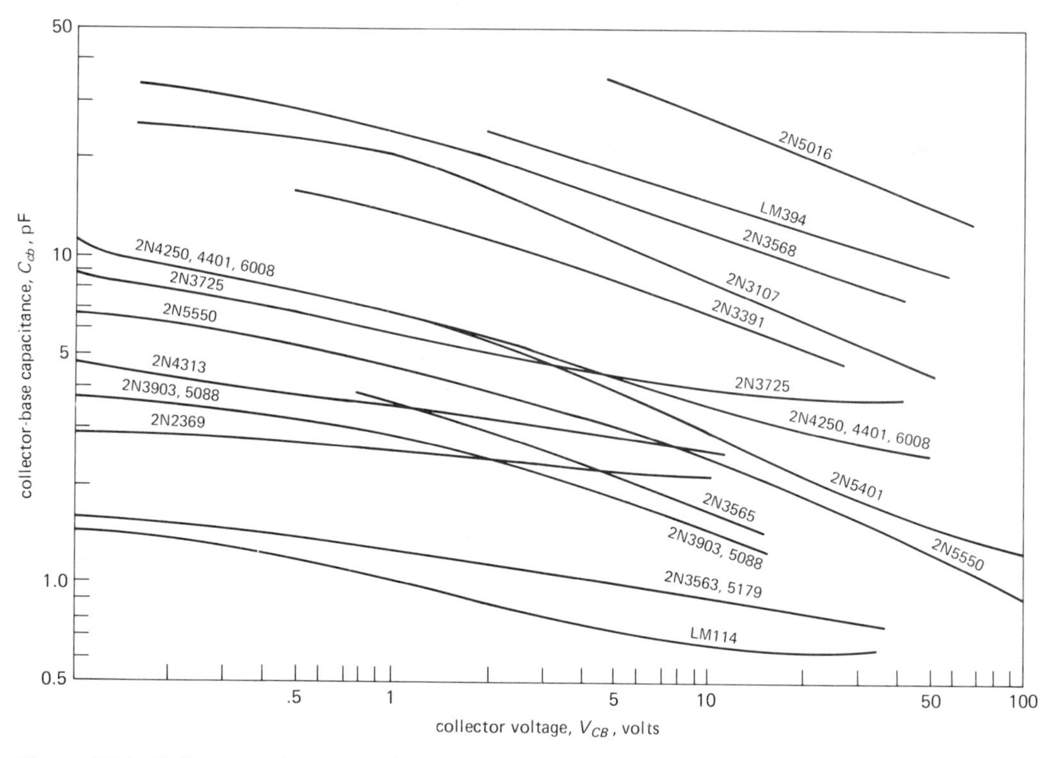

Figure 13.3. Collector-to-base capacitance versus voltage for some popular bipolar transistors.

remains roughly constant. As a result, the transistor's gain at a particular frequency depends primarily on the ratio between current lost into C_{ie} and current that actually "drives the base" and is not strongly dependent on collector current. Therefore, instead of attempting to specify C_{ie}, the transistor manufacturer usually specifies f_T, the frequency at which the current gain (h_{fe}) has dropped to unity. It is easy to show that f_T is given by

$$f_T = \frac{1}{2\pi C_{ie} r_e}$$

or, equivalently,

$$C_{ie} = \frac{1}{2\pi f_T r_e}$$

for particular values of C_{ie} and r_e at some collector current. Transistors intended for radiofrequency applications have f_T in the range of 500MHz to 10GHz whereas "general purpose" transistors have $f_T s$ in the range of 50MHz to 250MHz. Figure 13.4 shows the variation of f_T with collector current for typical transistors.

□ 13.03 A high-frequency calculation example

Let's apply our simple model to the design of a high-frequency broadband amplifier. We will show the driving stage also, so the driving (source) impedance is known. As it will turn out, the amplifier will exhibit poor performance and severe loading of the driving stage. The sort of performance problems you will see are characteristic of real-life circuit design, and we will talk about ways to improve performance by changes in circuit configuration and operating points. Figure 13.5 shows the circuit fragment. This subcircuit is assumed

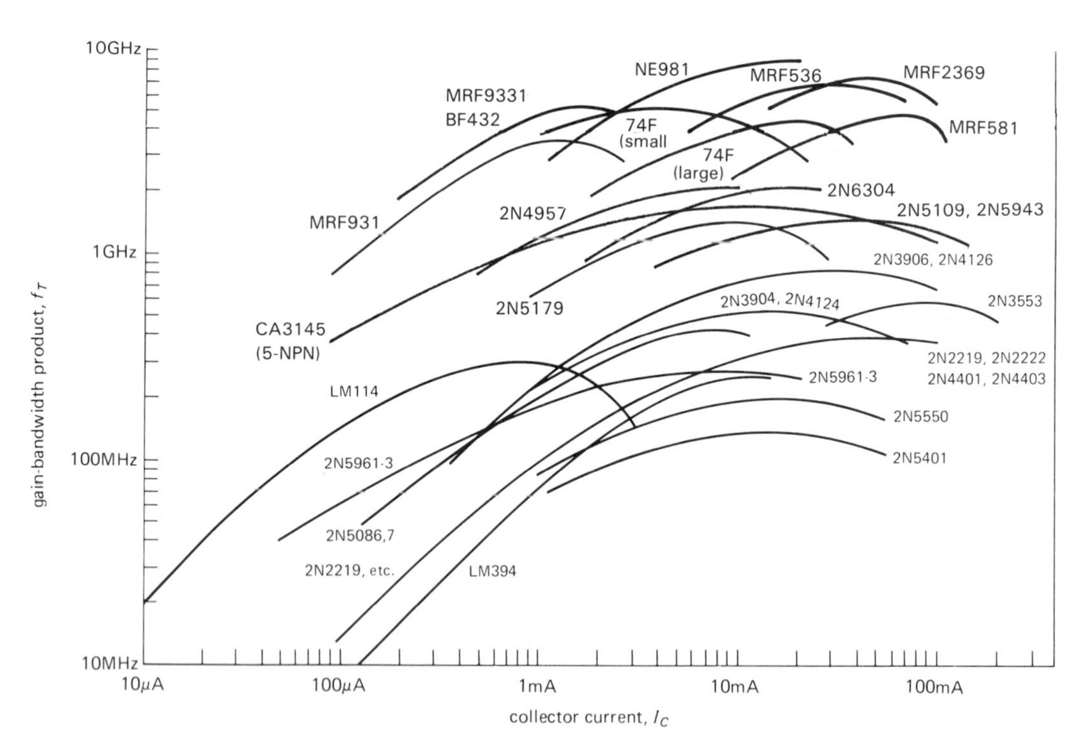

Figure 13.4. Gain-bandwidth product, f_T, versus collector current for some popular bipolar transistors.

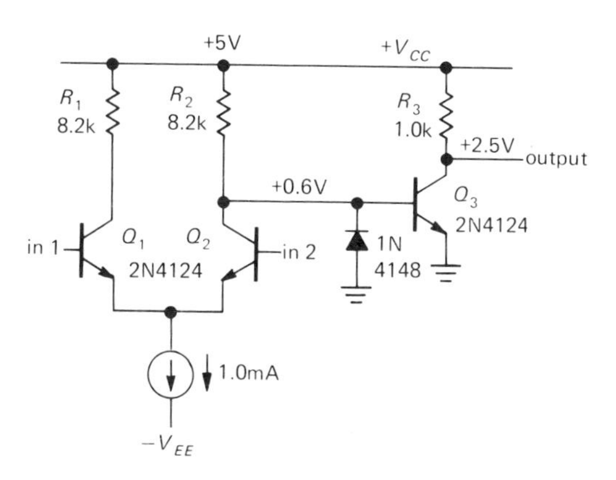

Figure 13.5

to lie within an overall amplifier circuit with feedback at dc to stabilize the quiescent point at $\frac{1}{2}V_{CC}$; it would not be biased stably as shown. Since we are interested in high-frequency performance, we won't worry further about how the biasing is accomplished. Note that the differential stage has very little common-mode input range, extending to perhaps +0.25 volt or so and limited in the negative direction by the compliance of the emitter current source.

Analysis of high-frequency rolloff

The differential stage has calculable gain and output impedance, allowing us to analyze the output-stage rolloff in detail. Our analysis of the gain of the Q_3 amplifier stage consists of the following:

1. Find the low frequency gain with zero source impedance. Then find the 3dB point, owing to output capacitance and feedback capacitance in combination with load resistance:

$$f_{-3\,\text{dB}} = \frac{1}{2\pi R_L(C_L + C_{cb})}$$

2. Find the input impedance, the combination of base input impedance (r_b and C_{ie}) and effective feedback capacitance ($G_V C_{cb}$).

3. Compute the 3dB point due to input loading of the source; compare with the "output 3dB point" calculated in step 1, to see where the high-frequency bottleneck is.

4. Improve performance, if necessary, by alleviating the problem that dominates the high-frequency rolloff.

Note that the feedback capacitance C_{cb} appears in both the output and input circuit calculations, multiplied by the voltage gain (Miller effect).

Let's try our simple method of analysis on this circuit, modeled in Figure 13.6. The 2N4124 is parametrized by C_{cb} = 2.4pF at 2.5 volts, $h_{fe} \approx 250$, and f_T =300MHz.

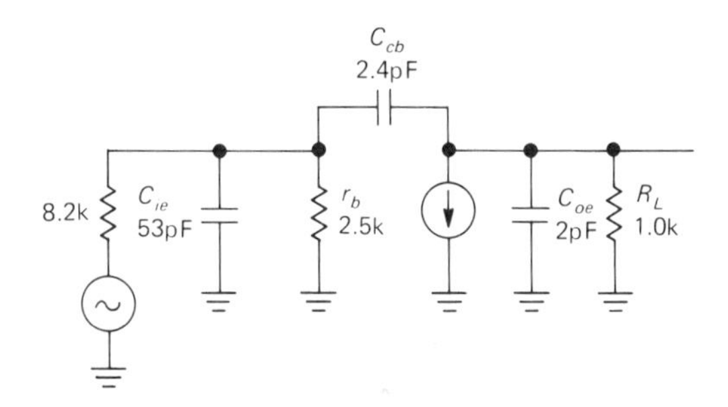

Figure 13.6

1. Assuming Q_3 is driven by a voltage source, its low-frequency voltage gain is 100, since r_e =10 ohms at 2.5mA collector current. The 3dB point set by output capacitance is at roughly 40MHz (2.4pF in parallel with 2pF, driven by 1.0k). Note that in this simple calculation we have ignored the load capacitance and stray wiring capacitance.

2. The input resistance is roughly 2.5k ($h_{fe}r_e$), paralleled by the Miller capacitance (240pF) and by C_{ie}; the latter works out to about 53pF, using the formula given earlier.

3. The 3dB point due to input capacitances comes out roughly at 280kHz (R = 8.2k paralleled with 2.5k; C = 240pF+53pF), dominated by the Miller-effect capacitance $C_{cb}G_V$ in combination with the relatively high impedance at the base. Note that the low-frequency gain is actually less than 100, considering the input signal to be the *unloaded* output if the differential stage, owing to the loading of the previous stage by the low input resistance; when you include this effect, the gain at low frequencies is actually $100\times2.5/(2.5+8.2)$, or about 23.

The combination of excessive previous stage loading and low 3dB rolloff frequency makes this a poor circuit, included here to illustrate the real-life problems of high-frequency amplifier design. In practice, you would improve performance by using greatly reduced collector impedances or by going to a different amplifier configuration intended to reduce or eliminate the effects of input capacitance (f_T) and feedback capacitance ($C_{cb}G_V$, Miller effect).

13.04 High-frequency amplifier configurations

As the preceding example illustrates, the Miller effect can dominate the high-frequency performance of an amplifier driven by moderately high source impedance. In that example, an f_T of 300MHz and an output time constant corresponding to 3dB point of 40MHz were swamped by an input time constant giving a 3dB point of 280kHz.

Three cures for Miller effect

Besides the brute-force approach of reducing collector resistances enormously, there are several interesting configurations that aim to reduce driving (source) impedance or reduce feedback capacitance or both. Figure 13.7 show these configurations, drawn in their simplest forms, without regard to bias or power supplies (i.e., the signal-frequency circuit alone is drawn).

In the first circuit, an emitter follower reduces the driving impedance seen at the input of a common emitter amplifier. This greatly reduces the degradation of high-frequency performance caused by f_T and $C_{cb}G_V$. The second circuit is the popular cascode, in which a common-emitter stage drives a common-base stage, eliminating $C_{cb}G_V$ Miller effect (Q_4's emitter is pinned by the fixed base voltage; it just passes Q_3's collector current through to R_L). In the third circuit a follower drives a common-base stage, eliminating Miller effect and reducing the driving impedance at the same time; this circuit is the familiar differential amplifier, with unbalanced collector resistors and one input grounded.

More techniques

In addition to these circuit configurations, there are two other approaches to the input and feedback capacitance problem, namely

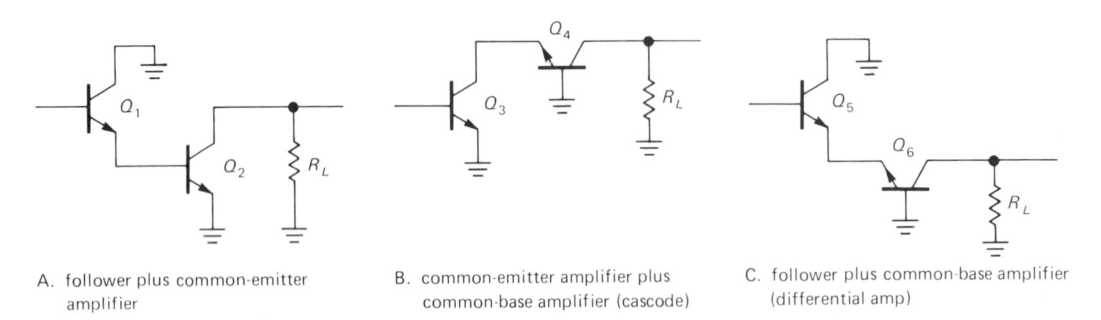

Figure 13.7. Simplified high-frequency amplifier configurations.

(a) the use of a simple grounded-base amplifier alone, if the driving impedance is low enough, and (b) the use of tuned circuits at the input and output of a common-emitter (or other) amplifier, to "tune away" the effects of interelectrode capacitance. Note that such a tuned amplifier does not have broadband response, but amplifies only a narrow range of frequencies (which may be an advantage, depending on the application). In addition, *neutralization* may be necessary. We will discuss narrowband tuned amplifiers in a later section of the chapter. An in-between approach involves the use of "peaking" inductances of a few microhenrys in series with collector load resistances to cancel some of the effects of capacitance and hold up the gain at frequencies somewhat above the normal high-frequency rolloff (Fig. 13.8).

In order to be able to estimate the high-frequency performance of circuits involving followers and grounded-base stages, we will need simple ac transistor models for these configurations (Fig. 13.9). Note that in the emitter follower model the impedances depend on source and load impedances (reactance as well as resistance). We will apply these models in the next example.

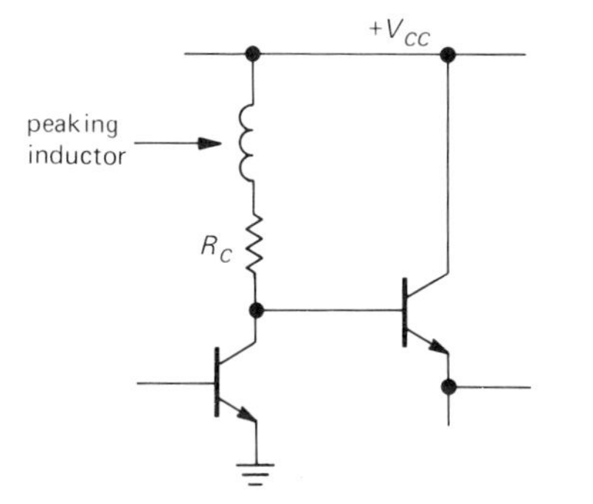

Figure 13.8

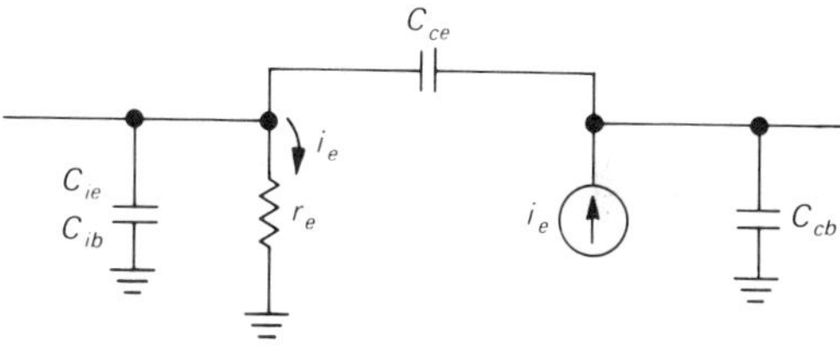

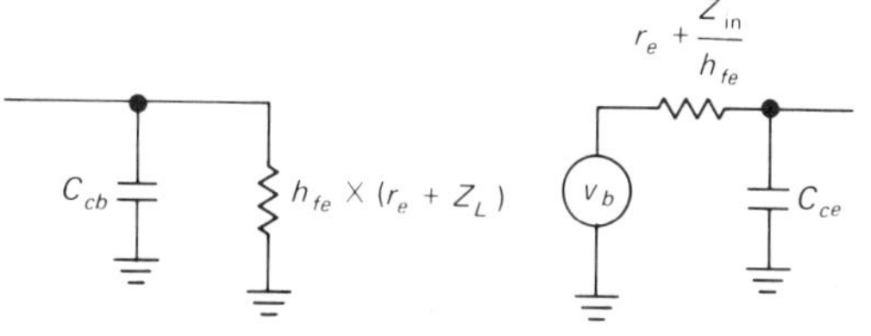

Figure 13.9

□ 13.05 A wideband design example

As an example of an improved broadband amplifier design, consider the circuit in Figure 13.10, configured to eliminate almost entirely the rolloff caused by Miller effect. It uses emitter follower inputs (high input impedance) to a differential amplifier; the output is isolated by an emitter follower from the cascode-connected output section of the differential amplifier. The design is based on the use of a good high-frequency transistor such as the 2N5179, with an f_T of 1000MHz (specified as $h_{fe} = 10$ at 100MHz) and a C_{cb} of 0.5pF at 2 volts. The approximate equivalent circuit, in terms of junction and stray capacitances and their shunt resistances, is shown in Figure 13.11.

To determine the high-frequency rolloff point of this amplifier, you have to go through each stage, analyzing the various RC's by substituting the appropriate equivalent circuits. There is usually one stage that sets the lowest limit, and with some intuition and guesswork you can often put your finger right on it. In this case the limiting performance is set by the finite driving impedance (300Ω) to Q_7's base, in combination with the capacitance of Q_7 and the load capacitance C_L as seen buffered at Q_7's base (remember that h_{fe} drops approximately as 1/f, so at very high frequencies the isolating effect of an emitter follower is seriously degraded).

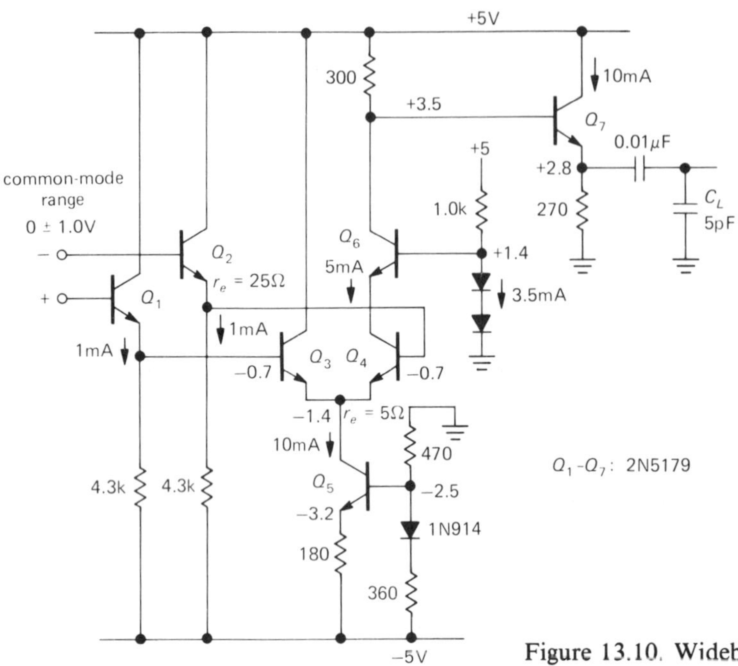

Figure 13.10. Wideband differential amplifier.

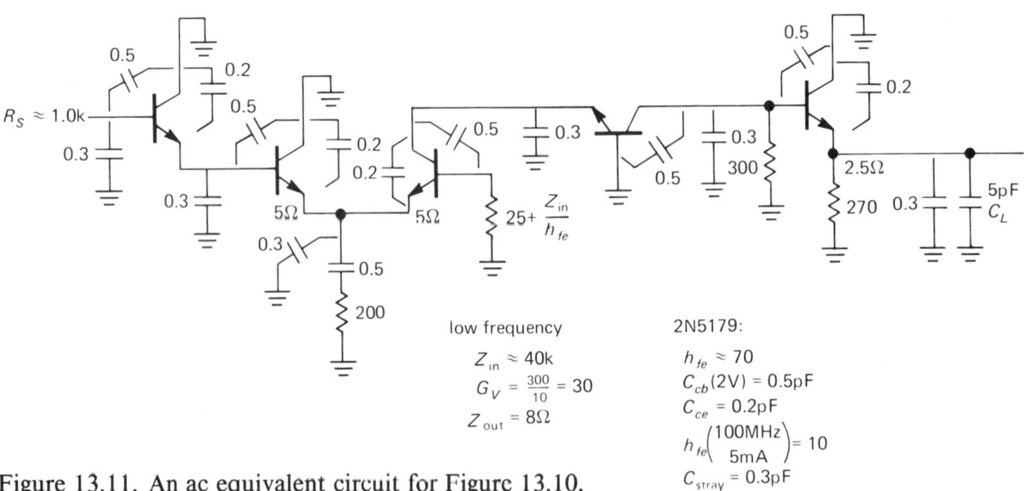

Figure 13.11. An ac equivalent circuit for Figure 13.10.

The simple method we used to figure the 3dB point goes something like this: Apply the emitter follower equivalent circuit to Q_7 to get the impedance looking into the base, knowing the load capacitance, junction capacitances, and wiring capacitance (we used $C_{cb} = 0.5$pF, $C_{ce} = 0.2$pF, and $C_{\text{stray}} = 0.3$pF). Since the impedance looking into the base depends on h_{fe} you have to do the calculation as a function of frequency (assuming $h_{fe} \approx 1/f$ at high frequency); we chose instead to do the calculation at a few high frequencies, guessing that the 3dB point would be in the neighborhood of a few hundred megahertz. Figure 13.12 summarizes the process. At frequencies of 100MHz, 200MHz, and 400MHz we took the load impedance, multiplied it by the transistor beta (of Q_7, assuming $h_{fe} \approx 1/f$), combined it with the other impedances already present at the base, and then calculated the resultant magnitude of impedance in order to get the relative output swing as a function of frequency. As can be seen, the output was down 3dB somewhere around 180MHz.

Now, using this as an estimate of the 3dB point, we went through the rest of the circuit, checking to see if other RCs gave significant attenuation at this particular frequency. As an example, Q_4's collector circuit would be down 3dB at about 1000MHz, *using the value of transistor beta at 180MHz* ($h_{fe} \approx 5$); in other words, the cascode portion of the circuit does not degrade overall performance.

In a similar manner, it is a relatively straightforward process to verify that no other portion of the circuit sets as low a 3dB point. When dealing with the input stage, you have to assume some value of driving (source) impedance. If you assume Z_s =1000 ohms (rather high for a video circuit like this), you find, finally, that the combination of source resistance in combination with input capacitance (1.0k, 0.8pF) contributes a 3dB point at about 200MHz. Thus, the overall circuit has good performance up to about 200MHz, for source impedances somewhat less than 1k, but this will be degraded for source impedances comparable to or exceeding 1k.

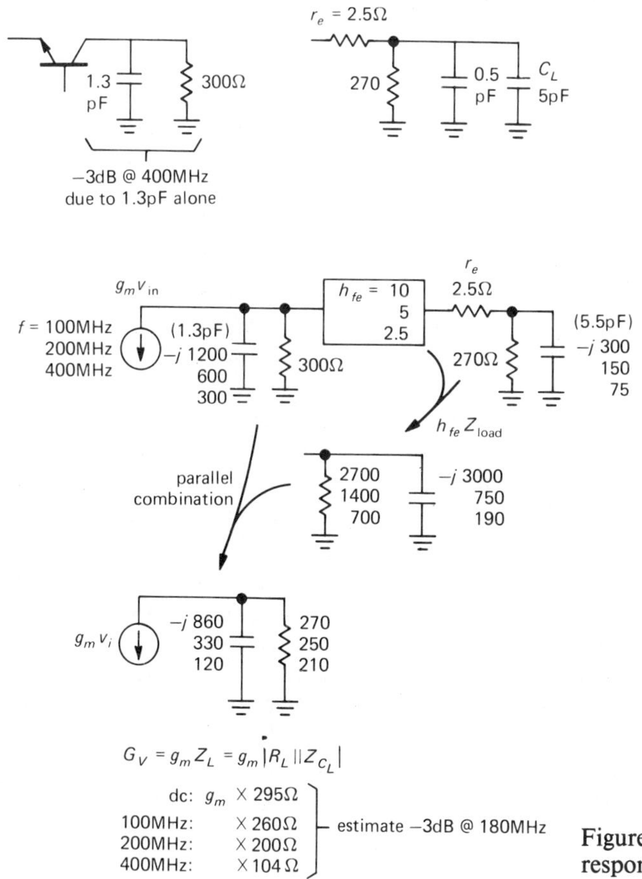

Figure 13.12. Calculating amplifier response rolloff.

This is a considerable improvement over the earlier circuit we analyzed.

□ 13.06 Some refinements to the ac model

□ *Base spreading resistance*

It is worth noting that the models we have been using are somewhat simplistic, and they neglect some important effects, e.g., the finite resistance r'_b of the base contact. Transistors intended for high-frequency use often specify the parameter $r'_b C_{cb}$, the "collector-base time constant." For the 2N5179 it is 3.5ps (typ), implying a base contact "spreading" resistance of about 7 ohms. When analyzing performance at extremely high frequencies, it is necessary to include such effects in the calculations; in this example it has no effect on the conclusions we reached earlier.

Pole splitting

Another simplification in the preceding treatment is the assumption that each RC rolloff acts independently of the others. It is intuitively easy to see that there must be some interaction, by the following argument: The Miller effect itself is a form of high-frequency negative feedback. Since it samples output *voltage*, it must therefore act to lower the output impedance of the transistor stage, particularly at higher frequencies, where its "loop gain" is high (of course, it also lowers the voltage gain, which is the whole problem). The resultant reduced impedance at the collector drives the $R_L C_L$ rolloff up to a higher frequency, since the collector impedance parallels R_L. Thus, lowering the frequency of the Miller-effect rolloff (by raising G_V or C_{cb}) raises the rolloff due to collector and load capacitance. This is known as "pole splitting."

□ 13.07 The shunt-series pair

A popular circuit for broadband low-gain amplifiers is the shunt-series pair (Fig. 13.13). The idea is to make an amplifier of relatively low gain (perhaps 10dB)

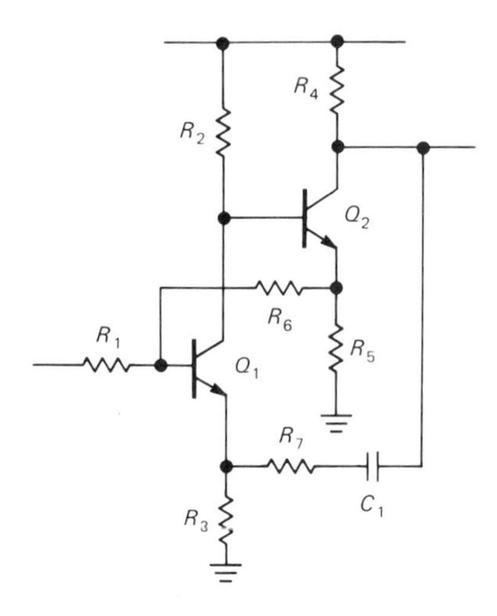

Figure 13.13. Shunt-series pair.

with flat response over a broad range of frequencies. That sounds like an application for negative feedback. However, negative feedback can be troublesome at radiofrequencies, owing to uncontrolled phase shifts in a high-loop-gain feedback path. The shunt-series pair overcomes this difficulty by having several feedback paths, each with relatively low loop gain.

In the preceding circuit, both Q_1 and Q_2 operate as low-gain voltage amplifiers, since their emitter resistors are not bypassed. R_6 provides feedback around Q_1 alone, since Q_2 is used as a follower for that loop. Once Q_1's overall voltage gain is set (R_6/R_1), R_4 is chosen to set Q_2's open-loop gain (R_4/R_5). Finally, the feedback to Q_1's emitter is added to reduce the gain to its design value.

The shunt-series pair is a convenient amplifier building block because it is extremely stable and easy to design. Amplifiers with bandwidth to 300MHz or so are easily constructed with this technique. You can get gains of 10dB to 20dB per amplifier, cascading several stages, if necessary, to obtain greater gain.

In Section 13.11 we will discuss the techniques used to construct tuned (narrowband) amplifiers, as contrasted with the broadband design we have been talking about so far. Since signals of interest are often confined to a narrow band of frequencies in instruments that operate at radiofrequencies, tuned amplifiers are extremely useful.

□ 13.08 Modular amplifiers

From the foregoing discussion of RF amplifiers it might seem that any project at high frequencies would become a formidable design effort, with messy calculations and numerous trial designs. Luckily, there are complete packaged amplifier *modules* available from more than a dozen suppliers, in configurations to meet almost any need. In fact nearly every RF component can be obtained as a module, including oscillators, mixers, modulators, voltage-controlled attenuators, power combiners and dividers, circulators, hybrids, directional couplers, etc. We will describe some of these other circuit elements in Section 13.12.

In its most basic form, the prepackaged RF amplifier comes as a thin-film hybrid circuit with gain over a wide band, packaged in a 4-pin transistor or surface-mount package (Fig. 13.14). Two of the pins are input and output terminals, with convenient 50 ohm impedance levels, and the remaining pins are for ground and the dc supply. There are dozens of different amplifiers available, some optimized for low noise and others for high power or large dynamic range. Individual amplifiers may be designed for operation over a very wide frequency range, or for a particular band of frequencies used in communications. As an example, the UTO-514 from Avantek has 15dB of gain over the frequency range of 30MHz to 200MHz, with a noise figure of 2dB (maximum) and a gain flatness of ±0.75dB. It is packaged in a 4-pin TO-8 transistor package.

The high performance Avantek UTO series and Watkins-Johnson A-series of modular amplifiers include almost a hundred models each, with bandwidths to 2GHz. We have found the less expensive Avantek GPD-series (or Watkins-Johnson EA-series) modules to be particularly affordable and handy. For example, the GPD-201 spans 5–200MHz with 30dB gain (min) and 3.5dB NF (typ), and costs $29.

These hybrid amplifiers can be used singly or in cascade, usually as part of a stripline (Section 13.21). To make life even easier, the amplifier manufacturers have thoughtfully provided completed amplifier building blocks as prepackaged modules. These beasts typically occupy a small metal box, perhaps 2 × 2 × 1 inch, with SMA-type RF coax connectors for input and output. You can choose from a list of these standard "connectored" amplifiers, or, if you prefer, you can specify a custom cascade of your choosing. Avantek will even sell you the little boxes and PC boards (that hold up to four modular amplifiers), if you want to make your own (Fig. 13.15).

To give an idea of what you can get, we have thumbed through the impressive Avantek catalog and come up with the following: The AMG-1020 is a nice low-noise amplifier with 34dB gain and a 2.7dB noise figure over the frequency range 50MHz to 1000MHz. For even wider bandwidth you might choose the UTC20-211 which spans 10MHz to 2000MHz with a 5dB noise figure and 26dB of gain. Wideband amplifiers are available to 18GHz and beyond, using GaAsFET (and HEMT) technology.

Amplifiers for use over a narrow band of frequencies can be optimized for low-noise performance; extremely good

A

B

Figure13.14. A. 10–200MHz cascadable amplifier.
B. Detail of hybrid construction technique on ceramic substrate, showing chip capacitors, thin-film inductors and resistors, transistors, and wire bonding. (Courtesy of Watkins-Johnson Company.)

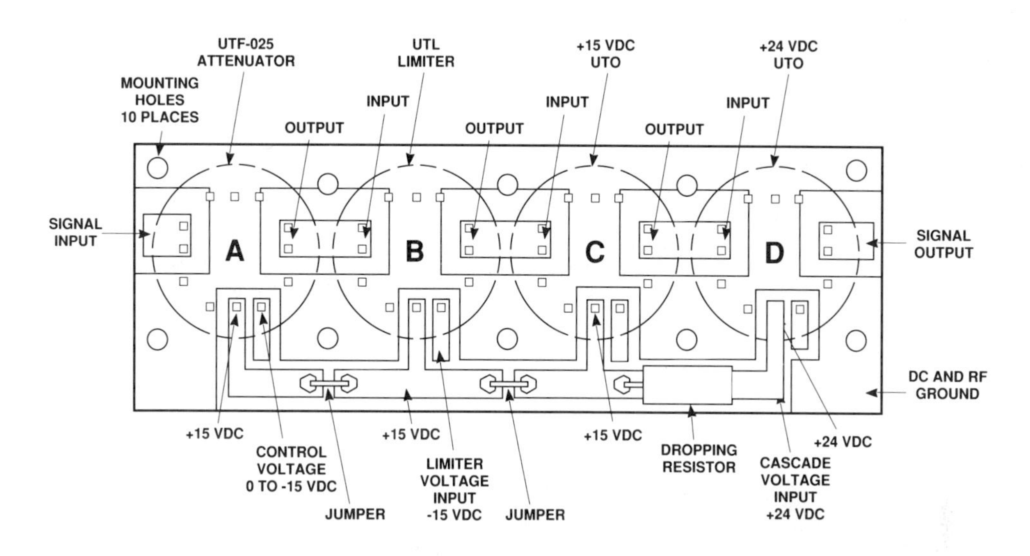

Figure 13.15. "Microstrip" (stripline) board for modular RF components. (Courtesy of Avantek, Inc.)

amplifiers are available for the communications bands. For example, for your own backyard satellite-downlink receiver, Avantek's AM-4285 has 50dB of gain (±0.5dB) in the 3.7–4.2GHz band, with an excellent 1.5dB noise figure (T_N = 120° K). In the 7.25–7.75GHz band, their AM-7724 delivers a gain of 35dB (±0.25dB) with an astounding 1.8dB noise figure.

With custom assembly and tuning it is possible to go considerably beyond the performance of these excellent commercial amplifiers. For example, radioastronomers routinely achieve noise figures of 0.7dB with custom L-band (1–2GHz) GaAs FET amplifiers, and 0.15dB noise figure when operated at liquid-nitrogen temperature (the corresponding noise temperatures are 50°K and 10°K; see Section 7.12). Recent designs employing high-electron-mobility FETs (HEMTs) have achieved a phenomenal 8°K noise temperature at 8.5GHz. The outstanding practitioner of this fine art is Sandy Weinreb at the National Radio Astronomy Observatory (Charlottesville, VA). And a small company known as Berkshire Technologies (Oakland, CA) will build you custom amplifiers with this sort of stunning performance; their cooled amplifiers are currently achieving 5°K noise temperature at L-band, and 15°K at 8.5GHz.

To measure noise figure (noise temperature) in microwave amplifiers, you invariably use the hot-load/cold-load method. Look back at Section 7.19 to remind yourself how that works.

There is plenty of commercial competition in these amplifier modules, as well as other RF modular components. For complete amplifier modules, some of the larger suppliers are Aertech/TRW, Avantek, Aydin Vector, Hewlett-Packard, Narda, Scientific Communications, and Watkins-Johnson. In practice, when designing an RF system you might well choose to thumb through catalogs of available (and custom) modules in order to assemble a system (Fig. 13.16). Screw them all down to a plate, connect them together with coaxial cable, and off you go!

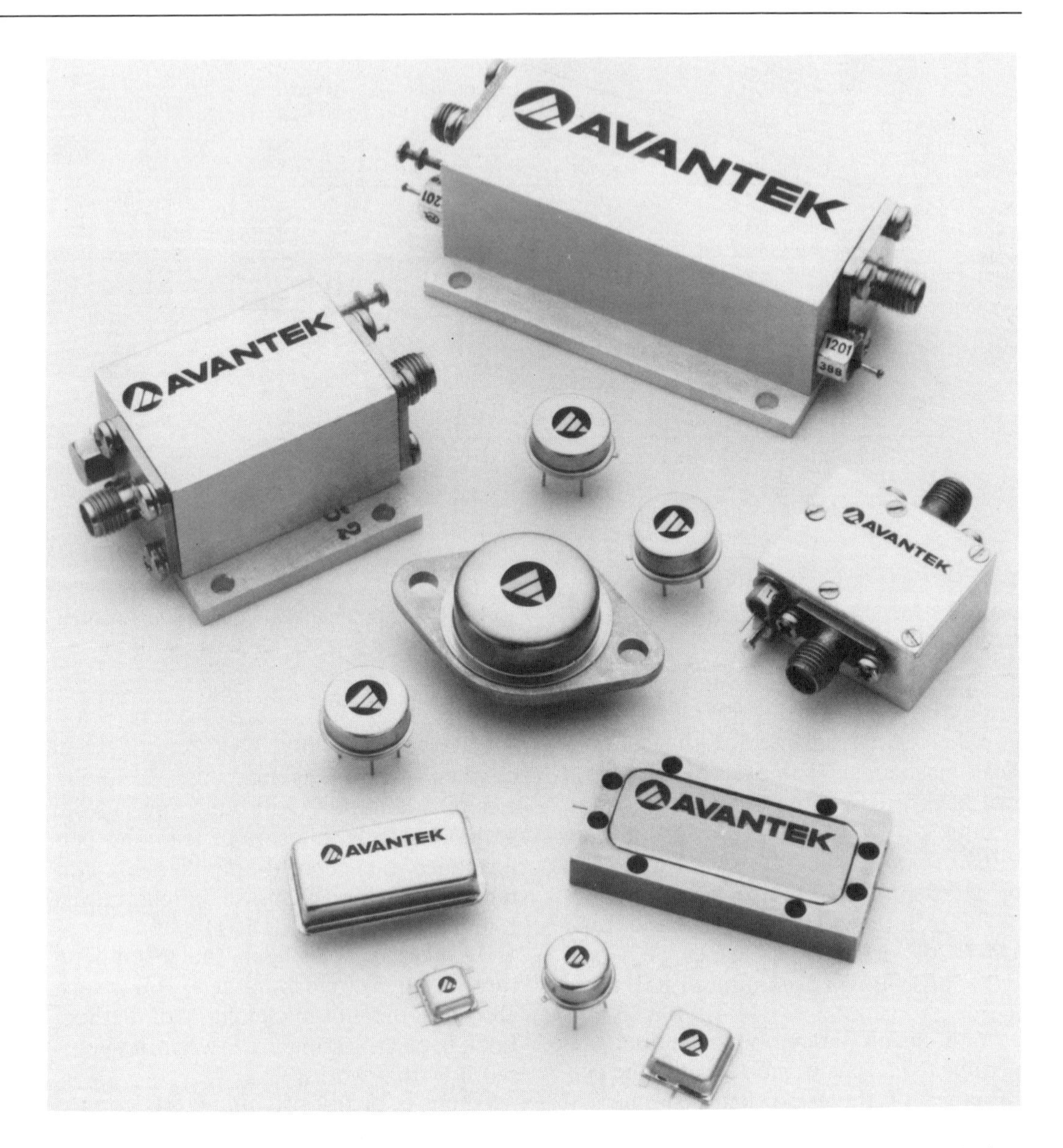

Figure 13.16. Radiofrequency modules are available with connectors, with through-board PC pins, and with surface-mount pads. (Courtesy of Avantek, Inc.)

Wide-bandwidth op-amps

If you are like most people, you probably think of op-amps as relatively low frequency amplifiers, not good for anything much above 100kHz, or perhaps 1MHz. That's certainly true for garden variety op-amps, with their f_T values typically in the range of 1MHz to 5MHz (see Table 4.1). However, as Table 7.3 shows, there is a class of accurate op-amps with gain bandwidth products up to 100MHz or so. In fact, if you can tolerate input offsets of 10mV or so, you can get op-amps with f_T values up to 1GHz. Unlike "video amplifier" ICs (which are single-ended,

TABLE 13.1. RF TRANSISTORS

Type	Case	P_{diss} T_C=25°C (W)	V_{CEO} (V)	V_{CBO}[a] (V)	I_C max (A)	h_{FE} typ	@ I_C (mA)	C_{cb} @10V (pF)	f_T (MHz)	@ I_C (mA)	Power gain (dB)	@ f (MHz)	Output power (W)	@ f (MHz)	Comments
2N3375	TO-60	12	40	65	1.5	10[m]	250	12	550	120	6	175	10	100	low cost, easy to mount
2N3553	TO-39	7	40	65	1.0	10[m]	250	7	600	100	15	175	7	100	low cost, popular
2N3866	TO-39	5	30	55	0.4	50	50	3	900	40	15	400	1[m]	400	oscillator
2N4427	TO-39	2	20	40	0.4	50	100	4	800	50	15	175	1[m]	175	
2N5016	TO-60	30	30	65	4.5	10[m]	500	25	600	500	8	400	30	100	
2N5109	TO-39	2.5	20	40	0.4	60	50	3	1500	70	15	200	–	–	low noise, popular
2N5179	TO-72	12	20	20	0.05	70	3	0.7	1500	10	20	200	0.02	500	low noise, $r'_b C_c$=7ps
2N5994	strip	35	30	65	5	–	–	70	–	–	10	100	35	175	VHF power
2N6267	strip	20	50	50	1.5	–	–	13	–	–	10	2000	10	2000	microwave power
2N6603	strip	0.5	15	25	0.03	80	15	0.5	5500	10	16	1000	–	–	microwave small signal
2N6679	strip	0.9	20	30	0.07	100	15	0.3	–	–	11	4000	0.07	4000	
MRF931	strip	0.05	5	10	0.005	70	0.25	0.4[b]	3000	1	12	1000	–	–	battery-powered, telemetry
NE981	strip	0.3	6	10	0.03	100	10	0.4	7000	20	14	1000	0.01	7000	NEC
MRF571	strip	1	10	20	0.07	100	30	0.7	8000	50	15	1000	0.1	1000	low noise, low cost
MRF941	strip	0.4	10	20	0.05	100	5	0.2	8000	15	13	2000	–	–	low noise
MRF951	strip	1	10	20	0.1	100	5	0.3	7500	30	13	2000	–	–	low noise
MRF9331	strip	0.05	8	15	0.001	80	0.5	0.2[b]	5000	1	20	1000	–	–	micropower
AT41485	strip	0.5	12	20	0.06	150	10	0.2	8000	25	12	2000	0.1	2000	Avantek, low noise
AT42085	strip	0.5	12	20	0.08	150	35	0.3	8000	35	10	4000	0.1	4000	Avantek, low noise
AT64020	strip	3	20	40	0.2	50	100	–	–	–	10	2000	0.5	4000	Avantek

(a) because the base is reverse-biased when the collector tuned circuit goes high, V_{CBO} is often the relevant breakdown voltage. (b) at V_{CB}=1V. (m) minimum.

TABLE 13.2. WIDEBAND OP-AMPS

Type	Mfg[a]	V_{os} max (mV)	I_{in} max (μA)	Supply: V_s max (V)	Supply: I_s typ (mA)	Curr fdbk[b]	Ext comp: Possible	Ext comp: Required	DIP pkg avail?	Input cap (pF)	e_n[c]	Bandwidth -3dB: G=1 (MHz)	Bandwidth -3dB: G=10 (MHz)	Bandwidth: large sig[d] (MHz)	Slew rate typ (V/μs)	Settling time to 1% (ns)	Small sig rise time (ns)	Output curr (mA)	V_{out} (peak) 50Ω (V)	V_{out} (peak) 150Ω (V)	# video loads[e]	Class A output?	Curr lim?[f]	Comments
AH0010	OE	20	20[t]	±18	5	–	–	–	•	–	–	30	–	24	1500	100[k]	–	100	5	10	3	–	–	buffer
CLC110	CL	8	50	±7	15	–	–	–	•	1.6	–	730	–	–	800	5	0.4	70	3	4	2	–	–	buffer
CLC400	CL	6	25	±7	15	25	–	–	•	0.5	2.7[f]	200	50	–	700	10	1.6	70	3	4	2	–	–	0.1% in 15ns
SL541	PL	5	25	+12, -6	16	–	•	–	•	–	–	150	100	–	175	50	–	6.5	–	–	–	•		
VA707	VT	6	1.1	±6	7	–	–	–	•	3	12[n]	–	20	5.6	105	150[k]	9	±50	±2.7	±4	5	–	•	decomp, G_V>12[p]
MSK737	KE	11	0.25	±18	37	–	–	–	–	–	65[g]	200	50	80[i]	2800	35	–	±120	±6	±12	10	–	–	
MSK738	KE	0.1	0.06	±18	37	–	–	–	–	3	4[g]	200	50	200[i]	3200	25	2.5	±120	±6	±12	10	–	–	precision, low I_b
MSK739	KE	0.03[t]	75pA[j]	±18	25	–	–	–	–	–	–	200	–	30	5500	8	–	±120	±6	±12	10	–	–	precision, FET
AD844A	AD	0.4	0.3	±18	7	0.3	•	–	•	2	2	67	43	10	2000	70[k]	–	±50	±2.5	±7.5	5	–	•	
AD846A	AD	0.2	0.5	±18	5	15	•	–	•	2	2	46	31	40[i]	450	50	–	50	5	10	3	–	•	0.01% in 80ns
1467	TP	0.1	25	±16	35	–	•	•	–	2	–	150	60	10	300	10	40	±10	–	–	–	–	–	precision
EL2003	EL	40	25	±15	10	–	–	–	•	–	–	110	–	10[i]	1000	–	–		6	12	2	–	•	buffer
EL2020C	EL	10	15	±18	9	40	–	–	•	–	7	60	40	30[i]	500	50	6	±30	±1.5	±5	3	–	–	good data sheet
EL2022	EL	2.5	20	±20	18	15	–	–	–	1.3	2	165	78	–	1900	22[k]	2.1	±100	±5	±11	5		E	good data sheet
SL2541	PL	10[t]	20	+12, -5	25	–	•	•	•	3.5	–	800[h]	220	40	1400	30	1.6	15	–	–	–	•	–	
CA3450	RC	15	0.35	±7	30	–	•	•	•	–	–	200	–	10[i]	400	35	–	75	±4	–	6	–	–	
AD5539S	AD	3	13	±8	14	–	•	•	•	–	4[g]	220	–	82	600	12	–	15	–	+2.3	1	•	–	
LM6364	NS	6	5	±18	5	–	–	–	•	3	8	–	20	4.5	300	100[k]	–	±30	±1.5	±5	3	–	–	decomp, G_V>5[r]
AD9610	AD	1	50	±18	21	15	–	–	–	2	0.7[o]	100	95	–	3500	18[k]	3.5	±50	±2.5	±8	5	–	–	
AD9611	AD	3	5	±6	74	5	–	–	–	3	1	280	270	210	1900	13	1.4	±50	±2.5	±3	5	–	–	
9826	OE	20	50	±15	15	–	–	–	–	2	20	200	10	30	1000	10	–	100	±2.5	±7.5	–		•	hybrid
SL9999	PL	15	18	+12, -5	35	–	•	•	•	–	–	400	200	–	1300	24	–	±50[l]	–	–	–	•	–	"A/D driver"

(a) see footnote to Table 4.1. (b) input current (μA) if current-feedback type. (c) nV/√Hz at 10kHz. (d) full-swing output. (e) number of 150Ω (double-terminated video) loads with video levels. (f) E - external. (g) at 1kHz. (h) G_V = +2, 50Ω input. (i) ±3V out. (j) typical, over full temp. (k) 0.1%. (l) programmable. (m) min/max. (n) μV rms, 10Hz to 100kHz. (o) above 5MHz. (p) VA706 for G = 1; VA708 for G > 3. (q) 0.5MHz. (r) 6361 for G=1; 6365 for G>25. (t) typical.

and operate at fixed gain), these are true op-amps (you use external feedback to determine the configuration and gain), and they are usable as closed-loop amplifiers to 100MHz or more. Many of these wide bandwidth op-amps use the vertical *pnp* process to get good performance. In a deviation from normal op-amp practice, you will often find unsymmetrical input impedances, with *current feedback*. Take a look at Table 13.2 for a sampling of these fast beasts.

RADIOFREQUENCY CIRCUIT ELEMENTS

13.09 Transmission lines

Before proceeding to the subject of communications circuits, it is necessary to deal briefly with the interesting subject of transmission lines. You have met these earlier in connection with digital signal communications in Chapter 9, where we introduced the ideas of characteristic impedance and line termination. Transmission lines play a central role in radiofrequency circuits, where they are used to pipe signals around from one place to another within a circuit, and often to an antenna system. Transmission lines provide one of the most important exceptions to the general principles (see Chapter 1) that a signal source ideally should have a source impedance small compared with the impedance of the load being driven and that the load should present an input impedance large compared with the source impedance driving it. The equivalent rule for transmission lines is that the load (and possibly the source) should present an impedance equal to the characteristic impedance of the line. The line is then "matched."

Transmission lines for signals of moderate frequency (up to 1000MHz, say) come in two major types: parallel conductors and coaxial line. The former is typified by the inexpensive molded 300 ohm "twin lead" used to bring the signal from a television antenna to the receiver, and the latter is widely used in short lengths with BNC fittings to carry signals between instruments. (Fig. 13.17).

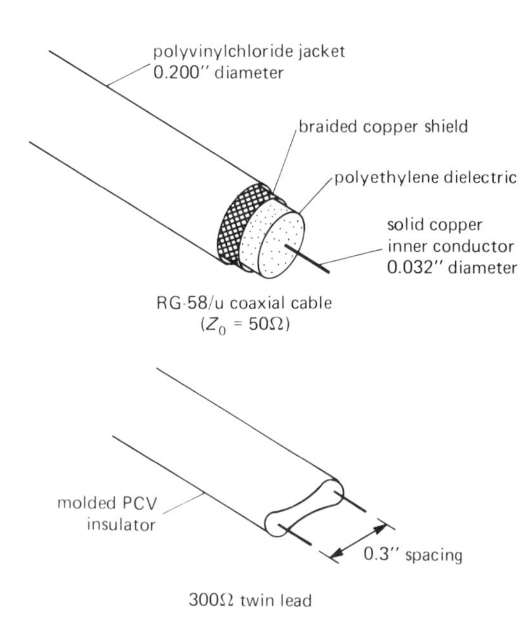

Figure 13.17

In the domain of ultra-high-frequency circuitry there are "stripline" techniques that involve parallel-conductor transmission lines as part of the actual circuit, and at the higher "microwave" frequencies (upwards of 2GHz, say) conventional lumped circuit elements and transmission lines are replaced by cavity and waveguide techniques, respectively. Except at these extremes of frequency, the familiar coaxial cable is probably the best choice for most radiofrequency applications. Compared with parallel conductor line, a properly matched coax line has the advantage of being totally shielded, i.e., there is no radiation or pickup of external signals.

Characteristic impedance and matching

A transmission line, whatever its form, has a "characteristic impedance" Z_0, meaning that a wave moving along the line has a ratio of voltage to current equal to Z_0. For a lossless line, Z_0 is resistive and equal to the square root of L/C, where L is the inductance per unit length and C is the capacitance per unit length. Typical coaxial lines have impedance in the range of 50 to 100 ohms, whereas parallel-conductor lines have impedances in the range of 300 to 1000 ohms.

When used with high-frequency (or short-rise-time) signals, it is important to "match" the load to the characteristic impedance of the line. The important facts are the following: (a) A transmission line terminated with a load equal to its characteristic impedance (resistance) will transfer an applied pulse to the termination without reflection. In that case all the power in the signal is transferred to the load. (b) The impedance looking into such a terminated line, at any frequency, is equal to its characteristic impedance (Fig. 13.18).

This is surprising at first, since at low frequencies you tend to think of a length of coax as a small capacitive load, generally a pretty high (capacitive) impedance. Also, at low frequencies (wavelength $\gg$ length of cable) there is no need to match the line's impedance, provided you can handle the capacitance (typically 30pF per foot). If the cable is terminated with a resistor, on the other hand, it magically becomes a pure resistance at all frequencies.

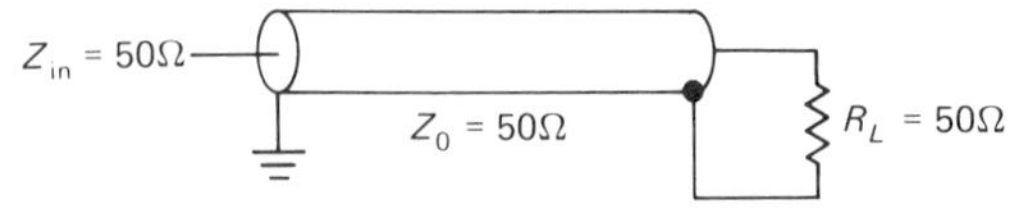

Figure 13.18

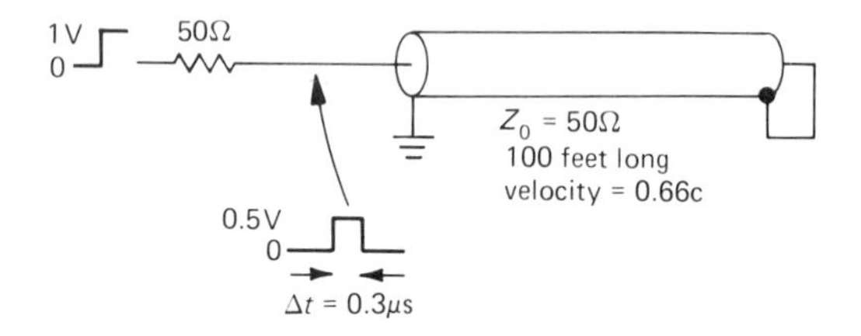

Figure 13.19. Pulse generation with shorted transmission line (inverted reflection).

Mismatched transmission lines

A mismatched transmission line has some interesting, and occasionally useful, properties. A line terminated in a short circuit produces a reflected wave of opposite polarity, with the delay time of the reflected wave determined by the electrical length of the line (the speed of wave propagation in coax lines is about two-thirds the speed of light, because of the solid dielectric spacing material). You can see the reason for this, since the short circuit enforces a point of zero voltage at the end; the cable produces this obligatory condition by creating a wave of opposite phase at the short. In similar manner, an open-circuited cable (boundary condition of zero *current* at the end) produces a noninverted reflection of amplitude equal to the applied signal.

This property of a shorted cable is sometimes exploited to generate a short pulse from a step waveform. The step input is applied to the cable input through a resistance equal to Z_0, with the other end of the cable shorted. The waveform at the input is a pulse of width equal to the round-trip travel time, since the reflected step cancels the input (Fig. 13.19).

Cables terminated with a resistance R unequal to Z_0 also produce reflections, although of lesser amplitude. The reflected wave is inverted if $R < Z_0$ and uninverted if $R > Z_0$. The ratio of reflected wave amplitude to incident wave amplitude is given by

$$A_r/A_i = (R - Z_0)/(R + Z_0)$$

Transmission lines in the frequency domain

Looked at in the frequency domain, a transmission line matched at the far end looks like a load of impedance Z_0, i.e., a pure resistance if line losses are neglected. That makes sense, because it just swallows any wave you apply, all the power going into the matching resistor. This is true independent of cable length or wavelength. It is when you deal with mismatched lines that things begin to get interesting in the frequency domain. Since, for a given line length, the reflected wave arrives back at the input with a phase (relative to the applied signal) that depends on applied frequency, the impedance seen looking into the input depends on the mismatch and on the electrical length of the transmission line, in wavelengths.

As an example, a line that is an odd number of quarter wavelengths long terminated in an impedance Z_{load} at the far end presents an input impedance $Z_{\text{in}} = Z_0^2/Z_{\text{load}}$. If the load is resistive, the input will look resistive. On the other hand, a line that is an integral number of half wavelengths long presents an input impedance equal to its terminating impedance (Fig. 13.20).

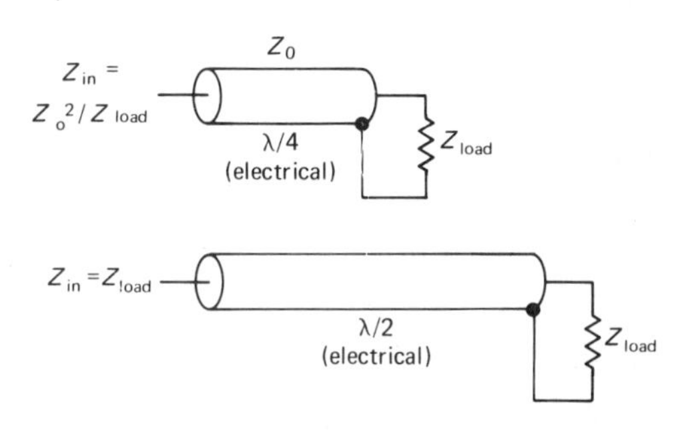

Figure 13.20

The presence of reflected signals on a transmission line is not necessarily bad. For operation at a single frequency, a mismatched line can be driven (through a line tuner) in such a way as to match its resultant input impedance, often with only negligibly greater line losses (due to higher voltages and currents for the same forward power) than with a matched load. But a mismatched line has different properties at different frequencies (the famous "Smith chart" can be used to determine transmission-line impedances and "standing-wave ratio," or SWR, a measure of the amplitude of reflected waves), making it undesirable for broadband or multifrequency use. In general, strive to terminate a transmission line in its characteristic impedance, at least at the receiving end.

13.10 Stubs, baluns, and transformers

There are some interesting applications of transmission lines that exploit the properties of mismatched sections or generally use sections of line in an unconventional way. The simplest is the quarter-wave matching section, which exploits the relationship $Z_{\text{in}} = Z_0^2/Z_{\text{load}}$. This can be rearranged to read $Z_0 = (Z_{\text{in}}Z_{\text{load}})^{1/2}$. In other words, a quarter-wave section can be used to match any two impedances by choosing the characteristic impedance of the matching section appropriately.

In a similar manner, a short length of transmission line (a "stub") can be used to "tune" a mismatched load by simply putting the stub across or in series with the mismatched line, choosing the stub length and termination (open or shorted) and its position along the mismatched line correctly. In this sort of application the stub is really functioning as a circuit element, not a transmission line. At very short wavelengths the use of sections of transmission line as circuit elements is common (Fig. 13.21).

Sections of transmission line (or a transformer made with several interconnected windings) can be used to construct a "balun," a device for matching an unbalanced line (coax) to a balanced load (e.g., an antenna). There are simple configurations for making fixed-impedance transformations at the same time (1:1 and 4:1 are common). Perhaps the nicest circuit

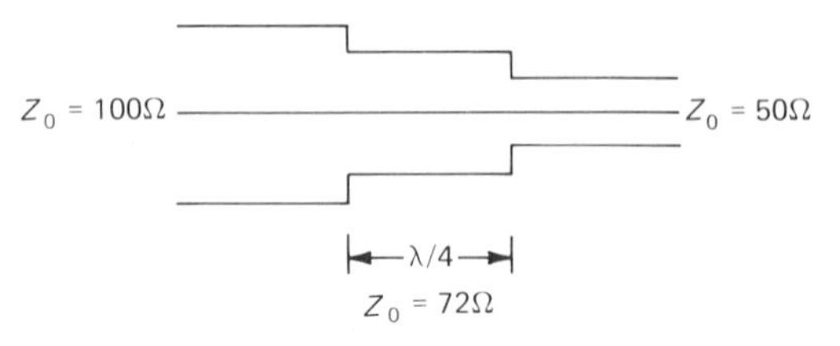

A. quarter-wave matching section

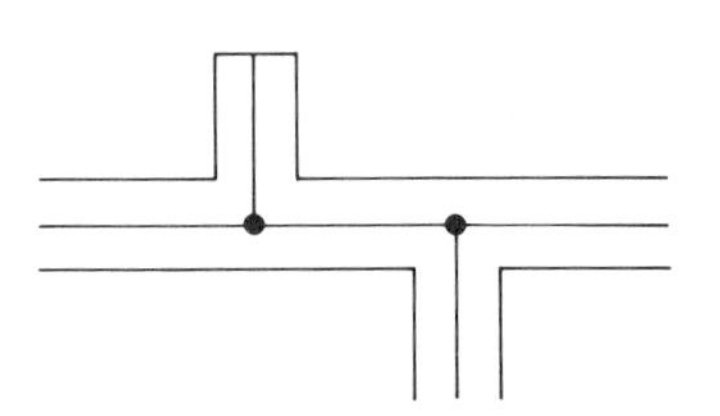

B. matching stubs

Figure 13.21

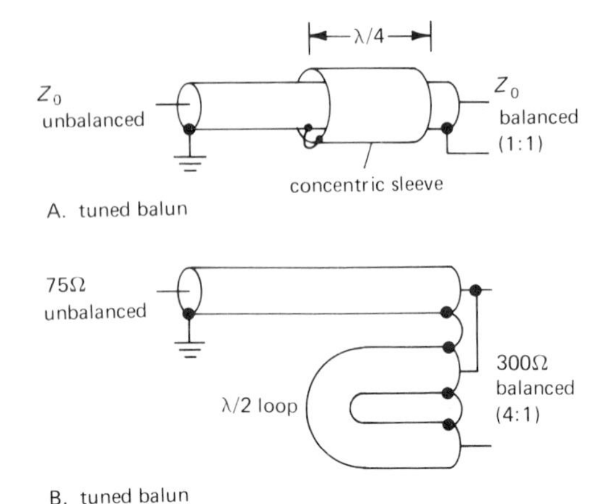

A. tuned balun

B. tuned balun

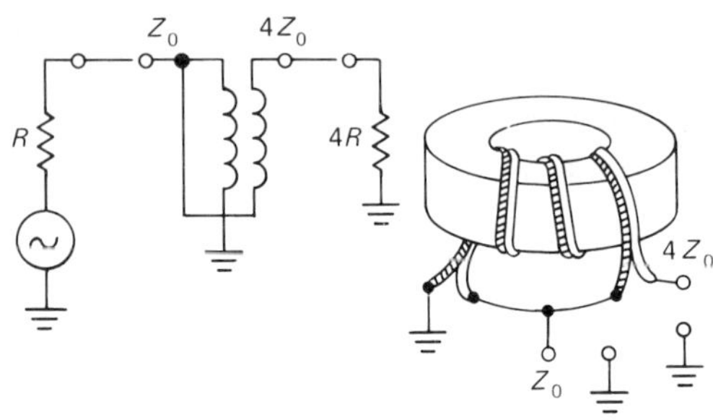

C. 4:1 unbalanced transmission-line transformer

Figure 13.22. Transmission-line transformers.

element made from transmission line is the broadband transmission-line transformer. These gadgets consist simply of a few turns of miniature coax or twisted-pair wound on a ferrite core, suitably interconnected. They avoid the high-frequency limitations of conventional transformers (caused by the resonant combination of "parasitic" winding capacitance and inductance) because the coils are arranged so that the winding capacitance and inductance form a transmission line, free of resonances. They can provide various impedance transformations with excellent broadband performance (e.g., less than 1dB loss from 0.1MHz to 500MHz), a property not shared by transformers constructed from simple coupled inductors. Transmission-line transformers are available from the Vari-L Co. and Mini-Circuits, among others, as packaged modules. Figure 13.22 shows a few examples of baluns and a transmission-line transformer.

13.11 Tuned amplifiers

In radiofrequency circuits intended for communications, or for other applications where the operating frequency is confined to a narrow range, it is common to use tuned LC circuits as collector or drain loads. This has several advantages: (a) higher single-stage gain, since the load presents a high impedance at the signal frequency ($G_V = g_m Z_{\text{load}}$) while allowing arbitrary quiescent current; (b) elimination of the undesirable loading effects of capacitance, since the LC circuit "tunes out" any capacitance by making it part of the tuned circuit capacitance; (c) simplified interstage coupling, since an LC circuit can be tapped or transformer-coupled (or even configured as a resonant matching network, as in the popular pi network) to achieve any desired impedance transformation; (d) elimination of out-of-band signals and noise owing to the frequency selectivity of the tuned circuits.

Examples of tuned RF circuits

You will see tuned RF amplifiers in their natural element when we discuss communications circuits shortly. At this point we would simply like to illustrate the use of tuned circuits in oscillators and amplifiers with a few examples. Figure 13.23 illustrates the classic tuned amplifier. A

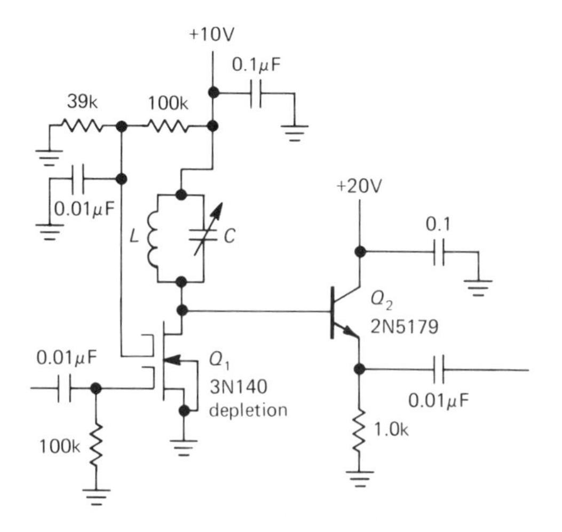

Figure 13.23. Dual-gate MOSFET (cascode) tuned amplifier.

dual-gate depletion-mode FET is used to eliminate the problems of Miller effect, since the input is untuned. By operating the lower gate at dc ground, the stage runs at I_{DSS}. The parallel tuned LC sets the center frequency of amplification, with output buffering via Q_2. Since the drain sits at +10 volts, the output follower requires a higher collector voltage. This sort of circuit has quite high voltage gain at resonance, limited by the LC circuit Q and loading by the follower.

In the circuit shown in Figure 13.24 a carefully constructed and tunable LC circuit is used to set the frequency of an oscillator. This is known as a VFO (variable-frequency oscillator); it is used as the tunable element in some transmitters and receivers, as well as variable-frequency RF signal sources. In this circuit a JFET provides the necessary power gain, with positive feedback from the source coupled into a "link" on L_1. The link has fewer turns than the inductor, providing voltage gain and therefore oscillation. By adding a varactor diode, which acts as a voltage-variable capacitor (see Fig. 5.44), you can make such an oscillator voltage-tunable. Note the use of a feed-through capacitor and decoupling RF chokes on the power-supply leads; this is nearly universal practice in radiofrequency circuits.

The circuit in Figure 13.25 is a 200MHz common-emitter transistor amplifier stage. This circuit illustrates neutralization, the technique of canceling out capacitively coupled signal from output to input by adding a current of the opposite phase. C_N is the neutralizing capacitor, driven from the bottom of the collector "tank" circuit where the phase is opposite to that at the collector. This circuit matches the output impedance to the line by tapping down on the "tank" (the collector LC circuit), a simple but inflexible method.

This last circuit (Fig. 13.26) is a 25kW RF amplifier, using a zero-bias grounded-grid triode. Vacuum tubes are still used in high-power frequency amplifiers, because no solid-state device can match their performance (for example, the 8973 power triode delivers 1.5MW at 50MHz!). The grounded-grid configuration requires no neutralization. The output circuit is the popular pi network, driven by blocking capacitor C_8. C_9, L_4, and C_{10} form the actual network, with their values determined by the desired resonant frequency, impedance transformation, and loaded Q (Q, or quality factor, is a measure of the sharpness of resonance, see Section 1.22). The RF choke at the output prevents dc voltage appearing there, and the plate RF choke is used to apply plate voltage while allowing signal swing at the operating frequency.

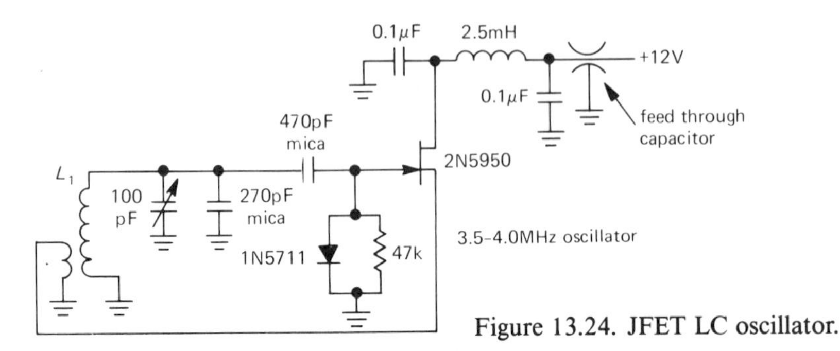

Figure 13.24. JFET LC oscillator.

13.12 Radiofrequency circuit elements

In radiofrequency circuits you meet several kinds of specialized circuit modules that have no equivalents in low-frequency circuitry. Before going on to consider RF communications circuits, we will mention some of these, since they enjoy widespread use in the generation and detection of radiofrequency signals.

Oscillators

If great stability is not important, a simple *LC* oscillator of the type just illustrated will generate a radiofrequency signal with adjustability over an octave or more by varying either *C* or *L* (the latter is somctimes called a "permeability-tuned oscillator," or PTO). With careful design and attention to detail during construction, such a VFO (variable-frequency oscillator) can have drifts of less than a few parts per million measured over hours and will be quite satisfactory for use in receivers and noncritical transmitting applications. *LC* oscillators can be operated at frequencies ranging from audiofrequencies to hundreds of megahertz.

Just as with the amplifier modules we mentioned in Section 13.08, prepackaged oscillator modules with excellent performance are readily available. Tunable oscillator modules use varactors (voltage-variable capacitors) to adjust the operating frequency of an *LC* oscillator in response to an externally applied voltage. A fancier version of tunable oscillator for gigahertz frequencies uses a yttrium/iron/garnet (YIG) sphere as a magnetically tunable resonant cavity; YIG-tuned oscillators provide high spectral purity and tuning linearity. A recent technique for making inexpensive oscillators of good stability

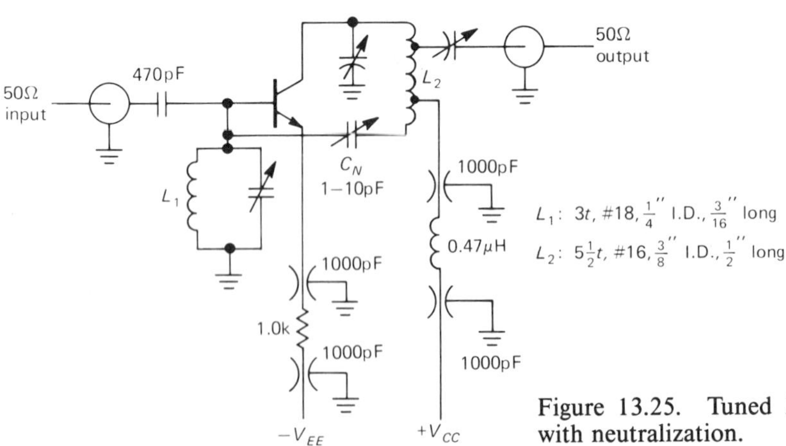

Figure 13.25. Tuned 200MHz RF amplifier with neutralization.

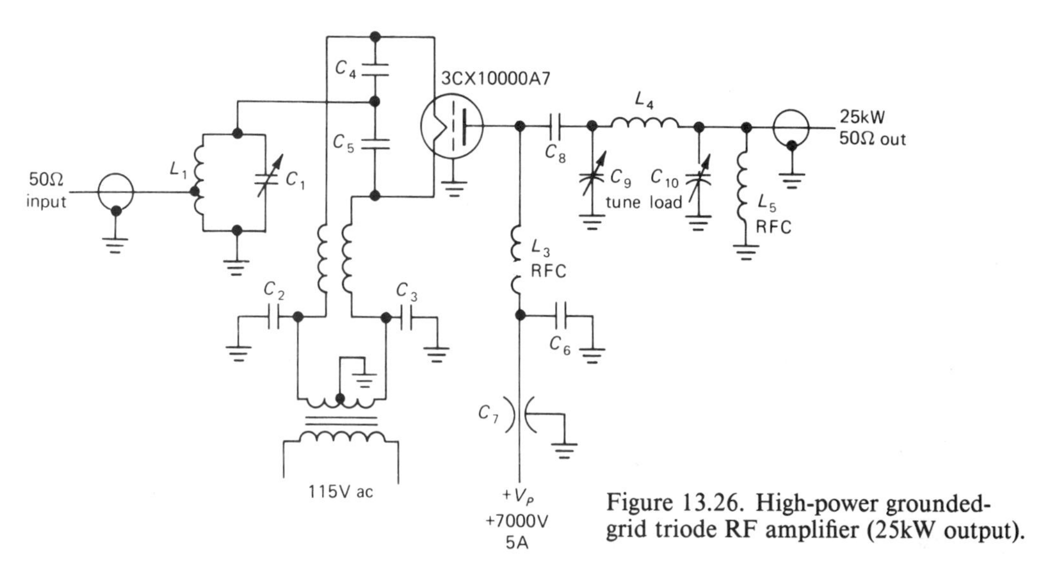

Figure 13.26. High-power grounded-grid triode RF amplifier (25kW output).

in the gigahertz range uses a dielectric "pill" resonator as the feedback element for a GaAs FET (or bipolar) oscillator. Oscillators using this "dielectrically stabilized" technique are simple and stable and have low noise.

For high stability, the best kind of oscillators use quartz crystals to set the operating frequency. With off-the-shelf garden-variety crystals, you can expect overall stabilities of a few parts per million, with tempco of order 1 ppm/degree or better. A temperature-compensated crystal oscillator (TCXO), which uses capacitors of controlled tempco to offset the crystal's frequency variation, can deliver frequency stability of 1 ppm over a temperature range of 0°C to 50°C or better. For the utmost in performance, oscillators with the crystal maintained in a constant-temperature "oven" are available, with stabilities of a few parts per billion over time and temperature. Even the so-called atomic oscillators (rubidium, cesium) actually use a high-stability quartz oscillator as the basic oscillating element, with frequency adjusted as necessary to agree with a particular atomic transition frequency.

Crystal oscillators are commercially available in frequencies ranging from about 10kHz to about 100MHz in all of the variations just mentioned. There are even little DIP and transistor can (TO-5) oscillators, with logic outputs. Only a slight electrical adjustment of frequency is possible, so the frequency must be specified when the oscillator or crystal is ordered.

To get both adjustability and high stability, a frequency synthesizer is the best choice. It uses tricks to generate any desired frequency from a single source of stable frequency, typically a 10MHz crystal oscillator. A synthesizer driven from a rubidium standard (stability of a few parts in 10^{12}) makes a nice signal source.

Mixers/modulators

A circuit that forms the product of two analog waveforms is used in a variety of radiofrequency applications and is called, variously, a modulator, mixer, synchronous detector, or phase detector. The simplest form of modulation, as you will see shortly, is amplitude modulation (AM), in which the high-frequency *carrier* signal is varied in amplitude according to a slowly varying *modulating* signal. A multiplier obviously performs the right function. Such a circuit can also be used as a variable gain control, thinking of one of the inputs as a dc voltage. There are convenient ICs to do this job, e.g., the MC1495 and MC1496.

A mixer is a circuit that accepts two signal inputs and forms an output signal at the sum and difference frequencies. From the trigonometric relationship

$$\cos\omega_1 t \cos\omega_2 t = \frac{1}{2}\cos(\omega_1+\omega_2)t + \frac{1}{2}\cos(\omega_1-\omega_2)t$$

it should be clear that a "four-quadrant multiplier," i.e., one that performs the product of two input signals of any polarity, is in fact a mixer. If you input two signals of frequency f_1 and f_2, you will get out signals at f_1+f_2 and f_1-f_2. A signal at frequency f_0 mixed with a band of signals near zero frequency (band-limited to a maximum frequency of f_{max}) will produce a symmetrical band of frequencies around f_0, extending from $f_0 - f_{max}$ to f_0+f_{max} (the spectrum of amplitude modulation, see Section 13.15).

It is not necessary to form an accurate analog product in order to mix two signals. In fact, any nonlinear combination of the two signals will produce sum and difference frequencies. Take, for instance, a "square-law" nonlinearity applied to the sum of two signals:

$$(\cos\omega_1 t + \cos\omega_2 t)^2 = 1 + \frac{1}{2}\cos 2\omega_1 t + \frac{1}{2}\cos 2\omega_2 t + \cos(\omega_1+\omega_2)t + \cos(\omega_1-\omega_2)t$$

This is the sort of nonlinearity you would gct (roughly) by applying two small signals to a forward-biased diode. Note that you get harmonics of the individual signals, as well as the sum and difference frequencies. The term "balanced mixer" is used to describe a circuit in which only the sum and difference signals, not the input signals and their harmonics, are passed through to the output. The four-quadrant multiplier is a balanced mixer, whereas the nonlinear diode is not.

Among the methods used to make mixers are the following: (a) simple nonlinear transistor or diode circuits, often using Schottky diodes; (b) dual-gate FETs, with one signal applied to each gate; (c) multiplier chips like the MC1495, MC1496, SL640, or AD630; (d) balanced mixers constructed from transformers and arrays of diodes, generally available as packaged "double-balanced mixers." The latter are typified by the popular M1 series of double-balanced mixers from Watkins-Johnson spanning the frequency range to 4000MHz with 20dB to 50dB of signal isolation, or the inexpensive SBL-1 mixer (1–500MHz) from Mini-Circuits Lab. Mixers are widely used in the generation of radiofrequency signals at arbitrary frequencies; they let you shift a signal up or down in frequency without changing its spectrum. You will see how it all works shortly.

The equations above show that the simple quadratic-law mixer produces outputs of equal amplitudes at both sum and difference frequencies. In communications applications (e.g., the "superheterodyne" receiver), where mixers are often used to shift frequency bands, it is sometimes desirable to suppress one of those mixer products. We'll see in Section 13.16 how to make such an *image-reject* mixer.

Frequency multipliers

A nonlinear circuit often is used to generate a signal at a multiple of the input signal's frequency. This is particularly handy if a signal of high stability is required at a very high frequency, above the range of good oscillators. One of the most common methods is to bias an amplifier stage for highly nonlinear operation, then use an LC output circuit tuned to some multiple of the input signal; this can be done with bipolar transistors, FETs, or even tunnel diodes. A multiplier like the 1496 can be used as an efficient doubler at low radiofrequencies by connecting the input signal to both inputs, thus forming the square of the input waveform. The square of a sine wave contains frequencies at the second harmonic only. Prepackaged frequency doublers that use balanced mixers are commercially available; they are very broadband (for example, the Watkins-Johnson FD25 accepts an input frequency from 5MHz to 2400MHz), with good suppression of both the input frequency (the "fundamental") and unwanted harmonics (typically 30dB). Exotic devices such as SNAP diodes and varactors are also used as multipliers. A frequency-multiplier circuit should include a tuned output circuit or should be followed by tuned amplifiers, since, in general, many harmonics of the input signal are generated in the nonlinear process.

Attenuators, hybrids, circulators

There are some fascinating passive devices that are used to control the amplitude and direction of radiofrequency signals passing between circuit modules. All of these are broadband transmission-line (or waveguide) components, meant to be inserted in a line of fixed impedance, usually 50 ohms. They are all widely available as modules.

The simplest is the attenuator, a device to reduce the amplitude of a signal. They come with a big knob and accurately calibrated steps of attenuation, or as voltage-controlled attenuators. The latter are simply balanced mixers with the control current serving as one of the multiplying inputs. Fixed attenuators (Fig. 13.27) are convenient for reducing signal levels between components, as you are assembling the modules of a 50 ohm radiofrequency system; they also reduce any impedance mismatch that may be present.

A hybrid (also known as a "rat race," magic T, 3dB coupler, or iso-T) is a clever transmission-line configuration with four ports. A signal fed into any port emerges from the two closest ports, with specific phase shifts (usually 0° or 180°). A hybrid that has one port terminated in its characteristic impedance is a 3-port "power splitter/combiner." Splitter/combiners can be cascaded to make multiport splitter/combiners. A close cousin of the hybrids is the directional coupler, a 3-port device that couples a small fraction of a passing wave out to a third port. Ideally there is no output at the third port for a wave passing through in the opposite direction.

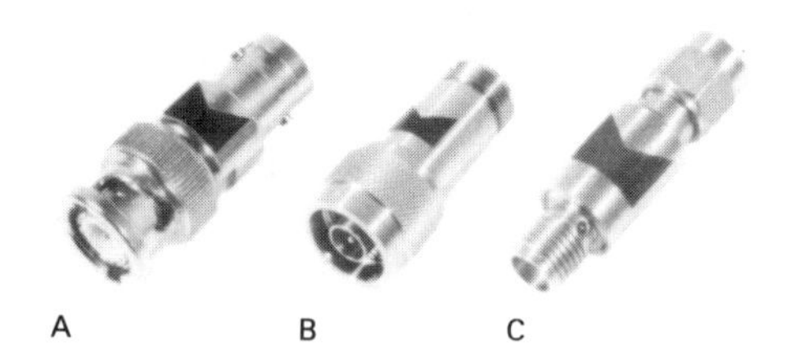

Figure 13.27. Fixed attenuators. (Courtesy of Merrimac Industries, Inc.)
A. BNC.
B. Type N.
C. SMA.

The most magical devices in this general category are the circulators and isolators. They employ exotic ferrite materials and magnetic fields to achieve the impossible: a device that will transmit waves in only one direction. The isolator has two ports and allows transmission in one direction only. Circulators have three or more ports, and they transmit an incoming signal at any port to the next port in succession.

Filters

As you will see, frequency selectivity is often needed in the design of radiofrequency circuits. The simple tuned LC amplifier provides a good measure of selectivity, with the peakiness of the response adjustable via the Q factor of the LC circuit. The latter depends on losses in the inductor and capacitor, as well as loading by the associated circuitry. Q values as high as several hundred can be easily obtained. At very high frequencies, lumped *LC* circuits are replaced by stripline techniques, and at microwave frequencies you use cavity resonators, but the basic idea remains the same. Tuned circuits can also be used to *reject* a particular frequency, if desired.

For applications where it is necessary to have a filter that passes a very narrow band of frequencies relatively unattenuated, with a sharp dropoff outside the limits of the band, a superior bandpass filter can be made from a set of piezoelectric (ceramic or quartz-crystal) or mechanical resonators. There are commercially available 8-pole and 16-pole crystal-lattice filters with center frequencies in the range of 1MHz to 50MHz and bandwidths ranging from as little as a few hundred hertz to several kilohertz. These filters are extremely important in setting receiver selectivity and in the generation of certain kinds of modulated signals. Recently, surface acoustic wave (SAW) filters have become popular and inexpensive; these, too, can have level passband characteristics with extremely steep skirts. This desirable characteristic is usually expressed as a "shape factor," e.g., the ratio of −3dB bandwidth to −40dB bandwidth, with values as small as 1.1. In a typical application SAW filters are used in television receivers and cable systems to limit the received passband.

Of course, in situations where such narrow passbands are not needed, filters can be designed with multiple resonant *LC* sections. Appendix H shows some low-pass and high-pass *LC* filter examples.

Detectors

The bottom line in the extraction of information from a modulated radiofrequency signal involves *detection*, the process of stripping the modulating signal from the "carrier." There are several methods, depending on the form of modulation (AM, FM, SSB, etc.), and we will discuss this important topic next, along with communications concepts.

13.13 Measuring amplitude or power

As we will soon see, detection of an AM signal is simply the generation of a voltage proportional to the instantaneous amplitude of the RF signal carrying the modulation. In many other applications as well (radioastronomy, laboratory RF measurements, signal generator "leveling," filter design, surveillance, etc.) it's important to be able to measure the amplitude or power of radiofrequency signals. Before moving on to communication, therefore, let's take a moment to look at these circuits and methods.

Signal rectification

In Section 1.30 we showed how to use a simple diode to derive an output voltage proportional to the amplitude of a signal. Since the diode's "knee" is somewhat soft, and offset by about 0.6 volt, we showed how to compensate the diode drop with a second diode. In Section 4.18 we showed how to circumvent the diode's nonlinearity and offset by putting the diode in the feedback path of an op-amp, thus forming a precision rectifier (or absolute-value) circuit.

Each of these circuits has problems. The simple diode detector has the advantage of working over an enormous range of frequencies (up to gigahertz, with the right kind of diodes), but it is nonlinear at low levels. The use of Schottky ("hot-carrier") diodes helps somewhat, because of their lower forward drop. You can improve the situation considerably by preamplifying the signal before rectification (this is used, for example, in the UTD-1000 amplifier/diode "level detector" from Avantek); but that limits dynamic range, owing to amplifier saturation (the UTD-1000 has a 30dB range and operates from 10 to 1000MHz). The active rectifier, by contrast, is highly linear; but it works well only at relatively low frequencies, in common with most op-amp circuits. You can use fast op-amps (Section 13.08) to improve the situation, but you will still be limited to frequencies below 10MHz or so.

Synchronous (homodyne) detection

An interesting method that combines dynamic range, accuracy, and speed is *synchronous detection*, also called "homodyne detection." In this method (Fig. 13.28) the output is rectified by being inverted during alternate half cycles. This obviously requires a clean signal at the same frequency as the signal being detected, either supplied externally or regenerated internally with a phase-locked loop (Section 9.27). Synchronous detection works well up to frequencies of a few megahertz, at least; the big disadvantage is the need for a coherent reference signal. We'll see this same circuit in Section 15.15, in the guise of a *phase detector* (in which form it also made a brief appearance in Section 9.27).

Current-source drive

Another solution to the problem of rectifier diode nonlinearity is to drive the rectifier circuit with a *current*, rather than a voltage; the output is then loaded resistively, to produce a proportional output voltage (Fig. 13.29). Figure 13.30 shows a good implementation with voltage-controlled transistor current source; the performance is shown in Figure 13.31. You can understand this circuit in another way: With no input signal, the amplifier's output is decoupled from the rectifier network, producing very high voltage gain (with its current-sink load); thus, only a very small input signal is needed to turn on the diodes. At that point the voltage gain drops to $G_V = R_L/(R_E + r_e)$ (in this case, $G_V \approx 3$), preventing saturation. With a wideband amplifier and fast diodes, this circuit design will operate to 100MHz or more.

Post-detection diode compensation

From Hewlett-Packard (HP Journal, 10/80) comes the circuit shown in Figure 13.32, which uses matched Schottky diodes in a clever arrangement in which each diode sees the same signal. Since the op-amps operate on the detected (low-frequency) signal, the bandwidth is limited only by the diode circuit. The designers of this circuit deserve a pat on the back (and three "attaboys").

Amplitude-tracking detector

Figure 13.33 shows another clever idea: Eliminate diode nonlinearities and offsets by using a locally generated signal, detected in a symmetrical circuit, to cancel the unknown current. Feedback adjusts

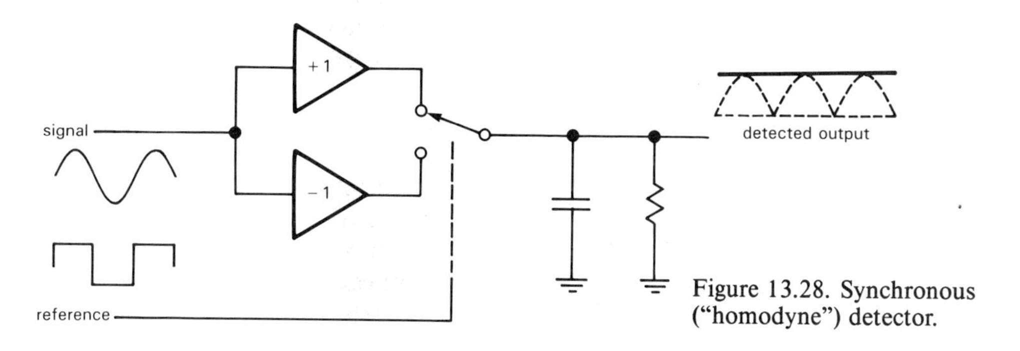

Figure 13.28. Synchronous ("homodyne") detector.

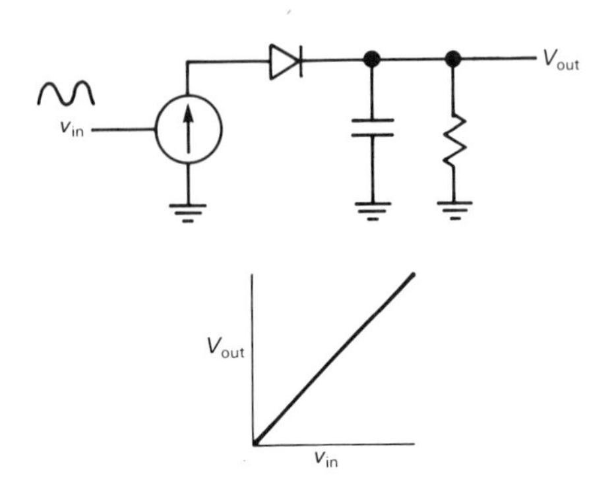

Figure 13.29

the amplitude of the local low-frequency signal until the rectified outputs balance. The frequency of the nulling signal is low enough so that its amplitude can be accurately measured with an op-amp precision rectifier. A good implementation of this circuit will operate linearly down to a few millivolts, and up to a gigahertz.

Power detection

The foregoing methods all measure the *amplitude* of a high-frequency signal. There are many occasions on which you really want to know *power*. Of course, for a sine wave, the two are simply related by $P = V_{\mathrm{rms}}^2/R_{\mathrm{load}}$, so you can simply convert a measured amplitude to a calculated power. For nonsinusoidal waveforms, however, a true power measurement can be made only by averaging the square of the actual voltage waveform. In the language of radiofrequency measurements, you need a "square-law detector."

There are numerous methods available. For signals of low to moderate frequency, you do well with a "function module," for example the AD637 monolithic rms-to-dc converter from Analog Devices. These devices use the exponential diode characteristic within feedback loops to form the square of the input signal, which is then low-pass filtered and fed to an analogous square-root circuit. You get excellent linearity and dynamic range, and good bandwidth, from these circuits. For example, AD637 has 8MHz bandwidth at full level, 0.02% nonlinearity, and 60dB dynamic range; it even has a logarithmic (dB) output.

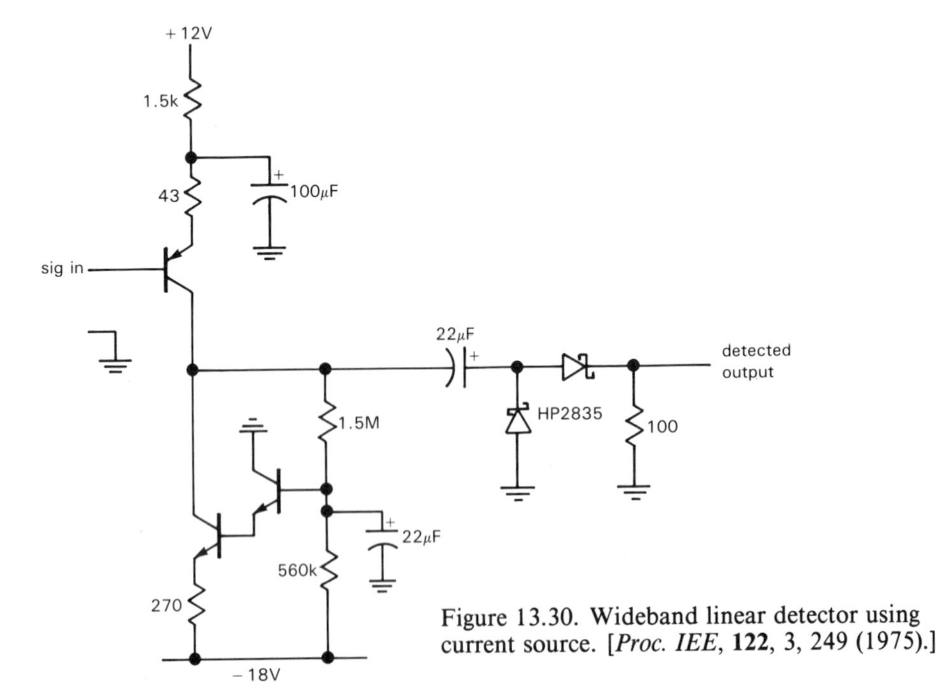

Figure 13.30. Wideband linear detector using current source. [*Proc. IEE*, **122**, 3, 249 (1975).]

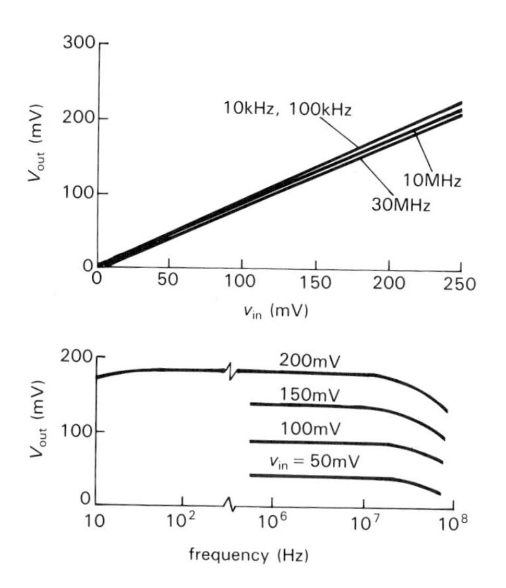

Figure 13.31. Performance curves for wideband detector.

For frequencies above a few megahertz these square/square-root methods of rms conversion fail, owing to inadequate op-amp loop bandwidth. However, there are other methods you can use. Figure 13.34 shows a simple square-law detector circuit that uses a *back diode*, which is nothing but a tunnel diode (Section 1.06) used in the nontunneling direction (where it has zero forward voltage drop). We got this circuit from the radioastronomers at the Haystack Observatory and were amazed at its extraordinary power linearity (Fig. 13.35).

An interesting square-law technique of considerable ancestry is the *bolometric* method, in which the input signal (perhaps amplified) is used to power a resistive heater, whose temperature is then measured. Since the heater power is accurately proportional to V^2, this method is intrinsically square-law. An example of a bolometric module is the LT1088 from Linear Technology. It uses a matched pair of resistive heaters coupled to a matched pair of temperature-measuring diodes. The input signal is applied to one heater, and feedback is applied to the reference heater to maintain the diodes at the same temperature. The reference heater's drive voltage is the output (Fig. 13.36).

The bolometric technique is intrinsically wideband, and accurately square-law. It has limited dynamic range, however, because it's hard to measure microscopic amounts of heat, and it's easy to blow it out with macroscopic amounts of heat! The exemplary LT1088, for example, works from dc to 300MHz, but has only 25dB of dynamic range. With careful bolometer design it is possible to extend the bandwidth to very high frequencies, and to wider dynamic range. The 432–438 series of bolometric power meters

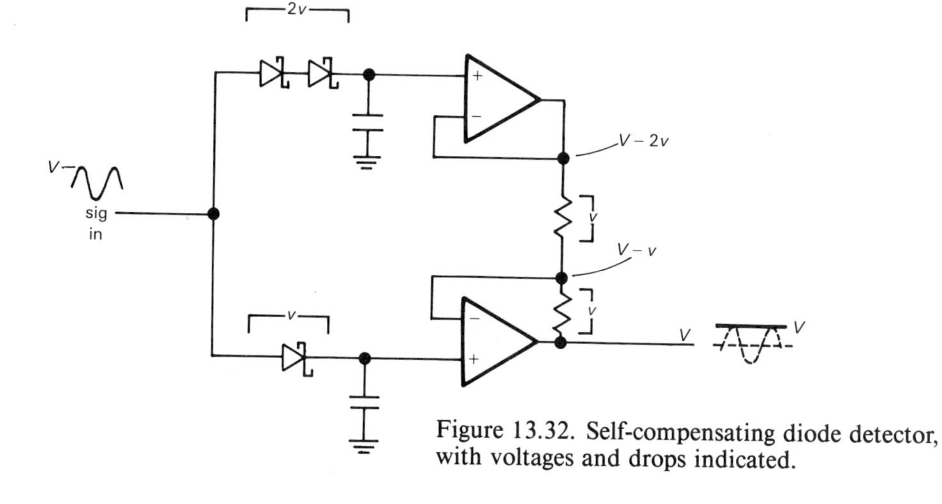

Figure 13.32. Self-compensating diode detector, with voltages and drops indicated.

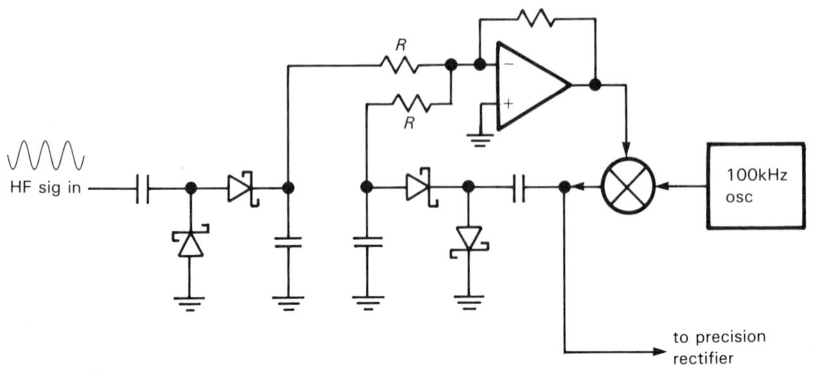

Figure 13.33. Amplitude-tracking detector.

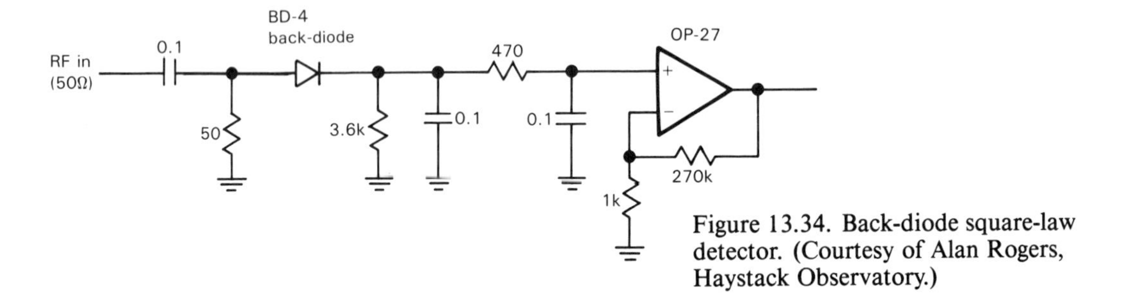

Figure 13.34. Back-diode square-law detector. (Courtesy of Alan Rogers, Haystack Observatory.)

from Hewlett-Packard span the frequency range from 100kHz to 50GHz, using a set of interchangeable bolometric power sensors. These cover the range +44 dBm (25W) to −70dBm (100pW), a total span of 114dB (factor of 2.5×10^{11} in power), although any one bolometer can accommodate at most a 50dB range.

RADIOFREQUENCY COMMUNICATIONS: AM

Since radiofrequency techniques find their greatest application in communications, it is important to understand how signals can be modulated and demodulated, i.e., how radiofrequencies are used to carry information from one place to another. Besides, how would you feel if, after taking a course in electronics, someone asked you how a radio works and you didn't know?

13.14 Some communications concepts

In communications theory we speak of a communications "channel," a means of conveying information from A to B. For example, the channel might consist of a cable or an optical-fiber link. In radiofrequency communications the channel is the electromagnetic frequency spectrum, which, roughly speaking, extends from very low frequencies (VLF) of a few kilohertz, through the "short waves" of a few megahertz to a few tens of megahertz, the very high frequencies (VHF) and ultrahigh frequencies (UHF) extending up to several hundred megahertz, and the microwave region beginning at about 1GHz.

A signal, consisting of speech, say, is sent on a radiofrequency channel by having it modulate a radiofrequency "carrier." It is important to understand why this is done at all, rather than transmitting the speech directly. There are basically two

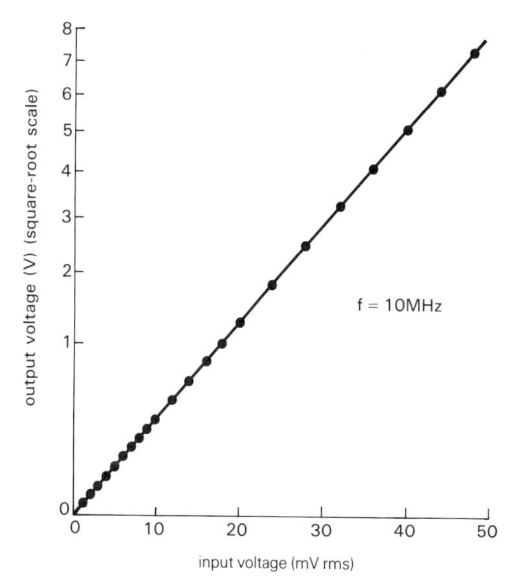

Figure 13.35. Performance of square-law detector.

reasons. First, if the information were transmitted at its natural band of frequencies with radio waves (in this case, in the VLF portion of the spectrum), any two signals would overlap and jam each other; i.e., by encoding the information onto carriers in separate portions of the spectrum, it is possible to "frequency-multiplex" the signal, and thereby maintain many channels simultaneously. Second, some wavelengths are more conveniently generated and propagated than others. For instance, in the region from 5MHz to 30MHz, signals can travel around the world by multiple reflections from ionosphere, and at microwave frequencies antennas of modest size can form narrow beams. Consequently, the HF (short-wave) region is used for over-the-horizon communication, whereas microwaves are used for line-of-sight repeaters and radar.

There are several ways to modulate a carrier. Roughly speaking, all methods have in common the property that the modulated signal occupies a bandwidth at least comparable to the bandwidth of the modulating signal, i.e., the bandwidth of the information being sent. Thus, a high-fidelity audio transmission will occupy 20–40kHz of spectrum, regardless of the carrier frequency. A perfect unmodulated carrier has zero bandwidth and conveys no information. A transmission of low information content, e.g., telegraphy, occupies a

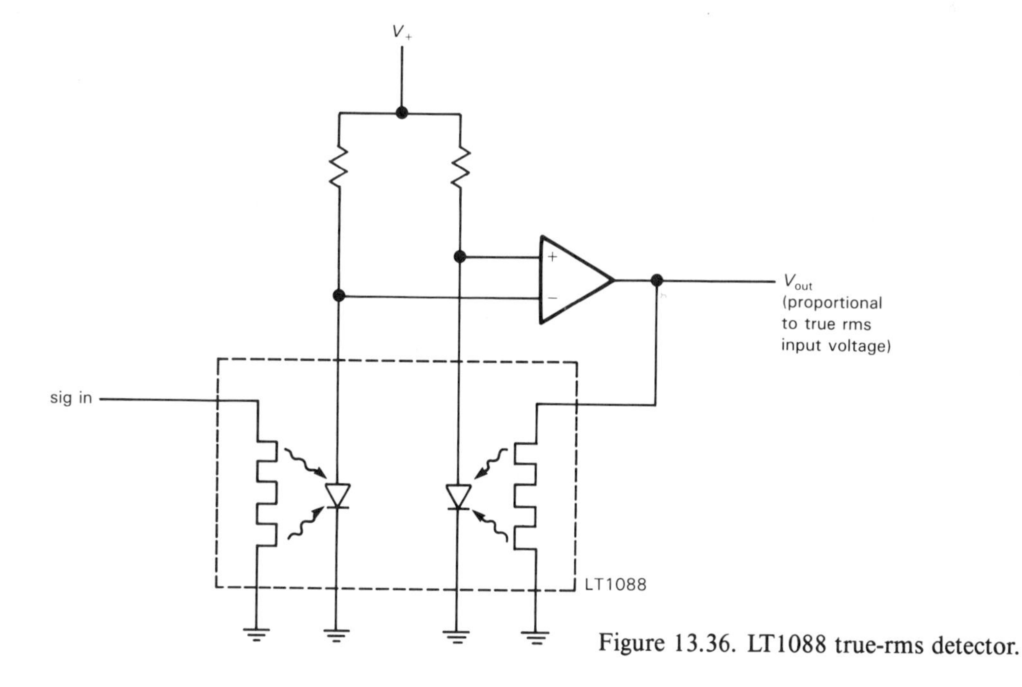

Figure 13.36. LT1088 true-rms detector.

relatively narrow slice of spectrum (perhaps 50–100Hz), whereas something like a television picture requires several megahertz. For completeness it should be pointed out that more information can be sent on a channel of given bandwidth if there is sufficiently high signal/noise ratio (SNR). Such "frequency compression" takes advantage of the fact that "channel capacity" equals bandwidth times $\log_2(1 + \text{SNR})$.

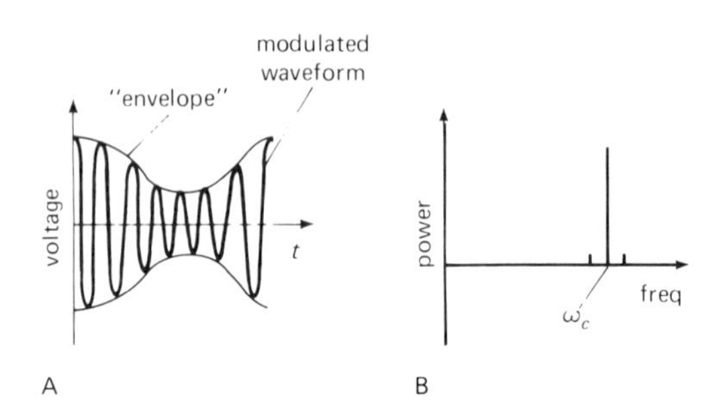

Figure 13.37. Amplitude modulation.

13.15 Amplitude modulation

Let's begin with the simplest form of modulation (AM), taking a look at its frequency spectrum and methods of detection. Imagine a simple carrier, $\cos \omega_c t$, varied in amplitude by a modulating signal of much lower frequency, $\cos \omega_m t$, in the following way:

$$\text{signal} = (1 + m \cos \omega_m t) \cos \omega_c t$$

with m, the "modulation index," less than or equal to 1. Expanding the product, you get

$$\text{signal} = \cos \omega_c t + \tfrac{1}{2} m \cos(\omega_c + \omega_m)t + \tfrac{1}{2} m \cos(\omega_c - \omega_m)t$$

i.e., the modulated carrier has power at frequency ω_c and at frequencies on either side ω_m away. Figure 13.37 shows the signal and its spectrum. In this case the modulation (m) is 50%, and the two "sidebands" each contain $\frac{1}{16}$ of the power contained in the carrier.

If the modulating signal is some complex waveform $[f(t)]$, like speech, the amplitude-modulated waveform is given by

$$\text{signal} = [A + f(t)] \cos \omega_c t$$

with the constant A large enough so that $A + f(t)$ is never negative. The resulting spectrum simply appears as symmetrical sidebands around the carrier (Fig. 13.38).

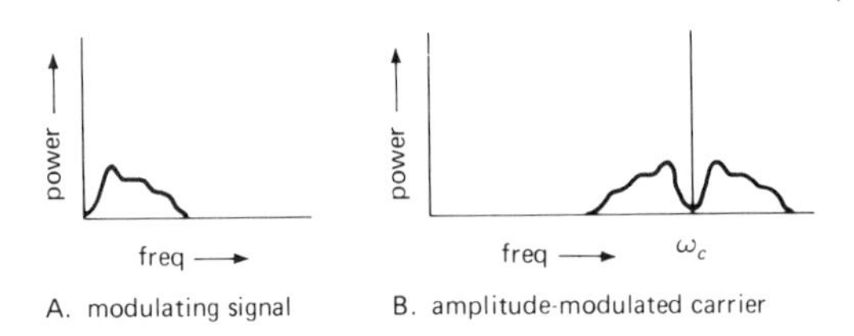

Figure 13.38. AM spectrum for a band of modulating frequencies (speech).

AM generation and detection

It is easy to generate amplitude-modulated RF. Any technique that lets you control the signal amplitude with a voltage in a linear manner will do. Common methods involve varying the RF amplifier supply voltage (if the modulation is done at the output stage) or using a multiplier chip such as the 1496. When the modulation is done at a low-level stage, all following stages of amplification must be linear. Note that in AM the modulating waveform must be biased up so that it never assumes negative values. Look at the graphs in Figure 13.39.

The simplest receiver of AM consists of several stages of tuned RF amplification, followed by a diode detector (Fig. 13.40). The amplifier stages provide selectivity against signals nearby in frequency, and they amplify the input signals (which may be at the microvolt level) for the detector. The latter simply rectifies the RF waveform, then recovers the smooth "envelope" with low-pass filtering. The

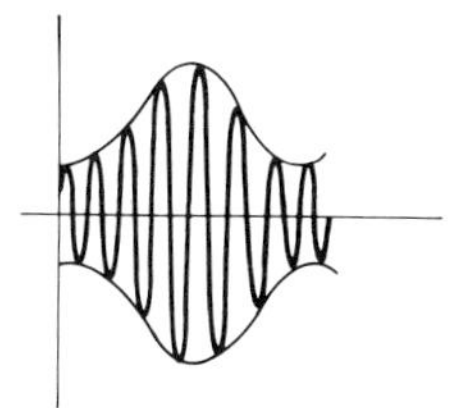

A. 50% modulation

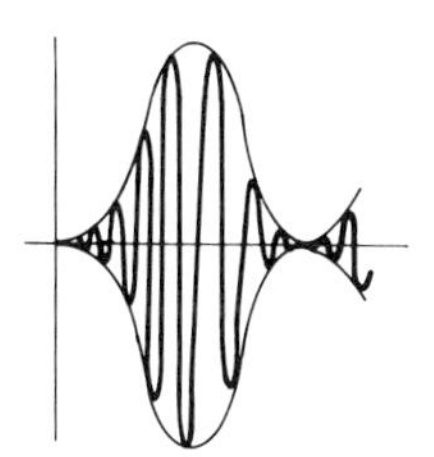

B. 100% modulation

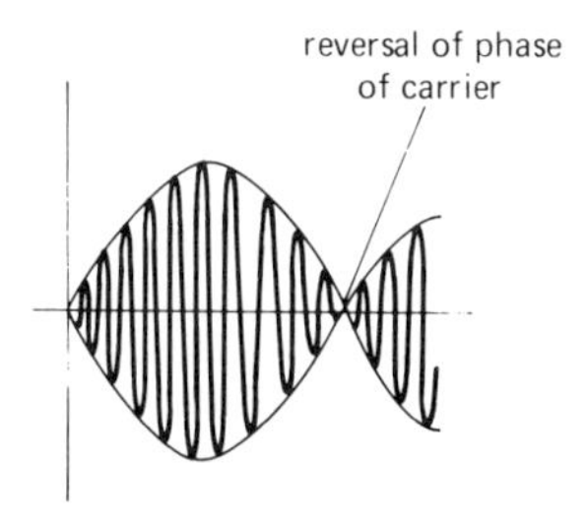

C. overmodulation

Figure 13.39

low-pass filter should reject RF while passing the audiofrequencies unattenuated. This simple scheme leaves much to be desired, as you will see. It is really just a glorified crystal set.

13.16 Superheterodyne receiver

A receiver consisting of a set of tuned RF amplifiers is undesirable for several reasons. First of all, the individual amplifiers must be tuned to the same frequency, requiring either great coordination by someone with a lot of hands or extremely good tracking of a set of simultaneously tuned LC circuits. Second, since the overall frequency selectivity is determined by the combined responses of the individual amplifiers, the shape of the passband will depend on the accuracy with which the individual amplifiers are tuned; the individual amplifiers cannot have as sharp a response as would be desirable, since tuning would then be practically impossible. And since the signal being received can be at any frequency within the tuning range of the amplifiers, it isn't possible to take advantage of crystal-lattice filters to generate a flat passband with steep falloff on either side (steep "skirts"), a very desirable passband characteristic.

A nice solution to these problems is the superheterodyne ("superhet") receiver shown in Figure 13.41. The incoming signal is amplified with a single stage of tuned RF amplification, then mixed with an adjustable local oscillator (LO) to produce a signal at a fixed intermediate frequency (IF), in this case 455kHz. From then on the receiver consists of a set of fixed-tuned IF amplifiers, including selective elements such as crystal or mechanical filters, finally terminating in a detector and audio amplifier. Changing the LO frequency tunes the receiver, since a different input frequency then gets mixed to the IF passband frequency. The input RF amplifier must be gang-tuned with the LO, but the alignment is not critical. Its purposes are (a) to

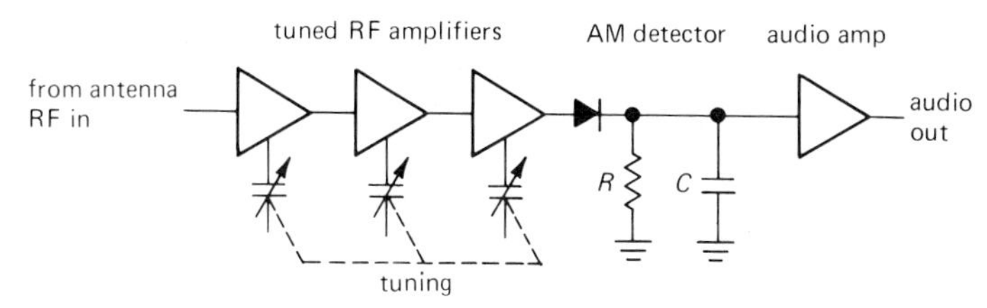

Figure 13.40

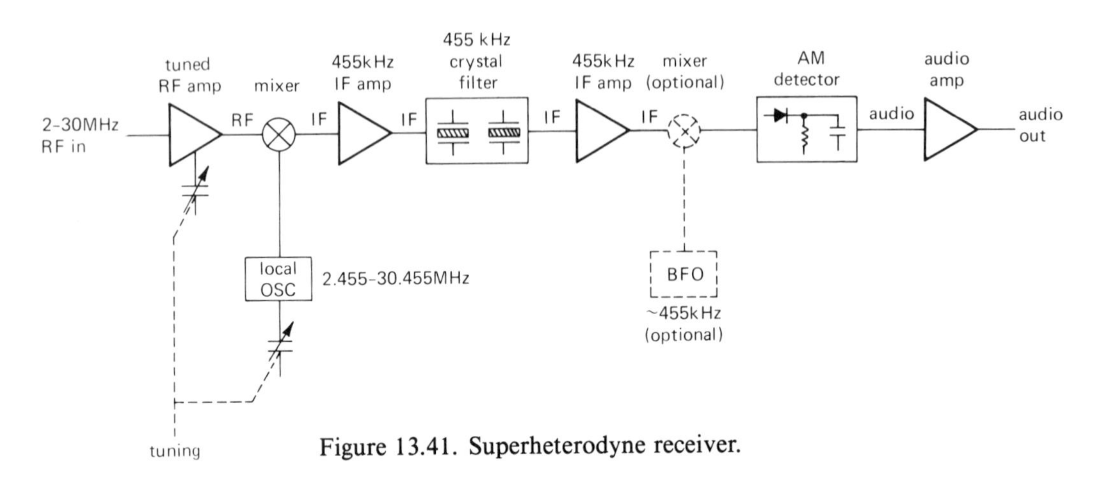

Figure 13.41. Superheterodyne receiver.

improve the sensitivity with a stage of low amplification prior to mixing and (b) to reject signals at the "image" frequency, in this case input signals at a frequency of 455kHz *above* the LO (remember that a mixer generates sum and difference frequencies). In other words, the superheterodyne receiver uses a mixer and local oscillator to shift a signal at the (variable) input frequency over to a fixed intermediate frequency where most of the gain and selectivity are concentrated.

Superhet potpourri

There are some additional features often added to a superheterodyne receiver. In this example a beat frequency oscillator (BFO) is shown; it is used in the detection of some signals with modulation other than AM (telegraphy, suppressed carrier telephony, frequency-shift keying, etc.). It can even be used for AM detection in what is known as a "homodyne" or "synchronous" detector. Receivers often have more than one mixing stage (they're called "multiple-conversion" receivers). By using a high first IF, image rejection is improved (the image is twice the IF frequency away from the actual received signal). A lower second IF makes it easier to use sharp-cutoff crystal filters, and a third IF is sometimes generated to allow the use of audio-type notch filters, low-frequency ceramic or mechanical filters, and "product detectors."

Recently, the use of direct up-conversion (an IF higher than the input signal frequency) in a front-end balanced mixer, with crystal filters at the ~40MHz IF, followed by detection with no further mixing, has become popular. Such a single-conversion scheme offers better performance in the presence of strong interfering signals, and it has become practical with the availability of good VHF crystal-lattice filters and low-distortion wide-range balanced mixers with good noise performance.

Image-reject mixers

The superhet receiver requires a tuned RF amplifier in order to reject the image band, which is separated by twice the IF frequency from the desired in-band RF signal frequency. The RF amplifier must be selective enough to reject the image band (i.e., its response to image-frequency signals must be much less than its response to in-band signals), and it must be tuned in order to keep its bandpass a constant (IF) frequency away from the LO, as the latter is adjusted to tune the receiver.

There is another way to suppress response at the image frequency, without using a tuned RF amplifier. Look at Figure

13.42, which shows an *image-reject* mixer: You begin with a pair of mixers, driven with quadrature LOs ("quadrature" means "differing in phase by 90°"); then combine the IF output signals, once again introducing a 90° phase shift in one path. The pair of 90° phase shifts adds for one sideband, and subtracts for the other, causing cancellation of the image band. Reversing the sign of the final 90° phase shift interchanges image band and signal band. In practice you usually use "4-port quadrature hybrids" to do the phase shifting, resistively terminating the unused output in each case. If you assemble an image-reject mixer from standard broadband components, you can expect something like 20dB suppression of the image sideband, with operation over one or two octaves of frequency. It's sometimes essential to be able to move around rapidly in frequency (called "frequency agility") without having to tune a tracking RF amplifier; in that case image-reject mixers are just what you want.

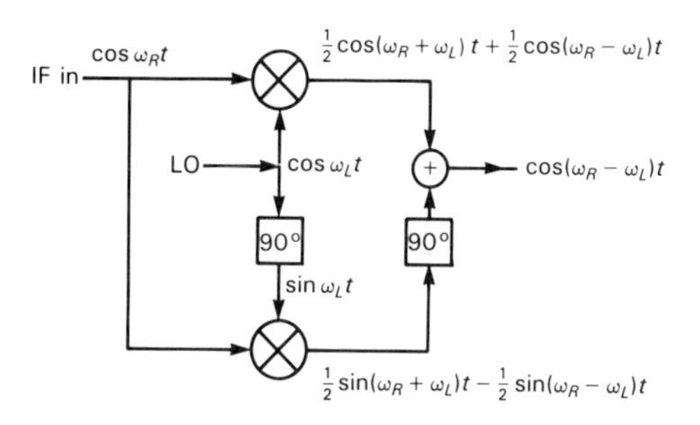

Figure 13.42. Image-reject mixer.

An interesting subtlety: As we remarked in Section 13.12, a mixer can be thought of as a modulator, and vice versa. The language you use to describe it depends on whether you are using the device to translate a low-frequency "baseband" of information up to high frequencies (in which case you call it a "modulator") or using it to translate a modulated RF band down to baseband [or perhaps an intermediate (IF) band along the way], where you demodulate it to extract the original modulating signal (in which case you call it a "mixer"). When you turn things around this way, what we called the image band becomes the other sideband. Our two methods of image rejection (RF filter, image-reject mixer) become the two classic methods of single-sideband modulation, namely the "filter" method and the "phasing" method. This may make more sense after you read the next section (but don't worry if it doesn't; we just couldn't resist trying to explain this unifying idea).

ADVANCED MODULATION METHODS

□ 13.17 Single sideband

From a glance at the spectrum of an AM signal it is obvious that things can be improved. Most of the power (67%, to be exact, at 100% modulation) is in the carrier, conveying no information. AM is at most 33% efficient, and that only when the modulation index is 100%. Since voice waveforms generally have a large ratio of peak amplitude to average amplitude, the modulation index of an AM signal carrying speech is generally considerably less than 100% (although speech-waveform "compression" can be used to get more power into the sidebands). Furthermore, the symmetrical sidebands, by conveying the identical information, cause the signal to occupy twice the bandwidth actually necessary.

With a bit of trickery it is possible to eliminate the carrier [a balanced mixer does the job; note that $\cos A \cos B = \frac{1}{2}\cos(A + B) + \frac{1}{2}\cos(A - B)$], creating what is known as "double-side-band-suppressed carrier," or DSBSC. (This is just what you will get if the audio signal multiplies the carrier directly, without first being biased so that the audio waveform is always positive, as in normal AM.) Then, either by using sharp crystal filters or by using a method known as "phasing," one of the remaining sidebands can be eliminated. The "single-sideband" (SSB) signal that remains forms a highly efficient mode of voice communication and is widely used by radio amateurs and commercial users for long-range high-frequency telephony channels. When you're not talking, there's nothing being transmitted. To receive SSB, you need a BFO and product detector, as shown in the last block diagram, to reinsert the missing carrier.

□ *Modulation spectra*

Figure 13.43 shows representative spectra of voice-modulated AM, DSBSC, and SSB. When transmitting SSB, either sideband can be used. Note that SSB consists simply of the audio spectrum translated upward in frequency by f_C. When SSB is being received, the BFO and mixer combine to translate the spectrum down to audiofrequencies again. If the BFO is slightly mistuned, all audiofrequencies will be offset by the amount of mistuning. This dictates good stability for the LO and BFO in a receiver used for single-sideband.

Note that a mixer (modulator) can always be thought of as a frequency translator, especially when combined with suitable filters to eliminate the undesired outputs: When used as a modulator, a low-frequency band of frequencies is shifted up by the carrier frequency, to form a band centered around f_C. When used as a mixer, a band of frequencies around f_C is shifted down to audiofrequencies ("baseband"), or to a band centered around the IF frequency, by the action of a high-frequency LO.

□ 13.18 Frequency modulation

Instead of modulating the amplitude of a carrier, as in AM, DSBSC, and SSB, it is possible to send information by modulating the frequency or phase of the carrier:

$$\text{signal} = \cos\left[\omega_c t + k\int f(t)\,dt\right] \quad \text{frequency modulation (FM)}$$

$$\text{signal} = \cos\left[\omega_c t + k f(t)\right] \quad \text{phase modulation (PM)}$$

FM and PM are closely related and are sometimes referred to as "angle modulation." FM is familiar as the mode used in the 88–108MHz VHF broadcast band, and AM is used in the 0.54–1.6MHz broadcast band. Anyone who has tuned an FM receiver has probably noticed the "quieting" of background noise characteristic of FM reception. It is this property (the steep rise of recovered SNR with increasing SNR of the channel) that makes wideband FM preferable to AM for high-quality transmission.

Some facts about FM: When the frequency *deviation* $kf(t)/2\pi$ is large compared with the modulating frequency [highest frequency present in $f(t)$], you have "wideband FM" as used in FM broadcasting. The modulation index, m_f, equals

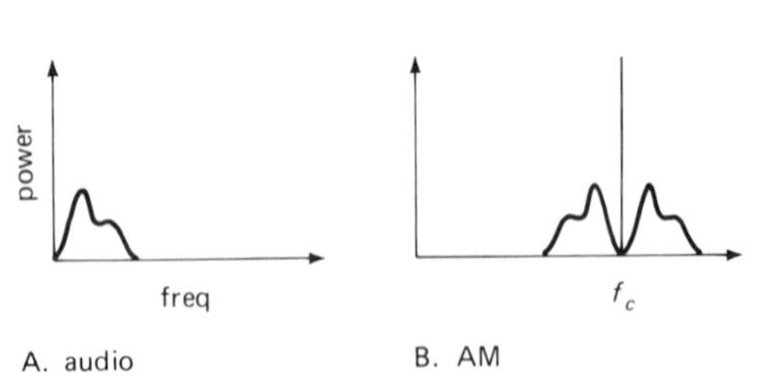

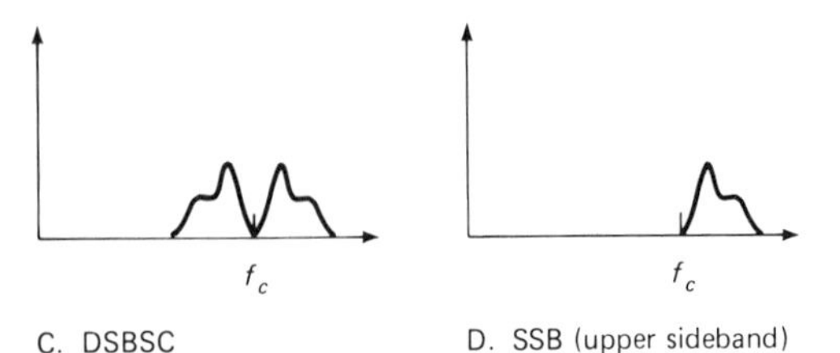

Figure 13.43. Suppressed-carrier spectra.

the ratio of frequency deviation to modulating frequency. Wideband FM is advantageous because under the right conditions the received SNR increases 6dB per doubling of FM deviation. The price you pay is increased channel bandwidth, since a wideband FM signal occupies approximately $2f_{\text{dev}}$ of bandwidth, where f_{dev} is the peak deviation of the carrier. FM broadcasting in the 88–108MHz band uses a peak deviation f_{dev} of 75kHz, i.e., each station uses about 150kHz of the band. This explains why wideband FM is not used in the AM band (0.54–1.6MHz): There would be room for only six stations in any broadcasting area.

FM spectrum

A carrier that is frequency-modulated by a sine wave has a spectrum similar to that shown in Figure 13.44. There are

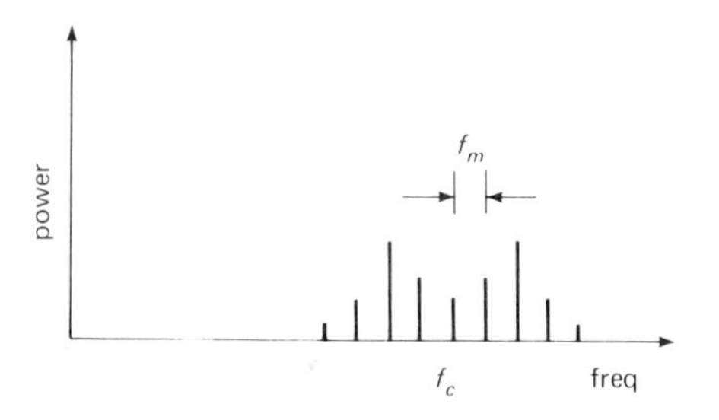

Figure 13.44. Wideband FM spectrum.

numerous sidebands spaced at multiples of the modulating frequency from the carrier, with amplitudes given by Bessel functions. The number of significant sidebands is roughly equal to the modulation index. For narrowband FM (modulation index <1), there is only one component on either side of the carrier. Superficially this looks the same as AM, but when the phase of the sidebands is taken into account, you have a waveform of constant amplitude and varying frequency (FM), rather than a waveform of varying amplitude and constant frequency (AM). With wideband FM the carrier amplitude may be very small, with correspondingly high efficiency, i.e., most of the transmitted power goes into the information-carrying sidebands.

□ Generation and detection

FM is easily produced by varying an element of a tuned circuit oscillator; a varactor (a diode used as a voltage-variable capacitor, Section 5.18) is ideal. Another technique involves integrating the modulating signal, then using the result to do phase modulation. In either case it is often best to modulate at low deviation, then use frequency multiplication to increase the modulation index. This works because the *rate* of frequency deviation is not changed by frequency multiplication, whereas the deviation is multiplied along with the carrier.

To detect FM, an ordinary superheterodyne receiver is used, with two differences. First, the final stage of IF amplification includes a "limiter," a stage run at constant (saturated) amplitude. Second, the subsequent detector (called a discriminator) has to convert frequency deviation into amplitude. There are several popular methods of detection:

1. A "slope detector," which is nothing more than a parallel *LC* circuit tuned off to one side of the IF frequency; as a result, it has a rising curve of response versus frequency across the IF bandwidth, thereby converting FM to AM. A standard envelope detector converts the AM to audio. There are improved versions of the slope detector involving a balanced pair of *LC* circuits tuned symmetrically to either side of the IF center frequency.

2. The Foster-Seely detector, or its variant, the "ratio detector," using a single tuned circuit in a fiendishly clever diode arrangement to give a linear curve of amplitude output versus frequency over the IF bandpass. These discriminators are superior to the simple slope detector (Fig. 13.45)

3. A "phase-locked loop" (PLL). This is a device that varies the frequency of a voltage-controlled oscillator to match an input frequency, as we discussed in Section 9.31. If the input is the IF signal, the control voltage generated by the PLL is linear in frequency, i.e., it is the audio output.

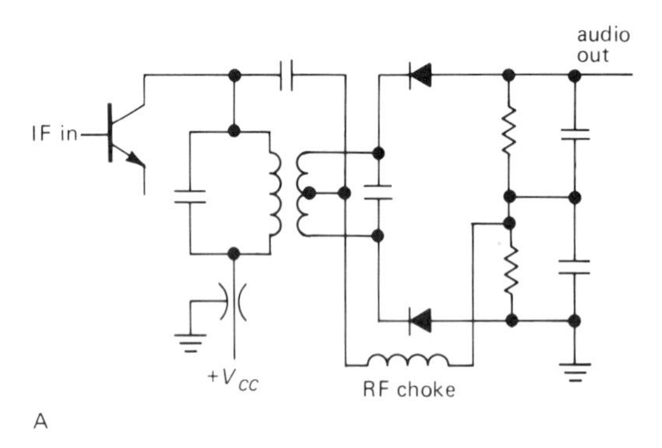

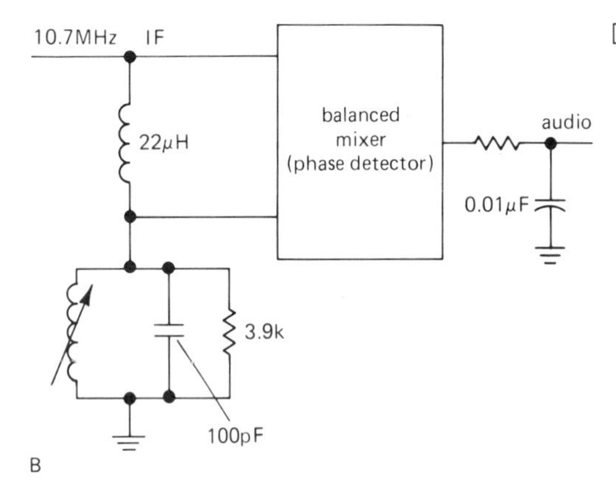

Figure 13.45. FM discriminators.
A. Foster-Seeley.
B. Balanced quadrature detector.

4. An averaging circuit, in which the IF signal is converted to a train of identical pulses at the same frequency. Averaging this pulse train generates an output proportional to IF frequency, i.e., the audio output plus some dc.

5. A "balanced quadrature detector," which is a combination of a phase detector (see Sections 9.27 and 9.31) and a phase-shifting network. The IF signal is passed through a network that produces a shift varying linearly with frequency across the IF passband (an *LC* circuit would do nicely). The resultant signal and the original signal are compared in a phase detector, giving an output that varies with relative phase. That output is the desired audio signal (Fig. 13.45).

It is often pointed out that FM provides essentially noise-free reception if the channel has sufficient SNR, as compared with AM, where the rejection of interference improves only gradually with increasing signal power. This makes sense when you remember that FM signals pass through a stage of amplitude limiting before detection. As a result, the system is relatively insensitive to interfering signals and noise, which appear as amplitude variations added to the transmitted signal.

□ 13.19 Frequency-shift keying

Transmission of digital signals (radioteletype, RTTY) is usually done by shifting a continuous-running carrier in frequency between two closely spaced frequencies according to the 1's and 0's being transmitted; 850Hz of shift is a typical value. The use of frequency-shift keying (FSK), rather than on/off modulation, is extremely effective in the presence of large signal fading from changing propagation conditions. To demodulate FSK, you simply use a differential amplifier looking at the outputs from a pair of filters set at the two detected audiofrequencies. You can think of FSK as digital FM. Narrow-shift FSK has been used to circumvent selective fading between the two signal frequencies. However, the shift cannot be reduced below the information bandwidth of the keyed signal itself, roughly the "baud" rate (number of bit cells per second), or about 100Hz for ordinary radioteletype.

□ 13.20 Pulse-modulation schemes

There are several methods whereby analog signals can be transmitted as pulses.

The basic fact that makes digital transmission of analog signals possible is expressed in the Shannon sampling theorem, which states that a band-limited waveform is fully described by sampling its amplitude at a rate equal to twice the highest frequency present. Thus, a method that conveys the amplitude of a waveform, by digital methods or whatever, at instants of time separated by $1/2f_{\max}$ can be used instead of a continuous modulation scheme. Several methods are shown in Figure 13.46.

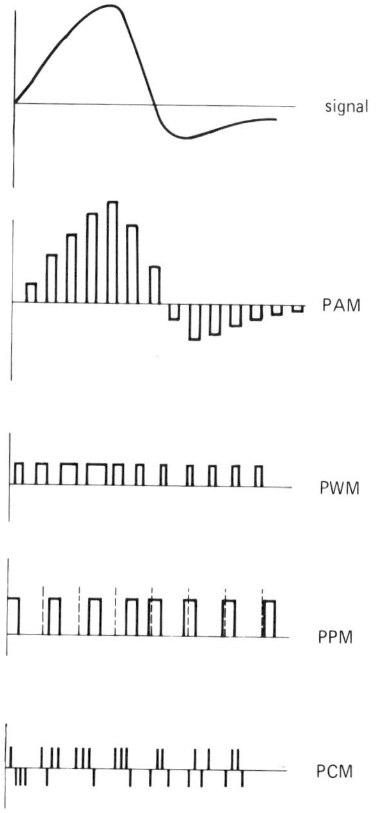

Figure 13.46. Pulse-modulation schemes.

In pulse-amplitude modulation (PAM), a train of pulses of amplitude equal to the signal is transmitted at regular intervals. This scheme is useful for *time multiplexing* of several signals on one information channel, since the time between samples can be used to transmit the samples of another signal (with an increase of bandwidth, of course). In pulse-width modulation (PWM), the width of fixed-amplitude pulses is proportional to the instantaneous signal amplitude. PWM is easy to decode, using simple averaging. In pulse-position modulation (PPM), pulses of fixed width and amplitude are delayed or advanced relative to a set of fixed times, according to the amplitude of the signal.

☐ *Pulse-code modulation*

Finally, in pulse-code modulation (PCM) the instantaneous amplitude of the signal is converted to a binary number and transmitted as a serial string of bits. In the illustration, a 4-bit offset binary code corresponding to 16 levels of quantization has been used. PCM excels when error-free transmission is required over noisy channels. As long as 1's and 0's can be identified unambiguously, the correct digital code, and hence a replica of the original signal, can be recovered. PCM is particularly useful in repeater application, e.g., transcontinental telephone channels, where the signal must pass through many stations and be amplified along the way. With any of the linear modulation schemes (AM, FM, SSB) noise accumulated in transit cannot be removed, but with PCM the digital code can be correctly regenerated at each station. Thus the signal starts anew at each station.

There are variations of PCM (known as coded PCM) in which techniques other than simple serial binary sequences are used to encode the quantized samples; for instance, a burst of one of 16 tones could be used in the preceding example. PCM is routinely used for telemetry of images from space vehicles, owing to its error-free properties. It is also used for "compact-disc" digital audio, in which each stereo channel is sampled and converted to a

16-bit number 44,100 times per second. In any PCM application the bit rate must be chosen low enough to ensure a low probability of error in bit recognition. In general, this limits transmission on a given channel to speeds much below what could be used with direct analog modulation techniques.

□ RADIOFREQUENCY CIRCUIT TRICKS

In this chapter we are attempting to highlight some of the motivation and techniques of circuitry operated at radiofrequencies. In such limited space it is not possible to consider circuit design and construction in as much detail as we have generally attempted in the other chapters, nor would that even be desirable in a book intended as a broad introduction to electronics. In keeping with this philosophy, we would like to give some idea of the techniques that are ordinarily used in RF circuits. These are generally aimed at reducing stray inductance and capacitance and coping with circuitry whose dimensions are often comparable with a wavelength. There will be no attempt to weave these together into a coherent methodology; just think of them as a bag of tricks.

□ 13.21 Special construction techniques

RF "chokes" (small inductors, in the range of microhenrys to millihenrys) are used extensively as signal-blocking elements. Power-supply voltages will usually be brought into a shielded enclosure with shielded "feedthrough capacitors" (bypass to ground combined with a mechanical feedthrough terminal), with an RF choke in series. A variation is to use ferrite beads on leads of transistors, FETs, etc. These are used because of the tendency of RF circuits toward "parasitic" oscillations, encouraged by unintentional tuned circuits at UHF formed by the wiring itself. Stringing a few beads on a base or collector lead here and there adds a lossy series inductance to prevent the oscillation (if you're lucky, that is!).

Inductors play a major role in RF design, and you see plenty of open coils and "slug-tuned" inductors and transformers (such as the little metal IF transformer cans you see everywhere in receiver circuits). Small-value air-variable capacitors are equally popular.

As suggested earlier, RF circuits are constructed in shielded enclosures, often with internal grounded partitions between sections of the circuit to prevent coupling. It is common to build circuits on double-sided PC board, with one side used as a ground-plane. Alternatively, a circuit may be constructed immediately adjacent to a shield or other grounded surface. Grounds can't be wishy-washy at RF; you've got to solder a shield along its whole length, and you have to use a lot of screws to mount a partition or cover.

When building circuits at higher radio frequencies, it is absolutely essential to keep component leads as short as possible. That means snipping off leads right at the resistor or capacitor and soldering them with no visible lead showing (the components get plenty hot, but they seem to survive). At VHF and UHF you often use ceramic capacitor "chips," soldered to PC strips, etc., without leads at all. Watch out when using conventional capacitors, because their internal series inductance can lead to self-resonance effects, sometimes at frequencies as low as a megahertz. The use of wide straps or metal ribbon, rather than ordinary wire, reduces inductance and is a favorite at UHF. At these frequencies you get into stripline and microstrip techniques, where every lead is itself a transmission line, complete with impedance matching. In fact, strips of sheet metal can be used as parts of tuned circuits; here's a specification for an inductor in a 440MHz circuit (ARRL handbook):

"$L_1 - L_3$, incl. – $2\frac{5}{8} \times \frac{1}{4}$-inch strip of brass, soldered to the enclosure on one end and to the capacitor at the other. Input and output taps are 1/2-inch up from the ground end." Of course, at microwave frequencies all such techniques give way to waveguide and cavity circuits, complete with exotica such as circulators and "magic T's" (Fig. 13.47).

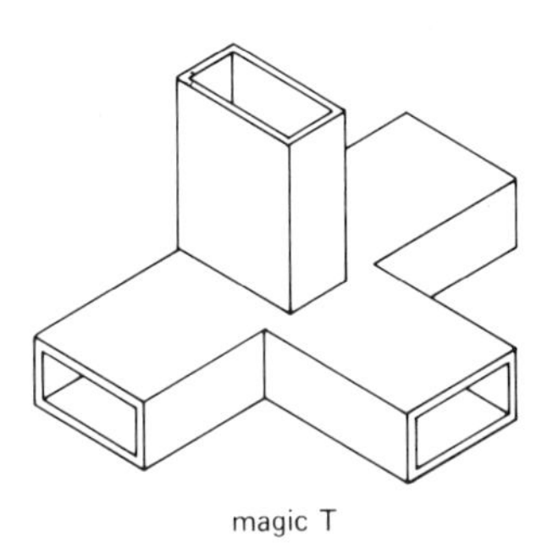

Figure 13.47

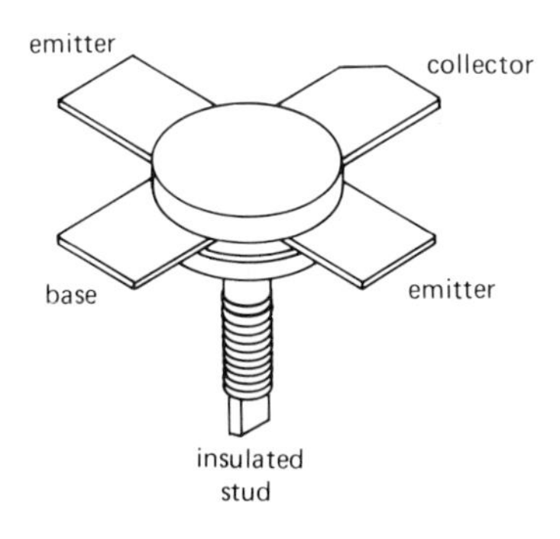

Figure 13.48

A facet of RF design that surprises beginners is the use of test instruments combined with "cut-and-try" techniques. You see widespread use of sweep generators (RF signal sources that sweep repetitively through a range of frequency), grid dip meters (for measuring resonances), SWR bridges, and spectrum analyzers, with plenty of circuit experimentation. At these frequencies you just can't predict everything; it takes some trial and error (and lots of experience) to make things work well.

□ 13.22 Exotic RF amplifiers and devices

Familiar devices such as bipolar transistors and FETs are used at radiofrequencies, although often in somewhat different incarnations. Transistors intended for use at VHF and above come in strange-looking packages, with flat strips radiating out from the center for connection to a stripline or PC board (Fig. 13.48). There are also devices and circuits with no low-frequency analog, such as the following.

Parametric amplifiers. These devices amplify by varying a *parameter* of a tuned circuit. An analogy is a pendulum formed by hanging a weight on a length of string. Imagine that the motion represents output signal. You can build up the swing by gently shoving the weight at the resonant frequency; this is analogous to an ordinary amplifier, with a transistor or other active device providing the "shove." But there's another completely different way to get the thing swinging, namely by pulling up and down on the string (varying its length, a parameter of the system) at *twice* the natural resonant frequency. Try it (Fig. 13.49). The pendulum is closely analogous

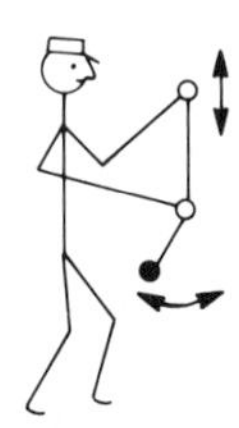

Figure 13.49. Pendulum analogy to the parametric amplifier.

to the Adler parametric amplifier. In a paramp you can vary the capacitance of a tuned circuit with a varactor (voltage-variable capacitor) by driving it with a "pump" signal. Paramps are used for low-noise amplification.

Masers. "Maser" is an acronym for microwaver amplification by stimulated emission of radiation. These things are basically atomic or molecular amplifiers, tricky to make and use but delivering the lowest noise of any amplifier.

GaAs FETs. The latest word in simple microwave amplifiers. Performance is comparable to that of paramps, without the fuss and bother. Present-day commercial GaAs FETs will deliver 28dB of gain at 10GHz with a 2dB noise figure. The latest in low-noise GaAs FETs are the so-called HEMTs – high-electron-mobility transistors – which can deliver astonishing noise figures [e.g., 0.12dB($T_N = 8°K$) at 8.5GHz in an experimental chilled amplifier].

Klystrons and traveling-wave tubes. Vacuum-tube amplifiers used at microwave frequencies, klystrons and TWTs take advantage of transit-time effects within the tube. A variation known as a reflex klystron works as an oscillator by bouncing its electron beam into its guts. There are klystrons available that can continuously deliver 0.5MW RF output at 2000MHz.

Magnetrons. The heart of radar and microwave ovens: a high-power oscillator tube, full of little resonant cavities, and operated in a large magnetic field to make the electrons spiral around inside.

Gunn diodes, IMPATT diodes, PIN diodes. These exotic devices are used extensively at UHF and microwave frequencies. Gunn diodes are used as low-power oscillators in the 5–100GHz range, delivering output powers of 100mW or so. IMPATT diodes are analogous to klystrons, with capabilities of a few watts at a few gigahertz. PIN diodes behave as voltage-variable resistances and are used to switch microwave signals on and off by becoming a short circuit across a waveguide. PIN diodes can be used as lumped attenuators or switching elements at lower radiofrequencies. A leader in this field is Unitrode Corp. (Lexington, MA), whose "PIN diode designers' handbook and catalogue" is indispensable.

Varactors, SNAP diodes. Varactors are reverse-biased diodes used as variable capacitances for tuning purposes, or in paramps. Because of their nonlinearity they are also used for harmonic generation, i.e., as frequency multipliers. SNAP diodes are also popular for harmonic generation, since they exhibit sub-picosecond rise times.

Schottky diodes, back-diodes. You have seen Schottky diodes earlier as high-speed diodes with low forward drop. They're often used as mixers, as are back-diodes, variations of tunnel diodes. Look at Section 13.13 for a nice back-diode application (square-law detector).

HIGH-SPEED SWITCHING

The same effects that limit linear amplifier performance at high frequencies (the combination of junction capacitance, feedback capacitance, with its Miller effect, and stray capacitance in combination with finite source and load resistance) impose speed limitations on high-speed digital circuits. Many of these problems don't affect the designer directly, since they've been handled well in the design of digital ICs themselves. The average circuit designer would have a difficult time even coming close to the performance of TTL circuits, for example, using discrete transistor design.

Nevertheless, there are plenty of occasions when you've got to know how to design fast switching circuits. For example, in driving some external high-voltage or high-current load (or a load of opposite polarity) from a logic output, it is quite easy to lose a factor of 100 in switching speed through careless design. Furthermore, there are situations in which no packaged digital logic is used at all, and you're on your own all the way.

In this section we will begin with a simple transistor model useful for switching-circuit calculations. We will apply it to a few example circuits to show how it goes (and how important the choice of transistor can be). We will conclude by illustrating transistor switching design with a complete high-speed circuit (a photomultiplier preamp/discriminator).

13.23 Transistor model and equations

Figure 13.50 shows a saturated transistor switch, connected as an inverter, driven

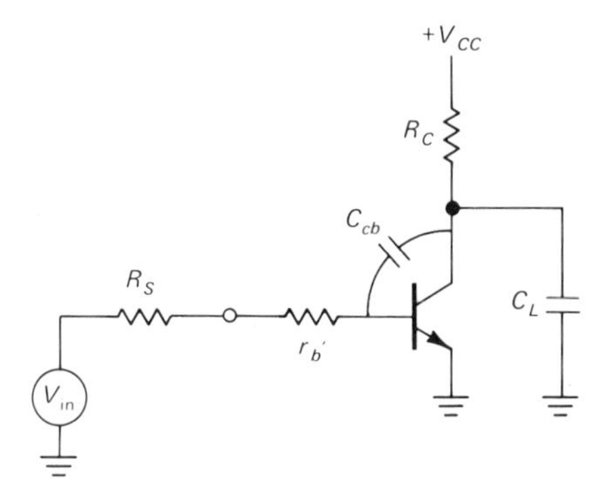

Figure 13.50

from a source of pulses with extremely fast rise and fall times. R_s represents the source impedance, r_b' is the relatively small intrinsic transistor base "spreading" resistance (of the order of 5Ω), C_{cb} is the all-important feedback capacitance, and R_C is the load resistance, paralleled by load capacitance C_L. The effects of finite load resistance can be included by letting R_C represent the Thévenin equivalent of the combined resistances, with V_{CC} suitably modified. The collector-to-emitter capacitance has been absorbed into C_L, and C_{be} has been ignored, since C_{cb} always dominates owing to Miller effect.

Figure 13.51 shows a typical output waveform from this circuit when driven

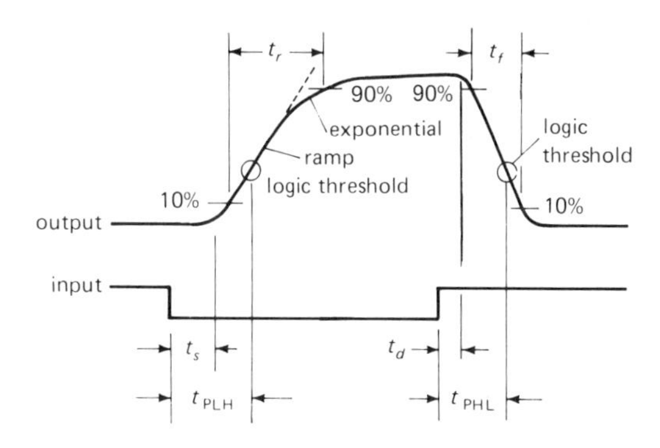

Figure 13.51. Waveform parameters of a transistor switch.

by a clean negative-going input pulse. The rise time, t_r, is usually defined as the time to go from 10% to 90% of the final value, with the corresponding definition of fall time, t_f. Note especially the relatively long storage time, t_s, that is required for the transistor to come out of saturation, compared with the correspondingly shorter delay time, t_d, to bring the transistor into conduction. These definitions are all conventionally taken between the 10% and 90% points. More useful for digital logic purposes are the propagation times, t_{PLH} and t_{PHL}, defined as the time from the input transition until the output passes through the logic threshold (rising or falling, respectively). Other symbols are in common use, e.g., t_{pd1} or t_{pr} is often used for what we've called t_{PLH}.

Let's use the circuit model to estimate rise and fall times for a given circuit. In the process you will even come to understand why the rising portion of the output waveform sometimes ends with an exponential.

□ *Estimation of rise time*

After the input signal has dropped to its LOW state and t_s has elapsed (more on that later), the collector begins to rise. Two effects limit the rate of rise: (a) R_C in combination with C_{cb} and C_L sets a time constant, generating an exponential rise toward V_{CC}, but (b) if that rate of collector rise is great enough, the resulting current through C_{cb} generates forward base bias across the source impedance $(R_s + r_b')$, and it can turn on the base, with the resultant effect of slowing the collector rise through negative feedback. What you have in the latter case is an integrator, and the collector waveform is a ramp. In general (depending on circuit values and transistor parameters), the collector waveform may begin as a ramp and change over to an exponential, as shown previously.

The scope photo in Figure 13.52 illustrates these effects. Instead of an *npn* transistor, however, we've used an *n*-channel enhancement-mode MOSFET. It behaves similarly, but the larger gate turn-on voltage makes the waveforms particularly clear. In addition, MOSFETs do not have storage or delay time effects, and they have no dc input current, which keeps things simple. Figure 13.53 shows the circuit, with our intentionally exaggerated signal source impedance. Note how feedback capacitance pins the gate voltage at the turn-on threshold while the drain switches. Note also that the rising drain waveform changes over to an exponential, when R_D is large.

A simple way to estimate circuit behavior is the following:

1. Compute the "integrator-limited" rate of rise of collector voltage, according to

$$\frac{dV_c}{dt} = \frac{V_{\rm BE} - V_{\rm in}({\rm LOW})}{C_{cb}(R_S + r_b')}$$

2. Find the collector voltage V_X at which the output waveform changes from a ramp to an exponential, according to

$$V_X = V_{CC} - \left(\frac{V_{\rm BE} - V_{\rm in}({\rm LOW})}{R_S + r_b'} + C_L \frac{dV_C}{dt} \right) R_C$$

This allows you to determine the collector waveform and rise time, as we will illustrate with examples presently. If V_X comes out negative, that means the entire collector ruse is exponential: The capacitive load dominates, and the base is never turned on via current through the feedback capacitor. The term r_b' is usually negligible.

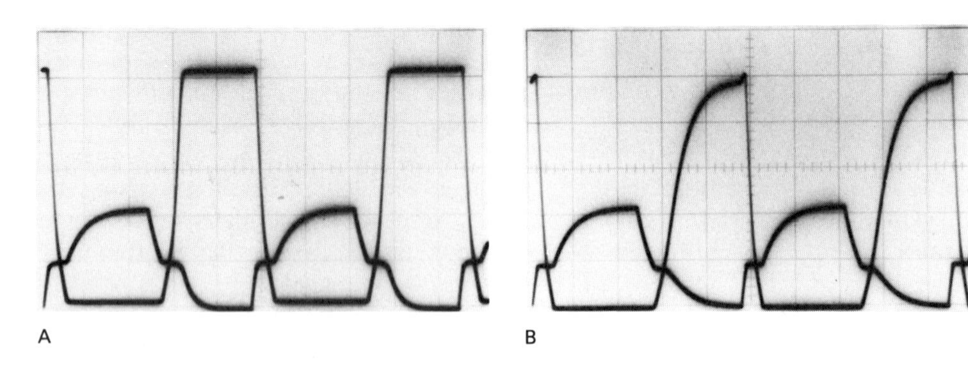

Figure 13.52. Gate and drain voltage switching waveform showing dynamic Miller clamping with (exaggerated) 100k source impedance. Vertical: 2V/div; signal frequency: 6kHz.
A. With 10k drain resistor.
B. With 200k drain resistor.

EXERCISE 13.1

Derive the two preceding formulas. Hint: For the second formula, equate the feedback current needed to bring the base into conduction with the available collector pull-up current less the current needed to derive the (capacitive) load.

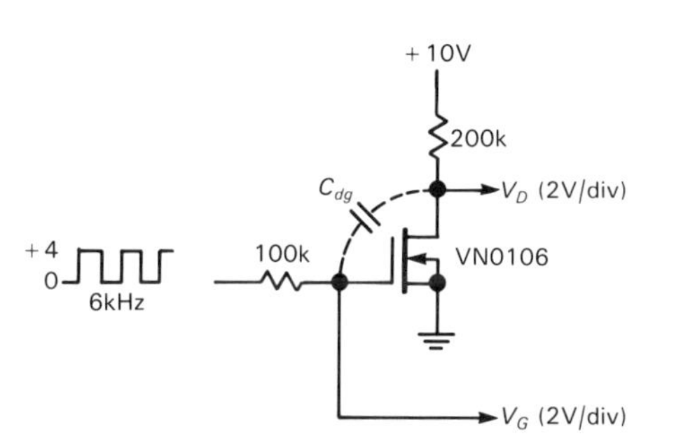

Figure 13.53

Estimation of fall time

Following the short delay time after the input has gone HIGH, the collector begins to drop, on its way to saturation. With a little bookkeeping of currents, it is easy enough to see that the collector current is given by

$$\begin{aligned} I_C &= \left(\frac{V_{\text{in}}(\text{HIGH}) - V_{\text{BE}}}{R_S + r_b'} + C_{cb}\frac{dV_c}{dt}\right) h_{fe} \\ &= \frac{V_{cc} - V_c}{R_c} - (C_L + C_{cb})\frac{dV_c}{dt} \end{aligned}$$

where the first line is the net base current multiplied by h_{fe} and the second line is the available collector current through R_C, less the current needed to drive the capacitance seen at the collector. Remember that dV_C/dt is negative. Rearranging, we get

$$-\frac{dV_c}{dt} = \frac{1}{C_L + (h_{fe}+1)C_{cb}} \times \left(\frac{V_{\text{in}}(\text{HIGH}) - V_{BE}}{R_S + r_b'} h_{fe} - \frac{V_{cc} - V_c}{R_c}\right)$$

where the first term in parentheses is recognizable as $h_{fe} i_{\text{drive}}$ and the second is $i_{\text{pull-up}}$. You are now licensed to try some circuits; you will be able to see what sort of rise times and fall times can be expected and which capacitance dominates. First, however, a word on delay and storage times.

Delay and storage times

In general, delay times are very short. The main effect is the time constant involved in moving the base capacitance up to V_{BE}, a time constant of order

$$T \sim (R_s + r_b)(C_{cb} + C_{be})$$

When working at extremely high speeds, transistor transit-time effects may also become important.

Storage times are another matter. A transistor in saturation has charge stored in the base region, and even after the base drive signal has gone close to ground (or even a bit negative), it requires a relatively long time for the extra injected minority carriers from the emitter to be swept from the base region by the collector current. Transistors differ widely in storage time; it can be shortened by using less base overdrive during saturation and by reverse-biasing the base to reverse the base current when switching the transistor OFF. This equation for storage time, t_s, makes these points:

$$t_s = K \ln \frac{I_B(\text{ON}) - I_B(\text{OFF})}{\frac{I_C}{h_{FE}} - I_B(\text{OFF})}$$

where $I_B(\text{OFF})$ is negative for "discharge" reverse base currents. The constant K includes a "minority-carrier lifetime" term, which is greatly reduced by gold doping. However, such doping reduces h_{FE} and increases the leakage current. This explains the good speed performance of TTL, along with its low breakdown voltage (7V).

Storage times can be as long as several hundred nanoseconds, and they are typically an order of magnitude longer than delay times. The popular general-purpose

2N3904, for instance, has a specified maximum delay time of 35ns and a storage time of 200ns under standardized test conditions, which include driving the base negative by two diode drops.

Since storage times can turn out to be a severe limitation on the performance of high-speed switching circuits, there are several measures that can be taken to circumvent the problem. One solution is to avoid saturation altogether. A Schottky clamping diode (a "Baker clamp") from base to collector will accomplish this by robbing current from the base when the collector is nearing saturation. It prevents transistor saturation, since its forward voltage drop is less than that of the collector-base junction. The Schottky families of TTL logic use this trick. A small "speedup" capacitor (25–100pF) across the base driving resistor is often a good idea in addition, since it can reduce storage time by providing a pulse of current to remove base charge at turn-off, and in addition it increases base drive current during turn-on transitions. Figure 13.54 illustrates these methods.

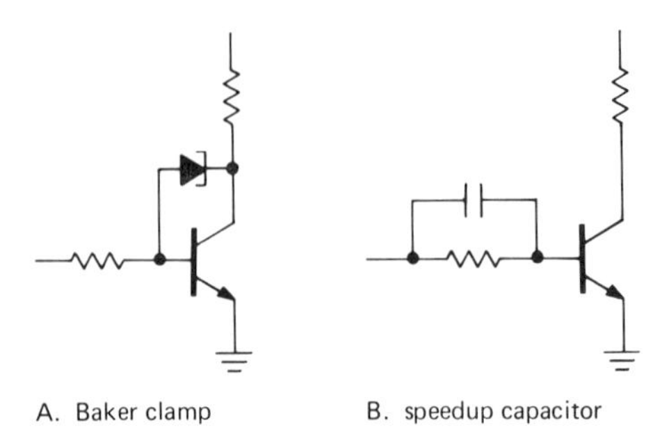

Figure 13.54

13.24 Analog modeling tools

It should be evident that designing both high frequency amplifiers and high speed switching circuits with discrete components is a complex subject, particularly when the full effects of parasitic capacitances and inductances are properly accounted for. Our simplified models provide good circuit intuition, but they may often be inadequate if you are trying to squeeze maximum performance from an amplifier in the gigahertz range of frequencies, for example. The traditional approach consisted of a combination of more complex modeling (plenty of calculation!) and lots of breadboarding.

A pleasant development is the maturation of computer-aided analog modeling, in particular a program called "SPICE" and its several commercial extensions. SPICE (originated by L. W. Nagel) models the operation of your trial circuit (using a library of sophisticated component models), predicting the gain, distortion, noise, frequency response, etc. You can ask it to show you the waveforms (voltage *and* current) at any point in the circuit – a computerized oscilloscope! Thus you can fiddle with your hypothetical circuit, trying speedup capacitors, exploring component and circuit trade-offs, and so on. In fact, good simulators even let you see the effects of component tolerance, via either conventional worst-case analysis or a more sophisticated "Monte Carlo" statistical treatment. Some modeling programs also perform a "sensitivity" analysis, telling you which components affect performance the most.

Computer-aided modeling programs are inexpensive and available for desktop computers as well as mainframes (e.g., IsSpice from Intusoft, or PSpice from Microsim). Most accept "netlists" as generated by schematic capture programs (Section 12.08). Modeling programs are not restricted to high frequency design. They can be used to model circuits with op-amps, digital circuits, and the innards of ICs themselves. Although SPICE dominates the current scene, there are at least a dozen competing modeling programs, optimized for specific applications.

SOME SWITCHING-SPEED EXAMPLES

In this section we will analyze the performance of a few simple circuits, based on the methods just discussed.

13.25 High-voltage driver

Let's begin with the circuit in Figure 13.55. It is a simple inverting stage intended for driving a piezoelectric crystal with 100 volt pulses, generated originally with TTL logic. The TTL output, and therefore the base driving signal, is roughly as indicated. In these calculations we will ignore r'_b, which is small compared with the source impedance.

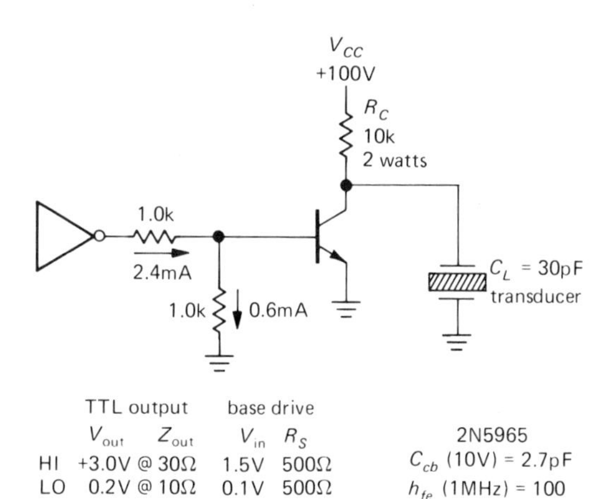

Figure 13.55

Rise time

We begin by calculating the integrator-limited collector rise:

$$\frac{dV_c}{dt} = \frac{V_{\mathrm{BE}} - V_{\mathrm{in}}(\mathrm{LOW})}{C_{cb}R_S} \approx 450\mathrm{V}/\mu\mathrm{s}$$

from which the estimated rise time will be

$$t_r = \frac{0.8V_{cc}}{dV_c/dt} \approx 180\mathrm{ns}$$

Now we find the collector voltage at which the rise changes from an integrator-limited ramp to an exponential:

$$V_X = V_{CC} - R_C\left(\frac{V_{\mathrm{BE}} - V_{\mathrm{in}}(\mathrm{LOW})}{R_S} + C_L\frac{dV_C}{dt}\right) \approx -50\mathrm{V}$$

This means that the collector rising waveform is exponential the whole way, with the feedback current $(C_{cb}dV_C/dt)$ insufficient to pull the base up into conduction, given the source impedance. The collector time constant is $R_C(C_L + C_{cb})$, or 0.33μs, with a rise time (10% to 90%) of 2.2 time constants, or 0.73μs. It is clear that the combination of collector resistor and load capacitance dominates the rise.

Fall time

To analyze the fall time, we use the formula derived earlier to find

$$-\frac{dV_c}{dt} = \frac{1}{C_L + (h_{fe}+1)C_{cb}} \times \left\{h_{fe}\left(\frac{V_{\mathrm{in}}(\mathrm{HIGH}) - V_{\mathrm{BE}}}{R_S}\right) - \frac{V_{CC} - V_C}{R_C}\right\} \approx 530\mathrm{V}/\mu\mathrm{s}$$

$$t_f = \frac{0.8V_{CC}}{dV_C/dt} \approx 0.15\mu\mathrm{s}$$

The last term depends on V_c, but is negligible compared with the first term in parentheses. If it weren't, you would have to evaluate it at several values of collector voltage to get a good picture of the falling waveform. At this point it should be noted that the calculated fall time corresponds to a frequency of about 3MHz, and therefore the value of h_{fe} =100 we used is realistic (f_T = 300). If a calculated rise time or fall time corresponds to a frequency much higher than originally assumed, it is generally necessary to go back and recompute the transition time, using a new h_{fe} based on a better estimate of the transition time. This iteration process will usually give a satisfactory answer on the second pass.

□ *Switching waveform*

For this circuit, then, the collector waveform is as shown in Figure 13.56. The rise is dominated by the time constant of the load capacitance and collector resistor, whereas the fall is dominated by the feedback capacitance in combination with the source impedance. To put it another way, the collector voltage falls at a rate such that the current through the feedback capacitance is almost sufficient to cancel the base drive current and bring the base out of conduction. Note that we have assumed throughout that the TTL output waveform is much faster than the output of our circuit. With typical rise and fall times of about 5ns, that is a good approximation.

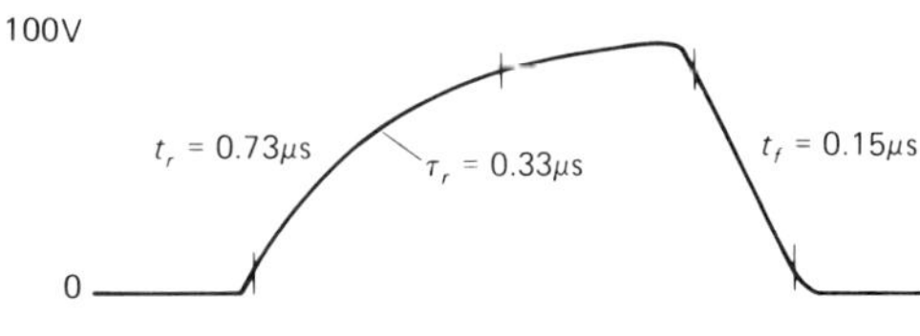

Figure 13.56

□ 13.26 Open-collector bus driver

Suppose we want to drive an open-collector TTL bus from the output of an NMOS circuit. We can do it by interposing an *npn* inverting stage, as in Figure 13.57. The base resistors are necessarily large because of the low output sourcing capability of NMOS operating from +5 volts (see Section 9.09). We have chosen two popular transistors in order to illustrate the effect that parameters like C_{cb} can have.

For rise time, we calculate as before, and we find the following integrator-limited rise times:

	2N5137	2N4124
dV_C/dt	8.5/μs	76V/μs
t_r	470ns	53ns

The crossover to exponential is calculated as:

	2N5137	2N4124
V_X	4.4V	1.1V
time const	66ns	52ns

The falling waveform comes out:

	2N5137	2N4124
dV_C/dt	−11V/μs	−78V/μs
t_f	360ns	51ns

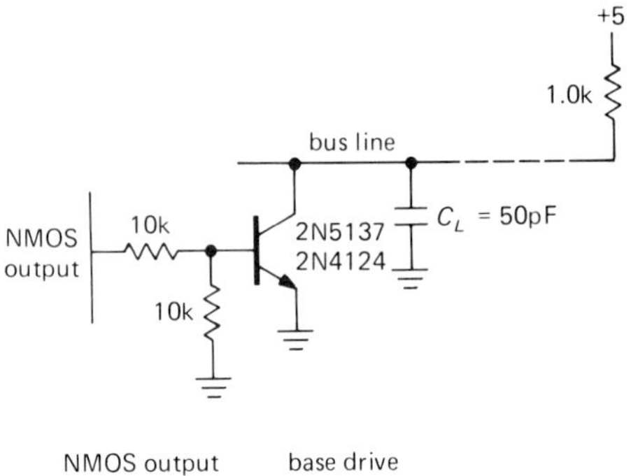

	NMOS output		base drive	
	V_{out}	Z_{out}	V_{in}	R_S
HI	+3.5	@ 1kΩ	1.7V	5.5k
LO	0.0V	@ 200Ω	0.0V	5.1k

	2N5137	2N4124
C_{cb} (10V)	16pF	1.8pF
h_{fe} (1MHz)	100	100

Figure 13.57

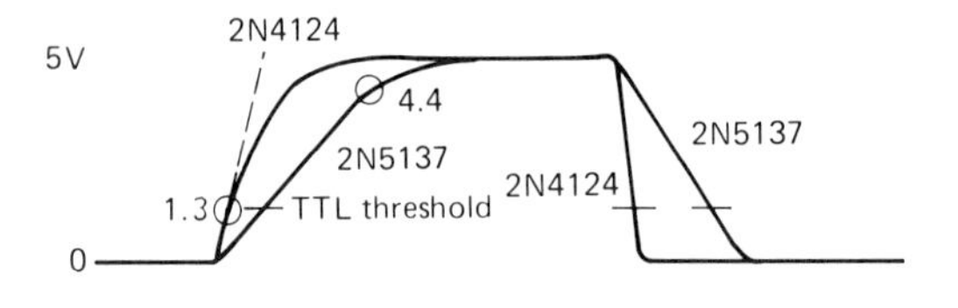

Figure 13.58

□ *Choice of transistor*

The situation is as shown in Figure 13.58. The inferior performance of the 2N5137 is due entirely to the effects of feedback capacitance, aggravated in this example by the relatively high value of source impedance. The transition times of the 2N4124 are probably a bit optimistic, since they correspond to a frequency of about 10MHz at which h_{fe} is somewhat lower than assumed.

It is interesting to look at the time required to reach the TTL threshold voltage of about 1.3 volts, the relevant parameter in a system in which TTL gates are driven by the bus signals. Ignoring storage and delay times, the times to reach the TTL threshold voltage are the following:

	2N5137		2N4124	
	calc	*meas*	*calc*	*meas*
rise(t_{PLH})	150ns	130ns	17ns	30ns
fall(t_{PHL})	340ns	360ns	47ns	52ns

The rise and fall times we measured are in reasonable agreement with the predictions of our somewhat simplistic model, except perhaps for the rise time of the 2N4124 circuit. There are a few possible explanations why the rise time we predicted is too small in that case: The calculation used the value for h_{fe} at 10MHz, whereas a 17ns rise time implies somewhat higher frequencies, and hence a lower value of h_{fe}. Also, by actual measurement the particular transistor has C_{cb} =2.2pF at 10 volts and C_{cb} =3pF at 2 volts. Curiously, the 2N5137 we used had a much lower C_{cb} (about 5pF) than specified on the data sheet, and so we added a small external capacitor to bring it up to specifications. This probably represents a change in the manufacturing process since the original data sheet was published.

EXERCISE 13.2

Verify the results just calculated for dV_C/dt (rise and fall) and V_X.

□ *Pull-up to +3 volts*

Note that the times to reach TTL threshold from a HIGH state are much longer than the times from a LOW state, even though the output slew rates (in the case of the 2N4124 circuits) are almost the same. That's because the TTL threshold voltage is not symmetrically positioned between +5 and ground, forcing the collector to slew through a larger voltage on the way down. For this reason TTL buses are often terminated to a source of +3 volts (a series pair of diodes tied to +5 is one trick sometimes used), or each line of the bus can be terminated with a voltage divider, as in Figure 13.59.

EXERCISE 13.3

Calculate the rise and fall times and the propagation delays for a 2N4124 driving the preceding bus with C_L =100pF. Show your work.

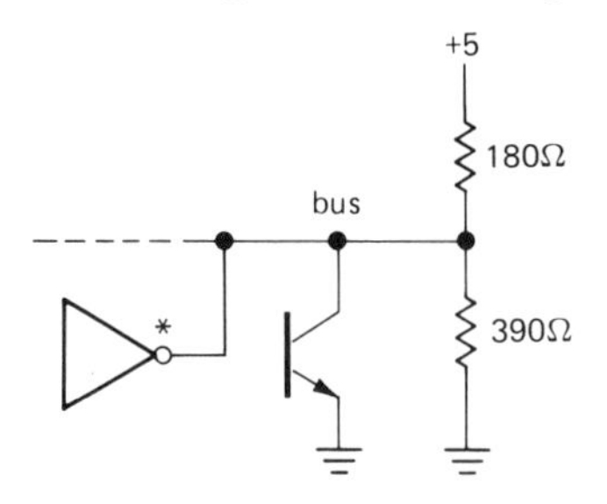

Figure 13.59

□ 13.27 Example: photomultiplier preamp

As we will discuss in Chapter 15, a device called a photomultiplier tube (PMT) is an extremely useful light detector, combining high sensitivity with high speed. Photomultipliers are also useful in applications where the quantity being directly measured isn't light, e.g., high-energy-particle detectors in which a scintillator crystal generates light flashes in response to particle bombardment. To take advantage of a photomultiplier's properties, it is necessary to use a charge-sensitive high-speed discriminator, a circuit that generates an output pulse when an input pulse of charge exceeds some threshold corresponding to the detection of a photon of light.

Figure 13.60 shows a circuit of a high-speed photomultiplier preamp and discriminator that illustrates the high-frequency and switching techniques discussed in this chapter. The output of the

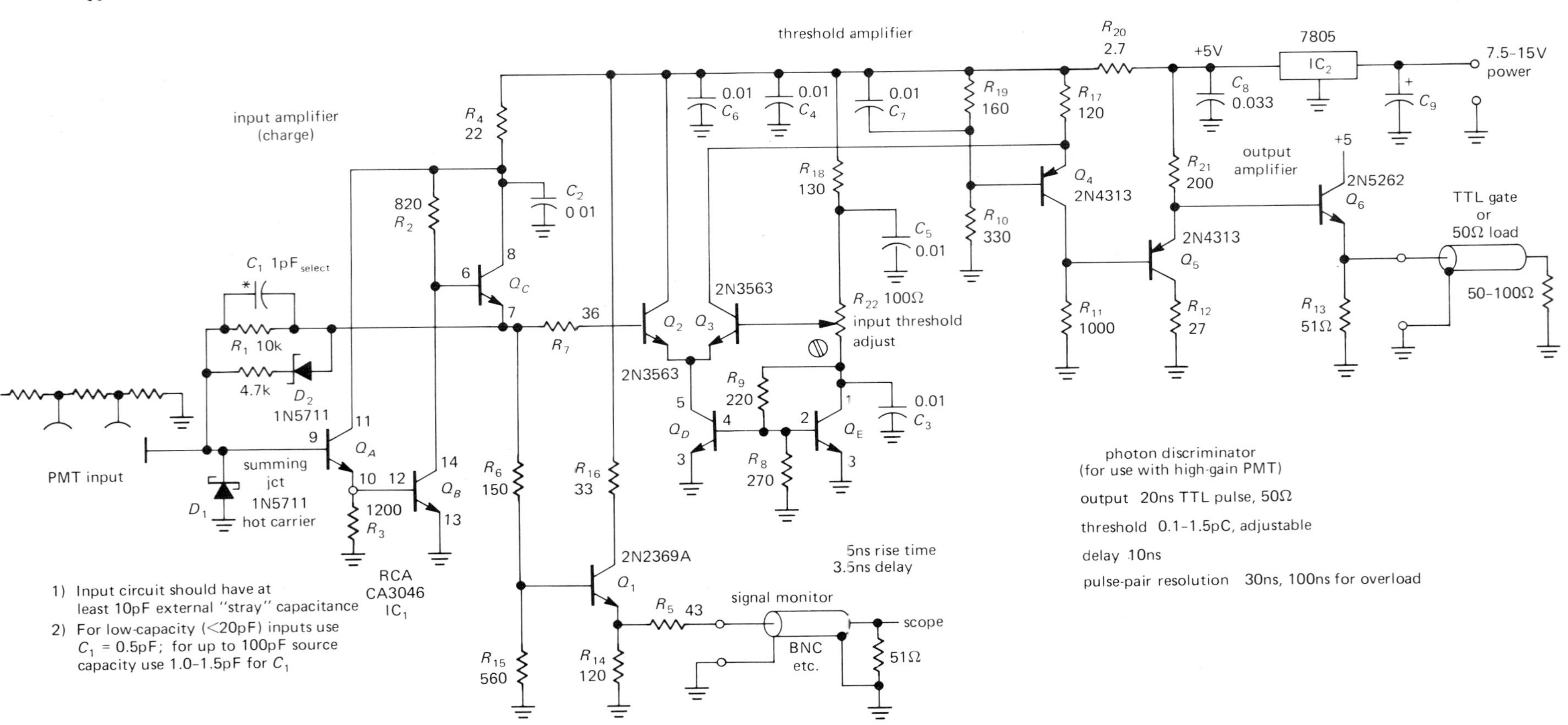

Figure 13.60. High-speed charge-sensitive amplifier for photon counting with photomultiplier tubes.

photomultiplier tube consists of negative pulses of charge (electrons are negative), each pulse having a width of perhaps 10–20ns. The larger pulses correspond to detected photons (quanta of light), but there are also lots of smaller pulses that arise from noise within the photomultiplier tube itself and that should be rejected by the discriminator.

□ *Circuit description*

The circuit begins with an inverting input amplifier (Q_A - Q_C) with current and charge feedback via C_1 and R_1. The input follower presents a low driving impedance to Q_B (which provides the voltage gain) to reduce the effects of feedback capacitance (C_{cb}). The follower at the output of the gain block, Q_C, provides a low output impedance while allowing Q_B to have a reasonable amount of gain. The signal at this point is a small positive pulse corresponding to the negative-charge input from the PMT; dc feedback stabilizes Q_C's output at about $2V_{BE}$. Q_1 is biased as a class A emitter follower, giving a low-impedance "monitor" output of the amplified photomultiplier pulses before discrimination.

Differential amplifier Q_2 and Q_3 form the discriminator; the threshold is set by R_{22} referenced to a voltage (set by Q_E, operating as an "adjustable diode") that tracks the input amplifier's $2V_{BE}$ quiescent point. This diode-drop tracking occurs because the transistors Q_A–Q_E are in a monolithic transistor array (CA3046) and are all at the same temperature. Q_4 forms an inverted cascode with Q_3, both for high speed and level shifting. Two stages of follower, arranged with opposite-polarity transistors to cancel V_{BE} offsets, complete the circuit.

There are several interesting points in this circuit. Transistor quiescent currents are set rather high (the differential pair Q_2Q_3 has 11mA emitter current, Q_5 idles at 20mA, and the output transistor has to source 120mA to drive a 50Ω load) in order to get a good high-speed performance. Note that the cascode base (Q_4) is bypassed to V_+, not ground, since its input signal is referenced to V_+ via R_{17}. The comparator's emitter current source is a current mirror, convenient since Q_E is already used for the threshold reference. D_1 and D_2 are used to improve overload performance. Although it complicates the circuit, clamping diode D_1 can be returned to Q_E's collector (bypassed to ground) to put a tighter limit on negative (overload) swings at the input.

□ *Performance*

Figure 13.61 shows a graph of output pulse shape and timing versus input pulse size (measured as quantity of charge). The output pulses are stretched by large input overloads, but the overall performance is quite good, measured by usual photomultiplier preamp standards.

SELF-EXPLANATORY CIRCUITS

13.28 Circuit ideas

Figure 13.62 shows a few wideband circuit ideas.

ADDITIONAL EXERCISES

(1) In this problem you are to work out in detail the high-frequency behavior of the circuit in Figure 13.10, summarized briefly in Section 13.05. ***(a)*** Begin by repeating the calculation of driver/output-stage rolloff diagrammed in Figure 13.12. Be careful as you combine complex impedances. Write to one of the authors if you find an error! ***(b)*** Now check to see that the high-frequency rolloffs of the

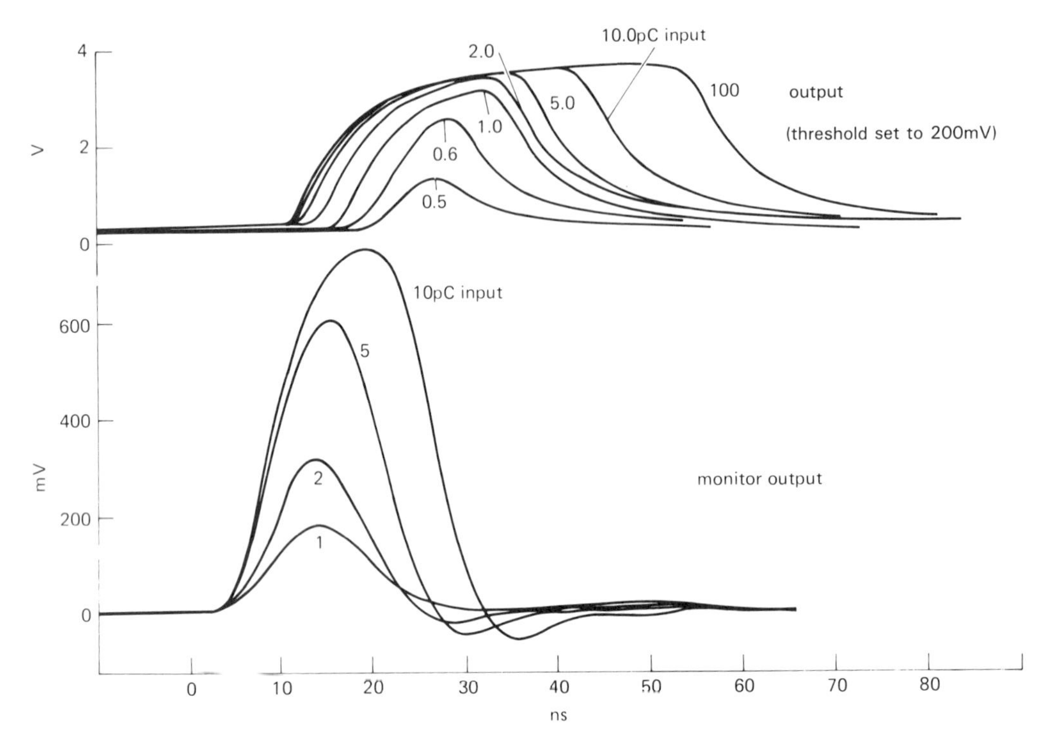

Figure 13.61. Pulse response of the photon-counting amplifier.

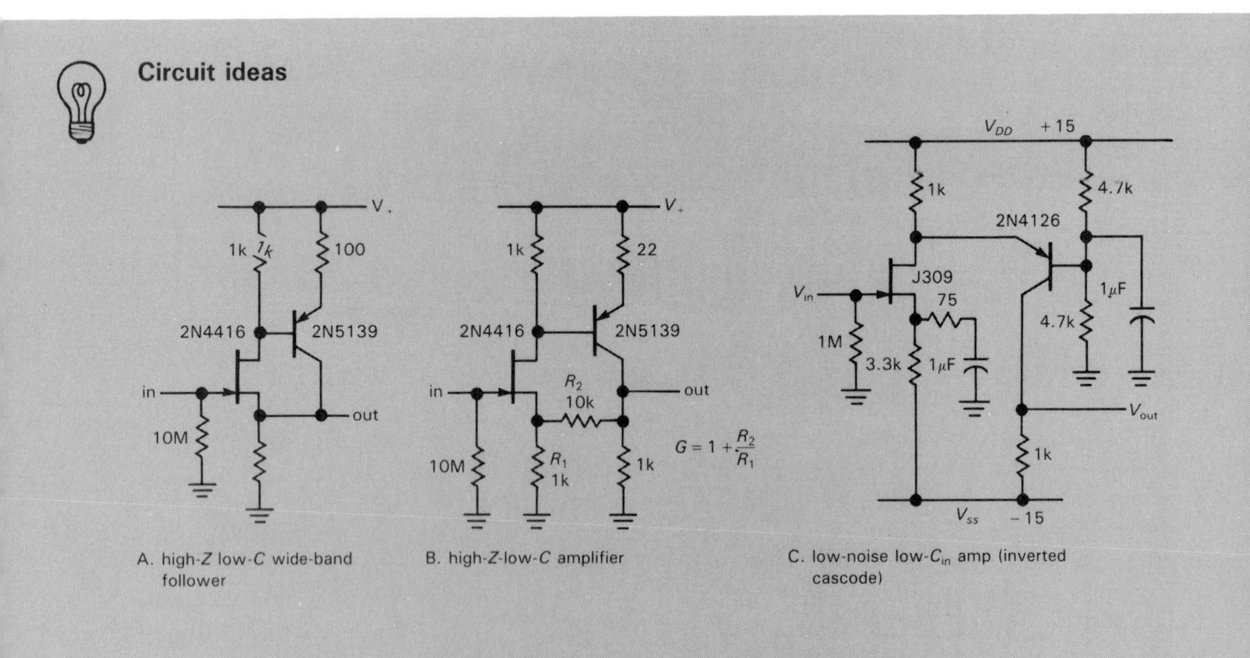

Figure 13.62

previous stages are significantly higher in frequency than the ~180MHz 3dB frequency of the output stage and its driver. In particular, check the following rolloffs: Q_1's output (emitter) impedance driving a capacitive load (see Fig. 13.11); Q_2's output driving a slightly different capacitive load (because Q_4's collector is not grounded); Q_3 and Q_4's emitters driving a capacitive load; Q_4's collector driving a capacitive load.

(2) What is the impedance looking into a length of coaxial cable that is ***(a)*** open-circuited at the far end, and a quarter wave long, electrically, at the frequency of interest, ***(b)*** short-circuited at the far end, and a quarter wave long, electrically, at the frequency of interest, ***(c)*** same as (a), but a half wave long, ***(d)*** same as (b), but a half wave long? The result in (d) is the basis of the so-called choke joint used in waveguides.

(3) Work out in detail the rise time and fall time of the high-voltage switching circuit in Figure 13.55, as summarized in Section 13.25. Use $V_{BE} = 0.7$V.

(4) "The Rise and Fall of a Bus Driver": Calculate the rise and fall times for the TTL bus driver circuit of Figure 13.57, as summarized in Section 13.25. Use $V_{BE} =$ 0.7V.

(5) Design a video amplifier with a gain of +5 and a rolloff of 20MHz or more. The input impedance should be 75 ohms, and the output should be able to drive a 75 ohm load with 1 volt pp output capability. A nice way to achieve the noninverting gain is to use a common-base input stage with an emitter follower output, as suggested in Figure 13.63. If you like the circuit, finish the design by choosing operating currents, resistor values, and biasing components. You can, of course, use something like a differential amplifier/cascode/follower combination, if you prefer. Note that the gain must be noninverting, or the image will be reversed.

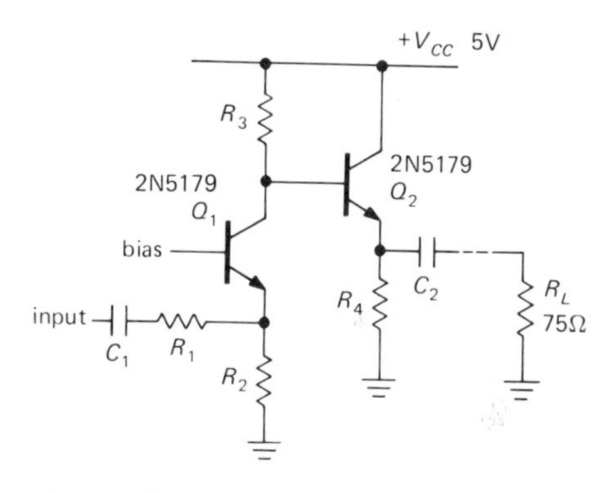

Figure 13.63

INTRODUCTION

Lightweight hand-held instruments, data loggers that make measurements at the ocean bottom, digital modems that power themselves from the "holding current" in the telephone line – these are just a few of the applications that invite (or demand) low-power electronic design techniques. Within such instruments you often find examples of all the areas of design already treated in this book – regulated supplies, linear circuits (both discrete and op-amp), digital circuits (almost invariably CMOS) and associated conversion techniques, and, increasingly, microprocessor circuits of considerable complexity. Although we have occasionally discussed power consumption and speed/power trade-offs (when comparing logic families, for example), the design of micropower electronic instruments involves special techniques and cautions throughout and requires a chapter all its own.

We will begin by considering a range of applications where low-power consumption is desirable or essential. They're not all of the exotic transmitter-attached-to-wild-moose variety, and in fact there are plenty of places where ac line power is plentiful, but where battery power is more convenient or performs better. Next we will review the power sources that make sense for low-power electronics, beginning with the ubiquitous "primary" (nonrechargeable) cells (alkaline, mercury, silver, lithium) and the closely related "secondary" (rechargeable) cells (nickel-cadmium, lead-acid, "gel" cells). We've hounded the manufacturers of batteries mercilessly for their latest technical literature, and so we've got really useful battery comparison data – energy content, discharge characteristics, effects of temperature, discharge rate, and storage conditions, etc. We'll help you pick the right battery for your application.

Batteries aren't the only power sources to consider, so we'll continue by talking about those little black plastic "wall-plug-in" modules you get with the consumer electronics gizmos you can buy. Wall plug-ins are remarkably inexpensive and are available as simple transformers, or unregulated dc supplies (transformer-rectifier-capacitor), or as complete regulated dc supplies; they also come in dual- or triple-supply varieties. Solar cells are useful in some unusual applications, so we'll discuss them too. Finally, we'll mention the use of signal currents, for example the dc current you find on the telephone line, or the ac relay energizing current in a thermostat or doorbell, to activate a micropower instrument.

We'll then talk about low-power design at the same level of detail that we've used in the rest of the book, mirroring a similar progression of subjects: regulators and references, then linear circuitry (discrete and op-amp), digital circuits and conversion techniques, and finally microprocessors and memories. There are, in addition, techniques that have no counterpart in ordinary design, for example "power switching," in which a normal circuit is rendered micropower by applying dc for extremely short intervals; for example, a microprocessor data logger might be powered for 20ms every 60 seconds. We'll talk about some of those tricks, and the curious pitfalls that await the uninitiated.

Finally, we'll talk about packaging – including the small plastic instrument cases that are widely available, complete with back door for access to the batteries. Micropower instruments are generally much easier to package than conventional instruments are, since they're usually lightweight, they don't produce heat, and they don't need the usual paraphernalia of power cords, line filters, and fuses. Micropower design is different and fun, and it presents new challenges for the jaded electronic designer. Read on to learn more about this exciting (but often neglected) subject!

We hate to say it, but this chapter could be passed over in a first reading of the book.

14.01 Low-power applications

We've collected together most of the reasons that might motivate you to do low-power circuit design. They're listed here, in no particular order.

Portability

You can't carry it around with you if it has a power cord running off to a wall outlet. "It" could be a commercial product such as a calculator, wristwatch, hearing aid, "walkabout" tape recorder or receiver, paging radio, or digital multimeter. Or it might be a custom portable instrument, for example a small transmitter used to study herd migration and physiology. Because batteries have a finite energy content, you've got to keep power consumption low in order to have reasonable battery life with acceptable battery weight. A multimeter that runs 1000 hours on a single 9 volt battery will outsell a competing unit that requires four D cells and gives only 100 hours of service. A portable transmitter for animal migration studies is useless if it runs only two days on a fresh set of batteries. Thus, low-power circuitry is at a premium in instruments designed for portability. In the special case of extremely small instruments (e.g., a wristwatch), the tiny energy content of the self-contained batteries dictates micropower design, with total current drains of just a few microamps.

Isolation

Instruments that are powered from ac line current are not suitable for some kinds of "floating" measurements at high potential. For example, you might want to measure microamp charged-particle-beam currents at the +100kV terminal of a particle accelerator. It may well be that you can't make the measurement by lifting the low-voltage end off ground (as in Fig. 4.79), because the power transformer of the high-voltage supply causes significant 60Hz currents to be capacitively coupled to the high-voltage supply through its transformer (or perhaps because of corona discharge and other high-voltage leakage effects, which add spurious current as measured at the ground return). If you try to build an ac-powered instrument to measure the current, say by using a differential amplifier connected across a precision resistor in the high-voltage lead, the power supply of your instrument will have to use a special power transformer rated at 100kV insulation breakdown so that the measuring circuit (op-amps, readouts) can float at 100kV. Since such a transformer is almost impossible to find, this is a good place for a circuit powered by a battery (or possibly by a solar cell, illuminated by light beam across the potential gap), intrinsically isolated from the power line and from earth ground.

In the foregoing example, an ac-powered circuit would have another problem, namely that it would impress some 60Hz ripple onto the circuit it was attempting to measure, due to capacitive coupling and leakage of 60Hz currents through its power transformer. So the power transformer would have to be of special design to ensure low inter-winding capacitance and low leakage currents. This problem of coupled 60Hz ripple can also crop up in conventional circuits that deal with signals at very low levels, for example weak audio signals. Although such problems can usually be solved by careful design, the isolation provided by a battery-powered preamp can be a real advantage in these situations.

Little power available

Dialers, modems, remote data-acquisition systems that send through the telephone lines – these are examples of instrument designs that can be powered by the holding current of the telephone line itself (it's about 50V dc open-circuit, driven from an impedance of about 600Ω; you must load it so that the dc voltage is below 6V, in order for the telephone company to think you're "off hook," and therefore to maintain the connection). Likewise, "smart" thermostats for heating systems often use NiCd rechargeable cells for their dc power, charged during intervals when the relay is not activated by the low-current ac then available (usually a 24V ac transformer in series with a relay coil of a few hundred ohms resistance).

This same trick of powering your circuit with ac signaling currents could also be used with doorbell circuits, and any other application where low-voltage ac relays are used. Another example of extracting power from a signaling current is the use of "industrial sensor current loops," in which a dc current in the range 4mA to 20mA (or, sometimes, 10mA to 50mA) is used to send analog sensor measurements over a two-wire system. Modules using this standard typically permit a voltage drop of 5 to 10 volts; hence the opportunity to power remote instrumentation from the signal current itself.

For these applications you have available a power supply delivering currents of the order of a few milliamps across a few volts, which is enough to power relatively complex low-power circuits. It is certainly attractive to attempt signal-current-driven low-power design, given the more cumbersome alternative of separate ac power sources for the same instrument.

A final example of a power source that limits you to a few milliwatts is the use of solar cells to power instruments (and/or charge their batteries). There are wristwatches and inexpensive pocket calculators built this way, and they have the advantage of (a) staying sealed and (b) remaining inexpensive, respectively.

No power available

Battery operation really becomes essential when there's nothing else available. Examples include physical oceanography, where you may wish to deploy a set of sensors on the ocean bottom for six months, quietly logging ocean currents, sediments, salinity, temperature, and pressure, as well as environmental studies, where remote measurements of pollutants at inaccessible sites are required. In these applications you usually want extended operation on a set of batteries, sometimes up to a year or more; hence the need for careful micropower design.

There are other situations where ac power is available, but not convenient. Household examples include smoke detectors and wall clocks.

Minimizing heat management

Digital circuitry constructed with ECL or Schottky technologies can easily consume 10 watts or more per board, and a system of several such boards requires forced air cooling. On the other hand, the newer high-speed CMOS logic families (with names such as 74AC*xx* and 74ACT*xx*) deliver performance comparable to that of their advanced Schottky cousins, with negligible static power consumption and greatly reduced dynamic power consumption (Figs. 8.18 and 9.2). That means smaller power supplies, closed dirt-free enclosures (no fan), and better long-term reliability.

The same considerations apply to linear design, making low power consumption a desirable objective to keep in mind in almost any application, even when plenty of power is available.

Uninterruptibility

Momentary power interruptions often cause microprocessor-based instruments to re-initialize themselves, computers to crash, etc. A nice solution is the use of an uninterruptible power supply (UPS), usually in the form of a battery-powered dc-to-ac inverter with 115 volt 60Hz output, able to switch on automatically within a few milliseconds of a power interruption. Uninterruptible power supplies are available with power ratings of many kilowatts. The big ones are expensive and bulky; however, there are compact units, powered by a small bank of lead-acid gel cells (see Section 14.02), for systems that use less than a kilowatt of ac power. For truly low-power systems, a small UPS inverter or direct dc battery backup (as in Fig. 1.83) is convenient and is a good reason to practice low-power design.

POWER SOURCES

14.02 Battery types

The Duracell "Comprehensive Battery Guide" lists 133 off-the-shelf batteries, with descriptions like zinc-carbon, alkaline manganese, lithium, mercury, silver, zinc-air, and nickel-cadmium. There are even subclasses, for example Li/FeS_2, Li/MnO_2, Li/SO_2, $Li/SOCl_2$, and "lithium solid state." And from other manufacturers you can get sealed lead-acid and gel-type batteries. For the truly exotic application you might even want to consider fuel cells or radioactive thermal generators. What are all these batteries? How do you choose what's best for your portable widget?

The foregoing list divides into so-called *primary* and *secondary* batteries. Primary batteries are designed for a single discharge cycle only, i.e., they're nonrechargeable. Secondary cells (NiCd, lead-acid, and gel-type in the foregoing list), by comparison, are designed to be recharged, typically from 200 to 1000 times. Among primary batteries, you usually make your choice of chemistry based on trade-offs among price, energy density, shelf life, constancy of voltage during discharge, peak current capability, temperature range, and availability.

Once you've picked the right battery chemistry, you figure out which battery (or series combination of batteries) has enough energy content for the job.

Fortunately, it's pretty easy to eliminate most of the batteries in the catalogs, if you follow our first suggestion: *Avoid hard-to-get batteries.* Besides being hard to find, they're usually not fresh. So it's usually better to stick with the varieties available at the drugstore, or perhaps photography store, even if it results in somewhat less than optimum design. We particularly recommend the use of commonly available batteries in the design of any consumer electronic device; as consumers ourselves, we shun those inexpensive marvels that use exotic and expensive batteries. (Remember those early smoke detectors that used an 11.2V mercury battery?)

□ *Primary batteries*

Now for details. Table 14.1 compares the characteristics of the various primary cells, and Table 14.2 and Figure 14.1 give actual numbers for the most popular cells.

The old-fashioned "dry cell" with a cat on the outside is a LeClanche cell. Inside it's as primitive as you might guess, with a carbon rod stuck down into a cathode mixture of manganese dioxide, carbon, and ammonium and zinc chloride electrolytes. There's a cylindrical separator made of flour-and-starch paste, then a zinc anode outer can. The top is sealed on with wax and asphalt seals, designed to vent the innards if too much pressure builds up. These cells are the cheapest you can buy, but you don't get too much for your money. In particular, the voltage drops and the impedance rises steadily as the battery is used; furthermore, the battery's capacity drops drastically if used at high currents.

The "heavy-duty" dry cells are similar, but with a higher proportion of zinc chloride and correspondingly different mechanical design to accommodate greater gassing. Although their total energy content is only slightly greater than that of LeClanche cells, these cells are considerably better in delivering most of their rated capacity even when operated at high currents. For example, a LeClanche D cell delivers 4.2 amp-hours (Ah) into a 150 ohm load, 1.2Ah into 15 ohms, and 0.15Ah into 1.5 ohms; the equivalent zinc chloride cell delivers 5.6, 5.4, and 1.4 amp-hours, respectively. The zinc chloride cell also shows less dropoff of capacity at low temperatures.

The alkaline manganese cell, generally sold as simply "alkaline," is better still in high-current-discharge and low-temperature operation. It is inside-out, compared with zinc-carbon, having the powdered-zinc negative anode and potassium hydroxide electrolyte in the middle, surrounded by a manganese-dioxide-and-carbon outer positive cathode. For comparison with the numbers above, an alkaline D cell delivers 10 amp-hours into 150 ohms, 8Ah into 15 ohms, and 4Ah into 1.5 ohms. Because of its particular chemistry, an alkaline battery maintains a low and slowly increasing internal resistance as it discharges, compared with the rapidly rising internal resistance of both types of zinc-carbon cells. It also works better at low temperatures. Alkaline batteries have a longer shelf life than LeClanche or zinc chloride. As Figure 14.1 suggests, the cell voltage-versus-discharge curve for all three types of batteries lets you easily estimate the condition of the battery. Figure 14.2 shows comparative performance for the three kinds of "dry cells."

Mercury, silver oxide, lithium – these are the real premium cells, with greatly superior performance compared with alkaline and zinc-carbon. The mercury cell uses an amalgamated zinc anode, mercuric-oxide-plus-carbon cathode, and sodium or potassium hydroxide electrolyte. It excels in constancy of open-circuit

TABLE 14.1. PRIMARY BATTERIES

Type	Advantages	Disadvantages
Zinc-carbon (LeClanche) (standard "dry cell")	least expensive widely available	lowest energy density (1–2Wh/in^3) sloping discharge curve poor high-current performance impedance increases as discharged poor low-temperature performance
Zinc-carbon (zinc chloride) ("heavy duty" dry cell)	less expensive than alkaline better than LeClanche at high curr and low temp	low energy density (2–2.5Wh/in^3) sloping discharge curve
Alkaline manganese ("alkaline" dry cell)	moderate cost better than zinc chloride at high curr and low temp maintains low impedance as discharged moderate energy density (3.5Wh/in^3) widely available	sloping discharge curve
Mercury	high energy density (7Wh/in^3) flat discharge curve good at high temperatures good shelf life low and constant impedance open-circuit voltage 1.35V±1%	expensive poor at low temp (<0°C)
Silver oxide	high energy density (6Wh/in^3) flat discharge curve good at high & low temp (to -20°C) excellent shelf life	expensive
Lithium oxyhalide	high energy density (8Wh/in^3) highest energy density per unit weight flat discharge curve excellent at high & low temp (to -55°C) extraordinary shelf life (5–10 yrs @70°C) light weight high cell voltage (3.0V)	expensive
Lithium solid-state	high energy density (5–8Wh/in^3) excellent at high & low temp (-40°C to 120°C) unbelievable shelf life (>20y @ 70°C) light weight	expensive low current drain only

TABLE 14.2. BATTERY CHARACTERISTICS

Type	R_{int} (Ω)	V_{oc} (V)	Capacity[a] continuous, to 1V/cell (mAh)	@ (mA)	(mAh)	@ (mA)	Size (in)	Weight (gm)	Connec[b]	Comments
9V "1604"										
Le Clanche	35	9	300	1	160	10	0.65x1x1.9	35	S	
Heavy Duty	35	9	400	1	180	10	"	40	S	
Alkaline	2	9	500	1	470	10	"	55	S	280mAh@100mA
Lithium	18	9	1000	25	950	80	"	38	S	Kodak Li-MnO_2
1.5V Alkaline										
D	0.1	1.5	10000	10	8000	100	1.3Dx2.2L	125	B	4000mAh @ 1A
C	0.2	1.5	4500	10	3200	100	1.0Dx1.8L	64	B	
AA	0.4	1.5	1400	10	1000	100	0.55Dx1.9L	22	B	
AAA	0.6	1.5	600	10	400	100	0.4Dx1.7L	12	B	
Mercury										
625	–	1.35	250[c]	1	250[c]	10	0.62Dx0.24L	4	B	
675	10	1.35	190[c]	0.2	–	–	0.64Dx0.21L	2.6	B	
431	–	11.2	1000[c]	25	–	–	1.0Dx2.9L	115	S	
Silver										
76	10	1.55	180	1	–	–	0.46Dx0.21L	2.2	B	
Li-Oxyhalide										
D	–	3.9	14000[d]	175	10500[d]	350	1.3Dx2.3L	113	B,T	$SOCl_2$/BrCl
D	–	3.95	14000[d]	175	12000[d]	1000	"	110	B,T	SO_2Cl_2/Cl_2
D	–	3.5	9500[d]	175	8500[d]	1000	"	120	B,T	$SOCl_2$
Li solid	–	4.0	350[d]	1μA	175[d]	0.1	1.2Dx0.23L	16	T	high impedance
Ni-Cd										
D	0.009	1.3	4000[c]	800	3500[c]	4000	1.3Dx2.3L	130	B	Saft/Powersonic
9V	0.84	8.1	100[c]	10			0.65x1x1.9	35	S	
Pb-acid										
D	0.006	2.0	2500[e]	25	2000[e]	1000	1.3Dx2.6L	180	T	

(a) see Fig. 14.1 for discharge curves. (b) B - button; S - snap; T - solder tabs. (c) to 0.9V/cell. (d) to 2.5V/cell. (e) to 1.75V/cell

voltage (1.35V, stable to 1%) as well as constancy of voltage during discharge (a "flat discharge curve"); see Figure 14.1. It performs well at temperatures up to 60°C, but performance is seriously degraded below −10°C.

The silver oxide cell is similar to the mercury cell, but with the mercuric oxide replaced by silver oxide. It, too, has a very flat discharge curve, but with higher open-circuit voltage (1.6V) and improved performance at low temperatures (to −20°C).

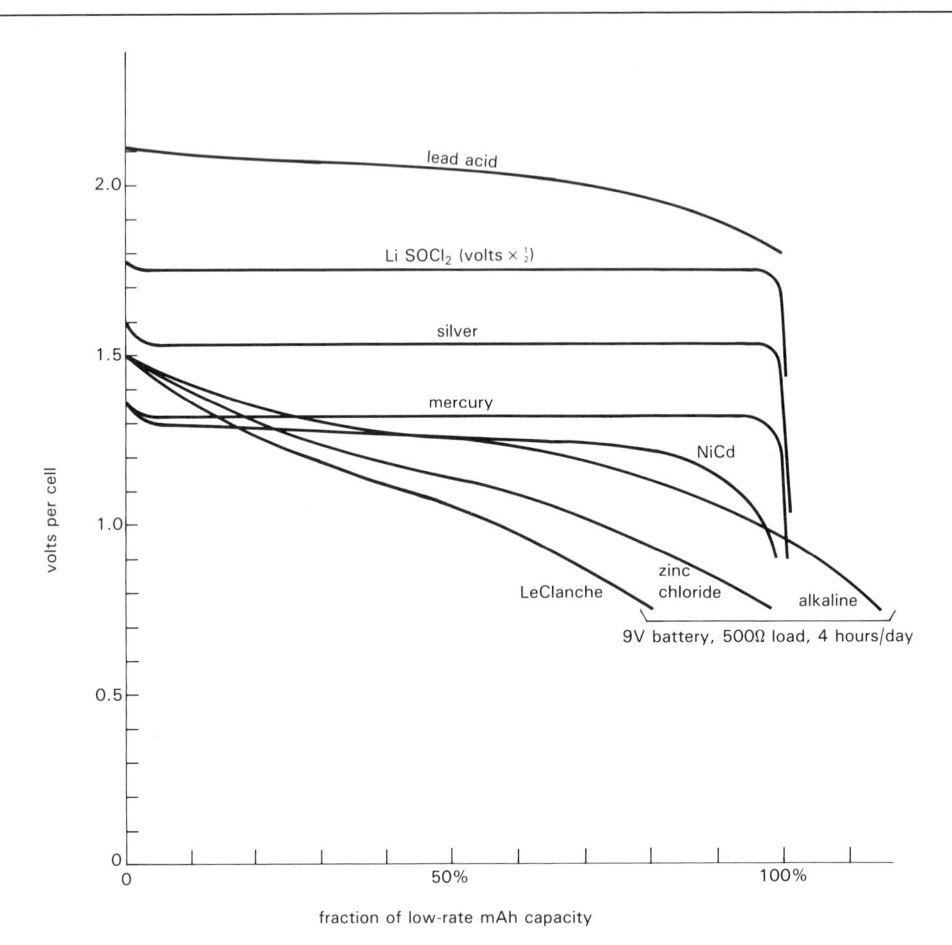

Figure 14.1. Discharge curves for primary batteries. (This and subsequent figures in this chapter are adapted from battery literature by Arco Solar, Duracell, Electrochem Industries, Eveready, Gates, Kodak, PowerSonic, Solavolt, and Yuasa.)

Lithium cells are the newest of the commercially available primary cells, with several different chemistries available. They have the highest energy density per unit weight. They are the best performers at very high and low temperatures, and they have extraordinary shelf life at all temperatures. For example, a D-size lithium thionyl chloride ($Li/SOCl_2$) cell delivers more than three times the energy (10Ah at a terminal voltage of 3.5V) of an alkaline D cell, with comparable size and weight. Lithium batteries will operate down to −50°C and up to 70°C (see Fig. 14.3), with 50% of their room-temperature service at −40°C, a temperature at which other primary batteries cease to function at all. Lithium batteries have shelf lives of 5 to 20 years at room-temperature and can be stored for 1–2 years at 70°C, a temperature that makes other batteries wither. They have a flat discharge curve. Their long shelf life and 3–3.5 volts make lithium batteries ideal for on-board CMOS memory backup.

Each lithium chemistry system has its own peculiarities. For example, lithium thionyl chloride batteries have a tendency to develop an electrode passivation that raises their internal resistance enormously; it can be "burned off" by momentary operation at high current. Lithium sulfur dioxide has been implicated in some battery explosions:

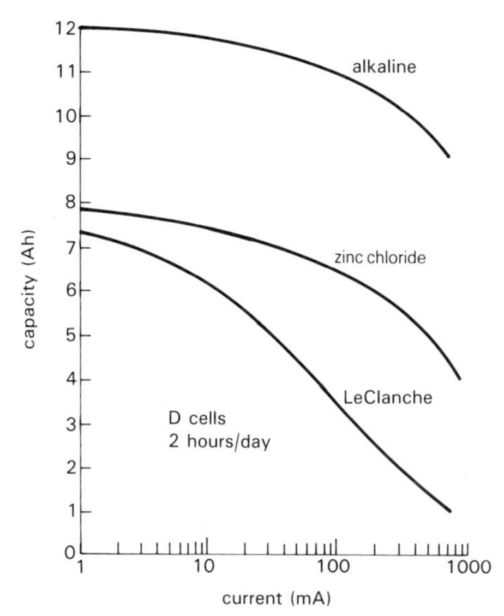

A. D-cell capacity versus load current

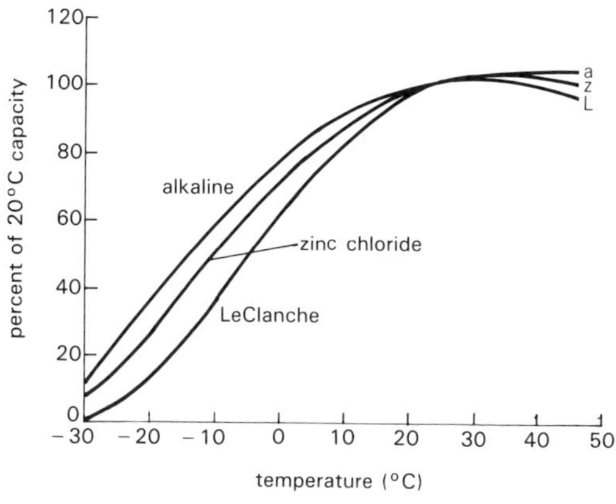

D. capacity versus temperature

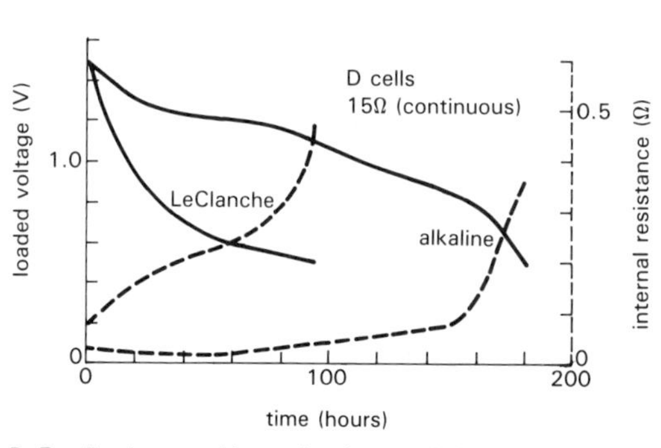

B. D-cell voltage and internal resistance during discharge

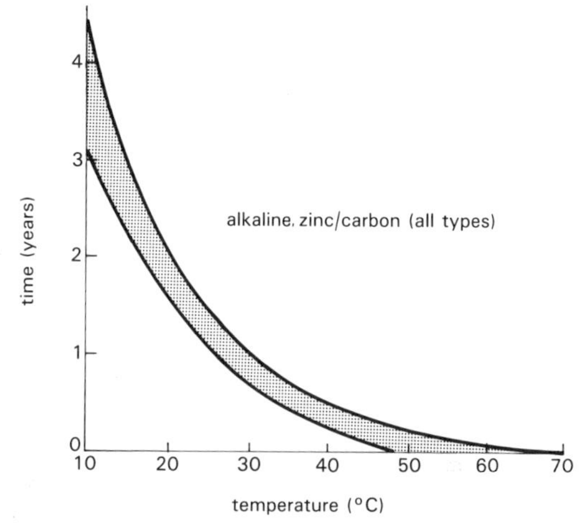

E. shelf life (10% loss of capacity) versus temperature

Figure 14.2. Zinc "dry-cell" performance comparison.

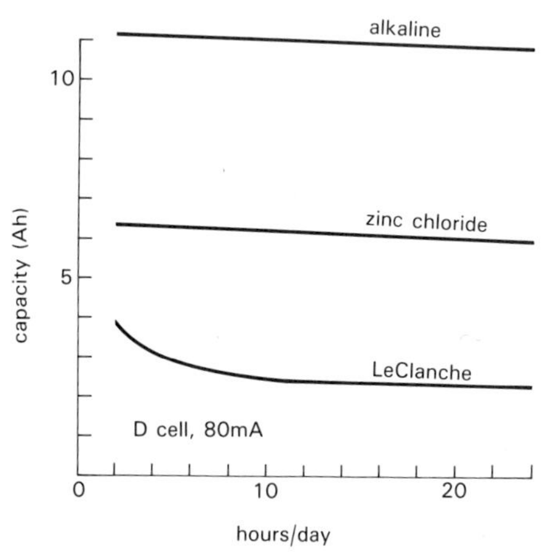

C. D-cell capacity versus duty cycle

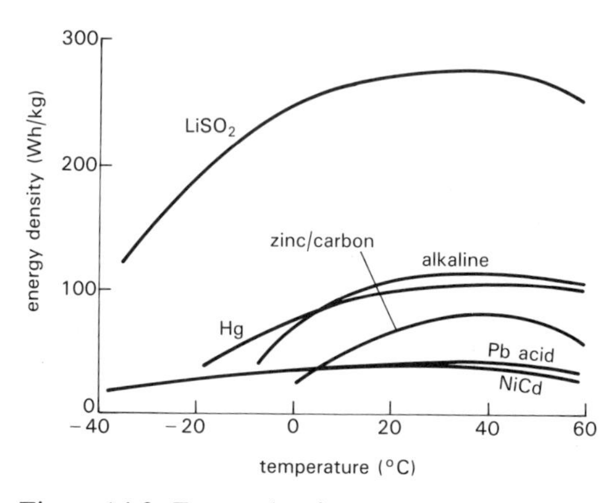

Figure 14.3. Energy density versus temperature for primary cells.

Warning: There have been incidents in which lithium batteries have exploded, in some cases causing severe personal injury. Having warned you about this, we will not be responsible for any calamities you might experience at the hands of a lithium battery.

□ *Secondary batteries*

For use in electronic equipment, your choices are (a) nickel-cadmium ("nicad") or (b) sealed lead-acid. Both have lower energy content than primary cells (Table 14.2), but they *are* rechargeable. Nicad cells provide 1.2 volts, are generally available in the 100mAh–5Ah range, and work down to −20°C (and up to +45°C); lead-acid batteries provide 2 volts per cell, are generally built to provide 1 to 20 amp-hours, and will work down to −65°C (and up to +65°C). Both types have relatively flat discharge curves. Lead-acid batteries have low self-discharge rates and are claimed to retain two-thirds of their charge after a year's storage at room temperature (though our experience leads us to be skeptical); nicad batteries have relatively poor charge retention, typically losing half their charge in 4 months (which we do believe!) (see Fig. 14.4). A nicad D cell provides 5Ah (at 1.2V), whereas a lead-acid D cell provides 2.5Ah (at 2V); the comparable alkaline cell provides 10Ah at 1.5 volts.

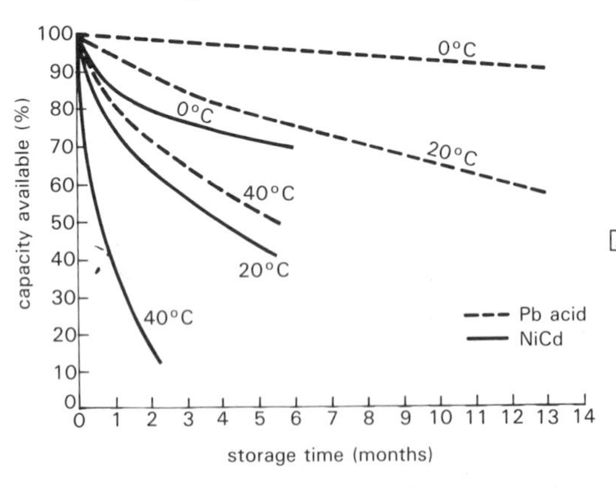

Figure 14.4. Secondary battery charge retention.

Both nicad and sealed lead-acid batteries claim to be good for 250–1000 charge/discharge cycles (more if they are only partially discharged each time; less if completely discharged, or charged/discharged rapidly). Nicads have an overall life expectancy of 2–4 years if held at a constant trickle charge current (see below); the comparable life for sealed lead-acid batteries held at constant "float" voltage is claimed to be 5–10 years.

It's worth pointing out that these rechargeable batteries really are sealed; they won't drip mysterious and terrible chemicals. In particular, although the name "lead-acid" conjures up images of husky car batteries with corroded terminals and leaking acid, the sealed types really are clean batteries: You can run them in any position, they don't drip or ooze, and they're generally well-behaved. In our experience you can design them into real electronic instruments without fear of your circuit boards dissolving into a white crusty plague, or the bottom of your expensive enclosure becoming awash with foul-smelling bilge juices.

Secondary batteries die young if they aren't charged correctly. The procedures are different for nicad and lead-acid. It's conventional to designate charging rates in terms of the ampere-hour capacity of the battery; for example, charging at "C/10" means applying a charging current equal to one-tenth of the ampere-hour capacity of the fully charged battery. For the nicad D cell above, that would be 500mA.

□ *Nickel-cadmium.* Nicad cells are designed to be charged at constant current and to withstand continuous charging at C/10. Because of inefficiencies in the charge/discharge cycle, you have to charge at this rate for 14 hours to guarantee a full charge; you can think of this as charging the battery 140%.

Although it's OK to overcharge nicad batteries indefinitely at C/10, it's better to switch over to a "trickle" charge, typically at C/30 to C/50. However, nicads are funny, with a "memory" effect, so that a trickle rate may fail to revive a fully discharged battery; a minimum of C/20 is recommended.

There are applications where you can't wait a whole day for nicads to recharge. Nicad literature gives you permission to charge normal cells at a "high rate" of C/3 to C/10, if you don't do it too much. Up to three days at C/3 is about the limit. There may be some venting of gases under these conditions, in contrast to "normal" charging at C/10 in which internally evolved oxygen gets recombined within the cell. There are special "fast-charge" nicad cells designed to be charged at C/1 to C/3 in a special charger that senses the fully charged condition by monitoring cell temperature (they have internal chemistry that makes them heat rapidly once they are charged). Unlike the situation with lead-acid batteries, you can't reliably determine when a nicad is fully charged by monitoring terminal voltage, because it changes with repetitive cycling, temperature, and rate. Nicads should not be charged by a constant voltage, nor held "floated" at a fixed voltage.

You can buy handy little nicad chargers from several companies, including the battery manufacturers themselves. They typically let you charge all the popular sizes (D, C, AA, and 9V).

Nicads have a pathology all their own. If you're like us, you probably take it personally when your rechargeable calculator dies during a tax audit. As the graphs show (Fig. 14.5), nicads have "memory" effects, so that the first discharge after a long period of charging may be poor. They are intolerant of reverse polarity; thus, the first cell to discharge suffers horribly if a series string of them is fully discharged. Likewise, nicads shouldn't be connected in parallel. You'll find people promoting various snake remedies, such as periodic "deep discharge," or shock therapy in the form of a substantial electrolytic capacitor discharged across a moribund nicad. Although we're skeptical of the latter, periodic deep discharge is important for nicad health.

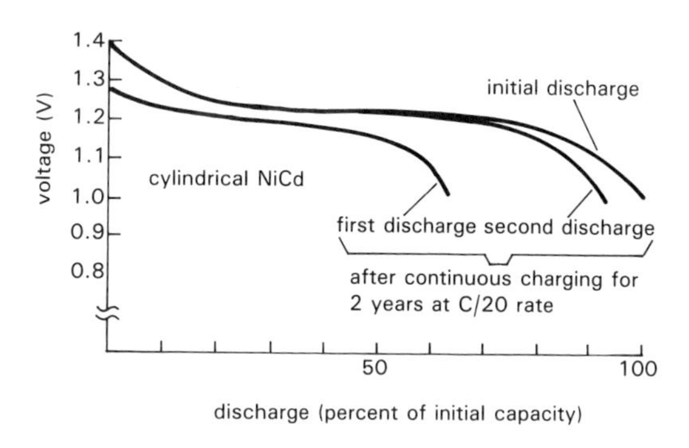

Figure 14.5. Nicad battery restored to good health by "deep discharge."

□ *Lead-acid.* These versatile batteries can be charged by applying a current-limited constant voltage, a constant current, or something in between. With current-limited constant-voltage charging, you apply a fixed voltage (typically between 2.3V and 2.6V per cell); the battery initially draws a high current (up to 2C), but tapers down as it charges, eventually leveling off to a trickle current that maintains the battery in a fully charged state. A higher applied voltage gives you a faster charge, at the expense of greater required charger current and reduced overall battery life. A simple implementation is to use a 3-terminal regulator like the 317 to supply a current-limited fixed voltage. The battery's charge can be held indefinitely by maintaining a fixed "float" voltage between 2.3 and 2.4 volts per cell (corresponding to a trickle current of C/1000 to C/500).

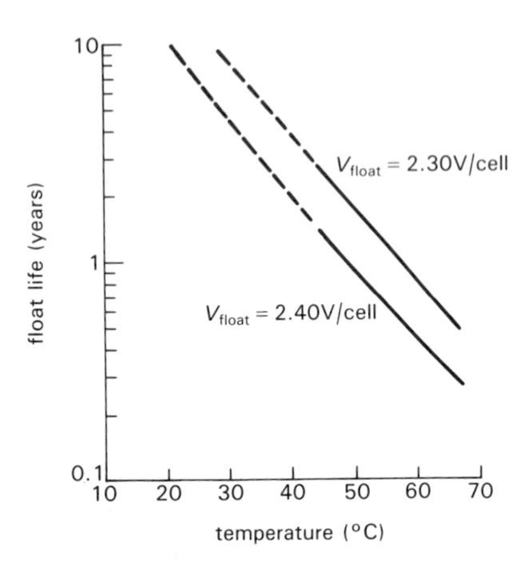

Figure 14.6. Excessive float voltage reduces life of lead-acid batteries.

Figure 14.6 shows the trade-offs. These charging and floating voltages are mildly temperature-dependent and should be adjusted by −4mV/°C for operation at extremes of temperature.

With constant-current charging (which is not often used) you apply a fixed current, typically C/5 to C/20; the battery voltage rises gradually as the battery charges, then increases dramatically as full charge is reached. At this point (indicated by a terminal voltage of 2.5V/cell) you must reduce the current, typically to a fixed C/500 rate, which will maintain full charge indefinitely. Sealed lead-acid batteries will give 8–10 years of service while being charged at a C/500 rate.

A nice lead-acid charging method is the so-called two-step technique (Fig. 14.7): After a preliminary "trickle" charge, you begin with a high-current "bulk-charge" phase, applying a fixed high current $I_{\max}$ until the battery reaches the "overcharge voltage," V_{OC}. You then hold the voltage constant at V_{OC}, monitoring the (dropping) current until it reaches the "overcharge transition current," I_{OCT}. You then hold a constant "float voltage," V_F, which is less than V_{OC}, across the battery. For a 12 volt 2.5Ah lead-acid battery, typical values are $I_{\max} = 0.5$ amp, $V_{OC} = 14.8$ volts, $I_{OCT} = 0.05$ amp, and $V_F = 14.0$ volts.

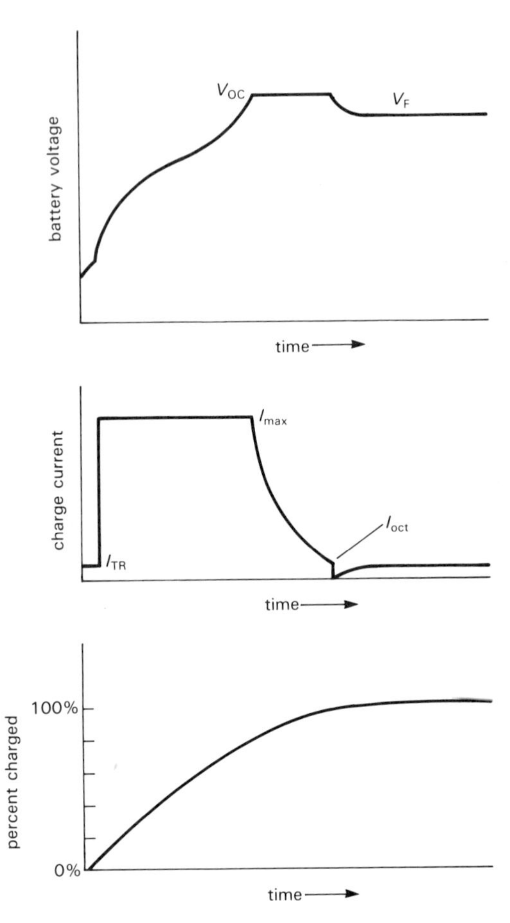

Figure 14.7. Recommended lead-acid battery charging cycle ("two step").

Although this all sounds rather complicated, it results in rapid recharge of the battery without damage. Unitrode makes a nice IC, the UC3906, that has just about everything you need to do the job. It even includes an internal voltage reference that tracks the temperature characteristics of lead-acid cells and requires only an external *pnp* pass transistor and four parameter-setting resistors.

Battery availability and recommendations

As we said at the outset, it's really a good idea to design your instrument to use a popular and readily available battery.

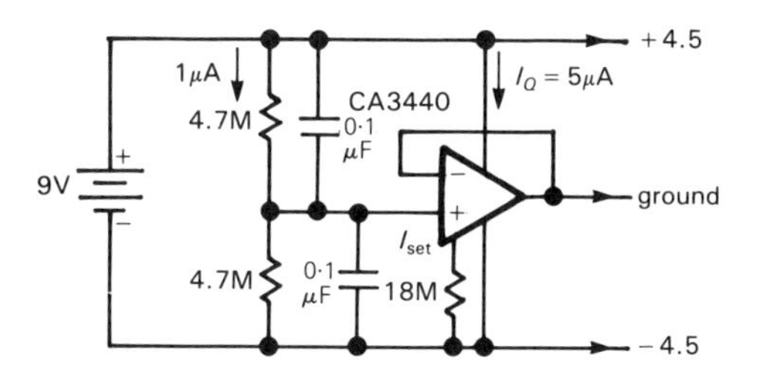

Figure 14.8. Creating a split supply from a single battery.

Tops on the list are the 9 volt "transistor" batteries, known generically as NEDA 1604 (1604, LeClanche; 1604D, heavy duty; 1604A, alkaline; 1604M, mercury; 1604LC, lithium; 1604NC, nicad). You can buy 9 volt alkaline batteries in every corner grocery store (or open-air market) in the world. Op-amps work well on ±9 volt supplies; you can even use ±4.5 volts if you use a resistive divider and a follower to generate a midpoint "ground" (Fig. 14.8; discussed further in Section 14.08). There are nice little plastic instrument cases, complete with 9 volt battery compartment underneath, available from many manufacturers at very attractive prices. We recommend using alkaline, rather than zinc-carbon, because of the improved electrical characteristics discussed earlier. Kodak's new "Ultralife" lithium 9 volt battery looks like a real winner, with 1000mAh capacity, long shelf life (80% retention after 10 years), and flat discharge curve (Fig.14.9); they wisely used 3 cells, not 2, so its terminal voltage

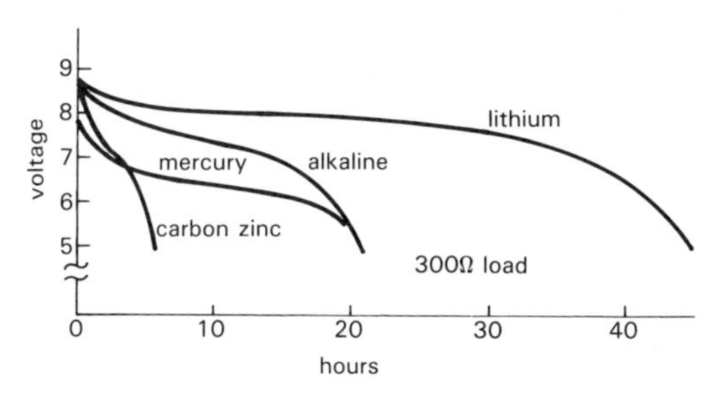

Figure 14.9. 9 volt battery discharge curves; "lithium" is the 3-cell Kodak "Ultralife."

is close to 9 volts, the same as alkaline. Our preliminary measurements on some early samples showed a rather high internal impedance, however.

The familiar alkaline AA, C, and D cells have more energy capacity (and correspondingly lower internal impedance) than the 500mAh of the 1604A (3, 9, and 20 times as much, respectively), and they're just as easy to get. But they are somewhat less convenient because of the problem of holding and connecting reliably to a group of series-connected cells. Everyone has noticed that if you shake a dim flashlight, it usually gets brighter. The problem is compounded by the tendency of some types (alkaline, mercury) to grow white deposits on the terminals (this is officially known as "salting").

Nicads are also available (though not in every drugstore) in the standard battery sizes (AA, C, D, and 9V), for applications where it makes sense to use secondary batteries. But you get only about 25%–50% the energy capacity, and reduced battery voltage (1.2V versus the alkaline's 1.5V per cell).

Lithium batteries are available in the same standard battery sizes, although they provide 3 volts or more per cell. Most manufacturers also provide them with solder tabs for more reliable connections; this makes good sense, considering their extended shelf life. Lithium cells are also available in flat "button" shapes with solder tabs, for use as CMOS memory backup, or to power calendar clock chips. Kodak's 9 volt lithium battery has a nice twist, namely gold-plated snap tabs, for more reliable connections. See our earlier warning on lithium-battery detonation.

Sticking to batteries you can get in any small town, in most photo stores you'll find a selection of mercury, silver, and lithium cells. They're meant to go into cameras (and calculators and watches) and are generally of the "button" variety. For

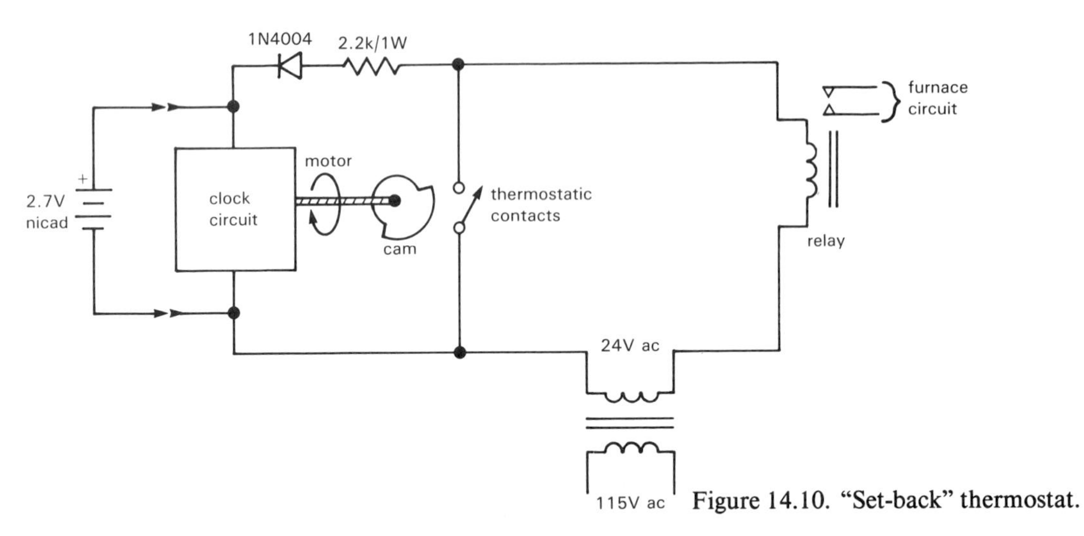

Figure 14.10. "Set-back" thermostat.

example, there's the popular 625 mercury button, hardly larger than a coat button and good for 250mAh. The smaller 76 silver oxide cell (and energetically equivalent type 675 mercury cell) has an interesting twist, namely an offspring lithium 3 volt cell (NEDA 5008L) of the same diameter and twice the height, intended to substitute for a pair of the 1.5 volt cells. At this voltage you can run CMOS logic directly, as well as low-voltage op-amps like the LM10 and the ICL7610 series and TI's versatile "LinCMOS"series of op-amps (the TLC251–254 series) and comparators (TLC372/4, TLC339/393, and TLC3702/4).

If your application requires the rechargeability and high peak currents of sealed lead-acid batteries, or some exotic form of primary battery, you'll generally have to deal with the battery manufacturers or their distributors. Look for names like Gates, Powersonic, and Yuasa for lead-acid. Duracell and Eveready dominate the primary-cell market. All of these companies have helpful and extensive data books on batteries and battery lore.

In the next few sections we will consider alternative power sources – wall-plug-in modules, solar cells, and signal currents – for low-power equipment. It's worth remembering that each of these power sources can be used to charge secondary batteries. For example, the popular "set-back" thermostats that turn down the heat at night use the high-impedance 24 volt ac relay signaling current to charge nicads and keep the clock running during periods of relay-ON (Fig. 14.10).

Table 14.3 summarizes our advice on the relative merits of various primary-cell battery types.

TABLE 14.3. PRIMARY-BATTERY ATTRIBUTES

	9V alkaline	1.5V alkaline	Mercury	Silver	Lithium
Properties					
Inexpensive	•	•	–	–	–
Available	•	•	–	–	–
Wide temp range	–	–	–	–	•
Stable voltage	–	–	•	•	•
Reliable contacts	•	–	–	–	•
Good at high current	•	•	–	–	•
Long shelf life	–	–	–	–	•
Miniature	–	–	•	•	•
Applications					
Linear circuits	•	–	–	–	–
Low voltage CMOS	–	•	•	•	•
4000-series CMOS	•	–	–	–	•
CMOS backup	–	–	–	–	•

14.03 Wall-plug-in units

Calculators, modems, tape recorders, telephone dialers, small measurement instruments – more and more low-power devices come with those familiar square black wall-plug-in power units (Fig. 14.11). Although they're usually labeled to match the instrument they power, you can easily get them in a variety of ratings, both in large quantities and small (distributors like Radio Shack and Digi-Key each stock a few types). The best news is the incredibly low price: A 9 volt 500mA (unregulated dc) wall plug-in costs about $2.50 in quantity.

Wall plug-ins are a good way to power small instruments that need more power than you can get from batteries, or that need to keep their rechargeable batteries charged. They're cheaper than internally mounted discrete or modular power supplies, and by using them you save space and keep heat (and high voltage) outside the instrument. Furthermore, they usually satisfy UL and CSA safety requirements, which is important if you want to market an instrument without the lengthy UL approval process.

Wall plug-ins come in three flavors: plain step-down transformers, filtered but unregulated dc supplies, and complete regulated supplies (both linear and switchers). All come in a variety of voltages and currents, and the regulated supplies even come in useful combinations like +5 volts at 1 amp and ±15 volts at 250mA. They have the usual features of IC regulators, namely current limiting and thermal shutdown, as well as optional overvoltage crowbar. You can get them with three-prong (grounding) wall plugs and with various output connectors; many of the larger units are also available as free-standing desk units, with an ac power cord. One word of caution: There is no standardization of connector type and voltage ratings. In fact, there is not even standardization of polarity! So it is effortless to blow out an instrument by plugging the wrong wall unit into it. Beware!

An extensive line of high-quality wall power units is manufactured by Ault (Minneapolis, MN). For inexpensive imported units, look at the catalogs of Condor (Sunnyvale, CA) or Multi Products

Figure 14.11. Wall-plug-in power supplies. (Courtesy of Ault, Inc.)

International (Cedar Grove, NJ). Check the EEM (see Bibliography) for addresses and additional manufacturers.

□ 14.04 Solar cells

A combination of lead-acid or nicad battery plus silicon solar cells forms a good power source for a moderate-power instrument that is to be deployed at a remote site for extended periods. For example, you might want to tether a buoy that makes ocean measurements and transmits them periodically. If the average power consumption is 1 watt, primary batteries become prohibitively bulky (you'd have to use 500 alkaline D cells to last 1 year). Full sunlight delivers about 1kW per square meter after traversing the atmosphere; after accounting for the inefficiencies of solar cells (they're about 10% efficient when operating into the correct load) and the daylight and weather cycle at middle northern latitudes (where you average 100W/m^2 in winter, 250W/m^2 in summer), you can average about 25 watts (in July) or 10 watts (in January) per square meter from good quality solar cells, which cost about $800 in 1986. In peak sunlight, such a solar module delivers 100 watts to a matched load.

With a bank of secondary cells for energy storage (lead-acid is better than nicad, because of its long life and wide operating temperature range), you can withdraw nearly this average power *continuously*; lead-acid cells are typically 70%–80% efficient, so, all factors (including weather) considered, you can withdraw something like 8 watts per square meter (winter) to 20 watts per square meter (summer), averaged over 24 hours.

For low-power instruments that only need to operate in bright light, you can omit the battery. Solar-powered CMOS calculators with liquid crystal displays are a boon to battery haters everywhere.

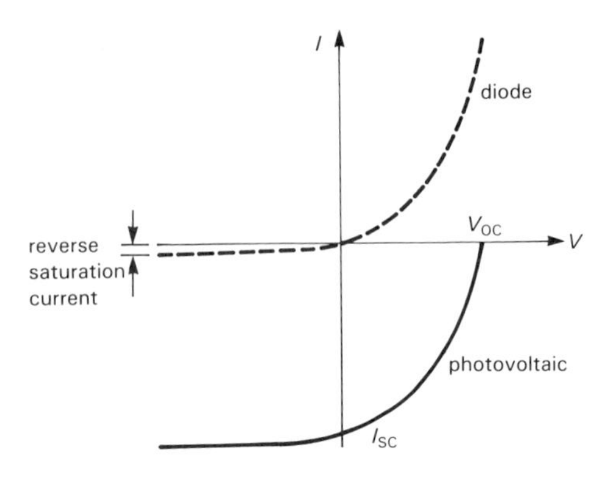

Figure 14.12. Solar-cell output voltage versus load current is simply a displaced diode V–I curve.

□ *VI characteristics*

Silicon solar cells have a simple and very useful volt–ampere characteristic. It turns out that the open-circuit voltage is almost independent of light level and averages about 0.5 volt per cell; the V–I curve is simply a displaced diode curve (Fig. 14.12). Typical solar panels consist of 36 series-connected cells, for an open-circuit voltage of about 18 volts. The terminal voltage stays nearly constant as load current is increased up to a maximum current, at which point the solar module becomes roughly a constant current source for further decreases in load impedance. The maximum current scales linearly with light level, giving a set of characteristic curves as shown in Figure 14.13. Solar cells work best when

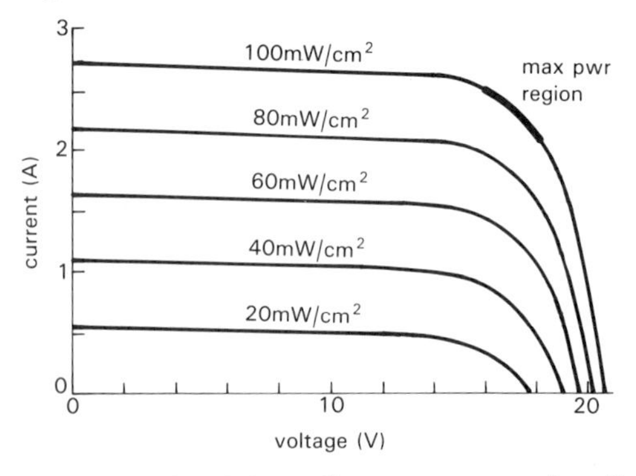

Figure 14.13. Solar-cell output versus irradiance (Solavolt MSVM4011).

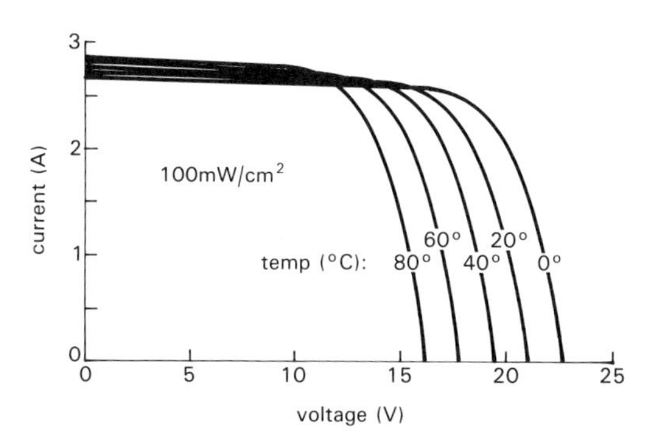

Figure 14.14. Solar-cell output versus temperature (Solavolt MSVM4011).

cold, since the open-circuit voltage drops with increasing temperature (Fig. 14.14).

For a given light level, maximum power is delivered when the operating point has maximum product *VI*; in other words, the point on the *V–I* curve that touches a family of hyperbolas (constant product *VI*) plotted on the same axes. Roughly speaking, that's at the knee of the *V–I* curve. Since the load impedance that corresponds to the knee changes rapidly with light level, you can't expect to maintain an optimum load (which would be a load impedance increasing inversely with light level, or, put another way, a load that draws a current proportional to light level, at roughly constant voltage). However, for low-power applications it isn't essential that the load extract maximum power – all that matters is that the load be powered under normal lighting conditions. That's the case for those solar-powered calculators, whose CMOS circuits draw so little current that there is plenty of reserve power except under very low light levels. Because of the wide voltage range of 74C/4000B "high-voltage" CMOS (3V to 18V), and the fact that solar cells have an open-circuit voltage that is relatively independent of light level, you don't need to use any voltage regulators; just power the CMOS directly from the module, with bypass capacitors, of course. A typical small module like the Solarex SX-2 provides 290mA at 8.5 volts in sunlight and has an open-circuit voltage of 11 volts; you could use it, unregulated, for high-voltage CMOS, or, with a regulator, for any +5 volt logic family.

For any application that uses secondary cells for energy storage, it's worth noting the rather good match of solar-cell *VI* characteristics to the charging requirements of lead-acid cells. A solar module provides roughly constant charging current into a discharged battery, changing over to a constant-voltage "float" as the battery voltage rises at the end of charging. The temperature coefficient of open-circuit voltage (−0.5%/°C) is a fair match to the recommended float-voltage tempco for lead-acid batteries (−0.18%/°C). So some suppliers make solar modules that are intended to charge lead-acid batteries directly, for example the Arco M65 (2.9A @ 14.5V). The more usual way to match solar modules to the charge/float characteristics of lead-acid batteries is with a series or shunt regulator circuit designed for the job. Many solar modules are designed to work this way, with 20 volts open-circuit voltage and matching regulator module for charging 12 volt batteries. The regulators switch over from charging to temperature-compensated floating, with automatic load disconnection if the battery voltage drops too low. These systems are available for multiples of 12 volt systems (24V, 36V, 48V, etc.), and you can get accessories such as 60Hz inverters (to make ac), or dc-operated refrigerators, attic ventilators, etc.

Some of the bigger names in solar modules and systems are Arco Solar (Chatsworth, CA), Mobil Solar (Waltham, MA), Solarex (Rockville, MD), and Solavolt (Phoenix, AZ).

14.05 Signal currents

Don't forget about the possibility of using signaling currents to power a micropower instrument. Four of the more

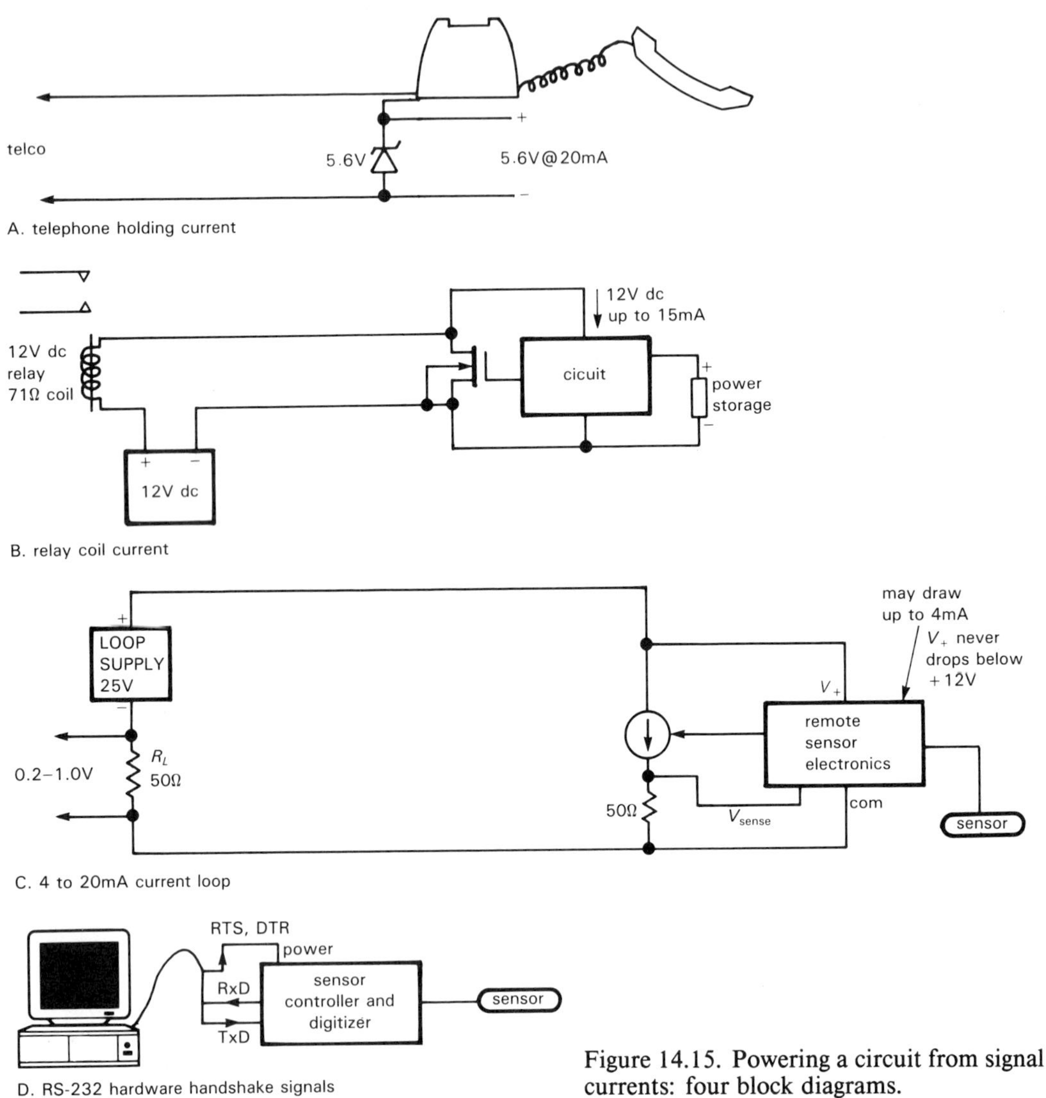

Figure 14.15. Powering a circuit from signal currents: four block diagrams.

common opportunities (Fig. 14.15) are (a) the dc holding current flowing through a telephone circuit that is "off hook," (b) the ac or dc voltage available from a relay circuit when it is not energized, (c) the 4–20mA dc current used for industrial-sensor current-loop signaling, and (d) the serial-port RS-232C bipolarity "handshake" signals (RTS, DSR, etc.). In the first two schemes, your source of power is available only part of the time – power disappears when the phone is hung up ("on hook") or when the relay circuit is energized by your instrument. If you need power continuously, you'll have to use rechargeable batteries, charged during periods of available power; for very low current loads, another possibility is a high-capacitance (up to 5 farads) "double-layer" capacitor, the same type used for CMOS memory retention.

Each of these power sources has rather strict limits on voltage compliance or maximum current. Here are their characteristics, and some hints on the parasitic use of these power sources.

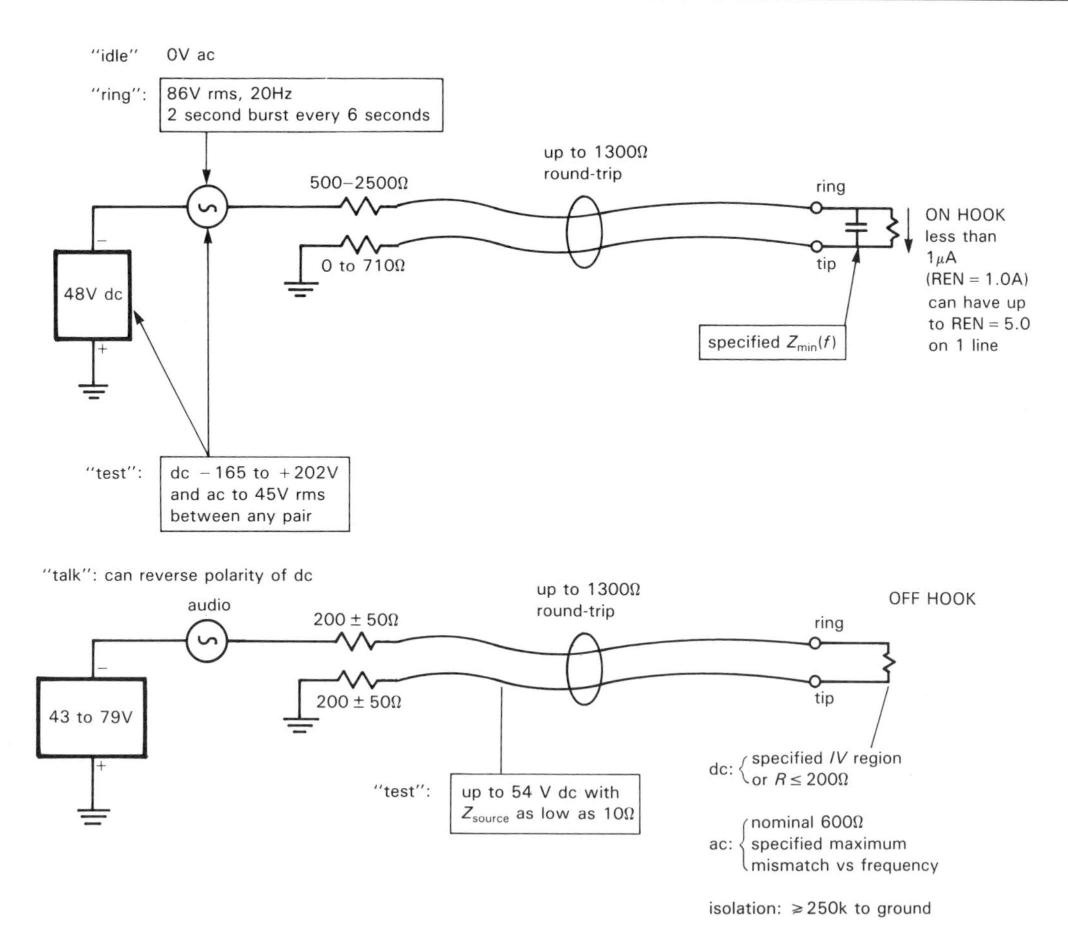

Figure 14.16. Telephone operating and test states.

□ *Telephone-line power*

There are several different states that the phone line can be in, depending upon what your phone is doing, and what the phone company is doing to you. The central office (or nearby equivalent) applies various dc (and ac) voltages to the two-wire phone loop (labeled "tip" and "ring") during these various stages of call progress (Fig. 14.16). In the *idle* state, the telephone company central office applies −48 (±6) volts dc in series with 500 to 2500 ohms to the "ring" line, and terminates the "tip" line to ground with 0 to 710 ohms. In addition, there is typically up to 1300 ohms of external line resistance between the central office and you (the "subscriber"). When you go off-hook, the central office goes into *dialing* mode, applying a dial tone and a dc level of −43 to −79 volts in series with 200 ohms (±50Ω) on "ring," and terminating "tip" with the same impedance to ground. The same dc voltage and source impedances are present in the *talking* state (after the connection is made), although the telephone company may, at its discretion, reverse the polarity of the dc voltage applied to "ring." Of course, in the talking state you also have audio signals superposed on the dc, which is the whole purpose of the telephone!

There are two other states. During *ring*, the phone company applies 86 (±2) volts rms, at 20Hz, on top of the usual −48 volt dc bias. As with the dc, the ac ringing signal is applied to the "ring" lead. The official ringing specification is 2 second bursts at 6 second intervals. During *test* mode, the phone company applies various ac and dc test signals to make sure that the network is working properly. They can apply dc voltages in the range −165 to +202 volts, and ac voltages up to 45 volts rms, between any pair of conductors (ring, tip, ground) in the on-hook state, and up to 54 volts dc with source impedance as low as 10 ohms in the off-hook state. The phone company also specifies a range of lightning-induced "high-voltage surges" that you may find on your phone lines. They're typically a few thousand volts, capable of sourcing a few hundred amps; equipment connected to the phone line is supposed to have transient suppressors so that it will survive such pulses. In addition, the phone company specifies "very high voltage surges" that may occur from a nearby lightning strike. These may reach 10kV and 1000 amps, and the idea is that even if your equipment gets fried, nobody should get hurt. So the specification says that the equipment shouldn't shoot out pieces of stuff, catch fire, or electrocute anyone.

Permissible loads are specified in terms of "ringer equivalence number" (REN). Typical phones have an REN of 1.0A, which corresponds to (a) an on-hook dc resistance of 50MΩ, and on-hook ac impedance that stays above a specified curve of impedance versus frequency (satisfied by keeping $|Z| > 125k$ from 4Hz to 3.2kHz, though it can be much lower over certain frequencies and voltages), (b) an off-hook dc characteristic that stays within the acceptable region of Figure 14.17 (or that measures 200Ω or less), and (c) an off-hook impedance that approximates 600 ohms from 200Hz to 3.2kHz (this is actually specified in terms of reflection when driven by a 600Ω signal source: at least 3.5dB from 200Hz to 3.2kHz and 7dB from 500Hz to 2.5kHz). Loads connected to the phone lines must be dc-isolated from ground (50MΩ on-hook, 250k off-hook). A total REN up to 5.0A is permissible, i.e., a load impedance as low as 1/5 the above values. The telephone company requires you to notify them of your total REN loading.

From the foregoing data it is obvious that the subscriber is not supposed to draw current in the on-hook state, and the 50MΩ minimum dc bridging resistance (for an REN of 1.0A) is really a leakage specification – 50MΩ corresponds to 1μA. Nevertheless, with careful design and component selection you can maintain CMOS circuitry (digital or analog) in a quiescent state with a few microamps, and have it "wake up" when the line goes off-hook. Use a small tantalum electrolytic (or "double-layer" memory-retention capacitor, available in tiny packages up to 5 farads!) to keep things going during dialing or other transients. In the off-hook talking state, you are guaranteed 6 volts dc (7.8V after a few seconds) at a minimum of 26mA (see Fig. 14.17), which is enough

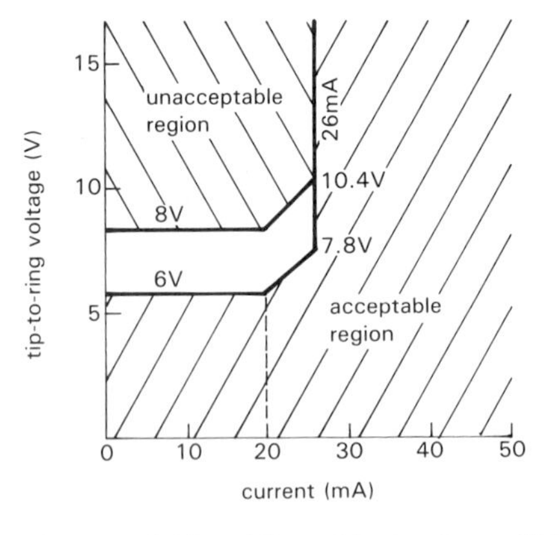

Figure 14.17. Allowable load conditions for "off-hook" telephone. The unshaded region is allowable beginning 1 second after going off-hook. (Adapted from Bell System Tech. Ref., Pub. 47001.)

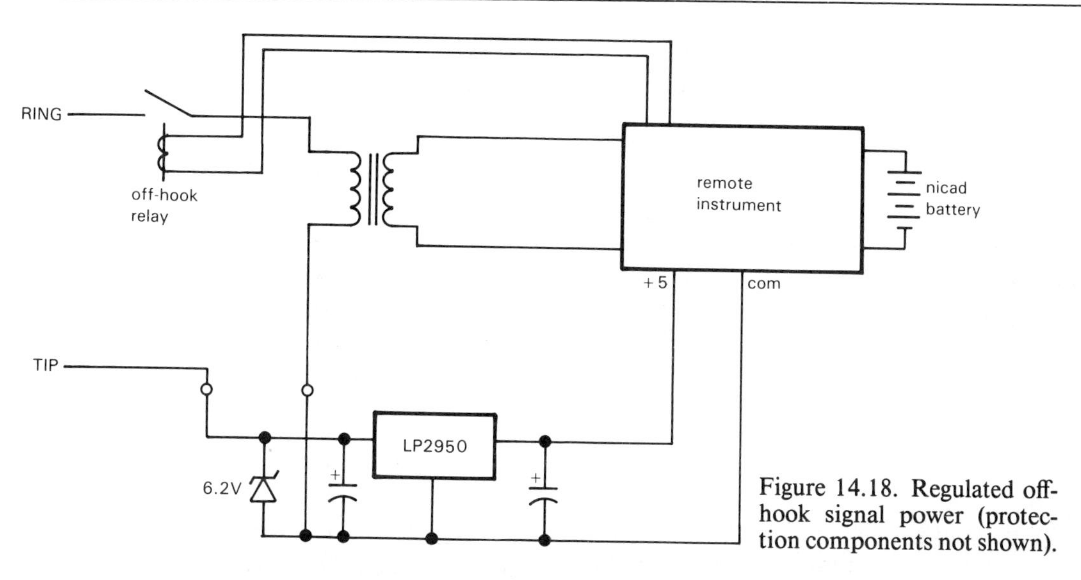

Figure 14.18. Regulated off-hook signal power (protection components not shown).

to operate lots of micropower circuitry; it's really comparable to the kind of power you get from a 9 volt battery. With a micropower low-dropout regulator like the LP2950 (80μA quiescent, 0.4V dropout at 100mA load) you can supply regulated 5 volts to digital circuitry, as in Figure 14.18. If you can be sure of having regular periods of off-hook operation, you can use the extra current then available to keep rechargeable batteries charged. For example, if you have an hour of off-hook operation per day, you can draw nearly a milliamp continuously.

Warning: Before designing any device for direct connection to the telephone system, be sure to get the relevant specifications. You must be in compliance with FCC regulations, which include test and approval procedures. Do not assume that the specifications quoted in this chapter are correct.

□ *Relay circuits*

Those popular "set-back" thermostats that turn down the heat at night, and turn it back up a half hour before you wake up, use the small ac power that can be extracted from the relay circuit without making the relay close. Typical mechanical control relays draw 100mA or more at the coil rating of 24 volts ac and can be relied on to stay open at 10% of the normal coil current. So you can have 10mA or so, at nearly 24 volts ac, to power your instrument. Be sure to include rechargeable cells (or perhaps a memory-retention capacitor, if that's all that's needed), because the power source goes away when you close the contacts to energize the relay. Figure 14.10 shows the idea.

Industrial current loops

In industrial environments there is a standard for current-loop signaling, in which a remote sensor (a thermocouple, say; see Section 15.01) sends back its measurements by converting them to an analog current, which then flows through a loop. The dc bias for the loop is usually provided at the receiving end (Fig. 14.15C). There are two standards, namely full-scale ranges of 4 to 20mA and 10 to 50mA. The 4–20mA standard is more popular and usually uses a dc bias of 24 volts dc (though sometimes higher). For simplicity it is often desirable to use the signaling current to power the electronics at the remote end.

For this purpose you can use part of the loop bias for power. Commercially available current-loop modules generally specify that the data recipient has to provide a maximum load resistance R and minimum dc bias V_s such that $(V_s - 12\text{V})/R_s$ is equal to the full-scale current. In other words, the remote module can drop up to 12 volts while still applying full-scale loop current. Of course, the module has to keep running when sending a loop current corresponding to minimum output. So the bottom line is that you always have available at least 12 volts at 4mA to power your equipment; you may have more, but don't count on it. That's plenty for even rather complex circuits, if you practice careful micropower design.

RS-232 serial-port signals

The RS-232C/D standard specifies bipolarity data and control signals of substantial drive capability (see Section 10.19); you can use one of the control signals (or even a data signal!) to run a low-power circuit. Officially an output must be able to assert ±5 volt to ±15 volt levels into a 3k to 7k load resistance. The RS-232 drivers in common use typically have an output impedance of a few hundred ohms, and current limit at 5 to 15mA. To become a parasite on this power source, you have to arrange your software to keep a known control line in a known (and stable) state. You can even use a *pair* of control lines, if available, to get split supply voltages (±5V, min). Remember that control signals (RTS, DTR, etc.) are asserted HIGH, which is the opposite of the data signals.

Since there's usually plenty of commercial ac power available around a computer, you aren't really doing anything miraculous by sucking the life forces from the 25-pin *D* connector. However, for a simple serial-port hang-on circuit it is an elegant source of power. You can get commercial network interfaces and modems that work this way.

POWER SWITCHING AND MICROPOWER REGULATORS

14.06 Power switching

You can tame your usual microprocessors, regulators, and other power-hungry components into a micropower application, if the design permits the circuit to be turned off (or put into a low-current standby state) most of the time, and only occasionally run at full current. For example, an oceanographic data logger might make a 10 second salvo of observations (temperature, pressure, salinity, ocean currents) once each hour for a 6 month period. Only the real-time clock need run continuously, with the analog signal conditioning circuitry, microprocessors, and data-recording media shut off except during actual data logging.

Even if you take pains to use micropower design techniques, you may still be forced to use some high-current devices, for example, if you need to use high-speed transducers or high-current actuators. You may need to use some specialized LSI digital circuits, op-amps, filters, or other circuits that are simply not available in low-power versions. In all these cases it is necessary to switch off power to the high-current portions of the circuit except when they must operate.

Such "power switching" can be the simplest form of micropower design, since ordinary design techniques with ordinary components can be used throughout. You've got to make sure the circuit "wakes up" gracefully (a linear circuit should be designed to avoid embarrassing momentary states, for example driving its outputs into saturation; a fully shut down microprocessor circuit would usually do a complete "cold boot"). Likewise, the circuit should be designed to shut down in an orderly manner.

There are several ways to do the power switching (Fig. 14.19):

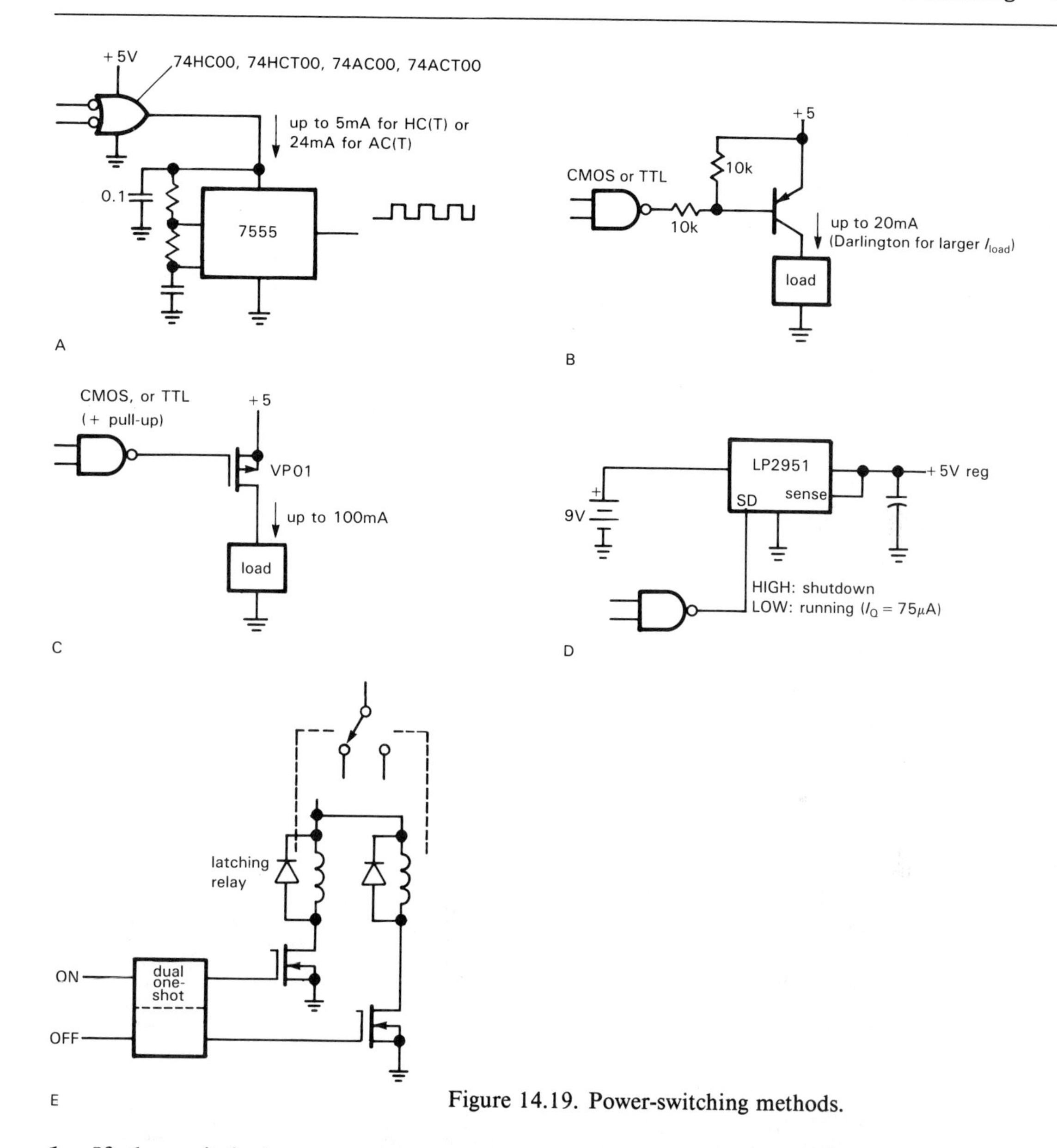

Figure 14.19. Power-switching methods.

1. If the switched components run at less than 5mA or so, you can power them directly from a CMOS logic output. The HC/HCT families can supply 5mA with only 0.5 volt drop below the positive rail; for higher current, several outputs can be used in parallel. The AC/ACT CMOS families are good for 24mA.

2. Use a power transistor, operating as a saturated switch (not a follower) to minimize forward drop (thus *pnp* for a positive supply). The necessary base drive, chosen conservatively large to guarantee saturation, is a disadvantage, though it will probably be smaller than the current used by the switched circuit.

3. Use a power MOSFET. As with bipolar transistors, use as a switch, not a follower (thus *p*-channel for a positive supply). MOSFETs are easy to drive and have no gate current in either state.

4. Many of the low-power regulators include a "shutdown" input, with very low quiescent current in the standby mode (see Section 14.07). You can do power switching by commanding such a regulator into the active state.
5. Use a mechanical relay, perhaps a *latching* relay. There's a good variety now available in DIPs and tiny metal cans, and they offer zero voltage drop, high overload capability, and the ability to switch bipolarity (or even ac) voltages. In addition, latching relays require no holding current. Be sure to use a diode to protect the relay driver from inductive spikes (Fig. 1.95).

Current limiting

It is essential to limit the inrush current in a power-switched circuit, for two reasons: The high peak currents that would result from switching a battery (bypassed with a capacitor) into a load (similarly bypassed) could destroy the switch; this is true even for a small mechanical relay, which is most likely to fail by having its contacts fuse shut. Furthermore, the momentary collapse of the battery voltage during a high current-switching transient can cause volatile memory and other circuitry being held in a standby state to lose information (Fig. 14.20).

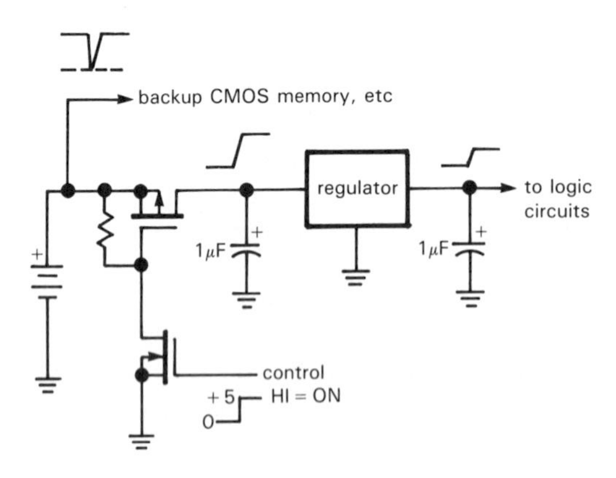

Figure 14.20. Inrush current can cause transient loss of battery voltage.

Several approaches are shown in Figure 14.21. As long as the switch can handle the transient, you can decouple the negative-going dip from the maintaining

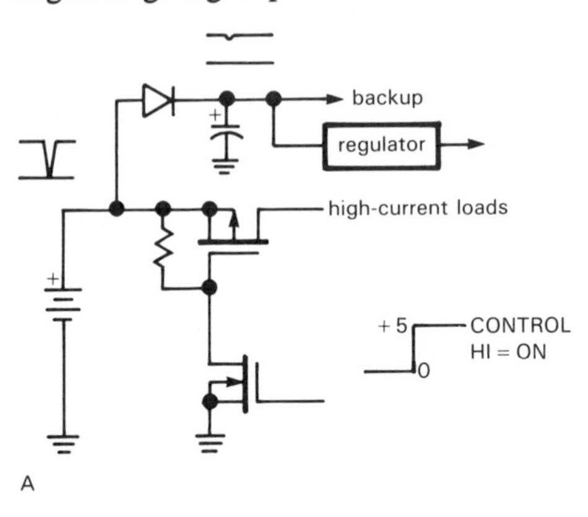

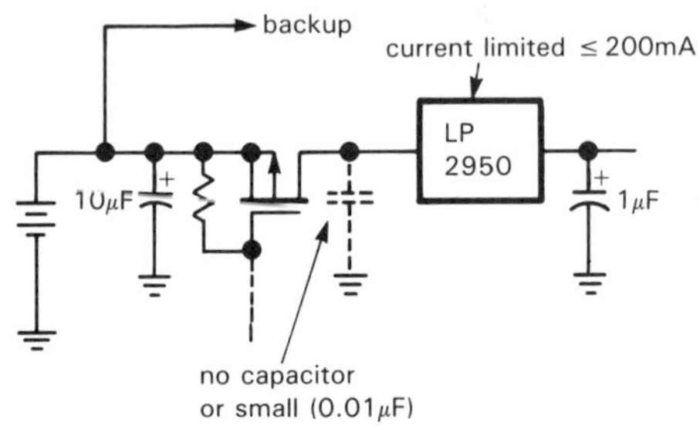

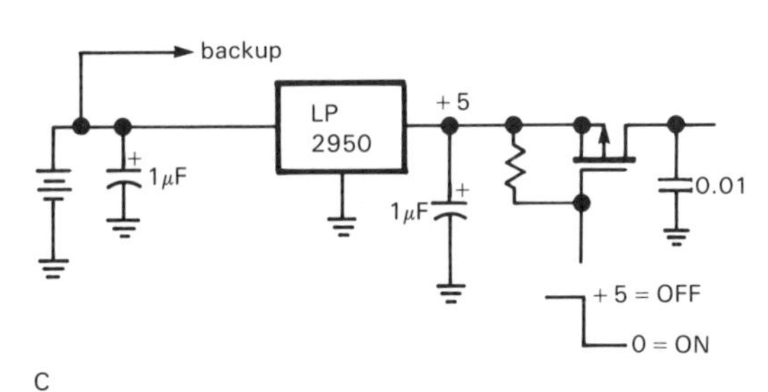

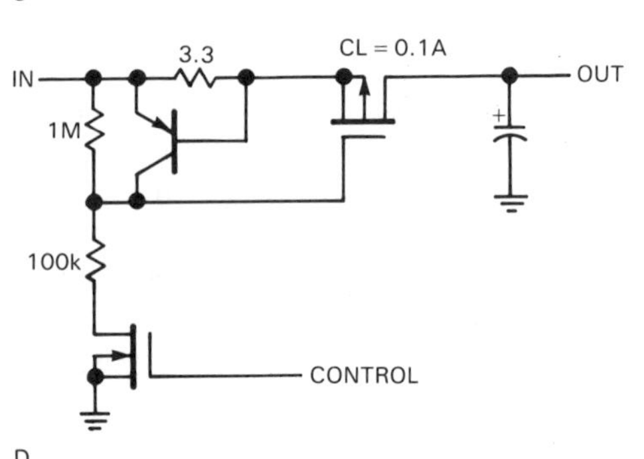

Figure 14.21. Four cures for inrush-current transients.

regulators with a diode, as in option 1. Alternatively, do the switching upstream of a current-limited regulator (keep its input bypass capacitor small), as in option 2, or put the switch after the regulator (option 3). The latter method isn't as good, because of degraded supply stiffness due to the switch's R_{ON}. Another method is to use upstream current-limited switching (option 4), in this case with 150mA current limit, to prevent collapse of V_{batt}.

14.07 Micropower regulators

Until recently it had been difficult to find voltage regulator ICs with microamp quiescent currents capable of substantial output currents. The choice was (a) the Intersil 7663/4 or (b) build your own! Fortunately the situation is improving. Here is the current selection:

ICL7663/4; MAX663/4/6 (Intersil; second-sourced by Maxim and others). These are multiterminal positive and negative regulators, with 1.5–16 volts operating range and maximum quiescent current of 10μA. The bad news is that they are slow (due to a "starved" servo amplifier; use lots of bypass capacitance) and only good for a few milliamps of load current (they're stiffer at higher input voltage, being CMOS devices); for example, with +9 volts input the output impedance is typically 70 ohms.

LP2950/1 (National). These are positive regulators, available as a 3-terminal +5 volt regulator (2950) and an 8-terminal adjustable regulator (2951). Quiescent current is 80μA at zero load current, rising to 8mA at 100mA load current. These regulators use *pnp* pass transistors for low dropout voltage (80mV max at 100μA, 450mV max at 100mA) and are designed so that the quiescent current does not soar when the input voltage dips below dropout (a common disease of bipolar transconductance regulators). This last feature is particularly useful for battery-powered instruments that can continue to function with a low battery. The 2951 includes a shutdown input and dropout-detector output.

LT1020 (Linear Technology). This is a multiterminal positive regulator with 40μA quiescent current, 2.5–35 volt output range, and 125mA maximum current. The *pnp* pass transistor gives low dropout voltage (20mV typ at 100μA, 500mV typ at 125mA). There is a shutdown input and dropout-detector output.

TL580C (Texas Instruments). This is a dual positive micropower *switching* regulator with 2.5–24 volt output range and 140μA quiescent current. Like all switching regulators, you get high efficiency (up to 80%) over a range of battery voltages and the flexibility to have output voltages greater than the unregulated input voltage.

MAX630 series (Maxim). These are micropower switching regulators, in a nice variety of options. The MAX630 is an adjustable (2V to 18V) positive step-up regulator (i.e., $V_{out} > V_{in}$), while the MAX634 is an inverting switcher (i.e., positive input, negative output). The MAX631–3 are fixed-voltage (5, 12, 15V) positive step-up switchers, with MAX635–7 being the inverting equivalents. The MAX638 is an adjustable positive step-down ($V_{out} < V_{in}$) switcher. All are capable of output currents of a few hundred milliamps, with quiescent currents around 100μA and efficiencies around 80%.

MAX644 series (Maxim). These micropower switching regulators are designed to generate a +5 volt output when powered from a one- or two-cell battery. This clever design uses a two-part switching upconverter: One section runs continually, providing a low-current (<0.5mA) +5 volt output; it also supplies +12 volts dc needed to switch the MOSFET used for the high-current (up to 50mA) +5 volt output. In standby mode (low-current output only) the quiescent current is 80μA. The MAX644 is designed for 1.5 volt nominal input and operates down to input voltages of 0.9 volt.

TABLE 14.4. LOW-POWER REGULATORS

Type	Mfg[a]	Pins	I_Q typ (μA)	Pol	V_{out} (V)	V_{in} (V)	Dropout V (V)	@ I (mA)	Shutdown	Tempco typ (ppm/°C)	Comments
ICL7663	IL+	8	4	+	1.5–16	1.5–16	0.8	20[b]	•	200	MAX663, 7663S improved
MAX664	MA+	8	6	–	1.3–16	2–16.5	0.2	20[b]	•	100	also ICL7664
MAX666	MA	8	6	+	1.3–16	2–16	0.9	40	•	100	MAX663 + dropout detec
LT1020	LT	14	40[c]	+	0–35	5–36	0.5	125	–	1%	dropout detector
LP2950	NS	3	75[c]	+	5	5–30	0.45	100	–	20	dropout I_Q = 110μA
LP2951	NS	8	75[c]	+	1.2–29	2–30	0.45	100	•	20	
MAX630	MA	8	70	+	V_{in} to 18	2–16.5	-	375	•	-	switching steput
MAX635-7	MA	8	80	±	–5,–12,–15	+2 to 16.5	-	375	–	-	switching inverter
MAX634	MA	8	100	±	to -20	2–16.5	-	375	•	-	switching inverter
MAX631-3	MA	8	135	+	5,12,15	1.5–V_{out}	-	325	–	-	switching stepup
MAX638	MA	8	135	+	< V_{in}	2–16.5	-	375	•	-	switching stepdown
TL580C	TI	8	140	+	2.5–24	2.4–30	-	100	–	-	
LM10	NS+	8	300	+	1–40	1.1–40	0.4	20	–	30	
LM2931	NS+	5	400	+	1.2–25	to 26V	0.2	150	•	-	TO-220
LM2931-5	NS+	3	400	+	5	5.2–26	0.2	150	–	-	TO-92; 2931CT is adjustable
TL750L05	TI	3	1000	+	5	5.2–26	0.6[d]	150	–	-	TO-92; TL751 has shutdown
LM317L	NS+	3	2500[e]	+	1.2–37	to 40V	2	100	–	0.7%	TO-92
LM337L	NS+	3	2200[e]	–	1.2–37	to –40V	2	100	–	0.7%	TO-92
78Lxx	FS+	3	3000	+	5,12,15	to 30V	2	100	–	-	TO-92
79Lxx	FS+	3	2000	–	–5,–12,–15	to –35V	2	100	–	-	TO-92; LM320L also
LM330	NS+	3	3000[c]	+	5	5.3–26	0.6	150	–	-	TO-220
LM2930	NS+	3	4000	+	5	5.3–26	0.6	150	–	-	TO-220; LM2935 also

(a) see footnote to Table 4.1. (b) for V_{in} = 9V. (c) no load. (d) over full temp range. (e) I_L (min).

In addition, there are several "low-power" regulators (78L05, LM330, LM317L, LM2930/1), characterized by quiescent currents of a few milliamps. These are useful for instruments with some external power source, for example solar cells or telephone holding current. Also, don't overlook the possibility of using a micropower voltage *reference*, rather than a regulator, if its voltage happens to be what you want. For example, the REF-43 from PMI is a 3-terminal 2.5 volt reference with 250μA maximum quiescent and excellent characteristics.

Look at Table 14.4 (which also includes the regulators above) for characteristics of most available micropower regulators.

Negative supplies

With the exception of the ICL7664/MAX664, all of the linear micropower regulators are positive polarity only (though the LT1020 can be used to make a dual supply). If you need negative supply voltages, there are (besides the feeble 7664) several possibilities, namely (a) a "flying-capacitor" voltage converter chip like the 7662, (b) a discrete realization of a flying-capacitor voltage converter, using complementary power MOS transistors, (c) a voltage converter using a CMOS oscillator chip like the 7555 (that's a CMOS 555) or the output of any CMOS logic gate that is driven by a square wave, (d) a switching supply, with inductive energy storage, or (e) the use of a single positive supply, with an op-amp-generated ground reference part way between ground and the positive rail. Let's take them in turn:

1. The 7662 (and its predecessor 7660) is a CMOS IC introduced by Intersil and widely second-sourced (see Section 6.22). It has an oscillator and CMOS switches (Fig. 6.58), and with a few external capacitors you can use it to generate either $-V_{supply}$ or $+2V_{supply}$, when powered by a positive voltage V_{supply}. Like most CMOS devices, it has a limited supply voltage range; for the 7662, V_{supply} can range only from 4.5 to 20 volts (1.5V to 10V for the 7660). The output is not regulated, and it drops significantly for load currents greater than a few milliamps. In spite of these drawbacks, it can be very useful in special circumstances, for example to power an RS-232C driver chip on a board that otherwise runs on a single +5 volt supply. The MAX680 and LT1026 are flying-capacitor dual supplies that generate ±10 volts (up to 10mA) from +5 volts (Fig. 6.60). There are also combination voltage converter and RS-232 driver/receivers available as single ICs, the LT1080 and MAX230–239 series. If your application requires RS-232 ports, you may be able to use the dual supply voltages generated by one of these RS-232 driver ICs to power your analog electronics.

2. To generate a larger negative voltage, you can use discrete MOS transistors in a flying-capacitor circuit (Fig. 14.22). The particular example shown idles at a few microamps and generates up to 30mA.

3. Figure 14.23 shows a simpler method, again somewhat limited in voltage range, using the CMOS 7555 timer chip. You can power the 7555 from a positive supply in the range 2 to 18 volts, thus generating up to −15 volts or so. With a voltage multiplier (see Section 1.28) you can, of course, generate higher voltages, with correspondingly poorer regulation. If you have some CMOS logic in your circuit, you can use the output of a CMOS gate instead of the 7555. However, if you're using a high-performance CMOS family such as HC/HCT or AC/ACT, then you are limited to 5 volt logic swings, whereas the older 4000 or 74C series permit 15 volt swings, albeit at lower current.

4. As we explained in Chapter 5, with inductive energy storage you can make switching supplies for which the output voltage is higher than the input, or much lower, or even negative, all with efficiencies of 75% or so, independent of input voltage. This is obviously useful in

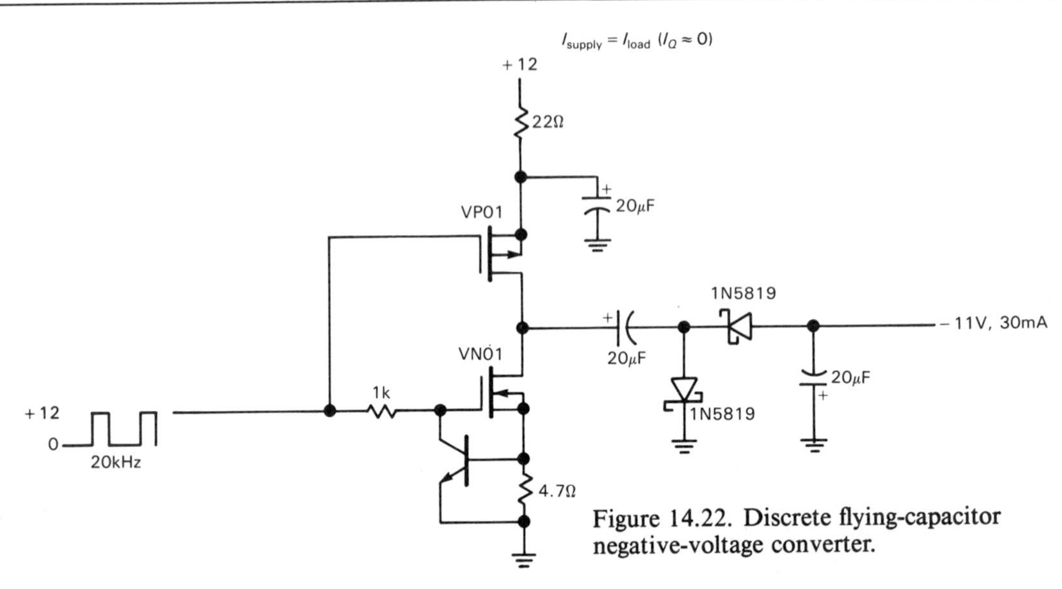

Figure 14.22. Discrete flying-capacitor negative-voltage converter.

micropower design, where the unregulated dc may be supplied by batteries whose voltage drops off with use. Switching supplies for micropower applications can be designed to maintain high efficiency even when unloaded (unlike ordinary high-current switchers), by using circuitry that shuts the oscillator off until the output drops, at which point it supplies a single charging pulse, then goes to sleep again. Figure 14.24 shows a +5 volt supply constructed with the low-power MAX631.

5. You may not need a separate negative supply, even if you are using op-amps with bipolarity output swings, etc. For example, you might generate a +4.5 volt ground reference (using a resistive divider and micropower op-amp follower) for an op-amp circuit running from a single 9 volt battery. Let's look at this method in some more detail.

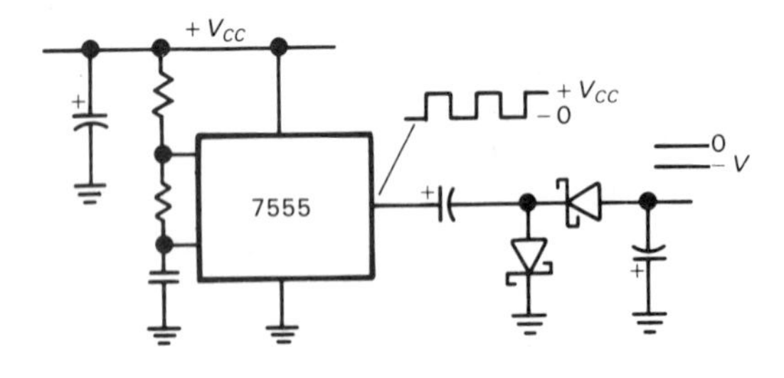

Figure 14.23. Negative-voltage generator from a positive square wave.

14.08 Ground reference

Most of our op-amp circuits in Chapter 3 used symmetrical power supplies, usually ±15 volts, because of the flexibility of dealing with signals near ground. As we mentioned in Section 4.22, however, it is possible to use only a single supply, by generating a reference voltage that substitutes for the ground potential of the usual bipolarity op-amp power supplies. When your power supply is a battery, there's an added incentive to keep things simple, preferably by using a single 9 volt battery.

The easiest way to generate an analog "common" is to split the battery voltage with a resistive voltage divider, then use a micropower op-amp follower to generate the low-impedance common. To the outside world that common voltage is "ground," with both ends of the battery floating; see Figure 14.8.

In the example circuit, we've chosen a 3440 CMOS programmable op-amp, biased to run at 5μA quiescent current. The divider's unusually large resistors keep its contribution to the current drain small, with capacitive bypassing to keep the impedance low at the midpoint, which otherwise would be susceptible to hum and pickup of other signal

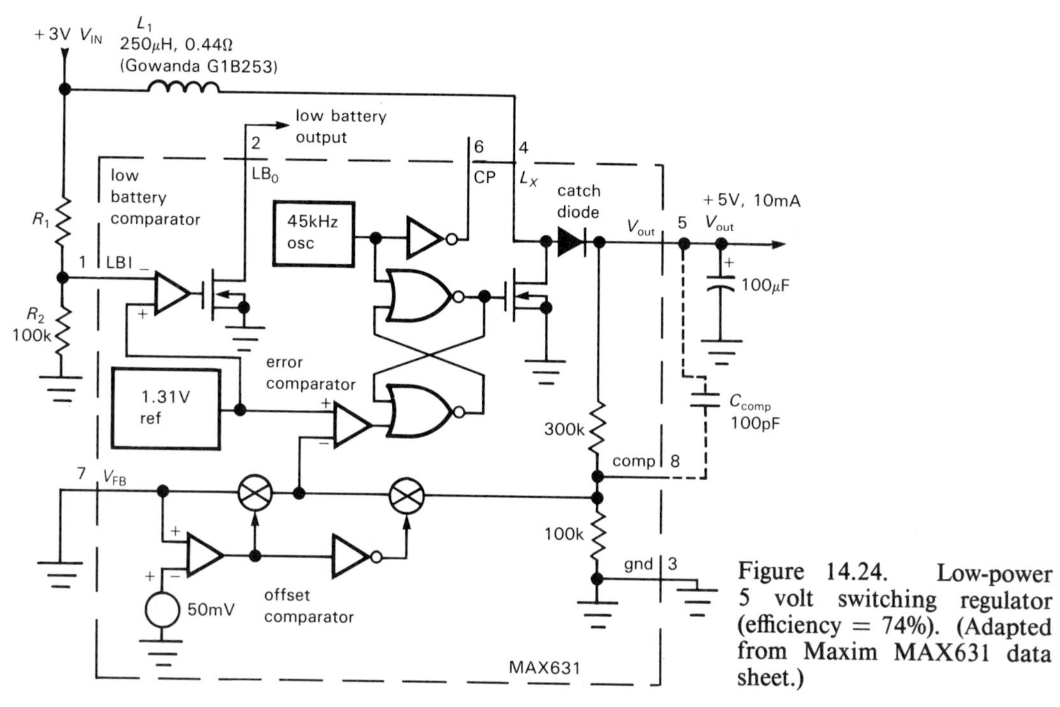

Figure 14.24. Low-power 5 volt switching regulator (efficiency = 74%). (Adapted from Maxim MAX631 data sheet.)

frequency interference. The 3440 is a good choice in this application, because it can sink or source substantial currents (up to a few milliamps) even when biased at 1μA; this property is not shared by all programmable op-amps, many of which have poor sourcing capability when operated at micropower levels. For example, the LM346 operating at 5μA can source only about 0.1mA, although it can sink 20mA (look ahead to Fig. 14.32).

Note that the reference voltage doesn't have to be half of the battery voltage; it may be best to split the supply unsymmetrically, to allow maximum signal swing. (We'll have an example in Section 14.12.) In some instances it may be preferable to put it at a fixed voltage from one rail, perhaps determined by a precision micropower voltage reference. That rail is then a regulated supply with respect to the common reference.

Output impedance

There are some situations in which you don't even need to use an op-amp to generate the ground reference. For example, if the reference voltage goes only to op-amp inputs (which would have been connected to ground in the usual split-supply configuration), then a high-impedance resistive divider, bypassed to maintain a low impedance at signal frequencies, will usually suffice.

In the more usual case, however, the ground reference must present a low impedance, both at dc and at signal frequencies. For example, some ICs may use it as their negative rail; it might be the common point for low-pass filters, biasing networks, loads, etc. Look at almost any normal split-supply circuit and you'll find dc and signal currents flowing into and out of ground. As in the example above, be sure the op-amp you choose to generate the ground reference has the source and sink capability the circuit needs. Micropower op-amps tend to have rather high open-loop output impedances (Fig. 7.16), so at high frequencies (where there isn't much loop gain) the impedance of the ground reference may rise to several thousand ohms.

The obvious cure is to bypass the ground reference (Fig. 14.25A), but this is likely to cause ringing or even oscillation because of the lagging phase shift of the bypass capacitor in combination with the op-amp's relatively large output impedance, all of which is inside the feedback loop. Figure 14.25B shows one cure – a decoupling resistor of a few hundred ohms, which, however, raises the impedance at dc since it is outside the feedback loop. With two more parts, Figure 14.25C does the trick, maintaining dc feedback (via R_2) and stability at the same time.

Whatever method you choose, make

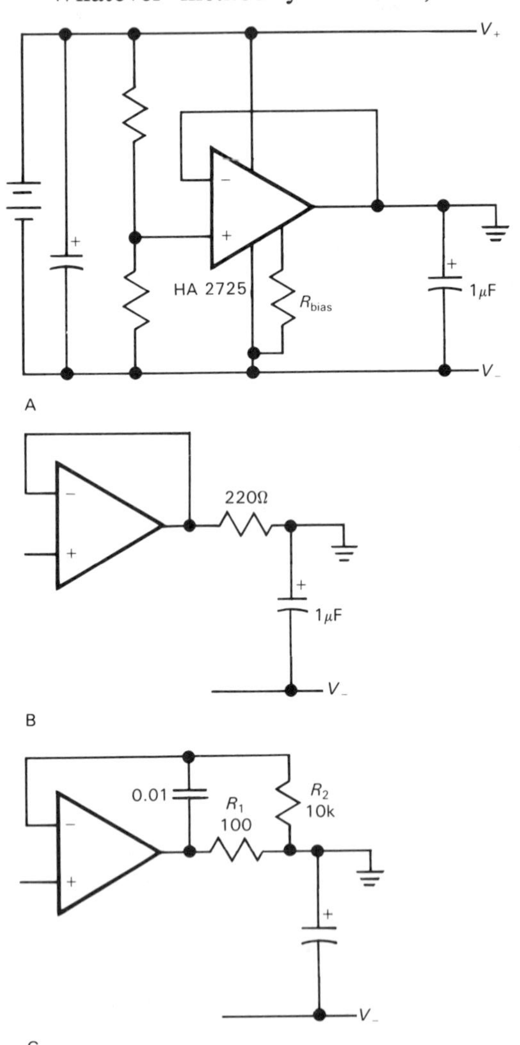

Figure 14.25. Bypassed split-supply generators.

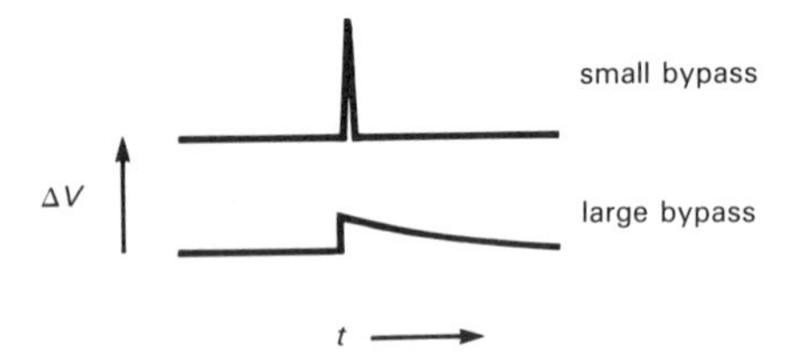

Figure 14.26

sure you test it under various steady-state and transient load conditions. A good way to test for transient behavior is to watch the voltage waveforms while applying a low-frequency square-wave load. There are some op-amps (e.g., the HA2725 and MC3476) that can drive directly into a large capacitive load without stability problems; apparently in these cases the external capacitor reflects back onto the internal compensation capacitor, pushing down the dominant pole in a "brute-force" compensation. In many cases, however, you're more likely to wind up with a pair of nearby lagging phase shifts, which spells trouble.

Note that the choice of bypass capacitor value may involve some subtlety: For a load-induced spike of fixed charge injection into the ground reference node (i.e., a fixed amp-second product), a larger bypass capacitor will keep transient ground noise smaller, but will have a longer recovery time than a small capacitor (Fig. 14.26). For a high-gain low-speed circuit that may be worse, perhaps producing slow exponential recovery instead of harmless little spikes at the output.

When designing ground reference circuits, don't overlook the reference voltage outputs that are sometimes provided on other ICs. For example, the LM322 timer provides a stable 3.15 volt output. Other chips that have external access to internal voltage references are A/D converters, V/F converters (e.g., the 331, with its 1.89V reference), and chips like the LM10, which has a 200mV reference and amplifier, in addition to an uncommitted op-amp. Figure 14.27 shows several buffered reference schemes.

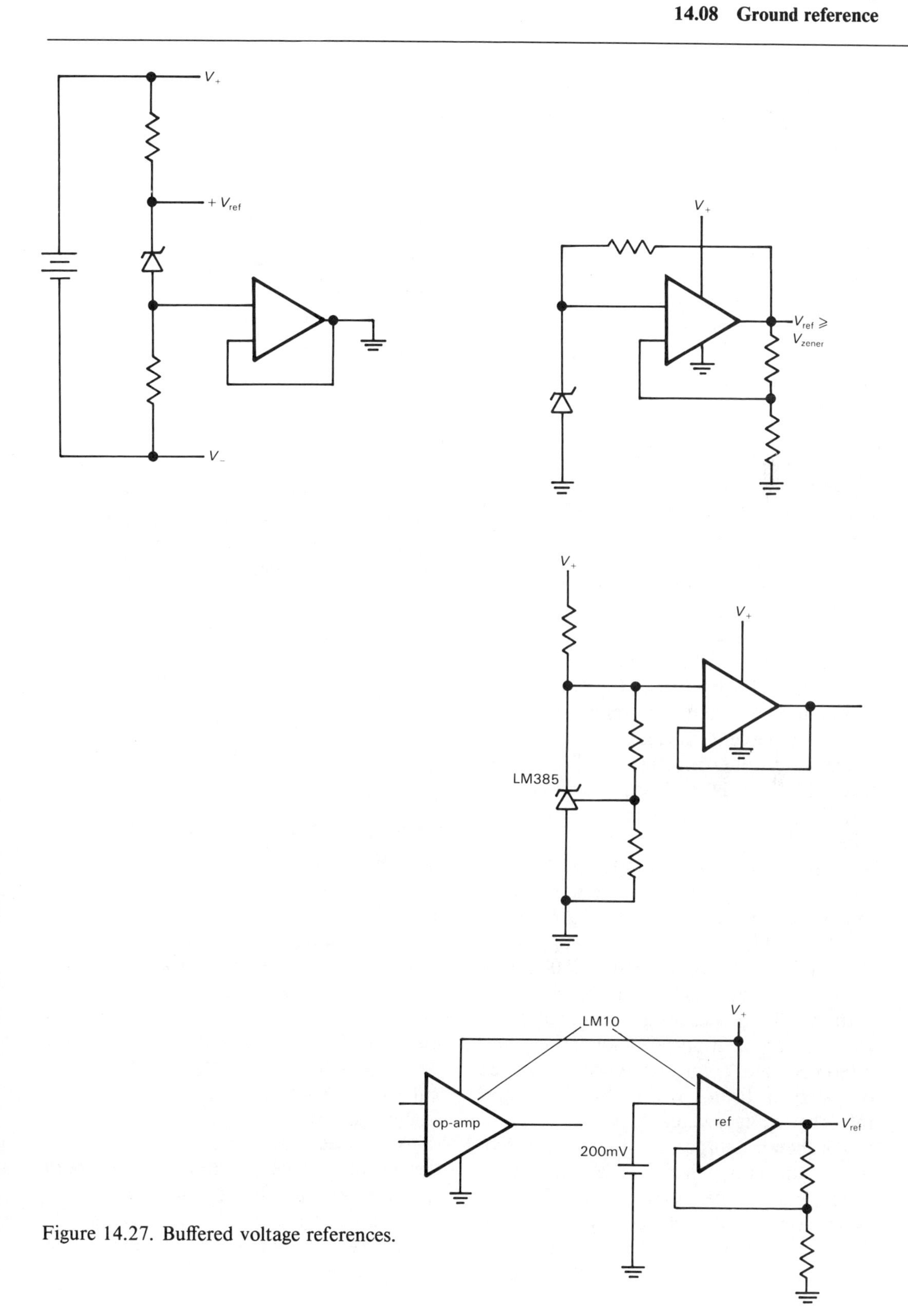

Figure 14.27. Buffered voltage references.

14.09 Micropower voltage references and temperature sensors

Most zener and bandgap references are relatively power-hungry and are not suitable for micropower circuits. As Table 6.7 demonstrates, most 3-terminal references run at about a milliamp, and most 2-terminal zener-like references are specified at similar operating currents.

Fortunately, there are several voltage references intended for micropower applications. The LM385 series includes a programmable 2-terminal bandgap reference (LM385, 1.24V–5.30V) and two fixed voltage references (LM385-1.2, 1.235V and LM385-2.5, 2.50V). The fixed-voltage models are specified to operate at currents down to 10μA, with dynamic impedances of 1 ohm at 40μA and 100μA, respectively. The minimum current of the programmable version goes from 10μA to 50μA, depending on voltage. All versions are available with tempcos down to 30ppm/°C. The ICL7663/4 regulators (Section 14.07) can be used as 3-terminal references, with typical quiescent current of 4μA and dynamic output impedance of 2 ohms. The ICL8069 is a 2-terminal bandgap reference that operates down to 50μA (where the dynamic impedance is 1Ω), with tempco down to 50ppm/°C. The AD589 has the same characteristics, with improved tempco (down to 10ppm/°C). The LT1004 from Linear Technology is like the LM385-1.2, while their LT1034 is a dual 2-terminal reference (1.2V and 7.0V) with minimum operating currents of 20μA and tempco of 20ppm/°C for its 1.2 volt reference; the 7 volt reference should be operated at 100μA (min), but is quieter than the bandgap references.

For better tempco at not-quite-micropower currents, there's the LM368 3-terminal reference, available in 5, 6.2, and 10 volt versions (0.05% accuracy). It draws 300μA, has low output impedance over frequency, and is available with tempcos down to 10ppm/°C. Even better is the REF-43, a 3-terminal 2.5 volt positive reference with 0.05% accuracy and 3ppm/°C tempco (max). It has low Z_{out} (0.1Ω), excellent regulation (2ppm/V_{in}, max), output current to 10mA, and quiescent current of 250μA, max.

Table 14.5 lists currently available micropower references.

Finally, there are micropower ICs that convert temperature to current or to voltage. The AD590 and AD592 are 2-terminal current sources that run on 4 volts to 30 volts, and give a current of 1μA/°K (e.g., 298.2μA at 0°C). The LM334 is similar, but with a programming pin to set the conversion factor; the operating range is 1μA to 10mA. The LM34 (Fahrenheit) and LM35 (Centigrade) are 3-terminal temperature sensors with voltage output (thus 0V at 0°F or 0°C, and 10mV/°F or °C, respectively) and quiescent current of 100μA. The LM335 is a 2-terminal IC zener with a breakdown voltage of 10mV/K (e.g., 2.982V at 0°C), operable down to 400μA. See Section 15.1 for additional information.

LINEAR MICROPOWER DESIGN TECHNIQUES

Thus far we have treated power sources, power-switching techniques, regulators, and references for the design of micropower instruments. Now, following the progression of topics in the rest of the book, we turn to the design of linear and digital circuits themselves. We will begin with a discrete linear circuit example (a high-gain micropower audio amplifier), then proceed to micropower op-amp design techniques. That will be followed by sections on digital and microprocessor design, and finally some comments on packaging techniques for low-power instruments.

TABLE 14.5. MICROPOWER VOLTAGE REFERENCES

Type	Mfg[a]	Term	B'gap/Zener	Pins	Trim	Voltage (V)	Acc'y (%)	Tempco max (ppm/°C)	Supply voltage min (V)	Supply curr[b] (μA)	Output curr max (mA)	Noise typ 10Hz-10kHz (μV,rms)	Drift typ (ppm/1kh)	Z_{out} typ (Ω) @	I_{ref} (μA)	Comments
TSC04A	TS	2	B	2	–	1.26	2	50	-	15	-	-	-	0.3	1000	
TSC05A	TS	2	B	2	–	2.5	3	50	-	20	-	-	-	0.3	1000	
REF25	FE	2	B	2	–	2.5	-	55[t]	-	60	-	-	-	1.5	-	
REF43F	PM	3	B	8	•	2.5	0.05	10	4.5	350	10	5[f]	-	0.03	-	2ppm/V_{in}, max
REF50	FE	2	B	2	–	5.0	-	55[t]	-	60	-	-	-	1.5	-	
LM385	NS	2	B	3	•	1.2-5.3	-	30	-	9	-	50[c]	20	0.4	100[c]	
LM385-1.2	NS	2	B	2	–	1.2	1	30	-	10	-	60	20	1	40	LT1004-1.2
LM385-2.5	NS	2	B	2	–	2.5	1	30	-	20	-	120	20	1	100	LT1004-2.5
AD589M	AD	2	B	2	–	1.24	2	10	-	50	-	5	-	0.6	500	MP5010
LT1034	LT	2	B	2	–	1.22	1	20	-	20	-	4[d]	20	0.3	100	dual ref
"	LT	2	Z	2	–	7	3	40[t]	-	100	-	-	20	-	-	
LP2950ACZ	NS	3	B	3	–	5.0	0.5	100	5.4	120	100	430[e]	-	0.02	100	regulator
ICL7663/4S	IL+	3	B	8	•	1.5-10	-	100[t]	1.5	10	40	-	-	2	-	MAX663/4
ICL8069	IL	2	B	2	–	1.23	2	50	-	50	-	5	1	1	50	
TSC9491A	TS	2	B	2	–	1.22	2	50	-	50	-	-	-	2	-	

(a) see footnote to Table 4.1. (b) min oper curr (for 2-term refs); max quiescent curr (for 3-term refs). (c) at V_{ref}. (d) 0.1–10Hz, pp.
(e) 10Hz–100kHz, 1μF bypass. (f) 0.1Hz–1kHz, max. (t) typical.

14.10 Problems of micropower linear design

In general, low-power linear design means low collector (or drain) currents and correspondingly large collector (drain) resistors. As a result, capacitance effects dominate, in the form of both Miller effect and normal *RC* rolloffs. You often resort to techniques that are ordinarily used only in radiofrequency design, for example the use of the cascode (Section 2.19), emitter followers, and the shunt-series pair (Section 13.07). RF transistors (with f_T of 1GHz or more) may be good choices even at low frequencies because of their extremely low feedback capacitance C_{ob}; for example, the MRF931 has $C_{ob} = 0.35\text{pF}$ at $V_{CE} = 1$ volt and is specified for use down to 1 volt and 0.1mA ($f_T = 3\text{GHz}$ at $I_C = 1\text{mA}$ and $V_{CE} = 1\text{V}$). In spite of measures like these, it still pays to choose low operating frequencies whenever possible, for example the clock rate for microprocessor or other CMOS digital systems.

Other undesirable effects of low-power operation are increased noise pickup (because of the relatively high signal source impedances), reduced drive capability (low operating currents, high impedances), and relatively high transistor noise voltage e_n (from Johnson noise in the relatively high r_e; see Section 7.13). This latter problem also afflicts micropower voltage references; be sure to check the noise specifications. Even with emitter followers, output impedances can be unreasonably large ($r_e = 25\text{k}\Omega$ at $I_C = 1\mu\text{A}$).

In general, operation at low voltage is desirable, because of the correspondingly reduced collector resistors for the same operating current. In addition, at the same collector current the *power* is reduced proportional to supply voltage.

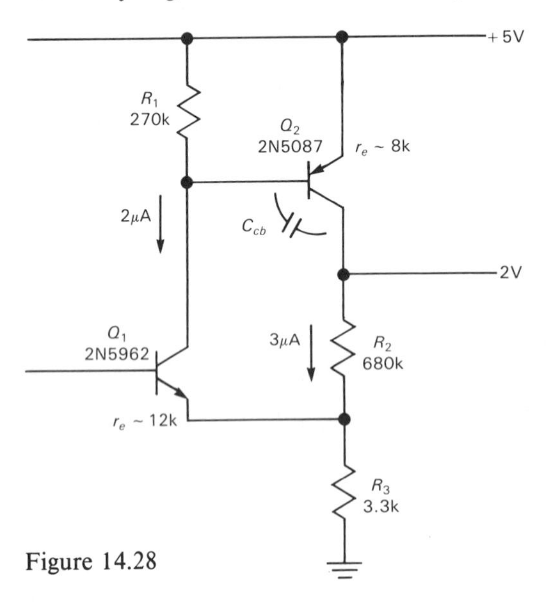

Figure 14.28

14.11 Discrete linear design example

Let us imagine we need a low-noise audio amplifier with lots of gain (at least 80dB) and low quiescent drain, to operate in some remote battery-powered application. Since the signal levels may vary over a wide range, it would be nice to include some provision for switching the gain over a range of, say, 60dB. To maintain shelf life from an alkaline 9 volt battery (500mAh) we should draw no more than 20μA total (corresponding to a 3 year life); since other circuits probably run from the same battery, we will budget 10μA for the amplifier.

The first thing to note is that a micropower op-amp cannot provide the performance we need. The exemplary "nanopower" CA3440, running at 10μA, has a dc gain of 80dB (min) and a gain-bandwidth product of 300kHz; i.e., at 20kHz its gain is only 15 (24dB). We will discuss micropower op-amp design, and its limitations, in the next section. For now, all we need to know is that the objectives of op-amp design (dc coupling, accuracy, unity-gain compensation) are quite different from what we need for this example, and that we can do better with discrete design.

Let's begin by trying a stage of "series-feedback pair" shown in Section 4.27. Figure 14.28 shows our first try, where we have run a pair of superbeta low-noise transistors at 5μA combined collector currents, with a design gain (R_2/R_3) of 200 (46dB). The bias arrangement is not shown. Q_1's quiescent current is set by a V_{BE} drop across R_1, and R_2 sets Q_2's collector voltage, once its quiescent current is chosen. The intrinsic emitter resistances r_e are quite high, 12k and 8k, respectively, owing to the low collector currents.

Two such stages, with their 90dB gain and 10μA quiescent current, would appear to do the job, perhaps with an emitter follower at the output. As we mentioned earlier, however, the effects of capacitance can be devastating in low-current high-impedance circuits. Let's see what capacitance does to this circuit's performance.

To evaluate Miller effect we need to know how the voltage gain is distributed between the two transistors. Q_2 has $r_e =$ 8k, so its voltage gain is about 85; the first stage, with feedback operating at the emitter, makes up the rest, with a gain of about 2.4. The high second-stage gain suggests that Miller effect there may dominate the amplifier's overall rolloff. In fact, that is the case: The 2N5087 has $C_{cb} = 6.5$pF at $V_{cb} = 2$ volts, which acts like 550pF input capacitance to ground at the base. The reactance of this Miller capacitance equals R_1 at 1kHz, which would cause 3dB of rolloff at that frequency in the absence of overall negative feedback. With feedback, the rolloff is higher, roughly 4.4kHz, but still far too low for an audio amplifier, which ideally should be flat to 20kHz.

The solution in this case is to notice that C_{cb} is the killer (via Miller) and use instead a transistor with very low C_{cb}. The *pnp* 2N4957 is a suitable choice, a low-noise high-gain UHF amplifier with $C_{cb} = 0.6$pF at 2 volts. Figure 14.29 shows the finished circuit, with a cascade of two series-feedback pairs and an output emitter follower. The second pair has a larger emitter resistor for proper biasing with dc coupling from the first stage. The gain switching is done with a 4066B CMOS switch array, which has low noise and good inter-switch isolation. Power-supply decoupling is a good idea in a high-gain amplifier circuit like this. With all switches closed, this amplifier has 90dB of gain (switchable down to 30dB with various combinations of switch closures), 27kHz of bandwidth, and input noise voltage of 12nV/$\sqrt{\text{Hz}}$; with 50k source impedance its noise figure is 1.1dB. It's worth noting by comparison that the CMOS CA3440 that we considered initially has 20dB more noise ($e_n =$ 110nV/$\sqrt{\text{Hz}}$); even the excellent bipolar OP-90 from PMI has $e_n = 40$nV/$\sqrt{\text{Hz}}$, and that with 20μA of supply current and a gain-bandwidth product of 25kHz (allowing a meager gain of only 2dB at 20kHz). For this application, discrete design techniques are mandatory.

It may seem paradoxical to be using UHF transistors for an audio application, but our example demonstrates that it makes good sense. There are some real gems in the RF data books. For example, the MRF9331 has $C_{cb} = 0.25$pF at 0 volts (!), and an f_T of 5GHz at 1mA. These devices are specified at low voltages and currents and are meant for battery-operated communications. For example, in actual measurements the MRF9331, operating at $V_{CE} = 1.5$ volts, had $h_{FE} =$ 30 at 10nA, and 60 at 1μA.

14.12 Micropower operational amplifiers

As with ordinary linear design techniques, it is tempting to abandon discrete low-power transistor design in favor of micropower op-amps, providing they can deliver the performance you need. Steady progress in bipolar linear ICs, along with recent improvements in CMOS IC fabrication, has led to a reasonable selection of micropower op-amps. There are some serious trade-offs and design constraints involved in the use of micropower op-amps, however. Let's look at some of these problems.

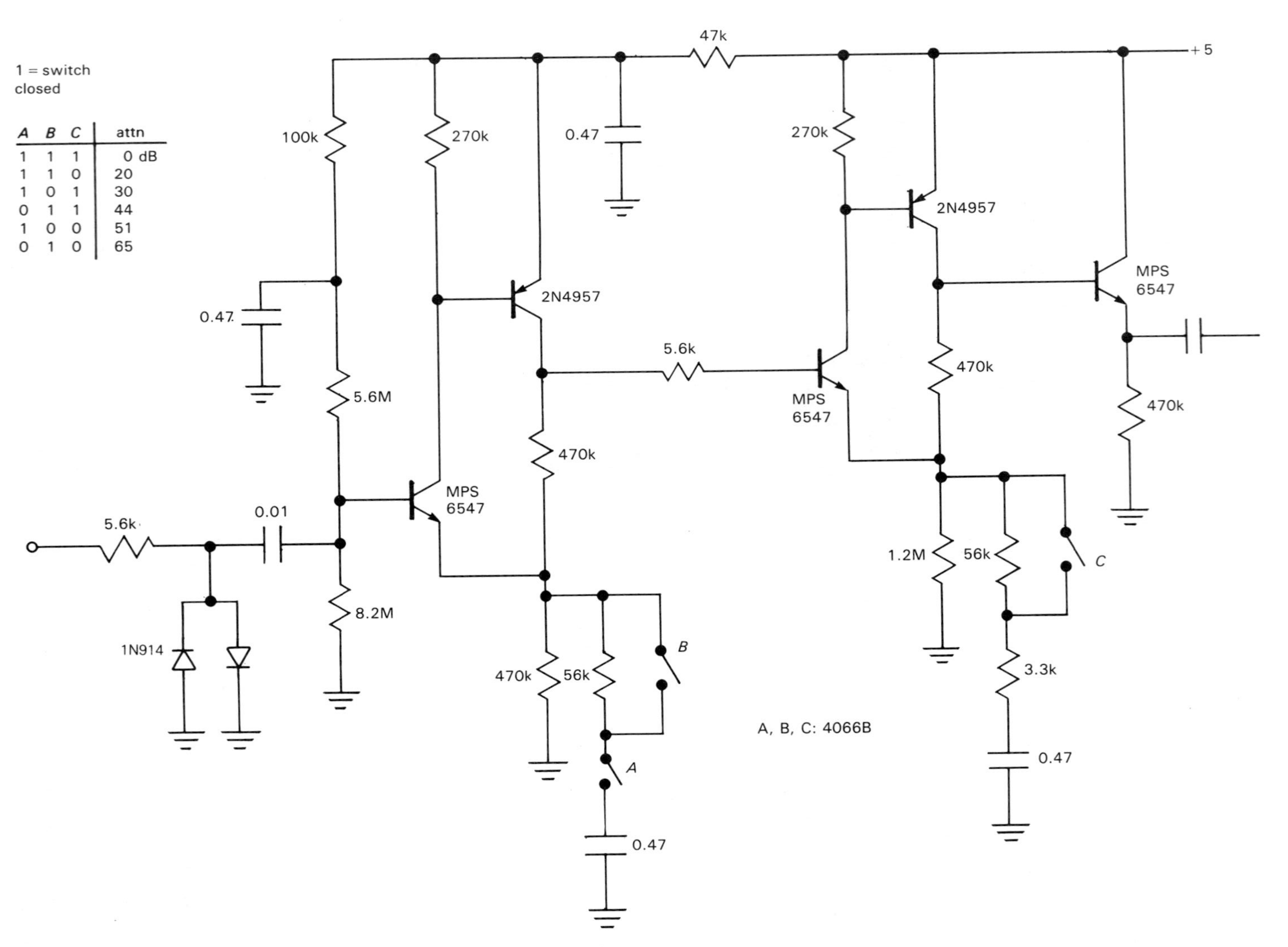

A	B	C	attn
1	1	1	0 dB
1	1	0	20
1	0	1	30
0	1	1	44
1	0	0	51
0	1	0	65

Figure 14.29. Micropower high-gain audio amplifier. The FET switches permit attenuation from maximum gain, as shown.

All other things being equal, reducing the operating current of an op-amp causes corresponding reductions in the unity-gain frequency f_T and the slew rate, and increases in the output impedance Z_{out}, crossover distortion, and input noise voltage e_n. In most cases it also tends to reduce the available output drive current I_{out}. In addition to these undesirable characteristics, the circuit compromises needed to design micropower op-amps can lead to additional pathology, for example the possibility of low-frequency oscillations (motorboating), latchup, or insufficient range of input offset trimming circuitry.

In many cases micropower design means battery-powered design, often with a single (and perhaps unregulated) dc supply voltage. An op-amp operating from a single low-voltage battery will have only limited output swing; in order to maintain good dynamic range and accuracy with these small signals, it is necessary to have smaller V_{os} compared with a conventional circuit using ±15 volt supplies. For example, an op-amp that can swing only to within 2 volts of the power-supply rails is limited to 3 volts pp maximum swing when operating from a single 9 volt battery (which has dropped to 7V at the end of its life), compared with 26 volts pp when operating from ±15 volt supplies. In this example you would need to keep op-amp input offsets nearly 10 times smaller in your battery-powered design to maintain equivalent accuracy.

Programmable op-amps and micropower op-amps

There are two categories of operational amplifiers that are suitable for micropower design: *programmable* op-amps, and *low-power* op-amps.

Programmable op-amps (Section 4.13) are like ordinary op-amps, but with an additional "programming" pin that lets you set the operating currents within the IC. Programmable op-amps typically use current mirrors in various ratios to set the operating currents of their internal stages, so that the total quiescent operating current is some multiple of the programming current, I_{set}. The usual way to set I_Q is to hook a resistor from the programming pin to one of the supplies (usually V_-), since the programming pin usually drives a current mirror directly. Quiescent currents down to a microamp or less are possible, though with correspondingly degraded dynamic performance (e.g., at $I_Q = 1\mu A$, a 4250 programmable op-amp has SR = 0.005V/μs and f_T = 0.01MHz). Some popular programmable op-amps are the LM346 and the 4250 (both bipolar) and the CMOS CA3440.

Low-power op-amps are simply op-amps internally designed to operate at low quiescent current, without a programming pin. Examples are the precision OP-20 (45μA) and OP-90 (12μA max) and TI's "LinCMOS" TLC27L2 (20μA). A variation on this theme provides pin-*selectable* operating current, according to whether the programming pin is tied to V_+ or V_- or is left open. Both the TLC271 and the ICL7612 work this way, with selectable operating currents of 10μA, 100μA, and 1mA.

Op-amp design example: stuck-node tracer

Let's start with a simple example, to see how micropower op-amp design goes. A tricky troubleshooting problem is a so-called stuck node, in which there is a short somewhere on a circuit board. It may be an actual short circuit in the wiring itself, or it may be that the output of some device (for example a digital three-state driver) is held in a fixed state. It's hard to find, because anywhere you look on that line you measure zero volts to ground.

A technique that does work, however, is to use a sensitive voltmeter to measure

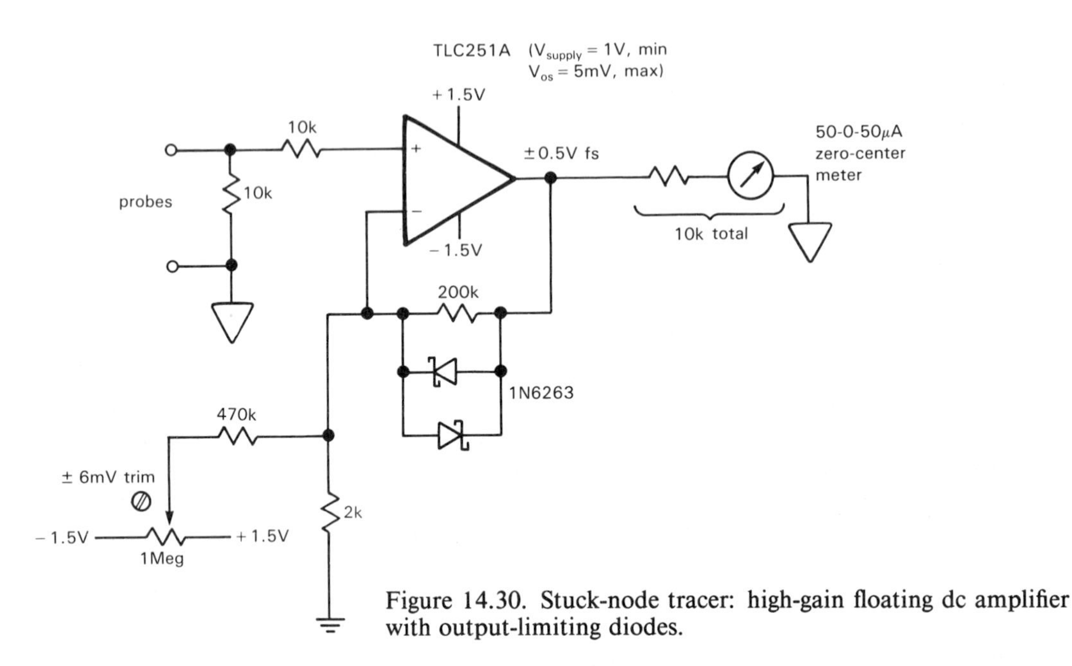

Figure 14.30. Stuck-node tracer: high-gain floating dc amplifier with output-limiting diodes.

voltage drops *along* the stuck trace. A typical signal trace on a printed-circuit board might be 0.012 inch wide and 0.0013 inch thick (1 ounce per square foot), which has a resistance along the trace of 44mΩ per inch. So if there's a device holding the line to ground somewhere, and you inject a diagnostic current of 10mA dc somewhere else, there will be a voltage drop of 440μV per inch in the direction of the stuck node.

Let's design a stuck-node tracer. It should be battery-powered so that it can float anywhere on the powered circuit under test. It should be sensitive enough to indicate a drop of as little as ±100μV on its zero-center meter, with larger meter deflections for larger drops. Ideally it should have a nonlinear scale, so that even for voltage drops of tens of millivolts the meter will not go off scale. With micropower design it should be possible to omit the on/off switch, since 9 volt batteries or AA-size cells give nearly their full shelf life (500mAh and 1400mAh, respectively) at drain currents less than 20μA.

With a floating supply provided by batteries, the simplest circuit is a high gain noninverting amplifier driving a zero-center meter (Fig. 14.30). Since the input and output are both intrinsically bipolarity, it's probably best to use a pair of AA cells, running the op-amp from ±1.5 volt unregulated supplies. The back-to-back Schottky diodes reduce the gain at large output swings and prevent pegging; Figure 14.31 plots the resulting response versus V_{in}. Note the input-protection

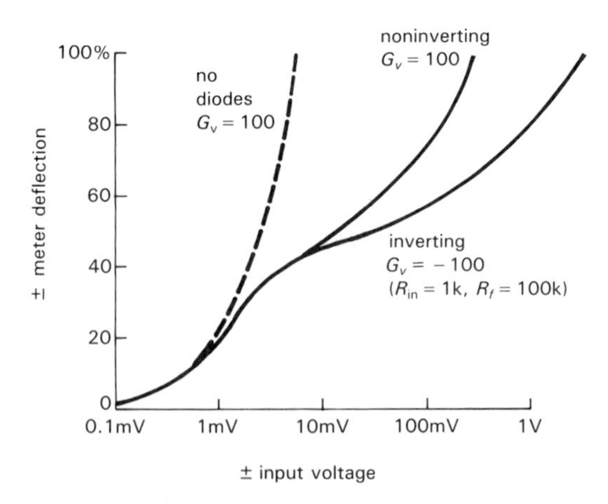

Figure 14.31. Stuck-node tracer achieves large dynamic range through nonlinear feedback.

resistor, in case voltages greater than ±1.5 volts are applied across the inputs. The 10k resistor across the input keeps the output zero when you're not connected to the circuit under test.

The major difficulty in this design is in achieving an input offset less than 100μV while maintaining micropower current drain, all with supply voltages of just ±1.5 volts. The TLC251A is specified to operate down to 1 volt total supply voltage, and its CMOS output stage gives rail-to-rail swings. It has selectable operating currents of 10μA, 150μA, or 1mA; naturally we choose 10μA (by tying pin 8 to V_+). This choice degrades slew rate and bandwidth, which we don't care about, but actually improves input offset drift (0.7μV/°C). The untrimmed input offset is 5mV, which obviously must be trimmed. However, the fine print on the spec sheet says that "the amount of nulling range varies with the bias selection .... In low bias or when the TLC251 is used below 4V, total nulling may not be possible on all units."

If the conventional offset trimming circuit won't work, design your own! Here we've resorted to the circuit shown. It is guaranteed to work, since it can introduce more than 5mV at the inverting input. And it adds only 3μA to the supply current. But it is a compromise, because the trim depends on battery voltage, which isn't regulated. The trimming current is proportional to battery voltage, so in the worst case (a full 5mV of initial input error) the offset voltage will drift 50μV per percent of battery voltage change.

Until recently there was no clean solution to this problem. However, the OP-90 from PMI happens to provide a perfect solution here. It is a micropower op-amp with 12μA of supply current and operation from supply voltages down to ±0.8 volt. In the best grade (OP-90E), it has V_{os} = 150μV max, untrimmed. Although it is bipolar, it swings clear down to the negative rail, and within a diode drop of the positive rail, which is good enough here. For this application it would probably make more sense to buy the cheap grade (OP-90G, 0.5mV) and trim the offset externally. One advantage of using a fixed-bias micropower op-amp, rather than a programmable op-amp, is that the offset trim is guaranteed to work.

Miscellaneous micropower op-amps

The first programmable op-amp (in fact, the first low-power op-amp) was the bipolar 4250, introduced in 1967 by Union Carbide, who subsequently sold their line of linear products to Solitron. In 1970 the 4250 cost $42.50. It became popular immediately (it still is) and is widely second-sourced. The 4250 is usable down to a microamp or so and will operate with 2 volts total supply voltage. It is cheap and delivers respectable performance; consider it an all-around "jellybean."

The 4250 does have one design peculiarity that can lead to problems when operated at low current. It has an amusing bias circuit that provides additional current to the output-stage drivers if the output load current to ground is large (compared with I_{set} multiplied by a pair of h_{FE}'s). This is supposed to help drive stiff loads, but the scheme can backfire if you overdo it, with the drivers robbing supply current from the rest of the op-amp. The op-amp then shuts off, relaxing via the compensation capacitor, then starts up again, etc., leading to a low-frequency oscillation at a few hundred hertz.

This problem was fixed in the quad bipolar LM346, which doesn't "motorboat," but which consequently has poor output-current sourcing capability at low supply current (see Fig. 14.32). The

TABLE 14.6. MICROPOWER OP-AMPS

Type	Mfg[a]	Supply current per amp (V_s=±5V) typ (μA)	#/pkg 1 2 4	FET[b]	CM to rails?[c]	Total supply voltage min (V)	Total supply voltage max (V)	V_{os} typ (mV)	I_b typ (nA)	e_n@10Hz typ (nV/√Hz)	Slew rate typ (V/μs)
CA3420	RC	400	• – –	M	N	2	20	2	0.05pA	350	0.5
AD821B	AD	400[m]	• • •	J	N	4	36	0.1	0.01	90	3
OP-97E	PM	380	• – –	–	–	4.5	40	0.01	0.03	17	0.2
LT1012C	LT+	380	• – –	–	–	4	40	0.01	0.03	17	0.2
LT1013C	LT	370	• • •	–	N	4	40	0.06	15	24	0.4
358/324	NS+	350	– • •	–	N	3	32	2	45	-	0.5
TSC911A	TS	350	• • •	M	N	4	16	0.005	0.07[m]	11[e]	2.5
TSC918	TS	300	• – –	M	N	4.5	16	0.005[m]	0.1[m]	4[e]	0.2
ALD1701	AL	300	• – –	M	B	2	12	2	0.001	-	1
LM10	NS+	270	• – –	–	N	1.1	40	0.3	10	50	0.1
312	NS+	240	• – –	–	–	4	40	2	1.5	45	0.1
MC34181	MO	210	• • •	J	–	3	36	0.5	0.003	65	10
OP-80E	PM	200[m]	• – –	M	N	4.5	16	1[m]	10fa	70[f]	0.2[m]
HA5151	HA	200	• • •	–	–	2	40	2	70	15[f]	4.5
TL031	TI	190	• • •	J	P	10	36	0.5	0.002	61	2.9
MC33171	MO	180	• • •	–	N	3	40	2	20	32[f]	2.1
OP-21	PM	170	• • •	–	–	5	36	0.04	50	20	0.3
TL061	TI+	170	• • •	J	–	4	30	3	0.03	85	3.5
MAX432	MA	170	• – –	M	N	6	32	0.001	0.01	1.2[e]	0.13
TLC25M2A	TI	150	– • •	M	N	1	16	5[m]	0.001	38[f]	0.6
LF441	NS	150	• • •	J	P[g]	6	36	0.3	0.01	50	1
AD548C	AD	150	• • –	J	–	9	36	0.25[m]	0.005	35[f]	1.8
TSC900	TS	140	• – –	M	N	4.5	16	0.005[m]	0.07[m]	4[e]	0.2
ICL7621	IL	100	• • •	M	–	2	18	5[m]	0.001	100[f]	0.2
LT1006	LT	90	• – –	–	N	4	40	0.03	10	24	-
TL022	TI	65	– • •	–	–	4	36	1	100	50[f]	0.5
HA5141A	HA	45	• • •	–	N	2	40	0.5	45	35	1.5
OP-20	PM	40	• • •	–	N	4	36	0.06	15	60	0.03
LT1078A	LT	40	– • •	–	N	2.2	44	0.03	6	29	0.07
LP324	NS	20	– – •	–	N	3	32	1	1	-	0.05
TLC1078C	TI	15	– • •	M	N	1.4	16	0.18	0.7pA	68[f]	0.05
LT1178A	LT	15	– • •	–	N	2	44	0.03	3	50	0.03
OP-90	PM	12	• • •	–	N	1.6	36	0.05	4	35[h]	0.01
TLC25L2A	TI	10	– • •	M	N	1	16	5[m]	0.001	70[f]	0.04

(a) see footnote to Table 4.1. (b) J - JFET; M - MOSFET. (c) input operating common-mode range: B - to both rails; N - to negative rail; P - to positive rail. (d) zero when *sourcing*. (e) μV pp, 0–10Hz (typ). (f) at 1kHz. (g) degraded SR and f_T. (h) at 30Hz. (m) min/max.

Type	f_T typ (kHz)	G_{OL} typ (dB)	Output current source (mA)	Output current sink (mA)	Output ΔV from rails V_+ (V)	Output ΔV from rails V_- (V)	Comments
CA3420	500	100	1.5	1.5	0.1	0.1	low I_b
AD821B	1300	120	10	10	0	0	accurate, single-supply
OP-97E	900	126	10	10	1	1	low-power OP-77
LT1012C	1000	126	10	10	1	1	precision, low-noise, low I_b
LT1013C	1000	137	20	20	1	0	greatly improved 358/324
358/324	500	115	20	20	1.5	0.5[d]	popular single-supply
TSC911A	1500	120	3.5	3.5	0.7	0	chopper; int cap
TSC918	700	130	-	-	1	0	chopper; inexpensive
ALD1701	1000	108	0.5	0.5	0	0	input, output to both rails
LM10	300	120	20	20	0.01	0.01	output to rails; int. ref.
312	400	110	5	5	1	1	original low I_b precision
MC34181	4000	88	8	11	1	0.5	fast, low distortion
OP-80E	300	100	10	10	1.5	0	ultra-low I_b, 5pA(max)@85°C
HA5151	1300	100	3	3	1	0.7[d]	fast
TL031	1000	83	8	20	1	1.1	improved TL061
MC33171	1800	114	4	15	1	1	
OP-21	600	120	-	-	1	1	precision, low-noise
TL061	1000	80	15	15	1.5	1.5	fast
MAX432	125	150	0.2	3	0	0	chopper; low clk noise; C_{int}
TLC25M2A	700	106	10	3	1.5	0	LinCMOS series
LF441	1000	100	4	6	2	2	favorite JFET
AD548C	1000	100[m]	5	5	2	2	improved LF441; dual=648
TSC900	700	130	2.5	2.5	1.5	0	chopper
ICL7621	500	102	20	0.6	0	0	output to both rails
LT1006	-	126	20	20	1	0	precision
TL022	500	80	2	2	2	2	
HA5141A	400	100	3	0.8	1	0.5[d]	fast
OP-20	100	120	0.5	0.5	0.5	0.5	precision
LT1078A	200	120	10	10	1	0	precision, recommended
LP324	100	100	10	5	1.5	0.5[d]	favorite bipolar
TLC1078C	110	118	15	15	1	0	LinCMOS, low offset
LT1178A	60	117	5	5	1	0	precision, recommended
OP-90	20	122	5	5	1	0.7[d]	precision, recommended
TLC25L2A	100	110	10	3	1.5	0	

TABLE 14.7. PROGRAMMABLE OP-AMPS

Type	Mfg[a]	Supply current		#/pkg			FET[b]	CM to rails?[c]	Total supply		I_s per amp[d]	I_{set}	V_{os} typ	I_b typ	e_n@10Hz typ	Slew rate typ	f_T typ	G_{OL} typ	Output current		Output ΔV from rails		Comments
		min (μA)	max (μA)	1	2	4			min (V)	max (V)	(μA)	(μA)	(mV)	(nA)	(nV/√Hz)	(V/μs)	(kHz)	(dB)	source (mA)	sink (mA)	V_+ (V)	V_- (V)	
OP-22	PM	1	400	•	–	–	–	N	3	30	10 100	1 10	0.1	3 20	90 40	0.008 0.08	20 200	125	0.7 5	0.4 2	0.8	0.8	
OP-32	PM	1	2000	•	–	–	–	N	3	30	10 100	0.6 6	0.1	2 15	120 50	0.03 0.3	100 1000	125	0.4 2	0.2 1	0.8	0.8	
XR094	XR		7000	–	–	•	J	–	-	36	10 100	2.3 23	3	0.08	18[ef]	0.1 1	20[g] 200	105	-	-	1.5	1.5	poor data sheet; I_{set} source
TLC251B	TI	10	1000	•	•	•	M	N	1	16	10 150	- -	2[m]	0.001	70[f] 38[f]	0.04 0.6	100 700		10	3	1.5	0	TLC 25xx, 27xx family
346	NS+	4	1000	–	–	•	–	–	3	40	10 100	0.3 3	0.5	3 20	150 70	0.01 0.1	30 300	120	0.5 5	20 20	1	1	3 + 1; I_{set} source
SL562	PL	10	3000	•	–	–	–	–	3	20	10 100	0.4 4	1	2 10	35[f]	0.01 0.3	50 400	90	-	-	0.8	0.8	I_{set} source
HA2725	HA	1	1500	•	–	–	–	–	2.4	36	10 100	1 9	2	2 8	150 125	0.05 0.5	60 600	92	1 5	1 5	1	1	
CA3078A	RC	0.1	1000	•	–	–	–	–	1.5	30	10 100	1 10	0.7 1.3	6 60	60 25	0.3[i] 1.5[i]	1000[i] 10000[i]	100	10	10	0.7	0.7	ext. comp.; I_{set} source
CA3440A	RC	0.02	10	•	–	–	M	N	4	15	10 100	1 10	2	0.01	250	0.03 0.3	50 300	100	1	2	2	2	lowest I_{supply}
MC3476	MO	0.1	1000	•	–	–	–	–	12	36	10 100	1 10	2	1 10	-	0.07 0.6	200 700	110 125	- 5	- 5	2	2	
XR4202	XR	-	-	–	–	•	–	–	3	36	10 100	0.3 3	0.5	200[m] -	300[j] 100[j]	0.006 0.06	15 150	80	-	-	1.4	1.4	poor data sheet
4250	all	0.5	300	•	–	–	–	–	2	36	10 100	1.6 16	3[m]	6 40	60	0.03 0.3	80 250	105	5	5	1	1	jellybean; first prog. op-amp
ICL7612	IL	10	1000	•	•	•	M	B	2	16	10 100	- -	2[m]	0.001	100[f]	0.02 0.2	40 500	102	20 20	0.07 0.7	0	0	76xx family
MC14573	MO	4	1500	–	–	•	M	N	3	15	10 100	5 50	8	0.001	1200 1600	0.2 2	800 2000	95 80	0.02 0.2	5 5	0.2	0.1	

(a) see footnote to Table 4.1. (b) J - JFET; M - MOSFET. (c) B - to both rails; N - to negative rail. (d) specs given at two values of I_s; I_{set} is sinking, unless stated otherwise. (e) at I_s=1.5mA per amp. (f) at 1kHz. (g) assuming $\propto I_s$. (i) uncomp/decomp. (j) 100Hz–10kHz. (m) min/max.

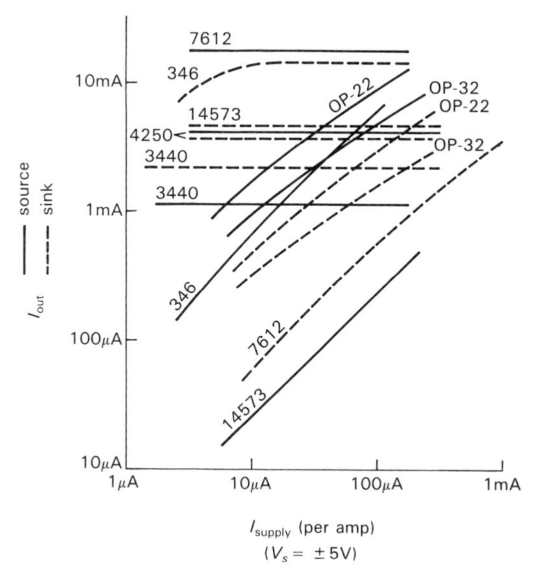

Figure 14.32. Output drive capability (source and sink) versus supply current for various programmable op-amps.

346 is otherwise a nice op-amp, organized as a single plus triple op-amp, with a corresponding pair of programming inputs.

One problem with bipolar programmable op-amps operated at very low currents is that the input bias current does not fall as fast as the supply current (i.e., the input-stage beta drops off at low collector current); thus, for example, the LM346 has a relatively large I_b(max) of 100nA when operating at 35μA per op-amp. This problem is often serious because most bipolar programmable op-amps do not use Darlington inputs or superbeta input transistors. More recent programmable op-amps have emphasized MOSFET design, for example the ICL761*x* series from Intersil, TI's "LinCMOS" TLC250/270 series, and the RCA CA3440. All have bias currents in the picoamp range, and operating common-mode range down to the negative rail. The ICL7612 has the additional wrinkle of operating with input common-mode swings beyond *both* rails. MOSFET outputs can swing to the supply rails; the 761*x*-series outputs can saturate all the way to both rails, while the TLC250/270 series are able to saturate only to the negative rail. Only the 3440 is continuously programmable (the others offer 3 supply current choices), and it is the undisputed champ for operation at extremely low currents. You can run it down to a few *nano*amps supply current, though it won't set any speed records: At 100nA supply current, the 3440's slew rate is 0.0004V/μs, and its f_T is 200Hz! However, because of MOS construction, it still delivers good output drive (±1mA at 2V from the rails). The 3440 is a very good choice for micropower design. (Warning: Note that Figures 7 and 8 in the data sheet should be labeled nA, not μA).

TI's LinCMOS line (TLC250/270 series) has some very nice features, including (like the 3440) good output drive at low supply current. It uses phosphorus-doped polysilicon gate technology to give extremely low offset drift with time (0.1μV/month), eliminating a traditional weakness of metal-gate MOSFET op-amps and comparators. TI has a real winner here, unfortunately poorly documented in their traditionally uninformative linear data sheets.

Most CMOS op-amps (including all those mentioned above) share the problem of limited total supply voltage (see Section 4.22), typically 16 volts (e.g., ±8V max). That's the bad news; the good news is that they can run on very low total supply voltages (2V for the 761x, 1V for the TLC250, 4V for the 3440).

We assembled in Tables 14.6 and 14.7 the low-power and programmable op-amps that we know about. If you compare them with Table 3.1, you'll see that micropower design is a specialty subject.

Micropower design example: integrating metronome

Figure 14.33 shows a micropower circuit that generates audible ticks at a rate proportional to the light intensity of a photographic-enlarger lamphead. Thus, if you time your enlargement exposures

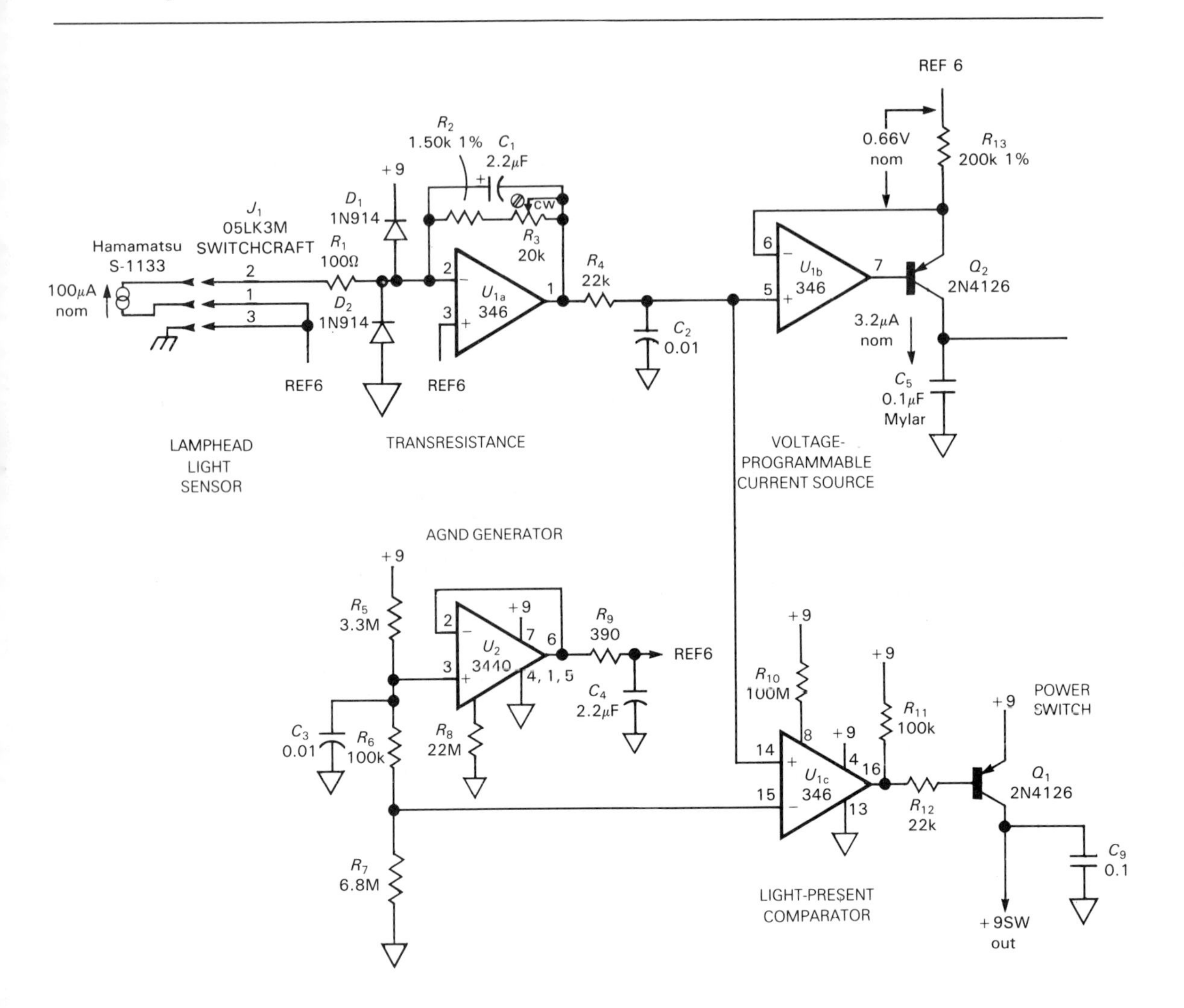

Figure 14.33. Micropower light-integrating darkroom timer.

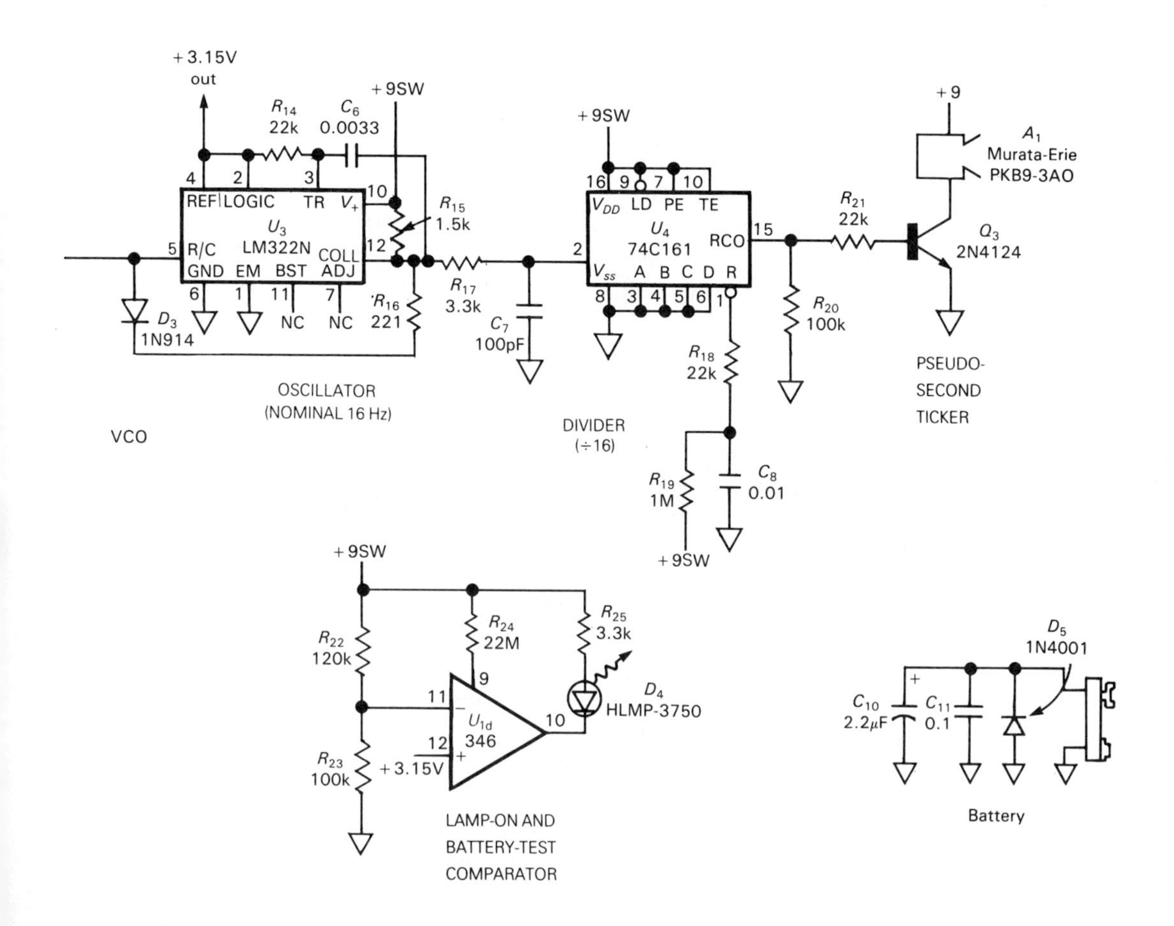
+3.15V
out
+9SW
R_{14}
22k
C_6
0.0033
REF
LOGIC
TR
V_+
U_3
LM322N
R/C
COLL
GND
EM
BST
ADJ
NC
NC
R_{15}
1.5k
R_{16}
221
R_{17}
3.3k
D_3
1N914
C_7
100pF
OSCILLATOR
(NOMINAL 16 Hz)
VCO
+9SW
V_{DD}
LD
PE
TE
U_4
74C161
RCO
V_{ss}
A B C D R
DIVIDER
(÷16)
R_{18}
22k
R_{19}
1M
C_8
0.01
+9SW
+9
A_1
Murata-Erie
PKB9-3AO
R_{21}
22k
Q_3
2N4124
R_{20}
100k
PSEUDO-
SECOND
TICKER
+9SW
R_{22}
120k
R_{23}
100k
R_{24}
22M
R_{25}
3.3k
U_{1d}
346
+3.15V
D_4
HLMP-3750
LAMP-ON AND
BATTERY-TEST
COMPARATOR
D_5
1N4001
C_{10}
2.2μF
C_{11}
0.1
Battery

by the clicks, the prints will be uniformly exposed even if the lamphead brightness is changing (perhaps due to power-line voltage variations, warmup of a fluorescent lamphead, etc.). Design objectives included 9 volt battery operation (simple, cheap), and no ON/OFF switch (people forget to turn it off). It also should signal (with an LED) that the battery is good.

Since battery drain has to be kept below 20μA to give 2 years of life with a 9 volt battery (500mAh, see Table 14.2), and since the LED and piezo beeper require a few milliamps each, the only way to eliminate the ON/OFF switch is to use power switching (see Section 14.17), turning on the power-hungry circuitry only when lamphead light is detected. This is a perfect application for programmable op-amps, which can be shut down via their programming pin. Let's walk through the circuit.

We needed a split supply, and we didn't want to use two batteries, so we began by using micropower follower U_2 to generate a "ground" rail at +6 volts (call it "REF6"). The divider (ignore R_6 for the moment) draws 1μA, and the 3440 is biased to operate at 2.5μA ($I_Q = 10I_{\text{set}}$) via R_8. The 3440 is a good choice because its CMOS design has negligible input current (50pA max) and maintains several milliamps of output drive capability (both source and sink) even at microamp quiescent currents. In fact, we could have run it at lower current; we picked this value because 22M is the largest standard resistor value, and the resulting current is within our budget! Note the bypass on the divider chain, to suppress capacitively coupled garbage (with megohm impedances you've got to be careful). The 2.2μF capacitor at the output keeps the rail a low impedance even at high frequencies, where U_2 has no gain (f_T = 0.01MHz at I_Q = 2.5μA), with R_9 for decoupling to prevent oscillation of U_2 into the capacitive load (see Section 7.07). U_2 is always powered.

The light detector is a photodiode, which generates a current (nominally 100μA, but proportional to lamphead intensity) into a short-circuit load, namely the virtual ground at U_{1a}. We want to generate a frequency accurately proportional to that current, which suggests a capacitor and relaxation oscillator. That won't work, though, because the photodiode, acting as a current source, has very little compliance (0.1V or less). In addition, we need a way to calibrate the instrument, i.e., an adjustment that makes the metronome tick at 1 second intervals when the particular lamphead is at normal brightness (which might be 50μA, or 200μA, rather than the nominal 100μA). Finally, we need a way to activate the power switching when light is detected.

For these reasons we began with a transresistance (current-to-voltage) stage, with gain adjustable over a 15:1 range via R_3. C_1 adds some smoothing for fluorescent light sources, which flicker at 120Hz. The photodiode is referenced to REF6, to keep it in the operating common-mode range of U_1. That makes U_{1a}'s output a voltage that is below REF6 by an amount proportional to light intensity, nominally 0.66 volt (for 1s ticks) when the calibration has been set. U_{1a}'s output drives two circuits: a comparator (U_{1c}), which controls the power switching, and a current source (U_{1b}), which drives the relaxation oscillator whose divided output will be the metronome output.

The comparator (U_{1c}) is one of the three sections of U_1 (a, b, and c) that run continually, biased via R_{10} at about 9μA total. We want to switch power on when U_{1a}'s output is slightly below REF6, so we pick off a reference from U_2's divider, via R_6, that is 0.1 volt below REF6. The output of comparator U_{1c} puts Q_1 into saturation to switch on the +9 volt power, called "+9SW," which therefore appears whenever the enlarger lamphead is turned on.

The current source (U_{1b}) runs continually. It is the standard op-amp plus *pnp* transistor configuration (see Section 4.07), and it sources 3.2μA into C_5 when its input is the nominal 0.66 volt below REF6. Note that one benefit of not using the photocurrent directly is that we can scale it to a convenient value; in this case, 3.2μA into 0.1μF (i.e., $dV/dt = I/C = 32$V/s) generates 16Hz in the 322 (which uses a precise internal reference to set its trigger point at 2.0V). The smorgasbord of resistors, capacitors, and diodes surrounding the 322 is, unfortunately, what it takes to form a relaxation oscillator with this sometimes awkward chip.

The 3.3k and 100pF network at the 322's output is what we have found necessary to cure its genetic predisposition toward double transitions (the 555 often has the same problem, with the same cure). The deglitched output drives a CMOS divide-by-16, whose output – ticks of $\frac{1}{16}$ second duration at 1 second intervals – saturates Q_3 to drive piezo beeper A_1. R_{19} – C_8 forces the divider to zero at the beginning of each exposure, so the first beep comes at the end of the first "pseudo-second."

The last section (U_{1d}) of the quad 346 is used, in comparator mode, as a combination "lamphead-on" and "battery-OK" indicator. Its quiescent current is separately programmable via R_{24}, hence is powered only when +9SW is on. D_4, a high-efficiency LED that we run at 2mA, is therefore ON when the lamphead is on, provided that voltage from divider R_{22} – R_{23} is at least 3.15 volts (a stable reference conveniently provided by the 322) above the negative rail. That will be true if the battery voltage is at least 7.0 volts, i.e., not near end of life.

Because of power switching, only $U_{1a\text{-}c}$ and U_2 run continually, with a combined drain of about 12μA. When U_{1c} senses lamphead photocurrent, it turns on 9SW, powering the 322 (2.5mA), the LED (2mA), and the piezo beeper ($\frac{1}{16} \times$ 8mA, i.e., 0.5mA average current). The battery life works out to about 5 years (or the "shelf life") in the quiescent state, and 100 hours operating. At 15 seconds exposure per average enlargement, that's 24,000 enlargements.

In the design phase, we chose the 3440 because of its good drive and low input current at low I_Q. We picked the 346 because of good overall characteristics and low price in a convenient quad package. The 322 was used because its internal reference meant that we didn't need regulated supplies (as we would have with something like a 555) when its timing capacitor was driven with an external current not referenced to the power supply. Its reference voltage output provided a nice bonus, in the form of a "battery-low" indication.

The "ground" rail (REF6) was put (asymmetrically, and as high as possible) at +6 volts to get maximum dynamic range of lamphead brightness: Since C_5 charges to 2 volts above the negative rail, the circuit will stop operating when the current programming voltage across R_{13} reaches nearly 4 volts (6 times nominal), since the current source will then run out of compliance. At the low end of the dynamic range, voltage offsets in U_{1a} and U_{1b} start to produce errors at about $\frac{1}{6}$ nominal brightness. Thus, the choices of ground-rail voltage (6V) and nominal programming voltage (0.66V) combine to give a dynamic range of $\frac{1}{6}$ to 6 times nominal, which is far more than any light source should ever fluctuate. For example, a fluorescent lamphead initially at room temperature produces about $\frac{1}{3}$ of its fully warmed-up brightness. We chose a rate of 16Hz because a single divide-by-16 can then provide the drive signal for the piezo beeper, without monostables.

Note the protective circuitry: R_1 prevents damage to the photodiode from the peak currents that could, under some unusual condition, flow from a charged C_1. Clamps D_1 and D_2 prevent damage to U_{1a} from something crazy plugged into the input. R_{18} prevents charged-up C_8 from putting U_4 into SCR latchup when +9SW turns off. Although these precautions may be unnecessary under most conditions, they were used anyway because this instrument is in commercial production, where a moderate field failure rate can wipe out all your profit (as well as your reputation!).

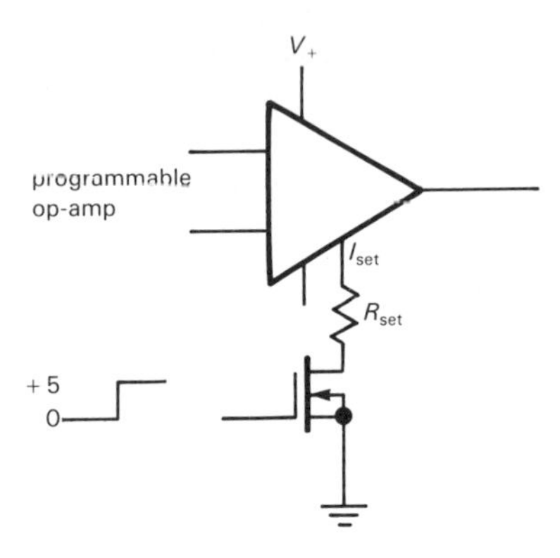

Figure 14.34. Using the programming pin as a power-down control.

Miscellaneous ideas: micropower op-amps

The programming pin can be used as a convenient on/off control for power switching, as in Figure 14.34. This is simpler than switching the op-amp's supply voltages, as we were forced to do for the other high-current loads in the metronome example. Most programmable op-amps (3440, 4250) work with an I_{set} *sink*, as in the figure, so you can use this simple circuit. It may be a good idea to use a high-value pullup resistor to the op-amp's positive supply, to bypass leakage currents and thus ensure complete turn-off.

Some op-amps are "quasi-programmable," in that they permit a choice of several fixed operating currents (typically 10μA, 100μA, and 1mA). Examples are the ICL7612 and the TLC251/271. The TLC250/270 series also includes multiply packaged op-amps with "low" (10μA/section), "medium" (150μA/section), and "high" (1mA/section) quiescent current, specified with part numbers like TLC27L2, TLC27M4, and TLC274 (low dual, medium quad, high quad, respectively).

When using CMOS op-amps as comparators, watch out for large drifts of V_{os} with time, an effect caused by sodium-ion migration in the gate region under long-term differential input voltage; this doesn't much affect their use as op-amps, where feedback enforces zero differential input voltage. The LinCMOS TLC270 series docs not suffer from this disease, because of its phosphorus-doped polysilicon gate.

Some CMOS op-amps exhibit a dramatic increase of input (leakage) current when operated at large supply voltage and nonzero input voltage; for example, the LinCMOS line from TI can have I_{in} as large as 20nA with $V_{in} = +2$ volts and $V_{supply} = \pm 9$ volts. Remember, too, that all FET op-amps (both JFET and MOSFET types) exhibit dramatic increases in input current with rising temperature, typically doubling each 10°C. At high temperatures, FET op-amps often have higher input currents than do good bipolar types; see Figure 3.30.

It's an unfortunate fact that most micropower op-amps are internally compensated for unity gain. At low quiescent current you need all the slew rate and bandwidth you can get, so it would be nice to have decompensated or uncompensated micropower op-amps for use in high-G_V applications. There is one such op-amp – the OP-32 from PMI – but its slew rate and f_T are only slightly improved over those for unity-gain compensated jellybeans like the 4250 and 346.

14.13 Micropower comparators

The same speed/power trade-off that limits micropower op-amp performance applies to comparators as well. As with conventional comparators and op-amps, however, you'll always get better speed with a comparator than with an op-amp of equivalent power; this is because comparator ICs, not intended for use with negative feedback, do not have speed-robbing frequency compensation. Of course, if you're not concerned with speed, an op-amp will often do the job nicely, as in the metronome circuit above.

As with op-amps, micropower comparators come in two varieties, namely *programmable* and *fixed* I_Q. An example of the former is the LP365, a quad programmable bipolar comparator that is specified as low as 10μA total supply current; it will run from 4 to 36 volts total supply voltage and has a separate output emitter terminal (like the 311) so that you can compare negative voltages while driving logic. An example of the latter is the LP339, a micropower (15μA/section) version of the popular low-power (200μA/section) LM339 quad comparator. TI makes CMOS versions of the 339/393 (TLC339/393), with even lower quiescent current, and with excellent speed/power performance; they also offer them with active pullups (TLC3702/4), so you don't waste precious current (and speed) with an external resistive pullup.

An unusual micropower comparator is the LT1040, which incorporates power switching to achieve an average quiescent current of 0.1μA when strobed externally once per second. Alternatively, you can use the internal-strobing oscillator, which uses an additional 0.5μA. A latency time of 1 second is perfectly adequate if you are monitoring slowly varying quantities, e.g., the level of fluid in a tank. The LT1040 is a dual comparator built with CMOS, and it has latched outputs. In addition, it provides a "pulsed-power" output pin, active for about 80μs during the conversion time, so you can drive a resistive network (e.g., a bridge with a thermistor in one leg) at the input with switched power also. This chip (or its close cousin the LT1041) would be a good choice simply as a low-frequency micropower oscillator (see next section), since microamp oscillators are not easy to make. Note, however, that it isn't a particularly stable oscillator.

You can use power switching with a conventional (i.e., fast) comparator if you want to do occasional fast comparisons at known times. For example, you might do acoustic ranging by sending short bursts of sound, then measuring the round-trip travel time while listening for the echo. By measuring the *difference* between travel times upstream and downstream, you can even measure a *velocity*. Speed is important here, and with CMOS logic you can do the job, but you probably want the performance of a full-power comparator (see Table 9.3). Power switching is a natural here, since you know when you want to make a measurement.

Table 14.8 lists most currently available low-power comparators.

14.14 Micropower timers and oscillators

In battery-operated instrumentation there is often a need to generate intervals of an hour or so. You may be making occasional measurements with various sensors, a power-switched microprocessor, and power-switched communications (or data logging). The whole system needs to "wake up" at predetermined intervals. An hour is far too long for *RC* timing, so you need a faster oscillator and a divider (programmable, perhaps, to set the interval to the next wake-up). Since the timer is the only part of the system that runs continuously, it needs to run at low current. What choices are available?

TABLE 14.8. LOW-POWER COMPARATORS

Type	Mfg[a]	#/pkg	Total supply min (V)	Total supply max (V)	I_s/unit typ (μA)	CM to V_-	V_{os} max (mV)	I_b max (nA)	Delay typ, @V_s=5V L→H (μs)	Delay typ, @V_s=5V H→L (μs)	Output[b]	Emitter[c]	I_{sink} typ (mA)	@ (V)
CMP-04F	PM	4	3	36	200	•	1	100	1.4	0.7	OC	–	12	1
CMP-404E	PM	4	3	36	55	•	1	50	3	4	OC	–	15	1
LP311	NS	1	3	36	150	–	7.5	25	1.2	1.2	OC	•	25	0.4
LM339	NS	4	2	36	200	•	5	250	1.3	0.75	OC	–	12	1
LP339	NS	4	2	36	15	•	5	25	13	7	OC	–	5	1
TLC339	TI	4	3	16	10	•	5	5pA[t]	2.5	2.1	OD	–	10	0.5
LP365[d]	NS	4	3	36	50	•	6	20	2	4	OC	–	2	1
"					5	•	6	5	20	40		–	0.2	1
TLC372C	TI	2	2	18	100	•	10	1pA[t]	0.65	0.65	OD	–	10	1
TLC374C	TI	4	2	18	100	•	10	1pA[t]	0.9	0.9	OD	e	10	1
LM393	NS	2	2	36	200	•	5	250	1.3	0.75	OC	–	12	1
TLC393	TI	2	3	16	10	•	5	5pA[t]	2.5	2.1	OD	–	10	0.5
LT1017	LT	2	1.1	40	30	•	1	15	18	25	TTL	–	10	0.15
LT1018	LT	2	1.1	40	110	•	1	75	6	6	TTL	–	10	0.15
LT1040	LT	2	2.8	16	$0.1f_s$[f]	g	0.5	0.3[t]	-	-	TTL	–	1.6	0.25
TLC3702	TI	2	3	16	10	•	5	5pA[t]	2.7	2.3	CMOS	–	10	0.5
TLC3704	TI	4	3	16	10	•	5	5pA[t]	2.7	2.3	CMOS	–	10	0.5
ICL7642C[h]	IL	4	2	16	10	•	10	0.05	150	300	CMOS	–	0.1	1
MC14574[i]	MO	4	3	15	45	•	30	0.05	10	5	CMOS	–	5	0.4
MC14578	MO	1	3.5	14	10[m]	–	50	1pA	-	-	CMOS	–	1	0.5

(a) see footnote to Table 4.1. (b) CMOS - CMOS output, swings to both rails; OC - open npn collector; OD - open n-channel drain. TTL - TTL active pullup output, for which the load may return to a more positive supply. (c) npn "open-emitter" pin. (d) prog oper curr; single prog pin for all 4 sections. (e) common to all 4 sections. (f) sampling comparator. (g) common-mode range extends to both supply rails. (h) CMOS low-power op-amp used as comp, with I_Q set to 10μA. (i) prog oper curr; prog pin for each pair. (t) typical.

CMOS relaxation oscillators

The first thing to note is that a conventional 4000-series CMOS relaxation oscillator (see Fig. 8.90) running at normal voltages draws a rather large class A current, due mostly to rail-to-rail conduction as the (relaxation) input approaches the CMOS threshold during each half cycle. This average current is in the neighborhood of 50μA when operated at 5 volts (rising rapidly at higher supply voltages) and is relatively independent of oscillation frequency. The problem is not helped by substituting the high-speed 74HC or 74AC series. However, if the oscillator is run from a 3 volt supply, for example from a lithium battery, the supply current will be in the microamp range. Figure 14.35 shows such a micropower oscillator and its current drain at several supply voltages. By substituting 74HC logic you get an oscillator with very low jitter, though the *voltage* stability of this type of circuit is poor (typically 10% change in frequency going from 1.0V to 1.6V).

□ *IC oscillators*

□ *Intersil ICM7242.* This is a CMOS *RC* oscillator plus 8-bit divider. It runs on 2–16 volts, and draws about 100μA at 5 volts. Supply current is not much reduced at low supply voltage, unfortunately. Typical temperature coefficient of frequency is 250ppm/°C.

□ *Intersil/Maxim ICM7240/50/60.* These are like the ICL7242, but with programmable digital dividers. They draw the same supply current.

□ *Intersil ICM7207/A.* These are CMOS crystal oscillators with dividers, intended to provide output frequencies of 100Hz/10Hz and 10Hz/1Hz, respectively; to get these frequencies you have to use 6.5536MHz and 5.24288MHz crystals. These chips draw 260μA at 5 volts, dropping to about 80μA at 3 volts. The data sheet claims they will operate down to 1 volt at lower frequencies, with supply currents of a few microamps.

□ *Intersil ICM7555/6 and others.* These are CMOS 555s, with generally improved characteristics (low supply current, higher maximum frequency, much smaller supply current transients). The quiescent current is 60μA at 5 volts, which is also the approximate operating current as an oscillator, if large timing resistors are used. Typical tempco is 150ppm/°C. The 7556 is a dual 7555. The similar National LMC555 draws 100μA and has tempco of 75ppm/°C. Look at Table 5.3 for other 555 lookalikes from TI, Advanced Linear Devices, and Exar.

□ *Op-amps.* An op-amp relaxation oscillator (Fig. 5.29), built with a micropower op-amp, makes a good low-frequency oscillator. Use an op-amp with true CMOS outputs for rail-to-rail swings, particularly at low supply voltages, in order to achieve low tempco and reliable oscillation. The 7611/2 types are good.

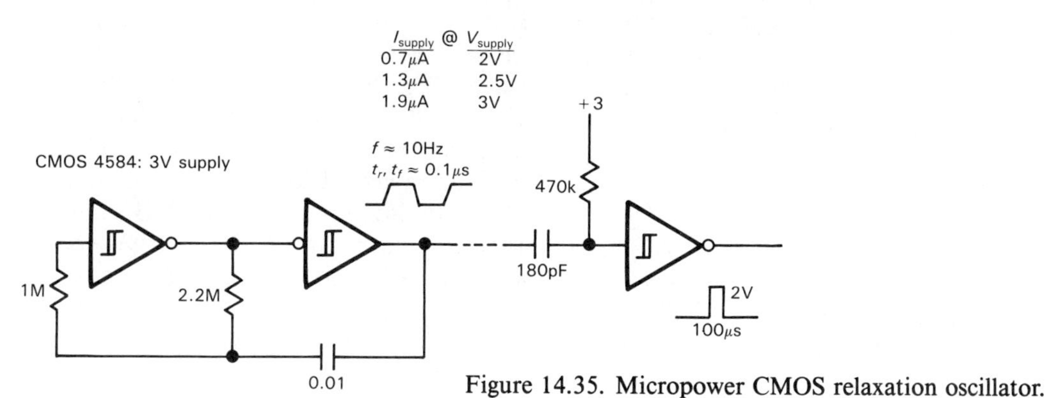

Figure 14.35. Micropower CMOS relaxation oscillator.

□ *LT1040.* This is the power-switched comparator from Linear Technology described above, with integral micropower oscillator (0.3μA at low frequency). The supply current scales linearly with frequency, given by $I(\mu A) = f(\text{Hz})/10$. The oscillator is not particularly stable (0.2%/°C, 10%/V), but it certainly is micropower at low frequencies.

□ *COPS timer.* The COP498 from National is a member of their COPS series of microcontrollers, with serial "MICROWIRE™" interface. The COP498 includes a 32.768kHz crystal oscillator circuit that runs at 20μA (max) at 2.4 volts. It must be programmed via the serial line, but then it can generate wake-up pulses at a 1Hz or 16Hz rate. Stabilities of a few ppm/°C are typical with "tuning-fork"-type crystals.

□ *Calendar clocks.* The MM58174/274 from NSC typifies timekeeping chips for microprocessors. They run from battery backup when the computer is off, and they maintain running time and date information, readable (and settable) via the computer's data bus as an I/O port. The 58174 idles (oscillator running) at 10μA (max) with 2.2 volts supply. Like the COPS chip, it can interrupt at periodic intervals, but only in its full-power 5 volt mode (1mA supply current); also, like the COPS chip, it requires programming via the bus to set up the interrupt interval. Other popular calendar clock chips are the DP8570 (National), the ICM7170 (Intersil), the MC146818 (Motorola), and the MSM5832 (Oki). Some of these allow the use of high-frequency (1MHz or above) crystals for better temperature stability. Epson makes a calendar chip with integral crystal (the RTC58321).

□ *Wristwatch circuits.* There are available low-voltage CMOS chips designed to run stepper motors in analog-display wristwatches. For example, the discontinued ICM7245 required a 32.768kHz crystal and ran on 1.5 volts (single-cell battery), drawing only 0.4μA; various versions provided output frequencies of 1Hz, 0.1Hz, or 0.05Hz. Since it was designed for timekeeping, it had very good stability, typically 0.1ppm over supply voltage. National offers the MM5368, a mini-DIP 32kHz oscillator that draws 50μA (max) at 3 volts and provides outputs at 1Hz, 10Hz, and 50/60Hz. Their MM53107 draws 75μA at 3 volts with a 1MHz crystal, providing ≈30Hz output.

□ *Programmable unijunction transistor.* A *unijunction transistor* (UJT) is a 3-terminal (emitter, base 1, base 2) negative-resistance device that enjoyed some popularity in the 1960s in triggering circuits and free-running oscillators. The device goes into heavy conduction from emitter to base 1 when the emitter terminal is brought more positive than a critical trigger voltage $V_t = \eta V_{BB} + 0.6$, i.e., a diode drop above a fixed fraction η (the "intrinsic standoff ratio," typically around 0.6) of the interbase voltage; it continues to conduct until the emitter current drops below some minimum ("valley current") value. The classic UJT oscillator is shown in Figure 14.36A, with the positive spikes at base 1 used to switch an *npn* transistor to generate full logic swings. You hardly see UJTs used anymore, because op-amps and ICs like the 7555 can do more, and do it better. However, there is an unusual UJT series known as *programmable* UJTs, in which the trigger parameters (η, peak and valley currents) can be set by an external divider. The 2N6028, in particular, is specified for peak currents as small as 0.1μA and can thus be operated as an oscillator at less than a microamp. Figure 14.36B shows a 10Hz oscillator with CMOS output swing and 1μA operating current; look also at Figure 6.57, where we used a PUJT in a micropower dc-dc converter.

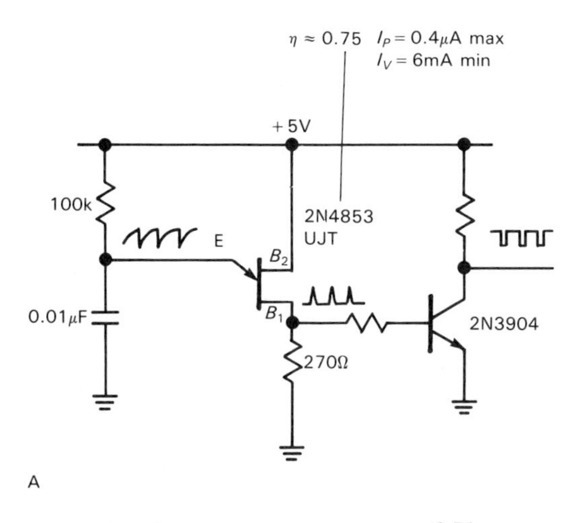

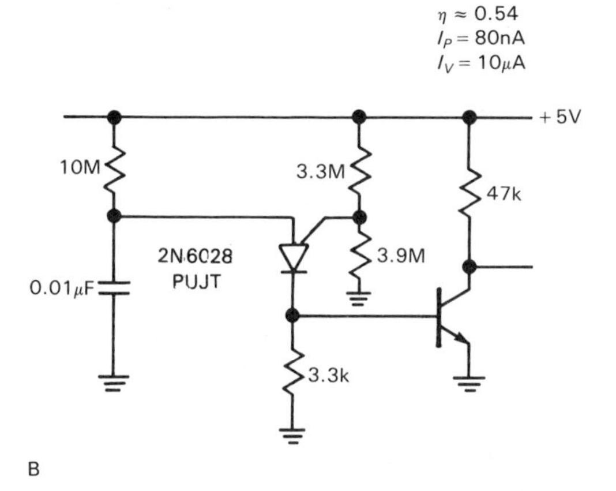

Figure 14.36. Unijunction transistor relaxation oscillators.

MICROPOWER DIGITAL DESIGN

At first glance, micropower digital design seems simple: Just use CMOS everywhere, including microprocessors and memory, right? Well, almost. CMOS is certainly the family of choice, although bipolar logic can be used with power switching. But there are now several CMOS variants, and there are many pitfalls that can render an unsuspecting CMOS design anything but micropower. In this last section of the chapter we'll review CMOS families and how to design with them so that you really achieve the low power you need for battery operation.

14.15 CMOS families

As we discussed at length in Chapter 9, there are several CMOS families; the best one to use depends on the application. They are summarized in Table 9.1.

4000B/74C series

This is the B-series enhancement of the original metal-gate CMOS, which the data sheets say can be operated with 3 to 15 volt supply. Operation is really marginal at 3 volts, however, with high Z_{out}, poor noise immunity, and low speed. The practical minimum supply voltage is 5 volts. At the high end of the supply voltage range there is considerable class A current during switching, and greater susceptibility to sudden death from spikes on the power supply. These CMOS families are pretty gutless in terms of output drive: 1mA or less when operating from 5 volts. 74C is electrically similar to 4000B, including supply voltage range, but with the functions and pinouts of 74 TTL. Fairchild offered an improved "Isoplanar C" family (faster, same voltage range), as does Philips/Signetics (called "LOCMOS"). These are the only series of CMOS that can be run over a large supply range; they are ideal for operation directly from a 9 volt battery.

74HC and 74AC series

Polysilicon-gate "high-speed" (HC=74LS speed, AC=74F/74AS speed) CMOS operates with 2 to 6 volts (or 1.5V to 5V) supply and is actually *specified* over that range. These devices are quite rugged, with good input protection and virtual freedom from SCR latchup. They have CMOS thresholds (i.e., half V_{DD}) and rail-to-rail output swings. The HC series includes many of the popular 4000B-series functions (e.g., 74HC4046), as well as the usual 74LS functions.

74HCT and 74ACT series

These are variants of the HC and AC series, with TTL input thresholds for compatibility when coexisting with bipolar TTL devices in the same circuit. Since micropower circuits don't generally have any bipolar TTL, you should always choose HC/AC, with its better noise immunity. HCT and ACT devices require 5 volts ±10%.

Low-threshold special function

As we will show below, the dynamic power consumption of CMOS is proportional to the square of the supply voltage. This is a powerful (!) incentive to operate at low supply voltage, and it is the reason for chips like the sub-microwatt (1.5V supply) wristwatch oscillator/dividers. These unusual ICs can be very useful, besides being inexpensive because of their large production runs.

M^2L (Mickey Mouse logic)

Don't overlook the possibility of using a few discrete components to make or invert a logic level; Figure 14.37 shows the idea. This can be particularly useful if you need to interface between different supply voltages. You can even create impromptu gates, etc., by adding diodes or paralleling transistor outputs.

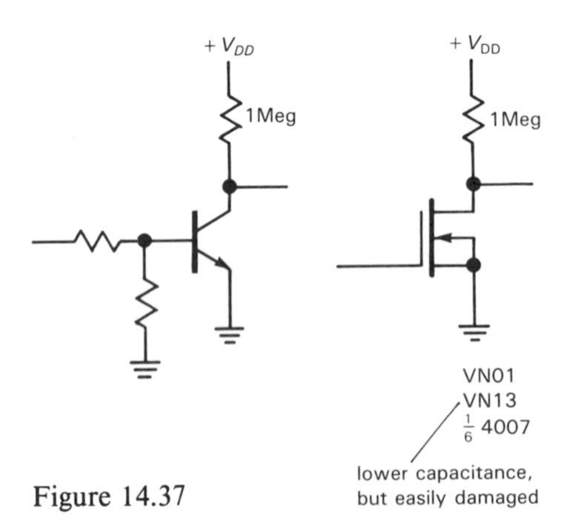

Figure 14.37

Choice of family

Use the 4000B/74C series if you don't need speed or high output current, or if you want to use unregulated or high supply voltages. Use HC (or AC) if you need the speed or output drive, but note the more limited supply voltage range. The AC (and ACT) families cause more problems from capacitive coupling and reflection (transmission-line effects), and also power-supply transients (crowbarring), because of their fast edge times and robust output drivers. They should be avoided unless absolutely needed. In general, avoid devices with TTL thresholds (HCT, ACT) unless you need them to interface to bipolar TTL, or to NMOS LSI circuitry with TTL logic levels.

14.16 Keeping CMOS low power

There are several routine measures you should take to achieve low-current CMOS operation. In addition, it's worthwhile raising your CMOS pathology awareness.

Routine design considerations

1. *Keep as few nodes as possible involved with high frequencies.* CMOS has no quiescent current (other than leakage), but current is required to charge internal (and load) capacitances during switching. Since the energy stored in a capacitor is $\frac{1}{2}CV^2$, and an equal amount of energy is dissipated by the resistive charging circuit, the power dissipated is

$$P = V_{DD}^2 fC$$

for a switching frequency f. Thus, CMOS devices consume power proportional to their switching frequency, as shown in Figure 14.38 (compare with Fig. 8.18). At their maximum operating frequency they may use more power than equivalent bipolar TTL logic. The effective capacitance C is often given on data sheets as the "power dissipation capacitance," to which you must add the load capacitance C_L before applying the formula above.

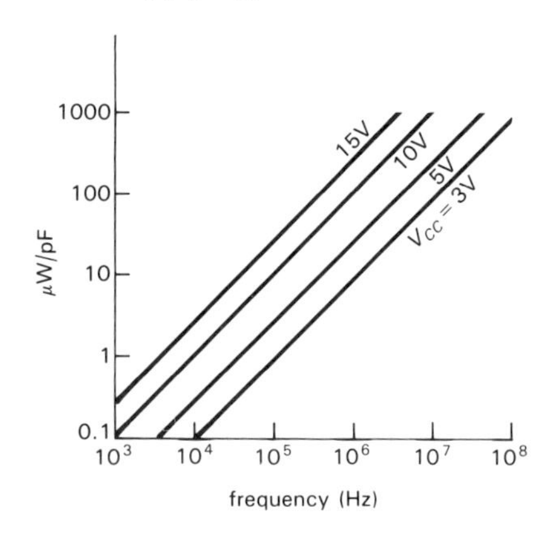

Figure 14.38. CMOS dynamic power consumption.

2. Within a circuit, keep all V_{DD}'s and V_{SS}'s the same. Otherwise you may have current flowing through input-protection diodes. Even worse, you may force a chip into SCR latchup (see pathologies, below).

3. Make sure logic swings go all the way to the rails. CMOS outputs swing rail-to-rail. Outputs from other devices – bipolar TTL, oscillators, NMOS chips – may hover in between, causing class A current and decreased noise immunity.

4. No open inputs. Open inputs are the enemy of micropower operation, since there may be considerable class A current (and even oscillation) as the input floats to the logic threshold. Tie all unused inputs to ground (or V_{DD}, if that disables something you don't want).

5. Arrange loads to keep normal-state drains low. Pullups, pulldowns, LEDs, and output drivers should be wired so that current is minimum in the usual state. Thus, for example, use an *npn* (not *pnp*) transistor to switch a high-voltage load from a node that spends most of its time low.

6. Avoid slow transitions. Again, class A current is the culprit. A sine-wave input driving a CMOS Schmitt trigger may cause a lot of supply current.

7. Put current-sensing resistors in the V_{DD} lead. In certain failure modes (see below), particularly those caused by static damage, a CMOS chip may draw excessive quiescent current; a 10 ohm resistor in series with V_{DD} on each board (bypass the load side) makes it easy to see if that is happening. Putting such a resistor on each chip (usually no need to bypass in this case) allows you to locate the bad chip quickly (Figure 14.39).

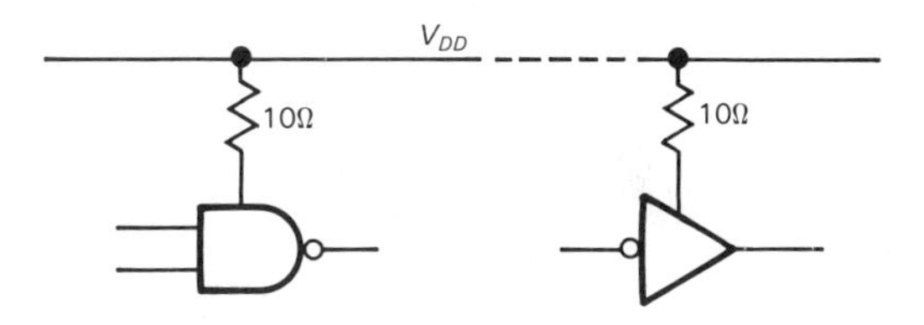

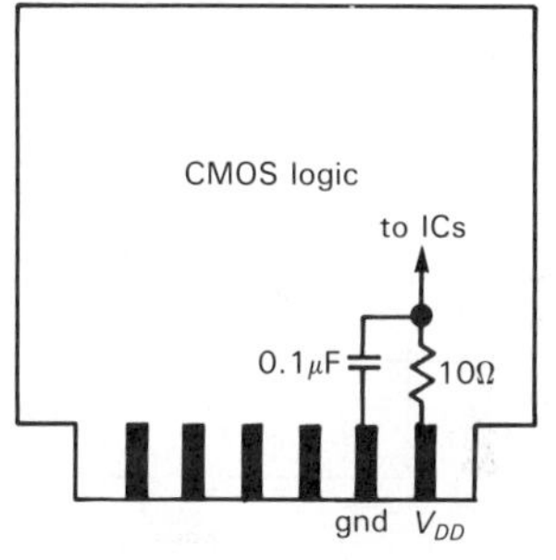

Figure 14.39. Supply current sensing ("current spy").

8. Quiescent current screening. A typical HC- or 4000B-series CMOS logic chip has a specified I_Q of 0.04µA (typ), 5µA (max). Most of the time it is rare to have a quiescent current anywhere near the maximum, but it can happen. If you are operating at low switching frequencies (therefore low dynamic current), and require comparably low quiescent current, you may need to screen incoming chips. The use of small series resistors as recommended above makes the job much easier. We've noticed that in the case of CMOS LSI chips (such as large memories) the typical quiescent current may be close to the manufacturer's maximum leakage specifications – beware!

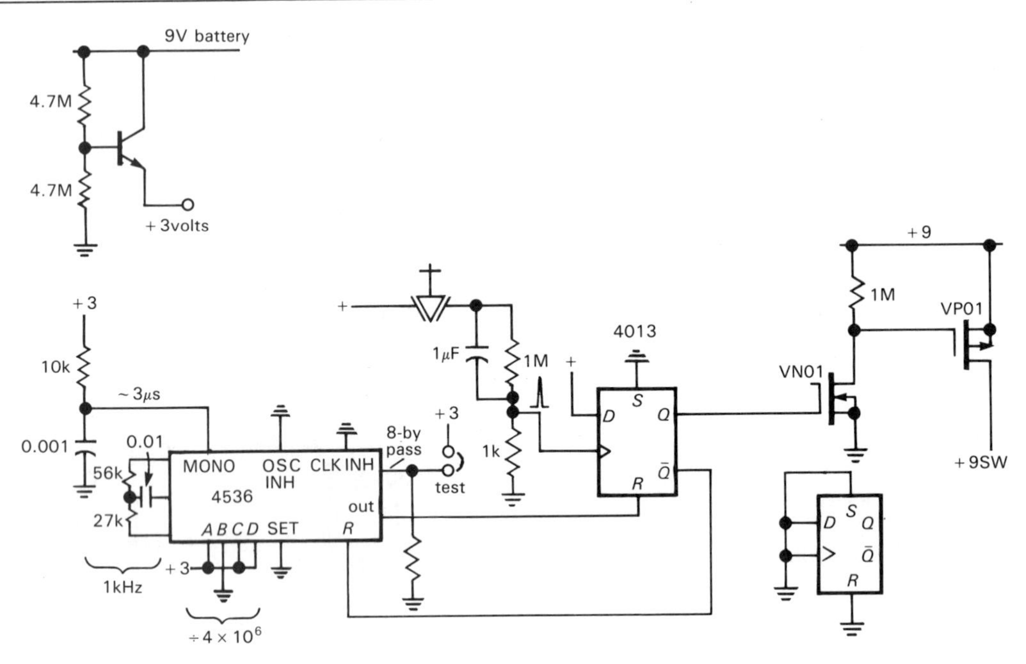

Figure 14.40. "An hour of power."

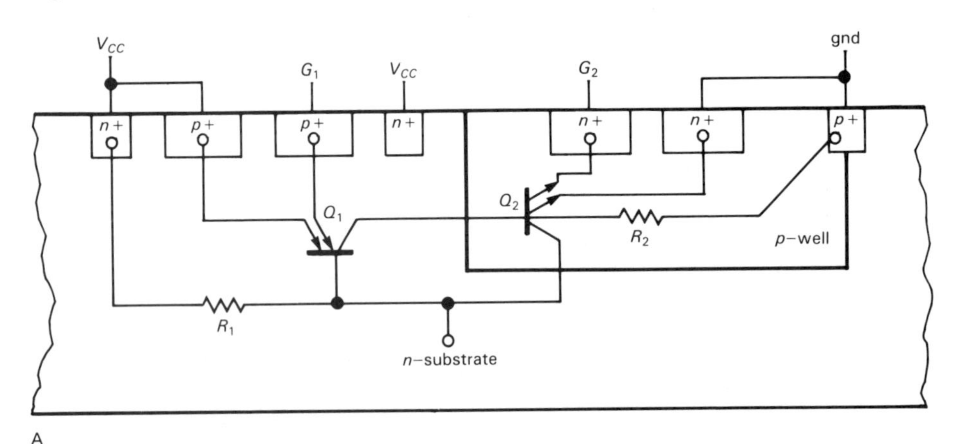

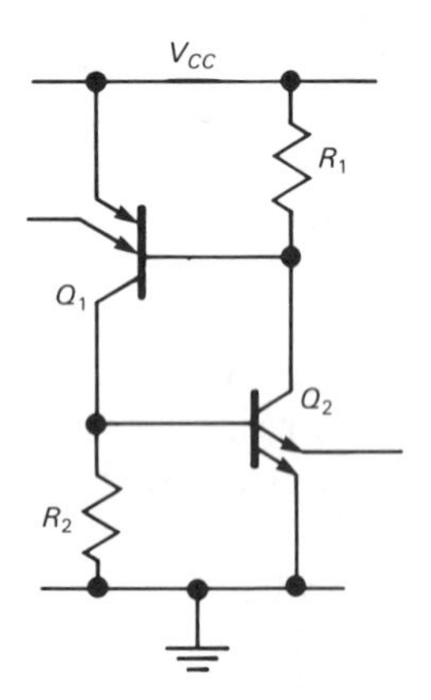

Figure 14.41. Parasitic 4-layer SCR lurking in CMOS ICs. (After App. Note 339, National Semiconductor Corp.)
A. CMOS IC cross section.
B. Equivalent circuit.

9. Time-out power switching. You can save a lot of power by making sure an instrument is turned off when no one is using it. Figure 14.40 shows a straightforward CMOS timeout circuit that turns off the switched +9 volt power an hour after the instrument has been turned on. You might build this into a manual instrument (e.g., a multimeter). It uses a 4536 oscillator/divider/monostable to reset a flip-flop that controls the instrument power. The circuit runs on +3 volts to keep current drain below 5μA. Using the monostable output prevents logic races and runt pulses, and the "8-bypass" line is used to test the circuit by shortening the delay to 15 seconds. MOS switches provide an easy interface with low quiescent power.

CMOS pathologies and failure modes

CMOS circuits exhibit strange behavior under some circumstances, and they can fail in weird ways. Some of these increase power dissipation dramatically. Here is a rundown.

1. SCR latchup. This is a major threat to low-power operation. The silicon substrate forms diode junctions with the elements of the CMOS circuit, producing a parasitic SCR-type circuit (Fig. 14.41) that can be triggered into heavy conduction. It takes typically 20mA–200mA (the larger values for the newer CMOS families) of current through the input-protection (or output-protection) diodes (Fig. 3.50) to turn on the parasitic SCR. Once on, it remains in heavy supply-to-supply conduction, with about 1 volt from supply to ground, often destroying the IC (or even the power supply!). To prevent SCR latchup, design your circuits with series-input current-limiting resistors in places such as external inputs, inputs that can be overdriven, outputs that go off-board, and signals between portions powered by separate supplies. Unfriendly loads that can drive currents into CMOS outputs are potential causes of latchup. For example, a current-output CMOS D/A converter (running from +5V and ground) driving the summing junction of an op-amp (powered from bipolarity supplies) seems reasonable enough; but on power-up you may get momentary current sinking at the op-amp's input, driving the DAC into latchup. A cure here is to add a Schottky diode to ground. (The newest D/A converters are designed to prevent this; look for a phrase like "protection Schottky not required.") Another place we've seen latchup is when switching large inductive loads with power MOSFETs, whose large feedback capacitances produce high dynamic currents at the gate driver (i.e., the CMOS logic output) during transitions. See "Signal coupling," below, for another latchup scenario.

An easy way to induce latchup is to plug a circuit board into a powered socket, since a signal line may connect before the power-supply line. (With any kind of circuitry it's a bad idea to plug or unplug boards or modules with power on.) When designing with CMOS, it's worth studying the official latchup specifications. The newer polysilicon-gate CMOS types have very effective protection circuits, and some manufacturers (e.g., National) claim that their HC/HCT or AC/ACT lines cannot be triggered into latchup.

2. Signal coupling. Because of the high impedances involved, CMOS is prone to capacitive coupling from nearby signals with fast edge times, producing logic spikes. For example, high-impedance pullups or pulldowns allow coupling of spikes through wiring capacitance from nearby lines carrying fast edges; to cure this, use a small (0.001μF) bypass capacitor. In general, wiring going off to panels can cause trouble by this mechanism. The relatively high capacitance can even couple to output lines, particularly with 4000B/74C-series CMOS running at 5 volts. In extreme cases (for example, switching high voltages with a relay in the same cable as logic levels) there can be enough coupling to induce SCR latchup.

3. Clock skew. As we mentioned in Chapter 9, the relatively high Z_{out} of 4000B/74C CMOS can lead to trouble in a synchronous system, particularly if the clock lines are heavily loaded capacitively, delaying the clock relative to the data. The relatively large scatter of CMOS logic thresholds only aggravates the situation. In an unregulated battery-operated system, it's important to check for reliable operation over the full range of supply voltage. Ironically, the problem tends to become worse at higher V_{DD}, where data delays and transition times become shorter. This is one argument for using regulated supplies in battery-operated CMOS systems.

4. Failure modes. Damage at the input can cause input leakage (or a short) to V_{SS} or to V_{DD}. Damage to the output stage can cause substantial quiescent current. It may cause one driver to open, so it cannot both sink and source. In such a case there may be quiescent current in one state only. Current-sensing resistors in the V_{DD} leads, as recommended above, make it easy to track down such problems. When using this method, note that it's easy to be fooled, since the symptom of a damaged input may be nonzero quiescent current in a healthy chip that drives the bad one.

A damaged CMOS chip may work only at very low speeds (faulty driver), or only at very high frequencies (faulty input stage, no dc connection, capacitively coupled). A similar symptom may occur if you've forgotten to connect an input: The circuit may "work" at high speeds due to capacitively coupled edges (Fig. 14.42).

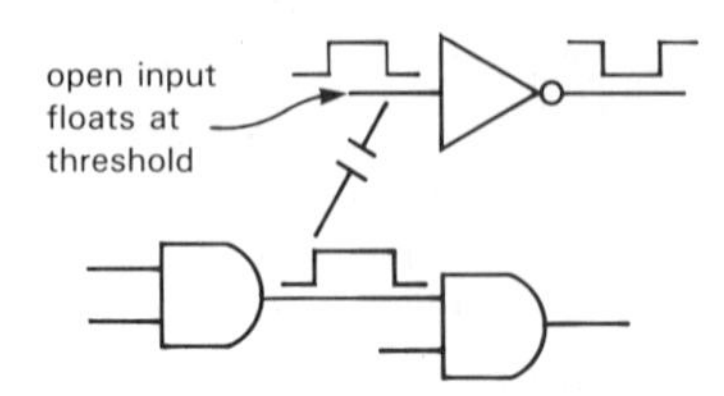

Figure 14.42

A pulldown will reveal this problem, by preventing the input from floating at the transition threshold. As mentioned in Section 8.35, forgetting to connect V_{DD} leads to weird symptoms, since chips can be powered through their logic inputs (via input-protection diodes); the power goes away, though, if all logic inputs go LOW simultaneously.

14.17 Micropower microprocessors and peripherals

There are available CMOS equivalents of many standard microprocessors. Thus, at first glance it would seem easy to design low-power microprocessor circuits. In fact, most of these CMOS microprocessors are simply pin-for-pin replacements of processors originally implemented in NMOS, and in many cases they do not have characteristics tailored to battery operation, e.g., the ability to run with a slow clock. Some aren't even low power – for example the 68020, which dissipates approximately 1 watt.

Since CMOS dissipation is usually proportional to clock frequency, the first question to ask is how much power is used when the clock is shut off. If the chip uses CMOS for its good electrical characteristics, but was not intended for very low power applications, it may include some non-CMOS circuits that result in significant static dissipation. Watch out, also, for chips that use the clock in mysterious ways: An example is a (discontinued) triple 16-bit counter that is microprocessor-compatible (six byte-wide three-state ports, plus control ports). It draws $10\mu A$ with the clock stopped; however, the surprise is that it needs the clock present even when it isn't talking to the processor. The data sheet doesn't tell you that the clock is used to synchronize the inputs and that its frequency must be at least three times the maximum counting rate. Thus, you need to keep it supplied with a fast clock for it to function as a counter, resulting in plenty of power dissipation.

The second question to ask is how low a clock frequency you can use while preserving reasonable operation. Some processors may have dynamic registers, mandating a fairly high minimum clock frequency. With a slow clock, some processors (particularly the "controller" types - see below) may respond *very* slowly; you may have a 10ms latency time from interrupts.

Computer versus controller

Microprocessors generally divide into two categories, according to their intended application. Computation-oriented types have at least a 64K address space and use only a few clock cycles per instruction (to achieve high speed). They require external peripherals for ports, timers, converters, etc. Controller-oriented types, meant for dedicated instrument use, usually have a small address space (2K or 4K is typical) and use many clock cycles per instruction. On the other hand, they tend to have lots of parallel ports built in, many interrupt pins, and an internal clock generator. They often have timers, UARTs, and even A/D converters and nonvolatile RAM.

Some examples

The 80C85, 80C86, and 80C88 are CMOS equivalents of popular NMOS computing processors. The 80C85 (Oki) draws 2mA with the clock stopped, and 10mA with a 6MHz clock. It may be unsuitable for micropower operation, unless you use power switching. The 80C86 and 80C88 (Harris, Intel) are better, with ≈0.1mA static current drain (though it is specified as 0.5mA max). With these you can stop the clock to conserve power, retaining register and program counter contents.

There is a subtlety when restoring clock input to a stopped processor. To keep standby power low, you would like to shut off the crystal oscillator, not just gate its output (with a synchronizer circuit, of course; see Section 8.19). However, because of its high-resonant Q, a crystal oscillator cannot start up instantaneously, and an oscillator in the megahertz range typically takes 5–20ms to start up; a 32kHz oscillator can take up to a *second* ($Q = 10^5$). Thus, to restore the clock signal, you can either wait for the oscillator to come up to speed before gating it through to the microprocessor, or you can hold the processor reset during that period. The first method is usually better, because you may not want to reboot each time you wake up (see "Power switching," below, for a "warm-boot" trick). The 82C85 (Harris) is a low-power clock generator intended for the 80C86/88, with internal circuitry supporting stop-clock, stop-oscillator ($\approx 25\mu A$), and low-frequency operation, with hooks to the microprocessor for software control.

The 80C31/51 are examples of popular controller-type processors designed with special features for battery operation. These processors include up to 32 programmable I/O lines, a pair of 16-bit timers, 128 bytes of on-chip RAM (they can use ports as multiplexed buses to address 64K of external memory; the 80C51 has in addition 4K bytes of mask-programmed ROM), 5 interrupts, and a programmable serial port. They can be put into a low-power "idle" mode (<1mA with 1MHz clock) in which interrupts, serial port, and timers continue to run while the processor clock is halted; all registers and on-chip RAM are preserved. They also can be put into a zero-current "power-down" mode, from which they can be revived only with a full reset, but which preserves on-chip RAM.

The 146805E2 is a similar controller, with the nice capability of waking up from

TABLE 14.9. CMOS MICROPROCESSOR ONE-CHIP CONTROLLERS

Type	Mfg[a]	Word size (bits)	Bus width (bits)	Adr bus (bits)	Pins	NMOS copy?	Number cruncher?	Curr @ $f_{clk}=0$?	Int oscillator?	Int RAM (words)	Int EPROM (words)	Timer?	Interrupts	Parallel port	Serial port?	Osc stop?	Comments
78C05	NE	8	8	16	64	–	•	0	•	128	–	•	2	•	–	•	SPI port
80C35	IN	8	8	12	40	•	–	0	•	64	–	•	•	•	–	•	8048 family (8051 family better)
80C39	IN	8	8	12	40	•	–	0	•	128	–	•	•	•	–	•	
80C31	IN	8	8	16	40	•	–	0	•	128	–	2	2	•	•	•	int baud rate gen; 8051 family
70C42	TI	8	–	int	40/44	–	–	0	•	256	4k	2	6	•	•	•	
146805E3	MO	8	8	16	40	•	–	0	•	112	–	•	•	•	–	•	
68HC704P3	MO	8	–	int	28	–	–	0	•	128	2k	•	•	•	–	•	
1468705G2	MO	8	–	int	40	–	–	0	•	112	2k	•	•	•	–	•	can use RC osc
HD63P05Y0	HI	8	–	int	64	–	–	0	•	256	8k	•	•	•	•	•	piggyback EPROM
HD6301V1	HI	8	8	16	40	•	•	0	•	128	–	•	6	•	•	•	a complete family
HD63P01M1	HI	8	–	int	40	–	•	0	•	128	8k	1	6	•	•	•	piggyback EPROM
HD6305Y2	HI	8	8	14	64	–	–	0	•	256	–	1	5	•	•	•	lots of I/O
HD6303R	HI	8	8	16	40	•	•	0	•	128	–	1	6	•	•	•	
HD6303Y	HI	8	8	16	64	–	•	0	•	256	–	2	8	•	•	•	3 volt version avail
HD647180	HI	8	–	int	84	–				512	16k	2		•	2	•	DMA
COP8788	NS	8	–	int	40/44	–	–	0	•	192	4k	2	13	•	•	•	has A/D; 8-bit internal Harv arch
16C54	GI	8/12	–	int	18	–	–	0	•	32	512	1	–	12	–	•	12-bit internal Harvard arch
16C55	GI	8/12	–	int	28	•	–	0	•	32	512	1	1	20	–	•	" "
87C51	IN	8	8	int	40	•	–	0	•	128	4k	2	5	32	•	•	CMOS vers of most pop controller
68HC11A8	MO	8	8	int	48/52	–	•		•	256	512[b]	4	2	28/32	•	•	6801 + 91 instr; A/D, SD1
80C196	IN	16	8/16	0/16	48/68	•	•		•	232	8k	2	8	•	•		twice performance of 8096
16003	NS	16	16	16	68	–	•	0	•	256	–	8	8	•	2	•	complete family
16084MH	NS	16	–	int	68	–	•	0	•	256	8k	8	8	•	2	•	A/D
78P312	NE	16	8/16	int	64	–	–		•	256	8k	4	15	32	•	•	new family, unique instr set

(a) see footnote to Table 4.1. (b) EEPROM.

zero-power "stop" mode via an interrupt. There's internal circuitry to wait for oscillator start-up before resuming processing. The specifications say that start-up time is typically 30ms (300ms max) at $f_{clk} = 1MHz$, $V_{DD} = 3$ volts. This is obviously unsuitable for an application where you must wake up frequently, say 10 times per second, but it would be fine in an application where you wake up once each minute. The 146805 series includes versions (specified by suffix) with different port and memory configurations, mask ROM, etc. We'll use it for a design example in the next section.

Table 14.9 lists most of the interesting low-power microprocessors available as this edition went to press.

Power switching

You can make any microprocessor operate at low average power, of course, if you switch the power on with a low duty cycle. This is actually easier with NMOS than with CMOS, because inputs such as WR′ can be held high during power-down (with CMOS, that usage would power the chip, through the input-protection diodes!), preventing spurious write cycles, etc. Thus, with CMOS you must use external logic to accomplish orderly shut-down. In either case you've got the clock start-up problem mentioned above, which you can solve with external delaying logic, or with a chip like the 82C85.

You usually don't want to go through the full "cold" reboot software on each restart. The best way to handle this is to let the CPU read a "power-on flag" flip-flop (CMOS, continuously on) at each restart, doing a cold boot the first time only, after which it sets the flip-flop.

With power switching of NMOS processors, you have to store edge-triggered interrupt requests in external CMOS logic during processor hibernation, servicing it when the processor is next restarted. You may need to use the same trick with some CMOS processors, those that lose edge-triggered interrupts during idle or stop modes.

When power switching NMOS processors, which draw typically 100mA or more of supply current, be sure to use a pretty hefty MOSFET, with $R_{ON} < 3$ ohms at 5 volts of gate drive.

CMOS peripherals

Many low-power peripheral chips are simply CMOS copies of NMOS parts, for example the 81C55 and 82C55 parallel ports. The data sheets are usually almost exact copies of the original NMOS data sheets, with a few changes. As such, the data sheets are often incorrect! For example, they may specify 2mA output sinking at 0.4 volt and 100μA sourcing at 2.4 volts, whereas their p-channel drivers can actually source 2mA, particularly at 2.4 volts. Input thresholds may also be misleading.

Another thing to worry about with CMOS is the disposition of floating inputs from a three-state bus. Harris and Intel have a "bus-hold" circuit that introduces a little positive feedback at the inputs to prevent class A current from floating inputs.

Watch out for CMOS peripherals that draw "quiescent" current because they have a clock. For example, UARTs like the 65C51 and 82C52 draw about 2mA at their recommended oscillator frequency (1.84MHz @ 1.4mA/MHz for the 65C51). You could imagine shutting off the oscillator, but then the UART can't receive data, for example a command to turn on! Other CMOS peripherals with quiescent currents of 1–5mA are A/D converters, modems, video drivers, EEPROMs, and keyboard encoders. A complex system with several of these devices may have a quiescent current of 25–50mA, giving a 9 volt battery life of only 10 hours. If that's OK, fine; otherwise, you have to power-switch. But be careful – inputs and outputs may misbehave; for example, the A/D's three-state bus drivers may pull LOW when the A/D is not powered (in which case use a separate CMOS three-state driver).

RS-232 drivers have traditionally been high-power devices: The classic 1488 quad driver draws ±20mA quiescent current, not including load current, and the 1489 quad receiver draws 15mA. Some recent RS-232 chips run at low power. Here are a few good choices:

Motorola MC145406. This is a triple CMOS driver/receiver that will work with supply voltages from ±5 volts to ±13 volts, with less than 15mW dissipation over that range. CMOS output drivers give rail-to-rail swings, so you can get RS-232 swings even with ±5 volts supplies. Various tricks were used in the design to permit CMOS to work with a total supply voltage up to 26 volts, and to permit receiver input swings 20 volts beyond the rails.

LT1032. This is a quad bipolar RS-232 driver with ±5 to ±15 volt supply range, and quiescent current of 0.5mA. It can be shut down (zero current) with a control pin; during shutdown, the outputs go to a high-impedance state.

LT1039. This is a triple bipolar driver/receiver with ±5 to ±15 volt supply range, and 4mA quiescent current. Like the LT1032, it has a shut-down pin. It also has a control line that allows one receiver to remain on while the rest of the chip is shut down; you could use that to power up the rest of the chip when something is received. The outputs go to a high-impedance state during shutdown.

MAX230–239/ICL232 series; LT1080/1. These devices from Maxim, Intersil, and Linear Technology are dual driver/receivers with on-chip voltage converters, so they run from a single +5 volt supply, producing ±9 volt output swings. All except the MAX233 and 235 (which have built-in capacitors) require four external tantalum capacitors for the voltage converters; their ±9 volt outputs are available externally to power low-current loads. Quiescent current is 5mA. The voltage-conversion portions of these chips are available separately as the MAX680 or LT1026, single + 5 volt to dual ±10 volt voltage converters; these could be used to power any of the other RS-232 chips above.

DS14C88/89. National's CMOS reworkings of the similarly named bipolar classics. The 14C88 driver works with supply voltages from ±4.5 volts to ±12 volts, producing the usual CMOS rail-to-rail swings. With ±5 volt supplies the quad driver draws 30μA max (unloaded), while the receiver draws 0.9mA max from its single +5 volt supply. (National also makes CMOS RS-422 chips (DS26C31-32).

14.18 Microprocessor design example: degree-day logger

Let's bring all these ideas together with an example. We'll design a small battery-powered data logger whose purpose is to monitor the ambient temperature once per minute, storing the "degree-day" averages in RAM for subsequent readout via serial communication. You might put such an instrument in a remote location, paying it twice-yearly visits during which you read out its data into a portable computer. Or you might prefer to "harvest" the loggers, then read them out when you've got them back home.

The instrument will operate on three alkaline C cells, with battery life of at least a year. To keep power consumption low, we'll use CMOS peripherals and a CMOS controller-type microprocessor with built-in shut-down modes. We'll use power switching of the CPU and front-end circuitry during the momentary data-collection intervals, with wake-up provided by a low-power calendar chip.

Since the serial port will be used only occasionally, we'll do power switching here, too. Our particular circuit is by no means unique; along the way we'll discuss alternative circuit solutions.

CPU

Figure 14.43 shows our design. We began by choosing the Motorola MC146805 CMOS controller, which is specified for operation down to 3 volts, and includes on-board circuitry to go into WAIT mode (low power, oscillator and timer running) or STOP mode (zero power, oscillator stopped; wake-up via interrupt or reset only). The -E2 suffix version uses external ROM and RAM, but includes 112 bytes of on-board RAM.

When run from 5 volts, the CPU typically uses 7mA running (5MHz clock), 1mA in WAIT mode, and 5μA in STOP mode. Since we need to gather data for only a few milliseconds once per minute, and a wake-up from STOP mode (due to crystal oscillator start-up) takes typically 30ms, we minimize power consumption by using an externally generated interrupt to wake up for each measurement. The alternative – using CPU timer-generated interrupts from WAIT mode – imposes an average CPU current drain of at least 1mA, corresponding to a C-cell battery life of only half a year. That could, of course, be extended to a year by using D cells; another solution would be to run at lower oscillator frequency (say 1MHz), where the WAIT-mode current is significantly lower. Still another possibility is to run at 3 volts, where the WAIT-mode current drain is about 150μA with a 1MHz clock. Any of these solutions is perfectly good. In this example, we'll stick with the power switching, because it illustrates additional techniques. It also provides convenient timekeeping via the calendar chip.

Calendar clock

For the calendar clock we needed a chip that not only can keep time at low current drain (all calendar chips do that) but also can make interrupts while in low-power mode. Since the primary use of calendar chips is for ac-line-operated computers, where there is plenty of power available when the CPU is running, many chips cannot interrupt in low-power (battery-backup) mode. We first looked at the ICM7170, a nice calendar chip from Intersil; it can interrupt in low-power mode, but it has an awkward power scheme for single-battery operation. The ever-popular MM58274 from National doesn't interrupt during backup. We finally settled on the Motorola MC146818, a popular part that is made by at least two other manufacturers and is intended for use with chips like the MC146805 CPU. It can be left running at full supply voltage, and it maintains low current drain (50μA max at 32kHz, external oscillator) while not enabled.

Motorola likes memory-mapped I/O, and their MC146805 is no exception. So you don't have the I/O strobes favored by Intel; instead, you decode some portion of memory space as "I/O space." With only a few I/O devices in the system you can be pretty sloppy about I/O decoding, as we explained in Section 10.06. In this case we put the UART at the bottom of external memory (the CPU chip monopolizes the bottom 80_H bytes for on-chip memory and ports), and the calendar clock at 80_H, using the same 'HC139 decoder that enables memory (see below).

Memory

For EPROM, we've used a standard 27C64, an 8K×8 part of which only the bottom quarter is addressable with the 11 address lines we've connected (we've use the top of address space for I/O). A smaller ROM would be fine, but memory manufacturers have generally been discontinuing the smaller memories as they are able to make larger ones. The 27C64 specifies

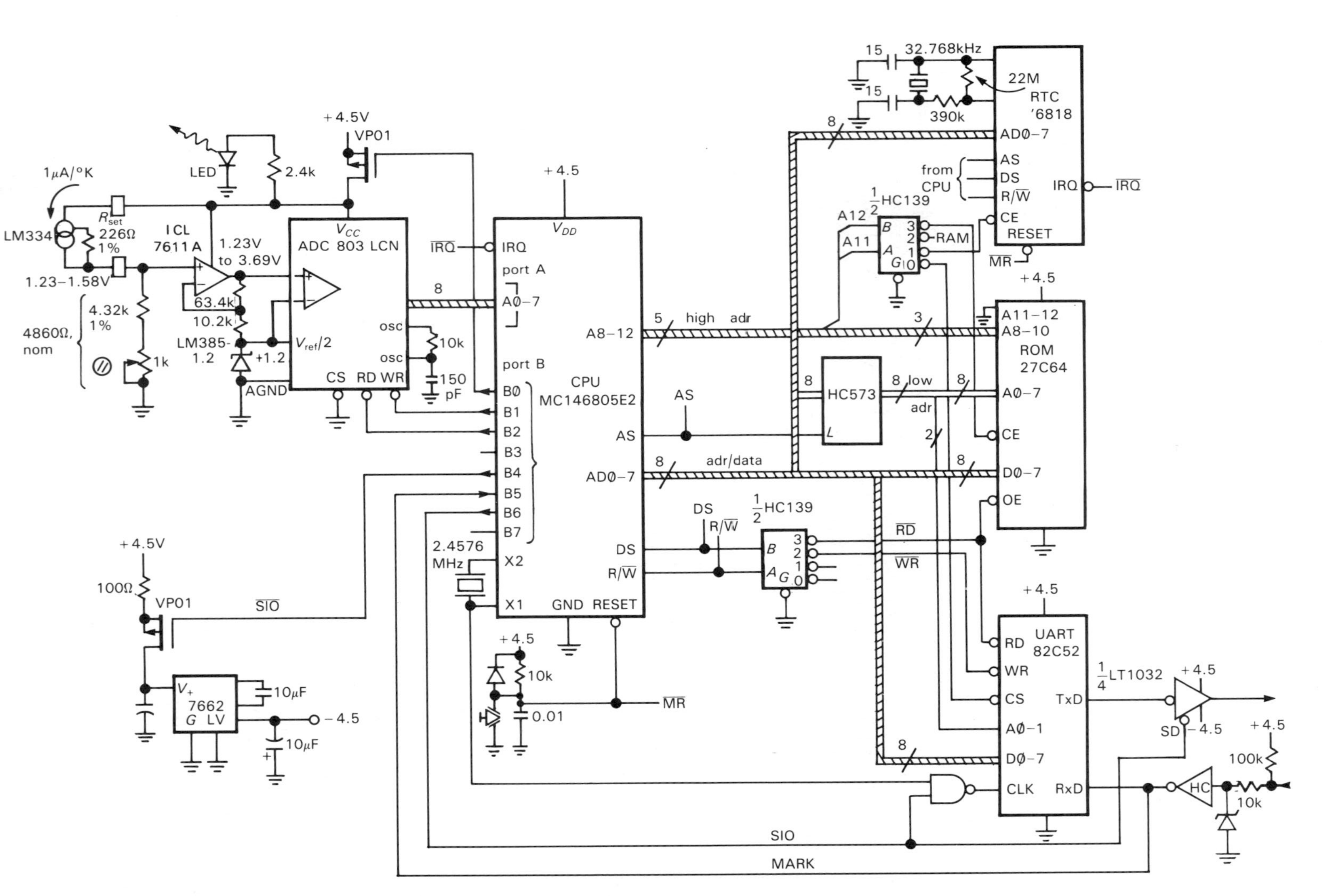

Figure 14.43. General-purpose low-power microprocessor example.

I_{DD} = 100μA, max, when deselected; in fact, the quiescent drain is likely to be less than 10μA. Note the low-order address latching from the multiplexed CPU bus, and the conversion of the Motorola pair (R/W′, DS) to the Intel pair (RD′, WR′), conveniently done with half of an 'HC139 decoder (the other half does address decoding).

The external CMOS RAM (optional, not shown) sits below ROM in address space (Fig. 14.44) and ties onto the same lines, with the exception of the address decoding. Once again, a smaller RAM would be fine, but the chip manufacturers haven't cooperated.

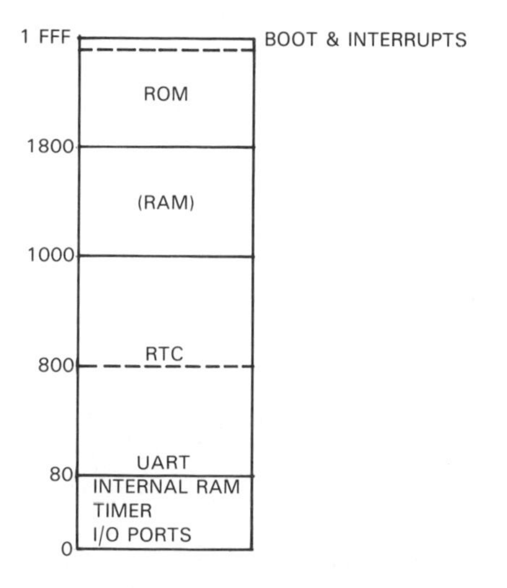

Figure 14.44. Memory map for microprocessor example.

Serial communication

For the serial communication (Section 10.19) we need a UART and bipolarity RS-232 driver/receiver. Since the serial port will be used only occasionally (during readout and initialization), it can be left in a power-down mode, with the CPU checking to see if something has been connected to it as part of its periodic once-per-minute duties. One possibility, then, is to use a conventional NMOS UART (e.g., an 8251), with power switching controlled by an output port bit of the CPU. This scheme is fine, although you have to be careful that the powered-down UART doesn't load the bus (or get powered by it, as a CMOS chip would): A good way to solve this problem is to use HC three-state bus buffers between the bus and the power-switched UART, putting them into the high-impedance state during power-down (Fig. 14.45).

In this example we took a simpler solution, namely the use of a CMOS UART that we leave powered at all times, switching the *oscillator* on only when the serial port is used; the resulting quiescent current, though not specified on the data sheet, is typically less than 20μA. For simplicity we also leave most of the RS-232 interface powered. The driver is one section of a low-power quad with a shutdown control; the specified quiescent current in the shutdown state is 1μA (typ), 10μA (max). The negative supply is provided by a 7662 voltage inverter, power-switched under control of one of the CPU's port bits; although the specifications give I_Q = 20μA (typ), the *maximum* I_Q is 150μA, enough reason to power-switch a portion of the circuit that might get used 10 minutes each year! Note the 100 ohm current-limiting resistor – the load looks like a short when power is first applied. For the RS-232 receiver, we've cheated by using an HC logic inverter with Schottky clamp and current limiter for the bipolarity input.

By leaving the receiver powered, we can detect when someone has connected to the serial port, since the RS-232 resting state ("mark," at least 3V negative) brings the inverter's output HIGH (note the pullup at the input). That is why the inverter's output, besides driving the UART, is tied to a CPU input port. Of course, the CPU could recognize that fact only during the short (<1s) interval each minute that it is awake. Thus, in practice, the serial-port user would connect to the port, then practice some patience until things begin to happen.

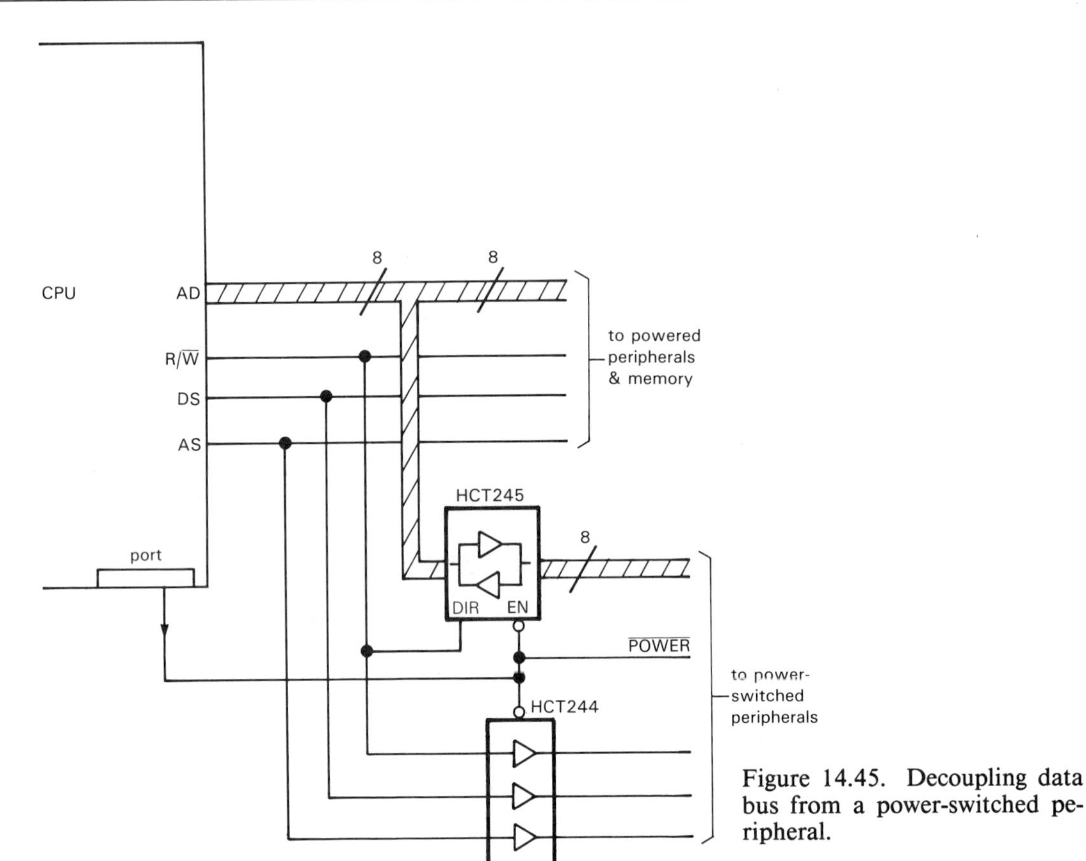

Figure 14.45. Decoupling data bus from a power-switched peripheral.

Front end

We chose the LM334 temperature-sensing current source as our sensor. The 2-terminal current is proportional to absolute temperature, set by a single resistor R_{set}: $I(\mu\,\text{A}) = 227\ T/R_{\text{set}}$, where T is in degrees Kelvin and R_{set} is in ohms. It has a voltage compliance from 1 to 40 volts, and initial accuracy of 6%. For the A/D, we picked the simple AD803, an inexpensive (less than $5) successive-approximation converter with internal clock generator, three-state outputs, and single +5 volt supply. It has a convenient differential input, with circuitry to set the full-scale range. Since it draws 1mA, we've used power switching, controlled by a CPU port bit.

In this implementation we chose $R_{\text{set}} =$ 226 ohms, i.e., 1μA/°K. A reasonable temperature range is −20°C to +50°C, with corresponding current range 254μA to 324μA. This has to be matched to the A/D input range. This particular A/D has a full-scale analog voltage span of $2V_{\text{ref}}$, when an externally supplied reference voltage V_{ref} is used; furthermore, the differential input structure lets you add an offset to the input. The simplest configuration is as shown, with an offset equal to V_{ref}; i.e., the analog input range is V_{ref} to $3V_{\text{ref}}$. With our +4.5 volt supply, the obvious reference is a 1.23 volt bandgap reference, say the micropower LM385-1.2. That makes the analog input span 1.23 to 3.69 volts. The rest is simple, namely choosing a load resistor to put the low end of the sensor output at 1.23 volts, then adding a stage of differential dc gain to put the high end of the sensor output at 3.69 volts. Accordingly, the load resistor for the sensor should be 4.84k, followed by a stage of voltage amplification of $G_V = 7.26$, as shown.

The ±10% trim adjustment is needed on the load resistor to accommodate initial errors (LM334, 6%; LM385, 2%; op-amp offset, 1%). Note the unusual piggyback configuration at the input, which makes the zero point of the dc amplifier equal to the reference, while simultaneously biasing the reference.

EXERCISE 14.1

Check our arithmetic by calculating for yourself the temperature range produced by the resistor values in Figure 14.43.

Note that we've interfaced the A/D via parallel port lines, rather than the more usual approach of using the CPU's data bus. We did this because the unpowered A/D would otherwise load the bus. Speed is unimportant in this application (especially since the CPU oscillator start-up can take as long as 250ms), and the port lines are available and unused.

TABLE 14.10. TEMPERATURE LOGGER CURRENT DRAIN[a]

	Idle		Data in		Serial comm	
Device	typ (μA)	max (μA)	typ (μA)	max (μA)	typ (μA)	max (μA)
CPU	10	175	4000	7500	4000	7500
ROM[b]	10	100	2500	5000	2500	5000
(RAM)	2	10	2500	5000	2500	5000
Sensor ckt	0	0	350	350	0	0
A/D	0	0	1100	1800	0	0
LED	0	0	1000	1000	0	0
UART	20	100[c]	20	100[c]	2500	3000
RS-232[d]	1	10	1	10	1600	2000
7662	0	0	0	0	20	150
Discrete	0	10	0	10	0	10
RTC	25	100	25	100	25	100
Totals	68μA	505μA	12mA	21mA	13mA	23mA

(a) V_{batt} = 4.5V, f_{CPU} = 2.5MHz. (b) assuming 50% ROM, 50% RAM accesses. (c) I_Q(max) not specified for 82C52; value for 82C51 used. (d) assuming 50% marking, and 5kΩ load impedance.

Power consumption

Table 14.10 shows how the current drain is proportioned, in each of the three operating states. Note the large spread between "typical" and "maximum" quiescent currents. If your ICs are "typical," the average current drain (assuming 500ms of wake-up time per minute) is 168μA, or 3 years with alkaline C cells (4500mAh). The worst-case average drain is 680μA (9 months), marginal for an unattended data logger. Of course," worst case" means that *every* IC's quiescent current is at the specified limit. There are two solutions: (a) Use much bigger batteries, to guarantee satisfactory worst-case battery life; or (b) pre-screen (by measuring I_Q) the ICs that dominate battery life, in this case the CPU. Most of the time you can probably get away with method (c): Live dangerously; most ICs have much lower leakage than the worst-case spec.

Programming

The ROM coding is straightforward and follows the general pattern of firmware coding that we talked about in Chapter 11. There are a few additional wrinkles, however.

When power-switching, be sure to allow time for the powered device to get going. It may include a crystal oscillator, in which case the required delay can be tens of milliseconds. The 7662 voltage inverter in this circuit has large capacitors and needs a few milliseconds to come up to voltage. If the switched device is connected to port bits (as the A/D has been), the port bits should either be set LOW or programmed as inputs before the device is switched off. If three-state drivers are used to isolate the bus, be sure to put them into the high impedance state *before* the device is powered down.

EXERCISE 14.2

Make a flow chart for the degree-day logger. Do the right things to the parallel port bits before going to sleep. Don't forget to initialize the calendar clock and UART. Be sure to check for serial port connection during each wake-up.

Design alternatives

As we mentioned at the outset, by operating at 3 volts instead of 4.5 volts (since battery voltage drops throughout its life, this is best done by regulating down from +4.5V with a micropower low-dropout regulator like the LP2951 or ICL7663), the CPU could be kept in WAIT mode, rather than STOP mode. In wait mode (200μA, max, at 3V and 1MHz), the internal oscillator runs, and supports both interrupts and the internal timer function. Thus, the external calendar clock could be replaced by a simple low-power 32kHz oscillator and divider (e.g., the inexpensive mini-DIP MM5368) to make 1pps interrupts; the internal timer would then wake up the CPU at a programmable interval, with everything else as before. Note that in the present design, most of the battery's energy is used by the calendar clock, a relatively expensive chip of excessive complexity.

EXERCISE 14.3

Draw a revised schematic for a temperature logger that sleeps in a WAIT state.

There are some recent chip offerings that allow you to simplify the serial port RS-232 circuitry. The LT1080 and the MAX230 series of RS-232 driver/receivers include on-chip voltage inverters, and, on some models, "shut-down" controls with $I_Q = 1\mu$A, typ (10μA, max). We could thus replace the 7662, its discrete power-switching circuitry, and the LT1032 with a single MAX235, which even includes built-in flying capacitors. Unfortunately, the latter's receiver sections go dead during shut-down, so we cannot use it to replace also our 'HC04 receiver. There are RS-232 driver/receivers available that do keep one receiver running ("hibernation"?) during micropower shut-down, for example the LT1039; however, this chip does not provide a solution here, because it requires continuous bipolarity power supplies during hibernation, which would require running the 7662 continuously.

While on the subject of serial port alternatives, it's worth noting that you can actually eliminate the UART, by using a pair of CPU parallel port bits for transmit and receive. To implement such a "software UART" you have to write software to generate and receive serial bit streams. The usual procedure is to use an internal CPU timer function, set to the appropriate baud rate. The transmit function is relatively straightforward, since you just generate 1's and 0's at each timer tick. The receive function turns out to be more of a challenge, since you have to sample the incoming port bit at a much higher clock rate (typically 8 times the baud rate) in order to sample near the middle of the bit cells. In spite of the programming problems, software UARTs are often used in small systems, because they eliminate a large chip.

Instead of the direct RS-232 outputs, the design could incorporate a power-switched modem for connection to the telephone line. The National 74HC943 would be good here, with 8mA quiescent current and single +5 volt supply. A passive "ring-detect" circuit (Fig. 14.46) would then substitute for the negative "mark-detect" circuit of Figure 14.43. It would be best to have the ring-detect trigger an interrupt, since no one likes waiting a minute for a phone to be answered. In the present design, the presence of a marking level (−3V or more) could also be used to trigger an interrupt.

EXERCISE 14.4

Show how to make a marking level trigger an interrupt. Be sure to provide a way to clear the interrupt via software.

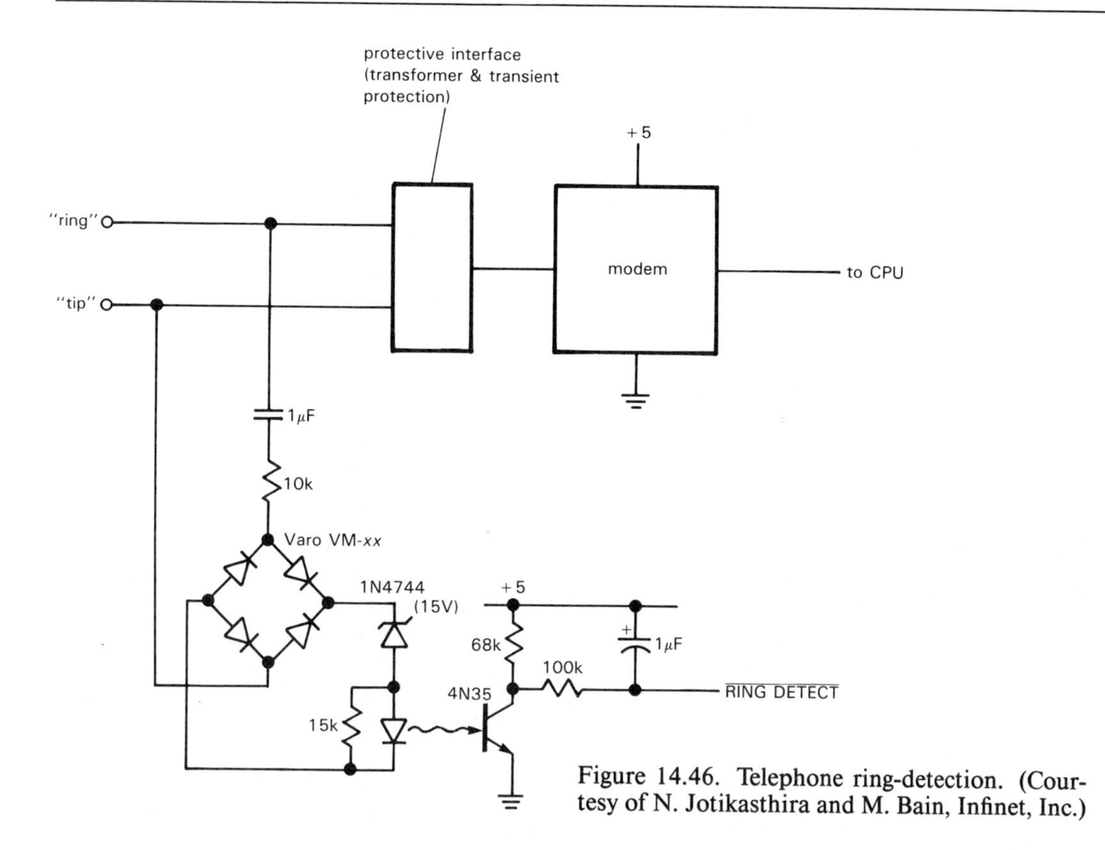

Figure 14.46. Telephone ring-detection. (Courtesy of N. Jotikasthira and M. Bain, Infinet, Inc.)

The A/D front end could tie into the CPU bus (instead of a port), providing that CMOS three-state drivers are used to isolate the powered-down device (Fig. 14.45). The same trick could be used to minimize power drain if a high-performance UART (e.g., an 85C30, as in Fig. 11.13) were substituted for the simple 85C52 used here.

A bit of philosophy: It's always tempting to try to squeeze the last ounce out of micropower design, by making diabolical use of the lowest-power devices, power switching, etc. For this circuit you could probably succeed in a design that used AA cells rather than C cells. But the extra effort (and cost) would not be worth it, because a reduction of 20% in size and weight wouldn't matter in this application. In fact, it would probably make more sense to simplify the design still further, for example by powering the 7662 from the same switched source as the A/D front end, or even leaving it powered continuously.

SELF-EXPLANATORY CIRCUITS

14.19 Circuit ideas

Figure 14.47 presents some low-power circuit ideas.

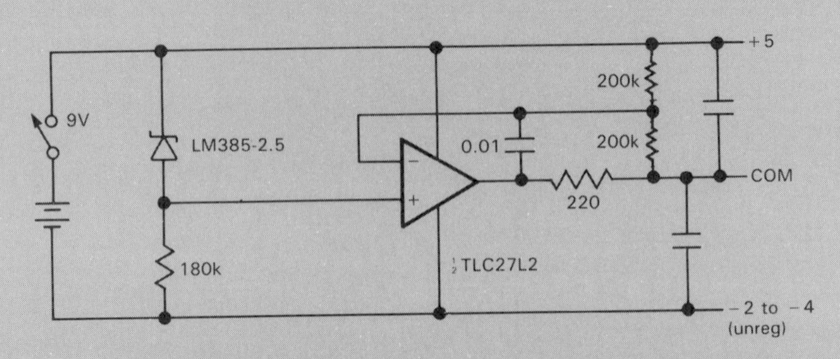

A. generating +5V and −2 to −4V from a single 9V battery

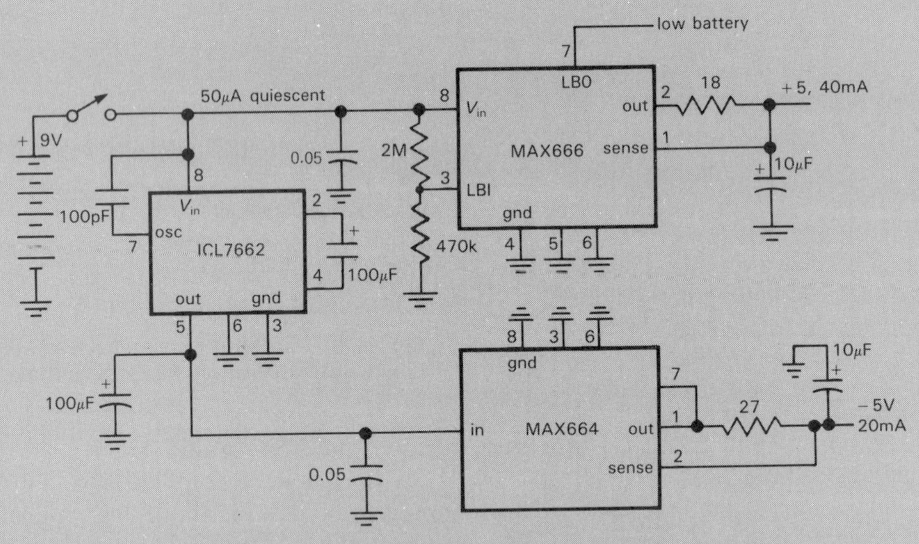

B. ±5 volt supply from single 9V battery

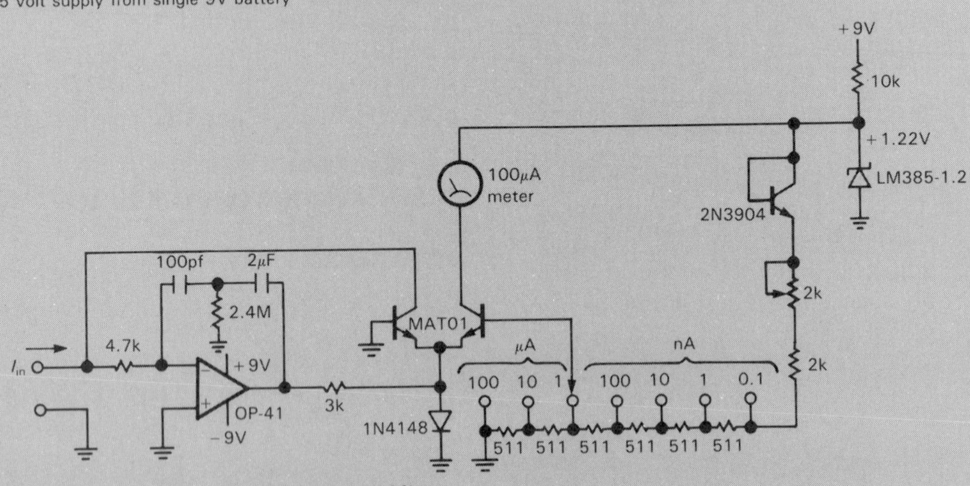

C. wide-range nano-ammeter (from PMI App. Note 106)

MEASUREMENTS AND SIGNAL PROCESSING

CHAPTER 15

OVERVIEW

Perhaps the most exciting (and most useful) area of electronics involves the gathering and manipulation of data from an industrial process or a scientific experiment. Generally speaking, *transducers* (devices that convert some physical quantity, such as temperature or light level, to a voltage or some other electrical quantity) are used to generate signals that can be manipulated by electronic circuits, quantified by analog-to-digital converters, and logged and analyzed by computers. If the signal you're looking for is masked by noise or interference, powerful "bandwidth-narrowing" techniques such as lock-in detection, signal averaging, multichannel scaling, and correlation and spectral analysis can magically retrieve the sought-after signal. Finally, the results of such physical measurements can be used to control the experiment or process itself, with "on-line" control usually provided by a small computer or micro-processor dedicated to the task. The recent development of powerful and inexpensive microprocessors and support chips has brought about an explosion in the use of electronics to control and log processes that would not have seemed likely candidates only a decade ago.

In this chapter we will begin with a sampling of quantities that can be measured and the transducers that are normally used for the job. In this area there is plenty of room for ingenuity, and the catalog of transducers we will describe should therefore be considered representative, not exhaustive. We will go into some detail describing the particular problems some of these measurement transducers present and the circuit solutions you might use with them. We will try to cover the most common difficulties, dealing with ultrahigh source impedances (hundreds of megohms in the case of microelectrodes or ion-specific probes), low-level low-impedance transducers (e.g., thermocouples, strain gauges, magnetic pickups), high-impedance ac sensors (capacitance transducers), and others.

The chapter will continue with a look at precision standards (standards of frequency and time, as well as voltage and resistance) and some of the techniques of precision measurements. We will then describe in some detail the whole business of bandwidth-narrowing, "pulling the signal from the noise." These techniques are extremely powerful, and they are mysterious to the uninitiated. Finally, we will conclude the chapter with a brief look at spectrum analysis and Fourier techniques. Readers interested primarily in electronic circuit design may wish to skip this chapter.

MEASUREMENT TRANSDUCERS

In some situations the quantity you want to measure is itself an electrical quantity. Examples might be nerve impulses (voltage), seawater conductivity (resistance), charged-particle fluxes (current), etc. In these cases measurement techniques tend to be relatively straightforward, with most of the difficulties centering around the kind of collection electrode to use and how to handle the signals once they've been collected. You might encounter very high impedances (e.g., with microelectrodes) or very small signals (e.g., a current generated from radioactive decay).

More often, a "transducer" of some sort is necessary to convert some physical quantity to a quantifiable electrical quantity. Examples are measurements of temperature, light level, magnetic field, strain, acceleration, sound intensity, etc. In the following sections we will take a look at some of the more common input transducers to give an idea of what can be measured and how accurately. We will go into somewhat greater detail when describing the more common measurements, such as heat and light, but in a book of this scope we can cover only a fraction of the measurement possibilities.

15.01 Temperature

Temperature transducers illustrate a nice variety of performance trade-offs. Temperature range, accuracy, repeatability, conformity to a universal curve, size, and price are all involved.

Thermocouples

A junction between two dissimilar metals generates a small voltage (with low source impedance!), typically in the millivolt range, with coefficients of about $50\mu V/°C$. This junction is called a *thermocouple*, and it is useful for measuring temperatures over a broad range. By using various pairs of alloys it is possible to span temperatures from −270°C to +2500°C with reasonable accuracy (0.5–2°C). The thermoelectric properties of different alloys are well known, so thermocouple probes in different formats (rods, washers, armored probes, etc.) made from the same alloys can be interchanged without affecting calibration.

The classic thermocouple circuit is shown in Figure 15.1. The particular choice of metals in this figure constitutes what is known as a type J thermocouple (look at Table 15.1 for a listing of the standard choices and their properties). Each couple is made by welding the two dissimilar metals together to make a small junction. (People have been known to get away with twisting the wires together, but not for very long!) The reference junction is absolutely necessary, since otherwise you wind up with additional dissimilar thermocouples where the dissimilar metals join the metal terminals. Those extra thermoelectric voltages produced at uncontrolled places in the circuit would result in erratic and inaccurate results. Even with a pair of thermocouple junctions, you still have thermocouples formed where the leads join the metal terminals. However, this seldom causes problems, since those junctions are at the same temperature.

The thermocouple circuit gives you a voltage that depends on the temperatures of both junctions. Roughly speaking, it is proportional to the *difference* between

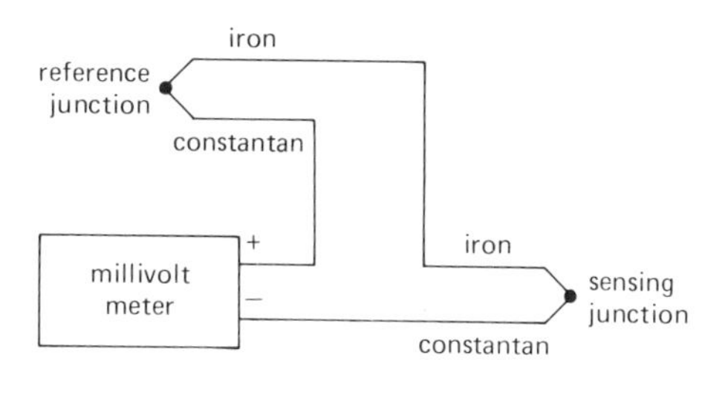

Figure 15.1. Classic thermocouple circuit.

the two junction temperatures. What you actually want is the temperature at the sensing junction. There are two ways to handle the problem of the reference: (a) Classically, you put the reference junction at a fixed temperature, usually 0°C. They used to use ice baths, and you still can, but you can also buy nice little stabilized cold boxes to do the same job. If you are measuring very high temperatures, you may not even care about small errors caused by having the reference junction at "room" temperature. (b) A more modern technique is to build a compensation circuit that corrects for the difference caused by having the reference junction at a temperature other than 0°C.

Figure 15.2 shows how this is done. The basic idea is to use a temperature-sensing chip and circuitry that adds in a voltage that makes up for the difference between the actual reference junction temperature and the standard 0°C. The AD590 (see the subsequent section on IC temperature sensors) produces an output current in microamps equal to the temperature in degrees Kelvin. R_1 is chosen according to the thermoelectric coefficient, in this case converting 1μA/°C to 51.5μV/°C (see Table 15.1), and the AD580 3-terminal reference (in combination with R_2 and R_3) is used to subtract the AD590 offset of 273μA at 0°C (273.15°K). Thus, there is no correction made when the reference junction is at 0°C, and 51.5μV/°C (the thermoelectric coefficient of a type J junction at room temperature) is added to the net output voltage of the pair of junctions when the reference junction is at some other temperature.

The metering circuit deserves a few words of comment. The circuit problems

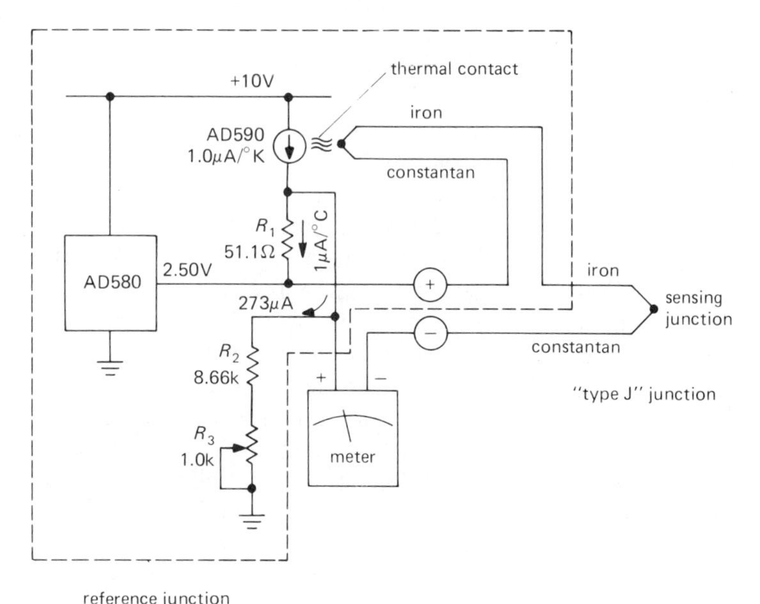

Figure 15.2. Thermocouple reference junction compensation.

TABLE 15.1. THERMOCOUPLES

Type	Alloy	Max temp[a] (°C)	Tempco @20°C (μV/°C)	Output voltage[b] 100°C (mV)	400°C (mV)	1000°C (mV)	30 gauge lead resistance[c] (Ω)
J	Iron / Constantan[d]	760	51.45	5.268	21.846	–	3.6
K	Chromel[e] / Alumel[f]	1370	40.28	4.095	16.395	41.269	6.0
T	Copper / Constantan[d]	400	40.28	4.277	20.869	–	3.0
E	Chromel[e] / Constantan[d]	1000	60.48	6.317	28.943	76.358	7.2
S	Platinum / 90%Pt–10%Rh	1750	5.88	0.645	3.260	9.585	1.9
R	Platinum / 87%Pt–13%Rh	1750	5.80	0.647	3.407	10.503	1.9
B	94%Pt–6%Rh / 70%Pt–30%Rh	1800	0.00	0.033	0.786	4.833	1.9

(a) thermocouple life is shortened by prolonged operation near maximum temperature. (b) reference junction at 0°C. (c) per double foot; for 24-gauge, multiply values by 0.25. (d) 55%Cu–45%Ni. (e) 90%Ni–10%Cr. (f) 96%Ni–2%Mn–2%Al.

you have with thermocouples stem from their low output voltage (50μV/°C or thereabouts), combined with large common-mode ac and radiofrequency interference. The amplifier must have good common-mode rejection at 60Hz and stable differential gain. In addition, the input impedance must be moderately high (of order 10k or more) in order to prevent error from loading, since the thermocouple leads do have some resistance (5 feet of 30 gauge type K junction wire has a resistance of 30Ω, for example).

The circuit shown in Figure 15.3 is a good solution. It is just the standard differencing amplifier with the T connection in the feedback path to get high voltage gain (200 in this case) while keeping the input impedance large enough so that loading of the source impedance doesn't contribute error. The op-amp is a precision low-offset type, with drift of less than 1μV/°C to keep its contribution to the measurement error much less than the 50μV that corresponds to a 1°C error. The input bypass capacitors are a good idea to reduce the common-mode interference at 60Hz and at radiofrequencies (thermocouples and their long connecting cables tend to behave like radio antennas). Since thermocouples respond slowly anyway, you can limit the bandwidth with capacitors across the feedback resistors, as shown. In cases of extreme radiofrequency problems, it may be necessary to shield the input leads and add RF chokes before the input bypass capacitors.

Note that the reference junction compensation circuit in Figure 15.3 acts as the *output*, rather than the usual method of compensating the voltage from the thermocouple at the input, as in Figure 15.2. This is done to keep the input truly differential, in order to preserve the advantages of the good common-mode rejection of the differencing amplifier. Since the amplifier has a voltage gain of approximately 200, the compensation circuit has to add 200×51.5μV/°C, or 10.3mV/°C at the output. Note that the OP-97E's input

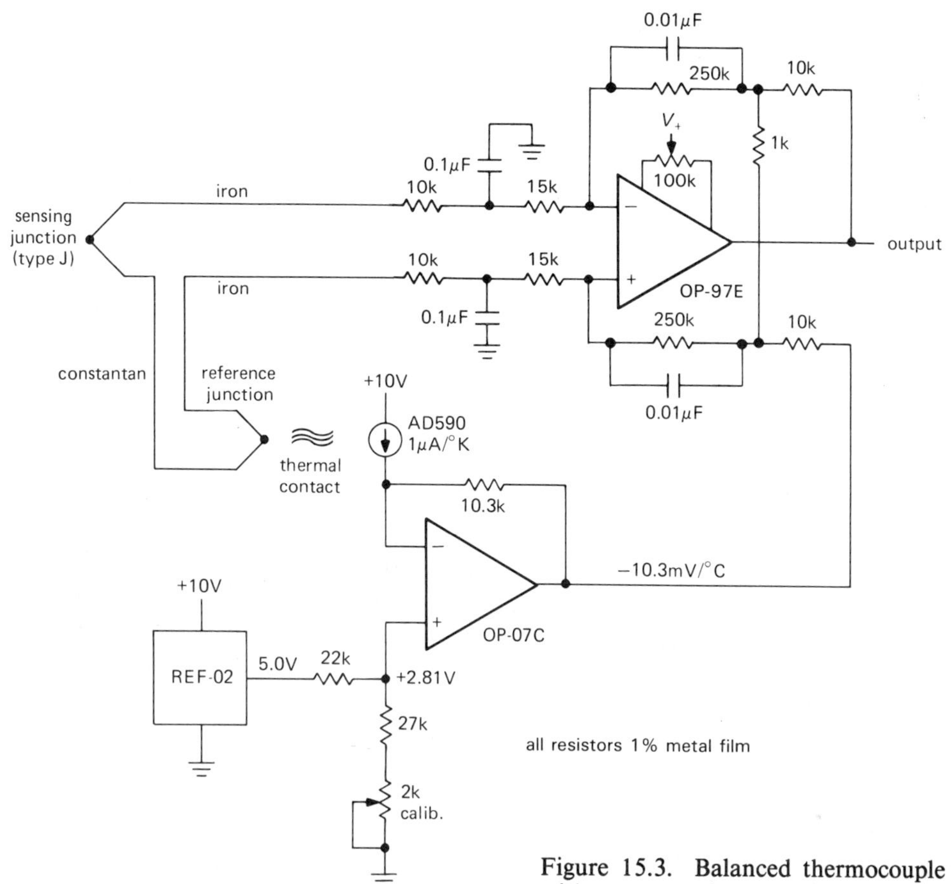

Figure 15.3. Balanced thermocouple amplifier with compensation at output.

offset current of 0.1nA (max) contributes 25μV of input error, which, along with the 25μV (max) of V_{os}, can be trimmed to zero. Alternatively, one could use a chopper-stabilized op-amp like the 7652 (V_{os} =5μV max, I_{os} = 40pA max).

An instrumentation amplifier, as in Figure 7.32, could be used instead of the differencing amplifier we've shown; in that case be sure to provide a dc bias path at the input.

Thermocouple users should be aware of Analog Devices' "thermocouple amplifier with cold junction compensation," the AD594 (type J) and AD595 (type K). These monolithic devices have everything you need (including an ice-point reference) to produce a temperature-proportional output voltage, or even programmable trip point, given a thermocouple input. The best grade is accurate (without trimming) to ±1°C at room temperature, increasing to ±3°C at −25°C and +75°C. Linear Technology makes the LT1025 "micropower thermocouple cold-junction compensator," designed to be used with an external precision op-amp. It includes compensation for all thermocouples in Table 15.1 (except type B), with second-order curvature correction to maintain accuracy over a wider temperature range. The best grade (LT1025A) is accurate (without trimming) to ±0.5°C at room temperature, increasing to ±2°C at −25°C and +80°C.

Complete "smart" temperature-measuring instruments configured for various thermocouple pairs are available commercially. These instruments include computational circuitry to convert the thermoelectric voltage to temperature. For instance, the digital thermometers manufactured by Analog Devices and Omega Engineering achieve an accuracy of about 0.4°C over a temperature range from −200°C to +1000°C and an accuracy of 1°C at temperatures up to +2300°C.

When compared with other methods of temperature measurement, thermocouples have the advantage of small size and wide temperature range, and they are particularly good for measuring high temperatures.

Thermistors

Thermistors are semiconductor devices that exhibit a negative coefficient of resistance with temperature, typically in the neighborhood of −4%/°C. They are available in all sorts of packages, ranging from tiny glass beads to armored probes. Thermistors intended for accurate temperature measurement (they can also be used as temperature-compensation elements in circuits, for instance) typically have a resistance of a few thousand ohms at room temperature, and they are available with tight conformity (0.1–0.2° C) to standard curves. Their large coefficient of resistance change makes them easy to use, and they are inexpensive and stable. Thermistors are a good choice for temperature measurement and control in the range of −50°C to +300°C. It is relatively easy to design a simple and effective circuit for "proportional temperature control" using a thermistor sensing element; see, for example, RCA application note ICAN-6158 or the Plessey SL445A data sheet.

Because of their large resistance change with temperature, thermistors make no great demands on the circuitry that follows. Some simple ways to generate an output voltage are shown in Figure 15.4. The circuit in part A expands the low-temperature end of the range because of the thermistor's exponential resistance change, whereas the circuit in part B produces a somewhat more linear variation of output voltage with temperature. For example, Figure 15.5 shows resistance versus temperature for two configurations – a bare 10k thermistor (Fenwal UUA41J1 "uni-curve"), and the same thermistor with a 10k resistor in series. The series pair is linear to within 3% from −10°C to +50°C, and better than 1% from 0°C to +40°C. The circuit of Figure 15.4B, sensing the drop across R, produces a voltage output of equivalent linearity.

Parts C and D of Figure 15.4 show elaborations of this linearizing idea, using composite matched "thermilinear" thermistors (and corresponding resistor pairs) from Yellow Springs Instrument Company. These 2-thermistor configurations achieve 0.2% linearity from 0°C to 100°C. YSI also makes modules with *three* thermistors (and three resistors), for even better linearity. The circuit in part E is the classic Wheatstone bridge, balanced when $R_T/R_2 = R_1/R_3$; since it is ratiometric, the null doesn't shift with variations in supply voltage. The bridge circuit, with a high-gain amplifier, is particularly good for detecting small changes about some reference temperature; for small deviations the (differential) output voltage is linear in the unbalance. With all the thermistor circuits you have to be careful about self-heating effects. A typical small thermistor probe might have a dissipation constant of 1mW/°C, meaning that I^2R heating should be kept well below 1mW if you would like your reading accurate to better than 1 degree.

Complete "smart" temperature-measuring instruments using curve-conforming thermistors are available commercially. These devices include internal computational circuitry to convert resistance readings directly to temperature. As an example, the Omega model 5800 digital thermometer covers −30°C to +100°C

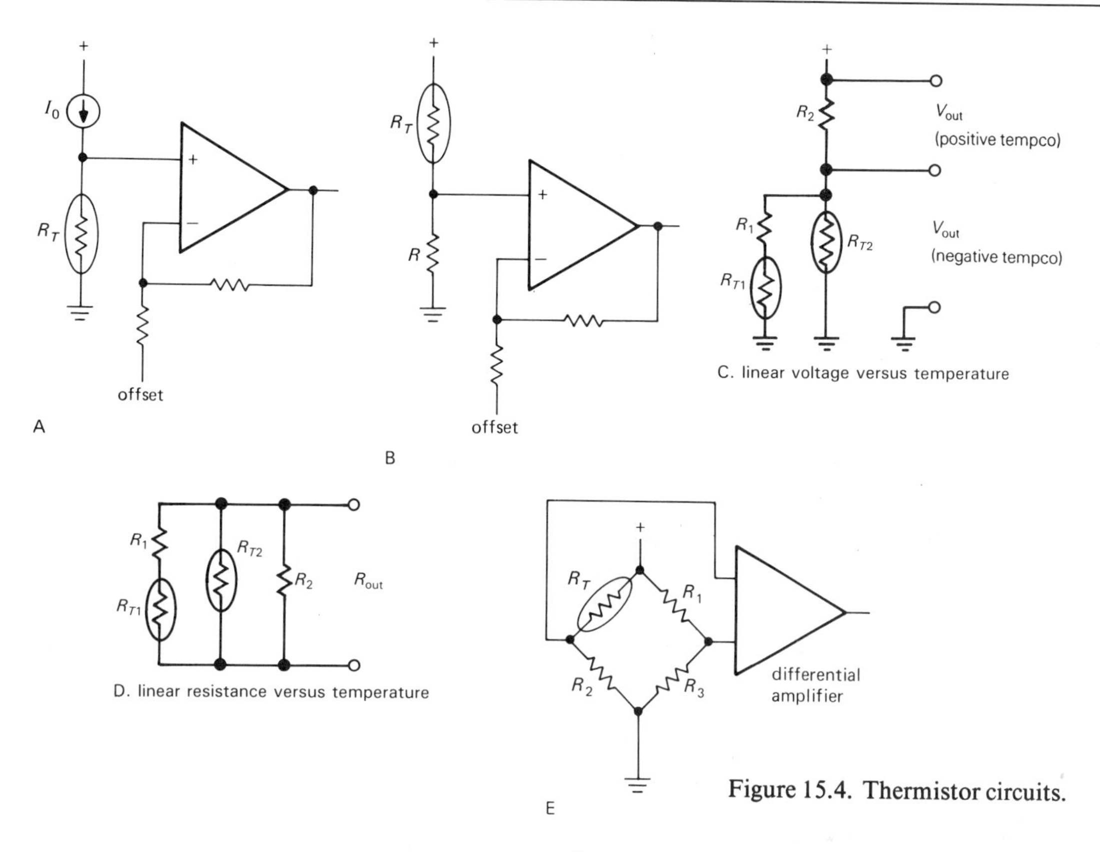

Figure 15.4. Thermistor circuits.

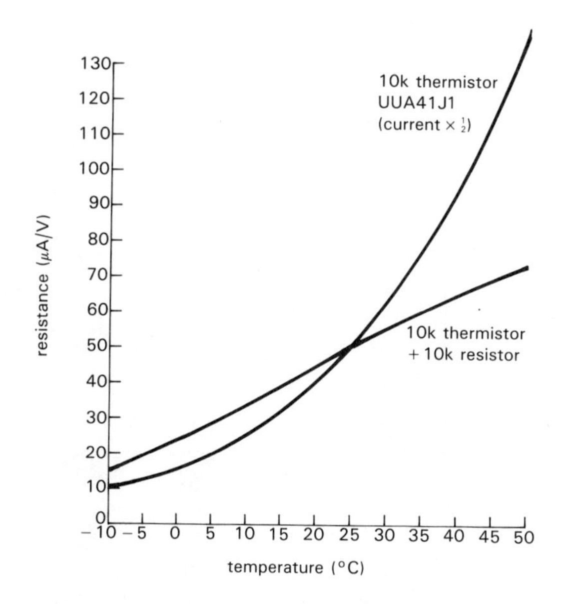

Figure 15.5 Resistance versus temperature for thermistor and thermistor-resistor pair.

in two ranges, reading out in either centigrade or Fahrenheit on a 4-digit LED display. It has 0.5°C accuracy over the full temperature range, with 0.1°C resolution.

When compared with other methods of temperature measurement, thermistors provide simplicity and accuracy, but they suffer from self-heating effects, fragility, and a narrow temperature range.

Platinum resistance thermometers

These devices consist simply of a coil of very pure platinum wire, which has a positive temperature coefficient of about 0.4%/°C. Platinum thermometers are extremely stable with time and conform very closely (0.02–0.2°C) to a standard curve. They are usable over a temperature range of −200°C to +1000°C. They aren't terribly cheap.

IC temperature sensors

As we remarked in Section 6.15, a bandgap voltage reference can be used to generate a temperature-sensing output voltage proportional to absolute temperature, as well as its usual stable zero-tempco reference output. The REF-02, for example, provides a "temp" output with a linear coefficient of +2.1mV/°C. If you buffer this output with an amplifier of adjustable gain and offset for calibration, you can achieve accuracies of about 0.5°C over the range −55°C to +125°C. The LM335 is a convenient 2-terminal temperature sensor that behaves like a zener diode with a voltage of +10mV/°K; e.g., at 25°C (298.2°K) it acts like a 2.982 volt zener (Fig. 15.6). It comes with an initial accuracy as good as ±1°C and it can be externally trimmed. A single point calibration can typically improve its accuracy to ±0.5°C max over a −55°C to +125°C range. After trimming, the output should be accurate to 0.1°C at the test temperature, with an accuracy budget increasing to ±0.5°C at the temperature extremes (Fig. 15.7).

The LM35 also provides a voltage output with +10mV/°C slope, but it behaves like a 3-terminal reference (rather than a 2-terminal zener), powered by +4 volts to +30 volts applied to a third terminal; it has an internal offset such that the output is 0 volts at 0°C. To operate near or below 0°C, you must use a pulldown resistor, as shown. The best grade (LM35A) has a maximum error of 0.5°C, but cannot be trimmed. Its cousin, the LM34A, works similarly, with readout in Fahrenheit (0 volts at 0°F).

Another approach to IC temperature sensors is the AD590, a two-terminal device that acts as a constant-current element, passing a current in microamps equal to the absolute temperature; e.g., at 25°C (298.2°K) it behaves like a constant-current regulator of 298.2μA (±0.5μA). With this simple device you get 1°C accuracy (best grade) over the range −55°C

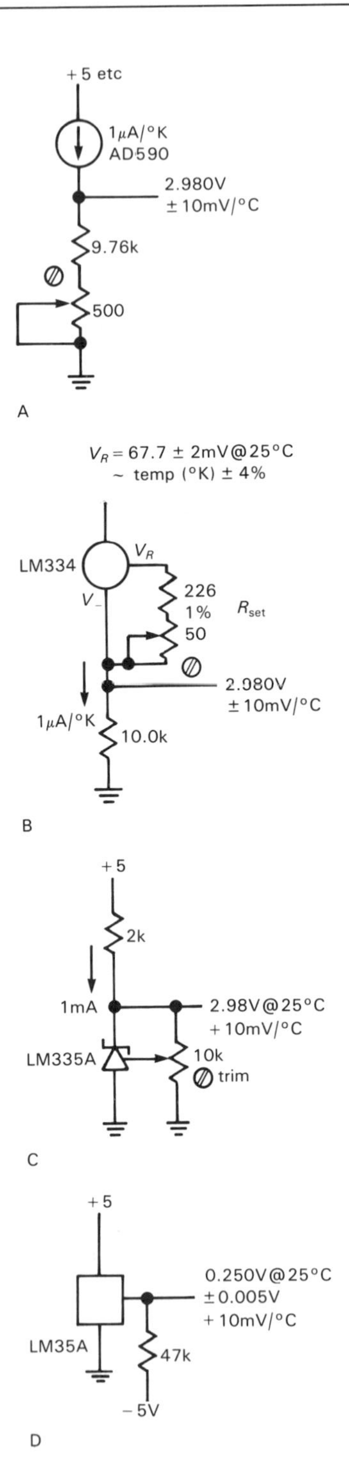

Figure 15.6 IC temperature sensors. Voltages are at 25°C. Methods A and B incur an additional 1% error per 33ppm/°C resistor tempco.

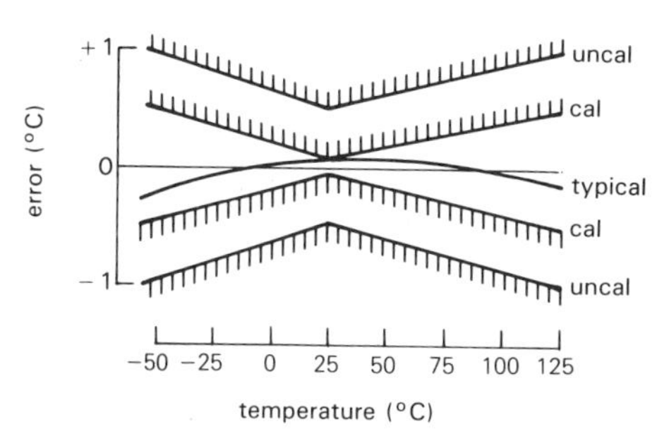

Figure 15.7. LM335 temperature error.

to +150°C. The simplicity of external circuitry required makes this a very attractive device. A plastic version, the AD592, has comparable performance over a smaller temperature range (−25°C to +105°C). The LM334 current-source IC (see Section 6.18) also has an output proportional to absolute temperature, settable with a single resistor according to $I_{out}(\mu A)=227T(°K)/R_{set}(\Omega)$ (see Fig. 15.6); this formula includes ≈5% correction for V_- current.

Quartz thermometer

The change of resonant frequency of a quartz crystal with temperature can be exploited to make an accurate and repeatable thermometer. Although the usual objective in quartz-crystal oscillator design is lowest possible temperature coefficient, in this case you choose a crystal cut with a large coefficient and take advantage of the high precision possible in frequency measurements. A good example of a commercial instrument is the Hewlett-Packard 2804A, a microprocessor-controlled thermometer with an absolute accuracy of 40 millidegrees over the range −50°C to +150°C (reduced accuracy over a wider range) and temperature *resolution* of 100 microdegrees. To get this kind of performance, the instrument contains calibration data for the individual sensor that it uses in the temperature calculation.

Pyrometers and thermographs

An interesting method of "noncontacting" temperature measurement is exemplified by the classic pyrometer, a gadget that lets you sight through a telescope at an incandescent object, comparing its glowing color with that of a filament inside the pyrometer. You adjust the filament's current until its brightness matches the object's, both being viewed through a red filter, then read off the temperature. This is a handy method of measuring the temperatures of very hot objects, objects in inaccessible places like ovens or vacuum chambers, or objects in oxidizing or reducing atmospheres where thermocouples cannot be used. Typical optical pyrometers cover a range of 750°C to 3000°C, with an accuracy of about 4°C at the low end and 20°C at the high end.

The development of good infrared detectors has extended this sort of measurement technique down to ordinary temperatures and for ordinary use. For example, Omega offers a line of digital-readout infrared pyrometers with a temperature range of −30°C to +5400°C. By measuring the intensity of infrared radiation, perhaps at several infrared wavelengths, you can determine with good accuracy the temperature of a remote object. Such "thermography" has recently become popular in quite diverse fields: in medicine, for the detection of tumors, and in the energy business, where a thermograph of your house can tell you where your energy dollars are evaporating.

Low-temperature measurements

Cryogenic (very cold) systems pose special problems when it comes to accurate temperature measurement. What matters there is how close to absolute zero (0°K=273.15°C) you are. Two popular methods involve measuring the resistance of ordinary carbon-composition resistors, which soars at low temperatures, and measuring the degree of paramagnetism of some salt. These are really specialty measurement techniques that will not be dealt with here.

Measurement allows control

If you have a way of adjusting some quantity, then the availability of a good measurement technique lets you control that quantity accurately. Thermistors, in particular, provide a nice method for controlling the temperature of a bath or oven.

15.02 Light level

The measurement, timing, and imaging of low light levels are parts of a well-developed field, thanks to the existence of amplification methods that do not rely on conventional circuit techniques. Photomultipliers, channel-plate intensifiers, CCDs (charge-coupled devices), and ISITs (intensified silicon intensifier target) are included in the catalog of high-performance optical detection devices. We will begin with the simplest detectors (photodiodes and phototransistors) and then go on to discuss the exotic and the wonderful.

Photodiodes and phototransistors

A diode junction acts as a photodetector: Light creates electron-hole pairs, and therefore a current through the external circuit. Diodes intended as photodetectors (photodiodes and PIN diodes) are packaged in a transparent case and are designed for high speed, high efficiency, low noise, and low leakage current. In the simplest mode of operation, a photodiode can be connected directly across a resistive load, or current/voltage converter, as shown in Figure 15.8. You get faster response (and the same photocurrent) by reverse-biasing the junction, as in Figure 15.9. High-speed PIN diodes have response times of a nanosecond or less (1GHz bandwidth) when loaded into a low impedance. It should be noted that the leakage current of good PIN diodes is so low (less than a nanoamp) that the Johnson noise in the load resistance dominates for load resistances less than 100MΩ or so, implying a speed/noise trade-off. An additional problem to be aware of is the error caused by the amplifier's input offset voltage, or the applied bias voltage, in combination with the photodiode's "dark resistance," when working at low light levels.

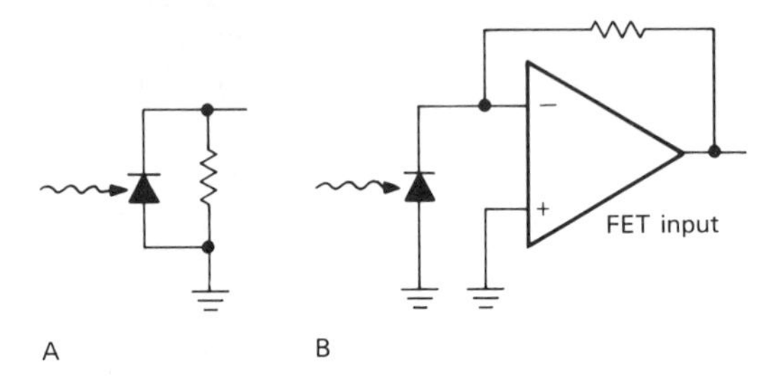

Figure 15.8

Photodiodes are pretty good light detectors when there is plenty of light around, but the output signal can be inconveniently small at low light levels. Typical sensitivities are of the order of 1 microamp per microwatt of incident light. A flux of 1000 photons per second, quite visible with the unaided eye, would cause a photocurrent of 4×10^{-16} amps when focused onto a PIN diode, totally undetectable when compared with the leakage current and noise. No silicon photodetectors are sensitive at the photon level (see the subsequent section on photomultipliers for that), but a device known as a *phototransistor* has considerably more output current than a photodiode at comparable light levels, bought at the expense of speed. It works like an ordinary transistor, with the base current provided by the photocurrent produced in the base-collector junction.

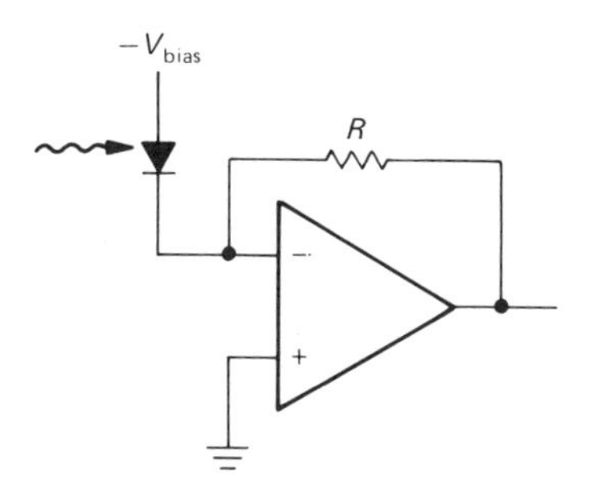

Figure 15.9

Inexpensive phototransistors like the MRD701 have output currents of a milliamp or so at an illumination of 1mW/cm^2, with rise and fall times of tens of microseconds, and photo-Darlingtons like the MRD711 have photocurrents roughly 50 times higher, but with rise times of 100μs or more. Note, however, that the additional current gain of a phototransistor or photo-Darlington doesn't improve its ability to detect extremely low light levels (its "detectivity"), since the ultimate limit is set by the detector diode's "dark current."

We have had excellent results with the photodiodes from UDT, Siemens, and Hamamatsu. The latter's catalog lists an impressive variety of detectors, including silicon *pn*-junction, silicon PIN diodes, GaAs (both diffusion and Schottky types) and GaP photodiodes, and avalanche detectors. It includes single detectors of many sizes and shapes, as well as linear photodiode arrays. UDT makes a series of detectors with integral MBC connectors.

Photomultipliers

For low-light-level detection and measurement (and, incidentally, for nanosecond resolution), you can't beat the photomultiplier. This clever device allows a photon (the smallest unit of light) to eject an electron from a photosensitive alkali metal "photocathode." The photomultiplier then amplifies this feeble photocurrent by accelerating the electron onto successive surfaces (dynodes), from which additional electrons are easily ejected. Figure 15.10 suggests the process. This use of "electron multiplication" yields extremely low noise amplification of the initial photocurrent signal. Typically, you use a voltage divider to put about 100 volts between successive dynodes, for a gain of about 10 per stage, or 1 million overall. The final current is collected by the anode, usually run near ground potential (look at Fig. 15.11), and is large enough so that subsequent amplifier noise is negligible.

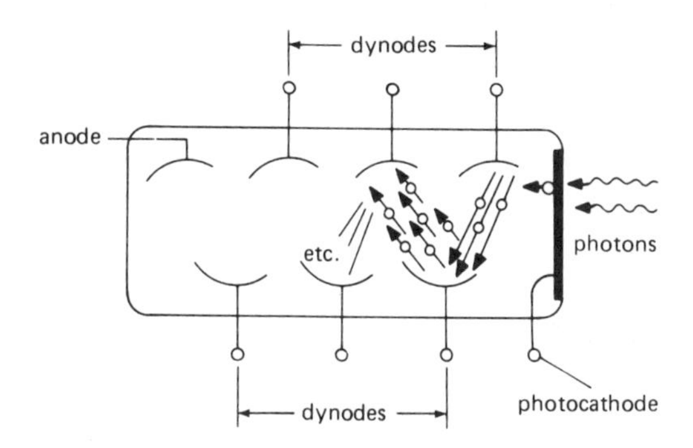

Figure 15.10. Photomultiplier multiplication process.

The most efficient photocathode materials have quantum efficiencies exceeding 25%, and with the large gain provided by the dynodes, individual photoelectron events are easily seen. At low light levels that's how you would use it, following the PMT (photomultiplier tube) with charge-integrating pulse amplifiers, discriminators (as detailed in Fig. 13.60), and counters. At higher levels the individual photoelectron count rate becomes too high, so you measure the anode *current* as a macroscopic quantity, instead. PMTs have typical sensitivities of an ampere per microwatt, although you could never operate them at such a high current; maximum PMT anode currents are limited to a milliamp or less. It should be noted that the practical limit to photon counting,

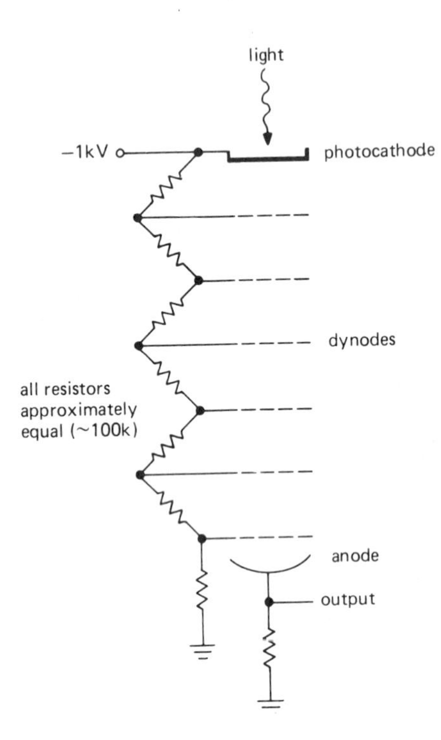

Figure 15.11. Photomultiplier biasing.

something like 1 million counts per second, corresponds to roughly 2 micro-micro-watts incident power!

Convenient electronic packages are available for both pulse-counting and output-current-measurement modes. An example is the "quantum photometer" from PAR, a handy gadget with built-in high-voltage supply and both pulse and current electronics. It has 11 ranges of pulse counting (10 pps to 10^6 pps, full scale) and 11 ranges of anode-current readout (10nA to 1mA, full scale).

Even in total darkness you get a small anode current from a photomultiplier. This is caused by electrons thermally excited from the photocathode and dynodes, and it can be reduced by cooling the PMT down to −25°C or so. Typical dark currents for a sensitive "bialkali" cathode PMT are in the neighborhood of 30 counts per second per square centimeter of cathode area, at room temperature. A cooled PMT with a small cathode can have dark currents of less than a count per second. It should be pointed out that a powered PMT should never be exposed to ordinary light levels; a PMT that has seen the light of day, even without power applied, may require 24 hours or more to "cool down" to normal dark-current levels. In some applications (e.g., fluorescence measurements) a PMT may be exposed to bright flashes of light at known times. In that case you can minimize the overload recovery time by disabling the accelerating voltage on the first few dynodes during the flash (some manufacturers offer this capability in their PMT/socket combinations).

When compared with photodiodes, PMTs have the advantage of high quantum efficiency while operating at high speed (2ns rise time, typically). They are bulky, though, and they require a stable source of high voltage, since the tube's gain rises exponentially with applied voltage.

It should be emphasized that PMTs are to be used with extremely low light levels. You run them with typical anode currents of a microamp or less, and they can easily see light that you cannot. Photomultipliers are used not only for the detection of light directly, as in astronomy (photometry) and biology (bioluminescence, fluorescence), but also in conjunction with scintillators as particle detectors and x-ray/gamma-ray detectors, as we will discuss in Section 15.07. Photomultipliers find wide use in spectrophotometry, where they are combined with prisms, gratings, or interferometers to make precise measurements of optical spectra. PMTs are manufactured by RCA (Burle), Hamamatsu, EMI, and EG&G.

CCDs, intensifiers, SITs, ISITs, and image dissectors

It is possible to do *imaging* at the light quantum level, thanks to some clever recent technology; i.e., you can form an image with the same sort of sensitivity to low light levels that you get with the (nonimaging) photomultiplier. These recent inventions are amazing to see. You can sit in what appears to be a completely dark room, then peer into a television monitor in which are imaged, albeit with plenty of "snow," all the objects in the room.

The key to all this is the image intensifier, an incredible device that produces as its output a brightened replica of an input image. You begin with either an ordinary silicon target vidicon (TV camera) or a CCD array. These are light-sensitive two-dimensional targets that accumulate an image and can be read out electronically by scanning with an electron beam or by shifting the image along as an analog shift register, respectively. At this point, all you have is a television camera whose sensitivity is far below the individual photon level; it is the two-dimensional analog of a photodiode. To bring about the miraculous, you simply put an imaging intensifier tube in front. Figure 15.12 suggests the process schematically.

Intensifiers come in two varieties. The first-generation type consists of a sensitive photocathode surface of the type used in photomultipliers, with electron-focusing optics and a phosphor screen arranged behind so that photoelectrons from the cathode are accelerated by high applied voltages and hit the phosphor with enough energy to give off a bright flash of light. With this kind of intensifier you can get single-stage light amplification of about 50, with resolution of about 50 lines/mm. Popular types cascade two, three, or four such stages of light amplification to achieve overall light amplifications of a million or more. The input and output may simply be glass surfaces with their internal photosensitive and phosphor coatings, or they may be paved with a dense fiber-optic bundle. Fiber-optics are nice because they let you match a perfectly flat entrance or exit surface to a curved tube surface, and they simplify the external optical system, since you can cascade these devices by just sticking things together, without any lenses.

Second-generation intensifiers using "microchannel plates" allow you to achieve much higher values of single-stage light amplification, and they are better at really low light levels because of fewer "ion events," the result of positive ions being ejected from the phosphor and returning to the cathode, where they make a big splash. In these channel-plate intensifiers the space from cathode to phosphor contains a bundle of microscopic hollow tubes whose insides are coated with a dynode-type multiplication surface. Photoelectrons from the cathode bounce their way down these channels, ejecting secondary

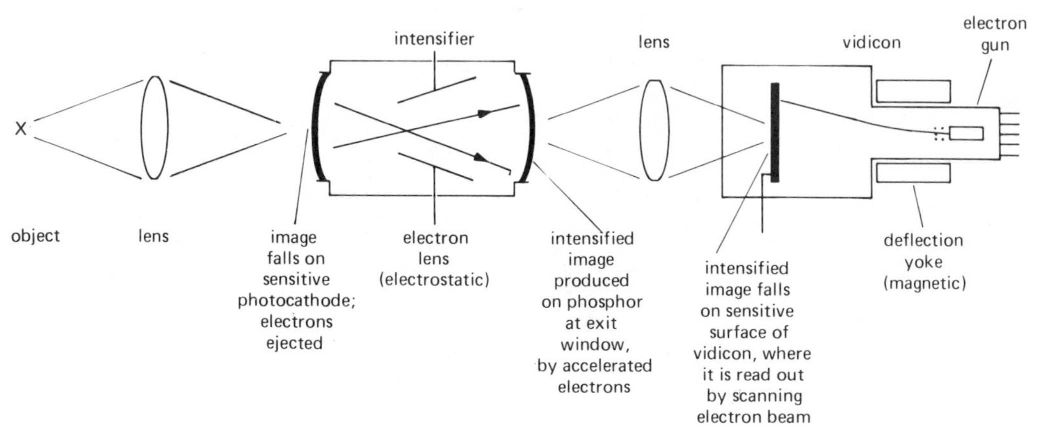

Figure 15.12. Vidicon with single-stage intensifier.

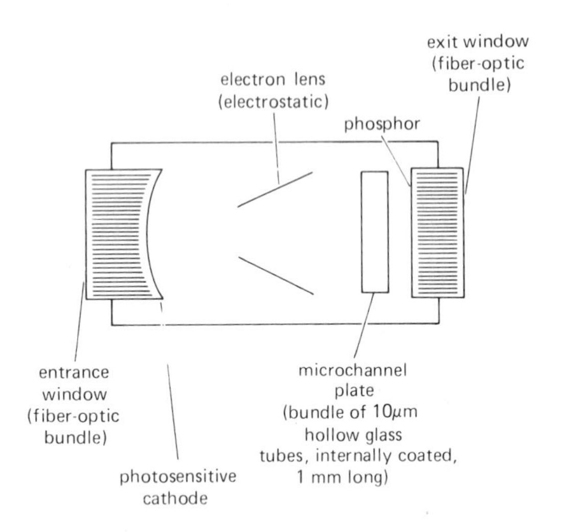

Figure 15.13. Electrostatically focused channel-plate intensifier.

electrons to give light amplification of about 10,000 (Fig. 15.13). You can get resolutions of about 20 lines/mm, and with special configurations ("J-channel," "C-channel," "chevron") the ion-event problem can be eliminated almost entirely. The result is an imaging intensifier with the same sort of quantum efficiency as photomultipliers (20%–30%). The use of nearly noiseless electron multiplication results in light amplification to a level that the vidicon or CCD can see.

Such an intensifier combined with a silicon-target vidicon in a single tube is called a "SIT" (silicon intensifier target). An ISIT is a SIT with an additional intensifier placed externally in front (Fig. 15.14); this is the sort of gadget that lets you see in the dark. These things are very popular with astronomers and with night-warfare people.

An interesting variation of the imaging intensifier is the so-called image dissector, a clever device that actually preceded the devices just described. It consists of a sensitive photocathode area, followed by the usual photomultiplier dynode chain. In between is a small aperture and some deflection electrodes, so that any spot on the photocathode can become the active area for electron multiplication by the dynode system. You can think of an image dissector as a photomultiplier with an electronically movable photocathode area. It has the quantum efficiency and gain of a conventional PMT, but it differs from the intensified vidicons, CCDs, and SITs

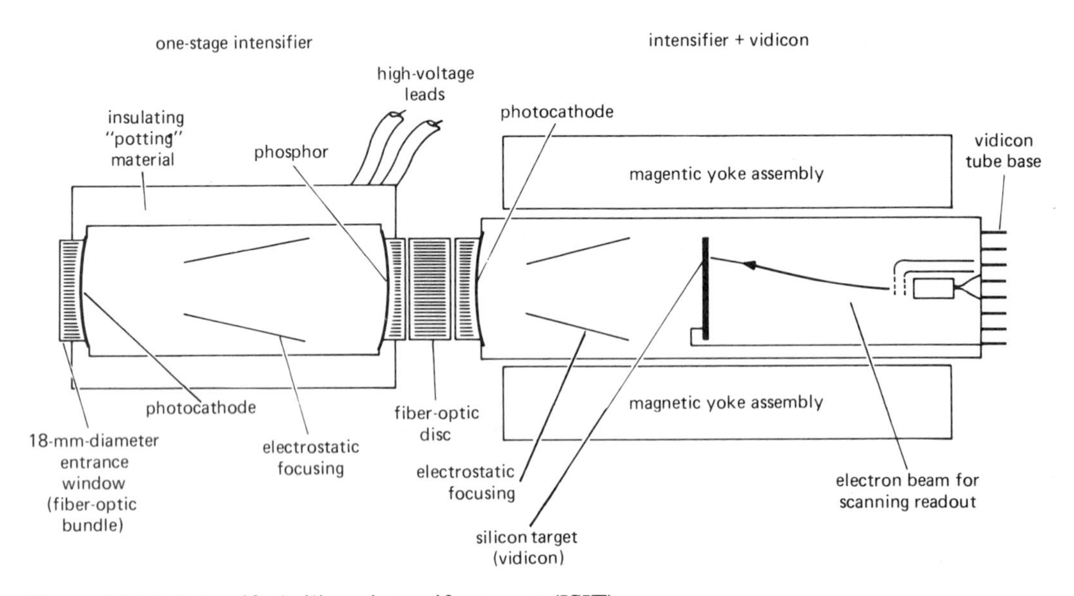

Figure 15.14. Intensified silicon intensifier target (ISIT).

(which are all image-integrating devices) in that it does not accumulate the image over the entire field in between readouts.

CCD detector arrays can, of course, be used without intensification. You can buy them from companies like EGG Reticon, Kodak, TI, Tektronix, Thomson, and Toshiba. They are available as 1-dimensional ("linear") arrays containing up to 4096 cells, or as 2-dimensional ("area") arrays containing 256K (512 × 512) or more cells (called "pixels," for "picture elements"); Toshiba has an area detector with an impressive 2 million pixels on a single chip. Linear arrays are handy as detectors for spectroscopy; area detectors, of course, are used for full 2-dimensional images, for example in television cameras.

All CCDs are light-*integrating* devices, which accumulate charge in each pixel until the array is read out. During readout the CCD becomes an analog shift register, with the image emerging, raster style, as a serial analog waveform at the single output line.

15.03 Strain and displacement

The field of measurements of physical variables such as position and force has its own bag of tricks, and any accomplished measurer should be aware of things like strain gauges, LVDTs, and the like. The key to all of these measurements is the measurement of displacement.

There are several nice ways to measure position, displacement (changes in position), and strain (relative elongation).

LVDT. A popular method is the LVDT (linear variable differential transformer), which is almost self-explanatory. You construct a transformer with a movable core, excite one set of coils with ac, and measure the induced voltage in the second set. The secondary is center-tapped (or brought out as two separate windings) and arranged symmetrically with respect to the primary, as shown in Figure 15.15. LVDTs come in an enormous variety of sizes, with full-scale displacements ranging from 0.005 inch to 25 inches, excitation frequencies from 50Hz to 30kHz, and accuracies of 1% down to 0.1% or better. A leader in this field is Schaevitz, whose catalog lists a broad selection of linear and angular ("RVDT") transducers, and measurement transducers utilizing LVDT sensors (e.g., pressure, force, acceleration, etc.), as well as LVDT readout electronics. If you become involved with LVDTs, you may want to build your own instrumentation, perhaps using the special ICs designed for the purpose. For example, the monolithic Signetics NE5520/1 "LVDT Signal Conditioner" provides sine-wave excitation signals and includes a synchronous demodulator to provide a voltage output proportional to the LVDT displacement. The 2S54/6 from Analog Devices are LVDT synchronous demodulators with excellent linearity (0.01%) and built-in A/D converters providing direct digital outputs (14 and 16 bits, respectively).

Strain gauge. A strain gauge measures elongation or flexure by subjecting an array of four metal thin-film resistors to deformation. They come as complete assemblies, in sizes from 1/64 of an inch to several inches, and they generally have impedances in the neighborhood of 350 ohms/leg. Electrically they look like a Wheatstone bridge; you apply dc across two of the terminals and look at the voltage difference between the other two, as discussed in Section 7.09. The output voltages are very small, typically 2mV per volt of excitation for full-scale deformation, with accuracies ranging from 1% down to 0.1% of full scale (see Fig. 15.15D).

It is not easy to measure small relative elongations, and strain-gauge specifications are notoriously unreliable. Small differences in the temperature coefficients

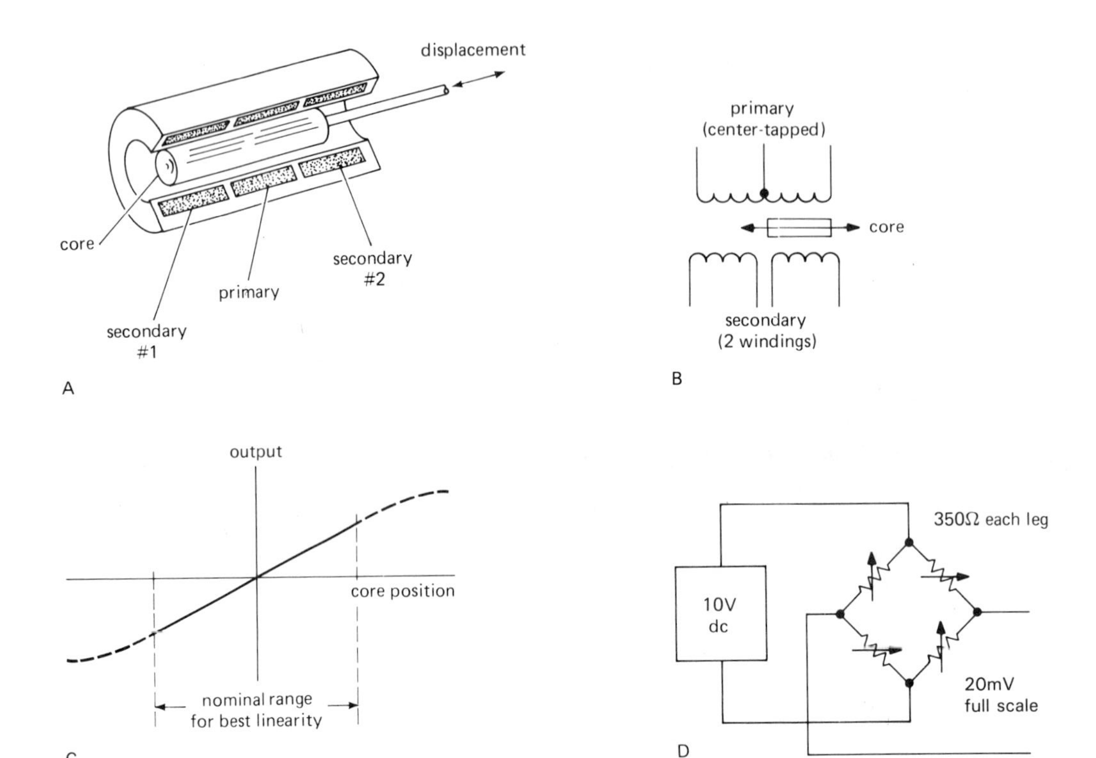

Figure 15.15. Displacement transducers.
A. Linear variable differential transformer (LVDT) cutaway view.
B. LVDT schematic.
C. LVDT output versus displacement.
D. Strain gauge schematic.

of the bridge elements are responsible for the temperature sensitivity, which limits the performance of the strain gauge. This is a problem, even in controlled-temperature environments, because of self-heating. For example, 10 volts of dc excitation on a 350 ohm bridge will produce 300mW dissipation in the sensor, with a temperature rise of 10°C (or more), causing errors corresponding to a real signal of 0.1% to 0.5% of full scale.

Recently, semiconductor strain gauges have become popular. They have outputs that are 10 times higher than those of the metal-film variety and impedances of a few thousand ohms. It is often necessary to use a current source as excitation, rather than a voltage source, to minimize temperature sensitivities.

Capacitance transducers. Very sensitive measurements of displacement can be made with a transducer consisting simply of two closely spaced plates, or a plate suspended between a pair of outer plates. By making the capacitor part of a resonant circuit, or by using a high-frequency ac bridge, you can sense or control very small changes in position. Capacitor microphones use this principle to convert acoustic pressure or velocity to an audio signal.

The amplifiers used with capacitor microphones illustrate some interesting circuit ideas, and they are of practical importance, since many of the best recording microphones are capacitive position transducers, made by supporting a thin metallized plastic foil in close proximity to a fixed plate. You charge the capacitor through a large resistor with a bias of 50 to 100 volts, and you look at the changes of voltage as the diaphragm moves in the sound field.

Capacitor microphones have enormously high source impedances (a typical capsule has about 20pF of capacitance, or a reactance of about 400M at 20Hz), which means that you don't have a chance of running the signal through any length of cable whatsoever without putting a preamp right at the capsule. Figure 15.16 shows two ways to buffer the voltage from the capsule, which might have an amplitude of 1mV to 100mV (rms) for typical ranges of program material. In the first circuit, a low-noise FET op-amp provides 20dB of gain and the low impedance necessary to drive a single-ended shielded line. Since the amplifier has to be located close to the microphone capsule (within a few inches), it is necessary to supply the operating voltages (bias for the capsule, as well as op-amp power) through the microphone cable, in this case on additional wires. Note the trick of floating the microphone capsule in order to simplify biasing of the op-amp. R_1 and C_1 filter the bias supply, and R_2 must be chosen to have a high impedance

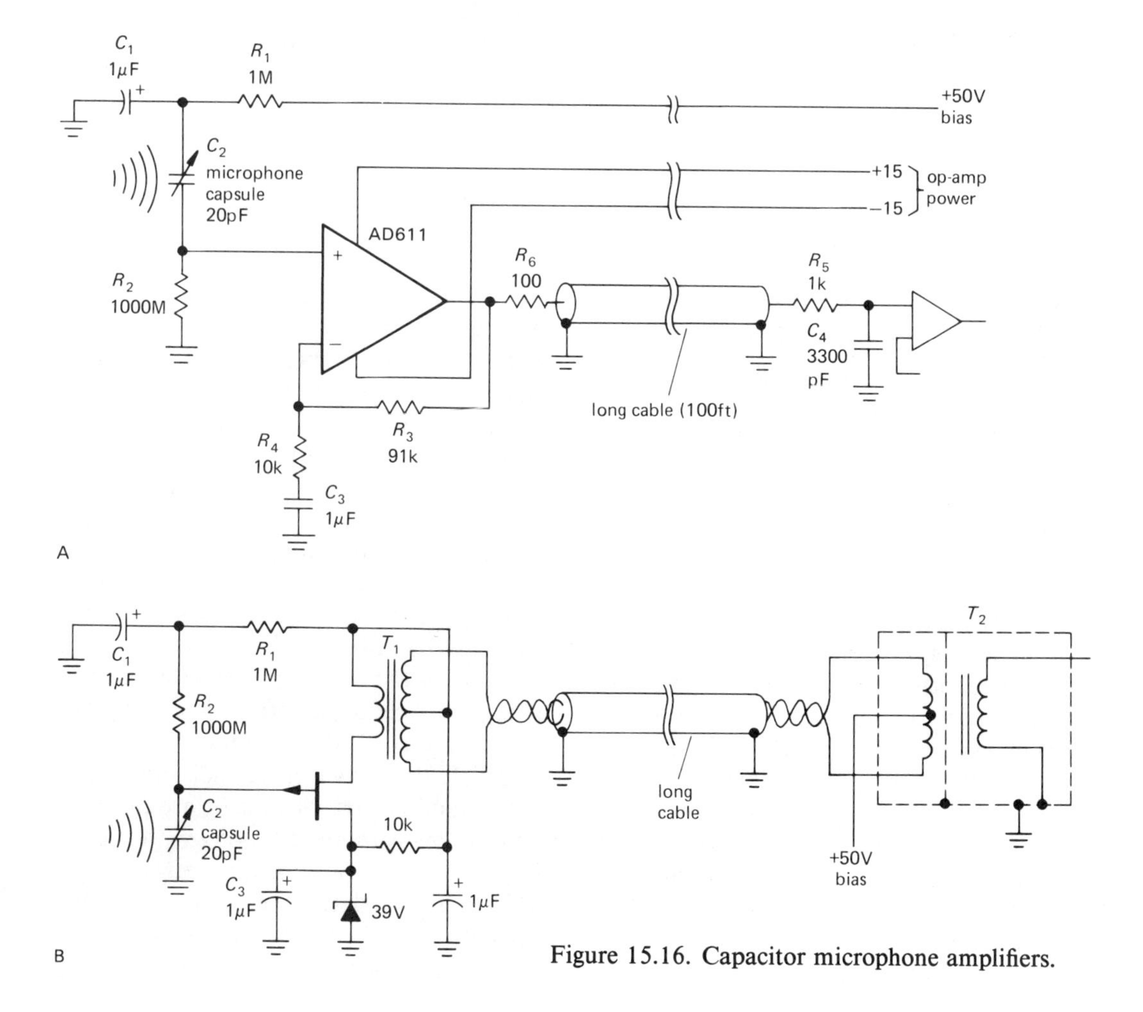

Figure 15.16. Capacitor microphone amplifiers.

compared with the capsule at all audio frequencies. R_5 and C_4 form an RF filter, since the line is unbalanced and therefore prone to radiofrequency interference.

There are a few bad features about this circuit. It requires a 4-wire cable, rather than the industry-standard shielded pair. Also, the floating capsule can create mechanical problems. These drawbacks are remedied in the second circuit, in which the capsule bias is sent on the same lines as the audio itself, which happens to be a balanced 200 ohm pair. One side of the capsule is grounded, and a p-channel JFET is used as a source follower to drive a small audio transformer. A single-ended output is recovered at the far end, where bias is applied via the transformer's center tap. Some would complain that the proliferation of transformers is a poor idea, but in practice they perform admirably.

□ *Angle.* It is possible to convert angles to electronic signals with pretty good precision. There are angular versions of the LVDT, for instance, and popular devices known as resolvers. In both cases you use an ac excitation, and you can easily measure angular position down to an arc-minute. With great care it is possible to measure angles at the arc-second level. There are other techniques, e.g., using light beams looking at a glass disc with gray coded radial stripes.

□ *Interferometry.* Highly accurate position measurements can be made by bouncing laser beams off mirrors attached to the object and counting interference fringes. The ultimate accuracy of such methods is set by the wavelength of light, so you have to work hard to do much better than a half micron (1 micron, or micrometer, is 1/1000 of a millimeter, or 1/25000 of an inch). An example of a commercial laser measurement instrument is the 5527A from Hewlett-Packard, which claims resolution of a microinch or better. Laser interferometer systems are now used routinely in surveying, in flatness measurements, and in various tasks around research laboratories.

The most precise distance measurements have been made interferometrically by Deslattes at the National Bureau of Standards. Deslattes is a real wizard when it comes to precise physical measurement, routinely measuring spacings to milliangstroms (one ten-millionth of a micron) and angles to milliseconds of arc.

□ *Quartz oscillators.* A quartz crystal responds to deformation with a change in its resonant frequency, thus providing a very accurate method for measuring small displacements or changes in pressure. Quartz-oscillator pressure transducers provide the highest resolution presently available (more about this later).

15.04 Acceleration, pressure, force, velocity

The techniques just mentioned allow you to measure acceleration, pressure, and force. Accelerometers consist of strain gauges attached to a test mass, or capacitance-sensing transducers that sense the change of position of the test mass. There are various tricks to damp the system to prevent oscillation, in accelerometers that simply measure the displacement of the test mass to provide an output signal; alternatively, some systems use feedback to prevent the test mass from being displaced relative to the body of the accelerometer, the amount of applied feedback force then being the accelerometer's output signal.

LVDT, strain gauges, capacitance transducers, and quartz oscillators are used for pressure measurements, along with special devices such as a Bourdon gauge, a spiral hollow quartz tube that unwinds when inflated. LVDT transducers, for example, are available with full-scale ranges going from 1 psi to 100,000 psi or more. Quartz-crystal oscillator types provide the highest resolution and accuracy. The types available from Paroscientific, for example, will deliver accuracies of 0.01% and stabilities of 0.001%. Hewlett-Packard has a quartz pressure gauge with 11,000 psi full-scale sensitivity and claimed resolution of 0.01 psi.

LVDT transducers are often used to measure force or weight, although any of the displacement techniques can be used. Full-scale sensitivities go from 10 grams to 250 tons for one popular series, with accuracies of 0.1%. For highly precise laboratory measurements of small forces, you will find quartz-fiber torsion balances, electrostatic balances, and the like. An interesting example of the latter is the clever gravimeter developed by Goodkind and Warburton. It uses a superconducting sphere levitated approximately to zero weight by a persistent magnetic field, then balanced the rest of the way by electrostatic sensing and levitating plates. It can measure changes in gravitational field of one part in a billion, and it easily sees barometric pressure variations because of the effect of the changing overhead air mass on local gravity!

Magnetic velocity transducers

The position transducers we have been talking about can also be used to keep track of velocity, which is just the time derivative of position. However, it is possible to make a direct velocity measurement by exploiting the fact that the voltage induced in a loop of wire moving through a magnetic field is proportional to the rate of change of magnetic flux linked by the loop. There are velocity-measurement gadgets available that consist of long coils of wire with magnetic rods moving through the central bore.

Much more prevalent are the magnetic velocity transducers used in the audio industry: microphones (and their inverse, loudspeakers), phono cartridges, and analog tape recorders. These devices typically generate signals at very low levels (a few millivolts is typical), and they present unique and interesting circuit challenges. For high-quality sound you have to keep noise and interference down 60dB or more, i.e., at the microvolt level. Since these signals get piped around over large distances in recording studios and radio stations, the problem can become serious.

Figure 15.17 shows how low-level signals from microphones and phono cartridges are usually handled. A dynamic microphone is a loudspeaker in reverse: A coil moves in a magnetic field, propelled by the sound pressure. Typically these things have output impedances of 200 ohms, with signals of 50μV to 5mV (rms) for quiet speech and concert-hall sound levels, respectively. For any significant length of connecting cable, you always use a balanced and shielded twisted-pair, terminated in the industry-standard Cannon XLR 3-pin audio connector.

At the far end you transform to a terminated impedance level of about 50k with a high-quality audio matching transformer, as indicated. Signal levels are then in the range of 1mV to 100mV (rms) and should be amplified by a low-noise preamp, as shown. Although you will see preamps with 40dB front-end gain, for good overload performance it is best to stick with a gain of 20dB. This is especially true for popular-music recording, where singers often wind up hollering into the microphone at close range.

The use of balanced 200 ohm microphone cable pretty much eliminates interference because of its good common-mode rejection. Good audio transformers for this kind of application have electrostatic shielding between the windings, which further reduces sensitivity to RF pickup. If radiofrequency interference is not suppressed enough with this scheme, as may be the case near transmitting

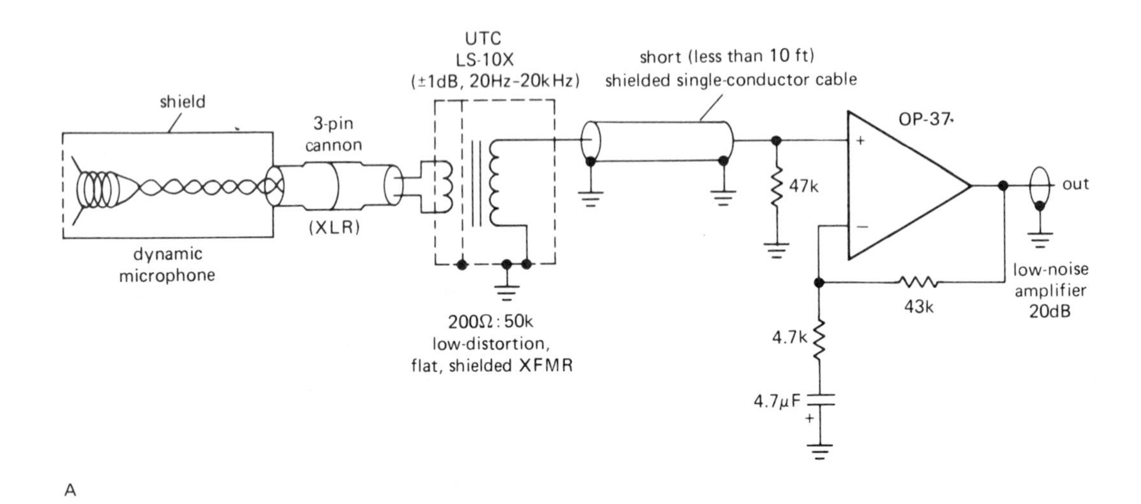

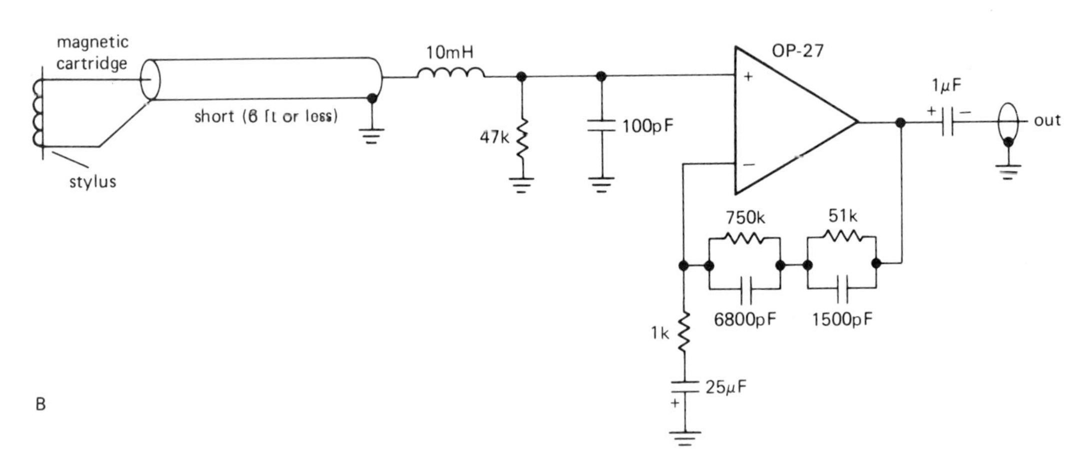

Figure 15.17. Dynamic microphone and phono cartridge amplifiers.

stations, you can add a low-pass filter at the preamp input. A 1k resistor or small RF choke in series at the input, followed by a 100pF capacitor to ground, will usually tame the beast.

Phono cartridges don't require balanced lines, because the cable run to the amplifier is usually very short. The standard method is simply to use single-conductor shielded cable, terminated with the 47k to ground that the cartridge requires for proper frequency response (Fig. 15.17B). We have also shown an input filter to reduce RF interference, because that is such a common problem in urban areas. RF signals at the inputs of audio equipment pose a particularly insidious problem, because the audio amplifier's nonlinearities at radiofrequencies produce rectification, with consequent interference (audio detection) and distortion. When designing RF filters, be sure to keep the load capacitance small (a maximum of 300pF, including cable capacitance), since otherwise the cartridge's frequency response is changed. The series impedance should not have a resistance greater than a few hundred ohms, to keep the noise low. Quite large values of inductance can be safely used, since the cartridge's inductance is typicallly 0.5 henry. The amplifier circuit shown has the standard RIAA response used for recording in the United States.

15.05 Magnetic field

Accurate magnetic-field measurements are important in the physical sciences, in connection with instrumentation that uses a magnetic field (magnetic resonance, magnetrons, magnetically focused electron devices, etc.), and in geology and prospecting. For measurements at the 1% level, a Hall-effect probe is adequate. The Hall effect is the production of a transverse voltage in a current-carrying conductor (usually a semiconductor) in a magnetic field, and commercial Hall-effect magnetometers cover a range of about 1 gauss to 10kG full-scale. To give an idea of scale, the earth's field is about 0.5 gauss, whereas the field of a strong permanent magnet is a few thousand gauss. Hall magnetometers are inexpensive, simple, small, and reliable. For example, the inexpensive TL173 from TI is a complete linear Hall-effect sensor in a 3-terminal plastic TO-92 package. You power it from +12 volts, and out comes a dc voltage that increases 1.5mV per gauss of applied magnetic field. Sprague also markets a line of linear Hall sensors (the UG3500/3600 series). The Hall effect is also used to make noncontacting keyboard and panel switches, as we remarked in Section 9.04.

A method with considerably greater roots into the past is the flip coil, a multiturn coil of wire that is either rotated in the magnetic field at some fixed speed or simply pulled out; you measure the induced ac voltage or the integrated current, respectively. A flip coil is simplicity itself, and it has the elegance of pure electromagnetic theory, but it tends to be a bit bulky and old-fashioned looking.

For measurements of minuscule magnetic fields you can't beat the exotic SQUID (superconducting quantum interference device), a clever arrangement of superconducting junctions that can easily measure a single quantum of magnetic flux (0.2 microgauss-cm^2). A SQUID can be used to measure the magnetic fields set up in your body when you drink a glass of water, for whatever that's worth. These are fancy devices that require a considerable investment in cryogenic hardware, liquid helium, etc., and shouldn't be considered ordinary circuit items.

For precision magnetic-field measurements in the kilogauss range you can't do better than an NMR (nuclear magnetic resonance) magnometer, a device that exploits the precession of nuclear (usually hydrogen) spins in an external magnetic field. This is the physicist's magnometer, and it effortlessly yields values of magnetic field accurate to a part in a million or better. Since the output is a *frequency*, all the precision of frequency/time measurements can be used (more about this later).

Devices known as flux-gate magnetometers and transductors provide yet another way to measure magnetic fields. They work by exciting a piece of ferrite with an ac excitation field, with the response, as modified by the ambient field, being observed.

15.06 Vacuum gauges

The measurement of vacuum presents no great obstacles, which is fortunate, since it is a quantity of great importance in processes such as transistor and IC manufacture, thin-film evaporation, and the preparation of freeze-dried coffee. The basic device here is the Bayard/Alpert ionization gauge, which looks like an inside-out vacuum tube (Fig. 15.18). A hot filament emits electrons that are collected at a positive wire anode. Along the way the electron beam scatters from residual gas molecules, creating positive ions that are

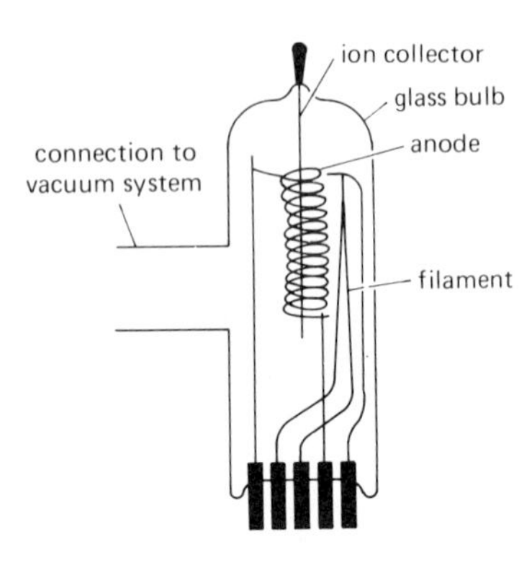

Figure 15.18. Ionization gauge.

collected at a central wire electrode held near ground. The ion current is accurately proportional to the density of gas molecules, i.e., the pressure. Ion gauges are usable at pressures (vacuums!) from about 10^{-3} to 10^{-11}mm Hg (1mm Hg is also known as 1 torr; atmospheric pressure is 760mm Hg). It takes great care to maintain a vacuum of 10^{-10}mm Hg; even a fingerprint on the side of the chamber will frustrate your efforts.

At more mundane levels of vacuum (1mm Hg down to 1μm Hg, which you get with mechanical "roughing" pumps) the popular choice of measurement device is the thermocouple bonded to a small heater; you run some current through the heater, then measure the temperature with the thermocouple. Residual gas cools the contraption, lowering the thermocouple's output voltage. Thermocouple gauges are usually used so that you know when it's safe to turn on the high-vacuum (diffusion or ion) pumps. Granville-Phillips has an improved heat-loss gauge that works on similar principles. It's called a "convectron" gauge and measures from 1μm Hg all the way up to atmospheric pressure.

15.07 Particle detectors

The detection, identification, spectroscopy, and imaging of charged particles and energetic photons (x rays, gamma rays) constitute an essential part of the fields of nuclear and particle physics, as well as numerous fields that make use of radioactivity (medical radiography tracers, forensic science, industrial inspection, etc.). We will treat x-ray and gamma-ray detectors first, then charged-particle detectors.

X-ray and gamma-ray detectors

The classic uranium prospector was a slightly grizzled and shriveled character who went poking around the desert with clicking Geiger counter in hand. The detector situation has now improved considerably. These detectors all have in common the property that they use the energy of an incoming photon to ionize an atom of something, giving off an electron via the photoelectric effect. What they do with the electron depends on the particular detector.

Ionization chamber, proportional counter, Geiger counter. These detectors consist simply of a cylindrical (usually) chamber, typically a few inches in size, with a thin wire running down the center. They are filled with some gas or mixture of gases. There's a thin "window" on one side, made of some material that the desired x rays can penetrate (plastic, beryllium, etc.). The central wire is held at a positive potential and connected to some electronics. Figure 15.19 shows a typical configuration.

When an x ray enters, it ionizes an atom by ejecting a photoelectron, which then loses energy by ionizing gas atoms until it is brought to rest. It turns out that the electron loses about 20 volts of energy per electron-ion pair it creates, so the total free charge left after the photoelectron is brought to rest is proportional to the x ray's initial energy. In an ionization chamber that charge is collected and amplified by a charge-sensitive (integrating) amplifier, just as with a photomultiplier. Thus, the output pulse is proportional to the x ray's energy. The proportional counter works the same way but with the central wire held at a higher voltage, so that electrons drawn toward it cause additional ionization, resulting in a larger signal. This charge-multiplication effect makes proportional counters useful at low x-ray energies (down to a kilovolt, or less) where an ionization counter would be useless. In a Geiger counter the central wire is at a high enough voltage that *any* amount of initial ionization causes a single large (fixed-size) output pulse. This gives a nice large output pulse, but in the process you lose all information about the x ray's energy.

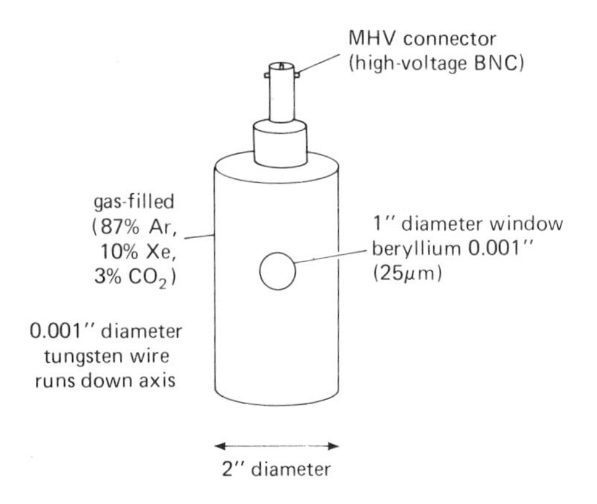

Figure 15.19. Proportional counter.

As you will see in Section 15.16, a clever device known as a pulse-height analyzer lets you convert an input stream of pulses of assorted heights into a histogram. If the pulse heights are a measure of particle energy, you wind up with an energy spectrum! Thus, with a proportional counter (but not with a Geiger counter) you are doing x-ray energy spectroscopy.

These gas-filled counters are usable in the energy range from about 1keV to about 100keV. Proportional counters have an energy resolution of about 15% at 5.9keV (a popular x-ray calibration energy provided by decay of iron 55). They're inexpensive and can be made in very large or very small sizes, but they require a well-regulated power supply (the multiplication rises exponentially with voltage) and are not terribly fast (25,000 count/s is a rough practical maximum counting rate).

Scintillators. Scintillators work by converting the energy of the photoelectron, Compton electron, or electron-positron pair to a pulse of light, which is then detected by an attached photomultiplier. A popular scintillator is crystalline sodium iodide (NaI) doped with thallium. As with proportional counters, the output pulse is proportional to the incoming x-ray (or gamma-ray) energy, which means that you can do spectroscopy, with the help of a pulse-height analyzer (see Section 15.16). Typically, an NaI crystal will give an energy resolution of about 6% at 1.3MeV (a popular gamma-ray calibration energy provided by decay of cobalt 60) and is usable in the energy range of 10keV to many GeV. The light pulse is about 1μs long, making these detectors reasonably fast. NaI crystals come in various sizes up to a few inches; they absorb water, though, so they have to be sealed. Since you must keep light out anyway, they're usually supplied in a metal package with thin aluminum or beryllium entrance window and integral photomultiplier tube.

Plastic (organic) scintillator materials are also popular, being very inexpensive. They have poorer resolution than sodium iodide and are used primarily at energies above 1MeV. Their light pulses are very short, roughly 10ns. Liquid scintillation "cocktails" are routinely used in biological studies. In such applications the material being examined for radioactivity is mixed into the scintillator cocktail, and the whole works is put into a dark chamber with a photomultiplier. You'll see handsome instruments in biology labs that automate the whole process, passing one vial after another through the counting chamber and recording the results.

Solid-state detectors. As with the rest of electronics, the great revolution in x-ray and gamma-ray detection has come about through advances in silicon and germanium technology. "Solid-state" detectors work just like the classic ionization chamber, but with the active volume filled with a nonconducting (intrinsic) semiconductor. An applied potential of about 1000 volts sweeps the ionization out, generating a pulse of charge. In silicon, an electron loses only about 2eV per electron-ion pair created, so many more ions are created for the same incident x-ray energy, as compared with a gas-filled proportional detector, giving better energy resolution through improved statistics. Other subtle effects also contribute to improved performance.

Solid-state detectors come in several varieties, Si(Li), Ge(Li), and intrinsic germanium, or IG (the first two are pronounced "silly" and "jelly"), according to the semiconductor material and dopants used to make it insulating. They are all operated at liquid-nitrogen temperature ($-196°C$), and the lithium-drifted types must be kept cold at all times (if allowed to warm up, they decay, permanently, with about the same time constant as fresh fish). Typical Si(Li) detectors come in diameters from 4mm to 16mm and are usable for x-ray energies from about 1keV to 50keV. Ge(Li) and IG detectors are used at higher energies, 10keV to 10MeV. Good Si(Li) detectors have energy resolutions of 150ev at 5.9keV (2.5%, six to eight times better than proportional counters), and the germanium detectors have energy resolutions of about 1.8keV at 1.3MeV (0.14%).

In order to illustrate what that extra resolution buys you, we bombarded a random hunk of stainless steel with 2MeV protons and measured the x-ray spectrum produced. This is called PIXE (proton-induced x-ray emission), and it is a powerful technique for determining spatially resolved trace-element distribution. Figure 15.20 shows the energy spectrum (made with a pulse-height analyzer), with two x-ray lines visible for each element, at least with the Si(Li) detector. You can see iron, chromium, and nickel. A few additional elements are visible if you expand the lower part of the graph. With the proportional counter, all you get is mush.

Figure 15.21 shows the same kind of comparison for gamma detectors. This time it's an NaI scintillator versus a Ge(Li). (We ran out of steam, so we cribbed this one from the friendly folks at Canberra Industries. Many thanks, Mr. Tench.) As before, solid-state detectors win, hands down, for resolution.

Solid-state detectors have the best energy resolution of all the x-ray and gamma-ray detectors, but they have the disadvantage of a small active area in a large clumsy package (see Fig. 15.22 for an example), relatively slow speed ($50\mu s$ or longer recovery time), high price, and high nuisance value (unless you enjoy being a full-time baby-sitter to a liquid-nitrogen guzzler).

Charged-particle detectors

The detectors we've just described are intended for energetic *photons* (x rays and gamma rays), not particles. Particle detectors have somewhat different incarnations; in addition, charged particles are deflected by electric and magnetic fields according to their charge, mass, and energy, making it much easier to measure particle energies.

Surface-barrier detectors. These germanium and silicon detectors are the analogs of the Ge(Li) and Si(Li) detectors. They don't have to be cooled, which simplifies the packaging enormously. (It also lets you take an occasional vacation!) Surface-barrier detectors are available in diameters from 3mm to 50mm. They are usable at particle energies of 1MeV to hundreds of MeV, and they have energy resolutions of 0.2% to 1% for 5.5MeV alpha particles (a

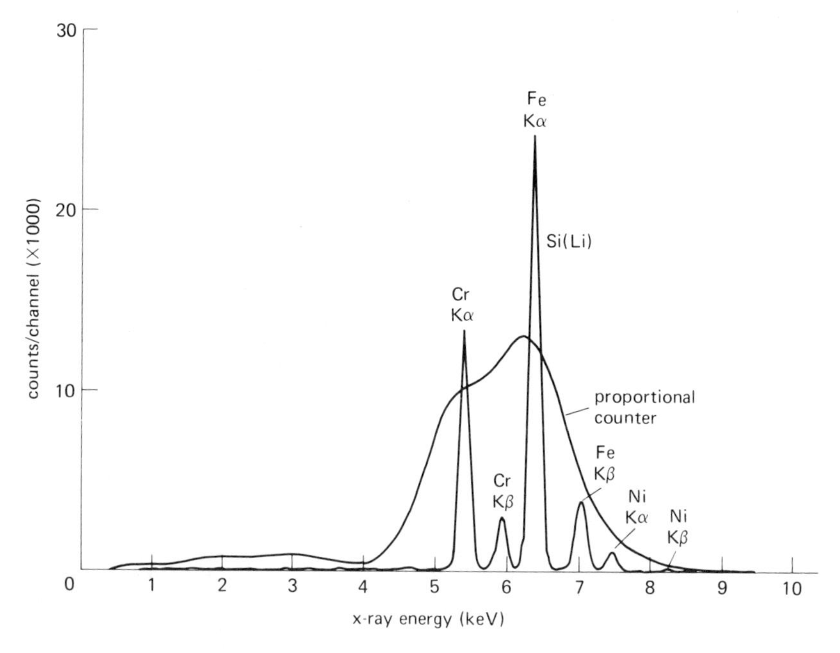

Figure 15.20. An x-ray spectrum from a piece of stainless steel, as seen by an argon proportional counter and a Si(Li) detector.

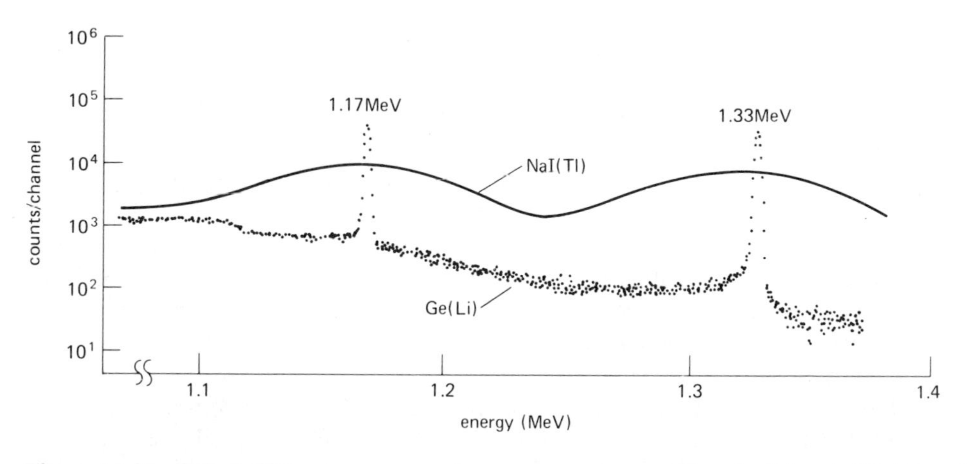

Figure 15.21. Cobalt-60 gamma-ray spectrum, as seen by a sodium iodide scintillator and a Ge(Li) detector. [From Canberra Ge(Li) Detector Systems brochure, Canberra Industries, Inc.]

popular alpha calibration energy provided by decay of americium 241).

□ *Cerenkov detectors.* At very high energies (1GeV and above) even a heavy charged particle can outrun light in material media, giving rise to Cerenkov radiation, a "visible sonic boom." They are used extensively in high-energy physics experiments.

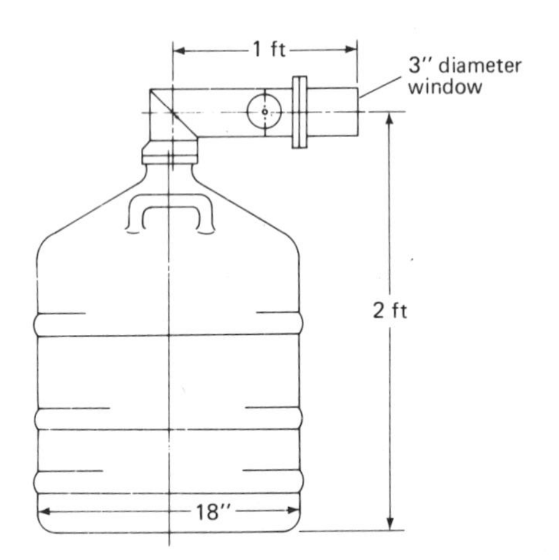

Figure 15.22. Ge(Li) cryostat. (Courtesy of Canberra Industries, Inc.)

□ *Ionization chambers.* The classic gas-filled ionization chamber previously described in connection with x-ray detection can also be used as a detector of energetic charged particles. In its simplest form it consists of a single collecting wire running the length of an argon-filled chamber. Depending on the particle energies involved, the chamber may range from inches to feet in size; variations include the use of multiple collecting wires or plates, and other filling gases.

□ *Shower chambers.* A shower chamber is the electron equivalent of an ionization chamber. An energetic electron enters a box of liquid argon, where it generates a "shower" of charged particles that are subsequently collected at charged plates. High-energy physicists like to call these things "calorimeters."

□ *Scintillation chambers.* A charged particle can be detected with very good energy resolution by using photomultipliers to detect the ultraviolet-rich scintillations caused by the particle's ionized path in a chamber filled with argon or xenon in gas or liquid form. Scintillation chambers are delightfully fast, in contrast with the more leisurely response of ionization and shower chambers.

□ *Drift chambers.* These are the latest rage in high-energy physics, and they are made possible by advances in high-speed on-line computing. They're simple in conception: a box filled with gas at atmospheric pressure (an argon-ethane mixture is typical) and crisscrossed by hundreds of wires with an applied voltage. The box is full of electric fields, and when a charged particle goes in and ionizes the gas, the ions are swept out by the array of wires. You keep track of the signal amplitudes and timing on all the wires (that's where the computer comes in), and from that information you deduce the particle's path. With an applied magnetic field, that also tells you the momentum.

The drift chamber has become the universal imaging charged-particle detector for high-energy physics. It can deliver spatial resolution of 0.2mm or better over a volume large enough for you to climb into.

15.08 Biological and chemical voltage probes

In the biological and chemical sciences there are many examples of measurement wizardry: electrochemical methods such as electrochemistry with ion-specific electrodes, electrophoresis, voltametry, and polarography, as well as techniques like chromatography, IR and visible spectroscopy, NMR, mass spectroscopy, x-ray spectroscopy, nuclear quadruple spectroscopy, ESCA, etc. It is hopeless in a volume this size to attempt any kind of comprehensive catalog of these sophisticated techniques. Furthermore, these techniques can be characterized as less fundamental than the direct physical measurements cataloged earlier in this chapter.

In order to give an idea of the special problems that arise in chemical and biological measurements, we will describe only the simplest sort of measurement: the determination of the potentials generated by a microelectrode (used to explore nerve and muscle signals in biological systems), by an ion-specific electrode (used to measure the concentration of some specific ionic species in solution), and by a voltametric electrochemical probe. As usual, there are some interesting electronic challenges you face, if you want to get anything meaningful out of your measurement.

Microelectrodes

In order to look at the voltage on nerves or in the interiors of cells, it is standard practice to make electrodes that are just a few hundred angstroms in tip diameter (1 angstrom $=10^{-8}$cm, approximately the size of a hydrogen atom). That turns out to be easily done by drawing a glass capillary, then filling it with a conductive solution. You wind up with a nice probe, but with interesting circuit problems arising from the electrode's source impedance of 100MΩ or more. Interference pickup, loading by the circuit, and high-frequency rolloffs of a few hertz due to cable and stray capacitances plague the unwary.

In order to see nerve or muscle signals you want to have decent high-frequency performance, at least out to a few kilohertz, or so (this isn't exactly high frequency in the sense of Chapter 13!). The amplifier must have very high input impedance, and preferably low input noise. In addition, it must be insensitive to common-mode interference.

The circuit in Figure 15.23 represents a good solution. The use of a reference electrode connected near the point of actual measurement keeps interference from appearing as normal (differential) mode signals. The inputs are buffered as close as possible to the actual microelectrode by low-noise FET-input op-amps IC_1 and IC_2, which also bootstrap the guard electrode in order to reduce the effective cable capacitance. Note that the guard is itself shielded. You've got to use FET amplifiers in order to get high input impedance and low input current noise; the particular types shown were chosen for their low input noise voltage (2.5μV pp max 0.1–10Hz), often a problem with FET or MOSFET amplifiers. The pair of buffered signals is applied to the standard differencing amplifier configuration with a low-noise-voltage low-drift op-amp, with 100mV of stable adjustable output offset added via IC_6.

At this point you have an amplifier with a differential gain of 10, suitable noise performance, good common-mode rejection, and low input current (<1pA). However, even with input guarding, the residual input capacitance at the input buffers and at the microelectrode tip will result in poor speed performance. For example, a 100MΩ source impedance driving 20pF has a high-frequency 3dB point of only 80Hz. The solution is active compensation via positive feedback, provided by IC_3 and IC_4 through C_1 and C_2. In practice, you adjust the voltage gain of amplifiers IC_3 and IC_4 for decent high-frequency performance (or transient response), with response to several kilohertz possible.

Ion-specific electrodes

The classic example of an ion-specific electrode is the pH meter, which measures the voltage developed between a reference electrode and a thin-walled glass electrode through which hydrogen ions can diffuse. Once again you're dealing with very high source impedances, although the problems here are less severe than with microelectrodes, because you don't often care about frequency response.

There are more than 20 kinds of ion-specific electrode systems available, e.g., to measure activities of K^+, Na^+, NH_4^+,

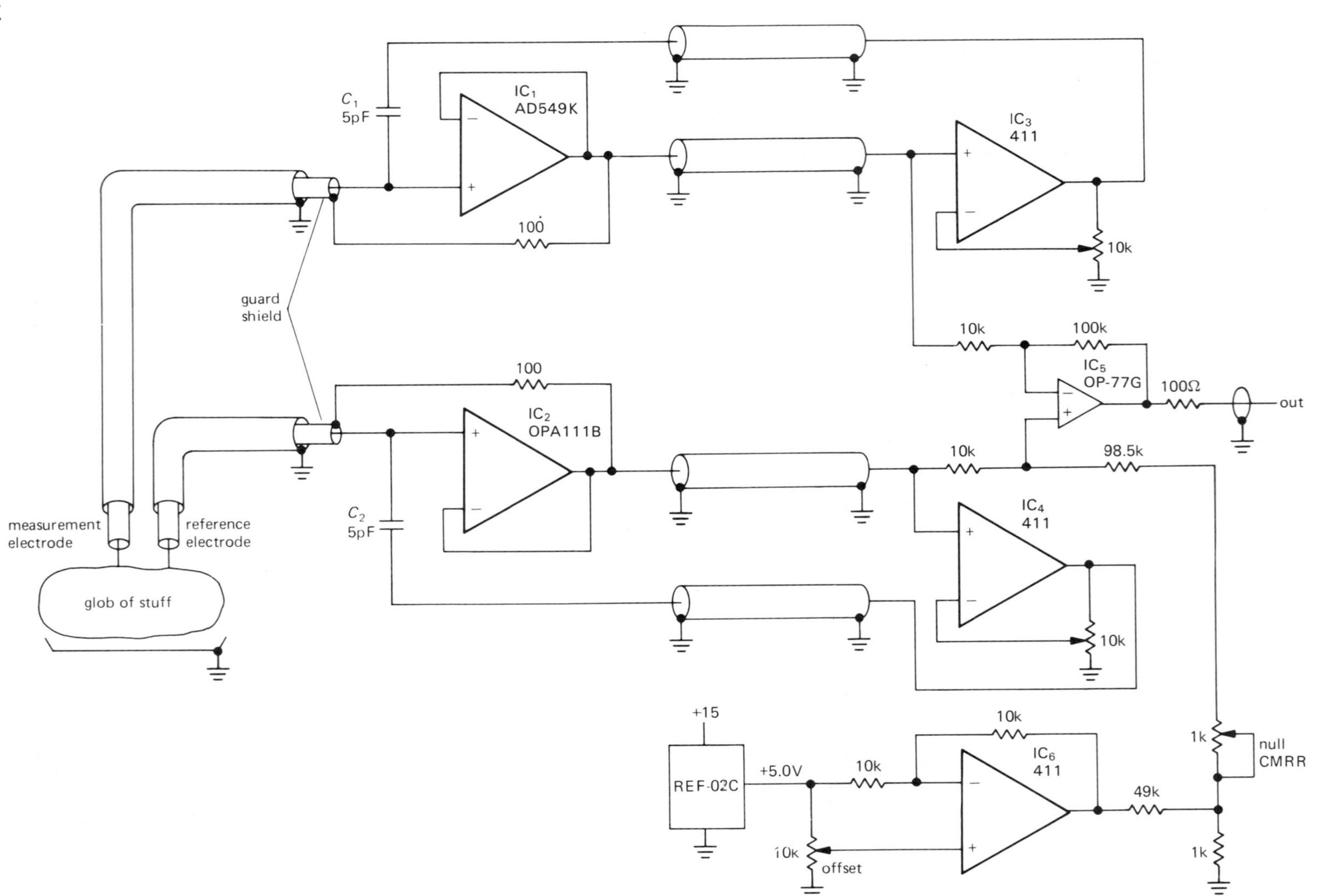

Figure 15.23. Compensating microelectrode amplifier with guarding and reference channel.

CN^-, Hg^{++}, SCN^-, Br^-, Cl^-, F^-, I^-, Ca^{++}, or Cu^{++}. In general, you have two electrodes: a reference electrode, typically silver-coated with silver chloride and immersed in a concentrated solution of potassium chloride that communicates with the solution you want to measure via a porous plug or gel, and an ion-specific electrode, typically consisting of an electrode immersed in a concentrated solution of the ion you're interested in and separated from the solution under test by a membrane that is selectively permeable to the ion of interest. The membrane is commonly an ion-selective glass or an organic liquid containing mobile ion-transporting organic molecules. Your task is to measure a voltage that is in the range of 0 to 2 volts, with an accuracy of a millivolt, while drawing less than 100pA. The situation is complicated by a temperature coefficient of as much as a few percent change in voltage per degree centigrade, which you can attempt to cancel automatically with thermistor-driven compensation circuitry. Conversion from measured ionic activity to concentration requires attention to the total ionic strength of the sample and to the crossover sensitivity of the ion-specific electrode to other ions present. In any case, chemists say you get best results with this sort of black art if you calibrate on some standard solutions just before and after making your measurements. With care, you can see concentrations of 0.1ppm and achieve measurement accuracies of about 1% in solutions of moderate concentration.

Electrochemical measurements

In the area of electrochemistry, it is possible to make very sensitive analytical measurements of the concentrations of specific ions by measuring electrode currents (reaction rates) versus applied voltage in a solution. By scanning the applied voltage, you pass through the potentials at which specific reactions occur, giving rise to steps or peaks. Terms such as cyclic voltametry, polarography, and anodic stripping voltametry (ASV) are used to describe various ways of doing such analytical measurements. Among the most sensitive of these techniques is ASV, which uses a hanging drop mercury (hdm) electrode, a renewable electrode onto which you electroplate at a relatively high potential for a while, then reverse the current and strip off each element sequentially. This technique can detect elements like lead and cadmium at the parts-per-billion level, and it should be considered on a par with other trace-element techniques such as neutron activation, flame spectroscopy, and x-ray and ion microprobes.

The technique of measuring a small current while subjecting a system to a fixed voltage is called a "voltage clamp," and it finds application also in nerve and cell physiology. Nerve membranes have voltage-dependent channels through which specific ions can diffuse, and nerve physiologists like to measure the voltages at which such channels open. Again, voltage clamps are used, this time with microelectrodes.

In preparative electrochemistry, the same techniques are used, but with currents measured in amperes rather than microamperes. Once again the idea is to drive a specific reaction product by applying the right voltage.

Figure 15.24 shows a simple potentiostat (or voltage clamp) circuit. The electrolytic cell consists of an electrode to inject current (the counter electrode), a common return electrode (the working electrode), and a small probe to measure the voltage in the solution near the working electrode (the reference electrode). IC_1 maintains a voltage equal to V_{ref} between the reference and working electrodes by varying the current into the counter electrode appropriately (in measurements of membrane potential, the upper two electrodes would be inside the cell, and the working electrode outside). IC_2 holds the working electrode at virtual ground, converting the current to an output voltage. The range of voltages encountered is typically ±1 volt; currents in the range of 1nA to 1mA are typical of analytical measurements, whereas currents of 1mA to 10 amps are used in preparative electrochemistry.

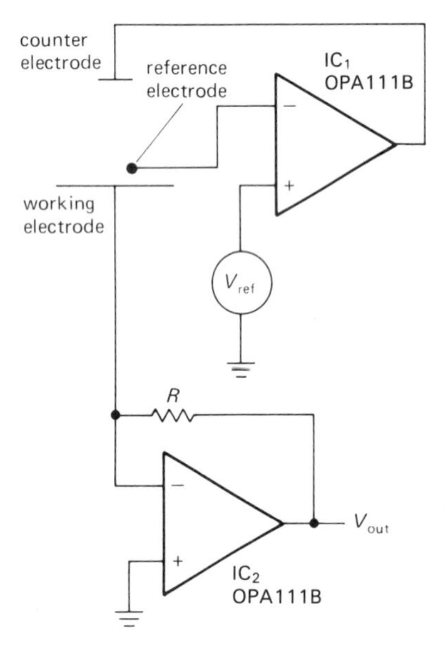

Figure 15.24. Potentiostatic electrochemistry circuit (voltage clamp).

In order to do scanning measurements, V_{ref} would be replaced with a ramp generator. For low-current membrane measurement you would have to shield the input leads carefully, and perhaps use guarding and positive feedback with a capacitor in the manner of Figure 15.23 to maintain some frequency response.

PRECISION STANDARDS AND PRECISION MEASUREMENTS

In Chapter 7 we talked about the circuit methods that are necessary in high-accuracy instruments to maintain small voltage offsets and drifts, e.g., when amplifying very small voltages. We dealt there only with analog electronics, the amplification of continuously varying voltages and currents. For a number of reasons, it turns out that measurements of digital quantities such as frequency, period, and time intervals can be made to far greater precision than any analog measurement. In the following sections, we will explore the accuracy of electronic standards (time, voltage, resistance), and you will see how to make analog measurements of high precision using such standards as references. We will devote the bulk of the discussion to time/frequency measurements, because of their greater inherent precision and because we have already treated precision analog circuitry in some detail in Chapter 7.

□ 15.09 Frequency standards

Let's take a look at the high-stability frequency standards you can get, and then discuss how you set and then maintain their frequency.

□ *Quartz-crystal oscillators*

Back in Sections 5.12–5.19, we described briefly the stabilities you can expect from frequency standards, going from the simplest *RC* relaxation oscillator to the atomic standards based on rubidium and cesium. For any serious timing you wouldn't consider anything less stable than a quartz-crystal oscillator. Fortunately, the cheapest crystal oscillators cost only a few dollars and can deliver stabilities of a few parts per million. For about $50 you can buy a good TCXO (temperature-compensated crystal oscillator), stable to one part per million from 0°C to 50°C. For better performance you need ovenized crystals, with price tags from a couple of hundred dollars to more than $1000. Once you begin talking about stabilities of a few parts per billion, you have to worry also about "aging," the tendency of crystal oscillators to drift in frequency at a more or less constant rate once they are initially "broken in." The 105B from Hewlett-Packard typifies good crystal oscillators with stability of better than 2 parts per billion over the full temperature range and aging rates of less than 0.5 part per billion per day.

Uncompensated crystal oscillators, and even TCXOs, are logical choices as part of a small instrument. The fancier ovenized oscillators are usually rack-mounted standards with an identity all their own.

□ *Atomic standards*

There are three standards in use today: rubidium, cesium, and hydrogen. Rubidium has a microwave absorption at 6,834,682,608Hz, cesium has an absorption at 9,192,631,770Hz, and hydrogen has an absorption at 1,420,405,751.768Hz. A frequency standard based on one of these is considerably more complicated (and expensive) than a good crystal oscillator.

□ *Rubidium.* In the rubidium standard you have a glass bulb with rubidium vapor, heated and contained in a microwave cavity with glass end windows. A rubidium lamp shines through the cavity, with a photocell detecting the transmitted light. Meanwhile, a modulated microwave signal referenced to a stable crystal oscillator is introduced into the cavity. By using lock-in detection (see Section 15.15) of the transmitted light, you can bring the microwave signal exactly to the rubidium resonance frequency, since the optical absorption of the rubidium gas is altered when its microwave resonance is excited. The crystal's frequency is then related in a known way to the rubidium resonance, so it is straightforward to generate a standard frequency like 10MHz. (There are actually several additional complications that we have glossed over.)

Rubidium standards have better stability than ovenized crystal oscillators, although they do exhibit a form of aging. Commercial units are available with stabilities of a few parts in 10^{11} over the full temperature range and long-term stabilities of 1 part in 10^{11} per month. Rubidium standards make sense in a laboratory situation, and you find them at observatories and other places where extremely accurate observations are made. It should by pointed out that a rubidium standard, just like a crystal oscillator, must be calibrated, because changing conditions within the resonance cell affect the frequency at the part-per-billion level.

□ *Cesium.* A cesium standard is practically a small atomic-beam laboratory, in which cesium atoms are launched from an oven into a vacuum chamber, where they pass through spin-state selector magnets and oscillatory electric fields before being detected with a hot-wire ionization detector. As with the rubidium standard, a microwave signal referenced to a stable crystal oscillator is locked to the resonance with feedback from a phase-sensitive detector, and the output frequency is synthesized from the crystal.

Cesium standards aren't small, and they aren't cheap. But they are *primary* standards; you don't have to calibrate them. In fact, by international agreement, cesium *defines* the second: "the duration of exactly 9192631770 periods of the radiation corresponding to the transition between the two hyperfine levels of the ground state of the cesium-133 atom." Cesium clocks are used to keep official time in this country and to calibrate time transmissions (more on this shortly). The cesium clocks used to keep time are elaborate devices, but even commercially built cesium standards keep exceptional time: long-term stability and reproducibility of 3 parts in 10^{12} for the model 5061B from Hewlett-Packard (priced at $32,500).

□ *Hydrogen.* Neutral hydrogen atoms have a hyperfine resonance at about 1420MHz, and in contrast to the situation with the other atomic standards, it is possible to make an actual oscillator with them. As with cesium, you make an atomic beam and run it through magnetic state selectors, then into a Teflon-coated quartz bulb in a microwave cavity. The atoms bounce around inside this "storage bulb" for about 1 second and give off enough radiofrequency energy to sustain an oscillation in the cavity. That makes it easy to lock a crystal oscillator, using PLLs and mixers. You call this object a hydrogen *maser* (microwave amplification by stimulated emission of radiation).

Hydrogen masers are extraordinarily stable over short times (up to a few hours), with stabilities of 1 part in 10^{15}. They have not replaced cesium-beam apparatus for primary timekeeping, however, because the problem of determining the frequency-pulling effect of the cavity has not been solved, and because of long-term drifts caused by the changing properties of the storage-bulb wall surface. R. Vessot at the Smithsonian Astrophysical Observatory (Cambridge, MA) is the world's leader in hydrogen clocks, having built more than two dozen; his timekeepers cost $0.5M apiece.

☐ *Methane laser.* A fourth atomic standard is used at infrared wavelengths, namely the methane-stabilized helium/neon laser. It has a frequency stability comparable to that of the other atomic standards, but at its frequency of 8.85×10^{13}Hz (3.39μm wavelength) it is not a usable radiofrequency standard.

☐ *Recent developments.* The latest research in stable frequency standards involves two promising areas: "trapped ions" and cryogenic hydrogen masers. Proponents of each talk about ultimate stabilities of parts in 10^{18}, if all goes well.

☐ *Calibrating a clock*

Unless you happen to own a cesium-beam standard, you've got to have access to a stable calibration signal to keep your oscillator on frequency. In addition, you may wish to keep accurate absolute time as well as frequency, i.e., you have to set your clock, after you have it running at the right rate. There are several services to help you keep time. On the East Coast of the United States, and in several other areas, you can receive Loran-C, a navigational signal at 100kHz, from which you can determine frequency and time. Loran-C is generated by cesium clocks and is compared with the cesium-beam master clock at the Naval Observatory, which publishes corrections each month. Another time service is WWVB, from the National Institute of Standards and Technology (formerly NBS) in Colorado. This is a 60kHz signal that you can receive most anywhere in the United States. For both these low-frequency transmissions you can synchronize to 1μs or better if you are within range of the "groundwave" signal (a few hundred miles), but ionospheric effects (day/night shifts, solar winds, etc.) make synchronization via the "sky wave" less accurate (10–50μs). A more recent network known as Omega transmits at very low frequencies (around 10kHz) and can be received anywhere, although with an accuracy of only 10μs or so. The geostationary weather satellites known as "GOES" transmit UHF timing signals (at 469MHz) that you can use for time synchronization to a millisecond, if you are in the "footprint" of one of them (all of North and South America).

If you can receive one of these time services, you can compare your oscillator frequency with the real thing. There are nice commercial gadgets that will take care of all the fuss and bother and even generate pretty graphs of the results. It is a bit more difficult to set your clock's time. The most reliable way is to carry it (or some portable clock) to one of the standards, set it, then carry it back. As soon as you get home you make observations of Loran-C, or whatever, to determine the time delay from the transmitter to you. Save that number! (We still remember the magic number 53,211μs for the delay from Loran-C in Nantucket to Harvard's 60 inch telescope dome.) As long as no one builds a new mountain between you and the transmitter, you're all set to tell time.

The recent Global Positioning System (GPS, or "NAVSTAR") is a constellation of 21 satellites, in 12 hour orbits of high inclination, containing on-board atomic clocks. When fully operational, it will permit the determination of time (to 20 nanoseconds) and location (to 10 meters) anywhere on earth, using a self-contained "smart" GPS receiver with a small L-band (1.2GHz and 1.6GHz) doorknob-shaped antenna. The incomplete GPS system has already been used to synchronize clocks worldwide to better than 50ns. The complete GPS system could ultimately be used to achieve time transfer at the 2ns level, if a set of proposed upgrades is adopted.

15.10 Frequency, period, and time-interval measurements

With an accurate reference oscillator and just a small amount of digital electronics, it is disarmingly easy to make frequency and period measurements of high precision.

Frequency

Figure 15.25 shows the basic circuit of a frequency counter. A Schmitt trigger converts the analog input signal to logic levels, at which point it is gated by an accurate 1 second pulse derived from a crystal oscillator. The frequency in hertz is the number of pulses counted by the multidigit BCD counter. It is best to latch the count and reset the counter between counting intervals.

In practice, you would arrange the clock circuit so that shorter or longer intervals can be selected, with a choice of 0.1 second, 1 second, and 10 seconds as a minimum. Also, you can eliminate the 1 second interval between measurements. Additional features might include the following: an adjustable preamp, with selectable trigger point and hysteresis, and perhaps a front-panel output from the discriminator so you can see the trigger point on an oscilloscope; BCD output for readout into a computer or logger; provision for an external oscillator, when a precision standard is available; a manual start/stop input for simple counting (totalization).

□ *Microwave counting.* You can go to frequencies of 3GHz with the digital ICs available today. In particular, GigaBit Logic manufactures a series of astoundingly fast GsAs ripple counters with guaranteed 3GHz clock rate. For higher frequencies you can use heterodyne techniques to mix the microwave input signal down to a directly countable frequency, or you can use a so-called transfer oscillator technique, in which you phase-lock the nth harmonic of a VCO to the input signal, then measure the VCO frequency and multiply the result by n.

□ *The ± 1 count ambiguity.* One disadvantage of this simple frequency counting scheme is that low frequencies cannot be measured to high precision, because of the ± 1 count error. For example, if you were to measure a signal near 10Hz with a gate time of 1 second, your answer would be accurate only to 10%, since the result would be either 9, 10, or 11. You could measure for a longer interval, but it would take a whole day's counting to get the relative accuracy (1 part per million) that you would get in 1 second when measuring

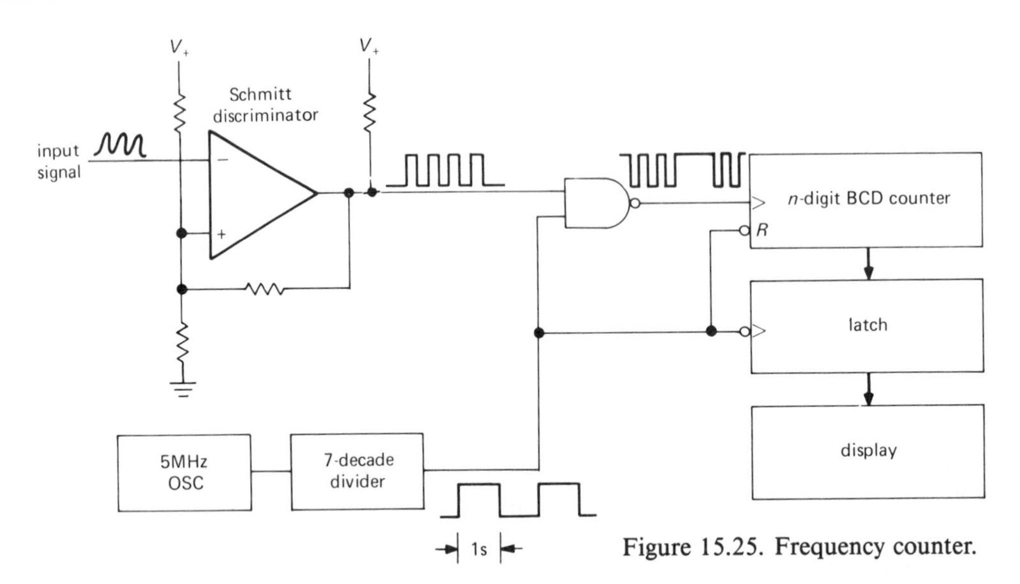

Figure 15.25. Frequency counter.

a 1MHz signal, for example. There are several solutions to this problem: period (or reciprocal) counting, interpolation methods, and phase-locked-loop frequency multiplication techniques. We will deal with the first two in the next sections, since they aren't really direct frequency measurements.

Figure 15.26 shows the PLL "resolution multiplication" technique. A standard phase-locked loop is used to synthesize a frequency of 1000 times the input signal, say, which is then counted as described earlier. The accuracy of this technique is limited by the phase jitter in the PLL phase detector and the loop compensation parameters. For example, if a 100Hz signal is multiplied by 1000 and counted for 1 second, and the jitter in the phase detector is 1% of a cycle (3.6°), or 100μs, then the *accuracy* of the measurement will be 1 part in 10,000, even though the *resolution* is 1 part in 100,000.

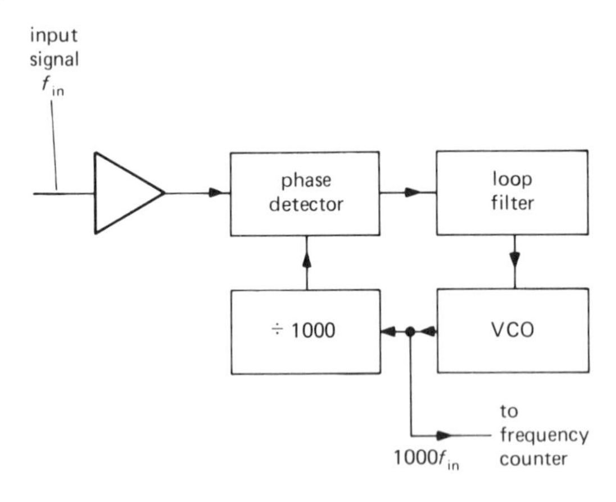

Figure 15.26. PLL resolution multiplication for low-frequency counting.

We will now mention two other ways to improve frequency-measurement accuracy: period measurement and interpolated time-interval measurement.

Period (reciprocal counting)

A good way to handle the problem of resolution when measuring low frequencies is to turn things around and use the input signal (or some subdivision of it) to gate the clock. Figure 15.27 shows the standard configuration for such a period counter. The number of periods measured is normally switch-selectable to some power of 10 (1, 10, 100, etc.). You will usually pick a number of periods such that the measurement takes a convenient length of time, typically a second, giving an answer to about seven significant figures. Of course,

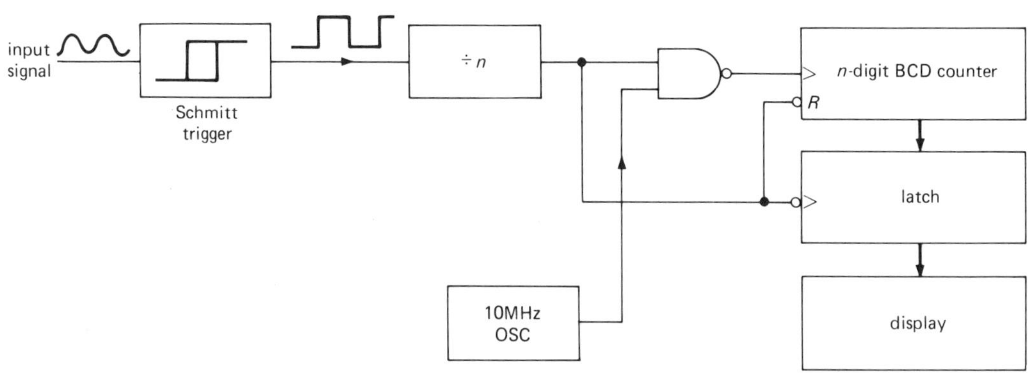

Figure 15.27. Period counter.

that answer is in units of time, not frequency, so you have to take the reciprocal to recover the frequency. Luckily, you soon won't even have to know how to divide, since modern counters use dedicated microprocessors to do the period-to-frequency conversion.

Note that the accuracy of period measurement is critically dependent on stable triggering and requires good signal/noise ratios. Figure 15.28 indicates the problem here.

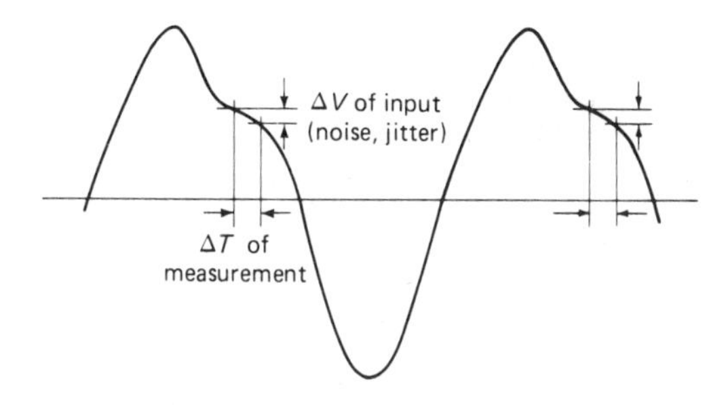

Figure 15.28

The main advantage of reciprocal counting is that you get a constant resolution $\Delta f/f$ for a given length of measurement, independent of the input frequency. The graph in Figure 15.29 compares the resolution of frequency and reciprocal frequency (period) measurements of duration 1 second, using a 10MHz clock. The period graph should actually be somewhat jagged, since you normally have to live with the closest power of 10 for the number of periods averaged. Even this restriction is evaporating, with the advent of "smart" microprocessor counters (e.g., the low-cost Hewlett-Packard 5315A) that have *continuous* adjustment of gate time; they know how many periods were averaged, and they divide the answer accordingly. They also oblige you by switching from period to frequency mode for input frequencies greater than the clock frequency, in order to get optimum resolution at any input frequency.

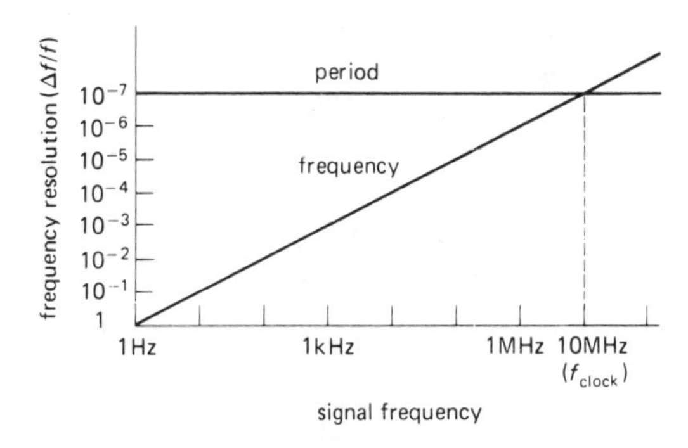

Figure 15.29. Fractional resolution of frequency and period counting.

A second advantage of reciprocal frequency measurement is the ability to control externally the time at which the gating occurs. This is advantageous if you wish to measure the frequency of a tone burst, for example, a situation in which a simple frequency counter would give incorrect results, since its internally controlled gating interval might not coincide with the burst. With period counting you can gate the measurement externally and can even make a set of measurements at various points along the burst, given the generally superior resolution of period measurement.

You might wonder if it is possible to do better than the "uncertainty principle" resolution limit of $\Delta f/f \approx 1/f_{\text{clock}}T$ (period measurement) or $1/f_{\text{input}}T$ (frequency counting), for the relative error $\Delta f/f$ of a frequency measurement made by counting for time T. The answer is yes. In fact, several clever schemes have been invented. We will discuss them in the next subsection (time-interval measurement), but just to show that it can be done, we've drawn in Figure 15.30 a method of measuring the frequency of a 1MHz oscillator to a resolution of 1 part in 10^{12} in 1 second of measurement time. The unknown oscillator is mixed with a stable reference offset slightly from 1.0MHz, say 1.000001MHz (this could be synthesized with a PLL). The mixer output contains the sum and difference frequencies. After low-pass filtering, you've got a 1Hz signal that is the difference between the two oscillators, and that can be easily measured with a period counter to one part per million in 1 second. In other words, you've measured 1MHz to 1μHz in 1 second.

This technique assumes that you have extremely good signal/noise ratios, and in practice you would have to worry about low-frequency noise, settling time of the filter, etc., so you might not do better than 1 part in 10^{10} in 1 second. Still, this is considerably better than simple frequency (or period) counting. In addition, the *accuracy* will be less than the resolution unless the reference oscillator is also accurate to 1 part in 10^{12} (possible, but not easy, with today's technology). You can think of this scheme as a way of comparing the relative frequencies of two oscillators, if you prefer.

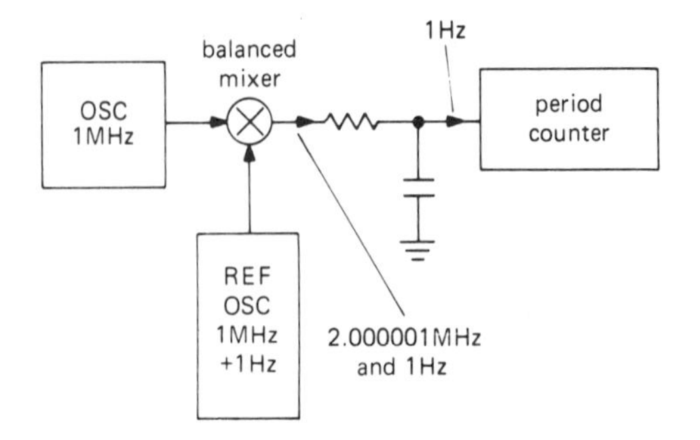

Figure 15.30. High-resolution frequency comparison.

Time-interval measurement

With a trivial change in the circuitry of the period counter, you can measure the time interval between two events. Figure 15.31 shows how. In practice it may be better to add a synchronizer, as shown in the second circuit, to prevent the generation of runt pulses. The best resolution is obviously obtained by running the oscillator at the highest possible frequency, and commercial counters use local oscillator references as high as 500MHz, phase-locked from a stable crystal at 5MHz or 10MHz. With a 500MHz reference, you have a resolution of 2ns.

As we hinted earlier, there are ways to beat the reciprocal frequency resolution limit when making time-interval measurements, essentially by exploiting the extra information you have about the position of zero crossings of the input signal relative to the reference. The oscillator comparison scheme we showed earlier really exploited that same information, but in a more subtle way. For these schemes you must have a clean signal with very low noise level. There are two interpolation methods in use in commercial instruments: linear interpolation and vernier interpolation.

□ *Linear interpolation.* Suppose you wish to measure the time interval between the start and stop pulses in Figure 15.32. You begin by measuring the number of

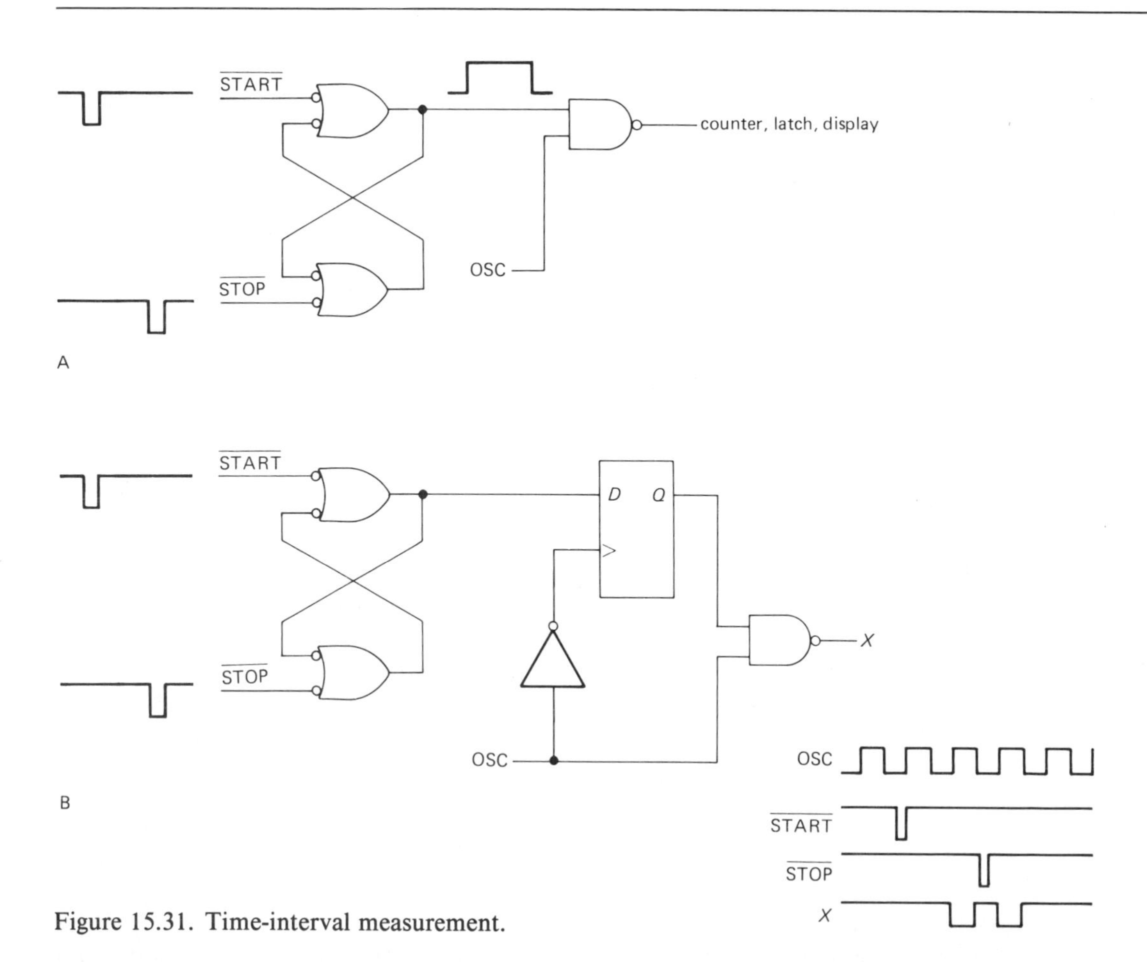

Figure 15.31. Time-interval measurement.

clock impulses, n, during the interval τ, as shown (with a synchronizer you would start and stop with the first clock pulse after the respective input signal, as shown). To improve the resolution of the measurement, all you need to know are the time intervals T_0 and T_1, the time elapsed from the occurrence of each input pulse to the next clock pulse. Assuming you're already running the system clock at the highest rate you can conveniently count, you have to expand those unknown intervals in order to measure them. A variation of the dual-slope principle works here: Integrate charge onto a capacitor during those intervals, then ramp down at a small fraction (say 1/1000th) of the charging rate, thus expanding the unknown time intervals by a factor of 1000. During those expanded intervals, you count the system clock, generating counts of n_0 and n_1. The unknown time interval is therefore given by

$$\tau = T_{\text{clock}} \times (n + n_0/1000 - n_1/1000)$$

with the obvious improvement in resolution. The ultimate accuracy of this method is limited by the accuracy of the interpolators and the system clock. An example of this kind of instrument is the Hewlett-Packard 5334B counter, which displays 9 digits (frequency or time) per second of gate time.

□ ***Vernier interpolation.*** Vernier interpolation is a digital technique that lets you

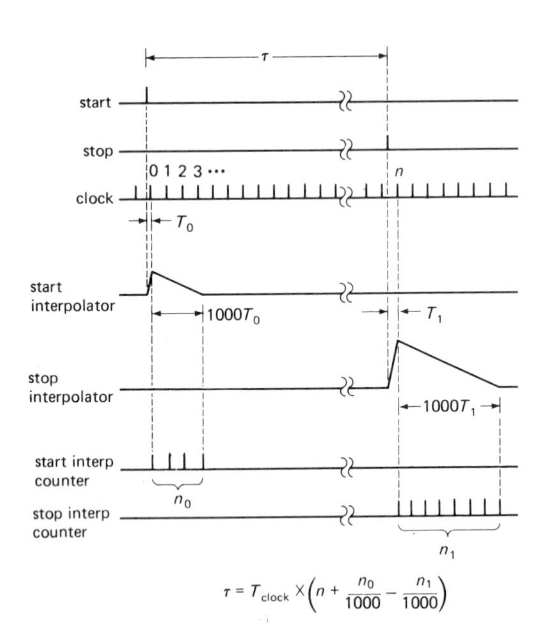

Figure 15.32. Linear interpolation (time-interval measurement).

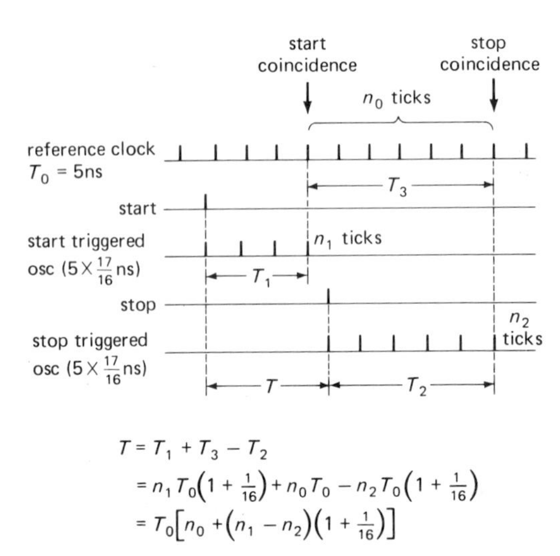

$$
\begin{aligned}
T &= T_1 + T_3 - T_2 \\
&= n_1 T_0\left(1 + \tfrac{1}{16}\right) + n_0 T_0 - n_2 T_0\left(1 + \tfrac{1}{16}\right) \\
&= T_0\left[n_0 + (n_1 - n_2)\left(1 + \tfrac{1}{16}\right)\right]
\end{aligned}
$$

Figure 15.33. Vernier interpolation (time-interval measurement).

find out where in the clock cycle the input pulse occurred. Figure 15.33 shows the method schematically. There are three clocks involved: The master reference clock runs continuously, with a period T_0 of, say, 5ns; the input START pulse triggers a second oscillator with a period greater than the reference by a factor $1+1/n$ (we've set n =16 for this example); the input STOP pulse triggers a third oscillator of the same period as the other triggered oscillator. Fast circuitry then looks for coincidences between the triggered oscillators and the master clock, while counting the number (n_1, n_2) of ticks of each before coincidence. The arithmetic is shown in the figure; the net result is to determine to within $1/n$th of a master clock pulse the duration between START and STOP.

The Hewlett-Packard 5370B uses this technique, with T_0 =5ns and n =256. The result is time-interval resolution of 20ps. The technique can clearly be used for period measurements, since period is simply the time interval for one cycle of an input wave. When used this way, the counter just mentioned determines frequency to 11-digit resolution in 1 second!

□ *Time-interval averaging.* There is a third way to improve the resolution of a time-interval measurement, namely by repeating the measurement many times and taking the average. The ±1 count ambiguity gets averaged that way, and the result converts to the true interval, provided only that the repetition rate of START pulses is not commensurate with the master clock. Some counters include a "jittered clock" to make sure this doesn't happen.

Spectrum analysis

A powerful technique that must be mentioned in connection with the measurement of frequency is *spectrum analysis*, looking at signals in the frequency domain. Spectrum analyzers can measure frequency (and in fact they are very useful when you need to know the frequency of a weak signal in the presence of other stronger signals), but in addition they can do a lot more. We will talk about them in Section 15.18.

□ 15.11 Voltage and resistance standards and measurements

As we hinted earlier, analog standards and measurements do not have anything like the precision we have just been talking about. Here you're lucky to get accuracy of a part per million. The analog standards are voltage and resistance; from them you can determine current, if need be.

The traditional standard of voltage is the Weston cell, an electrochemical device with reproducible output voltage, intended for use as a reference only (no more than $10\mu A$, and preferably no current at all, should be drawn from it). Its terminal voltage is 1.018636 volts at 20°C. Unfortunately, Weston cells are fussy gadgets. They have to be maintained at a precise temperature because of their large temperature coefficient ($40\mu V/°C$, far worse than good IC voltage references) and even larger sensitivity to temperature gradients (the individual "limbs" of the cell have tempcos of about $350\mu V/°C$). Standard cells are carefully maintained by the National Institute of Standards and Technology for comparison with secondary standards. Nowadays there are very stable solid-state references with controllable output voltage. They can be used to transfer a measurement from a finicky standard cell to an actual measurement situation. Typical specifications are 10 ppm stability for a month after calibration and 30 ppm stability in a year.

To make an accurate voltage measurement, you use precision voltage dividers (known as Kelvin-Varley dividers), available with linearities in the 0.1ppm range. The divider is used to generate a precise fraction of the unknown voltage, for comparison with the voltage standard. Accurate null detectors and instruments for compensation of wiring resistance are available for this task. Routine calibrations at accuracies of just a few parts per million are possible.

Recently a measurement based on a superconducting Josephson junction has replaced the standard cell as the definition of voltage. With care it is possible to measure voltages reproducibly to a few parts in 10^{10}. The method has the pleasant simplicity of requiring only a measurement of a frequency and knowledge of the physical constants h (Planck's constant) and e (the electron charge). Although Josephson junction technology has traditionally been considered too complicated for use as a voltage standard, the situation is changing: The National Institute of Standards and Technology (NIST, formerly NBS) has developed an affordable chip containing 19,000 series junctions, which can go to 10 volts and beyond. Serious users of voltage standards can now have their own precision Josephson standard, although at a cost of $100k. If recent breakthroughs in high-temperature superconductivity can be harnessed to make voltage standards, Josephson voltage references could become accessible to every laboratory.

As with voltage, standards of resistance are carefully maintained by NIST. By using such standards in a Wheatstone bridge circuit, you can calibrate a secondary standard and maintain accuracies of a few parts per million.

We should point out some of the limitations that prevent analog measurements from having the same high accuracy as time measurements. Analog measurements rely on physical properties such as electrochemical potentials, breakdown voltages, and resistances, and these all vary with temperature and time. Interfering effects such as Johnson and $1/f$ noise, leakage currents, and thermoelectric potentials (thermocouple effect) complicate any measurement. To measure a voltage with precision comparable to state-of-the-art time or frequency measurements would require a measurement accuracy of a picovolt at a voltage of 1 volt. Think of this not as an indictment of analog methods but merely as a celebration of the incredible precision attainable in the time/frequency domain. And, in practice, choose time/frequency transducers and measurements, rather than voltage/resistance measurements, whenever possible.

BANDWIDTH-NARROWING TECHNIQUES

15.12 The problem of signal-to-noise ratio

Up to this point we have been talking about the various experimental quantities that can be detected, how you might measure them, and what sort of tradeoffs you face. As luck would have it, the signals you often want to measure are buried in noise or interference, frequently to the extent that you can't even see them on an oscilloscope. Even when external noise isn't a problem, the statistics of the signal itself may make detection difficult, as, for example, when counting nuclear disintegrations from a weak source, with only a few counts detected per minute. Finally, even when the signal is detectable, you may wish to improve the detected signal strength in order to make a more accurate measurement. In all these cases some tricks are needed to improve the signal/noise ratio; as you will see, they all amount to a narrowing of the detection bandwidth in order to preserve the desired signal while reducing the total amount of (broadband) noise accepted.

The first thing you might be tempted to try when thinking of reducing the bandwidth of a measurement is to hang a simple low-pass filter on the output, in order to average out the noise. There are cases where that therapy will work, but most of the time it will do very little good, for a couple of reasons. First, the signal itself may have some high frequencies in it, or it may be centered at some high frequency. Second, even if the signal is in fact slowly varying or static, you invariably have to contend with the reality that the density of noise signal usually has a $1/f$ character, so as you squeeze the bandwidth down toward dc you gain very little. Electronic and physical systems are twitchy, so to speak.

In practice, there are a few basic techniques of bandwidth narrowing that are in widespread use. They go under names like signal averaging, transient averaging, boxcar integration, multichannel scaling, pulse-height analysis, lock-in detection, and phase-sensitive detection. All these methods assume that you have a repetitive signal; that's no real problem, since there is almost always a way to force the signal to be periodic, assuming it isn't already. Let's see what is going on.

15.13 Signal averaging and multichannel averaging

By forming a cumulative sum of a repetitive signal versus time, you can improve the signal/noise ratio enormously. This usually goes under the heading of "signal averaging," and it is often applied to analog signals. We will consider first what may seem to be an artificial situation, namely a signal consisting of pulses whose rate is proportional to the amplitude of some sought-after waveform versus time. We begin with this example because it makes our calculations easier. In reality, it isn't even an artificial situation, since it is the rule when using pulse-counting electronics such as particle detectors or photomultipliers at low light levels.

Multichannel scalers

We begin with multichannel scaling because it typifies all these techniques and, in addition, is easy to understand and quantify. The multichannel scaler (MCS) is a piece of hardware that contains a set of memory registers (typically 1024 or more), each of which can store a number up to 1 million (20 bits binary or 24 bits BCD) or so. The MCS accepts pulses (or continuous voltages, as will be described later) as its input; in addition, it accepts either a channel-advance signal (a pulse) or a parallel multibit channel address. Each time there is an input pulse, the MCS increments the count in the memory channel currently being addressed. Additional inputs let you reset the address to 0, clear the memory, etc.

To use an MCS you need a signal that repeats itself at some interval. Let's suppose for the time being that the phenomenon you're observing is itself periodic, with period T; although this is not the case most of the time (you usually have to coax the experiment into periodicity), there are good examples in the real world of strictly periodic phenomena, e.g., the light output of a pulsar. Let's suppose that the input consists of pulses, with rate proportional to the signal plus a large background rate of noise pulses, i.e., pulses randomly distributed in time (again, realistic for pulsars, where the actual signal is swamped by light from the night sky). By sending timing pulses to the channel advance and reset inputs, we arrange to sweep the MCS repetitively through its 1024 channels once every T seconds, accumulating additional input (signal plus background) counts into the memory channels each sweep. As time goes on, the signal will keep adding counts to the same subgroup of channels, with the background noise adding counts in all channels, because the sweep through the entire set of channels is timed to coincide with the signal's periodicity. Thus the signal keeps adding on top of itself, the accumulated sum getting larger after each repetition.

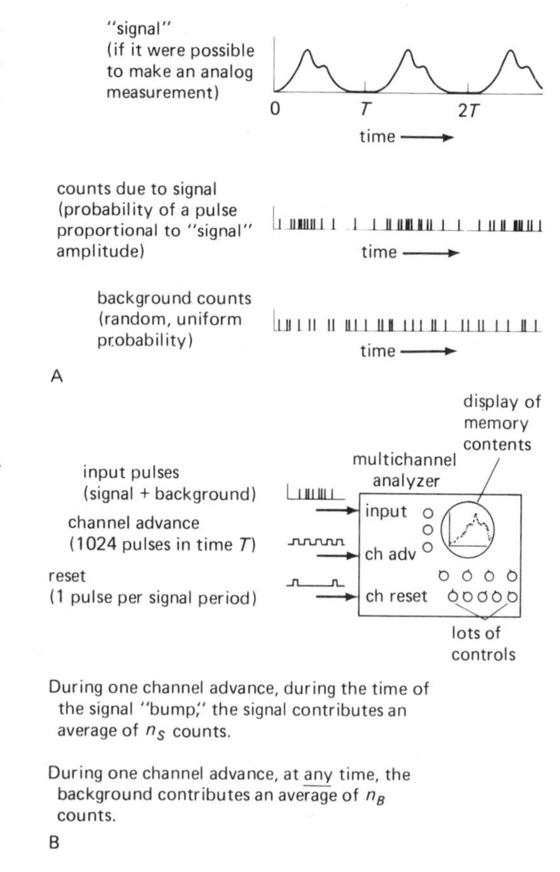

Figure 15.34. Multichannel signal averaging (pulse input).

Signal-to-noise computation

Let's see what happens. To be specific, let the background pulse rate have an average value that contributes n_b pulses per channel each sweep, with the signal contributing an additional n_s pulses into the channel where its peak lies (Fig. 15.34). Let's give ourselves a poor signal/background ratio, i.e., $n_s \ll n_b$, meaning that most of the counts added during each sweep through the memory are contributed by background, rather than signal. Now, when the memory contents are graphed, the signal should be recognizable as a bump above the background. You might think the criterion is that the number of signal counts in a channel with signal should be comparable with the number of counts contributed to that channel by the background noise. That would be wrong, since the *average* value contributed by noise is quite irrelevant; all that matters is the level of *fluctuations* of that average value about the mean.

Thus, a poor input signal/noise ratio is actually characterized by $n_s \ll \sqrt{n_b}$, meaning that in one sweep the signal will not be recognizable above the "noise" consisting of an undulating graph of accumulated random background pulses. For purposes of computation, let's let $n_s = 10$ and $n_b = 1000$. Therefore, in one sweep an initially cleared MCS will acquire an average of 1000 counts in each channel, with an additional 10 counts in the channels where the signal peaked. Since the fluctuations in the channel totals equal about 31 (square root of 1000), the actual signal bump is left pretty much buried in the noise after only one sweep. But after 1000 sweeps, say, the average count in any channel is about 1,000,000, with fluctuations of 1000. The channels where the signal peaks have an additional 10,000 counts (1000 sweeps × 10 counts/sweep), for a signal/noise ratio of 10. In other words, the signal has emerged from the background.

Example: Mössbauer resonance

Figure 15.35 shows the results of just such an analysis, in this case a Mössbauer resonance signal consisting of six dips in the transmission of an enriched iron-57 foil to gamma radiation from a cobalt-57 radioactive source. In this case $n_b = 0.4$ and $n_s = 0.1$, approximately, for a situation of poor signal/noise ratio. The Mössbauer signal is totally swamped by noise even after 10 or 100 sweeps; it becomes visible only after 1000 sweeps or so. The results are shown after 1000, 10,000, and 100,000 sweeps, with each graph scaled to keep this signal size the same. Note the rise of the "baseline" caused by the steady background, as well as the nice enhancement of SNR with time.

It is easy to see by what factor the ratio of signal amplitude to background fluctuation ("noise") increases as time goes on. The signal amplitude increases proportional to t; the average background count ("baseline") also increases proportional

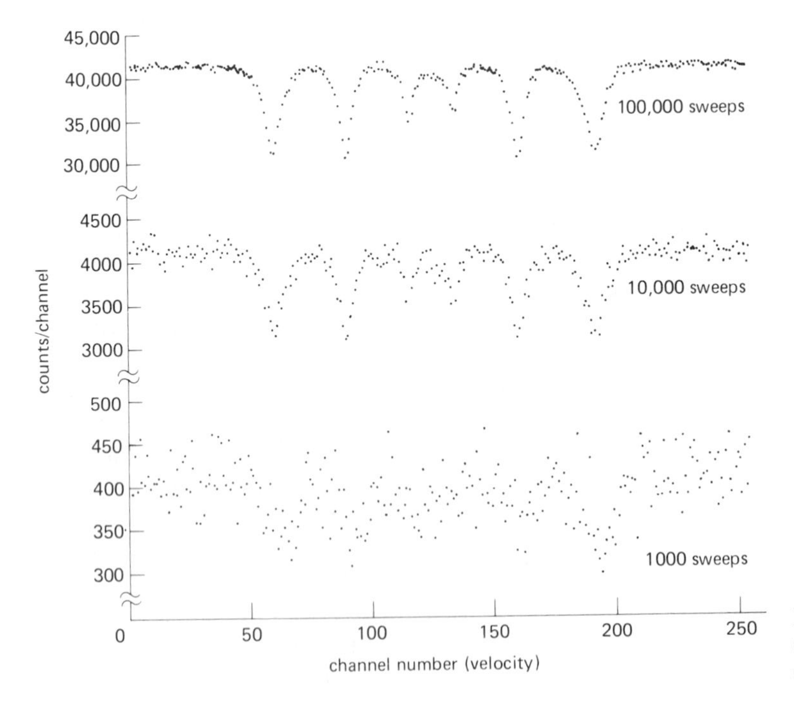

Figure 15.35. Mössbauer absorption spectrum, showing effect of signal averaging.

to t, but the *fluctuations* in the background count ("noise") rise only proportional to the square root of t. Therefore, the ratio between signal and fluctuations in background increases as t divided by the square root of t. In other words, the signal-to-noise ratio improves in proportion to the square root of time.

Multichannel analysis of analog signals (signal averaging)

You can play the same game with analog signals by simply using a voltage-to-frequency converter at the input. Commercial MCSs often provide the electronics for you, giving you a choice of analog or pulse input modes. In this form you often hear these gadgets called signal averagers or transient averagers. One company (TMC) called theirs a "CAT" (computer of averaged transients), and the name has stuck, in some circles at least.

It is possible to make a completely analog MCS by using a set of integrators to store the accumulated signal. A simpler device, known as a boxcar integrator, is an analog signal averager with a single "sliding channel." With the enormous reductions in digital memory prices that have taken place in the last decade, such analog signal averagers are becoming impractical, except perhaps for specialized applications.

Multichannel analysis as bandwidth narrowing

We suggested at the beginning of this discussion that there was an equivalence between the magical SNR-reduction methods and a reduction in effective measurement bandwidth. It is not hard to see how that goes in this case. Imagine another (interfering) signal added into the input, but with periodicity T' slightly different from the desired signal of period T. After just a few sweeps, its signal will also begin to accumulate, causing trouble. But wait – as time goes on, its "bump" will gradually drift along through the channels, successively contributing counts through all the channels. It will have drifted all the way around through all the channels once after a time.

$$t = 1/\Delta f$$

where Δf is the frequency difference $1/T - 1/T'$ between the desired signal and the interfering signal.

EXERCISE 15.1

Derive this result.

In other words, by accumulating data for a time t (as given in the preceding equation), the interfering signal has been spread equally through all the channels. Another way to say the same thing is that the measurement's bandwidth is reduced roughly to

$$\Delta f = 1/t$$

after accumulating data for time t. By running for a long time, you reduce the bandwidth and exclude nearby interfering signals! In fact, you also exclude most of the noise, since it is spread evenly in frequency. Viewed in this light, the effect of multichannel analysis is to narrow the accepted bandwidth, thereby accepting the signal power but squeezing down the amount of noise power.

Let's see how the calculation goes. After time t, the bandwidth is narrowed to $\Delta f = 1/t$. If the noise power density is p_n watts per hertz, and the signal power P_s stays within the measurement bandwidth, then the SNR after time t is

$$\mathrm{SNR} = 10\log(P_s t/p_n)$$

The signal amplitude improves proportionally to the square root of t (3dB for each doubling of t), just as we found in the analysis we did earlier by considering the number of counts per channel and its fluctuations.

15.14 Making a signal periodic

We mentioned initially that all signal-averaging schemes require a signal that repeats many times in order to realize significant reduction in signal/noise ratio. Since most measurements don't involve intrinsically periodic quantities, it is usually necessary to force the signal to repeat. There are many ways to do this, depending on the particular measurement. It is probably easiest to give a few examples, rather than attempt to set down rules.

A measurable quantity that depends on some external parameter can easily be made periodic – just vary the external parameter. In NMR (nuclear magnetic resonance) the resonance frequency varies linearly with the applied field, so it is standard to modulate the current in a small additional magnet winding. In Mössbauer studies you vary the source velocity. In quadrupole resonance you can sweep the oscillator.

In other cases an effect may have its own well-defined transient, but allow external triggering. A classic example is the pulse of depolarization in a nerve fiber. In order to generate a clean graph of the waveform of such a pulse, you can simply trigger the nerve with an externally applied voltage pulse, starting the MCS sweep at the same time (or even "anticipating" the trigger by starting the sweep, then triggering the nerve with a delayed pulse); in this case you would pick a repetition period long enough so that the nerve has fully recovered before the next pulse. This last case illustrates graphically the importance of a repeatable phenomenon as fodder for signal averaging; if the frog whose leg is twitching chances to expire, your experiment is over, whatever the signal/noise ratio!

It should be pointed out that cases where the phenomenon you're measuring has its own well-defined periodicity may in fact be the most difficult to work with, since you have to know the periodicity precisely. The graph of the "light curve" (brightness versus time) in Figure 15.36 is an example. We made this curve by using an MCS on the output of a photomultiplier stationed at the focus of a 60 inch telescope, run

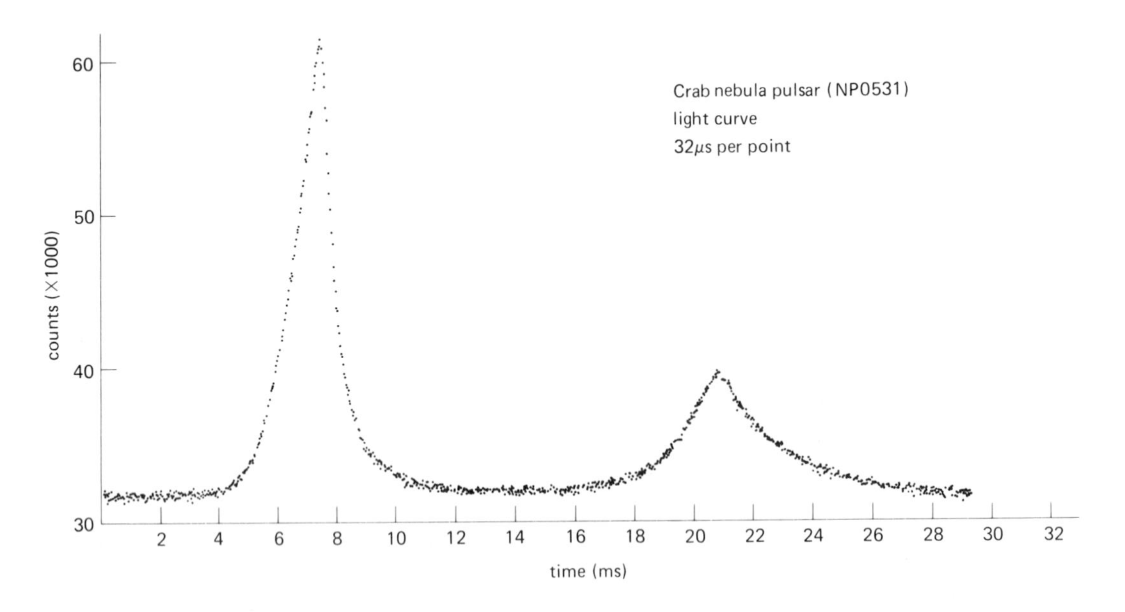

Figure 15.36. Crab nebula pulsar brightness versus time (light curve).

exactly in synchronism with the pulsar's rotation. Even with that size telescope it required an average of approximately 5 million sweeps to generate such a clean curve, since the average number of detected photons for each entire pulsar pulse was about 1. With such a short period, that puts enormous accuracy requirements on the MCS channel-advance circuitry, in this case requiring clocks of part-per-billion stability and frequent adjustment of the clock rate to compensate for the earth's motion.

It is worth saying again that the essence of signal averaging is a reduction in bandwidth, gained by running an experiment for a long period of time. The bottom line here is the total length of the experiment; the particular rate of scanning, or modulation, is usually not important, as long as it takes you far enough from the $1/f$ noise present near dc. You can think of the modulation as simply shifting the signal you wish to measure from dc up to the modulating frequency. The effect of the long data accumulation is then to center an effective bandwidth $\Delta f = 1/T$ at f_{mod}, rather than at dc.

15.15 Lock-in detection

This is a method of considerable subtlety. In order to understand the method, it is necessary to take a short detour into the phase detector, a subject we first took up in Section 9.27.

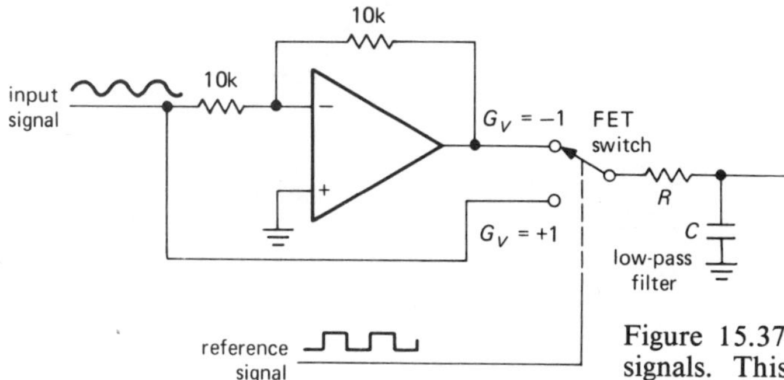

Figure 15.37. Phase detector for linear input signals. This scheme is used in the monolithic AD630.

Phase detectors

In Section 9.27 we described phase detectors that produce an output voltage proportional to the phase difference between two digital (logic-level) signals. For purposes of lock-in detection, you need to know about linear phase detectors, since you are nearly always dealing with analog voltage levels.

The basic circuit is shown in Figure 15.37. An analog signal passes through a linear amplifier whose gain is reversed by a square-wave "reference" signal controlling a FET switch. The output signal passes through a low-pass filter, RC. That's all there is to it. Let's see what you can do with it.

□ *Phase-detector output.* To analyze the phase-detector operation, let's assume we apply a signal

$$E_s \cos(\omega t + \phi)$$

to such a phase detector, whose reference signal is a square wave with transitions at the zeros of $\sin \omega t$, i.e., at $t = 0, \pi/\omega, 2\pi/\omega$, etc. Let us further assume that we average the output, V_{out}, by passing it through a low-pass filter whose time constant is longer than one period:

$$\tau = RC \gg T = 2\pi/\omega$$

Then the low-pass filter output is

$$\langle E_s \cos(\omega t + \phi)\rangle\big|_0^{\pi/\omega} - \langle E_s \cos(\omega t + \phi)\rangle\big|_{\pi/\omega}^{2\pi/\omega}$$

where the brackets represent averages, and the minus sign comes from the gain reversal over alternate half cycles of V_{ref}. As an exercise, you can show that

$$\langle V_{\mathrm{out}} \rangle = -(2E_s/\pi)\sin\phi$$

EXERCISE 15.2

Perform the indicated averages by explicit integration to obtain the preceding result for unity gain.

Our result shows that the averaged output, *for an input signal of the same frequency as the reference signal*, is proportional to the amplitude of V_s and sinusoidal in the relative phase.

We need one more result before going on: What is the output voltage for an input signal whose frequency is close to (but not equal to) the reference signal? This is easy, since in the preceding equations the quantity ϕ now varies slowly, at the difference frequency:

$$\cos(\omega + \Delta\omega)t = \cos(\omega t + \phi)$$
$$\text{with } \phi = t\Delta\omega$$

giving an output signal that is a slow sinusoid:

$$V_{\mathrm{out}} = (2E_s/\pi)\sin(\Delta\omega)t$$

which will pass through the low-pass filter relatively unscathed if $\Delta\omega < 1/\tau = 1/RC$ and will be heavily attenuated if $\Delta\omega > 1/\tau$.

The lock-in method

Now the so-called lock-in (or phase-sensitive) amplifier should make sense. First you make a weak signal periodic, as we've discussed, typically at a frequency in the neighborhood of 100Hz. The weak signal, contaminated by noise, is amplified and phase-detected relative to the modulating signal. Look at Figure 15.38. You need an experiment with two "knobs" on it, one for fast modulation in order to do phase detection and one for a slow sweep through the interesting features of the signal (in NMR, for example, the fast modulation might be a small 100Hz modulation of the magnetic field, and the slow modulation might be a frequency sweep 10 minutes in duration through the resonance). The phase shifter is adjusted to give maximum output signal, and the low-pass filter is set for a time constant long enough to give good signal/noise ratio. The low-pass-filter rolloff sets the bandwidth, so a 1Hz rolloff, for example, gives you sensitivity to spurious signals and noise only within 1Hz of the desired signal. The bandwidth also determines how fast you can adjust the "slow modulation," since now you must not sweep through any features of the signal faster than the filter can respond. People use time constants of fractions of a second up to tens of seconds and often do the slow modulation with a geared-down

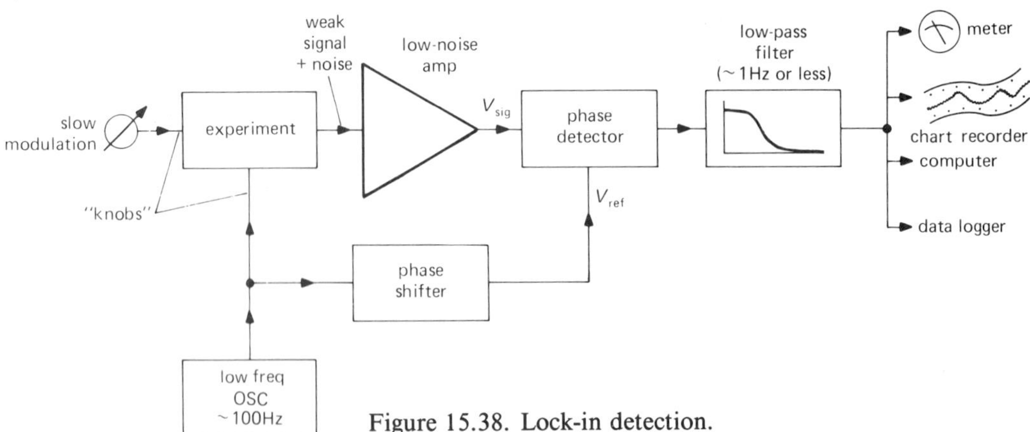

Figure 15.38. Lock-in detection.

clock motor turning an actual knob on something!

Note that lock-in detection amounts to bandwidth narrowing again, with the bandwidth set by the post-detection low-pass filter. As with signal averaging, the effect of the modulation is to center the signal at the fast modulation frequency, rather than at dc, in order to get away from $1/f$ noise (flicker noise, drifts, and the like).

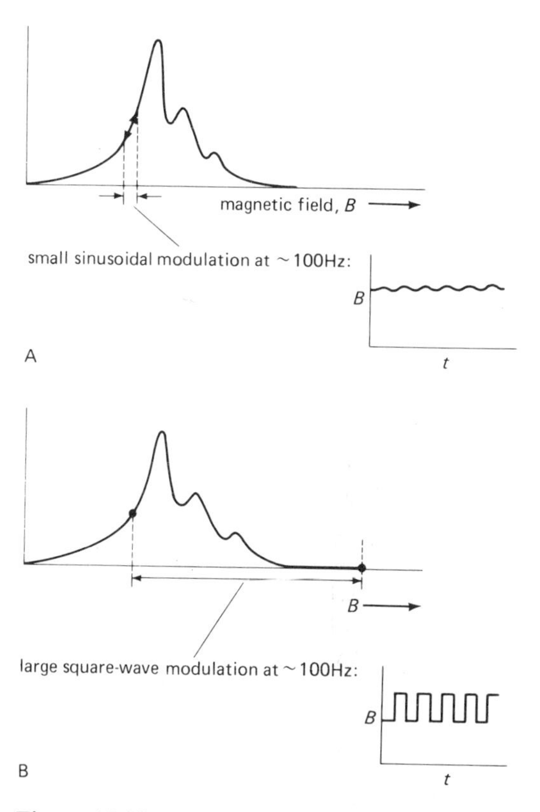

Figure 15.39. Lock-in modulation methods.
A. Small sinusoid.
B. Large square wave.

Two methods of "fast modulation"

There are some ways to do the fast modulation: The modulation waveform can be either a very small sine wave or a very large square wave compared with the features of the sought-after signal (line shape versus magnetic field, for example, in NMR), as sketched in Figure 15.39. In the first case the output signal from the phase-sensitive detector is proportional to the *slope* of the line shape (i.e., its derivative), whereas in the second case it is proportional to the line shape itself (providing there aren't any other lines out at the other endpoint of the modulation waveform). This is the reason all those simple NMR resonance lines come out looking like dispersion curves (Fig. 15.40).

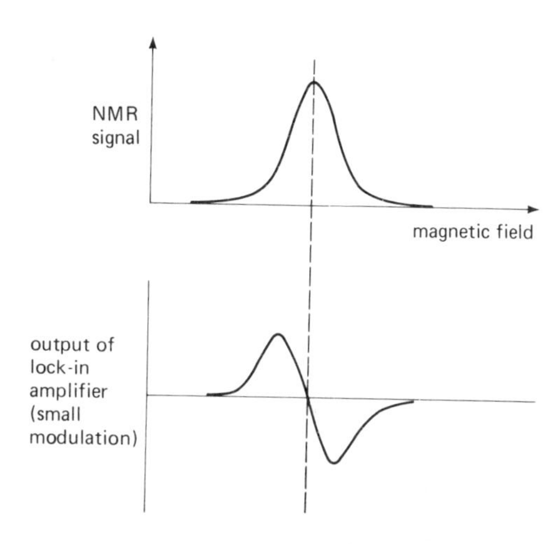

Figure 15.40. Line shape differentiation resulting from lock-in detection.

For large-shift square-wave modulation there's a clever method for suppressing modulation feedthrough, in cases where that is a problem. Figure 15.41 shows the modulation waveform. The offsets above and below the central value kill the signal, causing an on/off modulation of the signal at *twice* the fundamental of the modulating waveform. This is a method for use in special cases only; don't get carried away by the beauty of it all!

Large-amplitude square-wave modulation is a favorite with those dealing in infrared astronomy, where the telescope secondary mirrors are rocked to switch the image back and forth on an infrared source.

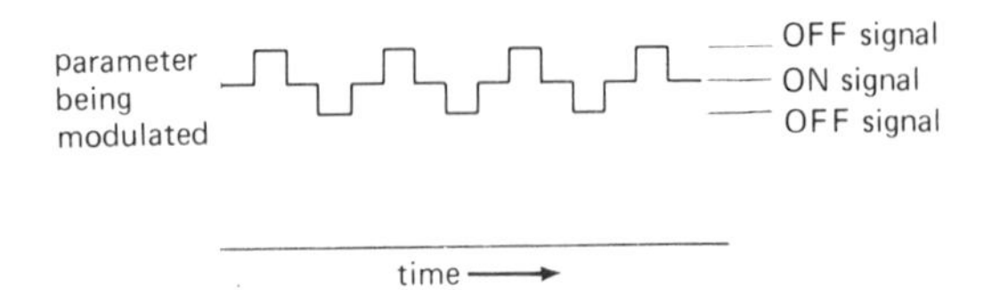

Figure 15.41. Modulation scheme for suppressing modulation feedthrough.

It is also popular in radioastronomy, where it's called a Dicke switch.

Commercial lock-in amplifiers have a variable-frequency modulating source and tracking filter, a switchable time-constant post-detection filter, a good low-noise wide-dynamic-range amplifier (you wouldn't be using lock-in detection if you weren't having noise problems), and a nice linear phase detector. They also let you use an external source of modulation. There's a knob that adjusts the phase shift, so you can maximize the detected signal. The whole item comes packed in a handsome cabinet, with a meter to read output signal. Typically these things cost a few thousand dollars and are manufactured by companies like EG&G Princeton Applied Research, Ithaco, and Stanford Research Systems. Board-level components are made by Evans Electronics, among others.

In order to illustrate the power of lock-in detection, we usually set up a small demonstration for our students. We use a lock-in to modulate a small LED of the kind used for panel indicators, with a modulation rate of a kilohertz or so. The current is very low, and you can hardly see the LED glowing in normal room light. Six feet away a phototransistor looks in the general direction of the LED, with its output fed to the lock-in. With the room lights out, there's a tiny signal from the phototransistor at the modulating frequency (mixed with plenty of noise), and the lock-in easily detects it, using a time constant of a few seconds. Then we turn the room lights on (fluorescent), at which point the signal from the phototransistor becomes just a huge messy 120Hz waveform, jumping in amplitude by 50dB or more. The situation looks hopeless on the oscilloscope, but the lock-in just sits there, unperturbed, calmly detecting the same LED signal at the same level. You can check that it's really working by sticking your hand in between the LED and the detector. It's darned impressive.

15.16 Pulse-height analysis

A pulse-height analyzer (PHA) is a simple extension of the multichannel scaler principle, and it is a very important instrument in nuclear and radiation physics. The idea is simplicity itself: Pulses with a range of amplitudes are input to a peak-detector/ADC circuit that converts the relative pulse height to a channel address. A multichannel scaler then increments the contents of the selected address. The result is a graph that is a histogram of pulse heights. That's all there is to it.

The enormous utility of pulse-height analyzers stems from the fact that many detectors of charged particles, x rays, and gamma rays have output pulse sizes proportional to the energy of the radiation detected (e.g., proportional counters, solid-state detectors, surface-barrier detectors, and scintillators, as we discussed in Section 15.07). Thus a pulse-height analyzer converts the detector's output to an energy spectrum.

Pulse-height analyzers used to be designed as dedicated hardware devices, with buckets of ICs and discrete components. Nowadays the standard method is to use an off-the-shelf microcomputer, preceded by a fast pulse-input ADC. That way you can build in all sorts of useful computational routines, e.g., background subtraction, energy calibration and line identification, disk and tape storage, and on-line control of the experiment. We have an apparatus that scans a proton microbeam over a specimen in a two-dimensional raster pattern, detects the emitted x rays, sorts them by chemical element, and stores a picture of the distribution of each element in the sample, all the while letting you view the x-ray spectrum and images as the picture accumulates. The whole operation is handled by a pulse-height analyzer that doesn't realize that it's really a computer.

There is an interesting subtlety involving the ADC front end of a pulse-height analyzer. It turns out that you can't use something like a successive-approximation A/D converter, in spite of its superior speed, because you wouldn't get exact equality of channel widths, with the disastrous effect of producing a lumpy baseline from a smooth continuum of input radiation. All PHAs use a so-called Wilkinson converter, a variation on single-slope conversion whereby an input pulse charges a capacitor, which is then discharged by a constant current while a fast counter (200MHz is typical) counts up the address. This has the disadvantage of giving an analyzer "dead time" that depends on the height of the last pulse, but it gives absolute equality of channel widths.

Most pulse-height analyzers provide inputs so that you can use them as multichannel scalers. Why shouldn't they? All the electronics are already there. Some big names in pulse-height analyzers are Canberra, EG&G, Nuclear Data, and Tracor-Northern.

15.17 Time-to-amplitude converters

In nuclear physics it is often important to know the distribution of decay times of some short-lived particle. This turns out to be easy to measure, by simply hooking a time-to-amplitude converter (TAC) in front of a pulse-height analyzer. The TAC starts a ramp when it receives a pulse at one input and stops it when it receives a pulse at a second input, discharging the ramp and generating an output pulse

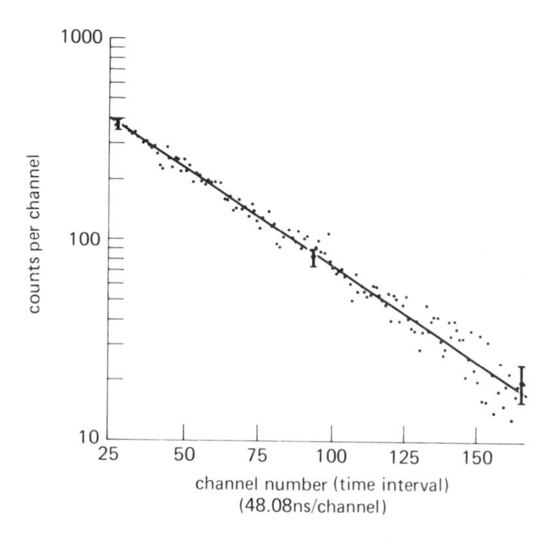

Figure 15.42. Muon lifetime measurement from time-interval spectrum (TAC + PHA).

proportional to the time interval between pulses. It is possible to build these things with resolution down in the picoseconds. Figure 15.42 shows a measurement of the muon lifetime made by a student by timing the delay between the capture of a cosmic-ray muon in a scintillator and its subsequent decay. Each event creates a flash of light, and a TAC is used to convert the intervals into pulses. A cosmic-ray muon decayed in this student's apparatus once a minute on the average, so he accumulated data for 18 days to determine a lifetime of 2.198 $\pm$ 0.02μs (accepted value is 2.197134 $\pm$ 0.00008μs). Note the use of log-lin axes to plot data that should be an exponential, and the systematic shift of $n^{1/2}$ (counting) error bars. The line plotted is the decay according to the accepted value, $n(t) = n_0 \exp(-t/\tau)$.

SPECTRUM ANALYSIS AND FOURIER TRANSFORMS

15.18 Spectrum analyzers

An instrument of considerable utility, particularly in radiofrequency work, is the spectrum analyzer. These devices generate an xy oscilloscope display, with y representing signal strength (usually logarithmic, i.e., in decibels), but with x representing frequency. In other words, a spectrum analyzer lets you look in the *frequency domain*, plotting the amount of input signal versus its frequency. You can think of it as a Fourier decomposition of the input waveform (if you know about such things), or as the response you would get as you tuned the dial of a broadrange high-performance (wide dynamic range, stable, sensitive) receiver through its frequency range. This ability can be very handy when analyzing modulated signals, looking for intermodulation products or distortion, analyzing noise and drift, trying to make accurate frequency measurements on weak signals in the presence of stronger signals, and making a host of other measurements.

Spectrum analyzers come in two basic varieties: swept-tuned and real-time. Swept analyzers are the most common variety, and they work as shown in Figure 15.43. What you have is basically a superheterodyne receiver (see Section 13.16), with a local oscillator (LO) that can be swept by an internally generated ramp waveform. As the LO is swept through its range of frequencies, different input frequencies are successively mixed to pass through the IF amplifier and filter. For example, suppose you have a spectrum analyzer with an IF of 200MHz and an LO that can sweep from 200MHz to 300MHz. When the LO is at 210MHz, input signals at 10MHz (± the IF filter bandwidth) pass through to the detector and produce vertical deflection on the scope. Signals at 410MHz (an "image" frequency) would also pass through, which is the reason for the low-pass filter at the input. At any given time, input frequencies 200MHz lower than the LO are detected.

Real spectrum analyzers allow lots of flexibility as to sweep range, center frequency, filter bandwidth, display scales, etc. Typical input frequency ranges go from hertz to gigahertz, with selectable bandwidths ranging from hertz to megahertz. A range of 10MHz to 22GHz is popular, with resolution bandwidths of 10Hz to 3MHz. In addition, sophisticated spectrum analyzers have convenience features such as absolute amplitude calibration, storage of spectra to prevent flicker during sweeping, additional storage for comparison and normalization, and display of digital information on the screen. Fancy spectrum analyzers let you analyze phase versus frequency, generate frequency markers, program the operation via the IEEE-488 bus, include tracking oscillators (for increased dynamic range), make precise frequency measurements of features in the spectrum, generate tracking noise voltages for system stimulus, and even do signal averaging (particularly useful for noisy signals).

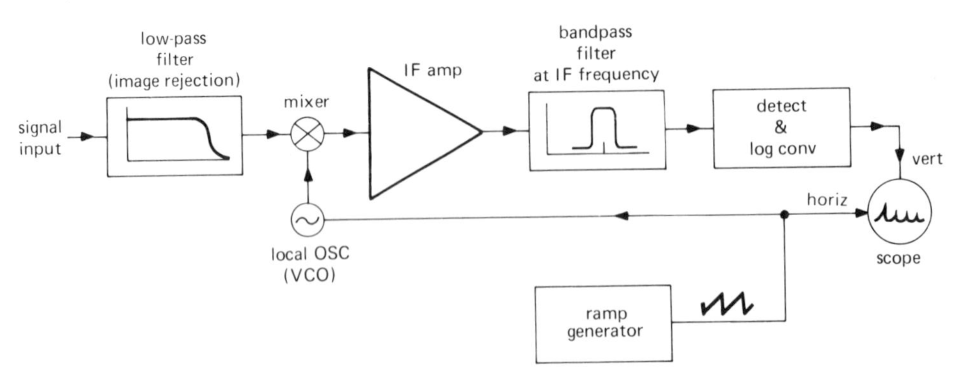

Figure 15.43. Swept-LO spectrum analyzer.

Note that this sort of swept spectrum analyzer looks at only one frequency at a time and generates a complete spectrum by sweeping in time. This can be a real disadvantage, since you can't look at transient events. In addition, when scanning with narrow bandwidth, the sweep rate must be kept slow. Finally, only a small portion of the input signal is being used at any one time.

These disadvantages of swept spectrum analysis are remedied in real-time spectrum analyzers. Again, there are several approaches. The clumsy method employs a set of narrow filters to look at a range of frequencies simultaneously. More recently, sophisticated analyzers based on digital Fourier analysis (in particular, the famous Cooley-Tukey fast Fourier transform, FFT for short) are becoming popular. These instruments convert the analog input signal (after mixing, etc.) to numbers, using a fast analog/digital converter. Then a special-purpose computer turns the crank on the FFT, generating a digital frequency spectrum. Since this method looks at all frequencies simultaneously, it has excellent sensitivity and speed, and it can be used for analysis of transients. It is particularly good for low-frequency signals, where swept analyzers are too slow. In addition, it can perform correlations between signals. Since the data comes out in digital form, it is natural to apply the full power of signal averaging, a feature available in some commercial instruments.

Note that these digital spectrum analyzers, being limited by computational speed, have *much* less bandwidth than the radiofrequency analog types (swept-LO or filter-bank). For example, the popular HP 3561A goes from 125μHz to 100kHz. You can, of course, use it to look at a 100kHz band centered at some higher frequency, by translating that band down in frequency with heterodyne techniques.

A clever real-time spectrum analyzer can also be constructed using the so-called chirp/Z transform. In this method a dispersive filter (delay time proportional to frequency) replaces the IF bandpass filter in the swept-LO analyzer (Fig. 15.43). By matching the LO sweep rate to the filter's dispersion, you get an output that superficially resembles the swept analyzer output, namely a linear scan of frequency versus time during each sweep. However, in contrast to the swept-LO analyzer, this scheme gathers signals from the entire band of frequencies continuously. Another interesting technique for real-time spectral analysis is the Bragg cell (or "acousto-optic spectrometer"), in which the IF signal is used to generate acoustic waves in a transparent crystal. These deformations diffract a laser beam, generating a real-time display of the frequency spectrum as light intensity versus position. An array of photodetectors completes the analyzer output. Bragg-cell spectrometers are used in radio astronomy. A typical unit has 2GHz instantaneous bandwidth, analyzed into 16,000 channels of 125kHz bandwidth each. When choosing a spectrum analyzer type, be sure to consider trade-offs among bandwidth, resolution, linearity, and dynamic range.

Figure 15.44 shows the sort of radiofrequency spectra that endear spectrum analyzers to people who earn their living above 1MHz. The first four spectra show oscillators: A is just a pure sine-wave oscillator, B is distorted (as indicated by its harmonics), C has noise sidebands, and D has some frequency instability (drifting or residual FM). You can measure amplifier intermodulation products, as in E, where second-, third-, and fourth-order intermodulation frequencies are visible in the output of an amplifier driven by a "two-tone" test signal consisting of pure sine waves at frequencies f_1 and f_2. Finally, in F you can see the uncouth behavior of a double-balanced mixer; there is feedthrough of both the LO and input signal, as well as distortion terms ($f_{LO} \pm 2f_{sig}, f_{LO} \pm 3f_{sig}$). This last spectrum may actually indicate quite respectable mixer performance, depending on the vertical scale shown. Spectrum analyzers are designed with enormous dynamic range (internally generated distortion products are typically down by 70dB or more; with a "tracking preselector" they're down by 100dB) so that you can see the failings of even a very good circuit.

The last graph G in Figure 15.44 shows what happens when you sweep the LO too fast in a swept analyzer. If the sweep causes a signal to pass through the filter bandwidth Δf in a time shorter than $\Delta t \approx 1/\Delta f$, it will be broadened, roughly to $\Delta f' \approx 1/\Delta t$.

15.19 Off-line spectrum analysis

The fast Fourier transform applied to digitized data from an experiment provides a very powerful method of signal analysis, particularly the recognition of weak signals of well-defined periodicity buried in interfering signals or noise, or the recognition of vibrations or oscillatory modes. For instance, we have used the FFT to search for pulsars, perform audio analysis, enhance the resolution of astronomical images (speckle imaging), and look for signals from intelligent life in space (SETI). In the last experiment, a GaAs FET amplifier connected to a receiving dish 84 feet in diameter drives a heterodyne receiver, with 400kHz of bandwidth analyzed (in real time) into 8 million simultaneous 0.05Hz channels. Our digital spectrum analyzer has 20,000 ICs and a half million solder joints (all done by hand!) and can detect narrowband signals 60dB below receiver noise in a 20 second integration. This corresponds to a radio flux of less than 1 nanowatt total over the entire earth's disk!

SELF-EXPLANATORY CIRCUITS

15.20 Circuit ideas

In Figure 15.45 we've collected some circuits that are useful in measurement and control applications.

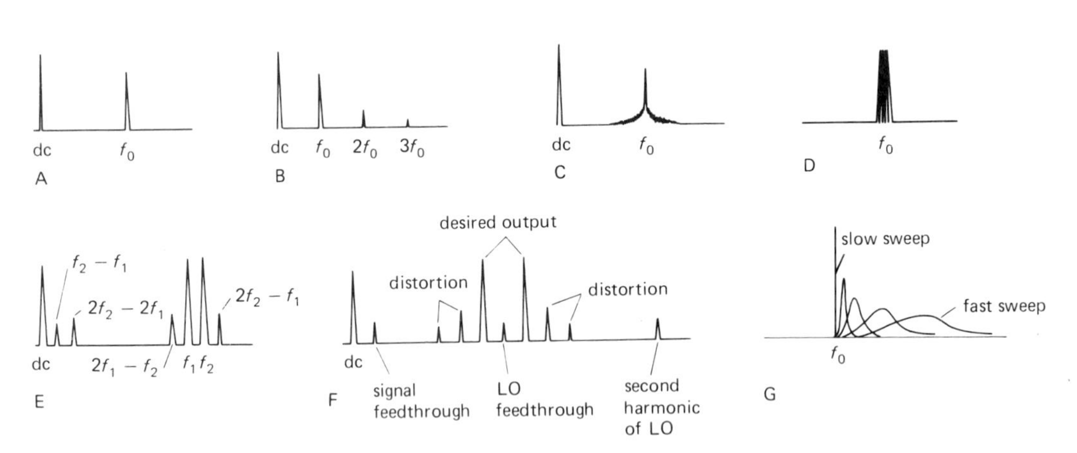

Figure 15.44. Spectrum analyzer displays.

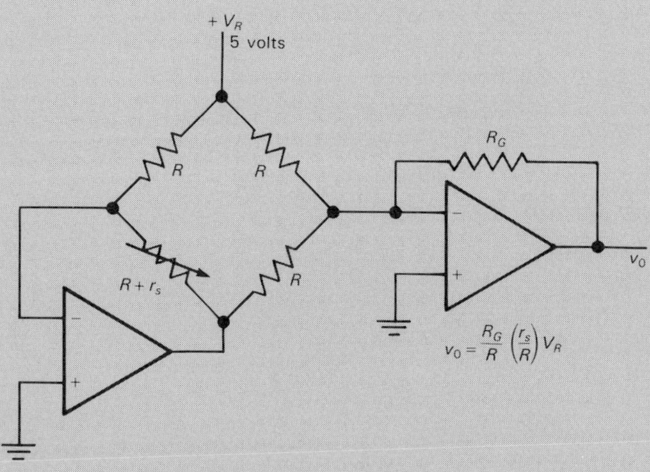

A. linearizes bridge response, with minimum V_{OS} effect; note that many bridges (e.g., semiconductor strain gauge) have high TCs of R.

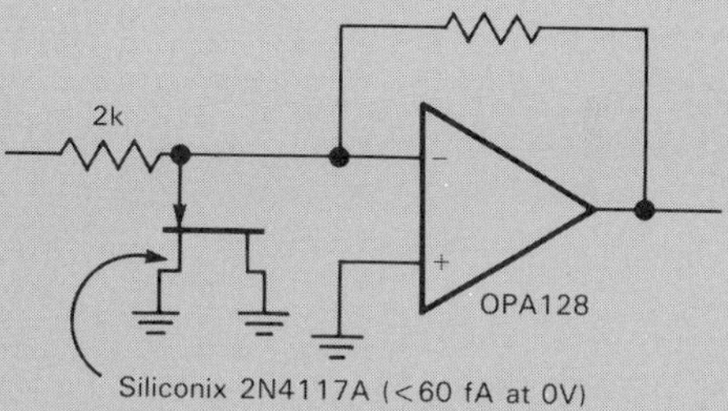

C. charge amplifier input protection with low leakage

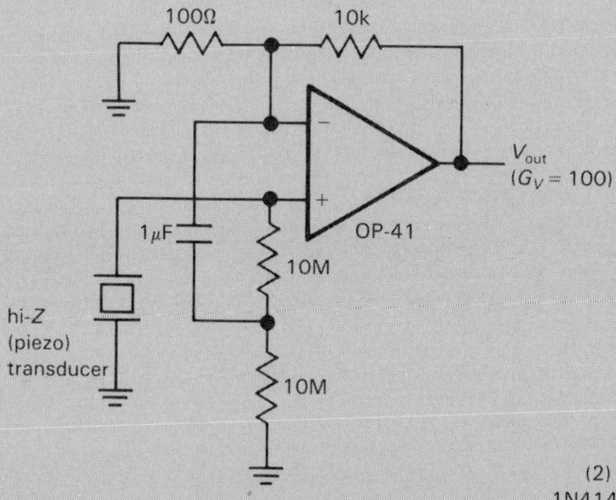

B. high-Z_{in} bootstrapped amplifier for piezo transducer

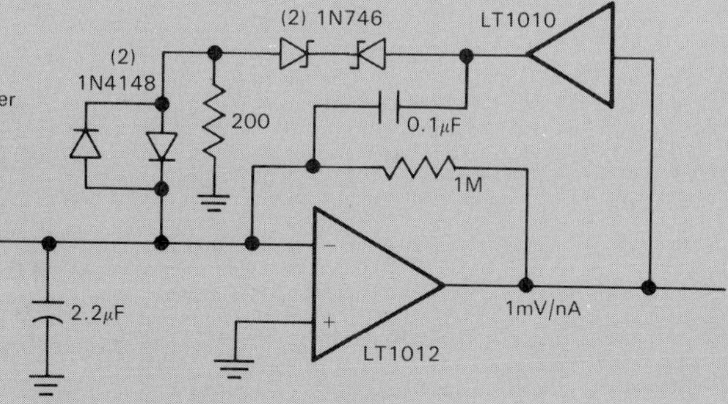

Figure 15.45

D. I-to-V converter, sensitive to 35pA, but able to keep summing junction under control to ±150mA

Circuit ideas (*cont.*)

photodiode amplifiers (adapted from Burr-Brown app. notes)

C_f

R_f

λ

I_D

C_D

detector cap

$V_0 = R_f I_D$

C_1 10pF

C_2 10pF

200pF C_3

for small $C_f = 0.5$–5pF

$$C_f = \frac{C_1 C_2}{C_1 + C_2 + C_3}$$

open-loop gain

I-to-*V* gain

$1/R_f C_f$

$\frac{C_f}{C_D} f_c$

$1/R_f C_f$

$1 + C_D/C_f$

noise gain

$1/R_f C_D$

$\omega \rightarrow$

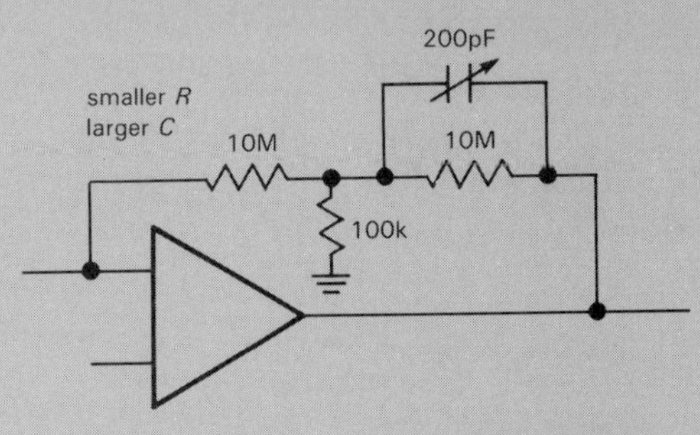

E. stabilizing photodiode amplifiers: contending with the necessary small C and large R

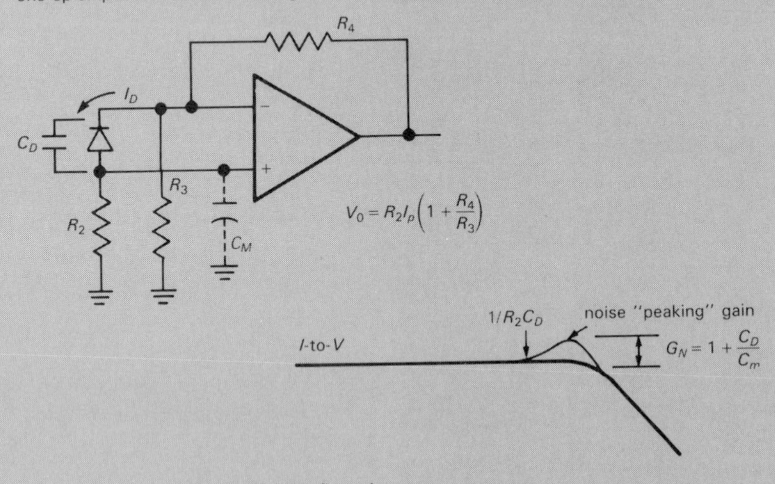

F. alternate photodiode amplifier configuration

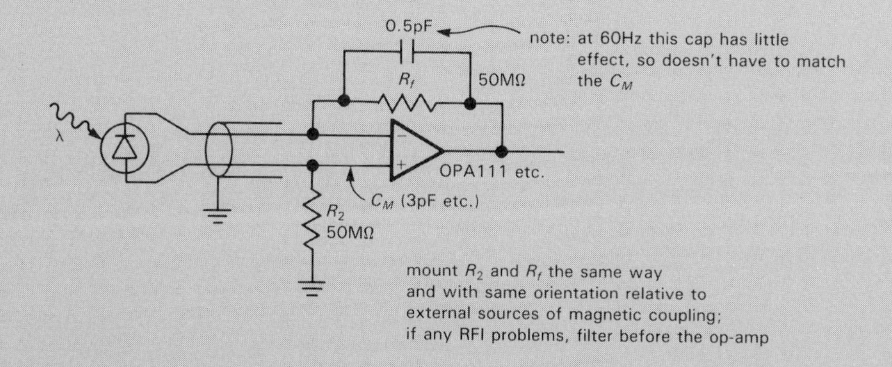

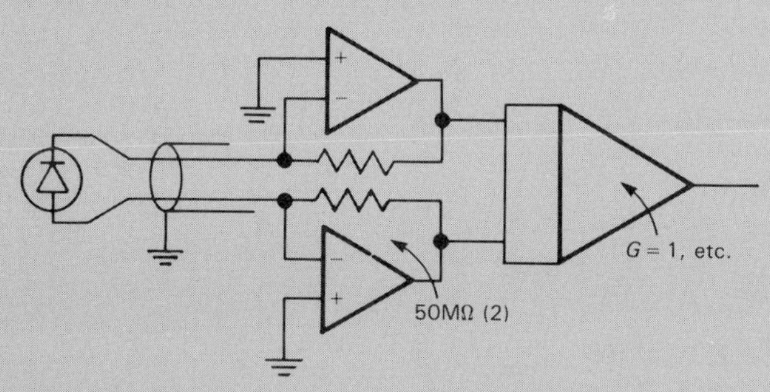

G. balanced photodiode amplifiers for long detector cables

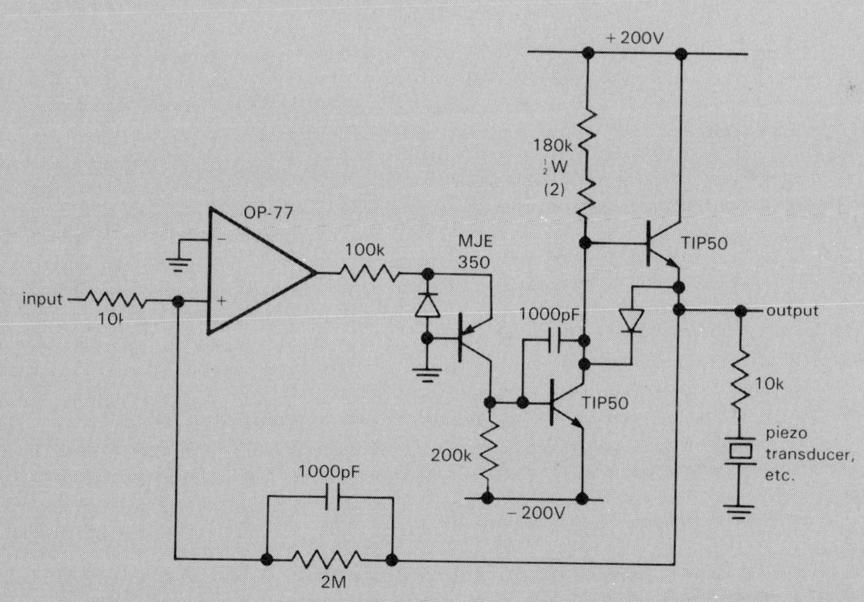

H. ±200V precision op-amp (from PMI App. Note 106)

APPENDIXES

APPENDIX A
THE OSCILLOSCOPE

The oscilloscope (scope for short) is the most useful and versatile electronic test instrument. As usually used, it lets you "see" voltages in a circuit as a function of time, triggering on a particular point of the waveform so that a stationary display results. We've drawn a block diagram (Fig. A1) and typical front panel (Fig. A2) to help explain how it works. The scope we will describe is usually called a dc-coupled dual-trace triggered scope. There are special-purpose scopes used for TV servicing and the like, and there are scopes of an older vintage that don't have the features needed for circuit testing.

VERTICAL

Beginning with the signal inputs, most scopes have two channels; that's very useful, since you often need to see the relationship between signals. Each channel has a calibrated gain switch, which sets the scale of VOLTS/DIVISION *on the screen.* There's also a VARIABLE gain knob (concentric with the gain switch) in case you want to set a given signal to a certain number of divisions. Warning: Be sure the variable gain knob is in the "calibrated" position when making voltage measurements! It's easy to forget. The better scopes have indicator lights to warn

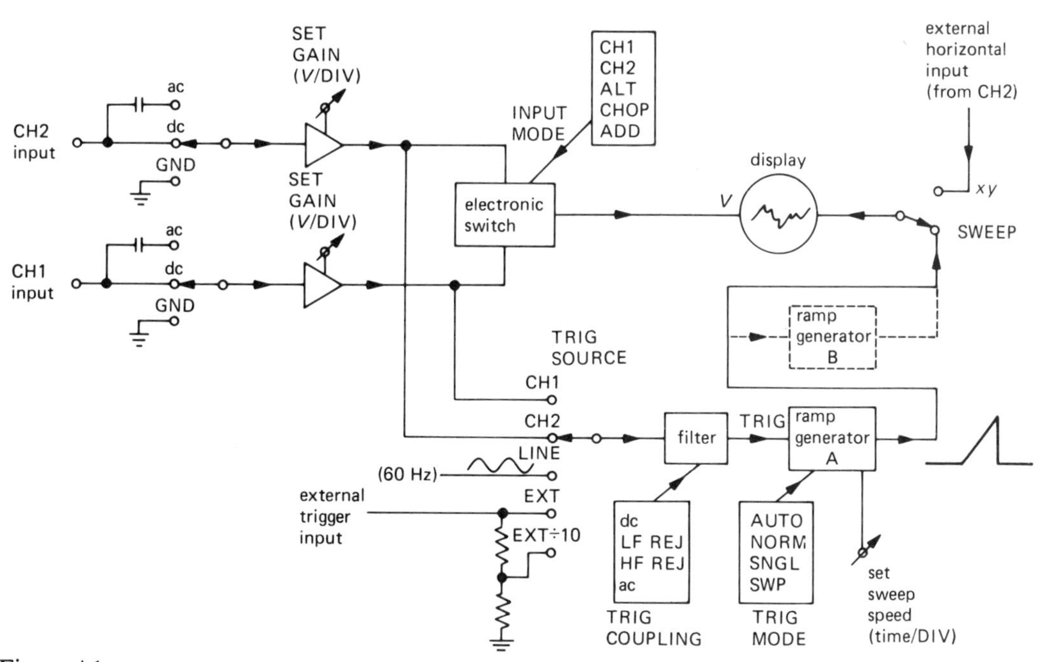

Figure A1

you if the variable gain knob is out of the calibrated position.

The scope is dc-coupled, an essential feature: What you see on the screen is the signal voltage, dc value and all. Sometimes you may want to see a small signal riding on a large dc voltage, though; in that case you can switch the input to ac coupling, which capacitively couples the input with a time constant of about 0.1 second. Most scopes also have a grounded input position, which lets you see where zero volts is on the screen. (In GND position the signal isn't shorted to ground, just disconnected from the scope, whose input is grounded.) Scope inputs are usually high-impedance (a megohm in parallel with about 20pF), as any good voltage-measuring instrument should be. The input resistance of 1.0 megohm is an

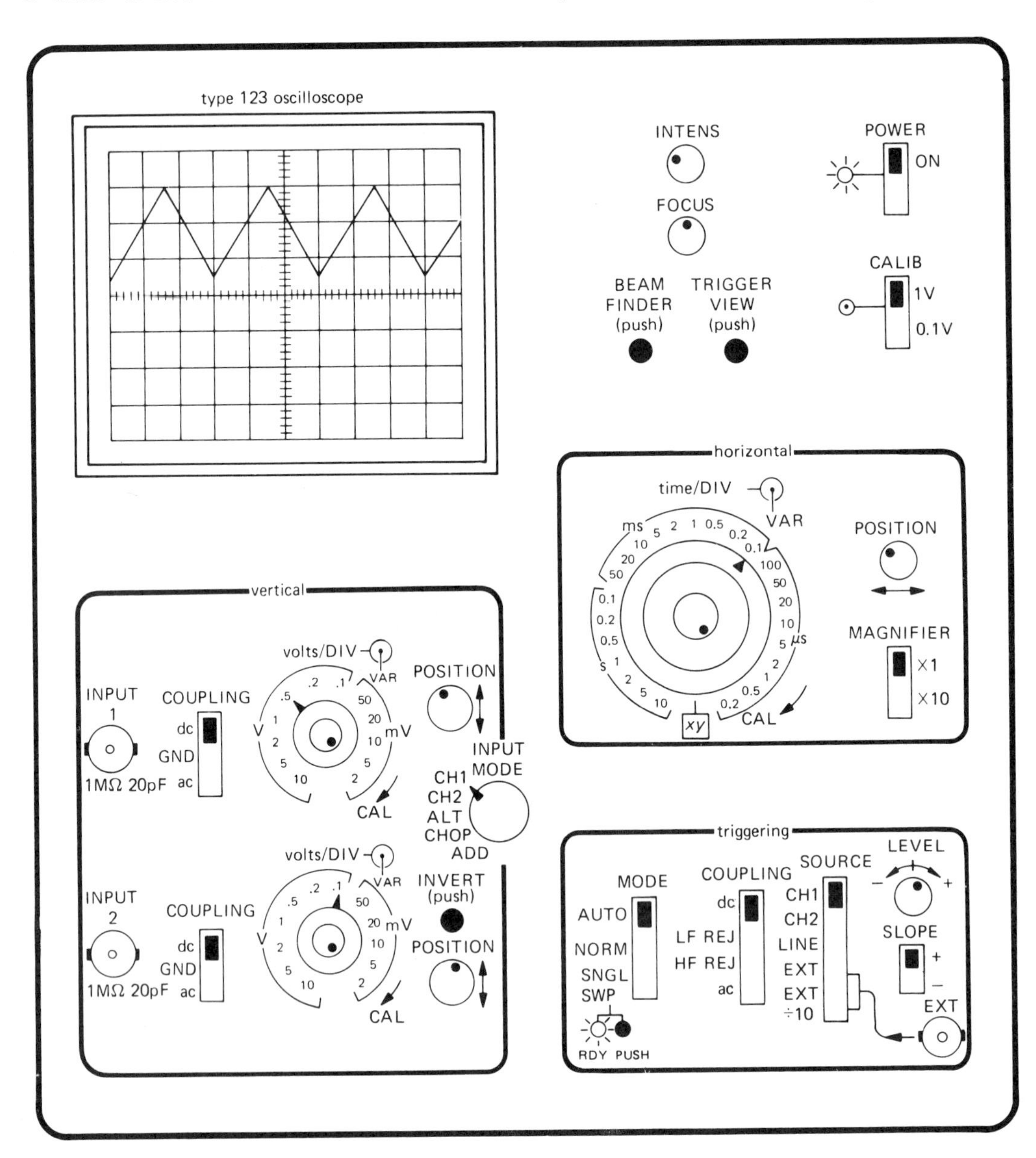

Figure A2

accurate and universal value, so that high-impedance attenuating probes can be used (as will be described later); unfortunately, the parallel capacitance is not standardized, which is a bit of a nuisance when changing probes.

The vertical amplifiers include a vertical POSITION control, an INVERT control on at least one of the channels, and an INPUT MODE switch. The latter lets you look at either channel, their sum (their difference, when one channel is inverted), or both. There are two ways to see both: ALTERNATE, in which alternate inputs are displayed on successive sweeps of the trace; CHOPPED, in which the trace jumps back and forth rapidly (0.1–1MHz) between the two signals. ALTERNATE mode is generally better, except for slow signals. It is often useful to view signals both ways, to make sure you're not being deceived.

HORIZONTAL

The vertical signal is applied to the vertical deflection electronics, moving the dot up and down on the screen. The horizontal sweep signal is generated by an internal ramp generator, giving deflection proportional to time. As with the vertical amplifiers, there's a calibrated TIME/DIVISION switch and a VARIABLE concentric knob; the same warning stated earlier applies here. Most scopes have a 10 × MAGNIFIER and also allow you to use one of the input channels for horizontal deflection (this lets you generate those beloved but generally useless "Lissajous figures" featured in elementary books and science fiction movies).

TRIGGERING

Now comes the trickiest part: triggering. We've got vertical signals and horizontal sweep; that's what's needed for a graph of voltage versus time. But if the horizontal sweep doesn't catch the input signal at the same point in its waveform each time (assuming the signal is repetitive), the display will be a mess – a picture of the input waveform superimposed over itself at different times. The trigger circuitry lets you select a LEVEL and SLOPE (+ or −) on the waveform at which to begin the sweep. You can see from the front panel that you have a number of choices about trigger sources and mode. NORMAL mode produces a sweep only when the source selected crosses through the trigger point you have set, moving in the direction (SLOPE) you have selected. In practice, you adjust the level control for a stable display. In AUTO the sweep will "free run" if no signal is present; this is good if the signal sometimes drops to small values, since the display won't disappear and make you think the signal has gone away. It's the best mode to use if you are looking at a bunch of different signals and don't want to bother setting the trigger each time. SINGLE SWEEP is used for nonrepetitive signals. LINE causes the sweep to trigger on the ac power line, handy if you're looking at hum or ripple in a circuit. The EXTERNAL trigger inputs are used if you have a clean signal available at the same rate as some "dirty" signal you're trying to see; it's often used in situations where you are driving some circuit with a test signal, or in digital circuits where some "clock" signal synchronizes circuit operations. The various coupling modes are useful when viewing composite signals; for instance, you may want to look at an audio signal of a few kilohertz that has some spikes on it. The HF REJ position (high-frequency reject) puts a low-pass filter in front of the trigger circuitry, preventing false triggering on the spikes. If the spikes happen to be of interest, you can trigger on them instead in LF REJ position.

Many scopes now have BEAM FINDER

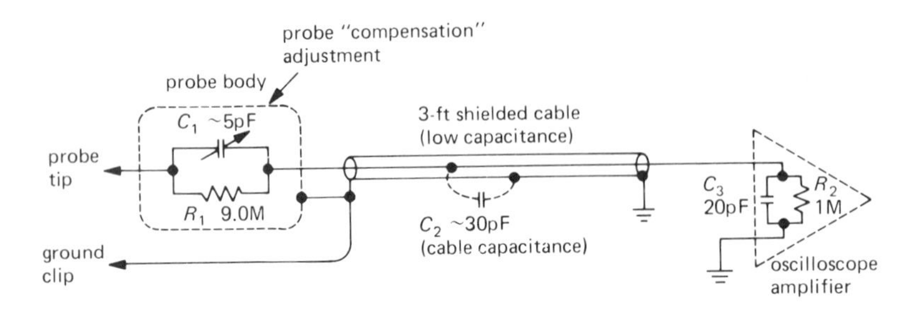

Figure A3

and TRIGGER VIEW controls. The beam finder is handy if you're lost and can't find the trace; it's a favorite of beginners. Trigger view displays the trigger signal; it's especially handy when triggering from external sources.

HINTS FOR BEGINNERS

Sometimes it's hard to get *anything* to show on the scope. Begin by turning the scope on; set triggering for AUTO, DC COUPLING, CH 1. Set sweep speed at 1ms/div, cal, and the magnifier off (×1). Ground the vertical inputs, turn up the intensity, and wiggle the vertical position control until a horizontal line appears (if you have trouble at this point, try the beam finder). Warning: Some scopes (the popular Tektronix 400 series, for example) don't sweep on AUTO unless the trigger level is adjusted correctly. Now you can apply a signal, unground the input, and fiddle with the trigger. Become familiar with the way things look when the vertical gain is far too high, when the sweep speed is too fast or slow, and when the trigger is adjusted incorrectly.

PROBES

The oscilloscope input capacitance seen by a circuit under test can be quite high, especially when the necessary shielded connecting cable is included. The resulting input impedance (1 megohm in parallel with 100 picofarads or so) is often too low for sensitive circuits and loads it by the usual voltage divider action. Worse yet, the capacitance may cause some circuits to misbehave, even to the point of going into oscillation! In such cases the scope obviously is not acting like the "low-profile" measurement instrument we expect; it's more like a bull in a china shop.

The usual solution is the use of high-impedance "probes." The popular 10× probe works as shown in Figure A3. At dc it's just a 10× voltage divider. By adjusting C_1 to 1/9th the parallel capacitance of C_2 and C_3, the circuit becomes a 10× divider at all frequencies, with input impedance of 10 megohms in parallel with a few picofarads. In practice, you adjust the probe by looking at a square wave of about 1kHz, available on all scopes as CALIB, or PROBE ADJ, setting the capacitor on the probe for a clean square wave without overshoot. Sometimes the adjustment is cleverly hidden; on some probes you twist the body of the probe and lock it by tightening a second threaded part. One drawback: A 10× probe makes it difficult to look at signals of only a few millivolts; for these situations use a "1× probe," which is simply a length of low-capacitance shielded cable with the usual probe hardware (wire "grabber," ground clip, handsome knurled handle, etc.). The 10× probe should be the standard probe, left connected to the scope, with the 1× probe used when necessary. Some probes feature a convenient choice of 1× or

10× attenuation, switchable at the probe tip.

GROUNDS

As with most test instruments, the oscilloscope input is referred to the instrument ground (the outer connection of the input BNC connectors), which is usually tied electrically to the case. That, in turn, connects to the ground lead of the ac power line, via the 3-wire power cord. This means that you cannot measure voltages between the two arbitrary points in a circuit, but are forced to measure signals relative to this universal ground.

An important caution is in order here: If you try to connect the ground clip of an oscilloscope probe to a point in the circuit that is at some voltage relative to ground, you will end up shorting it to ground. This can have disastrous consequences to the circuit under test; in addition, it can be downright dangerous with circuits that are "hot to ground" (transformerless consumer electronics like television sets, for example). If it is imperative to look at the signal between two points, you can either "float" the scope by lifting the ground lead (not recommended, unless you know what you're doing) or make a differential measurement by inverting one input channel and switching to ADD (some plug-in modules permit direct differential measurements).

Another caution about grounds when you're measuring weak signals or high frequencies: Be sure the oscilloscope ground is the same as the circuit ground where you're measuring. The best way to do this is by connecting the short ground wire on the probe body directly to the circuit ground, then checking by measuring the voltage of "ground" with the probe, observing no signal. One problem with this scheme is that those short ground clips are usually missing, lost! Keep your probe accessories in a drawer somewhere.

OTHER SCOPE FEATURES

Many scopes have a DELAYED SWEEP that lets you see a segment of a waveform occurring some time after the trigger point. You can dial the delay accurately with a multiturn adjustment and a second sweep-speed switch. A delay mode known as A INTENSIFIED BY B lets you display the whole waveform at the first sweep speed, with the delayed segment brightened; this is handy during setup. Scopes with delayed sweep sometimes have "mixed sweep," in which the trace begins at one sweep speed, then switches to a second (usually faster) speed after the selected delay. Another option is to begin the delayed sweep either immediately after the selected delay or at the next trigger point after the delay; there are two sets of trigger controls, so the two trigger points can be set individually. (Don't confuse delayed sweep with "signal delay." All good scopes have a delay in the signal channel, so you can display the event that caused the trigger; it lets you look a little bit backward in time!) Many scopes now have a TRIGGER HOLDOFF control; it inhibits triggering for an adjustable interval after each sweep, and it is very useful when viewing complicated waveforms without the simple periodicity of a sine wave, say. The usual case is a digital waveform with a complicated sequence of 1's and 0's, which won't generate a stable display otherwise (except by adjustment of the sweep-speed vernier, which means you don't get a calibrated sweep). There are also scopes with a "storage" that let you see a nonrepetitive event, and scopes that accept plug-in modules. These let you do just about anything, including display of eight simultaneous traces, spectrum analysis, accurate (digital) voltage and time measurements on waveforms, etc. Digital-storage analog oscilloscopes of a new generation are becoming popular; they let you catch a one-shot waveform, and even let you look backward in time (before the trigger event).

APPENDIX B
MATH REVIEW

Some knowledge of algebra and trigonometry is essential to understand this book. In addition, a limited ability to deal with complex numbers and derivatives (a part of calculus) is helpful, although not entirely essential. This appendix is meant as the briefest of summaries of complex numbers and differentiation. It is not meant as a textbook substitute. For a highly readable self-help book on calculus, we recommend *Quick Calculus*, by D. Kleppner and N. Ramsey (John Wiley & Sons, 1972).

COMPLEX NUMBERS

A complex number is an object of the form

$$\mathbf{N} = a + bi$$

where a and b are real numbers and i (called j in the rest of the book, to avoid confusion with small-signal currents) is the square root of -1; a is called the real part, and b is called the imaginary part. Boldface letters or squiggly underlines are sometimes used to denote complex numbers. At other times you're just supposed to *know*!

Complex numbers can be added, subtracted, multiplied, etc., just as real numbers:

$$(a + bi) + (c + di) = (a + c) + (b + d)i$$

$$(a + bi) - (c + di) = (a - c) + (b - d)i$$

$$(a + bi)(c + di) = (ac - bd) + (bc + ad)i$$

$$\frac{a + bi}{c + di} = \frac{(a + bi)(c - di)}{(c + di)(c - di)} = \frac{ac + bd}{c^2 + d^2} + \frac{bc - ad}{c^2 + d^2}i$$

All these operations are natural, in the sense that you just treat i as something that multiplies the imaginary part, and go ahead with ordinary arithmetic. Note that $i^2 = -1$ (used in the multiplication example) and that division is simplified by multiplying top and bottom by the *complex conjugate*, the number you get by changing the sign of the imaginary part. The complex conjugate is sometimes indicated with an asterisk. If

$$\mathbf{N} = a + bi$$

then

$$\mathbf{N}^* = a - bi$$

The magnitude (or *modulus*) of a complex number is

$$|\mathbf{N}| = |a + bi| = [(a + bi)(a - bi)]^{\frac{1}{2}} = (a^2 + b^2)^{\frac{1}{2}}$$

i.e.,

$$|\mathbf{N}| = (\mathbf{N}\mathbf{N}^*)^{\frac{1}{2}}$$

simply obtained by multiplying by the complex conjugate and taking the square root. The magnitude of the product (or quotient) of two complex numbers is simply the product (or quotient) of their magnitudes.

The real (or imaginary) part of a complex number is sometimes written

real part of $\mathbf{N} = \mathcal{R}e(\mathbf{N})$
imaginary part of $\mathbf{N} = \mathcal{I}m(\mathbf{N})$

You get them by writing out the number in the form $a + bi$, then taking either a or b. This may involve some multiplication or division, since the complex number may be a real mess.

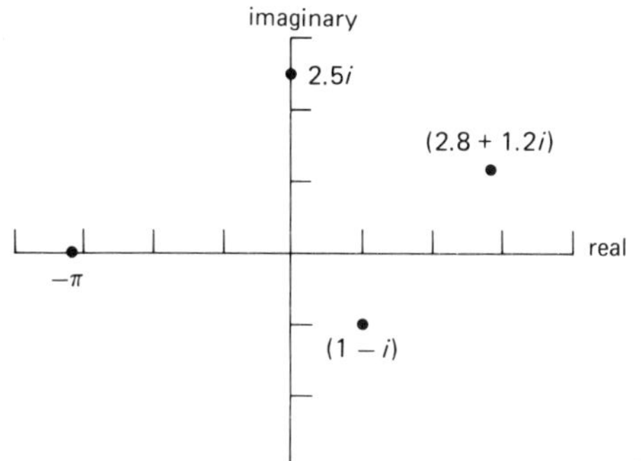

Figure B1

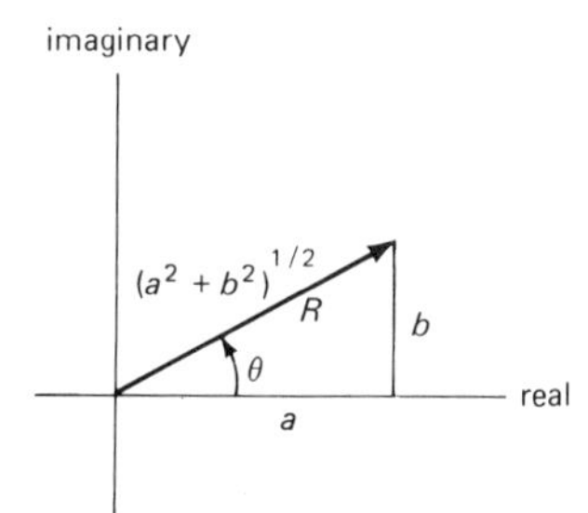

Figure B2

Complex numbers are sometimes represented on the complex plane. It looks just like an ordinary x, y graph, except that a complex number is plotted by taking its real part as x and its imaginary part as y; i.e., the axes represent REAL (x) and IMAGINARY (y), as shown in Figure B1. In keeping with this analogy, you sometimes see complex numbers written just like x, y coordinates:

$$a + bi \leftrightarrow (a, b)$$

Just as with ordinary x, y pairs, complex numbers can be represented in polar coordinates; that's known as "magnitude, angle" representation. For example, the number $a + bi$ can also be written (Fig. B2)

$$a + bi = (R, \theta)$$

where $R = (a^2+b^2)^{\frac{1}{2}}$ and $\theta = \tan^{-1}(b/a)$. This is usually written in a different way, using the fact that

$$e^{ix} = \cos x + i \sin x$$

(You can easily derive the preceding result, known as Euler's formula, by expanding the exponential in a Taylor series.) Thus we have the following equivalents:

$$\mathrm{N} = a + bi = Re^{i\theta}$$

$$R = |\mathrm{N}| = (\mathrm{NN}^*)^{\frac{1}{2}} = (a^2 + b^2)^{\frac{1}{2}}$$

$$\theta = \tan^{-1}(b/a)$$

i.e., the modulus R and angle θ are simply the polar coordinates of the point that represents the number in the complex plane. Polar form is handy when complex numbers have to be multiplied (or divided); you just multiply (divide) their magnitudes and add (subtract) their angles:

$$(ae^{ib})(ce^{id}) = ace^{i(b+d)}$$

Finally, to convert from polar to rectangular form, just use Euler's formula:

$$ae^{ib} = a \cos b + ia \sin b$$

i.e.,

$$\mathcal{R}e(ae^{ib}) = a \cos b$$

$$\mathcal{I}m(ae^{ib}) = a \sin b$$

If you have a complex number multiplying a complex exponential, just do the necessary multiplications. If

$$\mathrm{N} = a + bi$$

$$\begin{aligned} \mathrm{N}e^{ix} &= (a + bi)(\cos x + i \sin x) \\ &= (a \cos x - b \sin x) \\ &\quad + i(b \cos x + a \sin x) \end{aligned}$$

DIFFERENTIATION (CALCULUS)

We start with the concept of a *function* $f(x)$, i.e., a formula that gives a value $y = f(x)$ for each x. The function $f(x)$ should be *single-valued* i.e., it should give a single value of y for each x. You can think of $y = f(x)$ as a graph, as in Figure B3. The derivative of y with respect to x, written dy/dx ("dee y dee x"), is the *slope* of the graph of y versus x. If you draw a tangent to the curve at some point, its slope is dy/dx *at that point*; i.e., the derivative is itself a function, since it has a value at each point. In Figure B3, the slope at the point (1, 1) happens to be 2, whereas the slope at the origin is zero (you will see shortly how to compute the derivative).

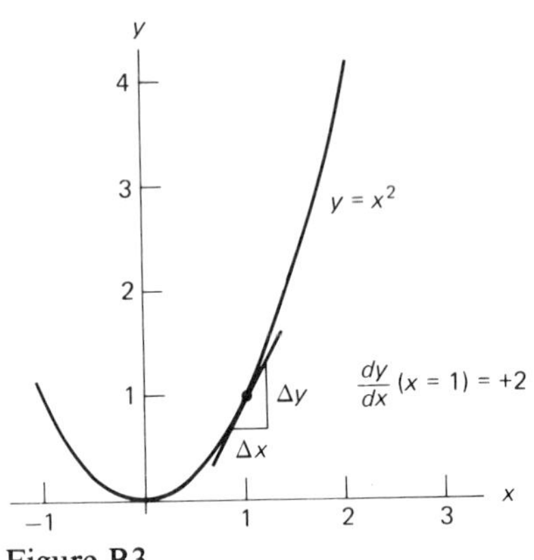

Figure B3

In mathematical terms, the derivative is the limiting value of the ratio of the change in y (Δy) to the change in x (Δx) corresponding to a small change in x (Δx), as Δx goes to zero. To quote a song once sung in the hallowed halls of Harvard (by Tom Lehrer and Lewis Branscomb),

You take a function of x, and you call it y
Take any x-nought that you care to try
Make a little change and call it delta x
The corresponding change in y is what you find nex'
And then you take the quotient, and now, carefully
Send delta x to zero, and I think you'll see
That what the limit gives us (if our work all checks)
Is what you call dy/dx
It's just dy/dx

(to the tune of "There'll Be Some Changes Made," W. Benton Overstreet).

Differentiation is a straightforward art, and the derivatives of many common functions are tabulated in standard tables. Here are some rules (u and v are arbitrary functions x):

Some derivatives

$$\frac{d}{dx}x^n = nx^{n-1}$$
$$\frac{d}{dx}\sin x = \cos x$$
$$\frac{d}{dx}e^x = e^x$$
$$\frac{d}{dx}au(x) = a\frac{d}{dx}u(x) \quad (a = \text{constant})$$
$$\frac{d}{dx}(u+v) = \frac{du}{dx} + \frac{dv}{dx}$$
$$\frac{d}{dx}\left(\frac{u}{v}\right) = \frac{v\frac{du}{dx} - u\frac{dv}{dx}}{v^2}$$
$$\frac{d}{dx}\{u[v(x)]\} = \frac{du}{dv}\frac{dv}{dx}$$

The last one is very useful and is called the chain rule.

Once you have differentiated a function, you often want to evaluate the value of the derivative at some point. Other times you may want to find a minimum or maximum of the function; that's the same thing as having a zero derivative, so you can just set the derivative to zero and solve for x. Here are some examples:

$$\frac{d}{dx}x^2 = 2x \quad \left(\text{Fig. B3: } \begin{matrix}\text{slope} = 2 \text{ at } x = 1, \\ \text{slope} = 0 \text{ at } x = 0\end{matrix}\right)$$
$$\frac{d}{dx}xe^x = xe^x + e^x \qquad \text{(product rule)}$$
$$\frac{d}{dx}\sin(ax) = a\cos(ax) \qquad \text{(chain rule)}$$
$$\frac{d}{dx}a^x = \frac{d}{dx}(e^{x\log a}) = a^x \log a \quad \text{(chain rule)}$$
$$\frac{d}{dx}\left(\frac{1}{x^{\frac{1}{2}}}\right) = -\frac{1}{2}x^{-\frac{3}{2}}$$

APPENDIX C
THE 5% RESISTOR COLOR CODE

Low-power axial-lead carbon-composition and film resistors with 2% to 20% tolerances have a standard set of values and a standard color-band marking scheme. Although it may seem diabolical to the beginner, the practice of color banding makes it easy to recognize resistor values in a circuit or parts bin, without having to search for a printed legend. The standard resistor values are chosen so that adjacent values have relative ratios of about 10% for the 2% and 5% tolerance types and 20% for the 10% and 20% tolerance types. Thus there are many values that could be described by the color code but that are not available.

Two digits and a multiplier digit determine the resistor value, and resistors are color-banded in that order starting from one end of the resistor (Fig. C1). A fourth tolerance band is usually present, and occasionally you'll see a fifth band for other parameters (such as yellow or orange band for MIL spec reliability rating).

Here is the set of standard values for the first two digits (lightface type indicates 2% and 5% only):

10	16	**27**	43	**68**
11	**18**	30	**47**	75
12	20	**33**	51	**82**
13	**22**	36	**56**	91
15	24	**39**	62	**100**

Carbon-composition resistors range in price from 3 cents each (in quantities of 1000) to 15 cents (quantities of 25). Distributors may be unwilling to sell less than 25 to 50 pieces of one value; thus an assortment box (made by Stackpole or Ohmite) may be a wise purchase.

digit	color	multiplier	number of zeros
	silver	0.01	−2
	gold	0.1	−1
0	black	1	0
1	brown	10	1
2	red	100	2
3	orange	1k	3
4	yellow	10k	4
5	green	100k	5
6	blue	1 M	6
7	violet	10 M	7
8	gray		
9	white		

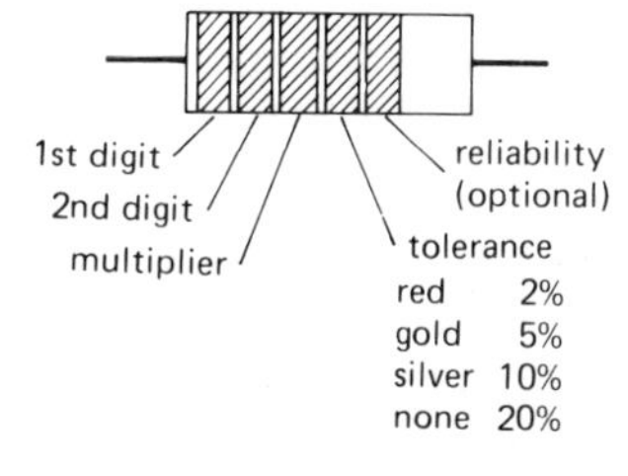

example: red-yellow-orange-gold is a 2, 4, and 3 zeros, or 24k 5%, resistor

Figure C1

1% PRECISION RESISTORS

APPENDIX D

Metal-film precision resistors with ½% and 1% tolerance ratings have seen sufficient use in the industry to have attractively low prices. In particular, the RN55D and RN60D resistors are often available for as little as 5 cents each in quantities of 100, and a distributor may be willing to sell an assortment of mixed values at a quantity discount. The RN55D resistors are the same size as ordinary 1/4 watt "composition" resistors (although their military rating will be 1/10 or 1/8 watt at 70°C ambient temperature), whereas the RN60D are the size of 1/2 watt composition resistors. The RN55D resistors have a temperature coefficient of 100ppm/°C, and RN55C resistors (same size) have a 50ppm/°C rating.

Metal-film precision resistors use a four-digit code printed on the resistor body, rather than the ordinary color-banding scheme. The first three digits denote a value, and the last digit is the "number-of-zeros" multiplier. For example, 1693 denotes a 169k resistor, and 1000 denotes a 100 ohm resistor. (Note that the color bands work the same way, but with only three digits altogether. Many capacitor types use this same printed number scheme.) If the resistor's value is too small to be described this way, an R will be used to indicate the decimal point; for example, 49R9 is a 49.9 ohm resistor, and 10R0 is 10.0 ohms.

The standard values range from 10.0 ohms to 301kΩ by approximately 2% ratios, although some companies may offer similar (non-MIL-spec) resistors with values from 4.99 ohms to 2.00 MΩ. Standard values in each decade are given in the list that follows.

One percent resistors are often used in applications that require excellent stability and accuracy; a small adjustable "trimmer" resistor may be connected in series to set a precise resistance value. But it's important to realize that from a worst-case standpoint, 1% resistors are only guaranteed to be within 1% of their rated value under a specified set of conditions. Resistance variation due to temperature change, high humidity, and operation at full rated power can easily exceed 1%. Resistance drift with time can approach 0.5%, particularly if the resistors are used at rated power. Circuits that require extremely accurate or stable performance (good to 0.1% or better, say) should use precision wire-wound resistors or some of the special metal-film resistors designed for such stability (for example, the Mepco 5023Z). This advice goes for composition resistors, as well. Resist, if you will, the temptation to regard the manufacturer's specifications as being overly conservative.

100	140	196	274	383	536	750
102	143	200	280	392	549	768
105	147	205	287	402	562	787
107	150	210	294	412	576	806
110	154	215	301	422	590	825
113	158	221	309	432	604	845
115	162	226	316	442	619	866
118	165	232	324	453	634	887
121	169	237	332	464	649	909
124	174	243	340	475	665	931
127	178	249	348	487	681	953
130	182	255	357	499	698	976
133	187	261	365	511	715	
137	191	267	374	523	732	

TABLE D.1. SELECTED RESISTOR TYPES

Property	Carbon comp (RCR-07)	Metal film standard (RN-55D)	Metal film precision (Mepco 5023Z)	Metal film miniature (Mepco 5063J)
Load life 1000hrs @85°C	10%	0.5%	0.01%	0.15%
Moisture Mil std 202	15%	0.5%	0.04%	1%
Temp cycle –65°C to +150°C	4%	0.25%	0.005%	0.25%
Low-temp operation –65°C	3%	0.25%	0.01%	0.25%
Short-term overload	2.5%	0.25%	0.01%	0.25%
Soldering @ 350°C	3%	0.25%	0.01%	0.25%
Shock 50G, 11ms	2%	0.25%	0.01%	0.25%
Vibration 10–2000Hz	2%	0.25%	0.01%	0.25%
Storage 1 year	–	–	0.003%	–
Tolerances avail	5%, 10%	0.1%–1%	0.025%–1%	1%, 5%
Tempco avail ppm/°C	5000	25–100	5–25	100
Voltage coef	–	5ppm/V	0.1ppm/V	–
Thermal emf	–	–	2μV/°C	–
Insulation resistance	–	–	10,000MΩ	1000MΩ

HOW TO DRAW SCHEMATIC DIAGRAMS
APPENDIX E

A well-drawn schematic makes it easy to understand how a circuit works and aids in troubleshooting; a poor schematic only creates confusion. By keeping a few rules and suggestions in mind, you can draw a good schematic in no more time than it takes to draw a poor one. In this appendix we dispense advice of three varieties: general principles, rules, and hints. We have also drawn some real knee-slappers to illustrate habits to avoid.

GENERAL PRINCIPLES

1. Schematics should be unambiguous. Therefore, pin numbers, parts values, polarities, etc., should be clearly labeled to avoid confusion.
2. A good schematic makes circuit functions clear. Therefore, keep functional areas distinct; don't be afraid to leave blank areas on the page, and don't try to fill the page. There are conventional ways to draw functional subunits; for instance, don't draw a differential amplifier as in Figure E1, because the function won't be easily recognized. Likewise, flip-flops are usually drawn with clock and inputs on the left, set and clear on top and bottom, and outputs on the right.

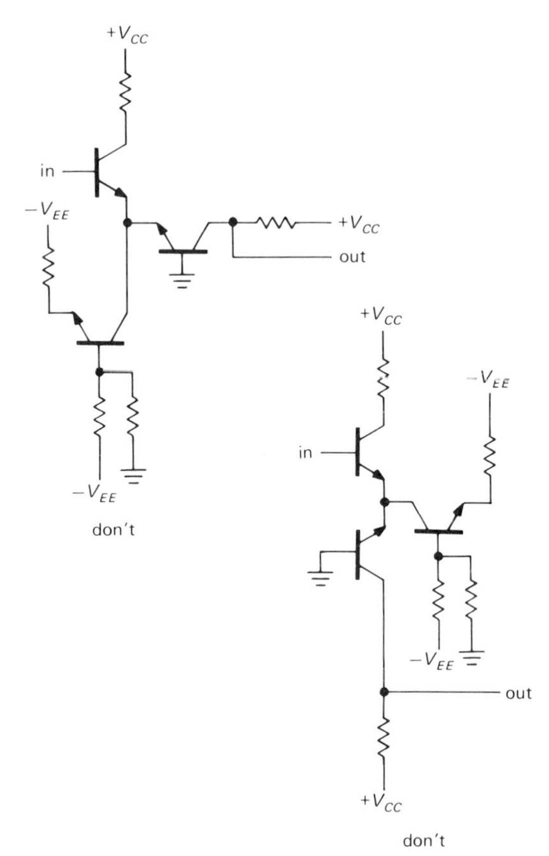

Figure E1

RULES

1. Wires connecting are indicated by a heavy black dot; wires crossing, but not connecting, have no dot (don't use a little half-circular "jog"; it went out in the 1950s).
2. Four wires must not connect at a point; i.e., wires must not cross *and* connect.
3. Always use the same symbol for the same device; e.g., don't draw flip-flops in two different ways (exception: assertion-level logic symbols show each gate in two possible ways).
4. Wires and components are aligned horizontally or vertically, unless there's a good reason to do otherwise.

5. Label pin numbers on the outside of a symbol, signal names on the inside.
6. All parts should have values or types indicated; it's best to give all parts a label, too, e.g., R_7 or IC_3.

HINTS

1. Identify parts immediately adjacent to the symbol, forming a distinct group giving symbol, label, and type or value.
2. In general, signals go from left to right; don't be dogmatic about this, though, if clarity is sacrificed.
3. Put positive supply voltages at the top of the page, negative at the bottom. Thus, *npn* transistors will usually have their emitter at the bottom, whereas *pnp*'s will have their emitter topmost.
4. Don't attempt to bring all wires around to the supply rails, or to a common ground wire. Instead, use the ground symbol(s) and labels like $+V_{CC}$ to indicate those voltages where needed.
5. It is helpful to label signals and functional blocks and show waveforms; in logic diagrams it is especially important to label signal lines, e.g., RESET′ or CLK.
6. It is helpful to bring leads away from components a short distance before making connections or jogs. For example, draw transistors as in Figure E2.
7. Leave some space around circuit symbols; e.g., don't draw components or wires too close to an op-amp symbol. This keeps the drawing uncluttered and leaves room for labels, pin numbers, etc.

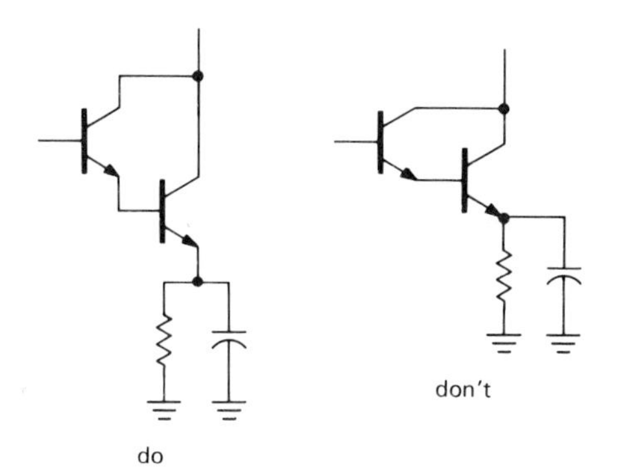

Figure E2

8. Label all boxes that aren't obvious: comparator versus op-amp, shift register versus counter, etc. Don't be afraid to invent a new symbol.
9. Use small rectangles, ovals, or circles to indicate card-edge connections, connector pins, etc. Be consistent.
10. The signal path through switches should be clear. Don't force the reader to follow wires all over the page to find out how a signal is switched.
11. Power-supply connections are normally assumed for op-amps and logic devices. However, show any unusual connections (e.g., an op-amp run from a single supply, where $V_- =$ ground) and the disposition of unused inputs.
12. It is very helpful to include a small table of IC numbers, types, and power-supply connections (pin numbers for V_{CC} and ground, for instance).
13. Include a title area near the bottom of the page, with name of circuit, name of instrument, by whom drawn, by whom designed or checked, date, and assembly number. Also include a revision area, with columns for revision number, date, and subject.
14. We recommend drawing schematics freehand on coarse graph paper (nonreproducing blue, 4 to 8 lines per inch) or on plain paper on top of graph paper. This is fast, and it gives very pleasing results. Use dark pencil or ink; avoid ball-point pen.
As an illustration, we've drawn a humble example (Fig. E3) showing "awful" and "good" schematics of the same circuit; the former violates nearly every rule and is almost impossible to understand. See how many bad habits you can find illustrated. We've seen all of them in professionally drawn schematics! (Drawing the "bad" schematic was an occasion of great hilarity; we laughed ourselves silly.)

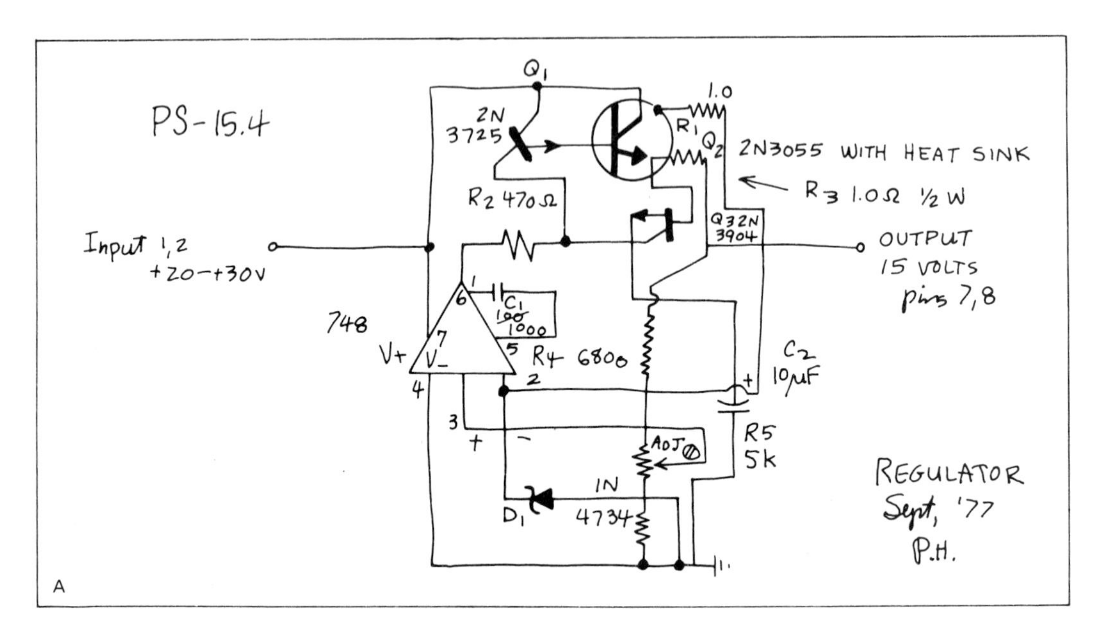
PS-15.4
Q1
2N 3725
1.0
R1
Q2 2N3055 WITH HEAT SINK
R3 1.0Ω ½ W
R2 470Ω
Q3 2N 3904
Input 1,2
+20–+30V
OUTPUT
15 VOLTS
pins 7,8
748
C1
1000
V+
V−
R4 6800
C2 10μF
R5 5k
ADJ
IN 4734
D1
REGULATOR
Sept, '77
P.H.
A

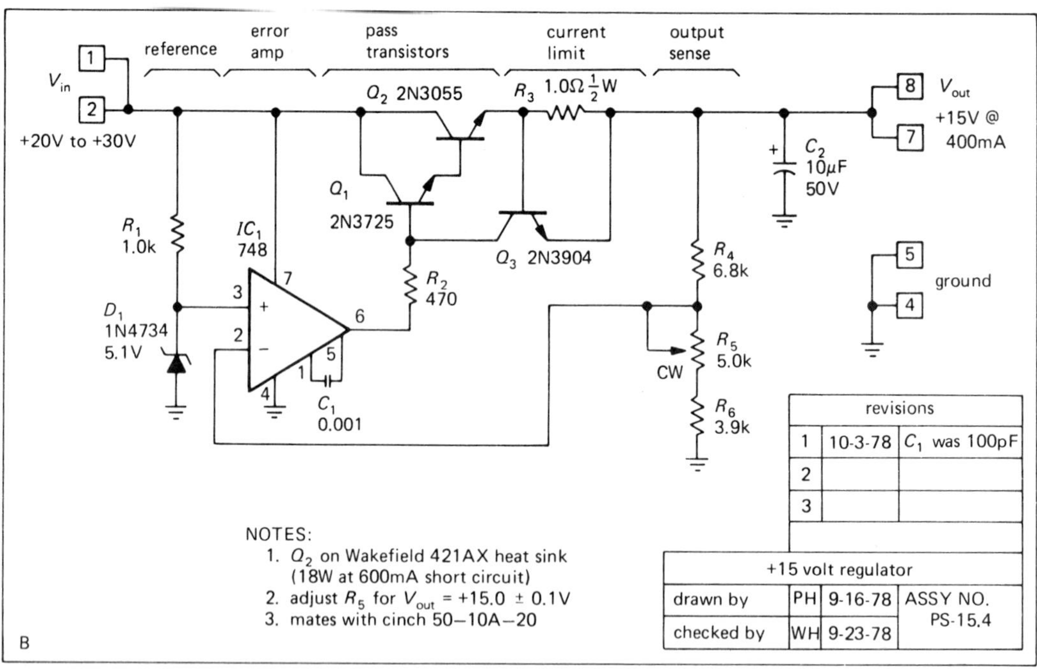
reference
error amp
pass transistors
current limit
output sense
Vin
+20V to +30V
Q2 2N3055
R3 1.0Ω ½W
Vout
+15V @ 400mA
C2 10μF 50V
Q1 2N3725
R1 1.0k
IC1 748
R2 470
Q3 2N3904
R4 6.8k
ground
D1 1N4734 5.1V
C1 0.001
R5 5.0k
CW
R6 3.9k
revisions
1 10-3-78 C1 was 100pF
2
3
NOTES:
1. Q2 on Wakefield 421AX heat sink (18W at 600mA short circuit)
2. adjust R5 for Vout = +15.0 ± 0.1V
3. mates with cinch 50–10A–20
+15 volt regulator
drawn by PH 9-16-78
checked by WH 9-23-78
ASSY NO. PS-15.4
B

Figure E3

LOAD LINES
APPENDIX F

The graphic method of "load lines" usually makes an early appearance in electronic textbooks. We have avoided it because, well, it just isn't useful in transistor design, the way it was in vacuum-tube circuit design. However, it is of use in dealing with some nonlinear devices (tunnel diodes, for example), and in any case it is a useful conceptual tool.

Let's start with an example. Suppose you want to know the voltage across the diode in Figure F1. Assume that you know the voltage-versus-current (V–I) curve of the particular diode (of course, it would have a manufacturing "spread," as well as depending on ambient temperature); it might look something like the curve drawn. How would you figure out the quiescent point?

One method might be to guess a rough value of current, say 0.6mA, then use the curve to get the drop across the resistor, from which you get a new guess for the current (in this case, 0.48mA). This iterative method is suggested in Figure F1. After a few iterations, this method will get you an answer, but it leaves a lot to be desired.

The method of load lines gets you the answer to this sort of problem immediately. Imagine *any* device connected in place of the diode; the 1.0k resistor is still the load. Now plot, on a V–I graph, the curve of resistor current versus device voltage. This turns out to be easy: at zero volts the current is just V_+/R (full drop across the resistor); at V_+ volts the current is zero; points in between fall on a straight line between the two. Now, on the same graph, plot the V–I curve of the device. The operating point lies on both curves, i.e., at the intersection, as shown in Figure F2.

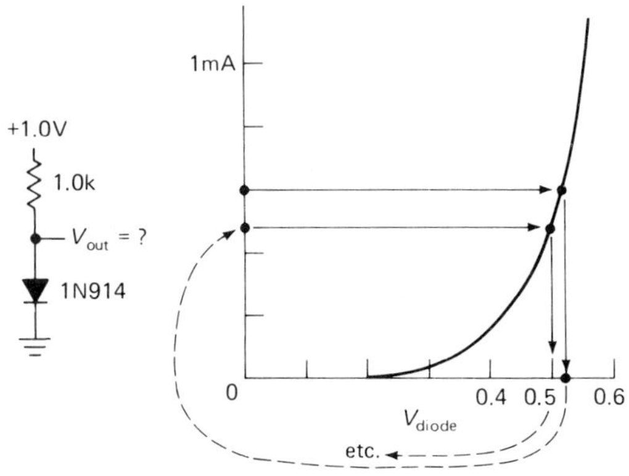

Figure F1

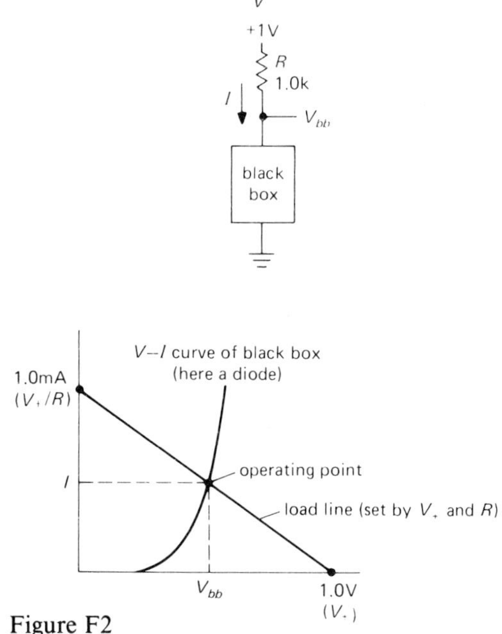

Figure F2

Load lines can be used with a 3-terminal device (tube or transistor, for example) by plotting a family of curves for the device. Figure F3 shows what such a thing would look like for a depletion-mode FET, with the curve family parameterized by gate-source voltage. You can read off the output for a given input by sliding along the load line between appropriate curves corresponding to the input you've got, then projecting onto the voltage axis. In this example we've done this, showing the drain voltage (output) for a gate swing (input) between ground and −2 volts.

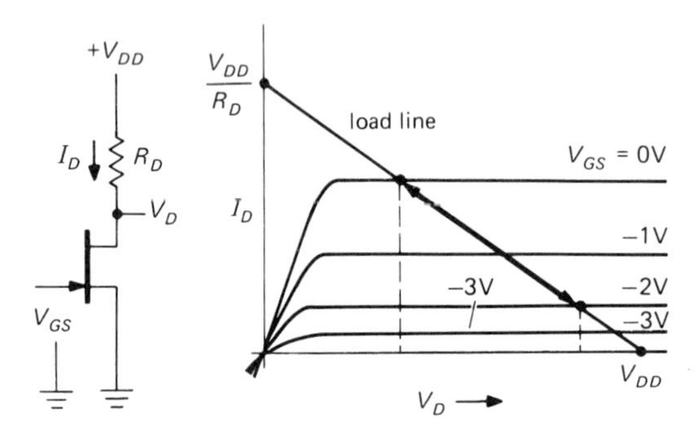

Figure F3

As nice as this method seems, it has very limited use for transistor or FET design, for a couple of reasons. For one thing, the curves published for semiconductor devices are "typical," with manufacturing spread that can be as large as a factor of 5. Imagine what would happen to those nice load-line solutions if all the curves shrank by a factor of 4! Another reason is that for an inherently logarithmic device like a diode junction, a linear load-line graph can be used to give accurate results only over a narrow region. Finally, the nongraphic methods we've used in this book are adequate to handle solid-state design. In particular, these methods emphasize the parameters you can count on (r_e, I_C vs. V_{BE} and T, etc.), rather than the ones that are highly variable (h_{FE}, V_P, etc.). If anything, the use of load lines on published curves for transistors only gives you a false sense of security, since the device spread isn't also shown.

Load lines turn out to be very useful in understanding the circuit behavior of highly nonlinear devices. The example of tunnel diodes illustrates a couple of interesting points. Let's analyze the circuit in Figure F4. Note that in this case, V_{in} takes the place of the supply voltage in the previous examples. So a signal swing will generate a family of parallel load lines intersecting with a single device V–I curve (Fig. F5A). The values shown are for a 100 ohm load resistor. As can be seen, the output varies most rapidly as the input swing takes the load line across the negative-resistance portion of the tunnel-diode curve. By reading off values of V_{out} (projection on the x axis) for various values of V_{in} (individual load lines), you get the "transfer" characteristics shown. This particular circuit has some voltage gain for input voltages near 0.2 volt.

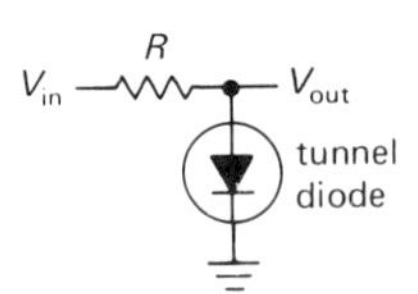

Figure F4

An interesting thing happens if the load lines become flatter than the middle section of the diode curve. That happens when the load resistance exceeds the magnitude of the diode's negative resistance. It is then possible to have *two* intersection points, as in Figure F6. A rising input signal carries the load lines up until the intersection point has nowhere to go and has to jump across to a higher V_{out} value. On returning, the load lines similarly carry the intersection point down until it must again jump back. The overall transfer characteristic has hysteresis, as shown. Tunnel diodes are used in this manner as fast switching devices (triggers).

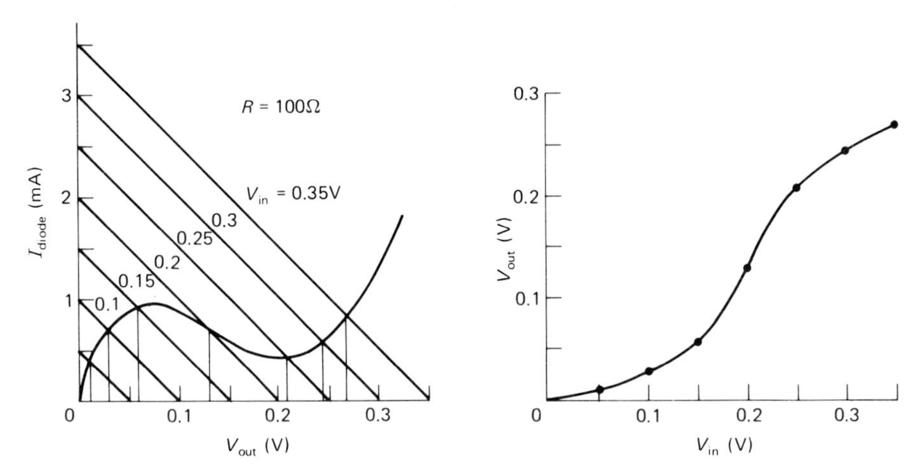

Figure F5

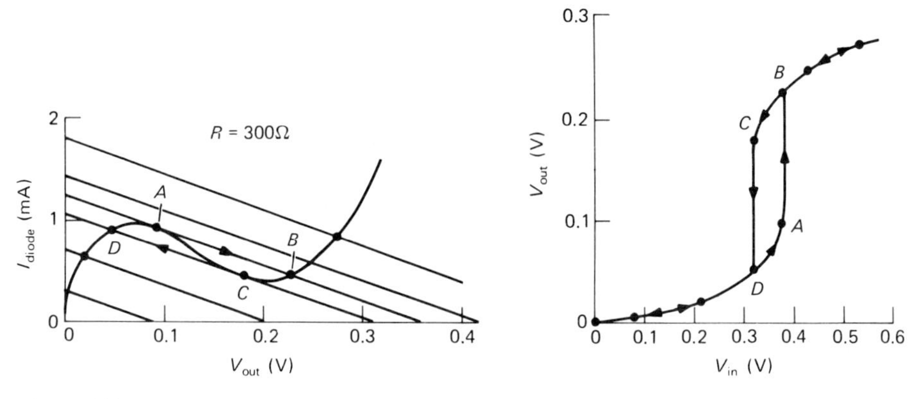

Figure F6

TRANSISTOR SATURATION
APPENDIX G

The subtitle of this appendix might be "transistor man defeated by the collector-base diode." With a simple model we can see the reason for the finite "saturation voltage" exhibited by bipolar transistors. The basic idea is that the collector-base junction is a big diode, with a high I_S (Ebers-Moll equation), so that it has a lower ON voltage for a given current than the base-emitter diode. Therefore, at small values of collector-to-emitter voltage (typically 0.25V or less), some of the base current will be "robbed" by conduction of the collector-base diode (Fig. G1). This lowers the effective h_{FE} and makes it necessary to supply relatively large base currents to bring the collector close to the emitter, as shown in the measured data of Figure G2.

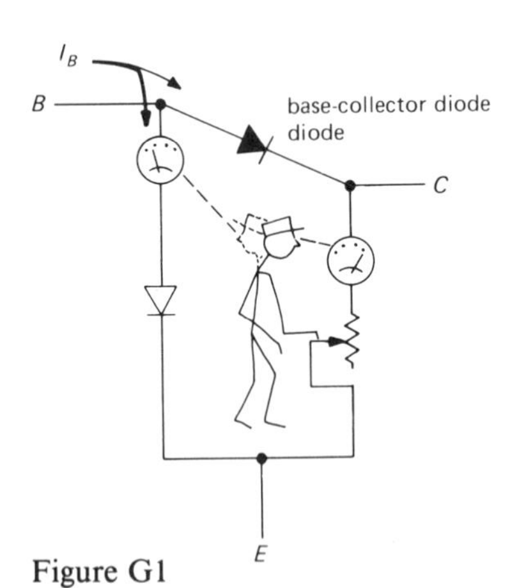

Figure G1

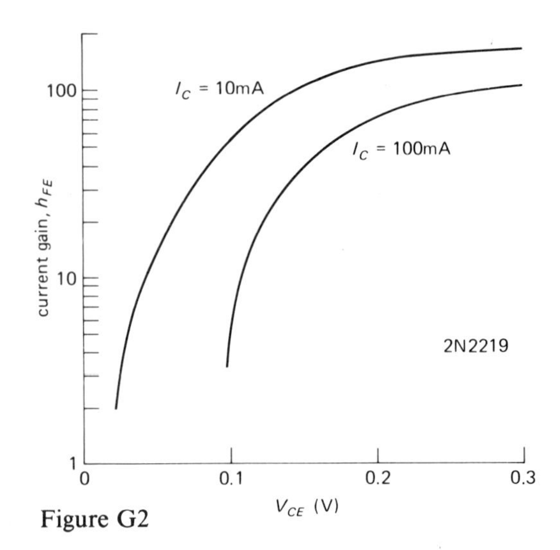

Figure G2

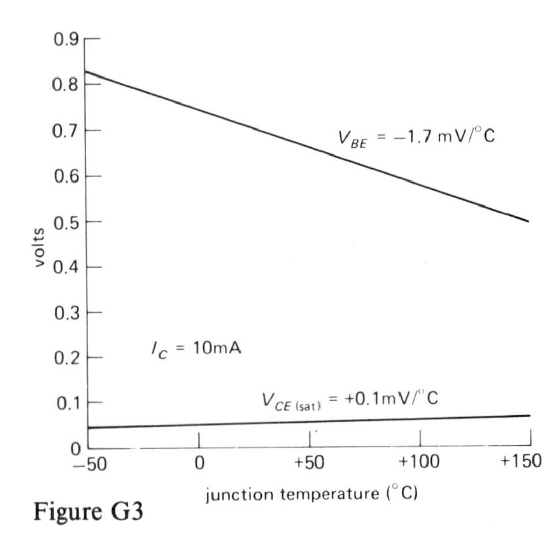

Figure G3

V_{CE}(sat), the collector saturation voltage at a particular value of base current and collector current, is also relatively independent of temperature because of cancellation of the temperature coefficients of the two diodes (Fig. G3). This is of interest because a saturated transistor is frequently used to switch large currents and may get hot (e.g., 10A at a saturation voltage of 0.5V is 5W, enough to bring the junction of a small power transistor to 100°C or more).

In saturated switching applications you usually provide generous amounts of base current (typically 1/10 or 1/20 of the collector current) to achieve values of V_{CE}(sat) of 0.05 to 0.2 volt. If the load inadvertently demands much greater collector currents, the transistor will go out of saturation, with greatly increased power dissipation. The measured data in Figure G4 show that it is hard to define exactly when a transistor is "saturated"; you might use some arbitrary criterion such as $I_C = 10I_B$.

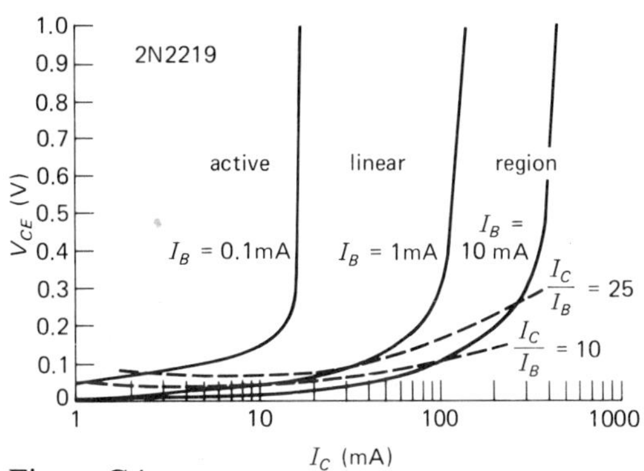

Figure G4

APPENDIX H

LC BUTTERWORTH FILTERS

Active filters, as discussed in Chapter 4, are very convenient at low frequencies, but they are impractical at radiofrequencies because of the slew-rate and bandwith requirements they impose on the operational amplifiers. At frequencies of 100kHz and above (and often at lower frequencies), the best approach is to design a passive filter with inductors and capacitors. (Of course, at UHF and microwave frequencies these "lumped-component" filters are replaced by stripline and cavity filters.)

As with active filters, there are many methods and filter characteristics possible with *LC* filters. For example, you can design the classic Butterworth, Chebyshev, and Bessel filters, each in low-pass, bandpass, high-pass, and band-reject varieties. It turns out that the Butterworth filter is particularly easy to design, and we can present in just a page or two all the essential design information for low-pass and high-pass Butterworth *LC* filters, and even a few examples. For further information we recommend the excellent handbook by Zverev cited in the Bibliography.

Table H1 gives the values of normalized inductances and capacitances for low-pass filters of various orders, from which actual circuit values are obtained by the frequency and impedance scaling rules:

Low-pass scaling rules:

$$L_n(\text{actual}) = \frac{R_L L_n(\text{table})}{\omega}$$

$$C_n(\text{actual}) = \frac{C_n(\text{table})}{\omega R_L}$$

TABLE H1. BUTTERWORTH LOW-PASS FILTERS[a] ($R_L = 1\Omega$)

π T	R_S $1/R_S$	C_1 L_1	L_2 C_2	C_3 L_3	L_4 C_4	C_5 L_5	L_6 C_6	C_7 L_7	L_8 C_8
$n = 2$	1	1.4142	1.4142						
	∞	1.4142	0.7071						
$n = 3$	1	1.0	2.0	1.0					
	∞	1.5	1.3333	0.5					
$n = 4$	1	0.7654	1.8478	1.8478	0.7654				
	∞	1.5307	1.5772	1.0824	0.3827				
$n = 5$	1	0.6180	1.6180	2.0	1.6180	0.6180			
	∞	1.5451	1.6944	1.3820	0.8944	0.3090			
$n = 6$	1	0.5176	1.4142	1.9319	1.9319	1.4142	0.5176		
	∞	1.5529	1.7593	1.5529	1.2016	0.7579	0.2588		
$n = 7$	1	0.4450	1.2470	1.8019	2.0	1.8019	1.2470	0.4450	
	∞	1.5576	1.7988	1.6588	1.3972	1.0550	0.6560	0.2225	
$n = 8$	1	0.3902	1.1111	1.6629	1.9616	1.9616	1.6629	1.1111	0.3902
	∞	1.5607	1.8246	1.7287	1.5283	1.2588	0.9371	0.5776	0.1951

[a] Values of L_n, C_n for 1Ω load resistance and cutoff frequency (−3dB) of 1 rad/s. See text for scaling rules.

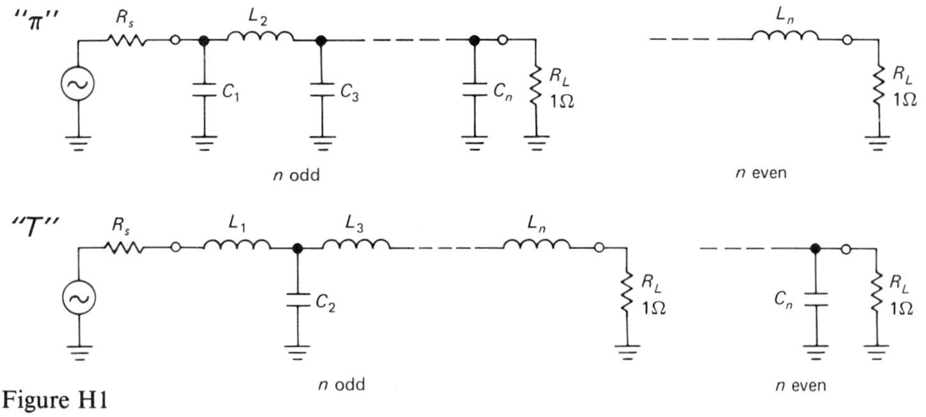

Figure H1

where R_L is the load impedance and ω is the angular frequency ($\omega = 2\pi f$).

Table H1 gives normalized values for 2-pole through 8-pole low-pass filters, for the two most common cases, namely (a) equal source and load impedances and (b) either source or load impedance much larger than the other. To use the table, first decide how many poles you need, based on the Butterworth response (graphs are plotted in Sections 5.05 and 5.07). Then use the preceding equations to determine the filter configuration (T or π; see Fig. H1) and component values. For equal source and load impedances, either configuration is OK; the π configuration may be preferable because it requires fewer inductors. For load impedance much higher (lower) than the source impedance, use the T (π) configuration.

To design a high-pass filter, follow the procedure outlined to determine which filter configuration to use and how many poles are necessary. Then do the universal low-pass to high-pass transformation shown in Figure H2, which consists simply of replacing inductors by capacitors, and vice versa. The actual component values are determined from the normalized values in Table H1 by the following frequency and impedance scaling rules:

High-pass scaling rules:

$$L_n(\text{actual}) = \frac{R_L}{\omega C_n(\text{table})}$$

$$C_n(\text{actual}) = \frac{1}{R_L \omega L_n(\text{table})}$$

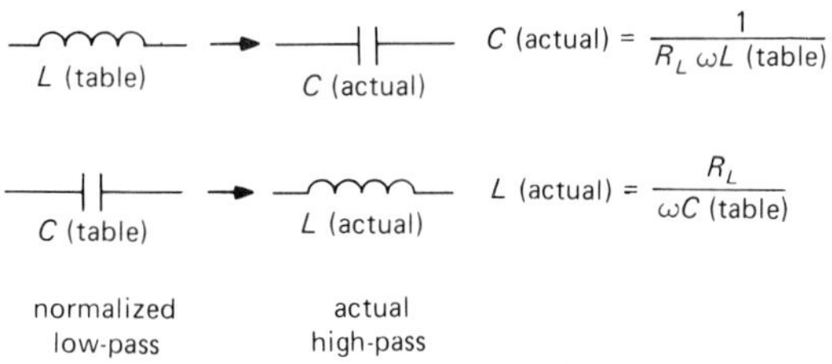

Figure H2

We will show how to use the table to design both low-pass and high-pass filters with a few examples.

EXAMPLE I

Design a 5-pole low-pass filter for source and load impedance of 75 ohms, with a cutoff frequency (−3dB) of 1MHz.

We use the π configuration to minimize the number of required inductors. The scaling rules give us

$$C_1 = C_5 = \frac{0.618}{2\pi \times 10^6 \times 75} = 1310\text{pF}$$
$$L_2 = L_4 = \frac{75 \times 1.618}{2\pi \times 10^6} = 19.3\mu\text{H}$$
$$C_3 = \frac{2}{2\pi \times 10^6 \times 75} = 4240\text{pF}$$

The complete filter is shown in Figure H3. Note that all filters with equal source and load impedances will be symmetrical.

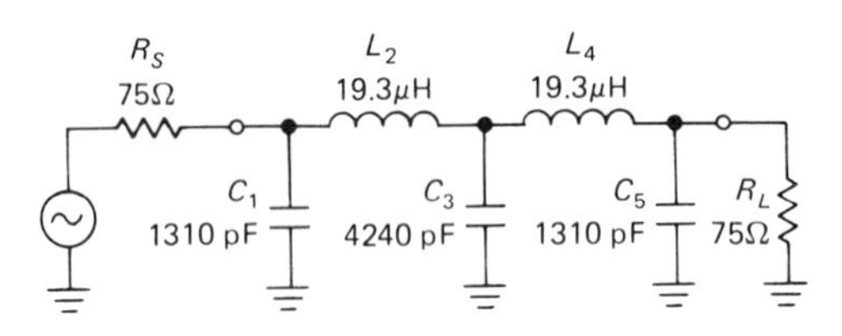

Figure H3

EXAMPLE II

Design a 3-pole low-pass filter for a source impedance of 50 ohms and a load impedance of 10k, with a cutoff frequency of 100kHz.

We use the T configuration, because $R_S \ll R_L$. For $R_L = 10\text{k}$, the scaling rules give

$$L_1 = \frac{10^4 \times 1.5}{2\pi \times 10^5} = 23.9\text{mH}$$
$$C_2 = \frac{1.3333}{2\pi \times 10^5 \times 10^4} = 212\text{pF}$$
$$L_3 = \frac{10^4 \times 0.5}{2\pi \times 10^5} = 7.96\text{mH}$$

The complete filter is shown in Figure H4.

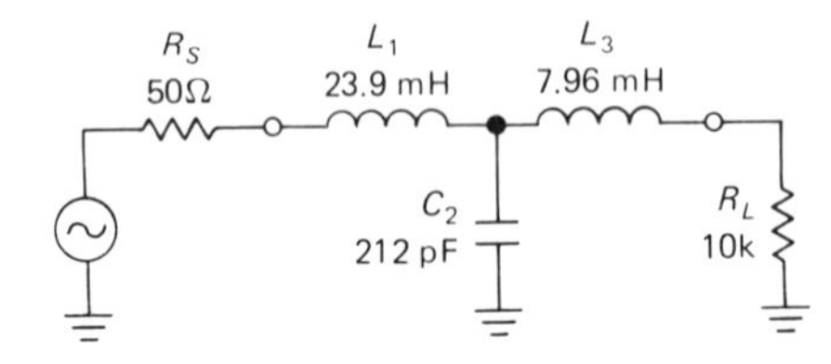

Figure H4

EXAMPLE III

Design a 4-pole low-pass filter for a zero-impedance source (voltage source) and a 75 ohm load, with cutoff frequency of 10MHz.

We use the T configuration, as in the previous example, because $R_S \ll R_L$. The scaling rules give

$$L_1 = \frac{75 \times 1.5307}{2\pi \times 10^7} = 1.83\mu\text{H}$$
$$C_2 = \frac{1.5772}{2\pi \times 10^7 \times 75} = 335\text{pF}$$
$$L_3 = \frac{75 \times 1.0824}{2\pi \times 10^7} = 1.29\mu\text{H}$$
$$C_4 = \frac{0.3827}{2\pi \times 10^7 \times 75} = 81.2\text{pF}$$

The complete filter is shown in Figure H5.

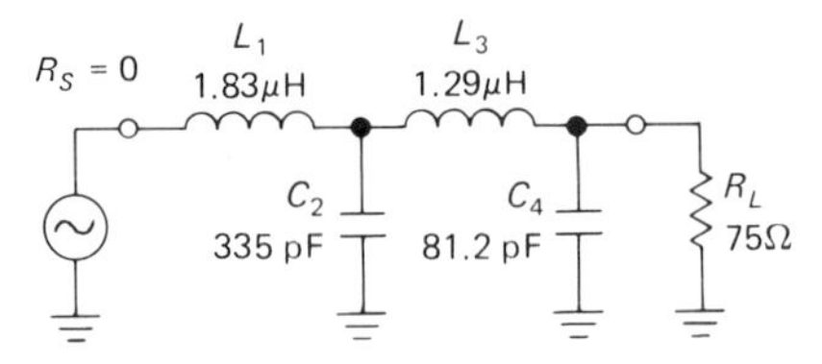

Figure H5

EXAMPLE IV

Design a 2-pole low-pass filter for current-source drive and 1k load impedance, with cutoff frequency of 10kHz.

We use the π configuration, because $R_S \gg R_L$. The scaling rules give

$$C_1 = \frac{1.4142}{2\pi \times 10^4 \times 10^3} = 0.0225\mu\text{F}$$
$$L_2 = \frac{10^3 \times 0.7071}{2\pi \times 10^4} = 11.3\text{mH}$$

The complete filter is shown in Figure H6.

EXAMPLE V

Design a 3-pole high-pass filter for 52 ohm source and load impedances, with cutoff frequency of 6MHz.

We begin with the T configuration, then transform inductors to capacitors, and vice versa, giving

$$C_1 = C_3 = \frac{1}{52 \times 2\pi \times 6 \times 10^6 \times 1.0}$$
$$= 510\text{pF}$$

$$L_2 = \frac{52}{2\pi \times 6 \times 10^6 \times 2.0} = 0.690\mu\text{H}$$

The complete filter is shown in Figure H7.

We could like to emphasize that the field of passive filter design is rich and varied and that this simple table of Butterworth filters doesn't even begin to scratch the surface.

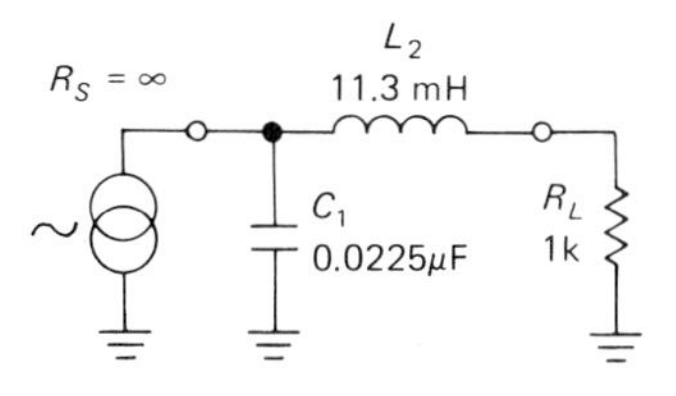

Figure H6

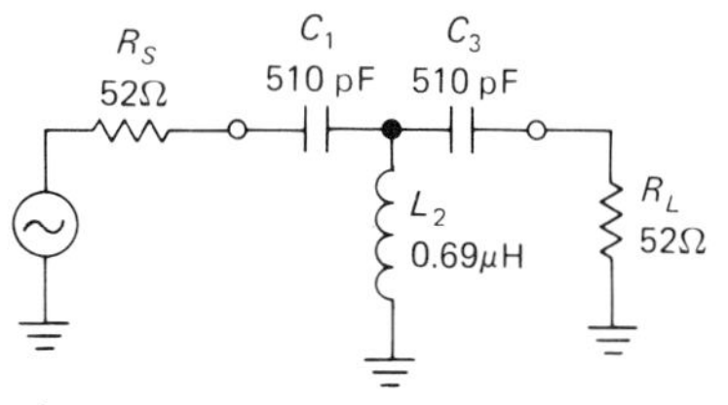

Figure H7

APPENDIX I
ELECTRONICS MAGAZINES AND JOURNALS

In this appendix we've gathered together a selection of magazines that are worth knowing about. Most are well supplied with advertisements proclaiming the specifications and virtues of new ICs, instruments, computers, etc., and these ads are a good source of information about new products; they don't have the nuisance value of ordinary newspaper and magazine ads. There is often a "reader information card" (or "bingo" card) at the back of the magazine on which you can circle the numbers corresponding to advertisements that you're interested in. More information then arrives in the mail within a few weeks. The system works well.

ELECTRONICS MAGAZINES

EDN; Electronic Design; Electronics; Electronic Products. At least one of these magazines should be considered required reading to keep up with new components and design ideas. The advertisements are as important as the articles.

EE Times; Electronic News. Newspapers for the electronics industry.

Electronics and Wireless World. British all-around electronics magazine for hobbyists and professionals.

Ham Radio. The most technical of the amateur radio publications.

Journal of Solid State Circuits (IEEE). Circuit design and new ICs.

QST. Amateur radio magazine from the official ARRL.

Spectrum (IEEE). General-interest electronics magazine, put out by IEEE. Good review articles covering broad range of subjects.

COMPUTER MAGAZINES

Byte. The first large-circulation personal computing magazine. Good tutorials on wide range of topics to which computers can be applied.

Computer Design. Definitive magazine on digital hardware and software techniques for large computer systems.

Computers in Physics. The title says it all.

Dr. Dobbs Journal. For programmers. Emphasis on software systems and design.

MacWorld; MacUser. End-user oriented evaluation of hardware and software; product reviews.

PC Magazine. Helpful magazine for the PC owner.

PC Tech Journal. Semi-technical magazine of PC computing.

PC Week; Infoworld; Macintosh Today. Newspaper-format trade weeklies, with the latest scuttlebutt.

OTHER MAGAZINES

Measurement and Control News. Includes biomedical and chemical instrumentation.

Nuclear Instruments and Methods; Review of Scientific Instruments. Scientific instrumentation.

IC PREFIXES
APPENDIX J

Typical problem: You need to replace an integrated circuit, or at least find some data on it. It says

DM8095N
7410 NS

and lives in a 16-pin DIP. What is it? The 7410 has a familiar sound, so you order a few. A week later they arrive, in 14-pin DIPS! Banging your head on the nearest wall (since you should have known this all along), you realize you're back where you were a week ago, but with a handful of spare 3-input NANDs for consolation.

What's needed is a master list of IC prefixes, from which you can quickly identify the manufacturer. This appendix is our attempt to bring some order out of chaos. We make no pretense of accuracy or completeness, especially since the list is proliferating daily. (That mystery chip is a National Semiconductor 8095 hex three-state TTL buffer, by the way, manufactured in the 10th week of 1974.)

PREFIXES

The various semiconductor manufacturers use distinctive (usually) prefixes in front of the IC number, even if it is an IC type made by many different companies. The DM in the preceding example indicates a digital monolithic IC made by National Semiconductor (also indicated by the NS logo). Here is a list of most of the prefixes now in use:

Prefixes	Manufacturer
ACF, AY, GIC, GP, SPR	General Instrument (GI)
AD, CAV, HAS, HDM	Analog Devices
ADC, DM, DS, LF LFT, LH, LM, NH	National Semiconductor (NSC)
AH	Optical Electronics Inc.
Am	Advanced Micro Devices (AMD)
AM	Datel
AN	Panasonic
Bt	Brooktree
BX, CX	Sony
C, I, i	Intel
CA, CD, CDP	GE/RCA
CA, TDC, MPY, THC, TMC	TRW
CM, HV	Supertex
CLC	Comlinear
CMP, DAC, MAT, OP, PM, REF, SSS	Precision Monolithics
CY	Cypress
D, DF, DG, SI	Siliconix
DS	Dallas Semiconductor
EF, ET, MK, SFC, TDF, TS,	Thomson/Mostek
EP, EPM, PL	Altera
F, μA, μL, U*nx*	Fairchild/NSC
FSS, ZLD	Ferranti
GA	Gazelle
GAL	Lattice
GEL	GE
HA, HI	Harris

HA, HD, HG, HL, HM, HN	Hitachi
HADC, HDAC	Honeywell
HEP, MC, MCC, MCM, MEC, MM, MWM	Motorola
ICH, ICL, ICM, IM	GE/Intersil
IDT	Integrated Device Technology Siemens
IMS	Inmos
INA, ISO, OPA, PWR	Burr-Brown
IR	Sharp
ITT, MIC	ITT
KA	Samsung
L	SGS
L, LD	Siliconix, Siltronics
L, UC	Unitrode
LA, LC	Sanyo
LS	LSI Computer Systems
LT, LTC, LTZ	Linear Technology Corp.
M	Mitsubishi
MA	Analog Systems, Marconi
MAX	Maxim
MB	Fujitsu
MCS	MOS Technology
MIL	Microsystems International
ML, MN, SL, SP, TAB	Plessey
ML, MT	Mitel
MM	Teledyne-Amelco, Monolithic Memories
MN	Micro Networks
MP	Micro Power Systems
MSM	Oki
N, NE, PLS, S, SE, SP	Signetics
*nn*G	Gigabit Logic
NC	Nitron
PA	Apex
PAL	AMD/MMI
R	Rockwell
R, Ray, RC, RM	Raytheon
RD, RF, RM, RT, RU	EG&G Reticon
S	AMI
SFC	ESMF
SG	Silicon General
SN, TL, TLC, TMS	Texas Instruments (TI)
SS	Silicon Systems
T, TA, TC, TD TMM, TMP	Toshiba
OM, PCD, PCF, SAA, SAB, SAF, SCB, SCN, TAA, TBA, TCA, TDA, TEA, U	AEG, Amperex, SGS, Siemens, Signetics, Telefunken
TML	Telmos
TP	Teledyne Philbrick
TPQ, UCN, UCS, UDN, UDS, UHP, ULN, ULS	Sprague
TSC	Teledyne Semiconductor
μPB, μPC, μPD	NEC
V	Amtel
VA, VC	VTC
VT	VLSI Technology Inc. (VTI)
X	Xicor
XC	Xilinx
XR	Exar
Z	Zilog
ZN	Ferranti
5082-*nnnn*	Hewlett-Packard (HP)

Suffixes

Suffix letters indicate package type and temperature range. There are three standard temperature ranges: "Military" (−55°C to +125°C), "Industrial" (−25°C to +85°C), and "Commercial" (0°C to +70°C). Commercial is adequate for anything intended for use in normal indoor environments. As luck would have it, each manufacturer has its own set of suffixes, subject to frequent modification. Be sure to look up the correct suffix before you order, or ask the distributor for assistance.

DATE CODES

Most ICs and transistors, and many other electronic components, are stamped with a simple four-digit code giving date of manufacture: the first two digits are the year, the last two are the week of the year. In the example given earlier, 7410 means the second week of March 1974. They're sometimes useful, for example to estimate the age of components that have a finite useful life (electrolytic capacitors, for instance); unfortunately the components with the shortest life (batteries) are often purposely coded so you can't figure out the date. If you get a batch of ICs with an abnormally high failure rate (most manufacturers test only a sample of each batch; typically 0.01% to 0.1% of the ICs you buy will not meet specifications), avoid replacements with the same date code. Date codes can also help you estimate the date of manufacture of commercial electronic equipment. Since ICs don't become stale, there's no reason to avoid an IC with an old date code.

APPENDIX K
DATA SHEETS

In this appendix we have reproduced three data sheets just as they were printed by the manufacturer. We chose representative or popular devices, looking especially for data sheets that were comprehensive and clear.

On the following pages you will find data sheets for these devices:

2N4400-4401 A popular signal transistor (from the *Motorola Semiconductor Library, Vol. 1*, 1974. (Courtesy of Motorola Semiconductor Products Inc.)

LF411-412 A popular series of JFET operational amplifiers (from the *National Semiconductor Linear Data Book*, Vol. 1, 1988. (Courtesy of National Semiconductor Corp.)

LM317 A popular 3-terminal adjustable positive voltage regulator (from the *National Semiconductor Linear Data Book*, 1978). (Courtesy of National Semiconductor Corp.)

2N4400
2N4401

NPN SILICON SWITCHING & AMPLIFIER TRANSISTORS

AUGUST 1966 — DS 5198

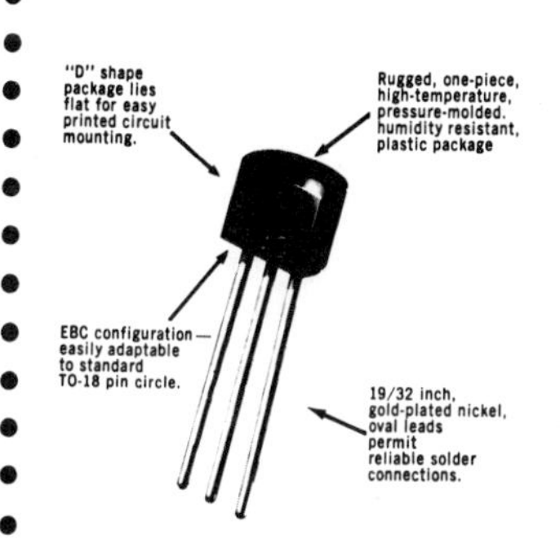

NPN SILICON ANNULAR* TRANSISTORS

. . . designed for general purpose switching and amplifier applications and for complementary circuitry with PNP types 2N4402 and 2N4403.

- High Voltage Ratings — BV_{CEO} = 40 V minimum
- Current Gain Specified from 0.1 mA to 500 mA
- Low Saturation Voltage
  $V_{CE(sat)}$ = 0.4 V maximum @ I_C = 150 mA
- Complete Switching and Amplifier Specifications
- One-Piece, Injection-Molded Unibloc† Package

MAXIMUM RATINGS

Characteristic	Symbol	Rating	Unit
Collector-Emitter Voltage	V_{CEO}	40	Vdc
Collector-Base Voltage	V_{CB}	60	Vdc
Emitter-Base Voltage	V_{EB}	6	Vdc
Collector Current - Continuous	I_C	600	mAdc
Total Device Dissipation T_A = 25°C	P_D	310	mW
Derate above 25°C		2.81	mW/°C
Operating & Storage Junction Temperature Range	T_J, T_{stg}	-55 to +135	°C

THERMAL CHARACTERISTICS

Characteristic	Symbol	Max	Unit
Thermal Resistance, Junction to Case	θ_{JC}	0.137	°C/mW
Thermal Resistance, Junction to Ambient	θ_{JA}	0.357	°C/mW

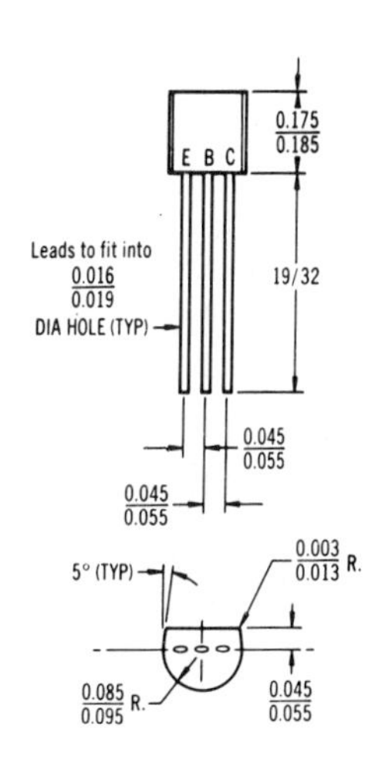

TO-92 OUTLINE

*Annular Semiconductors patented by Motorola Inc.
†Trademark of Motorola Inc.

ELECTRICAL CHARACTERISTICS (T_A = 25°C unless otherwise noted)

Characteristic	Fig. No.	Symbol	Min	Max	Unit
OFF CHARACTERISTICS					
Collector-Emitter Breakdown Voltage* (I_C = 1 mAdc, I_B = 0)		BV_{CEO}*	40	—	Vdc
Collector-Base Breakdown Voltage (I_C = 0.1 mAdc, I_E = 0)		BV_{CBO}	60	—	Vdc
Emitter-Base Breakdown Voltage (I_E = 0.1 mAdc, I_C = 0)		BV_{EBO}	6	—	Vdc
Collector Cutoff Current (V_{CE} = 35 Vdc, $V_{EB(off)}$ = 0.4 Vdc)		I_{CEX}	—	0.1	μAdc
Base Cutoff Current (V_{CE} = 35 Vdc, $V_{EB(off)}$ = 0.4 Vdc)		I_{BL}	—	0.1	μAdc
ON CHARACTERISTICS					
DC Current Gain	15	h_{FE}			—
(I_C = 0.1 mAdc, V_{CE} = 1 Vdc) 2N4401			20	—	
(I_C = 1 mAdc, V_{CE} = 1 Vdc) 2N4400			20	—	
2N4401			40	—	
(I_C = 10 mAdc, V_{CE} = 1 Vdc) 2N4400			40	—	
2N4401			80	—	
(I_C = 150 mAdc, V_{CE} = 1 Vdc)* 2N4400			50	150	
2N4401			100	300	
(I_C = 500 mAdc, V_{CE} = 2 Vdc)* 2N4400			20	—	
2N4401			40	—	
Collector-Emitter Saturation Voltage*	16, 17, 18	$V_{CE(sat)}$			Vdc
(I_C = 150 mAdc, I_B = 15 mAdc)			—	0.4	
(I_C = 500 mAdc, I_B = 50 mAdc)			—	0.75	
Base-Emitter Saturation Voltage*	17, 18	$V_{BE(sat)}$			Vdc
(I_C = 150 mAdc, I_B = 15 mAdc)			0.75	0.95	
(I_C = 500 mAdc, I_B = 50 mAdc)			—	1.2	
SMALL-SIGNAL CHARACTERISTICS					
Current-Gain — Bandwidth Product (I_C = 20 mAdc, V_{CE} = 10 Vdc, f = 100 MHz) 2N4400		f_T	200	—	MHz
2N4401			250	—	
Collector-Base Capacitance (V_{CB} = 5 Vdc, I_E = 0, f = 100 kHz, emitter guarded)	3	C_{cb}	—	6.5	pF
Emitter-Base Capacitance (V_{BE} = 0.5 Vdc, I_C = 0, f = 100 kHz, collector guarded)	3	C_{eb}	—	30	pF
Input Impedance (I_C = 1 mAdc, V_{CE} = 10 Vdc, f = 1 kHz) 2N4400	12	h_{ie}	500	7.5k	ohms
2N4401			1.0k	15k	
Voltage Feedback Ratio (I_C = 1 mAdc, V_{CE} = 10 Vdc, f = 1 kHz)	13	h_{re}	0.1	8	X 10^{-4}
Small-Signal Current Gain (I_C = 1 mAdc, V_{CE} = 10 Vdc, f = 1 kHz) 2N4400	11	h_{fe}	20	250	—
2N4401			40	500	
Output Admittance (I_C = 1 mAdc, V_{CE} = 10 Vdc, f = 1 kHz)	14	h_{oe}	1	30	μmhos
SWITCHING CHARACTERISTICS					
Delay Time — V_{CC} = 30 Vdc, $V_{EB(off)}$ = 2 Vdc,	1, 5	t_d	—	15	ns
Rise Time — I_C = 150 mAdc, I_{B1} = 15 mAdc	1, 5, 6	t_r	—	20	ns
Storage Time — V_{CC} = 30 Vdc, I_C = 150 mAdc,	2, 7	t_s	—	225	ns
Fall Time — I_{B1} = I_{B2} = 15 mAdc	2, 8	t_f	—	30	ns

*Pulse Test: Pulse Width ≤ 300 μs, Duty Cycle ≤ 2%

SWITCHING TIME EQUIVALENT TEST CIRCUITS

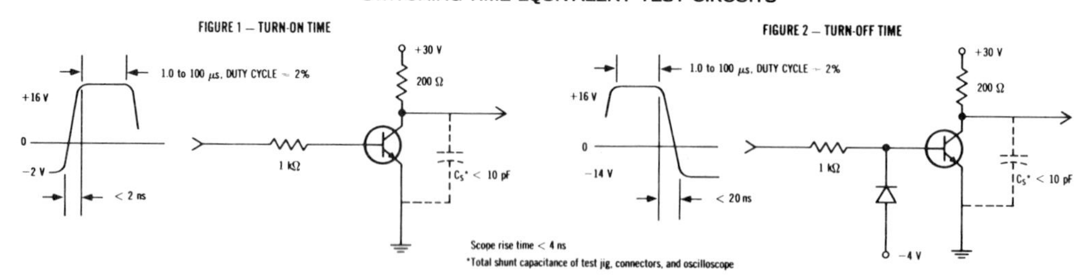

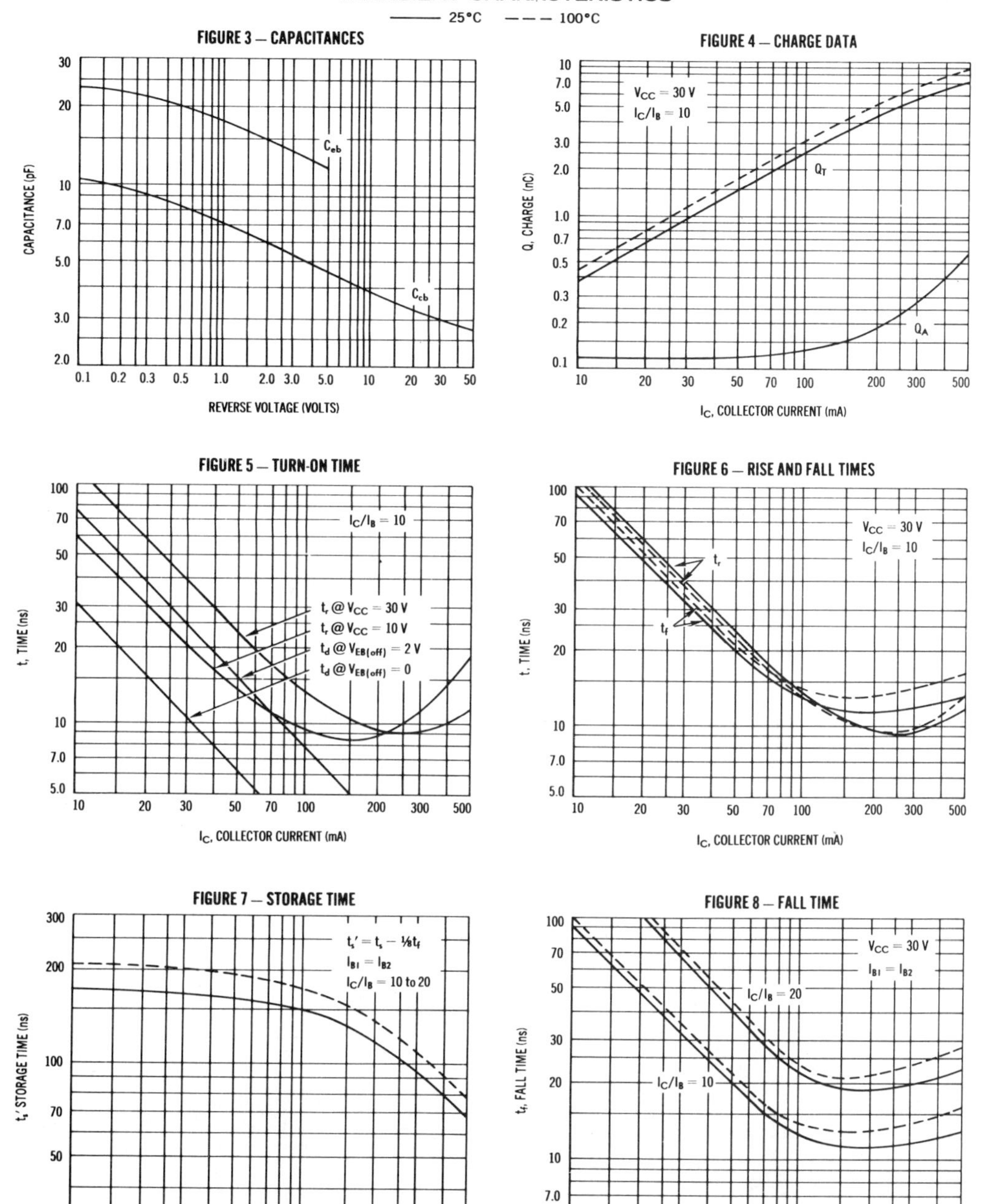
TRANSIENT CHARACTERISTICS
—— 25°C — — — 100°C
FIGURE 3 — CAPACITANCES
CAPACITANCE (pF)
C_eb
C_cb
REVERSE VOLTAGE (VOLTS)
FIGURE 4 — CHARGE DATA
Q, CHARGE (nC)
V_CC = 30 V
I_C/I_B = 10
Q_T
Q_A
I_C, COLLECTOR CURRENT (mA)
FIGURE 5 — TURN-ON TIME
t, TIME (ns)
I_C/I_B = 10
t_r @ V_CC = 30 V
t_r @ V_CC = 10 V
t_d @ V_EB(off) = 2 V
t_d @ V_EB(off) = 0
I_C, COLLECTOR CURRENT (mA)
FIGURE 6 — RISE AND FALL TIMES
t, TIME (ns)
V_CC = 30 V
I_C/I_B = 10
t_r
t_f
I_C, COLLECTOR CURRENT (mA)
FIGURE 7 — STORAGE TIME
t_s' STORAGE TIME (ns)
t_s' = t_s − 1/8 t_f
I_B1 = I_B2
I_C/I_B = 10 to 20
I_C, COLLECTOR CURRENT (mA)
FIGURE 8 — FALL TIME
t_f, FALL TIME (ns)
V_CC = 30 V
I_B1 = I_B2
I_C/I_B = 20
I_C/I_B = 10
I_C, COLLECTOR CURRENT (mA)

SMALL-SIGNAL CHARACTERISTICS

NOISE FIGURE

V_{CE} = 10 Vdc, T_A = 25°C

FIGURE 9 — FREQUENCY EFFECTS

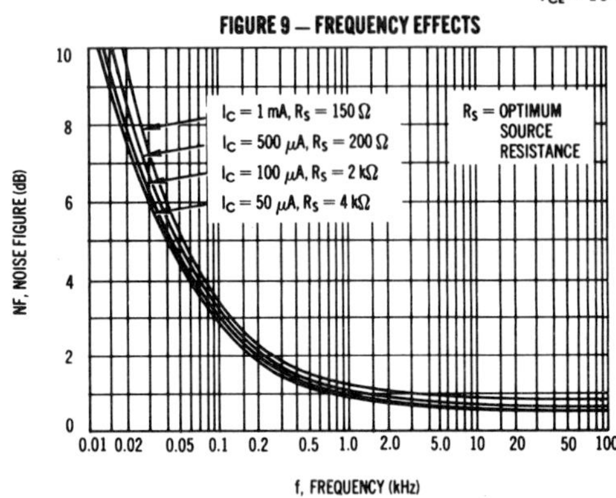

FIGURE 10 — SOURCE RESISTANCE EFFECTS

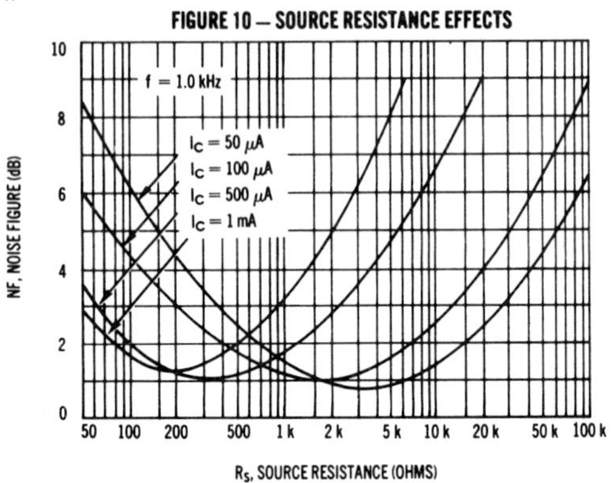

h PARAMETERS

V_{CE} = 10 Vdc, f = 1 kHz, T_A = 25°C

This group of graphs illustrates the relationship between h_{fe} and other "h" parameters for this series of transistors. To obtain these curves, a high-gain and a low-gain unit were selected from both the 2N4400 and 2N4401 lines, and the same units were used to develop the correspondingly-numbered curves on each graph.

FIGURE 11 — CURRENT GAIN

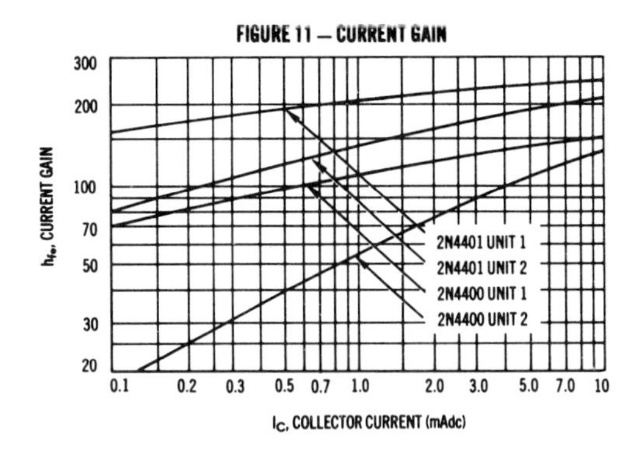

FIGURE 12 — INPUT IMPEDANCE

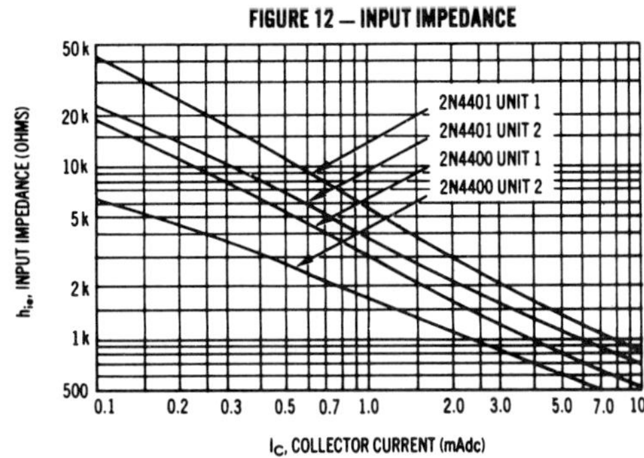

FIGURE 13 — VOLTAGE FEEDBACK RATIO

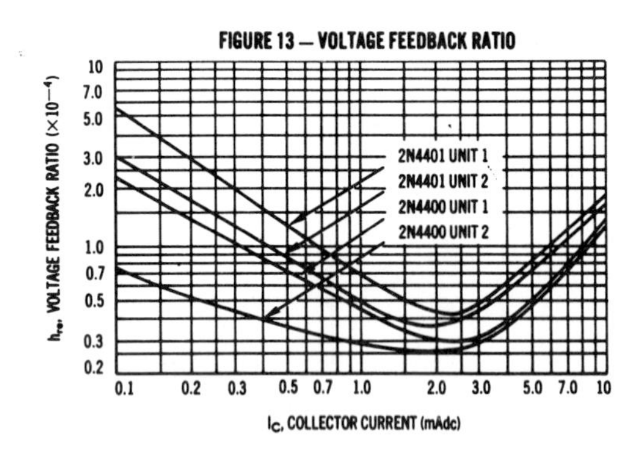

FIGURE 14 — OUTPUT ADMITTANCE

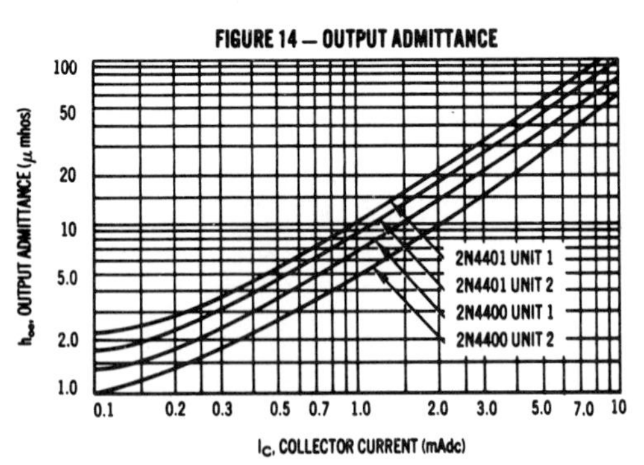

STATIC CHARACTERISTICS

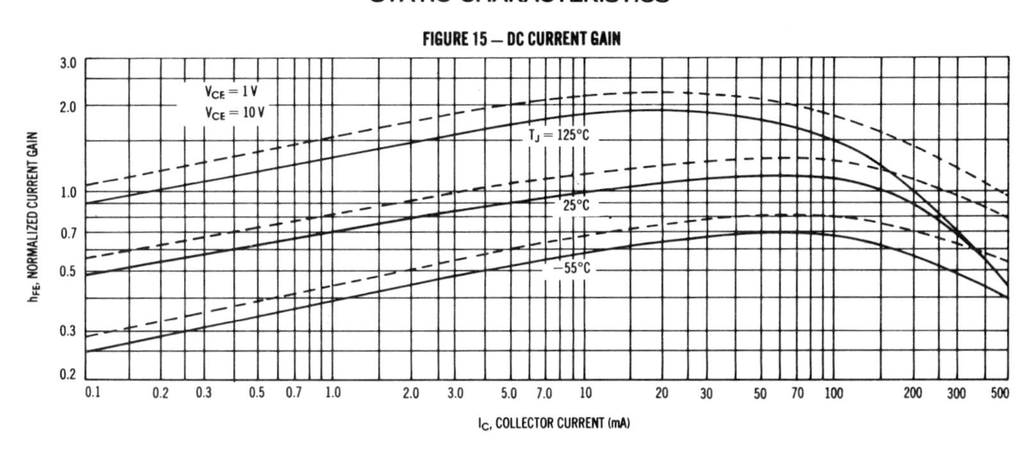

FIGURE 15 — DC CURRENT GAIN

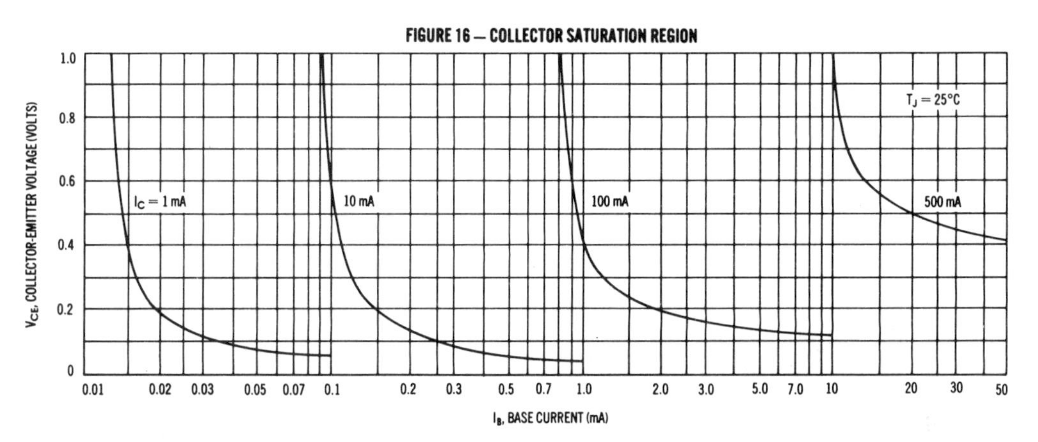

FIGURE 16 — COLLECTOR SATURATION REGION

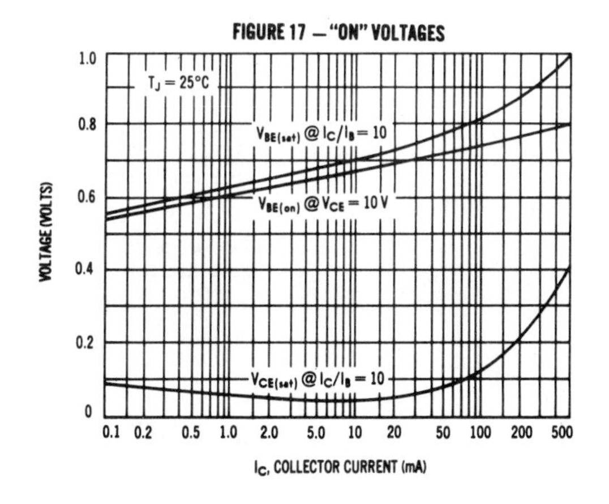

FIGURE 17 — "ON" VOLTAGES

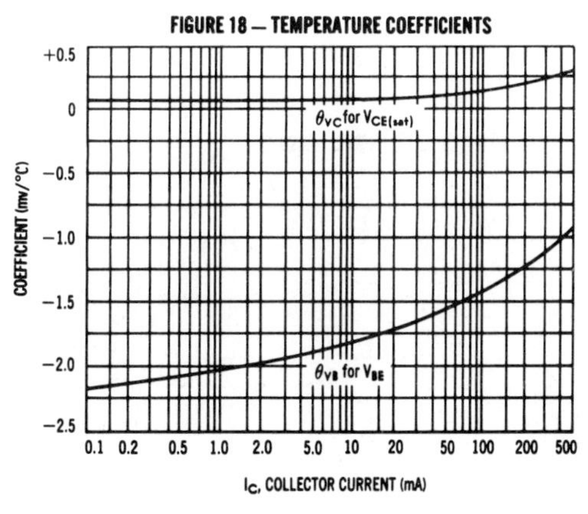

FIGURE 18 — TEMPERATURE COEFFICIENTS

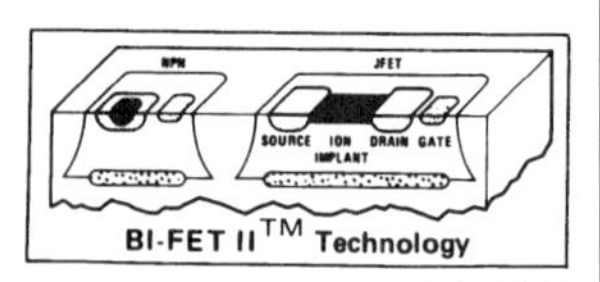

LF411A/LF411 Low Offset, Low Drift JFET Input Operational Amplifier

July 1987

General Description

These devices are low cost, high speed, JFET input operational amplifiers with very low input offset voltage and guaranteed input offset voltage drift. They require low supply current yet maintain a large gain bandwidth product and fast slew rate. In addition, well matched high voltage JFET input devices provide very low input bias and offset currents. The LF411 is pin compatible with the standard LM741 allowing designers to immediately upgrade the overall performance of existing designs.

These amplifiers may be used in applications such as high speed integrators, fast D/A converters, sample and hold circuits and many other circuits requiring low input offset voltage and drift, low input bias current, high input impedance, high slew rate and wide bandwidth.

Features

- Internally trimmed offset voltage — 0.5 mV(max)
- Input offset voltage drift — 10 μV/°C(max)
- Low input bias current — 50 pA
- Low input noise current — 0.01 pA/$\sqrt{Hz}$
- Wide gain bandwidth — 3 MHz(min)
- High slew rate — 10V/μs(min)
- Low supply current — 1.8 mA
- High input impedance — $10^{12}\Omega$
- Low total harmonic distortion $A_V = 10$, $R_L = 10k$, $V_O = 20$ Vp-p, BW = 20 Hz − 20 kHz — <0.02%
- Low 1/f noise corner — 50 Hz
- Fast settling time to 0.01% — 2 μs

Typical Connection

TL/H/5655-1

Ordering Information

LF411XYZ

X indicates electrical grade
Y indicates temperature range
"M" for military
"C" for commercial
Z indicates package type
"H" or "N"

Connection Diagrams

Metal Can Package

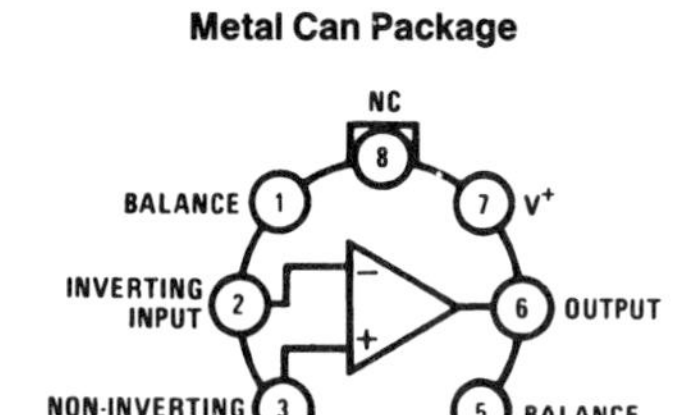

TL/H/5655-5

Top View

Note: Pin 4 connected to case.

Order Number LF411AMH, LF411MH, LF411ACH or LF411CH
See NS Package Number H08B

Dual-In-Line Package

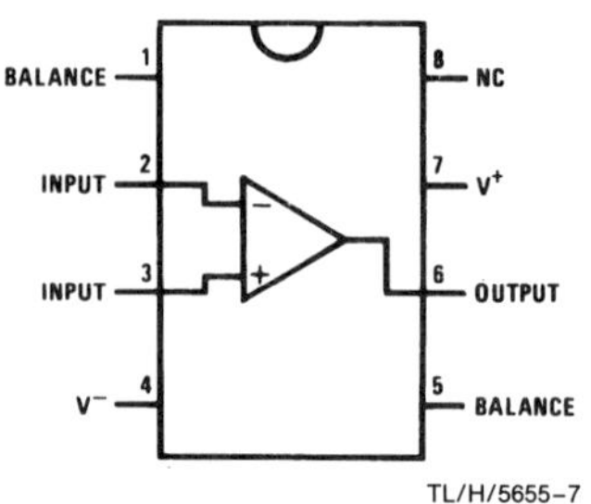

TL/H/5655-7

Top View
Order Number
LF411ACN or LF411CN
See NS Package Number N08E

Simplified Schematic

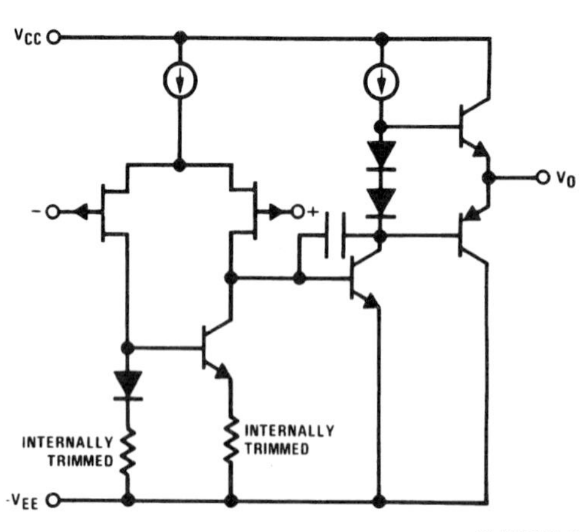

TL/H/5655-6

BI-FET II™ is a trademark of National Semiconductor Corporation.

Absolute Maximum Ratings

If Military/Aerospace specified devices are required, please contact the National Semiconductor Sales Office/Distributors for availability and specifications. (Note 8)

	LF411A	LF411
Supply Voltage	±22V	±18V
Differential Input Voltage	±38V	±30V
Input Voltage Range (Note 1)	±19V	±15V
Output Short Circuit Duration	Continuous	Continuous

	H Package	N Package
Power Dissipation (Notes 2 and 9)	670 mW	670 mW
$T_j$max	150°C	115°C
θ_jA	225°C/W (Still Air) 160°C/W (400 LF/min Air Flow)	120°C/W
θ_jC	25°C/W	
Operating Temp. Range	(Note 3)	(Note 3)
Storage Temp. Range	$-65°C \le T_A \le 150°C$	$-65°C \le T_A \le 150°C$
Lead Temp. (Soldering, 10 sec.)	260°C	260°C

ESD rating to be determined.

DC Electrical Characteristics (Note 4)

Symbol	Parameter	Conditions		LF411A			LF411			Units
				Min	Typ	Max	Min	Typ	Max	
V_{OS}	Input Offset Voltage	$R_S = 10\ k\Omega$, $T_A = 25°C$			0.3	0.5		0.8	2.0	mV
$\Delta V_{OS}/\Delta T$	Average TC of Input Offset Voltage	$R_S = 10\ k\Omega$ (Note 5)			7	10		7	20 (Note 5)	μV/°C
I_{OS}	Input Offset Current	$V_S = \pm 15V$ (Notes 4, 6)	$T_j = 25°C$		25	100		25	100	pA
			$T_j = 70°C$			2			2	nA
			$T_j = 125°C$			25			25	nA
I_B	Input Bias Current	$V_S = \pm 15V$ (Notes 4, 6)	$T_j = 25°C$		50	200		50	200	pA
			$T_j = 70°C$			4			4	nA
			$T_j = 125°C$			50			50	nA
R_{IN}	Input Resistance	$T_j = 25°C$			10^{12}			10^{12}		Ω
A_{VOL}	Large Signal Voltage Gain	$V_S = \pm 15V$, $V_O = \pm 10V$, $R_L = 2k$, $T_A = 25°C$		50	200		25	200		V/mV
		Over Temperature		25	200		15	200		V/mV
V_O	Output Voltage Swing	$V_S = \pm 15V$, $R_L = 10k$		±12	±13.5		±12	±13.5		V
V_{CM}	Input Common-Mode Voltage Range			±16	+19.5		±11	+14.5		V
					−16.5			−11.5		V
CMRR	Common-Mode Rejection Ratio	$R_S \le 10k$		80	100		70	100		dB
PSRR	Supply Voltage Rejection Ratio	(Note 7)		80	100		70	100		dB
I_S	Supply Current				1.8	2.8		1.8	3.4	mA

AC Electrical Characteristics (Note 4)

Symbol	Parameter	Conditions	LF411A			LF411			Units
			Min	Typ	Max	Min	Typ	Max	
SR	Slew Rate	$V_S = \pm 15V$, $T_A = 25°C$	10	15		8	15		V/μs
GBW	Gain-Bandwidth Product	$V_S = \pm 15V$, $T_A = 25°C$	3	4		2.7	4		MHz
e_n	Equivalent Input Noise Voltage	$T_A = 25°C$, $R_S = 100\Omega$, f = 1 kHz		25			25		$nV/\sqrt{Hz}$
i_n	Equivalent Input Noise Current	$T_A = 25°C$, f = 1 kHz		0.01			0.01		$pA/\sqrt{Hz}$

Note 1: Unless otherwise specified the absolute maximum negative input voltage is equal to the negative power supply voltage.

Note 2: For operating at elevated temperature, these devices must be derated based on a thermal resistance of θ_jA.

Note 3: These devices are available in both the commercial temperature range $0°C \leq T_A \leq 70°C$ and the military temperature range $-55°C \leq T_A \leq 125°C$. The temperature range is designated by the position just before the package type in the device number. A "C" indicates the commercial temperature range and an "M" indicates the military temperature range. The military temperature range is available in "H" package only.

Note 4: Unless otherwise specified, the specifications apply over the full temperature range and for $V_S = \pm 20V$ for the LF411A and for $V_S = \pm 15V$ for the LF411. V_{OS}, I_B, and I_{OS} are measured at $V_{CM} = 0$.

Note 5: The LF411A is 100% tested to this specification. The LF411 is sample tested to insure at least 90% of the units meet this specification.

Note 6: The input bias currents are junction leakage currents which approximately double for every 10°C increase in the junction temperature, T_j. Due to limited production test time, the input bias currents measured are correlated to junction temperature. In normal operation the junction temperature rises above the ambient temperature as a result of internal power dissipation, P_D. $T_j = T_A + \theta_{jA} P_D$ where θ_{jA} is the thermal resistance from junction to ambient. Use of a heat sink is recommended if input bias current is to be kept to a minimum.

Note 7: Supply voltage rejection ratio is measured for both supply magnitudes increasing or decreasing simultaneously in accordance with common practice, from $\pm 15V$ to $\pm 5V$ for the LF411 and from $\pm 20V$ to $\pm 5V$ for the LF411A.

Note 8: Refer to RETS 411AX for LF411AMH military specifications and to RETS 411X for LF411MH military specifications.

Note 9: Max. Power Dissipation is defined by the package characteristics. Operating the part near the Max. Power Dissipation may cause the part to operate outside guaranteed limits.

Typical Performance Characteristics

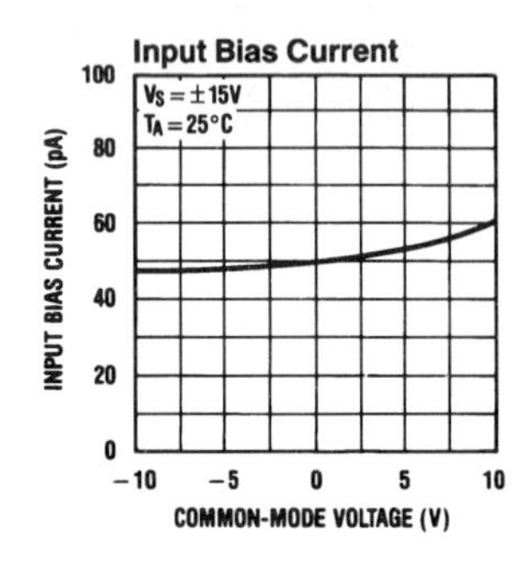

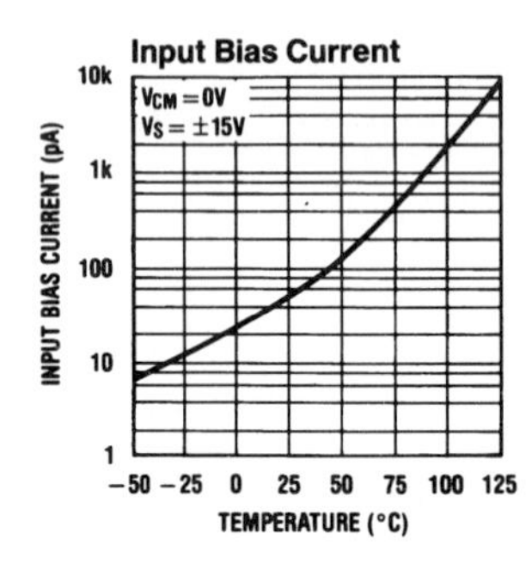

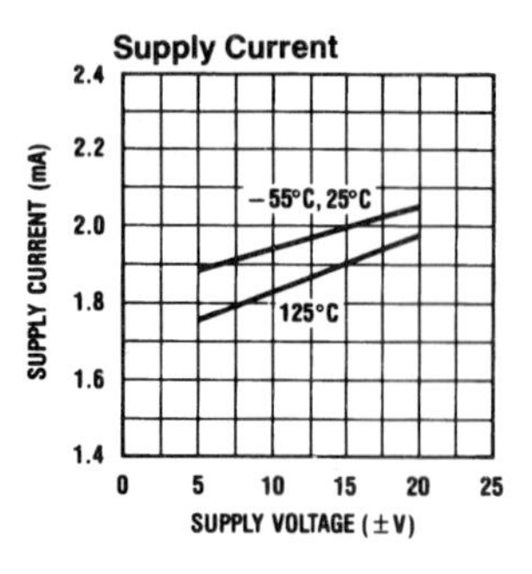

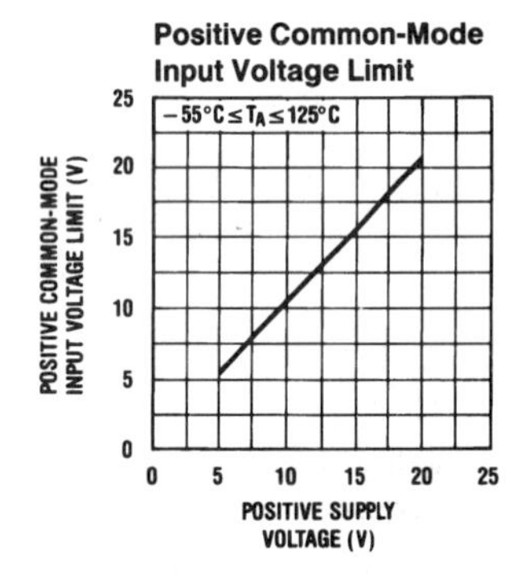

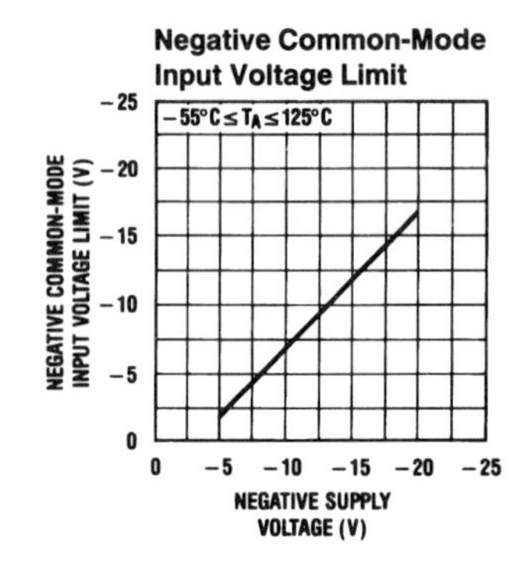

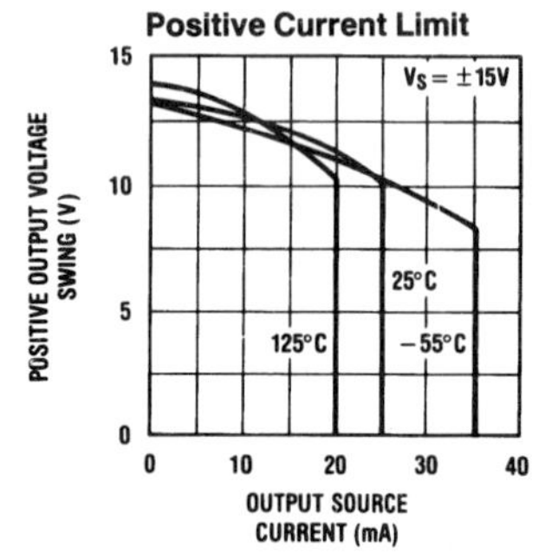

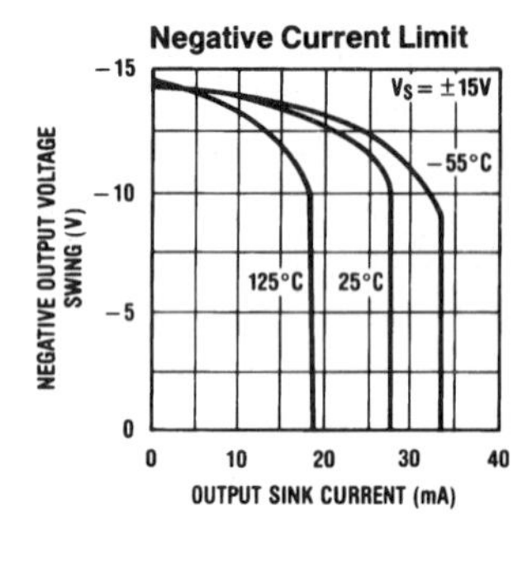

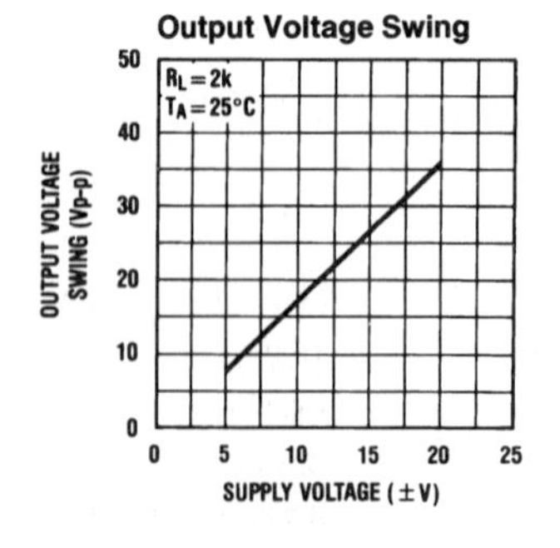

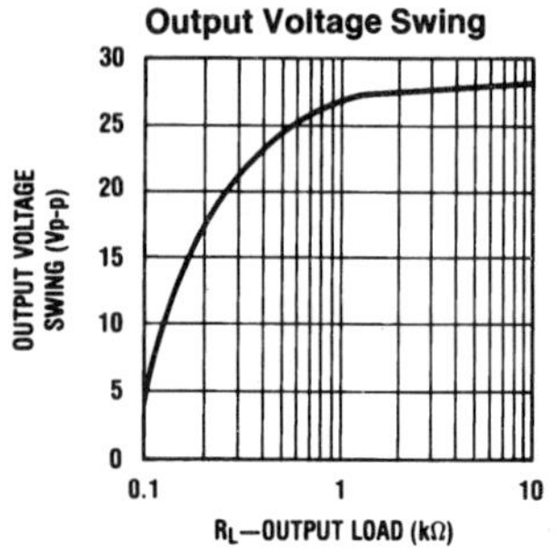

TL/H/5655–2

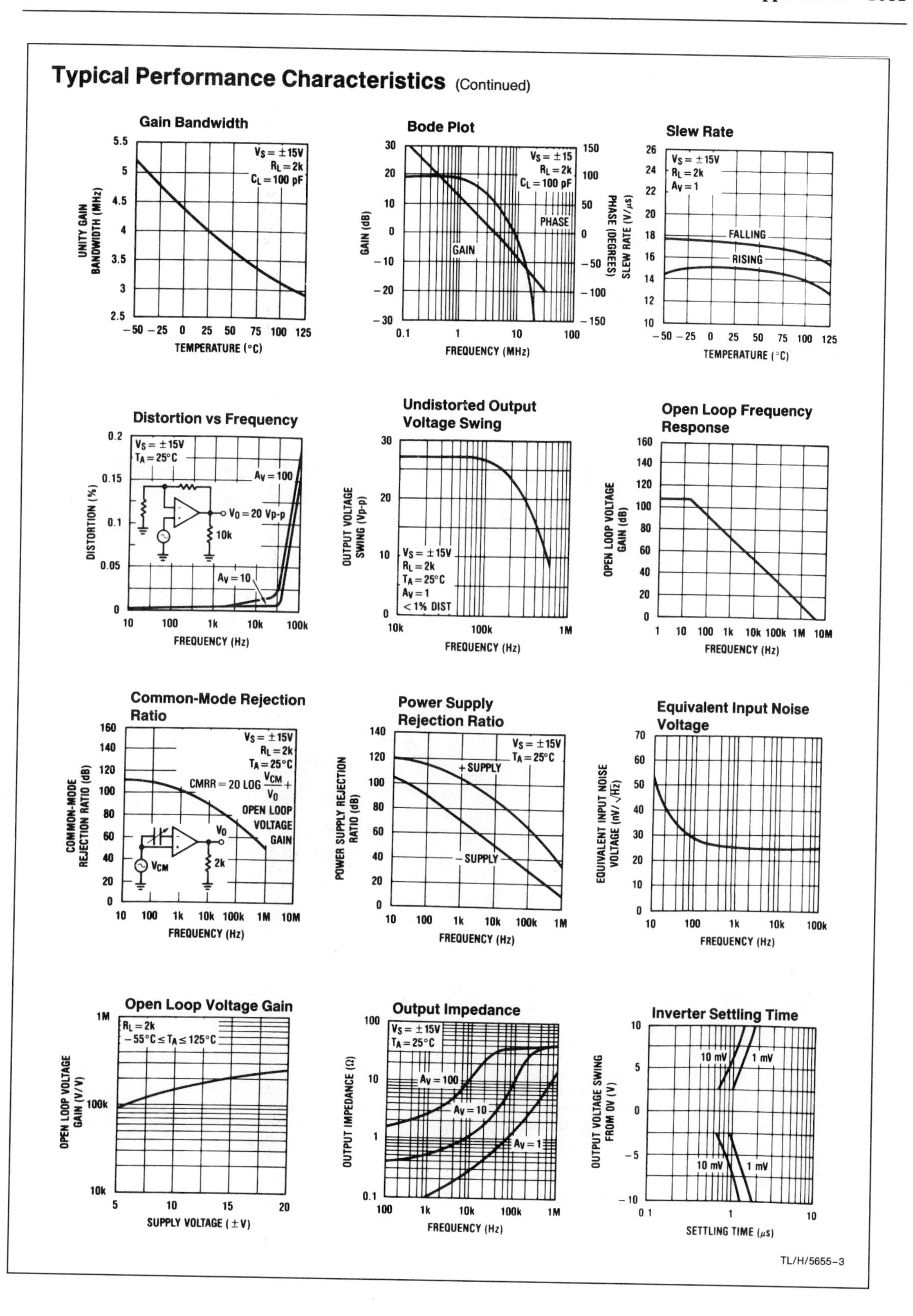
Typical Performance Characteristics (Continued)
Gain Bandwidth
UNITY GAIN BANDWIDTH (MHz)
TEMPERATURE (°C)
Vs = ±15V
RL = 2k
CL = 100 pF
Bode Plot
GAIN (dB)
PHASE (DEGREES)
FREQUENCY (MHz)
GAIN
PHASE
Slew Rate
SLEW RATE (V/μs)
TEMPERATURE (°C)
FALLING
RISING
Av = 1
Distortion vs Frequency
DISTORTION (%)
FREQUENCY (Hz)
TA = 25°C
Av = 100
Av = 10
V0 = 20 Vp-p
10k
Undistorted Output Voltage Swing
OUTPUT VOLTAGE SWING (Vp-p)
< 1% DIST
Open Loop Frequency Response
OPEN LOOP VOLTAGE GAIN (dB)
Common-Mode Rejection Ratio
COMMON-MODE REJECTION RATIO (dB)
CMRR = 20 LOG VCM/V0 + OPEN LOOP VOLTAGE GAIN
VCM
2k
Power Supply Rejection Ratio
POWER SUPPLY REJECTION RATIO (dB)
+ SUPPLY
− SUPPLY
Equivalent Input Noise Voltage
EQUIVALENT INPUT NOISE VOLTAGE (nV/√Hz)
Open Loop Voltage Gain
OPEN LOOP VOLTAGE GAIN (V/V)
SUPPLY VOLTAGE (±V)
−55°C ≤ TA ≤ 125°C
Output Impedance
OUTPUT IMPEDANCE (Ω)
Av = 100
Av = 10
Av = 1
Inverter Settling Time
OUTPUT VOLTAGE SWING FROM 0V (V)
SETTLING TIME (μs)
10 mV
1 mV
TL/H/5655–3

Pulse Response $R_L = 2\ k\Omega$, $C_L 10\ pF$

Small Signal Inverting

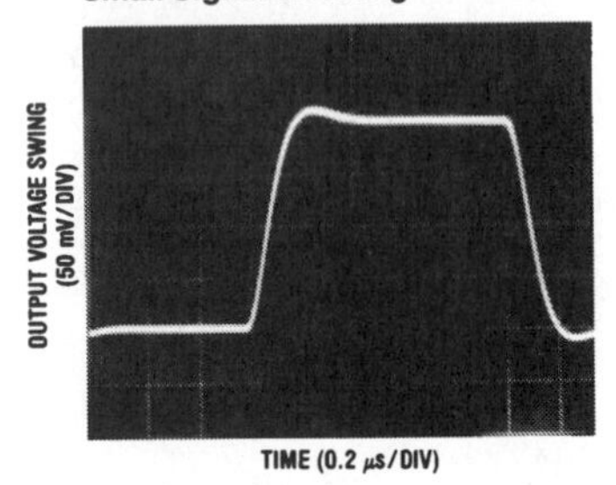

Small Signal Non-Inverting

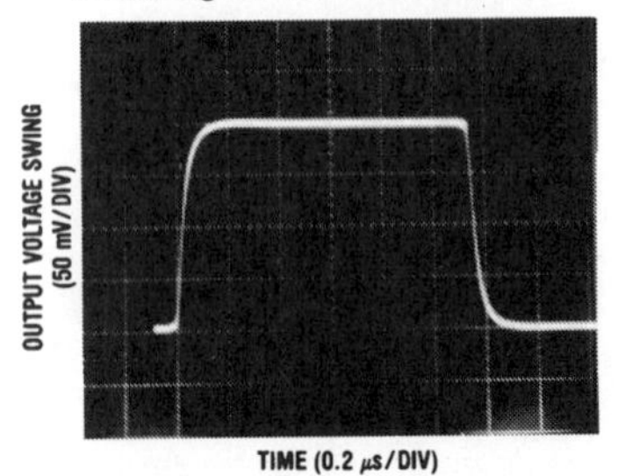

Large Signal Inverting

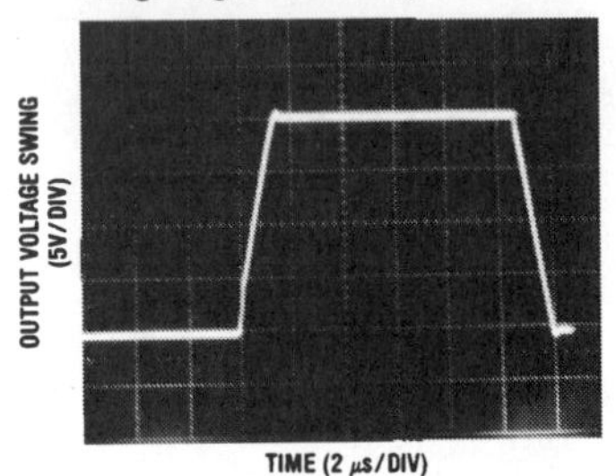

Large Signal Non-Inverting

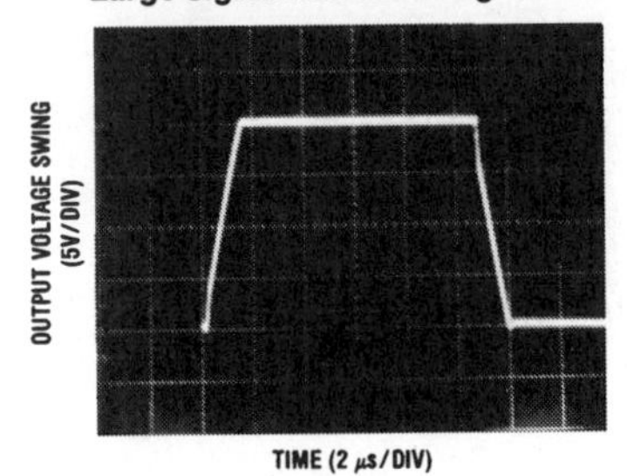

Current Limit ($R_L = 100\Omega$)

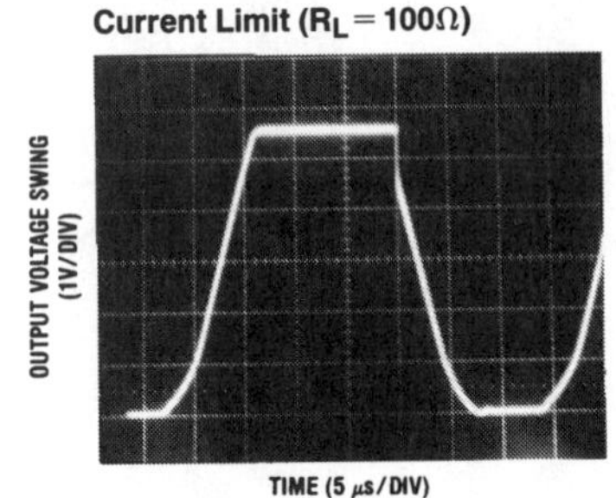

TL/H/5655–4

Application Hints

The LF411 series of internally trimmed JFET input op amps (BI-FET II™) provide very low input offset voltage and guaranteed input offset voltage drift. These JFETs have large reverse breakdown voltages from gate to source and drain eliminating the need for clamps across the inputs. Therefore, large differential input voltages can easily be accommodated without a large increase in input current. The maximum differential input voltage is independent of the supply voltages. However, neither of the input voltages should be allowed to exceed the negative supply as this will cause large currents to flow which can result in a destroyed unit.

Exceeding the negative common-mode limit on either input will force the output to a high state, potentially causing a reversal of phase to the output. Exceeding the negative common-mode limit on both inputs will force the amplifier output to a high state. In neither case does a latch occur since raising the input back within the common-mode range again puts the input stage and thus the amplifier in a normal operating mode.

Exceeding the positive common-mode limit on a single input will not change the phase of the output; however, if both inputs exceed the limit, the output of the amplifier may be forced to a high state.

Application Hints (Continued)

The amplifier will operate with a common-mode input voltage equal to the positive supply; however, the gain bandwidth and slew rate may be decreased in this condition. When the negative common-mode voltage swings to within 3V of the negative supply, an increase in input offset voltage may occur.

The LF411 is biased by a zener reference which allows normal circuit operation on ±4.5V power supplies. Supply voltages less than these may result in lower gain bandwidth and slew rate.

The LF411 will drive a 2 kΩ load resistance to ±10V over the full temperature range. If the amplifier is forced to drive heavier load currents, however, an increase in input offset voltage may occur on the negative voltage swing and finally reach an active current limit on both positive and negative swings.

Precautions should be taken to ensure that the power supply for the integrated circuit never becomes reversed in polarity or that the unit is not inadvertently installed backwards in a socket as an unlimited current surge through the resulting forward diode within the IC could cause fusing of the internal conductors and result in a destroyed unit.

Because these amplifiers are JFET rather than MOSFET input op amps they do not require special handling.

As with most amplifiers, care should be taken with lead dress, component placement and supply decoupling in order to ensure stability. For example, resistors from the output to an input should be placed with the body close to the input to minimize "pick-up" and maximize the frequency of the feedback pole by minimizing the capacitance from the input to ground.

A feedback pole is created when the feedback around any amplifier is resistive. The parallel resistance and capacitance from the input of the device (usually the inverting input) to AC ground set the frequency of the pole. In many instances the frequency of this pole is much greater than the expected 3 dB frequency of the closed loop gain and consequently there is negligible effect on stability margin. However, if the feedback pole is less than approximately 6 times the expected 3 dB frequency, a lead capacitor should be placed from the output to the input of the op amp. The value of the added capacitor should be such that the RC time constant of this capacitor and the resistance it parallels is greater than or equal to the original feedback pole time constant.

Typical Applications

Ultra High Speed Current Booster

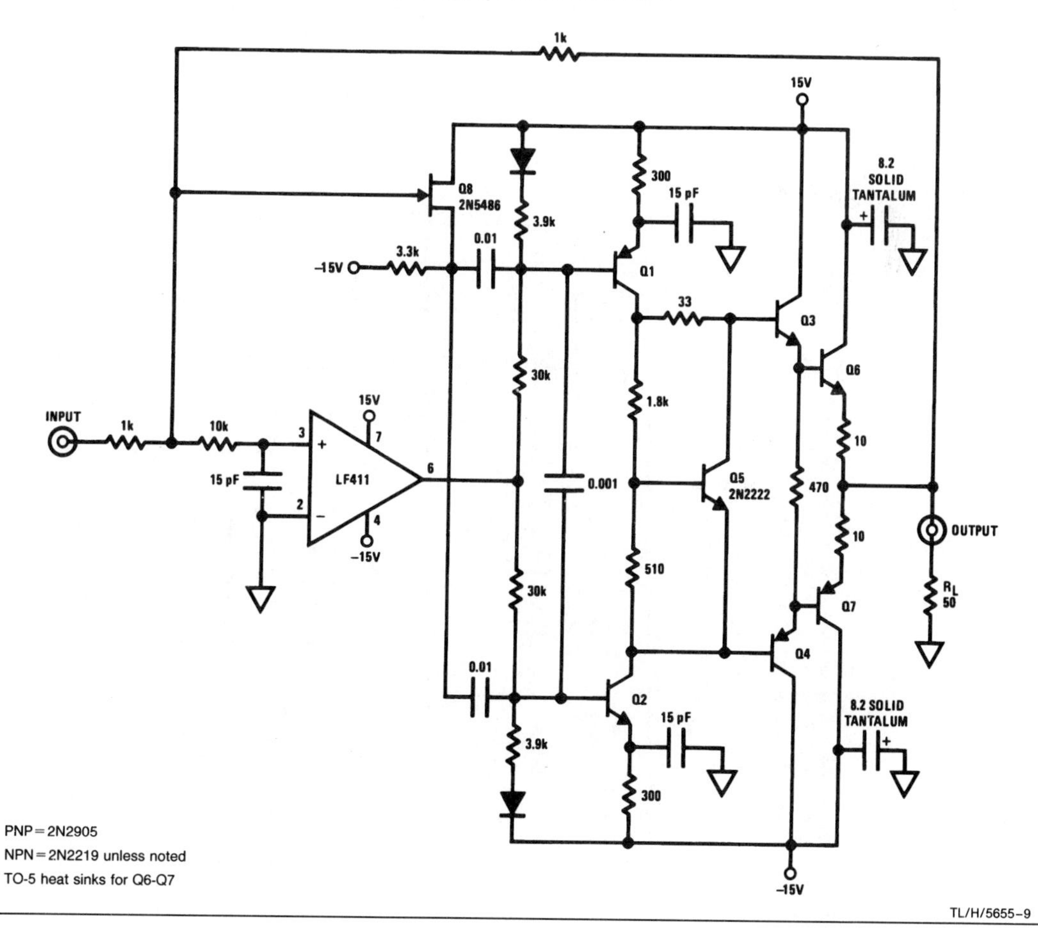

Typical Applications (Continued)

10-Bit Linear DAC with No V_{OS} Adjust

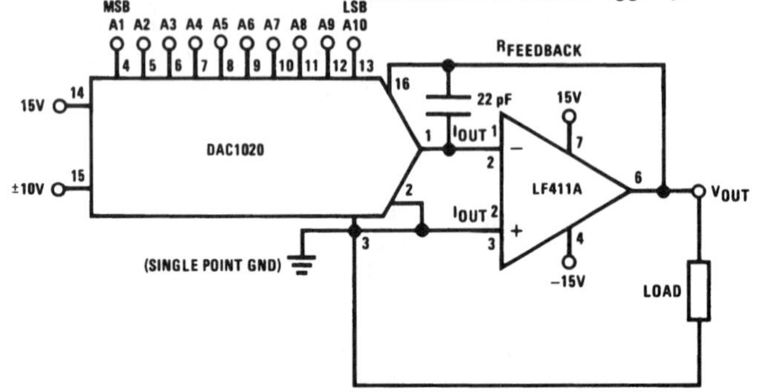

$$V_{OUT} = -V_{REF}\left(\frac{A1}{2} + \frac{A2}{4} + \frac{A3}{8} + \cdots \frac{A10}{1024}\right)$$

$$-10V \le V_{REF} \le 10V$$

$$0 \le V_{OUT} \le -\frac{1023}{1024} V_{REF}$$

where $A_N = 1$ if the A_N digital input is high

$A_N = 0$ if the A_N digital input is low

Single Supply Analog Switch with Buffered Output

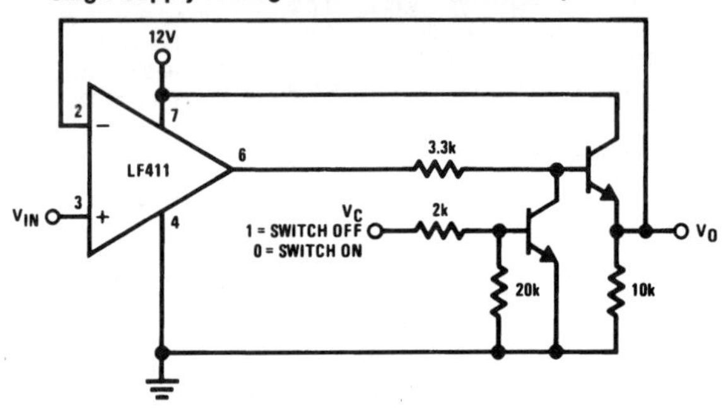

Detailed Schematic

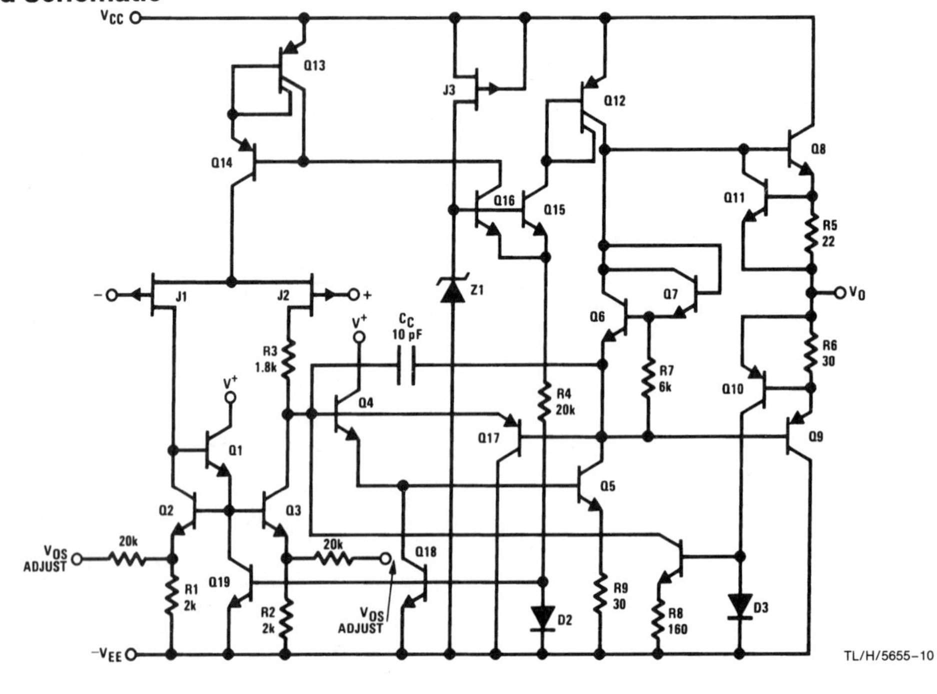

Physical Dimensions inches (millimeters)

Metal Can Package (H)
Order Number LF411AMH, LF411MH, LF411ACH or LF411CH
NS Package Number H08B

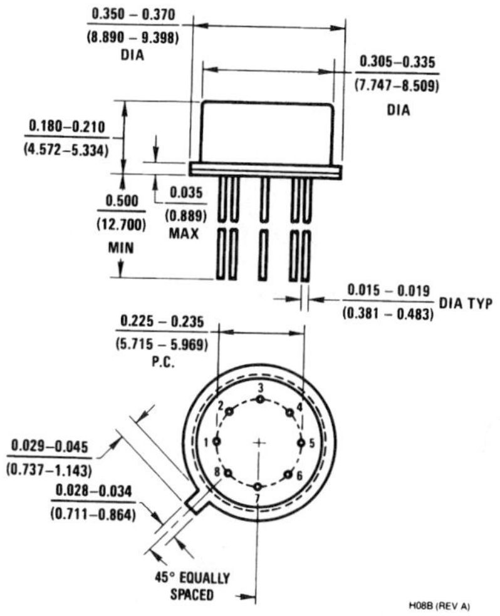

H08B (REV A)

Molded Dual-In-Line Package (N)
Order Number LF411ACN or LF411CN
NS Package Number N08E

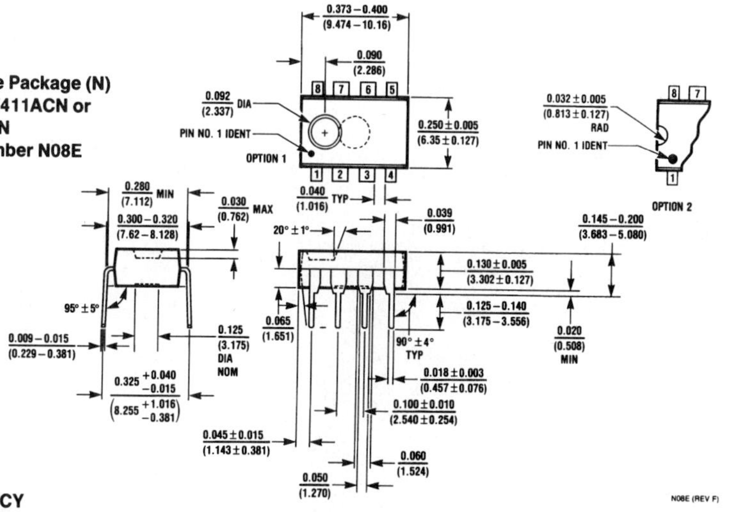

N08E (REV F)

LIFE SUPPORT POLICY

NATIONAL'S PRODUCTS ARE NOT AUTHORIZED FOR USE AS CRITICAL COMPONENTS IN LIFE SUPPORT DEVICES OR SYSTEMS WITHOUT THE EXPRESS WRITTEN APPROVAL OF THE PRESIDENT OF NATIONAL SEMICONDUCTOR CORPORATION. As used herein:

1. Life support devices or systems are devices or systems which, (a) are intended for surgical implant into the body, or (b) support or sustain life, and whose failure to perform, when properly used in accordance with instructions for use provided in the labeling, can be reasonably expected to result in a significant injury to the user.

2. A critical component is any component of a life support device or system whose failure to perform can be reasonably expected to cause the failure of the life support device or system, or to affect its safety or effectiveness.

National Semiconductor Corporation
2900 Semiconductor Drive
P.O. Box 58090
Santa Clara, CA 95052-8090
Tel: (408) 721-5000
TWX: (910) 339-9240

National Semiconductor GmbH
Westendstrasse 193-195
D-8000 Munchen 21
West Germany
Tel: (089) 5 70 95 01
Telex: 522772

NS Japan Ltd.
Sanseido Bldg. 5F
4-15 Nishi Shinjuku
Shinjuku-Ku,
Tokyo 160, Japan
Tel: 3-299-7001
FAX: 3-299-7000

National Semiconductor Hong Kong Ltd.
Southeast Asia Marketing
Austin Tower, 4th Floor
22-26A Austin Avenue
Tsimshatsui, Kowloon, H.K.
Tel: 3-7231290, 3-7243645
Cable: NSSEAMKTG
Telex: 52996 NSSEA HX

National Semicondutores Do Brasil Ltda.
Av. Brig. Faria Lima, 830
8 Andar
01452 Sao Paulo, SP. Brasil
Tel: (55/11) 212-5066
Telex: 391-1131931 NSBR BR

National Semiconductor (Australia) PTY, Ltd.
21/3 High Street
Bayswater, Victoria 3153
Australia
Tel: (03) 729-6333
Telex: AA32096

National Semiconductor — **Voltage Regulators**

LM117/LM217/LM317 3-terminal adjustable regulator

General Description

The LM117/LM217/LM317 are adjustable 3-terminal positive voltage regulators capable of supplying in excess of 1.5A over a 1.2V to 37V output range. They are exceptionally easy to use and require only two external resistors to set the output voltage. Further, both line and load regulation are better than standard fixed regulators. Also, the LM117 is packaged in standard transistor packages which are easily mounted and handled.

In addition to higher performance than fixed regulators, the LM117 series offers full overload protection available only in IC's. Included on the chip are current limit, thermal overload protection and safe area protection. All overload protection circuitry remains fully functional even if the adjustment terminal is disconnected.

Features

- Adjustable output down to 1.2V
- Guaranteed 1.5A output current
- Line regulation typically 0.01%/V
- Load regulation typically 0.1%
- Current limit constant with temperature
- **100% electrical burn-in**
- Eliminates the need to stock many voltages
- Standard 3-lead transistor package
- 80 dB ripple rejection

Normally, no capacitors are needed unless the device is situated far from the input filter capacitors in which case an input bypass is needed. An optional output capacitor can be added to improve transient response. The adjustment terminal can be bypassed to achieve very high ripple rejections ratios which are difficult to achieve with standard 3-terminal regulators.

Besides replacing fixed regulators, the LM117 is useful in a wide variety of other applications. Since the regulator is "floating" and sees only the input-to-output differential voltage, supplies of several hundred volts can be regulated as long as the maximum input to output differential is not exceeded.

Also, it makes an especially simple adjustable switching regulator, a programmable output regulator, or by connecting a fixed resistor between the adjustment and output, the LM117 can be used as a precision current regulator. Supplies with electronic shutdown can be achieved by clamping the adjustment terminal to ground which programs the output to 1.2V where most loads draw little current.

The LM117K, LM217K and LM317K are packaged in standard TO-3 transistor packages while the LM117H, LM217H and LM317H are packaged in a solid Kovar base TO-5 transistor package. The LM117 is rated for operation from −55°C to +150°C, the LM217 from −25°C to +150°C and the LM317 from 0°C to +125°C. The LM317T and LM317MP, rated for operation over a 0°C to +125°C range, are available in a TO-220 plastic package and a TO-202 package, respectively.

For applications requiring greater output current in excess of 3A and 5A, see LM150 series and LM138 series data sheets, respectively. For the negative complement, see LM137 series data sheet.

LM117 Series Packages and Power Capability

DEVICE	PACKAGE	RATED POWER DISSIPATION	DESIGN LOAD CURRENT
LM117	TO-3	20W	1.5A
LM217 LM317	TO-39	2W	0.5A
LM317T	TO-220	15W	1.5A
LM317M	TO-202	7.5W	0.5A

Typical Applications

1.2V–25V Adjustable Regulator

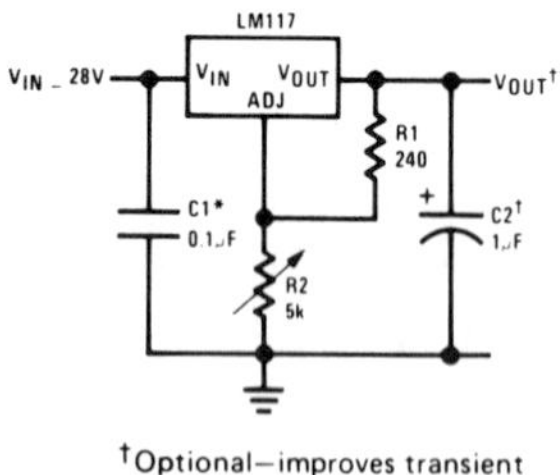

†Optional—improves transient response

*Needed if device is far from filter capacitors

$^{\dagger\dagger}V_{OUT} = 1.25V\left(1 + \frac{R2}{R1}\right)$

Digitally Selected Outputs

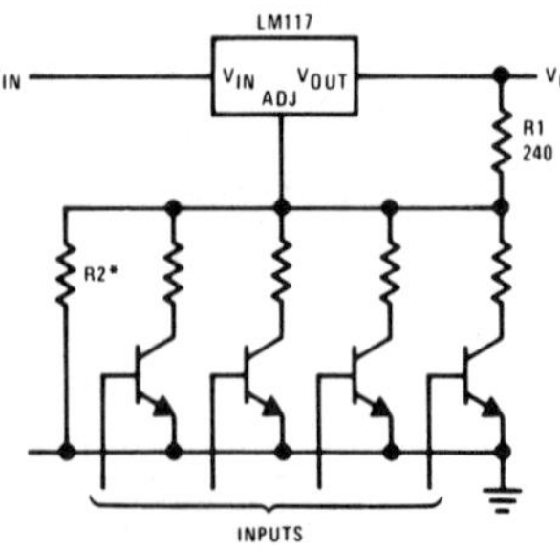

*Sets maximum V_{OUT}

5V Logic Regulator with Electronic Shutdown*

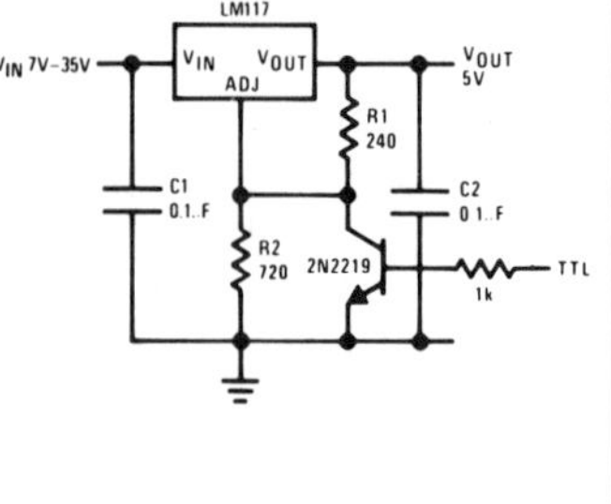

*Min output ≈ 1.2V

absolute maximum ratings

Power Dissipation	Internally limited
Input–Output Voltage Differential	40V
Operating Junction Temperature Range	
LM117	−55°C to +150°C
LM217	−25°C to +150°C
LM317	0°C to +125°C
Storage Temperature	−65°C to +150°C
Lead Temperature (Soldering, 10 seconds)	300°C

electrical characteristics (Note 1)

PARAMETER	CONDITIONS	LM117/217			LM317			UNITS
		MIN	TYP	MAX	MIN	TYP	MAX	
Line Regulation	$T_A = 25°C$, $3V \le V_{IN} - V_{OUT} \le 40V$ (Note 2)		0.01	0.02		0.01	0.04	%/V
Load Regulation	$T_A = 25°C$, $10\text{ mA} \le I_{OUT} \le I_{MAX}$							
	$V_{OUT} \le 5V$, (Note 2)		5	15		5	25	mV
	$V_{OUT} \ge 5V$, (Note 2)		0.1	0.3		0.1	0.5	%
Adjustment Pin Current			50	100		50	100	μA
Adjustment Pin Current Change	$10\text{ mA} \le I_L \le I_{MAX}$, $2.5V \le (V_{IN}-V_{OUT}) \le 40V$		0.2	5		0.2	5	μA
Reference Voltage	$3 \le (V_{IN}-V_{OUT}) \le 40V$, (Note 3) $10\text{ mA} \le I_{OUT} \le I_{MAX}$, $P \le P_{MAX}$	1.20	1.25	1.30	1.20	1.25	1.30	V
Line Regulation	$3V \le V_{IN} - V_{OUT} \le 40V$, (Note 2)		0.02	0.05		0.02	0.07	%/V
Load Regulation	$10\text{ mA} \le I_{OUT} \le I_{MAX}$, (Note 2)							
	$V_{OUT} \le 5V$		20	50		20	70	mV
	$V_{OUT} \ge 5V$		0.3	1		0.3	1.5	%
Temperature Stability	$T_{MIN} \le T_j \le T_{MAX}$		1			1		%
Minimum Load Current	$V_{IN}-V_{OUT} = 40V$		3.5	5		3.5	10	mA
Current Limit	$V_{IN}-V_{OUT} \le 15V$							
	K and T Package	1.5	2.2		1.5	2.2		A
	H and P Package	0.5	0.8		0.5	0.8		A
	$V_{IN}-V_{OUT} = 40V$							
	K and T Package		0.4			0.4		A
	H and P Package		0.07			0.07		A
RMS Output Noise, % of V_{OUT}	$T_A = 25°C$, $10\text{ Hz} \le f \le 10\text{ kHz}$		0.003			0.003		%
Ripple Rejection Ratio	$V_{OUT} = 10V$, $f = 120\text{ Hz}$		65			65		dB
	$C_{ADJ} = 10\mu F$	66	80		66	80		dB
Long-Term Stability	$T_A = 125°C$		0.3	1		0.3	1	%
Thermal Resistance, Junction to Case	H Package		12	15		12	15	°C/W
	K Package		2.3	3		2.3	3	°C/W
	T Package					5		°C/W
	P Package					12		°C/W

Note 1: Unless other wise specified, these specifications apply $-55°C \le T_j \le +150°C$ for the LM117, $-25°C \le T_j \le +150°C$ for the LM217 and $0°C \le T_j \le +125°C$ for the LM317; $V_{IN}-V_{OUT} = 5V$ and $I_{OUT} = 0.1A$ for the TO-5 package and $I_{OUT} = 0.5A$ for the TO-3 package and TO-220 package. Although power dissipation is internally limited, these specifications are applicable for power dissipations of 2W for the TO-5 and 20W for the TO-3 and TO-220. I_{MAX} is 1.5A for the TO-3 and TO-220 package and 0.5A for the TO-5 package.

Note 2: Regulation is measured at constant junction temperature. Changes in output voltage due to heating effects must be taken into account separately. Pulse testing with low duty cycle is used.

Note 3: Selected devices with tightend tolerance reference voltage available.

typical performance characteristics (K and T Packages)

Load Regulation

OUTPUT VOLTAGE DEVIATION (%)
$I_L = 0.5A$
$I_L = 1.5A$
$V_{IN} = 15V$
$V_{OUT} = 10V$
TEMPERATURE (C)

Current Limit

OUTPUT CURRENT (A)
$T_j = 150°C$
$T_j = -55°C$
$T_j = 25°C$
INPUT–OUTPUT DIFFERENTIAL (V)

Adjustment Current

ADJUSTMENT CURRENT (μA)
TEMPERATURE (C)

Dropout Voltage

INPUT–OUTPUT DIFFERENTIAL (V)
$\Delta V_{OUT} = 100$ mV
$I_L = 1.5A$
$I_L = 1A$
$I_L = 500$ mA
$I_L = 200$ mA
$I_L = 20$ mA
TEMPERATURE (C)

Temperature Stability

REFERENCE VOLTAGE (V)
TEMPERATURE (C)

Minimum Operating Current

QUIESCENT CURRENT (mA)
$T_j = -55$ C
$T_j = 150$ C
$T_j = 25$ C
INPUT–OUTPUT DIFFERENTIAL (V)

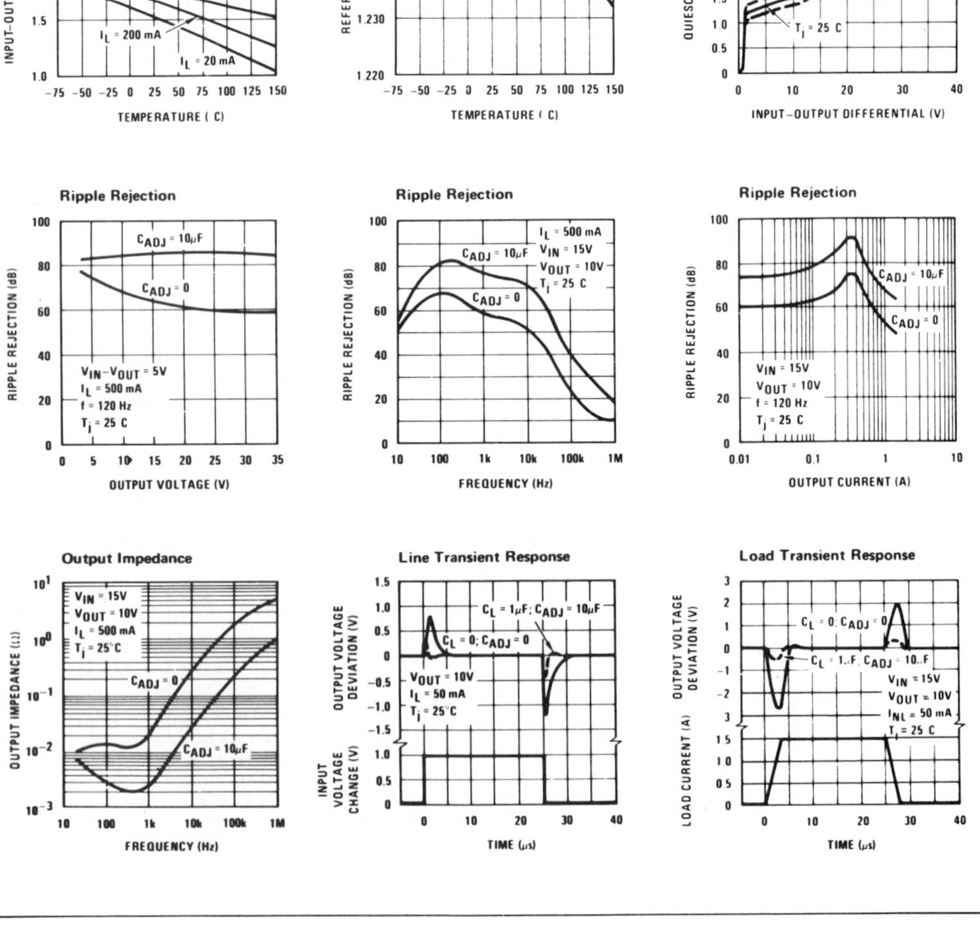

application hints

In operation, the LM117 develops a nominal 1.25V reference voltage, V_{REF}, between the output and adjustment terminal. The reference voltage is impressed across program resistor R1 and, since the voltage is constant, a constant current I_1 then flows through the output set resistor R2, giving an output voltage of

$$V_{OUT} = V_{REF}\left(1 + \frac{R2}{R1}\right) + I_{ADJ}R2$$

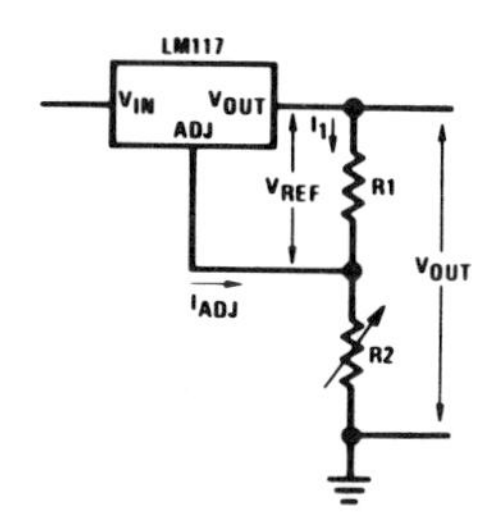

FIGURE 1.

Since the 100μA current from the adjustment terminal represents an error term, the LM117 was designed to minimize I_{ADJ} and make it very constant with line and load changes. To do this, all quiescent operating current is returned to the output establishing a minimum load current requirement. If there is insufficient load on the output, the output will rise.

External Capacitors

An input bypass capacitor is recommended. A 0.1μF disc or 1μF solid tantalum on the input is suitable input bypassing for almost all applications. The device is more sensitive to the absence of input bypassing when adjustment or output capacitors are used but the above values will eliminate the possibility of problems.

The adjustment terminal can be bypassed to ground on the LM117 to improve ripple rejection. This bypass capacitor prevents ripple from being amplified as the output voltage is increased. With a 10μF bypass capacitor 80 dB ripple rejection is obtainable at any output level. Increases over 10μF do not appreciably improve the ripple rejection at frequencies above 120 Hz. If the bypass capacitor is used, it is sometimes necessary to include protection diodes to prevent the capacitor from discharging through internal low current paths and damaging the device.

In general, the best type of capacitors to use are solid tantalum. Solid tantalum capacitors have low impedance even at high frequencies. Depending upon capacitor construction, it takes about 25μF in aluminum electrolytic to equal 1μF solid tantalum at high frequencies. Ceramic capacitors are also good at high frequencies; but some types have a large decrease in capacitance at frequencies around 0.5 MHz. For this reason, 0.01μF disc may seem to work better than a 0.1μF disc as a bypass.

Although the LM117 is stable with no output capacitors, like any feedback circuit, certain values of external capacitance can cause excessive ringing. This occurs with values between 500 pF and 5000 pF. A 1μF solid tantalum (or 25μF aluminum electrolytic) on the output swamps this effect and insures stability.

Load Regulation

The LM117 is capable of providing extremely good load regulation but a few precautions are needed to obtain maximum performance. The current set resistor connected between the adjustment terminal and the output terminal (usually 240Ω) should be tied directly to the output of the regulator rather than near the load. This eliminates line drops from appearing effectvely in series with the reference and degrading regulation. For example, a 15V regulator with 0.05Ω resistance between the regulator and load will have a load regulation due to line resistance of 0.05Ω x I_L. If the set resistor is connected near the load the effective line resistance will be 0.05Ω (1 + R2/R1) or in this case, 11.5 times worse.

Figure 2 shows the effect of resistance between the regulator and 240Ω set resistor.

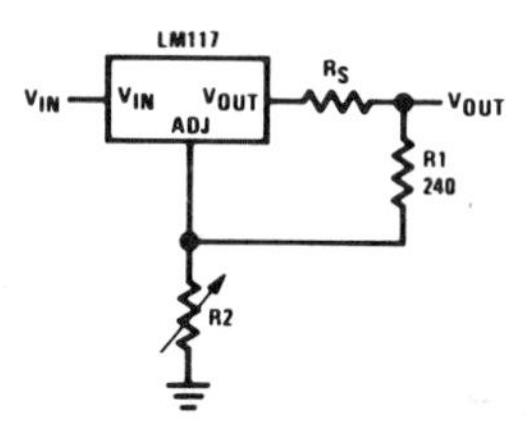

FIGURE 2. Regulator with Line Resistance in Output Lead

With the TO-3 package, it is easy to minimize the resistance from the case to the set resistor, by using two separate leads to the case. However, with the TO-5 package, care should be taken to minimize the wire length of the output lead. The ground of R2 can be returned near the ground of the load to provide remote ground sensing and improve load regulation.

Protection Diodes

When external capacitors are used with *any* IC regulator it is sometimes necessary to add protection diodes to prevent the capacitors from discharging through low current points into the regulator. Most 10μF capacitors have low enough internal series resistance to deliver 20A spikes when shorted. Although the surge is short, there is enough energy to damage parts of the IC.

When an output capacitor is connected to a regulator and the input is shorted, the output capacitor will discharge into the output of the regulator. The discharge current depends on the value of the capacitor, the output voltage of the regulator, and the rate of decrease of V_{IN}. In the LM117, this discharge path is through a large junction that is able to sustain 15A surge with no problem. This is not true of other types of positive regulators. For output capacitors of 25μF or less, there is no need to use diodes.

The bypass capacitor on the adjustment terminal can discharge through a low current junction. Discharge occurs when *either* the input or output is shorted. Internal to the LM117 is a 50Ω resistor which limits the peak discharge current. No protection is needed for output voltages of 25V or less and 10μF capacitance. *Figure 3* shows an LM117 with protection diodes included for use with outputs greater than 25V and high values of output capacitance.

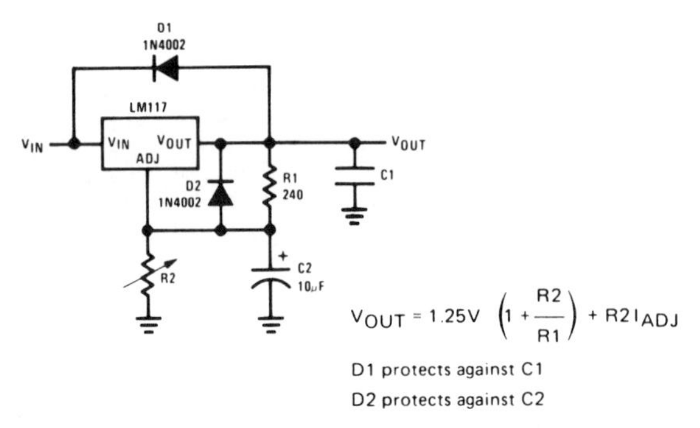

FIGURE 3. Regulator with Protection Diodes

schematic diagram

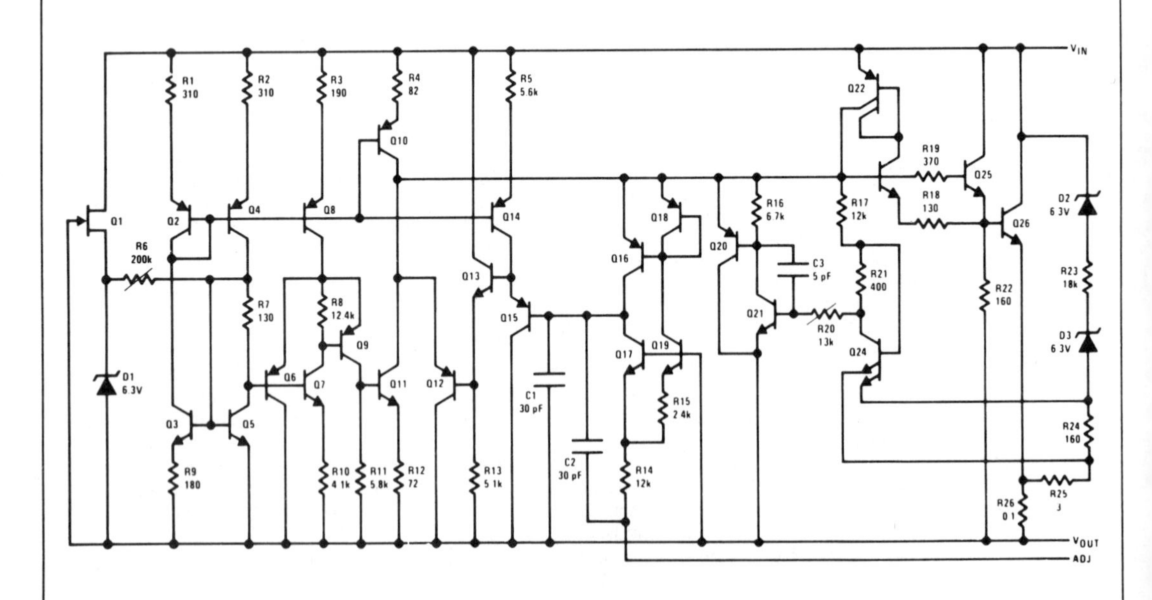

typical applications (con't)

Slow Turn-On 15V Regulator

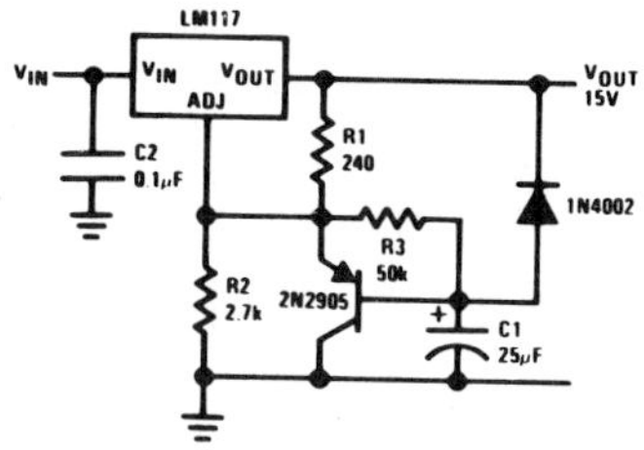

Adjustable Regulator with Improved Ripple Rejection

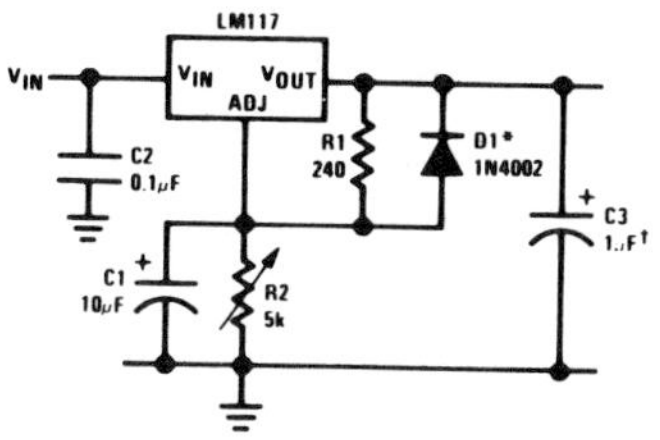

†Solid tantalum

*Discharges C1 if output is shorted to ground

High Stability 10V Regulator

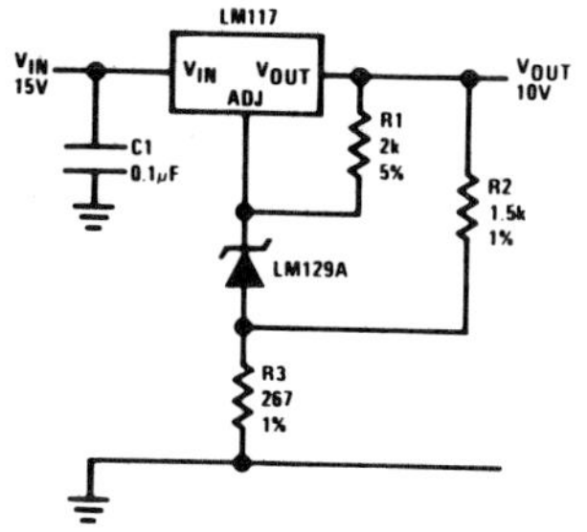

High Current Adjustable Regulator

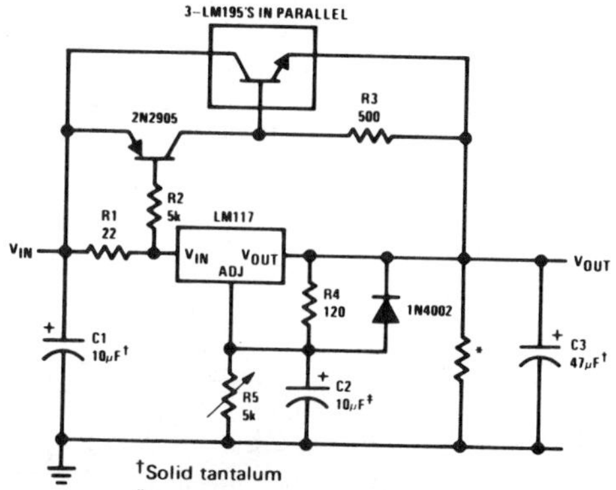

†Solid tantalum

*Minimum load current = 30 mA

‡Optional—improves ripple rejection

0 to 30V Regulator

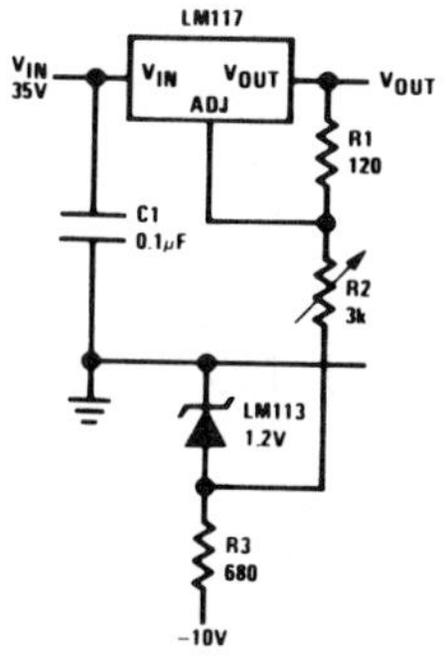

Power Follower

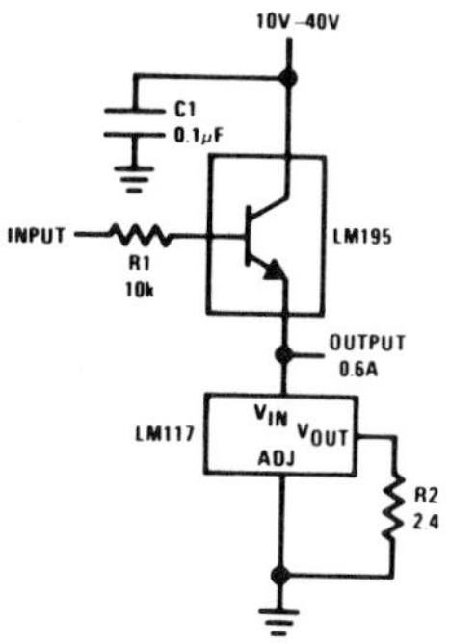

5A Constant Voltage/Constant Current Regulator

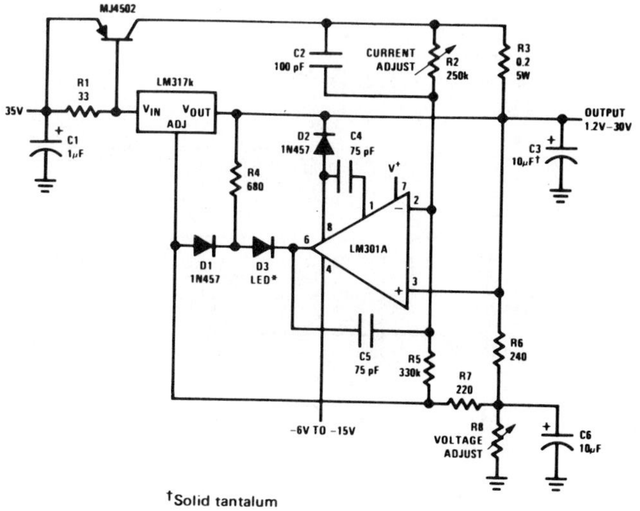

†Solid tantalum

*Lights in constant current mode

1A Current Regulator

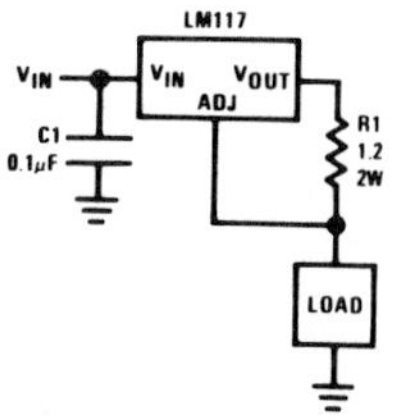

1.2V–20V Regulator with Minimum Program Current

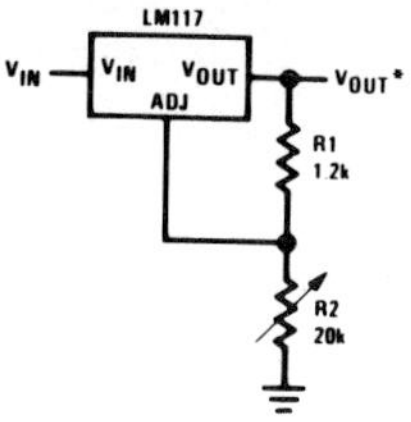

*Minimum load current ≈ 4 mA

typical applications (con't)

High Gain Amplifier

Low Cost 3A Switching Regulator

†Solid Tantalum

*Core—Arnold A-254168-2 60 turns

4A Switching Regulator with Overload Protection

†Solid Tantalum

*Core Arnold A-254168-2 60 turns

Precision Current Limiter

$I_{OUT} = \frac{1.2}{R1}$ *

$*0.8\Omega \le R1 \le 120\Omega$

Tracking Preregulator

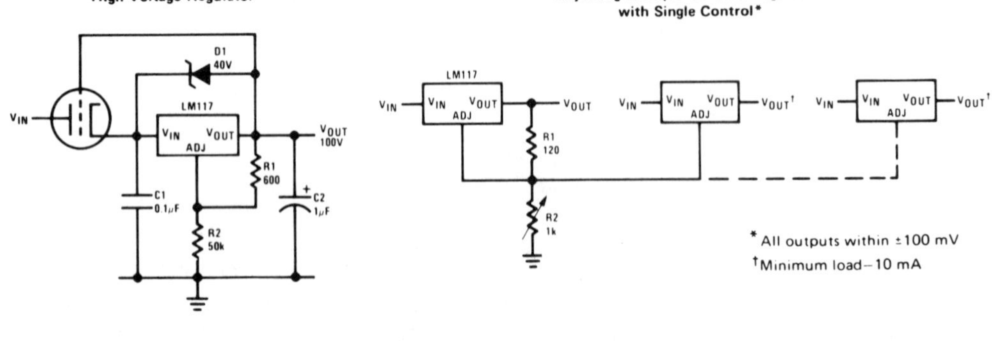

typical applications (con't)

AC Voltage Regulator

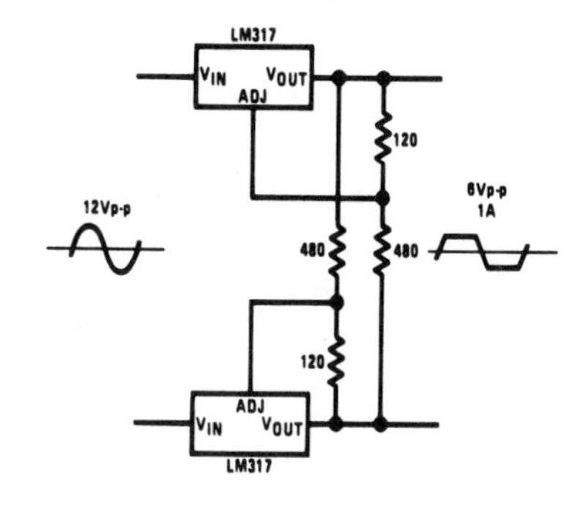

Adjustable 4A Regulator

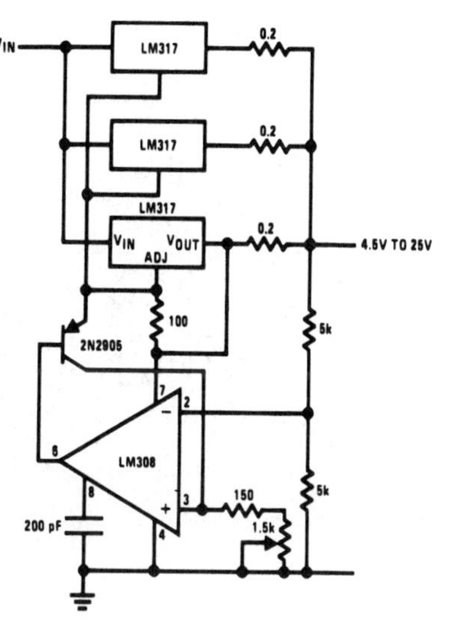

12V Battery Charger

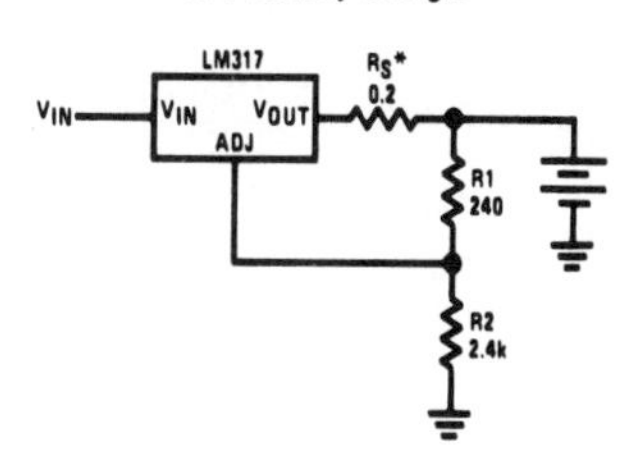

*R_S—sets output impedance of charger $Z_{OUT} = R_S \left(1 + \frac{R2}{R1}\right)$ Use of R_S allows low charging rates with fully charged battery.

50 mA Constant Current Battery Charger

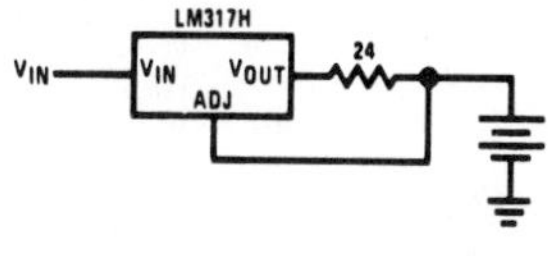

Current Limited 6V Charger

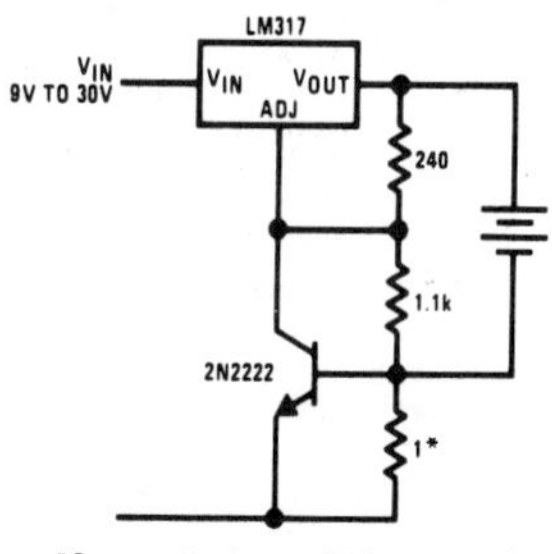

*Sets peak current (0.6A for 1Ω)

connection diagrams

Metal Can Package

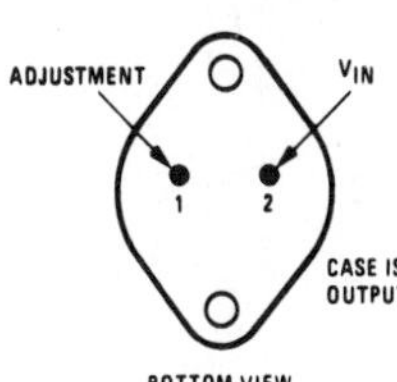

Order Number LM117K, LM217K or LM317K
See Package 18

Metal Can Package

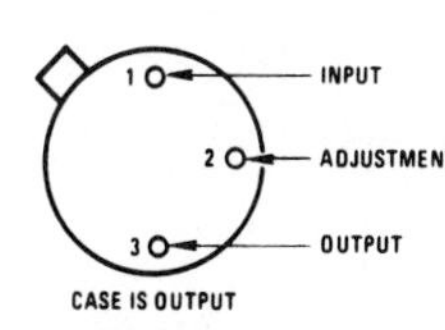
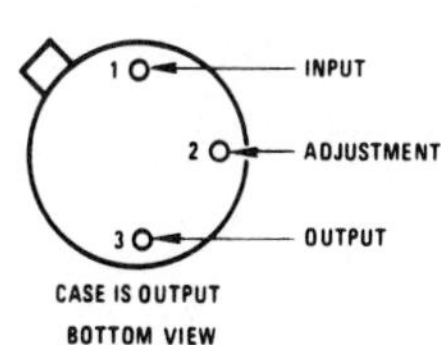

Order Number LM117H, LM217H or LM317H
See Package 9

Plastic Package

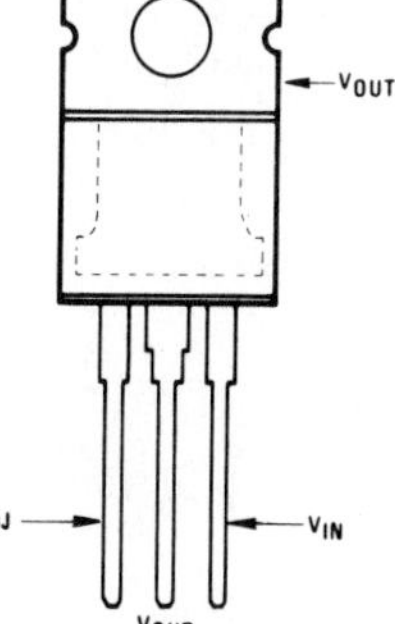

Order Number LM317T
See Package 26

Plastic Package

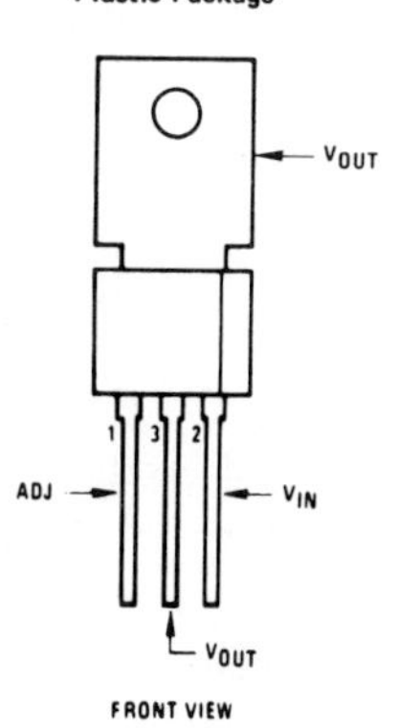

Order Number LM317P
See Package 37

BIBLIOGRAPHY

GENERAL

Handbooks

Fink, D. G., and Christiansen, D., eds. 1982. *Electronic engineers' handbook.* New York: McGraw-Hill. Encyclopedic.

Fink, D. G., and Beaty, H. W., eds. 1986. *Standard handbook for electrical engineers.* New York: McGraw-Hill. Tutorial articles on electrical engineering topics.

Giacoletto, L. J., ed. 1977. *Electronics designers' handbook.* New York: McGraw-Hill. Excellent tutorials and data.

Jordan, E., ed. 1985. *Reference data for engineers: radio, electronics, computer, and communications.* Indianapolis: Howard W. Sams & Co. General-purpose engineering data.

Master catalogs

EEM: Electronic engineers master catalog. Garden City, NY: Hearst Business Communications, Inc. Thousands of pages of manufacturers' data sheets, plus addresses of companies, their representatives, and local distributors. Extremely useful. Published annually.

IC master. Garden City, NY: Hearst Business Communications, Inc. Comprehensive selection guides and thousands of pages of data sheets. Extremely useful. Published annually.

Books

Bracewell, R. N. 1986. *The Fourier transform and its applications.* New York: McGraw-Hill. The classic in this field.

Brigham, E. O. 1973. *The fast Fourier transform.* Englewood Cliffs, NJ: Prentice-Hall. Highly readable.

Higgins, R. J. 1983. *Electronics with digital and analog integrated circuits.* Englewood Cliffs, NJ: Prentice-Hall. "The electronics book that's not Horowitz and Hill," according to its author.

Lathi, B. P. 1987. *Signals and systems.* Carmichael, CA: Berkeley-Cambridge Press. Network theory, transform methods, and communication.

Mead, C., and Conway, L. 1980. *Introduction to VLSI systems.* Reading, MA: Addison-Wesley. Device physics and circuit design; a classic.

Millman, J., and Grabel, A. 1987. *Microelectronics.* New York: McGraw-Hill. Highly recommended all-around text and reference.

Savant, C. J., Jr., Roden, M. S., and Carpenter, G. L. 1987. *Electronic circuit design.* Menlo Park, CA: Benjamin/Cummings. Good introduction to electronic circuits.

Senturia, S. D., and Wedlock, B. D. 1975. *Electronic circuits and applications.* New York: Wiley. Good introductory engineering textbook.

Siebert, W. M. 1986. *Circuits, signals, and systems.* Cambridge, MA: MIT Press. Network theory, transform methods, and signal processing.

Smith, R. J. 1984. *Circuits, devices, and systems.* New York: Wiley. Broad introductory engineering textbook.

Tietze, U., and Schenk, C. 1978. *Advanced electronic circuits.* Berlin: Springer-Verlag. Excellent all-around reference.

CHAPTER 1

Holbrook, J. G. 1966. *Laplace transforms for electronic engineers.* New York: Pergamon Press. Good for learning about the s-plane; out of print.

Johnson, D. E., Hilburn, J. L., and Johnson, J. R. 1986. *Basic electric circuit analysis.* Englewood Cliffs, NJ: Prentice-Hall. Passive circuit analysis.

Purcell, E. M. 1985. *Electricity and magnetism (Berkeley physics course, vol. 2).* New York: McGraw-Hill. Excellent textbook on electromagnetic theory. Relevant sections on electrical conduction and analysis of ac circuits with complex numbers.

CHAPTER 2

Ebers, J. J., and Moll, J. L. 1954. Large-signal behavior of junction transistors. *Proc. I.R.E.* **42**:1761–1772. The Ebers-Moll equation is born.

Grove, A. S. 1967. *Physics and technology of semiconductor devices.* New York: Wiley. Principles of fabrication and operation of bipolar and field-effect transistors.

Schilling, D. L., and Belove, C. 1979. *Electronic circuits: discrete and integrated.* New York: McGraw-Hill. Traditional h-parameter transistor treatment.

Searle, C. L., Boothroyd, A. R., Angelo, E. J., Jr., Gray, P. E., and Pederson, D. O. 1966. *Elementary circuit properties of transistors (semiconductor electronics education committee, vol. 3).* New York: Wiley. Physics of transistors.

Sze, S. M. 1981. *Physics of semiconductor devices.* New York: Wiley.

"Discrete products databook" and "Transistor databook." Soft-cover collections of transistor data sheets are published sporadically under these titles by all the transistor manufacturers, in particular GE, Motorola, National, and TI. Data sheets are essential for circuit design.

CHAPTER 3

Muller, R. S., and Kamins, T. I. 1986. *Device electronics for integrated circuits.* New York: Wiley. Transistor properties in ICs.

Richman, P. 1973. *MOS field-effect transistors and integrated circuits.* New York: Van Nostrand Reinhold. Recommended.

Tsividis, Y. P. 1987. *Operation and modeling of the MOS transistor.* New York: McGraw-Hill.

See also Grove, A. S., under Chapter 2 listings.

"FET databook," "Power MOSFET databook." Soft-cover collections of FET data sheets and applications notes are published every few years under these or similar titles by all the FET manufacturers, in particular GE (Intersil, RCA), Hitachi, IR, Motorola, National, Siemens, and TI. Data sheets are essential for design.

CHAPTER 4

Frederiksen, T. M. 1984. *Intuitive IC op-amps.* Santa Clara, CA: National Semiconductor Corp. Extremely good treatment at all levels.

Graeme, J. G. 1987. *Applications of operational amplifiers: third generation techniques.* New York: McGraw-Hill. One of the Burr-Brown series.

Jung, W. G. 1986. *IC op-amp cookbook.* Indianapolis: Howard W. Sams & Co. Lots of circuits, with explanations. See also Jung's *Audio IC op-amp applications.*

Meyer, R. G., ed. 1978. *Integrated circuit operational amplifiers.* New York: IEEE. Choice selection of reprints, somewhat dated.

Rosenstark, S. 1986. *Feedback amplifier principles.* New York: Macmillan. Design principles for discrete circuits.

Smith, J. I. 1971. *Modern operational circuit design.* New York: Wiley. A favorite, now out of print.

Soclof, S. 1985. *Analog integrated circuits.* Englewood Cliffs, NJ: Prentice-Hall. The design of linear ICs.

Stout, D. F., and Kaufman, M. 1976. *Handbook of operational amplifier circuit design.* New York: McGraw-Hill.

Explicit design procedures. See also their *Handbook of microcircuit design and application.*

Wait, J. V., Huelsman, L. P., and Korn, G. A. 1989. *Introduction to operational amplifier theory and applications.* New York: McGraw-Hill.

"Linear databook," "Analog databook," and "Op-amp databook." Soft-cover collections of linear IC data sheets and application notes are published approximately every two years under these titles by all the linear IC manufacturers, in particular Analog Devices, Burr-Brown, GE (RCA, Intersil), Linear Technology, Maxim, Motorola, National, Precision Monolithics, and TI. Data sheets are essential for circuit design.

CHAPTER 5

Bingham, J. A. C. 1988. *Theory and practice of modem design.* New York: Wiley. A good engineering guide; includes filters and oscillators.

Clarke, K. K., and Hess, D. T. 1971. *Communication circuits: analysis and design.* Reading, MA: Addison-Wesley. Good chapter on oscillators.

Hilburn, J. L., and Johnson, D. E. 1982. *Manual of active filter design.* New York: McGraw-Hill.

Jung, W. C. 1983. *IC timer handbook.* Indianapolis: Howard W. Sams & Co. All about 555s.

Lancaster, D. 1979. *Active filter cookbook.* Indianapolis: Howard W. Sams & Co. Explicit design procedure; easy to read.

Loy, N. J. 1988. *An engineer's guide to FIR digital filters.* Englewood Cliffs, NJ: Prentice-Hall. Design procedures and discussion.

Parzen, B. 1983. *Design of crystal and other harmonic oscillators.* New York: Wiley. Discrete oscillator circuits.

Zverev, A. I. 1967. *Handbook of filter synthesis.* New York: Wiley. Extensive tables for passive *LC* and crystal filter design.

See also Graeme, J. G., under Chapter 4 listings.

CHAPTER 6

Hnatek, E. R. 1981. *Design of solid-state power supplies.* New York: Van Nostrand Reinhold. Switching supplies.

Pressman, A. I. 1977. *Switching and linear power supply, power converter design.* Rochelle Park, NJ: Hayden Book Co; out of print.

"Voltage regulator databook," "Power databook." Soft-cover collections of voltage-regulator data sheets, power-component data sheets, and application notes are published sporadically under these and similar titles by Apex, Motorola, National, TI, and Unitrode. The "Linear databooks" referenced for Chapter 4 also contain regulator data sheets, which are essential for circuit design.

CHAPTER 7

Buckingham, M. J. 1983. *Noise in electronic devices and systems.* New York: Wiley.

Morrison, R. 1986. *Grounding and shielding techniques in instrumentation.* New York: Wiley.

Motchenbacher, C. D., and Fitchen, F. C. 1973. *Low-noise electronic design.* New York: Wiley. Recommended for low-noise amplifier design.

Netzer, Y. 1981. The design of low-noise amplifiers. *Proc. IEEE* **69**:728–741. Excellent review.

Ott, H. 1988. *Noise reduction techniques in electronic systems.* New York: Wiley. Shielding and low-noise design.

Sheingold, D. H., ed. 1976. *Nonlinear circuits handbook.* Norwood, MA: Analog Devices. Highly recommended.

Van Duzer, T. 1981. *Principles of superconductive devices and circuits.* New York: Elsevier. Overview of traditional superconductors and applications.

Wong, Y. J., and Ott, W. E. 1976. *Function circuits: design and applications.* New York: McGraw-Hill. Nonlinear circuits and op-amp exotica; out of print.

"Data acquisition databook" or "Linear databook." Soft-cover collections of data sheets and application notes relevant to precision design are published every few years under these or similar titles by many semiconductor manufacturers, in particular Analog Devices, Burr-Brown, Linear Technology, Maxim, National, Precision Monolithics, and Teledyne Semiconductor.

CHAPTER 8

Blakeslee, T. R. 1979. *Digital design with standard MSI and LSI.* New York: Wiley. Refreshing approach to practical logic design; includes two chapters of "nasty realities."

Hill, F. J., and Peterson, G. R. 1981. *Introduction to switching theory and logical design.* New York: Wiley. Classic logic design textbook.

Lancaster, D. 1979. *TTL cookbook.* Indianapolis: Howard W. Sams & Co. Practical circuits, good reading.

Lancaster, D. 1988. *CMOS cookbook.* Indianapolis: Howard W. Sams & Co. Good reading, down-to-earth applications. Includes widely used (but rarely mentioned) M^2L (Mickey Mouse logic) technique.

Wickles, W. E. 1968. *Logic design with integrated circuits.* New York: Wiley. Dated, but still good.

"TTL databook," "Logic databook," and "CMOS databook." Soft-cover collections of data sheets and applications notes are published approximately every two years under these and similar titles by most semiconductor manufacturers, in particular AMD/MMI, GE (RCA), Motorola, National, Signetecs, and TI. Look also for "Programmable logic databooks" (and similar titles) from manufacturers such as Altera, AMD/MMI, Cypress, Gazelle, Lattice, National, VTI, and Xicor. Data sheets are essential for design.

CHAPTER 9

Best, R. E. 1984. *Phase-locked loops.* New York: McGraw-Hill. Advanced techniques.

Davies, A. C. 1969. Digital generation of low-frequency sine waves. *IEEE Trans. Instr. Meas.* **18**:97. Digital sine-wave generation.

Gardner, F. M. 1979. *Phaselock techniques.* New York: Wiley. The classic PLL book: emphasis on fundamentals.

Hnatek, E. R. 1988. *A user's handbook of D/A and A/D converters.* New York: Wiley. Applications.

Jung, W. G. 1978. *IC converter handbook.* Indianapolis: Howard W. Sams & Co. Using modern converter ICs.

Sheingold, D. H., ed. 1976. *Nonlinear circuits handbook.* Norwood, MA: Analog Devices.

Sheingold, D. H., ed. 1980. *Transducer interfacing handbook.* Norwood, MA: Analog Devices.

Sheingold, D. H., ed. 1986. *Analog–digital conversion handbook.* Englewood Cliffs, NJ: Prentice-Hall. The A/D bible, from Analog Devices.

Yariv 1976. *Introduction to optical electronics.* New York: Rinehart & Winston. Physics of opto-electronics, lasers, and detection.

"Conversion products databooks," "Data acquisition databook." Soft-cover collections of data sheets and application notes are published periodically under these and similar titles by semiconductor manufacturers, in particular Analog Devices, Analogic, Brooktree, Burr-Brown, Crystal, Datel, Hybrid Systems, Teledyne Seniconductor, and Telmos. Data sheets are essential for design.

"Interface databook." Soft-cover collections of data sheets and application notes are published every few years under this and similar titles by semiconductor manufacturers, in particular Motorola, National, Sprague, and TI.

CHAPTER 10

Eggebrecht, L. C. 1986. *Interfacing to the IBM personal computer.* Indianapolis: Howard W. Sams & Co. By the PC system architect and design team leader.

Osborne, A. 1987. *An introduction to microcomputers. Vol. 1:*

Basic concepts. Berkeley, CA: Osborne/McGraw-Hill.

Sargent, M., III, and Shoemaker, R. L. 1986. *The IBM PC from the inside out.* Reading, MA: Addison-Wesley. Detailed guide to programming and hardware.

Sloan, M. E. 1980. *Introduction to minicomputers and microcomputers.* Reading, MA: Addison-Wesley. Enphasis on computing; software-oriented.

Sloan, M. E. 1983. *Computer hardware and organization.* Chicago: Science Research Assoc.

Tanenbaum, A. S. 1984. *Structured computer organization.* Englewood Cliffs, NJ: Prentice-Hall. Mainframes to micros to bit-slice.

See also Bingham, J. A. C., under Chapter 5 listings.

Also, manuals and data sheets on the 8086/8088 (Intel "MCS-86 user's manual," "iAPX 86,88 user's manual," "The 8086 family user's manual," etc.).

CHAPTER 11

Cramer, W., and Kane, G. 1986. 68000 microprocessor handbook. New York: McGraw-Hill. Introduction to 68000 hardware.

Eccles, W. J. 1985. *Microcomputer systems – a 16-bit approach.* Reading, MA: Addison-Wesley. Hardware and software, illustrated with the 68000.

Hancock, L., and Krieger, M. 1982. *The C primer.* New York: McGraw-Hill. Introduction for beginners.

Hansen, A. 1986. *Proficient C.* Bellevue, WA: Microsoft Press. Microsoft C on the IBM PC.

Harbison, S. P., and Steele, G. L., Jr., 1987. *C: a reference manual.* Englewood Cliffs, NJ: Prentice-Hall. Readable and definitive; has ANSI extensions.

Motorola, Inc. 1986. *M68000 programmer's reference manual.* Englewood Cliffs, NJ: Prentice-Hall. A must, if you plan to write 68000 code.

Peatman, J. B. 1977. *Microcomputer-based design.* New York: McGraw-Hill. Broad view of applying microprocessors.

Peatman, J. B. 1987. *Design with microcontrollers.* New York: McGraw-Hill.

Also, manuals and data sheets on the 68000/68008 (Motorola M68000 family reference book, P/N FR 68K/D).

CHAPTER 12

Coombs, C. F., Jr., ed. 1988. *Printed circuits handbook.* New York: McGraw-Hill. A wealth of information on the design, fabrication, and application of PC boards.

"Technical manual and catalog." Westlake Village, CA: Bishop Graphics, Inc. Frequently revised product catalog and information for PC layout.

CHAPTER 13

Carson, R. S. 1982. *High-frequency amplifiers.* New York: Wiley. RF transistor amplifiers.

DeMaw, D. 1982. *Practical RF design manual.* Englewood Cliffs, NJ: Prentice-Hall. Down-to-earth radiofrequency design.

Edwards, T. C. 1981. *Foundations for microstrip circuit design.* New York: Wiley.

Gonzalez, GH. 1984. *Microwave transistor amplifier analysis and design.* Englewood Cliffs, NJ: Prentice-Hall. Small-signal s-parameter design of oscillators and amplifiers.

Hayward, W. H. 1982. *Introduction to radiofrequency design.* Englewood Cliffs, NJ: Prentice-Hall. Design techniques; good on receiver systems.

Matick, R. E. 1969. *Transmission lines for digital and communication networks.* New York: McGraw-Hill.

Milligan, T. 1985. *Modern antenna design.* New York: McGraw-Hill. What you need to know about every antenna.

Rohde, U. L. 1983. *Digital PLL frequency synthesizers.* Englewood Cliffs, NJ: Prentice-Hall. Theory and lots of circuit detail.

Rohde, U. L. and Bucher, T. N. 1988. *Communications receivers.* New York: McGraw-Hill. Excellent compendium on receivers, mixers, modulation, and detection.

Skolnik, M. I., ed. 1979. *Radar handbook.* New York: McGraw-Hill. Incredible compendium of radar information.

Unitrode Corp. 1984. *Pin diode designers' handbook and catalog.* Lexington, MA: Unitrode Corporation. Theory, data sheets, and applications.

Viterbi, A. J. 1966. *Principles of coherent communication.* New York: McGraw-Hill. A classic; modulation theory; out of print.

Weinreb, S. 1980. Low-noise cooled GASFET amplifiers. *IEEE Trans. Microwave Theory and Techniques.* **MTT-28**, 10:1041–1054. Theory and practice of low-noise microwave amplifiers, by the world's expert.

"The radio amateur's handbook." Newington, CT: American Radio Relay League. Published annually, this is the standard handbook for radio amateurs.

"RF transistor data book." Soft-cover collections of data sheets and application notes are published sporadically under this and similar titles by the RF transistor manufacturers, in particular Avantek, GE (RCA), Mini-circuits, Mitsubishi, Motorola, Siliconix, and TRW.

CHAPTER 14

Meindl, J. D. 1969. *Micropower circuits.* New York: Wiley. Dated, but good for discrete design.

See also occasional appplication notes from Linear Technology, Maxim, and National. Extensive data books and applications notes are available from Duracell, Electrochem, Eveready (Union Carbide), Gates, Kodak, Power Conversion, Power Sonic, Saft, Tadiran, and Yuasa, among others. For information on solar cells contact manufacturers such as Arco Solar, Solarex, and Solavolt.

CHAPTER 15

Ferbal, T., ed. 1987. *Experimental techniques in high energy physics.* Reading, MA: Addison-Wesley.

Meade, M. L. 1983. *Lock-in amplifiers: principles and applications.* London: P. Peregrinus Ltd. How lock-in amplifiers work and how to design them.

Radeka, V. 1988. Low-noise techniques in detectors. *Ann. Rev. Nucl. and Part. Physics,* **38**:217–277. Amplifier design, signal processing, and fundamental limits in charge measurement.

Wobschall, D. 1987. *Circuit design for electronic instrumentation.* New York: McGraw-Hill. Sensors and associated electronics.

"Temperature measurement handbook." Stamford, CT: Omega Engineering Corp. (revised annually). Thermocouples, thermistors, pyrometers, resistance thermometers.

Hewlett-Packard application notes: AP52-2 ("Timekeeping and frequency calibration"), AP150 ("Spectrum analyzer basics"), and AP200 ("Fundamentals of quartz oscillators"). They are available without charge from the Hewlett-Packard Corp., Palo Alto, CA.

See also the annual product catalogs from the Hewlett-Packard Corp., EG&G Princeton Applied Research, Fluke/Phillips, and Tektronix.

INDEX

The letter "t" following a page number indicates a table.